Advances in Oil and Gas Exploration & Production

Series Editor

Rudy Swennen, Department of Earth and Environmental Sciences,
K.U. Leuven, Heverlee, Belgium

The book series "Advances in Oil and Gas Exploration & Production" focuses on publishing scientific monographs covering a wide spectrum of topics related to geophysical and geological research within both conventional and unconventional oil and gas systems. These topics are approached from both exploration and production perspectives. The series aims to establish a diverse library of reference works that describe the current state of research on selected themes, such as specific techniques used in the petroleum geoscience industry or regional aspects. Notably, all books in this series are authored and edited by leading experts actively engaged in their respective fields.

The "Advances in Oil and Gas Exploration & Production" series encompasses single and multi-authored books, edited volumes, as well as Conference Proceedings. You can obtain a Book Proposal Form from our website or request one directly from the Publishing Editor, Qiao Shu, at qiao.shu@springernature.com.

Akif Alizadeh · Ibrahim Guliyev ·
Parviz Mamedov · Elmira Aliyeva ·
Akper Feyzullayev · Dadash Huseynov ·
Lev Eppelbaum

Pliocene Hydrocarbon Sedimentary Series of Azerbaijan

Springer

Akif Alizadeh
Institute of Geology and Geophysics
of the Azerbaijan Academy
Baku, Azerbaijan

Parviz Mamedov
Department of Geophysics
Azerbaijan State Oil and Industry
University
Baku, Azerbaijan

Akper Feyzullayev
Department of Geochemistry and Fluid
Dynamics of Sedimentary Basins
Institute of Geology and Geophysics
of the Azerbaijan Academy
Baku, Azerbaijan

Lev Eppelbaum
Department of Geophysics
Faculty of Exact Sciences
Tel Aviv University
Tel Aviv, Israel

Azerbaijan State Oil
and Industry University
Baku, Azerbaijan

Ibrahim Guliyev
Institute of Geology and Geophysics
of the Azerbaijan Academy
Baku, Azerbaijan

Elmira Aliyeva
Department of Lithogenesis
of Oil and Gas Basins
Institute of Geology and Geophysics
of the Azerbaijan Academy
Baku, Azerbaijan

Dadash Huseynov
Institute of Geology and Geophysics
of the Azerbaijan Academy
Baku, Azerbaijan

ISSN 2509-372X ISSN 2509-3738 (electronic)
Advances in Oil and Gas Exploration & Production
ISBN 978-3-031-50437-2 ISBN 978-3-031-50438-9 (eBook)
https://doi.org/10.1007/978-3-031-50438-9

This Springer imprint is published by the registered company Springer Nature Switzerland AG
The registered company address is: Gewerbestrasse 11, 6330 Cham, Switzerland

Preface

Over the long history of the oil industry in the territory of the Azerbaijan Republic and adjacent waters of the Caspian Sea, a large complex of geological and geophysical research was carried out, a significant amount of structural mapping, search, support, and parametric drilling. Analysis and generalization of all accumulated geological-geophysical material determine the degree of knowledge and directions for further research. The accumulated vast archive of investigations in this region requires rethinking many studies and concepts. This book is suggested to review of the Pliocene hydrocarbon deposits make up most of Azerbaijan's explored deposits.

Oil and gas shows in Azerbaijan have been known since the 10th century and are mentioned in the works of famous scientists, historians, and travelers like Al-Mas'udi, El Istakhri, and Marco Polo. The works by M. Polo (the 13th century) contain data on oil production at Pirallakhi Island and morphological descriptions of oil wells and natural oil shows with indications of their production.

In 1654, in the Balakhani area, an Azerbaijani Allakhyar Mamed Nuri ogly dug the first 35-m-depth oil well. By 1735, 52 wells could be counted in Balakhani, and leather buckets bailed oil. In 1816, 116 wells were already found in Baku with maximal depth not exceeding 35 m. More than that, already in the early 19th century, a Bakuvian Kasymbeq was the first in the world to produce oil from the Caspian bottom from the wells constructed 20–30 m offshore.

The pioneering data on the western Absheron geological structure is met in the work by A. Humboldt (early 19th century) dedicated to the Azerbaijan oil fields. The 20-ies of the 19th century should be considered the commencement of scientific geological investigation of Azerbaijan. In 1827, N. I. Voskoboinikov was the first to give an in-depth description of oil wells of the Baku region, thus starting the research of the Azerbaijan Peninsula oil deposits.

A systematic study was started by the investigations of outstanding German (Russian) scientist G. W. Abich that began in the Caucasus in 1844 and continued until 1877. They were devoted to various issues of the Caucasian geology, specifically Azerbaijan oil-bearing regions.

It is necessary to note the contributions of Russian geologists in the late 19th century and the beginning of the 20th century: M. V. Abramovich, N. I. Andrusov, L. F. Batzevich, K. I. Bogdanovich, S. I., Charnotsky,

D. V. Golubyatnikov, I. M. Gubkin, K. P. Kalitsky, K. V. Kharichkov, A. Konshin, N. I. Lebedev, Ya. I. Lednev, D. V. Nalivkin, A. I. Sorokin, P. G. Volarovich.

It is essential to give prominence to Abdul Kadyr Efendiyev among these researchers, the first Azerbaijani mining engineer, who dealt with the geological structure study of the Fatmai oil-bearing region at the Absheron Peninsula (1910–1913). It is also necessary to mention Gasanbeq Zardabi, an Azerbaijan enlightener and natural scientist, and Farrukhbeq Javanshir, a mining engineer.

During the first part of the Soviet period (1920–1955), Azerbaijan was the main source of hydrocarbons in the former USSR. In searching for new hydrocarbon reserves and their best exploitation, the best scientists of that time were mobilized: M. V. Abramovich, M. M. Aliyev, L. M. Alpin, A. D. Arkhangelsky, Sh. A. Azizbekov, B. K. Babazade, B. V. Baturin, V. V. Fedynsky, T. A. Gasanov, V. A. Gorin, I. M. Gubkin, V. E. Khain, A. Yu. Khalilov, J. M. Khalilov, M. F. Mirchink, A. A. Melikov, A. B. Mamedov, and several other scientists. It is well-known that the victory in World War II was largely due to Baku oil.

In the second part of the Soviet period (1956–1991), a great contribution to Azerbaijan's hydrocarbon industry was introduced by R. A. Abdullayev, R. N. Abdullayev, F. S. Akhmedbeyli, A. A. Alizadeh, Ak. A. Alizadeh, K. A. Alizadeh, V. M. Babazade, T. G. Gadjiyev, G. A. Gamburtsev, A. D. Ismailzadeh, K. M. Kerimov, A. I. Mamedov, M. N. Mamedov, P. Z. Mamedov, A. K. Mirzajanzadeh, M. M. Radjabov, S. G. Salayev, N. I. Shapirovsky, Yu. A. Shikhaliyev, and Kh. B. Yusifzade. Reserves of many onshore hydrocarbon deposits were exhausted, and the direction of the search began to gradually shift to the Caspian Sea.

After Azerbaijan conquered its independence in 1991, International scientific relations were widely expanded, and a new stage of fundamental research, with novel modalities of engineering implementations, is underway. In a short period, the Institute was successfully integrated into world geological science. The process became even more active after the national leader of Azerbaijan, Heydar Aliyev signed in 1994 the "Oil Contract of the Century". Joint investigations with the largest world oil companies and visits to the research centers of these companies have radically altered the nature of international collaboration and the development pace.

Longstanding partnership with leading international oil companies such as "British Petroleum-Statoil", "Exxon", "Shell", "Amoco", "Mobil", "Yunocal", "Texaco", "Philips", "Total", "Elf", "Aqip", and some others (that started in the late 20th century and is successfully continuing at present) can be called a breakthrough in the international scientific space. During that period, more than 50 research projects in geology, paleontology and stratigraphy, oil and gas geochemistry, and geophysics have been prepared. Joint field research, as well as analysis and interpretation of research results that were later conducted at foreign research centers, permitted to adapt to the novel techniques applied in the research of Western scientists, accelerate the mastering of new software, on the one side, and speed up the training of new national scientific staffing, on the other side. The results of the studies are highly significant for predicting the structure and quality of reservoirs in unexplored areas, particularly the deep-water zone of the Southern Caspian Sea.

This book was developed based on the combined studies from the Institute of Geology and Geophysics (the Ministry of Science and Education of Azerbaijan) (M. T. Abasov, E. N. Alikhanov, A. A. Aliyev, G. A. Aliyev, E. G. Aliyeva, A. A. Alizadeh, Ak. A. Alizadeh, M.M. Alizadeh, F. G. Dadashev, A. A. Feizullayev, I. S. Guliyev, D. A. Huseynov, F. A. Kadirov, Yu. M. Kondryushkin, Sh. F. Mekhtiyev, A. A. Yakubov, a.o.) with the collaboration of P. Z. Mamedov (Azerbaijan State Oil and Industry University) and L. V. Eppelbaum (Tel Aviv University and Azerbaijan State Oil and Industry University).

Baku, Azerbaijan Akif Alizadeh
Baku, Azerbaijan Ibrahim Guliyev
Baku, Azerbaijan Parviz Mamedov
Baku, Azerbaijan Elmira Aliyeva
Baku, Azerbaijan Akper Feyzullayev
Baku, Azerbaijan Dadash Huseynov
Tel Aviv, Israel/Baku, Azerbaijan Lev Eppelbaum

Contents

About the Authors

Akif Alizadeh, Doctor of Geological and Mineralogical Sciences, Professor, Full Member of the Azerbaijan National Academy of Sciences (ANAS), Director of the Institute of Geology and Geophysics of ANAS, and Laureate of the State Prize of Azerbaijan for a series of works on paleontology and stratigraphy of Cretaceous deposits. He is one of the prominent geologists and significant science organizers. His scientific interests include regional geology and stratigraphy. Professor Akif A. Alizadeh authorized more than 270 scientific papers and ten monographs.

Ibrahim Guliyev, Doctor of Geological and Mineralogical Sciences, Full Member, Vice-president of the Azerbaijan National Academy of Sciences (ANAS), Editor-in-Chief of the Azerbaijan Oil Industry magazine, and Laureate of the Azerbaijan State Prize in the field of science. His areas of scientific research are geology and geochemistry of oil and gas, mud volcanism, and basin modeling. He is Author of about 220 scientific papers and eleven monographs.

Parviz Mamedov, Doctor of Geological and Mineralogical Sciences, Professor of the Azerbaijan State University of Oil and Industry (Head of the Department of Geophysics in 1991–2016), Full member of the National Academy of Sciences of Azerbaijan (ANAS), and Academician of the International Academy of Sciences (Austria). His areas of scientific research are seismic exploration, seismic stratigraphy, analysis of logging facies, sequence stratigraphy, integrated interpretation of geophysical data, and structural and formation analysis. He is Author of more than 240 publications and Author (Co-author) of six monographs, one atlas, eight textbooks, and ten teaching aids.

Elmira Aliyeva, Doctor of Geological and Mineralogical Sciences, Corresponding Member of the National Academy of Sciences of Azerbaijan (ANAS), and Laureate of the State Prize of the Azerbaijan Republic in the field of science. Her areas of interest are sedimentology of modern and ancient clastic sedimentary systems, paleogeography of the Meso-Cenozoic Caspian region, sequence stratigraphy, architecture of sedimentary strata, quality of hydrocarbon reservoirs, and analysis of petroleum systems.

Akper Feyzullayev, Doctor of Geological and Mineralogical Sciences, Professor, Full Member of the National Academy of Sciences of Azerbaijan (ANAS), Head of the Department of Geochemistry and Fluid Dynamics of Sedimentary Basins of the Institute of Geology and Geophysics of ANAS, and Laureate of the State Prize of Azerbaijan in science. He has 45 years of experience in the field of geology and geochemistry of oil and gas, mud volcanism, fluid dynamics, geochemical methods for searching for oil and gas fields, as well as geoecology.

Dadash Huseynov, Doctor of Geological and Mineralogical Sciences, Corresponding Member of the National Academy of Sciences of Azerbaijan (ANAS), Deputy Director and Head of the Department of Basin Modeling and Geotechnologies of the Institute of Geology and Geophysics of ANAS. He is Expert in geology and geochemistry of oil and gas, remote sensing methods of the Earth, and basin modeling. He is Author of about 150 scientific papers and five monographs.

Lev Eppelbaum, Research Professor (since 2005) at the Department of Geophysics, Tel Aviv University, and Honorary Professor (since 2020) at the Azerbaijan State Oil and Industry University. He received several scientific awards, including the Christian Huygens Medal of the European Geosciences Union (2019). L. Eppelbaum is Foreign Member of the Georgian Academy of Sciences. He is Author of about 440 publications, including 11 books, 190 articles, and about 80 proceedings. His scientific interests cover potential and quasi-potential geophysical field analysis, integrated interpretation of geological-geophysical data, nonlinear geophysics, and paleomagnetic, tectonic, and geodynamic reconstructions. Prof. Lev Eppelbaum is Editor-in-Chief of the International scientific journal "Geology, Geophysics, and Earth Sciences".

1 Productive Series of the South-Caspian Basin. The Modern Architecture of the South Caspian Megabasin Derived from Seismic and Other Methods

The South Caspian Megabasin (SCMB) is a significant tectonic element of the Earth's crust and a highly potential sedimentary basin within the central segment of the Alpine-Himalayan mobile belt (AHMB). The SCMB appears to have a distinct tectonic position and includes the most deep-seated depression of the South Caspian Basin (SCB) and its troughs: the Kur Trough (KT) from the west and the West-Turkmenian trough (WTT) from the east (Fig. 1.1). The two troughs being opened and widened appear to coalesce with the SCB. The megabasin trends in a sublatitudinal direction and is framed by the Greater Caucasus, Kopet-dag, Lesser Caucasus, Talysh, Elburs, and Aladag-Benaluda mountain structures. The SCMB modern structure has been formed due to the active tectonic movements and intensive sedimentation processes dominated during separate evolutional stages. All the SCB, KT, and WTT sections are distinguished by the presence of many types of structural elements and sedimentary complexes being different by their scales and preservation degree as well as by individual geometrical and morphological features from those developed within the other basins of the mobile belt and neighboring platform region.

High rates of subsidence and sedimentation characterize the SCB. During the Alpine cycle (~ 180 Ma), the total basin subsidence reaches 25–30 km. As a result, a vast sedimentary series has been accumulated equal to the thickness of the continental crust of the Scythian-Turanian platform in the Middle Caspian. It is significant that the shelf accumulation areal reduction and decrease of the basin sizes have taken place in the Cenozoic. This process was particularly marked in the Pliocene–Pleistocene time. The cause of the mentioned areal reduction is that the basin is located within the tectonic belt of compression, which has been influenced by the extreme crustal shortening in the Late Cenozoic.

1.1 The Notions of the Crustal Structure and SCMB Formation

Traditionally, the SCMB has been a striking representative of geosynclinal-folded sedimentary basins (SB). Several explanations have been advanced for the presence of the thinned continental crust under the SCB. The presence of old structures, massifs, and complexes (such as some Paleozoic structures of the Greater Caucasian major ridge; Dzirulian, Lockian, and Khramian massifs in the Transcaucasian and Nakhchivan block in the south), as well as Talysh-Vandam (on land) and South Caspian marine basement noses, being revealed by the gravity anomalies, appear to be grounds for fixists to classify all the South Caspian down warping region as the median mass.

The SCB has been considered in the modern structural plan as a Neogene-Anthropogenic depression superimposed on the mentioned

A. Alizadeh et al., *Pliocene Hydrocarbon Sedimentary Series of Azerbaijan*,
Advances in Oil and Gas Exploration & Production, https://doi.org/10.1007/978-3-031-50438-9_1

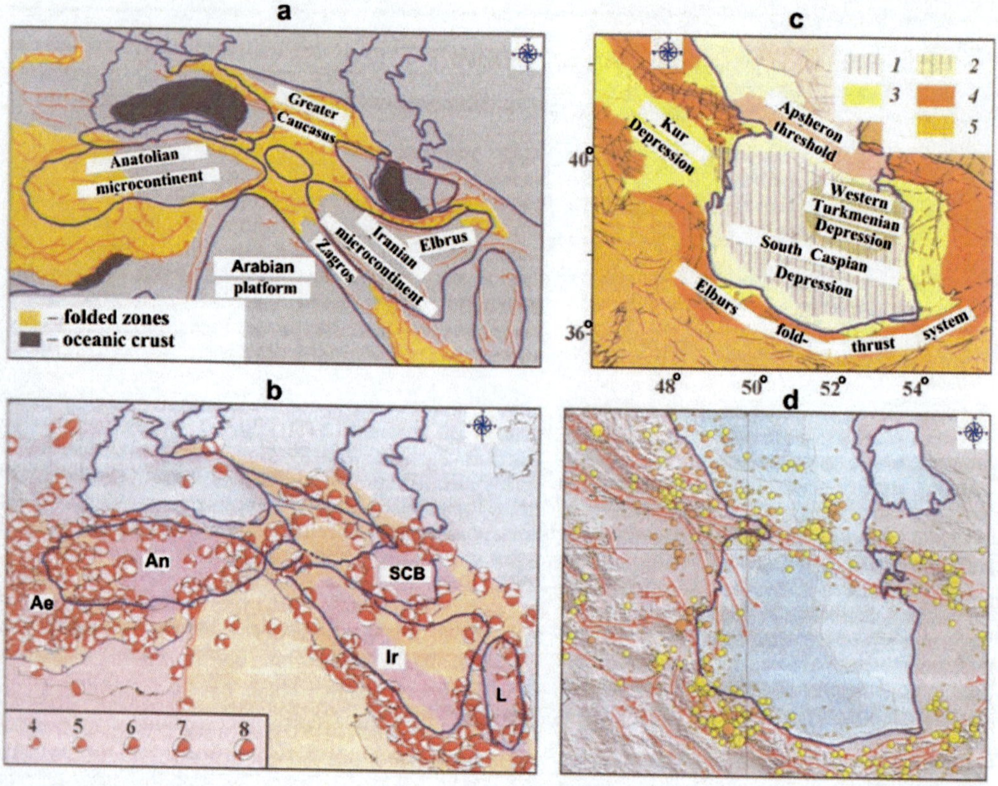

Fig. 1.1 **a** Tectonic plan of the AHMB central segment; **b** focal mechanisms of the earthquakes with their magnitudes; **c** tectonic map of the SCMB (after Khain, 2001); **d** earthquake epicenters in the SCMB

median mass. Although all the geophysical prospecting data have confirmed that the consolidated crust under the basin is of "non-granitic" (oceanic) type, the fixists assert up to now, without any proof and arguments, that it is of continental type of Paleozoic or even pre-Cambrian consolidation; "granitic" layer is supposed to be partly reworked and metamorphosed whereas "basaltic" layer became turned into eclogite. Meanwhile, many petrologists have been quite sure that the continental crust could never be turned into oceanic crust by disappearing the "granitic" layer. This contradicts the law of thermodynamics and may only be an inverse process (Uyeda, 1978).

Unfortunately, the fixist views on the formation (not only for the SCMB but also for all the mobile belts) have long been reflected in the minds of many geologists. An arbitrary and erroneous interpretation of the factual geological-geophysical, petrochemical, and paleomagnetic data led to the subjective and inconsistent tectonic notions of the SCMB formation and evolution. An interpretation of new geological-geophysical data and analyses of core samples obtained from the deep and superdeep drilling, as well as subsequent detailed reconstruction of the fixist's ideas related to the SCMB origin and evolution, leads us to believe that those views appeared to be the cause of an incorrect interpretation of the SCB evolution, its Kur and West-Turkmenian centroclines as well as bordered mountain structures. Most likely, all the mentioned misconceptions occurred because of essential factors such as horizontal transcurrent movements and rifting in the Late Mesozoic. The role of compressive tectonic movements in Miocene-Quaternary time has yet to be considered.

What is more, on the interpretation of natural features of the megabasin large structures and their morphological elements, it appears that among the factors which have been omitted are published data of geosynclines modern analogous such as active continental margins, their island arcs, backarcs, interares and so on. Many geologists have accepted the Lesser Caucasian ophiolitic sheets as the oceanic lithosphere remnants, determining the island-arched origin of the thick basaltic-andesite-rhyolitic series within the Kur depression and Talysh. Seismic surveys of the last decades revealed old extension structures and passive elements along the Scythian-Turanian plate margin as well as taking into consideration the new geophysical data led to the transformation of the existing views on regional tectonics and formation of the region's major geostructures (Alizadeh et al., 2016, 2017).

The investigators of the region understood that the SCMB formation and development history should be considered jointly with an evolution of the AHMB, whose formation process is closely connected with a complex history of the paleo-Tethys and neo-Tethys (or meso-Tethys) oceans. From the results of geophysical, structural, palaeogeographical, paleofacies, paleomagnetic, and petrological investigations carried out within this region, as well as from the detailed hydrodynamic reconstructions, it is inferred that in the geological past, the assemblage of structures and geological bodies typical to the active oceanic margins has been developed in the central segment of the AHMB, that is the Lesser Caucasian branch of the Meso-Tethys Ocean (which relics are ophiolites), volcanogenic island arc, back arched (marginal) Greater Caucasian Sea and passive elements on the Eurasian continental margin. All the mentioned structures' formation took place in the Meso-Tethys oceanic space where the Afro-Arabian and Eurasian platforms have interacted with the Anatolian, Iranian, South Caucasian, Nakhchivanian, Makerian microcontinents located between the mentioned platforms (Fig. 1.1).

In a few of the works published in the period of 1970–1980, the problems of the formation and evolution of deep-sea basins and intermountain troughs and bordered those mountain structures have been discussed from the standpoint of lithospheric plate tectonics (LPT). The South Caspian and the Black Sea deep basins have been the relics of the Meso-Tethys Ocean's Greater Caucasian branches on the active margin of the same ocean (Adamya et al., 1974; Mamedov, 2004; Zonenshain et al., 1990) or the relics of the Tethys Ocean (Dercourt et al., 1986; Nadirov et al., 1997). Other authors have brought up a proposal that the oceanic crust could have been formed in the "pull-apart" type structure originated after a slip (fault) in the Caucasian region during the Cretaceous time (Lomize and Khain, 1995; Sengor, 1984).

A large group of scientists has analyzed geological-geophysical data to substantiate the formation and evolution of geostructures in the Black Sea-Caspian region from the standpoint of the LPT. Numerous geodynamic, plate-tectonic, and palinspastic reconstructions, tectonic-structural maps, geological and seismogeological sections, and models have also been constructed. Besides, scientists have often lacked remarkable and goal-oriented investigation data in regional plans (such as necessary geochronological, petrological, isotopic, geomagnetic, geochemical, and geophysical investigations data). Therefore, it used only fragmental data obtained in the regions widely separated from each other and the data of single outcrops and core samples from the deep wells. Then, all these kinds of information have been correlated and synchronized. Some conclusions concerning the geostructure formation in marine conditions have been based on indirect appreciations of the materials obtained from the surrounding land or unambiguously interpreting data of the potential geophysical fields (e.g., Eppelbaum, 2019).

Naturally, it seems likely that many constructions and reconstructions have been considered too schematic, subjective, or even hypothetical. Such the final reconstruction vulnerability required an obligatory availability of the direct factual data, first of the type and parameters of consolidated crust under the SCMB and of the Moho discontinuity relief and of the erstwhile processes of the crust tension and subsidence, and the sedimentary basin expansion at the divergent stage as well as of subduction and collision processes at the convergent stage. That's why the mobilistic interpretation of the SCMB formation and evolution history needs confirmation by the proper, reliable, and objective factual data of structures typical to the marginal-marine paleo-basin. Information on the old buried continental slopes, known as direct indicators of the convergence of two different types of plates, is also required. The same is related to the sutural zone structure and mode of the lithospheric block collision. It is also essential in the information on the regional surfaces' discontinuity and the scale of erosion.

Thus, an eclectic conception of the TLP in the Caucasian-South Caspian region should be turned to the quest for the proofs, i.e., to seek out typical structures and morphoelements of the back arched sea, such as reefal tension and subsidence structures, relics of an oceanic crust, island areas, and volcanoes, passive elements, etc., or their buried fragments well preserved in the recent section as well as revealing of regional surfaces of unconformity and breaks related to the major stages of the basin evolution.

Besides, it seems most likely important to select factual and reliable information from the independent sources of different types, which correlation would have been explained within the framework of the mobilistic model foreseeing the presence of the marginal sea in the Late Mesozoic and Paleogene that turned out to be an intermountain basin with a relict deep-sea trough in the Late Cenozoic.

We believe that only combined employment of structural-geophysical, lithological-formation special features, geophysical, and petrophysical parameters and indications typical to the paleobasins of different genetic and tectonic types will make it possible to develop an actual tectonic-sedimentation model of the SCMB. It enabled a decrease of the quota of hypothetic elements in a notion of its formation mechanism and substantively interpreted the basin development history.

1.2 Geophysical Information of SCMB Lithosphere and Its Main Geodynamic Boundaries

One of the most critical problems of SCB geology is elucidating its earth crust type and structure. Traditionally, the two structural stages, genetically and substantively heterogeneous, have been distinguished in the basin earth's crust, namely: sedimentary cover (SC) and consolidated crust (CC), often being named as metamorphic (substratum) or crystalline basement (foundation), respectively. In recent years, some attempts have been made to single out one more stage, namely the subcrustal mantle down to the asthenosphere, which, together with SC and CC, makes up the lithosphere of this region. It is known that essential boundaries of the basin earth crust are the roof and foot of the consolidated crust, i.e., basement surface and Moho discontinuity. The importance of their revealing and tracing is caused by the fact that both are geodynamic boundaries.

Moreover, their position in the geological section indicates the time of the beginning and end of geological processes within which the CC has been formed. The Moho discontinuity indicates the time of the beginning and end of geological processes within which the CC has been developed because of the Greater Caucasus Marinal Sea (GCMS) rifting and opening in the Middle Jurassic time. A studied basin is an erosion remnant of the GCMS.

The basement surface's stratigraphic level corresponds to the completion time of the active magmatic, metamorphic, and tectonic processes within the basin, which led to the formation of a consolidated oceanic crust (Volozh & Leonov, 2004).

The other essential parameters of consolidated crust are its genetic type, thickness, and structure with which amplitude and tectonic down warping, sedimentary cover thickness, and sedimentation rate are connected. The deep fluid rising zones where an organic material was warming up and hydrocarbon "preparation" occurred. Such zones are concentrated, as a rule, within geodynamically strained, highly deformed, and warmed-up tension and shattering areas of consolidated crust (e.g., the regions of spreading, rifting, and plate subduction and obduction) (e.g., Kucheruk, 1990; Perrodon, 1985). Detailed studies of the mentioned parameters are required for further discoveries of potential earth degassing.

For a long time, the information source of the consolidated basin crust has been the deep seismic sounding data (DSS) (Aksenovich et al., 1962). Most of the 11 profiles have been worked out in the transitional zone between the Middle and South Caspian and within the Near Absheronian water area. Only two profiles (DSS-1 and DSS-9) crossed the middle South Caspian Basin (SCB) to the Iranian waters. As was established by the meeting travel time curves of the refracted waves, the cutoff velocities in the CC roof and foot are very high (6.4–6.7 km/s and 7.2–7.7 km/s, correspondingly), which is typical of the basaltic layer.

The Moho position has been established in separate profiles fragmentarily only. In the western SCB, the CC turned out to be thin (from 8 to 10 km), subsiding at great depths (lower than 22 km). The thickness of the SCB sedimentary cover appeared to be comparable with that of the adjacent Scythian-Turonian platform.

These data led us to believe that the SCB crust is characterized by anomalous structure. The first investigators of the DSS data (e.g., Aksenovich et al., 1962; Kosminskaya, 1968) concluded that the SCB crust is of (sub)-oceanic type and therefore, the South Caspian has been included in the tectonic map of the South of the USSR (1975) as a region of the absence of the crust' "granitic" layer. The conclusion of the subvolcanic type of CC in SCB has been of great scientific importance. It was rightly valued and accepted by most scientists studying this region's deep structure and evolution. In a few of the works published during 1970–1980, the deep-sea basins of the South Caspian and the Black Sea have been the relics of the GCBM-TO on the Meso-Tethys active margin (e.g., Adamiya et al., 1974; Gamkrelidze, 1984; Zonenshain et al., 1990), or the relics of Early Mesozoic Tethys Ocean (Amursky et al., 1968). The formation and evolution of the basin and its bordered mountain structures in the central part of the AHMB have been interpreted from the standpoint of lithospheric plate tectonics.

From the results of the DSS carried out in 1980, it was inferred that not all the data of those works had been derived and analyzed. That's why the data of profiles 1–2 and 9 have been reinterpreted with the employment of computational techniques, and new detailed and revised models have been constructed (Baranova et al., 1990). As a result, it was concluded that the Moho forms a mega-arch under a basin at 30–34 km. Besides, the CC within the largest part of the basin seems to be thin (Fig. 1.2). Its upper part is characterized by the cutoff velocity values (Vp = 6.5–6.8 km/s), typical to the oceanic crust (with the mafic composition of the rocks). At the CC lowermost strata, the mentioned values reach 6.8–7.9 km/s, which are typical of the paragenesis of basic and ultrabasic metamorphic rocks. The cutoff velocity along the Moho surface is high, reaching 8.0 km/s. CC in the eastern South Caspian appears to somewhat thicken (up to 12–14 km) at the expense of the basement raising to the day surface, where the upper part of the consolidated crust is characterized by the cutoff velocities estimated as Vp = 5.9–6.2 km/s. Important information on the high-speed model of SCB sedimentary cover has also been obtained. It is established that the sedimentary basin cover is characterized by the anomalous low average velocities of the longitudinal waves: $Vp < 4.8$ km/s, while within many other basins, this value is higher by the depths of 5.0–7.0 km being estimated as 5.5 km/s. It suggests a considerable porosity of the sedimentary cover.

Likely, the models with four boundaries are undoubtedly a gross idealization of SCB heterogeneous lithosphere. Moreover, because of the sparse network of the DSS profiles, the precise character of the obtained seismic information (through the hodograph), and the low resolving ability of this method at great depths, the mentioned models are incomplete and problematical.

The gravitational field special procedures have also been used to analyze the nature of the SCB

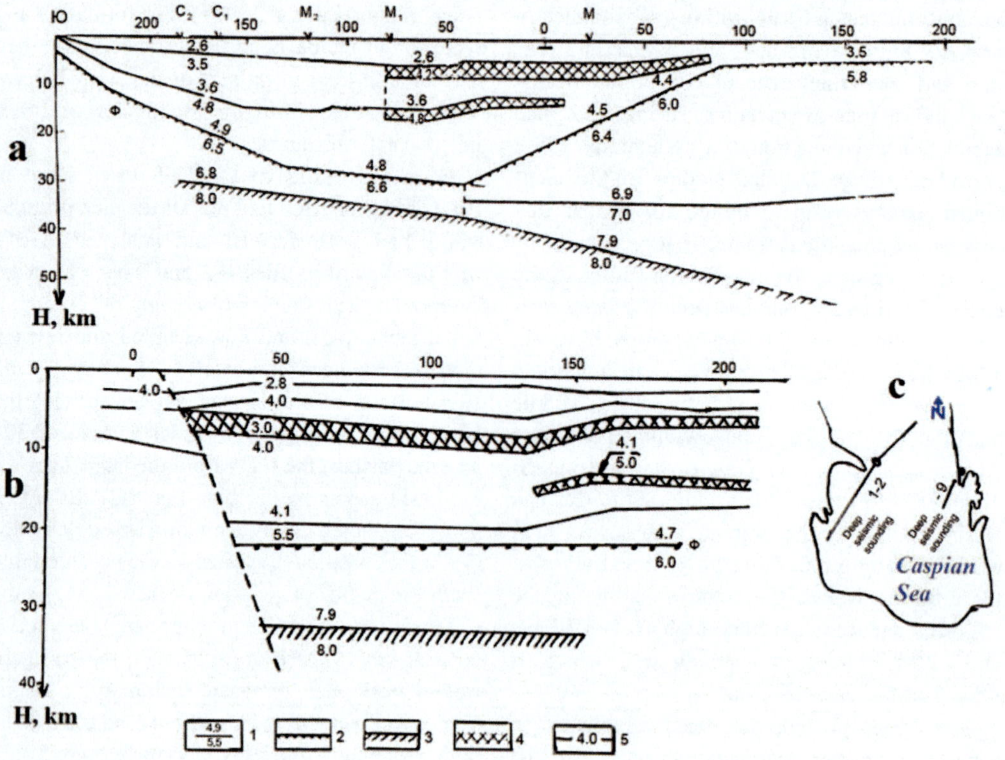

Fig. 1.2 High-speed models of the DSS profiles 1–2 and 9. (1) first-order boundary with the corresponding velocity jump, (2) basement surface; (3) Moho discontinuity, (4) high-speed series, (5) velocity isolines

lithosphere lower layers (Fig. 1.3). The marine and satellite gravity observations over the SCB and adjacent land (allowing for an altitude correction) show a sharp decrease in a gravitational field from Elburs to the Caspian Southern coast. The gravitational field in the western SCB is characterized by weakly manifested negative anomalies –(10/15 mGal). A significant negative linear anomaly (up to − 130 mGal) has been observed in the basin conjugation zone with a platform marked by comparative shallowness (from 20 to 100 m) and gently rolling bottom topography. Such an anomaly could not have been unexpected within the western and central SCB, where a thin and rigid CC lies under the water series (about 1 km) and thick (25–27 km) sedimentary cover. However, the mentioned anomaly is very weak, being close to the normal one.

The compact mass action in the upper mantle area has compensated for a significant negative anomaly. This action supports an isostatic condition in the "crust-mantle" contact zone, providing unusual crust stability in the SCB central part. The gravity gradient and a significant negative anomaly observed in the transitional zone relate to the mass deficit. The gravimetric data indicate that not only CC but also the top of the mantle is located much deeper than is required to isostasy. Therefore, it seems likely that some mechanism is drawing CC down into the mantle against Archimedes' principle. According to the seismic survey data, such a mechanism is likely CC subduction.

In the SCB density model of lithosphere developed by Granath et al. (2000) through the meridional profile, the negative anomaly is fixed under the CC subduction zone where the Moho discontinuity has deepened down to 60–70 km, causing a deficit of the dense mantle matter. The same negative anomalies are typical of the

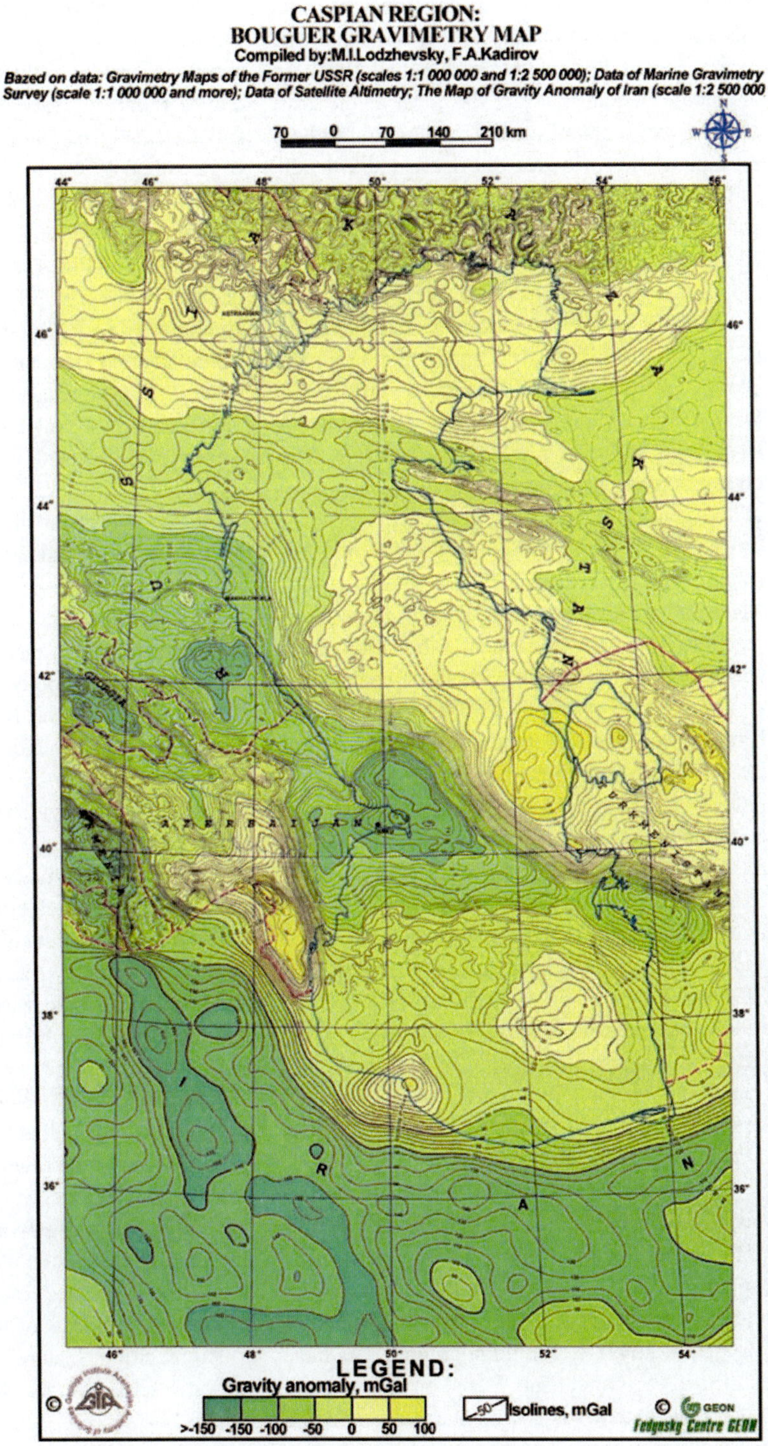

Fig. 1.3 Bouguer gravity map of the Caspian Sea and surrounding areas (in mGal) (compiled by M. I. Lodzhievsky and F. A. Kadirov)

trenches that occurred within the zones of oceanic crust subduction at the active margins of the recent oceans and in the backarched seas. An alternative model made in the form of the crust upthrow and underthrusted having a northern fault-plane dip provides the dense mass deficit. Still, it would have resulted in a relatively lesser negative anomaly. However, there are such additional factors in favor of the subduction model as the presence of a highly deformed sedimentary wedge (accretionary prism) over the CC within the region of its subduction at a great depth and high regional seismicity with deeply located hypocenters (from 60 to 75 km) which have been recorded north of the AHMB. This region is also characterized by increased values of the geoidal height field gradients, which are usually observed in the contact zones of lithospheric blocks different in their density. According to Uyeda (1978), Rodkin (1993), and other investigators, the recent subduction zones are marked by the same gradient anomalies.

The oceanic type of CC has also been confirmed by seismic tomography (Yakobson, 2000), which employs the Earth's crust transference with Rayleigh waves reflected from the earthquakes. The tomographic model represents three layers in sedimentary cover, the layer being related to the basement, i.e., CC (Fig. 1.5). The tomographic model represents three layers in sedimentary cover related to the basement, i.e., CC (Fig. 1.5).

The basement is mainly characterized by the "basaltic" velocities ($Vs > 3.9$–4.5 km/s) of transversal waves. Two zones with $Vs \geq 4.6$ km/s typical to the mantle matter have been revealed in the southern SCB.

The results of the DSS and seismological investigations led to a conclusion of the presence in the eastern basin of an elevated block of CC (up to 12–14 km); in which the upper part is marked by the spots with "cutoff" velocities ($Vs \leq 3.4$ km/s). Seismotomographic data have confirmed this conclusion. Based on these data, it is concluded that the sedimentary cover in the South Caspian more significant part is underlain by the "basaltic" substratum, which in places has burst by mantle matter.

The lithosphere model of the South Caspian region is also known, drawn up by latitudinal profile according to the functional analysis of the waves reflected from the earthquakes (receiver function analysis) (Jackson et al., 2002; Mangino & Priestley, 1998). In general outline, this model has 2–3 boundaries corresponding to our model constructed according to the superdeep seismometric data (Fig. 1.6), needing to be more detailed in the interpretation of the Earth's crust structure. The RFA model confirms the following conclusions: (1) greater thickness (more than 25 km) of sedimentary cover, (2) the presence of thin (about 6–8 km) consolidated crust, (3) the presence of elevated, 10–15 km thick, block of CC of the transitional type in the eastern basin.

More data is needed to determine geophysical parameters and substantial composition of the lithospheric lower level (sub-crustal mantle). Some data have been obtained from the DSS and seismotomographic methods.

The cutoff velocity of the refracted wave along (or below) the Moho defines elastic features of the upper mantle. Besides, the mentioned velocity appears to be changeably ranging from 7.9 to 8.3 km/s, evidence of the heterogeneous composition of the sub-crustal mantle substratum under the SCB. Some investigators believe that the highest cutoff velocities under CC are related to ultrabasic rocks such as peridotite, pyroxene, pyrolite, eclogite, etc. However, all these attempts to attribute various ultrabasic rocks to the upper mantle substratum seem most likely to be debatable since experimentally estimated longitudinal wave velocity in the mentioned rocks was not higher than 7.2 km/s (Kunin, 1989). At the same time, according to TRW and RFA analyses data, it is distinguished that a waveguide, i.e., a low-velocity layer ($Vs = 2.4$–2.9 km/s), is identified with the asthenosphere. It is mostly expressed in the Baku Archipelago region and southeast of this area.

An exciting feature is that relatively low-velocity values in the Jurassic-Cretaceous and Paleogene-Miocene sedimentary complexes are primarily confined to this waveguide rising zone, whose characteristic feature is that most submarine mud volcanoes are located here. Within

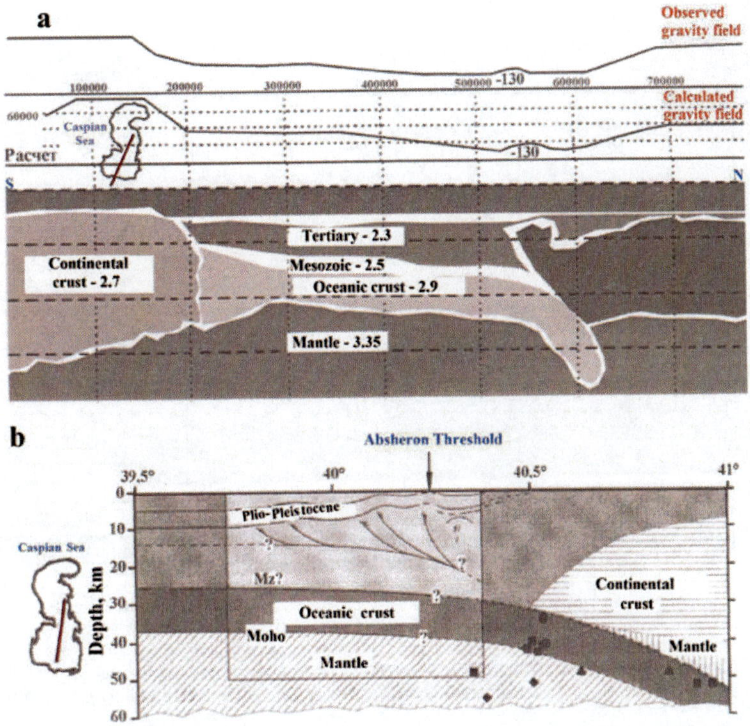

Fig. 1.4 The South Caspian lithosphere models constructed: **a** according to the satellite gravimetric data (Granath et al., 2000); **b** according to the superdeep seismic survey and seismogeological data (Knapp et al., 2004)

this Caspian area, the wave velocities' positive correlation and heat flow and estimated temperature values have also been established on the Earth's crust floor of the upper mantle (Rodkin, 1993). An interpretation of CC northwardly subsidence, taking into consideration the data of strong earthquake (M > 5) centers, is given using the subduction model (Fig. 1.4b) presented in Knapp et al. (2004). From the study of seismic wave energy, it is inferred that they are passing through the SCB crust. Those waves appeared highly absorbed (Knapp et al., 2004). Sharp seismic wave attenuation on the CC top level is fully attenuated below the Moho surface.

An anomalous wave absorption in the SCB crust may relate to increased content of the water fluid, which source is possibly a waveguide (i.e., asthenospheric layer) in the upper mantle. Because the western SCB mud volcanoes are mainly confined to the waveguide rising area, it is supposed that the fluids from the waveguide are partly squeezing—up the overlying beds of the Earth's crust. However, this supposition has not yet been supported by the results of the mud volcanic breccia analyses. Suppose the crystalline basement rocks inclusions would have been found in the breccia composition (apart from the remains of Cretaceous and Paleogene rocks). In that case, their finding might serve as a decisive argument in favor of the mentioned supposition. Such components in mud volcanic breccia composition through their "roots"—subvertical channels reach to the basement, and deeper levels are well seen in seismic sections under some mud volcanoes in southern Caspian (Mamedov, 1991; Mamedov & Guliyev, 2003) have not yet been found.

Up to 1990, the high-informative superdeep seismometric method was applied only in structural prospecting with a 6 to 8-s recording sweep in the Caspian Sea. The problems of SCB's deeply subsided area have been interpreted mainly in terms of the structure of Neogene-Quaternary sedimentary complexes down to 9–12 km. The absence of any information on more ancient sedimentary complexes and CC has

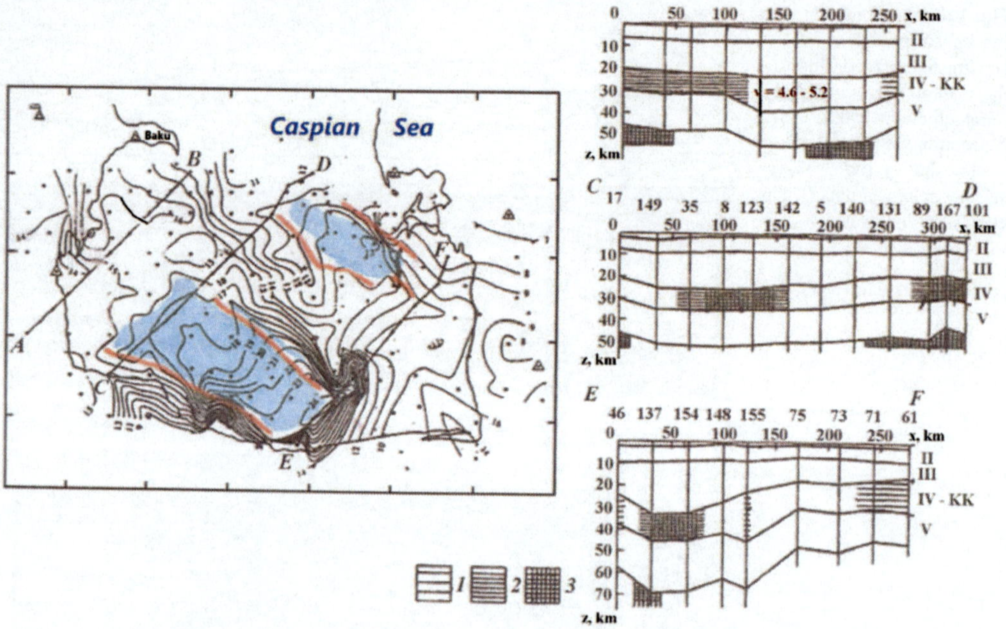

Fig. 1.5 Isopach map of the third sedimentary layer and the sections along profiles through the high-speed columns (after Yacobson, 1997). Transversal wave velocities: (1) Vs $\leq$ 3.8 km/s, (2) Vs > 4.6 < 5.0 km/s, (3) Vs > 5.0 km/s

puzzled many authors and led to an invention because of the deep basin structure as well as hypothetical conclusions of the age, type, and structure of its basement, thickness, stratigraphic range, and composition of the section "invisible" part. The regional sublatitudinal profile Byandovan-Okarem (B-O), about 300 km long, has been tried out by the deep seismic sounding method with 12-s scanning. The profile section has thrown light upon the Earth's crust structure down to 22 km. In this seismic section, we first saw basement protrusion in the eastern basin at a depth of 16–22 km, of which the wave field boundaries have stood out. Based on the section of the B-O regional profile and the sections connecting superdeep seismometric profiles tried out in the KT and WII using the DSS data, the author has compiled a composite seismostratigraphic section extended for about 560 km (Fig. 1.6a). This section, as the earth crust model of the South Caspian down-warping region, is widely used by oil companies and research centers to solve several problems of SCB geology and the presence of oil and gas (Brunet et al., 2003; Guliyev et al., 2003; Lerche et al., 1997; Tagiyev et al., 1997).

The second half of the 90th of the XX century has been marked by the expansion of technical and methodical abilities of the seismic survey by DSS methods used on land and sea. Such new methods of wave excitation, digital recording, detailed processing, high resolving ability, great deepness, and the latest 2D and 3D image methods made it possible to recognize comprehensively the Earth's crust's deeper layers. In the Southern and Middle Caspian, the regional seismic profiles have been worked out by the "Caspmorneftgeophysrazvedka" Assoc. and "Caspian Geophysical" joint enterprise using the DSS method with 16 to 20-s time-base recording.

Through the network of regional profiles, the DSS sections led to the volumetric examination of the SCB structure (Mamedov, 2006, 2008). It was obtained by special geological sections representing direct information about the interior part of the Earth up to 40–50 km deep. These

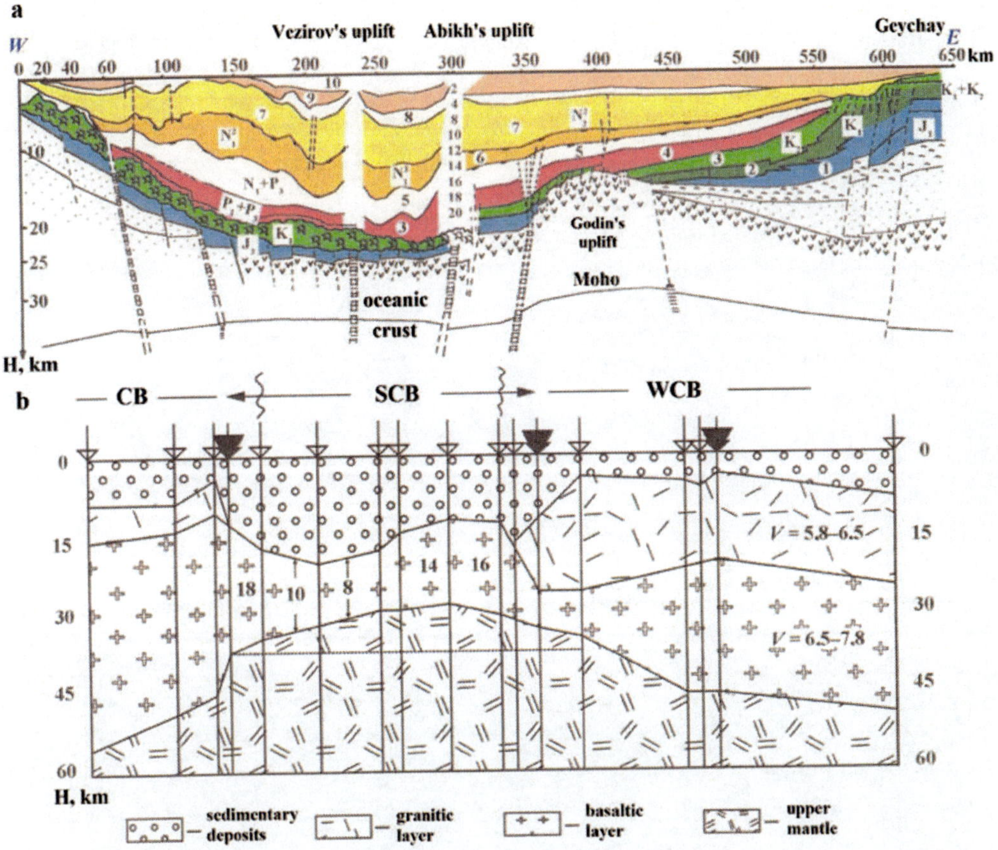

Fig. 1.6 **a** Regional seismostratigraphic section through the sublatitudinal geotraverse (Mamedov, 1991, 2006), **b** Lithosphere sequence compiled by the RFA method (Mangino & Priestley, 1998)

sections reflect large regional topographic forms, angular and stratigraphic discontinuity surfaces, stamp, and inversion (rootless) structural forms, faults, and ruptures. An interpretation of the regional profile data leads us to believe that the basement surface and the Moho (of CC roof and foot) seem to be traced by the wave field special features and, in places, using the reflection group correlation. The mantle-basement surface (at 23–28 km deep) is well distinguished at a low-frequency filtration (4–8–12–16 Hz) as a high amplitude section of the intermittent subparallel reflections (Fig. 1.7).

Below the 14–16 s (from 34 to 40 km), a crowding band of very weak low-frequency and intermittent axes of the phase coincidence is being well seen. This part of the section is known as the zone of "reflectivity" (Pavlenkova, 1996), by which the lower edge of the Moho discontinuity may have been traced.

From the sections of DSS profiles, it is inferred that the thickness of CC in the western basin is between 6 and 8 km and taking into consideration high seismic velocities in the crust (Vp > 6.5–7.8 km/s), which is typically exceptionally to the magmatic (basaltic) rocks, it becomes even more apparent that this crust is of oceanic type. The notable feature of CC is that it comprises separate "turbid" zones being changed by relatively transparent ones. Such seismic contrast range reflects the heterogeneous structure of CC, which is apparently connected with its block structure and substantial variegated composition. The fact that the mentioned structural data is indistinctly

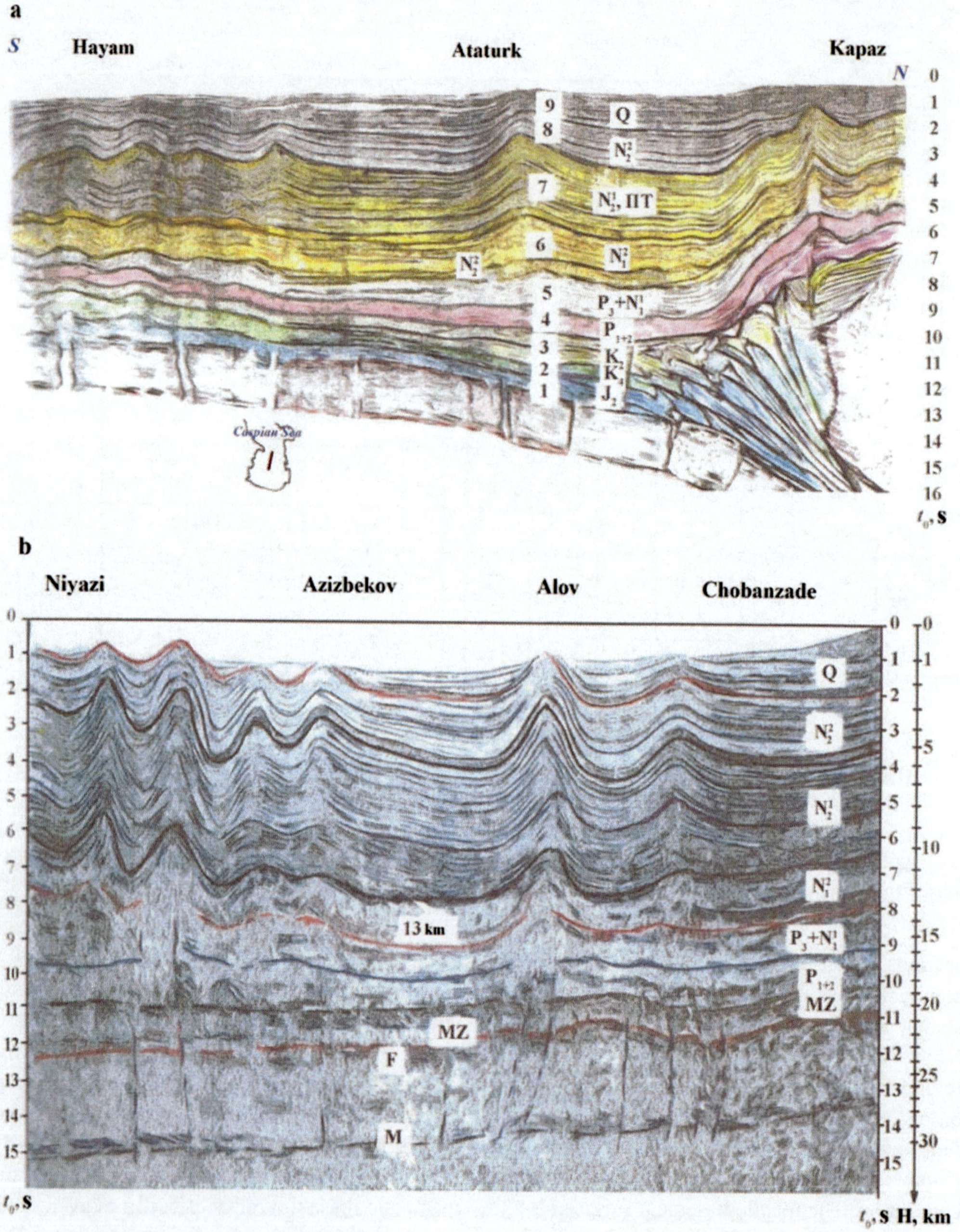

Fig. 1.7 The superdeep seismic time section along the meridional and sublatitudinal profiles displays the sedimentary cover's CC and sedimentation complexes

displayed in the wave field is caused by the relationship between their sites and the Fresnel disk sizes. From our calculations, it is concluded that in the SCB lithosphere, below 25 km, the heterogeneities of CC having a diameter less than 5–6 km and height of 1–2 km may not have been displayed on the reflected wavefield, while the heterogeneities which sites exceed the mentioned figures may have resulted in the "shaded" field of reflected waves.

Within the Absheron sill, seismic sections of submeridional profiles show a flexure and northwardly subduction of CC. Over the CC, it is marked by an unwarping of a highly deformed basal (Mesozoic-Paleogene) series of sedimentary cover (Fig. 1.7a). The fold-thrust, imbricate structure of sedimentary formations is a characteristic feature of the accretionary prisms over the submerging oceanic plates and marginal seas. An accretionary prism within the Absheron sill has been formed because of the rock scraping from the CC surface. The prism consists of post-rift (Cretaceous-Paleogenic) and syn-rift (from the Middle to Late Jurassic) rocks. The prism is uplifted and gently moved upon the Scythian-Turanian platform frontal region. The seismic time section (Fig. 1.7a) has been transformed into a deep section where CC occurs in the center of the SCB at a 23–25 km depth and smoothly subsided in a northerly direction at a depth of 30–32 km within an accretionary prism.

Taking as a basis a deep section (Fig. 1.7b) and using the DSS data from the Middle Caspian as well as the data of the Central Elburs tectonics (Nazari, 2006), the authors have compiled the earth crust regional section through the 500 km long submeridional geotraverse (Fig. 1.8). This section displays the deeply subsided structure with many kilometers of sedimentary covers and its joining with the continental blocks from the south and north. The SCB sedimentary cover is in contact with the metamorphic substratum of epi-Hercynian STP from the north and with block masses of the Iranian microcontinent of Riphean and pre-Devonian consolidation—from the south. The CC curve and subsidence within the Absheron sill have also been displayed on the models based on seismic data (Green et al., 2009; Knapp et al., 2004). From these models, it is inferred that the northward subsidence of the CC, considering the data of strong earthquake (M > 5) centers, is interpreted in the framework of the subduction model. As described by Allen et al. (2002) and Mamedov (2004), the basin's rapid down-warping is related to the regeneration of the CC subduction in the Late Miocene.

As mentioned above (Fig. 1.4a), CC subduction has also been reflected in the Earth's crust density model constructed according to the satellite data (Granath et al., 2000). A sizeable negative anomaly has been marked in the transitional zone close to the platform region. Here, the Moho appears to have deepened to 60–70 km, leading to a deficit of the mantle matter. Similar negative anomalies are typical to the trenches observed in the subduction zones of the oceanic crust at active margins of recent oceans and in the backarched seas.

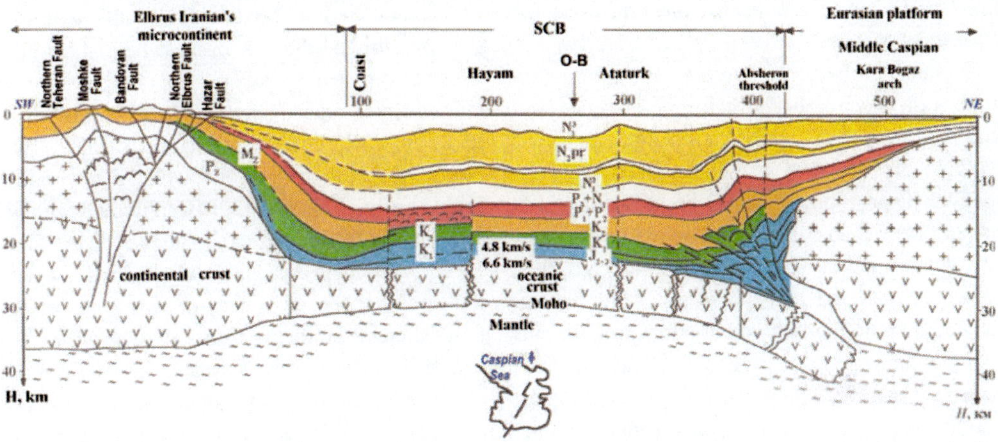

Fig. 1.8 Earth's crust regional section through the submeridional traverse along Central Elburs-SCB-Middle Caspian (after P. Z. Mamedov)

Such additional factors as the presence of a chain of positive magnetic anomalies in the transitional zone and high heat flow (even if fragmentarily), as well as regional high seismicity with the deep hypocenters (up to 60–70 km) in the northern AP region (Fig. 1.9), are agreed with the subduction model.

Detailed investigation of local mechanisms of earthquakes and deep epicenters is essential information on the CC subduction and the Benioff seismic zone in the transitional zone from the SCB to the platform. The SCB is surrounded from the west, south, and east by active seismic belts characterized by near-surface earthquakes with M > 5 at depths of 30 km. In comparison, the northern margin of the basin is remarkable for stronger and deep-focus earthquakes at depths from 45 to 80 km. The underthrusting of its oceanic crust causes these earthquakes in the SCB northern margin under a platform. At the same time, small-focus quakes manifested in its frame are considered a factor evidenced by the continuing convergence and collision of the plates.

Most deep-focus earthquakes have occurred near the folded zone of the Absheron sill at 40–80 km. The length of the CC subduction part, considering its curve, should be at least 80 km. A basin compression stage and an intake of its CC began at the end of the Oligocene and continued up to the Pliocene. The basal series of the sedimentary cover has been subjected to scraping from the CC surface and subsequent thrusting for almost 30 Ma.

A sharp basin deepening over the subduction zone took place between Miocene-Pliocene and was followed by a short dampening phase of tectonic activity that is evidenced by the subparallel disposition of the PS boundaries, which has not been disturbed even after their squeezing and the formation of anticlines in the Late Pliocene. An extremal lithosphere squeezing caused the regeneration of the subduction process and regional transverse compression of all the Neogene series. The presence of the Maikopian plastic clays under the mentioned series is a crucial factor of structurally localized folded zones, mainly the Absheron-Near Balkhan folded zone.

The linear folded zones traced over the subduction zone have mainly been formed because of mud volcanic activity and plastic mass injection in their cores. The buried mud volcanic vents are clearly revealed at the tops of many Pliocene brachyanticlines in the seismic sections.

Naturally, the large thrusts may not have been formed during a relatively short period except for upthrow overfaults that displaced the southern limbs of brachyanticlines. Fairly marked upthrow faults have been formed in the western sill, nearer to the northeastern Absheron Peninsula, where the accretionary prism is uplifted to 3–4 km. This phenomenon is mainly related to regional compression and possibly the catalytical role of the continual displacements in an underlying accretionary prism.

The seismic sections are believed that inserted significant corrections into a notion of the Absheron-Near-Balkhan folded zone (ANBFZ) formation. Traditionally, this zone morphology and genesis are related to the Greater Caucasus and Kopetdag, in which orogens began their growth in the Middle and Late Miocene due to the microcontinental block's collision with the platform at the place before the existent sedimentation basins. The ANBFZ has also been formed at the end of the Pliocene within the zone of convergence and contact of the two genetically different type plates over the subduction area of the oceanic basin crust under the platform continental crust. Tectonically, the ANBFZ origin is believed to be caused by the extremal compression of the sea basin where the linear anticlines appear to balance (or compensate) an uninterrupted crustal shortening and deform the 7 to 8-km thick Pliocene terrigenous rocks. The North-Absheron folded zone, which is the marine prolongation of the Tengi-Beshbarmag anticlinorium, is also related to the Greater Caucasus.

The SD-GDP data has also clarified the tectonics and mode of joining of two types of the Earth's crust in the transitional zone from the southern to the Middle Caspian. As shown from the time section, the marginal part of the STP basement is distinguished as an actual large continental structure in sharp contrast with the sedimentary cover and thin CC of the SCB (Fig. 1.10).

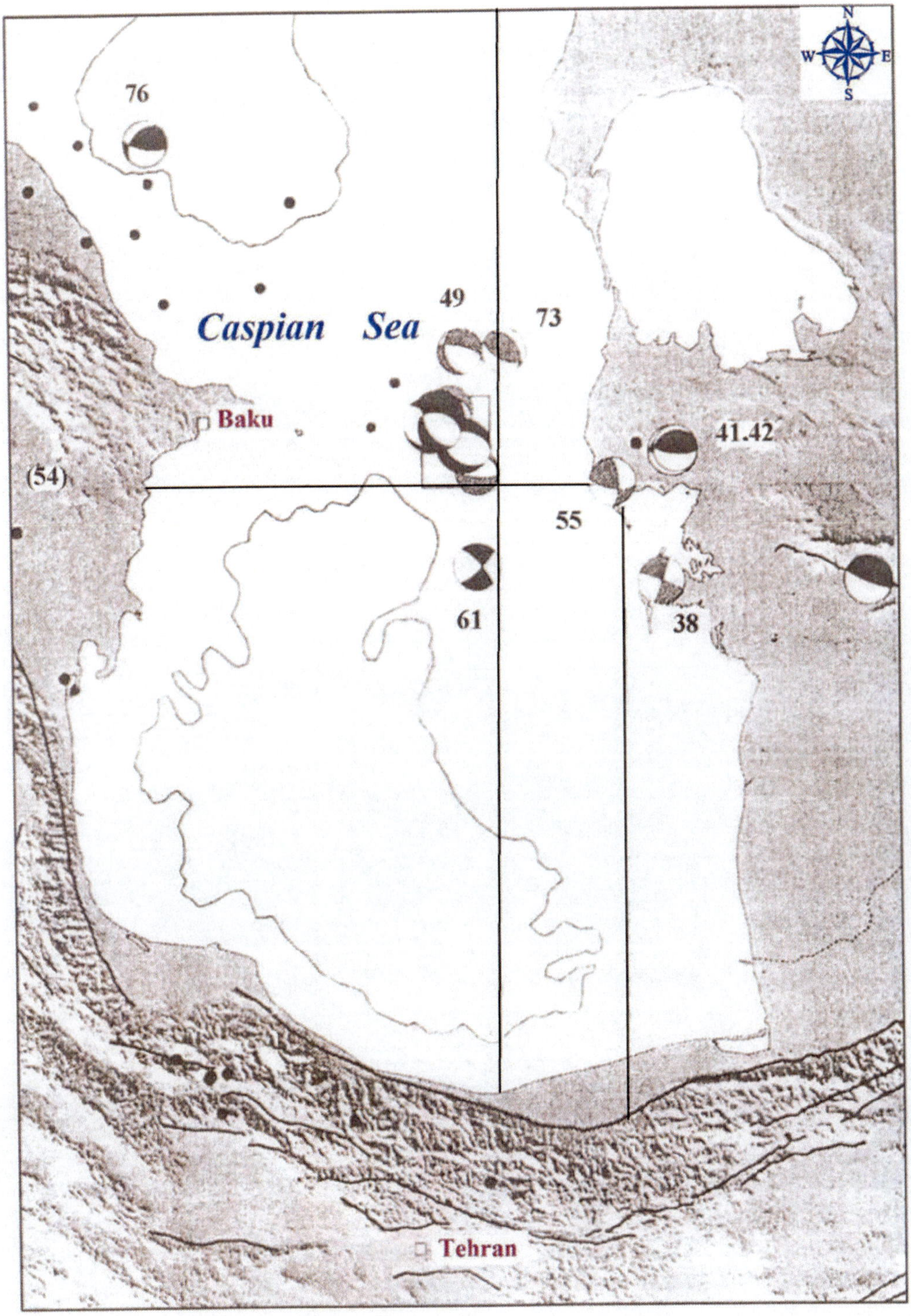

Fig. 1.9 Deep-focus earthquakes in the Caspian region

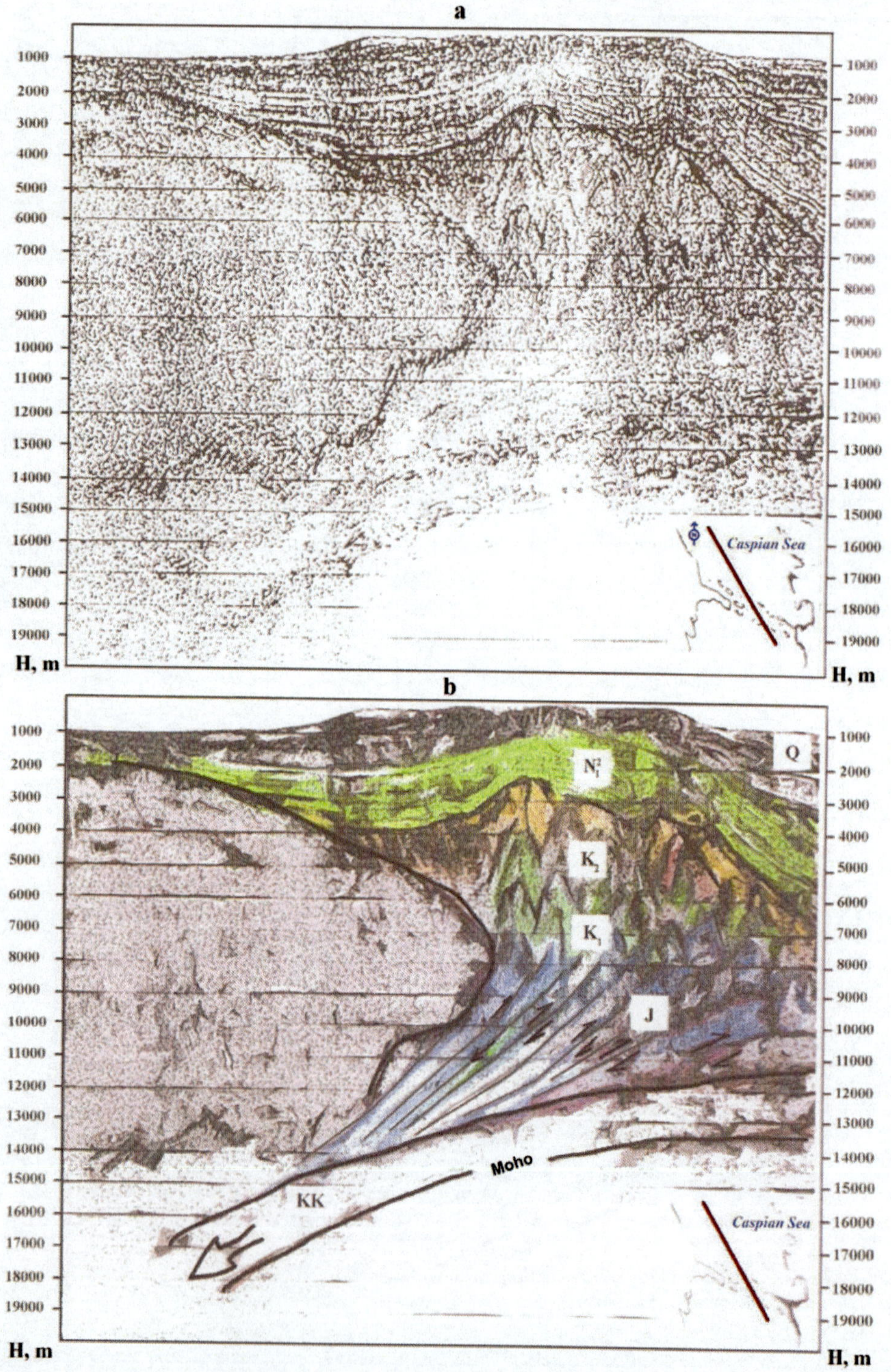

Fig. 1.10 Seismic images of the Scythian-Turanian platform southern termination in the Caspian and subduction zones of CC in SCB accretionary prism: **a** before, **b** after seismostatigraphic interpretation

Within the platform basement, the separate "turbid zones" appear to be distinguished, which could have been caused by the metamorphic substratum's heterogeneous and substantial variegated composition. Any extended (or intermittent) reflecting boundary surface dividing the CC into "granitic" and "basaltic" layers has not been observed inside this substratum. This conclusion conforms with seismic survey data obtained from many epicontinental basins. From the results of all these data studies, it is concluded that such a boundary does not exist in CC (Kunin, 1989). The contrast wave field boundary rather clearly marks the platform's frontal part. In the platform basement lowermost strata, an accumulation of short reflecting sections (by reflectivity), whose special ordering indicates a complex structure and configuration of the foot of metamorphic substratum, is observed. Such the basement foot shape may illustrate the non-mirror and unsharp mediums partition (viz., consolidated crust and upper mantle) characterized by different rheological and petrophysical properties. The broken and chaotic elements observed in the basement bottom half are undoubtedly small pieces of the rough surface over the Moho level.

According to the classical description of the basement of the young platforms, they represent the folded-crystalline foundation (substratum), which lived through the geosynclinal stage, followed by a calm platform regime. Judging by seismic sections, the Scythian-Turanian platform foundation has not been subjected to geosynclinal folding. The fold structures have only been revealed within the South-Mangyshlak and Karaudan regions of the graben-through systems, where the Permian–Triassic transitional complex is developed. These facts conform with Kunin's (1989) data, suggesting that most young platforms have yet to experience the geosynclinal stage of development. The stratified series are well distinguished over the platform's metamorphic basement, where a thin, 1.5–2 km thick platy complex consists of Jurassic-Cretaceous and Paleogene carbonate-terrigenous deposits.

More geophysical data is needed to determine the substantial composition and parameters of the lower lithosphere (subcrustal mantle of the SCB). Some information has been obtained from the DSS and seismotomographic methods. As this velocity is variable, ranging between 7.9 and 8.3 km/s, it is suggested that the subcrustal basement under the SCB is of heterogeneous composition. Some researchers believe that high boundary velocities under CC relate to ultrabasic rocks such as peridotite, pyroxene, pyrolith, eclogite, etc. However, attempts to convince that various ultrabasic rocks are related to the upper mantle substratum seem highly debatable since experimental values of longitudinal wave velocities in the rocks appear to be less than 7.2 km/s (Kunin, 1989). At the same time, according to the TRW and RFA data within the upper mantle series of SCB lithosphere, the waveguide is being marked, i.e., a layer of a low velocity ($Vs = 2.4$–2.9 km/s), which is identified with asthenosphere. At the modern level of geophysical science, the concept of an asthenosphere is not widely used because of its vagueness. Traditionally, it is used to motivate the conception of isostasy and magmatic "life" in the basins. According to the TRW data, a waveguide appears to be expressed most clearly within the Baku Archipelago and southeast of it (Yakobson, 2000). Relatively low-velocity values in the Jurassic-Cretaceous and Paleogene-Miocene sedimentary complexes are confined to the waveguide rising region. Most discovered submarine mud volcanoes are also located within this region. The relative correlation of the wave velocity values (Vs) in the upper mantle with the heat flow values and rated temperature values at the crustal foot has also been established within the Caspian area (Rodkin, 1993).

The data were obtained during the last years by classical seismic methods (DSS and SD-GDP), as by the other methods (seismotomography in TRW and RFA modifications, earthquake seismology, satellite altimetry), and the data of SCB lithosphere and framing geostructures,—all that have been exceptionally conducive to the deepening our notions of geophysical parameters and structural features of the Earth crust. Suppose all the mentioned data of CC would have been valued for their reliability and representativeness. In that case, information on the CC thickness and structure,

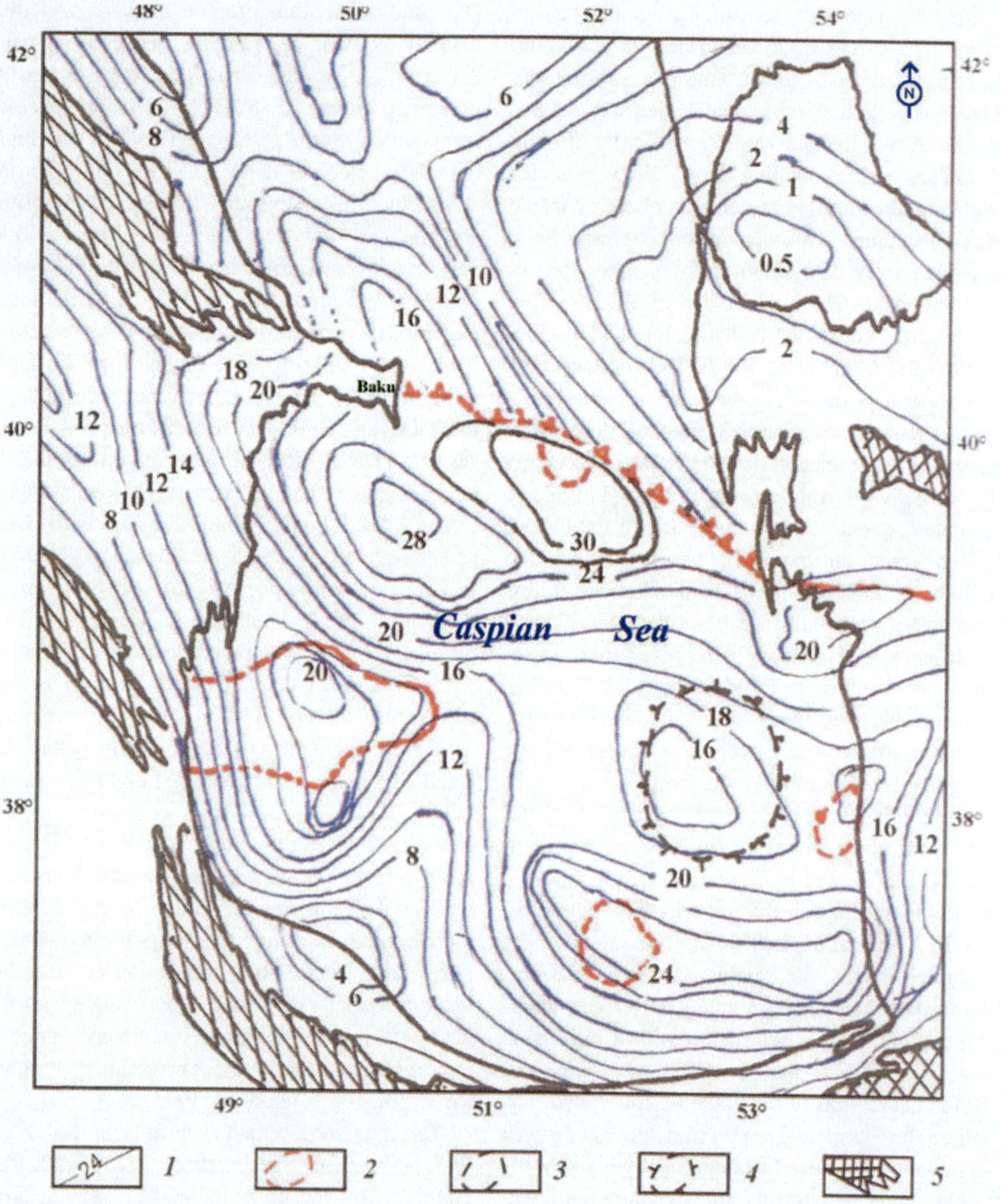

Fig. 1.11 Topographic scheme of the different-aged basement surface

clearly defined directly by seismic sections, should be put first. The SD-GDP seismic sections have also well reflected the buried old morphostructures and tectonic elements (or their fragments) typical to the basin widening stage, such as tension and block subsidence structures, continental slopes, and shelves in the northern edge zone, as well as buried structures of the volcanic island arch within the southern edge of the Kur trough (Figs. 1.14 and 1.15). The structures and geological bodies typical of the crustal shortening and compression stages have also been distinguished within the region. Those are the following: an oceanic crust curve and subduction, an overlying accretionary prism complicated by the thrusts and consisting of

Mesozoic-Paleogenic layers, compression structures in Neogene-Anthropogenic series (folding zones, thrusts, upthrusts), tectonic truncations, surfaces of unconformity, and erosion. Among the structures displayed by seismic sections, subvertical ruptures complicate mud volcano systems in folded series. All the mentioned data have been the factual fundamentals in constructing our objective tectonic-sedimentation model of the SCB and interpreting its real evolution history.

The SD-GDP high informative data are so vital that they may have been named the turning point in the SCMB investigation history. Thereby, it was as if finishing the stage in an indirect valuing of the deep basin structure based on insufficiently conditioned data of the other geophysical methods and the results of geological investigations on the surrounding land.

Determining a thin oceanic-type CC under sedimentary cover and its subduction under the platform is essential to interpreting the SD-GDP data. Up to the present, seismometry has disposed of the digital processing graph and well-developed interpretation technique. Therefore, it is very likely that the authenticity of the models is based on the same data and constructed by different authors (Green et al., 2009; Knapp et al., 2000, 2004; Mamedov, 2004, 2007) is beyond all manner of doubt.

These models are not a surmise or invention, as it was quite recently, and are not an abstract concept that can be refuted. The basin CC subduction under the platform is an immutable geological fact that may not have been ignored in studying tectonics and substantiation of the basin's evolutional development.

The two conceptually important special features are strongly marked in the new models of the consolidated part of the SCB lithosphere. Firstly, it is established that an oceanic-type CC lies at the base of the basin; thereby, a hypothesis of its development over the "median mass" has been disproved. Secondly, a thin CC under the SCB is structurally, substantially, and gravitationally heterogeneous; it is immersed smoothly in a northerly direction. The Absheron sill has subsided under the platform.

Based on the DS-GDP data interpretation and considering the data of other methods, we compiled the regional sections (2D models), topographic scheme of different-aged basement surfaces, and tectonic scheme of the basement. A subduction line represents the basin's northern boundary. It passes parallel to the Absheron sill. As shown on the basement surface typographic scheme (Fig. 1.11), within the SCB, there are outcrop zones of matter characterized by mantle velocities and a zone of relatively lower velocities typical to the sialic component of the crust (Yakobson, 2000).

1.3 Tectono-sedimentation Complexes of Sedimentary Cover as Megabasin Evolutionary Development Indicators

The SCMB development history abounds with significant geological events that chronicle appears to be most widely fixed in the structure and composition of sedimentary cover. During its formation, the geotectonic cycle stages were the most important factors. The epochs of alternation in tectonic activity and its relative rest have been important boundaries of large-scale geological events and reconstructions. It is natural that essential changes in the structural plan and formation-lithological composition of sedimentary complexes occurred during those epochs. Seismic sections drawn up by the SD-GDP method make it possible, as distinct from other geophysical methods, to obtain direct and continuous information of the basin sedimentary cover structure "from side to side". Essential features of these regional seismic sections are that they contain comprehensive information of geological evolution history; besides, deep negative tectonic relief with side structures is distinguished; some surfaces of unconformity and concordant boundaries dissect a sedimentary cover onto different-aged sedimentary complexes are also clearly traced. They are objective criteria and benchmarks conducive to fixing the megabasin development stages.

From the results of seismostratigraphic methods of tectonic, eustatic, and lithodynamic analyses, objective criteria are obtained to carry out the age correlation of sedimentary complexes and construct a reliable stratigraphic framework of the basin. Seismostratigraphic analysis (SSA) of time sections supplies invaluable data for perceiving the SCMB's deep structure and, at the same time, makes it possible to reveal new features and details of its crustal architecture.

In the SCMB sedimentary cover, according to some seismostratigraphic criteria, rather large and thick three-dimensional geological bodies (macrocomplexes) have been recognized. As concluded from the SSA results, a thick sedimentary series (up to a few km in range) formed during tens Ma have been united by the mentioned macrocomplexes. The unity of structural plan distinguishes them and are subdivisions of the regional stratigraphic scale. The critical seismic horizons represent the interfaces between these macrocomplexes, confined to the unconformities in the basin's peripheral zones. In the central megabasin, they are merged with concordant boundaries. The essence of these macrocomplexes picked out in the SCMB is not in the formation-rock composition and dislocation but in their historical-geological significance. Indeed, it happens in the consanguinity of the componential series, ensuring integrity and hierarchy with large stratigraphic units.

Like macrocomplexes, these subdivisions are the units of regional scale and tectonic-sedimentation specialization. Each of the macrocomplexes consists of the second-order space–time units-sequences, which in Russian publications is called "sedimentation seismocomplex" (SSC) (Peyton, 1977). The SSC represents a stratigraphic unit of the section composed of a sequence of relatively concordant and genetically interconnected beds. The unconformities or corresponding concordant surfaces limit its roof and foot. Anyone of SSC includes chronostratigraphic information since it has been deposited during some interval of geological time.

As for the stratigraphy, appropriately, the seismostratigraphic zonation acquires a real meaning when the sections are being put on the geological timetable that, in turn, requires that the biostratigraphic and, as far as possible, other absolute dating data would have been used. However, such data exists only for the SCMB edge zones where a thin sedimentary cover had been drilled, while there is no data for the central SCMB where the thickness of sedimentary reaches up to 25–30 km. Single wells 6–6.5 km deep in the sea have drilled the only upper section. The Cenozoic and Mesozoic underlying strata have only been studied by the seismic survey. In the absence of biostratigraphic data necessary for age and SSC boundaries determination, the fundamental conception, a global synchronism of the sea level relative variations (SLRV), may have been used. According to Vail et al. (1977), only one SSC is formed during every cycle of the second order (sea level elevation-lowering cycle). The authors have revealed fourteen second-order cycles on the SLRV cyclic recurrence charts made by seismic sections for the Phanerozoic. Nine of them are related to the Mesozoic and Cenozoic eras. As was established by investigations carried out in several regions of the former USSR (e.g., Kunin, 1989; Mamedov, 1991; Schlezinger, 1988), a decisive role in the formation of different aged units of sedimentary cover in the ranks of seismocomplexes and limited those concordant and discordant surfaces belong to the tectonic factor. SLRV may serve as an additional factor for the stratification of those parts of sedimentary cover that have not been drilled in.

Based on the SSA of the SCMB sections, we proved that the megabasin seismocomplexes had been formed under conditions of the Earth's crust tectonic expansion and compression in the regime as compensated by uncompensated sedimentation (Mamedov, 1991). The following nine SSC, which we named according to the objective seismostratigraphic criteria, have been distinguished in the SCMB sedimentary cover (Fig. 1.12):

- Jurassic, 1.0–5.0 km thick terrigenous-carbonate, mainly carbonate complex in the uppermost strata (SSC-1). Within the Greater

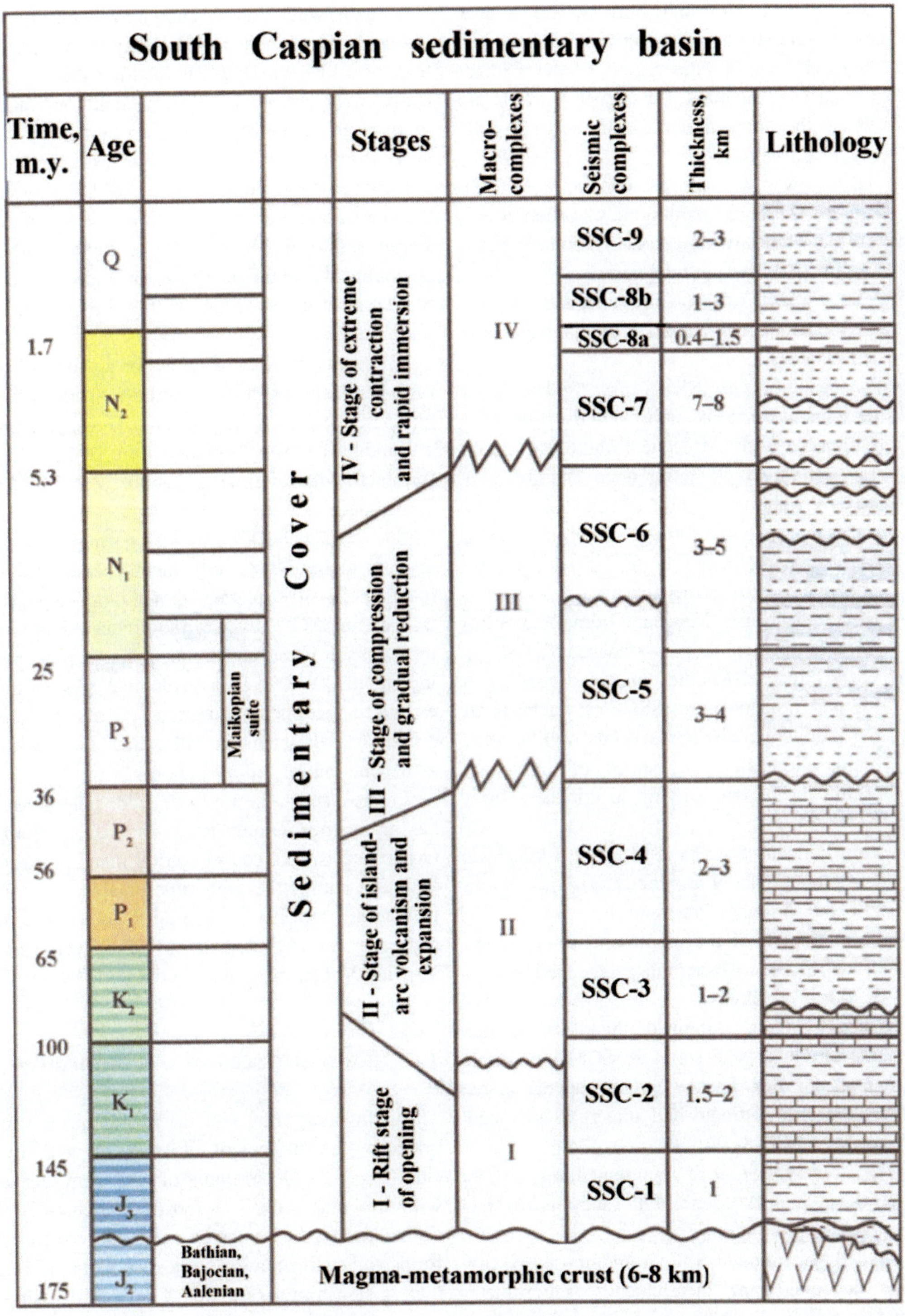

Fig. 1.12 Main tectonic-sedimentation seismic-stratigraphic complexes of the sedimentary cover

Caucasian flysch trough, Jurassic beds appear to be underlain by volcanogenic formations of 2–3 km thick. In the region of the Greater Caucasian structures' subsidence towards the SCB, the total thickness of Jurassic SSC reaches 5–6 km;

- The Lower Cretaceous complex (SSC-2), which is carbonate within the vast shelf zones, on the whole, is terrigenous-carbonate reaching up to 2.0 km in thickness;
- The Upper Cretaceous, occasionally upper-Cretaceous-Paleogene 2 km thick terrigenous-carbonate SSC-3;
- Paleogene-Eocene SSC-4 is 2–3 km thick terrigenous-carbonate, effusive in the southwest.
- Oligocene-Early Miocene (Maikopian suite), 3.0–4.0 km thick terrigenous mainly clay SSC-5. Maikopian suite is considered to be the primary source formation that is a source of hydrocarbons and a feeder of the mud volcanic system in megabasin;
- Middle-to Upper Miocene, from 2 to 3 km thick terrigenous (argillo-arenaceous) SSC-6;
- Early Pliocene (Middle Pliocene according to the past nomenclature) SSC-7. It encloses all the Productive-Red series, sandy-argillaceous, main oil&gas-bearing series of sediments within the region, varying in thickness from 7.0 to 8.0 km.
- Late Pliocene SSC-8a is of 0.4–1.5 km thick terrigenous (sandy-argillaceous) series of the Pliocene Akchagylian stage;
- Late Pliocene SSC-8b represents sandy series of the Pliocene Absheronian stage having 1.0–3.0 km in thickness;
- Because of the thinned thickness of the Absheronian stage series in SCMB in some areas, the SSC-8a and SSC-8b appear to be united and distinguished under an assumed name of SSC-8;
- Quaternary SSC-9 is represented by a 2.0–3.0 km thick clay series with sandy inclusions.

All the mentioned SSC are composed of the age succession of hierarchically subordinate middle and small-rank units (Mamedov, 2007).

Among sedimentary complexes, the Early Pliocene SSC-7 m, as a typical sequence (system tract of a low sea level), is distinguished by its enormous thickness and rhythmical alternation of sandy and clay deposits. Separating by an unconformity from the underlying system tract of a high sea level-Pontian stage, SSC-7, by its thickness and volume, excels in all the structural stages and even all the plate complex within epicontinental sedimentary basins adjacent to the SCMB from the north and within the piedmonts and foredeep.

This complex is remarkable for an abundance of dynamically expressed regular reflections and quasi-isotropic recording intervals (Figs. 1.7, 1.8, 1.9 and 1.10). The formation time of this SSC (productive-red series) is about 2.2–2.5 Ma, corresponding to the 1.5% geological time of the Alpine megacycle. Accumulated during this time appears to be 7.8–8.5 km thick, coarse detrital material. 25–30% of the SCMB total thickness falls on an SSC-7. Even without considering the rock consolidation and frequent faults (occupying up to 40–60% of a geological space), the estimated sedimentation rate is about 2.5–3.0 km/Ma. It is on an order that exceeds an avalanche rate of sedimentation.

In this respect, the Early Pliocene paleo-basin has no analogs among paleo- and recent World Ocean basins. The coarse detrital material transportation causes the high super-avalanche sedimentation rate into a closed Pliocene basin carried by numerous small and large rivers such as paleo-Volga, paleo-Kur, and paleo-Uzboy.

1.4 Main Surfaces of Unconformity

The sedimentary cover of the megabasin is saturated by the surfaces of unconformity and erosion (Fig. 1.8). Determining of long-term breaks with irregular scours of formerly accumulated sediments and the revelation of specific structural forms and sedimentation bodies led us to believe that seismostratigraphy was a powerful instrument for interpreting the region by stage development history. Within the SCMB lateral zones, according to the seismostratigraphic criteria and wavefield special features, there are mainly distinguished three regional (structural) surfaces of unconformity by which all the section is separated into four macrocomplexes. The main megabasin evolution stages have been marked by those macrocomplexes.

The lower unconformity surface (SU-1) in the northern lateral zone of marginal sea marks a surface of terrigenous-volcanogenic and red continental (or lacustrine) rocks of the rift complex, which is crushed by fault-block tectonics. The rift "shoulders" and interior horsts have served as a source of deposit ablation. The drilling data and seismofacial analysis show that over the SU-1, typically laminated terrigenous and carbonate rocks of the late Jurassic and early Cretaceous post-rift complex occur. This complex is characterized by the relatively unbroken bedding of its series since it was formed on the northern "passive" side of the extending sea basin. SU-1 marks a boundary between Jurassic and occasionally between early Cretaceous volcanogenic rocks and late Cretaceous carbonate rocks of the basin island arc side. In the central SCB, this boundary has not been traced because of the generally deep point method (GDPM) resolving power deterioration at great depths (22–24 km). A sharp unconformity separating the Lower Cretaceous limestones from the Upper Jurassic calc-alkali volcanites, which have been drilled in by the Saatly superdeep borehole (SD-1), seems to have been formed during arc splitting, and the initial stage of a new trough (graben-trough). This time corresponds to the extension within the Black Sea and South Caspian cells of the Greater Caucasian Sea (Mamedov, 2004, 2008; Shreider et al., 1997).

The stratigraphic cutoff accompanied by long breaks, age sliding, and structural changes appears to be the regional unconformity surface (SU-2) over which a megabasin is dismembered on the trough and uplift zones. From the above, carbonate-terrigenous beds of the Paleocene-Eocene sedimentation complex are truncated by the surface mentioned. These beds are well traced within the regions of old continental slopes of the marginal sea at the southern platform and not so well in the southern marginal zone of the Kur trough, where an island arc system has developed until the Oligocene. The SU-2 marks the border between the divergent and convergent stages of lithosphere evolution.

The results of the seismic survey carried out in the North Absheron folded zone trace the next surface of unconformity (SU-3), by which the beds of different ages belonging to the pre-Pliocene complexes are truncated (Fig. 1.13). The SU-3 was formed during a sharp fall of sea level at the end of Pontian and served as the surface of the Productive-Red series primary sedimentation. Mamedov (1984) has compiled a structural sketch map over the SU-3 surface reflecting PRS foot topography in northwestern Absheron sill for the first time. The SU-3 subsided step-like towards the topographic depression. Accordingly, this unconformity surface has the same structure within the Pliocene sea's marginal zones in the Kur and West Turkmenian troughs. The terrace scarps are remarkable for the apparent indications of tectonic down-warping and wave action. They are the first-rate markers of shorelines in transgressive seas of the Pliocene age. Obvious and hidden breaks have been an integral part of the sedimentary process in the SCMB.

The chronostratigraphic sections compiled by the Pliocene–Quaternary series indicate that more than 30–40% of geological time fell on the breaks in sedimentation or the "reverse sedimentation", on the erosions and washouts of formerly deposited sediments. The change in sedimentation regimes has occurred during tectonically unstable periods of basin development when the concerted course of tectonic, sedimentation, and eustatic processes have been disturbed.

1.5 Morphostructural Elements and Sedimentary Complexes in the SCMB Marginal Zones

The Pre-Caucasian and Middle-Caspian seismic sections reflect buried land slope morphostructures of the Jurassic-Eocene Sea. The traces of riftogenic extension and displacement of

Fig. 1.13 Unconformity surfaces of in the sedimentary cover: **a** Absheronian sill, **b** West Turkmenia trough

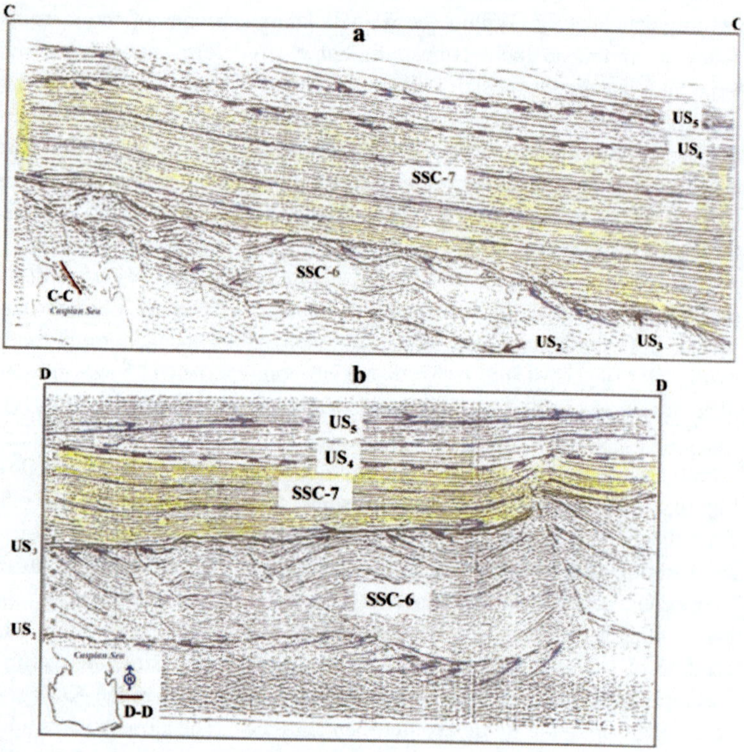

Paleozoic substratum blocks in a southerly direction appear to have been preserved on the marginal Scythian-Turanian platform (Fig. 1.14). During the post-rift stage of evolution, continental slopes and shelves formed over those blocks.

The most notable structures in the western land slope zone of the GCMS are thought to be the Paleozoic substratum cusps buried under the Mesozoic series. They are distinguished in seismic sections as an acoustic basement. Their surface is represented by a rough boundary having obvious indications of erosion and underground denudation. Such are the southern blocks of Karabogaz, Agzybirchala, and other cusps. The basement cusps played the role of the marginal uplifts separating a deep-sea trough of the marginal sea from the shallow epicontinental basins (SEB) on the platform. The latter has been formed within the vast shelf areas during the late Mesozoic and Early Paleogene transgressions.

From the seismic data analysis, it is inferred that continental slopes in the Mesozoic and a significant part of the Paleogene have been formed on the external limbs of marginal uplifts. Relatively steeply dipping and high (3–4 km) continental slopes are well readable in seismic sections of the Karabogaz southern limb (Fig. 1.14), where plane truncations ruptured clinoforms and collapse bodies are well recorded. Based on the interpretation of the seismic sections, it is suggested that the slope has been formed for a long time—beginning from the Late Jurassic up to the Oligocene. The slope surface has occasionally been eroded and furrowed by valleys and canyons. Among the borehole displayed seismic sections are also landslide seismofacies on the upper, comparatively gently sloping terrace of a slope (Fig. 1.14).

The same age complexes (sequences) are distinguished on two levels: at the shelf and the foot of the slope. The dominant controlling mechanism in transferring sedimentary cover is the cyclic falls of sea level. The Caspian buried continental slopes morphologically and by height (up to 3–4 km) are the same type as those

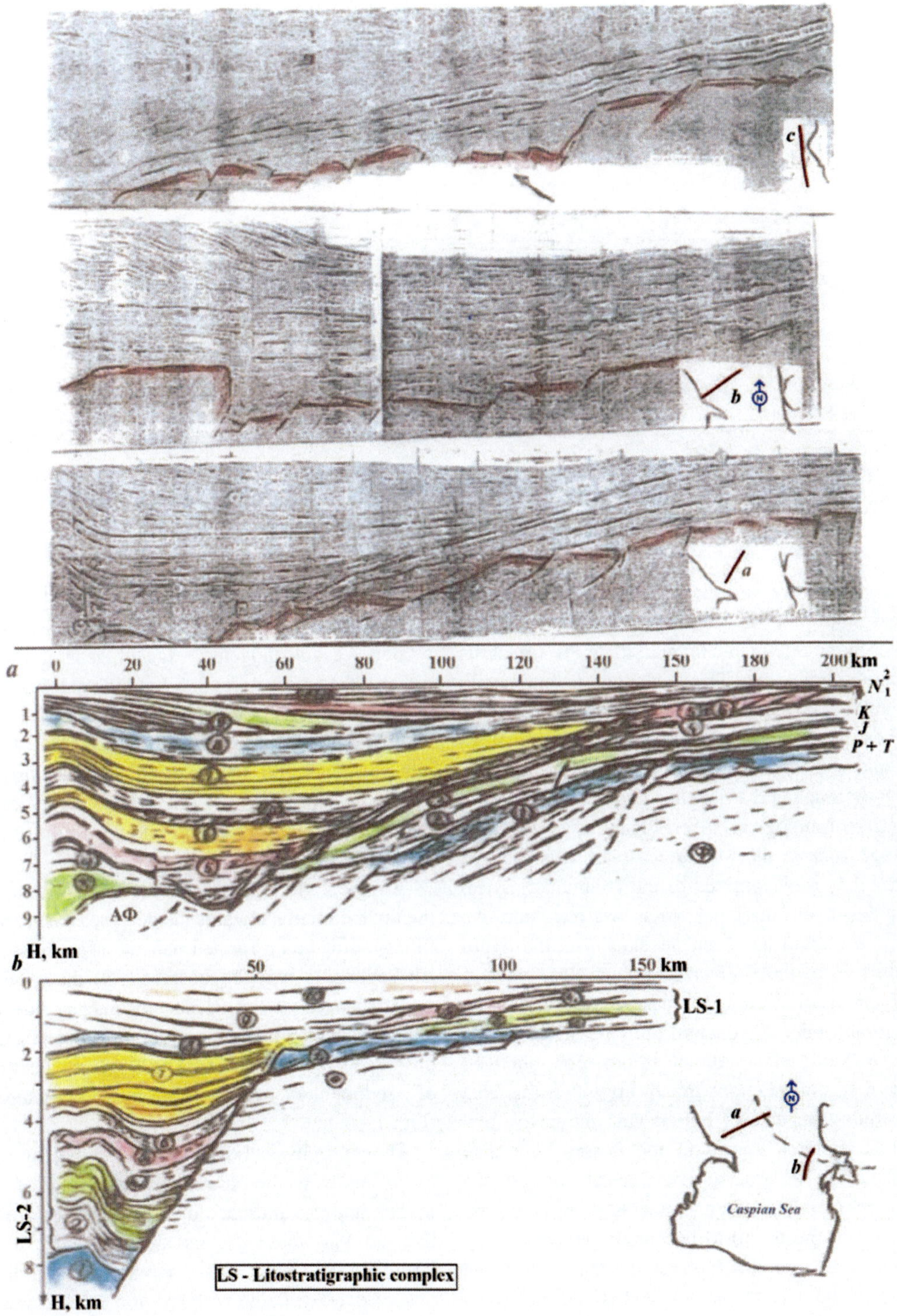

Fig. 1.14 Seismic sections displaying offset, tension, and subsidence structures of the basement blocks of the Scythian-Turanian platform and continental slopes

a b

Fig. 1.15 Seismic sections displaying the structures of extension and displacement of magmatic substratum blocks (southern continental slope of the Kur trough)

discovered by seismic survey in the southern marginal part of the Scythian-Turanian Plate within the Scythian-Turanian Plate Pre-Caucasus and the Black Sea. This fact suggests that the old continental slopes were large regional morphostructures. They extended along the northern flank of the joined marginal marine basin. From the thorough and strict consideration of structural and morphological special features of the near slope zone in the Middle Caspian, it is inferred that they are submitted to the common classical model of the marginal sea's "passive" side. An active margin has been developed on the opposite side of the Greater Caucasian Sea, where a volcanogenic island area has been formed in the northern microcontinental outlying areas.

A piece of important information obtained from seismic sections drawn up within this area includes the data of buried volcanic cusps, tectonic ruptures related to the Lesser Caucasian Island arc tension, and its breakdown into the frontal (southern)—Pre-Lesser Caucasian arc and inner (northern)-Mingechavir-Geychai-Saatly arc. The morphostructures of the pre-arched region, namely the trench and island arc external slope, are destroyed because of plate convergence. In contrast, the morphostructure of the island arc inner slope seems to be subjected to weak tectonic actions and is well preserved in the Kur trough section. As it is seen from the seismic sections, a volcanogenic ridge of general Caucasian orientation, up to 3–5 km high and from 2 to 70 km in width, is traced from the PreTalysh region almost up to the Dzirulian massif. The rare and rare-earth elements' content and distribution in volcanites of the Saatly superdeep borehole (CD-1) correspond entirely to those in volcanites of recent island arcs.

A typical gliding of magmatic blocks through the inclined faults towards the deep-sea basin and filling up of being formed narrow lakes by syn-rift sediments may be observed on the north-eastern slopes of the island arc inner segment. A fan-shaped divergence of carbonate beds observed on the block tops is a typical feature of sedimentary prisms in such structures (Fig. 1.10a).

The petrophysical data, such as the belonging of volcanites to the basalt-andesite-rhyolitic formation and the presence of rare-earth elements, indicate that these factors are typical for the island arc volcanism. The absolute identity of the observed wave fields with specific seismic images of sedimentary prisms in island arc systems, known from the publications, is also noted. The seismic images of the volcanic massif and petrophysical indicators lead us to believe there is volcanic cusps island arc genesis. Notably, they are close-fitted by up to 800 m thick Cretaceous carbonate series.

Owing to the sea level lowering at the end of the Cretaceous, the described area has been under continental conditions for almost 50 Ma. For a long time (Late Cretaceous-Paleogene), the carbonate sheet's washed-out surface served as the marginal sea's southern land slope. According to the adjoining layers' thickness, an insular slope's height is about 4 km, entirely corresponding to that of the GCMS northern passive margin.

1.6 The Results of Modeling of SCB Earth's Crust Subsidence

The reconstruction of the Earth's crust subsidence history is a necessary starting point in investigating any SB evolution, including SCMB. The SCMB evolution stages are well displayed by the Earth's crust subsidence history models and block diagrams of the basement tectonic down-warping (Brunet et al., 2003, 2005; Mamedov, 2008). Such processes as sea-level variations and crust subsidence provide additional space for accumulating sediments.

The numerical modeling has been done based on the deep drilling data and using "backstripping analysis" (Steckler and Watts, 1978). According to the generalized model (Fig. 1.16), during the Middle-Late Jurassic epoch, in the sea-spreading stage at the expense of thermal subsidence, CC has subsided to 1.2 km at the rate of 50–60 m/m y. By the beginning of the Cretaceous, the total tectonic setting of CC was 5–6 km, considering a load of sedimentary volcanogenic series and waterbed (2.5–3.5 km in height). The lower rate marked the same process in the Cretaceous (10–20 m/m y; in all 3 km).

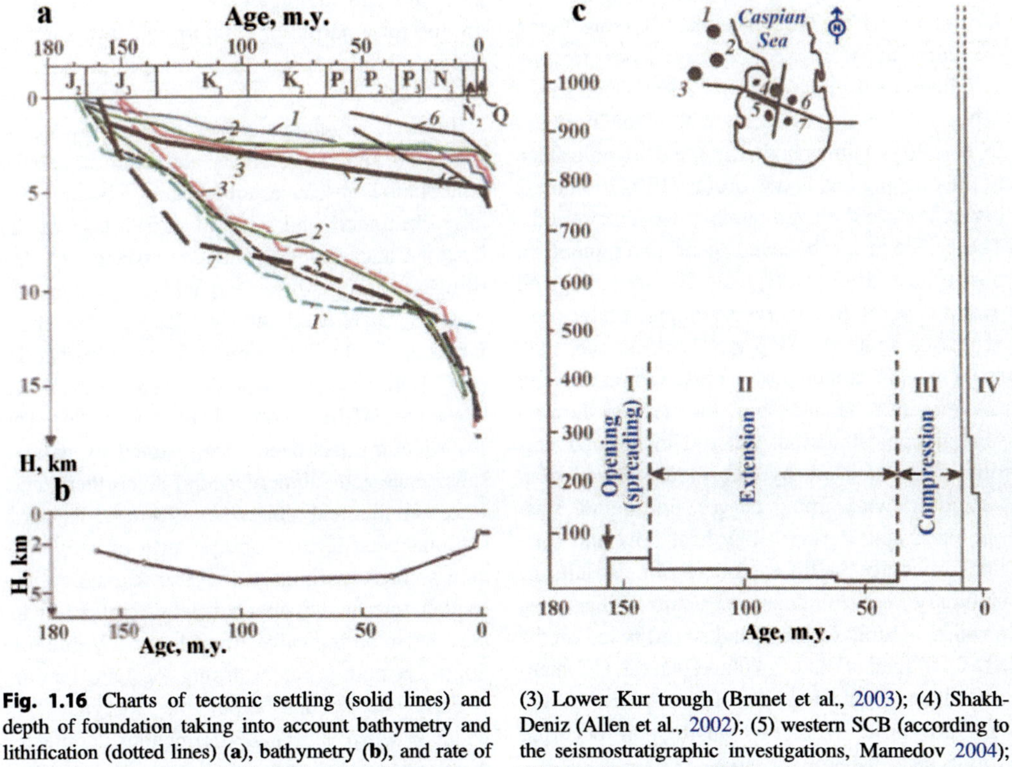

Fig. 1.16 Charts of tectonic settling (solid lines) and depth of foundation taking into account bathymetry and lithification (dotted lines) (**a**), bathymetry (**b**), and rate of tectonic down-warping (**c**). Numbers with primes refer to dashed lines. (1) and (2) Greater Caucasus SE subsidence; (3) Lower Kur trough (Brunet et al., 2003); (4) Shakh-Deniz (Allen et al., 2002); (5) western SCB (according to the seismostratigraphic investigations, Mamedov 2004); (6) SCB (Narimanov, 2003); (7) generalized charts of the SCB (Brunet et al., 2003)

Geophysical evidence suggests that the SCB crust in the Oligocene-Early Miocene age has tectonically subsided very slowly (Glumov et al., 2004). The basin was a deep-sea one in Paleocene-Eocene time, as well. By the beginning of the Oligocene, the sea spreading had practically been held up, accompanied by a sharp slowing down of the tectonic setting (5–8 m/m y). It is well known that the thermal subsidence of a newly formed crust in an ocean and marginal seas ended in 80–100 m y after their opening (Kukal, 1987). By the end of the Eocene, the CC had subsided to 1 km only by loading of sediments. The total subsidence amplitude reached 10.5 km. Only by increasing sedimentary series loading this reached 14–16 km in thickness.

Further, in the Middle and Late Miocene, the depth of the basin gradually decreased (Fig. 1.16) from 4.5–4 km to 2 km owing to the intense sedimentation process. The rate of tectonic down-warping in the Pliocene reached 1000–2000 m/m y. The calculated amplitude of subsidence within the Absheron sill, considering the corrections for sedimentary series loading and bathymetry, makes 10–12 km. The subsidence rate in the Pliocene within the northern SCB is 20–30 times higher in the riftogenic stage of the opening and is two orders (100–200 times) higher than in the Cretaceous Paleogene periods. These figures are believed to be preeminent. In this respect, to the SCB, there are no analogs all over the world. Many diverse opinions have been expressed about the SCB rapid subsidence, such as (1) "instantaneous" crust cave-in and (2) riftogenic subsidence of the crust in the present geological period. All the mentioned versions were related by subsidence simulation, tectonic down-warping curves, and actual seismic data, considering a low heat flow and seismic immunity in the central basin. Results of summary constructions (structural diagrams, profiles, seismic sections, and so on) based on the SD-GDP method data conclude that the CC sharp and deep subsidence has taken place in the northern basin where its subduction occurred within the Absheron sill area.

1.7 Main Stages and Substages of the Megabasin Evolution

Based on the actual seismic data and analysis of a series of representative features (including the type of CC, stratigraphic sequence, surfaces of unconformity, morphology, and genesis of continental slope structures, seismicity, and mud volcanism) as well as taking into account the results of geological-geophysical investigations, paleotectonic reconstructions carried out before; it is inferred that the two stages and six substages may have been distinguished in the region's earth crust development.

The divergent stage (Middle Jurassic-Eocene) began from the riftogenic substage (I). The continental crust splitting marked between the Early and Middle Jurassic (J_1-J_2) and the divergence of microcontinental blocks from the margin of the epi-Hercynian platform (which was attached in the Triassic) led to the opening of a system of echelon-like disposed of troughs. In the Bajocian-Bathonian, the latter served as a ground for a narrow marine trough with a newly formed basaltic crust and continental margins (Fig. 1.12).

Further, simultaneously with a narrowing of the Meso-Tethys, its crust subduction under the Anatolian-Transcaucasian-Iranian block of microcontinents and with an initial stage of the Lesser Caucasian volcanogenic island arc, the riftogenic trough has been widened within the systems mentioned above. During the Late Jurassic—Early Cretaceous interval (substage II) (J_3-C_1), the riftogenic trough turned out to evolve into a vast GCMS (Zonenshain et al., 1990). The sea widening has been accompanied by passive subsidence and submergence of its northern side, where continental slopes have formed. Its north side has been formed on the north of active (island arched) margins of the Transcaucasian and Iranian microcontinents where calc-alkali magmas have periodically flowed out. Within the South Caspian segment during the Late Jurassic—Early Cretaceous time, the marginal sea widened where carbonate-terrigenous complexes were accumulated under shallow conditions. As a result, a deep trough arose there, being highly stable over all the stages of evolution. An interval between the Late Jurassic and Eocene (J_3-Pg_2) has corresponded to the maximal widening and deepening stage (III). Before this stage, the main features seem to have been tectonic and volcanic activity slackening in the marginal parts of microcontinents, widening suspension, and a stable high level of the marginal sea (Fig. 1.17).

Thus, the geodynamic processes in the GCMS before the Oligocene have also been the same as those in the Cenozoic within the recent marginal (back-arc) seas. These processes conform entirely to a model of stage-by-stage development of the troughs in marginal seas (Maltman, 1994). In a region during the Oligocene—Early Miocene period, a basin partly isolated from an ocean exists and is characterized by stagnant

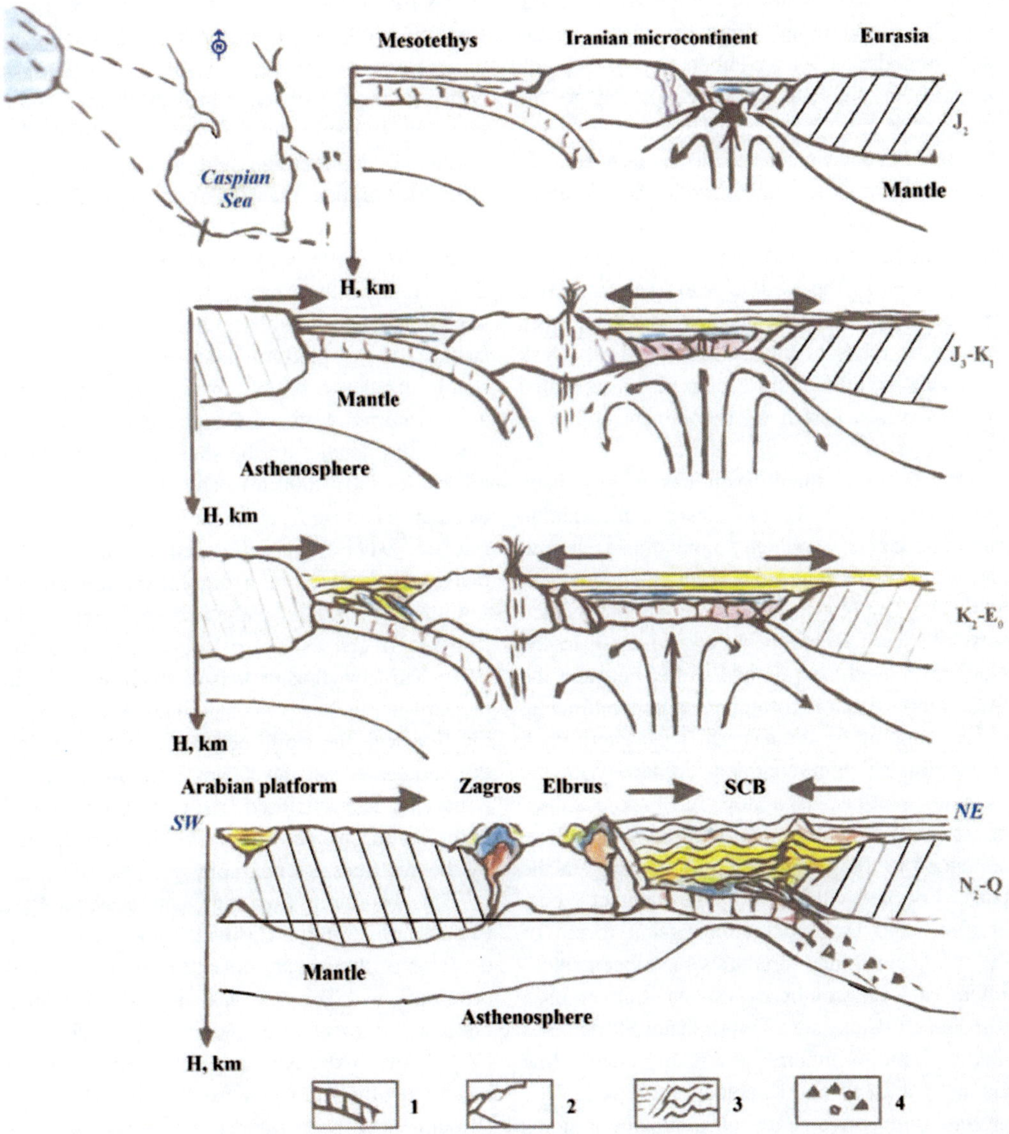

Fig. 1.17 The Earth's crust models indicate the following stages: **a** riftogenic trough opening, **b** formation, and **c** widening of the Transcaucasian marginal sea and extremal contraction and down-warping of its South-Caspian relic

conditions of sedimentation where high-bituminous black clays have accumulated. During the convergent stage, the microcontinental displacement to the north under Arabian wedge pressure led to the GCMS shortening and fragmentation and the consumption of the more significant part of its CC. Subsequently, each GCMS segment has been developed quite individually. Such fold belts as the subduction zones with an accretionary prism or the zones of continental collision with the gneissose granitic cusps have sprung up along the plate convergence boundaries. Active subduction of CC took place up to the Late Miocene, i.e., up to the collision of the Transcaucasian microcontinent with the Scythian plate and the eastern flank of the Iranian microcontinent with the Turanian plate. An intense uplifting of the framework mountain structures has begun in the plate collision areas. In the Middle Miocene, these two deep-sea basins with an oceanic crust have been isolated as relics of the GCMS: the East Black Sea basin and the South Caspian Basin, separated by the Transcaucasian microcontinent (substage IV) (Fig. 1.18).

The South Caspian basin has mainly been preserved due to Transcaucasian and Iranian microcontinents' geometry and those geodynamic processes that led to their collision with STP. The Transcaucasian microcontinent located west of the SCB and being forced close to the platform played the role of a buffer between the SCB and Anatolian-Iranian microcontinental block propping from the south. The eastern flank of the Iranian microcontinent collided with the Turanian platform is a stop that braked subsequent SCB reduction and, at the same time, conduced to the growth of the Kopetdag. In the late Miocene within the plate contact zone, uplifting and tangential compression forces of the mountain-folded structures have been intensified, and the southern systems' convergence (the Lesser Caucasus, Talysh, and Elbrus) has reduced the northern mountain chain (the Greater Caucasus and Kopetdag) took place. The growth and convergence of mountain systems have then been recovered by subsidence, filling with clastic product from the intermountain areas, and by contraction of sedimentary basins (Fig. 1.18).

The Miocene-Pliocene boundary is highly significant for the proto-Caspian and the arrangement of continental blocks all over the AHMB central segment. The Arabian wedge strutting off pressure onto neighboring microcontinents appears sharply increased in connection with the Red Sea opening and widening. It seems likely that substantial compressive forces from the south have been conducive to the underthrust-overthrust processes in the microcontinent collision zone, still more growth of the mountains, and extreme compression of the SCB. The horizontal compression and an enormous isostatic load of more than 15 km thick pre-Pliocene sedimentary series caused a sharp impulse to the SCB CC curve and subduction under the STP. As a result, a trench was formed—a deep trough in the northern basin in front of the frontal surface of the platform continental slope. The rapid subsidence of the crust in its subduction zone occurred between Pontian and early Pliocene times, leading to the sharp fall of sea level (down to 700–1,000 m). The waters of a vast Pontian basin moved towards the landlocked sea/lake where avalanche sedimentation took place, resulting from a significant amount of sedimentary product supplied by such full-flowing rivers as paleo-Volga, paleo-Uzboy, paleo-Kur as well as by tens of shallow rivers. In the Upper Pliocene (Akchagylian age), communication with the world ocean was renewed with the short-lived sea level elevation. In the Quaternary, a deep enclosed basin has been formed in the South Caspian—a relic kettle characterized by the shelf-basin topography.

Thus, beginning from the Quaternary, the East Caucasian, South Caspian, and Kopetdag domains of the sea are not at the same levels of development. The Caucasian and Kopetdag domains represent an orogenic (collisional) stage (VI). Under extremal lithospheric compression conditions, the Kur and West Turkmenian trough domains have been subjected to rising and closure. The complete draining of sedimentation

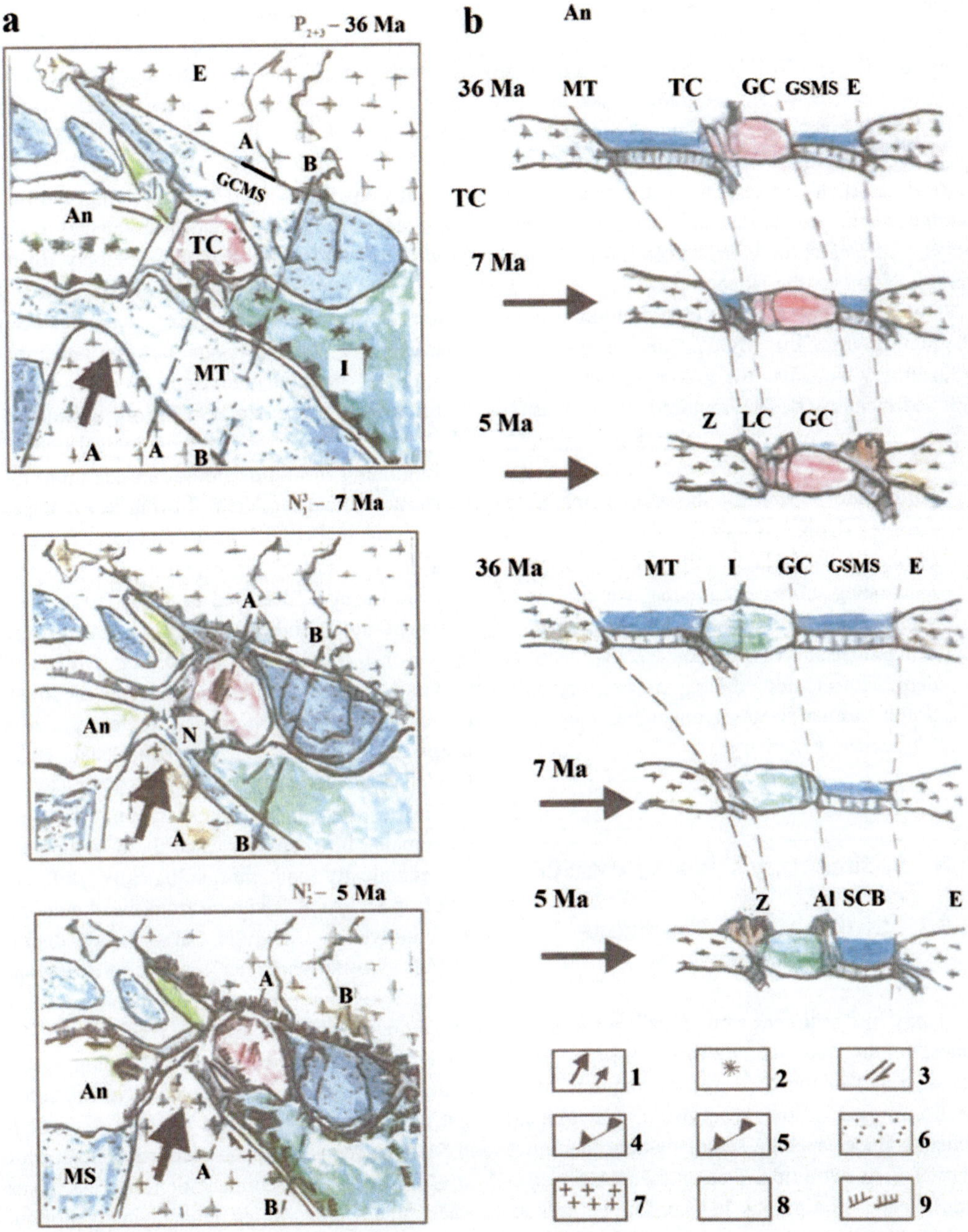

Fig. 1.18 Geodynamics of the Caucasian-South-Caspian region: 36.7, 7, and 5.0 Ma. Tectonic position of plates and microcontinents in the plane **a**; the same on the regional sections and profiles A–A and B–B **b**. (1) directions of plate movement, (2) island arc volcanoes, (3) displacement faults, (4) deep faults, (5) subduction lines, (6) sedimentary basin (SB) with the oceanic crust, (7) Eurasian (E) and Arabian (A) platforms, (8) microcontinents: Iranian (I), Anatolian (An), Transcaucasian (T-C), Nakhchivanian (N), Sh—Shatski, (9) folded zones: Greater Caucasian marginal sea (GCMS), MT (Meso-Tethys), Mediterranean Sea (MS); mountain systems: Zagros (Z), Lesser Caucasian (LC), Al (Alborz)

basins and transition to the continental regime (stage V) occurred. The sedimentary product from the KT and WTT has been extruded to the side of the SCB, where the Earth's crust appears to be sharply weighed. The geotectonic cycle that broke off in the IV stage is still going on, including such processes as the earth crust compression, CC subduction, and uncompensated sedimentation with terrigenous sediments progradation on the slopes.

To sum up, as the investigation shows, beginning from the Middle Jurassic up to the Quaternary inclusive, the following tectonotypes of sedimentary basins replaced each other in space and time have been existing in a region:

1. Riftogenic (Middle Jurassic), marginal-marine (Upper Jurassic-Eocene), deep-sea basins partly isolated from the world ocean and being under stagnating conditions of sedimentary (Oligocene-Early Miocene),
2. Molassic basins of intermontane and piedmont depressions (Middle Miocene-Quaternary),
3. South Caspian deep-sea uncompensated basin (Pliocene–Quaternary).

1.8 Sedimentary Cover of the SCB

1.8.1 Sedimentation Conditions in Paleobasins

A study on sedimentation complexes and the quest for oil and gas promising objects (structures, traps) in the SCB section requires being well-informed of the genotype of paleo-basins existing under specific sedimentation conditions during their evolution. During the last two decades, seismostratigraphy has become an influential factor, playing an essential role in geological-geophysical investigations in sedimentary basins.

The seismostratigraphic analysis (SSA) of the GDPM time section gives invaluable material to knowing a megabasin deep structure and its evolutionary development. The SSA and regional geodynamic analysis carried out from the PT standpoint, completing each other, leads us to believe that the SCMB modern structure may be interpreted as a total number of varieties of elementary basins changing each other in space and time in the form of revolutionary series within the framework of a single geotectonic (Alpine) cycle.

Mamedov (2004, 2007) believes that the two main stages may have been distinguished in the process of lithosphere evolution, namely divergent (J_2-P_2) and convergent (P_3-Q) evolution and several development stages. Each stage has been characterized by its inherent tectonic and morphological (lithodynamic) type of paleo-basin, distinguished by its structural form, sedimentation condition, and variety of geological bodies.

Beginning from the Middle Jurassic up to the Quaternary inclusive, the following tectonotypes of SB have existent in a region: riftogenic (Middle Jurassic), marginal-marine (Upper Jurassic-Eocene), deep-sea basin under stagnant sedimentation conditions, partly isolated from the world ocean (Pliocene-Early Miocene), molassic basins of intermontane and piedmont depressions (Middle Miocene-Quaternary) and a deep-sea South Caspian uncompensated basin (Pliocene–Quaternary).

The geodynamic processes and tectonic movements in a region caused the development of tectonically and morphologically different types of SB. As it is known from the standpoint of lithodynamic analysis, stratigraphy distinguishes two morphological types of SB: unconsolidated deep-sea basins (DUB) and shallow epicontinental basins (SECB) (Fig. 1.19). The latter are formed by descending sin-sedimentation tectonic movements typical to the platforms and intermountain troughs. According to Schlezinger (1988), the distinctive feature of SECB is that the sedimentation process takes place practically without change in its bottom depth. As to the DUB—those are types of structures. They are also called sedimentation traps or topographic depressions. DUB represents a morphologically expressed tectonic kettle basin, subjected to the expanding (horizontal) and descending (vertical) tectonic movements but is unfilled by sediments entirely. All the DUB has been formed just in such a way,

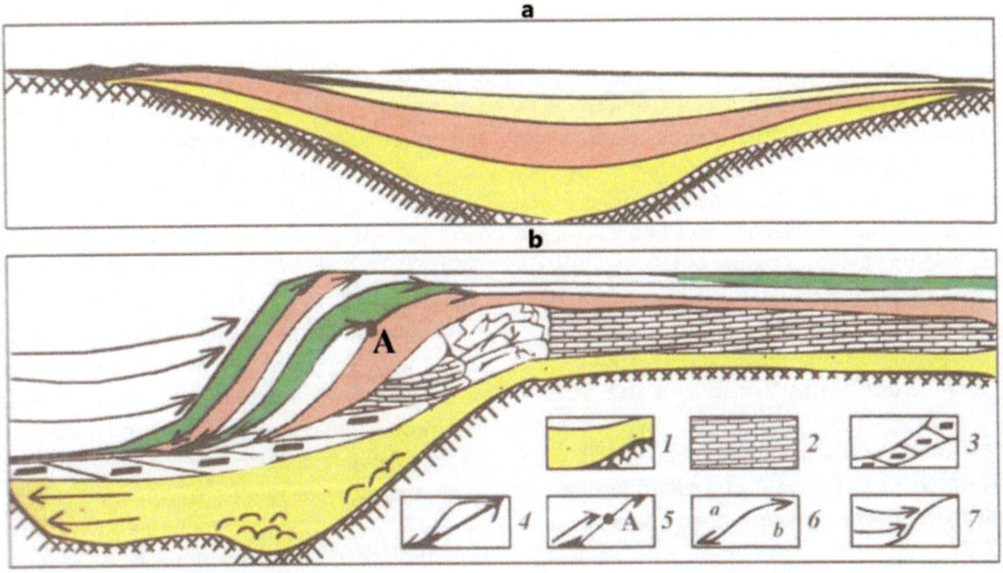

Fig. 1.19 SB tectonotypes: **a** shallow sin-sedimentation epicontinental basin (ECB), **b** deep-sea uncompensated basin (DUB). (1) lateral filling complex (reverse clinoform); (2) carbonate (straight) clinoform; (3) depression deposits; (4) progradation clinoforms; (5) adjacent point-downshift; (6) footing **(a)** and roofing **(b)** adjoining; (7) footing superposition (bedding)

including the Greater Caucasian riftogenic marginal-marine basin in the Middle Jurassic-Eocene epoch. The latter has been subjected to tectonic down-warping and uncompensated sedimentation since its space has been expanded up to the Oligocene. Intensive down-warping and uncompensated sedimentation are also believed to have continued in the late Cenozoic.

The modern relic of the marginal-marine basin—the SCB, is a stable region of the Earth's crust permanent downwarping during 170 Ma, where the sedimentary basin formation process (sedimentogenesis, folding, mud volcanism, and sea-level variation) is going on up to this day.

Specific sedimentation complexes and bodies such as lateral filling and side accretion complexes, carbonate scarps on the shelf brow, straight clinoforms, and condensed covers revealed by SCA in the SCMB section suggest that the paleobasins existed in the region at different periods, relate to the deep-sea and uncompensated type. The lateral filling complexes have been formed by terrigenous material derived from the shelf. With time, relatively gentle slopes (< 100) of paleobasins appear to be buried under these complexes, and as a result, a reverse clinoform has been formed (Fig. 1.2b). The complex mentioned above is composed of subparallel beds being onlap to the bottom relief roughness. An onlap is the most typical feature of DUB.

The lateral accretion complexes were formed under conditions of steeply dipping slopes (> 10 to 150) and an abnormal amount of river drift. In transit, through the shelf, a sedimentary product has been carried to paleobasin on which slopes the clinoform bodies were formed (Fig. 1.19). The clinoform progradation bodies have also been formed in avandeltas far from the deltas. The reconstruction of SCB subsidence history using backstripping analysis (Brunet et al., 2003; Mamedov, 2004, 2008) and seismostratigraphic analysis of the shape and configuration of sedimentation bodies (Mamedov, 1986, 1991) suggest that the old sedimentary basins have been of deep-sea (from 2 to 4.5 km in the divergent stage and from 2 to 0.9 km in the convergent stage) with developed continental margins. Before the results of seismostratigraphic investigations, even the Early Pliocene basin was considered a typical deep-sea basin and did not wholly compensate by sediments under avalanche sedimentation conditions. A narrow relict deep-sea basin has been preserved within the basin near the Iranian part that is confirmed by the presence of the bodies of the slope's lateral accretion and on lap seismofacies, the most important indicators of DUB.

1.8.2 Sediments Volume in the SCB

The SCB is the world's most deep-seated basin, whose sedimentary cover exceeds 25 km. One of the most critical problems of the investigation of SCB sedimentary cover to explain its unique characteristics is estimating sediment volume accumulated right from the start of the basin formation and estimating denudation for a region surrounding the SCB.

To carry out such estimation, Abdullayev (2015) used a 3D geological model of the SCB composed of more than 30 layers grouped in seven-time intervals (pre-Oligocene, Oligocene, Pontian, Early Pliocene, Akchagylian, Absheronian, and Pleistocene). As a result, it was established that the sedimentation rate before the Pliocene was wretched (5–35 m/m y) and then appeared to be sharply increased in the Early Pliocene (up to 300–600 m/m y and reached 1200–1500 m/m y. during Underkirmaki-Kala and Surakhani suites time). After a sharp decrease (to less than 70 m/m y) in the Akchagylian age, the sedimentation rate increased to 300–800 m/m y came to 200–400 m/m y in the Pleistocene. Besides, in the Early Pliocene lower division (Kala suite, Underkirmaki-Kirmaki suites, Superkirmaki sandy suite-Superkirmaki clay suite) and middle division (Break and Balakhani suites), the dominant role belonged to the paleo-Volga. In contrast, the role of paleo-Uzboy and Caucasian Mountain systems was prevalent in the Early Pliocene upper division (Sabunchi suite and Surakhani suite). The given numerical data are apparently averaged since, during the estimation of sediment volume and sedimentation rate, there was some vagueness connected with geochronological inaccuracy (in the absence of biostratigraphic data), as with the neglecting of the rock's seal failure. That is why the information on sedimentation rate in the Early Pliocene presented by different authors appeared to disagree: Mamedov (1991), 3000 m/m y; Van Baak (2010), 2000–4000 m/m y; Guliyev et al. (2003), 1000–1500 m/m y; Abdullayev (2015), 1500–2000 m/m y.

Estimated rates are undoubtedly averaged. In such a way, a super avalanche sedimentation rate of more than 3000 m/Ma in the Early Pliocene (Mamedov, 2006) seems possible to be too high for an average value. However, it may be actual as an instantaneous rate for some period or suite of the PS. The sedimentation process likely occurred unevenly depending on such factors as variations in denudation rates of drainage systems, slope gradients, rate of mountain folding within the surrounding area, and so on. Abdullayev (2015) estimated that SCB sediments volume is more than 775,000 km^3, of which 336,000 km^3 belong to the PS sediments. To analyze an irregularity in sediment inflow to the SCB, this investigator has broken up a sedimentary cover into four sedimentation stages.

(1) A slowed-up sedimentation stage, including the period from the beginning of the basin formation up to the Oligocene. Nearly 200,000 km^2 of sediments have slowly accumulated ($\sim$ 5 m/m y) under conditions of newly formed marginal sea remoteness from the provenance and the absence of mountain structures around this basin.
(2) A stage of increase in sedimentation at the rate of about 10 m/m.y. in the Oligocene and Early Miocene. More than 100,000 km^2 of rocks, 500 to 300 m in thickness, have accumulated during 30 m/y. Such an increase in the sedimentation rate has been related to mountain folding around this basin.
(3) A stage of avalanche sedimentation in the Early Pliocene (b-2.4 m y. ago) when more than 336,000 km^3 of PS rocks have been accumulated is equal to more than 45% of all the basin deposits. High sedimentation rates were caused by a sharp sea-level fall

during the Messinian time (Mediterranean Upper Miocene), paleo-Volga, and other rivers' delta integration. At the end of the Pliocene, the role of paleo-Uzboy increased, and as a result, sedimentation rates within the same suites were estimated to be 300 to 1400 m/m y.

(4) An avalanche deep-sea sedimentation stage, which began from the Akchagylian transgression, occurred after PS accumulation. Within several marine transgressions between 2.5–1.8 m/m y, a sedimentation rate has been estimated to be 75 m/m y, resulting in up to 19,000 km^3 of sediment accumulation. During the Absheron age of the Pliocene and in the Pleistocene (1.8 m y. ago), up to 40,000 km^3 of sediments have been accumulated at the rate of 300–500 m/m y.

Because of the absence of up-to-date seismic data on the Iranian sector of the Caspian Sea (20% of the SCB total area), it could not have been the estimated volume of sediments supplied by the rivers from the Elbrus; therefore, Abdullayev (2015) added more than 20% to the volume of the total deposits.

After the regression of marine waters of the Paratethys, which covered a more significant part of a region during the Neogene, the Caspian topographic basin appeared to be more continental by the Upper Miocene, and Pliocene was completely isolated in the Pontian age. The formation of the recent drainage system in a region may be attributed to the same time. The following paleo-rivers have been included in the drainage system: paleo-Volga, paleo-Kur, small rivers of the Greater and Lesser Caucasus, Elbrus, Kopetdag, and the relict system of paleo-Uzboy—"heiress of the Amudarya".

Based on the results of the South Caspian drainage system analysis and denudation rates of surrounding areas, Abdullayev (2015) determined the volume of sediments supplied into the SCB. According to the results of different estimation methods, the maximum rates of supply with sediments in the PS basin have been as follows: in the paleo-Amudarya drainage system (from 124,000 km^3/m y and over); within the Caucasian drainage system (800–320,000 km^3/m y), in paleo-Volga system (up to 125,000 km^3/m y), and least of all,—the Elbrus drainage system (up to 40,000 km^3/m y).

Abdullayev (2015) assumed the Caspian Sea level fell (after its isolation from the ocean 5.6 m y ago) down to 1500 m as the criterion for the vertical basin denudation. Figure 1.20 shows

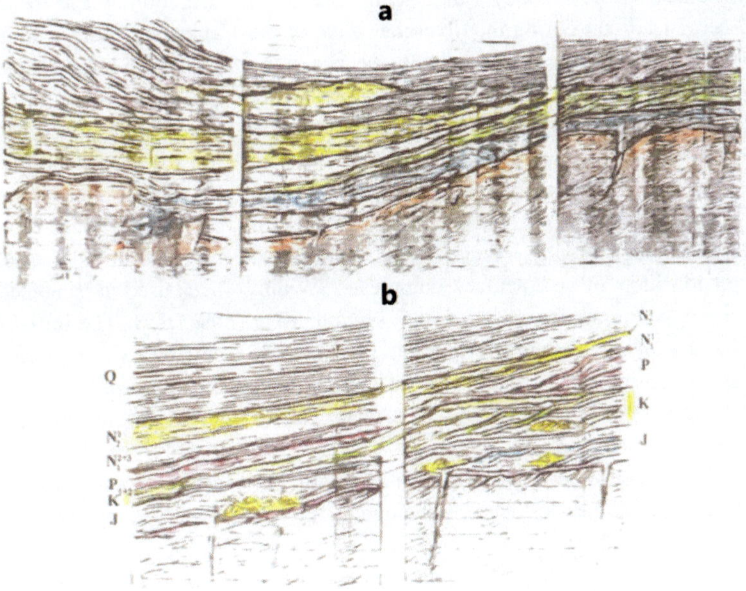

Fig. 1.20 Seismic time section of profile crossing the SCHB northern continental margin (**a**) and its fragment (**b**)

an evolution of the restored profile of the paleo-Volga basin in the line of "N-S" by each of the PS's suites. Of the paleo-Volga, an erosion profile (above an onlap point) has been changed from more steeply dipping (0.03) at the beginning of Kala suite time up to very gently sloping (0.01) in the Surakhani time, suggesting that to the end of the Early Pliocene, the paleo-Volga had played a decreased role in supply with sediments.

1.8.3 Seismostratigraphic Characteristics of Sedimentary Series

Tectonic and sedimentation factors are believed to have divided the megabasin development into separate stages and phases. As shown above, according to the vital seismic horizons and surfaces of unconformity in the SCMB sedimentary cover, the nine quasi-synchronous sedimentations (age) complexes are distinguished (Mamedov, 2007). Considering the biostratigraphic, petrological, well logging, and other data, the age succession, substantial composition, sedimentation condition, and other special features of the sedimentary cover structure, as well as a reliable megabasin framework has been drawn up (Mamedov, 1991, 2007). Based on the SSA, it is recognized that during the divergent stage of the thick sedimentary cover formation, the role of the sea-level eustatic changes (SLEC) has relatively been insignificant. However, they have partly been related to the tectonic factor (bottom depression, the rise of blocks, basin subsidence, and so on). However, the SLEC has influenced the paleogeography, facies, and substantial composition of sedimentary complexes. Within chronostratigraphic units of the section were formed deltaic, nearshore-marine, shelf, slope, and deep-sea facies zones, whose vertical succession determines a cyclic recurrence in the sedimentation process.

Thus, a eustatic factor is likely responsible for creating a vital facies class of geological objects, which may have been clearly distinguished on seismic sections by their configuration and dynamics of reflections, bedding mode, and sedimentation bodies' geometry.

One of the essential conclusions in paleogeographic and seismofacies analysis of SCMB is that the thicknesses of sediments in Meso-Paleogene seas have mainly been determined by the scale of erosion of their northern continental surroundings. The marginal marine basin has been filled primarily at the expense of clastic products removed from the platform by the rivers, mudflows, and dense underwater streams. Less of mainly pyroclastic material has been supplied from the island arc side. At the concurrent stage beginning from the Oligocene, the role of mountain systems of the Lesser Caucasus, Elbrus, Greater Caucasus, Kopetdag, and Talysh (as suppliers of sedimentary products) appeared to have sharply increased.

Modern seismostratigraphy is based on the careful and detailed examination of features of all the wave patterns, including an external shape of both substantial extent and weak interrupted reflections. It enables us to derive that the traditional flat-bedded sedimentation models are atypical and by no means dominant for the SCMB section, where diversified and multifaced sedimentation bodies of lateral filling and accretion appear to be predominant. The latter is characterized by an initial dip of beds and the sigmoidal configuration of their boundaries. The bed thicknesses have not represented the scales of synchronous vertical (adverse) tectonic movements. Therefore, such known methods as "thickness analysis", "unstripped thickness restoration", and other expedients of extrapolation and interpolation of surfaces may not have been used in paleotectonic reconstructions within the DUB. The initial flat-bedded series (leveling covers) indicate sedimentation on the shelves and the bottoms of deep-seated basins (condensed covers of quasi-liquid diffluence over the bottom). Even though the mentioned series have been subjected to tectonic actions (subsidence, rise, folding, etc.), they are recognized by subparallel reflection patterns. The thickness of the beds accumulated on the shelves is believed to depend on tectonic down-warping and may be considered their measures. This rule, with some assumptions, may also be applied to the carbonate plateau. At the same time, the thicknesses of beds on condensed covers at the depths appear to not depend on tectonic down-warping amplitude, mainly defined by supplied product quantity (Schlesinger, 1988).

Based on seismostratigraphic data obtained from regional profiles, it is concluded that uncompensated sedimentation has widely been developed in almost all the stages of basin development. It should be noted that synchronous sedimentation took place in almost all the paleogeographical zones depending on the tectonics, relief morphology, and hydrodynamics.

To single out and map sedimentation bodies being formed in the near slope zones of DUB, the seismostratigraphic fundamental principle is a principle of sedimentation bodies' three-dimensionality according to which subhorizontal, cross, and wedge-like laminations are indicators of sedimentation regimes.

The seismic sections show clear indications of DUB filling in the divergent stage, mainly from the north and the STP side, while in the convergent stage, the load sources have been mountain systems and hilly regions around the basins.

Within the basin joint zone with the epi-Hercynian platform or with mountain systems, the fluvial-deltaic formations shelves covers, lateral series of the slope terrigenous clinoforms, debris cones at the foot, and condensed covers at the abyssal zones appear to have formed a specific paragenesis highly typical to the DUB. During the periods when active removal sources and large deltaic systems have been at a great distance from the basin, the paragenesis of shelf carbonates (complicated by biogenic and reef knolls), slope depression deposits passing into thin covers close-fitting a bottom, have been formed on its continental margin. This paragenesis type is also one of the DUB's most important indicative features.

1.8.4 Sedimentation in Deep-Sea Basins of the Divergent Development Stage

Seismic data analysis shows an objective pattern of the old shelf and slope structure. It is concluded that a deep-sea basin of the GCMS has been separated from the shelf and epi-continental basins (ECB) (formed by sin-sedimentation down-warping at the platform) by the system of marginal uplifts (western cusp of Karabogaz arc (KBA), Agzybirchala horst (AH), and East Fore-Caucasian uplift). For a long time, these uplifts have existed as the Paleozoic basement's raised elements are subjected to erosion and washout.

Thus, they have played a significant role in forming the shelf and slope structures of the marginal sea and have been dominant factors controlling sedimentation conditions within the epi-continental basins on the platform. For instance, for a long time, almost up to the beginning of the Late Cretaceous, the KBA has served as provenance for both the GCMS, and epi-continental basins located on the platform. According to the drilling and seismic survey data, the Agzybirchala horst has been paleoland almost up to the Miocene. The basement other cusps occurred below sea level in the Mesozoic; they are believed to have existed in different periods as the islands are structurally elevated elements. The denudation surface of marginal uplifts and cusps indicates the erosional processes and the past dense streams' "action".

Subsequently, the mentioned cusps have been involved in subsidence because of the sedimentation basin's expansion and the platform margin's subsidence. Their eroded surface and external limbs turned out to be the continental shelves and slopes of the marginal sea, correspondingly.

1.8.4.1 Shelf Sedimentation Models

In connection with the platform margin epigenetic subsidence and sea-level elevation during the Jurassic and Cretaceous periods, an advance of the sea within the vast platform areas has taken place. A clear notion of the platform transgression and submergence may have been obtained from the high-information GDPM time sections (Fig. 1.20). Successive transgressive (basal) overlap of synphase axes upon the pre-Jurassic surface of unconformity (SU_1) is an essential and objective indicator of a shelf expansion in the Jurassic period. Upper Paleozoic metamorphic rocks (and locally Permian–Triassic transition complex rocks) have been cut by the SU_1, which settles the beginning of the STP development plate stage.

Subsequently, geological bodies of the shelves have been formed at the expense of uninterrupted (more often, differentiated) compensated downwarping. Compensation sedimentation took place only on active shelves subjected to the subsidence practically without change of the bottom depth and continued up to the opening of the equilibrium profile. The sedimentation surface on such shelves coincided with a terrestrial shelf erosion baseline. Subsequently, the bottom tectonic settling and hydrodynamics have controlled sedimentation conditions. Transgression and regional sedimentation on the shelf occurred during sea level (SL) raising. Subparallel beds have been superimposed onto the gently inclined bottom surfaces and appear pinched out by the shoreline (Fig. 1.20). On the passive shelves, which have not been involved in the subsidence process, the sediments have yet to be deposited and removed in transit to the deep-sea part of the basin. The waves and currents are believed to be responsible for forming the plane truncations. These truncations of the heads of underlying deposits are clearly seen on seismic time sections.

As shown from the Middle Caspian seismic section, the total thickness of the Jurassic shelf sediments (SSC-1) does not exceed 1.0–1.5 km, while the thickness of synchronous sediments is increased towards the Caucasian geosyncline where the flysch accumulated at the foot of the old slope, reaches 6–7 km in thickness. Further to the SW, within the Kur trough and in the central SCB, where an abyssal region of the marginal sea has been developing, the flysch thickness appears to have decreased to 1.5–2.0 km. As a result of sea-level lowering taking place between Jurassic and Cretaceous times (that is evident from the shoreline southwesterly displacement on the platform) (Fig. 1.3), the terrigenous sediments removed from the platform outcropping areas have been deposited on the shelf.

According to the drilling data, the Upper Cretaceous complex (SSC-3) comprises a thick carbonate series. As seen from time sections, several seaward accreted carbonate scarps may have been distinguished from the shelf brow. They form one large straight clinoform (Fig. 1.20). Using a particular method (Kunin, 1989), we estimated that the sea basin depth behind the shelf brow over the scarps' sigmoidal surfaces is between 300 and 400 m. These depths correspond entirely to the carbonate accumulation critical depth. Towards the sea, carbonates appear to be changed by depression formations. The facies transition from the shelf carbonates to the depression slope sediments may have been traced by the change for the worse of dynamic expressiveness and seismofacies appearance. The development of carbonate series on the external shelf indicates an uncompensated sedimentation regime and remoteness of the provenance. An accumulation of the mentioned series is typical for the shelf basin expansion phase, corresponding to the maximum transgression during the Earth crust extension damping stage. The Late Miocene is just the border between the GCMS extension "mature" stage and its gradual "extinction" stage (almost up to the Oligocene). At that time, a thin condensed cover had been formed in an abyssal zone of paleo-basin, with a small thickness (up to 0.5 km) and a wide age range ($K_2 \sim 30$ m y).

The carbonate shallow-marine formations are believed to be the best evidence of shelf sedimentation during global sea level rising and epeirogenetic subsidence of the platform margin. Almost monolithic calcareous series of up to 0.8 km thick Jurassic and up to 1.0 km thick Upper Cretaceous complexes are distinguished in seismic sections by their specific macro-laminated habit with carbonate clinoform thinning out behind a shelf brow. The same-age carbonate clinoforms framed in a deep-sea basin from the north have a similar pattern in pre-Caucasian seismic sections.

After warm sea transgression in K_2, a stable high sea level became settled at a low temperature of shelf subsidence. The carbonate scarps development zones on a shelf brow are favorable regions for reefing. We have distinguished the barrier reefs having high interval rates located on the platform margin (Fig. 1.4) and the northern terrace of the North-Absheron "Mesozoic cordillera" (Mamedov, 1991). Paleocene-Eocene times (SSC-4) have been marked by the fact that the slope regime has changed the shelf condition on the platform margin.

1.8.4.2 Slope Sedimentation Models

From an analysis of seismic data obtained in the transition zone between the South and Middle Caspian, it is inferred that several models of uncompensated slope sedimentation may have been distinguished. Among the others, there are distinguished bodies with original non-horizontal bedding and lateral thickness changes—so-named clinoforms (CF). Abundant terrigenous, mainly progradation clinoforms, have formed on continental slopes close to the provenance. Such terrigenous clinoforms of the Eocene–Paleocene seismocomplex (SSC-4) may have been recognized as an example of the slope setting. The slopes' lateral accretions have been caused by a sharp down-warping of the platform margin and the approach of the clastic product provenance in the late Cretaceous.

Farther south, the SSC-4 surface appears to have almost merged with the Mesozoic surface from a united unconformity surface. This surface is distinguished along the STP margin, separating the plate complex from the orogenic Oligocene-Anthropocene complex.

The gentle continental slopes have been developed in the zones of conjugation of active shelves with deep-sea troughs without marginal uplift. During the Cretaceous to Paleogene, within one of the same zones located between the KBA and Agzybirchala cusp, sharp transgressions took place that resulted in the formation of carbonate series with clinoform cusp on the vast shelf area remoted from the provenance. The KBA is relatively steeply dipping (Guliyev et al., 2003) and high (up to 3–4 km), and southern continental slopes are well-read in seismic sections where the platy truncations, broken clinoforms, and collapse bodies are also clearly recorded (Fig. 1.21).

Such slopes have been formed from the sections for a long time, beginning from the Middle-late Jurassic to the Oligocene. The fact that there are no deposits on the slope indicates that it was a

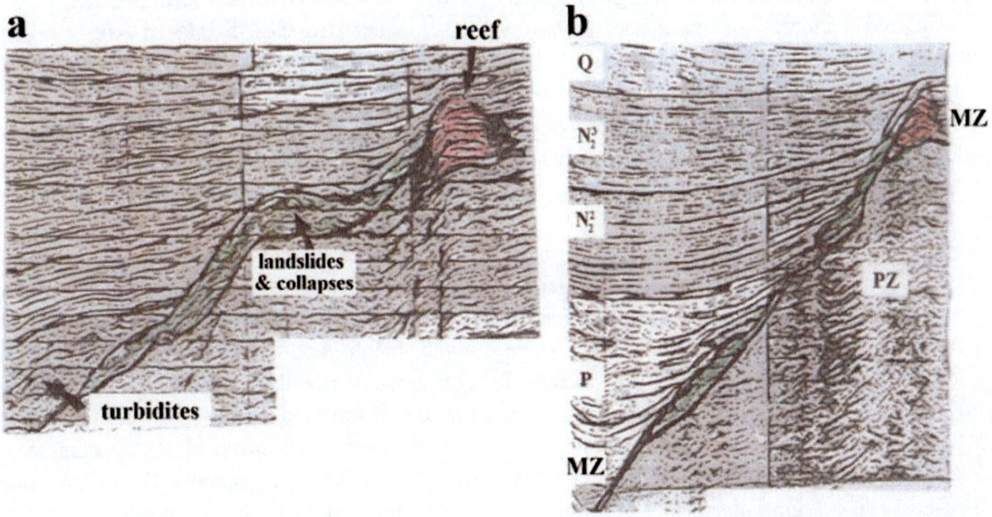

Fig. 1.21 Time-sections fragments illustrate the reef knolls on a shelf brow and landslide collapse (**a**), and turbidites on a slope (**b**)

transit region. A slope plane has, in places, eroded down to the basement and wrinkled by a system of canyons. The same age complexes (sequences) are distinguished at two levels: on a shelf and at the slope foot. The primary controlling mechanism of sedimentary cover transferring is believed to have been a cyclic sea level lowering.

The SSA shows that the upper canyons entering a shelf and cleaving aslope are movement channels for many sediments supplied from the shelf space. Benthic currents and turbidity flows have fulfilled this transportation. Submarine canyons are filled up by seismofacies characterized by hummocky and hilly overlapping of equi-phase axes. As time sections show, the "steamy" and landslide seismofacies are also well traced on the upper slope (Fig. 1.12).

From scrutinizing structural and morphological special features of the Mesozoic to Paleogene paleobasin edge zone, it is inferred that this zone is submitted to the general classical model of marginal seas "passive" side. Interestingly, the unique features of morphostructures, paleo-shelves, and paleo-slopes of the GCMS continental margin are quite like those of recent backarched seas. A specific character of their joining as with continent and with island arcs (microcontinents) is considered an important indicator to determine their marginal-marine genesis and tectonics of marine paleo-basins within the AHMB.

In summing up the seismostratigraphic analysis data, it can be distinguished between two following models of slope sedimentation within an old passive continental margin:

- Model I includes a gentle slope that has been periodically covered with the bodies of lateral accretion of clastic sediments and depression facies of carbonate clinoforms. Such a slope appeared to be a region of localized sedimentation,
- Model II includes a relatively steep slope on which the sediments have yet to accumulate. It has been a region of erosion and transit of deposits.

These two models are the typical structures of continental slopes within uncompensated deep-sea basins. Thus, from the above-described data, the basement border structures on the epi-Hercynian platform played an exceptional role in forming different age sedimentation complexes and bodies of specific forms within the SCMB. As displayed in seismic sections, the complexes of gravity, density flow, and collapses well preserved on a slope have consequently been buried under thick Cenozoic deposits.

All these exceptional data allow the outline of the density streams and gravitates distribution area with which non-structural hydrocarbon traps may be connected.

From the sections, such slopes have been formed for a long time, beginning from the Middle-late Jurassic up to the Oligocene. The fact that there are no deposits on the slope indicates that it was a transit region. A slope plane has in places, eroded down to the basement, and wrinkled by a system of canyons. The same age complexes (sequences) are distinguished at two levels: on a shelf and at the slope foot. The primary controlling mechanism of sedimentary cover transferring is believed to have been a cyclic sea level lowering.

1.8.5 Sedimentation Conditions on the GCMS Island Arc Margin

During the divergent stage, unique island arc system development features have predetermined sedimentation conditions and conditions (under which specific geological bodies have been formed on an active margin). It is known that from the seismic sections, detailed information on volcanic cusps may be obtained, such as the width, height, and complex configuration of the volcanogenic island arc surface. A relatively high (up to 4–5 km) and broad (30–70 km) island arc buried under a Cenozoic series appears to be traced as a ridge of generally Caucasian orientation for hundreds of kilometers, from the Talysh up to Dzirulian massif (Fig. 1.22). As

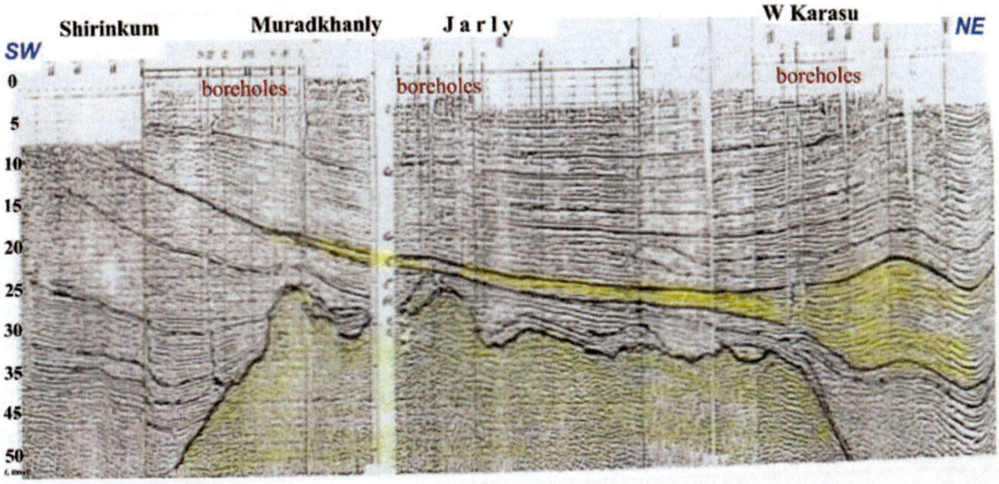

Fig. 1.22 The volcanogenic island arc imagery derived from seismic data

seismic sections show, a steeply dipping insular slope is well reflected on the island arc's northern flank. From the thickness of contiguous beds, it is estimated that the slope height has been at least 3.5–4 km, corresponding to the continental slope's height on the opposite northern seaside. Various geological accumulative bodies such as the paleorelief remnants, underwater canals, and erosional truncations appear to have been distinguished in the zones of underwater denudation and erosion of volcanogenic and sedimentary series within the island arc development area.

In the Late Cretaceous, subaerial volcanites being deprived of the provenance during an island arc subsidence have been embodied in carbonate formation complicated by biogenic and reefal buildups. The carbonate sheet thickness has reached 600–800 m.

Because of sea-level lowering in the Cretaceous, the upper island arc appeared to be exposed, and during almost 50 m/y, it was under continental conditions (up to the late Miocene). Marine sedimentation has been continued only in the northern island arc. A narrow sin-sedimentation basin is well distinguished from the volcanites shown in the seismic section of several profiles. The typical features of this basin are the Paleogene beds thinning out towards a periphery and following their fan-shaped divergence and widening. Such narrow basins, mainly in the northern island arc, are evidenced by the tension and subsidence of substrate blocks underlying volcanogenic massif. The stratigraphic range of the volcanogenic series (J_2 - K_1) affected by the faults and the age of terrigenous and tufogenic sediments filling up the trough suggests that the substrate blocks displacement and subsidence have most likely occurred in Paleogene time. The detection of fragments with a similar pattern in the sections of the Middle Kur trough and the southern (Iranian) part of the South Caspian would have been used as a documentary confirmation of the suppositions brought up by Schlezinger (1988) and Zonenshain et al. (1990) about a narrow trough opening on the GCMS northern periphery in the Eocene.

An onlap of the Eocene tufogenic rock layers onto volcanogenic massif surface has also been marked on the slopes of the Saatly-Geychai-Mingechavir segment of an island. Such facts as seismic horizons dip-slip (towards the MTO) and an ending of their tracing utilizing the foot onlap on a slope—indicate a sharp sea-level lowering (~ 300 m) in the late Miocene. In the joint zone of the island shelf with the island slope, the plane beveling formed by which a Mesozoic volcanogenic substrate appears to be exposed.

1.8.6 Sedimentation in Paleobasins of the Divergent Stage Deltaic Sedimentation Models

Among sedimentary tracts (systems), the fluvial-deltaic systems are most important because of an abundance of the rocks (lithofacies) having good reservoir properties. The deltaic Sedimentation is a considerable mechanism provided the Late Cenozoic paleobasins fill a region. Within the basin investigated, the paleo-Volga and paleo-Uzboy buried Pliocene deltaic systems have been studied in detail, considering the data gained from the natural exposures and core samples from the seismic survey and GTW data. We have used the synthesis of all the data of lithofacies composition and internal structure (texture) of the paleo-Volga exposed deltaic deposits in the Absheron Peninsula to study the SCB buried deltaic systems by the seismic reflection distinctive patterns as in drawing up seismostratigraphic models of sedimentation.

Pliocene time is highly significant to the beginning of SCB's sharp sea-level lowering, avalanche sedimentation, and rapid crust subduction. In the Pontian only (7.1–5.4 m/y), the amplitude of subduction came to more than 1 km (Mamedov, 2008; Volozh & Leonov, 2004). The vast areas of the Middle Caspian and recent intermontane troughs have been exposed. Such big rivers as paleo-Volga, paleo-Kur, paleo-Uzboy, and tens of small ones cut deep valleys in the surrounding land. They moved into the Pliocene lake, delivering many clastic products.

The paleo-Volga cutting valleys are well distinguished in the Middle Caspian seismic sections in the Cretaceous and Paleogene complexes. During the Kala time, the paleo-Volga delta occupied a vast area in the latitude of the recent Absheron Peninsula and the same name archipelago (Mamedov, 1991). Subaerial delta and prodelta have been moved far to the south (more than 50–100 km), throwing out on a shallow shelf a large volume of rudaceous deposits. The morphological and textural features of sandy-siltstone deltaic suites have been well studied in natural exposures of the Absheron Peninsula. As may be seen in several exposures (in the Kirmaki Valley, Zhiloy Island, Sumgait River Valley, and others), deltaic formations are represented by obliquely laminated partings whose upper parts are truncated by the sub-horizontal surface of erosion (Sultanov & Gorin, 1963).

Naturally, the mentioned features of the deltaic origin deposits appear to be out of the limits of medium frequency seismic survey resolving ability. However, their relatively large wedge-shaped and lenticular elements as well as streamy configuration and external morphological features, have been rather clearly manifested in high-frequency dynamic sections (Fig. 1.23b). From the drilling data, logging data analysis, and seismofacies analysis, it is concluded that the paleo-Volga River system and its deltaic and shelf complexes are genetically linked up lateral (mainly) and vertical series of accumulative bodies. The paleo-Volga delta height during Kala, Underkirmaki, and Kirmaki times has, on average, been several tens of meters, having a lobate form in a plan and wedge-shaped section. Along the strike, deltaic complexes are characterized by lenticular-bedded morphology: seismic sections and isopach maps trace stream courses. The scattered zones of anomalous thicknesses (more than 200–300 m) resembling recent delta lobes are distinguished on isopach maps of the PS foot suites (Mamedov, 1991).

1.9 Shelf Sedimentation Models

The paleo-Volga cutting valleys are well distinguished in the Middle Caspian seismic sections in the Cretaceous and Paleogene complexes. During the Kala time, the paleo-Volga delta occupied a vast area in the latitude of the recent Absheron Peninsula and the same name archipelago (Mamedov, 1991). Subaerial delta and prodelta have been moved far to the south (more than 50–100 km), throwing a large volume of rudaceous deposits out on a shallow shelf. The morphological and textural features of sandy-siltstone deltaic suites have been well studied in natural exposures of the Absheron Peninsula, as may be seen in several exposures (in the Kirmaki

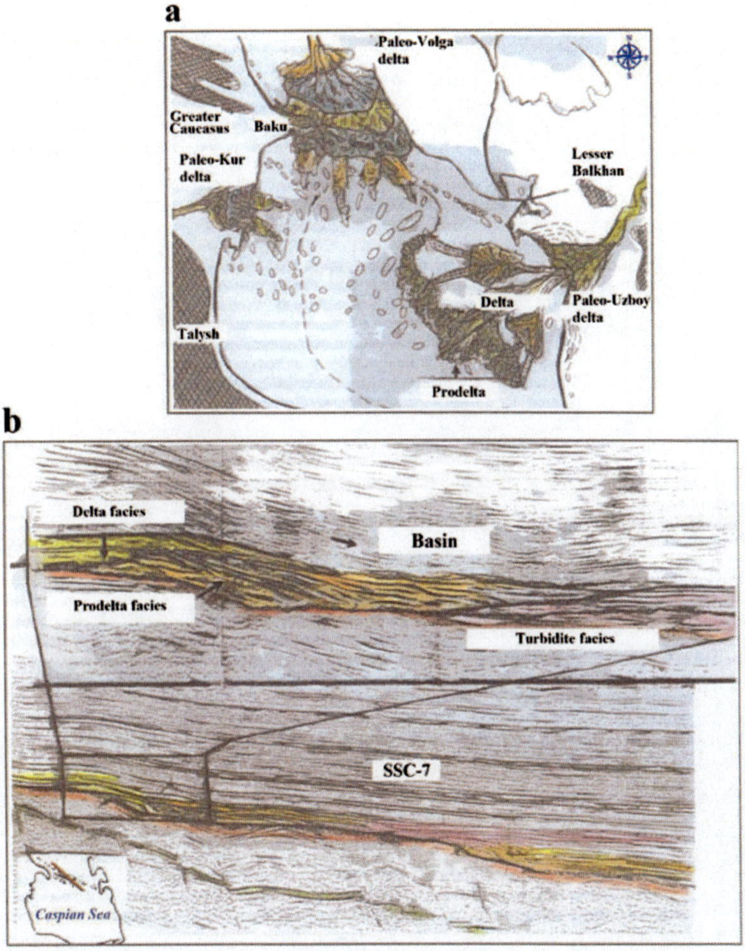

Fig. 1.23 Paleo-rivers deltas in the early Pliocene (**a**) and deltaic deposits displayed in the seismic section (**b**)

Valley, Zhiloy Island, Sumgait River Valley, and others), deltaic formations are represented by obliquely laminated partings whose upper parts are truncated by the sub-horizontal surface of erosion (Sultanov & Gorin, 1963).

Naturally, the mentioned features of the deltaic origin deposits are out of the limits of medium frequency seismic survey resolving ability. However, their relatively large wedge-shaped and lenticular elements, streamy configuration, and external morphological features have been rather clearly manifested in high-frequency dynamic sections (Fig. 1.23b). From the drilling data, logging data analysis, and seismofacies analysis, it is concluded that the paleo-Volga River system and its deltaic and shelf complexes are genetically linked up lateral (mainly) and vertical series of accumulative bodies. The paleo-Volga delta height during Kala, Underkirmaki, and Kirmaki times has, on average, been several tens of meters, having a lobate form in a plan and wedge-shaped section. Along the strike, deltaic complexes are characterized by lenticular-bedded morphology. The scattered zones of anomalous thicknesses (more than 200–300 m) resembling recent delta lobes are distinguished on isopach maps of the PS foot suites (Mamedov, 1991).

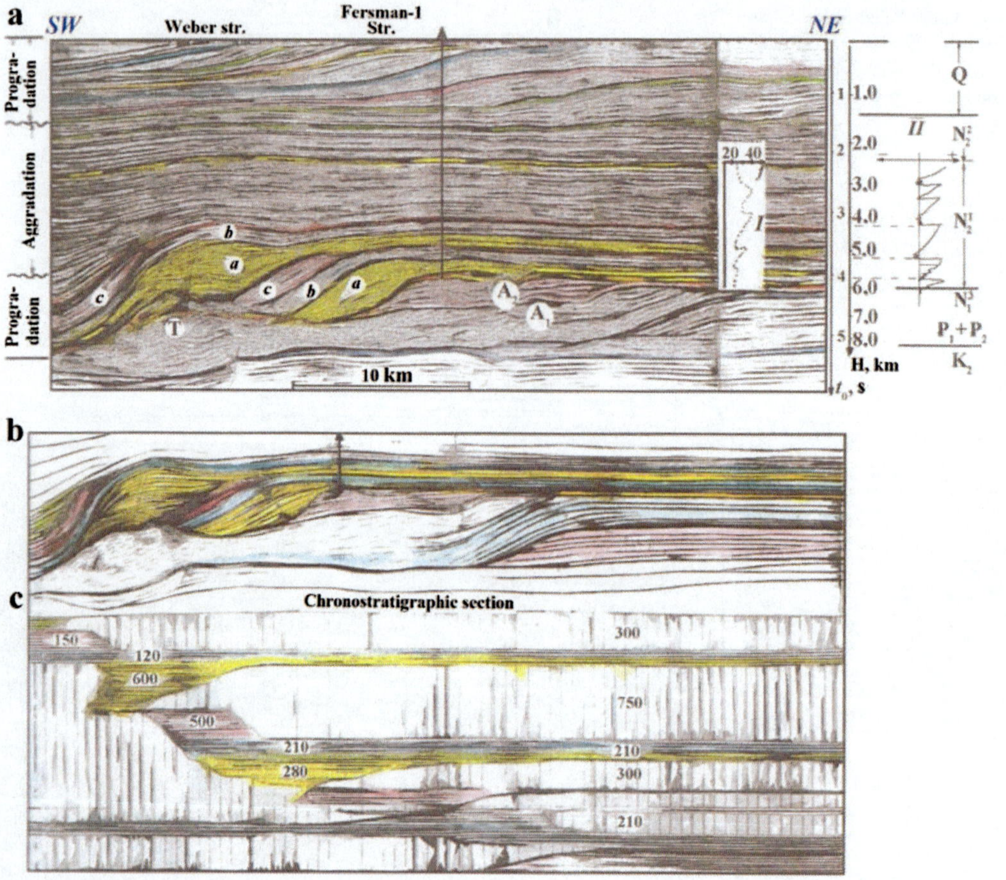

Fig. 1.24 Seismic time-section (**a**), seismic quantum (**b**), and chronostratigraphic (**c**) sections of the lateral accretion complex (the South Caspian Turkmenian shelf). A_1 and A_2 are the semifacies of the foredelta clinoform; (T) turbidites; a, b, c are the elementary bodies of the slope clinoforms (clinocycles); **a**—clinoform; **b**—clinosheet; **c**—obliquely laminated body; (I) Reflected wave visible frequency chart; (II) the SLRV chart

1.10 Foredeltaic and Slope Sedimentation in the Pliocene Basin

From the result of seismostratigraphic investigations of deltaic and shelf deposits in the SCB section, it is inferred that a great variety characterizes the two-dimensional models of deltaic sedimentary systems. Notably, the paleo-Volga deltas have been developed in the northwestern periphery of the Pliocene basin. In contrast, its northeastern periphery is marked by the development of fan-shaped delta-type sedimentary tracts having advanced far forward, up to 200 km (Fig. 1.24a). On investigating lateral accretion bodies in the Turkmenian shelf sections, a version of a relict deep-sea topographic depression in the southern periphery of the Pliocene basin (in the place of the modern Pre-Elbursian trough). The data of broad time-based regional profiles carried out in the nineties of the last century have confirmed this version (Mamedov, 2007).

Examples of foredeltaic clinoform advance for tens and hundreds of kilometers away from the mouth and delta are known from the published data (Lisitsyn, 1988; Shlezinger, 1988). Within the Turkmenian shelf, shallow incisions typical to the fan-shaped delta have been distinguished in lengthwise seismic sections drawn up through the Productive Series lowermost strata (red series). Here, the mentioned fan-shaped delta systems form alluvial fans accreted towards the topographic depression. The presence of subparallel beds marks an interior sea-side shelf zone. During a sharp sea-level lowering, the sediments transported from the shelf seem to have taken place over a topographic depression steep continental slope. Based on a few seismofacies and morphological features, the sizeable foredeltaic clinoform and two slope clinoforms have been distinguished (Fig. 1.23).

The foredeltaic clinoform has been formed during one full paracycle of SLRV (~ 500 t y). According to the seismic section data, the topographic depression's initial slope has laterally been accreted by the mentioned clinoform for about 10–12 km, while its vertical accretion appears to be only 100–200 m. As a result, a new steeper (~ 100) sedimentation slope has been formed, which during a long interval (~ 300 t y) has also been accreted with the slope clinoforms-clinocycles (Mamedov, 1991).

During the next SLRV paracycle (~ 800 t y), a slope has been laterally accreted by the first clinocycle for up to 8–10 km. It was formed when a sediment volume applied to the shelf appeared to have been overdrawn, the possibility of compensation accumulation. As a result, the deposits have been carried out of its edge. Each clinocycle has had an unmistakable cyclic character being controlled by SLRV; every stage of the clastic product delivery to a slope region is characterized by the concrete three-dimensional body, namely clinoform (lower SL), clinosheet (high SL) and obliquely laminated body (stable SL) (Fig. 1.23).

The next clinocycle was formed during the SLRV paracycle and lasted 840 thousand years. It is also composed of three elements. The high amplitude regular reflection from the clinoform frontal surface, bounding the northeastern slope of a relict deep-sea basin, inherited from the Pontian DUB. The studies on lateral accretion bodies in the PS section will be described in more detail in the next chapter.

References

Abdullayev, N. (2015). Evolution of the sedimentary cover and assessment of the prospects of the South Caspian Sea. Abstract of diss. Doc. Phil. in Earth Sciences. Acad. of Sci. of Azerbaijan, Baku.

Adamiya, S. A., Gamkrelidze, T. P., Zakhariadze, G. S., & Lordkipanidze, M. B. (1974). Adjara-Trialeti trough and the problem of formation of the Black Sea deep-water basin. *Geotectonics, 1*, 78–93.

Aksenovich, G. I. et al. (1962). Deep Seismic Sounding in the Central Part of the Caspian Sea. Acad. Sci. of the USSR, Moscow, 152 p. (in Russian).

Alizadeh, A. A., Guliyev, I. S., Kadirov, F. A., & Eppelbaum, L. V. (2016). *Geosciences in Azerbaijan. Volume I: Geology* (239 p.). Springer.

Alizadeh, A. A., Guliyev, I. S., Kadirov, F. A., & Eppelbaum, L. V. (2017). *Geosciences in Azerbaijan. Volume II: Economic minerals and applied geophysics* (340 p.). Springer.

Allen, M., Jones, S., Ismail-Zadeh, A., Simmons, M., & Anderson, L. (2002). Onset of subduction as the cause of rapid Pliocene-Quaternary subsidence in the South Caspian Basin. Geological Society of America. *Geology, 30*(9), 775–778.

Amursky, G. I., Tiunov, K. V., Kharikov, B. A., & Shlezinger, A. E. (1968). *Structure and tectonics of Bolshoi Balkhan* (p. 234). Nauka.

Baranova, E. P., Kosminskaya, N. P., & Pavlenkova, N. I. (1990). Results of reinterpreting of deep seismic sounding for Southern Caspian. *Geophysical Journal, 12*(5), 60–67. (in Russian).

Brunet, M. F., Korotaev, M. V., Ershov, A. V., & Nikishin, A. M. (2003). The South Caspian Basin: A review of its evolution from subsidence modeling. *Sedimentary Geology, 156*, 119–148.

Brunet, M.F., Barrier, E., Eban, S., Mammadov, P. Z., et al. (2005). Tectonic and subsidence evolutions of the South Caspian basin margins in Iran and Azerbaijan. Trans. of the V Intern. Conf. "Petroleum geology and hydrocarbon potential of Caspian and Black Seas region" (pp. 87–88). Baku.

Dercourt, J., Zonenshain, L. P., Ricou, L.-E., et al. (1986). Geological evolution of the Tethys belt from the Atlantic to the Pamirs since the LIAS. *Tectonophysics, 12*, 241–315.

Eppelbaum, L. V. (2019). *Geophysical potential fields: Geological and environmental applications* (467 p.). Elsevier.

Gamkrelidze, I. P. (1984). The tectonic structure and Alpine geodynamics of the Caucasus and adjacent areas. In Z. W. Otchmezuri (Ed.), *Tectonics and metallogeny of the Caucasus* (pp. 105–184). Mezniereba, Tbilisi (in Russian).

Glumov, I. F., Malovitsky, Y. P., Novikov, A. A., & Senin, B. V. (2004). *Regional geology and petroleum potential of the Caspian Sea* (342 p.). Nedra, Moscow (in Russian).

Granath, J. W., Soofi, K. A., Baganz, O. W., & Bagirov, E. (2000). Gravity modeling and its implications to the tectonics of the South Caspian Basin. *Trans. of the AAPG's international regional conference* (pp. 46–50), Turkey.

Green, T., Abdullayev, N., Hossack, J., Riley, G., & Roberts, A. (2009). Sedimentation and subsidence in the South Caspian Basin, Azerbaijan. In M.-F. Brunet, M. Wilmsen & J. W. Granath (Eds.), *South Caspian to central Iran Basins* (Vol. 312, pp. 241–260). Geol. Society, London, Spec. Public.

Guliyev, I. S., Levin, L. E., & Fyodorov, D. L. (2003). *Hydrocarbons potential of the Caspian region: System analysis*. Nafta Press.

Jackson, J., Priestley, K., Allen, M., & Berberian, M. (2002). Active tectonics of the South Caspian Basin. *Geophysical Journal International, 148*, 214–245.

Khain, V. E. (2001). *Tectonics of continents and oceans*. Scientific World, Moscow (in Russian).

Knapp, C. C., Knapp, J. H., & Connor, J. A. (2004). Crustal-scale structure of the South Caspian Basin revealed by deep seismic reflection profiling. *Marine and Petroleum Geology, 21*(8), 1073–1081.

Knapp, J. H., Diaconescu, C. C., & Connor, J. (2000). Crustal-scale imaging of the Absheron Ridge (South Caspian Sea) revealed by deep seismic reflection profiling. In *Trans. of the AAPG's Inaugural Region al International Conference* (pp. 153–154), Istanbul, Turkey.

Kosminskaya, N.P. (1968). Method of Deep Seismic Sounding of the Earth's Crust and Upper Mantle. *Development of the Fundamentals of the Method.* Nauka, Moscow, (227 p.). (in Russian).

Kunin, N. Y. (1989). *Structure of the lithosphere of continents and oceans* (286 p.). Nedra, Moscow (in Russian).

Kucheruk, E. V. (1990). Features of the exploration for hydrocarbon accumulations in basement rocks. *Geology of Oil and Gas, 10*, 8–9. (in Russian).

Kukal, Z. (1987). *Velocity of Geological Processes* (p. 300). Mir.

Lerche, I., Ali-Zade, A., Guliyev, I., Bagirov, E., Nadirov, R., Tagiyev, M., & Feizullayev, A. (1997). *South Caspian Basin: Stratigraphy* (p. 430). Nafta-Press, Baku.

Lisitsyn, A. P. (1988). *Avalanche sedimentation and breaks in sedimentation in the seas and oceans* (308 p.). Nauka, Moscow (in Russian).

Lomize, M. G., & Khain, V. E. (1995). *Geotectonics with fundamentals of geodynamics* (560 p.). KDU, Moscow (in Russian).

Maltman, A. (Ed.). (1994). *The geological deformation of sediments* (p. 359). Chapman & Hall.

Mamedov, P. Z. (1984). Some results of applying the principles of seismostratigraphy in the study of the lower boundary of the PT within the Absheron Archipelago. Oil-and-Gas Geology and Geophysics. *Reprint of Scient. Inst. of Economics of the Hydrocarbon Industry, 11*, 20–23. (in Russian).

Mamedov, P. Z. (1986). Identification of reef formations using seismostratigraphic studies. *Geology of Oil and Gas, 7*, 24–27. (in Russian).

Mamedov, P. Z. (1991). Seismostratigraphic studies of the geological structure of the sedimentary cover of the South Caspian megadepression in connection with the prospects for oil and gas. Abstract of a D.Sc. Thesis (50 p.) (in Russian).

Mamedov, P. Z. (2004). Genesis and Seismic stratigraphic model of the South Caspian Megabasin architecture. In A. A. Alizadeh (Ed.), *South Caspian Basin: Geology, geophysics, oil and gas content* (pp. 150–164). «Nafta-Press», Baku (in Russian).

Mamedov, P. Z. (2006). Peculiarities of the Earth's crust in the South Caucasus in the light of new geophysical data. *Izvestiya Acad. Sci. Azerb. Earth Sciences, 3*, 36–48.

Mamedov, P. Z. (2007). Seismostratigraphic subdivisions of the sedimentary cover of the SCB. *Jour. of stratigraphy and sedimentology of oil and gas basins of ANAS. Earth Sciences, 1*, 102–117.

Mamedov, P. Z. (2008). On the causes of rapid subsidence of the earth's crust in the South Caspian Basin. *Azerbaijan Oil Industry, 1*, 8–20. (in Russian).

Mamedov, P. Z., & Guliyev, I. S. (2003). Subvertical geological bodies in the sedimentary cover of the South Caspian Basin. *Izvestiya Acad. Sci. Azerb. Earth Sciences, 3*, 139–146. (in Russian).

Mangino, S., & Priestley, K. (1998). The crustal structure of the southern Caspian region. *Geophysical Journal International, 133*, 630–648.

Nadirov, R. S., Bagirov, E. B., Tagiyev, M. F., & Lerche, I. (1997). Flexural plate subsidence, sedimentation rates, and structural development of the super-deep South Caspian Basin. *Marine and Petroleum Geology, 14*(4), 383–400.

Narimanov, N. R. (2003). Geodynamic aspects of the sedimentary SCB's cover generation. *Geology of Oil and Gas, 6*, 26–31. (in Russian).

Nazari, H. (2006). Analise de la tectonic recente et active dans l'Alborz Central et la region de Teheran. Thes Doctoral, Montpeller.

Pavlenkova, N. I. (1996). Development of ideas about seismic models of the Earth's crust. *Geodynamics, 4*, 11–19. (in Russian).

Payton, C. E. (Ed.) (1977). Seismic stratigraphy—Applications to hydrocarbon exploration. *American Association of Petroleum Geologists Memoir, 26*, 516 p.

Perrodon, A. (1985). *Histore des Giandes Decouvertes Petrolieres* (p. 222). Edite par Dunod.

Rodkin, M. V. (1993). *The role of the deep fluid regime in geodynamics and seismotectonics* (189 p.). Nauka, Moscow (in Russian).

Schlezinger, A. E. (1998). *Regional seismostratigraphy.* Nauchnyi Mir, Moscow, 138 p. (in Russian).

Sengor, A. M. C. (1984). The Kimmeridge orogenic system and the tectonics of Eurasia. *Geological Society of America Special Paper, 195*, 181–241.

Shreider, A. A., Kazmin, V. G., & Ligin, V. G. (1997). Magnetic anomalies and problems of the Black Sea Basin age. *Geotectonics, 1*, 59–70.

Steckler, M. S., & Watts, A. B. (1978). Subsidence of the Atlantic-type continental margin off New York. *Earth and Planetary Science Letters, 41*, 1–13.

Sultanov, A. D., & Gorin, V. A. (1963). *Productive strata of the western side of the South Caspian depression* (290 p.). Azerneshr, Baku (in Russian).

Tagiyev, M. F., Nadirov, R. S., Bagirov, E. B., & Lerche, I. (1997). Geohistory, thermal history and hydrocarbon generation history of the north-west South Caspian basin. *Marine and Petroleum Geology, 14* (4), 363–382.

Uyeda, S. (1978). *The new view of the Earth: Moving continents and moving oceans* (122 p.). W. H. Freeman, San Francisco.

Vail, P. R., Mitchum, R. M., & Thompson, S. (1977). Seismic stratigraphy and global changes of sea level. Parts 3 and 4. Seismic stratigraphy. *Aapb Memoir, 26*, 83–97.

Van Baak, C. G. C. (2010). Glacio-Marine transgressions of the early and middle pleistocene Caspian Basin, Azerbaijan. M.Sc. thesis, Paleomagnetic Laboratory 'Fort Hoofddijk', Utrecht University, Faculty of Geosciences.

Volozh, Y.A., & Leonov, Y.G. (Eds.). (2004). *Sedimentary Basins: Methodology of investigation, structure and evolution* (526 p.). Nauchnyi Mir, Moscow (in Russian).

Yacobson, A. N. (1997). Seismic shear wave velocity in the Southern Caspian lithosphere. *Doklady Academy Science Russ., 353*(2), 258–260.

Yakobson, A. N. (2000). Lithosphere of South Caspian Sea. Tomographic model. In *Domestic geology* (pp. 34–45) (in Russian).

Zonenshain, L. P., Kuz'min, M. I., & Natapov, L. M. (1990). *Geology of the USSR: A plate-tectonic synthesis* (242 p.). American Geophysical Union, Geodyn. Ser. 21, Washington.

2 Lithological Composition, Lithostratigraphy, and Lithofacies Zonation of the PS Deposits

Productive Series (PS) is widely occurring within the western SCB, including the territories of the Absheron Peninsula, Absheron Archipelago, Baku Archipelago, Jeirankechmessian depression, Alyat ridge, Lower Kur depression, as well as in a sedimentary section of PreCaspian-Guba region within the western Middle Caspian basin. The following lithofacies types of PS sediments are distinguished: Volga (Absheron), Gobustan, Kur (including Geilyar (Lengebiz) sub-facies), Pre-Caspian, and South-Caspian. The latter unites the features of all the mentioned lithofacies zones (Fig. 2.1) and represents their continuation to the shelf zone of the basin (Aliyev et al., 2005; Alizadeh et al., 2017; Azizbekov et al., 1972; Mustafayev, 1964). At the same time, all the lithofacies zones are characterized by similar features as an accumulation of thick terrigenous sediments, lack of mollusk fauna, and scarcity of microfauna remains.

2.1 The Volga (Absheron) Type

This type has been identified by Baturin (1932) under the name of "disten-ilmenite province". It occupies all the territory of the Absheron Peninsula, Absheron Archipelago, and northern Baku Archipelago (Fig. 2.1). The depositional environment of the Productive Series has been the most debatable for a long time. Different theories were elaborated, such as sedimentation in lacustrine (Abramovich, 1921; Golubyatnikov, 1914) and shallow-marine environments (Abramovich, 1921). Because of the similarity in mineral composition of the Productive Series deposits of the Absheron oil and gas-bearing region and that one of the modern sediments of the delta of the Volga River, Baturin (1932, 1937) has developed a theory earlier announced by Kalitsky (1922) and Kovalevsky (1922), about the paleo-Volga River sedimentary input and deltaic environment of the Productive Series. The light mineral fraction was demonstrated to contain many quartz grains (sometimes up to 95%). According to Mirchink (1926a, 1926b, 1933), deposition of the Productive Series occurred in a closed basin affected by an ample sediment supply from big rivers flowing into this lake that caused accumulation of the Productive Series sediments in a vast depositional environment such as lacustrine, fluvial, and deltaic settings.

Many researchers studying the Productive Series paid particular attention to the problem of its depositional environment (Akhmedov et al., 1961; Aliyeva, 2004a, 2004b, 2008; Aliyeva et al., 2003; Babayev & Gadzhiev, 2003; Guliyev et al., 2003; Hinds et al., 2001, 2004, 2007; Kerimov & Averbuch, 1982; Khalifa-zade & Rustamova, 1999; Konukhov, 1951; Mustafayev, 1963; Pashaly et al., 1979, 1998; Potapov, 1954; Reynolds et al., 1998).

The results of the last decades' sedimentological investigations (Aliyeva, 2004a, 2004b, 2005, 2008; Aliyeva et al., 2003; Alizadeh et al., 2017; Babayev & Gadjiev, 2003; Feyzullayev &

A. Alizadeh et al., *Pliocene Hydrocarbon Sedimentary Series of Azerbaijan*,
Advances in Oil and Gas Exploration & Production, https://doi.org/10.1007/978-3-031-50438-9_2

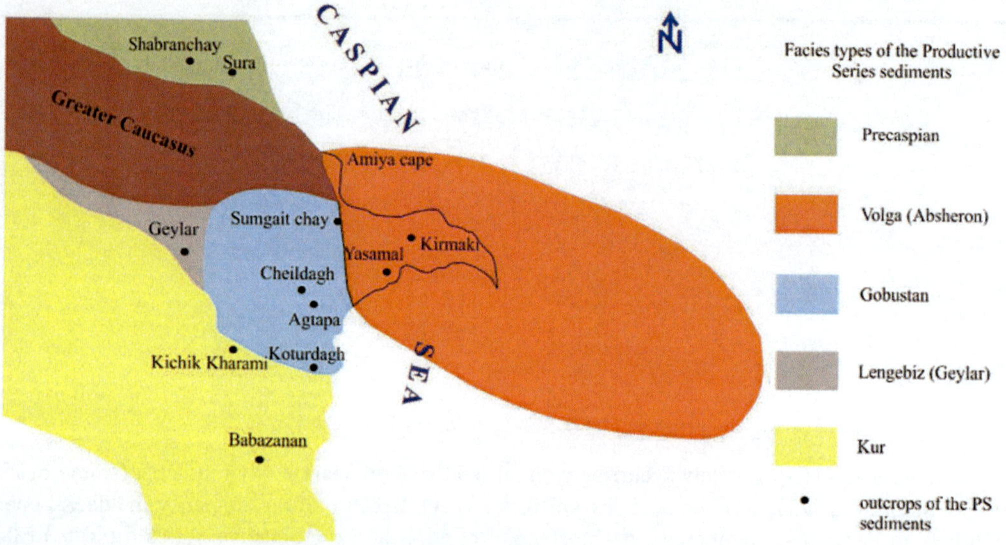

Fig. 2.1 Schematic facies zonation of the Lower Pliocene basin

Huseynov, 2002; Hinds et al., 2001, 2004, 2007) have confirmed the hypothesis proposed by Mirchink (1926a, 1933) regarding the accumulation of the Productive Series under different conditions including continental facies—fluvial, subaerial, lacustrine, and marginal settings in shallow marine and deltaic environments.

Together with the northern provenance (Russian Platform), there were other sources of clastic material of subordinate importance that is confirmed by the presence of Paleozoic pebbles in Absheron-type sediments (Vistelius and Miklukho-Maklay, 1951) which are believed to have been supplied from the Middle Caspian land (Aliyev, 1947, 1949; Aliyev & Daidbekova, 1955; Khain, 1950) and Cretaceous, Tertiary fauna from the Greater Caucasus (Aliyev, 1947, 1949; Aliyev & Daidbekova, 1955; Potapov, 1954; Putkaradze, 1958; Sultanov, 1949).

During PS time, the contours of the Volga-type sediments' occurrence zone have incredibly varied, probably due to the Caspian lake level and sediment supply changes. In response, the paleo-Volga delta would begin to retreat southward or backstep northward. The northern margin of the zone of Volga-type sediments appears to be around the cape of Amya while the western margin corresponds to the geographic boundary between the Absheron Peninsula and Gobustan area (Azizbekov et al., 1972; Mustafayev, 1964) (Fig. 2.1).

The thickness of rhythmically alternating sand, sandstones, and mudstones (Fig. 2.2) (the PS section of Kirmaki Valley is shown in Chap. 5) varies from several meters up to several hundred and thousand meters. In the eastern and central peninsula, sand beds have a much bigger thickness (especially in PS's lower portion) than in the western peninsula. The most significant thickness of the Volga-type sediments (up to 5000 m) is recorded in the offshore South Absheron trough.

The eastern sub-facies zone comprises nine hydrocarbon-bearing suites distinguished by their gross lithological characteristics (listed from older units to younger ones): the Gala, PreKirmaki, Kirmaki, PostKirmaki Sand, PostKirmaki Clay, Fasila, Balakhani, Sabunchi, Surakhani (Figs. 2.3 and 2.4). One of the most crucial production units, the Balakhani suite, is subdivided into six production horizons numbered from X to V. North and northwestwards onlap of the lower stratigraphic intervals is well defined in the seismic lines.

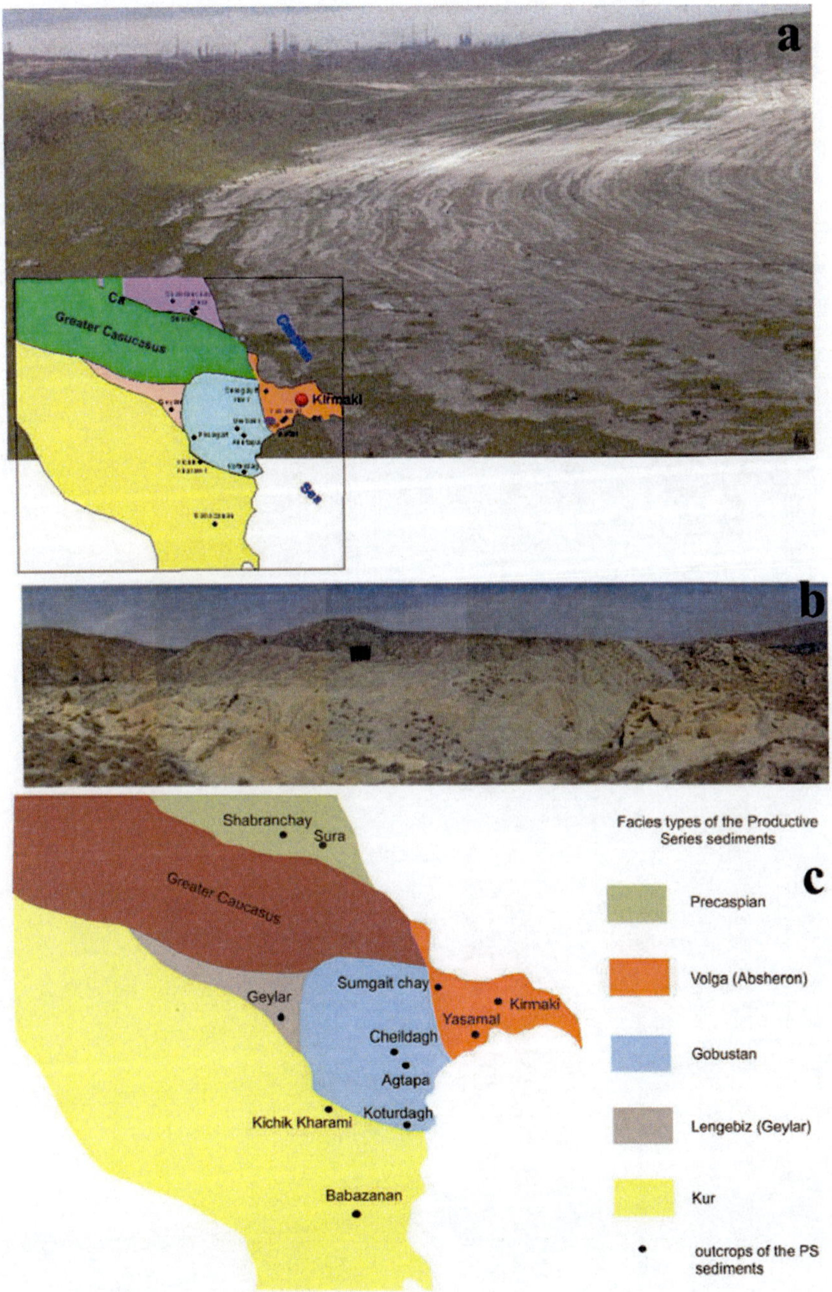

Fig. 2.2 Productive Series outcrop in the Kirmaki (**a**) and Yasamal (**b**) valleys within the Absheron Peninsula (**c**). Some of PS stratigraphic units may be observed for several kilometers

In the western peninsula, the stratigraphic scheme so named the Garadagh scheme contains the following seven horizons: Horizon VIII corresponds to the East Absheronian PostKirmaki Sand suite, horizon VII is an analog of the Fasila suite, the Horizons VI, V, and IV correspond to the Balakhani suite, the horizons III and II, which are the analogs of the Sabunchi suite and the

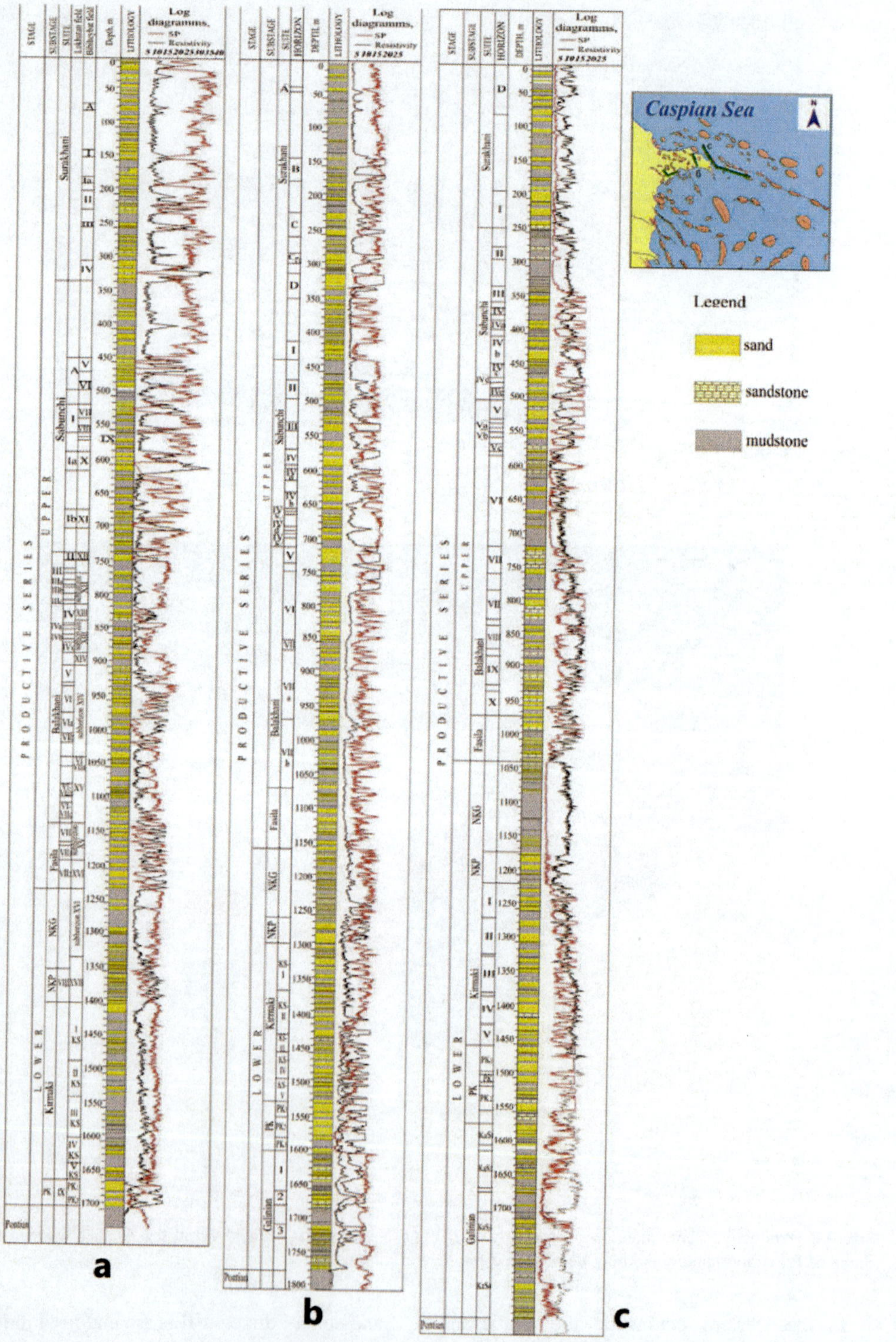

Fig. 2.3 PS composite sections of the southwestern, eastern Absheron, and Absheron-Pribalkhan zone. The following areas are shown in this figure: **a** Bibi-Heybat, Lockbatan, Sulu-Tepe, Puta, and Yassamal Valley; **b** Buzovna, Mashtaga, Gala, and Turkan; **c** Darwin bank, Pirallakhi, Chilov, Guyrgyan-deniz, and Neft Dashlari

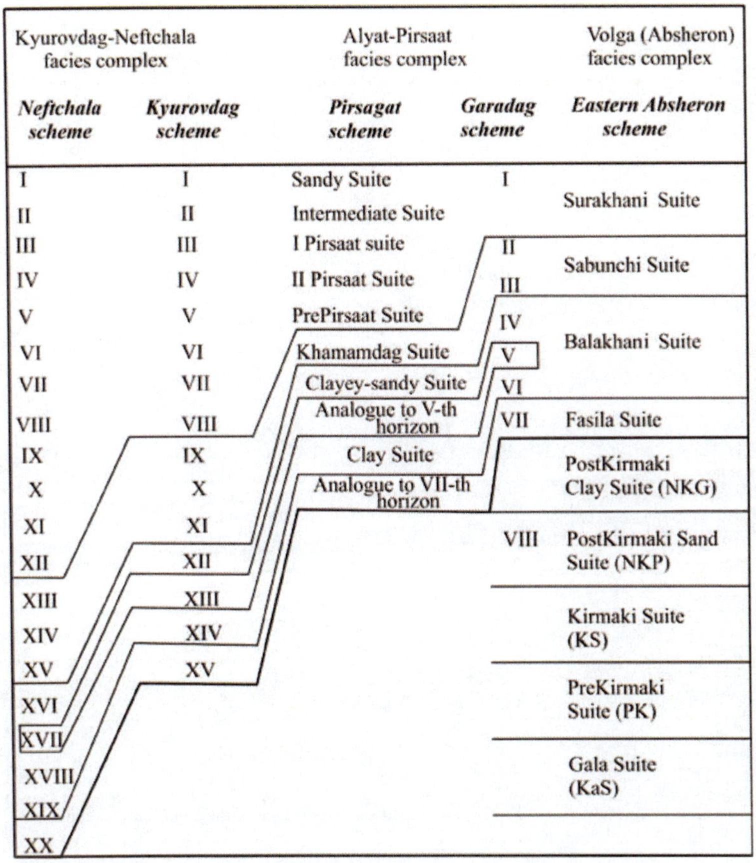

Fig. 2.4 Schematic correlation of the PS deposits within different facial zones

horizon I corresponding to the eastern Absheronian Surakhani suite (Fig. 2.4). Overall, the Productive Series succession in the western Absheron Peninsula is characterized by increased mudstone-dominated intervals.

The typical patterns of the Volga-type sediments, such as the high proportion of sand fraction, sand-dominated terrigenous successions, the high net-to-gross ratio in hydrocarbon reservoirs, and the occurrence of intraformational barriers to fluid flow comprising mudstone intervals are the result of the accumulation of these sedimentary successions under specific and highly varying depositional conditions and formation of relevant stratal architecture that will be discussed below.

PS's lower portion comprises the first five suites. The thickness of the PS lower portion is less compared with that of one of the upper portions. It is noted that PS in both divisions tends to increase in the southeastern and southern directions (Figs. 2.5 and 2.6). The proportion of sand fraction in the lower portion of the Productive Series is slightly higher than in the upper portion. Besides, the sandiest are sediments in the central Absheron Peninsula.

2.2 Productive Series' Lower Portion

Absheron Peninsula and Absheron Archipelago. It is absent in the northwestern Absheron Peninsula and some areas of the central and southern Absheron Archipelago.

The thickness of the Gala suite has increased from zero in the west-northwest of the Absheron

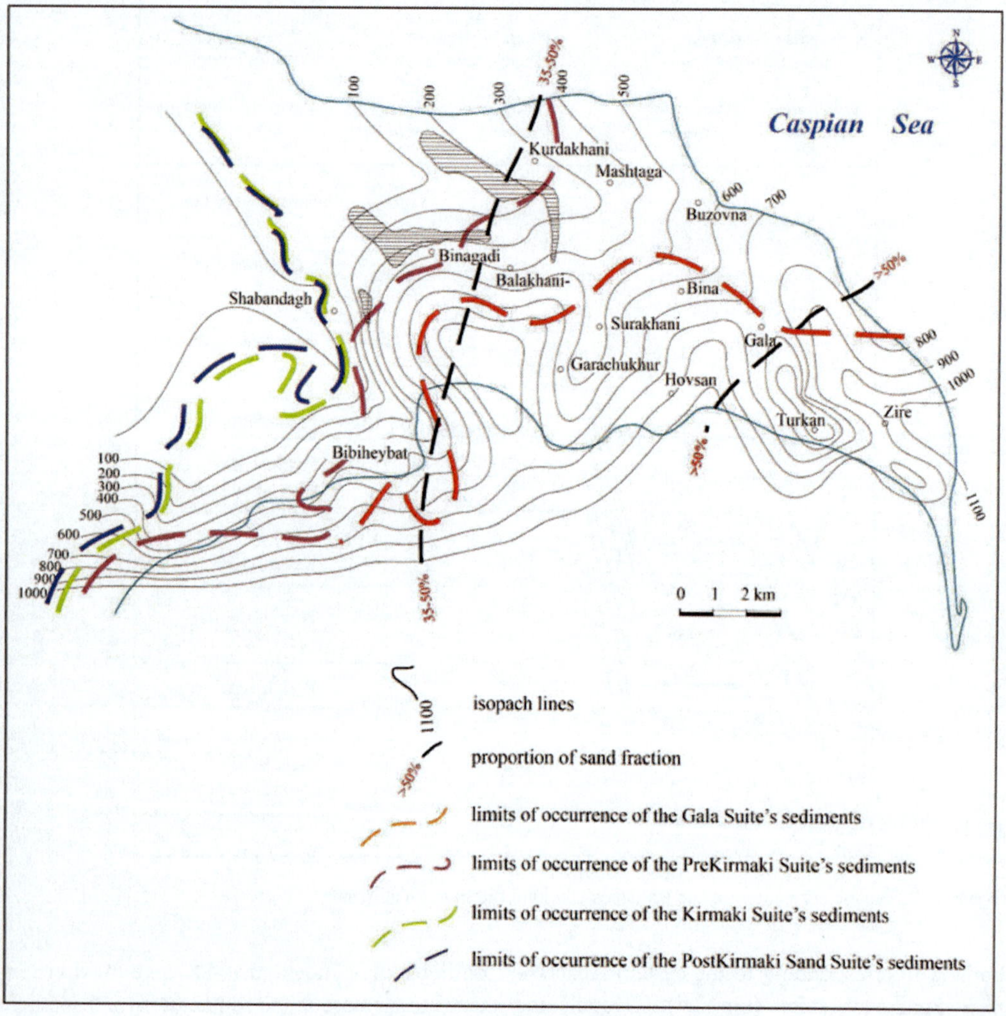

Fig. 2.5 The distribution of sand fraction and thickness of the PS's lower portion within the Absheron oil–gas–bearing region

Peninsula up to 400 m in the south-southeastern direction towards the deeply subsided part of the South Caspian Basin. The Gala suite sediments should be presented more in the sections of the South Caspian oil and gas fields. The drilling data on the onshore structures Bibi-Heybat, Garachukhur, Gala, Zirya, Neft Dashlari, Chilov Isl., Gum Isl., Gyurgyany-sea, and others display unconformable contact between the Gala suite on the Pontian successions.

The absence of these deposits in the crests of several structures (Surakhani, Zikh, Gala, Garachukhur, Neft Dashlari, and Janub) indicates their syn-tectonic sedimentation subsequent erosion in Upper Pliocene–Quaternary time. At the same time, the reduced thickness of the Gala suite in the crests of some anticlines suggests syn-sedimentary tectonics.

The sandy sediment proportion in the Gala suite section varies significantly within the area of their occurrence. A share of sand beds observed in the eastern Absheron structures (Neft Dashlari, Tyurkyan, Gyurgyan-deniz, and Hovsen) comes to 60–70% while in the southerly and

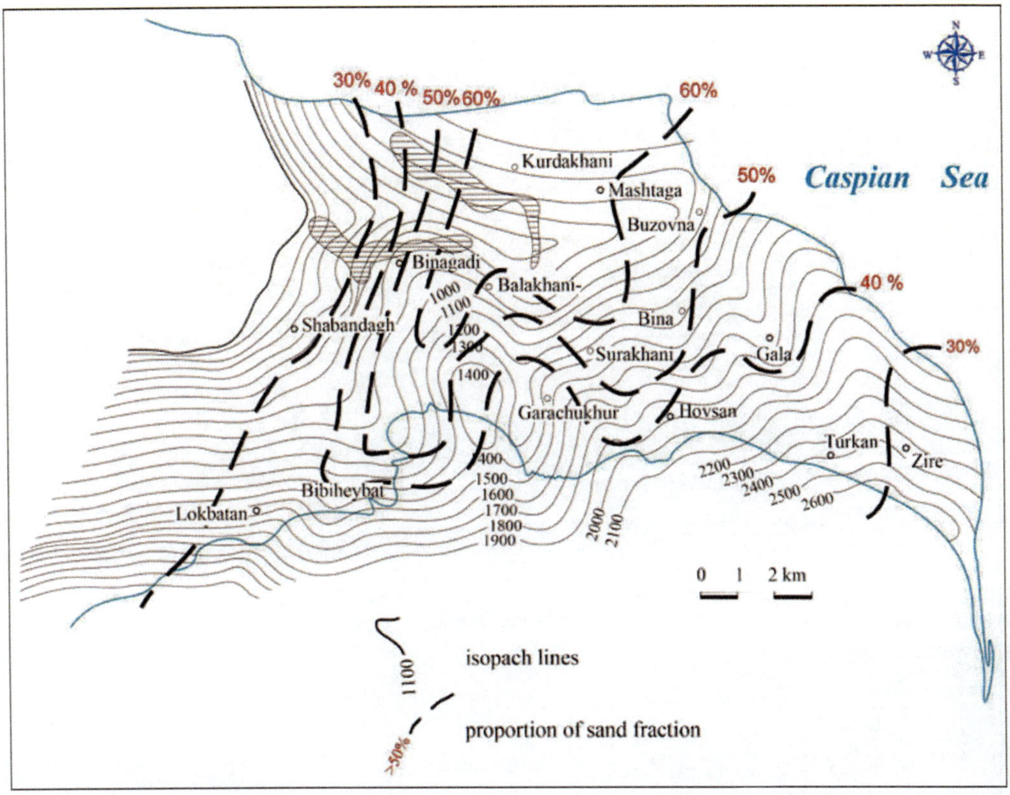

Fig. 2.6 The distribution of sand fraction and thickness of the PS's upper portion within the Absheron oil–gas–bearing region

southwesterly fields (Surakhani, Garachukhur, Bina, and Hovsen) it appears that mudstones dominate the section.

From the grain size analysis, it is estimated that the coarsest fraction > 0.25 mm reaches 15.7% on average; 0.25–0.1 mm gets 32.3–65.9% (39.2 on average); 0.1–0.01 mm—16–48% (24.9 on the average); < 0.01 mm makes 11–42% (23% on the average) (Table 2.1).

According to the log data, the Gala suite is subdivided into three portions: the lower portion is mud-prone, the middle comprises intercalating sandstones and mudstones, and the upper subsite is sand-dominated.

Such lithological variations result from the advance and retreat of the paleo-Volga delta across the Absheron Peninsula and adjacent offshore area, which should be the focus of a

Table 2.1 The grain size of the Gala suite sediments from the Absheron region (139 analyses)

Composition (%)	Content limits (%)	Average contents (%)
$CaCo_3$ content	5.5–34.0	18.5
Fraction > 0.25 mm	0.1–30.6	13.7
0.25–0.1 mm	32.3–65.4	39.2
0.1–0,01 mm	16.8–48.2	24.2
< 0.01 mm	11.6–42.0	22.9

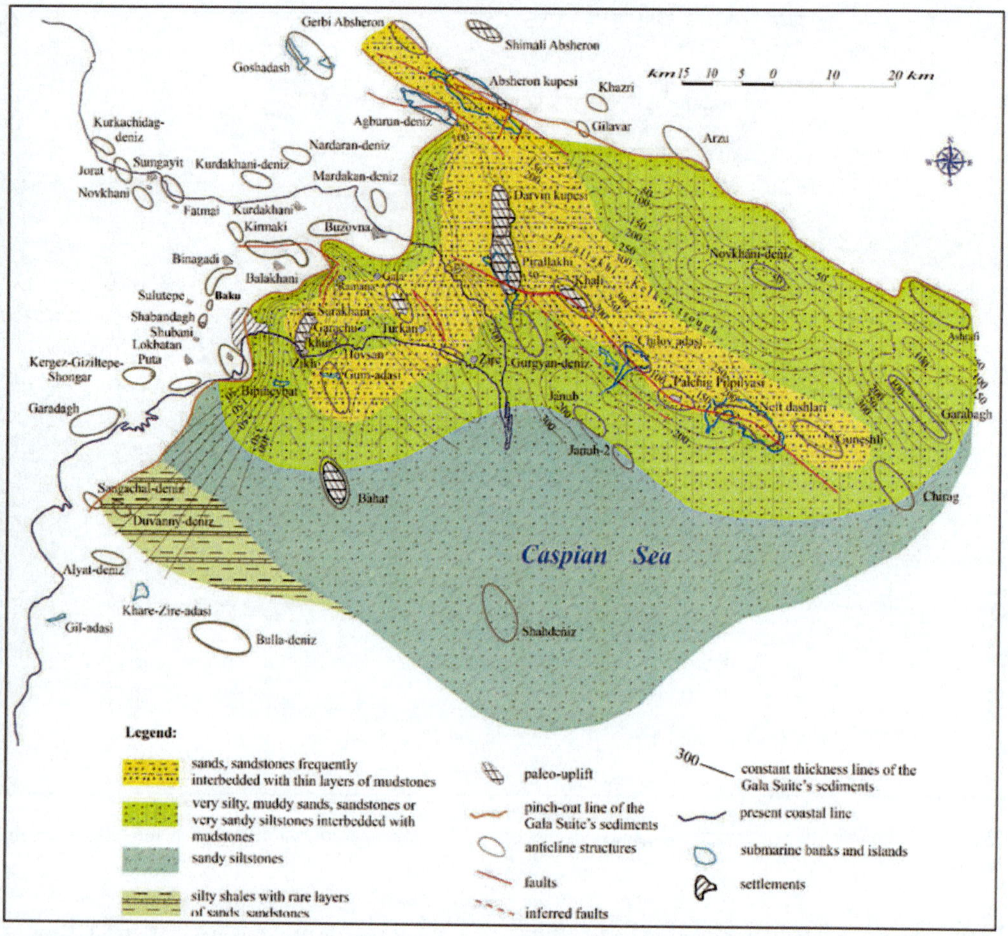

Fig. 2.7 Isopach and lithofacies map of the Gala suite within the South Caspian Basin's western flank (developed by E. Aliyeva)

particular study. The lithofacies map displays our sedimentary model of the upper sand-dominated sub-suite because of the interpretation of a large amount of the subsurface data. The map demonstrates the bifurcation of the paleo-Volga delta into two big channels (Fig. 2.7). The presumably main channel is southeast trending across the modern anticline structures of the Absheron sill and terminating near the Guneshli anticline. The terminal part of the southwest directed second big channel is recorded in the Gum Adasi and Zikh structures. Each such channel comprises a few distributive streams. The dominant type of sediments is channelized fine-grained sandstones separated by mudstone intervals. The maximal thickness of these channelized sandstones is 15–20 m.

A vast flood plain separating these main channels is moderately desiccated by the laterally interconnected distributive streams that led to poor preservation of the flood plain fines.

The southward transition to silty-muddy sandstones, frequently alternating with mudstone intervals, occurred due to environmental changes to delta front facies bounding the outer edge of the terrestrial part of the paleo-Volga delta. These sediments southward gradually pass into prodeltaic facies.

Fig. 2.8 The section of the PreKirmaki suite observed in the Kirmaki Valley. The lower photo (in the left side) shows the Pontian-PreKirmaki suite contact zone showing a sharp facies transition from mud-prone Pontian sediments to coarse sands of the PK suite. The upper photo (in the left side) shows a large mud clast found in the sandstone of the PK suite and indicates high-velocity flow

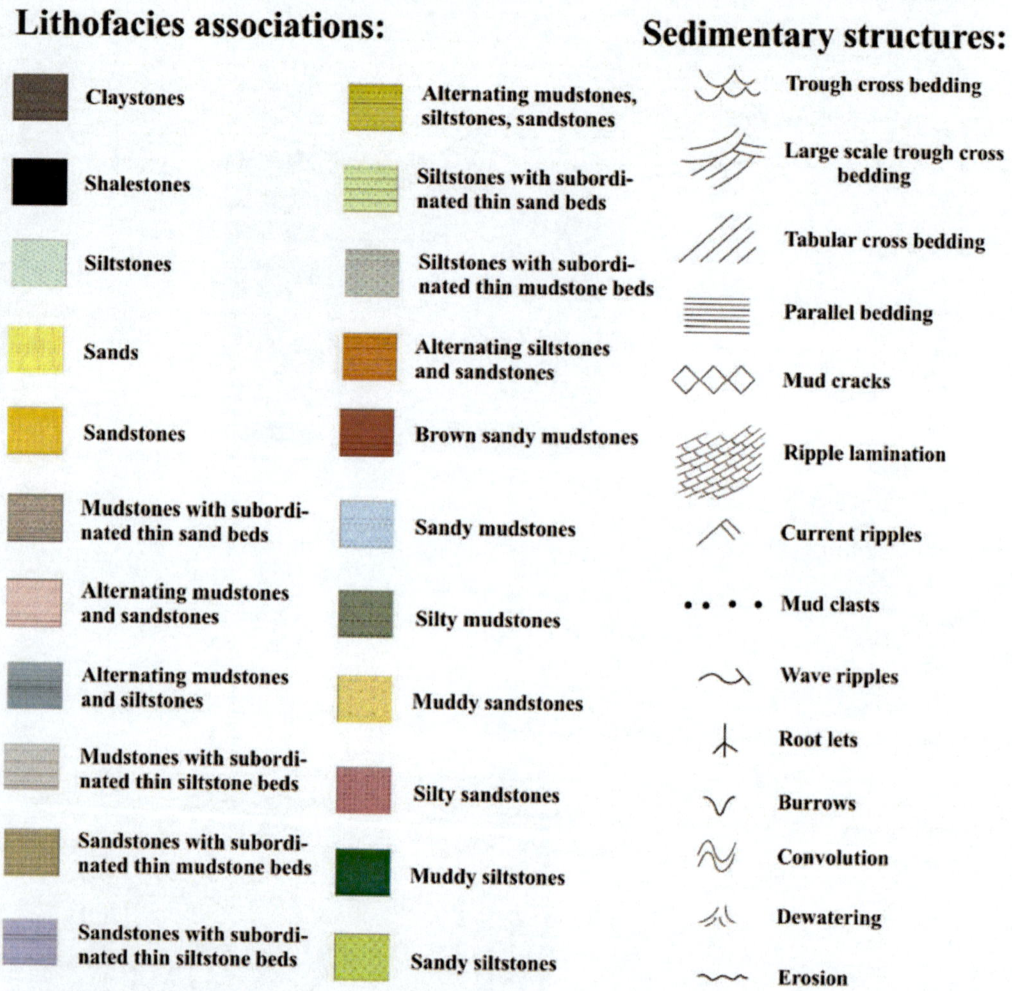

Fig. 2.8 (continued)

The colorization of the Gala suite's sandstones varies from grey, and dark grey to light-grey yellowish. The presence of small siliceous pebbles characterizes some sections. The proportion of the carbonate material in the sediments of the Guneshli, Chirag, and Azeri fields changes from 1 to 11%. The cement type is carbonate muddy. The maximum thickness of the Gala suite is 430 m.

Within the Absheron region, the next PreKirmaki suite comprises grey and light-grey sands, sometimes at the base of small, black, and angular pebbles (Fig. 2.8). The petrographic data points that dominate the mineralogical composition of the rocks are large and rounded quartz grains.

Shales dominate in the upper suite (PK_1, PK_2), and the lower portion (PK_4, PK_5) predominantly comprises sands and sandstones. Sandy shales are of subordinate importance, constituting the tenth part of the suite. The sandstones' thickness is decreasing northwest and westwards from the central Absheron Peninsula. Sandstones are mainly medium grained in the central and eastern Peninsula. The grain sizes decrease southeast and southwestwards. The suite thickness is increasing

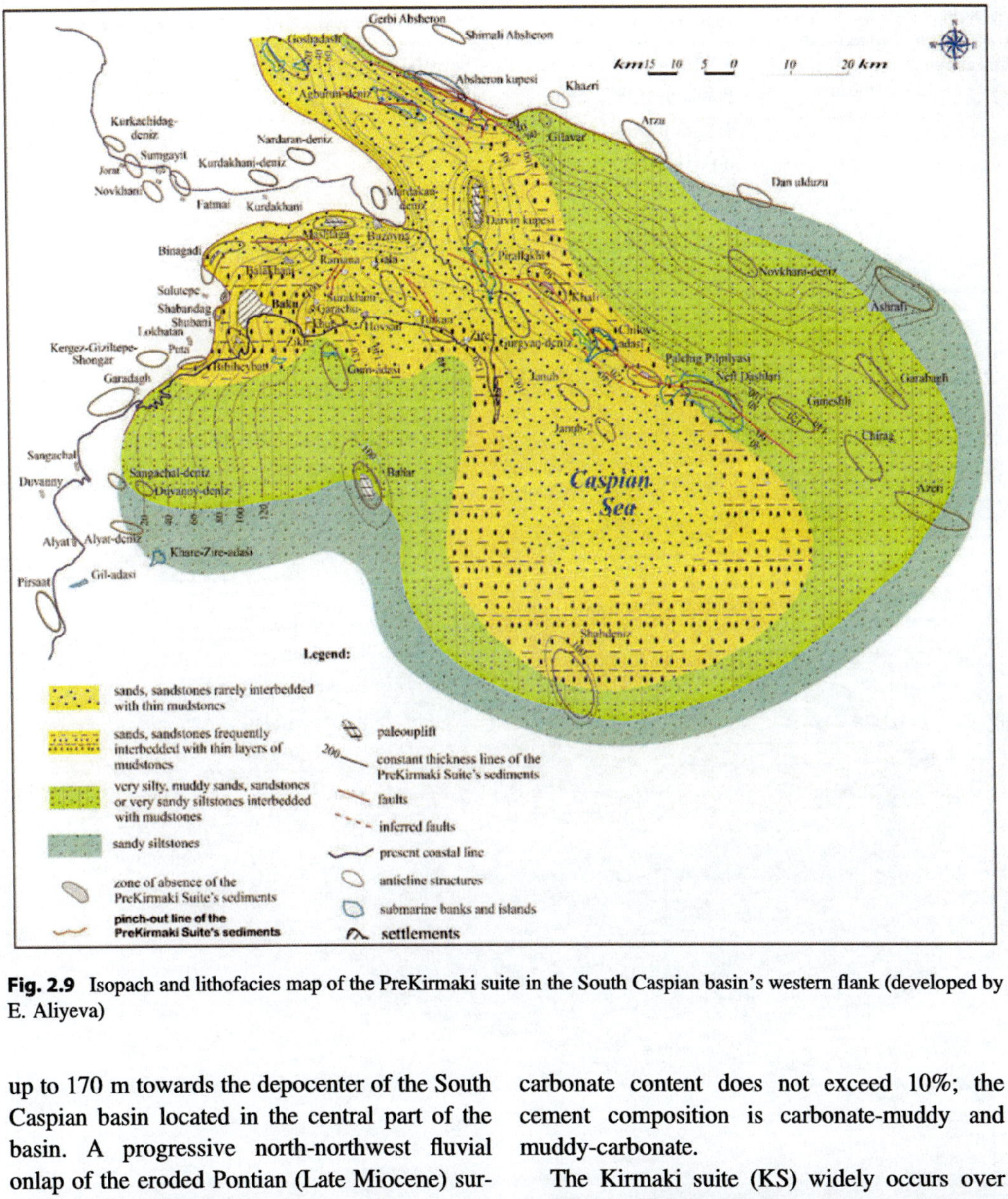

Fig. 2.9 Isopach and lithofacies map of the PreKirmaki suite in the South Caspian basin's western flank (developed by E. Aliyeva)

up to 170 m towards the depocenter of the South Caspian basin located in the central part of the basin. A progressive north-northwest fluvial onlap of the eroded Pontian (Late Miocene) surface by the PreKirmaki suite, as well as south-southwest paleo-Volga delta extension and progradation, probably, during normal regression that is confirmed by fluvial–deltaic facies in some offshore South Caspian fields (Fig. 2.9).

According to data from the fifteen fields of the Absheron Peninsula, the average grain size of the PK suite sediments is as follows: > 0.25 m—12.5%; 0.25 mm–0.1—34.7%; 0.1–0.01 mm—32%; < 0.01 mm—20.7% (Table 2.2). The carbonate content does not exceed 10%; the cement composition is carbonate-muddy and muddy-carbonate.

The Kirmaki suite (KS) widely occurs over the South Caspian Basin's western flank and pinches southwestwards (Fig. 2.10).

The increasing thickness of the Kirmaki suite from the west to the east, up to 270–300 m (Turkan, Zirya fields), and towards the central part of the basin testifies to two depocenters during the accumulation of KS sediments. On some marine structures (Darvin kupesi, Absheron kupesi, Mardakan-deniz), KS sediments have overlaid the Pontian and Diatom

Table 2.2 The grain size of the PreKirmaki suite sediments in the Absheron region (25 analyses)

Composition (%)	Content limits (%)	Average contents (%)
$CaCo_3$ content	3.5–40.0	16.5
Fractions > 0.25 mm	0.0–58.2	12.0
0.25–0.1 mm	13.3–69.3	34.8
0.1–0.01 mm	7.5–46.5	31.7
< 0.01 mm	2.8–32.2	21.5

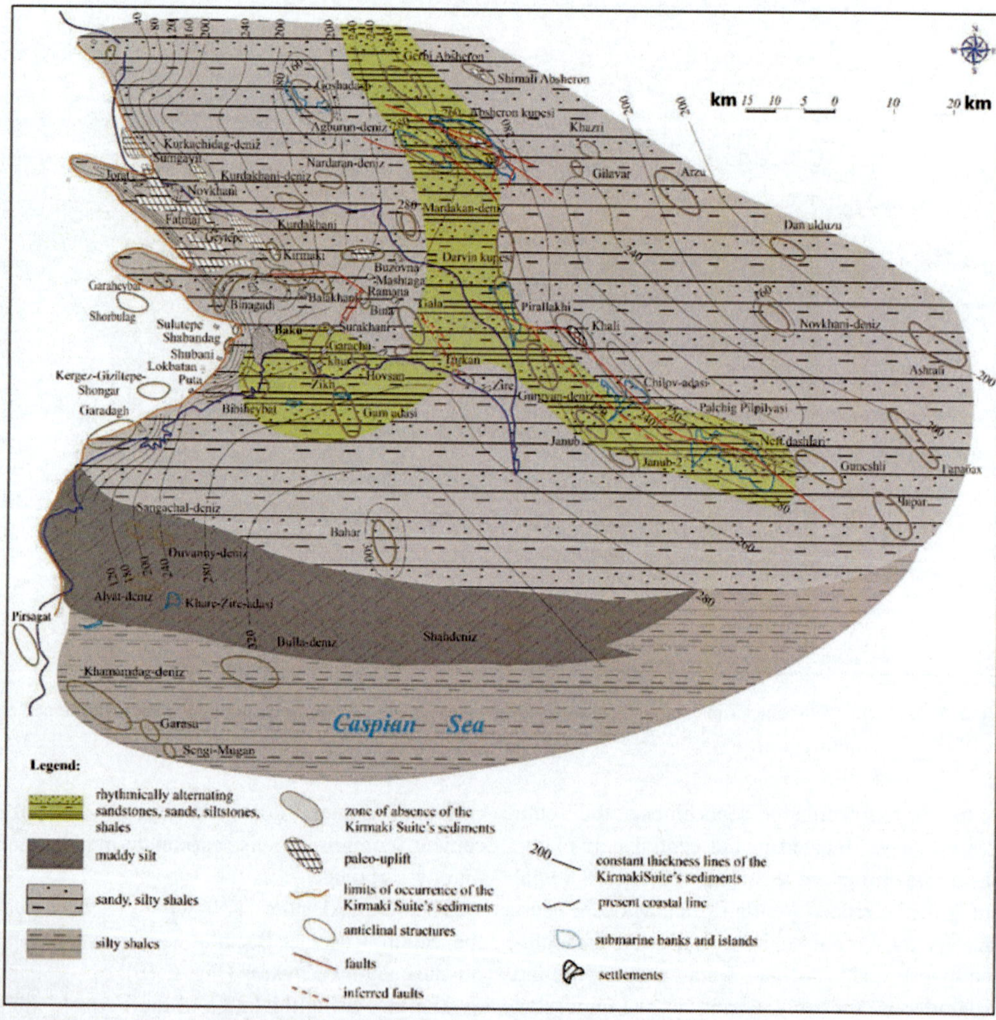

Fig. 2.10 Isopach and lithofacies map of the middle portion of the Kirmaki suite in the South Caspian Basin's western flank (developed by E. Aliyeva)

deposits. These sediments also rest on the old deposits on some local uplifts within the Absheron Peninsula (Fig. 2.10).

The Kirmaki Suite consists of alternating shale and sandstone beds having an average thickness of 0.2–0.3 m to 0.5–1.0 m (Fig. 2.11).

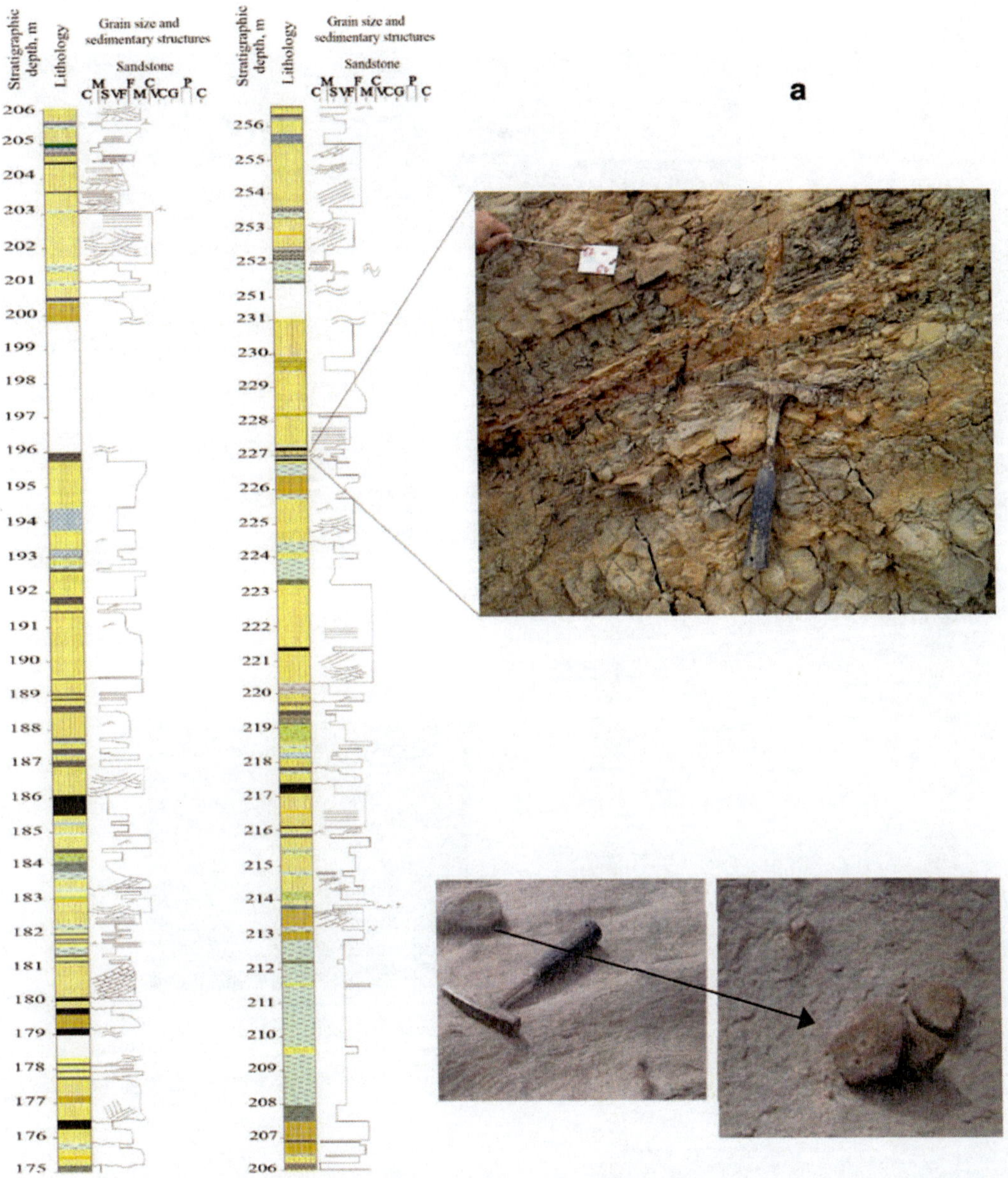

Fig. 2.11 The sedimentological log of the Kirmaki suite exposed in the Kirmaki Valley. **a** lowermost strata of the PK suite, desiccation cracks (upper photo) and root traces (lower photo) indicating that sedimentation occurred under continental conditions, possibly on the deltaic plain; **b** middle suite; sandy bed extended for several hundred meters (left photo) deposited within distal delta front environment (Reynolds et al., 1998); right photo—an example of frequently alternating beds of different lithology. For legend refer to Fig. 2.8

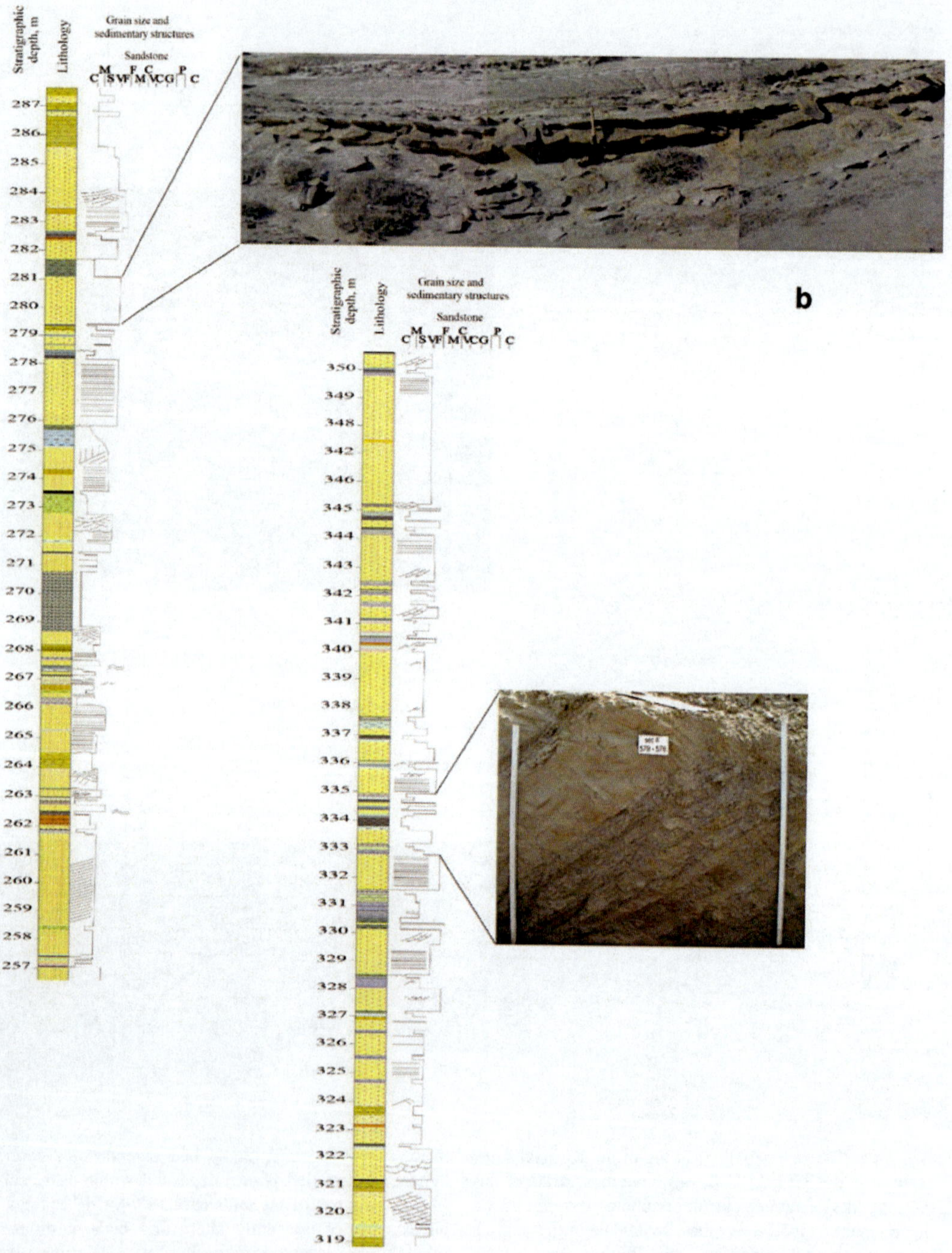

Fig. 2.11 (continued)

Table 2.3 The grain size of sediments of the Kirmaki suite in the Absheron region (26 analyses)

Contents (%)	Content limits (%)	Average content (%)
$CaCo_3$ content	5.0–8.3	6.6
Fractions: > 0.25 mm	0.0–8.0	1.3
0.25–0.1 mm	36.0–59.0	48.4
0.1–0.01 mm	4.0–38.0	19.9
< 0.01 mm	26.5–36.3	30.4

The suite may be divided into three portions: lower and upper, sandier, and middle mud-prone.

The sands are mainly fine-grained, sometimes strongly muddy, poor, or moderately sorted. Sandstone proportion in the section of the Kirmaki suite varies from 70 to 75% in the Fatmai, Binagadi, and Buzovni anticlines, which are probably a result of erosion of nearby local uplifts to 50–55% in Surakhani, Garachukhur, Bibi-Heybat, Gala, Pirallakhi, Gum adasi, Palchig Pilpilyasi, Neft Dashlari and other structures and 30–35% in the rest onshore fields (Table 2.3). The grain size analysis demonstrates that the sand fraction, on average, reaches 48.4%.

The lithofacies map of the middle portion of the Kirmaki suite (Fig. 2.10) demonstrates that the configuration of the area covered by the relatively coarse sediments resembles the two channels reported in the Gala suite (Fig. 2.7). However, both channels of the Kirmaki suite consist of alternating sandstones, siltstone, and mudstone beds in contrast to channels of the Gala suite, constituted by dominated sandstones. We assume such dramatic changes in lithology are due to the damming of the channels during the Caspian Lake transgression.

The silty, sandy rocks are grayish brown, while the shales appear to be dark grey and black. The suite's thickness is about 300 m. To the west of the Absheron Peninsula in the border zone between the Absheron and Gobustan, the Kirmaki suite falls out of the section, and the Productive Series starts from the PostKirmaki Sandy suite (Aliyev et al., 2005; Azizbekov et al., 1972).

The PostKirmaki Sandy suite (NKP) and PostKirmaki Clay suite (NKG) have been named based on the dominant lithology. The thickness of the NKP suite in the Absheron Peninsula varies from 30 to 40 m in the north and northeast of the peninsula up to 80 m in the southwestern part (Fig. 2.12). Significant changes in the suite's thickness are also reported in the local uplifts.

Based on our field observations, core descriptions, and log data interpretation, we can report a new episode in the South Caspian story of the paleo-Volga delta characterized by the delta retreat and south-southeast progradation. The data allow us to assume the further enlargement of the southeastern channel that covers a large area. In contrast to the PreKirmaki suite (Fig. 2.9), we observe the development of the southwestern channel and its southwards progradation. Generally, the NKP sands and sandstones are medium to coarse-grained, poorly sorted, and sometimes contain small black pebbles (Table 2.4; Fig. 2.13).

The base of the suite is constituted of a 36 m thick bed containing three coarsening upward from 9 to 15 m thick sand intervals thickness. Horizontally bedded coarse sands at the top of each interval have different thicknesses. A particular feature of this section is a sharp transition from the shales to sandstones. The 33 m thick sand bed, distinguished in the middle portion of the suite, is subdivided into two intervals resting on alternating shales and sands. The 6 m thick upper interval consists of more fine-grained sediments. Grey, brownish-grey, and yellowish are the typical colors of sandstones of PostKirmaki Sand suite. The thickness of individual beds varies from 0.2 to 4 m. The general paleo current direction is SSE (130°).

Sand rocks need to be better sorted, and carbonate cement prevails. The highest proportion of sand beds is recorded in the central Absheron Peninsula. Here, the NKP is subdivided into two portions—NKP1 and NKP2. The shale

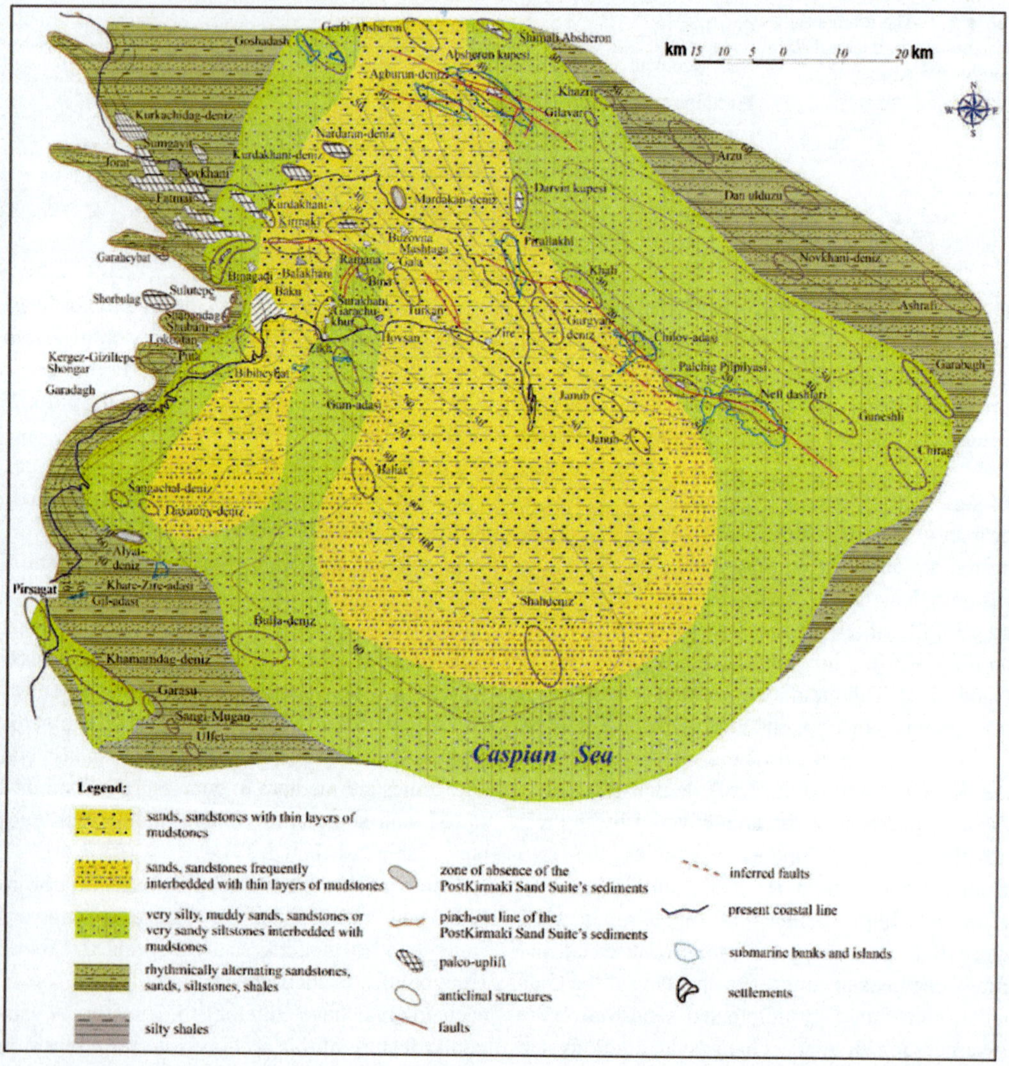

Fig. 2.12 Isopach and lithofacies map of the PostKirmaki Sand suite in the South Caspian Basin's western flank (developed by E. Aliyeva)

Table 2.4 The grain size of the NKP suite sediments in the Absheron region

Content (%)	Contents limits (%)	Average content (%)
$CaCo_3$ content	3.0–15.0	7.3
Fractions > 0.25 mm	0.2–10.0	0.2
0.25–0.1 mm	31.9–52.0	42.2
0.1–0.01 mm	16.0–45.0	38.3
< 0.01	4.0–30.0	19.3

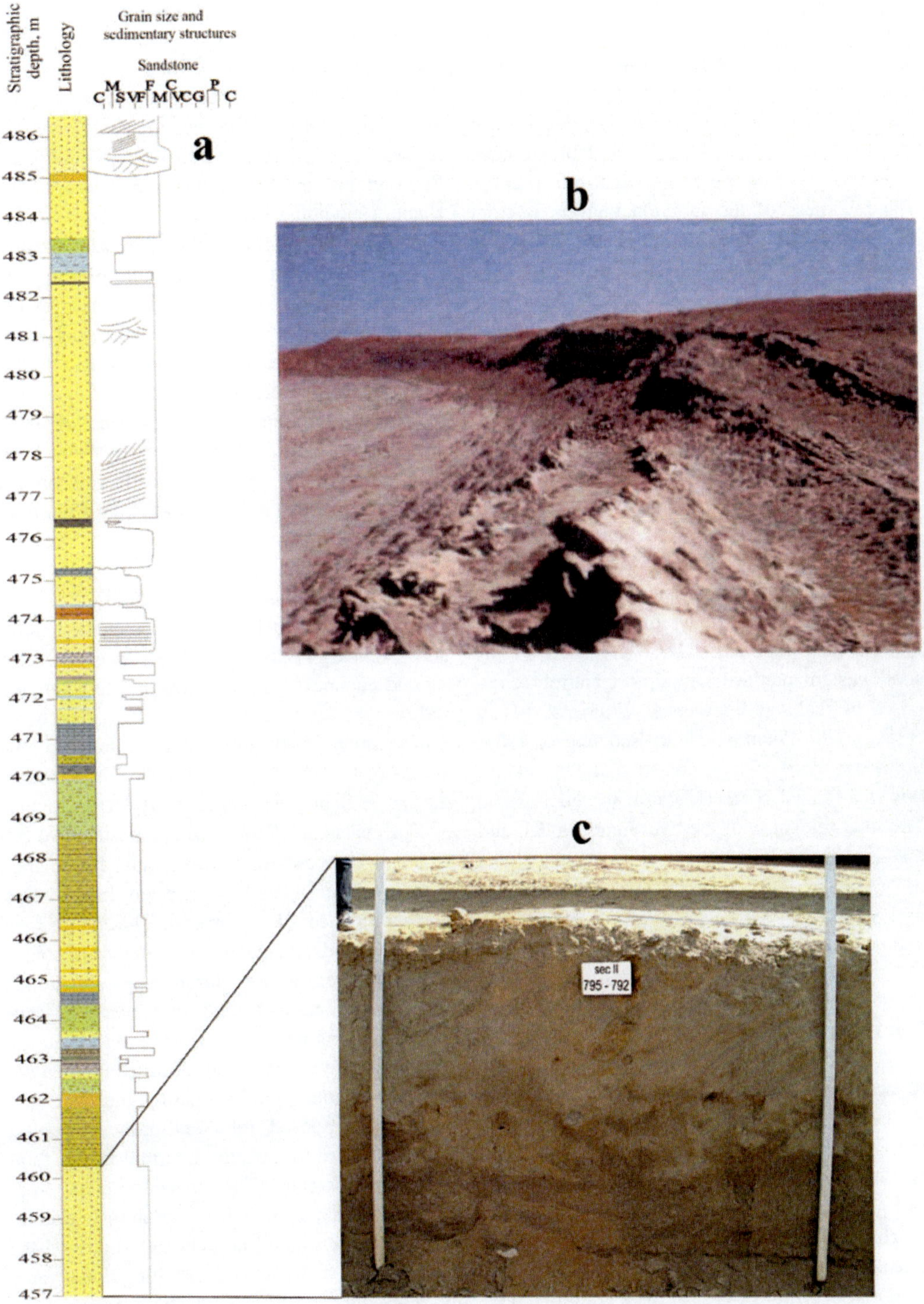

Fig. 2.13 The NKP suite section is exposed in the Kirmaki Valley (**a**), cropping out the basal part of the suite (**b**), **c** weakly cemented medium-grained fluvial sandstones with mud clasts

proportion gradually increases east southeastward. The thickness of shale beds changes from 0.1 m to 1.75 m. Shales, being rather monotonous, are dark-brown and black. Carbonate cementation predominates, and sideritic cement is a rare occurrence. Siltstones contain abundant mica and are of dark grey—to light-brown color. The thickness of the individual layers is from 0.1 m to 0.5 m. They are cemented by calcite cement.

The thickness of the PostKirmaki Clay suite (NKG) increases from southwest to southeast from 30–40 m to 250 m. At several of the crests within the Central Absheron, these deposits appear to have been eroded. Grayish-brown sandy shales with subordinate dominate the NKG suite in the central Absheron Peninsula interbeds of fine—to medium-grained, sometimes muddy sands. These sediments are passing southwards into silty shales and shales lacking sand intervals (Figs. 2.14 and 2.15).

A considerable quantity of loam and muddy loam is also reported. The NKG suite in the southwestern peninsula is almost entirely composed of shales (Aliyev et al., 2005; Azizbekov et al., 1972; Potapov, 1954; Putkaradze, 1958; Sultanov, 1949). Thus, coarse sediment starvation is a typical characteristic of the NKG suite. The total thickness of the two Suites—NKP and NKG, is about 250–280 m.

2.3 The Productive Series' Upper Portion

2.3.1 Fasila Suite

A succession of the PS' upper portion within the Absheron Peninsula and Absheron Archipelago begins from the Fasila suite. Its thickness changes from 80 to 130 m in the northwest to 200–240 m in the southeast (Fig. 2.16).

The lower Fasila suite in the Kirmaki Valley consists of three sandstone beds. The lowermost sand bed overlies a 1–1.5 m thick conglomeratic layer whose accumulation has been related to the paleo-Volga incision into underlying NKG deposits, i.e., it relates to the sequence boundary formation. However, this assumption is only sometimes supported by drilling data.

The lower sand bed containing conglomerates gets 8 m in thickness on average. Conglomerates include rock fragments up to 10 cm in diameter. Sand grains are of angular and semi-rounded shape. Cementing material is made of carbonate composition. The fining upward trend is clear.

The interpretation of the bulk of the data displays a general southern progradation of the paleo-Volga River delta through three big channels directed to the southeast, south, and southwest. Deeply penetrating the paleo-Volga delta inter distributary bays, locally named "kultuks," are typical features of the modern Volga river delta. We assume that these three large channels comprise the branching and merging of multiple small channels, creating a braided river system.

The biggest channel is southeast-directed and passes through the modern South Caspian structures of the Absheron sill—Khali, Chilov adasi, Palchig Pilpilyasi, and Neft Dashlari. The second channel was prograding to the south. Its sediments were recorded in the Gum adasi field. The sediments of the minor directed to the south-south-west channel compose a section of one of the oldest South Caspian oil fields, Bibi-Heybat.

The coarse, medium-grained sandstones of the channels pass downwards into delta front sandstones of the laterally connected mouth bars displaying sheet-like geometry. The mouth bars comprise sandstone bodies deposited from coarse sediment-laden waters during seasonal floods and interbedded with mudstone intervals accumulated during quiet periods (river out of flood). The pulse-like sedimentation resulted in the more complex stratal organization of mouth-bar succession, composed of sand-prone reservoir intervals more frequently interrupted by thin-bedded mudstones. The proximal delta front distributary-mouth bars form an almost continuous sheet-like sandstone body traced around the periphery of the delta and transiting downwards to more muddy distal delta front facies comprising silty sandstones and very sandy siltstones. These distal delta front sediments merge,

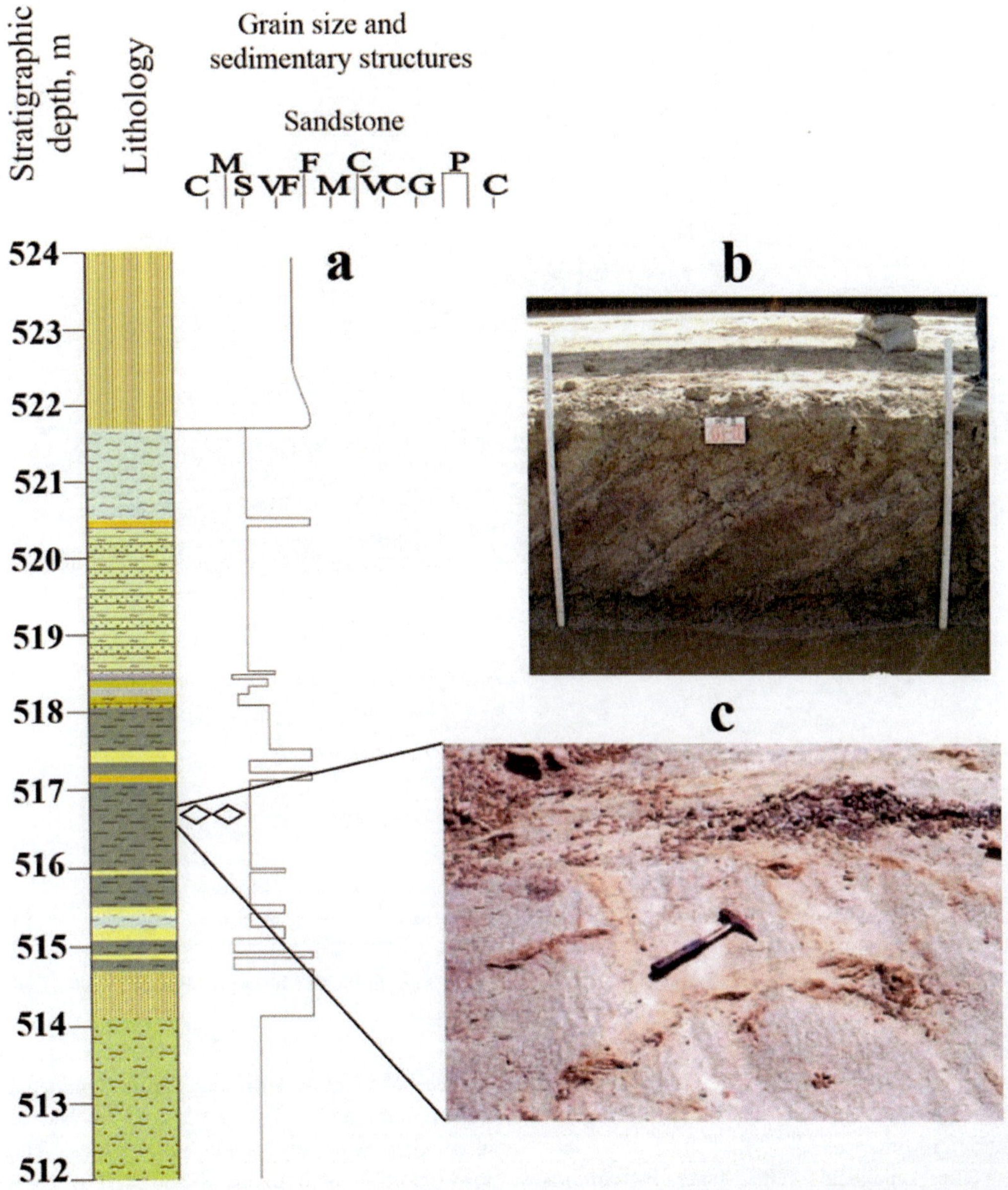

Fig. 2.14 The sedimentological log of the PostKirmaki Clay suite cropping out in the Kirmaki Valley. **a** large desiccation cracks testifying to a subaerial exposure environment and arid climate; **b & c** plastic structural shales interbedded with thin sandstone beds of less than 0.5 m thickness containing abundant current ripples

forming extended long-distance strips along the margin of the delta occasionally interrupted by the mudstone and siltstone-dominated intervals of the distributary bay fill facies. Interfingering sandstones here are a result of multiple splays.

We can also observe the progradation of the paleo-Pirsagat and paleo-Kur rivers deltas. The Fasila suite shales contain a high proportion of sand fraction and often rounded mud clasts formed due to erosion and redeposition of shales

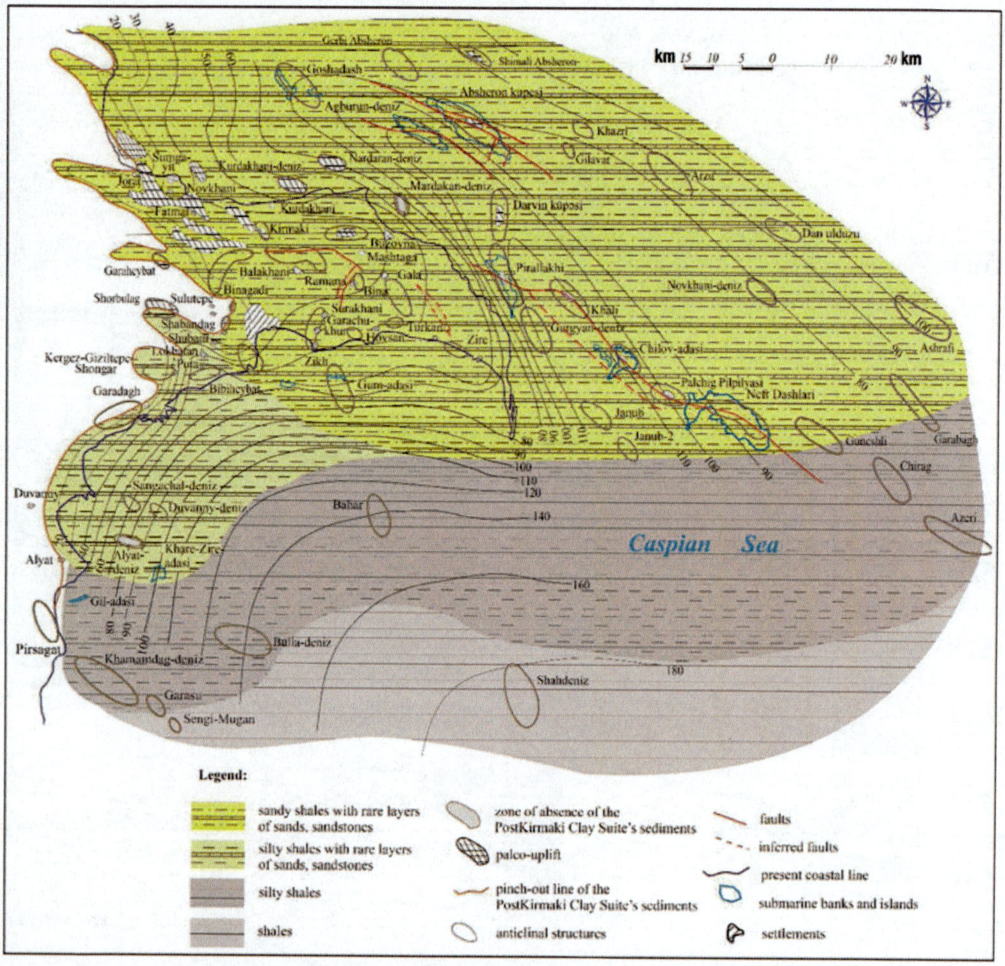

Fig. 2.15 Isopach and lithofacies map of the PostKirmaki Clay suite in the South Caspian Basin's western flank (developed by E. Aliyeva)

from the channel bed (Fig. 2.17). The suite average thickness in the Absheron region is about 100 m.

The Balakhani suite generally comprises medium-fine-grained sandstones and sands interbedded with low-thickness sandy, silty shale intervals. In the Absheron Peninsula, the suite thickness varies from 370 m in the crests of the folds up to 450 m on their franks. Based on the gross lithological characteristics, the Balakhani suite is subdivided into six horizons: V, VI, VII, VIII, IX, and X, which are distinguished in the log diagrams. All the mentioned horizons are well distinguished in the logs.

The horizons X, VIII, and VI are dominated by medium, fine-grained sands and sandstones alternating with sandy and silty shales. The lower portion of horizons X and VIII is composed of more coarse-grained sandstones and sands, while in the portion of the upper part, the sandstones are dominated by fine grains. The X horizon contains medium, coarse-grained, porous sandstones (Fig. 2.18). Their thickness is 75–98 m. The muddy horizon IX predominantly consists of shales containing thin, fine-grained sandstones and sands. Individual shale beds' thickness reaches from 8 to 15 m (Fig. 2.19).

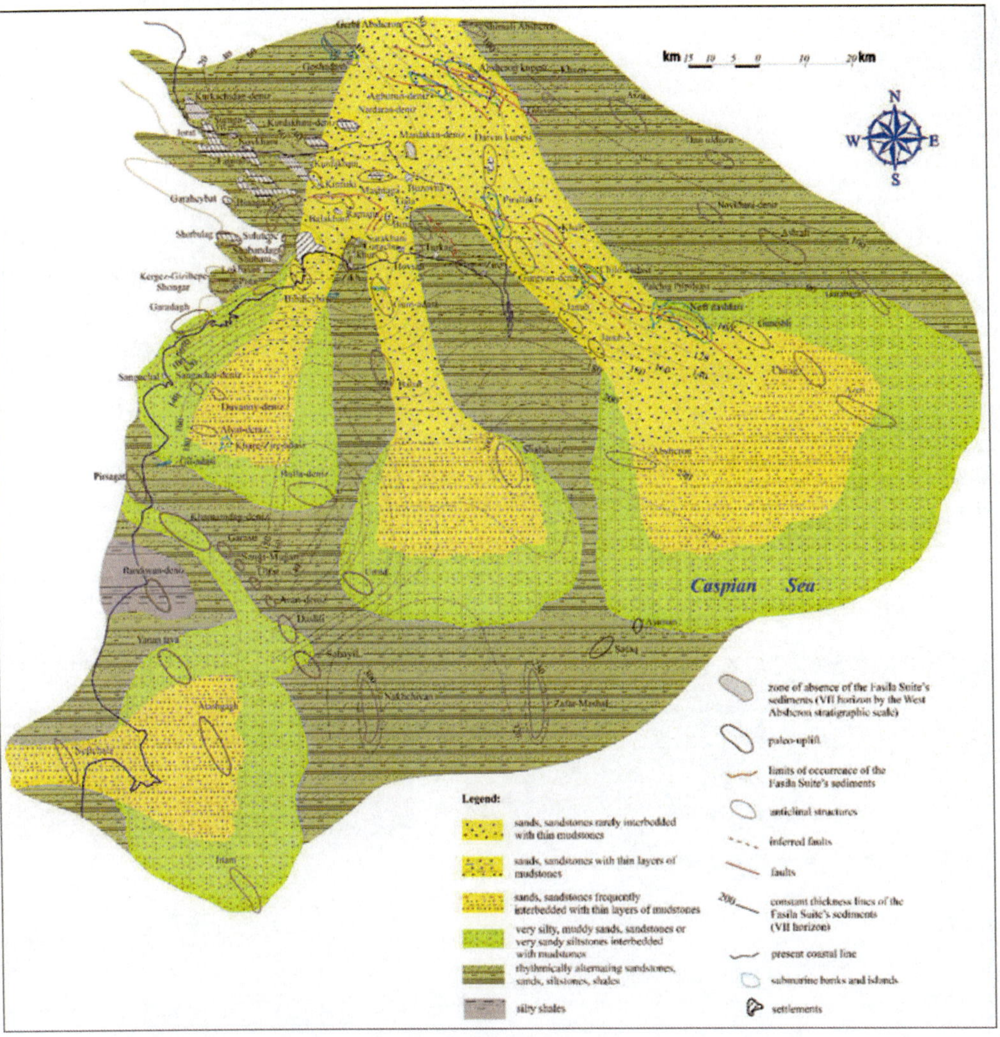

Fig. 2.16 Isopach and lithofacies map of the Fasila suite of the Absheron region and its stratigraphic analog horizon VII of the Baku Archipelago (developed by E. Aliyeva)

Transition to Balakhani horizon VIII is accompanied by the sharp facies' changes to coarse sediment sedimentation. The constructed lithofacies map of the Balakhani horizon VIII resembles the lithofacies map of the Fasila suite with less prograding paleo-Volga channels (Fig. 2.20). We assume simultaneous expansion of the paleo-Kur River delta towards the center of the South Caspian Lake.

A gradual progradation of the paleo-Kur River delta, together with the backstepping of the paleo-Volga River delta, had governed the depositional system of horizon VI of the Balakhani suite, the typical feature of which is the widening of the area covered by mud sedimentation. Deposition of coarse material is limited to three main paleo-Volga River channels that were significantly reduced in size and retrograded to the north. Sand sediments are also reported in the paleo-Volga delta front, the outer edge of which shifted to the north and is located close to the present-day Bakhar and Chirag anticline structures as well as within the paleo-Kur and paleo-Pirsagat rivers' channels (Fig. 2.21).

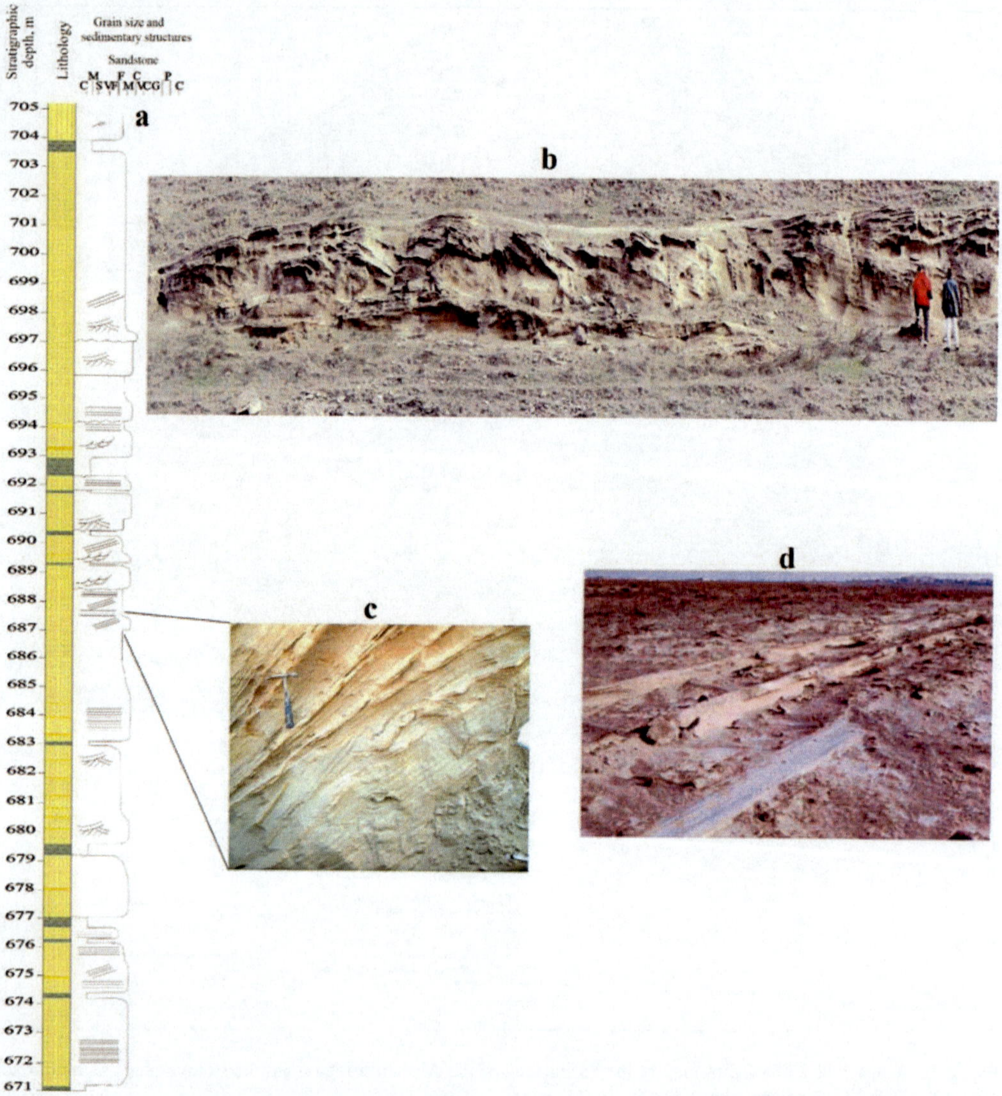

Fig. 2.17 Sedimentological log **(a)** of the Fasila suite observed in the Kirmaki Valley. **(b)** section of the lowermost strata of the Fasila suite shows several meters thick sandstones with southeast-oriented foresets; **(c)** parallel to cross-bedded coarse-grained sandstones; **(d)** continuous sand beds laterally trending for tens of meters and alternating with shales of small thickness that serve as a seal for hydrocarbon reservoirs. Sandstones are interpreted as the accreting mid-channel bar deposits in the river's major channels

The further backstepping of the paleo-Volga River delta continues in the next horizon V of the Balakhani suite (Fig. 2.22). We can report a limited occurrence of sand material in the northern Absheron region and within the channels of the paleo-Pirsagat and paleo-Kur rivers.

A comparison of the coarse sediments in the Fasila and Balakani suites demonstrates that the sands and sandstones of the Fasila suite are mostly coarse-grained, while in the Balakhani suite, the medium-grained sandstones predominate (Table 2.5).

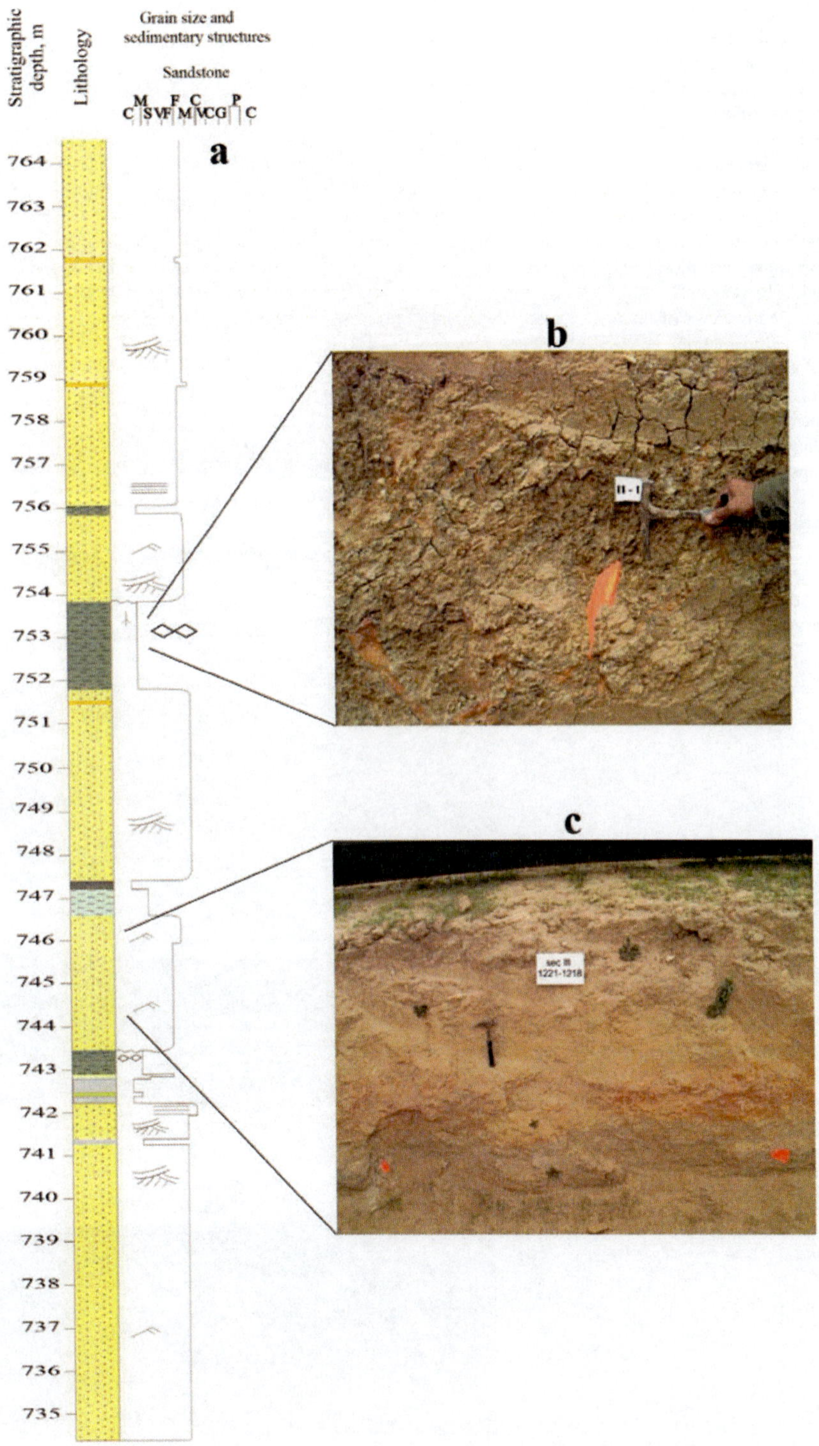

Fig. 2.18 Stratigraphic log (**a**) of the horizon X of the Balakhani suite in the Kirmaki Valley. **b**: mud cracks in shales of the horizon X; **c**: upward thickening sand beds accumulated in the prograding mouth bar environment

Fig. 2.19 Balakhani suite cropping out in the Yasamal Valley. **a** sands with current ripples structure, horizon VII of the Balakhani suite, **b** extended red shale suggesting sedimentation in the subaerially exposed environment, probably fluvial or deltaic plain horizon VII of the Balakhani suite. The most likely that sand beds have been accumulated because of the overbank sedimentation during the flood stages and deposition of coarse sediments on the muddy floodplain facies: **c** extended sand bed deposited in the river channel setting, horizon VI of the Balakhani suite, **d**: exaggerated part of the photo "c". A channel cutting into underlying reddish shales is clearly observed

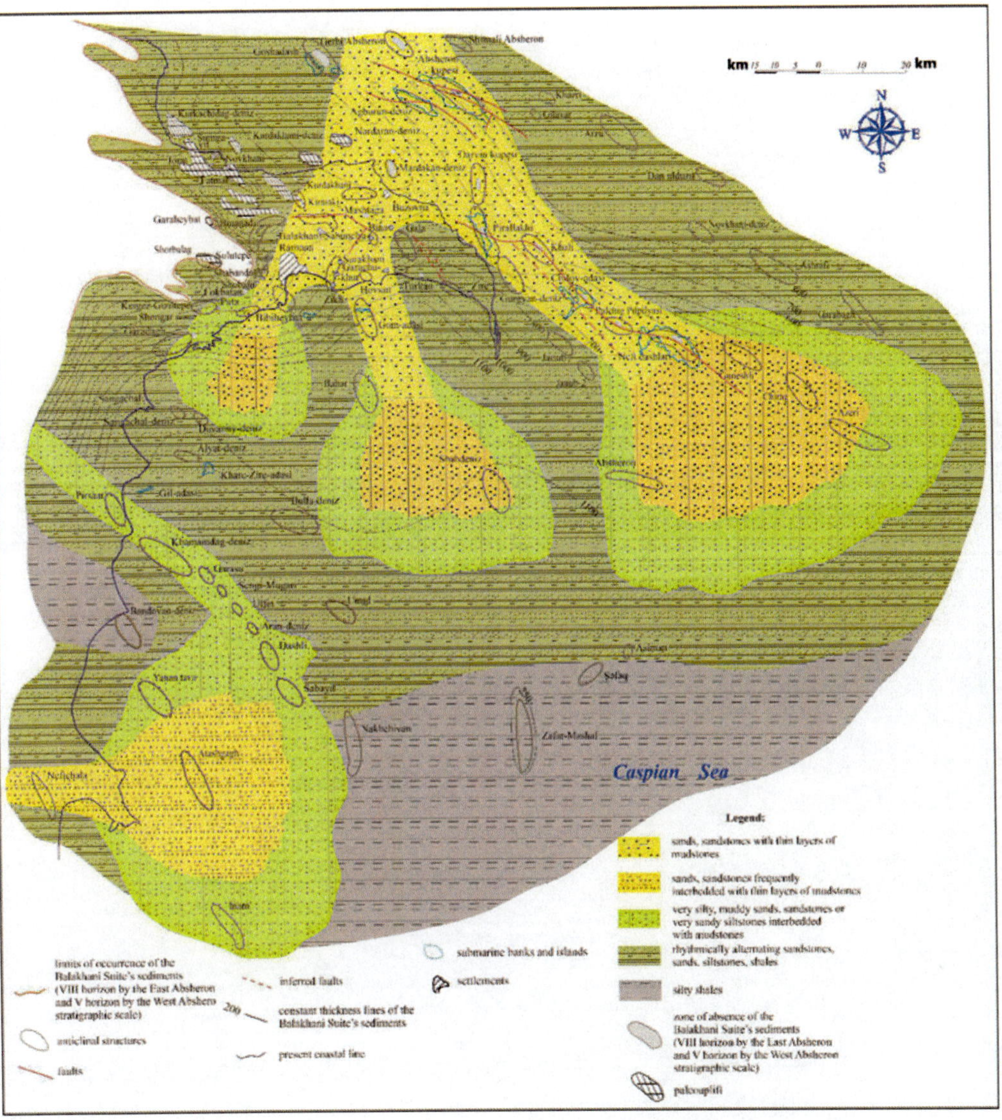

Fig. 2.20 Isopach map of the Balakhani suite and lithofacies map of the horizon VIII of the Balakhani suite (Absheron region), and its stratigraphic analog, horizon V of the Baku Archipelago (developed by E. Aliyeva)

A gradual widening of the sedimentation basin is observed in the youngest suites, i.e., in the Sabunchi and Surakhani times. It has been accompanied by a further northward shift of the paleo-Volga delta and expansion of the paleo-Kur River delta that sediments have been recorded in almost all fields of the Baku Archipelago and even in the westernmost area of the Absheron Peninsula in the Surakhani time (Fig. 2.23). These paleogeographic conditions were accompanied by increased sedimentation of muddy sediments.

The Sabunchi suite's succession lithology varies in different parts of the Absheron Peninsula. The proportion of sandstones and shales in the central peninsula is approximately equal. Sandstones of the Sabunchi suite are predominantly fine-grained or muddy, poorly sorted, and

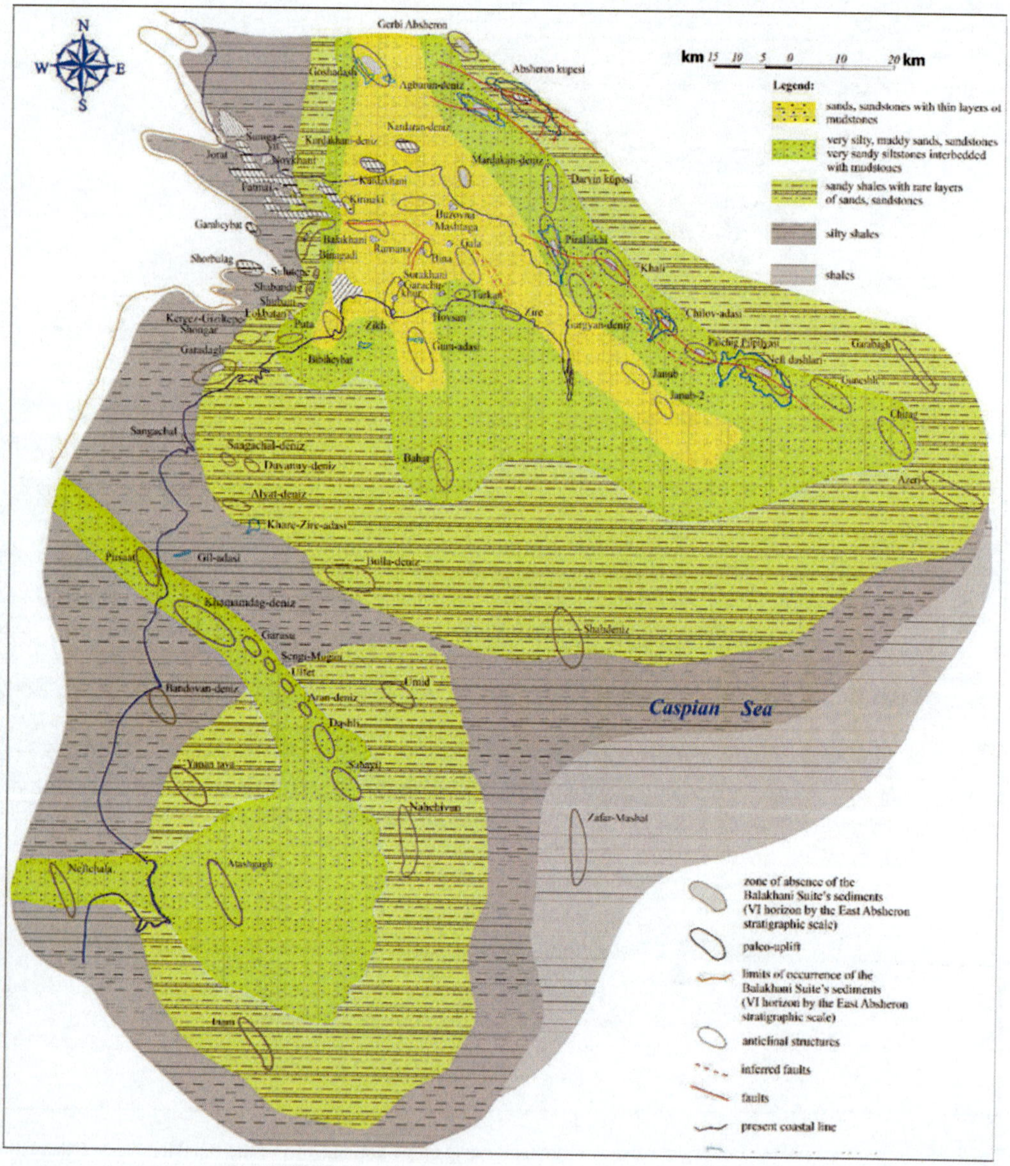

Fig. 2.21 Lithofacies map of the horizon VI of the Balakhani suite within the western flank of the South Caspian Basin (developed by E. Aliyeva)

cemented by carbonate material. A large amount of vein and interlayer types of gypsum is observed. The thickness of the Sabunchi suite varies from 450 m in the central Absheron Peninsula to 800 m northeast and southwards; in southeastern Gobustan, it changes from 0 to 800 m offshore.

The youngest stratigraphic unit of the Productive Series Surakhani suite is dominated by shale. Fine-grained thin sand beds are reported in the eastern Absheron Peninsula. An increase in the thickness of the Surakhani suite up to 2400 m towards the depocenter in the South Caspian offshore region is recorded.

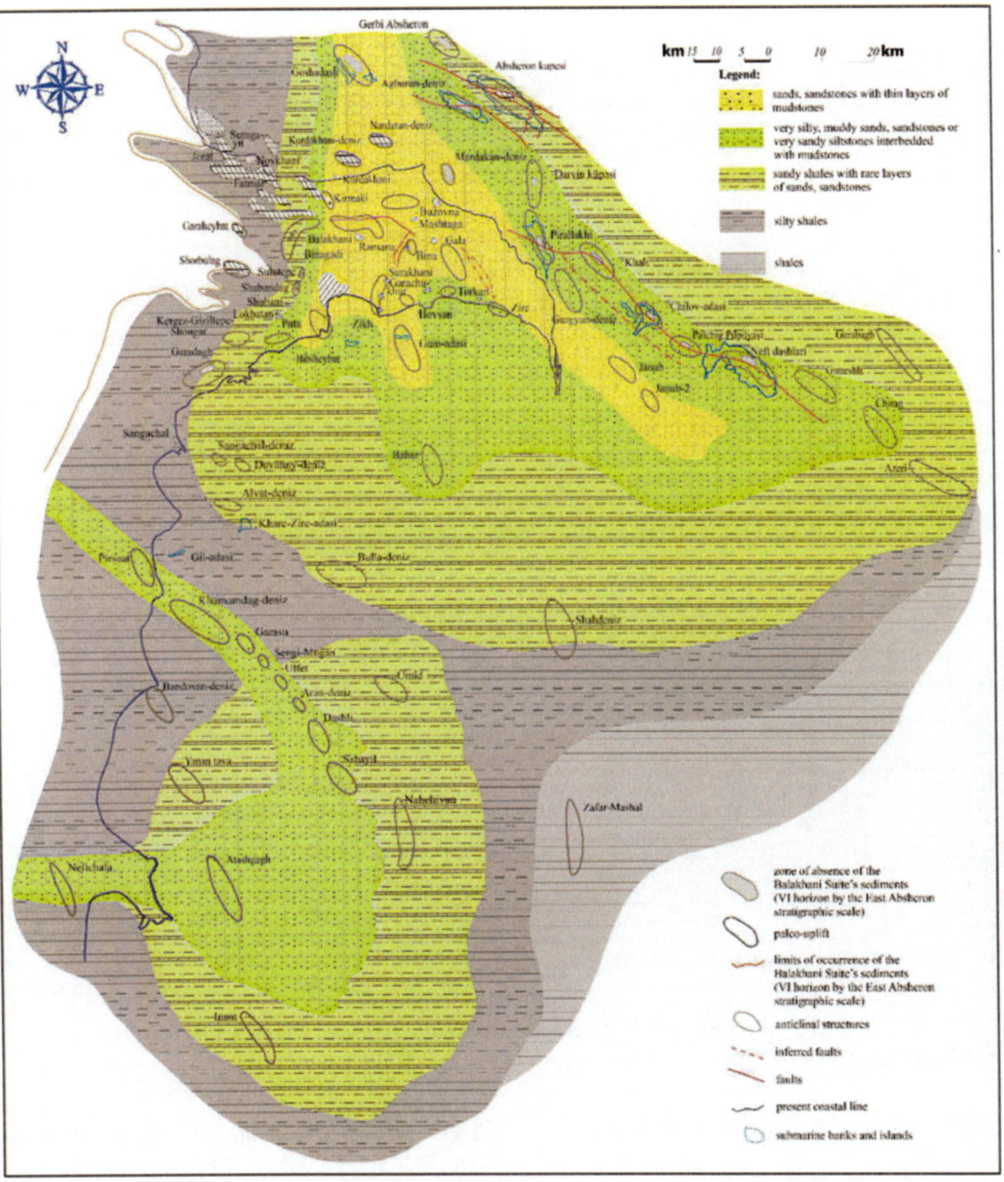

Fig. 2.22 Lithofacies map of the horizon V of the Balakhani suite within the western flank of the South Caspian Basin (developed by E. Aliyeva)

Table 2.5 The grain size of the Fasila and Balakhani suites sediments in the Absheron region (15 analyses)

Content (%)	Content limits (%)	Average contents (%)
$CaCo_3$ content	8.8–18.2	11.2
Fractions > 0.25 mm	27.2–45.8	35.6
0.25–0.1 mm	10.5–26.7	19.9
0.1–0.01 mm	15.5–41.9	28.4
< 0.01 mm	1.5–40.2	16.1

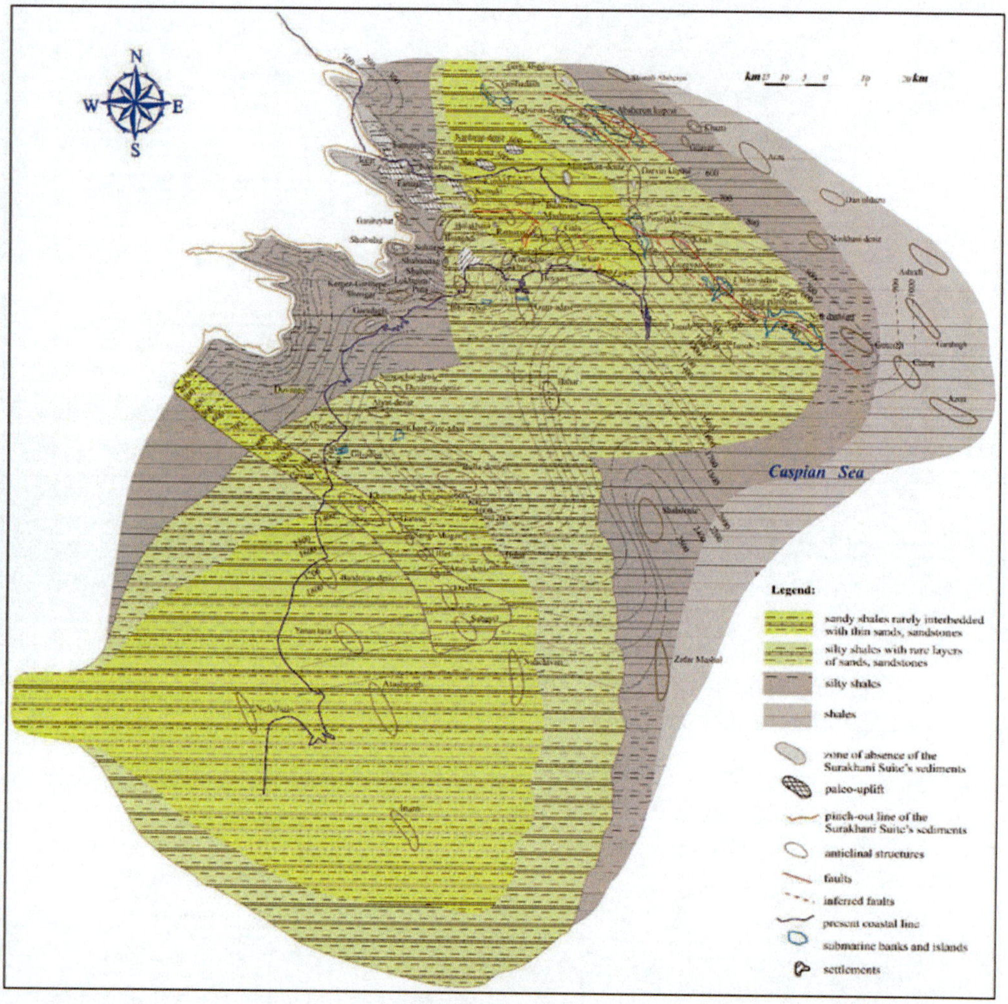

Fig. 2.23 Isopach and lithofacies map of the Surakhani suite on the western flank of the South Caspian Basin (developed by E. Aliyeva)

2.3.2 The Gobustan Facies Zone

The Gobustan facies zone is developed in the central and southeastern Gobustan. It is also named a "zircon-epidotic province" because of the predominance of these two minerals in sediments. The two subtypes, Donguzdyk and east-Gobustanian, are distinguished within the Gobustan type of PS deposits (Aliyev, 1947, 1949; Aliyev and Daidbekova, 1955; Alizade et al., 1967).

The PS deposits are represented by alternating shales (75%), silty (18%), and sandy (7%) rocks. The granulometric composition is highly varied over the zone area. A distinctive feature is the poor sorting of sedimentary material owing to the provenance proximity (Fig. 2.24). The most poorly sorted rocks were developed along the northwestern Djeirankechmes depression within the "Donguzdag series" (Kyrkyshlak, Kaftaran, and Donguzdyk areas). This zone's width is from 4 to 5 km, with an extent from 25 to 30 km. An alternation of carbonized sandy rocks, brown loams, and coarse sands containing pebbles, gravel, and conglomerates is a typical feature of this zone. All these deposits are divided into two

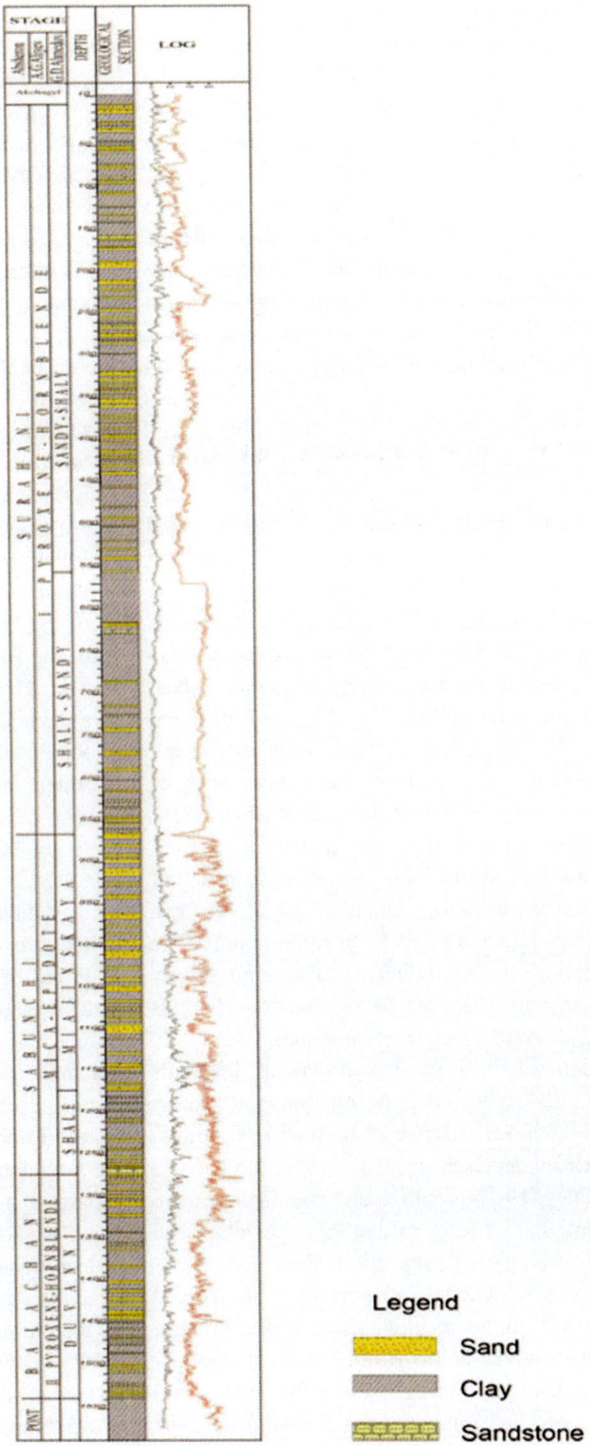

Fig. 2.24 The productive series composite section represents the following areas in the Gobustan facies zone: Cheildag, Ragim, Klych, Donguzduk, Anart, Shikhkaya, Cheilakhtarma, and Kaftaran

subdivisions. Coarse, pebbly sands represent the lower one, and an alternation of brown loams with conglomerates and shingles represents the upper one. Loam is of Miocene age. The thickness of the Donguzdyk series varies from 200 to 400 m.

Further south, a Donguzdag series appears to be changed to the east-Gobustanian subtype sediments characterized by decreasing grain size and increasing deposit sorting degree. In an easterly direction, the Donguzdyk series turns into the Absheron type (Azizbekov et al., 1972).

2.3.3 The East-Gobustanian Subtype

The East-Gobustanian subtype is distinguished noticeably by its more fine-grained composition compared with the Donguzdyk subtype. Being developed within the Djeirankechmes depression and Alyat-Pirsagat zone, these sediments also take part in marine structures located near Gobustan. They are poorly sorted and differentiated. The clean sand or shale beds are absent; dominating are mixed sediment types such as muddy-sandy sandy-silty sediments (Fig. 2.25).

A distinctive feature of sandstones is their poor sorting. Most of the section is composed of shales and siltstones, which in the Klych area come to 44.30 and 47.78%, correspondingly. The number of coarse sediments with grain size > 0.25 m averages 4%. Sandy fraction share is usually about 18–20%. Sometimes, the sandy bed riches 20 m thick, but in general, PS totals 2200–2500 m on land. Its maximum thickness (3200–3600 m) is fixed in the marine continuation of the depression.

Aliyev (1947) proposed an "East-Gobustanian stratigraphic scheme" of the PS sediments based on mineralogical composition. Some other dismemberment schemes of these deposits worked out based on their lithological characteristics, which are given below (from bottom to roof).

Grayish-brown shales with silty, sandy, and arenaceous beds represent the Solakhai suite. Its thickness is about 500 m. The Duvanni suite comprises sandstones and siltstones of 170–180 m thickness. The argillous-arenaceous suite consists of an alternation of shale marls with thick sandy and sandstone beds. The thickness is highly varied over an area from 500 to 700 m.

Apart from the generally accepted scheme of the PS division within Eastern Gobustan, there are also the stratigraphic schemes of PS deposits developed in separate areas where local stratigraphic complexes appear to be distinguished.

In such a way, the following stratigraphic scheme is used for the Pirsagat area (upwards):

- An analog of horizon VII (Fasila (Break) suite),
- Dividing shale suite (an analog of the VI horizon of the Garadagh scheme and X-IX horizons of the Balakhani suite),
- An analog of V horizon (horizon VIII of the Balakhani suite),
- Argillo-sandy suite (an analog of horizon IV of the Garadagh scheme and VII-V horizons of the Balakhani suite),
- The Hamamdag suite (an analog of the Sabunchi suite),
- SubPirsagat, Pirsagat, transitional and sandy suites (an analog of the Surakhani suite).

A somewhat simplified scheme is presented in Fig. 2.24, where the SubPirsagat, Pirsagat, Pirsagat-first, transitional, and sandy suites are united into Dashgil and Goturdag suites.

From the analyses of thicknesses and drilling data, it is concluded that the area of intensive sedimentation during PS lowermost strata accumulation (an analog of the Kirmaki suite) has occupied the southeast part of the Djeirankechmes depression only. After Kirmaki time, the boundaries of the sedimentation basin were expanded in northerly and northwesterly directions. As a result, the areas located in the north and northwest of the depression (Gyrdag, Kaftaran, Donguzdyk, and Glych) are characterized by the presence of weakly dislocated deposits of the PS upper division, which overlay the eroded surface of steeply dipping beds of the Maikopian, Chokrakian, and other upper Miocene stratigraphic units. Thus, a general tendency of the Lower Pliocene basin on the Absheron Peninsula as in Gobustan is a basin widening to the

Fig. 2.25 Outcrops of the Upper Productive Series: in southeastern Gobustan. **a** Panoramic photo of the Koturdag section comprising amalgamated fluvial sandstones; **b** Plane bedded sandstones (Koturdag outcrop); **c** Root traces in dark gray mudstones (Koturdag outcrop); **d** Sandstones displaying trough cross-stratification (Koturdag outcrop); **e** Current ripples, arrow shows the paleoflow direction (Agtapa outcrop); **f** Plane view of wave ripples, arrows indicate the reverse ripple trains (Agtapa outcrop); **g** Skilithos-type burrow (Agtapa outcrop); **h** Sand-filled desiccation cracks (Agtapa outcrop); **i** Paleosol horizon containing root traces and halos (Agtapa outcrop); **j** Current-rippled bioturbated sandstones containing paired burrows (Agtapa outcrop); **k** Panoramic photo of the Cheildag section; multiple sand beds were accumulated in the river channels; **l** Channelized sandstones with erosive base, I—areal map. All photos have been taken during a joint expedition of the Geology and Geophysics Institute of the Azerbaijan National Academy of Sciences (GIA) and CASP, Cambridge University (e.g., Vincent et al., 2001)

northwest during the second half of the PS accumulation period. It is most likely that contrasted tectonic movements on the boundary of the Lower Kur depression and southeastern Gobustan have predetermined an indicative thick variation of the PS deposits.

2.3.4 The Kur Type

This sediment type has accumulated in the Lower Kur intermontane trough limited by the Greater and Lesser Caucasian systems. A boundary between the Kur and east-Gobustanian sediment types passes between the Beyuk and Kichik Kharami ridges (Azizbekov et al., 1972).

The total thickness of the Kur lithofacies sediment type comes to 3500–4000 m, being increased in a southeasterly direction towards the regional plunge of the beds (Azizbekov et al., 1972; Ismailov et al., 1972). The clastic product has mainly been supplied to the basin by the most extensive river artery, paleo-Kur, and by paleo-Araz, sweeping a large quantity of erosion material from the Kur lowland, Greater and Lesser Caucasus, and Talysh.

The alternation of silt-argillo-sandy rocks represents this section. The essential feature of the Kur lithofacies type is that it differs from the Absheron type by its considerable argillization, comparative quartz content, and increased quantity of feldspars, pyroxene, and hornblendite that points to the predominance of igneous rocks in the erosion zone. In some cases, sandy beds come to 20 m thick (Figs. 2.26 and 2.27). In a northwesterly direction, the content of the sand fraction of these deposits appears to be increased due to the fact of being nearer to the provenance in the Lesser Caucasus. This section becomes more shaley towards the southeast (Neftchala) and the northeast (Pirsagat, Byandovan). From the analysis of thickness distribution in the PS

Fig. 2.26 Outcrop of the PS uppermost strata (analog to the Surakhani suite) Babazanan, the Lower Kur depression, **a** panoramic photo, **b** 3 m thick channel-fill sandstones, **c** wood debris in fine sandstones, **d** sand-filled mud cracks, **e** areal map (joint GIA and CASP expedition, 2001) (Vincent et al., 2001)

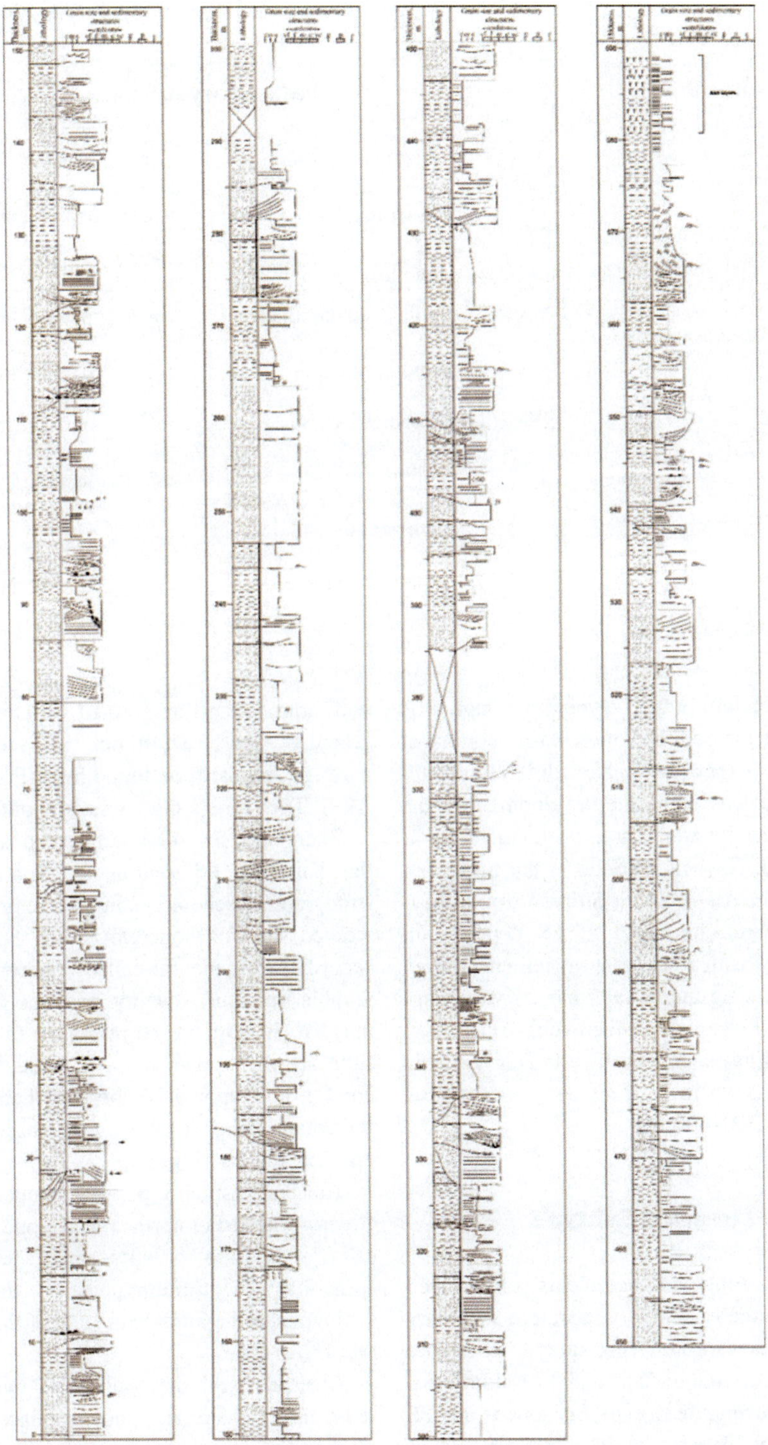

Fig. 2.27 The Babazanan outcrop, sedimentological log of the Upper Productive Series, an analog to the Surakhani suite (Vincent et al., 2001)

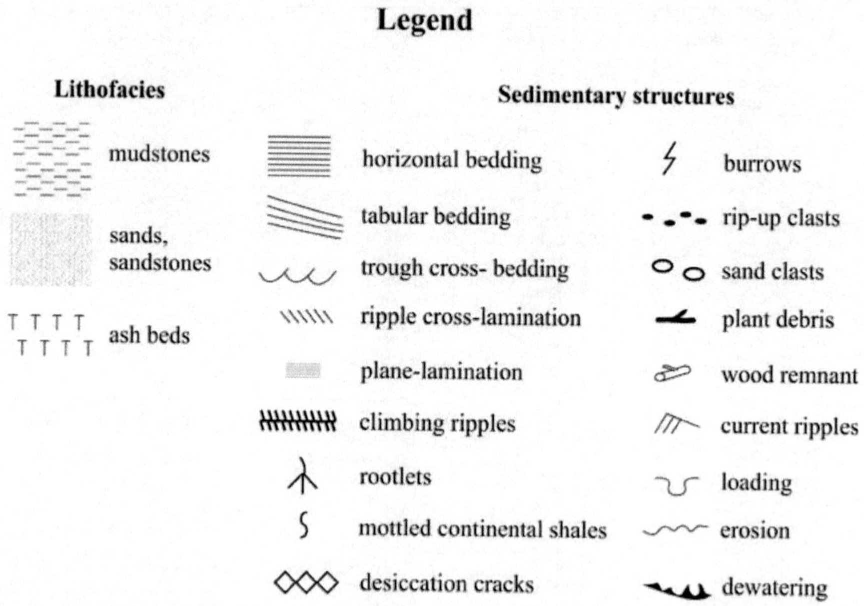

Fig. 2.27 (continued)

upper division within the Lower Kur trough, it is inferred that the sedimentation basin bottom at that time has experienced sharp differential subsidence accompanied by the development of the following positive and negative structures: up to 3000 m thick Navaga syncline in the northeastern part of the basin, South Shirvan trough having a maximum thickness of PS deposits up to 3300 m, Mugan-Salyan synclinal trough and the following anticlinal zones: Kyurovdag-Neftchala, Galamadyn-Mishovdag-Byandovan zones of uplifts in the southwest; Djarly-Saatly zone of uplifts where PS thickness appears to be decreased to 300–600 m.

2.3.5 The Lengebiz Subtype

The Lengebiz subtype of sediments is developed along the Lengebiz Ridge, a northern boundary of this lithofacies zone. This subtype combines Dongusdag, East-Gobustan, and Kur lithofacies types. The essential feature of this zone is that PS upper division deposits, jointly with a bed of conglomerates, appear to be immediately occurring on the Pontian eroded surface. On a level with dominating fine-grained varieties, inequity-granular sands, sandstones, and conglomerate partings, may also be found here (Figs. 2.28 and 2.29). This zone's thickness is about 1000 m.

There are several stratigraphic schemes for the Kur-type PS sediments. Apart from the stratigraphic scheme, some schemes are also worked out for concrete fields. For instance, according to the so-called "Neftchala stratigraphic division", twenty horizons from which the XX-th horizon corresponds to the Fasila suite are distinguished. In contrast, the XVI-th horizon corresponds to the VIII horizon of the Balakhani suite in the East Absheron stratigraphic scheme (Fig. 2.29).

The Pre-Caspian type is developed along the Greater Caucasian northeastern slope. Lithologically, it consists of silty sands, pebbles, and conglomerates. Stratigraphically, this type is analogous to the Surakhani suite of the Absheron lithofacies type.

A clear zonal distribution of two types of sedimentary material, sandy-argillaceous in the southeast (Gyzylburun type) and pebbly one in the northwest (Guba type), is well defined (Azizbekov et al., 1972). The sandy-argillaceous

Fig. 2.28 **a** Laterally continuous fluvial sandstones (sandstone is interbedded with shale layers), **b** scours at the base of channelized sandstones (overlying beds deeply truncate the underlying sandstones), **c** parallel bedded sandstones, **d** areal map (joint GIA and CASP expedition, 2001) (Vincent et al., 2001)

Gyzylburun complex mainly comprises fine-grained sediments; close-grained sands and sandstones are subordinate. Upwards the geological section, sandy material appears to be increased. Sandy's bed thickness is from a few cm to 1–5 m. This complex lower part is overlying by red gravel. Its maximum thickness is 2000 m (Fig. 2.30). On the whole, the thickness of this complex gradually decreased from the southeast towards the northwest, where the transition of the Gyzylburun type into the Guba lithofacies type is being noted.

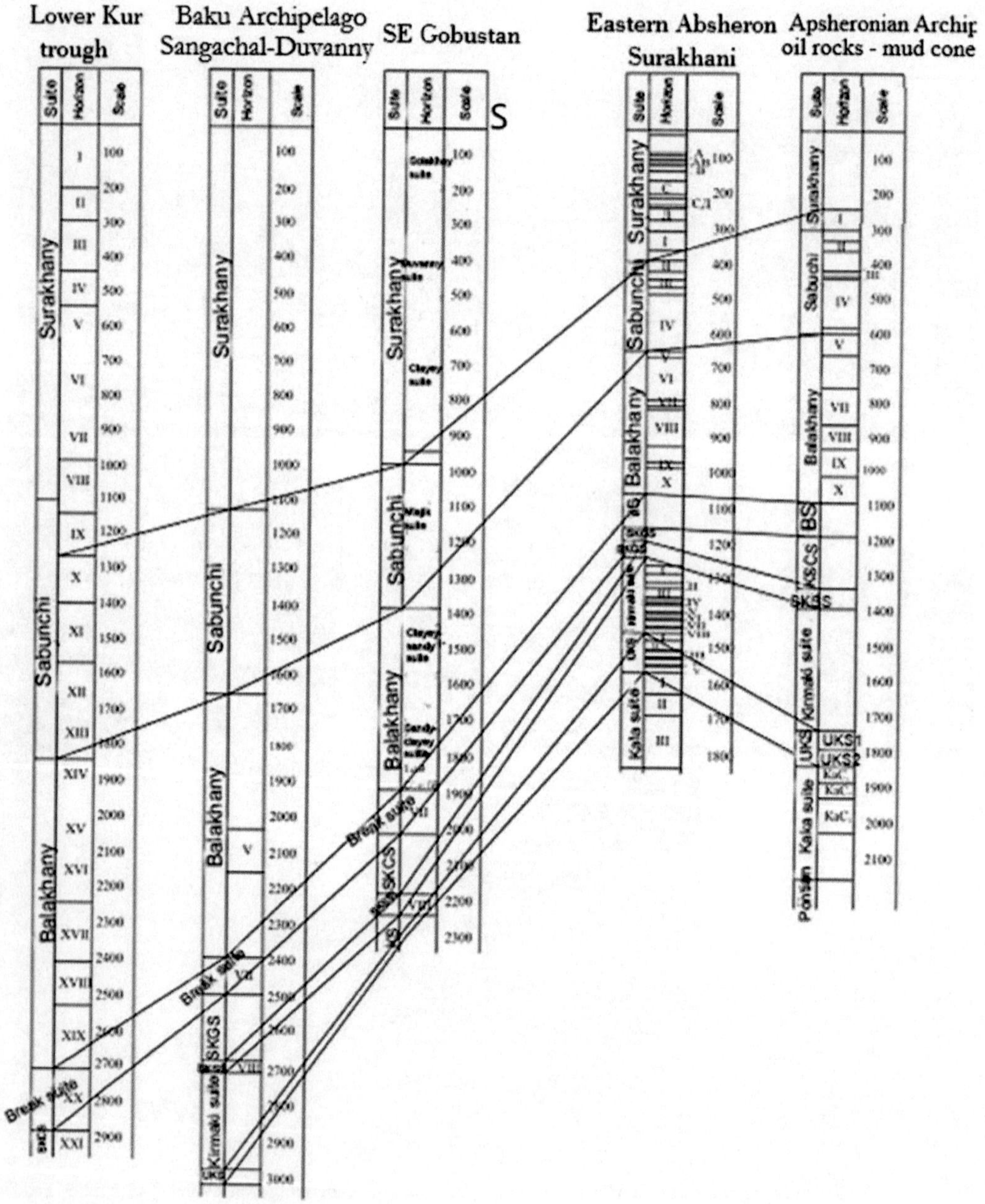

Fig. 2.29 Correlation scheme of the Productive Series sediments in different facies zones

The Guba lithofacies type is represented by poorly sorted pebbly beds alternating with coarse sands, sandstones, and conglomerates. The pebble sires are varied from small, < 2 cm, to isolated 1 m boulders of the Tithonian-Neokomian age. Petrographically, pebbles consist of Jurassic

Fig. 2.30 Outcrop of the PS uppermost strata (an analog to the Surakhani suite) in the Pre-Caspian region, Sura section. **a** Basal portion of the section; younging of sediments is to the left (thick shale succession containing fluvial sandstones), **b** current ripple lamination in very fine sandstones, **c** large desiccation cracks shown by arrows, **d** stacked channelized sandstones, **e** overview of the Sura section showing stacked highly cemented sand beds indicated by arrows, **f** areal map. The convening of sediments is to the left (joint GIA and CASP expeditions) (Vincent et al., 2001)

sandstones, limestones, shales, and Cretaceous to Eocene marls. Feldspars and rock waste mainly represent the pebbles' light fraction.

References

Abramovich, M. V. (1921). Section of productive series of the Surahan area. *Azerbaijan Oil Industry*, Nos. 4–5 (in Russian).

Akhmedov, G. A., Salayev, S. G., & Ismailov, K. A. (1961). *Prospects for the search for oil and gas in the Mesozoic deposits of the South-Eastern Caucasus* (283 p.). Azerneshr, Baku (in Russian).

Aliyev, A. G. (1947). *Petrography of the productive strata of Kabristan* (155 p.). Academy of Science of Azerbaijan, Baku (in Russian).

Aliyev, A. G. (1949). *Petrography of the tertiary deposits of Azerbaijan* (311 p.). Azneftizdat, Baku (in Russian).

Aliyev, A. G., & Daidbekova, E. A. (1955). *Sedimentary rocks of Azerbaijan* (331 p.). Aznefteizdat, Baku (in Russian).

Aliyev, I., Dadiyeva, T., Gasanov, F., Zokhrabova, V., Nabiyev, M., Pashaly, N., et al. (2005). In A. A. Alizadeh (Ed.), *Geology of Azerbaijan. Lithology* (Vol. II). Nafta-Press (in Russian).

Aliyeva, E. (2004a). Depositional environment and architecture of Productive Series reservoirs in the South Caspian basin. In: *South Caspian basin: Geology, geophysics, oil-gas content*. Special issue devoted to 32nd International Geological Congress in Florence, Italy (pp. 19–32). Naftha-Press, Baku.

Aliyeva, E. (2004b). Lithofacies zonation of Lower Pliocene Productive Series sediments, South Caspian basin. In *Proceedings of the AAPG Hedberg Conference "Sandstone Deposition in Lacustrine Environments: Implications for Exploration and Reservoir Development"* (pp. 72–74). Baku, Azerbaijan.

Aliyeva, E., Huseynov, D., & Kroonenberg, S. (2003). Depositional environment at rapid sea level fluctuations: implication to the South Caspian productive series architecture. In *Proceedings of the International Meeting "Dating Caspian Sea level Change", International Geological Correlation Programme* (pp. 4–5). UNESCO, Moscow.

Aliyeva, E. G. (2005). Reservoirs of the productive strata of the Lower Pliocene of the western side of the South Caspian depression. *Lithology and Minerals, 3*, 307–320. (in Russian).

Aliyeva, E. G. (2008). Sedimentation conditions, cyclicity, and architecture of reservoirs. In *Geology of Azerbaijan*, Vol. VIII, "Oil and Gas" (pp. 159–192) (in Russian).

Alizade, A. A., Alikhanov, E. N., & Shoikhet, P. A. (1967). *Investigation of the Conditions for the Transformation of Organic Matter in Modern Sediments of the South Caspian* (in Terms of the Oil Origin) (101 p.). Nedra, Moscow (in Russian).

Alizadeh, A. A., Guliyev, I. S., Kadirov, F. A., & Eppelbaum, L. V. (2017). *Geosciences in Azerbaijan. Volume II: Economic minerals and applied geophysics* (340 p.). Springer.

Azizbekov, S. R., Agakishibekova, R. R., Alizadeh, A. A., Alizadeh, K. A., Aliyev, M. M., Akhmedov, A. M., Akhmedov, G. A., Bairamov, A. S., Gadjiev, T. G., Zhouze, B. P., Zaitseva, L. V., Kashkay, M. A., Mekhtiyev, S. F., Sultanov, A. D., Khalilov, A. G., Shikhalibeyli, E. S., Efendiyev, G. H., & Yakubov, A. A. (Eds.). (1972). *Geology of the USSR*, Vol. 47, Azerbaijan Republic. Economic Deposits. Fossil Fuels (Oil and Gas). Moscow, Nedra (in Russian).

Babayev, D. K., & Gadzhiev, A. N. (2003). Paleogeography of the meso-cenozoic deposits and the role of the paleo-Volga in the evolution of sedimentary formations in the Caspian region. *Geologist of Azerbaijan. Scientific Bulletin, I*, 37–47.

Baturin, V. P. (1932). Petrography of Sands and Sandstones of productive series. *Proceed. of the Azerb. Scient. Inst., 4*, 1932. (in Russian).

Baturin, V. P. (1937). *Paleogeography by terrigenous components* (292 p.). AzONTI, Baku-Moscow (in Russian).

Feyzullayev, A. A., & Huseynov, D. A. (2002). Features of oil and gas formation within the Baku Archipelago. *Azerbaijan Oil Industry, 4*, 1–5. (in Russian).

Golubyatnikov, D. V. (1914). Detailed geological map of the Absheron Peninsula. Bibi-Heybat. *Transaction of the Russian Geology Communication* (216 p.), No. 106, New Ser. (in Russian).

Guliyev, I. S., Levin, L. E., & Fedorov, D. L. (2003). *Hydrocarbon potential of the Caspian region* (127 p.). Nafta-Press (in Russian).

Hinds, D. J., Aliyeva, E., Allen, M. B., Deavies, C. E., et al. (2004). Sedimentation in a discharge-dominated fluvial-lacustrine system: The Neogene Productive series of the South Caspian Basin, Azerbaijan. *Marine and Petroleum Geology, 21*, 113–138.

Hinds, D. J., et al. (2001). Depositional systems of the Neogene Productive Series, Apsheron Peninsula, Azerbaijan. *Azerbaijan project* (pp. 1–48), CASP-GIA, Report No. 8.

Hinds, D. J., Simmons, M. D., Allen, M. B., & Aliyeva, E. (2007). Architecture variability in the Pereriva and Balakhany suites of the neogene productive series, Azerbaijan: Implications for reservoir quality. In *Proceedings of the AAPG Studies in Geology "Oil and Gas of the Greater Caspian Area"* (Vol. 55, pp. 87–107).

Ismailov, K. A., Idrisov, V. G., Tagiev, E. A. (1972). *Productive Series of the Lower Kur Depression and the Baku Archipelago* (117 p.). Elm, Baku (in Russian).

Kalitsky, K. P. (1922). On the productive strata of the Absheron Peninsula. *Oil and Shale Industry, 3*(1–4), 29–35. (in Russian).

Kerimov, V. Y., & Averbukh, B. M. (1982). *Stratigraphic and lithological deposits of oil and gas in Azerbaijan* (139 p.). Elm, Baku (in Russian).

Khain, V. E. (1950). *Geotectonic development of the Southeast Caucasus* (225 p.). Azneftizdat, Baku (in Russian).

Khalifa-zade, C. M., & Rustamova, R. E. (1999). New data on the structure and conditions of formation of the Early Pliocene basin of the South Caucasus. In *Transaction of the International Conference on "Sedimentology of Sedimentary-Rock Basins of the Caspian-Black Seas Region"* (pp. 65–66). Baku (in Russian).

Konyukhov, I. A. (1951). The character of cross-bedding in the rocks of the productive strata of the Absheron Peninsula. *Geological Collection of Papers* 1 (IV) (pp. 63–72). Gostoptekhizdat, Moscow (in Russian).

Kovalevsky, S. A. (1922). On the parallelization of the Bibi-Heybat and Atashkya-Yasamal sections. *Azerbaijan Oil Industry, 3–4*, 65–71. (in Russian).

Mirchink, M. F. (1926). On the genesis of the productive series of the Apsheron Peninsula. *Azerbaijan Oil Industry, 1*, 23–37. (in Russian).

Mirchink, M. F. (1926). More about the genesis of the productive strata of Azerbaijan. *Azerbaijan Oil Industry, 10*, 15–25. (in Russian).

Mirchink, M. F. (1933). On the issue of the genesis of the productive series. *Azerbaijan Oil Industry, 2*, 10–13. (in Russian).

Mustafayev, I. S. (1963). *Lithofacies and paleogeography of the Middle Pliocene deposits of the Caspian Basin* (193 p.). Azerneshr, Baku (in Russian).

Mustafayev, I. S. (1964). Lithofacies and paleogeography of oil and gas bearing Middle Pliocene deposits of the Kura depression. In: *Essays on the Geology of Azerbaijan* (pp. 309–319). Academy Science of Azerbaijan, Baku (in Russian).

Pashaly, N. V., Kheyirov, M. B., & Saradzhalinskaya, T. M. (1998). Productive series, In: *Geology of Azerbaijan*, Vol. II, *Lithology* (pp. 186–229) (in Russian).

Pashaly, N. V., Saradzhalinskaya, T. M., & Katz, N. M. (1979). *Facies and paleogeography of the Pliocene and Quaternary lithogenesis and associated minerals (Baku Archipelago).* Report of the Inst. of Geology, Academy of Science of Azerbaijan (235 p.) (in Russian).

Potapov, I. I. (1954). *Absheron oil-bearing region (geological characteristics).* Academy Science of Azerbaijan, Baku, (539 p.) (in Russian).

Putkaradze, A. L. (1958). *Baku Archipelago.* Azerneftneshr, Baku, (335 p.) (in Russian).

Reynolds, A. D., Simmons, M. D., Bowman, M. B. J., Henton, J., Brayshaw, A. C., Ali-Zadeh, A. A., Guliyev, I. S., Suleymanova, S. F., Atayeva, E. Z., Mamedova, D. N., & Koshkarly, R. O. (1998). Implication of outcrop geology for reservoirs in the Neogene Productive Series. Apsheron Peninsula, Azrbaijan. *AAPG Bulletin, 82*, 25–49.

Sultanov, A. D. (1949). *The lithology of the productive series of Azerbaijan.* Academy of Science of Azerbaijan, Baku, (184 p.) (in Russian).

Vincent, S., Davies, C., & Aliyeva, E. (2001). Outcrop sedimentology of the Kura basin Upper Productive Series, Azerbaijan. CASP (Camridge Univ.)—GIA (Geol. Inst. of Azerbaijan) Report.

Vistelius, A. B., & Miklukho-Maklay, A. D. (1951). On Paleozoic pebbles from the productive strata of the Apsheron Peninsula. *Doklady Academy Science USSR, 79*(3), 15–21. (in Russian).

3 Bio-, Chemo-, and Magnetostratigraphy of the Productive Series

3.1 Biostratigraphic Characteristics of the PS Deposits

Biostratigraphic characteristics of the PS deposits in the Absheron oil/gas-bearing region are relatively well worked out. Faunal complexes established in the PS outcrop within the Kirmaki and Yassamal valleys of the Absheron Peninsula are described below.

According to Khalilov (1946) and Agalarova (1956), the Kala suite deposits contain, on a level with redeposited Miocene foraminifera *Nonion granosus Orb.*, *Elphidium macellum* (Ficht. et Moll.) and other, also the relict Pontian *Ostracoda Cytherissa bogatschovi Liv.*, *Cyprideis littoralis* (Brady), *Loxoconcha cf. djaffarovi* (Schneid.) and others (see Table 3.1). As mentioned above, these deposits have not been exposed anywhere and are known only by drilling data. In the Kirmaki valley, the PS section begins from the UKS deposits transgressively overlapping the Pontian regio-stage clays. Here, Baba-Zadeh (2011) has discovered 22 species of Pontian Ostracoda and a large quantity (more than 30) of bivalve mollusk shells *Corbicula fluminalis* (Muller) (Table 3.1; Fig. 3.1).

Four of the Ostracoda species from this complex, namely *Leptocytherelata* (Schneid.), *L. microlata Liv.*, *Pontoniella acuminata* (Zal.), and *P. Loczyi* (Zal), are the typical Pontian regio-stage forms which usually have not been found in more young deposits. The lower boundary of the PS is drawn by the disappearance of these forms from its section. The age of the PS uppermost strata is confirmed by the Ostracoda relict forms found in the rocks jointly with Miocene and Paleogene foraminifera. From time to time, found are the very small, thin-walled, transparent Ammonia *beccarii* (Linne), *Elphidium macellum* (*Ficht. et Moll.*), and *Nonion granosus* (Orb.) occurring "in situ" in the PS basin (Alizadeh et al., 2017). The mentioned forms are quite different from the shells of the same species, and genera redeposited from the Miocene by their depressed state connected with decreasing the water salinity of the PS accumulation basin.

The contact between the PS deposits and overlying Akchagylian deposits has been studied by Baba-Zadeh (2011) in the Yassamal Valley, about 300–400 km south of Gobu village. The Akchagylian age of overlying deposits is confirmed by the presence of the following typical Ostracoda fauna: *Candona combibo Liv.*, *C. Candida* (Muller), *Leptocythere litica Liv.*, *L. Verrucosa Liv.*, *L. palimpsesta Liv.* and by such foraminifera species as *Cassidulina ex gr. crassa Orb.*, *Discorbis multicameratus Chutz.* The presence of freshwater Ostracoda fauna confirms the affirmation that underlying deposits belong to the PS uppermost strata: *Ilyocypris brady sars*, *Cyprinotus triangularis Kasim.*, *C. Salinus* (Brady), *Candoniella Ibicans* (Brady), *Heterocypris incongruens* (Ramd.), *Darvinula stevensoni* (Bram. et Rob), *Xestoleberis chanakovi Lin.* and redeposited Cretaceous foraminifera.

A. Alizadeh et al., *Pliocene Hydrocarbon Sedimentary Series of Azerbaijan*,
Advances in Oil and Gas Exploration & Production, https://doi.org/10.1007/978-3-031-50438-9_3

Table 3.1 Comparison of schemes of the PS deposit dissection according to the Ostracoda fauna

Lithostratigraphy				Dissection by Ostracoda fauna	
Series	Subseries		Suites	Khalilov (1946)	Agalarova (1956)
	New division	Old division			
PS	Upper	Upper	Surakhani	Upper part	Freshwater Ostracoda stage
		Middle			
	Lower subseries	Lower subseries	Sabunchi	Middle part	Euryhaline Ostracoda stage
			Balakhani Break suite		
			Superkirmaki suite Clay suite	Lower part	Brackishwater Osracoda stage
			Superkirmaki clay suite		
			Kirmaki suite		
			Underkirmaki suite		
			Kala suite		

Agalarova (1956) concluded about the law-governed nature of redeposited foraminifera distribution in the PS section. In such a way, the Miocene and Paleogene forms have been discovered in the Lower suites, while Cretaceous forms—are in the Middle and Upper suites. The given fact is reflected in the paleogeography of the Lower Pliocene basin, a transgressive water body that reached its maximum at the end of the Lower Pliocene.

Based on the given data, the following scheme of the PS zonal dissection is proposed (Baba-Zadeh, 2011; Baba-Zadeh & Aliyeva, 2005) (Fig. 3.2):

1. A zone of in situ developed foraminifera (*Ammoniabeccarii*, *Elphidiummaccelum*, *Noniongnosus*, *N.sp.*), relict brackish water Ostracoda and redeposited Miocene, Paleogene, and Cretaceous foraminifera. It should be pointed out that redeposited Cretaceous forms are shown up in the uppermost strata of this zone that corresponded to the lithostratigraphic units of the Superkirmaki clay suite—the more significant part of the Balakhani suite. The lower part of this zone is characterized by the presence of only redeposited Paleogene and Miocene foraminifera and corresponds by its volume to the Underkirmaki-Superkirmaki suites.
2. This zone presents the brackish water Ostracoda development and the absence of in situ foraminifera. The foraminifera extinction and replacement of euryhaline Ostracoda by continently freshwater forms are observed between the Balakhani and Sabunchi suites. The appearance of brackish water Akchagylian fauna defines the upper PS boundary. At the end of Balakhani and the beginning of the Sabunchi times, the PS brackishwater sea has been gradually changed into a freshwater basin existent up to the end of the Surakhani time. From the beginning of the Akchagylian stage, this freshwater basin changed once again into a vast brackish water basin of the Akchgylian stage where such brackish water fauna as *Ostracoda*, mollusks, and other typical groups have been settled and widely developed.

3.2 Chemical-Stratigraphic Dismemberment of the PS Deposits

Although more than the age-long history of the PS investigation and detailed study of its macro- and microelement composition, no attempt was

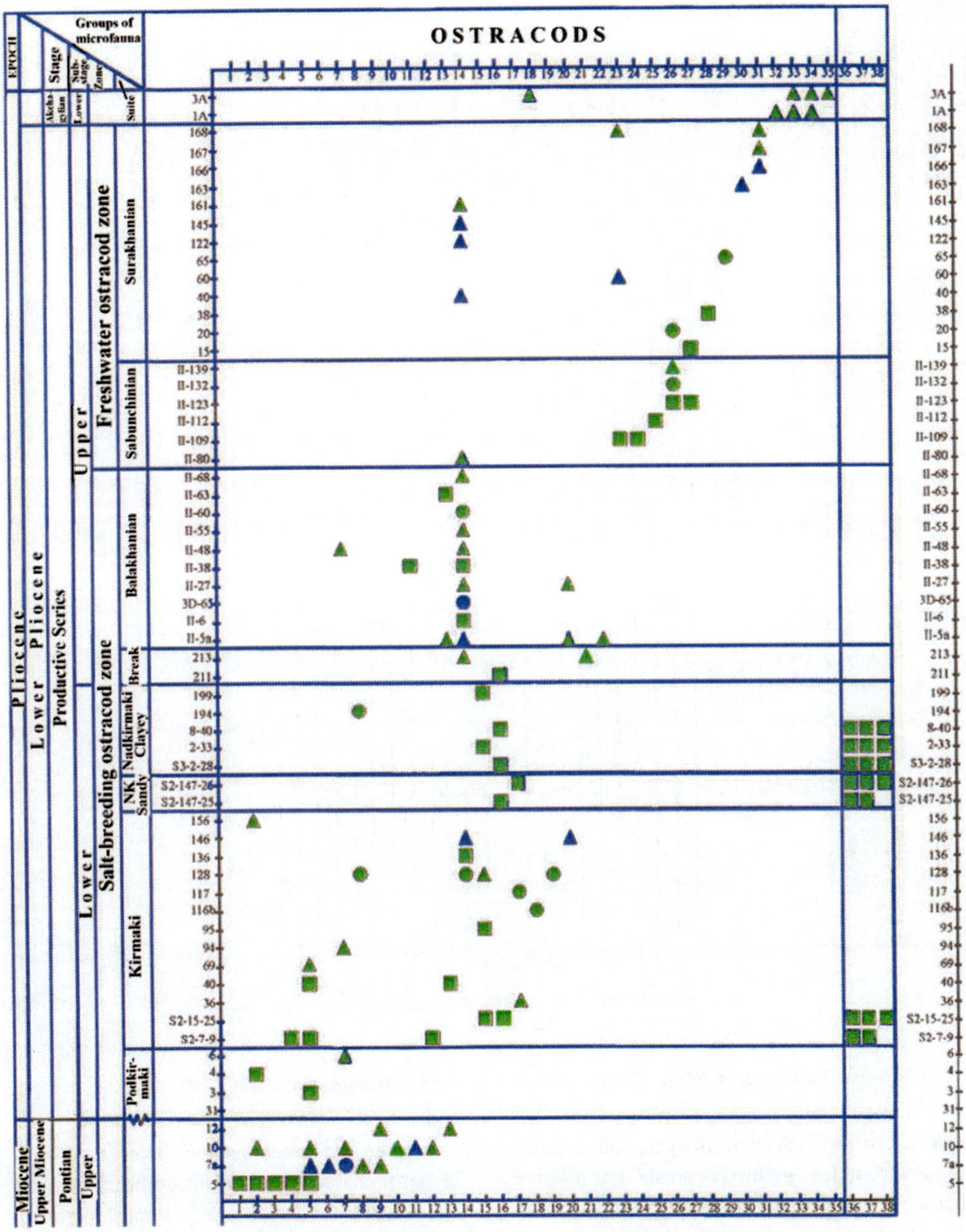

Fig. 3.1 Schematic diagram illustrating microfauna distribution in the PS deposits of the Absheron Peninsula (Kirmaki and Yassamal valleys) (Baba-Zadeh, 2011)

made to dismember these deposits based on chemical factors, and no geochemical explanation of the results obtained has been presented till quite recently. The work of Ali-Zadeh et al. (2013) played an essential role in this study.

The Kirmaki section's lowest interval, which will be described in detail in Chap. 5 (see Fig. 5.3), is characterized by anomalous values of almost all the elements within an interval of 128–134 m, distinguishing from all the rest part

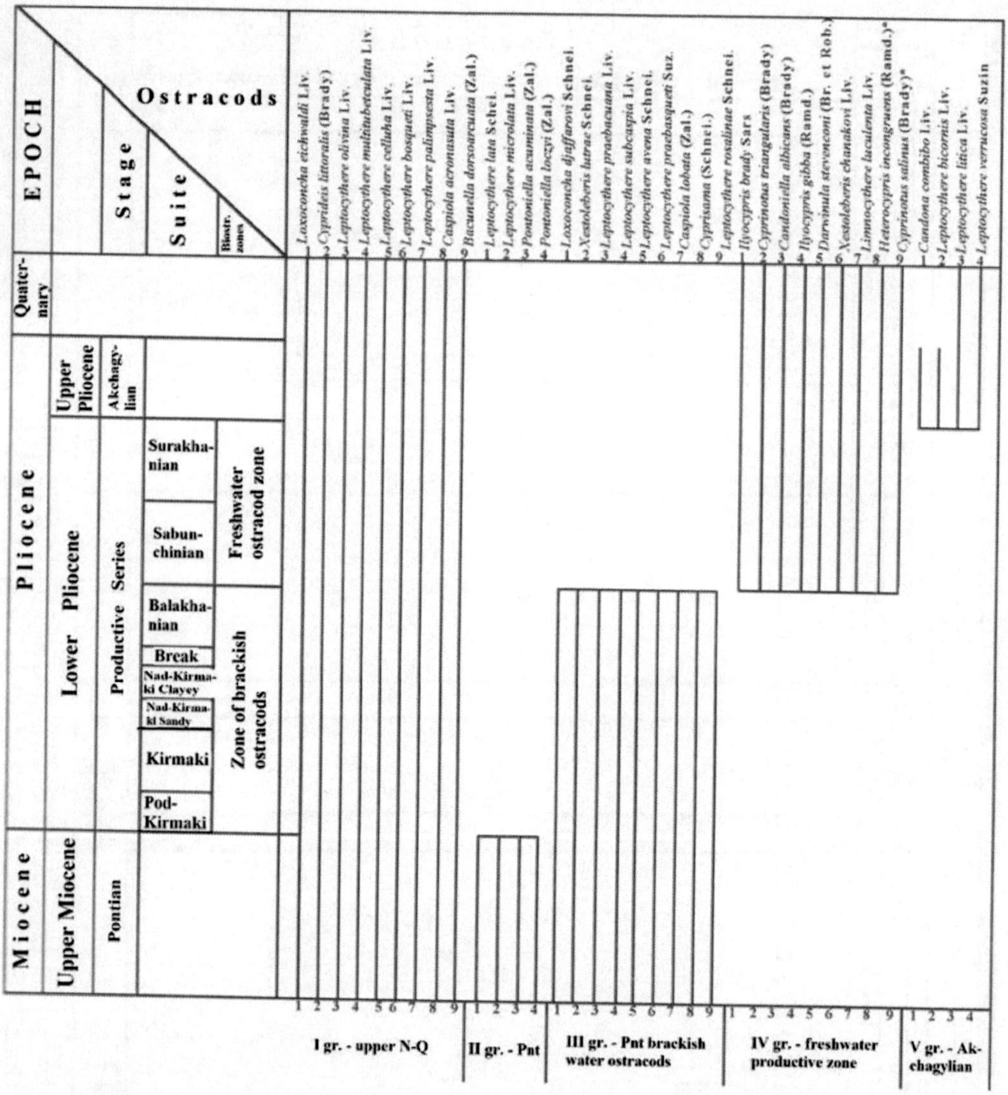

Fig. 3.2 Schematic illustration of the PS biostratigraphic complexes (Baba-Zadeh, 2011)

of the section. Such high element concentration takes place in fine-grained deposits. For instance, the concentration peak of such elements as Ca, Fe, Ba, Zn, Sr, Ni, Pb, and Cu is registered in sandy shales occurring in the interval of 128.5–128.8 (Fig. 3.3).

The mentioned elements' content in overlying coarse sands appears to be decreased except Sr. One peak of contents of practically all the studied elements has been noted some higher (at 152.5 m) within sandy shale partings. The same case is observed in the sandy-siltstone interval (171.3–171.7 m), characterized by insignificant comparative growth in the content of most elements except for Mg, Cu, and Zn. Thus, against the relatively monotonous distribution of microelements in Underkirmaki deposits, the mentioned peaks in the content of most microelements relate to pelitic and siltstone deposits. This outcome confirms that microelement concentration appears to be increased in fine-grained sediments. An exciting feature of rock-forming elements, which

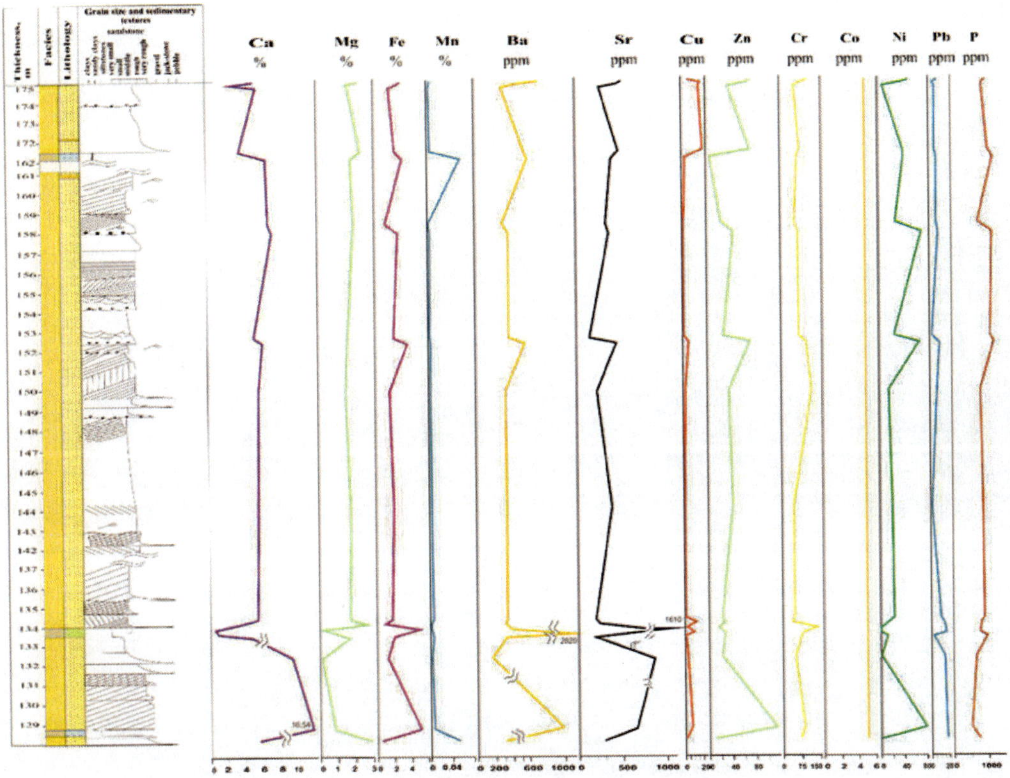

Fig. 3.3 Lithological-geochemical section of the Underkirmaki ('Podkirmakinskaya') suite

is hard to explain, is that some lithological regularities have been revealed in their behavior. In the PS Kirmaki suite, such regularity is displayed less clearly, which may be caused by a large quantity of clay and siltstone fraction in sandy sediments (Figs. 3.4, 3.5 and 3.6).

The distribution of elements over the PS lower-division section shows a relatively wide variation in their values. For instance, Ca content limits are highly variable, ranging from 0.2 to 16.54%, and the same values were found in the Underkirmaki suite.

The contents of other elements, being highly variable, are as follows: Ba—150–2820 ppm, Sr—20–1610 ppm, Cu—0.1–53 ppm, Zn—5–400 ppm, Ni—from less than 1 ppm to 108 ppm, Cr—from less than 1 ppm to 145 pm. Some elements show a different stratigraphic distribution relationship in the PS section. The Ca content of the Underkirmaki suite (US) is 1.5–2.5 times higher than those in different intervals of the K, SKS, and SKC suites (Fig. 3.7). Its concentration in the Middle KS is 1.5 times higher than in other suite parts. Besides, sedimentation conditions in the Middle KS are interpreted as the most distal within PS, changing from the delta front to the lacustrine conditions. It is hard to explain the high Ca content of UKS sediments by increased carbonate contents only since the studied section is composed of uncemented, coarse, and medium-grained sands. It is probably due to the presence of feldspars in these deposits, even if they are presented slightly. From our field observation, it is inferred that increased Ca content up to 3.7% is explained by predominant sandy varieties in the SKS section and, what is more, by the presence of many sandstones cemented with carbonate cement. Our results show that this element content of KS uppermost strata becomes relatively low and is estimated at 1.7%.

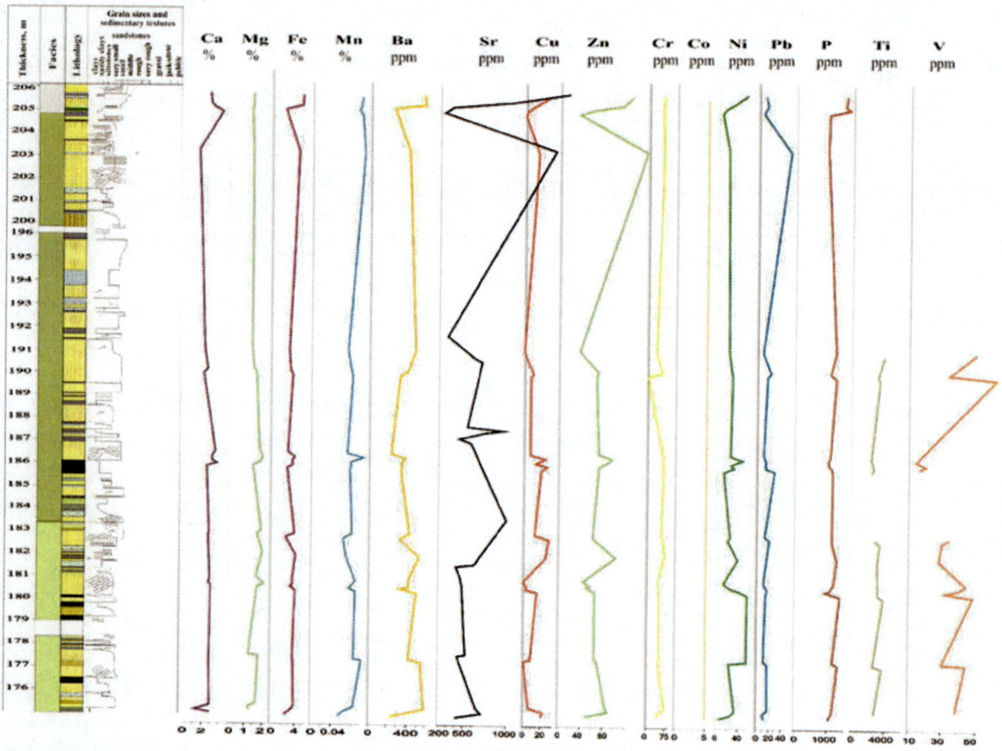

Fig. 3.4 Lithological-geochemical section of the lower Kirmaki suite

It seems likely that a sharp difference in lithological composition between the UKS and KS sediments apparently resulted in an essential Mg concentration. It appears that the average Mg content is increased from 1.7% in the UKS, composed of 98% of sand, and up to 1.8% in the lower KS section, composed of 50% fine-grained rocks as regarded as coarse-grained ones. In the Middle KS, where the sand-clay relation is approximately equal, an average Mg content appears to be sharply increased up to 2.5%, and then, in KS uppermost strata characterized by dominant sandy deposits, this value is decreased again to 1.6%. Lastly, in the SKSS, Mg percentage slightly increased to 2%.

The mentioned relationship resulted from the lithological control of the distribution of the elements. As it is known, an increase in Mg content in clay rocks is caused by the presence of Mg silicates in clay minerals.

The Fe distribution in the section is like that of Mg. Its smallest amount has accumulated in the UKS (on an average of 2.4%). Then its accumulation slightly increased—from 2.8% in the KS lowermost strata to 3.8% in the Middle KS, but in the KS uppermost strata, the Fe content decreased to 3.1%. This value is observed in SKSS and KSCS, as well.

The typical iron minerals are limonite, hematite, siderite, ilmenite, magnetite, pyrite, and chalcopyrite. Besides, a small amount of Fe is presented in epidote, chlorite, hornblende, biotite, pyroxene, and other minerals. Because some ferriferous minerals are easily decomposed, a certain iron amount may be passed into the solution. However, an essential part of these minerals is stable and may have been retained during weathering. From the study of the Kur and Terek rivers' water, it is inferred that 99.5–99.8% of Fe is migrating through bed loads (Konovalov et al., 1968).

As a rule, the most significant amount of Fe is accumulated in the littoral zones. Therefore, explaining its contents growing in the KS clay

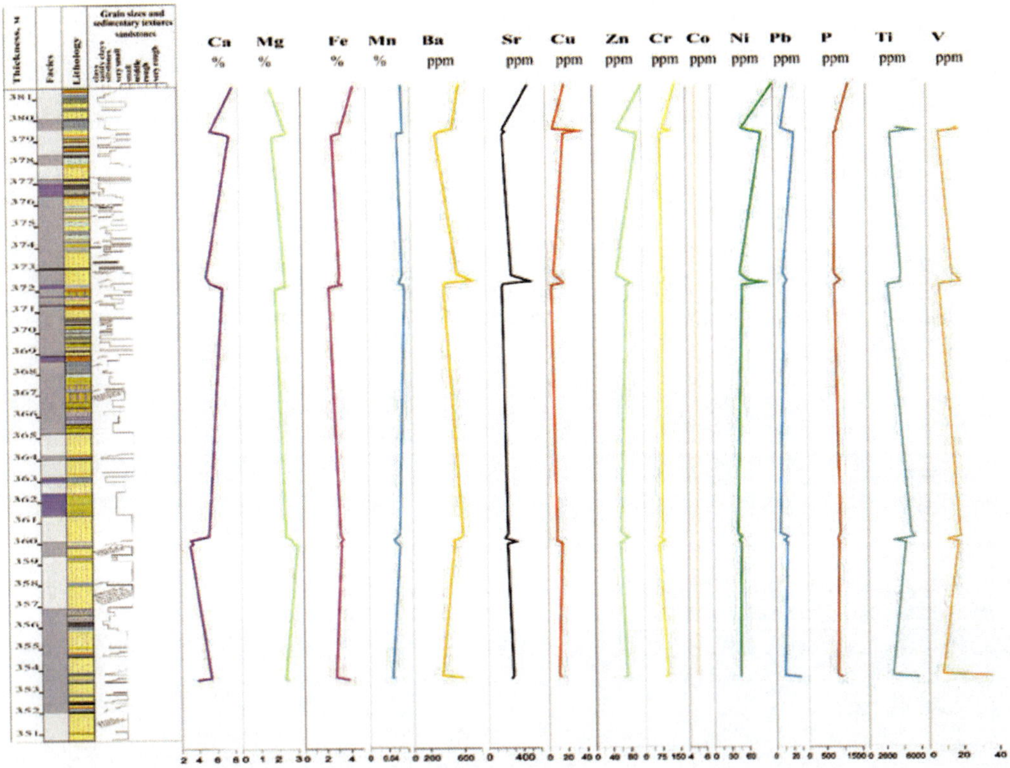

Fig. 3.5 Lithological-geochemical section of the middle Kirmaki suite

lacustrine-prodeltaic deposits is hard. In our understanding of this case, the obtained gross iron quantity dominates the lower iron oxide.

At the same time, these values are lower than the clark of this element given by Turekian and Wedepohl (1961) for fine-grained rocks (4.72%) as well as lower than the clark given by Vinogradov (1962).

The manganese content peak marked in the Middle KS sediments exceeds those in the UKS almost eight times, being 0.012 and 0.09 ppm. As it is known, the Mn geochemical distribution is characterized by its increased contents in the sediments of the basin's distal parts.

It is recognized that the PS lower division is characterized by a gradual increase in element content from the UKS to the more clay middle KS and further, their decrease in the uppermost strata of the KS, SKSS, and SKCS. Such geochemical distribution is typical for several elements: Zn contents are increased from the UKS to SKSS as follows: 40, 60, 72, 91, 69, 77%; Cr—75, 65, 88, 80, 71, 78%; Ni—30, 35, 58, 46,6, 44, 45% (all contents are given on an average over the suites). The same relationship is typical for vanadium. The distribution of copper and lead does not become well-defined.

As distinct from iron, the most incredible amounts of Mn, Ni, Co, Mo, Cu, and Zn are concentrated in unstable minerals like mica, amphibole, feldspars, etc. During the weathering, these minerals quickly decompose; consequently, an essential part of their elements appears to be passed into the solution. In the Caspian basin rivers (Volga, Ural, Kur, and Terek), the following elements are currently migrating in the form of solution: Fe (0.25%); Mn (1.7%); Ni (23%), and Cu (33.7%) (averaged percentage are presented by Lebedev et al. 1973).

Such geochemical features of the mentioned element and their high stability explain their location in the PS basin distal parts. Chrome,

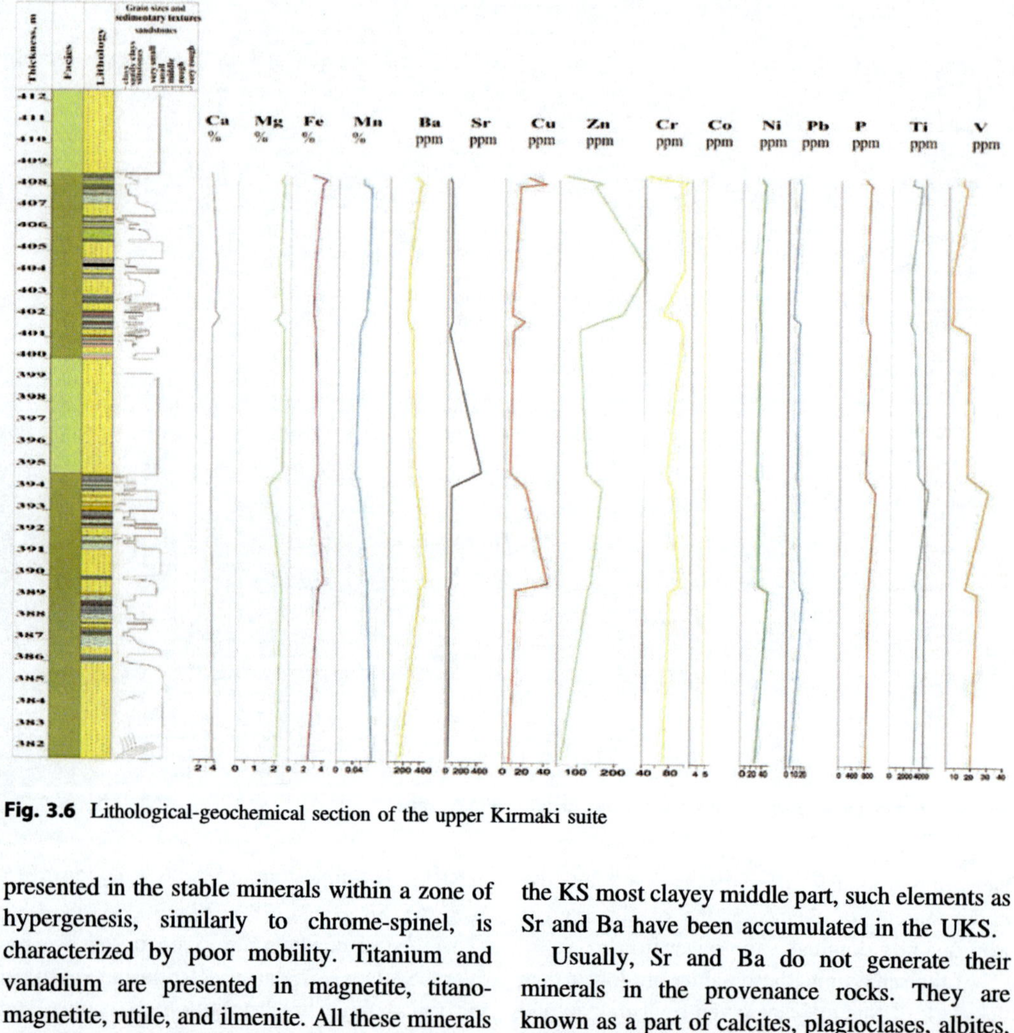

Fig. 3.6 Lithological-geochemical section of the upper Kirmaki suite

presented in the stable minerals within a zone of hypergenesis, similarly to chrome-spinel, is characterized by poor mobility. Titanium and vanadium are presented in magnetite, titanomagnetite, rutile, and ilmenite. All these minerals are stable against the action of weathering that results in their migration in the form of suspension.

It is known that Ni, Cr, and V are retarded by organic matter. It seems likely that this fact explains well an essential increase in their content in the Middle Kirmaki deposits accumulated in the distal part of the PS basin despite these elements' low migration ability. Our investigations show that the deposits in the distal part of the PS basin are characterized by the most incredible amount of organic matter (Aliyeva, 2005a), as distinct from the mentioned elements, whose maximum contents have been identified in the KS most clayey middle part, such elements as Sr and Ba have been accumulated in the UKS.

Usually, Sr and Ba do not generate their minerals in the provenance rocks. They are known as a part of calcites, plagioclases, albites, augites, and other minerals. Since these minerals were unstable in a hypergenesis zone, they have been decomposed during the weathering. Strontium and barium have passed into aqueous solutions and migrated as bicarbonate, chloride, and sulphide. The strontium content of the river waters is estimated at $1.3 \times 10^{-5}\%$.

In contrast, the content of sea waters is $1.3 \times 10^{-2}\%$. In other words, the river waters are considerably depleted in Sr. For sea waters to have been saturated with strontium, a salt concentration should be increased 4–5 times compared with the normal salt basin (3.5%); afterward, it will be followed by the strontium

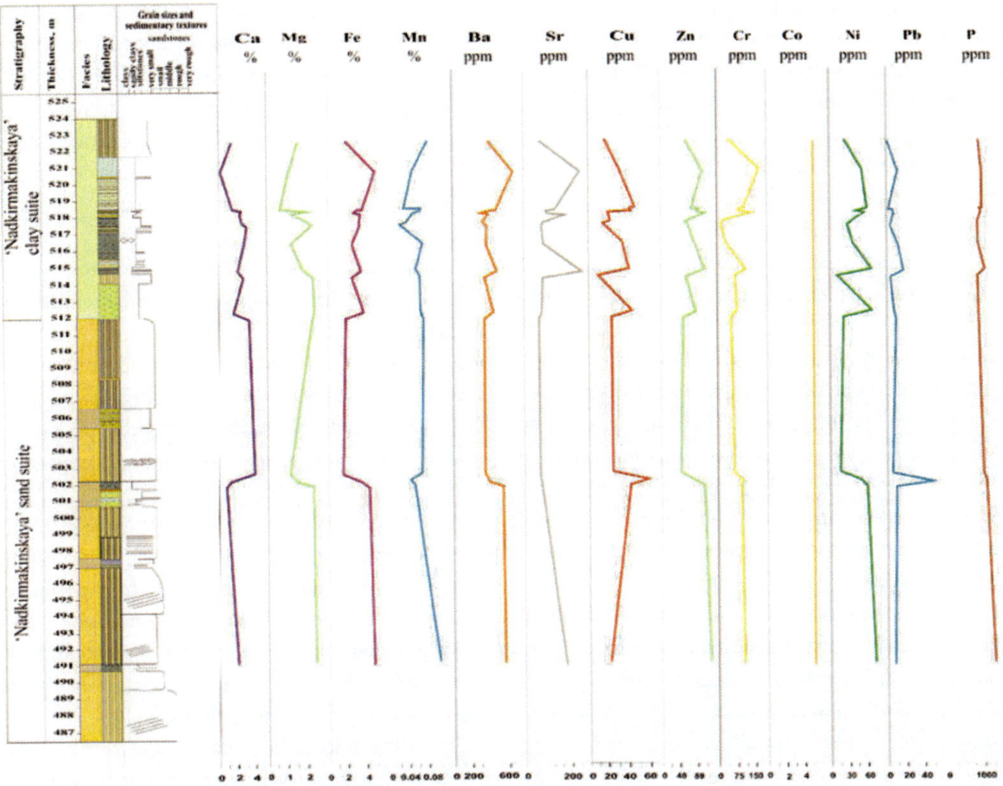

Fig. 3.7 Lithological-geochemical section of the SKSS and SKGS

settling in the form of celestine. So, the sediments accumulated in the freshwater basins are characterized by the greatest Sr content, mainly observed in alumosilicates (0.01–0.02%). The barium ions supplied with the river waters to the sea where SO_4 ions are abundantly present appear to be precipitated as a weakly soluble barite ($BaSO_4$), which is usually absorbed in near-shore deposits. The barium ions supplied with the river waters to the sea where SO_4 ions abundantly appear to be precipitated as weakly soluble barite ($BaSO_4$), usually absorbed in near-shore deposits.

The barium content of seawater is estimated as $5 \times 10^{-6}\%$, while the content of freshwater is much higher, being $1.7 \times 10^{-4}\%$. This element may accumulate in freshwater basins because of the clay particles' sorption. The Ba and Sr different geochemical behavior within a zone of hypergenesis allows the use of the Ba/Sr ratio for sedimentation medium reconstruction since this ratio in freshwater deposits is usually decreased compared with marine deposits.

The Sr/Ba ratio decreasing trend in the PS lower division is well-defined from the UKS to the SKCS. The distribution of this value within the SKCS itself, both in clay and sandy rocks, favors a more freshwater medium. The results of micropaleontological investigations confirm this conclusion. The foraminifer in this suite is highly stunted and displayed in its small sizes and transparent test walls. Based on the Ba/Sr ratio, it seems reasonable to suppose that the Caspian maximum salinity occurs during the SKSS accumulation. Thus, from the result of a few elements study and their distribution, it is inferred that the PS deposits are clearly divided into chemostratigraphic units, well correlated with their generally accepted lithostratigraphic scheme. In such a way, the UKS and KS deposits are clearly separated and differentiated. A well-defined trend in the distribution of macro-and microelements in the KS itself allowed us to single out this suite's lower, middle, and upper parts. It is noted that the SKSS and SKCS deposits need to be more clearly differentiated. It may be because the SKCS deposits are poorly stripped in the studied section.

It is known that the contents of uranium and thorium series elements and potassium-40 isotopes cause the rocks' natural radioactivity. A share of these elements in the total radioactivity of the rocks depends on their type, mode of occurrence, and formation. Therefore, a separate determination of radioactive elements' content and their share in total radioactivity could provide additional information on the rock's special features and origin (Kharitonova, 1964; Smyslov, 1974).

As it is well-known, the data of uranium, thorium, and potassium contents could play an essential role in sedimentation processes investigation. The notable feature of the potassium content data is that they allow the determination of the location and dominant effect of the provenance region and direction of terrigenous material evacuation to the basin and distinguish the zones of different clay minerals associations. Further from the provenance region, material sorting increases with decreasing feldspar and mica contents, and potassium concentration decreases from the littoral zones to the deep-sea area. In such a way, potassium distribution in sedimentary basins largely depends on the change in mineral composition over the basin area (Kovalev, 1965).

The distribution pattern of uranium in the sedimentation basin area is determined by an organic matter type and its accumulation level in different facies zones. The maximum uranium accumulation in the basins characterized by the organic matter contained within a clark's limits occurs under the most shallow-nearshore facies conditions. These zones are mainly characterized by an organic matter accumulation pattern and possibly uranium diagenetic redistribution in the shallow facies sediments, which are dominantly represented by sandy and sandy-silty deposits. Deep into the basin, uranium content becomes diminished, and the sediments of transitional facies appear to have differed by minimum uranium content only. The sediments formed under relatively deep-sea facies conditions show increased uranium but are dominantly in scattered form.

The thorium distribution over the area of the sedimentation basin appears to be less differentiated than those of potassium and uranium. When the rock formation occurred under one source area influence, the only marginal parts of the basin appear to be distinguished, characterized by relatively stable thorium accumulation observed in the rocks of different lithological compositions. It is also certain that the thorium distribution data allow us to determine the source area location, outline different transportation ways of supplying terrigenous material, and make out shallow and nearshore zones, which all may have been used as an influential factor in paleogeographic reconstructions.

The data gained from the relationships of the mentioned radioactive elements' behavior made it possible to carry out for the first time the PS deposits dissection based on their radioactive features. From the integral and spectral radioactivity tests we obtained for the PS deposits in the Kirmaki Valley (Fig. 5.3), there are widespread values and essential changes in the natural radioactivity of the PS different units. For instance, it appears that specific radioactivity values in the UKS deposits are usually marked in the interval from 0 to 21 Bk/kg, i.e., these are of low values, which, in our understanding, can be explained by dominating sandy fraction in this suite deposits (Fig. 3.8).

From the correlative connections pattern of radioactive elements in the UKS deposits, it is concluded that the mostly expressed correlative connection of integral radioactivity occurs with uranium (correlation coefficient r is 0.86) and favors this suite uranium nature. It is noted that correlative connections of integral radioactivity with thorium are also relatively high (correlation coefficient r is 0.76). Significant values have also

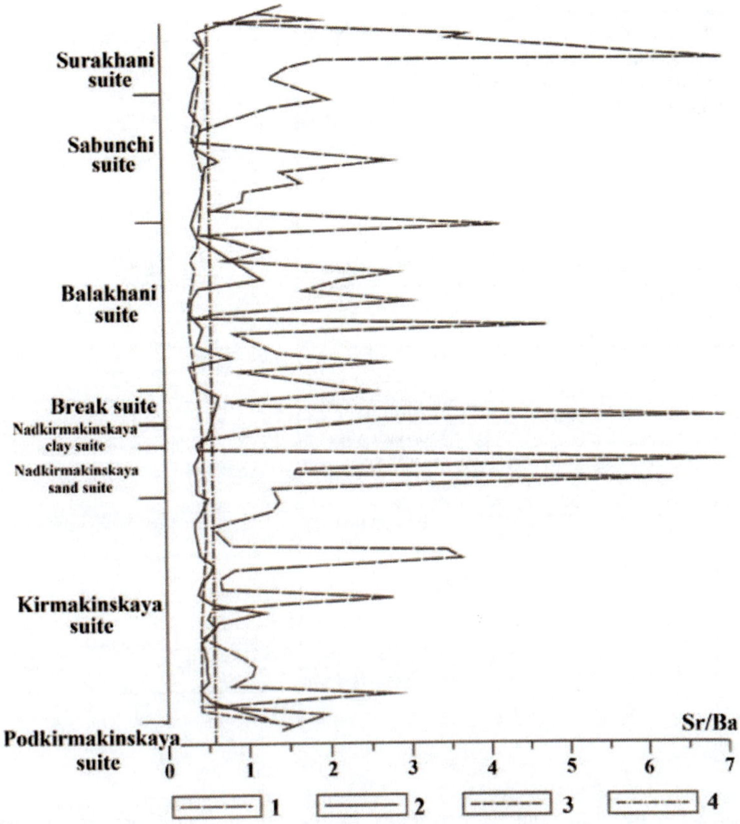

Fig. 3.8 Curves showing the variations of integral radioactivity values and radioactive elements contents through the UKS section

been obtained for the correlation of integral radioactivity with the potassium content (correlation coefficient *r* is 0.57).

Rather, high specific activity values characterize the Kirmaki suite. The samples most frequently have radioactivity values from 50 to 60 Bk/kg. The correlation connections pattern shows that the most integral radioactivity depends on the potassium content ($r = 0.67$). That is, from all the mentioned data, it is evident that the UKS deposits are characterized by natural radioactivity of the multi-elemental nature confirmed by similar relationships in the variations of integral radioactivity values and radioactive elements content all over the suite section (Fig. 3.8). Almost the same values are typical for the integral radioactivity correlation with uranium ($r = 0.57$).

An analysis of the variations in studied radioactivity values over the section has shown that its lowermost strata are characterized by a well-integral radioactivity correlation with potassium and thorium, while the upper part—is with potassium and uranium. The middle section is noted by the presence of natural uranium radioactivity only (Fig. 3.9). Finally, the uppermost suite is distinct in potassium effect, which becomes more perceptible.

Approximately the same integral radioactivity values have been noted in the SKSS deposits, from 59 to 80 Bk/kg. The correlation coefficient of integral radioactivity with thorium is 0.77. Those with uranium and potassium appear to be 0.14 and 0.12, respectively. Thus, the natural radioactivity type of the SKSS deposits is undoubtedly thorium, which is confirmed by the character in variations of radioactive elements contents and integral radioactivity values over the suite section (Fig. 3.10), showing the synchronism in variations of integral radioactivity

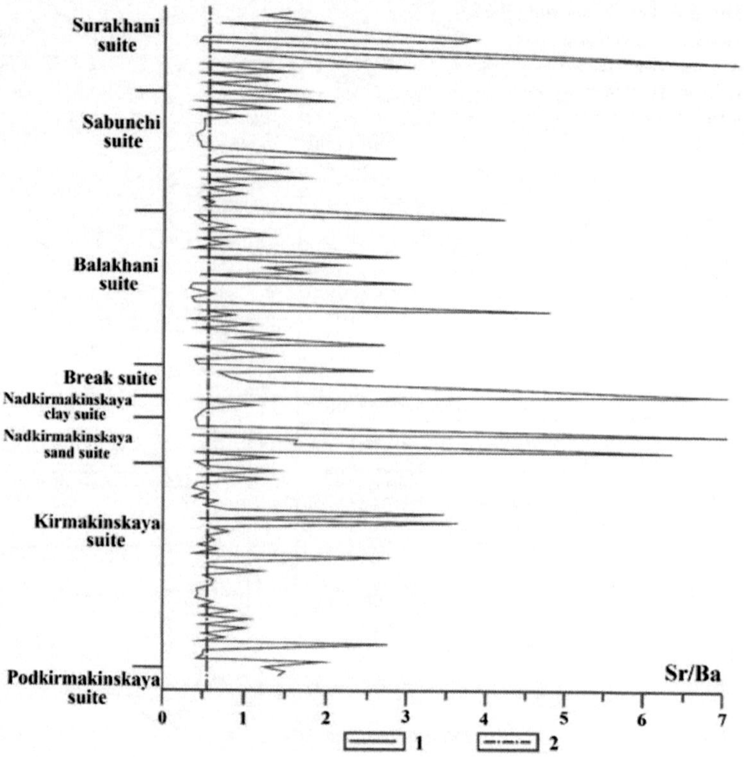

Fig. 3.9 Curves showing variations of the integral radioactivity values and radioactive elements contents through the KS section

and the thorium content, especially in the section's sandy part.

The highest integral radioactivity values within the studied section have been recognized in the Superkirmaki clay suite (SKCS) deposits, being estimated as from 63 to 104 Bk/kg. According to the correlation data, the radioactivity nature is of uranium (correlation coefficient r is 0.98). At the same time, from an analysis of the variations in the section, it becomes apparent that a good correlation with potassium also occurs (correlation coefficient is r 0.57) (Fig. 3.10).

Thus, an analysis of radioactive elements' distribution within the PS lower-division section shows certain relationships in these elements' distribution in the sediments of different lithological-facial compositions. In such a way, sandy deposits of the UKS and SKSS, deposited under river channel conditions, are distinguished by their natural thorium radioactivity. So, in most cases, natural thorium radioactivity has been dominant under proximal sedimentation conditions of the PS deposits. As mentioned above, the primary provenance of terrigenous material supplied to the Absheron facies zone, to which studied deposits have belonged, is the crystalline basement of the Russian platform. The thorium content of sediments was probably increased during this platform-intensive erosion and an accumulation of thick sandy series.

Similar sediments of the KS and SKCS lowermost and uppermost strata, which have been deposited, in our understanding, mainly under subaerial–deltaic conditions, are also characterized by similarity in their radioactivity having a dominant potassium nature. As known, potassium accumulation is mainly controlled by the feldspar content. It is suggested that the growth in the potassium content is caused by increases in the number of feldspar minerals during separate stages of the PS accumulation, which, in turn, may have related to an increased role of the other provenances among their number of the Lesser Caucasus.

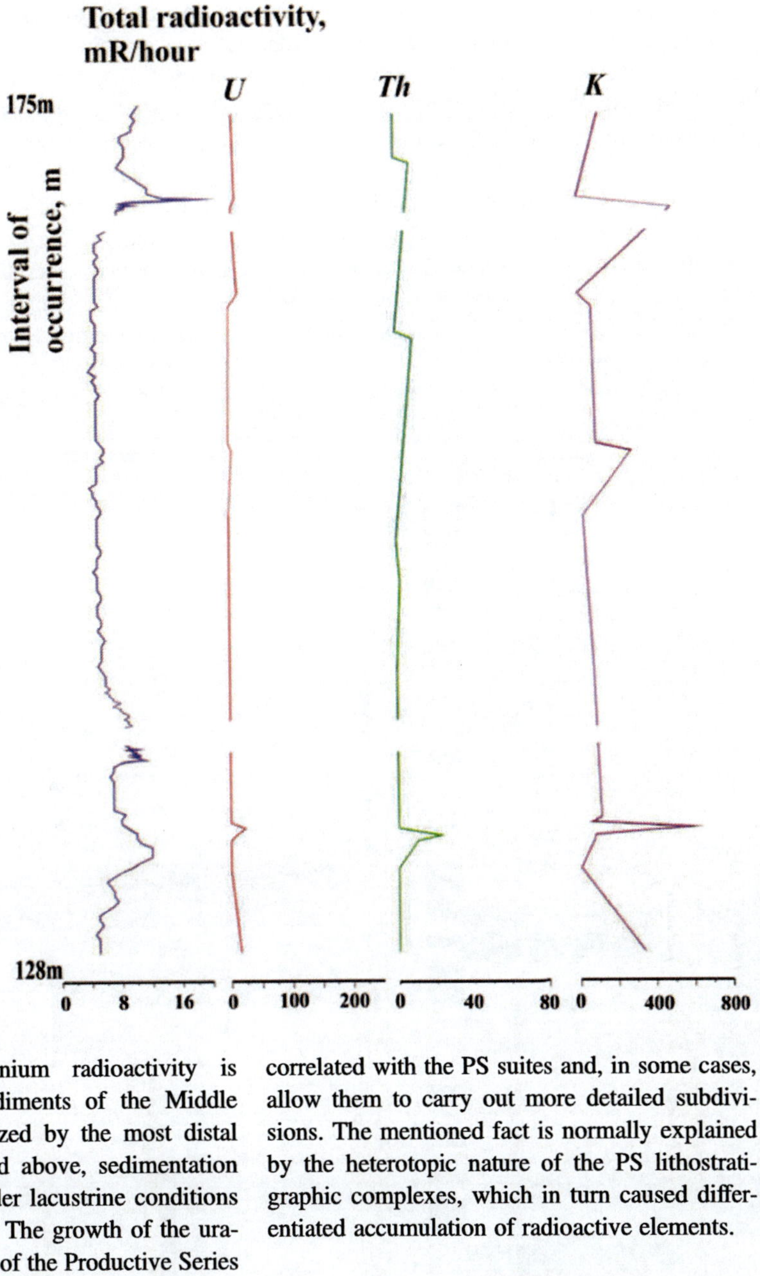

Fig. 3.10 Curves showing variations of the integral radioactivity values and radioactive elements contents through the SKSS and SKCS sections

Finally, natural uranium radioactivity is characteristic of the sediments of the Middle Kirmaki suite characterized by the most distal genesis. As it was noted above, sedimentation here has taken place under lacustrine conditions and in delta front facies. The growth of the uranium content in this part of the Productive Series (PS) section is probable because clay deposits having relatively high organic matter content within the Productive Series (Aliyeva, 2005b) are dominated in this section.

The data gained from distributing the radioactive elements is available to carry out the stratigraphic division of PS deposits in the Kirmaki Valley. These stratigraphic units are well correlated with the PS suites and, in some cases, allow them to carry out more detailed subdivisions. The mentioned fact is normally explained by the heterotopic nature of the PS lithostratigraphic complexes, which in turn caused differentiated accumulation of radioactive elements.

3.3 Magnetostratigraphy of the PS Deposits

This problem is one of the most debatable up to now. Many attempts have been undertaken to carry out detailed investigations into this trend, but the majority of them appeared unsuccessful because ferriferous minerals were lacking. Among those investigations, the most detailed and successful is the work of van Baak (2015) on the Productive Series of Azerbaijan, containing the data of the PS in several sections. The straight and reversed magnetization zones of the PS separate suites and horizons have been revealed by van Baak for the first time and then conjoined to the global paleomagnetic scale. van Baak (2015) has worked out the first magnetostratigraphic section of the Productive Series of Azerbaijan (Fig. 3.11).

According to van Baak (2015), the foot of the PS is "conjoined" to the uppermost strata of the chron (C3r), while the roof—to the C2An.1n. Besides, as it was mentioned above, the latest data indicate that the beginning and the end of PS age are identified by time-marks of 5.35 and 2.72 Rf BP, respectively. The fact that within the series, the 14 zones of direct and reverse magnetization have been distinguished by the author seems to offer a unique possibility for dating PS suites. In addition to the mentioned absolute age determination of the PS in some stratigraphic

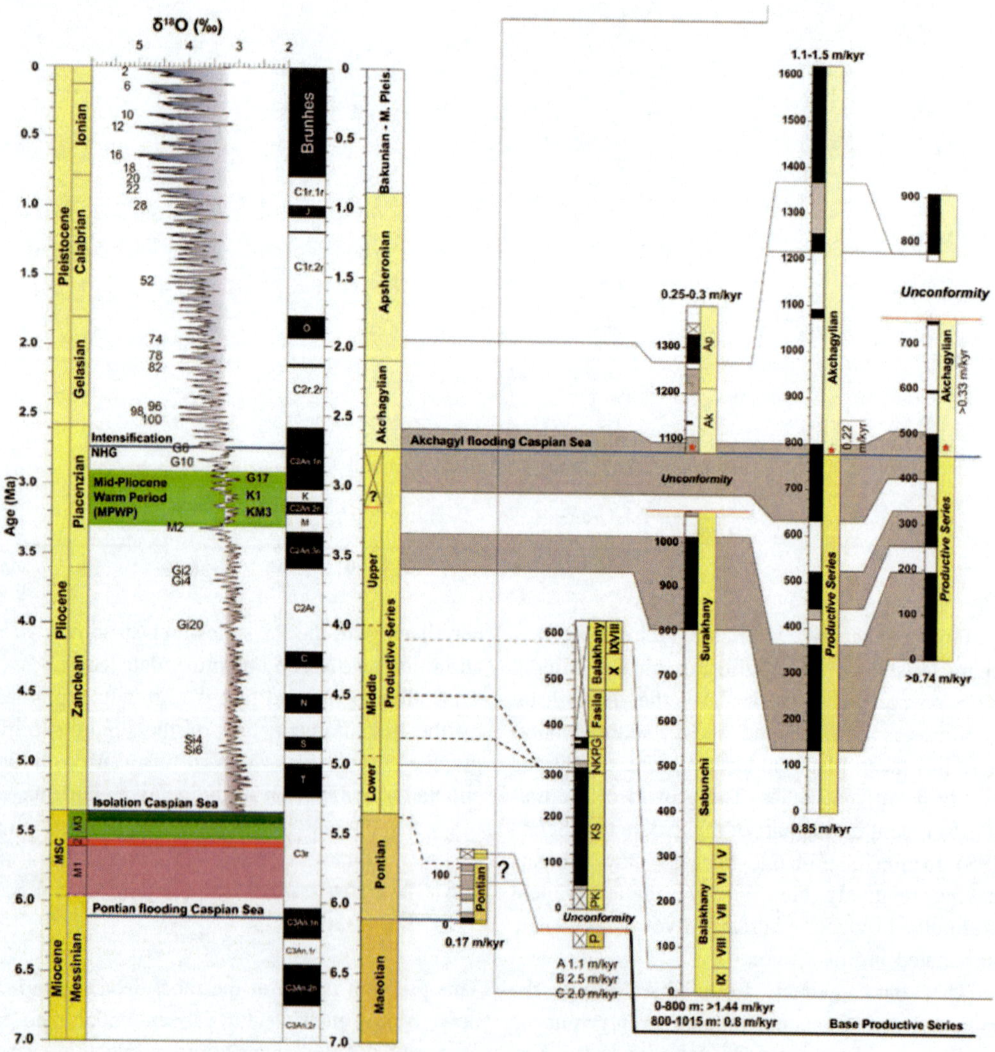

Fig. 3.11 Magnetostratigraphic section of the productive series of Azerbaijan (after van Baak, 2015)

complexes, the results gained by the author allow us to determine sedimentation rates during Lower Pliocene time intervals. A practical correlation of paleomagnetic data with hydrospheric disturbances and some other parameters for the second half of the Late Cenozoic was developed by Eppelbaum and Katz (2021).

References

Agalarova, D. A. (1956). *Microfauna of the productive series of azerbaijan and the red-colored Strata of Turkmenistan* (191 p.). Academy of Science of Turkmenia, Ashkhabad (in Russian).

Aliyeva, E. G. (2005a). Reservoirs of the productive stratum of the Lower Pliocene of the western side of the South Caspian depression. *Lithology and Minerals, 3*, 307–320.

Aliyeva, E. G. (2005b). Lithofacies zoning of the basin of the upper section of the productive strata. *Izvestiya Academy Science Azerbaijan, Earth Science, 1*, 35–42.

Ali-Zadeh, A., Aliyeva, E., Huseynov, D., & Guliyev, I. (2013). The elemental stratigraphy of the South Caspian lower Pliocene productive series. In *Proceedings of the First International Congress on Stratigraphy "At the Cutting Edge of Stratigraphy"* (pp. 827–831). Springer.

Alizadeh, A. A., Guliyev, I. S., Kadirov, F. A., & Eppelbaum, L. V. (2017). *Geosciences in Azerbaijan. Volume II: Economic minerals and applied geophysics* (340 p.). Springer.

Baba-Zadeh, A. D. (2011). Micropaleontology of the South Caspian Lower Pliocene productive series. *Stratigraphy and Sedimentology of Oil-Gas Basins, 2*, 3–14.

Baba-Zadeh, A. D., & Aliyeva, E. G. (2005). Microfaunal response to paleoenvironmental change in the Lower Pliocene productive series and Late Holocene Kura River delta in the South Caspian Sea. *Transaction of the IGCP-521 "Black Sea-Mediterranean Corridor during the last 30ky: Sea level change and human adaptation". First Plenary Meeting*, Istanbul, Turkey, October 8–15, 12–13.

Eppelbaum, L., & Katz, Y. (2021). Akchagylian hydrospheric phenomenon in aspects of deep geodynamics. *Stratigraphy and Sedimentation of Oil-Gas Basins, 2*, 8–26.

Khalilov, D. M. (1946). Микрофауна продуктивной толщи Апшеронского полуострова. *Transaction of the Academy Science of Azerbaijan, Baku, 6*, 4–6. (in Russian).

Kharitonova, R. S. (1964). On the content of uranium, thorium, and potassium in sedimentary rocks and their role in the overall gamma-activity. *Geochemistry, 8*, 831–835. (in Russian).

Konovalov, G. S., Ivanova, A. A., & Kolesnikova, T. K. (1968). Dispersed and rare elements dissolved in water and contained in suspended matter of the main rivers of the USSR. *Geochemistry of Sedimentary Rocks and Ores* (pp. 72–87). Nauka.

Kovalev, A. A. (1965). Geochemical aspects of studies of ratio Th/U in sedimentary deposits. *Geochemistry, 9*, 1171–1173.

Lebedev, L. I., Mayev, E. G., Bordovsky, O. K., & Kulakova, L. C. (1973). *Sediments of Caspian Sea* (119 p.). Nauka, Moscow (in Rissian).

Smyslov, A. A. (1974). *Uranium and thorium in the Earth's Crust* (231 p.). Nedra, Leningrad (in Russian).

Turekian, K. K., & Wedepohl, K. H. (1961). Distribution of the elements in some major units of the Earth's crust. *Geological Society of America Bulletin, 72*, 175–192.

van Baak, C. (2015). Mediterranean-Paratethys connectivity during the late Miocene to Recent. Unraveling geodynamic and paleoclimatic causes of sea-level change in semi-isolated basins. PhD Thesis, Utrecht University, The Netherlands, 275 p.

Vinogradov, A. P. (1962). The average content of chemical elements in the main types of igneous rocks of the Earth's crust. *Geochemistry, 7*, 555–571.

4 Mineralogical Composition and Provenances

4.1 Mineralogical Composition of the Productive Series

Based on generalizations of several works on the mineralogical composition of light and heavy fractions of the PS sandy rocks developed within the marine area of the SCB western slope and on land (Aliyev, 1947, 1949; Aliyev & Daidbekova, 1955; Alizadeh et al., 2016, 2017; Ali-Zade et al., 1985; Azizbekov et al., 1972; Davies et al., 2002; Guliyev et al., 2003; Pashaly and Suleimanova, 1983; Pashaly et al., 1979, 1988; Putkaradze, 1958; Suleimanova and Atayeva, 2002; Sultanov, 1949) (see Chap. 3, Fig. 3.2), it was compiled, if one may say so, the mineralogical "image" of sediments supplied from the various provenances to the basin by different water arteries—the Volga, Kur, and the Greater Caucasian mountain rivers.

Based on available data, the cyclic diagrams of the mineralogical composition of a light fraction within facies zones different stratigraphic intervals, i.e., in the Break suite, Balakhani, Sabunchi, and Surakhani suites. As it is known, the paleo-Volga sediments are characterized by a very high content of quartz washing out from the Russian platform's crystalline basement (Baturin, 1937).

The average quartz content is about 40–55%, up to 90% in some cases. The presence of the rest of the light fraction components—feldspars and rock waste—does not exceed 20% (Fig. 4.1).

As a distributive province, the Greater Caucasus plays an important role, though not as the Russian platform, in developing many SCB structures. The clastic material has been supplied from the northern slope by mountain rivers to the paleo-Volga itself. It merged in the total balance of solid runoff from the southern slope to the basin itself. The distinctive feature is that the rock debris is widely present in light fracture (up to 60–80%), while the rest of the components are lacking (Fig. 4.2).

4.2 Main Provenances of the Sediments

The sediments supplied by the paleo-Volga are believed to have been from the two provenances, the Greater and Lesser Caucasus, which accordingly have been reflected on approximately equal feldspars and rock debris contents (Fig. 4.3).

Exciting results have been gained from several areas where the light fraction component composition leads us to believe that the clastic material has been supplied from several sources. In such a way, some samples include identical amounts of quartz, field spars, and rock debris (Fig. 4.4), which, in our understanding, suggests that the Kur and Volga sediments are mixed in certain parts of the basin.

The strengthening of the Greater Caucasus runoff influence led to another facies zone, the shift zone of sediments of all three provenances, the Greater Caucasus, the Lesser Caucasus, and

A. Alizadeh et al., *Pliocene Hydrocarbon Sedimentary Series of Azerbaijan*,
Advances in Oil and Gas Exploration & Production, https://doi.org/10.1007/978-3-031-50438-9_4

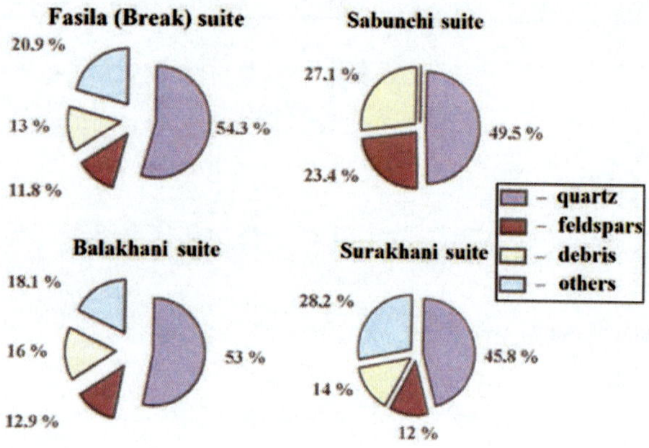

Fig. 4.1 Mineralogical composition of a light fraction of the upper PS sandstones within the Volga facies zone: quartz = 40–50%; feldspars = 10–20%; rock debris = 10–20%

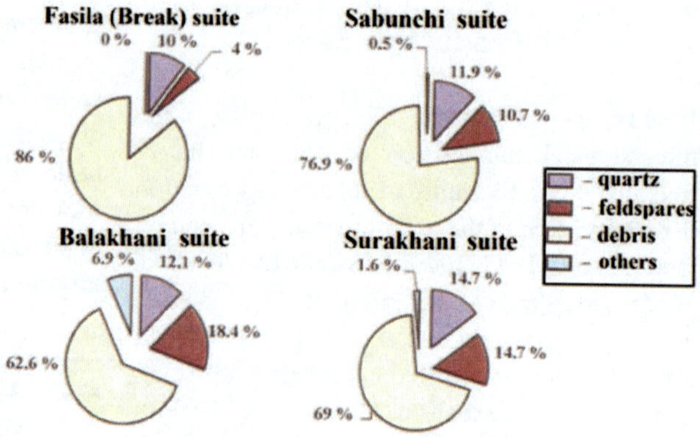

Fig. 4.2 Mineralogical composition of a light fraction of the PS upper division sandstones within the Gobustan facies zone: quartz = 10–15%, feldspars = 5–15%, rocks debris = 60–80%

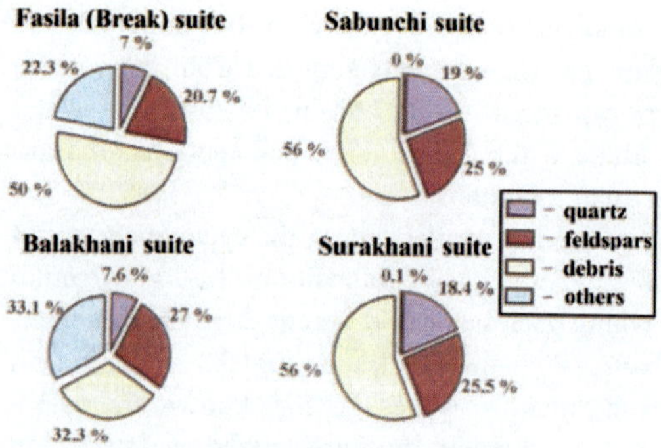

Fig. 4.3 Mineralogical composition of the upper PS sandstones in the Kur facies zone: quartz = 5–10%, feldspars = 20–30%, rock debris = 30–55%

the Russian platform. A typical feature of the mentioned sediments is high rock debris content (up to 50%) on a level with high quartz and feldspars, up to 30% and 20–25%, respectively (Fig. 4.5).

It is likely that within the uppermost suite on some structures of the Baku Archipelago, there is one more zone where Volga runoff has not been noted, and its deposits are composed of material supplied by the Greater Caucasian Mountain rivers as by the Kur River itself. The mentioned assumption is based on minimal quartz content and almost identical amounts of feldspars and lithic fragments (up to 40–45%).

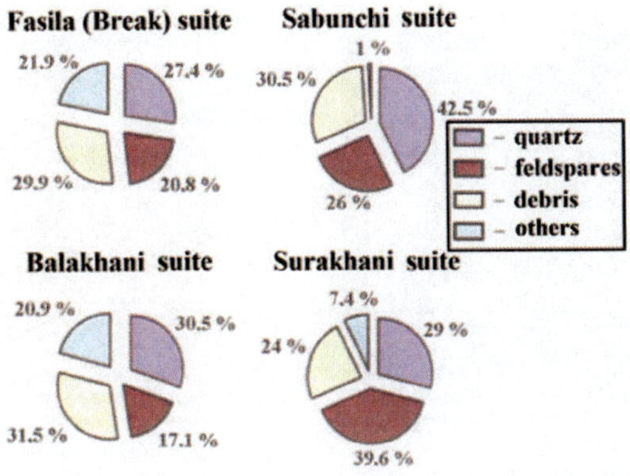

Fig. 4.4 Mineralogical composition of a light fraction of the upper PS sandstones within the Volga-Kur mixed facies zone

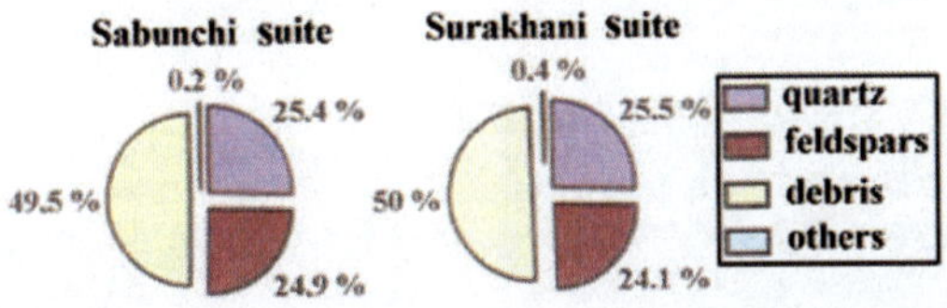

Fig. 4.5 Mineralogical composition of a light fraction of the upper PS sandstones within the Volga-Kur-Gobustan mixed facies zone: quartz = 25–30%, feldspars = 20–25%, rock debris = 45–50%

The mineralogical characteristic of the light fraction of the upper PS sandstones developed within different facies zones may be presented as shown in Table 4.1.

A series of maps were constructed to delineate the South Caspian Basin's different facies zones. These maps display information on the main components in the mineralogical composition of a light fraction of sandstones and disthen and staurolite. Unfortunately, the available data allowed the development of such maps for the PS upper division only.

From the maps of quartz, feldspars, and rock debris distribution in Break suite deposits (Fig. 4.6), it is concluded that these minerals' contents are highly varied through the areas that are most likely caused by sediment material delivery to the South Caspian from the different provenances.

The zone containing a tremendous amount of quartz in Break suite deposits extends strictly from the north to the south, including the following structures: Chirag, Gyuneshli, Azeri, Shakh-Deniz, Bakhar, and Nakhchivan (Fig. 4.6). Its

Table 4.1 Averaged mineralogical composition of a light fraction of the upper PS sandstones taken from the different facies' zones

Lithofacies zones	Quartz (%)	Feldspars (%)	Rock debris (%)
Absheron	40–50	20–10	20–10
Kur	5–20	20–30	30–55
Gobustan	10–15	5–15	60–80
Mixed Absheron—Kur	25–40	20–40	25–35
Mixed Absheron-Gobustan—Kur	25–30	20–25	45–50
Mixed Gobustan—Kur	10–15	25–30	45–50

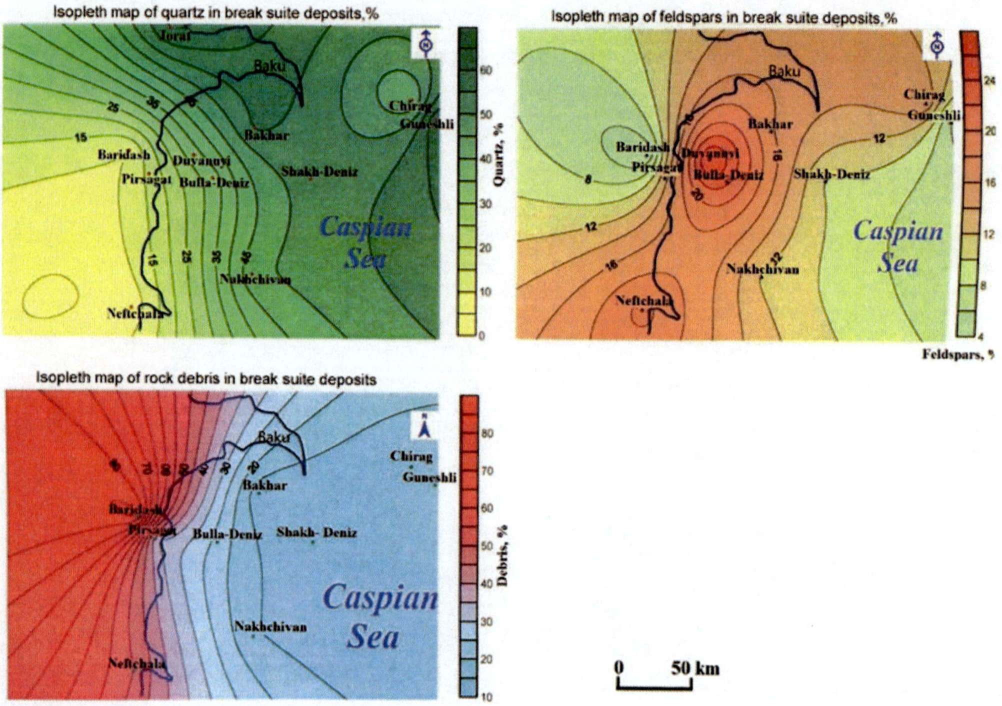

Fig. 4.6 Schematic maps showing the contents of light fraction minerals and disthen in the Break suite sandy deposits

contours almost entirely recurred those of the zones of lesser contents of feldspars and rock debris. This indicates that the Russian platform has been a significant provenance of terrigenous material for most areas of SCB.

It is noted that during the Balakhani suite accumulation, the zone of the most significant content of quartz appears to be moved to the north (Fig. 4.7). Notably, this zone covers the Absheron Peninsula and a region of Absheron sill, but its northern boundary is displaced to the north. It is situated slightly to the south of the Shakh-Deniz structure. Based on this fact, the Volga sediment propagation area gradually decreased.

An exciting feature is that the contours of the zone in Balakhani suite deposits where disthen presents have almost exactly recurred those of the zone with the most incredible quartz content; besides, the delivery trend of quartz and disthen to the Balakhani suite deposits is the same, from north to south, i.e., from the Russian platform to the SCB (Fig. 4.7). As it is known, the presence of disthen is one of the typical features of the Volga lithofacies sediments, and it was taken into consideration in the facial analysis given in Chap. 5. It is noted that a zone with rock debris predominance in a light fraction is somewhat reduced. It is known that the Greater Caucasus is a significant supplier of sedimentary products in this case.

At the same time, the sediments propagation area characterized by practically equal amounts of feldspars and rock debris and lower quartz content (Kur facies) appears to be increased to the east, is located around Hamamdag-Deniz, Yanan Tava, Byandovan-Deniz, Mugan-Deniz, and Inam structures. The same ascertainment is noted for a zone of approximately equal contents of all the light fraction components (Pirsagat,

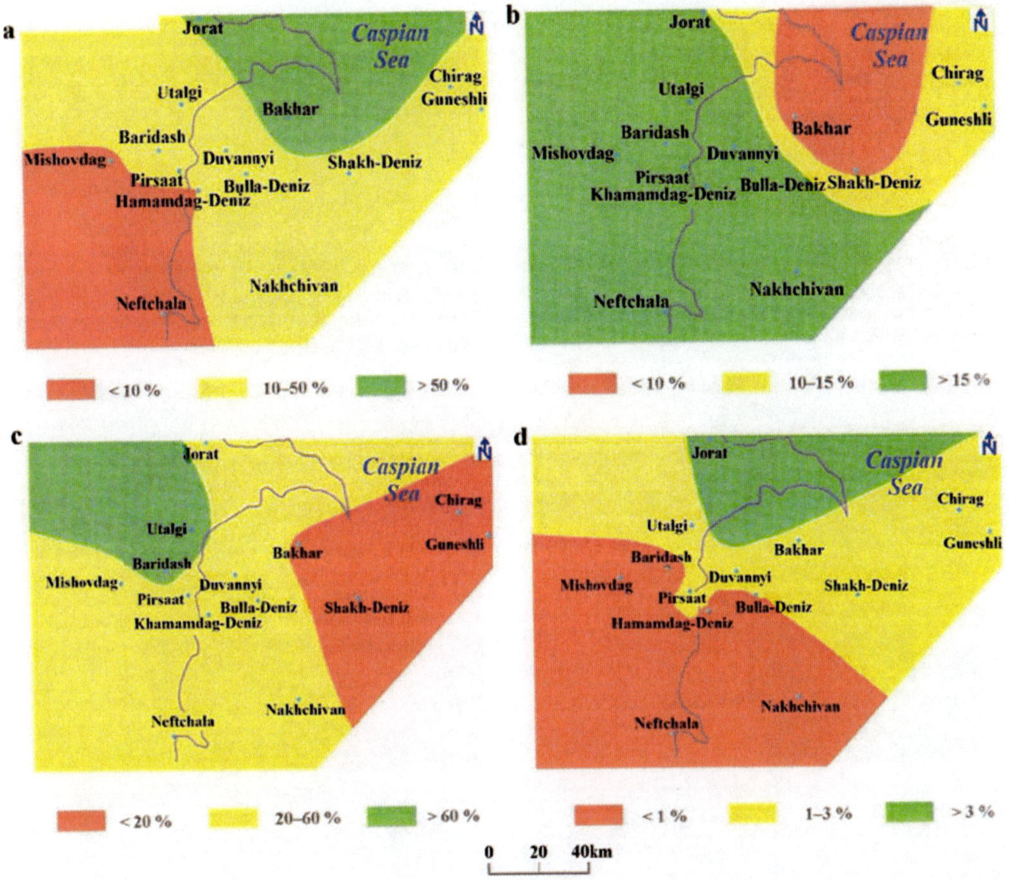

Fig. 4.7 Schematic maps of the content of light fraction minerals (quartz, feldspars, rock debris, and disthen) in sandy deposits of the Balakhani suite in the western SCB

Alyat-Deniz, Bulla-Deniz, Sangi-Mugan, Aran-Deniz, Sabail, and Nakhchivan structures).

In the Balakhani suite, the light fraction primary components delivery trends are as follows: quartz from the north to the south-southwest, rock debris from the northwest, and feldspars from the southwest. The transition to the Sabunchi suite has been accompanied by widening a zone of the greatest feldspar content, on an average of about 25% (Fig. 4.8).

A typical feature is that the rock debris fraction appears to be increased, averaging up to 56%, and quartz—up to 19%. This fact is believed to be evidence of a change in distributive provinces' importance in supplying sediment balance and increasing the share of the Kur solid runoff. The anomalous high rock debris contents (88.8%) in the Sangi-Mugan area and lower contents of quartz (9%) and feldspars (2.2%) point to the dominance of runoff from the Greater Caucasus may not be attributed to the common mineralogical pattern. A southern zone of high quartz contents and the presence of disthen (Volga facies) in the rocks of the Sabunchi suite has been located somewhat further south of the Bakhar area.

The mineralogical evidence of several provenances (the Greater Caucasus, Russian platform, and Lesser Caucasus) that may have

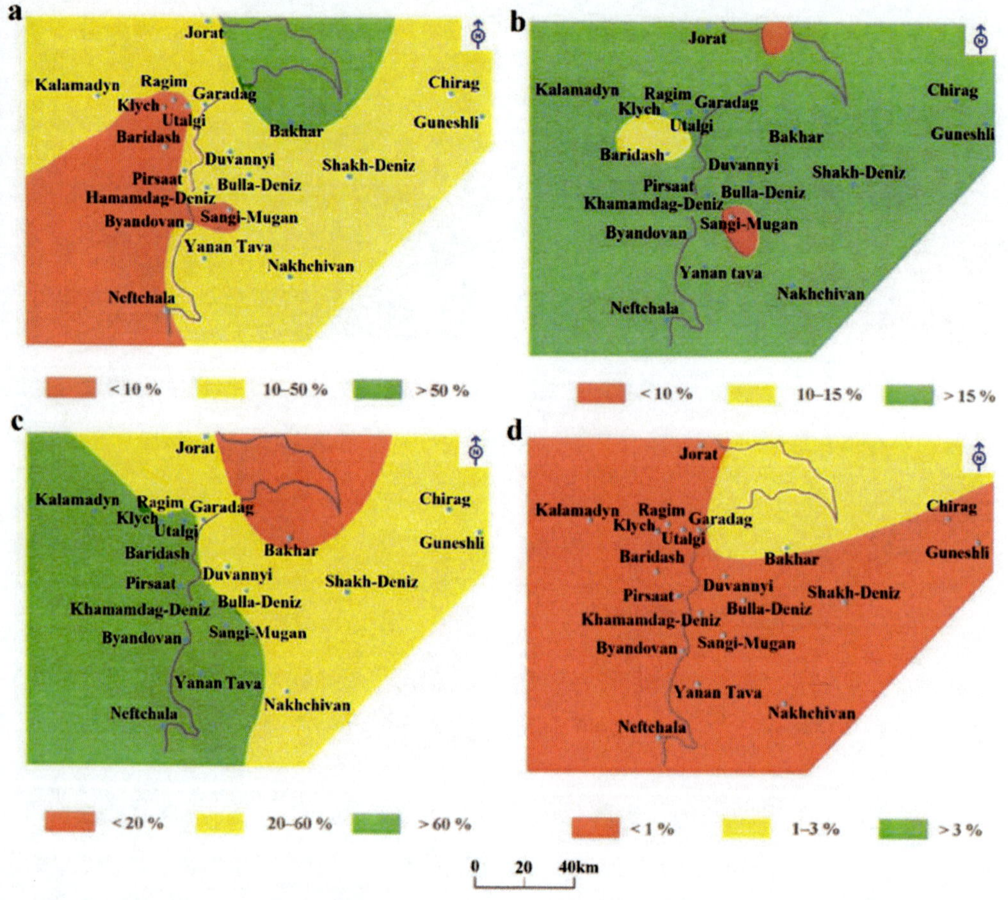

Fig. 4.8 Schematic maps showing the contents of light fraction minerals and disthen in sediments of the Sabunchi suite in the western SCB

taken place in the sedimentation of the Sabunchi suite have been obtained from several areas. For the first time, it is expressed in the development of the complex heterogenetic mixed-sediments zone. Besides, the mixed sediments are not of the Volga-Kur origin only but the Kur-Gobustan as well. Such mixed-sediment composition has been found in the Ragim and Klich areas where on a level with high rock debris content (Greater Caucasian provenance), an increased feldspars content up to 30% (Lesser Caucasian provenance) is noted. The mentioned tendencies are typical of the Surakhani suite, too. A zone of quartz domination and disthen in Surakhani suite sediments retreat to the north (Fig. 4.9).

A zone covered with sediments delivered by mountain river lets from the Greater Caucasus has become narrower and moved to the side of the basin Sabail structure. The sediments of Gobustan facies in Surakhani time are characterized by an average rock debris content of about 77%, feldspars, and quartz—of 15%. Like the light fraction composition in the Sangi-Mugan area of the Sabunchi suite, the deposits in the Sabail area of the Surakhani suite have very high rock debris content (up to 65%) and

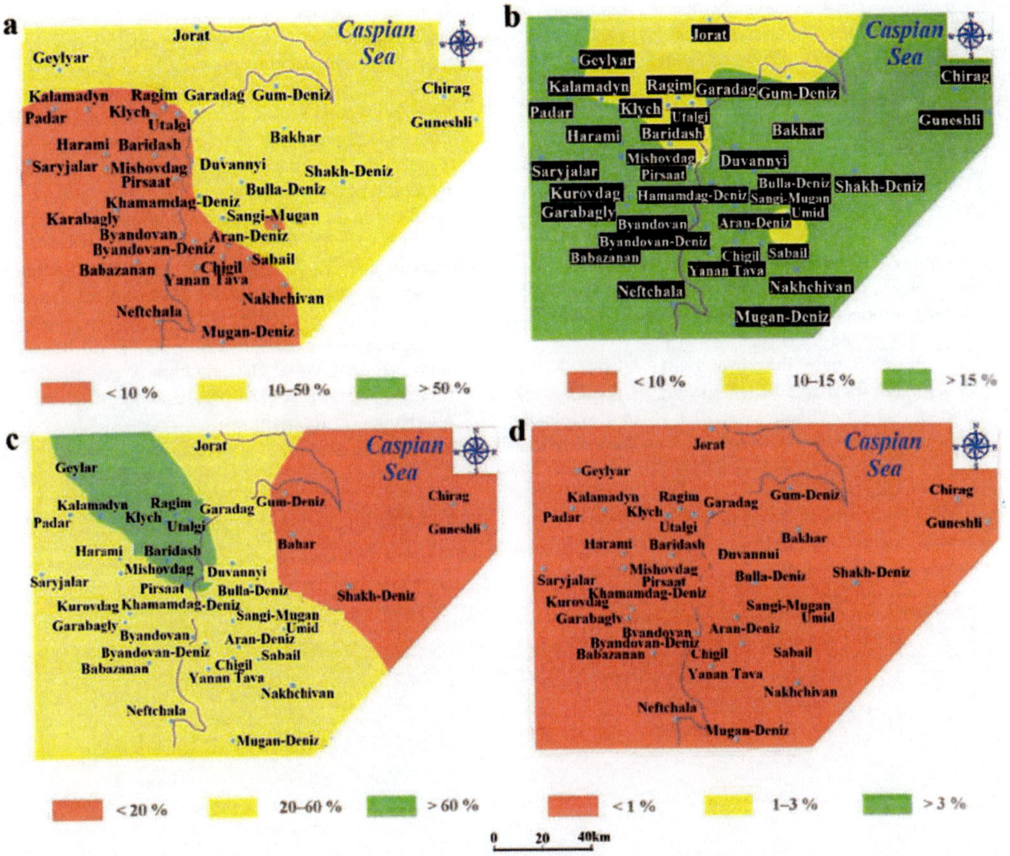

Fig. 4.9 Schematic maps showing the contents of light fraction minerals and disthen in sediments of the Surakhani suite in the western SCB

relatively lower feldspars and quartz contents, 15%, and 20%, respectively.

The paleo-Kur sediments range zone outlines are changed slightly against the background of essential expansion of mixed sediments from all three sources. This zone extends south and embraces the Hamamdag-Deniz and Sangi-Mugan areas (Fig. 4.9).

References

Aliyev, A. G. (1947). *Petrography of the productive strata of Kabristan* (155 p). Academy of Science of Azerbaijan (in Russian).

Aliyev, A. G. (1949). *Petrography of the tertiary deposits of Azerbaijan* (311 p). Azneftizdat (in Russian).

Aliyev, A. G., & Daidbekova, E. A. (1955). *Sedimentary rocks of Azerbaijan* (331 p). Azneftizdat (in Russian).

Alizadeh, A. A., Guliyev, I. S., Kadirov, F. A., & Eppelbaum, L. V. (2016). *Geosciences in Azerbaijan. Vol. I: Geology* (239 p). Springer.

Alizadeh, A. A., Guliyev, I. S., Kadirov, F. A., & Eppelbaum, L. V. (2017). *Geosciences in Azerbaijan. Vol. II. Economic minerals and applied geophysics* (340 p). Springer.

Ali-Zade, A. A., Salayev, S. G., & Aliyev, A. I. (1985). *Scientific assessment of the prospects for oil and gas potential in Azerbaijan and the South Caspian* (252 p). Elm (in Russian).

Azizbekov, Sh. R., Agakishibekova, R. R., Alizadeh, A. A., Alizadeh, K. A., Aliyev, M. M., Akhmedov, A. M., Akhmedov, G. A., Bairamov, A. S., Gadjiev, T. G., Zhouze, B. P., Zaitseva, L. V., Kashkay, M. A., Mekhtiyev, Sh. F., Sultanov, A. D., Khalilov, A. G., Shikhalibeyli, E. Sh., Efendiyev, G. H., & Yakubov, A. A. (Eds.). (1972). *Geology of the USSR. Vol. 47:*

Azerbaijan Republic economic deposits fossil fuels (oil and gas). Nedra (in Russian).

Baturin, V. P. (1937). *Paleogeography by terrigenous components* (292 p). AzONTI (in Russian)

Davies, C., Vincent, S., Hyden, F., & Aliyeva, E. (2002). *Petrographic and petrophysical analysis of the upper productive series, Kur Basin* (31 p). Azerbaijan Project, Joint CASP-GIA report.

Guliyev, I. S., Levin, L. E., & Fedorov, D. L. (2003). Hydrocarbon potential of the Caspian region (system analysis). Nafta-Press (in Russian).

Pashaly, N. V., Heirov, M. B., & Saradzhalinskaya, T. M. (1988). Productive series. In *Geology of Azerbaijan* (Vol. II, pp. 186–229) (in Russian).

Pashaly, N. V., Saradzhalinskaya, T. M., & Katz, N. M. (1979). *Facies and paleogeography of the Pliocene and quaternary lithogenesis and related economic minerals (Baku Archipelago)* (235 p). Report of the Institute of Geology, Azerbaijan Academy of Science (in Russian).

Pashaly, N. V., & Suleimanova, S. F. (1983). *Study of filtration properties and lithological and structural features of deep-seated reservoir rocks of the Middle Pliocene deposits of the South Caspian* (142 p). Report of the Institute of Geology, Azerbaijan Academy of Science (in Russian).

Putkaradze, A. L. (1958). *Baku Archipelago* (335 p). Azerneftneshr (in Russian).

Suleimanova, C. F., & Atayeva, E. Z. (2002). Sedimentology and stratigraphy of the Gunashli, Chirag, and Azeri oil & gas fields (Absheron Archipelago). *Azerbaijan Geologist Scient Bull, 7*, 109 (in Russian).

Sultanov, A. D. (1949). *The lithology of the productive series of Azerbaijan* (184 p). Academy of Science of Azerbaijan (in Russian).

5 Environmental Conditions, Sedimentation Cyclicity, and Architecture of the Productive Series Reservoirs

The orogenic processes took place at the end of the Pontian. They gave rise to the Caspian surrounding land areas, and enormous sea-level lowering, according to information available, from 600 to 1500 m (Reynolds et al., 1998), led to the complete isolation of the Caspian Sea in the Lower Pliocene. The sedimentation at that time took place under conditions of the South Caspian. This most isolated basin accumulated all the vast mass of terrigenous material supplied by the three largest river arteries: paleo-Volga, paleo-Amu-Darya, and paleo-Kur. Avalanche sedimentation at the rate of 2.5 mm/y and a high down-warping rate of the basin bottom led to the formation of a unique 7 km terrigenous Productive Series (Guliyev et al., 2003) containing up to 90% of all oil and gas reserves in South Caspian oil/gas—bearing basin (SCB). The question arises: within more than 30 km thick sedimentary series represented by a wide stratigraphic range—from the Middle Jurassic Aalenian stage up to the Holocene inclusive, the horizons possessing top-quality reservoir properties have been formed precisely in the Lower Pliocene deposits.

This investigation is intended to study sedimentation conditions and lithological composition of the Productive Series (PS) deposits, their change in vertical and lateral directions, and cyclicity in the PS sedimentation process. Besides this, we must understand how these factors affect the Pliocene hydrocarbon reservoir's structure and PS paleo-facies map the development of the western SCB. We aim to predict the structure and quality of reservoirs within unexplored areas of the South Caspian deep zone.

The rhythmic change in lithofacies composition of the PS deposits has been pointed out before by a member of researchers (Aliyev, 1947, 1949; Avdusin, 1952; Azizbekov et al., 1972; Golubyatnikov, 1914; Kalitsky, 1922; Kovalevsky, 1922; Mustafayev, 1963; Nikishin, 1974, 1981; Potapov, 1954; Sultanov, 1949). However, the event causality with sedimentation conditions change caused by the Caspian Sea level variations and change in volumes of solid runoff of the rivers flowing into the sea is examined for the first time. The problem of the role of the PS accumulation cyclicity in the reservoir's formation being examined in this chapter is also of high importance.

The following available data have been used in this investigation: the results of the sedimentological study of PS outcrops located on the Absheron Peninsula (Kirmaki and Yasamal valleys) in Gobustan (Akhtapa), within the Lower Kur depression (Babazanan), in Near-Caspian-Guba region (Sura) as well as the gamma-ray logs interpretation of many wells in the Bakhar oil–gas-condensate field and Shakh-Deniz gas-condensate field located within the Southeastern marine oil–gas-condensate region of the Baku Archipelago and also the data obtained from the Alyat-Deniz, Bulla-Deniz, Pirsagat, Aran-Deniz, Sabail fields located in the water area of the Baku Archipelago being adjacent to the Gobustan, and

A. Alizadeh et al., *Pliocene Hydrocarbon Sedimentary Series of Azerbaijan*,
Advances in Oil and Gas Exploration & Production, https://doi.org/10.1007/978-3-031-50438-9_5

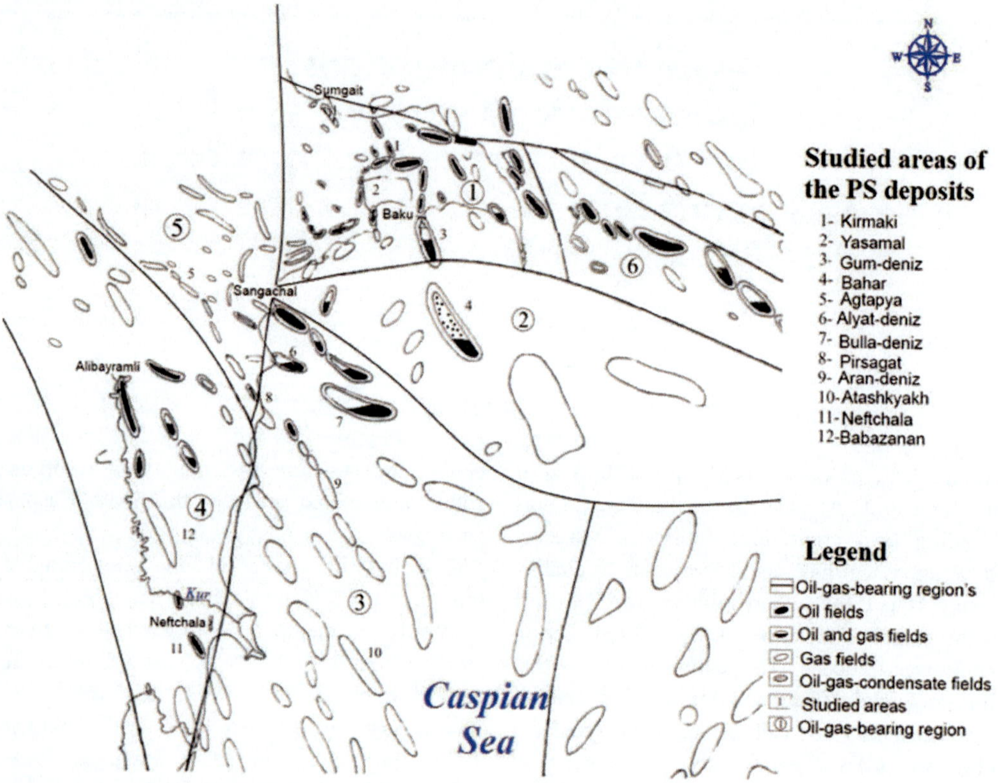

Fig. 5.1 Sketch map of the location of the western SCB local structures (after Guliyev et al. (2003), with modifications)

from the Atashgyakh field located in the Baku Archipelago water area adjacent to the Lower Kur depression. Besides, the Neftchala field in the Lower Kura depression has also been studied (Fig. 5.1).

The list of studied areas of the PS deposits widespread within the western SCB is given below.

I. The areas where sedimentological investigations and logging diagrams analyses have been carried out: 3—Kirmaki, 6—Yasamal, 7—Gum-Deniz, 8—Bakhar, 9—Shakh-Deniz, 10—Bulla-Deniz, 11—Alyat, 14—Aran-Deniz, 32—Pirsagat, 37—Babazanan, 38—Neftchala, 41—Atashkyakh, 43—Akhtapya.

II. The areas through which the PS deposits mineral composition has been studied: 1—Djorat, 2—Masazir, 4—Balahani, 5—Govsan, 7—Gum-Deniz, 8—Bakhar, 9—Shakh-Deniz, 10—Bulla-Deniz, 11—Alyat, 12—Hamamdag, 13—Sangi-Mugan, 14—Aran-Deniz, 15—Umid, 16—Bandovan-Deniz, 17—Yanan-Tava, 18—Chigil, 19—Sabail, 20—Garadagh, 21—Ragim, 22—Utalgi, 23—Klych, 24—Kyrlykh, 25—Baridash, 26—Harami, 27—Padar, 28—Kalamaddin, 29—Geoglyar, 30—Sarydjalyar, 31—Mishovdag, 32—Pirsagat, 33—Kyurovdag, 34—Khydyrly, 35—Byandavan, 36—Garabagly, 37—Babazanan, 38—Neftchala, 39—Inam, 40—Mugan-Deniz, 41—Atashkyakh, 42—Gyuneshli, 43—Akhtapya, 44—Sura.

The oil–gas-bearing regions (numbers in circles): 1—Absheron, 2—Southeastern Absheron water area, 3—Baku Archipelago, 4—Lower

Kur, 5—Djeirankechmez, 6—Adjichai-Alyat oil–gas-bearing zone of Shamakhi-Gobustan region, 7—Eastern Absheron water area.

During the last years, in connection with a new oil boom in the South Caspian, the problem of the PS sedimentation condition acquired special attention. Detailed investigations of PS outcrops in the Absheron Peninsula and the results of core sample analyses stimulate the development of various theories, including those opposed. In such a way, a theory of the distal deltaic-lacustrine genesis of PS deposits is maintained by Reynolds et al. (1998). Hinds et al. (2004) have concluded that the PS deposits are of proximal genesis, suggesting sedimentation conditions of all the PS sections, except some intervals of paleo-Caspian inundation. i.e., sea-level rising, have been varied within various facies settings. There is also a third theory propounded by Aliyeva (2008). The sedimentation in the Lower Pliocene time occurred under frequent and sharp paleo-Caspian Sea level variation accompanied by the change of environmental conditions from the proximal fluvial to the distal deltaic-lacustrine ones. Our understanding of this theory needs to be considered in more detail.

An outcrop stripped by a gas-pipeline trench in the Kirmaki Valley has been used as an object of investigation of the PS Absheron-type deposits. The trench passed through the strike practically all the PS suites from the Underkirmaki suite to Surakhani (Fig. 5.2).

5.1 Productive Series Lithological-Facies Characteristics

The sedimentological and lithological study compiled a composite section of all the PS suites (except the Kala and Surakhani suites), having an actual thickness of up to 1300 m and a total extent of 2500 m (Figs. 5.3a, 5.3b, 5.3c, 5.3d, 5.3e). It should be noted that the outcrops of the Balakhani, Sabunchi, and Surakhani suites in the Yasamal Valley have also been studied.

It seems likely that during this suite accumulation, the smooth facies transition from the upper deltaic plain to its lower part has occurred. The shoreline delimited the plain outer margin at the time. The mentioned facies change is indicated in the section by decreasing the share of the branched fluvial deposits and domination of argillic-silty lithofacies. Upward the section, an old shoreline stable advance to the land side appears to be preserved, and then, in the Kirmaki valley, the proximal delta front facies have been formed. Fine-grained, mainly parallel bedded sandy bands coarsen toward the roof and are interpreted as bay-mouth bar sediments. Though our interpretation of these sandy bands is generally in accord with earlier conclusions of Reynolds et al. (1998), their thickness does not exceed 1–2 m, i.e., much smaller than mentioned by those authors. Sandy band bottom is only sometimes well defined. Occasionally, undulation ripples may be observed.

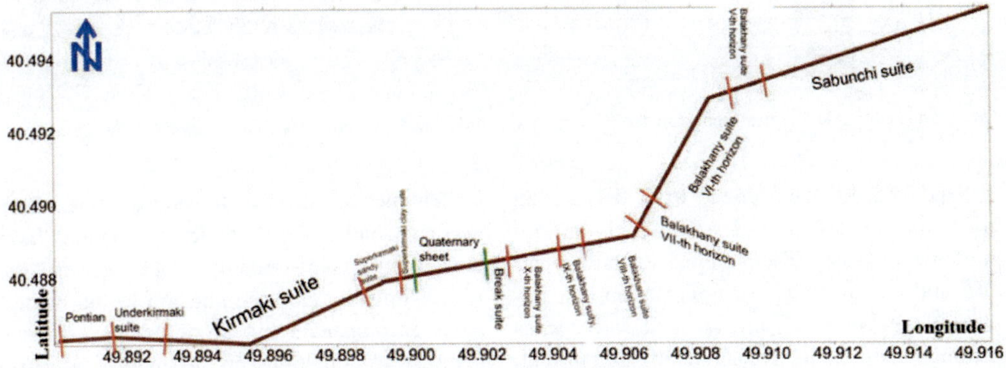

Fig. 5.2 Schematic representation of trench location in the Kirmaki Valley

Fig. 5.3a Lithological-facies section of the PS deposits in the Balakhani suite (after E.G. Aliyeva)

Sandy bands characterized by a decrease in grain size toward a roof and weakly eroded bottom also occur. Their typical feature is parallel and crossbedding changing by current ripples at the bottom. They are deltaic branches feeding a bay-mouth bar. Overall, summing up the results obtained, it is believable that within the Productive Series of lowermost strata, there was one semi-cycle of sea-level variations from the lower point taking place during the Underkirmaki suite accumulation to the highest point accompanied by the formation of distal delta front conditions in the middle Kirmaki suite. Afterward, in our understanding of this

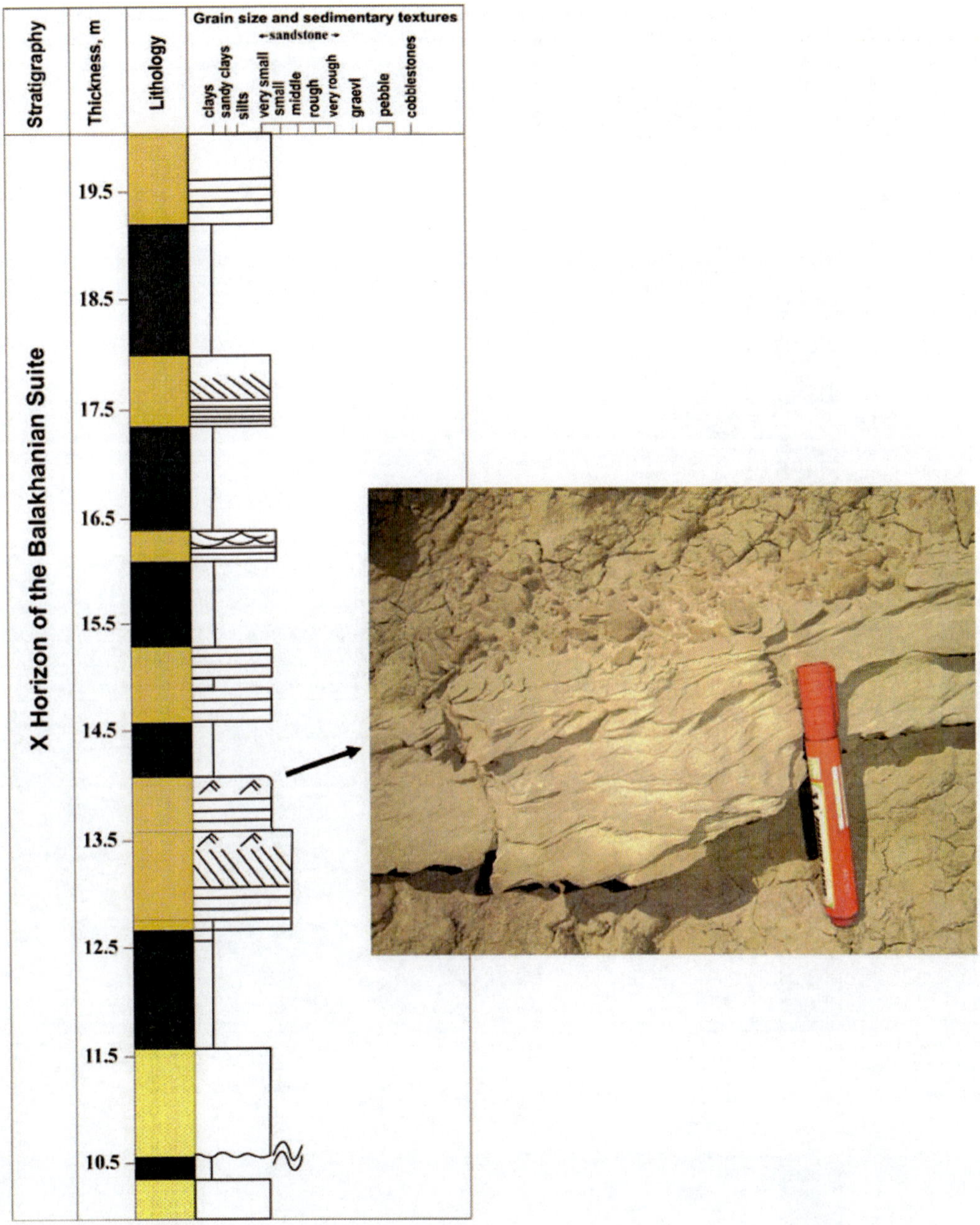

Fig. 5.3b Lithological-facies section of the PS deposits in the Balakhani suite, continuation (after E.G. Aliyeva)

process, a complete cycle of sea-level variations has taken place, including its lowering and inverse gradual rising by which the Kirmaki section's lower half has been completed. Overall, the thickness of the deposits corresponding to the first sedimentary cycle is 83 m, and those of the second complete cycle—56 m.

The transition to the middle suite has been marked by rising sea level to its highest elevation. This part of the suite consists dominantly of

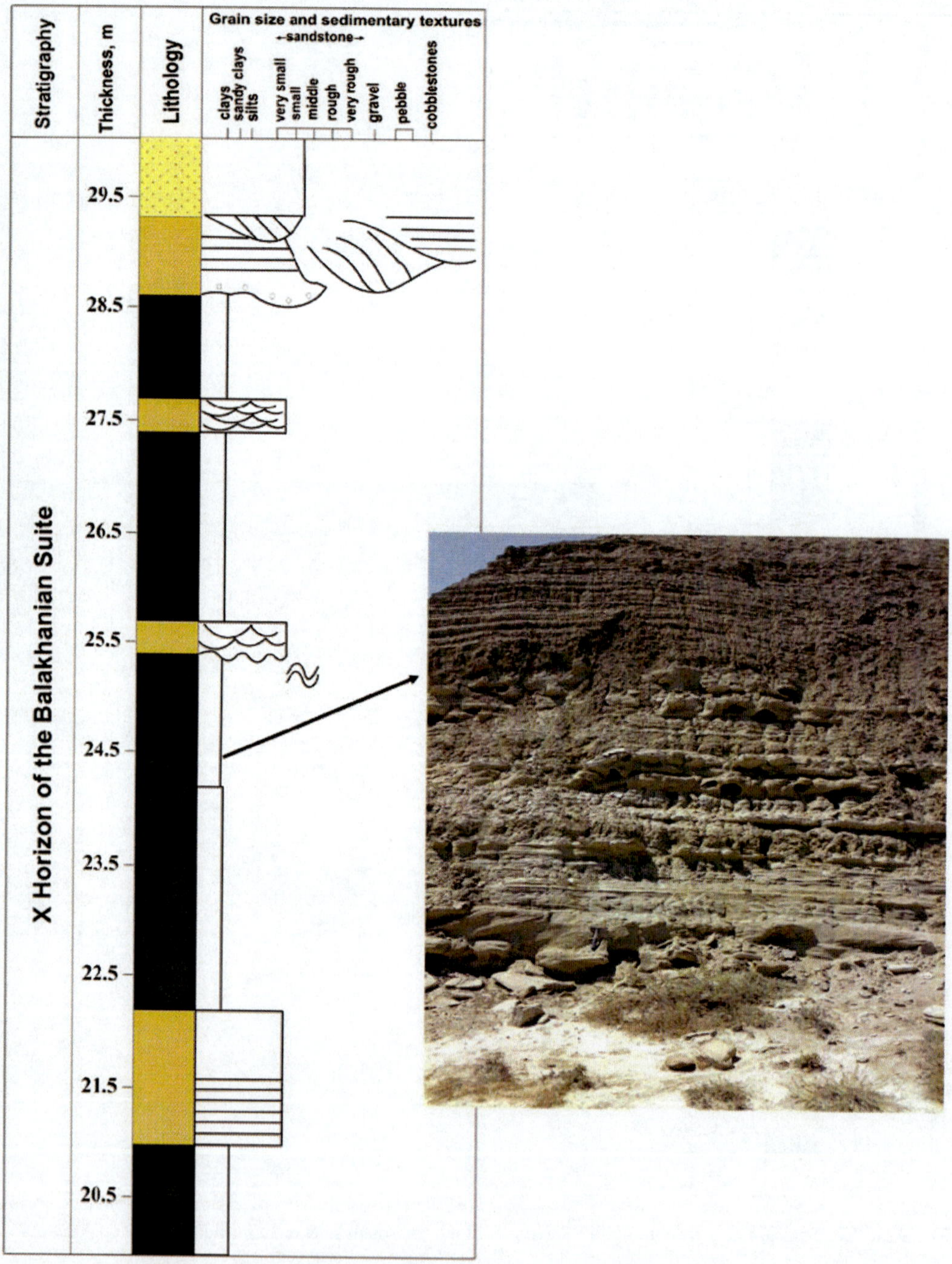

Fig. 5.3c Lithological-facies section of the PS deposits in the Balakhani suite, continuation (after E.G. Aliyeva)

fine-grained lithological varieties. In the section, alternating partings of argillaceous or fine-grained sands, siltstones, silty, or clean, very thin-bedded clay bands are repeatedly observed. Besides, variation in grain sizes (coarsening or reduction) may have taken the section upward.

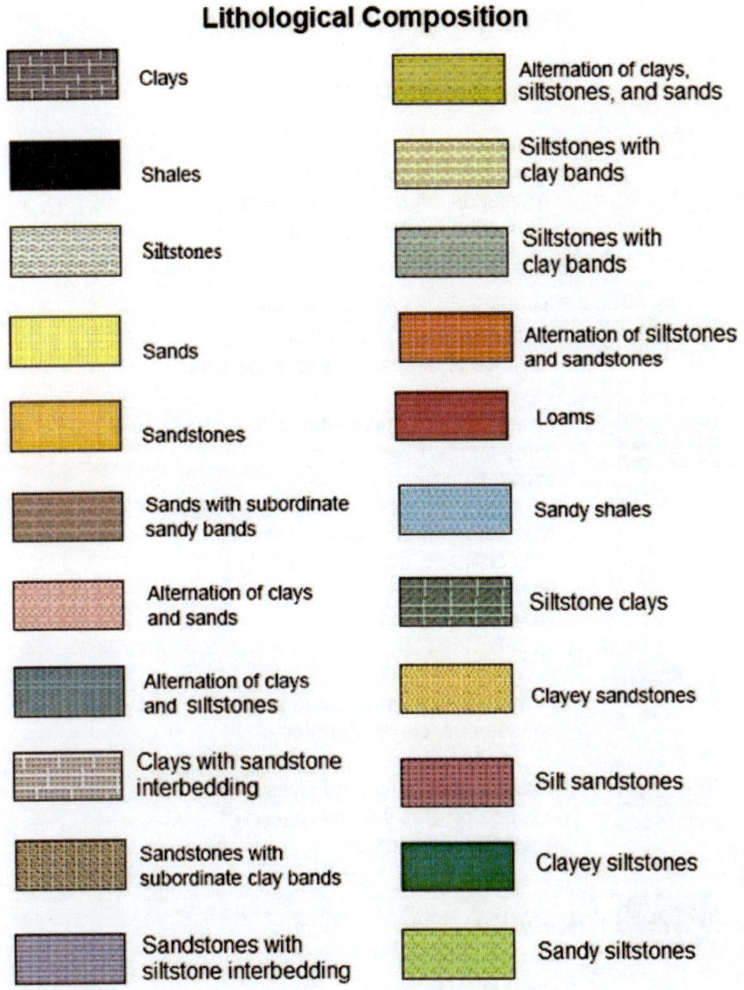

Fig. 5.3d Lithological composition (for Figs. 5.3a–5.33)

The thickness of sandy and siltstone bands varies from a few tens of cm to 2 m, and of clays is about a few cm. The band boundaries are well-defined, and even the transition from one to other lithological types is sharply marked. There is a clear difference in sedimentary textures between the middle and lower suites. The sands in the middle suite are mainly horizontally stratified or crisscross-bedded. Widely developed ripples characterize siltstones. Silty clay sediments are also characterized by convoluted texture resulting from the deformation during high-rate deposition. The sedimentation conditions may be interpreted as an accumulation on the delta front (sands) which has then been changed by both prodeltaic facies (siltstones with widely developed ripples) and typical lacustrine facies (thin-bedded clays). Such features characterize some sandy bands as a decrease in grain sizes, the sediment profile, and an unevenly eroded floor upward. According to our interpretation, these bands are considered most likely to represent the continuation of deltaic branches from the subaerial delta into its prodeltaic part.

The thickness of distal delta front sediments is considerably lesser than that of the proximal front.

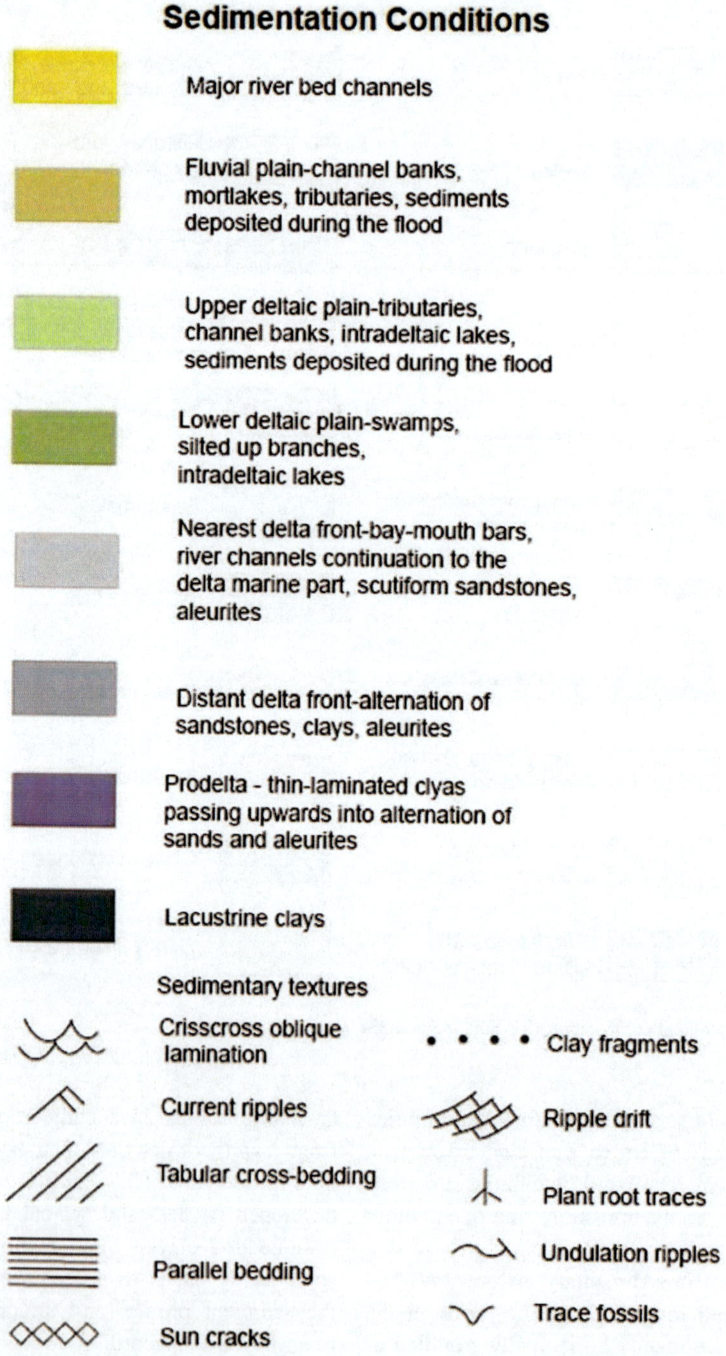

Fig. 5.3e Sedimentation conditions (for Figs. 5.3a–5.33)

The prodeltaic and lacustrine deposits are more thinly bedded. A special feature of this section interval is that it includes sedimentary cycles of different orders. On the background of relatively long-term cycles to which the sediments averaging about 30 m in thickness correspond, alternating deposits of different facies are frequently noted. It may suggest that sedimentation conditions within the section have frequently been changed. In such a way, sedimentation series, from 3 to 15 m in thickness, have been deposited in different sedimentation conditions: lacustrine, foredeltaic, and delta front facies. So, frequently and sharply changed variations in the paleo-Caspian fluctuation regime have also caused facies conditions. Thus, sedimentary series accumulated during the complete sedimentation cycles of the minor order are also noted. At the same time, it is noted that sedimentation conditions within even such small series take place more frequently. The thickness of the deposits within these series is, in places, a few cm. The mentioned frequently changed sedimentation conditions and an accumulation of lithological heterogeneous thin beds may be caused by sea-level variation rhythm and changes in the amount of sedimentary material supplied to the basin.

The thickness of the middle, more clay part of the Kirmaki suite is 114 m. As distinct from the lower suite accumulated mainly under aerial conditions, we believe that the middle suite has been formed under foredeltaic-lacustrine conditions.

The upper Kirmaki suite lithofacies composition is quite like the lower suite. Sandy varieties are dominant. The section's common feature is very fine sandy beds having an unevenly eroded bottom and passing upwards into more silty sediments. An interpretation of these sedimentary beds leads us to believe they were formed under deltaic branch conditions. The siltier varieties are distinguished by reddish color and by the presence of plant root traces. The dominant bedding type is parallel bedding, but in places, crisscross-bedding and current ripples may also be observed. It is suggested that a 75 m thick upper Kirmaki suite has been formed under aerial-deltaic conditions. The total thickness of the entire Kirmaki suite is determined as 282 m.

To sum up, the Kirmaki suite is clearly divided into three parts corresponding to the different sea-level stages and accompanied by changing environmental conditions from the continental in the lowermost and uppermost suite to the foredeltaic-lacustrine conditions in the middle suite.

An old shoreline advance has marked the transition from the Superkirmaki sandy suite to the seaside, accompanied by changing facial conditions from deltaic plain to the fluvial facies. It is noted that the section's intervals are composed of an alternation of thin interbeds of sandy, silty clays, sandy siltstones, and silty sands overlapping each other over clean and smooth contact. Predominant in the lithological composition are obliquely laminated medium sands. The thickness of sandy bands varies from 1 to 6 m. From the special features of these sandy bands' stratification in the section and their contact with underlying rocks, they may be considered river channel deposits. Like the Underkirmaki suite, the river channel bodies here appear entrenched into each other, forming a series of vertically well-communicating sandy bodies. As pointed out in a detailed description of the Underkirmaki suite, the deposits mentioned above have been formed under plain fluvial conditions.

Furthermore, the fact that more coarse varieties overlap the fluvial plain fine-grained sediments is probably connected with the riverbank's debacle during the freshets and the accumulation of sandy material on silty sediments. The formation of the entire Superkirmaki sandy suite took place under a low sea level stage. Essential facies' changes during the SKSS accumulation have yet to be observed.

Sometimes, transitions from the significant river channel conditions to the adjacent fluvial plain relate to the dynamics of all the river system development, including migration and silting or/and activation of the river channels that were mainly caused by the changes in the river solid runoff balance. However, there is one special feature distinguishing the SKSS from the UKS. With the single-type sedimentation conditions within the sizeable, branched river system of the paleo-Volga and at the same low base level of erosion existent during the sedimentation of the two suites (both were formed on a low sea level stage), the thickness and character of sandy bands appear to be unlike. The SKSS are relatively thin-bedded and represented by the shallow channel formations divided by more silty sediments of the channel banks or adjacent fluvial plains that are apparently caused by channel migration.

Thick bands represent Sandy bodies in the UKS, probably the major riverbed. Our understanding of the mentioned differences between the SKSS and UKS reservoir structure may have resulted from the changes in the amount of transported and deposited sedimentary material. The thickness of the SKSS is 55 m. The next Productive Series suite is the Superkirmaki clay suite (SKCS), partly stripped down to 12 m only. The Pleistocene deposits overlap its upper half. Of the SKCS, the sedimentation condition is the most debatable problem. The large sun cracks here suggest that these deposits have originated under subaerial conditions.

Overall, the SKCS stripped deposits, represented by sandy-silty clays with plant root traces and sandy siltstones with subordinate argillaceous sands, appear to be very close to the deltaic plain sediments that have been described in the Kirmaki suite, and therefore, we believe that the SKCS facies conditions may be interpreted in the same way. Thus, it is likely that a long-term period of sea-level lowering during the SKSS accumulation has been changed by its recurrent rise. Some researchers had considered that the SKCS lowermost strata sedimentation took place under highly aridic climate conditions when the paleo-Volga was regenerated into a so-named "ephemeral" river system with periodically drying channels and branches flowing into a shallow salt lake; no delta has been formed. Under these conditions, the dominant factor caused the changes in the lithological composition of sediments in the section and, accordingly, in reservoir structure appeared to have been a sharp variation in volumes of supplying water and sediments into the basin. The strong freshet stages have been accompanied by an increased volume of terrigenous material; particularly coarse sediments delivered by the river. The drying of branches and accumulation of clay varieties characterized the periods of the river activity decreasing. Therefore, the South Caspian basin was decreased, concentrated within its modern central part only. Of the suite, the uppermost strata have been accumulated on the Absheron under lacustrine conditions, which were formed because of the paleo-Caspian Sea level's sharp rise.

This hypothesis raises objections because it is hard to assume such sharp regeneration of that powerful branched river system as paleo-Volga into a drainage network of SKCS ephemeral channels during the SKSS accumulation. Such a fundamental reconstruction of all the river systems may be assumed in the case of climate changes during the Lower Pliocene within the drainage system, i.e., the Russian platform. However, there is no evidence of such sharp climatic changes, from the high humidity to the pronounced arid—on the Russian platform territory during the Lower Pliocene. From the result of a substantial composition study of carbonate taken from the Ostracod shell in this suite, it is also concluded that there were no arid conditions in the SKCS. It is also hard to understand what processes caused such sharp paleo-Caspian to rise in the suite uppermost strata.

Considering these arguments, the major factors controlling the SKCS sedimentation, and all the other suites of the Productive Series have been paleo-Caspian Sea-level variations, which led to the mentioned facies and lithological heterogeneity of the bottom beds. The contact of the SKCS with the Break suite has yet to be found in the studied section. However, as it is known from both suites outcropping in the Kirmaki Valley, basal conglomerates occurring on the bottom of the SKCS appear to be evidenced by deep erosion of SKCS deposits and paleo-Volga entrenchment of the Absheron Peninsula. The Break suite sediments represented the most coarse-grained part of the PS section, composed of coarse sands passing up into medium-grained varieties. The dominant bedding type is tabular crossbedding. Also observed are parallel bedded sandy bands smoothly passing from the coarse varieties on the bottom into more fine-grained ones in the roof, typical of river channel deposits. Some bands' thickness varies from 1 to a few meters. The sections overlap to form a series of sandy bodies with well-defined boundaries. A typical feature is that many clay-rounded fragments occur as partings on the bottom of sandy bands. The mentioned clay fragments likely have been eroded from the channel walls or adjacent fluvial plain and then re-deposited. Cretaceous and Jurassic large pebbles supplied from the Greater Caucasus, confirmed by petrographical data, represent conglomerates on the bottom of the suite. Clay fragments and pebbles found in the Break suite sands indicate the river stream's very high rate and energy. Furthermore, these clay bands crushing into separate fragments suggests they may not have been a barrier to the fluid stream.

Also noted are several red paleo-soils and possibly eolian deposits, which may serve as evidence of a highly arid climate during the Break suite accumulation that, in turn, may be the cause (or one of the causes) of retreat of the sea at that time. There was an intimate relationship between the paleo-Caspian Sea level variation and climatic events. The Break suite incomplete section (stripped by a trench) is 34 m.

Some authors have considered that the formation of such thick sandy bands in the Break suite is caused by the increase in water content and solid runoff of paleo-Volga when the river system advances deep into the basin, which is unrelated to its level fall. However, it occurred because of the increased volume of delivered clastic material (Nummedal et al., 2004).

However, the investigations have shown that the river channel facies are traced far to the south, making the paleo-Volga delta advance to the seaside during its rising questionable. Such high-scale changes in facies conditions and the formation of the branched river system may have been possible under conditions of essential changes in the paleo-Caspian Sea-level regime not only on the Absheron but also in the southern Baku Archipelago. The high arenosity of the PS section is well preserved in the next Balakhani suite, in which boundaries with the Break suite are marked by 1 m thick silt and sandy shale bands. A significant fine-granular composition of a sandy fraction compared with the Break suite is caused by current ripples, which establish the beginning of this stratigraphic interval. Sedimentation conditions of the Break suite are preserved in the Balakhani suite Horizon XI lowermost strata, where sandy bands are interpreted as the river channel deposits. However, the dominant deposits of the section appear to be relatively thick sandy bands with rightly sorted sediments typical to the deltaic deposits. Also noted is a 15 m sandy band passing up to sandy siltstones and clays with sun cracks. A typical feature of sandy deposits is the domination of crisscross lamination among other sedimentary textures. These deposits are interpreted as prograding shallow sandy bay-mouth bars and then being changed by plain deltaic facies, which upwardly pass into a fluvial setting.

Thus, an entire cycle of sea-level variation appears from its high stage during the formation of the SKCS deposits to its fall during an accumulation of the Break suite, the Balakhani suite X horizon's lowermost strata, and subsequent sea-level levels rise in the middle X horizon have taken place. Besides, sedimentation conditions were changed from the plain deltaic facies to the fluvial and delta front facies. Overall, it is noteworthy that the thickness of the deposits being formed under river channel conditions is much more than those of the deltaic and lacustrine sediments, which is explained by an increase in sedimentation rate in a depression of the base level of erosion during sea-level fall.

As mentioned above, the X horizon has ended with a return to the river channel conditions, including fine multilayer sand characterized by a cross and parallel bedding. Abundantly present clay fragments indicate a very high rate and stream energy. However, it is noteworthy that apart from the finer sandy material in the Balakhani suite compared with the SKSS, UKS, and Break suite, sandy bands in the Balakhani section occur monotonously, dividing between each other by silty clay bands. Furthermore, if sandy bodies entrenched into each other are repeatedly observed in SKSS, UKS, and Break suite sections, they have yet to be noted here. In our understanding, the sandy deposit sedimentation occurred in the paleo-Volga's numerous lateral branches. The thickness of Horizon X is 122 m.

The transition to the X horizon has been marked by the following change of facies conditions onto the deltaic plain with typical frequent alternation of thin, fine-grained sandy sediments. The contacts between these bands are mainly smooth. We also observed thicker, from 1 to 2 m, sandy bands and widely developed ripple texture. We interpreted them as deltaic branches. The thickness of the IX horizon is comparatively small, 26 m.

Horizon VIII, one of the most developing objects of the Balakhani suite, comprises 95% thick sandy bands with small pebbles on their bottom. Many rolled clay fragments indicate a high rate and energy of transporting streams. All types of cross-laminations, from tabular to crisscross ones, are present, and those result from the widely developed current ripple textures. Rarely noted are sandy bands with parallel bedding. A typical feature is decreasing sediment grain size from the roof to the bottom (reversed sorting), which is one more piece of evidence of these deposits' fluvial genesis. Some sandy bands representing fluvial deposits are up to 7 m in thickness. Also noted is a series of sandy bodies entrenched into each other, conducive to vertical fluid migration.

Furthermore, a many-meter extension provides favorable conditions for lateral migration. However, a few centimeters thick clay partings represented, as a rule, the transition to the floodplain facies because channels silting up also occur in the section. Such partings are caused by vertical lithological heterogeneity of the reservoir and may serve as a barrier to the fluid stream.

Overall, sedimentation conditions of the VIII horizon have been monotypic and have not been changed during its accumulation period. The low sea level stage conditions are preserved through the entire 80-m thick VIII horizon and have been changed by gradual sea level rising in the VII horizon's lowermost strata.

We believe sedimentation conditions in the VII horizon may have been considered a return to more distal delta front conditions, which were then changed by delta plain facies. Monotonous and extended sandy bands up to 10 m thick are characterized by a well-defined trend of increasing grain size upwards a section (correct sediment sorting) and mainly clear and even ground. As a rule, sands are obliquely laminated (tabular or crisscross lamination), rarely with current ripples, and represented by a fine-grained and very fine-grained fraction. Some sandy bands abundantly contain hard clay waste from adjacent delta plain or fluvial plain. All these deposits were interpreted as a proximal bay-mouth bar. It is noted that almost all sandy bands include 10–12 cm thick, hardly cemented sandstones. Cement is mainly carbonated. Also noted within the VII horizon of the Balakhani suite are sandy deposits characterized by an entirely reversed trend in sediment's grain size, namely their gradual reduction from the bottom to the roof (reversed sediment sorting). Sands are mostly obliquely laminated and are of delta branch deposits.

The sediments formed during the low sea level stage have been deposited under deltaic plain conditions, while bay-mouth bar deposits represent those formed during the high sea level stage. Overall, the transition to the delta front facies being changed by deltaic plain conditions has been twice noted in the VII horizon. A similar change in environmental conditions from the deltaic plain to the delta front is also typical to the VI horizon lowermost strata, within which one complete sedimentary cycle is developed. However, it is noted that sharper changes in environmental conditions from fluvial channel facies to the deltaic plain occur in the middle and upper VI horizon. A smooth return to the fluvial facies ends the VI horizon section.

Sedimentation conditions within the V horizon of Balakhani and Sabunchi suites we interpreted as more distal varying within the bounds of the delta plain and delta front. It is probable that during some very short time intervals, the paleo-Caspian transgression has occurred, accompanied by an accumulation of very thin lacustrine clay partings. However, the thickness and amount of such partings in the section are considerably less than those in the Kirmaki suite. During some periods, the transition to the plain fluvial facies has taken place. Because the latter is very close to the plain deltaic facies by their sedimentological features, it is hard to distinguish between them.

An analysis of paleo-streams direction shows that they are mainly unidirectionally extending in southerly and south-southeasterly directions. Detailed sedimentological investigations and facies interpretation have also been carried out in the Yasamal valley, where PS upper-division, specifically Sabunchi, Surakhani, and Balakhani suites (from IX to V horizons) are cropping out to the day. The thicknesses of the horizons in the Balakhani suite are somewhat increased compared with those in Kirmaki Valley, which relates to facies changes from typical channel facies in X, VIII, and V horizons in the Kirmaki Valley to the more flood plain conditions in the Yasamal Valley. The deposits of X horizon of the Balakhani suite outcropping in the Yasamal Valley may be roughly divided into two parts: the lower, relatively coarse-grained, and the upper, more fine-grained.

Clay partings of the lower interval are alternated with thin (1–3 m) sands representing laterally trending single channels having eroded bottom. An upper band is composed of isochronous upwardly coarsen fine-grained and silty sands up to 4 m in thickness and of brownish to reddish clay intervals. These sands' granularity and sedimentary textures are characterized by a typical gradual transition from crisscross laminated medium-grained sands in the lowermost parts of sandy bodies to their uppermost parts. Also noted are laterally trending 0.5–1.5 m thick sandy or siltstone bodies having clear and even bottoms and characterized by various sedimentary textures, from parallel bedding to small-scaled obliquely laminated and current ripples and convolutions.

In our view, facial conditions in the IX horizon of the Balakhani suite within the Yasamal Valley are like those in the Kirmaki Valley, suggesting that sedimentation has occurred under delta plain conditions. Numerous sun cracks observed in clay intervals indicate sedimentation under periodical flooding and dewatering conditions typical to delta plain conditions.

The transition to the next Horizon VIII has been marked by an essential change in the lithological composition of its component rocks. Dominating in the section are up to 25 m thick sandy bodies composed of the fluvial channel bodies entrenched into each other. Clay fragments occurring on their bottom have been eroded from the channel walls. The large and small-scale crisscross lamination and convolution present typical sedimentary textures. On the roof of sandy bodies, it may also be observed that current and "clambering" ripples indicate stream rate extinction and an increase in precipitation, respectively. The middle section is exclusively composed of fluvial sands, while an upper section is represented by clay and sandy intervals alternation as those in the Kirmaki suite. Environmental conditions here have been interpreted as those in the Kirmaki Valley. Sedimentation took place under significant channel conditions. However, the thickness of sandy formations and their transition to the argillic-sandy alternation suggests that sedimentation in the Yasamal Valley occurred under boundary conditions, i.e., between a fluvial plain and a significant channel.

The thickness of Horizon VIII is about 80 m. The VII horizon is lithologically very similar to the IX horizon and is also characterized by an alternation of fine-grained sandy beds having smooth and even bottoms with clay and silty intervals. It is noted that clay horizons are frequently covered with sun cracks, which have also been observed in the IX horizon. The dominant trend is that the sediments are coarsening the profile upwards, from clays to siltstones and fine-grained sands. Thus, sedimentation conditions in the VII horizon are very close to those in the IX horizon, i.e., they are presented by plain delta facies, which possibly have periodically been subjected to flooding by the Caspian Lake. Nevertheless, it is noticeable that the change of facies from the delta plain to the delta front is not so clearly manifested as those in the Yasamal Valley.

The Horizon VI lithologically and sedimentary is very similar to those in the Kirmaki Valley. They are laterally trending sandy bodies composed of separate up to 4 m thick channel bodies entrenched into each other. Like the other horizons, the reddish clay intervals are interpreted as plain deltaic formations. The thickness of the VI horizon here is much less than in Kirmaki (40 m and 100 m, respectively). The erosion of these sandy formations probably caused it. Sedimentological and lithological features of the V horizon are almost the same as in the Kirmaki Valley.

Its thickness is about 60 m. Overall, clay intervals of the Balakhani section suggest that sedimentation occurred under a decrease in the volume of supplied sedimentary material that is in accord with delta plain (or fluvial plain) conditions. As mentioned above, laterally trending sandy bodies observed in clay intervals corresponded to the stages of increase in the river water balance, making it possible to transport coarse deposits and their deposition on the deltaic plain.

The thickness of Horizon VIII and some sandy intervals of the X and VI horizons suggest that the sedimentation process took place under a considerable supply of terrigenous material. Besides, their bedding and extension mode indicate a significant riverbed condition and low base level of erosion, i.e., the most incredible sedimentation area. Such conditions may have been created during sea-level lowering and an advance of all the river systems deep into the basin. Moreover, at last, the thick sandy bands characterized by an increase in granularity upward of the sediment profile, which is clearly manifested in the Kirmaki Valley, and sporadic lenses in the form of the river channels being developed in clay intervals suggest that sedimentation took place under conditions of an increased volume of sedimentary material and enormous accommodation area (sedimentation space), i.e., under the foredeltaic conditions caused by the Caspian sea level rising.

The more significant part of the Sabunchi suite outcropping in the Yasamal Valley is represented by upwardly coarsen sandy-siltstone rocks, from 2 to 8 m in thickness. These bands are alternated with laminated grey, brown, and reddish clays. Some different sandy-silty intervals having reversed sorting of sediment are also noted. As a rule, such intervals are composed of sandy and silty bodies formed under river channel conditions. Clay fragments frequently occur on their bottom.

The thickness of these intervals varies from 1 to 4 m. A few up to 1 km laterally trending horizons of sandy rocks are noted in the middle suite. Their thickness is from 2 to 10 m. These heterogeneous bands are composed of sandy bodies interlocked and entrenched into each other and formed under river channel conditions. All mentioned above evidenced the development of the riverbed facies in the Yasamal Valley. It is suggested that the stream channel marginal part has taken place here.

Widely developed clay horizons have been subjected to the periodical outcropping that caused sun cracks. Also noted are sandy dykes, i.e., sandy bodies intrusion into overlying clay intervals. The development of such sedimentary textures is commonly connected with very rapid sedimentation, during which rapidly burying water-saturated sandy bands under the pressure of overlying plastic clayey sediments have injected the latter. Similar conditions are typical to the flood plain and plain delta facies.

Sedimentation conditions within the Sabunchi suite in the Yasamal Valley are interpreted as sedimentation processes mainly occurring in the delta plain condition. However, in our understanding, these relatively thick intervals of straightly sorted sands up to 8 m in thickness have been accumulated under near delta front conditions and are of bay-mouth bar formations.

The thickness of the Sabunchi suite is about 220 m. The Surakhani suite is the most clayey interval of the PS upper division. Its component rocks may conventionally be subdivided into three lithological types: sandy-siltstone bands, upwardly coarsen siltstone bands and clay bands. Sandy-siltstone intervals are a minor portion of the section and are represented by the thick laterally trending bands having clear and even bottoms. Typical sedimentary textures are cross-lamination, parallel bedding, current ripples, and "clambering" ripples.

The thickness of these bands varies between 1 and 6 m. Some of them are composed of sandy formations accumulated under river channel conditions. The more fine-grained, straightly sorted deposits are represented by alternating horizons of siltstone and clay rocks crowned with silty sands. As noted above, the same intervals have been observed in the Sabunchi suite. Parallel bedding current ripples characterize these intervals, and sun cracks parallelly occur on each other, forming a series of thickness bands up to 11 m. Clay intervals within the suite are, on average, 25 m thick. Clays are massive and very fine-grained. Their color varies from gray to green, brown, and reddish.

Widely developed are clays with oxidation spots. A typical feature of the Surakhani suite is an abundance of gypsum bands of epigenetic sedimentation origin, especially in the upper suite. Accumulating such gypsum bands and sun cracks indicates that the Surakhani suite has been formed under arid climatic conditions.

Overall, sedimentation conditions of the Surakhani suite are interpreted as an alternation of proximal fluvial plain or delta plain conditions, which have taken place during dry periods correspondent to the low sea level stages changed then by its rise and flooding. The development of grey lacustrine clays has marked these periods. Depending on their lithological character, sandy-siltstone bands overlying them may be Break facies of the riverbanks during freshet when the coarsest sedimentary varieties accumulated on the fluvial or delta plain or, on the other hand, they may have been interpreted because of the paleo-Volga delta propagation towards the Caspian lake during high river water when it carries much sedimentary material. Such fluvial streams saturated with sediments (so-named hyperpycnal inflows) have much more density than lacustrine waters, which leads to the fact that they appear to be diffluent over the floor of a shoaly lake in the place of the river mouth, in turn causing the development of enormous sandy-silty bodies.

The thickness of the Surakhani suite in Yasamal Valley is about 500 m. The Surakhani suite has been studied in another three outcrops located within other lithofacies zones of the PS, namely the Gobustan zone (Akhtapa outcrop) and the Lower Kur zone (Babazanan outcrop), and the near-Caspian zone (Sura outcrop).

The lithological composition of the Surakhani suite studied in the Akhtapa outcrop located not far from Umbaki village in southeastern Gobustan is very close to that in the Yasamal Valley (Vincent et al., 2001). The studied section is composed of clay rocks in which the paleo-soils with paleo root traces and organisms' activity are widely developed. Sometimes observed are thin-laminated clays. Many intervals of oxidized soils covered with sun cracks have also been noted.

Subordinate sandy-silty intervals composed of upwardly coarsen cross-laminated or parallelly bedded bands crowned with highly developed "clambering" ripples may also be present. A similar association of sedimentary textures is noted during stream flow gradual depletion that may have taken place under conditions of Break facies of the riverbanks and during deposition of coarser sandy-silty material on flood-plain clay deposits. However, it should be noted that the same sedimentary textures may have been developed under delta front conditions and during hyperpycnal inflows penetrated the basin. The origin of the mentioned sandy-silty intervals may be drawn only after analyzing all the features package, that is, the character of underlying clay bands, the organism's vital activity traces, and agitation ripple development. In such a way, agitation ripples observed in separate intervals indicate sedimentation has occurred under lacustrine conditions. Isolated single sandy bodies are typical fluvial channel formations up to 7 m thick. The trend of paleo-streams within the suite is north-northeast-southeast.

Overall, sedimentation conditions here, similarly to the Surakhani suite in the Yasamal valley, are interpreted as proximal fluvial-plain or delta front temporarily interrupted by the stages of the flooding and sedimentation under lacustrine conditions. It is noticed that the quality of reservoirs in the Akhtapa outcrop, which may be accepted as an example of the Gobustan facies, is essentially lower compared with the Absheron Peninsula since sandy bodies there are isolated, leads to the poor lateral and vertical communication.

The thickness of this section is 130 m. The other outcrop of the Surakhani suite, Babazanan, which we studied in detail, is located not far from Salyan town, adjacent to the Kur River within the Kur lithofacies zone (Vincent et al., 2001). According to the textural features of clay sediments, their color, and much facts evidenced the development of vital functions of organisms and plants. This section is very similar to those

Fig. 5.4 The complete cycle of the paleo-Caspian sea level fluctuations in the Lower Pliocene and corresponding sedimentation conditions

described in the Akhtapa outcrop. Sandy-silty intervals characterized by the grain size coarsening a profile upwards are less widely developed. Sandy bodies accumulated under fluvial conditions are observed to be a far large quantity. Typically, up to 20 m thick sandy bands are composed of numerous individual channel bodies considerably smaller in thickness than a band joining them. Such sandy intervals may be traced for many tens of meters.

Numerous channels transient into each other are noted within their limits. All that suggests, on the one hand, that a lateral channel migration occurs when a stream volume exceeds the volume of terrigenous material carried by this stream. On the other hand, it serves as evidence of a low base level of erosion, i.e., of a low level of the river basin. Sandy bodies' arrangement suggests they are well interrelated, which is very important from the standpoint of understanding the reservoir structure of the whole. At the same time, sandy bodies are more isolated in the vertical section, being divided by clay bands apparently of the food plain origin. Unlike the two described outcrops of the Surakhani suite, Yasamal Valley and Akhtapa, no evidence exists of the Caspian lake-sea rising and accumulation of lacustrine clays. It is noted that clay sediments are highly oxidized (except the uppermost strata of the suite, which is most likely due to the reduction processes) and characterized by many plant and animal traces. This leads to the fact that the vertical section reservoir appears non-uniform and is divided into separate sandy intervals disconnected from each other.

The stream's course is from the east to the south. Overall, features noted upward in the section decrease the quantity of the channel body as their horizontal connection slackens. It may be that it serves as evidence of the Caspian Sea level, which began at the end of the Surakhani suite and reached its maximum in Akchagilian time.

The thickness of this section is 580 m. The relatively small, thick Sura outcrop of the Surakhani suite in the Near-Caspian-Guba region is about 240 m. Its section is composed mainly of clay sediments, but sometimes observed are sandy bands composed of stratified beds that look like river channel bodies. The thickness of one such band reaches several meters. Abundantly present among sedimentological textures are current ripples. Also noted are sun cracks in clay partings. These sediments may be of fluvial origin. These sandy bands' accumulation is likely related to the rivers flowing down from the surrounding mountains to the seaside.

To sum up, all the above mentioned, it may be concluded that from the result of facies interpretation of the PS section, it is revealed that the sedimentation conditions of these deposits have repeatedly been changed. It is inferred that apart from the one long-term sedimentary cycle of the third order corresponding to the PS 2.8 Ma duration (according to the latest data of Abdullayev et al. (2018)) and accompanied by the change of environmental conditions from the proximal in the PS basal part to the more distal in the middle part of the uppermost Surakhani suite. It is reflected in a gradual reduction of grain sizes of the PS deposits from its bottom to the roof. The sedimentary cycles of the fourth and smaller orders may also be present. The studied section revealed that from the bottom Underkirmaki suite deposited under fluvial conditions, a gradual change to the delta plain and delta front sedimentation conditions occurs (Fig. 5.4). The most distal lacustrine-prodeltaic facies appear to be accumulated in the middle Kirmaki suite, caused by the highest sea-level position during this suite accumulation. From the KS to the SKSS, it is noted that a gradual shoreline advance toward the seaside is accompanied by the change of sedimentation conditions from the lacustrine to the delta plain and delta front facies. The latter has been widely developed during the Break suite deposition that, in our view, is related to the deep retreat of the sea and an advance of all the paleo-Volga River system deep into the basin and the formation of the high order sequent boundary (Fig. 5.4). It seems likely that within the period between the Break suite and the uppermost PS Surakhani suite, there was one more cycle of sea level, during which more distal facies were formed.

The abovementioned cycles corresponded to the fourth-order sedimentary cycles by their duration. As described above, the more high-order sedimentary cycles in this section have been accompanied by essential facies changes and significant changes in the stratigraphic architecture of the PS deposits directly influenced by reservoir structure, which will be shown below.

To carry out an analysis of environmental conditions in the western South Caspian Basin during Lower Pliocene time, as well as for correlation of reservoir structures within the different facies zones, the authors have interpreted the UKS and KS logs and also gamma-logging of a number of the water areas located in different parts of the Absheron and Baku archipelagos such as Bakhar oil and oil–gas condensate field in South Caspian depression 40 km of Baku (Alizadeh et al., 2017) Alyat-Deniz, Bulla-Deniz, Pirsagat, Aran-Deniz fields within the Baku Archipelago adjacent to the Lower Kur depression. The Neftchala field in the Lower Kur depression has also been studied (Fig. 5.1).

Several investigators proposed an interpretation of GTW data (geophysical testing of wells) to reconstruct paleogeographical conditions within PS in some stratigraphic complexes (Abdullayev, 2001; Abdullayev et al., 1998; Shilov & Bezmenov, 1994; Suleimanov, 2003). Described deposits have been studied in the profiles drawn up through the following suites: (a) KS, SKSS, and SKCS; (b) Break suite, Horizon X of the Balakhani suite; (c) IX to V horizons of the Balakhani suite; (d) Sabunchi suite.

The facies analysis of most of the Productive Series' suites (except the Kala suite, Underkirmaki suite, and Surakhani suite) and main oil-bearing complexes of the Baku Archipelago, i.e., Horizon VII and an analog of the Balakhani suite-argillic-arenaceous suite have been presented for the first time in this work. The logging data interpretation has been made, considering the result of sedimentological investigations of the PS outcrops. The profile through the Bakhar field has been chosen to be oriented in a south-southeasterly direction that reflects a trend of most streams during the PS formation stage (Hinds et al., 2004) (Fig. 5.5). The facies interpretation of the log diagrams of the Break and Balakhani suites over the Bakhar field has previously been undertaken by Abdullayev et al. (1998) and Abdullayev (2001).

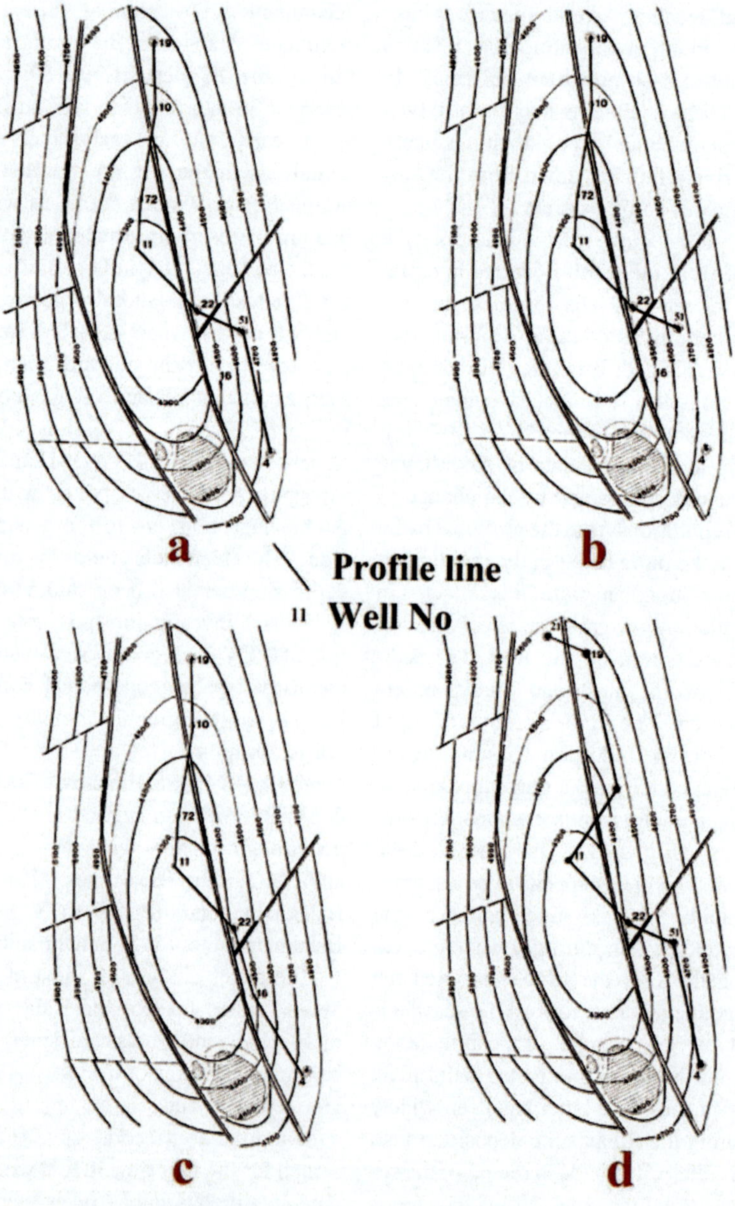

Fig. 5.5 Structural map through the X horizon's bottom within the Bakhar field

Overall, six suites, being the most productive part of the section, were studied. From the results of gamma-logging and UKS curves interpretation, it is inferred that the two trends are distinguished over the section reflecting either increasing or decreasing in grain sizes of sediment upward its profile, i.e., a gradual smooth transition from fine-grained to more coarse-grained varieties, and the other way round. The mentioned different paleoenvironmental conditions accordingly characterize two types of sediments. In the first case, as a rule, delta bars are formed on the delta front of the sediments deposited at the flood time when the riverbanks burst, and more coarse-grained varieties accumulate on silty river flats. The second event is typical of the river channels, delta branches, and punctate bars, which have been deposited on the riverbed (Reineck & Singh, 1980). In the Bakhar field, similarly to the Absheron Peninsula, a wide range of PS sedimentation conditions is noted—from the lacustrine to the fluvial conditions.

From the above–mentioned sedimentological analysis, it is inferred that the deposits in outcrops of the Kirmaki suite are mainly characterized by upward grain size, a sediment profile typical to the delta front facies (subfluvial delta facies). Similar facies are very clearly distributed onto the proximal and distal delta front. Under paleo-Volga near delta front conditions, sandy delta bars distinguished in the Bakhar section as up to several meters thick laterally trending sandy bodies (Fig. 5.6). One of the significant features is a gradual transition from the fine-grained sandy fraction at the bottom of the bar to the coarser in its roof. At the same time, typical channel bodies filled up with sandy material are characterized by quite the opposite trend in granularity regarding bar sediments. Such channels are interpreted as delta branches extending from the aerial delta into its underwater part and serving as main pathways for supplied terrigenous material.

The thickness of these sandstones is from several tens of cm to 1–2 m. The distal delta front conditions are presented lithologically by alternating several cm to several tens of cm thick clay, silty, and sandy bands. The deposits of delta branches are generally absent or may only be present sporadically. From the comparison of sedimentation conditions of the Kirmaki suite at Bakhar with its facies composition in the Kirmaki Valley, there are essential differences between them.

In such a way, sandy bodies entrenched into each other are absent at Bakhar. In contrast, they occur in the lowermost and uppermost strata of this suite in the Kirmaki Valley, which we interpreted as subaerial-deltaic (delta plain) sedimentation conditions. Widely developed at the Bakhar field are large laterally extending sandy bands and clay intercalations typical to the underwater delta facies (delta front) being changed in a southeasterly direction by monotonous clay bands formed under lacustrine conditions (Fig. 5.6).

Small river channels are rarely noted in the northwestern part of the field. It is interesting to note that environmental conditions appear to be frequently changed in a lateral direction, from the delta plain intersected by the branches in the Kirmaki Valley to the delta front deposits in northwestern Bakhar and typical lacustrine deposits in its southeastern part. The appearance of very thin-laminated clays defines the latter's presence in the section.

The middle Kirmaki suite is characterized by the transition to the delta front proximal and distal facies, which repeatedly alternated between each other and with lacustrine facies. Only two cycles of sea-level variation without any noticeable facies' changes may be noted in the Bakhar field.

In the Kirmaki valley, the change of environmental conditions repeatedly was caused by the frequent sea-level variations according to which several cycles may have been distinguished. In our opinion, such frequent changes in sedimentation conditions of the Kirmaki suite in the Kirmaki Valley compared with the Bakhar area are explained by the fact that the Kirmaki Valley at that time was in the basin's marginal part where the boundary conditions between the continental and lacustrine facies have been formed. Any slight sea-level change under similar conditions leads to material changes towards the continental or lacustrine facies.

As to the Bakhar field, because more deep-sea conditions of distal delta front and lacustrine facies have occurred there, sea level variation could not lead to significant facies transitions. It is most likely that the differences in the Kirmaki suite's environmental conditions in the Kirmaki Valley and the Bakhar caused the differences in reservoir structure. Within the Kirmaki Valley and adjacent oil and gas-bearing areas, lateral and vertical connections characterize multimeter sandy bodies of the delta plain and proximal delta front. In the Bakhar area, little thick sandy bands occurring in the distal delta front practically have no vertical communication, divided by impermeable silty deposits.

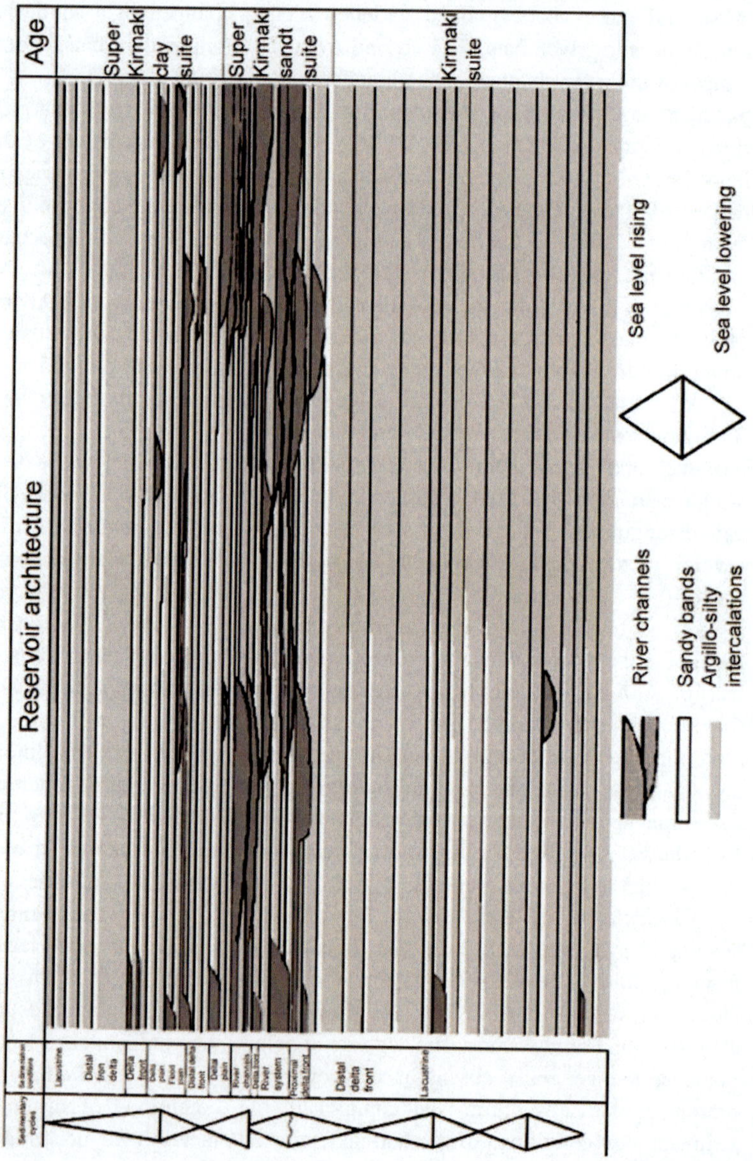

Fig. 5.6 Sedimentation conditions and reservoir architecture of the Kirmaki, Superkirmaki sandy, and Superkirmaki clay suites in the Bakhar area

The transition to the next Superkirmaki sandy suite (SKSS) took place under the conditions of an old shoreline stable advance towards the sea that led to the gradual progradation of the paleo-Volga delta, which during separate stages of the SKSS formation was located farther south of the Bakhar area (Fig. 5.6). Dominant in the SKSS lithological composition are obliquely laminated sands, which form just as in the Kirmaki Valley and Bakhar, multimeter and extended sandy bands entrenched into each other. From the special bedding features of these sandy bodies in the section and their contact with underlying rocks, it is inferred that they may be considered river channel deposits.

It is noted that there were two periods of the low sea-level stage during the SKSS sediment accumulation. A typical feature of these sediments is their forming erosional surfaces of sequential boundaries. The following sea level rise at the end of the first cycle has been accompanied by the transition to the delta front facies, where sporadic shallow sandy bodies separated from each other have been formed. The latter was the derivatives of delta branches continuing down the underwater delta. This conclusion is confirmed because very small pebbles have been found at the bottom of sandy bodies in the Kirmaki Valley's SKSS section.

A sharp change in environmental conditions within the SKSS proves that the Caspian Sea variation during the SKSS formation has been of wide amplitude, sharply, and frequently manifested. The transition to the PS most clay suite, Superkirmaki clay suite (SKCS), has been accompanied by the change in environmental conditions from the fluvial to the delta front and delta plain facies (Fig. 5.6). Compared with the SKSS deposits in the Kirmaki Valley, sandy bands of the same suite in Bakhar are represented by little thick derivates of shallow channels divided by more silty sediments of channel banks or adjacent fluvial plain sediments. Sometimes, separate channel bodies appear to be overlying each other, which results in their vertical interrelation but limited lateral connection.

Laterally trending thin sandy bands seen through the profile as slightly entrenched into underlying deposits and having neither lateral nor vertical connection seem most likely to be formed under plain fluvial conditions periodically downcutting by the same plain channels. Numerous sun cracks found in the SKCS deposits within the Kirmaki Valley confirm the conditions mentioned in the middle of this suite accumulation. However, in the author's opinion, it is hard to believe that the same conditions existed during the period of the SKCS formation on the Bakhar field territory. Many sandy partings and sizes appear to be reduced in this suite's lower and upper parts. Within these intervals, the SKCS deposits represented by sandy-silty clays with plant traces and subordinate intercalations of argillaceous sands are very close to the delta plain sediments of the Kirmaki suite.

The only thick argillic-silty sediments are noted in the uppermost strata, where they appeared to be either increased in grain sizes upward the profile or characterized by monotonous clay composition with relatively uniform granularity. The authors have interpreted similar changes in lithological composition as a smooth transition from the plain fluvial conditions to the delta front and typical lacustrine facies, which have not been noted in the Kirmaki Valley.

The Bakhar field is located within 50 km of the PS outcrop in the Kirmaki Valley, which suggests that abyssal environmental conditions have been formed here. Only one uninterrupted cycle of sea-level variation is noted within the SKCS, with its lowest stage occurring in the middle suite. The fluvial or delta plain deposits comprise approximately 1/3 of the total suite thickness, indicating that the low sea level stage was rather long. The reservoir structure is characterized by a complete absence of the vertically communicating sandy bodies, though they appear to be extended for a long distance. The formation of the next PS suite, the Break suite, as was mentioned above, took place under sea level deep lowering (Fig. 5.7).

From the facies interpretation of logging data, it is inferred that there were three stages of low sea level in the Bakhar area accompanied by considerable delta progradation, the development of a branched river system, and the formation of erosional surfaces may be sequential boundaries. Pebbly conglomerates confirm the latter in the Break suite bottom and well cropping out in the Kirmaki Valley. Repeatedly marked bands characterized by the coarse varieties in the roof are evidenced by the fact that these deposits have been formed during the flood and outburst of riverbanks. The first low-level stage on the basal part of the Break suite has been ended by its sharp rising and transition to the delta front facies accompanied by an extension by clay band formation. This event was followed by the subsequent sharp sea-level lowering and the formation of a highly branched river system with the channels incised into each other and overlapping each other. The transition to the delta plain facies with limitedly developed isolated sandy bodies of deltaic branches is noted in the section upwards. Relatively thick argillic-silty bands formed apparently in underwater delta conditions are noted to be still

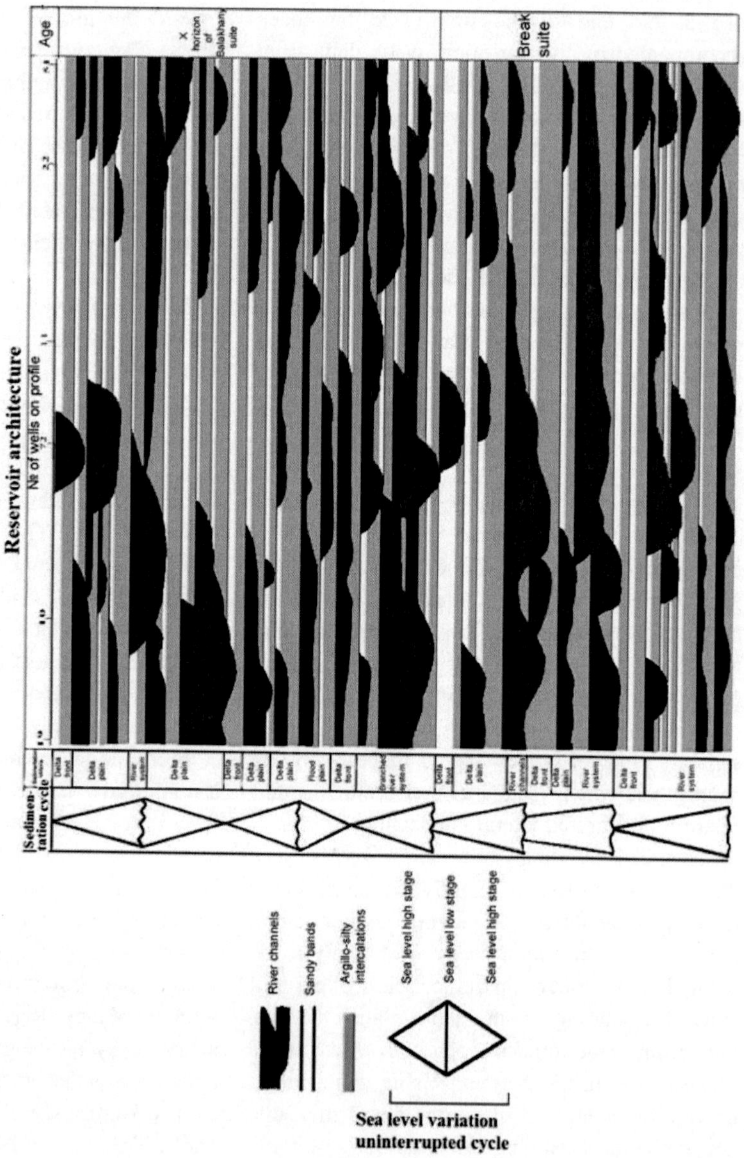

Fig. 5.7 Sedimentation conditions and reservoirs architecture of the Break ('Fasila') suite and X Horizon of the Balakhani suite

higher. The subsequent sea-level lowering has led to essential facies changes. A deep drainage system was developed at that time.

This low sea level stage was relatively short in a time interval and then changed by gradual rise with its smooth transition to the more distal conditions. The more significant part of the mentioned cycle is represented by delta plain facies composed mainly of thick laterally trending sandstones. The delta front clay series ends this cycle upwardly, passing into the more coarse sandy varieties.

Thus, relatively more distal facies compared with the Absheron Peninsula were formed in the Bakhar area. The three sedimentary cycles have been recognized by the Break suite total thickness of 135 m in the Bakhar field. They are remarkable for their very sharp character of sea-level variation with significant differences and amplitude.

As noted in the previous suites, specific differences in the reservoir structure of the Absheron Peninsula and Baku Archipelago are connected with heterogeneity of sedimentation conditions. As mentioned above, there are no essential barriers to the fluid flow in the Break suite deposits in the Kirmaki Valley. The homogenous lithological composition of sandy bodies vertically and horizontally characterizes the reservoir.

The delta plain and delta front clay formations are divided by sandy bodies into separated bands, thereby causing this reservoir's lateral and mostly vertical heterogeneity. The transition to the PS next suite, the Balakhani suite, took place under high sea-level conditions accompanied by the formation of delta front facies in the Bakhar area (Horizon X of the Balakhani suite) (Fig. 5.7).

The high sea level stage is characterized by its short duration being sharply changed by sea-level lowering that led to the following essential change in environmental conditions and the development of the branched river system at the beginning of the Balakhani suite accumulation time (Horizon X). Also observed is the river channels' well lateral connection while their vertical interrelation is limited. The transition from sandy deposits to silty varieties has resulted from the channel silting. At the same time, the bands characterized by the change in grain sizes from fine to more coarse-grained in the upper parts may be considered a result of the river-bank's outburst and fluvial plain silting deposition of more coarse varieties on silty-argillaceous flood-plain deposits.

The next band of laterally trending clays observed in the Balakhani suite section in the Bakhar area is the transition to the delta front or lacustrine deposits. Its analog in X horizon outcrop is up to 12 m thick sandy band with straightly sorting sediments typical to the deltaic deposits, which we interpreted as prograding shallow bay mouth bar. Shallow deltaic branches feeding this bar also take place.

The next turn to the fluvial conditions represented in the Kirmaki Valley and Bakhar, by the plain fluvial facies cut by shallow tributaries, is marked in the middle X horizon. In a south-southeasterly direction parallel to the chosen profile, the flood-plain facies appear to be passed into delta plain facies where the river channel sizes and their extension have been essentially reduced while the sediments with upwardly increased grain sizes become dominant.

As seen from the X horizon section of the Balakhani suite, plain delta facies dominate all the studied territory. On the next upper stage of the section, they are passed smoothly to the delta front conditions by which the second sedimentary cycle is ended with the beginning of the third cycle, where no sharp sea-level variations have been marked. Thus, the sea level variation during the formation of the Break suite and basal part of the X horizon of the Balakhani suite, the primary development objects, have been sensitive to frequent changes of movement and the development of incomplete cycles of sedimentation.

Sedimentation conditions in the middle and upper X horizon are not frequently changed, although sea level variation amplitude is usually tremendous, and a range of facies changing is relatively wide.

The following fourth sedimentary cycle, being marked in the lowermost X horizon, is characterized by the lower amplitude of sea-level variation and the development of delta front facies within this cycle during the high sea-level stage and plain delta facies during this cycle, the sea level low stage (Fig. 5.8). A sharp sea-level lowering has then retaken place, leading to the progradation of the paleo-Volga delta, which most likely has been located in the southeast of the Bakhar area. Thick sandy bands deposited within this area are overlapped by small intervals of silty rocks, which may apparently represent the bar deposits. The fluvial system formation time has not been extended. The facies change observed in the uppermost part of the section is a typical feature of all the field areas. If plain delta facies characterize the northeastern part of the studied area, the transition to the avandeltaic conditions occurs in a south-southeasterly direction.

The filth cycle (semi-cycle) is ended by a thick clay band deposited under delta front

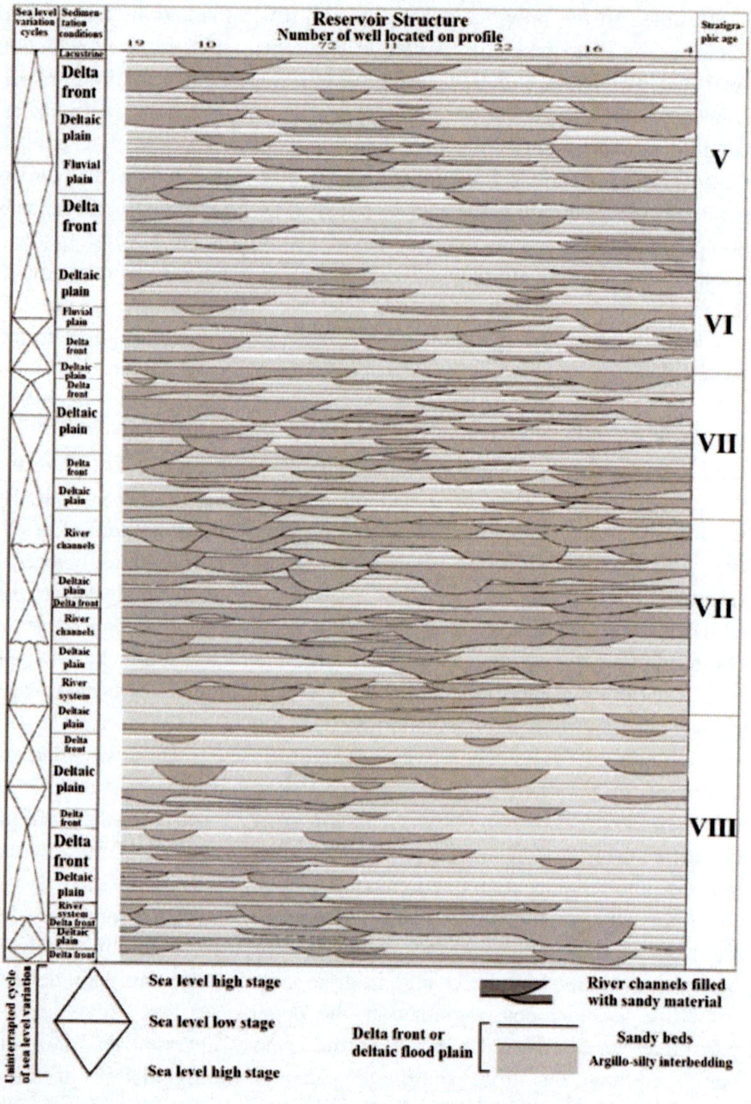

Fig. 5.8 Sedimentation conditions and reservoirs architecture of the IX-V horizons of the Balakhani suite in the Bakhar area

conditions. The X horizon section ends with a sixth short cycle, which is remarkable for its low-amplitude sea-level variation, and the beginning of the seventh cycle leads back to the regime of sharp, contrast sea level variations and the development of incomplete sedimentary cycles. For the mentioned regime, the typical feature is the absence of the deposits formed under transitional conditions, from the high sea level to the low stage. In the Bakhar area, three such sharp sea-level saltations led to the formation of a branched river system within the VIII horizon of the Balakhani suite (Fig. 5.8). Overall, the VIII horizon is characterized by the domination of fluvial facies compared with the rest of the horizons of the Balakhani suite.

Upwards profile, the rate of sea-level variation is slackened. The change of facies zones is less frequently observed. The two short-lived cycles marked in the VII and VI horizons did not lead to sharp changes in environmental conditions. According to the author's notion, it is likely that the plain fluvial facies take place in the uppermost V horizon of the Balakhani suite. The laterally trending thin bands having decreased grain sizes in the roof and deeply entrenched into underlying bands are interpreted as apparent offshoots from the primary channel.

In contrast, the bands coarsen in the roof are thought to be the flood-plain deposits. From the increasing share of red clay intercalations in the V horizon section within the Absheron Peninsula and considering that they have been formed in aerobic conditions, it is concluded that the flood-plain conditions were dominant there. Nevertheless, they are not observed generally in all areas. It is possible that in a south-southeasterly direction, the change of environmental conditions to the more distal deltaic plain or delta front facies occurred during a short-lived sea level rise.

In the Bakhar area, the dominant sedimentation conditions of the Sabunchi suite were plain deltaic facies, possibly changed in places by the plain fluvial facies (Fig. 5.9). The laterally trending isolated sandy bodies of the river channels are of branches or, in some cases, they present offshoots from the primary channel extended towards the fluvial plain.

In the middle of the fifth sedimentary cycle within the Sabunchi suite, a thick layer of reversed-sorted sandstones extended through the whole area and entrenched into underlying clay and sandy bands having an even bottom. The transition to the primary channel conditions may occur here, which we also observed in the Yasamal Valley.

The Caspian Sea level variation caused a sequential change of fluvial condition to the plain deltaic facies with typical isolated vertically bedded sandy-siltstone bodies divided by deltaic lake deposits. In the Absheron Peninsula during Sabunchi time, the high sea level stage corresponded to the formation of delta front facies. However, the time duration of these stages has been shorter, suggesting that an accumulation of the more significant part of the Sabunchi suite occurs during the sea-level lowering stages. By comparing the sedimentation conditions of the Sabunchi suite in the Bakhar area with those in the Yasamal and Kirmaki valleys, one may point to their identity and development of proximal facies.

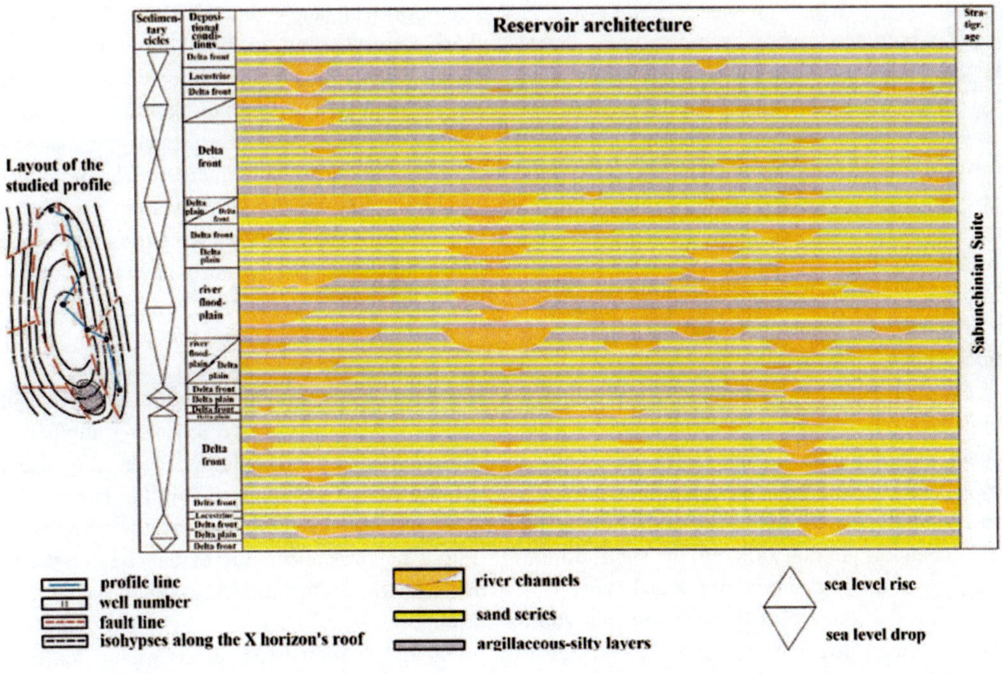

Fig. 5.9 Sedimentation conditions and reservoir architecture of the Sabunchi suite in the Bakhar area

Seven complete cycles of sea-level variation are distinguished within the Sabunchi suite. Furthermore, in its lower part, characterized by establishing more distal delta front facies in the Bakhar area, these cycles have been shorter in their extension.

Thus, numerous sedimentary cycles corresponding to the sea level variation cycles, from high to low and then the following high stages, are distinguished within the PS. According to the latest dating of volcanic ashes taken from the PS roof in the Lokbatan section of the Absheron Peninsula, the Productive Series formation time is 5.5–2.72 Ma (van Baak, 2015), i.e., about 2.8 Ma.

By their continuance in time, the mentioned sedimentary cycles come to about tens of thousands of years that, according to van Baak (2015), may have been interpreted as high-frequency (short-lived, small-scale) cycles. Different paleo-environmental types are formed within these cycles because of sea lever high-frequency variations, from a typical fluvial to the deltaic plain, delta front facies, and lacustrine deposits. Furthermore, the amplitude of environmental conditions change has been highly varied.

The presence of both sandy bodies characterizes the described short-lived sedimentary cycles as reservoir rocks and clay interbeds serving as cap deposits. In such a way, sea-level lowering has been accompanied by the progradation of the paleo-Volga delta, the branched river system formation, and sandy bodies well connected vertically and laterally. These stages in basin development have been characterized by a high rate of sedimentation and the formation of thick sandy bands serving as reservoirs.

Subsequent sea level led to delta retrogradation and the change of fluvial environmental conditions to the delta front facies characterized by laterally trending clay bands accumulation and limitedly developed sandy bodies being the river channels continuation into the shelf. Similar clay bands may be considered actual caps.

Thus, it is quite clearly observed that short-lived Caspian Sea level variation is reflected in the PS stratigraphic architecture that, in turn, caused structural heterogeneity of hydrocarbon reservoirs. Data obtained from the studies of some minor elements among their number of Sr and Ba, which are salt-basin-sensitive, show a good correlation of sedimentary cycles with content variations of these indicators.

All the analyzed lithological types of the PS rocks taken from the Kirmaki and Yasamal valleys are represented by authigenic minerals, in light fraction—by quartz whose content in sandstones is sometimes as high as 95%, and in heavy fraction—by disthen, staurolite, sillimanite and to a lesser extent—by magnetite, ilmenite, pyrite, garnet, zircon, tourmaline. It is noticeable that the authigenic minerals content is not high. However, certain mineralogical relationships appear to be outlined.

From the result of the Sr/Ba ratio study carried out on the samples, the curves showing average values of this index in various rock types (Fig. 5.10) and a cumulative curve showing all the lithological varieties and reflecting the Ba/Sr values variation over all the PS section (Fig. 5.11).

As shown from the analysis carried out, the most significant swing in these index values has been noted in sandy varieties (by a threshold of 0.56 corresponding to the Sr/Ba ratio in the recent Caspian sediments). In contrast, the Sr/Ba ratio curve in clays is less differentiated, whereas this value in siltstones is practically unchanged. Besides, single peaks of increased Sr/Ba values in clays are correlated with those in sandstones (Fig. 5.10). The cumulative curve reflects a Sr/Ba ratio mode of action over the section for sandy and pelitic fractions.

From the result obtained, it is inferred that the number of peaks of Sr/Ba ratio maximum values within some suites coincides with the number of cycles. In such a way, the following number of coincident peaks and cycles has been recognized: —in the Balakhani suite—13; in the Break suite —3; in the SKCS—1; in the SKSS—2 (Fig. 5.11). It is apparently evidenced in favor of the climatic factor affecting the Caspian Sea level.

An idea of short-lived paleo-Caspian Sea level variations accompanied by the change of

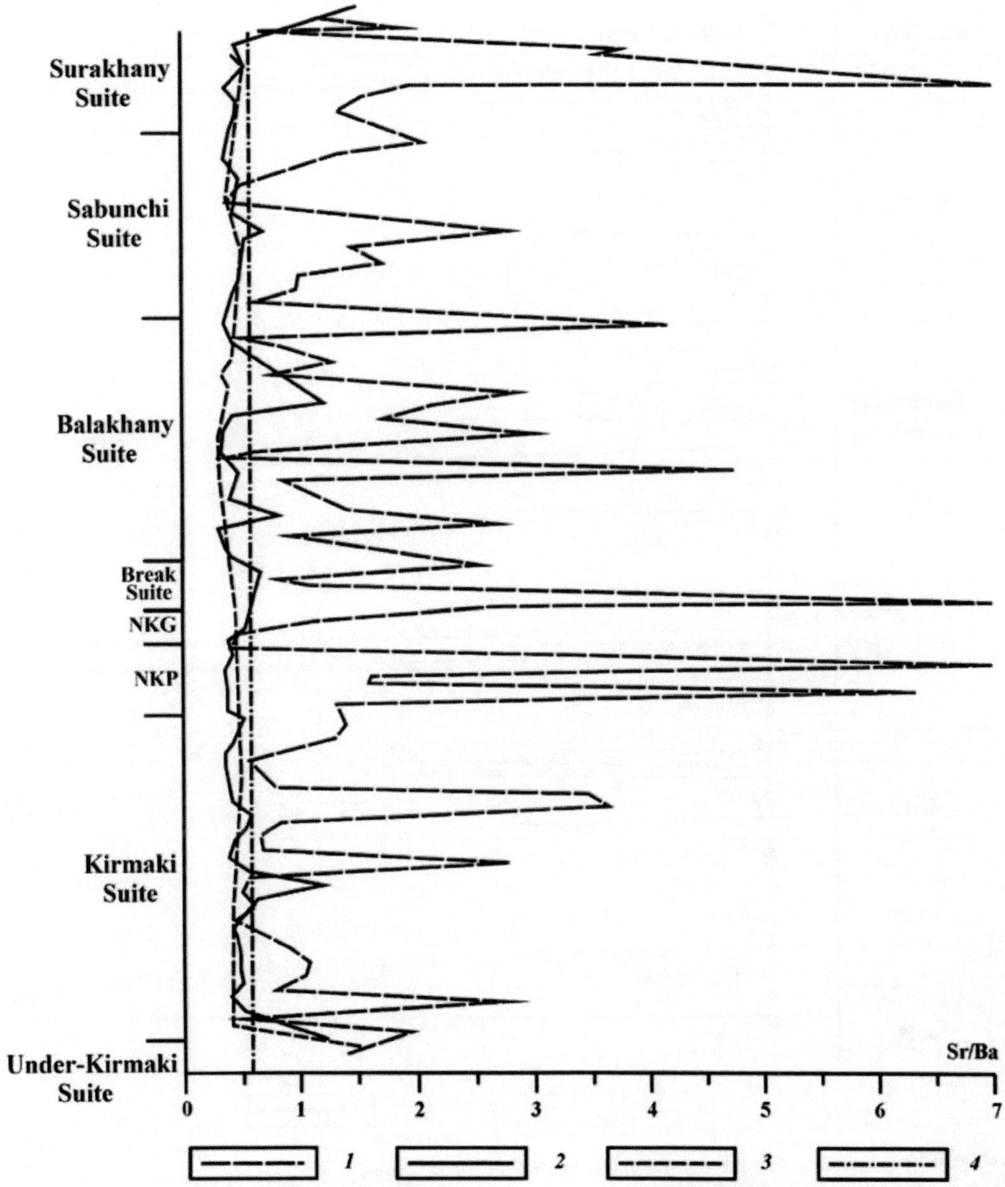

Fig. 5.10 Plots of the Sr/Ba ratios in different rock types

sedimentation conditions during PS accumulation has served as a base of the above-represented conception of the South Caspian sedimentation in the Lower Pliocene. However, to date, more information is needed to decide whether these frequently changed environmental conditions have resulted from the relative sea level or caused by the changes in the volume of supplying sedimentary material. For instance, it is debatable if an accumulation of the Break suite thick sandy bands tracing from the Absheron Peninsula far to the south on the SCB marine structures was caused by the shoreline progradation due to sea level lowering or it resulted from a sharply increased rate of erosion within the provenance and correspondingly from the

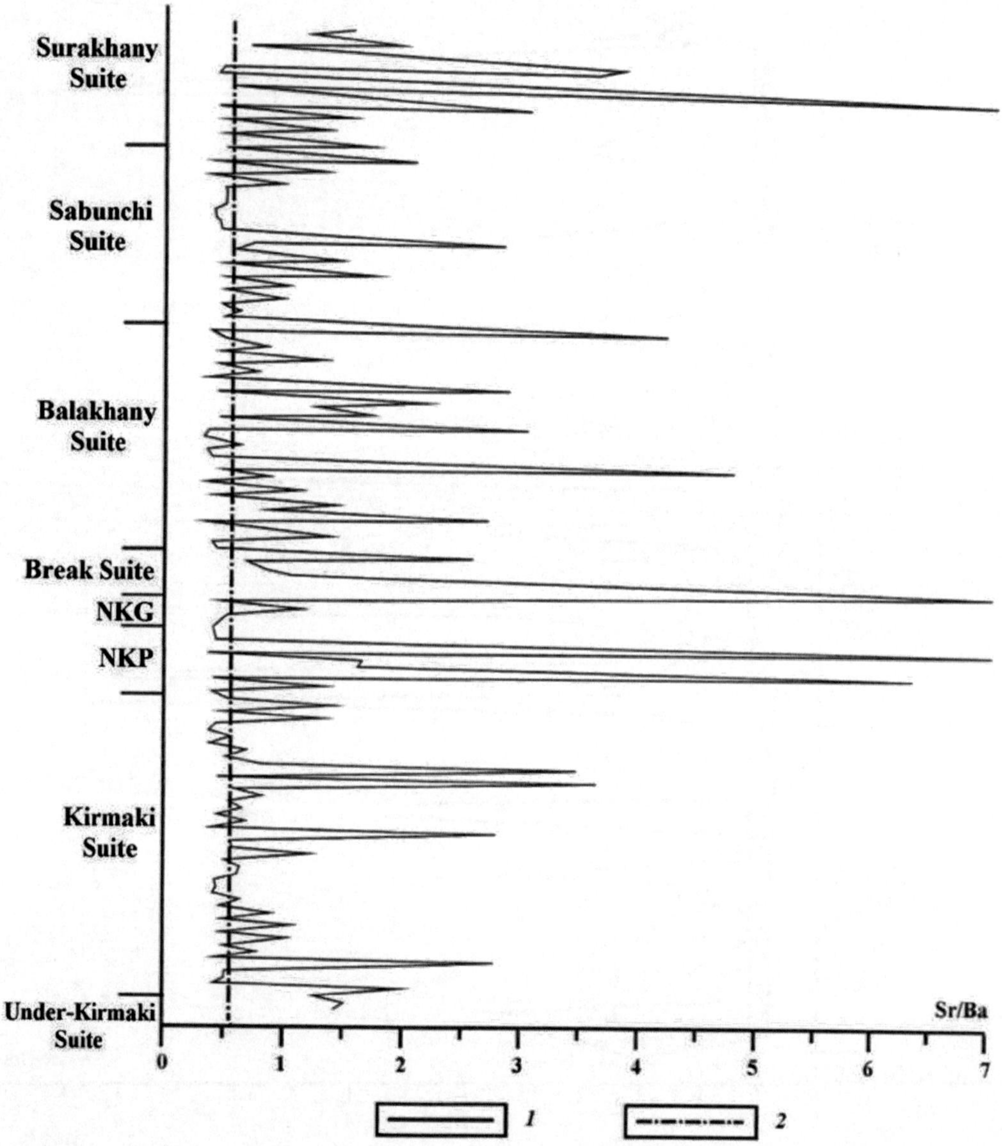

Fig. 5.11 Cumulative curve of the Sr/Ba values in the PS vertical section

increased volumes of sedimentary material. On the one hand, the deposition of thick terrigenous coarse-grained series suggests an increase in atmospheric precipitation and erosional process activity within distributive provinces resulting from the humidization of the local climate in the plain Russian region.

On the other hand, the geochemical data indicate an increase and frequent changes in paleo-Caspian water salinity during the Break suite and SKSS accumulation. It may have been caused by increased evaporation and sharp changes in the total water balance of this basin in the Lower Pliocene that may have taken place against the background of the desiccation in the Caspian region climate and paleo-Caspian Sea level lowering. Such contrastive processes within the provenances and depocenters may have resulted in the Caspian shoreline's fast-moving advance during the Break suite and SKSS accumulation due to the highly increased paleo-Volga solid run-off consequence of sharp sea-level lowering. In any case, climatic factors led to the Caspian sedimentation processes during the Lower Pliocene.

Nevertheless, the role of tectonic processes in Caspian-level high-frequency variations must not be denied. The correlation of gamma-logging carried out in some wells of the Gum-Deniz and Bakhar fields within the Kirmaki Valley shows that in the PS vertical section of the mentioned areas, the flooding surfaces occur in the SKCS and Kirmaki suites being noted every 8–9 m on average. The mentioned flooding surfaces are also noted in the SKSS and Break suite and in the lower horizons of the Balakhani suite, where they have been observed every 6–7 m, suggesting an accelerated regime of sea level variation over separate periods. Such a mechanism of the Caspian Basin development may be explained by climatic factors' effect and tectonic processes within this tectonically active region. Suppose tectonic movement marks have coincided with the direction of sea-level variations caused by climatic factors. In that case, the sharp transitions between different sea-level phases occur, accompanied by essential changes in sedimentation conditions. As observed from the PS section, when tectonic movement marks and climatically caused sea-level variations to have been equidirectional, the Caspian Sea development conditions were calmer with smooth transitions of one facies zone into the other.

Among the conceptions explaining the Caspian Sea level variation mechanism, the most ponderable is the climate-tectonic conception, considering both factors that align with Lilienberg's (1996, 2002) opinion.

As it follows from the analysis carried out, the factor of facies control of the PS reservoir structure is significant. It is quite normal that the question arises: firstly, what kind of sedimentary conditions had been in other parts of the basin? Secondly, what is the difference between the reservoir structure in SCB's different parts? The most significant interest is in the SCB marine area, which has become a sphere of intensive prospecting work. The authors have analyzed logging data from several marine areas within the Baku Archipelago, intending to carry out the paleo-reconstructions proceeding from the arrangement of chosen areas from the north to the south that reflects the paleo-Volga stream course and the change of facies zones. All chosen areas are located within several parallel anticlinal belts trending from NW to SE.

These are: Alyat-Seniz—Bulla-Deniz; Pirsagat-Hamamdag-Deniz—Aran-Deniz-Dashly-Sabail; Byandovan—Deniz—Atashgyakh—Inam; Kyurovdag—Neftchala. The major oil–gas-bearing complexes—the Break and Balakhani suites—have been investigated in detail. The Alyat-Deniz area is located within the Baku Archipelago water area adjacent to southeastern Gobystan, not far from the cape of Alyat southwest of the Sangachal-Deniz structure (Fig. 5.1).

This slightly asymmetrical fold with a steeply dipping northeastern limb and relatively gentle southwestern limb is located within the Alyat-Deniz—Bulla-Deniz anticlinal belt, which is the continuation of the Alyat ridge (Agabekov, 1963; Alizadeh et al., 1966, 1967). A slight rise separated from the Dashkil uplift by a shallow saddle is established here (Alizadeh et al., 1966). It was chosen a profile intersecting this area from the east to the west (Fig. 5.12).

From a comparison of the reservoir structure of Horizon VII (analogous to the Break suite) in the Alyat-Deniz area with the Break suite in the Bakhar area, it is inferred that they are similar. Widely spread here are typical river channel formations similar in geometry in both areas. As it is seen from the lower part of this section in the Alyat-Deniz and Bakhar areas, stream channels trending through these areas and filled with sandy material (according to the logging data) form a series of vertically bedded river channel bodies entrenched and transient into each other in a horizontal direction (Fig. 5.12).

A similar space setting is evidence of very good vertical and lateral communication of sandy bands in this section and, correspondingly, of the same well-fluid permeability. The mentioned interval is changed upwardly by a small horizon

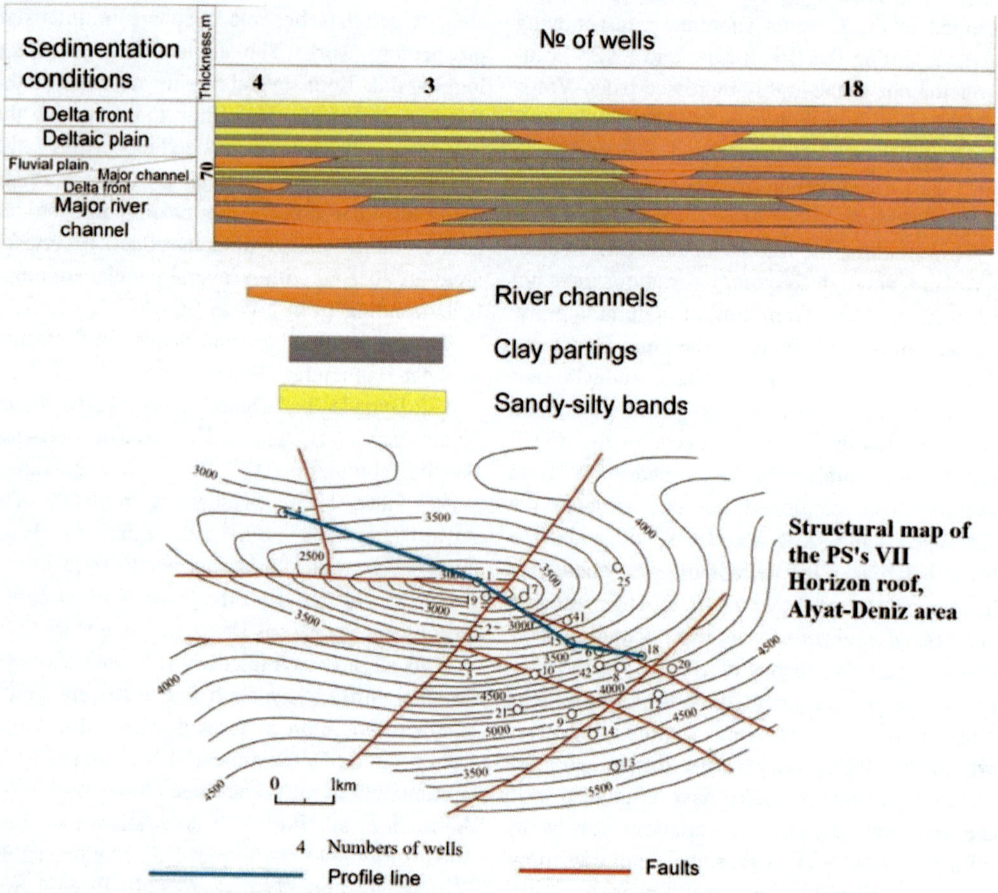

Fig. 5.12 Sedimentation conditions and architecture of the reservoirs of the PS VII horizon, Alayt-Deniz area

of mainly clay sediments with a single sandy band, which reminds it of a river channel body. A small, thick, sandy-silty band of straightly sorted rocks is also noted. This interval we interpreted as the transition to the delta front facies was also noted in the Bakhar area. In the middle section, the primary channel conditions of sedimentation are observed again, which is established by forming thick and extending sandy channel bodies entrenched in the western Alyat-Deniz area. However, unlike the Break suite in the Bakhar area, the Alyat-Deniz area is characterized by the transition to the plain fluvial facies indicated by the domination of more tinned sedimentary material and accumulation of upwardly coarsen sediments answered to the stages of the freshet and riverbanks debacle.

Thus, in the Alyat-Deniz area, the middle Break suite is marked by marginal facies of the paleo-Volga major fluvial plain passing into the floodplain conditions. As was mentioned in the Bakhar area, the Alyat-Deniz section's upper part is also characterized by the domination of upwardly coarsen and thick sandy-silty rocks. Besides, the river channel formations decreased in amount and thickness and appeared to be developed only in the western field. A similar space setting is typical to the delta front deposits, which, in our opinion, have existed at the end of Horizon VII accumulation in the Alyat-Deniz area.

There are some distinctions on a level with the similarity between the Break suite reservoir's structure observed in the Alyat-Deniz and

Bakhar areas. So, the Break suite thickness in Alyat-Deniz compared with Bakhar is reduced to about 70 m on average, reflected in the incomplete development of all sedimentary cycles in Bakhar. In the Alyat-Deniz area, only two stages of fluvial conditions have been established, whereas we believe that in Bakhar. There were three stages in the development of the significant channel facies and environmental conditions from the more distal to the proximal. The upper part of this suite in Alyat-Deniz has been washed out (Alizadeh et al., 1967). The existence of two transgressive and regressive cycles in the VII horizon within the Duvanni and Sangachal-Deniz areas has also been pointed out by Suleimanov (2003).

The transition to the next stratigraphic complex—the V-th division of the VII horizon in Alyat-Deniz is accompanied by essential changes in reservoir architecture (Fig. 5.13). The deposits of high sea levels and transgressive stages are becoming dominant. Also dominant in the section are clay intervals. The upwardly coarsen sandy-silty deposits take place but are essentially reduced in thickness. Sandy bodies of fluvial channels are also separately presented, being of small thickness, and are developed only in the eastern area. They are practically absent in the upper half of the section, where thick clay intervals are lacustrine in origin. The authors believe that the sedimentation during the mentioned stratigraphic interval occurred mainly on the delta front in the lower section and has been changed by prodeltaic-lacustrine conditions in the upper section. Besides, in our view, frequently changed environmental conditions have taken place in a lateral direction from the delta front to the prodeltaic facies, evidenced by the utter absence of the river channels with the domination of clay intervals.

We considered the sedimentation conditions within this horizon more distal than the IX horizon at Bakhar, where sedimentation occurred under largely deltaic plain-delta front conditions. Within the Bakhar area, also noted are the changes from more proximal environmental conditions in the lowermost strata of the IX horizon (river channel facies) to the more distal delta front facies in the uppermost strata of this horizon. The transition to the next V horizon, which may be an analog of the VIII horizon of the Balakhani suite within the Absheron facies zone, has been marked by essential changes in lithological composition and reservoir structure. Sandy deposits here are more widespread. Besides, the river channel deposits, like the VII horizon, dominate in the western area, whereas upwardly coarsen floodplain deposits dominate in the eastern part. Sedimentation conditions in the V horizon are interpreted similarly to the VII horizon—as marginal facies of the significant stream channel, which have been changed in the eastern area by the floodplain formations. The last episode is the only feature distinguishing this horizon from the VIII horizon in the Bakhar area, where dominant deposits are branched river system facies and sandy bodies have relatively good vertical and horizontal communication. Thus, the quality of the V horizon's reservoirs in the Alyat-Deniz area is not as good as that in the VIII horizon in the Bakhar area.

It is noticeable that the formation of the river system deposits corresponding to the sea level lowering has been accompanied by the development of the surface's unconformity, so-called "sequential boundaries," which appear to be coincident with flooding surfaces. Based on the PS outcrops investigation data, similar features have also been pointed out by Nummedal and Clifton (2002).

Furthermore, finally, the last described stratigraphic interval corresponding to the uppermost strata of the argillic-arenaceous suite (an analog to the Balakhani suite) is characterized by essential changes in reservoir architecture. Like the V-VII interval, the deposits of the high sea-level stage are dominant here. The sea level rise has been accompanied by changes in environmental conditions (Fig. 5.13). The river channel deposits are practically absent here. Upwardly, coarsen sandy-silty bands are obviously pinched out, passing into monotonous clay partings. Overall, sedimentation conditions here are interpreted as the most distal-prodeltaic-lacustrine. The existence of distal delta front facies may be in the uppermost profile only.

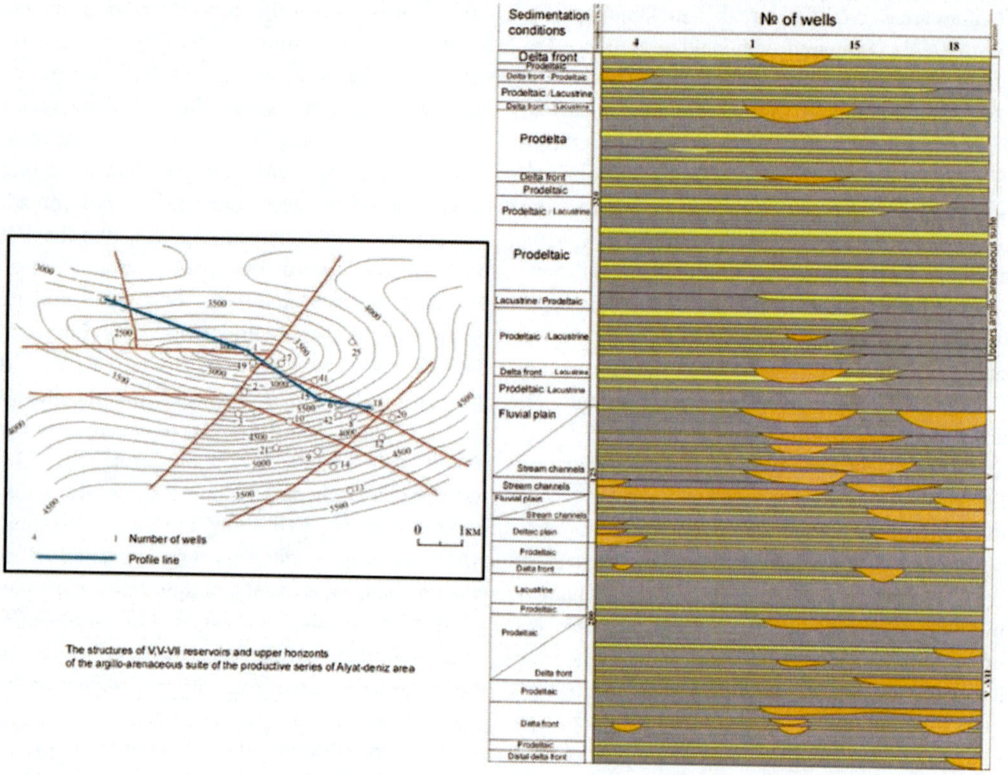

Fig. 5.13 Sedimentation conditions and architecture of reservoirs in the V horizon, within an interval of V-VIII horizons, and in the PS argillic-arenaceous suite upper-most strata; Alyat-Deniz area

The development of fluvial channel facies, even though reservoir geometry within these areas is quite different from that caused by the development of significant riverbed marginal facies in Alyat-Deniz. As in Bakhar, the lowest sea-level stages corresponding to the Break suite and VIII horizon formation time (accordingly, the VII and V horizons in Alyat) have been accompanied by the paleo-Volga River system progradation. Subsequent sea level rising led to the formation of distal delta front—prodeltaic—lacustrine facies here.

Among the notable features that must not be esteemed is the role of the changes in the volume of supplying material that led to the accumulation of monotonous laterally trending and upwardly coarsen sandy-silty series within individual sedimentary cycles. It is caused by an increase in sedimentary run-off and a break of the riverbanks accompanied by the accumulation of the coarser varieties on the fluvial plain clay sediments. Also, it may be caused by the penetration into a basin of more dense flows saturated with relatively coarse fractions.

The next area studied by the authors in the light of the reservoir structure is the Bulla-Deniz field (Fig. 5.1), located south-east of the Alyat-Deniz area and Khara-Zirya structure at the southeastern termination of the Alyat-Deniz—Bulla-Deniz anticlinal belt which is the Alyat ridge marine continuation (Guliyev et al., 2003). Tectonically, this area is a nearly symmetrical branchy-anticlinal fold (Alizadeh et al., 1966). A profile chosen in this area is oriented in a northwest-westerly-southeast-easterly direction (Fig. 5.14).

Horizon VII has been studied incompletely because this stratigraphic complex has yet to be stripped down to the bottom by the wells at our disposal. As seen in a constructed model,

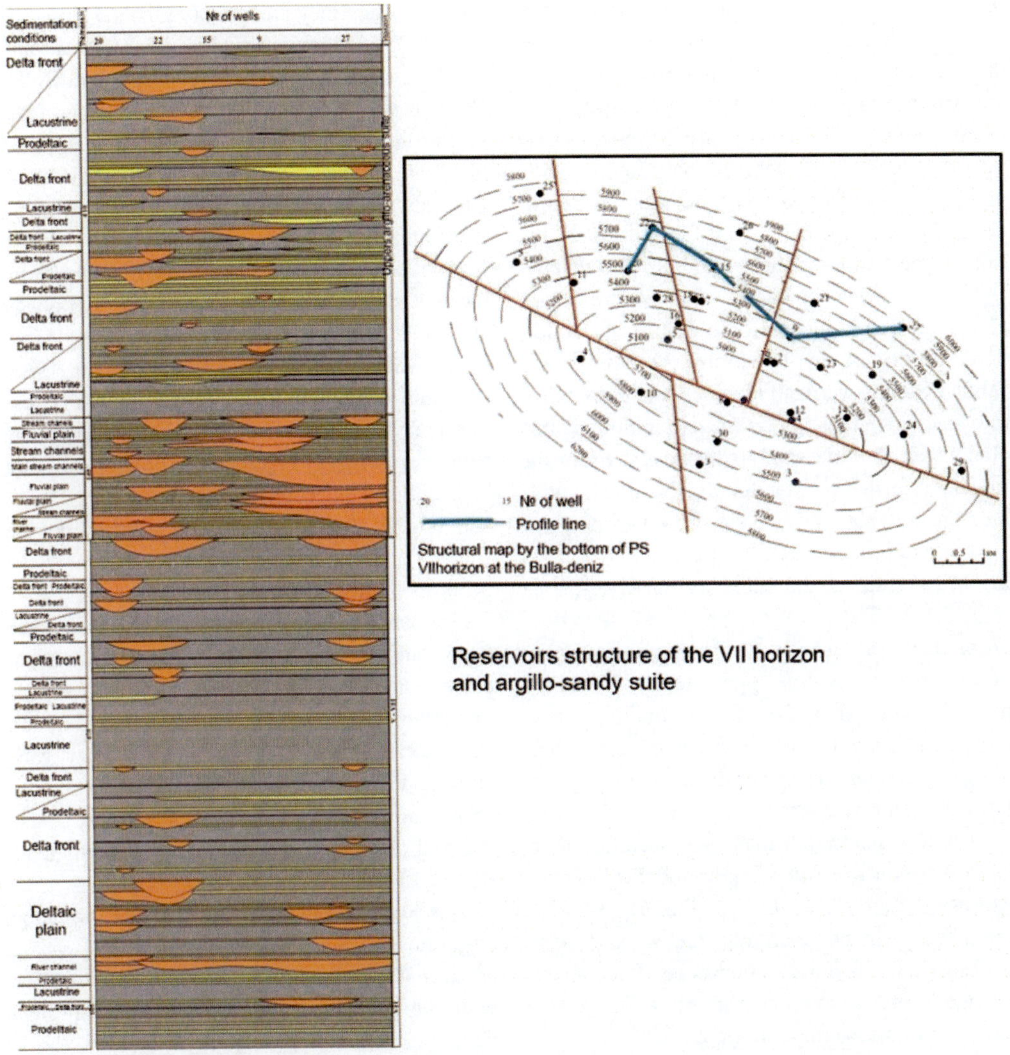

Fig. 5.14 Sedimentation conditions and architecture of the PS reservoirs within the Bulla-Deniz area

laterally trending thick sandy bands with relatively good vertical and horizontal communication are developed in the roof of the VII horizon (Fig. 5.14).

The described river complex may be correlated only with the third stage of fluvial deposit formation in the Bakhar area. The authors have established the same deposits in the Alyat-Deniz area. Below these deposits, the facies composition of roughly 2/3 of the Break suite total thickness has been cardinally changed, reflected immediately in the reservoir structure. It has formed thick clay horizons divided by small-thick intervals of upwardly coarsen sandy-silty rocks extending through this field. It is noticeable that the river channel deposits are practically absent here, except for a single case in the middle suite. This interval we interpreted as sediments being deposited in the prodeltaic-lacustrine facies.

In the light of time, they correspond to the uppermost interval of Horizon VII in the Alyat-Deniz area, where delta front facies have been established. An examination of overlying formations corroborated that more distal facies have been formed at Bulla-Deniz. In such a way, the lacustrine facies are dominated within an interval dividing the V and VII horizons.

As distinct from the Alyat-Deniz, sandy intervals must be developed here and quickly pinched along the profile. The more proximal delta plain or near delta front facies formed, the lower the stratigraphic interval. Prodeltaic-lacustrine facies upwardly change the latter. One more facies change onto delta front conditions occurs in the uppermost horizon where sandy channels vertically bedded and entrenched into each other are horizontally communicated along with the thin sandy-silty layers. From the reservoir formation standpoint, the most favorable are this horizon's upper and lower parts.

The stratigraphic architecture of the V horizon in Bulla-Deniz is the same as in Alyat-Deniz (Fig. 5.14). Thick sandy bands of the river channel origin, laterally well communicated but not so well vertically, appear to be developed here. As noted above, the facies conditions here have also been formed because of the paleo-Volga delta advance toward the basin during sea-level lowering and accumulation of Horizon V.

It is also noted that there is a specific difference between the Alyat-Deniz and Bulla-Deniz expressed in the fact that in the Alyat-Deniz, there were noted marginal facies that were passing into floodplain conditions. Such facies change is not marked in the Bulla-Deniz area except on its lowermost horizon.

Thus, the V horizon at Bulla-Deniz, compared with that at Alyat-Deniz, is characterized by rather well reservoir properties caused, apart from other reasons, by relatively vertical and horizontal communication of sandy bodies, which leads to the fluid migration and reservoir saturation. The development of the delta front-prodeltaic-lacustrine facies transient into each other along the profile is typical of the uppermost Balakhani suite (Fig. 5.14).

As in the Alyat-Deniz area, the river channels in the Balakhani suite are developed limitedly, and they are isolated from each other vertically. In places, they are connected by sandy-silty bands. The extension of sandy-silty bands decreases, and they are quickly pinched along the profile. The reservoir structure of the Break and Balakhani suites in both Alyat-Deniz and Bulla-Deniz is very similar, slightly distinguished from that in the Bakhar area. It should be mentioned that marginal facies were developed within some intervals that caused a sharp transition along these areas.

The next studied area is the Pirsagat field, located south-southwest of the Alyat railway station and located within the Pirsagat—Aran-Deniz—Sabail anticlinal belt (Agabekov, 1963; Alizadeh et al., 1966). A profile drawn across this area is oriented from the northwest to the southeast (Fig. 5.15). The VII horizon has not been entirely outcropped in wells No. 91, 93, 94, and 106 located along with this profile. Like the Bakhar area, the three river facies intervals have been distinguished here.

The large channel bodies formed within these facies are extended through the area and, in places, appear to be passed laterally to the floodplain facies. As mentioned above, they are characterized by the development of clay varieties and small-thick laterally trending sandy-silty bands generated during freshet and coarse material deposition on the floodplain formations (Fig. 5.15).

It is suggested that the fluvial facies formation stages in the Pirsagat area and the river system advance towards the sea have been alternated with the periods of distal delta front or plain deltaic facies. Within an interval dividing the V and VII horizons, gradual facies change upward in the section is observed upward as has been mentioned for above described Alyat-Deniz and Bulla-Deniz areas, from mainly delta front to more distal prodeltaic-lacustrine conditions (Fig. 5.15). It has been turned back directly on the reservoir structure.

Sedimentation occurred in the border zone between the Caspian Sea and the prodelta. The southeastern area, more distant from the deltaic region, is characterized by the development of typical lacustrine facies in the uppermost section. Suppose some isolated bodies formed under branch conditions and extended sandy-silty horizons uncommunicated vertically but being

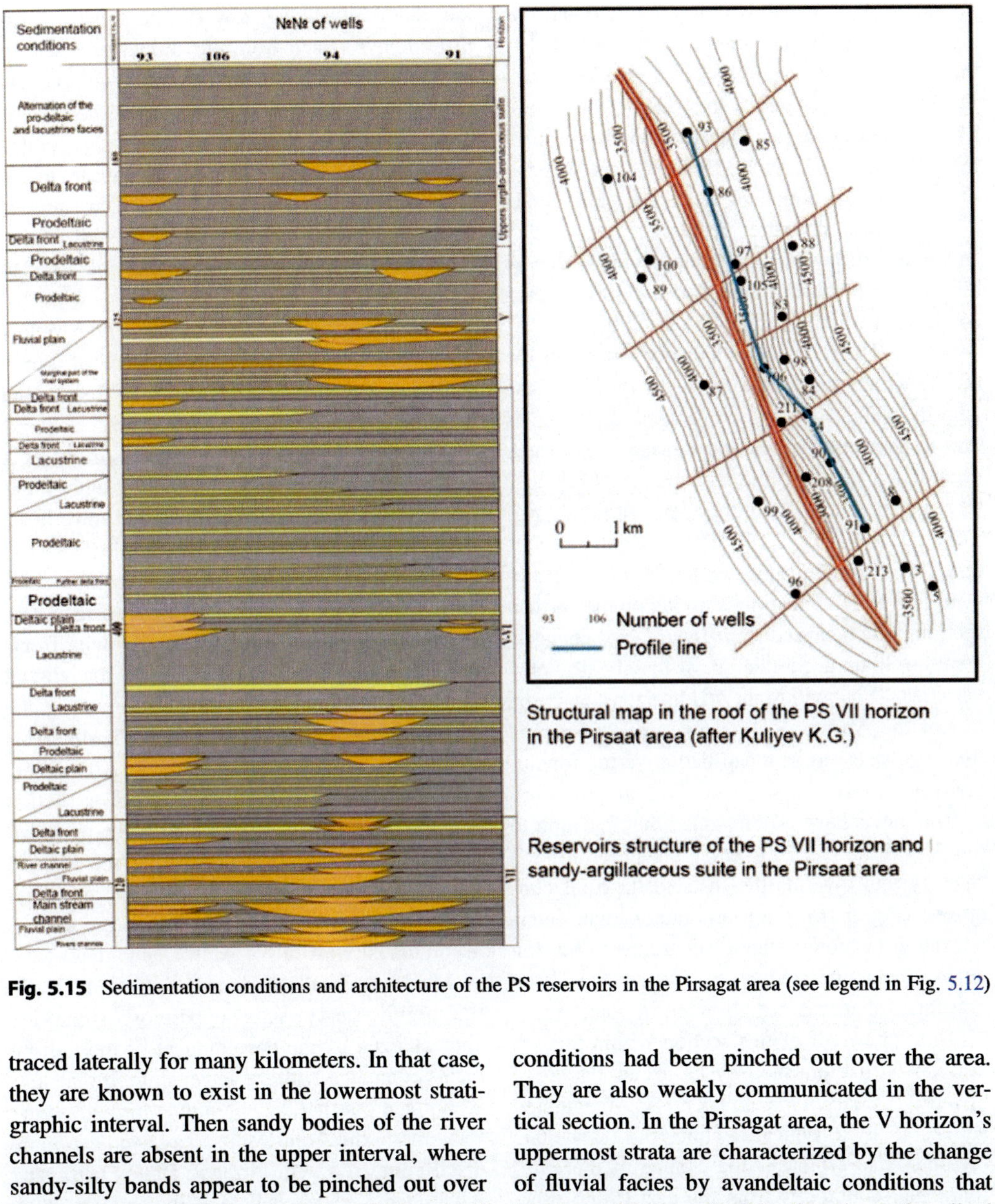

Fig. 5.15 Sedimentation conditions and architecture of the PS reservoirs in the Pirsagat area (see legend in Fig. 5.12)

traced laterally for many kilometers. In that case, they are known to exist in the lowermost stratigraphic interval. Then sandy bodies of the river channels are absent in the upper interval, where sandy-silty bands appear to be pinched out over the area, reflecting the change in environmental conditions from the prodeltaic to the lacustrine-argillaceous facies.

The return to fluvial conditions marks the V horizon. Here, the marginal facies are observed between the river channels in the Alyat-Deniz area. It is northwesterly seen to be changed by the flood-plain conditions (Fig. 5.15). The result was that sandy bands deposited under flood-plain conditions had been pinched out over the area. They are also weakly communicated in the vertical section. In the Pirsagat area, the V horizon's uppermost strata are characterized by the change of fluvial facies by avandeltaic conditions that have not been marked in the above-described areas. Regarding reservoir rocks, this interval is not so well compared with the horizon's lower half.

Argillic-silty lithofacies mainly represent the uppermost argillic-arenaceous suite (an analog to the Balakhani suite). The change of environmental conditions from mainly avandeltaic to the prodeltaic-lacustrine is observed within this interval. Like the V-VII horizon interval, the mentioned feature has directly been reflected in the reservoir structure (Fig. 5.15). The Aran-Deniz area is located 7 km SE of the Sangi-Mugan area at the southeastern continuation of the Pirsagat—Hamadag-Deniz—Aran-Deniz—Dashli—Sabail anticlinal belt (Alizadeh et al., 1966). This structure represents an axially asymmetrical branchy anticline having a steeply dipping northwestern limb and a gently sloping southeastern limb (Guliyev et al., 2003).

The VII horizon is completely tapped in the wells, which logs have been analyzed by the authors (Fig. 5.16). Its average thickness is 160 m. Like the Bakhar and Pirsagat areas, three dominative facies of the river channels (up to a few meters deep) are developed within its primary channel (Fig. 5.16). The channels are vast and well-communicated with each other. However, this communication is limited in the vertical section, which caused short-lived facies changes, resulting in an accumulation of mainly silty clay varieties. Thus, within the VII horizons, as in the rest of the Absheron Peninsula areas, so-named fluid stops served as a cap-like covering for fluvial sandy traps.

The facies have accompanied the transition to the V-VII horizon's interval change from the river system towards the distal delta front conditions (Fig. 5.16). Only one intermediate interval has been noted where, according to the log curves behavior, the river facies have been developed that was not observed in other areas.

The upper half of this section within the VII horizon is distinguished by an accumulation of upwardly coarsen sediments. As mentioned above, it is a typical feature of avandeltaic regions. Also typical is the change of the environmental conditions from the east to the west, specifically from the delta front to the prodelta and even to the lacustrine conditions. It points to the fact that during the Balakhani suite accumulation, this area was in the marginal zone between the delta front and lacustrine facies, leading to the frequent transition of one condition to another. As a rule, similar facies zones react sensitively to the river solid runoff volume changes or minor sea-level variations that lead to the delta advancing toward the basin or its recession. As seen from our model of reservoir architecture of the V-VII horizons section, the delta front facies development stages are accompanied by a single vertically separated river channel body. The latter appear to be changed by prodeltaic conditions characterized by forming clay varieties, thin sandy-silty partings, and lacustrine facies within which clay sediments have accumulated. Along with the profile, from the structure, the eastern pericline toward its western part, the change of sandy sediments by sandy-silty partings of distal delta front or prodeltaic facies are observed.

The Horizon V structure in the Aran-Deniz area is distinguished from that of the other areas by the river channel facies observed only in its marginal part and changed in profile from the east to the west by apparently argillic-silty sediments with thin sandy partings (Fig. 5.16). Upwards the V horizon section, the delta front-prodeltaic-lacustrine facies change the fluvial conditions. The same has been observed in the Pirsagat area. As distinct from the other intervals of the argillic-sandy section in the Aran-Deniz area, the typical feature of the V horizon section is the facies changes from the more proximal conditions in the western structure to the distal in the eastern part. This fact may be interpreted as the change in the source of the horizon sediments accumulated west of the Aran-Deniz area.

Moreover, finally, sedimentation conditions in the argillic-sandy suite uppermost strata are characterized by the development of more distal facies compared with the lower suite. It may also be pointed out that the conditions formed during this time in the Aran-Deniz area have been more distal compared with Pirsagat, Bulla-Deniz, and Alyat-Deniz areas, where sedimentation took place in the prodeltaic-lacustrine alternation conditions (Fig. 5.16). An interval of delta front facies with rather large filling sandy bodies is marked only in the middle stratigraphic complex. In the section upper strata, the repeated transitions between facies zones are noted, as was observed in the other areas.

Besides, facies change, and sandy-silty beds pinching out took place from the southeast to the

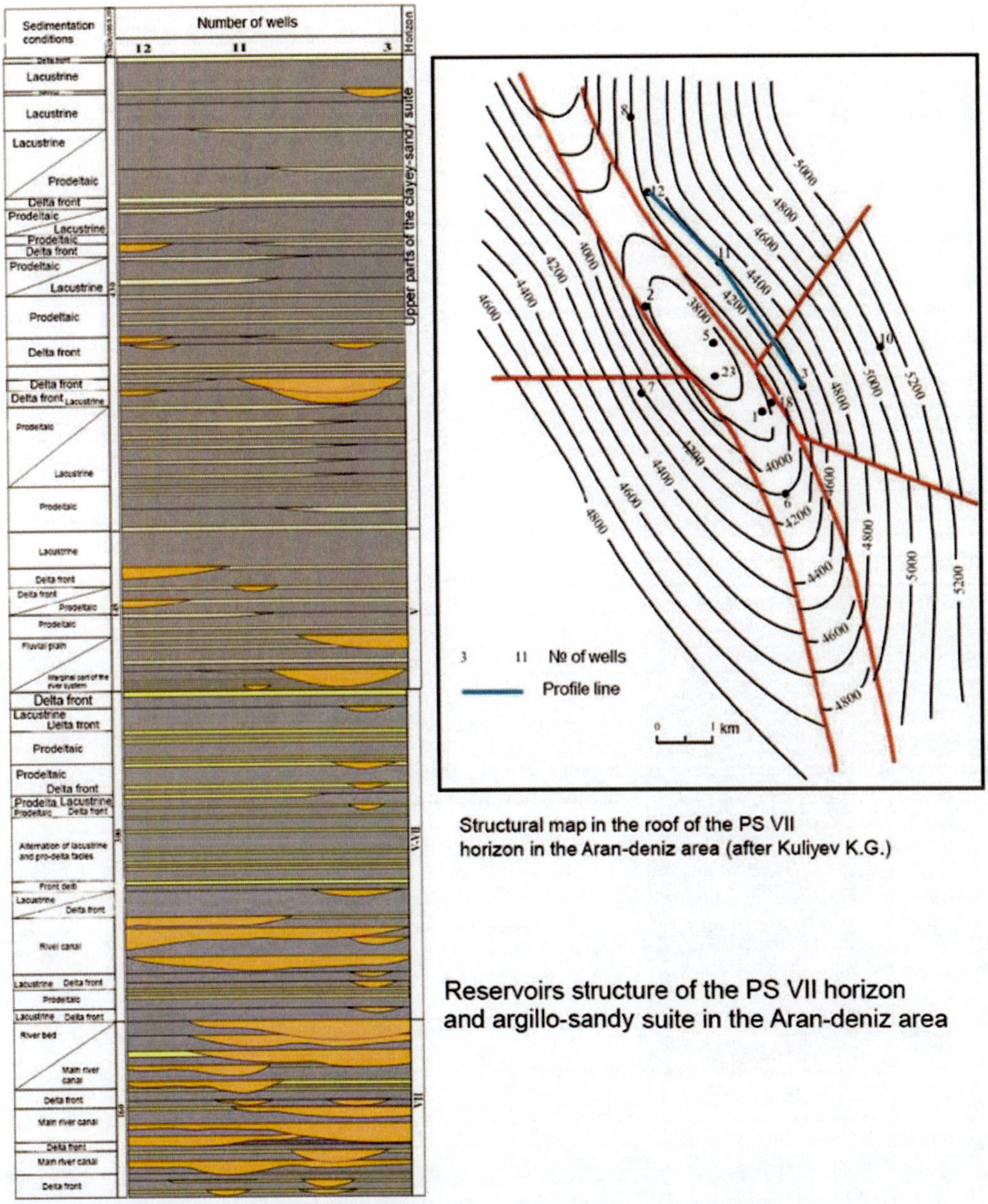

Fig. 5.16 Sedimentation conditions and architecture of the PS reservoirs in the Aran-Deniz area (see legend in Fig. 5.12)

northwest and the other way round—from the northwest to the southeast, suggesting the idea of a frequently changed direction of clastic material delivery to the area. The Atashkyakh area is located within the northwest-southeasterly oriented Byandovan-Deniz—Yanan-Deniz—Atashkyakh-Inam anticlinal belt, which is the continuation of Kalamadyn-Mishovdag-Byandovan anticlinal belt at the Lower Kur trough and is separated from the Pirsagat-Sabail

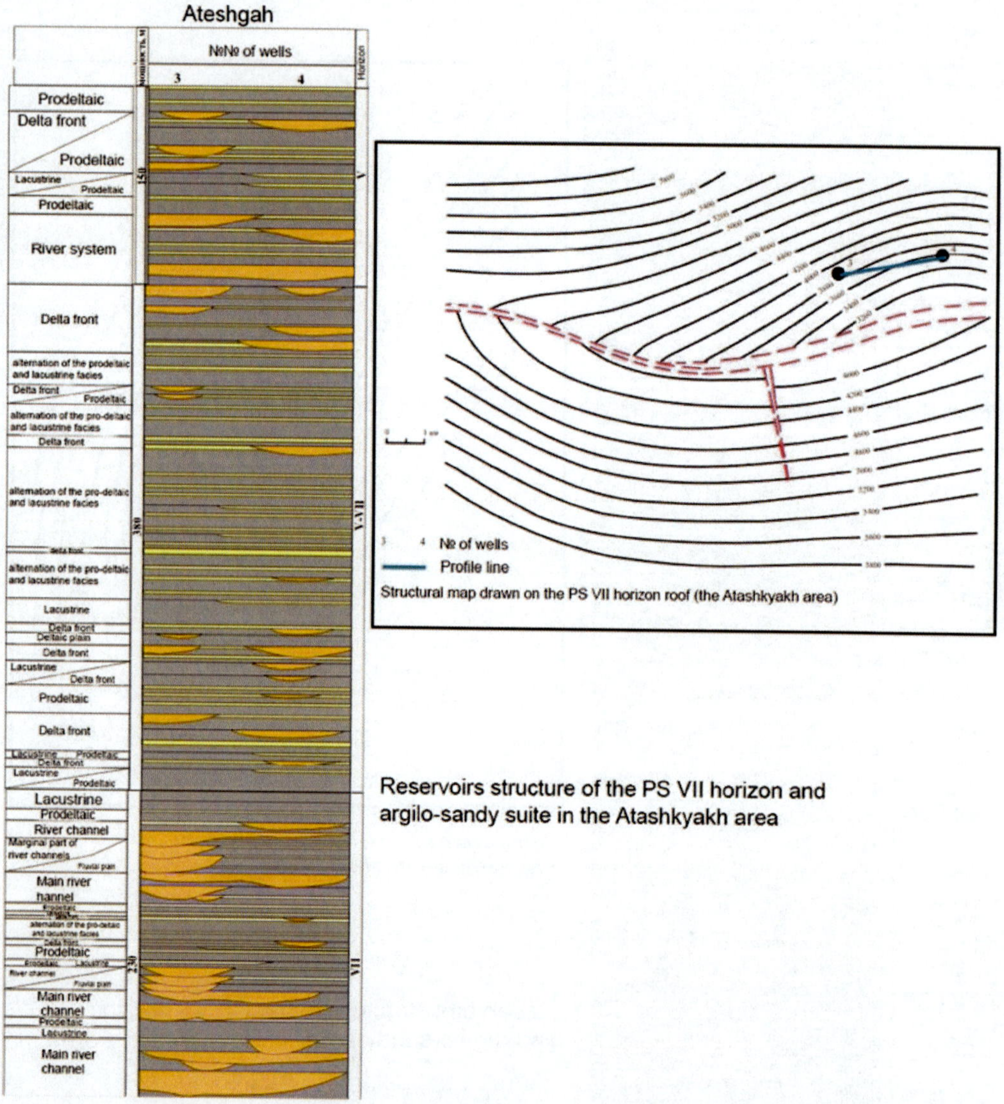

Fig. 5.17 Sedimentation conditions and reservoir architecture of the productive series in the Atashkyakh area (see legend in Fig. 5.12)

belt by a sizeable synclinal zone. The Atashkyakh area is located east of the Kur River delta, a large brachy-anticline (Alizadeh et al., 1966).

The VII horizon reservoirs architecture is characterized by all the features described for this stratigraphic complex in the Aran-Deniz area (Fig. 5.17). As was noted in other described areas, the VII horizon is also characterized by the three low sea-level stages with the development of a branched river system divided by the prodeltaic-lacustrine facies (Fig. 5.17). It is inferred that sandy bodies of fluvial origin are well-communicated vertically and horizontally. We may declare vehemently that during the Break suite formation, the paleo-Volga River system was moved far to the south up to the Atashkyakh structure.

The interval of these horizons by its reservoir structure is close to the synchronal Aran-Deniz interval. Some distinction is in the fact that during accumulation of the lower horizon, the Aran-Deniz area has been in the boundary zone between the delta front and prodeltaic facies, while the Atashkyakh has mainly been in a delta front zone, suggesting that these conditions have been moved far to the south (Fig. 5.17). Upwards the section, these conditions appear to be changed by the prodeltaic-lacustrine environment, pointing to the fact that the Atashkyakh area is in the marginal part of the prodeltaic-lacustrine zone.

As in the above-described areas, the V horizon is characterized by the development of large sandy bodies extended through all the areas that are probably connected with the river channels widespread here (Fig. 5.17). However, distinct from the Alyat-Deniz and Bulla-Deniz areas, the channel deposits in the vertical section are not bedded on each other, so they are poorly communicated. Overlapping clay partings are a barrier to the vertical migration of fluids that form the caps.

The last area studied considering the environmental conditions in the Balakhani and Break suites is the Neftchala structure. This asymmetrical branchy-anticline with a steeply dipping northeastern limb and a more gently sloping southwestern limb is confined to the northwest-southeasterly trending Kurovdag-Neftchala anticlinal belt (Alizadeh et al., 1966). Along with the areas, it was chosen to be profile-oriented, as in other studied areas, from the northwest to the southeast, crossing all the structures (Fig. 5.18).

From examining the XX horizon's reservoir architecture and facies composition, which is an analog to the Break suite, it is concluded that the fluvial facies are also developed here (Fig. 5.18). However, the geometry of sandy bodies representing the channel fillings is distinguished from that in the above-described areas. Here, they are of significantly lesser thickness. So, the depth of the river channels in the Neftchala area largely is less. The areal extent of the channels is also limited. It is also noted that the lateral connection of sandy bodies is limited, taking place only in sandy-silty bands deposited in fluvial plain. The river system has been formed mainly in the northwestern area. The profile shows the transition to the plain deltaic facies within which the few deltaic branches have been formed. Besides, the deposits of these branches are divided vertically by clay bands, while horizontally, they appear to be passed into thin sandy-silty bands.

From the reservoir structure model constructed for the XX horizon of the Neftchala area, it is inferred that the vertical communication of sandy bodies within the river facies is rather well caused by the fact that these channels in the vertical section are entrenched into each other, thereby forming a system of bedded sandy bands. In the southeastern structure, the river channels are limited and are isolated vertically, while horizontally, they appear to relate to the aid of sandy-silty bands. The changes in reservoir structure observed along the profile are caused by the transition of river facies to the deltaic plain conditions.

The formation of marginal river facies characterizes the upper section of the XX horizon sharply changed in profile by the delta front facies, which are passed to the lacustrine conditions. So, an analysis of the XX horizon's reservoir structure shows that they have not been formed by paleo-Volga as distinct from all the above-described areas but most likely by the paleo-Kur River system. It should also be pointed out that during an accumulation of the XX horizon sediments, this area has been in the marginal river system, sharply changed along the area by more distal conditions. It should be noted that the mentioned tendency is observed in the upper horizons as well.

The XIX-XVIII horizons, which may be considered the analogs to the lower horizons of the Balakhani suite, are, overall, characterized by the same environmental conditions passing over an area from the delta front to the lacustrine facies (Fig. 5.18).

Horizon XVII, an analog to the Balakhani suite's V horizon, is the transition to the proximal facies of the mainly deltaic plain. Besides, within a small interval of the section, an extended sandy band whose origin is most likely connected with the river delta advances to the basin and with short-lived conditions of the primary riverbed in

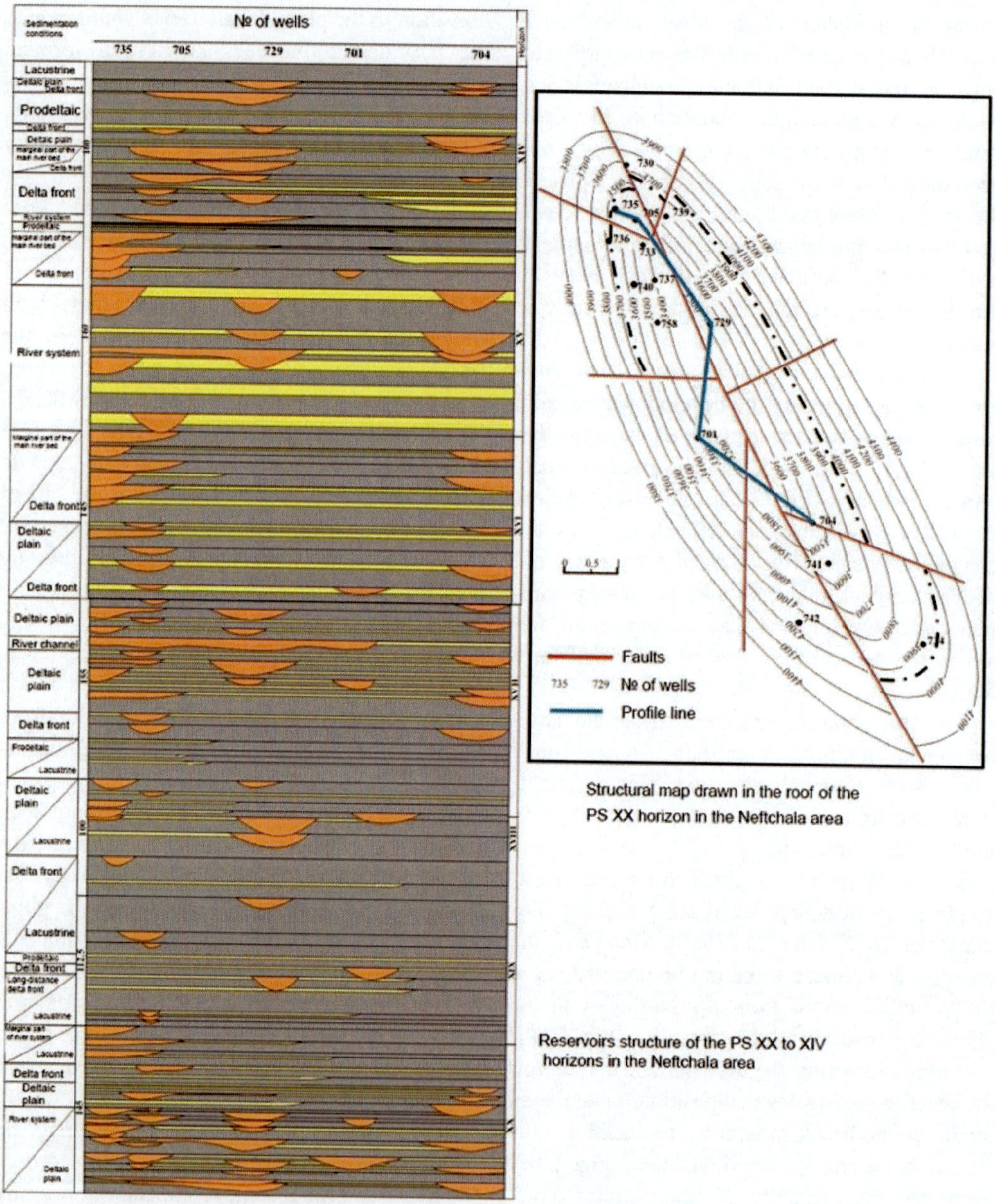

Fig. 5.18 Sedimentation conditions and PS reservoirs architecture in the Neftchala area (see legend in Fig. 5.12)

this area (Fig. 5.18). Reservoir structure is characterized by the development of isolated shallow lenses being formed due to the river channel activity. These deposits are not connected vertically, communicated laterally through the sandy-silty bands.

The plain deltaic facies are inherited in the XVII horizon, but they appear to be sharply changed in profile by the delta front conditions developed in the southeastern area (Fig. 5.18). It makes evident the paleo-Kur delta retrogradation and its advance to the west. The horizontal communication of sandy bodies here is also realized along with sandy-silty bands, which, unlike the XVII horizon, are of avandeltaic origin.

According to the sandy bodies' geometry and their arrangement in the section, the environmental conditions have been changed towards the most marginal part of the river system, which appears to be sharply changed in profile by more distal facies. This tendency of the transition to the river system facies is also observed in Horizon XV, where these sediments are developed completely (Fig. 5.18). A typical feature of this interval is an accumulation of relatively thick sandy-silty bands extended through the field and relatively deep river channels.

The reservoir structure in the Break and Balakhani suites of the Neftchala area is characterized by several features that have not been observed in other studied structures. These are: firstly—it is the nature of river channels that are shallow and non–extended in most cases. Secondly—a sharp change of environmental conditions over an area from the northwest to the southeast in the direction of the sea is practically observed in all stratigraphic intervals, especially in the transition of terrestrial facies of the river system deltaic plain to the avandeltaic conditions. Usually, such cases take place when bottom gradients are relatively high. The data obtained during the Kur River modern delta studies have confirmed this pattern, which will be shown below. Thirdly—as distinct from the paleo-Volga. According to the facial interpretation, its delta has gradually been recessed to the north (to the end of the Balakhani suite accumulation). The paleo-Kur has moved its delta to the basin, seizing more and more territory under the sphere of its influence. Such a situation may likely be explained only by the dynamics of the changes in both rivers' water balance that are directly dependent upon the climatic processes within their watersheds.

Summing up the facies mentioned above analysis of some areas within the SCB western slope, one may come to the following conclusions:

1. The Break suite and Horizon VII of the Balakhani suite formation time appeared to be a stage of the simultaneous advance of the paleo-Volga River system far to the south up to the Atashkyakh structure and paleo-Kur delta deep into a basin that apparently may be explained by paleo-Caspian significant sea-level fall.
2. In the Absheron oil-and-gas-bearing region, within the Balakhani suite and an analog to its argillic-sandy suite of the Baku Archipelago, the three following sedimentary cycles corresponding to the three stratigraphic intervals are distinguished: (a) The V to VII horizons section of the Baku Archipelago (an analog to the X and IX horizons of the Balakhani suite), (b) The V horizon (an analog to the VIII horizon of the Balakhani suite); (c) the uppermost strata of the argillic-sandy suite (an analog to the VI and V horizons of the Balakhani suite).
3. Each such cycle began with the advancement of the deltas of both paleorivers into the basin and the formation of river systems and deltaic plains on the side parts of the SCB sides and ended with the retreat of the delta systems and a change in sedimentation conditions.

Every of the mentioned cycles began from the advance of both paleo-river deltas to the basin and the formation of the deltaic plain and river system facies on the SCB slopes and ended with the regression of deltaic systems and the change of sedimentation conditions by the delta front-prodeltaic facies.

1. Beginning from the VIII horizon of the Balakhani suite, the paleo-Volga delta has progressively been recessed to the north. In contrast, the paleo-Kur delta has been moved towards the basin.
2. Against the background of those mentioned above, relatively large sedimentary cycles within the Break and Balakhani suites and no less than within all the PS suites, it is distinguished that numerous high-frequency sedimentary cycles caused by short-lived paleo-Caspian sea-level variations repeatedly led to

the progradation of the paleo-Volga and paleo-Kur deltas within the Absheron Peninsula and adjacent water area as well as in the Lower Kur depression resulted in the formation of high-order sequential boundaries.

3. The climate is essential in the Caspian Sea level's high-frequency variations. However, since the Caspian basin is an area of active tectonic manifestation, it is distinguished by its specific mechanism of development, where tectonic processes are also of great importance in forming sedimentary series.
4. The Caspian Sea-level short-lived variations in the Lower Pliocene have played a key role in infrequently changing sedimentation conditions that directly influenced reservoir formation and heterogeneity of some areas in the section. The not unimportant factor is the changes in the volume of clastic material supplied to the basin by the water arteries that influenced reservoir structure and caused their heterogeneity even within the mono-facies zones.
5. The location of some areas, such as Aran-Deniz, Atashkyakh, and Neftchala, in the marginal part of the facies zones caused the differences between their reservoir structure over an area.
6. Based on the data obtained from the Neftchala area, one may affirm that the reservoir structure in the areas located within the Kur facies zone is distinguished from those in the Absheron facies zone.

5.2 Environmental Dependence and Cyclicity of Organic Matter Accumulation in the PS Deposits

The pyrolysis results show that, overall, the organic carbon content of the PS deposits is not high. However, some authors have postulated good parental properties of PS separate intervals among their number of the Kala suite (Huseynov & Guliyev, 2004). The most significant analyses have been done on the samples from the Kirmaki suite. The maximum values have been noted in sediments taken from the middle Kirmaki suite (sample interval is 78–111) (Fig. 5.19). The average content of Corg is about 0.51%, while the limits of variations are from 0.035 to 0.7%. The maximum organic carbon contents (>0.55%) are noted in the samples No. 78, 81, 105, 108, and sedimentation conditions are interpreted as distal delta front-lacustrine facies. From the result of the facies interpretation of the Kirmaki Valley section, it is inferred that this place of the section is of the highest sea-level stage during the PS accumulation.

The next interval of the relatively high Corg contents in the studied section is between the samples from 112 to 149 taken from the same Kirmaki suite (Fig. 5.19). The Corg content limits are between 0.29 and 0.55%, and the average values are 0.4%. Sedimentation conditions here are mainly delta front facies that correspond to the gradual sea-level lowering regarding its highest level in the middle Kirmaki suite through sedimentation in somewhat distal facies. The same Corg contents are noted within an interval of samples 61–75 (Kirmaki suite), where their mean values are also equal to 0.44%, ranging between 0.28 and 0.55%. Sedimentation conditions in this interval have been changed within the limits of the lower delta plain and proximal delta front facies corresponding to the rising sea level.

The next interval of the Corg contents in decreasing order is marked in the section between the samples 153–170 (Kirmaki suite) (Fig. 5.19). Sedimentation here occurred within the delta plain facies that, in the light of prolonged-time answers to the successive sea-level lowering stage. Average Corg content is 0.27%, ranging between 0.12 and 0.39%. Furthermore, finally, the least values are noted within the intervals stratigraphically confined to the Underkirmaki suite to the lowermost strata of the Kirmaki suite and Superkirmaki sandy and

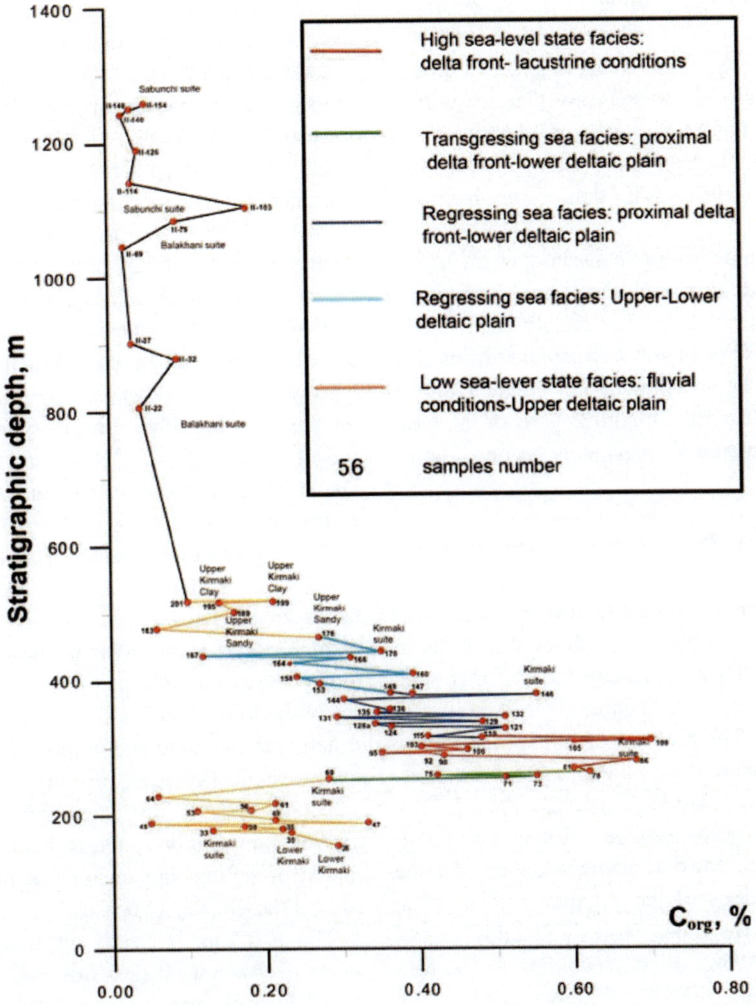

Fig. 5.19 The C_{org} content of the PS deposits

clay suites. The average values here are 0.19% and 0.18%, correspondingly. There is only one sample from the Superkirmaki suite with the Corg content equal to 0.42%.

Sedimentation occurred mostly under fluvial conditions and plain deltaic facies corresponding to the lowest sea-level stages. Besides, the Corg contents in the samples taken from the fluvial and deltaic plain are approximately equal. The samples taken from the rest suites-Balakhani and Sabunchi—have been analyzed selectively, and the Corg contents there were extremely low, mainly less than 0.1%. Thus, a specific pattern of the Corg content in the PS sediments depends on the conditions of their accumulation, where maximum Corg values are confined to the intervals of high sea-level state.

However, the PS deposits' oil-generation potential is very low, and the hydrogenous index (HI) values have reached no more than 100 except for a single sample with an HI of about 200 and one sample from the Kirmaki suite with an HI value of 585 (a sample from the sediments accumulated under lacustrine conditions). Besides, an oxygenous index is characterized by very high values. It suggests that an organic matter of a third type has been accumulated in the PS deposits, even in the sediments deposited under relatively distal delta front-lacustrine conditions.

As it is known, the PS basin is characterized by extremely low biological productivity. An organic matter balance has mainly been formed at the expense of this matter supplied from land, resulting in the accumulation of continental-type organic matter in the marginal parts of the basin during the middle Kirmaki suite accumulation.

5.3 Facies Zoning of the PS Basin

As it is known, the mineralogical composition of PS deposits in different parts of this basin is varied, essentially caused by the fact that sedimentary material is supplied from different provenances. The problem of an expanded range of the PS different facies zones has been interpreted in several works (Aliyeva, 2004a, 2004b; Khalilov, 2000; Mustafayev, 1963, 1964; Putkaradze, 1958). Most of them are dedicated to the range boundaries of the PS different facies on land except for those works (Khalilov, 2000; Putkaradze, 1958) where, according to the data obtained from the PS outcrops within the Absheron and Baku archipelagos and the results of the core material study, an approximate strike boundary of the Surakhani suite deposits of the Absheron lithofacies type is given. At the same time, delineating different fall zones, especially in the SCB marine part, is vital since it is directly influenced by reservoir prediction and the HC prospecting strategy.

Based on this factor, different lithofacies have been distinguished. This principle has been used as a basis in lithofacies delineation, which we carried out within the western SCB, mostly in its shelf zone and adjacent land area (Aliyeva, 2004b).

According to the isopleth maps data of the light fraction major components (quartz, feldspars, rock debris), the sediments of the Volzhsk (Absheron) lithofacies zone, which mainly were formed at the expense of terrigenous material have been extended far to the south up to the Nakhchivan structure during the Break suite formation (Fig. 5.20). It is in accord with the results of Davies et al. (2004). Based on some heavy mineral ratios, micro-admixtures in garnet, and light fraction composition, they concluded that the sediments of the Absheron lithofacies type within the Nakhchivan structure had accumulated during the Break suite formation. Besides, according to our data and the results of Davies et al. (2002), the Inam area is within the Kur zone.

The second zone clearly distinguished in the Break suite is, in our view, the Gobustan lithofacies zone, characterized by the predominance of rock debris in the light fraction composition (up to 80%) (see Chap. 4). It trends from the northwest to the southeast, i.e., from the Pirsagat structure to the north and northwest and encloses southwestern Gobustan territory. Such a trend reflecting rock debris decreasing in the light fraction composition suggests that the significant source of sedimentary material supplied to this zone is the Greater Caucasus.

The next zone is clearly distinguished by the minor quartz content (less than 8%) instead of a high feldspar content (up to 30%) and the presence of rock debris (about 30%). This zone encloses the upper Kur depression and the adjacent part of the Baku Archipelago. As it is known, a large quantity of feldspars has been supplied to the basin by the paleo-Kur River system, forming its watershed area at the expense of the Lesser and Greater Caucasian rivers. This fact likely explains well approximately equal quantities of feldspar and rock debris supplied from the Lesser and Greater Caucasus correspondingly.

We interpreted this zone as the Kur lithofacies propagation area. According to the feldspar's distribution map, the source from which they

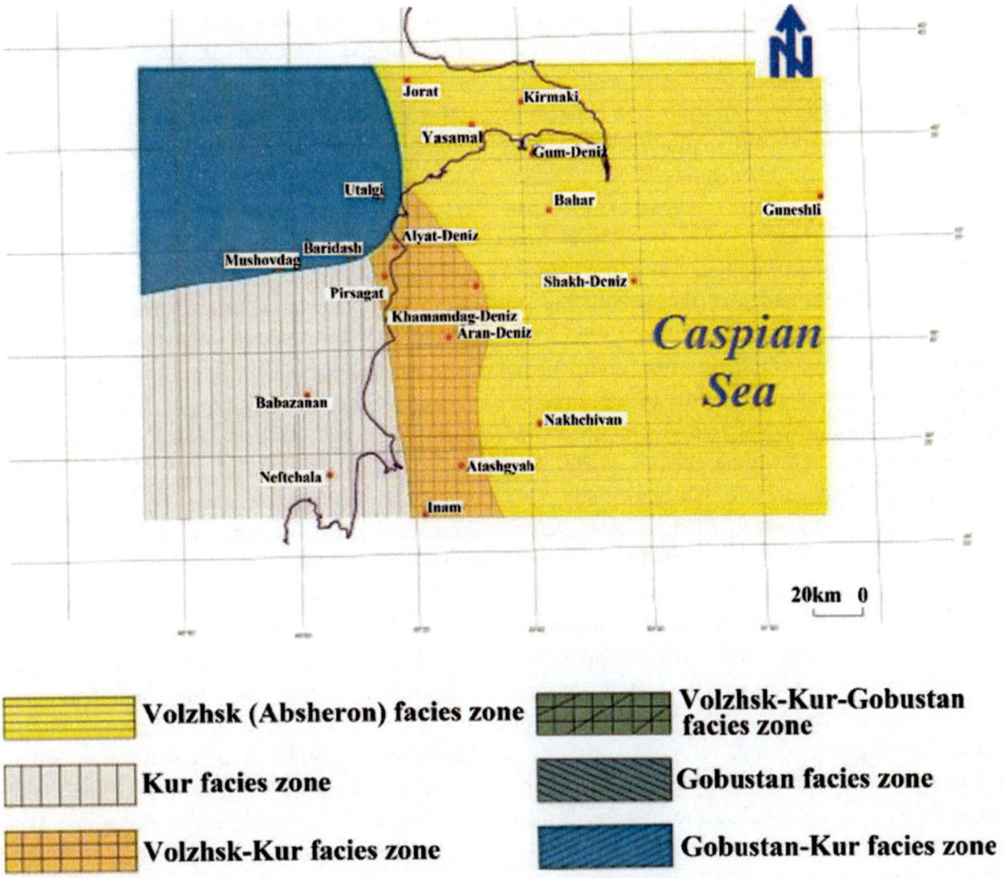

Fig. 5.20 Schematic distribution of the facies zones in the Break suite deposits

have been supplied to the basin is located southwest, i.e., it corresponds to the Lesser and Greater Caucasus disposition. The effusive fragments in the Kur lithofacies zone rock debris also suggest that the clastic material has been supplied from the regions mentioned above (Abdullayev et al., 2011; Davies et al., 2002).

As we see it, the one more clearly distinguished zone is a mixed Absheron-Kur lithofacies zone, which is noteworthy by its approximately equal contents of all the light fraction components, such as the quartz content varies within the limits of 25–40%; rock debris—25–35%; feldspars are of 20–40%. This is a region of the location of Bulla-Deniz, Pirsagat, Duvanni, Aran-Deniz, Yanan-Tava, and Atashkyakh structures. The Volzhsk facies delineation within the sediment's accumulation basin of the Balakhani suite has been carried out using the quartz and disthene distribution manner in the rocks of this suite, according to which its southern boundary is drawn farther south of the Shakhdeniz area.

The Gobustan lithofacies region is slightly reduced, covering the small territory of the Mishovdag structure and farther north of it than during the Break suite accumulation stage.

As it was mentioned above, this region is fixed according to the rock debris' high contents exceeding those of feldspar and quartz. The rock debris content of the Balakhani suite deposits is more than 60% on average, somewhat lower than those of the Break suite. At the same time, the Kur lithofacies zone propagation area appears to be increased insignificantly, moving forward in the direction of the Hamamdag-Deniz, Yanan-Tava, Byandovan-Deniz, Mugan-Deniz, and Inam structures. Here, as in the Break suite sediments, the quartz content does not exceed 8%, while feldspar and rock debris contents are approximately equal, being about 30%. The amount of the light fraction rest components seems to be increased. The area of the Volzhsk (Absheron)-Kur mixed sediments becomes wider-essentially. This zone extends in the form of a wedge between the Absheron and Kur lithofacies zones occupying Pirsagat, Alat-Deniz, Bulla-Deniz, Sangi-Mugan, Aran-Deniz Sabail, Nakhchivan structures locality. From about equal contents of all the light fraction components (quartz and rock debris about 30%; feldspars and the rest components about 20%), it is inferred that this zone has been under the influence of paleo-Kur and paleo-Volga.

The above-mentioned compositional relationships have been reflected in the lithofacial map compiled by the authors for deposits of the Balakhani suite in western SCB. The transition to the Sabunchi suite has been accompanied by the preservation of translational retrogradation of the paleo-Volga delta to the north and the widening of the Kur lithofacies zone to the east and north. The Sabail, Aran-Deniz, Chigil, and Umid areas were also within the Kur lithofacies zone (see Chap. 4). As was shown above, the same results have been obtained from the facies analysis data of the PS deposits in some SCB areas. We came to a similar conclusion based on widening a zone of the highest feldspar content, about 25% on average. Typically, the amounts of rock debris and quartz appear to be increased by an average of 56% and 19%, respectively.

The Absheron (Volzhsk) lithofacies' southern boundary was located slightly south of the Bakhar area. The quartz and disthene distribution data confirm that. We suggest that the contours of the Gobustan lithofacies' spreading zone are also changed. This zone is recessed in the west and, simultaneously, is moved southeasterly towards the basin. This conclusion is based on anomalous high rock debris content (88.8%) and low quartz (9%) and feldspar (2.2%) contents in the Sangi-Mugan area.

A similar tendency is preserved in the Surakhani suite, which serves as one more confirmation of this conclusion. The rock debris content in the Absheron (Volzhsk) lithofacies zone during Sabunchi time is stable, 77% on average, while the quartz and feldspar content does not exceed 10–12% (Fig. 5.21).

A typical feature of the Sabunchi suite deposits is the significant widening of the mixed lithofacies zone. It applies not only to the Volzhsk-Kur lithofacies but also to a mixing zone of the Kur and Gobustan sediments (Fig. 5.22). The latter is located within the Klich and Ragim structures, whose mineralogical composition differs from the other lithofacies zones. On a level with a high rock debris content that serves as the fact of the Gobustan lithofacies effect, it is noted an increase in feldspar content of up to 30% and quartz content of about 10%. No similar composition is typical to any of the above-described zones.

The paleo-Kur runoff zone widening is displayed, in our opinion, not only in an easterly advance of the Kur lithofacies itself and a mixed Volzhsk-Kur lithofacies to the north but also in the origin of a mixed zone of sedimentary material supplied from all the three primary sources, that is, from Russian platform, the Greater and Lesser Caucasus. We interpreted this zone as a mixed Absheron (Volzhsk)-Gobustan-Kur lithofacies zone, which relates to the river runoff effect from the Greater Caucasus and paleo-Kur. The light fraction composition here is characterized by a high rock debris content of 50% on average, distinguishing from that in the deposits of a mixed Volzhsk-Kur lithofacies zone and suggesting an essential influence of the Gobustan sediments. The quartz and feldspar percentage is also relatively high, about 25% on

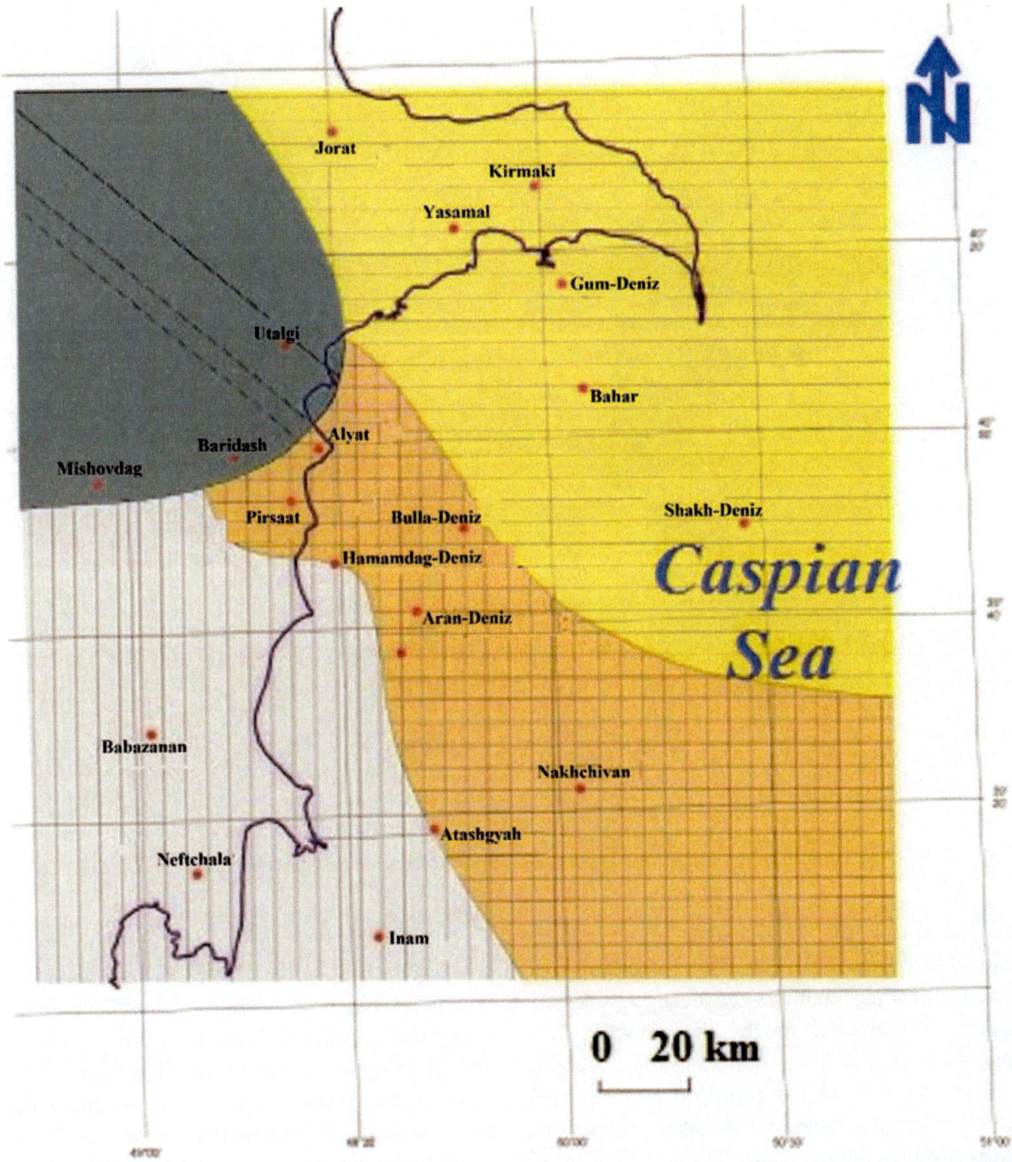

Fig. 5.21 Distribution of the facial zones in the Balakhani suite deposits (see legend in Fig. 5.20)

average, which indicates the simultaneous effect of the Kur and Volzhsk runoffs.

The given lithofacies zone extends from the northwest to the south and southeast, delineating all three lithofacies zones, the Volzhsk, Gobustan, and Kur zones, and enclosing the following areas: Pirsagat, Hamam-Deniz, Alyat, Alyat-Deniz, and Karadag.

The same main features characterize the Surakhani suite as those in the Sabunchi suite. From the quartz and disthene distribution data in the Surakhani suite, it is inferred that the Volzhsk lithofacies zone continues its recession to the north (Fig. 5.23).

The narrowed Gobustan lithofacies zone is moved towards the basin, which appears to be

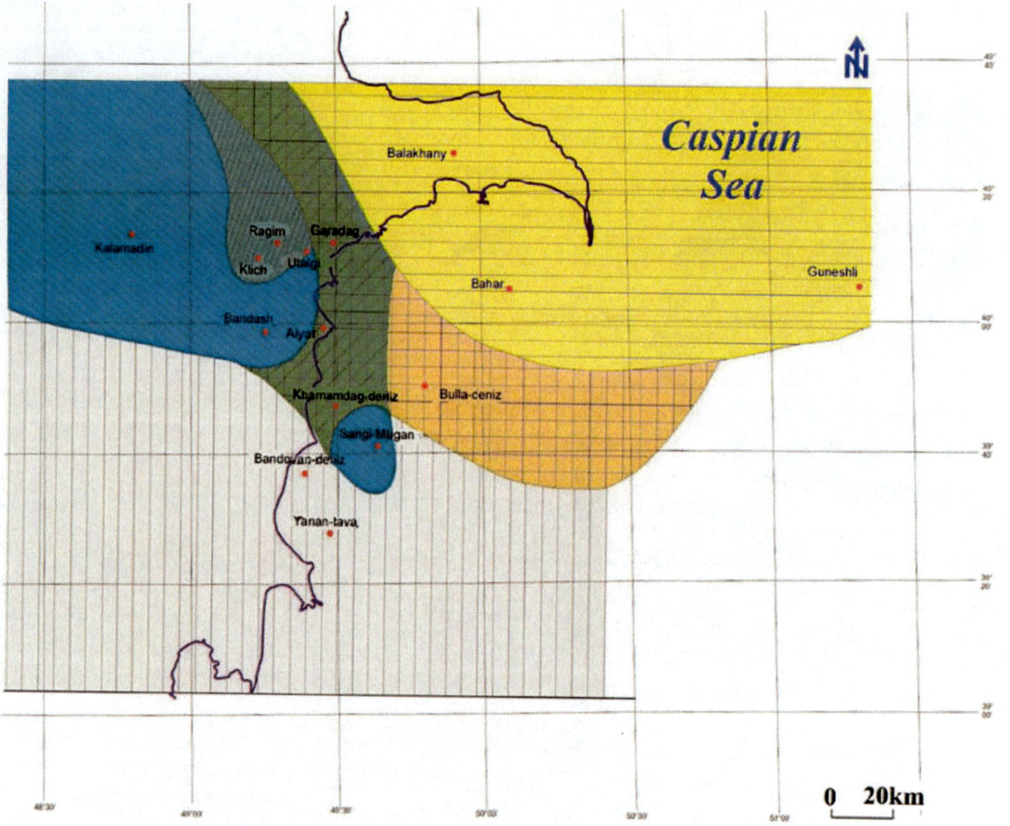

Fig. 5.22 Schematic map of the facial zones' distribution in the Sabunchi suite deposits (see legend in Fig. 5.20)

displaced from the Sangi-Mugan structure to the Sabail one (see Chap. 4). The paleo-Kur sediment spreading contours are changed unessentially. Furthermore, finally, the Volzhsk, Kur, and Gobustan sediments mixing zone becomes wider. Apart from the Hamamdag-Deniz area, this advanced southerly zone also encloses the Sangi-Mugan area.

According to the logging curves data, dominating facies within the Pirsagat structure have been the Volga fluvial plain facies, possibly trending south. However, the width of the fluvial plain development zone is essentially decreased due to the Kur River runoff effect and its deltaic formations spreading. The Kur runoff effect has been perceived both in a marginal part of the Volzhsk riverbed and within the floodplain facies since they are in a zone of both types of sediments—of Volzhsk and Kur types.

The lithofacies zonation carried out jointly with an interpretation of the PS sedimentation conditions has been taken as a basis for delineating different genetic sedimentation complexes and developing a series of facies maps by stratigraphic section of the Lower Pliocene of the western SCB. As mentioned above, the Break suite formation time is a stage of the critical advance of both river systems- paleo-Volga and paleo-Kur towards the basin (Fig. 5.24).

Besides, the paleo-Volga sediments have been spread far to the south up to the Atashkyakh structure and maybe some further covering a zone of the Kur and Absheron sediments mixing. Furthermore, since fluvial deposits have been formed in the Atashkyakh area, this may serve as grounds to predict the paleo-Volga more distal facies spreading further to the south up to the Alov and Araz structures.

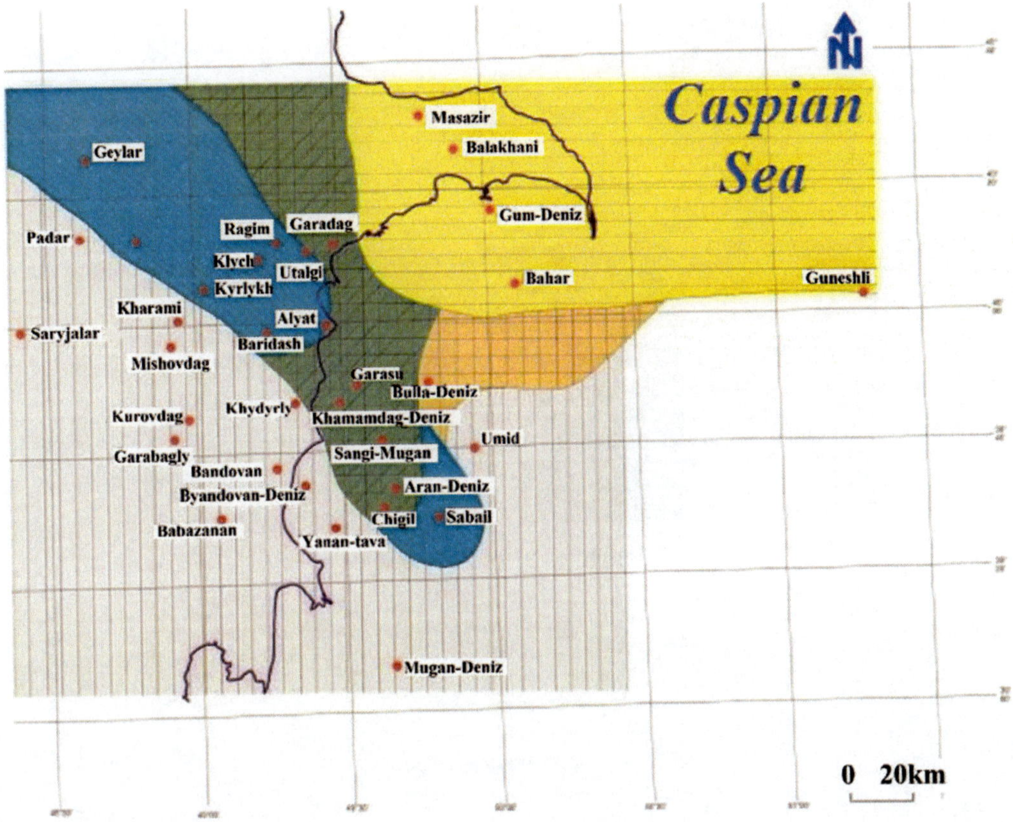

Fig. 5.23 Schematic map of the facial zones' distribution in the Surakhani suite deposits (see legend in Fig. 5.20)

Notably, in the Kur lithofacial zone, the fluvial facies appear to be advanced up to the Neftchala area, where deltaic deposits change them. By its dimensions, the paleo-Kur delta as of the present Kur River has not been as large as those of the paleo-Volga reflected in the Kur sediments areal extent. We predict the spreading of the Kur delta front facies up to the boundaries of the Kur sediments spreading zone delineated according to the mineral composition data.

The areal extent maps of facies zones in the Balakhani suite deposits and its analog, argillic-sandy suite of the Baku Archipelago, have been compiled for the three described sedimentary complexes formed during the following three complete sedimentary cycles:

- Horizons X and IX of the Balakhani suite (an analog to V-VII horizons' interval of the Baku Archipelago),
- Horizon VIII of the Balakhani suite (an analog to the V horizon of the Baku Archipelago),
- Horizons VI and V of the Balakhani suite (an analog of the uppermost argillic-sandy suite of the Baku Archipelago).

Facies maps have been compiled for the deposits of separate stages in each cycle, i.e., sedimentary complexes accumulated during paleo-Caspian low and high-level stages. The Balakhani age begins from the paleo-Volga system recession to the north (Fig. 5.25). Fluvial facies are developed only on the X horizon

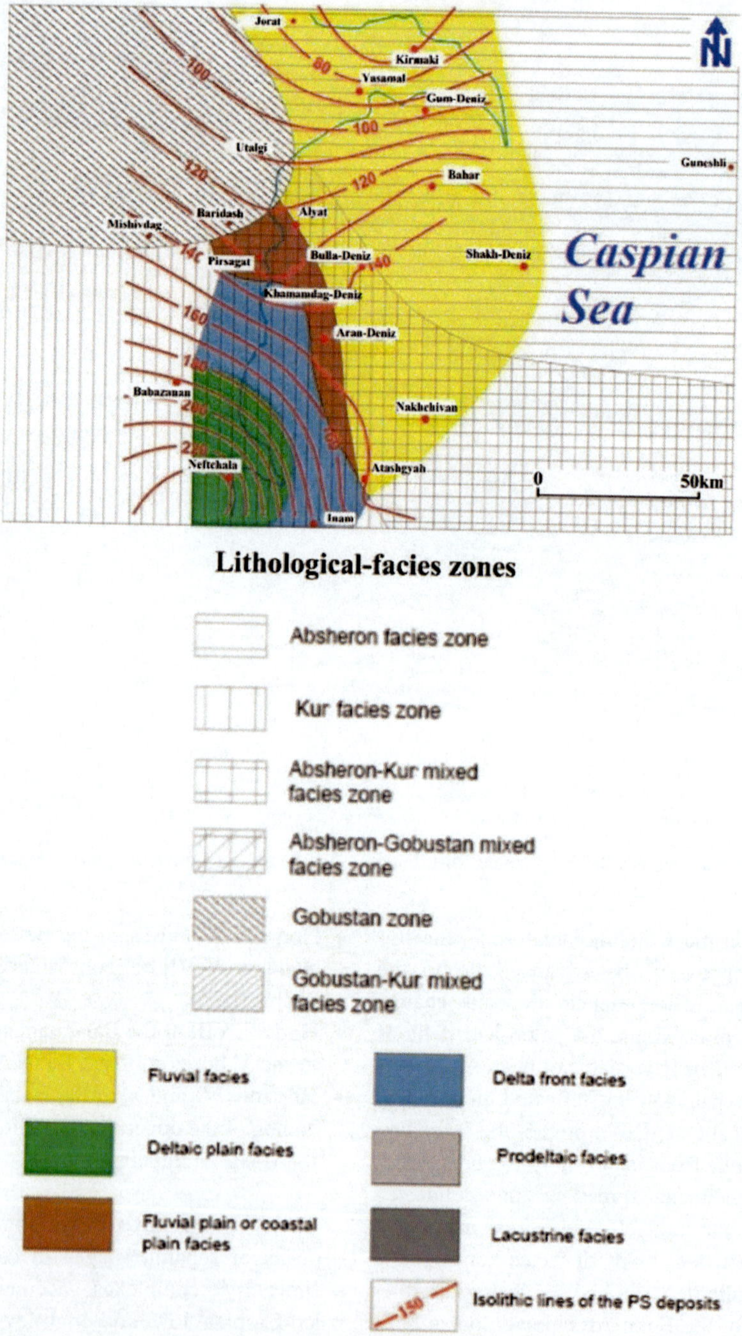

Fig. 5.24 Geological sketch map of the SCB's sedimentation conditions during the Break suite accumulation

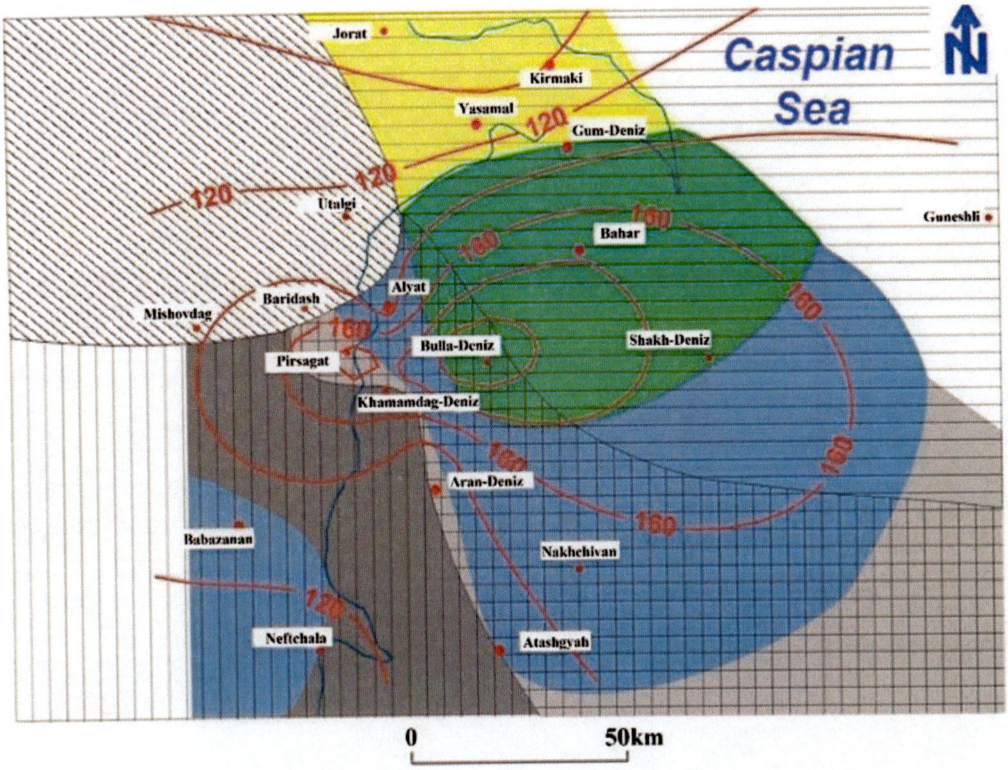

Fig. 5.25 Sketch map of the sedimentation conditions in the SCB during accumulating the Balakhani suite X horizon (see legend in Fig. 5.24)

of the Balakhani suite within the Absheron Peninsula. In our opinion, the Bakhar area dominated there have been terrestrial-deltaic sedimentation conditions extended up to the Shakh-Deniz area. However, because of the absence of available data from this area, it may not be drawn to outline the southern boundary of the paleo-Volga deltaic plain.

In the southwest, according to the facies interpretation of the V-VII horizons, the interval in the Bulla-Deniz area, a boundary between the deltaic plain and delta front, has been located precisely at this place. Numerous transitions from the terrestrial-deltaic to the avandeltaic conditions are observed along with this structure.

The sediments accumulated under paleo-Volga delta front conditions have been extended far to the south, replacing fluvial deposits of the Break suite on the Atashkyakh structure. The paleo-Volga delta front settings exist within all the mixing zones of the Absheron and Kur sediments. Their change to the prodeltaic facies took place in the Pirsagat and Aran-Deniz regions. However, as the Kur lithofacies zone has advanced to the east, the spreading area of the mentioned facies conditions in the western Absheron lithofacial zone has vastly decreased.

The facies' sharp change from the prodeltaic to the lacustrine deposits, under paleo-Volga runoff influence, occurs around the Aran-Deniz

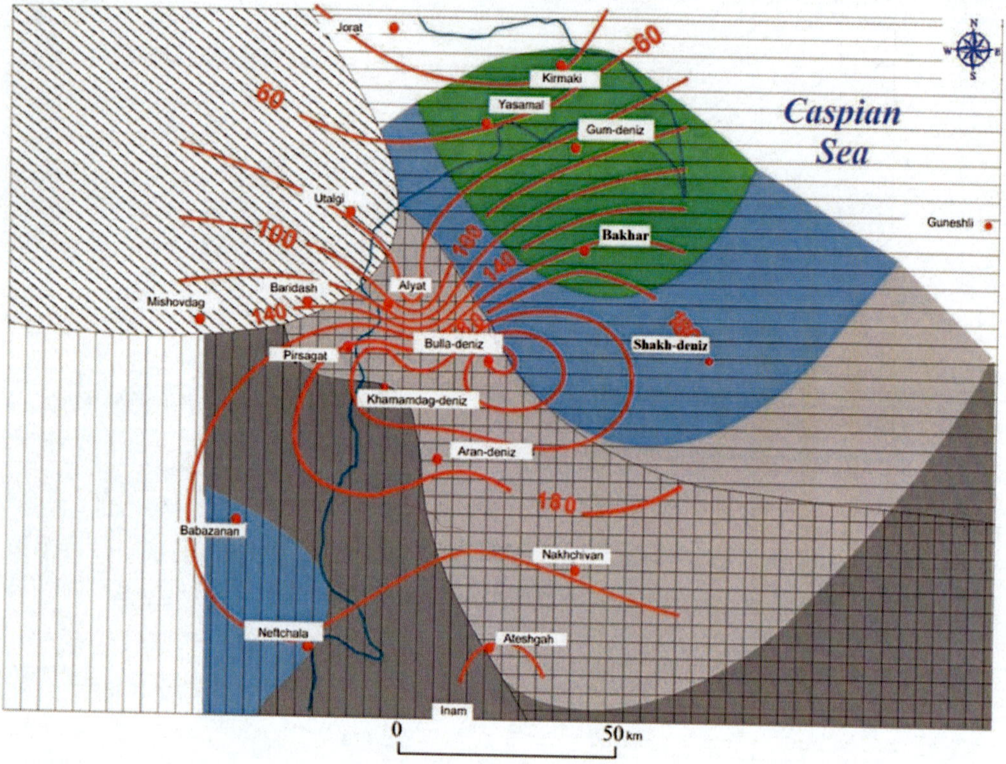

Fig. 5.26 Sketch-map of the sedimentation conditions in the SCB during the IX horizon accumulation in the Balakhani suite (see legend in Fig. 5.24)

structure. The lacustrine facies' eastern boundary passes along the Kur lithofacies boundary. At the beginning of the Balakhani age, the paleo-Volga retrogradation coincided with the paleo-Kur recession to the west. The Neftchala area at that time was in the border zone between the delta front and lacustrine facies.

In the upper half of the first sedimentary cycle, corresponding in our view to the IX horizon of the Absheron lithofacies zone and the V-VII horizon's interval, the deposits of the high sea-level stage have been formed. Within the Absheron Peninsula—in the Kirmaki and Yasamal valleys and Bakhar-terrestrial delta facies are observed passing farther south to delta front and prodelta conditions. The latter is near the Bulla-Deniz and Pirsagat areas (Fig. 5.26).

The Aran-Deniz and Atashkyakh structures have been located on the boundary of the transition of prodeltaic into lacustrine facies. Thus, during the X horizon sediment accumulation, the paleo-Volga delta has been displaced to the north and covered the Absheron Peninsula. This eastern delta boundary has been in the Neftchala structure area as in the sedimentary cycle's lower half. No movement dynamics of the paleo-Kur delta are observed within the Kur lithofacies zone.

The beginning of the V horizon of the Baku Archipelago and its analog—VIII horizon of the Balakhani suite—within the Absheron lithofacies zone has been marked by the beginning of a new sedimentary cycle. For the first time after the Break suite formation, the paleo-Volga River

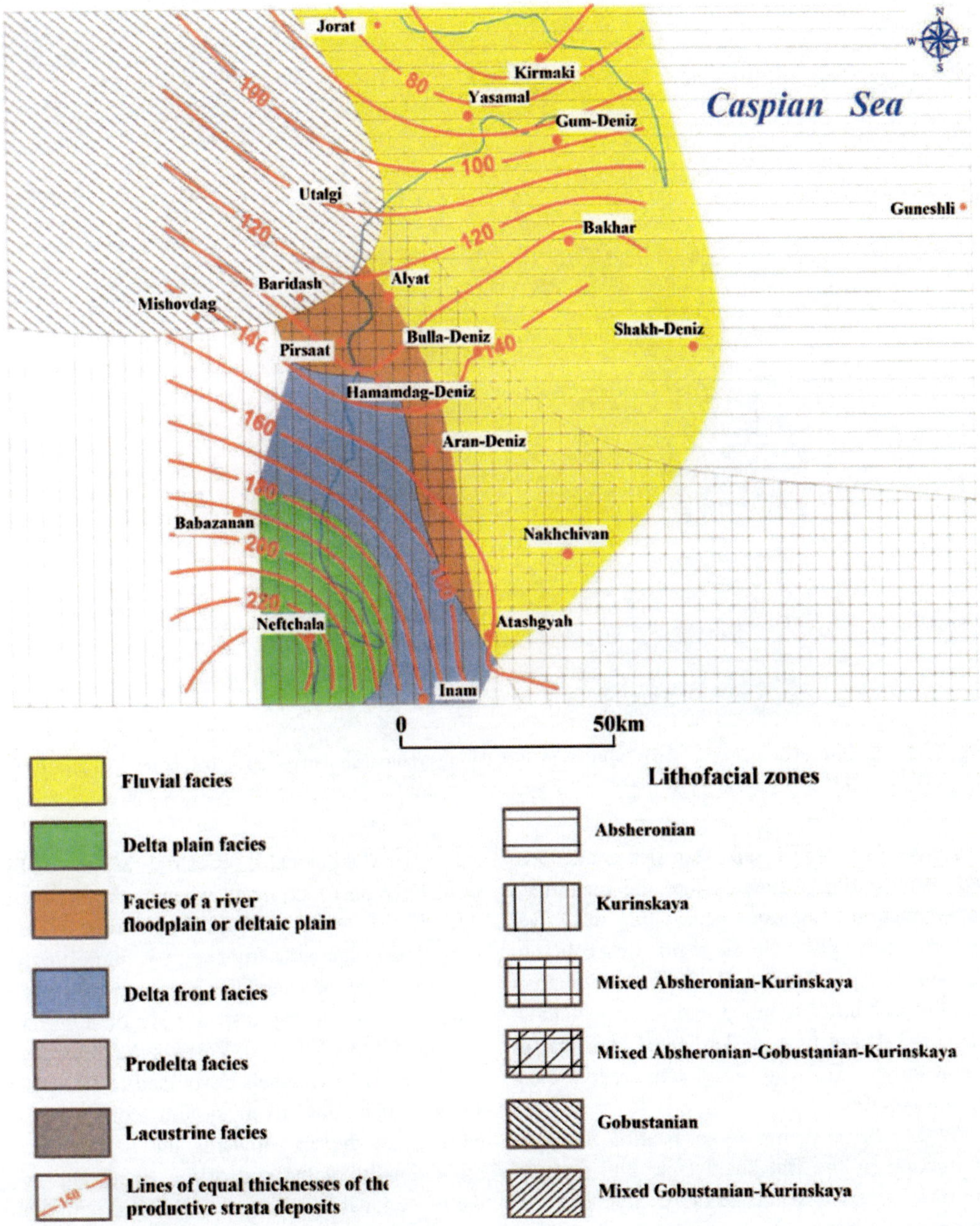

Fig. 5.27 Sketch-map of the sedimentation conditions in the SCB during accumulation of the VIII horizon lower half in the Balakhani suite (see legend in Fig. 5.24)

system advances far to the south up to the Atashkyakh area and has been marked again (Fig. 5.27).

However, the width of a zone of fluvial sediment development compared with the Break suite is slightly decreased, trending to the east up to the Kur lithofacies boundary. It is noted that between actual paleo-Volga channel formations and the Kur zone, there is a zone of paleo-Volga flood-plain conditions within which such

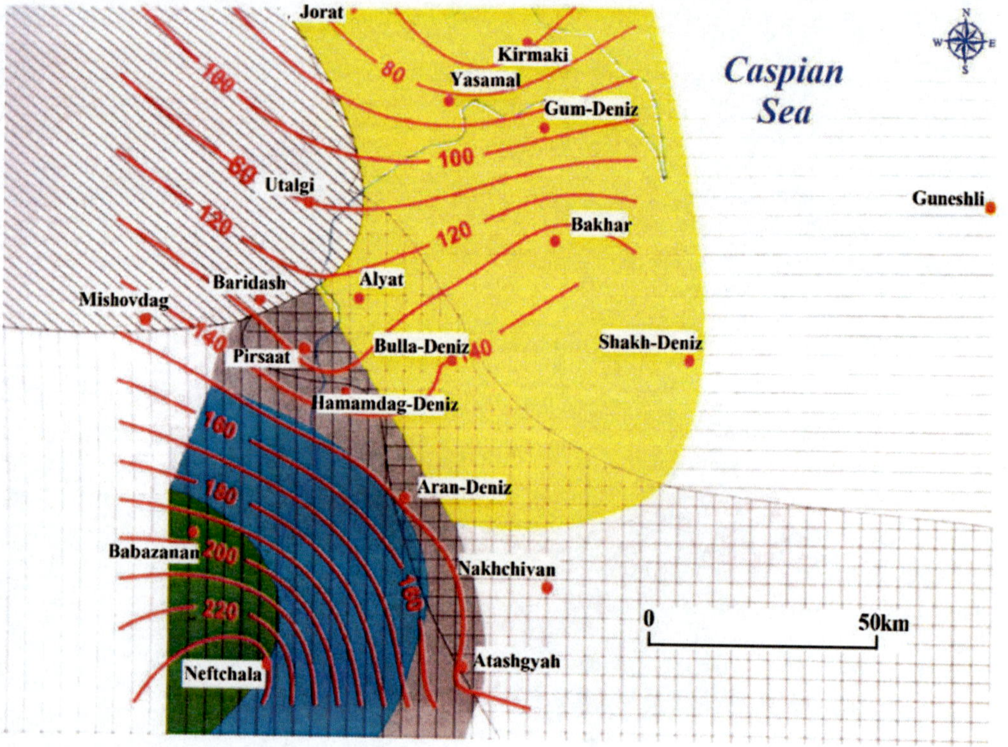

Fig. 5.28 Sketch-map of sedimentation conditions in SCB during accumulation of the upper half of the VIII horizon of the Balakhani suite (see legend in Fig. 5.24)

structures as Pirsagat and Aran-Deniz are located. An exciting feature is that the mentioned sedimentation conditions and partly the river system facies existent in the region of Absheron-Kur mixed type of sediments. It points to the fact that the amount of supplied sedimentary material of the Absheron type predominated over that of the Kur type. However, the latter's effect must not be ignored.

Within the Absheron oil/gas bearing zone, the V horizon of the Balakhani suite and its synchronous interval in the Baku Archipelago corresponded to the middle part of the upper half of the argillic-sandy suite are characterized by the continuing recession of paleo-Volga delta and widening of the Kur sediments spreading limits (Fig. 5.29). Fluvial deposits have accumulated only in the Absheron Peninsula region. The predominance of terrestrial-deltaic conditions has characterized the Bakhar area.

At the Bakhar field, terrestrial-deltaic conditions have been dominant, whereas the Pirsagat, Aran-Deniz, and Atashkyakh areas were within the paleo-Volga delta front setting. Interestingly, the boundary between the Kur and Absheron lithofacies zones was also a zone of a combination of both Kur and Volga delta front systems. The river channels developed at Neftchala suggest that this river system continues its advance to the east towards the sea. Furthermore, finally, at the end of the Balakhani age, the most distal facies were established within all the studied territories. Most of the Baku Archipelago is covered with lacustrine sediments passing into avandeltaic and terrestrial conditions toward the basin side (Fig. 5.30). This sedimentary complex may be considered a series of high sea-level stage deposits. Constructed thickness maps confirm the results of facies analyses.

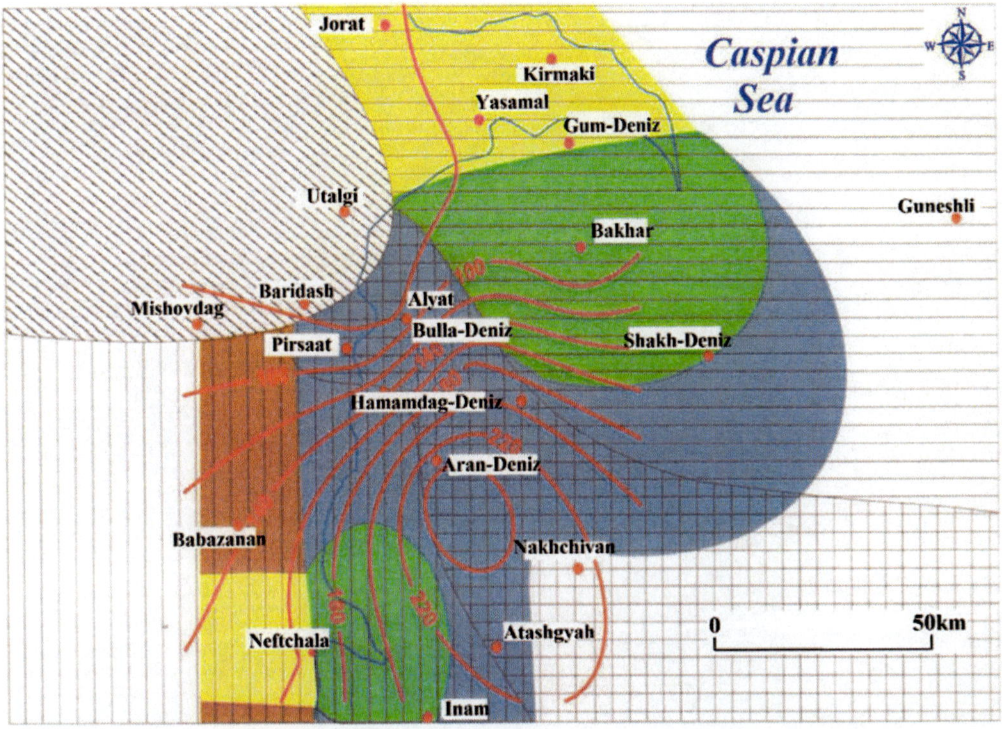

Fig. 5.29 Sketch-map of the sedimentation conditions in the SCB during accumulation of the upper half of the VI horizon in the Balakhani suite (see legend in Fig. 5.24)

Apparently, during the stage corresponding with the paleo-Volga River system advance towards the central paleo-Caspian (Break suite and Horizon VIII of the Balakhani suite), the thicknesses of the mentioned deposits are increased in the south-southeasterly direction (Figs. 5.24, 5.27 and 5.28). During the paleo-Volga delta recession, a depocenter has accordingly been displaced to the north, gradually covering a region of the Baku Archipelago (Figs. 5.25, 5.26, 5.29 and 5.30).

The paleogeographic map gives a true notion of the PS basin at the end of the Pliocene and of the location of deltas of the three main rivers flowing into the paleo-Caspian western shelf (Fig. 5.31). An analysis carried out shows that the facies composition of the PS deposits in the western SCB is extraordinarily complex and characterized by frequent transitions of one facies zone into others and an intersection of the spheres of influence of such large appliers of sedimentary material as paleo-Volga and paleo-Kur.

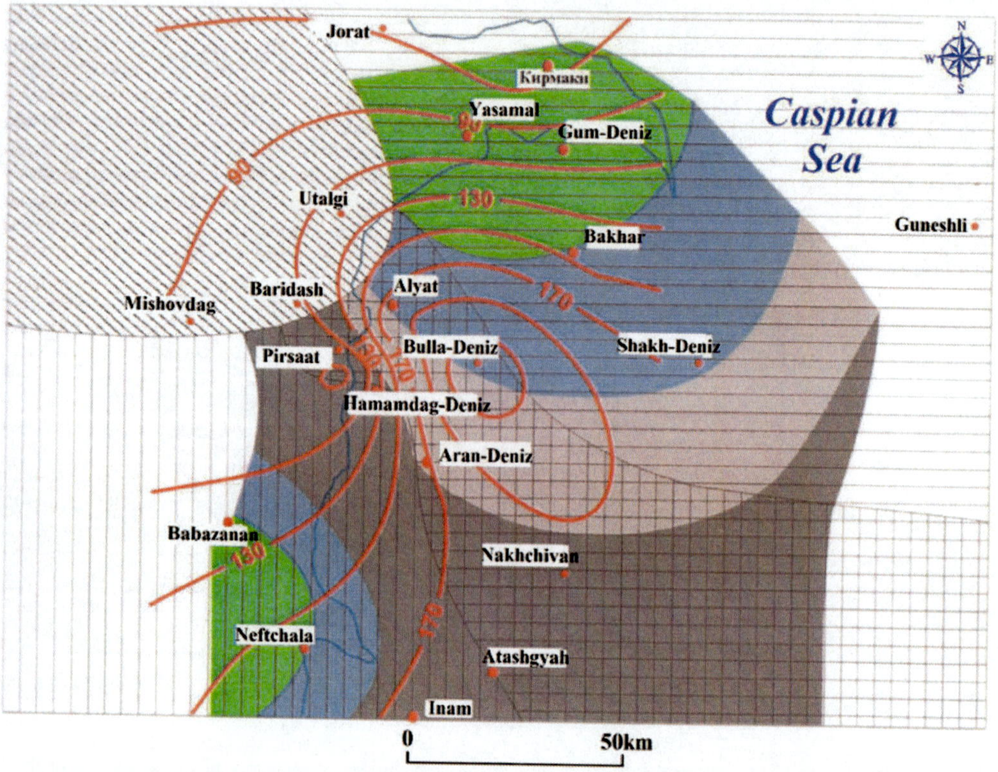

Fig. 5.30 Sketch-map of the sedimentation conditions during the accumulation of the upper half of the V horizon in the Balakhani suite (see legend in Fig. 5.24)

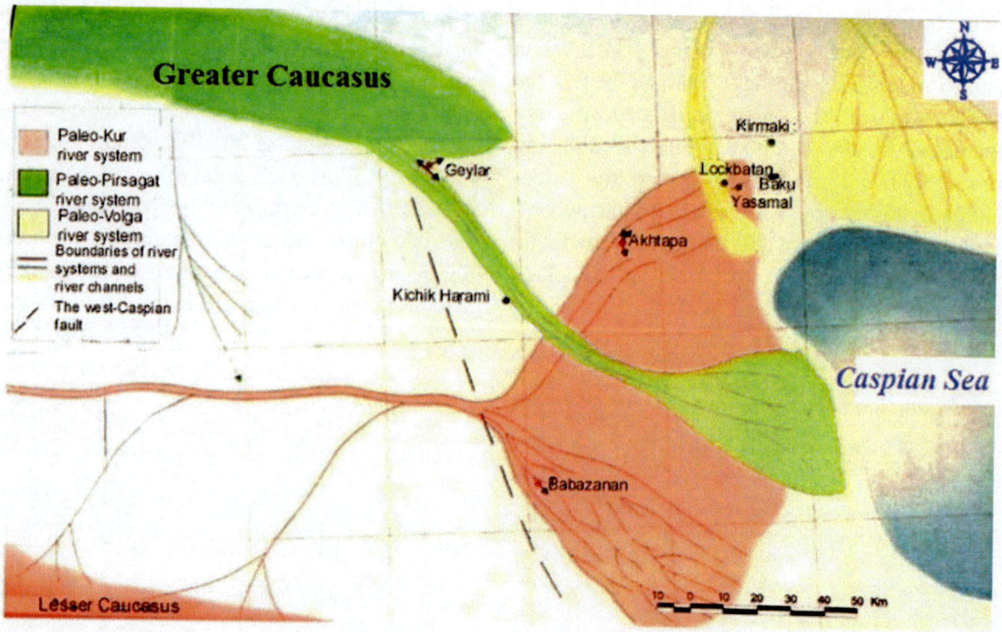

Fig. 5.31 A paleogeographic map of the SCB during the Surakhani suite accumulation

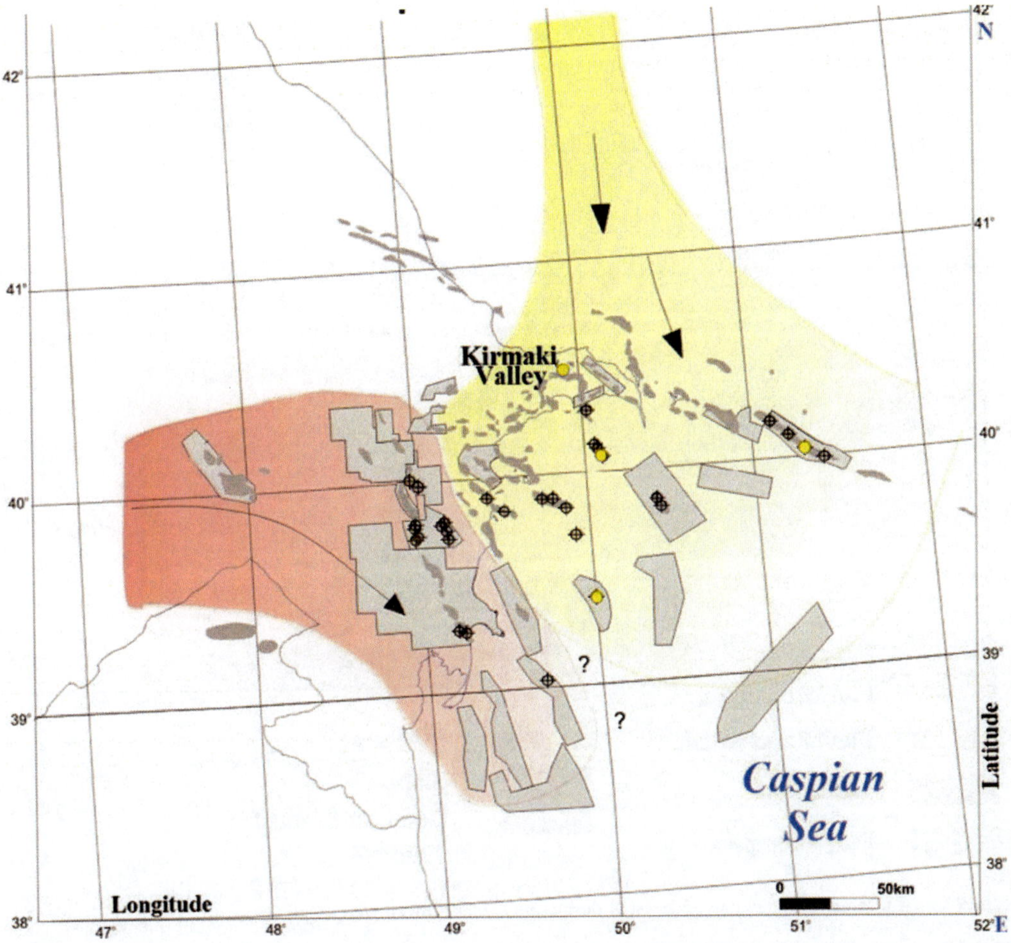

Fig. 5.32 A paleogeographic map of the SCB during the PS lower division accumulation

The PS upper division detailed facial analysis was carried out based on the abundant drilling, core, and log data. The lower division of the PS is also of some interest, especially considering its significance in subsequent prospecting works in SCB. However, the lack of geological material makes it unable to examine the problem of these deposit accumulations in detail as it was done for the PS upper division. Up-to-date isolated data on such areas as Shakh-Deniz, Zafar, Mashal, and Nakhchivan made it possible to construct a generalized map of sedimentation conditions in the lower division without detailed data on individual stratigraphic complexes (Fig. 5.32).

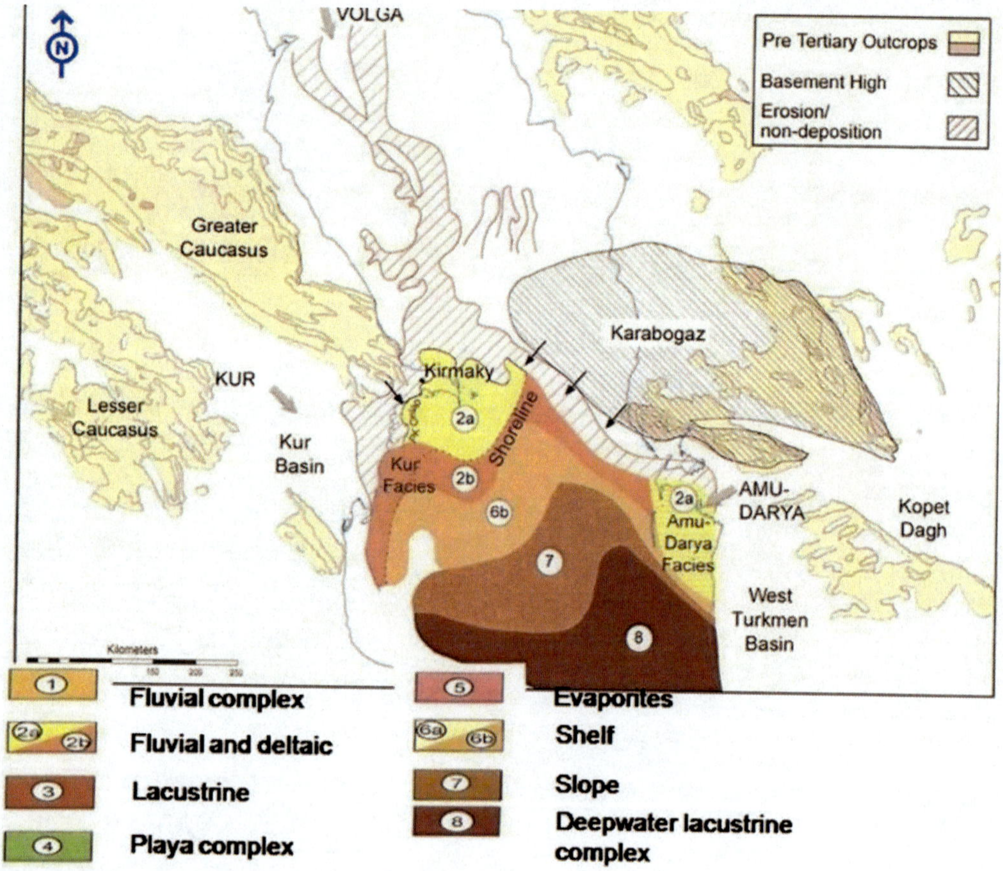

Fig. 5.33 A paleogeographic map of the SCB during the Underkirmaki suite accumulation according to seismic data (after Abdullayev, 2015)

As it is seen from the SCB paleogeographic map, the paleo-Volga delta was a major supplier of sedimentary material to the SCB during the lower division accumulation that is noted by quartz and other typical "Volzhsk" minerals with high content in the above-mentioned southern structures—Shakh-Deniz, Zafar, Mashal. The paleo-Kur effect was limited in the marine part of the SCB along the strand. This fact is essential for predicting the quality of reservoirs in SCB's southern unexplored structures.

An attempt to construct paleogeographic maps of some PS sections based on seismic data has been undertaken by Abdullayev (2015). Considering that a degree of seismic data settling does not allow us to carry out an interpretation with a high degree of accuracy, the maps mentioned are approximate (Figs. 5.33 and 5.34). Nonetheless, these maps also evidenced that the paleo-Volga played a dominant role in the sedimentation processes of the PS lower divisions.

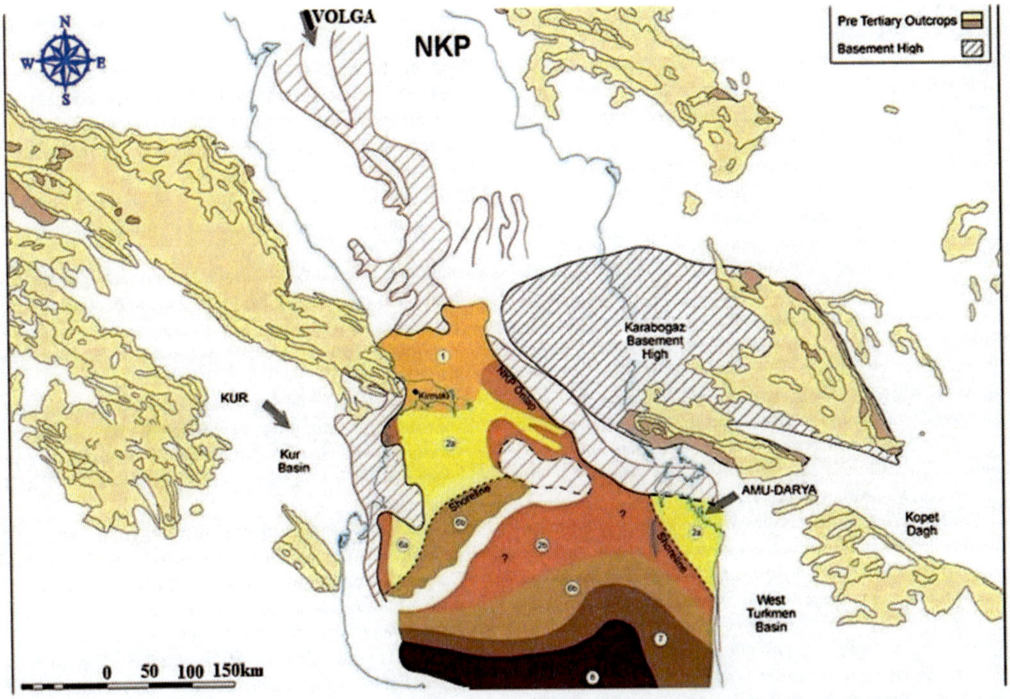

Fig. 5.34 A paleogeographic map of the SCB during the SKS accumulation according to the seismic data

References

Abdullayev, N., Riley, G., & Bowman, E. (2011). Sedimentation history of the productive sequence in the South Caspian considering basin subsidence. Part 1. *Azerbaijan Oil Industry, 5*, 20–30.

Abdullayev, N., Weber, J., Aliyeva, E., Leslie, C., Riley, G., O'Sullivan, P., van Baak, C. G. C., & Kislitsiyn, R. (2018). Detrital zircon and apatite constraints on depositional ages, sedimentation rates and provenance: Pliocene productive series, South Caspian Basin, Azerbaijan. *Basin Research, 30*(5), 835–862.

Abdullayev, T., et al. (1998). A reservoir model for the main Pliocene reservoirs of the Bahar field in the Caspian sea, Azerbaijan. *Petroleum Geoscience, 4*(3), 259–270.

Abdullayev, T. I. (2001). Stratigraphic architecture of horizon VIII of the Balakhani formation of the Bakhar deposit. Azerbaijan's geologist. *Scientific Bulletin, 6*, 66–75 (in Russian).

Abdullayev N. (2015). Evolution of the sedimentary cover and assessment of the prospects of the SCB. PhD Thesis, *Inst. of Geology and Geophysics, Azerbaijan Academy of Sciences*, Baku, Azerbaijan.

Agabekov, M. G. (1963). *Geological structure of Azerbaijan oil fields and their formation* (274 p). Azerneshr (in Russian).

Aliyev, A. G. (1947). *Petrography of the productive strata of Kabristan* (155 p). Academy of Science of Azerbaijan (in Russian).

Aliyev, A. G. (1949). *Petrography of the tertiary deposits of Azerbaijan* (311 p). Azneftizdat (in Russian).

Aliyeva, E. (2004a). Depositional environment and architecture of productive series reservoirs in the South Caspian basin. In: *South Caspian basin: Geology, geophysics, oil-gas content. Special issue devoted to 32nd international geological congress in Florence, Italy* (pp. 19–32). Naftha-Press.

Aliyeva, E. (2004b). Lithofacies zonation of Lower Pliocene productive series sediments, South Caspian basin. In *Proceedings of the AAPG Hedberg conference "sandstone deposition in Lacustrine environments: Implications for exploration and reservoir development"*. Baku, Azerbaijan (pp. 72–74).

Aliyeva, E. G. (2008). Sedimentation conditions, cyclicity, and architecture of reservoirs. In: *Geology of Azerbaijan*. Vol. VIII: "Oil and gas" (pp. 159–192) (in Russian).

Alizade, A. A., Akhmedov, G. A., Akhmedov, A. M., Aliyev, A. K., & Zeynalov, M. M. (1966). *Geology of oil and gas fields of Azerbaijan* (392 p). Nedra (in Russian).

Alizade, A. A., Alikhanov, E. N., & Shoikhet, P. A. (1967). *Investigation of the conditions for the transformation of organic matter in modern sediments of the South Caspian (in terms of the oil origin)* (101 p). Nedra (in Russian).

Alizadeh, A. A., Guliyev, I. S., Kadirov, F. A., & Eppelbaum, L. V. (2017). *Geosciences in Azerbaijan. Volume II: Economic minerals and applied geophysics*. (340 p). Springer.

Avdusin, P. P. (1952). *The structure of rocks and facies of the middle Pliocene of Eastern Transcaucasia* (72 p). Academy of Science (in Russian).

Azizbekov, Sh. R., Agakishibekova, R. R., Alizadeh, A. A., Alizadeh, K. A., Aliyev, M. M., Akhmedov, A. M., Akhmedov, G. A., Bairamov, A. S., Gadjiev, T. G., Zhouze, B. P., Zaitseva, L. V., Kashkay, M. A., Mekhtiyev, Sh. F., Sultanov, A. D., Khalilov, A. G., Shikhalibeyli, E. Sh., Efendiyev, G. H., & Yakubov, A. A. (Eds.). (1972). *Geology of the USSR. Vol. 47: Azerbaijan Republic. Economic deposits. Fossil fuels (oil and gas)*. Moscow, Nedra (in Russian).

Davies, C. E., Morton, A., Mekhdiyev, U., Aliyev, G.-M., Hyden, F., Aliyeva, E., Gasanov, F., & Allen, M. (2004). Provenance studies of the productive series, offshore Azerbaijan; petrographic and heavy mineral analyses. In *Joint report of CASP, HM research associates, Azerbaijan State Scientific Research Institute of the oil & gas industry, State Oil Company of Azerbaijan Republic, Oil Quest Ltd., Geology Institute, Academy of Sciences. Baku, Azerbaijan.*

Davies, C., Vincent, S., Hyden, F., & Aliyeva, E. (2002). *Petrographic and petrophysical analysis of the upper productive series, Kura basin* (pp. 1–31). Azerbaijan Project, Joint CASP-GIA Report.

Golubyatnikov, D. V. (1914). *Detailed geological map of the Absheron Peninsula. Bibi-Heybat* (216 p). Transmission of the Russian Geological Communication, No. 106, New Series (in Russian).

Guliyev, I. S., Levin, L. E., & Fedorov, D. L. (2003). *Hydrocarbon potential of the Caspian region* (127 p). Nafta-Press (in Russian).

Hinds, D. J., Aliyeva, E., Allen, M. B., Deavies, C. E., et al. (2004). Sedimentation in a discharge-dominated fluvial-lacustrine system: The Neogene productive series of the South Caspian basin, Azerbaijan. *Marine and Petroleum Geology, 21*, 113–138.

Huseynov, D. A., & Guliyev, I. S. (2004). Mud volcanic natural phenomena in the South Caspian basin: Geology, fluid dynamics and environmental impact. *Journal of Environmental Geology, 46*, 988–996.

Kalitsky, K. P. (1922). On the productive strata of the Absheron Peninsula. *Oil and Shale Industry, 3*(1–4), 29–35 (in Russian).

Khalilov, N. Yu. (2000). Geological structure and prospects of oil and gas potential of large structures Shah Deniz and Absheron located in the Caspian sea. *Azerbaijan Oil Industry*, 11–12, 9–14 (in Russian).

Kovalevsky, S. A. (1922). On the parallelization of the Bibi-Heybat and Atashkya-Yasamal sections. *Azerbaijan Oil Industry, 3–4*, 65–71 (in Russian).

Lilienberg, D. A. (1996). Problems of morphotectonics, geodynamics and geoecology of the Caspian. *Doklady Russian Academy of Sciences, Series: Geography, 6*, 140–146 (in Russian).

Lilienberg, D. A. (2002). The phenomenon of the Caspian sea and a new tectonic-hydroclimatic concept of fluctuations in the level of inland water bodies. *Izvest Azerbaijan Academy of Sciences, Series: Earth Science, 3*, 3–11 (in Russian).

Mustafayev, I. S. (1963). *Lithofacies and Paleogeography of the middle Pliocene deposits of the Caspian Basin* (193 p). Azerneshr (in Russian).

Mustafayev, I. S. (1964). Lithofacies and paleogeography of oil and gas bearing Middle Pliocene deposits of the Kura depression. In: *Essays on the geology of Azerbaijan* (pp. 309–319). Academy of Sciences (in Russian).

Nikishin, A. V. (1974). On the sedimentation rhythm of the oil and gas bearing strata of the Apsheron threshold. In *Book: "Patterns of formation and placement of oil and gas fields"* (pp. 45–48). IGIRGI (in Russian).

Nikishin, A. V. (1981). Sedimentary rhythm and comparison of middle Pliocene sections of the South Caspian Depression. In *Book: "Problems of geology and oil and gas content of the depressions of the Inland Seas"* (pp. 60–66). Nauka (in Russian).

Nummedal, D., & Clifton, H. E. (2002). The role of paleoclimate in unraveling the reservoir architecture of the Productive Series of the South Caspian Basin. In *Proceedings of the international conference* "petroleum geology of the Caspian basin".

Nummedal, D. et al. (2004). Productive series at Kirmaki Valley: 2 million years of fluvial-lacustrine interactions. In *Proceedings of the AAPG Hedberg conference "sandstone deposition in lacustrine environments: Implications for exploration and reservoir development". Guidebook for field trip* (pp. 1–25). Baku, Azerbaijan.

Potapov, I. I. (1954). *Absheron oil-bearing region (geological characteristics)* (539 p). Academy of Sciences (in Russian).

Putkaradze, A. L. (1958). *Baku Archipelago* (335 p). Azerneftneshr (in Russian).

Reineck, H.-E., & Singh, I. B. (1980). *Depositional sedimentary environments with reference to terrigenous clastics*. Springer-Verlag.

Reynolds, A. D., Simmons, M. D., Bowman, M. B. J., Henton, J., Brayshaw, A. C., Ali-Zadeh, A. A., Guliyev, I. S., Suleymanova, S. F., Atayeva, E. Z., Mamedova, D. N., & Koshkarly, R. O. (1998). Implication of outcrop geology for reservoirs in the Neogene Productive Series. Apsheron peninsula, Azerbaijan. *AAPG Bulletin, 82*, 25–49.

Shilov, G. Y., & Bezmenov, V. G. (1994). On the question of stratigraphic confinement of deposits of the VI, VII and VIII horizons of the PT in the Sangachaly-sea field. *Azerbaijan Oil Industry, 7–8*, 10–13. (in Russian).

Suleimanov, A. M. (2003). Paleogeographic conditions of the formation and oil and gas potential of the suite of the VII horizons of the productive strata (on the example of the Jeyrankechmez depression and neighboring areas of the Baku Archipelago). *Azerbaijan Oil Industry, 4*, 1–6. (in Russian).

Sultanov, A. D. (1949). *The lithology of the productive series of Azerbaijan* (184 p). Academy of Sciences (in Russian)

van Baak, C. (2015). Mediterranean-Paratethys connectivity during the late Miocene to Recent. Unravelling geodynamic and paleoclimatic causes of sea-level change in semi-isolated basins (275 p). Ph.D. Thesis, Utrecht University, The Netherlands.

Vincent, S., Davies, C., Aliyeva, E. (2001). Outcrop sedimentology of the Kura basin Upper Productive Series, Azerbaijan. *CASP (Cambridge Univ.) - GIA (Geol. Inst. of Israel)* Report.

6 Seismostratigraphic Analysis of the Early Pliocene Productive Redbed Series

The authors have presented the above-described special features of the SCMB basement, and the structural relationship of its sedimentary cover based on the results of regional seismostratigraphy of an area. The following problems have been examined in the previous chapter: an ascertainment of genetic types of paleobasins; a detailed study of characteristic features of consolidated crust (acoustic basement) and closer definition of crucial structural elements of sedimentary cover, determination of bedding types and spatial relationship of the large sedimentation complexes, clarification of common paleogeographic and paleotectonic setting and formation conditions of geological structures, determination of surfaces of unconformity and breaks, clarification of their genesis and creation of seismostratigraphic framework of the megabasin according to the regional profiles.

Using the general deep point method, the actual material to solve these problems was used seismic time and deep sections obtained from the Caspian offshore and adjacent land.

It is known that detailed (precise) seismostratigraphy includes an interpretation of particular features of different-age sedimentary series, their lithofacies and stratigraphic peculiarities, paleogeographic conditions of formation, clarification of their origin, and an assessment of lithofacies features of stratigraphic units (of medium and small ranks) of the section and also such of applied problems as detection and mapping of non-anticlinal traps, the discovery of oil-and gas-bearing series and so on. Some results of precise (detailed) seismostratigraphy of the Early Pliocene are given in the present chapter.

The Early Pliocene sedimentation seismocomplex SSC-7 deserves close attention as an essential part of the SCMB sedimentary cover that causes its structural characteristics and oil–gas content. These deposits present a thick terrigenous formation being formed relatively quickly (1.7–6.1 Ma) within an enclosed sea/lake isolated from the world ocean under rapid Earth's crust subsidence conditions, sharply contrasting tectonic movements and growth of surrounding mountain structures. As mentioned above (see Chap. 11), the Early Pliocene sedimentary complex is the primary oil and gas-bearing complex in the SCMB and named the Productive Series (PS) within the western Caspian and Azerbaijan. At the same time, it is called a red bed series (RBS) within the eastern Caspian and Western Turkmenistan. No unique fauna has been found in the PS and red-bed series terrigenous deposits. Therefore, the Early Pliocene age of these series is defined because they are underlaid by the Pontian clays having Upper Miocene fauna and overlapped with clay deposits of the Absheronian and Akchagylian stages of the Upper Pliocene.

The Early Pliocene deposits occur widely in the depression zones of the South Caspian regional down-warping area in the Middle Caspian structure (north-Absheron trough and Gusar-Devechi superimposed trough), Absheron

A. Alizadeh et al., *Pliocene Hydrocarbon Sedimentary Series of Azerbaijan*,
Advances in Oil and Gas Exploration & Production, https://doi.org/10.1007/978-3-031-50438-9_6

periclinal and Djeirankechmes superimposed marginal troughs, Lower Kur, and Middle-Kur depressions. Within the Absheron periclinal trough, the PS sections are exposed in the Kirmaki and Yasamal valleys. The oil fields of the Absheron Peninsula (and the same name sill) have been drilled through all their thicknesses. They represent an alternation of sandy, psammitic, siltstone, and clay bands. Notably, to the south of the Absheron Peninsula, PS thickness increases up to five times (from 1000 to 1200 m around the Kirmaki region to 1600 m on the Bakhar structure) with decreasing PS arenosity from 60 to 70% down to 30%. During the PS accumulation period, the basin's sequential widening took place. An areal extent of the Early Pliocene suites and horizons has been transgressive and regressive. The change of coastline in the Azerbaijan part of the Caspian took place only in a northwesterly and westerly direction. In contrast, in the Caspian Turkmenian part, this coastline travel took place, at first, to the west (RBS lower division) and then to the east. The general course and character of the accumulation process have been subjected to intense subsidence, and the rock lithological composition change has been connected first with terrigenous material provenance from the platform as from intermontane troughs.

The basin's bottom-down warping and sea-level variations caused interrupted-continuous filling of the basin by sedimentary material. The sedimentation occurred mainly under compensated conditions with short-lived uncompensated episodes and has been accompanied by breaks and erosions established by chronostratigraphic sections (Mamedov, 2004, 2006, 2007, 2008). The notable feature is that the PS and RBS, as synchronous constituent elements of a single sedimentary-petrean basin, are characterized by similar lithological composition represented with alternation of sandy-silty rocks and clays, in places (mainly in the RBS section) with anhydrite bands. It is noted that in the PS and RBS, clay material is increased the section downwards. Besides, if clay content in the PS is increased in the western SCB in a southerly direction, then in the Turkmenian shelf, the sandy-silty content is increased in the same direction. According to the mineralogical composition, quartz is predominant in the light fraction of the PS and RBS sandy-silty rocks. An average quartz value in PS rocks is 46–48%, and in RBS rocks, it is 38–39% (Pashaly et al., 1998). The rate of the river waters and solid runoff delivery has also influenced the lithofacies composition of the Early Pliocene rocks. Sandy suites and horizons have mainly been formed during the sediments supply to the basin and high-water mobility (in avalanche sedimentation). In decreasing the rate of solid runoff supply, argillaceous sediments have been deposited. There were different opinions on the formation of the Early Pliocene sedimentation complex.

In such a way, Khain and Shardanov (1954) believed that the PS age was a period of epigenetic movements with the maximum scope, and the sediment accumulation took place in systematically widening basin deep into the Kur and West-Turkmenian troughs. Kerimov et al. (1999), one of the authors of the published work, considered that the center of this basin generation had been a region of the Baku Archipelago, and sedimentation took place exclusively under conditions of the basin bottom compensated for down-warping. Besides, he considered that the basin contours widening is related to the crust subsidence in adjacent land areas. According to this author, several underwater island cups and even cordillera have existed within a shelf sea, which created an obstacle between the Absheron, South-Absheron, and Kur sub-basins. This researcher has assumed that the joined Pliocene basin has disintegrated into some independent basins. Each of these basins was characterized by its geological regime and feeding source, which caused the changes in their thickness and lithofacies. Kerimov et al. (1999) have considered that it was just a reason why numerous attempts by many researchers to correlate PS sections had no positive results.

According to one of the theses of this work (Bagirzade et al., 1988), PS sedimentation has been accompanied by tectonic movements and folding processes that caused the washout and erosion of most of the crests of the Pliocene structures. The mentioned authors suggested that the latter protruded from under sea level at the end of the Pliocene. The GDPM data does not confirm the "washout of the crests" thesis. Moreover, it seems especially dubious to affirm that most of the Pliocene anticlinal structures are characterized by a syn-sedimentation nature of development and that folding trends have repeatedly been changed in the different parts of the basin that led to the distinctions in thicknesses, facies composition, and degree of dislocation of the PS deposits. The authors of the mentioned work (Bagirzade et al., 1988) have considered that these reasons complicate a choice of keylogging benchmarks to carry out interareal correlations of the PS suites and horizons.

Thus, all the notions of the formation and development of the Early Pliocene paleobasin and tectonic-sedimentation conditions of the PS and RBS oil/gas bearing beds existent before the nineties of the twentieth century have been based on sparse data obtained from offshore and land drilling carried out mainly on the arched Pliocene structures and on the results of geophysical investigations carried out on land and in South and Middle Caspian shelf areas. As to features of structure and depth of subsidence of bottom deposits of the Early Pliocene lower suites in the central paleobasin, there was no seismic information on them up to the XX century.

Seismic data obtained by the geological-geophysical prospecting trust have had only a six-second sweep. Therefore, it could not light up the section below 6–8 km depth. The PS and RBS structure model presentations have been based on stereotyped notions of the sediments' primary horizontal bedding within syn-sedimentation basins.

The transition to the wave pattern logging up to 8–10 s and more made it possible to obtain direct information on the bottom and structure of the PS lower part and reveal its new formerly unknown features. Essential information has been obtained from seismostratigraphic interpretation of GDPM data from 1980 to 1990 (Mamedov, 1991, 2006).

6.1 Dissection and Layout of the Productive-Red Bed Series According to the Outcrop and GIW Data

For a long time, dissection and correlation of the Early Pliocene sections have been carried out mainly based on information obtained from drilling and geophysical investigation of wells (GIW). In connection with certain specific features of the lithological structure, mineralogical composition, and oil/gas saturation of PS and RBS deposits, individual layouts have been worked out for separate oil/gas-bearing regions. The following layout variants are known: Absheron, Garadagh, Pirsagat, and others compiled for the PS; Cheleken, Nebitdag, Chikishlyar, and others—for the RBS. Even though all these layouts are based on lithological-facial features and divide the section into separate sandy and clay bodies in discrete points, they are often called in geological publications as stratigraphic layouts. It relates to the fact that in the absence of biostratigraphic indicators in the PS and RBS barren beds and high–informative seismic data, correlation schemes of geophysical investigation of wells have been the major and most likely the only material allowed to dissect PS and RBS thin-laminated sections into separate fractional units.

In dissection and correlation of the Early Pliocene well logs, most of the research is based on the Absheron layout, according to which the following suites are distinguished: Kala (KaS), Underkirmaki (UKS), Kirmaki (KS), Superkirmaki sandy (SKS), Superkirmaki clay suite (SKCS), "Break" suite, Balakhani, Sabunchi, and Surakhani suites.

The Absheron layout is helpful in dissection within the regions adjacent to the Absheron Peninsula and the same name sill (Absheron Archipelago, South-Absheron zone, and partly the North of the Baku Archipelago). The typical feature of these regions is that terrigenous material supplied mainly from the north (Turanian platform) by the paleo-Volga River system and characterized by high quartz content (up to 50–80%) in its light fraction has been deposited here.

It seems likely that the enormous shelf areas slightly inclined to the south are characterized by their special sedimentation regime that caused similar lithology and bedding type of the section located south of the Absheron sill of the same name peninsula. At the same time, essential differences in lithological features are noted in a westerly direction within the Absheron—Pribalkhan folded zone and in the Turkmenian shelf area. The varied lithofacies composition in lateral series of the alluvial-deltaic, coastal-marine, and shallow-shelf complexes caused great difficulties in correlating PS and red-bed series by the GIW data not only in oil/gas-bearing regions but on local structures as well.

In an easterly direction (to the Pribalkhan region and Turkmenian structural terrace), essential changes are noted as lithofacies characteristics and the deposits' bedding character. The correlation of these sections becomes more complex in the transitional zone from the PS to the red-bed series. In such a variety of suites and horizons lithological characteristics, their interareal correlation by the GIW data appears impracticable. In this connection, several researchers (e.g., Nikishin, 1981; Potapov, 1954) have proposed to seek out similar features, not in separate suites or horizons but in their law-governed associations—sedimentation rhythms.

As it is known, rhythmic (cyclic) correlation methodology has assumed "rigorously" ordered succession in the vertical line. According to this principle, Potapov (1954) has picked out seven sandy-argillaceous rhythms. In his opinion, the boundaries between the rhythms correspond to the breaks in sedimentation. At the base of each rhythm are the traces of erosion of the previous regressive rhythm. Subsequent investigations show that the presence of sharp lithological boundaries is typical of the bottom of the Early Pliocene and the bottom of its middle parts (Break suite lowermost strata). In the rest of the cases, a gradual change from sandy to clay sediments and a sharper transition from sandstones to clays (Mamedov and Ragimkhanov, 1985; Nikishin, 1981).

Lack of geophysical testing of wells, particularly in offshore areas, and disjoined drilling information created an obstacle to the well-founded correlation of the sections according to the lithological character of the Early Pliocene on a regional scale. The point is that in most wells, except those in developing structures, geophysical testing is limited by standard electrical logging. Besides, natural potential diagrams (PD) allow for the determination of lithological special features objectively. In that case, apparent resistance diagrams (AR) are not the answer to this purpose because they react to the fluid saturation of the rocks. Moreover, as a rule, the wells appear to be drilled in highly dislocated arched parts of the Early-Pliocene structures widely separated from each other (up to 30–50 km). The interval between the wells within even well-studied structures is also considerable, reaching 2–3 km. Therefore, the interareal and inter-well correlation of the PS lithological units according to only one parameter (AR) may not be considered correct and unique. From the analysis of correlation diagrams carried out through the Absheron and Baku archipelagos structures and Kur trough, it is concluded that there is a certain periodicity (a steady rhythm) in the configurations of the AR averaged diagrams. As to the eastern SCB structures, AR diagrams have caught no rhythmic indications here. This rhythmicity in red-bed sections could be displayed more clearly, particularly within the Near-Balkhan zone and in the west Turkmenian trough. It is no mere chance that in the latest works of Turkmenian researchers dedicated to the lithologic features and oil/gas content of the red-bed series, its rhythmicity should be discussed.

Within such a tectonically mosaic area as the Early Pliocene basin, where different paleo-rivers have supplied deposits, general regional rhythmicity of the PS and red-bed series deposits should not be expected. Besides, it should be noted that from all the possible geological processes (tectonic down warping and compression, sea-level variations, sedimentation regime, and so on), only sea-level variations have been the general (regional and cyclic) for all the basins. Of course, the unidirectional tectonic movements of the different basement blocks were superimposed upon sea-level variations. Therefore, a "chronicle" of their total effect was preserved differently within different basin parts. In this connection, only shelf sheets of the Early Pliocene Sea may have been characterized by relatively persistent lithological composition and bedding type. The lithological boundaries, far from the shoreline, have coincided with or parallel to the chron boundaries. At a certain distance from shorelines, the boundaries of lithological rhythms have become diachronous and parallel to the waterline.

In such a case, a desire to synchronize different facies bodies according to the logging data obtained from the wells removed at a great distance (10–50 km and more) appears to be prevalent over common sense and may lead to severe misreckoning as in attempts to correlate and synchronize the suites (or their rhythms) as in prediction of oil/gas-bearing objects. As an example, it reminds us of the truth of the study of the PS zone of pinching out in the SW slope of the Mugan syncline.

In the late fifties, incorrect determination of the age of the PS underlying deposits in the "Daikend-1" well-log and the absence of typical Early Pliocene configurations on the logs have been served as grounds to draw a line of the PS pinching out east of the Daikend well. As a result of a seismic survey based on the GDPM data, it was revealed the same errors were assumed in the correlation of the PS sections according to electrical logging carried out within the Absheron Archipelago (Mamedov, 1984). Seismic prospecting carried out by extended horizons methods during 1962–63 (Mamedov, 1965) shows the fallibility of the above conclusion, and using a tracing of seismic horizons allowed to reveal a zone of PS pinching out within a region of the Novogolovka and Bilyasuvar uplifts, i.e., 80–100 km southwest of the Daikend structure. The mentioned seismic survey results established a 1.5-long Early Pliocene series in which thin layers have gradually been pinched out by the scheme of transgressive (bottom) overlap on the underlying deposits' erosion surface.

Since conceptual differences between the seismographic models and correlative diagrams of geophysical investigation of wells (GIW) have been established, the attempts undertaken in the quest for age analogs to the PS and RBS suites and horizons based exclusively on the logging should be considered erroneous. Meanwhile, despite the errors revealed in the mentioned premises, it should be noted that correlative electrical logging diagrams, apart from the solving of some oil/gas-prospecting problems and correlation of reservoirs and caps, have been used up to now in the PS stratigraphic dissection and even for external control of seismic prospecting. Over the last 15–20 years, deep offshore drilling with a combined geophysical investigation was carried out to revise the earlier accepted models. As a result, it was created an essential geological-geophysical database necessary for the detailed study of the PS structure and genesis, construction of its chronostratigraphic framework, and development of an objective model of its sedimentation in the Early Pliocene basin.

6.2 Pre-Pliocene Unconformity Surface and Its Geomorphological and Structural Features

One of the clean-cut and typical unconformity surfaces in the SCB section was formed between the Pontian and Early Pliocene. Within the paleobasin northern slope zone, this unconformity surface cut Pontian deposits to the Lower Pliocene in age (in the Scythian-Turanian platform mantle). At the same time, the break is increased at the expense of the PS lower suites and horizons omission and erosion of older underlying deposits.

A study and mapping of the Pre-Pliocene surface of unconformity are of the utmost practical interest since many lithological, and the mentioned breaks control stratigraphic traps in the Pliocene and its underlying deposits.

An SU was formed due to a fall in base level of erosion and South Caspian Lake's complete isolation from the World Ocean approximately 5.3–5.1 Mya. This event coincided in time with the Messinian salt crisis in the Mediterranean. SSA showed 300–700 m deep, and 10–20 km wide canyons were formed within the paleo-Volga and other extensive river systems (Fig. 6.1). Approximately 700–800 km to the south of the paleo-Volga canyon within the Middle Caspian and on a shelf, edge preceding the Pontian basin, an overlap of the earlier PS deposits on a surface of unconformity that marks a fall boundary of base level of erosion in Early Pliocene time. Basin isolation indicates that the balance between water, runoff supply, and evaporation has controlled the base level of erosion. This sea/lake was filled with sediments and water from the river systems.

6.2.1 The Nature of the Pre-Pliocene Unconformity Surface

From the tectonic-morphological point of view, the Early Pliocene basin is considered an intermountain basin enclosed between the Greater Caucasian—Bolshoi Balakhan—Kopetdag and Lesser Caucasian—Elbrussian—Aladag—Benaludian mountain-folded systems. It was formed on the tectonically mosaic geological basement. Separate blocks of the basement have been characterized by different movement and subsidence directivity and intensity reflected on different thicknesses of accumulating sediments. Within the Near-Caspian-Guba piedmont region, north of the Absheron sill (and the same name peninsula), all the sedimentary series was formed on the continental crust of the Scythian-Turanian platform while consolidated crust within the SCB was mainly of oceanic type.

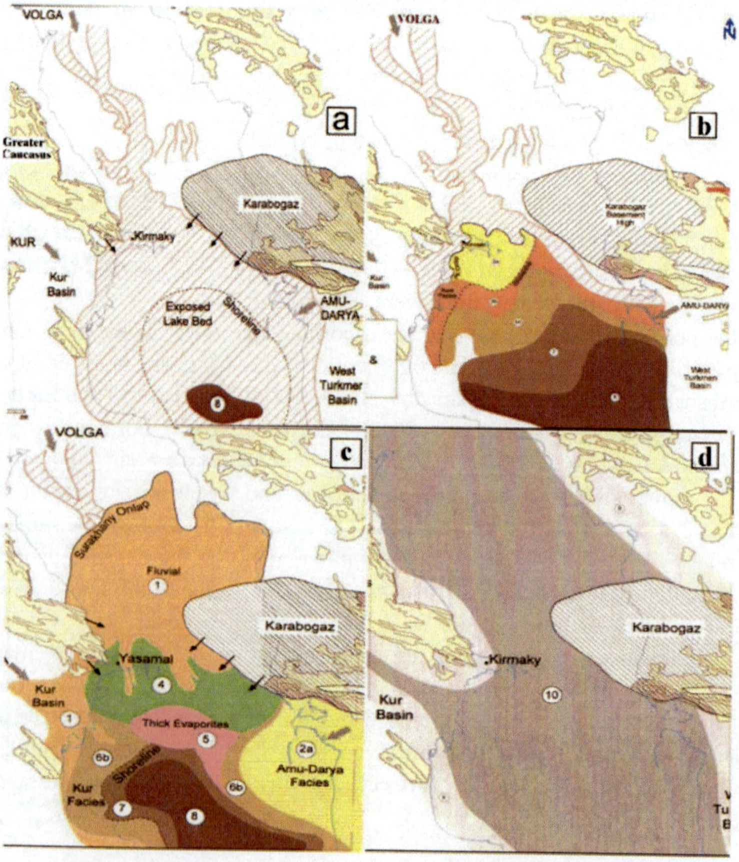

Fig. 6.1 Paleogeography in the Messinian time (the end of the Miocene). **a** During Underkirmaki suite time, **b** in Surakhani time, and **c** in the Upper Pliocene (Akchagylian time) (according to the BP company data)

Such a complex geodynamic setting was added and complicated by specific paleogeographic and sedimentation conditions connected with sea-level cyclic variations. According to Alizade (1960), the Pliocene sea-level variations have been connected not only with the change in the rate of tectonic down-warping but also with climatic changes. This researcher believed that because of the challenging cold climate and glacial period, which set in at the end of the Pontian age, the sea area was sharply reduced, being dissected into two basins (southern and northern). The northern lake-type basin has occupied a minimal area within the modern North-Absheron trough.

Some researchers (Mamedov, 1977; Sultanov and Gorin, 1963) have supposed that the Pontian Sea was shallowing and separated into several distinct basins due to a sharp Messinian sea-level lowering (down to 600–700 m) all over the Paratethys. It is confirmed by geological-geophysical data of sea-level relative changes investigated in the Mediterranean, the Black Sea, and the Caspian basin (Eppelbaum & Katz, 2021; Gromin et al., 1986; Lisitsyn, 1988; Mamedov, 1991).

Khain and Shardanov (1954) and Alizade et al. (1985) have considered that the basin widening is not connected with land subsidence and SCB marginal parts. This effect is associated with sea level and bottom elevation at the expense of increased inflow of the river waters supplying a large quantity of sedimentary material and the fact that the rate of the basin bottom subsidence has been lower than that of sedimentation. Such water ingression on land at the expense of abundant river water runoff (in the absence of descending tectonic movements) is named transfusion (but not as transgression) (Alizade, 1960).

Sultanov and Gorin (1963, p. 268) have been of another opinion, according to which "a rate of sedimentation basin subsidence in simultaneous delivery of an enormous quantity of solid runoff and the water (from the melting and downpours) appeared to be advanced as regards a rate of terrigenous material accumulation that led to the general water transgression".

The results of the SSA carried out by Mamedov (1984, 1986) indicate that denudation processes were widespread in the basin slope zones at the beginning of the Early Pliocene time. According to seismostratigraphic investigations, a primary sedimentation surface of the Pliocene paleo-basin has been an erosional and denudation surface (Fig. 6.2).

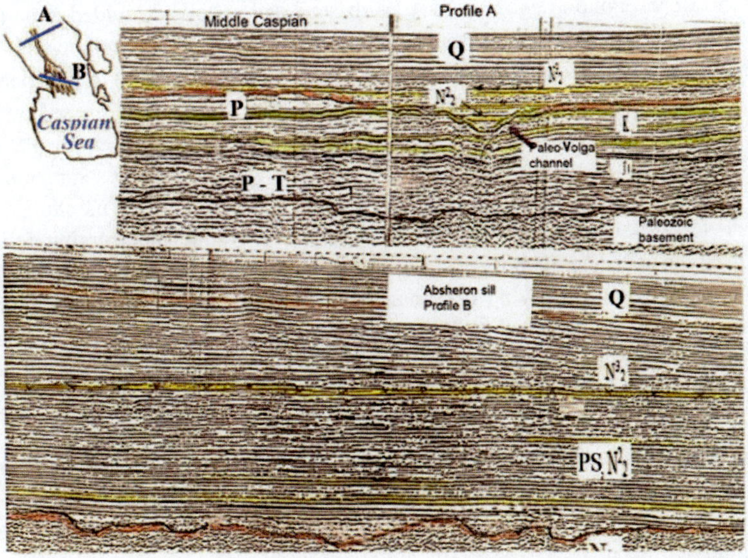

Fig. 6.2 Erosional downcutting of the paleo-Volga channel (**A**), downcuttigs, sub-fluvial bars, and grooves in the paleo-Volga deltaic system (**B**)

6.2.2 Tracing of Pre-Pliocene Unconformity Surface Within the SCB Slope Zone on the Absheron Sill According to Seismic Data

The surface of the unconformity between PS and underlying deposits on a vast area (≈3000 km^2) of the SCB northern slope has first been traced and mapped by the authors of this work based on seismostratigraphic interpretation data (Mamedov, 1984). It should be noted that the PS is well studied by logging and seismic surveys both on land and sea within the Absheron sill, the same-name peninsula, and in the Middle Caspian. It includes a 4–5 km thick deltaic-fluvial wedge, and lacustrine deposits thicken up to 7–8 km in the central basin. The PS appears thinned and progressively superposes the surface of unconformity in the Miocene deposits.

The unconformity surface on the PS bottom is well traced by truncation of underlying beds and bottomed overlap of the PS lower beds. As seen in the roof of the PS underlying complex, inclined beds are, in places, leaned against the unconformity surface. Such contact form between the beds of different ages suggests that considerable breaks in sedimentation and erosion processes in underlying deposits occurred at the end of the Pontian age due to a sharp sea-level lowering accompanied by the rise of mountain structures in adjacent areas.

As it is seen in seismic sections of the Absheron sill, an unconformity of interrupted elements is observed in some areas without any features of pinching out of the beds, making up the lower SSC-7. In some cases, a short way off it may be traced temporal lines that appear to be invisible where overlying beds meet the unconformity surface, as noted in the case of bottomed overlap. Overall, within the transition zone from the PS to the underlying complex, the recording quality in time sections is sharply better than before. That is why it becomes difficult to trace the surface of unconformity. We traced this surface in some places only due to seismostratigraphic detailed analysis of time sections revealing the typical seismofacies forms and using high-speed and energetic analyses of reflected waves.

According to the sections (Figs. 6.2 and 6.3) of NW–SE trending profiles through the Absheron sill, the unconformity surface is characterized by a rough shape with separate convexo-concave elements. An interrupted, largely low-amplitude reflection observed in lowland areas corresponds, according to seismofacial criteria, to lithologically homogenous deep-sea rocks. From the drilling data gained from the Absheron—Near-Balkhan folded zone, it is concluded that clay rocks represent the PS underlying deposits with scarce inclusions of sandy and limy varieties. PS terrigenous deposits lie on small, thick (70–120 m) Pontian deposits and places (when the latter are fully eroded) on the Miocene clays with angular unconformity and erosion traces. An analysis of time sections leads to a better understanding of morphological features of the surface of unconformity. Shallow valleys on the PS lowermost strata revealed in seismic sections (Fig. 6.2) we interpreted as erosional shears. Distinguishing in the middle seismic section is low-energy seismofacies consisting of low-amplitude, interrupted, and chaotically arranged reflections. The NW–SE trending seismoprofile traverses the northern slope of the southern anticlinal belt of the Absheron—Near-Balkhan folded zone (AN-BFZ).

The boundaries "P" and "a" shown in Fig. 6.4 correspond to the sheer erosional bottom and the roof of filling sediments. The "P" boundary is interpreted as the PS bottom since a typical PS comfortable subparallel bedding is observed above the mentioned bottom. Within a shear zone, interrupting reflections and their streamy shape seen in the sections of NW–SE trending profiles are typical of the homogenous fluvial rocks that filled up those shear zones at the primary stage of its filling.

It is known that the change of sedimentation basin boundaries took place because of oscillatory motions at the end of the Pontian age, led to the change of distribution area and accumulation rate of the PS lower suite,—Kala suite (KS) within the Absheron Archipelago (the Azi Aslanov, Gum Pilpilyasi, and Neft Dashlari

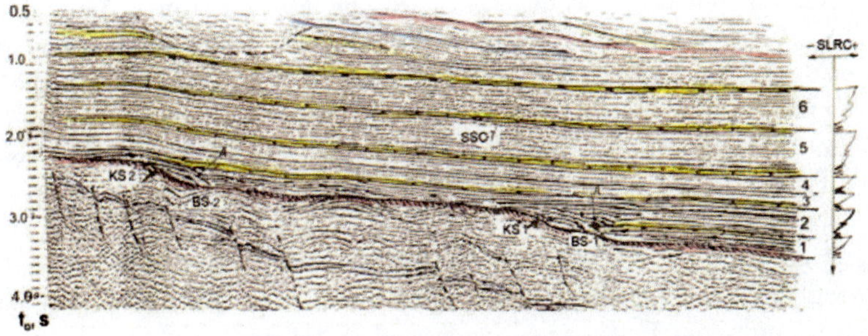

Fig. 6.3 Seismic time-section for profile No. 813455

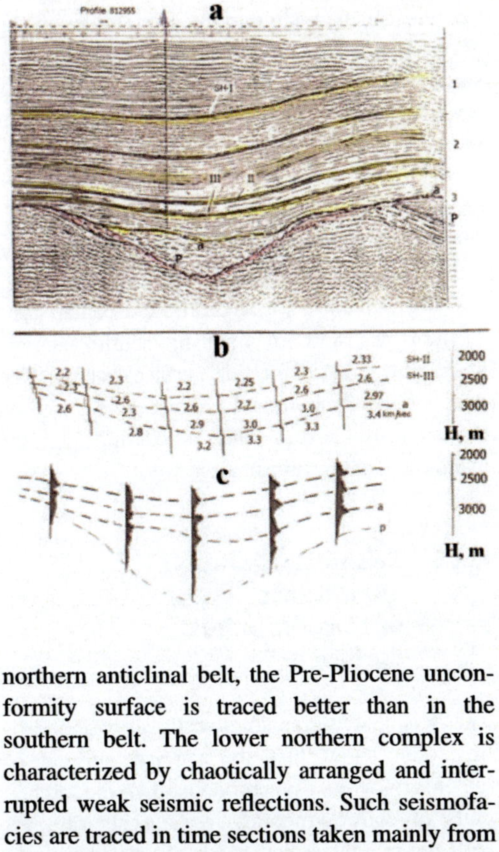

Fig. 6.4 Seismic time section (**a**), graphic representation of the interval rates (**b**), and energetic analysis (**c**)

structures). Figure 6.4 shows a seismic profile crossing the structures' northern periclines and pinching out of KS beds. PS thinning between SH-III and the unconformity boundary in the north-easterly direction is observed in the time section.

The KS thickness change suggests that submarine currents and subsequent tectonic movements have marked the sediment distribution and accumulation at the beginning of the PS age within the Absheron sill. In profiles crossing the northeastern slope of the northern anticlinal belt, the Pre-Pliocene unconformity surface is traced better than within the southern belt. In the profiles crossing the northeastern slope of the northern anticlinal belt, the Pre-Pliocene unconformity surface is traced better than in the southern belt. The lower northern complex is characterized by chaotically arranged and interrupted weak seismic reflections. Such seismofacies are traced in time sections taken mainly from the depressed structural areas and erosional downcutting of the bottom of the sedimentation basin.

Notably, the graphs of interval rates and energetic analysis carried out by digital processing of seismic records were very influential in tracing the Pre-Pliocene surface of unconformity. The result of high-speed series correlation along profiles allowed us to confidently trace the

PS lower boundary. Clay deposits underlying the PS are characterized by higher speeds than those of the PS lower suites (Fig. 6.4). It is explained by the high compaction of underlying clays and limy varieties in their composition. Energetic data analysis data controls the SU position, i.e., an energogram shows that an increase in speed corresponds to separate individual waves (Fig. 6.4b). Maximum values of signal-to-noise ratio correspond to the increasing waves' arrival time. The transition from the PS bottom to the underlying complex is marked by a sharp decrease in the energograms' amplitudes evaluation. It is likely connected to the absence of smooth and sharp boundaries. An application of energograms in an analysis of time sections guarantees a reliable definition of the sought surface position and then recommends that they be widely used. Thus, we revealed and traced the PS lower boundary within northeastern SCB due to the mentioned time section analysis.

The roof adjoining type, bottom onlap type seismofacies, and erosional downcutting served as reliable criteria for this surface tracing. The described analysis stage has been done based on the standard seismic data processing and the available drilling information.

6.2.3 Seismostratigraphic Significance of Unconformity Surface "P"

The seismostratigraphic investigation and drilling data show that the studied surface of unconformity is asynchronous. On the arched parts of some structures (e.g., Absheron), PS beds are in contact with more old deposits up to the Cretaceous. For example, within the Guneshli—Neft Dashlari structures region, this surface corresponds to the Pontian clays' erosional surface. Within the northeastern areas (Gilavar and Harbi Gilavar), PS is characterized by the absence of the lower Kala and Underkirmaki suites being underlaid by the diatomaceous suite (Azizbekov et al., 1972) of the Miocene. Consequently, in a north-westerly direction, the break-in sedimentation between the PS and underlying deposits appears to be increased at the expense of an omission of the PS lower suites and profound erosion of more old deposits.

It is recognized that the prevailing direction of underlying deposits' successive truncation and pinching out of the PS bottom bands and suites above the "P" unconformity surface is the trend of increasing break duration caused by an interval in sedimentation as well as by erosion along this surface. Most likely, northern slope zones composed of more old deposits have also been involved in denudation processes.

The mentioned "P" surface is asynchronous and characterized by different break amplitudes within its different parts. The chronostratigraphic significance of this surface is that it separates more old deposits from the recent deposits and reflects a paleorelief of the Pliocene basin bottom, i.e., a primary surface of the PS sedimentation. During the subsequently widening basin, this surface was overlapped by each following suite after the omission of the preceding one.

6.2.4 Geomorphological and Structural Special Features of "P" Unconformity Surface

Geoseismostratigraphic analysis of time sections and paleotectonic reconstruction indicates the step-like, terrace-like character of the PS subsidence (Fig. 6.5). Some of the almost horizontal areas of these terraces form a morphological plane trap for the sediments accumulated under shallow shelf conditions.

Thus, it becomes even more apparent that the terraces located on different hypsometric levels and overlapping seismobands are undoubted evidence of an interrupted-persistent history of the basin's tectonic development and sea-level variation. The SU scarps and bends have prominent geomorphological features of the wave action (coastal abrasion, down warping) and descending tectonic movements related to the bottom differentiated down warping and basin contours widening. The terrace scarps of the SU are not only the first-rate shoreline markers of transgressively widening sea, but they are indicators of gradual differentiated subsidence of the external relief as well. Near the first SU scarp, the KS lower sandy-silty band formed on a shallow shelf zone appears to be pinched out. The fact that subaerial and inclined axes of seismic band-1 phase coincidence are convergent suggests that shelf tectonic subsidence occurs.

As to the sigmoidal bodies accreted a narrow scarp zone, they should be considered indicators of sea-level variation. Its rising is marked by coastal overlap that was formed under the influence of abundantly supplied sediments and the water mass transgression ("transfusion" according to Alizade et al. (1985)). Their isostatic weight caused the bottom to differentiate downwarping and the Kala Sea to deepen from 3–10 m up to 30–50 m. During this period, the process of normally compensated sedimentation has been disturbed. At the same rate of sediment delivery, the near-shore (non-marine) and littoral sediments have formed a coastal overlap and accreted the terrace literally. At the same time, synchronic and genetically interconnected with them, littoral-marine (shelf) and marine sediments have formed a band of layers characterized by increasing thickness toward the basin. As seen in the time seismic section, this band corresponds to the transparent seismoband, composed of sandy-argillaceous deposits according to the Guneshli-1 well data.

The first elementary cycle of relative sea-level change has been completed by its sharp lowering at the end of Kala time, evidenced by the downwardly displaced reflecting boundary (Fig. 6.3; SF-2). During the second cycle, the UKS sandy-argillaceous deposits accumulated on a shallow shelf of the Underkirmaki Sea. Seismoband SB-3 and seismofacies SF-2 have been formed on a second gently sloping terrace being settled against the scarp of the "P" surface.

It should be noted that because of the small heights of terrace cusps and the same thick seismobands formed during the elementary cycles and owing to the post-Pliocene tectonic complications within this area, the scarps and bodies laterally accreted them may not always be revealed as sigmoidal and cross-laminated seismofacies units on time sections of standard processing.

The above-proposed model of the SU morphostructure formation and sedimentation conditions on gently sloping terraces has been constructed using time sections of several profiles passing through the relatively quiet zones not complicated with anticlinal folding in subsequent times. Folding movements have been superimposed on the bottom subsidence within the areas of active growth of the Pliocene anticlinal structures (Gilavar, Sevinge, Novkhany, Arzu, and others) observed are the "P" surface weakly bendings, or they are absent at all. In tectonically complicated areas, time sections taken from the PS lowermost strata show synsedimentation dips. An increase in dip angles may be observed on the limbs. The positive and negative structures are distinguished by the "P" surface more clearly than by overlying seismic horizons, SH-III (KS uppermost strata) and SH-II (Break suite lowermost strata), suggesting that folding movements are dying out in the Pliocene in due course. SSA shows that all the Pliocene folds located southeast of Gilavar reflect the folding of the Miocene deposits and the northwestern Pliocene structures (Absheron, Shimali-Absheron, Garbi-Absheron, Gilavar, and Shargi-Gilavar) they all are structures of the Greater Caucasian anticlinorium subsiding down into the sea. These structures were initiated by overlapped-overthrust processes within an accretional prism over the SCB subducting crust.

Analyzing seismic data from a seismostratigraphic standpoint allowed us to trace and map the SU ("P") relief in the Shimali-Absheron water area. A structural map over the surface of unconformity "P" is illustrated in Figs. 6.5 and 6.6.

A joint examination of structural maps over the surface of unconformity "P" and the seismic horizon–VI has revealed a distinction between their structural plans.

As seen on a map over the surface "P", against the background of the Absheron-Krasnovodsk anticlinal zone regional subsidence, there are noted anticlinal buried structures, expansive

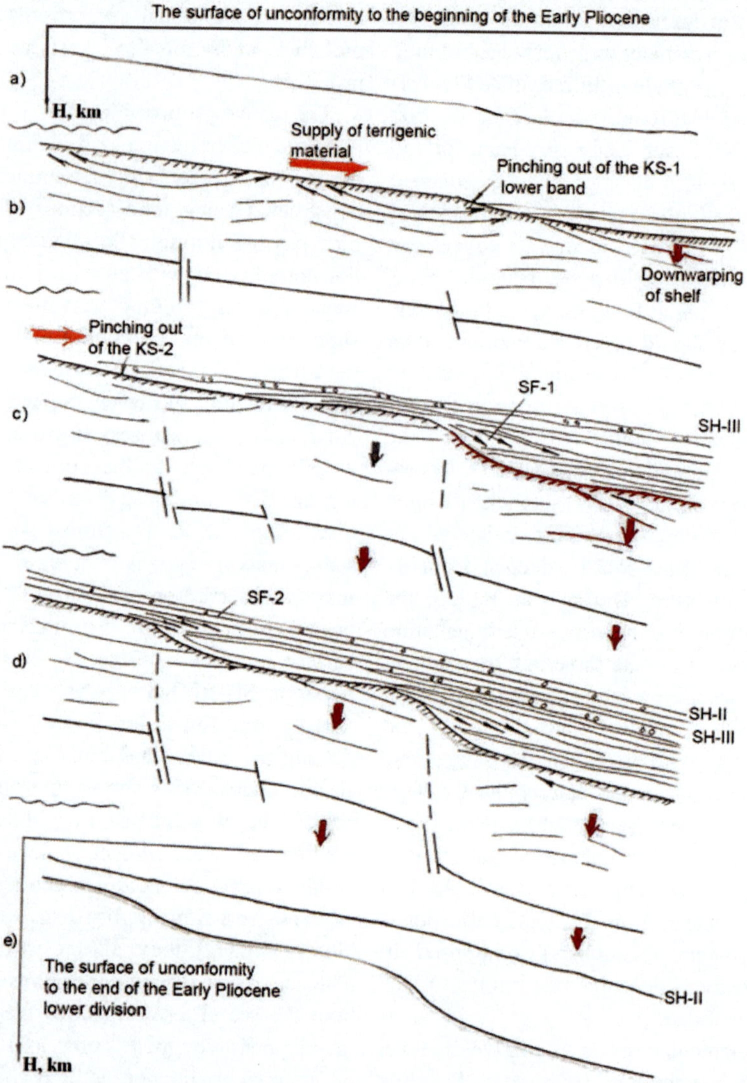

Fig. 6.5 Paleotectonic profile (**a**) and models of the PS lower part sedimentation under conditions of the bottom syn-sedimentation subsidence and the basin contours widening (**b–e**)

terraces, narrow troughs, and structural noses. The arches of the Sevindge, Novkhany, Arzu, and Gyandjlik structures are displaced to the south-east, having isometric forms. To the north of the Aipara structure (about 5–6 km), a brachy-anticline appears to be isolated and articulated with the Dan Ulduzu structure. A seismic survey has subsequently confirmed the presence of this fold in the Pliocene complex. It has been imaged on the structural map over seismic horizon II and named Ashrafi. The other fold, Shargi-Absheron, not isolated by seismic horizon-III, is articulated with the Shimali-Absheron fold, and is closed by an isoline of 2100 m. The Pirallakhi-Kelkor synclinal structure is also defined more precisely. At the beginning of the PS age, it was characterized by more steeply dipping slopes and a narrow trough-like form. Among the others illustrated in the map are the pinching-out zones of the PS lower bench, the Kala suite, and the

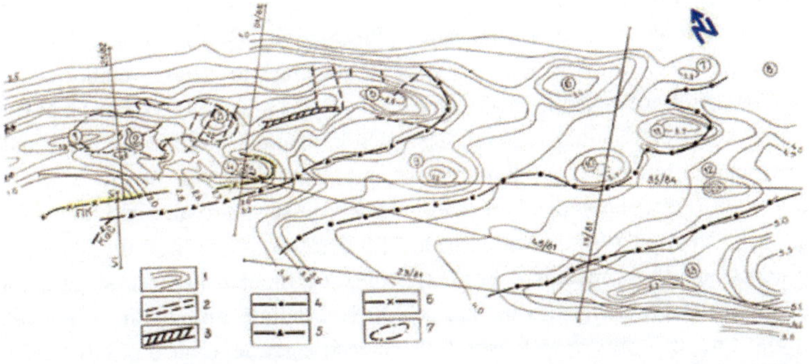

Fig. 6.6 Structural map over the unconformity surface "P". (1) isolines over the unconformity surface "P", (2) disturbances marked according to seismic data, (3) complex seismic zones, (4) KS pinching out zone, (5) pinching out zone of the KS lower sections, (6) UKS pinching out zone, (7) UKSS island (KS pinching out zone). Uplifts: 1—Shimali-Absheron, 2—Shargi Absheron, 3—Shargi-Gilavar, 4—Gilavar, 5—Arzu, 6—Dan Ulduzu, 7—Nameless, 8—Ashrafi, 9—Sevindge, 10—Novkhany, 11—Novkhany, 12—Gyandjlik, 13—Pirallakhi-Kelkor trough

Underkirmaki suite, which is of great practical interest in searching for lithologic-stratigraphic oil and gas pools.

Thus, seismostratigraphic investigations within the northern SCB slope-side zone prove that the unconformity surface may be traced in the RS lowermost strata by discordant relation synphase axes and wavefield borders. Occasionally occurring are erosional incisions, which subsequent burial leads to the generation of breaks and unconformities in sedimentary series. Among the studied problems are morphological features of the unconformity surface and its chronostratigraphic significance; all the mentioned pinching-out zones of the PS lower oil and gas-bearing suites are localized. A map compiled over the surface of unconformity "P" reflects today's most notable structural features of the NW Absheron sill. Many geologists have considered that the cores of some anticlines have already been formed at the beginning of the Pliocene. However, according to the seismic time sections, the PS layers' conformable accretion allows ascertaining that the NW Absheron sill structures have been formed after PS formation during the compression stage connected with regeneration of the subduction process.

So, the following problems have been discussed in this chapter: methodical ways of singling out and tracing the surface of unconformity, i.e., the lower PS boundary; a new model of sedimentation between the Pontian and Early Pliocene time has been proposed; it is motivated a way of tracing of this boundary in a regional plan; it is drawn up a block diagram over the "P" surface in a vast Absheron Archipelago territory.

Subsequent investigations of features of the mentioned boundary surface between the PS and underlying deposits and its regional mapping for all the SCB will make it possible to re-create a more reliable history of tectonic development beginning from the Early Pliocene.

6.3 Fluvial, Deltaic, and Lacustrine Complexes of the Pliocene

Significant sedimentation changes occurred at the end of the Miocene when Maikopian and Diatomic marine deposits were followed by the fluvial-lacustrine Productive Series (PS) and red bed series (RBS)—an equivalent of the PS in the Caspian Turkmenian part. This change over the basin is marked by a series of unconformities resulting from the violent fall on a base level of erosion and complete Caspian isolation from the world ocean (approximately 5.3–5.0 Mya). This event is considered coincident in time or immediately connected with the Messinian salt crisis in the Mediterranean. It is supposed that the fall on a base level of erosion led to the integration the hydraulic drainage systems of paleo-Volga and the Russian platform.

Detailed interpretation of seismic data over the central Caspian shows that a large canyon has been formed within the paleo-Volga River system in places up to 600 m deep and 20 km wide (Figs. 6.2 and 6.7). Nearly 700 km south from the shelf boundary of the preceding Pontian stage, an overlap of the PS earlier sediments is observed over the unconformity that marks fall boundaries on a base level of erosion. The Caspian isolation from the world's ocean indicates that the balance between water and sediment delivery and evaporation controlled its base level of erosion. In contrast, the pronounced displacement of a base level of erosion observed at the foot of the PS suggests that during its isolation, the Caspian has mainly been a lake much lower than sea level, being filled up with sediment and water from the river systems around.

The PS is well known from the drilling data, outcrops on land, the Absheron Peninsula, and the seismic and logging data obtained in the SCB marine part. This series includes deltaic-fluvial and lacustrine deposit complexes up to 5–7 km thick within the SCB. Toward the north of the Central Caspian, this series appears to be thinned and progressively overlaps the surface of unconformity of the Upper Miocene.

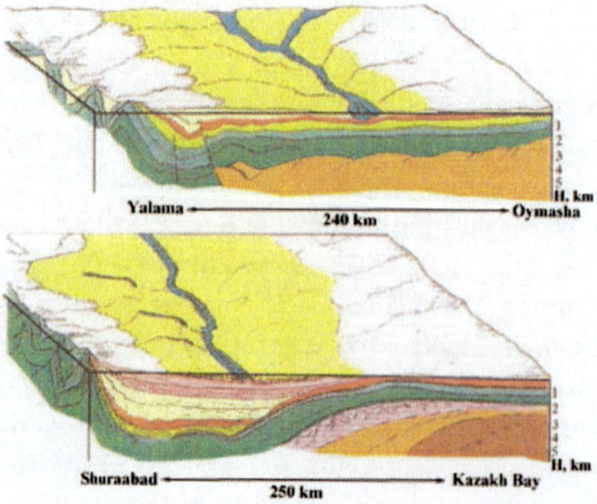

Fig. 6.7 The paleo-Volga River system in the Middle Caspian

Abdullayev et al. (2011) have divided all the Productive Series into three major units. According to these authors, the lower PS (Fig. 6.8) includes the Kala suite (KaS), Underkirmaki suite (UKS), Kirmaki suite (KS), and Underkirmaki sandy suite (UKSS). The lowermost interval, including UK and KaS sandstones, as well as the lowermost clay complex of the Kirmaki suite, are interpreted as a fluvial-deltaic unit of the sea-level lower stage; besides, sea level at first has been regressed and then transgressively recessed to the north end by the culmination, i.e., by maximum waterflood (clays in the uppermost Kirmaki suite). Overlying UKSS sandstones belong to the high sea level stage and are composed of extended fluvial sediments in the north, gradually passing to the deltaic sediments with prevailing fluvial elements in the central basin.

This lowermost PS unit is an interval of prograding development of the facies high in sand content all over the basin, then changed by the SKCS clay beds. An equivalent of the PS and RBS lowermost strata part is the Dagadjik suite.

The Middle PS consists of clay and sandy beds of the Balakhani and Break suites and is limited by the unconformity surface at its foot and considerable flooding in the roof. The mentioned PS middle part is mainly of the wave-built type, indicating a prolonged balance between the water/sediment-production rate and subsidence. This sedimentation takes place in a transgressive cycle with a tendency to be upwardly decreased. Lithologically, this unit consists of thick intervals of fluvial-deltaic sands alternating with lacustrine clays (in the Balakhani and Break suites). These sands correspond to the periods of large-scale growth of the river and plain deltaic facies during the lower sea-level stage. Horizon V of the upper RBS is equivalent to the PS Break suite.

The upper PS may be divided into two stratigraphic units. The lower part (Sabunchi suite) comprises fluvial sand beds alternating with relatively thick clay intervals. The baymouth bars and distributive canals have been found in the thick sandy beds. The offshore drilling data shows that the Sabunchi suite is almost wholly composed of clays and reflects the

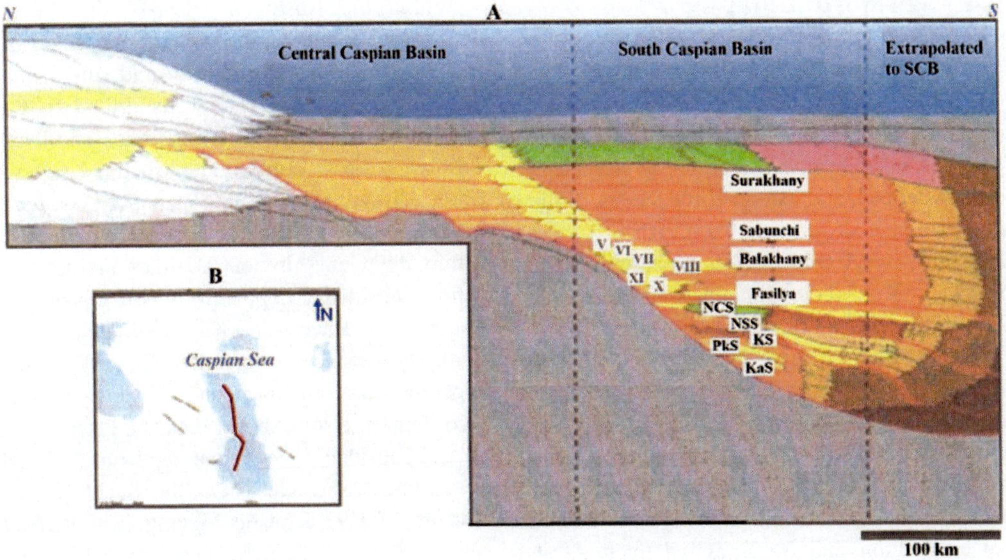

Fig. 6.8 Facies and sedimentation model in the early Pliocene basin (after Green et al., 2009, with modifications). *NCS* Nadkirmakinsky clayey suite, *NSS* Nadkirmakinsky sand suite, *PkS* Podkirmakinsky suite, *KS* Kirmakinsky suite, *KaS* Kalinsky suite

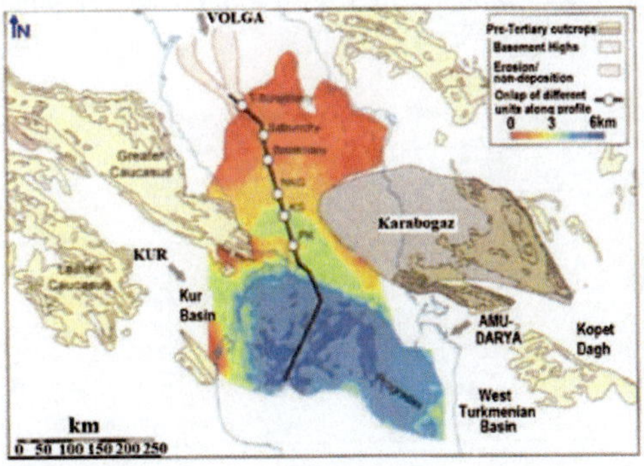

Fig. 6.9 Map of the isopach lines of the productive series indicating 6–8-km thickness in the center of the South Caspian depression and the places of pinching out of the productive series suites by a scheme of leaning to the unconformity surface, produced by a sharp reduction of the sea level

facies displacement towards the coast in respect of the underlying Balakhani suite. It is most likely connected with a decreased supply of sediments from the paleo-Volga during the general lowering of a natural accumulation gradient.

The overlying Surakhani suite's thickness in the Absheron Peninsula's deposits is about 500–600 m, while it reaches 2–2.5 km in the SCB offshore area. This is a sequence of mainly fine-grained sediments with thick clay beds and isolated fluvial sand bands. The uppermost Surakhani suite includes 20 m thick evaporite beds, which are considered to thicken up to a hundred meters in a southerly direction towards the middle part of the basin. As seen from the section, the Surakhani suite is a continuation of a rock cycle characterized by a decrease in arenosity connected with a decrease in the supply of coarse-grained material from the paleo-Volga system.

Moreover, this period was marked by increases in supply from the other point sources of clastic rocks. These processes occurred from the paleo-Kur in the western SCB and the Amu-Darya in the east, dominant in SCB (Fig. 6.9). The share of sediments supplied from the paleo-Volga was the least in the central SCB, while the thick evaporate beds appear to be preserved up to now.

6.4 Seismostratigraphic Characteristic of the Early Pliocene Seismocomplex

The Pliocene deposits are sharply distinguished from the other seismocomplexes by their dense, clearly defined, and extended reflections. Within the basin near-slope zones (North Absheron zone and Absheron Archipelago, southwestern Mugan syncline, Kur depression, Aladag-Messerinian step of the West Turkmenian trough), the SSC-7 roof, and bottom are well distinguished along the wavefield boundaries and by the ending of reflections along limiting surfaces of unconformity (Figs. 6.3 and 6.10). Within the mentioned zones, the lower SSC-7 is well isolated as an independent sequence with a system tract of lower position and transgressive tract. At the same time, in the central SCB, the seismocomplex sediments conform with Upper Pliocene and underlying Pontian-Miocene sediments, which have formed together with a single structural stage—i.e., macro-complex.

Within the Absheron—Near-Balkhan zone, the SSC-7 section has been studied in detail by drilling and GIW. The density and sound logs have been carried out in wells drilled in the Guneshli Guneshly, Azeri, Chirag, and Neft Dashlari-1 areas. Within the Absheron Kyupesi-Garbi Absheron uplifts, the PS bottom has been frilled in at 900–1000 m depths. In contrast, within the Guneshli-Neft Dashlari uplifts, this bottom subsided to depths from 3900 to 4200 m. Strong subsidence of SSC-7 took place in a southerly direction toward the SCB. Within the Shakhovo-sea structure and Vezirov's area, according to seismic survey data, the PS roof and bottom subsided down to the depths of 4–5 and 12–13 km, respectively. On the Shakhovo-sea structure, GIW, acoustic logging, and vertical seismic profiling methods have drilled in and studied the Upper PS deposits in detail. At first sight, based on the syn-sedimentation subparallel-bedded character of deposits of lower stratigraphic units traced in the PS section, everything seems to indicate that the Productive Series has

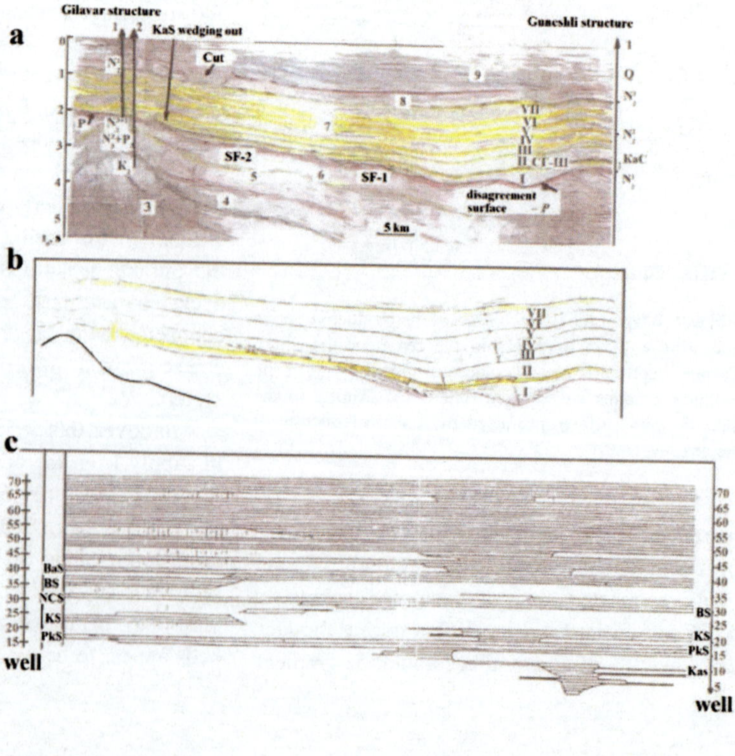

Fig. 6.10 Seismic section (**a**), seismic quantum (**b**), and chronostratigraphic section (**c**) of profile No. 814555. *BaS* Balakany suite, *BS* Break suite, *SF-1 and SF-2*, clinoform seismic facies, *NCS* Nadkirmakinsky clayey suite, *PkS* Podkirmakinsky suite, *KS* Kirmakinsky suite, *KaS* Kalinsky suite

been formed under compensation conditions within a vast shelf space. However, seismic survey data gained employing GDPM allowed the revealing of lateral accretion complexes in the SSC-7 lowermost strata in the Turkmenian structural terrace section and the paleo reconstruction of seismic utilizing straightening of seismic horizons. For instance, seismic horizon-II is confined to the Break suite. All indicators made it possible to find out continental slopes in the Pliocene basin. From the data of the Pliocene seas shelf edge displacement to the southwest in eastern SCB (Figs. 6.20–6.23), it is concluded that an accumulative-topographic noncompensated basin exists in eastern SCB. It was inherited from the deep-sea Pontian or Paleogene paleobasin. In the Absheron-Cheleken folded zone and adjacent areas to the south, the seismostratigraphic analysis of seismic sections carried out by a dense network of intercorrelated profiles allowed us to single out 6–7 covering chronostratigraphic (age) units within the SSC-7. These units are marked by the rank of seismoseries distinguished by density and dynamic expression of seismic reflections.

The seismo-series mentioned above (cycles) within the PS are rather thick (from 300–400 m up to 800–1000 m) cyclically bedded sedimentary bodies, which, per covering type of formation, were caused mainly by differentiated negative shelf movements and sea-level variations. In time sections of standard processing, the seismo-cycles are well dismembered onto relatively small-rank seismobands (Mamedov, 1991). The latter corresponded to the three-dimensional bedded bodies in which physical-geometrical and lithological parameters have been sustained in a vast shelf space of the Early Pliocene basins. In various paleogeographic zones within each seismo-cycle and seismoband, local seismofacies are composed of monotypic sediments corresponding to one elementary sea level cycle (where the relative variations are well distinguished). In seismic sections, they are clearly distinguished by a specific pattern of timelines group (synphase axes).

In the central basin within seismobands, some separated seismobeds describe the geological objects composed acoustically. They may be lithologically monotypic sediments during one SLRC elementary cycle. In seismic sections, they are distinguished by specific timeline patterns (synphase axes). The mentioned seismostratigraphic units form a hierarchical system of co-subordinate three-dimensional seismic bodies by which the Pliocene basins are filled up. The methods we used for the SSC-7 dissection of chronostratigraphic, hierarchical, and co-subordinate units are based on detailed investigations of the wave fields qualitative features and the mode of structural change in seismic sections in a vertical direction. Besides, quantitative valuation of the reflected waves' dynamic and high-speed parameters was employed. The qualitative features include density, configuration, and continuity of timelines, making it possible to determine the shapes and extension of seismic bodies (bands and beds). It is of some importance that confidently tracing timelines has created the PS stratigraphic framework. Within PS, most reflecting boundaries connect with acoustic stiffness shocks on sedimentation (bedded) surfaces and penetrate different facies strata on an isochronous level. Such reflections may also be found to unite special features of gradient mediums with weak acoustic hardness shocks.

The qualitative valuation of some parameters from seismic recording (amplitude, frequency, interval velocity, instant characteristics, etc.) enables us to gain information on acoustic differentiation, thickness, and bedding character of studying series and bands. The optional (by areas) qualitative valuation of the mentioned parameters improved the investigation of the sedimentation conditions.

Based on several seismic profile sections, the following seismostratigraphic and seismofacies characteristic features of hierarchical co-subordinate subdivisions in the SSC-7 section are described below. The profiles show many wells with initial acoustic models, lithologic logs, GIW diagrams, and other available data.

Figure 6.10 shows a basic profile's time and chronostratigraphic sections extending 75 km northward and crossing the Pirallakhi-Kalkor syncline. The profile southeastern end is allocated to the prospecting well No. 1 on the Guyneshli uplift's NE pericline. In contrast, its northwestern end is allocated to wells No. 1 and No. 2 on the Gilavar bank, drilled in all the PS sections and passed through underlying deposits at 4230 and 2700 m, accordingly. In a northwesterly direction, it marked an abrupt rise of observed horizons and a gradual shortening of the PS thickness. The profile chronostratigraphic section shows that in a northwesterly direction, seismic quantum quantity decreases with increasing sedimentation breaks. In a southeastern profile around the Guneshli structure, of 75 distinguished seismic quantums, 42 (about 66%) fall on sedimentation. A northwestern profile from 54 seismic quantum 28 falls on sedimentation equal to 52%. In a No. 1 well log (Guneshli) within the depths of from 4100 to 4222 m in the KS lowermost strata, it is distinguished as a thinly laminated band composed of alternating clays, siltstones, sands, and hard sandstones.

Seismoacoustic modeling has shown that dynamically expressed extendedly tracing wave (SH-III) is formed by a thin-laminated "b" band. Its two phases are complete, from the roof down to the bottom, formed by the deposit of two sandstone beds separated by clay bed. Some partings of this band are from 1 to 10 m in thickness. According to the acoustic log taken from well No. 5 (Guneshli), the mentioned band is acoustically well-differentiated and characterized by more often alternating beds with increased and decreased velocities in its lower part. There is no apparent periodicity in the beds' thicknesses and velocities. This band is also distinguished by the comparatively high values of its specific resistance ($\rho a = 200\ \Omega m$) and the negative anomaly of UKS. Nearly the same values but slightly other internal structure is typical to the band in the KS uppermost strata, which is shown in No. 1 and No. 5 well logs (Guneshli) and No. 1 and No. 2 well logs (Neft Dashlari-2).

A detailed SSA has shown that SH-III in the northwestern slope zones of the Pliocene basin in places marks a non-isochronous eroded surface of unconformity (Mamedov, 1991). When tracing SH-III within a station distance of 1440–1500, the two-time lines (seismoband) appear to be invisible by the frontal surface of SF-1, which laterally accretes peneplain scarp. The KS lower band's pinching out is likely to take place here.

Within a station distance of 1540–2000, it is observed time "steps" within which traced extremum gradually disappeared. Further, the other one, the higher extremum, is traced as being decreased in intensity up to its complete damping within a station distance of 1460–380 traced is a timeline that envelops hilly-like (basic) seismofacies of an "eye effect" type and then ended by the "P" surface of unconformity. A new horizon (SH-III-a) is traced up to the Gilavar bank, three "steps" higher than the first horizon traced in the profile right part. According to the Gilavar-1 well data, SH-III-a is confined to the thin-laminated band ($\Sigma h = 40$ m) observed in the UKS. Thus, the described interval is marked by lateral structural differentiation of the reflecting band (due to the lower beds pinching out) enveloped with an incident wave impulse.

As is seen in Figs. 6.3 and 6.10, the thickness of seismo-cyclite-1 enclosed between PH-P and SH-III is changeable. The obliquely laminated elements of footing and roofing overlap suggest that sea-level variations and connected sedimentation conditions at the Kala sea bottom terrace play an important part. After pinching out of lower beds at the frontal surface (SF-1), the seismo-cyclite surface has been planted by a clay band formed during stable high sea levels. A complete pinching out of seismoband-1 takes place at the southeastern pericline of the Gilavar bank uplift. Such tree-like spreading resembling a runoff system and an accumulative body widening in the direction of the basin reflected on an isopach map (Fig. 6.11) are typical of deltaic formations. In seismo-cyclite-1, reflections appear intermittent and uninterrupted, typical of the shoaly-shelf facies zone.

Seismocyclite-2, the second from below, is enclosed between SH-III and SH-II. According to the drilling data gained from the Absheron Archipelago and adjacent regions, the latter was formerly confined to the Break suite and even the Balakhani suite's X horizon. However, the new

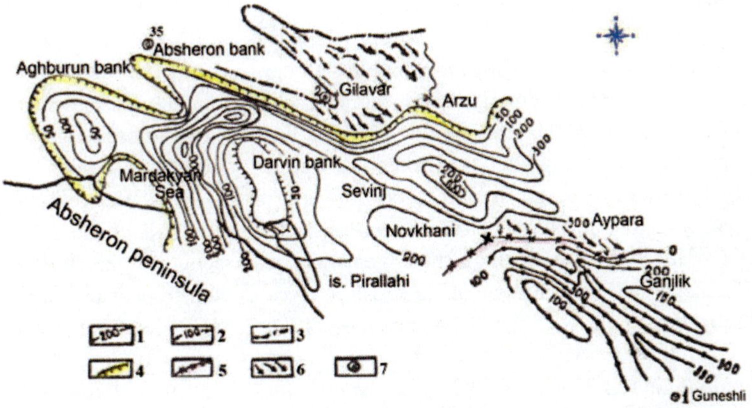

Fig. 6.11 Isopach map of the KS and its lower band (after Mamedov, 1984). (1) KS isopachous lines, (2) KS lower band's isopachous lines, pinching-out lines of (3) UK suite, (4) KS, (5) KS lower band, (6) fluvial-deltaic facies development zones, (7) prospecting wells

high-speed models based on the acoustic and seismic logging data made it possible to precisely define its conjunction with the SKCS roof.

Indeed, as shown in AL charts, the most significant wave velocity jump is in the SKCS roof between the PS lower and upper divisions. Within the paleo-basin sloping zones, SH-2 serves as a fixing element of both erosion and angular unconformity surfaces. Based on the data gained from the investigation of erosional cuts and washout of the SKCS clay sediments surface, as well as from the fact of coarse granularity of the following suite (Break suite) deposits in the Absheron Archipelago, it may have concluded that between the ages of the SKSS and SKCS formation, essential changes in sedimentation conditions have taken place. In bygone days, it was one of the main arguments for the PS subdivision into two separate divisions. According to several researchers (e.g., Aliyev, 1975; Potapov, 1954), the change from the shelf conditions in sedimentation to the continental within an enormous area of the Pliocene sea (lake) has corresponded to the mentioned age boundary.

So proceeding from the confinement of the mentioned seismo-cyclite, limited by SH-II and SH-III, to the UKS, KS, SKSS, and SKCS of the Productive Series, it may be affirmed that UKS and SKSS as the most sandy and oil/gas-bearing suites have been formed on a shallow shelf at a depth of only a few (1–5) m while clayey KS—at the depths of from 30 to 50 m. The SKCS has been formed roughly under the same conditions. In temporal sections, seismo-cyclite-II is represented by the relatively semitransparent, weakly differentiated wave field, including approximately 6–3 seismic quantum.

Seismocyclite temporal power within the Gilavar and Sevindge structures varies between 0.15 and 0.18 s; within the Guneshli and Chirag structures, it is up to 0.35–0.40 s. In a southerly direction, this value appears to be increased up to 0.5 s, corresponding to 550 m, while in sloping parts of many structures, this value is decreased almost doubly.

In the northwestern sloping zones of paleo-basin within seismo-cyclite-2 there are distinguished tile-shaped obliquely laminated seismofacies passing to the hummocky, chaotic seismofacies suggesting an active hydrodynamic regime with the intensive supply of terrigenous material. This section is composed of subparallel beds typical to the shelf facies zones in a southerly direction. South of the Shakhovo-sea and Absheron structures, seismo-cyclite dynamic definition is getting worse. Within the eastern structures of the Absheron sill, seismo-cyclite-2 is characterized by relatively low-speed values. In such a way, the KS low-speed beds are isolated in acoustic logging charts taken from wells No. 5 (Guneshli) and No. 1 (Neft Dashlari). A seismoband-2 is also characterized by low specific resistance from the KS clay seal failure.

It should be mentioned that the deltaic and underwater-deltaic sedimentation regime is also forecasted in the North-Absheron zone. An obliquely laminated seismofacies (SF-UKS) distinguished in several seismic sections between the Gilavar and East Gilavar structures is egg-shaped in form. Formerly, such complex obliquely laminated elements of the SF-UKS have been wrongly interpreted as tectonic dislocations that, in turn, caused incorrect conclusions of paleogeographic and paleotectonic character. The SF-UKS is brightly distinguished against a weak "transparent" recording of enclosing series reflection that is well displayed in high-resolution dynamic temporal sections, keeping a true amplitude relationship ("PAH-processing"). In a southeasterly direction, SF-UKS appears to be passed into intermittent-uninterrupted coastal plain seismofacies covered with semitransparent seismo-cyclite. The latter corresponds to the KS clay deposits of up to 200–250 m in thickness within the Absheron Archipelago. The thickness of seismo-cyclite-II is increased almost two times toward the south, indicating that the rate of syn-sedimentation subsidence of the Pliocene basin shelf is essentially increased.

The next seismo-cyclite-III is enclosed between SH-II and SH-1a, confined to the well-studied shelf deposits of the Balakhani suite within the Absheron sill. All the above-described Early Pliocene paleo-basins have been slightly overlapped by the Balakhani Sea, whose western coast on land is mapped near Akchagylian recent outcrops.

According to the GIW data, seismo-cyclite-III comprises a thick (up to 700–900 m) and the PS section's sandiest interval (more than 80% of quartz sands). Based on the acoustic logging data, the three thick (100–200 m) thin-bedded bands are distinguished within their limit. All of them are characterized by cyclic-layered structures: 20–30 m thick sandy beds are alternated with relatively thin-thick (5–10 m) clay beds. Notably, these bands with variable reflectance are suitable wave-generating objects. The waves reflected from these bands in inter-structural troughs are well resolved and correlated by two or three dynamically expressed phases. Vertically, the thickness of alternating beds decreases with the increase in the section's clayiness. It is noted that reflected waves on the crests are not resolvable because of the beds thinning out and are represented by continuous-wave train oscillations. In the central SCB, seismic section III is clearly distinguished against the background of "semitransparent" underlying and overlying sub-horizontal bodies.

The relatively thick (6–7 seismic quantum) and straticulate seismo-cyclite-IV seems to have acquired the features of anisotropic cover in coastal-marine and shelf areas of the central SCB. Its seismofacies characteristic change in a southerly direction is apparently connected with the argillization of the Sabunchi section. The notable feature of this series is that its thickness appears to be sharply increased (10–12 seismic quantum).

The fifth from the bottom cover body,—the seismocyclite-V, corresponds to the lower thin-laminated part of the Surakhani suite. South of the Absheron sill, it is distinguished clearly and brightly, just as seismo-series-III, among underlying and overlying "semitransparent" covers. The next seismo-cyclite-VI is the most "transparent" and slightly bedded body all over the basin.

In the central and Near-Absheron parts of the SCB, within almost all the thick seismo-cyclites, there were distinguished hierarchically subordinated stratified seismobands alternating with slightly bedded seismic bands. In submeridional profile sections, these seismobands' dynamic definition is changeable because SSC-7 is sharply subsided (down to 3–4 km). Moreover, some timelines may not be traced and may vanish from sight.

The PS section appears pressed within the folded zones. That's why it is hard to separate seismic bands on the crests. Nonetheless, they are correlated quite well in different structures, being distinguished confidently by dynamic features of reflected waves within inter-structural areas. The contrast range of the wave field and periodical alternation of two types of seismobands indicate the presence of quasi-synchronous geological bodies of small rank developed in a vast PS section and formed under similar sedimentation conditions. Each is characterized by a particular anomaly of the rocks' elastic properties and bedding type displayed in reflected waves' dynamic and speed parameters.

6.5 PS Dissection Using Analysis of Wave Velocities and Dynamic Parameters

The quantitative evaluation of the parameters derived from seismic data appears useful in PS dissection onto quasi-synchronous units. Derived data from velocity analysis, pseudo acoustic logging, and energetic and dynamic analysis are the sources of important information about the SSC-7 internal structure. It should be noted that combined analysis of all the derived data is of great importance because each of them characterizes one of the properties of reflected waves.

6.5.1 An Interval Speed Analysis

The initial data have been vertical spectrum velocities in a study of internal velocities. A high-quality and frequent carrying out of the ANVIT procedures over all the profiles (through every other 120 blasting points) made it possible to convert the $V_{g.d.p.}$ into V_{int} by the Urupov-Diks formula in the following circuit (Urupov, 1966):

$$V_{g.d.p.} \rightarrow V_{eff.} \rightarrow V_{d.t.} \rightarrow V_{int.}.$$

Besides, the Vint. determination error by $V_{g.d.p.}$ has been up to 5–10%. For safekeeping accuracy and because of the absence of effective velocity data, the interval velocities have been calculated for those series when the temporary thickness has been more than 0.20–0.30 s. Also, the $V_{int.} = f(t)$ plots were used in the process of digital processing in automatic operation through the profiles by a pitch of 200 m.

Within the NE Absheron Archipelago, "high-speed" bands are well correlated along with the profiles, although Vint. values within these bands are unstable and changeable. We believe that the mentioned instability relates to calculation errors and the lithological and acoustic properties of those seismobands. It should be noted that interval velocities are decreased in all clay bands (Fig. 6.12). In almost all the profiles, the best correlatability has been noted in a low-velocity band of the SSC-7 lowermost strata. According to the drilling data, this band is of Kirmaki clay deposits. It is noticeable that the $V_{int} = f(t)$ averaging lines and abnormally high reservoir pressure curves are very close or similar in their shape.

There is likely a correlation between interval velocity and electrical resistance in the KCS and possibly within all the PS sections. However, because of errors in Vint analytical calculation, it is hard to value this connection quantitatively by the GDPM data. This problem needs special investigations into this work.

6.5.2 Seismofacies Characteristics

Seismofacies characteristics were added by the interval velocities and reflection amplitude data

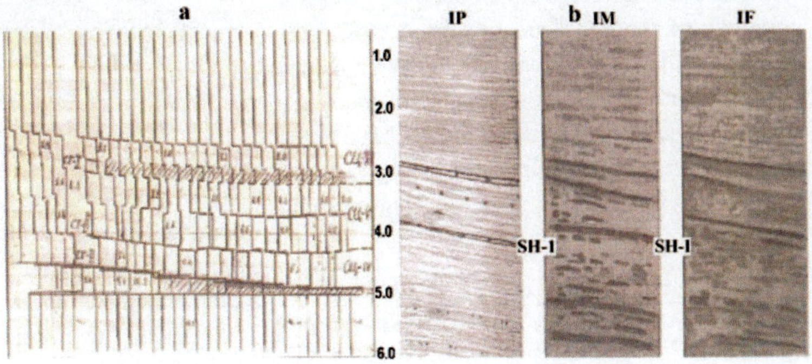

Fig. 6.12 V_{int} charts (**a**), dynamic time section fragments (**b**) (the Shakhovo-sea area)

that improved the reliability of seismostratigraphic interpretation. Within the PS, high-velocity bands are correlated more confidently in the central SCB, where four such bands are distinguished in the most subsided part of the PS upper division. Vint's most significant jump has been noted opposite a thick sandstone band in the Balakhani suite II horizon interlaying mainly between clay deposits. Based on the change in V_{int} and recording dynamic expression, we distinguished the transition zone from shallow-shelf genesis to the continental facies inside seismoband-4. The most brightly displayed sandy band in the seismic section corresponds to the seismoband-3 revealed by its high-speed anomaly values. Nevertheless, in the PS uppermost strata, the sandy band is not well characterized by abnormally high V_{int} values through correlation in seismic charts. Hence, we believe there is no direct connection between the lithology and interval velocity in the PS low-differentiated sections.

Because of its insufficiently precise determination, an interval velocity allows separating only those deposit associations sharply distinguished by their substantial composition. Besides, they reflect an average composition of more than 300–400 m in thickness vertically heterogeneous series. It is hard to conclude that they are of common lithofacies composition. At the same time, the PS dissection into high-speed series suggests the presence of genetically and lithologically isolated bedded geological bodies, i.e., mainly clay or sandy bands. In the V_{int} charts, sandy bands are distinguished by relatively high-speed values. From the AL and GIW joint analysis data, it is inferred that the boundaries of the high-speed series do not mark some concrete (sharp) acoustic boundary of the PS section in the central SCB. These boundaries are confined to the transition bedded intervals from primary clay to the sandy zones and vice versa. Therefore, they have not to be considered isochronous object boundaries, though they carried a particular chronostratigraphic load in cover complexes.

UKS and Sabunchi suite clay deposits. From the interval velocity analysis data, it is inferred that in the Absheron sill's eastern structures, the V_{int} is decreased by 300–400 m/s within the section intervals corresponding to the occurrence of mainly KS. Besides, the V_{int} absolute values for the PS suites are 2700–3200 m/s.

6.6 Dynamic Analysis of Seismic Data

One of the ways of the PS dissection onto objectively existing stratigraphic units is the employment of dynamic parameters derived from the observed wave field. Such values are instantaneous frequency, phase, and amplitude calculated through the Gilbert transformation.

Our investigations conclude that the most reliable separation of seismostratigraphic units in a rank of seismobands is reached in instantaneous amplitude sections. It should be remembered that instantaneous amplitude is the envelope of a complex interferential recording of the bands' reflection. Our modeling result shows that the dynamic (amplitude) expressiveness is displayed in IA sections when the thickness of the beds and bands having variable reflectance appear to be close to a quarter of the wavelength.

Figure 6.12 shows some fragments of Gilbert transformation in the time section, suggesting a stratified structure of the Shakhovo-sea uplift's pericline. The observed mosaic structure of the IF section represents lateral lithological differentiation of reflecting bands. As is seen in the instantaneous amplitude section, five contrast and "heavy" lines are distinguished against the background of the fluctuating amplitude values (spotted recording), causing an illusion of persistence in acoustic and geometrical parameters of seismobands. It should be noted that IP sections with many phase coincidence axes need to be adapted for seismostratigraphic analysis.

Increases in contrast range and instantaneous frequency between SH-1 and SH-1a and between SH-1a and SH-2 suggest an increase in very thin bedding within corresponding seismo-cyclites. Seismobands are likely synthetic outlines of thin-laminated beds characterized by slight AC differentials. The dominant phase of seismoband (by its amplitude) is formed by those beds whose thickness becomes close to ¼ of l.

Figure 6.13 shows an example of wave patterns in time sections representing instantaneous amplitudes in the Shakhovo-sea area (profile 4, UKS 2010-2130). Several reflections are distinguished in the mentioned section (Fig. 6.13) displayed at the medium-frequency filtration 7-14-20-40. This section is a real medium response to the air-operated wave transducer signal pulses, in which frequencies are decreased with depth. During this section reproduction in a low-frequency filtration (2-4-8-16) (Fig. 6.13), many of the axes of phase coincidence disappeared with the appearance of the new ones. In such a way, seismobands No. 2, 3, 4, 7, 8, 10, 11, and 12 are not characterized by well-defined reflection in this section.

Nevertheless, new seismobands such as A and B emerged. The fact that many reflections disappear in low-frequency filtration relates to the increased thin-lamination of the PS section. In a tremendous seismic wavelength at great depths, the thicknesses of separate beds appear to be much less than a quarter of the wavelength that led to the violation in the $h/l = 1/4$ condition. The appearance of new "A" and "B" reflections in the section indicates the presence of seismobands located lower than seismoband 3 and between seismobands 11 and 12, which are characterized by relatively "thick" beds.

From all the observed reflections, only two, namely No. 6 and No. 11, confined to the PS roof and Quaternary foot, keep up their fixed position and amplitude expressiveness in all the filtrations. This fact suggests that these reflections relate to frequency-independent objects, i.e., with sharp boundaries or surfaces of unconformity. The complete disappearance of seismic information in high-frequency filtration 18-37-71-142

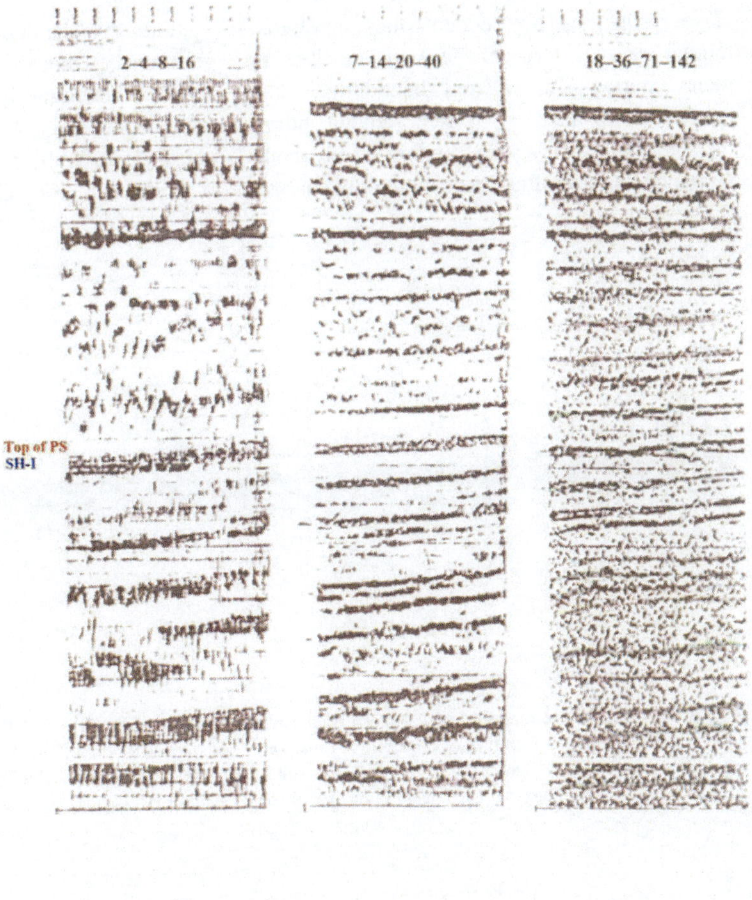

Fig. 6.13 Dissection of the PS section on different filtrations

is likely connected with the transmission band being outside the limits of incident wave spectrum impulse and those wave spectrums reflected from the deep levels.

6.7 Hierarchical PS Seismostratigraphic Units Derived from Acoustic Logging Data

Our investigations show that an acoustic logging chart and calculated pulse route of reflection factors, usually not used to reveal the PS certain relationships, may contain more detailed information about the section bedding nature and hierarchical units of medium and small ranks.

According to the acoustic logging data, a chance to do PS dissection into small stratigraphic units has been undertaken on an example of geoacoustic models of the Guneshli, Chirag, and Shakhovo-sea areas. From the Al charts, it is seen intervals of frequent change in deposits' elastic properties. So, bedded bands are characterized by spasmodic changes in velocities on their external and internal boundaries. Such bands are located, as a rule, among homo-bedded, mainly clay series, within which acoustic parameters are slightly and smoothly changed with depth. The lithofacies composition of surrounding sediments establishes the deep-sea formation conditions of these bands.

Klushin (1987, p. 41) believes that "any change (spasmodic or smooth) in acoustic parameters of the section reflects the changes in sedimentation regime parameters". According to this author, the mentioned sedimentation regime parameters are the totality of the section formation's paleogeographical and tectonic conditions. This thesis may likely be used for deep-sea basins with developed paleogeographical regions (shelf, continental slope, foot, and abyssal zone). Within the intracontinental shallow and shelf basins, the structure of very thin bedding and substantial composition of subparallel beds is controlled mainly by tectonic conditions (differentiated subsidence of the basin bottom) accompanied by an active effect of sea-level variation. Any acoustic (physical) boundary distinguished in AL charts records the basin bottom relief. Therefore, it is of chronostratigraphic significance.

From this standpoint, we investigate effective geoacoustic models given in Fig. 6.14. Any acoustic heterogeneity in the section is considered an elementary (small) seismostratigraphic unit (seismo-bed). Besides, these heterogeneities are reflected more clearly in the AS (acoustic

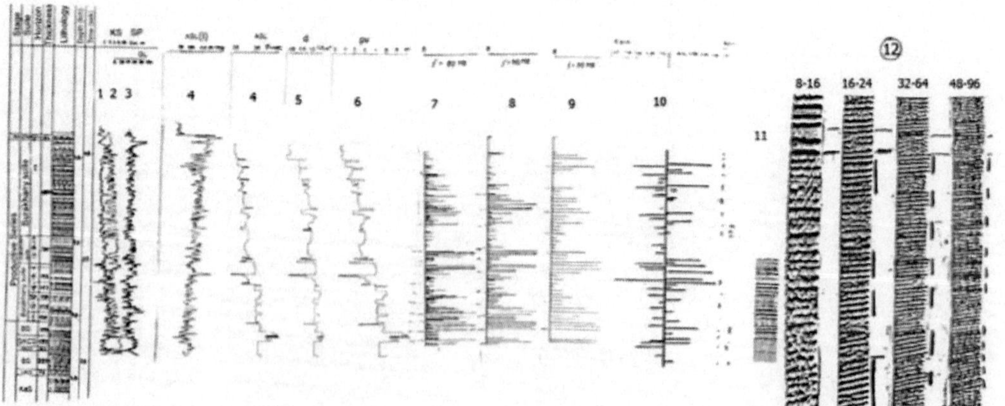

Fig. 6.14 Geoacoustic model of well No. 15 (Guneshli) (**a**), different-frequency filtration of an impulse route (**b**), and time section (**c**). Charts (1–3): (1) KS, (2) UKS, (3) CKS, (4) acoustic logging, (5) density lag, (6) acoustic stiffness, (7), (8), and (9) R_f absolute values on the frequencies 30, 60, and 90 Hz, respectively, (10) impulse route, (11) synthetic seismogram fragment, (12) time section's different-frequency filtration fragment

stiffness) charts than in the AL charts (Fig. 6.14). The AS chart describes that a bed model in the real medium may be considered interference with the many images of the bedded system elements.

The task of reconstructing such a system comes to selecting anomalies corresponding to the different SSUs. One of the simple dissection methods of the described section based on the AS and AL charts reveals slightly differentiated intervals and determination of background values employing acoustic parameters averaging. The three fragments are distinguished on graphs of acoustic parameters corresponding to the PS section gently bedded clay intervals (Fig. 6.14). According to the acoustic and dense logging data, a linear geoacoustic model is a detailed image of the PS bedding. The authors have constructed similar models describing total reflectance in the form of normal successions with practically unrestricted and uneven spectrums.

According to the reflectance chart, it has been proved (Mushin et al., 1990) that the PS dissection into cyclic units is possible by impulse route. Figure 6.15 shows an impulse route (well No. 5, Guneshli) and envelope lines indicating a weak acoustic dissection of quasi-homogeneous deposits formed in a stable (deep-sea) sedimentation condition except for separate rebound reflectance values, leaving out the limits of these lines. It is noted that on the background of very low reflectivity, the maximum reflectance values correspond, as a rule, to the acoustically sharp boundaries in sand bands connected with the breaks in sedimentation.

From the analyses of impulse routes taken from the other wells (well No. 6, Shakhovo-sea area; well No. 1, Girag), it is concluded that Rf (reflection factor) values are decreased with increasing clayiness of the rocks. It follows that an abnormal increase of amplitudes in AS (or AL) charts and Rf values in impulse routes indicate the basin shallowing periods. In contrast, those decreasing values indicate the periods of the basin deepening. Thus, the change in acoustic parameters indicates the changes in the bathymetry of the sedimentation basin during the Early Pliocene PRS formation.

Figure 6.14 shows the R_f moduli uneven succession connected with the PRS cyclic layered structure. It is shown that these R_f moduli are characterized by a broad spectrum (up to $f_{rp} = 90$ kHz). On the background of relatively weak and thinned-out lines, it is distinguished high-order intervals characterized by increased pulse duration lines. If R_f is decreased on the charts ($K = f(H)$) with limiting frequency $f_{rp} = 60$ and 30 Hz, it reveals approximately the same order units noted on effective seismic models—on time sections. An example of revealing seismo-cyclites 1–6 and of their dissection into more small units (seismobands) using $k = f(H)$ charts are given in Fig. 6.14 (7–9). Comparing the R_f model charts with time section fragments of different-frequency filtration shows that they are practically identical by a set of distinguished different order units and by line density. For example, the time section fragment by its timelines density on filtration 32–64 is identical to chart 8, while the 16–24 filtration fragments are identical to the 9th chart (Fig. 6.15).

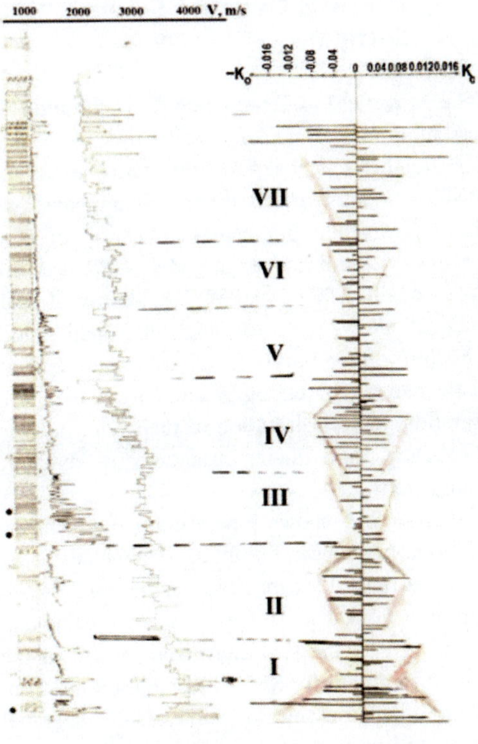

Fig. 6.15 Geoacoustic model of well 5 (Guneshli)

Geoacoustic model of well 5 (Gyuneshli)

Geological interpretation of /K/ = f(H) charts, taking into mind consideration lithological and bedding data of well No. 15 (Guneshli), shows that low-amplitude and thinned-out intervals in the time section are correlated with homogenous-bedded (clay) bands. In contrast, high-amplitude intervals are correlated with heterogeneous-bedded bands. The differentiation of this section based on other physical properties (according to the KS, UKS, and KCS charts) is directly connected with sedimentation conditions, resulting in the formation of different-ranked objects having their own substantial and acoustic characteristics. The example mentioned above shows that the effectiveness of dissection into hierarchical units may be raised using charts of acoustically active parameters (velocity, density, and acoustic stiffness) and reflectance moduli with different spectrums.

6.8 A Study of Very Thin Bedding and Cyclic Recurrence of PS Deposits Over the Seismic Sections

The wavefield different-frequency filtration is essential in sedimentary series dissection into hierarchical seismostratigraphic units. From the result of different filtrations comparisons, we have received a fragmental collection of time sections. Each describes a model of PS structure to a certain detailed extent. The totality of such models allows one to judge the hierarchical structure of the Early Pliocene complex and do their ranking based on reflections density and duration, the redistribution of their intensity and frequency, and their instantaneous dynamic parameters.

Figure 6.14 shows time section fragments in different-frequency filtration, distinguished by different-ranked seismostratigraphic units. When filtration from 8 to 16 Hz occurs, the PS is distinguished by the semitransparent record in the rank of sedimentary SSC-7. In the filtration from 16 to 32 Hz, this seismocomplex is disintegrated by its reflection density into two sub-complexes corresponding to the PS upper and lower division. According to the time section, the lower sub-complex is relatively more bedded than the upper one.

In most basin parts, the boundary between them is acoustically sharp, being connected with a long break in sedimentation. A thin sandstone bed in the bottom half of the Break suite suggests that on the border of the Early Pliocene lower and upper divisions, a deep-sea condition has been changed by shallow and maybe continental conditions. According to the AL data, the transition from the clays to close sandstones is sharp, while from sandstone to siltstones and argillites of the Break suite is gradual. Many subordinate sedimentation middle-ranked objects-seismo-cyclites are distinguished within the mentioned seismocomplexes under medium-frequency filtration from 16 to 32 Hz and 20 to 40 Hz. The latter comprises 400–500 to 1000 m thick, covering bodies zonally traced in different parts of the SCB. The seismo-cyclites revealed within the marine and shallow-marine complexes are characterized by progressive-regressive bedding type (prorecyclite) here. There correspond to the sedimentation rhythms distinguished before by the GIW data. Regressive parts of seismo-cyclites usually comprise the sequence of sandy-siltstone beds in the UKS, UKSS, and the bottom of the Surakhani, Sabunchi, and Balakhani suites. Their progressive parts are commonly presented by argillaceous and clayey beds (uppermost strata of KS, KaS, SKCS, and clayey parts of the PS upper-division suites). Thus, it is traced to a natural connection between reservoirs and regressive parts of seismo-cyclites, but clay covers relate to their progressive parts. This relationship is characteristic of the seismo-cyclites of marine facies complexes. At the same time, within the Gyuneshli, Shakhovo-sea, Umid, Dashly, Bulla-island, and other structures, a particular reduction in seismo-cyclite in some parts is noted. It relates to local changes in sedimentation conditions.

The distinctive feature of seismo-cyclites located in coastal-marine and deltaic complexes (basal suites within the paleo-seas peripheries) is the reverse directivities of the section regressive or regressive-transgressive bedding type, which is typical to the transition zones between continental and marine deposits. An interval without bedding characteristic to the sediments deposited under continental conditions is recognized in the border between the IV-th and V-th seismocyclites (between PS lower and upper divisions). It is inferred that during filtration from 32 to 64 Hz, these seismo-cyclites appear to disintegrate into seismobands limited in their expansion range. Because the PS lower-division suites have a small thickness (at great depths, less than the seismic wavelength), the wave packets tied—into them are very conventional. In the filtration from 36 to 72 Hz, dissecting some seismo-bands into two seismo-beds distinguished as a two-phase wave becomes possible. Synthetic seismogram calculations show that intensive quasi-regular reflections in the coastal–marine zones are tied to the relatively dense sandstone beds distinguished in the rank of seismic bed. They often proved to be the same as external seismo-cyclite boundaries. Besides, a decisive part in forming the reflected wave main extremum plays an acoustic boundary penetrating a different-facies medium on a level of the clay-sand interface (upwards).

All the III, IV, and V rank units examined are material pieces of evidence of successive natural geological processes (tectonic movements, SLRV, and the changes in paleogeography). Infiltration of the mentioned processes onto the sedimentation regime led to the formation of hierarchical co-subordinate units—sedimentation bodies limited by chronostratigraphic boundaries. Geological interpretation of $/K/ = f(H)$ charts, taking into mind consideration lithological and bedding data of well No. 15 (Guneshli), shows that low-amplitude and thinned-out intervals in the time section are correlated with homogenous-bedded (clay) bands. In contrast, high-amplitude intervals are correlated with heterogeneous-bedded bands. The differentiation of this section based on other physical properties (according to the KS, UKS, and KCS charts) is directly connected with sedimentation conditions, resulting in the formation of different-ranked objects having their own substantial and acoustic characteristics. The example mentioned above shows that the effectiveness of dissection into hierarchical units may be raised using charts of acoustically active parameters (velocity, density, and acoustic stiffness) and reflectance moduli with different spectrums.

6.9 Refinement of the Early Pliocene Deposit Genesis

The origin of the main oil-prospecting object in the SCMB is the Early Pliocene productive-red Series (PRS), which needs to be solved despite long-standing investigations. Baturin (1937), who studied the terrigenous-mineralogical zonation in the Near-Absheron region, concluded that alluvial-deltaic deposits of paleo-Volga and other rivers formed it. The essential role of paleo-rivers in the PS formation has also been noted in the works of Potapov (1954), Alizade (1960), Mustafayev (1963), Sultanov and Gorin (1963), Alizade et al. (1966) and others. The other group of researchers (e.g., Aliyev, 1947, 1949; Pustovalov, 1951; Shutov, 1962) believed that the source area feeding the Absheron terrigenous-mineralogical region may have been hypothetical land subsided at present under Middle Caspian waters.

Over the last years, the idea of the PS deltaic nature won new supporters and presented such essential data as lithofacies analyses of the rocks and geochemical and paleogeographic conditions of sedimentation confirming this conception. From the result of lithofacies analysis of the Absheron oil/gas-bearing region, it is established that the PS is an area of different-facies sedimentation conditions where the coastal-marine and lagoon conditions have changed the fluvial and deltaic environment. According to several researchers' data, on the Absheron sill and the same name archipelago, the deposits of the Kala suite (KaS), Underkirmaki suite (UKS), Superkirmaki sandy suite (SKSS), Break suite, and the lowermost strata of the Balakhani suite have been classified as mainly deltaic sediments. Only KS deposits have been classified as of typical lagoon or coastal-marine nature. In addition, the Kala suite (KaS) bottom bands occupied an intermediate position. The beginning of the PS sediments formation changed the Pontian Sea under the most expressed different-facies conditions.

Among the Absheron region, sandy-silty rocks dominate monomineral quartz varieties (90–80%). Within the Baku Archipelago, the quarts content in the PS sands and sandstones is increased, better sorted, and less clayey. A typical feature is that in the sections of the Baku and Absheron archipelagos, as well as of the Absheron Peninsula, sandy reservoirs with high filtration-capacity properties are extended up to 6–7 km in depth where non-lithified sands may also be found (Pashaly et al., 1985). According to many researchers, the major suppliers of a vast majority of sedimentary material were such big rivers as the paleo-Volga Plain River, paleo-Kur, paleo-Araz, paleo-Pirsagat, paleo-Uzboy, paleo-Tedzhen intermontane rivers and many other middle and small ones. Considering the presence of highly dissected topography on adjacent land in the Late Pontian and the spreading character of transported terrigenous material, the above suggests that broad deltaic zones have been developed in the regions of those paleo-rivers falling into the Early Pliocene paleobasin (Fig. 6.16).

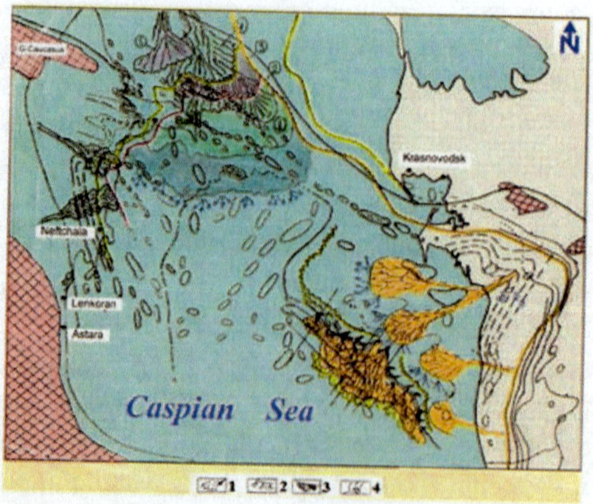

Fig. 6.16 Schematic representation of the spatial distribution of the Early Pliocene paleo deltaic shelf and slope complexes. (1) Paleodeltas, (2) foredeltas, (3) accretionary depocenters of clinoforms, (4) porosity and permeability equal value lines

From the information available on the lithology of PS sandy-argillaceous suites, it is inferred that they were formed during regressive cycles when relatively deep-sea (30–50 m) marine-shelf conditions of sedimentation have been changed by external subcontinental conditions. Several outcrops of thick obliquely laminated sands and sandstones in the section of the Absheron Peninsula indicate the river stream's activity. For example, lenticular sandy beds pinching out along the strike have been established in the UKS and SKSS sections. Since they have fibrous habit across the strike, Sultanov and Gorin (1963) believe they accrete the sedimentation slopes and are of deltaic origin.

The paleogeographic setting during the PS age was not single-valued and complex. Submarine valleys, erosional incisions, and Miocene folds dissected the basin bottom. Several bays have bordered shorelines. Among the others, widespread were clay diapirs connected with underlying deep-sea PS deposits of Oligocene–Miocene in age.

High sedimentation rate and the thickness of PRS sandy-argillaceous deposits (up to 6–7 km), cyclic structure recurrence, crossbedding, and progradation of sandy bodies with high quartz content. Finally, the complex's high oil and gas content are the typical features of deltaic sedimentary formations within the intermountain areas. In this connection, finding out and mapping the development zones of the PRS fluvial, deltaic, coastal-marine, and foredeltaic deposits acquire a tremendous national economic significance. In this respect, seismostratigraphic and seismofacies analyses effectively study the PS deposits' deltaic genesis and sedimentation conditions.

Some following concrete GDPM data show the effectiveness of these analyses. As mentioned above, the sharp fall of sea level in the Late Pontian caused an active process of erosional relief formation in adjacent mountain regions as in the plain areas of the Epi-Hercynian platform. The lowering of the sea level of erosion caused the process owing to which the paleo-Volga and other paleo-rivers appear to be filled up with clastic terrigenous deposits.

In time sections of latitudinal profiles drawn through the Middle Caspian fixed an erosional and truncated surface of different age deposits and up to 700 m deep paleo valley and incisions conditioned by the paleo-Volga and other paleo-river beds (Figs. 6.2, 6.7, and 6.17). The latter are filled up with terrigenous deposits of Upper Miocene-Pliocene in age. The paleo valley was composed of a wide (5–20 km) aggradation terrace and a gentle bed profile. It extends a long distance in the Middle Caspian platform part, reaching the northern slope of the North-Absheron trough (NAT). From the paleotectonic analysis, it is concluded that the paleo-Volga delta migrated from the north to the south, and its bed cut the Cretaceous and Paleogene-Miocene deposits.

Different provenances formed the Upper Miocene-Early Pliocene deposits within the NAT, but their formation was mainly caused by paleo-Volga deltaic mouth protrusion to the south. By the beginning of PS time, the NAT was a system of shallow lagoons and lakes on the platform edge. As a result of the Miocene sedimentary complex rise, caused by the formation of an accretionary prism in the Absheron sill, a natural barrier was formed to the water stream from the north. In that case, the NAT played the role of the trap for sedimentary material, decreasing its penetration to the SCB. The presence of filling seismofacies (Fig. 6.17) in the form of parallel and interrupted reflection with minor amplitude variations suggests that the deposits were formed at the beginning of Pliocene time under calm hydrodynamic conditions.

The large-scale regressions of the Pontian basin and ascending tectonic movements on its edge zones led to the perceptible paleogeographic reconstruction. All the territory around a tiny and closed relic in the South Caspian appears to be drained, proven by a long sedimentation break and by unconformity established by drilling and seismostratigraphic analysis data. The fact that PS underlying inclined beds and oil uplift arches of NW Absheron Archipelago are truncated is clearly fixed in the time section (IV in Fig. 6.17), which is objectively evidenced by ascending movements in the Pontian and

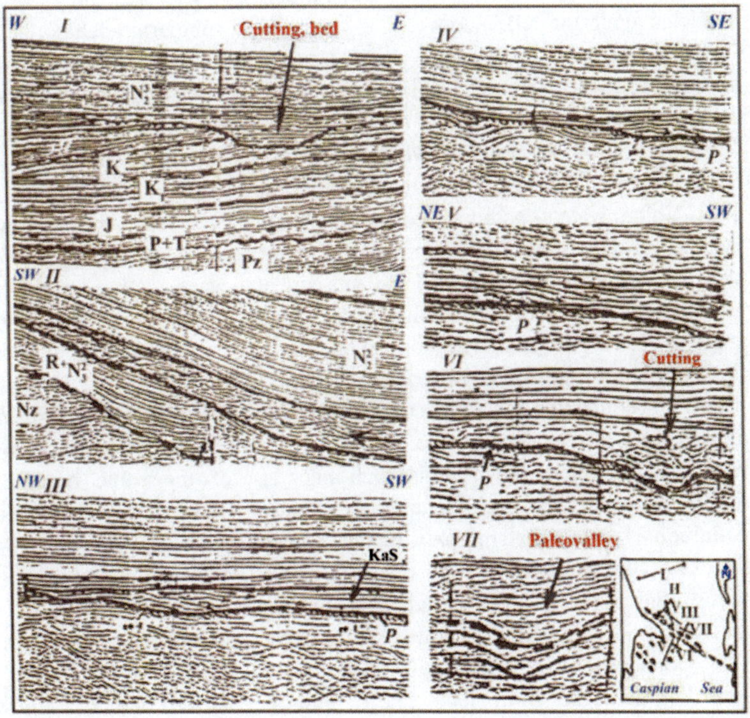

Fig. 6.17 Time section fragments illustrating exogenic cuttings (I), tectonic truncation (IV), paleo-valley (VII), stream seismofacies (III, VI), and microclinoforms on the terrace scarps (III)

subsequent rock scouring resulting from the transgression of PS basin. During the period described, when the lagoons and lakes in the NAT area have almost been filled up with northern terrigenous material and when the Caucasian-Balakhan isthmus draining has been completed, the basin has gone over to sub-meridional down warping (Potapov, 1954).

The power water streams have been moved from the north to the south through the narrow valley, paving the way over the most depressing parts of Pliocene paleorelief (IV in Fig. 6.17). The seismostratigraphic analysis and paleotectonic reconstructions to reveal erosional entrenchments in time sections drawn through the PS lowermost strata suggest that most of the river run-off has been delivered to the SCB in the locality of the Ashrafi structure.

The old KBA slope and Mesozoic cusps have screened the river stream from the NE. This paleo valley has been branched to form an enormous deltaic system in a southerly and south-westerly direction. The thickened fluvial and deltaic sands have accumulated along the KBA slope because of the fall in the flux rate. That explains the presence of hilly-like PS seismofacies along the KBA slope.

Deep erosion of Pontian and Miocene deposits up to the Maikopian suite established by drilling data in several areas (Neft Dashlari, Guneshli uplifts) suggests that vigorous streams and submarine flows have occurred there. The paleo-Volga waters outburst led to the sharp increase in the supply of clastic material to the closed Early Pliocene basin and the subsequent accumulation of thick terrigenous series. In a study of the PS genesis and formation conditions, the morphological features of paleorelief, i.e., primary sedimentation surfaces of the PS deposits, are of great importance.

The authors have drawn the isopach maps of the PS bottom set lithofacies to provide a closer definition of paleorelief morphostructure within the Kala Sea slope zones and reveal possible lithostratigraphic traps. The latter belongs to only the Kala suite (KaS) and its lower sandy band. The anomalous thick zones (250–300 m) with spreading-dendritic form look like a run-off system and widen in the basin direction are clearly distinguished in those maps (Figs. 6.11 and 6.18). Besides, some islets and small dome-shaped hills were distinguished, cut by separate ducts.

The mentioned ones are distinguished in time sections of sublatitudinal profiles as 2–3 km wide negative morphostructures, indicating the prominent deltaic distributive stream channels position. The paleo-Volga complicated sigmoidal-obliquely laminated deltaic deposits are clearly traced on the standard processing materials.

They turned out well in time sections of profound processing (in relatively high-frequency filtration with conservation of amplitude relations). In seismic sections oriented in the direction of prevailing sedimentary material supply, it well distinguished sigmoidal, interrupted-wavy synphase axes typical to the small-amplitude alluvial-deltaic sedimentation bodies of lateral accretion (III in Fig. 6.17). In cross-section, they are characterized by lenticular configuration (IV in Fig. 6.17). In the coastal marine zone, such bodies have a lobate shape in a plan but are lenticular bedded in the sections (Fig. 6.19). Therefore, they may be attributed to the submarine-deltaic facies belt within which sand and sandy-siltstone deposits are usually developed. The fact that the latter are pinched out up dip the slopes of the Kala and Underkirmaki seas should be considered a pre-condition for forming non-anticlinal traps.

Within the shallow-shelf basins, the main transporting agents of sedimentary material are wave action, alongshore, and drift currents. According to Strakhov (1963), the waves stirring up and deforming depths within the living Caspian and Black Sea basins correspondingly come to 25–40 m and 8–12 m. These parameters in the shallow Pliocene paleo-basin have probably been of the same order. The terrigenous material supplied by paleo-Volga and other streams (paleo-Kur, paleo-Uzboy) has periodically been downthrown on the shelf and transported by waves and currents to the central basin. Special sedimentation features were its cyclic nature dependent on SLRV and its very high (avalanche) rate (2–3 km/Ma).

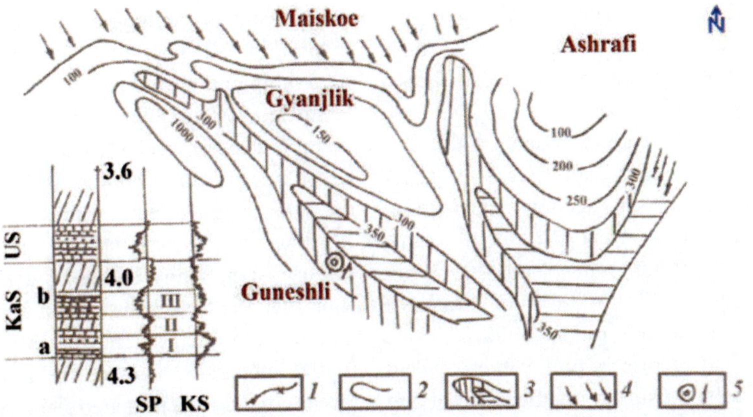

Fig. 6.18 Thickness map of the Kala suite lower pinching out and well No. 1 (Guneshli) electric logging curves. (1) pinching out boundaries of the band, (2) isopachs, (3) zones of abnormal thicknesses, (4) microclinoforms development zones, (5) exploratory drilling well, I, II, III are the micro-rhythms lithofacies of the Kala suite; anticlinal structures: a—Aipara, b—Ashrafi, c—Gyandjlik, d—Guneshli

The authors have used a method of logging facies analysis (Muromtzev, 1981) and lithofacies analysis data (Buryakovsky et al., 1995) for the unmistakable determination of terrigenous deposits origin of the PS bottom suites. Figure 6.18 shows the logging diagrams from the No. 1 well drilled on the Guneshli structure.

The SP curve of the No. 1 well log (Guneshli) shows that the highly serrated in-blocks and cylindrical forms shared by a positive anomaly of the natural potential correspond to the KaS lower band (Fig. 6.18). According to the logging facies genetic type and lithological analysis data, this band is characterized by favorable filtration-capacity properties and was formed under complicated hydrodynamic conditions when the joint effect of strong alongshore and fluvial streams has occurred. The lower coarse-grained band has been formed under intensive terrigenous material supply conditions during the low sea level. The obliquely laminated and lense-shaped partings of the deltaic origin found in the drill cores are confidently identified as the deposits of alongshore (regressive) and offshore bars. The basin transgression caused by the sea-level rising (1st cycle of SLRV) led to an advance of shoreline to the north deep into a land that resulted in coarse-grained bar deposits being overlapped by clay cover. During the period of sea-level stabilization, because of deltaic deposits reworking by alongshore currents in the coastal-marine zone, there was formed the upper coarse-grained part (with fine and median bedding) of the mentioned band (III-d microband). The UKS curve configuration indicates that this transitional type of band (coastal-marine) is located between the marine and continental facies. The grain sizing is changed from fine to coarse-grained varieties and vice versa (argillite-siltstone-argillite). As has been reported by Muromtzev (1981), such a band is related to the typical deltaic facies complex.

In the late Kala time, in connection with thick series formation within the deltaic system and sedimentation growth of anticlines, the main streams were displaced to the northwest. So, during the Underkirmaki and subsequent times, the paleo-river new delta developed within the area located west of the Gilavar and West-Absheron structures. From the time sections showing the area between the North Absheron and Gilavar structures, it is seen that on the background of partial reflections and in the conservation of proper amplitude relations, there are distinguished a few intensive obliquely laminated reflections with marked regions of their overlap on surface of unconformity and erosion. The latter, as it is well known, corresponded to the high-power sedimentation condition in the submarine-deltaic facies zone. In the plan, the mentioned seismofacies area has a ligulate shape (Fig. 6.14), while in a section, it is wedge-shaped with regressive stratification.

In the coastal marine zone (Gilavar—Guneshli—Neft Dashlari), the UKS is about 100–1300 m thick. Time sections obtained in 10–60 Hz filtration show that this suite (the same as the other small-thick suites of the PS lower division) comprises 0.04–0.06 s of the time interval that corresponds to the reflected wave one and one-half recording phase. It is evident that in such seismic survey methods, resolving power restrictions to make known the problem concerning the deposit genesis and sedimentation conditions (based on the SFA data) is almost out of the question. From the long and varied notable works in the region studied, it is inferred that the SFA may be effectively used if studying objects are represented at least by 3–4 timelines.

According to the GTW data obtained from the No. 4 well (Eastern Gilavar), the cross-laminated seismofacies distinguished on time sections of several profiles drawn through the northern Absheron—eastern Gilavar structures are confined to sandy beds of the Underkirmaki suite and lower regressive semifacies of the Kirmaki suite. The presence of alongshore (regressive) and coastal types logging facies on the UKS graph (Fig. 6.18) confirms our supposition about the submarine–deltaic genesis of the cross-laminated seismofacies (Fig. 6.19).

From the drill of the No. 39 well located close to the first pinching outline, it is concluded that the same logging facies represent the UKS and KS lowermost strata. During the Kirmaki Sea transgression, sandy bodies of deltaic and near-shore genesis have been buried under clay caps, and seismic images are represented by interrupted subparallel axes of the reflections' equ-phase condition. The paleo delta north-westerly recession occurred during the relatively long Kirmaki sea transgression.

Within the Middle Caspian in the Gyzylburun-Deniz and Agzybirchala-Deniz structures location area, we succeeded in tracing the two punching outlines related to seismobands, the first of which is indicates the Kirmaki Sea contours and the second—of the Break suite. Because of the Underkirmaki sandy suite (UKSS) and Superkirmaki clay suite (SKCS), we cannot determine whether it is pinching out exact points. From the seismostratigraphic and paleotectonic analyses, it is concluded that during the Early Pliocene, the terrace cusps leveling on the initial sedimentation surface were distinguished within the PS lower

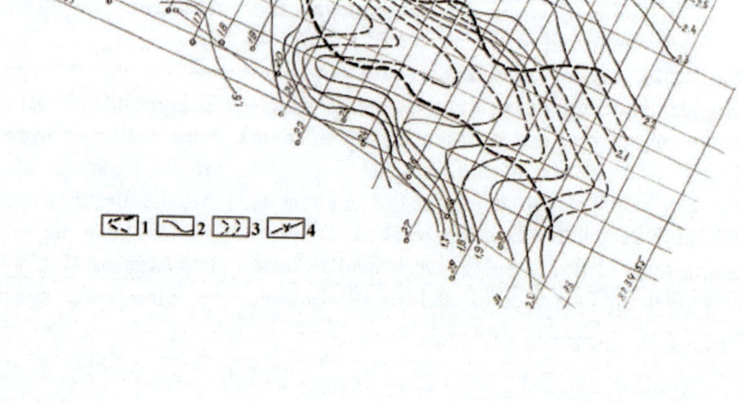

Fig. 6.19 Isopach map over the "SF-UKS" roofing on a scale of 1:100,000 (after P.Z. Mamedov). (1) seismofacies pinching out, (2) isolines over its roofing, (3) the surface of unconformity isolines, (4) seismofacies range boundary

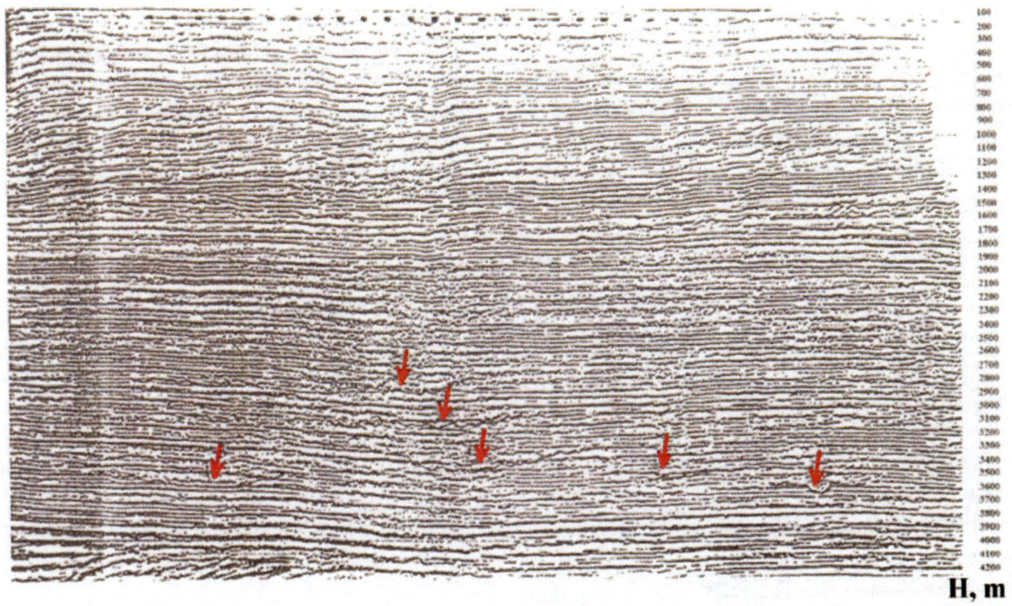

Fig. 6.20 The seismic time section shows erosional incisions and scours in the RBS section

division in STP peripheries. Correspondingly, seismobands have been formed over the continental blocks.

Within the NAT, they are composed of sub-parallel beds having sub-horizontal boundaries. Such seismobands' shape suggests that they have been formed only on the shelf zone at not large depths. Hence, the lower PS seismobands may be considered according to the law of the bed thicknesses perspective relationship when the sedimentary series lies on one united block. Thus, the northern Early Pliocene shelf zone has typically experienced its bottom compensated subsidence. This mechanism is hard to understand as nothing but a self-controlled process (subsidence amplitude has been dependent on sediment mass). Otherwise, two independent processes (sediment supply and block subsidence) should be coordinated.

Several rip currents and shallow shelf facies are distinguished on the PS diagram obtained from the No. 4 well drilled on the E. Gilavar structure and located 70–80 km from the Kirmaki seashore line. The latter is a distinguishing mark of high and median energetic sedimentation conditions. The lithologic characteristic of the core (fine-grained, poorly cemented sandstone) recovered from the depth of 2355–2360 m confirms the forecast on logging facies. The KaS is represented by shallow-marine facies passing up sections into the open sea facies (at 2720 m depth) within the Arzu seaside area.

From the analysis of "logging" facies data of the No. 4 well located in the North-Absheron area, it is inferred that the KS is mainly composed of coastal-marine facies of the north-westerly direction, the rip currents, and transgressive offshore bars that point out the nearness of a coast.

The alongshore transgressive bars-type coastal-marine facies with commercial oil reserves have been drilled in by the No. 35 well on the Western Absheron structure. In the Kirmaki lowermost strata, oil-bearing sandstones and sands are overlapped by a silty-clayey band. The sedimentation cyclicity within the Kirmaki Sea is expressed by the UKS curve, where six transgressive and three regressive bands are distinguished.

From the sea-level variation study using the lithofacies analysis and seismic material data, it is concluded that to the north of the Absheron sill, the UKS, UKSS, and Break suite sandy sediments containing a significant amount of quartz (up to 70–90%). The latter was formed during the low sea-level periods. The PS basin boundaries widened in a north-westerly direction due to the abundant fluvial run-off of paleo-Volga and other rivers and the intensive (super-compensation) supply of sedimentary material. The two significant factors—abundant solid run-off and sea-level variation, have been conducive to forming the PS cyclically bedded sandy-clayey deposits of deltaic origin. During the UK age, sedimentation cyclicity is reflected in seismic sections (by the cyclic repetition of similar seismobands) over enormous shelf areas. The seismobands with close-arranged axes of phase coincidence corresponding to sandy-silty bodies are shown in the time section as alternating semi-transparent intervals. The DIW and drilling data show that they tend to be mainly clayed series.

As the rate of subsidence of the Early Pliocene basin bottom has increased, its contours appear to be widened, and most of the land on the basin slopes has been filled up with seawater. The basin boundaries tend to be moved further to the northwest, with the paleo-Volga delta recession simultaneously (see Fig. 6.16). An analysis of the facts on the Absheron Archipelago shows that the most important oil and gas accumulations in the PS bottom suites are confined to the sandy-silty sediments of alluvial-deltaic and coastal-marine origin. In this connection, such not yet drilled uplifts of the northern anticlinal belt as Sevindje, Dan-Ulduzu, Novkhany, Aipara, Hamdam, and others may be classified as highly prospective structures containing great HC concentration. All the mentioned structures are located within estuarine and deltaic systems development zones during the Kala and Underkirmaki times and within alongshore and offshore (regressive and transgressive) bars during the first half of the Middle Pliocene.

Besides, the traps in the lagoon (lacustrine) and alluvial-fluvial facies of the Kirmaki and Superkirmaki clay suites within the North Absheron water area may be oil and gas-bearing structural elements. So, from the seismofacies analysis of seismic data and logging facies study (using lithofacies analysis of cores), it is closer to defined formation conditions of sedimentary series and revealed the lateral zones of alluvial, lagoon, deltaic, and coastal-marine deposits.

In the marginal part and not far off the PS sedimentation basin shorelines, it was revealed that the mapped zones of development of obliquely laminated and sigmoidal bodies of deltaic origin were hardly determinable in compensation basins. The seismostratigraphic analysis (using the data of paleorelief morphostructure study) and considering the suite and band thicknesses made it possible to interpret the problem of the genesis of the PS bottom suites deposits from an essentially new standpoint. As a result, the favorable zones for forming essential lithologic-stratigraphic traps for oil and gas accumulation were revealed.

6.10 Studies of the Genesis and Evolution Regime of Lateral Accretionary Sedimentation on the Turkmenian Shelf During the Early Pliocene and Their Stratigraphic Models

The GDPM seismic material indicates that the Early Pliocene and Quaternary lateral accretionary sedimentation layers are widely developed in the Turkmenian shelf (TS). It is well-known that lateral accretionary layers are generated in uncompensated down-warping basins (UDWB) in sea-level variation and sedimentation rate aggregate. These bodies, weakly controlled by tectonic movements, are mainly generated during relatively low sea-level stages when shelf regions appear to be subjected to erosion resulting from the strengthening of paleo-rivers and sub-fluvial currents activity when the base-level of erosion becomes sharply depressed. Sedimentation bodies are generated by the river and submarine fans on the inclined (slope) surface.

A generally accepted thesis of a weak influence of tectonic movements on clinoform shaping is based on the data obtained from the platform regions or those far away from the mobile belts. Such bodies are characterized by the beds' initial inclination and specific clinoform configuration of their boundaries. This thesis does not reflect sedimentation conditions in uncompensated paleobasin of the Alpine mobile belt during an orogenic stage.

It is known that the tectonic regime in mobile belts is characterized by instability and repetition of uplifting cycles in denudation regions and down-warping cycles taking place in accumulation regions. We believe that in fundamental studies of oil and gas-bearing PS and RBS structures, it should be rejected all the hackneyed model notions of original horizontally stratified series under the condition of basin bottom compensated down warping; what is more, all existing views of the basin structural history, as well as oil and gas formation mechanism, should be revised. Therefore, a study on the genesis and development regime of clinoform bodies in SCB sedimentary cover, their synchronization with erosion surfaces, and the construction of cyclic sedimentation models should be based on the argumentative geological conceptions and fundamental seismostratigraphic principles. In addition, the following immutable geological factors confirmed by seismostratigraphic analysis data should be accepted:

1. The Early Pliocene basin was formed at the base of the Pontian deep-sea basin. Consequently, an uncompensated trough existent in its central and southeastern parts, whose slopes appeared to be accreted by lateral-deltaic and avandeltaic bodies having clinoform shapes,
2. The primary mechanism provided for filling in SCB shelves and slope areas is believed to have been fluvial-deltaic and avalanche sedimentation, correspondingly.

With the purpose of the PS and RBS dissection and synchronization of their suites and horizons by seismostratigraphic methods, the authors have used such fundamental principles as the three-dimensionality of sedimentation bodies, chronological significance of their boundaries, and the age assessment of seismostratigraphic units by sea-level eustatic variation-scale (Wayl scale) or by RSLV local scale. As a basis for these postulates, sub-horizontal and cross-macrobedding indicators are considered sedimentation conditions in the SCB basin slope zones. Besides, the leading indicators mapped over the vertical and lateral series of quasi-synchronous bodies are the parageneses of the deltaic sigmoidal bodies, shelf compensation covers, foredeltaic and slope terrigenous clinoforms as well as deep-sea condensed covers.

6.10.1 Lateral Series of Red Sedimentation Beds and Eastern Pliocene Basin

The formation of thick clinoform red beds in the eastern Pliocene basin was caused by the following factors:

1. The nearness of source areas (Kopetdag mountain system and plain territories),
2. Widely spread river drainage and deltaic systems caused by an abundant supply of coarse waste into a sedimentation basin,
3. High energetic level of sedimentation,
4. Cyclic change of relative sea level,
5. Redeposition of the previously formed deposits on the shelf and slope.

These factors' combination appears conducive to forming a sedimentary series of deltaic genesis in the coastal zone and to the widespread temporal and turbidity streams and landslide events. During the sea-level cyclic lowering, intense storms, and undercurrents, the surplus sedimentary material has been downthrown from the shelf platform onto the slope area, forming foredeltaic and slope clinoform bodies.

Based on SSA of abundant seismic data and lithofacies analysis of cores obtained from the SCB, we concluded that during the Early Pliocene time, especially in its first half, a fan delta-shelf type was developed in the low sea-level stage with the development of the continental slope system (Mamedov, 1986). Abundant rutaceous material has been supplied from the raised areas of the basin's eastern frame (Bolshoi Balkhan, Malyi Balkhan, and the Kopetdag mountain system), being delivered by the branched water streams and numerous river systems to the sedimentary basin up to the fan deltas.

The paleo-Volga fan deltas have primarily been advanced from the northwest along an axis of Kyzyl-Kum trough's bay-shaped embryo and along with Alandag-Messerian and Gograndag-Chekishlyar tectonic benches in the region of Kamyshdja-Okarem structures—from the southeast (Fig. 6.16).

According to the paleogeographic reconstructions and seismofacies analysis, an initially sizeable deltaic system has been developed in the region southwest of the junction between the Bolshoi Balkhan and Malyi Balkhan and then reached the Uzun-ada and Erdekly structures location area. The granulometric analysis data shows that the zone containing the maximum number of sandy-silty rocks (up to 80%) was in the region of the Boyadag, Kumdag, Karadashly, Kamyshldja, and Gograndag structures (Alizade et al., 1985). In mountain environments, sandy rock amounts reach 90% of the lower red bed (LRB) total thickness.

Porosity in the LRB reservoirs is increased toward the basin with decreasing clayiness and carbonate content. In the region of Malyi Balkhan, the isopleths of porosity, clayiness, and carbonate content look like they broke off, having fan-shaped outlines (Fig. 6.16) that point out the river valley provenance location.

The RBS lower part is characterized by delta front and fan delta distal facies developed in its foreshore line. Timelines on NE-SW-oriented seismic sections around the Ogurdja island and Sharg Ulduz structure form a cross-laminated pattern. Oblique axes of the reflection phase coincidence are convergent dips toward the basin. On the sections of latitudinally oriented profiles, reflections from the lower RBS have upwardly curved, hilly-shaped, and hummocky configurations. The signs of the presence of erosional channels and scours are also noted (Fig. 6.20).

The mentioned seismofacies units correspond to the coastal-marine sedimentation condition typical to the proximal and medial fan delta. In the lower RBS, reflection tracing and dynamic expressiveness improve in a southwesterly direction. A macro-bedded sub-horizontal reflection type typical to the shelf planation cover has been noted in the Khanlar, Fersman, and Fedynsky mud volcanic structures region. The time section also notes the foredeltaic and clinoform bodies of lateral accretion. A wide foredelta beyond the shelf brow indicates the unerring existence of a deep-sea tectonic basin on its west-southwestern margin.

As mentioned above, the basin has been inherited from the Pontian relic of the Paratethys, which by the end of the Early Pliocene was a closed basin bordered from all sides by immense mountain structures and abrasive-denudation plains. The latter has been the shelves of the Pontian basin, widespread in an enormous area (Semenenko, 1987). By the beginning of the Pliocene time, this basin had played the natural sedimentation trap in connection with a wide river system formation and transportation of bulk clastic sediments. During the Early Pliocene transgression initial phase, a thick foredeltaic complex was formed on steeply dipping slopes of this basin, which is distantly located opposite the deltaic system. The time section revealed this complex by its specific seismostratigraphic features and then mapped in southwestern shelf margins. Based on the thickness of the foredeltaic complex at the initial phase of the slope accretion, the basin depth came to 500–700 m and more. Subsequently, a primary slope has been accreted by this complex up to 10–15 km (Fig. 6.21).

The next episode occurred when the sea basin was filling, evidenced by sedimentation area displacement toward the slope. At that time, the avalanche mentioned above the sedimentation mechanism was controlled by sea-level variations. Many clastic materials that omit shelves have been thrown out to the slope area.

A series of large clinoform-lenticular bodies represent the lateral accretion complex elongated archwise to 100–200 km (approximately) along a slope. During the relatively short geological time (about 0.7–1.0 Ma), these clinoforms have successively accreted a slope that resulted in the formation of a new sedimentation slope being displaced towards the basin for more than 25–30 km.

6.10.2 Seismostratigraphic Model of the Shelf-Slope with Its Lateral Accretion

The seismostratigraphic model of the shelf-slope with its lateral accretion elements developed in the southeastern basin during the first half of the Early Pliocene is given below.

1. The shelf cover. The RBS bottom shelf cover transgressively overlies the surface of the primary sedimentation, which is eroded and tectonically dislocated. The latter appears to be a surface of unconformity (SU) and, as it was mentioned above, is step-like subsided landward. The bottom shelf cover has been formed during high sea levels and low energetic conditions. In seismic sections, it is distinguished as up to 0.7–1.0 km thick seismoband (SB-1) that flattens out an SU scarp. According to Alizade et al. (1985), the lower Dagadjik suite of RBS, the most widespread within the WTB, is characterized by a thickness of 500 m in the littoral zone (Karadashly, Kamyshldja, and other structures). Because of the absence of seismic data on the shallow littoral zone (about 70–80 km wide), we have not been able to correlate SB-1 with the Dagadjik suite. Nevertheless, their synchroneity has no doubt.

2. Foredeltaic clinoforms. The transition from the shelf cover to the slope corresponds to the transition from the shallow to the relatively deep-sea conditions. The sediments accrete a relatively steeply dipping slope (up to 160) in a southeasterly lateral direction. The sizeable rear clinoform represented this complex at the early stages of its formation with many specific structural features. The most distinctive are cross- and hummocky-bedding and general elongation along a slope (Figs. 6.21 and 6.22). The cross-bedding feature in the upper clinoform (A_1) indicates a steady-state sedimentation condition during relatively low sea levels. The fact that a single roofing surface truncates this slope suggests an effect of submarine erosion and multiple reworking of sediments. The bedding feature is expressed weakly in the middle and

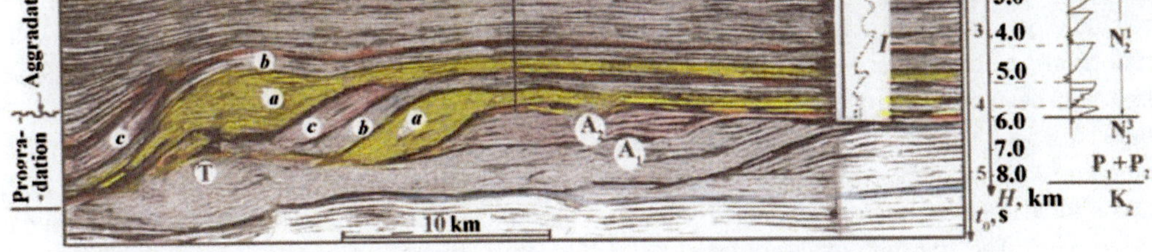

Fig. 6.21 The seismic time section shows laterally accreted bodies in the RBS section (Turkmenian shelf). A_1 and A_2 are the foredeltaic clinoform's semifacies; T designates turbidites; a, b, c are the elementary bodies of the slope clinoforms (clinocycles); a—clinoform, b—clinocover, c—cross-bedded bodies. I: reflected waves visible frequency chart

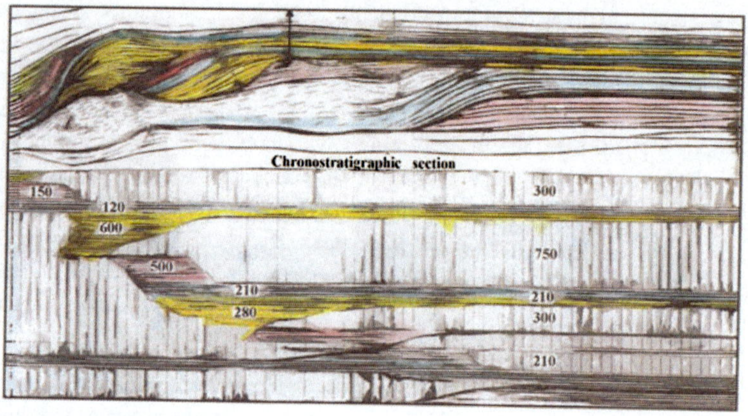

Fig. 6.22 The seismic quantum and chronostratigraphic sections of the lateral accretion complex in the Turkmenian shelf

lower clinoforms, and prevailing reflections are intermittent and hummocky configurations. This reflection pattern corresponds to the foredeltaic sediments deposited in changeable conditions (Payton, 1977). The beds are crumpled into fine folding because sedimentation slumps on the slope. Sedimentation faults typical to the foredeltaic sediments are dipping toward SB subsidence. The "hummocky" relief is well expressed in the lower part of the clinoform.

Transparent seismofacies appear to be formed under a system of slope clinoforms. The lower foredeltaic clinoform "A_1" comprises intermittent chaotically arranged axes of phase coincidence of very weak reflections. The deep-sea sediments of the lower slope clinoform-1 have undoubtedly accumulated in this part of the basin. Usually, similar seismofacies are composed of foredeltaic plastic clays. As to the sedimentation faults in overlying clinoform-1, we propose that they have been formed due to sedimentary load quick-accumulating deep-sea sediments.

Such specific features as semi-transparent seismofacies habit, weakly expressed inclined reflections, a large cone volume, and infinitesimal thickness of synchronous shelf deposits (or their absence) suggest that the seismofacies as mentioned above may confidently be attributed to the foredeltaic regressive semifacies of the big paleo-river. The transgressive semifacies A_2 of this complex are relative to less volume. It is represented by sub-horizontal long-periodic reflections overlapping the sigmoidal-corrugated surface of the regressive semifacies A_1. Towards the basin, reflections appear to be confluences or discontinued dependent on the beds' successive pinching out resulting from the sediments increasing the deficit. With increasing vertical accretion of foredeltaic sediments, the depth of the basin's uncompensated part has increased, and a new sedimentation slope was formed.

So, a thick foredeltaic complex has been formed for one uninterrupted cycle of sea-level lowering-rising. Besides, a lateral accretion of the primary paleo-slope up to 15 km with insignificant vertical accretion of the shelf (about 100–200 m) has occurred. The steepness of a newly formed sedimentation slope in its most steeply dipping part reaches 20–250.

3. Slope (cyclic) clinoforms. As noted above, this clinoform type is formed on uncompensated basin slopes where the predominantly gravitational genesis sediment volume is greater than the compensatory aggradation potentiality. Morphologically, slope clinoforms differ from the preceding foredeltaic complex, indicating sedimentation changes. These factors are conducive to the avalanche-like "sweeping" of sedimentary material from the shelf and the formation of thick progradational clinoforms on the slopes. The sedimentation process is cyclic. Each stage in the mobile supply of terrigenous material and its passing to the shelf edge is characterized by specific and concrete surfaces of unconformity. The latter are separated from each other and

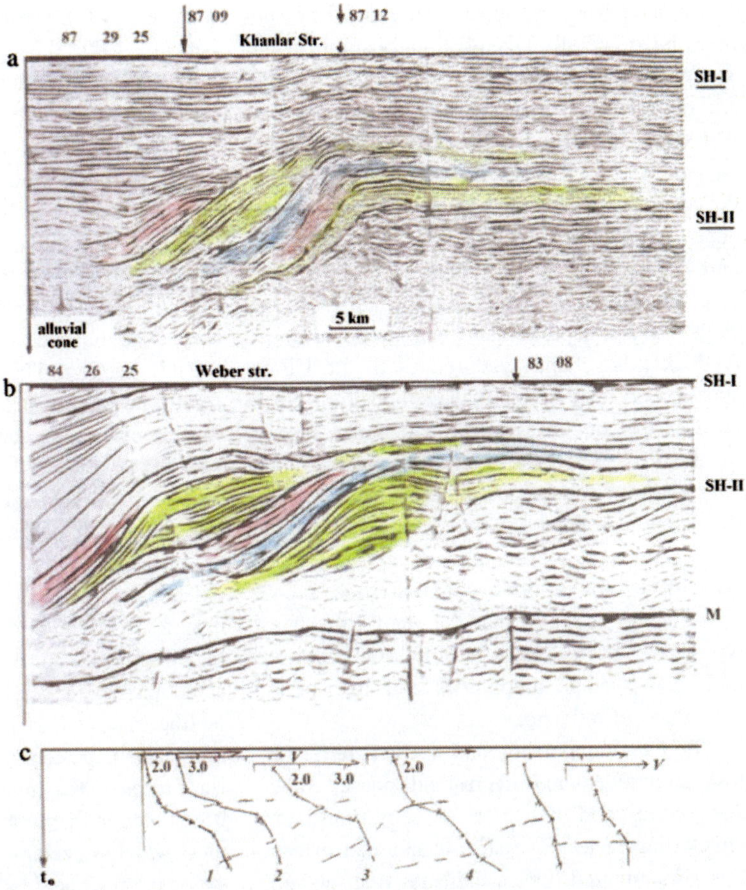

Fig. 6.23 Time sections showing the lateral accretion bodies in the early Pliocene complex (**a**, **b**), **c** is the charts of $V_{ort} = F(f_0)$

synchronous cross-laminated-lenticular sedimentation bodies. The latter are distinguished in the seismic section by their specific seismofacies features.

6.11 Seismostratigraphic Model of Slope Clinoform Sedimentation

The time section fragment defines two large slope clinoforms limited by high-amplitude reflections. Each of them is dismembered into three distinguished elementary bodies. The first elementary body is clinoform-1a (CF-1a), whose internal reflections have a sigmoidal configuration. The seismic section also shows a particular law-governed nature in their alternation.

The CF-1a formation began from the low sea-level stage, evidenced by the denudation-erosional sedimentation surface overlapped by the lower reflections. At that time, sedimentary material was supplied on the deep-sea slope through the dried shelf. Calculations carried out by methods described in Kunin and Kucheruk (1984) allowed estimating the basin depth to be equal to 200–300 m.

Sigmoidal beds accreted the clinoform mentioned above laterally formed during subsequent SL rising. According to the drilling data (No. 1 and No. 2 wells drilled on the Fersman structure), this clinoform corresponds to the clay band from which a two-phase well-defined seismic oscillation is reflected (Fig. 6.21). The angle of the beds' slope within the thickest part of CF-1a reaches 6–80.

As seen from Fig. 6.21, sigmoidal bed surfaces have laterally taken a high position in the section that testifies to successive RSL increases and retreat of the shelf far to the east. It is known that subsea depth may be increased due to either eustatic sea-level rise or the basin bottom subsidence. In absolute value, an increase of subsea depth has occurred only in an uncompensated part of the basin since lateral accretion of the slope has taken place only in this area. As to the shelf zone, its depth remains in a low-level stage. A thick sandy band synchronous to the CF-1a speaks well for the mentioned supposition. Such facts as the roughness of the paleorefief surface of the clinoform 1-a underform roofing part and roofing truncation of the beds suggest that erosional processes have occurred on the shelf. Accretion of the underform part up to 300–400 m for a brief geologic time (about 0.1–0.2 Ma) suggests that the shelf subsidence was active. Typical features of the shelf zone are high and average amplitudes and long-period oscillation of reflections.

The CF-1a underform part is characterized by low amplitudes and interrupted phase coincidence axes and transition to semi-transparent enveloping facies. The steeply inclined orientation of sigmoidal bodies indicates that the sediments of predominantly gravitational genesis have been laterally accreted. As is known, the latter is more highly saturated with terrigenous material. Therefore, steeply dipping angles of the natural inclination of high-density sediments are preserved in the seismic section. The length of the clinoform-1a body itself is about 8 km by 1.5 km maximum thick.

During a high sea-level stage, the process of terrigenous material supplied by density streams has been discontinued, and clay material has vertically been deposited, enveloping this clinoform. As a result, a poorly bedded covering seismic body of regional expansion was formed, which is believed to be a component element of the slope clinoform-1. In the publication, several researchers have named this body differently, such as clino-blanket (Gogonenkov et al., 1988), transgressive semifacies (Kunin, 1989), and progressive part of seismofacies (Sosedkov & Surkov, 1989). From our investigation, it is concluded that this seismic blanket body is not only a component element of the lateral accretion complex but of the shelf complex as well. Therefore, we believe that the "clinoform" term is the most acceptable in the described seismic body.

A clino-blanket-1b infills the clinoform accumulative slope in the low-energetic condition of sedimentogenesis. This unit's roofing surface is located at a higher horizon of the seismic section. This clinoform is traced as a semi-transparent recording related to the homogenous lithological composition in the time section.

The next element of slope clinoform-1 is tangentially cross-bedding seismofacies (SF-1b). A series of steeply dipping axes of phase coincidence is up-dip-ended in the roof adjoining design. It is inferred that the recording cross-bedding pattern corresponds to the section lateral accretion stage in a relatively stable low sea level on the shelf when all volume of sedimentary material is deposited on the slope area, omitting a shelf terrace. The highest amplitude values have been noted in the SF upper part. This area has the most widely developed sands and siltstones with clays in a slope-depressional zone. Reflection phase coincidence axes in seismofacies' lower part are sharply broken at the bottom surface, which relates to insufficient GDPM resolving power.

As distinct from the clino-blanket and clinoform accreting a slope to a certain extent, a cross-bedded geological body in SF-1b accretes accumulative slope only. Many fans and turbidites form a vast alluvial fan (Fig. 6.16). In space, this fan-shaped body is located quite near the dislocation facies and mass-turbidites wasting by underwater currents to the slope foot. It should be noted that almost all the notions existing about a eustasy (or SLRV) are supposed geologically quick, momentary sea-level lowering after its stabilization stage. However, a slow SL lowering has been reported in several regions (Kunin, 1989).

SSA of the data on the southern Caspian testifies that sea-level lowering in the Early

Pliocene time took place instantly after each SLRV paracycle. A new stage of lateral accretion and the second clinoform body (CF-2a) formation began from the low sea-level phase. It is evidenced by the downward shifting of the oblique seismofacies frontal surface reflections from the highest position of the roofing adjoining (C point) down to the bottom, overlapping the lowest position (D point) inside of the CF-2a body. Erosional shearing in SF-1a upper parts favors a sharp SL lowering after its formation. A clinoform body (CF-2a) is almost half as much as CF-1a. Post-sedimentation fractures and intrusion of plastic clays from the underlying complex deformed it. According to the borehole No. 1 drilling data (Fersman structure), the CF-2a underform part comprises sandy deposits from the RBS 2-d horizon.

The clinocover II-b flatting out the clinoform II-a, as it was noted in the case of CC-Ib, is represented by finely dispersed clays accumulated during the high sea-level stage. During the period of stabilization in the relatively low sea-level stage, the sedimentary material transported by underwater currents has been deposited on the slope, resulting in the formation of a cross-laminated body (clinoform-IIb) that is seismofacies-IIb. The body's roofing erosional surface delineates the sedimentation slope of the relic basin. The latter had the form of a relatively narrow bay-like basin within the Pre-Elbursian trough being merged with a vast basin of the PS accumulation. Because of a sharp drop in carrier force of underwater gravitational streams, the lateral accretion appears to end with the beginning of the basin filling.

The detailed structural analysis of the sedimentation bodies revealed that lateral accretion of the sedimentation prism is controlled by the SLRV cycles. So, based on an analysis of a large number of time sections, a seismostratigraphic model of the Early Pliocene complex within the South Caspian Turkmenian shelf. It presumes the presence of two large cyclically formed slope clinoforms. Each consists of three elements: clinoform, clinocover, and cross-laminated body. The latter are the elements by which a primary tectonic and erosional paleo-slope of the uncompensated basin has been achieved up to 25–30 km. Simultaneously, synchronous shelf covers have been vertically accreted only to 1.5 km.

The most terrigenous material volume has been supplied to the slope area of an uncompensated basin during sea-level lowering when the shelf served as a region of erosion and the passage of the sediments. The quiescent interval in the sedimentation process marked during the rising sea-level stage has been ended by the accumulation of finely dispersed material under low-energetic conditions.

Each clinoform is a lens-shaped or hilly geological body elongated along the strike. The available seismic data indicate that the combination of seismofacies within the same complex is essentially changed from profile to profile, suggesting that seismofacies relate to a locally developed separate elementary body. The complex lens-like-hilly internal structure of lateral accretion bodies is most clearly visible on the sections of profiles oriented along the strike (Fig. 6.23).

6.12 The Employment of Velocities and Dynamic Analyses for the Study of Lateral Accretion Complex

The available data of longitudinal wave velocity in the deposits are known from SL and VSP in the Fersman-1 well. The speed analysis of digital machining data is obtained from one profile. The mentioned well was drilled in only a separate part of the clinoform (CF-1a and CF-2a), both clino-covers and upper parts of the foredeltaic complex. The density data are absent. The GGL (gamma-gamma log) has been carried out neither in the No. 1 nor the No. 2 well. The available core data answering this purpose are lacking.

Speed analysis of the GDPM data carried out for the No. 842,622 profile showed that the 3-speed bands (series) are distinguished in the terrigenous composition of the shelf complex within the upper RBS, which are well correlated with those distinguished by seismic logging in the No. 1 well.

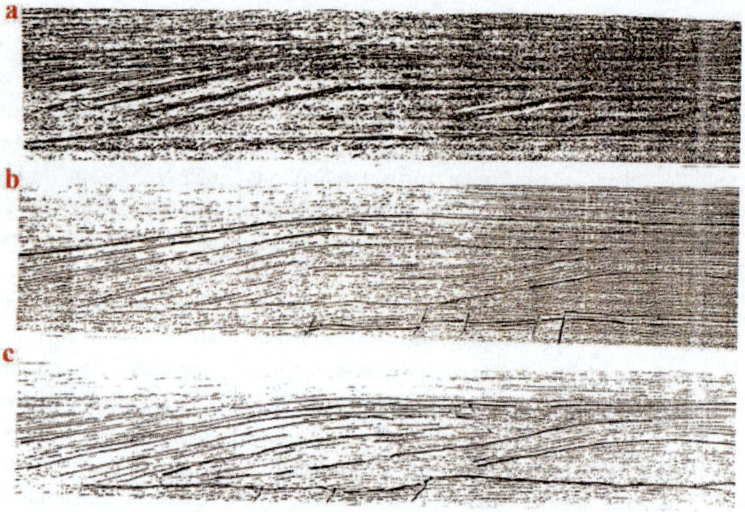

Fig. 6.24 Results of dynamic analysis along profile 842,625 in the Turkmenian shelf. **a** instantaneous amplitudes, **b** instantaneous phases, **c** initial time section

Although the average speeds usually are and sustainedly changed with the depth, a sharp increase in these values is marked within the intervals of the clinoform complex development. An exciting feature is that the V_{av}. increasing interval coincides with an area of the shelf edge rise and displacement in a south-westerly direction that is undoubtedly caused by relatively highly compact sandstones in the clinoform body.

Without drilling data and based on typomorphic seismofacies (sigmoidal-cross-laminated bodies), we forecasted that CF-1a had been developed under low—and mid-energetic conditions sedimentogenesis during SL relative rising. This forecast was confirmed by drilling data and geophysical testing in the Fersman No. 1 well, which stripped an underform part of the CF-1a. The 10–15 m thick seven sandstone beds have been revealed among the thicker (50–70 m) clay bands of the open-sea shelf. In addition, it was confirmed the fact that the clinocover is mainly of clay composition (Fig. 6.23). It is inferred that the dynamic sections with restored amplitude relationship and transformed into complex traces of instantaneous amplitudes appeared to be the most effective in mapping lithofacies zones in the clinoform bodies. The clinoform-lens-like bodies are contrastively distinguished in instantaneous amplitude sections (Fig. 6.24). From the PS section stratification, it appears that these instantaneous (envelope) amplitudes correspond to the sandy-siltstone bands in the terrigenous section. At the same time, light intervals between them correspond to the homogeneous clay bands of the same section. The molted and interrupted pattern of the section is thought to relate to the lithofacies variability of the clinoform objects.

All types of elementary clinoform-lenticular bodies are inwardly heterogeneous due to changes in their facies composition, porosity, etc. Their total combination in the form of the large lateral accretion bodies is more heterogeneous.

Such features as a change in their bedding, fallout of some elements from the section, and an outgrowth of a new one caused a space variation of their reflectance. An analysis of different-frequency profiling data showed that the sigmoidal configuration of the phase coincidence axes describing sedimentation bends are well expressed in relatively high-frequency filtrations. In a specific section interval at low-frequency filtration, some clinoform elements appear disguised, making it possible to form a proper notion of an object structure by subparallel reflections.

The other positive feature of the different-frequency filtration in the LA bodies development is recognizing the reflection from the PS surface

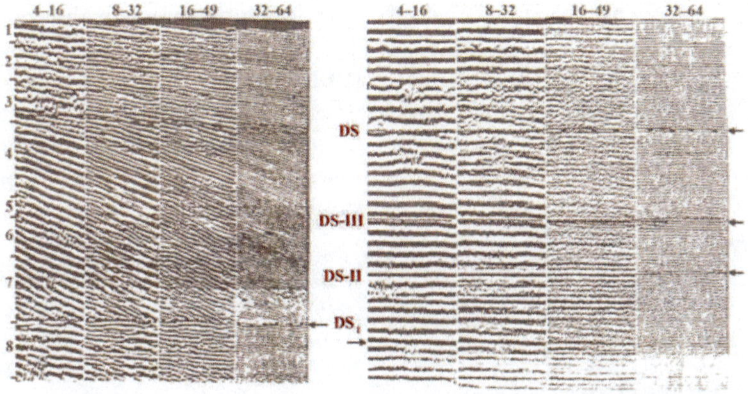

Fig. 6.25 The results of dynamic analysis carried out through the profile 842,625 (Turkmenian shelf). **a** instantaneous amplitude section, **b** instantaneous phase section, **c** initial section

of unconformity (erosion, washout). During records reproduction at different-frequency filtrations, most reflections preserve their fixed position in TS. The three surfaces of unconformity formed during SL lowering have been revealed within the complex investigated (Fig. 6.25).

The instantaneous phase sections are not well adapted for SFA within an area of LA bodies' development. Besides, these sections include a lot of false axes of phase coincidence unrelated to the natural bodies. That is why these instantaneous phase sections appear to be ineffective. On the other hand, tectonic dislocations in the underlying complex are well traced in these sections (Fig. 6.26). Thus, based on seismostratigraphic analysis of the GDPM data and using the results of speed and dynamic analyses, some facies zones were revealed, and a scheme of the spatial distribution of deltaic, shelf foredeltaic, and slope complexes. Accretion depocenters of clinoform and cross-laminated bodies have been outlined.

From the result of SSA, it is concluded that the following uplifts—Fersman, Fedynsky, Weber, Gasan-Kuli-sea, Dmitriyev, and Western

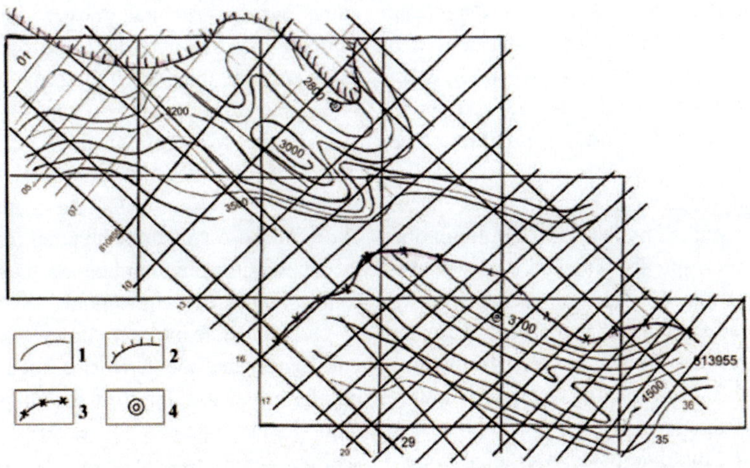

Fig. 6.26 A structural map drawn over the PS pinching out horizons (Absheron Archipelago) (after P.Z. Mamedov). (1) isolines along pinching out horizons, (2) pinching outlines of the lower KaS, (3) the same for the upper KaS, (4) recommended wells

—all revealed before by the lower redbed deposits—are sedimentation structures composed of the clinoform-lenticular bodies. The morphostructure of clinoform bodies is complicated by post-sedimentation faults, a typical feature of avalanche sedimentation bodies. It seems likely that irregular flow and uprise of plastic clays, as in the PS underlying complex in depressional facies of the clinoform complex, played a particular part in forming the mentioned structures.

6.13 Non-anticlinal Oil and Gas Traps in the Early Pliocene Deposits of the Northwestern Absheron Archipelago

The SCB northwestern slope zone is a promising area. The essential features conducive to the formation of non-anticlinal oil and gas traps are the following: (1) the suites and series of regional pinching out along the natural boundaries of the sedimentation basin and in the slope zones, (2) the presence of the breaks and unconformities caused erosional and tectonic truncation of the beds, (3) constructive topographic forms, and (4) exogenic processes. Within the northwestern Absheron Archipelago, where the Mesozoic floor lies at a shallow depth, it is observed that such geologic factors as regional and local reduced thicknesses of Pliocene and Miocene-Oligocene deposits. These deposits are truncated and transgressively overlapped by more young complexes and marked contact of the mentioned deposits with old (Mesozoic) formations. The most widely spread are lithologic-stratigraphic traps confined to the PS bottom suites. Their formation is caused by the PS accumulation within an uninterruptedly widened basin, resulting from the basin contours of each following suite or band appearing wider than before. After this, there were conditions for trap formations connecting with the pinching-out zones and a stratigraphic disconformity.

Analysis of lithologic-paleogeographic conditions and studies of a law-governed range of facies and thicknesses show that the most promising prospecting objects for lithologic-stratigraphic traps within northwestern Absheron Archipelago are the Kala suite (KaS), Underkirmaki suite (UKS) and partly Kirmaki suite (KS). Many workers have conducted field-oriented studies of the paleogeographic boundaries of the PS bottom suites. It was used mainly for drilling and seismic survey (GDPM) data until recently. It turned out well to determine the geographical range of the KaS and partly UKS near the Gosha dash, Agburun Absheron, and the Darvin banks. However, because of the absence of deep drilling data and insufficient seismic survey data of former times, it could not determine the KaS and UKS boundaries to the northeast of the Absheron bank.

Tectonically, the northwestern Absheron Archipelago is of sea continuation of the Greater Caucasian meganticlinorium southeastern plunge. Within the region investigated, the following local uplifts with common structural features are distinguished: deeply washed-out arch (up to the PS upper-division inclusive), asymmetrical structure, and the presence of dislocations with a break in continuity. A correspondence between structural plans is observed in the zones of reduced Cenozoic thickness (on local uplift crests). Some tectonic differences in the structural plans of the Paleogene-Neogene and Mesozoic structural stages are noted with the increase in the thickness of the Paleogene-Neogene complex that is usually observed in depression zones. For instance, in some cases, structural "noses" are visible on Pliocene deposits over the Mesozoic local uplifts. In other cases, the axes of Pliocene folds appear to be displaced southeasterly regarding the structural plan of the Mesozoic surface. SCB explains it down warping intensification at the beginning of Pliocene time. As a result, the southeastern periclines of those structures inverted to the center of the depression have undergone the most significant influence of the mentioned down-warping.

Overall, SCB intensive down warping, on the one hand, and the Greater Caucasus intensive bulging up, beginning from the Oligocene, on the other hand, appeared to be decisive factors of sedimentation that caused the presence (or absence) of any types of deposits. Such tectonic movement combination has been manifested in the fact that on the crests of most structures, the

Mesozoic deposits are not deep-seated, occurring at a depth of about 1000–1500 m. The result was the development and forming of the Tengiz-Beshbarmak anticlinorium of the Greater Caucasus. Therefore, the change of the old (Mesozoic) deposits by younger (Oligocene) deposits is observed from northwest to southeast on the crests of the mentioned anticlinorium and its marine continuation. Based on the exploratory drilling data, it was established that significant breaks in sedimentation within the Agburun, Absheron, and Gilavar banks resulted in the fact that many lithologic-stratigraphic intervals are omitted here. In such a way, within the Agburun bank, the Upper Cretaceous (Datonian) deposits have been stripped under Pliocene deposits. However, there were stripped Barremian (Lower Cretaceous) deposits under the same deposits. In the Gilavar bank, Miocene-Oligocene deposits of diatomaceous and Maikopian suites occur under the PS Kirmaki suite, underlain by the Upper Cretaceous (Campanian and Santonian) deposits. The mentioned breaks are well fixed by truncating underlying deposits by the unconformity surface (P) and contiguous covering beds.

As was noted above, a break in sedimentation between PS and underlying deposits is increased in a north-westerly direction at the expense of the PS lower suites omission from the section and deeper erosion of the old deposits. The break duration is increased in the direction of successive truncation of the PS underlying deposits and pinching out its bottom bands and suites above the unconformity surface. During the basin expansion, this surface has been overlapped immediately after omission from the section of the lower suite deposits.

Typical seismofacies on the surface (P) bench have been used as an indicator of pinching out of the PS lower bands and suites from the section in the northern tectonic zone of the Absheron Archipelago (Mamedov, 1986). Within the Absheron and Agburun banks, seismic surveys and drilling data from numerous wells have determined the KS basin boundaries. From the drilling results of wells No. 4, 5, 7, 8, 14, it is concluded that during all the Kala time, a southeastern pericline of the fold has been subjected to the plunge that caused an accumulation of up to 300 thick KaS sediments. During the PS bottom suites accumulation, a northeastern limb of the structure was a land composed of diatomaceous deposits. Then, at the time of the basin's subsequent expansion, they were overlapped by the KS beds. The drilling data suggests that during Underkirmaki time, the Agburun Bank began its subsidence while the Absheron Bank underwent a rise. This fact explains the presence of UKS sediments on the crest of the Agburun bank and, the other way around, their absence on the crest and northeastern limb of the Absheron bank. We have established the UKS boundary within the studied region based on the wavefield seismofacies analysis (Figs. 6.2 and 6.6).

Within the northwestern Absheron Archipelago, the Kala Sea has formed a bay between the Iki Gardash Dashlari and Agburun structures—in the north and the Nardaran structural nose—in the south (Fig. 6.7). In our view, the KaS paleobasin western boundary has been the Pirshagi transversal flexure, which is clearly traced within an area of the cape Bezymyanny (Nameless) and Two Brothers Rock. As it is seen on the isopach map, the KaS thickness is increased in a southerly and southeasterly direction, reaching the enormous thickness on the southeastern pericline of the Arzu structure in a zone of its articulation with the Sevindj structure (300 m), on southeastern pericline of the Absheron bank, and in the place of its contiguity to the Darwin bank. Besides, the Kas thickness also increased between the structures of Darwin bank, Agburun, and Mardakan Sea banks, where KaS's thickness reaches 280 m.

Such essential factors as the above-described geotectonic development conditions at the beginning of Pliocene, pinching out of the PS bottom bands and suites by the "P" surface, the closing of the traps by terrace scarps of this surface, the presence of screening clay mass over the KaS upper strata, this suite pinching out established by our seismostratigraphic analyses, an inversion of traps by their open sides to the deep-seated Artyom-Kelkor trough and the SCB as well as a regional dip of beds to the southeast

—all that have created favorable conditions for the formation of non-anticlinal hydrocarbon traps. Besides, the latter may relate to the zones of stratigraphic pinching out of all the suites, as with the zone of pinching out of its separate sandy horizons up-dip to the crests of local syn-sedimentation uplifts.

All the mentioned data allowed us to conclude that the KaS basin is lithologically stratigraphically enclosed from the north and northwest. The slope zones of the Bilgya and Shuvalan troughs may likely be considered coffer folds that form a reservoir having a smoothing bottom and elevated sealed slopes. They may serve as a lithologic-stratigraphic trap on hydrocarbon migration from SCB-subsided zones to the northwest. Gas and gas-condensate inflow from the KaS deposits in the No. 4 prospecting well and No. 7 structural-prospecting well on the Absheron bank (within an interval of 1650–1675 m) suggests that the KaS in the northern Absheron Archipelago possesses oil-generating potential.

From the studies of the PS contact with underlying deposits and the KaS thick and lithofacies analyses data, it is inferred that the outlooks for oil and content in the northwestern centrocline of the Artyom-Kelkor trough are related to this suite. Besides, both sides of the trough likely contain stratigraphic and lithologic traps in the PS lowermost strata. In this respect, the most promising traps are those mapped, based on seismostratigraphic investigations, at southeastern periclines of the Arzu, May, and Nakhichevan structures.

The regional beds dip towards a deep trough and trap closure by the initial sedimentation surface are reasonable factors to believe that the two structural-prospecting wells are advisable for stripping the stratigraphic-type deposits (see Fig. 6.11). In addition to the factors mentioned above, the KaS, as PS bottom suite, is remarkable for its lithologic variability reflected in its sporadic oil and gas content. Therefore, apart from the stratigraphic oil pools, this suite is expected to contain lithologically limited pools. Considering the presence of fractured dislocations,—tectonically screened oil accumulations.

Analysis of geological-geophysical conditions of non-anticlinal trap formation in the region examined. The results of seismic and seismostratigraphic investigations show that oil and gas pool discoveries are expected in the zones with a shallow attitude of productive beds, which may serve as objects for exploratory drilling. It is advisable to carry out well spacing in an inevitable succession in the profiles parallel to the beds' pinching outlines and the suites arranged in 200–250 m isopachous lines. Based on the above, prospecting works for NAT within KaS and UKS.

References

Abdullayev, N., Riley, G., & Bowman, E. (2011). Sedimentation history of the productive sequence in the South Caspian taking into account basin subsidence. Part I. *Azerbaijan Oil Industry, 5*, 20–30.

Aliyev, A. G. (1947). *Petrography of the productive series of Kabristan* (155 p). Academy of Sciences (in Russian).

Aliyev, A. G. (1949). *Petrography of tertiary deposits of Azerbaijan* (311 p). Aznefteizdat (in Russian).

Aliyev, A. I. (1975). Zoning of oil and gas distribution in the South Caspian depression in connection with the prospects for prospecting for large gas and (gas condensate) deposits at great depths. In *Search and exploration of gas fields, No. 47/55, Nedra, Moscow* (pp. 160–169) (in Russian).

Alizade, A. A. (1960). *Paleogeography of the Balakhani Stage Basin* (67 p). AzINTI (in Russian).

Alizade, A. A., Akhmedov, G. A., Akhmedov, A. M., Aliyev, A. K., & Zeynalov, M. M. (1966). *Geology of oil and gas fields of Azerbaijan* (392 p). Nedra (in Russian).

Alizade, A. A., Salayev, S. G., & Aliyev, A. I. (1985). *Scientific assessment of the prospects for oil and gas potential in Azerbaijan and the South Caspian* (252 p). Elm (in Russian).

Green, T., Abdullayev, N., Hossack, J., Riley, G., & Roberts, A. (2009). Sedimentation and subsidence in the south Caspian Basin, Azerbaijan. In: (Brunet, M.-F., Wilmsen, M., Granath, J.W., Eds.), South Caspian to CentralIran Basins. *The Geological Society, London, Special Publications*, 312, 2241–2260.

Azizbekov, Sh.A., Alizadeh, K.A., Shikalibeyli, E.Sh. and Gadjiev, T.G. (Eds.), (1972). *Geology of the USSR,Azerbaijan, Volume XLVII*, Nedra, Moscow (in Russian).

Bagirzade, F. M., Kerimov, K. M., & Salayev, S. G. (1988). *Deep structure and oil and gas potential of the South Caspian Megadepression* (304 p). Azgosizdat (in Russian).

Baturin, V. P. (1937). *Paleogeography on terrigenous components* (292 p). AzONTI (in Russian).

Buryakovsky, L. A., Djevanshir, R. D., & Chilingar, G. V. (1995). Abnormally-high formation pressures in Azerbaijan and the South Caspian Basin (as related to smectite/illite transformations during diagenesis and catagenesis). Journal of Petroleum Science and Engineering, 13(3-4), 203-218.

Eppelbaum, L., & Katz, Y. (2021). Akchagylian hydrospheric phenomenon in aspects of deep geodynamics. *Stratigraphy and Sedimentation of Oil-Gas Basins, 2,* 8–26.

Gogonenkov, G. N., Mikhailov, Y. A., & Elmanovich, S. S. (1988). Development of seismostratigraphic forecasting of oil prospects. In: *Modern geophysical methods in solving problems of petroleum geology* (pp. 5–12). Nedra (in Russian).

Gromin, V. I., et al. (1986). Clinoforms of the northwestern shelf of the Black Sea, their genesis and conditions of oil and gas potential. In *Geology of oil and gas, No. 10* (in Russian).

Kerimov, K. M., Gadzhiev, F. M., & Gasanov, I. S. (1999). Hydrocarbon resources of the Kur-South Caspian megadepression. *Azerbaijan Oil Industry, 7,* 1–11 (in Russian).

Khain, V. E., & Shardanov, A. N. (1954). *Geological history and structure of the Kur depression* (357 p). Academy Sciences of Azerbaijan (in Russian).

Klushin, S. V. (1987). Study of sedimentation cyclicity by dynamic parameters of reflected waves. In: *Applied issues of sedimentation cyclicity and oil and gas potential* (pp. 38–52). Nauka (in Russian).

Kunin, N. Y. (1989). *Structure of the lithosphere of continents and oceans* (286 p). Nedra (in Russian).

Kunin, N. Ya., & Kucheruk, E. V. (1984). Seismostratigraphy in solving problems of prospecting and exploration of oil and gas fields. In *Science and technology results* (Vol. 13, 195 p). VNIITN (in Russian).

Lisitsyn, A. P. (1988). *Avalanche sedimentation and breaks in sedimentation in the seas and oceans* (308 p). Nauka (in Russian).

Mamedov, A. V. (1977). *History of geological development and paleogeography of the middle Kur depression in connection with oil and gas potential* (212 p). Elm (in Russian).

Mamedov, P. Z. (1965). Study of the spectral features of reflected waves in the Lower Kur lowland. *Izvestiya Vuzov, Series: Oil and Gas, 10* (in Russian).

Mamedov, P. Z. (1984). Some results of applying the principles of seismostratigraphy in the study of the lower boundary of the PS within the Absheron Archipelago. In: *Oil and gas geology and geophysics II. All-Union Institute of Oil & Gas Economics, Moscow, No. 11, 20–23* (in Russian).

Mamedov, P. Z. (1986). Identification of reef formations with the help of seismostratigraphic studies. *Geology of Oil and Gas, 7,* 24–27 (in Russian).

Mamedov, P. Z. (1991). *Seismostratigraphic studies of the geological structure of the sedimentary cover of the South Caspian megadepression in connection with the prospects for oil and gas potential* (50 p). Doctor of Science Thesis, Oil and Chemistry Institute (in Russian).

Mamedov, P. Z. (2004). Genesis and seismic stratigraphic model of the South Caspian Megabasin architecture. In A. A. Alizadeh (Ed.), *South Caspian Basin: Geology, geophysics, oil and gas content* (pp. 150–164). Nafta-Press (in Russian).

Mamedov, P. Z. (2006). Peculiarities of the Earth's crust in the South Caucasus in the light of new geophysical data. Izvestiya Academy of Sciences Azerbaijan. *Earth Sciences, 3,* 36–48 (in Russian).

Mamedov, P. Z. (2007). Seismostratigraphic subdivisions of the sedimentary cover of the SCB. *Journal of Stratigraphy and Sedimentology of Oil and Gas Basins of ANAS, Earth Sciences, 1,* 102–117 (in Russian).

Mamedov, P. Z. (2008). On the causes of rapid subsidence of the earth's crust in the South Caspian Basin. *Azerbaijan Oil Industry, 1,* 8–20 (in Russian).

Mammadov, P. Z., & Ragimkhanov, F. G. (1985). The study of the surface of unconformity in the lower middle Pliocene in the northwestern part of the Absheron threshold based on the results of seismostratigraphic studies. *Izvestiya Vuzov, Series: Oil and Gas, 12,* 14–19 (in Russian).

Muromtzev, V. S. (1981). Methods of local forecasting of sandy beds—Lithological oil and gas traps based on electrometric facies. In *Proceedings of all-union oil scientific Institute* (pp. 7–24) (in Russian).

Mushin, I. A., Brodovoy, L. Y., Kozlov, E. A., & Khatyanov, F. I. (1990). *Formation-structural interpretation of seismic data* (299 p). Nedra (in Russian).

Mustafayev, I. S. (1963). *Lithofacies and paleogeography of the middle Pliocene deposits of the Caspian Basin* (193 p). Azerneshr (in Russian).

Nikishin, A. V. (1981). Sedimentary rhythm and comparison of sections of the middle Pliocene of the South Caspian depression. In *Problems of geology and oil and gas content of the depressions of the Inland Seas* (pp. 60–66). Nauka (in Russian).

Pashaly, N. V., Kheirov, M. B., & Khairulin, R. K. et al. (1985). The lithology of the middle Pliocene deposits of the Western and Eastern Shelves of the South Caspian. In *Trans. of the All-Union Geological Conference* (pp. 151–152) (in Russian).

Pashaly, N. V., Kheirov, M. B., & Saradzhalinskaya, T. M. (1998). Productive series. In *Geology of Azerbaijan. Vol. II: Lithology* (pp. 186–229) (in Russian).

Payton, C. E. (Ed.). (1977). *Seismic stratigraphy-applications to hydrocarbon exploration* (516 p). American Association of Petroleum Geologists, Memoir 26. Tulsa.

Potapov, I. I. (1954). *Absheron oil-bearing region (geological characteristics)* (539 p). Academy Sciences of Azerbaijan (in Russian).

Pustovalov, L. V. (1951). *On detrital quartz from the productive strata of the Apsheron Peninsula. Series: Geological, No. 4.* Izvestiya USSR Academy of Sciences (in Russian).

Semenenko, V. N. (1987). *Stratigraphic correlation of the upper Miocene-Pliocene of the Eastern Paratethys and Tethys* (232 p). Naukova Dumka (in Russian).

Shutov, V. D. (1962). *Epigenesis zones in terrigenous deposits of the platform cover (on the example of the study of Riphean and Paleozoic deposits in the southeastern part of the Russian Platform). Series: Geological, No. 3* (pp. 30–44). Izvestiya USSR Academy of Sciences (in Russian).

Sosedkov, V. S., & Surkov, Y. N. (1989). Regional seismostratigraphy of the Mesozoic of Northwestern Siberia. In *Seismic exploration for lithology and stratigraphy* (pp. 35–43). Western Siber Oil & Gas Institute (in Russian).

Strakhov, N. M. (1963). *Fundamentals of the theory of lithogenesis* (Vol. I, 210 p, Vol. II, 574 p). Academy of Sciences (in Russian).

Sultanov, A. D., & Gorin V. A. (1963). *Productive series of the Western Side of the South Caspian Depression* (290 p). Azerneshr (in Russian).

Urupov, A. K. (1966). *The study of velocities in seismic exploration* (225 p). Nedra (in Russian).

7 Reservoir and Screening Properties of the Productive Series' Sediments

7.1 The Productive Series Succession: Main Peculiarities

The PS succession comprises terrigenous sediments and rocks containing grains of an extensive range from very coarse to very fine. A summary of the mineralogical composition and cementation of the Productive Series rocks is given below. This issue is partially described in Chap. 4.

Sandy-silty rocks make up 20–50% of the PS succession. The Lower Kur trough consists of polymictic (quartzose-feldspathic-graywacke) varieties containing abundant rock fragments (see Chap. 4). The proportion gradually increases towards the Baku Archipelago and becomes predominant there. As a result, polymictic rocks appear to be turned into oligomictic feldspathic-quartzose rocks. The amount of quartz also increases in the Absheron Peninsula and Absheron Archipelago, where monomictic quartz-rich sandstones and siltstones have a significant occurrence, particularly in the PS lower portion. The sorting of the PS sandstones varies from very well-sorted to poorly sorted varieties. Such heterogeneity of mineral distribution in the PS sediments is due to different provenance terrigenous material composing PS succession as well as post-sedimentary processes influencing the number of clay minerals such as montmorillonite, kaolinite, and chlorite in a matrix and cementing material that strongly affects reservoir quality.

The amount of cementing material changes between 12 and 35%. We know several cement types: carbonate-muddy, muddy-carbonate, muddy, gypsum, chloritic, and pyritic (Fig. 7.1). The first three types are common, and the last two are of infrequent occurrence. The cementation type is pore filling, contact, basal, and mixed.

The Productive Series in the Lower Kur depression is characterized by a high proportion of clay fraction and abundant authigenic montmorillonite, which are the primary reasons for the frequent occurrence of low-permeability reservoirs here (Fig. 7.2).

In the Baku Archipelago, the rocks are characterized by a reduced share of clay fraction and little occurrence of authigenic montmorillonite, which produce better conditions for increased reservoir quality. The data presented here indicates the high variability of clay mineral content in the Productive Series reservoir rocks in the different parts of the South Caspian Basin. Besides the role mentioned above of the multiplicity of parent rocks and diagenetic processes in the heterogeneity of the PS sediments' mineralogical composition, the type of rock-saturating fluids can also be one of the crucial factors affecting the clay mineralogy of these sediments (Alizadeh et al., 2017).

The composition of cement in reservoir rocks is changed in space. For instance, oil prevents the transformation of clay minerals while water and

A. Alizadeh et al., *Pliocene Hydrocarbon Sedimentary Series of Azerbaijan*,
Advances in Oil and Gas Exploration & Production, https://doi.org/10.1007/978-3-031-50438-9_7

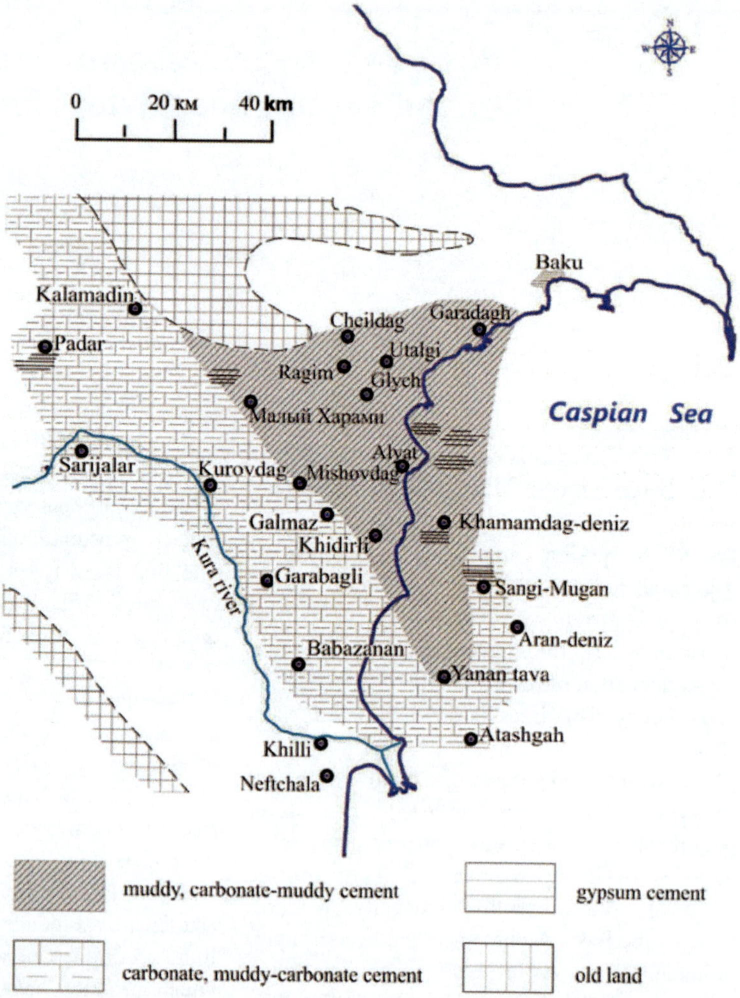

Fig. 7.1 Distribution of different cement types in the Productive Series rocks within the western flank of the South Caspian basin

gas promote the formation of such authigenic minerals as montmorillonite and kaolinite.

7.2 Reservoir Properties of the PS Deposits in the Absheron Peninsula and Absheron Archipelago

7.2.1 The Lower Portion of the Productive Series

This study is based on the use of analytical and log data to estimate the reservoir properties of the Productive Series deposits. The objective is to establish the best reservoir units and reservoir distribution within the South Caspian Basin and delineate and map them. Many published and unpublished sources have been involved in summarizing and integrating the Productive Series reservoir properties data.

7.2.2 The Gala Suite

This lowermost unit of the Productive Series has limited occurrence in the South Caspian basin and was recorded only in the subsurface of the southern and southeastern Absheron Peninsula, North Absheron uplift zone, southern Absheron

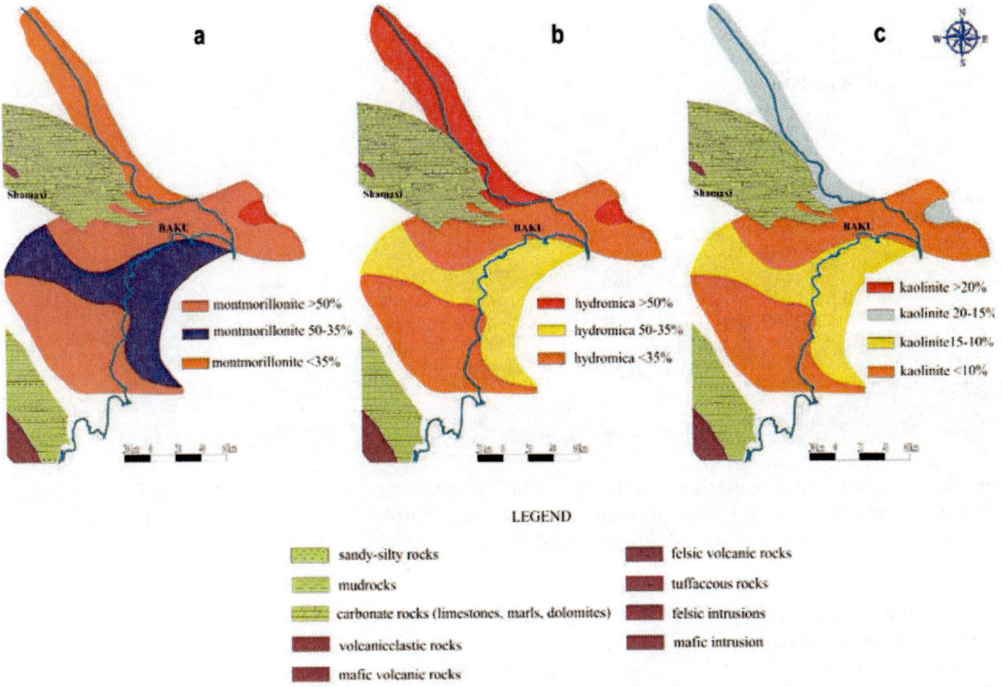

Fig. 7.2 The authigenic clay minerals in the cementing material of the PS reservoir rocks: **a** montmorillonite, **b** hydromica, **c** kaolinite

Archipelago, and northern Baku Archipelago (Fig. 2.7). Only one outcrop of the Gala suite was reported on Chilov Island.

The maximum thickness of the Gala suite is 430 m. The stratigraphy of the suite is based on the log data that display clear separation of the Gala succession into three units: the lower mud-dominated sub-suite, the middle one containing alternating sandstone and mudstone intervals, and the upper sand-dominated sub-suite. The Gala suite sandstones are poorly sorted. Most sandstones can be classified as fine to medium-grained.

The fine-grained sandstones dominate coarse sediments (the median grain size is 0.154 mm), which is also confirmed by the results of SEM analyses of the Gala suite sediments in several fields. Below, we present the SEM image of the Gala suite rocks in the South Caspian Chirag field as an example (Fig. 7.3).

In the mineralogical composition of the light fraction of sands and sandstones, the quartz amount varies from 30 to 87% and is equal to 62% on average (feldspar content—9.5% (from 5 to 38%)). Reported disthen and staurolite testify to the granite source of the Gala suite sediments transported, most likely from the Russian platform. Besides the mentioned minerals, a few other minerals (marcasite, celestite, anhydride, etc.) were recorded in the thin sections pared from the Gala suite sediments.

The map of sandiness of the Gala suite sediments (Fig. 7.4) shows that the contours of the area containing the rocks with the highest proportion of sand material very well correspond to the major channels of the paleo-Volga River recorded in the South Caspian basin in the Gala time (Fig. 2.7). This result clearly demonstrates the degree of the paleo-Volga River's impact on sedimentation and its role as a principal coarse sediment supplier to the Pliocene Caspian Lake. The clay fraction displays a clear trend of southward increase of the fine material proportion (Fig. 7.5).

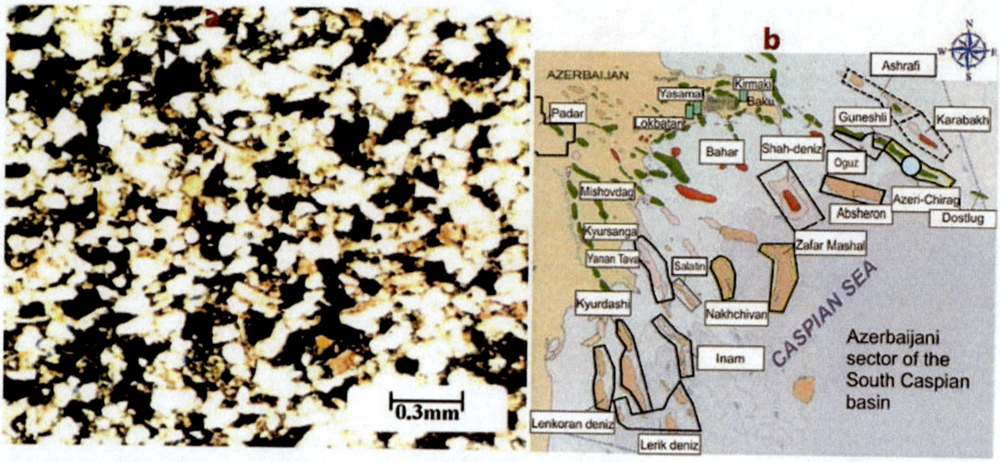

Fig. 7.3 **a** An SEM image of the Gala suite sandstones in the Chirag field, well No 6, (3671–3676 m), **b** areal map. Reservoir properties: porosity—2.5%; permeability—0.01 × 10^{-15} m^2; density—2.41 g/cm^3; carbonate content—20%; grain size—0.154 mm; sorting—good

Fig. 7.4 The average percentage of sand fraction in the Gala suite sediments of the South Caspian basin western flank

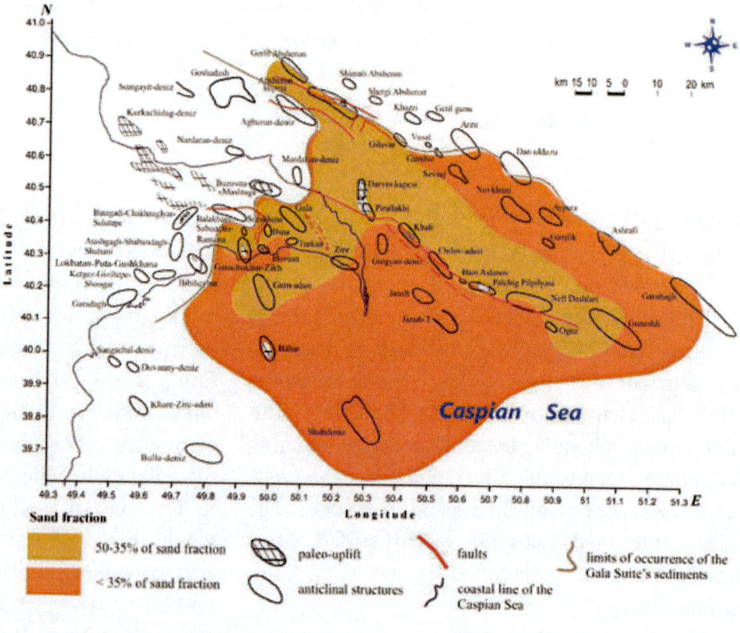

The constructed reservoir properties maps demonstrate the Gala suite reservoir rocks' good to medium reservoir quality. Porosity changes from lower values of less than 10 to 22%, and permeability—from negative to around 250 × 10^{-15} m^2 (Fig. 7.6). Most areas of the Gala suite occurrence are covered by rocks displaying good fluid filtration properties (Fig. 7.7).

In most rocks, a low amount of carbonate material is recorded (Fig. 7.8), significantly improving the reservoir quality of the Gala suite sediments and making a good prognosis for this succession.

The maximal oil production from the Gala suite succession was reported in the Gala and Zirya fields. Within the Gala field, the share of a

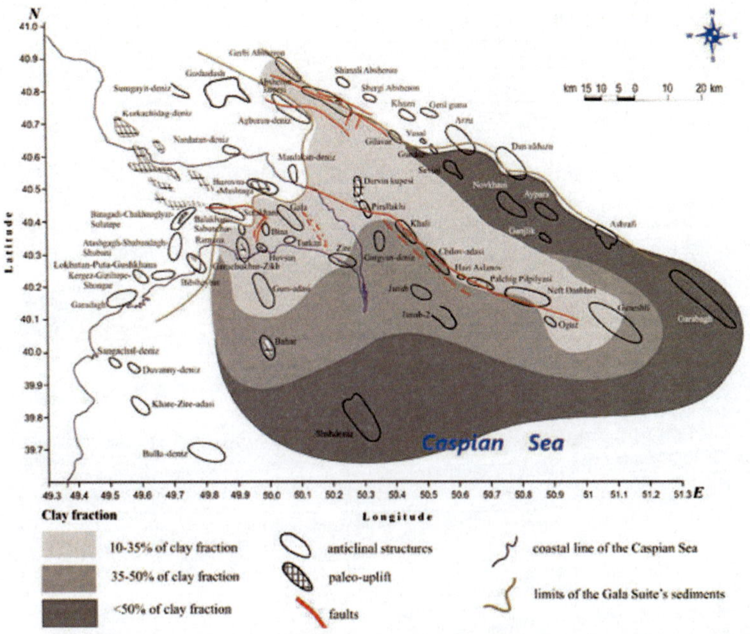

Fig. 7.5 The average percentage of the clay fraction in the Gala suite sediments of the South Caspian basin western flank

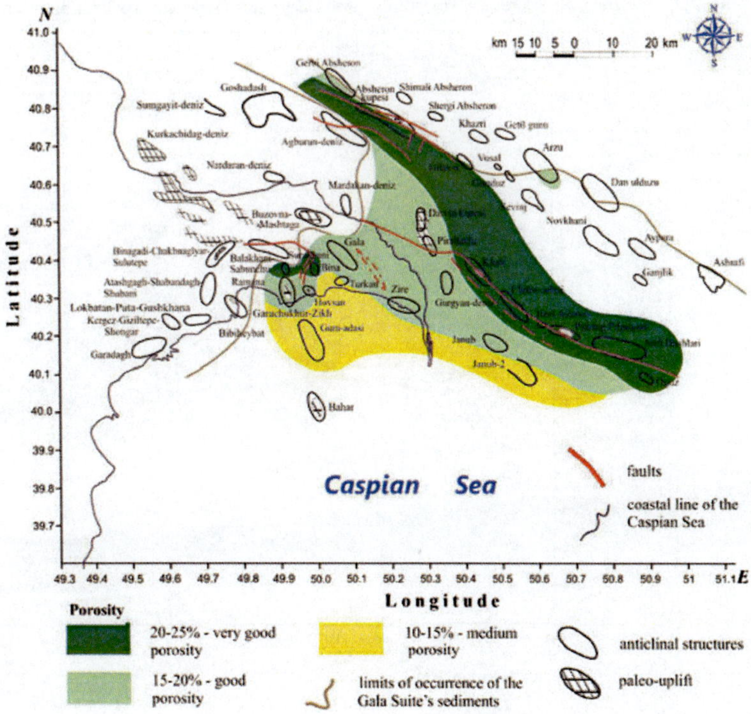

Fig. 7.6 The mapped average porosity (%) of the Gala suite reservoir rocks in the South Caspian basin western flank

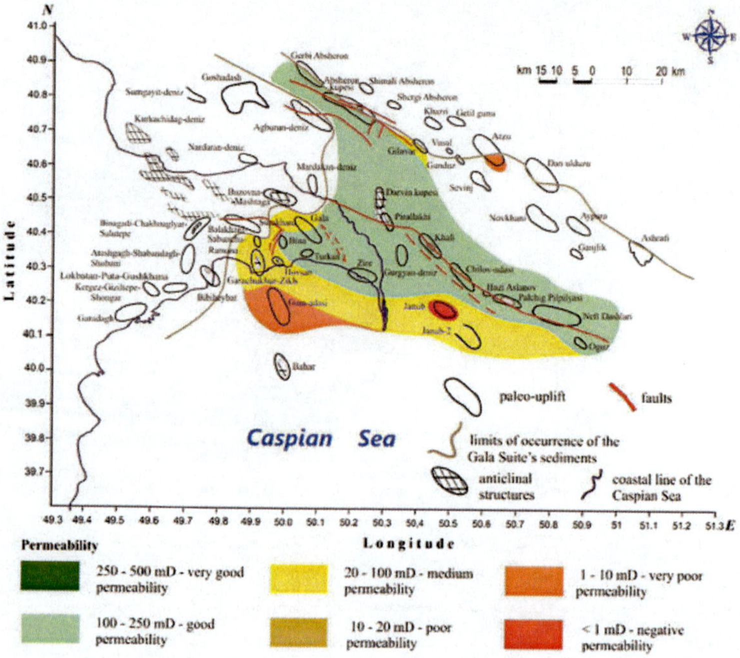

Fig. 7.7 The mapped average permeability of the Gala suite reservoir rocks in the South Caspian basin western flank, 10^{-15} m^2

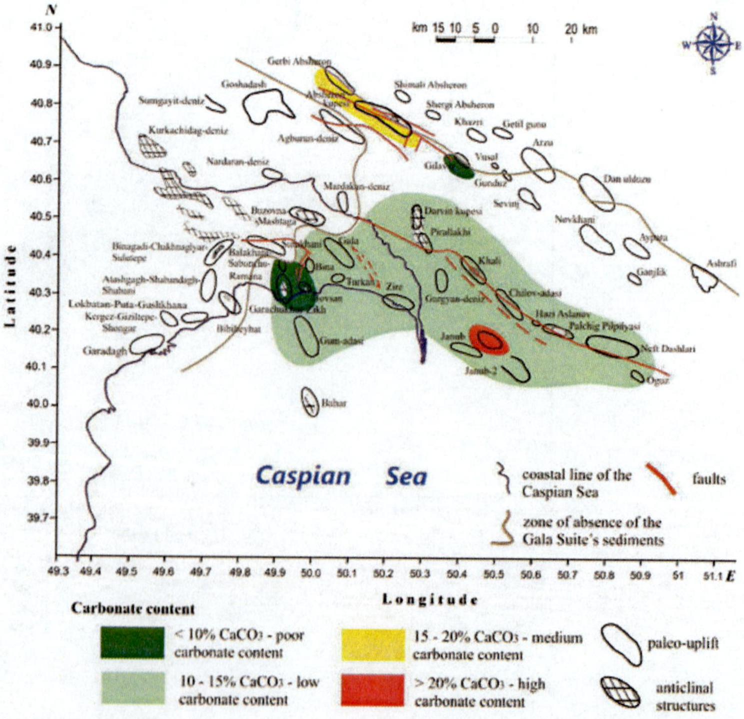

Fig. 7.8 The map of variations of the carbonate material's average amount (%) in the Gala suite sediments of the South Caspian basin western flank

sand fraction varies from 12.7 to 26%; clay fraction—19.3 to 30.4%; carbonate content—9.2 to 29%; porosity—5.4 to 16.5%; permeability—20 to 132 × 10^{-15} m^2.

In the Surakhani field, oil-saturated Gala suite sediments were encountered in the southern limb of the fold. The average reservoir properties are as follows: sand fraction—39.3%; clay fraction—13.7%; carbonate content—8.5%; porosity—23.1%.

In the Garachukhur field, the Gala suite deposits are widely occurring on the eastern limb. Reservoir properties are as follows: sand fraction—23.1 to 48.8%; clay fraction—14.6 to 29.3%; carbonate content—12.7 to 26.25%; porosity—10.7 to 20.3%; permeability—3.8 to 24.7 × 10^{-15} m^2.

Within the Zikh field, the Gala suite succession contains several productive horizons: the Ka_1 horizon, which has the following characteristics: share of sand fraction—23.4 to 70.9%; clay fraction—8.3 to 36.4%; carbonate content—6.2 to 10.5%; porosity—16.9 to 22.2%; permeability—20.6 to 354 × 10^{-15} m^2.

The Ka_2 horizon: sand fraction—11.9%; clay fraction—16.5%; carbonate content—13.9%; porosity—22.2%; permeability—297 × 10^{-15} m^2. The Ka_3 horizon: sand fraction—40.8%; clay fraction—7.5%; carbonate content—8.2%; porosity—24.1%; permeability—167 × 10^{-15} m^2.

In the Hovsan field, three productive intervals are distinguished: the Ka_1 horizon: sand fraction—21.9 to 44.7%; clay fraction—14.5 to 26.6%; carbonate content—13.0 to 19.5%; porosity—9.8 to 27.6%; permeability—13.6 to 589 × 10^{-15} m^2.

The Ka_2 horizon: sand fraction—8.1 to 44.4%; clay fraction—11.9 to 34.4%; carbonate content—7.4 to 21.9%; porosity—11.6 to 18.9%; permeability—9.4 to 147.1 × 10^{-15} m^2.

The Ka_3 horizon: sand fraction—32.1 to 52.9%; clay fraction—12.8 to 26.4%; carbonate content—8.4 to 17.3%; porosity—9.1 to 18.9%; permeability—10 to 512.3 × 10^{-15} m^2.

We have not recorded depth-dependent variations in the proportion of the sand-sized grains (Fig. 7.9a). However, the porosity and permeability display well-pronounced changes with the depth expressed in the higher maximal values of these parameters in the upper intervals of the Gala suite section (Fig. 7.9b, c).

More insight into controls on the reservoir properties of the Gala suite's rocks is demonstrated by two plots reflecting the link between the amounts of carbonate and clay material and permeability values of studied samples (Fig. 7.10a, b). These graphs show a clear grouping of the permeability data based on the quantities of carbonates and clay fraction. The first cluster (group) unites samples with poor to negative permeability not exceeding 20 mD. In such samples, we have recorded the highest share of carbonates—up to 38%, and clay size—a maximum of 50%. In the second cluster, the maximum amounts of carbonate and clay material decrease to less than 35% and 45%, respectively, which is accompanied by the permeability variations from minimal 20 mD to maximal 100–150 mD that testifies to medium permeability of the rocks.

In the third cluster, we have reported well-permeable samples with permeability values varying from 100–150 to 250–300 mD. The maximal share of carbonates and clay-size particles in these samples falls to less than 25% and 35%, respectively. Very well-permeable rocks display the lowest percentage of carbonates—less than 20%, and fine terrigenous material—less than 30%.

7.2.3 The PostKirmaki Suite (PK)

The PostKirmaki (PK) suite is dominated by grey and light-grey sands and sandstones containing abundant black angular pebbles and large rounded quartz grains. Sands in the central and eastern Absheron Peninsula are medium to coarse-grained, while they become more fine-grained in the western peninsula. Quartz is the predominant mineral in the light fraction. Its content is varied within 45–95% limits. The rest of the light fraction comprises feldspar (10–30%), rock fragments (0–2%), glauconite, volcanic glass, and analcime. The thickness of sand beds decreases in the north, northwest, and westwards.

The lower portion of the suite (PK_4; PK_5) is dominated by sands (Fig. 7.11), while in the

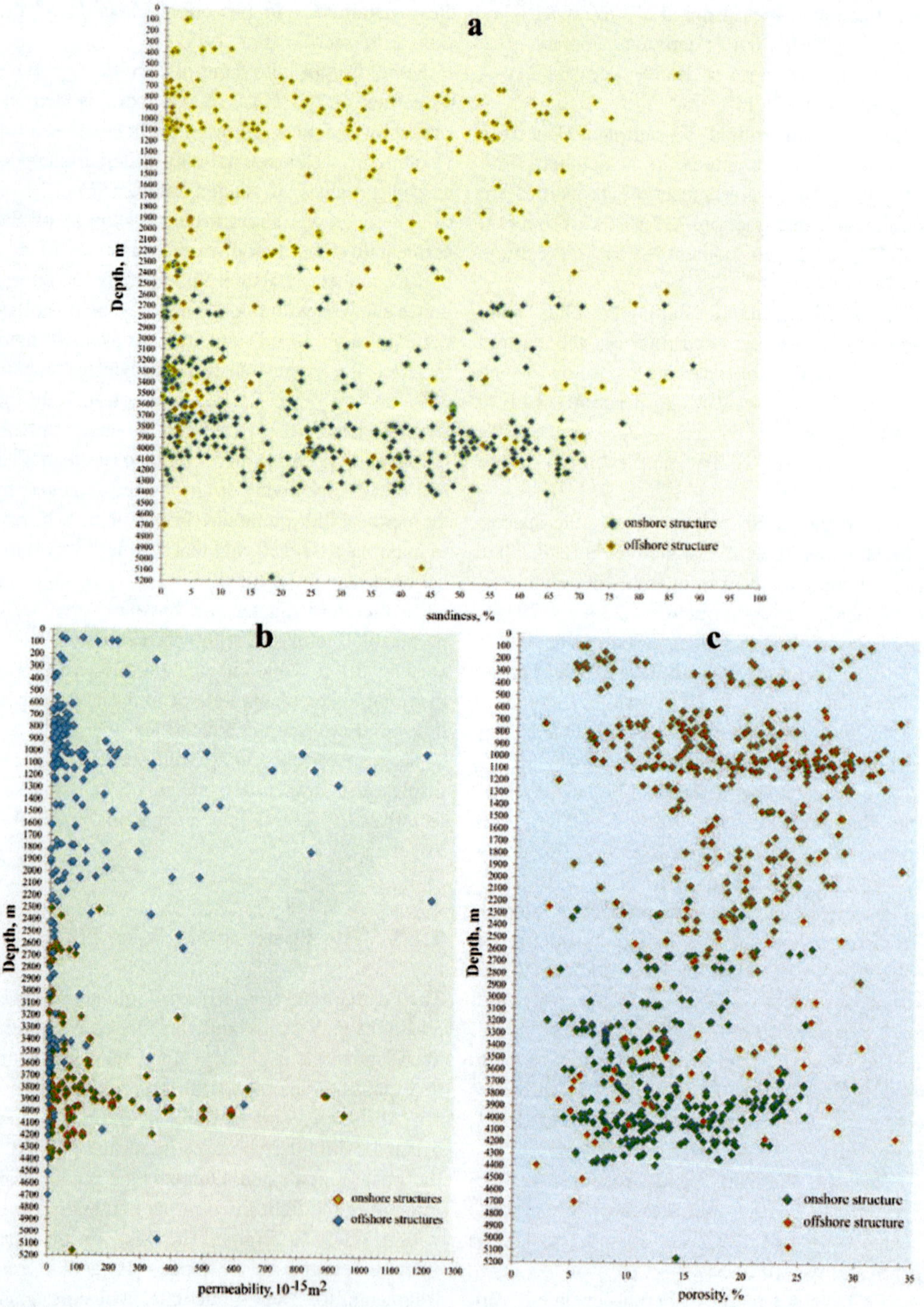

Fig. 7.9 Plots (**a**, **b**, and **c**) display the vertical alignment of the reservoir properties' parameters of the Gala suite reservoir rocks

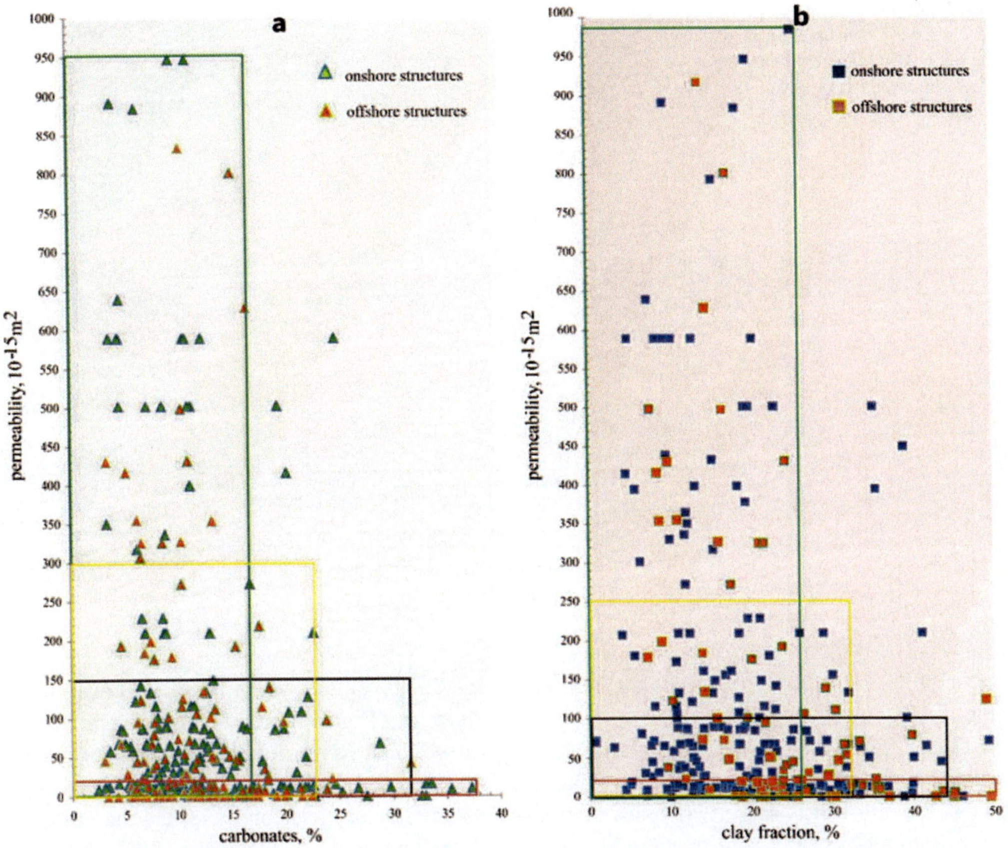

Fig. 7.10 Plots (**a** and **b**) display carbonate material and clay fraction-dependent alignment of the Gala suite reservoir rocks' permeability values

Fig. 7.11 The average percentage of sand fraction in the lower PK suite sediments of the South Caspian basin western flank

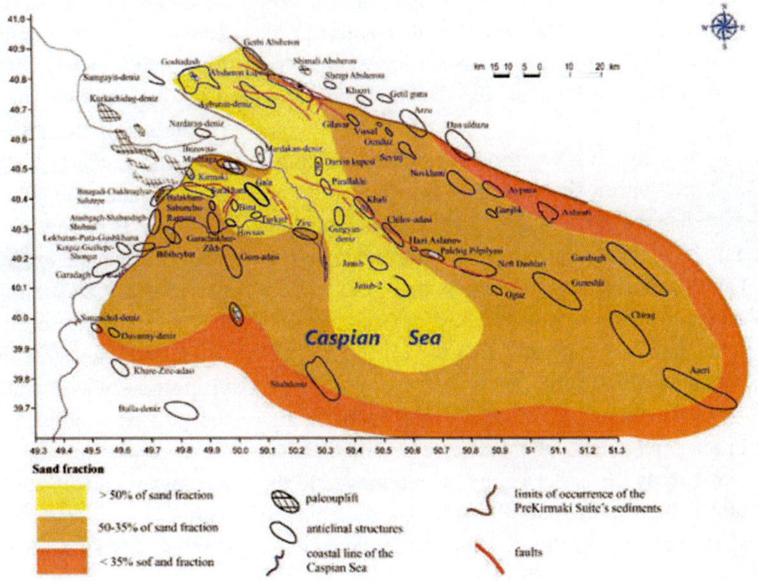

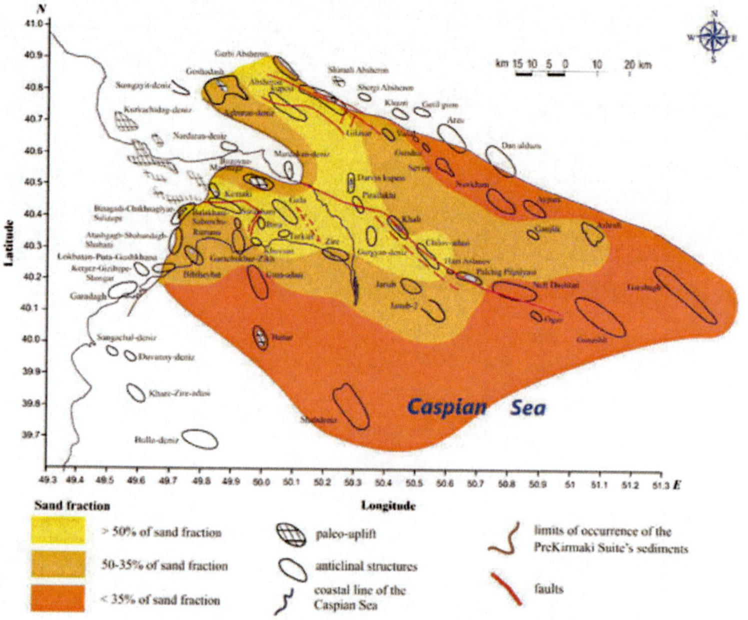

Fig. 7.12 The average percentage of sand fraction in the upper PK suite sediments of the South Caspian basin western flank

upper portion, the share of mudrocks increases (Fig. 7.12).

These two maps provide insight into the sedimentation history of the South Caspian basin. We still can observe two zones of maximum occurrence of the sand fraction, like in the Gala suite, but with some differences. Here, at the beginning of the PK suite, we can report a coarser fraction compared with the Gala suite. The zone of sand fraction content varying from 50 to 35% covers a vast area that includes the Absheron Peninsula, the Absheron Archipelago, and a big portion of the Baku Archipelago (Fig. 7.11).

Also, the eastern zone of maximum coarse material with a share of sand fraction exceeding 50% has shifted west and occupied the Janub and Janub-2 structures. This eastern zone significantly increased in size. In contrast, the western coarsest sediment zone reduced in size compared with the Gala suite and focused on the western Absheron Peninsula.

In the upper PK suite, we can report a significant reduction of the coarse sediment zone, particularly the eastern zone, which moved to the north (Fig. 7.12). This PK suite's sedimentation story clearly reflects the paleogeographic conditions in the South Caspian Lake in the Pliocene. The southward progradation of the paleo-Volga River delta at the onset of the PK suite brought the coarse material to the southern South Caspian fields and widened the sand sedimentation area. The upward decrease in the proportion of coarse sediments is a result of the paleo-Volga River delta backstepping at the end of PK time.

Sandstone porosity changes from 9 to 25% (Fig. 7.13), and permeability—from 26 to very high values close to 1000×10^{-15} m^2 (Fig. 7.14). This data allows us to classify the reservoir properties of the PK suite sediments as good—very good, which is also confirmed by the results of SEM analyses (Fig. 7.15).

In the Bibi-Heybat field, the rocks of the PK suite are characterized by high porosity and permeability, reaching, on average 20.9% and 179×10^{-15} m^2, correspondingly (Fig. 7.14).

Despite the stratigraphic control and higher proportion of sand fraction in the lower PK suite, the plot of sandiness variations with the depth demonstrates an equal share of coarse grains in the lower and upper intervals (Fig. 7.16a). We assume that is due to the different stratigraphic depths of PK suite occurrence in the various

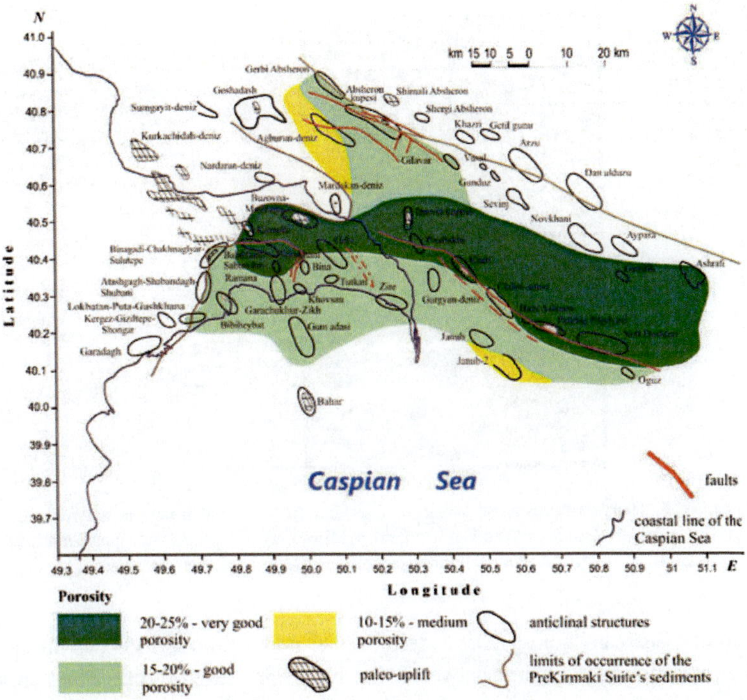

Fig. 7.13 The mapped average porosity (%) of the lower PK suite reservoir rocks in the South Caspian basin western flank

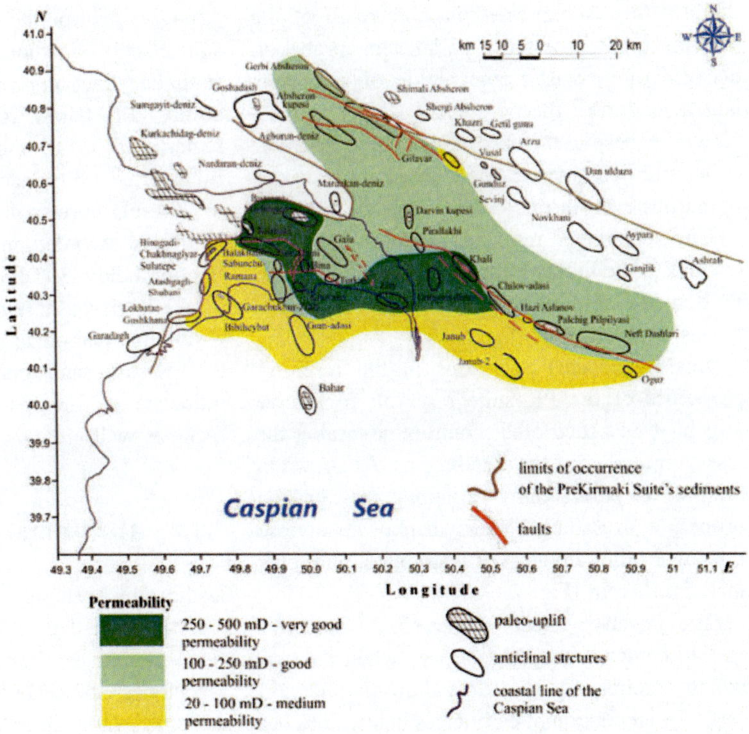

Fig. 7.14 The mapped average permeability (10^{-15} m^2) of the lower PK suite reservoir rocks in the South Caspian basin western flank

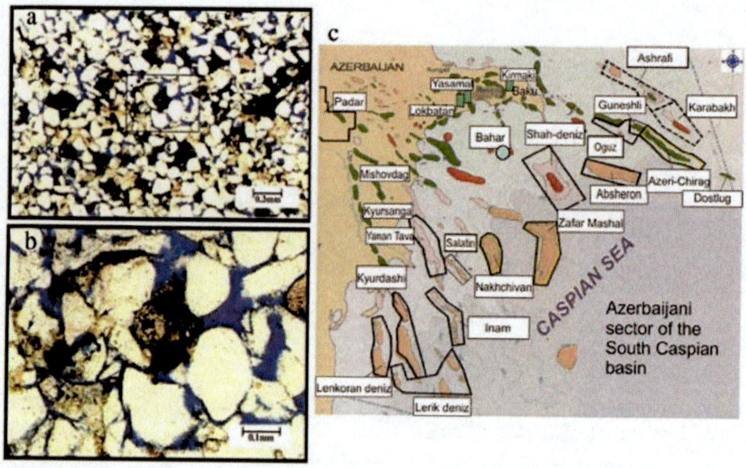

Fig. 7.15 Exaggerated image of the thin section (**a**) of the PK suite sandstones in the Bakhar field, well No. 19, 5033–5040 m depth interval. **b** Increased middle part of the image **a**. **c** Areal map. Reservoir properties: porosity—20.2%, permeability—52.8×10^{-15} m^2, density—2.4 g/cm^3, carbonate content—7.3%

South Caspian fields. Also, despite the non-monotonic relationship between the reservoir properties of the PK Suite rocks and the depth of their occurrence, we report porosity and permeability reduction in the lower intervals below 2500 m that is expressed in a decrease of the maximal values of these parameters, probably, because of increased compaction of sediments and reduction of the pore space (Fig. 7.16b, c). However, despite this fact, the PK suite reservoir rocks still demonstrate good- excellent reservoir properties even in the lower intervals.

The constructed reservoir properties plot also testifies to the higher hypsometric position of the PreKirmaki suite in the offshore fields compared to onshore structures (Fig. 7.16).

Besides vertical variations in the reservoir properties of the PK suite reservoir rocks, we also have recorded other controls governing the reservoir quality of these sediments. As shown in the plots of permeability variations, the filtration properties of sampled rocks display an inverse link with the amount of carbonate and clay material therein (Fig. 7.17a, b).

The negative effect of these both reservoir quality controls sharply increases when the carbonate amount exceeds 30% and the clay fraction 40%. The reduction of carbonates below 30% but above 25% has brought permeability increase up to 100 mD. The permeability gets a good value exceeding 300 mD when carbonate content is less than 20%.

The amount of clay material displays an even stronger effect on the reservoir quality of the PK suite rocks. In samples with the quantity of fine sediment fraction exceeding 45%, the permeability falls below 20 mD. A decrease of fine material to 35–40 and 30–35% is accompanied by a permeability increase to 50–60 mD, and up to 200 mD correspondingly. Very good permeability is recorded in the rocks with the clay fraction below 30%.

We interpret these results as testifying to both—possibly carbonate and muddy cementation of the PK suite sandstones and the effect of matrix influence on the reservoir properties of these rocks as well.

7.2.4 The Kirmaki Suite

Within the Absheron OGR, the Kirmaki suite is reported in all fields. The Kirmaki suite succession comprises frequently alternating sand, sandstones, and shale intervals. The thickness of the suite varies from 0 to 175–270 m in the

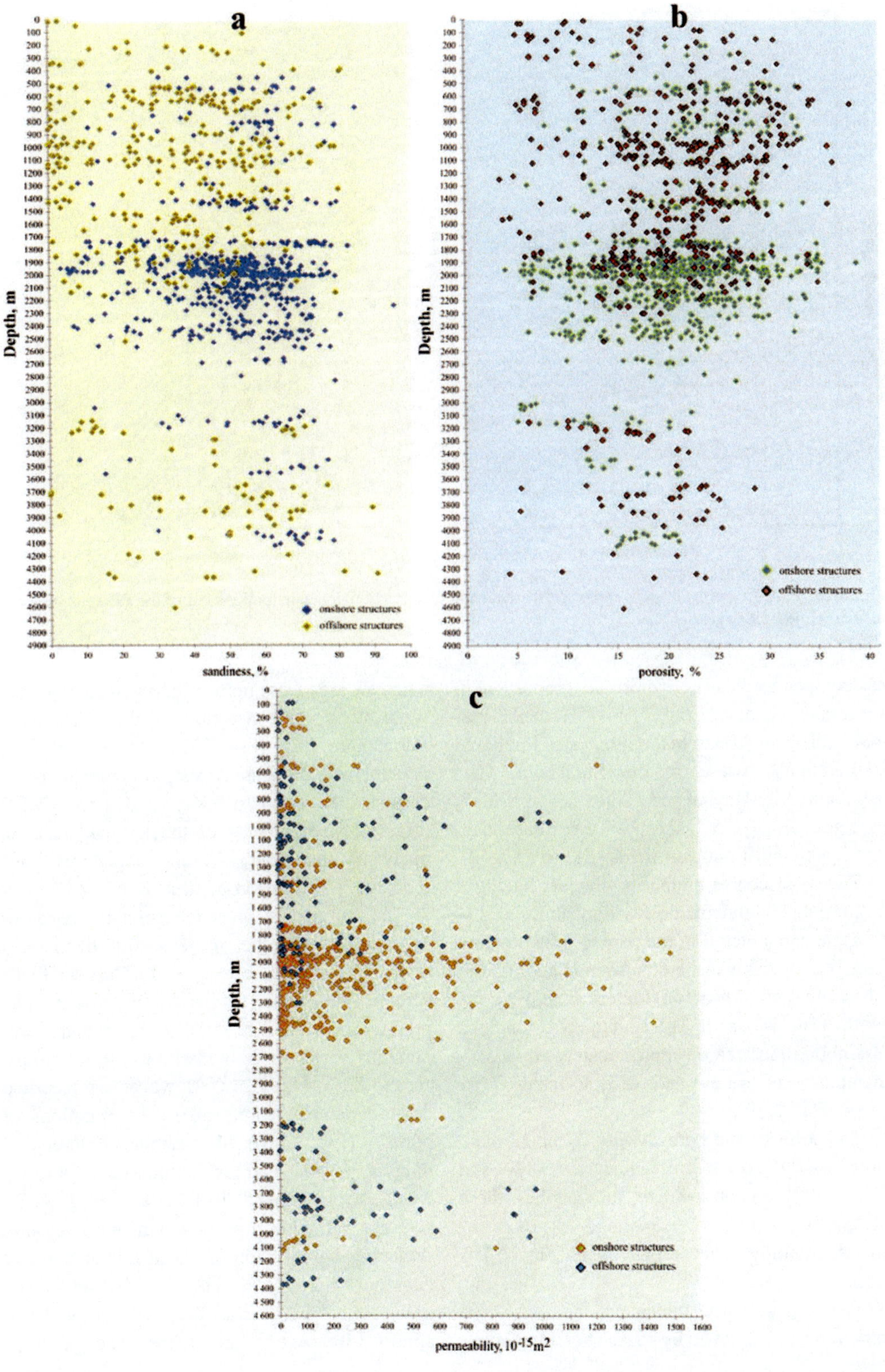

Fig. 7.16 Plots (**a**, **b**, and **c**) display vertical alignment of the reservoir properties' parameters for the PK suite rocks

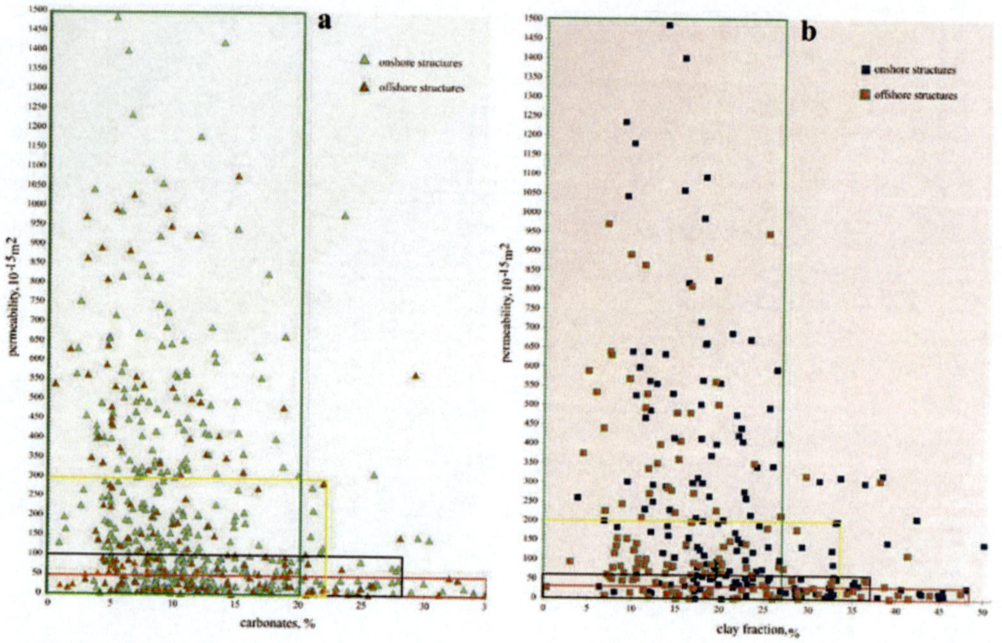

Fig. 7.17 Plots (**a** and **b**) display carbonate material and clay fraction-dependent alignment of the PK suite reservoir rocks' permeability values

eastern peninsula. The sandstone proportion in succession changes from 70 to 75% in the Fatmai, Binagadi, Buzovna, Gala, and Pirallakhi Island fields, while in the Surakhani, Garachukhur, Bibi-Heybat and Chilov island fields it is equal to 50–55%, and in the Shabandag, Atashgakh and Lokbatan fields—30 to 35%.

The most coarse-grained sediments having a proportion of sand fraction within limits of 20–35% are encountered in the narrow area trending from the North Absheron Archipelago (Gerbi Absheron and Absheron kupesi fields) to the south (Fig. 7.18). It splits into two portions resembling the area of approximately equal sedimentation of sandy and muddy material by shape (Fig. 2.10).

The porosity and permeability of the Kirmaki suite sandy rocks are equal to 18% and 100×10^{-15} m^2 on average; in siltstones, these parameters reduce to 7% and 34×10^{-15} m^2 correspondingly. The sorting coefficient (S_o) is 1.65.

The best porosity, which can be considered very good, is displayed by rocks stretching out as a narrow area from north-north-west to east-east-south along some structures of the Absheron—Pribalkhan Sill (Fig. 7.19). Most Absheron Archipelago and East Absheron Peninsula fields demonstrate good porosity and permeability (Fig. 7.20). Sediments of the Kirmaki suite in many of the South Caspian structures have moderate reservoir properties.

The cumulative plots reflecting the behavior of reservoir properties in the section of the Kirmaki suite of the several South Caspian fields demonstrate some regularity (Fig. 7.21). The proportion of sand material in upper and lower horizons varies in the wide limits—from almost zero to 80% (Fig. 7.21a). However, we recorded depth-dependent alignment in the variations of porosity (Fig. 7.21b). Most samples in the upper interval above 2500 m demonstrate porosity change from around 10% to almost 40%. In contrast, in the lower portion of the Kirmaki suite occurring below 2500 m, the maximal porosity does not exceed 27%. The same is true for the permeability that reaches the maximum value of almost 1300×10^{-15} m^2 in the upper horizons

Fig. 7.18 The average percentage of sand fraction in the Kirmaki suite sediments of the South Caspian basin western flank

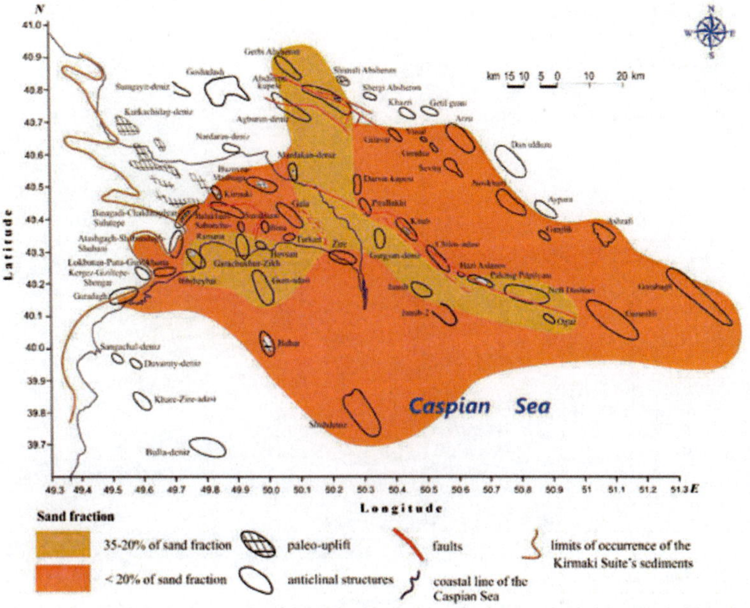

Fig. 7.19 The mapped average porosity (%) of the Kirmaki suite reservoir rocks in the South Caspian basin western flank

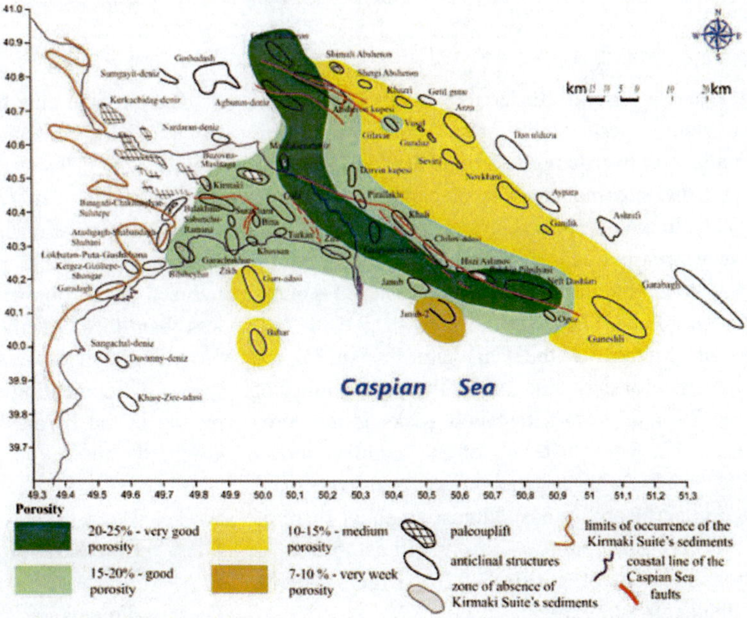

and twice decreases to almost 600 × 10^{-15} m^2 in the lower portion of the suite below 2500 m (Fig. 7.21c).

However, these data indicate still excellent reservoir properties of the Kirmaki suite reservoir rocks in the whole section of the suite. The lowest permeability changing from almost zero values to 20 × 10^{-15} m^2 has been found in samples with carbonate content up to 40% (Fig. 7.22a).

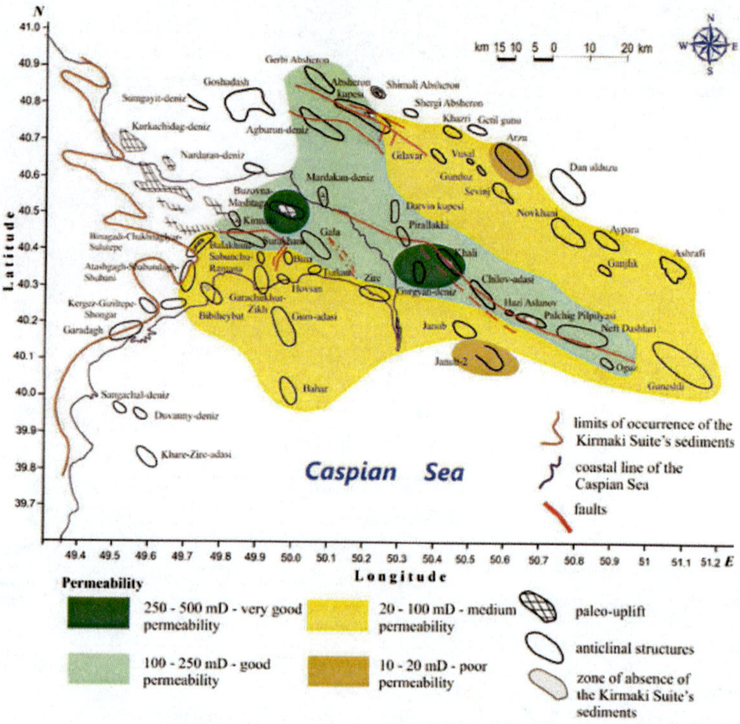

Fig. 7.20 The mapped average permeability (in 10^{-15} m^2) of the Kirmaki suite reservoir rocks in the South Caspian basin western flank

In the second cluster that unites samples of moderate permeability, the maximum of this parameter is growing to around 100×10^{-15} m^2, and the maximal number of carbonates falls to 30%. In the other two clusters, we have well, and very well-permeable samples distinguished by the low amount of carbonate material—maximum 20 and 17.5%.

In contrast to the Gala and PK suites, the influence of clay material on the permeability of the Kirmaki suite's reservoir rocks is less pronounced. The clustering of the samples shows that rocks with negative, poor, moderate, and good permeability have almost an equal share of the clay size particles—from almost 5% to a maximum of 50% (Fig. 7.22b), which is quite challenging to explain.

We assume that these data demonstrate a good indication of the carbonate material's negative influence on the fluid filtration properties of the Kirmaki suite's rocks and dominating carbonate cementation in contrast to the Gala and PK suites that are characterized by both types of cementing material—carbonate and muddy.

The effect of clay fraction on the permeability of the Kirmaki suite rocks is even clearer (Fig. 7.22b). The weakly permeable rocks (less than 20×10^{-15} m^2) show the highest values of clayiness, almost reaching 50%. The moderately and well-permeable rocks stand out, with the reduced proportion of clay-size particles getting less than 45%. All these data demonstrate well-pronounced evidence of the occurrence of both types of cementing material—carbonate and muddy in the Kirmaki suite reservoir rocks and the matrix effect.

7.2.5 PostKirmaki Sand Suite (NKP)

These sediments are widely occurring within the Absheron Peninsula and Absheron and Baku archipelagos. A light fraction of sand and sandstone's mineralogical composition is dominated by quartz (up to 80%). The highest quartz percentage has been noted in northeastern fields. To the southwest, the quartz content is decreasing and not exceeding 50% (Bibi-Heybat, Lokbatan,

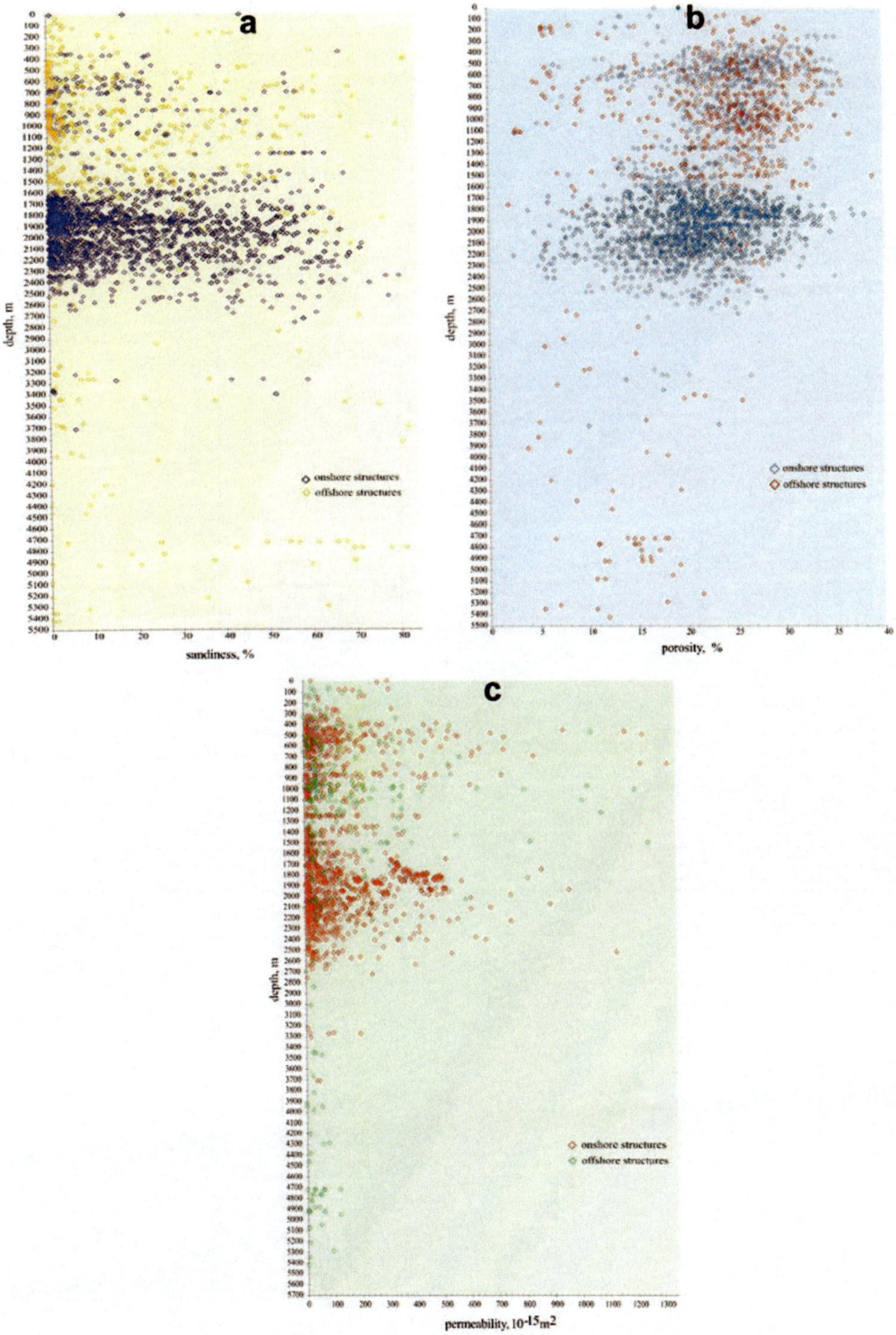

Fig. 7.21 The reservoir properties change of the Kirmaki suite's sandstones (**a**, **b**, and **c**) and siltstones with the depth of their occurrence

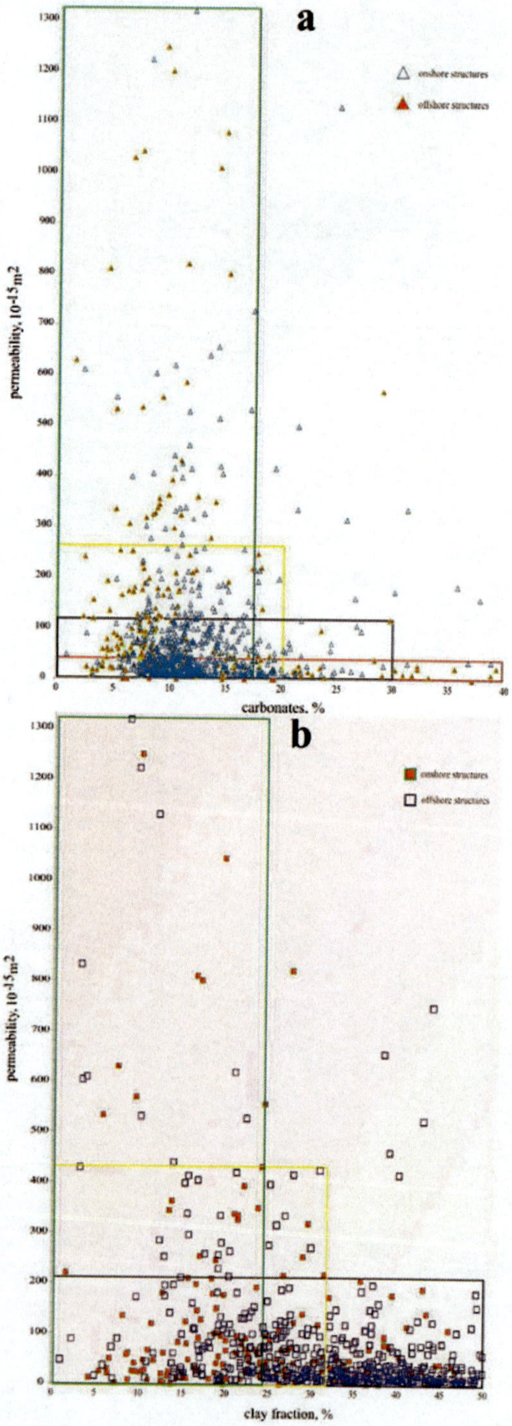

Fig. 7.22 Plots displaying carbonate (**a**) and clayiness-dependent (**b**) alignment of the Kirmaki suite reservoir rocks' permeability values

and Gushkhana). Other components of a light fraction have approximately an equal share—feldspars up to 26% and rock fragments up to 25%.

In heavy fractions, both authigenic such as pyrite, hydrous ferric oxides, glauconite, and allothigenic—magnetite, ilmenite, muscovite, chlorite, disthen, staurolite, garnet, and other minerals are detected. In the mineralogical composition of shales, the chlorite-montmorillonite-kaolinite-hydromica and chlorite-kaolinite-montmorillonite-hydromica complexes are reported.

The PostKirmaki Sand suite is composed of alternating sands, sandstones, silts, siltstones, and shales. The lithology of this suite in the Absheron area and northern Baku Archipelago is dominated by coarse sediments (Fig. 7.23). Here, the share of a sand fraction exceeds 50%. In some sections, conglomerates containing very small pebble-sized clasts are encountered. The shape of an area of coarse sedimentation exactly resembles the area covered by the paleo-Volga primary channel and delta. Two areas dominated by sand sedimentation—the large south-south-eastern directed and minor south-southwestern trending, are separated by an area of reduced coarse material where the share of sand fraction falls to 35–50%. We assume that these two areas are covered by sedimentation in the multiple splits within the vast paleo-Volga delta encountered in the large portion of the South Caspian Lake. The outer edge of this delta is outlined by an area of little deposition of sand material not exceeding 35%.

The proportion of fine fraction does not change much within the South Caspian Basin and is recorded within the 10–35% limits (Fig. 7.24). Based on the number of shale horizons, two distinguished (in the Surakhani, Sabunchi, Balakhani, and Binagadi areas) or three (in the Bibi-Heybat field) subsites in the NKP suite section. To the southwest, the amount of shale beds and proportion of clay fraction gradually increase. The colorization of sandy, silty rocks is primarily dark grey, sometimes greyish brown.

The section of NKP suite in the most significant portion of the South Caspian basin western flank comprises reservoir rocks containing low carbonate material (Fig. 7.25), not getting more than 15% on average. It is slightly higher in the Garadagh, Lokbatan, Gushkhana, and Shubani fields. Such a low number of carbonates positively affects the reservoir quality of the NKP

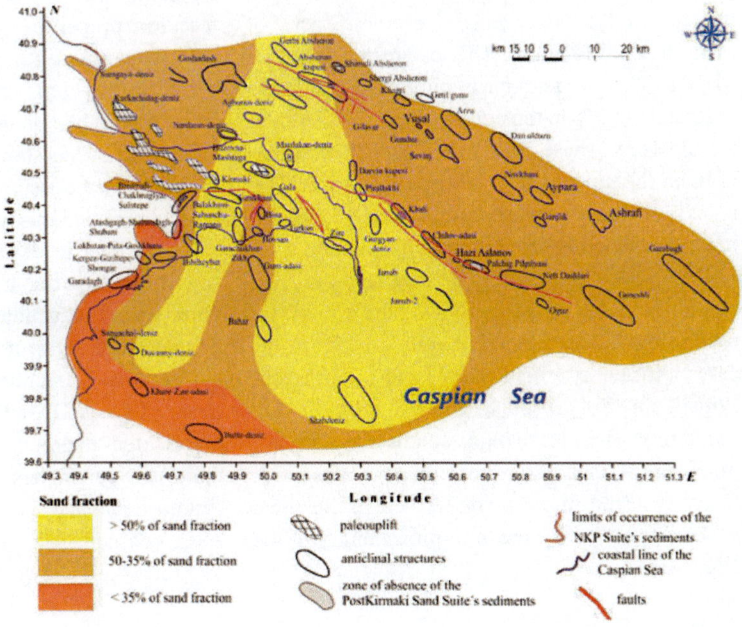

Fig. 7.23 The average percentage of sand fraction in the PostKirmaki Sandy suite sediments of the western flank of the South Caspian basin

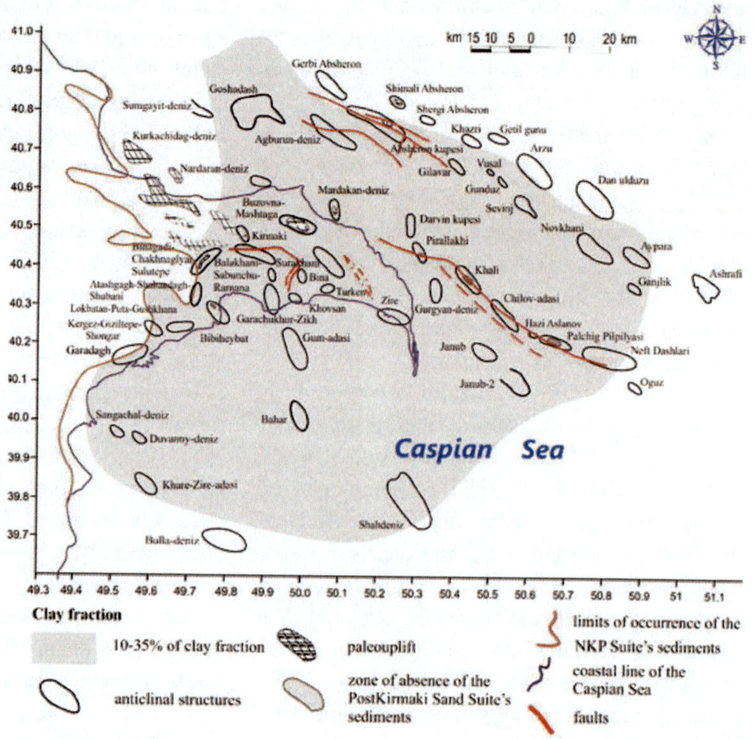

Fig. 7.24 The distribution of the average amount of clay fraction (in %) in the PostKirmaki Sandy suite sediments of the South Caspian basin western flank

suite sediments. It manifests in the good reservoir characteristics of these rocks in most South Caspian fields.

The porosity reaches a minimum of 20% on average in the NKP suite's reservoir rocks of many fields (Fig. 7.26). This parameter decreases to the southwest, getting a minimum of 10% and a maximum of 5% in the northwest part of the Baku Archipelago. These data fully agree with variations in the amount of sand material in the NKP suite's sandstones and testify to these sediments' homogenous grain size and porous medium.

The same regularities are reported for permeability variations in the NKP suite's reservoir rocks (Fig. 7.27). The NKP suite's rocks in the east Absheron Peninsula and Absheron-Pribalkhan Sill display good fluid filtration properties. The clear south, southwest, and east trend of permeability worsening is observed. However, even in the fields located in this area, we can report on the medium permeability of the NKP suite's rocks.

This data testifies to the south, southwest, and east trends of the degradation of the NKP suite's reservoir properties; however, we can classify the quality of these reservoirs as medium. Describing reservoir properties changes with depth, and we can state that both parameters—porosity, and permeability, are decreasing below 2500 m (Fig. 7.28a, b), predominantly in the shallow NKP suite succession of the offshore fields.

This raised questions about the controls on the reservoir properties—whether these characteristics are worsening in the lower intervals of the NKP suite succession due to sediments compaction and pore space reduction or the facies variations governed by reservoir quality. The plot demonstrating sand fraction amount changes versus depth of the NKP suite's rock sample occurrence reveals the nonmonotonic relationship between these two parameters that allows us to conclude that no significant coarse grains' quantity changes, and, consequently, facies variations between hypsometrically lower intervals penetrated in the

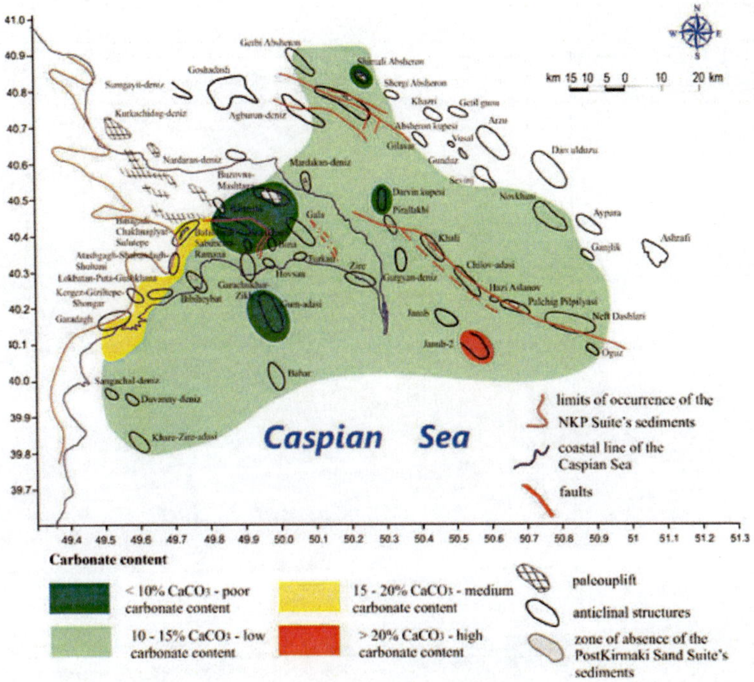

Fig. 7.25 Variations of the carbonate material's average quantity (in %) in the PostKirmaki Sand suite's reservoir rocks of the South Caspian basin western flank

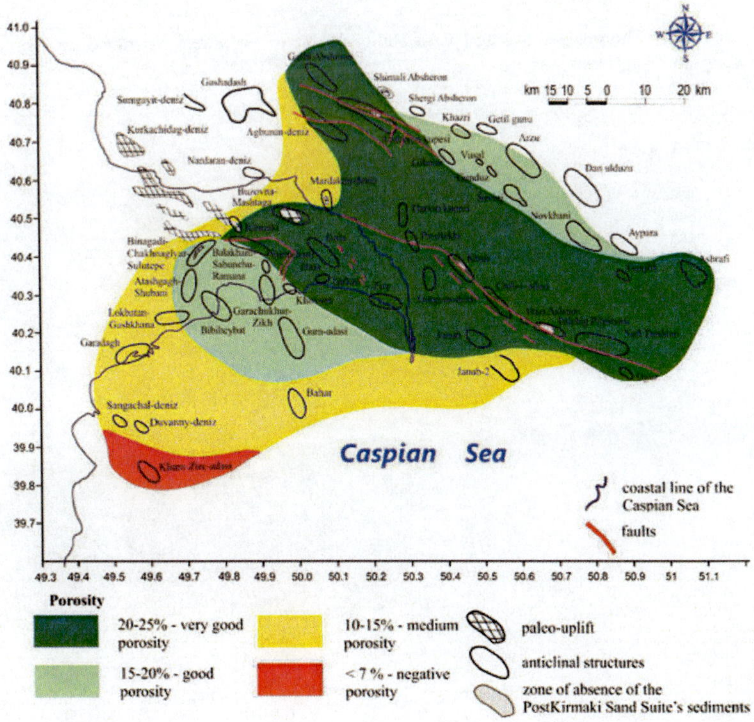

Fig. 7.26 The mapped average porosity (%) of the PostKirmaki Sand suite's reservoir rocks of the South Caspian basin western flank

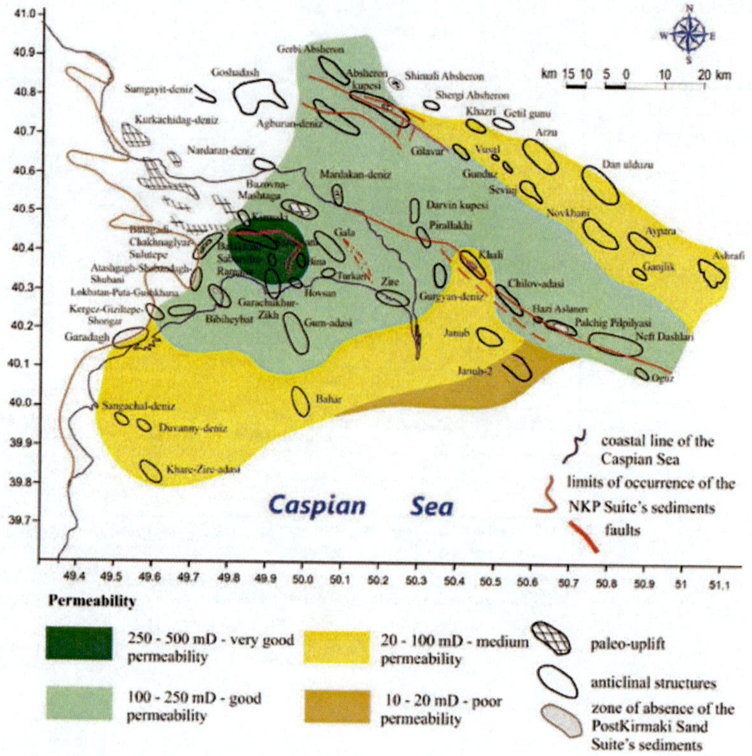

Fig. 7.27 The mapped average permeability (10^{-15} m^2) of the PostKirmaki sand suite's reservoir rocks in the South Caspian basin western flank

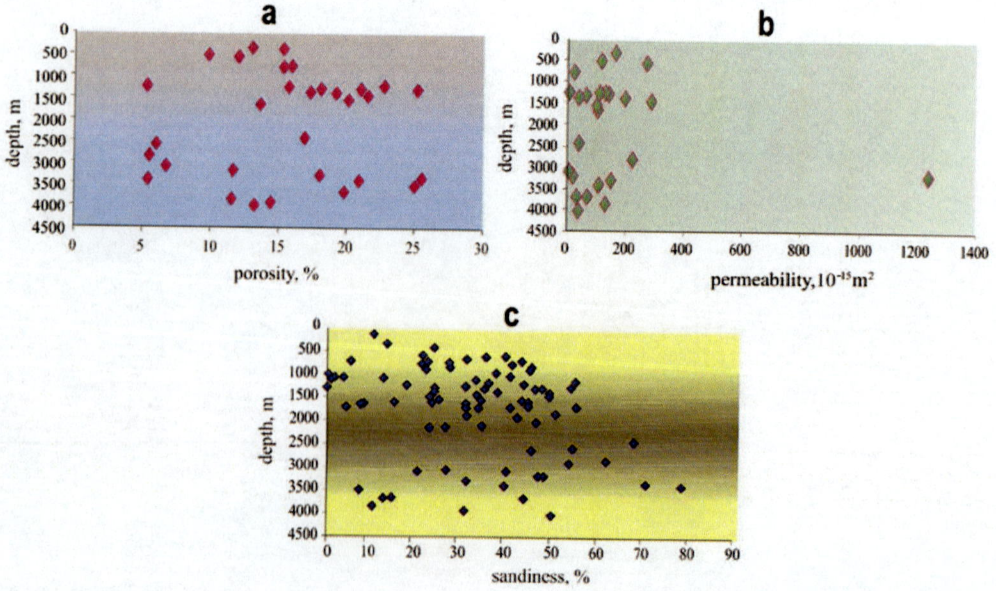

Fig. 7.28 Plots (**a**, **b**, and **c**) demonstrate the reservoir properties changes of the PostKirmaki Sandy suite's sandstones with the depth of their occurrence

deeply subsided offshore fields, and elevated onshore succession (Fig. 7.28c).

Many samples display chaotic distribution of the sand fraction proportion in the NKP suite reservoir rocks, varying from shallow values to more than 85% in the same depth interval, including the lower horizons. However, the analysis of rock sandiness by fields demonstrates the complex stratigraphic regularity (Fig. 7.28d, e). We have recorded in many onshore fields (Gushkhana, Garachukhur, Balakhani-Sabunchu-Ramana, Shubandagh, and Bibi-Heybat) that the mid interval of the NKP suite succession is distinguished by the lower proportion of coarse clastic material on average comparing with the upper and lower portions of the suite.

This trend is not reported in some onshore structures, standing out by the short sampling interval that probably resulted in the incomplete stratigraphic records achieved in these fields and is less pronounced in the offshore fields.

Two plots constructed for permeability—carbonate material and clay fraction data series also display scattered arrangement (Fig. 7.29a, b).

The non-linear link between the quantity of carbonates and the permeability of the NKP suite's reservoir rocks can be explained by the analysis of correlation coefficients. It equals − 0.235 under carbonate content below 25% and sharply increases, reaching − 0.35 when the carbonate amount exceeds 25%. This data points to a weak inverse correlation between the filtration properties of the rocks and the amount of carbonate material under conditions of its high content exceeding 25%. The absolute majority of PostKirmaki suite reservoir rocks have low amounts of carbonate material of less than 25%, which is a reason for the insignificant correlation between carbonate content and permeability of the PostKirmaki Sand suite reservoir rocks. However, we could record some regularity. It is clear from the permeability values versus carbonates amount plot (Fig. 7.29a) that the impermeable—low-permeable rocks exhibit a wide range of carbonate material quantity variations—from 5% to more than 35%, and in most samples, the highest content exceeds 25% but is below 30% (red square in the plot). Increasing permeability is accompanied by a progressive fall in the number of carbonates. The highest permeability, more than 250×10^{-15} m^2 is reported in the samples, with the lowest number of carbonates is around 12%.

Another phenomenon is the need for a significant effect of clay fraction on the permeability of the NKP suite rocks, which is quite challenging to explain. One possible explanation is the low amount of swelling mineral montmorillonite, at most 5%, that enables preserving reservoir properties. On the other hand, our statistical processing of analytical data also shows that the permeability of the NKP suite reservoir rocks significantly falls when clay fraction content is around 10%. In this case, a correlation between these two parameters gets − 0.49. Increasing the number of clay-size particles to

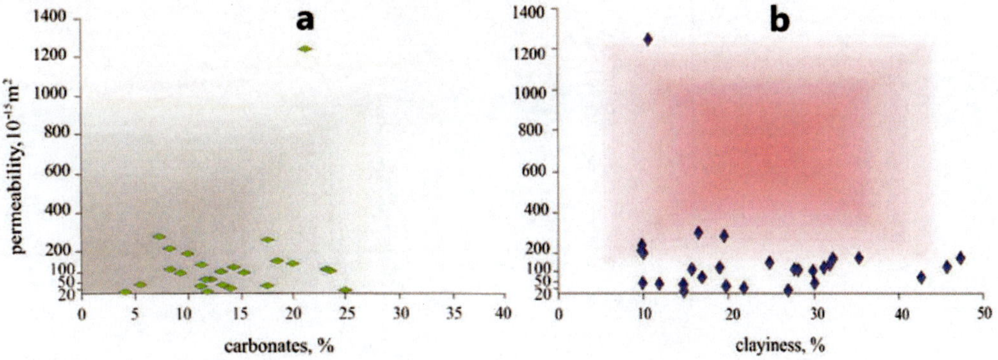

Fig. 7.29 The arrangement of carbonates (**a**) and clay fraction (**b**) versus permeability values for the PostKirmaki Sandy suite's reservoir rocks

20% reduced the correlation coefficient to − 0.16. In the rocks with the amount of clay fraction exceeding 30%, this correlation coefficient is recorded as − 0.14.

One possible explanation for this quite exciting regularity is the following. Pure sandstones are the most sensitive to occlusion of the pore space by clay particles, resulting in a strong reaction of the reservoir properties to even minor variations in the clay material amount. In the muddy sandstones and siltstones containing more than 30% of the clay size grains, such small changes in their amount only affect the filtration properties of these rocks.

The described reservoir characteristics of the PostKirmaki Sand suite support a conclusion on the low heterogeneity of these reservoirs within the Absheron Peninsula and adjacent onshore area. The reservoir quality here is good. It is slightly worsening towards the Baku Archipelago and North Absheron uplift zone, where the reservoirs can be classified as moderate quality.

7.2.6 The PostKirmaki Clay Suite (NKG)

These sediments are widely occurring within the Absheron Peninsula and Absheron and Baku archipelagos. The thickness of the NKG succession is varied from 15 to 20 m in the northern Absheron Peninsula (the Kurdakhani area) to 40–50 m in the southern peninsula (the Zikh field) and from 10 to 120 m in the western (the Sulutepe field) and eastern (the Gala field) parts of the peninsula. As is seen from the suite's name, it is mainly composed of fine-grained sediments.

However, the share of sand-size grains in the NKG suite succession is stable and does not exceed 35% in the South Caspian Basin western flank (Fig. 7.30). An exception is a small area covering Bibi-Heybat, Garachukhur-Zikh, Atashgagh-Shubani structures, and the southern part of the Surakhani field distinguished for the higher proportion of the sand fraction in the NKG suite sediments varying from 35 to 50%.

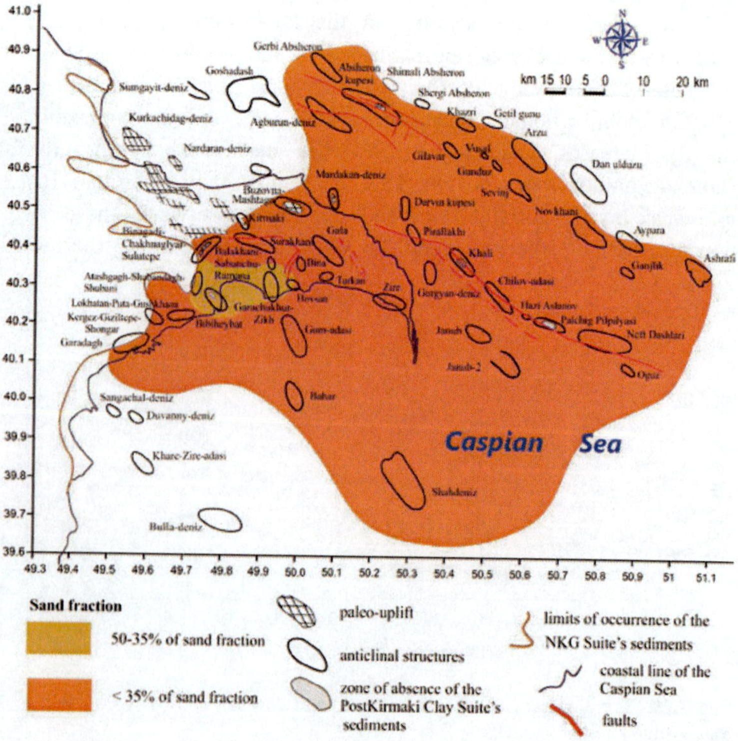

Fig. 7.30 The average percentage of sand fraction (%) in the PostKirmaki Clay suite sediments of the South Caspian basin western flank

The highest proportion of sand beds is recorded in the middle portion of the PostKirmaki Clay suite. Quartz is a predominant mineral in a light fraction of sandy and silty rocks. Its contents are estimated to have been 45–80%, and the average is 60%. Feldspar's content varied widely from 12 to 70%, averaging about 18%. The heavy fraction contains mica, amphibole, garnet, tourmaline, disthen, staurolite, and other minerals.

A clear southward trend of clay fraction growth in the NKG suite succession is observed (Fig. 7.31).

Its share in the Baku Archipelago's southern fields exceeds 50%. In the Absheron Peninsula and adjacent offshore areas, including Absheron-Pribalkhan Sill, the content of the finest sediments does not exceed 35%. However, the clay fraction share is at most 20% in most South Caspian fields. Considering a small proportion of coarse sediments here (Fig. 7.30), we can conclude about the prevalence of silt fraction in the NKG suite succession.

Despite the dominance of the fine-grained sediments in the NKG suite section, this succession is one of the hydrocarbon production units in the Absheron oil–gas region. Occurring here, thin sandstone and siltstone beds display good reservoir properties that have been widely studied in several fields (Sabunchi-Balakhani-Ramani, Bibi-Heybat, Mashtaga, Gum adasi, Garachukhur, Buzovna, Atashgagh, Shubani, Surakhani, Garadagh, Zikh, and Mardakan-deniz), the Darvin bank, Neft Dashlari. For instance, the amount of carbonate material in the reservoir rocks in most areas of the Absheron Peninsula and Absheron Archipelago does not exceed 15% on average with some exceptions (Garachukhur-Zikh, Lokbatan-Puta-Gushkhana, Kergez-Giziltepe, and Garadagh fields) (Fig. 7.32) that is a good indication of weak cementation of NKG suite sandstones, siltstones by carbonate material. The less

Fig. 7.31 The distribution of the average amount of clay fraction (%) in the PostKirmaki Clay suite sediments of the South Caspian basin western flank

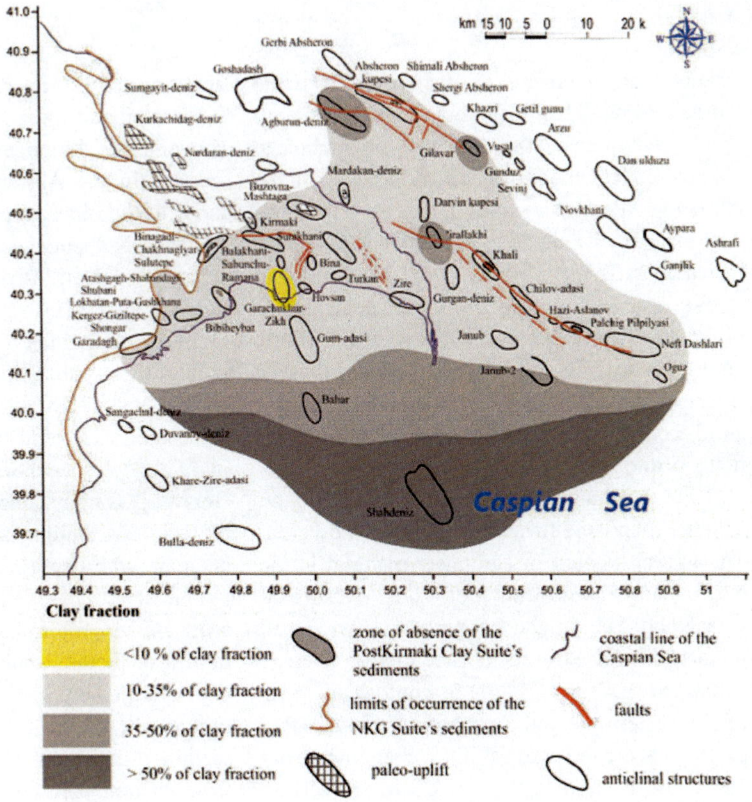

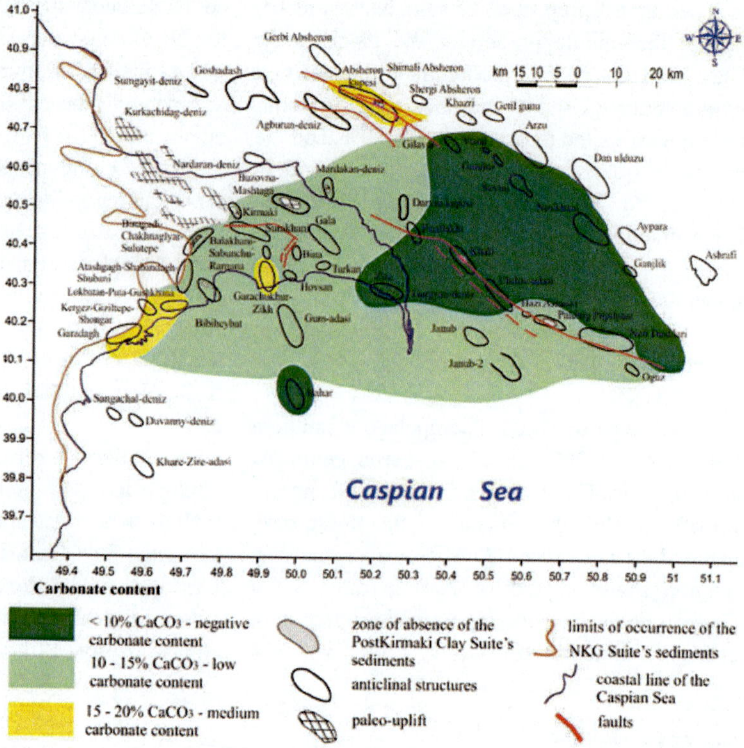

Fig. 7.32 Variations of the carbonate material's average quantity (%) in the PostKirmaki Clay suite's reservoir rocks of the South Caspian Basin western flank

cemented are sediments in offshore fields of the Absheron-Pribalkhan Sill and North Absheron uplift zone. Here, the quantity of carbonate material is less than 10%. In most areas of the Absheron Peninsula and northern Absheron Archipelago, the number of carbonates slightly increased to 15%.

We assume that weak carbonate cementation of the NKG suite reservoir rocks resulted in their good porosity (Fig. 7.33), which changes in most Absheron fields within the limits of 20–28%, and good-moderate permeability (Fig. 7.34) in many fields of the Absheron oil–gas region.

The worsening of reservoir properties' south, southeast, and southwestward trend is observed. The lowest amount of carbonate material in the NKG suite sediments of the Absheron-Pribalkhan Sill on the background of reduced permeability of these rocks here that on average equals 20–100 × 10^{-15} m^2 in contrast to rocks of the eastern, central Absheron Peninsula displaying average permeability changing from 100 × 10 to 500 × 10^{-15} m^2, arose question about reservoir properties controls. This question is underpinned by the clayiness map (Fig. 7.31) showing the homogenous distribution of clay fraction in the Absheron oil–gas region. Similarly, to the older suites, the reservoir properties of the NKG suite sediments demonstrate two types of regularities of their distribution with a depth (Fig. 7.35).

In the upper intervals above 2000 m, the data points representing porosity and permeability values in the NKG suite's reservoir rocks are distributed in a wide range of values, not showing any depth-dependent manner (Fig. 7.35a, b). Below 2000 m, we observe a monotonic function of their distribution characterized by the value decrease with depth.

The share of sand-size grains in the total sediment volume indicates the higher sandiness of the onshore samples (Fig. 7.35c) compared with the offshore sediments. We also noted that samples in the mid-interval of the NKG suite section differ by the higher percentage of the sand fraction. The same interval is distinguished

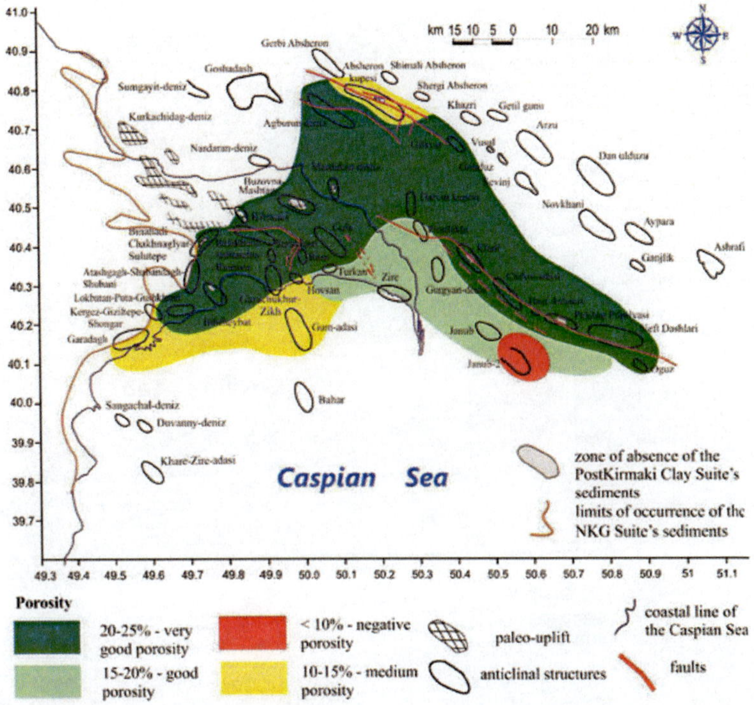

Fig. 7.33 Variations of the average porosity (%) of the PostKirmaki Clay suite's reservoir rocks of the South Caspian basin western flank

by the better parameters of the reservoir properties (Fig. 7.35a, b) that raised a question on the NKG suite reservoir quality control—whether reservoir properties change along the section due to stratigraphic control more sandiness in the mid-suite interval or because of their degradation because of depth-dependent compaction.

In some fields, the mid-interval of the NKG suite section stands out by the highest proportion of sand fraction that points to stratigraphic control on this parameter. These results indicate the multiple controls affecting the reservoir quality of the NKG suite sediments, including facies distribution, sandiness variability along the section, and compaction due to overburden pressure. We assume that the last one was the main factor causing the rapid degradation of the reservoir properties of the NKG suite sediments in the lower horizons.

However, we cannot entirely exclude the effect of other factors, such as carbonates and clay fraction content, on the reservoir properties. Although two plots reflecting permeability dependence from the amount of carbonate and clay material exhibit chaotic arrangement, we can state that the lowest permeability below 20×10^{-15} m^2 is recorded in the samples with the highest carbonate content exceeding 20% in the cluster outlined by the red line (Fig. 7.36a). Such low-permeable rocks demonstrate the highest clay material (Fig. 7.36b).

In other clusters of samples demonstrating permeability variations in the value ranges of 20–50×10^{-15} m^2; 50–100×10^{-15} m^2; 100–250×10^{-15} m^2; above 250×10^{-15} m^2 we observe the gradual fall of the carbonate content to below 10% in the most-permeable rocks. The share of clay fraction does not change so dramatically, only in the rocks displaying permeability of more than 250×10^{-15} m^2. This parameter has the lowest value below 20%.

7.2.7 The Fasila Suite

The Fasila suite is a principal production unit in the South Caspian basin. It contains abundant coarse material, including different sizes and

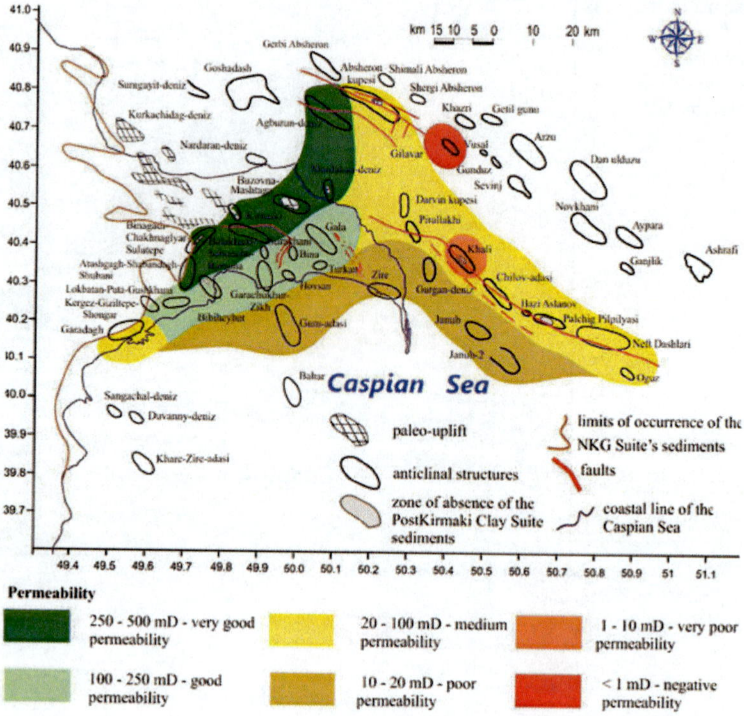

Fig. 7.34 The map of average permeability variations (10^{-15} m^2) of the PostKirmaki Clay suite's reservoir rocks of the South Caspian basin western flank

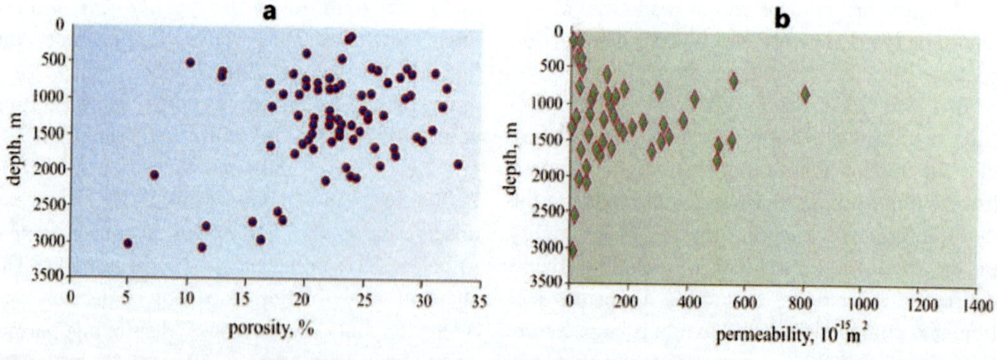

Fig. 7.35 The reservoir properties changes of the PostKirmaki Clay suite's sandstones and siltstones with the depth of their occurrence: **a** porosity, %, **b** permeability, 10^{-15} m^2

morphology pebbles, recorded in conglomeratic beds in the base of the Fasila suite in the Kirmaki Valley and other fields.

The 80–90% of the Kirmaki suite section in the central and eastern Absheron Peninsula and Absheron-Pribalkhan Sill comprises course, medium, less fine-grained sands, and sandstones. A thorough analysis of an extensive data set on the grain size of the Fasila suite sediments shows clear evidence of channel shape morphology of

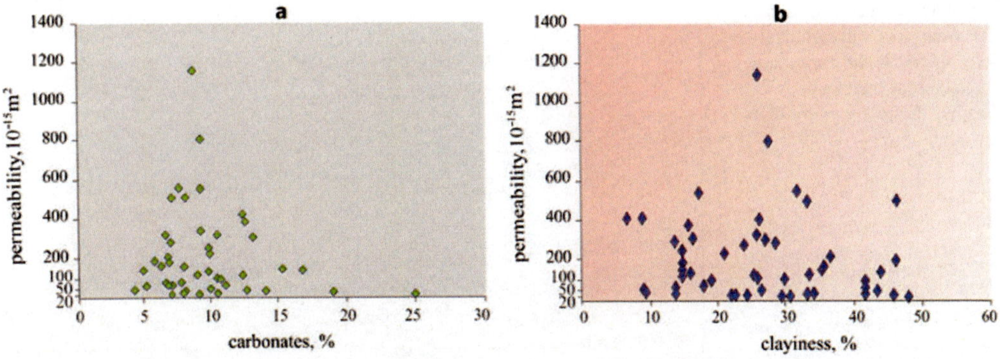

Fig. 7.36 The arrangement of carbonates (**a**) and clay fraction (**b**), versus the permeability values in the PostKirmaki Clay suite reservoir rocks

sand-dominated areas containing more than 50% of sand fraction in the Fasila suite section (Fig. 7.37). We have found three such areas trending from north to southeast, south, and southwest, resembling three paleo-Volga delta branches. The biggest one covers the structures of the Absheron-Pribalkhan sill. The smallest one is reported in the western Absheron Peninsula and extends to the structures Sangachal-Deniz, Duvanny-Deniz, Alyat-Deniz, and Khara-Zirya in the Baku Archipelago.

Addressing back to the sandiness map of the older suites, we can conclude about the episode of the most considerable accumulation of coarse material since the beginning of the Pliocene time that is, obviously, the result of the biggest progradation of the paleo-Volga River delta. Based on the distribution of sand-prone sediments, we can state further development of the middle branch, which originated in the PK suite (Fig. 7.10), and the western branch, which was already well developed in the NKP suite (Fig. 7.23). We also record the split of the vast eastern portion of the delta (Figs. 7.10 and 7.23) into the largest eastern and middle branches. The adjacent area has a good proportion of sand material varying from 35 to 50% that is probably distributed in the extensive network of crevassing channels desiccating the mud-dominated delta plain and in the inter-distributary bay environment.

These three sand-dominated areas are separated by the areas containing less proportion of sand material, and the one between the western and middle channels is the most sand-starved. Here, the share of clay fraction reaches 39% on average. The lithofacies (Fig. 2.16) and sandiness (Fig. 7.37) maps of the Fasila suite sediments indicate the significant lithological changes that brought reservoir heterogeneity and spatial variations of the reservoir properties. For instance, the porosity and permeability of sandstones are estimated to be equal to 10%–28% and 12–94 $\times$ 10^{-15} m^2, respectively. In siltstones, these parameters vary porosity—from 7 to 26%, permeability—from 6 $\times$ 10^{-15} m^2 to 82 $\times$ 10^{-15} m^2; in sandy loams—13 to 16%, and 4–30 $\times$ 10^{-15} m^2.

Thus, the porosity changes in a wide range from very good values exceeding 20% to very small values in several fields in the Absheron Peninsula (Fig. 7.38). The porosity map demonstrates the confinement of the most porous rocks of the Fasila suite to the biggest eastern branch of the paleo-Volga delta and the adjacent eastern portion of the delta plain. Another such area of good porosity of the Fasila suite rocks is reported in the western branch site. The same regularities were recorded in permeability variations (Fig. 7.39). The permeability changes from medium values in the Absheron—Pribalkhan sill (the maximum value of 357.8 $\times$ 10^{-15} m^2 is

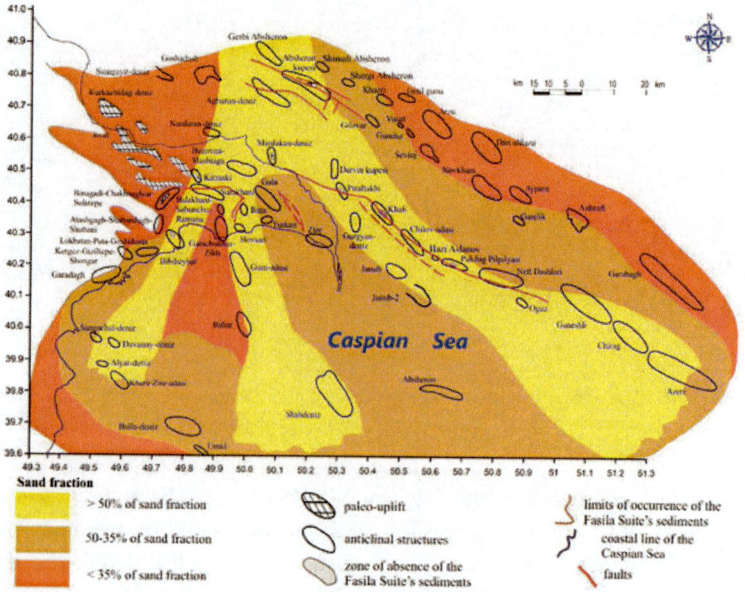

Fig. 7.37 The distribution of the average amount of sand fraction (%) in the Fasila suite sediments of the South Caspian Basin's western flank

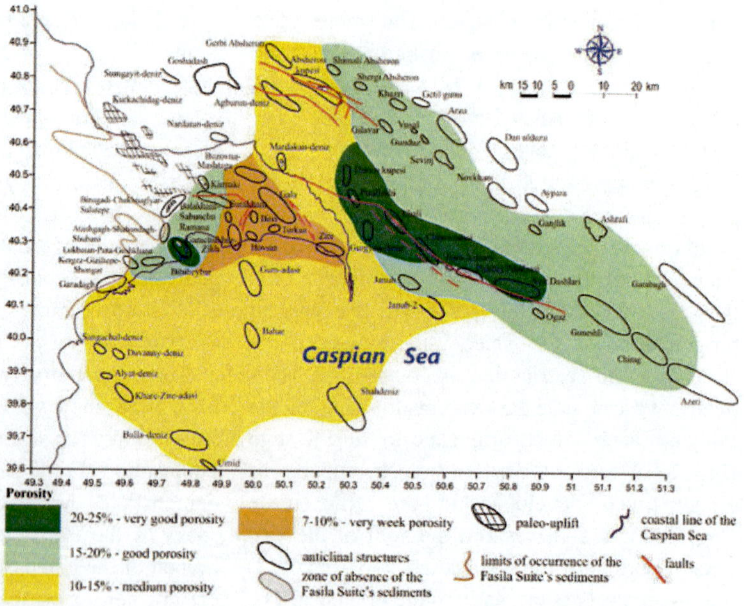

Fig. 7.38 The map of average porosity variations (%) of the Fasila suite's reservoir rocks of the South Caspian basin western flank

recorded in the Darvin bank field, then it drops towards Neft Dashlari structure) to medium—very good permeability in some fields of the Absheron Peninsula located within the western branch area. For instance, in the Gushkhana field the permeability is as high as 288.3×10^{-15} m^2. Within the Baku Archipelago, the data testifies to moderate reservoir properties of the Fasila suite rocks (Fig. 7.40). Thus, the obtained results show a significant dependence on the reservoir properties of the Fasila suite rocks from their spatial location.

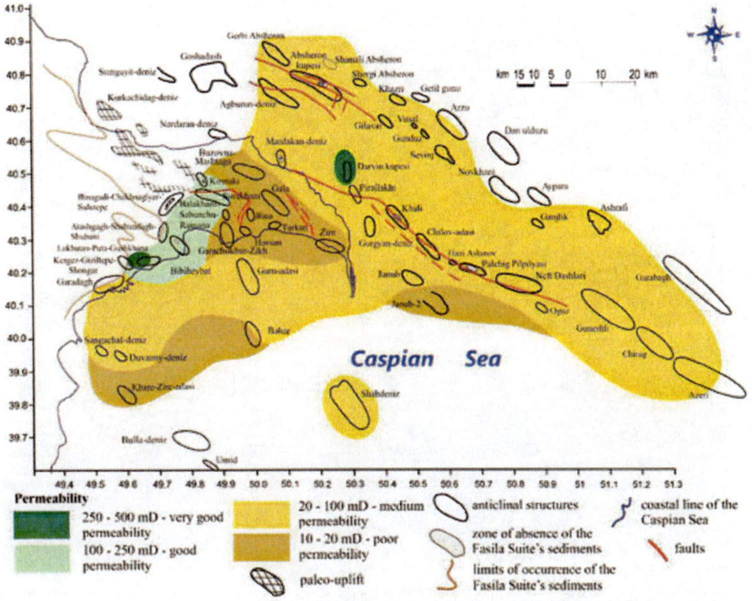

Fig. 7.39 The average permeability variations of the Fasila suite's reservoir rocks of the South Caspian basin western flank, 10^{-15} m^2

At the same time, in other fields of the Absheron Peninsula (Buzovna-Mashataga, Bina, Turkan, Zirya, Garachukhur-Zikh, etc.) the reservoir quality of the Fasila suite rocks significantly reduces. Unexpectedly, here, in the middle branch site, we have recorded very weak porosity and poor permeability (Figs. 7.38 and 7.39).

The data on the carbonate cementation testifies to medium carbonate content in the fields of the Absheron Peninsula (Fig. 7.41) that could not negatively affect reservoir properties of the Fasila suite's rocks and so dramatically reduce them. We assume that such worsening of the reservoir properties is due to a very low proportion of sandy sediments in these fields (Fig. 7.37).

Two plots displaying a link between the permeability of the Fasila suite's reservoir rocks and the amount of carbonate material (Fig. 7.42a) and clay fraction (Fig. 7.42b) have been constructed.

Both plots show no strong correlation between the studied parameters. The low-permeable rocks with permeability records less than 20×10^{-15} m^2 have very variable amounts of carbonates and clay fraction, changing from low values around 5% to high ones exceeding 35% for carbonates, and 50% for clay fraction (Fig. 7.42). However, the low-permeable rocks display that most samples create clusters exhibiting a percentage of carbonate material below 25% as the highest value (Fig. 7.42a). A similar number of carbonates is recorded in the moderate-permeable rocks with permeability ranging from 50×10^{-15} m^2 to 100×10^{-15} m^2. Here, the highest carbonate material content is around 20%.

The rocks displaying good permeability varying in the range of 100×10^{-15} m^2 to 250×10^{-15} m^2 have a medium quantity of carbonate material and a fine-grained fraction (from 3 to 13% and 10% to 20% correspondingly). The highly permeable rocks display a scattered arrangement that makes it difficult to catch any patterns.

These results raise a question about the principal factors affecting the reservoir quality of the Fasila suite's rocks. The porosity plot reflecting the vertical distribution of this parameter in the sediments collected from different South Caspian structures shows several clusters (Fig. 7.43a).

The lowest porosity, less than 7%, is reported in the group of offshore core samples from the uppermost interval above 1000 m (red square). Below 2500 m, we recorded another closely

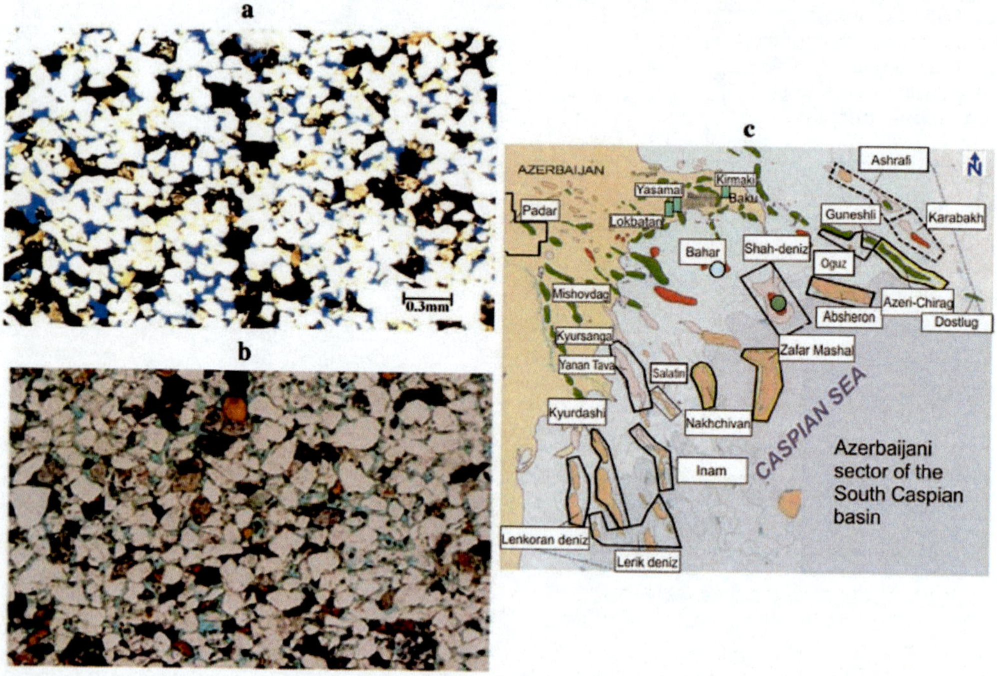

Fig. 7.40 SEM image of the Fasila suite sandstones. **a** the Bakhar field; well No. 73, depth 4533–4539 m; reservoir properties: porosity—11.3%; permeability—71.4 × 10^{-15} m^2; density—2.4 g/cm^3; carbonate content—7.5%; grain size—0.194 mm; sorting is well. **b** the Shakh-Deniz field, depth 5613.5–5613.7 m; **c** areal map. Reservoir properties: porosity—18.7% (primary intergranular porosity—15.7%, secondary, mostly, intragranular porosity—3%); pore size—10–200 μm); permeability—from moderate to a good; pore size—50 to 550 μm; sorting is well

spaced sample set demonstrating medium to very good porosity values (black square) except several separately standing scattered samples (Fig. 7.43a). Above 2500 m, another cluster of highly porous sediments showing porosity change in the 16–32% limits, excluding several standings beyond the group samples, is reported (green square). These data display no significant difference between onshore and offshore samples. Both data sets demonstrate equal porosity values.

The permeability plot shows a wide depth range of equal permeability values. Only in the interval above 4000 m do we record the highest permeability of the Fasila suite sediments. This analysis allows us to conclude the nonlinear relationship between reservoir characteristics and the depth of their occurrence in contrast to NKP suite reservoirs.

The carbonate content in the clustered samples displays equal values in the upper and lower intervals (Fig. 7.43c) and is around 20% as the highest. As described above, the negative influence of carbonate material on the filtration properties of the rocks becomes significant under conditions of its highest content exceeding 25%, which was also proved for the Fasila suite's reservoir rocks. We tried to find the reasons for reservoir characteristics variations in the grain size versus depth plots (Fig. 7.43d, e). In both plots reflecting the distribution of shares of sand and clay fractions, the data have scattered arrangement and does not display any regularity of their arrangement with the depth. The onshore and offshore sediments show the same values, and the values vary in a wide range even within the same depth intervals of the same field, which, we assume, indicates the strong lateral

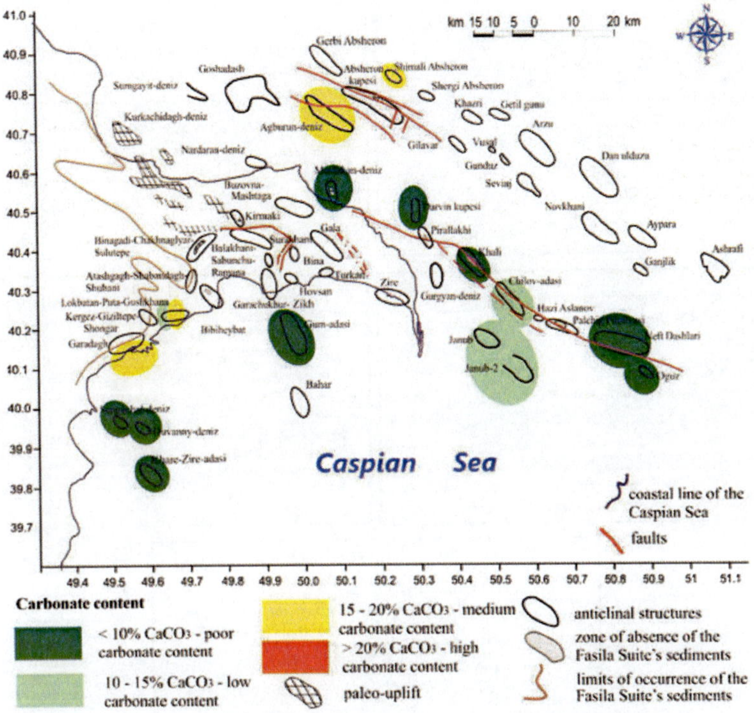

Fig. 7.41 The variations of the average amount of carbonate material (%) in the Fasila suite's reservoir rocks of the South Caspian basin western flank

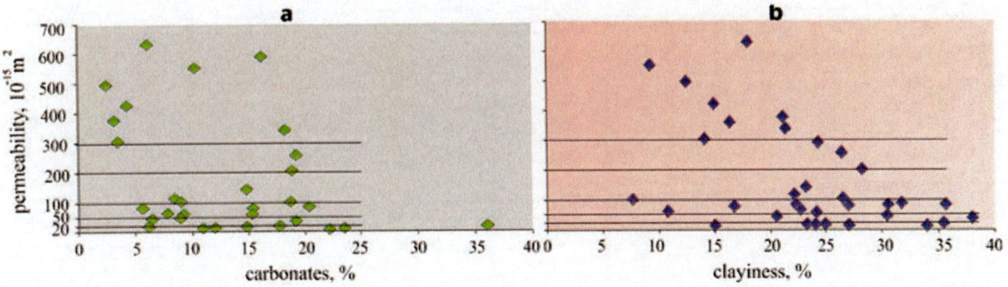

Fig. 7.42 The arrangement of carbonates (**a**) and clay fraction (**b**) versus the permeability values of the Fasila suite reservoir rocks

environmental variations. However, two other plots constructed for sandiness values of different offshore and onshore fields (Figs. 7.43f, 7.43g) demonstrate two types of data distribution. In the first type, data vary in a wide range, and the highest percentage of coarse-grained sediments exceeds 70%, recorded in all studied offshore fields and some onshore structures. Other data types display the lower maximum of the sand fraction proportion, below 50%, in most samples, which does not exceed 40%. It was reported in specific onshore fields such as Surakhani, Garachukhur, Puta, and Lokbatan. Here, in these structures, we have recorded the lowering of reservoir properties of the Fasila suite sediments (Figs. 7.38 and 7.39).

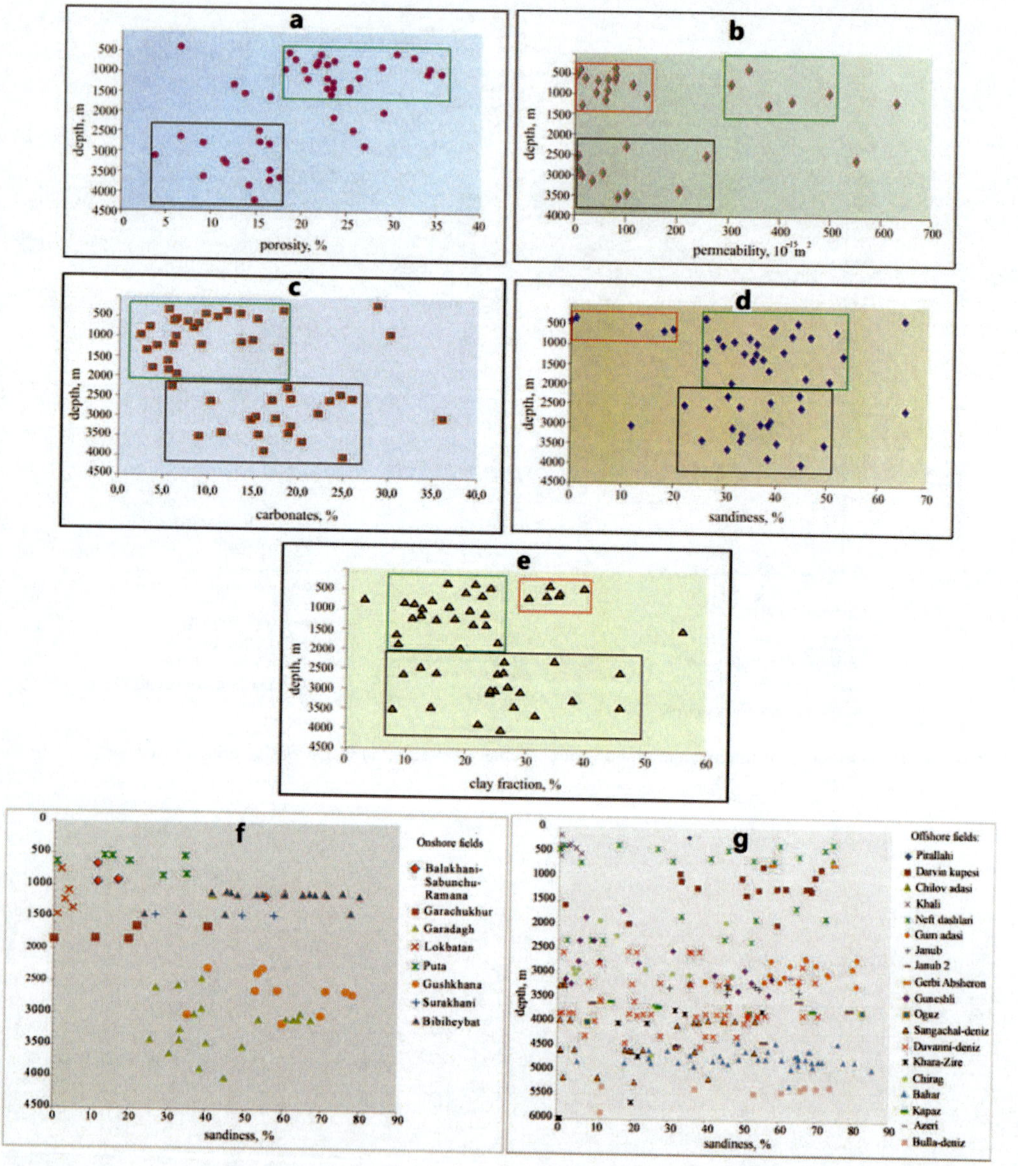

Fig. 7.43 Reservoir characteristics of the Fasila suite rocks (**a**–**g**) versus the depth of their occurrence

These results testify to the significant facies control on the reservoir parameters of the Fasila suite rocks. The data in diagrams 7.43f and 7.43g also display that they are aligned to certain intervals separated by the intervals of lack of data. We record four such data intervals referring to reservoir rocks in the offshore fields and three in the onshore ones. It gives us proof to speak about the stratigraphic controls for the reservoir rocks accumulation replaced in the vertical section by the periods of coarse clastic material's starvation intervals and accumulation of fine-grained sediments, which served as intraformational seals for the underlying sandstones.

The statistical analysis of the Fasila suite sandstones showed a very weak inverse correlation between porosity and clay fraction amount when the latter varies between 15 and 30%. We did not record a significant link between the quantity of clay-size particles and the porosity and permeability when the clay fraction content exceeds 30%.

These results allow us to assume that, like the NKP suite's sandstones, the pore space of the Fasila suite's coarse sediments containing a small amount of clay fraction react more strongly to even insignificant growth of the clay size particles' share in the matrix and cementing material as well. Besides that, the Fasila suite sandstones also display a clear downward porosity–permeability reduction profile that is better pronounced in the porosity depth-dependent variations.

The performed analysis of the reservoir characteristics data of the rocks of the Fasila and other Productive Series suites illuminate the effect of many controls on reservoir quality, such as facies-dependent grain size variations, matrix influence, depth of occurrence, and stratigraphic control, type of cementation, mineralogy of cementing material that all together provide the nonlinear, multifunctional distribution of reservoir properties of these sediments.

7.2.8 The Balakhani Suite

The Horizon X

It is the lowermost horizon of the Balakhani suite. Its maximum thickness of about 550 m has been recorded in the Gum adasi, Garachukhur, Surakhani, and Balakhani-Sabunchi-Ramana fields.

In the area under study, reservoir rocks are composed of sands, sandstones, sometimes poorly sorted, and siltstones. The typical feature is color variations—from grey, greyish-yellow, yellow to reddish grey. The grain size of particles constituting sandy rocks ranges from typical fine to coarse sand. Shale intervals are also reported in the horizon X section.

The lithology of these sediments is significantly changing. Some fields of the central Absheron (Garachukhur, Surakhani, and Buzovna-Mashtaga) display a moderate proportion of the coarse-grained clastic material within the horizon X section. Here, reservoir rocks are dominated by coarse material (Fig. 7.44). The share of the sand fraction ranges from 32.1 to 69.6%. The highest percentage of coarse sediments was reported in the Garachukhur area, the lowest—in the Balakhani-Sabunchi-Ramana field. A fining southwest is observed.

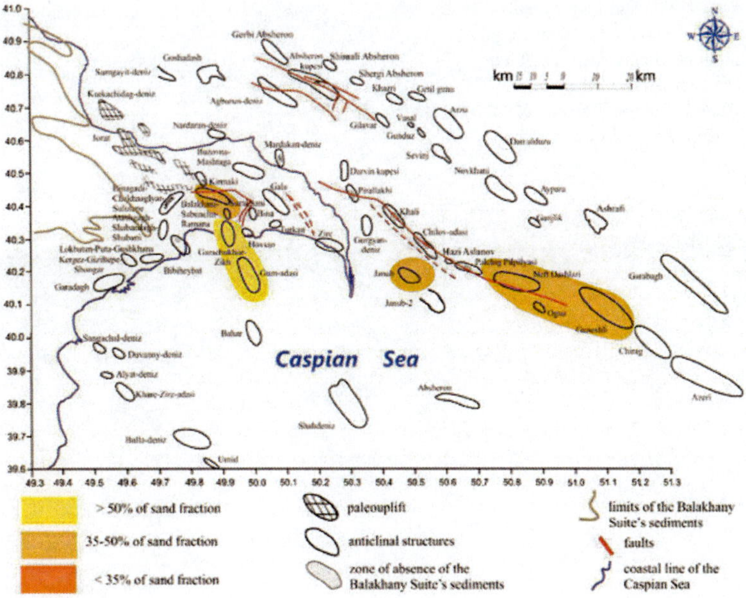

Fig. 7.44 The distribution of the average amount of sand fraction in sediments (%) of the Balakhani suite horizon X within the Absheron OGR

Quartz is recognized as a dominant mineral of the light fraction (up to 80%). Feldspar and rock fragments are of subordinate importance. At the same time, quartz content appears to be decreasing on the background of southward increasing feldspar amount.

Heavy fraction is composed of allogeneic and authigenic minerals. The common minerals are magnetite, ilmenite, disthen, staurolite mica, epidote, zoisite, hornblende, and pyroxene. Authigenic minerals are represented by pyrite, limonite, and dolomite. The share of clay fraction in the Balakhani X sediments does not change significantly, displaying a monotonous distribution over the Absheron OGR (Fig. 7.45).

The most porous horizon X sediments are reported in the Surakhani and Garachukhur fields. The average porosity here reaches 26.9% and 26.5%, respectively (Fig. 7.46). South, southeast, and northwards, this parameter is decreasing to 17.8% and 17.6% on average in the Gum Adasi and Bakhar fields; 20.7%, 23%, and 25.5% in the Janub, Neft Dashlari, and Guneshli fields respectively; 20.7% in the Balakhani–Sabunchu–Ramana field. Overall, the porosity widely ranges from 15 to 30% and gets an average value equaling 24% over the horizon X of the Balakhani suite. The effective porosity of reservoir rocks has been estimated in the Chakhnaglar area and Yasamal Valley, where it is 11.7% and 10.1%, respectively.

The permeability variations over the studied area demonstrate similar patterns. It gets the maximal values of 519×10^{-15} m^2 in the Garachukhur field; 316×10^{-15} m^2 in the Balakhani-Mashtaga field; 339×10^{-15} m^2 in the Balakhani-Sabunchi-Ramana field; 337×10^{-15} m^2 in the Puta field; 330×10^{-15} m^2 in the Surakhani fields; 399×10^{-15} m^2 in the Gum adasi and 436×10^{-15} m^2 in the Janub fields (Fig. 7.47). Thus, in these areas, the rocks of the Balakhani horizon X display good and very good fluid filtration properties southwards. The permeability decreases and reaches values not exceeding 100×10^{-15} m^2 in the Bakhar field.

Such porosity and permeability lateral variation trend is explained by the behavior of other reservoir characteristics affecting reservoir properties. The limited grain size data demonstrates no apparent lateral changes in the sand fraction percentage over the Absheron onshore and offshore areas. Moreover, the share of sand particles is slightly higher in the offshore fields, exhibiting a little southward-directed trend of sand fraction

Fig. 7.45 The map reflects the distribution of the average amount of clay fraction (%) in the sediments of the Balakhani suite horizon X within the Absheron OGR

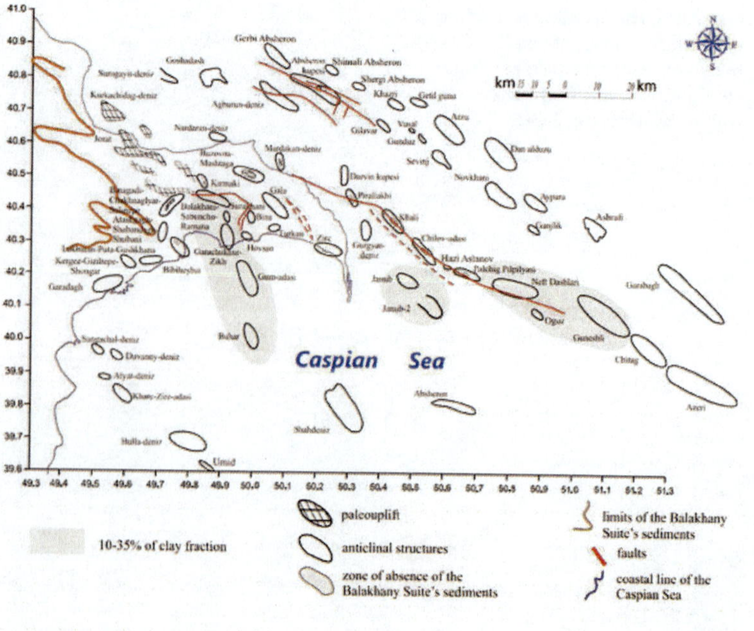

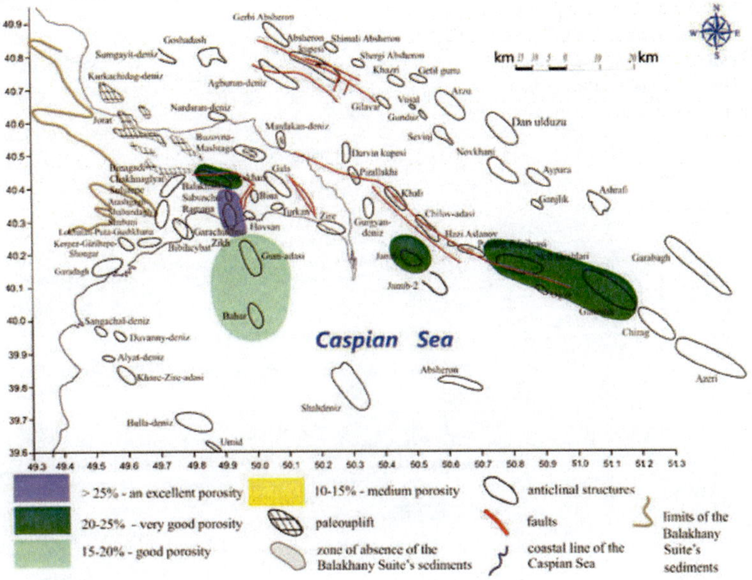

Fig. 7.46 The porosity variations (%) in the reservoir rocks of the horizon X of the Balakhani suite within the Absheron OGR

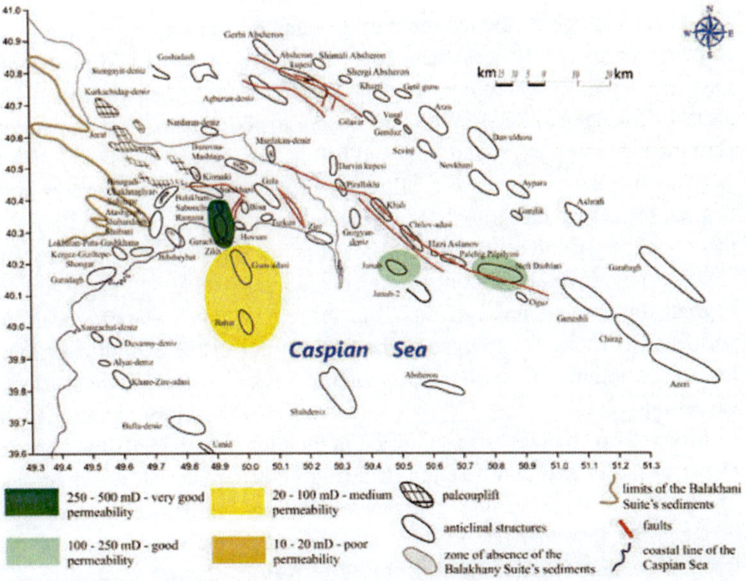

Fig. 7.47 The permeability variations (10^{-15} m^2) in the reservoir rocks of horizon X of the Balakhani suite within the Absheron OGR

amount increase (Fig. 7.44). It is natural to assume that the main effect on the reservoir properties over the studied area is exerted by the degree of cementation of the described rocks.

The thin-section studies show that cement content in the sandy-silty reservoir rocks varies from 3 to 38%. Cement is silica or carbonate. Our analytical data demonstrate a low amount of carbonate material, at most 15%, which testifies to the minor role of carbonate cementation in the Balakhani horizon X rocks (Fig. 7.48).

However, we could record some regularity. In the plot reflecting the link between permeability and carbonate material amount in the rocks of the

Fig. 7.48 Variations of the average amount of carbonate material (%) in the rocks of horizon X of the Balakhani suite within the Absheron OGR

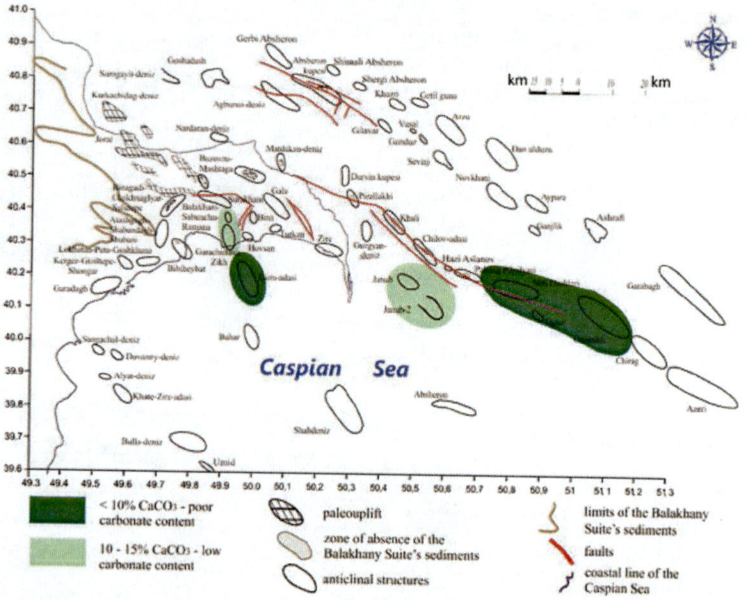

Balakhani horizon X, the less permeable samples showing permeability less than 20×10^{-15} m^2 have the highest carbonate content, reaching almost 15% (Fig. 7.49a). With the decrease of carbonate quantity to around 12% and below, the samples display an increase of filtration properties confirmed by the growth of permeability of the rocks from 50×10 to 100×10^{-15} m^2. Except for several discrete dots positioned beyond the area of most of the data, the well-permeability rocks are grouped in the field of the lowest amount of carbonate material not exceeding 10%.

More sharp changes are manifested in the plot of permeability and clay fraction data (Fig. 7.49 b). In the low permeable samples exhibiting permeability less than 20×10^{-15} m^2, the clay fraction content changes from 10 to 37%. Nevertheless, in the moderate and well-permeability rocks (permeability exceeds 20×10^{-15} m^2, but below 250×10^{-15} m^2), the share of fine material decreases below 20% (except in a single sample). Again, excluding two separately standing samples, we can conclude that in the very well-permeable rocks displaying permeability of more than 250×10^{-15} m^2 the percentage of clay fraction falls below 12%. Such permeability changes are due to variations in the muddy cement and matrix proportion.

The weakly pronounced effect of carbonate material on the reservoir quality of the studied rocks is observed in the plot, reflecting the link between porosity and carbonate content variations (Fig. 7.50a). We have recorded a wide range of porosity changes under the samples' relatively high and low numbers of carbonates. Only samples with carbonate amounts less than 12% exhibit excellent porosity exceeding 25%.

Such a weak effect of carbonate cement on the reservoir quality of the described rocks can be explained by the low amount of carbonate material in the studied samples, whose negative influence is enormously multiplying when the carbonate content exceeds 20%, as was shown above.

The porosity values are even less dependent on variations in the amount of fine-grained material (Fig. 7.50b). The onshore samples display a better porosity value compared with the offshore samples. They also demonstrate a scattered distribution in the plot in contrast to offshore data showing a cluster alignment excepting separate samples. Most offshore samples have porosity changing from 15 to 22% under clay

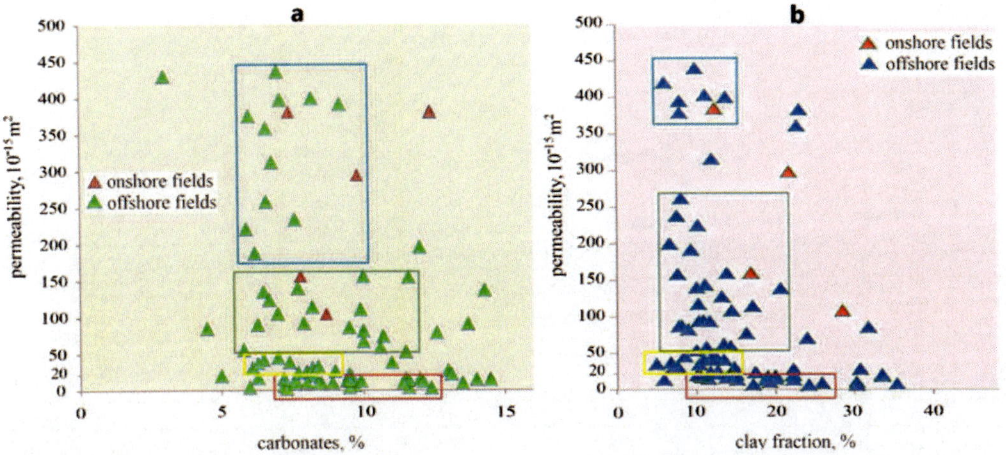

Fig. 7.49 The permeability values of the carbonates (a) and clay fraction (b) of horizon X of the Balakhani suite within the Absheron OGR

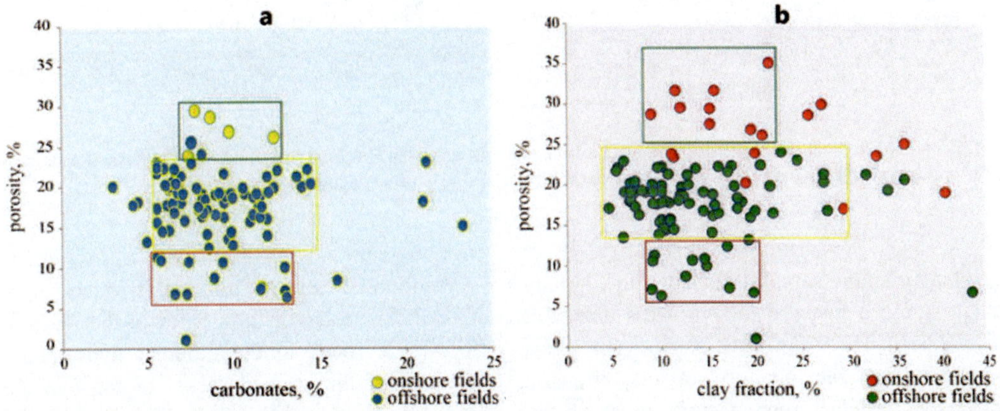

Fig. 7.50 The arrangement of the porosity values of the carbonates (a) and clay fraction (b) of horizon X of the Balakhani suite within the Absheron OGR

fraction amounts varying from 5 to 15%. However, some offshore and onshore samples of the same porosity have a higher proportion of clay-size particles reaching 40%. Moreover, the excellent—porosity rocks display a wide range variation of the clay fraction amount from 10 to 35%.

The statistical analysis also indicates no correlation between porosity and clay fraction amount and a weak negative relationship between porosity and carbonate content in the Balakhani horizon X rocks (the correlation coefficient is − 0.33). Because of these results, the question arises: What principal factors affect these rocks' reservoir quality? The reservoir parameters' values versus depth are summarized in Fig. 7.51.

We did not record any vertical variations in the sand fraction amount in the Balakhani horizon X rocks (Fig. 7.51a). We can note that offshore samples demonstrate more variations of the sand fraction content and contain more coarse clastic material, confirmed by a high proportion of sand fraction therein, up to 80%.

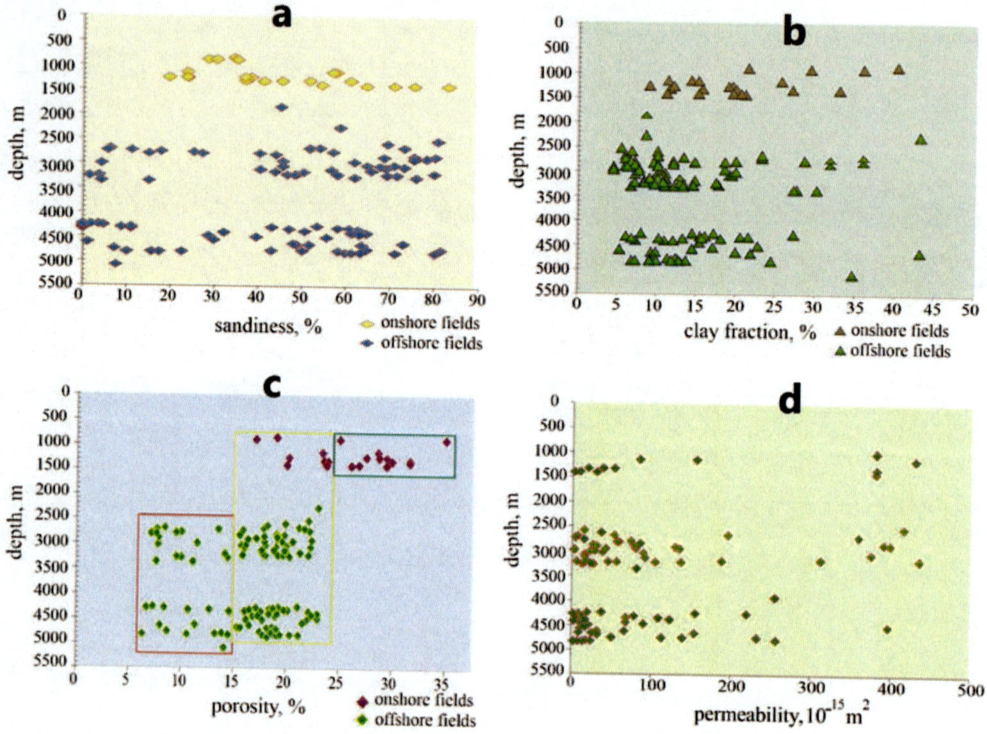

Fig. 7.51 The vertical distribution of the reservoir characteristics of the Balakhani suite horizon X rocks: **a** sandiness (%), **b** clay fraction (%), **c** porosity (%), **d** permeability (10^{-15} m^2) versus depth (m)

Many offshore samples display a low amount of fine-grained material therein—less than 20% (Fig. 7.51b). The rocks of the onshore fields have a widely variable quantity of clay fraction. However, despite a more significant proportion of coarser sediments and a lower share of fine fraction, the offshore samples show less porosity (Fig. 7.51c). The porosity plot demonstrates that samples are well clustered based on porosity data: the lowest porosity, not exceeding 15%, is specific for the offshore rocks (outlined by the red square in Fig. 7.51c); the good and excellent porosity is demonstrated by both offshore and onshore samples (yellow square in Fig. 7.51c); the excellent porosity is characteristic of the onshore samples except one single offshore data (green square in Fig. 7.51c). It is difficult to conclude the permeability difference between offshore and onshore rocks as we have only a few data on onshore fields (Fig. 7.51d).

We found the explanation of this phenomenon of the better reservoir properties of the onshore Horizon X rocks in the different hypsometric positions of the studied rocks. In the onshore fields, the horizon X of the Balakhani suite occurs above 1500 m; many offshore samples lie below 2500 m, which testifies to well-pronounced depth control and associated compaction effect on the reservoir properties of the studied rocks. The correlation coefficient between the rocks' occurrence depth and porosity equaling − 0.588 is further evidence.

The Horizon IX

Mudstones dominate this horizon. However, in some fields in the Absheron Peninsula and adjacent southern offshore, the fine-grained sandstones and siltstones provide suitable hydrocarbon reservoirs of commercial importance. There is not much data on the reservoir properties of the Horizon IX sediments. However, the existing few analytical results allow us to draw some conclusions.

The highest percentage of sand fraction of the horizon IX sediments has been recorded in the Gum Adasi field—up to 74%. In the central Absheron Peninsula in the Surakhani and Zikh fields, this parameter is as high as 64%. Southeastwards from the Gum adasi field, the share of sand fraction gradually decreased to 24%.

The highest clay fraction proportion, estimated to be more than 46%, has been reported in the Gum adasi field. This proportion is slightly falling to 43% in the Garachukhur and Zikh fields.

The maximum carbonate content in the Balakhani horizon IX sediments is reported in the Gum adasi field (up to 26%). However, it should be mentioned that it is a single case. This parameter exceeds 10% in most samples but is less than 15%. North and north-westwards, the carbonate material content decreases to 4.2% minimally, averaging 7% on average. Excellent-porosity rocks (more than 30%) have been found in the Surakhani and Zikh fields. South and southeastwards, this parameter decreases to 18% on average.

The best-permeable rocks have been reported in the fields of Garachukhur (around 600 × 10^{-15} m^2) and Gum adasi (around 1050 × 10^{-15} m^2). However, this value was registered in one sample. On most of the Absheron OGR, the permeability is estimated as less than 150 × 10^{-15} m^2. It has been recorded that the filtration properties of the rocks are worsening southwards. Plots illustrating reservoir properties of IX horizon reservoir rocks as a function of different factors are given below (Fig. 7.52).

Like Horizon X, the porosity of the Horizon IX sediments demonstrates higher values in the upper interval lying above 2000 m (Figs. 7.52a). Below this level, the porosity decreases and does not display any vertical variations. However, it still has high values exceeding 20% in most offshore and almost all onshore samples. The average value exceeding 10% indicates a good void space in the most analyzed samples.

In contrast to porosity, the sandiness of the reservoir rocks does not differ between the upper and lower intervals of the Horizon IX section (Figs. 7.52b). It changes in a wide range from a deficient percentage of less than 1 to almost 80%. The permeability data testify to dominating moderate—well-permeable rocks in the offshore fields and the very well and excellent permeability sediments in the onshore samples set (Figs. 7.52c).

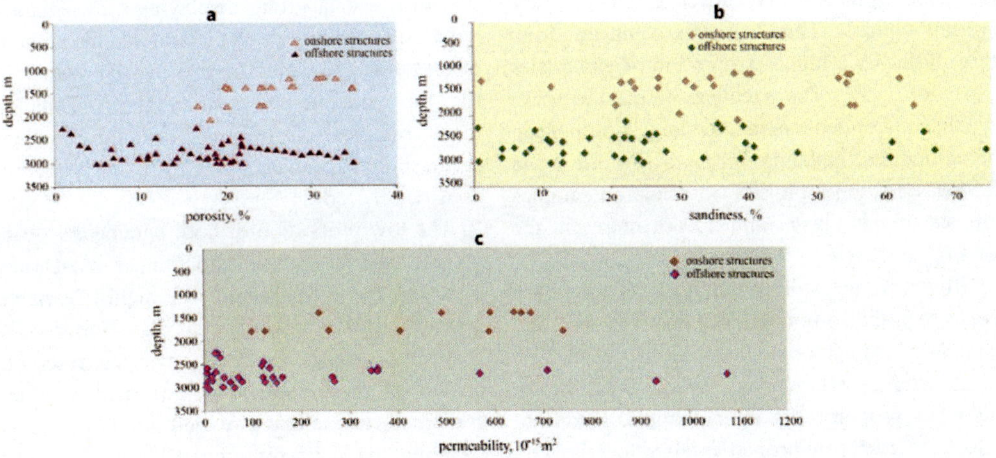

Fig. 7.52 The reservoir characteristics (**a** porosity, **b** sandiness, and **c** permeability, versus depth) of the Balakhani suite horizon IX

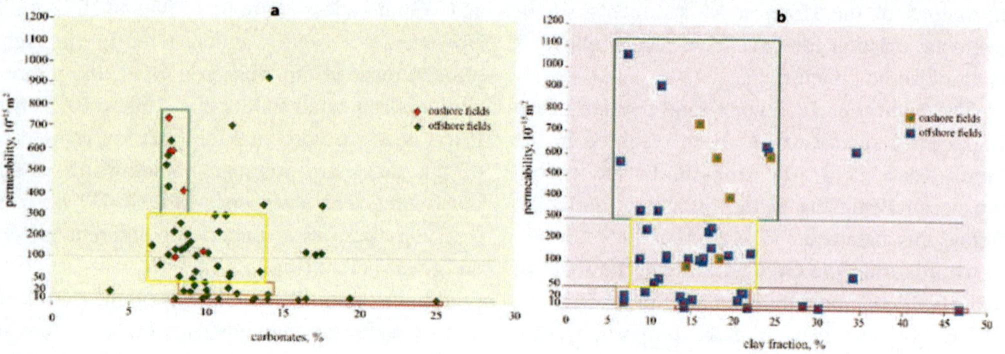

Fig. 7.53 The arrangement of the carbonates (**a**) and clay fraction (**b**) versus the permeability values for the horizon IX of the Balakhani suite within the Absheron OGR

The values of reservoir properties affecting parameters are presented in Fig. 7.53. As it follows from the constructed plot, most samples demonstrate the low-medium proportion of carbonates getting more than 7% but less than 15% (Fig. 7.53a). It can be mentioned that the poor-permeability rocks (less than 10×10^{-15} m^2) stand out by the highest percentage of carbonate material reaching 25% (red square). In the rocks of permeability ranging from 20×10 to 50×10^{-15} m^2 the carbonate content varies in the 8–14.5% limits (orange square).

The rocks of moderate, good, and very good permeability varying from 50 to 300×10^{-15} m^2 are clustered in a group distinguished by the fall of the lower limit of carbonates amount to 6%; the upper limit in this cluster is around 14% (yellow square). Three samples stand out from this cluster by a higher proportion of carbonates —16 to 17.5%. The excellent-permeable rocks (green square) demonstrate the lowest proportion of carbonates, equaling 7% as a low limit and 8.5% as an upper limit. Two scattered samples are out of this cluster and do not represent the general tendency.

However, we still cannot discuss the sharp contrast in the carbonate amounts. The recorded content of the carbonate material is at most 17% in all samples except single cases. Such a low-moderate proportion of the carbonates could not affect the reservoir properties significantly, i.e., this parameter cannot be considered a principal reservoir quality control.

The proportion of the clay fraction demonstrates the high changeability. Here, it is even more difficult to catch any tendency. The lower and high-permeable rocks show the same amount of the fine-grained fraction (Fig. 7.53b). In the poor-permeable rocks displaying permeability less than 10×10^{-15} m^2, the percentage of fine-grained particles varies in a wide range from 14 to 46% (Fig. 7.53b). There is not a big difference in the clay fraction variations between the two groups of samples showing permeability values of 10–50×10^{-15} m^2 and 50–100×10^{-15} m^2. In these samples, the share of clay fraction changes from around 7 to 22%.

The high-permeable rocks are scattered and do not form a data group displaying close characteristics. However, we can record that the amount of clay fraction varies from 6 to 24%, which is like the moderate-permeability group. Thus, in most samples manifesting medium-excellent permeability, the share of clay fraction therein does not exceed 24%.

The low proportion of both parameters—carbonate material and clay fraction, in most samples on the background of highly variable porosity and permeability of the Horizon IX sediments, allows assuming that the reservoir quality of these rocks was governed by their combined impact that provided a multiple effect reducing the reservoir properties. At the same time, we observe strong depth control on the reservoir quality.

The Horizon VIII

The median grain size of the Horizon VIII sediments varies between 0.078 and 0.153 mm. The sandy fraction content is also widely changed from 31.9 to 44.5%. The maximum proportion of sand fraction is recorded in several fields in the Absheron Peninsula and Absheron Archipelago (Fig. 7.54). In the fields of the Baku Archipelago, the number of sand-size grains is less than 35%.

The proportion of clay fraction varies insignificantly in the limits of 10–35% (Fig. 7.55). The highest amount of pelitic sediments is recorded in the Surakhani field. The carbonate content in the Balakhani suite horizon VIII sediments in the Absheron Peninsula displays minor variations between 4.3 and 10.0% that testify to low carbonate cementation of these rocks (Fig. 7.56) To the south and southeast, the amount of carbonate material in the sediments of Absheron—Pribalkhan Sill and Baku Archipelago increases to 15% that is an indication of reservoir quality south, so south-eastwards creasing trends.

Reservoir rocks of this horizon are generally characterized by high porosity estimated to be 20.8–29.8%. The best capacity properties are typical of the rocks from the Absheron Peninsula, where these values are estimated to have been about 30% (Fig. 7.57). To the south, southwest, southeast, and north of the Absheron Peninsula within the Absheron OGR, the porosity decreases significantly to about 20%. In sediments of the Baku Archipelago, the values of this parameter vary in the range of 10–15%.

In contrast to their porosity, the filtration properties of these rocks display significantly higher variations. They can take values within one field that are characteristics of very poor permeability to very good permeability rocks.

Overall, based on the average for each field value, the horizon VIII sediments' permeability may be considered satisfactory—very good. The highest values are recorded in the Absheron Peninsula (Fig. 7.58). Here, the permeability of the discussed sediments may vary from 10×10^{-15} m^2 to more than 800×10^{-15} m^2. To the south, southeast, and northeast of this area, the permeability decreases to values typical for medium permeability rocks.

The proportion of sand fraction does not display any vertical changes and has a scattered distribution (Fig. 7.59a). We also do not see any geographical variations in this parameter,

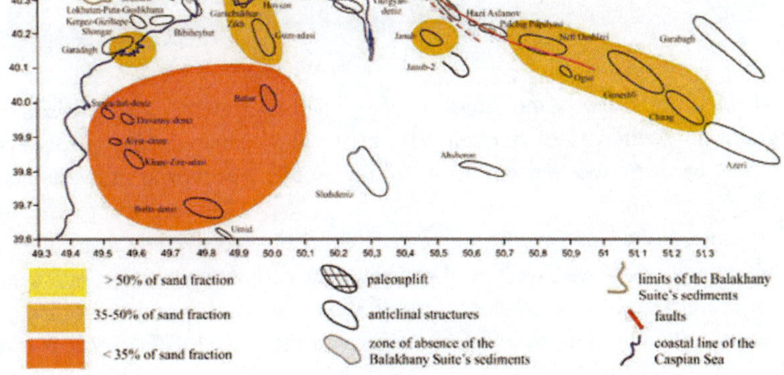

Fig. 7.54 The distribution of the average amount of the sand fraction in the sediments (%) of the Balakhani suite horizon VIII within the western flank of the South Caspian basin

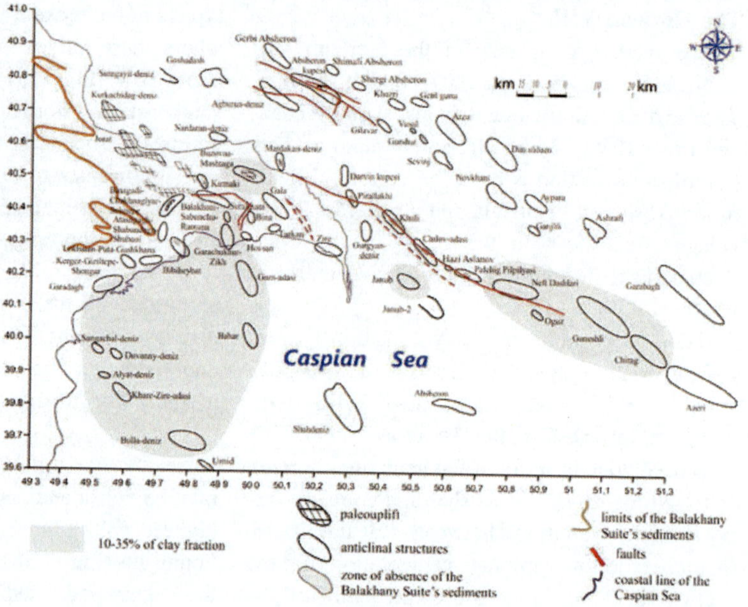

Fig. 7.55 The distribution of the clay fraction content (%) in the reservoir rocks of the Balakhani suite horizon VIII within the western flank of the South Caspian basin

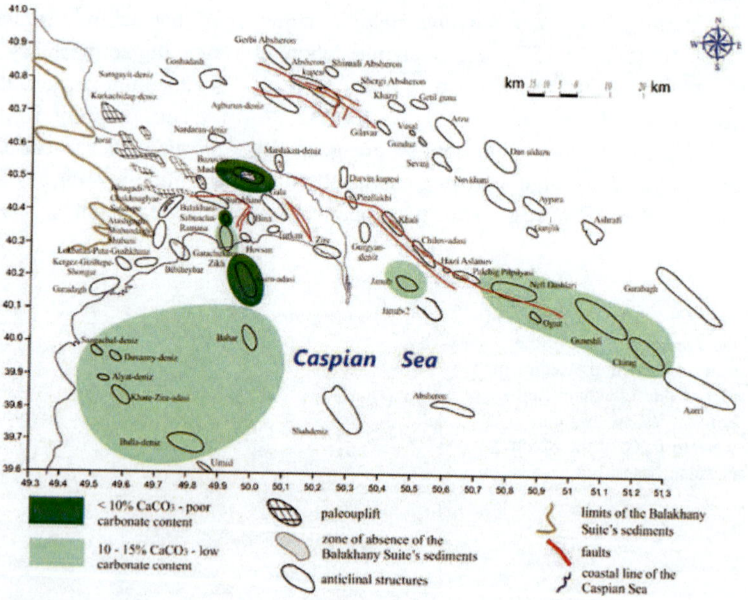

Fig. 7.56 The variations of the average amount of the carbonate material in rocks (%) of the horizon VIII of the Balakhani suite within the western flank of the South Caspian basin

which means the same number of sand-size grains in sediments of horizon VIII in the different parts of the western flank of the South Caspian Basin.

As described above, the downward trend of the porosity decrease is a typical characteristic of the Productive Series sediments (Fig. 7.59b). We record a gradual decrease of pore space with the depth. As a result, the highest porosity is encountered in rocks of the Absheron Peninsula, occurring in the depth interval of 1000–2000 m. The sediments of the Absheron Archipelago recovered from the depth interval of 2000–4500 m have intermediate porosity. The lowest porosity is recorded in sediments of the Baku Archipelago that is a result, as we assume, of at

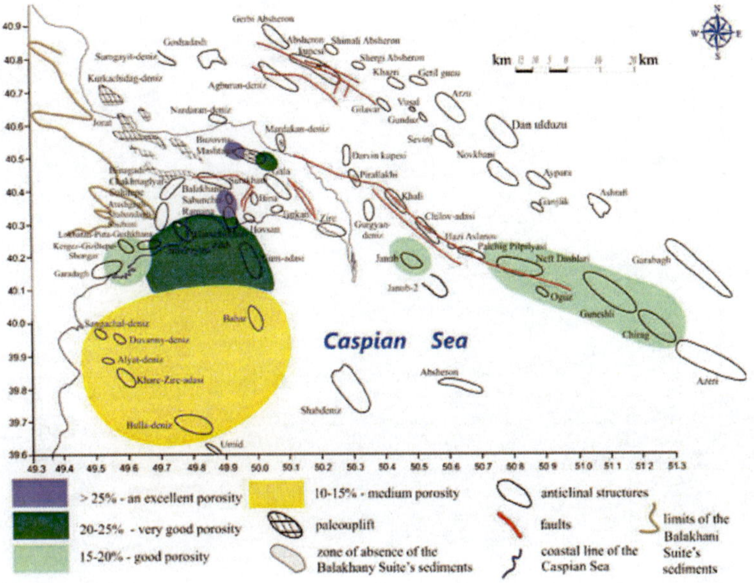

Fig. 7.57 The porosity map of the Balakhani suite horizon VIII reservoir rocks (%) within the western flank of the South Caspian basin

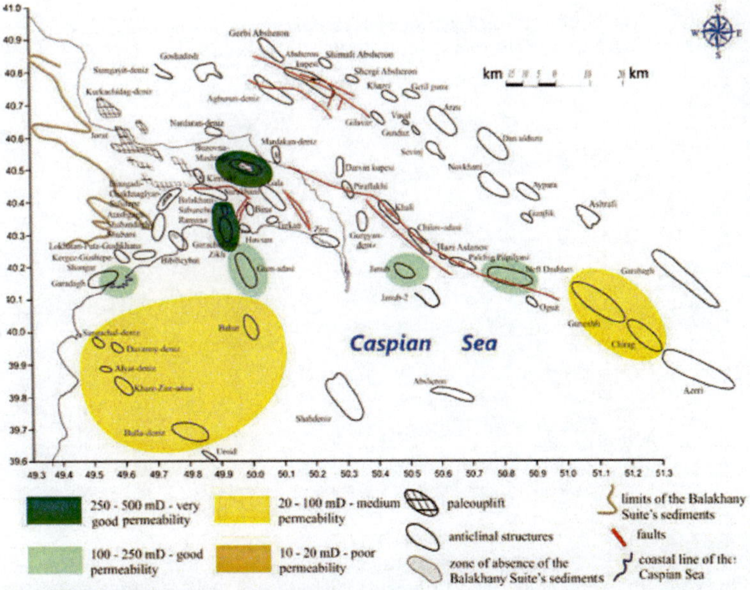

Fig. 7.58 The permeability map (in %) of the Balakhani suite horizon VIII reservoir rocks within the western flank of the South Caspian basin

least two affecting factors—the significant depth of these sediments occurrence and consequent reduction of the pore space and increasing carbonate cementation (Fig. 7.59d). In the same manner, and even more sharply, the permeability changes versus depth (Fig. 7.59c).

Analogously to sediments of the older stratigraphic units of the Productive Series, the permeability of the horizon VIII sediments demonstrates similar changes in response to variations of carbonate material and clay fraction amounts (Fig. 7.60).

However, the low-permeably rocks, in contrast to previously described sediments, display considerable variations in the amount of clay fraction and carbonate material, i.e.,

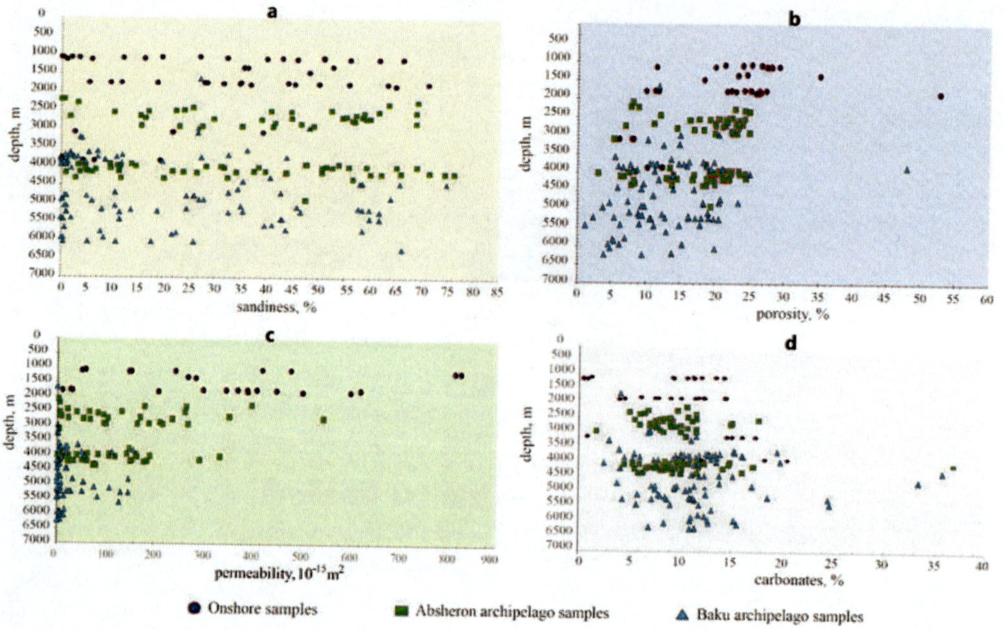

Fig. 7.59 Values of reservoir characteristics (**a**—sandiness, **b**—porosity, **c**—permeability, and **d**—carbonates) of the Balakhani suite horizon VIII rocks versus depth of their occurrence

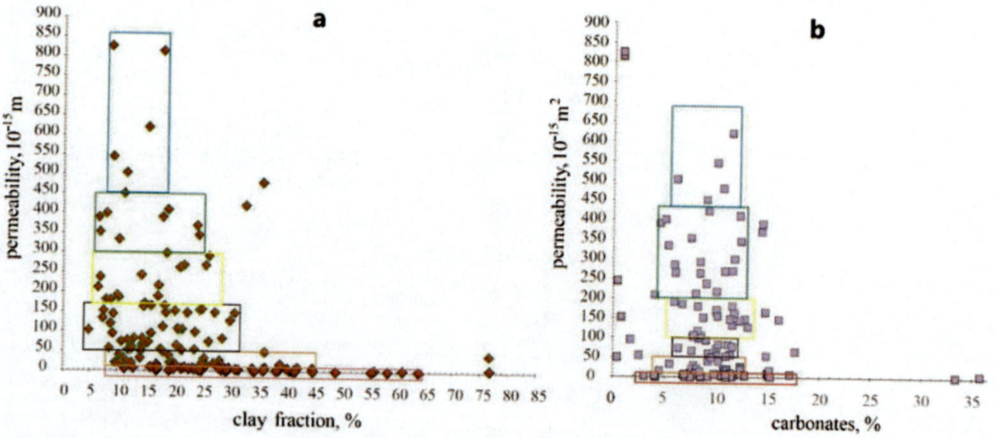

Fig. 7.60 The permeability of VIII horizon deposits as a function of the carbonate material (**a**) and clay fraction (**b**) content

demonstrating both high and low values of these reservoir characteristics. We assume that is evidence of a cumulative effect of many factors affecting reservoir properties of the horizon VIII rocks.

The high capacitive properties of the Balakhani suite horizon VIII rocks are confirmed by SEM analysis (Fig. 7.61). Below, we submit an SEM image of the core sample recovered from the Bakhar field.

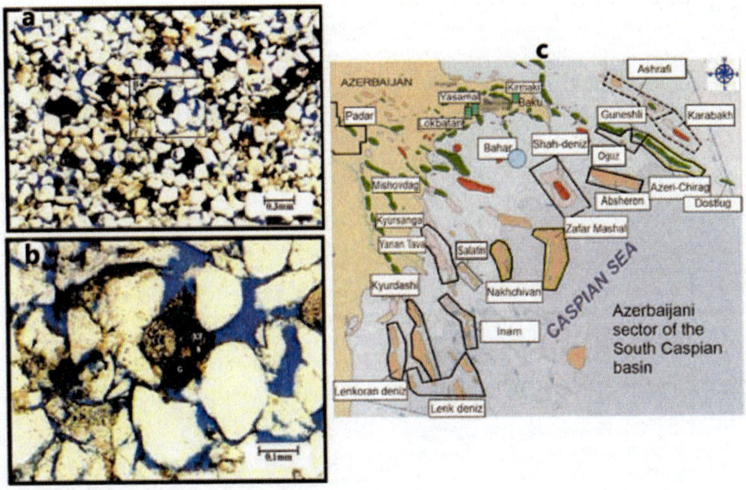

Fig. 7.61 SEM photographs (**a** and **b**) of the sandstone from the Balakhani suite horizon VIII from the Bakhar field, depth interval of 3996–3999 m; **c** areal map. The reservoir properties: porosity—20.1%, permeability—21.9 × 10^{-15} m^2, grain size—0.224 mm, well sorted

7.2.9 The Surakhani Suite

The Surakhani suite is the uppermost and most mud-prone unit in the Productive Series section. However, it contains several sand-silt horizons, which display moderate-excellent reservoir properties (Fig. 7.62).

As it is clear from the plot demonstrated in Fig. 7.62, the best reservoir characteristics are recorded in the Surakhani suite sediments recovered from the onshore wells in the Absheron oil–gas region. Here, most samples have permeability exceeding 50 × 10^{-15} m^2, and porosity of more than 12%. In the Absheron Peninsula, the Surakhani suite contains several production horizons in many fields.

In contrast to onshore samples, the rocks from the offshore wells of the Absheron Archipelago demonstrate low reservoir properties. The bulk of these sediments have negative permeability despite good porosity. We assume that is a result of the increased proportion of muddy sediments towards the Caspian Sea due to the change to more distal facies of the paleo-Volga prodeltaic portion in the Absheron Archipelago.

More promising look data from the Baku Archipelago. Most sediments display very low permeability, but many samples have moderate to excellent reservoir properties.

The environmental investigations show that the western part of the Baku Archipelago in Surakhani time turned into a depocenter of terrigenous material supplied from the Greater Caucasus by several rivers such as paleo-Pirsagat, paleo-Djeirankechmez that resulted in deltaic conditions and deposition some amount of coarse material here in contrast to paleo-Volga affected part of the Baku Archipelago. Such environmental heterogeneity and lithofacies variability brought very changeable reservoir properties of the Surakhani suite sediments.

Our data set also testifies to reservoir properties preservation of the Productive Series principal reservoir rocks at large depths. Below, we present the basin modeling results and the SEM analysis data proving good reservoir quality of the Fasila suite and Balakhani suite horizon VIII sediments occurring below 5000 m (Figs. 7.63, 7.64 and 7.65).

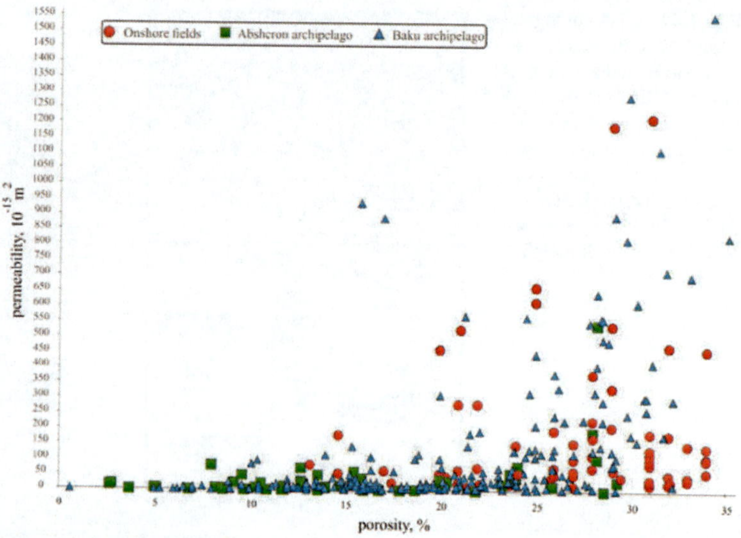

Fig. 7.62 The reservoir properties of the Surakhani suite rocks in the different parts of the South Caspian basin's western flank

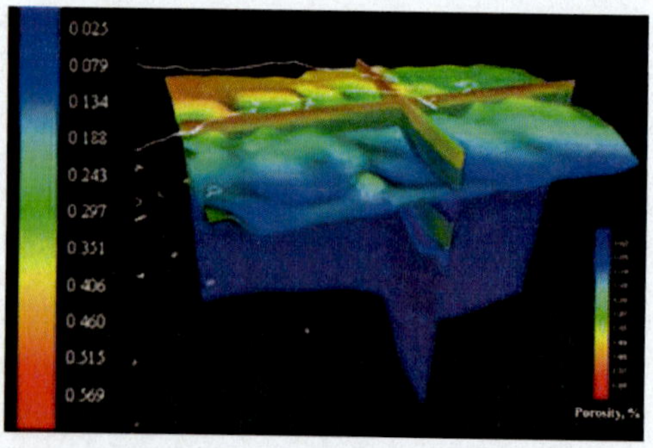

Fig. 7.63 A porosity model of the South Caspian Fasila suite rocks. Note the good porosity of these sediments in the lower interval of the suite

7.3 Reservoir Properties of the PS Deposits Within the Lower Kur Depression and Southeastern Gobustan (Onshore the South Caspian Basin Western Flank)

Within these regions, the reservoir properties of the Productive Series sediments occurring in the upper portion of the Series have been extensively studied. Our routine analysis of the porosity and permeability of the core samples from several Lower Kur depression fields revealed six clusters (rock groups) distinguished by different reservoir characteristics. The data are summarized in Fig. 7.66, testifying to the wide variations of the porosity and permeability of the Productive Series sediments. Cluster 1 unites the rocks of negligible permeability. Such rocks can be considered non-reservoir rocks. This cluster includes several fields, such as Janubi Kyursanga, Neftchala, Kurovdag, Kyursanga, Kalamadin, and Galmas.

Poor reservoir properties distinguish cluster 2: porosity—5 to 10%; permeability—from 1 to 10×10^{-15} m^2. Such rocks have been recorded in the following fields: Kyursanga, Neftchala, Janubi Kyursanga, Kurovdag, Kalamadin, Mishovdag, Kichik Harami, Galmas, and Garabagli. Cluster 3 contains rocks with moderate to poor reservoir properties. These rocks are characterized by from good to excellent porosity—> 10% and weak permeability estimated to be from 1 to 10×10^{-15} m^2. It should be noted that this cluster contains the most studied samples. Cluster 4 represents the moderate-highly porous (the porosity > 12%) and moderate permeability (10–50 $\times 10^{-15}$ m^2) sediments. The sediments of cluster 5 are characterized by excellent porosity exceeding 20% and permeability variating within the interval from 50 to 250×10^{-15} m^2. Cluster 6 contains excellent reservoir rocks: porosity is > 20%, permeability—more than 250×10^{-15} m^2.

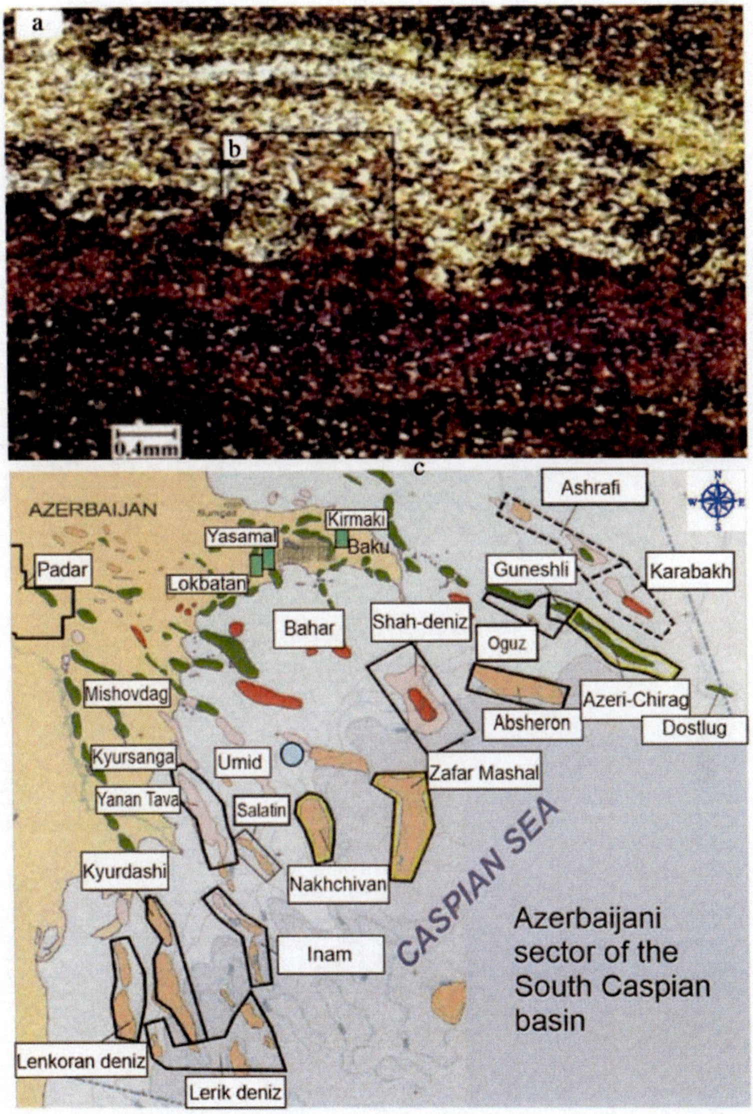

Fig. 7.64 SEM images (**a** and **b**) demonstrate the Fasila suite sandstones in the Umid field, depth interval is 5455–5460 m; **c** areal map. Note the retained reservoir properties at this depth: porosity—11.3%, permeability—71.4×10^{-15} m^2, grain size—0.032 mm, well sorted

The good and excellent reservoir parameters have been revealed in the PS rocks of Kurovdag, Kyursanga, Kalamadin, Mishovdag, Garabagli, and Neftchala fields. Figure 7.66 indicates that many samples have poor—moderate reservoir properties. Figure 7.67 displays variations in the amount of sand and clay-size grains. Most samples are concentrated in sediments, which do not demonstrate the prevalence of any fraction and have approximately equal proportions of sand, silt, and clay-sized particles.

An approximately equal number of samples have good and excellent reservoir properties. The same number of samples are distinguished by a high content of sand (red square) and silt (blue square) fractions that defined such good reservoir potential of these rocks. Consequently, the non-reservoir rocks have the highest percentage of clay fraction (black square in Fig. 7.67). The dominating montmorillonite in the cementing material also negatively affects reservoir properties.

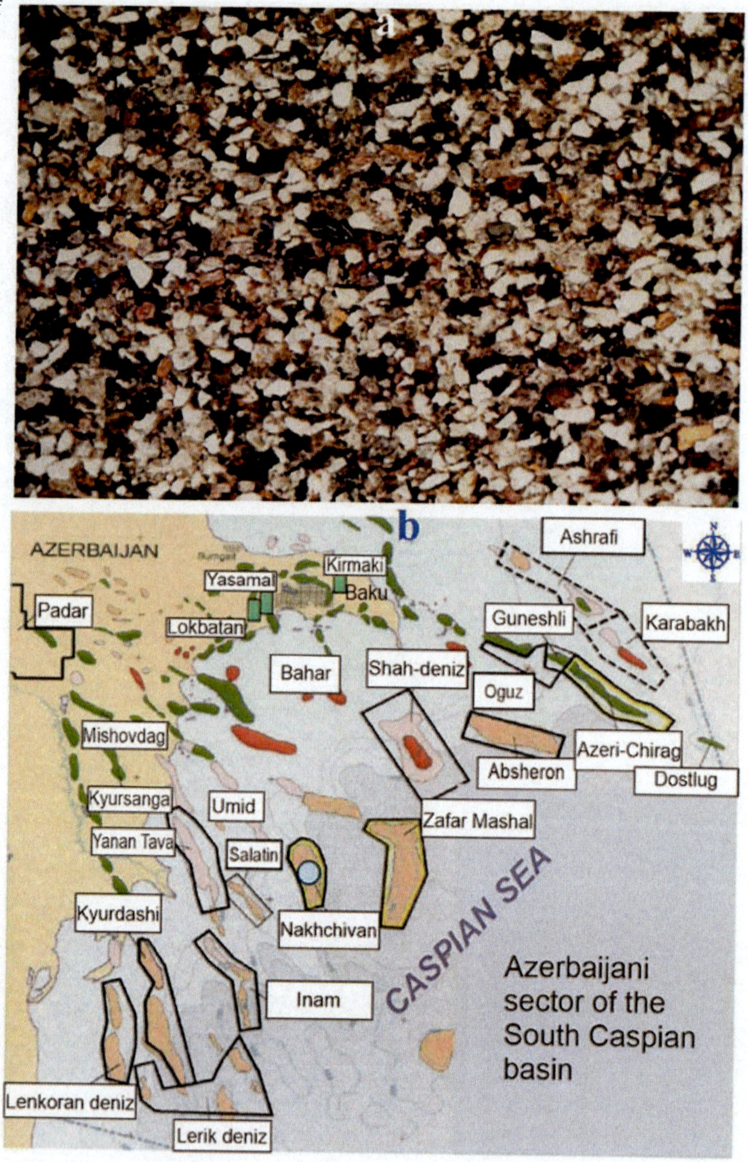

Fig. 7.65 An SEM image of sandstone: **a** from the Balakhani suite in the Nakhchivan field recovered from a depth of 6262.4 m, **b** areal map. The rocks display the moderate reservoir properties: porosity—12%, grain size—0.04 to 0.065 mm, sorting—from poor to moderate

The mapped reservoir properties within the Low Kur depression confirm their significant variability (Fig. 7.68). We can conclude that the best permeability—the Productive Series rocks display porosity within the narrow stripe along the Kurovdagh—Garabagli—Babazanan anticlinal line corresponding to the paleo-Kur river channel. The reservoir properties appear to be sharply worsening outside of this area, which, in our view, relates to the facies change from the significant river channel to the mud-rich flood plain.

We assume that well-pronounced variations of reservoir properties over the sections of some fields result from frequent avulsion of the meandering paleo-Kur River accompanied by rapid lithofacies changes in the Productive Series succession of the Lower Kur depression.

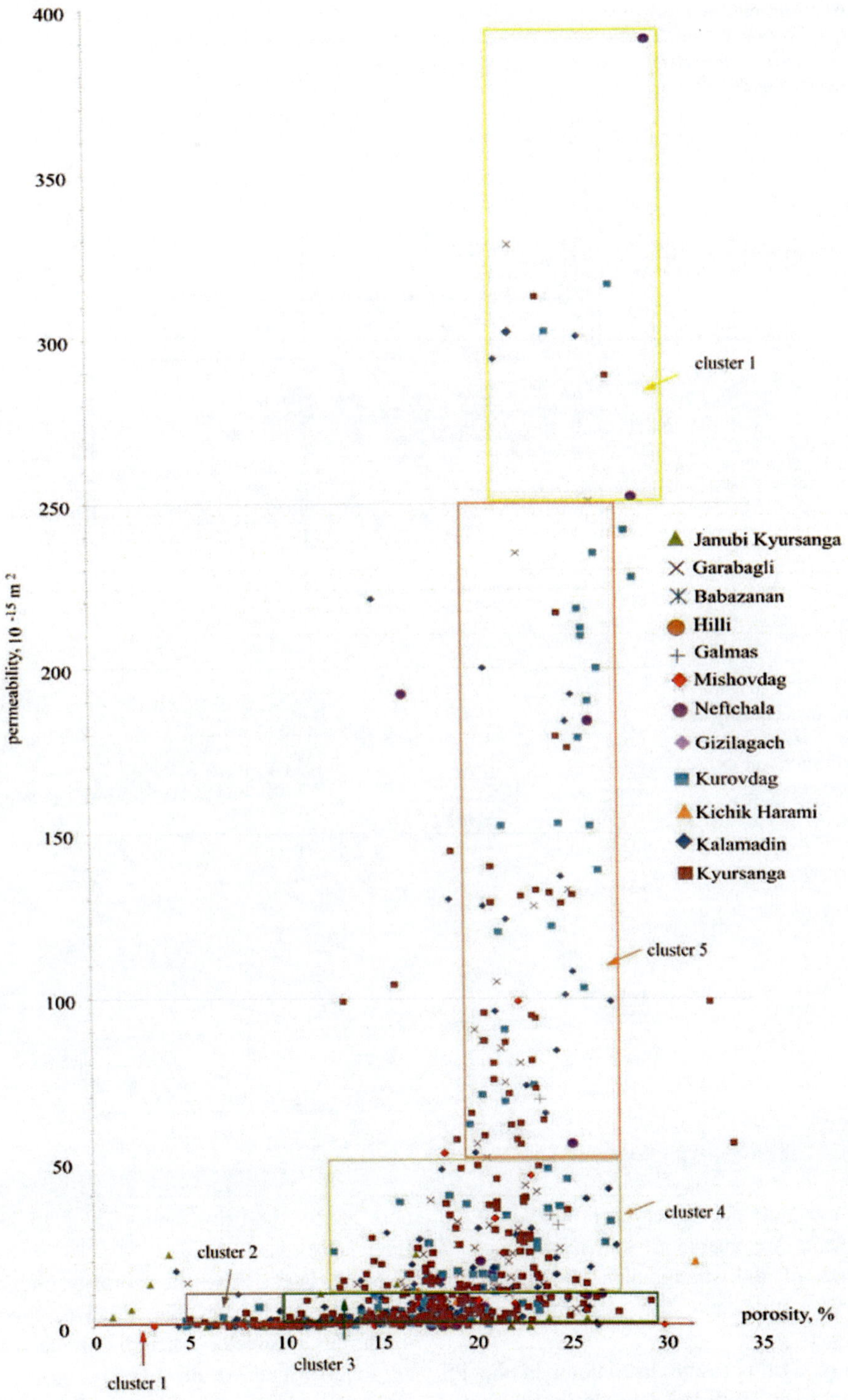

Fig. 7.66 The reservoir properties of the Productive Series rocks in the Low Kur depression

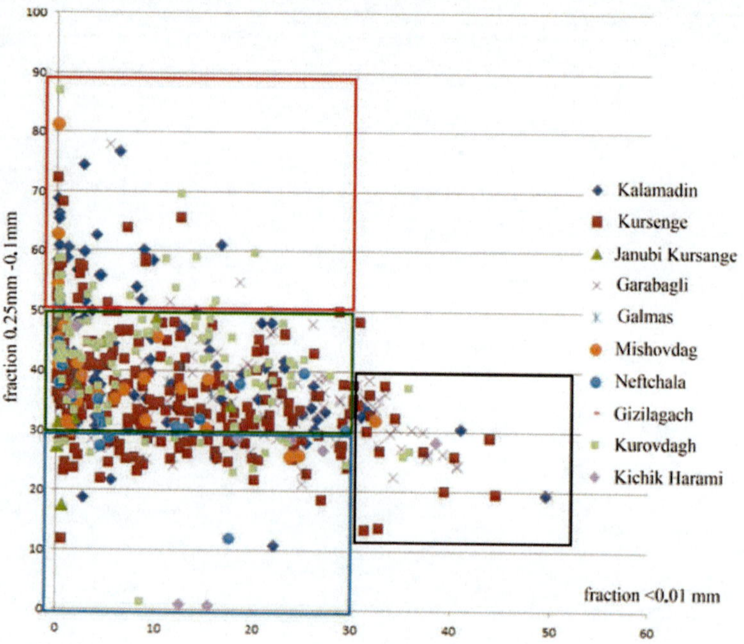

Fig. 7.67 Properties of sand and clay-size grains in the Productive Series sediments of the Lower Kur depression

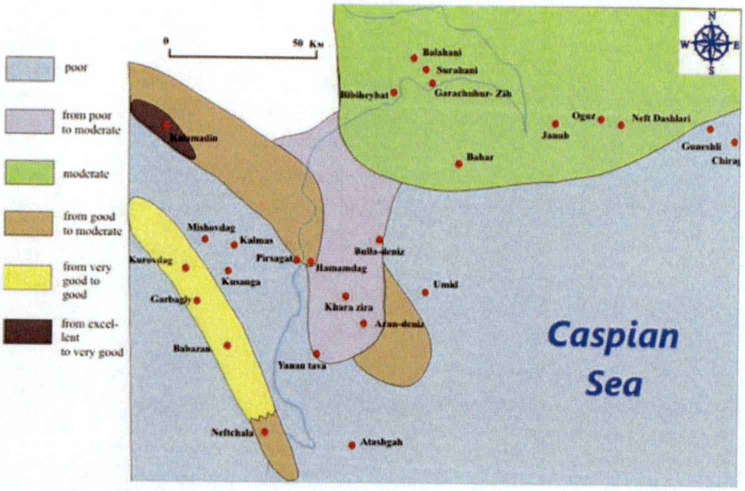

Fig. 7.68 SCB zonation according to the quality of the reservoir rocks of the Surakhani suite and its stratigraphic analogs

The moderate-poor reservoir properties of the PS deposits are visible in the SEM image of sediments of the stratigraphic analog of the Balakhani suite in the Kalamadin field (Fig. 7.69).

The plot of porosity distribution in the PS rocks versus the depths of their occurrence shows that most samples occur within the depth interval of 3000–4000 m and are characterized by good pore space—porosity ranges from 15 to 25% (Fig. 7.70).

Most intensively, the porosity reduces in intervals of 2.5–3 km (Fig. 7.71a–c). From 3 to 4 km, this parameter either slightly decreases or an inverse process of porosity growth takes place, as we assume, due to fracturing and decompaction of sediments as well as mineral dissolution.

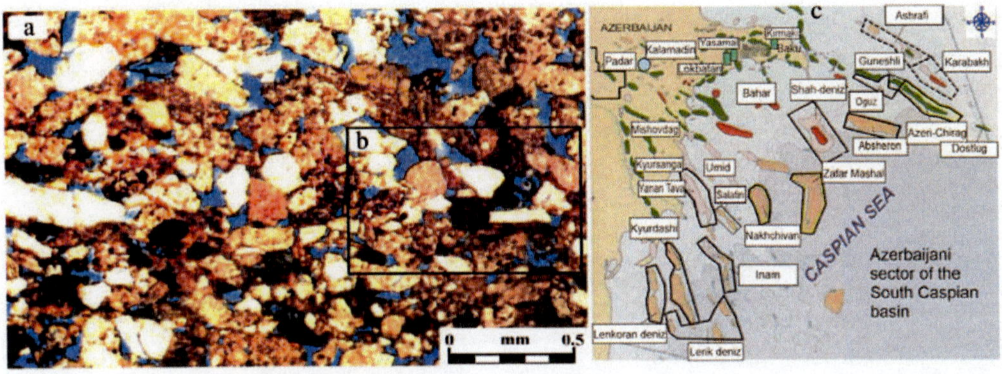

Fig. 7.69 SEM images of sandstones (**a** and **b**) from the stratigraphic analog of the Balakhani suite in the Kalamadin field; **c** areal map. The reservoir characteristics are as follows: porosity—7.0%, permeability—21 × 10^{-15} m^2, grain size—0.310 mm, moderate sorting

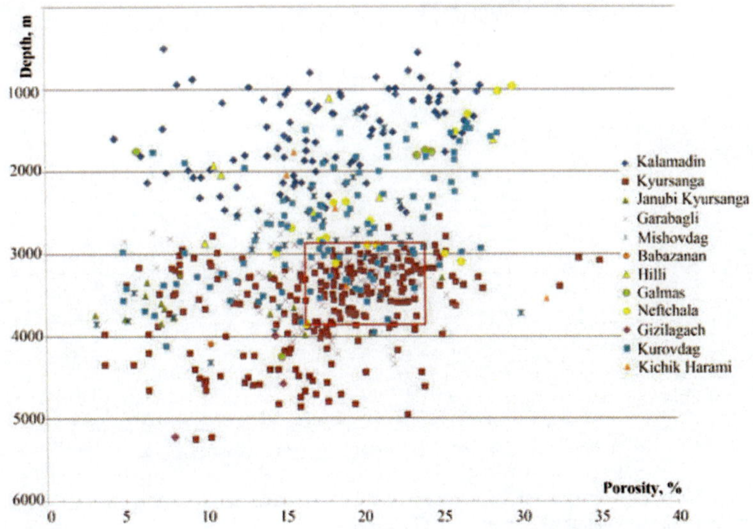

Fig. 7.70 The distribution of porosity of the Productive Series sediments in the Low Kur depression versus the depth

The fact that reservoir properties of sediments are improved at large depths because of post-sedimentary processes is confirmed by SEM images of rocks of the horizon XX (an analog of the Fasila suite) in the Neftchala field (Fig. 7.72).

The primary porosity—7% of these sediments, is partially preserved during compaction, and because of complete or partial grain dissolution, the initial porosity increased to 13.7%. A typical feature of the PS reservoir properties in southeastern Gobustan is the predominance of moderate-poor quality reservoir rocks (Fig. 7.73) like these sediments in the Low Kur depression.

7.3.1 Evaluation of Reservoir and Screening Properties of the South Caspian Basin Productive Series Rocks Recovered from the Large Depth

The evaluation of reservoir properties of PS rocks occurring lower than 0.5 km depth, a boundary below which core recovery is absent, is carried out based on an extrapolation of the core data or geophysical parameters of the section stripping part.

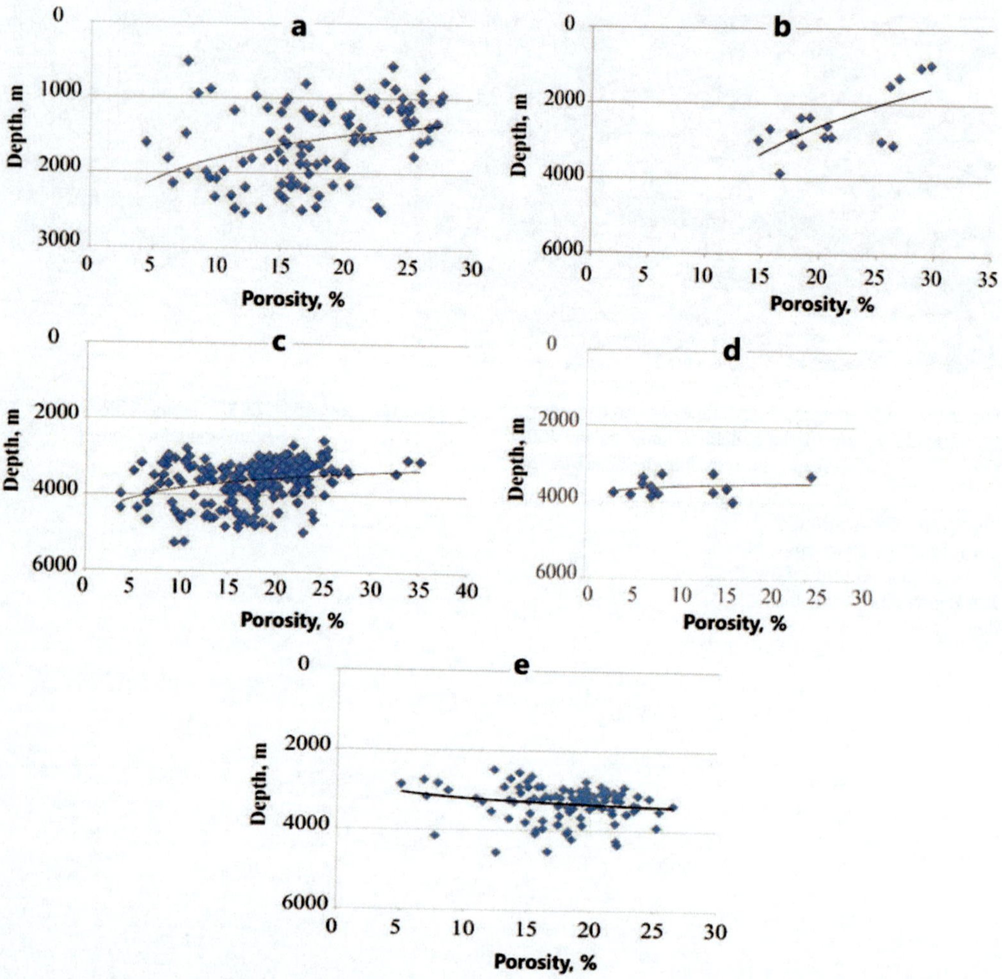

Fig. 7.71 The distribution of porosity of the Productive Series sediments versus depth: **a** Kalamadin, **b** Neftchala, **c** Kyursanga, **d** Janubi Kyursanga, **e** Garabagli

This section has partly discussed the problem of the rock's reservoir properties conservation or, in some cases, their improvement at a great depth. It will be described below in more detail and based on the data obtained from more than 100 developing and prospecting areas of the Absheron and Baku archipelagos and Lower Kur, Pre-Balkhan, Keimir-Chikishlyar regions (mainly to the depth of 4.5 km), Aliyev et al. (1973) presented porosity change curves which extrapolation based on the regression equation made it possible to value the rocks reservoir properties at 5–7 km depths. In such a way, according to the prognostic valuation within the SCB at a depth of 7000 m, the porosity of sands and sandstones in Pliocene deposits has been estimated as 12.2–20.2% and siltstones—7.0 to 18.8%. Geophysical forecasting shows that the porosity at a depth of 8–9 km should be about 7–14% and permeability from 2 to 3 × 10^{-15} m^2.

Reservoir properties of the rocks depend on such factors as mineralogical, granulometric, qualitative, and quantitative contents of cement, degree of grain roundness, secondary transformation of rocks, and so on. The mentioned transformation changes during increases in

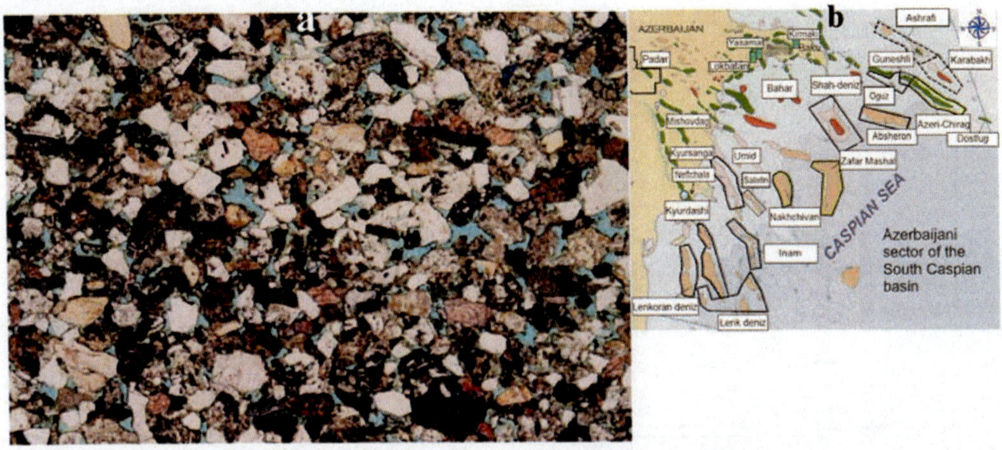

Fig. 7.72 An SEM image of the sandstone of the horizon XX in the Neftchala field recovered from a depth interval of 5455–5460 m. Note green spots indicating the secondary porosity. The reservoir properties: porosity—13.7%, medium permeability, grain sizes—30 to 950 μm, poorly sorted

depth, pressure, and temperature occur with different intensities depending on concrete geological-geochemical conditions. The variation in these processes may be evaluated based on the general laws of thermodynamics.

It is most likely that the totality of physical–chemical processes that leads to the change in reservoir properties of the rocks is a gradual successive transition of sedimentary formations from metastable conditions to the free energy lower-level conditions. It is suggested that the processes mentioned above include the following:

- mechanical rock compaction when grains' relative position appears to be ordered and changed (close packing mode),
- the allothigenic minerals dissolution and authigenic minerals formation as well as reduction, recrystallization, and other physical–chemical processes. The stability ratio of the system (rocks-fluids-organic matter) is dia-catagenetic transformation stages valued by several indexes, first, reflection and refraction indexes of organic matter (vitrinite, kaolinite) and metamorphization ratio. It appears that there is a specific connection between some indexes.

For instance, it is suggested (Chernikov & Kurenkov, 1977) that the distribution of coefficient ratios of silty-sandy rocks metamorphization is in accord with those of coaly inclusions defined by reflectance value ($10{\cdot}R$) which serves as an index of the rock's lithification degree. Of sandstone and siltstone, open porosity values are in inverse relationship with their transformation degree.

Correlative relations between open porosity, metamorphization coefficients, and reflection (refraction) indexes may be used to forecast reservoir properties of the rocks (of the same formation complexes).

It is inferred that the most stable reservoir rocks—pure quartz sandstones, have been compacted, resulting in their low porosity (less than 10%) and permeability. The lower of this gradation reservoir properties of sandstones are estimated to be positive only when jointing (secondary porosity) originates. Thus, the degree of catagenesis of the rocks is one of the major factors controlling the changes in porosity and permeability. In this connection, the primary goal in evaluating reservoir properties at great depths within the unhidden part of the section is the forecasting of zoning of the rock and OM catagenesis under SCB conditions. It may be done

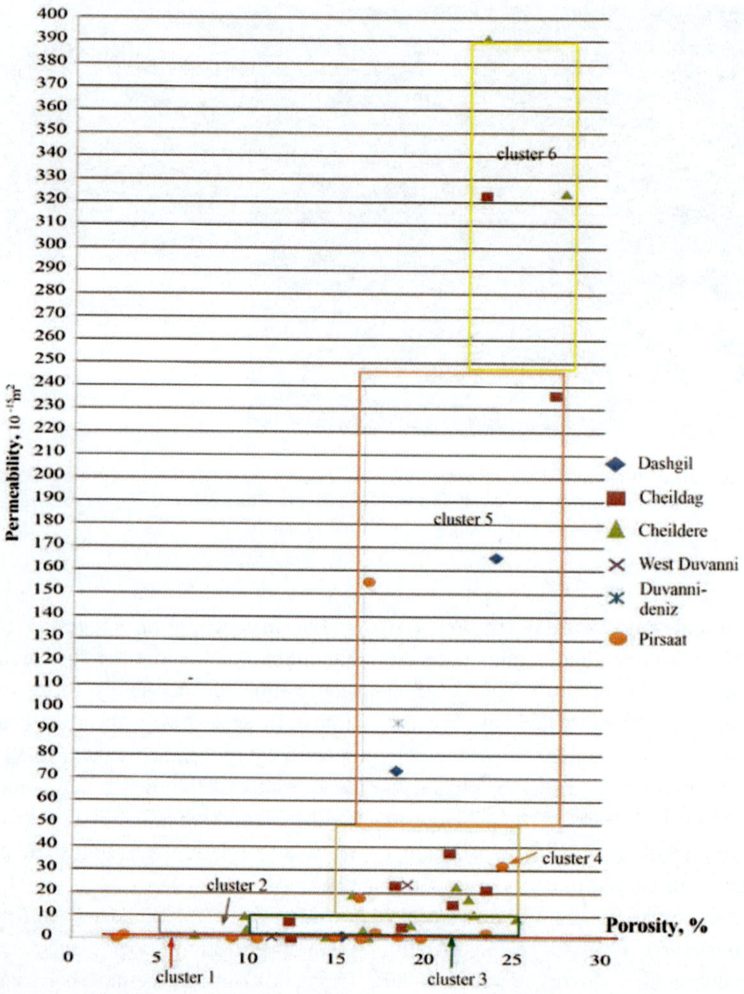

Fig. 7.73 The reservoir properties diagram of the PS deposits from southeastern Gobustan

using the widely applied total heat pulse calculation method. An essential feature of this method is that it accounts for such highly significant factors of catagenesis as time and temperature (Lopatin, 1983).

Using this method for the Pliocene deposits in the Baku Archipelago, it was revealed that they are in the protogenesis—early catagenesis stage. Theoretical calculation carried out by this method follows experimental data of organic matter samples taken from the drilled wells as with outburst rocks of mud volcanoes. As is seen from Figs. 7.72 to 7.73, organic matter catagenetic transformation in the Eocene-Upper Pliocene stratigraphic interval, according to the vitrinite reflectance data measured in the Baku Archipelago and Absheron water areas, is in the initial stages. A specific extension of contacts between the grains serves as a ketogenesis stages index (Chernikov & Kurenkov, 1977).

From the analysis of rock samples picked out from the depths of 1553 to 5417 m in the Bakhar and Bulla-sea areas, it is concluded that the specific extension values of the contacts and

Table 7.1 Specific contact extension and rock transformation stage (after Guliyev et al., 1991)

Structure name	Average grain diameter	Specific contacts extension	Transformation stage	Well no	Depth, m	Nearness to the dislocation line
1	2	3	4	5	6	7
Bulla-deniz	0.105	7.05	Disgenesis	3	1499–4502	Far
	0.166	4.22	Disgenesis	4	1553–4502	Far
	0.150	14.32	Catagenesis	25	5693–5699	Far
	0.101	3.76	Disgenesis	29	3566–3573	Close
	0.142	8.58	Disgenesis	30	5540–5545	Far
	0.175	6.78	Disgenesis	31	2590–2608	Far
	0.167	7.08	Disgenesis	31	4782–4783	Far
	0.235	8.28	Disgenesis	31	5437–5441	Far
	0.156	7.29	Disgenesis	32	5071–5072	Far
	0.117	15.38	Disgenesis	42	5390–5393	Far
Bakhar	0.177	16.3	Catagenesis	10	3844–3847	Very close
	0.151	15.3	Catagenesis	10	4132–4137	Very close
	0.191	11.08	Disgenesis	19	4821–4827	Close
	0.189	13.43	Catagenesis	19	4834–4835	Close
	0.236	13.6	Catagenesis	19	5033–5040	Close
	0.142	15.55	Catagenesis	49	4671–4673	Very close
	0.184	14.36	Catagenesis	49	5284–5289	Very close
	0.232	7.13	Disgenesis	56	3809–3810	Close
	0.268	7.38	Catagenesis	59	5262–5270	Close
	0.268	6.8	Catagenesis	64	5043–5046	Very close
	0.291	7.84	Catagenesis	64	5072–5079	Very close
	0.279	9.42	catagenesis	64	5417–5420	Very close
	0.261	9.42	Catagenesis	65	4732–4737	Close
	0.204	7.27	Disgenesis	66	4018–4924	Close
	0.212	5.87	Disgenesis	71	4215–4220	Far
	0.212	8.24	Catagenesis	72	3996–3999	Close
	0.243	9.23	Catagenesis	73	3765–3773	Close
	0.171	4.78	Disgenesis	75	4243–4245	Far

average grain diameter in studied sandstones correspond to the diagenesis—early catagenesis stage (Table 7.1). It is noted that the core located close to the fractured zone is characterized by marked post-sedimentation changes that are due to the tectonic movements.

The authigenic mineral formation processes in Lower Pliocene rocks at the early catagenetic stage are characterized by the fact that pore space is generally not subjected to essential changes and preserves its primary capacitive properties (Table 7.1).

What is more, in many cases, such processes as dissolution (of quartz, feldspar, carbonate, and sulphate) and recrystallization of carbonate and clayey material facilitate the development of secondary porosity and microfracturing that is also a positive factor for reservoir rock preservation at great depths. The same is related to the salt leaching. Rock's carbonate and clayey

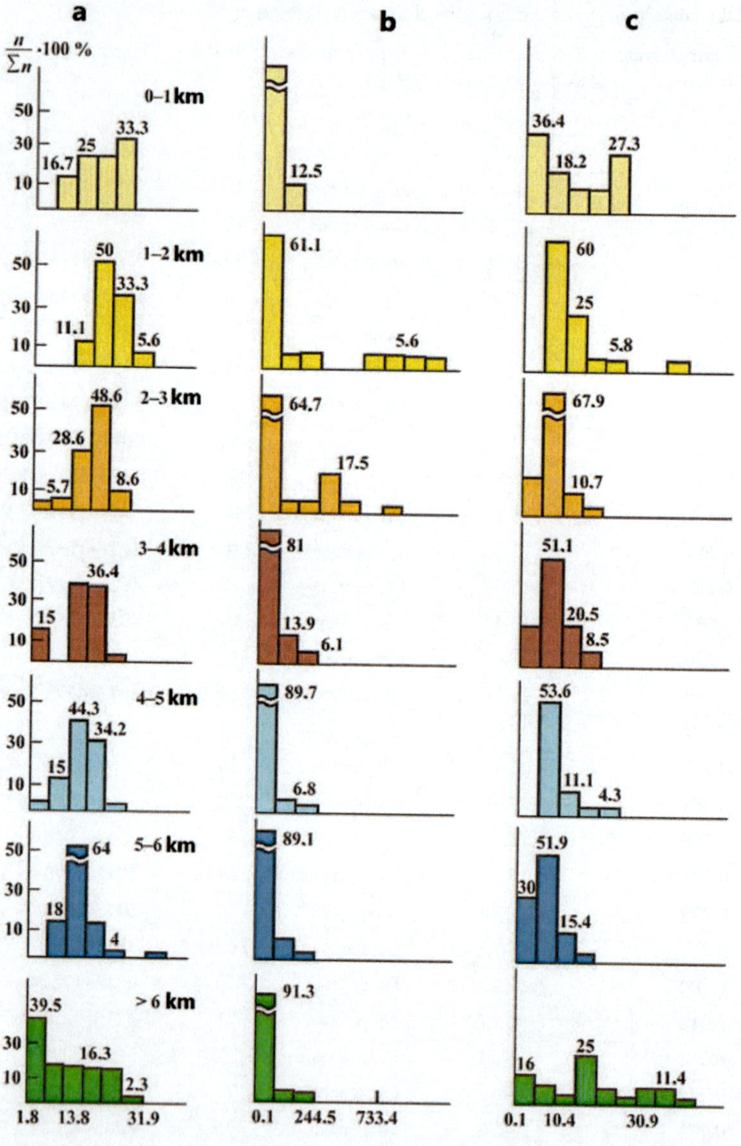

Fig. 7.74 Histograms showing the change in the reservoir parameters of sand rocks with depth in the South Caspian basin. **a** porosity, **b** permeability, **c** carbonate content (after Guliyev et al. (1991), with modifications)

content exert an essential effect on capacitive properties. Figures 7.74, 7.75 and 7.76 noticeably show the changing pattern of reservoir parameters of different rocks through the SCB sequence.

From the histogram of porosity change in sandy reservoir rocks, it is seen that the percentage of porosity with 20–26% in value is decreased interval-wise to 4 km depth. The same rocks' permeability analysis shows its change in a great range from 0.1 to 1000 × 10^{-15} m^2 in the interval of 0.3 km, suggesting that these reservoir rocks belong to all five classes that are especially marked at 1–2 km depth. Generally, these rocks are of a third class below, with sandstone porosity decreasing to 14–20%.

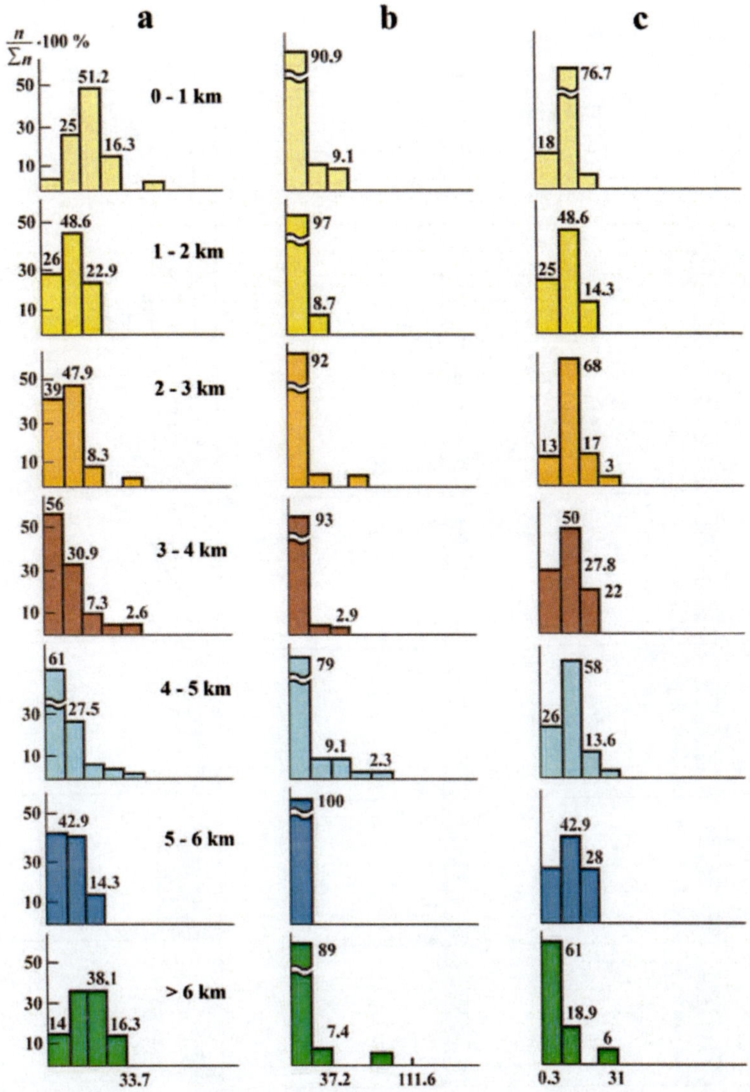

Fig. 7.75 The histograms show the change in the reservoir parameters of mudrocks with depth in the South Caspian basin. **a** porosity, **b** permeability, **c** carbonate content (after Guliyev et al. (1991), with modifications)

A notable feature is that at a depth of more than 6 km, this value varies widely, ranging from 2 to 37.9%, suggesting that favorable capacitive properties are preserved in sandy reservoir rocks occurring lower than 4 km and characterized by the presence of primary porosity and formation of its secondary type. Permeability stabilization in the range of 0.6–366 × 10^{-15} m^2 occurs at the same depths where 80–90% of sandstones appear to be reservoir rocks of III-II classes.

Analysis of carbonate material change in sandy reservoir rocks depending on the depths suggests that its quantity (no more than 10%) has not negatively affected the preservation of favorable properties at 6 km and more depths. 68% of silty reservoir rocks preserve 15–28% porosity at a depth of from 0 to 3 km.

At a depth of from 0 to 6 km, the rocks mentioned have approximately permeability as in sandstone being slightly decreased to

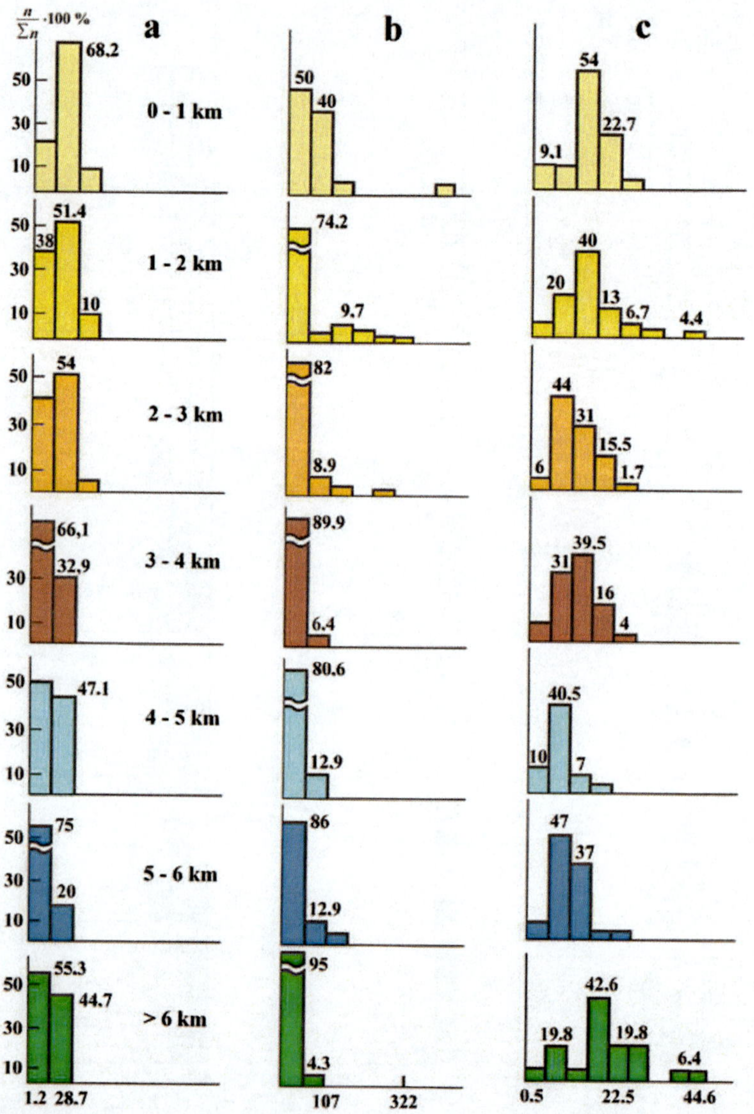

Fig. 7.76 The histograms show the change in the reservoir parameters of silt rocks with depth in the South Caspian basin. **a** porosity, **b** permeability, **c** carbonate content (after Guliyev et al. (1991), with modifications)

(0–429) × 10^{-15} m^2. Generally, siltstones have a permeability of (0–53.5) × 10^{-15} m^2, i.e., they belong to the III-IV classes.

At a depth of more than 6 km, these siltstones become more porous; 45% of their total quantity has up to 28% porosity, and permeability of (0–53) × 10^{-15} m^2 suggests their favorable reservoir qualities. Carbonate material in described rocks is slightly increased compared with sandstones and stably changed from 5.8 to 17.2% over the 0–6 km sequence.

However, at great depths (more than 6 km), reservoir properties have yet to be essentially subjected to the carbonate material action because post-sedimentation processes occur at these depths conducive to forming additional pore space.

Porosity in clay rocks varies widely at a depth of 0–1 km. 50% of these rocks have 16–29% porosity, and from 1 to 3 km, this value is changed from 8 to 17%. More porous varieties may be observed in a deeper zone. As to the permeability, the interval of its change is narrowed to be (0–116) × 10^{-15} m^2, and the filtration property of most clays varies from 0 to 18.5 × 10^{-15} m^2.

At 3–5 km depths, clays appear less porous, although porosity value may sometimes be up to 35%. It is noted that an increase in porosity is also observed in the rocks occurring at depths of more than 6 km. The quantity of clay rocks has permeability up to 92 × 10^{-1} m^2, confirming our standpoint of the secondary alteration effect on the unsealing process at these depths. It is noted that the carbonate content of clay rocks tends to decrease for a depth. In such a way, its content of groundmass does not exceed 15% all over the sequence, and spreading in large values is decreased with increasing depth estimated to be less than 8% lower than 6 km. The same tendency has been marked in nongraded rocks, i.e., porosity is decreased with increasing depth (0–3 km) ranging from 10 to 20%. At a 5–6 km depth, porosity values seem to be slightly stabilized (10–15%). Permeability varies from 0 to 665 × 10^{-15} m^2, reaching its highest values at 2–4 km depths. Favorable filtration properties characterize the most described rocks and belong to reservoir rocks' IV-III and IInd classes. In a zone deeper than 6 km, the nongraded rocks become more porous; 41% of these varieties have permeability from 19 to 24% and capacity characteristics at this depth range as widely as at shallow depths (0.6–34%).

Even though the carbonate interval is increased to a certain extent (7–14%), the permeability in III and IInd classes reservoir rocks is preserved. All the above mentioned corroborates an opinion that the secondary processes are another major factor affecting the pore volume formation at great depths.

The capacitive-filtration characteristics in the interval from 6.5 to 10 km have been investigated using the data of the solid product of the Upper Cretaceous-Miocene mud volcanic outburst (Guliyev et al., 1991). The sandy-silty rocks occurring below 6 km depth are characterized by porosity of 19% and permeability—70 × 10^{-15} m^2. Most parts of clay rocks below 5–6 km depth have a porosity of 7–29%, and permeability averages 8 × 10^{-15} m^2. Rarely, this value reaches 92 × 10^{-15} m^2.

From the above, it follows that deep-seated clays in some sequence intervals are uncompacted, and reservoir properties of sandy-silty rocks appear to be slightly worsened. It is known that there are a significant number of theoretical curves showing a change in the physical properties of rocks. Figure 7.77 illustrates the curves of a change in porosity with depth drawn for young basins. The high sediment subsidence rate is the most distinctive feature of the Alpine basins compared with the platforms and the more significant part of the foredeep.

Because low rates characterize the Earth's crust subsidence and settling on platform areas, the reworking process of sedimentary material and its adaptation to new environmental conditions take a long time. Within the geosynclinal zones, crustal subsidence amplitude and depositional rate are known to be high. Besides, these deposits frequently appear to be buried in an

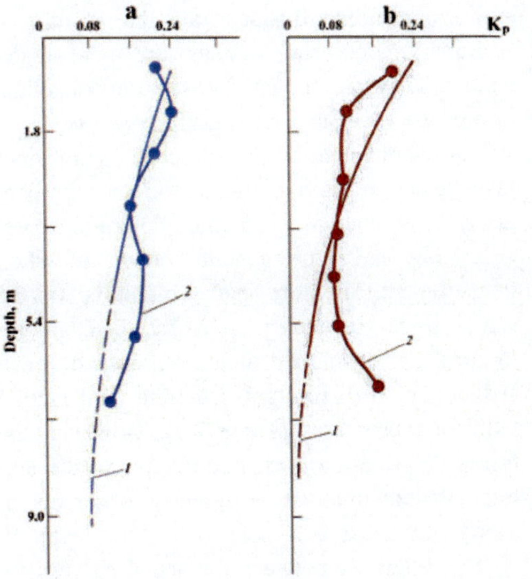

Fig. 7.77 Changes in the porosity with a depth for young basins: **a** sandstone, **b** clay. (1) modeled, (2) actual

under-consolidated state. It seems likely that such distinctive features of the mentioned curves relate to the accumulation mode. The more reliable pattern of a change in porosity with depth in sandy-silty and clay rocks is created by real curves plotted based on the physical measurements (Fig. 7.77). These curves, as distinct from the model curves, reflect the physical properties of rocks occurring at SCB great depths. The total combination of physical–chemical processes leads to the change of properties with depth successive transition of sedimentary formations from free energy high-level state (metastable state) to the lower states. The stages of dia determine the degree of the system stability (rock-fluid-organic matter)—and catagenetic transformations estimated by lithologic indexes of reflection and refraction of OM material (e.g., vitrinite) and metamorphization ratio. The post-sedimentation processes in the deep-seated Neogene-Paleogene deposits of different lithotypes are mainly predetermined by the conditions of sedimentation and manifested in mineral neogenesis (calcitization, limonitization, cericitization, kaolinzation, and chloritization) as well as in dissolution of carbonates, sulfates, feldspars, quartz and carrying away both these components and easily soluble salt from the rocks and then recrystallization of carbonate and clayey cement and compaction. Besides, all the mentioned secondary processes' actions are not single-valued. They may be conducive to unconsolidation of the rocks and restrain this process.

The authigenesis of Pliocene sandstones investigated in thin sections and closely connected with abyssal thermobaric alterations suggest the early catagenetic stage of their development. The pore space volume is still the same since secondary porosity and microjointing are developed at the expense of post-sedimentation dissolution, leaching, and recrystallization processes (Table 7.2). However, the brittle deformation traces, conducive to the textural disjunction, are frequently observed in sandy-silty rocks at a more than 4 km depth.

The difference between the actual curves of a change in porosity with depth in clays and sandstones is explained by different degrees of compaction or noncompaction of these rocks. It depends directly on mineral and granulometric composition and structural-textural peculiarities. An exciting feature is the change of mineral composition in shielded clays and clayey fraction of reservoir rocks, as well as the noticeable trend in their catagenetic transformation.

The sediments of the western and eastern shelf zones have accumulated within a single basin but in different tectonic and climatic conditions, being recharged from different provenances. This is a prominent feature of studied deposits.

Clays and clay fractions of the Lower Pliocene other rock types are polymineralic with the same association of rock-forming clay minerals (hydromica, montmorillonite, and kaolinite), which seem to be changed quantitatively in the western shelf (from the Absheron to the Baku archipelagos) and more markedly from the eastern shelf to the western one. In such a way, the montmorillonite content of the Baku Archipelago deposits is relatively low at a depth of 2–7 km, but it decreases with depth from 60 to 28%, hydromica—from 45 to 28% with increasing of the content of such minerals as kaolinite from 10 to 34%, chlorite from 5 to 7% as well as of mixed-layered minerals of hydromica-montmorillonite and chlorite-montmorillonite composition. In the rocks of the Absheron Archipelago, the kaolinite content decreases with depth (39–20%), while the value of hydromica appears to be increased up to 30–45%, and the content of montmorillonite remains in a stable state (20–25%).

The amount of mixed-layered minerals of hydromicaceous-montmorillonitic composition is estimated to be from 5 to 15%. Within the South Caspian eastern shelf, a finely dispersed fraction of clays and other rock types is characterized by predominant hydromica (50–60%), and subordinate montmorillonite (traces-17%), kaolinite (10–25%) and chlorite (5–10%). The mixed-layered hydromicaceous-chloritic, montmorillonite-hydromicaceous, and vermiculite-montmorillonitic minerals may also be present (5–10%).

The mineral composition of clays in the Lower Kur depression is specific. The

Table 7.2 Reservoir properties of the deep-seated Pliocene deposits

Area	Well no.	Depth interval, m	Porosity, %	Permeability, 10^{-15} m^2
SCB western coast				
Bulla	527	5416–5421	18.3	82.1
	549	5589–5582	13	128
Duvanni-sea	505	5305–5507	14.2	93.5
	565	5589–5590	18.6	58.2
	533	5228–5231	18.4	94
	1	5444–5449	12.3	111.1
	1	5951–5954	10.1	129.9
	1	5778–5783	16.6	121.8
Southern-2	3	4991–4996	28.9	152.1
	7	5171–5176	19.9	116.2
The Samedov area		6120–6121	14.8	7.1
SCB eastern slope				
Ogurchinsky	3	4929–4934	16.4	48
	2	5148–5153	24.4	12.5
West Erdeklin	11	5412–5414	18.1	126
	1	5470–5480	21.8	32.8
	1	5670–5675	19.2	48.9
	3	5301–5303	24.4	286.8
	3	5565–5570	18.5	77
Bakhar		4533–4539	11.3	71.4
		3996–3999	20.1	21.9
Shakh-deniz		5613.48–5613.7	18.7	
		5241–5241.22	17.3	
Umid		5455–5460	15.8	13.8
Nakhchivan		6596.03–6597	11.7	

montmorillonite decreases with depth from 55 to 25% while hydromica increases from 30 to 45%; the kaolinite content of all the sections is almost stable (10–25%).

Typically, the mineral composition of more old clay Paleogene deposits (Limanov bank—SCB Eastern shelf), according to the mud volcanic outburst data (Proshlyakov et al., 1987) is almost the same as those of Pliocene clays (montmorillonite —5 to 15%; hydromica—55 to 65%; kaolinite—45 to 30%). Mineral composition of Oligocene—Lower Pliocene clays (Agzybir in the Lower Kur depression) is also like those in the Pliocene: hydromica—20 to 40%; montmorillonite—40 to 60%; kaolinite—20%. Hence, kata genetic transformation with depth seems not to be increased. Thus, the quantitative distinction of clay minerals relates to qualitative features of distribution provinces supplying the eastern and western shelf with clastic products and in turn, with different intensities of post-sedimentation changes.

The Talysh mountains and the Greater and Lesser Caucasus were the major provenances during the Pliocene deposits' formation within the western shelf. The fact that the Mesozoic and Cenozoic volcanogenic fragments are widely developed in the mentioned provenances caused the high montmorillonite content of these deposits. The most notable features of the South Caspian Basin—Baku Archipelago sequence are the volcanic nature of its source material and the exceptional preservation of swelling minerals at such great depth (6 km).

Several researchers believe that when saturated with potassium, low order dioctahedral montmorillonite of volcanic origin, no irreversible compression of this mineral lattice takes place. This mineral may, in fact, not be transformed. Within the eastern shelf, these deposit formations relate to removal material from the Kopetdag, Malyi and Bolshoi Balkhan, and Guba-Dag, where a large quantity of muscovite, biotite, and altered plagioclase are presented at the expense of hydromica has been formed.

Thus, it was recognized that different source materials resulted in certain conditions under which secondary processes were formed. The fact that these rocks are deeply seated and consider specific features of primary mineral formation causes some more transformation of clay minerals, which may be seen in thin sections and photos of clay samples made by scanning electron microscope. In addition, as seen from the studied sections, rigid clay minerals appear to be increased with depth conducive to the formation of local micro-jointing both longitudinal and transverse. As a rule, it is mostly displayed at deeper parts of the section in clayey lithotypes of Paleogene in age within the Southeast Gobustan and Lower Kur depressions. It is known that a distinctive feature of clayey oil reservoirs is their mineral composition, represented mainly by rigid clay minerals, such as chlorite, hydromica, and kaolinite (Abdullayev & Leroy, 2016). This feature is just like the SCB deep-seated deposits as well. The mineralogical factor and the process of authigenic clayey mineral formation that influences reservoir properties can promote a decrease in pore space. If clay caps and clays occurring as integral parts of the rocks (cement) are compared, then it becomes apparent that clay minerals appear to be changed earlier in sandstone cement. A favorable relation between swelling and rigid clay minerals is created at the expense of forestalling transformation in reservoir rocks. It is explained by the more rapid and active interbedded solution of the clastic product in sandstones. Hence, the development of the secondary processes in clay caps seems to be slowing down and accelerated by such deep factors as temperature and pressure.

The rising of brittle deformation in all lithological varieties is dependent mainly upon mineralogical and tectonic factors. The results of our investigation of pore space structure transformation and analysis of the specific extent of contacts and median grain sizes show that the most secondary variation is observed in the rocks stripped close to the dislocation and at great depths (Table 7.3).

The South Caspian is recognized for its differentiation in down-warping intensity both in time and space and by significant crushing in blocks of deep tectonic fractures, and Neogene fractures developed as thrust, fault, and upthrust. This is just one more argument of an inequivalent manifestation of secondary variations in the deposits of the western and eastern shelves.

Klubova (1985) believes that the specific tectonic features of the region, where future clay reservoirs are being formed, create conditions conducive to the noncompactness of surrounding rocks (in that case, these are clays).

Thus, within the Meso-Cenozoic complex, the rocks in the interval 0–9 km are characterized by pore and fissure healing with newly formed authigenic material, and increasing their void volume caused new capacities. As a result, the secondary elements appeared to exist in the primary granular deep-seated reservoirs, suggesting that the latter may be preserved at great depths. However, in such a case, the reservoir type will be changed from the pore to the pore fractured and jointly vascular types. The data gained during the investigation of secondary alteration in deep-seated clays indicates that the Absheron—Near Balkhan tectonic zone reservoirs may be found within these deposits.

An analysis of reservoir properties and general trend in secondary alteration, considering the depth and subsidence longevity, confirms that deep-seated fractured-vesicular clay reservoir

Table 7.3 Structural parameters showing rock transformation stages

Area	Wells	Depth, m	Median grain diameter, mm	The specific extent of contacts, 1/mm	Rock transformation stages
Bulla Sea	3	4499–4502	0.105	7.05	Diagenesis
	4	1553–1560	0.166	4.22	Diagenesis
	25	5692–5699	0.15	14.32	Catagenesis
	23	3566–3573	0.101	3.76	Diagenesis
	30	5540–5545	0.142	8.58	Diagenesis
	31	2590–2603	0.175	6.78	Diagenesis
	31	4782–4783	0.167	7.08	Diagenesis
	31	5437–5441	0.235	8.28	Diagenesis
	32	5071–5072	0.156	7.29	Diagenesis
	42	5390–5393	0.117	15.38	Catagenesis
Bakhar	10	3844–3847	0.177	16.3	Catagenesis
	10	4132–4137	0.151	15.3	Catagenesis
	19	4821–4827	0.191	11.08	Diagenesis
	19	4834–4835	0.189	13.43	Catagenesis
	19	5033–5040	0.236	13.6	Catagenesis
	49	4671–4673	0.142	15.55	Catagenesis
	49	5284–5285	0.184	14.36	Catagenesis
	56	3809–3810	0.232	7.13	Diagenesis
	59	5262–5270	0.268	12.38	Catagenesis
	64	5043–5046	0.268	6.9	Catagenesis
	64	5072–5079	0.291	7.84	Catagenesis
	64	5417–5420	0.279	9.42	Catagenesis
	65	4732–4732	0.261	9.42	Catagenesis
	66	4918–4924	0.204	7.27	Diagenesis
	71	4215–4220	0.212	5.87	Diagenesis
	72	3996–3999	0.212	8.24	Catagenesis
	73	3765–3773	0.243	9.23	Catagenesis
	75	4243–4245	0.171	4.78	Diagenesis
Zhdanov	13	2305–2310	0.111	10.47	Catagenesis
	21	3600–3605	0.209	3.77	Diagenesis
	34	4371–4376	0.143	13.09	Catagenesis

stripping is possible, i.e., the rocks in association "shield-reservoir rock" may be qualitatively changed with depth.

Thus, we can conclude that:

- An analysis of the regional distribution of studied parameters shows that each suite and horizon are characterized by their typical features over an area. There is no common trend in the variation of these parameters within the SCB. Also, it shows a strong dependence of such parameters as clay fraction and sand fraction on sedimentation conditions. Environmental conditions are believed to control the clastic material distribution within the SCB firmly.

- If the distribution of granulometric composition of the deposits, in particular sandy fraction, is to be compared with porosity data through the studied separate stratigraphic sections, then it becomes evident that there is no clear correlation between the mentioned two values.
- From the analysis of variation in porosity values over some studied areas, it is concluded that porosity is strongly dependent on structural plan, and as a rule, it decreases in the direction of the plunge. Some increase of this value is noted in the shatter zone. A strong dependence of the deposit's porosity on their depth of occurrence has also been established. At a 2.5–3.0 km depth, porosity appears to be significantly reduced because of sediment compaction. This regularity is most clearly displayed in the lower horizons of the studied sequence, i.e., Superkirmaki sandy suite, Superkirmaki clay suite, Break suite, and lowermost strata of the Balakhani suite. Some increase of porosity noted in the samples taken out from a depth below 3000 m is likely caused by secondary processes in these rocks.
- It is established that the carbonate content of deposits depends on the pelitic fraction content, i.e., on the clay fraction of these deposits, suggesting that carbonates in the studied section are of sedimentation origin. The carbonate content of the section is increased with the subsiding of sediment, which points to the fact that carbonates have been formed in the process of diagenesis. It is known that the carbonate precipitation from pore water takes place under the influence of pressure and temperature. The previous arguments conclude that the deposits in the studied suites include primary (sedimentation) carbonates and secondary (diagenetic and epigenetic) carbonates.
- The correlation analysis of porosity, permeability, clay fraction, and the carbonate content value in studied deposits, as well as regularities in their variation over the SCB area, show the presence of negative dependence of permeability on the pelitic fraction and carbonate contents. Hence, it follows that both clayey and carbonate cement are present in the SKSS, SKGS, and Break suite deposits.

7.3.2 Characteristics of Screening Rocks

The most widespread rocks in the PS of Azerbaijan are clayey rocks averaging 60% of the total thickness, reaching 75–80% in the Lower Kur depression and Baku Archipelago areas. In the Baku Archipelago sections, these rocks form up to 70–80 m thick overall bands. Clays are mostly compact without bedding, but sometimes, horizontally, and obliquely laminated varieties may also be present. Their structures are silty-pelitic, pelitic, and psammo-silty-pelitic; microstructures are irregular, massive, and thinly lamellar. The pelitic fraction content of clays ranges from 51 to 90%. The most elutriated varieties have been noted in the Western Absheron, Southeastern Gobustan, and Baku Archipelago sections. The carbonate content of clays is from 10 to 25%.

Clays contain grains of quartz, feldspar, magnetite, ilmenite, chlorite, glauconite, pyrite, limonite, and other minerals. Dominating in clays of the PS lower division is hydromica, which constitutes more than 40% of the pelitic mass.

Besides, its content decreases upwards the section with an increase in montmorillonite content. The kaolinite content is 10–20%, and the mixed-layered varieties content is from traces to 10%. The clays from the Absheron Archipelago are markedly distinguished by their composition from those of the Baku Archipelago. The montmorillonite content of clays in the Absheron Archipelago is 15–30%, while hydromica content is more than 50%. In addition to the essential clayey minerals, the Break suite clays contain corrensite.

From the qualitative valuation of clay minerals, it is inferred that the PS upper division clays, occurring above the Break suite (horizon VII according to the Garadagh nomenclature), are markedly distinguished from the Break suite clays. The dominant clay mineral here, except in the Absheron Archipelago and Near-Caspian-Guba region, is montmorillonite, unlike the Break suite, where the dominant clay mineral is hydromica.

A particular law-governed nature is noted in the spatial distribution of clays within the PS upper division. Quantitatively, the majority of montmorillonite, mostly of non-micaceous origin (more than 50% of its fraction is smaller than 1 mm), occurs in the southeastern Kur depression and within the Shamakhi-Gobustan and Absheron oil–gas-bearing regions.

Its content appears to be slightly decreased from 50 to 30% in the northern and northeastern Kur depression, in eastern Gobustan and southern Absheron region, as well as seaward east of the Alyat locality. At the same time, the most increased content of hydromica with a decrease in montmorillonite content is noted within the Baku Archipelago in northerly and northeasterly directions.

It is recognized that clay rocks serve as a fluidal stop to the PS oil and gas accumulation. The best of them are PS clays of the Baku Archipelago and Lower Kur depression, which are characterized by marked thin-pelitic fraction, mainly swelling clay minerals, significant total thickness of clay bands, high values of the section clay fraction ratio (more than 0.8), high degree of dispersion and plasticity.

Besides, the rock insulating properties improvement is noted east of the Lower Kur depression towards the Baku Archipelago, from which this process is continued in a southeasterly direction resulting from the fact that the rocks' dispersion and plasticity appear to be increased in the same direction. The Lower Kur depression PS clays are not as fluid as those from the Baku Archipelago due to sandy-silty interbeds alternating with clay bands.

It is also noted that from the Baku Archipelago towards the Absheron Archipelago, the clay cap-rock insulating properties become less well, which is explained by increased sandy-silty content of clays and by decreased quantity of clay minerals. Relatively improved insulating properties of clayey rock-caps are marked west of the Absheron Archipelago towards the Absheron island and further in a southwesterly direction up to the Shamakhi-Gobustan oil/gas-bearing region.

References

Abdullayev, E., & Leroy, S. A. G. (2016). Provenance of clay minerals in the sediments from the Pliocene Productive Series, western South Caspian Basin. *Marine and Petroleum Geology, 73*, 517–527.

Aliyev, A.I., et al. (1973). Recommendations for the direction of prospecting for large gas (gas condensate) deposits in deep-immersed zones of the South Caspian depression. In *VNIIGaz* (52p), Moscow. (in Russian).

Alizadeh, A. A., Guliyev, I. S., Kadirov, F. A., & Eppelbaum, L. V. (2017). *Geosciences in Azerbaijan. Volume II: Economic minerals and applied geophysics* (340p). Springer.

Chernikov, O. A., & Kurenkov, A. I. (1977). *Lithological studies of sandy productive reservoirs (oilfield lithology)* (110p). Nauka. (in Russian).

Guliyev, I. S., Suleymanova, S. F., & Klyatsko, N. V. (1991). Exploring the reservoir properties of rocks from the sedimentary cover of the southern Caspian Basin. *Soviet Geology, 7*, 7–15. (in Russian).

Klubova, T. T. (1985). Lithological and geochemical factors of formation and preservation of reservoir properties of rocks at great depths. In: *Reservoir properties of rocks at great depths* (pp. 59–68). Nauka. (in Russian).

Lopatin, N. V. (1983). *The formation of fossil fuels* (192p). Nedra. (in Russian).

Proshlyakov, B. K., Galyanova, T. I., & Pimenov, Yu. G. (1987). *Reservoir properties of sedimentary rocks at great depths* (201p). Nedra. (in Russian).

8 Thermobaric Conditions in the South Caspian Basin

8.1 Geothermal Characteristics of the South Caspian Basin

The South Caspian Basin is characterized by the variegated thermal regime of its tectonic elements and the low warming—to the extent of the Pliocene–Quaternary deposits. Analysis and generalization of many temperature measurements in wells show that the temperature distribution in sections reflects, without going into detail, the essential tectonic features of the South Caspian Basin.

As it is seen from the sections, the geoisotherms correspond to the depression contour lines and, in broad outline, to the basement surface contour lines (Fig. 8.1). This Figure shows that temperature fall is observed at a horizontal section level of 2000 m from the basin slopes to its center.

The same event is marked at a depth of 5000 m. Within the marginal zones of the Gabyrry-Adjinour and Evlakh-Agdjabedi troughs, the temperatures at the same depth are estimated to have been in the range of 160–200 °C (Gadzhiev, 2008).

In the Absheron and Lower Kur troughs, reservoir pressure values average 1250 °C, respectively. According to the borehole measurements carried out within the Baku Archipelago, the temperatures towards the regional basement subsidence with increasing sedimentary thickness at a depth of 5000 m do not exceed 100 °C.

Mukhtarov et al. (2003) have reported that the temperature on the PS roof ranges from 17–20 °C (in the PS pinching out boundaries) to 80–85 °C (at the most subsided parts of the PS roof) (Fig. 8.2). The distribution of temperatures has corresponded to the PS roof structure being increased from the depression slope parts to its center. The same is observed in a southeasterly direction towards the SCB deep-seated zone. For the structures located in SCB's deeply subsided zone, the authors have used the temperatures forecasted by Rustamov (2001).

Local temperature maximums are observed along with the Kur depression (Fig. 8.3) in the Adjinour area, where they are estimated to have been in the range of 80 °C; in the Sor-Sor and Amirarkh areas, these temperatures vary between 65 and 70 °C in the Sarkhanbeili-Garabagli-Kurovdag areas those are 60–65 °C. Local temperature minimums (30–35 °C) have been marked in the Neftchala area. The other vaster minimum (35–40 °C) comprises the Duvanny-Deniz and Khara-Zirya structures and widens along the zone of Khamam-Deniz, Garasu, Sangi-Mugan, Aran-Deniz areas. The lowest temperature (13.5 °C) on the PS roof has been found within the Palchig Pilpilyasi structure of the Absheron OGR. The maximum temperature (84.5 °C) in the horizontal section has been marked within the SCB deep-seated Djanubi Sabukhi area. Local maximums (82 °C) have also been marked in the Sabukhi and Alesker

A. Alizadeh et al., *Pliocene Hydrocarbon Sedimentary Series of Azerbaijan*,
Advances in Oil and Gas Exploration & Production, https://doi.org/10.1007/978-3-031-50438-9_8

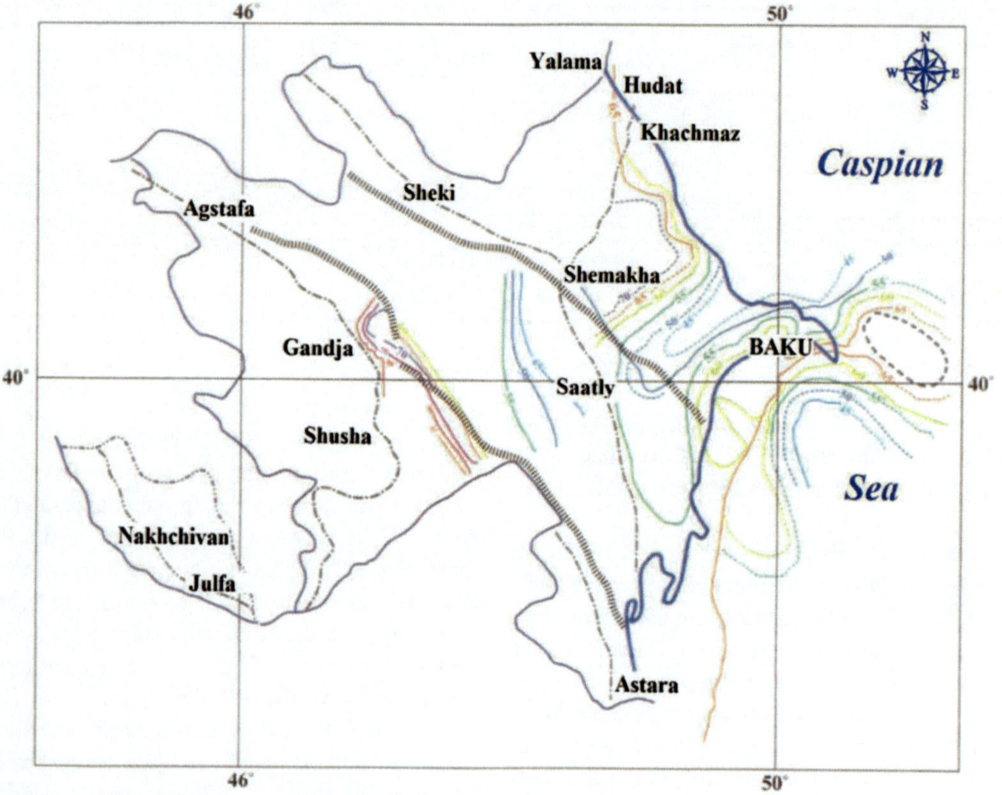

Fig. 8.1 Temperature distribution in the western SCB at a section level of 2000 m (after Aliyev S. & Aliev A., 1995)

Alekperov areas and the Adjinour area, where it reaches 83.5 °C.

A low-temperature zone extends from the northwest to the southeast, embracing some digressional zones. According to the same authors, the temperature in the PS bottom varies from 20 °C up to 130 °C in the SCB subsided areas (Fig. 8.3).

It is noted that an isotherm corresponding to 50 °C passes close near the pinching-out boundaries. In contrast, an isotherm corresponding to the temperature of 100 °C embraces the more significant part of the Lower Kur and South Caspian depressions. Within the SCB deeply seated areas (PS lowermost strata), the temperature conditions corresponding to the beginning of hydrocarbon generation are present.

Figure 8.4 illustrates generalized graphs of temperature change with depth for the fields of the Absheron Island, Baku Archipelago, Lower Kur depression, and Absheron Archipelago. An approximation of available data of measured temperature changes with depth has been carried out using linear, power, and polynomial (second and third orders polynomials) functions. These functional dependencies are given in Table 8.1.

This table illustrates the temperature-depth dependence in the Absheron Peninsula. The second and third-order polynomials best describe the Absheron and Baku archipelagos. As to the Lower Kur depression, the power function approximation has gained the best results.

The above follows that the Pliocene–Quaternary interval in SCB sedimentary cover is characterized by a very low geothermal gradient ranging from 12 to 27 °C/km with anomalous low values marked within the basin water area (Fig. 8.5). Gradients higher than 22–35 °C have

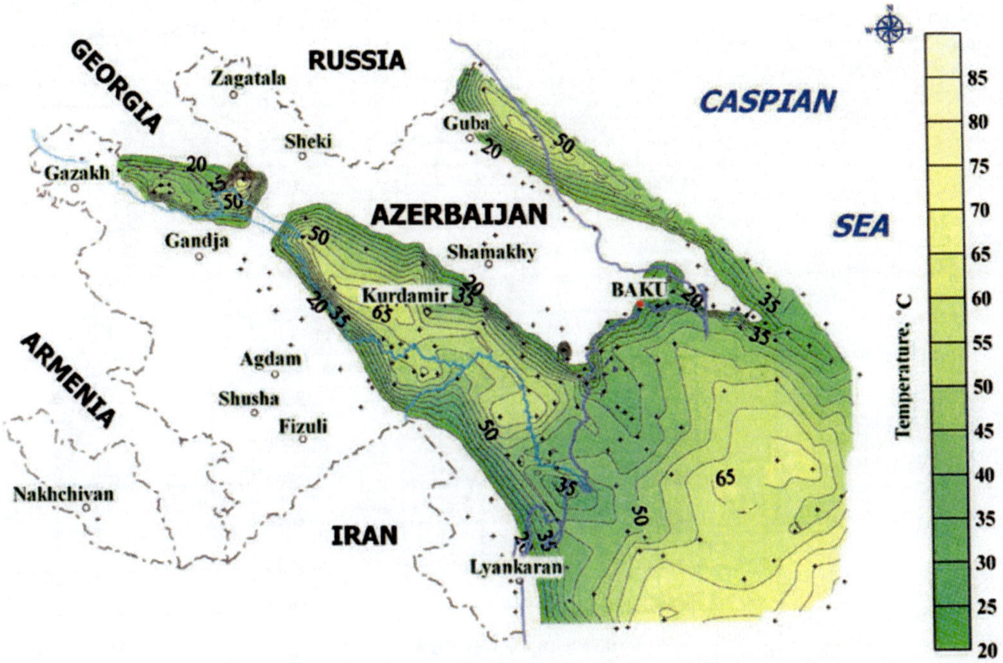

Fig. 8.2 The temperature distribution in the Productive series roof

Fig. 8.3 The temperature distribution map (°C) is drawn up by the PS bottom

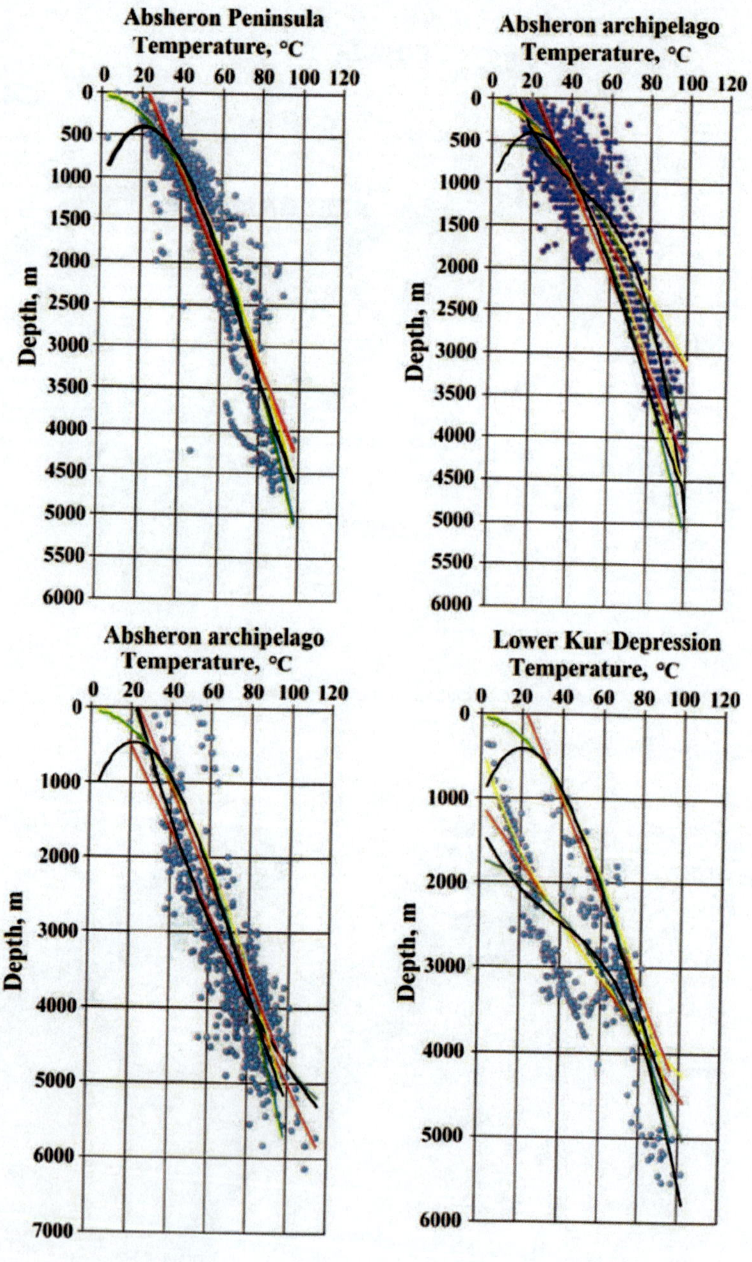

Fig. 8.4 The reservoir temperature changes with depth in the SCB

been marked within the Absheron Peninsula, Absheron Archipelago, and Shemakha-Gobustan trough.

Mukhtarov (2011) estimated a complete horizontal gradient of geothermal field $G(x, y)$, which usually is determined by the following formula:

$$G(x,y) = \left[\left(\frac{\partial t(x,y)}{\partial x}\right)^2 + \left(\frac{\partial t(x,y)}{\partial y}\right)^2\right]^{\frac{1}{2}}, \tag{8.1}$$

where $t(x, y)$ are the temperature values.

Table 8.1 Temperature (T)—depth (*H*) dependence functions for the different regions of SCB

Region	Approximate dependence	Certainty factor (R^2)
Absheron Archipelago	$T = 0.017 * H + 25.10$	$R^2 = 0.634$
	$T = 1E{-}06 * H^2 + 0.013 * H + 27.33$	$R^2 = 0.637$
	$T = 1E{-}09*H^3 - 5E{-}06 * H^2 + 0.022 * H + 24.39$	$R^2 = 0.640$
	$T = 3.327 * H^{0.372}$	$R^2 = 0.524$
Absheron Peninsula	$T = 0.014 * H + 28.67$	$R^2 = 0.800$
	$T = 2.237 * H^{0.430}$	$R^2 = 0.795$
	$T = - 3E{-}06 * H^2 + 0.025 * H + 20.37$	$R^2 = 0.841$
	$T = 7E{-}10 * H^3 - 7E{-}06 * H^2 + 0.033 * H + 16.78$	$R^2 = 0.845$
Baku Achipelago	$T = 0.0125 * H + 28.553$	$R^2 = 0.707$
	$T = 7E{-}07 * H^2 + 0.0087 * H + 33.187$	$R^2 = 0.711$
	$T = - 9E{-}10 * H^3 + 8E{-}06 * H^2 - 0.0107 * H + 45.003$	$R^2 = 0.731$
Lower Kur depression	$T = 0.0175 * H + 1.2023$	$R^2 = 0.607$
	$T = 0.0072 * H^{1.108}$	$R^2 = 0.667$
	$T = - 4E{-}07 * H^2 + 0.0199 * H - 1.8496$	$R^2 = 0.608$
	$T = 3E{-}10 * H^3 - 3E{-}06 * H^2 + 0.0274 * H - 7.476$	$R^2 = 0.609$

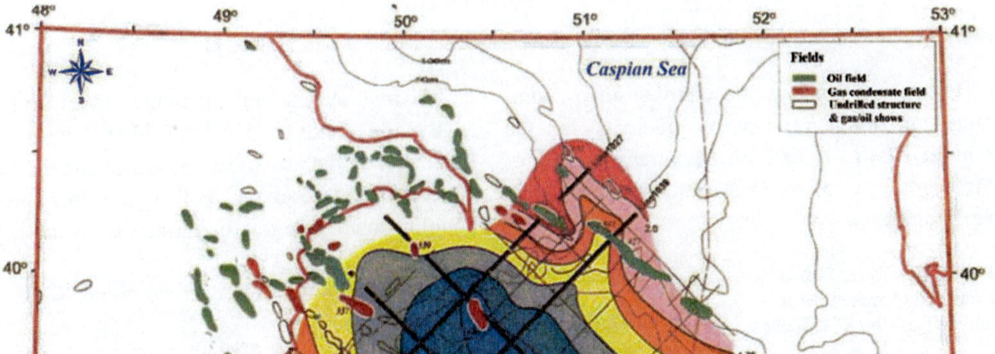

Fig. 8.5 A sketch map of the temperature gradients spatial change in the SCB (generalized from the data of BP Azerbaijan) (Feyzullayev, 2011, 2014)

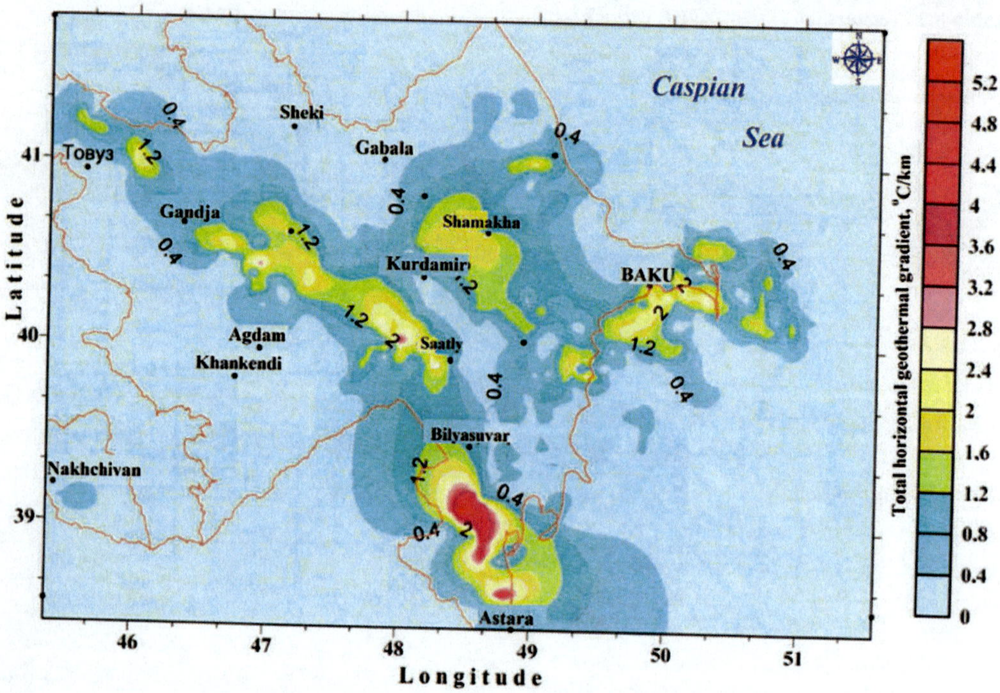

Fig. 8.6 The distribution of the total horizontal gradients for the section depths of 5000 m

Figure 8.6 displays the distribution of complete horizontal geothermal gradients for the section depth of 5000 m. Maximum values of these gradients correspond to the active regional fault zones.

Maximal measured temperature values for the SCB marine fields are listed in Table 8.2. It should be noted that the actual data do not contradict the conception of reservoir temperature functional coupling with a degree of proximity to the day

Table 8.2 The maximal measured temperatures in the wells of the SCB fields

Structure	Well no	Depth, m	Temperature, °C
Baku Archipelago			
Shakh-Deniz	4	6500	122
Bulla-Deniz	46	5730	115
Bulla-Deniz	38	6150	110
Bulla-Deniz	42	5850	110
Sangachal-Deniz	550	5770	113
Garasu	28	5650	112
Garasu	30	5683	106
Duvanny-Deniz	39	4450	111
Absheron and Absheron Archipelago			
Absheron-Deniz	3	5000	110
Arzu	2	4708	105
Janub	2	4710	102
Janub-2	12	4127	100
Bakhar	19	5450	99

surface of basement rocks warmed up by deep generation heat. According to this conception, the sections warming up relate to the hypsometric position of the crustal basement. Where it lies closer to the surface, a deep conductive heat transfer to the upper parts of the sedimentary series is carried out with lesser scattering than within the regions of thicker sedimentary infilling.

At the same time, the low-temperature conditions of the SCB Pliocene–Quaternary deposits are also associated with abnormally high rates of basin subsidence, accumulation of thick sedimentary series, and their properties under compaction, which strongly prevents the warming-up of all the series. An unsteady geothermal field caused by avalanche sedimentation takes place. The other factor of anomalous low warping-up of these deposits is the presence of up to 6 km thick clay series of underlying Paleogene-Miocene complex characterized by below-average heat conduction and recognized as the regional heat insulator of conductive heat transfer upwards the section that causes a decrease in total heat supplying to the upper intervals. The hypsometric position of the Paleogene-Quaternary main top controls the Lower Pliocene–Quaternary heat field intensity. It has resulted in the fact that the most heat activity is manifested within the areas where Lower Pliocene foot occurs at relatively no great depth.

The distribution of temperature patterns in Mesozoic deposits of the Evlakh-Agdjabedi trough is clearly evidenced by the Paleogene-Miocene clay series' critical thermal insulating role. In such a way, in the Sovetlyar area at a depth of 3800 m, the temperature in the Upper Cretaceous deposits reaches 121 °C, increasing to 135 °C at 4500 m depth. We believe that the significant factors limiting a heat transfer intensity upward the section and decrease in the quantity of heat transmitted from the Paleogene-Miocene complex to the upper intervals are (a) seal failure and fluid saturation due to the intensive hydrocarbon formation and (b) endothermic metamorphic process or dehydration of containing swelling modifications of clay minerals accompanied by partial deep heat discharge that leads to reduce in its flux to the Earth's surface. The dehydration processes are widely developed in Paleogene-Miocene deposits within the SCB, which is proved by the results of isotopic-geochemical investigations of mud volcanic water fluids that will be described in detail in the following chapters.

A study of temperature distribution patterns in some anticlinal structures shows that their crests are much warmer than arch limbs (Eppelbaum et al., 2014; Mekhtiyev et al., 1960). In all cases, the geothermal gradient on the limbs appeared to be lower than in their periphery. Each field reflects the local temperature maximum on a plan: a positive geothermal gradient. This feature is most distinctly manifested in the same horizontal sections within some fields where several fractures, mud volcanoes, diapirism, and other discontinuities complicate crystal parts. For instance, within the Balakhani-Sabunchi-Ramani field, the temperature difference at the same depths in the Bog-Boga mud volcano (crest line) region and the depression area is estimated to be 4–5 °C. Within the Lokbatan, Zikh, Gum Island, and Bibi-Heybat fields in the region of the same name volcanoes, the temperature is 5–10 °C higher than on the limbs and periclines. In the Neft Dashlari field, apart from the temperature anomaly marked around the mud volcano area, the temperature values estimated close to the tectonic disturbance areas are 6–8 °C higher than those in more tranquil structural zones. Within the Pirallakhi and Sulutepe fields, the increased temperature zones are also passed over the disjunctive areas with a difference of more than 18 °C (Mekhtiyev et al., 1968, 1973; Rachinsky & Kuliev, 1984). Thus, in the formation of temperature fields in local uplifts, a vital role plays both conductive and convective components of a deep heat flow formed by migratory thermal fluids from the SCB deep zones.

The fact that all the SCB oil fields are confined to the local positive heat anomalies is one of the convincing arguments favoring a fluid-dynamic conception of hydrocarbon deposit formation in the Lower Pliocene productive and redbed series.

The most important feature of the processes in the bowels is the heat flow density. The quantity of heat energy Q passing through a unit of the Earth's surface area in time unit and expressed mathematically by the following formula:

$$Q = \lambda \text{ grad } T,$$

where λ is the thermal conduction medium depending on the lithological composition of the section deposits and on the reservoir temperature, pressure, and humidity, grad T is the geothermal gradient.

From the functional relation of heat flow with geothermal gradient and hence with the section temperature characteristics, it is concluded that very low heat flow values characterize the South Caspian basin. According to Aliyev S. & Aliyev A. (1995), Lebedev & Tomara (1981), heat flow values in this region range widely from less than 20 mW/m^2 up to 120 mW/m^2, creating a specific mosaic pattern. The following special features of heat flow distribution are distinguished: low heat flow values to 30 mW/m^2, are typical to all the western SCB (Fig. 8.7). Average values from 30–40 to 50–70 mW/m^2 are predominant in the Lower Kur depression on the Absheron Peninsula and adjacent water area of the Absheron Archipelago (Fig. 8.7).

Most parts of the Baku Archipelago and deep-sea South Caspian are characterized by relatively low heat flow values from 20 to 50–60 mW/m^2 (Fig. 8.7). Against this background, the areas with increased heat flow values (from 50–100 to 400 mW/m^2) are confined to the western and northwestern South Caspian periphery and locally to the narrow linear zones characterized by abnormal geological conditions caused by the influence of faults and confined mud volcanoes. The fact that most of the increased heat flow

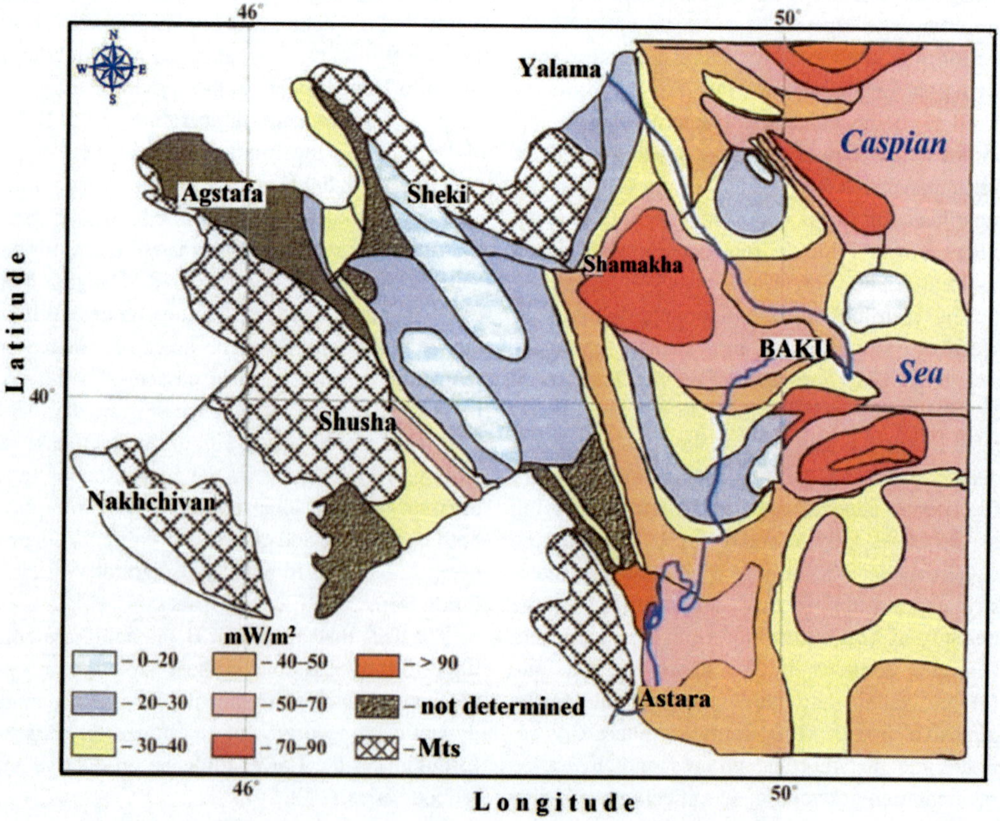

Fig. 8.7 A schematic map showing the heat flow distribution in the western South Caspian basin

values are marked along the northwestern South Caspian Basin may be explained by up-to-date geological processes (Eppelbaum et al., 2014). This region has a high sedimentation rate, intensive fluid generation, mud volcanism, increased seismic activity, and related large-scale displacement of sedimentary material up to 300 m (Lebedev et al., 1973; Lebedev, 1994).

It is noted that the zones having low heat flow values connected with eastern and southeastern South Caspian have well coincided with the tectonically stable region where mainly tranquil brachy-anticlines are developed. The zones of high heat flows correspond to the regions of large oil/gas fields and mud volcano locations.

8.2 Geobaric Characteristic Under Conditions of the Anomalous High-Pressure Formation in the South Caspian Basin

The most important feature of the South Caspian basin is that its section contains intervals with anomalous high pores and reservoir pressures. In some cases, these values may be much higher than hydrostatic pressure, reaching the lithologic ones. For example, according to Khalilov and Imanov (1979) a blowout of mud 2.2 t/m^2 in density from the well at a depth of 3600 m has been marked in the Khara-Zirya field. Assuming the volume weight equal to 2.3 t/m^2 we will obtain the lithologic (geostatistic) pressure as P_g = 3600 m × 2.3 t/m^2 = 8280 t/m^2 and reservoir pressure as P_f = 3600 m × 2.2 t/m^2 = 7920 t/m^2. Hence, in this case, the effective stress will come to P_f/P_g = 0.96, indicating that the fluid enclosing system is on the boundary or in hydroexplosion state. An abnormality coefficient ($C_{AHFP,AHPP}$ = P_f/Phydrostatic head pressure) comes to 2.2. The same is typical for other SCB fields (Djanub, Shakh-Deniz, Kyursyangya). Table 8.3. shows the interval of abnormal ratio changes in virgin pressure for the SCB fields. It is well seen that abnormal maximum and minimum values are mostly typical to the Baku Archipelago and Lower Kur depression fields. Figure 8.8 displays a distribution of virgin pressure in the SCB fields.

An interesting feature in these gradients' distribution has been revealed within the Baku Archipelago fields. So, in the VII horizon of the Sangachal-Deniz, Duvanny-Deniz, and Khara-Zirya fields, the highest reservoir pressure gradient (RPG) areas are confined to the fractured zones delimiting the structure into separate tectonic blocks (Fig. 8.9). It is also noticeable that anomalous high reservoir pressure areas appear localized within the same zones (Fig. 8.10).

It confirms splendidly the active fluid distributive role of these dislocations in the oil pool formation and the mechanism of traps filling at the expense of vertical fluidal streams.

Regionally, it is well outlined an increase of reservoir pressure gradients from the Absheron Peninsula and Absheron Archipelago in the southwesterly, southerly, and southeasterly directions towards the Baku Archipelago, Lower Kur depression, and the central SCB (Fig. 8.11). The data mentioned above indicates that a depth decrease of the abnormally high reservoir pressure (AHRP) zone occurrence in the SCB is to be expected.

As shown by pore pressure calculation data based on the field-geophysical investigation of wells and their correlation with actual gradients of mud pressure in the Bulla-Deniz area (Fig. 8.12), the upper line of the AHRP zone begins from 500 to 650 m depth independently of the investigated wells located on the structure. Stratigraphically, this 600–6230 m thick depth interval includes the Lower Pleistocene (Old Caspian folds and Absheron suite), Upper Pliocene (Akchagylian suite), and in some wells, the Lower Pliocene roof—Productive Series (PS) (Yusufzade et al., 1976; Khalilov & Imanov, 1979) (Fig. 8.12). In this interval, pore pressure gradients are high and estimated to be 23 MPa/km.

The second sharp increase in pressure along the columnar section in the SCB has been marked at a depth below 5 km (Fig. 8.13). This is the most remarkable zone where AHRP correlates with the geostatic pressure (Buryakovsky et al., 1982; Khalilov & Imanov, 1979).

Table 8.3 The distribution of reservoir pressure abnormality coefficients in the SCB fields

Area	Interval of depths, m		Anomalous coefficient	
	Since	Till	Since	Till
Absheron Archipelago				
Gum-Deniz	1938	3466	0.96	1.02
Bakhar	4044	4920	0.95	1.26
Janub	814	4292	1.092	2.32
Palchig Pilpilyasi	571	1774	1.14	1.716
Chilov	564	1487	0.44	0.92
Baku Archipelago				
Sangachal-Deniz	2250	5600	1.196	1.414
Duvanny-Deniz	2100	5113	1.037	1.67
Khara-Zirya	3337	5850	1.031	2.2
Bulla-Deniz	4300	6600	1.029	1.574
Alat-Deniz	2900	3927	1.16	1.7
Baligli	4090	4377	1.196	1.414
Garasu	4860	5000	1.38	1.41
Sangi-Mugan	4000	4940	1.127	1.45
Shakh-Deniz		5068	1.471	2.026
Lower Kur depression				
Kurovdag	373	3374	0.87	1.735
Garabagli	2552.5	3820	0.66	1.39
Kursangi	2657	4550	1.125	2.31
Mishovdag	891	1626	1.06	1.379
Galmaz	1115	2399	0.97	1.260
Kalamadin	1010	1475	0.63	1.696
Pirsagat	1090	3890	0.949	1.669

The mechanism of AHRP formation in the SCB has yet to be appropriately studied and remains a disputable problem. Thus, the totality of previous investigation results on fluid pressure distribution within the SCB based on geophysical testing of wells and the measurements carried out down to a depth of 7 km made it possible to determine two major AHRP zones within the Baku Archipelago. The upper boundary of the first zone occurs in the interval of 500–1200 m. Approximately 4 km below the first zone, the pressure gradients appear to be stable and high. A new, more marked AHRP zone begins approximately at a depth of 5 km (Fig. 8.12). The most important works in this field were carried out during the period of the seventies to the beginning of the 1990s (Buryakovsky et al., 1982, 1986, 1991; Khalilov & Imanov, 1979; Yusufzade et al., 1976).

It is suggested that the abnormal pressure in sedimentary series may have arisen under the following conditions: (1) as a result of a decrease in a system volume including gas, oil, and water resulting from the rock consolidation and fluid embarrassed withdrawal; (2) when fluids appear to be increased in volume resulted from (a) phase transformations in the fields depended on temperature rising during submergence; (b) hydrocarbon generation; (c) clayey minerals degradation accompanied by a release of

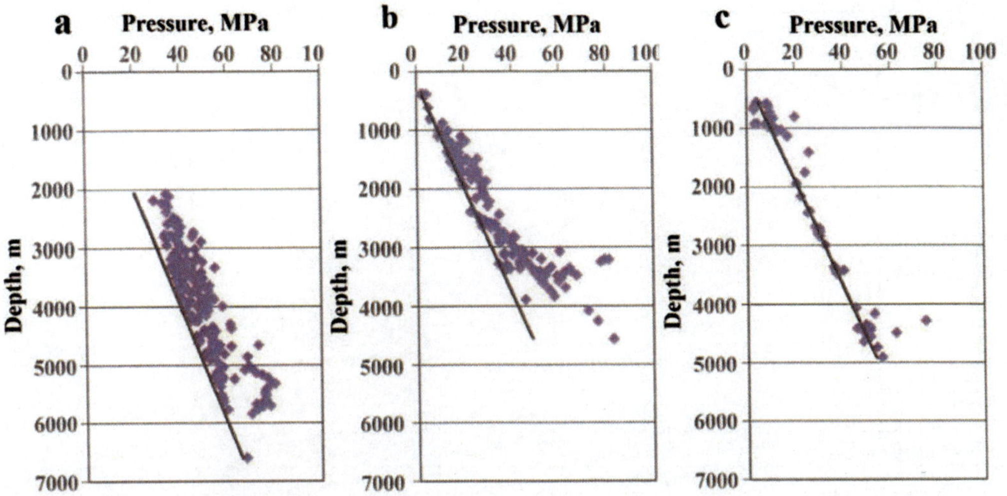

Fig. 8.8 The distribution of the virgin pressure in the SCB fields (drawn line indicates the arbitrary static pressure)

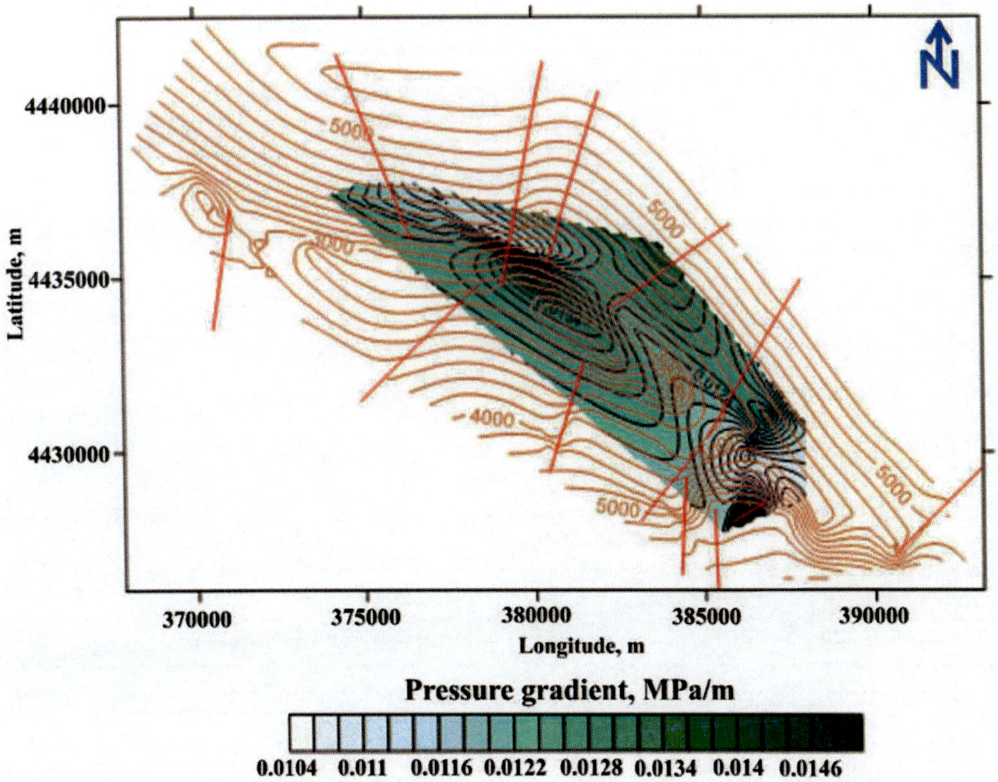

Fig. 8.9 A schematic map showing the distribution of the reservoir pressure gradients in the VII horizon of the Sangachal-Deniz—Duvanny-deniz—Khara-Zirya fields

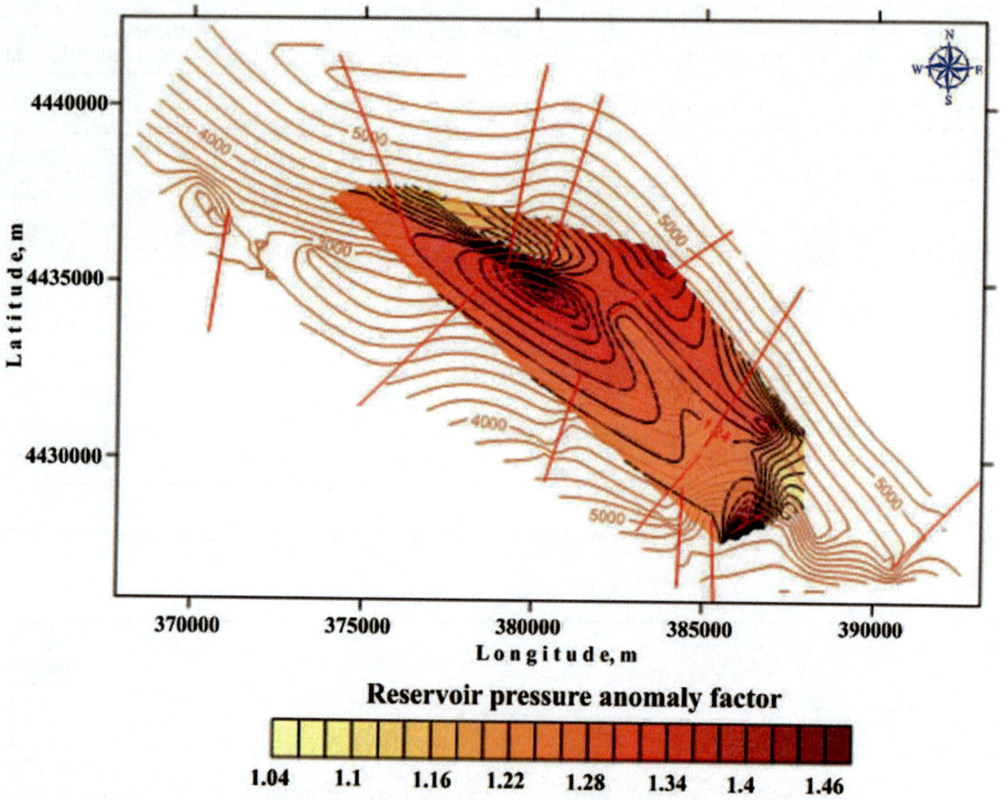

Fig. 8.10 A schematic map showing the distribution of the reservoir pressure abnormality coefficients in the VII horizon of the Sangachal-Deniz—Duvanny-Deniz—Khara-Zirya fields

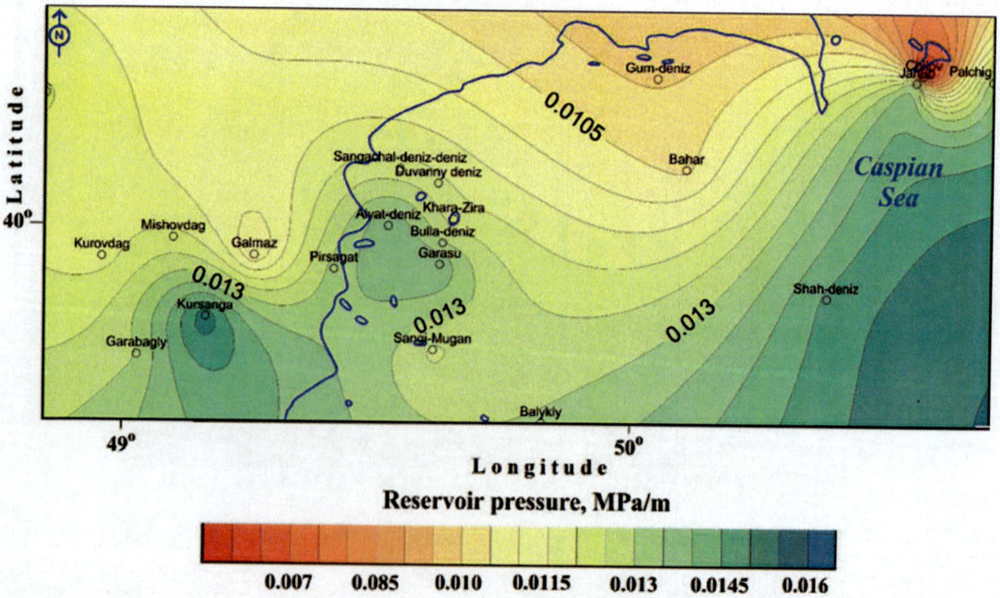

Fig. 8.11 The distribution of the reservoir pressure gradients in the SCB fields

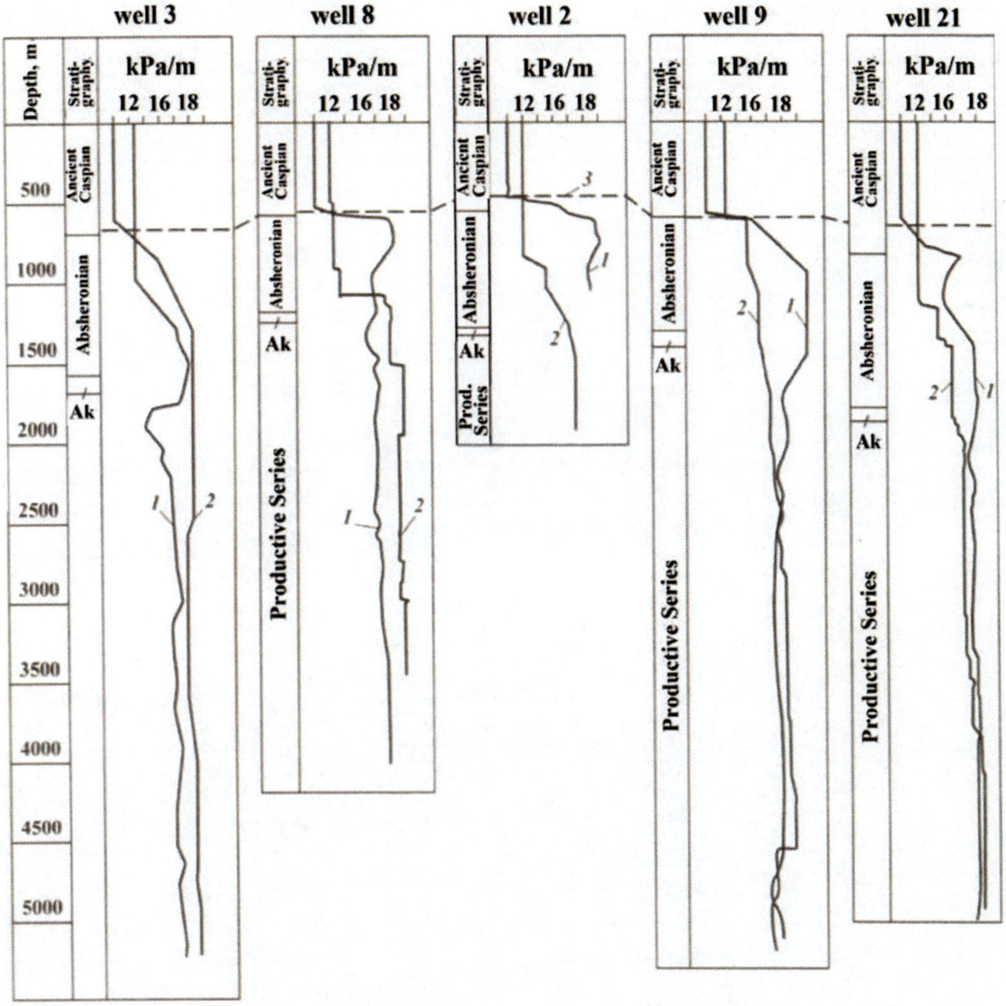

Fig. 8.12 Change of the pressure gradients along the section in the Bulla-Deniz-1 field. (1) Calculated gradient, (2) measured gradient (by drilling mud density), (3) upper border of the overpressure (Yusufzade et al., 1976)

significant volumes of combined and interlayer waters into the free phase; (d) high-head fluids migration from the high-pressure zones to the low-pressure intervals.

What kind of mechanism is prevalent in the abnormal fluid pressure generation in the SCB? Most recent workers believe that abnormal pressure generation relates to a decrease in rock volume. This conclusion is mainly based on two compaction hypotheses: gravitational compaction of clays and tectonic movements. There is another point of view, too, according to which a significant factor in the reservoir excess pressure in the SCB at the beginning of an intensive generation of hydrocarbons, especially gas (in the oil and gas windows) (Feyzullayev & Lerche, 2009; Shykhaliyev et al., 2010).

As mentioned above, pressure gradients estimated at the fields located in the SCB Productive Series tend to be increased from the Absheron Peninsula in southerly and southwesterly directions (Fig. 8.11). Buryakovsky et al. (1986) have served this factor as a basis to separate the three zones shown in Fig. 8.14 and Table 8.4. It agrees

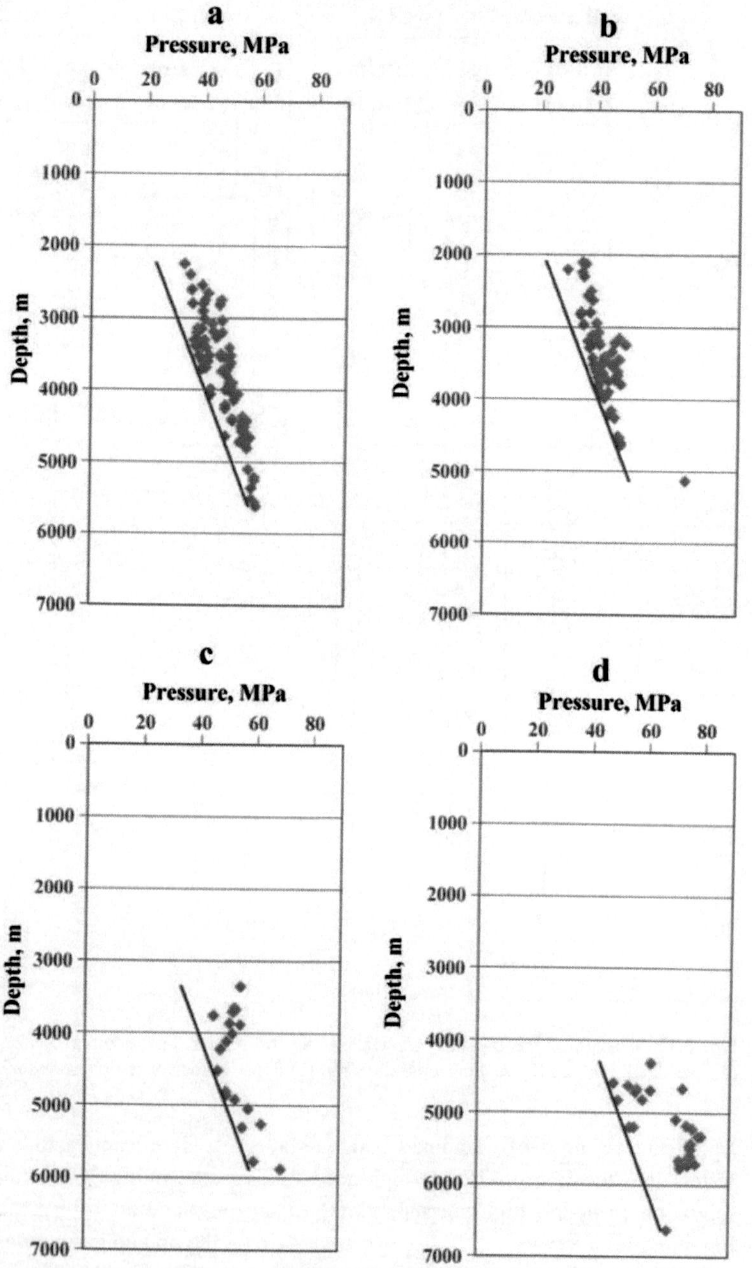

Fig. 8.13 The distribution of reservoir pressures in the Baku Archipelago fields

with an increase in the amount of the PS clayey rocks and the thickness of clay horizons in the PS section.

Table 8.4 illustrates a distinct connection between fluidal pressure abnormality and clayness in the sedimentary section. At the same time, the mentioned relationship reflects different dynamics in the compaction processes of the rocks and, hence, fluid withdrawal dynamics.

From the change in porosity of the rocks with depth in the Baku Archipelago relative to that in

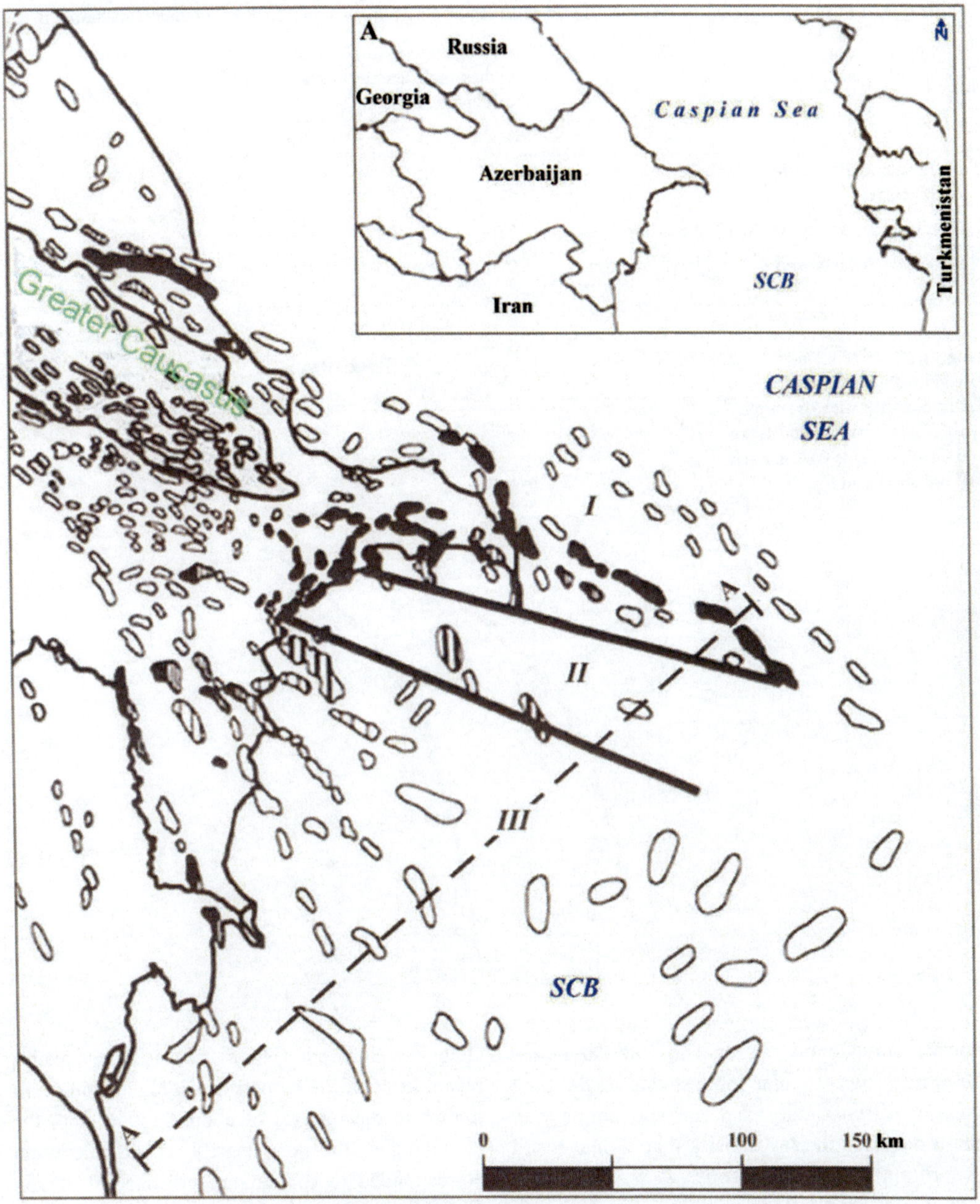

Fig. 8.14 The distribution of the reservoir pressure zones in the SCB. I—Absheron Peninsula and Absheron Archipelago, II—South Absheron water area; III—Baku Archipelago and position of the A-A regional profile (after Buryakovsky et al., 1986)

the Absheron Archipelago characterized by moderate fluid pressure, it becomes evident that AHRP decelerates the consolidation of the rock in the Baku Archipelago (Fig. 8.15).

It is also confirmed by the density variation curve of clayey rocks versus depth in the Bulla-Deniz field (Fig. 8.16a). Figure 8.15 shows that this curve is sharply distinguished from the

Table 8.4 The spatial relationship between the thicknesses of clay horizons and pressure gradients in the SCB

Zones	Average power (m) of shale bodies in different intervals of depths, km				Barometric gradient, MPa/km
	1–2	2–3	3–4	4–5	
I—Apsheron Peninsula and Absheron Archipelago	50	40	30	20	13.5
II—Offshore area of the South Absheron	750	235	185	150	16.3
III—Baku Archipelago	900	725	460	350	18.0

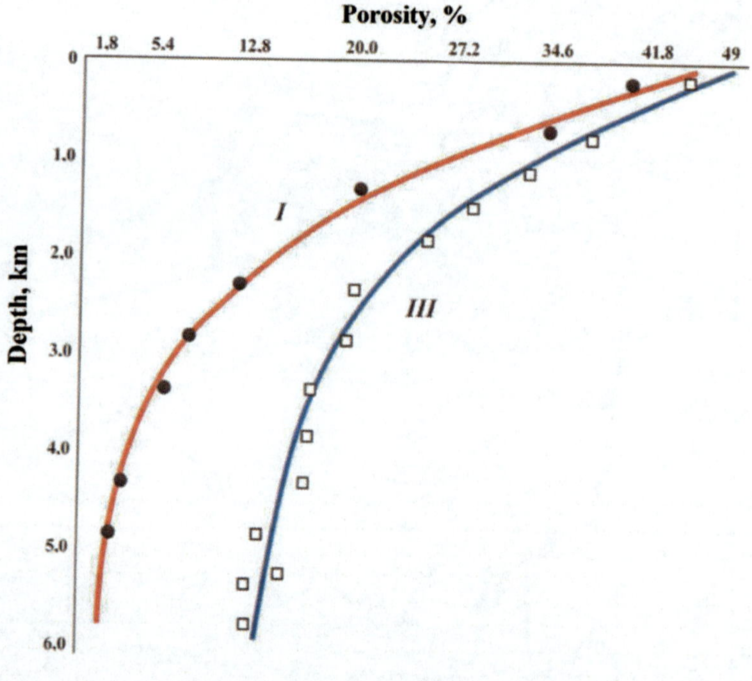

Fig. 8.15 A schematic diagram showing variations in rock porosity with the depth in the Absheron (I) and Baku (III) archipelagos (zones I and III are shown in Fig. 8.14)

normal compaction curve. The interval below 5000 m is of particular interest due to the pore pressure discontinuity and, on the contrary, a sharp decrease in clay density that should result in forming a seal failure zone.

Suppose gravitation compaction is a significant factor in forming the AHRP upper interval. In that case, it is unacceptable for the lower zone of ultrahigh abnormal reservoir and pore pressures. Due to a buffering effect in the rock matrix, the overlying deposit load may not create pressure exceeding lithostatic or equal pressures.

From the comparison of density and pore pressure curves of clays (Fig. 8.16a) and temperature change with depth (Fig. 8.16b), it is revealed a coincidence of the second-density inversion (minimum) in the interval of about 3500 m with the beginning of the pore density second maximum corresponded to the temperature interval of 75–80 °C. It is known that this interval is the beginning of the transformation of clay minerals of the smectite group, mainly montmorillonite, when essential volumes of constitutional water appear to be escaped from their lattice and the interpacket space of clay minerals with a corresponding increase in pore pressure in their matrix.

At first, the generation of abnormal pressure in clay rocks increases progressively during their

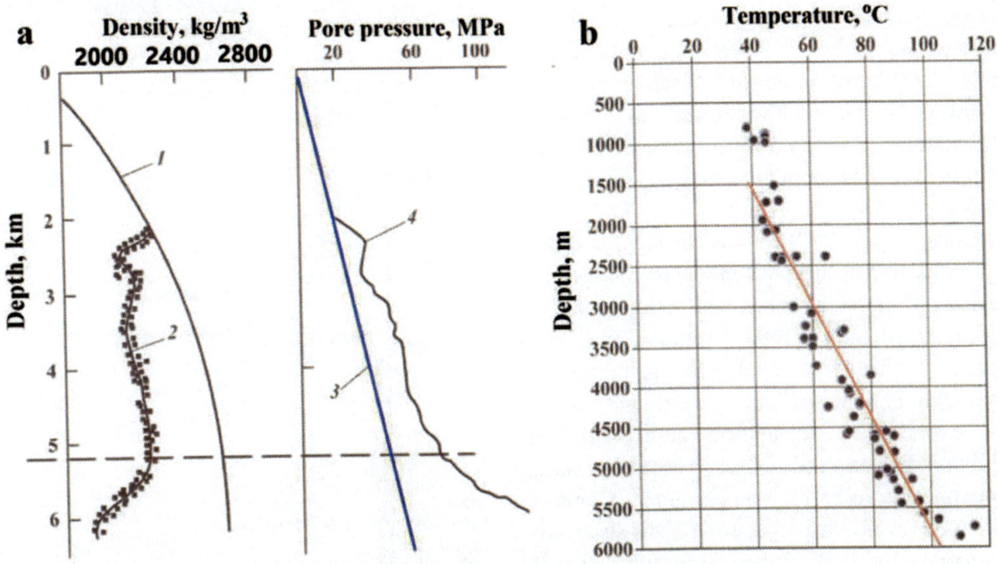

Fig. 8.16 **a** Schematic diagram showing variations in the clay density with depth in the Bulla-Deniz field. (1) and (2) are the curves of the normal and actual compaction, respectively, and (3) and (4) are the hydrostatic and actual compaction curves, respectively. **b** Schematic diagram showing the temperature change with the depth in the Bulla-Deniz field (both **a** and **b** after Feyzullayev & Lerche, 2009)

submergence, reaching its maximum at a temperature of 105–110 °C. Then, as montmorillonite transformation to hydromica appears to be near completion, this process gradually dies down between 150 and 170 °C. Under the mentioned conditions, swelling components of clay rocks are transformed into unswelling ones, leading to the release of 10–15% water of consolidated sediments' total volume (e.g., Blokh, 1977; Burst, 1969; Kartsev & Vagin, 1973; Kholodov, 1983; Magara, 1982; Perry & Hower, 1972; Powers, 1967). A sharp inflection of the density curve in 5500 m with negative density inversion (Fig. 8.16a), points to the fact that the pore pressure in the temperature interval of about 100 °C is highly increased (Fig. 8.16b). This tendency continues multi-directionally down to a depth of 6100 m with increases in temperature up to 115 °C that point to a significant phase in the dehydration of clays and formation of "regenerated" waters. The process of clay unsealing at the expense of intrapore space saturation by surplus water volume may be accompanied by increasing the clay's pore space up to 30–40% (Bro, 1980; Kartsev & Vagin, 1973). The most important feature is that interlayer water has a higher density than free water. Therefore, the removal of interlayer water is accompanied by its density inversion and spasmodic growth of its volume that causes the same sharp growth of fluidal pressure in pore space.

From results of the detailed studies of the SCB Cenozoic deposits, it is concluded that relatively recent Lower Pliocene deposits of the Lower Kur depression and Baku Archipelago are characterized by the absence of visible catagenetic transformations despite great depths of occurrence. At depths exceeding 6000 m, these deposits, without appreciable catagenetic transformation, contain such highly dispersed swelling minerals as montmorillonite (Kheirov, 1987). Our investigations show that the oil–gas-forming zone is confined to this deep temperature interval in the Baku Archipelago. The clay series has yet to be realized completely and preserves its potential to generate dehydration water.

A significant feature of the oil–gas-forming process is an increase in a matter volume that is shown from a simple calculation: oil having a density of 600–700 kg/m³ and hydrocarbon gases—100 to 200 kg/m^3 (under reservoir conditions) are formed from the organic matter of 1400–1500 kg/m^3 in density. Since a void space is always filled up with water, such an increase in the volume of newly formed product from the OM thermolysis should cause repeated intrapore pressure in clays.

Thus, within the Baku Archipelago, at depths of 5000 m and deeper, the essential part of abnormal pressure generation belongs to the processes of hydrocarbon formation and dehydration. Nonetheless, it must be agreed that the formation of abnormal pressure surficial zones at the expense of biochemical gases and their accumulation in sandy lenses limited by plastic impermeable rocks may be permitted. Commercial reserves of such a gas have been discovered on the Bulgarian Black Sea coast, within the Po River valley (Haly), and in Japan.

It is known that the incidents when marine platforms in the SCB collapsed due to the mentioned gas accumulation. Besides, within the coastal South Caspian (Neftchala area), during an irrigation canal sinking at a depth of 6 m, natural gas accumulation was found with the well-defined methane-carbonic biochemical isotopic marker of δ^{13}C—81.5 to 82.3‰ (Feyzullayev & Lerche, 2009).

According to the mud logging data from the Shakh-Deniz field, highly gas-saturated zones have been marked in the Pleistocene deposits between 700 and 1200 m (Fig. 8.17). Also noted are two large zones of near-surface gas accumulation on the Absheron structure in the interval between the Quaternary and Upper Absheron deposits. The data obtained from the "Chevron" company confirm that significant gas-hydrate accumulations have been discovered in Quaternary deposits.

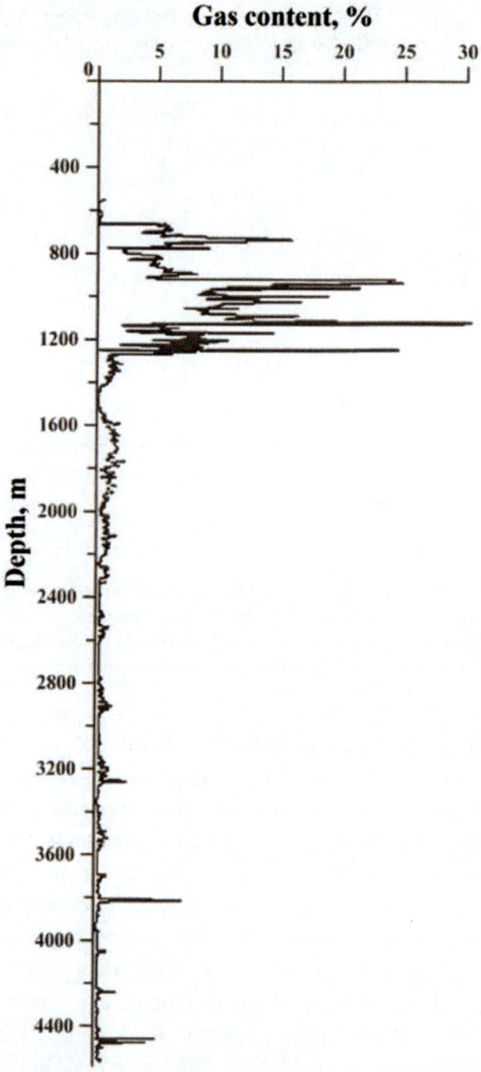

Fig. 8.17 The gas logging analysis in the Shakh-Deniz field with the Pleistocene minimum hydrocarbon gas content (after Feyzullayev & Lerche, 2009)

The first zone prevailing in AHRP may be attributed to the rocks' disequilibrium consolidation due to high sedimentation rates. It is easy to forecast such a zone, and usually, no problem arises during the drilling process there.

The lower AHRP zone is of significant risk. Its intensity prediction plays a vital role from the standpoint of the right choice of mud density. This zone extension is not known at great depths (which have not yet been drilled up). As mentioned above, this zone of AHRP relates to intensive fluid generation. According to the isotopic-geochemical investigation of HC fluids, organic matter maturity, and basin modeling, the highest-pressure gradients, as well as the most dangerous drilling hazards, should relate to the depths below 9–10 km where intensive gas generation is in progress (Figs. 8.18 and 8.19) (Feyzullayev & Lerche, 2009; Shykhaliyev et al., 2010).

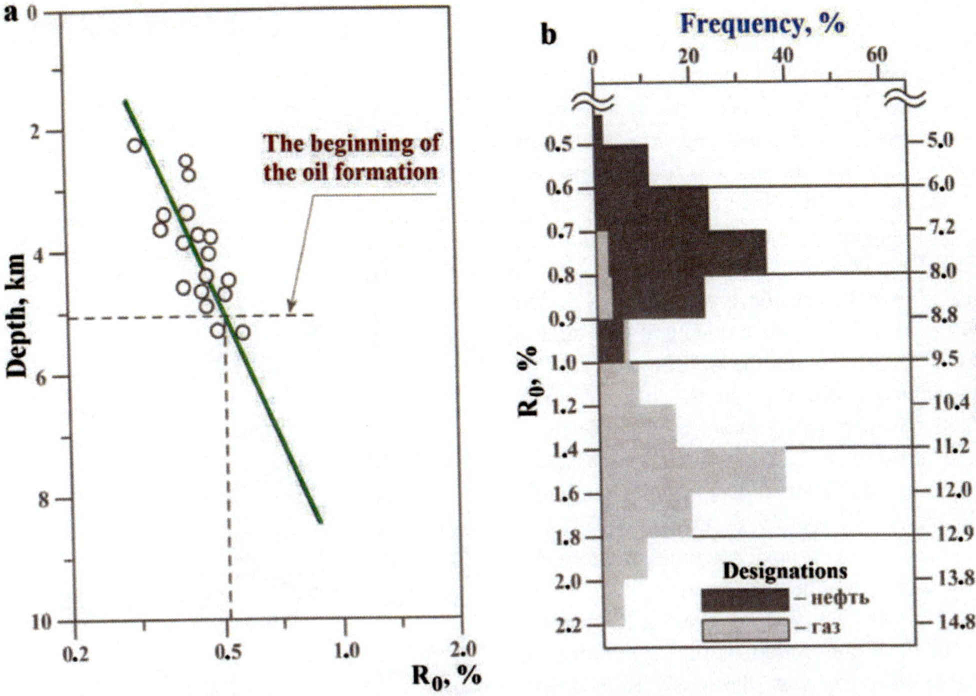

Fig. 8.18 Schematic diagrams show vitrinite reflectance as a function of depth (**a**) and prediction of depth intervals of oil and gas generation in the SCB based on their maturity according to the isotopic-biomarker data (**b**) (after Feyzullayev & Lerche, 2009)

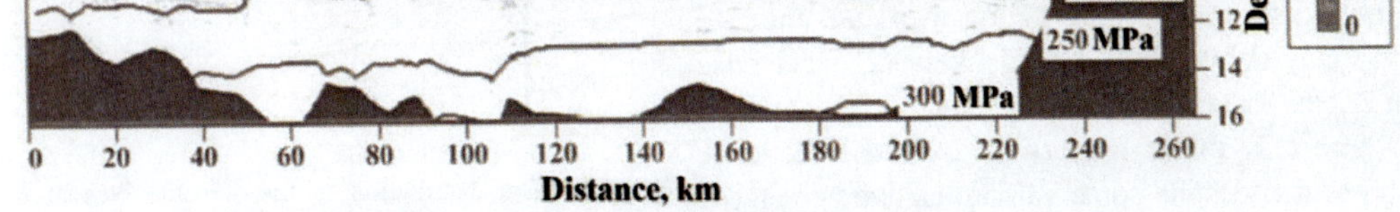

Fig. 8.19 The distribution of the fluid pressure in the SCB submerged area. Data are from the conducted 2-D modeling along with profile A-A (the location of this profile is shown in Fig. 8.14) (after Feyzullayev & Lerche, 2009)

The highest anomalous reservoir pressure values characterize the Baku Archipelago. Initially, it was a relatively closed system where generated HC-fluid discharge was embarrassed by thick clay beds rich in organic matter. As is known from physical chemistry, preventing reaction product discharge from the system leads to a slowdown of the reaction velocity. From this point of view, the successive elevation of temperature caused by the basin subsidence and thermal decomposition of organic matter is reflected in the stable increase in hydrocarbon volumes generated in the massive clayey series. At the same time, the process of organic matter decomposition is delayed by low migration ability. Comparative analyses of this process have defined this deceleration feature in the oil and gas generation process in three different basins (Fang et al., 2007; Helgeson, 1985). It was revealed that the organic matter maturity in the wells located close to each other and with similar thermal histories within the AHRP zones appears to be much lower than in the intervals with normal pressures (Fang et al., 2007). The retardation in the clay dehydration process has also been observed in interval depths with abnormally high pressures (Dódony & Lovas, 2003). The rates of thermochemical reactions and organic matter transformation may sharply be increased if, because of such factors as tectonic, thermal, and chemical convection or uprise of fluidized unstable clayey masses, the newly formed product's removal from the system takes place (for example during mud volcano eruption).

In most parts of the SCB, the clays are approximately 40–50% (and sometimes more) and consist of smectite. At the same time, the SCB is characterized by low-temperature conditions. For instance, in the Baku Archipelago, characterized by the most widely developed AHRP, the temperature gradient ranges between 15 and 18 °C per kilometer. As mentioned above, the temperature required for smectite transformation into illite must be in the interval from 75 to 150 °C (110 °C on average). Intensive clay dehydration should likely be expected at depths exceeding 7 km.

The clay mineralogy study over the section shows that despite some variations, average clay content ranges insignificantly down to a depth of 6200 m (Fig. 8.20). The persistence of this mineral content is caused by the formation of secondary montmorillonite from hydromica.

According to Buryakovsky et al. (1986), maximum depths of montmorillonite distribution in the SCB range from 15 to 17 km. The mud-volcanic outburst analyses have proved that smectite is preserved at the great depth in the SCB.

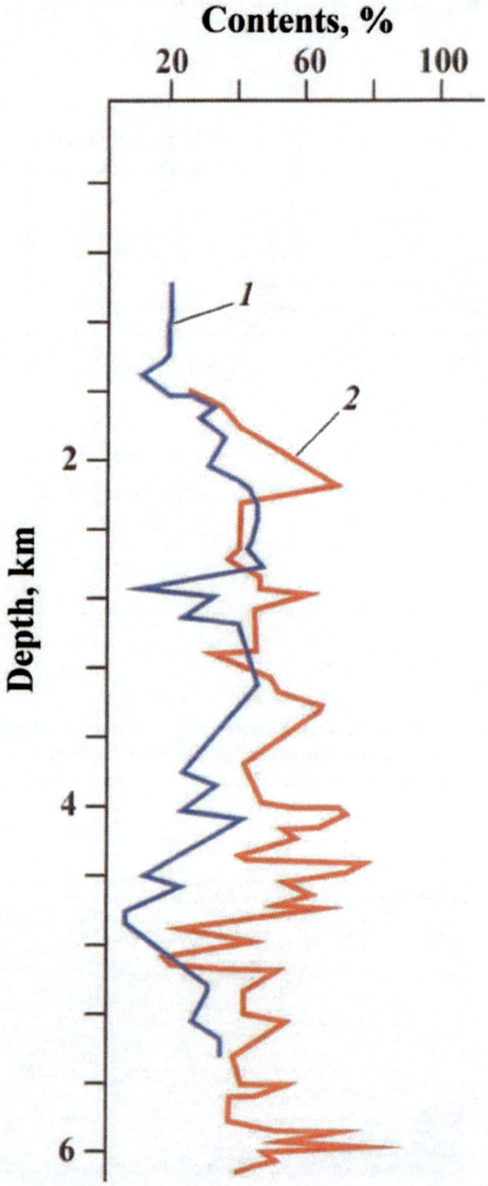

Fig. 8.20 The distribution of the smectite content in clays as a function of depth. (1) Absheron Archipelago, (2) Baku Archipelago (after Kheirov, 1979)

The authors have tried to evaluate the contribution to the AHRP creation at the expense of increasing volume during gas generation. It was required to define a methanee volume obtained from the carbon-containing in a prescribed volume of the rock at the appointed depth. It is known that the amount of generated methane mole is equal to the same of carbon. Therefore, in all reasonings, carbon takes place instead of methane.

According to the Mendeleyev-Clapeyron equation, it is obtained:

$$\frac{P \times V_g}{T} = \frac{M_c}{12} \times R,$$

where P is the pressure at the given depth, V_g is the gas volume at the given depth, T is the temperature at the appointed depth, M_c is the reacted carbon mass, '12' is the carbon atomic mass, and R is the gas constant.

The carbon mass is calculated as follows. A body of taken rock, V_r (cm^3), by the density of ρ (g/cm^3), contains an amount of organic matter equal to K_{org} (%). In turn, organic matter contains a fixed carbon of K_c (%). Besides all, the percentage is shown by mass.

As the earliest investigation established, the K_{org}: K_c relation for the South Caspian has been estimated to be equal to 1.7 (Geodakyan, 1968).

Thus, a finite formula is expressed as:

$$\frac{V_g}{V_r} = \frac{\rho \times K_{org} \times K_c \times R \times T}{12 \times P \times 10}$$

For convenience in the calculation, all the values in this formula should be expressed as follows: the pressure P—in atmospheres; K_{org}—in percents (organic matter content of rock), and K_c—in fractions (the carbon content of organic matter).

The relation obtained from the given formula shows an amount of gas volume (V_g) liberated from the carbon in the taken rock volume (V_r). We carried out the mentioned calculations for the depths of 2000, 4000, and 5860 m within the Bulla-Deniz field and the depths of 2000, 4000, and 5000 m in the Sangachal-Deniz area. The temperature values measured for the mentioned depths have been used as well. Geostatic pressure is estimated to equal the product of average rock density (2.3 g/cm^3) into depth. The organic carbon content of this part of the Lower Pliocene does not exceed 0.4%. The results of these calculations are given in Table 8.5.

An increment in the volume in both areas tends to be markedly decreased with depth. This process at 2000 m is almost two times more than at 5000 and 5850 m depths. Overall, an increment in the volume during gas generation does not exceed 2.65% at a depth of 2000 m and 1.52% at a depth of 5850 m. Thus, gas generation intensity weakens with increasing depth and consequently in the temperature and pressure.

As mentioned above, the temperature interval between 100 and 110 °C within this part of the SCB corresponds to the intensive degradation depth zone. It is known that the liberating volume of interlayer water ranges from 10 to 15% of the rock volume. Even if the liberating water is accepted to be half as much, then an increment in the volume at the expense of dehydration at a depth of 5000–5850 m will be at least 3–5 times more.

The overpressure water thermal expansion creates is considerably less than the other factors, particularly in the SCB characterized by anomalous low temperatures (Feyzullayev & Lerche, 2009). One of the probable and vital factors of anomalous pressure creation is tectonic movement.

Table 8.5 An increment in the volume during gas generation within the SCB

Area	Depth, m	Temperature, °C	V_g/V_r
Sangachal-Deniz	2000	52	0.0265
	4000	83	0.0145
	5000	100	0.0122
Bulla-Deniz	200	45	0.0259
	4000	80	0.0144
	5850	110	0.0152

The modern structure of the SCB located within the mobile Alpine-Himalayan tectonic belt is controlled by the collision of the Arabian and Eurasian plates, which is still going on (Axen et al., 2001; Jackson et al., 2002; Philip et al., 1989). Hence, all the pre-conditions exist for AHRP creation along the zones of lateral tectonic stresses.

From the results of GPS modeling within the SCB (Guliyev et al., 2003), it is revealed that there is a noticeable decrease in horizontal movement velocity from 12–14 mm/y (in the Talysh mountains region) to 0–2 mm/y around the foothills of the Greater Caucasian southeastern termination where mud volcanic and diapiric processes are widely developed.

The vertical movements are also of some importance. All the southeastern Caucasus, South Caspian, and Western Turkmenistan regions have experienced differential vertical movements, which are of different marks in adjacent areas.

On a level of the SCB general subsidence, an essential factor is that uplift occurs, especially in the areas where mud volcanoes are developed. The GSP modeling results show that the rates of uplift on different structures are: + 4.7 mm/y (Sengi-Mugan), + 2.5 mm/y (Lokbatan), + 3.5 mm/y (Zenbil); + 1.2 mm/y (Duzdag) and even + 90 mm/y in Balakhani. At the same time, the subsidence rates in adjacent areas have been determined as − 18 mm/y (Surakhani), − 12 mm/y (Bibi-Heybat), and so on. Moreover, the subsidence rate may come to − 25 mm/y and even − 50 mm/y (Lilienberg, 2002; Pobedonostsev, 1972; Shikhalibeyli, 1996).

The rate of change in reservoir and pore pressures during tectonic movements is much greater than during sedimentation. The relative rate of vertical movements in the SCB may reach 100–140 mm/y. It exceeds, to a great extent, the rate of sedimentation within the Caspian, estimated as 0.2–6 mm/y (Mayev, 1961).

Each of the mentioned geologic processes leads to an increase in pore pressure in clays that, in turn, results in an increase in pressure in reservoirs. Accordingly, commercial hydrocarbon accumulations are often connected with the AHRP zones.

References

Aliyev, S.A., & Aliyev, A.S. (1995). Heat flow in Depression of Azerbaijan. Proceed. of the Conf. "Terrestrial Heat Flow and Geothermal Energy in Asia". New Delhi-Bombay-Calkutta, 295–309.

Axen, G., Lam, P., Crove, M., et al. (2001). Exhumation of the west-central Alborz Mountains, Iran, Caspian subsidence, and collision-related tectonics. *Journal of Geology, 29*(6), 559–562.

Blokh, A. M. (1977). On the universality of the Powers and Burst model of dehydration of sedimentary strata. *Izvestiya of the Acadamy of Sciences USSR, Geological Series*, (6), 119–124. (in Russian).

Bro, E. G. (1980). *The influence of catagenesis on the physical properties of terrigenous rocks and groundwater mineralization* (152p). Nedra. (in Russian).

Burst, J. (1969). Diagenesis of Gulf Cost clays sediments and its possible relation to petroleum migration. *Bulletin of the American Association of Petroleum Geologists, 53*(1), 73–93.

Buryakovsky, L. A., & Dzhevanshir, R. D. (1986). Interrelation and mutual influence of the transformation of clay minerals with thermobaric conditions of the subsurface. *Geochemistry*, (4), 512–521. (in Russian).

Buryakovsky, L. A., Jafarov, I. S., & Dzhevanshir, R. D. (1982). *Prediction of physical properties of oil and gas reservoirs and tires* (200p). Nedra. (in Russian).

Buryakovsky, L. A., Jafarov, I. S., & Kerimov, V. Yu. (1991). *Search and exploration of offshore oil and gas fields* (232p). Nedra. (in Russian).

Dódony, I., & Lovas, G. (2003). Crystalchemistry of clay-minerals around the border of an overpressure zone in one of the deep sub-basins of the southern part of the great Hungarian plain. In *Acta Mineralogica-Petrographica, Abstract Series, Szeged* (Vol. 1, pp. 26–27).

Eppelbaum, L. V., Kutasov, I. M., & Pilchin, A. N. (2014). *Applied geothermics* (751p). Springer.

Fang, H., Zou, H., Gong, Z., et al. (2007). Hierarchies of overpressure retardation of organic matter maturation: Case studies from petroleum basins in China. *AAPG Bulletin, 91*(10), 1467–1498.

Feyzullayev, A. A. (2011). About retardation of physicochemical processes in over pressured sediments, South-Caspian basin, Azerbaijan. *Natural Science, 3* (5), 359–364.

Feyzullayev, A. A. (2014). Spatial heterogeneity of the South Caspian Basin in the context of its oil and gas potential. *Oil and Gas Journal* (Russia), (6), 42–50. (in Russian).

Feyzullayev, A. A., & Lerche, I. (2009). Occurrence and nature of overpressure in the sedimentary section of the South Caspian Basin, Azerbaijan. *Energy Exploration & Exploitation, 27*(5), 345–366.

Gadzhiev, F. M. (2008). Thermobaric conditions, hydrogeological and hydrochemical characteristics of oil and gas complexes. In: *Geology of Azerbaijan* (Vol. b, pp. 336–378). Nafta-Press. (in Russian).

Geodakyan, A. A. (1968). *Geochemical features of oil and gas formation in the South Caspian depression* (152p). Nedra. (in Russian).

Guliyev I.S., Levin L.E., Fedorov, D.L. (2003). *Hydrocarbon Potential of the Caspian Region*. Nafta-Press, Baku, (127p). (in Russian).

Helgeson, H. C. (1985). Adjective—Diffusive/dispersive transport of chemically reacting species in hydrothermal system. *Grant U.S. Department of Energy: DE-FG03–85ER13419*.

Jackson, J., Priestey, K., Allen, M., & Berberian, M. (2002). Active tectonics of the South Caspian Basin. *Geophysical Journal of International, 148*, 214–245.

Kartsev, A. A., & Vagin, S. B. (1973). On the role of "interlayer" waters of clay minerals in the formation of groundwater. *Izvestiya Vuzov, Geology and Exploration*, (3), 64–67. (in Russian).

Khalilov, N. Yu., & Imanov, A. A. (1979). The influence of abnormally high reservoir pressures on drilling performance. *Azerbaijan Oil Industry*, (10), 9–13. (in Russian).

Kheirov, M. B. (1987). Catagenesis of clay deposits and forecasting of abnormally high reservoir pressure before drilling wells. *Azerbaijan Oil Industry*, (4), 5–10. (in Russian).

Kheirov, M. B. (1979). Influence of the depth of sedimentary rocks on the transformation of clay minerals. *Izvestiya Academy of Science, Series: Earth Science*, (8), 144–151. (in Russian).

Kholodov, V. N. (1983). Formation of gas-water solutions in sandy-clay strata of elision basins. In: *Sedimentary basins and their oil and gas potential* (pp. 28–44). Nauka. (in Russian).

Lebedev, L.I., Tomara G.A. (1981). On some features of the distribution of heat flow in the Southern Caspian. In: Geothermometers and paleotemperature gradients. Nauka, Moscow, 156–161 (in Russian).

Lebedev, L. I. (1994). The influence of tectonic factors on the oil and gas content of inland seas. *Geology of Oil and Gas*, (7), 1–6. (in Russian).

Lebedev, L. I., Mayev, E. G., Bordovsky, O.- K. et al. (1973). *The sediments of the Caspian Sea* (119p). Nauka. (in Russian).

Lilienberg, D. A. (2002). The phenomenon of the Caspian Sea and the new tectonic-hydroclimatic concept of fluctuations in the level of inland water bodies. *Izvestiya Earth Science of Academy of Sciences of Azerbaijan*, (3), 3–11. (in Russian).

Magara, K. (1982). *Rock compaction and fluid migration* (295p). Nedra. (in Russian).

Mayev, E. G. (1961). The thickness of modern sediments and sedimentation rates in the Southern Caspian. *Oceanology, 1*(4), 658–664. (in Russian).

Mekhtiyev, Sh.F., Geodekyan, A.A., Rachinsky M.Z. (1973). Geothermal regime of the South Caspian depression. *Soviet Geology, 3* (in Russian).

Mekhtiyev, Sh. F., Mirzajanzade, A. Kh., Aliyev S. A., et al. (1960). Thermal regime of oil and gas fields. *Azerneftneshr, 384*. (in Russian).

Mekhtiyev, Sh. F., Yakubov, A. A., & Rachinsky, M. Z. (1968). Geothermal indicators of migration of hydrocarbons and movement of formation waters in the productive strata of the Apsheronsk oil and gas region. *Geology of Oil and Gas*, (6). (in Russian).

Mukhtarov, A. Sh. (2011). The structure of the thermal field of the sedimentary complex of the South Caspian basin. *Dissertation Doctor of Geological Sciences—Minimum Institute of Geology*, Baku. (in Russian).

Mukhtarov, A. Sh., Tagiev, M. F., & Imamverdiev, R. A. (2003). Models of oil and gas formation and forecast of the phase state of hydrocarbons in the Baku Archipelago. *Izvestiya Academy of Sciences of Azerbaijan, Series Earth Sciences*, (2), 17–25 (in Russian).

Perry, E., & Hower, J. (1972). Late-stage dehydration in deeply buried pelitic sediments. *AAPG Bulletin, 56* (10), 2013–2021.

Philip, H., Cisternas, A., Gvishiani, A., et al. (1989). The Caucasus: An actual example of the initial stages of continental collision. *Tectonophysics, 161*, 1–21.

Pobedonostsev, S. V. (1972). Modern vertical movements of the coasts of the seas washing the European territory of the USSR. *Oceanology, 12*(4), 741–745. (in Russian).

Powers, M. (1967). Fluid-release mechanisms in compacting marine mudrocks and their importance in oil exploration. *AAPG Bulletin, 51*(7), 1240–1254.

Rachinsky, M. Z., & Kuliev, L. G. (1984). On the nature of hydrochemical inversion in the South Caspian depression. *Izvestiya Vuzov. Geology and Exploration*, (12), 25–32. (in Russian).

Rustamov, R. I. (2001). Geothermal regime of the Southern Caspian. *Azerbaijan Oil Industry*, (4–5), 28–32. (in Russian).

Shikhalibeyli, E. Sh. (1996). *Some problematic issues of geological structure and tectonics of Azerbaijan*. Elm. (in Russian).

Shykhaliyev, Y. A., Feyzullayev, A. A., & Lerche, I. (2010). Pre-drill overpressure prediction in the South Caspian Basin using seismic data. *Energy Exploration & Exploitation, 28*(5), 397–410.

Yusufzade, H. B., Kasumov, K. A., Alexandrov, B. A., & Dragunov, E. N. (1976). Studying and prognosing of abnormal formation pressure based on logging geophysics data. *Azerbaijan Oil Industry*, (5), 1–8.

9 Oil and Gas Content of the Productive Series and Analysis of Geological-Prospecting Efficiency

9.1 Productive Red Series: Some Characteristics

The main oil and gas bearing complex in the South Caspian Basin (SCB) is the Productive Red Series of the Lower Pliocene age, with which the most important oil and gas fields in Azerbaijan and Southwestern Turkmenistan are connected. This complex is characterized by a higher specific density of the explored (proved) reserves and prospective and expected oil and gas resources. The region of this series' areal extent, with its proven HC potential resources, is related to the high-class perspective area.

Within the SCB, the total local structures of the productive red beds series proved by drilling and geological-geophysical data have been estimated to be 531 uplifts, including 274 at sea and 257—on land. Of those, 206 have been prospected (86 at sea and 120—on land) by drilling. During the period of prospecting works carried out in Azerbaijanian and Turkmenian parts of the SCB, there were discovered 763 oil and gas pools joined into 83 fields (Table 9.1), from which 674 or 88% are confined to the productive red beds series. Besides, more than 70% (487) of oil and gas pools discovered in the Lower Pliocene deposits are confined to the Productive Series of Azerbaijan. Of those discovered in the SCB, 411, or 53.8% appeared to be oil pools, 158, or 20.7%—were oil and gas pools, 142, or 18.6%—were gas condensate, and 52, or 6.9%—were oil and gas—condensate pools (Aliyev, 2004).

9.2 Productive Series Efficience Estimation

The currently estimated exploration and efficiency coefficients by sea are 0.314 and 0.477, respectively. By land, those are 0.467 and 0.600, 0.388 and 0.548 all over the basin. The exploration coefficient estimated for the Caspian water area of Azerbaijan is 0.456, and for the Turkmenian part of the Caspian Sea, it is 0.173 (Aliyev, 2004).

For a long time, prospecting-exploratory drilling in Azerbaijan and southwestern Turkmenistan has mainly been carried out within the slope parts of depressional zones where many oil fields were discovered and put into development. This fact served as a reason to recognize the SCB as a mainly oil-bearing basin. However, from the beginning of the sixties of the last century, several gas condensate and oil–gas condensate fields within the SCB were discovered (Shakh-Deniz, Bulla-Deniz, Bakhar, 8 March, Umid, Zirya, and Southern) (Fig. 9.1). Besides, such factors as a low-governed increase in gas content toward regional folding subsidence and periodic mud volcanic gas outbursts to the atmosphere have

A. Alizadeh et al., *Pliocene Hydrocarbon Sedimentary Series of Azerbaijan*,
Advances in Oil and Gas Exploration & Production, https://doi.org/10.1007/978-3-031-50438-9_9

Table 9.1 Location of the oil and gas pools in the South Caspian Basin (Aliyev, 2004)

Period	Epoch	Stage	Pool types	large structural element						Quantity of pools in stratigraphic complexes				
				Lower Kur depression	Shamakhi-Gobustan trough	Baku Archipelago	Absheron periclinal trough	Near-Balkhan uplift zone	Gograndag-Chikishlyar step	Total	Including			
											O	OG	GC	OGC
Neogene	Pliocene	Absheronian	O	2			2	18		47	22			
			OG	3				6				9		
			GC					6					6	
			OGC					10						10
		Akchagylian	O	2				17		36	19			
			OG					2				2		
			GC	1				9	4				14	
			OGC					1						1
		Productive Series	O	26		3	268	67		674	364			
			OG	11	3	4	111	10	8			147		
			GC	6	1	3	39	44	29				122	
			OGC			2	10	20	9					41
	Miocene	Chokrak-Diatom	O		1		3			4	4			
			OG									0		
			GC										0	
			OGC											0
Paleogene	Oligocene	Maykop	O		2					2	2			
			OG									0		
			GC										0	
			OGC											0
Quantity of fields				9	5	7	37	17	8	83				
Quantity of pools				51	7	12	433	210	50	763				
Oil (O)				30	3	3	273	102			411			
Oil and gas (OG)				14	3	4	111	18	8			158		
Gas condensate (GC)				7	1	3	39	59	33				142	
Oil and gas cond. (OGC)						2	10	31	9					52

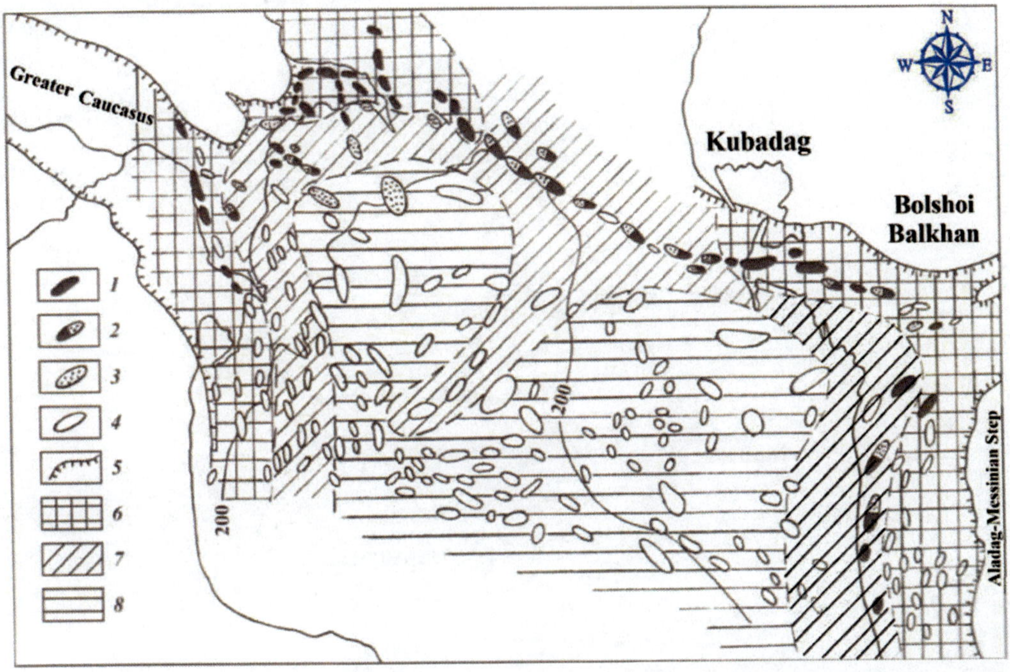

Fig. 9.1 Classification of the oil and gas fields in the South Caspian Basin (after Aliyev, 2004). Deposits: (1) oil, (2) oil and gas, (3) gas condensate, (4) promising structures, (5) the boundary of the deposits of the Productive Red rock formation. Zones, predominantly: (6) oil accumulation, (7) oil and gas accumulation, (8) gas accumulation

changed formerly existent opinion. Now, the SCB may be classified as an oil and gas-bearing basin. In the most subsided traps, the condensate content of the gas phase reaches 0.3–0.35 g/cm^3. In some cases, gas-condensate pools contain pay oil fringes.

9.3 Zonation Scheme of the SCB Region

The current existing oil and gas zonation scheme of the SCB region is given in Fig. 9.2. The productive Red Series section is characterized by clearly manifested oil and zonal gas distribution depending on the depth and thermobaric parameters of the pools. Special features of the distribution of revealed oil and gas geological reserves by depth intervals indicate that oil and gas explored reserves within the SCB predominantly occur at depths from 0 to 3500 m and deeper (Aliyev, 1975; Aliyev et al., 1973).

The hydrocarbon deposits within the SCB relate to the conditions of their formation, reformation, and destruction during the Neogene-Quaternary development history of the basin as with the change in the thermobaric regime of the depths (Alizadeh et al., 2017). The areal

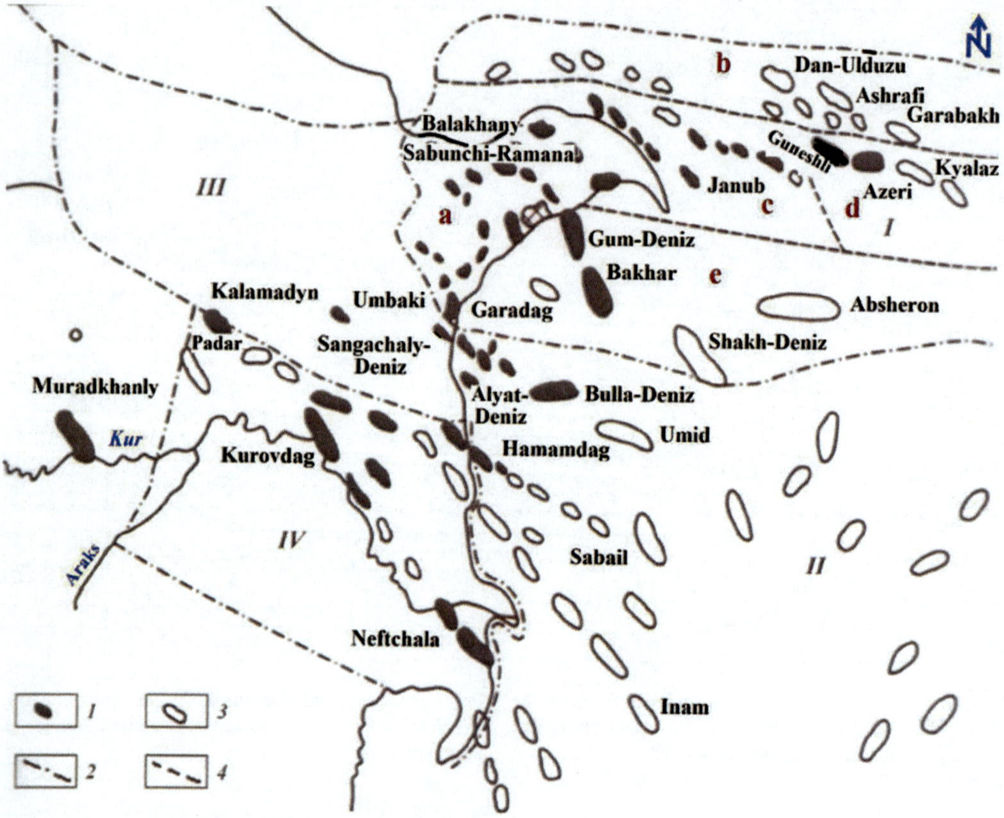

Fig. 9.2 The oil-and-gas bearing regions: I—Absheron, zones of (a) Absheron Peninsula, (b) North Absheron, (c) Absheron Archipelago; (d) South Absheron water area; II—Baku Archipelago; III—Shamakhi-Gobustan region; IV—Lower Kur region. (1) oil-and-gas fields, (2) structures' boundaries, (3) oil and gas bearing regions, (4) zones of oil and gas content

distribution features of initial total HC resources in the region are mainly defined by differences in the gas generation potential of source rocks and trap and reservoir parameters controlling oil and gas accumulation volumes.

Most oil and gas fields in Azerbaijan are confined to the Lower Pliocene deposits. Meanwhile, the outlook for discovery of a new hydrocarbon accumulation, especially in the Caspian deep sea part, is good. According to Guliyev et al. (2003), the Pliocene deposits in Azerbaijan part of the Caspian Sea contain hydrocarbon sums of 15.2 billion t, i.e., 2.5 times more than those on land (5.8 billion t) (Fig. 9.3).

The most important feature of the SCB fields is that they are multilayer. For example, in the Surakhani field located in the Absheron

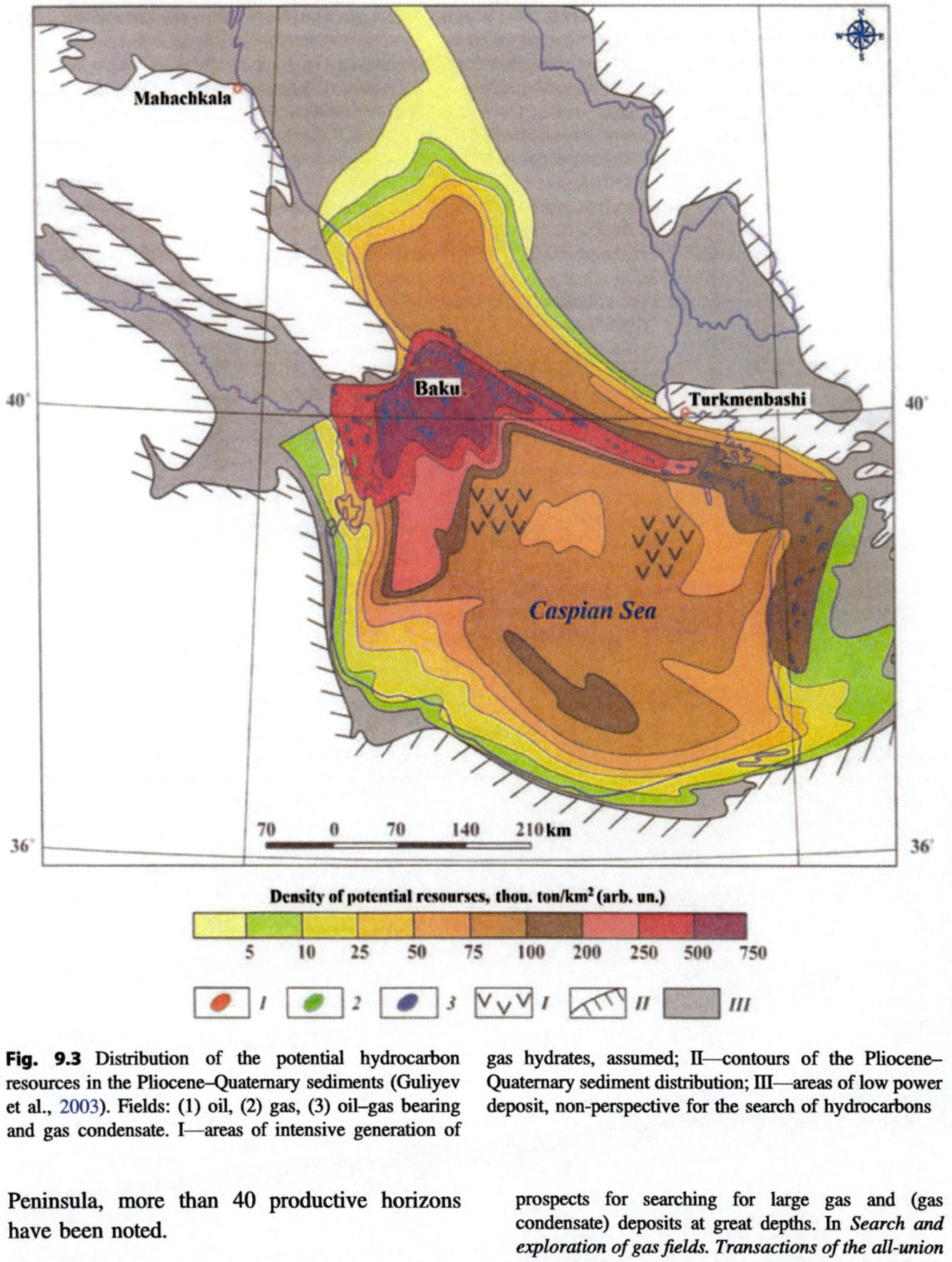

Fig. 9.3 Distribution of the potential hydrocarbon resources in the Pliocene–Quaternary sediments (Guliyev et al., 2003). Fields: (1) oil, (2) gas, (3) oil–gas bearing and gas condensate. I—areas of intensive generation of gas hydrates, assumed; II—contours of the Pliocene–Quaternary sediment distribution; III—areas of low power deposit, non-perspective for the search of hydrocarbons

Peninsula, more than 40 productive horizons have been noted.

References

Aliyev, A. I. (1975). Zoning of oil and gas distribution in the South Caspian Basin in connection with the prospects for searching for large gas and (gas condensate) deposits at great depths. In *Search and exploration of gas fields. Transactions of the all-union gas scientific institutes No. 47/55* (pp. 160–169). Nedra.

Aliyev, A. I. (2004). Conditions for the formation of oil and gas deposits in the South Caspian Basin. *Izvestiya Academy Sciences of Azerbaijan, Earth Sciences*, (4), 13–24. (in Russian).

Aliyev, A. I., et al. (1973). *Recommendations for the direction of searching for large gas (gas condensate) deposits in deep-submerged zones of the South Caspian Basin* (52p). All-Union Gas Scientific Institutes. (in Russian).

Alizadeh, A. A., Guliyev, I. S., Kadirov, F. A., & Eppelbaum, L. V. (2017). *Geosciences in Azerbaijan. Volume II: Economic minerals and applied geophysics* (340p). Springer.

Guliyev, I. S., Levin, L. E., & Fedorov, D. L. (2003). *Hydrocarbon potential of the caspian region* (127p). Nafta-Press. (in Russian).

10 Generation Potential of the Lower Pliocene Deposits and Its Importance (Contribution) to Hydrocarbon Generation in the South Caspian Basin

10.1 The Origin of Hydrocarbons Relating to the Productive Series

One of the fundamental problems of SCB geology and geochemistry is to clear up the origin of hydrocarbon (HC) payable accumulations in the Productive Series. Many scientists (e.g., Agabekov, 1963; Aliyev, 2004; Alizadeh, 1980; Alizadeh et al., 1975; Dadashev et al., 2006) consider that oil in the PS is syngenetic. The other more numerous group (e.g., Abrams & Narimanov, 1997; Alizadeh et al., 2017; Feyzullayev et al., 2001, 2015; Guliyev & Feyzullayev, 1996; Guliyev et al., 2004; Gurgey, 2003; Huseynov, 2003a; Inan et al., 1997; Katz et al., 2000; Wavrek et al., 1996) believes that hydrocarbons in the PS are epigenetic.

The results of the latest investigation of sedimentary rocks in the SCB based on the up-to-date laboratory methods of analyses allow a correct estimate of HC potential in the PS and, hence, its role in forming discovered oil and gas commercial accumulations. This estimation has been carried out based on the pyrolysis of core samples from the wells drilled in on potential structures and fields on land and sea of the SCB.

Selective sampling has been carried out according to their lithological composition (picked samples were of clay intervals in well logs). Many laboratory analyses of organic matter in clay rocks have been done, including optical investigation, pyrolysis, and total organic carbon determination. In its turn, optical data include vitrinite reflectance values (R%). The quantity and quality characteristics of organic matter in the PS rocks based on the results of the pyrolytic investigation are given in Table 10.1 and reflected in Figs. 10.1 and 10.2.

The data in Table 10.1 and Figs. 10.1 and 10.2 indicate that organic matter in the PS deposits is mainly represented by 3-d type kerogen and rarely by that of 2-d type, consisting of reworked vegetable material with trace amounts of amorphous and algal matter.

Therefore, PS deposits are characterized by low oil-yielding properties.

From the oxygen index (01) values shown in Table 10.1, it is inferred that the sedimentation process during the Lower Pliocene occurred under geochemically unfavorable (mostly oxidizing) conditions.

Overall, the data presented in Table 10.2 form an accurate notion of the geochemical properties of the PS rocks. The data shown in this table shows that the Kala bottom suite (KaS) has relatively more favorable hydrocarbon-producing properties than the other PS stratigraphic units (Huseynov, 2003a; Huseynov et al., 2004).

An exciting feature in the spatial distribution of the OM content and type is that within the Kala suite, the highest OM concentration is

A. Alizadeh et al., *Pliocene Hydrocarbon Sedimentary Series of Azerbaijan*,
Advances in Oil and Gas Exploration & Production, https://doi.org/10.1007/978-3-031-50438-9_10

Table 10.1 Statistical data of the pyrolysis in the PS rocks

Parameters	TOC (%)	$S_1 + S_2$ (mgHC/g rock)	S_3 (mgCO_2/g rock)	HI (mgHC/g TOC)	OI (mgCO_2/g TOC)	T_{max} (°C)	Ro(%)
Minimum	0.14	0.16	0.02	17	2	315	0.25
Maximum	2.38	2.28	2.33	318	1031	446	0.48
Average	0.47	0.57	1.02	84	269	389	0.33

typical to the North-Absheron and Absheron banks (1.2–2.38%) while in westerly and southeasterly directions, towards the West-Absheron and Gyuneshli structures, its concentration appears to be decreased to 0.4–0.68%. It is noted that OM quality changes from type II to type III in the same directions (Fig. 10.3).

10.2 Organic Matter Maturity

To realize the hydrocarbon potential of rocks and organic matter transformation into hydrocarbons, adequate temperature conditions are needed. Therefore, a degree of OM maturity is the most critical parameter, which, together with other geological-geochemical parameters, allows obtaining certain information on the depth of the oil–gas generation zone, stratigraphic confinement of source beds, and the condition and direction of HC fluids migration.

The data from the experimental investigation suggest that the up-to-date temperature value of the depths at which oil generation begins is approximately 1000 °C. At the same time, the paleo-temperature index is about 0.55% (Peters & Moldovan, 1993).

Measured vitrinite reflectance and T_{max} show a low level of OM maturity in the Lower Pliocene rocks (lower of the threshold of HC generation onset) (see Table 10.1). As is known from the SCB's present geological structure, the PS roof reaches a depth of oil generation onset only at separate areas of the deep-seated central basin. This event took place during the last stage of the basin's development history. In other words, considering subsidence history and recent occurrence of the PS deposits as their warm-up duration, it may be concluded that temperature conditions for OM transformation into HC have been lacking. Thus, the HC potential study of the PS rocks shows that OM quantity, quality, and maturity level need to follow the conception that the Productive Series possesses oil-and-gas source properties. The PS generation potential is insufficient to form any known or possible hydrocarbon resources.

That is why it is reasonable to assert that oil fields in the PS have been formed due to allochthonous hydrocarbon inflow from the underlying deposits. This conclusion is confirmed by the low-governed nature of spatial distribution and in the vertical section of isotopic-geochemical features in oil composition, thermobaric parameters, and gas factors revealed at the Gyuneshli, Garadagh, Kyurovdag, Kyursyangya, Neftchala, and other fields. All the mentioned features indicate that the HC feeding source is present in the underlying deposits (Feyzullayev, 2013; Feyzullayev et al., 2005; Huseynov, 2003b, 2012).

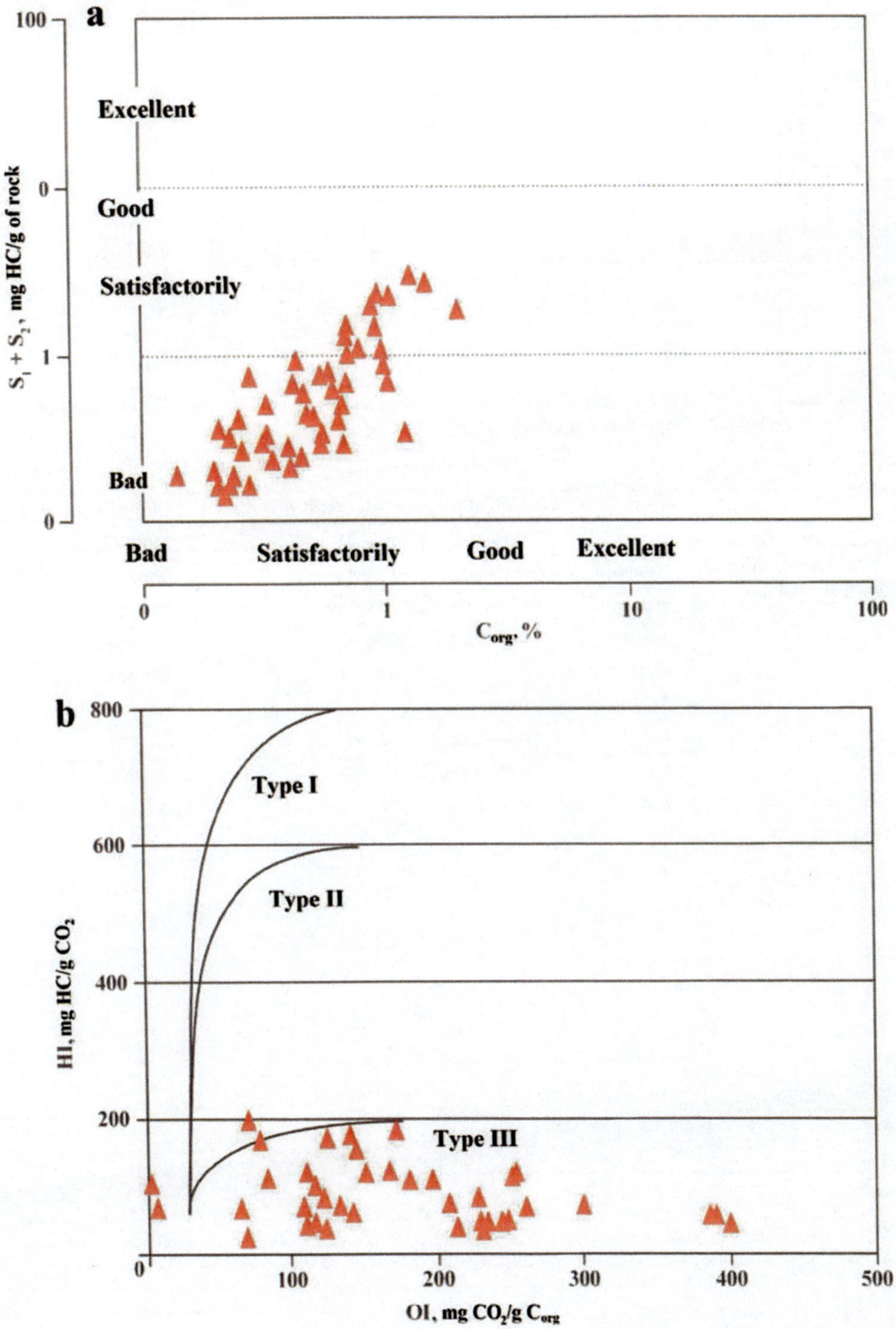

Fig. 10.1 The generalized diagrams showing the organic matter quantitative (**a**) and qualitative (**b**) characteristics of the PS rocks

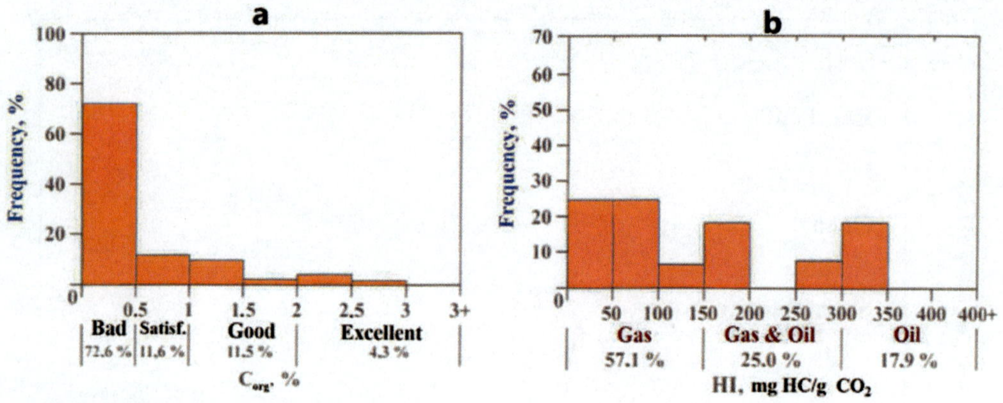

Fig. 10.2 Histograms of the distribution of the total organic carbon and oxygen index values. The classification of OM is from Peters (1986) (after Feyzullayev et al., 2001)

Table 10.2 The quantitative and qualitative characteristics of OM in the Productive Series different stratigraphic units

Suite	TOC (%)	HI, (mgHC/gTOC)	OI ($mgCO_2$/gTOC)
Surakhani	0.16–0.48	27–31	346–1031
Sabunchi	0.19–0.53	17–43	296–609
Balakhani	0.11–1.31	24–169	97–638
Fasila	0.23–0.33	26–48	161–263
SKG	0.32–0.80	25–114	157–451
SKS	0.64	25	258
KS	0.48–0.82	27–67	151–381
UK	0.65–0.81	45–84	112–263
KaS	0.40–2.38	47–318	98–188

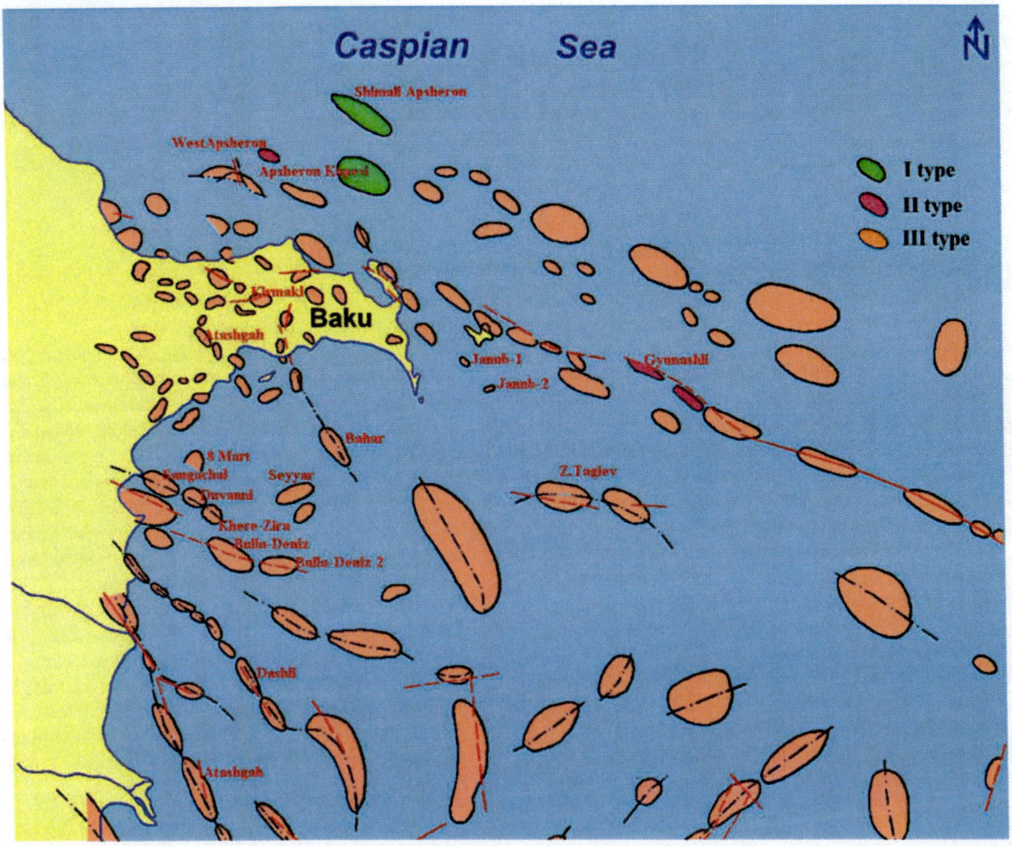

Fig. 10.3 The distribution of the organic material types in the PS Kala suite deposits

References

Abrams, M. A., & Narimanov, A. A. (1997). Geochemical evaluation of hydrocarbons and their potential sources in the western South Caspian depression, Republic of Azerbaijan. *Marine and Petroleum Geology, 14*(4), 451–468.

Agabekov, M. G. (1963). *Geological structure of oil fields of Azerbaijan and their formation* (274p). Azerneshr. (in Russian).

Aliyev, A. I. (2004). Conditions for the formation of oil and gas deposits in the South Caspian Basin. *Izvestiya Academy of Science Azerbaijan, Earth Sciences,* (4), 13–24.

Alizadeh, A. A. (1980). On the so-called "mother strata". In *Proceedings of the Azerbaijan Oil Institute* (172p). Nedra (in Russian).

Alizadeh, A. A., Akhmedov, G. A., Aliyev, G.-M. A. et al. (1975). *Assessment of oil-producing properties of meso-cenozoic deposits of Azerbaijan* (140p). Elm. (in Russian).

Alizadeh, A. A., Guliyev, I. S., Kadirov, F. A., & Eppelbaum, L. V. (2017). *Geosciences in Azerbaijan. Volume II: Economic minerals and applied geophysics* (340p). Springer.

Dadashev, F. H., Mammadova, P. A., & Alakbarov, E. F. (2006). *The Moughan monocline* (p. 183). Nafta-Press Publication.

Feyzullayev, A. A. (2013). Migration pathways of hydrocarbons in the South-Caspian basin. *Geology and Geosciences, 2*(3), 1–6.

Feyzullayev, A. A., Guliyev, I. S., & Tagiyev, M. F. (2001). Source potential of the Mesozoic-Cenozoic rocks in the South Caspian Basin and their role in forming the oil accumulations in the Lower Pliocene reservoirs. *Petroleum Geoscience, 7*(4), 409–417.

Feyzullayev, A., Sadykh-zade, L., & Hasanov, A. (2005). On some aspects of the formation of oil and gas deposits in the Productive stratum of the South Caspian basin. *Azerbaijan Oil Industry*, (4), 13–18.

Feyzullayev, A. A., Tagiyev, M. F., & Lerche, I. (2015). On the origin of hydrocarbons in the main Lower Pliocene reservoirs of the South Caspian Basin,

Azerbaijan. *Energy Exploration & Exploitation, 33*(1), 1–14.

Guliyev, I. S., & Feyzullayev, A. A. (1996). Geochemistry of hydrocarbon seepages in Azerbaijan. In D. Shumacher & M. Abrams (Eds.), *Hydrocarbon migration and its near-surface expression* (Vol. 66, pp. 63–70). AAPG Memoir.

Guliyev, I. S., Huseynov, D. A., & Feyzullayev, A. A. (2004). Geochemical features and fluid sources of mud volcanoes in the South Caspian sedimentary basin in the light of new data on carbon, hydrogen, and oxygen isotopes. *Geochemistry,* (7), 792–800. (in Russian).

Gurgey, K. (2003). Correlation, alteration, and origin of hydrocarbons in the GCA, Bahar, and Gum Adasi fields, western South Caspian Basin: Geochemical and multivariate statistical assessments. *Marine and Petroleum Geology, 20*(10), 1119–1139.

Huseynov, D. A. (2003a). New data on oil source rocks in Pliocene sediments of the South Caspian petroleum system. In *Proceedings of the 1st North Africa/ Mediterranean Petroleum and Geoscience EAGE Conference, Tunis* (pp. 56–60).

Huseynov, D. A. (2003b). Integrated geochemical analysis of oils—Application to reservoir infill on Caspian shelf. In *Proceedings of the EAGE 65th Conference and Exhibition, Stavanger, Norway, on CD.*

Huseynov, D. A. (2012). Application of the oil maturity data to migration and reservoir infill in the Low Kura depression, South Caspian basin. In *Proceedings of the 74th EAGE Conference and Exhibition incorporating SPE EUROPEC, Copenhagen, Denmark* (pp. 1–4).

Huseynov, D. A., Aliyeva, E., Nummedal, D., Guliyev, I., Riley, G., & Friedmann, J. (2004). Oil source rocks in the lower Pliocene deltaic-lacustrine successions in the South Caspian basin. In *Proceedings of the AAPG Hedberg Conference "Sandstone Deposition in Lacustrine Environments: Implications for Exploration and Reservoir Development", May 21–28, Baku, Azerbaijan* (pp. 1–3).

Inan, S., Yalchin, N., Guliyev, I., Feyzullayev, A., & Kuliyev, K. (1997). Deep petroleum occurrences in the Lower Kura depression, South Caspian basin, Azerbaijan. *Marine and Petroleum Geology, 14*(7/8), 731–762.

Katz, K. J., Richards, D., Long, D., & Lawrence, W. (2000). A new look at the components of the petroleum system of the South Caspian basin. *Journal of Petroleum Science and Engineering, 28*, 161–182.

Peters, K. E. (1986). Guidelines for evaluating petroleum source rock using programmed pyrolysis. *AmericanAssociation of Petroleum Geologists, 70*, 318–329.

Peters, K., & Moldovan, J. (1993). *The biomarker guide: Interpreting molecular fossils in petroleum and ancient sediments* (p. 363). Prentice Hall.

Wavrek, D., Collister, J., Curtiss, D., Quick, J., Guliyev, I., & Feyzullayev, A. (1996). Novel application of geochemical inversion to derive generation/expulsion kinetic parameters for the South Caspian petroleum system (Azerbaijan). In *Proceedings of the AAPG/ASPG Research Symposium "Oil and Gas Petroleum Systems in Rapidly Subsiding Basins", 6–9 October, Baku, Azerbaijan* (pp. 1–4).

11 Isotopic-Geochemical Characteristics of Organic Matter and Hydrocarbon Fluids from the SCB Productive Series. Oil-Rock Relationships

The oil fields in the SCB are confined to the reservoir and widely ranged in age from the Upper Cretaceous to the Upper Pliocene–Quaternary. Besides this, the exceptionally thick sedimentary infilling of the basin that has no analog all over the world and the presence of several oil-generating complexes—all that is of significant difficulty to ascertain an actual contribution of the mentioned factors to the formation of hydrocarbon accumulations in the Productive Series, the main oil-bearing object of the Lower Pliocene age. Some well-known researchers (e.g., Alizade et al., 1967, 1985; Bagirzade et al., 1987; Weber, 1978, 1983) believe that oil in the PS is primary, i.e., it is produced by an organic matter of the PS itself (or of its lower division) and then it was literally, laterally-step-like and vertically migrated to the place of an accumulation from the generation areas located in the SCB deep-seated zones. The other group of scientists is devoted to the idea of the secondaries of oil and its migration to the PS reservoir rocks from the underlying Paleogene-Miocene and older deposits along the faults and fractures (e.g., Abrams & Narimanov, 1997; Feyzullayev & Tagiyev, 2008; Feyzullayev et al., 2015; Gorin & Buniyatzade, 1971; Mekhtiyev, 1956, 2010).

Combining investigations, including geochemical studies of organic matter of rocks, oil and gas, can clear up this problem. The chemical, gas-chromatographic, and infrared-spectrometric methods in studies of oils and organic matter of rocks, before being applied in Azerbaijan, allowed us to obtain a particular notion of their geochemical properties. However, these methods appeared ineffective enough to solve this problem because of a need for more informative data for detailed classification according to the OM genetic type, paleoecological, and facies-geochemical formation conditions of oil-producing deposits. We believe this problem may be solved by utilizing up-to-date high-technological methods based on the results of an analysis of variations in stable carbon isotopes and studies of biomarkers that react sensitively to the fluctuation in sedimentation conditions paleoclimate, genetic type of initial organic material, and degree of oil maturity. An impressive analytical material on carbon isotopic composition and oil biomarkers accumulated by the authors to date from practically all fields of Azerbaijan, together with the available geochemical data, made it possible to interpret this problem at a qualitatively new level and obtain specific results described in a number published works (e.g., Feyzullayev, 2019; Feyzullayev et al., 2001, 2005; Guliyev et al., 1999a, 1999b, 2001a, 2001b, 2004, 2005; Huseynov, 2000, 2003; Huseynov & Guliyev, 2004).

A. Alizadeh et al., *Pliocene Hydrocarbon Sedimentary Series of Azerbaijan*,
Advances in Oil and Gas Exploration & Production, https://doi.org/10.1007/978-3-031-50438-9_11

11.1 Isotopic Composition of the OM Carbon in the SCB Mesozoic-Cenozoic Deposits

Detailed investigations of Mesozoic-Cenozoic bituminous extracts have revealed a well-defined distinction in carbon isotopic composition between separate stratigraphic complexes. According to the kerogen's $\delta^{13}C$ distribution, the Meso-Cenozoic sedimentary section is delimited into pre-diatomic and post-diatomic sections. The $\delta^{13}C$ content of the Jurassic kerogen ranges from 27.35 to − 25.27‰ being on an average 26.33‰. The organic matter of the Cretaceous deposits is characterized by almost the same $\delta^{13}C$ values, from − 27.22 to − 24.05‰ on the average − 25.63‰.

The $\delta^{13}C$ content of the kerogen in Paleogene-Lower Miocene deposits varies from 28.24 to − 24.15‰, being on average − 26.48‰. The abovementioned data shows that the highest $\delta^{13}C$ content of organic matter is typical of the Paleogene and Middle to Upper Miocene (diatomic) deposits (Fig. 11.1).

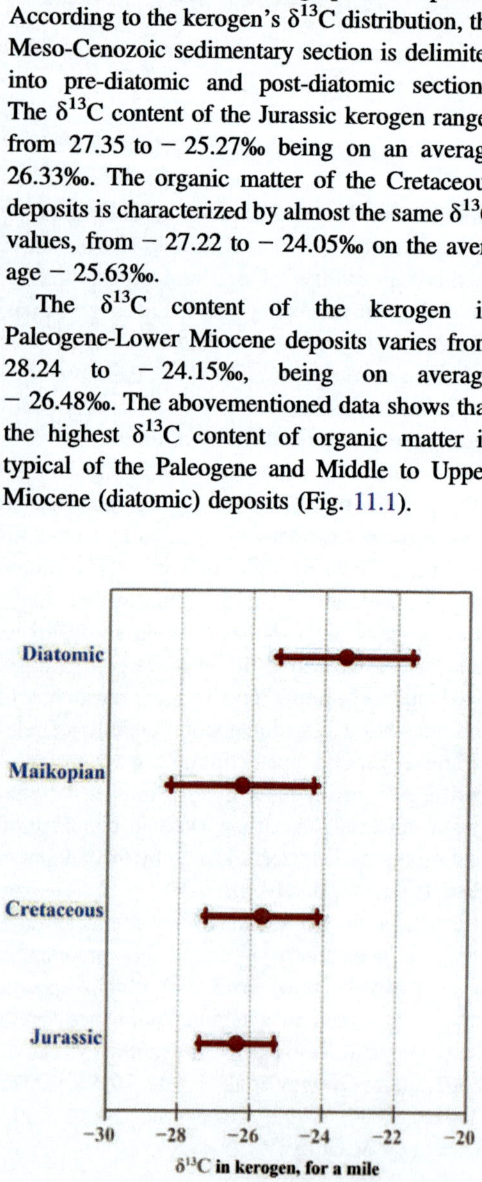

Fig. 11.1 The distribution of $\delta^{13}C$ in the kerogen of the Mesozoic-Cenozoic deposits within the SCB and adjacent areas (after Feyzullayev et al., 2001)

The oil generated by the organic matter of these deposits and characterized by inherited isotopic values will also be distinguished by the $\delta^{13}C$ isotopic composition. Hence, the $\delta^{13}C$ isotopic composition may be used in the SCB as the most important diagnostic feature for stratigraphic and genetic typification of oils—on the one hand, and the rock-to-oil and oil-to-oil correlation on the other hand.

11.2 The Isotopic Composition of Oil Carbon from the SCB

This section represents the stable isotopic results of total oil carbon and its alkanoic and aromatic fractions based on 152 samples of wells from 38 fields located in the Absheron, Evlakh-Agdjabedi, Shemakha-Gobustan, Lower Kur oil-gas-bearing regions, within the Baku and Absheron archipelagos and the Kur and Gabyrry interfluves, i.e., from oil reservoirs aged from the Upper Cretaceous to the Upper Absheron inclusive. Of all studied samples, 67 are of crude oil, and 85—are of alkanoic and aromatic fractions.

Based on these results, there were constructed and interpreted the graphs and histograms of isotopic ratios for oils sum from reservoirs of Upper Cretaceous to Upper Pliocene in age and for oils from a reservoir of concrete age as well as for oils taken from each oil-gas-bearing region. The $\delta^{13}C$ isotopic values of oils from the SCMB vary widely, ranging from − 28.0 to − 24.34‰ for total carbon and from − 29.1 to − 24.8‰ for oil's alkanoic fraction.

Besides, it is markedly noted that the isotopic ratio of the alkanoic fraction is characterized by bimodal frequency distribution. In contrast, the polymodal frequency distribution is typical to the total carbon value (Fig. 11.2a, b), according to which the SCMB oils are grouped into two classes: (1) isotopically light (high gravity) oils having $\delta^{13}C$ values from − 28.0 to − 27.0‰ of total carbon and − 29.1 to − 27.0‰ of the carbon alkanoic fraction and (2) isotopically weighted (low gravity) oils having $\delta^{13}C$ values from − 26.5 to − 24.0‰ and − 26.5 to − 24.0‰, respectively, that confirms the data defined previously

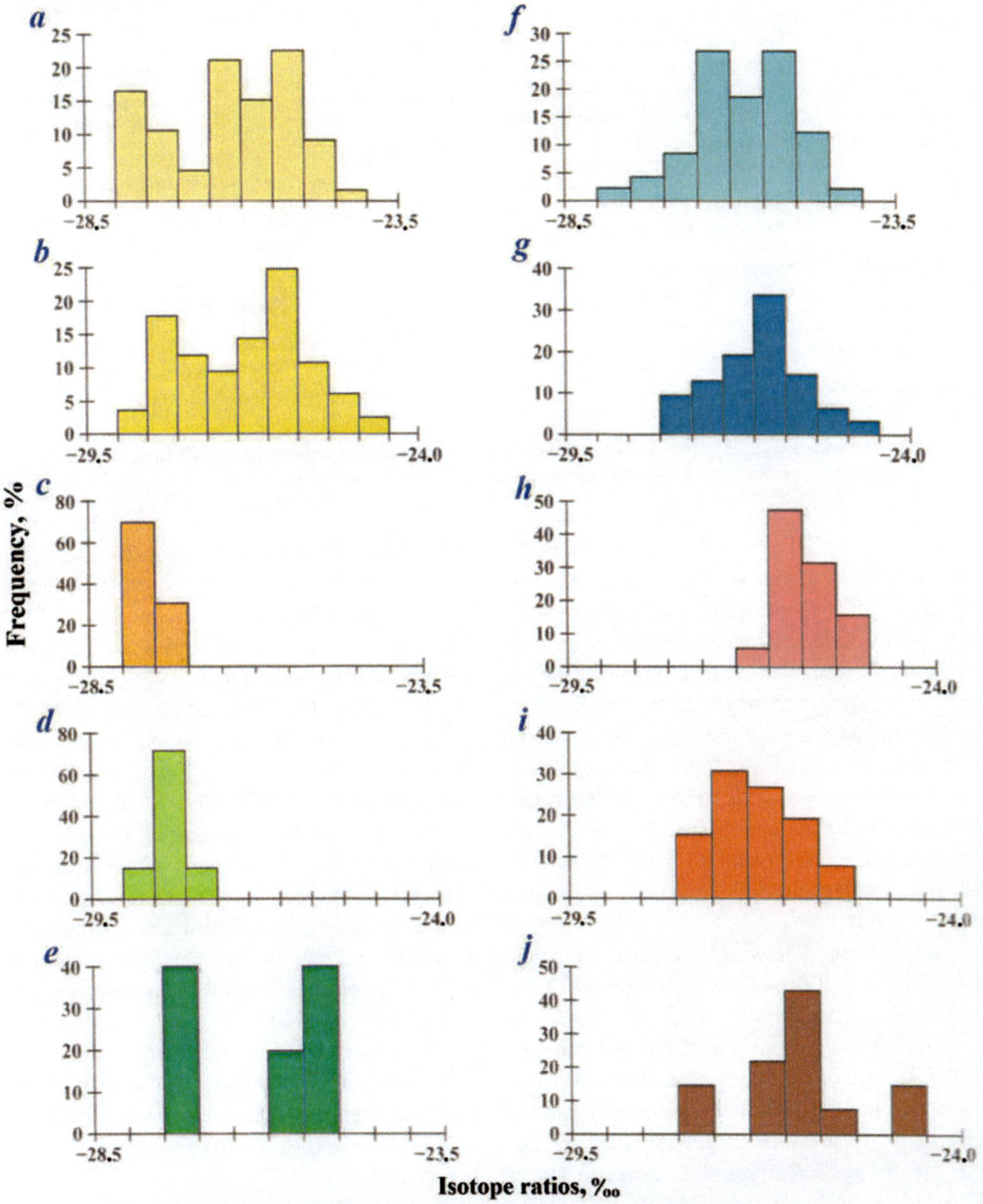

Fig. 11.2 Frequency distribution of the isotopic ratio values of the total carbon and its alkanoic fraction in oils from the South Caspian megabasin. **a, b** In oils from the reservoirs of the Upper Cretaceous, Eocene, Maikopian, Diatomic, and Pliocene complexes; **a** in total carbon of oils; **b** in oils alkanoic fraction; **c, d** in oils from the Upper Cretaceous, Eocene, Maikopian and Chokrakian complexes; **c** in total carbon of oils; **d** in oils alkanoic fraction; **e** in total carbon of oils from the Diatomic complex; **f, g** in oils from the Pliocene complex; **f** in total carbon of oils; **g** in oils alkanoic fraction; **h** in alkanoic fraction of oils from the Pliocene reservoir in the Baku and Absheron archipelagos; **i** in alkanoic fraction of oils from the Pliocene reservoir in Lower Kur oil gas-bearing region; **j** in alkanoic fraction of oils from the Pliocene reservoir in the Absheron Peninsula

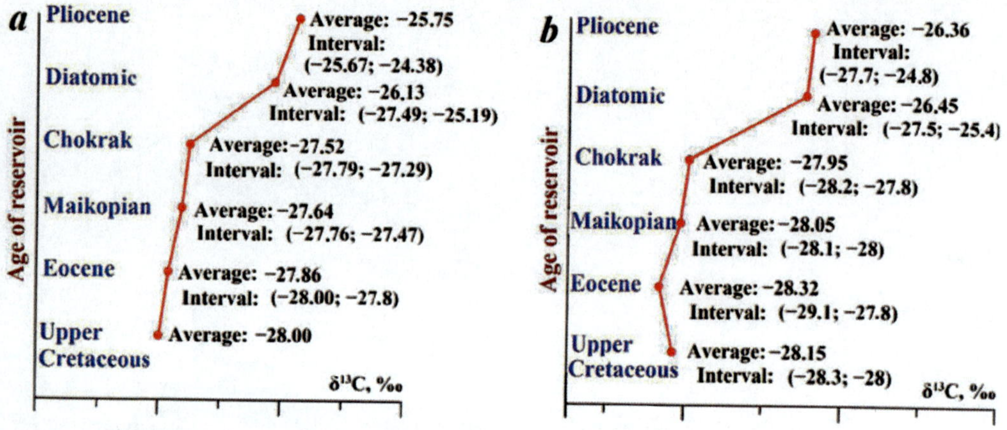

Fig. 11.3 The variations in isotopic ratio values of the total carbon and alkanoic fraction of oils from the reservoirs of the South Caspian megabasin of: **a** oil total carbon, **b** alkanoic fraction carbon

(Guliyev & Feizullayev, 1996). In the SCB, the bulk of oils is of the second group, i.e., from 57.64 to 68.66% of analyzed samples, while isotopically light oils are estimated to be from 31.35 to 42.35%. The most important feature is the marked regular change of isotopic ratio values in the stratigraphic section (Fig. 11.3). In such a way, the most isotopically light oils are typical to the Upper Cretaceous reservoirs (δ^{13}C values range from − 28.15 to − 28.0‰; here and below, the first figure corresponds to the δ^{13}C value of alkanoic fraction and the second one—to the total carbon of oil).

Successively following are oils from the reservoirs of Eocene in age (− 28.32 to − 27.86‰); Maikopian oils (− 28.05 to − 27.64‰); oils from the Chokrak complex (− 27.95; − 27.5‰). It appears that sharp isotopic weighting of oils takes place in Diatomic reservoirs (− 26.45; − 26.13‰). The heaviest oils are from the reservoirs of Lower to Upper Pliocene in age (− 26.35; − 25.75‰).

At the same time, a spread between the upper and lower limits of the δ^{13}C values is increased in the same direction. If isotopically light oils from the Upper Cretaceous, Eocene, Maikopian, and Chokrakian reservoirs are characterized by intervals in variation from − 29.1; − 28.0 to − 27.8; − 27.2‰ with the difference of 0.1 and 0.5‰, then in oils from the Diatomic reservoirs these intervals are: − 27.5; − 27.79 to − 25.4; − 25.19‰ with the difference of 2.1 and 2.3‰. Maximal variation in the carbon ratio values is marked in the Pliocene reservoirs (from − 27.7; 27.67 to − 24.8; − 24.38‰) with a difference of 2.9 and 3.29‰. As is seen from Fig. 11.2e, 40% of oils from the Diatomic reservoirs in the Absheron oil/gas-bearing region are isotopic light oils (− 27.49 to − 27.01‰), and 40% are heavy isotopic oils (− 25.67 to − 25.19‰).

According to the δ^{13}C data of alkanoic fraction, in the Pliocene reservoirs of the Absheron, Shemakha-Gobustan, Lower Kur oil/gas-bearing region, and the Baku and Absheron archipelagos, 22.6% of oils are of isotopic light (− 28.0 to − 27.0‰) and 9.68% are of heavy isotopic oils (− 25.5 to − 24.5‰).

The bulk of oils is of intermediate class with δ^{13}C values as − 27.0 to − 25.5‰ (Fig. 11.2f). The same regularity is noted in the distribution of total oil carbon (Fig. 11.2d).

11.3 Isotopic-Geochemical Factors of the OM Sedimentation Conditions, Oil Formation, and Oil-Rock Correlation

It is generally recognized that high concentrations of normal alkanes n-C15, n-C17, n-C19 at incomparable low n-C27, n-C29, and n-C31 values, as well as low pristan-phytane ratios and

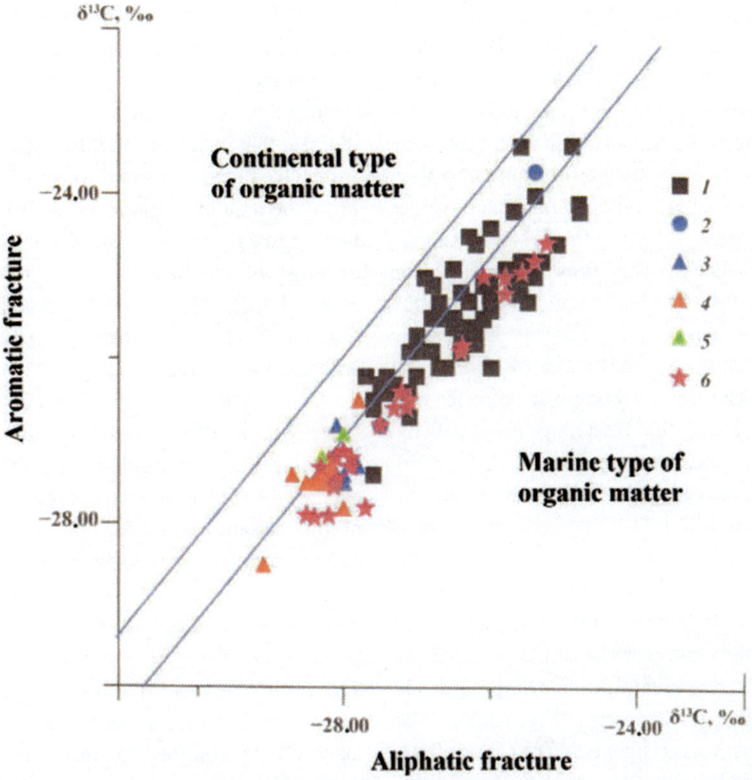

Fig. 11.4 Diagram illustrating the relationship between the values of carbon isotopic ratios in alkanoic and aromatic fractions of oils in the South Caspian basin. Oils from the following reservoirs (1–5): (1) Pliocene, (2) Diatomic, (3) Maikopian, (4) Eocene, (5) Upper Cretaceous complexes, (6) from the mud volcanoes

the content of sulfur rarely reaching 1.4% and 0.4%, respectively, are typical features of oils from the SCB, that is in according with Chung et al. (1992), Collister and Wavrek (1996), Sofer (1984), Peters et al. (1986) who have postulated that the same features are typical to the oils being formed under marine delta conditions. It is also evidenced by the manner of the carbon isotopes' relationship in alkanoic, and aromatic fractions of oils expressed in their figurative points on the plots of δ^{13}Calk-δ^{13}Carom, which are arranged in the field of the primary marine organic medium. Graphically, these isotopes form a stretched-out and ascending trend along the boundary line of the "Continental-Marine OM types" (Fig. 11.4). It should be specially pointed out that this conclusion is more evidence of the fact that the oils from the Pliocene reservoir are epigenetic (Feyzullayev & Tagiyev, 2008; Feyzullayev et al., 2015) regarding the environmental deposits because the latter, according to the paleogeographic construction, have been formed in closed freshen basin where on a level with terrigenous clastic material the continental organic material has intensively been supplied. The mentioned factor should likely be reflected in the isotopic composition of carbon generated by these oils (Aliyeva, 2002; Hinds et al., 2004; Reynolds et al., 1998).

The preceding statement is also confirmed by biomarker parameters of oils from the Pliocene reservoirs of all oil and gas-bearing regions. This data is based on the results of several analyzed samples.

Chromato-spectrometric studies have established the oils' affinity and uniformity in all the PS stratigraphic ranges. As shown in all fragmentograms of hydrocarbons having mass number 191, the peaks of the same biomarker complex-triterpanes: *18a(H)-*, *18b(H)*-oleanane, gammacerane, *17a(H)-22,29,30-trisnorhopane*, *18a(H)-trisnorneohopane* appear to be clearly outlined with the same value and intensity. As a result, these oils have very close values to their indexes: the oleanane and gammacerane values vary in a narrow range: 0.05–0.1 and 0.036–0.046, respectively. The presence of oleane molecules in oil composition indicates that the age of oil-producing complexes is mostly the Upper Cretaceous. A small share of continental organic material in oil composition also evidences this. From the moderate value of the gammacerane index and established Pr/Ph ratio, it is inferred that the paleobasin was of normal salinity developing under sub-reducing conditions of OM fossilization.

A typical feature of oils from the PS on all horizons is equal distribution and the same ratios of normal sterans and isosterans (C27-C28-C29), which make approximately 33%:35%:32% and 31%:36%:33%, respectively (Fig. 11.5). The C29/C28 ratio varies between 0.92 and 0.98, indicating that OM marine geochemical facies are predominant. The (C29/C28 + C27) ratio comes to 0.497–0.522, being displaced toward the relict organic algal facies. The C29 increased content and its slightly predominant over the C27 content (C29/C27 = 1.06–1.15) are caused, in our opinion, by the presence of dolomites in the composition of oil-generating complexes and, in turn, indicate that the deposits of the Miocene Diatonic suite have taken part in the oil generation process. From the diatomic culture study, it is inferred that predominant steroids in the oil composition are an ancestor of stigmastane—C29-sterol (Volkman et al., 1981). Thus, isotopic-biomarker characteristics of oils show that they are not syngenetic with the Productive Series.

An experience of isotopic–geochemical investigations of oils based on the study of different–age objects from the world's various oil and gas-bearing provinces shows that the value of the carbon stable isotopic ratio reaches its maximum in those oils generated from the middle—to Upper Miocene deposits. It is considered that this event took place owing to the global decrease in CO_2 concentration in an atmosphere that resulted in CO_2 solution in seawater, causing a decrease in isotopic fractionation of light isotopes C^{12} by marine plankton during photosynthesis. This interpretation seems to adequately explain the process of oil isotopic weighting according to the average $\delta^{13}C$ values in SCB Diatomic and Pliocene reservoirs.

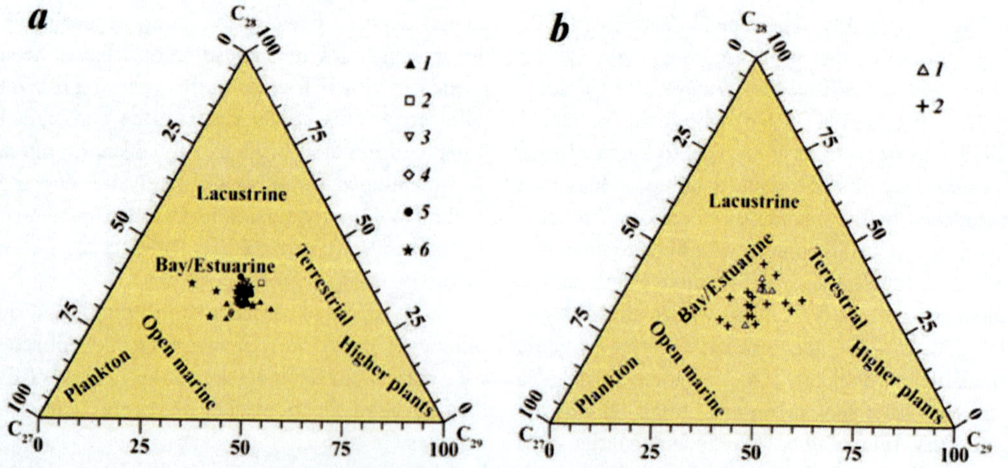

Fig. 11.5 Distribution of normal sterans in oils and organic matter in the SCB different age deposits: **a** oils from the following reservoirs: (1) Pliocene, (2) Diatomic, (3) Chokrakian and Maikopian, (4) Eocene, (5) Upper Cretaceous, (6) oil manifestations in mud volcanoes; **b** Bitumoidal extracts from (1) Pliocene deposits, and (2) Oligocene–Miocene deposits

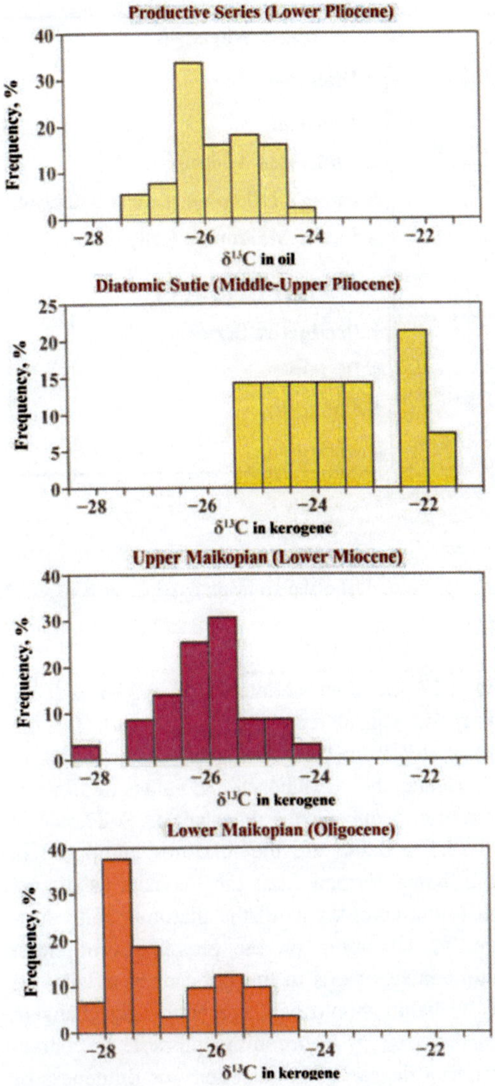

Fig. 11.6 Rock-oil correlation using the carbon isotopic ratios in kerogen of different age deposits and PS' oils (Feyzullayev et al., 2001)

Hence, the localization in the SCB Diatomic-Pliocene reservoirs of both facies-uniform isotopically-light and isotopically heavy oils with the difference in the $\delta^{13}C$ values from 2.1 to 3.29‰ is evidenced by their generation at least by two different—age series of deposit—pre-Diatomic (Cretaceous-Maikopian) and Diatomic (Karagan-Konk-Sarmatian, Meotian) (Chung, 1992; Tissot & Welte, 1984). The isotopic-geochemical correlation of the PS oils and kerogen from the different intervals Paleogene-Miocene in age (Fig. 11.6) shows that they are a mixture of oils supplied from the pre-Pliocene deposits. Considering that oils are, as a rule, depleted in $\delta^{13}C$ by 0.5–1.5% compared with source kerogen (Omokawa, 1985; Peters & Moldovan, 1993), we believe that OM plays a leading part in oil generation in the PS reservoirs. An affinity of kerogen from the Miocene deposits (Upper Maikopian, Chokrakian, and Diatomic suites) with oils from the PS reservoirs has been recognized by the correlation of $\delta^{13}C$ ratios in their alkanoic and aromatic fractions (Fig. 11.7).

11.4 Assessing of Contribution of Different Stratigraphic Intervals to the Process of Oil Generation in the Productive and Red Series

The $\delta^{13}C$ values of alkanoic fraction in oils of the Pliocene reservoirs distinctly differentiate oil and gas-bearing regions. According to the $\delta^{13}C$ average values, these oil and gas-bearing regions are arranged in the following succession: Lower Kur region (− 26.8‰) → Absheron region (− 26.29‰) → Shemakha-Gobustan region (− 26.1‰) → Baku Archipelago (− 26.04‰) → Absheron Archipelago (− 25.87‰) (Fig. 11.8).

As the above plot shows, the Pliocene reservoir oils are distinguished by a particular weighting in their isotopic composition. However, this distinction appears to be more intensified and displayed more contrast when examining $\delta^{13}C$ values variation. In such a way, for oil of the Lower Kur oil and gas-bearing region and Absheron Peninsula, the $\delta^{13}C$ upper and lower intervals are, respectively: − 27.7‰; − 25.4‰ with the difference of 2.3‰ and − 27.6‰; − 24.8‰ with the difference of 2.8‰ while those of the Absheron and Baku archipelagos are − 26.3‰; − 25.4‰ and − 26.5‰: − 25.0‰ with a difference of 0.9‰ and 1.9‰, respectively. This fact points out that oils from marine Pliocene reservoirs, as distinct from the reservoirs of the Lower Kur oil and gas-bearing region and Absheron Peninsula, are more

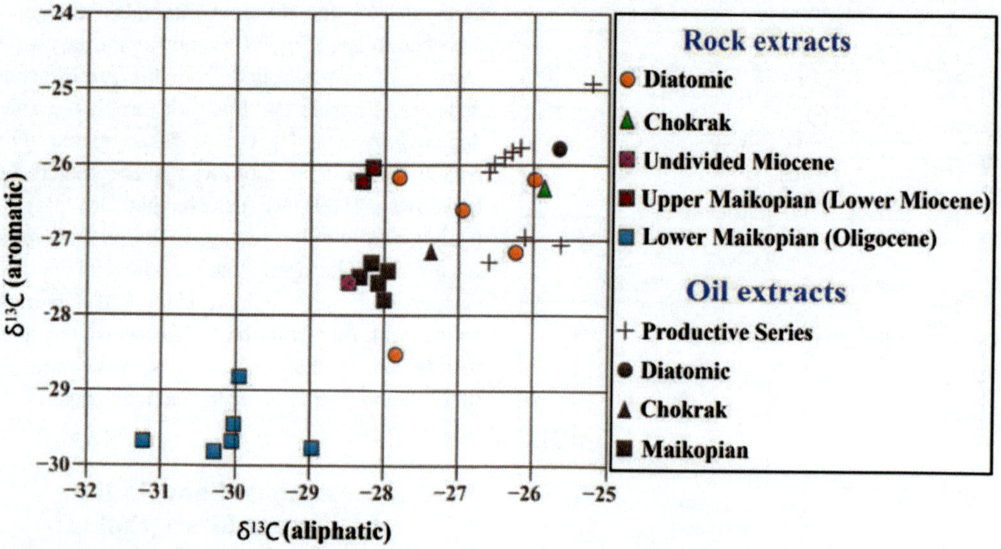

Fig. 11.7 Rock-oil correlation using the carbon isotopic ratios of the aliphatic and aromatic fractions in kerogen of different age deposits and PS' oils (Feyzullayev et al., 2001)

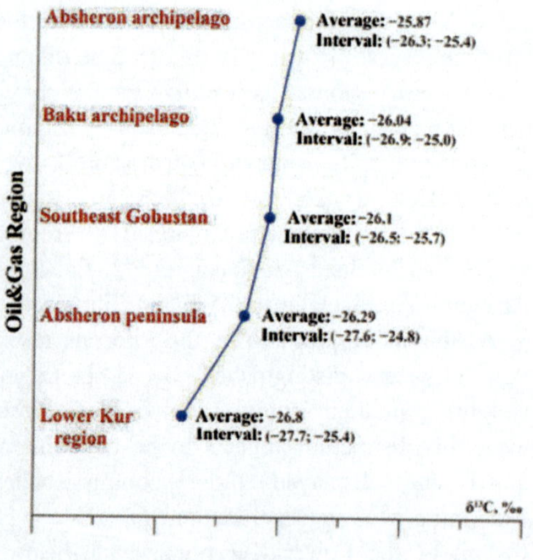

Fig. 11.8 Diagrams of variations in values of carbon isotopic ratios of the alkane fraction of the oils of the Pliocene reservoir of the South Caspian megadepression

homogenous. If it is taken into account that maximum $\delta^{13}C$ values of alkanoic fraction in oils of diatomic reservoir is equal to − 25.4‰ and those of the Miocene reservoir is − 24.8‰, at the upper limit of this value for oils of stratigraphic range from the Upper Cretaceous to the Middle Miocene lowermost strata (Chokrak) being equal to − 27.5‰, then the interval of − 25.1 ± 0.3‰ may be considered as a datum mark for oils generated from the diatomic parent deposits.

Hence, the oils whose $\delta^{13}C$ values of alkanoic fraction occur in the interval of − 27.5‰ to − 25.1 ± 0.3‰ are the mixtures composed of oils being formed from the deposits as of prediatomic complex as of the diatomic suite. Naturally, depending on the prevalence of either component of oils in the Pliocene reservoir, the $\delta^{13}C$ value should be correspondingly changed. In that case, it is becoming possible to characterize a degree of homogeneity or diluteness of oils in Pliocene reservoirs located in SCB oil and gas-bearing areas, to correlate them with each other, and, most importantky, to form a correct quantitative estimate of real participation of diatomic suite and underlying deposits in the process of oil and gas fields generation in the Pliocene complex as well as taking into account oil generating potential of these deposits, to estimate the possible hydrocarbon resources on a level of the Miocene-Paleogene stage. Histograms of oil distribution frequencies constructed by the $\delta^{13}C$ values of alkanoic fraction for each oil and gas-bearing region visually illustrate the relationship of heterogeneous oils distinguished by isotopic values (Fig. 11.2 g–i). These graphs show that marine fields mainly comprise mixed Paleogene-Lower Miocene and diatomic oils. Within the Absheron Peninsula, the volume of oils having diatomic isotopic marks remains the same. The Absheron Peninsula and SCMB water area are characterized approximately by the same volume of modal values (− 26.5 to − 26.0‰), which is equal to 43.36 and 42.85%. Within the Lower Kur oil and gas bearing region, the Pliocene reservoir varies strikingly by its mixed content of Paleogene-Lower Miocene and Diatomic oils from the above-described reservoirs by the absence of mainly diatomic oils, an increased volume of Paleogene-Lower Miocene oils, and sharply displaced modal values of $\delta^{13}C$ in the limits of − 27.5 to − 27.0‰ (isotopic light oils) in a volume of 30.77% and $\delta^{13}C = -27.0$ to − 26.5‰ in a volume of 26.92% which form the sum of 57.69%. From the above, it is shown that Paleogene-Lower Miocene oils dominate in the oil composition of the Pliocene reservoir of the Lower Kur oil and gas-bearing region.

Determining the $\delta^{13}C$ boundary values typifying pre-Diatomic and diatomic oils allows quantitatively evaluating a particular oil-producing complex's contribution to forming hydrocarbon accumulation in Pliocene reservoirs in the SCMB, overall, and in the field taken separately.

From the calculation based on the defined isotopic mark of oils, it is inferred that the role of different oil-generating intervals of Paleogene-Lower Miocene and Diatomic complexes in the generation of oil pools in the Pliocene reservoir of the Absheron Peninsula is roughly the same. The mentioned feature is also typical of the Pliocene reservoirs of southeastern Gobustan and Baku Archipelago, although the Diatomic complex is slightly dominant here. About 3/4 of oils in the Lower Kur oil and gas-bearing region were formed in Paleogene-Lower Miocene deposits. In contrast, the Diatomic deposits played the leading role in oil pool formation in the PS of the Absheron Archipelago. These deposits are estimated to contain up to 2/3 of all PS oils. The fact that the carbon isotopic composition of oils is estimated to be quite similar in the Baku Archipelago and southeastern Gobustan confirms the formerly drawn conclusion that the northern Baku Archipelago is, as a matter of fact, a marine continuation of southeastern Gobustan (Djeirankechmes depression).

The isotopic weighting of oils is traced from land to sea (Fig. 11.9), which is explained by the entrainment of more young deposits in the "oil window". It seems natural since younger Pliocene-Quaternary deposits thicken in the same direction, and underlying older formations have subsided considerably.

Determination of oil composition carried out for the Pliocene reservoir of the central SCMB shows that it is composed of roughly equal relation between the Paleogene-Lower Miocene and Diatomic oils. Assuming the fact that the thickness of Oligocene-Lower Miocene deposits in the central megabasin is roughly twice exceeded that of Diatomic deposits by the average organic carbon content being equal (0.68% and 0.63% respectively for each series) (Alizade et al., 1975), it is inferred that about a half of the realized hydrocarbon potential of the pre-Diatomic complex is concentrated in the Pliocene reservoirs. Hence, at least 50% of hydrocarbon resources generated by Paleogene-Lower Miocene deposits remain unrevealed.

11.5 Carbon Isotopic Composition of Oils from the SCB Mud Volcanic Oil Manifestations

It is known that oils emitted by mud volcanoes of naphthene-aromatic and methane composition are highly oxidized and bio-degraded. The isotopic composition of carbon in oils varies from − 28.5 to − 25.4‰ (in saturated fraction). Based on the isotopic mark revealed in bituminous extracts of Paleogene-Miocene deposits and oils of SCB reservoirs different in age and because of the "rock-oil" and "oil-oil" correlation, it turned out to determine a contribution of various stratigraphic complexes to the feed of mud volcanoes with oils (Guliyev et al., 2003; Alizadeh et al., 2017). It should be mentioned that the oils generated by Paleogene-Lower Miocene

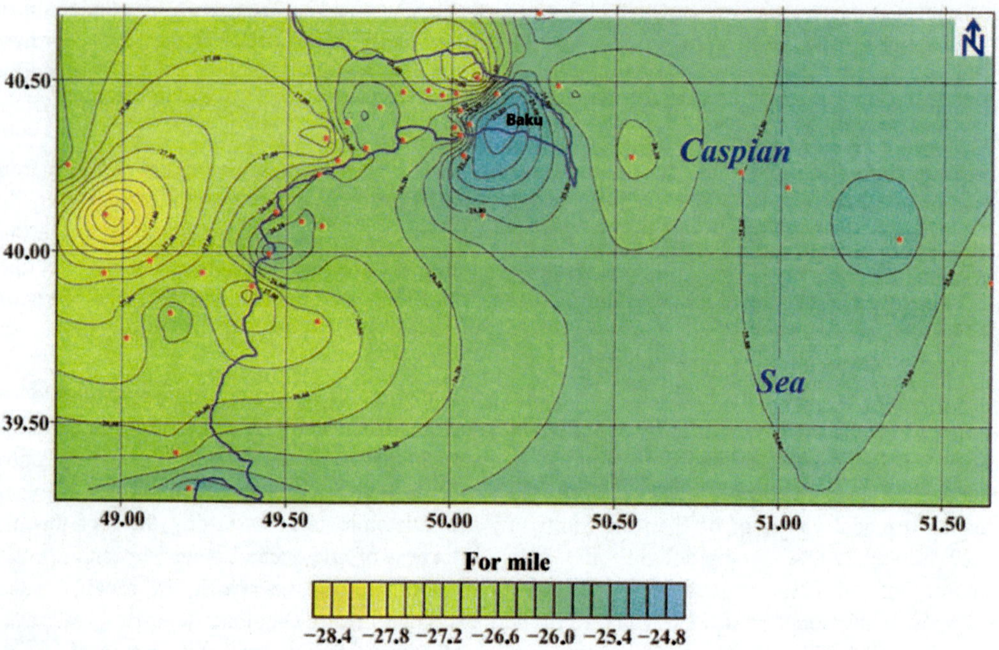

Fig. 11.9 The distribution of oils according to the carbon stable isotopes relationship in the SCB Pliocene reservoirs

(Eocene, Maikopian) complexes are isotopically light ($\delta^{13}C = -28.5$ to -27‰), whereas Middle to Upper Miocene (diatomic) oils are isotopically heavy ($\delta^{13}C = 26$‰ to -24.5‰).

From the result of isotopic-geochemical analyses of oils picked out from oil pools related to the mud volcanoes, it is inferred that two types of oils are distinguished here, namely oils distinguished by typical Paleogene-Lower Miocene $\delta^{13}C$, and mixed oils produced by Paleogene-Lower Miocene and Diatomic deposits. Nearly 50% of mud volcanoes are characterized by sweeping exclusively Paleogene-Lower Miocene oils. 17% of mud volcanic oils are mostly of Diatomic complex, and 33% are mixed of Paleogene-Lower Miocene and Diatomic complexes (Fig. 11.10).

It is of some importance in the fact of the law-governed spatial arrangement of mud volcanic oil pools according to which those of mostly diatomic components in oil appear to be separated in the highly distant northwestern SCB, i.e., within the zone of conjugation of the Lower Kur and Shemakha-Gobustan troughs (Fig. 11.11).

Tectonically, these structures are divided by the Adjichai-Alyat deep fault, along which the Paleogene-Lower Miocene deposits of the southwestern slope of the Shemakha-Gobustan tectonic zone are moved upon the Middle-Upper Miocene and Pliocene complexes of the northeastern Lower Kur basin. The fact that in the overthrust plate and overlapped interval, the $\delta^{13}C$ of oils is the same suggests that the oil source is mostly located within the Diatomic complex of the Lower Kur basin. This conclusion is confirmed by the lowest degree of oil maturity in this mud volcanic cluster and the same oil maturity in the northern Lower Kur basin (Kalamadyan, Small Kharami).

Thus, the following principal conclusions may be drawn:

- The isotopically light oils have been formed by the Paleogene-Lower Miocene intervals, whereas isotopically heavy oils—mostly by Diatomic deposits. It is not out of the question that the PS itself (largely its lower division) takes a slight part in the oil-producing process

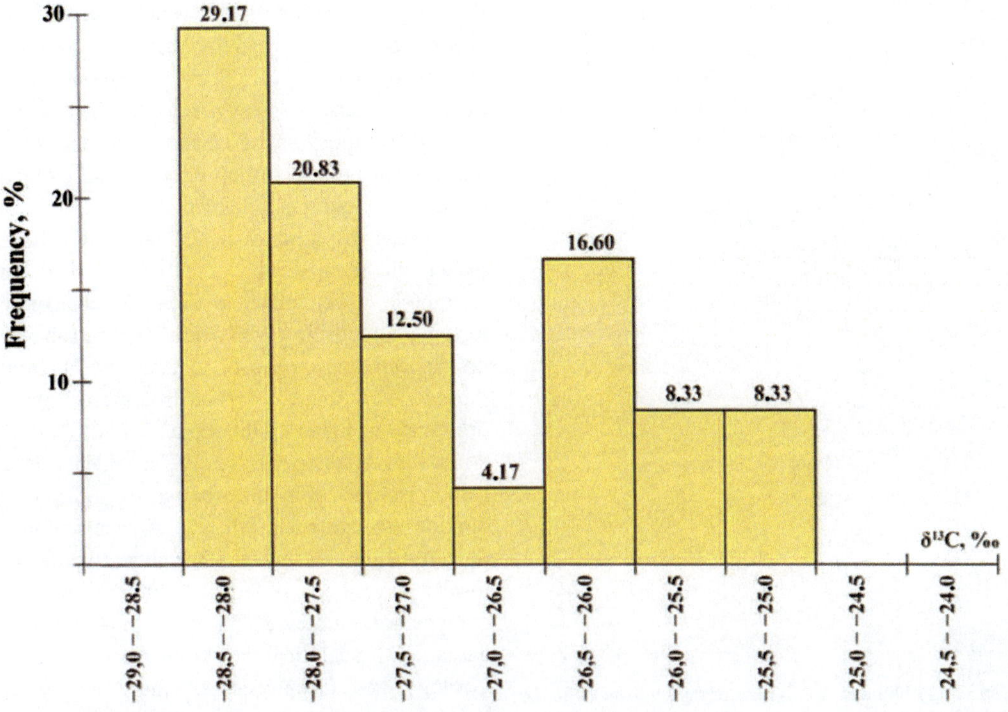

Fig. 11.10 The frequencies of the $\delta^{13}C$ distribution in the SCB mud volcanic oil

(dependent on its low hydrocarbon potential), but this suggestion demands additional investigations and realization of the “rock-oil” correlation,

- Pliocene reservoir contains a mixture of oils from different source series, whose importance varies depending on concrete geologic conditions (as on a regional scale, as within separate oil and gas-bearing regions),
- Based on the revealed regularities, the values of different source complexes in oil pool generation were determined within different oil and gas-bearing regions.

11.6 The Isotopic Composition of Carbon and Hydrogen in the Gases of Azerbaijan

A study of the carbon isotope of methane and its analogs in gases from 29 oil, gas-condensate, and gas fields of Azerbaijan has been carried out on 45 free gas samples taken from the PS different horizons and suites (38 samples) and underlying Chokrak and Maikopian deposits (7 samples). The results of these analyses indicate that the $\delta^{13}C$ value of methane varies from − 37.2 to − 60.3‰ (on an average − 45.0‰), the $\delta^{13}C$ of ethane from − 21.0 to − 40.3‰ (on an average − 28.9‰), of propane from − 10.5 to − 33.7‰ (on an average − 23.7‰), of isobutane from − 21.5 to − 35.8‰ (on an average − 27.0‰), of normal butane from − 14.8 to − 30.8‰ (on an average − 23.5‰); of butane—from − 18.0 to − 32.6‰ (on an average − 25.3‰); the hydrogen isotopic composition of methane varies from − 101.0 to − 277‰ (on an average − 207‰). The hydrogen dioxide in laboratory samples corresponds to the hydrocarbon gases. The $\delta^{13}C$ values of the carbon dioxide vary from − 13.2 to + 21.5‰ (on an average + 3.6‰).

The oxygen isotopic composition of the carbon dioxide also varies widely from + 2.5‰ to − 12.5‰ (on an average − 2.9‰). The result of gas isotopic composition studies allows for a

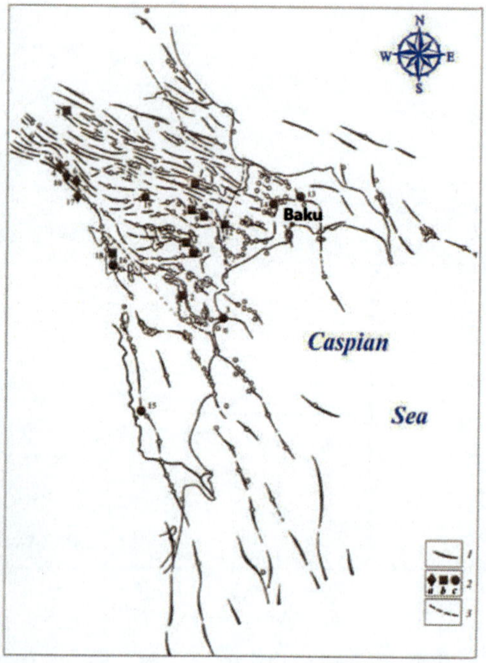

Fig. 11.11 Spatial distribution of mud volcanoes in the SCB according to the $\delta^{13}C$ mark of oils. (1) anticlines, (2) mud volcanoes emitting the following types of oil: (a) with diatomic $\delta^{13}C$ mark, (b) with Paleogene $\delta^{13}C$ mark; (c) mixed Diatomic and Paleogene oils, (3) boundaries between the oil and gas-bearing regions

correlation of this distribution over the section by depth and types of oil pools. Such a wide range in variation of the carbon dioxide isotopes compared with carbon isotopes of hydrocarbons is reasonably explained by heterogeneous sources of CO_2, their origin, and fractionation.

It appears unable to reveal an apparent regularity in the variation of $\delta^{13}C$ values of hydrocarbon gases and carbon dioxide and hydrogen isotopes of methane and oxygen. Isotopically, the most marked distinction is noted in gases from the Chokrak horizon and Maikopian suite compared with those of the Productive Series.

The correlation of the $\delta^{13}C$ average values of hydrocarbon gases and carbon dioxide from different types of oil and gas pools shows that according to all hydrocarbon and carbon dioxide components, the carbon isotopic composition is enriched with heavy isotopes. In contrast, hydrogen isotopic composition is lightened, but the oxygen of carbon dioxide is not changed. It is noticed that carbon and hydrogen of methane from both oil and gas pools to the gas-condensate ones are enriched in heavy isotopes. The carbon isotopic composition of ethane, propane, iso— and normal butane, carbon dioxide, and oxygen of carbonic acid gas is lightened.

Because the samples investigated have been picked out from depths of 230 to 5754 m, an attempt has been made to trace the change in isotopic composition with depth. The plots based on the carbon isotopes of methane and carbon dioxide show that the carbon isotopic composition tends to lighten with depth.

In the direction of the PS subsidence, the carbon isotopes of hydrocarbon gases and carbon dioxide are lightened, whereas oxygen isotopes of carbon dioxide become weighty. Significant results have been obtained from the isotopic composition studies of the SCB mud volcanic gases. With analytical data from the Institute of Geology and Geophysics of the Ministry of Science and Education of Azerbaijan, the results of published works have also been used (e.g., Dadashev, 1985; Dadashev et al., 1982, 1986; Valyayev et al., 1978, 1980, 1982, 1985).

The carbon isotopic composition of methane in the SCB mud volcanoes varies from − 50 to − 40‰, which corresponds to the methane maturity average stage. Nearly 15% of these volcanoes have methane with heavy carbon isotopic composition (from − 40 to − 36‰) that characterizes a late stage of methane maturity (Fig. 11.12a).

Overall, well-defined zoning is observed in the spatial distribution of mud volcanoes by the isotopic composition of methane (Fig. 11.13). Isotopically heaviest, then catagenetically mature gases are typical to the mud volcanoes in the Shemakha-Gobystan zone (the $\delta^{13}C$ on average is − 40‰ by greasy gas content of 0.1‰). Toward the Lower Kur basin and South Caspian water area, the carbon isotopic composition of methane is markedly lightened (on an average $\delta^{13}C = -47$‰ by the greasy gas content of 2.1‰), corresponding to its maturity degree to the early and late stage of catagenesis of its producing organic matter. Such zoning relates to various geological

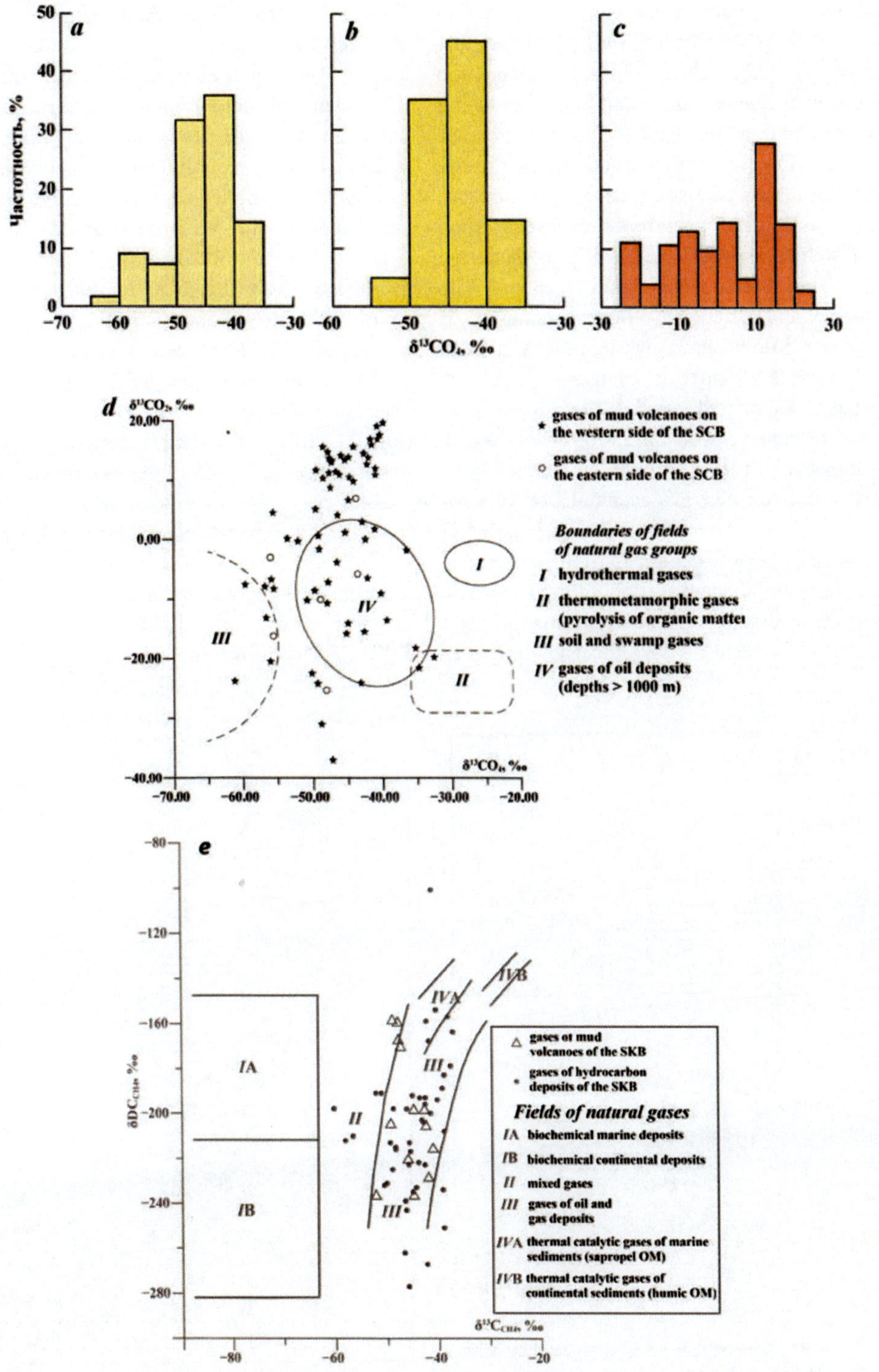

Fig. 11.12 Distribution of the $\delta^{13}C$ values of methane in mud volcanoes (**a**), in oil and gas fields (**b**), $\delta^{13}C$ of carbon dioxide in the mud volcanoes (**c**), $\delta^{13}C$ of CH_3–CO_2 relation (**d**). The hydrogen isotopic composition of the CH_4 in SCB mud volcanoes (**e**)

conditions of hydrocarbon gas formation and conservation in sedimentary series. The fact that within the Shemakha-Gobustan zone, the Paleogene-Miocene oil and gas-producing deposits occur at no great depth and are outcropped on its enormous area, and seismic activity of this zone caused a loss of young, immature gas.

The widespread occurrence of hydrocarbon manifestation confirms an exposing and intensive degassing of these deposits within this zone. The Mesozoic and Paleogene-Miocene complex of the Lower Kur basin is overlapped by a thick series (up to 6 km) of Pliocene–Quaternary deposits. Compared with the Shemakha-Gobustan zone, fewer mud volcanoes and a low frequency of their eruption are marked here. A significant oil and gas accumulation in Pliocene–Quaternary reservoirs indicates a well HC preservation and low degassing of these deposits. Thus, the isotopic analysis data of mud volcanic gases show that as terrestrial as marine mud volcanoes connected with Pliocene structures (Lower Kur basin, Baku Archipelago) are characterized by lighter $\delta^{13}C$ of methane than those located on the Paleogene-Miocene outcrops (Gobustan, Absheron Peninsula). The features noted on mud volcanoes are likely like those revealed in the gas of oil fields. Namely, the $\delta^{13}C$ appears to be weighed depending on the depth of occurrence of the Mesozoic surface. The data obtained from the $\delta^{13}C$ study of mud volcanic CO_2 gases show that $\delta^{13}C$ values vary widely from − 49‰ to + 25‰, indicating the presence of CO_2 of different genesis (Fig. 11.12b), which is also confirmed by the $\delta^{13}C$ CH_4–$\delta^{13}C$ CO_2 relation (Gutsalo & Plotnikov, 1980) (Fig. 11.12d), including metamorphogenetic (from + 8‰ to − 4‰), thermocatalytic (from − 16‰ to + 2‰), biochemical (< − 16‰) (Fig. 11.12d). The principal maximum comprises an interval from + 16‰ to + 10‰, whereas subordinate intervals are from + 2‰ to − 2‰ and from − 6‰ to − 10‰ (Fig. 11.12c). The $\delta^{13}C$ CH_4 confirms the mixed nature of mud volcanic gases—$\delta^{13}C$ C_2H_6 relation (Fig. 11.14) (Feyzullayev, 2012; Feyzullayev et al., 2022).

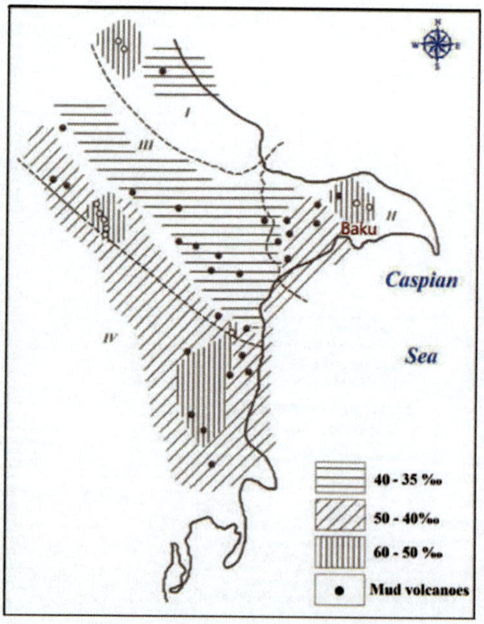

Fig. 11.13 Zonation of the carbon isotopic composition of methane in the mud volcanoes of the western flange of the SCB (Dadashev et al., 1982, 1986). Oil and gas bearing regions (downfolds): I—Pre-Caspian-Guba; II—Absheron; III—Shamakhi-Gobustan; IV—Lower Kur

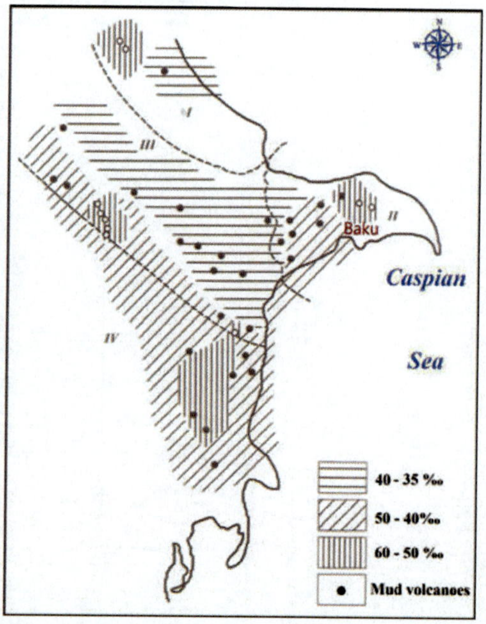

Fig. 11.14 A relationship between the $\delta^{13}C$ of methane and ethane in oil&gas fields and mud volcanoes (Feyzullayev, 2012)

Nearly half of studied mud volcanoes have ultra-heavy $\delta^{13}C$ of CO_2 ($\delta^{13}C > +8‰$), and naturally, they are of special interest. For a long time, a source such as CO_2 has not been revealed. From the result of our investigation of $\delta^{13}C$ in gases of oil and gas fields on the SCB western slope, it is inferred that ultra-heavy $\delta^{13}C$ of CO_2 is typical to the fields located at no great depths where the reservoir temperature is lower than 70 °C, and oils are oxidized and largely biodegraded.

Features typical of these gases are a positive correlation between the $\delta^{13}C$ of carbon dioxide and total CO_2 content. It is well-known that the liquid hydrocarbon oxidization is accompanied by the abundant emission of Neogene CO_2, which is ultra-heavy (Dimitrakopolos & Muehlenbachs, 1987).

The same feature characterizes both mud volcanic gases and those from oil and gas fields, that is, an interrelation between the $\delta^{13}C$ of carbon dioxide and total CO_2 content. Thus, the presence of ultra-heavy $\delta^{13}C$ of CO_2 in mud volcanic gases suggests that liquid hydrocarbon accumulations experienced intense bacterial-oxidizing destruction may be present in the section of mud volcanic structures (Feyzullayev & Movsumova, 2010). It should be emphasized that the revealed isotopic-geochemical relationship plays an essential role in the identification of mud volcanoes and may be successfully used as a reliable geochemical indicator in the prospecting of concealed liquid hydrocarbon accumulation.

References

Abrams, M. A., & Narimanov, A. A. (1997). Geochemical evaluation of hydrocarbons and their potential sources in the western South Caspian depression, Republic of Azerbaijan. *Marine and Petroleum Geology, 14*(4), 451–468.

Aliyeva, E. (2002). Depositional environment of hydrocarbon-bearing Lower Pliocene Productive Series in Southwest Caspian basin. In *Proceedings of the 16th Sedimentological Congress, South Africa* (pp. 9–12).

Alizade, A. A., Akhmedov, G. A., & Aliyev, G.-M. A. (1975). *Evaluation of the oil-producing properties of the Meso-Cenozoic deposits of Azerbaijan* (140p). Elm. (in Russian).

Alizade, A. A., Alikhanov, E. N., & Shoikhet, P. A. (1967). *Investigation of the conditions for the transformation of organic matter in modern sediments of the South Caspian (in terms of the oil origin)* (101p). Nedra. (in Russian).

Alizade, A. A., Salayev, S. G., & Aliyev, A. I. (1985). *Scientific assessment of the prospects for oil and gas potential in Azerbaijan and the South Caspian* (252p). Elm. (in Russian).

Alizadeh, A. A., Guliyev, I. S., Kadirov, F. A., & Eppelbaum, L. V. (2017). *Geosciences in Azerbaijan. Volume II: Economic minerals and applied geophysics* (340p). Springer.

Bagirzade, F. M., Kerimov, K. M., & Salayev, S. G. (1987). *Deep structure and oil and gas potential of the South Caspian Megadepression* (304p). Azgosizdat. (in Russian).

Chung, H., Rooney, M., Toon, M., & Claypool, G. (1992). Carbon isotope composition of marine oils. *The American Association of Petroleum Geologists Bulletin, 76*(7), 1000–1007.

Collister, J., & Wavrek, D. (1996). $\delta^{13}C$ composition of saturate and aromatic fractions of lacustrine oils and bitumens evidence for water column stratification. *Organic Geochemisry*, 1–8

Dadashev, A. A. (1985). *Features of the isotopic composition of natural hydrocarbon gases of the Western side of the South Caspian depression in connection with the assessment of the prospects for the oil and gas potential of its deep zones* (21p). [Extended Abstract for PhD Thesis, Moscow State University, Moscow]. (in Russian).

Dadashev, A. A., Feyzulayev, A. A., & Guliyev, I. S. (1986). Vertical zonation of oil and gas formation according to the carbon isotopic composition of methane from mud volcanoes in Azerbaijan. *Express Information, Series: Oil&Gas Geology and Geophysics,* (6), 24–26. (in Russian).

Dadashev, A. A., Zorkin, L. M., & Blokhina, G. G. (1982). New data on the carbon isotopic composition of natural gas methane from Azerbaijan mud volcanoes. *Doklady Academy Science USSR, 262*(2), 399–401. (in Russian).

Dimitrakopolos, R., & Muehlenbachs, K. (1987). Biodegradation of petroleum as a source of 13C-enriched carbon dioxide in the formation of the carbonate cement. *Chemical Geology (Isotope Geosciences Section), 65*, 283–291.

Feyzullayev, A. A. (2012). Mud volcanoes in the South Caspian basin: Nature and estimated depth of its products. *Natural Science, 4*(7), 445–453.

Feyzullayev, A. A. (2019). Isotope-geochemical characteristics of hydrocarbons on the north-western flank of the South Caspian basin. *ANAS Transactions, Earth Sciences,* (1), 3–10.

Feyzullayev, A. A., Huseynov, D. A., & Rashidov, T. M. (2022). Isotopic composition of the products of activity of mud volcanoes in the South Caspian basin in connection with the oil and gas content of deep-seated sediments. *ANAS Transactions, Earth Sciences,* (1), 68–80.

Feyzullayev, A. A., & Movsumova, U. A. (2010). The nature of the isotopically heavy carbon composition of carbon dioxide and bicarbonates in the waters of mud volcanoes in Azerbaijan. *Geochemistry International, 48*(5), 517–522.

Feyzullayev, A. A., Tagiyev, M. F., & Lerche, I. (2015). On the origin of hydrocarbons in the main Lower Pliocene reservoirs of the South Caspian Basin, Azerbaijan. *Energy Exploration & Exploitation, 33* (1), 1–14.

Feyzullayev, A., Guliyev, I., & Tagiyev, M. (2001). Source potential of the Mesozoic-Cenozoic rocks in the South Caspian Basin and their role in forming the oil accumulations in the Lower Pliocene reservoirs. *Petroleum Geoscience, 7*(4), 409–417.

Feyzullayev, A., Sadykh-zade, L., & Gasanov, A. (2005). On some aspects of the formation of oil and gas deposits in the Productive stratum of the South Caspian basin. *Azerbaijan Oil Industry*, (4), 13–18. (in Russian).

Feyzullayev, A., & Tagiyev, M. (2008). Some aspects of formation of oil fields in the Productive Series, the South Caspian basin. In *Proceedings of the EAGE International Conference on Petroleum Geology and the Hydrocarbon Potential of the Caspian and Black Sea Regions*. Baku, 6–8 October 2008.

Gorin, V. A., & Buniyat-zade, Z. A. (1971). *Deep faults, gas-oil volcanism and oil and gas deposits of the western side of the South Caspian depression* (192p). Azerneshr. (in Russian).

Guliyev, I. S., & Feizullayev, A. A. (1996). Natural hydrocarbon seepages in Azerbaijan/Near surface expression of hydrocarbon migration. In *Proceedings of the AAPG Hedberg Research Conference* (pp. 63–70). Vancouver.

Guliyev, I. S., Aliyeva, E. G., & Huseynov, D. A. (2001). Deep foci of hydrocarbon formation in the South Caspian oil and gas basin. In *Proceedings of the Institute of Geology of the National Academy of Science of Azerbaijan* (No. 29, pp. 79–99). Nafta-Press. (in Russian).

Guliyev, I. S., Aliyeva, E. G., & Huseynov, D. A. (2005). Hydrocarbon systems of non-equilibrium basins: opportunities for improving the search for oil and gas deposits. *Izvestiya Academy Science Azerbaijan, Series: Earth Sciences*, (2), 3–23. (in Russian).

Guliyev, I. S., Feyzullayev, A. A., & Huseynov, D. A. (1999a). Carbon isotopic composition of oils in the South Caspian megadepression. *Azerbaijan Oil Industry*, (6), 3–13. (in Russian).

Guliyev, I. S., Feyzullayev, A. A., & Huseynov, D. A. (1999b). Genetic typing of oils from reservoirs of different ages in the western part of the South Caspian oil and gas basin in the light of new isotope-geochemical data. In *Proceedings of the 3rd International Conference "New Ideas in Geology and Geochemistry of Oil and Gas"* (pp. 74–77). Moscow State University. (in Russian).

Guliyev, I. S., Feyzullayev, A. A., & Huseynov, D. A. (2001). Isotopic composition of carbon in hydrocarbon fluids of the South Caspian megadepression. *Geochemistry,* (3), 271–278. (in Russian).

Guliyev, I. S., Huseynov, D. A., & Feyzullayev, A. A. (2004). Geochemical features and fluid sources of mud volcanoes in the South Caspian Sedimentary basin in the light of new data on carbon, hydrogen, and oxygen isotopes. *Geochemistry,* (7), 792–800. (in Russian).

Guliyev, I. S., Levin, L. E., & Fedorov, D. L. (2003). *Hydrocarbon potential of the Caspian region* (127p). Nafta-Press. (in Russian).

Gutsalo, L. K., & Plotnikov, A. M. (1980). The ratio of carbon isotopes in the system as an indicator of the origin of methane and carbon dioxide in natural gases. In *Proceedings of the VIII All-Union Symposium on Stable Isotopes in Geochemistry* (pp. 241–244). Nauka. (in Russian).

Hinds, D. J., Aliyeva, E., Allen, M. B., Deavies, C. E., et al. (2004). Sedimentation in a discharge-dominated fluvial-lacustrine system: The Neogene Productive Series of the South Caspian basin, Azerbaijan. *Marine and Petroleum Geology, 21*, 113–138.

Huseynov, D. A. (2000). Origin of oils in the western part of the Kura South Caspian oil-gas bearing basins. In *Proceedings of the 62nd EAGE Conference, Glasgow, UK* (p. 20).

Huseynov, D. A. (2003). New data on oil source-rocks in Pliocene sediments of the South Caspian petroleum system. In *Proceedings of the 1st North Africa/ Mediterranean Petroleum and Geoscience EAGE Conference Tunis* (pp. 56–60).

Huseynov, D. A., & Guliyev, I. S. (2004). Mud volcanic natural phenomena in the South Caspian basin: Geology, fluid dynamics and environmental impact. *Journal of Environmental Geology, 46*, 988–996.

Mekhtiyev, Sh. F. (1956). *Issues of the origin of oil and the formation of oil deposits in Azerbaijan* (320p). Azerbaijan National Academy of Sciences. (in Russian).

Mekhtiyev, Sh. F. (2010). *Selected publications* (474p). Nafta-Press. (in Russian).

Omokawa, M. (1985). Source rock—Oil correlation using stable carbon isotopes. The case of Niigata basin. *Journal of Japanese Association of Petroleum Technology, 50*, 9–16

Peters, K., & Moldovan, J. (1993). *The biomarker guide: Interpreting molecular fossils in petroleum and ancient sediments*. Prentice Hall.

Peters, K., Moldowan, J., Shoell, M., & Hempkins, W. (1986). Petroleum isotopic and biomarker composition related to source rock organic matter and depositional environment. *Organic Geochemistry, 10*, 17–27.

Reynolds, A. D., Simmons, M. D., Bowman, M. B. J., Henton, J., Brayshaw, A. C., Ali-Zadeh, A. A., Guliyev, I. S., Suleymanova, S. F., Atayeva, E. Z., Mamedova, D. N., & Koshkarly, R. O. (1998). Implication of outcrop geology for reservoirs in the Neogene Productive Series. Apsheron Peninsula, Azerbaijan. *AAPG Bulletin, 82*, 25–49.

Sofer, Z. (1984). Stable carbon isotope composition of crude oils: Application to source depositional environments and petroleum alteration. *The American Association of Petroleum Geologists Bulletin, 68*(1), 31–49.

Tissot, B., & Welte, D. (1984). *Petroleum formation and occurrence*. Springer-Verlag.

Valyayev, B. M., Grinchenko, Yu. I., Erokhin, V. E., Prokhorov, V. S., & Titkov, G. A. (1985). Isotopic appearance of mud volcano gases. *Lithology and Mineral Resources,* (1), 72–87. (in Russian).

Valyayev, B. M., Grinchenko, Yu. I., Prokhorov, V. S., & Titkov, G. A. (1982). On the zonality of the carbon isotopic composition of mud volcano gases and its tectonic control. *Doklady Academy of Science USSR, 267*(5), 1222–1225. (in Russian).

Valyayev, B. M., Prokhorov, V. S., & Grinchenko, Yu. I. (1978). New data on the carbon isotopic composition of mud volcanoes in the south of the USSR. In *Proceedings of the VII All-Union Symposium on Stable Isotopes and Geochemistry* (pp. 67–69). Nauka. (in Russian).

Valyayev, B. M., Prokhorov, V. S., & Grinchenko, Yu. I. (1980). Carbon isotopic composition of gases from mud volcanoes in the south of the USSR in connection with their genesis. *Doklady Academy of Science USSR, 254*(6), 1459–1461. (in Russian).

Volkman, J., Gilan, F., Jones, R., & Eglinton, G. (1981). Sources of neutral lipids in a temperate interidal sediment. *Geochimica et Cosmochimica Acta, 45*, 1817–1828.

Weber, V. V. (1978). *Diagenetic stage of oil and gas formation* (253p). Nedra. (in Russian).

Weber, V. V. (1983). *Conditions for the formation and occurrence of oil* (367p). Nedra. (in Russian).

12 The Maturity of Hydrocarbon Fluids in the PS Reservoirs and Deep-Stratigraphic Confining of Their Formation Sources

12.1 Oil Maturity in the Productive Series Reservoirs of the South Caspian Depression

Oil maturity level is the most critical parameter, which in combination with other geologic-geochemical features, allows gaining information on the depth of the gas and oil formation zone, stratigraphic confining of source bed, the trend and migration conditions of hydrocarbon fluids, and the result, to form a correct estimate of oil and gas-bearing basin potential. Detailed investigations of oil maturity levels in oil and gas-bearing basins, which include several fluid-generating complexes involved in the process of HC accumulation, are of particular importance. It is also noteworthy that several fluid-generating complexes are distinguished in its Cenozoic section, that is, the Eocene, Oligocene-Lower Miocene (Maikopian), Middle Miocene (Chokrak horizon), and Middle-Upper Miocene (Diatom suite). An example of the latter may be the intensely developing South Caspian megabasin, whose uniqueness is in its total thickness, which comes close to 30–32 km.

Tectonically, the South Caspian regional down-warping area being conjugated with the most geostructural elements of the Caucasus, Kopetdag, and Elburs appeared to be broken onto several depressions such as intermountain areas and troughs different in their geologic and tectonic structure (Alizadeh et al., 2017). The geological and tectonic structure of the SCB troughs affects the geochemical features of included oils. The most important feature of the megabasin is a wide qualitative geochemical variety of oils in the Pliocene reservoir, which produces different types of oil such as light, heavy, low sour, resinous, methane, naphthene, and other oil compositional types.

From the result of geochemical investigations carried out by the authors, it is concluded that the influence of several thermocatalytic, migration, biodegradation, and genetic factors caused a wide geochemical variety of oils in this reservoir. The last one means that fluid-generating complexes of different ages have occurred in the oil formation process. An essential feature of the SCB is a low maturity degree of oils in the Upper Cretaceous, Oligocene-Lower Miocene (Maikopian), and Miocene reservoirs of the Shemakha-Gobustan oil and gas-bearing region compared with those from the Pliocene reservoir in adjacent regions. It is known that the oil maturity degree is a function of the bowel's temperature regime and the longevity of oil-producing deposits in the oil formation zone. It is interrelated with the age of the oil-generating series. Oil generated by underlying series should be more mature than those from the overlapping complexes. The following very remarkable fact is the extremely low oil maturity within known oil and gas condensate fields that is not tied with the existing notion of vertical zoning in oil and gas formation, according to which condensates are the products of organic matter's high catagenetic destruction.

A. Alizadeh et al., *Pliocene Hydrocarbon Sedimentary Series of Azerbaijan*,
Advances in Oil and Gas Exploration & Production, https://doi.org/10.1007/978-3-031-50438-9_12

The high-technological methods widely applied in the study of organic matter and oil on a level of molecular fossils (biomarkers) made it possible to carry out "oil-oil" and "oil-source rocks" correlation and, what is most important, to determine the stratigraphic age of oils and their maturity degree that exactly means a level of catagenetic transformation of kerogen from which these oils are generated. With that end in view, such parameters as isomerization degrees of terpanes, steranes, their aromatization degrees, aromatic steranes, and steroids relation have been used.

The authors have also used such highly informative and widely applied biomarker parameters as isomerization degree of sterane {αααC29 (20S/S + R)} and aromatization degree of monoaromatic sterane {C28 triarom. sterane/C28triarom. + C29monoarom.sterane}.

The first parameter's essence is that the living organisms synthesize only an epimere predecessor of 20R sterane C27, C28, C29. With increasing kerogen maturity, the 20R epimere is exposed to an isomerization reaction, leading to the formation of 20S epimere. Hence, the 20R epimere appears to be always dominant in slightly mature source beds and oils generated from them. As kerogen maturity has increased, a share of the 20S epimere increases in steranes composition, reflected in the increase of C29(20S/S + R) relation in oils. Mature oils are characterized by high epimerization values of a C29(20S/S + R) steranes estimated to be from 0.50 to 0.54, which corresponds to the equilibrium between biological (R) and geological (S) epimeres (Peters et al., 1986; Petrov, 1994). These values are characteristic of the intervals of the oil window, where source beds of organic matter generate the most significant volumes of liquid hydrocarbons. A peak in oil generation sets in. These intervals corresponded to the kerogen maturity level equal to the equivalent of the vitrinite reflectance, that is (EVR)R_0 = 0.8–0.9%.

The essence of the C28triarom.sterane/C28triarom + C29 monoarom.sterane parameter is that it is increased from 0 to 1 with increasing the thermal maturity level of oils. This is explained by the fact that under challenging thermocatalytic conditions, monoaromatic sterane C29 of organic matter and oils lose its methyl group, being transformed into triaromatic sterane C28 (Mackenzie, 1984; Peters et al., 1986; Requejo, 1992). The maximal coefficient of this process is equal to EVR R_0 = 0.8–0.9%.

The data obtained from the fields of all oil and gas-bearing regions in the South Caspian megabasin show that oil maturity estimated by sterane isomerization varies from 0.155 to 0.49. The most mature oils, having values from 0.35 to 0.49, are found within the oil and gas-bearing region of the Middle Kur trough. As shown in histograms of the frequency allocation, nearly 50% of oil objects have maturity values estimated by sterane isomerization as 0.45–0.5, which is the equivalent of the vitrinite reflectance (EVR), R_0 = 0.68–0.73% and in 40% of the same objects this value varies from 0.4 to 0.5 or R_0 = 0.63–0.68%, i.e., on the whole oils investigated indicate the low or average transformation of kerogen in source beds.

Within the stratigraphic interval comprising the Upper Cretaceous-Eocene-Oligocene-Lower Miocene section, the mentioned values are roughly the same and estimated to have been 0.43–0.45 on average (by a degree of sterane isomerization).

Based on such factors as all studied fields from the Upper Cretaceous section within the Evlakh-Agjabedi trough are confined to the effusive formations (Muradkhanly field), their overlapping by Eocene oil-bearing sandy-clayey deposits and the same catagenetic transformation of oils from the two reservoirs, it becomes apparent that interformational migration of hydrocarbon fluids from the Paleogene complex into Upper Cretaceous volcanogenic unit took place here. An identity of oils from the studied reservoirs is confirmed by such genetic biomarker parameters as normal steranes αααC27:C28:C29 relation, isosteranes αββC27:C28:C29 relation, isoprenoids relation (Pr/Ph), and so on (Guliyev et al., 1999). The fact that an oleanolic molecule is generally present in oils is the most crucial indicator of their similarity and belonging to the Cenozoic complex. At the same time, the two groups of fields are clearly distinguished in oil and gas-bearing regions within the Kur and Gabyrry interfluve. One of the groups is characterized by the least mature values of oils in the Middle Kur basin, which is $R_0 = 0.58–0.63\%$.

Genetic biomarker parameters, jointly with the carbon isotopic relation, indicate that the two Eocene lithofacies complexes have occurred in oil generation within this basin. The other group of more mature oils is characterized by the carbon isotopic relation of 28.0–27.8% (estimated by an alkane fraction), predominant C27 over C28 and C29, and roughly equal relationship of isosteres. The carbon low isotopic values (from − 29.1 to − 28.7‰) are typical of more low-mature oils, where C28 predominates over C27 and C29 among the normal steranes and isosteres. However, it appears that oleanane has been found in both the first and second groups of oils. The isotopic-biomarker characteristics indicate that oil-generating deposits of the first group are related to the marine facies, whereas those of the second group are of coastal-marine and deltaic facies of the bays and lagoons (Peters & Moldovan, 1993; Peters et al., 1986).

The relatively low catagenetic transformation degree of oil within the Middle Kur basin ($R_0 = 0.58–0.75\%$) is well correlated with a level of OM mature in enclosing Paleogene deposits within an interval of catagenetic gradation ($R_0 = 0.65–0.85\%$) (Aliyev, 1993). It is convincingly evidenced that studied oils are syngenetic with the Paleogene complex and as a weighty argument for the excellent prospect of underlying Mesozoic oil-producing and oil-accumulating deposits (Fig. 12.1).

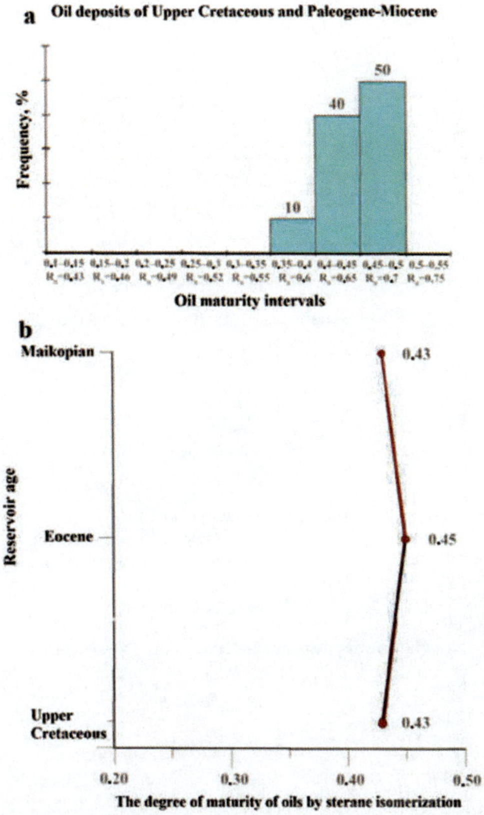

Fig. 12.1 Middle Kur depression: frequency of deposits distribution by a degree of oil maturity (**a**) (in the upper row along the abscissa axis, the intervals of the values of the degree of maturity of oils for the isomerization of sterane, in the bottom—EVR (R_0, %)) and change of the degree of maturity of the oils (by sterane isomerization) in stratigraphic section (**b**)

The Pliocene fields located within the Lower Kur, Shemakha-Gobustan, Absheron troughs, and Baku and Absheron archipelagos of the SCMB are characterized by widely varying degrees of oil maturity, from $R_0 = 0.45$ to 0.67%. Most Pliocene fields (which means the age of reservoirs) in the SCB are objects (81.58%) having an average R_0 value in an interval of 0.53–0.63%, which suggests a low degree of OM transformation in source beds. Against this background, there are fields (about 15%) with very low values of oil maturity ($R_0 = 0.5–0.53\%$) and a small number (5.2%) of those with relatively mature oils ($R_0 = 0.7\%$).

Histograms of frequency allocation and plots (Figs. 12.2 and 12.3) clearly illustrate quantitative relations of Pliocene objects according to their oil maturity. As it follows from the given plots, the slightest degree of transformation is a typical feature of oils in the Absheron Peninsula and Shemakha-Gobustan trough ($R_0 = 0.43–0.56\%$). It is noticed that the maturity-increased values are typical of oils in the Lower Kur basin and Absheron Archipelago ($R_0 = 0.62\%$).

At the same time, there is great diversity between the fields of the mentioned areas. It is

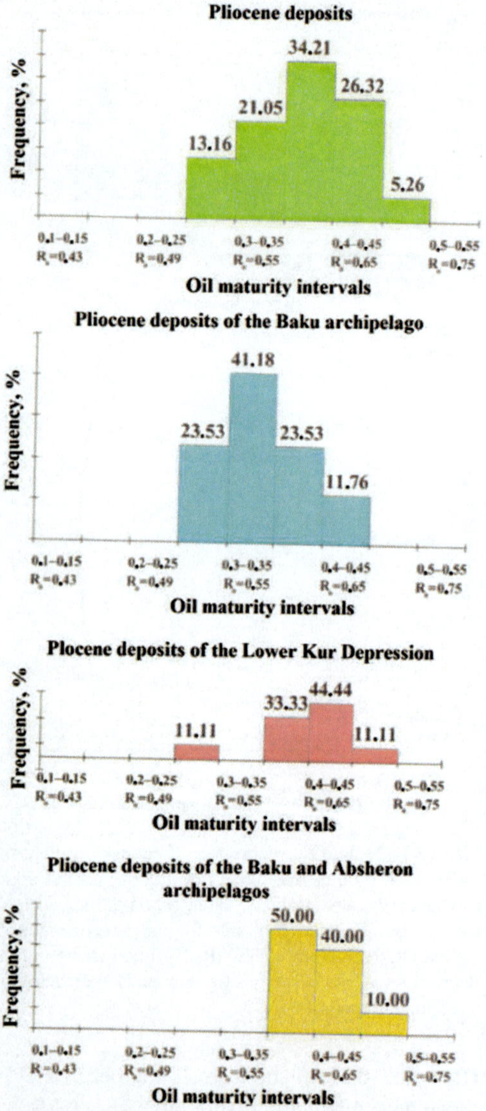

Fig. 12.2 Frequency distribution of oil fields in the SCMB Pliocene reservoirs according to the oil maturity degree (abscissa axis of the upper row shows the intervals of oil maturity values by the sterane isomerization, a lower row presents EVR (R_0, %))

expressed in markedly different shares of the Paleogene-Lower Miocene and Diatomic complexes in the saturation of oil reservoirs. This conclusion is based on the result of our detailed investigation of the stable carbon isotopes of oils and their alkane fraction from the reservoirs different in age (from the Upper Cretaceous to the Upper Pliocene inclusive) of all the SCMB oil and gas-bearing regions. As a result, a distinct differentiation of oils in a stratigraphic section has been revealed (Fig. 12.3a, b).

The carbon isotopic composition of oils appears to be weighted with the rejuvenation of reservoir geologic age (Guliyev et al., 1999, 2000, 2001). In such a way, the most isotopically light oils are typical of the Lower Cretaceous reservoirs (− 28.15; − 28.0‰) (here and then, the first figure corresponds to $\delta^{13}C$ in alkanoic fraction, the second one—to the total carbon content of oil). The following isotopic composition values are oils from the reservoirs of the Eocene (− 28.32; − 27.86‰), Oligocene-Lower Miocene (Maikopian) (− 28.05; − 27.64‰), and Middle Miocene (Chokrak) (− 27.95; − 27.5‰) complexes. A sharp isotopic weighting of oils occurs in the Diatomic reservoir (Middle to Upper Miocene) (− 26.45; − 26.13‰). The most isotopic-heavy oils are confined to the Pliocene reservoirs (− 26.35; − 25.75‰).

At the same time, a spread between the upper and lower limits of $\delta^{13}C$ values occurs in the same direction. In the previous chapter, the authors have explained this tendency by a generation of oils from the rock series different in age: (1) from the pre-Diatomic complex (Cretaceous-Chokrak) generating oils with carbon isotopic mark of alkane fraction as − 29.1, − 28.8‰ and of all oil as 28.0, − 27.2‰; (2) from the Diatomic (Karagan, Konk, Sarmatian, Meotian) complex generating oils with carbon isotopic mark of alkane fraction as − 25.4, − 24.8% and of all oil as − 25.19, − 24.38‰. This tendency is clearly traced in histograms of oil distribution frequency constructed according to $\delta^{13}C$ values separately for each reservoir (Fig. 11.2a–e). From the given graph, it is shown that oils from the pre-Diatomic reservoirs (Cretaceous-Chokrak) are characterized by single modal distribution with maximal $\delta^{13}C$ values not exceeding − 27.5, − 27.0‰, whereas Diatomic and Pliocene reservoirs are characterized by a bimodal distribution of oils with widely varying intervals from $\delta^{13}C$ − 28.0 to − 24.0‰.

As it is seen on histograms of oils from the Diatomic reservoirs (Fig. 11.2d), a frequency

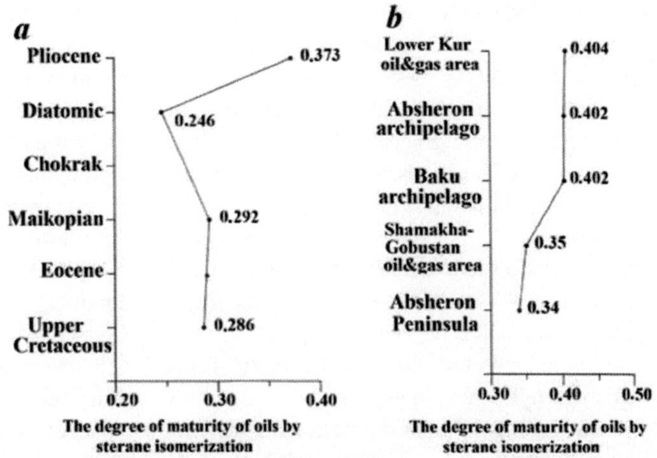

Fig. 12.3 Variations of the oil maturity values in stratigraphic section (**a**) and in the SCB oil and gas-bearing regions (**b**) estimated by sterane isomerization data

class of δ^{13}C values (− 27.5‰ to 27.0‰) occupies an isolated position comprising 40% of all objects; it is also noticed that the objects of intermediate classes (− 27.0‰ to − 26.5‰) and (− 26.5‰ to − 26.0‰) are absent. At the same time, the frequency classes of the δ^{13}C values of (− 26.0‰ to 25.5‰) and (− 25.5‰ to − 25.0‰) comprising 20% and 40%, respectively, are well distinguished on histograms with a marked predominance of the latter class. All the mentioned above lead us to believe that isotopically light oils are uncharacteristic of this complex. They likely have migrated from the underlying Oligocene-Paleogene and probably older deposits.

From those mentioned above, it is concluded that the younger the reservoir, the more likely its saturation with oils from different stratigraphic levels and the higher their heterogeneity. Histograms of frequency allocation of oils constructed according to the δ13C values of alkanoic fraction for each oil and gas-bearing region separately clearly illustrate a relation between isotopically heterogeneous oils (Fig. 11.2 h, i, k). From the given graphs, it follows that marine fields mainly consist of diatomic oils. In the areas of the Absheron Peninsula, the volume of mixed oils is decreased by separating mainly Paleogene-Lower Miocene oils. In contrast, oils with diatomic isotopic marks remain approximately the same. A common feature of the megabasin marine part and the Absheron Peninsula is nearly the same volume of modal values (− 26.5, − 26.0‰), equal to 47.36% and 42.85%, respectively. The Pliocene reservoir of the Lower Kur OG-BR by its mixed Paleogene-Lower Miocene and Diatomic oils content is cardinally distinguished from the above-described reservoirs by the absence of mainly diatomic oils and increased volume of Paleogene-Lower Miocene oils that are expressed in sharply displaced modal δ^{13}C values toward the isotopically-light (− 27.5 to − 27.0‰) areas. It should be noted that overall, the isotopic composition of oils in the Pliocene reservoir tends to be weighed toward the South Caspian Basin (see Chap. 9), is in accord with the increasing thickness of young deposits in the same direction and more entrainment of diatomic complex and possibly Pliocene lower suites into the oil window.

It is highly noteworthy that it was established that the oil maturity degree in the Lower Pliocene reservoir (Productive Series) appears to change with the increasing thickness of sedimentary deposits and depending on the relation between the Paleogene-Lower Miocene and diatomic complexes.

The postulated dependence may be confirmed by an example of the Lower Kur basin, one of the deepest SCMB depressions where the pre-alpine basement subsided down to 30 km. Its intensive down warping from the beginning of Cenozoic time predetermined an accumulation of more than 10 km thick sedimentary series, including

4000 m Paleogene-Lower Miocene complex, up to 1000–1500 m thick Middle to Upper Miocene and about 6000 m thick Pliocene-Quaternary complex. Such the relation between the Paleogene-Lower Miocene and Middle to Upper Miocene complexes explains the isotopically light-weight oil composition.

However, in the general light, the isotopic background, a sharp isotopic weighting of oils, is traced in a southeasterly direction with distinguished anomalies having a typical diatomic mark in oils of the most southeasterly distant Neftchala field. A unique feature of this field is that its oils are the most matured within the Lower Kur basin, as in the central SCMB. In the seismologic profile through the cross-section of the Neftchala fold (Fig. 12.4), it is well seen that west of its axial line within the Lower Kur synclinorium, the Paleogene-Miocene complex is sharply decreased in its thickness. As a result, the Oligocene–Miocene deposits at a depth of 7000 m are superimposed on the Mesozoic surface (Bagirzade et al., 1988). The oil pools in the Neftchala field are confined to the near-crest part of the southwestern limb of SE-NW trending anticlinal branchy-fold. This fold is complicated by numerous cross and longitudinal faults, among which two large longitudinal are distinguished by cross-cutting the near-crest part throughout its extension. The fault surface falls to the southwest, toward the Lower Kur synclinorium, where the significant hydrocarbon hearth is supposedly located. Mud volcanic manifestations are also connected with these major faults.

Within the northwestern anticlinal zone at the Kyurovdag and Garabagly fields complicated by a southwesterly dipping fault system confined to the northeastern Lower Kur synclinorium characterized by a full section of Oligocene-Miocene deposits, a share of oils with Paleogene-Oligocene isotopic mark is markedly increased with the same maturity values (Fig. 12.5). The same interrelation is also observed at those fields adjacent to the inner Kargala trough (Mishovdag filed) as located within its limits (the Kyursyangya filed). It should be specially pointed out that there is no relation between the degree of maturity of SCMB oils and their isotopic composition. In such a way, relatively highly matured oils are characterized by both isotopically light and isotopically heavy oils, independently of the age of the reservoir. The same applies with low matured oils.

It is crucial to reveal the causes of oils' low maturity with the Paleogene-Maikopian isotopic mark in the Pliocene reservoirs (PS) of the Absheron trough and oils from the Oligocene-Miocene fields of the Shemakha-Gobustan trough. This problem may be cleared up in its joint investigations with anomalies manifested in oil-gas-condensate and gas-condensate fields

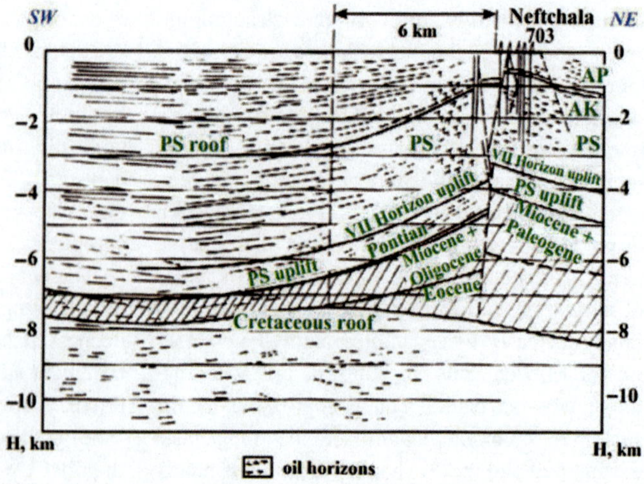

Fig. 12.4 The seismo-geological cross-section through the Neftchala field

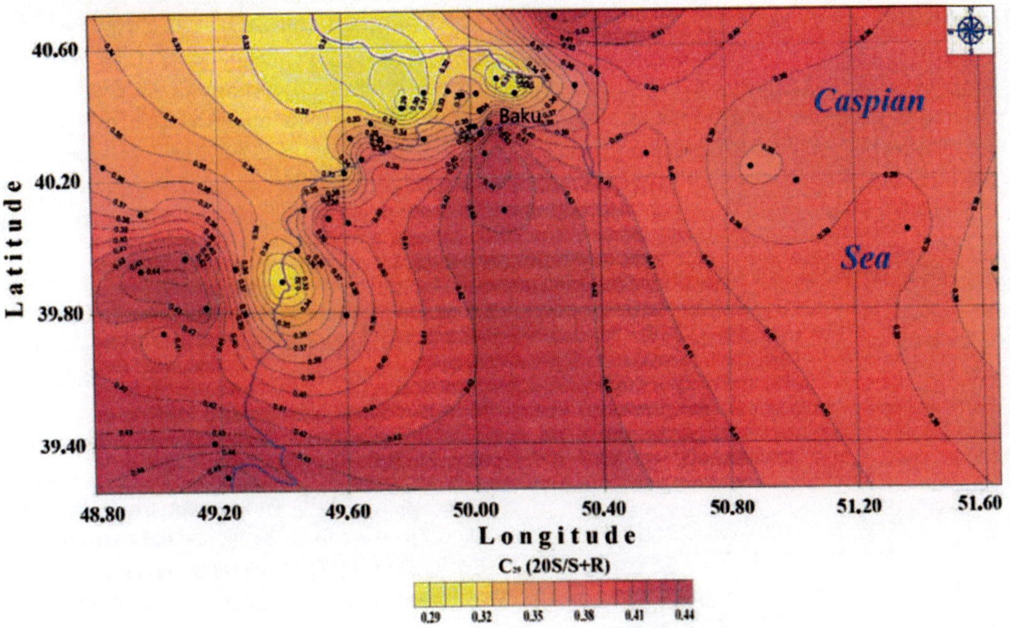

Fig. 12.5 Oil distribution in the SCMB Pliocene reservoir according to the degree of maturity

since we believe these problems are interconnected. One of the anomalous oil-gas-condensate objects, the Garadagh field, is in the extreme southwestern sea-shore part of the Absheron Peninsula. This problem is discussed in detail in the following section dedicated to reconstructing reservoir filling mechanisms using isotopic-geochemical features of hydrocarbon fluids.

Thus, in summing up all the mentioned above, it is drawn the following principal conclusions:

- Overall, in the SCMB, a low degree of oil maturity not exceeding $R_0 = 0.73\%$ (the Middle Kur basin) essentially increases the prospects for discovering more matured oils.
- Proceeding from the fact that more matured oils of known fields have been generated in a diatomic suite of the central SCB and taking into consideration its geological structure, it becomes apparent that the oil generation peak ($R_0 = 0.8$–0.9%) is confined to the Paleogene and Lower Miocene deposits.

The isotopic weighting of the Productive Series oils simultaneously increasing their maturity in the direction of the South Caspian Basin is not connected with the catagenetic transformation of organic matter. However, it has resulted from the entrainment into the oil generation zone diatomic producing isotopically heavy oils.

12.2 Maturity of Hydrocarbon Gases in the Productive Series

Hydrocarbon gas's degree of thermal maturity is clearly reflected in the relationship between hydrocarbon components. It is often expressed in the humidity coefficient representing methane—its homologs ratio. However, in practice, for example, in the case of mud volcanoes or shallow oil and gas pools, the system described appears to be enriched in biogenic methane due to intense biochemical gas generation that increases gas dryness.

The other significant feature of hydrocarbon gas thermal transformation is the carbon isotopic composition of methane and its homologs, which become weighed with increasing their maturity (Galimov, 1968; Prosolov et al., 1980; Schoell, 1980). At the same time, biological methane admixture characterized by highly light carbon isotopic composition ranging from $\delta^{13}C$ – 70 to 75‰ may also distort an actual level of hydrocarbon gas catagenetic transformation. We have used the carbon isotopic composition of ethanee and other homologs to rule out such mistakes.

The isotopic composition of methane from the SCB oil and gas fields varies depending on the age of reservoirs from the Upper Cretaceous to the Quaternary, ranging widely from – 35 to 61‰ with modal values from – 35 to 50‰ (Fig. 12.6).

At a depth interval of 300–4500 m, where most oil and gas pools are localized, the average carbon isotopic composition of ethane comes to 45.0‰. It is noted that heavy ($\delta^{13}C$) isotope content is normally increased with depth. At the same time, there is no correlation between the $\delta^{13}C$ of ethane and the age of these pools enclosing rocks. The carbon isotopic composition of ethane in oil and gas fields varies from – 21.1 to – 40.3‰ (Fig. 12.7).

The quantitative calculations based on the experimental data of the relationship between the carbon isotopic composition of gas and the level of its catagenetic maturity (R_0) (Faber, 1987) as of ethane: R_0 ($\delta^{13}C_{C2H6}$(o/oo) = 22.61 g R_0(%) —32.2 show that the maturity of hydrocarbon gases in the SCB oil and gas fields varies widely from 0.4 to 3.1 R_0 (%) among which the most widespread are those in the interval of from 1.3 to 1.6 R_0 (Fig. 12.8). It points to the fact that gases and condensates have been generated in the zone of catagenesis and corresponded to the peak of a "gas window". At the same time, highly matured gases are found in some fields (Absheron bank, Neftchala), corresponding to the zone of thermogenesis characterized by R_0 more than 2.3–3%. Overall, an increase in gas maturity is typical of those fields localized within the regions of Mesozoic deposits' shallow occurrence (Fig. 12.9). As is shown in the same figure, the fields with highly matured hydrocarbon gases are localized within the Absheron Peninsula and Absheron Archipelago.

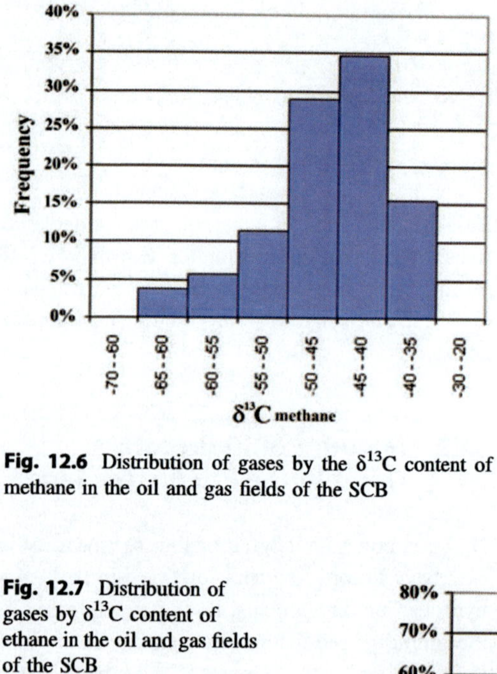

Fig. 12.6 Distribution of gases by the $\delta^{13}C$ content of methane in the oil and gas fields of the SCB

Fig. 12.7 Distribution of gases by $\delta^{13}C$ content of ethane in the oil and gas fields of the SCB

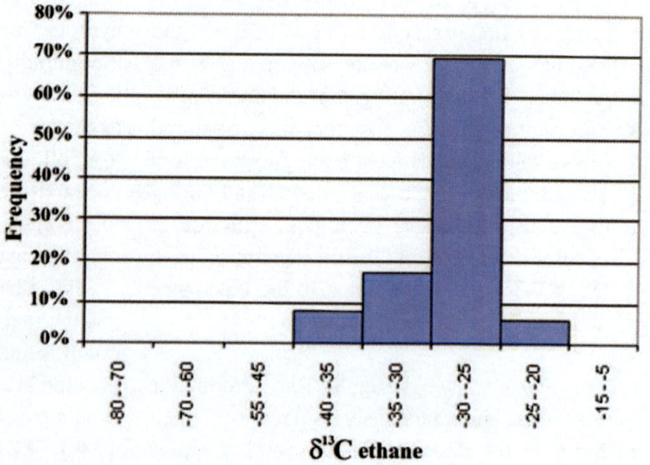

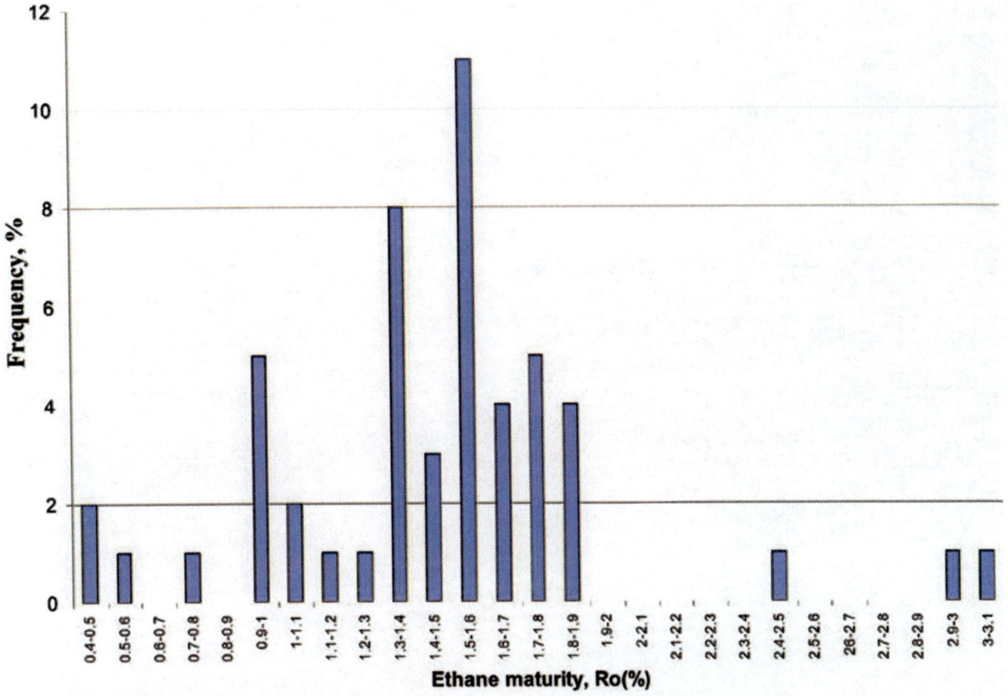

Fig. 12.8 The distribution of gases by the ethane maturity degree in the oil and gas fields of the SCB

The least maturity values ($R_0 = 0.4$–0.5) are typical of the Umbaki field from the Mesozoic and Chokrak reservoirs.

12.3 Correlation of Oil and Gas Maturity in the Productive Series and Prognostic Valuation of Deep-Stratigraphic Confining of Their Sources

From the result of the oil and gas correlation in the SCB fields, it is inferred that the oil maturity degree is between 0.65 and 0.7, whereas this value for gas is from 1.4 to 1.6 (Fig. 12.10). Remember that the most matured oils with maximum Ro values are localized within the Baku and Absheron archipelagos. Considering that the peak of oil generation is in the interval of $R_0 = 0.8$–0.9, it becomes evident that the more significant part of hydrocarbons has R_0 from 0.7 (maximally matured oils) to 1.1 (minimally matured hydrocarbon gases) has still not been discovered in the SCB. It serves as evidence of the SCB's high potential from the standpoint of future prospecting of more mature oils in the lower Productive Series and underlying deposits that have not been drilled to date within the SCB studied area.

The given plots show that the difference between oil and gas maturity is reduced toward the deep-sea part of the basin in the region of the Zafar Mashal and Alov structures. Hydrocarbons characterized by $R_0 = 0.8$–1.3 may be found within this area. This is also in accord with the spatial correlation of oil and gas maturity degree changes (Fig. 12.11).

The correctness of these conclusions is confirmed by the fact that the significant gas-hydrate accumulations having the typical gas composition (Table 12.1) and the carbon isotopic composition are confined to the mentioned areas.

Table 12.1 shows that hydrated gases are characterized by high humidity with heavy hydrocarbons summing up to 38%, which

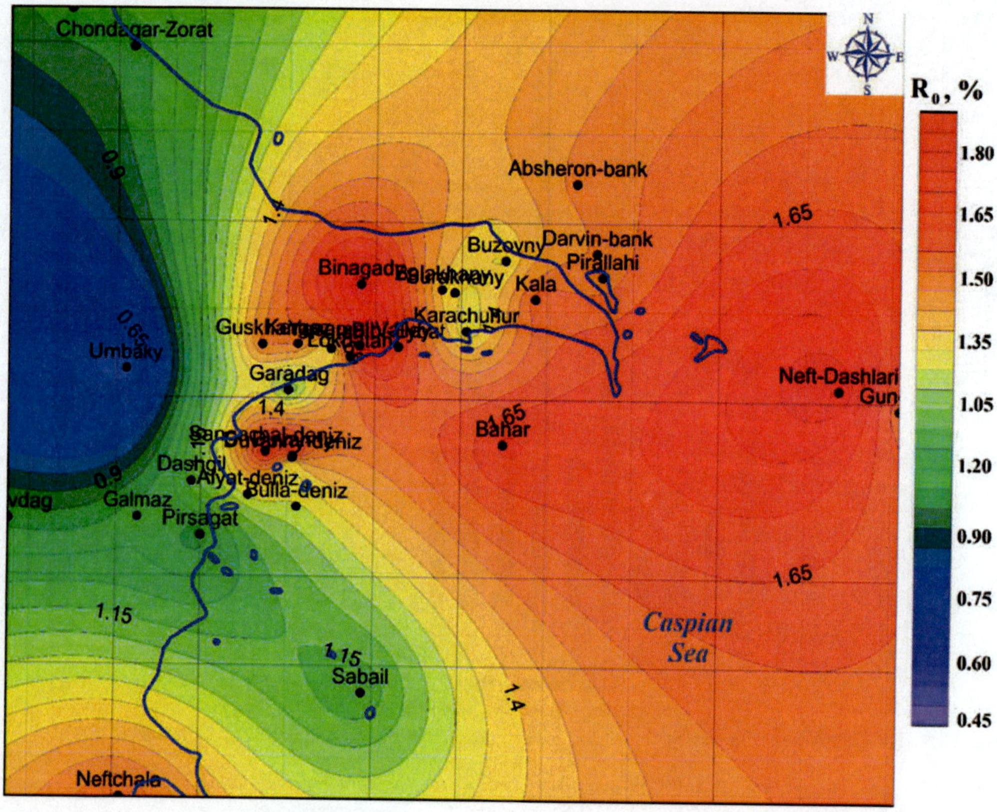

Fig. 12.9 The distribution of oil fields by the gas maturity degree (by R_0 of ethane)

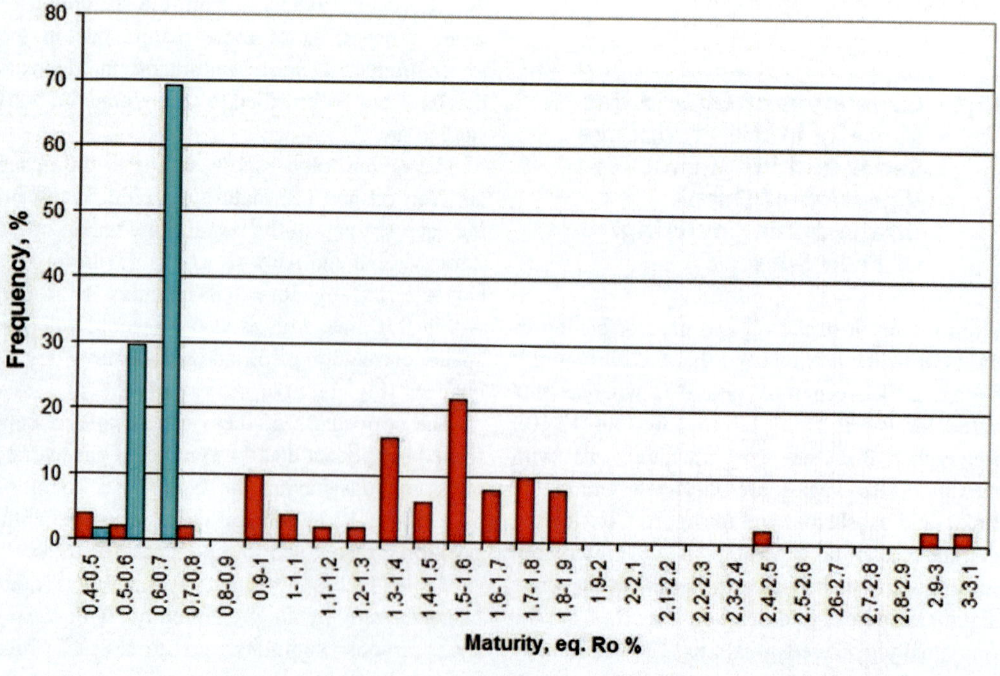

Fig. 12.10 Correlation of the hydrocarbon maturity in the South Caspian oil and gas fields (blue—oil, red—gas)

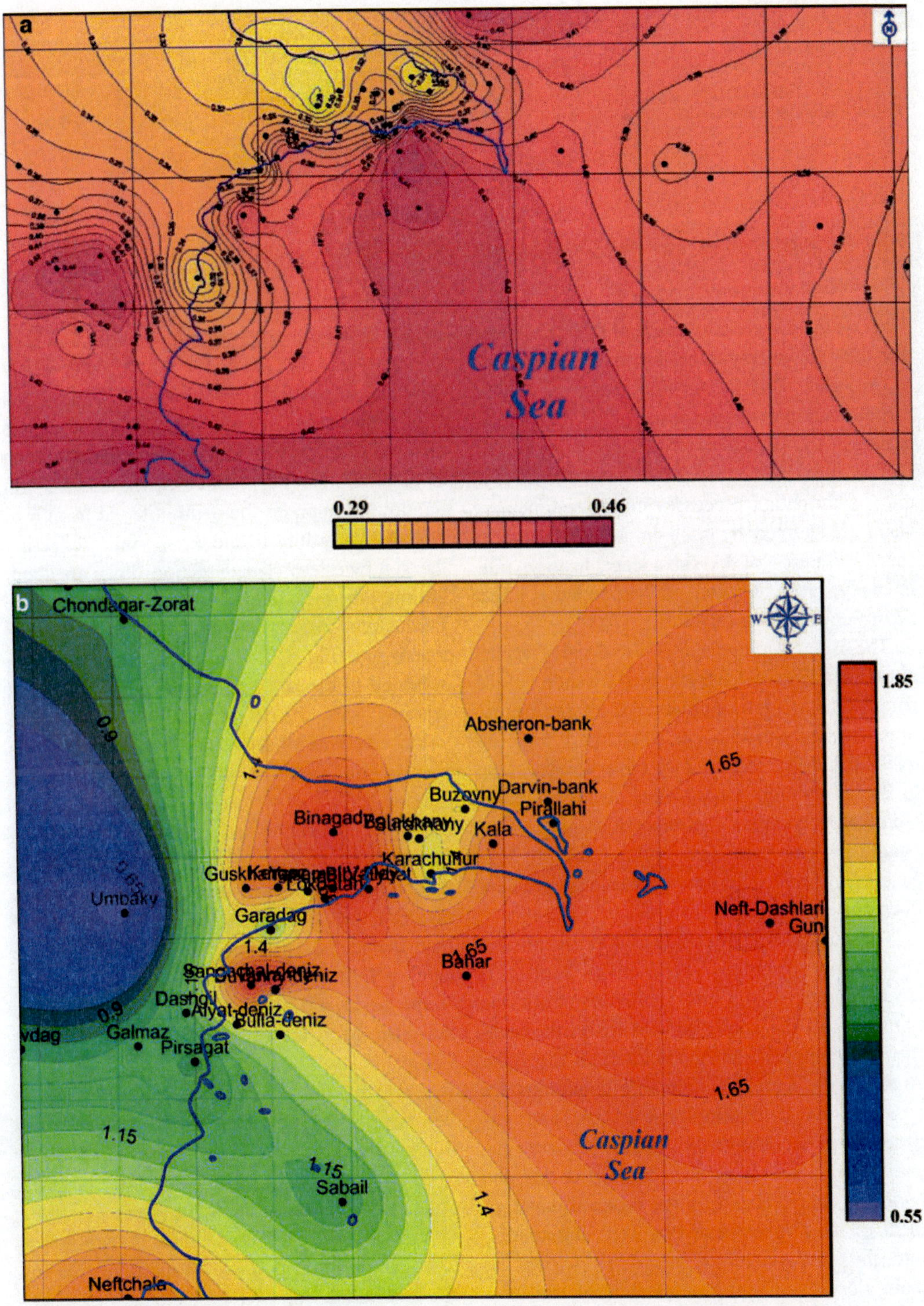

Fig. 12.11 Spatial correlation of the of the hydrocarbon's maturity peculiarity changes in space. South Caspian oil and gas fields (**a** oil, **b** gas)

Table 12.1 The hydrocarbon gas content of gas-hydrate aggregates (after Ginzburg et al., 1988)

Mud volcano	Interval	The components content, %					
		CH_4	C_2H_6	C_3H_8	iC_4H_{10}	nC_4H_{10}	C_5H_{12}
Zafar-Mashal structure (D-9)	0–0.2 m	80.8	13.6	4.2	0.3	0.4	0.02
		87.8	10.4	1.8	0.1	0.4	0.06
		74.2	17.0	6.0	0.7	0.9	0.11
		58.7	19.4	15.8	2.5	2.0	0.68
Azizbekov structure (D-6)	0–0.4 m	81.4	15.3	1.6	0.2	0.7	N.O

suggests the catagenetic nature of these gases and that they are weakly transformed and correspond to oil gases. This fact is also confirmed by the carbon isotopic composition of methanee and ethanee: $\delta^{13}C_{CH4} = -44.8$ to $-57.3‰$, $\delta^{13}C_{C2H6} = -25.7$ to $-28.4‰$. As it is known, the gas formation processes in any sedimentary basin are defined by such important factors as special features of its subsidence history, rate, intensity, and duration of temperature effect taking place within this basin.

The Baku Archipelago is the South Caspian most subsided area characterized by such unique parameters as high anomaly rate of sedimentation, especially in Pliocene–Quaternary time (up to 3500 m/Ma), sedimentary cover thickness up to 25 km including 10–11 km of the Pliocene-Quaternary complex, anomaly low-temperature conditions (its heat flow comes close to 25–51 mW/m^2 and average temperature gradient does not exceed 2 °C/100 m).

The thickness of the SCB significant oil and gas-bearing reservoir, Productive Series, is reached here at 4.5 km, increasing from the northwest to the southeast. Owing to the sedimentary complex's great thickness in the Baku Archipelago, there are favorable conditions for complete zonal development of oil and gas generation, including the lower primary gas generation zone. It is also confirmed that the region is characterized by the rampant development of mud volcanoes every year, outbursting into the atmosphere an incredible amount of mainly dry (methane) gas, which is typical for a high stage of gas generation.

The parameters mentioned above indicate that the zone of oil and gas generation is subsided to great depths being extended by an area (Feyzullayev & Huseynov, 2002). The reconstruction of the Baku Archipelago paleotemperatures during Miocene time according to the vitrinite reflectance data and their extrapolation at great depths (Fig. 12.12) shows that the "oil window" comprises the depth interval from 6 to 11 km. According to the 12-s profile reflecting the southwestern Sabail section down to about 20 km, the "oil window" zone is confined to the Productive Series lowermost stratum to the Paleogene (Fig. 12.13). Considering that rocks subsided to the adjacent synclinal zone, the "oil window" should stratigraphically comprise more young deposits of Miocene-Pliocene in an age that is in accord with the abovementioned factors.

In this connection, the results of the valuation of hydrocarbon gases maturity based on experimental data obtained from the relationship between the carbon isotopic composition of methane and its gaseous homologs on the one hand and the factors of paleotemperature regime in the basin and vitrinite reflectance (R_0), from the other hand are of some importance (Faber, 1987).

Applying the mentioned relationship allows determining, with a particular share of probability, stratigraphic and depth interval of either hydrocarbon gas (methane, ethane, propane, butane) generation in the SCB.

The authors have used ethane's carbon isotopic composition (CIC) to be more reliable. An application of the CIC of methanee has been recognized as unadvisable since it is not out of the question that thermocatalytic methane (with relatively heavy CIC) would be mixed with

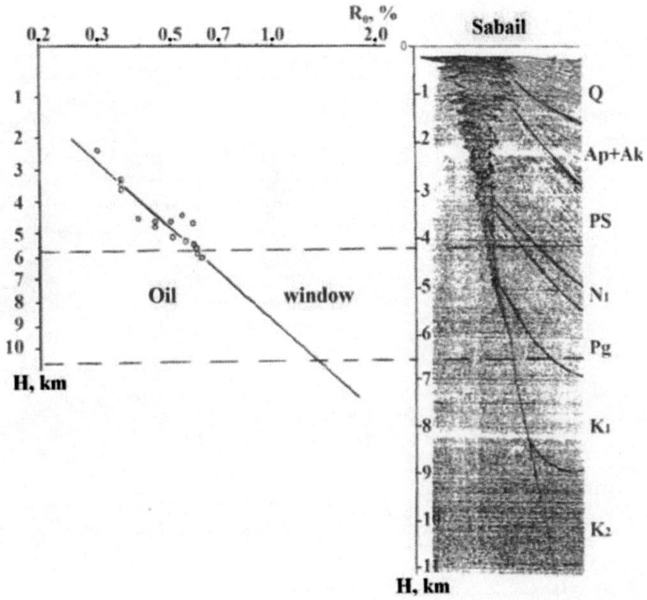

Fig. 12.12 Model of the depth and stratigraphic confinement of the "oil window" in the Baku Archipelago (according to the paleotemperature data and seismic section of the Sabail area)

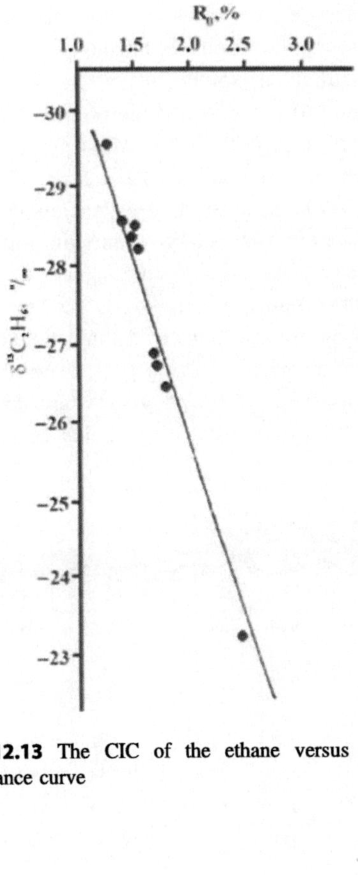

Fig. 12.13 The CIC of the ethane versus vitrinite reflectance curve

biochemical methane (with light CIC), which may lead to the distortion of its primary values. Figure 12.13 shows graphical experimental dependence of $\delta^{13}C_{C2H6}$–R_0(%); its empiric expression is described as: 22.6 lg R_0(%)–32.2.

These calculations have been done according to the CIC data of ethane in gases from the Sangachal-Deniz, Alyat-Deniz, Bulla-Deniz, Duvanny-Deniz, and Sabail-Deniz fields. As it is seen from the curve and empiric dependence of $\delta^{13}C_{C2H6}$–R_0(%) (Fig. 12.13), hydrocarbon gas maturity within these fields is in the interval of R_0 = 1.1–1.75, which corresponds to the late stage of oil generation (condensate/grease gas stage, that is methane generation stage).

Calculations using data from oil biomarkers, the isotopic composition of carbon in ethane, and the thermal history of the SCB are allowed to establish that the interval of oil and gas generation occupies a depth range of 5–15 km, with a hypsometric shift between the zones of liquid and gas HC. Accordingly, the oil generation interval is confined to 5–9 km (peak generation at 7–8 km), and for gas lies between 7 and

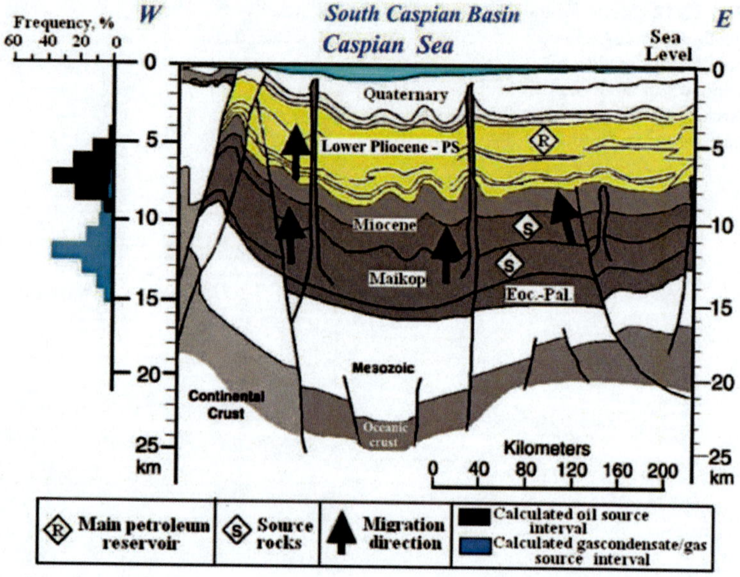

Fig. 12.14 A schematic geological profile directed across the SCB with the location of calculated "oil window" and "gas window" depth intervals

15 km (peak generation at 11–12 km) depth (Fig. 12.14).

From the result of the R_0 measurements carried out to 6 km depth, where this value becomes changed from 0.33 to 0.61 and its subsequent extrapolation on the greater depths, it follows that a depth of ethane generation in the Baku Archipelago fields is in the interval of from 9–10 to 13 km (Fig. 12.12).

The most notable feature is the least ethane maturity in the most subsided reservoir of the Sabail structure (5972–5975 m), the most profound and most extreme eastern drilled structure of the Baku Archipelago. The highest degree of ethane transmutation is a typical feature of hydrocarbon gas from the extreme northwestern Sangachal-Deniz structure. So, it becomes apparent that the depths of ethane generation are decreased in a south-southeasterly direction. According to the 12-s seismic profile, relatively more recent deposits of the Miocene-Paleogene complex are noted in the same direction (Fig. 12.12). In contrast, gas generation in the northwestern Archipelago occurs in Mesozoic deposits.

Hydrocarbon gas composition conforms with the carbon isotopic ethane composition of HC gas thermal maturity. Table 12.2 shows the HC gas composition in studied areas, where gas from the northwestern Archipelago fields consists of

Table 12.2 The gas composition of the oil and gas fields in the Baku Archipelago

Deposits	Average depth, m	Horizon	C_1, %	C_2, %	C_3, %	i-C_4, %	n-C_4, %	i-C_5, %	n-C_5, %	Heavy HC_5 (C_2–C_5), %	Amount of C_1–C_5, %
Sangachal-Deniz	5029	VII, PS	99.8	0.0	0.0	0.1	0.04	0.03	0.03	0.11	100.0
Duvanny-Deniz	3833	VII, PS	92.1	3.8	1.9	0.37	0.63	0.24	0.23	7.17	99.2
Bulla-Deniz	5738	VII, PS	91.3	4.7	2.0	0.39	0.73	0.24	0.25	8.31	99.6
Alat-Deniz	4260	VII, PS	91.9	4.8	1.7	0.38	0.65	0.17	0.16	7.48	99.9
Alat-Deniz	3609	VII, PS	91.6	5.3	1.8	0.37	0.61	0.11	0.1	8.29	99.8
Sabail	5974	VII, PS	77.1	18.1	4.1	0.6		0.1		22.9	100

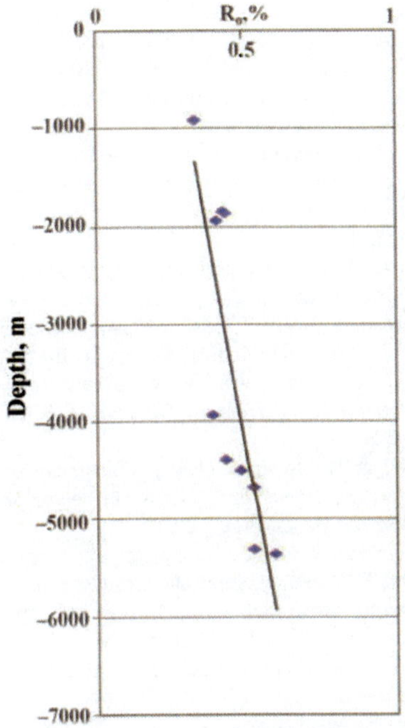

Fig. 12.15 The vitrinite reflectance values measured in boreholes' cores in the Bakhar and Gum-deniz fields

91.6–99.8% methane (dry gas). The share of heavy HC gases (methane homologues) comes from 0.11 to 8.31%. It is noted that studied gases in the Sabail field are essential for heavy hydrocarbons (23%).

From land toward the central South Caspian Basin, it is clearly observed that thermocatalytic transmutation of HC gases has decreased reasonably, suggesting that the same should result in a depth of their generation and confining to the heavy hydrocarbon gas stage.

The presence of oil and gas having essentially different maturity degrees within the northwestern Baku Archipelago's fields, as well as of two groups of HC gases distinguishing from each other in chemical and isotopic composition and a level of their thermocatalytic transmutation, suggest firstly a significantly dragged interval of hydrocarbons generation and secondly that at least two stratigraphic complexes, namely gas-generating Mesozoic and oil and gas-generating Paleogene-Miocene complexes (possibly the Pliocene lowermost strata) have taken place in this process within the Baku Archipelago.

From the results of vitrinite reflectance measurements in the Bakhar and Gum-Deniz fields within the northern SCB (carried out to the depth of 5500 m and then extrapolated on a deeper level (Fig. 12.15)), it is inferred that according to the dependence of

$$H = -9757.1 * R_0, \% + 396.71,$$

where H is the depth, the focuses of HC gas generation correspond to the hypsometric depth's interval from 13 to 15 km. These depths, considering the available geological data, correspond to the interval of the Paleogene deposits.

The oil and gas accumulation area (PS reservoir) is located at comparatively no great depths (2600–4700 m) (Tables 12.3 and 12.4). It follows that vertical migration took place at least 1000–4000 m for oil and 9000–10,000 m for

Table 12.3 Calculated depths of ethane generation in the Bakhar field

Well	Top	Base	$\delta^{13}C_{C2H6}$	R_0, %	Depth, m
Bakhar-178	4036	4083	– 28.2	1.5	14,239
Bakhar-198	4253	4258	– 28.0	1.5	14,239
Bakhar-208	4348	4397	– 28.7	1.4	13,263

Table 12.4 The calculated depths of oil generation in the Bakhar and Gum-Deniz fields

Well	Back	Base	20S/S + R	R_0, %	Depth, m
Bakhar-135			0.419	0.65	5946
Bakhar-148			0.447	0.70	6434
Bakhar-182	4660	4700	0.4	0.62	5653
Gum-348	2656	2686	0.453	0.72	6630

gas. From the calculation carried out for this area, it is concluded that the "oil window" corresponds to hypsometric depths from 5600 to 6700 m (Table 12.4) and the Miocene deposits. It is well correlated and confirmed by the results of isotopic-geochemical studies of oils and kerogen.

References

Aliyev, G.-M. A. (1993). Forecast of the quality and genetic types of oils on the promising structures of the oil and gas-bearing regions of Azerbaijan. *Azerbaijan Oil Industry,* (7–8), 16–20.

Alizadeh, A. A., Guliyev, I. S., Kadirov, F. A., & Eppelbaum, L. V. (2017). *Geosciences in Azerbaijan. Volume II: Economic minerals and applied geophysics* (340p). Springer.

Bagirzade, F. M., Kerimov, K. M., & Salayev, S. G. (1988). *Deep structure and oil and gas potential of the South Caspian Megadepression* (304p). Azgosizdat.

Faber, E. Z. (1987). Isotopengeochemie gasformiger Kohlenwasserstoffe. *Erdole, Erdgas and Kohle, 103,* 210–218. (in German).

Feyzullayev, A. A., & Huseynov, D. A. (2002). Features of oil and gas formation within the Baku Archipelago. *Azerbaijan Oil Industry,* (4), 1–5.

Galimov, E. M. (1968). *Geochemistry of stable carbon isotopes* (p. 198). Nedra.

Ginzburg, G. D., Muradov, Ch. S., & Dadashev, A. A. (1988). Underwater-mud volcanic type of accumulation of gas hydrates. *Doklady Academy of Sciences USSR, 300*(2), 253–258.

Guliyev, I. S., Aliyeva, E. G., & Huseynov, D. A. (2001). Deep foci of hydrocarbon formation in the South Caspian oil and gas basin. In *Proceeding of the Institute of Geology of the National Academy of Sciences of Azerbaijan* (No. 29, pp. 79–99). Nafta-Press.

Guliyev, I. S., Aliyev, G.-M. A., Aliyeva, E. G., & Muradov, Ch. S. (2000). Multicomponent anomaly in bottom sediments and seawater in the central part of the South Caspian depression. *Geochemistry,* (9), 1010–1017. (in Russian).

Guliyev, I. S., Feyzullayev, A. A., & Huseynov, D. A. (1999). Carbon isotopic composition of oils in the South Caspian megadepression. *Azerbaijan Oil Industry,* (6), 3–13.

Mackenzie, A. (1984). Application of biological markers in petroleum geochemistry. In *Advances in Petroleum Geochemistry* (Vol. 1, pp. 115–214). Academic Press.

Peters, K. (1986). Guidelines for evaluating petroleum source rock using programmed pyrolysis. *American Association of Petroleum Geological Bulletin, 70,* 318–329.

Peters, K., & Moldovan, J. (1993). *The biomarker guide: Interpreting molecular fossils in petroleum and ancient sediments.* Prentice Hall.

Peters, K., Moldowan, J., Shoell, M., & Hempkins, W. (1986). Petroleum isotopic and biomarker composition related to source rock organic matter and depositional environment. *Organic Geochemistry, 10,* 17–27.

Petrov A. A. (1994). Geochemical typification of oils. *Geochemistry,* (6), 876–891.

Prosolov, E. M., Lobkov, V. A., & Yakutseny, V. P. (1980). Intensity and depth of methane generation in the earth's crust (according to isotopic data). *Doklady Academy of Sciences USSR, 252*(6), 1476–1479.

Requejo, A. G. (1992). Quantitative analysis of triterpane and sterane biomarkers: Methodology and applications in molecular maturity studies. In J. M. Moldowan et al. (Eds.), *Biological markers in sediments and petroleum* (pp. 223–240). Printed Hall.

Schoell, M. (1980). The hydrogen and carbon isotope composition of methane from natural gases of various origins. *Geochemica et Cosmochimica Acta, 44*(5), 649–661.

13 Special Features of Hydrocarbon Migration and the Mechanism of Oil Trap Filling in the Productive Series

13.1 Conditions and Special Features of Hydrocarbon Migration in the South Caspian Basin and Their Spatial Variation

One of the essential processes in oil and gas ontogeny is hydrocarbon migration. That is why the problem of oil migration conditions in connection with oil field formation is an essential fundamental and, simultaneously, the most complex problem in oil–gas geology.

The mentioned complexity is wholly related to the oil pool formation in the Productive Series, mainly if to stick to the dominant opinion of their formation at the expense of migration from the underlying thick Oligocene–Miocene source deposits. The notions and ideas of HC migration in the SCB have been developed as the knowledge of their evolution and modern structure has been widened and as isotopic-geochemical analyses of OM in rocks, oil, and gas have been improved. A significant volume of fundamental investigations served as a basis for analysis of conditions and special features of HC migration in the SCB have been carried out over the past two decades.

The SCB geologic history and geological structure define special features of HC migration processes here. These are the following factors:

1. An avalanche rate of sedimentation resulted in the squeezing rate of sedimentation waters appearing considerably lower than the sediment down-warping and compaction rate. In such a way, if compaction and squeezing of the bulk of primary pore water in ordinary basins take place to the depths of 2–3 km, then this process in the SCB is extended down to 4 km,
2. When the "oil window" appears to be displaced at great depths (to 5–9 km in the central basin) under anomalous-low temperature conditions,
3. Formation of massive Oligocene–Miocene source series exceeding 5000 m thick in the central basin,
4. Dominant (up to 80–90%) fine-pelitic fraction in source rock without visible sandy-silty partings in the more significant part of the section (according to the well logging).

In correspondence with the factors mentioned in points "a" and "b", the processes of intense water squeezing and mass oil generation in the section are displaced in respect of each other, and that is why oil transfer in the form of a molecular solution in the SCB interstitial waters may not be considered as a potential mechanism of primary oil migration (Fig. 13.1).

From the result of the experimental investigation, it is inferred that if the thickness of the clayey source bed is relatively high, the oils being formed in its central part due to HC kerogen cracking appear to be practically locked.

A. Alizadeh et al., *Pliocene Hydrocarbon Sedimentary Series of Azerbaijan*,
Advances in Oil and Gas Exploration & Production, https://doi.org/10.1007/978-3-031-50438-9_13

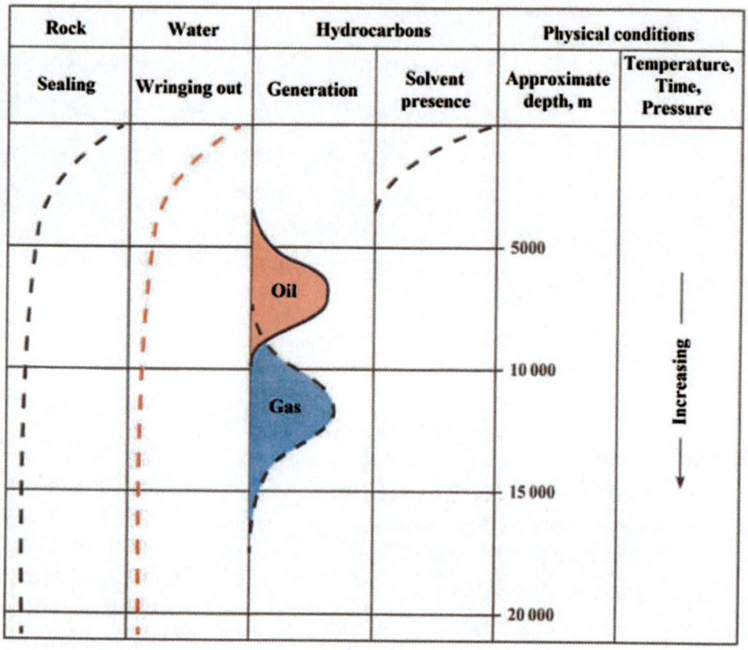

Fig. 13.1 The relationships between the peaks of sedimentation water squeezing and oil-and-gas generation in the SCB section (Feyzullayev, 2013)

They cannot be forced up into the reservoir. It is defined by anomaly high pore pressures provoking diapirism and mud volcanism.

Taking into consideration such factors as the uneven character of commercial oil and gas saturation in the SCB, the concentration of primary HC resources in the Absheron region (on land and sea), the law-governed nature of spatial change in lithofacies characteristics of rocks and anomalous high fluid pressure, it is supposed that HC displacement from source bed into reservoirs is characterized by different efficiency in different parts of the basin.

For instance, abnormal high reservoir pressure (AHRP) in the Baku Archipelago indicates HC-impeded discharge from the source bed. It is not out of the question that the problem of commercial saturation of surrounding structures and low efficiency in discovering commercial HC accumulation relates to AHRP. HC migration under these conditions should likely be impulse explosive.

Anomalous high reservoir pressure is being formed here due to HC-impeded discharge. On reaching a critical level exceeding a threshold of the rock strength, should impulsively eject hydrocarbons through the system of horizontal and vertical joints up to the carrier beds or press out fluidized clayey mass through the fault zone up to the reservoirs in the form of the diapir. In that case, the mentioned process should be continuous-discontinuous, mainly defining the Lower Miocene (explosive) nature of HC migration in the gas–gas/condensate phase (Feyzullayev, 2013).

Such a form of migration suggests that the mentioned process should be repeated in the source bed more than once during the geologic time to ensure HC migration in considerable volumes necessary for commercial HC accumulation. However, it should be noted that the HC migration process in the SCB is characterized by a short time since it is supposed that effective primary-secondary oil migration here began in the Late Pliocene. In the Absheron OGR (on land and sea), where major HC resources are localized, and frequently alternated clayey and sandy-silty rocks characterize the most favorable lithofacies conditions, the secondary HC migration is highly effective. It is believed that the factors that could be conducive to this process are subhorizontal HC migration through the thin (millimeter and

Fig. 13.2 A part of the section of the Maikopian series natural outcrop in the SE Gobustan (Feyzullayev, 2013)

Fig. 13.3 Microphoto of thin section (×90) reflecting the typical stratified structure of the Miocene clays in the Masazir area (well No. 132, an interval between 239 and 248 m) of the Absheron Peninsula (Feyzullayev, 2013)

centimeter thick) microlayers, which cannot be fixed by borehole logging but are visible in natural outcrops (Fig. 13.2), in an interlayer space (layered character of Miocene clays in the Absheron region is established, Fig. 13.3) and in horizontal micro joints up to the hearth of HC vertical discharge (tectonic disturbance) (Feyzullayev, 2013).

Since the water column pressure on subsided beds is decreased at great depths from the central sedimentation basin to its marginal parts, the fluid movement in reservoirs and in massive source series will be decreased in the same direction up to reaching highly permeable sub-vertical joints. The fact that at great depths, adequate horizontal pressure is lower than that of vertical pressure is conducive to the mentioned process (Fig. 13.4) (Feyzullayev, 2013).

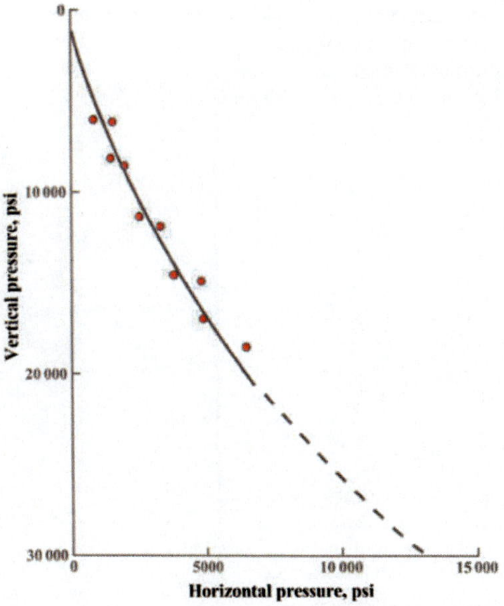

Fig. 13.4 A relationship between the vertical and horizontal pressures in the sedimentary section (after Price, 1979)

Currently, the prevailing opinion concerning the problem of oil genesis in the Productive Series (PS—Lower Pliocene) is that most of the oil resources here are epigenetic. This standpoint is based on many isotopic-geochemical analyses of organic matter in oil and gas, the correlation of their parameters, and basin modeling carried out over the last 15 years.

The mentioned dominant epigenetic conception of HC deposits in the SCB Productive Series and their youthful stage (their formation time does not exceed 1 Ma) suggest that the bulk of HC transportation from the generation source to the reservoirs seems to most likely have taken place at the expense of filtration form of mass-transportation (sub-vertical migration through the deep faults, tectonic breaks, and mud volcanic channels). According to the data given in previous subchapters, the South Caspian northern edge zone, the Absheron Archipelago region, has been an area with the most favorable geological conditions for such mass transportation. This affirmation is because the subduction zone characterized by high seismicity and dislocated rocks is confined to this region. This, in turn, caused continuous intensive HC migration and a high rate of trap filling.

At the same time, it is noted that farther south (for example, the central and southern Baku Archipelago areas), this basin is aseismic and composed of a thick series of notably less dislocated plastic rocks. It is therefore suggested that the migration process under these troublesome conditions has been impulsive, particularly in the form of gas blow-out through mud volcanic vents. That explains the fact that mud volcanoes are more widespread and more active here than in the Absheron Archipelago (Fig. 13.5).

13.2 Mechanism of Trap Filling in the Productive Series of the South Caspian Basin (According to Isotopic-Geochemical Data)

The reservoir filling process modeling is the most crucial factor in determining the mechanism of oil field formation, with which correct exploration and prospecting are closely connected. This is considered particularly important concerning the complex oil and gas-bearing systems. A good example is provided by the South Caspian oil and gas-bearing basin characterized by stressed tectonics, mud volcanism, and thick Paleozoic sedimentary series (up to 30–35 km)

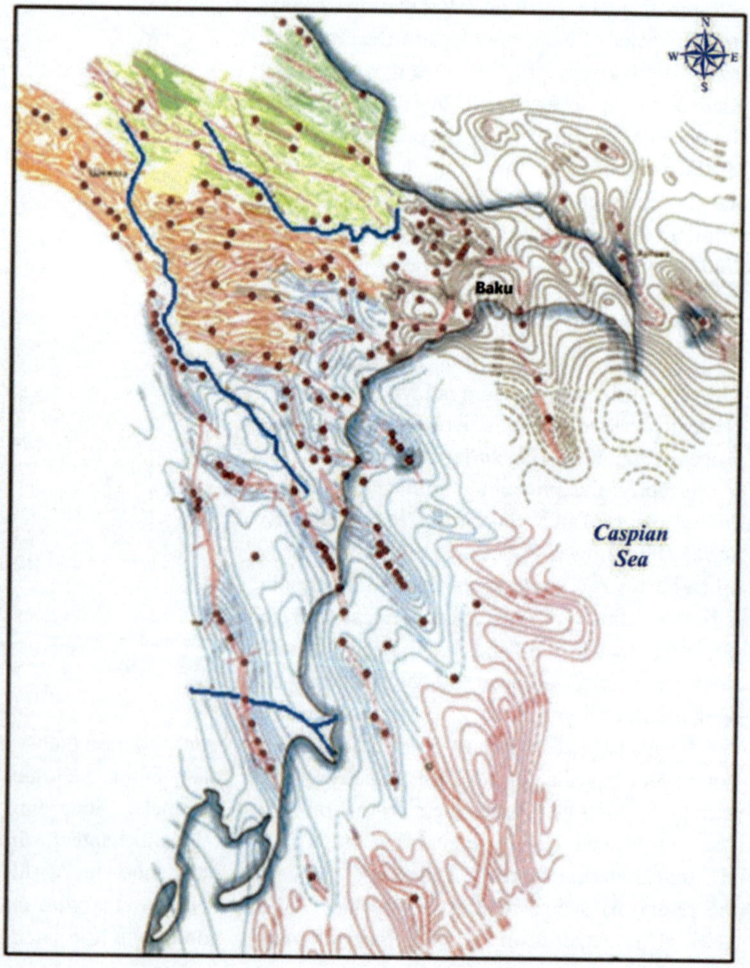

Fig. 13.5 Schematic sketch-map of the mud volcanoes distribution in Azerbaijan (on the land and adjacent offshore area)

enclosing several oil-producing complexes aged from the Jurassic to the Miocene where multilayer oil reservoirs are developed.

The following oil fields have been chosen as objects for the model geochemical construction: the Guneshli multilayer marine deposit located within the Caspian Azerbaijan shelf, the Garadagh oil field in the south-eastern Absheron Peninsula and much of the Lower Kur trough fields as Kyurovdag, Garabagli, and Neftchala.

Guneshli deposit. A topic of interest investigation carried out in this deposit is that the results obtained may be used in planning research works in adjacent Chirag and Azeri structures. The Guneshli deposit occupies 48 km^2 (12 km × 4 km) and is confined to the Absheron-Pribalkhan zone of uplifts. It is the northwestern end of a sizeable NW–SE trending anticline, including the Chirag and Azeri deposits. The Gunashli anticlinal uplift is an asymmetrical fold with a relatively gently dipping (at 150–170) limb displaced along an upthrust. Its southwestern limb is steeply dipping from 170–190 to 300–350. The deposit structure is highly complicated by a series of longitudinal (NW–SE trending) and transverse faults dividing this deposit into more than 15 blocks (Fig. 13.6).

Geologically, this oil field is composed of Pliocene and Quaternary deposits. Stratigraphic units of the Lower Pliocene (PS) serve as oil reservoirs. The thickness of the PS here reaches 3500 m. It is represented by alternating sand, sandstone, clay, and siltstone. The Productive Series is subdivided into lower division (PS$_1$) and upper division (PS$_2$). The lower division consists of the following five suites: Kala suite (KS), Underkirmaki suite (UKS), Kirmaki suite (KirS), Superkirmaki suite (SKS), and Superkirmaki clayey suite (SKCS). The Break, Balakhani, Sabunchi, and Surakhani suites represent the upper division. All the lower division suites (PS$_1$), Break suite, and Balakhani suite of the upper division (PS$_2$) are oil-bearing. The most oil-saturated are the Break and Balakhani suites.

Determining a source of oil generation in these deposits is necessary to introduce a correction in the geochemical modeling of reservoirs in multilayer deposits (enclosing several oil-producing complexes). With that end in view, the geochemical model constructions in the Guneshli multilayer reservoir have been carried out in two stages. The first stage studied the geochemical features of oils from the PS multilayer reservoirs and their genetic and geochemical correlation. Within the second stage, a spaced relationship in changes in basic geochemical parameters has been studied, and proper model constructions have been done.

According to more than 100 density tests, oils from all the PS stratigraphic horizons have an average density varying from 0.830 to 0.874 g/cm^3 (on average, 0.866 g/cm^3). A unique feature of oils from all PS suites is their similar hydrocarbon group and continental compositions.

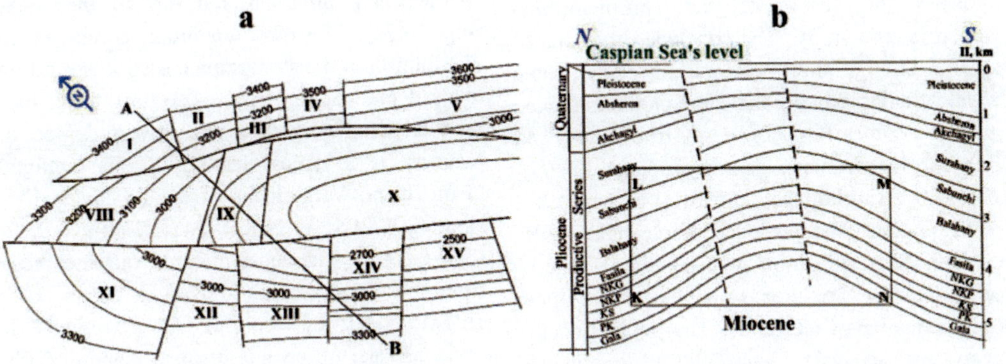

Fig. 13.6 Structural sketch-map of the Guneshli oil field over the break suite roof (**a**) and profile along the line of A-B (**b**). The KLMN block limits the studied interval

The group hydrocarbon composition of the fraction <2000 C is composed of alkanes (49.40–55.2%), cyclanes (35.30–37.62%), and aromatic hydrocarbons (7.14–13.54% of the fraction). It is noted that studied oils are characterized by high resin content and extremely low asphaltene content, estimated from traces to 0.3%. Detailed gas-chromatographic investigation of oils sampled from all horizons of PS_1 and PS_2 revealed that all of them contain normal alcanes with carbonic atoms numbering up to C_{36} inclusive and isoprenoids up to C_{23}. Besides, all the analyzed oils mostly comprise n-alcanes (up to C_{18}). A typical feature of oils from all horizons is a roughly equal number of normal alcanes, isoprenoids, and predominance of the isoprenoids over the methanee hydrocarbons and the even alcanes (n-C_{10}-n-C_{30}) over uneven alcanes (n-C_{11}-n-C_{29}) varied in ratios within the interval of from 0.670 to 0.879. An absolute similarity in the distribution and relationship of alcanes in oils of all PS horizons indicates that these oils' source organic matter composition and formation conditions have been the same. In this connection, it should be specially pointed out the fact that the sum of n-C_{15} + n-C_{19} repeatedly exceeds the sum of n-C_{27} + n-C_{29} + n-C_{33} (that is 1.07–2.8 as against 0.00–0.26) which is joint with a low isoprenoids ration (Pr/Ph = 1.16–1.36) compared with normal alcanes (Pr/n-C_{17} = 0.42–0.43 and Ph/n-C_{18} = 0.34–0.36) indicates the mostly algal marine type of oil-generating organic matter (Gelpi et al., 1970; Kenneth & Moldowan, 1993; Tissot & Welte, 1984).

From the result of the chromatomass-spectrometric study, it is concluded that oils of all PS stratigraphic horizons have the same characteristics, namely the same biomarkers, i.e., normal steranes (C_{27}–C_{28}–C_{29}), triterpanes: 18α(H),-18b(H)-oleanane, gamma-serpane, 17α(H)-22,29,30-trisnorhopane, 18α(H)-trisnorneohopane relation. The presence of oleanane molecules in oil composition suggests that the age of oil-generating complexes is like the Upper Cretaceous-past-Cretaceous. This fact also suggests that an insignificant part of continental organic matter is present in oil composition. These homogenous and isotopically heavy oils we have formerly studied within the SCB are typical of those generated in the diatomic suite (Guliyev et al., 1999a, 1999b, 2001).

The isotopic ratio values vary in very narrow limits, from − 25.56 to 24.65%, which suggests their common source. Also, this conclusion follows from the carbon stable isotope distribution in oil fractions (saturated, asphaltene, heterocyclic, and in resins). The $\delta^{13}C$ values of the carbon stable isotopic ratio in alcanic and aromatic oil fractions indicate that organic matter is mostly of marine type. Thus, a detailed isotopic-geochemical investigation of oils from the Guneshli field shows their complete uniformity. This is because they have been generated from a single complex of oil-producing deposits of the Middle-Upper Miocene Diatomic suite.

The oil maturity defined by normal sterane isomers relation 24-ethllcholestane (20S/20S +20R), comes to 0.410–0.460 that, which indicates a high degree of transformation. The same maturity is defined by pentacyclic triterpenes. For example, according to the ratio of 17α(H)-22,29,30-trisnorneohopane and 18α(H)-22,29, 30-trisnorneohopane isomers, the absolute values of maturity are changed from 0.45 to 0.48. Empirically, vitrinite reflectance equivalent (R_0) estimated by sterane C_{29} is:

$$R_0 = 0.59 - 0.66\%.$$

An exciting feature is observed in directivity in the change of oil maturity degree over all the PS section, marked in its increase in a south-southeasterly direction and dip of the rocks (Fig. 13.7a). The most important feature is oil distribution in the stratigraphic section according to its degree of maturity. According to all biomarker parameters data, the lowest degree of maturity is a typical feature of oils confined to the I and VIII blocks of the PS lower division (20S/20S + 20R = 0.41; Ts/Tm = 0.82; R_0 = 0.59%). The most matured oils are those from the Break suite (20S/20S + 20R = 0.45; Ts/Tm = 0.87; R_0 = 0.66%). Oils from the IX-X horizons take up an intermediate position (20S/20S + 20R = 0.44; Ts/Tm = 0.855; R_0 = 0.63%). An analysis of oil distribution within the Break

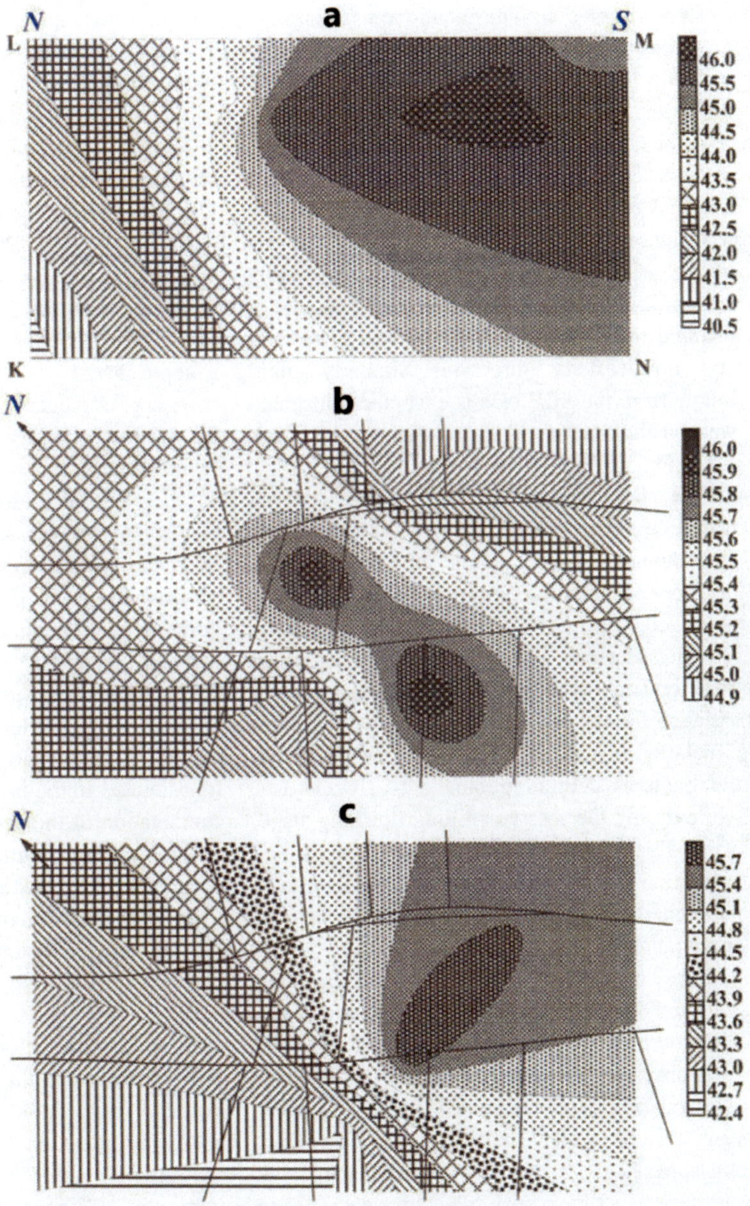

Fig. 13.7 The changes in oil maturity according to the normal sterane C_{29} (2OS/2OS + 2OR) isomerization in the Guneshli field. **a** across the field within the block KLMN in the profile A-B, **b** in the Break suite, **c** in the Balakhani suite

suite and in IX-X horizons shows that the change in their maturity degree generally occurs in the same direction (Fig. 13.7b, c).

Besides, it is noted that maximally matured oils are confined to the X block and the fractured zone delimiting the XIII and XIV blocks and the strip that delineates these maximums and is placed at an angle to the anticline trend. This northwesterly trending strip cuts the field into two parts. The maximum in Horizon X gravitates towards the crystal fault zone of conjugation of blocks X and XIII, and it is quite easy to note that areas of maximally matured oils in IX-X horizons appear to be imaged on the same areas in the Break suite unchanging its orientation. At the same time, it is remarkable that the least matured oil fields appear to be utterly coincident within both horizons of block XIII, acquiring the western orientation. The mentioned factor is of great importance from the standpoint of reconstruction of the process and mechanism of HC migration and subsequent filling of the Guneshli reservoirs. This fact favors the vertical migration of hydrocarbons in which streams have been arranged crosswise to the structure trend. Otherwise, i.e., in lateral reservoir filling from the side of the SCB significant oil and gas-generating source, it was to be expected reversed change of matured oils by less matured ones in an easterly and northeasterly direction. Similarly, fluid inflow from the SCB side is excepted, illustrated well on the map of oil distribution in the Break suite (Fig. 13.7a).

The field block structure causes highly favorable conditions for vertical fluid migration from the underlying Diatomic suite of the Middle to Upper Miocene age. From the interpretation of space relationships in oil distribution, it is inferred that tectonic dislocations displayed varied fluid-dynamic activity during different stages of anticline formation. It may be assumed that fluid circulation in the setting stage occurred through the fractures delimiting blocks I-VIII and I-II, while during the reservoir filling finishing stage, active fractures were limiting the IX block and delimiting blocks XIII-XIV. Also, the longitudinal arched rupture was active within the zone of conjugation of X and XIII blocks.

Based on all biomarker parameters suggesting that the PS different-aged complexes contain oils varied in degrees of maturity, it is supposed to be an approximate mechanism of time sequence in reservoir filling. We believe that the PS_1 reservoirs have been filled in the first part of the time, and some later—of the PS_2. Besides, the Break suite reservoirs have been filled considerably and more intensely than reservoirs in the IX and X horizons. Vertically migrating, a late portion of fluids pressed back less matured oils in a southwesterly direction (Fig. 13.7).

Garadagh field. As was noted in the previous chapter, the least matured oils in the SCB are confined to the Paleogene-Lower Miocene reservoirs. The notable feature is an increase in oil maturity towards the central basin in the direction of regional subsidence of the beds. A spatial correlation of thermal and genetic indexes (increase in oil maturity and weighting of the isotopic composition of oils, respectively) is observed. Naturally, the question arises: why oils from the Oligocene-Lower Miocene reservoir (isotopically light oils) are less matured than oils from younger Pliocene reservoir (PS), and why oils from the PS reservoir having older Paleogene isotopic marks are less matured than those from the same PS reservoir with diatomic isotopic mark?

A separate question arises about the reasons for the low maturity of oils having Paleogene-Maikopian isotopic mark in the Pliocene reservoir of the Absheron trough and oils from the Oligocene—Lower Miocene oil fields in the Shemakha-Gobustan trough. Here, it is necessary to solve this problem jointly with the disputable question of anomalies taking place in some oil-and-gas condensate and gas condensate fields. One anomalous oil-and-gas condensate object is the Garadagh field located in the extreme southwestern maritime part of the Absheron Peninsula. It is situated in the latitudinal limb of the marine continuation of the deep fault zone and confined to the anticlinal uplift, having a steeply dipping southern limb (350–600) and a slightly gentle (300–350) northern limb (Fig. 13.8). A significant 200–300 m tectonic break in shift amplitude passes along the fold axis. It is known that oil-bearing objects here are confined to the Miocene and Lower Miocene deposits.

The isotopic composition of oils from the Miocene reservoir is estimated to be as light as $\delta^{13}C = 27.5‰$, characterized by very low maturity ($R_0 = 0.5\%$). These parameters in oils from the Pliocene reservoir appear to be increased $\delta^{13}C$ till 26.25‰ and $R_0 = 0.6\%$. The share of the diatomic component in oil composition has increased.

The facts that the low matured Miocene pools are confined to the northern limb, whereas the Pliocene pools are confined to the southern limb turned upon the Caspian Sea (marine continuation of the Djeirankechmes depression) (Fig. 13.8) and that the southern limb as regards the northern one is subsided along the southerly

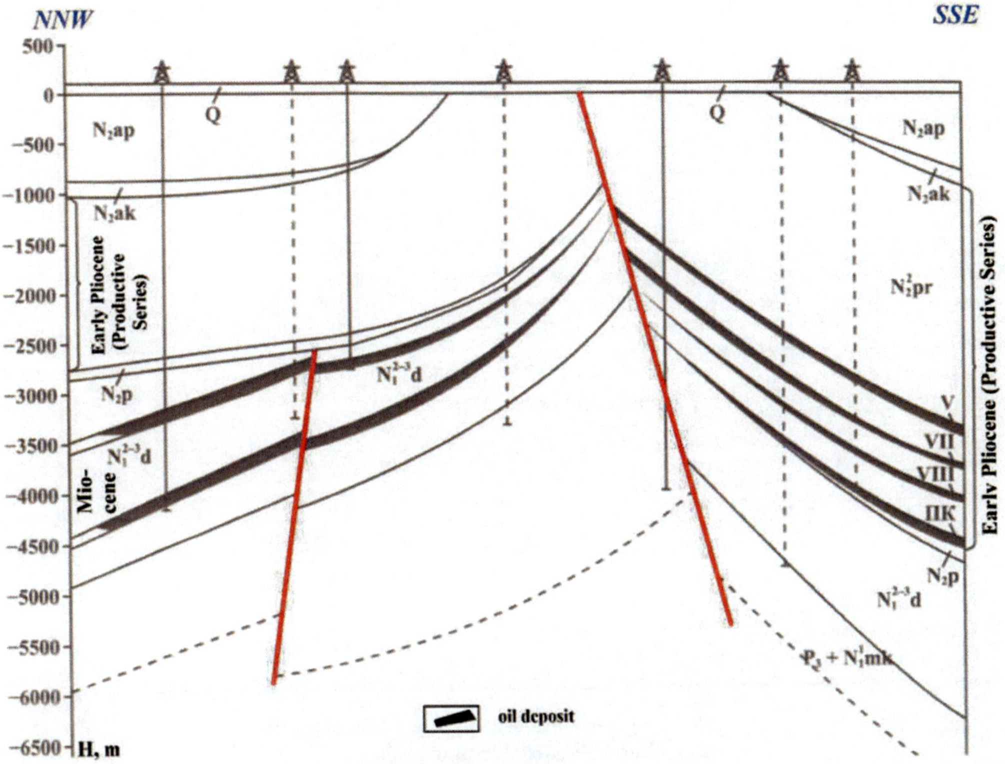

Fig. 13.8 A cross section through the Garadagh oil-and-gas field

dipping fault with an amplitude of 1000–1200 m together with the fact that the low matured oils of the Paleogene-Miocene fields (Umbaki, Adjiveli, Cheildag, and other fields) appear to be regionally arranged around more highly matured oils of the PS suggests that the reservoirs of the Pliocene fields, on the whole, and the Garadagh field in particular, have been filled in two stages (Alizadeh et al., 2017). We believe that this is caused by the SCB intensive down warping during Upper Pliocene–Quaternary time along such limiting deep faults as the mentioned longitudinal Adjichai-Alyat fault and transverse Yashma and regional West-Absheron faults (Fig. 13.9). As a result, it was disturbed the Paleogene-Lower Miocene fluid guiders served as a pathway for regional lateral fluid migration from the central Djeirankechmes—South Absheron trough toward its slopes, in the direction of Gobustan, Absheron Peninsula, northern uplifts of the Baku Archipelago and Absheron sill (Fig. 13.9). Because of this, almost all the formed pools appear to be isolated from the Paleogene Lower Miocene feeder source. The Prealpine basement fast-moved subsidence in this part of the South Caspian resulted in the fact that a thick sedimentary series (up to 9000–10,000 m) was formed during a short time (about five Ma), causing accelerated involvement of the younger Middle-Upper Miocene (Diatomic) deposits into a zone of oil formation that becomes a significant factor of the predominance of isotopically heavy and more matured oils in marine fields.

This example leads us to conclude that the isotopic composition and maturity of oils in the SCB directly depend on its tectonic history, causing multiphase fluid migration and trap filling.

The Lower-Kur trough's oil fields. An analysis of a map of oil distribution in the SCB according to varying degrees of maturity shows that the most transformed are oil fields in the Lower Kur trough.

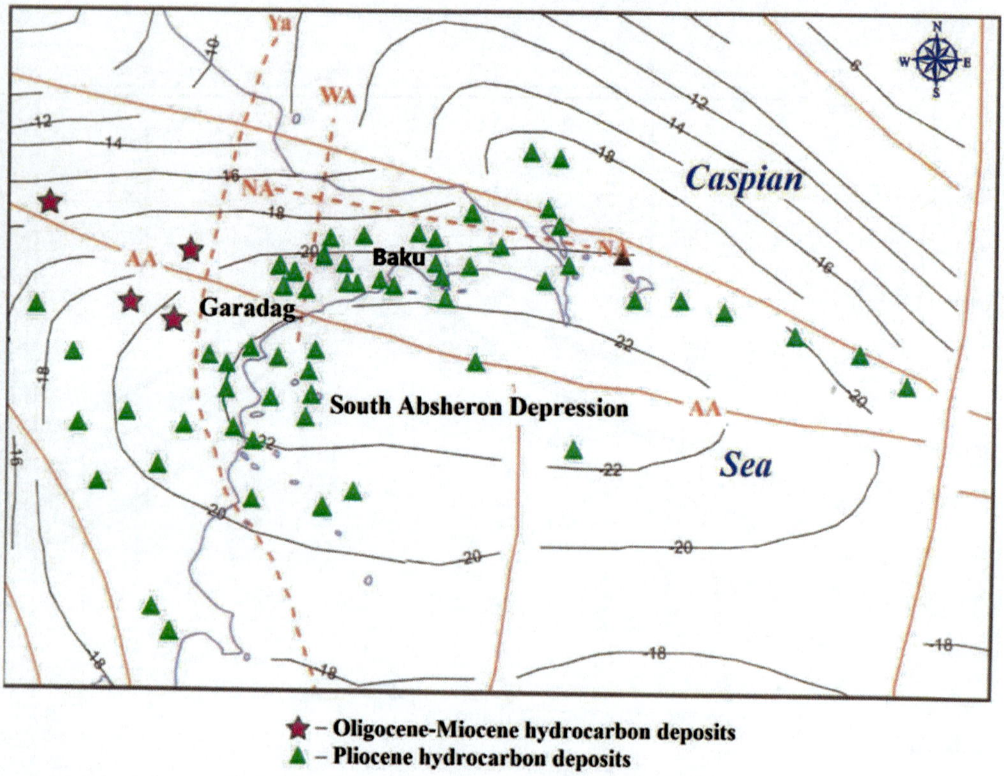

Fig. 13.9 A generalized sketch map of the surface of the SCB crystalline basement (in km). AA—Adjichai-Alyat deep fault; YA—Yashma deep fault (flexure); WA—Gyarbi-Absheron regional fault; NA—Shimali-Absheron fault. Stars are Pliocene oil, oil–gas, and oil–gas condensate fields. A fast-moved subsidence along the Yashma, West, and North Absheron faults in Late Pliocene–Quaternary time broke off the hydrocarbons' lateral migration from the Maikopian (Paleogene-Lower Miocene) fluid-generating source to the SCB towards the Gobustan (to the east) and Absheron (to the north)

The fact that oils in this region are highly matured does not make sense in a notion of regional lateral oil migration from the South Caspian trough up dip the beds in the PS reservoir.

From the results of detailed investigations on oil fields of the Kyurovdag-Garabagli-Neftchala anticlinal zone, it is inferred that oils in these fields are characterized by different trends in the increase of their maturity (Fig. 13.10) from an arch of anticlinal traps toward the adjacent Mugan-Salyan and Kyursyangya troughs. It serves as a reason to believe that reservoirs of this anticlinal zone have been filled with hydrocarbons from local sources located in the mentioned troughs. For example, a reservoir on the Kyurovdag field has been filled at the expense of backward fluid migration from the two troughs. As a result, low-matured oils of the early migration phases have been driven out toward an arch and appeared bordered by more matured oils of the late phases. The traps of the Garabagly and Neftchala fields have been filled with fluids migrating from the Mugan-Salyan trough. The more so that geological-geophysical data confirm HC generation in described troughs. According to the paleotectonic and paleogeographic reconstructions, an initial subsidence stage began in the Paleogene. By interpreting seismic data (Fig. 13.11), the thickness of Paleogene-Miocene deposits in these troughs reaches 5 km, and they overlie the Mesozoic volcanogenic-carbonate formations, subsiding to more than 7 km. The deep wells testing data estimates the temperature at a horizontal section depth of 6000 m exceeds 110–1200°C.

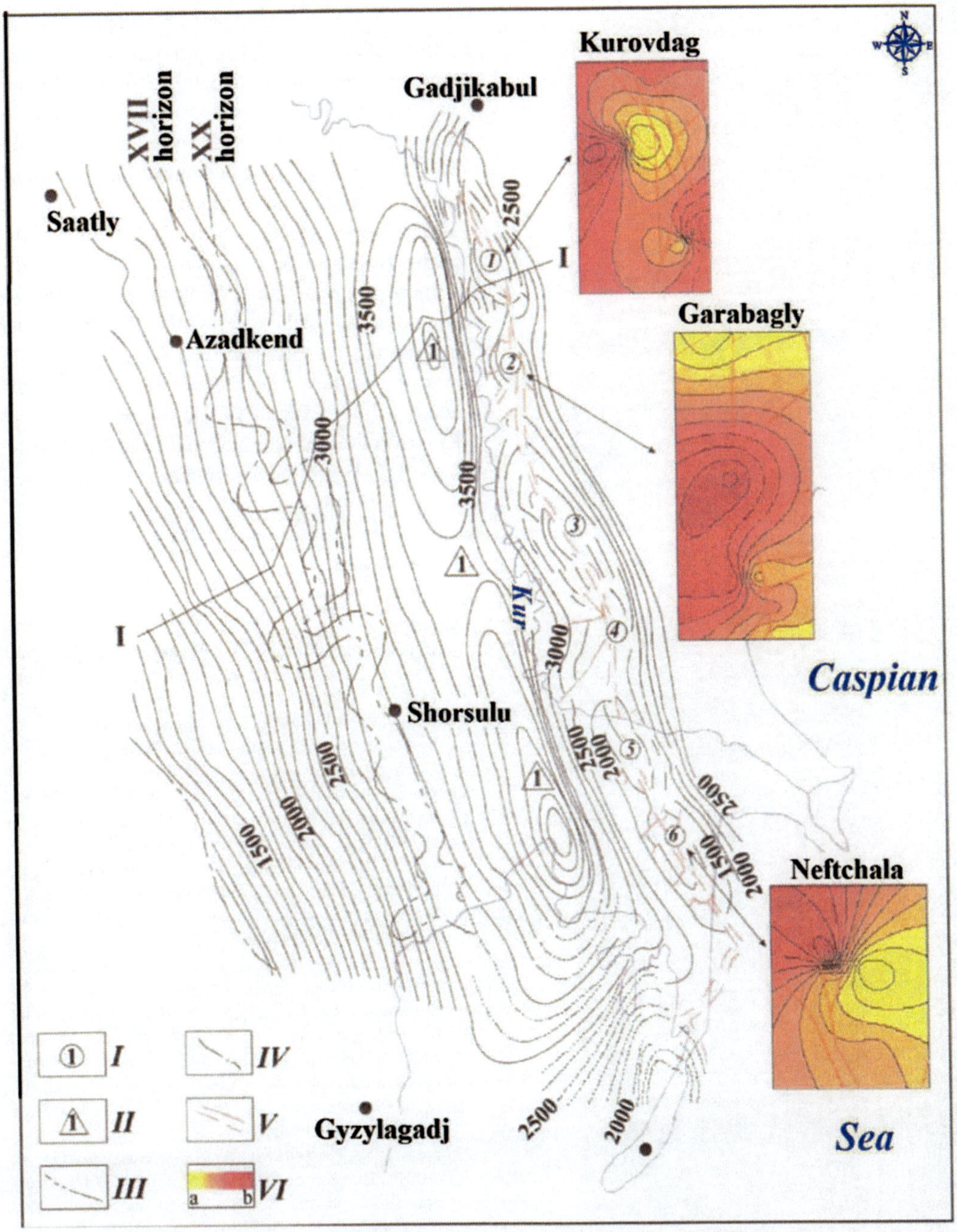

Fig. 13.10 Structural map of the southwestern Lower Kur trough over PS roof and oil distribution scheme drawn according to the degrees of maturity. I—anticlinal structures (oil fields): 1—Kyurovdag, 2—Garabagly, 3—Babazanan, 4—Durovdag, 5—Hilly, 6—Neftchala; II—Mugan-Salyan trough; III—the 17th and 20th horizons pinching out zone; IV—the PS pinching out zone; V—regional and local tectonic disturbances; VI—oil maturity degrees estimated by isomerization of normal sterane C_{29} (20S/20S+20R) (a—minimum, b—maximum)

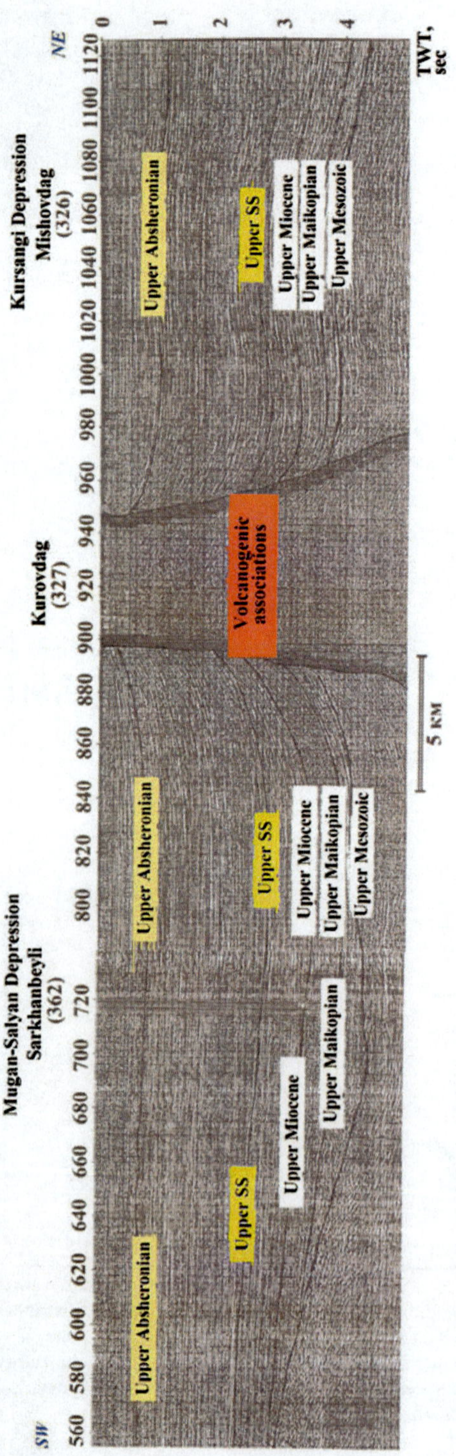

Fig. 13.11 Seismic profile through the Lower Kur trough (along the line of I-I shown in Fig. 13.10)

Consequently, Paleogene-Miocene deposits occur within a zone of the oil–gas window. The temperatures at all horizontal sections fall sharply toward the South Caspian Basin. It means that Paleogene-Miocene deposits in the Mugan-Salyan and Kyursyangya troughs have subsided to a zone of oil window much earlier, reflected in a high degree of oil maturity in the Lower Kur basin. It should be emphasized that the southwestern limbs of the Kyurovdag and Garabagli structures (turned towards the Mugan-Salyan trough) are oil-saturated to a far greater extent than the northeastern limbs, confirming preferential fluid migration into reservoirs of Kyurovdag-Garabagli-Neftchala anticlinal zone from the Mugan-Salyan trough.

References

Alizadeh, A. A., Guliyev, I. S., Kadirov, F. A., & Eppelbaum, L. V. (2017). *Geosciences in Azerbaijan. Vol. II: Economic minerals and applied geophysics* (340 p). Springer.

Feyzullayev, A. A. (2013). Migration pathways of hydrocarbons in South-Caspian Basin. *Journal of Geology and Geosciences, 2*(3), 1–6.

Gelpi, E., Schneider, H., Mann, J., & Oro, J. (1970). Hydrocarbons of geochemical significance in microscopic algae. *Phytochemistry, 9*, 603–612.

Guliyev, I. S., Feyzullayev, A. A., & Huseynov, D. A. (1999a). Carbon isotopic composition of oils in the South Caspian megadepression. *Azerbaijan Oil Industry, 6*, 3–13. (in Russian).

Guliyev, I. S., Feyzullayev, A. A., & Huseynov, D. A. (1999b). Genetic typing of oils from reservoirs of different ages in the western part of the South Caspian oil and gas basin in the light of new isotope-geochemical data. In *Proceedings of the 3rd international conference "new ideas in geology and geochemistry of oil and gas"* (pp. 74–77). Moscow State University (in Russian).

Guliyev, I. S., Feyzullayev, A. A., & Huseynov, D. A. (2001). Isotopic composition of carbon in hydrocarbon fluids of the South Caspian megadepression. *Geochemistry, 3*, 271–278 (in Russian).

Kenneth, P., & Moldowan, J. (1993). *The biomarker guide: Interpreting molecular fossils in petroleum and ancient sediments* (365 p). web.

Tissot, B., & Welte, D. (1984). *Petroleum formation and occurrence*. Springer-Verlag.

14 Hydrocarbon Preservation Conditions in the Productive Series

One of the obligatory conditions for forming and preserving the minable oil and gas accumulation in the earth's crust is the presence of reservoir rocks and cap rocks. It is well-known that the SCB is confined to highly dislocated and mobile folded regions characterized by such widespread processes as orogenesis, fault tectonics, seismicity, anomalous high pressure, diapirism, and mud volcanism. All that creates favorable conditions for subvertical HC migration, playing positive and negative roles in the HC ontogeny.

14.1 Some Physical–chemical Conditions of Hydrocarbon Preservation Within PS

The quality of fluidal caps playing an essential role in the location of oil and gas accumulation zones is defined by the following features: lithologic-petrographic, granulometric and mineral composition, structural, textural, physical–chemical properties, the degree of discontinuity (the presence of dislocation with a break in continuity and jointing) and so on. It is also known that clayey rocks and evaporites (salt, gypsum, and anhydrite) are the most reliable fluidal caps. The best among them are montmorillonite varieties, which appear to be swelled in the presence of moisture and become practically impermeable.

The thickness and extent of these caps are also essential features. It is supposed that potentially significant oil fields may only be formed under thick caps. However, this affirmation may not be recognized as acceptable for the frequent and rhythmical alternation of sandy-silty and clayey rocks, for example, in the section of the PS lower division in the Absheron oil-and gas-bearing region. It is believed that under these conditions, the regionally persistent thick clayey caps, immediately overlapping reservoir rocks, and small thick clayey interbeds may play the role of reliable fluidal caps (intraformational caps). This relates to the fact that the pressure gradient P/H in such small-thick interbeds is insufficient. In this connection, small-thick clayey interbeds separating sandy-silty rocks in the PS lower division of the Absheron oil and gas–bearing region are classified as intraformational cap rocks. It is confirmed by a distinction between the oil composition in different horizons and the absence of oil in water-bearing reservoir beds.

As is seen from the SCB sections, the thicknesses of the clay series are changed from fractions of a centimeter to a few tens of meters. The carbonate content of studied clays is estimated to be in the range of 10–24%. Clays in the PS are mainly plastic. Their structure is politic, silty, and psammo-siltstone, while their microstructure is irregular, mainly massive, and rarely stratified. Also, the basin's age and longevity of the pool's destruction process play an essential role in preserving hydrocarbon accumulation since oil and gas pool formation, reformation, and destruction are united and interconnected.

A. Alizadeh et al., *Pliocene Hydrocarbon Sedimentary Series of Azerbaijan*,
Advances in Oil and Gas Exploration & Production, https://doi.org/10.1007/978-3-031-50438-9_14

Continuous processes of oil pool formation may simultaneously be accompanied by its destruction, which in turn may be replenished by reserves at the expense of steady lateral and vertical migration. The last feature is clearly manifested in young basins such as the SCB, where the young age of pools (for example, their formation time in the PS does not exceed 1 Ma) destruction process is not long (in geologic measurements) (Alizadeh et al., 2017).

The chemical and biochemical transformation scale appears insignificant if there is no reliable fluidal cap.

Based on the data obtained from the deep control of hypergene processes, all the SCB fields have been classified according to their involvement in these processes using the data of the PS horizons' minimal depth of occurrence. It was determined that about 70% of the SCB fields, one way or another, occur in a zone of oxidation and biodegradation of oils (Table 14.1), being confined to the basin slope zones.

A depth of pool (or, in other words, the thickness of overlapping deposits) as one of the significant factors determining a scale of the HC devolatilization is clearly reflected on a series of histograms of the distribution of gas factor values made for different depth intervals all over the SCB (Fig. 14.3).

Figure 14.4 shows a marked tendency toward an increase in gas factor with depth. At the same time, specific zoning in the change of this parameter with depth is observed, and two zones (I and II) and two zonules (Ia and IIa) are distinguished. A zone II (at a depth of more than 4 km) may be attributed to the interval of the beginning of gas generation (gas factor is > 300m^3/t).

Zone I reflects an interval of sub-vertical gas migration from zone II (gas factor is 100–200 m^3/t). In the near-surface interval (at 0–2 km), this migration process appears to be disturbed because of faulty preservation conditions (zonule Ia, a gas factor is < 100 m^3/t).

An analysis allows the conclusion that at the end of Absheron time, particularly in the Quaternary, the SCB near-slope structures have been exposed to the most significant uplift and erosion jointly with numerous dislocations caused that they have been highly degassed.

These structures are known in the northern and northwestern Absheron Peninsula (Kirmaki, Balakhani-Sabunchi-Ramany, Binagadi, Atashkyakh, and others), in the Absheron Archipelago (Chilov Island, Absheron Peninsula, and Arsu) and Baku Archipelago (Sangachal-Deniz, Duvanny-Deniz, Hamamdag-Deniz and others), in Gobustan (Toragay, Kyanizadag and others) and within the Lower Kur basin (Babazanan).

By granulometric composition, the clayey rocks of deep-sea SCB are commonly represented by well-elutriated varieties containing about 30% of sandy-silty material. The granulometric composition of the PS clays is controlled by the paleogeographical condition of sedimentation and the distance from the feeding source of terrigenous material. The most dispersive clays are from the Baku Archipelago, with more than 80% pelitic fraction content. It is noted that more "pure" clayey varieties have also been found. Certain regularity characterizes their areal distribution according to which PS clays in northwestern and western SCB contain more siltic admixture than those in the southeastern and eastern basins. PS clays in the Baku Archipelago

Table 14.1 Statistical analysis of the minimal depth of occurrence of the PS horizons in the SCB oil and gas-bearing areas

Pool type	Minimal depth of occurrence (km)/Quantity of areas					
	< 1	1–2	2–3	3–4	4–5	5–6
Oil pool	15	6	2	2	1	–
Oil and gas pool	27	13	5	5	3	–
Gas field	6	1	2	–	–	–
Gas-condensate field	–	–	–	2	–	1
In all	48	20	9	9	4	1

and southeastern Lower Kur trough are better elutriated and more plastic than in northwestern and western areas.

The Break suite (Fasila) deposits are distinguished from the PS overlying and underlying deposits by widespread sandy-silty formations. This suite mainly comprises sandstones, poorly graded rocks, and siltstones. Rarely observed are silty clays. The carbonate content of these rocks does not exceed 10%. Consequently, there are no reliable caps in the Break suite. In that case, the clay rocks of the PS upper division serve as the caps. According to the quantitative X-ray diffractometric analysis, clays of the PS upper interval are markedly distinguished by increased smectite mineral association from those of the Break suite. It is noted that the smectite content in the Break suite does not exceed 20%, whereas the content of hydromica exceeds 60%. Another minor factor is that, unlike the clays of the PS upper interval, the Break suite clays contain a mixed-layered chlorite-smectite mineral.

In the northern and northwestern Lower Kur trough, the content of smectite decreases to 30–40%; the same smectite content (not exceeding 30%) is noted in the northern and northwestern Baku Archipelago, where clays from the PS lower and upper divisions are relatively consolidated, calcic ($CaCO_3$ = 10–12%) and polymictic. The latter includes such clay minerals as hydromica, smectite, kaolinite, chlorite, sepiolite, palygorskite, and various clayey mixed-layered minerals. The content of smectite and hydromica is changed as in section as an area, decreasing in a southeasterly direction with increasing hydromica content in the same direction. The predominance of hydromica mineral association characterizes the clays of the PS upper division within the Absheron-Pribalkhan tectonic zone. This caused a change for the worse in the insulating properties of clayey caps. Clays of the PS lower division, like those from the Break suite, are characterized by slightly increased hydromica and low montmorillonite content. It is caused by changing environmental conditions in sedimentation basins and clay minerals delivery at different times. In this connection, it is interesting that montmorillonite is absent in the clay of the Sabunchi suite of the PS upper division at the Umid area (№2 well within the depth interval from 4636 to 4631 m).

In contrast, it is present in a sufficient quantity at other intervals. A degree of dislocation and modern geodynamic activity of the basin is believed to be essential in determining HC preservation conditions. As it is well-known, the SCB is confined to the highly dislocated and mobile folded region where processes such as orogenesis, fault tectonics, seismicity, abnormally high pressure, diapirism, and mud volcanism are widespread. All that creates favorable conditions for sub-vertical HC migration, which plays positive and negative roles in hydrocarbon ontogeny. Since HC resources in the SCB are mainly concentrated in Lower Pliocene deposits, the preservation conditions of HC pools are defined by folding, which began at the end of the Pliocene and reached its maximum in the Middle Anthropogenic (Khain & Shardanov, 1954).

From the result of recent crustal movement monitoring in Azerbaijan, it is inferred that this process (folding) has been going on until now (Kadirov et al., 2008). The development of ruptures in the near-crest parts of anticlines caused the formation of discharge sites for underground fluids, which have been found on the Gyuneshli, Binagadi, and other structures. The geological failure intensity is defined by the scales of HC vertical migration from the pools to the surface and controlled by the hypsometric position of a pool, by the density of tectonic fractures and joints developed in its overlying deposits, and by the presence of mud volcanic vents and their activity. The presence of numerous fractures, mainly formed after oil and gas pool formation as well as seismicity, might provide significant evidence illustrating the increased probability of the failure in the integrity of fluid stop sealing up oil reservoir. After all, they may lead to the reformation and even destruction of HC accumulation.

An analysis of the SCB geological structure and the results of previous investigations shows that of all the studied areas, the Shamakhi-Gobustan oil and gas-bearing region (OGR) is the most dislocated area (widespread mud

Table 14.2 Folding tension data for the SCB (after Feyzullayev & Tagiyev, 2007)

Region	Area (km^2)	The total extent of anticlinal lines (km)	The specific extent of anticlinal axes (km/100 km^2)
Absheron	7060	491	6.9
Shamakhi-Gobustan	5830	1150	19.7
Lower Kur	5300	270	5.1
Baku Archipelago	10,300	480	4.7
Western Turkmenistan (SCB eastern slope)	15,300	920	6.0

volcanoes, folds, and tectonic breaks). Hence, this region is characterized by the worst conditions for preserving HC accumulation. The fact that the rocks here are highly dislocated compared with those from other OGRs is seen by the parameter proposed by Sultanov et al. (1983) (Table 14.2).

Most oil shows coming out to the day in Gobustan are confined to the zones of tectonic breaking and related mud cones, gryphous, and salses. In such a way, oil and gas shows confined to the ruptured zones having high vertical slips (up to 2000 m) in southwestern Gobustan appear to be more intensive than those in Jeirankechmes depression where ruptures localized in the Pliocene deposits have relatively low slips (in order of 100–200 m) (Salayev & Kastryulin, 1974).

An analysis of the relationship between the specific extent of tectonic dislocations and the quantity of oil and gas shows that these features are interconnected (Fig. 14.4). Salayev and Kastryulin (1974) believe that the significant role in destroying oil and gas accumulations in Paleogene-Miocene deposits in Gobustan may belong to sublatitudinal axially longitudinal ruptures and their derivatives—local cross faults. No visible connections of natural oil and gas shows with submeridional ruptures have been observed.

Widespread large longitudinal ruptures extending at a great distance along anticlinal belts and numerous local cross faults are typical features of the SCB. An interpretation of such significant features of oil and gas, showing their development history, depth and time of location, activity, strike, dip angles, and so on, is critical to clarify the role of the mentioned dislocations in preserving oil and gas accumulations.

From the result of laser measurement of HC concentration in terrestrial atmosphere carried out along regional profiles, it was estimated that the highest density of superphones sub-vertical HC streams and hence highest density of anomalous oil and gas shows in terrestrial atmosphere per unit of profile length is marked in the Shamakhi-Gobustan region (2.0) as compared with Absheron and Lower Kur regions estimated to be (0.4) and (0.2), respectively (Guliyev et al., 1985).

From the result of the valuation of specific streams in different methanee micro shows of Azerbaijan carried out by some researchers, it is concluded that the most intensive degassing occurs on mud volcanoes and within the Greater Caucasian southeastern termination. Relatively high values of these streams are noted in the Absheron Peninsula; the lowermost values have been marked within the Lesser Caucasian carbonated province and the Kur basin (Table 14.3).

14.2 Gas Pool Hypsometric Position Value

Depending on its modern depth, the intensity of gas component loss in HC accumulation is clearly seen in an example of the change in gas factor value within the Balakhani-Gum Island anticlinal line. In the Balakhani-Sabunchi-Ramany field, the minimum depth of the producing bed roof is 8 m, whereas, in the Gum Island field, its depth of occurrence exceeds 1800 m.

Table 14.3 Quantitative valuation of the methane micro-shows in Azerbaijan (after Feyzullayev & Tagiyev, 2007)

Region	Micro-shows (methane-specific stream) (t. km^{-2} $year^{-1}$)	Quantity of measurements	Source
On mud volcanoes	250	87	Etiope et al. (2004)
The Greater Caucasian southeastern termination including the mud volcanic region	158	15	Balakin et al., 1981
The territory of Azerbaijan out of mud volcanoes			
The Kur basin	4.4	10	Balakin et al. (1981)
The Lesser Caucasus	3.0	8	Balakin et al. (1981)
The Absheron Peninsula	35.0	–	Sokolov (1971)
Azerbaijan (measurement limits)	10.2–73.0	–	Voitov (1975)

As shown in Figs. 14.1, 14.2 and 14.3, gas factor values, in conformity with the above-mentioned geological structure, are the least in relatively shallow Balakhani-Sabunchi-Ramany and Surakhani fields.

The fact that the minable oil and gas content has been established in the Absheronian and Akchagylian suites (in the Quaternary and Pliocene deposits, respectively) within some areas of the Lower Kur (Fig. 14.5a) and Absheron (Fig. 14.5b) oil and gas-bearing regions which have not to possess necessary conditions for the syngenetic HC generation. It is unmistakable evidence that they have been formed at the expense of the reformation of HC accumulation from the underlying Productive Series. It is also certain that mud volcanoes are complicating most of the SCB structures and play a part in the geological destruction of oil and gas fields.

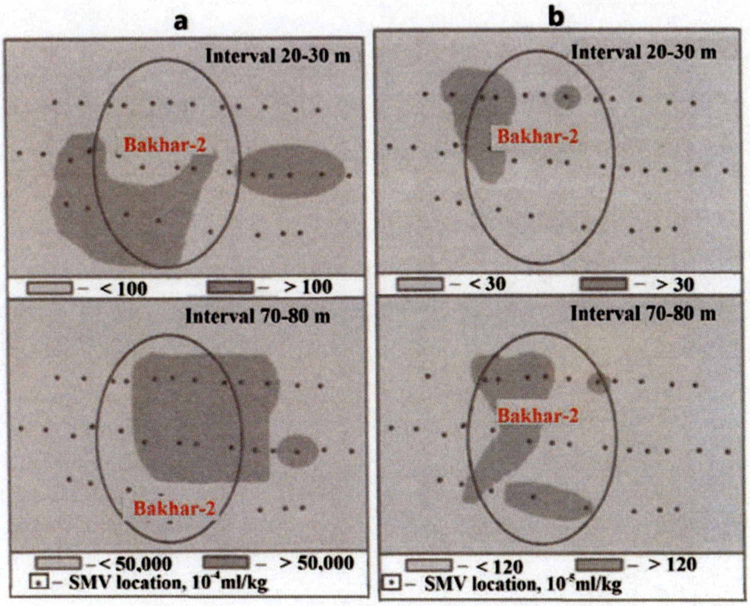

Fig. 14.1 Gas factors change and oil resources of the fields in the Balakhani-Gum Island anticline line (Feyzullayev, 2013)

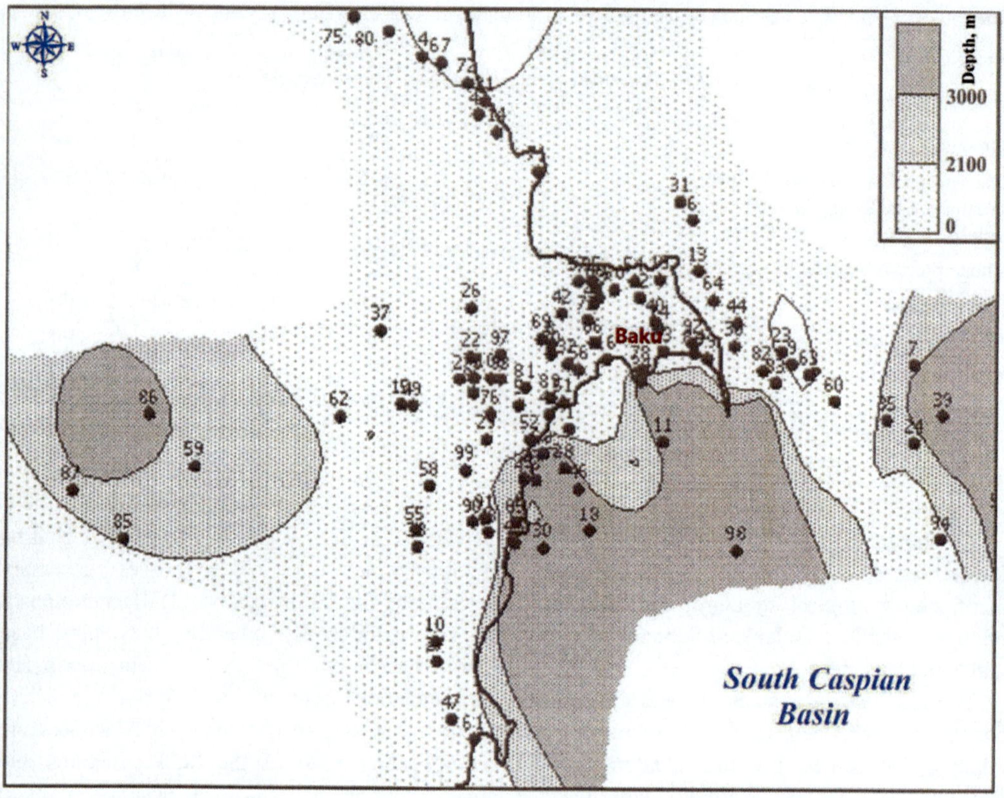

Fig. 14.2 Schematic map showing the distribution of the SCB fields according to the minimal depth of occurrence of the productive horizon (Feyzullayev & Tagiyev, 2007)

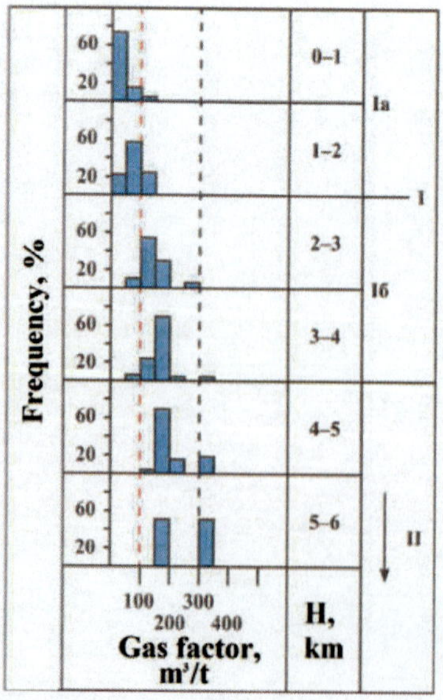

Fig. 14.3 General changes in the gas factor with the depth in the SCB (after Feyzullayev & Tagiyev, 2007)

An explosive primary/secondary HC migration in the form of fluidized clayey mass moving through the fracture up to the reservoir (in the form of diapir), on reaching a surface, appears to be transformed into a mud volcano. From that moment, this migration is characterized as a tertiary process since the HC gases (sometimes oil) are continuously and explosively discharged into the atmosphere (hydrosphere). This process destroys HC accumulations at depths (Fig. 14.6).

An important feature is that the participation of HC gaseous micro streams in the tertiary migration process is reflected in the increased background of these gases over regional oil and gas-bearing series and increasing gas anomalies over local minable HC accumulations (Fig. 14.7).

Based on an analysis of the relationship between the change in lithofacies characteristics and mineral composition of the PS clays, the present study may be summarized as follows:

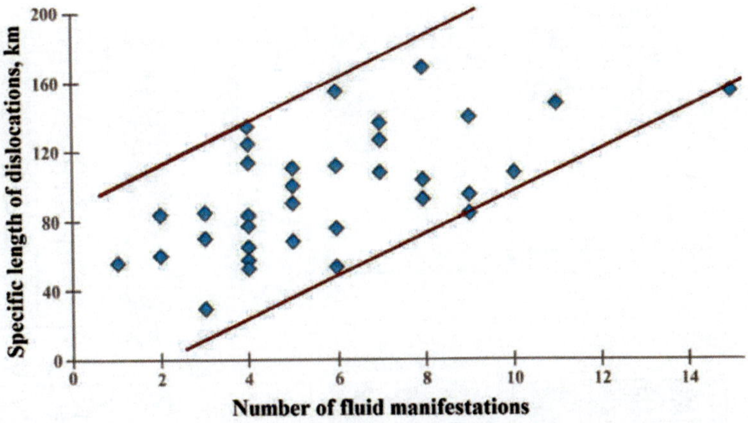

Fig. 14.4 The relation of dislocations' specific extent to the density of fluidal shows in the SCB western slope (after Feyzullayev & Tagiyev, 2007)

- Clay rocks of the Lower Pliocene (PS and RBS) deposits in the Absheron-PriBalkhan structural zone are more compacted than those in the Lower Kur trough and Baku Archipelago at the same depths that is caused, on the one hand, by relatively high geothermic gradient value (24–28 °C/km) and tectonic activity of this zone and, on the other hand, by less dispersion and heterogeneity of clays and also by the fact that they are mainly expressed by upswelling clayey minerals.
- In the Absheron-PriBalkhan structural zone, clays of the PS are presented mainly by minerals of the hydromica group, which form more than 50% of the fine politic fraction. It is also confirmed by chemical analysis data of this fraction, that is, by increased values of K_2O (2.01–3.65%) and Al_2O_3 (23.1–28.3%) and low water content (3.6–8.6%).
- From the Absheron OGR towards the Baku Archipelago, it is noted that hydromica content increases with decreasing montmorillonite content. This process continues in a southeasterly direction within the Sangachal-Deniz–Khara-Zirya anticlinal zone. The same is noted within the Pirsagat-Sangi-Mugan zone.
- To the north and northwest of the Baku Archipelago, it is noted that relative change for the worse in clay leak proofness takes place due to increases in sandy fraction. From the detailed analysis, it is inferred that a low minable oil and gas content of the central and southern Baku Archipelago may not relate to bad conditions of HC preservation since the quality of clay caps here is comparatively higher than in the Absheron Archipelago and Peninsula. At the same time, low permeability, and a great thickness of HC-generating rocks lead to the loss and discharge of liquid HC in the form of focused and impulsive streams through mud volcanic vents. These continuous—intermittent outbursts reach several million cubic meters of gas with dissolved liquid hydrocarbons. This fact and the high density of the currently active mud volcanoes within the Baku Archipelago suggest no good condition for HC preservation.

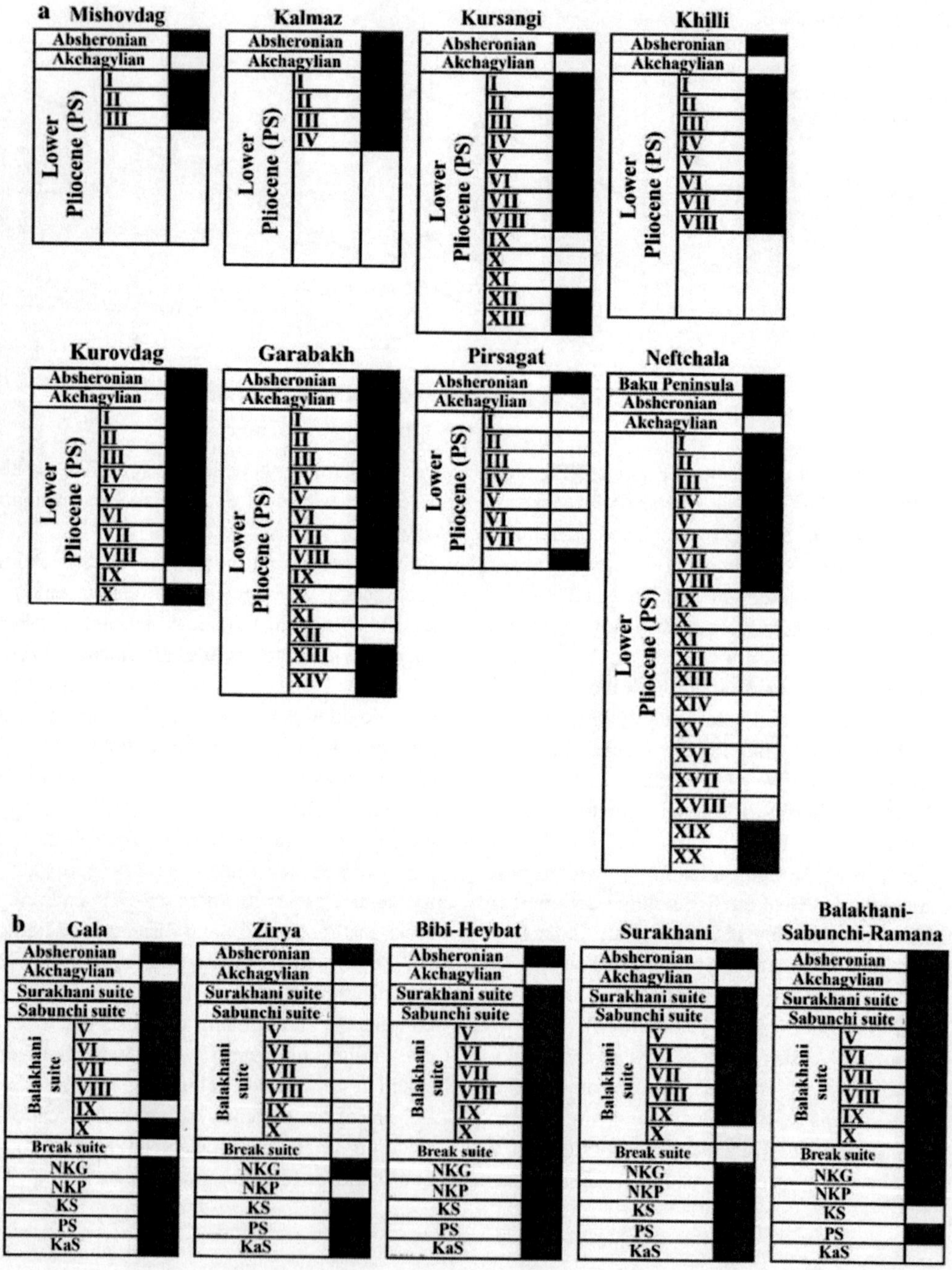

Fig. 14.5 Oil and gas-bearing intervals in the Absheronian and Akchagylian deposits within the Lower Kur trough (**a**) and Absheron Peninsula (**b**). NKG, OverKirmaki Clayey suite ('NadKirmakindskaya Glinistaya' suite), NKP, Overkirmaki Sand suite ('NadKirmakinskaya Peschanaya' suite), KS, Kirmaki suite ('Kirmakinskaya' suite), PS, Underkirmaki suite ('PodKirmakinskaya' suite), KaS, Kalinsky suite ('Kalinskaya' suite) (after Feyzullayev, 2013)

Fig. 14.6 Two examples of gas and oil discharge to the surface through the mud volcanic vent (after Feyzullayev, 2013)

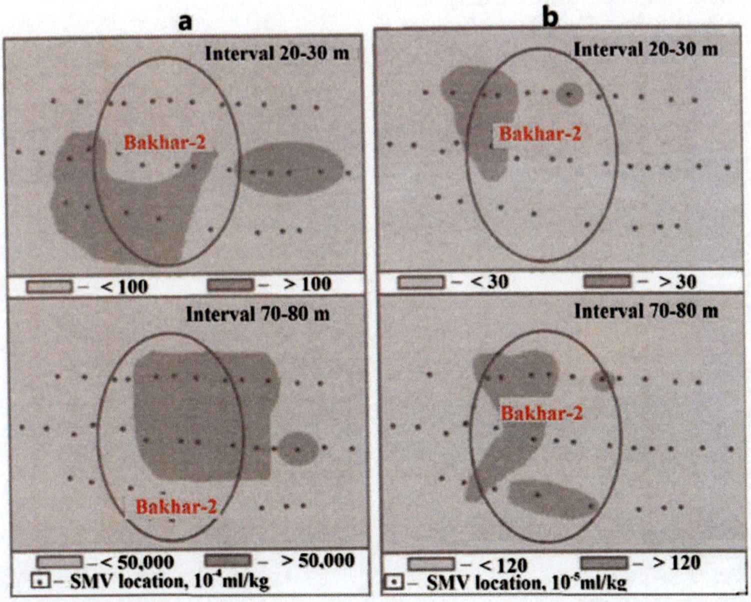

Fig. 14.7 Schematic figures display the distribution of methane (**a**) and its total homologs (**b**) at different depths in the Bakhar area (after Feyzullayev, 2013)

References

Alizadeh, A. A., Guliyev, I. S., Kadirov, F. A., & Eppelbaum, L. V. (2017). *Geosciences in Azerbaijan. Volume II: Economic minerals and applied geophysics* (340 p.). Springer.

Balakin, V. A., Gabrielants, G. A., Guliyev, I. S., Dadashev, F. G., Kolobashkin, V. M., Popov, A. I., & Feyzullayev, A. A. (1981). Test of experimental study of hydrocarbon degassing of the lithosphere of South Caspian Basin and adjacent mountains systems, using laser gas-analyzer "Iskatel-2." *Doklady Academy Science USSR, 260*(1), 154–156. (in Russian).

Etiope, G., Feyzullayev, A., Baciu, C. L., & Milkov, A. V. (2004). Methane emission from mud volcanoes in eastern Azerbaijan. *Geology, 32*(6), 465–468.

Feyzullayev, A. A. (2013). Migration pathways of hydrocarbons in South-Caspian Basin. *Journal of Geology & Geosciences, 2*(3), 1–6.

Feyzullayev, A. A., & Tagiyev, M. F. (2007). Conditions for preserving hydrocarbons in the South Caspian Basin. *Proceedings of the Institute of Geology of Azerbaijan National Academy Science, 35*, 150–167.

Guliyev, I. S., Feyzullayev, A. A., & Dadashev, A. A. (1985). Experience in Quantitative Estimation of Degassing of Subsoil by a Complex of Geological and Geochemical Parameters. In *Proceedings of the All-Union Meet. "Earth Degassing and Geotectonics"* (pp. 93–94). Nedra, Moscow (in Russian).

Kadirov, F., Mammadov, S., Reilinger, R. E., & McClusky, S. (2008). Some new data on modern tectonic deformation and active faulting in Azerbaijan (according to global positioning system measurements). *Proceedings of the Azerbajian National Academy Science Earth Sciences, 1*, 82–88.

Khain, V. E., & Shardanov, A. N. (1954). *Geological history and structure of the Kur depression* (357 p.). Academy Science of Azerbaijan, Baku (in Russian).

Salayev, S. G., & Kastryulin, N. S. (1974). The role of tectonic faults in the formation, preservation, and destruction of oil and gas deposits in southern Gobystan. *Materials on the geology of oil and gas in Azerbaijan* (pp. 72–88). Elm, Baku (in Russian).

Sokolov, V. A. (1971). *Geochemistry of natural gases* (336 p.). Nedra, Moscow (in Russian).

Sultanov, A. D., Salayev, S. G., & Ovnatanov, S. T. (1983). Correlation between the frequency of folding, the mass of sedimentary deposits, and hydrocarbons in the South Caspian depression. *Doklady Academy Science USSR, 272*(4), 937–941. (in Russian).

Voitov, G. I. (1975). Gas breath of the Earth. *Nature, 3*, 91–98. (in Russian).

15 Tectonic-Geodynamic Model of Oil and Gas Field Formation in the South Caspian's Productive Series

15.1 Tectonic-geodynamic Model of Oil and Gas Field Formation: An Introduction

The young though relatively rigid South Caspian microplate moving under the Arabian Plate head in a northeasterly direction formed the Absheron-Pribalkhan uplift, having bedrock outcrop above the sea level in the contact zone with the Scythian-Turanian Plate. During the Pliocene–Quaternary time, the Caspian microplate exercised fast-moving subsidence accompanied by avalanche sedimentation (up to 3 km/Ma). The total thickness of the sedimentary series is estimated to be 25–30 km.

As a result, during the relatively short geologic time (about 5 Ma), it formed a thick (up to 10 km) Pliocene–Quaternary series that increased the total thickness of sedimentary cover up to 25–30 km. The fact that sedimentary rocks saturated with organic matter have subsided at great depths caused the generation, migration, and formation of minable HC accumulation. The sublatitudinal zone of significant HC resources revealed in northern SCB is spatially well coincided with the accretionary prism formed in the frontal zone of the underthrusting fault (Fig. 1.10). An important feature is that all gas and oil fields are related to brachy-anticlinal uplifts complicated by longitudinal and cross faults, which were actively developed during the Late Pliocene and Anthropocene. That is why the age of these structures is very young and is at most 2–3 Ma. The HC generation and migration processes preceding these structures' formation are commensurable with the beginning of the South Caspian subduction under the Middle Caspian, evaluated as 5.5 Ma (Allen et al., 2002).

As mentioned above, several kilometers thick Mesozoic rocks have been involved in the SCB underthrust zone, which formed an accretionary prism with almost vertical tectonic faults immediately in the frontal part of the prism and slightly gentle faults in its rear part (see Fig. 1.10). All these features predetermine special features of HC migration and the formation of their commercial accumulation (Alizadeh et al., 2017).

First, it should be noted that gas plays a vital role in oil transportation to the SCB reservoirs (Feyzullayev, 2013; Feyzullayev et al., 2016). It is because Oligocene–Miocene source beds contain the II-III type organic matter (OM), generating a tremendous amount of gas on a level with oil. Another factor is that considerable volumes of HC gases are formed in Mesozoic deposits (mainly in the Jurassic deposits rich in gas-generating organic matter) involved in the underthrust zone.

Taking into consideration the young age of these fields and the difficulty of liquid hydrocarbon migration from the low-permeable massive Oligocene–Miocene source beds (that is confirmed by widespread mud volcanoes, clay diapirs, anomalous high fluid pressure, and their impulsive discharge), it seems most likely that

A. Alizadeh et al., *Pliocene Hydrocarbon Sedimentary Series of Azerbaijan*,
Advances in Oil and Gas Exploration & Production, https://doi.org/10.1007/978-3-031-50438-9_15

the factors mentioned above are the most real mechanisms of primary HC migration and subsequent transportation to the leader-rocks or immediately to the reservoirs.

The compression structures (Feyzullayev, 2013) formed within buried Absheron-PriBalkhan uplift subsided under recent deposits immediately in the frontal zone of the accretionary prism may be classified as the "through" structures (Lebedev, 1994) and complicated by dislocations with a break in continuity. These structures filling with hydrocarbons took place at the expense of their subvertical stream and subordinate lateral inflow marked on a short distance. It is believed that the conditions mentioned above are favorable for more intense HC delivery into anticlinal traps located immediately over the frontal accretionary prism (that is, along the Absheron-Pribalkhan sill). The following facts confirm this affirmation: *firstly*, the most significant HC accumulations within the SCB (the Balakhani-Sabunchi-Ramany, Neft Dashlari, Gyuneshli-Chirag-Azeri, and other fields) have been discovered in its sublatitudinal zone and the northern part (on the Absheron Peninsula and Absheron-Pribalkhan uplift); *secondly*, all the structures of this belt are saturated with hydrocarbons on both limbs (Fig. 15.5), and reservoir rocks comprise a wide stratigraphic interval (Fig. 15.6); *thirdly*, based on thermodynamic factors (reservoir pressure, recent and paleotemperatures) and geochemical features (gas factor, carbon isotopic composition of oil, the hydrochemical water composition, and so on), it was revealed the fact that the HC feeder source of the Pliocene structures is from underlying deposits (on an example of the Guneshli field) (Feyzullayev, 2013).

15.2 Results of Basin Modeling

In this connection, fascinating and important results of the basin modeling carried out for this part of the SCB from which it follows that down warping here after Maikopian suite accumulation took place highly uneven. Naturally, oil formation in local troughs during the late Miocene (5.5 Mya) also occurred unevenly (Fig. 15.1a). Accordingly, pore pressures in clays are characterized by minimal values on the crests of uplifts estimated by the Maikopian roof (Fig. 15.1f). In this case, the phenomenon is that against the background of comparatively low pore pressure, the fluidal stress loads appear to be sharply increased, resulting in the formation of local areas with anomalous low practical stress values. Pore pressure within these areas reaches lithostatic loads and exceeds them, leading to hydrofracturing formation (Fig. 15.2a). Synchronously, the Maikopian clay's porosity in the zones of hydrofracturing appears to be sharply increased, reaching the reservoir rock values (Fig. 15.2f). It should be noted that hydrofracturing processes began taking the time of an areal oil formation to create the ways for fluid withdrawal.

Between the 4.5–4.0 Ma time intervals approximately corresponding to the PS lower division suites, the geographical range of oil generation is slightly increased (Fig. 15.1b, c). In contrast, the former tendency is preserved in the distribution of pore pressures, effective stresses, and hydro-fracture. That is to say, the system continues to be de-fluidized in formerly mentioned areas.

To the end of the PS, 3.5 Mya, the oil generation processes have intensified (Fig. 15.1e). A critical moment in the evolution of the fluidal system was the end of the Akchagylian –beginning of Absheronian times (1.8 Ma) when the Maikopian suite was practically located in the oil window of the SCB. In contrast, this suite in the eastern basin has been in the depletion stage of oil potential and entered the gas and condensate generation zone (Fig. 15.1d). Although this complex realized its oil-generating potential for about 45–50% of the sources of effective stress, hydro-fracture and high porosity of rocks are invariably localized in the same places, being manifested in more contrast (Fig. 15.2e).

All the territories have been under stress conditions, and the hydro-fracture and permeable rock sources inherit their position (Fig. 15.2d). It may be presumed that the present stage corresponds to the master phase of HC migration.

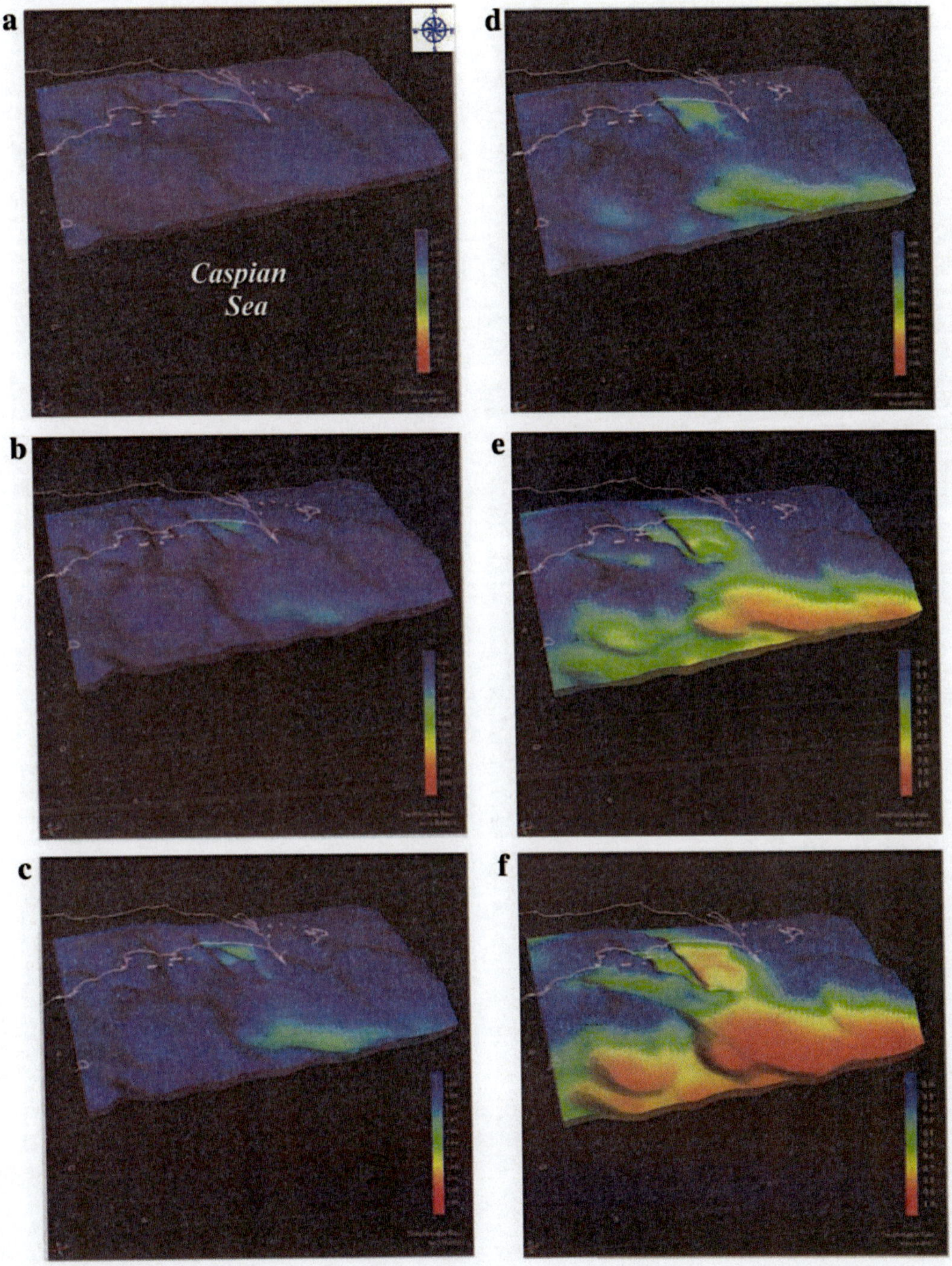

Fig. 15.1 Three-dimensional evolutionary model of oil and gas generation (**a–f**) and pore pressure (**g–l**) change in the Maikopian deposits: **a, g** = 5.5 Ma, **b, h** = 4.5 Ma, **c, i** = 4.0 Ma, **d, j** = 3.5 Ma, **e, k** = 1.8 Ma, **f, l** = nowadays

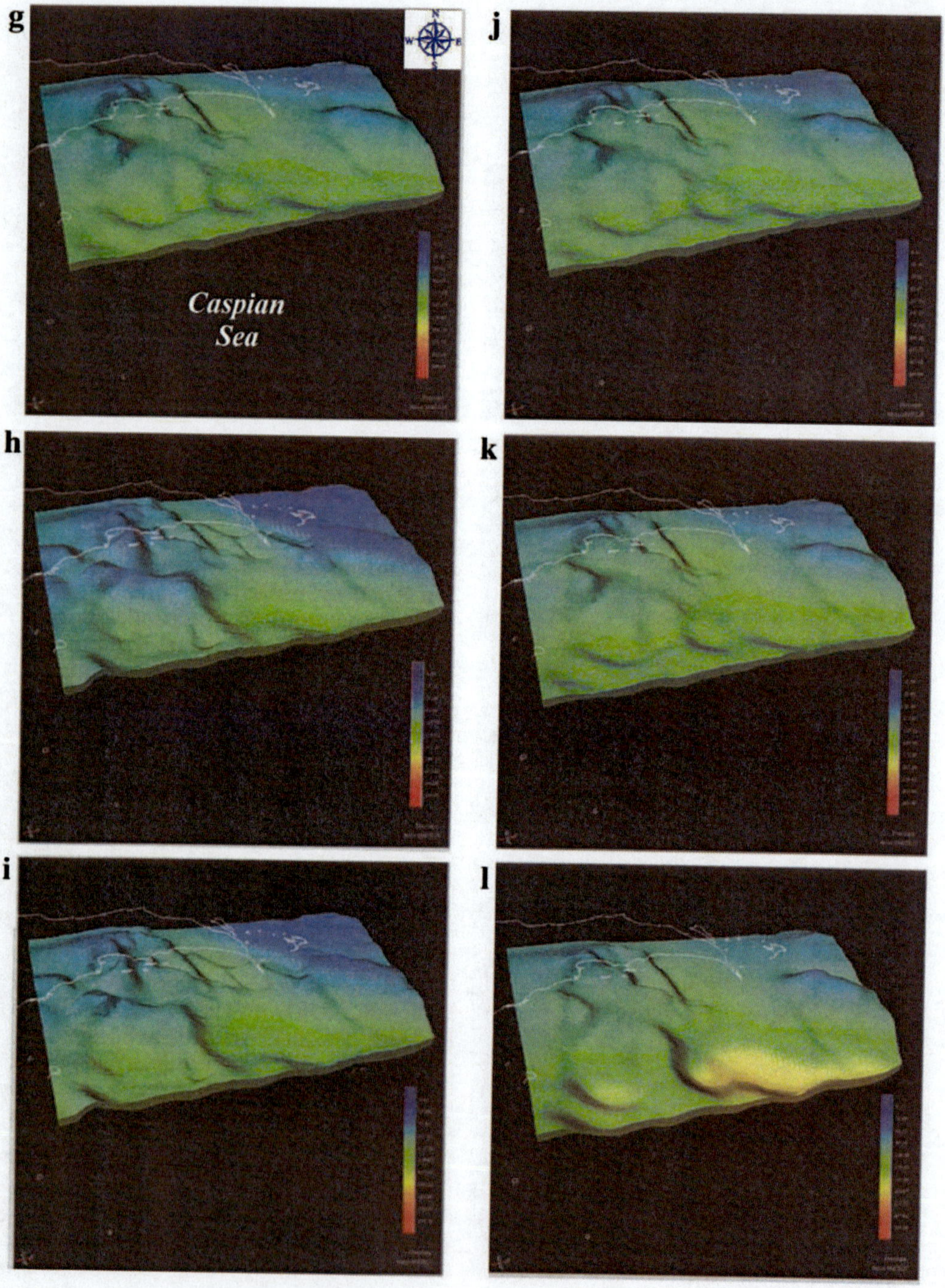

Fig. 15.1 (continued)

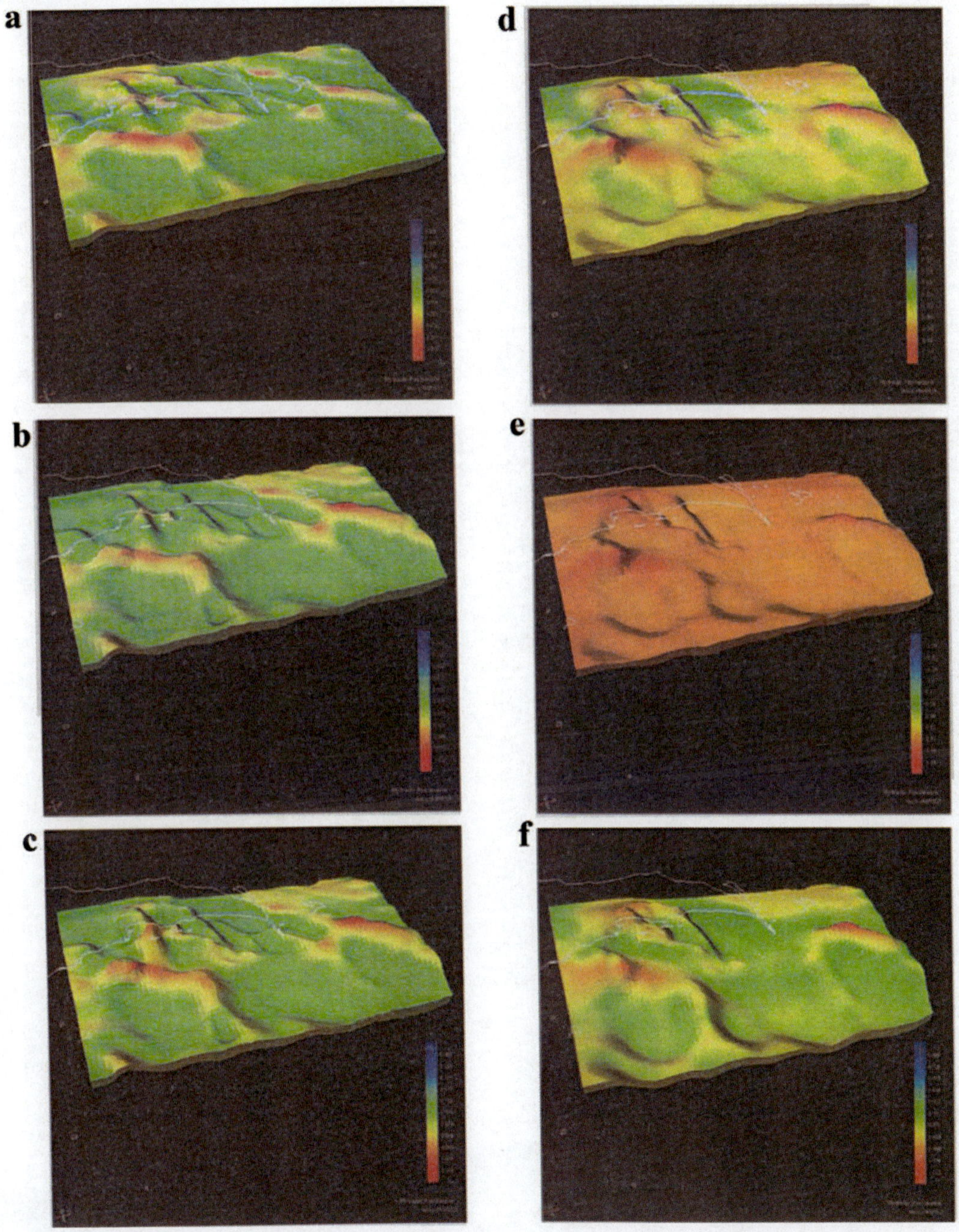

Fig. 15.2 Three-dimensional model of hydro-ruptures formation (**a–f**) and pore pressures change (**g–m**) in the Mesozoic deposits: **a, g** = 5.5 Ma, **b, h** = 4.5 Ma, **c,** i = 4.0 Ma, **d, k** = 3.5 Ma, **e, l** = 1.8 Ma, **f, m** = nowadays

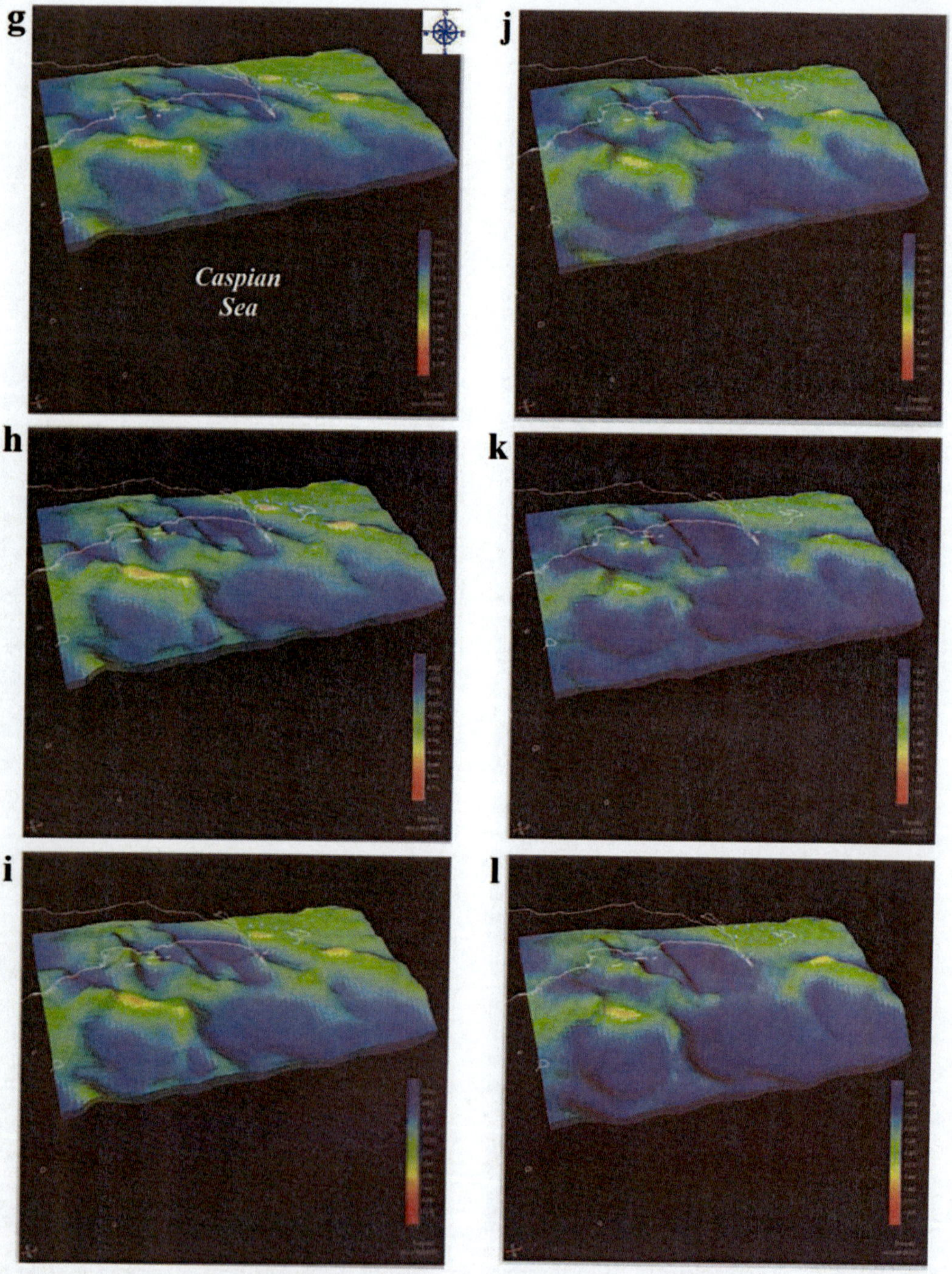

Fig. 15.2 (continued)

Maikopian deposits widespread in the eastern South-Absheron trough have essentially realized their oil potential. They are relatively mature in the western trough (Fig. 15.1e). However, in most parts of the Absheron Peninsula and Absheron Archipelago, these deposits are in the middle and early stages of maturity. Even though stress conditions within these troughs have been essentially reduced (Fig. 15.2e, m), the states of hydro-fracture and high porous mediums have been preserved, being confined exclusively to the domes inherited from Miocene time and which

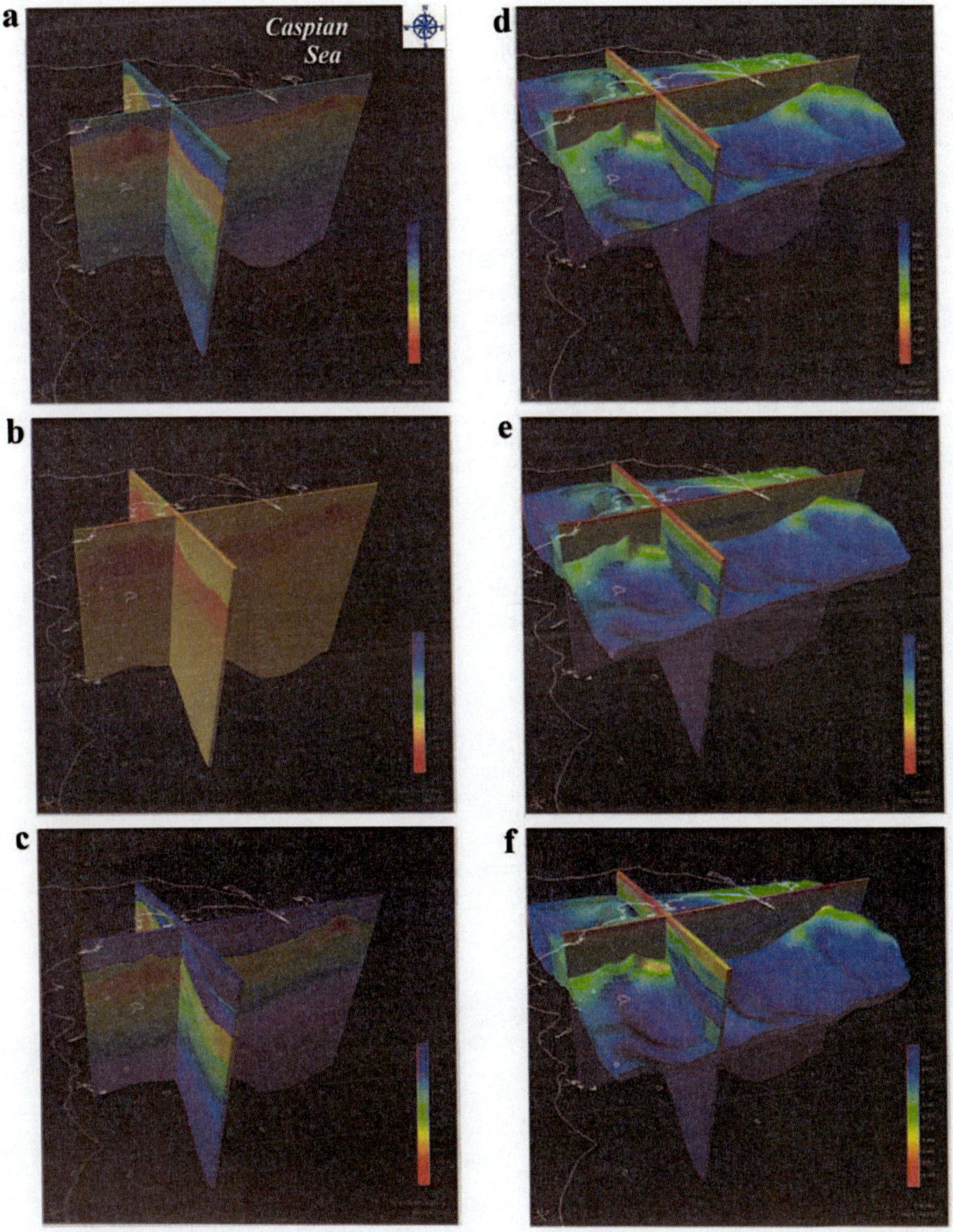

Fig. 15.3. 3D model of the reservoir fracturing development (**a, b, c**) and porosity changes (**d, e, f**) in the Paleogene-Miocene and Pliocene sediments: **a, d** 3.5 Ma, **b, e** 1.8 Ma, **c, d** nowadays

are something like "windows" or "holes" involving fluids from the Miocene up to the present. This tendency is repeated by modeling the Middle-Upper Miocene fluid generating system. From the detailed analysis of all the system functioning, a columnar-vertical character and dome-shaped intrusion of unsealed fluid-generating and fluid-conductive mass (Fig. 15.3), and most importantly, a vertically inherited nature of the fluid stream.

It is imperative to note that the Paleogene-Miocene vertical zones of a fluid stream are exactly imaged on the richest and oldest Paleocene structures as Bibi-Heybat, Balakhani-Sabunchi-Ramany, Surakhani, Kala, Neft Dashlari-Palchyg Pilpilyasi. Considering that all the oil and gas-bearing structures of the studied territory are complicated to a considerable extent by steeply dipping ruptures reaching PS's lowermost strata (Fig. 15.4), it may be concluded that the mechanism of HC vertical migration during these structures' infilling is manifested highly distinct.

It explains that these fields have underplated during more than a century of their development and serve as highly ponderable evidence of their replenishment primarily by oil from the adjacent local troughs. Also, it makes an affirmation of distal-lateral fluid migration from the central SCB an unfounded conclusion.

The mechanism of vertical migration through the ruptures and reservoir infilling has been shown in an example of the Guneshli field using high-informative isotopic-biomarker technologies and an analysis of primary conditions.

Thus, three-dimensional basin modeling shows different conditions and their deep-lateral expansion range under which Paleogene-Miocene impermeable clayey source beds in the SCB have very high fluid-conductive and reservoir potential. Hence, we believe that the following investigations are reasonably warranted: (1) a particular geological-geophysical exploration of vertically oriented zones in Paleogene-Miocene deposits on the mentioned structures to possibly draw them into development, and (2) there may be the possibility of oil and gas accumulation discovery in a zone of the dynamic effect of subvertical geological bodies (SGB) revealed by a seismic survey. These SGBs are likely the zones of increased permeability, porosity, and low pressures (compared with environmental rocks) (Figs. 15.2 and 15.3).

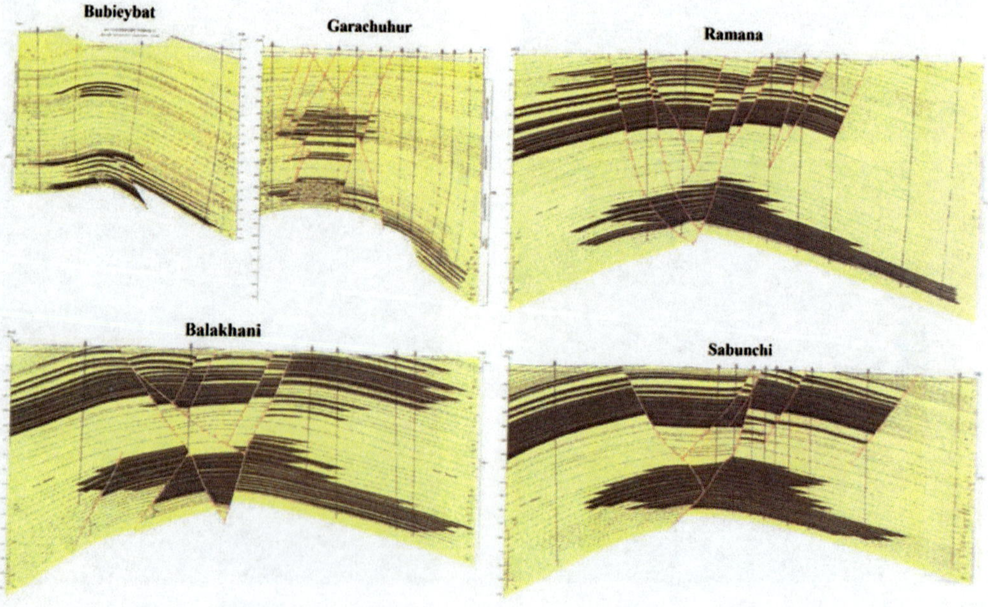

Fig. 15.4 Geological sections of the Bibi-Heybat, Garachukhur, Ramana, Balalkani, and Sabunchi fields

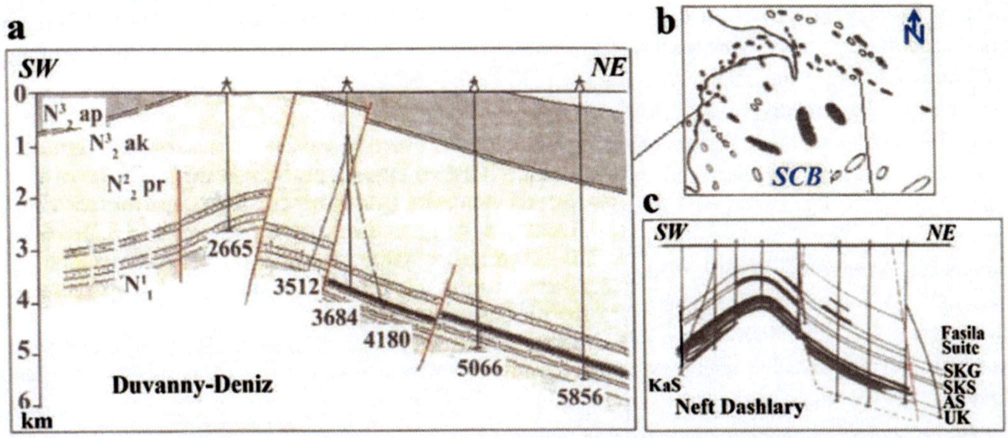

Fig. 15.5 Special features of hydrocarbon filling of the frontal (**a**) and rear zones (**b**) of subduction, on an example of the Neft Dashlari and Duvanny fields, correspondingly, **c** schematic location of these fields (after Feyzullayev et al., 2016)

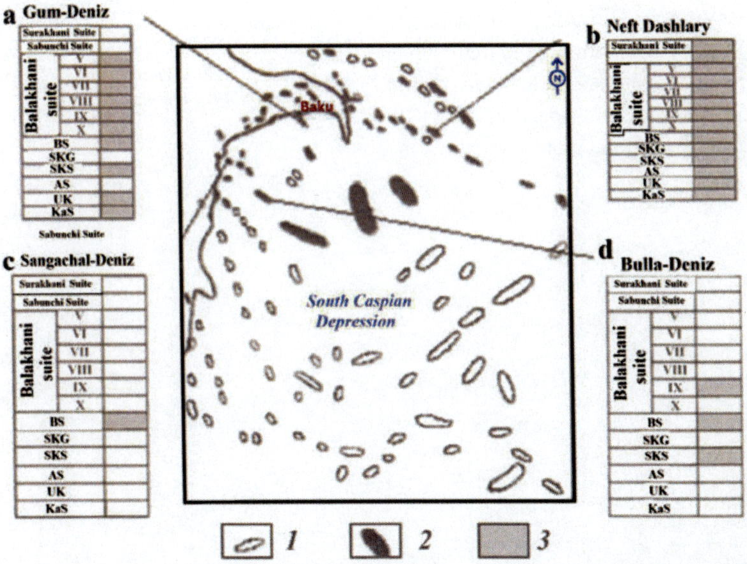

Fig. 15.6 The distribution of HC pools in the SCB stratigraphic section within the frontal zone of the Absheron-Pribalkhan trough (on an example of the Gum-Deniz (**a**) and Neft Dashlari fields (**b**) and in rear parts of this trough (Sangachal-Deniz (**c**) and Bulla-Deniz (**d**) fields). (1) anticlinal structures, (2) oil and gas fields, (3) oil and gas-bearing intervals. *BS* Break suite, *UKC* Upper-Kirmaki ('Nadkirmaki') clayed suite, *UKA* Upper-Kirmaki ('Nadkirmaki') arenaceous suite, *KS* Kirmaki suite, *SK* Subkirmaki ('Podkirmaki') suite, *KaS* Kala suite (productive series of the lower pliocene) (after Feyzullayev et al., 2016)

It provides constant HC inflow from environmental space. Moreover, if mud volcanoes appear confined to these zones, periodic eruptions may activate the inflow process. Thus, the processes of forming and evolution of the subvertical geological bodies can be compared with the giant pump forcing out the hydrocarbons from the sedimentary rocks.

First, employing subvertical migration through the faults, these hydrocarbons reached the prominent reservoir leader and filled structural traps. It has resulted from relatively distant lateral up-dip migration with subordinate subvertical HC inflow. Somewhat another formation and the migration pattern of HC fields are marked in the opposite rear part of the South Absheron trough, where all structures (Sangachal-Deniz, Duvanny-Deniz, Khara-Zyrya, Bulla-Deniz) have been filled due to the step-like HC migration. Such a mechanism is confirmed by the fact that within this region, the hydrothermal filling is marked only by the limbs facing the frontal subduction zone. What is more, oil reserves in these structures are less than in the above-mentioned structures of the Absheron-Pribalkhan region because the stratigraphic interval of the HC saturation here is markedly less (Feyzullayev et al., 2016) (Fig. 15.5).

The phase state of HC accumulations is mainly controlled by their preservation conditions which are worse in the Absheron-Prirbalkhan region, where oil and related gas fields dominate. The gas/gas condensate accumulations (Shakh-Deniz, Bulla-Deniz, Umid, Absheron-Deniz fields) are dominant in the more subsided part of this trough (Fig. 15.6).

References

Alizadeh, A. A., Guliyev, I. S., Kadirov, F. A., & Eppelbaum, L. V. (2017). *Geosciences in Azerbaijan. Volume II: Economic minerals and applied geophysics* (340 p.). Springer.

Allen, M., Jones, S., Ismail-zadeh, A., Simmons, M., & Anderson, L. (2002). Onset of subduction as the cause of rapid pliocene-quaternary subsidence in the South Caspian basin. Geological society of America. *Geology, 30*(9), 775–778.

Feyzullayev, A. A. (2013). Migration pathways of hydrocarbons in the South-Caspian basin. *Geology and Geosciences, 2*(3), 1–6.

Feyzullayev, A. A., Kadirov, F. A., & Kadyrov, A. G. (2016). Tectono-geophysical model of the Southern Caspian in the context of the presence of oil and gas. *Izvestiya, Physics of the Solid Earth, 52*(6), 913–923.

Lebedev, L. I. (1994). The influence of tectonic factors on the oil and gas content of inland seas. *Geology of Oil and Gas, 7* (in Russian).

16 Akchagylian Hydrospheric Phenomenon and Its Connection with Deep Geodynamics

16.1 Essence of the Problem

The relationship of most commercial hydrocarbon deposits with the structural features of the studied region is well known (e.g., Eppelbaum & Khesin, 2012; Alizadeh et al., 2016, 2017). Nevertheless, there is a basin factor in which hydrospheric disturbances determine the sedimentation nature of the rocks hosting hydrocarbon deposits. The most known among them are the Jurassic-Early Cretaceous oil-bearing formations of Western Siberia (Eder et al., 2020), formed under the conditions of the boreal transgression, the Late Jurassic salt-bearing formations of the Mexican and northern parts of the Persian Gulf (Khain & Polyakova, 2008), formed under the conditions of the equatorial regression, hydrocarbon deposits of the Mediterranean Sea associated with both the Middle Cretaceous equatorial transgression (Eppelbaum & Katz, 2011, 2015) and the Late Cenozoic Messinian hydrospheric crisis (Dalla et al., 1997; Kontakiotis et al., 2020; Roberts & Peace, 2007), and other events (Eppelbaum et al., 2012).

The Pliocene basin of the Paratethys region (including the Akchagylian Stage), lying to the north of the Mediterranean Sea (the Tethyan Basin), is a paleogeographic region, the deposits of which contain numerous and varied hydrocarbon deposits used for millennia and actively exploited today. The influence of the Akchagylian hydrospheric phenomenon on the ancient hominin dispersal is studied in detail by Eppelbaum and Katz (2022).

16.2 The Akchagylian Phenomenon Against the Background of the Hydrospheric Disturbances of the Late Cenozoic

The hydrospheric disturbances are one of the significant elements of the development of both the stratigraphic scale (Andrussov, 1902, 1918) and the scale of planetary periodization (Balukhovsky, 1966; Katz, 1986; Milankovitsch, 1941; Sobolev, 1926; Stovas, 1963; Veronnet, 1912). During the last 12 million years of the Late Cenozoic (Fig. 16.1), there were three hydrospheric maxima with a height of about + 200 m—Early Sarmatian, Early Pontic, and Middle Akchagylian, and two minima—the Late Messinian-Early Pliocene (5.7–5.5 Mya)—up to − 300 m, and Early Pleistocene (0.84–0.85 Mya)—up to − 200 m (Cassian-Tiurkian regression).

After this regressive event, the epoch of equatorial transgression (Fig. 1) gave way to the epoch of polar transgression due to the long-term change in the shape and speed of the Earth's rotation (Katz, 1986). The tectonic-eustatic regime of hydrospheric disturbances was replaced by the glacioeustatic regime, which is a multiple of the period of change in the eccentricity of the Earth's orbit (100,000 years)

A. Alizadeh et al., *Pliocene Hydrocarbon Sedimentary Series of Azerbaijan*,
Advances in Oil and Gas Exploration & Production, https://doi.org/10.1007/978-3-031-50438-9_16

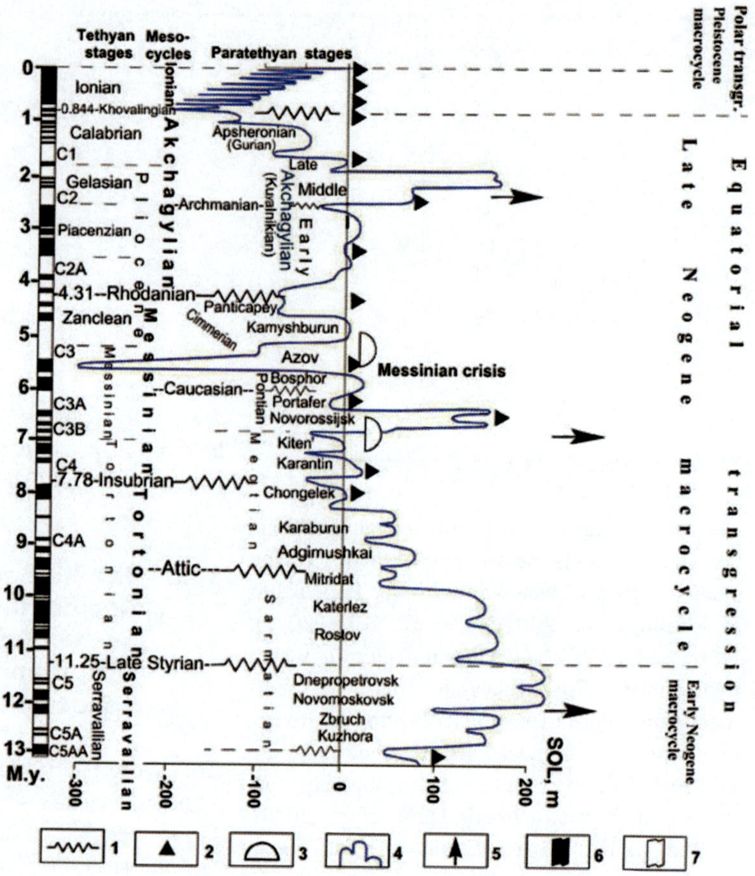

Fig. 16.1 The scheme of hydrospheric disturbances in the basins of the Mediterranean Sea region—Paratethys in the second half of the Late Cenozoic, and their paleomagnetic correlation (after Eppelbaum and Katz (2021), with modifications). (1) tectonic phases, (2) glacial epochs, (3) main stages of evaporite sedimentation, (4) 'Standard Ocean Level (SOL) curve, in meters, (5) peaks of the positive hydrospheric disturbances, (6) magnetic field of direct polarity, (7) magnetic field of reverse polarity

(during which the epochs of continental glaciations and interglacials alternate (Milankovitsch, 1941).

Among the three hydrospheric maxima of the Late Cenozoic described above, the Akchagylian basin is the most exceptional paleogeographic and sedimentation-biotic terms (Nevesskaya et al., 1986). According to the comparative paleogeography of the Tethys and Paratethys basins in the Late Cenozoic, presented in this work (see Fig. 4 in Nevesskaya et al. 1986), the development in the Akchagylian time of the anomalously elongated Caspian-Turanian basin, isolated from the Pontus, with a typology of transition from a brackish water regime to a semi-marine and marine, is phenomenal one (see Fig. 5 in Nevesskaya et al. 1986).

Notably that the number of mollusks arising in this basin is abnormal for the entire Late Cenozoic and makes up 94% of the total species composition (see Table 12 in Nevesskaya et al. 1986). This analysis from the work of Nevesskaya et al. (1986) is very important for the determination of the Akchagylian hydrospheric phenomenon in terms of the paleogeography, the hydrochemical regime, and the biotic characteristics of the vast Caspian-Turanian basin, which stretches far to the north in comparison with other Paratethys basins.

Further analysis of the Akchagylian hydrospheric phenomenon will be carried out below using detailed comparative analysis. In the section below, we examined only the content and essence of this phenomenon against the background of other historical and geological stages of the Late Cenozoic.

16.3 Stratotypic Area of the Akchagylian Stage and Its Distinctive Features from Other Basins of the Paratethys

The Caspian-Turanian region, where the Akchagylian Stage is developed, is unique in terms of the sharp facies contrast of the deposits corresponding to its chronostratigraphic volume (Naidin & Naidina, 1992). The lectostratotype located in the Turan region (the Ushak section on the Krasnovodsk Peninsula in West Turkmenistan) is composed of the marine formations (Mironova, 1982), relating, according to the paleomagnetic studies (Trubikhin, 1977), to the Gauss zone and the lower parts of the Matuyama zone. Moreover, the same sequence is developed throughout the Turan region (Ali-zade, 1961, 1967; Danukalova, 1990; Popov, 1969). At the same time, two branches of sea straits are outlined (Richards et al., 2018). One of them stretches through the Turgai trough to the Western Siberia Basin, and the second—through the Caspian Sea to the Pechora-Barents Sea Basin. The vast extent of the Turanian Akchagylian, the development of carbonate facies in it, and the distribution of the remnants of marine mollusks throughout the section—from the Gauss zone to the Matuyama zone (Trubikhin, 1977), contributed to the concept of the "wide Akchagylian". The latter is contrasted with the "narrow Akchagylian" of Azerbaijan and other more western areas of the Caspian region (Krijgsman et al., 2019), which is identified only with the analogs of the lower parts of the Matuyama zone.

The South Caspian-Caucasian region itself, where the relatively thin formations of the marine Akchagylian sharply contrast with the fluvial terrigenous strata of the Productive Series, has turned out in recent years to be a key factor for understanding the Akchagylian Phenomenon. First, the regional tectonic, deep geophysical, geodynamic, and paleogeographic conditions of the formation of this oil&gas region in the Late Cenozoic were thoroughly studied (Abdullayev et al., 2015), and the unity of the Productive Series within the framework of a single regressive stage of the Early Pliocene was established (Richards et al., 2021). The facial studies of the uppermost Productive Series (Surakhani Formation) (Vincent et al., 2010) showed that they represent the final regressive stage of the Early Pliocene in the form of an alluvial incision of the paleo-Volga–paleo-Kur river system. In the Akchagylian complex of the South Caspian itself, various remnants of the marine microorganisms were found, unambiguously confirming the hypotheses made in the middle of the twentieth century (Alizade, 1954; Ali-zade, 1961) about a more transgressive stage in the development of the second half of the Akchagylian Basin. In recent years, just on the sections of the Azerbaijani Akchagylian, the previously unexplained problem (Nevesskaya et al., 1986) of the sharp renewal of the Akchagylian mollusk fauna was solved. The data about the possible connections of the Akchagylian Basin with the Mediterranean and the Polar Basins were not paleogeographically confirmed.

The remnants of dinoflagellates of the *Algidasphaeridium cf capillatum*, first described in the Miocene-Pliocene of the Bering Sea islands (Matsuoka & Bujak, 1988), were found in several sections of the Azerbaijani Akchagylian (Richards et al., 2018).

Tectonic-geophysical instability at the Early and Middle Pliocene boundary in the Caucasian segment of the Akchagylian Basin (Abdullayev et al., 2015) led to the development of breaks at the boundary of the decisive fluvial stage of the Productive Series and the Akchagylian-Anthropogenic transgressive complex. As a result, the paleomagnetic zones' age at the Early and Middle Pliocene boundary was interpreted ambiguously in many sections. This concerns the data on the reference section Jeyrankechmez (Gurary, 1988; Van Baak et al., 2019), wherein the Surakhani Formation (according to the data of Gurary's (1988), in the Balakhani Formation), the change of relatively stable complexes of the direct and reverse magnetized rocks is interpreted as a sign of their belonging to the Gauss paleomagnetic epoch. Nevertheless, they most likely correspond to the more ancient paleomagnetic Gilbert epoch.

In this case, the broader correlations (Lazarev et al., 2021) are unreliable since any metric analysis should be based on the tectonic-geophysical fundamental data sedimentation-paleogeographic analysis of sedimentary strata belonging to different structures and regions. The separation of part of the regressive Productive Series of the Early Pliocene due to an arbitrary interpretation of the results of specific studies contradicts the data of both the regional geology of the Eastern Caucasus [the difference of which from the Hercynides of the Turan plate is obvious (Fig. 16.2)], and the data of the historical planetology (Fig. 16.1). This testifies to the sharply regressive nature of the Late Messinian-Early Pliocene formations: Lago Mare, Zanklian, Cimmerian, Torongly red flowers, and the Productive Series compared to the Akchagylian transgressive maximum.

The data from regional geological mapping and studies of sections in the southern part of Azerbaijan (Gasanov & Aliyeva, 2003) significantly supplement the paleogeographic and facial concepts of the nature of the southern framing of the Akchagylian Caspian Basin. In the Azerbaijani part of the Lesser Caucasus, within the Garabakh region, the Akchagylian sediments transgressively overlap the various age complexes of the Cenozoic.

At the same time, the stage itself decreases in thickness compared with the sections of the northern regions and is composed of shallower facies, among which the basal conglomerates, limestones, and sandy rocks should be noted. The deposits themselves correspond to the Middle-Upper Akchagylian. Facies of this kind are relatively close not to the sections of the Greater Caucasus but to the Epigercynian Turanian plate. Thus, the southern edge of the Caspian Basin is uplifted, and it is unlikely that the Tethys and Paratethys basins could be connected through the uplift of a block of the Lesser Caucasus (Naidin & Naidina, 1992). Moreover, this is explained by the recently obtained regional tectonic-geophysical data on the influence of a giant mantle rotating structure on near-surface geological objects (Eppelbaum et al., 2021).

16.4 The Meridional Basin of the Volga-Ural Region

This extended basin, stretching from the Caspian Sea to the vast coasts of the Polar Basin, is vital for understanding the Akchagylian problem in terms of the paleontological, paleomagnetic, and paleogeographic correlations of sections of the stratotype region and areas where the influence of the hydrospheric disturbances and the glacial-climatic phenomena of the Earth's polar segment are developed.

On the other hand, on the border with the Caspian Sea, within the relatively stable East European Precambrian platform (Fig. 16.2), the maximum hypsometric levels of the Akchagylian marine transgressions can be estimated, and in the north of the region—the maximum depth of the regressive levels of the Pliocene and Early Pleistocene. This region has been well and reasonably comprehensively studied with the help of lithological-sedimentation, paleontological, paleogeographic, and paleomagnetic studies, with the extensive use of methods for analyzing the natural outcrops and numerous sections obtained because of reference drilling. It is stratigraphically compared with the more westerly sections adjacent to the Black Sea basin (Kamaletdinov & Yakhimovich, 1981). Due to the detailed facies-biostratigraphic division and paleomagnetic correlation, the region under consideration ultimately makes it possible to unambiguously estimate the volume and content of the Akchagylian Stage of the Azerbaijan-Turkmenian stratotype region. First, it should be noted that at the base of the section lies the Late Miocene-Early Pliocene strata of continental formations containing the remains of the Pont-Cimmerian type flora and ostracods occurring in the Productive Series of Azerbaijan and the red-colored Torongly Formation of Turkmenistan (Kamaletdinov & Yakhimovich, 1981), underlying the Akchagylian marine formations.

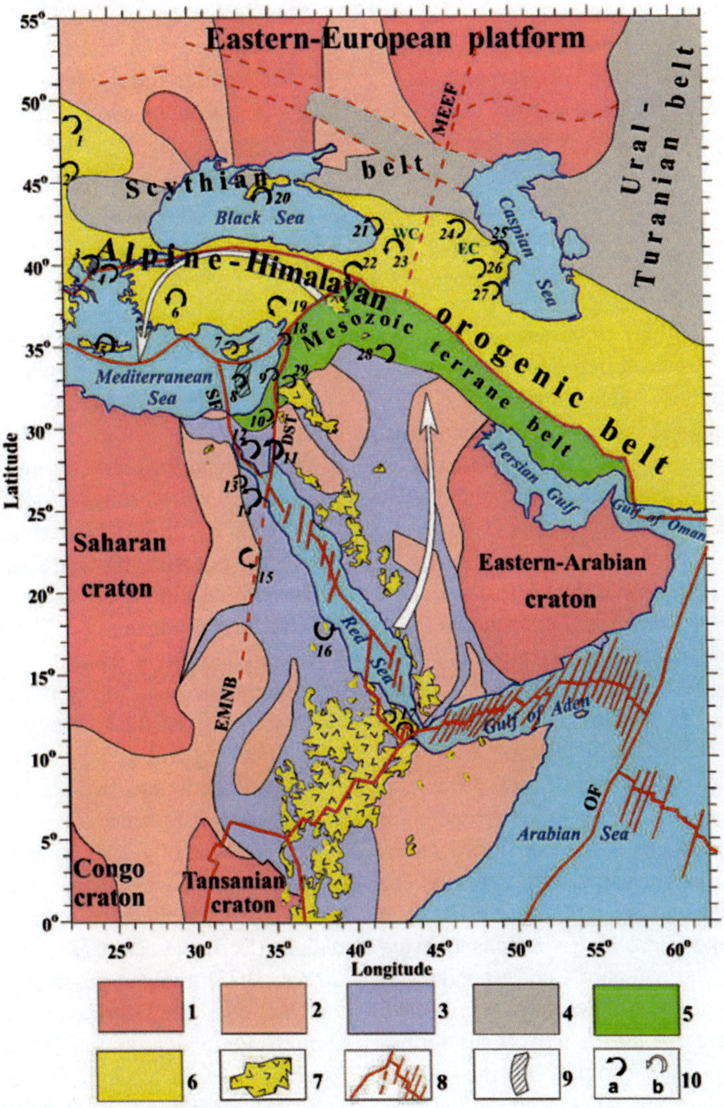

Fig. 16.2 An integrated tectonic-geodynamic-geophysical scheme of the region under study (after Eppelbaum et al. (2021), with modifications). (1) Archean cratons, (2–4) fold belts: (2) Paleo–Middle-Proterozoic, (3) Neoproterozoic, (4) Late Paleozoic (Hercynian), (5) Mesozoic terrane belt, (6) Alpine–Himalayan orogenic belt, (7) Cenozoic traps of the African–Arabian rift belt, (8) main fault systems, (9) Kiama paleomagnetic hyperzone of inverse polarity, (10) rotational geodynamic elements derived from: paleomagnetic (major) and tectonic (minor) data: 10a: (*1*—Piennine klippen belt (West Carpathians), *2*—Getic Basin (South Carpathians–Moesia), *3*—Athos Peninsula (Greece), *4*—Samothraki (Greece), *5*—Crete (Greece), *6*—Menderes massif (western Anatolia, Turkey), *7*—Cyprus, *8*—Kiama paleomagnetic hyperzone (Eastern Mediterranean), *9*—Galilee magmatic and block systems (Israel), *10*—dikes of Makhtesh Ramon (Negev terrane, Israel), *11*—block systems of Gulf of Aqaba (northern Egypt) and Midian fault zone (far west of Saudi Arabia), *12*—dikes of Sinai Peninsula, *13*—Sinai Bay block system, *14*—northern Red Sea bottom deposits, *15*—Mansouri ring complex (SE Desert, Egypt), *16*—Barake suture zone (Red Sea), *17*—northeast of the Afar Depression (Eritrea), *18*—central part of the Galilee–Lebanon terrane, *19*—Eastern Taurides (Anatolia, Turkey), *20*—seismotectonic shear zone (Black Sea, southern Crimean Peninsula), *21*—Rioni Basin (Georgia), 22—Eastern Pontides, *23*—Achara–Trialet belt (Georgia), *24*—Dagestan Mountains (Eastern Caucasus, Russia), *25*—Kur Depression (Azerbaijan), *26*—Somkheto-Garabakh belt, *27*—Nakhichevan and Talysh regions (Azerbaijan), *28*—Kata-Rash area (Iraq), *29*—Mt. Hermon foot (northern Israel)), 10b: GPS constructions, with generalization (after Reilinger et al. 2006). *SF* sinai fault, *DST* dead sea transform, *MEEF* main Eastern-European fault, *EMNB* Eastern-Mediterranean-Nubian belt, *OF* owen fault, *WC* Western Caucasus, *EC* Eastern Caucasus

According to the paleomagnetic data, this continental sequence, crowned by the lower part of the Karlaman Formation, corresponds to the Gilbert paleomagnetic zone and the underlying C3r zone. The first signs of the transgressive marine Akchagylian in the Bashkir Urals are developed in the upper part of the Karlaman Formation, where the seaweed *Zostera* and boreal foraminifera are present. The paleomagnetic data indicates that the age of these layers corresponds to the first half of the Gauss paleomagnetic epoch (Kamaletdinov & Yakhimovich, 1981). The marine Akchagylian of the Turkmenistan stratotype region (e.g., Danukalova, 1990; Trubikhin, 1977) and Azerbaijan (e.g., Alizade et al., 1987) also correspond to this sharp ecostratigraphic and paleogeographic boundary. It also manifests itself in other areas, primarily in the well-studied sections of Iceland, where at the base of the Gauss zone, there is a sequence of the most transgressive terrigenous formations with the mollusks *Serripes groenlandicus*, above which the glacial formations (Furuvik tillites) are developed in the upper part of the Gauss zone (Einarsson, 1957; Gladenkov, 1978). This glacioeustatic stage corresponds to the formation of the regressive Kumurly Formation in the Urals, composed of lacustrine and alluvial deposits. The overlying middle and upper marine Akchagylian of the Urals and Trans-Volga regions (Akchagylian in the "narrow sense", corresponding to the lower part of the Matuyama paleomagnetic epoch) is represented by the terrigenous formations with a diverse complex of the marine mollusks (Danukalova, 1990; Kamaletdinov & Yakhimovich, 1981). The presented materials unambiguously indicate that the revision of the boundaries and volume of the Akchagylian Stage is incorrect from carefully and comprehensively studied Pliocene sections and the history of the development of structures and movements in the East Paratethys Basin. The arid regressive stage of the Late Messinian, Cimmerian, Zanclian, Torongly, and Productive Series differ sharply from the Akchagylian transgressive-glacial complex; these two geological events are sharply opposite stages in the Earth's development (Alizade et al., 1987; Naidina & Richards, 2020).

This region contains a significant amount of data concerning both the hypsometric levels of the Akchagylian transgression's maxima and the regressive incisions corresponding to the Messinian crisis and other hydrospheric minima. These data are crucial for analyzing hydrospheric disturbances and understanding their planetary nature. In this case, the Caspian-Ural region is a bridge connecting the seas of Eurasia and Gondwana.

It is imperative to attract data on the Timan-Pechora northern region, where, because of extensive prospecting-research studies, the deep incisions (up to − 300 m) of the marine Pliocene of the Prosundui Formation into the Mesozoic sediments were discovered (Kamaletdinov & Yakhimovich, 1981). The overlying Kolvinskaya Formation contains the remains of numerous mollusks of the Boreal-Atlantic type, like both the overlying Padimeiskaya Formation and the Pliocene complex of Icelandic mollusks. These biostratigraphic data do not make it possible to identify analogs of the Akchagylian Stage in the Timan-Pechora Basin. Considering the absence of paleomagnetic data, we can use a method of hydrospheric disturbances.

The presence of a deep (up to − 180 m) post-Kolvinsky erosional incision of the Paleo-Pechora with the presence of sandy-boulder formations (Belkin et al., 1972; Zarkhidze & Krasnozhen, 1973) makes it possible to compare this complex with the alluvial formations of the Kumurli Formation in the Bashkir Urals and Furuvik tillites in Iceland. The overlying Padimeyskaya Formation, which contains mollusks of the Kolva type, can be compared with the Middle Akchagylian of the Urals and the Breidavik Beds of Iceland.

These data indicate the fundamental unity of the structure of the Pliocene deposits of the Eastern Paratethys mid-latitude basin and the high-latitude subpolar Timan-Pechora Basin. Such coincidences are due to the unity of the hydrospheric disturbances and confirm the fundamental regional geological-geophysical constructions of the Ural-Caspian region, which do not need a radical revision. On the other hand, the detection of the Messinian incision (up to

– 300 m) in the polar segment explains the Akchagylian paradox of the radical renewal (up to 94%) of the malacofauna (Nevesskaya et al., 1986). The phenomena of a sharp change in the hydrosphere levels of a non-glacial nature, tectonophysically and planetary-ecologically, have been described and studied on the example of the history of the Earth's development in the Mesozoic and Paleozoic (Katz, 1986; Lapkin & Кац, 1990) and are called the "Messinian-like" events.

Interlayer lithospheric disruptions, a sharp change in the ratio between the areas of the shelf and oceanic depressions, and a radical transformation of the shelf biotas accompany them. The paradox is that the Messinian event itself, compared to other geodynamically similar events, needs to be sufficiently comprehended biotically. A detailed analysis of the Akchagylian Phenomenon contributes to eliminating this biohistorical gap. Furthermore, the Akchagylian Turano-Caspian basin, where an endemic complex of the Sarmatian mollusks appears in the post-Messinian period (Danukalova, 1990; Nevesskaya et al., 1986), is a unique one. It makes it possible to rethink transformations in planetary ecosystems at the boundary of the Miocene and Pliocene when various climatic, hydrospheric, and tectonothermal events influenced the nature of biotic transformations.

One of the most critical problems of the Akchagylian Stage of geological history is the determination of the maximum heights of the level of its hydrosphere compared to the modern one. Theoretically, it is convenient to consider this level close to the mid-latitude level, which does not influence the meridian distortions of the Earth's figure and the maximum extrema of the geoid (Katz, 1986). In addition, the tectonic activity of the region should also be considered. Within the Eastern Paratethys Basin, the regions of the East European platform located on the outskirts of the Caspian Basin have long been used to analyze the hypsometry of the alignment surfaces.

The highest elevations of the Akchagylian transgression, within 180–186 m, were obtained for the Southern Trans-Volga and Obshchy Syrt (Vostryakov, 1964). In other mid-latitude European regions, the marine Pliocene deposits are not developed, except for the areas adjacent to the North Atlantic Basin. In this regard, the classic is the marine Pliocene of southeastern England—the Coral Crag and the Lenham Beds—ferruginous sands with mollusks' remnants brachiopods, bryozoans, and corals occurring at altitudes up to 180 m (Bennison & Wright, 1969). Thus, within the vast Eurasian tectonic plate for the Akchagylian and Atlantic basins, there is a similarity in the hypsometric level of the maximum Pliocene transgression.

The task of comparing this level for the plates belonging to the southern segment—Central Gondwana—has not been posed by anyone. However, this area has been comprehensively studied for a long time in connection with the broad development of various minerals and other natural resources.

16.5 Paleogeographic Mapping of the Akhchagylian Basins of Eurasia and Gondwana

Since our geological and geophysical studies for many years have been concentrated in the southern regions of Eurasia and the northern regions of Gondwana, we attempted to synthesize numerous studies for the paleogeographic mapping of the Akchagylian basins to clarify their relationship and the degree of difference in the manifestation here the Akchagylian hydrospheric phenomenon. As a geographical base, we used the contours of the coastlines of the modern sea basins, and the geodynamic bases were oriented to the lithospheric plates developed in the region and mainly interplate deep faults (Fig. 16.3). In this case, the maximum amplitude of manifestations of the Akchagylian transgression was considered. This study created a somewhat bizarre picture when comparing the coastline contours of the Akchagylian seas with the modern ones.

Consolidated paleogeographic maps of Akchagylian were developed for the Caspian-Turan region (Nevesskaya et al., 1986; Richards

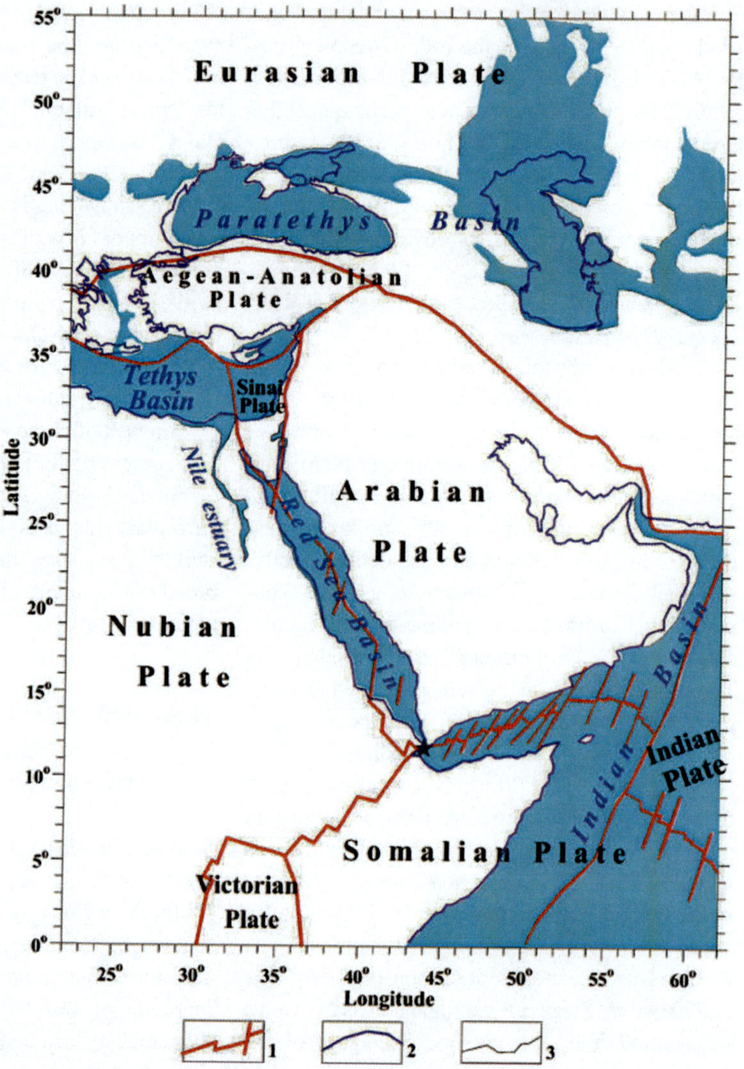

Fig. 16.3 The Akchagylian paleogeographic map of the Eurasian-Arabian-Nubian region with tectonic elements. (1) interplate faults, (2) modern land-sea boundaries, (3) land-sea boundaries in the period of the maximal Middle-Pliocene (Akchagylian) transgression

et al., 2018). They are based on numerous data from the past years. However, no special paleogeographic studies have been carried out in recent years, indicating the links between the Turano-Caspian Basin and the Timan-Pechora Basin. So, the paradox of the fauna separation of the Akchagylian basin still needs to be resolved.

Some part of the Akchagylian fauna of mollusks through the Manych Strait reaches the Taman coast of the Azov Sea (Danukalova, 1990). However, overall, the same-aged Akchagylian Euxinian kuyalnik is represented by a set of malacofauna of a substantially desalinated basin isolated from the Caspian Akchagylian (Nevesskaya et al., 1986). According to modern paleogeographic and paleotectonic data, the relationship between the Euxine Basin at the Kuyalnikian epoch and the marine Tethys-Levant Basin is problematic.

At this time, the rhomb-shaped graben (pull-apart basin) of the Sea of Marmara was a zone of the fold uplift (Karakaş et al., 2018), and the Aegean Sea was almost drained and highly narrow (Clauzon et al., 2008; Mandic et al., 2015;

Neubauer et al., 2015). This area is characterized by a highly mosaic structure and complexity of the block tectonics (Le Pichon et al., 1995).

Paradoxical concerning the deep and extended Akchagylian Basin of the Caspian-Turan, is the Tethyan Basin (Fig. 16.3), limited in area due to the drainage of the Aegean Sea. The southern part of the Aegean-Anatolian lithospheric plate is currently occupied by the sea, except for the large islands—of Crete and Cyprus. However, the narrow bands of the Marine Pliocene covering their plateaus indicate that under conditions of a high transgression of the middle Pliocene (Bellas & Keupp, 2010; Koskeridou et al., 2009; Moskovits, 2012), significant tectonic uplifts occurred simultaneously in the conditions of the boundary structures of the Aegean-Anatolian lithospheric plate (Faccenna et al., 2016). This area requires a more detailed study.

On the other hand, within the southern flank of the Tethys basin, in the Levant and Nubia (Fig. 16.4), the Pliocene basin boundary extends beyond the current coastline and is located at different levels of the adjacent high plateau at the elevations of 50–100 m and 180–210 m (Chumakov, 1967; Gvirtzman & Buchbinder, 1969; Raynaud et al., 2010; Said, 1962). Due to the tectonic and geodynamic features and, first, manifestations of plate geodynamics, two paleogeographic regions can be considered here, where the Pliocene formations are quite different in the composition and origin of rocks.

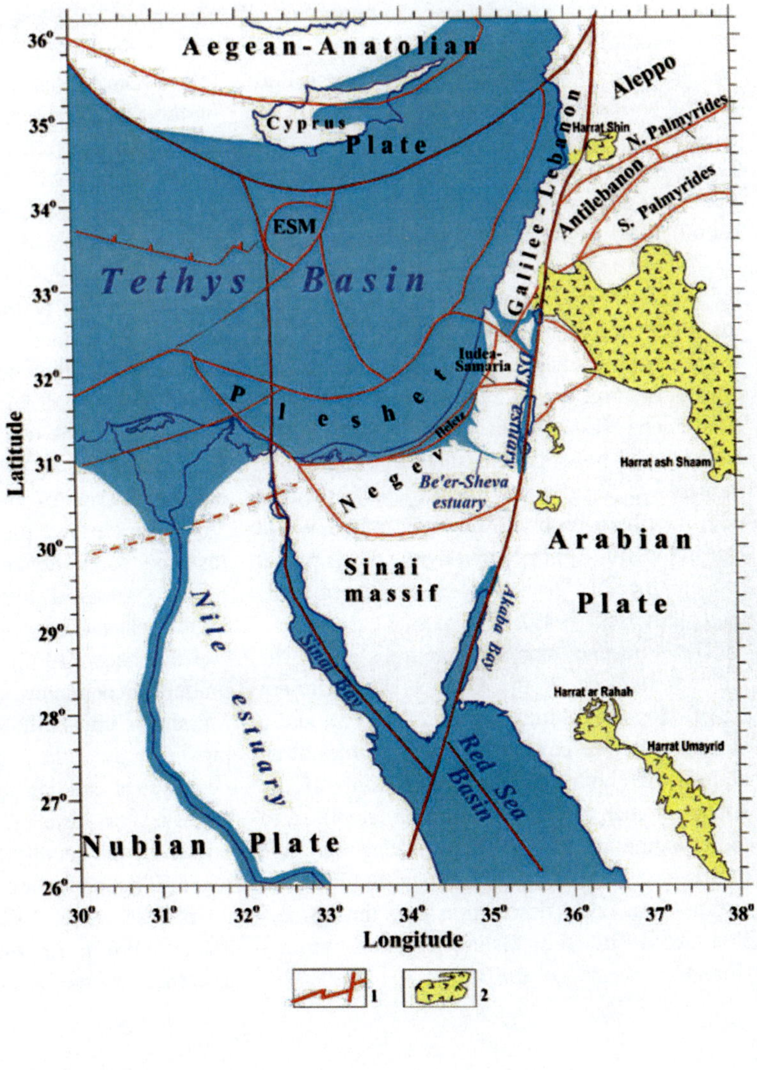

Fig. 16.4 The structural-paleogeographical map of the region under study. (1) interplate faults, (2) trap fields. ESM—Eratosthenes seamount, DST—Dead Sea Transform

The first belongs to the Sinai Plate, and the second to the Nubian Plate (Fig. 16.3). On the Sinai Plate, outcrops of the Pliocene formations are exposed mainly in Israel, in the coastal zone of the Eastern Levant, where the Mesozoic Terrane Belt is developed (Eppelbaum & Katz, 2015). The Pliocene marine and brackish water formations here occur in the Galilee-Lebanon, Judea-Samaria, Pleshet, Heletz, and Negev terranes, mainly on the rocks of the Upper Cretaceous and, less often, of the Paleogene.

They often underlie or alternate with the basaltic bodies, which are a branch of the most extensive trap field in Arabia–Harrat ash Shaam (Fig. 16.3). At the very border with the Mediterranean Sea, the outcrops of the transgressive Pliocene are generally parallel to the modern coastline and compose the upper part of the coastal plain of Israel, the Shfela uplift, and the Beer Sheva valley. The heights of the Pliocene outcrops, as they approach the uplifts of the Negev and Shomron terranes, reach 300 m. However, in general, the middle part of the Pliocene outcrops does not exceed 100–200 m. It should be borne in mind that this is an area of not only active tectonic movements associated with the presence of an active interplate deep shear—the Dead Sea Transform (TMM), and no less active regional diagonal structures and assemblies of rotation caused by the deep geodynamics phenomena (Eppelbaum et al., 2020, 2021). Here, in the post-Pliocene time, the hydrostatic movements were widely manifested, associated with the sharp drops in the level of the World Ocean by 100–120 m caused by the development of continental glaciations and interglacials (Milankovitsch, 1941).

The Pliocene marine formations form the residual terraces of the Carmel Mt. (northern Israel) slopes at altitudes of about 100 m, and a structural terrace covered with the pebbles at an altitude of about 200 m. To the north of the Kishon graben in Galilee (northern Israel) up to the coastline of the Mediterranean Sea, outcrops of this type are developed, forming two Pliocene terraces, smoothly descending into the immersion zone of the Galilee-Lebanon terrane (Fig. 16.4) to the Levant Basin.

These deposits are generally located in the trough zone and have an estuary-like contour. The Marine Pliocene was not found in the TMM troughs. Here, the alluvial and brackish-water Pliocene terrigenous formations are developed, among which the foraminifera remain *Ammonia beccarii*, *A. tepida,* and *Elphidium sp.* were discovered (Almogi-Labin et al., 1995; Hall et al., 2005). The Late Messinian Beer Sheva—Gaza Canyon incised into the Miocene and older formations and composed of the marine Pliocene (Buchbinder & Zilberman, 1997), is located at the same latitude as the southernmost outcrops of the DST brackish Pliocene estuary (Fig. 16.4). Thus, within the limits of Gondwana in the coastal plain of Israel, near the Tethys Basin are reliably present as the elements of the Messinian crisis itself [first discovered by A.N. Mazarovich in 1927 in the Volga-Caspian region (Vostryakov, 1964)], and the ingression complex of the Mediterranean analogs of Akchagylian, filling the Late Miocene canyon. The latter's height is about 200 m, which corresponds to the Caspian hydrospheric maximum. These data were expressed and described earlier, after analyzing the surveys in the Nile Valley in Nubia (Egypt), to clarify the geodynamic stability of the site in connection with the choice of the alignment and the construction of the high dam of the Aswan hydroelectric power station—Sadd el-Aali (Chumakov, 1967). This work determined not only the engineering and energy goals and objectives of two far-flung regions (the Volga region and the Nubian Desert) but also revealed some fundamental similarities in the recent geological history of the Caspian and Mediterranean seas.

Chumakov (1967) was the discoverer of a unique phenomenon in geological history—the Messinian crisis, the reliability of which was proved by the painstaking geological-geophysical studies of the international community. For instance, it has been irrefutably proven by the geophysical surveys and the deep-sea drilling under the Glomar Challenger program (Hsü et al., 1972). Simultaneously, Chumakov (1967), for the first time, proved the existence of the Akchagylian analogs in the

Mediterranean and concluded that the ingression of this sea level should reach the Nile second rapids—to an altitude of about 200 m. Chumakov was the first to link the Mediterranean Messinian crisis with the tectonic-eustatic processes in the Red Sea region.

Preliminary materials on the analysis of the heights of occurrence of the Pliocene marine complexes on the Red Sea islands indicate their excess in places by tens of meters above sea level. However, an interpretation of these data can be genetically diverse. This problem requires a particular study (primarily to assess these prospects for the occurrence of commercial hydrocarbon deposits), which is not included in the research program in the present work.

Much more urgent is the problem of the absence of the marine Pliocene on the Arabian Plate, which immediately makes it unrealistic to search for a connection between the nearby occurring Caspian Akchagylian and the "dry" Persian Gulf. The absence in this bay of deposits associated with the Pliocene hydrospheric maximum is theoretically explainable from the standpoint of a geodynamic active deep structure recently discovered in the studied region (Eppelbaum et al., 2020, 2021). However, to explain the real structural and geodynamic reasons for the drainage of the Persian Gulf in Pliocene at the maximum values of transgression, an analysis of empirical data is significant. Such data indicate the structural-tectonic mapping and the analysis of the movements in the ophiolite belt of the Mushandam Peninsula (Searle, 2019), which testify that the closure of the Strait of Hormuz was caused by the powerful Late Cenozoic movements in the zone of development of the Semail ophiolite belt.

To the southeast of this region, the obduction phenomena occur on the Indian Ocean coast of Arabia, around the younger Cretaceous Masirah ophiolite belt. The consequence of this process was that eastern Arabia became an obstacle to the Pliocene seas (Schreurs, 1999).

16.6 Expanding Future Searches for Hydrocarbons in the Region

Thus, the data from the study of the Pliocene deposits of remote water areas at the boundary of Eurasia and Gondwana indicate the commonality and the difference in their properties, structure, and development history. These facts require an explanation from the standpoint of the latest regional deep geophysics and geodynamics data.

For this purpose, the paleogeographic map was combined with a deep zonation scheme containing the elements of plate geodynamics, the lines of conjugate deformation of the Earth's figure, and the geoid's stepped zone, and isolines of the projection of the deep mantle structure onto the Earth's surface (Fig. 16.5). This data provides a real opportunity to verify the obtained conclusions and the formulations in their assessment. Based on this, the "Akchagylian Phenomenon" concept receives a planetological and geophysical assessment: the Turan-Caspian Basin is topologically separated from the adjacent basins of the Arabian-Caucasian syntaxis.

Three considered basins, the Red Sea, the Tethys, and the Euxine, are in the zone of influence of the deep (lower mantle) structure (Eppelbaum et al., 2021). The data from these studies (Fig. 16.5) testify that the marginal northwestern contour of this structure projection coincides with the Caspian Basin boundaries. Its geodynamics (primarily counterclockwise rotation of the deep structure) influenced the basin's southern and western parts. In the south, the Alpine stage's final development stage was expressed in the uplift of folded structures and the Arabian Plateau. In the west, the Main East European fault divided the Eastern and Western Caucasus zones, which differed in the intensity of uplift and the direction of the tectonic block rotation (Figs. 16.2 and 16.5).

It should be noted that such deep structures are the most unstable and often include mantle

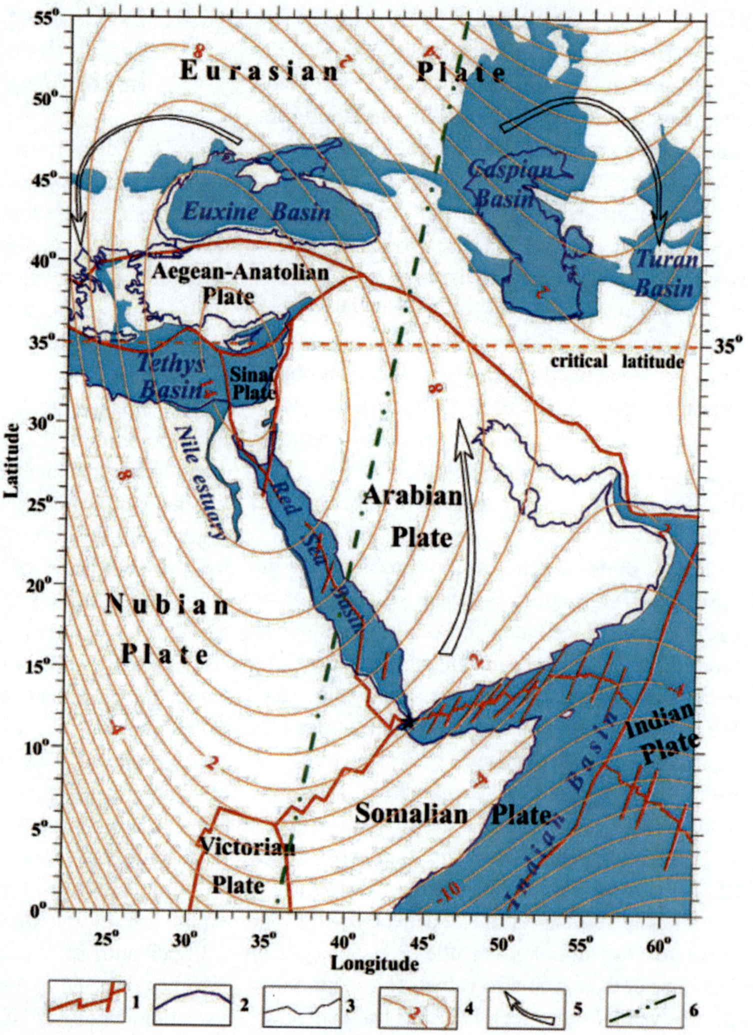

Fig. 16.5 The Pliocene paleogeographical map of the region under study joined with the integrated geophysical-geodynamical scheme (after Eppelbaum and Katz (2021), with modifications). (1) interplate faults, (2) modern land-sea boundaries, (3) land-sea boundaries during the maximum Middle Pliocene (Akchagylian) transgression, (4) residual regional gravity map, (5) rotation of the Earth's crust according to the GPS observations, paleomagnetic and structural data, (6) the generalized position of the planetary Ural-African step [(4) and (5) according to Eppelbaum et al. (2020), (2021), respectively]

plumes, which affect the geodynamics and magmatism of the upper mantle and crust layers and the formation of hydrocarbon reservoirs in the near-surface layers (Trubitsyn, 2012).

The Indian and Turano-Caspian basins are located outside the deep mantle structure but also belong to the Ural-African zone of the negative geoid anomalies (Eppelbaum et al., 2020). These two basins are separated by a critical parallel of 35° (Veronnet, 1912), a zone of the conjugate deformation of the Earth's rotation ellipsoid, where anomalously high zones of the Earth's surface relief are formed. Therefore, the assumptions made so far by many researchers about a possible marine connection of the Caspian Pliocene through the Oman and Persian gulfs seem to be theoretically untenable.

Another important aspect is the development of the Messinian crisis in the Tethys and Paratethys basins. We will not consider the global aspects of this phenomenon associated with hydrospheric disturbances. In our regional case, it is vital to consider the identified geological and geophysical factors of the influence of the deep mantle structure on the near-surface geodynamics (Eppelbaum et al., 2021).

The explanation of the traditional local property—the drainage of the entire Tethys water area due to the closure of the Rif corridor—does not seem convincing enough (Bellas & Keupp, 2010; Krijgsman et al., 2018) since the reasons for the isolation of the basin could be multifaceted ones.

In addition to the plate geodynamics, other deep-seated factors could also have influenced, including an extended geodynamical active deep mantle structure, the top of which is in the zone of the most incredible Earth's tensions, on the critical latitude of 35°. At the same time, the highest movements in this apical zone counteract the development of spreading in the Red Sea zone, and it can be assumed that in the absence of this deep structure, the active rifting would lead to the unification of the Levant and the Red Sea. Furthermore, the influx of water flows from the southeast would have offset the Messinian crisis.

The most significant conclusions of the conducted research are as follows.

- A comprehensive analysis of the paleomagnetic and biostratigraphic data, biogeographical and facial characteristics for the Akchagylian Stage of the Eurasian-Arabian-Nubian region was carried out.
- A generalized paleogeographic map of the studied region has been constructed for the first time, indicating the similarities and differences of the Akchagylian water basins in different paleotectonic zones.
- The planetary nature of the Akchagylian Phenomenon is shown by comparing the Caspian basin with the far-distant vast basins of the Atlantic Ocean, the Polar Ocean, and the Tethys Ocean.
- Explained the effect of the influence of the rotating counterclockwise deep mantle structure on the near-surface and surface geodynamic processes, including the Akchagylian Phenomenon.

The research can be used to develop an enhanced strategy for prospecting hydrocarbon deposits in the region under study.

References

Abdullayev, N. A., Kadirov, F. A., & Guliev, I. (2015). Subsidence history and basin-fill evolution in the South Caspian Basin from geophysical mapping, flexural backstripping, forward lithospheric modelling and gravity modelling. *Geological Society, London, Special Publications, 427*(1), 175–196.

Ali-zade, A. A. (1961). *Akchagylian of Turkmenistan* (Vol. 1, 405 p.). Gosgeoltekhizdat, Moscow (in Russian).

Ali-zade, A. A. (1967). *Akchagylian of Turkmenistan* (Vol. 2, 300 p.). Gosgeoltekhizdat, Moscow (in Russian).

Alizade, A. A., Pashaly, N. V., Mamedov, A. V., Fedorov, P. V., Saradzhalinskaya, T. M., & Ibarov, T. R. (1987). Azerbaijan. In *The boundary between the neogene and quaternary systems in the USSR* (pp. 89–94). Academy of Science of the USSR. Nauka, Moscow (in Russian).

Alizadeh, A. A., Guliyev, I. S., Kadirov, F. A., & Eppelbaum, L. V. (2016). *Geosciences in Azerbaijan. Volume I: Geology* (239 p.). Springer.

Alizadeh, A. A., Guliyev, I. S., Kadirov, F. A., & Eppelbaum, L. V. (2017). *Geosciences in Azerbaijan. Volume II: Economic minerals and applied geophysics* (340 p.). Springer, Heidelberg.

Alizade, K. A. (1954). *Akchagylian stage of Azerbaijan.* Acad. Sci. Azerb. SSR. Inst. of Geology. Baku, Azerbaijan, 344 p. (in Russian).

Almogi-Labin, A., SimaneTov, R., Rosenfeld, A., & Debard, E. (1995). Occurrences and distribution of the foraminifer Ammonia beccari tepida (Cushman) in water bodies, recent and quaternary, of the Dead Sea rift, Israel. *Marine Micropaleontology, 26*, 153–159.

Andrussov N. I. (1902). Materials for the knowledge of the Near-Caspian Neogene. Akchagylian beds. *Transactions of the Russian Geological Committee* (Vol. 15, 154 p.) (in Russian).

Andrussov, N. I. (1918). The relationship between the Euxine and Caspian basins in the Neogene Epoch. *Izvestiya Russian Academy of Science, 8*, 749–760. (in Russian).

Balukhovsky, N. F. (1966). *Geological cycles* (169 p.). Institute of Geological Sciences of Ukrain. Academy of Science, Naukova Dumka, Kiev (in Russian).

Belkin, V. I., Zarkhidze, V. S., & Semenov, I. N. (1972). Stratotypic section of the Kolvinskaya suite. In *Questions of stratigraphy and correlation of the Pliocene and Pleistocene deposits of the Northern and Southern Parts of the Cis-Urals*. Bashkir. Branch of the USSR Acad. Sci., No. 1, Ufa, 5–8 (in Russian).

Bellas, S., & Keupp, H. (2010). *Contribution to the Late Neogene stratigraphy of the ancient Gortys area (southern central Crete, Greece)* (pp. 579–593). Bulletin of the Geology Society of Greece.

Bennison, G. M., & Wright, A. E. (1969). *The geological history of the British Isles* (p. 143). Geol., Univ. Birmingham, London.

Buchbinder, B., & Zilberman, E. (1997). Sequence stratigraphy of Miocene-Pliocene carbonate-siliciclastic shelf deposits in the eastern Mediterranean margin (Israel): Effects of eustasy and tectonics. *Sedimentary Geology, 112*, 7–32.

Chumakov, I. S. (1967). Pliocene and pleistocene deposits of the Nile Valley in Nubia and Upper Egypt. In *Transactions of the Geology Institute, Russian Academy of Science* (Vol. 170, 115 p.), Nauka, Moscow (in Russian).

Clauzon, G., Suc, J.-P., Dumurdžanov, N., Melinte-Dobrinescu, M. C., & Zagorchev, I. (2008). The Pliocene Gilbert-type fan delta of Draĉevo (Skopje area, Republic of Macedonia): Paleogeographic inference. *Geologica Macedonica, 2*, 21–28.

Dalla, S., Harby, H., & Serazz, M. (1997). Hydrocarbon exploration in a complex incised valley fill: An example from the late Messinian Abu Madi Formation (Nile Delta Basin, Egypt). *The Leading Edge, 12*, 1820–1824.

Danukalova, G. A. (1990). *Bivalve mollusks and the Akchagylian stratigraphy* (PhD thesis). Paleontological Institute of the USSR Academy of Science (192 p.), Moscow (in Russian).

Eder, V. G., Ryzhkova, S. V., Kostyreva, E. A., Pavlova, M. A., Sotnich, I. S., Zamirailova, A. G., & Ponomareva, E. V. (2020). Lithologic, geochemical, and geophysical characteristics of the boundary Strata of the Bazhenov and Kulomza Horizons (Lower Cretaceous Base) in the central regions of the West Siberia. *Geology and Geophysics, 61*(7), 767–782.

Einarsson, T. (1957). Magneto-geological mapping in Iceland with the use of a compass. *Philosophical Magazine*, Suppl. 6, No. 2, 232–239.

Eppelbaum, L. V., Ben-Avraham, Z., Katz, Y. I., Cloetingh, S., & Kaban, M. (2020). Combined multifactor evidence of a giant lower-mantle ring structure below the Eastern Mediterranean. *Positioning, 11*, 11–32.

Eppelbaum, L. V., Ben-Avraham, Z., Katz, Y., Cloetingh, S., & Kaban, M. (2021). Giant quasi-ring mantle structure in the African-Arabian junction: Results derived from the geological-geophysical data integration. *Geotectonics, 55*(1), 67–93.

Eppelbaum, L. V., & Katz, Y. I. (2011). Tectonic-geophysical mapping of Israel and eastern Mediterranean: Implication for hydrocarbon prospecting. *Positioning, 2*(1), 36–54.

Eppelbaum, L. V., & Katz, Y. I. (2015). Eastern Mediterranean: Combined geological-geophysical zonation and paleogeodynamics of the Mesozoic and Cenozoic structural-sedimentation stages. *Marine and Petroleum Geology, 65*, 198–216.

Eppelbaum, L., & Katz, Y. (2021). Akchagylian hydrospheric phenomenon in aspects of deep geodynamics. *Stratigraphy and Sedimentation of Oil-Gas Basins, 2*, 8–26.

Eppelbaum, L. V., & Katz, Y. I. (2022). Combined zonation of the African-Levantine-Caucasian areal of ancient hominin: Review and integrated analysis of paleogeographical, stratigraphic, and geophysical-geodynamical data. *Geosciences (Switzerland), 27* (1), 1–23.

Eppelbaum, L. V., Katz, Y. I., & Ben-Avraham, Z. (2012). Israel–petroleum geology and prospective provinces. *AAPG European Newsletter, 4*, 4–9.

Eppelbaum, L. V., & Khesin, B. E. (2012). *Geophysical studies in the Caucasus*. Springer.

Faccenna, C., Gueydan, F., Sokoutis, D., Philippon, M., Kydonakis, K., et al. (2016). The two-stage Aegean extension, from localized to distributed, a result of slab rollback acceleration. *Canadian Journal of Earth Sciences, 53*(11), 1142–1157.

Gasanov, T. A., & Aliyeva, S. G. (2003). Stratigraphy of sediments of akchagylian and apsheronian in foothills of southeast part of Lesser Caucasus. *Geology of Oil, 37*(1), 23–30.

Gladenkov, Y. B. (1978). Pliozen-antropogen (Pliocene-Anthropogen. In *Iceland and the mid-ocean ridge. Stratigraphy and lithology* (pp. 62–85). Nauka, Moscow (in Russian).

Gurary, G. Z., 1988. *Geomagnetic field during reversals in the late Cenozoic* (Vol. 415, 207 p.). Transaction of the Geological Institute Academy of Science USSR (in Russian).

Gvirtzman, G., & Buchbinder, B. (1969). Outcrops of Neogene formation in the central and southern coastal plain Hashefela and Beer Sheva regions, Israel. *Geological Survey of Israel Bulletin, 50*, 1–76.

Hall, J. K., Krasheninnikov, V. A., Hirsch, F., Benjamini, C., & Flexer, A. (2005). *Geological framework of the Levant. Vol. II: The Levantine Basin and Israel* (826 p). Historical Productions-Hall, Jerusalem, Israel.

Hsü, K. J., Citta, M. B., & Ryan, W. B. F. (1972). The origin of the Mediterranean evaporites. In A. G. Kaneps (Ed.), *Initial reports of the deep-sea drilling project*, XIII (Part 2) (pp. 1203–1231). U.S. Government Printing Office, Washington, D.C.

Kamaletdinov, M. A., & Yakhimovich, V. L. (Responsible Editors). (1981). Pliocene and pleistocene of the Volga-Ural region. *Bashkir Branch of the USSR Academy of Science* (176 p.), Institute of Geology, Nauka, Moscow (in Russian)

Karakaş, Ç., Armijo, R., Lacassin, R., Suc, J.-P., & Melinte-Dobrinescu, M. C. (2018). Crustal strain in the Marmara Pull-Apart region associated with the propagation process of the North Anatolian fault. *Tectonics, 37*(5), 1507–1523.

Katz, Y. I. (1986). Cretaceous Thalassocratic Maximum and Planetary Movements of the Hydrosphere. In D. P Naidin (Ed.), *Cretaceous period. Paleogeography and paleooceanology* (pp. 191–237). Nauka, Moscow (in Russian).

Khain, V. E., & Polyakova, I. D. (2008). Large and giant hydrocarbon accumulations in the transitional continent-ocean zone. *Geotectonics, 3*, 3–17.

Kontakiotis, G., Karakitsios, V., Cornée, J.-J., Moissette, P., Zarkogiannis, S. D., Pasadakis, N., Koskeridou, E., Manoutsoglou, E., Drinia, H., & Antonarakou, A. (2020). Preliminary results based on geochemical sedimentary constraints on the hydrocarbon potential and depositional environment of a Messinian sub-salt mixed siliciclastic-carbonate succession onshore Crete (Plouti section, eastern Mediterranean). *Mediterranean Geoscience Reviews, 2*, 247–265.

Koskeridou, E., Vardala-Theodorou, E., & Moissette, P. (2009). Pliocene and Pleistocene shallow-water chitons (Mollusca) from Rhodes Island, Greece. *Neues Jahrbuch fur Geologie und Palaontologie-Abhandlungen, 251*(3), 303–330.

Krijgsman, W., Capella, W., Simon, D., Hilgen, F. J., Kouwenhoven, T. J., Meijer, P. T., Sierro, F. J., Tulbure, M. A., van den Berg, B. C. J., van der Schee, M., & Flecker, R. (2018). The Gibraltar Corridor: Watergate of the Messinian Salinity Crisis. *Marine Geology, 403*, 238–246.

Krijgsman, W., Tesakov, A., Yanina, T., Lazarev, S., Danukalova, G., van Baak, C. G. C., Agusti, J., Alçiçek, M. C., Aliyeva, E., Bista, D., Bruch, A., Büyükmeriç, Y., Bukhsianidze, M., Flecker, R., Frolov, P., Hoyle, T. M., Jorissen, E. L., Kirscher, U., Koriche, S. A., ... Wesselingh, F. P. (2019). Quaternary time scales for the Pontocaspian domain: Interbasinal connectivity and faunal evolution. *Earth-Science Reviews, 188*, 1–40.

Lapkin, I. Y., & Кац, Y. I. (1990). Geological events at the carboniferous and permian boundary. *Izvestiya Academy of Science USSR, 8*, 45–58. (in Russian).

Lazarev, S., Kuiper, K. F., Oms, O., Bukhsianidze, M., Vasilyan, D., Jorissen, E. L., Bouwmeester, M. J., Aghayeva, V., van Amerongen, A. J., Agustí, J., Lordkipanidze, D., & Krijgsman, W. (2021). Five-fold expansion of the Caspian Sea in the late Pliocene: New and revised magnetostratigraphic and 40Ar/39Ar age constraints on the Akchagylian Stage. *Global and Planetary Change, 206*, 1–19.

Le Pichon, X., Chamot-Rooke, N., & Lallema, S. (1995). Geodetic determination of the kinematics of central Greece with respect to Europe: Implications for eastern Mediterranean tectonics. *Journal of Geophysical Research, 100*(B7), 12675–12690.

Mandic, O., Kurečić, T., Neubauer, T. A., & Harzhauer, M. (2015). Stratigraphic and paleogeographic significance of lacustrine mollusks from the Pliocene Viviparus beds in central Croatia. *Geologia Croatica, 68*(3), 179–207.

Matsuoka, K., & Bujak, J. P. (1988). Cenozoic dinoflagellate cysts from the Navarin Basin, Norton Sound and St. George Basin, Bering Sea. *Bulletin of the Faculty of Liberal Arts, Nagasaki University, Natural Science, 29*(1), 1–147.

Milankovitsch, M. (1941). *Canon of insolation and the ice-age problem* (Vol. 132, 634 p.). Special Publication of the Royal Serbian Academy, Serbia.

Mironova, L. V. (Chief Editor) (1982). *Stratigraphic dictionary of the USSR* (611 p.). Paleogene, Neogene, Quaternary System. Nedra, Leningrad (in Russian).

Moskovits, S. (2012). The mollusca in the marine Pliocene and Pleistocene sediments of the south-eastern Medititerranean Basin (Cyprus–Israel). *The Geological Survey of Israel, Report GSI/25/2012*, 1–159.

Naidin, D. P., & Naidina, O. D. (1992). On the so-called Great Akchagylian transgression. *Izvestiya Vyzov. Geologiya i Razvedka, 6*, 3–18. (in Russian).

Naidina, O. D., & Richards, K. (2020). The Akchagylian stage (Late Pliocene-Early Pleistocene) in the North Caspian Region: Pollen evidence for vegetation and climate change in the Urals-Emba region. *Quaternary International, 540*, 22–37.

Neubauer, T. A., Harzhauser, M., Kron, A., Georgopoulou, E., & Mandic, O. (2015). A gastropod-based biogeographic scheme for the European Neogene freshwater systems. *Earth Science Reviews, 143*, 98–116.

Nevesskaya, L. A., Goncharova, I. A., Il'ina, L. B., Paramonova, N. P., Popov, S. V., Babak, E. V., Bagdasaryan, K. G., & Voronina, A. A. (1986). History of the Neogene mollusks of the Paratethys. *Transactions of the Paleontology Institute of the Academy Science USSR* (Vol. 200, 208 p.). Nauka, Moscow (in Russian).

Popov, G. I. (1969). Scheme of the Akchagylian stage stratification. In *Geology and Economic Deposits of Turkmenistan (Stratigraphy)* (pp. 179–223). Ashkhabad (in Russian).

Raynaud, S., de la Boisse, H., Makroum, F. M., & Bertho, J. (2010). Geological and topographical study of the original hills at the base of Fourth Dynasty Egyptian monuments of the Memphite plateau. *Bulletin de la Société géologique de France, 181*(3), 279–290.

Reilinger, R. E., McClusky, S., Vernant, P., Lawrence, S., Ergintav, S., Cakmak, R., Ozener, H., Kadirov, F., Guliev, I., Stepanyan, R., Nadariya, M., Hahubia, G., Mahmoud, S., Sakr, K., ArRajehi, A. et al. (2006). GPS constraints on continental deformation in the Africa–Arabia–Eurasia continental collision zone and implications for the dynamics of plate interactions. *Journal of Geophysical Research: Solid Earth, 111* (BO5411), 1–26.

Richards, K., van Baak, C. G. C., Athersuch, J., Hoyle, T. M., Stoica, M., William. E. N., Austin, W. E. H., Cage, A. G., Wonders, A. A. H., Marreti, F., & Pinnington, C. A. (2018). Palynology and micropalaeontology of the Pliocene—Pleistocene transition in outcrop from the western Caspian Sea, Azerbaijan: Potential links with the Mediterranean, Black Sea and the Arctic Ocean? *Palaeogeography, Palaeoclimatology, Palaeoecology, 511*, 119–143.

Richards, K., Vincent, S. J., Davies, C. E., Hinds, D. J., & Aliyeva, E. (2021). Palynology and sedimentology of the Pliocene productive series from Eastern Azerbaijan. *Palynology, 45*(4), 569–598.

Roberts, G., & Peace, D. (2007). Hydrocarbon plays and prospectivity of the Levantine Basin, offshore Lebanon and Syria from modern seismic data. *GeoArabia, 12*, 99–124.

Said, R. (1962). *The geology of Egypt* (p. 734). Elsevier Publ. Co.

Schreurs, G. (1999). West-northwest directed obduction of the Batain Group on the eastern Oman continental margin at the cretaceous-tertiary boundary. *Tectonics, 18*(1), 148–160.

Searle, M. (2019). *Geology of the Oman Mountains, Eastern Arabia* (478 p.). GeoGuide, Springer Nature Switzerland AG.

Sobolev, D. N. (1926). *Earth and Life. I. The geological cycles* (60 p.). Kiev (in Russian).

Stovas, M. V. (1963). Some questions about tectogenesis. In *Problems of planetary geology* (pp. 222–274). Gosgeoltekhizdat, Moscow (in Russian).

Trubikhin, V. M. (1977). Paleomagnetism and stratigraphy of the Agchagylian deposits of West Turkmenistan. In *Transactions of the Geology Institute, Russian Academy of Science* (Vol. 301, 79 p.). Nauka, Moscow (in Russian).

Trubitsyn, V. P. (2012). Passage of oceanic plates through the upper and lower mantle boundary. *Doklady Russian Academy Science, 446*(6), 677–679. (in Russian).

Van Baak, C. G. C., Grothe, A., Richards, K., Marius, S. M., Aliyeva, E., Daviesh, G. R., Kuiper, K. F., & Krijgsman, W. (2019). Flooding of the Caspian Sea at the intensification of Northern Hemisphere Glaciations. *Global and Planetary Change, 174*, 153–163.

Veronnet, A. A. (1912). Rotation de l'Ellipsoide Heterogene et Figure Exacte de la Terre. *Journal de Mathematiques Pures et Appliquees, 6-me ser., 8*, 331–463 (in French).

Vincent, S. J., Davies, C. E., Richards, K., & Aliyeva, E. (2010). Contrasting Pliocene fluvial depositional systems within the rapidly subsiding South Caspian Basin; a case study of the palaeo-Volga and palaeo-Kura River systems in the Surakhany Suite, Upper Productive Series, onshore Azerbaijan. *Marine and Petroleum Geology, 27*(10), 2079–2106.

Vostryakov, A. P. (1964). Buried and accumulative surfaces of the Southern Trans-Volga region. In *Alignment surface problems* (pp. 107–115). Nauka, Moscow (in Russian).

Zarkhidze, V. S., & Krasnozhen, A. S. (1973). Experience in the selection of polygenetic surfaces of alignment in the north of the Timan-Ural region. *Izvestiya Komi Branch of the Geographical Society of USSR* (Vol. 16, pp. 97–105). Syktyvkar (in Russian).

17 Estimating Informational Content for Hydrocarbons Searching in Azerbaijan

17.1 An Essence of the Informational Approach

It is well-known that most inverse problem solutions in geophysics are ill-posed (e.g., Zhdanov, 2002). It means that the solution does not exist, is not unique, or is not a continuous function of observed geophysical data (when a small perturbation in the observations will cause an arbitrary mistake in the solution). This fact calls for the broad application of informational and probabilistic methodologies in applied geophysics and geology.

Geophysical observations in searching for useful minerals are notoriously complicated by numerous factors. Numerous interpretational methodologies have been developed to eliminate these disturbances (Eppelbaum, 2019; Eppelbaum & Khesin, 2012; Khesin et al., 1996). However, at times, the complexity of the geological environment (variability in lateral and vertical physical properties), significant depth of anomalous targets (**AT**) occurrence, and additional disturbances make it impossible or unfeasible to apply these methodologies. In such cases, information-probabilistic methods are effective tools for recognizing and classifying targets (e.g., Duda & Hart, 1973), estimating the potential information value of geophysical methods, and deciding upon a workable solution. Geophysical surveys aim to obtain qualitative and quantitative information about the geometric and physical characteristics of buried **AT**, e.g., to develop physical-geological models (***PGM***) of target objects. ***PGM***s of varying degrees of complexity (the simplest ***PGM***s are purely target identification, and complex ***PGM***s can be 3D models of buried **AT**) can be used to justify drilling (excavations) in defined areas and generate future strategies for geophysical-geological searches in areas where **AT** are known to exist.

This Chapter briefly describes some of the critical features of the informational approach to geophysical methods applied for searching useful minerals. Here, simple informational and probabilistic criteria can be applied to formalize the relative value of geophysical methods, geophysical information, financial cost, and time factors to generate integrated parameters.

Estimating the information value of geophysical, geochemical, and other means can be formalized based on the following criteria (after Eppelbaum et al., 2003) (Fig. 17.1):

1. Informativeness of the application (informational criterion Γ);
2. Cost of implementing the method (cost criterion C);
3. Time required to carry out the method (time criterion T).

Criteria C and T are easy to calculate directly, but criterion Γ is a non-trivial research problem. A simplified algorithm can be written as (Eppelbaum et al., 2003):

A. Alizadeh et al., *Pliocene Hydrocarbon Sedimentary Series of Azerbaijan*,
Advances in Oil and Gas Exploration & Production, https://doi.org/10.1007/978-3-031-50438-9_17

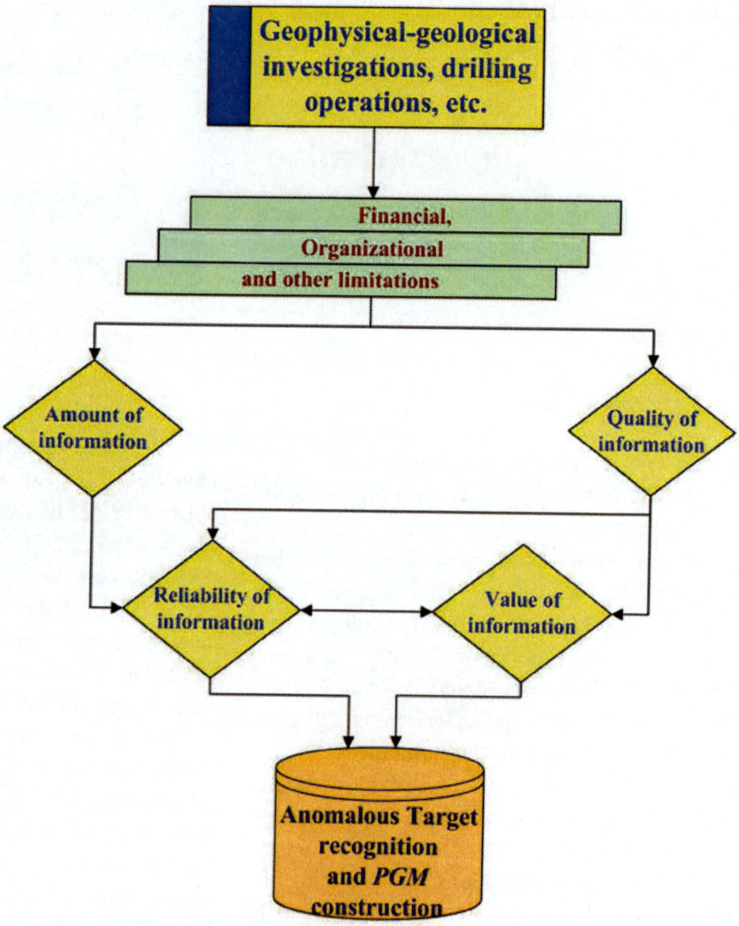

Fig. 17.1 A scheme of anomalous target recognition and *PGM* construction with elements of information theory

$$\Omega = \Gamma \cup C \cup T, \quad (17.1)$$

where $\cup$ is the symbol of unification.

All the available geological information can be represented as the classic three-level model (Fig. 17.1): (a) syntactic—the quantity of information, (b) semantic—the content of information, and (c) pragmatic—the value of information. The logical-heuristic model for describing environmental information thus takes the following form (Eppelbaum, 2014):

$$\Gamma = I \cup R \cup V, \quad (17.2)$$

where I is the quantitative estimation of information, R is the estimation of the informational reliability corresponding to the semantic criterion, and V is the estimation of the informational value in terms of feasibility according to the pragmatic criterion.

Algorithm (17.2) is based on the fundamental terms of information theory and is combined with the structural (hierarchical) approach. This approach defines each indicator as a structure reflecting a set of typical situations and is then used to calculate the value of each estimator using the informational measure. Parameters V and R should be estimated geologically, but this is beyond the scope of this study. Here, parameters V and R will be neglected, and it is assumed that $\Gamma = I$.

A variant of the possible analysis of geological-geophysical information is presented in Fig. 17.2.

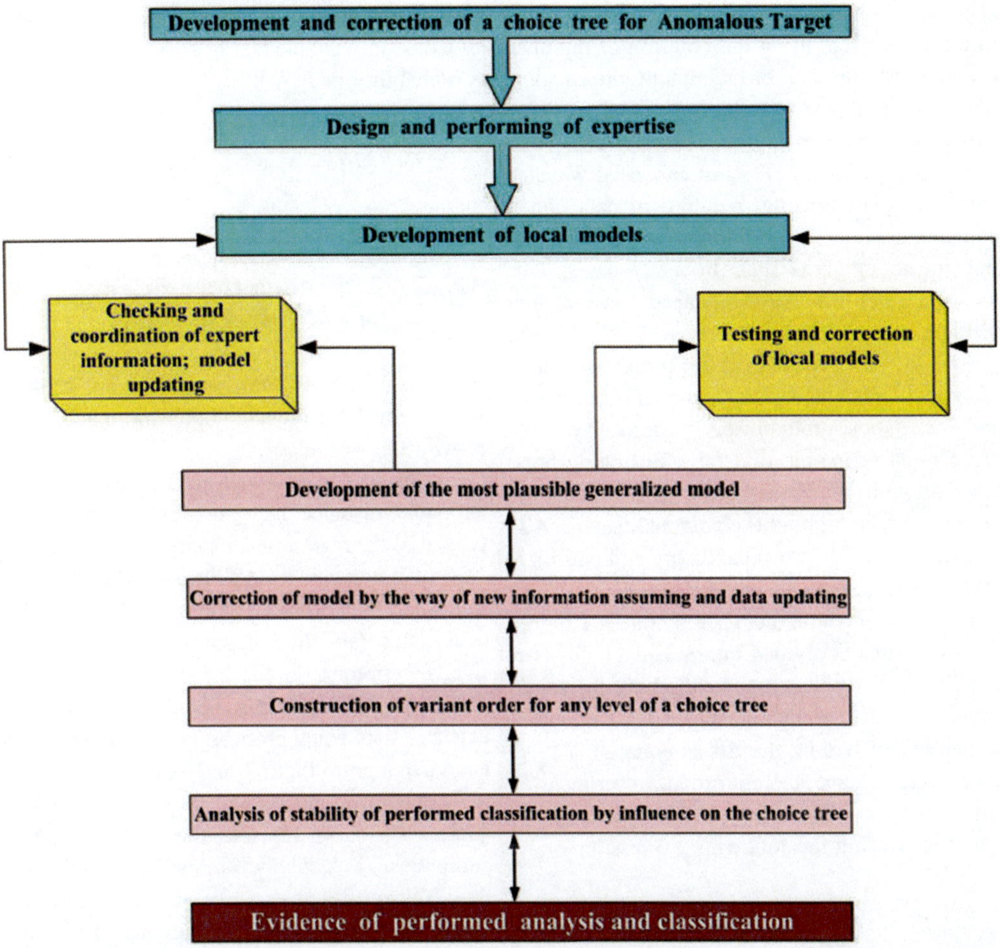

Fig. 17.2 Flow-chart of integrated geological-geophysical analysis of combined data obtained in geological-geophysical prospecting using the comprehensive informational approach

17.2 Evaluating the Efficiency of Geophysical Methods with Informational-Statistical Procedures

17.2.1 General Considerations

Choosing the suitable method (or several methods) can be based on a qualitative-quantitative estimate (Figs. 17.1 and 17.2). For this purpose, reliable informational and statistical criteria are needed. The first issue is the quantity of information that can be obtained by a single method or a set of methods (geophysical, geochemical, and other data). The second is to define a criterion to express the decision-making risk as a function of the obtained data. Nevertheless, informational criteria are preferable because geological-geophysical prospecting is a permanent process of acquiring and analyzing different kinds of information.

The classic information-theoretic works by Shannon (1948) and Brillouin (1962) prompted Khalfin (1958) to apply these criteria to geology and geophysics first. It has been shown (Khesin et al., 1996) that informational and statistical approaches represent two aspects of a shared approach. For instance, the solution to an identification problem using the criterion of minimal average risk or that of maximum information (minimal residual uncertainty) under definite conditions results in the same expressions (Nikitin, 1993; Svetov, 1992). An enhanced wavelet approach to integrating geophysical data under complex geological conditions was proposed by Eppelbaum et al. (2011). In many cases, this approach is like the enhanced information method.

This lays the groundwork for presenting some of the key elements of the theory of information and statistical-probabilistic approaches to prospecting economic minerals. Illustrating how these straightforward methods can yield valuable results should highlight commonalities in **AT** localization and geophysical theory and optimize ***PGM*** development and successful **AT** discovery.

Let us assume that there is only one **AT** in the area. This area is divided into N equal cells. For simplicity, each cell is assumed to have the same probability of containing the **AT**. Thus, the probability of finding the **AT** is equal $P = 1/N$ in each cell. Hence, the entropy of experiment β (discovery of **AT**) is $\log_2 N$. The entropy is determined using the following expression:

$$H(\beta) = -\sum_{i=1}^{N} P(B_i) \cdot \log_2 P(B_i), \qquad (17.3)$$

where $P(B_i)$ is the probability of a B_i outcome (B is the range of outcomes of experiment β).

In this situation $H(\beta)$ value ($\log_2 N$) is the maximum possible uncertainty. Experiment α (geophysical observation) gives additional information. The A value is a range of outcomes for the experiment α.

The difference between uncertainties in the β results before and after experiment α serves to estimate the information in α as related to β:

$$I(\alpha, \beta) = H(\beta) - H(\beta|\alpha), \qquad (17.4)$$

where $H(\beta|\alpha)$ is the conditional entropy for the experiment β (provided that the experiment α has been conducted).

The conditional entropy is the average value of a random variable taking a $H(\beta_i|\alpha_i)$ value with a probability of $P(A_i)$:

$$H(\beta_i|\alpha_i) = \sum_{i=1}^{N} P(A_i) H(B_i|A_i). \qquad (17.5)$$

17.2.2 Estimating the Efficiency of Individual Methods

When selecting methods for integration, it makes sense to evaluate the amount of information each provides.

Let us start from a well-investigated area with an equal distance between observation points, typical of the region under study. We assume that 1/50 of it contains an **AT**. It is a priori assumption that the Bouguer gravity field over **AT** occurrence (anticline structure) is always positive. In contrast, in the area's empty part (not containing **AT**), it may be either positive or negative with equal probability.In other words, it is known a priori that 2% and 98% of the area are **AT**-containing and empty, respectively, and in 49% and 51% of this area, the Bouguer gravity anomalies, respectively, are negative and positive. The results of experiment β can be designated as follows: B—**AT** occupying part of the area, $\overline{B}$—empty part of the area. Thus, $P(B) = 0.02$, $P(\overline{B}) = 0.98$. According to expression (17.3) $H(\beta) \cong 0.14$.

The result of experiment α is expressed as follows: A is a positive field, $\overline{A}$ is a negative field. The relative partial entropy (after recording the positive Bouguer gravity at the measurement point) can be calculated in the following way: $P(A) = 0.51; P(\overline{A}) = 0.49$; $P(B|A) = 2/51$, $P(\overline{B}|A) = 1 - P(B|A) = 49/51$. Thus, $H(\beta|\alpha) \cong 0.24$. According to the recorded negative Bouguer gravity, the area is certainly empty: $H(B|\overline{A}) = 0$. Consequently, the partial entropy for experiment β under the conditions of α, as stated in Eq. (17.5), is $H(\beta|\alpha) \cong 0.12$. Thus, the uncertainty of the **AT** determination decreased after Bouguer gravity observation/computation from 0.14 to 0.12.

By considering the financial cost of each measurement, the effectiveness of the respective methods can be estimated (e.g., Eppelbaum et al., 2003). For this purpose, it is sufficient to compare the values (costs) of the information units provided by each method. On this basis, a valid solution for a reasonable integration of the methods can be made.

17.2.3 Estimation of Information by Indicator (Field) Gradations

The applied geophysical fields usually have maximal and minimal intensity within the studied areas. The difference between the maximal and minimal intensities may be subdivided into some intervals (gradations). The gradations of indicators can also be used to obtain information about the types of **AT**. Physical fields, geochemical analyses, **AT** features, etc. can serve as indicators. Let $P(A_i|B)$ denote the posterior probability of finding the *i*-th gradation of indicator *A* (e.g., magnetic field intensity) over the targets *B*, $P(A_i)$ is the prior probability of finding the same gradation in a survey area. Thus, the partial information on the presence of the target $I_{A_i \to B}$ contained in the recorded A_i gradation takes the following form:

$$I_{A_i \to B} = \log_2 \left[\frac{P(A_i|B)}{P(A_i)} \right], \qquad (17.6a)$$

as the uncertainty in the recording of A_i before the survey was $\log_2 P(A_i)$ and after the survey —$\log_2 P(A_i|B)$.

Similarly,

$$I_{A_i \to \overline{B}} = \log_2 \left[\frac{P(A_i|\overline{B})}{P(A_i)} \right]. \qquad (17.6b)$$

Statistically, the probabilities (or, to be more precise, relative frequencies) are expressed by the following ratios: *S* is the total number of values, S_i is the number of values occupied by the *i*-th interval of the A_i values, S_p is the total number of points reflecting the **AT** projections to earth's surface, and S_{pi} is the number of points common for interval A_i and the **AT** projections.

Then:

$$P(A_i) = \frac{S_i}{S}, \quad P(A_i|B) = \frac{S_{pi}}{S_p},$$
$$P(A_i|\overline{B}) = \frac{S_i - S_{pi}}{S - S_p}.$$

The increment in the information contained about the presence of the object

$$\Delta I_i = I_{A_i \to B} - I_{A_i \to \overline{B}} = \log_2 \left[\frac{P(A_i|B)}{P(A_i|\overline{B})} \right] \qquad (17.7)$$

are summed up in each elementary cell of the studied geological site.

To estimate the informational significance, the following expression can be used (Eppelbaum, 2020):

$$I'_{A \to B} = \sum_i \left[P(A_i) \cdot I_{A_i \to B} \right]. \qquad (17.8)$$

The estimates served to substantiate the ratio between the informational significance, financial cost, and time (see Eq. (17.1)). To normalize the results, the following semi-empirical expressions (obtained based on an informational approach and long-term experience in the field of applied geophysics) can be used (Expressions 17.9a and 17.9b). It is explained in the example of geophysical methods I and II. At the same time, it can be geochemical methods or other geoscientific tools:

$$\Omega_I = \frac{I'_{A \to B(I)}}{C_{(I)}/C_{(II)} \cdot T_{(I)}/T_{(II)}}, \qquad (17.9a)$$

$$\Omega_{II} = \frac{I'_{A \to B(II)}}{C_{(II)}/C_{(I)} \cdot T_{(II)}/T_{(I)}}. \qquad (17.9b)$$

A comparison of the corresponding factors will enable us to estimate the geophysical method's effectiveness.

17.2.4 Estimates of the Efficiency of Geophysical Integration Based on the Probability of Type I and II Errors

Classification efficiency can be estimated quantitatively for separate methods and geophysical integration by calculating the reliability of **AT** revealing. Let us assume that 20 geophysical anomalies have been contoured and localized in some regions by a set of geophysical methods. The revealed anomalies were divided into three groups with various degrees of desired **AT** discovery. Under the assumption that the drilling results are reliable, classification reliability can be assessed by calculating the probability of type I and type II errors.

The probability of a type I error (M_1) is expressed as the relative frequency of an erroneous diagnosis for objects from the sampling $\overline{B}$ (the remainder of the objects). The probability of a type II error (M_2) is expressed as the relative frequency of an erroneous diagnosis for objects from sampling B (**AT**). These errors are used to determine the total unconditional error of separation q between the classes B and $\overline{B}$ (i.e., the risk of an erroneous solution):

$$q = M_2P(B) + M_1P(\overline{B}), \tag{17.10}$$

where $P(B)$ and $P(\overline{B})$ are the prior probabilities of the appearance of objects of the first and second classes, respectively.

If $P(B) = P(\overline{B}) = 0.5$, then q value corresponds to the intersection area of distribution densities $P(X|B)$ and $P(X|\overline{B})$. Here X is the separation index. It can represent geophysical field amplitude or gradient, the value of the integrated indicator, etc.

The separation reliability (γ) is:

$$\gamma = 1 - q. \tag{17.11}$$

The total empirical error should be compared to the theoretical error. The approximation of the errors can confirm a correct assumption and provide high reliability of identification. Using logical-informational methods (Khesin et al., 1996), the classification reliability is estimated solely from empirical errors.

The errors due to assigning observation results to a class (with or without the **AT**) can be determined as follows. The absence of anomalies for the complex indicator in a known target-containing area is a type II error, or "omission of target". These anomalies in the empty part of this area are a type I error or a "false alarm".

Comparing these rapid results with those of a more complex integration can also be used to estimate the respective errors and the classification reliability. From the previous experience was recognized that the hydrocarbon nature of 14 out of 20 revealed gravity-magnetic anomalies in the Middle Kur depression was confirmed (Khesin et al., 1985). New **ATs** were not revealed at the areas where geophysical methods were applied. Thus, $M_1 = 0$, $M_2 = 14/20$. If $P(B) = P(\overline{B}) = 0.5$ and considering Eqs. (17.10) and (17.11), then $q = 0.70$ and $\gamma = 0.30$.

17.2.5 Calculation of Information Parameters

Typically, the first step in the qualitative interpretation involves visually inspecting profile observations or compiled maps. However, more than this simple analysis is often required to locate buried objects.

Detecting any geophysical anomalies caused by **AT** is often difficult because the influence of surrounding inhomogeneities of different sizes and intensities may mask the effect. Various transformation and filtering procedures have been developed for more effective localization of finite objects (Parasnis, 1997; Telford et al., 1990, etc.). We have shown elsewhere (Khesin & Eppelbaum, 1997; Eppelbaum, 2013, 2014, 2020) that applying information parameters often overcomes these problems. The method involves considering that the noise component—the combined effect of the surrounding inhomogeneities that are not associated with the target objects—is random. The amount of information (I_i) at each observation points by application of the i-th method is

$$I_i = -\log_2 P_j \tag{17.12a}$$

or

$$I_i \cong \log_2\left(\frac{H_i}{\Delta H_i}\right), \tag{17.12b}$$

where P_j is the frequency rate of the j-th interval of the i-th indicator (field) in the histogram (more precisely, Bayes' evaluation of the probability that the results will end up in the j-th interval) and H_i and ΔH_i are the amplitude and determination error of this indicator, respectively.

Eppelbaum et al. (2008) proposed the following expression:

$$I_i = U_i\left(\sum_i^n U_i\right)^{-1} \log_{10}\left[U_i\left(\sum_i^n U_i\right)^{-1}\right]K, \tag{17.12c}$$

where U_i is the geophysical observation at i-th point ($U_i > 0$) in the area under study, n is the total number of observations, and K is some coefficient.

After summing the information elements that suggest a priori that a target object is present, the random noise and components caused by dissimilar geological features are suppressed. To avoid singling out fictitious objects by the plot of $\frac{1}{n}\sum_{i=1}^{n} I_i$, which can occur when a large amount of information is contained in the data of only one or two methods. An additional complex criterion that depends on the number of significant indicators can be calculated while adjusting for their relative influence:

$$I_{\text{integr}} = \sum_{k=1}^{\frac{n(n-1)}{2}} \frac{(I_p)_k}{(I_p)_{\max}}, \tag{17.13}$$

where I_p is determined from the formula

$$I_p = (I_1 + I_2)\frac{I_1}{I_2}, \quad (I_1 \le I_2),$$

by using pairwise combinations of the n methods employed.

The results of the parameter calculation may be compiled as maps of I_i and I_{integr}. To avoid missing deeply embedded objects, in some cases it is better to use frequency rates of average values or average field estimates on a sliding scale instead of the P_j and H_i values, respectively. The correlation of I_{integr} with the sum of information elements makes it possible to avoid missing an object, that for these or other reasons, was not revealed by other indicators. The combination of indices permits certain interpretative conclusions.

In practice, J_i is usually replaced by the relative amount of information, also known as the coefficient of informativity (Khesin et al., 1996):

$$K_i = \frac{J_i}{\overline{J}_i}. \tag{17.14}$$

The value of $\overline{J}_i$ determines the information obtained when the result of U_j falls into the x_j interval of the histogram with an equal probability of falling into any of the R intervals. According to probability principles (Daston, 1988; Ventsel, 1969), it is equal to the average (complete) information obtained when using a single method:

$$\overline{J}_i = \log_2 R. \tag{17.15}$$

The application of K_i considers differences in the ranges of different fields. However, the application of expressions (17.11) and (17.13)–(17.15) may not be practical for sparse sampling.

17.2.6 Estimating Integration Efficiency by Localization of Weak Anomalies

Suppose a set of methods is focused on investigating independent indicators of equal value. In that case, the anomaly detection reliability γ can be described by an error function (probability integral) (Khesin & Eppelbaum, 1997) as:

$$\gamma = F\left(\frac{\sqrt{\sum_i \upsilon_i}}{2}\right), \tag{17.16}$$

where υ is the ratio of the anomaly squared to the noise dispersion for each i-th geophysical field, and F is the probability integral of type

$$F(t) = \frac{1}{\sqrt{2\pi}} \int_{-\infty}^{t} \mathrm{e}^{\left(\frac{-x^2}{2}\right)} \mathrm{d}x.$$

Now, let us assume that three points indicate the anomaly and that the mean square of the anomaly for each field is equal to the noise dispersion. For a single method, the reliability of detecting an anomaly of a known form and intensity by Kotelnikov's criterion (e.g., Borda, 2011) is expressed by $F(t)\left(\sqrt{\upsilon_i}/2\right)$. Hence the reliability for individual methods is 0.61 and 0.77 and 0.87 for a set of two or three methods, respectively. This means that the q value (risk of an erroneous solution) when integrating two or three methods decreases by factors of 1.7 and 3.0, respectively (according to Eq. (17.11)). A comparison of the

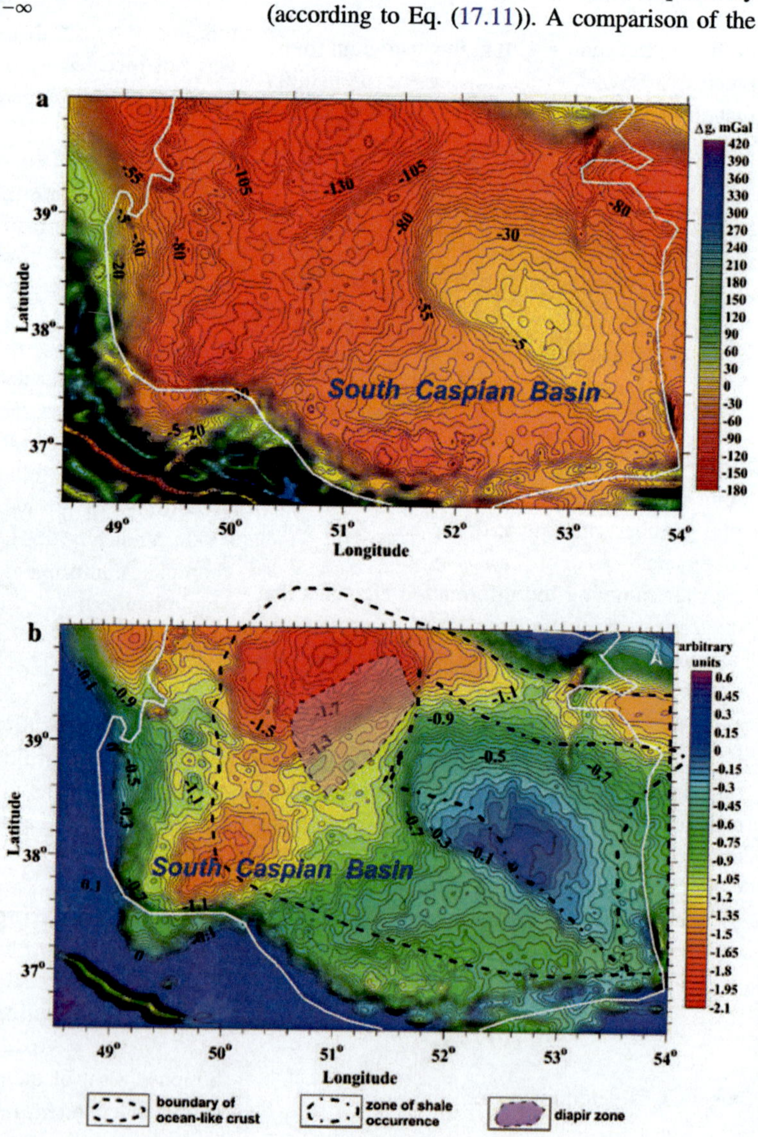

Fig. 17.3 **a** Satellite gravity map of the South Caspian Basin and its SW border compiled based on Geosat and ERS-1 altimetry data. **b** Map of the informational parameter I_i transformed from the map presented in Fig. 17.3a

risk with financial costs C yields an optimum set of methods.

17.3 Application of the Information Methodologies On-Field Examples

Let us begin to consider the application of the informational parameter on the regional gravity field processing. It is well-known that the South Caspian Basin (*SCB*) is one of the main hydrocarbon provinces in the world (Alizadeh et al., 2016). The satellite gravity data of the *SCB* were obtained from the World Gravity DB as retracked from Geosat and ERS-1 altimetry (e.g., Sandwell & Smith, 2009). A highly positive factor is that these observations were made with regular global 1-min grids, and the error of gravity data computation for the newly arrived data is estimated at 1.0–1.5 mGals (Sandwell et al., 2013). The gravity data retracked to the Earth's surface are close to the free air gravity and can be effectively applied to the tectonic-geophysical zonation (Eppelbaum & Katz, 2011). The compiled gravity map (Fig. 17.3a) shows the gravity pattern within the *SCB* and southwest land bordering. The gravity data range from − 180 up to + 420 mGals, and the anomalies in the center of the *SCB* are not brightly expressed. After the application of the informational approach (Eq. 17.12c), the image was significantly transformed (Fig. 17.3b), and the gravity anomalies in the *SCB* can be effectively recognized (it should be noted that in both maps, the same color pattern was applied). Geological-geophysical examination of the transformed anomalies may be an exciting problem, but it is beyond the scope of this study.

A Bulla-Deniz gas deposit located in the Bay of Baku (Azerbaijan) is one of the rich deep hydrocarbon accumulations occurring below

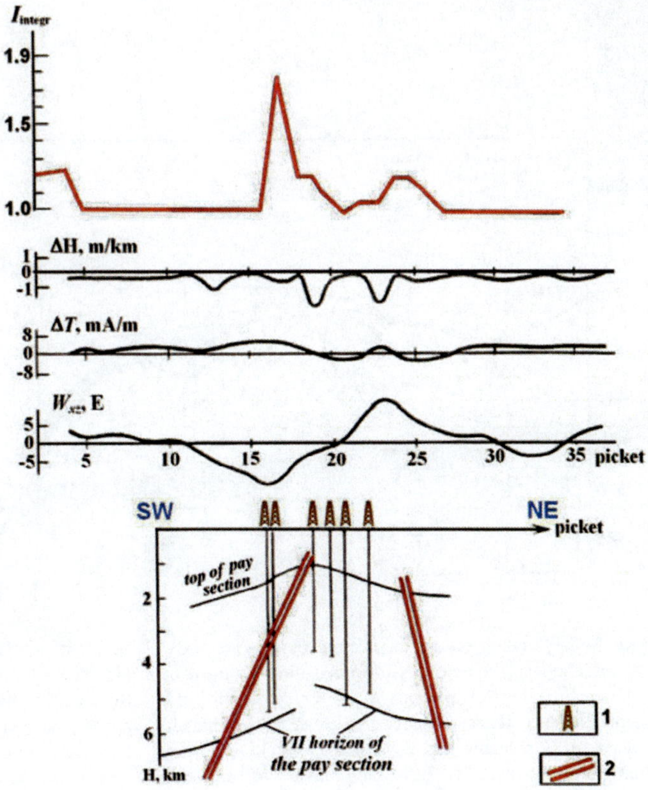

Fig. 17.4 Qualitative delineation of faults by computing parameter I_{integr} at the Bulla area (Baku Archipelago, Caspian Sea). (1) deep wells, (2) faults delineated by combined analysis of geological and geophysical data

5000 m within the *SCB* (Guliyev et al., 2011). To calculate the parameter I_{integr} [see Eqs. (17.12a) and (17.13)], three different fields were employed: the local magnetic anomalies ΔT (marine survey data), the second horizontal derivative of the gravity potential W_{xz} (data from the bottom gravity survey were utilized), and $\Delta H/H$ (relative changes of the sea bottom

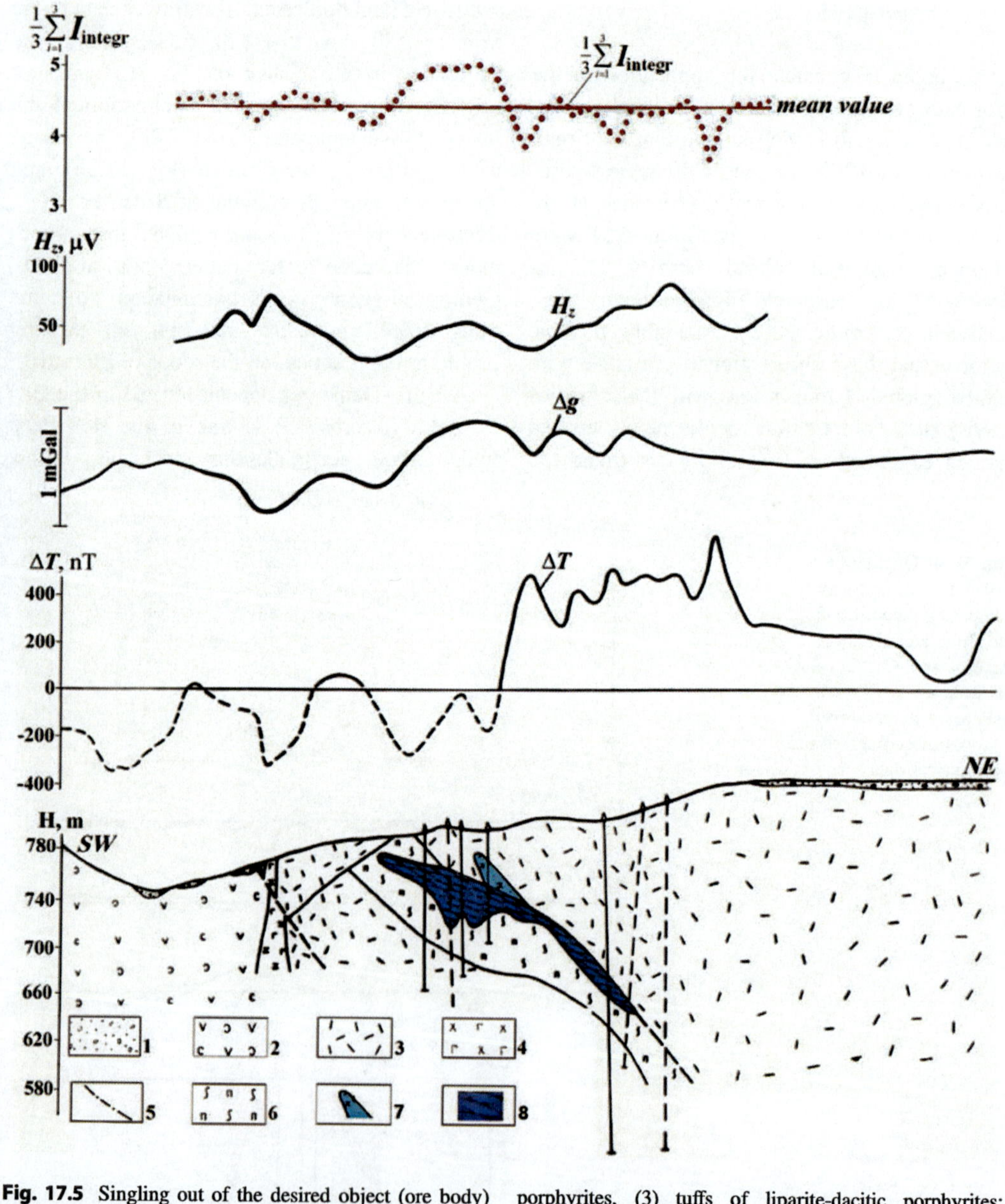

Fig. 17.5 Singling out of the desired object (ore body) by summing up the amounts of information obtained by different geophysical methods in the Gyzylbulagh gold-pyrite deposit (Lesser Caucasus, western Azerbaijan) (based on Eppelbaum and Khesin 2012). (1) Quaternary deluvial deposits; (2) tuffs and lavas of andesitic porphyrites, (3) tuffs of liparite-dacitic porphyrites; (4) dikes of andesite-basalts; (5) disjunctive dislocations; (6) zones of brecciation, crush and boudinage with lean pyrite-chalcopyrite ore; (7) zone of brecciation, crush and boudinage with rich impregnating mineralization; (8) massive pyrite-chalcopyrite ore

topography) (Fig. 17.4) (modified after Eppelbaum & Khesin, 2012). As can be seen from this figure, both faults associated with the pay section are reflected in the graph of I_{integr}. The large amplitude of the SW anomaly of the I_{integr} parameter compared to NE may be explained by the SW fault's proximity to the sea bottom's surface.

A practical example of applying this methodology is given in the example of one of the ore deposits in Azerbaijan. A gold-pyrite deposit Gyzylbulagh is situated in the Agdara ore district (Lesser Caucasus, western Azerbaijan) (Alizadeh et al., 2016). In this deposit, gravity field (Δg_B), magnetic field (ΔT), and *VLF* (Very Low Frequency) electromagnetic field were employed (Fig. 17.5). In gravity and *VLF* (vertical component H_z) fields, minor positive anomalies over the ore body were observed, whereas, in the magnetic field, a negative anomaly was registered (in this deposit, the pyrite body with gold mineralization does not contain magnetic minerals and occurs in a magnetic medium). Integration of these geophysical methods [Eqs. (17.12b) and (17.13) were used] enabled us to obtain a significant anomaly I_{integr} indicating the position of the **AT** (ore body) upper edge.

Effective integration of gravity, VLF, and temperature data in the mine of Katsdag polymetallic deposit (southern slope of the Greater Caucasus, north-west Azerbaijan) was shown in Eppelbaum et al. (2014). However, the **ATs** in mines (a significant difference from land or satellite surveys) can occur below and in all surrounding space, which complicates employment of the informational algorithms.

Thus, the information value of applying geophysical (geochemical, geological) methods for searching oil-and-gas deposits can be estimated using various probabilistic-statistical procedures. As shown in this Chapter, the components of these procedures can serve to assess the informativity of not only a single geophysical method but also of any geophysical integration. Elements of information theory can be easily applied to optimize the process of target recognition and help reveal anomalous effects from an **AT** against the significant noise background (*low signal-to-noise ratio*). The simple convolution of geophysical information, financial cost, and time suggested here serves to compare the parameters.

References

Alizadeh, A. M., Guliyev, I. S., Kadirov, F. A., & Eppelbaum, L. V. (2016). *Geosciences in Azerbaijan. Volume I: Geology*. Springer.

Borda, M. (2011). *Fundamentals in information theory and coding*. Springer.

Brillouin, L. N. (1962). *Science and information theory* (2nd ed.). Academic Press.

Daston, L. (1988). *Classical probability in the enlightenment*. Princeton Univ. Press.

Duda, R. O., & Hart, P. E. (1973). *Pattern classification and scene analysis*. Wiley.

Eppelbaum, L. (2020). Theories of probability, information and graphs in applied geophysics. In K. Kyamakya (Ed.), *Prime archives in applied mathematics* (pp. 1–35). Vide Leaf.

Eppelbaum, L., & Katz, Y. (2011). Tectonic-geophysical mapping of Israel and eastern Mediterranean: Implication for hydrocarbon prospecting. *Positioning, 2*(1), 36–54.

Eppelbaum, L. V. (2013). Geophysical observations at archaeological sites: Estimating informational content. *Archaeological Prospection, 21*(2), 25–38.

Eppelbaum, L. V. (2014). Estimating informational content in geophysical observations on example of searching economic minerals in Azerbaijan. *Izvestiya Academy Science of Azerbaijan Report Series: Earth Sciences, 3–4*, 31–40.

Eppelbaum, L. V. (2019). *Geophysical potential fields: Geological and environmental applications*. Elsevier.

Eppelbaum, L. V., Alperovich, L., Zheludev, V., & Pechersky, A. (2011). Application of informational and wavelet approaches for integrated processing of geophysical data in complex environments. In *Proceedings of the 2011 SAGEEP Conference* (Vol. 24, pp. 24–60), Charleston, South Carolina, USA.

Eppelbaum, L. V., Eppelbaum, V. M., & Ben-Avraham, Z. (2003). Formalization and estimation of integrated geological investigations: Informational approach. *Geoinformatics, 14*(3), 233–240.

Eppelbaum, L. V., Ezersky, M. G., Al-Zoubi, A. S., Goldshmidt, V. I., & Legchenko, A. (2008). Study of the factors affecting the karst volume assessment in the Dead Sea sinkhole problem using microgravity field analysis and 3D modeling. *Advances in GeoSciences, 19*, 97–115.

Eppelbaum, L. V., & Khesin, B. E. (2012). *Geophysical studies in the Caucasus*. Springer.

Eppelbaum, L. V., Kutasov, I. M., & Pilchin, A. N. (2014). *Applied geothermics*. Springer.

Guliyev, I., Aliyeva, E., Huseynov, D., Feyzullayev, A., & Mamedov, P. (2011). Hydrocarbon potential of ultra-deep deposits in the South Caspian Basin. Adapted from oral presentation at the *AAPG European Region Annual Conference*, Kiev, Ukraine, October 17–19, 2010. Search and Discovery Article #10312.

Khalfin, L. A. (1958). Information theory of geophysical interpretation. *Doklady AN USSR, 122*(6), 1007–1010.

Khesin, B. E., Alexeyev, V. V., & Eppelbaum, L. V. (1996). *Interpretation of geophysical fields in complicated environments*. Modern Approaches in Geophysics, Kluwer Academic Publ.

Khesin, B. E., Alexeyev, V. V., Eppelbaum, L. V. et al., (1985). Combined analysis of potential geophysical fields in the Middle Kura depression. *Research Report*, YuzhVNIIGOEFIZIKA (Baku, Azerbaijan).

Khesin, B. E., & Eppelbaum, L. V. (1997). The number of geophysical methods required for target classification: Quantitative estimation. *Geoinformatics, 8*(1), 31–39.

Nikitin, A. A. (1993). *Statistical processing of geophysical data*. Bookseries: Advanced Geophysics. Russian Experience, Issue No.2, Electromagnetic Research Centre, Moscow.

Parasnis, D. S. (1997). *Principles of applied geophysics*. Chapman & Hall.

Sandwell, D. T., Garcia, E., Soofi, K., Wessel, P., & Smith, W. H. F. (2013). *Towards 1 mGal global marine gravity from CryoSat-2, Envisat, and Jason-1. The leading edge* (Vol. 32, No. 8, pp. 892–899).

Sandwell, D. T., & Smith, W. H. F. (2009). Global marine gravity from retracked Geosat and ERS-1 altimetry: Ridge Segmentation versus spreading rate. *Journal of Geophysical Research, 114*(B01411), 1–18. https://doi.org/10.1029/2008JB006008

Shannon, C. E. (1948). A mathematical theory of communication. *Bell System Technical Journal, 27* (3–4), 379–432, 623–656.

Svetov, B. S. (1992). *Information theory basis of geophysics*. Bookseries: Advanced Geophysics. Russian Experience, Issue No.1, Electromagnetic Research Centre, Moscow.

Telford, W. M., Geldart, L. P., & Sheriff, R. E. (1990). *Applied geophysics* (2nd ed.). Cambridge University Press.

Ventsel, E. S. (1969). *The probability theory*. Nauka Publ.

Zhdanov, M. S. (2002). *Geophysical inverse theory and regularization problems*. Ser.: Methods in geochemistry and geophysics (Vol. 36). Elsevier.

18 Deep Structure of Azerbaijan and Its Relationship with Hydrocarbon Reserves

World experience indicates that combined geophysical studies of the deep structure opens new prospects for the search and exploration of hydrocarbon deposits (Eppelbaum, 2019; Gasi & Hachay, 2017; Gupta, 2021).

18.1 Brief Geological-Geophysical Background

The complexity of Azerbaijan's territory's geological structure stems from its location in the Alpine-Himalayan tectonic belt (AHTB) (e.g., Aliyev et al., 2005a, 2005b; Alizadeh, 2012; Eppelbaum, 2015; Kadirov et al., 2015; Khain, 2000; Leonov, 2008). The NE part of Azerbaijan is a fragment of the Pre-Caucasian foreland filled with Cenozoic terrigenous sediments. A heterogenic Nakhchivan folding system is in the SW part, where Paleozoic carbonate strata and Cenozoic magmatic formations are mixed. At the mega-anticlinorium of the Greater Caucasus, stratified Cenozoic and Mesozoic thick (predominantly sedimentary) strata are presented. The prevalence of Mesozoic magmatic formations is typical of the mega-anticlinorium of the Lesser Caucasus. The Kur mega-synclinorium, dividing the Greater and Lesser Caucasus, is characterized by an accumulation of thick (up to several kilometers) Cenozoic terrigenous sediments. The Talysh anticlinorium is located on the SE flank of the Kur depression, where Paleogene magmatic associations are widely distributed (Aliyev et al., 2005a; Ismail-zadeh et al., 2005).

According to Khain (2000), the most ancient Pre-Baikalian [this complex is termed as Cadomian in Western publications (Khain, 2007)] structural complex is characterized by a sub-meridional strike. A less metamorphosed Baikalian complex is rumpled to latitudinal folds in separate areas. The Caledonian complex is practically unknown. The Hercynian complex is characterized by a Caucasian strike identical to the overlying Mesozoic rocks.

More complete geological data characterize the Alpine tectonic–magmatic cycle. The frequently changing geological associations on the vertical and lateral axes for Azerbaijan territory are typical. The geological structure complicates multifarious fold and fault structures and regional and local metamorphism (Alizadeh, 2012). All these factors make the development of reliable models of these media sufficiently complex.

The first models of the Earth's crust of Azerbaijan were put forward in the mid-1960s (Gadjiev, 1965; Shekinsky et al., 1967; Tzimelzon, 1965). The first generalized velocity models of the Azerbaijanian Earth's crust were developed by Radjabov (1977). These models were subsequently evaluated in the works of Tzimelzon (1970), Azizbekov et al. (1972), Shikhalibeyli (1972), Shikhalibeyli and Grigoriants (1980), Alexeyev et al. (1988), and later by Khesin et al., (1993, 1996). Ruppel and McNutt (1990) studied the regional compensation of the

A. Alizadeh et al., *Pliocene Hydrocarbon Sedimentary Series of Azerbaijan*,
Advances in Oil and Gas Exploration & Production, https://doi.org/10.1007/978-3-031-50438-9_18

Greater Caucasus using the Bouguer gravity. Sarker and Abers (1998) attempted to apply *P* and *S* wave tomography to examine the Greater Caucasus deep structure. After that, modified models based on gravity field analysis were presented in Kadirov (2000a, 2000b).

Some significant regional peculiarities of the deep structure of Azerbaijan and adjacent Caspian area are reflected in: Krasnopevtseva (1984), Rappel and Mcnutt, 1990, Pilchin and Eppelbaum (1997), Mangino and Priestley (1998), Gasanov (2001), Jackson et al. (2002), Kaban (2002), Allen et al. (2003), Brunet et al. (2003), Ulomov (2003), Glumov et al. (2004), Guliyev and Panachi (2004), Ismailzadeh et al. (2004), Mamedov (2004), Mukhtarov (2004), Knapp et al. (2004), Aliyev et al. (2005a, 2005b), Babayev and Gadjiev (2006), Eppelbaum and Pilchin (2006), Kadirov (2006), Khalafly (2006), Reilinger et al. (2006), Saintot et al. (2006), Artyushkov (2007), Khain (2007), Leonov (2008), Ricketts et al. (2008), Mamedov (2008, 2009), Mamedov et al. (2008), Egan et al. (2009), Green et al. (2009), Mosar et al. (2010), Yetirmishli et al. (2011), Alizadeh et al. (2012), Eppelbaum and Khesin (2012), Koulakov et al. (2012), Pavlenkova (2012), Yetirmishli and Kazimova (2013), Forte et al. (2013), Mederer et al. (2013), Ruban (2013), Kadirov and Gadirov (2014), Abdullayev et al. (2015), Eppelbaum (2015), Kadirov et al. (2015), Sharkov et al. (2015), Alizadeh et al. (2017), Eppelbaum (2017), Eppelbaum et al. (2021), Abdullayev et al. (2022), Kadirov et al. (2023), and others.

We will present several examples of geophysical investigation separately for the Azerbaijanian land and sea.

18.2 Azerbaijan's Land

18.2.1 Seismic and Seismotomography Data Analysis

Seismic data analysis and seismotomography play an essential role in developing physical-geological models of deep structures.

18.2.1.1 Deep Seismic Sounding

A review of the main recent publications in this field was presented in Sect. 18.1. Here we will show the last examples of re-interpretation of deep seismic profiles crossing Azerbaijan territory (Pavlenkova, 2012) (Figs. 18.1, 18.2 and 18.3).

Profiles of deep seismic sounding (DSS) 'Stepnoe—Bakuriani' and Volgograd—Nakhchivan were carried out by the Ministry of Geology of the former USSR in the 60s of the XX century (Krasnopevtseva et al., 1970; Yurov, 1963). These profiles cross the Greater Caucasus (Fig. 18.1) and, till the present time, are sole DSS profiles (Figs. 18.2 and 18.3) intersecting across the strike of the Greater Caucasus and surrounding regions of Azerbaijan. Therefore, the re-interpretation of these profiles is of great interest.

The new velocity sections indicate that the tectonic movements generating the Caucasian orogen were a complex combination of horizontal and vertical movements covering all the Earth's crust and upper mantle. The boundary M1 in the upper mantle and ramps in the southern part of the orogen are relics of the subhorizontal movements (Pavlenkova, 2012).

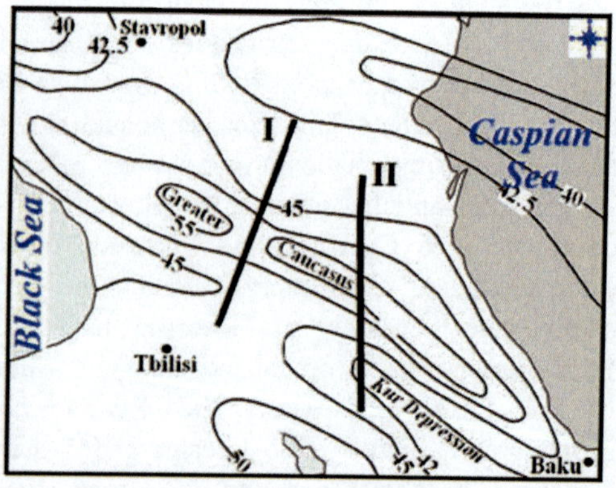

Fig. 18.1 Scheme of the deep seismic sounding profiles location: Stepnoe–Bakuriani profile (I) and southern part of the Volgograd–Nakhchivan profile (II) and map of the Moho discontinuity of the Greater Caucasus (after Pavlenkova, 2012; Rezanov & Shevchenko, 1978, with minor modifications)

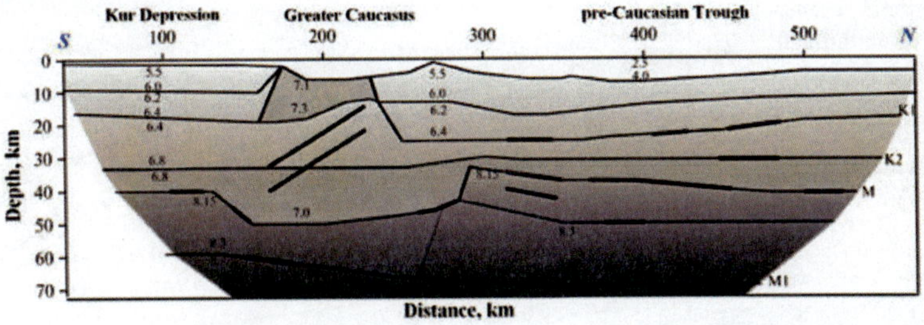

Fig. 18.2 New velocity section along the profile Stepnoe-Bakuriani (I). Other explanations are presented in Fig. 18.1. M boundary shows the Earth's crust bottom, M1 is the wave reflected from the mantle boundary; K1 and K2 are the boundaries inside the Earth's crust. Thin lines represent the boundaries between layers with different seismic velocities (km/s), and thick lines represent the reflecting areas (after Pavlenkova, 2012)

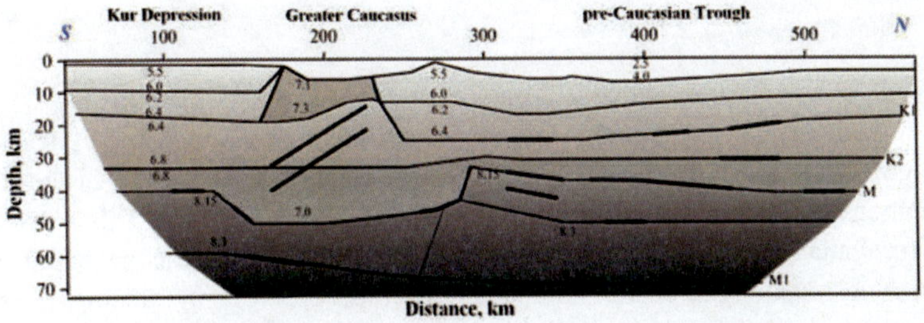

Fig. 18.3 New velocity section along with the southern part of profile Volgograd-Nakhchivan (II). Other captions are presented in Figs. 18.1 and 18.2 (after Pavlenkova, 2012)

The newly obtained data were utilized to reconstruct Azerbaijan's 3D deep physical-geological models.

Seismic sections of regional and composite profiles provide enough space for reconstructing the main features of the tectonic development and sedimentation conditions. Information obtained from seismic sections and updated by available geological-geophysical and borehole data was used by us for compiling some seismostratigraphic sections (Fig. 18.4). The tracing of regional surfaces of unconformity on seismic sections has a principal meaning not only for deep structure analysis but also for historical-genetic analysis. The unconformity surfaces between sequences control most revealed and forecasted non-anticlinal traps in the region, outlining the applied aspect of their study (Guliyev et al., 2003).

18.2.1.2 Seismotomography of the Azerbaijanian Earth's Crust

As one of the methods for Earth's deep structure analysis, seismic tomography allows (based on the travel time of elastic waves from earthquakes) to obtain independent information about the structure and physical properties of Earth's crust and mantle. The tomographic model is used to construct an integrated geological-geophysical model with geological and geophysical (magnetic, gravity, and thermal data and magnetotelluric sounding) data. Just the velocity model is the tool that can be used for revealing interrelations between geophysical field parameters and concrete depth intervals (Koronovsky, 2000). The development of new algorithms and methodologies, the use of the new recording equipment, and the improvement of computational methods allow

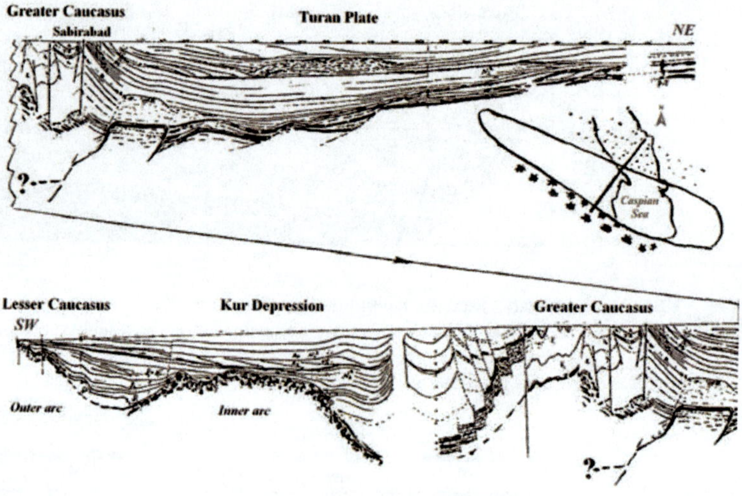

Fig. 18.4 Regional seismic-stratigraphic section through the Turan Plate, southeastern troughs of the Greater Caucasus, Kur Depression, and Lesser Caucasus (after Mamedov, 1991)

more efficient and accurate restoration of the velocity structure of the upper layers of the Earth. Novel information obtained based on cooperative analysis of recalculated velocity models with geological-tectonic maps significantly improves the present notions about deep geological structure peculiarities and geodynamic processes (Tarakanov, 2006).

18.2.1.3 Methodology of Investigation: A Brief Description

The aim of the tomographic inversion consists of a detailed analysis of the medium's velocities and absorbing properties. Such investigations are based on the travel time of the first arrivals or observed amplitudes for a set of pairs 'receivers-sources'. Any geometry of sources and receivers can be applied. The sole limitation is that the seismic rays should create a common net; seismic rays must cross each point of the observed medium in all directions (Fig. 18.5). At the first step of the tomographic inversion selection, an initial velocity model (and absorbing model for amplitude inversion) must be created (Adamova & Sabitova, 2004). Further interpretation is realized in two stages: (1) solving the direct problem and (2) solving the inverse problem (Firbas, 1987; Richter, 1958).

In this investigation, a leading role played "TomoTetraFD" software—a program for computing 3D velocity models based on the seismotomography method developed at the Inst. Geosphere Dynamics of the Russ. Acad. of Sci. (Usoltseva, 2004). This software allows for restoring velocity structure based on analysis of discrepancies of ray tracing containing information about anomalous zones to wave propagation.

For sorting of geodynamic events, Kissling's and Roecker's criteria were used (Abers & Roecker, 1991; Barber et al., 1996). In the investigation, seismological parameters about the characteristics of local earthquakes and the travel time of *P* and *S* waves registered by the telemetric station net for the period of 2004–2014 (Fig. 18.6). Parameters of hypocenters (coordinates and depth of focus, time of event generation, and travel times of *P* and *S* waves) were received from the Republic Center of Seismological Service catalogs. For event classification, Kissling's and Roecker's criteria were used (Abers & Roecker, 1991; Barber et al., 1996). Based on the geometry of earthquake focus location and seismological station disposition, the depth of the investigation may consist of up to 60 km (Kazymova & Kazymov, 2010; Yetirmishli et al., 2011).

In the process of velocity model computation, four input files were utilized: (1) coordinates of usable seismological stations; (2) 1D velocity model; (3) travel time of seismic waves from focus to the station; (4) main file jointing all

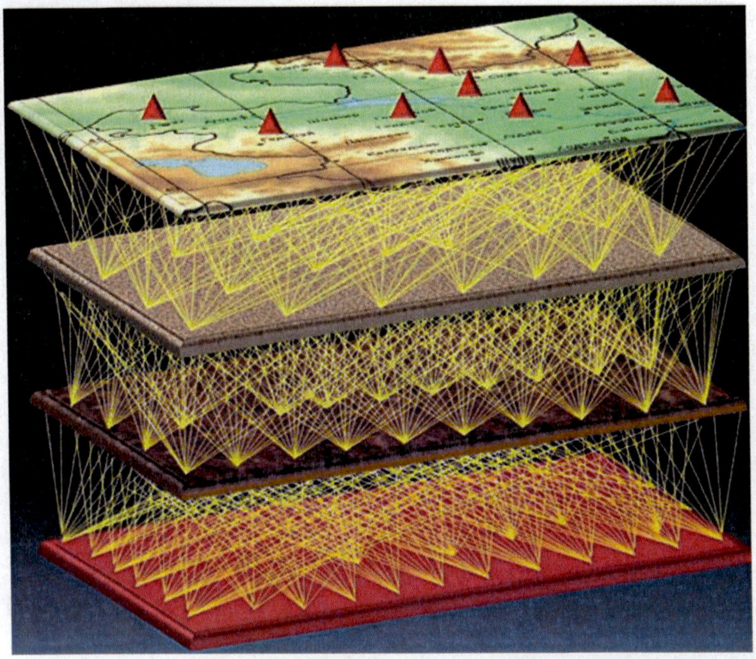

Fig. 18.5 3D scheme of seismic ray distribution in tomographic inversion (after Yetirmishli & Kazimova, 2013)

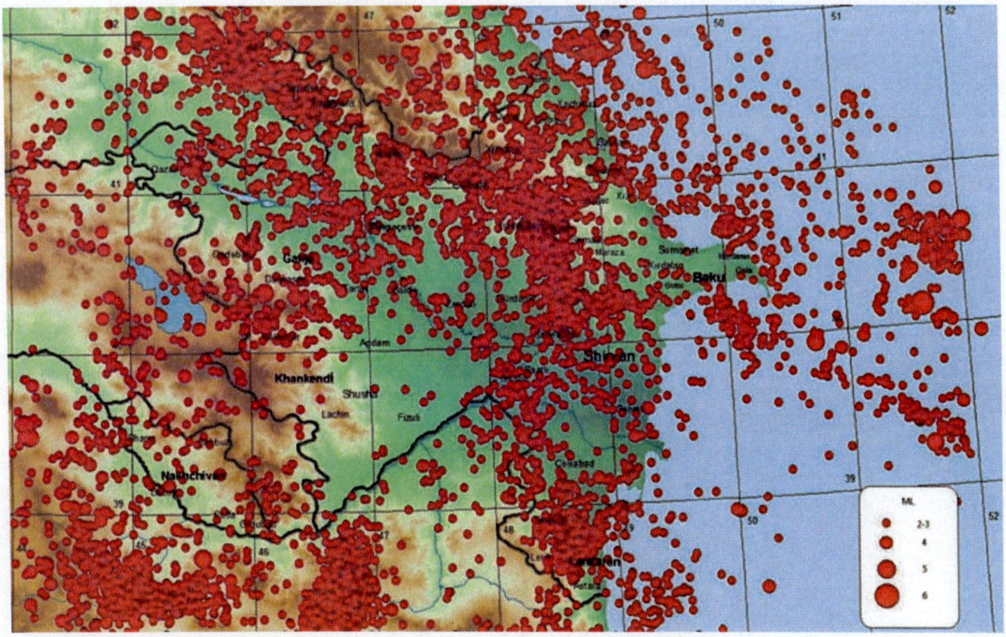

Fig. 18.6 Map of earthquake epicenters for the region under study for the period of 2004–2014 with ml $\geq$ 2.0 (from materials of Yetirmishli, F. Kadirov, and S. Kazymova; Alizadeh et al., 2017)

input and output arrays. The main criterion of velocity model optimization has selected the closeness of RMS residuals (deviation of travel wave times from the used velocity model) for all rays to zero.

The investigated volume up to the depth of 60 km was divided into layers with a thickness of 2 km in the depth interval of 0–10 km and a thickness of 5–10 km in the depth interval of 10–60 km. Data from 35 seismological stations in Azerbaijan covering all the regions under study (Fig. 18.6) were analyzed.

After numerous iterations were received, the thick red line shows 1D models for P and S-waves in Fig. 18.7. The dash black line shows a current velocity model for calculating earthquake hypocenters.

Seismotomographic investigation has been performed in three stages. In the first stage, an initial data set was analyzed. In the second stage, the optimal 1D velocity layered model was calculated using 'Velest' software, and the stability of the obtained model was checked. Simultaneously, the selection of the 1D model has been performed by re-estimating hypocenter parameters and time in the source in the 1D model and calculating temporary corrections.

The third stage (most important) consisted of checking the resolving capacity of the available data set and constructing a 3D velocity model. For the 3D velocity field computation, the program TomotetraFD was used. In this program, a classic seismotomographic method for the case when sources and receivers are within the investigating region is realized. The finite difference method has been used to calculate seismic ray trajectories.

The obtained 3D model is a linear quasi-continuous function—velocity function discontinues at concrete depths. In the horizontal direction, this function is continuous (Usoltseva et al., 2010). Figures 18.8, 18.9, 18.10, 18.11, 18.12, and 18.13 display the horizontal slices of 3D velocity models at different depths in the region under study.

Analysis of the geological-geophysical data reveals at the depth interval of 1–3 km a clear boundary between the sedimentary rocks and volcanogenic associations. This depth interval relates to Lower and Middle Jurassic rocks, presented by limestone with clay interbedding, marl, sandstone, tuffs, mudstone, and dolomites. In the 4–6 km depth interval, a transition from porphyry olivine-pyroxene-plagioclase basalts to

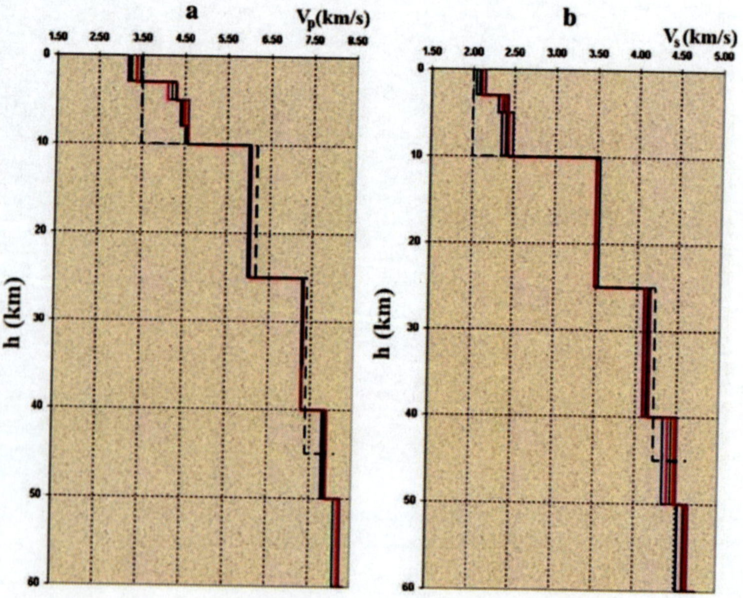

Fig. 18.7 1D velocity model for the compressional (**a**) and shear (**b**) waves (after Yetirmishli & Kazimova, 2013)

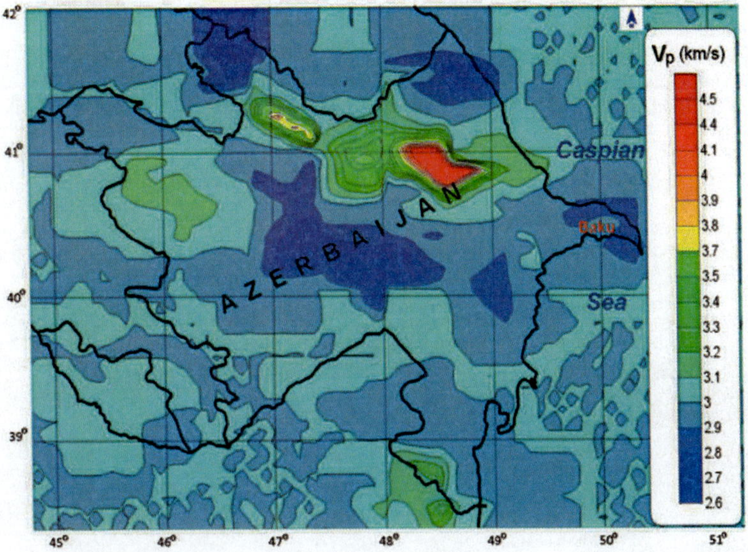

Fig. 18.8 The horizontal slice of velocity model for the territory of Azerbaijan at a depth of 1 km (after Yetirmishli & Kazimova, 2013)

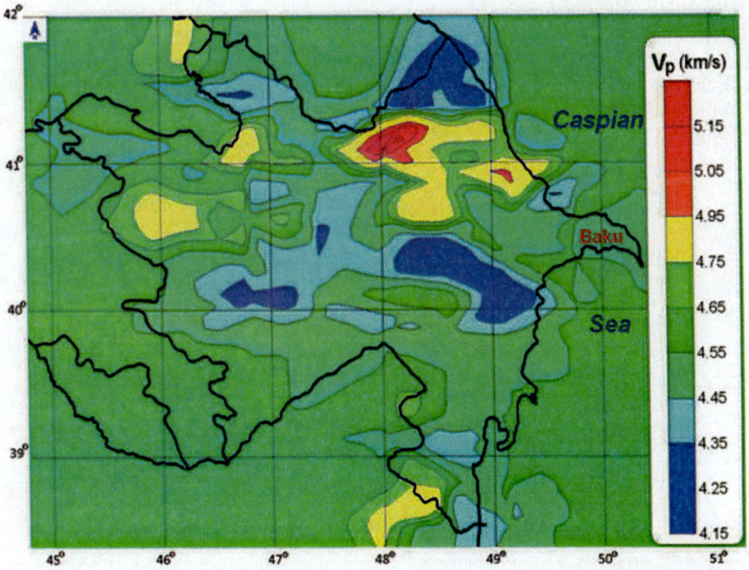

Fig. 18.9 The horizontal slice of velocity model for the territory of Azerbaijan at a depth of 5 km (after Yetirmishli & Kazimova, 2013)

andesite-basalts of the Lower Cretaceous is observed (Radjabov, 1974).

At a depth of 7 km, a boundary—transition from andesites to dacites and plagiorhyolites, their tuffs, and pumice breccia. The 7–10 km depth reveals a roof of a pre-Alpine basement. This interval observes seismic velocity decreasing in the Evlakh-Agjabedi and Kurdamir-Saatly zones of the Middle Kur depression (this fact confirms the presence of rock fracturing and decompression zone). At a depth of 15 km also, a geological boundary was detected. Here, seismic velocities increase from 5.9 to 6.4 km/s (it is known that velocities of 6.0–6.2 km/s correspond to granites and 6.5–6.7 km/s—to the "basaltic" layer).

It is essential to underline that not all velocity boundaries in the volcanogenic strata are associated with changing rock composition. Some

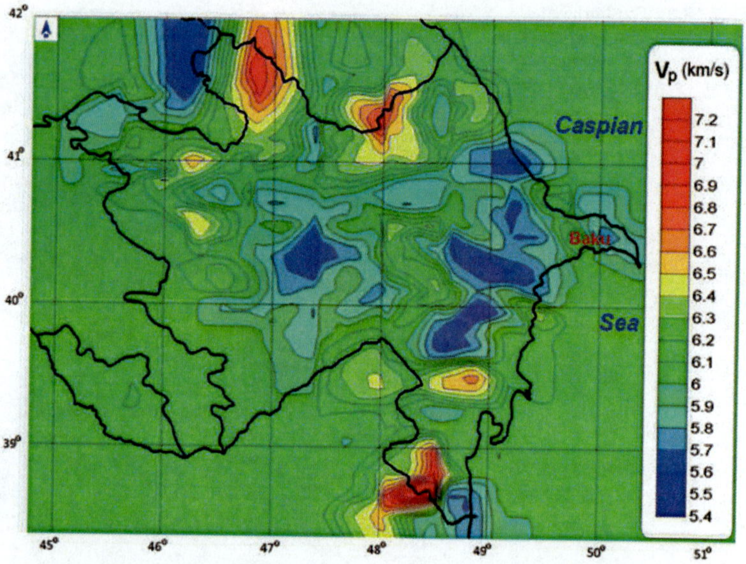

Fig. 18.10 The horizontal slice of velocity model for the territory of Azerbaijan at a depth of 10 km (after Yetirmishli & Kazimova, 2013)

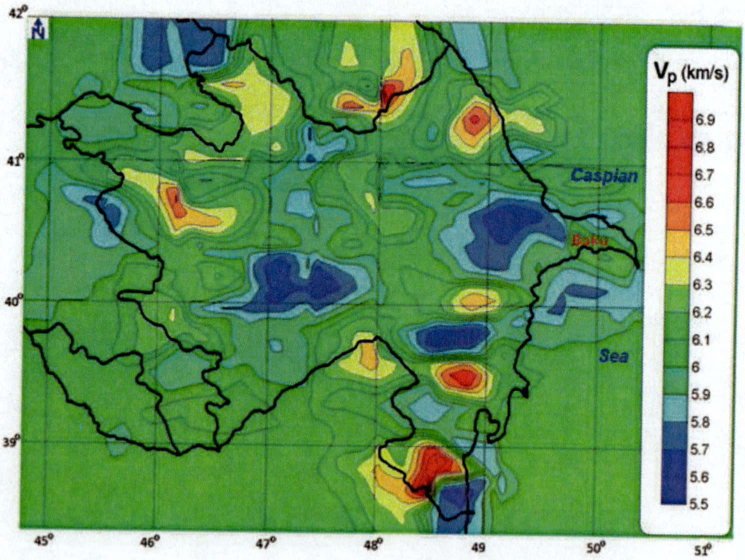

Fig. 18.11 The horizontal slice of velocity model for the territory of Azerbaijan at a depth of 15 km (after Yetirmishli & Kazimova, 2013)

boundaries are induced by different stress conditions of geological matter at the depth, with the superposition of metamorphic processes, rheological fracturing, and some other processes. We can suggest that these depths reflect an upper surface of the substrate from metamorphized rocks of the pre-Alpine basement and in separate zones—from consolidated volcanogenic and metamorphized associations of the Mesozoic age.

Based on the obtained data, we can note that the first interval (0–10 km) relates to the Cenozoic-Mesozoic boundary in sedimentary cover and is characterized by velocities of 2.8–5.9 km/s, the second interval (10–25 km) associated with a roof of consolidated Earth's crust ("granitic" layer) with velocities of 6.0–7.3 km/s, the third (25–40 km) with velocities of 7.4–7.8 km relates to "basaltic" layer, and the fourth

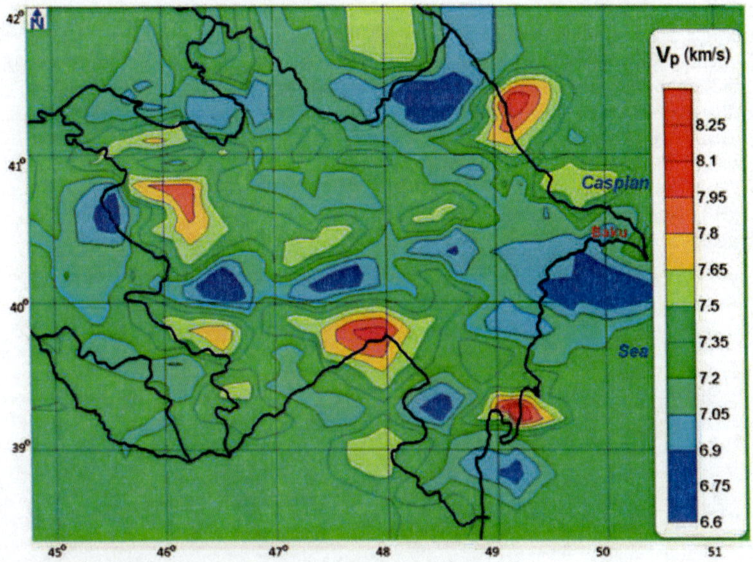

Fig. 18.12 The horizontal slice of velocity model for the territory of Azerbaijan at a depth of 25 km (after Yetirmishli & Kazimova, 2013)

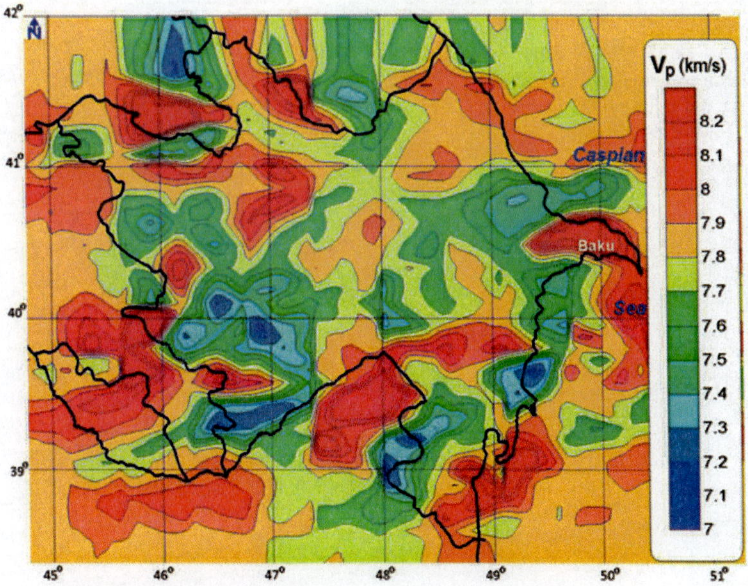

Fig. 18.13 The horizontal slice of velocity model for the territory of Azerbaijan and adjacen areas at a depth of 40 km (after Yetirmishli & Kazimova, 2013)

(40 km and deeper) with velocities of 8.0–8.2 km associates with the Moho discontinuity. The abovementioned calculations were used for the development of 3D maps of "granitic" (Fig. 18.14a), "basaltic" (Fig. 18.14b), and Moho (Fig. 18.14c) discontinuities.

18.2.2 Thermal Data Analysis

Thermal data analysis has a vital role in the deep structure unmasking in different world regions (e.g., Eppelbaum et al., 2014; Lyubimova, 1968; Turcotte & Schubert, 1982).

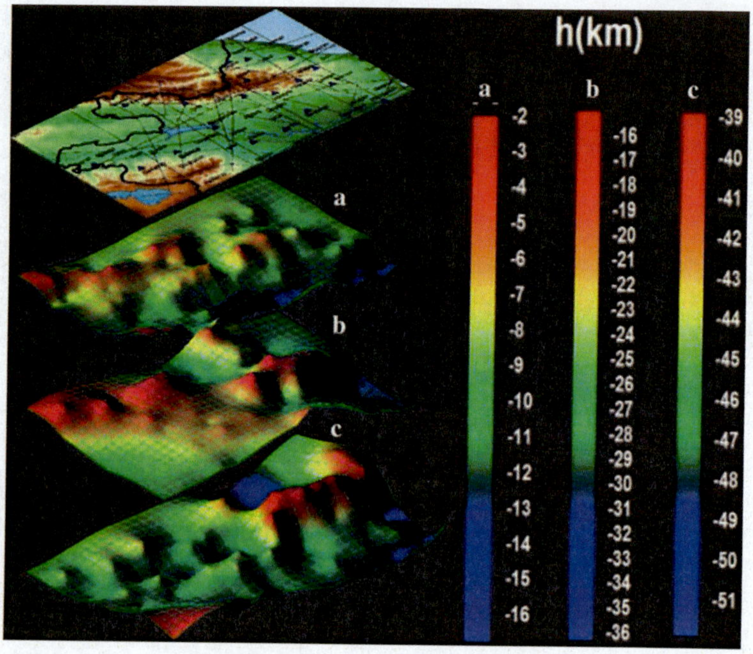

Fig. 18.14 3D scheme of depths distribution for the "granitic" (**a**), "basaltic" (**b**), and Moho (**c**) boundaries obtained from the seismological data interpretation (after Yetirmishli & Kazimova, 2013, with minor modifications)

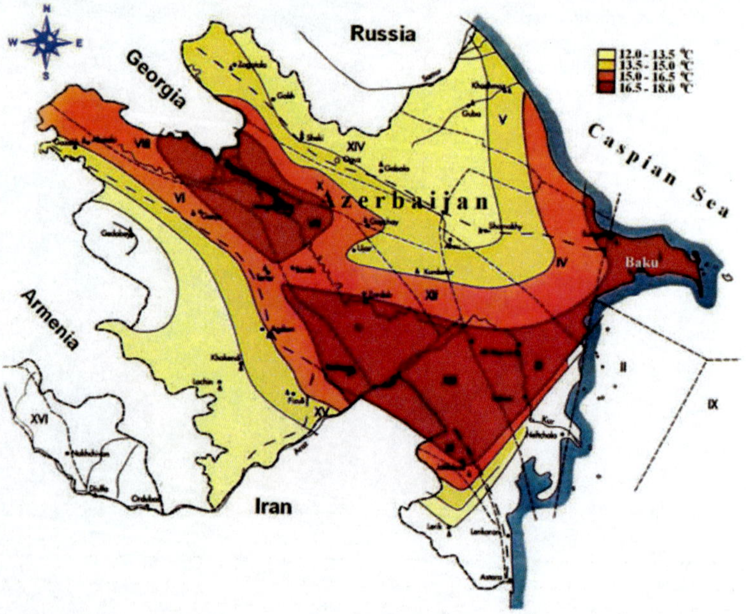

Fig. 18.15 The neutral temperature layer distribution for the territory of Azerbaijan (Mukhtarov, 2011)

It is well-known that all deep structure analyses of thermal data began by developing a map of the neutral layer distribution of the region under study. Such a map was composed by Mukhtarov (2011) (Fig. 18.15).

The analysis of soil temperature data recorded by meteorological stations in Azerbaijan revealed that the temperature recorded at a depth of 3.2 m is very close to the temperature of the neutral layer. Within the limits of measurement error, it can be assumed that it is equal to the temperature of the neutral layer. The neutral layer temperature in Azerbaijan varies from 5–6 to 10–12 °C. The observed values in the Caspian Sea water and high-altitude areas were below 12 °C (Fig. 18.15). The high-temperature area of the neutral layer corresponds to the Kur Depression. In this area, the temperature ranges from 16 to 19 °C.

Its footwall lies at a depth of around 20 m, where temperature stabilizes and does not change (Mukhtarov, 2011). At this depth, the amplitude of annual temperature variations is reduced to a minimum, and the yearly average temperature is close to the temperature of the neutral layer. The water masses' impact explains the neutral layer's lower temperatures.

On the horizontal section 2000 m, the lowest temperature was observed at the Garamaryam area (38 °C) (Ajinour oil and gasbearing region), Agamamedli (39 °C), Yevlakh-Agjabedi oil-and-gas bearing region, Nakhchivan (39 °C), Absheron oil and gas bearing region and Bulla Deniz (40 °C), and oil and gas bearing area of the Baku Archipelago (Fig. 18.16).

This section's maximum temperature (110 ° C) is observed in the Arkevan-Lenkaran-Astara region. Local temperature maximums (80.4 °C) are also observed in the Bibi-Heybat and Garachukhur-Zikh areas of the Absheron oil-and-gas-bearing region.

The low-temperature area extends from the northwest to the southeast and covers depression areas of the considered region. This section's maximum temperature (110 °C) is observed in the Arkevan-Lenkaran-Astara region. Local temperature maximums (80.4 °C) are also observed in the Bibi-Heybat and Garachukhur-Zikh areas of the Absheron oil and gas-bearing region. At the same local temperature, maximums (70–95 °C) are observed in the deepest part of the depression in the deep-water part of the Caspian Sea (Mukhtarov, 2012).

Temperature field distribution is generally similar at 4000 m (Fig. 18.17) and 6000 m (Fig. 19.18) horizontal sections. To a certain degree, the computed isotherms correspond to structural contours and reflect the main tectonic peculiarities of the region. At the same time, it is noticeable that regional structures are well-defined at deeper levels (4000–6000 m), while local structural elements become distinct closer to the surface (2000 m) (Fig. 18.16).

Temperature change is determined by structural and lithological, hydrogeological, and other factors. For example, the increase of shale

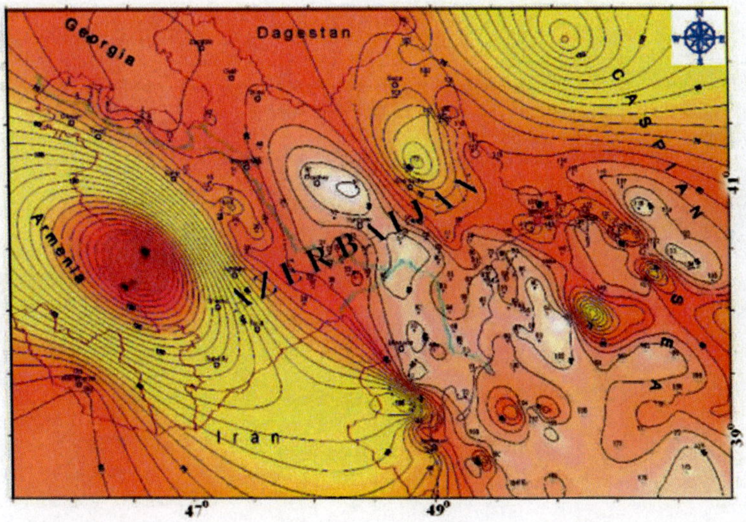

Fig. 18.16 The temperature field distribution at a depth of 2000 m (after Mukhtarov, 2011)

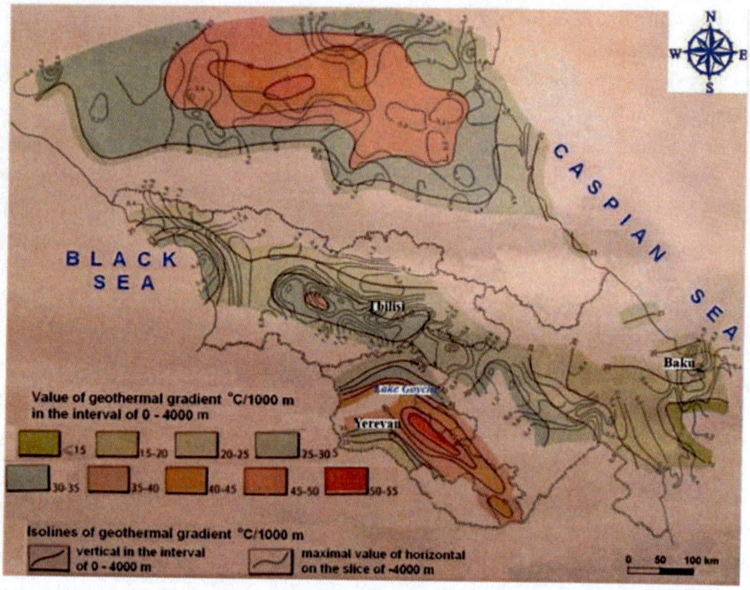

Fig. 18.17 Values of vertical and horizontal geothermal gradients at depths of 0–4000 m in the Caucasian region (modified after Kerimov et al., 1989, Eppelbaum et al., 2014)

content in the section results in a drop in temperature. In contrast, the rise of sand content has an opposite effect related to their thermal conductivity (an ability to conduct heat). Filtration of surface water also results in a temperature decrease, and the rise of groundwater increases the temperature (Mekhtiyev et al., 1960, 1971).

At the 6000 m elevation section (Fig. 18.18), temperature also decreases from the sides of the depression (180–160 °C) towards deeper parts (110–100 °C). In the center of the depression, minimum temperatures extend in the southeast direction from 120 to 100 °C (Mukhtarov, 2012).

Comparison of vertical and horizontal geothermal gradients at depths of 0–4 km is shown in Figure 18.18.

All the thermal maps (here, only a tiny part of the developed thermal maps is presented) play an essential role in the deep structure analysis and hydrocarbon reservoir delineation.

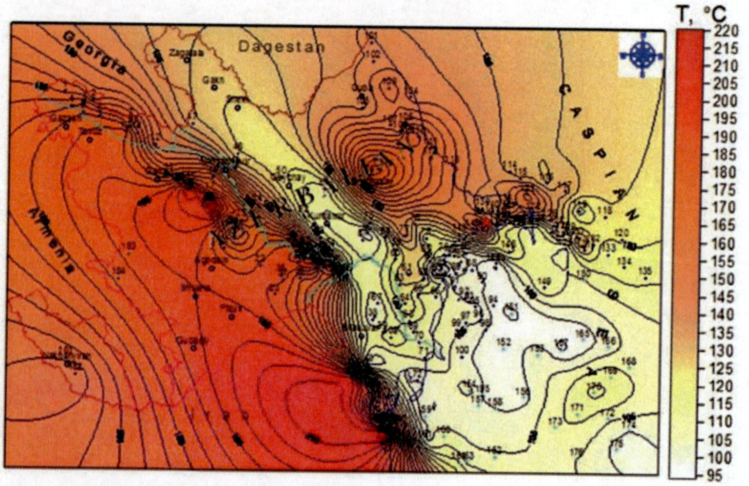

Fig. 18.18 The temperature field distribution at a depth of 6000 m (after Mukhtarov, 2011)

18.2.3 Magnetotelluric Data Analysis

H.D. Jafarov, with collaborators, made a significant contribution to studying Azerbaijan's deep structure using magnetotelluric sounding (e.g., Jafarov et al., 1967).

From 1988 to 1991, telluric and magnetic field variations were observed at three points on a profile perpendicular to the Greater Caucasus (Trofimov, 1995). One of these sites, KSN, is located on the northern slope of the ridge in the vicinity of Kusnet village (Guba region, Azerbaijan). These observations were utilized to study the Greater Caucasus's deep structure and analyze geodynamic activity. An example of such investigations is shown in Fig. 18.19.

These observations were utilized for deep structure studies of the Greater Caucasus and geodynamic activity analysis. An example of such investigations is shown in Fig. 18.19.

According to Gugunava (1988), the Caucasian crust contains several electrically conducting units: (1) the sedimentary cover (up to 15–20 km thick), (2) relicts of magma chambers within mountainous regions, and (3) the asthenosphere. This model was later confirmed by Spichak (1999). Chamber relicts were detected as an oblate ellipsoid at a depth of about 20 km (Greater Caucasus) and isolated lenses at a depth of 10–20 km (Lesser Caucasus). A crustal asthenosphere underlies the Transcaucasus (0–20 km thick) and attains maximal thickness beneath the Greater and Lesser Caucasus. High conductivity reflects the increased temperature at its depth; its highest values were detected in the southeastern prolongation of the Absheron Peninsula (Fig. 18.20), where increased geothermal gradients were measured (e.g., Kerimov et al., 1989; Mukhtarov, 2011).

18.2.4 Remote Sensing

Today, the origin and evolution of the Caucasus are mainly explained in terms of plate tectonics (e.g., Khain & Ryabukhin, 2002; Pilchin & Eppelbaum, 2020). In a brief review, Kopf et al. (2003) noted that the present-day Caucasus is dominated by thrust faulting due to continental collision. From the Jurassic to the Paleogene eras, subduction of the Tethyan seafloor occurred along the southern margin of the Turkish and

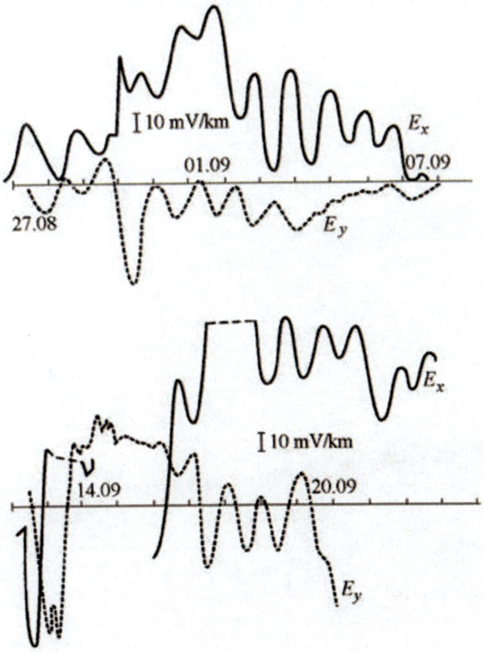

Fig. 18.19 Long-period variations in the telluric field at the KSN site in the Guba region of Azerbaijan (Trofimov, 1995)

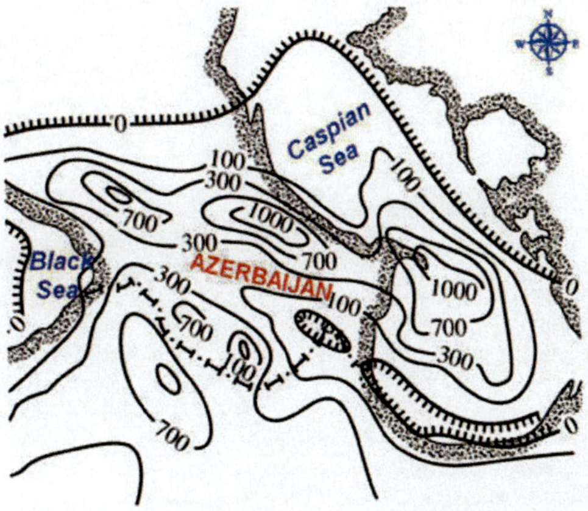

Fig. 18.20 Map showing the longitudinal conductance (in Siemens) of the inverted crustal layer in the Caucasus (after Gugunava, 1988, with minor modifications)

Iranian blocks, resulting in calc-alkaline arc volcanism and a wide backarcs basin system. The spread of the Red Sea began during the Early Miocene, and the Arabian Plate migrated northward, accompanied by a reduction in the width of the Tethys. Subduction shifted to the north after its closure (~ 20 Ma). As a result of the indentation of the Arabian block, the continuous backarc basin was separated, and the oceanic crust they have only remained in the Black Sea and the southern Caspian Sea. The continuous northward drift of the Anatolian Plate led to initial continental collision expressed by the formation of the Lesser Caucasus and the subsequent resurrection of the Greater Caucasus during the Middle Pliocene. Continental convergence continues at a rate of up to ~ 20–30 mm/y (e.g., Reilinger et al., 2006) along strike-slip faults, where most of the modern tectonic activity is localized (Alizadeh et al., 2017). Reilinger et al. (2006) note the tendency for motion rates to increase from west to east along the strike of the Main Caucasus thrust (MCT) (120° azimuth). Besides this, there is little change in the magnitude of velocity estimates across the Lesser Caucasus along transects perpendicular to the Caucasus range. Considering the low level of seismic activity within the Lesser Caucasus, Reilinger et al. (2006) propose to imply a block-like, counterclockwise rotation of the Lesser Caucasus, resulting in increased convergence from west to east along the MCT.

Without hesitation, the GPS data constructions (e.g., Kadirov et al., 2008; Reilinger et al., 2006) must be coordinated with other remote sensing data or regional geophysical field distribution. For example, the map of deep-seated lineaments (Fig. 18.21) (undeservedly little used) shows fascinating tectonic patterns for Azerbaijan and adjacent regions. Besides other peculiarities, the intersection of deep faults was detected at the northern point of a sharp tectonic wedge where the destructive Spitak earthquake of 1988 (coordinates: 40°49′ N, 44°15′ E) occurred.

Gadjiev et al. (1989) identified this feature from this map of the Caucasus (Fig. 18.21) by calculating the lithospheric heterogeneity at different depths. The total length of the lineaments within square cells with different sides (corresponding to cubic blocks of different depths)

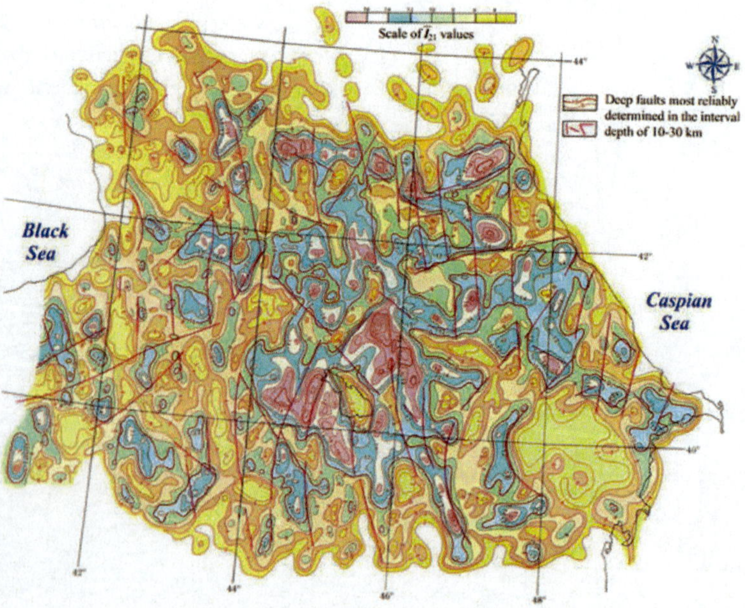

Fig. 18.21 Deep structure of Azerbaijan and adjacent regions (the cut at ≈10 km in depth) according to the satellite data analysis (after Eppelbaum & Khesin, 2012; Gadjiev et al., 1989)

served as a measure of heterogeneity (Khesin et al., 1996).

Note that this map of the deep structure of the Caucasian lithosphere is consistent with the depiction of horizontal gradients of isostatic and lithospheric gravity anomalies (Gorshkov & Niauri, 1984).

18.2.5 Gravity Data Analysis

The method of iterative selection is widely applied in gravity field analysis. This method is used when a researcher can construct an initial model of the geological section. A selection criterion is a coincidence of observed and calculated fields (if the developed geophysical model corresponds to available geological criteria). Different approaches in terms of the realization of the selection method exist (e.g., Bulakh, 2000; Bulakh & Markova, 1994; Tikhonov & Arsenin, 1977). Mathematically, the applied methodologies correspond to the problem of minimization of multi-parametric functional.

18.2.5.1 Physical-Geological Model Along with Profile Masally—Poilu

The gravity model along the studied profile was constructed applying a forward modeling process, including the fit of the initial model to the observed gravity profile, re-calculation of the anomaly, and comparison of the modeled and observed anomalies. This procedure was repeated until the calculated and observed anomalies were considered sufficiently alike (based on data uncertainties and model resolution), adjusting the model parameters in such a way as to improve the matching between the observed and modeled anomalies.

To study the deep structure of the Kur Depression, a gravity field interactive modeling has been conducted along the profile DSS-3 (Masally—Poilu) (Aliyev et al., 2005a; Kadirov, 2000a; Shikhalibeyli et al., 1990) (Fig. 18.22).

In the SE part of the studied region, the interpreting profile crosses the Mughan monocline, and in the NW part—the pre-Lesser Caucasian trough. The geological-geophysical sections based on the DSS were used to compose the initial model of this profile (Gadjiev, 1965; Krasnopevsteva, 1978; Pavlenkova et al., 1991).

Figure 18.22 shows gravity graphs of observed anomaly (solid line), computed from the initial model (dotted line) and computed from the final model (point line). The right part of this profile shows sufficient discrepancy between the observed and computed (from the initial model) gravity fields. The difference in gravity effect here consists of 70 mGal, which indicates the deficit of masses in the initial model.

Based on the analysis of observed and computed gravity curves in the right part of this profile, some additional contact boundaries and mantle bodies were introduced. The model includes the upper edges of the mantle bodies with a redundancy density of 0.3 g/cm^3 (300 kg/m^3), which occur here at the depths of 12.5 and 11 km (with the corresponding width of 12.5 and 11 km) (Kadirov, 2000a).

In the left part of the profile, non-significant disagreement between the observed and computed graphs was detected. This disagreement was removed by changing the configuration of the consolidated crust of the roof of the Tovuz-Ganja uplift of the crystalline basement.

Based on the results of gravity modeling, the geological section of the Earth's crust within the DSS profile No. 3 is revised as follows. On the surface of the consolidated crust and the profile are two large zones of uplifts corresponding to the gravity maximums Tovuz-Ganja and Bilasuvar-Karadonly. The depths of the boundary of consolidated crust in axial parts are 7 and 4 km, respectively. In the Evlakh-Agjabedi trough area, the consolidated crust's computed depth is 18.5 km. The Evlakh-Agjabedi trough appears in the subsurface layers: troughs are found on the bottom of the Productive Series and the Cenozoic bottom. The boundary "basalt" along the profile repeated the form of

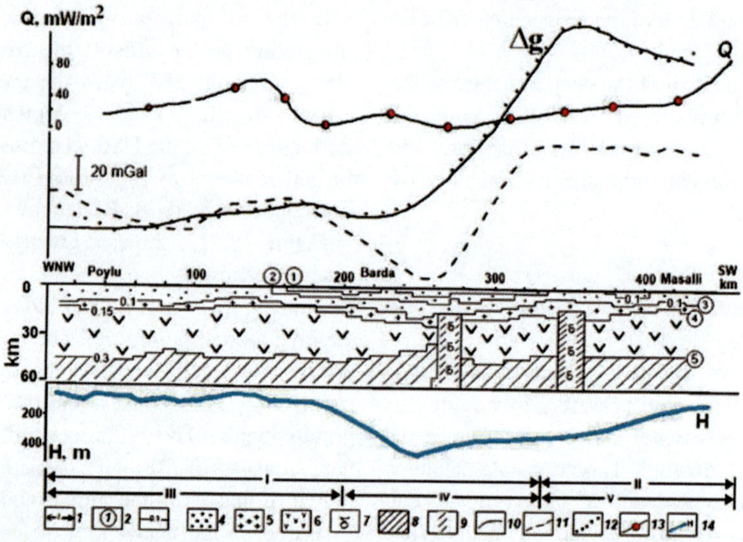

Fig. 18.22 Gravity model along with profile Poylu-Masalli. The values of density differences are in g/cm^3. (1) zonation carried out by geological data: I—pre-Lesser Caucasian Depression; II—Mughan monocline; for deep horizons: III—Tovuz-Ganja uplift, IV—Evlakh—Agjavedi depression, V—Bilyasyvar—Karadonlin uplift; (2) contact boundaries in the earth's crust and upper mantle (digits in circles): 1, 2—in sedimentary strata, (3) surface of a consolidated crust, (4) "basaltic" layer, (5) Moho discontinuity; (6 and 7) density boundaries in the upper mantle; 3—redundant densities at contact boundaries; (4) Meso-Cenozoic rocks; (5) rocks of consolidated crust; (6) basic rocks; (7) mantle's matter; (8) mantle; (9) deep faults; gravity field: (10) observed, (11) computed from the initial model, (12) computed from the selected model; (13) graph of thermal flow; (14) thickness of the Quaternary deposits (modified after Kadirov, 2000a)

consolidated crust, and its depth in axial parts is the following: Tovuz-Ganja zone—14 km, Evlakh-Agjabedi—22 km, and Bilasuvar-Karadonly—12 km. At the same time, the Moho boundary occurs reversely with two superincumbent boundaries in most parts of this profile. The Moho boundary rises only in the zone of the Bilasuvar-Karadonly maximum. In the Tovuz-Ganja zone, the depth of the Moho discontinuity is about 47 km (to NW along with profile, the depth increases up to 52 km), in Evlakh-Agjabedi—39 km and Bilasuvar-Karadonly—41.5 km.

Figure 18.22 also displays graphs of the thermal flow distribution (*Q*) (Aliyev, 1982) and changing thickness of the Quaternary layer (*H*) deposits along with the profile (Mamedov, 1984).

The thermal flow distribution along the profile (Aliyev, 1985) displays a change in its level by a transition from the Tovuz-Ganja zone to the Evlakh-Agjabedi zone. That testifies to differences in geotectonic evolution. The highest thermal flow values (50–70 mW/m^2) along the profile are typical for the comparatively young Tovuz-Ganja tectonic–magmatic zone, structures of which were formed in the Meso-Cenozoic. The low and middle thermal flow values (20–40 mW/m^2) were observed in the Evlakh-Agjabedi and Bilasuvar-Karandonlin zones and indicated the ancient consolidation of the region.

The developed physical-geological section along the profile DSS No. 3 indicates that the region beginning from the axis of the Evlakh-Agjabedi trough and to the east, including the Bilasuvar-Karandonlin zone, subjected to a joint by-genesis appearance of magmatism in the pre-Baikalian time. The presence of a residual gravity anomaly of 70 mGal (in the initial model) in the right part of the profile and the change in the observed anomaly amplitude compared with the left part of the profile may be explained by the presence of ancient massif embedded from the mantle. Revealing by gravity modeling the additional boundaries of mantle-lithospheric bodies and their small areal distribution testifies that deep faults destroyed the crystalline basement (Shikhalibeyli et al., 1990).

The distribution of thicknesses of Quaternary deposits along this profile (Mamedov, 1984) indicates the contrasting movement at the neotectonic stage. It continues to rapidly raise the Bilasuvar-Karandonlin zone, on both sides—to the Caspian Sea in the east and to the Evlakh-Agjabedi trough in the intense west subsidence is observed.

18.2.5.2 Physical-Geological Model Along with Profile Agjabedi—Zardab—Mususli—Karamaryam

Various researchers studied the deep structure of the Kur Basin (e.g., Abbasov, 2002; Azizbekov et al., 1972; Gadjiev, 1965; Krasnopevtseva, 1984; Mamedov et al., 2008; Radjabov, 1977). However, most of these works were based on the classic fixist theory of basin formation. According to the plate tectonics conception (Khain, 2005; Pilchin & Eppelbaum, 2020), the Kur Depression and framing mountain massifs of the Greater and Lesser Caucasus result from a complex combination of the lithosphere's horizontal and vertical motions. Thus, we can propose that the formation of the subsided (Kur Depression) and uplifted (Greater and Lesser Caucasus) regions of the Earth's crust are the results of the horizontal displacements and collision of the Arabian and Eastern-European plates (Reilinger et al., 2006).

An attempt was made to construct the gravity model of the Kur Basin deep structure based on these aforementioned ideas. A particular field study and the earlier compiled seismic profile of Agjabedi—Zardab—Myusyusli—Karamariyam was carried out (Fig. 18.23). Gravimetric and geodetic measurements on this profile were made every 250 m at 350 points.

The sedimentary section along the investigated profile is composed of Mesozoic, Paleogene, Neogene, and Quaternary rocks. The first 45 km of profile (from Agjabedi to Zardab area) in the regional gravity field is influenced by the SE part of Alazan-Middle Kur minimum (Yevlakh-Agjabedi Trough of the crystalline basement) and the residual part of the profile—by the western termination of Azerbaijan gravity

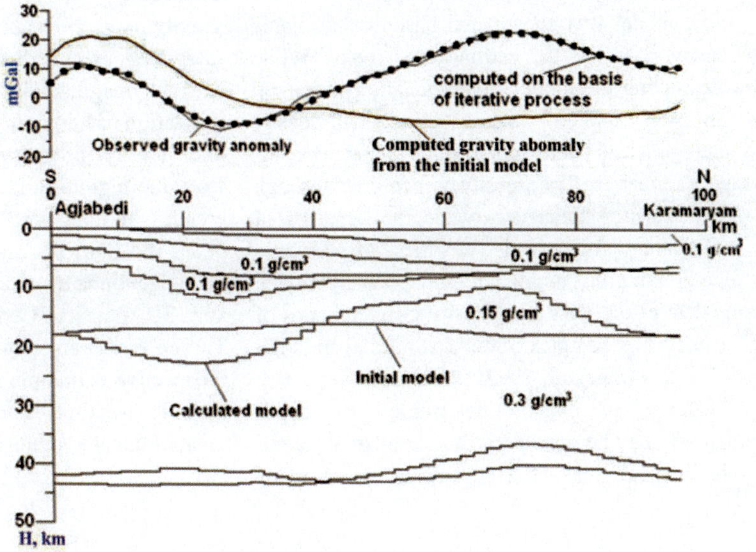

Fig. 18.23 Gravity model along the meridional profile Agjabedi—Karamaryam (Central Azerbaijan). The values of density differences are given in g/cm³ (Kadirov, 2000a)

maximum (combined influence of the Geokchay-Mingechavir and Saatly-Kurdamir projections of the crystalline basement) (Azizbekov et al., 1972; Gadjiev, 1965).

An initial gravity model was constructed based on deep seismic sounding. In this model, the geological-geophysical section along with the seismic profile 'Agjabedi—Zardab—Mususli—Karamaryam' is represented by six contact petrophysical boundaries: foot of the Quaternary deposits (Absheron), the boundary between the Neogene and Paleogene deposits foot of the Eocene deposits, foot of the Mesozoic deposits, the surface of the "basaltic" layer and Moho discontinuity. Such boundaries as the "basaltic" layer and Moho discontinuity are initially compiled by gravity data (Azizbekov et al., 1972; Gadjiev, 1965; Tzimelzon, 1965). The average seismic velocity in the Absheron stage strata is 2.5 km/s, in the other three boundaries—3.5 km/s, in the Mesozoic—4 km/s. At the section interval between the Mesozoic complex and consolidated crust, the seismic velocities increase to 5.2 km/s. Most Mesozoic discontinuity along the profile has a boundary velocity of 6.2 km/s. The boundary rate on the "basaltic" surface varies from 6.5 to 7.3 km/s. Boundary velocities weakly change for the "Moho" discontinuity and consist of 8.1 km/s (Krasnopevsteva, 1984; Pavlenkova et al., 1991; Radjabov, 1977).

The density model for the upper part of the initial model was constructed based on available density data for the sedimentary rocks. While constructing the model, some density generalization was applied, i.e., a few small-density heterogeneities were replaced by standard average parameters. The density is acquired through the correlation linkage between the densities of rocks and propagation velocities of *P*-waves therein. For constructing a regional density model, the data of the DSS and density characteristics of the enclosing complex of rocks at normal and high *P–T* terms (Salekhli, 1993) were used.

Change of density determined by depth increase may be considered linear at small depth intervals, whereas at a significant variation of depth change, this parameter corresponds to the exponential law. The density of deeper layers was determined by the P-wave velocities (determined experimentally) at high pressures and temperatures (Balakishibeyli et al., 1996). Density contrast for the "basaltic" boundary and the Moho discontinuity is assumed as + 0.15 g/cm^3 and + 0.3 g/cm^3, respectively (Aliyev et al., 2005a; Balakishibeyli et al., 1996).

To select the geometrical parameters, about 60 iterations were used, and to choose the density parameters, 20 iterations were done. The difference between the observed and calculated fields is about three mGal at 38 km.

Figure 18.23 demonstrates gravity curves of the observed anomaly (solid line) calculated by the initial model of the section (dotted line) and the final model—from the presented geological-geophysical section (dotted line).

18.2.6 Combined Gravity-Magnetic Analysis

Based on our experience (Eppelbaum & Khesin, 2012; Kadirov, 2000b), it was recognized that gravity, magnetic, and especially combined gravity-magnetic modeling is a powerful tool for studying the deep variable structure of Azerbaijan. This study must be preceded by a combined qualitative and advanced quantitative gravity/magnetic data analysis supported by an integrated examination of available geological, seismic, magnetotelluric, and thermal data and utilization of numerous magnetic, paleomagnetic, and density properties of geological samples from the region under study. The practical preferences of integrated analysis are presented in detail in Chap. 17. The final product of such an investigation is the development of a series of 2.5 and 3D Physical-Geological Models (*PGMs*). These *PGMs* can to substantiate various types of prospective economic deposits and delineate the tectonic-structural factors affecting the long-term seismological prognosis (Eppelbaum & Khesin, 2012).

18.2.6.1 Preferences for Integrated Interpretation

Gravity-magnetic data processing is generally intended to reduce and eliminate noise factors of different origins and intensities. The main problem faced by qualitative interpretation is to single out the desired target, whereas quantitative interpretation needs to determine and refine the target parameters. Thus, geological issues need to be resolved in terms of (1) the capabilities of the geophysical method selected for measurements of the field containing the information required, (2) the physical properties of the medium under study and their capability to generate detectable signals (anomalies), (3) the methods for data processing and interpretation; namely, their ability to extract information from geophysical fields and reveal the effects from the geological targets. Figure 18.24 presents a general flowchart for analyzing and synthesizing geophysical data for complex regions. Each step in this flowchart is divided into sub-steps with a more concrete formulation (Eppelbaum & Khesin, 2012).

The estimation of the efficiency of integrated interpretation from the probabilistic and information points of view is considered in detail in Eppelbaum (2014a). Interestingly, from analysis of a classical "Four Color Problem", two independent geophysical method applications theoretically are sufficient for successive mapping of any area of any geological complexity (Eppelbaum, 2014b). Undoubtedly, this fact supports the theoretical basement of 3D combined gravity-magnetic field modeling.

18.2.6.2 Short Description of the Employed Algorithm

The *GSFC* (Geological Space Field Calculation) program was developed to solve a direct 3-D gravity and magnetic prospecting problem under complex geological conditions (Eppelbaum & Khesin, 2004; Khesin et al., 1996). This program has been designed for computing the field of Δg (Bouguer, free-air or observed value anomalies), magnetic field components ΔZ, ΔX, ΔY, total magnetic field ΔT, as well as second derivatives of the gravitational potential under conditions of rugged relief and oblique magnetization. The geological space is approximated by (1) three-dimensional, (2) semi-infinite bodies, and (3) those infinite along the strike closed, left-hand (LH) non-closed, right-hand (RH) non-closed, and open bodies.

The program has the following main advantages (besides the above-mentioned ones): (1) Simultaneous computing of gravity and magnetic fields; (2) Description of the terrain

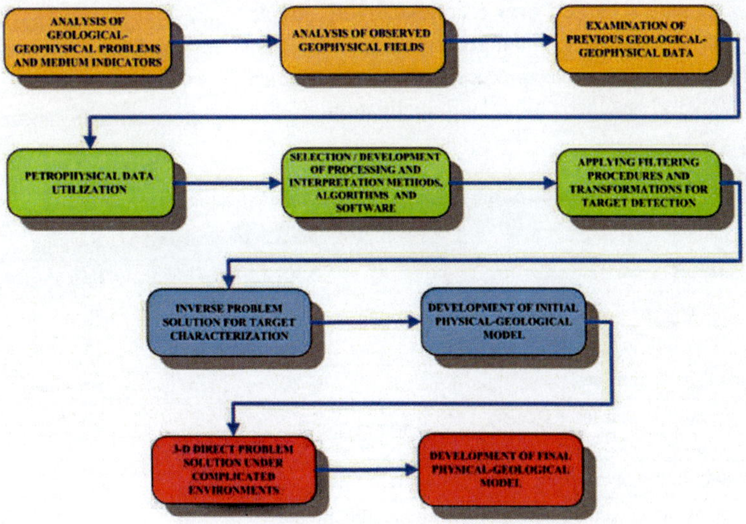

Fig. 18.24 Interpretation of potential geophysical fields under complex environments: A general scheme (revised after Eppelbaum & Khesin, 2011)

relief by irregularly placed characteristic points; (3) Computation of the effect of the earth-air boundary directly in the process of interpretation; (4) Modeling of the interpreting profiles draping over rugged relief or at various arbitrary levels (using characteristic point description); (5) Simultaneous modeling of several profiles; (6) Description of a large number of geological bodies and fragments (up to 1000). The basic algorithm realized in the GSFC program is the computation of the 3-D combined gravity and magnetic effects for the horizontal limited in the polygonal strike prism (Eppelbaum & Khesin, 2004). In the developed algorithm, integration over a volume is realized on the surface, limiting the anomalous body.

The advantage of the interactive computer selection system is that it eliminates the need for many iterations, which are inevitable in automated selection algorithms that use successive parameter variation searches within a specified range.

The automatic selection of geophysical fields has certain limitations. Goldshmidt et al. (1981) and Bulakh et al. (1984) developed such initial gravity field selection. First, the selection must be meaningful, e.g., changes in physical properties and 3-D geometrical parameters should be carried out in defined numerical and space intervals. These limitations stem primarily from the program. However, there are more severe issues. Take the example of a 3-D integrated gravity and magnetic selection over a comparatively simple section consisting of ten geological bodies. Each geological body has three petrophysical variables (density, magnetization, and inclination of the magnetization vector), geometric variables (left-hand (y_1) and right-hand (y_2) end faces of the body), and finally, its geometric parameters in the plane of geological section.

The number of points (variables) to describe bodies in the plane of section a priori is unknown. For simplicity and given that many of these points are calculated twice by the contouring objects, the number of these points can be assumed to be ten. To calculate the number of possible combinations of all variables by combined 3-D modeling, we need to calculate the approximate ranges of the variables (Table 18.1). These ranges are relative and only estimate the number of combinations.

Applying known combinatorial analysis (e.g., Riordan, 2014), for one body, we have the number of combinations $C^1_{30} \cdot C^1_{60} \cdot C^1_{24} \cdot C^1_{30} \cdot C^1_{30} \cdot C^1_{100} \approx 4 \cdot 10^9$ correspondingly, for ten bodies, we have $\approx 4 \cdot 10^{90}$ combinations. Such combinations considerably complicate automatic 3-D integrated gravity-magnetic modeling, even using supercomputers.

18.2.6.3 Description of the Interpretation Methodology

The complete description of the interpretative process structure was given by Strakhov (1976). An interpretation process may be roughly subdivided into the following stages: (1) summarizing prior information, (2) sequential analysis, and (3) geological synthesis.

The development of 3-D ***PGM*** is usually performed using these three stages.

Table 18.1 Calculation of the number of possible combinations of variables (Eppelbaum, 2015)

Variable	Interval of changing	Ranging	Number of combinations
Density, g/cm^3	2.30–2.60	0.01	30
Magnetization, mA/m	0–3000	50	60
Inclination of magnetization, degree	0–360	15	24
Left-hand end face, y_1/x_m*	0–20	non-linear	30
Right-hand end face, y_2/x_m	0–20	non-linear	30
Geometrical coordinates of a geological body in the plane of the geological section	min 10 points	–	min 10 · 10

* x_m is the maximum length of the interpreting profile

The first stage (summarizing prior information) is as follows:

(A) First of all, the main geological-geophysical conception of tectonic development of the region under study must be analyzed and adopted (without it, the process of ***PGM*** construction will not be successive and logical). Construction of the geological section includes a compilation of all intrusive, effusive, and other associations, as well as faults and the surface of folded foundation based on geological data within a strip of 15–20 km (in some cases—20–40 km) wide. The interpreting section is in the middle of this strip. Undoubtedly, a geophysicist-interpreter must have a good knowledge of the region under study.

Such a section characterizes the upper portion of the Earth's crust with a thickness from 2–3 to 5–8 km from the Earth's surface to the Baikalian basement. Deeper parts of the intrusive bodies and certain faults are formed by extrapolating the available constructions, general geological considerations, and the results of previous geophysical analyses.

(B) A preliminary petrophysical model of the section is developed. Here, all the geological bodies acquire density and magnetization values according to the preceding petrophysical data analysis and results of geophysical field interpretation. When no data are available on the magnetization direction, it is assumed to be parallel to the normal geomagnetic field in the region under study. Further, the magnetization direction is refined in the process of physical-geological modeling. Density properties are received from the borehole sample examination and converted from the seismic data by the known method (e.g., Barton, 1986). The petrophysical model includes deep-seated layers (slabs) of the Earth's crust: (1) the "basaltic", (2) the intermediate between the crust and the upper mantle, and (3) the upper mantle. Their surfaces are constructed, and physical properties are associated with them according to the data from previous seismic, magnetotelluric, thermal, and other geophysical studies. Paleomagnetic data examination may be of high significance.

(C) The initial (preliminary) petrophysical model also includes hidden bodies. Their location, thickness, depth, density, and magnetization are obtained from the quantitative analysis of magnetic and gravity fields and seismic data examination.

The second stage (sequential analysis) includes applying combined gravity and magnetic field modeling along the interpreting profiles using the 3D *GSFC* program. Each time, the gravity-magnetic effects from different bodies, groups of bodies, and the total computed model are displayed and compared to the observed gravity and magnetic fields. Using the results of this comparison, the changes that match the gravity and magnetic effects in the model of the medium are introduced. The computations and comparisons of fields and model modifications are repeated until the desired fits between the computed and observed fields are obtained.

Then, a regional gravity (and sometimes magnetic) field is roughly selected. Next, the geophysical fields of local bodies are fixed. If necessary, this is followed by a verification of the regional fields and the fields of the local bodies. As a rule, the densities of deep-seated complexes are not changed; the modifications only affect the shape of their roof.

A separate analysis of gravity and magnetic fields is carried out at each computational step. Geometrical coordinates of geological bodies are verified in the subsequent steps and then introduced into the model. This procedure integrates qualitative and quantitative interpretations for anomalous gravity and magnetic fields. The modeling is completed when the computed gravity and magnetic fields coincide accurately with the observed fields. All this modeling process must be carried out in full compliance with the known geological principles.

The third stage (geological synthesis) involves a detailed geological interpretation of these models. A 3-D *PGM* of the area under investigation is developed based on the available geological data obtained at the previous stages and qualitative and quantitative geophysical data examination. This yields the final physical-geological sections, and the models are characterized by a complete rendering of the geological targets, including crustal blocks, intrusions, faults, and economic deposits.

The geological interpretation of the geological associations, complexes, and local bodies of the constructed (final) petrophysical model usually has no complex problem.

The geological interpretation of the geological associations, complexes, and local bodies of the constructed (final) petrophysical model usually has no a complex problem. In implementing the interactive selection system, almost all the bodies in the PGM acquire some specific geological content. The geological nature of new sources introduced into the model during the iterative modeling and reflected either in the initial geological section or in the initial PGM is determined according to the similarity of their physical properties, dimensions, and depth of occurrence concerning the known targets. The bodies' age is determined according to their interrelations with the surrounding (host) rocks.

18.2.6.4 Application of 3D Combined Gravity-Magnetic Field Modeling in Azerbaijan (Land)

A simplified tectonic map of Azerbaijan with interpreting magnetic-gravity profiles is shown in Fig. 18.25. Advanced interpretation methods (improved modifications of tangents, characteristic point methods, and areal method) were applied to study magnetic and gravity anomalies along all profiles surrounding SuperDeep borehole SD-1 in the Saatly area of Azerbaijan. A fragment of this interpretation along Profile 18 is shown in Fig. 18.26. First, note that the behavior of the magnetic ΔZ curve and graph $\partial \Delta g_B / \partial x$ is very similar, which testifies that these anomalies are due to the same geological objects. A quantitative analysis of the magnetic curve allowed us to delineate two magnetic targets. The main target is a source of the Talysh-Vandam gravity anomaly (its upper edge coincides with the data obtained by SD-1 drilling) (Aliyev et al., 2005b). The obtained data were utilized by constructing a PGM of first approximation for 3D combined physical-geological modeling (Eppelbaum & Khesin, 2012).

18.2.6.5 Integrated Physical-Geological Model of the Saatly Super-Deep Borehole

For many years, the dominant point of view was that in the Kur Depression separating the mega-anticlinoria of the Greater and the Lesser Caucasus, thick sedimentary deposits were present in the crystalline pre-Alpine basement structures divided by sub-vertical deep faults. On the buried uplift of the basement, hypothesized based on high densities and velocities of elastic waves, the Saatly superdeep borehole SD-1 was designed in 1965; the drilling began in 1977 and stopped in 1991 (Popov & Kremenetsky, 1999). The drilling area was selected using the analysis of seismic profiles 9, 16, and 18 (see Fig. 18.25) and regional gravity field analysis (Gadjiev, 1965; Tzimelzon, 1965).

However, analysis of the magnetic properties of rocks and the magnetic survey results showed that the basement was not magnetized, and a large part of the geological section of the Middle Kur depression was occupied by Mesozoic magmatic associations of basic and intermediate composition with high magnetization (Eppelbaum & Khesin, 2012). These mainly Jurassic associations are widely distributed in the north-eastern part of the Lesser Caucasus. They have a deep-seated, gently sloping underthrust under the sand-shale thick series of the Greater Caucasus Jurassic rocks.

The validity of the interpretation was fully confirmed by the results of the SD-1 drilling (Fig. 18.27). The borehole exposed Mesozoic volcanogenic rocks at a depth of 3.6 km and did not reach their bottom even at 8.2 km (Eppelbaum & Khesin, 2012). Given the location of the lower edge of the magnetized masses and by

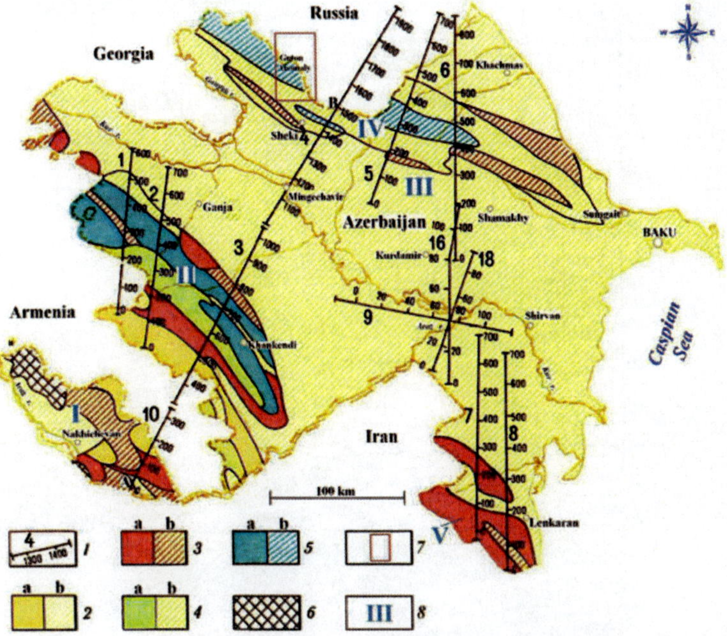

Fig. 18.25 Areal map of the main profiles used for physical-geological modeling in Azerbaijan and adjacent regions. (1) profiles and pickets, (2) Pg_3-Q: (a) orogenic magmatic associations, (b) background sedimentary deposits, (3) K_2-Pg_2: (a) pre-orogenic magmatic associations, (b) background sedimentary deposits, (4) J_3-K_1: (a) magmatic associations of the Late Alpine sub-stage, (b) background sedimentary deposits, (5) J_1-J_2: (a) magmatic associations of the Early Alpine sub-stage, (b) background sedimentary deposits; (6) Pz deposits, (7) contour of the Guton magnetic anomaly, (8) tectonic regions: I—Nakhchivan folding region, II—SE part of the Lesser Caucasus mega-anticlinorium, III—central and SE parts of the Kur mega-synclinorium, IV—SE part of the Greater Caucasus mega-anticlinorium, V—Talysh anticlinorium

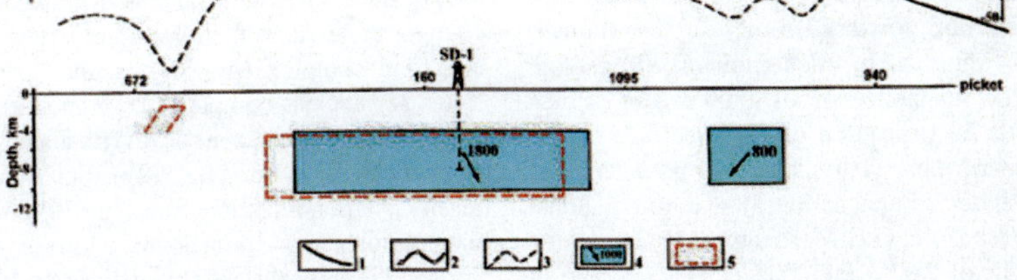

Fig. 18.26 Fragment of the gravity and magnetic fields analysis along with profile 18 (see Fig. 18.25) (after Eppelbaum & Khesin, 2011). (1) gravity field Δg_B, (2) magnetic field ΔZ, (3) first horizontal derivative of the gravity field $\partial \Delta g/\partial x$, (4) contour of the determined magnetized body and position of magnetization vector derived from the magnetic field analysis, (5) contour of the body determined by analysis of $\partial \Delta g/\partial x$

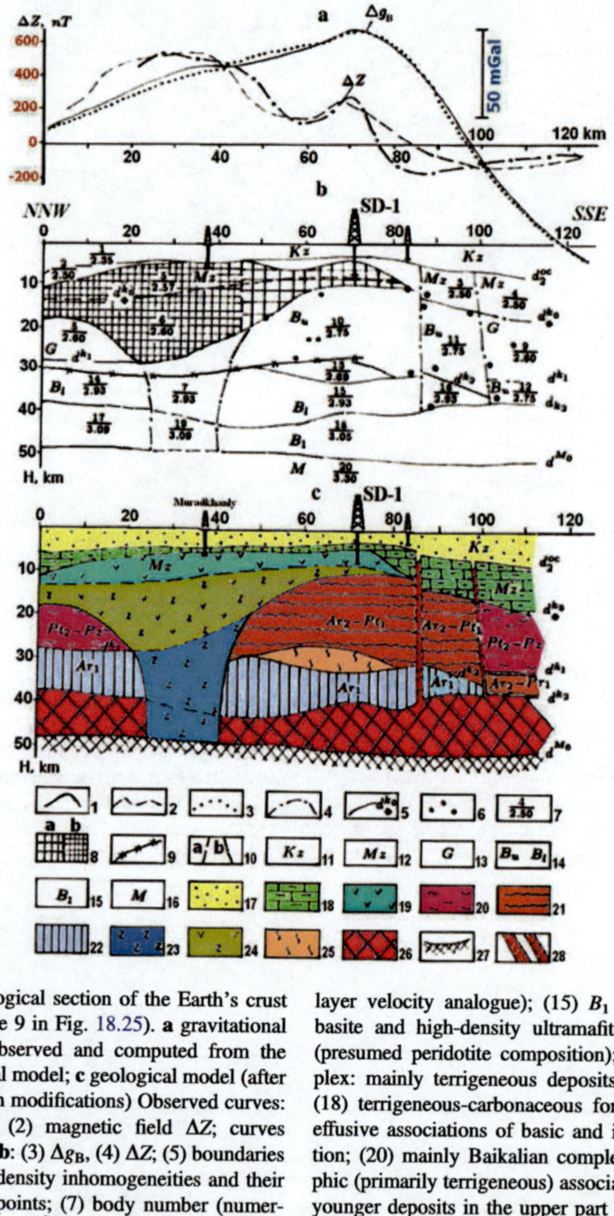

Fig. 18.27 Deep geological section of the Earth's crust in the SD-1 area (profile 9 in Fig. 18.25). **a** gravitational and magnetic fields, observed and computed from the model **b**; **b** petrophysical model; **c** geological model (after Khesin et al., 1996, with modifications) Observed curves: (1) gravity field $\Delta_B g$, (2) magnetic field ΔZ; curves computed by the model **b**: (3) Δg_B, (4) ΔZ; (5) boundaries of the velocity and the density inhomogeneities and their indices; (6) diffraction points; (7) body number (numerator) and density value, g/cm^3 (denominator); (8) geological bodies with a magnetization of 2500 mA/m (**a**) and 2,800 mA/m (**b**); (9) projection of Curie surface on the basis of geothermal data; (10) subvertical boundaries of bodies on the basis of magnetic (a) and gravitational (b) fields; (11) Cenozoic; (12) Mesozoic; (13) *G* complex (velocity analogue of the "granitic" layer); (14) B_u and B_l subcomplexes of *B* complex (complex *B* is the "basaltic" layer velocity analogue); (15) B_1 complex (apparently, basite and high-density ultramafites); (16) *M* complex (presumed peridotite composition); (17) Cenozoic complex: mainly terrigeneous deposits; Mesozoic complex: (18) terrigeneous-carbonaceous formations, (19) mainly effusive associations of basic and intermediate composition; (20) mainly Baikalian complex (Pt_2-Pz): metamorphic (primarily terrigeneous) associations (the presence of younger deposits in the upper part is possible); (21) Pre-Baikalian complex (Ar_2-Pt_1): mainly gneisses and marbles; (22) ancient complex (Ar_1): gneisses and amphibolites; (23) root of the basic magmatism; (24) undivided effusive-intrusive complex; (25) rock complex of a low density (serpentinization zone ?); (26) complex of magmatic associations corresponding to crust-to-mantle transition; (27) upper mantle roof position; (28) large fault zones

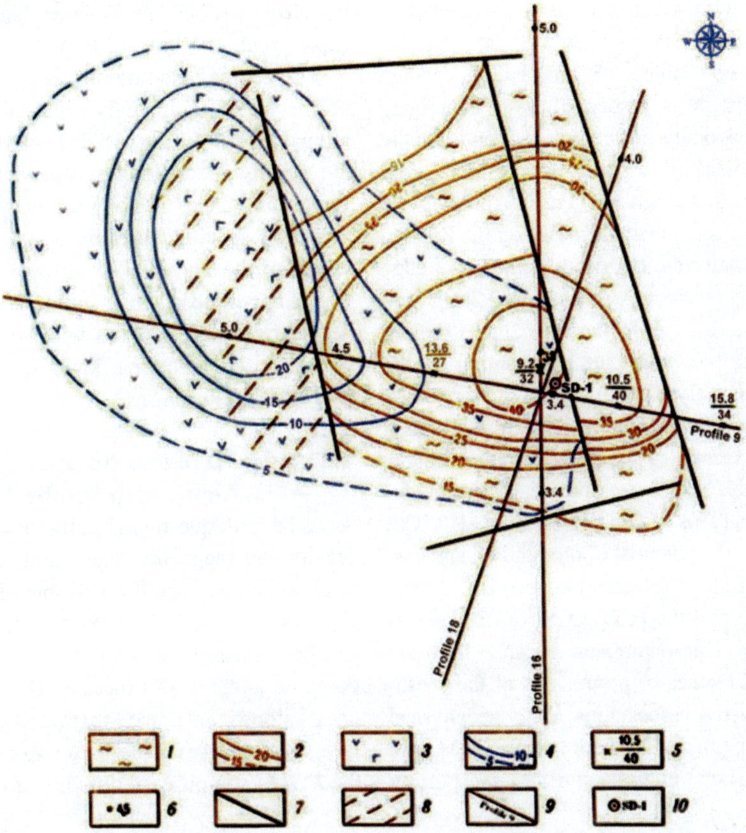

Fig. 18.28 Main elements in the 3-D physical-geological model for the SD-1 area (location of profiles 9, 16 and 18 is shown in Fig. 18.25) (Khesin & Eppelbaum, 2007) (1) Pre-Baikalian (Ar_2-Pr_1?) rock associations of heightened density (initially mainly sedimentary), (2) total isopachs of Pre-Baikalian associations including Ar_2-Pr_1 (solid line is reliable contour, dashed line is proposed contour), (3) highly magnetized Mesozoic magmatic rocks of basic and intermediate composition, (4) isopachs (km) of Mesozoic magmatic rocks (solid line—reliable, dash line—proposed), (5) depth of occurrence (km) of roof (numerator) and lower boundary (denominator) of Ar_2-Pr_1uplift, (6) depth of occurrence (km) of Mesozoic magmatic rocks, (7) major faults—the boundary of blocks, (8) projection of root zone of basic magmatism, (9) interpretation profiles, (10) SD-1 location

analogy of this geological section to ones in the Lesser Caucasian, it is likely that magmatic rocks occur down to 10 km (compatible with the Lower Bajocian rocks of the Lesser Caucasus).

The gravity and magnetic fields were examined for profiles 9, 16, and 18 (with seismic and thermal data). The results determined the main features of a spatial geological-geophysical model of the Earth's crust for the SD-1 area (Fig. 18.28). The diagram shows the primary sources of gravity-magnetic anomalies in the eastern part of the area where SD-1 is located and the western part where the Zardab magnetic maximum was found. The magnetic and gravity field analysis (including 3D modeling of these fields) and petrophysical data provided additional evidence to account for the Ganja regional magnetic maximum (Eppelbaum & Khesin, 2012).

The SD-1 stopped at a depth of 8.2 km. Given the location of the lower edge of the magnetized masses and by the analogy of this geological section to ones in the Lesser Caucasian, it is likely that magmatic rocks occur down to 10 km (compatible with the Lower Bajocian rocks of the Lesser Caucasus).

The primary source of the Talysh-Vandam gravity maximum is associated with underlying high-density, strongly metamorphized, and initially mainly sedimentary associations of the non-magnetic (or low-magnetic) Pre-Baikalian floor. This floor has a submeridional strike on the Russian and African platforms. The depth of the upper edge of these highly dense rocks is estimated at 9.5 km.

Thus, SD-1 has yet to discover (these drilling operations may or may not be continued) the source of the Talysh-Vandam gravity anomaly. Still, it has identified the origin of the Ganja magnetic maximum. This has enormous importance for analyzing the tectonic–magmatic evolution of the Caucasus region (Alizadeh et al., 2000) and evaluating the potential of ore- and oil and gas-bearing prospecting. For example, the geomagnetic models can extend the prospecting of oil deposits in the Middle Kur depression because they are associated with zones of protrusions of the Mesozoic magnetoactive associations. Many geophysical methods have mapped these dense Mesozoic associations, but only magnetic prospecting revealed basic and middle-consistency magmatites (Eppelbaum & Khesin, 2011).

The geomagnetic model along the meridional profile 16 is significantly different from sublatitudinal profile 9 (Fig. 18.28). At the same time, the gravity field along profile 16 is smoother. These differences in potential geophysical fields reflect the specificities of the deep structure of the source of the Talysh-Vandam gravity maximum, which is composed of a submeridional strike of deeply occurring massses with superfluous density and a sublatitudinal strike of younger magnetoactive associations.

18.2.6.6 3D Magnetic-Gravity Modeling Along with Profile 1

A visual example of 3D combined modeling of gravity and magnetic fields along with Profile 1 (see a scheme of profiles presented in Fig. 18.25) is shown in Fig. 18.29. Profile 1, crossing the Lesser Caucasus, illustrates this region's very complex geological structure. This profile stretches along a line Mez-Mazra—Gedabey—Dzegam-Jirdykhan. The Late-Alpine effusives in the *PGM*s compose an ophiolite zone (a relic of

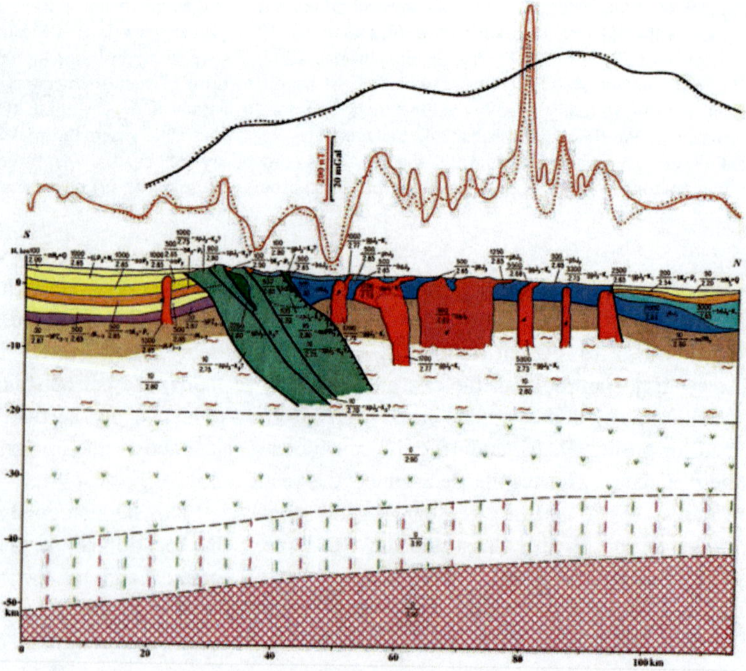

Fig. 18.29 3D modeling results along with profile 1 (location of this profile is shown in Fig. 18.25) (Eppelbaum & Khesin, 2011)

the ocean crust). The same rocks are thought to occur in the NE immersion of the Lesser Caucasus. Pre-orogenic and orogenic intrusive and effusive rocks are fixed in the southern parts of the ***PGM***.

Thick sedimentary deposits are developed in the northern part of this profile. A smooth high of the Moho discontinuity is observed from south to north from a depth of 52 km up to 42 km (Eppelbaum & Khesin, 2012). Such a Moho boundary behavior agrees with the latest data of deep seismic profile re-interpretation (Pavlenkova, 2012). Depths of the magnetized body's lower edges were estimated based on various geophysical field analyses but mainly based on thermal and magnetic data examination (Pilchin & Eppelbaum, 1997). Here were revealed the acid intrusions of lower density and magnetization, basic magmatic rocks of increased density and magnetization, and fault zones. It was determined that the most evident density boundaries were associated with the base of the Cenozoic sedimentary strata and, to a lesser degree, with the base of the Alpine complexes. According to the performed modeling, the geomagnetic boundaries are associated mainly with the roof and bottom of the Mesozoic floor of heightened magnetization.

The geomagnetic model along the meridional profile 16 significantly differs from the sublatitudinal profile 9 (see locations of profiles in Fig. 18.25). At the same time, the gravity field along profile 16 is smoother. These differences in potential geophysical fields reflect the specificities of the deep structure of the source of the Talysh-Vandam gravity maximum, which is composed of a submeridional strike of deeply occurring masses with superfluous density and a sublatitudinal strike of younger magnetoactive associations.

The examination indicates that the area of the Talysh-Vandam gravity maximum is highly inhomogeneous in terms of the geological structure, and individual elements of the gravity maximum—anomalies of the second-order—reflect a different genesis in areas of Earth's crust and suggest that they developed apart from one another (Eppelbaum & Khesin, 2011).

A visual example of 3D combined interactive modeling of gravity and magnetic fields along stakes 700–1350 of profiles 3–4 (see scheme presented in Fig. 18.25) is presented in Fig. 18.30. This profile crosses the Lesser-Caucasus, the Middle Kur Depression, and ends at the submountainous zone of the Greater Caucasus. An initial ***PGM*** was constructed based on drilling data, analysis of seismic data and petrophysical material, and an advanced examination of magnetic and gravity anomalies. An important role played here in determining the values of the Curie Point Depth and lower edges of magnetized bodies (Eppelbaum et al., 2014). It explains, for instance, the comparatively shallow position of nonmagnetic rocks in the SW part of the investigated profile.

As seen from this figure, despite the complex geological section, there was obtained a good agreement between the observed and computed magnetic and gravity fields.

18.2.6.7 Results of 3D Combined Interactive Modeling Along Profile A-B (Profiles 10, 3, and 4)

Regional profile A–B (its location is shown in Fig. 18.25) is composed of three profiles (10, 3, and part of profile 4) and starts near the town of Djulfa (Nakhchivan Autonomic Republic (AR) of Azerbaijan). It crosses Armenian territory and ends near the town of Sheki (near the border with Georgia) (Fig. 18.31). It is the most important profile, and this ***PGM*** will be described in detail.

In this ***PGM***, the Earth's crust decrease is observed within the Kur Depression. The surface of the "basaltic" layer displays the most uplift in the area of the core of the Lesser Caucasus mega-anticlinorium, where associations of the pre-collisional effusive formations are found.

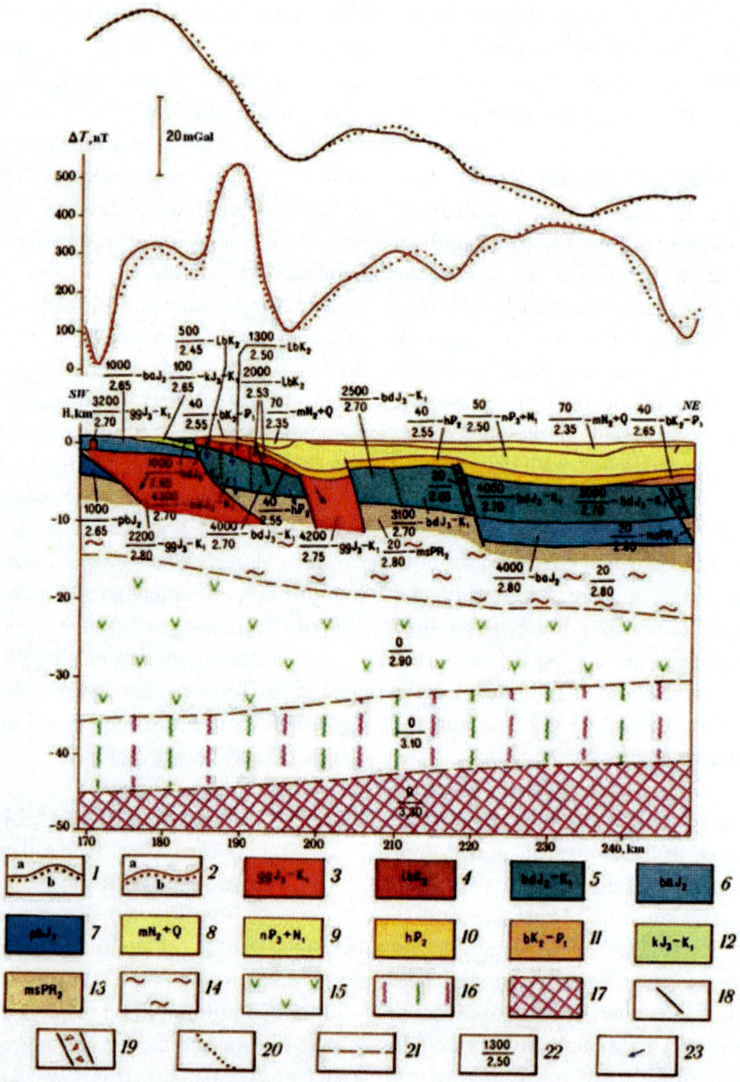

Fig. 18.30 Construction of a physical-geological model (for stakes 700–1350 of profiles 3–4; location of profiles is shown in Fig. 18.25) (after Khesin et al., 1996, with modifications). (1) gravity field Δg_B: a—observed, b—computed; (2) magnetic field ΔT: a—observed, b—computed; (3) intrusive gabbro-diorite-granodioritic association; (4–6) effusive associations: (4) liparite-basaltic, (5) basalt-andesite-dacitic, (6) basalt-andesitic; (7) basalt-andesite-plagioliparitic; (8–13) background sedimentary deposits: (8) upper molassic, (9) lower molassic, (10) terrigeneous, (11) terrigeneous-carbonaceous, in some places flyschoid, (12) carbonaceous-sandy, reef rocks, (13) metamorphic schists and other metamorphites; (14–16) deep-seated complexes: (14) granites and gneisses, in some places amphibolites, (15) basic rocks, (16) basic rocks—eclogites (ultramafic-mafic complexes ?); (17) upper-mantle peridotites (ultramafic-mafic complexes ?); (18) faults, upthrusts; (19) crush zones; (20) boundaries of physical properties changing within the same geological association; (21) boundaries of physical properties changing along all geological section; (22) physical properties (numerator = magnetization, mA/m, denominator = density, g/cm^3); (23) direction of the magnetization vector other than the averaged geomagnetic field inclination for the region

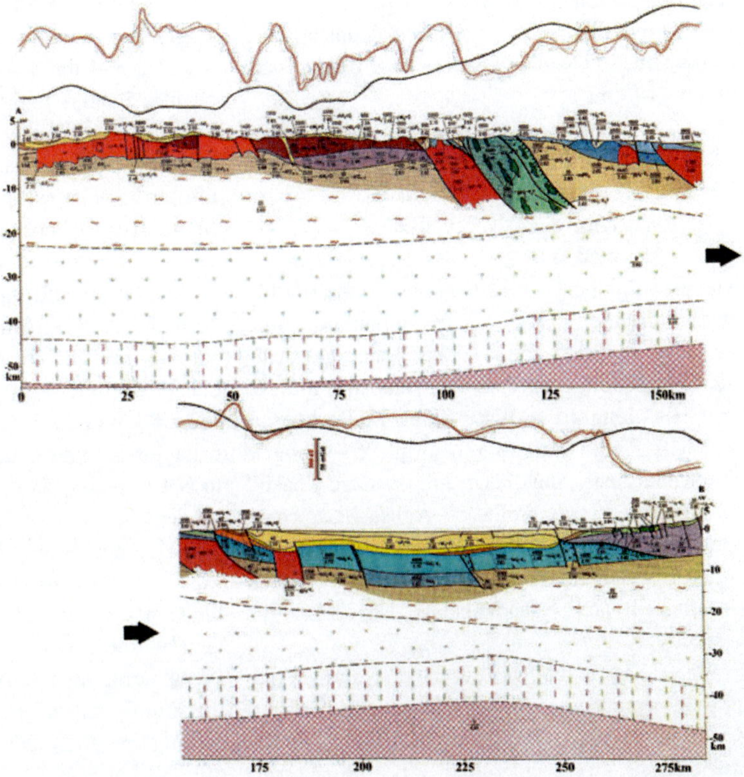

Fig. 18.31 3D modeling results along profile A–B (location of the profile is shown in Fig. 18.25) (Eppelbaum & Khesin, 2011)

The upper part of the geological section visually reflects Azerbaijan's geological formation stages. Orogenic granodiorite-porphyric and pre-orogenic gabbro-monzonite-dioritic intrusive formations are developed at the beginning of the SW profile (Nakhchivan region). Orogenic effusive dazite-andesite-basaltic formations are found on the SE of the town of Istisu and orogenic sedimentary deposits (molasses) in the Kur Depression (with a thickness of 4–5 km).

Associations of pre-orogenic effusive formations are present in (1) the SW part of this profile, where they occur on sub-platform terrigenous-carbonate deposits, (2) in the transfer area from the Lesser Caucasus to the Kur Depression, (3) in the southern slope of the Greater Casuscas immersion.

The post-collisional effusive formations are outcropped in the Lesser Caucasus's middle part of the profile. These formations create a so-called "ophiolitic zone". Physical-geological modeling indicates that these formations can occur under young sedimentary deposits in the Kur Depression and under Lower-Middle-Jurassic sand-shale associations in the Greater Caucasus pre-collisional stage (where these formations were underthrusted as a result of tectonic processes).

In the Lesser Caucasus, there are presumed to be thick buried intrusives of post-collisional gabbro-diorite-granodiorite formations. Intrusives of these formations are usually metalliferous. Sedimentary deposits of the post-collisional stage are exposed in separate areas of the Lesser Caucasus and are more widely present on the southern slopes of the Greater Caucasus.

Finally, effusive associations of the pre-collisional stage are (as was mentioned earlier) found in the core of the Lesser Caucasus mega-anticlinorium. The presence of these associations was predicted as well in the middle part of the

Kur Depression below the post-collisional effusives. Pre-collisional sedimentary deposits exist in the Greater Caucasus, where their thickness is about 7–8 km.

18.2.6.8 Results of 3D Combined Interactive Modeling Along with Profiles 2, 5, and 6

The pre-Alpine basement is closest to the Earth's surface in the vicinity of the ophiolitic zone, where its depth is only 1–1.5 km. Other physical-geological models have been developed. Profile 2 stretches along a line Lake Karagel—Dashkesan—Beyuk-Kasik (Fig. 18.32), profile 5—along a line Kutkashen—Shakhdagh Mt.—village Chakh-Chakh (Fig. 18.33), profile 6—Aghsu—Lagich—Gusar (Fig. 18.34). Profiles 7 and 8 (not presented in the Chapter) stretch along Kyalvazchay—Lerik—Jalilabad and Shandankalasi Mt.—Masally, respectively.

The ***PGMs*** in models 1, 2, 5, and 6 show fixed pre-collisional cores of the Lesser and Greater Caucasus with Jurassic associations. These associations were identified in the Lesser Caucasus as effusive basaltic-andesite-plagioliparite and basaltic-andesite formations and in the Greater Caucasus—Lower-Middle-Jurassic associations—as sand-shale associations. The surface of the pre-Alpine foundation occupies the highest position in these cores.

The post-collisional effusive formations in the ***PGMs*** 1 and 2 (Lesser Caucasus), as in the ***PGM*** along with the A–B line, compose an ophiolitic zone. The same formations are presumed to occur in the NE submersion of the Lesser Caucasus. In ***PGMs*** 5 and 6 (Greater Caucasus), their broad evolution can be predicted by 3D combined gravity-magnetic modeling under the pre-collisional Lower-Middle-Jurassic sand-shale deposits of the Greater Caucasus (occupying this position because of overthrust), as well as under young deposits of the Kur Depression.

Pre-orogenic and orogenic intrusive and effusive formations are in the southern parts of ***PGMs*** 1 and 2. Sedimentary deposits of these stages developed in the northern parts of profiles 1 and 2 and the southern and northern parts of profiles 5 and 6. In the northern parts of profiles 5 and 6, at a depth of 9–10 km, are associations of significant magnetization, presumably related to tuff-terrigenous sub-platform associations of the Middle-Late Paleozoic.

The behavior of the Moho discontinuity in the ***PGMs*** overall agrees with the ***PGM*** along with the A–B direction. In ***PGMs*** 1 and 2, a smooth uplift of the Moho discontinuity can be observed from south to north at a depth of − 52–54 km up to − 42 km. In ***PGMs*** 5 and 6, its behavior is more complex: in ***PGM*** 5 (from SE to NW), there is an uplift from − 44 to − 41 km (under the area of the abovementioned overthrust). Then is a submersion to a depth of − 58 km (below the conjunction of anticlinoria of the Caucasian Main Ridge and Side Ridge) and then an uplift to a value of − 52 km. In PGM 6 (from north to south), there is a stepped submersion from − 47 to − 60 km (under the same conjunction area) and a smooth uplift to a depth of − 58 km.

An interesting peculiarity in the southern part of ***PGM*** 2 is a zone of strong rock decompaction in the roots of the Dalidagh intrusive massif. Its depths range from − 4 to − 17 km, and a horizontal thickness of 30 km. This decompaction zone produces the greatest Kelbadzhar-Dalidagh gravity minimum. This large "granite room" was a source of pre-orogenic and orogenic granitoid magmatism.

PGM 7 and 8 are in the Talysh geostructural zone (see scheme in Fig. 18.25). There is a Moho uplift from south to north (from − 52 km to − 42 km and from − 44 km to − 31 km, respectively) and to the east (on the side of the Caspian Sea).

Within the Talysh zone, intrusive bodies from ultrabasic to acid consistency were delineated. Based on the 3D gravity-magnetic modeling, surface trachite-trachiandesite-trachibasalt effusive formations widely presented at the Earth's

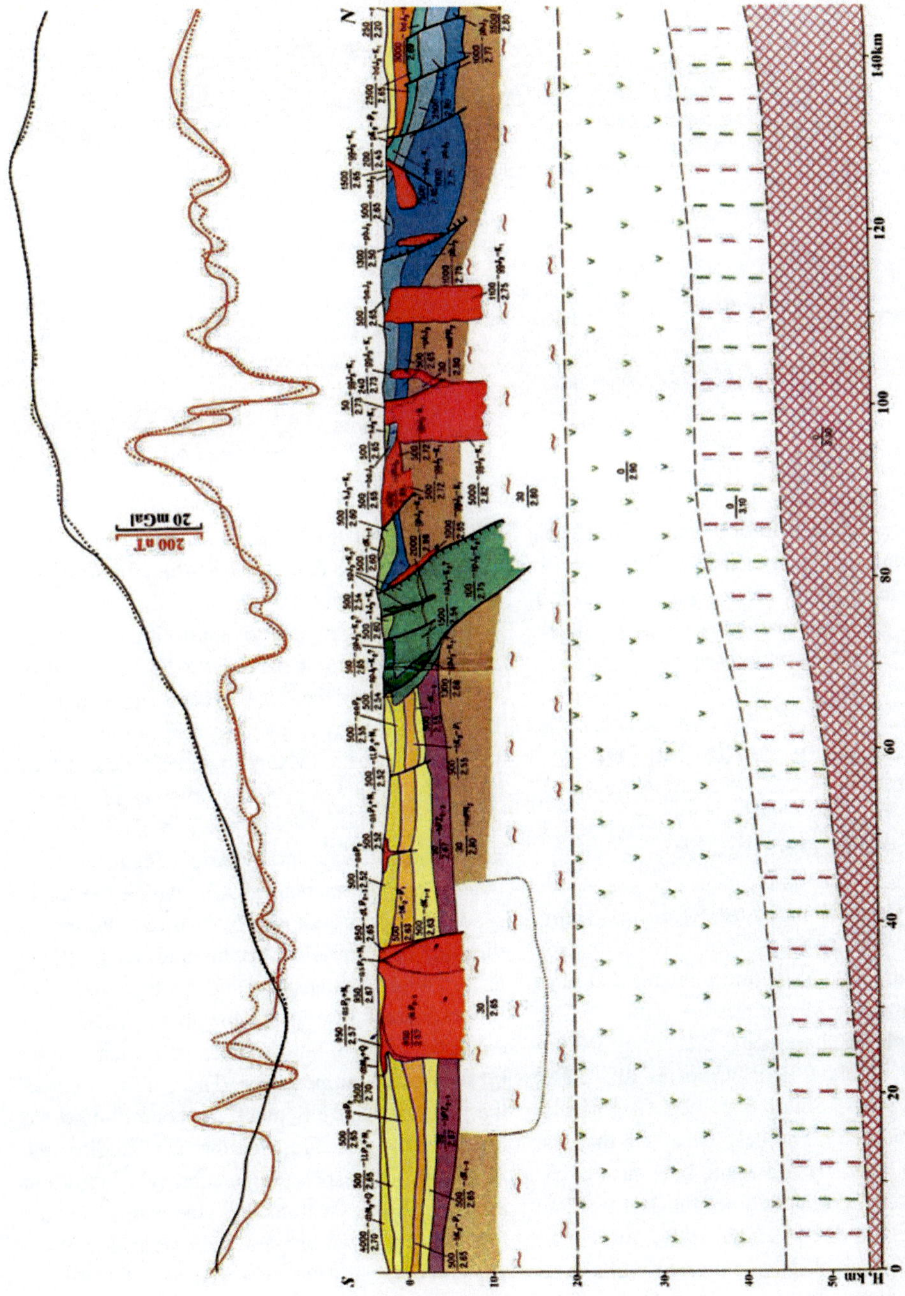

Fig. 18.32 3D modeling results along with profile 2 (location of the profile is shown in Fig. 18.25) (Eppelbaum & Khesin, 2011)

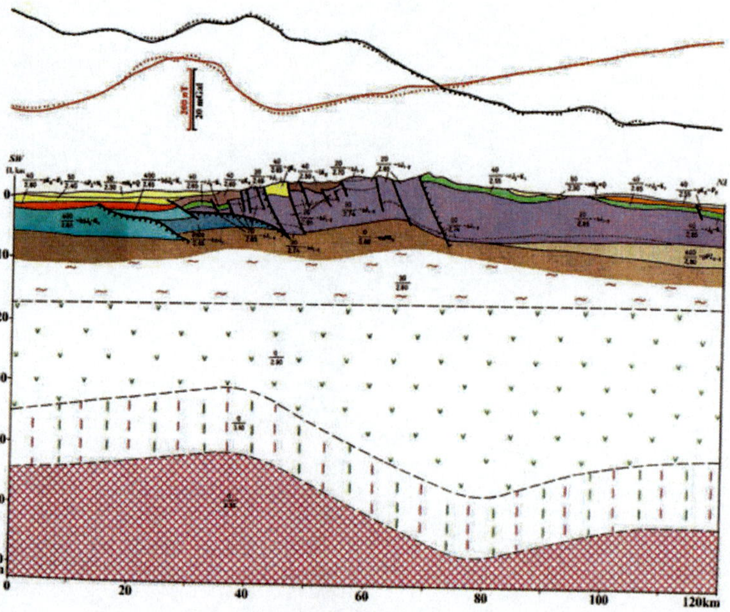

Fig. 18.33 3D modeling results along with profile 5 (location of the profile is shown in Fig. 18.25) (Khesin & Eppelbaum, 2007)

surface were divided into blocks of differing magnetization and (or) density. These blocks have tectonic contacts. The bodies of gabbro-syenitic and gabbro-monzonite-dioritic formation were characterized quantitatively.

18.3 Azerbaijan: South Caspian Basin

18.3.1 Seismic Data Analysis

The South Caspian Basin (SCB) occurs within intra-continental settings. Still, it is poorly understood in terms of the mechanisms that have controlled its subsidence history (Fig. 18.35), with a depth to basement exceeding 20 km (Abdullayev et al., 2015; Brunet et al., 2003; Egan et al., 2009; Shikhalibeyli & Grigoriants, 1980). Although it is widely accepted that the SCB was initiated by Mesozoic back-arc extension related to the subduction of the Tethys Plate (e.g., Zonenshain & Le Pichon, 1986), more than half of the 20 km subsidence presently observed occurred within the tectonic framework of the evolution of the Alpine–Himalayan orogenic belt (Egan et al., 2009).

Analysis of the PGM presented in Fig. 18.36 indicates that the depth of the Moho discontinuity varies from about 60 km beneath the onshore South Caspian region to approximately 30 km beneath the deepest part of the SCB. Based on seismic velocity data, it has been suggested that the 'granitic' crustal layer is absent in the central part of the basin (Mangino & Priestley, 1998). This could be because the crust is of oceanic origin, although thicker than typical oceanic crust. Another appropriate model suggests that it could be a continental crust in which the upper section has been removed by erosion, faulting, or thinned and intruded continental crust (Egan et al., 2009). Zonenshain and Le Pichon (1986) proposed that the crust beneath the SCB is of continental origin but has been subjected to high pressures and temperatures. The granite rocks of the continental basement metamorphosed to become eclogite. From earthquake studies, the SCB crust appears to be a relatively rigid and aseismic block within the framework of the Alpine-Himalayan orogenic belt (Priestley et al., 1994). This is evidence of a rheological difference between the crust of the SCB and that of the surrounding region (Egan et al., 2009; Mangino & Priestley, 1998).

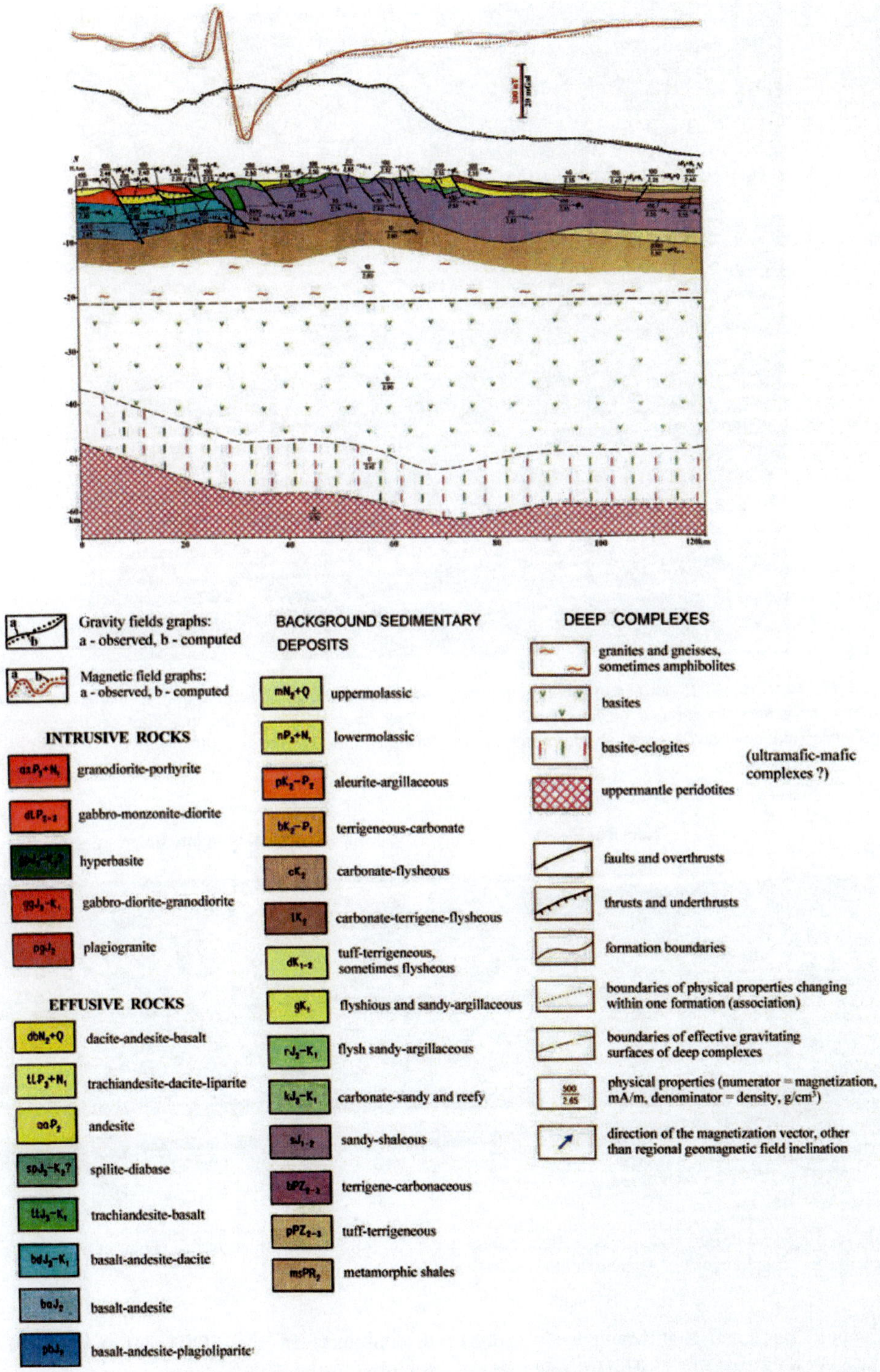

Fig. 18.34 3D modeling results along with profile 6 (location of the profile is shown in Fig. 18.25) (after Khesin et al. (1996), with modifications)

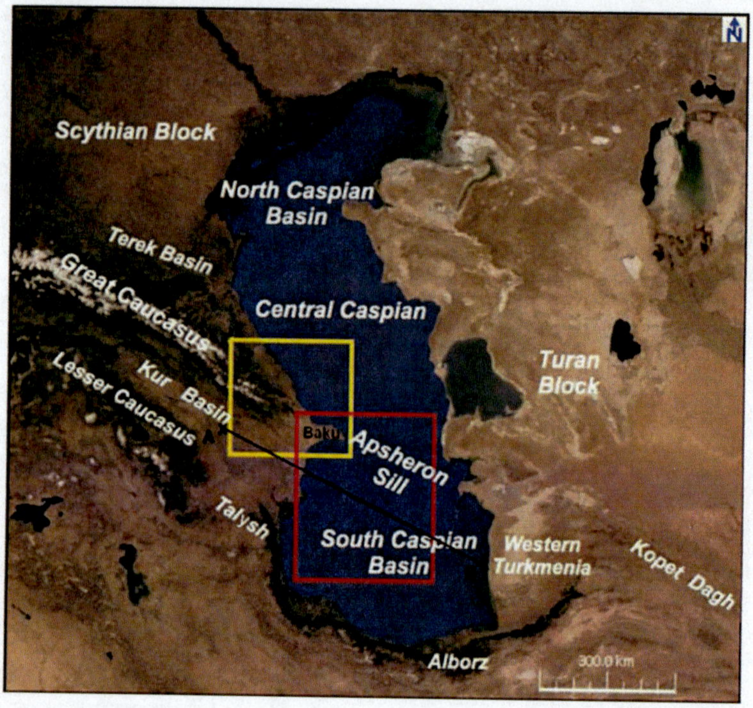

Fig. 18.35 Map of SCB and surrounding regions illustrating the general location of onshore (yellow box) and offshore (red box) study areas (background image courtesy of NASA World Wind). Points A and B and the black line designate the position of the investigated profile (after Egan et al. (2009), with minor modifications)

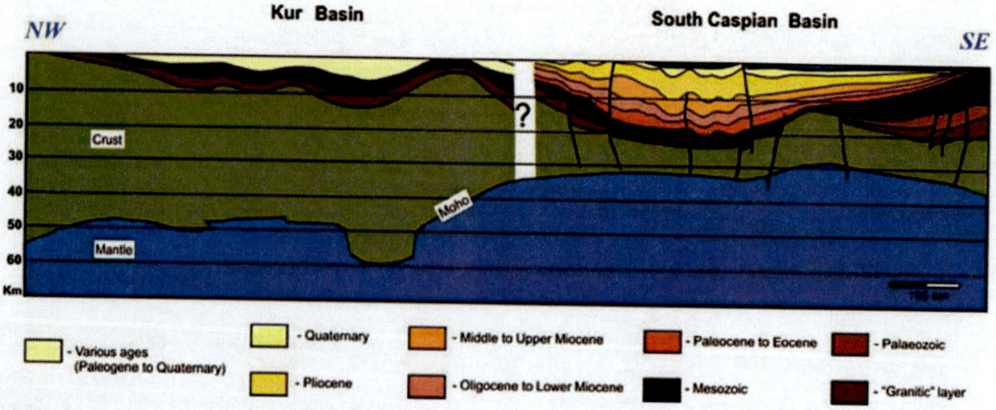

Fig. 18.36 Regional crustal-scale cross-section through the Kur and South Caspian basins (Egan et al. (2009) with adapted seismic constructions of Baranova et al. (1991) and Mamedov (1992)

It should be noted that Mamedov's (2008) seismo-geological profile (SW–NE orientation) (Fig. 18.37) is a basis for many tectonic and geophysical reconstructions. The depth of Moho discontinuity in the SCB central part also is found at about 30 km.

Alizadeh et al. (2017) present a more detailed seismic time section. Analysis of this section

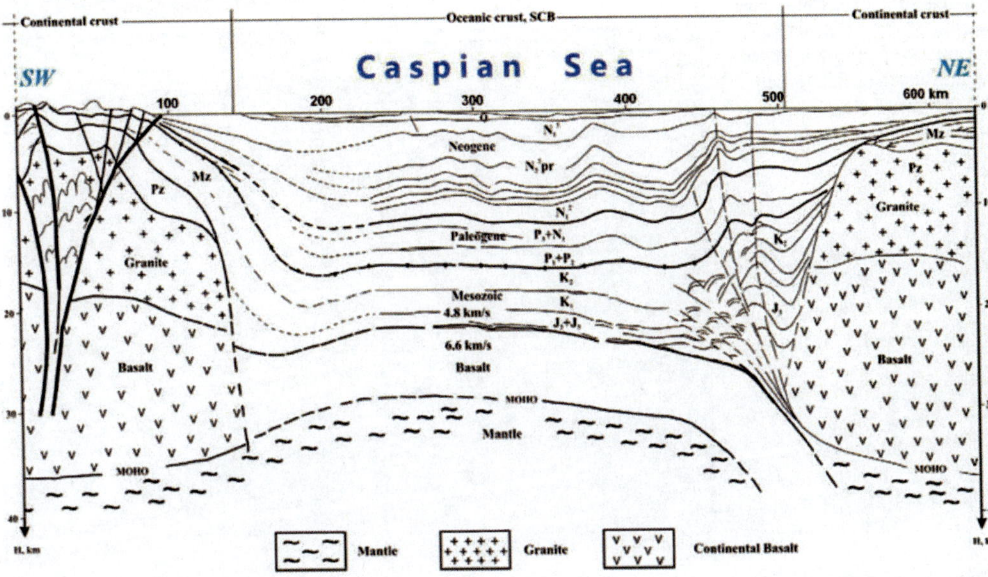

Fig. 18.37 The seismic-geologic cross-section along with a regional profile from the Alborz (SW) to the Absheron sill (NE) (after Mamedov, 2008)

indicates that the Caspian Sea level is lowering on the Jurassic and Cretaceous boundaries, confirmed by the shoreline displacement to the south and southwest that caused the deposition of terrigenous sediments on the shelf area.

Figure 18.38 shows the time section and the submeridional profile in the South Caspian Basin. In this section, the consolidated crust is characterized by a thickness of 8–10 km, and it is fluently submersed to the side of the Turan platform. According to Mamedov (2004), interpretation in Absheron ridge over the consolidated crust is observed tectonic congestion of Mesozoic and Palaeogene strata like accretional prism image in subduction zones.

Abdullayev et al. (2015) presented an effective seismic model (Fig. 18.39). This profile runs from north to south and is developed by available surface sets of various 2D reflection seismic profiles, some of which are ultra-deep 2D TWT profiles. It is broadly like the profile modeled in Green et al. (2009) but differs by being constructed to avoid structurally complex parts of the Absheron ridge. The model consists of nine layers corresponding to seismically mappable intervals covering the Jurassic to the present (Fig. 18.39). A portion of the profile north of the Absheron ridge represents the continental crust of the SCB, while the southern part is located over the SCB crust. The profile is characterized by an increase in thickness, from south to north, in the SCB portion of the Mesozoic–Lower Palaeogene age. The complete profile was subjected to restoration with 'FlexDecomp' software initially with constant *b*-factors, based on the assumption of changing crustal types across the profile. The northern portion of the profile includes significant crustal shortening under the Absheron ridge and cannot be confidently restored using this methodology.

Abdullayev et al. (2015), based on analysis of reflection seismic and some other data, have developed a map of the crystalline basement (Fig. 18.40).

18.3.2 Gravity Field Analysis

The main structural elements of the SCB are demonstrated in Fig. 18.41. The central part of the basin, unlike its periphery, is practically aseismic, with earthquakes primarily confined to

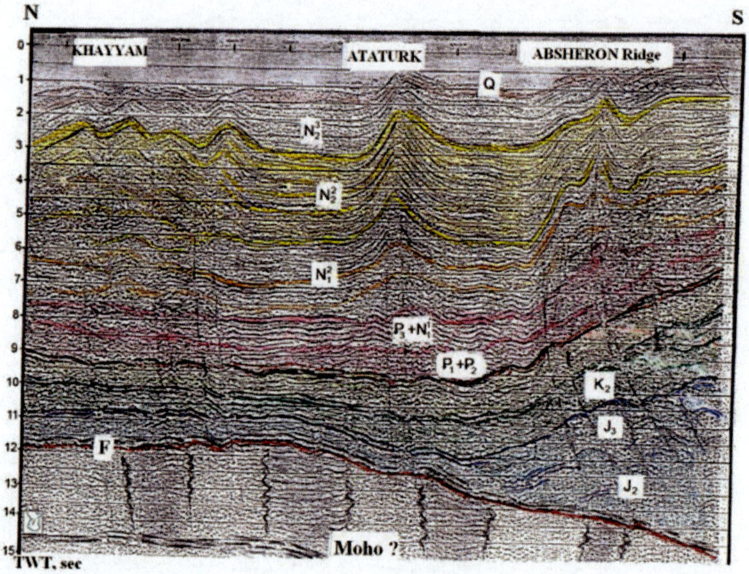

Fig. 18.38 The seismic time section of the profile crosses the western part of the SCB (after Mamedov, 2004). Letter 'F' designates a top of crystalline basement

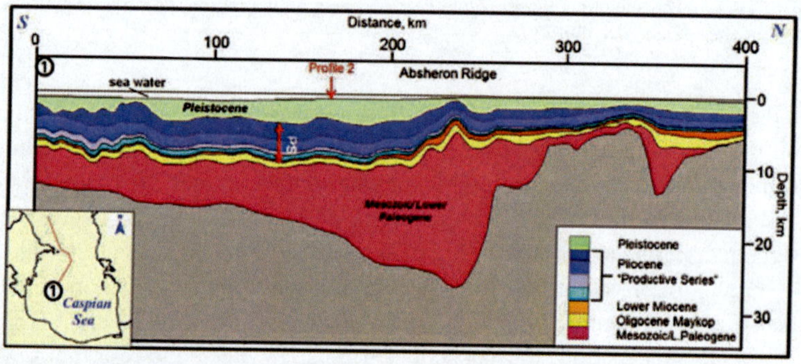

Fig. 18.39 Geoseismic profile 1 through the SCB and the Absheron ridge shows the variable thicknesses across the profile (Abdullayev et al., 2015)

the borders (Ambraseys & Melville, 1982; Jackson et al., 2002; Mangino & Priestley, 1998; Priestley et al., 1994; Ulomov, 2003). Despite the observed extremely low heat flow values in the SCB, some small areas of higher flow are likely related to the activity of mud volcanoes (Mukhtarov, 2004; Yakubov et al., 1971).

New seismic data have recently been acquired, providing new constraints on basin structure and formation history. However, the problem of the origin of the basin as a unified structure (Artyushkov, 1993, 2007; Khain, 2005) and assumptions about the subduction of the lithosphere of the South Caspian underneath the mid-Caspian (Allen et al., 2003; Egan et al., 2009; Granath et al., 2000; Green et al., 2009; Khalilov et al., 1987; Knapp et al., 2004; Mamedov, 2008) remain disputable. The purpose of this study has been to propose a crustal model and describe some characteristics of the crustal structure of the SCB along with a regional profile from the Alborz to the Absheron sill, integrating available seismic data with the Bouguer gravity anomaly along with the profile. The deep structure analysis along the selected profile has been combined with an analysis of earthquake focal mechanisms and GPS velocity data to evaluate the basin's active tectonics and geodynamic evolution.

The tectonic-structural scheme of the SCB region and its surroundings are of high complexity. Its generalization is presented in Fig. 18.41, where the investigated profile A—A is shown.

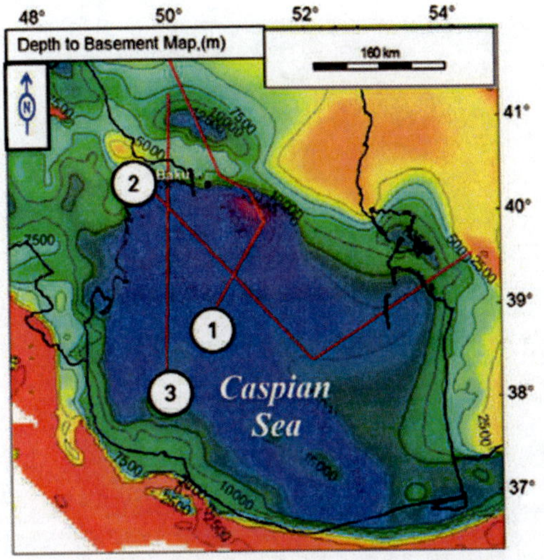

Fig. 18.40 Depth to crystalline basement map integrating reflection seismic results and published data (Abdullayev et al., 2015; Glumov et al., 2004)

A new map of the Bouguer gravity field of the SCB has been developed by Kadirov and Gadirov (2014) based on Dehghani and Makris (1983), Gravity map of the USSR (1990), and Kadirov (2000a, 2000b) (Fig. 18.42).

Anomalies of various geometries and amplitudes characterize the gravity pattern. The area of increased horizontal gradients sub-parallel to the strike of the Caucasus represents the anomaly field of the northern part of the basin. In the northwestern part of the SCB, there is a vast gravity minimum with an amplitude reaching 125 mGal. The central part of the basin is represented by a gravity minimum and an isometric maximum of the gravity field in the southwest and southeast (Safidrud and Godin uplifts). The area of increased gradients from the Alborz Mountains to the south up to the deep-water part of the basin is typical for the southern part of the SCB. The area of the Central Alborz is characterized by a negative anomaly (− 120 mGal).

A model profile has been generated extending almost 650 km from the southwest Central Alborz range towards the northeast in the mid-Caspian (see Fig. 18.41). The initial model was created by Mamedov (2008) and later utilized in models of Brunet et al. (2003) based on the recent ultra-deep 20 s TWT cross-sections integrated with published data on deep seismic sounding and earthquake analysis (Aksenovich et al., 1962; Allen et al., 2003; Babayev & Gadjiev, 2006; Baranova et al., 1991; Brunet et al., 2003; Jackson et al., 2002; Knapp et al., 2004).

The density values used in our model for the crust and upper mantle were selected from published data, while the mantle density was set at 3300 kg/m^3. The developed initial model demonstrated a sequence of zones with excessive mass (0–62 km and 220–500 km) along with the profile. Calculated values of the gravity field from the initial geological and geophysical models could have explained the observed gravity field better. There is a vast gravity minimum in the northwestern part of the SCB with an amplitude reaching 125 mGal. The most significant discrepancies appeared to be in the intervals of 10–65 km (54–100 mGal), 260–485 km (40–55 mGal), and 605–630 km (35–54 mGal). For compensation for the observed gravity field, a selection of density boundaries (upper boundary of "basaltic" layer of oceanic and continental crust and Moho surface) was conducted until the slightest discrepancy between the observed and selected values of the gravity field. The final gravity model, topography, and observed and calculated Bouguer gravity anomalies along the studied profile are demonstrated in Fig. 18.43. There is a general correlation between the geometry of layers' boundaries and topography along with the profile. For instance, we can note the boundaries between the Neogene and Paleogene and between the Paleogene and Mesozoic.

The new model best fit (Fig. 18.43) introduces a dense "basaltic" type crust boundary only 6 km thick in places close to Absheron Sill (location about 480 km) and draws it as down-going below the Absheron Sill, therefore suppressing the crust-mantle boundary. Isostatic compensation in this model, therefore, will occur in the mantle. Unlike previous models by Granath et al. (2000), the thickness of this thin, dense crust is not kept constant but is variable and increases to 10–12 km to the south around the middle of SCB. The best fit also substitutes a very thick sediment pile from the Mesozoic

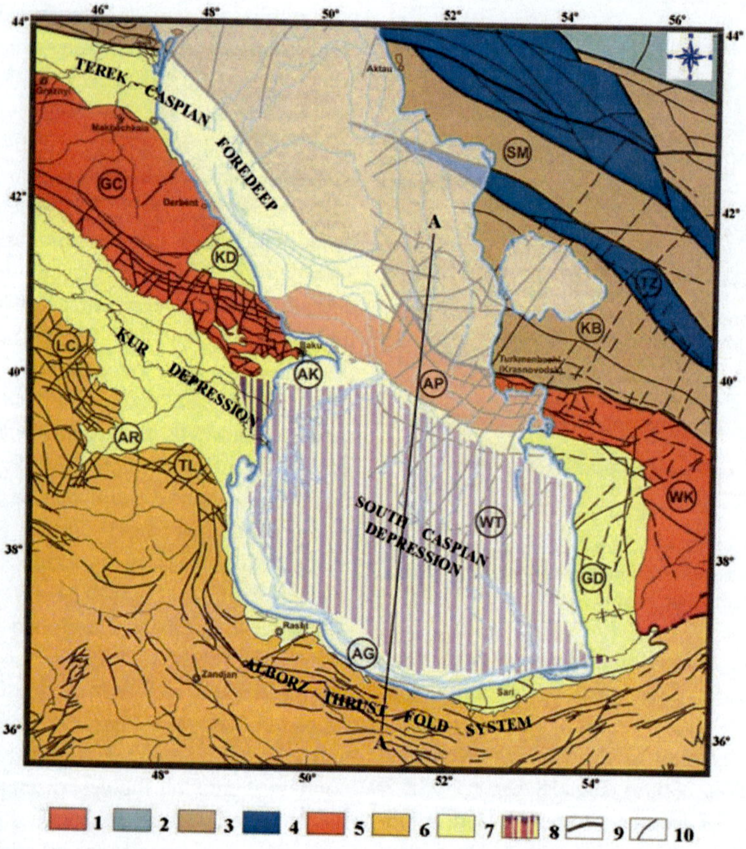

Fig. 18.41 Major structural elements of the SCB (based on Khain, 2005). The basement of platform areas (1–4): (1) Early Precambrian, (2) Baikalian, (3) Hercynian, (4) Early Cimmerian; the Alpine folding-cover systems (5, 6): (5) Greater Caucasus and Kopetdagh, (6) Lesser Caucasus, Talysh, Alborz; (7) foredeep and troughs; (8) troughs with a crust of an oceanic type; (9) fault corresponding borders of large structures; (10) other relevant faults. The major structures (letters in circles): *TZ* Tuarkyr zone, *KB* Middle-Caspian-Karabogaz anticline, *KD* Gusar-Devechi trough, *AK* Absheron–Kobustan trough, *AP* Absheron–Pribalkhan Sill, *WK* West Kopetdagh zone, *LC* fold system of the Lesser Caucasus, *AR* Lower-Aras flexure, *TL* Talysh zone, *AG* Alborz-Gogran foredeep, *WT* West-Turkmen trough, *GD* Gogran Dagh-Okarem zone, *GC* Greater Caucasus folds system, *SM* South Mangyshlak-Ustyurt system of troughs; *AA* Location of profile forming the basis of this research investigation (after Kadirov & Gadirov, 2014)

wedge with an upper crustal "granitic layer", supported by seismic observations. Our best-fit basement surface is considered the deepest interpreted seismic reflector in Green et al. (2009), which is generally shallower than in the initial model by about 5 km across all the profiles and ranges between 16 and 20 km. Our best-fit model also puts maximum depth to the sediments layer underneath Absheron Sill to about 22–25 km instead of 30 km from some places in the initial model.

Modeled "bending" of the oceanic crust, in both initial and best-fit models, can explain large gravity north of 480 km. Values of gravity must have a less dense downgoing slab and thickened continental crust north of Absheron Sill to achieve a fit and, therefore, may be less reasonable.

The thickness of the Mesozoic layers in the middle part of the profile is ~ 8 km, which increases northward, reaching the peak value (~ 20 km) in the Absheron-Pribalkhan Sill

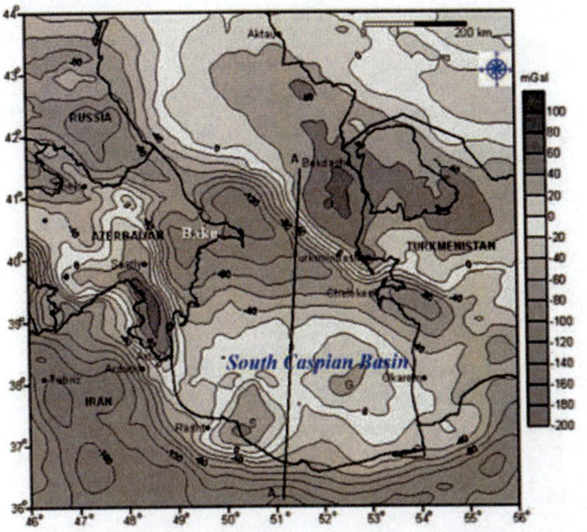

Fig. 18.42 Map of the Bouguer gravity anomalies of the SCB. *S* Safidrud uplift, *G* Godin uplift (after Kadirov & Gadirov, 2014)

rather stay than constant as in Granath et al. (2000). Subduction underneath the Absheron-Pribalkhan Sill is accompanied by sliding the sediments from the basaltic layer, forming an accretionary wedge, which can be observed on deep seismic profiles (Knapp et al., 2004). Some of the increased thickness is the result of shortening.

Green et al. (2009) also state that while some anomalous thickenings result from shortening, incredibly close to the Absheron Sill, a geographically larger part of the thick is believed to be an original feature. According to the authors, this feature is caused by Mesozoic rifting followed by thermal subsidence and sediment loading in the post-rift phase following the basin's formation in the Late Jurassic. Green et al. (2009) performed forward lithospheric modeling, which also supports the existence of a thin (6 km) oceanic crust with high values of stretching *b*-factor south of Absheron Sill, increasing in thickness to the south. This crustal variation has been modeled in Fig. 18.43 to achieve the best fit.

There are, of course, other possible combinations of densities and depths to achieve the best fit. However, the proposed crustal model is supported by modeling and observation from Green et al. (2009) and better describes a geological scenario than the initial model.

The gravity model matches the observed gravity field of Profile 1 (Fig. 18.44b), which supports the presence of denser; 'oceanic-type' crust (5–7 km) underneath the northern portion of the SCB, below and to the north of the Absheron ridge. This gravity observation is also well supported by reflection seismic and earthquake observations (Kadirov & Gadirov, 2014; Kadirov et al., 2008). This area corresponds to a large negative Bouguer anomaly (Fig. 18.44a). It is best modeled as a root of the South Caspian crust that has displaced the standard lithospheric mantle, creating a sizeable isostatic anomaly, as shown in Granath et al. (2000).

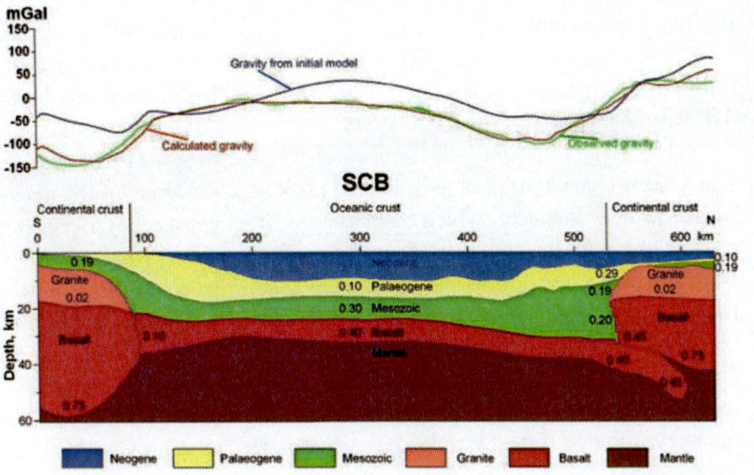

Fig. 18.43 Gravity model and topography along with meridional profile Alborz–Absheron Sill. The values of density differences are in g/cm^3 (after Kadirov & Gadirov, 2014)

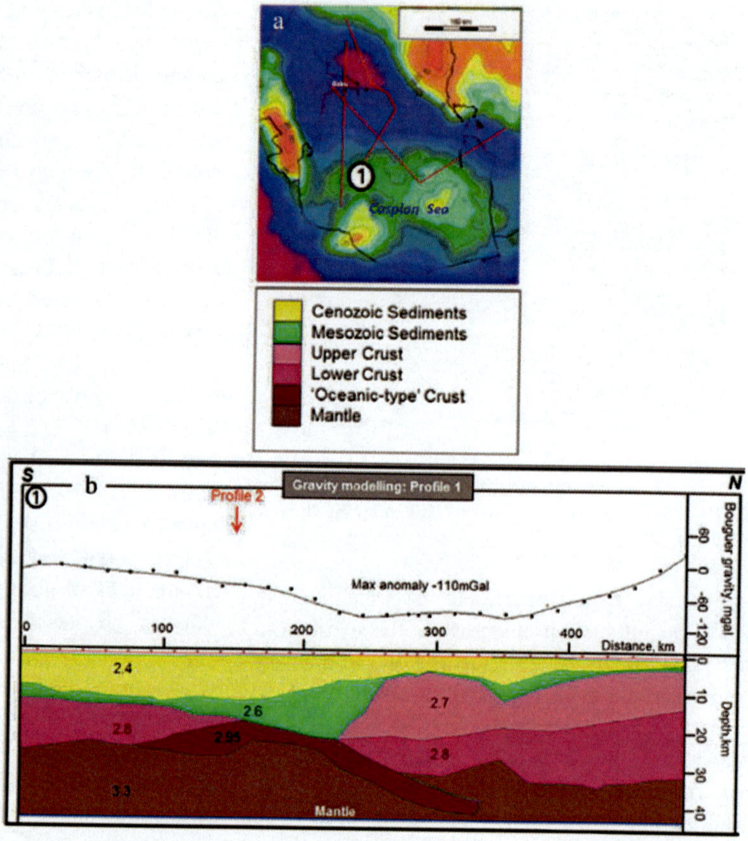

Fig. 18.44 Gravity map in the Bouguer reduction of the SCB and surrounding areas (**a**) (after Kadirov & Gadirov, 2014) and gravity modeling results of Profile 1 (**b**) showing the observed (dotted line) and modeled (solid line) Bouguer gravity anomaly (after Abdullayev et al. (2015), with minor modifications)

The southern part of the profile is characterized by a small negative Bouguer anomaly and can be modeled with thicker (+ 10 km) and less dense crust (we used lower continental crust densities of 2800 kg/m^3). The values of the Bouguer gravity field slowly increase towards positive values south of the SCB.

18.3.3 Magnetic Field Analysis

The SCB is characterized by a smoothed positive magnetic field intensity, which decreases to the south along with the common abrupt subsidence of the surface of the crustal basement and a sharp increase in sedimentary Mesozoic-Cenozoic complexes. In the western part of the SCB (area of the Baku Archipelago), there are many local positive and negative anomalies of small intensity (up to 100 nT) reflecting geological peculiarities of the sedimentary association structure (Fig. 18.45).

The geological nature of the regional magnetic maximum in the central part of the SCB is still being debated. According to Dzabayev (1969) and Khesin et al. (1996), the depth to the upper surface of the anomalous body is 17–20 km, which coincides with the depth to the crystalline basement found by deep seismic sounding (e.g., Abdullayev et al., 2015; Mamedov, 2008). Dzabayev (1969) connected this anomaly to the influence of the magmatic rock complex of basic consistency occurring in the body of the basement. This point of view was confirmed by the behavior of the upward continued magnetic field to a level of 4 and 10 km: parameters of maximum magnetic change weakly with the level of upward continuation. At the same time, a careful analysis of this anomaly indicates that it is composed of several different anomalies. It was

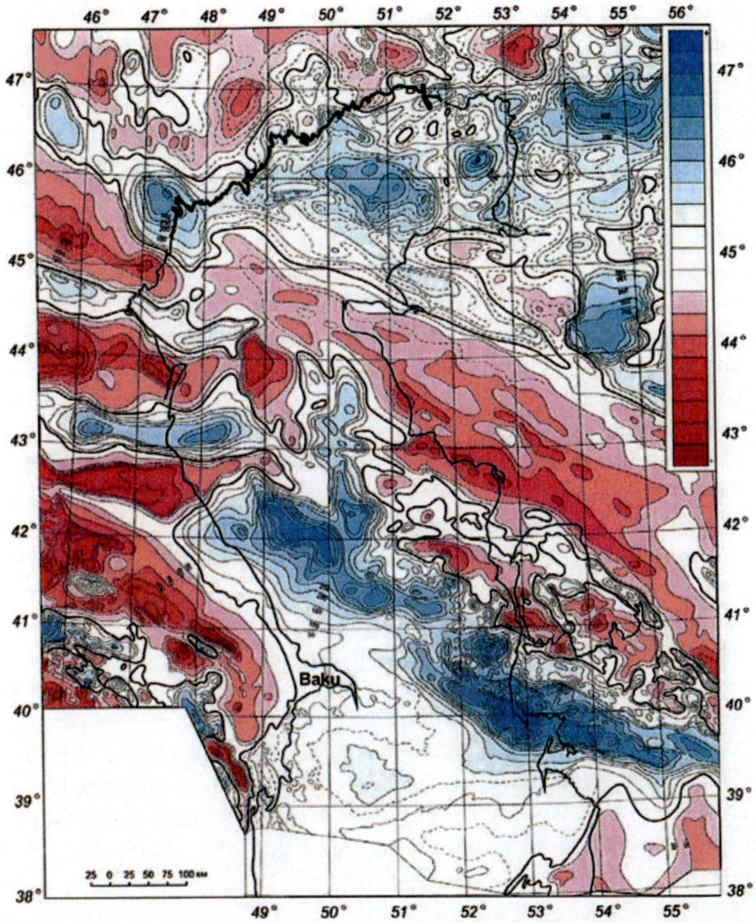

Fig. 18.45 Map of the anomalous magnetic field ΔT_a (after Glumov et al., 2004)

clearly detected from a combined interpretation of residual magnetic anomalies (0–10 km—the difference between the land observed magnetic field and analytically continued to the level of 10 km). It should be noted that the magnetic field maximum is localized in the northern area of a large gravity maximum corresponding to the western part of the SCB.

A similar interrelation of gravity and magnetic anomalies was studied in the Middle Kur Depression. It was explained by the physical-geological model when younger (Mesozoic) magmatic rocks were introduced into a weakened zone on the peripheral part of the uplift of the pre-Alpine basement (Eppelbaum & Khesin, 2012).

The Pliocene-Anthropogenic deposits in the SCB are the thickest. The paleomagnetic characteristics of these deposits on the eastern and western sides of the SCB are identical. This fact enables interpretation by applying the principles of directly magnetized deposits (Baku, a small interval of Absheron and the lower part of Akchagylian) and inversely magnetized deposits (Absheron, upper part of Akchagylian and productive red bed for Western Turkmenistan) [Ismailzadeh (1983) and his unpublished reports (1972–1984)]. The average magnetic susceptibility values usually range from 1000–2000 (up to 5000) $\cdot 10^{-6}$ SI in Eastern Azerbaijan and $1000 \cdot 10^{-6}$ SI in Western Turkmenistan.

Within the eastern part of the Caspian Sea and Gulf of Garabogaz-Gol is selected a central Caspian—Garabogaz-Gol zone of the alternating mosaic field divided from the Mangyshlak zone of maximums by the Kendyrly zone of the normal magnetic field.

Analysis of the anomalous magnetic field of the folded zones extending to the South Caspian in the west (Caucasus) and the east (Kopetdag) does not show a continuation of these zones far out to sea. The zone of magnetic minimums of the Greater Caucasus and located to the south Transcaucasian zone of the intensive alternating mosaic field by extending to the seacoast sharply changes their orientation to the south, in the direction of the Alborz folded construction (Glumov et al., 2004).

A magnetostratigraphic analysis of Plio-Pleistocene transgressions in the South Caspian Basin combined with paleo-environmental reconstructions enables the development of the following conclusions (Van Baak et al., 2013). Rock magnetic analyses combined with thermal demagnetization data indicate that the magnetic signal is carried dominantly by the iron oxide magnetite in the Productive Series of the Lokbatan and Xocashen sections. The marine deposits Akchagylian and Apsheronian of Lokbatan are characterized by the iron sulphide greigite, which is of (near-) primary origin. The most logical correlation of the magnetic polarity patterns to the Geomagnetic Polarity Time Scale dates to the Akchagylian transgression at ~ 3.2 Ma, a significant transgression during the Apsheronian at ~ 2.0 Ma, and the Bakunian transgression at 0.85–0.89 Ma (Fig. 18.46) (Van Baak et al., 2013). The conclusions are essential for the quantitative interpretation of magnetic anomalies and 3D magnetic field modeling in this region.

An effective potential field generalization has been realized in Fig. 18.47, where gravity-magnetic field intensities were ranged for the Caspian Sea and some adjacent areas.

The development of combined seismic-gravity-magnetic profiles crossing the SCB (with the thermal and paleomagnetic data attraction) will help to unmask the deep structure of this region [similarly to unmasking the deep structure of the Easternmost Mediterranean (Eppelbaum & Katz, 2015)].

18.3.4 Thermal Field Analysis

According to Mukhtarov's (2004) calculations, the sedimentation rate in the Jurassic in the SCB was 120–180 m/My if the maximum thickness of the sedimentary cover was about 30 km. It

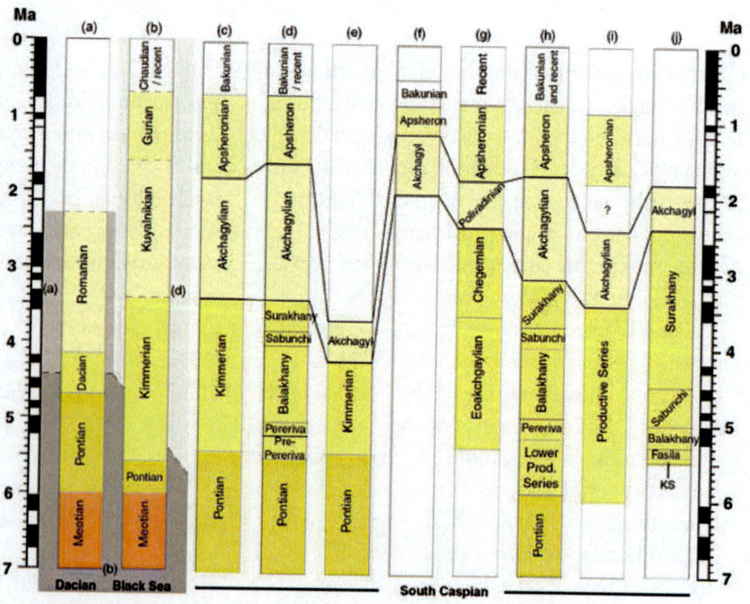

Fig. 18.46 Existing Caspian Basin and Eastern Paratethys timescales (Van Baak et al., 2013). Lines connect the upper and lower boundaries of the Akchagylian regional stage. Authors of the works (a)–(j) are given in Van Baak et al. (2013)

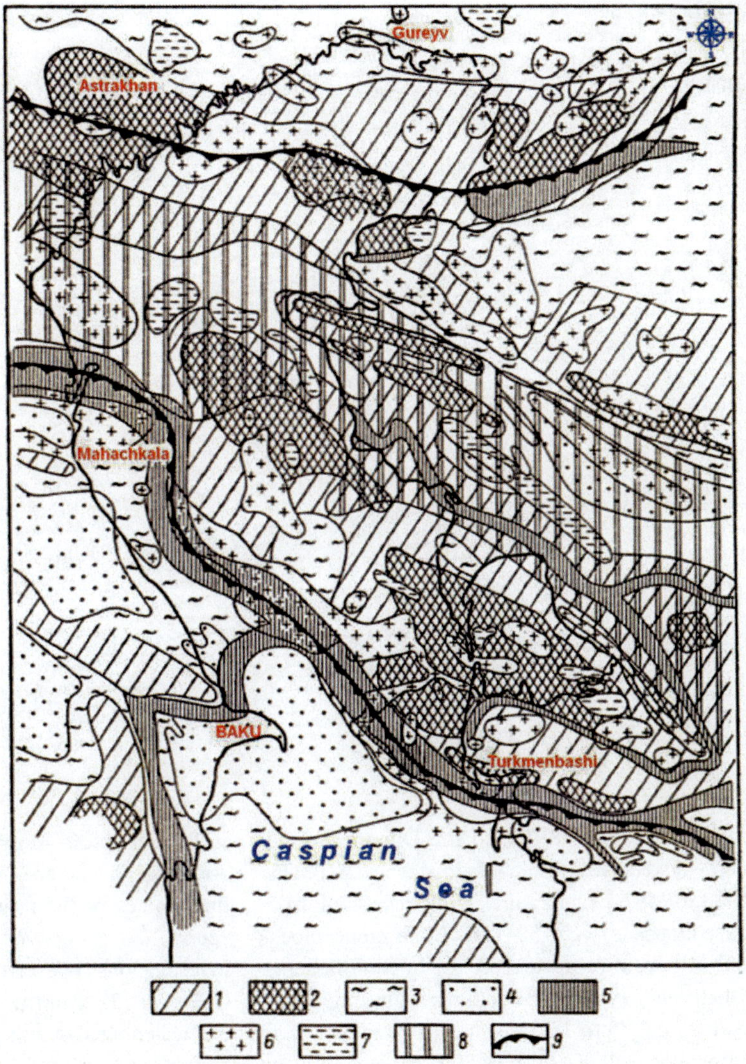

Fig. 18.47 Scheme of tectonic-structural elements of the Caspian Sea and adjacent areas derived from potential geophysical fields (after Malovitsky et al. (1977), with minor modifications). Gravity anomalies: (1) relatively positive, (2) the same, most intensive, (3) relatively negative, (4) the same, most intensive, (5) gravity steps; Magnetic anomalies: (6) maximums, (7) minimums, (8) zone of regional strip depression, (9) boundaries between the main different age tectonic elements

became lower in the Cretaceous and Paleogene, and in the Pliocene, it reached avalanche values —1.8 km/My. Results of modeling the thermal evolution of the basin with the account for non-stability of the heat field demonstrated that the temperature in the base of the sedimentary layer changed 400–500 °C (Mukhtarov and Adigezalov, 1999; Mukhtarov et al., 2003). Lately, opinions have appeared that the total thickness of the sedimentary cover in the SCB is up to 30 km. This may result in the sedimentation rate at the early stages of sedimentation being more than was calculated for the 20 km thickness of sediments. As a result, the share of deep heat flow in the sedimentary thickness will be much lower.

Levin and Viskovski (2000) proposed that in the Jurassic, the sedimentation rate in the SCB varied from 10–25 to 50 m/My. Temperatures in the base of the system were 150–200 °C. They increased in deeper blocks up to 300–450 °C. In the Cretaceous, the sedimentation rate was 2.5–10 m/My. Temperatures in the base were from

Table 18.2 Borehole maximum temperature data in SCB (after Mukhtarov, 2004)

Structure	No. of well	Depth, m	Temperature, °C
Baku Archipelago			
Shakh Deniz	4	6500	122
Bulla Deniz	46	5730	115
Bulla Deniz	38	6150	110
Bulla Deniz	42	5850	110
Sangachal Deniz	550	5770	113
Garasu	28	5650	112
Garasu	30	5683	106
Duvanny Deniz	39	4450	111
Absheron and the Absheron Archipelago			
Absheron Deniz	3	5000	110
Arzu	2	4708	105
Jenub	2	4710	102
Jenub	12	4127	100
Bakhar	19	5450	99
The South–West Caspian			
Gubkin uplift		3485	74.5
Barinov uplift		4420	91.5
Zhdanov uplift	24	3993	88
Lam uplift	1	4353	94

50–150 to 250–300 °C. In the Oligocene–Miocene system, the sedimentation rate was 0.025–0.4 km/My. Temperatures in the base of Miocene deposits were 50–100 °C; in some blocks only, they were up to 200 °C. In the Pliocene–Quaternary system, the sedimentation rate changed from 0.75 to 1.75 m/My. Temperatures were from 100–150 °C to 200–300 °C. It should be considered that in the SCB there, a lot of wells were drilled. While considering temperature data from the wells, one can be sure that the temperature regime of sedimentary layers uncovered by the wells is moderate enough (Table 18.2).

The sedimentation rate is one of the factors affecting the heat regime in sedimentary basins. With avalanche rates of sedimentation, intensive decay of deep heat flow occurs. For this reason, one can observe relatively low heat flow density values throughout the Caspian region. It should be noted that values of the heat flows determined by a well method (20–40 mW/m^2 in wells of the Baku Archipelago and the Absheron Sill) were lower than values of the heat flows determined by sea sounds (30–50 mW/m^2 and more). With special investigations, one can judge the reason for this difference. To construct the map of heat flows in the water area of the Caspian Sea (Fig. 18.48), there were used results of the determination of the heat flow by marine thermal probes (Catalogue of data on heat flow in the territory of the USSR, 1973; Lyubimova et al., 1976; Tomara, 1979; Aliyev et al., 1979; Lebedev & Tomara, 1981). Moreover, there were used data obtained by the well method in the shelf zone and the onshore territory (Aliyev, 1988; Ashirov, 1984; Catalogue of data on heat flow in the territory of the USSR, 1973; Kashkay & Aliyev, 1974).

The main features of the heat flow distribution in the Caspian Sea agree with the tectonic peculiarities of its deep structure (Eppelbaum

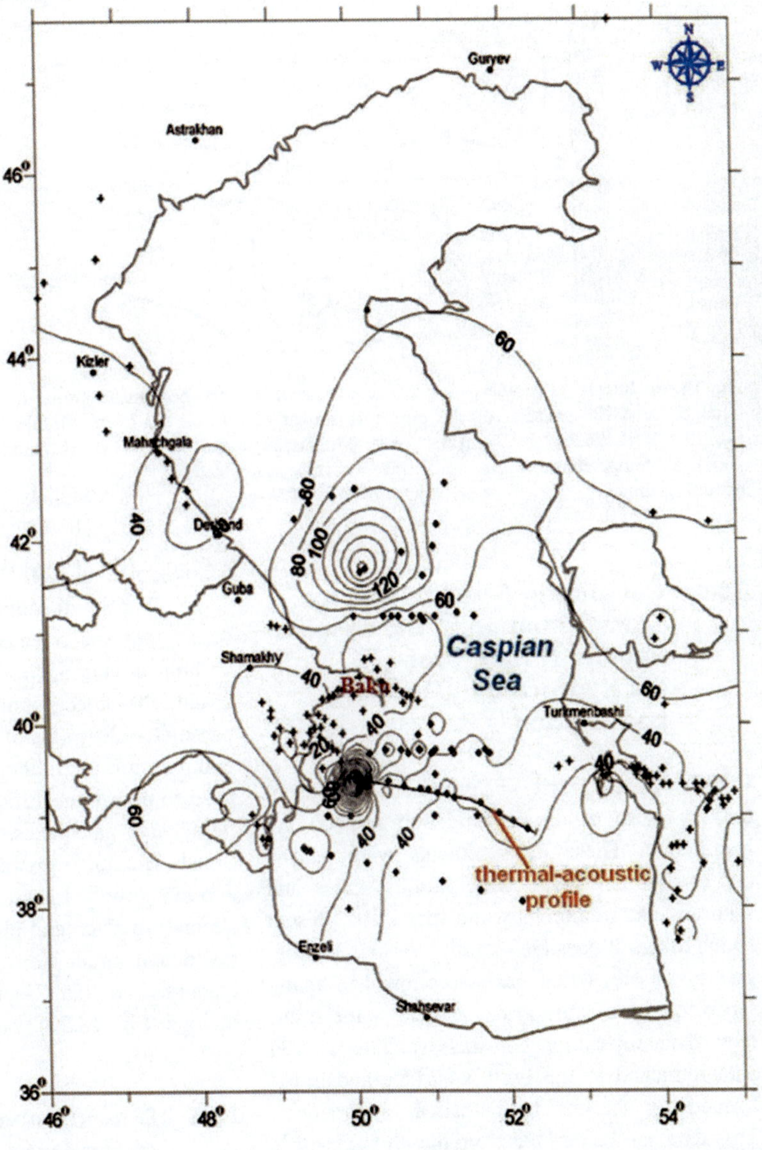

Fig. 18.48 Heat flow distribution map in the Caspian Sea. The crosses show the points of determination of the heat flow density in wells, rhombs—points of determination of the heat flow density by sea thermal soundings (Mukhtarov, 2004)

and Pilchin, 2006). For instance, increased heat flow values are conditioned by the impact of such tectonic structures as faults and mud volcanoes. One of the stations west of the SCB recorded an abnormally high value (480 mW/m^2) of heat flow (Fig. 18.49). The available geophysical data do not explain this anomaly as the impact of the intrusive body (Lebedev & Tomara, 1981). That is due to the evacuation of fluids along the active dislocations (faults or vents of mud volcanoes).

Measuring the heat flow is of a point character, and main geologic factors affecting the density may reflect some local areal peculiarities. This causes the necessity of performing special investigations in areas that cover local zones of active faults, mud volcanoes, and remote calm areas. The distribution of mud volcanoes and heat flows in the SCB proves the abovementioned. The mud volcanoes are mainly spread in the contoured isolines of the heat flow 40 mW/m^2.

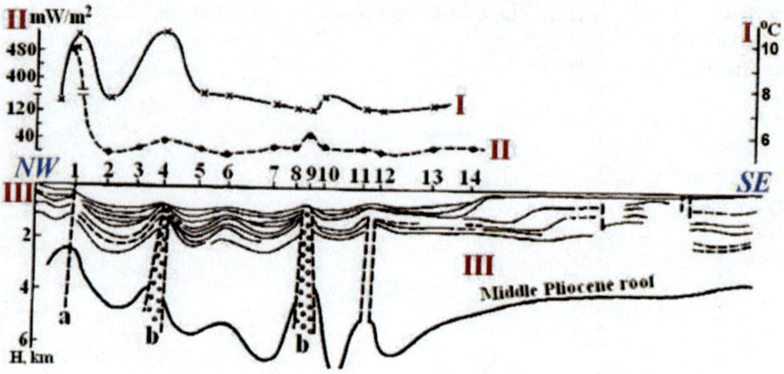

Fig. 18.49 Integrated geothermal-seismoacoustic profile through the SCB (position of the profile is shown in Fig. 19.48) (Lebedev & Tomara, 1981; Mukhtarov, 2004). I—temperature of the upper layer of the sea bottom sediments, °C; II—heat flow, mW/m^2; III—seismoacoustic profile of the sedimentary series as deep as the top of the Middle Pliocene; (a) zone of faults near the west coast; (b) mud volcano; 1–14—numbers of stations

18.4 Computing Satellite Gravity Transformation in the Caspian Region: A New Tool for Hydrocarbon Deposit Localization

Earth's surface gravity measurements are essential (as close to the investigation targets) but insufficient. These measurements were carried out at different years, with various scales and accuracy and numerous white spots. The present epoch makes it possible to utilize various satellite gravity missions that have accomplished many repetitions, the same grids, and the same accuracy. This subsection considers satellite-derived data retracked to the Earth's surface and transformed by various mathematical apparatuses. This data can be derived from the global Earth's satellite data, mainly from the GRACE and GRACE-FO missions. The gravity gradient tensor Γ (the Marussi tensor) is a tensor of the second derivatives of the disturbing potential T of the gravity field model. This tensor was considered the centerpiece of traditional differential geodesy. It is analogous to the tidal deformation from geodesy and geophysics; one can imagine the direction of such a deformation due to "erosion" brought about solely by gravity (Klokočníket al., 2014). The strike angles usually show chaotic directions. We aimed to detect where they are oriented dominantly in one prevailing direction (linearly or creating a halo around the object). Another applied gravitational parameter allows us to obtain the distribution of compressions and dilatations. The values may be used to detect subsurface structures, i.e., oil–gas fields and paleolakes. Integrating the conventional Bouguer gravity maps with satellite-derived gravity transformations will enable the generation of crucial physical-geodynamical and geological conclusions. One example has been computed for the Caspian Sea and surrounding regions (Fig. 18.50).

18.5 Development of the Map of the Deep Structure of Azerbaijan with Adjacent Regions

Accumulated experience shows the importance of performing independent zonation using initial field and regional anomaly maps. Comparing the results in a unified scheme makes it possible to determine common features and differences in deep structure elements and the upper portion of the section and reveal the elements

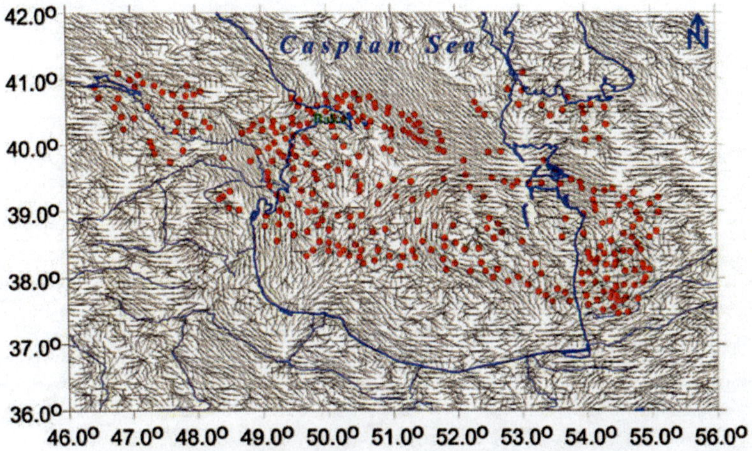

Fig. 18.50 The "combed" strike angles (black dashes) in the South Caspian Basin accompanied by the hydrocarbon deposit location (red dots) (after Kadirov et al., 2023)

(primarily, disjunctive dislocations and blocks) manifesting themselves at the different structural stages.

The transforms reflecting the degree of field heterogeneity, such as anomalous coefficients, dispersion, entropy, etc., are the most useful for singling out field areas in open regions. In different areas, these parameters can take on different geological meanings. They are also appropriate for solving more specific problems in the zonation of separate portions by the degree of heterogeneity of their lithological complexes. Here, it makes sense to compute the specific sinuosity of the isolines.

The map of gravity-magnetic anomalies (Fig. 18.51) shows both these minima since the problem of their nature can be solved at the stage of anomaly examination (Alexeyev et al., 1988; Eppelbaum and Khesin, 2012).

A description of gravity anomalies was based on: (1) difference fields ("ring" anomalies) $\Delta g_{B(8-20)}$ and $\Delta g_{B(4-10)}$ computed as the difference between upward continued fields in the Bouguer reduction (terrain correction was computed with a radius of 200 km and the density of the intermediate layer was defined as 2.67 g/cm^3) at heights of 8 and 20, 4 and 10 km, respectively, (2) the residual field $\Delta g_{B(0-4)}$ (local anomalies were calculated as the difference between the Earth's surface Bouguer gravity and this field continued upward to a height of 4 km), (3) horizontal gravity gradient field at a height of 2 km ($\Delta g_{x(2)}$).

These features were analyzed based on the following descriptors.

I. The residual field $\Delta g_{B(0-4)}$ in the mountain-folded regions reflects the influence of near-surface Alpine structures composed of volcanogenic and volcanogenic-sedimentary associations and intrusions (both outcropping to the earth's surface and near-surface), and in the Kur Depression as regards the distribution of dense inhomogeneties of the upper part of the sedimentary floor.
II. The peculiarities of difference field ("ring" anomalies) $\Delta g_{B(8-20)}$ were described in Eppelbaum and Khesin (2012).
III. Significant anomalies of $\Delta g_{x(2)}$ should be taken into consideration by delineation and examination of regional gravity anomalies.

These descriptors were determined based on analysis of petrophysical parameters of the medium as well as data on density of metamorphic foundation outcropped in Azerbaijan and surrounding territories.

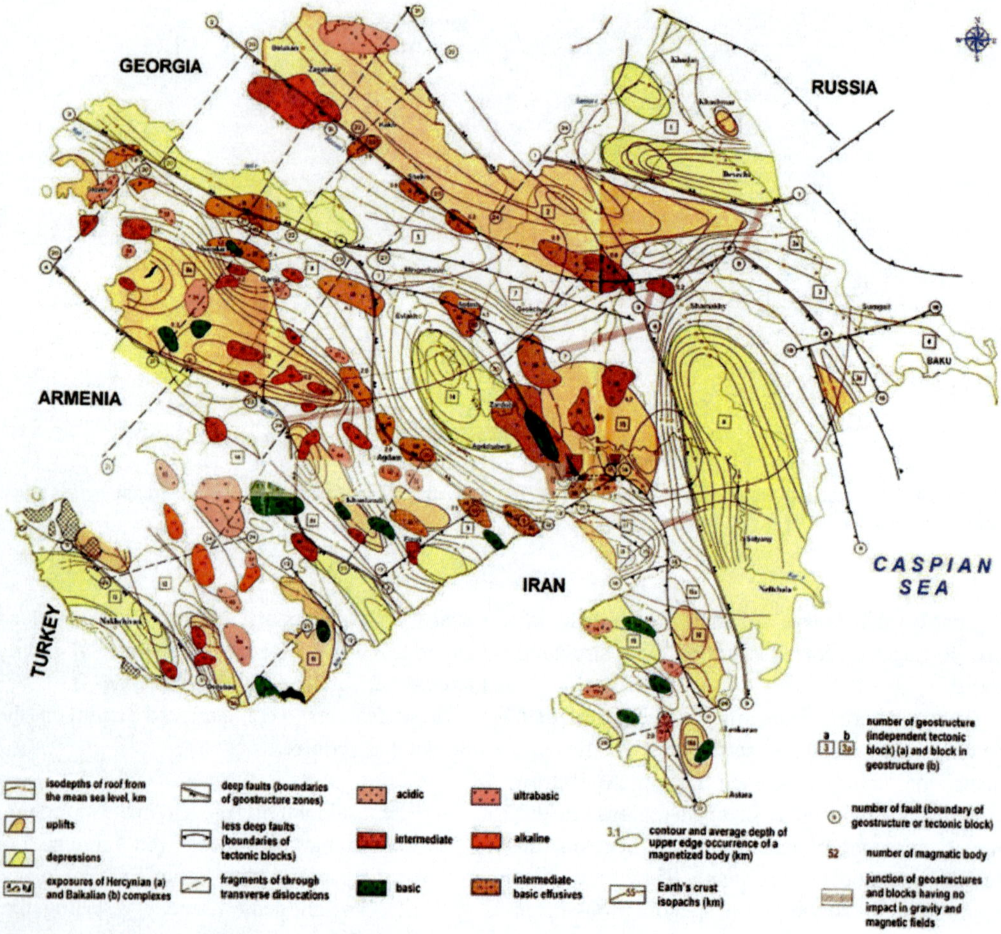

Fig. 18.51 Map of the deep structure of Azerbaijan with its adjacent regions derived from combined analysis of gravity and magnetic data [after Eppelbaum and Khesin (2012) with minor modifications; geology after Azizbekov et al. (1972)]

References

Abbasov, A. B. (2002). Yevlakh-Agjabedi and South Kur troughs paleogeography based on seismostratigraphic studies. *Geophysics News in Azerbaijan, 1–2*, 11–15. (in Russian).

Abdullayev, N. R., Huseynova, S., Javadova, A., Kadirov, F., Maharamov, B., & Mukhtarov, A. S. (2022). An interpretation of crustal structure and hydrocarbon evolution of the South Caspian and Kura Basins, Azerbaijan. *ResearchSquare reprint*, 1–15.

Abdullayev, N. R., Kadirov, F., & Guliyev, I. S. (2015). Subsidence history and basin-fill evolution in the South Caspian Basin from geophysical mapping, flexural backstripping, forward lithospheric modelling and gravity modelling. In: M.-F. Brunet, T. McCann, & E.R. Sobel (Eds.), *Geological evolution of Central Asian Basins and the Western Tien Shan Range. Geological Society, London, Special Publications* (vol. 427, pp. 1–22).

Abers, G. A., & Roecker, S. (1991). Structure of an arc-continent collision: Earthquake relocation and inversion for upper mantle P and S wave velocities beneath Papua New Guinea. *Journal of Geophysical Research, 96*(B4), 6379–6401.

Adamova, A. A., & Sabitova, T. M. (2004). 3D velocity model of Tyan-Shan earth's crust. *Izvestiya, Academy of Sciences, Russian, Physics of the Earth, 5*, 58–67.

Aksenovich, G. I., Aronov, L. E., Gagelgants, A. A., Galperin, E. I., Kayonchkovskiy, M. A., Kosminskaya, I. P., & Krakshina, P. M. (1962). *Deep seismic sounding in the central part of the Caspian Sea*. Academy of Sciences, USSR, Moscow (in Russian).

Alexeyev, V. V., Gadjiev, T. G., Karkoshkin, A. I., Khesin, B. E., & Ismailzadeh, T. A. (1988). *Gravity and magnetic anomalies in Azerbaijan and their geological interpretation*. Map Printing Factory (in Russian).

Aliyev, A. S. (1985). In T. A. Ismailzadeh, & T. G. Gadjiev (Eds.), *Map of heat flows in the depression zones of Azerbaijan. Scale 1: 500,000*. Ministry of Geology of the USSR.

Aliyev, G. A., Akhmedbeyli, F. S., Ismailzade, A. D., Kangarli, T. H., & Rustamov, M. I. (2005). In V. E. Khain, & A.A. Alizadeh (Eds.), *Geology of Azerbaijan* (vol. IV). Tectonics. Nafta-Press (in Russian).

Aliyev, S. A. (1988) Geothermal fields of depression zones in the SCB and their relationship with oil-gas potential. *D.Sci. Thesis*, Institute of Geology, Baku (in Russian).

Aliyev, S. A., Aliyev, C. S., Askerhanova, H. Q., Balakishibeyli, S. A., Qasanov, A. O., Qasanov, A. B., Zolotovichaya, T. A., Ismailzade A. D., Isayeva, M. I., Kadirov, F. A., Mukhtarov, A. S., Rzayev, A. Q., Cafarov, I. B., & Khalafly, A. A. (2005). In A. A. Alizadeh (Ed.), *Geology of Azerbaijan: Physics of the Earth* (vol. V) (in Russian).

Aliyev, S. A., Ashirov, T., & Sudakov, N. P. (1979). New data on the heat flow through the Caspian Sea floor. *Izvestiya, Academy of Sciences, Turkmenia, SSR, Series: Physics, Technology and Geological Science, 2*, 124–126. (in Russian).

Alizadeh, A. A. (Ed.). (2012). *Modern problems of geology and geophysics of the Caucasus*. Nafta-Press.

Alizadeh, A. A., Guliyev, I. S., Kadirov, F. A., & Eppelbaum, L. V. (2017). *Geosciences in Azerbaijan. Volume II: Economic minerals and applied geophysics* (340 p.). Springer.

Alizadeh, A. A., Khain, V. E., & Ismailzadeh A. D. (Eds.) (2000). *Saatly Superdeep Well. Studies of Deep Structure of the Kur Intermountain Depression Based on Materials of Drilling of Saatly Superdeep Well SG-1*. Nafta-Press (in Russian).

Allen, M. B., Vincent, S. J., Alsop, G. I., Ismailzadeh, A., & Flecker, R. (2003). Late Cenozoic deformation in the South Caspian region: Effects of a rigid basement block within a collision zone. *Tectonophysics, 366*, 223–239.

Ambraseys, N. N., & Melville, C. P. (1982). *A history of Persian earthquakes*. Cambridge University Press.

Artyushkov, E. V. (1993). *Physical tectonics*. Nauka.

Artyushkov, E. V. (2007). Formation of ultra-deep basin in the South Caspian because of phase transition in the continental crust. *Geology and Geophysics, 48*(12), 1289–1307. (in Russian).

Ashirov, T. (1984). *Geothermal field of Turkmenia*. Nauka.

Azizbekov, S. A., Alizadeh, K. A., Shikalibeyli, E. S., & Gadjiev, T. G. (Eds.) (1972). *Geology of the USSR, Azerbaijan* (vol. XLVII). Nedra (in Russian).

Babayev, D. K., & Gadjiyev, A. N. (2006). *Deep structure and prospects oil-and-gas content basin of Caspian Sea*. Geological Institute, Azerbaijan, Academy of Sciences, Baku (in Russian).

Balakishibeyli, S. A., Salekhli, T. M., Gasanov, A. B., & Guliyev, R. D. (1996). Results of petrophysical research. In: *Geophysical investigations in Azerbaijan* (pp. 352–381). Nafta-Press (in Russian).

Baranova, E. P., Kosminskaya, I. P., & Pavlenkova, N. I. (1991). A reinterpretation of south Caspian DSS data. *Geophysical Journal, 10*, 666–677.

Barber, C. B., Dobkin, D. P., & Huhdanpaa, H. T. (1996). The Quickhull algorithm for Convex Hulls. *ACM Transactions on Mathematical Software, 22*(4), 469–483.

Barton, P. J. (1986). The relationship between seismic velocity and density in the continental crust—A useful constraint? *Geophysical Journal International, 87*, 195–208.

Brunet, M.-F., Korotayev, M. V., Ershov, A. V., & Nikishin, A. M. (2003). The South Caspian Basin: A review of its evolution from subsidence modelling. *Sedimentary Geology, 156*, 119–148.

Bulakh, E. G. (2000). Review of publication on minimization methods used in inverse gravimetry and magnetometry problems. *Geophysical Journal, 19*, 709–733.

Bulakh, Y. G., Markova, M. N., Timoshenko, V. I., & Boiko, P. D. (1984). *Mathematical support of automated system of gravity anomaly interpretation*. Naukova Dumka.

Catalogue of data on heat flow in the territory of the USSR. (1973). Nauka (in Russian).

Dehghani, G. A., & Makris, J. (1983). The gravity field and crustal structure of Iran. *Geodynamic Project (Geotraverse) in Iran* (pp. 50–68). Final Report. 51. Geological Survey of Iran.

Dzabayev, A. A. (1969). *Principles of the searching and study of oil-and-gas bearing structures by aeromagnetic method (South Caspian Basin)*. Statistika (in Russian).

Egan, S. S., Mosar, J., Brunet, M. F., & Kangarli, T. (2009). Subsidence and uplift mechanisms within the South Caspian Basin: Insights from the onshore and offshore Azerbaijan region. *Geological Society, London, Special Publications, 312*, 219–240.

Eppelbaum, L. V. (2014a). Estimating informational content in geophysical observations on example of searching economic minerals in Azerbaijan. *Izvestiya (Proceedings). Academy of Sciences, Azerbaijan, Reports Series: Earth Sciences, 3–4*, 31–40.

Eppelbaum, L. V. (2014b). Four color theorem and applied geophysics. *Applied Mathematics, 5*, 358–366.

Eppelbaum, L. V. (2015). Comparison of 3D integrated geophysical modeling in the south Caucasian and eastern Mediterranean Segments of the Alpine-Himalayan tectonic belt. *Izvestiya (Proceedings), Academy of Sciences, Azerbaijan, Series: Earth Sciences, 3*, 25–42.

Eppelbaum, L. V. (2017). Satellite gravimetry ('Big Data'): A powerful tool for regional tectonic examination and reconstructions. In B. Veress, & J. Szigethy

(Eds.), *Horizons in earth science research* (vol. 17, pp. 54–86). Nova Science Publishers.

Eppelbaum, L. V. (2019). *Geophysical potential fields: Geological and environmental applications* (467 p.). Elsevier.

Eppelbaum, L. V., Ben-Avraham, Z., Katz, Y., Cloetingh, S., & Kaban, M. (2021). Giant quasi-ring mantle structure in the African-Arabian junction: Results derived from the geological-geophysical data integration. *Geotectonics, 55*(1), 67–93.

Eppelbaum, L. V., & Katz, Y. (2015). Eastern Mediterranean: Combined geological-geophysical zonation and paleogeodynamics of the Mesozoic and Cenozoic structural-sedimentation stages. *Marine and Petroleum Geology, 65*, 198–216.

Eppelbaum, L. V., & Khesin, B. E. (2004). Advanced 3-D modelling of gravity field unmasks reserves of a pyrite-polymetallic deposit: A case study from the Greater Caucasus. *First Break, 22*(11), 53–56.

Eppelbaum, L. V., & Khesin, B. E. (2011). Development of 3-D gravity-magnetic models of Earth's crust of Azerbaijan and adjacent areas: A generalized review. *Positioning, 2*(2), 84–102.

Eppelbaum, L. V., & Khesin, B. E. (2012). *Geophysical studies in the Caucasus*. Springer.

Eppelbaum, L. V., Kutasov, I. M., & Pilchin, A. N. (2014). *Applied geothermics*. Springer.

Eppelbaum, L. V., & Pilchin, A. N. (2006). Methodology of Curie discontinuity map development for regions with low thermal characteristics: An example from Israel. *Earth and Planetary Sciences Letters, 243*(3–4), 536–551.

Firbas, P. (1987). Tomography from seismic profiles. In G. Nolet (Ed.), *Seismic tomography, with applications in global seismology and exploration geophysics* (pp. 189–202). Springer.

Gadjiev, R. M. (1965). *Deep geological structure of Azerbaijan.* Azgosizdat (in Russian).

Gadjiev, T. G., Nechayev, Yu. V., Potapova, E. I., & Sattarova, V. M. (1989). *The map of deep structure of the Caucasus according to cosmic data.* Map Printing Factory of the Ministry of Geology of the USSR.

Gasanov, A. G. (2001). *Deep structure and seismicity of Azerbaijan in Relation to Hydrocarbons Prognosis.* Elm (in Russian).

Gasi, S., & Hachay, O. (2017). *Oil and gas exploration: Methods and application.* Wiley & American Geophysical Union.

Glumov, I. F., Malovitsky, Y. P., Novikov, A. A., & Senin, B. V. (2004). *Regional geology and oil and gas bearing of the Caspian Sea.* Nedra (in Russian).

Goldshmidt, V. I., Kuzmin, Y. I., & Shabaldin, V. N. (1981). An automated system of gravity data interpretation. In *Theory and methods for interpretation of gravity and magnetic fields* (pp. 58–70). Naukova Dumka (in Russian).

Gorshkov, A. I., & Niauri, G. A. (1984). Reflection of fault tectonics of the Caucasus in zones of horizontal gravity anomaly gradients. *Izvestiya, Academy of Sciences, USSR, Physics of the Earth, 20*(9), 699–704 (in Russian).

Granath, J. W., Soofi, K. A., Baganz, O. W., & Bagirov, E. (2000). Gravity modelling and its implications to the tectonics of the south Caspian basin. In *AAPG inaugural regional international conference* (pp. 46–49). Extended Abstracts.

Gravity map of the USSR. (1990). *Scale 1:2,500,000.* Ministry of Geology.

Green, T., Abdullayev, N., Hossack, J., Riley, G., & Roberts, A. (2009). Sedimentation and subsidence in the South Caspian Basin, Azerbaijan. In: M. F. Brunet, M. Wilmsen, J. W. Granath (Eds.), *South Caspian to Central Iran Basins* (vol. 312, pp. 2241–2260). The Geological Society of London, Special Publication.

Gugunava, G. E. (1988). The structure of Caucasus inferred from the analysis of some geophysical fields. *D. Sci. Thesis.* Institute of Geophysics of the Georgian Academy of Sciences, Institute of Geology of the Azerbaijan Academy of Sciences (in Russian).

Guliyev, I., & Panachi, B. (2004). Geodynamics of the deep sedimentary basin the Caspian Sea region: Paragenetic correlation of the seismicity and mud volcanism. *Geo-Marine Letters, 24*(3), 169–176.

Guliyev, I. S., Mamedov, P. Z., Feyzullayev, A. A., Huseynov, D. A., Kadirov, F. A., Aliyeva, E. H. M., Tagiyev, M. F. (2003). In I. S. Guliyev (Ed.), *Hydrocarbon systems of the South Caspian Basin.* Nafta-Press.

Gupta, H. K. (2021). *Encyclopedia of solid geophysics.* Springer Cham.

Ismailzadeh, A. D., Alizadeh, A. A., Guliyev, I. S., Babayev, D. K., & Hajiyev, A. H. (2004). On the possible mantle character of hydrocarbon fluids in the South Caspian Depression. *Izvestiya, Academy of Sciences, Azerbaijan, Series: Earth Sciences, 3*, 6–11 (in Russian).

Ismailzadeh, A. D., Mustafayev, G. V., & Rustamov, M. I. (2005). In A. A. Alizadeh (Ed.), *Geology of Azerbaijan, Vol. III, Magmatism.* Nafta-Press (in Russian).

Ismailzadeh, T. A. (1983). Paleomagnetic investigations of Meso-Cenozoic of Azerbaijan. *D.Sci. Thesis*, Institute of the Physics of the Earth, Moscow (in Russian).

Jackson, J., Priestly, K., Allen, M., & Berberian, M. (2002). Active tectonics of the South Caspian Basin. *Geophysical Journal International, 148*, 214–245.

Jafarov, H. D., Kerimov, A. A., & Solomonov, B. I. (1967). Deep structure of the Ajinohur region by data of exploration geophysics. *Izvestiya VUZov, 9*, 11–14. (in Russian).

Kaban, M. K. (2002). A gravity model of the north Eurasia crust and upper mantle: 2. The Alpine-Mediterranean fold belt and adjacent structures of the southern former USSR. *Russian Journal of Earth Sciences, 4*(1), 19–33.

Kadirov, F. A. (2000a). *Gravity field and models of deep structure of Azerbaijan.* Nafta-Press.

Kadirov, F. A. (2000b). Application of the Hartley transform for interpretation of gravity anomalies in the Shamakhy-Gobustan and Absheron oil- and gas-bearing regions Azerbaijan. *Journal of Applied Geophysics, 45*, 49–61.

Kadirov, F. A. (2006). Nature of gravitational anomaly in the South Caspian Basin. *Transaction of the EAGE 68th conference* (pp. 1–4), Vienna, P325.

Kadirov, F. A., Floyd, M., Reilinger, R., Alizadeh, A. A., Guliyev, I. S., Mammadov, S. G., & Safarov, R. T. (2015). Active geodynamics of the Caucasus Region: Implications For earthquake hazards in Azerbaijan. *Izvestiya (Proceedings), Academy of Sciences, Azerbaijan, Series: Earth Sciences, 3*, 3–17.

Kadirov, F. A., & Gadirov, A. H. (2014). A gravity model of the deep structure of South Caspian Basin along submeridional profile Alborz-Absheron Sill. *Global and Planetary Change, 114*, 66–74.

Kadirov, F. A., Klokočník, J., Eppelbaum, L. V., Kostelecký, J., & Bezděk, A. (2023). Integration of the Bouguer gravity data with satellite gravity transformation integration in the Caspian region: An introduction. *ANAS Transactions, Earth Sciences, 1*, 11–16.

Kadirov, F. A., Mammadov, S. A., Reilinger, R., & McClusky, S. (2008). Some new data on modern tectonic deformation and active faulting in Azerbaijan (according to global positioning system measurements). *Izvestiya (Proceedings) Academy of Sciences, Azerbaijan, Series: Earth Sciences, 1*, 82–88.

Kashkay, M. A., & Aliyev, S. A. (1974). Heat flow in the Kur depression. In *Deep heat flow in the European Part of the USSR* (pp. 95–108). Naukova Dumka (in Russian).

Kazymova, S. E., & Kazymov, I. E. (2010). 1D velocity model of Kur depression from the data of telemetric stations of Azerbaijan. In *Science 21 century, "Volgapromstroibezopasnost"*, Saratov, No. 2 (in Russian).

Kerimov, K. M., Pilchin, A. N., Gadzhiev, T. G., & Buachidze, G. Y. (1989). *Geothermal map of the Caucasus, Scale* 1:1,000,000 Baku, Cartographic Plant No. 11 (in Russian).

Khain, V. E. (2000). *Tectonics of continents and oceans*. Nauchnyi Mir (in Russian).

Khain, V. E. (2005). The problem of origin and age of South Caspian Basin and its probable solutions. *Geotectonics, 1*, 40–44. (in Russian).

Khain, V. E., & Ryabukhin, A. G. (2002). Russian geology and the plate tectonics revolution. *Geological Society, London, Special Publications, 192*, 185–198.

Khalafly, A. A. (2006). *Paleomagnetism and some problems of shear deformations of the Lesser Caucasus*. Takhsil (in Russian).

Khalilov, E. N., Mekhtiyev, Sh. F., & Khain, V. E. (1987). On some geophysical data confirming the collision origin of the Greater Caucasus. *Geotectonics*, 2, 54–60. (in Russian).

Khesin, B. E., Alexeyev, V. V., & Eppelbaum, L. V. (1993). 3-D modelling of gravity and magnetic fields as a final stage of application of effective interpretation system of geophysical data under difficult geological conditions. *Geoinformatics, 4*(3), 177–188.

Khesin, B. E., Alexeyev, V. V., & Eppelbaum, L. V. (1996). *Interpretation of geophysical fields in complicated environments*. Ser: Modern Approaches in Geophysics, Kluwer Academic Publications (Springer).

Khesin, B. E., & Eppelbaum, L. V. (2007). Development of 3-D gravity/magnetic models of Earth's crust in complicated regions of Azerbaijan. In *Proceedings of the 69th EAGE conference* (5 p.), P343, London, Great Britain.

Klokočník, J., Kostelecký, J., Eppelbaum, L., & Bezděk, A. (2014). Gravity disturbances, the Marussi tensor, invariants and other functions of the geopotential represented by EGM 2008. *Journal of Earth Science Research, 2*(3), 88–101.

Knapp, C. C., Knapp, J. H., & Connor, J. A. (2004). Crustal-scale structure of the South Caspian Basin revealed by deep seismic reflection profiling. *Marine and Petroleum Geology, 21*, 1073–1081.

Kopf, A., Deyhle, A., Lavrushin, V. Y., Polyak, B. G., Gieskes, J. M., Buachidze, G. I., Wallmann, K., & Eisenhauer, A. (2003). Isotopic evidence (He, B, C) for deep fluid and mud mobilization from mud volcanoes in the Caucasus continental collision zone. *International Journal of Earth Sciences, 92*, 407–425.

Koronovsky, N. V. (2000). Seismic tomography. *Earth Sciences (Soros' Russian Journal*), *6*(11), 63–68.

Koulakov, I., Zabelina, I., Amanatashvili, I., & Meshkina, V. (2012). The nature of orogenesis and volcanism in the Caucasus region based on results of regional tomography. *Solid Earth, 3*, 327–337.

Krasnopevtseva, G. V. (1984). *Deep structure of the Caucasian seismoactive region*. Nauka.

Krasnopevtseva, G. V., Matushkin, B. A., & Shevchenko, V. I. (1970). Novel interpretation of the DSS data along the profile Stepnoe—Bakuriani in the Caucasus. *Soviet Geology, 8*, 113–120.

Lebedev, L. I., & Tomara, G. A. (1981). Some peculiarities of distribution of heat flow in the South Caspian. In: *Geothermometers and paleotemperature gradients* (pp. 156–161). Nauka (in Russian).

Leonov, Y. G. (Ed.). (2008). *The Greater Caucasus in the Alpine Epoch*. Geos.

Levin, L. E., & Viskovski, Y. A. (2000). The SCB: Evolution and thermal regime of oil gas systems. In *Proceedings of the III Azerbaijan international geophysics conference* (p. 239). Baku.

Lyubimova, E. A. (1968). Thermal history of the Earth. In: *The Earth's Crust and Upper Mantle* (vol. 13, pp. 63–77). American Geophysical Union, Geophysics Monograph Series.

Lyubimova, E. A., Nikitina, V. N., & Tomara, G. A. (1976). *Thermal fields of the inner and marginal seas in the USSR*. Nauka.

Malovitsky, Y. P., Kogan, L. I., Mistrukov, Y. M., et al. (1977). *Marine geophysical investigations*. Nedra (in Russian).

Mamedov, A. V. (1984). Tectonics of quaternary deposits of Azerbaijan. In: *Essays of geology of Azerbaijan* (pp. 256–270). Academy of Sciences, Azerbaijan (in Russian).

Mamedov, P. Z. (1991). Paleo-deltaic complexes in the north of the South Caspian depression. *Petroleum Geology, 225*, 9–10.

Mamedov, P. Z. (1992). Seismostratigraphical investigations of geological structure of sedimentary cover of South Caspian superdepression and perspectives of oil–gas productivity. *D.Sci. Thesis*, National Academy of Sciences, Baku, Azerbaijan (in Russian).

Mamedov, P. Z. (2004). Genesis and seismic stratigraphic model of the South Caspian Megabasin architecture. In: A. A. Alizadeh (Ed.), *South Caspian Basin: Geology, geophysics, oil and gas content* (pp. 150–164). Nafta-Press.

Mamedov, P. Z. (2008). The subsidence evolution of the South Caspian Basin. In *Proceedeedings of the EAGE conference "Petroleum Geology & Hydrocarbon Potential of Caspian and Black Sea Regions"*, 6–8 October 2008, Baku, Azerbaijan.

Mamedov, P. Z. (2009). Sedimentation models in the SCMB and their seismostratigraphic characteristics. *Stratigraphy and Sedimentology of Oil-Gas Basins, 3*, 70–85.

Mamedov, P. Z., Guliyev, Q. Q., Kadirov, F. A., Shykhalyev, Y. A., Guliyev, I. S., Aliyeva, E. H., Feyzullayev, A. A., Dadashov, F. Q., Kheirov, M. B., & Tagiyev, M. F. (2008). In A. A. Alizadeh (Ed.), *Geology of Azerbaijan, vol. VII: Oil and gas*. Nafta-Press (in Russian).

Mangino, S., & Priestley, K. (1998). The crustal structure of the southern Caspian region. *Geophysical Journal International, 133*, 630–648.

Mekhtiyev, S. F., Mirzadjanzade, A. H., & Aliyev, S. A. (1971). *Geothermal studies of oil and gas fields*. Nedra.

Mekhtiyev, S. F., Mirzadjanzade, A. H., Aliyev, S. A., et al. (1960). *Thermal regime of oil and gas fields*. Azerneftneshr (in Russian).

Mosar, J., Kangarli, T., Bochud, M., Glasmacher, U. A., Rast, A., Brunet, M.-F., & Sosson, M. (2010). Cenozoic-recent tectonics and uplift in the Greater Caucasus: A perspective from Azerbaijan. *Geological Society, London, Special Publications, 340*, 261–280.

Mukhtarov, A. S. (2004). Heat flow distribution and some aspects of thermal field formation in the Caspian region. In: A. A. Alizadeh (Ed.), *South Caspian Basin: Geology, geophysics, oil and gas content* (pp. 165–172). Nafta-Press.

Mukhtarov, A. S. (2011). Structure of thermal flow of sedimentary complex of the Southern Caspian Basin. *D.Sci Thesis*, Institute of Geology, Baku (in Russian).

Mukhtarov, A. S. (2012). Models of distribution of temperatures in sedimentary section of the South Caspian Basin. In: A. A. Alizadeh (Ed.), *Modern problems of geology and geophysics of the Caucasus* (pp. 214–230). Nafta-Press.

Mukhtarov, A. S., Tagiyev, M. F., & Imamverdiyev, R. A. (2003). Models of oil-gas generation and prediction of the phase state of hydrocarbons in the Baku Archipelago. *Izvestiya, Academy of Sciences Azerbaijan, Series: Earth Sciences, 2*, 17–25. (in Russian).

Pavlenkova, G. A. (2012). Structure of the Caucasus' earth crust along the deep seismic sounding profiles Stepnoe-Bakuriani and Volgograd-Nakhchivan (results of initial data re-interpretation). *Izvestiya, Physics of the Earth, 5*, 16–25.

Pilchin, A. N., & Eppelbaum, L. V. (1997). Determination of magnetized bodies lower edges by using geothermal data. *Geophysical Journal International, 128*(1), 167–174.

Pilchin, A. N., & Eppelbaum, L. V. (2020). Plate tectonics and Earth evolution: A conceptual review. *ANAS Proceedings, Earth Sciences, 2*, 3–32.

Popov, V. S., & Kremenetsky, A. A. (1999). Deep and superdeep scientific drilling on continents. *Earth Sciences, 11*, 61–68. (in Russian).

Priestley, K., Baker, C., & Jackson, J. (1994). Implications of earthquake focal mechanism data for the active tectonics of the South Caspian Basin and surrounding regions. *Geophysical Journal International, 118*, 111–141.

Radjabov, M. M. (1974). Seismic models of the Earth's crust of Azerbaijan. *D. Sci. Thesis*. Institute of the Physics of the Earth, Moscow (in Russian).

Radjabov, M. M. (1977). Compiling of velocity model of the Earth crust on example of the Kur Depression. *Izvestiya, Academy of Sciences, Azerbaijan, Series: Earth Sci, 6*, 13–17. (in Russian).

Rappel, C., & Mcnutt, M. (1990). Regional compensation of the Greater Caucasus mountains based on an analysis of Bouguer gravity data. *Earth and Planetary Science Letters, 98*, 360–379.

Reilinger, R., McClusky, S., Vernant, P., Lawrence, S., Ergintav, S., Cakmak, R., Ozener, H., Kadirov, F., Guliyev, I., Stepanyan, R., Nadariya, M., Hahubia, G., Mahmoud, S., Sakr, K., ArRajehi, A., Paradissis, D., Al-Aydrus, A., Prilepin, M., Guseva, T., ... Karam, G. (2006). GPS constraints on continental deformation in the Africa-Arabia-Eurasia continental collision zone and implications for the dynamics of plate interactions. *Journal of Geophysical Research, 111*(B05411), 1–26.

Rezanov, I. A., & Shevchenko, V. I. (1978). *Structure and evolution of Earth's crust of geosynclines*. Nedra.

Richter, C. F. (1958). *Elementary seismology*. W. H. Freeman & Co.

Ricketts, J. W., Takedatsu, R., Mellors, R. J., Sandwell, D., Gok, R., & Yetirmishli, G. (2008). Crustal structure of the Caucasus and Caspian region using gravity and receiver functions. In *Proceed. of the American Geophysical Union Fall Meet.*, abstract #T21A-1922.

Riordan, J. (2014). *An introduction to combinatorial analysis*. Princeton University.

Ruban, D. A. (2013). The Greater Caucasus—A Galatian or Hanseatic terrane? Comment on "The formation of Pangea" by G. M. Stampfli, C. Hochard, C. Vérard, C.

Wilhem, & J. von Raumer [*Tectonophysics*, 593 (2013), 1–19]. *Tectonophysics*, 608, 1442–1444.

Saintot, A., Brunet, M.-F., Yakovlev, F., Sebrier, M., Stephenson, R., Ershov, A., Chalot-Prat, F., & McCann, T. (2006). The Mesozoic-Cenozoic tectonic evolution of the Greater Caucasus. *The Geological Society of London, Memoirs, 32*, 277–289.

Salekhli, T. M. (1993). Facial alternation of Cenozoic deposits on the geotraverse Kyurdamir-Byandovan and its reflection on petrophysical characteristics. In *Structural-Formational and Seismostratigraphic Investigation of Sedimentary Strata of South Caspian Mega-Depression* (pp. 12–23). AzVNIIGEOFIZIKA.

Sarker, G., & Abers, G. A. (1998). Deep structures along the boundary of a collision belt: Attenuation tomography of *P* and *S* waves in the Greater Caucasus. *Geophysical Journal International, 133*, 326–340.

Sharkov, E., Lebedev, V., Chugayev, A., Zabarinskaya, L., Rodnikov, A., Sergeyeva, N., & Safonova, I. (2015). The Caucasian-Arabian segment of the Alpine-Himalayan collisional belt: Geology, volcanism and neotectonics. *Geosciences Frontiers, 6*(4), 513–522.

Shekinsky, E., Radjabov, M., Timukyan, G., Levi, V., & Riger, R. (1967). Study of the Earth's crust structure in Azerbaijan using deep seismic sounding. *Izvestiya, Academy of Sciences, Azerbaijan, Series: Earth Scienes, 5*, 41–50. (in Russian).

Shikhalibeyli, E. S. (1972). Location of Azerbaijan in general structure of the Caucasus and surrounding folded region. In S. A. Azizbekov, K. A. Alizadeh, E. S. Shikalibeyli, & T. G. Gadjiev (Eds.), *Geology of the USSR, Azerbaijan, vol. XLVII* (pp. 286–290). Nedra (in Russian).

Shikhalibeyli, E. S., & Grigoriants, B. V. (1980). Principal features of the crustal structure of the South-Caspian Basin and the conditions of its formation. *Tectonophysics, 69*, 113–121.

Shikhalibeyli, E. S., Askerhanova, N. G., Kadirov, F. A., & Gairov, A. G. (1990). Gravity modeling along the GSZ No. 3, Poylu-Masalli profile. *Izvestiya, Academy of Sciences, Azerbaijan, Series: Earth Sciences, 2*, 107–110. (in Russian).

Spichak, V. V. (1999). *Magnetotelluric fields in three-dimensional models of geoelectrics*. Nauchnyi Mir (in Russian).

Tarakanov, R. Z. (2006). Velocity models and P-waves hodographs for the Far East region. *Herald of the Far East Branch of the Russian Academy of Sciences, 1*, 81–94. (in Russian).

Tikhonov, A. N., & Arsenin, V. Y. (1977). *Solutions of Ill-posed problems*. V. H. Winston and Sons (distributed by Wiley).

Tomara, G. A. (1979). Thermal stream of deepwater trough of Caspian Sea. In: *Experimental and theoretical study of hot streams* (pp. 99–112). Nauka (in Russian).

Trofimov, I. L. (1995). Observations of telluric field variations in the seismic region of Azerbaijan. *Izvestiya, Academy of Sciences Russian, Physics of the Solid Earth, 30*(7–8), 672–678.

Turcotte, D. L., & Schubert, G. (1982). *Geodynamics: Applications of continuum physics to geological problems*. Wiley.

Tzimelzon, I. O. (1965). Deep structure of the Earth's crust and tectonics of Azerbaijan on the geophysical data analysis. *Soviet Geology, 4*, 103–111. (in Russian).

Tzimelzon, I. O. (1970). Relation between the tectonics of sedimentary deposits of Azerbaijan and earth's deep structure. *Geotectonics, 5*, 69–81. (in Russian).

Ulomov, V. I. (2003). Voluminous model of dynamics of lithosphere, seismicity structure and variations of the level of Caspian Sea. *Izvestiya, Academy of Sciences Russian, Physics of the Earth, 5*, 5–17. (in Russian).

Usoltseva, O. A. (2004). *3D velocity models of the Tyan-Shan Earth's crust based on the bi-spline parametrization and delone triangulation*. Nauka.

Usoltseva, O. A., Kazymova, S. E., & Kazymov, I. E. (2010). Seismotomographic investigation of crust of south-eastern Caucasus using of Delone triangulation by data of P-wave traveltime. In *Dynamic Processes in Geospheres* (pp. 140–147). Russian Academy of Sciences (in Russian).

Van Baak, C. G. C., Vasiliev, I., Stoica, M., Kuiper, K. F., Forte, A. M., Aliyeva, E., & Krijgsman, W. (2013). A magnetostratigraphic time frame for Plio-Pleistocene transgressions in the South Caspian Basin, Azerbaijan. *Global and Planetary Change, 103*, 119–134.

Yakubov, A. A., Alizade, A. A., & Zeynalov, M. M. (1971). *Mud volcanoes of Azerbaijan Rep., Atlas*. Academy of Sciences of the Azerbaijan SSR. Academy of Sciences of Azerbaijan (in Russian).

Yetirmishli, G. C., Kazimova, S. E., & Kazimov, I. E. (2011). One-dimensional velocity model of the Middle Kur Depression from the local earthquakes data of Azerbaijan. *Izvestiya, Russian Academy of Sciences, Physics of the Solid Earth, 47*(9), 847–856.

Yetirmishli, G. D., & Kazimova, S. E. (2013). Velocity model of crust of Azerbaijan from the data of digital seismic stations. *Geology and Geophysics of South of Russia, 1*, 59–74.

Yurov, Yu. G. (1963). Structure of Earth's crust in the Caucasus and isostasy. *Soviet Geology, 9*, 42–47. (in Russian).

Zonenshain, L. P., & Le Pichon, X. (1986). Deep basins of the Black Sea and Caspian Sea as remnants of Mesozoic back-arc basins. *Tectonophysics, 123*, 181–211.

19 An Analysis of Geological Studies and Recommendations for the Near Future

19.1 Strategy for Development of Hydrocarbon Resources in the South Caspian Basin

In the nineties, the South Caspian attracted Western oil companies due to the high effectiveness of geological prospecting works primarily carried out in the Absheron-Pribalkhan zone and because PS potential resources have been evaluated as much as from 20–30 to 50 billion tons in oil equivalent. The mentioned factors and the outlook for the discovery of new gigantic and essential oil and gas fields have been followed up by the conclusion of 15 agreements (of "production-division" type) with foreign companies, including the fact of signing a "Century Contract" in 1994.

These agreements include 26 structures in different parts of the Caspian Azerbaijan sector. The Western companies' strategy in the development of South Caspian resources was based on the following principles: (1) first of all to draw into prospecting those structures whose resources have been evaluated as more than 50 mln. ton, (2) to analyze the results of geological-geophysical investigations in the Caspian offshore area which have been carried out by insufficient employment of new seismic survey methods, (3) to reveal oil and oil-and-gas condensate fields in the Caspian shallow zone and analyze existing notions of geological conditions under which these fields seem to have been roughly the same, that is to say, the prospects for oil and gas content in different parts of the Caspian basin have exemplarily been correlated, (4) to analyze existing notions that all structures located at deep sea parts have to be filled with hydrocarbons up to the trap hinge.

However, many of the existing notions have been refuted by prospecting works carried out by the world's different oil companies. Even though it was discovered gigantic gas-condensate Shakh-Deniz field and some small hydrocarbon accumulations on the Garabakh and Ashrafi structures, borehole drilling was carried out on the Talysh-Deniz, Kyurdashi, Araz-Deniz, Oguz, Nakhchivan, and Zafar-Mashal structures appeared to be ineffective. As a result, the Penzoil, Adjip, Total-Fina-Elf, Vintershell, and Chevron-Texaco operating companies have closed their offices in Azerbaijan.

Based on the prospecting technology data, the four following primary possible reasons for their insufficient effectiveness in the SCB have been distinguished:

1. Incorrect prediction of available commercial accumulation and HC phase state in an object of study,
2. Erroneous valuation of HC probable reserves,
3. False choice of the first prospecting well location,
4. Incorrect choice of technical and technological parameters of drilling and well testing and their incongruity with concrete geological conditions.

A. Alizadeh et al., *Pliocene Hydrocarbon Sedimentary Series of Azerbaijan*,
Advances in Oil and Gas Exploration & Production, https://doi.org/10.1007/978-3-031-50438-9_19

That is why an analysis of prospecting data for each concrete is given below. It should be noted that all contract field geological and geophysical materials are confidential, making it challenging to perform this analysis in detail. Therefore, there are given only generalized opinions (Alizadeh et al., 2017; Guliyev et al., 2005) and the statements of senior officials from many oil companies and our scientific investigations.

The first prospecting wells were drilled in 1997 and 1998 on the three contract structures (Garabakh, Ashrafi, and Dan Ulduzu) in the northeastern Absheron Archipelago. Commercial HC accumulations have been revealed on the Garabakh and Ashrafi structures, whereas Dan Ulduzu appeared barren.

From well No. 2, drilled in the PS Under Kirmaki suite on the Garabakh structure, it was discharged about 846,000 m^3/day of gas and 24.6 m^3/day of condensate; a gusher of about 300 t/day discharged from well No. 3, drilled in the PS Superkirmaki sandy suite.

From the first prospecting well drilled on the Ashrafi structure, commercial gas inflow from the PS Under-Kirmaki suite of about 640,000 and 70 m^3/day of condensate; from the PS Superkirmaki sandy suite, oil discharge has been estimated as 556 t/day and gas discharge as 27,000 m^3/day.

The valued liquid HC reserves revealed on the Garabakh and Ashrafi structures in 1997–1998 being under conditions of meager price for oil existing at that time may not ensure their profitable development. However, a sharp rise in oil prices (almost 5–6 times) during the last years makes it possible to suppose their profitable development under present conditions.

Regarding the outlook for the Dan Ulduzu structure, it is challenging to draw well-founded conclusions based on the data of only two wells located on the distant NW flank of this structure and characterized by the fact that sampling works have yet to be carried out in them. An analysis of logging diagrams indicates the presence of potentially producing objects here. A detailed investigation should be continued into this structure.

Prospecting works have also been carried out in the Baku Archipelago and adjacent areas. In the Kyurdashi block of the Araz-Deniz structure, the first drilled prospecting well appeared to be barren and was liquidated for geological reasons. The prospecting well drilled in the adjacent Lyankaran-Deniz area appeared to be also barren. Because of the rhythmic character of sedimentation in this area, it is not as stable as in the Absheron oil/gas-bearing region. All drill logs here are represented mainly by clayey lithofacies. The results of prospecting works have confirmed the existing opinion of most Azerbaijan geologists that the outlook for the Pliocene deposits in this region could be better. That is why further prospecting has been stopped and displaced to the Absheron-Pribalkhan area, namely, the Oguz zone. As is known, this zone was considered a highly promising and potentially oil- and gas-bearing area but prospecting well drilled after a detailed seismic survey appears to be weakly oil- and gas-saturated. It seems most likely that it was caused by the late formation time of this deformational trap. Therefore, the bulk of HC has been accumulated in neighboring uplifted Neft Dashlari structures.

After Oguz, the quest continued in the highly promising Shakh-Deniz structure revealed by a seismic survey as long ago as 1967. This uplift is the continuation of the known Fatmai-Zikh-Gum-Deniz-Bakhar anticlinal zone. In 1974 "Aznipineft" (Azerbaijan scientific-research and Design Oil Institute) drew up a prospecting project in the Shakh-Deniz area and evaluated its gas and condensate reserves. According to this project, three prospecting wells have been drilled. Two of them, on reaching the depths of 2172 and 906, respectively, have been liquidated for technical causes. The next well was drilled in the VII horizon of the Balakhani suite at a depth of 5500 m. Because of technical and pecuniary embarrassment, further drilling on this structure has not been continued. Detailed seismic survey (2D and 3D) and drilling of the first prospecting well (SDX-1) on the northeastern limb began after signing a "Century contract". At the bottom hole (6316 m), this well was drilled in horizons VIII and X of the Balakhani and Break suites.

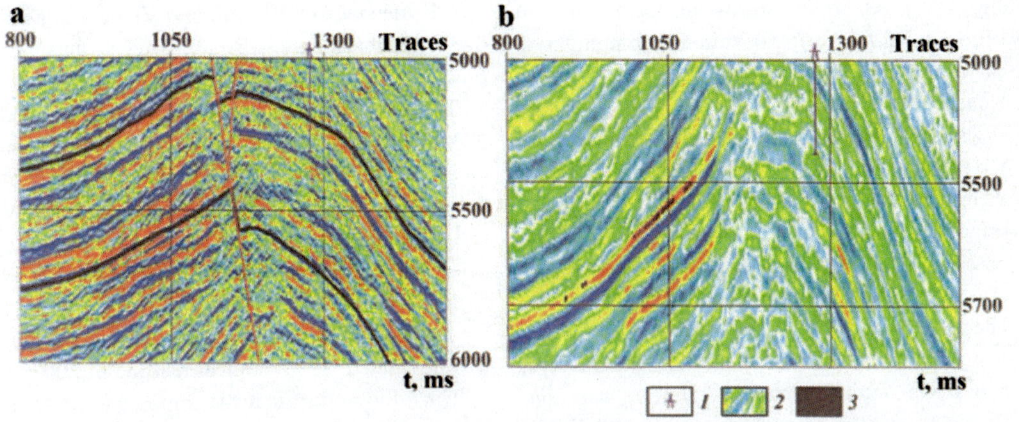

Fig. 19.1 The Absheron structure: Time section fragment (on the left) and productivity forecast (on the right). (1) prospecting well, (2) seismic time section, (3) productive horizon (Shikhaliyev and Feyzullayev, 2010)

From the last suite interval of 6225–6209 m, it yielded 1.5 mln. m^3 of gas and about 400 t/day of condensate. The VIII and X horizons of the Balakhani suite have not been tested for technical causes. The second well (SDX-2) is located near the arch zone drilled in the Break suite at the bottom hole of 5892 m. Testing an interval of 5730–5815 m yielded 1.8 mln. m^3/day of gas and 400 t condensate. The third well (SDX-3) located on the far northern pericline was behind the gas-bearing outline. The northwestern pericline is more complex than it was supposed to be before. Therefore, a well SDX-3 was in a water-saturated tectonic block.

Based on the drilling data, the preliminary counting of reserves on the Shakh-Deniz structure was estimated at 625 bln. m^3 of gas and 101 t of condensate. Later, these digits were increased. It is suggested that the oil forecast be valued at 190 mln. tons have not been confirmed. Prospecting works followed the discovery of the gigantic Shakh-Deniz gas-condensate field greater in size. Absheron structure, 30–35 km long, 7.7 km wide, which is two times as much as the Shakh-Deniz structure. It is known that the Absheron structure has been well studied by a seismic survey carried out by methodology and program pack worked out by Shykhaliyev and Gauzer (2006) that made it possible to put a direct HC diagnosis into practice. With that end in view, it was used the time section of one of the 30 profiles passing through the only prospecting well ABX-1 drilled here. The seismic horizons of the Productive Series Balakhani and underlying suites have been distinguished and traced in this profile (Shykhaliyev & Feyzullayev, 2010) (Fig. 19.1).

Of the Absheron structure, the southeastern pericline is complicated by a currently active mud volcano. Several longitudinal and transverse dislocations and the zones dangerous for a borehole sinking have been distinguished here by a 3D seismic survey. Oil reserves have initially been estimated as 100–150 mln. ton and more than 2.1 trillion m^3 of gas. According to the GTW data, the prospecting well drilled down to 6506 m has not stripped productive objects of the Break suite. A gas-saturated band found at the bottom hole is unproductive. Therefore, it has yet to be tested and liquidated for geological reasons.

The distribution of reservoir speeds over the section resulting from the inversion of the seismic wave field in the time section is highly differentiated. High-speed horizons are alternated with low-speed ones.

From the result of analyses of all sections reflecting the distribution of petrophysical parameters, it became possible to forecast the main probably productive objects whose wave pattern is displayed more intensively (see

Fig. 19.1). All these objects are located on the NE limb below the Break suite in a time interval of 5.5–5.7 s. The depth of occurrence of those forecasted productive beds is about 6.7–7.0 km. Figure 16.1 shows that a place of unproductive well location has been chosen wrongly. According to seismic data, it was drilled on the opposite limb, where productive horizons are absent. It is possible that on the Absheron structure, PS beds are produced mainly on the one (northern) limb by analogy with the Sangachal-Deniz, Duvanny-Deniz, Khara-Zyrya, and Bulla-Deniz structures. Later, it was confirmed by drilling data of the second well, ABX-2, at a depth of about 6550 m. In September 2011, the French company represented by "Total" declared a discovery of the gas-condensate field "Absheron", having reserves estimated as 300 bln. m^3 of gas and 45 mln. ton of condensate.

Somewhat earlier, the fact of the discovery of gas-condensate field "Umid" in the Baku Archipelago, located slightly farther south of the Bulla-Deniz oil–gas structure, was declared by the leadership of Azerbaijan State Oil Company. Six wells drilled here before appeared to be ineffectual. From the forecast carried out before based on seismic data (Fig. 19.2), it is inferred that the deposits of VI, VII, and VIII horizons of the Balakhani and Break suites may be productive.

The first prospecting wells on the Inam and Nakhchyvan structures have been located on their periclines. In the Nakhchivan area, one of the wells (No. 1) drilled in the potential object (V and VII horizons) has been stopped at a depth of 6747 m (SKGS). As it was supposed, this well turned out to be behind the gas-oil pool outline of the mentioned objects, and without sampling, it was liquidated for geological reasons in 2002.

Thus, the notion that the trap has been filled with hydrocarbons up to its end and drilling the first prospecting wells on its "critical" points based on the mentioned notion appeared to be an erroneous idea. In the Inam area, a well No. 1 boring clashed with severe complications because of abnormally high pressure. As a result, this borehole was stopped for technical reasons at a depth of 4442 m in 2001. No drilling is renewed here connected with the overestimation of existing notions about expecting an HC phase state at this location as well as due to the complex technological conditions of drilling requiring an additional outlay. Now, this well is in dead storage, and in all probability, a new well drilling very soon is not contemplated.

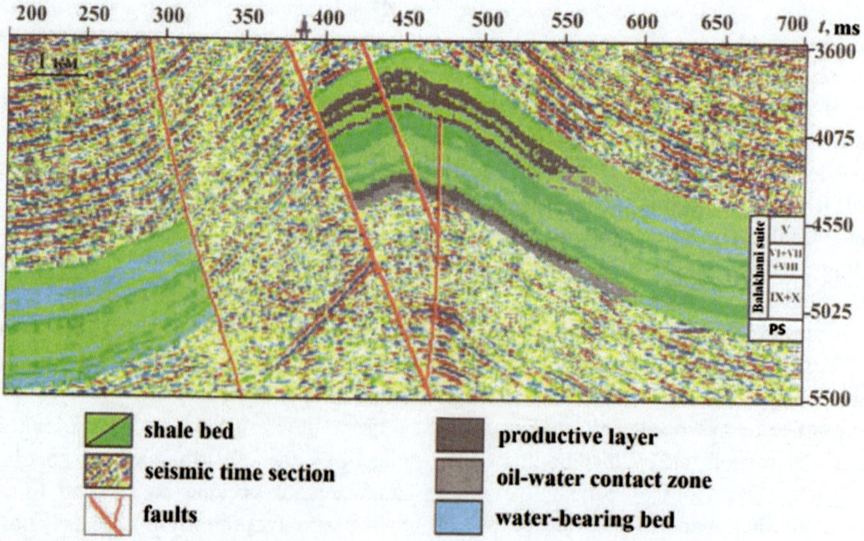

Fig. 19.2 The "Umid" area. A section productivity forecast (Shikhaliyev and Feyzullayev, 2010)

In 2001, the well JTX-1 boring at the Yanan Tava area was stopped at a depth of 4715 m for technical reasons, not stripped of a total thickness of the significant VIIth horizon. Although AT-1 stripped horizon VII well in the Atashkyakh area, the presence of oil and gas has not been confirmed. It was liquidated in 2003. After the JAOC company had fulfilled its contractual obligations, the project closed.

In November 2003, a super-deep well drilled in the Zafar Mashal area. Its frilling has been carried out from the most powerful rig, "Heydar Aliyev", however, because of the high-power waterspout from the Absheron deposits. This well has been liquidated for technical reasons.

In 2004, a new prospecting well was located on the structure's steeply dipping south-eastern limb. It was contemplated that this well should be drilled—in the Break suite objects at a depth of 7088 m. Nevertheless, the authors had no information on these borehole drilling conditions. That is why it is impossible to assert unequivocally if those objects have been tapped. Project operators have closed this contract ("Exxonmobil") because of the lack of oil and gas reserves. The generalized results of prospecting drilling in the contract areas are given in Table 19.1.

During a period of development of the South Caspian resources by western oil companies, prospecting works were carried out in 15 areas, from which only five areas (Garabakh, Ashrafi, Shakh-Deniz, Umid, and Absheron) appeared to be commercial oil and gas-bearing. The Shakh-Deniz area, instead of expecting oil–gas-condensate field, had been recognized as a gas-condensate deposit. The number of discovered productive traps is about 36% of all prospecting areas. From the 27 wells drilled in the mentioned areas, 23 wells have reached the projected objects. Commercial output has been recognized in 8 wells. Proceeding from the mentioned facts, the efficiency of prospecting works according to the number of productive wells is about 34%. At first sight, a low efficiency, high degree of economic hazard depending on high prospecting outlay (sometimes up to 75 mln. for one marine area) may not be conducive to optimistic conclusions about this region.

However, in our opinion, the present situation does not refute the high HC potential of the Caspian region, overall, and South Caspian—in particular. In Azerbaijan (on land and sea), more than 500 bill. m^3 of gas and one mln. ton of oil had been recovered. Besides, several gigantic fields had been discovered. The conventional fuel reserves evaluation was estimated to be from 4 to 8–10 milliard tons (Alizadeh et al., 2017). The last evaluations of initial potential HC resources carried out jointly with Russian experts include more than 20 billion tons of conventional fuel (Guliyev et al., 2003). Nearly the same values have been estimated by a scientific team from the state oil company of Azerbaijan (Kerimov et al., 1999).

19.2 Former Strategic and Methodological Miscalculations

As to some adverse outcomes of oil search in SCB, they have mainly resulted from strategic, technological, and possibly geological miscalculations, such as the following:

1. An analysis of the SCB reserves structure according to the field size classes and phase structure of reserves (oil–gas relation) is absent. An account of these factors for oil–gas-bearing regions must be taken as a basin in the choice of prospecting works and forecast their results.
2. There needs to be a proper notion of oil and gas content in different facial types of sediments composing the primary reservoir and South Caspian source series. As it is known, the high productivity of structures and low degree of hazard are mainly connected with the Absheron type of PS sediments characterized by highly favorable capacity-filtration properties and very weakly developed abnormal high pressures. The second one, by the areal extent, is the Kur type of sediments confined to the southern and southeastern subsided areas of the Baku Archipelago. This sediment type is mainly represented by clays

Table 19.1 Prospecting drilling data from the contract structures (O-oil; G-gas; C-condensate)

Structures	Forecast reserves: (Oil, mln. ton, gas, bil. m^3)	Drilled wells quantity	Drilling results	Oil reserves valuation according to BP data, mln. ton	Failure reasons	Current state on 01.01.2005
Garabakh	130/–	3	Oil, gas, and condensate are quarried from 3 wells	Oil, 20–40	+	Contract is closed because of low profitableness
Ashrafi	150/30–50	1	Oil, Gas, Condensate	Oil, 20–40	+	«-»
Dan Ulduzu	«-»	1	Negative	–	An absence of HC accumula-tion conditions	The contract is closed because oil reserves are lacking
Lankaran-Deniz	80–100/–	1	Negative	–	An absence of favorable lithofacies for HC accumulation	«-»
Shakh-Deniz	190/500	3	Gas and condensate quarried from 2 wells	Gas, 625 bln. m^3; Condensate, 110 mln. т	+	Beginning of the development first stage
Araz-Deniz	95/–	1	Negative	–	An absence of favorable lithofacies for HC accumulation	The contract is closed because of the oil reserves' absence
Nakhchivan	110/85	1	Negative	–	No favorable conditions for HC accumulation	«-»
Inam	170/100	1	–	–	The well is stopped due to complications related to anomalous high reservoir pressure	New well boring is not planned
Oguz	100/–	1	Negative	–	This trap formed after HC migration	The contract is closed because of the absence of oil reserves
Absheron	120/300	2	Positive	–	–	Prospecting works are continued
Umid	/300	6	Positive	–	–	Prospecting works are continued
Yanan Tava	100–120/–	1	Negative	–	This well was not drilled—in all objects of VII and VIII horizons	The contract is closed because of the oil reserves' absence
Atashkakh	«-»/–	2	Negative	–	No favorable conditions for HC accumulation	«-»
Zafar Mashal	140/–	1	«-»	–	The well is supposedly not drilled—in the Break suite	«-»

with high anomalous pressures ($K_{an} = 1.5 - 1.9$) and is characterized by a higher degree of hazard. The geographical range of facies changes in sources series must be established as trustworthy.

3. There needs to be more analysis of laws governing anomalous high-pressure development and the reasons for sharp differential pressures within the zones of high abnormality. Significant pressure drops relate to different types of barriers (lithologic, tectonic, hydrodynamic, etc.), which play an essential role in the hydrocarbon filling of structures. It only sometimes turns out well to reveal these barriers using a seismic survey. Therefore, they are usually established based on drilling data. The presence of the mentioned barriers may be considered one of the reasons for the incomplete HC filling of traps (on the South Caspian structures, this filling coefficient does not exceed 0.7).
4. The role of faults, diapirs, and mud volcanoes in oil and gas formation and accumulation should be re-estimated. Many of the South Caspian oil and gas fields are confined to the local and regional stress nodes (zones of discharge) formed by tectonic faults action. These nodes caused stream course migration and the mode of traps filling with hydrocarbons. Also, the block structure should be taken into consideration in the choice of the borehole location place.
5. There needs to be more studies of the problem of trap formation time and the beginning of HC generation and migration processes relation.

An essential factor determining the reserve amount and the outlook of the structure is its location and distance from the nearest regions of HC generation. An analysis of the field's areal extent in the South Caspian shows that HC reserves are more in case the hydrocarbon field is confined to a significant structure surrounded by several troughs containing oil and gas-generating series.

As to the problem of decreased economic risk and unfounded outlays in further prospecting works, all previous data must be revised by skilled experts on a level with additional, more detailed, and complex purposive investigations (not excluding the fact re-comprehension of theoretical fundamentals of oil and gas fields generation and their reflection in geophysical and geochemical fields applicable to the quick-subsided and extremely unstable young basins). From the standpoint of the prospects of the new hydrocarbon pools discovery in the SCB, it is essential to note that they are inadequate for oil and gas.

19.3 Prospects for Identifying New Hydrocarbon Deposits

The outlook for an increase in oil reserves with the purpose of partial compensation of oil. The outlook for an increase in oil reserves for partial compensation of oil production volume over Azerbaijan continues to be related to the PS deposits and, to a lesser extent, to Paleogene-Miocene and Mesozoic deposits. Besides, prospecting objects in the PS are mainly connected with shallow slope areas in the northern SCB, where all the fund of promising structures is exhausted. Therefore, the search for new promising areas of oil accumulation should be concentrated in depositional and stratigraphic traps.

The problem of increment of gas and gas-condensate reserves is solved more optimistically at the expense of the SCB deep-subsided structures, which are characterized by significant technological hazards. The SCB major reservoir with which more than 90% of oil and gas production is connected in Azerbaijan continues to be the Lower Pliocene complex (Productive Series), which is the main prospecting object very soon.

The choice of prospecting areas in the PS falls on the most promising South Caspian northern zone comprising the Absheron Archipelago and the South Caspian deep-sea part adjacent to the south.

The following objects may be recommended for oil prospecting:

1. The Garabakh-Ashrafi structural block is in the northwestern Absheron Archipelago within the anticlinal zone, being the marine continuation of southeastern periclinal subsidence of the Greater Caucasian Meganticlinorium.
2. Such local uplifts as Absheron Archipelago, Gilavar, Dan Ulduzu, and Ashrafi are located within the mentioned NW–SE-trending anticlinal zone (Fig. 19.3).

The geological structure of the northwestern anticlinal zone comprised of the uplifts mentioned above (Absheron Archipelago, northern and western Absheron, Hazri, and Gilavar) is characterized by such general features as the deeply washed crown (up to the PS upper division, inclusive), asymmetrical structure, the present of dislocations with a break in continuity. The key factors defining special sedimentation features, the presence or absence of either deposit series, are the SCB extreme down warping, on the one hand, and the Greater Caucasus rising (beginning from the Middle Jurassic), on the other hand. As a result of the mentioned tectonic movement combination, the Mesozoic deposits appeared to be at a shallow depth (of about 1000–1500 m).

Essential breaks in sedimentation established by prospecting drilling on the Absheron-Deniz, Absheron Archipelago, and Gilavar structures have resulted in several lithostratigraphic intervals omitted. Deep erosion of the arch zone on northwestern structures is seen in an example of

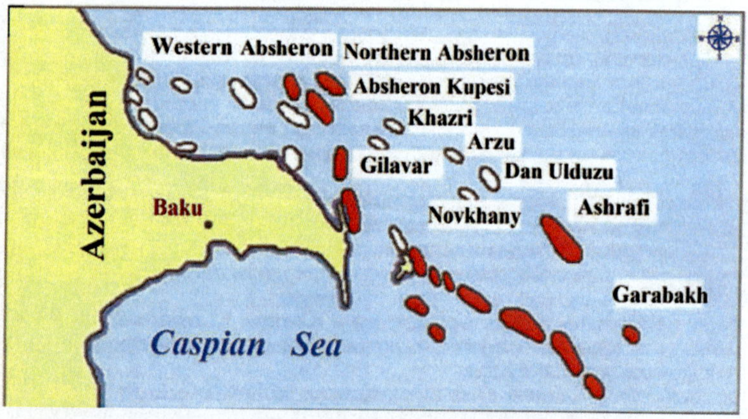

Fig. 19.3 Schematic location of structures within the anticlinal zone in the NE Absheron Archipelago

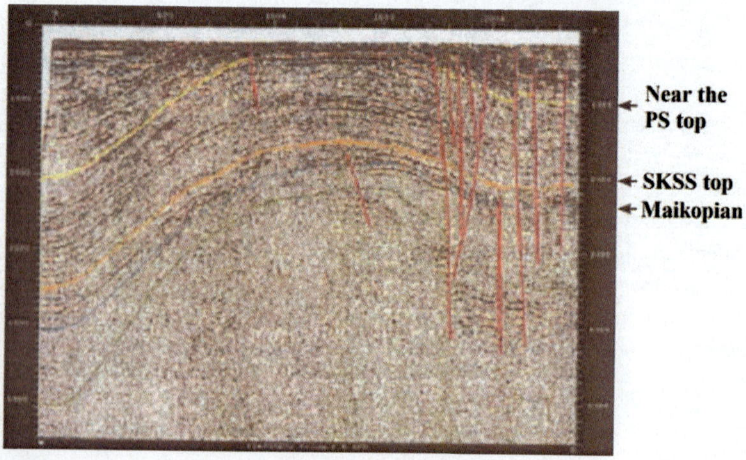

Fig. 19.4 Seismic time section through the northern Absheron structure

the seismic time section of the northern Absheron area (Fig. 19.4). In such a way, on the Agburun structure, the Danian deposits are stripped under the Pliocene formation and Barremian deposits—on the Absheron Kyupesi. In the Gilavar area, the Miocene-Oligocene deposits (Diatomic and Maikopian suites) occur under the PS Kirmaki suite and are underlaid by the Upper Cretaceous (Campanian, Santonian) deposits.

It is noted that on the crests of the mentioned anticlinorium and its southeasterly trending continuation, the old (Mesozoic) deposits are replaced by younger (Pliocene) ones that increased in thickness in the same direction. The latter are overlapped by the Quaternary sediments (Table 19.2). The first prospecting wells on the Garabakh, Ashrafi, and Dan Ulduzu contract structures were drilled in 1997 and 1998. The payable HC accumulation has been discovered on the first two structures, whereas a well drilled in the Dan Ulduzu appears barren (Table 19.3).

Estimated liquid HC reserves discovered on the Garabakh and Ashrafi structures in 1997–80 under conditions of the meager price of oil at that time (about 10–12 USD/barrel) may not be considered profitable for their development. However, the rise in oil prices over the last years requires a revaluation of the profitableness of these oil fields' development under present conditions. With that end of view, it is necessary to reinterpret the seismic survey results considering the drilled wells data to delimit oil pools discovered in the PS and prospect new accumulations in the Miocene-Oligocene and possibly Mesozoic sections. It is also recommended to include in the Garabakh-Ashrafi block such adjacent promising structures as Dan Ulduzu, Ufug, and Vurgun (Fig. 19.5).

Table 19.2 The NW–SE change in thickness of drilled deposits within the NE Absheron Archipelago

Area	Minimum depth of the Pre-Pliocene sediments deroofing, m	Minimum depth of the Mesozoic sediments deroofing, m
Absheron Archipelago	625	1900–2300
W. Absheron	973	
N. Absheron	1912	
Khazri	2647	
Gilavar	2325	
Novkhani	3564	3500–4300
Dan Ulduzu	3078	
Ashrafi	> 3650	

Table 19.3 Sampling data of the prospecting wells drilled in the Ashrafi and Garabakh areas

Well No.	Sampling interval, m	Horizon	Sampling data		
			Oil, t/day	Condensate, t/day	Gas, thous. m^3/day
Ashrafi					
1	3524.5–3536	PK		70	640
1	3274–3283	NKP	556		27
Garabakh					
2	3665–3681	PK		24.6	846
2	3414–3433	NKP		83	700
3	3723–3750	PK			Gas
3		NKP	300		

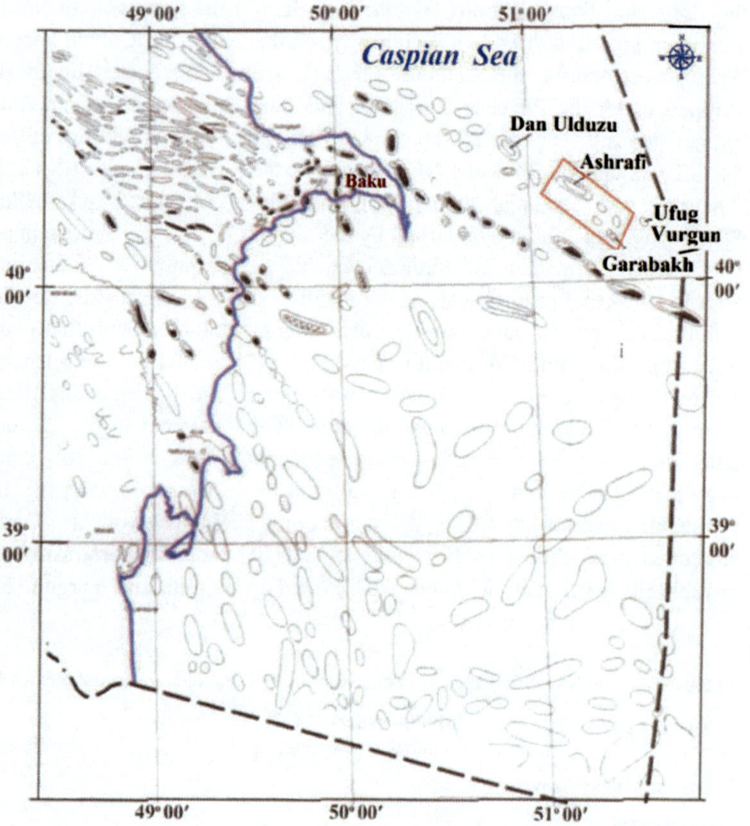

Fig. 19.5 A schematic representation of the Garabakh-Ashrafi block location

However, to draw unequivocal and well-founded conclusions about the outlook for oil and gas content in the Dan Ulduzu structure based only on two incomplete data is incorrect. It is believed that a detailed investigation should be continued if log analyses show the presence of potentially productive objects here.

Prospect drilling and contour mapping revealed the series of perspective structures in a shallow, 20–30 m transition zone (with a width of about 20 km and primarily outlining the Absheron Peninsula) (Fig. 19.6). These structures are either (a) the continuation of onshore structures, (b) the continuation of the anticlinal zones onshore-offshore, or (c) the individual structures.

No seismic survey has been undertaken in the transitional zone for technological reasons. Therefore, essential information about its geological structure is absent. Besides, the distribution density of anticlines revealed by prospecting drilling and structural construction in the mentioned zone is noticeably lower than on coastal land, which is opposed to the law-governed nature of the basin's geological-tectonic evolution.

At present, seismic survey technologies applicable in the offshore areas have been worked out, and they may be used in the described zone to verify the location of before revealed and new structures as well as non-anticlinal oil and gas-bearing traps (first, those related to the regional pinching out of the PS Underkirmaki suite). Regional pinching out zone of the PS suites/horizons on SW Absheron Near-Balkhan sill where all the following revealed anticlinal structures are characterized by commercial oil and gas content.

It is known that prospecting works have mainly concentrated on discovering HC accumulation of anticlinal type, whereas possible

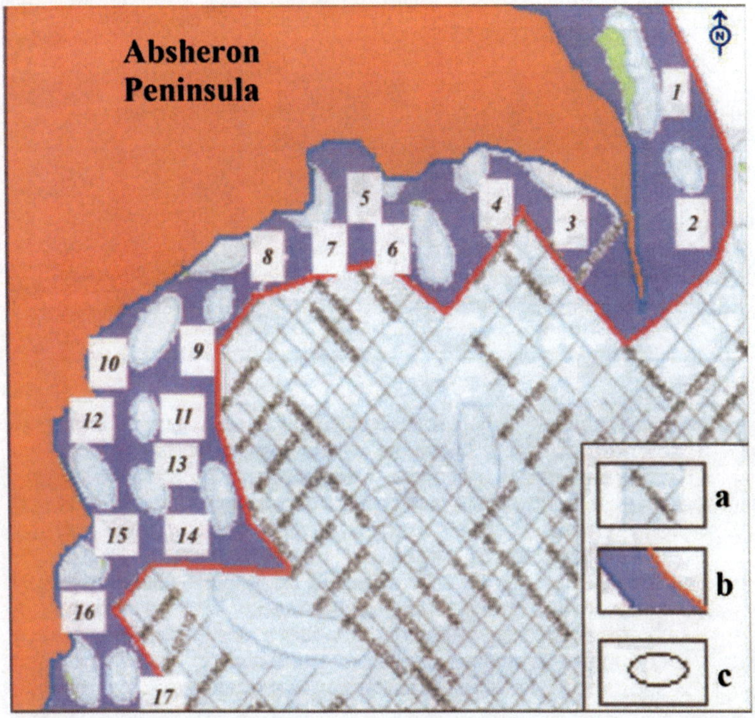

Fig. 19.6 Schematic representation of a shallow (down to 20–30 m) transition zone about 20 km wide bordering the Absheron Peninsula on the South Caspian: **a** seismic profile lines used up to the present, **b** transitional zone, **c** geological structures: (1) Pirallakhi, (2) Gyurgyan-Deniz, (3) Turkan-Deniz, (4) Govsany, (5) Zikh-Deniz, (6) Gum-Deniz, (7) Bibi-Heybat-Deniz, (8) Lockbatan-Deniz, (9) Shakh-Deniz, (10) Garadagh-Deniz, (11) March, (12) Sangachal-Deniz, (13) Duvanny-Deniz, (14) Khara-Zyrya, (15) Alyat-Deniz, (16) Hamamdag-Deniz, (17) Garasu

non-anticlinal HC accumulations have been out of attention. However, according to the existing expert estimation, about 40–70% of the world's HC resources may be related to non-anticlinal traps. They proceeded from the above estimation, exploration for oil and gas accumulation in such non-anticlinal conditions as sand bars and lenses in the Volga riverbed which is a principally new object of investigation. In this connection, the regional pinching out zone of some suites/horizons (for example, of the Balakhani suite, Fig. 19.7) is recommended as a prospecting object. First, distant SW limbs of commercially oil/gas-bearing structures of the Absheron-PriBalkhan sill. It should be noted that the mentioned SW limbs turned into a deep South-Absheron trough, the primary center of HC generation in the SCB. Such necessary gas-condensate fields as Shakh-Deniz, Umid, and Absheron have been discovered over the last years.

Currently, prospecting in the South Caspian is concentrated chiefly within its deep-sea zone, considered the most worthwhile area. Considering this fact, several prospecting objects have been recommended in the South Caspian deep-sea. All of them are evaluated as potential areas according to the results of the processing and interpretation of seismic survey data and geophysical testing of wells (GTW), "REZAYR" with the employment of methods, technology, and a packet of programs developed in the SOCAR Inst. (Baku) in 2006. In the mentioned methods, all seismic survey and GIW data are considered a unified complex being worked out and interpreted jointly. The significant features of

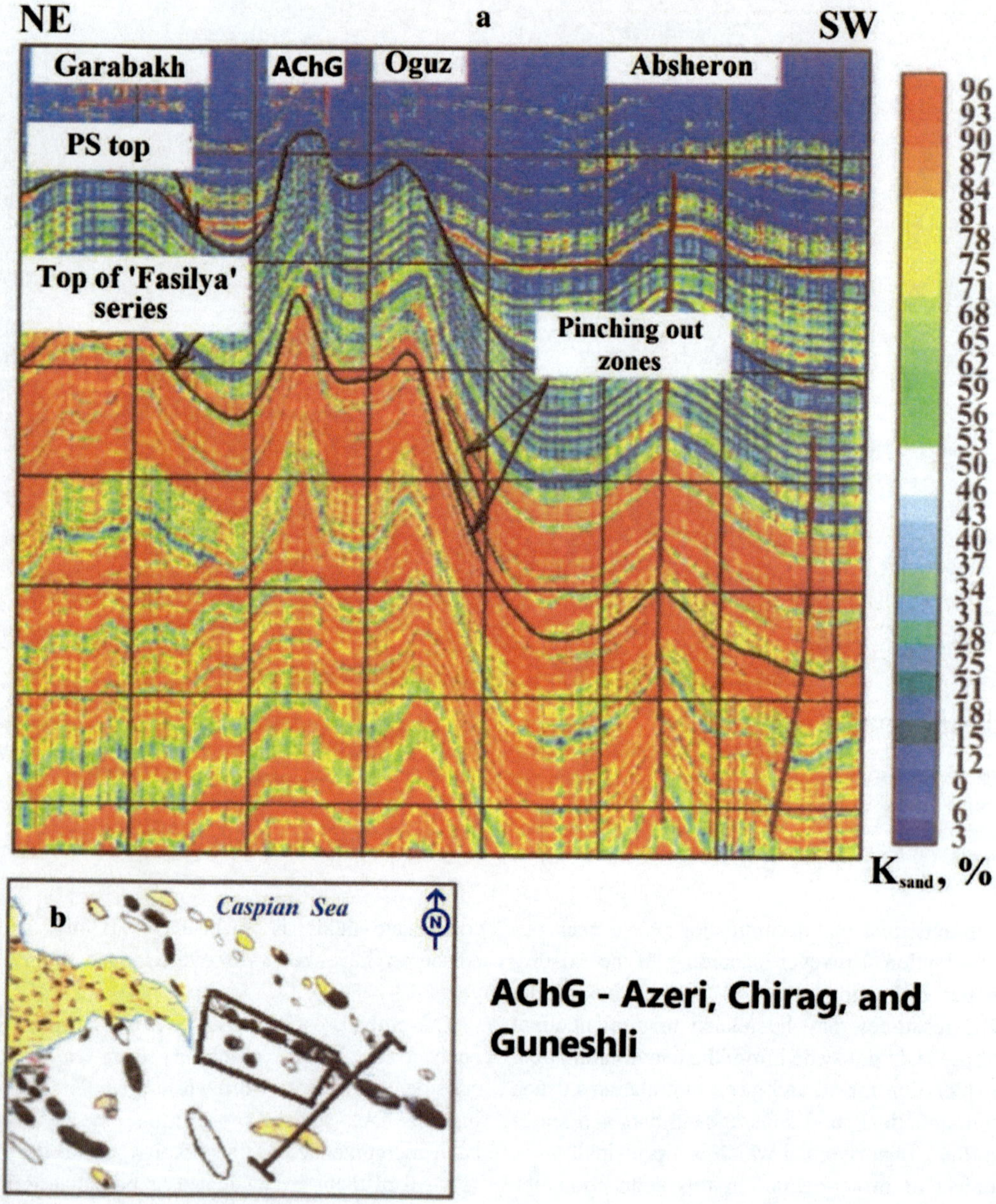

Fig. 19.7 Seismo-time section **a** showing arenosity percentage and the PS pinching out on the SW slope of the Absheron-Pribalkhan sill, Azerbaijan sector of the SCB; **b** areal map (after Feizullayev & Shykhaliyev, 2011)

this complex are a wide range of GTW data compared with seismic survey data and different scales of GTW and seismic survey data. GIW data are presented on a scale of depths, whereas seismic survey data is on a time scale.

The applied methodology is based on mathematical methods of analysis and interpretation of the available data realized in an interactive regime that allows determining such important geologic-geophysical features as:

- stratigraphic boundaries,
- average speed values by each route,
- practical porosity values through the profile,
- volumetric clayiness values in every point of the profile,
- mean-grain diameter values through the section and determination of lithofacies boundaries,
- determination of sedimentation cyclicity boundaries,
- construction of the effective geological model of the section,
- determination of the beds' reservoir properties and their productivity.

An essential feature of this methodology is that having a priori knowledge of the section speed characteristics required is not obligatory. However, its efficiency may be increased in the presence of such data as acoustic logging, vertical seismic profile, or high-speed logging. An excellent example of this method's high efficiency is the prediction of the Umid structure productivity (Fig. 19.8).

The present figure shows that the deposits of VI, VII, and VIII horizons of the Balakhani suite are very likely productive. The oil–water surface position is traced only from the anticlinal side since its other side may not be predicted because of seismic data quality.

The prediction was then confirmed by the discovery of gas-condensate commercial accumulations, whose geologically inferred reserves are about 200 billion m^3. A unique feature of the proposed methods is that the GTW and seismic survey data processing is realized in a united interpretative cycle using feedback at its different stages. In addition, it is realized by employing separate relatively short graphs made up of procedures, some of which realize their algorithms, and the others are of the processing standard procedures.

The processing efficiency and reliability of the results depend on the integration being used and the fact that the mentioned graphs allow attaining an optimal result utilizing successive approximation in an interactive regime. It answers for the GTW complex as seismic survey demands. Within the South Caspian deep-sea area, the following promising structures are recommended to search for new commercial HC (mostly gas-condensate) accumulations: Babek (in the Umid-Babek block), Nakhchivan, Zafar-Mashal. The information on these structures is based on the above-described processing methods and interpretation of seismic data.

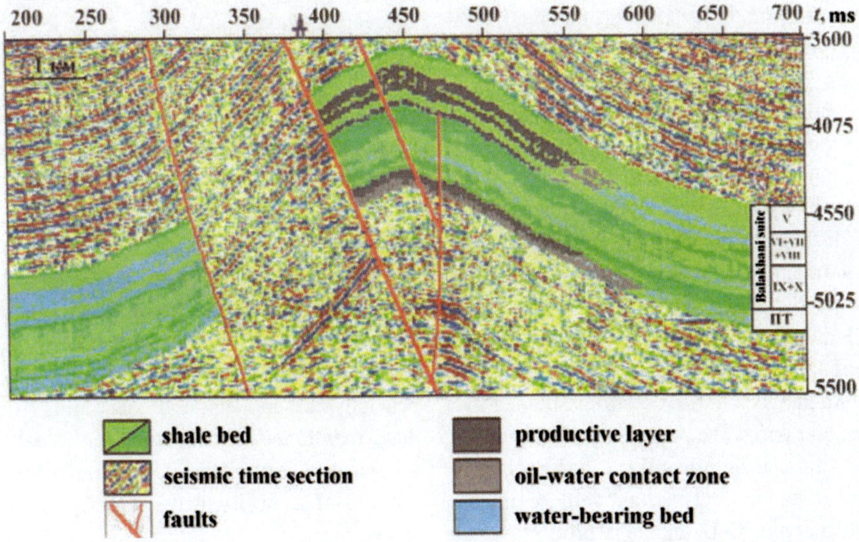

Fig. 19.8 Prediction of the Umid structure productivity by seismic data (after Shykhaliyev & Feyzullayev, 2010)

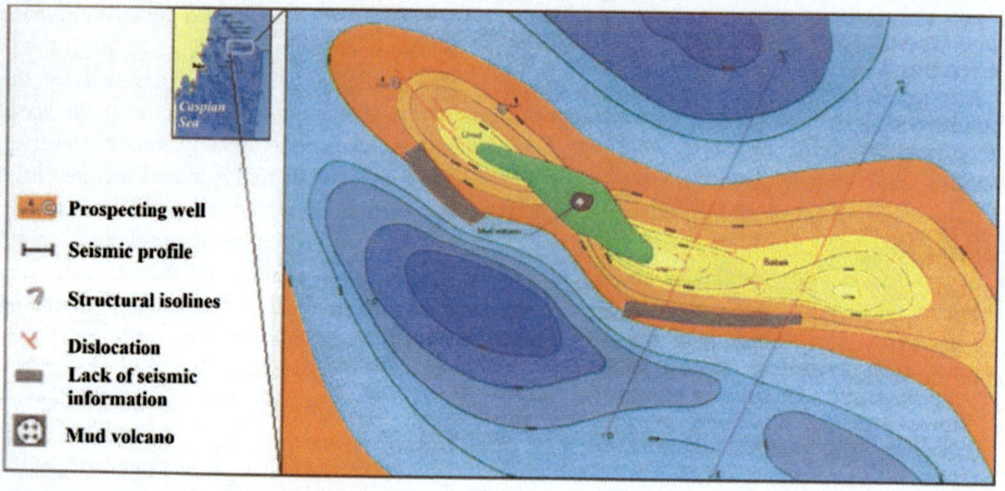

Fig. 19.9 A structural scheme was drawn over the Break suite roof and seismic profiles located on the Umid-Babek block

The Babek structure is within the Umid-Babek block. The Umid-Babek megastructure was discovered by a seismic survey in 1953 (Fig. 19.9). Ten prospecting wells have been drilled in on the NW Umid structure. In 2012, a 3D seismic survey was carried out in the Umid-Babek area. At the same time, it established its commercial productivity, totaling about 250 bln. m^3 of gas-condensate reserves. Therefore, a second uplift (Babek) within this megastructure is also potential.

The results of 2D seismic profile interpretation are given below (Figs. 19.10 and 19.11).

Various geological objects confined to the definite stratigraphic ages are depicted in seismic-time sections of these profiles. Special attention has been paid to the Break and Balakhani suites as possible promising objects.

The distinguished objects' reliability and boundary tracing are controlled by connecting profiles and the Umid wells (Nos. 4, 6, 8, 10) data. All distinguished objects are identified in profiles of D-D and E-E through the connecting profiles. An analysis of time sections showed that their character and wave pattern are distinguished from each other insignificantly. It has been confirmed by dynamic analysis, as well. A notable feature is that the D-D and E-E profiles' wave fields intersecting the Babek structure are more dynamically marked.

It should be noted that the wave pattern of time sections is complex on the crest, being more tranquil on the flank where seismic horizons are indeed correlated. Some dislocations revealed in time sections resemble one another, and two are traced through all sections up to the Productive Series. An area between these dislocations looks like a shutter zone. The wave pattern within this zone is chaotic and complicated by multiple and diffracted waves and other interferences. To eliminate erroneous results, these areas in seismic sections have not been analyzed. An anticlinal part of a seismic section in a time interval of from 3600 to 5500 m has been interpreted in detail to embrace all over the Balakhani and Break suites of the Productive Series. The time section transformation into a section of geophysical parameters is a rather complex problem. Its solution depends on factors such as a priori knowledge of the section structure's noteworthy features, the rocks' lithofacies composition, petrophysical models, and constraints. Using the data mentioned above, the wave patterns of seismic sections have been transformed into sections of arenosity, wave velocity, and porosity ratios.

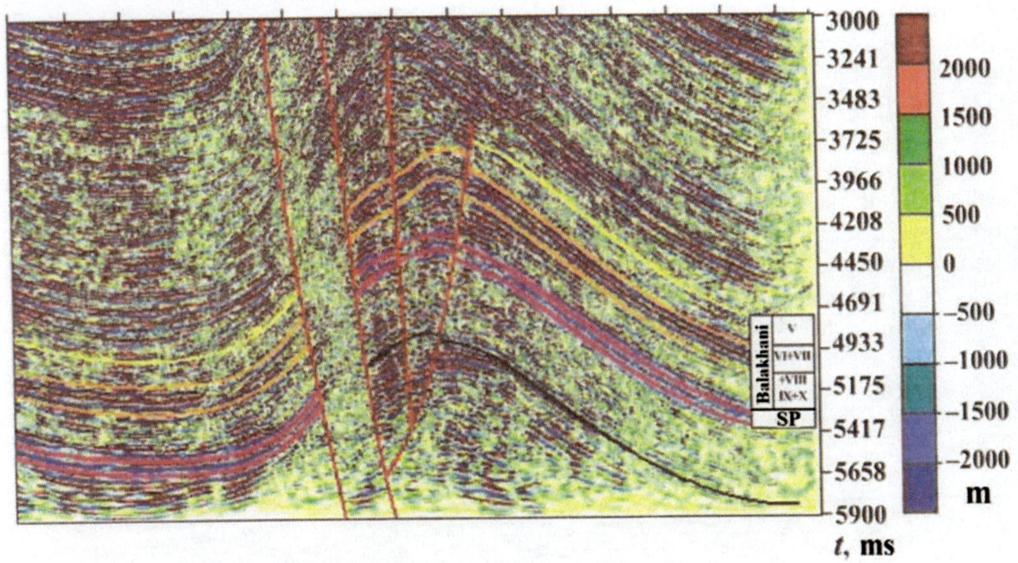

Fig. 19.10 Babek area: Seismic-time section through the profile of D-D

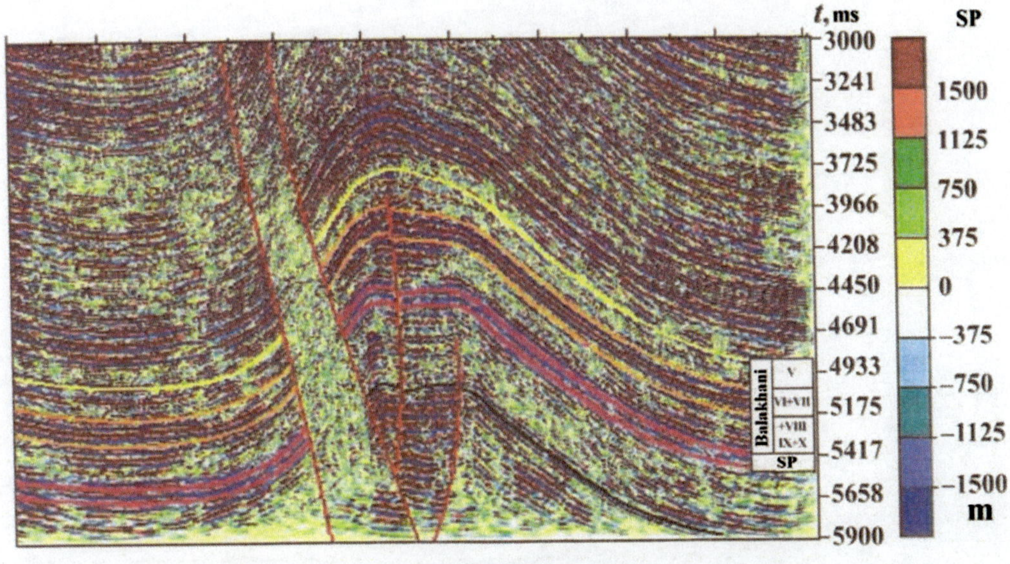

Fig. 19.11 Babek area: Seismic time section through the E-E profile

To reach a maximum effect in geophysical forecasting parameters and to achieve coupling between seismic parameters and drilling data, seismic profiles have been transformed into values of other characteristics such as instantaneous parameters, energy, etc. Such constraints may be determined by the knowledge of a priori information or the available statistical and analytical dependences for the concrete deposit type (in the present instance, for the Balakhani and Break suites). The problem of forecasting based on the profiles of D-D and E-E in the absence of drilling data is solved by compensating these profiles' characteristics with comparable data of the profiles intersecting the Umid structure. As a result, it was interpreted the distribution of geophysical parameters in the D-D and E-E profiles. The results obtained allow for making forecasting geophysical parameters more precise and reliable. The authenticity of the obtained information is confirmed by an identity of synthetic paths calculated by forecasted wave velocities and seismic paths in critical stations.

Based on analyzed profiles, schematic sections reflect the distribution of various geophysical parameters. An analysis of arenosity ratio distribution in the sections constructed for the D-D and E-E profiles (Figs. 19.12 and 19.13) showed that the quantity of sandy beds in the D-D profile is markedly less than those in the E-E profile. It relates to such factors as the geotectonic condition of the area development and the composition and location of the provenance of sedimentary terrigenous material. The beds characterized by arenosity lower than 40% are predominant here. However, there are more high-grade sandstones in the Break suite and in the VI, VII, and VIII horizons of the Balakhani suite (Productive Series).

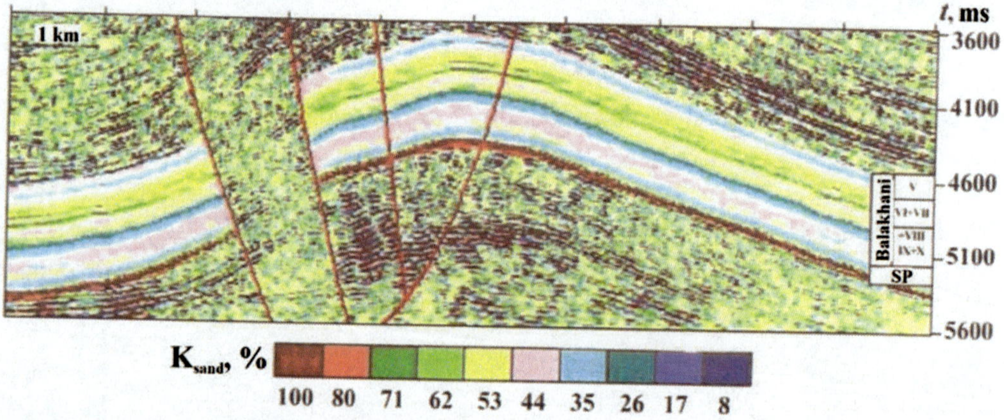

Fig. 19.12 The Babek area: distribution of arenosity ratios along the D-D profile

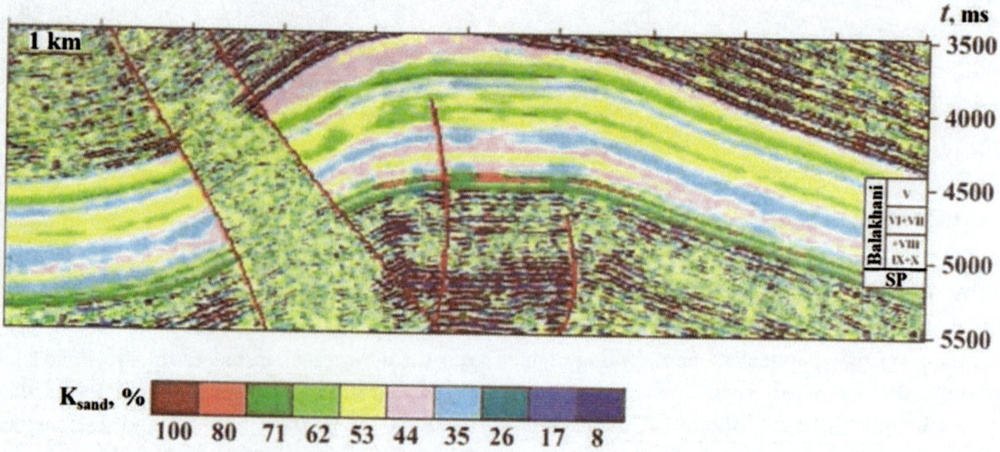

Fig. 19.13 The Babek area: distribution of the arenosity ratios along the E-E profile

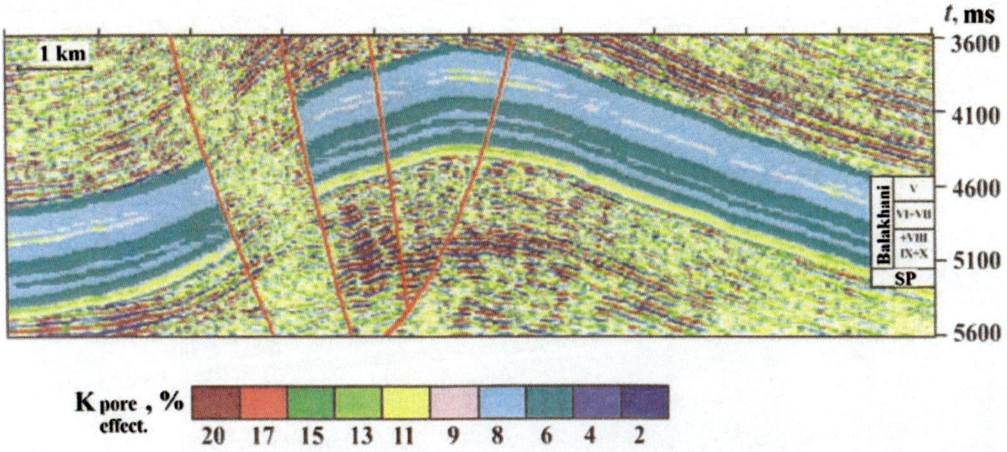

Fig. 19.14 The Babek area: distribution of the effective porosity ratios along the D-D profile

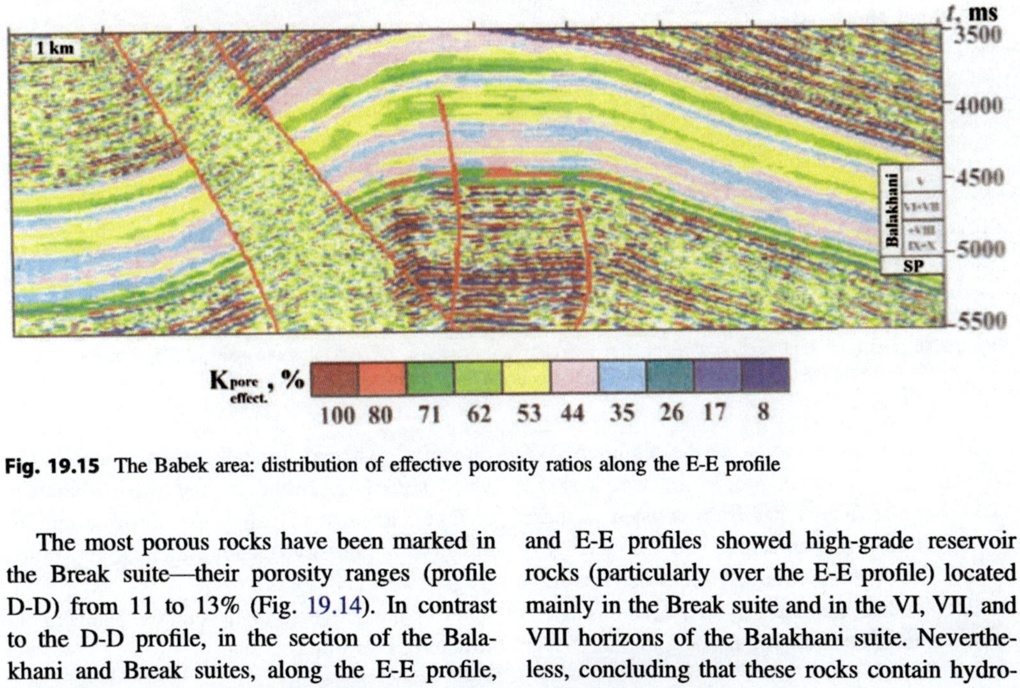

Fig. 19.15 The Babek area: distribution of effective porosity ratios along the E-E profile

The most porous rocks have been marked in the Break suite—their porosity ranges (profile D-D) from 11 to 13% (Fig. 19.14). In contrast to the D-D profile, in the section of the Balakhani and Break suites, along the E-E profile, there are rocks with porosity up to 15% (Fig. 19.15).

The distribution of geophysical parameters allows for estimating the total volume of reservoir rocks in the section, which is necessary information on total reservoir volume and hydrocarbon reserves and estimating their productivity. Seismic data processing over the D-D and E-E profiles showed high-grade reservoir rocks (particularly over the E-E profile) located mainly in the Break suite and in the VI, VII, and VIII horizons of the Balakhani suite. Nevertheless, concluding that these rocks contain hydrocarbons may be done only from the result of additional investigations.

The Nakhchivan structure is in the central part of the SCB Azerbaijan sector. A 3D seismic survey was carried out a 3D seismic survey, and a deep prospecting well (about 6800 m deep) was drilled. Moreover, this area has been worked out by numerous 2D seismic profiles.

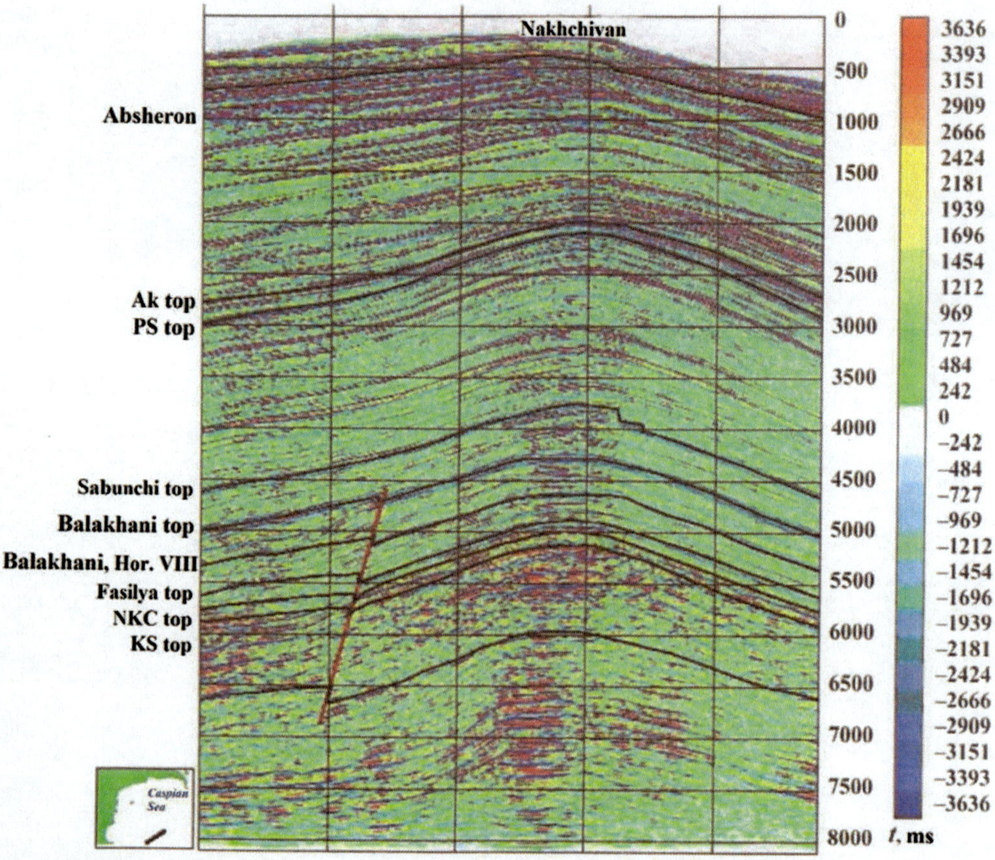

Fig. 19.16 Seismic time section through the Nakhchivan structure. *Ak* Akchagylian suite, *PS* Productive Series, Fasilya, Break suite, *NKS* 'Nadkirmakinskaya' sandy suite, *KS* Kirmaki suite

In addition, a fragment of the regional SW–NE trending seismic profile intersecting the Nakhchivan structure has been enlisted in these investigations (Feyzullayev and Shykhaliyev, 2011). The dynamic time section of this profile is remarkable for its intensity. Also, ten seismic horizons have been distinguished and traced over this section (Fig. 19.16).

From the result of wave field inversion in this profile, it was constructed a section of wave velocities distribution (Fig. 19.17). The wave velocity change is clearly seen in the vertical and horizontal directions. Besides, these velocities appear to be distinctly changed in a vertical direction, whereas laterally, the change takes place smoothly. Based on the wave velocities change, it was revealed the distribution of effective porosity and arenosity ratios (Figs. 19.18 and 19.19).

From the above-given data on the Nakhchivan structure, it is inferred that:

- Horizon VIII of the Balakhani and the Break suites contains high-grade reservoir rocks,
- The PS lower division (particularly SKS) also contains high-grade reservoir rocks, which may be productive,
- Zafar-Mashal structure

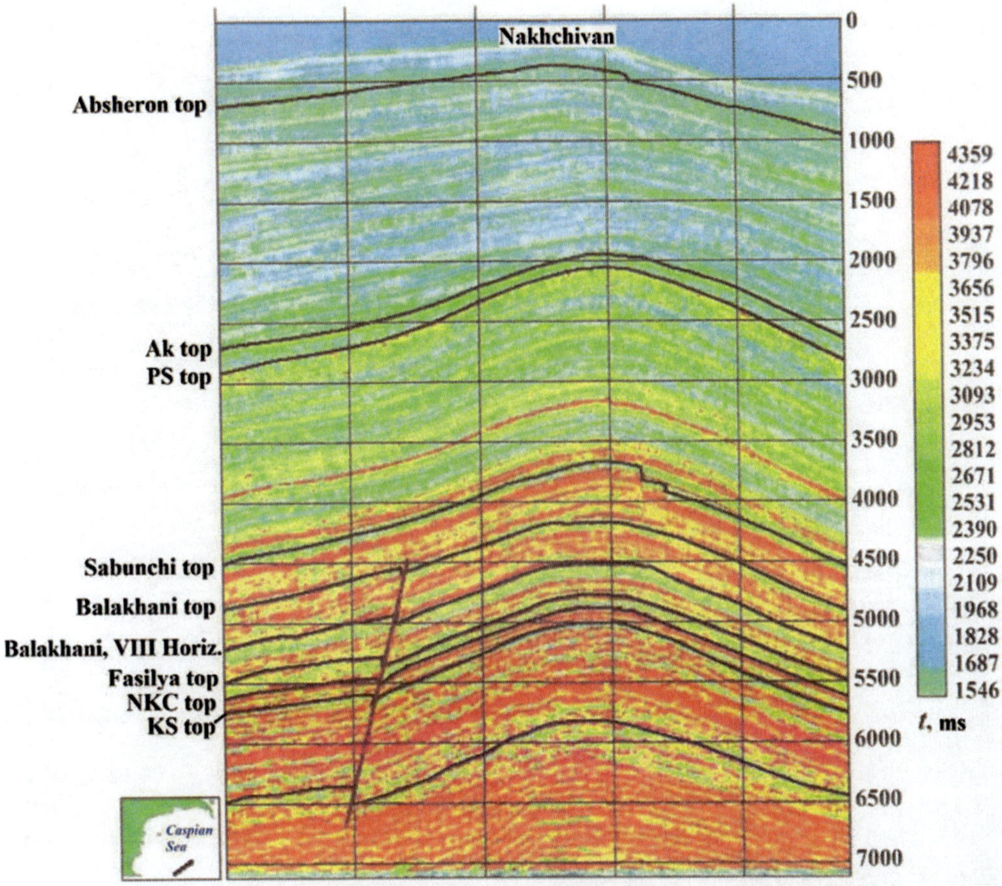

Fig. 19.17 Distribution of wave velocities along the profile through the Nakhchivan structure

It is also located in the central Azerbaijan sector of the SCB beside the Nakhchivan structure. Here a 3D seismic survey and deep exploratory drilling (drilled well depth was about 7050 m) have been carried out. In addition, this area is intersected by numerous 2D seismic profiles. An analyzed profile intersects the Zafar-Mashal structure in a sublatitudinal direction. It should be noted that the seismic time section of this profile is remarkable for its wave field dynamic expressiveness (Fig. 19.20).

Forecasting reservoir velocity values over this profile have been corrected using vertical seismic profile data obtained from the well ZAFX-1H1.

Figure 19.21 shows the distribution of values of the reservoir wave velocity forecasting over the seismic paths. As shown in this figure, the reservoir velocity section over an analyzed profile is not in contrast. As a result of the inversion, were constructed sections reflecting the coefficients of sandiness (Fig. 19.22) and effective porosity (Fig. 19.23) of sediments.

From the analyzed results obtained from the Zafar-Mashal area, it is inferred that the deposits of horizon VIII of the Balakhani suite and those

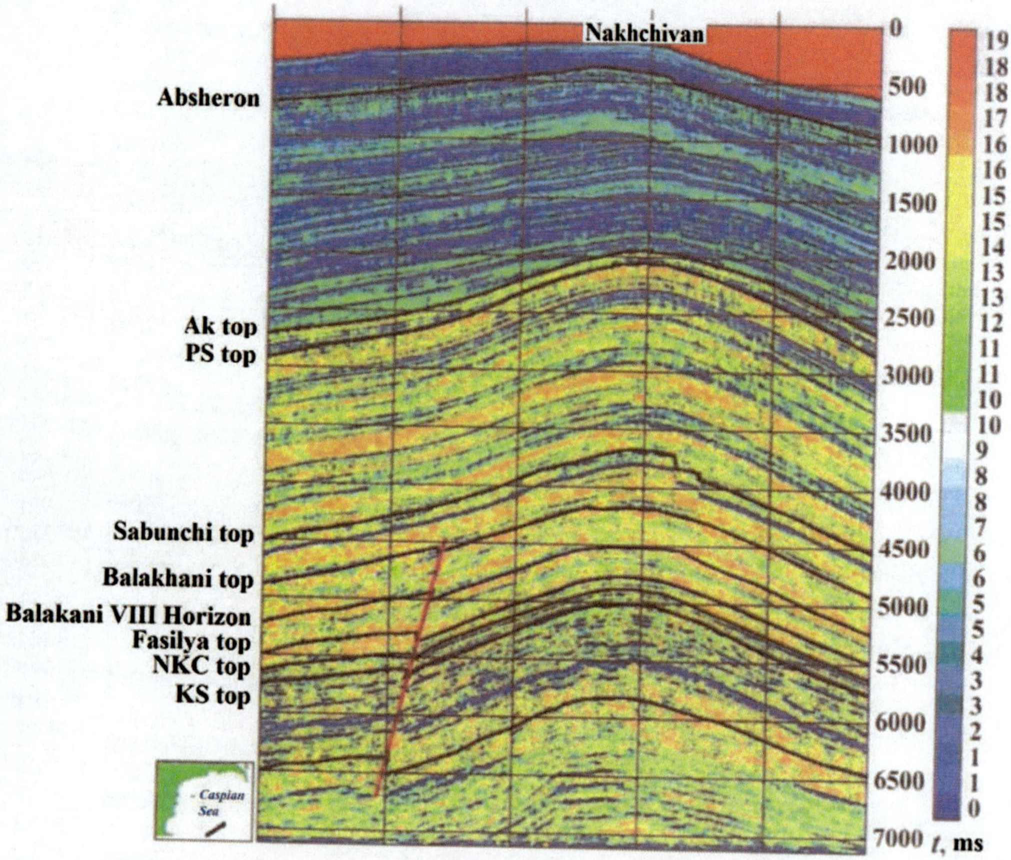

Fig. 19.18 Distribution of effective porosity ratios along the profile through the Nakhchivan structure

of the Break suite, SKS, and KaS are characterized by satisfactory reservoir properties.

With the prospect of a new oil and gas accumulation, oil and gas reserves increment may be fulfilled at the expense of a re-survey within the developed areas.

<u>On the SCB land area, these are:</u>

- The XX horizon and the Kursanga field's lower PS horizons are within the Lower Kur depression. The wells drilled—here before having stripped the deposits down to the PS XVII to XVIII horizons, but the lower horizons have not been studied until now. The depths here are about 6000–6200 m, and the expected production is gas-condensate-oil—oil products. Geological hazards here are low, whereas technical and technological risks are high. It is believable that the generalization of the available geological-geophysical and drilling data should precede the drilling of prospecting wells, and in the case of need, the 3D seismic survey should be carried out,
- The XX horizon and the PS lower horizons at the Khydyrly area in the Lower Kur depression. The Khydyrly structure is the only one in the SCB where PS lower horizons have yet to be drilled. The depths of the aimed horizons here are from 4500 to 5500 m.

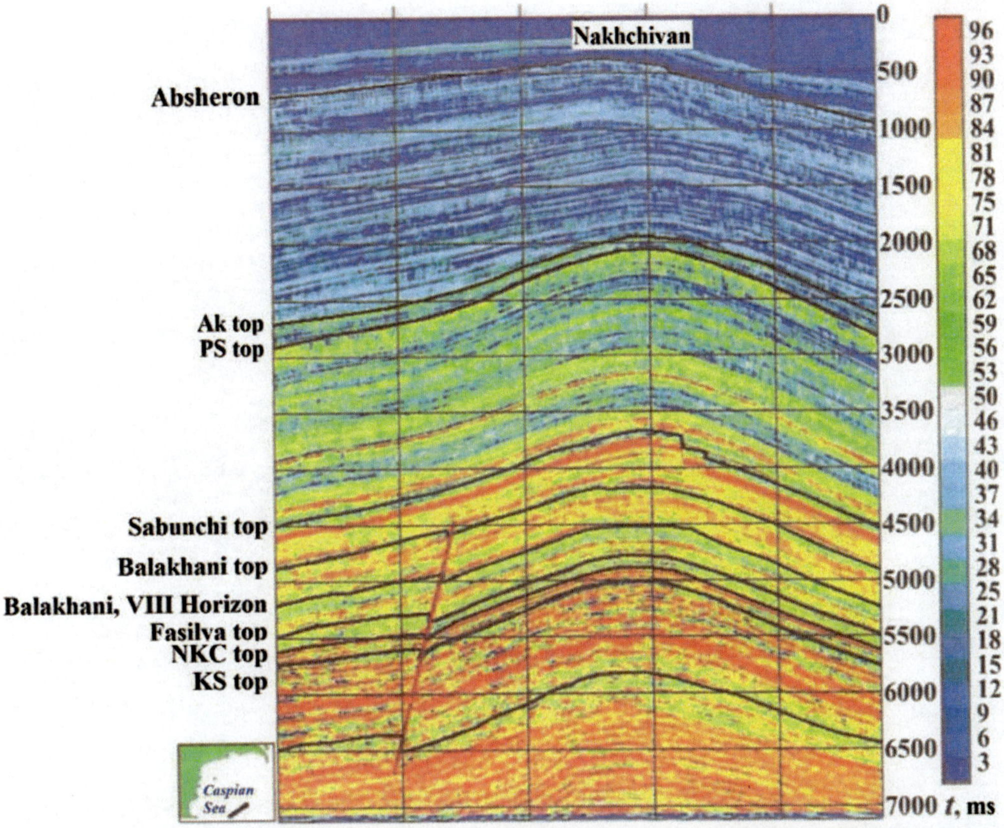

Fig. 19.19 Distribution of effective arenosity ratios along the profile through the Nakhchivan structure

Using a 3D seismic survey to study this area's geological structure and choose an optimal location for prospecting wells is advisable. These works' geological and technological risks have been estimated as average.

In the basin marine area, these are:

- The PS lower horizons (from the VIII horizon to the UKS) within the Bulla-Deniz, Alyat-Deniz, and other areas. A 3D seismic survey should be carried out before prospecting drilling to subsequently sink a well in optimal conditions. Besides, expected production over these areas is gas and gas-condensate, but the presence of oil pools is possible. Geologic hazards here are minimal, whereas technological risks are high,
- Only partially proved areas of the PS bottom suites, first, the Kala suite with which both anticlinal and non-anticlinal (related to the pinching out zone and the structural-stratigraphic and structural-lithological traps) oil and gas accumulations are connected.

On a level with those mentioned above, an extremely perspective object for prospecting works is those structures located within the Alov-Araz blocks in the Caspian deep-sea Azerbaijan sector (Guliyev et al., 2000, 2001).

The most important index of oil-generating processes in the mentioned South Caspian area is a large multi-component anomaly discovered by the authors on the strength of complex geochemical and hydrochemical investigations and gas indexes of the bottom sediments ooze

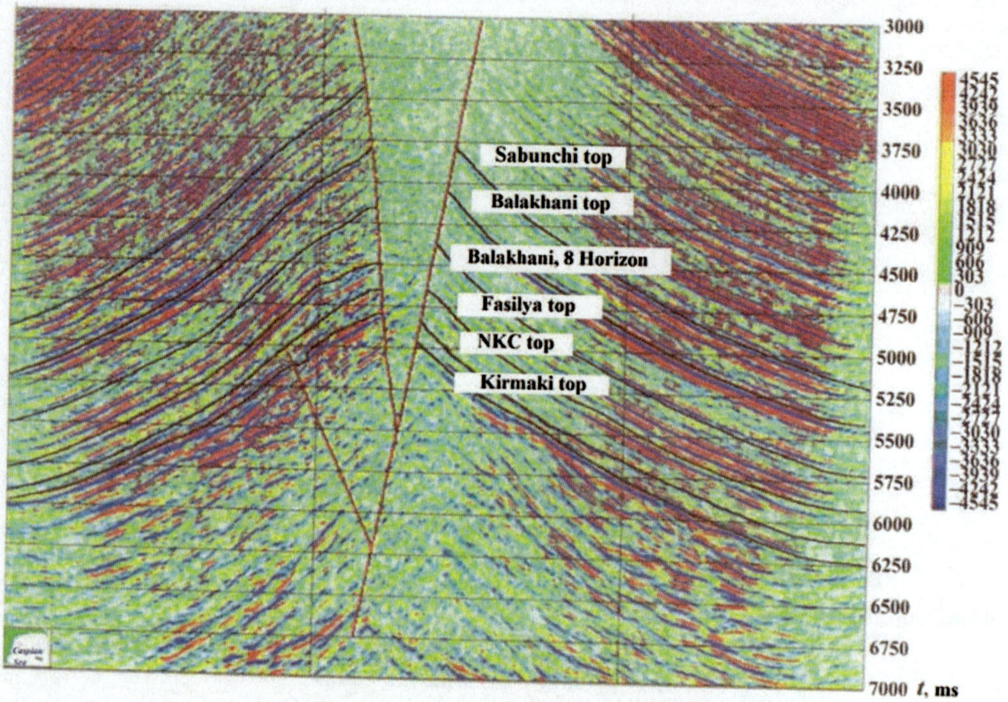

Fig. 19.20 Zafar-Mashal area: Seismic time section (after Shykhaliyev & Feyzullayev, 2010)

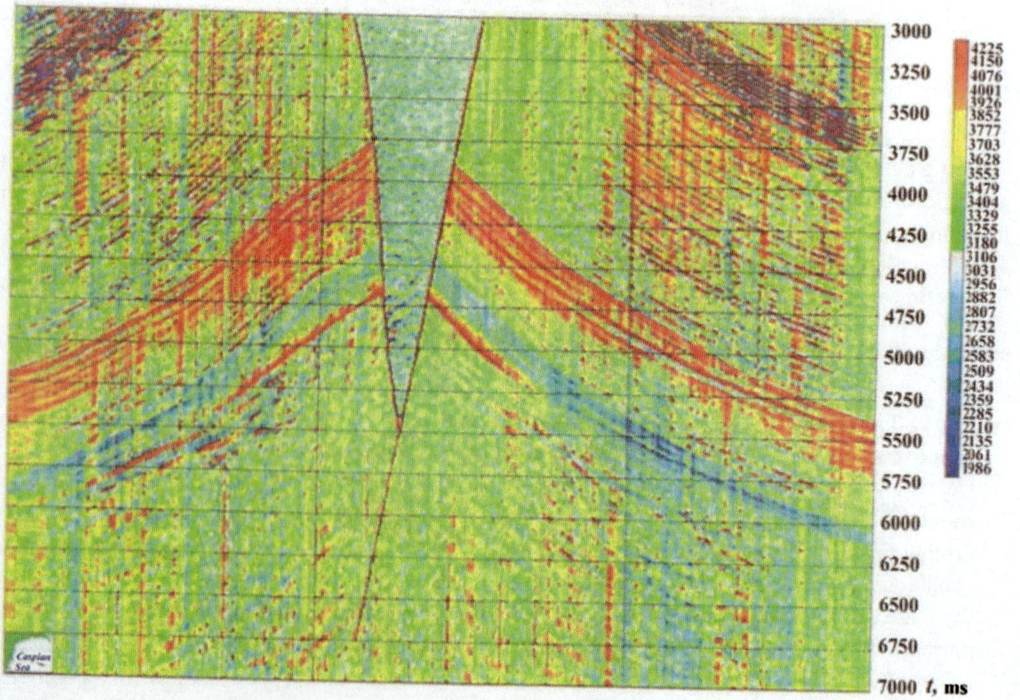

Fig. 19.21 Zafar-Mashal area: distribution of the reservoir wave velocity values

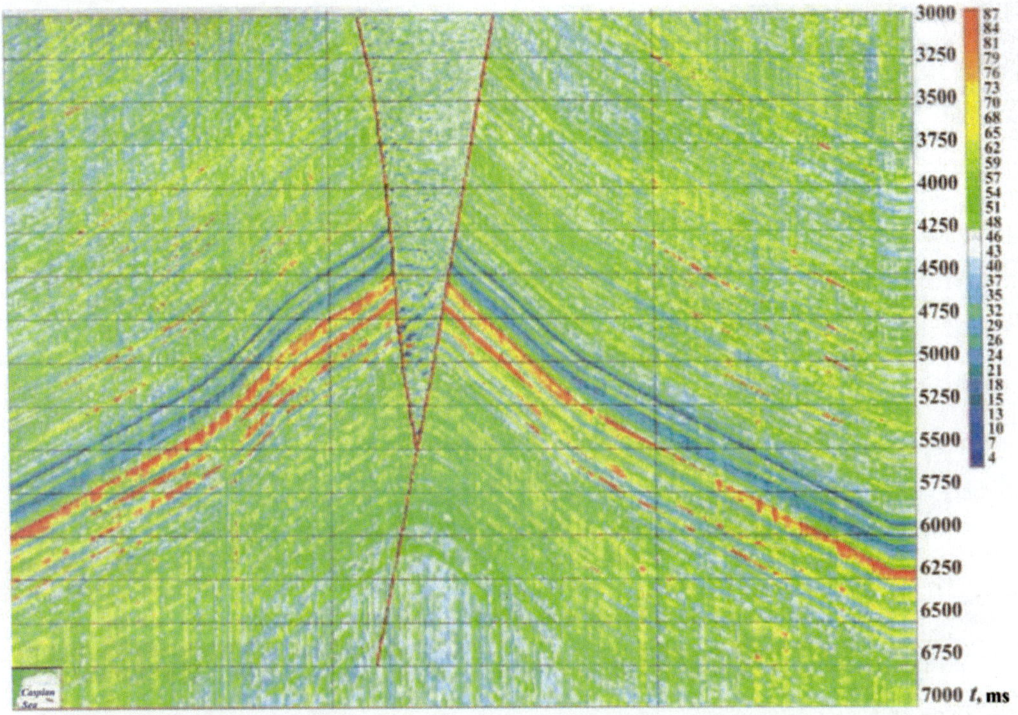

Fig. 19.22 Zafar-Mashal area: distribution of the arenosity values

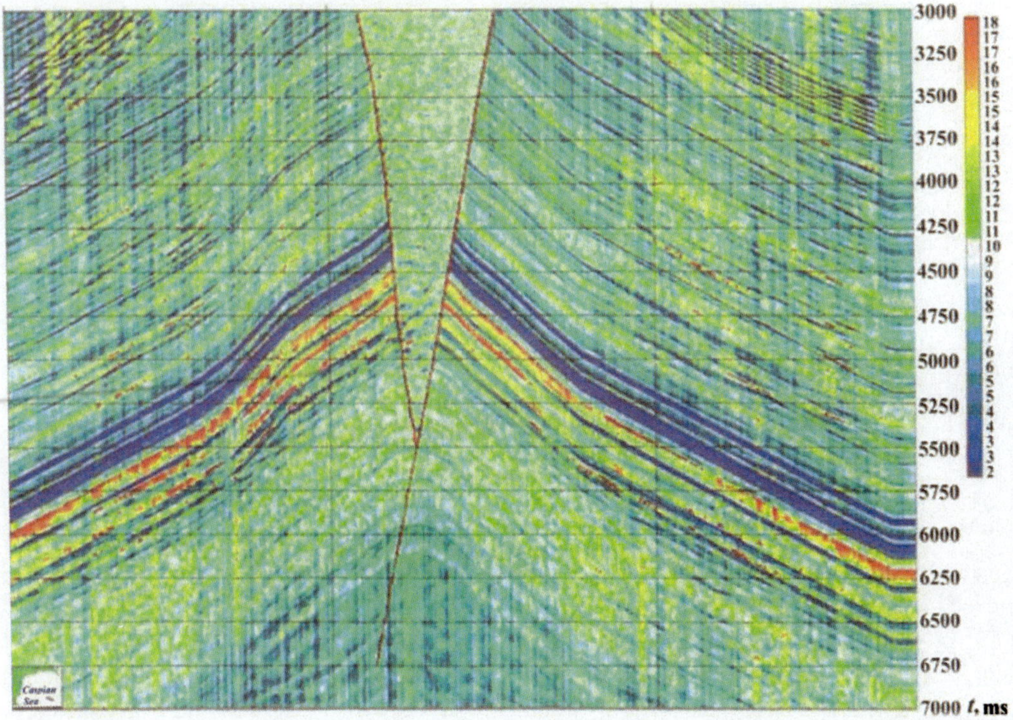

Fig. 19.23 Zafar-Mashal area: distribution of effective porosity values

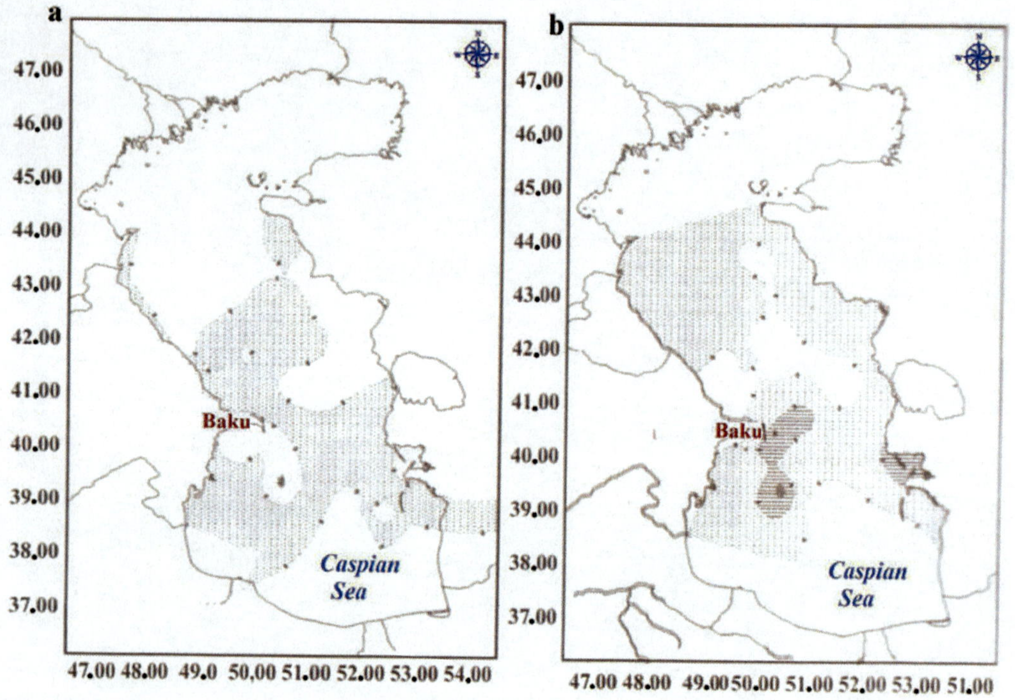

Fig. 19.24 The distribution of bitumen ratio in the SCB bottom set beds (to the left—vertical section of from 0 to 50 cm to the right—from 0 to 150 cm)

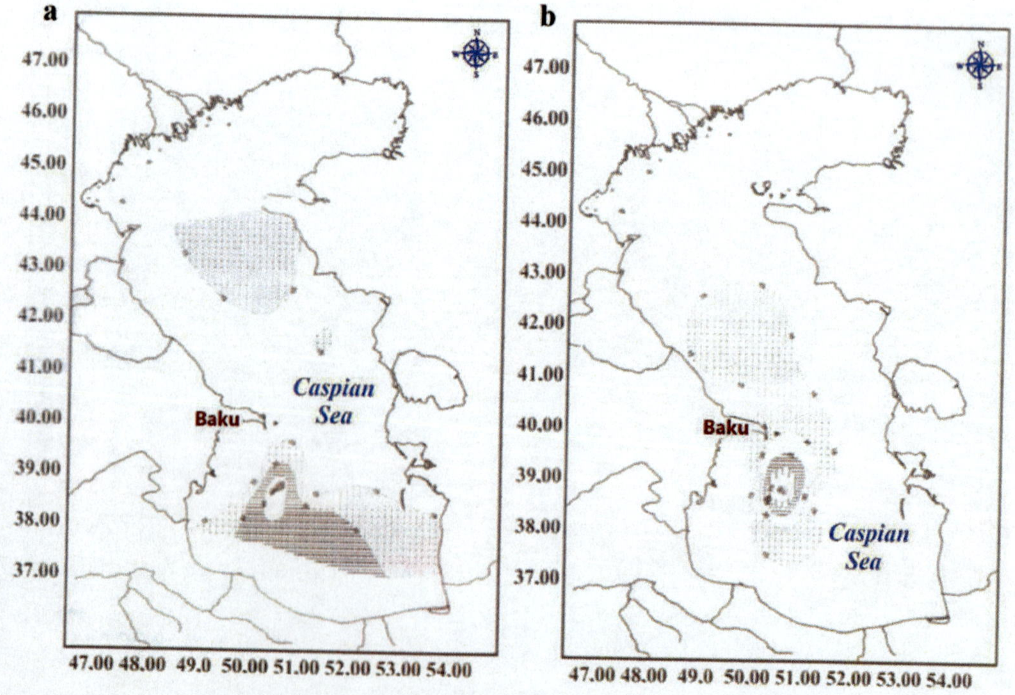

Fig. 19.25 The distribution of NH_4^-/J^- in the SCB bottom set beds (to the left—vertical section of from 0–50 cm to the right—from 0 to 150 cm)

solutions and the Caspian Sea water. Anomalous high values of geochemical indexes are marked all over the sequence of sediments and seawater. In such a way, this zone is characterized by the degree of bituminization in the organic matter, which appears to be increased, the component composition of the bitumen is changed, and the share of its oil fraction is also increased to 45% (Fig. 19.24). High values of anomalous direct indicators of the presence of oil and gas are established in ooze solutions. For instance, an ammonium-to-iodine ratio exceeds background values four times in the upper seven-meter bed and seven times in the interval of from 0.6 to 1.5 m (Fig. 19.25). The content of hydrocarbon gases in this zone is also increasing (Fig. 19.26).

An integrated analysis of the results of geological-geophysical investigations and a detailed interpretation of modern fluid dynamic processes makes it possible to carry out a regional ranking of the SCB water area according to the degree of outlook and risk.

Based on this range (Fig. 19.27), the following zones are distinguished within the SCB water area:

I High-potential zone of oil and gas accumulation with minimal geological, technological, and economic risks, including the Near-Absheron water area (transitional zone),

IIa Promising zone located mainly in the PS lower division characterized by some technological risk connected with AHRP,

IIb Unpromising zone,

III Potential zone for gas and condensate accumulation with increased economic and technological risks related to the large reservoir and sea depths.

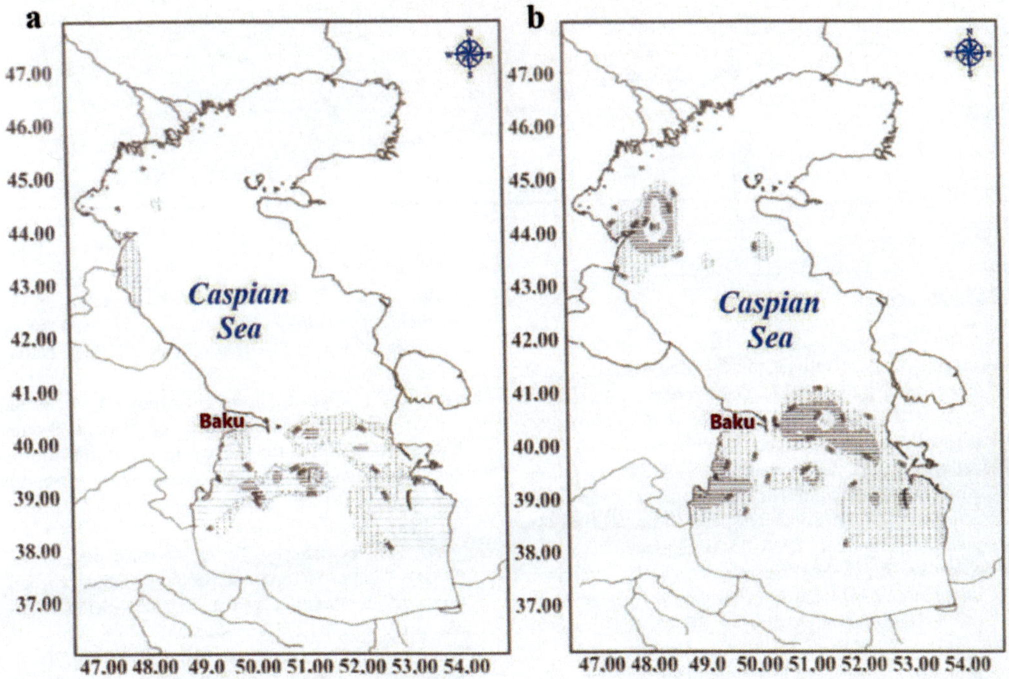

Fig. 19.26 The distribution of HC gases in the SCB bottom set beds (to the left—CH_4, to the right—heavy HC gases)

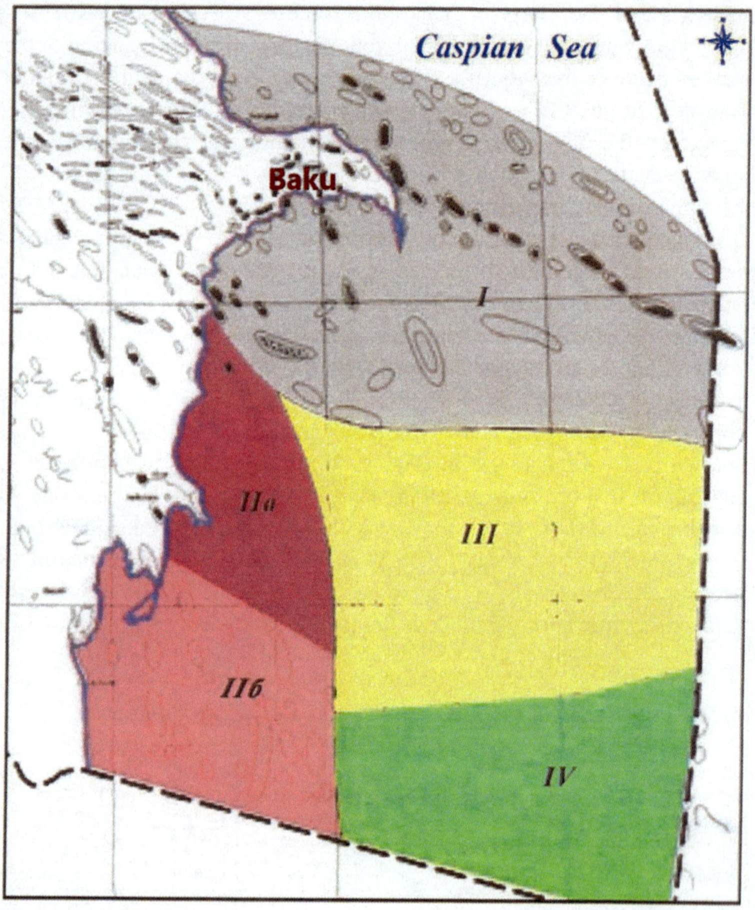

Fig. 19.27 A schematic map showing the SCB water area ranking according to the outlook degree

References

Alizadeh, A. A., Guliyev, I. S., Kadirov, F. A., & Eppelbaum, L. V. (2017). *Geosciences in Azerbaijan. Volume II: Economic minerals and applied geophysics* (340 p.). Springer.

Feizullayev, A. A., & Shykhaliyev, Y. A. (2011). On the strategy for searching for oil and gas in deep-seated sediments of the South Caspian Basin. *Oil & Gas Journal (Russia), 1–2*, 30–35. (in Russian).

Feyzullayev, A., & Shykhaliyev, Y. (2010). Promising directions of exploration in the South Caspian basin. *Oil & Gas Journal (Russia), 9*, 38–41. (in Russian).

Guliyev, I. S., Aliyev, G.-M.A., Aliyeva, E. G., & Muradov, C. S. (2000). Multicomponent anomaly in bottom sediments and seawater in the central part of the South Caspian basin. *Geochemistry, 9*, 1010–1017. (in Russian).

Guliyev, I. S., Aliyeva, E. G., & Huseynov, D. A. (2001). Deep sources of hydrocarbon formation in the South Caspian oil and gas basin. In *Proceedings of the Institute of Geology of the National Academy of Sciences of Azerbaijan* (No. 29, pp. 79–99). Nafta-Press (in Russian).

Guliyev, I. S., Aliyeva, E. G., Huseynov, D. A., et al. (2005). Hydrocarbon systems of nonequilibrium basins: Possibilities for improving the search for oil and gas deposits. *Izvestiya Academy of Sciences Azerbaijan, Series: Earth Sciences, 2*, 3–23. (in Russian).

Gulyiev, I. S., Feizullayev, A. A., & Huseynov, D. A. (2001). Carbon isotope composition of hydrocarbon fluids of the South Caspian megadepression. *Geochemistry, 3*, 271–278. (in Russian).

Guliyev, I. S., Levin, L. E., & Fedorov, D. L. (2003). *Hydrocarbon potential of the Caspian region* (120 p.). Nafta-Press (in Russian).

Kerimov, K. M., Gadzhiev, F. M., & Gasanov, I. S. (1999). Hydrocarbon resources of the Kura-South Caspian megadepression. *Azerbaijan Oil Industry, 7*, 1–11.

Shykhaliyev, Y. A., & Feyzullayev, A. A. (2010). On the oil&gas-bearing of the Absheron structure in the South Caspian depression: Pros and cons. *News of the Azerbaijan Academy of Sciences and Earth Sciences, 2*, 3–6.

Shykhaliyev, Y. A., & Gauzer, G. E. (2006). On the issue of forecasting zones of abnormally high reservoir pressures based on seismic data. *Geophysics (Azerbaijan), 1*, 21–25.

Conclusions

For about 150 years (most intensely in the first part of the Soviet period, 1920–1955), Azerbaijan has been a crucial source for hydrocarbon deposit exploration worldwide. However, this famous oil province has not lost its importance today. Since the search for significant hydrocarbon deposits has shifted mainly to the Caspian Sea, our book is primarily devoted to this region.

The productive, red-colored strata of the Lower Pliocene of the South Caspian basin is a unique geological phenomenon with no analogs. Orogenic processes that took place at the end of the Pontus and led to the uplift of land areas surrounding the Caspian Sea and a considerable drop in its level, which, according to some sources, reached from 600 m to 1500 m, led to the complete isolation of the Caspian Sea in the Lower Pliocene. Sedimentation at this time took place in the conditions of a small basin isolated in the South Caspian region, which accumulated the entire huge mass of terrigenous material supplied by the three largest river arteries - PaleoVolga, PaleoAmu-Darya, PaleoKur. Avalanche sedimentation at a rate reaching 2.5 mm/year and high subsidence rates of the basin bottom led to the formation of a unique 7-kilometer terrigenous Productive Series. This stratum contains up to 90% of all oil and gas reserves of Azerbaijan and South-Western Turkmenistan. This complex is characterized by the highest specific density of explored (proven) reserves and promising and forecasted oil and gas resources. The area of development of these deposits with established oil and gas potential according to the specificdensities of potential hydrocarbon resources belongs to the highest category of promising territories. The question arises: why, in more than 30 km of sedimentary sequence, represented by a wide stratigraphic range of depositsfrom the Aalenian stage of the Middle Jurassic to the Holocene inclusive (lasting about 174 Ma), it was precisely in the Lower Pliocene for an extremely short time (within 22.5 Ma) a complex of sediments was formed, accounting for about 25% of the total thickness of the sedimentary fill of the basin and having excellent reservoir properties. In this book, an attempt was made to collect some impressions about the Pliocene hydrocarbon deposits of Azerbaijan, including some authors developments. It is challenging to realize in one volume book of a middle format. In addition, much of the data is proprietary. Therefore, many of the best geological-geophysical results are beyond the scope of this book.

The geological part of this book presents in detail the following main items:

- Geological structure and architecture of the South Caspian Basin (SCB),
- Lithological composition of the Productive Series,
- PS' lithofacial mapping,
- PS' mineralogical composition and demolition sources,
- PS' reservoir properties,

A. Alizadeh et al., *Pliocene Hydrocarbon Sedimentary Series of Azerbaijan*,
Advances in Oil and Gas Exploration & Production, https://doi.org/10.1007/978-3-031-50438-9

- PS' oil and gas-bearing,
- Geochemical characteristics of the organic matter,
- Peculiarities of the migration process in the SCB,
- Maturity of the hydrocarbon fluids,
- Conditions of the hydrocarbon's preservation,
- Tectonic-dynamic models of deposit generation within the SCB,
- Recommendation for future investigations.

The geophysical part of this book mainly contains seismic data (the traditional main geophysical instrument for hydrocarbon studies). Meanwhile, potential geophysical fields (gravity, magnetic, and thermal) and paleomagnetic data are also presented. A separate Chapter applies informational and probabilistic-statistical approaches in oil and gas geophysics. Since searching for hydrocarbons is impossible without knowledge of deep structure, independent attention is paid to the integrated investigations of Azerbaijan's Earth crust (both on land and sea). A novel geophysical-geodynamic approach was employed for unmasking the total area of the Akchagylian stage in the Caucasian and other regions.

In conclusion, the book's authors want to thank thousands of ordinary geologists and geophysicists who have worked in Azerbaijan for different periods and are working in this region now. The comprehensive geological-geophysical studies for discovering new Azerbaijan hydrocarbon reserves are continuing…

Index

A. Alizadeh et al., *Pliocene Hydrocarbon Sedimentary Series of Azerbaijan*,
Advances in Oil and Gas Exploration & Production, https://doi.org/10.1007/978-3-031-50438-9

F

G

H

I

K

L

M

N

Advances in Experimental Medicine and Biology

For further volumes:
http://www.springer.com/series/5584

Michael G. Rossmann • Venigalla B. Rao
Editors

Viral Molecular Machines

Springer

Editors
Michael G. Rossmann
Department of Biological Sciences
Purdue University
West Lafayette, IN 47907-2032, USA
mr@purdue.edu

Venigalla B. Rao
Department of Biology
The Catholic University of America
Washington, DC 20064, USA
rao@cua.edu

ISSN 0065-2598
ISBN 978-1-4614-0979-3 e-ISBN 978-1-4614-0980-9
DOI 10.1007/978-1-4614-0980-9
Springer New York Dordrecht Heidelberg London

Library of Congress Control Number: 2011941434

Printed on acid-free paper

Springer is part of Springer Science+Business Media (www.springer.com)

Acknowledgment

We thank Ms. Alice Kuaban for her assistance in the production of this book. We are grateful to all the authors who have freely given their time to make this book possible while having busy schedules to maintain their own research programs. Special thanks to our present and former lab members and collaborators for their contributions over the years. The research in our laboratories has been funded for many years by the National Science Foundation (MGR, MCB-0443899; VBR, MCB-0923873) and the National Institutes of Health (MGR and VBR, NIAID-R01AI081726).

Contents

Contributors

Matthew L. Baker National Center for Macromolecular Imaging, Verna and Marrs McLean Department of Biochemistry and Molecular Biology, Baylor College of Medicine, Houston, TX, USA

Dennis H. Bamford Department of Biosciences and Institute of Biotechnology, University of Helsinki, Viikki Biocenter, Helsinki, Finland

Ricardo A. Bernal Chemistry Department, University of Texas at El Paso, El Paso, TX, USA

Lindsay W. Black Department of Biochemistry and Molecular Biology, University of Maryland School of Medicine, Baltimore, MD, USA

Dennis T. Brown Department of Molecular and Structural Biochemistry, North Carolina State University, Raleigh, NC, USA

Christopher B. Buck Laboratory of Cellular Oncology, National Cancer Institute, National Institutes of Health, Bethesda, MD, USA

Sarah J. Butcher Structural Biology and Biophysics Programme, Institute of Biotechnology, University of Helsinki, Helsinki, Finland

Juan Chang National Center for Macromolecular Imaging, Verna and Marrs McLean Department of Biochemistry and Molecular Biology, Baylor College of Medicine, Houston, TX, USA

Lia Cardarelli Department of Biochemistry, University of Toronto, Toronto, ON, Canada

Giovanni Cardone Laboratory of Structural Biology, National Institute for Arthritis, Musculoskeletal and Skin Diseases, National Institutes of Health, Bethesda, MD, USA

Sherwood R. Casjens Division of Microbiology and Immunology, Pathology Department, University of Utah School of Medicine, Salt Lake City, UT, USA

Yann R. Chemla Department of Physics, Center for the Physics of Living Cells, Center for Biophysics and Computational Biology, University of Illinois, Urbana-Champaign, Urbana, IL, USA

Naiqian Cheng Laboratory of Structural Biology, National Institute for Arthritis, Musculoskeletal and Skin Diseases, National Institutes of Health, Bethesda, MD, USA

Wah Chiu National Center for Macromolecular Imaging, Verna and Marrs McLean Department of Biochemistry and Molecular Biology, Baylor College of Medicine, Houston, TX, USA

Graduate Program in Structural and Computational Biology and Molecular Biophysics, Baylor College of Medicine, Houston, TX, USA

Kyung H. Choi Department of Biochemistry and Molecular Biology, University of Texas Medical Branch, Galveston, TX, USA

Alan R. Davidson Department of Molecular Genetics, University of Toronto, Toronto, ON, Canada

Department of Biochemistry, University of Toronto, Toronto, ON, Canada

Robert W. Doms Department of Microbiology, University of Pennsylvania, Philadelphia, PA, USA

Bentley A. Fane Department of Plant Sciences, The BIO5 Institute, University of Arizona, Tucson, AZ, USA

Michael Feiss Department of Microbiology, Roy J. and Lucille A. Carver College of Medicine, University of Iowa, Iowa City, IA, USA

Barbie K. Ganser-Pornillos Department of Molecular Physiology and Biological Physics, University of Virginia School of Medicine, Charlottesville, VA, USA

Department of Cell Biology , The Scripps Research Institute, La Jolla, CA, USA

Tom S.Y. Guu Department of Biochemistry and Cell Biology, Rice University, Houston, TX, USA

Roger W. Hendrix Department of Biological Sciences, Pittsburgh Bacteriophage Institute, University of Pittsburgh, Pittsburgh, PA, USA

Raquel Hernandez Department of Molecular and Structural Biochemistry, North Carolina State University, Raleigh, NC, USA

J. Bernard Heymann Laboratory of Structural Biology, National Institute for Arthritis, Musculoskeletal and Skin Diseases, National Institutes of Health, Bethesda, MD, USA

Zacariah L. Hildenbrand Chemistry Department, University of Texas at El Paso, El Paso, TX, USA

John E. Johnson Department of Molecular Biology, MB31, The Scripps Research Institute, La Jolla, CA, USA

Nelli J. Karhu Next Biomed Technologies NBT Oy, Helsinki, Finland

Amy Kendall Department of Biological Sciences, Vanderbilt University, Nashville, TN, USA

Kazuhiko Kinosita, Jr. Department of Physics, Waseda University, Okubo, Shinjuku-ku, Tokyo, Japan

Petr G. Leiman Institut de physique des systèmes biologiques, Ecole Polytechnique Fédérale de Lausanne (EPFL), Lausanne, Switzerland

Xiangan Liu National Center for Macromolecular Imaging, Verna and Marrs McLean Department of Biochemistry and Molecular Biology, Baylor College of Medicine, Houston, TX, USA

Ming Luo Department of Microbiology, University of Alabama at Birmingham, Birmingham, AL, USA

Erika J. Mancini Division of Structural Biology, Wellcome Trust Centre for Human Genetics, Oxford University, Oxford, UK

Violeta Manole Structural Biology and Biophysics Programme, Institute of Biotechnology, University of Helsinki, Helsinki, Finland

Karen L. Maxwell The Donnelly Centre for Cellular and Biomolecular Research, Department of Molecular Genetics, University of Toronto, Toronto, ON, Canada

Leonard Mindich Public Health Research Institute Center, UMDNJ – New Jersey Medical School, Newark, NJ, USA

Ian J. Molineux Section of Molecular Genetics and Microbiology, Institute for Cellular and Molecular Biology, University of Texas at Austin, Austin, TX, USA

Marc C. Morais Department of Biochemistry and Molecular Biology, Sealy Center for Structural and Computational Biology, University of Texas Medical Branch at Galveston, Galveston, TX, USA

Elena V. Orlova Crystallography and Institute of Structural and Molecular Biology, Birkbeck College, London, UK

Lisa G. Pell Department of Molecular Genetics, University of Toronto, Toronto, ON, Canada

Department of Biochemistry, University of Toronto, Toronto, ON, Canada

Minna M. Poranen Department of Biosciences and Institute of Biotechnology, University of Helsinki, Viikki Biocenter, Helsinki, Finland

Owen Pornillos Department of Molecular Physiology and Biological Physics, University of Virginia School of Medicine, Charlottesville, VA, USA

Department of Cell Biology, The Scripps Research Institute, La Jolla, CA, USA

B.V. Venkataram Prasad Department of Biochemistry and Molecular Biology, Keck Center for Computational Biology, Baylor College of Medicine, Houston, TX, USA

Peter E. Prevelige Department of Microbiology, University of Alabama at Birmingham, Birmingham, AL, USA

Devon R. Radford Department of Molecular Genetics, University of Toronto, Toronto, ON, Canada

Venigalla B. Rao Department of Biology, The Catholic University of America, Washington, DC, USA

Ryan H. Rochat National Center for Macromolecular Imaging, Verna and Marrs McLean Department of Biochemistry and Molecular Biology, Baylor College of Medicine, Houston, TX, USA

Graduate Program in Structural and Computational Biology and Molecular Biophysics, Baylor College of Medicine, Houston, TX, USA

Michael G. Rossmann Department of Biological Sciences, Purdue University, West Lafayette, IN, USA

Michael F. Schmid Department of Biochemistry and Molecular Biology, Keck Center for Computational Biology, Baylor College of Medicine, Houston, TX, USA

Mikhail M. Shneider Shemyakin-Ovchinnikov Institute of Bioorganic Chemistry, Moscow, Russia

Douglas E. Smith Department of Physics, University of California at San Diego, La Jolla, San Diego, CA, USA

Alasdair C. Steven Laboratory of Structural Biology, National Institute for Arthritis, Musculoskeletal and Skin Diseases, National Institutes of Health, Bethesda, MD, USA

Gerald Stubbs Department of Biological Sciences, Vanderbilt University, Nashville, TN, USA

Yizhi J. Tao Department of Biochemistry and Cell Biology, Rice University, Houston, TX, USA

Paulo Tavares Unité de Virologie Moléculaire et Structurale, UPR CNRS 3296 and IFR 115, Gif-sur-Yvette, France

Julie A. Thomas Department of Biochemistry and Molecular Biology, University of Maryland School of Medicine, Baltimore, MD, USA

John C. Tilton Department of Microbiology, University of Pennsylvania, Philadelphia, PA, USA

Benes L. Trus Laboratory of Structural Biology, National Institute for Arthritis, Musculoskeletal and Skin Diseases, National Institutes of Health, Bethesda, MD, USA

Imaging Sciences Laboratory, Center for Information Technology, National Institutes of Health, Bethesda, MD, USA

Roman Tuma The Astbury Centre for Structural Molecular Biology, Institute of Cellular and Molecular Biology, University of Leeds, Leeds, UK

Craig B. Wilen Department of Microbiology, University of Pennsylvania, Philadelphia, PA, USA

Mark Yeager Department of Molecular Physiology and Biological Physics, University of Virginia School of Medicine, Charlottesville, VA, USA

Department of Cell Biology, The Scripps Research Institute, La Jolla, CA, USA

Wenjie Zheng Department of Biochemistry and Cell Biology, Rice University, Houston, TX, USA

Sophie Zinn-Justin Laboratoire de Biologie Structurale et Radiobiologie, iBiTec-S and URA CNRS 2096, Gif-sur-Yvette, France

Chapter 1
Viruses: Sophisticated Biological Machines

Michael G. Rossmann and Venigalla B. Rao

Abstract Virus infection involves coordination of a series of molecular machines, including entry machines, replication machines, assembly machines, and genome packaging machines, leading to the production of infectious virions. This chapter provides an introduction to various viral molecular machines described in this book.

Ever since the observations by Dimitri Ivanovsky in 1892 and 6 years later by Martinus Beijerinck that a "filterable" virus causes tobacco mosaic disease, viruses had captured the interest of biologists, chemists, and physicists alike. It took nearly 30–40 years to resolve the controversy about the true nature of viruses, whether they are products of the host or whether they are "minute living organisms" distinct from the host. Worth mentioning are the contributions by Félix d'Herelle, who in 1917 coined the term "phage" and described the nature of bacteriophages; Wendell Stanley, who in 1935 showed that viruses have a discrete size and mass, containing roughly equal amounts of nucleic acids and proteins; and Max Delbrück, Salvador Luria, and Alfred Hershey, who in the 1940s gave a detailed description of a typical bacteriophage life cycle that paved the way for modern virology. Finally, viruses are now established as being unique "organisms," which do not multiply by "fission" like the rest of the biological world, but instead use the host cell as a factory to assemble hundreds to thousands of virus "particles."

In the late 1950s, the determination of the three-dimensional structures of myoglobin and hemoglobin was a milestone in the development of structural biology. For the subsequent two decades the primary focus was on understanding the utilization of prosthetic groups by proteins like hemes for oxygen transport and the mechanisms of simple enzymes like glycosidases, proteases, and nucleases. At that time protein structures at near atomic resolution was a novelty. Gradually it became apparent that proteins were built of functional domains. These domains could be integrated into single polypeptide chains in such a way that a substrate bound between the domains, each with a specific function. Furthermore, regulation of these processes could be obtained by associating the

M.G. Rossmann (✉)
Department of Biological Sciences, Purdue University, 240 S. Martin Jischke Drive, West Lafayette, IN 47907-2032, USA
e-mail: mr@purdue.edu

V.B. Rao (✉)
Department of Biology, The Catholic University of America, 620 Michigan Avenue NE, Washington, DC 20064, USA

M.G. Rossmann and V.B. Rao (eds.), *Viral Molecular Machines*,
Advances in Experimental Medicine and Biology 726, DOI 10.1007/978-1-4614-0980-9_1,

polypeptide subunits into oligomers. Thus knowledge of these, relatively simple, small molecular machines accumulated. At the same time the techniques for the study of these machines grew in sophistication and power to consider structure determination of much larger assemblies such as viruses.

The first structural information on viruses was based on X-ray diffraction analyses in the classical work on the rod-shaped Tobacco Mosaic Virus by Bernal and Fankuchen in the 1930s and later by Franklin, Klug, and Holmes in the 1950s. The first three-dimensional, near atomic resolution, structures of simple, spherically shaped, icosahedral viruses started to emerge in the late 1970s and was based on X-ray diffraction from single crystals. However, complexes of these viruses with their receptors or neutralizing antibodies could not be crystalized. Furthermore, larger viruses, particularly those with lipid envelopes, were often too heterogenous or are too large to form sufficiently strong lattice contacts to produce useful crystals. Fortunately 3D electron microscopy, using initially negative staining but later more productive freezing techniques, became available to study viruses at moderate resolution supplemented by crystallography of the component parts.

Although viruses had been considered as merely dull, static containers, and protectors of genomes, this false concept was replaced by the realization that viruses are beautiful intricate machines, essential to biological evolution, capable of invading cells, stealthily avoiding the protective barriers of the host, usurping the host's synthetic machinery for their own survival and able to self assemble into complex molecular machines. Indeed it has become apparent that the capabilities of viral machines far exceed those of the simple enzymes first studied in the mid-twentieth century. This book is a partial description of some of the amazing things accomplished by viruses in infecting a host and replicating themselves.

The first essay (Chap. 2) is not about viruses at all but about the F_1-ATPase rotary motor that is part of the F_1 F_o-ATP synthase machine that makes ATP using the energy of an electrochemical gradient to pump protons from one side of a membrane to the other. It is the prototypical machine that has aspects common to many of the viral molecular machines. For instance, the ATP binds to a protein fold that is probably the most common fold in all proteins and forms the engine of nearly all ATP-driven molecular devices. To understand the mechanism of this motor is an entry into understanding other biological motors.

The third chapter deals with the cryo microscopy, a major new tool that has been developed to aid in the determination of virus structure. The original techniques for the study of virus structure were based fairly exclusively on X-ray diffraction techniques that have been described in numerous text books. However, the more recently developed cryo-electron microscopy techniques are vital for the study of the huge conformational changes that occur during the life cycle of viruses. Thus this chapter describes single particle cryo-electron microscopy and cryo-electron tomography techniques as used to determine virus structure.

The fourth chapter describes the principles of virus structure, clearly an essential topic for the study of any virus. This topic grew up alongside the expanding knowledge of protein structure, but was initially merely abstract ideas based on symmetry. With time these have been translated into detailed structural information for many viruses.

The subsequent chapters deal with the consecutive events in the life cycle of a virus. Part I examines the processes that occur when a virus recognizes a molecule that is likely to yield to the virus' desire to enter the host, either by inserting only its genome into the host's cytoplasm (Chaps. 5–7) or by encouraging the host cell to envelope the entering virus into an endocytotic vesicle (Chaps. 8–10). Having entered the cell it is then necessary for the virus to have its genome replicated numerous times and its structural proteins and replication-aiding enzymes synthesized as described in Part II, Chaps 11 and 12. However, at least some of the new proteins need viral- or host-coded chaperons to complete their reproduction. This is accomplished by a variety of better- (like GroEL of *E. coli*) and lesser-known chaperone molecular machines (Part 3, Chap. 13). With that task done, the different, newly formed, viral components need to be assembled into progeny virions as described in Chaps.

14–20, Part III. Most viruses code for one or more types of scaffolding proteins to aid the assembly process (Chap. 14). The virus assembly process is perhaps the most amazing event of all of these functions and has been extensively studied. During this process the partially assembled virions go from immature noninfectious objects to mature infectious particles. The process of arming the virus ("maturation") needs to be accomplished carefully in order not to damage the host cell that is harboring the new crop of virions. Thus maturation must be timed to occur when the new particles are at the point of leaving their host, often in a budding process. The assembly for six very different types of viruses is then examined in Chaps. 15–20. Finally some of the bigger assembled parts need to be put together. This especially relates to bacteriophages and herpes viruses that assemble empty procapsids which are then packaged with their genome. This process requires recognition of the correct nucleic acid molecule for pumping into and concentrating in the head using energy supplied by an ATPase motor. Once the head has been filled, it is sealed by plugging the portal with a class of proteins that form the "neck" in tailed bacteriophages. These aspects are discussed in a series of Chaps. 21–27, Part IV. The final chapter (Chap. 28) describes genome packaging in helical viruses where the capsid protein subunits and the nucleic acid coassemble to form the infectious virion.

After nearly a century of virus research, we can now not only visualize the beauty of many viral molecular machines in amazing detail but also begin to understand their dynamic mechanisms. We hope that this collection of essays describing some of these machines will be useful to students as well as postdoctoral and senior scientists. We are grateful to all the authors who have freely given their time to make this book possible while having busy schedules required to maintain their own research objectives.

Chapter 2
F_1-ATPase: A Prototypical Rotary Molecular Motor

Kazuhiko Kinosita, Jr.

Abstract F_1-ATPase, the soluble portion of ATP synthase, has been shown to be a rotary molecular motor in which the central γ subunit rotates inside the cylinder made of $\alpha_3\beta_3$ subunits. The rotation is powered by ATP hydrolysis in three catalytic sites, and reverse rotation of the γ subunit by an external force leads to ATP synthesis in the catalytic sites. Here I look back how our lab became involved in the study of this marvelous rotary machine, and discuss some aspects of its rotary mechanism while confessing we are far from understanding. This article is a very personal essay, not a scientific review, for this otherwise viral machines book.

2.1 A Rotary Molecular Machine

To my common sense (in a bad sense), rotation within a molecule seemed impossible, because I was taught that a protein molecule is made on the basis of the lock-and-key principle, or that any part of a molecule is complementary to the other (Fig. 2.1). A distortion is alright, or even a shift by one tooth unit or two along a well-designed interface may be possible. Rotation, however, would require continuous shift of one part against the other, over 360°.

So, at about the year 1980 when Oosawa began to preach that F_1-ATPase must rotate and begged young Japanese biophysicists to show him the rotation experimentally, I did not listen to him. I was not aware that Boyer (Boyer and Kohlbrenner 1981) and other eminent scientists also proposed that F_1, or the ATP synthase of which F_1 is a part, may well rotate. Oosawa's theory was published much later (Oosawa and Hayashi 1986), to which I did not pay much attention at the time. In retrospect, I could have been more imaginative, already knowing that myosin slides past actin for many reaction cycles (Huxley 1957; Huxley 1969) and that bacterial flagellar motor, a gigantic machinery, rotates (a review by Berg 2003). If a linear molecular motor such as myosin can shift on a substrate track continually, all one needs to make a rotary motor is to roll the track into a ring (Kinosita et al. 1998).

Earlier in 1977, I happened to listen to a lecture by Kagawa, who showed that active F_1-ATPase can be reconstituted from purified subunits of thermophilic origin (Yoshida et al. 1977). I was deeply impressed, and dreamed of future collaboration. Oosawa's proposal in 1980, however, failed to push

K. Kinosita, Jr. (✉)
Department of Physics, Waseda University, Okubo, Shinjuku-ku, Tokyo 169-8555, Japan
e-mail: kazuhiko@waseda.jp

M.G. Rossmann and V.B. Rao (eds.), *Viral Molecular Machines*,
Advances in Experimental Medicine and Biology 726, DOI 10.1007/978-1-4614-0980-9_2,

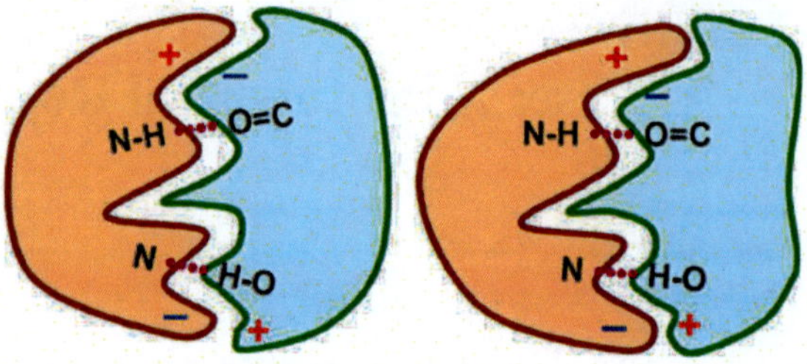

Fig. 2.1 Lock-and-key principle of protein construction

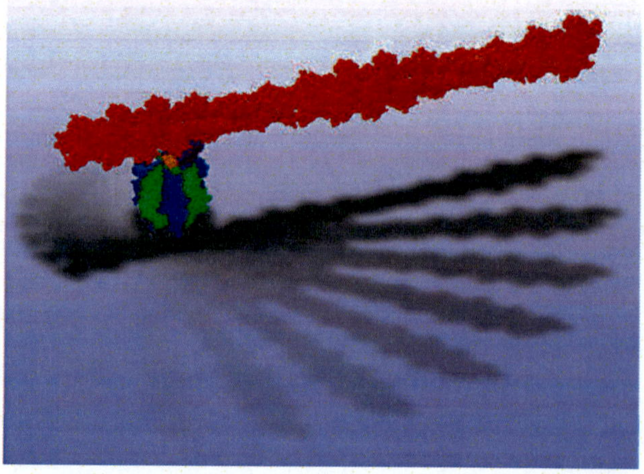

Fig. 2.2 Rotation of F_1-ATPase, a part of the enzyme ATP synthase. The rotary molecular motor consists of six stator subunits (three β in *green* and three α in *blue*) and a central rotor subunit γ (*orange*). Its rotation, powered by ATP hydrolysis, has been videotaped under an optical microscope by attaching a long actin filament (*red*; actual length far exceeds the image size) to the rotor through streptavidin (*purple*)

me into real collaboration. I was simply full of timid common sense. It was only after young associates in our lab, excluding myself, successfully began single-molecule work that the long-dreamed collaboration materialized (Noji et al. 1997), with the team of Yoshida, a former associate of Kagawa. People involved in the collaboration have different stories of how the collaboration began. Indeed, several members of both teams had thought about this collaboration, including those who eventually could not join. From my standpoint it was my admiration for the work of Kagawa's team that led me to talk to Yoshida. More important, it was no longer timid myself who would venture into the rotation experiment.

The collaboration started in 1996 and immediately bore fruit. By that time, a crystal structure of F_1 had been solved (Abrahams et al. 1994), showing that the putative rotor, the γ subunit, deeply penetrates the stator cylinder made of $\alpha_3\beta_3$ subunits. The tip of the rotor was made mostly of hydrophobic residues and the portion of the stator that surrounds the tip was also hydrophobic. The oily, or waxy, residues would act as a "molecular bearing" (Abrahams et al. 1994), strongly suggesting rotation. Both Yoshida and I were still highly suspicious, but were willing to bet on young associates. The actual bet we agreed upon, with young Noji and Yasuda, was not whether F_1 would rotate. We betted upon the sense of rotation, right or left. That is the way single-molecule physiologists must adopt.

Yasuda found a rotating F_1 molecule (actually a rotating actin filament that was supposedly attached to the rotor subunit) in the very first observation chamber he made for microscopic observation (Fig. 2.2). This sample was the first mutant that Noji prepared, among many he had designed. These two were the luckiest people whom I have known personally. I cannot understand why, but I am certain that being loved by Lady Luck is a talent. The talent I desperately lack.

Another rotating filament was soon found, and we were ready to toast after a third. But we had to wait for 2 months for the third. That's our single-molecule physiology.

2.2 Rotary Mechanism

Various aspects of the rotary mechanism of F_1 and experimental evidence leading to our tentative views have been discussed (Adachi et al. 2010, 2011). Below I present my current, personal views on some topics.

2.2.1 Symmetry Considerations

The primary reason why rotation in F_1-ATPase was proposed in early days was that there are three catalytic sites that are equivalent in function and there is only one γ subunit that was known to be crucial to catalysis and to be devoid of a threefold symmetry. The only way for the asymmetric γ to interact with the three catalytic sites impartially would be to rotate (not necessarily in a unique direction).

Physicists tend to believe that Nature likes symmetry, or at least physicists are the people who love the idea of symmetry. The Monod–Wyman–Changeux model of allosteric transitions (Monod et al. 1965), for example, was very popular among biophysicists of which I was one. At present, though, I think that the experimental distinction between this concerted model and the induced-fit theory (Koshland 1958) is very difficult, or rather, the truths are somewhere in between.

Shortly after I was shown the F_1 rotation, I began to think about how it may rotate. The crystal structure solved by Abrahams et al. (1994) had shown that the three catalytic sites reside at β–α interfaces, hosted primarily by β (Fig. 2.3a, b). As seen in Fig. 2.3a, the upper portion of the β subunit is bent toward γ when ATP (analog) or ADP is bound, whereas it is unbent in the nucleotide-free state, suggesting that these nucleotide-dependent bending and unbending drive γ rotation (Wang and Oster 1998). To correlate the nucleotide states and γ rotation, I drew a diagram like one in Fig. 2.3c, where γ was represented by a triangle. The simplification itself should have been okay, but somehow the triangle fooled me into thinking that the interaction of γ with a β subunit would basically (though not exactly) be threefold symmetric (Fig. 2.3d). This led me nowhere into rotation, which requires broken symmetry. Soon I learned that previous workers (Oosawa and Hayashi 1986; Wang and Oster 1998) correctly assumed non-threefold interactions such as one in Fig. 2.3e.

Symmetry consideration may turn out to be useful in predictions and explanations, but Nature does not always respect it. In vivo, the F_1 motor is connected to another rotary motor F_o, which is driven by proton flow, to constitute the ATP synthase (Fig. 2.4). The proton motive force under physiological conditions operates in the direction from top to bottom in Fig. 2.4, which drives F_o clockwise when viewed from above. This rotary direction is opposite to that of F_1 which rotates counterclockwise when it hydrolyzes ATP. The rotors of the two motors are supposed to be fused to each other, and hence clockwise rotation of F_o drives the F_1 motor in its reverse direction, resulting in the reversal of the ATP hydrolysis reaction in the three catalytic sites of F_1. This is the accepted view of the mechanism of ATP synthesis. A small problem from a physicist's view is that the rotor of the F_o motor comprises a ring of several *c* subunits, to which the asymmetric γ rotor of F_1 is attached via ε. Symmetry would predict that $\gamma\varepsilon$ would rotate against the *c* ring, but then ATP synthesis would fail. I think it is still possible that the interface between $\gamma\varepsilon$ and the *c* ring serves as a kind of clutch, which may disengage when sufficient ATP has been accumulated. Another small uneasiness that I have been having is that the basically symmetric stator of the F_1 motor ($\alpha_3\beta_3$ ring) is

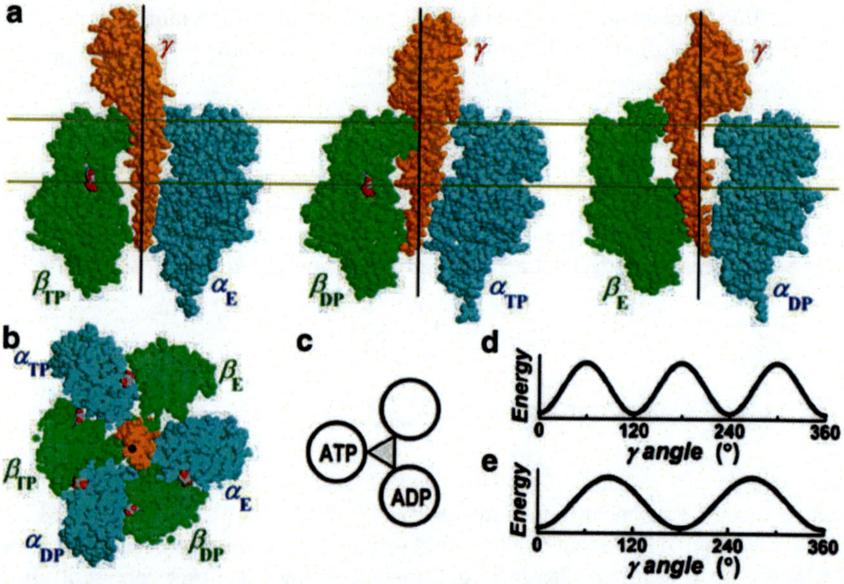

Fig. 2.3 Structure of F_1-ATPase and schematic diagrams. (**a, b**) An atomic structure of bovine mitochondrial F_1-ATPase (Gibbons et al. 2000). The γ rotor and an opposing pair of α and β subunits are shown in (**a**), and the bottom view of the section between the *gold lines* is shown in (**b**). Nucleotides are shown in CPK colors. The stator subunits are designated according to the nucleotides in catalytic sites found in the original structure (Abrahams et al. 1994): TP site between β_{TP} and α_{TP}, largely in β_{TP}, bound an ATP analog (AMPPNP), DP site ADP, and E site none, while non-catalytic sites in the other three interfaces bound the ATP analog. The *vertical line* in (**a**) and the *black dot* in (**b**) represent the putative rotation axis (Wang and Oster 1998). (**c**) A schematic diagram showing the relationship between catalytic nucleotides and γ orientation. (**d**) A possible diagram showing the interaction energy between γ and a β subunit. (**e**) Another possible energy diagram

connected to the single-unit F_o stator (*a*) via δb_2. In the whole ATP synthase, the three catalytic sites of F_1 may not be entirely equivalent.

Linear molecular motors such as kinesin and myosin have two identical legs (no distinction between right and left feet) that are joined in twofold symmetry as opposed to the mirror symmetry in humans (Kinosita et al. 2005). The most natural way for these molecular motors to walk is to rotate 180° every step, as in a dance (Howard 1996). Several attempts to observe this all failed (Kinosita et al. 2005) showing that the motors, under the experimental constraints, can walk disregarding the twofold symmetry. But recently Komori et al. (2007) have succeeded in showing rotation in either way randomly, indicating that molecular motors want to observe symmetry principles. I still hope to observe rotation in a unique direction.

Beauty is the prime goal of (theoretical) physics. In the biological world of individualism, there are many different beauties, and a beauty is not always a charm. C'est la vie, which I enjoy.

2.2.2 *Kinetics of Chemo-Mechanical Coupling*

The kinetic scheme we had arrived at by 2007 is summarized in Fig. 2.5, and evidence supporting the scheme has been discussed (Adachi et al. 2010). In brief, ATP binding initiates and drives rotation from 0° (an ATP-waiting angle; the angle descriptions below refer to the pink, or the lower left, catalytic site in Fig. 2.5b or c) to 80°. After 200° of rotation since the ATP binding, that ATP is split into ADP and

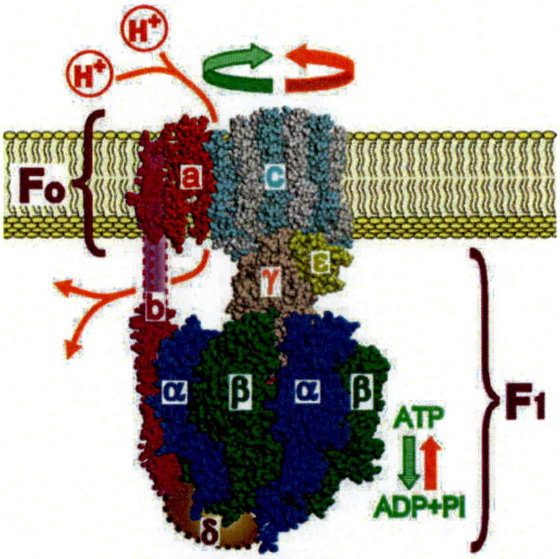

Fig. 2.4 Subunit assembly of ATP synthase. The minimal composition inferred from bacterial enzymes is $\alpha_3\beta_3\gamma\delta\varepsilon$ for F_1 and ab_2c_{8-15} for F_o (Yoshida et al. 2001; Junge et al. 2009; Watt et al. 2010). F_1 is a rotary motor driven by ATP hydrolysis (the subcomplex $\alpha_3\beta_3\gamma$ suffices for rotation in isolation), and F_o is another rotary motor driven by proton flow (Diez et al. 2004). The two motors are connected and coupled, such that the common rotor consists of $\gamma\varepsilon c_{8-15}$ and the stator assembly of $\alpha_3\beta_3\delta ab_2$. When the free energy obtained from proton flow (*curved orange arrows*) is greater than that from ATP hydrolysis, the F_o motor gains control and the rotor rotates in the *orange* direction, resulting in ATP synthesis in F_1. When F_1 wins, rotation in the *green* direction occurs by ATP hydrolysis and protons are pumped back. The atomic structures shown, arbitrarily arranged, are from Gibbons et al. (2000), Stock et al. (1999), Rastogi and Girvin (1999), Dmitriev et al. (1999), and del Rizzo et al. (2002)

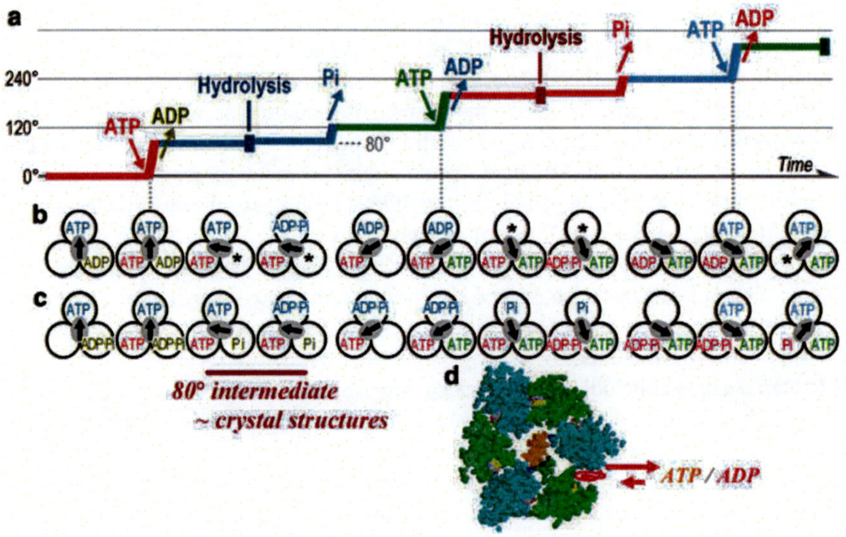

Fig. 2.5 Proposed scheme for coupling between catalysis and rotation. (**a**) Schematic time course of rotation. Colors indicate the site at which the rate-limiting reaction is to occur in (**b**). Angles quoted in the text refer to events in the *pink* (*lower left*) site. (**b, c**) Alternative schemes suggested by Adachi et al. (2007), differing in the timing of phosphate release by 120°. Chemical states of three catalytic sites (*circles*) and γ orientation (*central arrows*) are shown. (**d**) Filling of the third catalytic site by a medium nucleotide (Shimo-Kon et al. 2010). The *asterisked* site in (**b**) is freely accessible to medium nucleotide, which, upon entry, quenches the fluorescence of the reporter tryptophan (*yellow*). The crystal structure shown is from Gibbons et al. (2000), where one catalytic site is empty and is fully open

inorganic phosphate. The phosphate is either immediately released to drive rotation from 200° to 240° (Fig. 2.5b), or remains bound for another 120° rotation to drive the last rotation from 320° to 360° (Fig. 2.5c). ADP is released after a third ATP is bound, during rotation from 240° to 320°. The ADP release likely contributes to the 80° rotation in addition to the third ATP binding (Adachi et al. 2007).

We still had a few problems. The timing of phosphate release, either Fig. 2.5b or c, had to be solved. Also, either of the schemes indicates that the site occupancy, the number of nucleotides bound to the three catalytic sites, remains two for most of the time except for the brief moment of rotation over 80°. The occupancy number of two was at odds over tryptophan quenching studies including our own. Earlier, Weber et al. (1993) have introduced a reporter tryptophan residue in each catalytic site (yellow in Fig. 2.5d) to show that the tryptophan fluorescence is quenched when a nucleotide is bound. With *Escherichia coli* F_1, they have shown that the time-averaged site occupancy goes up to three when the ATP concentration is high and the enzyme is exerting its full activity. Dou et al. (1998) have confirmed the results with the thermophilic F_1 that we work with, and we were also getting similar results. Because the nucleotide scheme in Fig. 2.5 was derived from experiments using a fluorescent ATP analog (Cy3-ATP), we thought that the behavior of the analog may be different from that of unlabeled ATP used in the tryptophan studies.

The discrepancy in the site occupancy was resolved when Shimo-Kon noticed that active F_1 undergoing catalysis and inactive F_1 either in an inhibited state or binding ADP alone show the same occupancy pattern for the third site, as probed by tryptophan quenching (Shimo-Kon et al. 2010). The implication is that the third nucleotide is not necessarily the nucleotide being catalyzed; it may come from the medium. In Fig. 2.5b, the asterisked site that has been vacated by leaving ADP can be filled by a medium nucleotide, ATP or ADP (Fig. 2.5d). The Cy3-ATP experiments that led to Fig. 2.5b or c were made at very low (nanomolar) concentrations of the nucleotide, and thus the third site was not filled. This view that the third nucleotide comes mainly from the medium is consistent with the crystal structures solved by the Walker group. Of the many structures solved to date, only one structure binds three catalytic nucleotides (Menz et al. 2001), another one no nucleotide (Kabaleeswaran et al. 2009), and all the rest bind two nucleotides and the third site is fully open. We think that the two-nucleotide crystal structures resemble the 80° intermediate in the kinetic scheme (Fig. 2.5) and thus the intermediate would readily accommodate a medium nucleotide in the open site.

I have to confess that I liked the bi-site mechanism of rotation that Boyer has been championing (Boyer 1998, 2002). The site occupancy alternates between one and two in the bi-site mode, warranting maximal asymmetry that, I thought, would assure powerful rotation. So, it was not without my personal regret that we eliminated almost completely the possibility of bi-site catalysis in our enzyme working with unlabeled ATP (Shimo-Kon et al. 2010).

An added bonus of the tryptophan study (Shimo-Kon et al. 2010) was the settlement of the timing of phosphate release. Because the asterisked site in Fig. 2.5b allowed binding of a medium ATP, the site cannot hold phosphate as in Fig. 2.5c. When we added phosphate in the medium, binding of ATP to the asterisked site was hindered.

I was to stop here, to conclude that the kinetic coupling scheme is now complete. The site occupancy problem that had been lingering in my mind for years is finally gone. Watanabe et al. (2010), however, have recently reported intricate experiments that indicate, at least for the slow-hydrolysis mutant they used, phosphate release at 200° is too slow to be compatible with the scheme in Fig. 2.5b. The authors suggest Fig. 2.5c. To reconcile with the observation above that the third site can accommodate a medium ATP, the authors also suggest that the phosphate retained by 320° would immediately be released upon reaching 320°, leaving the 320° site open for most of the 320° dwell (the asterisked site in Fig. 2.5 will also be open). But then the 40° rotation from 320° to 360° cannot be driven by phosphate release, as opposed to our contention that phosphate release confers driving torque for the 40° rotation, whether from 200° to 240° or from 320° to 360° (Adachi et al. 2007). Something, or somethings, must be wrong, or the F_1-ATPase allows different kinetic pathways and chooses one (or more) depending on the reaction conditions and/or mutations.

Thus, at this moment, I no longer have a clear-cut view of F_1 kinetics. All aspects of the kinetics may have to be re-examined. In this regard, I note that, under a certain circumstance, ATP hydrolysis may take place at 120° (Shimabukuro et al. 2006), not at 200° as shown in Fig. 2.5b or c. Also, in a related enzyme V_1-ATPase of *Thermus thermophilus*, rotation proceeds in steps of 120°, without the 80° and 40° substeps as in F_1; all chemical reaction steps, at least ATP binding, hydrolysis, and another rate-limiting reaction likely ADP or phosphate release, take place at the same angle (Furuike et al. 2011). Why, then, does F_1 bother to divide them into two angles? Possibly to produce torque over a wide range of angles and to synthesize ATP efficiently? If so, why does the *Thermus* V_1, which also participates in efficient ATP synthesis when combined with V_o (Toei et al. 2007), not adopt the same strategy?

2.2.3 Binding Changes and ATP Synthesis

Boyer has proposed that binding changes are the essence of the function of ATP synthase (Boyer 1998). The final step of the synthesis, for example, is the release of a tightly bound ATP, or a decrease in the binding constant. The binding change results from a conformational change of the catalytic site, and the conformational change is driven by mechanical rotation. Other chemical steps, bindings of ADP and phosphate to the catalytic site and linking of phosphate to ADP in the catalytic site, all occur through rotation-driven binding changes.

Isolated F_1 alone has been shown to catalyze ATP synthesis when the γ rotor is forced to rotate in reverse by an external, artificial force (Itoh et al. 2004; Rondelez et al. 2005), implying that the binding changes can be effected by γ rotation alone, or that the γ angle determines the equilibrium constants of each catalytic site for ADP binding, phosphate binding, ATP synthesis/hydrolysis, and ATP unbinding (a "γ-dictator," or "γ-controlled," mechanism). In the rotation driven by ATP hydrolysis, the central γ will coordinate the chemical steps in the three catalytic sites.

A general framework of how the mechanical work of rotation (the mechanical energy liberated in the hydrolysis-driven rotation, or the energy input required for synthesis rotation) is coupled to the binding changes (shifts of the equilibrium constants of the catalytic sites) has been discussed and used to build models of F_1 rotation (Oosawa and Hayashi 1986; Wang and Oster 1998; Kinosita et al. 2004). All models treat the equilibrium constants as a continuous function of the γ angle, and hence the energy involved, or the torque as the derivative of the energy, is also a continuous function of the γ angle. This is in accord with the observation that, apparently, F_1 generates torque at all angles except when it is waiting for ATP binding and phosphate release (Kinosita et al. 2000; Pänke et al. 2001; Palanisami and Okamoto 2010). (This "power stroke" view, though, represents only one side of a coin; see Adachi et al. 2010, and discussion below).

Angle-dependent binding change has been quantified for phosphate around the phosphate release angle: the association constant for phosphate decreases by a factor well above 10^4 upon rotation from 200° to 240° (Adachi et al. 2007). Experiments have also indicated that the association constant for ATP (or the rate of ATP binding) increases around the ATP-binding angle (Watanabe-Nakayama et al. 2008; Iko et al. 2009). Using Cy3-ATP and Cy3-ADP, we have been measuring the binding changes for ATP and ADP over all angles and under various conditions. Years of analyses are now almost complete, which we will report in the near future. One hint from the analyses is that ATP synthesis in F_1 appears to be basically the reverse of hydrolysis: when forced to rotate in the reverse direction, catalytic events seem to occur from right to left in Fig. 2.5a.

Our current view, yet abstract, is that binding changes and protein conformational changes (including γ rotation) are two sides of the same coin (Fig. 2.6). ATP, for example, binds through the process of "induced fit" (Koshland 1958), whereby the catalytic site adapts itself to better accommodate ATP, increasing the number of weak bonds (blue dots in Fig. 2.6) that hold ATP. Thus, if ATP

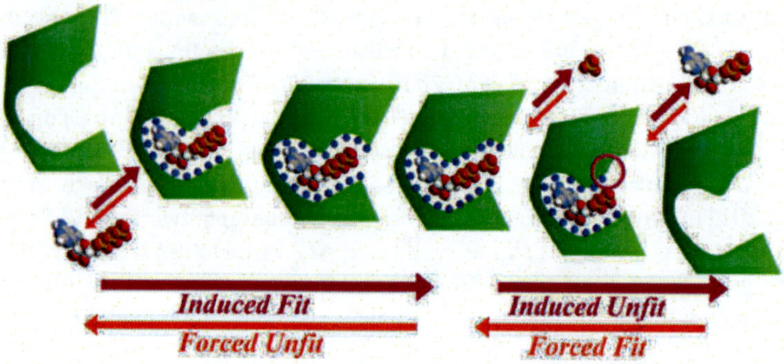

Fig. 2.6 Chemo-mechanical and mechano-chemical energy conversion through the processes of "induced fit" and "induced unfit." *Violet arrows* show reactions driven by free energy liberated by ATP hydrolysis; *red arrows*, reverse reactions driven by an external force. *Green blocks* show conformations of the protein part in a highly schematized fashion, their vertical locations representing the free energy level in an environment favoring hydrolysis and in the absence of an external force. *Small blue dots* surrounding a nucleotide represent weak bonds (mainly hydrogen bonds) that hold the nucleotide in the catalytic site. Through progressive formation of these bonds a binding ligand (ATP) pulls the protein toward a conformation that better fits the ligand (induced fit). A leaving nucleotide promotes a reverse process, which we call "induced unfit." Both induced fit and induced unfit can be reversed by an external force on the protein part, resulting in forced binding changes. The *pink circle* highlights the small protrusion that would hinder ATP binding (but not ADP and phosphate separately); ATP binding should somehow be hindered for efficient synthesis

binding drives a conformational change, that particular conformational change, when induced thermally or by an external force, will increase the affinity of the catalytic site for ATP. Likewise, phosphate or ADP release goes through "induced unfit," and the conformational change accompanying the release process will decrease the affinity. To state that binding/unbinding drives a conformational change is the so-called power stroke view. An alternative view is that a thermal fluctuation into a proper conformation induces and stabilizes binding/unbinding ("conformational selection"). Again, truths are always somewhere in between, I suppose (Adachi et al. 2010).

The induced fit and induced unfit are both basically downhill, toward lower free energy. An external force applied to the protein part can reverse these processes, by forcing fit and unfit. This is how we view ATP synthesis on the protein machine (Fig. 2.6, from right to left). For efficient synthesis, forced fit for ADP must not allow ATP binding, and the little projection in the pink circle in Fig. 2.6 is a highly schematic representation of this distinction mechanism, one possibility being preoccupation by phosphate which must be present in the medium for ATP syntheses. In the last step of synthesis, it must be ATP and not ADP that is released into the medium. For this, we conjecture that a small clockwise rotation of γ would shift the equilibrium between ADP + phosphate and ATP toward the synthesis side. A corollary, then, is that ATP hydrolysis in the catalytic site would accompany a small counterclockwise rotation of γ, which seems natural to me.

2.2.4 Structural Basis of Rotation

Wang and Oster (1998) made movies of rotating F_1 by interpolating the crystal structure (Abrahams et al. 1994) in which the three β subunits adopt different conformations as though the conformational changes directly drive γ rotation. The movies were pretty much impressive to me, and I thought that the basic mechanism would be the push and pull of bending and unbending β against γ (Fig. 2.3a) as these authors suggested.

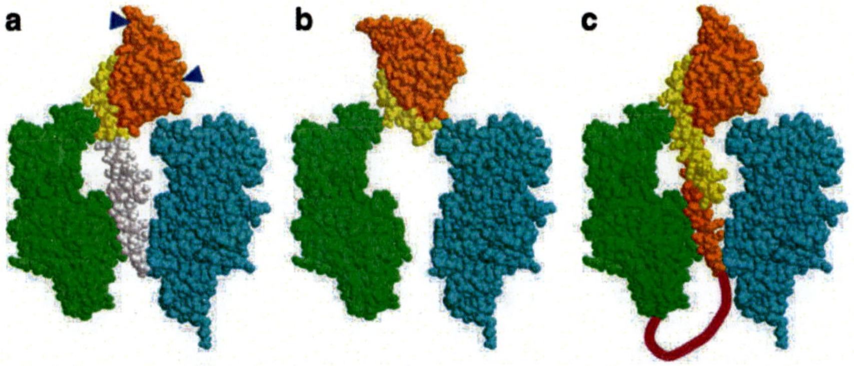

Fig. 2.7 Truncations of the γ rotor. (**a**) An axle-less construct in which the *white* portion is deleted (Furuike et al. 2008). *Arrowheads* indicate approximate positions of the cysteine residues introduced for the attachment of a rotation marker. (**b**) The rotor head in (**a**) appears to occasionally adopt such orientations where the marker (a *gold bead*) would be right above the rotor. (**c**) A construct (Kohori et al. 2011) devoid of the entire amino-terminal α helix (*yellow*). The carboxyl terminus is connected to β via a short peptide linker (*magenta*)

When Hossain joined us as a postdoc of physics background, I suggested if he might wish to redo the γ truncation work of Müller et al. (2002) as biology training. Thus started a never-ending story. Truncation of γ in the direction from bottom to top led him to the final product shown in Fig. 2.7a. This axle-less construct rotated in the correct direction for >100 revolutions, albeit slowly (Furuike et al. 2008). The rotation was visualized by attaching a gold bead to the two cysteine residues near the arrowheads in Fig. 2.7a, and the obliquely attached bead moved in a circle. Occasionally, though, the bead moved to the center of the circular trace, suggesting an upright orientation as in Fig. 2.7b. That the remaining rotor head in Fig. 2.7a continued to rotate without being carried away by thermal diffusion is already a surprise, and the orientation in Fig. 2.7b still clinging is more than amazing. The push–pull action requires a relatively rigid axle pivoted at the bottom. Obviously, the axle-less construct rotates by a different mechanism, which is totally a mystery to me.

An undergraduate student Kohori attempted a sideways truncation, removing the entire amino-terminal α helix of γ (yellow in Fig. 2.7c). The remaining carboxyl terminal helix was connected to a β subunit via a short peptide liner. This construct rotated at a quarter of the wild-type speed and produced approximately half the wild-type torque (Kohori et al. 2011). Truncation from top is now under way, and the preliminary indication is that the entire γ head is dispensable. If we take AND of all constructs, a conclusion seems inevitable that none of the γ residues are needed for rotation.

Is this the end of the story? That rotation does not depend on specific interactions between γ and the stator has led us to propose that F_1 may have evolved from a motor that can rotate anything (other than an object with threefold symmetry) albeit inefficiently. Evolution, though, cannot be proved, being one-time, irreproducible experiment that God is playing at. So, I am suggesting young colleagues to resort to a "creationist approach," to create F_1 that indeed rotates an unrelated object such as DNA.

2.3 Spiral Rotations and Viral Machines

Now I come to the viral packaging machine, and my interest is whether it rotates.

God did not invent too many different kinds of proteins and nucleotides, and he often resorts to symmetry for building larger structures, ending in helical filaments, disks, and spheres. Repetitive interactions with a helix (or a disk) should lead to rotation, as indeed observed. A dynein and some kinesins spirally rotate a microtubule (Vale and Toyoshima 1988; Walker et al. 1990; Yajima et al. 2008), myosin or formin rotates an actin filament (Nishizaka et al. 1993; Sase et al. 1997; Mizuno et al. 2011), and RNA polymerase rotates DNA (Harada et al. 2001). Conversely, myosin, when unconstrained, spirals around an actin filament (Ali et al. 2002, 2004).

These spiral motions often disregard the helical pitch of the filament. The pitch of spiral motion and the structural pitch of the helix will agree when movement during one interaction cycle is shorter than half the helical pitch, as for DNA polymerase and formin above and regular kinesin (Howard 1996). Linear molecular motors that walk with long strides disobey the helical pitch, and some appear to produce lateral force to go sideways.

There seems to be no a priori reason why the helical DNA being actively packaged into a viral capsid should not rotate. Although rotation of the portal motor has been denied (Hugel et al. 2007), the stepwise interaction of DNA with the packaging motor (Moffitt et al. 2009) suggests rotation. Extensive DNA rotation, however, would hinder its packaging, and thus if DNA does rotate, its torsion would have to be relaxed by occasional slippage in the motor or, say, by a topoisomerase action. As an ex-physicist, I do not care if DNA really rotates during packaging in vivo. If the machine has the potential of rotation, however, I would like to see it. If the machine is not willing, why not let it go, by artificially preventing slippage if needed. By the time this book appears, someone hopefully will have shown me a movie of rotation.

Acknowledgments I thank members of Kinosita and Yoshida labs for collaboration and discussion, R. Kanda-Terada for technical support, and S. Takahashi, K. Sakamaki, M. Fukatsu, and H. Umezawa for encouragement and lab management. This work was supported by Grants-in-Aids for Specially Promoted Research from the Ministry of Education, Sports, Culture, Science and Technology, Japan.

References

Abrahams JP, Leslie AGW, Lutter R, Walker JE (1994) Structure at 2.8Å resolution of F_1-ATPase from bovine heart mitochondria. Nature 370:621–628

Adachi K, Oiwa K, Nishizaka T, Furuike S, Noji H, Itoh H, Yoshida M, Kinosita K Jr (2007) Coupling of rotation and catalysis in F_1-ATPase revealed by single-molecule imaging and manipulation. Cell 130:309–321

Adachi K, Furuike S, Hossain MD, Itoh H, Kinosita K Jr, Onoue Y, Shimo-Kon R (2010) Chemo-mechanical coupling in the rotary molecular motor F_1-ATPase. In: Gräslund A, Rigler R, Widengren J (eds) Single molecule spectroscopy in chemistry, physics and biology – Nobel symposium, vol 96, Chemical physics. Springer, Heidelberg, pp 271–288

Adachi K, Nishizaka T, Kinosita K Jr (2011) Rotational catalysis by F_1-ATPase. In: Ferguson S (ed) Comprehensive biophysics, vol 9. Elsevier, Amsterdam, in press

Ali MY, Uemura S, Adachi K, Itoh H, Kinosita K Jr, Ishiwata S (2002) Myosin V is a left-handed spiral motor on the right-handed actin helix. Nat Struct Biol 9:464–467

Ali MY, Homma K, Iwane AH, Adachi K, Itoh H, Kinosita K Jr, Yanagida T, Ikebe M (2004) Unconstrained steps of myosin VI appear longest among known molecular motors. Biophys J 86:3804–3810

Berg HC (2003) The rotary motor of bacterial flagella. Annu Rev Biochem 72:19–54

Boyer PD (1998) Energy, life, and ATP (Nobel lecture). Angew Chem Int Ed 37:2296–2307

Boyer PD (2002) Catalytic site occupancy during ATP synthase catalysis. FEBS Lett 512:29–32

Boyer PD, Kohlbrenner WE (1981) The present status of the binding-change mechanism and its relation to ATP formation by chloroplasts. In: Selman BR, Selman-Reimer S (eds) Energy coupling in photosynthesis. Elsevier, Amsterdam, pp 231–240

del Rizzo PA, Bi Y, Dunn SD, Shilton BH (2002) The "second stalk" of *Escherichia coli* ATP synthase: structure of the isolated dimerization domain. Biochemistry 41:6875–6884

Diez M, Zimmermann B, Börsch M, König M, Schweinberger E, Steigmiller S, Reuter R, Felekyan S, Kudryavtsev V, Seidel CAM, Gräber P (2004) Proton-powered subunit rotation in single membrane-bound F_0F_1-ATP synthase. Nat Struct Mol Biol 11:135–141

Dmitriev O, Jones PC, Jiang W, Fillingame RH (1999) Structure of the membrane domain of subunit *b* of the *Escherichia coli* F_0F_1 ATP synthase. J Biol Chem 274:15598–15604

Dou C, Fortes PAG, Allison WS (1998) The $\alpha_3(\beta Y341W)_3\gamma$ subcomplex of the F_1-ATPase from the thermophilic *Bacillus* PS3 fails to dissociate ADP when MgATP is hydrolyzed at a single catalytic site and attains maximal velocity when three catalytic sites are saturated with MgATP. Biochemistry 37:16757–16764

Furuike S, Hossain MD, Maki Y, Adachi K, Suzuki T, Kohori A, Itoh H, Yoshida M, Kinosita K Jr (2008) Axle-less F_1-ATPase rotates in the correct direction. Science 319:955–958

Furuike S, Nakano M, Adachi K, Noji H, Kinosita K Jr, Yokoyama K (2011) Resolving stepping rotation in *Thermus thermophilus* H^+-ATPase/synthase with an essentially drag free probe. Nat Commun 2:233

Gibbons C, Montgomery MG, Leslie AGW, Walker JE (2000) The structure of the central stalk in bovine F_1-ATPase at 2.4Å resolution. Nat Struct Biol 7:1055–1061

Harada Y, Ohara O, Takatsuki A, Itoh H, Shimamoto N, Kinosita K Jr (2001) Direct observation of DNA rotation during transcription by *Escherichia coli* RNA polymerase. Nature 409:113–115

Howard J (1996) The movement of kinesin along microtubules. Annu Rev Physiol 58:703–729

Hugel T, Michaelis J, Hetherington CL, Jardine PJ, Grimes S, Walter JM, Falk W, Anderson DL, Bustamante C (2007) Experimental test of connector rotation during DNA packaging into bacteriophage ϕ29 capsids. PLoS Biol 5:e59

Huxley AF (1957) Muscle structure and theories of contraction. Prog Biophys Biophys Chem 7:255–318

Huxley HE (1969) The mechanism of muscular contraction. Science 164:1356–1366

Iko Y, Tabata KV, Sakakihara S, Nakashima T, Noji H (2009) Acceleration of the ATP-binding rate of F_1-ATPase by forcible forward rotation. FEBS Lett 583:3187–3191

Itoh H, Takahashi A, Adachi K, Noji H, Yasuda R, Yoshida M, Kinosita K Jr (2004) Mechanically driven ATP synthesis by F_1-ATPase. Nature 427:465–468

Junge W, Sielaff H, Engelbrecht S (2009) Torque generation and elastic power transmission in the rotary F_0F_1-ATPase. Nature 459:364–370

Kabaleeswaran V, Shen H, Symersky J, Walker JE, Leslie AGW, Mueller DM (2009) Asymmetric structure of the yeast F_1 ATPase in the absence of bound nucleotides. J Biol Chem 284:10546–10551

Kinosita K Jr, Yasuda R, Noji H, Ishiwata S, Yoshida M (1998) F_1-ATPase: a rotary motor made of a single molecule. Cell 93:21–24

Kinosita K Jr, Yasuda R, Noji H, Adachi K (2000) A rotary molecular motor that can work at near 100% efficiency. Philos Trans R Soc Lond B 355:473–489

Kinosita K Jr, Adachi K, Itoh H (2004) Rotation of F_1-ATPase: how an ATP-driven molecular machine may work. Annu Rev Biophys Biomol Struct 33:245–268

Kinosita K Jr, Ali MY, Adachi K, Shiroguchi K, Itoh H (2005) How two-foot molecular motors may walk. Adv Exp Med Biol 565:205–219

Kohori A, Chiwata R, Hossain MD, Furuike S, Shiroguchi K, Adachi K, Yoshida M, Kinosita K Jr (2011) Torque generation in F_1-ATPase devoid of the entire amino-terminal helix of the rotor that fills half of the stator orifice. Biophys J 101:188–195

Komori Y, Iwane AH, Yanagida T (2007) Myosin-V makes two Brownian 90° rotations per 36-nm step. Nat Struct Mol Biol 14:968–973

Koshland DE (1958) Application of a theory of enzyme specificity to protein synthesis. Proc Natl Acad Sci USA 44:98–104

Menz RI, Walker JE, Leslie AGW (2001) Structure of bovine mitochondrial F_1-ATPase with nucleotide bound to all three catalytic sites: implications for the mechanism of rotary catalysis. Cell 106:331–341

Mizuno H, Higashida C, Yuan Y, Ishizaki T, Narumiya S, Watanabe N (2011) Rotational movement of the formin mDia1 along the double helical strand of an actin filament. Science 331:80–83

Moffitt JR, Chemla YR, Aathavan K, Grimes S, Jardine PJ, Anderson DL, Bustamante C (2009) Intersubunit coordination in a homomeric ring ATPase. Nature 457:446–451

Monod J, Wyman J, Changeux J-P (1965) On the nature of allosteric transitions: a plausible model. J Mol Biol 12:88–118

Müller M, Pänke O, Junge W, Engelbrecht S (2002) F_1-ATPase, the C-terminal end of subunit γ is not required for ATP hydrolysis-driven rotation. J Biol Chem 277:23308–23313

Nishizaka T, Yagi T, Tanaka Y, Ishiwata S (1993) Right-handed rotation of an actin filament in an in vitro motile system. Nature 361:269–271

Noji H, Yasuda R, Yoshida M, Kinosita K Jr (1997) Direct observation of the rotation of F_1-ATPase. Nature 386:299–302

Oosawa F, Hayashi S (1986) The loose coupling mechanism in molecular machines of living cells. Adv Biophys 22:151–183

Palanisami A, Okamoto T (2010) Torque-induced slip of the rotary motor F_1-ATPase. Nano Lett 10:4146–4149

Pänke O, Cherepanov DA, Gumbiowski K, Engelbrecht S, Junge W (2001) Visco-elastic dynamics of actin filaments coupled to rotary F-ATPase: angular torque profile of the enzyme. Biophys J 81:1220–1233

Rastogi VK, Girvin ME (1999) Structural changes linked to proton translocation by subunit *c* of the ATP synthase. Nature 402:263–268

Rondelez Y, Tresset G, Nakashima T, Kato-Yamada Y, Fujita H, Takeuchi S, Noji H (2005) Highly coupled ATP synthesis by F_1-ATPase single molecules. Nature 433:773–777

Sase I, Miyata H, Ishiwata S, Kinosita K Jr (1997) Axial rotation of sliding actin filaments revealed by single-fluorophore imaging. Proc Natl Acad Sci USA 94:5646–5650

Shimabukuro K, Muneyuki E, Yoshida M (2006) An alternative reaction pathway of F_1-ATPase suggested by rotation without 80°/40° substeps of a sluggish mutant at low ATP. Biophys J 90:1028–1032

Shimo-Kon R, Muneyuki E, Sakai H, Adachi K, Yoshida M, Kinosita K Jr (2010) Chemo-mechanical coupling in F_1-ATPase revealed by catalytic site occupancy during catalysis. Biophys J 98:1227–1236

Stock D, Leslie AGW, Walker JE (1999) Molecular architecture of the rotary motor in ATP synthase. Science 286:1700–1705

Toei M, Gerle C, Nakano M, Tani K, Gyobu N, Tamakoshi M, Sone N, Yoshida M, Fujiyoshi Y, Mitsuoka K, Yokoyama K (2007) Dodecamer rotor ring defines H^+/ATP ratio for ATP synthesis of prokaryotic V-ATPase from *Thermus thermophilus*. Proc Natl Acad Sci USA 104:20256–20261

Vale RD, Toyoshima YY (1988) Rotation and translocation of microtubules in vitro induced by dyneins from Tetrahymena cilia. Cell 52:459–469

Walker RA, Salmon ED, Endow SA (1990) The *Drosophila claret* segregation protein is a minus-end directed motor molecule. Nature 347:780–782

Wang H, Oster G (1998) Energy transduction in the F_1 motor of ATP synthase. Nature 396:279–282

Watanabe R, Iino R, Noji H (2010) Phosphate release in F_1-ATPase catalytic cycle follows ADP release. Nat Chem Biol 6:814–820

Watanabe-Nakayama T, Toyabe S, Kudo S, Sugiyama S, Yoshida M, Muneyuki E (2008) Effect of external torque on the ATP-driven rotation of F_1-ATPase. Biochem Biophys Res Commun 366:951–957

Watt IN, Montgomery MG, Runswick MJ, Leslie AGW, Walker JE (2010) Bioenergetic cost of making an adenosine triphosphate molecule in animal mitochondria. Proc Natl Acad Sci USA 107:16823–16827

Weber J, Wilke-Mounts S, Lee RS, Grell E, Senior AE (1993) Specific placement of tryptophan in the catalytic sites of the *Escherichia coli* F_1-ATPase provides a direct probe of nucleotide binding: maximal ATP hydrolysis occurs with three sites occupied. J Biol Chem 268:20126–20133

Yajima J, Mizutani K, Nishizaka T (2008) A torque component present in mitotic kinesin Eg5 revealed by three-dimensional tracking. Nat Struct Mol Biol 15:1119–1121

Yoshida M, Sone N, Hirata H, Kagawa Y (1977) Reconstitution of adenosine triphosphatase of thermophilic bacterium from purified individual subunits. J Biol Chem 252:3480–3485

Yoshida M, Muneyuki E, Hisabori T (2001) ATP synthase – a marvelous rotary engine of the cell. Nat Rev Mol Cell Biol 2:669–677

Chapter 3
Principles of Virus Structural Organization

B.V. Venkataram Prasad and Michael F. Schmid

Abstract Viruses, the molecular nanomachines infecting hosts ranging from prokaryotes to eukaryotes, come in different sizes, shapes, and symmetries. Questions such as what principles govern their structural organization, what factors guide their assembly, how these viruses integrate multifarious functions into one unique structure have enamored researchers for years. In the last five decades, following Caspar and Klug's elegant conceptualization of how viruses are constructed, high-resolution structural studies using X-ray crystallography and more recently cryo-EM techniques have provided a wealth of information on structures of a variety of viruses. These studies have significantly furthered our understanding of the principles that underlie structural organization in viruses. Such an understanding has practical impact in providing a rational basis for the design and development of antiviral strategies. In this chapter, we review principles underlying capsid formation in a variety of viruses, emphasizing the recent developments along with some historical perspective.

3.1 Introduction

Viruses are metastable macromolecular assemblies composed of the viral genome enclosed within a proteinaceous capsid. They come in variety of sizes, shapes, and forms. Some are large, and some are small; some are spherical, and some are rod-like; some have lipid envelopes. Many of these viruses exhibit exquisitely symmetric organization. Irrespective of their shape and size, the underlying theme in all these viruses is that the virus structure is designed to contain and protect the viral genome and deliver it to a specific host cell for subsequent replication of the virus. Viruses are also distinguished based on the type of the genome that they contain: single-stranded or double-stranded RNA or DNA. The viral genome, in addition to encoding the proteins that constitute the capsid, also encodes other proteins referred to as nonstructural proteins, so called because they are not part of the final capsid's organization. These nonstructural proteins are essential for viral replication inside the host cell. In some viruses, particularly of bacterial origin, viral genome encodes a protein called scaffolding protein that may not be part of the mature capsid but may be a critical factor in facilitating the capsid assembly.

B.V.V. Prasad (✉) • M.F. Schmid
Department of Biochemistry and Molecular Biology, Keck Center for Computational Biology, Baylor College of Medicine, Houston, TX 77030, USA
e-mail: vprasad@bcm.tmc.edu

M.G. Rossmann and V.B. Rao (eds.), *Viral Molecular Machines*,
Advances in Experimental Medicine and Biology 726, DOI 10.1007/978-1-4614-0980-9_3,

Often, the size of the virus is proportional to the size of the genome. However, the viral genome contributes far less to the total mass of the virion than the capsid proteins. It was this observation that prompted Watson and Crick to suggest that the capsid has to be formed by the association of multiple copies of the capsid protein(s) (Crick and Watson 1956, 1957). Such an assembly with repeating subunits then greatly reduces the amount of genetic information required. In some viruses, the capsid formation involves a single gene product, whereas in other viruses which are more complex, it involves multiple gene products. Such an assembly involving repeating subunits raises several interesting questions. How do these subunits interact with one another with high fidelity and specificity to form the capsid architecture? This question becomes even more interesting in complex viruses in which the capsid formation involves multiple gene products. Are there any specific structural properties that these proteins should have for the capsid formation? How is capsid assembly directed and controlled? How is the genome encapsidated? In addition to containing and protecting the genome, the capsid architecture must also be conducive for interactions with the host cell for entry; how is this process coordinated? Given that capsid has to disassemble to make the genome available for replication, what are the cues for disassembly? How does the capsid organization respond to and evade the antiviral response mounted by the host?

In the last half century, structural studies on a variety of viruses have provided a wealth of information regarding some of the questions listed above. In addition to providing insight into the fundamental principles underlying various aspects of capsid assembly, more importantly, such studies have had practical impact in providing a rational basis for the design and development of antiviral strategies. Several excellent reviews on virus structures and principles underlying capsid formation have been published periodically over years (Klug and Caspar 1960; Caspar and Klug 1962; Rossmann and Johnson 1989; Johnson and Speir 1997; Harrison 2007); we will emphasize here the recent developments along with the some historical perspective.

3.2 Structural Techniques

Two principal techniques used in the structural studies on viruses are electron microscopy and X-ray crystallography. Contributions from other elegant studies using a variety of biochemical and biophysical techniques and theoretical modeling have been crucial in providing a more complete understanding of the capsid construction and assembly pathways. Electron microscopy of negatively stained virus specimens provided the first glimpse of viruses and led to early classification of viruses based on shape and form (Green et al. 1956; Brenner and Horne 1959; Horne and Wildy 1962, 1979; Wildy and Horne 1963). Even today, this technique is used as a diagnostic tool in identifying clinical virus samples. Subsequently, the discovery that EM images of virus particles, which are essentially projection images, can be used to reconstruct the three-dimensional structure of the virus using computer image analysis protocols (Crowther et al. 1970) paved the way for spectacular advances in specimen preparation (Knapek and Dubochet 1980; Dubochet et al. 1988), electron imaging, and computer image reconstructions. In the last two decades, this exciting new technology called three-dimensional cryo-electron microscopy (cryo-EM) has revolutionized the structure analysis of a variety of viruses (Baker et al. 2010; Crowther 2010; Grigorieff and Harrison 2011).

Much of our understanding of subunit interactions in a viral capsid at the atomic level has come from X-ray crystallographic structure of spherical viruses. Beginning with the structures of three small spherical plant viruses in early 1980s (Harrison et al. 1978; Abad-Zapatero et al. 1980; Liljas et al. 1982), over the last three decades, X-ray crystallography has been successfully applied to study a variety of larger and more complex spherical viruses including human viruses (Hogle et al. 1985; Rossmann et al. 1985; Liddington et al. 1991; Grimes et al. 1998; Reinisch et al. 2000; Wikoff et al. 2000; Reddy et al. 2010). The closely related technique of X-ray fiber diffraction has been used to study viruses that have helical symmetry (Namba and Stubbs 1986; Namba et al. 1989). In recent years, cryo-EM technique has allowed visualization of a variety of spherical viruses at subnanometer (Bottcher et al. 1997; Conway et al. 1997; Jiang et al. 2003; Zhang et al. 2003; Saban et al. 2006; Li et al. 2009) to near-atomic resolutions (Yu et al. 2008a, b; Baker et al. 2010; Liu et al. 2010; Wolf et al. 2010; Settembre et al. 2011, see also Chap. 4). For some viruses that are not amenable for high-resolution structural analysis by these techniques, complementarity between cryo-EM and X-ray crystallography has been exploited in deriving the pseudoatomic models of the capsid (Grimes et al. 1997; Mathieu et al. 2001; Zhang et al. 2002, 2007; Settembre et al. 2011). In these studies, when the virus capsid could not be crystallized, but a lower resolution structure could be determined by cryo-EM, as this technique does not require the specimen in a crystalline form, independently determined X-ray crystallographic structures of the capsid components are fitted into lower-resolution cryo-EM map of the capsid. Such a hybrid technique has been most useful in studying capsid–receptor, capsid–antibody interactions and in studying capsid-associated structural dynamics (Rossmann et al. 1994; Ilag et al. 1995; Smith et al. 1996; Stewart et al. 1997; Belnap et al. 2000; Conway et al. 2001; Martin et al. 2001; Nason et al. 2001; Dormitzer et al. 2004; Gan et al. 2006). Structure determination of spherical viruses either by X-ray crystallography or cryo-EM techniques relies implicitly on the symmetry of the capsid. As a result, the structural organization of the encapsidated genome is amenable to these structural techniques only when the genome follows the capsid symmetry. However, in recent years, there are several examples in which the entire genome or a significant portion of it is observed to follow the capsid symmetry and visualized in the structural analysis (Chen et al. 1989; Namba et al. 1989; Fisher and Johnson 1993; Larson et al. 1993). A detailed discussion on the genome organization in viruses is provided in a review by Prasad and Prevelige (2003). In this review, we mainly focus on the capsid organization. In addition to the X-ray crystallographic and cryo-EM structural techniques, other diffraction techniques such as neutron diffraction (Bentley et al. 1987), low-angle X-ray scattering (Tsuruta et al. 1998), and spectroscopic techniques (Tuma and Thomas 1997; Benevides et al. 2002) have been useful in understanding the capsid organization in viruses.

3.3 Capsid Organization in Spherical Viruses

3.3.1 Cubic Symmetry

It was Watson and Crick (1956) who first proposed that spherical viruses exhibit cubic symmetry involving at least four threefold rotational symmetry axes. They argued that only such symmetry in an isometric capsid allows close packing of repeating subunits such that each subunit has identical environment. The cubic symmetry is inherent in the platonic solids, the tetrahedron, cube, octahedron, dodecahedron, and icosahedron. The distinguishing characteristics of these polyhedral solids are that each polyhedron consists of identical faces, in the shape of a regular polygon; identical vertices, which are meeting points of the same number of faces; and identical edges, which are lines joining the adjacent vertices (Fig. 3.1). For example, a cube consists of six identical faces, each of which is a square; eight identical vertices, which are meeting points of three squares; and 24 edges, each joining adjacent vertices. These highly symmetric polyhedra can also be described in terms of the inherent rotational symmetry axes, wherein rotation of $360/n°$ about an n-fold symmetry axis produces n invariant views of the polyhedron. For instance, a cube has twofold rotational symmetry axes passing through the centers of the opposite faces and threefold axes along the diagonals passing through each of the vertices. The combination of these two rotational symmetry elements gives rise to additional rotational symmetry axes: three fourfold symmetry axes going through centers of the

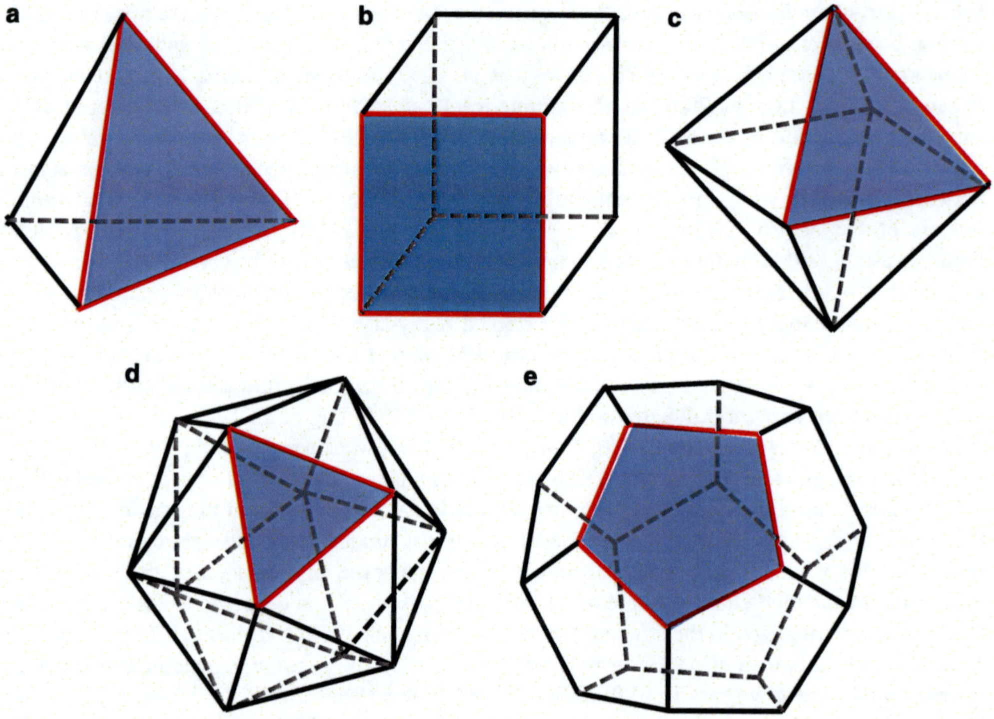

Fig. 3.1 The five platonic solids. One of the faces in each of these solids is colored in *pale blue*. The number of vertices (*V*), faces (*F*), and edges in each of these solids follows Euler formula, $F+V=E+2$. (**a**) Tetrahedron, which displays two- and threefold symmetry axes, has 4 triangular faces, 4 vertices, and 6 edges. (**b**) Cube, with four-, three-, and twofold symmetry, has 6 square faces, 8 vertices, and 12 edges. (**c**) Octahedron which has two- and threefold symmetry has 8 triangular faces, 8 vertices, and 12 edges. (**d**) Icosahedron with five-, three-, and twofold symmetry axes has 20 triangular faces, 12 vertices, and 30 edges. (**e**) Dodecahedron, which also displays the same 5-3-2 symmetry elements as an icosahedron, has 12 faces, 20 vertices, and 30 edges. In contrast to an icosahedron in which three triangular faces meet at a vertex, in the dodecahedron, three pentagonal faces meet at each of the vertices

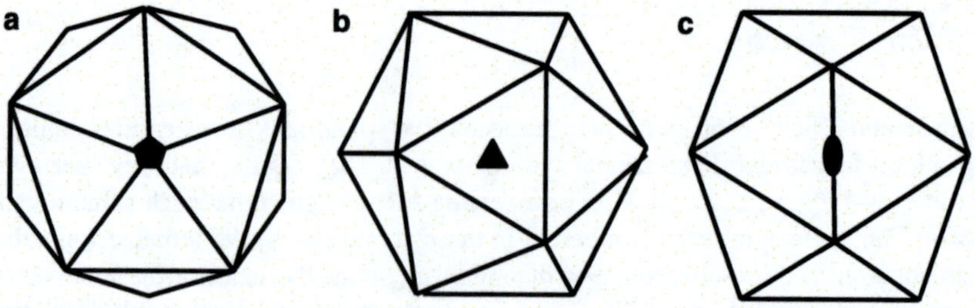

Fig. 3.2 Icosahedral axes of symmetry. An icosahedron displayed along the (**a**) five-, (**b**) three-, and (**c**) twofold symmetry axes. The fivefold rotation axis passes through the vertices of the icosahedron; the threefold axis passes through the middle of each triangular face; and the twofold axis through the center of each edge

opposite faces and six twofold symmetry axes going through the midpoints of opposite edges. Similarly, in an icosahedron, there are 12 vertices with fivefold rotational symmetry, 20 triangular faces with threefold symmetry, and 30 edges with twofold symmetry (Fig. 3.2). Among the platonic solids, fivefold rotational symmetry is present only in the icosahedron and its dual, the dodecahedron (Fig. 3.1). By analyzing the X-ray diffraction patterns obtained from the crystals of a spherical plant

Fig. 3.3 Construction of an icosahedron from a template based on a planar hexagonal net. (**a**) The template consists of 20 equilateral triangles cut from a hexagonal net. Each triangle has a threefold axis at the center, indicated by a small triangle symbol, and a twofold axis at the center of the line that joins the adjacent triangles, indicated by an *oval* symbol. Application of the threefold rotational symmetry to a motif (*quotes*) placed in one of the triangles generates two other motifs in the same triangle, and successive application of the twofold rotational symmetry between the adjacent triangles results in each of the 20 triangles having three motifs as shown. To construct an icosahedron from this planar net, first, the two ends of the central body consisting of ten equilateral triangles are folded and glued at the edge shown as a *red band*. Each of the five triangular faces at the top is then joined with one another at the *blue bands* to form a single vertex, and a similar operation with the five triangular faces at the bottom results in an icosahedron with 12 fivefold vertices. (**b**) Hexagonal net in (**a**) folded into an icosahedron. The 60 motifs (*quotes*) in the icosahedron are now related by five-, three-, and twofold symmetry, indicated by a *pentagon*, *triangle*, and *oval* symbols, respectively

virus, tomato bushy stunt virus (TBSV), which showed characteristics spikes that only arise because of fivefold symmetry, Caspar provided the first evidence for the presence of icosahedral (5-3-2) symmetry in spherical viruses (Caspar 1956). Soon after, Klug, by analyzing the X-ray diffraction patterns from another spherical plant virus called tomato yellow mosaic virus (TYMV), provided further evidence for icosahedral symmetry in spherical viruses (Klug et al. 1957). At that time in the 1950s when these fundamental discoveries were made, the X-ray diffraction methods and the computer software were not adequately developed for determining the three-dimensional structures. However, later structural studies overcoming these technical limitations clearly established icosahedral capsid organization in a variety of spherical viruses.

3.3.2 *Icosahedral Capsid Organization*

The icosahedron with five-, three-, and twofold rotational symmetry axes allows placement of 60 identical units with equivalent contacts between each of them. In contrast, a cube with four-, three-, and twofold symmetry axes, as mentioned above, allows placement of 12 identical units. Placement of 60 units in an icosahedron can be illustrated by considering the construction of an icosahedron based on a hexagonal net (Fig. 3.3a). By placing a subunit at one corner of one of the 20 triangular facets, as a result of the threefold symmetry axis, passing through the center of the triangle and perpendicular to the plane of the triangle, the other two corners of this facet will have subunits, and then, as a result of repeated application of the twofold axis of symmetry, at the center of the edge joining the adjacent triangular facets, we generate 60 subunits that are related by 5-3-2 symmetry, with every subunit having an identical environment (Fig. 3.3b). A pertinent question at this point is whether all the spherical viruses with icosahedral symmetric capsids have only 60 identical subunits? Although there are a number of viruses in which capsid consists of 60 identical subunits, such as satellite plant viruses (Ban et al. 1995), there are many spherical viruses which consist of more than 60 subunits, as was first shown by Brenner and Horne using EM of negatively stained virus specimens (Brenner and Horne 1959). From later studies, we know that not only do these viruses have more than 60 identical subunits but also, in several cases, the capsid is formed by subunits of different gene products. For simplicity, we will consider first how spherical viruses with more than 60 identical subunits, i.e., single gene product, conform to icosahedrally symmetric organization. Caper and Klug (1962) provided an elegant conceptual framework that specifically addressed this question.

3.3.3 Triangulation Numbers

It is clear from symmetry considerations that more than 60 identical subunits cannot be placed in such a way that they have strictly equivalent positions on an icosahedral surface. However, Casper and Klug (1962) argued that this can be achieved by allowing minimal distortions from the strict equivalence. To achieve this, with an assumption that size of the subunit remains the same and only the number of subunits varies, the basic triangular facet of the icosahedron has to be first enlarged and then subdivided into smaller triangles. Such a subdivision, which is referred to as triangulation for convenience in this discussion, cannot be arbitrary, and because of geometrical considerations, it is dictated by the equation $T=H^2+HK+K^2$, where T is the triangulation number, and h and k are 0 or positive integers (Fig. 3.4). That is, the T number takes discrete values such as 1, 3, 4, 7, etc., and subdividing an equilateral triangle into, say, five or eight smaller triangles is not allowed.

The concept of triangulation and how it allows for more than 60 subunits with consequent increase in the size of the icosahedron can be illustrated using a hexagonal lattice with H- and K-axes crossing at 60° angle (Fig. 3.4a). By arbitrarily choosing a lattice point as the origin (0, 0) and considering it as the position of a fivefold vertex of an icosahedron, the position (H, K) of the neighboring fivefold vertex that is closest to the origin signifies the T number of that icosahedron. The equilateral triangle with the length of each side equal to the distance between the origin and the position (H, K) corresponds to one of the 20 triangular faces of such an icosahedron with each of its corners representing a fivefold vertex. The icosahedron with no triangulation can be described as having a triangulation number of 1 ($T=1$) in which the closest fivefold vertex is positioned at ($H=1$, $K=0$). By defining the equilateral triangle representing each face in such a $T=1$ icosahedron as a unit triangle, in a $T=4$ icosahedron, for example, with closest fivefold vertex positioned at ($H=2$, $K=0$), the equilateral triangle describing each facet consists of four unit triangles (Fig. 3.4a). That means, in each triangular facet of the $T=4$ icosahedron, 12 subunits can be placed, in contrast to three subunits in the facet of the $T=1$ icosahedron. With 12 subunits in each of the 20 facets, a $T=4$ icosahedron then will accommodate 240 ($60T$) subunits in contrast to 60 subunits in the $T=1$ icosahedron (Fig. 3.4b). Also, as can be seen, the size of the $T=4$ icosahedron compared to the size of the $T=1$ icosahedron has proportionally increased. It should be pointed out here that subdividing the triangular facet resulting in icosahedra with $T>1$ need not necessarily increase the size and that increased size is only with the assumption that molecular mass of the subunit remains approximately the same. Generally, however, spherical viruses with $T>1$ tend to be of larger size.

3.3.4 Classes of Icosahedra

Based on their T numbers, icosahedra can be divided into three classes (Fig. 3.5). In the first class, icosahedra have T numbers based on the H-axis (1, 4, 9, etc.) with $H\geq 1$ and $K=0$. In these icosahedra, lattice lines run parallel to the edges of the triangular icosahedral facet. In the second class, the icosahedra have T numbers along the line that bisects H- and K-axes (3, 12, 27, etc.) with $H=K$. In this class of icosahedra, the lattice lines bisect the angles between the edges of the icosahedral

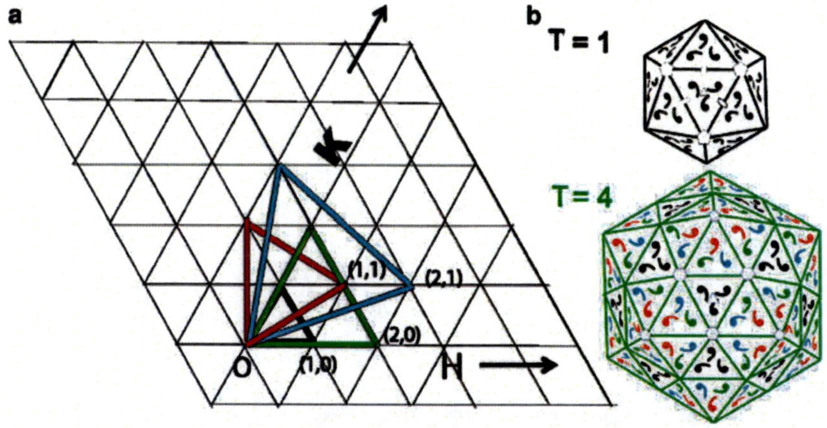

Fig. 3.4 Triangulated icosahedral lattices. (**a**) The hexagonal net used for constructing an icosahedron in Fig. 3.3 also serves to illustrate the formation of equilateral facets in the triangulated icosahedral lattices. The facet of the triangulated lattice of an icosahedron of a given *T* number is specified by a vector **T** (*H*, *K*) from the origin (**O**). The unit vector **T** (1, 0) (*black*) in this representation specifies the facet of the basic *T*=1 icosahedron. Other vectors such as **T** (1, 1), **T** (2, 0), and **T** (1, 2), for example, form the basis for successively larger *T*=3 (*red*), *T*=4 (*green*), and *T*=7 (*blue*) icosahedral facets, respectively. To construct a triangulated icosahedral lattice of a particular *T* number ($=H^2+HK+K^2$), its facet, which is an equilateral triangle, is generated, as specified by the vector **T** (*H*, *K*), using the underlying planar hexagonal net. Nineteen other triangles are then generated to give a template, as in Fig. 3.3, for constructing an icosahedron with a given *T* number. The *black*, *red*, *green*, and *blue triangles* represent facets for the *T*=1, 3, 4, and 7 icosahedra, respectively. Each vertex of the facet triangle becomes fivefold when the template is folded into an icosahedron as shown in Fig. 3.3. (**b**) The folded *T*=1 (*H*=1, *K*=0) and *T*=4 (*H*=2, *K*=0) templates, in *black* and *green*, respectively, are shown for comparison. The subunit organization in these icosahedra is shown using *quotes* to represent a "subunit" as in Fig. 3.3. In contrast to the *T*=1 icosahedron, which has 20 unit triangles, specified by the vector **T** (1, 0), with 60 subunits, the *T*=4 lattice, specified by **T** (2, 0), consists of 80 (20*T*) unit triangles with 240 (60*T*) subunits. The folded *T*=4 icosahedron clearly illustrates how the nonvertex sixfold positions in the net remain hexavalent (indicated by *hexagon* symbols) whereas the vertex sixfold positions in the net become pentavalent (indicated by *pentagon* symbol) upon folding. As in any triangulated icosahedron, with the exception of the unit triangle at the icosahedral threefold position, the other triangles exhibit local (or quasi) threefold symmetry; similarly, the adjacent unit triangles are related by local (or quasi) twofold axes, unless the midpoint of the edge coincides with icosahedral twofold axis. Each facet in the *T*=4 icosahedron is comprised of four triangles. The icosahedral threefold axis, at the center of the facet, divides the facet into three equivalent parts. Each part, consisting of one-third of the central triangle and one of the remaining three triangles in the facet, constitutes the icosahedral asymmetric unit. The four "subunits" in the asymmetric unit, one from the central triangle and three from one of the other triangles in facet, are shown in different colors to indicate their quasi-equivalent environment. The subunit in the central triangle is colored in *black*, while the subunits in the other triangle are colored in *red*, *blue*, and *green*. Application of the strict five-, three-, and twofold rotations to these quasi-equivalent subunits in the asymmetric unit generates rest of the 240 subunits. The arrangement of the subunits clearly illustrates how the triangulation leads to the formation of the rings of 5 (*red* subunits) and 6 (*black*, *green*, and *blue* around the hexavalent points) because of the triangulation. Furthermore, it also illustrates that although the formation of rings of 5 and 6 in a triangulated lattice is a geometrical necessity, clustering of the subunits into pentamers and hexamers is not obligatory. In the *T*=4 icosahedron shown here, either each subunit can be considered separately or as a cluster of three subunits (trimer). With a trimer as the building block, one trimer (*three black* subunits) occupies the central triangle of the facet, at the icosahedral threefold position, and another "quasi-equivalent" trimer (*red*, *green*, and *blue* subunits) occupies the other triangle in the facet. Such a *T*=4 organization with trimers is observed in the case of alphaviruses (see Fig. 3.12). See Figs. 3.7a and 3.8a for folded *T*=3 (*red* facet) and *T*=7 (*blue* facet) icosahedral lattices, respectively

facet. The third class, called the skew class, includes icosahedra with other allowed *T* numbers (7, 13, 19, 21, etc.), in which the lattice lines are not symmetrically disposed with respect to the edges of the icosahedral facet. In the first two classes, the lattice points that signify the *T* number are uniquely defined, whereas in the skewed class, there are two possibilities for the lattice points that define their *T* numbers. For example, for *T*=7 (*H*=2, *K*=1) and (*H*=1, *K*=2) are equally possible.

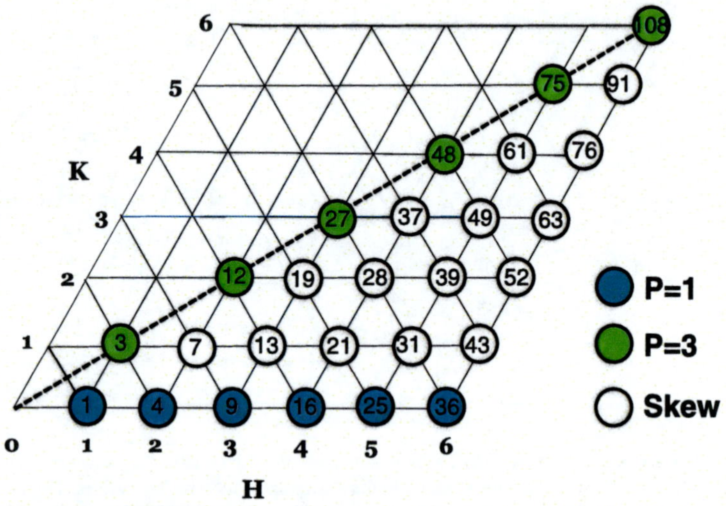

Fig. 3.5 Different classes of icosahedra. The origin ($H=0$, $K=0$) in the equilateral triangular net represents the position of a fivefold vertex of an icosahedron, and the position of the neighboring fivefold vertex in that icosahedron is represented at a position (H, K) based on its T number (shown inside the *circles*). The icosahedra with $K=0$ (or $H=0$) belong to $P=1$ class (*blue-filled circles*). The icosahedra with $H=K$ (on the *dashed line*) belong to $P=3$ class (*green-filled circles*). All other icosahedra belong to the skew class (*open circles*). These icosahedra exhibit lattices with handedness. Skewed icosahedra with $H>K$, below the *dashed* $H=K$ *line*, are left-handed (*laevo*) lattices, and those with $H<K$, above the *dashed line*, are right-handed lattices (*dextro*)

These two represent left-handed (*leavo*) and right-handed (*dextro*) enantiomorphic configurations. Examples of virus structures with skew icosahedral lattice include polyomaviruses ($T=7d$), bacteriophage ($T=7d$), and rotavirus ($T=13l$). In the literature, the first and second classes of icosahedra are also referred to as $P=1$ and $P=3$ classes, respectively, by defining the T number as $T=Pf^2$. In this equation, f is the largest common divisor between H and K, and $P=h^2+hk+k^2$, where h and k are integers without any common factor. All icosahedra with $P\geq7$ belong to skew class.

Another form of icosahedron that is found in some viruses such as phi29 (Tao et al. 1998), T-even bacteriophages (Fokine et al. 2004; Rao and Black 2010), and aberrant flock house virus particles (Dong et al. 1998; Wikoff and Johnson 1999) is the prolate icosahedron. Here, the icosahedron is stretched along one of the icosahedral axes. A typical icosahedron can be considered as having a central cylindrical body, consisting of ten triangles, with two caps, each with five triangles, at its bottom and the top (Fig. 3.6a). The prolate icosahedra are characterized by a combination of T number and an elongation number called Q according to the formula $n=30(T+Q)$ (Fig. 3.6b), where T is the triangulation number as defined previously, Q is any positive integer, that defines the triangulation number for the central body, and n refers to number of subunits (Fokine et al. 2004). In the triangles that constitute the caps, the triangulation follows the T number, whereas in the triangles that constitute the main cylindrical body, it follows the Q number. As can be seen, when $T=Q$, then the icosahedra become isometric with $n=60T$, as discussed above. When $Q>T$, the resulting icosahedron is referred to as a prolate icosahedron, and when $Q<T$, it is referred to as an oblate icosahedron. The structure of T4 bacteriophage exhibits a prolate icosahedron with $T=13$ and $Q=21$ (Fokine et al. 2004) (Fig. 3.6c), whereas the major capsid of the phi29 exhibits a prolate icosahedron with $T=3$ and $Q=5$ (Tao et al. 1998).

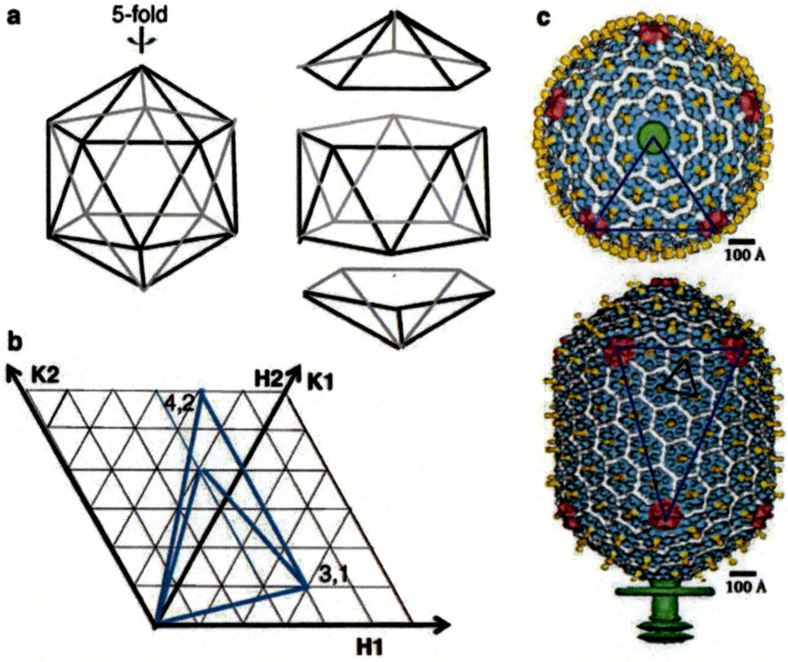

Fig. 3.6 Prolate icosahedra. (**a**) Illustration of how an icosahedron (*left*) can be considered as having two end caps, each with five triangular faces separated by a central cylindrical body consisting of ten triangular faces right. (**b**) Representation of *T* and *Q* numbers on an equilateral triangular net in a prolate icosahedron. The prolate icosahedra have two types of facets, one in the end caps and the other in the central body with unequal sides. The triangulation in these two facets can be represented on a hexagonal net by considering two coordinate points ($H1$, $K1$) and ($H2$, $K2$) (Moody 1999; Fokine et al. 2004). The vector from the origin to the point ($H1$, $K1$) specifies the equilateral facet for the end caps as in Fig. 3.4, whereas the facet for the central body with unequal sides is specified by two vectors, one from the origin to the point ($H1$, $K1$) and the other from the origin to the point ($H2$, $K2$). The triangulation in the equilateral facets of the end caps is defined as $T\ (H1,\ K1) = H1^2 + H1K1 + K1^2$ as illustrated in Fig. 3.4, whereas the triangulation in the facet, with unequal sides, of the central body is defined by $Q = H1H2 + H1K2 + K1K2$. As an example, the triangulation in the prolate icosahedron with $T = 13l$ ($H1 = 3$, $K1 = 1$) in the end caps and $Q = 20$ ($H2 = 4$, $K2 = 2$) in the central body is illustrated here. (**c**) Surface representation of the cryo-EM reconstruction (adapted from Fokine et al. 2004, with permission from Dr. M.G. Rossmann) of the T4 bacteriophage head illustrating subunit arrangement in the $T = 13$ lattice in the end caps (*above*), shown along the fivefold axis, and in the $Q = 20$ lattice in the central body (*below*). The facets in both the lattices are indicated

3.3.5 Icosahedral Asymmetric Unit and Quasi-Equivalent Subunits

An important concept that emerges from the triangulation is the icosahedral asymmetric unit. The threefold rotational symmetry axis passing through the center of the triangular icosahedral facet divides the subunits in the facet into three symmetrically equivalent sets. Each set of the subunits is defined as one icosahedral asymmetric unit, and application of the 5-3-2 rotational symmetry to this asymmetric unit produces all the $60T$ subunits in the icosahedron. In a $T = 1$ icosahedron, the asymmetric unit consists of one subunit, whereas in a $T = 4$ icosahedron, for example, the asymmetric unit consists of four subunits (Fig. 3.4b). As the interacting environment between these subunits in the icosahedron with $T > 1$ cannot be strictly equivalent, they are termed as "quasi-equivalent" subunits (black, blue, red, and green commas in Fig. 3.4b). Caspar and Klug proposed that these quasi-equivalent subunits in the icosahedral shell retain similar bonding interactions with minor distortions in their intersubunit interactions in order to adapt to the nonsymmetry-related environments.

3.3.6 Pentavalent and Hexavalent Positions

Another important consequence of the "quasi-equivalence" theory developed by Caspar and Klug is that triangulation necessarily results in the generation of hexavalent (sixfold) locations in addition to pentavalent (fivefold) positions on the icosahedral lattice (Fig. 3.4b). Irrespective of the T number, all icosahedra necessarily have 12 pentavalent positions. In the icosahedra with $T>1$, because of the triangulation, $10(T-1)$ hexavalent positions are generated. As a result, in icosahedra with $T>1$, the $60T$ subunits could cluster into 12 pentamers around the fivefold vertices and $10(T-1)$ hexamers at the hexavalent lattice points. Caspar and Klug defined these subunit clusters as morphological units. An icosahedron with a specific T number will then have $10(T+2)$ morphological units [i.e., $10(T-1)$ hexamers + 12 pentamers]. Although the arrangement of $60T$ subunits into rings of 5 and 6 is a geometrical necessity, clustering of these subunits into pentamers and hexamers is not; clustering into $20T$ trimers, $30T$ dimers, or $60T$ monomers is possible (Fig. 3.4b). Casper and Klug argued that if a particular oligomeric state, for example, hexamers, if they are particularly stable, they might be preformed, but when they are assembled into the shell, these conceptually planar units would have to be transformed into convex pentamers, resulting from the removal of a subunit, to occupy the "domed" pentavalent positions in the icosahedral structure. They suggested that such transformation would only require minor alterations in the dihedral angles between the subunits but essentially maintaining similar intersubunit contacts.

3.4 High-Resolution Structures of Spherical Viruses and Quasi-Equivalence Theory

Two basic tenets of the quasi-equivalence theory as discussed above are that (1) the icosahedral lattice with the possibility of triangulation presents the most efficient geometrical design for close packing of identical subunits in a spherical shell and (2) that the structural organization involves quasi-equivalent interactions requiring minimal distortions in the subunit bonding. Since the proposal of this theory in 1962, in the last three decades, high-resolution structures of several spherical viruses of different sizes have been determined. These structures clearly established that the capsid organization in spherical viruses follows icosahedral symmetry and that in viruses with subunits greater than 60, capsid organization is based on a triangulated icosahedral lattice as suggested from the quasi-equivalence theory. For example, several ssRNA plant viruses and some human viruses such as noroviruses, with capsid composed of 180 copies of the capsid protein, exhibit the expected $T=3$ icosahedral organization with rings of five and six subunits. In dsDNA bacterial viruses, such as P22 (Jiang et al. 2003), HK97 (Wikoff et al. 2000), with 420 copies of the capsid protein, the capsid organization is based on the expected $T=7$ icosahedral lattice with 60 hexamers and 12 pentamers. However, there are major surprises as well. The capsid organization in papovaviruses (papilloma- and polyomaviruses) represents a stunning departure from the quasi-equivalence theory (Liddington et al. 1991; Wolf et al. 2010). In these dsDNA spherical viruses, the capsid consists of 360 copies of the major capsid protein VP1. Such a number cannot be accommodated on a triangulated icosahedral lattice because $T=6$ (360/60) is forbidden as it does not adhere H^2+HK+K^2 rule (see above). Initially, although in conflict with the biochemical analysis which indicated 360 copies of the capsid protein in these virus, based on EM images of negatively stained specimens and computer reconstruction, an icosahedral structure with 420 copies organized as 60 hexamers and 12 pentamers on a $T=7$ icosahedral lattice, as expected from the quasi-equivalence theory, was proposed (Klug and Finch 1968). However, subsequent high-resolution structures of these viruses unambiguously revealed that the capsid indeed consists of 360 subunits and that these subunits, instead of pentamer–hexamer clustering, are organized as 72 pentamers at 60 hexavalent and 12 pentavalent location on a triangulated $T=7d$ icosahedral lattice. (Liddington et al. 1991; Wolf et al. 2010) Other examples include adenovirus, a dsDNA virus, in which 240 trimers of the major capsid protein (hexons) occupy hexavalent positions on a $T=25$ (pseudo) icosahedral lattice (Roberts et al. 1986; Liu et al. 2010; Reddy et al. 2010) and a unique icosahedral organization with 120 (forbidden $T=2$) subunits which is a recurring theme in dsRNA viruses including fungal L-A virus (Naitow et al. 2002), partitivirus (Ochoa et al. 2008), and inner shells of bluetongue virus (Grimes et al. 1998), rotavirus (Lawton et al. 1997; Chen et al. 2006; Settembre et al. 2011), and reovirus (Reinisch et al. 2000).

3.4.1 Conformational Switching

Although the high-resolution structures of spherical viruses show general concordance with the quasi-equivalent theory in terms of subunit arrangement on triangulated icosahedral lattices, they exposed limitations of the theory particularly in regard to the concept of quasi-equivalent interactions that bond the subunits in the context of icosahedral lattice. These structures showed that formation of the icosahedral capsid can be governed by nonequivalent interactions involving internally located conformationally "flexible" arms of the capsid protein subunits. Except for these conformationally flexible arms, which function as molecular switches to allow the subunits to adapt to quasi-equivalent environments of the triangulated icosahedral lattice, the majority of the capsid protein remains structurally invariant.

Capsid organization in many small plant and animal viruses follows $T=3$ icosahedral symmetry with 180 chemically identical subunits (Fig. 3.7). In a $T=3$ lattice (Fig. 3.7a), the icosahedral asymmetric unit consists of three quasi-equivalent subunits conventionally referred to as A, B, and C (Rossmann and Johnson 1989). These subunits form two types of quasi-equivalent dimers: A/B dimers, located at quasi-twofold axis of the $T=3$ lattice, and C/C dimers, located at the icosahedral twofold axis (Fig. 3.7b). Thus, the capsid consists of 60 A/B dimers and 30 C/C dimers. The A/B dimers surround the fivefold vertex forming rings of 5, and at the six-coordinated positions, A/B and C/C dimers alternate to form rings of 6. Despite lacking any sequence homology, the domain that participates in the formation of the icosahedral shell in these viruses exhibits a highly conserved polypeptide fold. This fold with two flanking twisted β-sheets, each consisting of four antiparallel β-strands, is called the eight-stranded jelly roll β-barrel motif (Fig. 3.7c). How do the dimers conform to the quasi-equivalent environment of the $T=3$ lattice and form a closed spherical shell? In conforming to the curvature of the spherical shell, A/B dimer exhibits a bent conformation whereas the C/C dimer is flat (Fig. 3.7d). Despite some thematic variations between the viruses, the internally located flexible N-terminal arm (NTA) is implicated in providing a switch to facilitate bent A/B and flat C/C conformations during $T=3$ capsid assembly (Fig. 3.7e). In the plant tombus viruses (Harrison et al. 1978; Hogle et al. 1986) and sobamoviruses (Abad-Zapatero et al. 1980), the NTA of the C subunit is ordered, whereas the equivalent regions in the A and B subunits are disordered, providing a switch to allow bent and flat conformations of the A/B and C/C dimers, respectively. In nodaviruses, an ordered arm of the C subunit and a piece of genomic RNA are implicated in keeping the flat conformation of the C/C dimers (Fisher and Johnson 1993). In human noroviruses, the ordered NTA of the B subunit, which interacts with the base of the shell domain of the neighboring C subunit, is suggested to provide such a switch (Prasad et al. 1999). Thus, in these viruses, at the level of the NTA, the $T=3$ icosahedral symmetry is reduced to $T=1$ with only one of the three NTAs of the quasi-equivalent subunits being ordered. The structures of animal caliciviruses exhibit a novel and distinct variation from any of these viruses (Chen et al. 2006; Ossiboff et al. 2010). The NTAs of A, B, and C in these viruses are equally ordered and provide a network of interactions essentially maintaining the $T=3$ symmetry at this level. Instead of an order-to-disorder transition as described

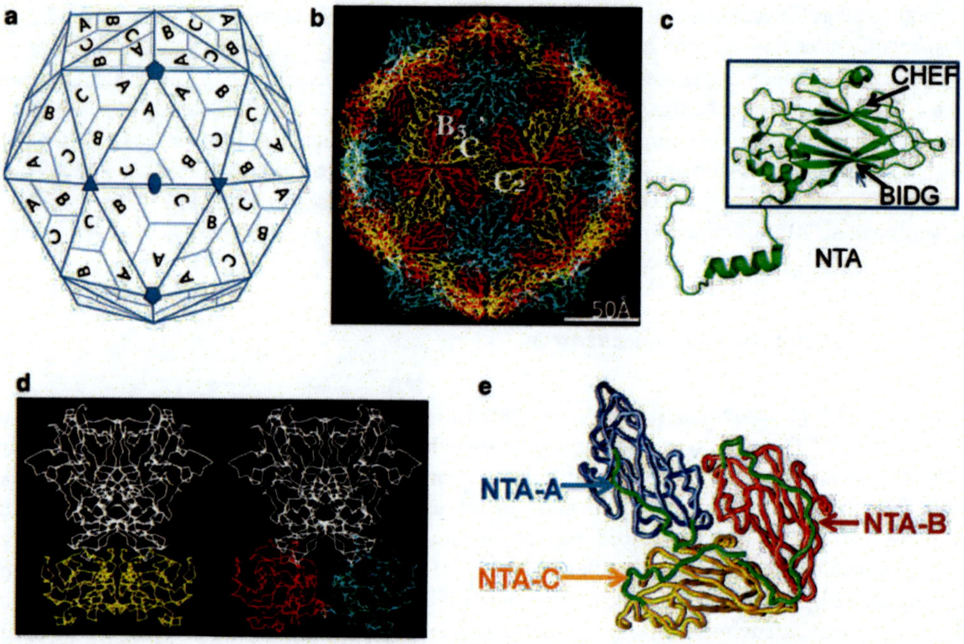

Fig. 3.7 Subunit organization and conformational switching in $T=3$ icosahedral viruses. (**a**) $T=3$ lattice as viewed along the icosahedral twofold axis. A set of five-, three-, and twofold axes of the $T=3$ icosahedron are denoted. The three quasi-equivalent "subunits" in the asymmetric unit of the $T=3$ lattice are indicated by A, B, and C. In a $T=3$ structure, these subunits are chemically equivalent. Application of the icosahedral symmetry generates 60 sets of these quasi-equivalent subunits. Disposition of the symmetry-related subunits is also indicated. The formation of the rings of 5 and 6 is clearly seen. The A subunits cluster around the fivefold axis, whereas B and C alternate around the six-coordinated positions. (**b**) The packing of the canonical trapezoid-shaped β-barrel domains as typically observed in a $T=3$ icosahedral virus structures. The view is along the icosahedral twofold axis, same as in (**a**). The β-barrels corresponding to the quasi-equivalent A, B, and C in (**a**) are colored in *cyan*, *red*, and *yellow*, respectively. The icosahedral shell can be considered as built from 60 A/B to 30 C/C dimers. The A/B dimers are related by local twofold symmetry (AB_5 in the figure), and C/C dimers are related by icosahedral symmetry (C/C_2 in the figure). (**c**) A typical jelly roll β-barrel domain [taken from the structure of San Miguel sea lion virus (Chen et al. 2006), an animal calicivirus] that participates in the $T=3$ icosahedral shell. The β-barrel motif (inside the *box*) consists of eight β-strands organized into two twisted antiparallel β-sheets generally referred to as BIDG and CHEF (Rossmann and Johnson 1989). The letters refer to the position of β-strands in the primary sequence: B is the most N-terminal, and H is the most C-terminal. An N-terminal arm that projects inward from the β-barrel participates in the intersubunit interactions. (**d**) Comparison of the C/C2 (*left*) and A/B5 (*right*) dimers in the Norwalk virus $T=3$ structure (Prasad et al. 1999) showing the flat and bent conformations, respectively. In Norwalk virus, the capsid protein consists of two domains: the shell domain (S), with a β-barrel motif, and a protruding domain (P). Only the S domain participates in the shell contacts. The β-barrel domains of A, B, and C are colored as in (**b**). The P domain which participates in the dimeric interactions is colored in *white*. Although structurally not similar, capsid protein of tombus viruses, such as TBSV, also has a protruding (P) domain (Harrison et al. 1978). (**e**) Nonequivalent interactions, as viewed from inside of the capsid, between the NTAs of the ABC subunits that are observed in the Norwalk virus structure, shown here for example (see text). The NTAs (*green*) from different subunits are marked. The ABC subunits are shown using the same color scheme as in (**b**)

above, a distinct conformational change involving a Pro residue in the B subunit that leads to the formation of a ring-like structure around the fivefold axis appears to provide a switch. In several of these viruses, the role of NTA in conferring correct capsid assembly is substantiated by mutational studies. Deletion of NTA in the sobamovirus results in formation of a $T=1$ structure (Erickson et al. 1985), mutation of NTA residues involved in the RNA interactions in nodaviruses results in aberrant particles (Dong et al. 1998; Wikoff and Johnson 1999), and deletion of N-terminal 35 residues in norovirus results in abrogation of capsid assembly (Bertolotti-Ciarlet et al. 2002).

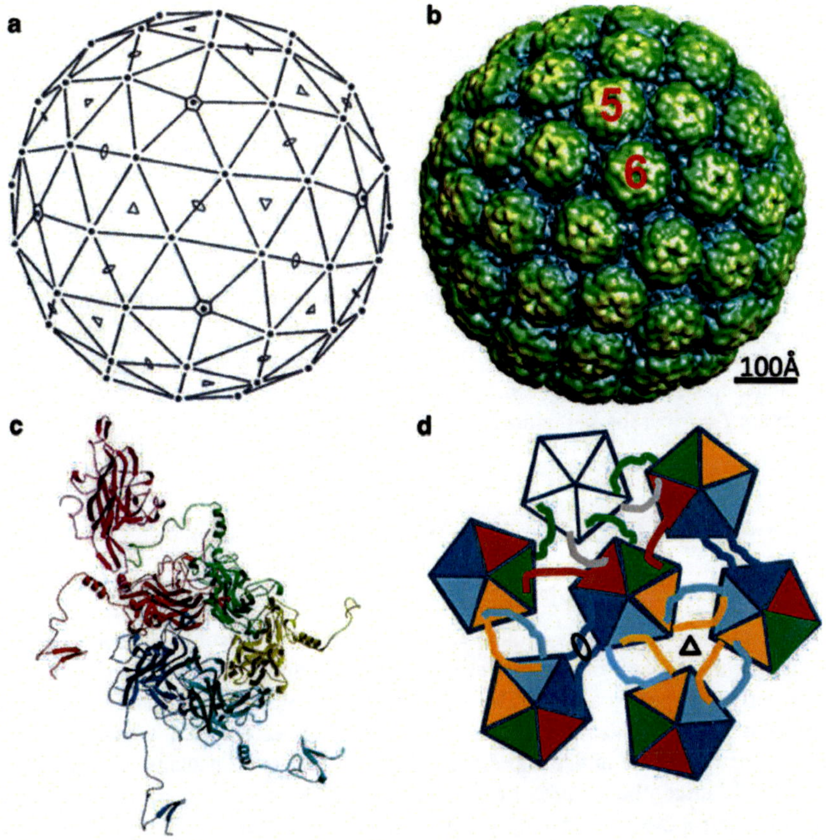

Fig. 3.8 Subunit packing and intersubunit interactions in the all-pentamer $T=7$ structure of polyomavirus. (**a**) $T=7$ icosahedral lattice as viewed along the icosahedral twofold axis. The five-, three-, and twofold symmetry elements are denoted by *pentagon*, *triangle*, and *oval* symbols. (**b**) Crystallographic structure of polyomavirus, viewed along the twofold axis as in (**a**), showing the close packing of pentamers on a $T=7$ icosahedral lattice (obtained from VIPER, Reddy et al. 2001). Pentamers sit on the five- and six-coordinated positions of the lattice. Two of these locations are denoted. (**c**) The icosahedral asymmetric unit consisting of a pentamer at the six-coordinated position and a subunit (shown in *pink*) from the pentamer at the fivefold axis. The subunits in the pentamer were colored differently (obtained from VIPER, Reddy et al. 2001). Application of the five-, three-, and twofold symmetry generates the structures shown in (**b**). The CTAs emerging from each subunit are clearly seen. (**d**) A schematic representation of the interchange of CTA arms between the pentamers that hold the capsid together (Liddington et al. 1991). The pentameric subunits are depicted as *triangles* inside a *pentagon*. The subunits in the pentamers at the sixfold node are colored in different colors as in (**c**). The pentamer at the fivefold axis is shown in *white*. The CTAs emerging from each subunit is shown in the same color as the subunit. The CTA from the pentavalent pentamer is shown in *gray*. Three distinct types of inter-pentameric interactions involving interchange of CTAs are observed in the structure. First type, between the pentavalent pentamer and two hexavalent pentamers related by the local threefold axis (*gray*, *red*, and *green* CTAs), between the two hexavalent pentamers related by the local twofold axis (*yellow* and *cyan*), and the third type, between the hexavalent pentamers related by the icosahedral twofold axis (*blue–blue*)

Conformational switching, as observed in the case of $T=3$ icosahedral viruses, is also evident in the capsid structure of in papovaviruses with all-pentamer arrangement on a $T=7$ icosahedral lattice (Liddington et al. 1991; Wolf et al. 2010), as mentioned earlier (Fig. 3.8a, b). In this structure, there are two types of pentamers, one located at the pentavalent lattice points of the $T=7$ lattice and the other located at the hexavalent lattice points (Fig. 3.8b). Each pentamer exhibits a roughly cylindrical shape with a hollow conical interior (Fig. 3.8c). Except for the flexible C-terminal arm (CTA), the subunits that compose both types of pentamers exhibit a similar structure. Interestingly, the main body exhibits the same canonical eight-stranded jelly roll β-barrel fold as the subunits in the $T=3$ icosahedral viruses except that the long axis of this fold is oriented radially in the polyomavirus in contrast to a tangential orientation in the $T=3$ viruses. The subunits in the pentavalent pentamer surrounded by five hexavalent pentamers have identical environments. However, the local environment for the subunits in the hexavalent pentamer, surrounded by a pentavalent pentamer and five other hexavalent pentamers, varies. How do these subunits conform to these local variations? The situation is analogous to the two quasi-equivalent dimers in the $T=3$ structures. Instead of a flexible NTA as in $T=3$ viruses, in papovaviruses, the flexible CTA of the capsid protein subunit provides the molecular switch to adapt to the local environment. The CTA extending away from the main body of each pentameric subunit invades the subunit of an adjacent pentamer such that each pentamer receives and donates five arms (Fig. 3.8d). The required variability for the capsid formation comes from the manner in which the conformationally flexible CTA are exchanged between the adjacent pentamers in the $T=7$ icosahedral lattice.

3.4.2 *Triangulated Lattice and Subunit Packing*

One of the main tenets of the quasi-equivalence theory that triangulated icosahedral lattices allow for efficient close packing of subunits is clearly evident in the high-resolution structures of spherical viruses including the papovaviruses. However, there are distinct variations in how triangulated lattices confer optimal subunit packing. The all-pentamer papovavirus structure represents a unique variation in which subunit packing involves a triangulated lattice despite the mismatch between the molecular symmetry and the lattice coordination. Except for the 12 pentamers at the fivefold positions of the $T=7$ lattice, the location of other pentamers, at the six-coordinated positions, is not consistent with their molecular symmetry. The advantage, however, is clearly evident considering that a triangulated lattice intrinsically provides locations for efficient hexagonal close packing of the subunits. The papovavirus structures elegantly demonstrate how the six-coordinated positions in the $T=7$ lattice allow for a favorable close packing of the roughly cylindrical pentamers between the fivefold positions requiring only three different types of interpentamer contacts (Fig. 3.8d).

In the $T=3$ structures with 180 identical subunits, as discussed previously, the subunit packing is as per the quasi-equivalent theory. The trapezoidal-shaped β-barrel domains of the capsid protein pack closely into rings of 6 at the hexavalent lattice points and rings of 5 around the pentavalent positions. Structures of picornaviruses (Hogle et al. 1985; Rossmann et al. 1985; Luo et al. 1987; Acharya et al. 1989) present an interesting variation depicting how a $T=3$ lattice allows similar packing but with nonidentical subunits. In picornaviruses, the icosahedral asymmetric unit consists of chemically nonidentical VP1, VP2, and VP3. Despite not having any sequence homology with capsid proteins of $T=3$ viruses or with themselves, VP1, VP2, and VP3 exhibit the same canonical β-barrel fold. The arrangement of 180 β-barrels, from 60 copies of each of the VP1, VP2, and VP3, is strikingly similar to that observed in the $T=3$ icosahedral lattices formed by 180 identical copies of a capsid protein. In the picornavirus structures, VP1, VP3, and VP2 occupy the same positions as chemically equivalent A, B, and C, respectively, in the $T=3$ icosahedral viruses. Because VP1, VP2, and VP3 are chemically nonidentical, the picornavirus structures are described as a pseudo $T=3$ lattice. The pseudo $T=3$ organization in the comovirus provides yet another interesting variation with two chemically distinct polypeptide chains, the small (S) protein with one β-barrel domain and the large (L) protein with two β-barrel domains (Lin et al. 1999). The packing of the β-barrels from these two proteins follows $T=3$ lattice and is similar to that observed in the $T=3$ icosahedral structures. The β-barrel of the S protein occupies the position corresponding to A in the $T=3$ lattice, whereas the two β-barrels of the L protein occupy positions corresponding to B and C.

3.4.3 Capsid Assembly

In an attempt to answer what dictates appropriate conformational switching in the formation of icosahedral shells, Berger et al. (1994) has proposed a local rule-based theory suggesting that protein subunits make use of local information to guide the capsid assembly and that the choice for a particular interaction is dictated by its immediate neighbors. In the $T=3$ icosahedral viruses, dimer of the capsid protein is thought to be the building block for the assembly. In solution, these dimers are perhaps in a dynamic equilibrium between the "bent" and "flat" conformations, and during the assembly, these dimers adopt appropriate conformational states. Two different assembly pathways have been proposed for the $T=3$ icosahedral viruses. In the case of sobamoviruses, such as southern bean mosaic virus (SBMV), it is suggested that assembly involves the formation of pentamers of dimers as an intermediate step followed by the association of other dimers resulting in the formation of a $T=3$ shell (Erickson et al. 1985). Such a pathway is consistent with the observation that deletion of NTA in the SBMV results in a $T=1$ structure formed by the association of 12 pentamers of dimers. A similar pathway involving pentamers of dimers as an assembly intermediate is also proposed for noroviruses (Prasad et al. 1999), which is supported by recent mass spectroscopic analysis of recombinant Norwalk virus particles (Shoemaker et al. 2010). In the case of tombus viruses, such as TBSV, it is suggested that the assembly intermediate involves trimers of dimers, consistent with the observation that the ordered NTAs of the C subunits form a stable internal structure at the icosahedral threefold axes of the $T=3$ shell (Sorger et al. 1986). These viruses have a stretch of basic residues at the N terminus (R arm) of the capsid protein that can interact with the RNA. The intermediate assembly unit, either pentamers of dimers or trimers of dimers, is thought to be nucleated by interaction with the packaging signal in the genomic RNA (Sorger et al. 1986; Harrison 2007). In the case of noroviruses, the capsid protein, lacking the basic R arm, itself has all the determinants for the formation of the $T=3$ structure because the recombinant capsid protein of noroviruses readily assembles into $T=3$ structures (Prasad et al. 1999). The role of RNA in directing the assembly pathway is readily apparent in the structure for flock house virus (a nodavirus) (Fisher and Johnson 1993). For picornaviruses, the assembly pathway is somewhat better characterized. The 5S structural unit consisting of one copy of VP0, VP3, and VP1 and 14S pentameric caps of VP1–VP2–VP3 are known to be the intermediates. In the case of papovaviruses, in which preformed stable pentamers are the building blocks, the correct assembly of the pentamers onto a $T=7$ lattice is likely dictated by interactions with the viral minichromosome. Initial interactions between the pentamer and the DNA, involving the N-terminal arms of the subunits, may serve as a nucleation center for further stepwise addition of individual pentamers or a cluster of pentamers, consisting of one pentavalent pentamer surrounded by five other pentamers (1+5 cluster), to form the T=7 capsid structure with the encapsidated genome (Stehle et al. 1996; Mukherjee et al. 2007). During this process, the curvature is appropriately modulated by local alterations in the "bonding" between the pentamers involving the CTAs.

As can be seen from the above discussion, the interplay between the global restraints, which allow optimal packing of the subunits to form a closed shell of proper size to accommodate the genome, and local conformational variability, which allows, necessary flexibility in the intersubunit contacts for modulating the curvature, dictates structural realization in icosahedral viruses.

3.4.4 Scaffolds, Glue Proteins, and Cores

In more complex icosahedral viruses such as dsDNA bacteriophages, herpes simplex viruses, and adenovirus, the establishment of the icosahedral lattice involves several other factors like scaffolding proteins, accessory proteins, maturation-dependent proteolysis, and even larger-scale conformational changes in their major capsid proteins than is observed in simpler icosahedral viruses.

In contrast to assembly based on the encapsidation of the nucleic acid concurrently with the formation of the capsid shell by stepwise addition of smaller assembly intermediates as discussed in the previous section, in some of these viruses, capsid assembly is accomplished by the formation of a complete capsid shell, followed by the insertion of the nucleic acid. Such a mechanism avoids some of the problems associated with the former assembly mechanism, such as the requirement that both the capsid proteins and the nucleic acid be brought to a common assembly point and properly staged for assembly, and the necessity of using a nonicosahedral component (the RNA) to help build an icosahedral capsid. However, this mechanism introduces other problems that must be solved. First, since most dsDNA phages and viruses packaged by this mechanism have a single genome segment, it is absolutely required that only one copy of the nucleic acid be inserted into each capsid shell. This is accomplished by having a DNA packaging machine, called the portal, located at only one of the 12 vertices of the capsid (see also Chap. 22). Second, the capsid shell must be of the correct size to fit the nucleic acid genome, and it must be empty of cellular proteins, which could interfere with DNA insertion. This is accomplished by assembling the capsid around a set of virally encoded scaffolding proteins. Thus, the shell is guaranteed to be the right size and *T* number, and the scaffolding proteins are removed either during DNA insertion or by proteolysis after capsid shell assembly. Third, in the cases studied so far, the capsid proteins of the procapsid have a different conformation that they do in the mature virion. This could serve as a signal that the DNA has been packaged, and is ready for the final steps of virus maturation, like the addition of the phage tail. Finally, the DNA insertion is accompanied by an increase in internal pressure generated by the repulsion of the charged nucleic acid. This pressure may be as high as 40 atm and has been suggested to facilitate or power DNA release. Most dsDNA bacteriophages have mechanisms, such as glue proteins in epsilon 15 or chemical crosslinking of the capsid shell proteins as in HK97, to strengthen the capsid shell against this pressure. So how do the capsid proteins, the scaffolding proteins, the glue, and the portal interact in this system of virus assembly?

3.4.4.1 "True" T=7 Capsid Organization

Structures of procapsids and mature capsids of dsDNA bacteriophages such as P22 (Prasad et al. 1993; Thuman-Commike et al. 1996; Jiang et al. 2003; Chang et al. 2006; Chen et al. 2011), ε15 (Jiang et al. 2006, 2008) and HK97 (Wikoff et al. 2000, 2006; Huang et al. 2011) have been well characterized. In these viruses, the icosahedral lattice is established according to the classical Casper and Klug quasi-equivalence theory, i.e., with 415 identical subunits arranged as 11 pentamers at the pentavalent positions and 60 hexamers at the hexavalent lattice points of the $T=7l$ icosahedral lattice (Fig. 3.9). The true $T=7$ symmetry is locally broken at one of the fivefold vertices because of the presence of the portal protein complex and sometimes a tail; as a result, there are only 11 pentamers instead of 12. The subunits at the hexavalent positions in the procapsids of these viruses show a skewed arrangement and only adopt a near sixfold symmetric organization in the mature capsids (Fig. 3.9a, b). Interestingly, as in the case of $T=3$ viruses, which all exhibit a canonical jelly roll β-barrel structure, these viruses and other tailed bacteriophages (Baker et al. 2005; Fokine et al. 2005) also exhibit a common fold that was first described in the high-resolution crystallographic structure of HK97 and hence is called HK97-like fold (Wikoff et al. 2000, see also Chap. 15) (Fig. 3.9c). Unlike in the case of papovaviruses, in which the capsid protein is preformed into pentamers (one species), the capsid protein subunits in these viruses have to switch between hexameric and pentameric clustering on the $T=7$ lattice.

3.4.4.2 Scaffolding Protein

To ensure accurate formation of the capsid shell, these viruses use an elaborate mechanism involving a scaffold which is either provided by a separate virally encoded protein as in the case of P22 (Fane

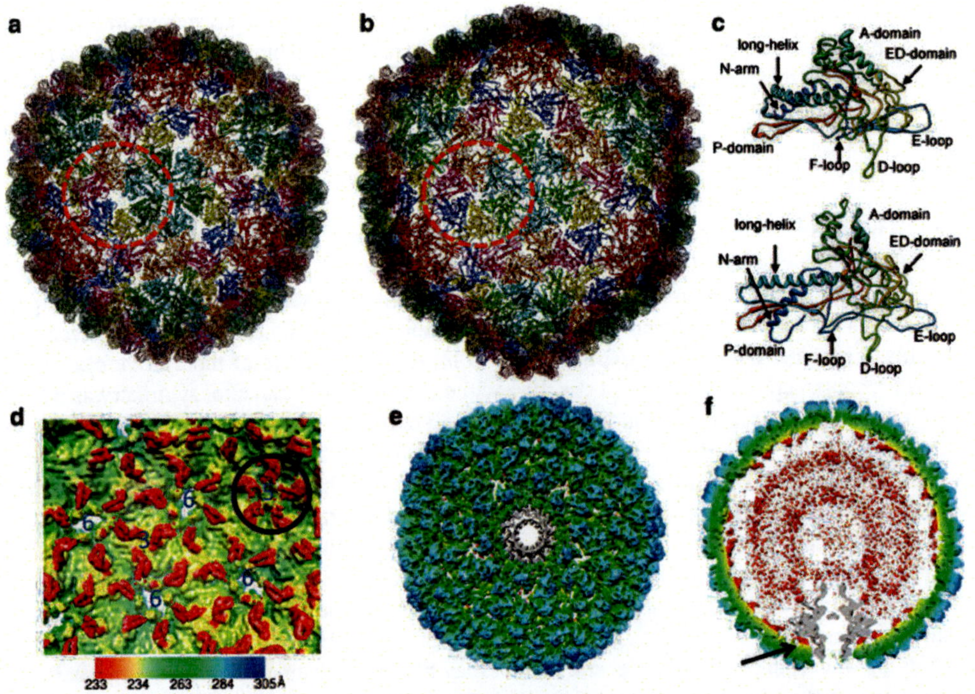

Fig. 3.9 Structural organization in the procapsid and mature capsid of P22. (**a**) Subunit packing in the procapsid showing the hexameric and pentameric clustering of the capsid protein subunits on a $T=7l$ lattice. The backbone structure of the subunits shown is derived from the ~4 Å cryo-EM structure. The subunits surrounding the fivefold axes are colored in *red*, and the six quasi-equivalent subunits at the hexavalent nodes (*circled* in *red*) are shown in different colors. Notice the spherical shape of the capsid, skewed clustering of the subunits around a hole at the hexavalent positions. These holes may be the exit points for the scaffolding proteins. (**b**) Mature capsid structure at ~4 Å determined using cryo-EM technique. The subunits are colored as in (**a**). Notice the angular shape, symmetric organization of the hexavalent subunits, and closure of the hole (*red circle*). (**c**) Conformational changes in the capsid subunit. *Above*: procapsid subunit. *Below*: Mature capsid subunit. The domains and the loops in the HK97-like fold are shown. (**d**) Inside surface of the procapsid showing the density due to scaffolding protein (*red*). (**e**) Asymmetric reconstruction of the procapsid, viewed along the 12-fold axis of the portal. (**f**) Cross-section showing the inward extension of the portal density. Figures from Donghua Chen and Wah Chiu

and Prevelige 2003, see also Chap. 14) or by a part of the capsid protein itself as in the case of HK97 (Huang et al. 2011). In P22, the scaffolding protein exits signaling for the maturation of the capsid, and then, it is recycled for the further rounds procapsid assembly. In the case HK97, the reordering of the N-terminal region of the capsid protein is suggested to provide a trigger for the maturation (Huang et al. 2011).

Recent cryo-EM reconstructions of P22 procapsid have provided some insights into the possible role of scaffolding proteins in directing the $T=7$ procapsid assembly (Parent et al. 2010a, b; Chen et al. 2011). In the reconstructions of the P22 procapsid, imposing icosahedral symmetry, the major portion of the scaffolding protein density is not clearly observed, suggesting that the majority of the scaffolding protein does not conform to the icosahedral symmetry of the capsid. However, the portion that interacts with the inside surface of the capsid shell is visible (Fig. 3.9d). These interactions are observed to be stronger with the hexavalent subunits than with the pentavalent subunits, suggesting that scaffolding protein may promote or stabilize the hexamer of the capsid protein relative to the pentamer. It is clear that the scaffolding proteins influence the icosahedral lattice because P22 mutants with altered or missing scaffolding proteins make a significant number of $T=4$ capsid shells with half the number of hexavalently disposed units, i.e., 30 versus 60

(Thuman-Commike et al. 1998). Thus, the scaffolding protein may influence the formation of the hexameric and pentameric clustering of the capsid protein subunits by altering the ratio of hexamers to pentamers during the assembly process.

3.4.4.3 The Portal

There is no evidence that the capsid proteins of P22 oligomerize into either hexamers or pentamers before being incorporated into the capsid shell. Thus, the shell appears to be built monomer by monomer, without the participation of preformed pentamers or hexamers. It is likely that the portal protein together with scaffolding protein guides the assembly of capsid protein into the correct shell. Visualization of the portal complex, because it is located at only one of the 12 fivefold vertices, requires reconstructions of the capsid structure without imposing icosahedral symmetry as was first done in the case of phi29 (Tao et al. 1998) and subsequently with several other dsDNA bacteriophages including P22 (Jiang et al. 2006; Chen et al. 2011). In the cryo-EM reconstructions of P22 procapsid, without imposing the icosahedral symmetry, the portal protein complex is clearly seen and sits as a dodecamer at one of the fivefold vertices (Chen et al. 2011) (Fig. 3.9e, f). Such a reconstruction shows that the portal interacts more directly with scaffolding protein than with the capsid shell in the procapsid consistent with the notion that scaffolding protein recruits the portal protein complex (Fane and Prevelige 2003). This suggests that the portal protein is a key participant in the capsid assembly (Chen et al. 2011). If the level of portal protein is kept well below that of the scaffolding or capsid proteins, the portal protein may nucleate the assembly of a capsid shell by its interactions with scaffolding and capsid shell proteins. This would in turn assure that there is only one portal per capsid, which is a requirement for proper DNA packaging. Capsids assembled in the absence of portal proteins are aberrant in their size, form, and/or symmetry.

3.4.4.4 Maturation and Expansion

During maturation of P22, the capsid protein undergoes a major conformational change, making the capsid shell thinner and the hexon interactions more symmetric. The shell becomes more angular from its round shape in the procapsid, is thinner by about 40 Å, and expands in diameter by about 100 Å (Prasad et al. 1993; Chen et al. 2011). The process of DNA packaging and concomitant release of the scaffolding proteins, through the small holes at the hexavalent positions of the procapsid, may trigger the conformational changes in the subunits and subunit contacts resulting in a dramatically expanded capsid. Similar expansion accompanied by large-scale conformational change upon maturation is also observed in HK97 and ε15 (Jiang et al. 2006; Wikoff et al. 2006). In the HK97, maturation is accompanied by the formation of an intersubunit isopeptide linkage that cross-links the entire capsid conferring stability to the capsid to withstand the pressure created by DNA packaging. In P22, which does not form a cross-linked capsid like HK97, strong protein–protein interactions among its capsid shell subunits appear to render sufficient structural rigidity in the mature virion (Parent et al. 2010a, b; Chen et al. 2011).

3.4.4.5 Glue Proteins

The nucleocapsid of herpesvirus, a dsDNA virus of eukaryotic origin with a distinctly different structural organization, nevertheless shares several characteristics with dsDNA bacteriophages (Heymann et al. 2003). It exhibits a $T = 16$ icosahedral structure (~1,250 Å in diameter) with pentamer and hexamer clustering of the major capsid protein VP5. 960 copies of VP5 forms 12 pentamers and 150 (10T−1) hexamers as expected from the quasi-equivalent symmetry (Schrag et al. 1989; Zhou et al. 1994, 2000). In addition to VP5, capsid also consists of what are known as triplex proteins, VP19c and VP23 (in a stoichiometric ratio of 1:2), that are positioned at the local and strict threefold axes of the T=16 lattice. These triplexes function as glue holding together the surrounding VP5 capsomeres. Similar to dsDNA bacteriophages, capsid assembly is assisted by scaffolding protein which is proteolytically degraded prior to the packaging of the genomic DNA (Nicholson et al. 1994; Rixon et al. 1996). Recombinant baculovirus expression of VP5 and VP19c alone, i.e., in the absence of VP23 and the scaffolding protein, results in smaller-sized particles with T=7 symmetry suggesting the role of VP23 and the scaffolding protein in ensuring the correct capsid assembly (Saad et al. 1999). Another interesting similarity with dsDNA bacteriophages is that VP5 exhibits the HK97-like fold (Baker et al. 2003; Bowman et al. 2003). In addition to these components, as revealed by the electron cryo-tomographic reconstructions, the nucleocapsid has a unique fivefold vertex with a 12-fold symmetric portal for inserting the genomic dsDNA (Cardone et al. 2007; Chang et al. 2007; Deng et al. 2007). The nucleocapsid capsid also exhibits maturation-dependent morphological alterations (Heymann et al. 2003). These observations clearly indicate that DNA packaging mechanism in herpes virus resembles that in the tailed dsDNA bacteriophages.

Adenovirus, another dsDNA virus, represents more intricate and complex architecture (~1,000 Å in diameter) involving the arrangement of multiple proteins on a triangulated icosahedral lattice. It shows a different solution for the problem of pentamer and hexamer clustering on such a lattice. In this virus, the capsid organization is based on a pseudo T=25 lattice (Stewart et al. 1991; Liu et al. 2010; Reddy et al. 2010). Instead of the same protein forming a pentameric and hexameric clusters as in the case of dsDNA bacteriophages discussed above, two different proteins occupy the pentavalent and hexavalent positions on the T=25 lattice (Fig. 3.10a). The protein called penton base forms a penton at the 12 vertices incorporating a trimer of the fiber protein, and another protein called hexon occupy the 260 hexavalent positions on the T=25 lattice, however, as trimers. Crystallographic structures of penton (Zubieta et al. 2005), hexon (Roberts et al. 1986; Athappilly et al. 1994), and the fiber protein (van Raaij et al. 1999) have been determined. Previous cryo-EM studies provided some insights into the overall capsid organization in this complex virus (Stewart et al. 1991; Saban et al. 2006). Recently, the structure of the adenovirus capsid has been determined to near 3 Å resolution using X-ray crystallographic (Reddy et al. 2010) as well as cryo-EM techniques (Liu et al. 2010) providing a clearer picture of the intricate interactions between various components in the capsid organization (Fig. 3.10). The hexon protein, which forms stable trimers in solution, is incorporated in this oligomeric state into the capsid structure. Each hexon subunit exhibits an elaborate structure with two jelly roll β-barrels. In the hexon trimer, the disposition of these six β-barrels, two from each subunit, is such that it gives a pseudohexagonal contour to the trimer base appropriate for close packing of hexons at the hexavalent positions in the T=25 lattice. The pentons and hexons are held together in the capsid structure by an intricate network of glue (or cement) proteins (Fig. 3.10b). A similar pseudo T=25 capsid organization is also observed in PRD1, an enveloped dsDNA bacteriophage (Abrescia et al. 2004, 2008; Cockburn et al. 2004, see also Chap. 16). The P3 protein of PRD1 positioned at the hexavalent positions of the T=25 lattice also exhibits a similar "double β-barrel trimer" suggesting such a trimeric organization is evolutionarily conserved (Benson et al. 1999, 2002, 2004; Saren et al. 2005).

3.4.4.6 Cores

The dsRNA viruses particularly in the family Reoviridae such as rotavirus, bluetongue virus, and reovirus with 10–12 dsRNA genomic segments, which encode multiple structural proteins, exhibit even more elaborate and complex icosahedral capsid organization with multiple concentric capsid layers (Fig. 3.11) (Lawton et al. 2000; Jayaram et al. 2004; Harrison 2007). The complex structural

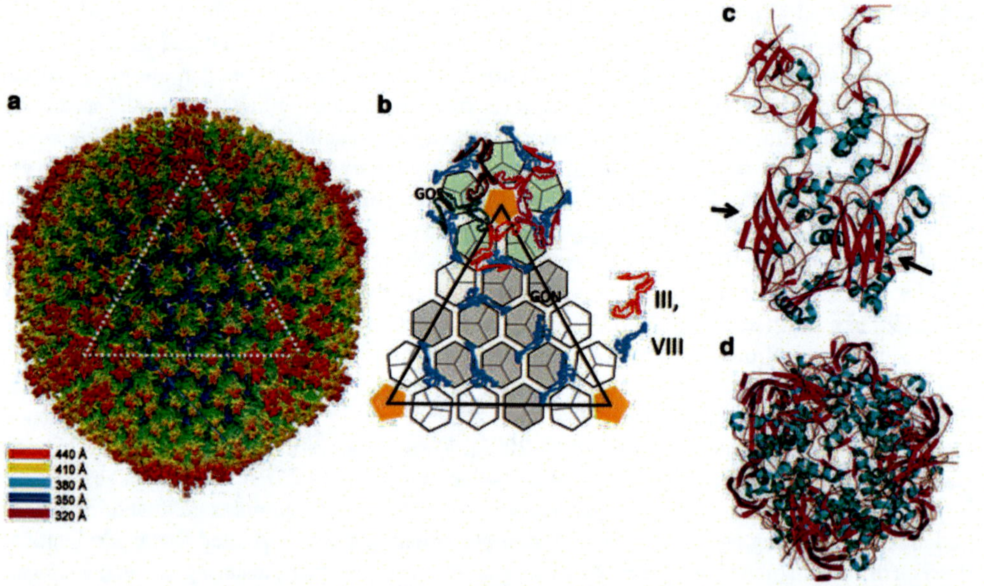

Fig. 3.10 Adenovirus capsid organization. (**a**) 3.4 Å cryo-EM structure of adenovirus capsid as viewed along the icosahedral threefold axis. One of the icosahedral facets is indicated by *dashed white lines*. The structure is radially colored according to the chart shown. (**b**) Schematic representation of structural organization of pentons, hexons, and the cement proteins (VIII, IIIA) in the $T=25$ (pseudo) icosahedral facet [same as indicated in (**a**)]. The hexon trimers are shown as *hexagons* and pentons as *pentagons* in *golden yellow*. The color coding for the cement proteins are shown in the side. The so-called GON (group of nine) hexons, because they remain associated following the capsid disassembly, are shaded in *gray*. The fiber protein which insets in to the penton is not observed in the cryo-EM structure as they are flexible. (**c**) Crystallographic structure of the hexon subunit. The orientation of the subunit structure is chosen such that top region is at the virion exterior. The structure is colored according to the secondary structural elements; α-helices in *cyan*, β-sheets in *pink*, and loops in *brown*. The radially oriented β-barrels are shown by *black arrows*. (**d**) Hexon trimer as viewed from the top, corresponding to the view of the trimer in the center of virion structure (**a**) along the icosahedral threefold axis. The structure is colored following the same scheme as in (**c**). The trimer has a striking hexagonal contour at its base, because of the orientation of the two β-barrels in each subunit, which is appropriate for its location at the 6-fold positions of the T=25 lattice. Fig. 3.10a, b from Dr. Hong Zhou

organization in these viruses appears to have evolved to compartmentalize the requirement of host cell interaction and the necessity to transcribe multiple genome segments within the capsid layers. Because the host cells do not have enzymatic machinery to convert dsRNA into mRNA, these viruses carry within the capsid interior of the enzymes required for transcribing the genome segments into mRNA and also capping the transcripts. It is to the advantage of the virus to carry out transcription endogenously not only to avoid any degradation of the genome by cellular nucleases but also to prevent cellular antiviral interferon response triggered by increased concentrations of dsRNA during replication, particularly in mammalian hosts.

One common theme in these dsRNA viruses is the assembly of the outer capsid layer(s), typically based on $T=13l$ icosahedral symmetry, and is assembled on the innermost layer that surrounds the dsRNA genome (see also Chap. 17). This innermost layer exhibits unusual icosahedral organization consisting of 120 subunits (forbidden "$T=2$") arranged as 60 asymmetric dimers on a $T=1$ icosahedral lattice (Lawton et al. 1997; Grimes et al. 1998; Reinisch et al. 2000; Chen et al. 2006) (Fig. 3.11e). Such an organization appears to be highly conserved in all the dsRNA viruses including dsRNA viruses of bacterial and fungal origins (Naitow et al. 2002; Ochoa et al. 2008), and the available structural data also indicate that the polypeptide folds of the proteins that form this layer are similar, despite lacking any noticeable sequence similarity.

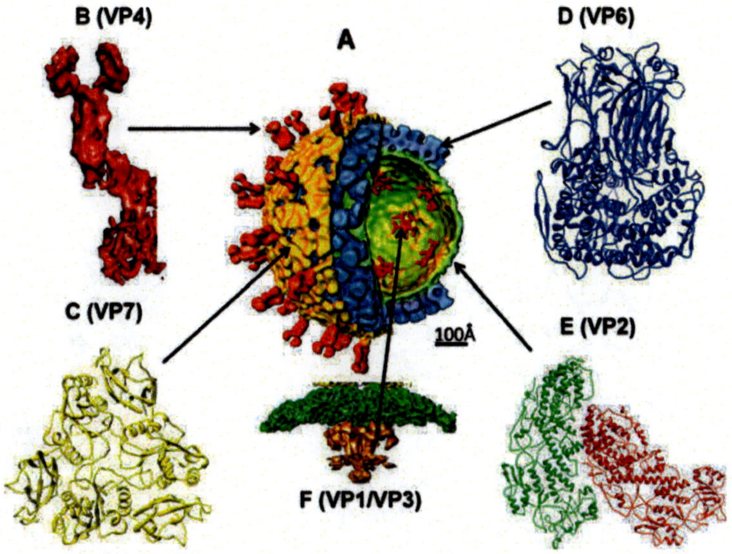

Fig. 3.11 Structure and location of various protein components in the triple-layered architecture of rotavirus are shown using a cutaway view of a lower-resolution cryo-EM structure of the TLP as the reference. (**a**) The mature rotavirus particle consisting of VP4 spikes (*red*) that emanate from the outer capsid layer, VP7 layer which exhibits a $T=13l$ icosahedral organization with 260 VP7 trimers located at all the icosahedral and local threefold axes (*yellow*), VP6 layer (*blue*), which also exhibits a $T=13$ icosahedral organization with 260 VP6 trimers, and the innermost VP2 layer consisting of 120 subunits of VP2 (*green*). The transcriptional enzymes VP1/VP3 anchored to the VP2 layer and projecting inward from the fivefold axes are shown in *red*. (**b**) A VP4 spike extracted from the cryo-EM map (Li et al. 2009). The foot domain is also indicated. (**c**) X-ray structure of VP7 trimer (PDB ID:3FMG, Aoki et al. 2009). (**d**) X-ray structure of VP6 trimer (PDB ID:1QHD, Mathieu et al. 2001). (**e**) Structure of the VP2 dimer that constitutes the icosahedral asymmetric unit of the innermost capsid layer (PDB ID:3KZ4, McClain et al. 2010) (**f**) Transcription enzymes VP1/VP3 (shown in *gold*) inside the VP2 layer in *green*

It is possible that this unique organization of the core capsid layer in these viruses has evolved to serve a dual purpose: to properly position the transcription enzyme complex and organize the genome to facilitate endogenous transcription. In many of these viruses, the virally encoded enzymes required for the transcription (RNA-dependent RNA polymerase) and capping intimately associate with this layer at the fivefold positions (Prasad et al. 1996; Zhang et al. 2003; Nason et al. 2004). One model is that pentamers of dimers of the subunits in the core layer associate along with the transcription enzyme complex, and 12 of these pentamers further associate to form the core layer with 120 subunits, which then serves as a platform for the assembly of the outer layers perhaps by sequential addition oligomeric (typically trimers) of capsid protein that constitutes the outer layer(s).

3.5 Helical Viruses

In addition to icosahedral symmetry that is commonly found in the spherical viruses, another symmetry that is prevalent among viruses is the helical symmetry. Many rod-shaped viruses such as viruses belonging to family Tobamoviruses (Namba and Stubbs 1986, see also Chap. 28), potyviruses (Kendall et al. 2008), rhabdoviruses (Ge et al. 2010), and nucleocapsids of several animal viruses such as Sendai viruses (Egelman et al. 1989) exhibit this kind of symmetry. In contrast to icosahedral symmetry which involves only rotational symmetry, helical symmetry involves both

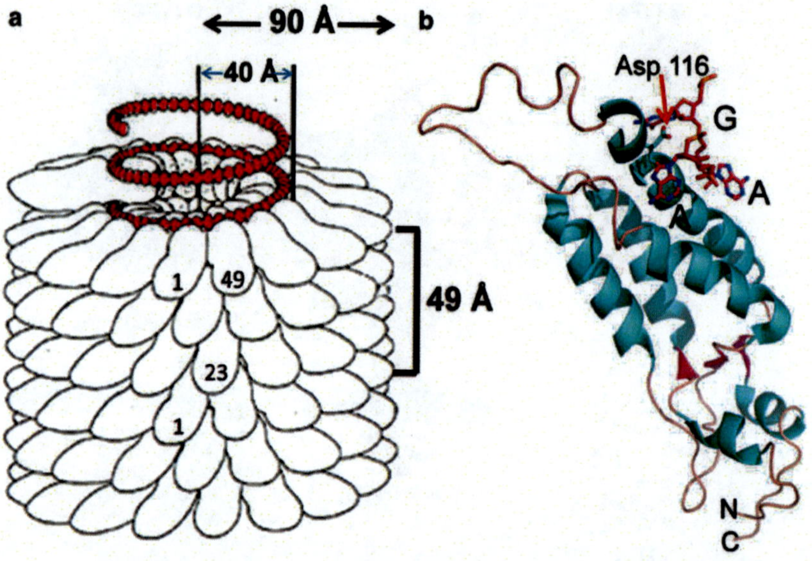

Fig. 3.12 The structure of TMV, an example of a helical virus. (**a**) Cartoon representation of the helical arrangement of the subunits in the TMV. The shoe-shaped capsid protein subunits wrap around the genomic RNA (*red*) following helical symmetry. Some of the subunits are numbered to indicate number of subunits in a turn of the helix. The radius of the helical assembly, pitch of the helix, along with inside radius at which protein interacts with the genomic RNA is indicated. (**b**) A cartoon representation of the crystallographic structure of the TMV capsid protein subunit with three nucleotides of the genomic RNA (PDB ID:2TMV, Namba et al. 1989). The N- and C-termini are denoted. The structure is colored according to the secondary structural elements (*cyan* – α-helices, *red* – β-strands, and *brown* – loops). The RNA bases are indicated along with the location of the residues Asp 116 of the capsid protein

rotational and translational components, which when combined, give a screw axis. Repeated application of rotation to a motif followed by translation along an axis gives rise to a structure with helical symmetry with a defined pitch (P). Helical structures are typically characterized by the radial location of the subunit with respect to the helix axis; the rotation per subunit, which is equivalent to the number of subunits per turn (n) in the helix; and the axial rise per subunit (h).

The best-characterized example of a helical virus is tobacco mosaic virus (TMV). It is a rod-shaped virus ~3,000 Å long and 180 Å in diameter, with a central hole of 40 Å in diameter. The structure of TMV has been determined to 2.8 Å resolution using X-ray fiber diffraction studies (Namba and Stubbs 1986; Namba et al. 1989) and most recently to a similar resolution using cryo-EM techniques (Ge and Hong Zhou 2011). The capsid protein of TMV forms a one-start, right-handed helix of pitch 23 Å, with 16-1/3 subunits in each turn (Fig. 3.12a). Interestingly, such a capsid organization leads to hexagonal close packing of the subunits, which is a common theme in many of icosahedral virus structures as discussed above. In icosahedral viruses, the encapsidated genome often does not obey the symmetry of the capsid; as a result, in the structural studies of these viruses which implicitly make use of the icosahedral symmetry, information about the structure of the genome and how it is organized is lost. However, in the case of TMV, the entire genome is well ordered and follows the helical symmetry of the capsid, thus providing details of not only how the capsid subunits interact with the genome but also how these interactions lead to virus assembly. The capsid subunits in the TMV structure wrap around the genomic RNA such that the RNA lies inside a groove, at a radius of 40 Å, between successive helical turns. Each subunit interacts with three nucleotides. The subunit structure is predominantly alpha-helical with a core consisting of a right-handed four antiparallel α-helix bundle (Fig. 3.12b). One of the helices from this bundle makes extensive contacts with the genomic RNA.

The interactions between the TMV capsid protein and the genomic RNA in general are nonspecific involving the basic residues of the protein and the phosphate groups of the RNA. One exception, however, is the anomalous repulsive interaction between the carboxylate group of Asp116 and a phosphate group of the RNA (Namba et al. 1989; Ge and Hong Zhou 2011) (Fig. 3.12b). Considering that TMV, and in general any viral assembly, has to assemble and disassemble during its infectious cycle, it is suggested that such a repulsive interaction, which may be required to maintain an energy balance, confers metastable nature to the TMV and functions as a trigger for driving viral disassembly (Namba et al. 1989; Culver et al. 1995; Stubbs 1999).

The assembly process in TMV has been studied extensively, and it serves as the best-characterized example of cocondensation of capsid protein and genomic RNA. In this process, the viral RNA interacts with a 20S aggregate, a two-turn helix of the capsid protein, and the assembly proceeds by addition of 20S aggregates through a highly cooperative process pulling the 5′ end of the RNA through the central hole of the growing TMV rod (Caspar and Namba 1990; Butler 1999; Klug 1999). It is suggested that a disorder-to-order transition of a loop in the capsid protein, analogous to conformational switching in the icosahedral viruses, that is induced by the binding of RNA may play an important role in this process (Namba et al. 1989; Culver et al. 1995). In the X-ray structure of the 20S aggregate, this particular loop is disordered, whereas in the TMV structure, it is ordered. The TMV assembly is initiated by the specific recognition of the sequence AAGAAGUCG in the viral RNA by the TMV capsid protein (Zimmern 1976; Butler 1999). The high-resolution structure of the fully assembled TMV has provided some insights in to how the capsid protein recognizes this sequence (Namba et al. 1989). Although the three RNA-binding sites in each subunit can accommodate any base, one of the binding sites is particularly suitable for G and allows favorable hydrogen bond interactions. In the initiation sequence, every third nucleotide being G thus may provide a strong discrimination for the higher affinity binding of the packaging signal over the rest of the sequence in which the XXG motif does not occur in phase with a statistically significant frequency.

3.6 Enveloped Viruses

During their morphogenesis, several viruses acquire a lipid envelope derived from a cellular organelle which remains part of their capsid organization. In some of the viruses such as influenza viruses, herpes viruses, coronaviruses, bunya viruses, and HIV, the lipid envelop is externally located and is studded by various viral proteins, whereas in others such as alphaviruses and flaviviruses, the lipid envelop is internally located underneath the outer proteinaceous capsid layer. With the exception of alphaviruses and flaviviruses, many of the enveloped viruses lack highly symmetric organization and are less amenable for high-resolution structural analysis. However, structures of the individual protein components of these enveloped viruses, e.g., hemagglutinin (Wilson et al. 1981; Bullough et al. 1994), neuraminidase (Varghese et al. 1983; Xu et al. 2008), and M2 protein of influenza virus (Acharya et al. 2010); glycoprotein of mouse hepatitis virus (Xu et al. 2004); gp120 and gp41 of retroviruses (Chan et al. 1997; Chen et al. 2005); envelop protein of a flavivirus (Rey et al. 1995; Rey 2003; Li et al. 2008); and most recently, E1/E2 complex of alphaviruses, have been well characterized (Li et al. 2010; Voss et al. 2010). These structures have provided significant insight into the mechanism of how the lipid bilayers in these viruses fuse with the membrane of the target cell (Rey 2006; Lamb and Jardetzky 2007; Harrison 2008). In addition, the highly symmetric parts of enveloped viruses such as the nucleocapsid of herpes virus that exhibit $T=16$ icosahedral organization (as discussed above) and the core of hepatitis B virus, which can form either a $T=3$ or a $T=4$ icosahedral structure, have been well characterized (Crowther et al. 1994; Conway et al. 1997; Wynne et al. 1999).

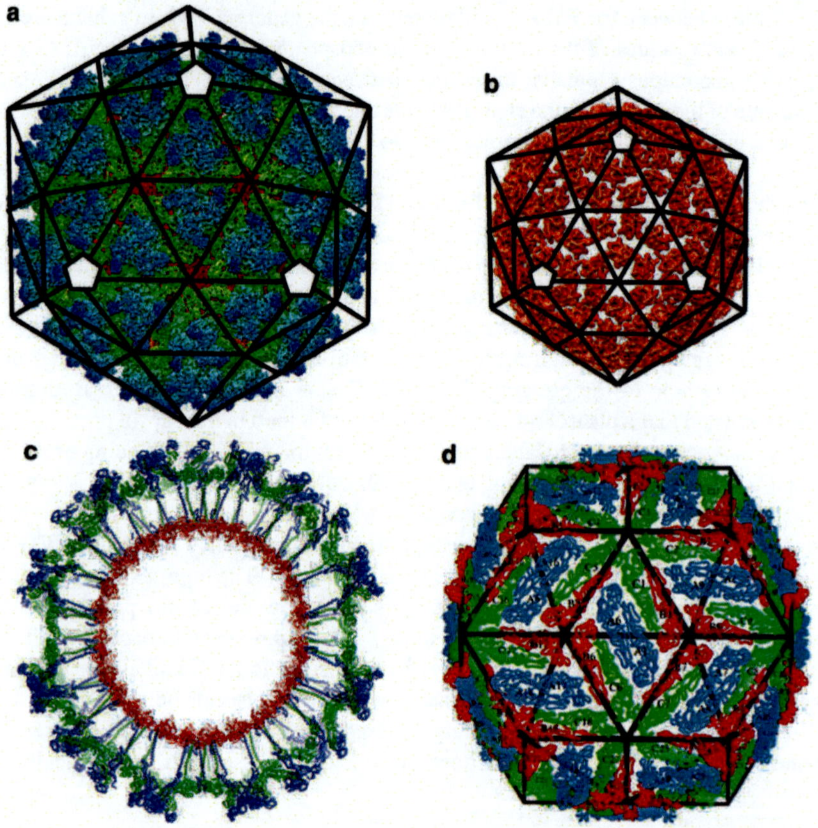

Fig. 3.13 Icosahedral organization in the enveloped viruses. (**a**) Cryo-EM structure of Venezuelan encephalitis virus (VEE), an alphavirus, determined to ~4.5 Å resolution showing the overall organization of the outer capsid composed of 80 trimers of E1 (*green*) and E2 (*blue*) heterodimers (Zhang et al. 2011). These trimers occupy local and strict threefold axes of the $T=4$ icosahedral lattice (*black lines*). (**b**) Structure of the VEE inner capsid extracted from the cryo-EM of the virion shown in (**a**). 240 subunits of the capsid protein are organized on a $T=4$ lattice (*black lines*) in register with the outer $T=4$ E1/E2 capsid layer. (**c**) An equatorial section showing the interactions between the outer E1/E2 proteins (*green* and *blue*) with the inner capsid protein (*red*) through the membrane bilayer. (**d**) Structural organization in the dengue virus, a flavivirus (obtained from VIPER, Reddy et al. 2001). The 180 copies of the E protein are organized as 90 dimers on the surface of the virion. The three subunits in the icosahedral asymmetric unit are colored in *blue*, *green*, and *red*. Despite having 180 subunits, as can be seen, the packing of the 90 dimers in a herringbone-like pattern is not quasi-equivalent and does not follow quasi-equivalent $T=3$ icosahedral lattice (shown in *black lines* for reference) (Fig. 3.13**a–c**, courtesy Corey Hryc and W. Chiu)

The only enveloped viruses that have been structurally characterized to high resolution are the internally enveloped alphaviruses such as Sindbis virus (Paredes et al. 1993, 1998; Zhang et al. 2002), Venezuelan encephalitis virus (Zhang et al. 2011), and Semliki Forest virus (Mancini et al. 2000) and flaviviruses such as dengue virus (Kuhn et al. 2002; Kaufmann et al. 2006; Pokidysheva et al. 2006; Li et al. 2008; Yu et al. 2008a, b). These viruses exhibit icosahedral organization. Alphaviruses consist of two concentric $T=4$ capsid layers that bracket the lipid bilayer (Figs. 3.13a, b). The outer layer is formed by heterotrimers of E1 and E2 glycoproteins, whereas the inner layer is formed by the core protein (Fig. 3.13b). The N-terminal transmembrane helices of the E1 and E2 protein subunits penetrate across the underlying lipid layer and interact one-to-one with the core protein subunits enabling proper registry of the outer and inner $T=4$ layers (Fig. 3.13c).

Although the subunit organization in the capsid layers of the alphaviruses is generally in accordance with the principles of quasi-equivalence theory, the capsid of flaviviruses presents an entirely novel organization (Kuhn et al. 2002). The flavivirus outer capsid, which sits above the lipid bilayer, is composed of 90 dimers of the major capsid protein, E glycoprotein, arranged as sets of three nearly parallel dimers that lie tangentially on the icosahedral surface (Fig. 3.13d). Although the capsid consists of 90 dimers, their arrangement does not conform to a classical $T=3$ quasi-equivalent lattice. This structure thus can be best described as a $T=1$ icosahedral structure with three molecules in the asymmetric unit, analogous to the situation in the core particles of rotavirus with 120 subunits which is described as having a $T=1$ structure with two molecules in the asymmetric unit. In addition to the E protein, flavivirus capsid also consists of M protein whose precise location is not definitively assigned. In contrast to the alphaviruses, which have well-organized core structure, the nucleocapsid of the flavivirus formed by the C protein does not exhibit any ordered organization.

3.7 Conclusion

The quasi-equivalence theory proposed almost half a century ago has been remarkably useful in describing icosahedral virus structure. The generality of its predictions arises from the simplicity afforded by the assumption that groups of subunits packing in a plane are energetically favored to have six neighbors (i.e., are hexagonally close-packed) and that curvature can be generated by introducing nodes (vertices) with only five such neighbors. However, such generality and adaptability can come at the cost of specificity. The propensity for subunits to form quasi-equivalent interactions leading to a wrong (usually smaller and simpler) final capsid structure indicates that more information is sometimes required. So we observe that in some viruses with $T=7$ symmetry, such as P22, scaffolding proteins are used either as a template core around which a proper-sized capsid can form, or, more subtly, to control the relative proportion of subunits which adopt hexameric interactions to those which form pentamers. This idea of altering the stability of one conformational form over another is also exemplified in the RNA viruses, where the RNA plays a role in stabilizing capsid proteins in the proper ratio of conformations. In the members of *Reoviridae*, the inner core layer forms a simple $T=1$ "permanent scaffold" from 60 dimers, upon which the $T=13$ outer layers are built. In adenovirus and herpesvirus, the role of scaffolding is augmented by the incorporation of accessory proteins in the capsid shell, which also appear to function to influence the curvature, and therefore the triangulation number of the capsid shell. Finally, the tendency for hexagonal close packing to be a controlling influence in virus assembly appears to be so robust that in the case of polyomavirus, where pentamers seem to be the only oligomeric species present, pentamers come to occupy the hexavalent positions. The exceptions to the quasi-equivalence theory prove the rule.

Acknowledgments We acknowledge the support from NIH grants R37AI36040 (BVVP), PO1AI057788 (BVVP), P41RR02250 (MFS) and a grant from Robert Welch Foundation (Q1279) to BVVP. We are grateful to Donghua Chen, Corey Hryc, Liya Hu, Wah Chiu, and Hong Zhou for the help in the figures.

References

Abad-Zapatero C, Abdel-Meguid SS, Johnson JE, Leslie AG, Rayment I, Rossmann MG et al (1980) Structure of southern bean mosaic virus at 2.8 A resolution. Nature 286:33–39

Abrescia NG, Cockburn JJ, Grimes JM, Sutton GC, Diprose JM, Butcher SJ et al (2004) Insights into assembly from structural analysis of bacteriophage PRD1. Nature 432:68–74

Abrescia NG, Grimes JM, Kivela HM, Assenberg R, Sutton GC, Butcher SJ et al (2008) Insights into virus evolution and membrane biogenesis from the structure of the marine lipid-containing bacteriophage PM2. Mol Cell 31:749–761

Acharya R, Fry E, Stuart D, Fox G, Rowlands D, Brown F (1989) The three-dimensional structure of foot-and-mouth disease virus at 2.9 A resolution. Nature 337:709–716

Acharya R, Carnevale V, Fiorin G, Levine BG, Polishchuk AL, Balannik V et al (2010) Structure and mechanism of proton transport through the transmembrane tetrameric M2 protein bundle of the influenza A virus. Proc Natl Acad Sci USA 107:15075–15080

Aoki ST, Settembre EC, Trask SD, Greenberg HB, Harrison SC, Dormitzer PR (2009) Structure of rotavirus outer-layer protein VP7 bound with a neutralizing Fab. Science 324:1444–1447

Athappilly FK, Murali R, Rux JJ, Cai Z, Burnett RM (1994) The refined crystal structure of hexon, the major coat protein of adenovirus type 2, at 2.9 A resolution. J Mol Biol 242:430–455

Baker ML, Jiang W, Bowman BR, Zhou ZH, Quiocho FA, Rixon FJ et al (2003) Architecture of the herpes simplex virus major capsid protein derived from structural bioinformatics. J Mol Biol 331:447–456

Baker ML, Jiang W, Rixon FJ, Chiu W (2005) Common ancestry of herpesviruses and tailed DNA bacteriophages. J Virol 79:14967–14970

Baker ML, Zhang J, Ludtke SJ, Chiu W (2010) Cryo-EM of macromolecular assemblies at near-atomic resolution. Nat Protoc 5:1697–1708

Ban N, Larson SB, McPherson A (1995) Structural comparison of the plant satellite viruses. Virology 214:571–583

Belnap DM, McDermott BM Jr, Filman DJ, Cheng N, Trus BL, Zuccola HJ et al (2000) Three-dimensional structure of poliovirus receptor bound to poliovirus. Proc Natl Acad Sci USA 97:73–78

Benevides JM, Juuti JT, Tuma R, Bamford DH, Thomas GJ Jr (2002) Characterization of subunit-specific interactions in a double-stranded RNA virus: Raman difference spectroscopy of the phi6 procapsid. Biochemistry 41:11946–11953

Benson SD, Bamford JK, Bamford DH, Burnett RM (1999) Viral evolution revealed by bacteriophage PRD1 and human adenovirus coat protein structures. Cell 98:825–833

Benson SD, Bamford JK, Bamford DH, Burnett RM (2002) The X-ray crystal structure of P3, the major coat protein of the lipid-containing bacteriophage PRD1, at 1.65 A resolution. Acta Crystallogr D Biol Crystallogr 58:39–59

Benson SD, Bamford JK, Bamford DH, Burnett RM (2004) Does common architecture reveal a viral lineage spanning all three domains of life? Mol Cell 16:673–685

Bentley GA, Lewit-Bentley A, Liljas L, Skoglund U, Roth M, Unge T (1987) Structure of RNA in satellite tobacco necrosis virus. A low resolution neutron diffraction study using $1H_2O/2H_2O$ solvent contrast variation. J Mol Biol 194:129–141

Berger B, Shor PW, Tucker-Kellogg L, King J (1994) Local rule-based theory of virus shell assembly. Proc Natl Acad Sci USA 91:7732–7736

Bertolotti-Ciarlet A, White LJ, Chen R, Prasad BV, Estes MK (2002) Structural requirements for the assembly of Norwalk virus-like particles. J Virol 76:4044–4055

Bottcher B, Wynne SA, Crowther RA (1997) Determination of the fold of the core protein of hepatitis B virus by electron cryomicroscopy. Nature 386:88–91

Bowman BR, Baker ML, Rixon FJ, Chiu W, Quiocho FA (2003) Structure of the herpesvirus major capsid protein. EMBO J 22:757–765

Brenner S, Horne RW (1959) A negative staining method for high resolution electron microscopy of viruses. Biochim Biophys Acta 34:103–110

Bullough PA, Hughson FM, Skehel JJ, Wiley DC (1994) Structure of influenza haemagglutinin at the pH of membrane fusion. Nature 371:37–43

Butler PJ (1999) Self-assembly of tobacco mosaic virus: the role of an intermediate aggregate in generating both specificity and speed. Philos Trans R Soc Lond B Biol Sci 354:537–550

Cardone G, Winkler DC, Trus BL, Cheng N, Heuser JE, Newcomb WW et al (2007) Visualization of the herpes simplex virus portal in situ by cryo-electron tomography. Virology 361:426–434

Caspar DL (1956) Structure of bushy stunt virus. Nature 177:475–476

Caspar DL, Klug A (1962) Physical principles in the construction of regular viruses. Cold Spring Harb Symp Quant Biol 27:1–24

Caspar DL, Namba K (1990) Switching in the self-assembly of tobacco mosaic virus. Adv Biophys 26:157–185

Chan DC, Fass D, Berger JM, Kim PS (1997) Core structure of gp41 from the HIV envelope glycoprotein. Cell 89:263–273

Chang J, Weigele P, King J, Chiu W, Jiang W (2006) Cryo-EM asymmetric reconstruction of bacteriophage P22 reveals organization of its DNA packaging and infecting machinery. Structure 14:1073–1082

Chang JT, Schmid MF, Rixon FJ, Chiu W (2007) Electron cryotomography reveals the portal in the herpesvirus capsid. J Virol 81:2065–2068

Chen ZG, Stauffacher C, Li Y, Schmidt T, Bomu W, Kamer G et al (1989) Protein-RNA interactions in an icosahedral virus at 3.0 A resolution. Science 245:154–159

Chen B, Vogan EM, Gong H, Skehel JJ, Wiley DC, Harrison SC (2005) Structure of an unliganded simian immunodeficiency virus gp120 core. Nature 433:834–841

Chen R, Neill JD, Estes MK, Prasad BV (2006) X-ray structure of a native calicivirus: structural insights into antigenic diversity and host specificity. Proc Natl Acad Sci USA 103:8048–8053
Chen DH, Baker ML, Hryc CF, DiMaio F, Jakana J, Wu W et al (2011) Structural basis for scaffolding-mediated assembly and maturation of a dsDNA virus. Proc Natl Acad Sci USA 108:1355–1360
Cockburn JJ, Abrescia NG, Grimes JM, Sutton GC, Diprose JM, Benevides JM et al (2004) Membrane structure and interactions with protein and DNA in bacteriophage PRD1. Nature 432:122–125
Conway JF, Cheng N, Zlotnick A, Wingfield PT, Stahl SJ, Steven AC (1997) Visualization of a 4-helix bundle in the hepatitis B virus capsid by cryo-electron microscopy. Nature 386:91–94
Conway JF, Wikoff WR, Cheng N, Duda RL, Hendrix RW, Johnson JE et al (2001) Virus maturation involving large subunit rotations and local refolding. Science 292:744–748
Crick FH, Watson JD (1956) Structure of small viruses. Nature 177:473–475
Crick FHC, Watson JD (1957) Virus structure: general principles. In: Wolstenholme GEW, Millar ECP (eds) CIBA foundation symposium on the nature of viruses. Little Brown and Co, Boston
Crowther, RA (2010) From envelops to atoms: The remarkable progress of biological electron microscopy. Advances in Protein Chemistry and Structural Biology Vol 81 "Recent Advances in Electron cryomicroscopy" Part A, (eds, Ludtke S, and Prasad, BVV), pp 2–33.
Crowther RA, Amos LA, Finch JT, De Rosier DJ, Klug A (1970) Three dimensional reconstructions of spherical viruses by fourier synthesis from electron micrographs. Nature 226:421–425
Crowther RA, Kiselev NA, Bottcher B, Berriman JA, Borisova GP, Ose V et al (1994) Three-dimensional structure of hepatitis B virus core particles determined by electron cryomicroscopy. Cell 77:943–950
Culver JN, Dawson WO, Plonk K, Stubbs G (1995) Site-directed mutagenesis confirms the involvement of carboxylate groups in the disassembly of tobacco mosaic virus. Virology 206:724–730
Deng B, O'Connor CM, Kedes DH, Zhou ZH (2007) Direct visualization of the putative portal in the Kaposi's sarcoma-associated herpesvirus capsid by cryoelectron tomography. J Virol 81:3640–3644
Dong XF, Natarajan P, Tihova M, Johnson JE, Schneemann A (1998) Particle polymorphism caused by deletion of a peptide molecular switch in a quasiequivalent icosahedral virus. J Virol 72:6024–6033
Dormitzer PR, Nason EB, Prasad BV, Harrison SC (2004) Structural rearrangements in the membrane penetration protein of a non-enveloped virus. Nature 430:1053–1058
Dubochet J, Adrian M, Chang JJ, Homo JC, Lepault J, McDowall AW et al (1988) Cryo-electron microscopy of vitrified specimens. Q Rev Biophys 21:129–228
Egelman EH, Wu SS, Amrein M, Portner A, Murti G (1989) The Sendai virus nucleocapsid exists in at least four different helical states. J Virol 63:2233–2243
Erickson JW, Silva AM, Murthy MR, Fita I, Rossmann MG (1985) The structure of a T = 1 icosahedral empty particle from southern bean mosaic virus. Science 229:625–629
Fane BA, Prevelige PE (2003) Mechanism of scaffold-assisted viral assembly. Adv Protein Chem 64:259–299
Fisher AJ, Johnson JE (1993) Ordered duplex RNA controls capsid architecture in an icosahedral animal virus. Nature 361:176–179
Fokine A, Chipman PR, Leiman PG, Mesyanzhinov VV, Rao VB, Rossmann MG (2004) Molecular architecture of the prolate head of bacteriophage T4. Proc Natl Acad Sci USA 101:6003–6008
Fokine A, Leiman PG, Shneider MM, Ahvazi B, Boeshans KM, Steven AC et al (2005) Structural and functional similarities between the capsid proteins of bacteriophages T4 and HK97 point to a common ancestry. Proc Natl Acad Sci USA 102:7163–7168
Gan L, Speir JA, Conway JF, Lander G, Cheng N, Firek BA et al (2006) Capsid conformational sampling in HK97 maturation visualized by X-ray crystallography and cryo-EM. Structure 14:1655–1665
Ge P, Hong Zhou Z (2011) Hydrogen-bonding networks and RNA bases revealed by cryo electron microscopy suggest a triggering mechanism for calcium switches. Proc Natl Acad Sci USA 108:9637–9642
Ge P, Tsao J, Schein S, Green TJ, Luo M, Zhou ZH (2010) Cryo-EM model of the bullet-shaped vesicular stomatitis virus. Science 327:689–693
Green EM, Horne RW, Matthews RE (1956) Electron microscope observations of periodicities in the surface structure of tobacco mosaic virus. Nature 178:635–636
Grigorieff N, Harrison SC (2011) Near-atomic resolution reconstructions of icosahedral viruses from electron cryomicroscopy. Curr Opin Struct Biol 21:265–273
Grimes JM, Jakana J, Ghosh M, Basak AK, Roy P, Chiu W et al (1997) An atomic model of the outer layer of the bluetongue virus core derived from X-ray crystallography and electron cryomicroscopy. Structure 5:885–893
Grimes JM, Burroughs JN, Gouet P, Diprose JM, Malby R, Zientara S et al (1998) The atomic structure of the bluetongue virus core. Nature 395:470–478
Harrison SE (2007) Principles of virus structure. Lippincott Williams and Wilkins, Philadelphia, PA
Harrison SC (2008) Viral membrane fusion. Nat Struct Mol Biol 15:690–698
Harrison SC, Olson AJ, Schutt CE, Winkler FK, Bricogne G (1978) Tomato bushy stunt virus at 2.9 A resolution. Nature 276:368–373

Heymann JB, Cheng N, Newcomb WW, Trus BL, Brown JC, Steven AC (2003) Dynamics of herpes simplex virus capsid maturation visualized by time-lapse cryo-electron microscopy. Nat Struct Biol 10:334–341
Hogle JM, Chow M, Filman DJ (1985) Three-dimensional structure of poliovirus at 2.9 A resolution. Science 229:1358–1365
Hogle JM, Maeda A, Harrison SC (1986) Structure and assembly of turnip crinkle virus. I. X-ray crystallographic structure analysis at 3.2 A resolution. J Mol Biol 191:625–638
Horne RW, Wildy P (1962) Recent studies on the fine structure of viruses by electron microscopy, using negative-staining techniques. Br Med Bull 18:199–204
Horne RW, Wildy P (1979) An historical account of the development and applications of the negative staining technique to the electron microscopy of viruses. J Microsc 117:103–122
Huang RK, Khayat R, Lee KK, Gertsman I, Duda RL, Hendrix RW et al (2011) The prohead-i structure of bacteriophage HK97: implications for scaffold-mediated control of particle assembly and maturation. J Mol Biol 408:541–554
Ilag LL, Olson NH, Dokland T, Music CL, Cheng RH, Bowen Z et al (1995) DNA packaging intermediates of bacteriophage phi X174. Structure 3:353–363
Jayaram H, Estes MK, Prasad BV (2004) Emerging themes in rotavirus cell entry, genome organization, transcription and replication. Virus Res 101:67–81
Jiang W, Li Z, Zhang Z, Baker ML, Prevelige PE Jr, Chiu W (2003) Coat protein fold and maturation transition of bacteriophage P22 seen at subnanometer resolutions. Nat Struct Biol 10:131–135
Jiang W, Chang J, Jakana J, Weigele P, King J, Chiu W (2006) Structure of epsilon15 bacteriophage reveals genome organization and DNA packaging/injection apparatus. Nature 439:612–616
Jiang W, Baker ML, Jakana J, Weigele PR, King J, Chiu W (2008) Backbone structure of the infectious epsilon15 virus capsid revealed by electron cryomicroscopy. Nature 451:1130–1134
Johnson JE, Speir JA (1997) Quasi-equivalent viruses: a paradigm for protein assemblies. J Mol Biol 269:665–675
Kaufmann B, Nybakken GE, Chipman PR, Zhang W, Diamond MS, Fremont DH et al (2006) West Nile virus in complex with the Fab fragment of a neutralizing monoclonal antibody. Proc Natl Acad Sci USA 103: 12400–12404
Kendall A, McDonald M, Bian W, Bowles T, Baumgarten SC, Shi J et al (2008) Structure of flexible filamentous plant viruses. J Virol 82:9546–9554
Klug A (1999) The tobacco mosaic virus particle: structure and assembly. Philos Trans R Soc Lond B Biol Sci 354:531–535
Klug A, Caspar DL (1960) The structure of small viruses. Adv Virus Res 7:225–325
Klug A, Finch JT (1968) Structure of viruses of the papilloma-polyoma type. IV. Analysis of tilting experiments in the electron microscope. J Mol Biol 31:1–12
Klug A, Finch JT, Franklin RE (1957) The structure of turnip yellow mosaic virus; x-ray diffraction studies. Biochim Biophys Acta 25:242–252
Knapek E, Dubochet J (1980) Beam damage to organic material is considerably reduced in cryo-electron microscopy. J Mol Biol 141:147–161
Kuhn RJ, Zhang W, Rossmann MG, Pletnev SV, Corver J, Lenches E et al (2002) Structure of dengue virus: implications for flavivirus organization, maturation, and fusion. Cell 108:717–725
Lamb RA, Jardetzky TS (2007) Structural basis of viral invasion: lessons from paramyxovirus F. Curr Opin Struct Biol 17:427–436
Larson SB, Koszelak S, Day J, Greenwood A, Dodds JA, McPherson A (1993) Double-helical RNA in satellite tobacco mosaic virus. Nature 361:179–182
Lawton JA, Zeng CQ, Mukherjee SK, Cohen J, Estes MK, Prasad BV (1997) Three-dimensional structural analysis of recombinant rotavirus-like particles with intact and amino-terminal-deleted VP2: implications for the architecture of the VP2 capsid layer. J Virol 71:7353–7360
Lawton JA, Estes MK, Prasad BV (2000) Mechanism of genome transcription in segmented dsRNA viruses. Adv Virus Res 55:185–229
Li L, Lok SM, Yu IM, Zhang Y, Kuhn RJ, Chen J et al (2008) The flavivirus precursor membrane-envelope protein complex: structure and maturation. Science 319:1830–1834
Li Z, Baker ML, Jiang W, Estes MK, Prasad BV (2009) Rotavirus architecture at subnanometer resolution. J Virol 83:1754–1766
Li L, Jose J, Xiang Y, Kuhn RJ, Rossmann MG (2010) Structural changes of envelope proteins during alphavirus fusion. Nature 468:705–708
Liddington RC, Yan Y, Moulai J, Sahli R, Benjamin TL, Harrison SC (1991) Structure of simian virus 40 at 3.8-A resolution. Nature 354:278–284
Liljas L, Unge T, Jones TA, Fridborg K, Lovgren S, Skoglund U et al (1982) Structure of satellite tobacco necrosis virus at 3.0 A resolution. J Mol Biol 159:93–108
Lin T, Chen Z, Usha R, Stauffacher CV, Dai JB, Schmidt T et al (1999) The refined crystal structure of cowpea mosaic virus at 2.8 A resolution. Virology 265:20–34

Liu H, Jin L, Koh SB, Atanasov I, Schein S, Wu L et al (2010) Atomic structure of human adenovirus by cryo-EM reveals interactions among protein networks. Science 329:1038–1043

Luo M, Vriend G, Kamer G, Minor I, Arnold E, Rossmann MG et al (1987) The atomic structure of Mengo virus at 3.0 A resolution. Science 235:182–191

Mancini EJ, Clarke M, Gowen BE, Rutten T, Fuller SD (2000) Cryo-electron microscopy reveals the functional organization of an enveloped virus, Semliki Forest virus. Mol Cell 5:255–266

Martin CS, Burnett RM, de Haas F, Heinkel R, Rutten T, Fuller SD et al (2001) Combined EM/X-ray imaging yields a quasi-atomic model of the adenovirus-related bacteriophage PRD1 and shows key capsid and membrane interactions. Structure 9:917–930

Mathieu M, Petitpas I, Navaza J, Lepault J, Kohli E, Pothier P et al (2001) Atomic structure of the major capsid protein of rotavirus: implications for the architecture of the virion. EMBO J 20:1485–1497

McClain B, Settembre E, Temple BR, Bellamy AR, Harrison SC (2010) X-ray crystal structure of the rotavirus inner capsid particle at 3.8 A resolution. J Mol Biol 397:587–599

Moody MF (1999) Geometry of phage head construction. J Mol Biol 293:401–433

Mukherjee S, Abd-El-Latif M, Bronstein M, Ben-nun-Shaul O, Kler S et al (2007) High Cooperativity of the SV40 Major Capsid Protein VP1 in Virus Assembly. PLoS ONE 2(8): e765. doi:10.1371/journal.pone.0000765

Naitow H, Tang J, Canady M, Wickner RB, Johnson JE (2002) L-A virus at 3.4 A resolution reveals particle architecture and mRNA decapping mechanism. Nat Struct Biol 9:725–728

Namba K, Stubbs G (1986) Structure of tobacco mosaic virus at 3.6 A resolution: implications for assembly. Science 231:1401–1406

Namba K, Pattanayek R, Stubbs G (1989) Visualization of protein-nucleic acid interactions in a virus. Refined structure of intact tobacco mosaic virus at 2.9 A resolution by X-ray fiber diffraction. J Mol Biol 208:307–325

Nason EL, Wetzel JD, Mukherjee SK, Barton ES, Prasad BV, Dermody TS (2001) A monoclonal antibody specific for reovirus outer-capsid protein sigma3 inhibits sigma1-mediated hemagglutination by steric hindrance. J Virol 75:6625–6634

Nason EL, Rothagel R, Mukherjee SK, Kar AK, Forzan M, Prasad BV et al (2004) Interactions between the inner and outer capsids of bluetongue virus. J Virol 78:8059–8067

Nicholson P, Addison C, Cross AM, Kennard J, Preston VG, Rixon FJ (1994) Localization of the herpes simplex virus type 1 major capsid protein VP5 to the cell nucleus requires the abundant scaffolding protein VP22a. J Gen Virol 75(Pt 5):1091–1099

Ochoa WF, Havens WM, Sinkovits RS, Nibert ML, Ghabrial SA, Baker TS (2008) Partitivirus structure reveals a 120-subunit, helix-rich capsid with distinctive surface arches formed by quasisymmetric coat-protein dimers. Structure 16:776–786

Ossiboff RJ, Zhou Y, Lightfoot PJ, Prasad BV, Parker JS (2010) Conformational changes in the capsid of a calicivirus upon interaction with its functional receptor. J Virol 84:5550–5564

Paredes AM, Brown DT, Rothnagel R, Chiu W, Schoepp RJ, Johnston RE et al (1993) Three-dimensional structure of a membrane-containing virus. Proc Natl Acad Sci USA 90:9095–9099

Paredes AM, Heidner H, Thuman-Commike P, Prasad BV, Johnston RE, Chiu W (1998) Structural localization of the E3 glycoprotein in attenuated Sindbis virus mutants. J Virol 72:1534–1541

Parent KN, Khayat R, Tu LH, Suhanovsky MM, Cortines JR, Teschke CM et al (2010a) P22 coat protein structures reveal a novel mechanism for capsid maturation: stability without auxiliary proteins or chemical crosslinks. Structure 18:390–401

Parent KN, Sinkovits RS, Suhanovsky MM, Teschke CM, Egelman EH, Baker TS (2010b) Cryo-reconstructions of P22 polyheads suggest that phage assembly is nucleated by trimeric interactions among coat proteins. Phys Biol 7:045004

Pokidysheva E, Zhang Y, Battisti AJ, Bator-Kelly CM, Chipman PR, Xiao C et al (2006) Cryo-EM reconstruction of dengue virus in complex with the carbohydrate recognition domain of DC-SIGN. Cell 124:485–493

Prasad BV, Prevelige PE (2003) Viral genome organization. Adv Protein Chem 64:219–258

Prasad BV, Prevelige PE, Marietta E, Chen RO, Thomas D, King J et al (1993) Three-dimensional transformation of capsids associated with genome packaging in a bacterial virus. J Mol Biol 231:65–74

Prasad BV, Rothnagel R, Zeng CQ, Jakana J, Lawton JA, Chiu W et al (1996) Visualization of ordered genomic RNA and localization of transcriptional complexes in rotavirus. Nature 382:471–473

Prasad BV, Hardy ME, Dokland T, Bella J, Rossmann MG, Estes MK (1999) X-ray crystallographic structure of the Norwalk virus capsid. Science 286:287–290

Rao VB, Black LW (2010) Structure and assembly of bacteriophage T4 head. Virol J 7:356

Reddy VS, Natarajan P, Okerberg B, Li K, Damodaran KV, Morton RT et al (2001) Virus Particle Explorer (VIPER), a website for virus capsid structures and their computational analyses. J Virol 75:11943–11947

Reddy VS, Natchiar SK, Stewart PL, Nemerow GR (2010) Crystal structure of human adenovirus at 3.5 A resolution. Science 329:1071–1075

Reinisch KM, Nibert ML, Harrison SC (2000) Structure of the reovirus core at 3.6 A resolution. Nature 404:960–967

Rey FA (2003) Dengue virus envelope glycoprotein structure: new insight into its interactions during viral entry. Proc Natl Acad Sci USA 100:6899–6901

Rey FA (2006) Molecular gymnastics at the herpesvirus surface. EMBO Rep 7:1000–1005

Rey FA, Heinz FX, Mandl C, Kunz C, Harrison SC (1995) The envelope glycoprotein from tick-borne encephalitis virus at 2 A resolution. Nature 375:291–298

Rixon FJ, Addison C, McGregor A, Macnab SJ, Nicholson P, Preston VG et al (1996) Multiple interactions control the intracellular localization of the herpes simplex virus type 1 capsid proteins. J Gen Virol 77(Pt 9):2251–2260

Roberts MM, White JL, Grutter MG, Burnett RM (1986) Three-dimensional structure of the adenovirus major coat protein hexon. Science 232:1148–1151

Rossmann MG, Johnson JE (1989) Icosahedral RNA virus structure. Annu Rev Biochem 58:533–573

Rossmann MG, Arnold E, Erickson JW, Frankenberger EA, Griffith JP, Hecht HJ et al (1985) Structure of a human common cold virus and functional relationship to other picornaviruses. Nature 317:145–153

Rossmann MG, Olson NH, Kolatkar PR, Oliveira MA, Cheng RH, Greve JM et al (1994) Crystallographic and cryo EM analysis of virion-receptor interactions. Arch Virol Suppl 9:531–541

Saad A, Zhou ZH, Jakana J, Chiu W, Rixon FJ (1999) Roles of triplex and scaffolding proteins in herpes simplex virus type 1 capsid formation suggested by structures of recombinant particles. J Virol 73:6821–6830

Saban SD, Silvestry M, Nemerow GR, Stewart PL (2006) Visualization of alpha-helices in a 6-angstrom resolution cryoelectron microscopy structure of adenovirus allows refinement of capsid protein assignments. J Virol 80: 12049–12059

Saren AM, Ravantti JJ, Benson SD, Burnett RM, Paulin L, Bamford DH et al (2005) A snapshot of viral evolution from genome analysis of the tectiviridae family. J Mol Biol 350:427–440

Schrag JD, Prasad BV, Rixon FJ, Chiu W (1989) Three-dimensional structure of the HSV1 nucleocapsid. Cell 56:651–660

Settembre EC, Chen JZ, Dormitzer PR, Grigorieff N, Harrison SC (2011) Atomic model of an infectious rotavirus particle. EMBO J 30:408–416

Shoemaker GK, van Duijn E, Crawford SE, Uetrecht C, Baclayon M, Roos WH et al (2010) Norwalk virus assembly and stability monitored by mass spectrometry. Mol Cell Proteomics 9:1742–1751

Smith TJ, Chase ES, Schmidt TJ, Olson NH, Baker TS (1996) Neutralizing antibody to human rhinovirus 14 penetrates the receptor-binding canyon. Nature 383:350–354

Sorger PK, Stockley PG, Harrison SC (1986) Structure and assembly of turnip crinkle virus. II. Mechanism of reassembly in vitro. J Mol Biol 191:639–658

Stehle T, Gamblin SJ, Yan Y, Harrison SC (1996) The structure of simian virus 40 refined at 3.1 A resolution. Structure 4:165–182

Stewart PL, Burnett RM, Cyrklaff M, Fuller SD (1991) Image reconstruction reveals the complex molecular organization of adenovirus. Cell 67:145–154

Stewart PL, Chiu CY, Huang S, Muir T, Zhao Y, Chait B et al (1997) Cryo-EM visualization of an exposed RGD epitope on adenovirus that escapes antibody neutralization. EMBO J 16:1189–1198

Stubbs G (1999) Tobacco mosaic virus particle structure and the initiation of disassembly. Philos Trans R Soc Lond B Biol Sci 354:551–557

Tao Y, Olson NH, Xu W, Anderson DL, Rossmann MG, Baker TS (1998) Assembly of a tailed bacterial virus and its genome release studied in three dimensions. Cell 95:431–437

Thuman-Commike PA, Greene B, Jakana J, Prasad BV, King J, Prevelige PE Jr et al (1996) Three-dimensional structure of scaffolding-containing phage p22 procapsids by electron cryo-microscopy. J Mol Biol 260:85–98

Thuman-Commike PA, Greene B, Malinski JA, King J, Chiu W (1998) Role of the scaffolding protein in P22 procapsid size determination suggested by T = 4 and T = 7 procapsid structures. Biophys J 74:559–568

Tsuruta H, Reddy VS, Wikoff WR, Johnson JE (1998) Imaging RNA and dynamic protein segments with low-resolution virus crystallography: experimental design, data processing and implications of electron density maps. J Mol Biol 284:1439–1452

Tuma R, Thomas GJ Jr (1997) Mechanisms of virus assembly probed by Raman spectroscopy: the icosahedral bacteriophage P22. Biophys Chem 68:17–31

van Raaij MJ, Mitraki A, Lavigne G, Cusack S (1999) A triple beta-spiral in the adenovirus fibre shaft reveals a new structural motif for a fibrous protein. Nature 401:935–938

Varghese JN, Laver WG, Colman PM (1983) Structure of the influenza virus glycoprotein antigen neuraminidase at 2.9 A resolution. Nature 303:35–40

Voss JE, Vaney MC, Duquerroy S, Vonrhein C, Girard-Blanc C, Crublet E et al (2010) Glycoprotein organization of Chikungunya virus particles revealed by X-ray crystallography. Nature 468:709–712

Wikoff WR, Johnson JE (1999) Virus assembly: imaging a molecular machine. Curr Biol 9:R296–R300

Wikoff WR, Liljas L, Duda RL, Tsuruta H, Hendrix RW, Johnson JE (2000) Topologically linked protein rings in the bacteriophage HK97 capsid. Science 289:2129–2133

Wikoff WR, Conway JF, Tang J, Lee KK, Gan L, Cheng N et al (2006) Time-resolved molecular dynamics of bacteriophage HK97 capsid maturation interpreted by electron cryo-microscopy and X-ray crystallography. J Struct Biol 153:300–306

Wildy P, Horne RW (1963) Structure of animal virus particles. Prog Med Virol 5:1–42

Wilson IA, Skehel JJ, Wiley DC (1981) Structure of the haemagglutinin membrane glycoprotein of influenza virus at 3 A resolution. Nature 289:366–373

Wolf M, Garcea RL, Grigorieff N, Harrison SC (2010) Subunit interactions in bovine papillomavirus. Proc Natl Acad Sci USA 107:6298–6303

Wynne SA, Crowther RA, Leslie AG (1999) The crystal structure of the human hepatitis B virus capsid. Mol Cell 3:771–780

Xu Y, Liu Y, Lou Z, Qin L, Li X, Bai Z et al (2004) Structural basis for coronavirus-mediated membrane fusion. Crystal structure of mouse hepatitis virus spike protein fusion core. J Biol Chem 279:30514–30522

Xu X, Zhu X, Dwek RA, Stevens J, Wilson IA (2008) Structural characterization of the 1918 influenza virus H1N1 neuraminidase. J Virol 82:10493–10501

Yu IM, Zhang W, Holdaway HA, Li L, Kostyuchenko VA, Chipman PR et al (2008a) Structure of the immature dengue virus at low pH primes proteolytic maturation. Science 319:1834–1837

Yu X, Jin L, Zhou ZH (2008b) 3.88 A structure of cytoplasmic polyhedrosis virus by cryo-electron microscopy. Nature 453:415–419

Zhang R, Hryc CF, Cong Y, Liu X, Jakana J, Gorchkov R, et al (2011) Cryo-EM structure of an enveloped alphavirus venezuelan equine encephalitis virus. submitted for publication. EMBO J 30:3854–3863

Zhang W, Mukhopadhyay S, Pletnev SV, Baker TS, Kuhn RJ, Rossmann MG (2002) Placement of the structural proteins in Sindbis virus. J Virol 76:11645–11658

Zhang X, Walker SB, Chipman PR, Nibert ML, Baker TS (2003) Reovirus polymerase lambda 3 localized by cryo-electron microscopy of virions at a resolution of 7.6 A. Nat Struct Biol 10:1011–1018

Zhang Y, Kaufmann B, Chipman PR, Kuhn RJ, Rossmann MG (2007) Structure of immature West Nile virus. J Virol 81:6141–6145

Zhou ZH, Prasad BV, Jakana J, Rixon FJ, Chiu W (1994) Protein subunit structures in the herpes simplex virus A-capsid determined from 400 kV spot-scan electron cryomicroscopy. J Mol Biol 242:456–469

Zhou ZH, Dougherty M, Jakana J, He J, Rixon FJ, Chiu W (2000) Seeing the herpesvirus capsid at 8.5 A. Science 288:877–880

Zimmern D (1976) The region of tobacco mosaic virus RNA involved in the nucleation of assembly. Philos Trans R Soc Lond B Biol Sci 276:189–204

Zubieta C, Schoehn G, Chroboczek J, Cusack S (2005) The structure of the human adenovirus 2 penton. Mol Cell 17:121–135

Chapter 4
Reconstructing Virus Structures from Nanometer to Near-Atomic Resolutions with Cryo-Electron Microscopy and Tomography

Juan Chang*, Xiangan Liu*, Ryan H. Rochat*, Matthew L. Baker*, and Wah Chiu

Abstract The past few decades have seen tremendous advances in single-particle electron cryo-microscopy (cryo-EM). The field has matured to the point that near-atomic resolution density maps can be generated for icosahedral viruses without the need for crystallization. In parallel, substantial progress has been made in determining the structures of nonicosahedrally arranged proteins in viruses by employing either single-particle cryo-EM or cryo-electron tomography (cryo-ET). Implicit in this course have been the availability of a new generation of electron cryo-microscopes and the development of the computational tools that are essential for generating these maps and models. This methodology has enabled structural biologists to analyze structures in increasing detail for virus particles that are in different morphogenetic states. Furthermore, electron imaging of frozen, hydrated cells, in the process of being infected by viruses, has also opened up a new avenue for studying virus structures "in situ". Here we present the common techniques used to acquire and process cryo-EM and cryo-ET data and discuss their implications for structural virology both now and in the future.

Abbreviations

1-D	One dimension; one dimensional
2-D	Two dimensions; two dimensional
3-D	Three dimensions; three dimensional
CCD	Charge-coupled device
Cryo-EM	Electron cryo-microscopy

* Juan Chang, Xiangan Liu, Ryan H. Rochat and Matthew L. Baker contributed equally.

J. Chang • X. Liu • M.L. Baker
National Center for Macromolecular Imaging, Verna and Marrs McLean Department of Biochemistry and Molecular Biology, Baylor College of Medicine, Houston, TX 77030, USA
e-mail: jc131361@bcm.edu; xiangan.liu@bcm.edu; mbaker@bcm.edu

R.H. Rochat • W. Chiu (✉)
National Center for Macromolecular Imaging, Verna and Marrs McLean Department of Biochemistry and Molecular Biology, Baylor College of Medicine, Houston, TX 77030, USA

Graduate Program in Structural and Computational Biology and Molecular Biophysics, Baylor College of Medicine, Houston, TX 77030, USA
e-mail: rochat@bcm.edu; wah@bcm.edu

M.G. Rossmann and V.B. Rao (eds.), *Viral Molecular Machines*,
Advances in Experimental Medicine and Biology 726, DOI 10.1007/978-1-4614-0980-9_4,

Cryo-ET	Electron cryo-tomography
CT	Computed tomography
CTF	Contrast transfer function
EM	Electron microscope; electron microscopy
EMDB	Electron Microscopy Data Bank (http://www.emdatabank.org)
FSC	Fourier shell correlation
FT	Fourier transform
GUI	Graphical user interface
SNR	Signal-to-noise ratio
SSE	Secondary structural elements
WPOA	Weak phase object approximation
ZPC	Zernike phase contrast

4.1 Introduction

Many viruses are composed of a highly symmetric capsid shell that serves as a container for their genome. In the past few decades, structures of icosahedrally arranged capsid shells solved by electron cryo-microscopy (cryo-EM) have progressed from low-resolution (30–50 Å) to well beyond the subnanometer resolution threshold (<10 Å). However, the imposition of icosahedral symmetry to reconstruct the capsid shell structure conceals the structures of the protein components that do not have such an arrangement (i.e., nonicosahedral). These protein components can play key roles in a virus' life cycle, from packaging its genome, to infecting its host cell, and even viral genome transcription. To recover these features, it is necessary to reconstruct the virus structure without any imposed symmetry. While it may seem conceptually easy to identify these special features of a virus from raw images, in reality this process is extremely difficult and relies on specialized software that is powerful enough to detect these features. Once such nonicosahedrally organized components can be resolved, their structures can also be determined at subnanometer resolutions.

The single-particle approach to virus reconstruction usually assumes a high degree of structural homogeneity in the virus sample, which may not be true in some cases [e.g., the tegument proteins of a herpesvirus or the unevenly distributed envelope spikes of human immunodeficiency virus (HIV)]. To address this issue, electron cryo-tomography (cryo-ET) has found a niche for visualizing extremely heterogeneous samples. Unlike the single-particle approach, cryo-ET obtains structural information by collecting a series of images of the same specimen area at different tilt angles and reconstructing them into a 3-D map, much in the same way that modern medical X-ray computed tomography (CT) scanner provides a 3-D view of the human body. Cryo-ET followed by posttomographic averaging of computationally extracted tomographic subvolumes can provide a 3-D density map of homogeneous molecular components in a heterogeneous virus structure. This approach can also provide snapshots of a biological process in situ. For example, it is possible to study the structural changes in viruses that accompany infection, capsid assembly, capsid maturation, and other processes in the virus life cycle. While the level of detail generated from this approach is far lower than that possible with cryo-EM in the current state of the technology, there is still much that can be learned about virus–host cell interactions even at these low resolutions.

Regardless of the tools used to generate a 3-D virus reconstruction, the ultimate goal is to produce an atomic model of the protein components of the virus. The structural detail of the reconstruction is a function of the resolution of the map (Fig. 4.1). In a low-resolution map (20–40 Å), only shape and overall arrangement of the virus capsomeres can be determined. At subnanometer resolutions (<10 Å), α-helices and β-sheets become visible in the map, and as the cryo-EM reconstructions push toward near-atomic resolutions, Cα backbones and some bulky side chains can be delineated in the map (beyond 4.5 Å).

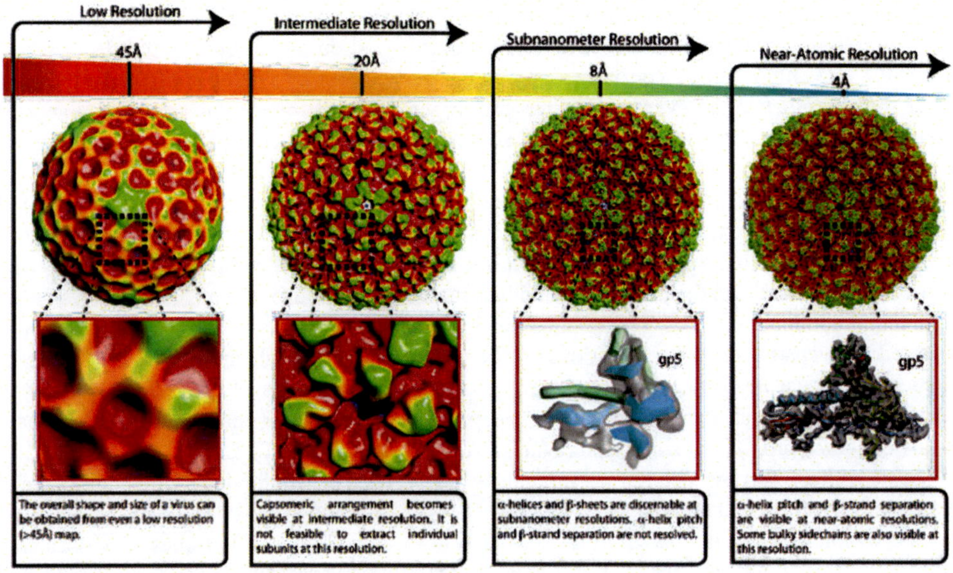

Fig. 4.1 Structural features of bacteriophage P22 procapsid [EMD-1824] that are visible as a function of resolution. The map density is radially colored

In this chapter, we will introduce both cryo-EM and cryo-ET, with a focus on how to process the data generated by these techniques. Specifically, we will discuss single-particle reconstructions of both icosahedral and nonicosahedral components of virus particle, and tomography of pleomorphic specimens containing homogeneous components (e.g., a virus envelope with protein spikes and cells infected with viruses). We will also address how these results can be validated and interpreted, including annotation and model building. Additionally, in line with such a rapidly evolving field, we will cover selected emerging techniques that will likely play a role in structural virology in years to come.

4.2 Cryo-Electron Microscopy and Tomography

Light microscopes have been used for centuries to observe objects that are too small for the human eye to resolve. In the 1930s, with the advent of the electron microscope, it became possible to visualize objects that were beyond the range of resolvability by conventional light microscopes. While the principles that dictate the behavior of these two instruments are similar, the main difference lies in the fact that electron microscopes use electrons (not light) as an illumination source. As resolvability is inversely proportional to the wavelength of the illumination source (4,000–7,000 Å for light, 0.025 Å for 200 KeV electrons), electron microscopes have the distinct advantage that they can potentially resolve features tens of thousands of times smaller than even the best light microscope.

Electron microscopy has been implemented for a variety of applications, but here we focus on the use of transmission electron microscopy (TEM) in cryo-EM for reconstructing virus particles. Cryo-EM first started in the mid-1970s (Taylor and Glaeser 1974, 1976), and during the 1980s, it was refined using a rapid plunge-freezing procedure (Adrian et al. 1984; Chiu 1986; Dubochet et al. 1988; McDowall et al. 1983). This rapid freezing procedure has the advantage that the sample is frozen quickly enough that ice crystals do not have time to form, resulting in a thin vitrified ice layer embedding the sample of interest. These ice-embedded specimens can be readily imaged in an electron microscope as long as the sample is kept at a temperature that is below the phase transition temperature of the vitreous ice (~ −145°C). However, while it is possible to preserve the specimen at these low temperatures, vitrified specimens are extremely sensitive to the electron beam and can tolerate only limited electron exposure. It has been shown by a number of radiation damage studies that the practical temperature for imaging frozen, hydrated biological specimen is approximately −170°C (Bammes et al. 2010; Comolli and Downing 2005; Iancu et al. 2006), and in general, the electron exposure used for single-particle cryo-EM ranges from 15 to 25 e/Å².

In biological cryo-EM, the images collected on the microscope can be interpreted in terms of the weak phase object approximation (WPOA) (Erickson and Klug 1971; Unwin and Henderson 1975). The WPOA makes two assumptions: the electrons are elastically scattered by the specimen (they suffer no energy loss), and the phase change of the scattered electrons is slight. The phase variations of the scattered electrons in the object plane, which carry the specimen's density information, are converted into spatial intensity variations in the image plane through interference. As a result, the specimen must be very thin in order to satisfy the assumptions of the WPOA. Under the WPOA, the 2-D image can be treated as the projection of the original 3-D object's density, modulated (convoluted) by a sinusoidal contrast transfer function (CTF). In practice, in addition to this phase contrast, there is also a small amount (~10% or less) of amplitude modification, which is expressed in the form of a co-sinusoidal function.

If the sample is not thin enough to satisfy the WPOA assumptions, it becomes very difficult to restore the original object's density function from the collected 2-D particle images. Even though there have been theoretical predictions, based on the WPOA, for the maximum attainable resolution as a function of specimen thickness (DeRosier 2000; Ho et al. 1988; Leong et al. 2010; Schmid et al. 1994; Zhou and Chiu 2003), no systematic experimental verification has yet been reported. Interestingly, the 3.6-Å resolution map of human adenovirus, reconstructed by cryo-EM, is almost identical to the structure determined by X-ray crystallography (Liu et al. 2010a; Reddy et al. 2010), demonstrating that with the 900-Å-thick virus particle, the WPOA assumptions have not been violated.

In addition to the CTF, there are numerous instrument factors that affect the image, including the spatial and temporal coherence of the electron source, instability of the lens and specimen holder, and the point spread function of the recording media (Chiu and Glaeser 1977). All of these factors act to dampen the high-frequency information in the images and thus the likelihood that the data will be useful for a high-resolution reconstruction. This dampening can be expressed analytically as the product of a series of envelope functions which can be approximated simply as a Gaussian function $E(s)=\exp(-Bs^2)$, with its width at half maximum set to the experimental *B*-factor (Saad et al. 2001). In the course of image reconstruction, all of the parameters inherent in the CTF and the envelope function should be determined on a per particle basis (Jiang and Chiu 2007).

4.3 Single-Particle Reconstruction Methods

The goal of any single-particle virus reconstruction is the generation of a 3-D density map of the virus from the 2-D projection images produced by the electron microscope. The ability to start with 2-D projections of an object and then work back to a full 3-D density function is entirely based on the central section theorem (DeRosier and Klug 1968). This theorem states that the Fourier transform (FT) of a 2-D projection of a 3-D object is identical to a central section of the FT of the 3-D object. The section is perpendicular to the direction of projection and passes through the origin. Thus, the original 3-D object can be recovered by filling a volume in Fourier space with many of these central sections and then computing the inverse FT of this volume.

The theory behind the single-particle approach for cryo-EM is that by collecting thousands of images of a structural homogeneous virus particle in different orientations, the 2-D particle images can, with sufficient angular sampling, be used to reconstruct its 3-D structure. In order to obtain a

3-D reconstruction from 2-D images, views of the specimen in sufficiently different directions are required. In single-particle imaging, multiple copies of a randomly oriented specimen are distributed throughout the vitreous ice layer. Thus, imaging any area will provide different views of the specimen. The assumption of homogeneity predicts that the differences that are observed in the raw images are due to projections of the particles in these different orientations. In the case of a specimen that has a preferred orientation on the grid (e.g., only top views), tilting the grid to a fixed angle during imaging (random conical tilt) can circumvent the problem by generating views from different directions. A few fixed angles can be used in order to increase the number of views from different directions. Unfortunately, as a consequence of both these assumptions, by combining multiple images together on the basis of structural homogeneity, any structural heterogeneity in the data will be averaged out and not appear in the final reconstruction.

As mentioned above, the particle images are convoluted with the CTF and envelope function of the electron microscope. As such, the computational tasks required for this process are to determine the orientation of each of the individual particle images and correct any image defects prior to reconstructing the data. Since a virus particle is typically made up of both icosahedrally and nonicosahedrally arranged components, the methods of reconstruction differ depending on whether icosahedral symmetry is assumed. We will discuss the various methods used to determine both icosahedrally and nonicosahedrally arranged structures in the virus particles.

4.3.1 Reconstructing Icosahedral Viruses

In the 1970s, DeRosier, Crowther, and Klug developed algorithms that we still use today for reconstructing 3-D maps of icosahedral virus particles from electron micrographs (Crowther 1971; Crowther et al. 1970a; DeRosier and Klug 1968). These methods use common lines to find the orientations of the 2-D particles and Fourier–Bessel inversion to compute the 3-D map as described below. This approach was first applied to tomato bushy stunt virus (TBSV), which was prepared using the traditional negative stain procedure. From just six particles, even though the particles had shrunk by 10% as a result of the negative staining procedure, the overall morphology of TBSV could still be determined from the reconstruction. In the 40 years following these first reconstructions, the resolutions of icosahedral virus reconstructions have improved to the point that the once lumpy densities representing virus capsomeres can now be resolved at near-atomic resolutions, detailing the intricate structures that comprise them (Fig. 4.1) (Chen et al. 2011). These advances were made possible by numerous improvements in cryo-specimen preservation, the coherence of the electron beam, the stability of the cryo-specimen holder, and particularly in the software used to process the data. While there are many options when it comes to choosing a software package for virus reconstruction, all of them follow a common path of particle image preparation, 2-D image orientation determination, and 3-D map synthesis. The differences in these programs lie in the algorithms they use for alignment and reconstruction, which will be described below.

4.3.1.1 Initial Data Preparation

The first step in any single-particle reconstruction is determining the quality of the individual micrographs or CCD frames (Fig. 4.2a) and selecting the best ones for further processing. The quality of a micrograph can be judged visually by an experienced microscopist or through inspection of the CTF rings in the power spectrum (Fourier transform) of the micrograph (Fig. 4.2b). It is a common practice to exclude images that contain drift and astigmatism as seen in Fig. 4.2d.

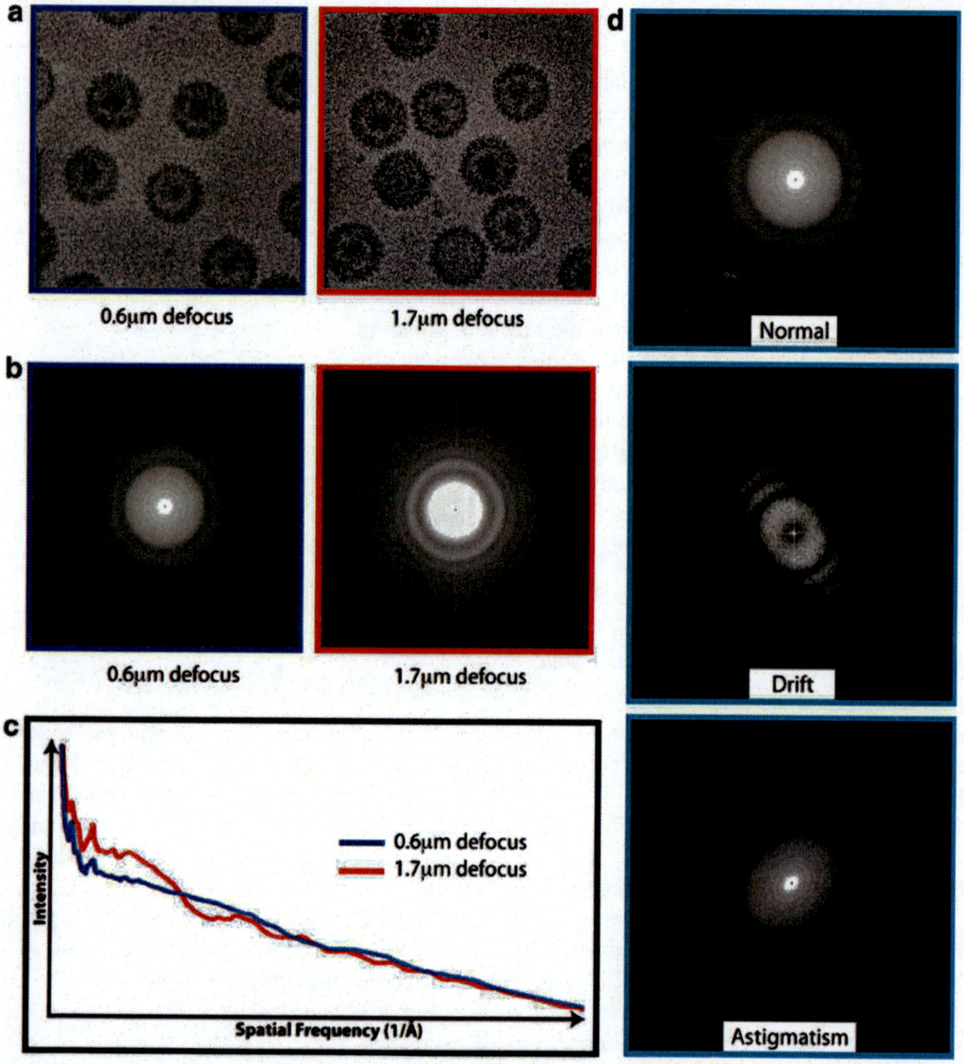

Fig. 4.2 Image preprocessing and data assessment. (**a**) Typical electron micrographs of herpes simplex virus type I collected on a 300-kV electron cryo-microscope. Higher contrast (*right side* of **a**) as a result of defocus is visible in the raw images. (**b**) Differences in defocus are also visualized in Fourier space with different CTF oscillating patterns. (**c**) Power spectrum curves from boxed herpes particles at different defocuses, which can be used to fit CTF curves of the corresponding micrographs. (**d**) Comparison between a normal image and images with aberrations (e.g., drift, astigmatism). For the purposes of quality control, images that contain these aberrations are typically excluded prior to processing the data for a 3-D reconstruction

Once a subset of good micrographs has been selected, particles in these micrographs are individually selected (boxed) and stored as individual particle images. Next, the CTF parameters and the envelope function for each micrograph can be determined from the circularly averaged power spectrum (Fig. 4.2c) of the boxed particles from the micrograph. From such a power spectrum, it is possible to estimate the signal-to-noise ratio (SNR) at any resolution (spatial frequency) of the data. Images that have a poor SNR at the desired resolution are typically removed from further consideration. While there are programs that can automatically or semiautomatically determine these CTF parameters (*fitctf* in EMAN, ctffind3/ctftilt, and *tf ed* in SPIDER),

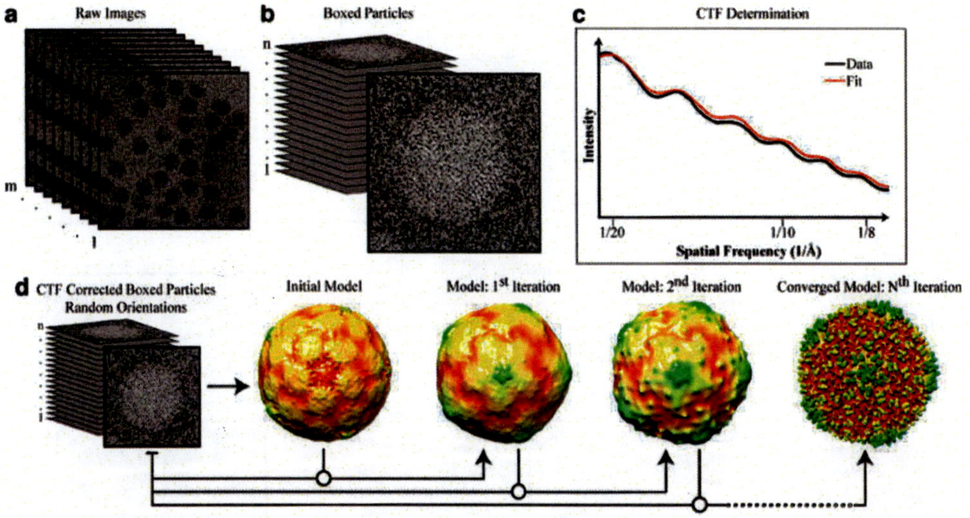

Fig. 4.3 Generalized schematic for single-particle reconstruction. (**a**) A single cryo-EM micrograph or CCD frame can contain anywhere from just a few to hundreds of particles. (**b**) The first step in reconstruction is to box the individual particles from all micrographs. Depending upon the software used, it may be necessary to invert the particles' contrast prior to alignment and reconstruction. (**c**) Once boxed, the power spectrum of the boxed particles of a micrograph can be used to determine the particles' CTF. (**d**) Using the determined CTF parameters, the particles can be aligned to a model and then reconstructed to a better-resolved map. The first step in this process is to build an initial model, as discussed in text. While there are a variety of different alignment schemes used in single-particle virus reconstruction, the general approach is to align the single-particle images to references projected from either an initial model or a previously reconstructed model, and then iteratively refine this model until it converges

many users often choose to visually inspect the output and/or manually refine the parameters using a GUI program (*ctfit* in EMAN, *robem* in AUTO3DEM, *bshow* in BSOFT, and *ctfmatch* in SPIDER) (Heymann et al. 2008a; Ludtke et al. 1999; Mindell and Grigorieff 2003; Shaikh et al. 2008; Yan et al. 2007).

4.3.1.2 Building an Initial Model

After the CTF parameters of the boxed particles have been determined, the next stage of the process is to start the iterative refinement for each of the five spatial parameters (2 for center, 3 for orientation) of each particle image (Fig. 4.3). The first step of this process generally requires an initial 3-D model as a reference. One way to obtain an initial model is to download an icosahedral virus density map from the electron microscopy data bank (EMDB: http://www.emdatabank.org), scale its dimensions to approximately match the current dataset, and then low pass filter the map to keep only the icosahedral shape and size information. It is important when selecting a virus from the EMDB to choose a different virus than the one under study as it virtually eliminates all possibility of model bias in the reconstruction. Alternatively, an initial model can be generated from the raw data by randomly assigning icosahedral orientations to a few hundred particles and reconstructing a 3-D density that has approximately the right size for the target particle (Liu et al. 2007). A more sophisticated approach involves computing the self-common lines of a handful of particle images, thereby determining their icosahedral orientations without using any reference model (the particle itself is the reference; see below) from which an adequate initial model can be built (Fuller et al. 1996). In the early days of cryo-EM, model bias was a concern for 3-D virus reconstruction as refining a

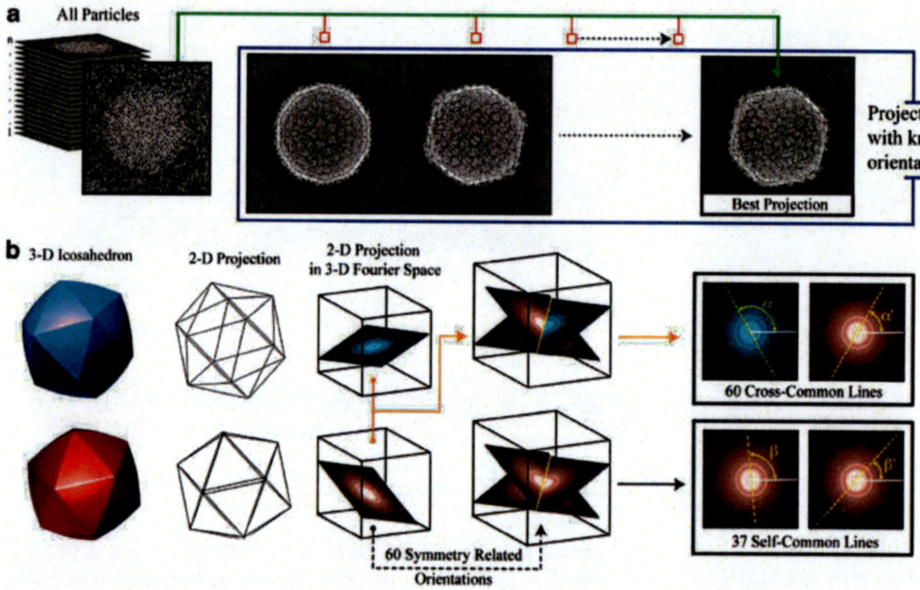

Fig. 4.4 Single-particle 2-D alignment methods. (**a**) Schematic of projection matching method. By comparing individual raw single-particle images to projections made of a model at known orientations, it is possible to define classes of particles that can then be averaged together to obtain a 3-D density map. (**b**) Cross-common line method. According to central section theorem, the Fourier transforms of any two images/projections intersect along a line where the values are identical. If the two images are generated from two totally different orientations, the intersection line is called cross-common line (right *top* of **b**). Considering the icosahedral symmetry of a virus particle, there are 60 pairs of cross-common lines between the two images. If the orientations of the two images are only related to its icosahedral symmetry, they are actually the same image. The *intersection line* is called self-common line. There are 37 unique pairs of self-common lines for an icosahedral particle image

model from an inaccurate starting model could provide erroneous results. However, with improvements in both image quality and alignment/reconstruction software, researchers can routinely achieve subnanometer resolution maps for most virus capsids. As secondary structural elements (SSEs) in the reconstruction become apparent in these subnanometer resolution reconstructions, the question of model bias is practically inexistent because it is unlikely for a map suffering from model bias to reach such a resolution.

4.3.1.3 Particle Orientation and Center Determination

The most time-consuming aspect of generating a 3-D virus map is the process of iteratively refining the orientation and center parameters (Fig. 4.3d). This procedure involves comparing the particles images with some 2-D references projected from a 3-D model (either an initial model or a model computed from a previous iteration) to determine their centers and orientations. Accuracy in determining the center and orientation of each particle image is an important factor in the quality of the 3-D reconstruction. While there are a variety of ways to determine these parameters, the most frequently used methods are projection matching and common lines (Fig. 4.4).

Projection Matching

In the projection matching method, unique 2-D projections are generated from the 3-D model for all possible views at a given angular step size. The initial model can be obtained as described in Sect. 4.3.1.2 and can be updated using a reconstruction produced at the end of each refinement cycle during this iterative process. These projections are compared against the pre-centered raw particle images in order to determine the orientations of the raw particle data. The orientation of the particle is assumed to be the same as the orientation of the projection it matches most closely. However, before this comparison can be made, in order to fully utilize the information in a raw image, the CTF must be deconvoluted from the raw particle images (equivalently, the CTF could be applied to the projection images). As the resolution of the map is dependent on how accurate the particle orientations are, a finer angular step size is required in order to refine the model to higher resolutions. Unfortunately, as these projections are sampled in 3-D space, the number of projections increases cubically with finer step size. In the process of refining a model to higher resolution, there is a trade-off between resolution and computation time as the requirement for precise orientation determination requires an ever-increasing number of comparisons to be made.

Projection matching can be carried out either in real or Fourier space. In order to speed up the computation, the alignment is typically performed in both Fourier and real space and involves determining a particle's projection angles (θ, ϕ) and in-plane rotation angle (ω) separately and refining the predetermined center (x and y). Different software packages implement these three steps in a variety of ways, each using different mathematical principles.

One approach to projection matching is to find the best in-plane rotation angle (ω) for a given projection, use this angle to refine the center, and then use these two results to find the best match among all the projections. To find the in-plane rotation angle, both the raw particle image and the model projection are converted to polar coordinates. Then a 1-D FT of an annulus of the data is computed, and the in-plane rotation is determined by the highest correlation between the 1-D FTs of the raw particle image and model projection. After in-plane alignment, the center of the particle is refined using the correlation between the image and projection. Given the in-plane rotation and center between the image and the projection, a correlation score can be computed. Applying these steps to a single raw particle image and each of the model projections will result in a set of scores. The particle orientation is assigned to the in-plane rotation, center, and projection with the best score (Ludtke et al. 1999).

In order to minimize the error propagation implicit in the process of sequential parameter search, another approach is to perform the local center refinement and global in-plane rotation search simultaneously. As a result, this method calculates the rotational correlation between an image and a projection for each of the possible center positions in the neighborhood of the image's predetermined center (Joyeux and Penczek 2002). This simultaneous search is time-consuming (for example, a 10-pixel neighborhood will increase calculation by 10 times). To speed up computation, FRM2D (fast rotational matching for 2-D images) converts the alignment problem (usually formulated by one in-plane rotation and two translations) into one translational and two rotational searches (Cong et al. 2003). It takes advantage of fast Fourier transforms (FFTs) in rotational space to speed up the search for the two rotational parameters, while the one translational parameter is explored by limited exhaustive search. FRM2D can search more efficiently, depending on the fineness of the angular sampling and linear search range (Cong and Ludtke 2010).

The polar Fourier transform (PFT) is another approach that uses a different sequence of steps for determining the orientation parameters (Baker and Cheng 1996). First, the precentered particle image is converted to polar coordinates (Γ, r), and a 1-D FT for each annulus of data is computed. The values in an annulus have a fixed radius r in the Cartesian image but different Γ angles, and the amplitudes of the 1-D FTs of these data are rotationally invariant (invariant in the in-plane rotation ω in the Cartesian image). As a result, the best matching projection can be determined first, and this provides the two Euler angles (θ, ϕ) associated with the projection. Next the in-plane rotation ω can be determined by rotational correlation between the best matching projection and the raw particle image. Given the three Euler angles, cross-correlation can be used to refine the image center. This method is very fast, but may suffer from interpolation and propagation errors because of the three separate searches. Furthermore, as this method only uses amplitude information from the 1-D FTs (phase information is ignored), it may not be very sensitive at high resolution.

Once the centers and orientations of the individual particles have been determined, all the particles that matched the same projection can be averaged together. These 2-D class averages have the advantage of increasing the SNR above that of a single image. It is important to note that prior to averaging, the CTF must be deconvoluted in the individual single-particle images. Otherwise, the average of images with different CTF parameters would result in the loss of signal at specific spatial frequencies. These class averages can then be used for reconstructing a 3-D volume (Sect. 4.3.1.4).

Common Lines

Another approach to particle orientation determination relies on the central section theorem, which assumes that planes in Fourier space intersect at the origin (DeRosier and Klug 1968). In the case of cryo-EM, the individual images recorded in a microscope are projections of randomly oriented objects, and the Fourier transform of these individual particle images exists as 2-D planes in a 3-D Fourier volume. For any two planes in Fourier space, the line at which the planes intersect is known as a common line (Fig. 4.4b). For icosahedral viruses, each of the 60 equivalent icosahedral orientations defines a plane in Fourier space that is related to each other through this symmetry (Fig. 4.4b). The line at which any two of these planes intersect is known as a self-common line. For any icosahedral particle, there are 37 self-common lines, including 12 from the fivefold symmetries, 10 from the threefolds, and 15 from the twofolds (Crowther et al. 1970b; Glaeser et al. 2007). Therefore, there is a great deal of information inherent in a single icosahedral particle with which the Fourier amplitudes and phases of two particles can be matched to each other (provided the particle's center is correctly identified).

The self-common line method works best for icosahedral viruses as it relies on the symmetry inherent in the particle. Additionally, as a result of the icosahedral symmetry in these particles, it is possible to determine the particle's center either from a center of mass calculation or from the cross-correlation of an image and the same image rotated by 180°. To determine the individual particles' orientations, an exhaustive search is performed for all possible orientations. The accuracy of an orientation is judged by the phase residual, which is a score that measures the discrepancy of the self-common lines for the given orientation. The orientation with the smallest phase residual is assumed to be the correct one. Subsequently, the predetermined center of the particle can be further refined based on this orientation (Fuller et al. 1996). Since the self-common line method uses the individual raw image itself as a reference image, it does not require an initial model, thus freeing it of model bias. Unfortunately, the accuracy of the self-common line method for determining particle centers and orientations is limited by the small number of self-common lines (37 of them) and error propagation from one step to the next, which precludes its use for studies with excessively noisy images.

In the same vein as the self-common line approach are cross-common lines (Fig. 4.4b). Unlike the self-common line method, which uses the particle itself as a reference, cross-common lines occur between a particle and another particle. The references in the cross-common line method can be either projections from a 3-D model/map or other single-particle images whose orientations are known. Cross-common lines can use multiple references instead of only one reference (the image itself, as in self-common lines), which makes it possible to use all of the information in the 2-D Fourier transform.

To determine a particle's orientation, the cross-common line method also uses the phase residuals of the cross-common lines to judge if an orientation is correct. As more references are included in the comparison, the phase residual will be more sensitive to variation in the orientation when searching near the true orientation. While this approach allows for the orientation to be determined very accurately, it also increases the possibility of inaccurately determining the true orientation (i.e., very small radius of convergence) when a large angular step size is used in an exhaustive search. This small radius of convergence requires a very exhaustive and simultaneous search of the center and orientation parameters, a process that was once nearly impossible to achieve (Fuller et al. 1996). Fortunately, as more computational resources have become available, the cross-common line method can be exploited for accurate single-particle orientation determination (Zhou et al. 1998).

The cross-common line method was implemented in the refinement loop of several software packages: SAVR (Jiang et al. 2001b), IMIRS (Liang et al. 2002), and SPIDER (Shaikh et al. 2008). However, it requires an accurate determination of the particle's center in the image, so that the search can be mostly limited to orientation space (θ, ϕ, ω). The multi-path simulated annealing (MPSA) optimization algorithm was later developed to use cross-common lines to globally determine both center and orientation simultaneously by minimizing the phase residual of the cross-common lines between the raw image and reference projections (Liu et al. 2007). The core of the MPSA algorithm relies on a Monte Carlo optimization scheme, which is formulated to optimally search a very small portion of the 5-D (2-D for center, 3-D for orientation) space, saving a tremendous amount of CPU time as compared with the exhaustive search methods. However, as a result of using Monte Carlo, there is no guarantee that the search result represents the true global minimum (i.e., the correct center and orientation). Nevertheless, the speed of the MPSA algorithm enables it to perform the same search multiple times, and the consistency of these results/solutions (reaching the same orientation) can confirm the determined orientation as a global minimum.

Since both the center and orientation parameters are determined simultaneously in the common line approach, there is no error propagation as compared with the other sequential methods. This characteristic of the cross-common line approach permits the accuracy of the orientation search to extend far beyond the resolution range used in the orientation search, resulting in a converged 3-D volume sooner. However, as most of cross-common lines do not overlap with the 2-D Fourier-space grid points, the method would suffer from interpolation errors in the high-resolution data. While it may be possible to overcome these interpolation errors by oversampling the 2-D Fourier transform of the images, this process will inevitably have larger memory demands.

4.3.1.4 Reconstructing a 3-D Volume

The only requirement for reconstructing a 3-D volume is a set of single-particle images (or class averages) and their corresponding orientations. The resolution of the 3-D map will depend on the total number of particles, accuracy of the orientations, the quality of the data, and many other factors (e.g., weighting scheme, interpolation method, CTF correction). Discussed below are the methods by which these maps are synthesized, either in Fourier or real space.

Fourier-Space Reconstructions

The most direct way to build a 3-D model from 2-D projections is through an inverse Fourier transform. Here, as in common lines, the central section theorem is used, but this time as a straightforward application for computing a particle's 3-D volume from a set of 2-D projections. The direct Fourier inversion technique takes the FTs of the particle images, and inserts these 2-D FTs into an empty 3-D matrix, according to the determined particle centers and orientations. Once all the 2-D FTs have been inserted into this volume, it is possible to generate a real-space 3-D map by computing the inverse FT of this 3-D Fourier volume. In this method, CTF correction must be applied to the individual particles before inserting the 2-D FTs into the 3-D Fourier volume (class averages must already have corrected CTFs as well). As with many of the particle alignment protocols, errors in interpolation can limit the pursuit of high resolution. Since this procedure requires a 2-D section to be placed in a 3-D volume, interpolation is inevitable because the grid points (voxels) do not overlap. While 3-D oversampling can potentially reduce the effect of interpolation errors, the massive memory and computational resources required for large virus particles often constrain the use of this approach for high-resolution reconstructions.

An alternative method, Fourier–Bessel synthesis, can partially address the memory and interpolation issues that arise from the direct Fourier inversion approach. This method uses cylindrical coordinates $\rho(r, \phi, z)$ in real space or $F(R, \Phi, Z)$ in Fourier space (Crowther 1971). Each of the inserted FT of the particle images $F(R, \Phi, Z)$, at a given cylindrical annulus (R and Z are constant in an annulus), can be expanded into a Φ-related Fourier–Bessel series, which overcomes the data interpolation problem in the Φ direction. The expansion coefficients can be determined by least squares fitting, so long as the total number of particle images is greater than the number of the expansion coefficients, as determined by targeted reconstruction resolution. Once the coefficients are determined, the real-space density $\rho(r, \phi, z)$ can be synthesized in two sequential steps of 1-D integration/summation that requires only a very small amount of memory. Even though Fourier–Bessel synthesis is still susceptible to interpolation errors, the memory requirements of this approach are significantly lower than for direct Fourier inversion.

A third method for reconstruction is spherical harmonics, which is similar to the Fourier–Bessel method but uses spherical coordinates $\rho(r, \theta, \phi)$ and/or $F(R, \Theta, \Phi)$ instead of cylindrical coordinates (Liu et al. 2008a). The advantage of spherical coordinates is that the method is less susceptible to interpolation errors. However this comes at the cost of significantly increased CPU time for the reconstruction. This method expands each of the inserted 2-D Fourier transforms into a series of spherical harmonic coefficients at a given Fourier shell radius (R is held constant). The two-dimensional expansion coefficients can also be obtained through least squares fitting, and then the density map $\rho(r, \theta, \phi)$ can be synthesized from these coefficients. Just as with Fourier–Bessel synthesis, the Cartesian coordinates for the real-space 3-D map are interpolated from the spherical coordinates of the spherical harmonic method.

Real-Space Reconstruction (Back Projection)

Conceptually, the images collected on the microscope can be visualized as a flattening process by which a 3-D object becomes a 2-D image. If this process were directly reversible, then the 2-D image could be "stretched" back into 3-D space by recovering the height information, which was lost due to this flattening, from any nonparallel projection of a similar object. While the process described above appears to be relatively straightforward, nonuniform sampling caused by the conversion from polar to Cartesian coordinates when the 2-D images are "stretched" back into 3-D space overweights the low-resolution components, resulting in a blurred 3-D map. This overweighting can be corrected using FTs to apply a windowed ramp filter to the 2-D images prior to back projecting. As this filtering process requires both forward and backward FTs, back projection requires one extra Fourier inversion for each image compared to the direct Fourier inversion method described above, which quickly increases computation times. However, the back projection method requires half the memory needed by direct Fourier inversion, making it desirable for problems that are intractable due to memory constraints. Furthermore, and most importantly, simple linear interpolation is often sufficient for the back projection approach, while more complicated approaches are needed for interpolating high-resolution data in Fourier space reconstruction methods above (Stark et al. 1981).

4.3.1.5 Processing and Refining a 3-D Map

Once a reconstruction cycle is complete, the 3-D map may need additional processing before the next iteration of refinement. In some cases, to enhance the visibility of structural features, it may be necessary to boost the high-resolution components and/or filter out noise at specific spatial frequencies of the map. The real-space noise outside the capsid and any density within the capsid (if not of interest) can be removed by masking the 3-D volume. As masking can cause edge artifacts, a soft mask is preferred over a sharp mask. In addition, normalizing the density map can help reduce any artifacts that may result from masking the 3-D volume. These steps are important for minimizing any noise in the reference projections used in the subsequent refinement iterations. This iterative refinement process stops when the 3-D map no longer shows any improvement.

4.3.1.6 Postprocessing of 3-D Volumes

Map Sharpening

Upon inspection of a refined 3-D map, it may appear as if certain features are not adequately resolved in the volume. While this can have many potential causes (e.g., preferred orientations in the data, uneven weighting, interpolation errors), it is possible to recover much of the missing information by "sharpening" the map. The process of map sharpening scales specific spatial frequencies (resolutions) above others. One such sharpening procedure is accomplished by multiplying by an appropriate *B*-factor curve (i.e., $\exp(Bs^2)$) in Fourier space to preferentially scale up the high-resolution components (Fernandez et al. 2008; Rosenthal and Henderson 2003; Zhang et al. 2010a). As there is no precise criterion for choosing the proper *B*-factor, a more objective way to sharpen the map is to scale the power spectrum of the map to match the solution scattering of the virus (Schmid et al. 1999).

Noncrystallographic Symmetry Averaging

The icosahedral symmetry in many virus capsids means there are 60 copies of the asymmetric unit in each particle. Thus, this symmetry can be used not only to determine orientation but also to increase the SNR of the data, thereby improving the resolution of the reconstruction. Additionally, each of the 60 icosahedral asymmetric units may be comprised of several protein subunits (typically given by the virus T-number) (Glaeser et al. 2007; Johnson et al. 1994). Provided that these subunits are found in similar conformations, it is possible to average them together to further increase the SNR of the data, thus improving the map's resolution. As this local averaging does not follow the overall icosahedral symmetry of the capsid, it cannot be performed during reconstruction and must be done afterward. The process of extracting and averaging similar subunits is called "noncrystallographic symmetry (NCS) averaging," which is borrowed from a term in X-ray crystallography. NCS averaging was first used in cryo-EM to improve a subnanometer resolution reconstruction of herpes simplex virus capsid (Zhou et al. 2000) and has since been used to enhance several near-atomic resolution reconstructions (Settembre et al. 2011; Zhang et al. 2010b) (see Table 4.1).

Visualization and Segmentation

Once a finalized 3-D volume has been obtained, using any of the methods described above (Sect. 4.3.1.4), the data can be visualized with a variety of programs, including Chimera (Pettersen et al. 2004), Amira, and Avizo. These programs allow the user to manipulate the volume in 3-D, and some even support the use of stereographic presentation so that the 3-D maps can be visualized in a "true 3-D" environment.

Table 4.1 Selected examples of near-atomic resolution structures

Virus	EMDB/PDB IDs	Particle numbers	Reported resolution (Å) (resolution criterion)	Program (algorithm)	Symmetry
Epsilon-15 phage (Jiang et al. 2008)	5003/3C5B	36,259	4.5 (FSC 0.5)	EMAN (projection matching, direct Fourier inversion)	Icosahedral
P-SSP7 phage (Liu et al. 2010c)	1713/2XD8	36,000	4.6 (FSC 0.5)	MPSA (cross-common line with optimization), EMAN (direct Fourier inversion)	Icosahedral
Human adenovirus (Liu et al. 2010a)	5172/3IYN	31,815	3.6 (FSC 0.5)	IMIRS (common lines, astigmatism correction, spherical harmonics reconstruction)	Icosahedral
Bovine papilloma virus (Wolf et al. 2010)	5155/3IYJ, 5156 (with NCS)	3,977	4.9 (FSC 0.5), 4.2 (NCS, FSC 0.5)	Frealign (projection matching, per particle defocus refinement, interpolation in Fourier space) Multiple programs for NCS	Icosahedral NCS
Aquareovirus infectious subvirion particle (VP3, VP6) (Zhang et al. 2010b)	5160/3IYL	18,646	3.3 (FSC 0.143)	CTFFIND, IMIRS, Frealign	Icosahedral NCS
P22 phage procapsid (Chen et al. 2011)	1824/2XYY	23,400	3.8 (FSC 0.5)	EMAN (projection matching, per particle defocus refinement)	Icosahedral
P22 phage (Chen et al. 2011)	1826/2XYZ	18,300	4.0 (FSC 0.5)	EMAN(projection matching, per particle defocus refinement)	Icosahedral
Cytoplasmic polyhedrosis virus (Cheng et al. 2011)	5233/3IZ3	29,000	3.9 (FSC 0.143)	EMAN, IMIRS, ISAF	Icosahedral
Rotavirus (Settembre et al. 2011)	5199/3IYU (scaffold, VP6), 3N09 (VP7, VP4)	4,187	6.5 (FSC 0.5), 4.3 (FSC 0.143), 3.8 (NCS, FSC 0.143)	Frealign (projection matching, per particle defocus refinement)	Icosahedral NCS

One of the most important steps in interpreting and annotating a map is segmentation, a process that enhances visualization of the density maps. While it is possible in some cases to directly segment the data based on the discontinuities observed between densities, for the majority of cases, segmentation is performed based on X-ray data of the molecule or homologous structures, biochemical experiments, biological information, or other supporting experimental evidence. Amira and Avizo are two commercially available programs that are designed for manual segmentation. Alternatively, densities can be segmented manually using "color zone" tools, or semiautomatically using Segger (Pintilie et al. 2010) that is part of the Chimera suite.

For high-resolution maps (<5 Å), further interpretation and validation are often needed but have the distinct advantage of being able to produce Cα backbone traces of the proteins in the map. As the resolution of a reconstruction approaches atomic resolution, it becomes possible to visualize and model the complex side-chain interactions in these proteins (see Sect. 4.5).

4.3.1.7 Selected Examples of Near-Atomic Resolution Structures

Table 4.1 and Fig. 4.5 are selected compilations of the near-atomic resolution virus structures solved by single-particle cryo-EM. The structures of these viruses were solved by a variety of computational procedures and software packages. While crystal structures exist for some of the structures, in other cases they have yet to be solved. Where crystal structures are available, they demonstrate remarkable structural similarities both in their backbone traces and in some of their side-chain densities (Chen et al. 2009; Reddy et al. 2010). In the case of adenovirus, the cryo-EM map resolves more of the connective density representing the cementing capsid proteins than is visible in the corresponding crystal map (Liu et al. 2010a; Reddy et al. 2010). In the case of P22, the detailed structural changes between the procapsid, which contains scaffolding proteins, and the mature phage, which contains the viral DNA, clearly illustrate the structural mechanism of phage maturation (Chen et al. 2011). These are just a few examples of the knowledge gained about the structural features and assembly of virus capsids; while only a few such structures have been solved to date (Hryc et al. 2011; Zhou 2011), more are expected to come in the future and likely at higher resolutions.

4.3.2 Reconstructing Particles Without Imposing Symmetry

In the traditional application of cryo-EM for studying virus particles, any features/proteins that do not satisfy the condition of being icosahedrally arranged will be lost. However, there are a variety of situations where the assumption of icosahedral symmetry prevents the visualization of certain virus features such as the genome packaging complex of some dsDNA viruses. Furthermore, these nonicosahedrally ordered features may have their own symmetry which is obscured by the symmetry mismatch with the icosahedral capsid shell. Symmetry mismatch in complex macromolecular assemblies has been recognized for over 30 years (Hendrix 1978). This problem arises when two objects that have different symmetries (e.g., 5- and 12-fold) are part of the same macromolecular complex. Essentially, the symmetry averaging for either of the objects must be decoupled because assigning one symmetry to the whole complex would obscure the subunit arrangement of the other component. The problems in directly visualizing such symmetry mismatches have recently been solved and is attributable to the tremendous advancements in both instrument and the software used to acquire and process the cryo-EM data. Unfortunately, in order to achieve a resolution equivalent to an icosahedral map, a symmetry-free map requires at least 60 times as much data.

In general, to correctly determine a particle's symmetry-free orientation, the nonicosahedrally arranged components need to have enough signals to overcome both the noise in cryo-EM images as

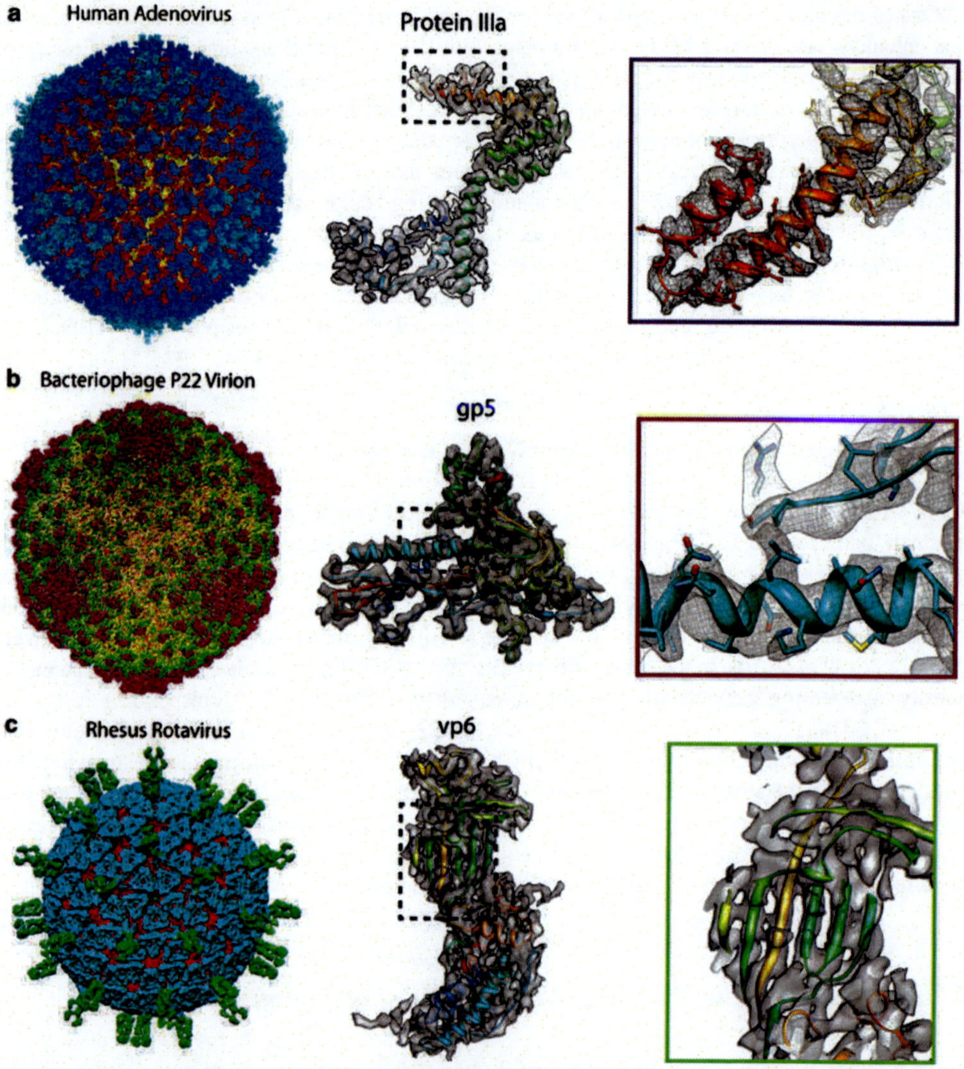

Fig. 4.5 Selected near-atomic resolution cryo-EM density maps. The *left column* of the figure is the whole virus density map, the middle is the segmented subunit density map of the virus, and the *right column* is a close-up that shows their detailed features. (**a**) Human adenovirus [EMD-5172]. The protein IIIa density shows more connecting density than the X-ray density map. (**b**) Bacteriophage P22 [EMD-1826]. Individual bulky side chains are visible in the gp5 subunit of P22. (**c**) Rotavirus [EMD-5199]. At near-atomic resolution, the separations of beta strands in VP6 [EMD-1461] can be clearly seen

well as the signal from the other components in the particle (e.g., capsid proteins, nucleic acids). Ideally, if these nonicosahedral components have a large mass, their signal is sufficient for them to be identified in the particle images. When this is not the case, the use of defocus adjustment to increase contrast in the micrographs can enhance the ability to identify these features in the projection images. Unfortunately, as discussed above, one potential downside of defocus-based contrast enhancement is the loss of high-resolution information. Although solutions may be forthcoming from a variety of alternative electron optics methodologies (e.g., a Zernike phase plate), their application in low-contrast cryo-EM situations is still in its infancy. Such methodology is discussed in Sect. 4.6.5.

When available, the crystal structures of viral proteins are often docked into the cryo-EM density map to paint a picture of how these complex macromolecular machines function. In the case of the T4 phage, the capsid and tail were reconstructed separately by cryo-EM, and then the crystal structures of their component proteins were docked into the density maps to gain an understanding of how the individual parts contribute to the function of this complex molecular machine (Leiman et al. 2010). However, new developments in data processing have now made it possible to resolve the structure of complex viruses as a single entity without imposing any symmetry, removing the need to separate the components prior to cryo-EM imaging. The following section describes how these advancements are implemented for symmetry-free subnanometer resolution reconstructions and how they have the potential to be extended to even higher resolutions.

4.3.2.1 Initial Data Preparation

The same data that are used for high-resolution icosahedral reconstructions can also be used for a symmetry-free reconstruction. For viruses that have extended asymmetric protrusions, such as a bacteriophage tail, the particle images may need to be adjusted prior to processing. Specifically, the box with which the particle image was initially selected may need to be enlarged to accommodate the tail structure. Furthermore, as a result of this larger box, it may be necessary to down-sample the images if memory and computation power are limited.

For any data collection scheme designed to see asymmetric features, to enhance contrast, the individual images are typically acquired in a range of 3–5 μm defocus. However, there is no specific defocus value that is optimum for every specimen, so the defocus value is often determined empirically. It is good practice to collect a series of images of a new sample across a broad range of defocus values and choose a defocus value that accentuates the asymmetric feature (as judged by visual inspection). However, discerning the asymmetric components in raw images is not always necessary because some symmetry-free reconstruction programs can identify the true asymmetric orientation even when the feature is not readily apparent in the raw images (Liu et al. 2010c). Thus, the defocus target for acquiring the images should also depend on the desired resolution because of the trade-off between visualizing the feature in the raw micrographs and retaining the high-resolution information which is harder to retrieve at higher defocus. For example, a subnanometer resolution, symmetry-free reconstruction would preferentially record data closer to focus than if the target is at 25 Å resolution.

4.3.2.2 Building an Initial Model

Just as for an icosahedral reconstruction, a symmetry-free reconstruction requires an initial model for starting the orientation search. This initial model must simulate the nonicosahedral components and comes with the same caveat as before of not biasing the final 3-D reconstruction. There are three common ways to create an initial model for symmetry-free virus reconstructions. The first involves creating a chimeric model comprised of the icosahedral particle reconstruction combined with a model of the nonicosahedral feature produced by low pass filtering a known structure for the isolated component itself or an equivalent (Agirrezabala et al. 2005). The second approach uses a simple geometric shape that mimics the approximate shape of the nonicosahedral feature [e.g., a cylinder for the tail of P22 and epsilon-15 (Chang et al. 2006; Jiang et al. 2006)]. The third, and most direct, method uses an icosahedral reconstruction map to find the general shape of the nonicosahedrally arranged components (Liu et al. 2010c). By lowering the isosurface threshold (generally 10- to 20-fold), the map's non-icosahedral features become faintly visible at each of their icosahedrally related positions and can be extracted as an initial model (Fig. 4.6). The lower display threshold is needed because imposing icosahedral symmetry to the nonicosahedrally symmetric features reduces their visibility in the map.

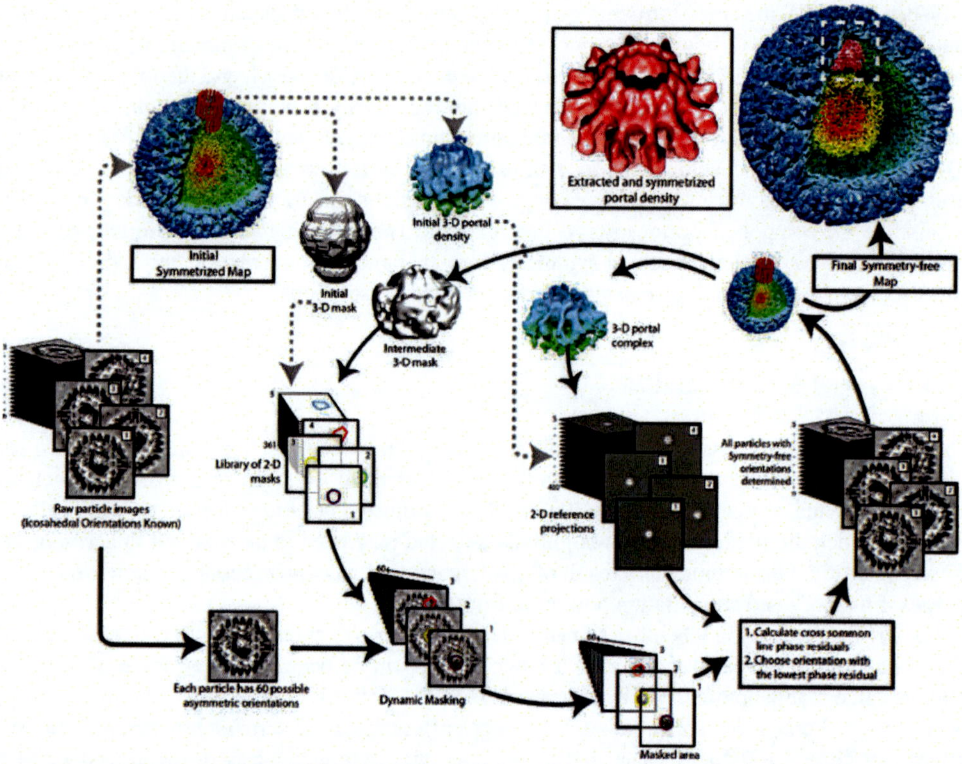

Fig. 4.6 A flowchart for symmetry-free reconstruction using the dynamic masking technique developed in MPSA. Starting with a collection of single-particle images (whose icosahedral orientations are known) and the icosahedrally reconstructed map, a 3-D initial model and mask of the interested asymmetric features can be extracted and generated (*red cylindrical area*). The 3-D mask is used to generate a series of 2-D masks for the raw particle data. After these 2-D masks are dynamically applied to a particle image to extract the interested asymmetric features, the resulting masked images are aligned to reference projections of the 3-D model. This process is iterated, each time generating a new 3-D mask and model from the intermediate 3-D symmetry-free reconstruction, until the symmetry-free map converges

4.3.2.3 Determining the Asymmetric Orientation

Currently available software can determine the asymmetric orientation of a particle in two ways. The first method, which does not make any assumptions about the symmetry of the particle, finds the orientation by exhaustively searching every possible orientation. While this approach is conceptually easy to understand, the exhaustive approach to orientation determination necessitates a large, homogeneous search space that is very time-consuming to process. Furthermore, it is more susceptible to model bias. The second method takes advantage of prior knowledge about the icosahedral organization present in many viruses, and splits the process of asymmetric orientation determination into two steps. The first step determines the particle's icosahedral orientation, at which point the asymmetric orientation search is limited to the 60 equivalent icosahedral orientations. The problem of symmetry-free alignment then becomes a process of determining which one of the 60 possible choices is the "true" asymmetric orientation of the particle. Care must be exercised with this second approach for two reasons: it assumes that the icosahedral orientation is correctly assigned, and the second step of this alignment procedure does not correct for errors in the first step.

To accurately determine the asymmetric orientation of a single-particle image, there needs to be enough contrast in the image to distinguish the nonicosahedral component from the other densities/features in the image. If the component contributes a significant part of the molecular mass of the whole particle (e.g., the tail in bacteriophage epsilon-15 (Fig. 4.7)), it is relatively easy to find these orientations. In the event that this does not hold true, it is necessary to increase the signal of the nonicosahedrally arranged component relative to the other components. One way to enhance this contrast is to subtract the known icosahedrally symmetric components (i.e., the capsid shell) from the image. This is accomplished by projecting the 3-D icosahedral map in the same orientation as a given raw particle image. This projection, which, as a result of icosahedral averaging, does not display the nonicosahedrally symmetric feature, can be subtracted from the raw particle image. Conceptually, the density that remains should correspond to the nonicosahedrally symmetric component alone. In practice, there are many factors that can reduce the effectiveness of this procedure, notably inaccuracy in the icosahedral orientation for projecting the image to be subtracted and improper scaling of its density to match that of the raw particle image.

Another way to enhance the contrast is by using a dynamic 2-D mask to remove portions of the raw particle image that do not contain the nonicosahedrally symmetric component (Liu et al. 2010b) (Fig. 4.6). This 2-D mask is generated by projecting a 3-D mask encompassing a volume that is believed to contain the nonicosahedrally symmetric feature, at each of the possible icosahedral positions and orientations previously determined for a given single-particle image. Depending upon the orientation with which the 3-D mask is projected, the resulting 2-D mask will take on a variety of positions and shapes, hence a dynamic mask. To facilitate this process, a library of all possible 2-D masks is generated before beginning the orientation search process. Typically these masks are projected with an angular spacing of 5–10°. While it is possible to use a smaller angular step size, which would remove more of the undesirable signals, this would require more masks and would take longer to process. In contrast to the subtraction technique, this dynamic masking technique does not remove any of the undesirable signals (i.e., capsid proteins, nucleic acids) in the projection that overlaps with the nonicosahedrally symmetric components in the 2-D projection image.

The choice of which of the above approaches to use depends upon the scientific question at hand, and whether the 2-D images would benefit from contrast enhancement. Once the 2-D raw particle images have been modified, the process of alignment and refinement can proceed. Just as for an icosahedral orientation search, alignment for symmetry-free reconstructions can either be in real space by projection matching or in Fourier space by common lines but without imposing any icosahedral symmetry in the alignment procedure.

4.3.2.4 Reconstructing Asymmetric Density Maps

Once the asymmetric particle orientations have been determined, the reconstruction can be performed either in real space (back projection) or in Fourier space (direct Fourier inversion, Fourier–Bessel method, etc.). The reconstruction procedure is basically the same as for icosahedral reconstruction (see Sect. 4.3.1), but since no symmetry is imposed in the process, it is more sensitive to preferred orientations. Since only one of the 60 symmetry-related planes is used during reconstruction, it is possible to sample this volume non-uniformly, resulting in artifacts in the 3-D map. As before, these symmetry-free reconstructions are refined until no further improvement is seen in the 3-D volume. Similarly, masking, normalizing, and the other processing steps described for icosahedral reconstructions can be applied to these symmetry-free 3-D maps.

4.3.2.5 Selected Examples of Symmetry-Free Structures

Spherical viruses are composed of both icosahedrally and nonicosahedrally arranged proteins and nucleic acids. In the past few years, developments in image processing software have enabled cryo-EM to solve the structures of these nonicosahedral components (Table 4.2 and Fig. 4.7). These

Table 4.2 Selected examples of symmetry-free structures

Virus	EMDB IDs	Particle numbers	Reported resolution (Å) (resolution criterion)	Program (algorithm)
T7 phage procapsid (Agirrezabala et al. 2005)	1161	4,460	24 (FSC 0.5)	SPIDER
T7 phage (Agirrezabala et al. 2005)	1163	4,785	25 (FSC 0.5)	SPIDER
Epsilon-15 phage (Jiang et al. 2006)	1175	15,000	20 (FSC 0.5)	EMAN, SAVR
Epsilon-15 phage (Murata et al. 2010)	5207	5,600	13(FSC 0.5)	MPSA, EMAN
P22 phage (Chang et al. 2006)	1222	16,000	20 (FSC 0.5)	EMAN
P22 phage (Lander et al. 2006)	1220	20,267	17 (FSC 0.5)	EMAN, SPIDER
K1E phage (Leiman et al. 2007)	1333	6,105	16.6 (FSC 0.4)	EMAN, SPIDER
Phi29 phage (Tang et al. 2008)	1420	12,682	7.8 (FSC 0.5)	EMAN, Frealign, AUTO3DEM
P-SSP7 phage (Liu et al. 2010c)	1715	15,000	9.2 (FSC 0.5)	MPSA, EMAN
P22 phage procapsid (Chen et al. 2011)	1827	43,850	8.7 (FSC 0.5)	MPSA, EMAN
HSV-1 B-capsid (Rochat et al. 2011)	5259	2,308	25 (FSC 0.5)	MPSA, EMAN

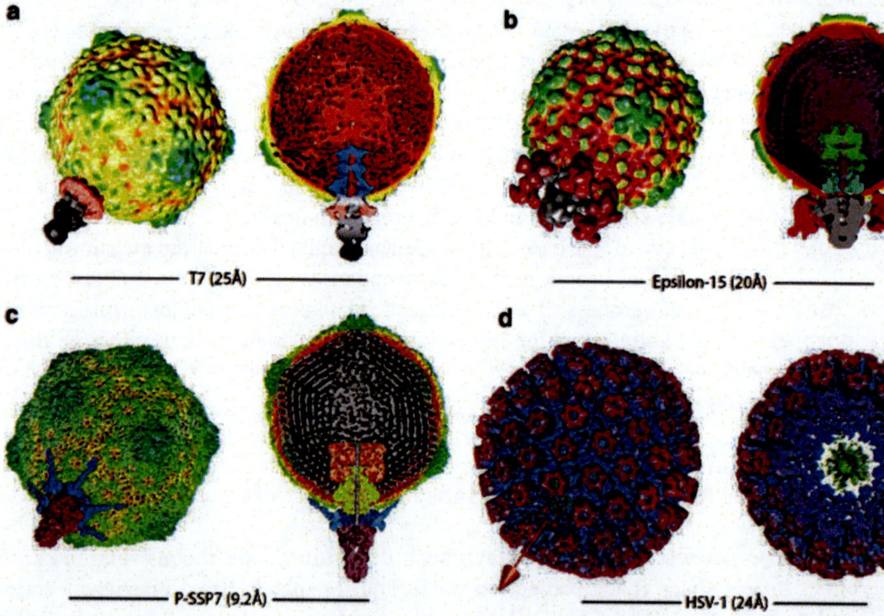

Fig. 4.7 Selected symmetry-free cryo-EM density maps. (**a**) Bacteriophage T7 [EMD-1163]. This is the first symmetry-free reconstruction of a whole virus. An oblique view (*left*) and coronal slice (*right*) are shown. (**b**) Bacteriophage epsilon-15 [EMD-1175]. This is the first reconstruction to clearly show the genome organization and DNA packaging/injection apparatus in the context of the virion. (**c**) Cyanophage P-SSP7 [EMD-1715]. This symmetry-free reconstruction is the first to show how the symmetry mismatch between 5-fold capsid and 12-fold portal structure is accommodated. (**d**) Human herpes virus type I B-capsid [EMD-5259]. This symmetry-free reconstruction represents the first such reconstruction for a nontailed virus particle with images recorded from an electron microscope equipped with the Zernike phase contrast optics. The green portal density can be easily visualized in the cutaway map and is located in one of the fivefold vertices (*arrow*)

advances have opened up a new area of structural virology where it is possible to understand the complex structural organization of all the viral components regardless of their underlying symmetry. Furthermore, the resolutions of these studies are gradually moving toward the subnanometer mark, where it becomes possible to detect α-helices and β-sheets. An example of the value of these maps is the light they have cast on the longstanding puzzle as to how the differing symmetries of the DNA packaging machinery and the capsid shell accommodate each other at a single unique vertex (Chang et al. 2006; Chen et al. 2011; Jiang et al. 2006; Lander et al. 2006; Liu et al. 2010c; Murata et al. 2010; Tang et al. 2008). Furthermore, the maps of the full and empty capsids of P-SSP7 phages have revealed the structural changes that occur in the nozzle proteins to trigger genome release upon infecting its host cell (Liu et al. 2010c).

4.4 Tomographic Reconstruction Methods

The two methods described in Sect. 4.3 assume that the particles under study are structurally homogeneous. Therefore, any differences in the 2-D images are assumed to be due to different particle orientations. However, these methods are not suitable to study viruses which do not have uniform structures, or viruses which are infecting the host cells.

In reality, no virus is made up of icosahedrally symmetric components only, and many have no icosahedrally symmetric components at all. For example, influenza (*Orthomyxoviridae*) surrounds its genome with a flexible helical capsid that is enclosed by a pleomorphic envelope, and HIV has an envelope that encircles its nonicosahedrally symmetric nucleocapsid. As highlighted in Sect. 4.3.2, even icosahedral viruses have nonsymmetric components. For example, herpesviruses have an icosahedral capsid inside irregular tegument and envelope layers. In all these cases, the shapes and sizes of the membranes, their relationship to the capsids, and the distribution of the glycoproteins may vary from particle to particle. Similarly, viruses infecting their host cells will exhibit nonuniform properties as each virus is in a different step of infection. Fortunately, heterogeneity, polymorphism, flexibility, and labile structures pose less of a problem for cryo-ET. Using posttomographic averaging to enhance the contrast of the subvolumes of identical features, one can retrieve 3-D structures of individual protein components in a structurally heterogeneous virus or viruses in different stages of infecting a cell.

4.4.1 Specimen Preparation Considerations

Electron microscopy offers the opportunity to study a whole cell being infected by viruses. However, specimen thickness becomes an important consideration as specimens can become impenetrable to the electron beam if they are too thick (e.g., over 1 μm). For very thick specimens, one solution is cryo-sectioning (Al-Amoudi et al. 2004; Hoenger and Bouchet-Marquis 2011). In this procedure, the sample of interest is cooled very rapidly to liquid nitrogen temperature at very high pressures and usually in the presence of a cryoprotectant. Freezing at high pressure slows the formation of ice crystals and allows the formation of vitreous ice, which preserves structural features in the specimen. Ultra-thin cryo-sections are cut from the block of frozen material and transferred to a specimen grid for imaging in the EM. While this approach may seem ideal, the sections invariably contain many artifacts that are caused by the sectioning process and can complicate data interpretation (Dubochet et al. 1988). Furthermore, as the optimal sectioning thickness is less than the diameter of many viruses (~750 Å), this approach may not provide the full picture of the virus structure in the cells. The following section describes the use of cryo-ET of specimens including isolated viruses and cells infected with viruses, which are thin enough to be prepared by the plunge freeze method as described above.

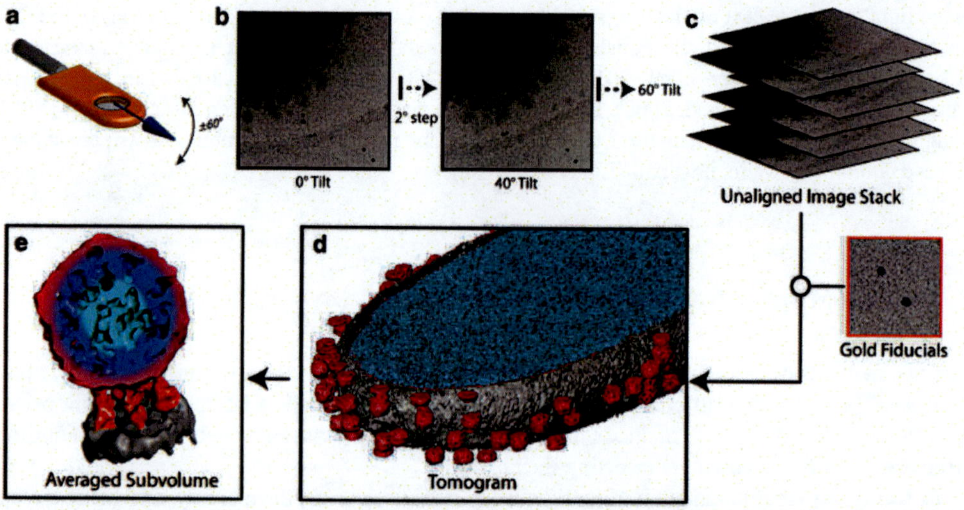

Fig. 4.8 Generalized schematic for a tomographic reconstruction. (**a**) Schematic of a cryo-tomographic holder, which enables the sample to be tilted at a specified angle. Most holders allow for the sample to be tilted in the range of ±60° to ±70°. (**b**) A series of tilted images are collected with a step size of 2°. (**c**) Images from a typical tilt series are not aligned with respect to each other due to mechanical imperfections in the TEM. Fiducial markers, typically gold nanoparticles, can be used to aid in alignment as they appear as high-density dots in micrographs. The aligned image stack can be used to reconstruct a tomogram (**d**). From the tomogram, the individual virus particles attached to the cell can be computationally extracted into subvolumes, aligned, and averaged. Shown here is bacteriophage epsilon-15 attached to its *Salmonella* host cell [EMD-5216] (**e**)

4.4.2 Cryo-ET Data Collection

In contrast to single-particle data collection, where the sample is imaged at a fixed angular position, the sample in tomography is rotated around an axis during imaging (Fig. 4.8). This allows for images of the same specimen area to be acquired at different tilt angles. Usually the microscope is mechanically limited to a tilt range of ±70°, but specimen thickness or other factors may further constrain this range to ±60°. Within this range, the specimen can be tilted by a constant angular increment (typically 1° or 2°) or by a variable step that is proportional to the cosine of the current tilt angle (Saxton scheme) (Saxton et al. 1984). In the Saxton scheme, the angular step is larger at low tilt angles and smaller at high tilt angles, which results in more even sampling in Fourier space.

In order to minimize damage to the specimen, the total electron exposure must be kept as low as possible. It has been established that low-resolution structures can be preserved up to an electron exposure of ~100 e/Å^2 (Bammes et al. 2010). Therefore, this total electron dose must be divided among all the images (usually 60–150) in a tomogram. Consequently, each image will receive an exposure of only ~1 e/Å^2, which is less than 5% of the exposure normally used for single-particle cryo-EM. As a result of this low exposure, each image is very noisy and has low contrast, which makes alignment and CTF correction very difficult. One way to improve the contrast is to image at a higher defocus (typically at least 3–10 μm). This facilitates alignment of the images but does not necessarily help with CTF correction. As a result, CTF correction is usually not performed for tomography, which limits the resolution to the first zero of the CTF. Another significant limitation to the resolution is the angular step that is used in collecting the tilt series (Crowther et al. 1970b). Ideally, a small angular step would be used for finer sampling (higher-resolution data collection), but this is not possible because of the limited total exposure that must be partitioned among all the images.

4.4.3 Processing the 2-D Data (Tilt Series)

Subtle mechanical imperfections in the microscope stage can cause the specimen to shift as the stage is tilted during tomography. Thus, before a volume can be reconstructed, the individual images in a tilt series must be aligned to each other. To facilitate this alignment, gold particles (fiducials), which appear as high-density dots in the images, can be applied to the specimen (Fig. 4.8). As the fiducial markers are fixed relative to the specimen, tracking their position across a tilt series allows the individual images to be shifted into alignment with each other (Amat et al. 2008; Brandt et al. 2001b; Heymann et al. 2008b; Kremer et al. 1996; Lawrence et al. 2006; Mastronarde 1997; Nickell et al. 2005). The gold particles must be large enough for them to be visible under the imaging conditions used for tomography but not too large, as it is harder to accurately determine the center of a larger, nonspherical gold fiducial, potentially introducing errors in alignment. As thicker specimens may need larger gold particles in order for them to be visible (especially at higher tilt), a reasonable starting point is to use gold particles that are ~15 pixels wide (usually 100–200 Å, depending on imaging conditions) and make adjustments as dictated by the specimen. Alternatively, a mixture of sizes can be used.

Fiducial-based alignment algorithms perform an initial coarse alignment using cross-correlation between the images of tilted specimen, but they can also be aligned by visual inspection. This simplifies the process of locating the corresponding fiducials in each image. The more widely distributed the fiducial markers are throughout the volume of the specimen, the better the global alignment of the images will be. Even though three points are sufficient to define a plane, selecting more fiducials will allow the software to provide a better solution. Once individual fiducials have been selected in the 0°-tilt image, the software can use these coordinates and the initial tilt angles for each image supplied in the readout from the electron microscope stage to predict the positions of these same fiducials in the other images. However, when the software fails to locate and center a fiducial accurately, user intervention is necessary. The user should examine the software solution and adjust the outliers (visually inspect the fiducials with large errors and make sure they are accurately centered on the gold particle). Once all the selected fiducials have been accurately positioned, their coordinates are used to transform the 2-D images (translate, rotate, scale, distort) before reconstructing the 3-D volume (i.e., a tomogram).

While it is relatively easy to use gold fiducials to align tilt series, there are algorithms for marker-free alignment (Brandt et al. 2001a; Castano-Diez et al. 2007, 2010; Liu et al. 1995; Sorzano et al. 2009; Winkler 2007; Winkler and Taylor 2006; Zheng et al. 2007). These methods can align the images by cross-correlation or by tracking identifiable features in the specimen itself. For example, membranes, which contain lipids with large positive head groups, tend to have good contrast in the images. Indeed, any prominent features in a sample that are evenly distributed in the tilt-series images can act as markers for fiducial-free alignment. Aligning the images in this way is essentially the same process as is used for fiducial-based alignment, but an advantage of this method is that the resulting 3-D reconstructions do not have the "shadows" that are a characteristic of gold fiducials and have the potential to obscure regions of interest (Fig. 4.8). In addition, marker-free alignment does not suffer from possible movement of the gold beads under the electron beam (Wright et al. 2006), which would cause inaccurate tilt-series alignment.

The aligned images are typically reconstructed into a volume by using weighted back projection rather than the Fourier-based methods (Sect. 4.3.1.4). In general, reconstructed volumes from cryo-ET are much larger than those from single-particle cryo-EM, so back projection, which requires half the computer memory as Fourier methods, is preferred. The weighting function used for this back projection is as previously described in Sect. 4.3.1.4, but the ramp function in this case is the radius in Fourier space at a defined cutoff (Gilbert 1972). Thus, the Fourier amplitudes are weighted by their distance from the origin, and then zeroed after a certain resolution. Other methods to obtain the 3-D density include iterative and algebraic algorithms, such as algebraic reconstruction technique

Table 4.3 Selected examples of cryo-ET reconstructions

Virus	EMDB IDs	Subvolume numbers	Reported resolution (Å) (resolution criterion)	Program (algorithm)
Vaccinia virus (Cyrklaff et al. 2007)	n/a	n/a	n/a	IMOD
Human immunodeficiency virus Env (Liu et al. 2008b)	5019	4,741	19 (FSC 0.5)	Inspect3D, Protomo, IMOD
Human immunodeficiency virus Env + antibody (Liu et al. 2008b)	5018	4,323	22 (FSC 0.5)	Inspect3D, Protomo, IMOD
Human immunodeficiency virus Env + antibody + CD4 (Liu et al. 2008b)	5020	4,900	23 (FSC 0.5)	Inspect3D, Protomo, IMOD
Epsilon-15 phage (Chang et al. 2010)	5216, 5217, 5218, 5219	83, 85, 44, 10	n/a	IMOD, EMAN (Tomohunter)
Sulfolobus turreted icosahedral virus (Fu et al. 2010)	n/a	n/a	n/a	IMOD
P-SSP7 phage (Liu et al. 2010c)	1714	29	24 (FSC 0.5)	IMOD, EMAN (Tomohunter)
Maloney murine leukemia virus RNA (Miyazaki et al. 2010)	1806	38	n/a	IMOD, EMAN (Tomohunter)
Epsilon-15 phage (Murata et al. 2010)	5203, 5204, 5208	132	32, 38, 26 (FSC 0.5)	IMOD, EMAN (Tomohunter)
Poliovirus (Bostina et al. 2011)	n/a	n/a	n/a	SPIDER, IMOD

(ART) and simultaneous iterative reconstruction technique (SIRT) (Lakshminarayanan and Lent 1979; Marabini et al. 1998). ART and SIRT iteratively minimize the differences between the projections from 3-D and the 2-D data, similar to the iterations of projection matching as discussed in Sect. 4.3.1 (Fig. 4.3). These iterative methods have higher computational demands, so they have been used less frequently than the weighted back projection approach.

All 3-D volumes derived by cryo-ET suffer from missing wedge artifacts. This is due to the lack of tilted views from 70° to 90° (or even from 60° to 90°, as mentioned above). The missing views result in a missing wedge of data in 3-D Fourier space. As a consequence, the densities in the final tomogram are elongated in the direction parallel to the electron beam in real space. Nonetheless, a great deal of insight can still be gained from the 3-D map even in the presence of these artifacts. While there is a great deal of information that can be directly extracted from an individual tomogram, for the purposes of this chapter, we will focus exclusively on computationally extracting subvolumes from this 3-D map for further processing. Various implementations of this approach are available and have been applied to different specimens (Table 4.3) (Bartesaghi et al. 2008; Bostina et al. 2011; Nicastro et al. 2006; Schmid 2011).

After reconstructing a tomogram of a polymorphic specimen, identical or homogeneous components can be recognized, such as virus glycoproteins or capsids. In contrast to overlapping densities in the 2-D images for single-particle cryo-EM, the different densities are discernible in 3-D, which simplifies processing these data. These components are candidates for extraction into subvolumes for 3-D alignment, classification, and averaging to improve the visibility of features in the volume. Each subvolume should contain one copy of the component of interest, and the subvolumes can originate from a single tomogram or from several tomograms.

Processing the subvolumes to enhance the structural features normally requires 3-D alignment and averaging. However, if the particles are not identical, classification is necessary. This case would require iterative refinement by alternating the alignment and classification steps because of the interdependence between these two steps. Accurate alignment needs homogeneous structures, while accurate classification needs accurate alignment to determine differences due to structure instead of orientation. To improve the 3-D alignment, an initial visual inspection can roughly classify the subvolumes (e.g., on the basis of particle diameter, presence of DNA in the capsid, or other coarse features) before aligning. An example of this classification is illustrated for bacteriophage epsilon-15 attached to ghost cells that were visually separated on the basis of the presence of DNA in the capsids prior to alignment (Chang et al. 2010).

4.4.4 Aligning and Averaging Subvolumes

One approach to 3-D alignment is an exhaustive search of all rotations. The subvolumes are rotated around three axes, and each transformation is compared against a fixed (not rotated) reference subvolume (e.g., an initial or intermediate model). The reference can be a 3-D map derived from the single-particle method (Sect. 4.3.2), downloaded from EMDB, or another subvolume. The rotation that results in the best match between the two subvolumes is retained as the correct alignment. Once all the subvolumes have been aligned, they can be averaged together to increase the SNR of the data, reduce the effect of the missing wedge, and increase the resolution of the total aggregate subvolume. Reduction in the missing wedge is accomplished by averaging particles that have their missing wedges in different orientations so that the gaps in the structural information are effectively filled when these particles are combined. Similarly, the angular step between each of the tilted images, which is fixed by the angular step size at which the tilt series was collected, can be reduced by combining particles derived from tilt series which have images taken at different tilt angles or rotations.

Cross-correlation is frequently used to align objects in 3-D (Schmid 2011). This is accomplished by determining the orientation that maximizes the product from multiplying the 3-D Fourier transforms of the two subvolumes. However, each subvolume extracted from the tomogram has a wedge of missing data in Fourier space, and multiplication by this artifact will ablate any signal (zero times anything is zero). This produces a very strong directional signal, and, if left unaccounted for, the process of maximizing the cross correlations will preferentially align the missing wedges rather than the two objects themselves. In order to compensate for this misleading scoring function, the cross-correlation has to be scaled by the fraction of data that is not eliminated by the missing wedge (Schmid and Booth 2008). Counting the number of nonzero amplitudes is easy to understand but not straightforward to accomplish computationally. Fortunately, it is possible to simplify this problem using Parseval's theorem so that instead of counting the number of nonzero amplitudes in Fourier space, the cross-correlation map in real space is normalized (Schmid 2011). By Parseval's theorem, scaling the cross-correlation map in real space is equivalent to scaling the intensity in Fourier space. Hence, the magnitudes of the data voxels in the product (representing cross-correlation in Fourier space) will be scaled up when there are fewer nonzero voxels and scaled down when there are more nonzero voxels. This approach reduces the tendency to align the missing wedges of the particles, thus producing more reliable results.

A similar algorithm minimizes the dissimilarities between the two subvolumes and uses masks to restrict the comparison to only the region where the data in the two subvolumes overlap (Bartesaghi et al. 2008). This general dissimilarity function can take the form of a cross-correlation or other scoring functions. Changing the Cartesian coordinates into spherical coordinates allows the use of the convolution theorem (spherical harmonics) to compute the rotational transformation that produces the best match, similar to the PFT for 2-D alignment (Sect. 4.3.1.3). The same is performed in Cartesian coordinates to find the best translational transformation. The main advantage of this approach is that by using spherical harmonics, the slower exhaustive search can be avoided.

Another method for determining subvolume orientations is similar to the projection-matching scheme applied to 2-D images for single-particle reconstruction (Sect. 4.3.1.3), and requires an initial reference model (Förster et al. 2005). The reference model is rotated around all three axes to generate a set of 3-D references. Next, a missing wedge is applied to these 3-D references, by multiplying a 3-D mask in Fourier space, so that the missing wedge effect is the same as that in the actual data. Then the individual subvolumes are compared with all the 3-D references, and the alignment having the highest cross-correlation coefficient is assigned to the subvolume. This is analogous to comparing the 2-D particle images with the 3-D model projections to find their orientations for single-particle cryo-EM.

As with the normalized cross-correlation map method, comparing the subvolume to all the references represents an exhaustive search. However, prior knowledge about the specific sample can be used to optimize the computation by limiting the search space. For example, searching for the icosahedral orientation of a virus capsid can be limited to one of the 60 equivalent rotations rather than to a complete sphere (Murata et al. 2010; Schmid et al. 2006a). Similarly, aligning the glycoproteins on the envelope of a virus may only require an exhaustive search in the neighborhood of the glycoprotein's axis (Förster et al. 2005). An exhaustive search can also be hastened by beginning with down-sampled volumes and a large search step size. Once the approximate orientations are known, the original data can be processed with a finer step size to improve its accuracy.

Subvolumes that are aligned in 3-D can be averaged together. The algorithm to combine them must take into account the missing wedge. In the 3-D FT of subvolumes, voxels in the missing wedge have a value of 0, and combining these voxels with data voxels would reduce the average, resulting in an improper weighting of the data. One implementation uses the Fourier amplitudes to weight the sum of the real and imaginary voxels in Fourier space (Schmid and Booth 2008). Another method keeps track of all the transformations required to align the subvolumes, applies the transformations to the corresponding subareas extracted from the raw tilt series, and reconstructs a 3-D subvolume (Walz et al. 1997).

4.4.5 Classification of Subvolumes

Many tomograms consist of a mixture of particles with different structures. Therefore, separation of the structural forms is necessary prior to performing subvolume averaging as described above. Classification relies on accurate alignment, so after a preliminary visual classification, alignment, and averaging, a more objective and accurate classification can be used for refinement. For example, a modified maximum likelihood algorithm separated newly assembled *Sulfolobus* turreted icosahedral viruses (STIVs) inside the cell into empty and full capsids (Fu et al. 2010). Another algorithm uses the cross-correlation peak scaled by the number of nonzero amplitudes, which can also be used for alignment (above), to separate subvolumes into different groups (Schmid et al. 2006b). In addition, an all-vs-all dissimilarity function between two subvolumes (above) can be used in a hierarchical ascending classification technique (Bartesaghi et al. 2008). After classifying the subvolumes into new groups with any of these algorithms, the next iteration would refine the alignment. As in the single-particle cryo-EM, iteration stops when there are no further improvements.

Iterative processing of carboxysomes is one such example of particle classification (Schmid et al. 2006b). Selected subvolumes were grossly classified into nine arbitrary classes based on diameter, as determined by a 1-D radial density plot. Then all-vs-all alignment and averaging was performed on the subvolumes in each class, resulting in nine averaged class subvolumes. Next the subvolumes were compared against these averaged class subvolumes to generate a new classification of the subvolumes. At the end, the initial nine classes were reduced to seven stable classes by alternating alignment and classification in an iterative refinement. Compared to the original subvolumes, the seven averaged subvolumes provided more interpretable information.

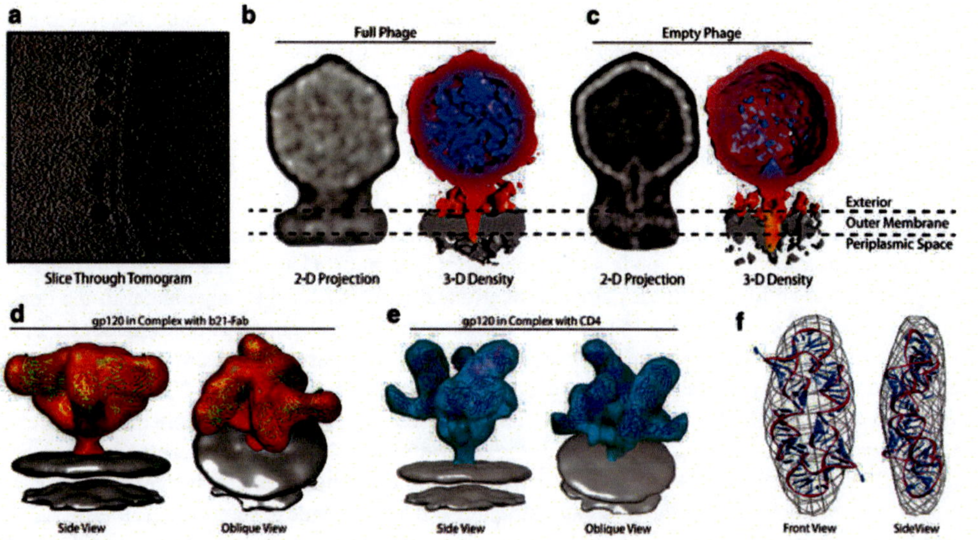

Fig. 4.9 Selected tomographic subvolume averages. (**a**) Slice through a reconstructed tomogram showing the bacteriophage epsilon-15 at the surface of its *Salmonella* host cell. (**b, c**) Shown are averaged 3-D maps of epsilon-15 attached to the cell before release of its genome (**b**) and after genome release (**c**). (**d, e**) Structure of the HIV gp120 surface glycoproteins: (**d**) in complex with b12-Fab (*green*, PDB ID: 2NY7), EMD-5018 (*red*), and EMD-5022 (*gray*) and (**e**) in complex with CD4 and 17b-Fab (*blue*, PDB ID: 1GC1), EMD-5020 (*cyan*), and EMD-5022 (*gray*). (**f**) Orthogonal views of the averaged RNA density of Moloney murine leukemia virus fitted to the ensemble of [ΨCD]2 NMR structures. This average represents the smallest structure solved to date by cryo-ET (43 kDa)

4.4.6 Selected Examples of Cryo-ET Reconstructions

Cryo-ET is the method of choice for studying heterogeneous samples. In the case of the *Archaea* bacteria *Sulfolobus* infected by STIV, cryo-ET showed a novel pyramid structure and distribution of capsids in different assembly states in the cytoplasm (Fu et al. 2010). Cryo-ET with subvolume averaging has emerged as a powerful method for studying the structures of virus particles and infected cells. When applied to the study of the trimeric HIV glycoprotein (gp120) (Fig. 4.9d), the averaged densities showed how gp120 changes conformation in order to interact with cellular receptors, subsequently leading to membrane fusion and infection (Liu et al. 2008b). Furthermore, the study challenged the proposed structural conformation based on X-ray crystallography (Chen et al. 2005). Another example of subvolume averaging focused on a virus–cell interaction, where snapshots of epsilon-15 bacteriophages infecting their bacterium host (Fig. 4.9a–c) revealed a tubular density spanning the bacterium periplasmic space (Chang et al. 2010). For the first time, this density has shown what is believed to be the conduit for viral DNA to enter the cell.

4.5 Data Validation and Interpretation

No matter what their resolution is, both cryo-EM and cryo-ET reconstructions allow researchers unparalleled access to virus structure and function. However, as these reconstructions have not yet reached atomic resolutions, it is often necessary to use a variety of field-specific computational tools for analyzing these density maps. Interpreting and validating a 3-D density map depend on a variety of factors: the resolution of the map, availability of related structures, biochemical data, and other experimental information. Here we discuss the tools used and process behind interpreting and validating the 3-D structural information that results from these studies.

4.5.1 *Map Validation*

Conceptually, perhaps the most direct method for validating a cryo-EM reconstruction is to compare the structure to known atomic models or homologous structures. Generally, structures of individual components can be fit visually or computationally and assessed for their agreement to the density map (see Sect. 4.5.4 for details). In a number of examples, such as herpes simplex, hepatitis B, and rice dwarf viruses, the subnanometer resolution cryo-EM density maps were actually published before the availability of their corresponding crystal structures (Baker et al. 2003; Böttcher et al. 1997; Conway et al. 1997; Zhou et al. 2001). These subsequent atomic models bore out the structural model determined from the density map for the individual capsid proteins. Recently, work on rotavirus and adenovirus has also clearly demonstrated the outstanding match between the cryo-EM density map, model, and available crystal structures at near-atomic resolutions (Liu et al. 2010a; Zhang et al. 2008).

While no validation methods are rigorously implemented in cryo-EM, several techniques have begun to emerge. One method uses tilted pairs, in which the same specimen area is imaged at 0° tilt and 10° tilt (Rosenthal and Henderson 2003). Once the orientations of the particles in this tilt pair have been determined, the accuracy of the alignment can be assessed by verifying that the orientations differ by 10°. In addition to assessing the accuracy of the reconstruction, this method can also provide the handedness of the virus particle. An alternative approach uses a range of resolutions to determine the orientations of the particles (e.g., 15–1,000 Å) before reconstructing them using the entire range of spatial frequencies in the data. If the resolution of the reconstruction is found to be higher than what was used to orient the particles (e.g., 10 Å), one can be assured that the higher-resolution information in the map is real and not due to alignment of noise by the software.

Bootstrapping has also been used to measure the validity of a 3-D reconstruction by generating a 3-D variance map, which can approximate the potential error at each voxel in the reconstruction (Penczek et al. 2006). Using the same particles and particle orientations that go into the final reconstruction, a new set of 2-D images is created in which a fraction of the particles are replaced by duplicates of the other images. This process is repeated multiple times (tens to thousands), reconstructing a 3-D model for each iteration. These reconstructions are combined to produce a 3-D density representative of the error at each voxel in the volume (a variance map). The variance map generated through this process shows the error present in each voxel of the final reconstruction. This error/variance can prove useful in interpreting the density map as regions of high variance are likely to correspond to flexible regions in the structure (Chen et al. 2008; Ludtke et al. 2011; Zhang et al. 2010a).

4.5.2 *Segmenting Cryo-EM and Cryo-ET Maps*

Segmentation is the process of isolating specific regions of density in a 3-D map so that they can be extracted and visualized/processed separately. This process can be performed on any volume, and the features that can be segmented depend on the resolution of the data (Fig. 4.1). While it may only be possible to segment the capsid shell at low resolutions (~40 Å), at subnanometer resolutions, the protein subunits in the capsid shell can be approximately identified and extracted for further analysis. As the data reach near-atomic resolutions, it becomes possible to unambiguously identify specific protein subunits in the virus and extract them for further processing (Sect. 4.3.1.6).

Segmentation can be accomplished with completely interactive tools like Amira and UCSF Chimera (Pettersen et al. 2004) or automated routines like Segment3D (Ludtke et al. 1999). Semiautomated tools like VolRover (Baker et al. 2006b) and Segger (Pintilie et al. 2010) can take advantage of symmetry inherent in the sample, while other methods can be used to segment, average, and classify multiple copies of a subunit within a tomogram (Forster et al. 2008). Regardless of technique, a mask is generally created for one or more subunits that define its boundaries and can be used to separate it from the reconstruction. It is important to note that this process of segmentation is often imprecise and can lead to erroneous results, particularly in lower-resolution reconstruction. In the case of epsilon-15 phage, the initial 9-Å map failed to resolve the globular density on top of the major capsid protein as a second capsid protein, which was subsequently identified in the 4.5-Å resolution structure of the virus (Jiang et al. 2008; Jiang et al. 2006).

4.5.3 Identifying Structural Features in a Density Map

The availability of high-resolution cryo-EM density has stimulated the development of a number of specialized tools for structural analysis and model building (Table 4.4). In principle, many of these

Table 4.4 Computational tools used in the validation and analysis of cryo-EM density maps

Function	Program name (reference)
Segmentation	Amira (Visage Imaging, GmbH)
	Avizo (VSG, France)
	Chimera (Pettersen et al. 2004)
	CoDiv (Volkmann 2002)
	EMAN (Ludtke et al. 1999; Tang et al. 2007)
	Segger (Pintilie et al. 2010)
	VolRover (Baker et al. 2006b)
Rigid body fitting	Chimera (Pettersen et al. 2004)
	CoFi (Volkmann and Hanein 1999)
	DockEM (Roseman 2000)
	EMFit (Rossmann 2000)
	Foldhunter (Jiang et al. 2001a)
	Mod-EM (Topf and Sali 2005)
	Situs (Wriggers et al. 1999)
	UROX (Siebert and Navaza 2009)
Flexible fitting	DireX (Schröder et al. 2007)
	Flex-EM (Topf et al. 2008)
	MDFF (Trabuco et al. 2009)
	NMFF (Tama et al. 2004)
	NORMA (Suhre et al. 2006)
	Yup.scx (Tan et al. 2008)
	Situs (Rusu et al. 2008)
SSE identification	*Helixhunter* (Jiang et al. 2001a)
	Sheetminer/Sheetracer (Kong and Ma 2003; Kong et al. 2004)
	SSEHunter (Baker et al. 2007)
Modeling	Coot (Emsley and Cowtan 2004)
	EM-IMO (Zhu et al. 2010)
	Gorgon (Baker et al. 2011)
	Modeller (Topf et al. 2006; Topf and Sali 2005)
	O (Jones et al. 1991)
	Rosetta (Baker 2006; DiMaio et al. 2009)

(continued)

Table 4.4 (continued)

Function	Program name (reference)
Visualization	Amira (Visage Imaging, GmbH) Avizo (VSG, France) Chimera (Pettersen et al. 2004) PyMol (DeLano Scientific LLC, USA) VMD (Humphrey et al. 1996)

tools are not specific for subnanometer resolution density maps and, while the results may not be as accurate, can be readily applied to lower-resolution data.

In subnanometer resolution density maps, SSEs begin to be resolved (Baker et al. 2010a; Chiu et al. 2005). In general, α-helices appear as long rod-like densities, and β-sheets appear as thin, flat sheets at ~8 Å resolution. Around 7 Å resolution, connections between SSEs begin to be resolved, although not to the point to unambiguously identify the protein backbone trace. At ~5 Å resolution, the pitch of α-helices begin to become visible, and β-sheets transform from flat densities to a series of interconnected strands, and when pushed to near-atomic resolutions, large charged side chains become visible. Based on these observations, a number of tools have been developed to directly analyze and model protein structure at subnanometer resolutions. Helixhunter (Jiang et al. 2001a) was the first computational tool to detect α-helices in a density map and was later followed by SheetMiner and SheetTracer (Kong and Ma 2003; Kong et al. 2004) for detecting β-sheets. SSEHunter offers a comprehensive solution to detect α-helices and β-sheets (Baker et al. 2007), as well as the connections between the secondary structure elements (Ju et al. 2007).

While SSEs can be used to describe gross protein structure at subnanometer resolutions, they can also be important in identifying structural and functional domains within a protein. In adapting fold recognition software, such as DejaVu (Kleywegt and Jones 1997) and COSEC (Mizuguchi and Go 1995), a set of SSEs can be queried against a structural database to identify potential structural homologues in the absence of structural similarity. This is particularly useful in identifying common structural motifs in seemingly unrelated proteins, as in the case of the capsid proteins from herpes simplex virus type I and bacteriophages (Baker et al. 2005). Furthermore, these SSEs can be used as landmarks for fitting atomic models to the density map (Abeysinghe et al. 2010).

4.5.4 Model Fitting to Cryo-EM Density Maps

When a known atomic model is available, fitting it to a cryo-EM density map provides a valuable means for analyzing structure and function in a macromolecular assembly [reviewed in (Rossmann et al. 2005)]. Fitting known structures is not restricted by the resolution of the density map and has been used both in the highest resolution single-particle reconstructions (Zhang et al. 2008) and in relatively low-resolution cryo-ET reconstructions (Zanetti et al. 2006). Over the past two decades, a number of fitting routines have evolved for use in cryo-EM, including Situs (Wriggers et al. 1999), Foldhunter (Jiang et al. 2001a), CoFi (Volkmann and Hanein 1999), UROX (Siebert and Navaza 2009), and EMFit (Rossmann 2000). Generally, these tools use a rotational and translational search of a given model within the density map in which a scoring metric is used to measure the agreement of the model with the density. However, each program utilizes a slightly different metric and reports a variety of fitting scores, so it is important to visually inspect the fit and assess the results in the context of all available structural information.

Traditionally, the aforementioned fitting methods treat an atomic model as a rigid body when fitting it to the density map. However, with flexible fitting methods, such as DireX (Schröder et al.

2007), Flex-EM (Topf et al. 2008), MDFF (Trabuco et al. 2009), NMDF (Tama et al. 2004), Norma (Suhre et al. 2006), and Yup (Tan et al. 2008), atomic models are morphed to best fit the density map. This type of fitting is particularly useful when fitting homologous structures, or structures in different conformations to the density map (Trabuco et al. 2010). In these flexible fitting methods, an atomic model is generally first fit to a density map as a rigid body and then allowed to deform over time to better fit the density map. The flexible deformation can be accomplished through a variety of algorithms, including normal modes, molecular dynamics, and geometric constraints. Again, flexible fitting can be used at any resolution, though higher-resolution density maps offer more well-resolved features that can increase the accuracy of the fit.

4.5.5 Deriving Protein Models from Cryo-EM Density Maps

In addition to model fitting, computational modeling tools can be used to construct a structural model for a protein or domain in a density map (Table 4.4). In this context, a cryo-EM density map becomes a constraint for constructing and evaluating models generated by computational tools such as Modeller and Rosetta (Baker et al. 2006a; DiMaio et al. 2009; Topf et al. 2005, 2006). If a structural homologue is known, constrained comparative modeling creates an initial sequence–structure alignment and model (Topf et al. 2005, 2006; Zhu et al. 2010). The alignment and model are then allowed to "evolve" in an effort to improve the model's fit to the cryo-EM density map. An example of this was in the near-atomic resolution structure of grass carp reovirus, where EM-IMO was used to build and refine protein structures in the virus based on the mammalian reovirus crystal structures and using the cryo-EM density map as a constraint (Zhu et al. 2010).

In the absence of a structural homologue, a structural model for a small protein or domain may be constructed using a similar constrained ab initio modeling approach. Rather than using an initial structural template, ab initio modeling constructs a gallery of potential models, typically with Rosetta (Bradley et al. 2005), and then evaluates their fit to the density map along with a variety of other model-based scores (Baker et al. 2006a). This procedure was used to construct a model for HSV-1 VP26, a small capsid protein bound to the tips of hexon subunits (Baker et al. 2006a). However, it should be noted that this procedure only works on relatively small proteins or domains (<200 amino acids). A recent extension of this procedure has been used to build and refine full atomic models of entire macromolecular assemblies using density constraints, though an initial structural template of all of the components is required (DiMaio et al. 2009; Ludtke et al. 2008).

In both of the above density-constrained modeling approaches, resolution is a key determinant in of the accuracy of the models (Topf et al. 2005). At lower resolutions, a density map provides only basic information about shape and may only be able to discriminate between possible protein folds. At higher resolutions, accuracy increases and even allows for optimal side-chain fitting at near-atomic resolutions.

4.5.6 De Novo Modeling

Single-particle cryo-EM reached another milestone in 2008 when the first near-atomic resolution structures of rotavirus (3.88 Å) (Zhang et al. 2008), GroEL (4.2 Å) (Ludtke et al. 2008), cytoplasmic polyhedrosis virus (4.0 Å) (Yu et al. 2008), and bacteriophage epsilon-15 (4.5 Å) (Jiang et al. 2008) were reported. At this resolution, the pitch of the α-helices and separation of the β-strands could be seen unambiguously, and when combined with computational techniques, Cα backbone models could be built directly from the cryo-EM density maps (Baker et al. 2010b; Jiang et al. 2008;

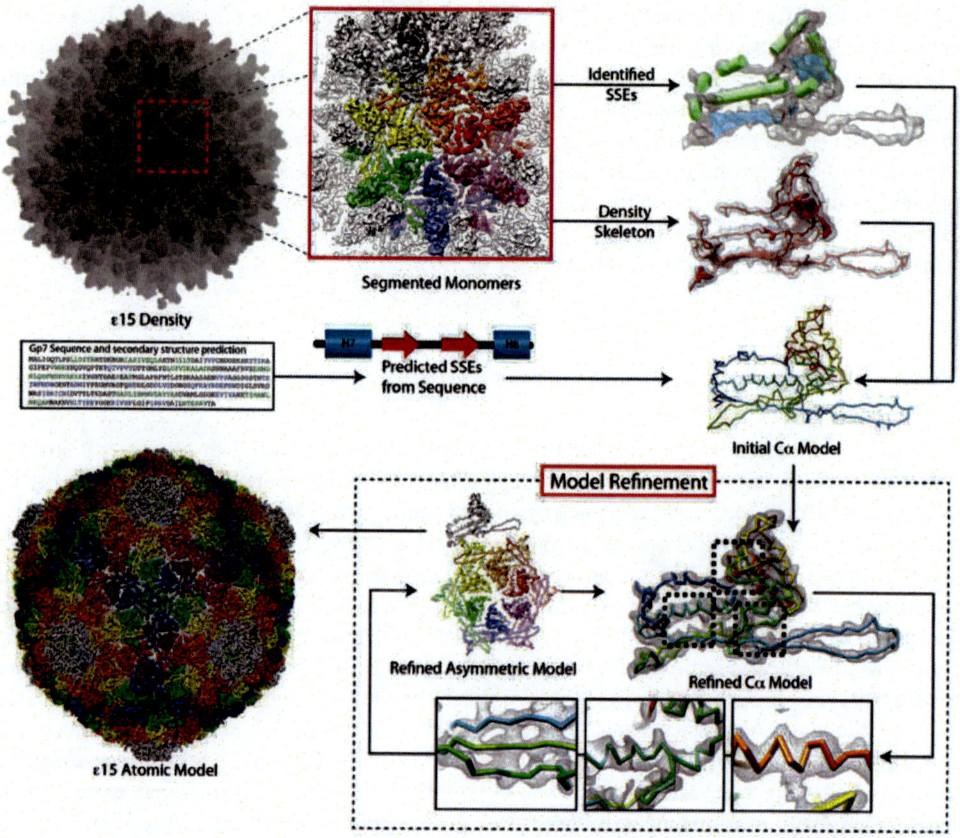

Fig. 4.10 De novo modeling. A schematic diagram illustrates the process of building a de novo backbone model directly from the cryo-EM density map of bacteriophage epsilon-15 without the aid of a structural template. Once an individual protein has been segmented from the density map, SSEs are identified along with a corresponding density skeleton. This density skeleton provides a set of possible topological paths that link the observed SSEs. From the sequence of the protein in question, secondary structure prediction yields a linear array of SSEs. Using a modified graph matching approach, SSEs from the density map are correlated to those in the sequence based on relative size, position, and connectivity. From this, an initial topological model can be created by placing Cα atoms along the density skeleton of the specified path. This model is then iteratively optimized with respect to visible density features and interactions with neighboring subunits in order to produce a final model

Ludtke et al. 2008). However, these de novo models relied almost entirely on visual interpretation of the density and manual structure assignment as they did not have the resolution to utilize standard X-ray crystallographic methods for model construction. Current state-of-the-art reconstructions have continued to push toward higher resolutions and have now made it possible to resolve atomic details (Baker et al. 2010b; Grigorieff and Harrison 2011; Zhou 2011).

At these near-atomic resolutions, SSE detection coupled with density skeletonization (Abeysinghe et al. 2008; Ju et al. 2007) and sequence analysis can be used for constructing de novo structural models directly from a density map without the aid of an existing structural template [described in (Baker et al. 2010a) and (Baker et al. 2010b)]. In de novo modeling, SSEs are used as anchors for establishing a topological model for a protein (Fig. 4.10). Once a topology has been established, individual atoms can be placed on annotated SSEs and along the density skeleton, in a similar manner to the techniques used in X-ray crystallography. While many model building toolkits, such as Coot (Emsley and Cowtan 2004) and O (Jones et al. 1991), offer a mechanism for placing atoms within a density map, none are specifically tailored to cryo-EM or near-atomic resolution density maps. Gorgon, a molecular modeling toolkit designed for subnanometer resolution density maps from cryo-EM and X-ray crystallography, provides a comprehensive suite of utilities that implement a complete protocol for Cα model construction (Baker et al. 2011). The streamlined procedure provided by Gorgon has made it possible to construct complete backbone traces in a fraction of the time required by the original de novo methods (Chen et al. 2011).

At resolutions where side-chain density becomes visible, computational and de novo modeling protocols can be extended to compute full atomic models (Zhang et al. 2010a). Essentially, a Cα backbone trace is first transformed into a full model and then iteratively refined to fit the density and satisfy biochemical and geometric constraints. This process can either be done manually using interactive software like Coot and O or completely automated with refinement routines in Rosetta (DiMaio et al. 2011). However, a density map may not necessarily have uniform resolution throughout the map, and thus only partial full atom maps can be generated for the protein (Chen et al. 2011). Regardless of the technique used to construct the model, model validation techniques at this resolution (4.5–3.0 Å) are still limited and will require future efforts to determine their reliability and limit.

4.6 Conclusion and Future Directions

The ability to study large macromolecular complexes in near-native environments at subnanometer resolutions is one of the driving forces behind the development and growth of cryo-EM in the past decade. The success of cryo-EM for homogeneous samples spurred the development of cryo-ET for studying heterogeneous samples where the assumptions of single-particle cryo-EM do not hold. Both of these techniques have opened the door to the study of the structures and conformational changes that occur in the life cycle of viruses (Chen et al. 2011; Liu et al. 2010c). While it took nearly 40 years to see the first near-atomic resolution virus structure from cryo-EM, the technology has advanced to the point that subnanometer resolution reconstructions are now routine for most virus capsid structures (Grigorieff and Harrison 2011; Hryc et al. 2011; Zhou 2011). Nevertheless, even though the field has matured dramatically in the past few years, there is still a great deal of knowledge to be gained by pushing the resolution of these techniques even further. History has shown that these incremental improvements in resolution are the result of years of technological advances in both the microscope and in the software used to acquire and process the data. However, as the scientific community begins to embrace the utility of cryo-EM and cryo-ET as tools for structural biology, these advancements are likely to come at a hastened pace. Below we discuss what is on the horizon for virus reconstruction in an era of high-resolution electron microscopy imaging and reconstruction.

4.6.1 Single-Particle Reconstruction at Atomic Resolutions

From a simplistic statistical perspective, pushing these reconstructions to higher resolutions would only require more data. Most of the near-atomic resolution maps discussed were reconstructed using 20,000–30,000 single-particle images, and the assumption of icosahedral symmetry equates this to roughly 1.2–1.8 million asymmetric projections that go into these single reconstruction. From a theoretical standpoint (Rosenthal and Henderson 2003), to improve the resolution of these maps from near-atomic to atomic resolutions (4–3 Å), more than ten times the current number of single-particle images will be needed (assuming a crystallographic *B*-factor of 100 Å^2). Collecting such massive, high-quality datasets can be facilitated by an automated imaging system that can operate continuously and efficiently (Carragher et al. 2000; Nakamura et al. 2010; Potter et al. 1999; Stagg et al. 2006; Zhang et al. 2009). The sheer magnitude of these datasets will require tapping into large-scale computing resources (CPU power, memory, and storage) to align and reconstruct the 3-D density maps. While it may be possible to process these data using the hundreds of CPUs available on modern-day graphics cards (GPUs), doing so will require the current software to be updated accordingly (Zhou 2011).

Not only will atomic resolution cryo-EM reconstructions require magnitudes more data, but these images must be of high quality as well. Specifically, the individual images must have detectable signal up to and beyond 3.0 Å so that the alignment software can extract this information for a reconstruction. There are numerous factors that need to be considered when targeting a 3.0-Å resolution reconstruction (e.g., microscope, image recording medium, alignment/reconstruction algorithm). On the instrument side, the electron optical system and the cryo-specimen holder have to be very stable, with coherent illumination but without drift or lens aberration. The current generation of microscopes appear to perform well, out to ~2.4 Å, and substantial improvement seems unlikely in the near future. Improvements can also be anticipated at the detector level. The CCDs currently in use do not directly record electrons but employ a scintillator to convert the electrons into photons, which are then recorded by the CCD. Unfortunately this process blurs the electron signal into an area of photons, effectively increasing background noise. The modulation transfer function of the CCD reduces the signal transfer substantially and loses practically all high-resolution information beyond the 0.5–0.6 Nyquist frequency of the detector. A CMOS-based direct electron detection device has recently shown the potential to improve the signal transfer efficiency substantially (Deptuch et al. 2007; McMullan et al. 2009; Milazzo et al. 2005, 2010). This offers the possibility of enhancing the high-resolution signal recorded on the detector by a factor of 5 or more. Such an improvement will in turn reduce the number of particle images needed to achieve a specific resolution by the same factor. Furthermore, the enhanced high-resolution contrast it provides may also improve the accuracy of the particle center and orientation determination. All these benefits will contribute to pushing the resolution of the reconstruction of biological specimens toward the resolving limit of the microscope.

It is unlikely that hardware improvements alone will solve all high-resolution reconstruction issues. Other parameters like defocus variation within a micrograph, magnification scaling for different particle images, depth of focus, etc., must also be considered in order to achieve these resolutions. In the few near-atomic resolution virus maps solved to date, attempts have been made to address these issues in the software used for alignment and reconstruction. Achieving atomic resolutions may also necessitate avoiding even the slightest error propagation in alignment, so all of these parameters may need to be refined at the same time, together with the five orientation parameters as discussed above. To handle this simultaneous, high dimensional refinement, optimized search algorithms will be required for processing the data.

4.6.2 High-Resolution Symmetry-Free Reconstruction

Current technology can generate subnanometer resolution symmetry-free reconstructions of virus particles from single-particle cryo-EM images. In terms of resolution, symmetry-free reconstructions of viruses are lagging far behind those where icosahedral symmetry is assumed. The reason for this is the sheer magnitude of data needed (60 times more than for an equivalent resolution in an icosahedral reconstruction), together with ambiguity in correctly determining the true asymmetric orientation. As for icosahedral reconstructions, part of this problem will be solved as the systems that automate image acquisition become mainstream. However, while it has been assumed that software can always determine the true asymmetric orientation of each particle image for symmetry-free reconstructions, there is no guarantee that this orientation is correct. Maximizing the number of images where a true asymmetric orientation can be unambiguously identified will require more sensitive algorithms and possibly advanced EM technology. Nevertheless, regardless of any improvement made in data collection and image processing, structural homogeneity is a requirement, a reality that should be considered for any specimen before attempting high-resolution asymmetric reconstruction.

4.6.3 CTF Correction in Cryo-ET

Some viruses may contain structurally homogeneous components that are not uniformly distributed, such as the glycoproteins on the envelopes of HSV and HIV. While these components have been reconstructed using cryo-ET, these results have been limited to ~20 Å resolution (Liu et al. 2008b), due largely in part to the defocus gradient in tilted images as well as limitations in electron exposure. For cryo-ET, the difficulty in determining the defocus of the images (a major contributor to the CTF) is a result of the low dose per image, making correction of the CTF difficult. While potential solutions to this problem have been proposed by correcting the CTF along strips of data parallel to the tilt axis (Fernandez et al. 2006; Winkler and Taylor 2003; Xiong et al. 2009), these methods have not yet been widely adopted in the field. In the event that the CTF can be accurately corrected for cryo-ET, it will become possible to process the data extracted from these volumes to higher resolutions and potentially into the subnanometer range.

4.6.4 Model Validation

Modeling cryo-EM or cryo-ET maps is an essential pipeline in the methodology described here. When used properly, these tools can provide amazing insight into the structure and function of individual proteins, macromolecular complexes, organelles, and even cells as illustrated in recent literature. However, the accuracy with which one can analyze a density map is a function of the map itself. Size, complexity, and quality all play critical roles in the ability to analyze a density map regardless of resolution. Another important consideration comes from the need to understand the limitations imposed by resolution when annotating structural features. While one would not expect to see side-chain density or strand separation at 9 Å resolution, the absence of such features at 3.5 Å resolution may indicate potential problems in the reconstruction. It is certain that the development of computational tools for modeling and annotating remains a necessity in order to make this process faster and the results more accurate.

4.6.5 Phase Plate Electron Microscopy

With current technology, there is a trade-off between high-contrast and high-resolution information. One promising technology to achieve both is phase plate electron microscopy, which was initially proposed nearly 65 years ago (Boersch 1947). One advantage of a phase plate electron microscope is that it dramatically enhances the low-resolution contrast of cryo-EM images because of the conversion of the CTF from a sine to a cosine function. The theory behind phase plate electron microscopy is that, by altering the electron beam, it is possible to shift the phase of the scattered electron beam, relative to the unscattered beam, thus dramatically enhancing the image contrast. Many devices have been proposed to induce this phase shift (Cambie et al. 2007; Danev and Nagayama 2001; Thon 1971), and Zernike-type phase plates have already begun to show very promising results for cryo-EM (Danev and Nagayama 2008; Murata et al. 2010; Rochat 2011). While there are many technical challenges to implementing this type of device, preliminary results suggest that it may be possible to reduce the data required for a subnanometer reconstruction by a factor of 3 (Murata et al. 2010). Zernike phase contrast cryo-EM has also proven to be valuable for cryo-ET. For components that have proven difficult to study by conventional cryo-EM due to the weak signal from the low-resolution component of the particle image, phase contrast microscopy may provide the necessary means to resolve their structures at resolutions beyond the current limit.

4.7 Conclusion

Cryo-EM, cryo-ET, and image processing software have been used to determine the structures of many viruses. This includes not only the icosahedral capsids but also portals, tails, glycoproteins, and other components without icosahedral symmetry. Analyses of these results have provided virologists and structural biologists with valuable insight into how viruses work. These structural studies have been enormously enriched by the use of the techniques described above that have developed over the past 40 years. Furthermore, the future holds enormous potential for visualizing viruses interacting with their host cells, which will extend our picture of the life cycles of these fascinating biological systems and make their structural investigation of interest to an even wider community.

Acknowledgments JC, XL, RHR and MLB contributed equally to this work. This work has been supported by grants from NIH (P41RR002250, R01AI0175208, and R01GM079429) and Robert Welch Foundation (Q1242). RHR is supported by the NIH training grants (GM07330 through the MSTP, T15LM007093 through the Gulf Coast Consortia). We thank Dr. Frazer Rixon for assistance with manuscript preparation.

References

Abeysinghe SS, Baker ML, Chiu W, Ju T (2008) Segmentation-free skeletonization of grayscale volumes for shape understanding. In: IEEE international conference on shape modeling and applications, Stony Brook, NY

Abeysinghe S, Baker ML, Chiu W, Ju T (2010) Semi-isometric registration of line features for flexible fitting of protein structures. Comput Graph Forum 29:2243–2252

Adrian M, Dubochet J, Lepault J, McDowall AW (1984) Cryo-electron microscopy of viruses. Nature 308:32–36

Agirrezabala X, Martín-Benito J, Castón JR, Miranda R, Valpuesta JM, Carrascosa JL (2005) Maturation of phage T7 involves structural modification of both shell and inner core components. EMBO J 24:3820–3829

Al-Amoudi A, Chang J-J, Leforestier Al, McDowall A, Salamin LM, Norlén LPO, Richter K, Blanc NS, Studer D, Dubochet J (2004) Cryo-electron microscopy of vitreous sections. EMBO J 23:3583–3588

Amat F, Moussavi F, Comolli LR, Elidan G, Downing KH, Horowitz M (2008) Markov random field based automatic image alignment for electron tomography. J Struct Biol 161:260–275

Baker D (2006) Prediction and design of macromolecular structures and interactions. Philos Trans R Soc Lond B Biol Sci 361:459–463

Baker TS, Cheng RH (1996) A model-based approach for determining orientations of biological macromolecules imaged by cryoelectron microscopy. J Struct Biol 116:120–130

Baker ML, Jiang W, Bowman BR, Zhou ZH, Quiocho FA, Rixon FJ, Chiu W (2003) Architecture of the herpes simplex virus major capsid protein derived from structural bioinformatics. J Mol Biol 331:447–456

Baker ML, Jiang W, Rixon FJ, Chiu W (2005) Common ancestry of herpesviruses and tailed DNA bacteriophages. J Virol 79:14967–14970

Baker ML, Jiang W, Wedemeyer WJ, Rixon FJ, Baker D, Chiu W (2006a) Ab initio modeling of the herpesvirus VP26 core domain assessed by CryoEM density. PLoS Comput Biol 2:e146

Baker ML, Yu Z, Chiu W, Bajaj C (2006b) Automated segmentation of molecular subunits in electron cryomicroscopy density maps. J Struct Biol 156:432–441

Baker ML, Ju T, Chiu W (2007) Identification of secondary structure elements in intermediate-resolution density maps. Structure 15:7–19

Baker ML, Baker MR, Hryc CF, Dimaio F (2010a) Analyses of subnanometer resolution cryo-EM density maps. Methods Enzymol 483:1–29

Baker ML, Zhang J, Ludtke SJ, Chiu W (2010b) Cryo-EM of macromolecular assemblies at near-atomic resolution. Nat Protoc 5:1697–1708

Baker ML, Abeysinghe SS, Schuh S, Coleman RA, Abrams A, Marsh MP, Hryc CF, Ruths T, Chiu W, Ju T (2011) Modeling protein structure at near atomic resolutions with Gorgon. J Struct Biol 174:360–373

Bammes BE, Jakana J, Schmid MF, Chiu W (2010) Radiation damage effects at four specimen temperatures from 4 to 100 K. J Struct Biol 169:331–341

Bartesaghi A, Sprechmann P, Liu J, Randall G, Sapiro G, Subramaniam S (2008) Classification and 3D averaging with missing wedge correction in biological electron tomography. J Struct Biol 162:436–450

Boersch H (1947) Über die Kontraste von Atomen im Elektronenmikroskop. Z Naturforsch 2a:615–633

Bostina M, Levy H, Filman DJ, Hogle JM (2011) Poliovirus RNA is released from the capsid near a twofold symmetry axis. J Virol 85:776–783

Böttcher B, Wynne SA, Crowther RA (1997) Determination of the fold of the core protein of hepatitis B virus by electron cryomicroscopy. Nature 386:88–91

Bradley P, Misura KM, Baker D (2005) Toward high-resolution de novo structure prediction for small proteins. Science 309:1868–1871

Brandt S, Heikkonen J, Engelhardt P (2001a) Automatic alignment of transmission electron microscope tilt series without fiducial markers. J Struct Biol 136:201–213

Brandt S, Heikkonen J, Engelhardt P (2001b) Multiphase method for automatic alignment of transmission electron microscope images using markers. J Struct Biol 133:10–22

Cambie R, Downing KH, Typke D, Glaeser RM, Jin J (2007) Design of a microfabricated, two-electrode phase-contrast element suitable for electron microscopy. Ultramicroscopy 107:329–339

Carragher B, Kisseberth N, Kriegman D, Milligan RA, Potter CS, Pulokas J, Reilein A (2000) Leginon: an automated system for acquisition of images from vitreous ice specimens. J Struct Biol 132:33–45

Castano-Diez D, Al-Amoudi A, Glynn AM, Seybert A, Frangakis AS (2007) Fiducial-less alignment of cryo-sections. J Struct Biol 159:413–423

Castano-Diez D, Scheffer M, Al-Amoudi A, Frangakis AS (2010) Alignator: a GPU powered software package for robust fiducial-less alignment of cryo tilt-series. J Struct Biol 170:117–126

Chang J, Weigele P, King J, Chiu W, Jiang W (2006) Cryo-EM asymmetric reconstruction of bacteriophage P22 reveals organization of its DNA packaging and infecting machinery. Structure 14:1073–1082

Chang JT, Schmid MF, Haase-Pettingell C, Weigele PR, King JA, Chiu W (2010) Visualizing the structural changes of bacteriophage epsilon15 and its Salmonella host during infection. J Mol Biol 402:731–740

Chen B, Vogan EM, Gong H, Skehel JJ, Wiley DC, Harrison SC (2005) Structure of an unliganded simian immunodeficiency virus gp120 core. Nature 433:834–841

Chen D-H, Luke K, Zhang J, Chiu W, Wittung-Stafshede P (2008) Location and flexibility of the unique C-terminal tail of Aquifex aeolicus co-chaperonin protein 10 as derived by cryo-electron microscopy and biophysical techniques. J Mol Biol 381:707–717

Chen JZ, Settembre EC, Aoki ST, Zhang X, Bellamy AR, Dormitzer PR, Harrison SC, Grigorieff N (2009) Molecular interactions in rotavirus assembly and uncoating seen by high-resolution cryo-EM. Proc Natl Acad Sci USA 106:10644–10648

Chen D-H, Baker ML, Hryc CF, DiMaio F, Jakana J, Wu W, Dougherty M, Haase-Pettingell C, Schmid MF, Jiang W et al (2011) Structural basis for scaffolding-mediated assembly and maturation of a dsDNA virus. Proc Natl Acad Sci USA 108:1355–1360

Cheng L, Sun J, Zhang K, Mou Z, Huang X, Ji G, Sun F, Zhang J, Zhu P (2011) Atomic model of a cypovirus built from cryo-EM structure provides insight into the mechanism of mRNA capping. Proc Natl Acad Sci USA 108:1373–1378

Chiu W (1986) Electron microscopy of frozen, hydrated biological specimens. Ann Rev Biophys Biophys Chem 15:237–257

Chiu W, Glaeser RM (1977) Factors affecting high resolution fixed-beam transmission electron microscopy. Ultramicroscopy 2:207–217

Chiu W, Baker ML, Jiang W, Dougherty M, Schmid MF (2005) Electron cryomicroscopy of biological machines at subnanometer resolution. Structure 13:363–372

Comolli LR, Downing KH (2005) Dose tolerance at helium and nitrogen temperatures for whole cell electron tomography. J Struct Biol 152:149–156

Cong Y, Ludtke SJ (2010) Single particle analysis at high resolution. Methods Enzymol 482:211–235

Cong Y, Kovacs JA, Wriggers W (2003) 2D fast rotational matching for image processing of biophysical data. J Struct Biol 144:51–60

Conway JF, Cheng N, Zlotnick A, Wingfield PT, Stahl SJ, Steven AC (1997) Visualization of a 4-helix bundle in the hepatitis B virus capsid by cryo-electron microscopy. Nature 386:91–94

Crowther RA (1971) Procedures for three-dimensional reconstruction of spherical viruses by Fourier synthesis from electron micrographs. Philos Trans R Soc Lond B Biol Sci 261:221–230

Crowther RA, Amos LA, Finch JT, De Rosier DJ, Klug A (1970a) Three dimensional reconstructions of spherical viruses by Fourier synthesis from electron micrographs. Nature 226:421–425

Crowther RA, DeRosier DJ, Klug A (1970b) The reconstruction of a three-dimensional structure from projections and its application to electron microscopy. Proc R Soc Lond A 317:319–340

Cyrklaff M, Linaroudis A, Boicu M, Chlanda P, Baumeister W, Griffiths G, Krijnse-Locker J (2007) Whole cell cryo-electron tomography reveals distinct disassembly intermediates of vaccinia virus. PLoS One 2:e420

Danev R, Nagayama K (2001) Transmission electron microscopy with Zernike phase plate. Ultramicroscopy 88:243–252

Danev R, Nagayama K (2008) Single particle analysis based on Zernike phase contrast transmission electron microscopy. J Struct Biol 161:211–218

Deptuch G, Besson A, Rehak P, Szelezniak M, Wall J, Winter M, Zhu Y (2007) Direct electron imaging in electron microscopy with monolithic active pixel sensors. Ultramicroscopy 107:674–684

DeRosier DJ (2000) Correction of high-resolution data for curvature of the Ewald sphere. Ultramicroscopy 81:83–98

DeRosier DL, Klug A (1968) Reconstruction of three-dimensional structures from electron micrographs. Nature 217:130–134

DiMaio F, Tyka MD, Baker ML, Chiu W, Baker D (2009) Refinement of protein structures into low-resolution density maps using rosetta. J Mol Biol 392:181–190

DiMaio F, Terwilliger TC, Read RJ, Wlodawer A, Oberdorfer G, Wagner U, Valkov E, Alon A, Fass D, Axelrod HL et al (2011) Improved molecular replacement by density- and energy-guided protein structure optimization. Nature 473(7348):540–543

Dubochet J, Adrian M, Chang JJ, Homo JC, Lepault J, McDowall AW, Schultz P (1988) Cryo-electron microscopy of vitrified specimens. Q Rev Biophys 21:129–228

Emsley P, Cowtan K (2004) Coot: model-building tools for molecular graphics. Acta Crystallogr D Biol Crystallogr 60:2126–2132

Erickson HP, Klug A (1971) Measurement and compensation of defocusing and aberrations by Fourier processing of electron micrographs. Phil Trans R Soc Lond B 261:105–118

Fernandez JJ, Li S, Crowther RA (2006) CTF determination and correction in electron cryotomography. Ultramicroscopy 106:587–596

Fernandez JJ, Luque D, Caston JR, Carrascosa JL (2008) Sharpening high resolution information in single particle electron cryomicroscopy. J Struct Biol 164:170–175

Forster F, Pruggnaller S, Seybert A, Frangakis AS (2008) Classification of cryo-electron sub-tomograms using constrained correlation. J Struct Biol 161:276–286

Förster F, Medalia O, Zauberman N, Baumeister W, Fass D (2005) Retrovirus envelope protein complex structure in situ studied by cryo-electron tomography. Proc Natl Acad Sci USA 102:4729–4734

Fu C-Y, Wang K, Gan L, Lanman J, Khayat R, Young MJ, Jensen GJ, Doerschuk PC, Johnson JE (2010) In vivo assembly of an archaeal virus studied with whole-cell electron cryotomography. Structure 18:1579–1586

Fuller SD, Butcher SJ, Cheng RH, Baker TS (1996) Three-dimensional reconstruction of icosahedral particles – the uncommon line. J Struct Biol 116:48–55

Gilbert PF (1972) The reconstruction of a three-dimensional structure from projections and its application to electron microscopy. II. Direct methods. Proc R Soc Lond B Biol Sci 182:89–102

Glaeser RM, Downing KH, DeRosier D, Chiu W, Frank J (2007) Electron crystallography of biological macromolecules. Oxford University Press, New York

Grigorieff N, Harrison SC (2011) Near-atomic resolution reconstructions of icosahedral viruses from electron cryo-microscopy. Curr Opin Struct Biol 21:265–273

Hendrix RW (1978) Symmetry mismatch and DNA packaging in large bacteriophages. Proc Natl Acad Sci USA 75:4779–4783

Heymann JB, Cardone G, Winkler DC, Steven AC (2008) Computational resources for cryo-electron tomography in Bsoft. J Struct Biol 161:232–242

Ho MH, Jap BK, Glaeser RM (1988) Validity domain of the weak-phase-object approximation for electron diffraction of thin protein crystals. Acta Crystallogr A 44(Pt 6):878–884

Hoenger A, Bouchet-Marquis C (2011) Cellular tomography. Adv Protein Chem Struct Biol 82:67–90

Hryc CF, Chen DH, Chiu W (2011) Near-atomic-resolution cryo-EM for molecular virology. Curr Opin Virol 1(2):110–117

Humphrey W, Dalke A, Schulten K (1996) VMD: visual molecular dynamics. J Mol Graph 14:33–38, 27–38
Iancu CV, Wright ER, Heymann JB, Jensen GJ (2006) A comparison of liquid nitrogen and liquid helium as cryogens for electron cryotomography. J Struct Biol 153:231–240
Jiang W, Chiu W (2007) Cryoelectron microscopy of icosahedral virus particles. Methods Mol Biol 369:345–363
Jiang W, Baker ML, Ludtke SJ, Chiu W (2001a) Bridging the information gap: computational tools for intermediate resolution structure interpretation. J Mol Biol 308:1033–1044
Jiang W, Li Z, Zhang Z, Booth CR, Baker ML, Chiu W (2001b) Semi-automated icosahedral particle reconstruction at sub-nanometer resolution. J Struct Biol 136:214–225
Jiang W, Chang J, Jakana J, Weigele P, King J, Chiu W (2006) Structure of epsilon15 bacteriophage reveals genome organization and DNA packaging/injection apparatus. Nature 439:612–616
Jiang W, Baker ML, Jakana J, Weigele PR, King J, Chiu W (2008) Backbone structure of the infectious epsilon15 virus capsid revealed by electron cryomicroscopy. Nature 451:1130–1134
Johnson, J.E. & Fisher, AJ (1994) Principles of virus structure. In Encyclopedia of Virology (Webster, R.G. & Granoff, A., eds.), London, Academic Press, 1573–1586
Jones TA, Zou JY, Cowan SW, Kjeldgaard M (1991) Improved methods for building protein models in electron density maps and the location of errors in these models. Acta Crystallogr A 47(Pt 2):110–119
Joyeux L, Penczek PA (2002) Efficiency of 2D alignment methods. Ultramicroscopy 92:33–46
Ju T, Baker ML, Chiu W (2007) Computing a family of skeletons of volumetric models for shape description. Comput Aided Des 39:352–360
Kleywegt GJ, Jones TA (1997) Detecting folding motifs and similarities in protein structures. Methods Enzymol 277:525–545
Kong Y, Ma J (2003) A structural-informatics approach for mining beta-sheets: locating sheets in intermediate-resolution density maps. J Mol Biol 332:399–413
Kong Y, Zhang X, Baker TS, Ma J (2004) A structural-informatics approach for tracing beta-sheets: building pseudo-C(alpha) traces for beta-strands in intermediate-resolution density maps. J Mol Biol 339:117–130
Kremer JR, Mastronarde DN, McIntosh JR (1996) Computer visualization of three-dimensional image data using IMOD. J Struct Biol 116:71–76
Lakshminarayanan AV, Lent A (1979) Methods of least squares and SIRT in reconstruction. J Theor Biol 76:267–295
Lander GC, Tang L, Casjens SR, Gilcrease EB, Prevelige P, Poliakov A, Potter CS, Carragher B, Johnson JE (2006) The structure of an infectious P22 virion shows the signal for headful DNA packaging. Science 312:1791–1795
Lawrence A, Bouwer JC, Perkins G, Ellisman MH (2006) Transform-based backprojection for volume reconstruction of large format electron microscope tilt series. J Struct Biol 154:144–167
Leiman PG, Battisti AJ, Bowman VD, Stummeyer K, Muhlenhoff M, Gerardy-Schahn R, Scholl D, Molineux IJ (2007) The structures of bacteriophages K1E and K1-5 explain processive degradation of polysaccharide capsules and evolution of new host specificities. J Mol Biol 371:836–849
Leiman PG, Arisaka F, van Raaij MJ, Kostyuchenko VA, Aksyuk AA, Kanamaru S, Rossmann MG (2010) Morphogenesis of the T4 tail and tail fibers. Virol J 7:355
Leong PA, Yu X, Zhou ZH, Jensen GJ (2010) Correcting for the Ewald sphere in high-resolution single-particle reconstructions. Methods Enzymol 482:369–380
Liang Y, Ke EY, Zhou ZH (2002) IMIRS: a high-resolution 3D reconstruction package integrated with a relational image database. J Struct Biol 137:292–304
Liu Y, Penczek PA, McEwen BF, Frank J (1995) A marker-free alignment method for electron tomography. Ultramicroscopy 58:393–402
Liu X, Jiang W, Jakana J, Chiu W (2007) Averaging tens to hundreds of icosahedral particle images to resolve protein secondary structure elements using a Multi-Path Simulated Annealing optimization algorithm. J Struct Biol 160:11–27
Liu H, Cheng L, Zeng S, Cai C, Zhou ZH, Yang Q (2008a) Symmetry-adapted spherical harmonics method for high-resolution 3D single-particle reconstructions. J Struct Biol 161:64–73
Liu J, Bartesaghi A, Borgnia MJ, Sapiro G, Subramaniam S (2008b) Molecular architecture of native HIV-1 gp120 trimers. Nature 455:109–113
Liu H, Jin L, Koh SB, Atanasov I, Schein S, Wu L, Zhou ZH (2010a) Atomic structure of human adenovirus by cryo-EM reveals interactions among protein networks. Science 329:1038–1043
Liu X, Rochat RH, Chiu W (2010b) Reconstructing cyano-bacteriophage P-SSP7 structure without imposing symmetry. Nat Protoc. doi:10.1038/nprot.2010.1096
Liu X, Zhang Q, Murata K, Baker ML, Sullivan MB, Fu C, Dougherty MT, Schmid MF, Osburne MS, Chisholm SW, Chiu W (2010c) Structural changes in a marine podovirus associated with release of its genome into Prochlorococcus. Nat Struct Mol Biol 17:830–836
Ludtke SJ, Baldwin PR, Chiu W (1999) EMAN: semiautomated software for high-resolution single-particle reconstructions. J Struct Biol 128:82–97

Ludtke SJ, Baker ML, Chen DH, Song JL, Chuang DT, Chiu W (2008) De Novo backbone trace of GroEL from single particle electron cryomicroscopy. Structure 16:441–448

Ludtke SJ, Tran TP, Ngo QT, Moiseenkova-Bell VY, Chiu W, Serysheva I (2011) Flexible architecture of IP3R1 by cryo-EM. Structure 19(8):1192–1199

Marabini R, Herman GT, Carazo JM (1998) 3D reconstruction in electron microscopy using ART with smooth spherically symmetric volume elements (blobs). Ultramicroscopy 72:53–65

Mastronarde DN (1997) Dual-axis tomography: an approach with alignment methods that preserve resolution. J Struct Biol 120:343–352

McDowall AW, Chang JJ, Freeman R, Lepault J, Walter CA, Dubochet J (1983) Electron microscopy of frozen hydrated sections of vitreous ice and vitrified biological samples. J Microsc 131:1–9

McMullan G, Chen S, Henderson R, Faruqi AR (2009) Detective quantum efficiency of electron area detectors in electron microscopy. Ultramicroscopy 109:1126–1143

Milazzo AC, Leblanc P, Duttweiler F, Jin L, Bouwer JC, Peltier S, Ellisman M, Bieser F, Matis HS, Wieman H et al (2005) Active pixel sensor array as a detector for electron microscopy. Ultramicroscopy 104:152–159

Milazzo AC, Moldovan G, Lanman J, Jin L, Bouwer JC, Klienfelder S, Peltier ST, Ellisman MH, Kirkland AI, Xuong NH (2010) Characterization of a direct detection device imaging camera for transmission electron microscopy. Ultramicroscopy 110:744–747

Mindell JA, Grigorieff N (2003) Accurate determination of local defocus and specimen tilt in electron microscopy. J Struct Biol 142:334–347

Miyazaki Y, Irobalieva RN, Tolbert BS, Smalls-Mantey A, Iyalla K, Loeliger K, D'Souza V, Khant H, Schmid MF, Garcia EL et al (2010) Structure of a conserved retroviral RNA packaging element by NMR spectroscopy and cryo-electron tomography. J Mol Biol 404:751–772

Mizuguchi K, Go N (1995) Comparison of spatial arrangements of secondary structural elements in proteins. Protein Eng 8:353–362

Murata K, Liu X, Danev R, Jakana J, Schmid MF, King J, Nagayama K, Chiu W (2010) Zernike phase contrast cryo-electron microscopy and tomography for structure determination at nanometer and subnanometer resolutions. Structure 18:903–912

Nakamura N, Shimizu Y, Shinkawa T, Nakata M, Bammes B, Zhang J, Chiu W (2010) Automated specimen search in cryo-TEM observation with DIFF-defocus imaging. J Electron Microsc 59:299–310

Nicastro D, Schwartz C, Pierson J, Gaudette R, Porter ME, McIntosh JR (2006) The molecular architecture of axonemes revealed by cryoelectron tomography. Science 313:944–948

Nickell S, Förster F, Linaroudis A, Net WD, Beck F, Hegerl R, Baumeister W, Plitzko JM (2005) TOM software toolbox: acquisition and analysis for electron tomography. J Struct Biol 149:227–234

Penczek PA, Yang C, Frank J, Spahn CM (2006) Estimation of variance in single-particle reconstruction using the bootstrap technique. J Struct Biol 154:168–183

Pettersen EF, Goddard TD, Huang CC, Couch GS, Greenblatt DM, Meng EC, Ferrin TE (2004) UCSF Chimera – a visualization system for exploratory research and analysis. J Comput Chem 25:1605–1612

Pintilie GD, Zhang J, Goddard TD, Chiu W, Gossard DC (2010) Quantitative analysis of cryo-EM density map segmentation by watershed and scale-space filtering, and fitting of structures by alignment to regions. J Struct Biol 170:427–438

Potter CS, Chu H, Frey B, Green C, Kisseberth N, Madden TJ, Miller KL, Nahrstedt K, Pulokas J, Reilein A et al (1999) Leginon: a system for fully automated acquisition of 1000 electron micrographs a day. Ultramicroscopy 77:153–161

Reddy VS, Natchiar SK, Stewart PL, Nemerow GR (2010) Crystal structure of human adenovirus at 3.5 A resolution. Science 329:1071–1075

Rochat RH, Liu X, Murata K, Nagayama K, Rixon FJ, Chiu W (2011) Seeing the portal in herpes simplex virus type 1 B capsids. J Virol 85:1871–1874

Roseman AM (2000) Docking structures of domains into maps from cryo-electron microscopy using local correlation. Acta Crystallogr D Biol Crystallogr 56:1332–1340

Rosenthal PB, Henderson R (2003) Optimal determination of particle orientation, absolute hand, and contrast loss in single-particle electron cryomicroscopy. J Mol Biol 333:721–745

Rossmann MG (2000) Fitting atomic models into electron-microscopy maps. Acta Crystallogr D Biol Crystallogr 56:1341–1349

Rossmann MG, Morais MC, Leiman PG, Zhang W (2005) Combining X-ray crystallography and electron microscopy. Structure 13:355–362

Rusu M, Birmanns S, Wriggers W (2008) Biomolecular pleiomorphism probed by spatial interpolation of coarse models. Bioinformatics 24:2460–2466

Saad A, Ludtke SJ, Jakana J, Rixon FJ, Tsuruta H, Chiu W (2001) Fourier amplitude decay of electron cryomicroscopic images of single particles and effects on structure determination. J Struct Biol 133:32–42

Saxton WO, Baumeister W, Hahn M (1984) Three-dimensional reconstruction of imperfect two-dimensional crystals. Ultramicroscopy 13:57–70
Schmid MF (2011) Single-particle electron cryotomography (cryoET). Adv Protein Chem Struct Biol 82:37–65
Schmid MF, Booth CR (2008) Methods for aligning and for averaging 3D volumes with missing data. J Struct Biol 161:243–248
Schmid MF, Prasad BV, Chiu W (1994) Structural studies of viruses by electron cryomicroscopy. Arch Virol Suppl 9:523–529
Schmid MF, Sherman MB, Matsudaira P, Tsuruta H, Chiu W (1999) Scaling structure factor amplitudes in electron cryomicroscopy using X-Ray solution scattering. J Struct Biol 128:51–57
Schmid MF, Paredes AM, Khant HA, Soyer F, Aldrich HC, Chiu W, Shively JM (2006) Structure of Halothiobacillus neapolitanus carboxysomes by cryo-electron tomography. J Mol Biol 364:526–535
Schröder GF, Brunger AT, Levitt M (2007) Combining efficient conformational sampling with a deformable elastic network model facilitates structure refinement at low resolution. Structure 15:1630–1641
Settembre EC, Chen JZ, Dormitzer PR, Grigorieff N, Harrison SC (2011) Atomic model of an infectious rotavirus particle. EMBO J 30:408–416
Shaikh TR, Gao H, Baxter WT, Asturias FJ, Boisset N, Leith A, Frank J (2008) SPIDER image processing for single-particle reconstruction of biological macromolecules from electron micrographs. Nat Protoc 3:1941–1974
Siebert X, Navaza J (2009) UROX 2.0: an interactive tool for fitting atomic models into electron-microscopy reconstructions. Acta Crystallogr D Biol Crystallogr 65:651–658
Sorzano CO, Messaoudi C, Eibauer M, Bilbao-Castro JR, Hegerl R, Nickell S, Marco S, Carazo JM (2009) Marker-free image registration of electron tomography tilt-series. BMC Bioinformatics 10:124
Stagg SM, Lander GC, Pulokas J, Fellmann D, Cheng A, Quispe JD, Mallick SP, Avila RM, Carragher B, Potter CS (2006) Automated cryoEM data acquisition and analysis of 284742 particles of GroEL. J Struct Biol 155:470–481
Stark H, Woods J, Paul I, Hingorania R (1981) Direct Fourier reconstruction in computer tomography. IEEE Trans Acoust Speech Signal Process 29:237–245
Suhre K, Navaza J, Sanejouand YH (2006) NORMA: a tool for flexible fitting of high-resolution protein structures into low-resolution electron-microscopy-derived density maps. Acta Crystallogr D Biol Crystallogr 62:1098–1100
Tama F, Miyashita O, Brooks CL (2004) Normal mode based flexible fitting of high-resolution structure into low-resolution experimental data from cryo-EM. J Struct Biol 147:315–326
Tan RKZ, Devkota B, Harvey SC (2008) YUP.SCX: coaxing atomic models into medium resolution electron density maps. J Struct Biol 163:163–174
Tang G, Peng L, Baldwin PR, Mann DS, Jiang W, Rees I, Ludtke SJ (2007) EMAN2: an extensible image processing suite for electron microscopy. J Struct Biol 157:38–46
Tang J, Olson N, Jardine PJ, Grimes S, Anderson DL, Baker TS (2008) DNA poised for release in bacteriophage phi29. Structure 16:935–943
Taylor KA, Glaeser RM (1974) Electron diffraction of frozen, hydrated protein crystals. Science 186:1036–1037
Taylor KA, Glaeser RM (1976) Electron microscopy of frozen hydrated biological specimens. J Ultrastruct Res 55:448–456
Thon F (1971) Phase contrast electron microscopy. In: Valdre U (ed) Electron microscopy in material sciences. New York, Academic, pp 571–625
Topf M, Sali A (2005) Combining electron microscopy and comparative protein structure modeling. Curr Opin Struct Biol 15:578–585
Topf M, Baker ML, John B, Chiu W, Sali A (2005) Structural characterization of components of protein assemblies by comparative modeling and electron cryo-microscopy. J Struct Biol 149:191–203
Topf M, Baker ML, Marti-Renom MA, Chiu W, Sali A (2006) Refinement of protein structures by iterative comparative modeling and cryoEM density fitting. J Mol Biol 357:1655–1668
Topf M, Lasker K, Webb B, Wolfson H, Chiu W, Sali A (2008) Protein structure fitting and refinement guided by cryo-EM density. Structure 16:295–307
Trabuco LG, Villa E, Schreiner E, Harrison CB, Schulten K (2009) Molecular dynamics flexible fitting: a practical guide to combine cryo-electron microscopy and X-ray crystallography. Methods 49:174–180
Trabuco LG, Schreiner E, Eargle J, Cornish P, Ha T, Luthey-Schulten Z, Schulten K (2010) The role of L1 stalk-tRNA interaction in the ribosome elongation cycle. J Mol Biol 402:741–760
Unwin PN, Henderson R (1975) Molecular structure determination by electron microscopy of unstained crystalline specimens. J Mol Biol 94:425–440
Volkmann N (2002) A novel three-dimensional variant of the watershed transform for segmentation of electron density maps. J Struct Biol 138:123–129
Volkmann N, Hanein D (1999) Quantitative fitting of atomic models into observed densities derived by electron microscopy. J Struct Biol 125:176–184

Walz J, Typke D, Nitsch M, Koster AJ, Hegerl R, Baumeister W (1997) Electron tomography of single ice-embedded macromolecules: three-dimensional alignment and classification. J Struct Biol 120:387–395
Winkler H (2007) 3D reconstruction and processing of volumetric data in cryo-electron tomography. J Struct Biol 157:126–137
Winkler H, Taylor KA (2003) Focus gradient correction applied to tilt series image data used in electron tomography. J Struct Biol 143:24–32
Winkler H, Taylor KA (2006) Accurate marker-free alignment with simultaneous geometry determination and reconstruction of tilt series in electron tomography. Ultramicroscopy 106:240–254
Wolf M, Garcea RL, Grigorieff N, Harrison SC (2010) Subunit interactions in bovine papillomavirus. Proc Natl Acad Sci USA 107:6298–6303
Wriggers W, Milligan RA, McCammon JA (1999) Situs: A package for docking crystal structures into low-resolution maps from electron microscopy. J Struct Biol 125:185–195
Wright ER, Iancu CV, Tivol WF, Jensen GJ (2006) Observations on the behavior of vitreous ice at approximately 82 and approximately 12 K. J Struct Biol 153:241–252
Xiong Q, Morphew MK, Schwartz CL, Hoenger AH, Mastronarde DN (2009) CTF determination and correction for low dose tomographic tilt series. J Struct Biol 168:378–387
Yan X, Sinkovits RS, Baker TS (2007) AUTO3DEM – an automated and high throughput program for image reconstruction of icosahedral particles. J Struct Biol 157:73–82
Yu X, Jin L, Zhou ZH (2008) 3.88 A structure of cytoplasmic polyhedrosis virus by cryo-electron microscopy. Nature 453:415–419
Zanetti G, Briggs JA, Grunewald K, Sattentau QJ, Fuller SD (2006) Cryo-electron tomographic structure of an immunodeficiency virus envelope complex in situ. PLoS Pathog 2:e83
Zhang X, Settembre E, Xu C, Dormitzer PR, Bellamy R, Harrison SC, Grigorieff N (2008) Near-atomic resolution using electron cryomicroscopy and single-particle reconstruction. Proc Natl Acad Sci USA 105:1867–1872
Zhang J, Nakamura N, Shimizu Y, Liang N, Liu X, Jakana J, Marsh MP, Booth CR, Shinkawa T, Nakata M, Chiu W (2009) JADAS: a customizable automated data acquisition system and its application to ice-embedded single particles. J Struct Biol 165:1–9
Zhang J, Baker ML, Schröder GF, Douglas NR, Reissmann S, Jakana J, Dougherty M, Fu CJ, Levitt M, Ludtke SJ et al (2010a) Mechanism of folding chamber closure in a group II chaperonin. Nature 463:379–383
Zhang X, Jin L, Fang Q, Hui WH, Zhou ZH (2010b) 3.3 A cryo-EM structure of a nonenveloped virus reveals a priming mechanism for cell entry. Cell 141:472–482
Zheng SQ, Keszthelyi B, Branlund E, Lyle JM, Braunfeld MB, Sedat JW, Agard DA (2007) UCSF tomography: an integrated software suite for real-time electron microscopic tomographic data collection, alignment, and reconstruction. J Struct Biol 157:138–147
Zhou ZH (2011) Atomic resolution cryo electron microscopy of macromolecular complexes. Adv Protein Chem Struct Biol 82:1–35
Zhou ZH, Chiu W (2003) Determination of icosahedral virus structures by electron cryomicroscopy at subnanometer resolution. Adv Protein Chem 64:93–124
Zhou ZH, Chiu W, Haskell K, Spears H Jr, Jakana J, Rixon FJ, Scott LR (1998) Refinement of herpesvirus B-capsid structure on parallel supercomputers. Biophys J 74:576–588
Zhou ZH, Dougherty M, Jakana J, He J, Rixon FJ, Chiu W (2000) Seeing the herpesvirus capsid at 8.5 Å. Science 288:877–880
Zhou ZH, Baker ML, Jiang W, Dougherty M, Jakana J, Dong G, Lu G, Chiu W (2001) Electron cryomicroscopy and bioinformatics suggest protein fold models for rice dwarf virus. Nat Struct Biol 8:868–873
Zhu J, Cheng L, Fang Q, Zhou ZH, Honig B (2010) Building and refining protein models within cryo-electron microscopy density maps based on homology modeling and multiscale structure refinement. J Mol Biol 397:835–851

Part I
Viral Entry Machines

Chapter 5
Contractile Tail Machines of Bacteriophages

Petr G. Leiman and Mikhail M. Shneider

Abstract Bacteriophages with contractile tails epitomize the concepts of "virus" and "phage" for many because the tails of these phages undergo a large conformational change – resembling the action of a syringe – upon the attachment to the host cell. The contractile tails belong to the recently recognized class of "contractile systems," which includes phage tails, their close relatives R-type pyocins, the bacterial type VI secretion system, and the virulence cassette of *Photorhabdus*. Their function is to deliver large proteins and/or DNA into the cytoplasm of a bacterial or eukaryotic cell. The structure of the core components of all contractile tail-like systems is conserved, but the corresponding genes have diverged to such a degree that the common ancestry can no longer be easily detected at the level of amino acid sequence. At present, it is unclear, whether the contractile systems originated in bacteria or in phages. This chapter describes the structure and function of phage contractile tails and compares them with other phage tails and with other known contractile systems.

5.1 Introduction

The host-cell attachment organelle of bacteriophages – the tail – is a complex macromolecular machine, which is responsible for host-cell recognition, attachment, and cell envelope penetration. The tail forms a channel connecting the phage capsid with the host cell during infection. Phage genomic DNA and proteins, which are packaged in the capsid, are translocated into the host-cell cytoplasm through this channel.

At present, all tailed phages (the order *Caudovirales*) are divided into three families based on the tail morphology: *Myoviridae* (long contractile tail), *Siphoviridae* (long noncontractile tail), and *Podoviridae* (short noncontractile tail) (Ackermann 2003). This classification is based on the visual appearance of the tail and does not take into account the life cycle of the phage, its genome replication strategy, or any other phage–host interactions, which occur inside the cell. Nevertheless,

P.G. Leiman (✉)
Institut de physique des systèmes biologiques, Ecole Polytechnique Fédérale de Lausanne (EPFL), BSP-415, CH-1015 Lausanne, Switzerland
e-mail: petr.leiman@epfl.ch

M.M. Shneider
Shemyakin-Ovchinnikov Institute of Bioorganic Chemistry, 117997, Moscow, Russia

M.G. Rossmann and V.B. Rao (eds.), *Viral Molecular Machines*, Advances in Experimental Medicine and Biology 726, DOI 10.1007/978-1-4614-0980-9_5,

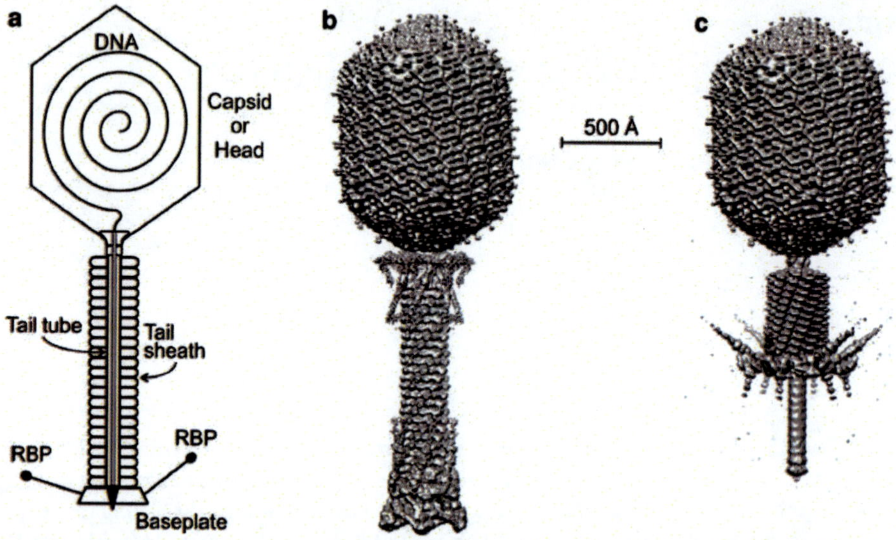

Fig. 5.1 Structure of a phage with a contractile tail. (**a**) Schematic showing major components of a contractile tail phage. RBP stands for receptor-binding proteins. (**b, c**) CryoEM reconstructions of phage T4 with the tail in the extended and contracted conformations (Fokine et al. 2004; Leiman et al. 2004; Kostyuchenko et al. 2005)

bacteriophages with contractile tails on average have larger genomes than other bacteriophages. In fact, giant *Myoviridae* bacteriophages represent some of the most complex viruses with genomes as large as ~700 kbp (Serwer et al. 2007). Phages with contractile tails and large genomes are strictly lytic, i.e., infection leads to host-cell lysis and phage genome does not integrate into the host-cell chromosome. All known phages with contractile tails have double-stranded DNA genome.

A contractile tail consists of the baseplate, the central tube (or the core), the external contractile sheath, and the terminator complex (Fig. 5.1) (Leiman et al. 2010). The baseplate carries host-cell-binding proteins [also called receptor-binding proteins (RBPs)] and coordinates host-cell attachment with sheath contraction. The tail terminator complex attaches the tail to the phage capsid via a head-to-tail connector protein, a cone- or mushroom-shaped dodecamer, which is positioned in one of the 12 pentameric vertices of the capsid (Fig. 5.1).

Upon binding to the host-cell surface with the RBPs, the baseplate changes its conformation triggering sheath contraction. The sheath contracts to about half or more of its original length, which moves the entire phage particle closer to the cell surface and drives the rigid internal tail tube through the cell envelope (Fig. 5.1). Depending on the phage species, the tube might extend to several hundreds of angstroms from the plane of the baseplate.

Phage tails have evolved to effectively breach all the constituent layers of the bacterial host-cell envelope. The extended external polysaccharides (if present) are both recognized and locally digested by RBPs, which possess an enzymatic activity specific to these sugars (Leiman et al. 2007; Walter et al. 2008). This allows the phage to bind to the cell's outer membrane, which is breached by the tail tube upon contraction of the tail sheath. The peptidoglycan layer is disrupted with the help of a glycosidase, which is also part of the tail, but its location in the tail and, possibly, number of such molecules in the tail varies in different phages. The tube is then able to interact with the cytoplasmic membrane.

The final step of the infection is the translocation of DNA and proteins from the phage capsid into the host-cell cytoplasm. The tail is thought of as a passive conduit in this process because it does not appear to contain a source of energy necessary to make it an active participant of this translocation event. A very detailed description of phage genome delivery into the host cell is given in a different chapter of this book (see Chap. 7 in this book).

The tail of bacteriophage T4 is by far the best-studied contractile system (Coombs and Arisaka 1994; Leiman et al. 2010). The T4 tail composition, assembly, structure, and conformational changes on attachment to the host cell are fairly well understood (Coombs and Arisaka 1994; Leiman et al. 2010). The structure and function of phage P2 tail (the other "model" contractile tail) have been studied to a much lesser extent (Kahn et al. 1991). These model systems are used for gene function prediction and assignment in all newly characterized contractile tail-like systems.

The accumulating genomic and protein structure data suggest that the contractile tail systems contain a set of conserved genes and have a common ancestor. Furthermore, many genes are shared with noncontractile tails also pointing to the common ancestry. A very compelling and plausible hypothesis stating that the contractile tails have likely evolved from noncontractile tails is beautifully presented in Chap. 6. Thus, this chapter will not discuss such evolutionary relationships, but instead will summarize the currently available data and emphasize common features of contractile tail-like systems.

5.2 Contractile Tail Structure

Contractile tails have an average diameter of ~220 Å but vary greatly in length from ~1,000 to ~4,500 Å. Despite its relatively large size, the organization of a contractile tail is fairly simple because its major part is a repetitive structure composed of two proteins – the internal tube and the external contractile sheath. Most of the tail's complexity is concentrated in the baseplate, which is the control center coordinating the host attachment process. The sheath and the tube represent two major bands on sodium dodecyl sulfate polyacrylamide gel electrophoresis (SDS PAGE) and serve as the definitive markers of a contractile sheath-like structure.

All known contractile phages have sheaths with molecular weights (MWs) between ~40 and ~80 kDa with the mean shifted to the lower values. A sheath with a MW of ~45 kDa is the most widespread among currently sequenced phages. The sheath proteins of two model phages, P2 and T4, have MWs of 43 and 71 kDa, respectively. On the other side, known tube proteins display less variation in their size. Their MWs are within the 15–21-kDa interval (18.5 kDa for T4 and 19.1 kDa for P2). The pair of sheath and tail tube genes also serves as a genetic marker of a contractile tail because the two genes are adjacent to each other with the sheath gene preceding the tube one and are transcribed from the sheath gene promoter.

5.2.1 *Structure of the Sheath*

The structure of the T4 sheath, the model object for all contractile sheath-like systems, was studied by several generations of structural biologists (reviewed in Coombs and Arisaka 1994; Leiman et al. 2010). However, a major breakthrough happened only recently, when the crystal structure of the T4 sheath protein gp18 (gene product 18) was solved (Aksyuk et al. 2009a). Full-length recombinant gp18 forms polysheaths of various lengths, which are not suitable for crystallization (King 1968). These polysheaths contain defects (missing subunits and other irregularities) making them difficult to analyze to high resolution because averaging techniques cannot be fully exploited. A large body of mutagenesis work went into producing a gp18 mutant, which would not form a polymeric structure (Kuznetsova et al. 1998; Poglazov et al. 1999). The result of this work is a mutant called gp18M constituting about three-fourth of the full-length molecule (residues 1–510 out of 659) with one amino acid substitution Arg510→Pro.

Gp18 consists of four domains. Twenty N-terminal and ~160 C-terminal residues form Domain I, which is not present in the gp18M crystal structure (the domains are renumbered compared to Aksyuk et al. 2009a) to make the nomenclature consistent with other sheath proteins). Domain II

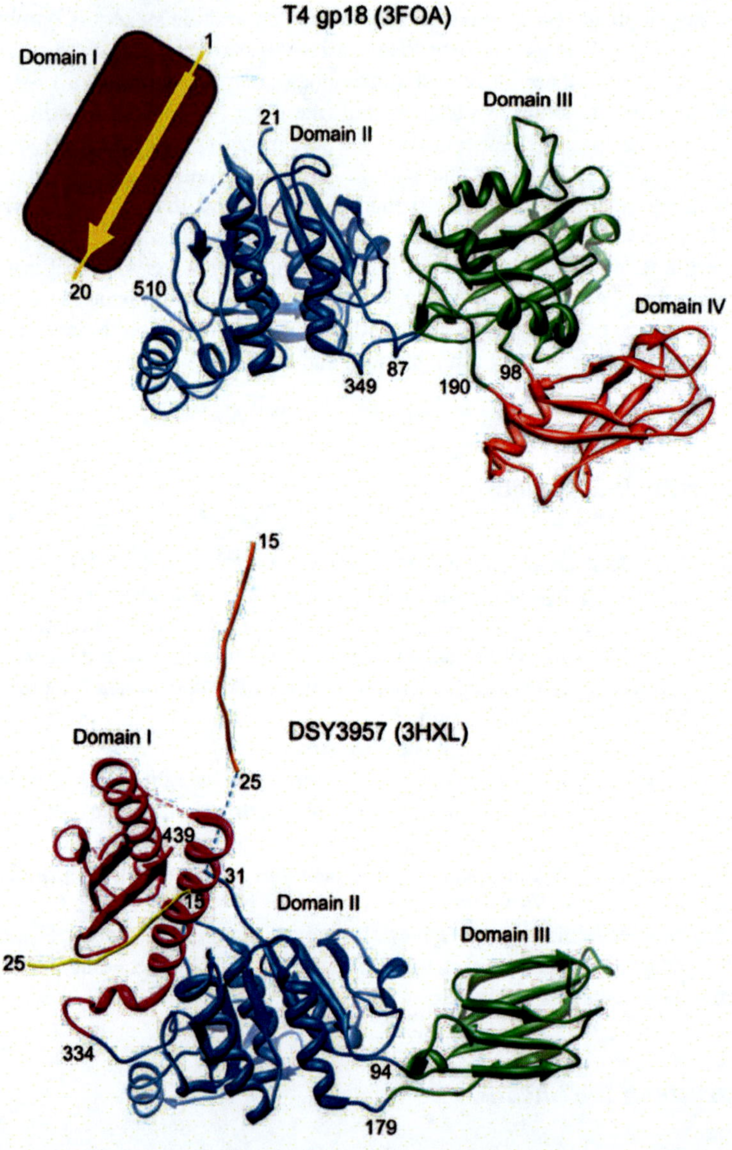

Fig. 5.2 Structure of the sheath protein. T4 gp18 (PDB ID 3FOA) and DSY3957 (PDB ID 3HXL) are shown as ribbon diagrams in the *top* and *bottom* panel, respectively. Domains I, II, III, and IV are colored *purple*, *blue*, *green*, and *red*, respectively. Domain I is not present in the crystals structure of T4 sheath protein gp18 and is shown as a *purple rectangle*. The domain-swapped N-terminal arm, which is donated to Domain I by a different subunit belonging to the same strand of the sheath, is colored *yellow*. The N-terminal arm belonging to the polypeptide chain of DSY3957 (PDB ID 3HXL) shown in the *lower* panel is colored in *orange*. In the crystal structure, it is donated to Domain I of a molecule related by crystal symmetry (not shown)

(residues 21–87 and 346–510) consists of a β-sheet with five parallel and one antiparallel β-strands plus six α-helices, which surround the β-sheet. Domain III (residues 88–97 and 189–345) is a two-layer β-sandwich, flanked by four small α-helices. Domain IV (residues 98–188) is a six-stranded β-barrel plus an α-helix. Together domains III and IV form the protease resistant fragment of the sheath protein (Fig. 5.2).

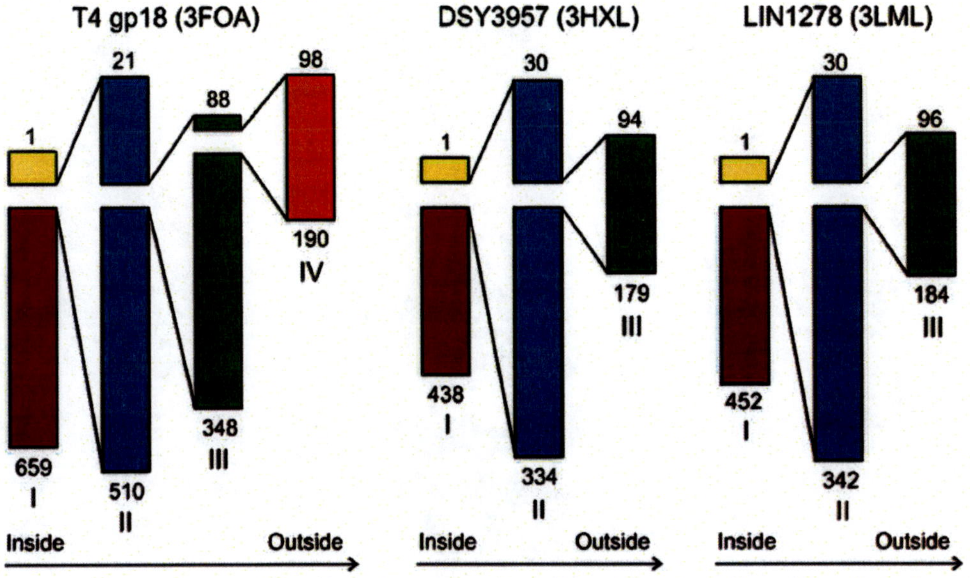

Fig. 5.3 Domainal organization of sheath proteins. The color code as in Fig. 5.2 Domains I, II, III, and IV are colored *purple*, *blue*, *green*, and *red*, respectively. The domain-swapped N-terminal arm is colored *yellow*. The orientation of the sheath subunit relative to the tail axis is shown below each diagram

The overall topology of the gp18 polypeptide chain is quite remarkable. Domain IV is "inserted" between residues 97 and 189, which belong to a loop connecting two β-strands of Domain III (Figs. 5.2 and 5.3). The latter is inserted between residues 87 and 346 into a loop of Domain II. Domain II, in turn, is an insertion into Domain I.

T4 tail has been studied by cryo-electron microscopy (cryoEM) in the extended and contracted conformations (Leiman et al. 2004; Kostyuchenko et al. 2005). The crystal structure of gp18M fits (as a rigid body) into the cryoEM map of the sheath in both conformations. Domain I interacts with the tail tube and forms the internal core of the sheath. Domains II and III are partially exposed to solution. Domain IV forms the sheath's external surface.

T4 sheath is composed of 138 gp18 subunits, which are arranged in a six-start helix (Leiman et al. 2004). Each of the six strands comprising the helix contains 23 gp18 subunits (Leiman et al. 2004). The sheath is 240 Å wide and 925 Å long. The pitch and twist of the helix are 40.6 and 17.2 Å, respectively (Kostyuchenko et al. 2005) (Fig. 5.4). Each gp18 subunit interacts with four neighbors: two related by the sixfold symmetry (interstrand contacts) and two within the same strand (intrastrand contacts). The interactions within the strand govern the connectivity and helical interpretation of the sheath structure because the surface of the intrastrand subunit contact is five times that of the interstrand (Kostyuchenko et al. 2005).

The contracted T4 sheath is 330 Å wide and 420 Å long. It is also a six-start helix with a 16.4 Å pitch and 32.9° twist (Leiman et al. 2004). Compared to the side-by-side arrangement of the gp18 strands in the extended sheath, the strands in the contracted sheath are interdigitated (Fig. 5.4). The axial compression or contraction is accomplished by sliding one layer of the sheath subunits into the next one, which is accompanied by the increase in the sheath's diameter. Upon sheath contraction, each of gp18 subunits moves by ~50 Å away from the tail's axis and tilts by ~45° (Aksyuk et al. 2009a).

In the process of sheath contraction, gp18 subunits move as rigid bodies without refolding or significant domain shifts (Aksyuk et al. 2009a). Domains I of the gp18 subunits belonging to the same strand maintain the connectivity throughout the contraction process, thus preserving the helical

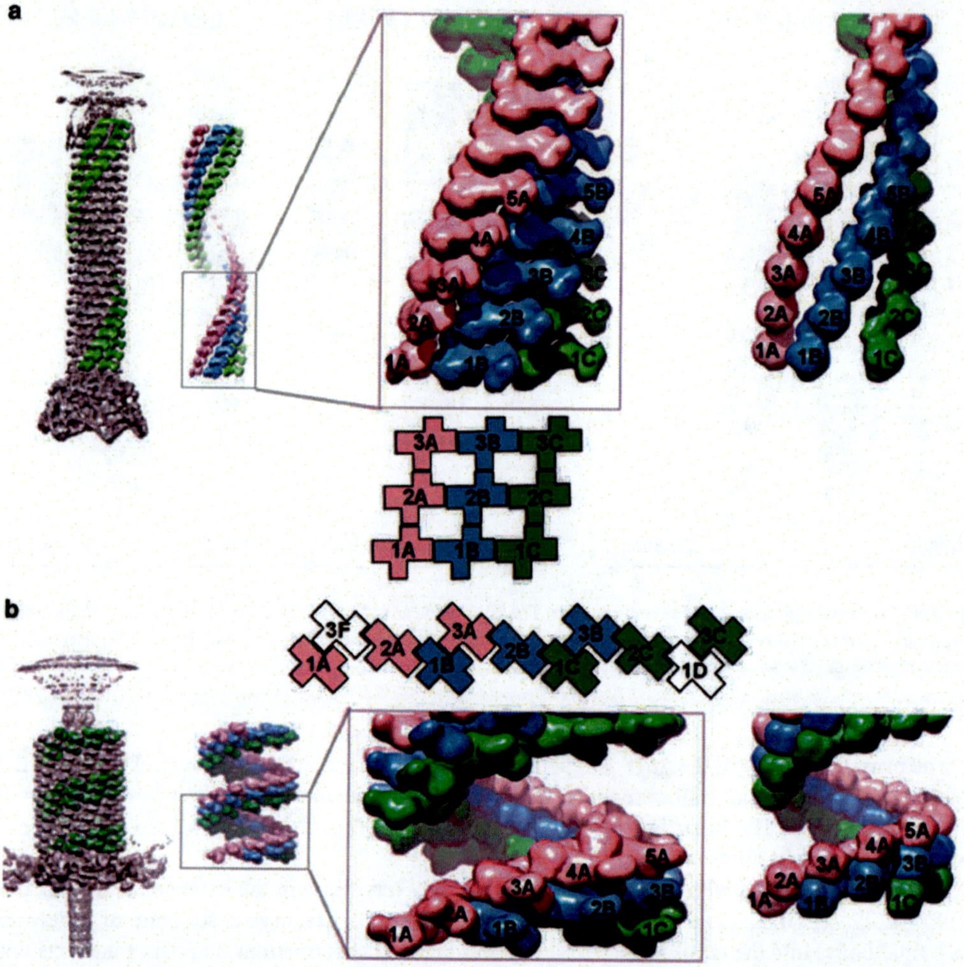

Fig. 5.4 Structure of T4 sheath in the extended and contracted states as determined by cryoEM (Leiman et al. 2004; Kostyuchenko et al. 2005; Aksyuk et al. 2009a). Three of the six gp18 helices are extracted from the complete tail structure, which is shown in the extended (**a**) and contracted (**b**) conformations. Each strand is colored in a distinct color (*pink*, *blue*, and *green*). The successive layers of gp18 subunits are numbered 1, 2, 3, 4, and 5 with the layer numbered 1 being closest to the baseplate. The middle panels of (**a**) and (**b**) show the arrangement of Domains II trough IV in three neighboring strands. The *rightmost* panels show the interactions of Domains I from the same three strands (note, that these domains maintain connectivity in the extended and contracted states). A simplistic representation of how Domains II through IV belonging to the three neighboring strands change their interactions in the extended and contracted states is also shown in the *middle* panels

structure of the sheath, whereas Domains II through IV form new interactions in the contracted state (Fig. 5.4). Upon contraction, the contact area between gp18 molecules increases by about four times (Aksyuk et al. 2009a).

5.2.2 *Evolution of Sheath Proteins*

The topology of gp18 polypeptide chain alone suggests that Domain I is likely the oldest part of the sheath structure (Figs. 5.2 and 5.3). Domains II, III, and IV were "picked up" over the period of

Table 5.1 Comparison of the T4 sheath protein gp18M fragment crystal structure with crystal structures of two sheath proteins currently available at the PDB: DSY3957 (PDB ID 3HXL) and LIN1278 (PBD ID 3LML). The number of aligned residues, the RMSD between the equivalent C_α atoms and the sequence identity percentage, is given for each compared pair

Superimposed pairs	DSY3957 (3HXL)	LIN1278 (3LML)
T4 gp18 Domains II+III	227 residues/3.5 Å/10%	230 residues/2.6 Å/17%
T4 gp18 Domain II	155 residues/2.3 Å/12%	154 residues/2.1 Å/16%
T4 gp18 Domain III	84 residues/2.6 Å/12%	80 residues/2.5 Å/25%

Table 5.2 Gp18 Domain I (residues 1–20 plus 511–659) structure prediction using HHpred

Matching structure	Probability (%)	*E*-value	*P*-value	Sequence identity (%)
DSY3957 (residues 308–438)	100	2.50E-29	1.00E-33	19
LIN1278 (residues 314–452)	99.9	1.80E-24	7.30E-29	14

evolution of the sheath to its current structure. This assumption is strengthened by comparing the primary and tertiary structure of gp18 with its homologs.

Due to the diversity of phage tail sheath sequences, only relatively close relatives of phage T4 can be identified using the gp18 amino acid sequence. Mapping the amino acid sequence conservation onto the gp18M structure shows that Domains I and II are the most conserved parts of the structure.

In addition to gp18, crystal structures of two other sheath proteins are currently available at the Protein Data Bank (PDB, http://www.pdb.org) under the deposition codes (IDs) 3HXL and 3LML (Figs. 5.2 and 5.3). They represent the complete products of genes *dsy3957* and *lin1278* from two different prophages infecting Gram-positive *Desulfitobacterium hafniense* and *Listeria innocua*, respectively (note that T4 is a Gram-negative *Escherichia coli* lytic phage). DSY3957 (3HXL) and LIN1278 (3LML) have very similar folds and show 43% sequence identity and a root mean square deviation (RMSD) of 2.8 Å upon superposition with 81% of residues participating in the alignment. The proteins consist of three domains connected by flexible linkers. The domains are shifted and rotated relative to each other, thus elevating the RMSD value, but the matching domains have virtually identical folds.

A careful inspection of the structure of DSY3957 and LIN1278 shows that folds of two of their three domains are similar to those of gp18 Domains II and III even though there is no significant sequence similarity (17% or lower overall) (Table 5.1) (Fig. 5.3). Domains II of the three proteins are very similar in size (~220 residues), whereas Domain III of gp18 is almost twice as big as that in the other two proteins (~160 residues vs. ~90 residues) because it contains many small and large insertions, one of which is Domain IV (90 additional residues). The third domain of DSY3957 and LIN1278 corresponds to Domain I of gp18, which is not present in the gp18M structure. HHpred analysis shows that Domains I of the three proteins have a 100% probability to have the same fold (Table 5.2).

This analysis shows that gp18, DSY3957, and LIN1278 have a common ancestor. Furthermore, we can propose that Domains I, II, and III likely represent the smallest unit, which can form a contractile sheath-like structure. The common ancestry of the sheath proteins also suggests that all sheaths undergo similar reorganization during the contraction event with Domains I maintaining the connectivity throughout the process and Domains II and III forming new and more extensive intersubunit interactions in the contracted state compared to the extended one.

A possible explanation of how Domains I maintain their connectivity during the contraction event can be given with the help of the DSY3957 crystal structure and the T4 sheath cryoEM map. Domain I of DSY3957 is formed by residues 308–438 of one polypeptide chain and residues 1–25 of the symmetry-related molecule. In other words, one molecule donates its N-terminal arm to its crystallographic neighbor, where these residues become a part of a β-sheet within Domain I. By superimposing DSY3957 onto the structure of gp18 fitted into the T4 sheath cryoEM map, we find that the length of the donated N-terminal arm and the spatial orientation of Domains I stacked onto each other should allow such a domain swapping to take place within the sheath structure. Therefore, the entire inner part of the sheath is "cross-linked" by the chain-swapped Domains I. This property is likely universal for all contractile tail-like structures.

5.2.3 Structure of the Tube

The accumulated body of data does not allow to give a fully consistent description of the tail tube structure. Earlier studies of isolated T4 tubes suggested that their helical structure is identical to that of the sheath in the extended state. Furthermore, the sheath and the tube appear to contain the same number of subunits. The sheath polymerizes into the extended conformation only in the presence of both the baseplate and the tube. It was proposed that during tail assembly the tube serves as a scaffold for the sheath and that the latter adapts the helical symmetry of the tube.

These conclusions are difficult to reconcile considering the latest structural data. Subunits, comprising the tube, are not well resolved in the T4 tail cryoEM maps, making symmetry determination impossible. If the tube and the sheath had the same symmetry, tube subunits should be well resolved, similar to the sheath subunits, but this is not the case. Furthermore, as it is shown in Chap. 6 in this book, the contractile tails, the long noncontractile tails, and the bacterial type VI secretion system (T6SS) have a common ancestor. The T6SS tube protein (called Hcp) forms tubes with sixfold symmetry in the crystalline form and, possibly, in solution. However, in these tubes, Hcp hexamers are stacked onto each other without any twist, which is inconsistent with the helical symmetry of the sheath.

It is nevertheless possible that the tube and the sheath have the same helical symmetry. The subunits of the tube might be unresolved in the cryoEM maps due to the powerful Fourier ripple of negative density caused by the sheath. This ripple overlaps with the tube density and significantly deteriorates it.

An explanation – consistent with current data – for the tube and the sheath to have different symmetries also exists. The subunits of the sheath and the tube might come in register only at certain points of the elongated structure, and the tube would still serve as a scaffold for sheath polymerization. Yet another possibility is that sheath subunit polymerization into the extended conformation is determined and driven by the baseplate, and the tube per se is not required for polymerization. Instead, only the proteins associated with the tube's end are necessary for stabilization of the sheath, which is known to depolymerize quickly without the tail capping complex.

5.2.4 The Simplest Contractile Tail-Like Structure

In phages with large genomes, the baseplate is a very complex organelle. For example, the baseplate of phage T4 consists of 12 proteins. Two additional proteins form a cylindrical structure, onto which the tail tube subunits are assembled and polymerized into the tube. The structure and position of nine constituent proteins (or their major domains) in T4 baseplate are known. This information is used to understand the function of proteins in other baseplates.

Phage genomes are diverse in general, but even close relatives display large variations in their RBPs, which emanate from the baseplate. Since the baseplate is responsible for interaction with the host-cell surface, it evolves faster compared to the rest of the genome because a small change in the baseplate structure might cause the phage to switch host specificity, and such an adaptability is in general beneficial for phage survival.

The diversity of baseplate genes of closely related phages increases toward the periphery culminating in the host-cell-binding regions of RBPs. This diversity pattern is even more pronounced in phages, which are related very distantly. Often, homologous proteins show no similarity at the amino acid sequence level, but the common ancestry could still be found at the level of protein structure. Unfortunately, structures of very few non-T4 baseplate proteins are currently available. One such pair of previously unknown homologs is gp27 (the central hub protein of the T4 baseplate) and gp44 of phage Mu. The two proteins have very similar folds but show only 9.6% sequence identity (Kondou et al. 2005). It is clear now that the baseplate hub proteins of contractile and noncontractile tails have a common ancestor (Sciara et al. 2010). This is described in detail in Chap. 6 in this book.

Despite the limited information on the structure of non-T4 baseplate proteins, it is possible to establish the relationship between the T4 proteins with known functions and their phage P2 and phage Mu homologs and extend this information to all other contractile tail-like systems. Furthermore, we can even attempt to describe the composition of a simplest contractile tail-like structure. The HHpred Web server (Soding 2005; Soding et al. 2005) and several crystal structures solved in the framework of the structural genomics initiative are very useful for this analysis.

The hub structure. The central part of a simplest contractile tail-like structure is formed by the homologs of T4 gp27 and gp5 proteins. The orthologs of these and other proteins mentioned below, which comprise the particles of P2 and Mu phages, are listed in Table 5.3 and shown in Fig. 5.5. Gp5 forms the central membrane-puncturing needle or spike in T4 baseplate. The N-terminal domain of gp5 is an OB-fold, and HHpred clearly shows that the N-terminal domain of P2 gpV and Mu gp45 is a similar OB-fold. The three OB-fold domains of gp5 trimer are inserted into the channel formed by the trimeric gp27 donut-like structure, whereas the C-terminal domain comprises the membrane-attacking spike. In T4 gp5, this domain is a very stable, intertwined trimeric β-helix, and a high β-structure content is predicted for the C-terminal domains of P2 gpV and Mu gp45 suggesting that they also might form a β-helix. In general, the N-terminal OB-fold domain and the C-terminal β-structure-rich domain is a common feature of all membrane-puncturing needles or spikes in phages of Gram-negative bacteria.

The trimeric gp5–gp27 hub forms the centerpiece of the otherwise sixfold symmetric baseplate. The gp27 trimer acts as a three to sixfold symmetry adaptor. One part of the gp27 trimer tightly wraps around the three N-terminal domains of gp5 trimer, whereas the other part is composed of six tandem domains with two domains per monomer, which results in $2\times3=$sixfold symmetry. The six tandem tube domains of gp27 trimer serve as a starting point for polymerization of the tail initiator complex and the rest of the tail tube. The structure of gp27 trimer is conserved not only in contractile but also in noncontractile tails making the three to sixfold adjustment principle universal for all long tails.

One feature of T4 tail not found in other contractile tails is a lysozyme domain in their gp5–gp27 complex orthologs. In T4, the lysozyme domain is "inserted" between the OB-fold and the C-terminal β-helical domain of gp5. The gp5 lysozyme domain shows 43% sequence identity to T4L, the cytoplasmic T4 lysozyme encoded by gene *e* (Mosig et al. 1989). T4L and the gp5 lysozyme domain have very similar structures with conserved catalytic residues suggesting their common ancestry (Kanamaru et al. 2002). The likely function of gp5 lysozyme domain is to locally digest the peptidoglycan layer during infection to make an opening for the tube (Kanamaru et al. 2002). However, a phage carrying a mutant gp5 with only 10% of the wild-type activity is almost as infective as the wild-type phage, suggesting that the residual activity is likely sufficient for infection in the laboratory conditions (Kanamaru et al. 2005).

Table 5.3 The minimal set of proteins comprising a functional contracted tail

	Number of polypeptide chains per tail	Phage T4	Phage P2	Phage Mu	Relevant PDB entries
Baseplate hub	3	gp27	gpD	gp44	1K28, 1WRU, 3CDD, 3D37, 3GS9, 2P5Z
Spike/needle	3	gp5	gpV	gp45	1K28, 2P5Z
Baseplate	6	gp25	gpW	gp46	2IA7
	12	gp6	gpJ	gp47	3H2T
	6	gp53 or gp7[a]	gpI	gp48	
Receptor-binding protein	18 (6 trimers)	gp12	gpH	gp49	1OCY, 1H6W, 2V5I
Tail tube initiator	6[b]	gp48/gp54	gpU[b]	gp43[b]	
Tape measure	3–6	gp29	gpT	gp42	
Tail tube	90+	gp19	gpFII	gp40	1Y12, 2GUJ
Tail sheath	90+	gp18	gpFI	gp39	3HXL, 3LML, 3FOA
Tube terminator	6	gp3	gpS[b]	gp37	2GJV
Sheath terminator	6	gp15	gpR[b]	gp38[b]	

[a]At present, there is not enough data to confirm T4 gp53 as P2 gpI or Mu gp48 ortholog. It is possible that in T4 the ortholog of P2 gpI or Mu gp48 is gp7
[b]Hypothetical assignment

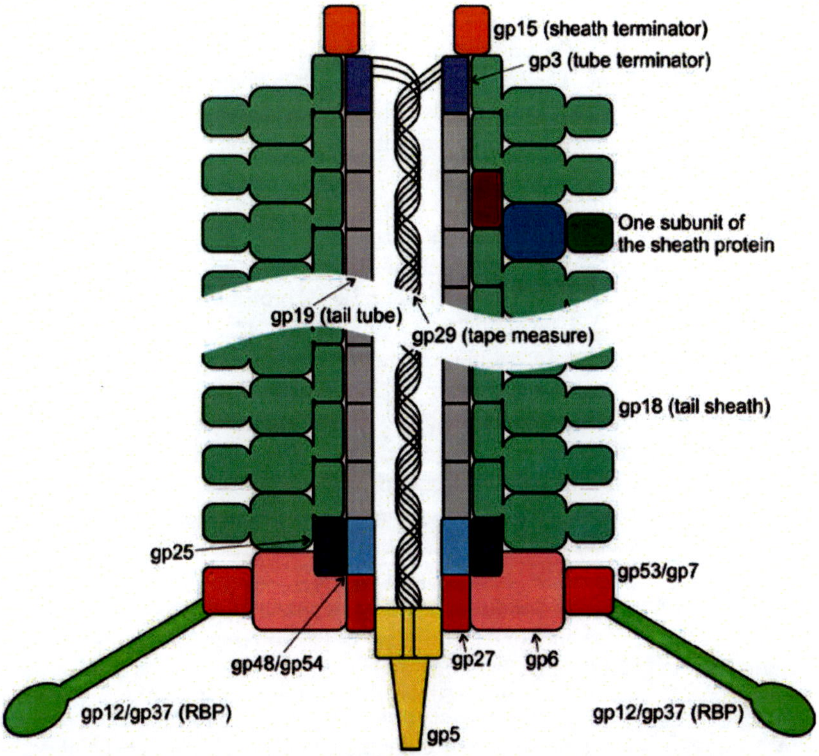

Fig. 5.5 Structure of the simplest contractile tail. The proteins are labeled with the names of their T4 phage orthologs (Table 5.3) and are colored in distinct colors. The three domains of one sheath protein subunit are colored using the color scheme in Figs. 5.2 and 5.3

Nevertheless, the presence of the lysozyme domain in T4 gp5–gp27 complex remains somewhat enigmatic because even in close relatives of T4, such as, for example, in *Pseudomonas aeruginosa* phage KVP40, the gp5 ortholog does not contain a lysozyme domain (Rossmann et al. 2004). Gp5 orthologs of P2 and Mu do not contain a lysozyme domain either (Table 5.3). However, a lysozyme or glycosidase domain can be detected at the C-terminus of tape measure proteins in some contractile and noncontractile phages (Piuri and Hatfull 2006; Boulanger et al. 2008). Most probably, the C-terminal domain of the tape measure protein interacts with the ortholog of gp27 hub protein and is translocated into the periplasm during infection. In summary, a functional glycoside does not appear to be required for contractile tails under all infection conditions.

It is worth to mention that the gp27 ortholog of the contractile tail iodobacteriophage φPLPE (Leblanc et al. 2009) encoded by gene 58 contains a lysozyme domain at its C-terminus (goose-type lysozyme), whereas its gp5 ortholog, encoded by gene 59, is lysozyme domain-free. Assuming that gp58 and gp59 of φPLPE form a complex similar to T4 gp5–gp27, the position of the lysozyme domain of φPLPE gp58 in the complex and in the baseplate would match that of the gp5 lysozyme domain.

Orthologs of T4 gp5–gp27 complex are also found in phages infecting Gram-positive bacteria, but the membrane-interacting C-terminal domain of gp5 proteins in these phages is a coiled coil instead of the β-helix (e.g., gp98 and gp99 from the *Listeria* phage A511 are homologs of T4 gp27 and gp5, respectively).

The minimal baseplate. The structure of T4 baseplate shows that the hub is surrounded by three proteins – gp6, gp25, and gp53. The ortholog of gp25 can be identified in the genomes of P2 and Mu using sequence similarity alone. The crystal structure of a gp25 homolog from a *Geobacter sulfurreducens* prophage (GenBank NP_952040.1) was solved by structural genomics (PDB ID 2IA7). The fold of this protein is very similar to that of Domain I of the sheath protein, which forms the inner part of the sheath. It is likely that gp25 and Domain I interact in the baseplate and that Domain I and, possibly, the rest of the sheath originates from a gp25-like protein. In general, the gp25 ortholog is the most conserved protein (in terms of the amino acid sequence) in the baseplate.

The orthologs of T4 gp6 in the baseplates of P2 and Mu (gpJ and gp47, respectively) are predicted by HHpred with a very high confidence (97% probability), but the "fifth" baseplate protein (the other four being gp5, gp27, gp25, and gp6) is somewhat enigmatic. In P2 and Mu, this protein is gpI and gp48, respectively, but none of them shows any sequence similarity to each other or T4 gp53. Their only common trait is a similar size (~180 residues). It is possible that the T4 ortholog of the "fifth" protein is a domain of gp7.

The position and function of the three conserved baseplate proteins (orthologs of T4 gp25, gp6, and gp53/gp7) can be derived from the structure of T4 baseplate (Kostyuchenko et al. 2003; Aksyuk et al. 2009b). The ortholog of T4 gp25 protein is positioned near the hub and might interact with the sheath. There are six copies of gp25 in the baseplate in agreement with the sheath symmetry. The gp6 ortholog forms a major part of the "minimal" baseplate. Six gp6 ortholog dimers form a continuous ring enveloping the hub and the gp25 ortholog protein. Six copies of the "fifth" baseplate protein (T4 gp53/gp7 ortholog) bind to the periphery of the gp6 ring. Tail fibers bind to the "fifth" protein and, possibly, to gp6 ortholog.

RBPs: The tail fiber/tailspike and tail fiber chaperone. The tail fiber/tailspike is responsible for host-cell receptor binding. It is an essential part of a functional tail. All known phage fibers are trimers having folds with very complex topologies often dominated by intertwined β-helices and, less often, α-helical coiled coils (Steven et al. 1988; van Raaij et al. 2001; Thomassen et al. 2003; Walter et al. 2008; Bartual et al. 2010). The C-terminal domain of the fiber is responsible for receptor binding, and its N-terminal domains attach the fiber to the baseplate.

The folding and attachment of the fiber to the baseplate is governed by a phage-encoded chaperone, which is always positioned downstream to the fiber. In some phages, the chaperone is part of the same gene or might remain bound to the fiber after its attachment to the baseplate (Riede et al. 1985; Tetart et al. 1996). In the latter case, the chaperone participates in host-cell receptor recognition and binding. The chaperones show a certain level of specificity in attachment of the fiber to the baseplate, suggesting that the chaperone–fiber complex binds to the tail during the fiber attachment (Williams et al. 2008).

The tail tube initiator. T4 clearly possesses the tail tube initiator complex, which serves as a starting point for polymerization of the tail tube. The initiator complex is built onto the gp27 trimer. In T4, this complex contains gp48 and/or gp54 (Kostyuchenko et al. 2003). Gp54 shows a significant sequence similarity to T4 tail tube protein gp19 and tube terminator protein gp3. There is not enough data to identify tube initiator proteins in P2 or Mu, but since P2 gpU and Mu gp43 are tail components with unknown function and location, we can tentatively assign them the tube initiator function.

The tape measure protein. Similar to noncontractile tails, the length of the tail is controlled by the tape measure protein (Abuladze et al. 1994). In T4, the tape measure protein gp29 appears to participate in the formation of the hub and the gp48/gp54 tube initiation complex (Kikuchi and King 1975a, b, c, d). The tape measure proteins of contractile and noncontractile tails are most likely homologous and thus function in a similar manner. The structure and function of tape measure proteins are described in detail in Chap. 6 in this book. Briefly, several copies of the tape measure protein extend through the central channel of the tail tube as it is being built. In the fully assembled tube, the tape measure protein adapts an extended α-helical conformation, and its N-terminus interacts with the tail tube terminator and, possibly, with the sheath terminator protein. Some tape measure proteins contain an identifiable glycosidase domain at the C-terminus. High-resolution studies of T4 tail did not provide sufficient detail on whether the tube channel is occupied by the tape measure protein or phage genomic DNA extending from the capsid.

Interestingly, the contractile tail of iodobacteriophage φPLPE mentioned earlier is one of the shortest known. The sheath part of the tail is only 600 Å (Leblanc et al. 2009), and its putative tape measure protein of φPLPE (gp55) is only 392 residues long. It does not contain an identifiable peptidoglycan hydrolase domain and is likely to have an α-helical structure throughout the sequence (McGuffin et al. 2000). In the extended α-helical conformation, it can form a fiber with a length of ~600 Å, which roughly corresponds to the length of the tail, but unlike other tape measure proteins φPLPE gp55 does not appear to form even a small globular domain at its termini. As the sheath proteins of all phages have a common core structure (see below), the pitch of the sheath helix is likely to be conserved and approximately equal to that found in T4, i.e., ~40 Å. Thus, the sheath and, consequently, the tube of φPLPE contain approximately $15 \times 6 = 90$ subunits each.

The tail tube and the sheath. The structure of the tube and the sheath of phage T4 is described above. The helical parameters and the length of the sheath vary in different phages even though the core structure of the sheath protein (the three inner domains) is conserved. The length of the tail in different phages varies, and phages with larger genomes have longer tails on average.

The tail terminator complex. In contractile tail phages, the tail is capped by two proteins, one of which interacts with the tail tube, whereas the other one most probably interacts with both the tube and the sheath (Coombs and Arisaka 1994; Leiman et al. 2010). The tail tube capping protein is a homolog of the noncontractile tail terminator protein. The tube and sheath terminators in T4 are encoded by gp3 and gp15, respectively. It is not possible to identify the terminator proteins in neither P2 nor Mu using the sequence data alone, except for Mu gp37, which is clearly related to the noncontractile tail terminator proteins (see Chap. 6).

In phage T4, the tail terminator complex can adapt two conformations – closed and open – associated with the extended and contracted sheath conformations, respectively (Kostyuchenko et al. 2005). The conformation of the tail terminator complex controls the location of the leading end of DNA molecule in the tail. In the extended tail, the central channel of the terminator complex is closed, and DNA cannot exit the capsid. Sheath contraction causes the channel to open up, allowing the leading end of phage DNA to exit from the capsid and descent down the tail tube in preparation for ejection into the host cell.

5.3 Assembly of a Contractile Tail

Extensive studies of T4 tail morphogenesis showed that the baseplate is assembled first (Kikuchi and King 1975a, b, c, d). The tail tube and sheath are built starting from the baseplate. Without the tube or the baseplate, T4 sheath assembles into a long polymeric structure called polysheath, whose structure is similar to that of the contracted sheath (Moody 1967a, b, 1973). The tube and the sheath must be capped with the tail terminator complex, or they would depolymerize (King 1968; Coombs and Eiserling 1977; Tschopp et al. 1979). The fibers are the last proteins, which attach to the baseplate. T4, like many large phages, contains two set of tail fibers. The short tail fibers attach to the baseplate directly, whereas attachment of the long tail fibers requires the phage particle to assemble fully: the tail must be sheathed and bound to the DNA-filled capsid. Another set of fibers, called whiskers or fibritins, emanating from the head-to-tail junction, participates in the long tail fiber assembly. No other phage tail assembly was studied to such an extent, but because all contractile tails have a common ancestor, their assembly pathway is likely to resemble that of T4.

The assembly process is strictly ordered. If a protein is excluded from the process, the assembly intermediate preceding the missing protein will accumulate in the cell (Kikuchi and King 1975a, b, c, d). Association of baseplate, tube, or sheath subunits into the complete structure does not appear to require any chemical energy. Except for gp5, no protein in the tail appears to undergo maturation proteolysis upon incorporation into the tail complex (Coombs and Arisaka 1994; Leiman et al. 2010).

It is apparent, however, that the tail is originally assembled as a high-energy metastable structure, akin to an extended spring, which is locked in this conformation by noncovalent interactions and can contract when those interactions are disturbed. Exposure of tails to 3 M urea and heat, and abrupt pH changes disturb these interactions causing contraction (Coombs and Arisaka 1994). Furthermore, the sheath protein assembles into a contracted sheath-like structure (the polysheath) in the absence of the baseplate and/or the tail tube. The contracted conformation of the sheath must be a very low-energy configuration because it is resistant to 6 M urea and detergents.

Still, there must be a source of energy, which is used to build the extended sheath structure in the first place. It is possible that the process is kinetically driven. The formation of the sheath in the extended conformation occurs about 50 times faster than the formation of polysheaths and, therefore, is a preferred pathway due to a lower activation energy (Arisaka et al. 1979; Tschopp et al. 1979). The integration of the N-terminal arm of the sheath protein into the C-terminal domain of the next layer subunit in the sheath (Sect. 5.2.2) might also be an important factor driving the polymerization of the sheath. Because the arm is evolved to interact with the C-terminal domain, its free energy is higher when it is extended into the solvent. This characterizes the sheath subunit before it becomes part of the polymeric structure. Thus, sheath subunits have a high intersubunit affinity, and the baseplate–tail tube complex shifts the assembly process toward the extended structure.

T4 baseplate can exist in one of two conformations, which are coupled to two sheath states: the "hexagonal" conformation is associated with the extended sheath, whereas the "star" conformation is associated with the contracted sheath, which is found in T4 particle after attachment to the host cell (Coombs and Arisaka 1994; Kostyuchenko et al. 2003; Leiman et al. 2004). The free energy of the hexagonal conformation is higher than that of the star because hexagonal baseplates switch to stars after prolonged storage. There is no published data about the structure or conformational changes of simpler baseplates (e.g., from phages P2 or Mu), but the structure of a noncontractile phage baseplate in two conformations is available (Sciara et al. 2010). Upon host-cell binding, this baseplate undergoes a rearrangement resembling that of T4 baseplate, from which we can conclude that the conformation of the baseplate is coupled to host-cell binding and determines sheath conformation in all contractile tail phages.

5.4 Interaction of the Contractile Tail with the Host-Cell Envelope

The baseplate coordinates host recognition and attachment with sheath contraction, which is initiated at the baseplate and is propagated through its entire length of the sheath in a wavelike fashion (Simon and Anderson 1967a, b; Moody 1973; Aksyuk et al. 2009a). Tail fibers, which emanate from the baseplate, are the primary determinants of host-cell specificity (Tetart et al. 1998; Williams et al. 2008). The host-cell binding parts of the fibers (their C-terminal domains) are positioned as far as 1,000 Å from the baseplate in some phages. Successful receptor binding triggers the conformational change of the baseplate, which then leads to sheath contraction. Thus, the host-cell-binding signal must be transmitted through the length of the fiber to the baseplate. The transmission of the signal occurs by way of a change in the orientation of the fibers relative to the baseplate because the fibers are not known to undergo refolding upon receptor binding. When the phage particle is free in solution, the fibers do not have a fixed orientation and point roughly sideways or even toward the capsid.

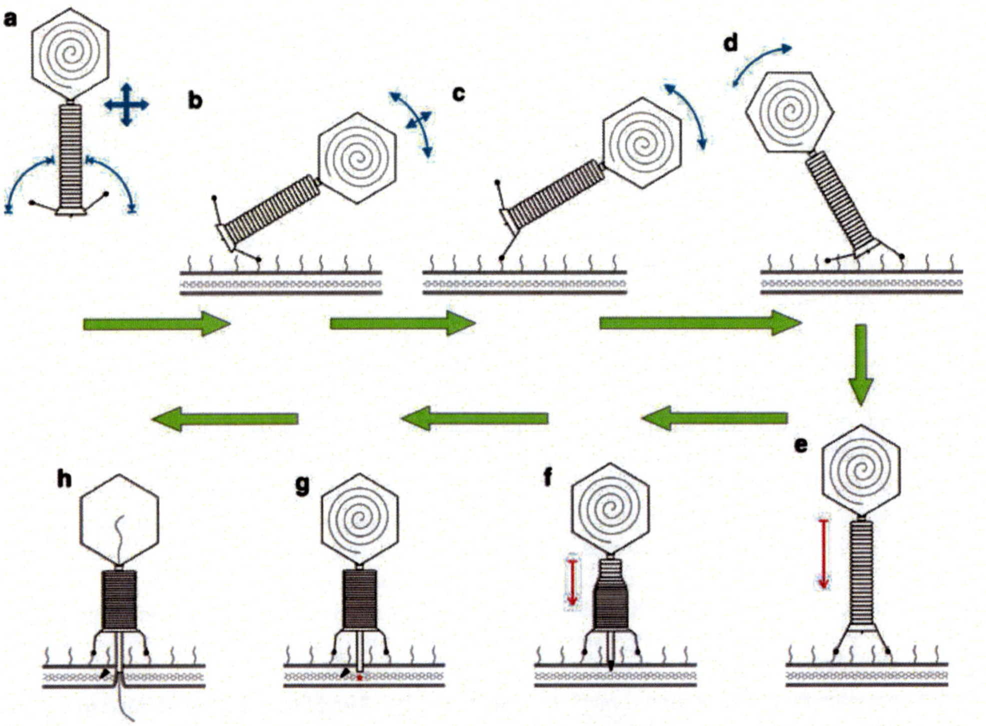

Fig. 5.6 Breaching the host-cell envelope. **(a–h)** Show stages and substages of the infection process. *Blue arrows* indicate motions caused by solvent movement (Brownian, thermal, convection, etc.). *Magenta arrows* show the movement of the phage capsid towards the host-cell surface, which is driven by sheath contraction. **(a)** The phage is free in solution, and its fibers move freely around their most favorable position. **(b)** The phage binds to the host cell with one or two of its fibers, which serve(s) as a tether restricting the movements of the phage particle. **(c)** The phage is moved in such a way that the fiber is brought into a conformation from which it cannot switch back to the "free phage" fiber configuration. The fiber forms new interaction with the baseplate proteins, and the baseplate initiates its conformational change. **(d)** The phage continues its hula dance on the cell surface while being attached to it with the tail fiber tether. Eventually, the other fibers find their binding partners on the cell surface. **(e)** As in **(c)**, the phage is moved in such a way that all of its fibers, which have bound to the cell surface receptors, point toward the host cell. The conformational change of the baseplate now proceeds. **(f)** The baseplate conformational change initiates contraction of the sheath, which drives the capsid and the tube toward the cell membrane. The outer cell membrane is punctured with the help of the baseplate central spike protein. **(g)** The spike dissociates from the tail tube tip, thus opening the tail tube channel. The tail-associated glycosidase creates a small opening in the peptidoglycan layer. **(h)** The tail tube interacts with the inner membrane, which is pushed toward it by the osmotic pressure of the cytoplasm in small region now lacking the peptidoglycan layer. Phage DNA is released into the cytoplasm

The fibers of the phage bound to the host-cell surface point toward it. We will describe two possible mechanisms, which explain how and why this change in fiber orientation happens only on the cell surface, but not when the phage is free in solution (Fig. 5.6). Notably, no chemical energy is used in fiber reorientation and baseplate triggering.

The first mechanism, which we favor, can be described as follows (Fig. 5.6). When the phage comes across a susceptible host cell, most probably only one or two of the fibers bind to the host-cell receptors. Thus, attached phage is now under the influence of all the solvent movements originating from cell swimming and other molecular motions of various sorts (Brownian, thermal, convective currents, etc.). The phage is tethered to the cell with its fiber(s), while it is constantly being shaken by the solvent. Eventually, the other fibers bind to the host-cell receptors, which lead to both, orienting the phage perpendicular to the host-cell membrane and, at the same time, causing all the fibers to be pointed toward the host-cell surface. When the fibers are in this configuration, new interactions are formed in the baseplate causing it to switch to a lower energy conformation, which unlocks the sheath allowing it to contract. The spontaneous tail contraction does not occur because the orientations of the fibers associated with the extended and contracted sheath conformations are separated by a significant free energy peak. This peak can be overcome only if the fibers have become immobilized on the cell surface and the phage particle is being jiggled around by the solvent. This mechanism is supported by structural investigations of phage T4 baseplate (Leiman et al. 2004; Kostyuchenko et al. 2005) and R-type pyocin (P. Leiman, unpublished data).

The other possible mechanism of how fiber reorientation can be coupled to sheath contraction upon attachment to the host-cell surface comes from the observation that many phages require divalent cations (most often calcium) for infection. Cell surface molecules, which include phage receptors (polysaccharides and proteins), bind those cations making their concentrations near the cell surface very high. The high concentration of the ions can cause the baseplate to change its conformation and extend the fibers to bind to the host-cell surface. The baseplate will then change its conformation and unlock the sheath. This mechanism is similar to that proposed for the Gram-positive *Lactococcus lactis* bacteriophage p2 (Sciara et al. 2010), which requires calcium for infection. However, many contractile tail phages have very week requirements for medium composition for successful infection and can infect without calcium or any other divalent ions. Furthermore, phage tails contract in 3 M urea but stay intact in calcium or other divalent cations (Leiman et al. 2004; Leblanc et al. 2009).

It is possible that both proposed mechanisms are in play. The fibers become pointed to the host-cell surface as a result of solvent-induced phage movement relative to the host-cell surface, but, at the same time, the baseplate is more predisposed to changing its conformation in the presence of "surface-associated" ions. It is possible that binding of fibers to the host receptors releases additional ions, thus increasing the local concentration even further.

As the sheath contracts, it moves the capsid closer to the cell surface making the tail tube protrude from below the plane of the baseplate. The baseplate central hub complex, onto which the tube is assembled initially and which consists of T4 gp5 and gp27 orthologs, is dislodged from the baseplate and now forms the tip of the tube. This complex is the membrane-piercing "device," which is driven through the cell outer membrane by the energy of the sheath. The C-terminal domain of the T4 gp5 ortholog forms a membrane-piercing spike or needle, and its N-terminal domain plugs the opening of the gp27 ortholog trimer, which forms the rim of the tube. The gp5 ortholog must dissociate from the gp27 ortholog to open the tube channel to allow the phage DNA to exit through the tube. It is unclear whether the gp5 ortholog is used to disrupt both membranes and it eventually leaves the gp27 ortholog in the cytoplasm or it comes off in the periplasm. Furthermore, there is little data about the interaction of the tube with the inner membrane.

The large energy, which is stored in the condensed conformation of phage DNA packaged into the capsid, is not used in membrane penetration because the pyocins (DNA-free complexes) are able to disrupt the host-cell envelope and kill the cells (Scholl et al. 2009). Furthermore, sheath contraction and DNA ejection are not linked (Leiman et al. 2004). The contraction can be induced by subjecting the phage to moderate concentrations of urea or pH and/or osmotic shock. Phages with contracted tails are, however, fairly stable and do not release their DNA. Therefore, the cytoplasmic membrane must contain a specific receptor, which opens up the tube and triggers the release of DNA. This receptor must be at least as abundant as the phage receptor on the cell surface. It is possible that the tube channel opening might be triggered by certain lipids comprising the cytoplasmic membrane. Translocation of DNA into the host cell through the tube is the next step in the infection process. The physics of this process is a hot topic of debate in today's biophysics. This is covered in Chap. 7 of this book.

5.5 Structure and Function of Other Contractile Tail-Like Systems

In the past few years, several systems, which appear to share their ancestry with contractile tail phages, have been characterized. Among those are the R-type pyocins, the bacterial type VI secretion system (T6SS), the *Photorhabdus* virulence cassette (PVC), and rhapidosomes.

5.5.1 R-Type Pyocins

Pyocins are proteinaceous (DNA-free) bactericidal agents of *P. aeruginosa*. They were first described by Francois Jacob as high molecular weight bacteriocins of *P. aeruginosa* (Jacob 1954), but similar bacteriocins were found in other Gram-negative and Gram-positive bacteria (Williams et al. 2008). Pyocin DNA resides silently in the bacterial genome until it is activated with a UV or mitomycin C treatment (Matsui et al. 1993). This treatment causes DNA damage and activates RecA, which degrades the PrtR protein, the repressor of PrtN, a positive transcription regulator of the pyocin gene cluster.

R-type pyocins are close relatives of phage P2 tail (Nakayama et al. 2000). The R-type pyocin particle consists of at least 11 proteins all of which are related to P2 tail genes listed in Table 5.3. R-type pyocin represents one of the smallest contractile tail-like systems. The tube/sheath capping proteins of R-type pyocins have significantly diverged (in terms of the primary structure) from their P2 orthologs.

A pyocin-lysogenized bacterium can produce up to 200 R-type pyocin particles that leave the host by causing its lysis. Each of the newly released particles can bind to a bacterium and form a pore in its envelope causing depolarization of the cytoplasmic membrane and resulting in host death. Like phage, pyocins are very potent bacteriocins, and one pyocin particle is enough to kill a bacterium. A pyocin particle can strike only once. Being DNA-free, it does not multiply. Unlike that for a phage, the "infection" of a cell by R-type pyocin does not cause cell lysis, which means that the cytoplasmic content is not spilled out into the milieu immediately upon cell death. The host specificity of pyocins and phages is determined by the tail fibers, and *P. aeruginosa* pyocins can be retargeted to kill *E. coli* by equipping the pyocin with a fiber from an *E. coli* phage (Scholl et al. 2009).

R-type pyocins are divided into five subtypes (R1 through R5), each of which has narrow host range (up to a single *P. aeruginosa* strain). *P. aeruginosa* cells, which carry a particular pyocin, are resistant to it. Thus, *P. aeruginosa* cells carrying a pyocin can suppress growth or completely eliminate other susceptible *P. aeruginosa* cells in a given environmental niche (Kohler et al. 2010).

5.5.2 Type VI Secretion System

The type VI secretion system (T6SS) has been characterized as a secretion system of Gram-negative bacteria relatively recently compared to the other five types (Mougous et al. 2006; Pukatzki et al. 2006). It is a remarkably versatile system: T6SS was shown to mediate the interaction between a T6SS-expressing bacterial cell, its eukaryotic host and other bacteria (reviewed in (Records 2011). T6SS manifests itself in an increased pathogenicity or determines it for some strains of *Vibrio cholerae*, *P. aeruginosa*, *Edwardsiella tarda*, *E. coli*, and other bacteria. T6SS can function as a mediator of competitive interbacterial interactions and as a regulator of a cooperative social behavior. T6SS is found in plant-associated bacteria, including both pathogens and symbionts (Records 2011).

T6SS genes are clustered in pathogenicity islands containing 15–20 open-reading frames and can be easily detected using a set of conserved genes as markers. It is found in about one quarter of all bacterial genomes currently present in the database, often in several copies per genome. In addition to the contiguous cluster, homologs of certain T6SS genes could be found in as many as 25 copies in different genomic loci. The T6SS genes are predicted to encode cytoplasmic, periplasmic, and membrane-associated proteins, ATPases, lipoproteins, and various substrates, which are typically recognized by virtue of their extracellular secretion (reviewed in Filloux et al. 2008). Secreted substrates lack recognizable signal sequences and appear in culture supernatant as unprocessed polypeptides. Inactivation of individual T6SS genes causes the secreted proteins to be trapped in either the bacterial cytoplasm or periplasm, depending on the mutation (Raskin et al. 2006). One of the curious characteristics of the T6SS is that some of the proteins required for the function of the apparatus are also extracellularly secreted by the system (Mougous et al. 2006; Raskin et al. 2006).

Several T6SS proteins show either structural or sequence similarity to phage tail proteins (Leiman et al. 2009). In particular, T6SS contains an ortholog of the tail tube protein called Hcp (the *h*emolysin-*c*oregulated *p*rotein), an ortholog of the gp5–gp27 complex called VgrG (*v*aline-*g*lycine *r*epeat protein *G*), and a T4 gp25-like protein. Furthermore, two T6SS proteins with MWs of ~19 and ~55 kDa form a structure virtually identical to T4 polysheath in low-resolution electron microscopy images. This structure can be disassembled by the T6SS-encoded AAA+ ATPase in the presence of ATP and Mg ions. The presence of Hcp and VgrG in the external medium is the hallmark of active T6SS. VgrG often carries the "effector" or the pathogenicity domain at its C-terminus. Using T6SS, bacteria are able to deliver this large (1,000+ residues) trimeric protein directly into the cytoplasm of the host cell. Considering the information outlined above, it was proposed that T6SS forms a large phage tail-like structure, which is able to extend from the cell surface to attack the host cell. Furthermore, this complex uses a contractile tail-like mechanism for translocation of large molecules, such as a VgrG trimer, into the host-cell cytoplasm.

There are several features of T6SS, which set it apart from contractile phage tails. First, only three of the five baseplate proteins listed in Table 5.3 are found in T6SS. The orthologs of T4 gp6 and gp53 (or P2 gpI) are not found in the T6SS gene cluster. Second, there is no tape measure protein. Third, T6SS sheath is formed by two proteins. Fourth, phage tail contraction is a one-time irreversible event, but T6SS sheath must act continuously, and thus, it must be able to reset itself.

Nevertheless, T6SS is clearly related to phage contractile tails (Leiman et al. 2009). Because the structure of T6SS machine is unknown, it is very difficult to create a description of its mechanism, which would explain all the peculiarities of its structure and function.

5.5.3 *Photorhabdus* Virulence Cassette and Rhapidosomes

The cytotoxic activity toward eukaryotic cells by *Serratia entomophila* and *Photorhabdus* species is attributed to the expression of a set of genes from the so-called PVC (Hurst et al. 2004; Yang et al. 2006; Hurst et al. 2007). These genes assemble into structures resembling R-type pyocin particles. Bioinformatic analysis of the PCV gene cluster shows that it contains two orthologs of the tail tube protein or Hcp (*afp1* and *afp5*), three sheath orthologs (*afp2*, *afp3*, and *afp4*), a protein with a glycosidase activity (*afp7*), the T4 gp5–gp27 complex ortholog (*afp8*), a T4 gp25-like protein (*afp9*), a T4 gp6 ortholog (*afp11*), and an AAA+ ATPase (*afp15*). The PVC gene cluster appears to represent a link between the pyocins and T6SS. On the one hand, it contains several tube and sheath proteins and the T6SS-like ATPase, but on the other hand, PVC appears to contain the complete minimal baseplate and, most importantly, is secreted into the medium, a property, which has never been seen for T6SS, but instead is a common feature of R-type pyocins and phages. Thus, the PVC is likely a pyocin-like structure, adapted to kill eukaryotic cells in a similar one-off stroke.

The function and structure of rhapidosomes are quite unclear due to limited data (Yamamoto 1967). Rhapidosomes appear to represent pyocin-like structures produced by *Photobacterium*, *Proteus*, and *Saprospira*. Apparently, many environmental isolates use such objects to inhibit growth of other strains or closely related species in their ecological niche.

5.6 Conclusion

The analysis presented in this chapter shows that the presence of just two proteins, orthologs of T4 gp25 and gp27, in a phage-like cluster of genes is a required and sufficient condition to define this cluster as a contractile tail relative. Orthologs of T4 gp25, but not the sheath or tube proteins, are better conserved and can often be detected using sequence alignment alone. For example, T4 gp25 and P2 gpW are the only two proteins from T4 and P2 tails with a sequence identity of 26%, which is slightly above the "gray area" threshold (Haggard-Ljungquist et al. 1995). The orthologs of T4 gp27 are more diverse, but because of their characteristic and unique secondary structure and domain organization, they can be identified with help of HMM algorithms (Soding 2005; Soding et al. 2005) with a high degree of confidence even if they display no significant sequence identity to known gp27 orthologs.

Contractile tail-like systems are widespread in the biosphere. Remarkably, the contractile tail-like secretion systems can penetrate through both, the bacterial and eukaryotic cell envelopes, suggesting that the underlying principle of creating a membrane channel for subsequent protein/DNA translocation must be universal for bacteria and eukaryotes. However, the structure of many proteins, which are involved in creating this channel, and their fate during membrane penetration remain unknown at present.

Acknowledgments We would like to express our sincere thank you to Prof. Michael Rossmann and Prof. Venigalla Rao, the editors of this book, for choosing the perfect timing in organizing this volume. This book was a unique opportunity for many scientists to summarize and publish the information, which was understood by many in the field because it was presented at conferences, but has never been published. We are very grateful to Prof. Rao for his great efforts in persuading the contributors to produce interesting chapters and, at the same time, keeping the entire project to a reasonable deadline. Discussions with Prof. Alan Davidson, Prof. Ian Molineux, and Dr. Anastasia Aksyuk were extremely useful in preparation of this manuscript. We are very grateful to Prof. Liang Tong for sharing the coordinates of one of the sheath proteins prior to the publication of the paper describing the crystal structure.

References

Abuladze NK, Gingery M, Tsai J, Eiserling FA (1994) Tail length determination in bacteriophage T4. Virology 199:301–310

Ackermann HW (2003) Bacteriophage observations and evolution. Res Microbiol 154:245–251

Aksyuk AA, Leiman PG, Kurochkina LP, Shneider MM, Kostyuchenko VA, Mesyanzhinov VV, Rossmann MG (2009a) The tail sheath structure of bacteriophage T4: a molecular machine for infecting bacteria. EMBO J 28:821–829

Aksyuk AA, Leiman PG, Shneider MM, Mesyanzhinov VV, Rossmann MG (2009b) The structure of gene product 6 of bacteriophage T4, the hinge-pin of the baseplate. Structure 17:800–808

Arisaka F, Tschopp J, Van Driel R, Engel J (1979) Reassembly of the bacteriophage T4 tail from the core-baseplate and the monomeric sheath protein P18: a co-operative association process. J Mol Biol 132:369–386

Bartual SG, Otero JM, Garcia-Doval C, Llamas-Saiz AL, Kahn R, Fox GC, van Raaij MJ (2010) Structure of the bacteriophage T4 long tail fiber receptor-binding tip. Proc Natl Acad Sci USA 107:20287–20292

Boulanger P, Jacquot P, Plancon L, Chami M, Engel A, Parquet C, Herbeuval C, Letellier L (2008) Phage T5 straight tail fiber is a multifunctional protein acting as a tape measure and carrying fusogenic and muralytic activities. J Biol Chem 283:13556–13564

Coombs DH, Arisaka F (1994) T4 tail structure and function. In: Karam JD (ed) Molecular biology of bacteriophage T4. American Society for Microbiology, Washington, DC, pp 259–281

Coombs DH, Eiserling FA (1977) Studies on the structure, protein composition and assembly of the neck of bacteriophage T4. J Mol Biol 116:375–405

Filloux A, Hachani A, Bleves S (2008) The bacterial type VI secretion machine: yet another player for protein transport across membranes. Microbiology 154:1570–1583

Fokine A, Chipman PR, Leiman PG, Mesyanzhinov VV, Rao VB, Rossmann MG (2004) Molecular architecture of the prolate head of bacteriophage T4. Proc Natl Acad Sci USA 101:6003–6008

Haggard-Ljungquist E, Jacobsen E, Rishovd S, Six EW, Nilssen O, Sunshine MG, Lindqvist BH, Kim KJ, Barreiro V, Koonin EV, Calendar R (1995) Bacteriophage P2: genes involved in baseplate assembly. Virology 213:109–121

Hurst MR, Beard SS, Jackson TA, Jones SM (2007) Isolation and characterization of the *Serratia entomophila* antifeeding prophage. FEMS Microbiol Lett 270:42–48

Hurst MR, Glare TR, Jackson TA (2004) Cloning *Serratia entomophila* antifeeding genes – a putative defective prophage active against the grass grub *Costelytra zealandica*. J Bacteriol 186:5116–5128

Jacob F (1954) Biosynthèse induite et mode d'action d'une pyocin, antibiotique de *Pseudomonas pyocyanea*. Ann Inst Pasteur (Paris) 86:149–160

Kahn ML, Ziermann R, Deho G, Ow DW, Sunshine MG, Calendar R (1991) Bacteriophage P2 and P4. Methods Enzymol 204:264–280

Kanamaru S, Ishiwata Y, Suzuki T, Rossmann MG, Arisaka F (2005) Control of bacteriophage T4 tail lysozyme activity during the infection process. J Mol Biol 346:1013–1020

Kanamaru S, Leiman PG, Kostyuchenko VA, Chipman PR, Mesyanzhinov VV, Arisaka F, Rossmann MG (2002) Structure of the cell-puncturing device of bacteriophage T4. Nature 415:553–557

Kikuchi Y, King J (1975a) Assembly of the tail of bacteriophage T4. J Supramol Struct 3:24–38

Kikuchi Y, King J (1975b) Genetic control of bacteriophage T4 baseplate morphogenesis. I. Sequential assembly of the major precursor, in vivo and in vitro. J Mol Biol 99:645–672

Kikuchi Y, King J (1975c) Genetic control of bacteriophage T4 baseplate morphogenesis. II. Mutants unable to form the central part of the baseplate. J Mol Biol 99:673–694

Kikuchi Y, King J (1975d) Genetic control of bacteriophage T4 baseplate morphogenesis. III. Formation of the central plug and overall assembly pathway. J Mol Biol 99:695–716

King J (1968) Assembly of the tail of bacteriophage T4. J Mol Biol 32:231–262

Kohler T, Donner V, van Delden C (2010) Lipopolysaccharide as shield and receptor for R-pyocin-mediated killing in *Pseudomonas aeruginosa*. J Bacteriol 192:1921–1928

Kondou Y, Kitazawa D, Takeda S, Tsuchiya Y, Yamashita E, Mizuguchi M, Kawano K, Tsukihara T (2005) Structure of the central hub of bacteriophage Mu baseplate determined by X-ray crystallography of gp44. J Mol Biol 352:976–985

Kostyuchenko VA, Chipman PR, Leiman PG, Arisaka F, Mesyanzhinov VV, Rossmann MG (2005) The tail structure of bacteriophage T4 and its mechanism of contraction. Nat Struct Mol Biol 12:810–813

Kostyuchenko VA, Leiman PG, Chipman PR, Kanamaru S, van Raaij MJ, Arisaka F, Mesyanzhinov VV, Rossmann MG (2003) Three-dimensional structure of bacteriophage T4 baseplate. Nat Struct Biol 10:688–693

Kuznetsova TA, Efimov AV, Aijrich LG, Kireeva IY, Marusich EI, Cappuccinelli P, Fiori P, Rappelli P, Kurochkina LP, Poglazov BF, Mesyanzhinov VV (1998) Properties of recombinant bacteriophage T4 tail sheath protein and its deletion fragments. Biochemistry (Mosc) 63:702–709

Leblanc C, Caumont-Sarcos A, Comeau AM, Krisch HM (2009) Isolation and genomic characterization of the first phage infecting Iodobacteria: phi PLPE, a myovirus having a novel set of features. Environ Microbiol Rep 1:499–509

Leiman PG, Arisaka F, van Raaij MJ, Kostyuchenko VA, Aksyuk AA, Kanamaru S, Rossmann MG (2010) Morphogenesis of the T4 tail and tail fibers. Virol J 7:355

Leiman PG, Basler M, Ramagopal UA, Bonanno JB, Sauder JM, Pukatzki S, Burley SK, Almo SC, Mekalanos JJ (2009) Type VI secretion apparatus and phage tail-associated protein complexes share a common evolutionary origin. Proc Natl Acad Sci USA 106:4154–4159

Leiman PG, Battisti AJ, Bowman VD, Stummeyer K, Muhlenhoff M, Gerardy-Schahn R, Scholl D, Molineux IJ (2007) The structures of bacteriophages K1E and K1-5 explain processive degradation of polysaccharide capsules and evolution of new host specificities. J Mol Biol 371:836–849

Leiman PG, Chipman PR, Kostyuchenko VA, Mesyanzhinov VV, Rossmann MG (2004) Three-dimensional rearrangement of proteins in the tail of bacteriophage T4 on infection of its host. Cell 118:419–429

Matsui H, Sano Y, Ishihara H, Shinomiya T (1993) Regulation of pyocin genes in *Pseudomonas aeruginosa* by positive (prtN) and negative (prtR) regulatory genes. J Bacteriol 175:1257–1263

McGuffin LJ, Bryson K, Jones DT (2000) The PSIPRED protein structure prediction server. Bioinformatics 16:404–405

Moody MF (1967a) Structure of the sheath of bacteriophage T4. I. Structure of the contracted sheath and polysheath. J Mol Biol 25:167–200

Moody MF (1967b) Structure of the sheath of bacteriophage T4. II. Rearrangement of the sheath subunits during contraction. J Mol Biol 25:201–208

Moody MF (1973) Sheath of bacteriophage T4. 3. Contraction mechanism deduced from partially contracted sheaths. J Mol Biol 80:613–635

Mosig G, Lin GW, Franklin J, Fan WH (1989) Functional relationships and structural determinants of two bacteriophage T4 lysozymes: a soluble (gene e) and a baseplate-associated (gene 5) protein. New Biol 1:171–179

Mougous JD, Cuff ME, Raunser S, Shen A, Zhou M, Gifford CA, Goodman AL, Joachimiak G, Ordonez CL, Lory S, Walz T, Joachimiak A, Mekalanos JJ (2006) A virulence locus of *Pseudomonas aeruginosa* encodes a protein secretion apparatus. Science 312:1526–1530

Nakayama K, Takashima K, Ishihara H, Shinomiya T, Kageyama M, Kanaya S, Ohnishi M, Murata T, Mori H, Hayashi T (2000) The R-type pyocin of *Pseudomonas aeruginosa* is related to P2 phage, and the F-type is related to lambda phage. Mol Microbiol 38:213–231

Piuri M, Hatfull GF (2006) A peptidoglycan hydrolase motif within the mycobacteriophage TM4 tape measure protein promotes efficient infection of stationary phase cells. Mol Microbiol 62:1569–1585

Poglazov BF, Efimov AV, Marco S, Carrascosa J, Kuznetsova TA, Aijrich LG, Kurochkina LP, Mesyanzhinov VV (1999) Polymerization of bacteriophage T4 tail sheath protein mutants truncated at the C-termini. J Struct Biol 127:224–230

Pukatzki S, Ma AT, Sturtevant D, Krastins B, Sarracino D, Nelson WC, Heidelberg JF, Mekalanos JJ (2006) Identification of a conserved bacterial protein secretion system in *Vibrio cholerae* using the Dictyostelium host model system. Proc Natl Acad Sci USA 103:1528–1533

Raskin DM, Seshadri R, Pukatzki SU, Mekalanos JJ (2006) Bacterial genomics and pathogen evolution. Cell 124:703–714

Records AR (2011) The type VI secretion system: a multi-purpose delivery system with a phage-like machinery. Mol Plant Microbe Interact 24(7):751–757

Riede I, Degen M, Henning U (1985) The receptor specificity of bacteriophages can be determined by a tail fiber modifying protein. EMBO J 4:2343–2346

Rossmann MG, Mesyanzhinov VV, Arisaka F, Leiman PG (2004) The bacteriophage T4 DNA injection machine. Curr Opin Struct Biol 14:171–180

Scholl D, Cooley M, Williams SR, Gebhart D, Martin D, Bates A, Mandrell R (2009) An engineered R-type pyocin is a highly specific and sensitive bactericidal agent for the food-borne pathogen *Escherichia coli* O157:H7. Antimicrob Agents Chemother 53:3074–3080

Sciara G, Bebeacua C, Bron P, Tremblay D, Ortiz-Lombardia M, Lichiere J, van Heel M, Campanacci V, Moineau S, Cambillau C (2010) Structure of lactococcal phage p2 baseplate and its mechanism of activation. Proc Natl Acad Sci USA 107:6852–6857

Serwer P, Hayes SJ, Thomas JA, Hardies SC (2007) Propagating the missing bacteriophages: a large bacteriophage in a new class. Virol J 4:21

Simon LD, Anderson TF (1967a) The infection of *Escherichia coli* by T2 and T4 bacteriophages as seen in the electron microscope. I. Attachment and penetration. Virology 32:279–297

Simon LD, Anderson TF (1967b) The infection of *Escherichia coli* by T2 and T4 bacteriophages as seen in the electron microscope. II. Structure and function of the baseplate. Virology 32:298–305

Soding J (2005) Protein homology detection by HMM-HMM comparison. Bioinformatics 21:951–960

Soding J, Biegert A, Lupas AN (2005) The HHpred interactive server for protein homology detection and structure prediction. Nucleic Acids Res 33:W244–W248

Steven AC, Trus BL, Maizel JV, Unser M, Parry DA, Wall JS, Hainfeld JF, Studier FW (1988) Molecular substructure of a viral receptor-recognition protein. The gp17 tail-fiber of bacteriophage T7. J Mol Biol 200:351–365

Tetart F, Desplats C, Krisch HM (1998) Genome plasticity in the distal tail fiber locus of the T-even bacteriophage: recombination between conserved motifs swaps adhesin specificity. J Mol Biol 282:543–556

Tetart F, Repoila F, Monod C, Krisch HM (1996) Bacteriophage T4 host range is expanded by duplications of a small domain of the tail fiber adhesin. J Mol Biol 258:726–731

Thomassen E, Gielen G, Schutz M, Schoehn G, Abrahams JP, Miller S, van Raaij MJ (2003) The structure of the receptor-binding domain of the bacteriophage T4 short tail fibre reveals a knitted trimeric metal-binding fold. J Mol Biol 331:361–373

Tschopp J, Arisaka F, van Driel R, Engel J (1979) Purification, characterization and reassembly of the bacteriophage T4D tail sheath protein P18. J Mol Biol 128:247–258

van Raaij MJ, Schoehn G, Burda MR, Miller S (2001) Crystal structure of a heat and protease-stable part of the bacteriophage T4 short tail fibre. J Mol Biol 314:1137–1146

Walter M, Fiedler C, Grassl R, Biebl M, Rachel R, Hermo-Parrado XL, Llamas-Saiz AL, Seckler R, Miller S, van Raaij MJ (2008) Structure of the receptor-binding protein of bacteriophage det7: a podoviral tail spike in a myovirus. J Virol 82:2265–2273

Williams SR, Gebhart D, Martin DW, Scholl D (2008) Retargeting R-type pyocins to generate novel bactericidal protein complexes. Appl Environ Microbiol 74:3868–3876

Yamamoto T (1967) Presence of Rhapidosomes in various species of bacteria and their morphological characteristics. J Bacteriol 94:1746–1756

Yang G, Dowling AJ, Gerike U, ffrench-Constant RH, Waterfield NR (2006) Photorhabdus virulence cassettes confer injectable insecticidal activity against the wax moth. J Bacteriol 188:2254–2261

Chapter 6
Long Noncontractile Tail Machines of Bacteriophages

Alan R. Davidson, Lia Cardarelli, Lisa G. Pell, Devon R. Radford, and Karen L. Maxwell

Abstract In this chapter, we describe the structure, assembly, function, and evolution of the long, noncontractile tail of the siphophages, which comprise ~60% of the phages on earth. We place particular emphasis on features that are conserved among all siphophages, and trace evolutionary connections between these phages and myophages, which possess long contractile tails. The large number of high-resolution structures of tail proteins solved recently coupled to studies of tail-related complexes by electron microscopy have provided many new insights in this area. In addition, the availability of thousands of phage and prophage genome sequences has allowed the delineation of several large families of tail proteins that were previously unrecognized. We also summarize current knowledge pertaining to the mechanisms by which siphophage tails recognize the bacterial cell surface and mediate DNA injection through the cell envelope. We show that phages infecting Gram-positive and Gram-negative bacteria possess distinct families of proteins at their tail tips that are involved in this process. Finally, we speculate on the evolutionary advantages provided by long phage tails.

6.1 Introduction

Tailed phages are the predominant type of phage on earth, comprising 95% of the total (Ackermann 2007). These phages are divided into three classes based on morphology: long noncontractile (*Siphoviridae* or siphophages), long contractile (*Myoviridae* or myophages), or short (*Podoviridae* or podophages). These tails fulfill two essential functions in phage propagation (1) attachment of the phage particle to the outer surface of the bacterium, and (2) injection of the phage DNA into the host cell. The construction of a phage tail involves many different phage

---

A.R. Davidson (✉) • L.G. Pell
Department of Molecular Genetics, University of Toronto, Toronto, ON, Canada

Department of Biochemistry, University of Toronto, Toronto, ON, Canada
e-mail: alan.davidson@utoronto.ca

L. Cardarelli
Department of Biochemistry, University of Toronto, Toronto, ON, Canada

D.R. Radford
Department of Molecular Genetics, University of Toronto, Toronto, ON, Canada

K.L. Maxwell
The Donnelly Centre for Cellular and Biomolecular Research, Department of Molecular Genetics, University of Toronto, Toronto, ON, Canada

M.G. Rossmann and V.B. Rao (eds.), *Viral Molecular Machines*,
Advances in Experimental Medicine and Biology 726, DOI 10.1007/978-1-4614-0980-9_6,

gene products and requires a carefully regulated multistep assembly pathway. The prevalence of tailed phages in spite of the complexity and metabolic cost of building these structures implies that tails must provide phages with a tremendous evolutionary advantage. Although attachment to the cell surface could likely be achieved through many mechanisms, injection of the genome through the thick peptidoglycan layer of the Gram-positive bacterium, or the labyrinthine double membrane of the Gram-negative bacterium may present the most difficult challenge in the life cycle of a phage. A tail appears to provide the best solution to this problem, and may represent one of nature's best designed machines for the transfer of macromolecules into bacteria.

In this chapter, we describe the long noncontractile tail of the siphophages, which are the most abundant class of tailed phages. A description of the contractile tail of myophages is provided in Chapter 5. We first provide an overview of the genomic and structural aspects of noncontractile tails, placing particular emphasis on features that are most conserved and on current knowledge pertaining to tail assembly. The second section explains how the tail components are utilized during the process of cell surface attachment and DNA injection. Finally, we discuss issues related to the evolution of long phage tails. In presenting this material, we include new insight gleaned from our own unpublished sequence and structural analyses, and some speculation that we hope will not seem too far-fetched. Remarkably, this chapter represents the first ever general review on siphophage tails; thus, we had to cover a broad range of topics and many decades worth of research.

6.2 The Structure of the Noncontractile Tail

6.2.1 General Architecture and Assembly of a Noncontractile Phage Tail

As alluded to above, long phage tails can be divided into two distinctive classes: *noncontractile*, which are found in the siphophages, and *contractile*, which are found in the myophages. While this chapter is focused on noncontractile tails, definite evolutionary connections between these two types of long tails have become apparent in the past few years, and this issue is raised at many points in the chapter. Figure 6.1 displays the conserved features and protein components of the noncontractile tail. As depicted in this figure, we have adopted generic names and abbreviations for these proteins. The use of these names is aimed at clarifying our discussion as most of these proteins are confusingly designated by either numbers or letters that are not consistent among different phages.

As is depicted in Fig. 6.1, the "top" surface of the tail, which interacts with the head, is formed by the tail terminator protein (TrP). The TrP attaches to the top of the tail tube by binding to the tail-tube protein (TTP), which is the most abundant protein in the *Siphoviridae* tail and comprises the bulk of the tubular portion. Tail tubes are usually composed of a stack of hexameric rings of the TTP. The tape measure protein (TMP) is found within the tail tube and its size determines the tube length by precisely regulating the number of tail tube rings that can polymerize into the tail structure. The tail tip complex (TTC) is positioned at the "bottom" of the tail and is responsible for interaction with the host cell. The TTC varies in structure and protein composition among different groups of siphophages with some displaying a narrow tapered tip while others possess diverse "baseplate" assemblies. Cell surface attachment is usually mediated by tail fibers, which project from the TTC. We have divided tail fibers into two groups depending on their position within the TTC, with central fiber proteins (CFPs) being attached to the center of the TTC and side fiber proteins (SFPs) emanating from the side of the TTC. The folding of SFPs is aided by the side fiber chaperones (SFCs). Two more families of proteins, which we refer to as the tail completion proteins (TCPs) and tail assembly chaperones (TACs), are also highly conserved in tail operons yet their functions are poorly defined.

Although the siphophage tail assembly pathway has been extensively characterized only for the phage λ system, the basic assembly steps are very likely to be the same in all of these phages. As will

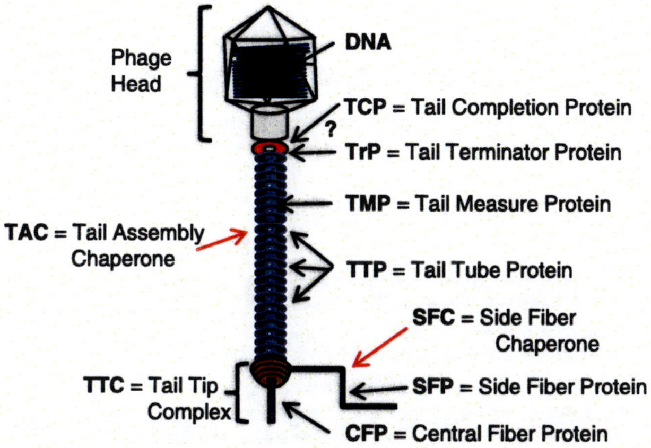

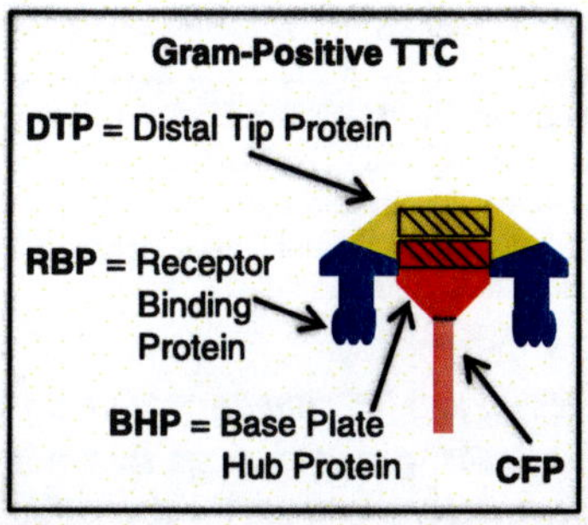

Fig. 6.1 The conserved features and proteins of the siphophage tail. A schematic of the typical siphophage tail is shown. The names and abbreviations used in this chapter for conserved tail proteins are indicated. Proteins designated with *red arrows* are not found in the final tail structure, but likely serve as chaperones. The "?" indicates that the position of the TCP within the phage tail is not known. The *inset* shows schematic of features of the tail tip complex of a Gram-positive infecting siphophage. RBPs and CFPs are not always both present. The CFP may or may not be covalently joined to the BHP. The *hatched boxes* in the Gram-positive TTC represent regions of the indicated proteins that display the tail-tube fold

be described below, assembly initiates with formation of the TTC. The TTC serves as a platform upon which TTP polymerization occurs. The TMP is attached to the TTC prior to TTP polymerization, and this reaction proceeds until the end of the TMP is reached. At this point, the TrP is added to permanently halt polymerization.

6.2.2 The Conserved Order of Siphophage Tail Genes

In general, the genes encoding proteins comprising the tails of diverse siphophages are arranged in a similar order. In light of the extreme sequence diversity seen among siphophage tail proteins, this conserved gene order provides crucial information for the prediction of gene function. The tail gene order usually follows the "body plan" of the phage tail with genes encoding proteins at the top of the tail (e.g., the TrP) appearing at the 5′-end of the operon and genes encoding proteins at the bottom of the tail being positioned at the 3′-end. Landmarks distinguishing most tail operons include large open-reading frames encoding the TMP and tail fibers (CFPs and/or SFPs) and a programmed translational frameshift involving the TAC, the gene of which is almost always located between the TTP and TMP genes. In Fig. 6.2, we show the genomic layout of well-studied siphophages that will be highlighted

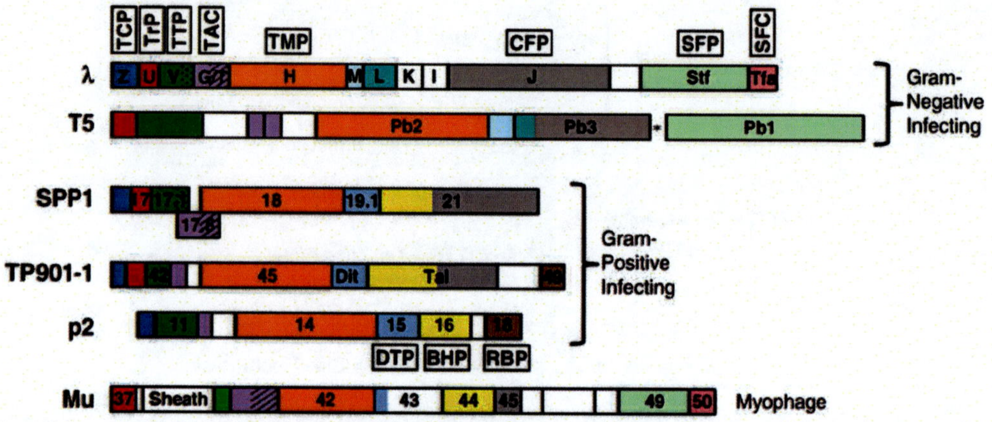

Fig. 6.2 The conserved genome arrangements of siphophage tail-encoding regions. Tail-encoding regions from siphophages, *E. coli* phages λ, and T5, *B. subtilis* phage SPP1, *L. lactis* phages TP901-1 and p2, and the *E. coli* myophage Mu are shown. Genes encoding proteins that are likely to be homologous are depicted in the same color. Homology was established through searches using PSI-BLAST or HHPred. The generic abbreviations we have used for phage tail proteins are shown above the genes. Certain abbreviations are used only for Gram-positive infecting phages. Genes colored in *gray* all encode spike or fiber proteins positioned at the center of the tail. However, homology has not been established for all of these proteins. Ig-like domains attached to TTPs are shaded as are experimentally established frameshifts in TAC genes

in this chapter: *Escherichia coli* phages λ, and T5, *Bacillus subtilis* phage SPP1, and *Lactococcus lactis* phages TP901-1 and p2. In some cases, gene functions shown in Fig. 6.2 have been assigned through our own bioinformatic analyses performed in the process of preparing this chapter.

6.2.3 The Tail Terminator Protein

Although tail length is determined by the TMP, as will be discussed in Sect. 6.2.5, addition of a TrP to the top of the tail is required to permanently halt tail polymerization, and form the interface for head attachment. Phage bearing nonsense mutations in TrP-encoding genes are unable to assemble viable phage particles, and lysates of such mutants display aberrantly elongated tails that are not attached to heads. Mutations affecting likely TrPs have been described in several myophages [Mu (Grundy and Howe 1985), P2 (Lengyel et al. 1974), T4 (Vianelli et al. 2000), and SPO1 (Parker and Eiserling 1983)]. Phage λ gpU is the only characterized TrP from a siphophage, but it has been studied in much more detail than any other TrP.

GpU is monomeric at high concentration (~1 mM) on its own in solution, and its structure in this state was determined using NMR spectroscopy (Edmonds et al. 2007). Upon contact with the hexameric ring of the TTP at the top of the tail, gpU hexamerizes into a ring of similar dimension to the tail tube (Katsura and Tsugita 1977). The biologically relevant hexameric structure of gpU was solved by X-ray crystallography, and mutagenesis studies based on this structure allowed identification of the top (i.e., interacting with the head) and bottom surfaces of the gpU ring (Fig. 6.3) (Pell et al. 2009a). It was found that amino acid substitutions at crucial positions on the top surface resulted in the production of normal length tails that could not attach to heads, presumably because these mutants were still able to interact with the tail tube and prevent aberrant polymerization. These mutants, when expressed from plasmids within infected cells, also inhibited the growth of wild-type λ phage ("dominant negative" phenotype). On the other hand, mutants bearing substitutions at the bottom surface of gpU were not dominant negative, and elongated tails were produced in

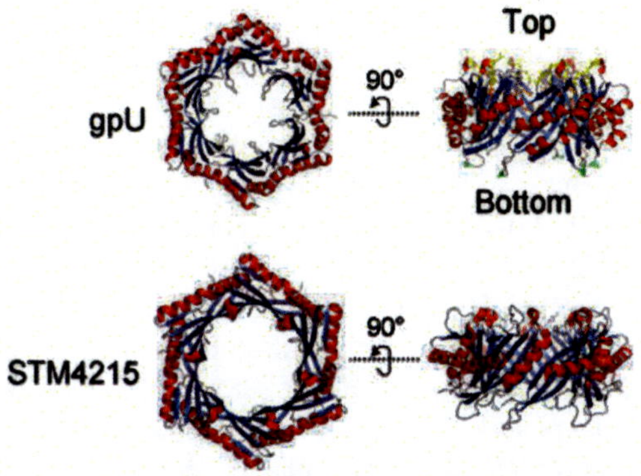

Fig. 6.3 The crystal structures of the TrP of phage λ (gpU) and a prophage-encoded homologue. The structure of the biologically relevant hexameric form of gpU is shown (PDB ID: 3FZ2). The experimentally determined top (head binding) and bottom (tail binding) surfaces of the gpU hexamer are indicated (Pell et al. 2009a). Specific residues implicated in tail (Asp54) and head binding (Arg111 and residues 29–33) are shown in stick representation, and are colored *green* and *yellow*, respectively. The structure of gpU is compared to that of STM4215 (PDB ID: 2GJV), which is encoded in a prophage-derived region of *Salmonella*. STM4215 is a component of a myophage tail

their presence, indicating that they were unable to interact with the tail tube and halt polymerization. Comparison of the NMR and crystal structures of gpU showed that regions on the bottom surface of the ring undergo large conformational transitions upon hexamerization, which involve unstructured regions becoming ordered. The unstructured regions in monomeric gpU were hypothesized to prevent self-hexamerization until the tail tube is contacted. These studies on gpU illustrate some recurring themes in investigations on morphogenetic proteins; namely, that unstructured regions often play crucial roles in regulating assembly, and that the isolation of dominant negative mutations can lead to the visualization of informative assembly intermediates. In addition, insight into the structural transitions occurring upon assembly can be obtained by comparing monomeric NMR structures with oligomeric crystal structures of the same protein.

Since all long-tailed phages must avoid uncontrolled tail-tube polymerization and create a head-joining interface at the top of the tail tube, they are all expected to possess a TrP. However, λ is the only siphophage with a proven TrP, and extensive PSI-BLAST (Altschul and Koonin 1998) searching for sequences similar to gpU yielded very few hits to other siphophages (Pell et al. 2009a). By assuming that the TrP-encoding genes of siphophages generally lie adjacent to the TTP-encoding gene as is seen for λ (Fig. 6.2), we identified putative TrPs in diverse siphophages and found that the predicted secondary structure of these proteins was similar to the known secondary structure of gpU, despite the lack of sequence similarity between gpU and these other proteins (Edmonds et al. 2007). This observation suggests that these proteins may be TrPs. It is notable that many putative TrPs from diverse siphophages, which we have identified based on genome position, can be connected by PSI-BLAST iterative sequence searches (our own unpublished results). For example, a PSI-BLAST search using the putative TrP of HK97 as a query revealed sequence similarity to putative TrPs in *B. subtilis* phage SPP1, *Pseudomonas aeruginosa* phage D3, and a variety of *Staphylococcus* and *Streptococcus* phages. Recently, a putative siphophage TrP from a prophage of *B. bronchisepta* was solved by a structural genomics group (PDB ID: 2L25). Its structure closely resembles that of gpU and its sequence is similar to the putative TrP of *E. coli* phage T1, and those of many other siphophages. This structure further supports the common evolutionary origins of TrPs with highly diverged sequences.

Although the sequence of gpU could not be linked by sequence alone to putative TrPs of most siphophages, gpU is clearly evolutionarily related to the TrPs of a large number of myophages. Through a database search, the structure of gpU was found to be very similar to that of protein

STM4215 (PDB ID: 2GJV), which is encoded in a prophage-related region of *Salmonella* (Fig. 6.3). Although this prophage element is uncharacterized, the gene encoding STM4215 lies adjacent to other genes encoding components of a contractile tail and the gene order resembles the tail region of the myophage Mu (Fig. 6.2). STM4215 was crystallized as a hexameric ring with very similar tertiary and quaternary structures to that observed for the hexameric crystal structure of λ gpU (Pell et al. 2009a). Through PSI-BLAST searches and detection of transitive homology, the sequences of gpU and STM4215 were shown to be related to one another and to a diverse group of putative myophage TrPs (Pell et al. 2009a). These related proteins include phage Mu gp37 (gpK), which is almost certainly a TrP based on its genomic position (Fig. 6.2) and the phenotype of mutants in the gene encoding it (Grundy and Howe 1985; Smith et al. 2010). The conservation of structure and sequence among putative TrPs from diverse myo- and siphophages suggests that TrPs may be a conserved feature of all long-tailed phages that may have arisen from a common ancestor.

6.2.4 The Tail-Tube Protein

The TTP is the most abundant protein within siphophage tails, and is also referred to as the major tail protein in many publications. Here we use the term "tail-tube protein" to differentiate this protein from the tail sheath protein of myophages, which has also been referred to as a major tail protein, and to emphasize the probable common evolutionary origin of TTPs of both myo- and siphophages. The TTP of λ, gpV, is the only TTP that has been studied intensively. The N-terminal domain of gpV (gpV_N, residues 1–160) is necessary and sufficient for the formation of a tail tube, while the C-terminal domain (gpV_C, residues 161–246), which is a member of the Ig-like domain family, plays an accessory role (see Sect. 6.3.1) (Katsura 1981). The λ tail tube is comprised of a stack of 32 hexameric rings of gpV. Purified unassembled gpV exists in a monomer–dimer equilibrium (Katsura and Tsugita 1977), but when the TTC is encountered, it readily hexamerizes and polymerizes into a tube (for a review of λ tail assembly see Katsura 1983). The TMP (gpH in phage λ), extends inside the tail tube, and determines the number of gpV hexamers that are incorporated.

Determination of the solution structure of gpV_N revealed a strong structural similarity with the TTP of the well-characterized PBSX prophage of *B. subtilis* (Fig. 6.4a) (Pell et al. 2009b), which was solved by a structural genomics group. This was a striking finding because the particles produced by PBSX possess a contractile tail; thus demonstrating that myo- and siphophage tail tubes are constructed with the same building block. Although the TTP of PBSX and gpV_N display no detectable sequence similarity, this structural similarity, combined with the gene order conservation observed in regions encoding these proteins in most sipho- and myophage genomes (Fig. 6.2), provides strong evidence of an evolutionary link between the tubes of contractile and noncontractile tails. Due to the extreme sequence divergence among TTPs and other conserved tail proteins, the detection of structural similarities between these distantly related proteins is crucial for tracing their evolution.

The structure of gpV_N was also discovered to be very similar to the putative tube protein of the Type VI secretion system, Hcp1 (Fig. 6.4b) (Pell et al. 2009b). This structural similarity provided additional support to previous observations that this secretion system is evolutionarily related to phage tails (Leiman et al. 2009). Remarkably, Hcp1 is assembled into hexameric rings within its crystal lattice that stack to form tubes (Fig. 6.4c). The dimensions of the Hcp1 hexameric rings (90 Å, with a 40-Å central hole) are strikingly similar to the dimensions of the hexameric rings of gpV in tails (90 Å, with a 30-Å central hole; Katsura 1981). Modeling of gpV_N into the hexameric structure of Hcp1 led to a prediction of the assembled structure of gpV_N in the tail and mutagenesis provided biological support for this model. These data provided the first molecular details of how a TTP assembles within the tail tube.

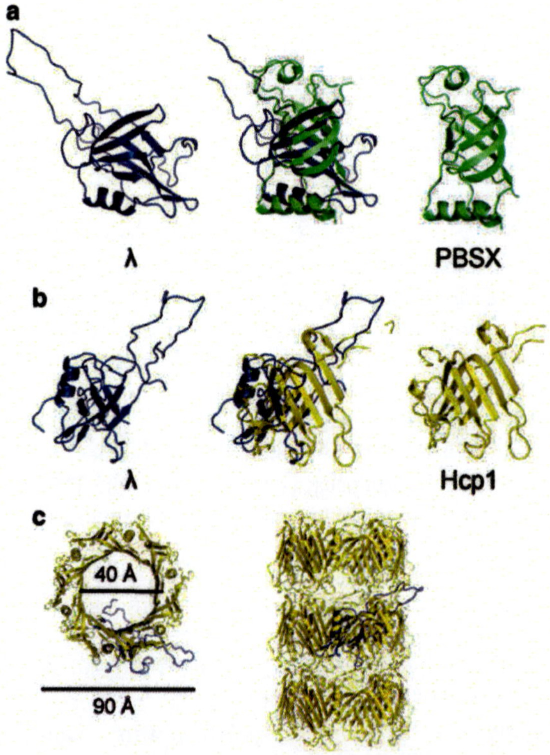

Fig. 6.4 Comparison of the structure of the N-terminal domain of the TTP of λ (gpV_N) to related structures. (**a**) The structure of gpV_N (PDB ID: 2K4Q) is superimposed on the TTP structure (PDB ID: 2GUJ) from the *B. subtilis* prophage-encoded myophage-like particle, PBSX. These structures can be overlaid with an RMSD of 1.9 Å over 57 equivalent C_α positions. (**b**) The structure of gpV_N is superimposed on the structure of Hcp1 (PDB ID: 1Y12), the likely tube protein from the Type VI secretion system. These structures can be overlaid with an RMSD of 2.1 Å over 74 equivalent C_α positions. (**c**) Hcp1 forms a hexamer which is of the same size as the rings of gpV found within λ tails (Pell et al. 2009b). As shown on the *right*, Hcp1 also formed tubes upon crystallization that resemble phage tails. One molecule of gpV_N (*blue*) is overlaid on the Hcp1 hexamer (*yellow*) in this figure

Given the structural conservation of TTPs in both a contractile and noncontractile tail, as well as the Type VI secretion system, we predict that this conservation will be seen in all long-tailed phages. One apparent outlier among TTPs is that of phage T5, which exhibits threefold symmetry rather than sixfold (Effantin et al. 2006). Interestingly, this TTP, which has few identifiable homologues, is approximately twice the size of other TTPs. It is possible that it possesses two copies of a TTP-like fold within its sequence, such that a pseudohexameric structure might exist. This phenomenon occurs in baseplate hub proteins (BHPs) and will be discussed in Sect. 6.2.7.1.

6.2.5 *The Tape Measure Protein*

The TMP is a conserved feature of all long phage tails, and it plays key roles in both tail assembly and injection of the phage DNA into the host cell. The TMP attained its name because its length controls the length of the phage tail. Within any particular species of phage, virtually every particle possesses a tail with precisely the same length. It has been shown for both sipho- and myophages that this precisely determined tail length can be shortened or elongated by a proportional change in the length of the TMP (Abuladze et al. 1994; Katsura and Hendrix 1984; Pedersen et al. 2000). Katsura most elegantly illustrated the role of the TMP in tail length determination by generating a series of deletions and one duplication in the λ TMP that resulted in tails ranging from less than half the size of wild-type to one more than 20% longer (Katsura 1987). The tail lengths observed for the 12 different mutants correlated almost perfectly with the length of the mutant TMPs.

The position of the TMP within the tail as well as its precise mechanism of action have not been unequivocally determined. However, it has been shown for both λ (Katsura and Kuhl 1975) and *L. lactis* phage TP901-1 (Vegge et al. 2005) that the TMP is a component of the precursor complex upon which TTP polymerization is initiated. Since all TMPs are predicted to be highly helical over most of their length, they are surmised to form extended helices that are encased within the tail tube. Supporting this idea, the amino acid lengths of most TMPs modeled as a continuous helix are consistent with the observed length of their associated phage tail. For example, the 853 amino acid TMP of λ would form a single continuous helix approximately 128 nm in length (~1.5 Å translation per residue in an α-helix), which is similar to the ~135 nm length of the λ tail tube. Given that the TMP forms a long helix that is attached to the tail tube initiator complex, the TTP likely polymerizes around the TMP and this polymerization pauses once the end of the TMP has been reached. At this point, the TrP must be incorporated or polymerization will continue, resulting in aberrantly long tails. The TMP may function by simply allowing TTP polymerization, while at the same time preventing the interaction of the TrP with the growing tail tube. This possibility is supported by the observation that the λ TrP will prevent the formation of any tail tube–related structures when TTP polymerization is initiated by aberrant initiator complexes lacking the TMP (Katsura 1976). Furthermore, the λ TrP is expressed at a higher level than the TTP (Murialdo and Siminovitch 1972), which would facilitate its rapid addition to the growing tail tube once the end of the TMP was reached. The TMP is clearly not essential for TTP polymerization as this reaction can occur in the absence of the TMP in several different mutant backgrounds of phage λ (Katsura 1976).

The stoichiometry of the TMP of λ was estimated at 6–7 per phage particle from microdensitometer analysis of protein gels (Casjens and Hendrix 1974a), which fulfills the expectation that six copies of the TMP per phage are required to match the sixfold symmetry of the tail tube. Since a six-helix bundle could fit inside the confines of a tail tube (Fig. 6.5), this stoichiometry is reasonable. Interestingly, many TMPs possess recognizable conserved folded domains within their sequences. These domains often have high similarity to peptidase (e.g., pfam family: Peptidase_M23, PF01551; Finn et al. 2010) or peptidoglycan-degrading (e.g., pfam family: SLT, PF01464) domains. They are usually positioned near the C-terminal end of TMPs and might protrude from the bottom of the tail tube. However, deletion of one of these domains from the TMP of Mycobacteriophage TM4 did not completely abrogate function. This deletion did result in a shorter tail (Piuri and Hatfull 2006), strongly implying that this folded domain, which is functionally important under some conditions (see Sect. 6.3.3), must be squeezed within the interior of the tail tube. The means by which TTP polymerization on the TMP could cause deformation of a folded domain is intriguing and has not been investigated.

Another interesting feature of TMPs is that several, including those of phages T5 (Zweig and Cummings 1973), TP901-1 (Pedersen et al. 2000), and PSA (Zimmer et al. 2003), are found within phage particles in a proteolytically processed form. Studies on phage λ TMP showed that approximately 10 kDa are cleaved from the C terminus (Walker et al. 1982), and that this cleavage occurs during tail assembly only after TTP polymerization is complete and the TrP has been incorporated (Tsui and Hendrix 1983). Since the C terminus of the TMP lies at the bottom of the tail, there must be a means by which completion of tail polymerization is sensed at the bottom of the tail so that cleavage is brought about at that location. The purpose for TMP cleavage and whether TMP cleavage occurs by the same mechanism and for the same purpose in different phages is not known.

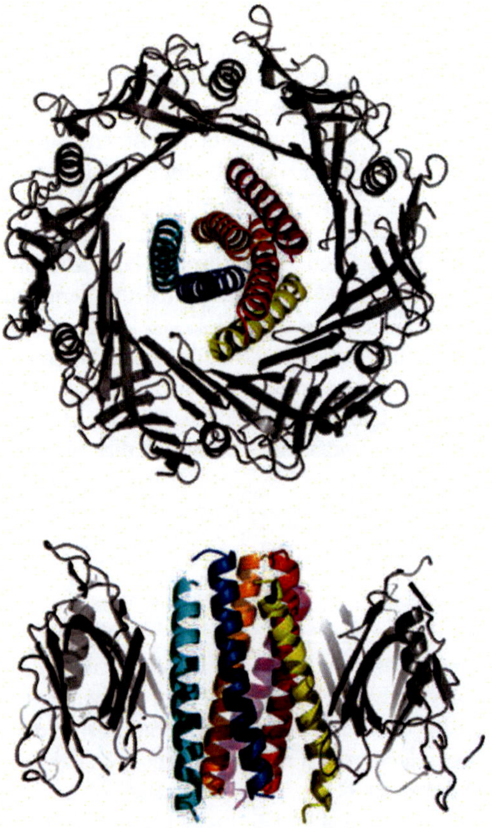

Fig. 6.5 A six-helix bundle, as may be formed by the TMP, can fit inside a phage tail tube. The six-helix bundle formed by gp41 of HIV (PDB ID: 1K33) was positioned inside the hexameric ring formed by the tube protein of the Type VI Secretion System (PDB ID: 1Y12). This ring possesses approximately the same inner diameter as a phage tail tube. The *lower* picture is a sideview with the front monomers removed

6.2.6 The Tail Assembly Chaperone

Another conserved feature of the genomic regions encoding long phage tails is the presence of two ORFs between those encoding the TTP and the TMP. These ORFs correspond to the *G* and *T* genes in λ and have often been referred to by these names. In λ, gpT is never made on its own, but is only synthesized in a form where it is fused to the C terminus of gpG to form a protein known as gpGT (Levin et al. 1993). The gpGT fusion protein is formed through a −1 translational frameshift, which occurs at the end of the *G* ORF. This frameshift occurs at a frequency of approximately 4% resulting in a fixed ratio between the abundance of gpG and gpGT during infection. Remarkably, the presence of a predicted frameshift mechanism is highly conserved across very diverse long-tailed phages in genes encoding the likely homologues of gpG and gpGT (Xu 2001). Frameshifting in these genes has been experimentally proven for a number of different phages, including myophages [e.g., *E. coli* phages Mu (Xu et al. 2004) and P2 (Christie et al. 2002)] and siphophages infecting diverse bacterial species (e.g., *Streptomyces* phage φC31, *Mycobacteria* phage L5 and *E. coli* phage HK022; Xu et al. 2004).

Although the functions of gpG and gpGT have not been fully elucidated, these proteins and their homologues have been referred to as TACs. This designation is based on the observation that gpG and gpGT are required for tail assembly and gpG is very highly expressed during a λ infection at a level similar to the TTP, yet neither protein is found within assembled λ particles (Levin et al. 1993). The gpG to gpGT ratio is crucial for proper tail assembly, which explains the high conservation of the frameshift between these two ORFs as frameshifting ensures a fixed ratio between the two proteins regardless of transcription or translation level (Xu 2001). The covalent attachment of the gpT moiety to gpG is also essential for function. Crystallographic studies of a putative TAC (as deduced from its genomic position and presence of a frameshifted ORF) from *L. lactis* phage p2 revealed a spiral structure with an internal diameter of 36 Å, which is similar to the internal diameter of a tail tube (Siponen et al. 2009). The authors speculated that the TAC may form a spiral around the long helical portion of the TMP and keep this protein in solution. A putative interaction between the TAC and TMP is consistent with experiments showing that gpG of λ binds to the TMP and increases its solubility (Xu 2001), and with the observation that in the absence of the putative TAC of *E. coli* myophage P2, its TMP is proteolyzed at a rapid rate (Lengyel et al. 1974). The biological relevance of the spiral seen for the phage p2 TAC is supported by other unpublished structures present in the PDB (3KLU and 3FGX) that we have identified as likely prophage-encoded TACs. These proteins also form spirals in the crystal with internal diameters of approximately 40 Å. The TAC of phage HK97 (2OB9), which was solved by X-ray crystallography in our laboratory, was found to form an octameric ring with a 40-Å diameter. All these structures support the hypothesis that the TAC forms a spiral or rings around the TMP, and may carry out its chaperone function through this mechanism. The function of the C-terminal extension placed onto the TAC by frameshifting is not known, but some experiments have suggested that this region of λ gpGT binds to the TTP (Xu 2001).

6.2.7 The Tail Tip Complex

While the upper portions of the tails of all siphophages and myophages are likely constructed with related components assembling through similar mechanisms, tail tip structures are more diverse. EM studies have revealed a variety of structures at the tips of siphophages including tapered conical tips with a long straight fiber at the end as seen on phages λ and SPP1, baseplate-like structures as seen on *L. lactis* phages TP901-1 and p2, and even rosettes as seen at the tip of *E. coli* phage RTP (Wietzorrek et al. 2006). The nature of the receptor-binding proteins found at siphophage tail tips can vary markedly with many phages, such as λ, SPP1, and T5 utilizing fibers for interaction with their hosts while other phages, such as p2, use only a globular receptor-binding protein (RBP). TP901-1 displays both globular RBPs and a fiber on its tail. Despite this apparent diversity, recently solved structures of siphophage tail tip proteins point to evolutionarily conserved features among diverse tail tips of sipho- and myophages.

6.2.7.1 The Baseplate Hub Protein

In myophages, the TTC is generally referred to as the "baseplate" and a component known as the baseplate hub protein (BHP) lies at the center of this structure (see Chapter 5). In the *E. coli* myophage T4, the six wedges of the baseplate are connected to the BHP, gp27, which forms a central ring-shaped trimer. The BHP serves as the adaptor to connect the sixfold symmetric tail tube to the threefold symmetric tail spike (gp5), which protrudes from the bottom of the tail. This adaptor function is possible because the two domains comprising the top of the BHP possess the same tertiary structure; thus they form a pseudo-hexameric ring within the trimeric BHP structure that matches the sixfold symmetric components of the tail tube (Fig. 6.6a). The BHP of *E. coli* myophage Mu, known as gp44 or gpP, possesses the same tertiary and quaternary structure as T4 gp27 despite a lack of detectable sequence similarity between these two proteins (Kondou et al. 2005). The conservation of this structure in these diverse phages and the membership of Mu gp44 in a very large sequence family (pfam family: PF05954) imply that the baseplate hub is a widely conserved feature of myophages. A very similar structure is also found as a component of the Type VI bacterial secretion system, which is related to myophage tails (Leiman et al. 2009).

The crystal structure of the complete TTC of the *L. lactis* siphophage, p2, revealed the presence of a trimeric central component, composed of ORF16, with strong structural similarity to the myophage

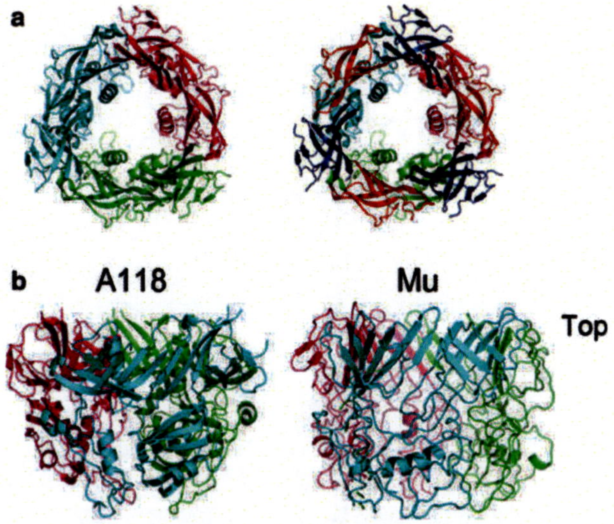

Fig. 6.6 Structures of BHPs. (**a**) A *top view* of the *Listeria* prophage BHP shows the trimeric structure (*left*) and the pseudohexameric structure (*right*). In the *right side* depiction, domain 1 (residues 3–90) is colored *blue* and domain 4 (residues 178–201, 268–322) is colored *red*. These domains possess a very similar tertiary structure. (**b**) The structure of a BHP encoded by a *Listeria* prophage (PDB ID: 3GS9) that is 95% identical to the BHP of *Listeria* phage A118 is shown next to the BHP from *E. coli* myophage Mu (PDB ID: 1WRU). The domains comprising the upper ring of these structures (domains 1 and 4) can be superimposed on each other with an RMSD of 1.6 Å over 93 equivalent C_α positions. The second domains of these structures can be superimposed on each other with an RMSD of 1.6 Å over 62 equivalent C_α positions, and the third domains display an RMSD of 1.7 Å over 39 equivalent C_α positions

BHPs (Sciara et al. 2010). ORF16 occupies a similar position as the myophage baseplate hubs at the bottom of the tail tube; however, it does not connect to a tail spike or to baseplate wedge subunits. Since ORF16 possesses very few homologues detectable through sequence comparisons, the question of whether BHP-like proteins are a widespread component of siphophage tails was not clear upon publication of this structure. However, an unpublished structure (3GS9) of a protein encoded by a *Listeria* prophage possesses strong structural similarity to BHPs (the individual domains of this protein can be overlaid on those from other BHPs with RMSD values of 1.8 Å or less, Fig. 6.6b). This protein is 95% identical to gp18 of the characterized *Listeria* siphophage A118, and is encoded in a genomic position consistent with its performing the function of a BHP (i.e., its gene is located 2 ORFs downstream of the TMP, Fig. 6.2). Our own bioinformatic studies focusing on gp18 indicate that homologues are found in almost all siphophages infecting Gram-positive organisms, including well-characterized phages such as SPP1 and TP901-1. However, no BHP-related sequences were detected in Gram-negative-infecting siphophages. Since there are currently no structures available for any TTC protein of a Gram-negative infecting siphophage, whether a new family of BHPs will be found in this class of phages remains an open question.

A distinctive feature of the BHP-fold-containing proteins of siphophages is their highly variable length. While gp18 of phage A118 is only 341 amino acids and is entirely encompassed by the BHP structure, phage tail proteins with strong similarity to gp18 at their N-termini range in length to well over 1,000 residues. Many of these proteins, such as Tal of phage TP901-1 and gp21 of phage SPP1, are known to encode the CFPs of these phages. While in myophages the trimeric tail spike is attached to the BHP in a noncovalent manner, it appears that in many siphophages, the likely trimeric tail fiber is attached covalently to the BHP. Interestingly, some of these BHP-fiber fusion proteins are known to be proteolytically cleaved during tail assembly (Vegge et al. 2005; Kenny et al. 2004), and it is possible that cleavage of the fiber from the hub must occur at some point in the infection cycle for all of these phages. In some cases, this cleavage may occur after adsorption onto the cell surface.

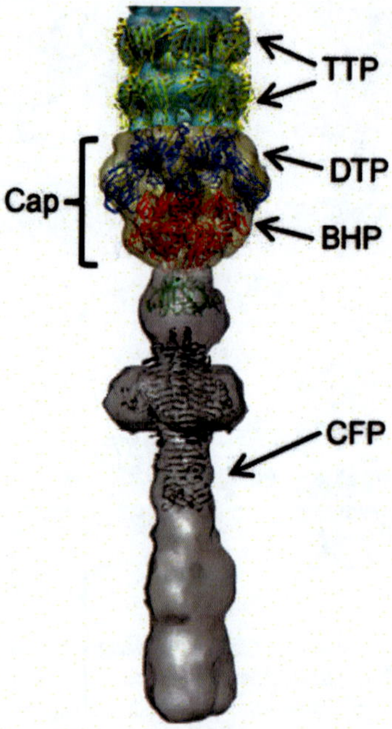

Fig. 6.7 An EM-based model of a Gram-positive infecting phage tail. The electron density outline derived from a three-dimensional reconstruction of the phage SPP1 tail based on negatively stained images is shown (Plisson et al. 2007). Two hexamers of Hcp1 (*yellow*) were fit into the region corresponding to the tail tube. The "cap" region was filled with one hexamer of the DTP from phage SPP1 (*blue*) (PDB ID: 2X8K), and one trimer of the BHP from phage p2 (*red*) (PDB ID: 2WZP). The head-binding domain (PDB ID: 1LKT) and main domain (PDB ID: 1TSP) from *E. coli* podophage P22 tail spike were previously fitted (Plisson et al. 2007) into the EM density corresponding to the CFP (gp21). This fitting shows that a protein with a Tail Spike structure can fit into this density. There is no evidence of homology between the SPP1 and P22 proteins. This figure was supplied through the courtesy of D. Veesler, C. Cambillau, and coworkers and a manuscript describing this has been submitted

6.2.7.2 The Distal Tail Protein

Another highly conserved feature of the TTCs of siphophages infecting Gram-positive bacteria is the distal tip protein (DTP), which is also known as Dit. PSI-BLAST searches reveal sequence similarity between the DTP homologues of most Gram-positive siphophages, including SPP1, TP901-1, and A118 (Veesler et al. 2010). Recent X-ray crystallographic and cryoEM studies on the DTP of phage SPP1 indicated that the DTP forms a hexameric ring that lies between the bottom of the tail tube and the top of the BHP. The likely arrangement of the BHP and TTP within a reconstruction of the intact SPP1 tail obtained using contrast EM (Plisson et al. 2007) is shown in Fig. 6.7. The DTP of phage SPP1 is very similar in structure to the DTP of phage p2 (ORF15) despite a lack of detectable sequence similarity between these proteins. The DTP of phage p2 was solved in the context of the complete TTC and it was seen to interact with the BHP in the same manner as that postulated for the SPP1 proteins in Fig. 6.7 (Sciara et al. 2010). The DTP is a two-domain protein with the N-terminal domain comprising the central ring and the C-terminal domains protruding from this ring. In phage p2, the C-terminal domain interacts with proteins comprising the baseplate to which the RBPs are attached, while the tapered tip of SPP1 does not require this interaction.

It should be noted that the placement of the DTP and BHP of SPP1 in the tail EM density, as shown in Fig. 6.7, contradicts the previous interpretation of this EM reconstruction (Plisson et al. 2007), which had assigned the conical "cap" region at the bottom of the tail as being filled by a C-terminal domain of the TMP. This assignment was based on previous work showing that antibodies reactive against the TMP of TP901-1 labeled the tip of the tail as visualized by immunogold TEM (Mc Grath et al. 2006). However, to achieve this labeling of the TMP, a baseplate protein (BppU or ORF48) mutant had to be used, which might have destabilized the tail and allowed exposure of the TMP during preparation of the EM micrographs. The fitting of the DTP and BHP shown in Fig. 6.7 is probably correct because there are no other regions of EM density into which these proteins can be fit in any satisfactory manner (the fitting of the DTP into this density is described in (Veesler et al. 2010), and fitting of the BHP will be described in a future manuscript; C. Cambillau, personal communication). This interpretation of the SPP1 electron density agrees with a recent EM analysis of the TTC of phage TP901-1 where a similarly positioned, cone-shaped electron density at the bottom of the tail tube was assigned as a BHP structure comprised of the N-terminal domain of the Tal protein (ORF47) (Bebeacua et al. 2010). This arrangement of the tail tube, DTP, and BHP is likely conserved among most Gram-positive infecting siphophages.

Although DTP and BHP sequences are highly conserved among phages infecting Gram-positive bacteria, extensive PSI-BLAST searching performed in our laboratory failed to detect any similar sequences among tail proteins of phages infecting Gram-negative hosts. This result demonstrates that certain proteins in the TTCs of siphophages are distinctly conserved among phages infecting Gram-positive bacteria. This situation contrasts sharply with conservation patterns seen among proteins in other locations of the tail (e.g., the TTPs, see Sect. 6.4.1) where proteins from phages infecting both types of bacteria are found within the same sequence family. Since some proteins in the TTC must interact directly with the bacterial cell envelope, this sequence divergence of TTC components found in Gram-positive- and Gram-negative-infecting phages is explainable if these proteins interact with the cell surface.

6.2.7.3 The TTC of Gram-Negative Infecting Siphophages and Initiation of Tail Assembly

Although no structural data is available for proteins comprising the TTCs of Gram-negative-infecting siphophages, the TTC of phage λ is the best characterized of all siphophage TTCs with respect to assembly (Katsura 1983). The assembly of the λ tail initiates at the CFP, gpJ, the activity of which could be isolated on its own by sucrose-gradient ultracentrifugation (Katsura and Kuhl 1975). Another assembly intermediate was isolated in the same manner after addition of gpI, gpK, and gpL. The TMP (gpH) assembles on to this complex in the presence of gpG/GT to form another discrete complex, and then gpM incorporates to complete the "initiation complex" onto which the TTP, gpV, is assembled. Although the biological activity of each of these tail assembly intermediates was detected after separation on sucrose gradients, none were isolated in quantities large enough for subsequent characterization with respect to protein composition or structure. However, based on both experimental evidence (Casjens and Hendrix 1974a) and comparison with other phage tails we expect that the λ CFP is trimeric and that one of the TTC proteins must serve as a trimer-to-hexamer adaptor in a manner analogous to the BHPs of myophages and Gram-positive infecting siphophages. In negatively stained EM micrographs of the λ tail tip, there are protein rings clearly visible below the rings of the TTP (Katsura 1976); thus, some of the proteins comprising the TTC likely oligomerize into ring-like structures upon assembly. In the Gram-positive infecting siphophage, TP901-1, the Tal protein, which is a fusion of the BHP and CFP, binds to the DTP and TMP to form the initiator of tail-tube polymerization (Vegge et al. 2005). The possible analogy between this system and phage λ has been pointed out, but the evolutionary relationships between the TTC components in these two types of phages are not yet clear. We expect that the determination of the structures of TTC proteins of Gram-negative-infecting siphophages will aid in clarifying this issue.

The appearance of the tail tips of most Gram-negative-infecting siphophages is similar to that of λ suggesting that these structures may be assembled from building blocks that are related to the λ proteins mentioned above. This idea is supported by the presence of proteins related to gpM, gpL, gpK, and gpI in diverse members of this phage family. For example, using extensive PSI-BLAST

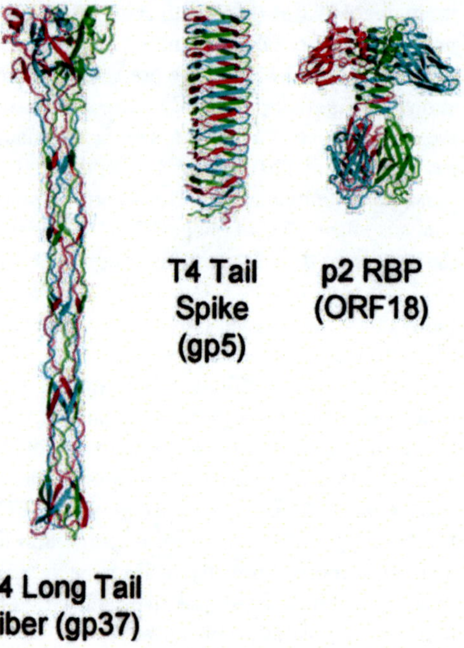

Fig. 6.8 Structures of tail fibers, spikes, and receptors. The structures of the myophage T4 long tail fiber (gp37, PDB ID: 2XGF), the T4 tail spike (gp5, PDB ID: 1K28), and the *L. lactis* phage p2 RBP are shown (ORF18, PDB ID: 1ZRU)

searches with a variety of starting queries, we have been able to identify clear homologues to λ gpL in almost every Gram-negative-infecting siphophage (unpublished results). In most cases, the genomic location of the genes encoding these putative homologues was consistent with gpL (i.e., the gene was located between the TMP and CFP genes, Fig. 6.2). Interestingly, no cross-hits were observed in these searches to TTC proteins of Gram-positive infecting siphophages. This result parallels our observation that no homologues of certain Gram-positive TTC components (i.e., the BHP and DTP) could be detected in Gram-negative-infecting siphophages. It appears that a small subset of proteins at siphophage tail tips are highly adapted to the cell envelope type of the host.

6.2.7.4 Tail Fibers and Receptor-Binding Proteins

Siphophages bind to a variety of cell-surface molecules, including specific proteinaceous receptors, lipopolysaccharide (LPS), and teichoic acid. These interactions are most often mediated by long fibers (~1,000 amino acids) situated at or near the tip of the tail. Although no high-resolution structure of a siphophage tail fiber has been determined, they are likely at least partially composed of a trimeric intertwined β-strand structure. We make this supposition because this property is shared among all of the solved structures of phage "fibers" or "spikes" that are involved in cell-surface interaction. For example, the tail spikes of podophage P22 and myophage T4 (gp5) form a trimeric β-helical structure (Fig. 6.8). The T4 tail spike is thought to be a membrane-penetrating apparatus and the β-helical structure may be important for this function (Kanamaru et al. 2002). A similar β-helical structure is formed by the short tail fibers of T4 (gp12) (van Raaij et al. 2001). By contrast, the long tail fiber of T4 (gp37) forms a much more extended intertwined β-strand structure (Fig. 6.8). The presence of some repetitive β-strand structure in most tail fibers is supported by our observation that most are predicted to possess β-helical structure by the program β-wrap (Bradley et al. 2001), which was designed to predict the occurrence of this type of structure.

The Gram-negative-infecting siphophages λ and T5 possess fibers at the bottom and sides of their tail tips (Effantin et al. 2006; Hendrix and Duda 1992). The SFPs of λ (gpStf) and T5 (pb1, L-shaped tail fiber) can be linked by PSI-BLAST searches to the long tail fiber of T4, gp37. Interestingly, a portion of the λ SFP can functionally replace the C-terminal region of gp37 (Montag et al. 1989). Another commonality between the λ SFP and T4 gp37 is that they both require a chaperone for their proper folding, which we have designated the SFC. The λ SFC, gpTfa, can functionally substitute for the chaperone required for gp37 assembly, gp38 (Montag and Henning 1987). This example is striking because it is the only case of such close functional and sequence (26% identical) similarity between tail proteins from a sipho- and myophage. The SFCs comprise a large-sequence family (pfam ID: Caudo_TAP, PF02413), the members of which are found in diverse siphophages (e.g., *E. coli* phage HK97 and *Yersinia* phage PY54) and myophages (e.g., *E. coli* phages Mu and P2). The genes encoding SFPs are almost invariably located immediately 3′ to genes encoding a tail fiber, supporting the close functional relationship between SFPs and tail fibers.

In myophages, such as T4, Mu, and P2, the SFC-associated tail fibers are attached to the outer edge of the baseplate, while the "tail spike," a prism-like β-helical structure, descends from the center of the tail tube where it is attached to the BHP. We speculate that the CFPs of siphophages may be more closely related functionally and structurally to the tail spikes of myophages based on their common locations and apparent ability to fold without a chaperone. In Gram-positive infecting siphophages, BHPs and CFPs are often encoded within the same ORF, which demonstrates a common arrangement of these proteins in myophages and siphophages. The CFPs and SFPs may represent two fundamentally different classes of tail fibers with the CFPs possibly playing a more active role in membrane penetration in a manner similar to the myophage tail spikes (Kanamaru et al. 2002; Kageyama et al. 2009; Suzuki et al. 2010), while the SFPs are involved only in the recognition of cell surface molecules. The distinctive structure of phage T4 gp37 in comparison with the tail spike (Fig. 6.8) may be a general reflection of the structural differences between the two types of fiber. That said, it must be emphasized that the delineation of families of fiber proteins through sequence comparisons is difficult because they are very long and repetitive, and also can contain additional domains, such as Ig-like domains (Fraser et al. 2006), that result in cross-hits with many different proteins. Given that proteins related to SFCs are found almost exclusively in Gram-negative infecting phages, the SFP class of fiber, which seems to always be associated with an SFC, may have evolved primarily to negotiate the distinctive outer membrane architecture of these bacteria.

In contrast to most siphophages, some lactococcal phages possess globular RBPs at their tail tips. For example, the *L. lactis* phage p2 possesses a baseplate-like structure at its tail tip to which six trimeric RBP complexes are attached (Fig. 6.8) (Sciara et al. 2010). These receptor complexes can bind various saccharidic moieties and likely interact with teichoic or lipoteichoic acid on the cell surface. RBPs with similar structure are found on other lactococcal phages, such as TP901-1 and bIL170 (Spinelli et al. 2006; Ricagno et al. 2006). Although RBPs are different in structure from tail fibers, they still share a trimeric structure and also possess a short region of β-helix as is seen in tail-spike proteins (Fig. 6.8). Interestingly, phage p2 appears to bind to the cell surface using only its RBPs, while phage TP901-1 possesses both RBPs and a CFP, which may also be involved in cell-surface recognition and/or cell-wall degradation (see Sect. 6.3.3).

6.2.8 The Tail Completion Protein

Phage λ gpZ is the prototype for a highly conserved family of tail proteins that we have designated the TCPs. In the absence of gpZ, complete phage particles are produced, but they have low infectivity (Casjens and Hendrix 1974b). Neither the location of gpZ within phage particles nor its mechanism of action are known, but the genomic location of the *Z* gene between regions encoding the head and tail genes suggests that it may assemble near the head–tail connector regions (Fig. 6.2). GpZ is believed to be a component of phage particles because its biological activity was detected in dissociated purified phage particles (Katsura and Tsugita 1977). Through sequence-based searches, we have identified homologues of gpZ in more than 170 long-tailed phage genomes and more than 1,500 prophage genomes (unpublished results), making this one of the most highly conserved phage structural proteins. Most siphophage genomes possess a gpZ homologue that is encoded 2 ORFs upstream from the TTP-encoding gene as is seen in phage λ (Fig. 6.2). Remarkably, gpZ homologues are also found in a wide range of myophages and include gpS and gpG (Mup31) of *E. coli* phages P2 and Mu, respectively. Similar to λ, phage Mu mutants lacking gpG are able to form complete phage particles of normal appearance that possess low infectivity (Grundy and Howe 1985). Phage P2 *S* amber mutants predominantly form normal-looking tails that are unable to attach to heads (Lengyel et al. 1974), which is consistent with a role for this protein in the head–tail joining region. A small number of tails with vastly elongated sheaths were also seen in lysates of this mutant.

Insight into the function of TCPs was provided by studies on λ gpZ. In mature λ particles, the right end of the genomic DNA is found within the tail tube and penetrates approximately one-third of the way down the tube. This DNA positioning was detected by treating phage with a chemical cross-linker and then dissociating heads and tails with formamide, revealing DNA molecules attached to the top surface of tails (Thomas 1974). DNA penetration into the tail was not observed in particles lacking gpZ (Thomas 1978). Given that interactions between phage genomic DNA and the tail have been seen in mature particles of three diverse siphophages (SPP1 of *B. subtilis*, T5 of *E. coli*, and A118 of *Listeria monocytogenes*) (Tavares et al. 1996; Saigo 1975; Loessner et al. 2000), we hypothesize that the last step in the morphogenesis of many long-tailed phages may be the opening of the head–tail connector complex to allow one end of genome to occupy a position within the tail. In this way, the genome becomes poised for ejection upon contact with a host cell. The general function of TCPs may be to facilitate this reaction.

6.2.9 Observations Pertaining to the Tail Proteins of Phage T5

Phage T5 is an unusual siphophage in that its genome is relatively large (122 kbp, Sect. 6.4.5), and most of its tail proteins are not easily connected by sequence similarity to components of other characterized siphophages. Since T5 is a well-studied siphophage, we felt that it would be useful to investigate the relationships of T5 tail proteins to those from other phages. Using HHpred (Soding et al. 2005), we were able to assign T5 ORFs corresponding to the TrP and TAC families, and we also found that the ORF immediately downstream of the TMP gene encoded a protein in the same family as λ gpM (Fig. 6.2). PSI-BLAST searches revealed that the N-terminal region of pb3 is similar to λ gpL and the C-terminal region of this protein is similar to the λ CFP (gpJ). These sequence-based connections reveal a previously unrecognized gene order similarity between T5 and λ (Fig. 6.2), and the similarity between gpJ and pb3 implies that the long fiber seen at the center of the T5 tail is composed of pb3. However, this conclusion contradicts the previous designation of this fiber as being composed of pb2 (Heller and Schwarz 1985), which is the T5 TMP. We believe that this previous assignment of pb2 as the straight tail fiber is incorrect because at 1,219 residues, pb2 is simply not long enough to form an extended helix inside the ~170 nm T5 tail tube (Effantin et al. 2006) (Sect. 6.2.5), and extend for more than 50 nm below the tube. It should also be mentioned that T5 encodes another protein, pb5, that is encoded in the head gene region yet is required for binding to the T5 receptor, FhuA (Plancon et al. 2002). We were unable to connect this protein to other siphophage proteins, and its location within the T5 particle is not known.

6.3 How Does the Tail Machine Work? The DNA Injection Process

6.3.1 Binding to the Cell Surface

A common theme of cell-surface recognition by siphophages is a multistep recognition process. For example, early studies on phage λ showed that interaction with its cell-surface receptor, LamB, could be reversible or irreversible depending on conditions (Schwartz 1975). Irreversible binding resulted in inactivation of the phage in vitro due to ejection of the phage genome. Visualization of complexes between λ phage particles and LamB embedded in liposomes indicated that the reversible complex involved an interaction between the tip of the CFP (gpJ) and the receptor while a direct interaction between the tail tube and the liposome surface was seen in irreversible complexes (Roessner et al. 1983). DNA was also injected into the liposomes under these conditions. While the two steps of λ interaction with the cell surface involve only the LamB receptor, many phages interact with more than one cell surface molecule. The phage T5 tail first interacts nonspecifically with the O-antigen of *E. coli* LPS through its L-shaped tail fibers (pb1) (Heller and Braun 1982) and then makes strong specific interactions with cell-surface receptor, FhuA through another putative TTC protein, pb5 (Heller and Schwarz 1985; Plancon et al. 2002; Heller 1984). In an analogous manner, the Gram-positive infecting phage SPP1 binds reversibly to cell-wall teichoic acids (Baptista et al. 2008) and irreversibly to a specific cell-surface receptor, YueB (Sao-Jose et al. 2004). Interaction of SPP1 particles with the purified YueB ectodomain causes ejection of the genomic DNA (Sao-Jose et al. 2006). In keeping with the dual molecule recognition strategy of T5 and SPP1, the original isolate of phage λ (known as Ur-λ) possesses SFPs that aid in phage adsorption, likely through a nonspecific interaction with an abundant cell-surface molecule, such as LPS (Hendrix and Duda 1992; Shao and Wang 2008). These SFPs are not essential for λ propagation and were lost from many lab strains through a frameshift mutation (Hendrix and Duda 1992).

In contrast to the phages mentioned above, the well-characterized lactococcal phages p2, TP901-1, and Tuc2009 all have large baseplate assemblies onto which are arrayed 6 or 18 copies of the trimeric RBP complex (Sciara et al. 2010; Bebeacua et al. 2010). The RBPs of these phages bind to saccharidic moieties on the cell surface and are believed to achieve high-affinity binding through avidity resulting from the large number of RBPs displayed on the baseplate. Interestingly, the phage p2 baseplate appears to undergo a large conformational change upon contacting the cell surface, which is brought about by Ca^{2+}-binding (Sciara et al. 2010). One unusual *L. lactis* phage, Q54, utilizes a frameshifting mechanism to attach its RBP to the TTP, such that ~25% of the TTP molecules are fused to the RBP (Fortier et al. 2006). This phenomenon also serves to place a large number of copies of this RBP domain, which appears by sequence to be a carbohydrate binding moiety, onto the tail such that avidity effects could also come into play.

In a manner similar to phage Q54, many phages append accessory domains onto various tail proteins, possibly as a means to aid in cell-surface recognition. For example, approximately 25% of sipho- and myophage virions have been recognized to display Ig-like domains somewhere on their surface, often in a tail protein (Fraser et al. 2006). These domains are usually positioned at the C terminus of a virion protein and are often added using a frameshifting mechanism. Due to their sequence similarity to Ig-like domains in carbohydrate-metabolizing enzymes, we have speculated that these domains on phages are involved in binding interactions with cell-surface carbohydrates, such as LPS or teichoic acid (Fraser et al. 2006, 2007). In phage λ, deletion of the C-terminal Ig-like domain from the TTP causes a 100-fold reduction in plating efficiency (Pell et al. 2010), but the cause is not yet known.

Although too few systems have been analyzed to derive strong general statements about the process, cell-surface binding by phage tails appears to typically involve recognition of multiple cell-surface molecules by more than one tail protein. One might wonder why phages have evolved these multipronged systems. Is this just a fail-safe mechanism providing back-up binding modes in case one fails? Maybe, but we tend to think that there are deeper explanations. One consideration is the imperative for phage to avoid "premature ejection" of DNA that could be caused by interaction with cellular fragments, nonhost cells, or nonviable cells. By utilizing multiple receptors for binding to several different cell-surface molecules, the chance of DNA being ejected into an appropriate and intact cell may be increased. In the usual laboratory setting where single pure phage stocks are tested with single bacterial strains, competitive advantages provided by the complete set of tail fibers and RBPs may not be manifested as they would be in a more "natural" environment where various phages are competing for the same hosts and phages must also avoid wasting their DNA in futile interactions with isolated proteins, membrane fragments, and nonpermissive hosts.

6.3.2 The Mechanism of DNA Injection

The mechanism by which phage DNA moves from the interior of the phage head to the cytoplasm of the host cell remains as one of the least understood aspects of phage biology. While much research has been done in recent years on the physical aspects of DNA ejection from phage in vitro (Inamdar et al. 2006; Jeembaeva et al. 2008), there is much controversy in this field and the connection between in vitro results and in vivo mechanisms is not clear (Panja and Molineux 2010). We will focus here only on the in vivo data pertaining to DNA ejection. In this area, only two siphophage systems, λ and T5, have been examined in any detail.

In Gram-negative bacteria, after the initial interaction of the phage and cell surface, a channel must be created to traverse the two membranes and periplasmic space to allow the DNA to pass through efficiently. By investigating the injection of λ DNA into liposomes into which the LamB receptor has been incorporated, the formation of transmembrane channels as a result of phage interaction with LamB has been clearly demonstrated (Roessner and Ihler 1986). These channels allowed the escape of small molecules, but not protein. Furthermore, their conductance properties were very different from that of the LamB pore, implying that phage proteins mediated the formation of this channel and that DNA does not move through the LamB pore (Berrier et al. 2000). Since the TMP (gpH) or CFP (gpJ) of λ are the only individual TTC proteins long enough to form a continuous channel through both membranes of *E. coli*, one of these likely forms the channel for DNA injection. Measuring the changes in proteolytic susceptibility of gpJ and gpH during the interaction of λ with LamB-containing liposomes showed that gpJ became more susceptible to proteolysis upon formation of irreversibly bound complexes (see Sect. 6.3.1). Conversely, gpH became more protease resistant upon irreversible binding when phage particles interacted with LamB embedded in liposomes. This increased protease resistance did not result when particles interacted with soluble LamB. These results implied that gpH becomes buried within the liposome when phage-induced channels are formed and that gpH must be the tail constituent of this channel (Roessner and Ihler 1984).

Studies performed on phage T5 also concluded that the TMP (pb2) is able to form pores through lipid membranes and liposomes (Feucht et al. 1990; Boulanger et al. 2008). A region within the C-terminal 184 residues of pb2 formed oligomeric structures and was able to insert into liposomal membranes, eliciting fusogenic effects (Boulanger et al. 2008). Visualization using cryoelectron tomography of T5-mediated transfer of DNA into liposomes directly demonstrated that pb2 forms a channel through the membrane, and that this channel is distinct from that formed by the T5 cell surface receptor FhuA (Boulanger et al. 2008). The membrane insertion of pb2 was postulated to be mediated by two predicted membrane spanning helices in the C-terminal region.

The existing data for phages λ and T5 indicate a crucial role for the TMP in DNA injection. The general model that emerges from these studies is that after initial contact between the phage tail and the cell surface, which is mediated by tail fibers or RBPs, a significant conformational change must occur to bring the bottom of the tail tube into direct contact with the cell surface. At this point, release of DNA from the phage head, triggered by the binding reaction with the tail tip, would force the TMP out of the tail tube (we assume that DNA egress would be impossible without prior removal of the TMP) and it would subsequently penetrate the cell membrane and form the channel for DNA passage into the cell interior. The tail tubes of most siphophages are plugged at the bottom by the CFP. Since DNA is unlikely to be able to pass through these structures, removal of the CFPs prior to TMP ejection appears to be necessary. Consistent with this model, the CFP of λ has been shown to fall off when DNA is released from particles by osmotic shock (Konopa and Taylor 1979). In addition, structural data derived from EM studies showed directly that the SPP1 tail tip does not have a channel for DNA egress and that interaction of the tip with the bacterial receptor causes release of the tip from the tail (Plisson et al. 2007). In some cases, the CFP may be proteolytically cleaved, as are those of TP901-1 and Tuc2009 (Vegge et al. 2005; Kenny et al. 2004), to aid in its removal from the tail tip when necessary. Given the common occurrence of predicted protease-like domains in tail proteins (e.g., gpK of λ and its homologues possess a metalloprotease domain; Ambroggio et al. 2004), the hypothesis that CFPs may often be cleaved is reasonable. Some phages, such as *L. lactis* phage p2, completely lack a CFP; thus, the BHPs of these phages, which can adopt an open conformation to release the TMP and DNA (Sciara et al. 2010), may directly contact the cell surface without removal or cleavage of any tail proteins.

An outstanding mystery in the DNA injection process is how events occurring at the tail tip cause DNA to emerge from the head which is generally situated at least 100 nm away. EM studies on the SPP1 tail before and after receptor-mediated DNA ejection in vitro showed a conformational change in the TTP subunits of the tail. It was postulated that this conformational change in the tail tube carried a domino-like signal to the head–tail junction leading to DNA release (Plisson et al. 2007). Another model, which we favor, posits that the opening of the tail at the bottom allows the TMP to emerge and the DNA follows this protein. The penetration of the phage genome into the tail at the final step of particle assembly, which may be a universal feature of siphophages (see Sect. 6.2.8), could lead to a stable interaction between the DNA and the TMP that facilitates DNA egress. Given that the TMP also forms the channel for DNA entry, an association between these molecules within the phage particle could be advantageous.

A final issue that has been investigated little is the role of host proteins in the phage injection process. *E. coli* mutants, known as *pel*⁻, were shown to resist the injection of λ DNA even though λ particles could adsorb to the outer membrane of these mutants normally (Scandella and Arber 1974). These mutations were found to lie in the gene encoding ManY, which is an inner membrane component of the mannose transport system (Elliott and Arber 1978; Williams et al. 1986). The role of ManY in the injection process has not been characterized. Oddly, other members of the PtsM complex, of which ManY is a subunit, are not required for injection. Mutations that restored the ability of λ to grow on *pel*⁻ strains were found to lie in the TTP and TMP genes (Scandella and Arber 1976), highlighting a role for these proteins in the injection process. The only other host genes that have been shown to be specifically involved in DNA ejection are *E. coli dcrA* and *dcrB*. Mutations in these genes, which also encode inner membrane proteins, inhibit DNA ejection of phages C1 and C6 (Samsonov and Sineoky 2002). Several other tested phages were able to grow on the *dcrA*, *dcrB*, and *pel*⁻ strains, implying that different phages utilize distinct pathways for DNA injection even though some of the tested phages bound to the same outer membrane receptors. We expect that a variety of inner membrane and periplasmic proteins may be involved in the DNA injection process and identifying these proteins will be a fruitful area for future research.

6.3.3 The Role of Tail-Mediated Peptidoglycan Degradation

Since injected phage DNA must pass through a layer of cell-wall peptidoglycan in both Gram-negative and Gram-positive bacteria, the placement of enzymes on phage tails to degrade this barrier may be essential. Consistent with this idea, a screen of virion constituent proteins of diverse tailed phages using zymogram analysis revealed that many different phage particles do carry peptidoglycan hydrolytic activities (Moak and Molineux 2004). Oddly, a hydrolytic activity was not detected for the Gram-negative-infecting siphophages λ and HK022 despite them both possessing a protein (gpK in λ and gp19 in HK022) with a region that is clearly related to a domain implicated in peptidoglycan cleavage (pfam family: NlpC/P60, PF00877) (Rigden et al. 2003). The failure to detect activity in these particles may be due to a failure of the relevant proteins to renature after separation in SDS PAGE gels or the activity may only be manifested in intact phage particles during the infection process. Since motifs associated with peptidoglycan hydrolysis can be recognized in at least one tail protein in most siphophages (our own observations), there is likely a conserved role for this activity.

The peptidoglycan-degrading activities of domains located in the TMPs of phage T5 (Boulanger et al. 2008) and mycobacteriophage TM4 (Piuri and Hatfull 2006) have been characterized as having that of the Tal protein of *L. lactis* phage Tuc2009. Interestingly, the M23 family peptidase (pfam ID: PF01551) motif found in Tal mediates both peptidoglycan hydrolysis and self-cleavage of Tal itself (Kenny et al. 2004). The cleavage of Tal, which possesses a long CFP-like region, may be required for the proper positioning of the tail-tube bottom onto the outer surface of the cell for carrying out the injection process (see Sect. 6.3.2). A likely TMP encoded in the SPβ prophage of *B. subtilis* possesses two different peptidoglycan hydrolase domains (Sudiarta et al. 2010). This protein was not recognized as a TMP in the paper describing its cell-wall-degrading activities.

Despite the ubiquity of peptidoglycan-degrading domains in phage tails, the function and importance of these domains have been studied very little. Deletion of a peptidoglycan hydrolase motif in the TMP of mycobacteriophage TM4 did not affect the ability of this phage to grow on exponentially dividing cells, but did abrogate growth on stationary-phase cells (Piuri and Hatfull 2006). This result was consistent with experiments on *E. coli* myophage T4 (Kanamaru et al. 2005) and podophage T7 (Moak and Molineux 2000) showing that the peptidoglycan hydrolase activity present on the tails of these phages was necessary for infection only under certain conditions. At present, a plausible hypothesis is that peptidoglycan-degrading domains on phage may not be always required, but that they provide enough of an advantage under some conditions to make their presence on most siphophages advantageous.

6.4 The Evolution of Long Phage Tails

6.4.1 The Diversity of Tail Proteins

A striking feature of families of tail proteins performing the same function is their tremendous sequence diversity. For example, the sequences of the known TTPs from a group of characterized siphophages, Gram-negative-infecting phages λ and HK97, and Gram-positive infecting phages SPP1, A118, A2, and Sfi19, differ so markedly in their sequences that meaningful pairwise alignments cannot be made between any of them. However, extensive PSI-BLAST searches using these sequences as queries show that the TTPs of λ, HK97, A118, and SPP1 can be connected to one another after many iterations (our own unpublished results). Furthermore, the λ, HK97, and A118 sequences can be seen to be much more closely related to one another because they cross-hit many of the same sequences in PSI-BLAST searches. The cross-hit frequency was much lower between these sequences and the TTP of SPP1. The TTPs from phages A2 and Sfi19 are members of a distinct sequence family, possessing many cross-hits between them, but showing no cross-hits to the other TTPs. It is interesting that members of the λ group of TTPs are found in phages infecting a wide variety of Gram-positive and Gram-negative organisms while the A2 group are found only in phages infecting Gram-positive bacteria. Overall, these findings highlight the difficulties in tracing the evolutionary relationships between phage tail proteins. The importance of solved structures in this effort is also emphasized as structures remain conserved even after sequences diverge beyond recognizable similarity. It is important to note that phage tail protein sequence relationships that are impossible to detect using PSI-BLAST can often be uncovered using HHPred (Soding et al. 2005), which generates a hidden Markov model (HMM) from a PSI-BLAST search of the input sequence and compares it to databases of HMMs, such as Pfam.

A further challenge in comparing the tail-encoding regions of various phages is the often mentioned mosaic nature of phage genomes (Hendrix et al. 1999). For example, while the TTPs of λ and HK97 can be linked easily through sequence comparisons, their TrPs, encoded immediately upstream of the TTP in each genome, cannot be connected through PSI-BLAST searches. By contrast, the sequences of the TrPs of λ and myophage Mu can be connected through PSI-BLAST searches (Pell et al. 2009a) even though their TTPs cannot. Dramatic variations are often observed in the relatedness of proteins in different regions of phage tails. For example, phages λ and N15 have very similar TrPs and TTPs (69 and 61% identical, respectively) while the proteins comprising their tail tips range from 31 to 36% identity. Conversely, the TrP and TTP of N15 display almost no detectable similarity to those of HK97, yet proteins at the tail tip of N15 are very similar to HK97, ranging from 69 to 93% identity. It is clear that this phenomenon has occurred through a recombination event within the TMP gene of the progenitors of these phages. Other similar examples of recombination within morphogenetic regions have been reported (Juhala et al. 2000). These findings indicate that certain components of the tail are malleable enough to adapt to dramatic changes in the sequences of interacting proteins. Elucidating the mechanisms by which these recombination breaks are tolerated will provide new insight into the morphogenetic process.

6.4.2 The Common Ancestry of Long Phage Tails

Given that myophage and siphophage tails perform the same function and have a similar appearance, a common evolutionary origin for these structures is likely. There are now many lines of evidence supporting this common evolution that were alluded to in other sections of this review, and will be summarized here. TTPs (Pell et al. 2009b), TrPs (Pell et al. 2009a), and BHPs (Sciara et al. 2010) from sipho- and myophages have been shown to adopt very similar structures (Figs. 6.3, 6.4, and 6.6), and sequence similarity has been detected among some TTPs (our own unpublished results) and TrPs (Pell et al. 2009a) from the two types of tails using PSI-BLAST. HMM-based comparisons using HHPred support the evolutionary connection between several other groups of sipho- and myophage protein families (our own unpublished results). For example, an HHPred search using the sequence of the *Listeria* siphophage A118 BHP as a query yields high probability hits to the BHPs of the myophages P2 and Mu. Hits were also obtained to the BHP-like proteins of the type VI secretion system (Vgr protein family). The common ancestry of sipho- and myophage tails is also reflected in the conserved gene order among many genomic regions encoding the two tail types (Fig. 6.2). Particularly striking is the conservation of the frameshift involving the TAC genes, which always lie between the genes encoding the TTP and TMP. It is probable that the simpler siphophages were the first long-tailed phages, and that sheaths were later added to the original design to make myophage.

6.4.3 The Tail-Tube Fold: The Primordial Building Block of Long Phage Tails

Structural comparisons by our group led to the surprising discovery that gpFII, the head–tail connector protein of λ, possesses the same tertiary structure as the TTP of λ. In addition, functional studies indicated that gpFII and the TTP oligomerized into the same quaternary structure within the phage tail (Cardarelli et al. 2010). Further structural comparisons revealed that the same "tail-tube fold"

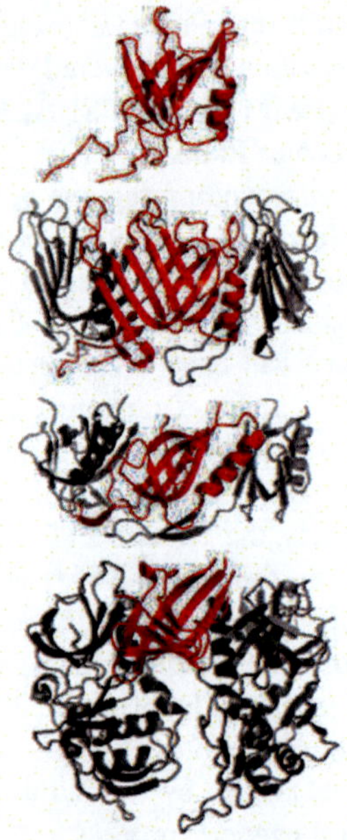

Fig. 6.9 The tail-tube fold. The monomeric structure of the TTP of λ is shown along with the hexameric structures of Type VI secretion system tube protein, Hcp1 (PDB ID: 2Y12), the DTP of phage SPP1 (PDB ID: 2X8K), and the trimeric structure of the *Listeria* prophage BHP (PDB ID: 3GS9). One domain of each structure that displays the tail-tube fold is shown in *red* in each structure. These domains can be superimposed with RMSD values of ~2 Å and possess essentially the same topology. The tail-tube fold segments are oriented the same way in the hexamer or pseudohexamer quaternary structure of each protein. The C-terminal domains of the DTP are omitted for clarity as are some of the subunits of the DTP and BHP rings

was also seen in the structures of several DTPs and BHPs, and that the quaternary arrangement of the tail-tube fold was the same in these structures (Fig. 6.9). We also detected a similar portion of tertiary structure to the tail-tube fold in TrP structures even though the overall fold of TrPs is somewhat different (Cardarelli et al. 2010). We concluded that all of these proteins derived from an ancestral TTP through the mechanism of gene duplication followed by differentiation (Cardarelli et al. 2010). The same mechanism has been suggested for the evolution of myophage T4 proteins gp10, gp11, and gp12, which are all situated in the baseplate and display the same protein fold (Leiman et al. 2006). We have also observed that the phage T4 TrP (gp3) and TTP (gp19) are similar enough in sequence to hit each other in PSI-BLAST searches, and that these searches also hit gp54, which forms a ring at the bottom of the tail tube. These proteins are clearly homologous as judged by sequence similarity yet perform distinct functions within the phage tail. These data provide unique insight into the mechanisms by which very elaborate multiprotein complexes may arise by iteration of simpler building blocks, and emphasize the great utility of phage systems for elucidating evolutionary mechanisms.

6.4.4 The Benefit of Having a Tail

According to the most recent comprehensive survey of 5,500 phages that have been observed by EM, approximately 95% possess either long tails (sipho- or myophages) or short tails (podophages) even though more than 15 visibly distinguishable types of phages have been identified (Ackermann 2007). While a survey like this could be somewhat biased because tailed phages may be generally easier to spot in EM micrographs, we nevertheless conclude that for a phage, having a tail likely provides a considerable evolutionary advantage. What might this advantage be? Consideration of the sizes of phage genomes provides a likely answer to this question. By analyzing the complete list of sequenced phage genomes available at NCBI, we discovered that the average size of tailed phage genomes is 62,000 bp ($n=471$) while the average size of nontailed phage genomes is only 7,300 bp ($n=57$). No class of nontailed phage possesses a genome of greater than 15,000 bp. These data imply that possession of a tail allows phages to utilize a much larger genome, and that having a larger genome provides an evolutionary advantage. A tail appears to be the only apparatus that is robust enough to transfer a large phage genome across the heavily armored bacterial cell surface.

While some might view it as obvious that a bigger genome is better, in the world of phages, one could have also concluded that being small and thus being able to replicate more quickly would be an equally effective approach. However, the available data lead us to conclude that even though making a tail requires a larger genome and a much more complicated morphogenetic pathway, this investment pays off for phages in allowing their genomes to attain sizes that are far in excess of what is required to encode the information needed for a simple lytic lifecycle. It is intriguing to consider that the bulk of the "extra" DNA carried by most tailed phages encodes "accessory" proteins, the functions of which are often not obviously connectable to the survival of the phage. We suspect that many genes in the regions of phage genomes that are dispensable for plaque formation in the lab provide significant advantages for survival of the phage under natural conditions or increase the fitness of the host when a lysogen is formed.

6.4.5 The Relationship Between Tail Type and Genome Size

Further analysis of the sequenced phages at NCBI reveals distinct relationships between genome size and tail type. A striking feature of siphophages is their narrow range of genome lengths. Among 241 examples, 80% possess genomes with lengths between 30,000 and 60,000 bp, with an average length of approximately 50,000 bp. Oddly, siphophages infecting mycobacteria are outliers in the distribution of genome lengths, with greater than 45% of these phages possessing genomes of greater than 60,000 bp. If mycobacterial phages are excluded, fewer than 10% of siphophages possess genomes of greater than 60,000 bp. By contrast, myophage genomes are much larger with the 101 examples averaging 110,000 bp in length. Greater than 60% of myophage genomes are larger than 60,000 bp. Although the heads of many myophages are larger than those of siphophages as is required to encapsulate their larger genomes, these heads are nonetheless essentially the same in structure and assembly as siphophages. Thus, the ability of myophages to generally bear considerably larger genomes is likely attributable to the superiority of the contractile tail design for delivering a large genome into bacteria. Since myophages and siphophage tails are clearly related (see Sect. 6.4.2) and myophages most likely evolved from the siphophages, we speculate that the addition of a tail sheath and contractile mechanism to the original noncontractile tail design provided a significant improvement in DNA injection performance. Indeed, the myophages may be in the process of overtaking the siphoviridae and will someday become the dominant family of phage on our planet. Fortunately for those of us who study siphophages, this is unlikely to happen in our lifetimes.

If one is willing to accept that myophages evolved from siphophages, it is entertaining to speculate as to when this event occurred. We have observed that among sequenced phages infecting Gram-positive bacteria ($n=210$), only 11% are myophages, while among phages infecting Gram-negative bacteria ($n=124$), 56% are myophages. If this trend is generally representative of the frequencies of these types of phages found in nature, then perhaps myophages emerged and spread first among Gram-negative bacteria, and subsequently spread to the Gram-positive bacteria through horizontal gene transfers. The improvement of phage tails seen in the myophages may have been driven by the added difficulty of penetrating the double membrane of the Gram-negative cell.

6.4.6 The Mystery of Tail Length

Although it is well established that the length of the TMP determines the length of phage tails in both types of long-tailed phages, a little discussed mystery is why phage tails are so long. Tail lengths of siphophages vary widely from ~80 nm to greater than 800 nm, with an average tail length of ~190 nm (Ackermann 2007). Through sequence searches, we have identified the TMPs from close to 200 different siphophages and 50 different myophages. As expected given the wide range in tail lengths, TMPs range from 600 to greater than 5,000 residues in length. On average, siphophages have longer tails than myophages with the average TMP length of siphophages being 1,200 residues and 900 residues for myophages. Remarkably, 60 siphophages and 3 myophages possess TMPs of 1,400 residues or longer. Why has this seemingly excessive tail length been tolerated in the fast-paced milieu of phage evolution? Given that the construction of these vastly elongated tails must expend considerable metabolic energy, we surmise that some advantage must be provided by increased tail length. At this point we can only speculate. One possibility is that extra-long tails increase the chance of a phage contacting a suitable host cell in conditions where cell concentration is low.

Another puzzling aspect of tail length is that even the shortest siphophage tails (~90 nm corresponding to a TMP of ~600 residues) seem longer than necessary for the job at hand. If the TMP must form a channel through the cell envelope, then the width of the cell envelope could place a lower limit on TMP length. However, the total width of a Gram-negative cell envelope is only 30–40 nm, while Gram-positive cell walls generally range from 20 to 80 nm. Additional insight into this problem can be gleaned by considering gene transfer agents (GTAs), which are widely observed phage-like particles in α-proteobacteria. Although GTAs are undoubtedly related to siphophages (i.e., their protein sequences are similar to phage proteins; Lang et al. 2002), these particles contain very small pieces of bacterial DNA (usually ~5,000 bp) that they transfer to other bacteria of the same species. GTAs possess small heads, as is expected due to the small size of their genomes, and also possess very small tails (~50 nm, TMP lengths of ~200 residues). In spite of their small tails, GTAs are still able to inject the DNA contained within their heads into host cells, implying that a TMP of 200 residues is sufficient for crossing a Gram-negative cell envelope. The existence of GTAs and their short tails leads us to hypothesize that the minimum length of the TMP is not set by the width of the cell envelope, but by the length of the phage genome. In other words, injection of a longer genome requires a longer TMP. Although there are currently no direct experimental data to support this hypothesis, siphophages with TMP lengths of less than 800 residues generally do possess considerably shorter genomes with an average length of 37,000 bp compared to an average length of 49,000 bp for phages with TMPs greater than 800 residues. A similar trend was seen in the myophages. These data suggest that the size of the genome may place the lower limit on the size of the TMP. However, within any given genome size range, the observed length of TMPs can vary widely, indicating that the upper limit for TMP size is very flexible.

6.5 Conclusion

The siphophage tail has been an intriguing object of study for 40 years. We hope that this chapter has convincingly demonstrated that much progress has been made in this time. Structural studies performed on individual phage tail proteins and tail complexes coupled with the increased power of bioinformatics as the sequence database grows are providing a powerful basis for future progress in understanding the function and evolution of phage tails. Nevertheless, many fundamental questions remain unanswered, particularly in the areas of tail assembly and mechanisms of DNA injection. In the future, an increased number of detailed mechanistic studies on tails from a variety of siphophages involving mutagenesis and functional assays will be required to gain a full understanding of these fascinating molecular machines.

Acknowledgments The authors thank Christian Cambillau for supplying the image used in Fig. 6.7. We also thank Petr Leiman for his encouragement during the writing of this chapter and for reading the final version. Work in our labs is supported by operating grants from the Canadian Institutes of Health Research to A.R.D. (Fund No. MOP-77680) and to K.L.M. (Fund No. MOP-6279).

References

Abuladze NK, Gingery M, Tsai J et al (1994) Tail length determination in bacteriophage T4. Virology 199:301–310

Ackermann HW (2007a) 5500 Phages examined in the electron microscope. Arch Virol 152:227–243

Ackermann HW (2007b) Bacteriophages: tailed, Encyclopedia of life sciences. Wiley, New York

Altschul SF, Koonin EV (1998) Iterated profile searches with PSI-BLAST – a tool for discovery in protein databases. Trends Biochem Sci 23:444–447

Ambroggio XI, Rees DC, Deshaies RJ (2004) JAMM: a metalloprotease-like zinc site in the proteasome and signalosome. PLoS Biol 2:E2

Baptista C, Santos MA, Sao-Jose C (2008) Phage SPP1 reversible adsorption to *Bacillus subtilis* cell wall teichoic acids accelerates virus recognition of membrane receptor YueB. J Bacteriol 190:4989–4996

Bebeacua C, Bron P, Lai L et al (2010) Structure and molecular assignment of lactococcal phage TP901-1 baseplate. J Biol Chem 285:39079–39086

Berrier C, Bonhivers M, Letellier L et al (2000) High-conductance channel induced by the interaction of phage lambda with its receptor maltoporin. FEBS Lett 476:129–133

Boulanger P, Jacquot P, Plancon L et al (2008) Phage T5 straight tail fiber is a multifunctional protein acting as a tape measure and carrying fusogenic and muralytic activities. J Biol Chem 283:13556–13564

Bradley P, Cowen L, Menke M et al (2001) BETAWRAP: successful prediction of parallel beta-helices from primary sequence reveals an association with many microbial pathogens. Proc Natl Acad Sci USA 98:14819–14824

Cardarelli L, Pell LG, Neudecker P et al (2010) Phages have adapted the same protein fold to fulfill multiple functions in virion assembly. Proc Natl Acad Sci USA 107:14384–14389

Casjens S, Hendrix R (1974b) Comments on the arrangement of the morphogenetic genes of bacteriophage lambda. J Mol Biol 90:20–25

Casjens SR, Hendrix RW (1974a) Locations and amounts of major structural proteins in bacteriophage lambda. J Mol Biol 88:535–545

Christie GE, Temple LM, Bartlett BA et al (2002) Programmed translational frameshift in the bacteriophage P2 FETUD tail gene operon. J Bacteriol 184:6522–6531

Edmonds L, Liu A, Kwan JJ et al (2007) The NMR structure of the gpU tail-terminator protein from bacteriophage lambda: identification of sites contributing to Mg(II)-mediated oligomerization and biological function. J Mol Biol 365:175–186

Effantin G, Boulanger P, Neumann E et al (2006) Bacteriophage T5 structure reveals similarities with HK97 and T4 suggesting evolutionary relationships. J Mol Biol 361:993–1002

Elliott J, Arber W (1978) *E. coli* K-12 pel mutants, which block phage lambda DNA injection, coincide with ptsM, which determines a component of a sugar transport system. Mol Gen Genet 161:1–8

Feucht A, Schmid A, Benz R et al (1990) Pore formation associated with the tail-tip protein pb2 of bacteriophage T5. J Biol Chem 265:18561–18567

Finn RD, Mistry J, Tate J et al (2010) The Pfam protein families database. Nucleic Acids Res 38:D211–D222

Fortier LC, Bransi A, Moineau S (2006) Genome sequence and global gene expression of Q54, a new phage species linking the 936 and c2 phage species of *Lactococcus lactis*. J Bacteriol 188:6101–6114

Fraser JS, Maxwell KL, Davidson AR (2007) Immunoglobulin-like domains on bacteriophage: weapons of modest damage? Curr Opin Microbiol 10:382–387

Fraser JS, Yu Z, Maxwell KL et al (2006) Ig-like domains on bacteriophages: a tale of promiscuity and deceit. J Mol Biol 359:496–507

Grundy FJ, Howe MM (1985) Morphogenetic structures present in lysates of amber mutants of bacteriophage Mu. Virology 143:485–504

Heller K, Braun V (1982) Polymannose O-antigens of *Escherichia coli*, the binding sites for the reversible adsorption of bacteriophage T5+ via the L-shaped tail fibers. J Virol 41:222–227

Heller KJ, Schwarz H (1985) Irreversible binding to the receptor of bacteriophages T5 and BF23 does not occur with the tip of the tail. J Bacteriol 162:621–625

Heller KJ (1984) Identification of the phage gene for host receptor specificity by analyzing hybrid phages of T5 and BF23. Virology 139:11–21

Hendrix RW, Duda RL (1992) Bacteriophage lambda PaPa: not the mother of all lambda phages. Science 258:1145–1148

Hendrix RW, Smith MC, Burns RN et al (1999) Evolutionary relationships among diverse bacteriophages and prophages: all the world's a phage. Proc Natl Acad Sci USA 96:2192–2197

Inamdar MM, Gelbart WM, Phillips R (2006) Dynamics of DNA ejection from bacteriophage. Biophys J 91:411–420

Jeembaeva M, Castelnovo M, Larsson F et al (2008) Osmotic pressure: resisting or promoting DNA ejection from phage? J Mol Biol 381:310–323

Juhala RJ, Ford ME, Duda RL et al (2000) Genomic sequences of bacteriophages HK97 and HK022: pervasive genetic mosaicism in the lambdoid bacteriophages. J Mol Biol 299:27–51

Kageyama Y, Murayama M, Onodera T et al (2009) Observation of the membrane binding activity and domain structure of gpV, which comprises the tail spike of bacteriophage P2. Biochemistry 48:10129–10135

Katsura I (1976) Morphogenesis of bacteriophage lambda tail. Polymorphism in the assembly of the major tail protein. J Mol Biol 107:307–326

Katsura I (1981) Structure and function of the major tail protein of bacteriophage lambda. Mutants having small major tail protein molecules in their virion. J Mol Biol 146:493–512

Katsura I (1983) Tail assembly and injection. In: Hendrix RW et al (eds) Lambda II. Cold Spring Harbor, New York

Katsura I (1987) Determination of bacteriophage lambda tail length by a protein ruler. Nature 327:73–75

Katsura I, Kuhl PW (1975) Morphogenesis of the tail of bacteriophage lambda. III. Morphogenetic pathway. J Mol Biol 91:257–273

Katsura I, Hendrix RW (1984) Length determination in bacteriophage lambda tails. Cell 39:691–698

Katsura I, Tsugita A (1977) Purification and characterization of the major protein and the terminator protein of the bacteriophage lambda tail. Virology 76:129–145

Kanamaru S, Ishiwata Y, Suzuki T et al (2005) Control of bacteriophage T4 tail lysozyme activity during the infection process. J Mol Biol 346:1013–1020

Kenny JG, McGrath S, Fitzgerald GF et al (2004) Bacteriophage Tuc 2009 encodes a tail-associated cell wall-degrading activity. J Bacteriol 186:3480–3491

Kondou Y, Kitazawa D, Takeda S et al (2005) Structure of the central hub of bacteriophage Mu baseplate determined by X-ray crystallography of gp44. J Mol Biol 352:976–985

Konopa G, Taylor K (1979) Coliphage lambda ghosts obtained by osmotic shock or LiCl treatment are devoid of J- and H-gene products. J Gen Virol 43:729–733

Lang AS, Taylor TA, Beatty JT (2002) Evolutionary implications of phylogenetic analyses of the gene transfer agent (GTA) of Rhodobacter capsulatus. J Mol Evol 55:534–543

Lengyel JA, Goldstein RN, Marsh M et al (1974) Structure of the bacteriophage P2 tail. Virology 62:161–174

Leiman PG, Basler M, Ramagopal UA et al (2009) Type VI secretion apparatus and phage tail-associated protein complexes share a common evolutionary origin. Proc Natl Acad Sci USA 106:4154–4159

Leiman PG, Shneider MM, Mesyanzhinov VV et al (2006) Evolution of bacteriophage tails: structure of T4 gene product 10. J Mol Biol 358:912–921

Levin ME, Hendrix RW, Casjens SR (1993) A programmed translational frameshift is required for the synthesis of a bacteriophage lambda tail assembly protein. J Mol Biol 234:124–139

Loessner MJ, Inman RB, Lauer P et al (2000) Complete nucleotide sequence, molecular analysis and genome structure of bacteriophage A118 of Listeria monocytogenes: implications for phage evolution. Mol Microbiol 35:324–340

Mc Grath S, Neve H, Seegers JF et al (2006) Anatomy of a lactococcal phage tail. J Bacteriol 188:3972–3982

Moak M, Molineux IJ (2000) Role of the Gp16 lytic transglycosylase motif in bacteriophage T7 virions at the initiation of infection. Mol Microbiol 37:345–355

Moak M, Molineux IJ (2004) Peptidoglycan hydrolytic activities associated with bacteriophage virions. Mol Microbiol 51:1169–1183

Montag D, Henning U (1987) An open reading frame in the *Escherichia coli* bacteriophage lambda genome encodes a protein that functions in assembly of the long tail fibers of bacteriophage T4. J Bacteriol 169:5884–5886

Montag D, Schwarz H, Henning U (1989) A component of the side tail fiber of *Escherichia coli* bacteriophage lambda can functionally replace the receptor-recognizing part of a long tail fiber protein of the unrelated bacteriophage T4. J Bacteriol 171:4378–4384

Murialdo H, Siminovitch L (1972) The morphogenesis of bacteriophage lambda. IV. Identification of gene products and control of the expression of the morphogenetic information. Virology 48:785–823
Panja D, Molineux IJ (2010) Dynamics of bacteriophage genome ejection in vitro and in vivo. Phys Biol 7:045006
Parker ML, Eiserling FA (1983) Bacteriophage SPO1 structure and morphogenesis. I. Tail structure and length regulation. J Virol 46:239–249
Pedersen M, Ostergaard S, Bresciani J et al (2000) Mutational analysis of two structural genes of the temperate lactococcal bacteriophage TP901-1 involved in tail length determination and baseplate assembly. Virology 276:315–328
Pell LG, Gasmi-Seabrook GM, Morais M et al (2010) The solution structure of the C-terminal Ig-like domain of the bacteriophage λ tail tube protein. J Mol Biol 403:468–479
Pell LG, Kanelis V, Donaldson LW et al (2009b) The phage lambda major tail protein structure reveals a common evolution for long-tailed phages and the type VI bacterial secretion system. Proc Natl Acad Sci USA 106:4160–4165
Pell LG, Liu A, Edmonds L et al (2009a) The X-ray crystal structure of the phage lambda tail terminator protein reveals the biologically relevant hexameric ring structure and demonstrates a conserved mechanism of tail termination among diverse long-tailed phages. J Mol Biol 389:938–951
Plancon L, Janmot C, le Maire M et al (2002) Characterization of a high-affinity complex between the bacterial outer membrane protein FhuA and the phage T5 protein pb5. J Mol Biol 318:557–569
Plisson C, White HE, Auzat I et al (2007) Structure of bacteriophage SPP1 tail reveals trigger for DNA ejection. EMBO J 26:3720–3728
Piuri M, Hatfull GF (2006) A peptidoglycan hydrolase motif within the mycobacteriophage TM4 tape measure protein promotes efficient infection of stationary phase cells. Mol Microbiol 62:1569–1585
Roessner CA, Struck DK, Ihler GM (1983) Morphology of complexes formed between bacteriophage lambda and structures containing the lambda receptor. J Bacteriol 153:1528–1534
Roessner CA, Ihler GM (1986) Formation of transmembrane channels in liposomes during injection of lambda DNA. J Biol Chem 261:386–390
Roessner CA, Ihler GM (1984) Proteinase sensitivity of bacteriophage lambda tail proteins gpJ and pH in complexes with the lambda receptor. J Bacteriol 157:165–170
Ricagno S, Campanacci V, Blangy S et al (2006) Crystal structure of the receptor-binding protein head domain from *Lactococcus lactis* phage bIL170. J Virol 80:9331–9335
Rigden DJ, Jedrzejas MJ, Galperin MY (2003) Amidase domains from bacterial and phage autolysins define a family of gamma-D, L-glutamate-specific amidohydrolases. Trends Biochem Sci 28:230–234
Saigo K (1975) Tail–DNA connection and chromosome structure in bacteriophage T5. Virology 68:154–165
Samsonov VV, Sineoky SP (2002) DcrA and dcrB *Escherichia coli* genes can control DNA injection by phages specific for BtuB and FhuA receptors. Res Microbiol 153:639–646
Sao-Jose C, Baptista C, Santos MA (2004) *Bacillus subtilis* operon encoding a membrane receptor for bacteriophage SPP1. J Bacteriol 186:8337–8346
Sao-Jose C, Lhuillier S, Lurz R et al (2006) The ectodomain of the viral receptor YueB forms a fiber that triggers ejection of bacteriophage SPP1 DNA. J Biol Chem 281:11464–11470
Scandella D, Arber W (1974) An *Escherichia coli* mutant which inhibits the injection of phage lambda DNA. Virology 58:504–513
Schwartz M (1975) Reversible interaction between coliphage lambda and its receptor protein. J Mol Biol 99:185–201
Sciara G, Bebeacua C, Bron P et al (2010) Structure of lactococcal phage p2 baseplate and its mechanism of activation. Proc Natl Acad Sci USA 107:6852–6857
Scandella D, Arber W (1976) Phage lambda DNA injection into *Escherichia coli* pel-mutants is restored by mutations in phage genes V or H. Virology 69:206–215
Shao Y, Wang IN (2008) Bacteriophage adsorption rate and optimal lysis time. Genetics 180:471–482
Siponen M, Sciara G, Villion M et al (2009) Crystal structure of ORF12 from *Lactococcus lactis* phage p2 identifies a tape measure protein chaperone. J Bacteriol 191:728–734
Smith ML, Avanigadda LN, Liddell PW et al (2010) Identification of the J and K genes in the bacteriophage Mu genome sequence. FEMS Microbiol Lett 313:29–32
Soding J, Biegert A, Lupas AN (2005) The HHpred interactive server for protein homology detection and structure prediction. Nucleic Acids Res 33:W244–W248
Spinelli S, Campanacci V, Blangy S et al (2006) Modular structure of the receptor binding proteins of *Lactococcus lactis* phages. The RBP structure of the temperate phage TP901-1. J Biol Chem 281:14256–14262
Sudiarta IP, Fukushima T, Sekiguchi J (2010) *Bacillus subtilis* CwlP of the SP-{beta} prophage has two novel peptidoglycan hydrolase domains, muramidase and cross-linkage digesting DD-endopeptidase. J Biol Chem 285:41232–41243

Suzuki H, Yamada S, Toyama Y et al (2010) The C-terminal domain is sufficient for host-binding activity of the Mu phage tail-spike protein. Biochim Biophys Acta 1804:1738–1742

Tavares P, Lurz R, Stiege A et al (1996) Sequential headful packaging and fate of the cleaved DNA ends in bacteriophage SPP1. J Mol Biol 264:954–967

Thomas JO (1974) Chemical linkage of the tail to the right-end of bacteriophage lambda DNA. J Mol Biol 87:1–10

Thomas JO (1978) Altered arrangement of the DNA in injection-defective lambda bacteriophage. J Mol Biol 123:149–161

Tsui LC, Hendrix RW (1983) Proteolytic processing of phage lambda tail protein gpH: timing of the cleavage. Virology 125:257–264

van Raaij MJ, Schoehn G, Burda MR et al (2001) Crystal structure of a heat and protease-stable part of the bacteriophage T4 short tail fibre. J Mol Biol 314:1137–1146

Veesler D, Robin G, Lichiere J et al (2010) Crystal structure of bacteriophage SPP1 distal tail protein (gp19.1): a baseplate hub paradigm in gram-positive infecting phages. J Biol Chem 285:36666–36673

Vegge CS, Brondsted L, Neve H et al (2005) Structural characterization and assembly of the distal tail structure of the temperate lactococcal bacteriophage TP901-1. J Bacteriol 187:4187–4197

Vianelli A, Wang GR, Gingery M et al (2000) Bacteriophage T4 self-assembly: localization of gp3 and its role in determining tail length. J Bacteriol 182:680–688

Walker JE, Auffret AD, Carne A et al (1982) Solid-phase sequence analysis of polypeptides eluted from polyacrylamide gels. An aid to interpretation of DNA sequences exemplified by the *Escherichia coli* unc operon and bacteriophage lambda. Eur J Biochem 123:253–260

Wietzorrek A, Schwarz H, Herrmann C et al (2006) The genome of the novel phage Rtp, with a rosette-like tail tip, is homologous to the genome of phage T1. J Bacteriol 188:1419–1436

Williams N, Fox DK, Shea C et al (1986) Pel, the protein that permits lambda DNA penetration of *Escherichia coli*, is encoded by a gene in ptsM and is required for mannose utilization by the phosphotransferase system. Proc Natl Acad Sci USA 83:8934–8938

Xu J (2001) A conserved frameshift strategy in dsDNA long tailed bacteriophages. University of Pittsburgh, Pittsburgh, PA

Xu J, Hendrix RW, Duda RL (2004) Conserved translational frameshift in dsDNA bacteriophage tail assembly genes. Mol Cell 16:11–21

Zimmer M, Sattelberger E, Inman RB et al (2003) Genome and proteome of Listeria monocytogenes phage PSA: an unusual case for programmed + 1 translational frameshifting in structural protein synthesis. Mol Microbiol 50:303–317

Zweig M, Cummings DJ (1973) Cleavage of head and tail proteins during bacteriophage T5 assembly: selective host involvement in the cleavage of a tail protein. J Mol Biol 80:505–518

Chapter 7
Short Noncontractile Tail Machines: Adsorption and DNA Delivery by Podoviruses

Sherwood R. Casjens and Ian J. Molineux

Abstract Tailed dsDNA bacteriophage virions bind to susceptible cells with the tips of their tails and then deliver their DNA through the tail into the cells to initiate infection. This chapter discusses what is known about this process in the short-tailed phages (*Podoviridae*). Their short tails require that many of these virions adsorb to the outer layers of the cell and work their way down to the outer membrane surface before releasing their DNA. Interestingly, the receptor-binding protein of many short-tailed phages (and some with long tails) has an enzymatic activity that cleaves their polysaccharide receptors. Reversible adsorption and irreversible adsorption to primary and secondary receptors are discussed, including how sequence divergence in tail fiber and tailspike proteins leads to different host specificities. Upon reaching the outer membrane of Gram-negative cells, some podoviral tail machines release virion proteins into the cell that help the DNA efficiently traverse the outer layers of the cell and/or prepare the cell cytoplasm for phage genome arrival. Podoviruses utilize several rather different variations on this theme. The virion DNA is then released into the cell; the energetics of this process is discussed. Phages like T7 and N4 deliver their DNA relatively slowly, using enzymes to pull the genome into the cell. At least in part this mechanism ensures that genes in late-entering DNA are not expressed at early times. On the other hand, phages like P22 probably deliver their DNA more rapidly so that it can be circularized before the cascade of gene expression begins.

7.1 Introduction

All tailed phages face similar obstacles to successful delivery of their DNA into host cells, but because of their very different tail types, the *Myoviridae*, *Siphoviridae*, and *Podoviridae* appear to eject their DNA in different ways. This review focuses on the *Podoviridae*, which are the tailed bacteriophage morphotype that has short tails (Hendrix and Casjens 2005; Lavigne et al. 2008). [These phages are still sometimes referred to by the outdated term "Bradley type C" morphology (Bradley 1967).] Although they have been lumped together in current classification schemes, different

S.R. Casjens (✉)
Division of Microbiology and Immunology, Pathology Department,
University of Utah School of Medicine, Salt Lake City, UT 84112, USA
e-mail: sherwood.casjen@path.utah.edu

I.J. Molineux
Section of Molecular Genetics and Microbiology, Institute for Cellular and Molecular Biology,
University of Texas at Austin, Austin, TX 78712, USA

M.G. Rossmann and V.B. Rao (eds.), *Viral Molecular Machines*,
Advances in Experimental Medicine and Biology 726, DOI 10.1007/978-1-4614-0980-9_7,

short-tailed phages often appear to have rather little in common. For example, the best-studied podoviruses, ϕ29, T7, and P22, have essentially no nucleotide sequence that is recognizably similar, and only a small number of their encoded proteins have been demonstrated to be homologues that are very divergent in sequence. Nonetheless, this review considers these viruses together since they may have overcome the barriers faced during DNA delivery in similar ways. Bacteriophage "tail machines" have been defined as the complex of proteins in tailed phage virions that is involved in DNA packaging and delivery of DNA into cells (Johnson and Chiu 2007). These machines are present as a tail structure that projects from one icosahedral vertex, called the portal vertex, of the virion. Some of the proteins in the tail machine, in particular the portal protein (below), can have important roles in DNA packaging into the virion as well as in DNA exit during ejection, but only the latter will be considered in this discussion.

DNA ejection by tailed phages is one of the least understood aspects of their life cycles (we use the term "ejection" rather than the historically popular term "injection," in order to avoid the macroscopic syringe analogy, which is not appropriate). There are several reasons for the poor state of our current knowledge. These include the facts that ejection usually occurs rapidly, data are limited to very few phages, and ejection often appears to be an all-or-none process without easily distinguishable intermediate states. Furthermore, genetic analysis is seriously hindered by the fact that tail proteins also have critical roles in tail assembly and must function there before they perform their roles during DNA delivery. Hence, mutations that debilitate tail proteins usually disable tail assembly, and without a tail, there is no infection to study. Informative mutations that affect tail function without affecting tail assembly can be very useful but are not always simple to obtain. Finally, because they have such small tails, any protein movements that might occur in podoviral tails during adsorption and ejection are particularly difficult to observe, so their tail machines are perhaps the most incompletely understood among the tailed phages.

DNA ejection by all tailed phages is generally believed to have several stages. First, the virion binds reversibly to a specific "primary receptor" on the surface of the target cell and then a second, usually irreversible binding event to a "secondary receptor" takes place; this multistep process of binding to a target cell is called adsorption. Although we will use this description of reversible adsorption here, it does not necessarily describe all events leading to infection. Phage particles are as likely to first encounter a potential host cell with their heads as with their tails, and maintaining contact with this bacterium is a potentially useful property. Some candidates for this type of initial interactions are the head fibers of ϕ29, the Ig-like domains found in structural proteins of many phages (Fraser et al. 2006), and the Dec protein present on the outside surface of podovirus L's head that has amino acid sequence similarity with myovirus T4 tail fibers (Tang et al. 2006). Such presumably weak and nonspecific binding may help preserve cell association, and allow additional time for the tail fibers or tailspikes to recognize the primary receptor on the cell surface.

It should be remembered that reversible adsorption is an operational definition and is dependent on the methods used; it is not therefore defining a specific event. In contrast, the meaning of irreversible adsorption is clearer – *infective* particles cannot be recovered from the infected cell–phage complex. Upon completing adsorption, the virion is committed to ejecting its DNA and releases the DNA from the virion head through a channel in the tail, past the cell wall and membrane(s) into the host cell's cytoplasm. A rudimentary understanding of some of these processes has been attained with the myovirus T4, in part because of its large size and the massive protein movements (tail contraction) that occur during the process. Binding of the T4 secondary receptor is thought to trigger the tail contraction protein rearrangements that release the DNA from the head and form a conduit for DNA to pass into the cell (Kanamaru et al. 2002, see chapter 22 of this book). Such protein movements during siphovirus and podovirus ejection have been much more difficult to observe. Certainly they are not so great as those in T4, but some movements must occur; at the *very least* the proteins that plug the exit channel must get out of the way so that DNA can be released from the virion. Furthermore, the tails of podoviruses are usually too short to span the entire cell envelope; either the

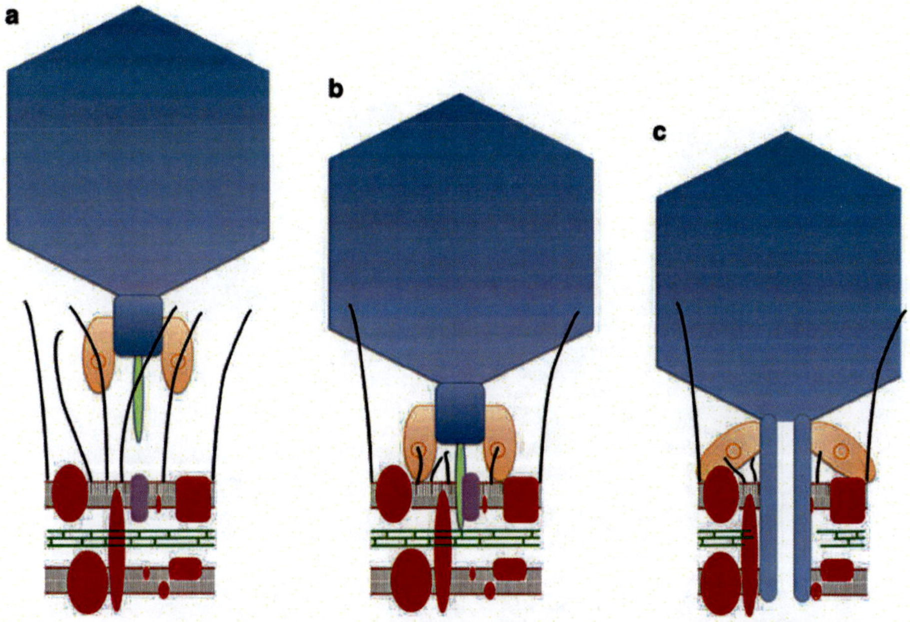

Fig. 7.1 Podovirus DNA delivery into a target cell. **(a)** A virion binds to its cell surface primary receptor (*black curved lines*, in this example depicting a polysaccharide) through its tailspikes (*orange*). **(b)** Tailspikes cleave the polysaccharide, the virion approaches the cell surface, and secondary receptor contact is made; this is shown as contact of the tail needle (*green*) with an outer membrane protein (*purple*). **(c)** Tail machine proteins (*blue*) rearrange to allow local enzymatic cleavage of the peptidoglycan (*green lines*) and form a passage for the DNA through the cell envelope. There are many variations in how podoviruses bind to cells and deliver their DNA; see text for more information

tail must be functionally extended to traverse the cytoplasmic membrane or that membrane must be induced to approach the tip of the tail at the site of adsorption. Among the podoviruses, limited information on such protein rearrangements is only available for T7, and it is discussed below.

Below we consider each step of podovirus DNA delivery in general and then examine the differences and similarities among podovirus subtypes in specific discussions of the better studied of these, the P22, ϕ29, T7, N4, ε15, and BPP-1 type phages. Figure 7.1 shows podovirus DNA delivery diagrammatically.

7.2 Adsorption

7.2.1 Initial or Reversible Binding to a Host Cell

The initial specific binding of any tailed phage virion to a host cell typically occurs through the tail fibers or tailspikes (the term fiber is used for long thin tail appendages, whereas spike is used for shorter and thicker appendages) but can also occur through less extended tail proteins. These virion proteins target host surface proteins or polysaccharides, and examples of podoviral fibers and tailspikes are shown in Fig. 7.2. The host molecule that is initially bound by the phage is referred to here as the primary receptor. Among the podoviruses, some like P22 are well known for using extended, long carbohydrate chains as their primary receptor (Lindberg et al. 1970), while others like N4 use surface protein receptors (McPartland and Rothman-Denes 2009). Binding of primary host receptors

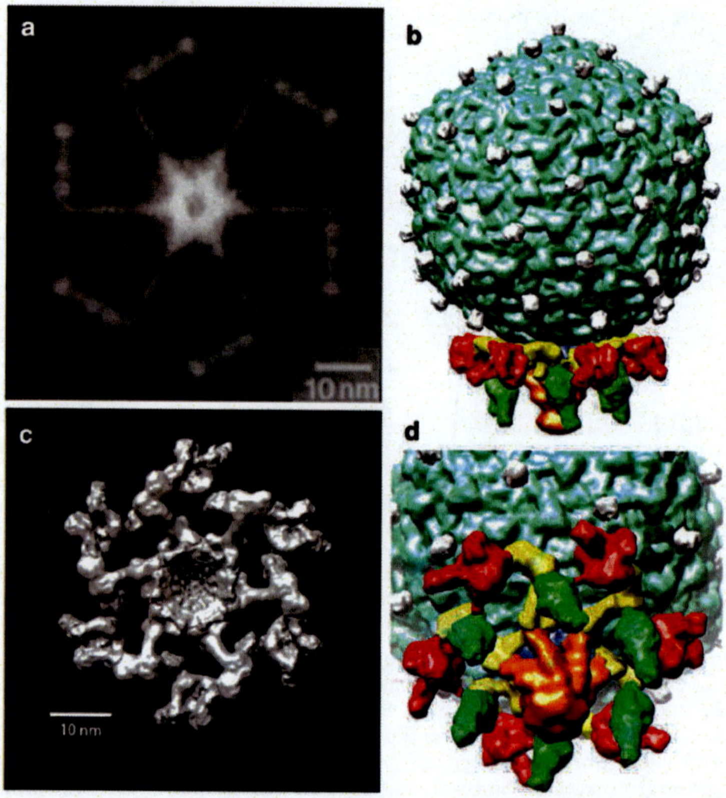

Fig. 7.2 Phage T7 tail fibers and phage K1-5 tailspikes. (**a**) Two dimensionally reconstructed image made using piecewise correlation averaging of electron micrographs of negatively stained tails showing an end-on view of the T7 tail illustrating the six thin, trimeric tail fibers emanating from the central hub (reproduced from Steven et al. 1988 with permission). (**b**) Three-dimensional reconstruction of the K1-5 virion. (**c**) The K1-5 tail from below. (**d**) Angled view of tail at the "bottom" of the virion. The products of K1-5 genes *33, 37, 46*, and *47* in (**b**) and (**d**) are shown in *orange*, *yellow*, *green*, and *red*, respectively. The gp46 and gp47 proteins have cleavage activity for, respectively, glucosamine-*N*-acetyl-glucosamine polymer (the *E. coli* K5 capsular polysaccharide) and polysialic acid (K1 polysaccharide) (Petter and Vimr 1993; Hanfling et al. 1996). (**b–d**) Modified from Leiman et al. (2007) and P. Leiman unpublished with permission. The reader is also directed to the recent cryo-EM reconstruction of P-SSP7 virions (Liu et al. 2010), which was published after this review was completed

mediates initial contact of the virion with the target cell, but does not trigger DNA release from the virion. In addition to conferring the specificity (at least in part) of adsorption, this step helps reduce the search space for a productive interaction of the phage with a bacterium (i.e., infection) from three to two dimensions. This idea is formally described as the strategy of reduction of dimensionality (RD) (Adam and Delbrück 1968). It is hypothesized that release and rapid rebinding of individual receptors allows the phage to explore a two-dimensional surface in a search for a productive site of adsorption, and the role of primary receptor binding is to speed the process of finding the secondary receptor. Experiments with myovirus T4 have lead to a model in which the phage uses this reversible binding of its tail fibers to recognize the correct type of cell and at the same time allows a flexible and rapid search for the secondary receptor (Goldberg et al. 1994). However, RD can only be useful if the reversible adsorption is sufficiently weak to allow desorption and rebinding at alternative neighboring sites, and it still requires the initial three-dimensional search, and a phage just released has a very good chance of diffusing back and adsorbing to the same cell (Berg and Purcell 1977). Furthermore, the well-studied siphoviruses, λ, T5, and SPP1, can infect perfectly well in the laboratory without their long tail fibers, utilizing only their secondary receptor proteins LamB, FhuA, and YueB, respectively,

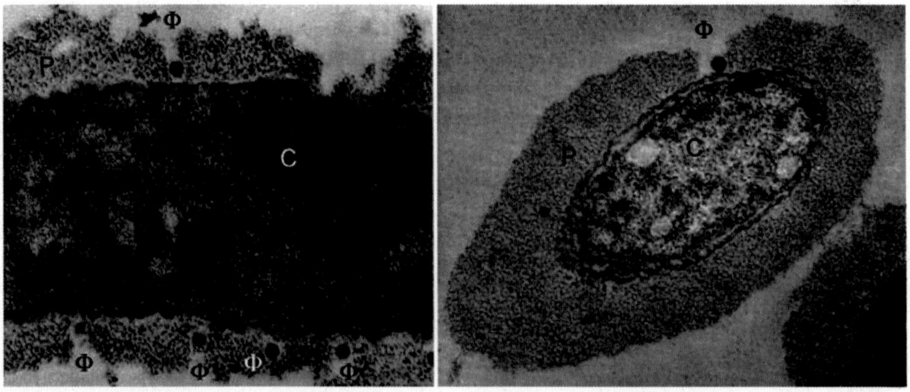

Fig. 7.3 Two views of phage K29 adsorbing to *E. coli*. Phage K29 is a podovirus with polysaccharide depolymerizing tailspikes. Symbols in the figure are C cell cytoplasm, P polysaccharide capsule, Φ phage virion. Figure is modified from Bayer et al. (1979) with permission

even though the rate of irreversible adsorption of the phage to its host cell is substantially reduced (Saigo 1978; Heller and Braun 1979; Hendrix and Duda 1992; Baptista et al. 2008). Regardless of the detailed mechanism(s) of reversible adsorption, the dispensability of the proteins supports the idea that primary receptor binding is not an obligatory step in the virion's route to the cell surface.

Phages are not restricted to having only one type of primary receptor-binding protein, and several mechanisms to increase the number of hosts that can be infected are known. In the podoviruses, the simplest of these are phages that carry two types of tailspikes that have different target specificities, either of which can function in primary receptor binding. For example, phage K1-5 is a T7-like phage that has two different tailspikes in its virion that can bind *Escherichia coli* α-polysialic acid (K1) or poly-glucuronic acid-*N*-acetylglucosamine (K5) capsular polysaccharides (Scholl et al. 2001, 2004; Leiman et al. 2007) (Fig. 7.2c, d). Other phages have complex mechanisms that vary the amino acid sequence of the tail fiber or spike, and one of these is mentioned below in more detail with the podovirus BPP-1. In addition, a related group of podoviruses is not limited to possessing long thin fibers or thick shorter spikes: T7 has long fibers, whereas its relatives K1F, K1-5, and K1E have shorter spikes (Steven et al. 1988; Petter and Vimr 1993; Scholl et al. 2004; Leiman et al. 2007) (Fig. 7.2). The K1F tailspike retains the tail-binding domain of the T7 tail fiber, fusing it to the enzymatic domain of the sialidase. In contrast, the K1-5 and SP6 tailspike proteins lack a tail-binding domain but encode a separate adapter protein that can link different tailspike enzymes to the tail (Stummeyer et al. 2006; Leiman et al. 2007).

7.2.2 *Getting Down to the Cell Surface*

When some podoviruses, for example, T7, bind to primary receptors that are near the cell surface, the act of binding may correctly position the virion for irreversible binding and ejection; however, other phages like P22 and K1F bind to bacterial surface-bound polysaccharides that extend beyond the core lipopolysaccharide (LPS), so that initial binding may not be near the surface. In these latter cases, an additional process must allow these virions to work their way down to the outer cell membrane surface. To date all phages that utilize extended polymeric primary receptors (O-antigen or capsules) have thick tailspikes, and in nearly all of the cases where information is available, these spikes have an enzymatic activity that cleaves the receptor polymer. Known phages have 6 or 12 tailspikes, each of which is a homotrimer with three polysaccharide-binding sites. Figure 7.3 shows striking electron micrographs in which the receptor-degrading short-tailed phage K29 appears to

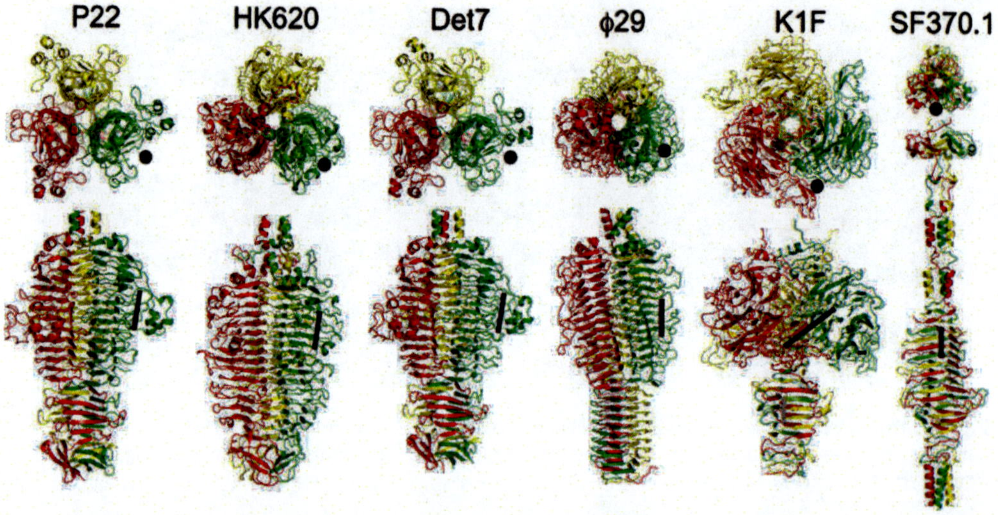

Fig. 7.4 Tailspike receptor-binding domains. Ribbon diagrams of the receptor binding domains of trimeric tailspikes from six tailed phages are shown with the three identical subunits shown in *red*, *green*, and *yellow* [*top* views (*upper*) are from the N-terminus and *side* views (*lower*) have N-termini at the top]; the N-terminal region is bound to the virion in all cases (except SF370.1 where the virion orientation is not known). The approximate locations of the substrate binding sites on one subunit are indicated by *black* and *gray circles* (*upper*) and *bars* (*lower*). The diagrams were produced from the coordinates in the following references: P22 amino acids 109–666 (Steinbacher et al. 1996), HK620 amino acids 113–709 (Barbirz et al. 2008), Det7 amino acids 152–708 (Walter et al. 2008), ϕ29 amino acids 89–691 (Xiang et al. 2009), K1F amino acids 246–911 (Stummeyer et al. 2005), and SF370.1 amino acids 7–336 (Smith et al. 2005). The P22, HK620, and Det7 proteins have very similar, largely β-helix folds and bind and cleave LPS polysaccharides. The HK620 protein has a different binding specificity and a completely different amino acid sequence from those of P22 and Det7, which have similar amino acid sequences and binding specificities. The ϕ29 protein is also largely β-helix, but it cleaves teichoic acid polymers. The K1F and SF370.1 proteins are different and more complex, but they are largely β-structure, and both include triple-strand β-helices. The K1F protein cleaves polysialic acid, and the SF370.1 protein cleaves hyaluronic acid polymers. Phage SF370.1 is an inducible siphovirus prophage in the genome of *Streptococcus pyogenes* strain SF370 (Canchaya et al. 2002), whereas Det7 is a myovirus (Walter et al. 2008). The amino acids that form the capsid-binding domain are not shown for the *left* five proteins, and the ϕ29 and K1F proteins do not show their C-terminal domains that are important for trimerization but that are removed by autocatalytic proteolysis during maturation (Xiang et al. 2009; Schulz et al. 2010)

have burrowed its way through its *E. coli* host's capsule to the cell surface. Bayer and Bayer (1981) presented evidence that suggests that receptor-degrading phages take longer (about 1 min) to complete the adsorption process than phages that do not degrade their receptors. It appears that this cleavage activity can "clear a path" through the polysaccharide for the virion to reach the cell surface. Exactly how this occurs is not well understood, but it can be important because an α-polysialic acid capsule on the surface can largely shield *E. coli* from infection by T7, which has no such enzyme activity (Scholl et al. 2005). However, by expressing sialidase in infected cells, it was shown that the released enzyme from a few infected cells allowed T7 to grow. It seems possible that such cooperation could extend to natural phages as follows: degradation of surface capsular polysaccharide by the tailspike of one phage, which may or may not be able to productively infect the underlying cell, can render that same bacterial cell susceptible to infection by a fibered phage.

In the few cases that have been studied, there are differences among the properties of the receptor-cleaving tailspikes that make construction of a universal model for their action during adsorption difficult. The phage K1F tailspike protein has been shown to act processively with the sequential release of trisaccharides, and the processivity is thought to be mediated by a second substrate-binding site in addition to the cleavage site (Schulz et al. 2010) (Fig. 7.4). This second binding site directs the enzyme toward the reducing end of the polysaccharide, which is anchored to the cell surface

(Stummeyer et al. 2005). The rate of degradation by K1F tailspike on a minimal substrate is 250/min (Schwarzer et al. 2009); as K1 capsule chains have an average length of about 200 trisaccharide units (Rohr and Troy 1980), chains could be fully processively degraded in the 1 min estimated to be necessary for receptor-degrading phages to reach the cell surface (Bayer and Bayer 1981). However, the rate of O-antigen cleavage by P22 tailspike was estimated at only 2/min (Baxa et al. 1996), which is too slow for an efficient infection if chains consisting of 20–35 repeating subunits (Palva and Mäkelä 1980) need to be degraded prior to cell surface binding. No second substrate-binding site has been identified in the P22 tailspike; however, the single known binding site does bind the polysaccharide in such a way that it could in theory correctly orient the virion for tail tip contact with the outer membrane. It has been suggested that rapid receptor release and rebinding by the P22 tailspike (with or without cleavage) might facilitate the phage finding a productive site for infection without dissociation from the cell (Baxa et al. 1996). At least three tailspikes are necessary for successful infection (Israel 1976, 1978), suggesting that the processivity of a single tailspike with its three binding sites is insufficient to carry the virion to the surface. However, the requirement for multiple tailspikes (or tail fibers) could also reflect the necessity of maintaining the tail-containing vertex in the correct orientation with respect to the cell surface. Finally, the *Salmonella anatum* phage g341 (also known as "c341," Uetake and Hagiwara 1969) tailspike does not degrade the O-antigen chain, but merely catalyzes its deacetylation (Iwashita and Kanegasaki 1976; Bayer et al. 1980). The P22 and g341 findings might suggest how a bound virion could move between O-antigen chains on the same bacterium as well as become properly oriented with their tail closest to the outer membrane. However, they do not address the question of whether there is an active directionality of movement toward the cell surface, and if it exists, how it is achieved.

Since a phage tail has no obvious source of internal energy, it is unlikely that its movement toward the cell surface could be made directional through the expenditure of stored energy. Theoretical means of achieving a directionally biased movement of virions toward the cell surface include several, perhaps not mutually exclusive, possibilities of which we mention three (in these the actions of the virion tailspikes are uncoordinated – coordination of tailspike actions would add another level of sophistication to the process, but there is no evidence for this at present). First, polysaccharide cleavage is an energy-releasing process, and this could be the underlying source of energy that drives processive movement of the tailspikes along the O-antigen polysaccharide toward the cell (Schulz et al. 2010). Second, a Brownian ratchet (Feynman 1963; Astumian 1997) type mechanism might be used. A tailspike could bind, cleave, and release its substrate polysaccharide, while other tailspikes remain bound. Then Brownian motion forces would buffet the bound virion, and if the forces happen to be "right" it would be pushed closer to the surface, where an unbound tailspike could bind for another round binding–cleavage–dissociation. Directionality could be achieved because cleavage removes distal O-antigen binding sites, and eventually the remaining ones in the vicinity are all closer to the cell surface. Third, it is possible that phages with tailspikes that do not cleave the O-antigen (like g341) or do so slowly (like P22) might only use the polysaccharide as a means of holding the virion close to the surface, while a random diffusional search for the putative secondary receptor on the cell surface occurs (see above). Note that this search is not perfectly random in the sense that, when successful, the phage becomes committed to further steps in the infection process.

7.2.3 Irreversible Binding and the Secondary Receptor

As mentioned above, the interaction of a phage with its primary receptor does not usually cause the virion to release its DNA. It is generally thought that another type of binding to a "secondary receptor" at the cell surface signals the virion to release its DNA into the cell (reviewed by Vinga et al. 2006). This second type of binding can cause the virion to be irreversibly bound to the cell and/or causes the virion to become noninfective; in theory, this could be mediated by the primary receptor-binding protein or another virion protein interacting with a different receptor or in a new way with the primary receptor.

This part of podovirus adsorption is particularly poorly understood, but current evidence suggests that they do have secondary receptor interactions. For example, T7-like phage SP6 grows well on *Salmonella enterica* strains that lack the O-antigen polysaccharide that is normally bound and degraded by its tailspike (Scholl et al. 2004), and SP6 mutants with mutant tail proteins, but with a wild-type tailspike, grow well in *E. coli* which does not have the SP6 tailspike receptor (I. J. Molineux, unpublished). Furthermore, mutants of K1F have been isolated that grow on noncapsulated rough *E. coli* strains (I. J. Molineux, unpublished), and the *E. coli* host range of T7 is primarily determined by its tail proteins, not its tail fibers (W. P. Robins and I. J. Molineux, unpublished). These observations clearly show that cleavage of surface polysaccharide is not obligatory for initiating host-specific infection by these podoviruses, and thus strongly suggest that there must be a secondary receptor for these phages. On the other hand, searches for conditions or mutations that allow phage P22 to grow on hosts that lack O-antigen have not been successful.

7.2.4 Diversity and Evolution of Tail Fibers and Tailspikes

Phages can change or expand their host range by mutation of their tail fibers or tailspikes and in so doing increase the extant diversity of these proteins. Bacterial mutants that are resistant to infection by a particular bacteriophage have often lost or modified the primary receptor, and the mutations that have been studied that alter the host range of a phage usually modify its tail *fibers*. The most dramatic example is the T2-like myovirus Ox2. In a series of selections, tail fiber mutants of Ox2 were shown to change their receptor from OmpA to OmpC to OmpX to the innermost glucose residue of *E. coli* B LPS (Drexler et al. 1991). The same phage protein can thus be changed by mutation to use one of several outer membrane proteins and even carbohydrate as its receptor, thereby providing the potential for extending host range without the necessity for recombination with host or prophage sequences. This approach has provided important avenues for experimental identification both of the host receptor and virus primary receptor, but these are not reviewed here.

Seckler and coworkers have determined the crystal structures of the receptor-binding domains of the tailspikes of P22-like phages P22, HK620, and Sf6 (Steinbacher et al. 1994, 1997; Barbirz et al. 2008; Muller et al. 2008) and found that although these have no recognizable amino acid sequence similarity (their capsid-binding domains do have sequence similarity), all three have the same polypeptide fold. These proteins have clearly been diverging for a very long time. Tailspike diversity appears to be exceptionally high among the P22-like phages. Among the 57 currently known P22-like phage genomes, there are 28 extremely different versions of the tailspike receptor-binding domain. These versions are not recognizably similar in amino acid sequence (below 15% amino acid sequence identity), and in all cases where the receptor is known, each sequence type corresponds to a structurally different receptor polysaccharide (Casjens and Thuman-Commike 2011). These are by far the most diverse protein parts of the P22-like phage virions, and their genome sequences suggest that there has been substantial horizontal exchange of these receptor-binding domains among these phages (Casjens and Thuman-Commike 2011). Such diversity may be in part a result of lysogenic conversion in which a prophage encodes genes that alter the surface polysaccharide of its bacterial host. Lysogenic conversion is common in many P22-like (and other) phages, and the changes in the cell envelope in turn select for new phages that have acquired an altered tailspike that can recognize the changed cell surface receptor.

Recombinational shuffling of mutationally diverged primary receptor-binding domains that have different host specificities among different phages is undoubtedly a source of diversity in the evolution of tailed phages. It was noticed a number of years ago that evolutionary horizontal exchange of genes and parts of genes that encode receptor-binding thin fibers is frequent, and these include examples of exchange among the podo-, sipho-, and myoviruses (George et al. 1983; Haggard-Ljungquist et al. 1992; Sandmeier et al. 1992; Walter et al. 2008; Bull et al. 2010). The polymer-cleaving tailspikes, were discovered in, and are very common in podoviruses, and they appear to have exchanged among the podoviruses quite extensively. For example, the P22 and SP6 tailspikes are clear homologues and bind the same *S. enterica* serovar Typhimurium O-antigen primary receptor (Dobbins et al. 2004; Scholl et al. 2004); similarly, P22-like phage CUS-3 and T7-like phage K1F have homologous tailspikes that bind the *E. coli* K1 polysialic acid capsular polysaccharide (King et al. 2007 see also phage P22 section below). In addition, tailspikes homologous to those of podoviruses have also been found in myo- and siphoviruses (Machida et al. 2000; Leiman and Molineux 2008; Walter et al. 2008; S. Casjens and R. Hendrix, unpublished data), suggesting that they have undergone horizontal exchange among very distantly related phages. Tailspikes do not all have the same polypeptide fold, but crystal structures of a number of such tailspikes show that they are all very high in β-helix structure (Steinbacher et al. 1994, 1997; Smith et al. 2005; Stummeyer et al. 2005; Mishra et al. 2009; Xiang et al. 2009). Figure 7.4 shows structural diagrams of tailspikes from phages P22, HK20, Det7, ϕ29, K1F0, and SF370.1. These phages infect *S. enterica*, *E. coli*, *Bacillus subtilis*, and *Streptococcus pyogenes*, which have O-antigen polysaccharide, polysialic acid, teichoic acid, and hyaluronic acid surface polymers, respectively. Thus, phages with such tailspikes are not limited to Gram-negative or Gram-positive bacteria, and a wide variety of surface polymers can be utilized for primary adsorption.

A satisfying explanation for the exceptional diversity and high level of horizontal exchange of tail fibers and tailspikes is that evolutionary sparring between the host and phage at other points in the life cycle eventually limits the competitiveness of a particular phage. Thus, acquisition (by diversification or horizontal tail fiber/tailspike exchange) of the ability to deliver DNA into a new host that has not seen a particular phage before may give it access to bacteria that are not prepared for its particular style of intracellular development. A phage with new receptor specificity could therefore have an immediate competitive advantage.

7.3 DNA Release and Delivery

7.3.1 Packaged DNA Structure

In tailed DNA phages, the packaged genome is maintained in a condensed, dehydrated B-form at a density of ~500 mg/ml. In order to maintain electroneutrality, the interior of the capsid must also contain ~1.5 M counterion charges. The DNA is very tightly bent and so is trapped in an energetically unfavorable state that favors DNA release (Jeembaeva et al. 2010 and references therein). This energetically unfavorable state, often referred to as the "DNA pressure" within the phage virion, was originally estimated from single molecule DNA packaging reactions to reach ~60 atm (Smith et al. 2001), a value later revised to ~100 atm (Rickgauer et al. 2008). It has commonly been assumed that this "pressure" is stored in mature virions and is ultimately used to eject DNA into a cell in the next infection cycle. During ejection by all tailed phages, DNA passes out of the virion through the channel in the portal vertex and the tail. In mature particles, this channel is plugged. By this model, when the plug is removed, DNA and counterions spontaneously escape the energetically unfavorable confines of the capsid. Indeed, in some studied cases, binding of the virion to purified receptor in vitro with no external energy source can result in rapid and complete DNA release. Phages λ and T5 have been reported to release their genomes in response to binding their purified secondary receptors at rates up to 60,000–75,000 bp/s (Mangenot et al. 2005; Grayson et al. 2007).

However, DNA ejection is certainly not this simple. Rates and extents of DNA ejection are highly dependent on the solution conditions, which alter the stability of the packaged DNA, and thus alter the free energy change associated with ejection. In addition, biophysical calculations suggest that spontaneous DNA exit through the tail should slow down significantly as release proceeds. In elegant in vitro experiments, suppression of genome ejection from the siphoviruses λ, T5, and SPP1 has been achieved by the presence of high concentrations of polyethylene glycol (PEG) in the external environment (e.g., Evilevitch et al. 2003; Grayson et al. 2006; Sao-Jose et al. 2007; Leforestier et al. 2008; Jeembaeva et al. 2010). For λ and SPP1, though notably not T5, the length of the DNA remaining in the capsid depends on the concentration of the PEG, which provides an osmotic force opposing ejection. At high external osmotic pressure, T5 responds "normally," but at lower levels multiple populations of T5 virions containing varying amounts of DNA coexist (Leforestier et al. 2008). The theoretical principles underlying these experiments were first presented 40 years ago (Zarybnicky 1969) and have recently been extended (e.g., Tzlil et al. 2003; Purohit et al. 2005 and references therein). Significantly, in the presence of a concentration of PEG that provides an osmotic pressure comparable to that in the bacterial cytoplasm (turgor pressure), in vitro ejection of λ and SPP1 is limited to about half the phage genome (Grayson et al. 2006; Sao-Jose et al. 2007). Therefore, regardless of any pressure or internal forces within the mature phage capsid, turgor necessitates that during natural infection there be mechanisms that complete cytoplasmic internalization of a phage genome (Molineux 2001, 2006; Grayson and Molineux 2007).

Tailless virions of T7 can be purified and are stable in solutions of moderate ionic strength. These properties allowed the first high-resolution cryoEM images of DNA spooled around the internal core within a capsid (Cerritelli et al. 1997). Viewed along the portal vertex (tail) axis, the DNA showed patterns of circular striations, about 2.5 nm apart, an average dsDNA strand separation which agrees with low-angle X-ray diffraction analysis of other phage virions (Earnshaw and Casjens 1980). These data led to a model in which T7 DNA is wrapped axially in concentric shells of an inverse spool (first layer laid down is on the outside, next layer immediately inside that, and so on toward the center). Subsequent cryoEM reconstructions of several tailed phages have revealed comparable toroidal structures in all three tailed phage morphotypes (Hud and Downing 2001; Olson et al. 2001; Fokine et al. 2004; Agirrezabala et al. 2005; Chang et al. 2006; Effantin et al. 2006; Jiang et al. 2006; Lander et al. 2006; Xiang et al. 2006; Leiman et al. 2007; Choi et al. 2008; Comolli et al. 2008; Tang et al. 2008). Only a few outer layers of DNA are clearly visible, and the central portion is much less ordered; this is usually explained by the necessity to bend the DNA more sharply as the inner radius of the toroid becomes smaller. However, it should be remembered that all these reconstructions are averaged structures, and a recent cryoEM analysis of T5 DNA ejection in vitro showed a different pattern inside individual virions that is not compatible with the inverse spool model (Leforestier and Livolant 2009, 2010). The mature T5 capsid appears full of hexagonally packed DNA, even in the central region, but many defect walls are visible that effectively partition the DNA into domains.

7.3.2 DNA Release

The plug that maintains stability of a mature virion must be removed for successful DNA exit; presumably binding to the secondary receptor is the trigger that normally initiates this process. In the podoviruses that have been examined, DNA exit during ejection is oriented with a particular end of the viral chromosome (likely the last end packaged) leaving the virion first (Saigo 1975; Krawiec et al. 1981; Choi et al. 2008). Also, in ϕ29, ε15, K1E, and P-SSP7 (and presumably their relatives, including T7), and perhaps P22, one end of the DNA (the "first out" end?) appears to be tethered within the portal channel in three-dimensional virion reconstructions (Jiang et al. 2006; Lander et al. 2006; Leiman et al. 2007; Tang et al. 2008; Liu et al. 2010). As mentioned above, even if forces internal to the virion initiate genome ejection, the obligatory turgor pressure of growing bacterial cells requires that other mechanisms complete transport of the phage genome into the cell cytoplasm. Indeed, there appear to be several rather different paradigms for podovirus DNA ejection into cells (see below).

Podoviruses that must circularize their DNA for virus-specific transcription and replication, like P22, probably eject their DNA into the host cell rapidly, although it has not been measured directly. In contrast, phages like T7, which does not circularize its DNA, eject their DNA much more slowly. In the T7 phage family, genome entry into the cell is coupled to transcription of its genome, and this may contribute to controlling the timing of phage gene expression (Molineux 2001). These differences are biologically consistent with the fact that the T7 chromosome has defined ends and every virion chromosome is identical, whereas P22 intravirion molecules are circularly permuted, so that in each virion the phage genes have a different relationship to the chromosome ends. The differences in DNA release mechanisms from these two kinds of virions are not understood, but could be reconciled if the "slow ejectors" have an intravirion mechanism, perhaps as simple as a channel with a diameter close to that of dsDNA, that impedes rapid DNA release such that external energy must be supplied to pull the DNA out. It has been noted with some tailed phages that during DNA ejection there is a temporary cellular membrane potential depolarization and/or leakage of cytoplasmic ions. Exactly what this corresponds to in physical or molecular terms is unknown, and we note that podovirus P22 (Rao et al. 1972; Bandyopadhyay et al. 1979; Ter-Nikogosian et al. 1991) and several other phages cause such an effect, whereas T7 does not (Kuhn and Kellenberger 1985; E. Ramanculov and I. J. Molineux, unpublished data).

7.3.3 *Periplasm Transit by DNA, Protein Ejection, and Peptidoglycan Cleavage*

The implications of the seminal "Hershey–Chase experiment" (Hershey and Chase 1952) have obscured to most molecular biologists, though not to Hershey himself (Hershey 1955), the fact that proteins are ejected into the host cell along with the DNA by all tailed phage virions. In all cases where it has been examined, some protein molecules are released with the phage DNA, and these can have important functions in the target cell. This is usually more than simple release of the protein that plugs the DNA exit channel, and, for example, in T7 and P22 60–80 individual protein molecules of five and four types, respectively, are ejected with the DNA (Casjens and King 1974; Israel 1977; Kemp et al. 2005; Chang et al. 2010). The exact roles of ejected proteins are in general not well understood; however, some of these proteins have important roles in DNA delivery, and others, such as inhibitors of restriction enzymes or RNA polymerase (RNAP) effectors, do not. The few specific cases where some information is available on their roles in DNA delivery are discussed below.

Several barriers need to be overcome to successfully deliver DNA into the cytoplasm in concert with its release from the virion. It may be important to (1) clear a path through the bacterial cell wall by enzymatic cleavage of the peptidoglycan at the point of entry, (2) build a conduit or path for the DNA to pass through the periplasmic space and/or host lipid membranes, and (3) protect the ends of the DNA in the cytoplasm. It has been known for some time that some tailed-phages release enzymes with the DNA that are capable of creating holes in the peptidoglycan cell wall. The best studied of these is the contractile-tailed phage T4 where a lysozyme domain in the tail protein gp5 is cleaved from the tail by an unknown protease, and three such molecules are released into the periplasm upon tail contraction (Kanamaru et al. 1999, 2002). Although in the two cases examined, phages T4 and T7, the activity is only conditionally essential (Moak and Molineux 2000; Kanamaru et al. 2002), it seems likely that most if not all tailed phages do benefit from such a muralytic enzyme in the virion, and such an activity has been reported in diverse phages (Moak and Molineux 2004). Proteins with muralytic activity have been identified in virions of the podoviruses T7 (and its relatives), N4, and ϕ29 (Moak and Molineux 2004; Stojković and Rothman-Denes 2007; Xiang et al. 2008). P22 gp4 was mistakenly concluded by (Moak and Molineux 2004) to contain muralytic activity, but this has subsequently been found to be due to the P22 endolysin gp19, which appears to be packaged at low levels in virions (P. Kemp and I. J. Molineux, unpublished). The T4 gene *e*-encoded lysozyme is also found at low levels in virions (Emrich and Streisinger 1968). Any biological significance of the last two observations is not known. Some P22 relatives may have such an activity in their ejected gp16, but this has not been shown experimentally (see below).

The precise mechanism by which proteins are ejected from phage heads remains unknown, and many questions remain. What drives their release? If they have to exit in a coordinated manner in order, for example, to functionally extend the short podovirus tail so that it can span the cell envelope, how is coordination achieved? Further, since proteins have to pass through the narrow portal channel, are they unfolded during ejection and if so, are they unfolded in the virion? If an unfolded state does exist, how does refolding occur in the outer layers of the cell where there are no general chaperone proteins and no source of energy?

7.3.4 Superinfection Exclusion and DNA Delivery

A number of phages express genes whose products help protect the host bacterium from superinfecting phages, and several of these are known to affect adsorption and/or DNA delivery. Some of these are expressed from the prophages of temperate phages, and a common mechanism is modification of receptors on the bacterial surface so that phages with the same receptor specificity as the prophage cannot adsorb. This was discovered in podovirus ε15 and typically takes the form of the expression of enzymes that alter the polysaccharide portion of LPS (Uetake et al. 1955; Uetake and Hagiwara 1969; Wright 1971; Allison and Verma 2000; Kropinski 2000; Villafane et al. 2008 and references therein). In addition to such a system, phage P22 has two additional excluding mechanisms mediated by the SieA and SieB proteins (Rao 1968; Susskind et al. 1971). The first of these may affect DNA delivery. The product of the *sieA* gene is expressed from the P22 prophage and appears to block all functioning of the DNA of superinfecting virions of phages P22, L, MG178, and perhaps MB78 at a stage after DNA release from the virion (Susskind et al. 1971; Joshi et al. 1982). Since generalized transduction fails in the presence of the SieA system (Ebel-Tsipis and Botstein 1971; Susskind et al. 1971), the superinfection block is not related to the sequence of the incoming DNA. SieA protein likely blocks successful DNA delivery by the superinfecting phage (Susskind and Botstein 1978), and the observation that the SieA protein is found in the cell membrane is compatible with such an idea (Hofer et al. 1995). Proteins with SieA-like functions may be widespread in phages. The *imm* gene product, a membrane protein encoded by the myovirus T4, prevents successful DNA ejection by superinfecting phage (Lu and Henning 1994), and membrane proteins encoded by the *Lactococcus lactis* temperate phage Tuc2009 (and other members of the P335 group of siphoviruses) prevent successful genome delivery by lytic phages belonging to the 936 siphovirus group (McGrath et al. 2002; Mahony et al. 2008). The mechanism(s) by which such proteins prevent successful ejection is unknown.

7.4 Podovirus Tail Machines

7.4.1 Universal and Variable Tail Machine Features

The short-tailed phages that are discussed below were chosen because, among the podoviruses, their DNA delivery processes are best understood. The structures of their virions are quite well understood, and P22, N4, ϕ29, and T7 have been studied intensely for many years, so a significant amount of information is available about many aspects of their life cycles. Phages BPP-1, K1F, K1E, and ε15 are less well understood, but they provide a glimpse into the wide diversity that exists even within

Fig. 7.5 Podovirus tail structures. Three-dimensional reconstructions of the virions of six podoviruses are shown, with surface renderings above, and slices through the center of the virion along the tail axis below (the ϕ29 head fibers and T7 tail fibers are not shown). The virions are not shown precisely to scale with respect to each other, but they are all of rather similar size; e.g., ϕ29 and P22 are about 900 Å from the top of the head to the tip of the tail, and N4 is the largest and is about 1,200 Å. The coat protein shell and interior DNA loops are shown in *gray*, the approximate position of the portal protein ring is *green*, and the rest of the tail machines are *orange*. The figure was inspired by Fig. 2 of Johnson and Chiu (2007), and the three-dimensional reconstructions are from the following sources: T7 (Agirrezabala et al. 2005), 24 Å resolution, European Bioinformatics Institute accession No. em1164; K1E (Leiman et al. 2007), 16 Å, em1336; ε15 (Jiang et al. 2006), 20 Å, em1175; P22 (Lander et al. 2006), 17 Å, em1220; ϕ29 (Tang et al. 2008), 7.8 Å, em1420; and N4 (Choi et al. 2008), 29 Å, em1472. The reader is also directed to the cryo-EM reconstruction of P-SSP7 virions (Liu et al. 2010), which was published after this review was completed

this narrow aspect of the podoviruses. The currently known structures of six of these phage virions, with their tail machines highlighted, are shown in Fig. 7.5. Among those shown, only T7 and K1E are of the same "phage type," but they have very different primary receptor binding proteins; T7 has long thin fibers (not shown in Fig. 7.5, see Fig. 7.2), and K1E has thick tailspikes. Although the tails of these phages are all short, the overall morphology of their tails is quite variable even at the resolution shown in Fig. 7.5. The variable sizes, shapes, and numbers of the tail proteins from the different podovirus types (e.g., T7, P22, ϕ29, ε15, and N4 types), and the almost universal lack of amino acid sequence similarity among the proteins involved, make it likely that many of their tail proteins are not homologous across phage types. Each of these tail structures shows some density within the portal vertex channel that is usually attributed to DNA; however, some of this density could also be due to proteins that are ejected. The podovirus types we discuss below represent only a small fraction of this group's total diversity; for example, the genomes of *E. coli* phage ϕEco32 (Savalia et al. 2008), *Pseudomonas aeruginosa* phage PaP3 (Tan et al. 2007), *L. lactis* phage KSY1 (Chopin et al. 2007), and *Prochlorococcus* phage Syn5 (Pope et al. 2007) have been sequenced, and the predicted amino acid sequences of the proteins in their short tails appear to have little in common with those discussed here.

Definitive answers to the question of homology across phage types for most of the podovirus tail proteins must await high-resolution structure determination; however, one tail machine protein, the portal protein (green in Fig. 7.5), is universally present and homologous throughout all tailed phages. Where information is available, all tailed phage virions contain a 12-subunit ring of portal protein that replaces a pentamer of coat proteins at one vertex (fivefold symmetry axis) of the virion's icosahedral shell. The portal protein-coat contacts are complex, since a 12- to 5-fold symmetry mismatch occurs, but nonetheless, these form the connection between the tail machine and the coat protein shell of the virion. Existing evidence suggests that the tail machine/portal protein cannot rotate freely relative to the protein shell in the virion (Baumann et al. 2006; Lander et al. 2006; Hugel et al. 2007). Crystal structures have been obtained for the portal proteins of the podoviruses ϕ29 and P22, siphoviruses SPP1 and HK97, and the myovirus *B. subtilis* prophage Skin (Simpson et al. 2000; Lebedev et al. 2007; Cardarelli et al. 2010; Olia et al. 2011). In spite of a lack of recognizable sequence similarity, they all share a major domain with the same polypeptide fold. It is thus believed that all tailed phage portal proteins are very ancient homologues with a similar function in DNA packaging (the DNA packaging motor is reviewed elsewhere in this book). The portal has an active role in DNA packaging (Casjens et al. 1992; Isidro et al. 2004), although it is not the actual ATPase motor. It is not known if the portal serves any active role in DNA and protein ejection beyond forming part of the exit channel.

Where the process has been examined, the podovirus tail machines have been found to assemble stepwise on the virion capsid, and not as free tails. This is in contrast to the myo- and siphoviruses, which assemble free tails that then join to heads to form the complete virion. Podoviral tail assembly is perhaps best studied in P22, where the order of assembly clearly shows that the tail grows outward from the portal ring during its assembly (details below). Although the non-portal protein parts of the tail machines may perform similar roles in the different phages, they appear on the surface to have rather little in common among podoviruses, except that they all have 12-fold symmetry at the portal that steps down to six- and/or three- or twofold in the more distal parts. A few clear exceptions to the lack of sequence similarity and/or homology among non-portal tail proteins are cases where the tailspike/fiber proteins have been exchanged among phage types (above).

7.4.2 Phage T7 and Relatives

E. coli phage T7 is the only tailed phage where mechanistic details of genome ejection are moderately well understood. This is largely a consequence of the process being slow, occupying about one-third of the latent period. Furthermore, both adsorption and the initial steps where the trans-envelope channel is made are highly synchronous, allowing accurate measurements of the rates of genome internalization.

T7 tail fibers are encoded by gene *17*, and, by extension from studies of the related phage T3 (Prehm et al. 1976), they interact with the inner core of the LPS of *E. coli* B, C, and K-12 strains, and many rough *Shigella* spp. and *S. enterica* strains, but probably without high specificity. The T7 fiberless tail has also been found to interact with the LPS inner core in *E. coli* K-12 (W. P. Robins and I. J. Molineux, unpublished); the tail fibers may thus only serve to accelerate tail binding. Following productive T7 adsorption, about 30 copies of the tail protein gp7.3, and about 18 copies of the head protein gp6.7, are ejected from the virion and become associated with the outer membrane of the infected cell. Both of these small proteins are normally degraded by an unknown outer membrane or periplasmic protease, although degradation is not important for a successful infection (Kemp et al. 2005; Chang et al. 2010). Gene *7.3* is essential, and virions made in its absence are unable to successfully adsorb to cells, even though they contain DNA and a normal complement of tail fibers. Cell–phage complexes cannot be isolated by centrifugation even without dilution (Kemp et al. 2005), showing directly that tail fiber binding is not strong. Despite gp7.3 being part of the virion structure, its essential role is probably during tail assembly, where it may have a chaperone-like activity that organizes the correct assembly of the two major tail proteins gp11 and gp12. Supporting this idea are the following observations: Gene *17* mutants produce virions lacking both tail fibers (gp17) and gp7.3, but these can be complemented in vitro by tail fiber addition. However,

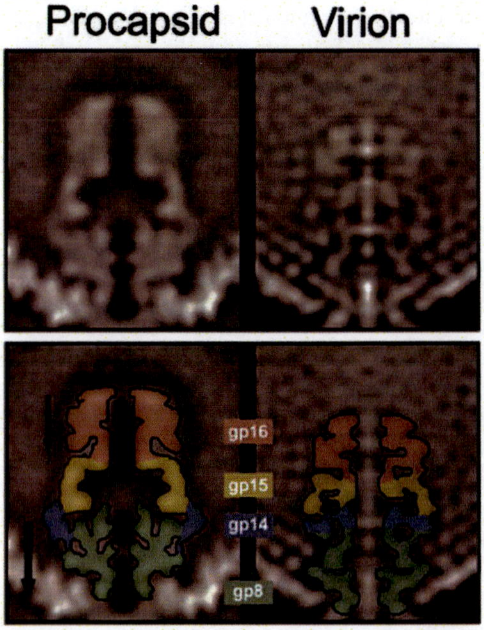

Fig. 7.6 The phage T7 virion internal core. Slices through the portal vertex regions of three-dimensional reconstructions of T7 procapsids and virions are shown above, and the putative locations of the T7 internal core proteins are indicated below; the external part of the tail extends below the panels. *Black arrows* note apparent protein movements during procapsid maturation. After infection, the three internal core proteins (gp14, gp15, and gp16) are ejected into the infected cell, where they are thought to functionally extend the T7 tail so that it spans the cell envelope, thereby providing a conduit for DNA to enter the cytoplasm (see text for details). Figure is adapted from Agirrezabala et al. (2005) with permission. The reader is also directed to the cryo-EM reconstruction of P-SSP7 virions both before and after infection (Liu et al. 2010), which was published after this review was completed

complementation fails if the phage is defective in both genes *17* and *7.3* (Kemp et al. 2005). Furthermore, viable second site suppressors of gene *7.3* deletion mutants lie in genes *11* and/or *12* (O. Perng, P. Kemp and I. J. Molineux, unpublished). The precise location of gp7.3 in the tail is not known. Gene 6.7 is also essential; virions prepared in its absence infect only at a low, strain-dependent efficiency, and the burst of viable progeny per infected cell is reduced. Gp6.7 is present in tailless virions and is therefore a head protein. It is possible that gp6.7 is a head-completion protein that serves as a plug to prevent premature exit of DNA into the tail into the portal vertex channel (Kemp et al. 2005).

Irreversible adsorption of T7 corresponds to the ejection of gp7.3 and gp6.7 into the infected cell, and these are followed by ejection of the internal core proteins, gp14, gp15, and gp16 (Kemp et al. 2005; Chang et al. 2010). In the core, gp14 is closest to the portal, with gp16 at the distal end (Agirrezabala et al. 2005) (Fig. 7.6). The conformation of the proteins in the core is not known, but on geometric grounds alone, it is likely that at least the 82 kDa gp15 and the 144 kDa gp16 must leave the head and pass through the portal channel at least partially unfolded. Both proteins are predicted to be largely α-helical, separated by unstructured regions, and it has been suggested that the α-helices might be preserved during ejection through the portal and tail, allowing for a rapid refolding in the periplasm (Molineux 2001; Chang et al. 2010). The approximately ten molecules of ejected gp14 are found stably associated with the outer membrane, whereas the eight molecules of gp15 and four of gp16 both apparently span the periplasm and associate with the inner membrane (Molineux 2001; Chang et al. 2010). Contact between gp15 and gp16 with gp14 is preserved in the infected cell, and in the absence of specific structural information, it is currently assumed that the proteins functionally extend the length of the short stubby T7 tail to span the entire cell envelope, thereby providing a conduit through which the genome can penetrate into the cell cytoplasm. In support of this idea, mature T7 virions can be induced in vitro to eject their internal core proteins, which remain attached to the tail. Furthermore, incomplete particles found in lysates display a 40–55-nm tail extension, thought to consist of the core proteins, penetrating membrane vesicles (Serwer et al. 2008).

The N-terminal domain of gp16 must rapidly refold in the periplasm of the infected cell. This domain has lytic transglycosylase activity, which is essential for degrading the cell wall under sub-optimal conditions of growth (Moak and Molineux 2000). With cells in the exponential phase of growth at 30°C or above, there is no defect in phage growth when this activity is inactivated. However, as cells approach stationary phase or are grown at low temperatures, a lack of muralytic activity delays penetration of the infecting phage genome into the cytoplasm. Interestingly, although all close relatives of T7 harbor lytic transglycosylase activity at the N-terminus of gp16, more distant relatives, for example, SP6 (Dobbins et al. 2004; Scholl et al. 2004) and the *P. aeruginosa* phage øKMV (Lavigne et al. 2004), have a lysozyme activity associated with a C-terminal domain of their gp15 counterpart (Moak and Molineux 2004). Perhaps the addition of a cell wall-degrading activity is evolutionarily relatively new, and DNA encoding the enzymatic domain independently inserted into the same general region of T7 and SP6 or øKMV, but fused to adjacent, but different genes. This enzymatic activity suggests that in the infected cell, the N-terminal region of T7 gp16 and the C-terminal region of the gp15 homologues in other phages are both located near the peptidoglycan layer after ejection from the virion.

The extreme C-terminus of T7 gp16 is essential for the phage genome to enter the cell cytoplasm. Removing the last 4 of its 1,318 amino acids has no major phenotype, but removing five results in noninfective virions (Chang et al. 2010). Mutant virions adsorb normally and eject proteins into the cell, however, but no T7 DNA can be detected in the bacterial cytoplasm. It is not yet known whether the phage genome remains entirely in the head in this nonproductive infection, or whether it enters the channel across the periplasm but cannot cross the cytoplasmic membrane. Extragenic suppressors of the gene *16* defect have been isolated that alter gene *15*, indicating that gp15 and gp16 cooperate in translocating the phage genome into the cell. It is not yet known which part(s) of gp15 and/or gp16 forms the channel across the cytoplasmic membrane. Computer algorithms that predict membrane-spanning domains indicate that gp15 is not a membrane protein, and in sandwich alkaline phosphatase fusion assays neither individual predicted transmembrane domains of gp16 nor the entire protein prevents enzyme localization to the periplasm (Y-H. Kim, J-H. Lee, P. Kemp, and I. J. Molineux, unpublished). We note, however, that staphylococcal α-hemolysin is also not predicted to be a membrane protein by those same computer algorithms, as they primarily search for potential amphipathic helices; in addition, both α-hemolysin and the T7 proteins normally approach the cytoplasmic membrane from the opposite face than the proteins used to train the prediction algorithms. It is also possible that both gp15 and gp16 need to be present, or both in concert with DNA, in order to be inserted into the membrane.

In a normal infection, after formation of a channel across the cytoplasmic membrane, about 1 of the 40 kbp T7 genome spontaneously enters the cell (Zavriev and Shemyakin 1982; Moffatt and Studier 1988; Garcia and Molineux 1995). The remainder of the genome is internalized by transcription, first by the host RNAP, and then by T7 RNAP. Blocking host transcription by treatment of cells with rifampicin prior to infection prevents further genome entry, although the 1 kbp inside the cytoplasm is not degraded. T7 encodes an anti-RecBCD protein (gp5.9) that is functionally similar to λ gamma protein, but gp5.9 is not a component of the virion and is in any case nonessential, even in *recBCD*$^{+}$ strains (Lin 1992). Gp15 and/or gp16 presumably protects the entering genome end from degradation as no specific end-binding protein, equivalent to T4 gp2 (Lipinska et al. 1989), has been identified.

The first experiments showing the stepwise mode of T7 DNA transfer into the cell used a modified Hershey–Chase protocol (Zavriev and Shemyakin 1982). Cells are infected with ^{32}P-labeled phage, and at various times, complexes are sonicated to desorb attached particles. Cells are then removed by centrifugation, and the phage DNA remaining in the supernatant fractions quantified. In the presence of rifampicin, most phage DNA remains outside the cell; by restriction enzyme digestion, it was concluded that only the leading ~10% of the genome entered the cell. This segment includes promoters for *E. coli* RNAP that program transcription toward DNA that has not yet entered the cell. Allowing host RNAP-catalyzed transcription, but preventing protein synthesis, results in ~60% of the genome being pulled into the cell cytoplasm. It was proposed that a T7 protein, likely T7 RNAP, was responsible for completion of genome entry. However, late T7 genes (located on the last part of the genome to be internalized) are expressed, albeit at low levels, in the absence of T7 RNAP (Studier 1972), and thus they must be in the cell.

A comparable assay was employed with the cyanophage Pf-WMP4 with similar results; the leading end of the genome enters the cell in a transcription-independent manner; the host RNAP transcribes – and thus pulls in – the following ~40% of the genome (Liu et al. 2007). Complete genome internalization requires synthesis of an unidentified Pf-WMP4 encoded protein. It is not obvious why the above "desorption by sonication" assay shows that the host RNAP is insufficient to internalize the complete T7 and Pf-WMP4 genomes, whereas a Dam-methylation assay (below) with T7 clearly indicates otherwise. This discrepancy warrants further investigation.

Understanding of the mechanism of T7 genome internalization was greatly facilitated by the development of a very sensitive assay that indirectly measures the time after infection that GATC nucleotide sequence sites on the genome get methylated inside the cell by Dam methylase (Garcia and Molineux 1995). At various times after infection, DNA is isolated from infected cells and cut to completion with the methylation-dependent enzyme DpnI. A Southern hybridization then reveals those parts of the phage genome that have entered the cell and been methylated by Dam. DNA still in the phage head is unmethylated and can be visualized as bands whose sizes are generally larger than any Sau3AI fragment (Sau3AI is insensitive to the state of Dam methylation at the GATC sequence). In order to ensure rapid methylation of DNA entering the cell, the *dam* gene is overexpressed in these experiments, inducing a partial SOS response. Although cells, therefore, grow more slowly, the phage latent period is not significantly altered, and there is only an approximately twofold reduction in burst size. The 40 kbp T7 genome only contains six GATC sites, so additional sites were introduced to improve the accuracy of the measurements. A paucity of GATC sites is common among close relatives of the T7 family. Statistically about 150 sites would be expected in a 40-kbp genome, but the actual numbers for enteric T7-likes range from 2 to 13. Presumably, there has been selection against the GATC sequence in these phage genomes, probably because it is a major regulatory sequence for the host bacterium. Unfortunately, this assay is not generally applicable, partly because many phage genomes are circularly permuted and partly because of the frequency of GATC sites in most Gram-negative phage genomes (and too few, or even none, in *Bacillus* phages). In addition, DpnI is the only known restriction enzyme that is completely dependent on a methylated substrate. The converse assay, methylating entering DNA with the cognate methylase of a restriction enzyme, and then cutting isolated DNA with that nuclease, is both less sensitive and subject to error. Entering genomes become resistant to the restriction enzyme after methylation in the cell cytoplasm and thus produce bands of longer DNA on a Southern blot that cannot readily be distinguished from incomplete digestion.

The leading ~1 kbp of the T7 genome enters into a rifampicin-treated cell at a constant rate of 70 bp/s at 30°C (Garcia and Molineux 1996; Struthers-Schlinke et al. 2000; Kemp et al. 2004). It is not known what makes this internalization process stop although it appears to be a measurement of length; deletion mutants lacking most of the leading wild-type DNA sequence still stop entering after ~1 kbp of the mutant genome is inside the cytoplasm. A selection for mutants that allow more DNA to enter the cell before the transcription phase resulted only in gene *16* mutants (Struthers-Schlinke et al. 2000). Different mutants define two important regions near the middle of gp16 (residues 733–761 and 809–860 of the 1,318 amino acid polypeptide); all have the same phenotype where the entire genome enters into a rifampicin-treated cell and at the same constant rate. The mutants are thus defective in stopping DNA translocation after ~1 kbp, but no mechanistic information could be gleaned from the mutational spectrum. Attempts to select for faster rates of DNA translocation have not been fruitful.

The constant rate of DNA internalization for the entire phage genome using these gene *16* mutants directly proves that this process is not driven by any pressure associated with the densely packed and tightly wound DNA in the virion (Kemp et al. 2004; Molineux 2006). One would naively expect from a virion-pressure driven reaction that the initial rate of genome ejection would be fast, and would decline to zero as the head empties. Furthermore, the rate of transcription-independent T7 genome internalization from virions containing a mutant gp16 varies with temperature (Kemp et al. 2004), a result not expected from a pressure-based model of genome ejection. The actual rate data can be fitted to an Arrhenius plot, with an apparent activation energy of 19 kcal/mol. Extrapolating the rate data to the point at which T7 begins to enter the infected cell cytoplasm allows an estimate of the time taken to form the trans-envelope channel. This value is also temperature dependent; an Arrhenius plot provides an apparent activation energy of 28 kcal/mol, significantly higher than that for the DNA translocation process. The initial stages of infection, those leading to the formation of a channel from the phage head into the cell cytoplasm, are therefore energetically infection determining. Interestingly, the activation energy of 28 kcal/mol for T7 is comparable to that determined for in vitro ejection of SPP1, λ, and T5 DNAs (de Frutos et al. 2005; Raspaud et al. 2007), suggesting that the bacterial cell is not resisting infection, and that after adsorption phages "control their own fate." The protein conformational changes that allow genome ejection from different mature phage virions may thus be comparable. Furthermore, T5 ejects its genome in discrete steps, both in vivo and in vitro (Lanni 1968; Mangenot et al. 2005); the activation energy required for each step is the same (de Frutos et al. 2005) and thus is independent of the length of DNA remaining in the virion. The internal pressure of encapsidated DNA does not therefore appear to provide the driving force for the conformational changes in the protein(s) controlling DNA ejection.

Three promoters (A1, A2, A3) for *E. coli* RNAP and one for T7 RNAP lie within the leading 1 kbp of the T7 genome that is internalized by the transcription-independent mechanism. If T7 RNAP is provided in the host cell prior to infection, the T7 promoter is recognized, and transcription internalizes the complete genome at a 30°C rate of about 250 bp/s (Garcia and Molineux 1995), in good agreement with optimal rates of T7 RNAP transcription in vitro (Anand and Patel 2006). Although one normally pictures enzymes moving along a large stationary DNA molecule, topologically there is no difference if the enzyme is stationary and DNA is threaded through the enzyme. It can be imagined that an RNAP moves along T7 DNA until it encounters the cytoplasmic membrane but continues to transcribe, thereby pulling DNA from the phage capsid into the cell.

The A1, A2, and A3 *E. coli* promoters on T7 DNA are among the strongest unregulated promoters known; this is not a requirement for genome internalization, but transcription from these promoters normally provides all early T7 mRNAs. Any of the three can be utilized to bring the infecting T7 genome into the cell, as can the minor *E. coli* promoters on the T7 genome if the three A promoters are deleted (Moffatt and Studier 1988; Garcia and Molineux 1995; Kemp et al. 2004). At 30°C, a wild-type T7 genome is internalized at a constant 40 bp/s across the genome (Garcia and Molineux 1995; Kemp et al. 2004), about twice the rate of host mRNA synthesis in the cell, but the same rate as transcription of the rRNA operons. This is because transcription from the natural A1, A2, and A3 promoters is subject to antitermination at a *boxA* element that is identical in sequence to *boxA* of the ribosomal operons. Transcription antitermination of *E. coli* RNAP on the T7 genome also requires the same host proteins (e.g., NusA, NusB, NusG) that are used by the host cell for rRNA synthesis (W. P. Robins and I. J. Molineux, unpublished). Antitermination reduces the extent of pausing by RNAP, leading to an increased overall rate of transcription, and thus an increased rate of genome entry. This may confer a small fitness benefit to the phage by decreasing the time of early gene expression and, as a consequence, the phage latent period. As would therefore be expected, at least two early promoters and a *boxA* sequence are found near the leading genome end of all close relatives of T7, although, among them, transcription has been only shown experimentally to be required for genome internalization for T3 (M. Pajunen and I. J. Molineux, unpublished) and the cyanophage Pf-WMP4 (Liu et al. 2007). However, the phage RNAP gene in the T7-like øKMV family is so far from the genome end that the time taken for RNAP to transcribe from promoters near the genome end across that gene would effectively be the entire latent period (Lavigne et al. 2003). Thus, øKMV may use the T7 transcription-independent mode of DNA translocation (about twice as fast as transcription mediated) for most of or the entire genome – it does encode counterparts of T7 gp15 and gp16 – or its mode of genome internalization may be completely different.

E. coli RNAP is one of the strongest motor proteins known (Yin et al. 1995; Wang et al. 1998a, b), and it is therefore not surprising that it can pull a DNA molecule into the cell cytoplasm. As an enzyme, RNAP is expected to follow Arrhenius kinetics. The apparent activation energy required for formation of an RNAP elongation complex (27.2 kcal/mol) corresponds closely to the activation energy for trans-envelope channel formation discussed previously. However, the apparent activation energy for transcription elongation itself is biphasic: ~10 kcal/mol at ≥30°C but ~25 kcal/mol at ≤30°C. The reason(s) underlying the difference is unclear, but may reflect the effects of temperature on the mechanism of transcription antitermination.

The overall similarity of the energetics of both the transcription-independent and transcription-dependent modes of T7 genome internalization led to the idea that the former, which utilizes the two ejected virion proteins gp15 and gp16, might also be enzyme catalyzed. If so, the enzyme needs an energy source as it is doing work in moving the phage genome from the virion into the cell. Under conditions where intracellular ATP levels were uncoupled from the membrane potential, collapsing the latter as T7 DNA was known to be translocating into the cell, led to an immediate cessation of the gp15/gp16-mediated process (Kemp et al. 2004). In contrast, *E. coli* RNAP could complete genome internalization, albeit at a much-reduced rate. This reduced rate suggests that a membrane potential is required to maintain a functional channel across the cell envelope. T4 infection also requires a membrane potential (Labedan and Goldberg 1979; Labedan et al. 1980), but it is not yet clear whether it is necessary for DNA transport or for maintaining the channel across the cytoplasmic membrane. Neither gp15 nor gp16 contains ATP or other nucleotide-binding motifs in their primary sequence, but it has not been shown directly that high-energy phosphate is *not* required. One technical difficulty is that the methylation-dependent assay of genome entry requires *S*-adenosyl methionine (as the methyl donor), but this compound turns over rapidly, and its synthesis requires ATP. Thus, any perturbations in T7 genome entry that might be observed after depleting intracellular ATP levels would be difficult to evaluate. It should be noted that not all phages require a membrane potential for DNA ejection. Both first- and second-step transfer of the siphovirus T5 DNA occur in the absence of cellular energy (allowing for the obligatory protein synthesis between those steps) (Filali Maltouf and Labedan 1983), and λ also has been shown to eject its DNA into a completely energy-poisoned cell (Filali Maltouf and Labedan 1985).

Another incompletely resolved question that results from a transcription-mediated T7 DNA internalization process is the fate of the positive supercoils that accumulate ahead of the transcription complex. In principle, the supercoils could be removed either by a cytoplasmic topoisomerase or they could be translated along the DNA helix into that part of the genome still in the phage head. The rate of T7 DNA internalization catalyzed by *E. coli* or T7 RNAP is not changed significantly by inhibition of DNA gyrase and/or topoisomerase IV (P. Kemp and I. J. Molineux, unpublished), implying that relaxation of the transcription-induced supercoils occurs within the phage head. This could be accomplished either by rotation of the entire DNA structure remaining in the capsid or by transmission of the supercoils along the helical axis where they can be dissipated at the trailing genome end.

The rates of T7 genome internalization by *E. coli* and T7 RNAP are in excellent agreement with optimal rates of transcription by those enzymes in vitro; the extended tail of T7 that spans the cell envelope does not therefore resist transcription-mediated DNA translocation significantly. A third enzyme has also been shown to internalize the genome. Type I restriction enzymes bind to their cognate recognition site and then, while still bound to that site, spool duplex DNA through themselves in an ATP-dependent process. Restriction enzyme cutting occurs when spooling is blocked, either because of a collision with a second restriction enzyme or because of topological constraints in a circular DNA. A linear DNA containing only a single recognition site is therefore not subject to restriction (Murray 2000). A unique EcoKI site was introduced within the initial 1 kbp of the genome of a T7 mutant that lacked its four natural sites. In the absence of transcription, the genome of the phage was internalized by cells harboring an active EcoKI enzyme, but not by cells lacking the enzyme (Garcia and Molineux 1999). In the presence of EcoKI, genome translocation at 30°C occurred at 100 bp/s, again in good agreement with the in vitro EcoKI translocation rate (Studier and Bandyopadhyay 1988).

The biology of T7 provides a rationale for its tripartite mode of genome internalization. First, slow entry could provide a simple but effective means of gene regulation. Genes expressed early are the first to enter the cell, whereas the late, virion structural gene cluster is prevented from being expressed as they are not in the cell at early times. Second, the first gene to be expressed on the T7 genome, gene *0.3*, encodes an antirestriction protein whose structure resembles duplex DNA (Walkinshaw et al. 2002). Gp0.3 binds to the DNA sequence specificity subunit of type I restriction enzymes, thereby inhibiting their activity. Unlike λ, for example, T7 grows normally on *E. coli* B and K-12 strains, regardless of its previous host, even though it contains unmethylated sites for restriction enzymes in those cells. By demanding that transcription internalizes the T7 genome, there is obligatory gene expression, in particular here of the antirestriction function, and thus, by the time cognate sites for a restriction enzyme enter the cell, gp0.3 should have inactivated the enzyme. This idea was tested directly by packaging gene 0.3^{+} T7 DNA into λ and allowing the λ ejection machinery to deliver T7 DNA into the cell cytoplasm. Ejection occurs rapidly into K-12 cells lacking EcoKI, but the T7 DNA is immediately degraded in cells containing the enzyme (Garcia and Molineux 1999). A reduction in plating efficiency, albeit less dramatic than when T7 DNA is packaged in a λ virion, is observed when T7 RNAP, which transcribes about sixfold faster than the *E. coli* enzyme, is used for genome internalization in restriction-proficient cells.

7.4.3 Phage N4

The tail of *E. coli* phage N4 contains 12 flexible, apparently trimeric "appendages" (made of gp66) that may be comparable to tailspikes in other podoviruses. The role of these appendages in infection is not known, but by analogy with other podoviruses, they could have a (nonessential?) role in adsorption. However, they have not yet been implicated in receptor binding, whereas the nonfibrous sheath protein has been identified as a host receptor-binding protein (McPartland and Rothman-Denes 2009). Figure 7.7 shows a three-dimensional reconstruction of the N4 tail. Note that the N4 "sheath" is not a clear analog of the sheath of the contractile tails of myoviruses like T4. Two host receptors have been identified for N4: the outer membrane protein NfrA, which interacts with the virion sheath protein (McPartland and Rothman-Denes 2009), and a cytoplasmic membrane protein NfrB, proposed to provide the transmembrane channel for cellular internalization of the virion RNAP and the phage genome (Kiino and Rothman-Denes 1989; Kiino et al. 1993a, b). The roles of two other genes involved in N4 adsorption are *nfrD* and *nfrC* (= *wecB, rffe*) are not yet understood (Kiino and Rothman-Denes 1989; Kiino et al. 1993a, b); the latter affects biosynthesis of enterobacterial common antigen (ECA) (Kuhn et al. 1988). An altered ECA could prevent access of N4 to NfrA, or perhaps ECA has a coreceptor function.

Phage N4 is unique among the well-studied phages in that its virions contain about four molecules of a huge protein, the 3,350-amino-acid, single-polypeptide, phage-encoded RNAP (vRNAP).

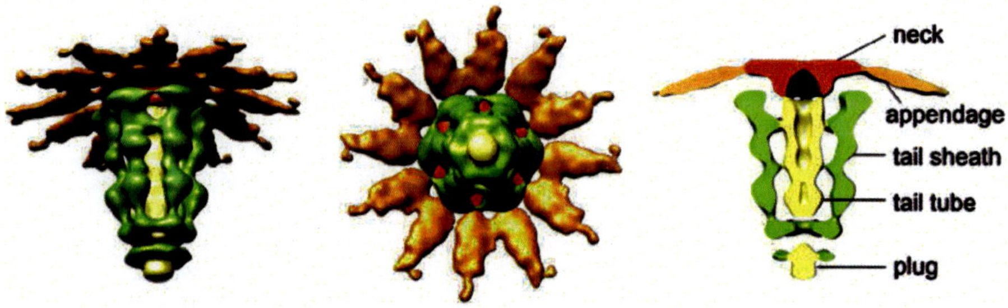

Fig. 7.7 The phage N4 tail. A three-dimensional reconstruction of the sixfold averaged N4 tail, extracted virtually from the virion structure, is shown with *side* (*left*), *bottom* (*middle*), and *slice* (*right*) views. The neck or portal protein is encoded by N4 gene *59*, and the appendage and sheath proteins are encoded by genes *66* and *65*, respectively. The genes that encode the tube and plug proteins have not been identified. The figure is modified from Choi et al. (2008) with permission

The vRNAP recognizes promoters found only on the N4 genome and is required for expression of the phage early genes (Falco et al. 1977, 1980; Choi et al. 2008). The three-dimensional reconstruction of the N4 virion shows internal core density above the portal ring (Fig. 7.5), and similar reconstruction of a virion that lacks only vRNAP has a smaller core and is missing density immediately above the portal, indicating this as a possible location for at least part of the vRNAP. The authors suggest that such a location could position vRNAP for ejection before DNA exit (Choi et al. 2008). It is not known if any other N4 proteins are ejected with the DNA.

The 72-kbp genome of phage N4 has, like T7, been shown to enter the cell in a stepwise manner (Kazmierczak and Rothman-Denes 2006). When infected cells, but not the phage virion, are permeabilized with chloroform, limited DNase I digestion produces 1–2-kbp fragments of N4 DNA inside the cell, whereas DNA inside the virion remains uncut. PCR reactions on these fragments using a bank of N4-specific primers can then identify how much of the genome has entered the cell (A. Demidenko and L. Rothman-Denes, personal communication). The leading end of the viral DNA carries several N4-specific promoters, the first of which is within the ~400-bp long terminal direct repeat (Haynes and Rothman-Denes 1985). The N-terminal domain of vRNAP is essential for internalization of ~500 bp of the N4 genome (Falco et al. 1977, 1980; Choi et al. 2008), and the vRNAP must be ejected into the cell either before or concomitant with this DNA. Transcription from the first vRNAP promoter to exit the virion directs synthesis of a small protein (gp1) that is necessary for internalization beyond the leading 1 kbp. Genome entry continues, powered by the vRNAP–gp1 complex, for about 40 kbp, and the last part of the genome is internalized via transcription by the N4-encoded middle RNAP (RNAP II) and the DNA-binding protein gp2 (Kazmierczak and Rothman-Denes 2006). Both the ejected vRNAP and RNAP II are related to T7 RNAP (Kazmierczak et al. 2002; Willis et al. 2002) and may, therefore, transcribe at comparable rates. DNA replication begins by 10–12 min after infection at 37°C, suggesting that internalization of the 70-kbp genome should be complete by that time; the current estimate is within 5–6 min after infection (A. Demidenko and L. Rothman-Denes, personal communication).

7.4.4 Phage ϕ29

Phage ϕ29 infects *B. subtilis*, which, unlike the other hosts discussed here, is a Gram-positive bacterium that has a thicker cell wall than Gram-negative bacteria and no outer membrane. In the virion, its tail machine has 12 radially arranged trimeric gp12 tailspikes (called "appendages" in this phage)

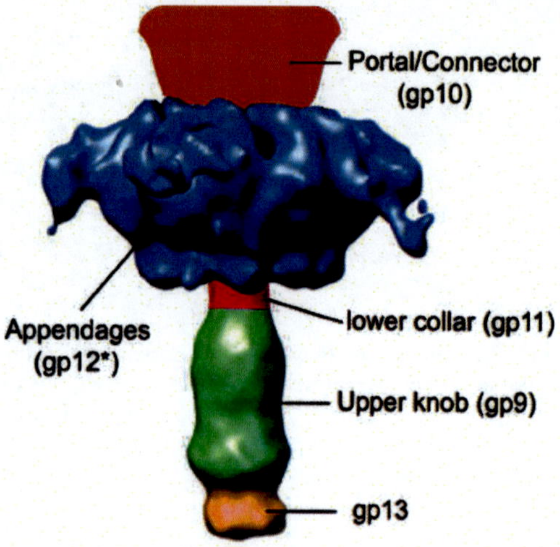

Fig. 7.8 The phage ϕ29 tail machine. Five protein components of the ϕ29 tail are shown in different colors. The approximate location of the portal ring is indicated above in *brown*. The figure is modified from Xiang et al. (2008)

which can degrade the acidic polymer, polyglycerolphosphate teichoic acid found on the surface of *B. subtilis* (Xiang et al. 2008), a dodecameric portal ring (Simpson et al. 2000) (called the "connector" in this paper), as well as three additional proteins (gp9, gp11, and gp13) that make up the central protruding "knob" (Cohen et al. 2008 and references therein) (Fig. 7.8). The latter three proteins are not recognizably similar in amino acid sequence to proteins in the tail machines of other tailed phages. The knob projects several hundred Angstroms beyond the tailspikes, positioning it for first contact with the cell surface during infection. The ϕ29 tailspikes/appendages, like other known tailspikes (Fig. 7.4), are built with largely β-helical secondary structure. Like phage K1F tailspikes (Schwarzer et al. 2007, 2009), they have a C-terminal "autochaperone" domain that helps trimerize the molecule; this domain is subsequently autoproteolytically removed so that it is not present in mature virion tailspikes (Xiang et al. 2009). Asymmetric three-dimensional reconstructions of virions and virions from which DNA has been artificially released show that after DNA exits, the portal vertex channel remains open, and that a small extra density appears at the tail tip after DNA release suggesting some protein rearrangement (Xiang et al. 2006). The ϕ29 virion reconstruction shows density attributed to DNA filling the portal channel from the interior about half way to the tip. Curiously, at a wide spot in the channel where the portal protein (gp10) and the tube protein (gp11) meet, the DNA adopts an *extremely* distorted conformation that appears as a 20 Å thick toroid with an inner diameter of only 18 Å (Tang et al. 2008). The biological role, if any, of this DNA deformation is mysterious. The ϕ29 virion DNA also has a 29-kDa terminal protein, gp3, that is covalently attached to the 5′ end of each strand. Terminal protein is used in priming DNA replication (reviewed by Meijer et al. 2001a). This protein has not been clearly identified in virion reconstructions, and any active role of gp3 in DNA delivery is unknown.

It is not known if any proteins (other than the terminal protein gp3) are released from the ϕ29 virion during natural DNA ejection into the host bacterium, but gp13 is at the tip of the knob and appears to be the portal channel plug (Cohen et al. 2008; Xiang et al. 2008). Its crystal structure shows that it has two separate domains with interesting structures. The N-terminal region has a transglycosylase fold that is similar to the λ endolysin, an enzyme that cleaves the polysaccharide portion of peptidoglycan (Bienkowska-Szewczyk and Taylor 1980). The C-terminal domain has a metalloendoprotease fold that is similar to the enzyme LytM, which is known to cleave the peptide cross-links in *Staphylococcus* peptidoglycan (Cohen et al. 2008; Xiang et al. 2008). Thus, gp13 seems well positioned to cleave both the polysaccharide backbone and the peptide cross-links of the peptidoglycan in preparation for DNA passage into the cell.

Salas and coworkers have studied DNA entry into the cell during ϕ29 ejection and concluded that it uses a strategy that is so far unique among podoviruses, termed a "push-pull" mechanism by

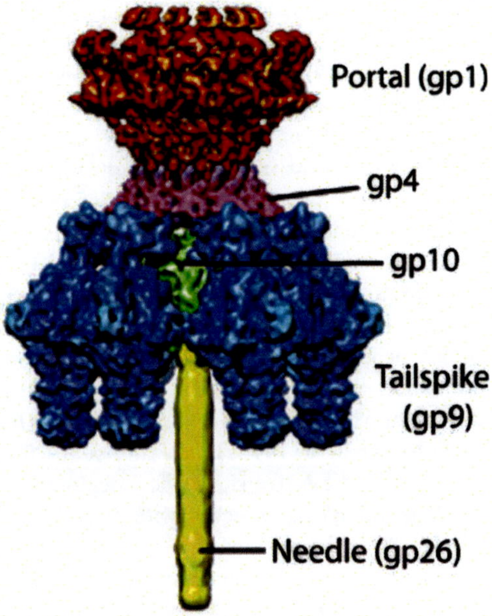

Fig. 7.9 The phage P22 tail machine. A three-dimensional reconstruction of the P22 tail at 9.4 Å resolution after its chemical extraction from the virion. The five protein components are shown in different colors as indicated in the figure. Figure is modified from Lander et al. (2009) with permission

these authors. In this phage, whose virion DNA is not circularly permuted, the genome enters the bacterium from right to left with respect to the conventional genetic map (Krawiec et al. 1981). Using a chromatin immunoprecipitation (CHIP)–based assay with the ϕ29 genome architectural protein p6 (Gonzalez-Huici et al. 2004a), it was shown that the leading ~65% enters the cell in the absence of protein synthesis (Gonzalez-Huici et al. 2004b, 2006). The pressure associated with the tightly packaged genome in a mature phage (see above) was assumed responsible for this "push" phase of DNA entry.

The last 35% of the DNA is then described as being "pulled" from the virion by an unknown mechanism that appears not to directly involve transcription, but does require energy, which is supplied by the membrane potential, and the ϕ29 proteins gp17 and gp16.7 (Gonzalez-Huici et al. 2004b, 2006; Alcorlo et al. 2007). These two proteins are not present in the infecting virion and are expressed from the first part of the DNA that is ejected into the cytoplasm. The "pull" phase requires energy, in addition to that required for synthesis of gp16.7 and gp17, and it is thought that the latter has a role in transducing energy from the electrochemical gradient into the cell. Gp16.7 and the membrane-associated gp17 also have poorly understood roles in phage DNA replication in vivo (Meijer et al. 2001b; Serna-Rico et al. 2002). The "pull" step of genome entry also requires that the ϕ29 DNA be supercoiled, and is blocked by the addition of novobiocin, an inhibitor of the negative supercoiling activity of DNA gyrase (Gonzalez-Huici et al. 2004b, 2006). DNA gyrase is not thought to be directly pulling the ϕ29 genome into the cell, as entry is not affected by nalidixic acid. Overall, ϕ29 DNA delivery is different from that of T7 and more similar to that of the siphovirus T5, in the respect that phage proteins expressed from the first part of the ejected DNA are essential for the completion of genome internalization.

7.4.5 *Phage P22*

Phage P22 infects *S. enterica* and adsorbs to the LPS O-antigen polysaccharide through its tailspikes. The structure of its tail machine is exceptionally well understood. Without the three ejection proteins discussed below, it contains 51 protein molecules of five different types. Johnson and coworkers

(Tang et al. 2005; Lander et al. 2006, 2009) have determined a 17 Å resolution structure of the whole P22 virion (Fig. 7.5) and a 9.4 Å resolution structure of the tail machine after its extraction from the virion (Fig. 7.9). In addition, work on the stepwise assembly of this structure (Strauss and King 1984; Olia et al. 2006, 2007a; Bhardwaj et al. 2007) and crystal structures of the tailspike and needle (Steinbacher et al. 1994, 1997; Olia et al. 2007b) allows an accurate estimation of the positions of most of the intersubunit boundaries in the tail machine electron density.

The homotrimeric tailspike contains two domains with a hinge between the receptor binding/cleavage domain and the domain that binds to the virion (Steinbacher et al. 1994, 1997). The hinge is bent in the virion so that the elongated C-terminal domain points approximately orthogonally away from the head. On one side of this domain a protruding "fin" appears to contact gp10 in the virion tail, but the resolution of the structure is such that the exact nature of this contact cannot be determined, and its importance remains unknown (Lander et al. 2006). The receptor-binding domain has mainly β-helical secondary structure, a very common theme in tailspikes (see above). The crystal structure of the P22 tailspike bound to its O-antigen polysaccharide receptor has been determined (Steinbacher et al. 1996; Landstrom et al. 2008), and each of the six tailspikes in P22 virions has three enzymatic active sites. The receptor polysaccharide is bound in these sites so that the virion would be oriented "correctly" with its tail closest to the bacterial surface if the substrate is stretched outward from the cell surface.

The P22 tail also contains a single "needle" that projects outward from the center of the tail (Figs. 7.5 and 7.9). The X-ray structure of this gp26 trimer has been determined; it has several domains: a virion proximal N-terminal domain that binds to the body of the tail, a long central coiled-coil region, and a virion distal C-terminal region with a partly β-helix and partly complex fold (Bhardwaj et al. 2007; Olia et al. 2007b). The latter domain is hinged relative to the rest of the molecule; it is bent at different angles in the two molecules in the unit cell of the crystals used for structure determination, and it does not appear in three-dimensional virion reconstructions, presumably due to its flexibility (Olia et al. 2009). The distal tip of the needle projects about 130 Å beyond the distal ends of the tailspikes and so is well positioned for first contact with the target cell's outer membrane surface. There are three tyrosine side chains lying on the needle's distal tip that could interact with some surface feature, and there is a band of high positive charge around the needle about 20 Å from the tip that in theory could interact with the negatively charged core region at the base of the LPS or with phospholipid head groups. The virion structure clearly shows that previous analyses, which indicated that gp26 is the protein that plugs the tail machine channel, are correct, since the N-terminal domain of the needle fills the distal opening in the tail's central channel, and particles with packaged DNA are very unstable until gp26 is added to the structure (Lenk et al. 1975; Berget and Poteete 1980; Strauss and King 1984; Lander et al. 2006; Bhardwaj et al. 2007).

It is attractive to speculate that the P22 tailspikes bind and cleave the O-antigen until the virion works its way to the surface (see above) and the tip of the needle contacts the cell surface. The (rather slow) release of DNA from P22 virions by purified LPS-O-antigen suggests that P22's secondary receptor might also be part of the LPS molecule (Andres et al. 2010). The P22 virion structure suggests two possible models for the transmission of this signal to the rest of the tail. (1) The distal tip of the gp26 needle might contact the putative secondary receptor, and this contact could cause a conformational change in gp26 that propagates to the proximal end of the needle to release the needle (the portal channel plug) from the virion and initiate the process of ejection. However, to date no secondary receptor has been identified for P22. (2) An alternative model is that the tip of the needle binds something on the bacterial surface that mechanically pulls the virion closer to the surface. This action could cause the tailspikes to splay out and lose their "fin contacts" with gp10. Since the needle binds to gp10 in the tail, loss of contact with the tailspike could induce a conformational change in gp10 that releases the needle. In agreement with both models, gp26 is indeed released from the virion during DNA delivery (Israel 1977); however, the high thermal stability of the folded needle trimer makes the required conformational change in the first model somewhat difficult to imagine (Bhardwaj et al. 2007).

P22 DNA is circularized by the host homologous recombination machinery, so the 9–15-min postinfection reported for the time circles appear (Rhoades and Thomas 1968) is the maximum time required for complete DNA entry, although it could be considerably less. Aside from this, little is known about the actual entry of P22 DNA into cells during ejection. Unlike T7, N4, and ϕ29, the DNA in P22 virions is circularly permuted (Tye et al. 1974). Thus, the sequence of the first part of the DNA to be ejected is different in different individual virions. This in turn precludes slow, ordered DNA entry mechanisms that help control the temporal pattern of gene expression, since different genes enter first from different virions. It also seems to preclude the necessity for synthesis of any P22-encoded proteins to allow complete genome internalization. The circular permutation also makes determination of the direction of ejection along the genome sequence difficult to determine, and this is unknown for P22.

The P22 virion contains three "ejection proteins," gp7, gp20, and gp16, which are released from the virion during the DNA ejection process (Israel 1977). Phages with null mutations in any or all of these genes assemble normal-appearing "virions" that do not deliver their DNA properly (Botstein et al. 1973; King et al. 1973 E. Gilcrease and S. Casjens, unpublished data). At least 60% or more of the DNA is released from the virion in the absence of gp16 function, but no genes are expressed from the released DNA. The precise fate of this DNA is not known; however, it appears not to enter the cell cytoplasm (Hoffman and Levine 1975b). It has been shown that gp16 from a wild-type virion can complement a particle that lacks gp16, suggesting that it is at least partly diffusible upon release from the complete virion; however, gp20 and gp7 are not able to function in *trans* in this way (Hoffman and Levine 1975a, b; Susskind and Botstein 1978). The three ejection proteins assemble into the virion independently from one another (Botstein et al. 1973; King et al. 1973 E. Gilcrease and S. Casjens, unpublished data), and gp16 and gp20 eject in the absence of the other two proteins, but gp7 appears to require the presence of gp16 in the virion to be ejected (Israel 1977). The location of the ejection proteins in the P22 virion is unknown [the density tentatively attributed to the ejection proteins above the portal in the virion structure by Lander et al. (2006) is extremely tentative]. The P22 ejection proteins have no homology to other proteins that might shed light on their functions, and their cellular location after release from the virion is not known. They are nonetheless prime candidates for proteins that help DNA to transit the cell membranes and periplasm. They could also have a role in the cytoplasm (Benson and Roth 1997), but gp16 expressed inside cells does not rescue infection by particles missing gp16 (Umlauf and Dreiseikelmann 1992). Purified gp16 can cause the release of a dye from, and entry of DNA into, liposomes in the presence of an artificially produced membrane potential, but the biological relevance of this tantalizing observation remains unclear (Perez et al. 2009). Interestingly, gp16 is extremely variable among the P22-like phages, and a convincing lysozyme-like domain is present in the gp16 of a defective P22-like *Shigella* prophage that is not present in P22 gp16 (Casjens and Thuman-Commike 2011, unpublished). These findings are all consistent with gp16 playing a role in DNA transport across the cytoplasmic membrane and periplasm during ejection, but the mechanism remains unknown.

7.4.6 Phages BPP-1, ε15, and Other Less-Studied Podoviruses

The more recently discovered phage BPP-1 infects *Bordetella* and is not a close relative of other known podoviruses. Most of its tail proteins have not been unambiguously identified genetically (Liu et al. 2002, 2004), but it has a short tail with six long, flexible tail fibers extending radially that have large knobs at their distal ends, and six short "spikes" that project down from the bottom of the tail (Dai et al. 2010) (Fig. 7.10). An especially interesting feature of BPP-1 is that it encodes a directed, error-prone reverse transcriptase-mediated enzymatic machinery that generates limited nucleotide sequence diversity at specific sites in the *mtd* gene (Liu et al. 2002; Doulatov et al. 2004;

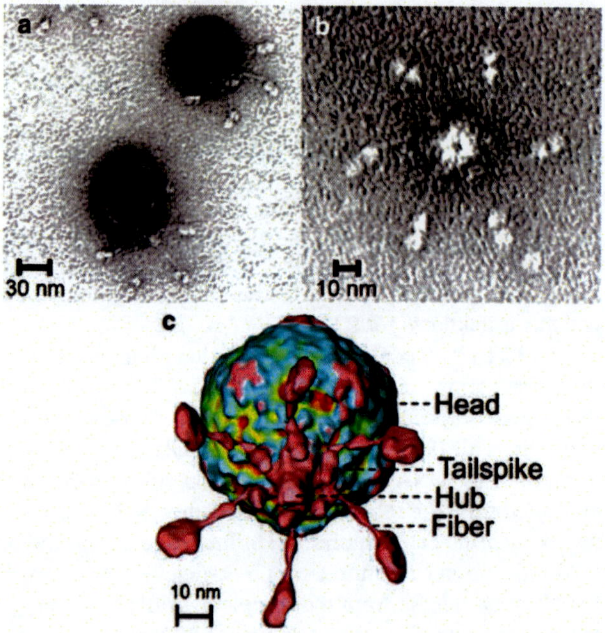

Fig. 7.10 Phage BPP-1. Negatively strained transmission electron micrographs of (**a**) phage BPP-1 virions, (**b**) a fortuitously released tail with six fibers, and (**c**) an electron microscopic tomogram with averaged tail fibers. (**a, b**) Modified from Liu et al. (2002) and (**c**) modified from Dai et al. (2010) with permission

Miller et al. 2008). The Mtd (major tropism determinant) protein appears to be present as the distal half of the knob at the end of the tail fiber, and the crystal structure of this protein bound to its primary host receptor protein, pertactin, has been determined. The conformation of Mtd does not appear to be altered by receptor binding (Miller et al. 2008).

Bordetella changes its surface proteins when environmental conditions are altered, and BPP-1's programmed variation of Mtd allows a small fraction of a phage population to utilize alternative receptors in *Bordetella* that are in a different state (Liu et al. 2004). Mtd binding to the receptor is not very strong, in the micromolar K_d range, suggesting that multiple fiber knobs may need to contact their receptor for tight virion binding (Miller et al. 2008). However, binding of one Mtd to pertactin positions the phage in an orientation where additional copies of Mtd can bind readily and the K_d of phage binding is in the picomolar range. Electron micrographic tomograms of individual virions adsorbed to the surface of the host suggest that the fibers are splayed upward when the bottom of the tail hub is in contact with the cell membrane surface, and that the tailspikes may also tilt about 30° inward upon adsorption (Dai et al. 2010). The authors suggest that these apparent changes could be the signal for the virion to release its DNA upon binding to a secondary receptor. It is not known if BPP-1 ejects any virion proteins into the host cell, and, unlike T7, ε15, and P22, it appears to lack an internal core structure above the portal vertex. Thus, its mode of genome delivery into the host cell could be different.

S. enterica phage ε15 is an example of yet another type of podovirus that is very different from those discussed above. Although few details of its molecular life cycle have been studied, a striking three-dimensional reconstruction of whole virions has been reported (Jiang et al. 2006) (Fig. 7.11). The virion reconstruction shows that its tail has six large tailspikes, which are presumably responsible for the observed cleavage the O-antigen polysaccharide of its *S. enterica* host (Kanegasaki and Wright 1973; Takeda and Uetake 1973). A substantial internal core above the portal ring is reminiscent of that of phage T7, and there is no central needle as is found in P22. Neither the ε15 tailspike nor the other putative tail proteins (except portal protein) have any recognizable sequence similarity to tail proteins in other podoviruses. The ε15 genome sequence shows nine genes between the coat protein and tailspike genes in the position whether other podoviruses often encode tail structural proteins and proteins that are ejected with the DNA; to date four of these have been shown to be present in relatively small numbers in ε15 virions (Kropinski et al. 2007).

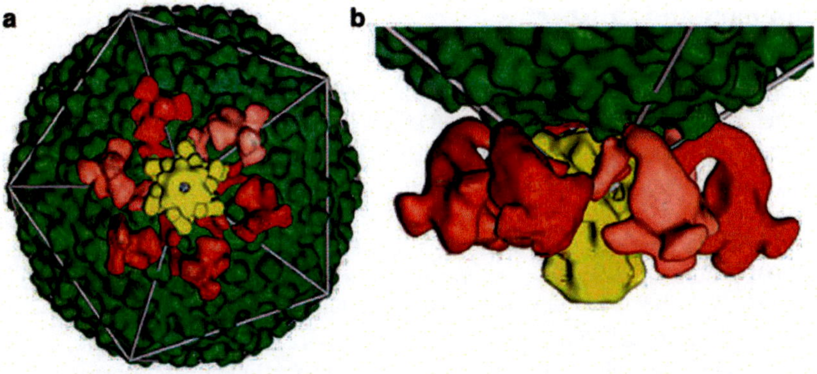

Fig. 7.11 The phage ε15 tail. Three-dimensional reconstruction of (**a**) the virion viewed along the tail axis and (**b**) enlarged side view of the tail. *Green* indicates the coat protein, and white wires outline the faces of the icosahedral head; *yellow* indicates the tail hub, and the tailspikes are shown in *red*. Figure is modified from Jiang et al. (2006) with permission

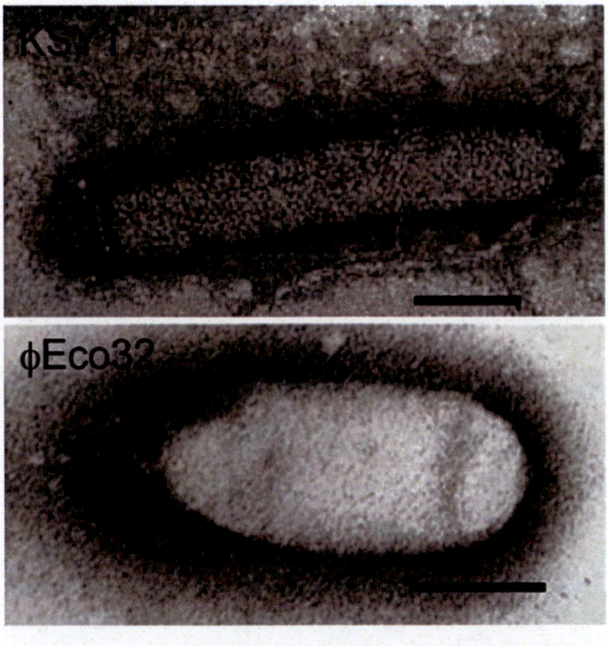

Fig. 7.12 Negatively stained electron micrographs of podoviruses KSY1 and ϕEco32. The KSY1 micrograph is from Chopin et al. (2007) and reproduced with permission. The ϕEco32 micrograph was produced by Dr. H. W. Ackermann as described in Savalia et al. (2008). The *black scale bars* are 50 nm

Some other podoviruses whose life cycles have not yet been studied in detail, for example, *Lactobacillus* phage KSY1 (Chopin et al. 2007) and *E. coli* phage ϕEco32 (Savalia et al. 2008), have unusually long heads, and KSY1 tails appear to be more elaborate than those discussed here. These two virions are shown in Fig. 7.12. Several of KSY1's tail proteins have sequence similarity to tail proteins of Lactococcal siphoviruses r1t and TP901-1. These include similarities to siphovirus receptor-binding proteins, so the proteins that encode the body of the KSY1 tail are likely different from those of other podoviruses. Interestingly, one of the putative ϕEco32 tail proteins has weak sequence similarity to a T4-baseplate protein, the biological significance of which is unknown. Putative tail fiber-encoding genes are present, but again the body of the tail is likely very different from the better-studied podoviruses.

7.5 Summary and Prospectus

The *Podoviridae* encompass a wide variety of short-tailed phages that share few features not also found in other morphotypes of tailed phages. It seems likely that all portal proteins will have similar dodecameric structures, and the symmetry mismatch with the fivefold vertex of the icosahedral capsid shell will likely be accommodated in comparable ways. The major coat protein fold of the siphovirus HK97 is also represented in phages of other morphotypes, including the podoviruses T7, ϕ29, ε15, P22, and BPP-1 (Jiang et al. 2003, 2008; Morais et al. 2005; Agirrezabala et al. 2007; Dai et al. 2010). Further, although most commonly described in podoviruses, structurally related tailspikes with enzymatic activity have also been found in both myo- and siphoviruses. The largest podovirus genome described to date is about 90 kbp, severalfold smaller than the largest myo- or siphovirus genomes, although any significance of this apparent limitation is obscure.

There are, however, some problems confronted by podoviruses that, if not unique, are more acute than with long-tailed phages. At least in principle, some long-tailed phages may be able to span the distance from their site of adsorption on the cell surface into the cytoplasm. Thus, aside from opening a channel through the phage tail by ejecting their tape measure and plug proteins (although the latter may simply rotate out of the way), the myo- and siphoviruses may not require complex mechanisms to allow their genomes access to the cell cytoplasm. However, it should be remembered that the T-even phages, from which the idea developed that phages are like nano-syringes, may not directly eject their genomes into the cell. On topological grounds, after tail contraction the tip of the T4 tail tube only reaches the outer surface of the cytoplasmic membrane and does not directly extend into the cytoplasm itself (Simon and Anderson 1967). Furthermore, some siphoviruses may also need to assemble a conduit through the periplasmic space and cytoplasmic membrane for DNA transport, similar to what is likely an essential step for all podoviruses.

It can be imagined that podoviruses infecting Gram-positive hosts approach the cytoplasmic membrane by tunneling their way through the thick cell wall, and the lack of an outer membrane suggests that these phages may not require additional specialized transit machinery. On the other hand, podoviruses infecting Gram-negative hosts must functionally extend their tails or otherwise generate a channel across the cell envelope. In particular, although porins could, in principle, provide a route through the outer membrane, it seems unlikely that the capsids of phages that adsorb to LPS could rapidly and efficiently penetrate the hydrophobic lipid layer of the outer membrane in order for their tail to access the cytoplasmic membrane directly. Indeed, in P22 virions that have ejected their DNA, the capsid appears well outside the outer membrane, and much of the tail is visible (Fig. 4b in Lenk et al. 1975; Casjens, unpublished data). It has been suggested that T7 uses its internal core proteins to functionally extend its tail across the cell envelope (Molineux 2001; Kemp et al. 2005; Chang et al. 2010), and P22 also ejects several tail machine proteins into the cell (Israel 1977). Although the final localization of the P22 proteins has not yet been determined, they might function in a similar structural way to the T7 core proteins. Alternatively, they may make the cytoplasmic membrane bleb out toward the adsorbed phage and thereby reduce the distance between the tail knob and the cytoplasm. A precedent for the latter idea comes from the conjugal plasmid R1, where the P19 protein makes small lesions in the cell wall, causing intracellular material to efflux from the cell (Bayer et al. 2001). Finally, the N-terminal domain of the N4 vRNAP is essential for internalization of the leading end of the genome (A. Demidenko, personal communication, cited as unpublished data in Kazmierczak and Rothman-Denes 2006) and may either escort the DNA or interact with the cytoplasmic membrane protein NfrB to form a complete trans-envelope channel. It is possible that most podoviruses will follow these or possess alternative strategies, both to overcome the "tail size" problem and to ensure that the infecting genome is protected from periplasmic nucleases.

The precise mechanism of DNA transfer into the cell cytoplasm is not well understood in any phage system, although podoviruses provide three of the better-studied examples. T7 uses energy-requiring molecular motors for all DNA translocation; likewise, more than 98% of the N4 genome is internalized by transcription; how the leading 500 bp enters the cell is not yet known. In contrast, it is suggested that the leading ~65% of the ϕ29 genome enters the cell using the pressure associated with the tightly packed DNA. The remaining 35% is internalized by an energy-requiring process, where the energy is supplied by the membrane potential. However, it is not known, like the situation in T4, whether the membrane potential provides energy that directly moves DNA or whether that energy is only required to maintain an open channel across the cytoplasmic membrane. These ideas are not mutually exclusive and must become a focus of future study. It may be significant that the only two phage systems where it is known that cellular energy is *not* required for DNA transport are the siphoviruses λ and T5 (Filali Maltouf and Labedan 1983, 1985), whose tail tips are long enough to potentially span the cell envelope.

An alternative model to explain phage DNA ejection was proposed some 5 years or so ago, but it has not yet received much attention. It was suggested that when a phage breaches the cell envelope, a hydrodynamic flow is initiated whereby water and small ions pass from the environment into the phage head, down through the tail and into the cytoplasm (Molineux 2006). Flow exerts a force on the phage DNA that is inserted into the portal (or tail tube) that then drags the genome into the cell. This model requires no cellular energy for DNA movement; the overall osmotic gradient between the environment and the cell cytoplasm ensures water flow. Importantly, this model is totally independent of the pressure inside the capsid that is usually attributed to the close-packed, condensed DNA. The hydrodynamic model was shown to be feasible (Grayson and Molineux 2007) and has recently also been shown to explain all the in vitro DNA ejection data (Panja and Molineux 2010). Some of the main attributes of this model are that it can explain how the entire phage genome is ejected into the cell, how internal virion proteins are also ejected, and why T7 requires molecular motors to transport its genome. It also explains why the only known electron micrograph that captures phage DNA entering the cell shows that the double helix may be stretched beyond its normal helical repeat of 3.4 nm (Simon and Anderson 1967). It is to be hoped that critical experimental tests of the hydrodynamic model for phage DNA ejection will be part of the future studies.

Fully understanding the role of the tail machines and the mechanisms of DNA transport into the cell will obviously require a deeper understanding of the structure of the packaged genome. Although advances in techniques of cryo-electron microscopy in the past decade or so have provided much support for the inverse toroidal structure of DNA inside the capsid, this model is inexact. It is also not in agreement with coarse-grained molecular mechanics models, which apply principles of polymer physics in simulations of DNA packaging (e.g., Comolli et al. 2008; Petrov et al. 2009). Perhaps most importantly, the inverse spool structure of packaged DNA is inconsistent with the structure of DNA molecules in individual mature T5 virions as evaluated from cryoEM micrographs (Leforestier and Livolant 2010), and these authors suggest that other phage genomes may be packaged in a comparable way to T5. This suggestion warrants further analysis and perhaps a careful reanalysis of the original micrographs that led to the averaged structures depicting an inverse spool arrangement of the genome. Clearly, DNA delivery by the podoviruses (and other tailed phages) is far from being well understood and will remain fertile research ground for some time to come.

Acknowledgments We thank all our colleagues in this unusually collegial research field for many years of open and insightful discussions. We especially thank our own coworkers for their contributions to our respective research programs, which are supported by NIH grants GM32095 to IJM and AI074825 to SRC.

References

Adam G, Delbrück M (1968) Reduction in dimensionality in biological diffusion processes. In: Rich A, Davidson N (eds) Structural chemisty and molecular biology. W. H. Freeman and Company, San Francisco, pp 198–215

Agirrezabala X, Martin-Benito J, Caston JR, Miranda R, Valpuesta JM, Carrascosa JL (2005) Maturation of phage T7 involves structural modification of both shell and inner core components. EMBO J 24:3820–3829

Agirrezabala X, Velazquez-Muriel JA, Gomez-Puertas P, Scheres SH, Carazo JM, Carrascosa JL (2007) Quasi-atomic model of bacteriophage T7 procapsid shell: insights into the structure and evolution of a basic fold. Structure 15:461–472

Alcorlo M, Gonzalez-Huici V, Hermoso JM, Meijer WJ, Salas M (2007) The phage ϕ29 membrane protein p16.7, involved in DNA replication, is required for efficient ejection of the viral genome. J Bacteriol 189:5542–5549

Allison GE, Verma NK (2000) Serotype-converting bacteriophages and O-antigen modification in *Shigella flexneri*. Trends Microbiol 8:17–23

Anand VS, Patel SS (2006) Transient state kinetics of transcription elongation by T7 RNA polymerase. J Biol Chem 281:35677–35685

Andres D, Hanke C, Baxa U, Seul A, Barbirz S, Seckler R. (2010) Tailspike interactions with lipopolysaccharide effect DNA ejection from phage P22 particles in vitro. J Biol Chem 285:36768–36775

Astumian RD (1997) Thermodynamics and kinetics of a Brownian motor. Science 276:917–922

Bandyopadhyay PN, Das Gupta B, Joshi A, Chakravorty M (1979) Is the injection of DNA enough to cause bacteriophage P22-induced changes in the cellular transport process of *Salmonella typhimurium*? J Virol 32:98–101

Baptista C, Santos MA, Sao-Jose C (2008) Phage SPP1 reversible adsorption to *Bacillus subtilis* cell wall teichoic acids accelerates virus recognition of membrane receptor YueB. J Bacteriol 190:4989–4996

Barbirz S, Muller JJ, Uetrecht C, Clark AJ, Heinemann U, Seckler R (2008) Crystal structure of *Escherichia coli* phage HK620 tailspike: podoviral tailspike endoglycosidase modules are evolutionarily related. Mol Microbiol 69:303–316

Baumann RG, Mullaney J, Black LW (2006) Portal fusion protein constraints on function in DNA packaging of bacteriophage T4. Mol Microbiol 61:16–32

Baxa U, Steinbacher S, Miller S, Weintraub A, Huber R, Seckler R (1996) Interactions of phage P22 tails with their cellular receptor, *Salmonella* O-antigen polysaccharide. Biophys J 71:2040–2048

Bayer M, Iberer R, Bischof K, Rassi E, Stabentheiner E, Zellnig G, Koraimann G (2001) Functional and mutational analysis of p19, a DNA transfer protein with muramidase activity. J Bacteriol 183:3176–3183

Bayer ME, Thurow H, Bayer MH (1979) Penetration of the polysaccharide capsule of *Escherichia coli* (Bi161/42) by bacteriophage K29. Virology 94:95–118

Bayer ME, Takeda K, Uetake H (1980) Effects of receptor destruction by *Salmonella* bacteriophages epsilon 15 and c341. Virology 105:328–337

Bayer ME, Bayer MH (1981) Fast responses of bacterial membranes to virus adsorption: a fluorescence study. Proc Natl Acad Sci USA 78:5618–5622

Benson NR, Roth J (1997) A *Salmonella* phage-P22 mutant defective in abortive transduction. Genetics 145:17–27

Berg HC, Purcell EM (1977) Physics of chemoreception. Biophys J 20:193–219

Berget PB, Poteete AR (1980) Structure and functions of the bacteriophage P22 tail protein. J Virol 34:234–243

Bhardwaj A, Olia AS, Walker-Kopp N, Cingolani G (2007) Domain organization and polarity of tail needle gp26 in the portal vertex structure of bacteriophage P22. J Mol Biol 371:374–387

Bienkowska-Szewczyk K, Taylor A (1980) Murein transglycosylase from phage lambda lysate. Purification and properties. Biochim Biophys Acta 615:489–496

Botstein D, Waddell CH, King J (1973) Mechanism of head assembly and DNA encapsulation in *Salmonella* phage P22. I. Genes, proteins, structures and DNA maturation. J Mol Biol 80:669–695

Bradley DE (1967) Ultrastructure of bacteriophage and bacteriocins. Bacteriol Rev 31:230–314

Bull JJ, Vimr ER, Molineux IJ (2010) A tale of tails: Sialidase is key to success in a model of phage therapy against K1-capsulated *Escherichia coli*. Virology 398:79–86

Canchaya C, Desiere F, McShan WM, Ferretti JJ, Parkhill J, Brussow H (2002) Genome analysis of an inducible prophage and prophage remnants integrated in the *Streptococcus pyogenes* strain SF370. Virology 302:245–258

Cardarelli L, Lam R, Tuite A, Baker LA, Sadowski PD, Radford DR, Rubinstein JL, Battaile KP, Chirgadze N, Maxwell KL, Davidson AR (2010) The crystal structure of bacteriophage HK97 gp6: defining a large family of head-tail connector proteins. J Mol Biol 395:754–768

Casjens S, King J (1974) P22 morphogenesis. I: Catalytic scaffolding protein in capsid assembly. J Supramol Struct 2:202–224

Casjens S, Wyckoff E, Hayden M, Sampson L, Eppler K, Randall S, Moreno E, Serwer P (1992) Bacteriophage P22 portal protein is part of the gauge that regulates packing density of intravirion DNA. J Mol Biol 224:1055–1074

Casjens S, Thuman-Commike P (2011) Evolution of mosaically related tailed phage genomes seen through the lens of phage P22 virion assembly. Virology 411:393–415

Cerritelli ME, Cheng N, Rosenberg AH, McPherson CE, Booy FP, Steven AC (1997) Encapsidated conformation of bacteriophage T7 DNA. Cell 91:271–280

Chang CY, Kemp P, Molineux IJ (2010) Gp15 and gp16 cooperate in translocating bacteriophage T7 DNA into the infected cell. Virology 398:176–186

Chang J, Weigele P, King J, Chiu W, Jiang W (2006) Cryo-EM asymmetric reconstruction of bacteriophage P22 reveals organization of its DNA packaging and infecting machinery. Structure 14:1073–1082

Choi KH, McPartland J, Kaganman I, Bowman VD, Rothman-Denes LB, Rossmann MG (2008) Insight into DNA and protein transport in double-stranded DNA viruses: the structure of bacteriophage N4. J Mol Biol 378:726–736

Chopin A, Deveau H, Ehrlich SD, Moineau S, Chopin MC (2007) KSY1, a lactococcal phage with a T7-like transcription. Virology 365:1–9

Cohen DN, Erickson SE, Xiang Y, Rossmann MG, Anderson DL (2008) Multifunctional roles of a bacteriophage ϕ29 morphogenetic factor in assembly and infection. J Mol Biol 378:804–817

Comolli LR, Spakowitz AJ, Siegerist CE, Jardine PJ, Grimes S, Anderson DL, Bustamante C, Downing KH (2008) Three-dimensional architecture of the bacteriophage ϕ29 packaged genome and elucidation of its packaging process. Virology 371:267–277

Dai W, Hodes A, Hui WH, Gingery M, Miller JF, Zhou ZH (2010) Three-dimensional structure of tropism-switching *Bordetella* bacteriophage. Proc Natl Acad Sci USA 107:4347–4352

de Frutos M, Letellier L, Raspaud E (2005) DNA ejection from bacteriophage T5: analysis of the kinetics and energetics. Biophys J 88:1364–1370

Dobbins AT, George M Jr, Basham DA, Ford ME, Houtz JM, Pedulla ML, Lawrence JG, Hatfull GF, Hendrix RW (2004) Complete genomic sequence of the virulent *Salmonella* bacteriophage SP6. J Bacteriol 186:1933–1944

Doulatov S, Hodes A, Dai L, Mandhana N, Liu M, Deora R, Simons RW, Zimmerly S, Miller JF (2004) Tropism switching in *Bordetella* bacteriophage defines a family of diversity-generating retroelements. Nature 431:476–481

Drexler K, Dannull J, Hindennach I, Mutschler B, Henning U (1991) Single mutations in a gene for a tail fiber component of an *Escherichia coli* phage can cause an extension from a protein to a carbohydrate as a receptor. J Mol Biol 219:655–663

Earnshaw W, Casjens S (1980) DNA packaging by the double-stranded DNA bacteriophages. Cell 21:319–331

Ebel-Tsipis J, Botstein D (1971) Superinfection exclusion by P22 prophage in lysogens of *Salmonella typhimurium*. 1. Exclusion of generalized transducing particles. Virology 45:629–637

Effantin G, Boulanger P, Neumann E, Letellier L, Conway JF (2006) Bacteriophage T5 structure reveals similarities with HK97 and T4 suggesting evolutionary relationships. J Mol Biol 361:993–1002

Emrich J, Streisinger G (1968) The role of phage lysozyme in the life cycle of phage T4. Virology 36:387–391

Evilevitch A, Lavelle L, Knobler CM, Raspaud E, Gelbart WM (2003) Osmotic pressure inhibition of DNA ejection from phage. Proc Natl Acad Sci USA 100:9292–9295

Falco SC, Laan KV, Rothman-Denes LB (1977) Virion-associated RNA polymerase required for bacteriophage N4 development. Proc Natl Acad Sci USA 74:520–523

Falco SC, Zehring W, Rothman-Denes LB (1980) DNA-dependent RNA polymerase from bacteriophage N4 virions. Purification and characterization. J Biol Chem 255:4339–4347

Feynman RP (1963) The Feynman lectures on physics. Addison-Wesley, Reading, MA

Filali Maltouf A, Labedan B (1983) Host cell metabolic energy is not required for injection of bacteriophage T5 DNA. J Bacteriol 153:124–133

Filali Maltouf AK, Labedan B (1985) The energetics of the injection process of bacteriophage lambda DNA and the role of the ptsM/*pel*-encoded protein. Biochem Biophys Res Commun 130:1093–1101

Fokine A, Chipman PR, Leiman PG, Mesyanzhinov VV, Rao VB, Rossmann MG (2004) Molecular architecture of the prolate head of bacteriophage T4. Proc Natl Acad Sci USA 101:6003–6008

Fraser JS, Yu Z, Maxwell KL, Davidson AR (2006) Ig-like domains on bacteriophages: a tale of promiscuity and deceit. J Mol Biol 359:496–507

Garcia L, Molineux I (1995) Incomplete entry of bacteriophage T7 DNA into F plasmid-containing *Escherichia coli*. J Bacteriol 177:4077–4083

Garcia LR, Molineux IJ (1996) Transcription-independent DNA translocation of bacteriophage T7 DNA into *Escherichia coli*. J Bacteriol 178:6921–6929

Garcia LR, Molineux IJ (1999) Translocation and specific cleavage of bacteriophage T7 DNA *in vivo* by EcoKI. Proc Natl Acad Sci USA 96:12430–12435

George DG, Yeh LS, Barker WC (1983) Unexpected relationships between bacteriophage lambda hypothetical proteins and bacteriophage T4 tail-fiber proteins. Biochem Biophys Res Commun 115:1061–1068

Goldberg EB, Grinius L, Letellier L (1994) Recognition, attachment and injection. In: Karam J (ed) The molecular biology of bacteriophage T4. ASM Press, Washiington, DC, pp 347–356

Gonzalez-Huici V, Alcorlo M, Salas M, Hermoso JM (2004a) Binding of phage ϕ29 architectural protein p6 to the viral genome: evidence for topological restriction of the phage linear DNA. Nucleic Acids Res 32:3493–3502

Gonzalez-Huici V, Salas M, Hermoso JM (2004b) The push-pull mechanism of bacteriophage ϕ29 DNA injection. Mol Microbiol 52:529–540

Gonzalez-Huici V, Salas M, Hermoso JM (2006) Requirements for *Bacillus subtilis* bacteriophage ϕ29 DNA ejection. Gene 374:19–25

Grayson P, Evilevitch A, Inamdar MM, Purohit PK, Gelbart WM, Knobler CM, Phillips R (2006) The effect of genome length on ejection forces in bacteriophage lambda. Virology 348:430–436

Grayson P, Han L, Winther T, Phillips R (2007) Real-time observations of single bacteriophage lambda DNA ejections *in vitro*. Proc Natl Acad Sci USA 104:14652–14657

Grayson P, Molineux IJ (2007) Is phage DNA 'injected' into cells – biologists and physicists can agree. Curr Opin Microbiol 10:401–409

Haggard-Ljungquist E, Halling C, Calendar R (1992) DNA sequences of the tail fiber genes of bacteriophage P2: evidence for horizontal transfer of tail fiber genes among unrelated bacteriophages. J Bacteriol 174:1462–1477

Hanfling P, Shashkov AS, Jann B, Jann K (1996) Analysis of the enzymatic cleavage (beta elimination) of the capsular K5 polysaccharide of *Escherichia coli* by the K5-specific coliphage: reexamination. J Bacteriol 178:4747–4750

Haynes LL, Rothman-Denes LB (1985) N4 virion RNA polymerase sites of transcription initiation. Cell 41:597–605

Heller K, Braun V (1979) Accelerated adsorption of bacteriophage T5 to *Escherichia coli* F, resulting from reversible tail fiber-lipopolysaccharide binding. J Bacteriol 139:32–38

Hendrix R, Casjens S (2005) Podoviridae. In: Fauquet C, Mayo M, Maniloff J, Desselberger U, Ball A (eds) Virus taxonomy. Elsevier, Amsterdam, pp 71–79

Hendrix RW, Duda RL (1992) Bacteriophage lambda PaPa: not the mother of all lambda phages. Science 258: 1145–1148

Hershey AD, Chase M (1952) Independent functions of viral protein and nucleic acid in growth of bacteriophge. J Gen Physiol 36:39–56

Hershey AD (1955) An upper limit to the protein content of the germinal substance of bacteriophage T2. Virology 1:108–127

Hofer B, Ruge M, Dreiseikelmann B (1995) The superinfection exclusion gene (*sieA*) of bacteriophage P22: identification and overexpression of the gene and localization of the gene product. J Bacteriol 177:3080–3086

Hoffman B, Levine M (1975a) Bacteriophage P22 virion protein which performs an essential early function. I. Analysis of *16-ts* mutants. J Virol 16:1536–1546

Hoffman B, Levine M (1975b) Bacteriophage P22 virion protein which performs an essential early function. II. Characterization of the gene *16* function. J Virol 16:1547–1559

Hud NV, Downing KH (2001) Cryoelectron microscopy of lambda phage DNA condensates in vitreous ice: the fine structure of DNA toroids. Proc Natl Acad Sci USA 98:14925–14930

Hugel T, Michaelis J, Hetherington CL, Jardine PJ, Grimes S, Walter JM, Falk W, Anderson DL, Bustamante C (2007) Experimental test of connector rotation during DNA packaging into bacteriophage ϕ29 capsids. PLoS Biol 5:e59

Isidro A, Santos MA, Henriques AO, Tavares P (2004) The high-resolution functional map of bacteriophage SPP1 portal protein. Mol Microbiol 51:949–962

Israel V (1976) Role of the bacteriophage P22 tail in the early stages of infection. J Virol 18:361–364

Israel V (1977) E proteins of bacteriophage P22. I. Identification and ejection from wild-type and defective particles. J Virol 23:91–97

Israel V (1978) A model for the adsorption of phage P22 to *Salmonella typhimurium*. J Gen Virol 40:669–673

Iwashita S, Kanegasaki S (1976) Deacetylation reaction catalyzed by *Salmonella* phage c341 and its baseplate parts. J Biol Chem 251:5361–5365

Jeembaeva M, Jonsson B, Castelnovo M, Evilevitch A (2010) DNA heats up: energetics of genome ejection from phage revealed by isothermal titration calorimetry. J Mol Biol 395:1079–1087

Jiang W, Li Z, Zhang Z, Baker ML, Prevelige PE Jr, Chiu W (2003) Coat protein fold and maturation transition of bacteriophage P22 seen at subnanometer resolutions. Nat Struct Biol 10:131–135

Jiang W, Chang J, Jakana J, Weigele P, King J, Chiu W (2006) Structure of epsilon15 bacteriophage reveals genome organization and DNA packaging/injection apparatus. Nature 439:612–616

Jiang W, Baker ML, Jakana J, Weigele PR, King J, Chiu W (2008) Backbone structure of the infectious epsilon15 virus capsid revealed by electron cryomicroscopy. Nature 451:1130–1134

Johnson JE, Chiu W (2007) DNA packaging and delivery machines in tailed bacteriophages. Curr Opin Struct Biol 17:237–243

Joshi A, Siddiqi JZ, Rao GR, Chakravorty M (1982) MB78, a virulent bacteriophage of *Salmonella typhimurium*. J Virol 41:1038–1043

Kanamaru S, Gassner NC, Ye N, Takeda S, Arisaka F (1999) The C-terminal fragment of the precursor tail lysozyme of bacteriophage T4 stays as a structural component of the baseplate after cleavage. J Bacteriol 181:2739–2744

Kanamaru S, Leiman PG, Kostyuchenko VA, Chipman PR, Mesyanzhinov VV, Arisaka F, Rossmann MG (2002) Structure of the cell-puncturing device of bacteriophage T4. Nature 415:553–557

Kanegasaki S, Wright A (1973) Studies on the mechanism of phage adsorption: interaction between phage epsilon15 and its cellular receptor. Virology 52:160–173

Kazmierczak K, Rothman-Denes L (2006) Bacteriophage N4. In: Calendar R (ed) The bacteriophages. Oxford University Press, Oxford, pp 302–314

Kazmierczak KM, Davydova EK, Mustaev AA, Rothman-Denes LB (2002) The phage N4 virion RNA polymerase catalytic domain is related to single-subunit RNA polymerases. EMBO J 21:5815–5823

Kemp P, Gupta M, Molineux IJ (2004) Bacteriophage T7 DNA ejection into cells is initiated by an enzyme-like mechanism. Mol Microbiol 53:1251–1265

Kemp P, Garcia LR, Molineux IJ (2005) Changes in bacteriophage T7 virion structure at the initiation of infection. Virology 340:307–317

Kiino DR, Rothman-Denes LB (1989) Genetic analysis of bacteriophage N4 adsorption. J Bacteriol 171:4595–4602

Kiino DR, Licudine R, Wilt K, Yang DH, Rothman-Denes LB (1993a) A cytoplasmic protein, NfrC, is required for bacteriophage N4 adsorption. J Bacteriol 175:7074–7080

Kiino DR, Singer MS, Rothman-Denes LB (1993b) Two overlapping genes encoding membrane proteins required for bacteriophage N4 adsorption. J Bacteriol 175:7081–7085

King J, Lenk EV, Botstein D (1973) Mechanism of head assembly and DNA encapsulation in *Salmonella* phage P22. II. Morphogenetic pathway. J Mol Biol 80:697–731

King MR, Vimr RP, Steenbergen SM, Spanjaard L, Plunkett G 3rd, Blattner FR, Vimr ER (2007) *Escherichia coli* K1-specific bacteriophage CUS-3 distribution and function in phase-variable capsular polysialic acid O acetylation. J Bacteriol 189:6447–6456

Krawiec S, Jimenez F, Garcia JA, Villanueva N, Sogo J, Salas M (1981) The orderly, in vitro emergence of DNA from bacteriophage ϕ29 particles. Virology 111:440–454

Kropinski AM (2000) Sequence of the genome of the temperate, serotype-converting, *Pseudomonas aeruginosa* bacteriophage D3. J Bacteriol 182:6066–6074

Kropinski AM, Kovalyova IV, Billington SJ, Patrick AN, Butts BD, Guichard JA, Pitcher TJ, Guthrie CC, Sydlaske AD, Barnhill LM, Havens KA, Day KR, Falk DR, McConnell MR (2007) The genome of epsilon15, a serotype-converting, Group E1 *Salmonella enterica*-specific bacteriophage. Virology 369:234–244

Kuhn A, Kellenberger E (1985) Productive phage infection in *Escherichia coli* with reduced internal levels of the major cations. J Bacteriol 163:906–912

Kuhn HM, Meier-Dieter U, Mayer H (1988) ECA, the enterobacterial common antigen. FEMS Microbiol Rev 4:195–222

Labedan B, Goldberg EB (1979) Requirement for membrane potential in injection of phage T4 DNA. Proc Natl Acad Sci USA 76:4669–4673

Labedan B, Heller KB, Jasaitis AA, Wilson TH, Goldberg EB (1980) A membrane potential threshold for phage T4 DNA injection. Biochem Biophys Res Commun 93:625–630

Lander GC, Tang L, Casjens SR, Gilcrease EB, Prevelige P, Poliakov A, Potter CS, Carragher B, Johnson JE (2006) The structure of an infectious P22 virion shows the signal for headful DNA packaging. Science 312: 1791–1795

Lander GC, Khayat R, Li R, Prevelige PE, Potter CS, Carragher B, Johnson JE (2009) The P22 tail machine at subnanometer resolution reveals the architecture of an infection conduit. Structure 17:789–799

Landstrom J, Nordmark EL, Eklund R, Weintraub A, Seckler R, Widmalm G (2008) Interaction of a *Salmonella enteritidis* O-antigen octasaccharide with the phage P22 tailspike protein by NMR spectroscopy and docking studies. Glycoconj J 25:137–143

Lanni YT (1968) First-step-transfer deoxyribonucleic acid of bacteriophage T5. Bacteriol Rev 32:227–242

Lavigne R, Burkal'tseva MV, Robben J, Sykilinda NN, Kurochkina LP, Grymonprez B, Jonckx B, Krylov VN, Mesyanzhinov VV, Volckaert G (2003) The genome of bacteriophage ϕKMV, a T7-like virus infecting *Pseudomonas aeruginosa*. Virology 312:49–59

Lavigne R, Briers Y, Hertveldt K, Robben J, Volckaert G (2004) Identification and characterization of a highly thermostable bacteriophage lysozyme. Cell Mol Life Sci 61:2753–2759

Lavigne R, Seto D, Mahadevan P, Ackermann HW, Kropinski AM (2008) Unifying classical and molecular taxonomic classification: analysis of the *Podoviridae* using BLASTP-based tools. Res Microbiol 159:406–414

Lebedev AA, Krause MH, Isidro AL, Vagin AA, Orlova EV, Turner J, Dodson EJ, Tavares P, Antson AA (2007) Structural framework for DNA translocation via the viral portal protein. EMBO J 26:1984–1994

Leforestier A, Brasiles S, de Frutos M, Raspaud E, Letellier L, Tavares P, Livolant F (2008) Bacteriophage T5 DNA ejection under pressure. J Mol Biol 384:730–739

Leforestier A, Livolant F (2009) Structure of toroidal DNA collapsed inside the phage capsid. Proc Natl Acad Sci USA 106:9157–9162

Leforestier A, Livolant F (2010) The bacteriophage genome undergoes a succession of intracapsid phase transitions upon DNA ejection. J Mol Biol 396:384–395

Leiman PG, Battisti AJ, Bowman VD, Stummeyer K, Muhlenhoff M, Gerardy-Schahn R, Scholl D, Molineux IJ (2007) The structures of bacteriophages K1E and K1-5 explain processive degradation of polysaccharide capsules and evolution of new host specificities. J Mol Biol 371:836–849

Leiman PG, Molineux IJ (2008) Evolution of a new enzyme activity from the same motif fold. Mol Microbiol 69:287–290

Lenk E, Casjens S, Weeks J, King J (1975) Intracellular visualization of precursor capsids in phage P22 mutant infected cells. Virology 68:182–199

Lin L (1992) Study of bacteriophage T7 gene 5.9 and gene 5.5. Ph.D. dissertation. SUNY, Sonybrook, New York

Lindberg AA, Sarvas M, Makela PH (1970) Bacteriophage attachment to the somatic antigen of *Salmonella*: Effect of O-specific structures in leaky R mutants and S, T1 Hybrids. Infect Immun 1:88–97

Lipinska B, Rao AS, Bolten BM, Balakrishnan R, Goldberg EB (1989) Cloning and identification of bacteriophage T4 gene *2* product gp2 and action of gp2 on infecting DNA *in vivo*. J Bacteriol 171:488–497

Liu M, Deora R, Doulatov SR, Gingery M, Eiserling FA, Preston A, Maskell DJ, Simons RW, Cotter PA, Parkhill J, Miller JF (2002) Reverse transcriptase-mediated tropism switching in Bordetella bacteriophage. Science 295:2091–2094

Liu M, Gingery M, Doulatov SR, Liu Y, Hodes A, Baker S, Davis P, Simmonds M, Churcher C, Mungall K, Quail MA, Preston A, Harvill ET, Maskell DJ, Eiserling FA, Parkhill J, Miller JF (2004) Genomic and genetic analysis of *Bordetella* bacteriophages encoding reverse transcriptase-mediated tropism-switching cassettes. J Bacteriol 186:1503–1517

Liu X, Shi M, Kong S, Gao Y, An C (2007) Cyanophage Pf-WMP4, a T7-like phage infecting the freshwater cyanobacterium *Phormidium foveolarum*: complete genome sequence and DNA translocation. Virology 366:28–39

Liu X, Zhang Q, Murata K, Baker ML, Sullivan MB, Fu C, Dougherty MT, Schmid MF, Osburne MS, Chisholm SW, Chiu W (2010) Structural changes in a marine podovirus associated with release of its genome into *Prochlorococcus*. Nat Struct Mol Biol 17:830–836

Lu MJ, Henning U (1994) Superinfection exclusion by T-even-type coliphages. Trends Microbiol 2:137–139

Machida Y, Miyake K, Hattori K, Yamamoto S, Kawase M, Iijima S (2000) Structure and function of a novel coliphage-associated sialidase. FEMS Microbiol Lett 182:333–337

Mahony J, McGrath S, Fitzgerald GF, van Sinderen D (2008) Identification and characterization of lactococcal-prophage-carried superinfection exclusion genes. Appl Environ Microbiol 74:6206–6215

Mangenot S, Hochrein M, Radler J, Letellier L (2005) Real-time imaging of DNA ejection from single phage particles. Curr Biol 15:430–435

McGrath S, Fitzgerald GF, van Sinderen D (2002) Identification and characterization of phage-resistance genes in temperate lactococcal bacteriophages. Mol Microbiol 43:509–520

McPartland J, Rothman-Denes LB (2009) The tail sheath of bacteriophage N4 interacts with the *Escherichia coli* receptor. J Bacteriol 191:525–532

Meijer WJ, Horcajadas JA, Salas M (2001a) ϕ29 family of phages. Microbiol Mol Biol Rev 65:261–287

Meijer WJ, Serna-Rico A, Salas M (2001b) Characterization of the bacteriophage ϕ29-encoded protein p16.7: a membrane protein involved in phage DNA replication. Mol Microbiol 39:731–746

Miller JL, Le Coq J, Hodes A, Barbalat R, Miller JF, Ghosh P (2008) Selective ligand recognition by a diversity-generating retroelement variable protein. PLoS Biol 6:e131

Mishra P, Prem Kumar R, Ethayathulla AS, Singh N, Sharma S, Perbandt M, Betzel C, Kaur P, Srinivasan A, Bhakuni V, Singh TP (2009) Polysaccharide binding sites in hyaluronate lyase – crystal structures of native phage-encoded hyaluronate lyase and its complexes with ascorbic acid and lactose. FEBS J 276:3392–3402

Moak M, Molineux IJ (2000) Role of the Gp16 lytic transglycosylase motif in bacteriophage T7 virions at the initiation of infection. Mol Microbiol 37:345–355

Moak M, Molineux IJ (2004) Peptidoglycan hydrolytic activities associated with bacteriophage virions. Mol Microbiol 51:1169–1183

Moffatt BA, Studier FW (1988) Entry of bacteriophage T7 DNA into the cell and escape from host restriction. J Bacteriol 170:2095–2105

Molineux IJ (2001) No syringes please, ejection of phage T7 DNA from the virion is enzyme driven. Mol Microbiol 40:1–8

Molineux IJ (2006) Fifty-three years since Hershey and Chase; much ado about pressure but which pressure is it? Virology 344:221–229

Morais MC, Choi KH, Koti JS, Chipman PR, Anderson DL, Rossmann MG (2005) Conservation of the capsid structure in tailed dsDNA bacteriophages: the pseudoatomic structure of ϕ29. Mol Cell 18:149–159

Muller JJ, Barbirz S, Heinle K, Freiberg A, Seckler R, Heinemann U (2008) An intersubunit active site between supercoiled parallel beta helices in the trimeric tailspike endorhamnosidase of *Shigella flexneri* phage Sf6. Structure 16:766–775

Murray NE (2000) Type I restriction systems: sophisticated molecular machines (a legacy of Bertani and Weigle). Microbiol Mol Biol Rev 64:412–434

Olia AS, Al-Bassam J, Winn-Stapley DA, Joss L, Casjens SR, Cingolani G (2006) Binding-induced stabilization and assembly of the phage P22 tail accessory factor gp4. J Mol Biol 363:558–576

Olia AS, Bhardwaj A, Joss L, Casjens S, Cingolani G (2007a) Role of gene *10* protein in the hierarchical assembly of the bacteriophage P22 portal vertex structure. Biochemistry 46:8776–8784

Olia AS, Casjens S, Cingolani G (2007b) Structure of phage P22 cell envelope-penetrating needle. Nat Struct Mol Biol 14:1221–1227

Olia A, Prevelige P, Johnson J, Cingolani G (2011) Nat Struct Mol Biol 18:567–603

Olia AS, Casjens S, Cingolani G (2009) Structural plasticity of the phage P22 tail needle gp26 probed with xenon gas. Protein Sci 18:537–548
Olson NH, Gingery M, Eiserling FA, Baker TS (2001) The structure of isometric capsids of bacteriophage T4. Virology 279:385–391
Palva ET, Mäkelä PH (1980) Lipopolysaccharide heterogeneity in *Salmonella typhimurium* analyzed by sodium dodecyl sulfate polyacrylamide gel electrophoresis. Eur J Biochem 107:137–143
Panja D, Molineux IJ (2010) Dynamics of bacteriophage genome ejection in vitro and in vivo. Phys Biol 7:045006
Perez GL, Huynh B, Slater M, Maloy S (2009) Transport of phage P22 DNA across the cytoplasmic membrane. J Bacteriol 191:135–140
Petrov AS, Locker CR, Harvey SC (2009) Characterization of DNA conformation inside bacterial viruses. Phys Rev E Stat Nonlin Soft Matter Phys 80:021914
Petter JG, Vimr ER (1993) Complete nucleotide sequence of the bacteriophage K1F tail gene encoding endo-N-acylneuraminidase (endo-N) and comparison to an endo-N homolog in bacteriophage PK1E. J Bacteriol 175:4354–4363
Pope WH, Weigele PR, Chang J, Pedulla ML, Ford ME, Houtz JM, Jiang W, Chiu W, Hatfull GF, Hendrix RW, King J (2007) Genome sequence, structural proteins, and capsid organization of the cyanophage Syn5: a "horned" bacteriophage of marine *synechococcus*. J Mol Biol 368:966–981
Prehm P, Jann B, Jann K, Schmidt G, Stirm S (1976) On a bacteriophage T3 and T4 receptor region within the cell wall lipopolysaccharide of *Escherichia coli* B. J Mol Biol 101:277–281
Purohit PK, Inamdar MM, Grayson PD, Squires TM, Kondev J, Phillips R (2005) Forces during bacteriophage DNA packaging and ejection. Biophys J 88:851–866
Rao GR, Chakravorty-Burma M, Burma DP (1972) Transient depression in the active transport across the membrane of *Salmonella typhimurium*, after infection with bacteriophage P22. Virology 49:811–814
Rao RN (1968) Bacteriophage P22 controlled exclusion in *Salmonella typhimurium*. J Mol Biol 35:607–622
Raspaud E, Forth T, Sao-Jose C, Tavares P, de Frutos M (2007) A kinetic analysis of DNA ejection from tailed phages revealing the prerequisite activation energy. Biophys J 93:3999–4005
Rhoades M, Thomas CA Jr (1968) The P22 bacteriophage DNA molecule. II. Circular intracellular forms. J Mol Biol 37:41–61
Rickgauer JP, Fuller DN, Grimes S, Jardine PJ, Anderson DL, Smith DE (2008) Portal motor velocity and internal force resisting viral DNA packaging in bacteriophage ϕ29. Biophys J 94:159–167
Rohr TE, Troy FA (1980) Structure and biosynthesis of surface polymers containing polysialic acid in *Escherichia coli*. J Biol Chem 255:2332–2342
Saigo K (1975) Polar DNA ejection in bacteriophage T7. Virology 65:120–127
Saigo K (1978) Isolation of high-density mutants and identification of nonessential structural proteins in bacteriophage T5; dispensability of L-shaped tail fibers and a secondary major head protein. Virology 85:422–433
Sandmeier H, Iida S, Arber W (1992) DNA inversion regions Min of plasmid p15B and Cin of bacteriophage P1: evolution of bacteriophage tail fiber genes. J Bacteriol 174:3936–3944
Sao-Jose C, de Frutos M, Raspaud E, Santos MA, Tavares P (2007) Pressure built by DNA packing inside virions: enough to drive DNA ejection in vitro, largely insufficient for delivery into the bacterial cytoplasm. J Mol Biol 374:346–355
Savalia D, Westblade LF, Goel M, Florens L, Kemp P, Akulenko N, Pavlova O, Padovan JC, Chait BT, Washburn MP, Ackermann HW, Mushegian A, Gabisonia T, Molineux I, Severinov K (2008) Genomic and proteomic analysis of ϕEco32, a novel *Escherichia coli* bacteriophage. J Mol Biol 377:774–789
Scholl D, Rogers S, Adhya S, Merril CR (2001) Bacteriophage K1-5 encodes two different tail fiber proteins, allowing it to infect and replicate on both K1 and K5 strains of *Escherichia coli*. J Virol 75:2509–2515
Scholl D, Kieleczawa J, Kemp P, Rush J, Richardson CC, Merril C, Adhya S, Molineux IJ (2004) Genomic analysis of bacteriophages SP6 and K1-5, an estranged subgroup of the T7 supergroup. J Mol Biol 335:1151–1171
Scholl D, Adhya S, Merril C (2005) *Escherichia coli* K1's capsule is a barrier to bacteriophage T7. Appl Environ Microbiol 71:4872–4874
Schulz EC, Schwarzer D, Frank M, Stummeyer K, Muhlenhoff M, Dickmanns A, Gerardy-Schahn R, Ficner R (2010) Structural basis for the recognition and cleavage of polysialic acid by the bacteriophage K1F tailspike protein EndoNF. J Mol Biol 397:341–351
Schwarzer D, Stummeyer K, Gerardy-Schahn R, Muhlenhoff M (2007) Characterization of a novel intramolecular chaperone domain conserved in endosialidases and other bacteriophage tail spike and fiber proteins. J Biol Chem 282:2821–2831
Schwarzer D, Stummeyer K, Haselhorst T, Freiberger F, Rode B, Grove M, Scheper T, von Itzstein M, Muhlenhoff M, Gerardy-Schahn R (2009) Proteolytic release of the intramolecular chaperone domain confers processivity to endosialidase F. J Biol Chem 284:9465–9474
Serna-Rico A, Salas M, Meijer WJ (2002) The *Bacillus subtilis* phage ϕ29 protein p16.7, involved in ϕ29 DNA replication, is a membrane-localized single-stranded DNA-binding protein. J Biol Chem 277:6733–6742

Serwer P, Wright ET, Hakala KW, Weintraub ST (2008) Evidence for bacteriophage T7 tail extension during DNA injection. BMC Res Notes 1:36

Simon LD, Anderson TF (1967) The infection of *Escherichia coli* by T2 and T4 bacteriophages as seen in the electron microscope. I. Attachment and penetration. Virology 32:279–297

Simpson AA, Tao Y, Leiman PG, Badasso MO, He Y, Jardine PJ, Olson NH, Morais MC, Grimes S, Anderson DL, Baker TS, Rossmann MG (2000) Structure of the bacteriophage ϕ29 DNA packaging motor. Nature 408:745–750

Smith D, Tans J, Smith S, Grimes S, Anderson D, Bustamante C (2001) The bacteriophage ϕ29 portal motor can package DNA against a large internal force. Nature 413:748–752

Smith NL, Taylor EJ, Lindsay AM, Charnock SJ, Turkenburg JP, Dodson EJ, Davies GJ, Black GW (2005) Structure of a group A streptococcal phage-encoded virulence factor reveals a catalytically active triple-stranded beta-helix. Proc Natl Acad Sci USA 102:17652–17657

Steinbacher S, Seckler R, Miller S, Steipe B, Huber R, Reinemer P (1994) Crystal structure of P22 tailspike protein: interdigitated subunits in a thermostable trimer. Science 265:383–386

Steinbacher S, Baxa U, Miller S, Weintraub A, Seckler R, Huber R (1996) Crystal structure of phage P22 tailspike protein complexed with *Salmonella* sp. O-antigen receptors. Proc Natl Acad Sci USA 93:10584–10588

Steinbacher S, Miller S, Baxa U, Budisa N, Weintraub A, Seckler R, Huber R (1997) Phage P22 tailspike protein: crystal structure of the head-binding domain at 2.3 A, fully refined structure of the endorhamnosidase at 1.56 A resolution, and the molecular basis of O-antigen recognition and cleavage. J Mol Biol 267:865–880

Steven AC, Trus BL, Maizel JV, Unser M, Parry DA, Wall JS, Hainfeld JF, Studier FW (1988) Molecular substructure of a viral receptor-recognition protein. The gp17 tail-fiber of bacteriophage T7. J Mol Biol 200:351–365

Stojković E, Rothman-Denes L (2007) Coliphage N4 N-acetylmuramidase defines a new family of murien hydrolases. J Mol Biol 366:406–419

Strauss H, King J (1984) Steps in the stabilization of newly packaged DNA during phage P22 morphogenesis. J Mol Biol 172:523–543

Struthers-Schlinke JS, Robins WP, Kemp P, Molineux IJ (2000) The internal head protein Gp16 controls DNA ejection from the bacteriophage T7 virion. J Mol Biol 301:35–45

Studier FW (1972) Bacteriophage T7. Science 176:367–376

Studier FW, Bandyopadhyay PK (1988) Model for how type I restriction enzymes select cleavage sites in DNA. Proc Natl Acad Sci USA 85:4677–4681

Stummeyer K, Dickmanns A, Muhlenhoff M, Gerardy-Schahn R, Ficner R (2005) Crystal structure of the polysialic acid-degrading endosialidase of bacteriophage K1F. Nat Struct Mol Biol 12:90–96

Stummeyer K, Schwarzer D, Claus H, Vogel U, Gerardy-Schahn R, Muhlenhoff M (2006) Evolution of bacteriophages infecting encapsulated bacteria: lessons from *Escherichia coli* K1-specific phages. Mol Microbiol 60:1123–1135

Susskind MM, Wright A, Botstein D (1971) Superinfection exclusion by P22 prophage in lysogens of *Salmonella typhimurium*. II. Genetic evidence for two exclusion systems. Virology 45:638–652

Susskind MM, Botstein D (1978) Molecular genetics of bacteriophage P22. Microbiol Rev 42:385–413

Takeda K, Uetake H (1973) In vitro interaction between phage and receptor lipopolysaccharide: a novel glycosidase associated with *Salmonella* phage epsilon15. Virology 52:148–159

Tan Y, Zhang K, Rao X, Jin X, Huang J, Zhu J, Chen Z, Hu X, Shen X, Wang L, Hu F (2007) Whole genome sequencing of a novel temperate bacteriophage of *P. aeruginosa*: evidence of tRNA gene mediating integration of the phage genome into the host bacterial chromosome. Cell Microbiol 9:479–491

Tang J, Olson N, Jardine PJ, Grimes S, Anderson DL, Baker TS (2008) DNA poised for release in bacteriophage ϕ29. Structure 16:935–943

Tang L, Marion WR, Cingolani G, Prevelige PE, Johnson JE (2005) Three-dimensional structure of the bacteriophage P22 tail machine. EMBO J 24:2087–2095

Tang L, Gilcrease EB, Casjens SR, Johnson JE (2006) Highly discriminatory binding of capsid-cementing proteins in bacteriophage L. Structure 14:837–845

Ter-Nikogosian VA, Vartanian MK, Trchunian AA (1991) Changes in membrane potential and transport of ions through the *S. typhimurium* LT2 membrane induced by bacteriophages. Biofizika 36:281–285

Tye BK, Huberman JA, Botstein D (1974) Non-random circular permutation of phage P22 DNA. J Mol Biol 85:501–528

Tzlil S, Kindt JT, Gelbart WM, Ben-Shaul A (2003) Forces and pressures in DNA packaging and release from viral capsids. Biophys J 84:1616–1627

Uetake H, Nakagawa T, Akiba T (1955) The relationship of bacteriophage to antigenic changes in Group E salmonellas. J Bacteriol 69:571–579

Uetake H, Hagiwara S (1969) Transfer of conversion gene(s) between different *Salmonella* phages g341 and epsilon-15. Virology 37:8–14

Umlauf B, Dreiseikelmann B (1992) Cloning, sequencing, and overexpression of gene *16* of *Salmonella* bacteriophage P22. Virology 188:495–501

Villafane R, Zayaz M, Gilcrease E, Kropinski A, Casjens S (2008) Genomic analysis of bacteriophage epsilon34 of *Salmonella enterica* serovar Anatum (15+). BMC Microbiol 8:e227
Vinga I, Sao-Jose C, Traveres P, Santos M (2006) Bacteriophage entry into the host cell. In: Wegrzyn G (ed) Modern bacteriophage biology and biotechnology. Research Signpost, Kerala, India, pp 165–205
Walkinshaw MD, Taylor P, Sturrock SS, Atanasiu C, Berge T, Henderson RM, Edwardson JM, Dryden DT (2002) Structure of Ocr from bacteriophage T7, a protein that mimics B-form DNA. Mol Cell 9:187–194
Walter M, Fiedler C, Grassl R, Biebl M, Rachel R, Hermo-Parrado XL, Llamas-Saiz AL, Seckler R, Miller S, van Raaij MJ (2008) Structure of the receptor-binding protein of bacteriophage Det7: a podoviral tail spike in a myovirus. J Virol 82:2265–2273
Wang HY, Elston T, Mogilner A, Oster G (1998a) Force generation in RNA polymerase. Biophys J 74:1186–1202
Wang MD, Schnitzer MJ, Yin H, Landick R, Gelles J, Block SM (1998b) Force and velocity measured for single molecules of RNA polymerase. Science 282:902–907
Willis SH, Kazmierczak KM, Carter RH, Rothman-Denes LB (2002) N4 RNA polymerase II, a heterodimeric RNA polymerase with homology to the single-subunit family of RNA polymerases. J Bacteriol 184:4952–4961
Wright A (1971) Mechanism of conversion of *Salmonella* O-antigen by bacteriophage epsilon 34. J Bacteriol 105:927–936
Xiang Y, Morais MC, Battisti AJ, Grimes S, Jardine PJ, Anderson DL, Rossmann MG (2006) Structural changes of bacteriophage ϕ29 upon DNA packaging and release. EMBO J 25:5229–5239
Xiang Y, Morais MC, Cohen DN, Bowman VD, Anderson DL, Rossmann MG (2008) Crystal and cryoEM structural studies of a cell wall degrading enzyme in the bacteriophage ϕ29 tail. Proc Natl Acad Sci USA 105:9552–9557
Xiang Y, Leiman PG, Li L, Grimes S, Anderson DL, Rossmann MG (2009) Crystallographic insights into the autocatalytic assembly mechanism of a bacteriophage tail spike. Mol Cell 34:375–386
Yin H, Wang MD, Svoboda K, Landick R, Block SM, Gelles J (1995) Transcription against an applied force. Science 270:1653–1657
Zarybnicky V (1969) Mechanism of T-even DNA ejection. J Theor Biol 22:33–42
Zavriev SK, Shemyakin MF (1982) RNA polymerase-dependent mechanism for the stepwise T7 phage DNA transport from the virion into *E. coli*. Nucleic Acids Res 10:1635–1652

Chapter 8
Infection of Cells by Alphaviruses

Dennis T. Brown and Raquel Hernandez

Abstract It is widely accepted that alphaviruses enter cells by a process involving endocytosis and low-pH-mediated virus membrane–cell membrane fusion. This model and the data supporting it have received extensive and numerous reviews. The major points presented in support of this model are summarized briefly herein. It is the primary objective of this review to present an alternative mechanism describing the penetration of cells by alphaviruses which does not involve endocytosis or exposure to acid environment. The data supporting this model are summarized in detail.

8.1 The Alphaviruses

The alphaviruses were at one time designated the "group A arboviruses." Indeed, the term arbovirus (ARthropod BOrne virus) reflects in part the enormous host range of these agents which bridges taxonomic phyla (Brown and Condreay 1986; Strauss and Strauss 1994). Alphaviruses are vectored by hematophagous, poikilothermic insects and are transmitted primarily to birds, reptiles, and small mammals which maintain the enzootic cycle. Alphaviruses are agents of significant disease of man and domestic animals. Several alphaviruses such as Western, Eastern, and Venezuelan Equine Encephalitis viruses cause encephalitis in human and domestic animals. Recently, Chikungunya virus, an alphavirus which produces arthralgia, has reemerged as a new and significant threat to human health.

8.2 The Study of Animal Virus Penetration of Host Cells: Some General Considerations for Membrane Containing Viruses

Elucidating the mechanism by which animal viruses penetrate cell membranes to initiate infection has proven to be one of the most difficult areas of virus research. Technology involving direct and indirect observation of this process has been employed with both approaches making assumptions

D.T. Brown (✉) • R. Hernandez
Department of Molecular and Structural Biochemistry, North Carolina State University, 128 Polk Hall, Campus, Box 7622, Raleigh, NC 27695, USA
e-mail: dennis_brown@ncsu.edu

M.G. Rossmann and V.B. Rao (eds.), *Viral Molecular Machines*, Advances in Experimental Medicine and Biology 726, DOI 10.1007/978-1-4614-0980-9_8,

difficult to prove. Using indirect analysis, the ability of the virus to initiate infection and synthesize virus-specific RNA or protein is used as an indication that cell penetration has taken place (Helenius et al. 1982; Kielian et al. 1984; Marsh et al. 1982, 1983). Synthesis of virus RNA and protein, however, occurs after attachment and cell penetration is completed, and failure to detect these events may or may not be related to the penetration process. Analysis by direct observation employs microscopy to view the interaction of virus with host cells (Helenius 1984; Marsh et al. 1983). While this approach holds great potential for visualizing the process of cell penetration, it is incapable of distinguishing a virus which is engaged in a productive process of infection and a virus involved in a dead-end interaction with the cell. This problem is exacerbated by the fact that for many animal viruses, the ratio of noninfectious to infectious virus particles can be very high. If this value is as low as 10 to 1 (which is exceptional for some viruses), only one in ten particles observed has the possibility of creating a productive infection.

These caveats cause considerable difficulty for experiments in which cells are treated with a chemical or biological agent and are then tested for virus RNA synthesis. If synthesis is detected, then most assuredly penetration has occurred. If it is not detected, penetration may have been blocked or RNA synthesis may have been inhibited by some direct or indirect effect of the agent. Likewise, if the infectivity ratio is poor, the observation that, for example, particles are entering endosomes or phagosomes may be misinterpreted to suggest that engulfment is an essential step in the penetration process. Evidence that a specific endocytic compartment is employed by specific cells for viral internalization, e.g., clathrin-coated pits, presupposes that endocytosis is the route of infection. These are all examples of indirect evidence of a specific penetration event.

Similar problems arise when artificial membranes, or liposomes, are employed as models to study the interaction of virus with living cells (Kielian and Helenius 1984). In these model systems, the membranes do not contain the necessary receptor proteins or other receptor-related structures. A second problem with these model systems is that the lipid composition of the artificial membrane may not reflect that of the living cell at the point of penetration. An additional complication is that the lipid and protein compositions of the mosquito cell membranes are vastly different from those of a mammalian host cell (Clayton 1964; Cleverley et al. 1997; Hafer et al. 2009; Mitsuhashi et al. 1983). The artificial membrane systems employed are formulated to maximally interact with the virus, not to reflect the composition of the mammalian or insect host membrane.

8.3 The Alphaviruses as Infectious Agents

The alphaviruses are uniquely suited for the study of virus penetration of a membrane containing virus. The structure of the alphaviruses is highly ordered and stable. Many strains of Sindbis virus have a particle to pfu ratio of 100/1 which is low for an enveloped virus. One strain of Sindbis virus, the heat-resistant or HR strain, is stable upon heating and produces virus stocks with a particle to pfu ratio of 1. Using this strain, SVHR, as a model system ensures that the starting material for any experiment involving functional evaluation of the infection begins with 100% infectious virus (Hernandez et al. 2006).

8.4 Alphavirus Structure

A discussion of alphavirus structure is essential for discussing the function of the virions. Many membrane-containing viruses are pleomorphic in morphology consisting of an enclosing membrane bilayer in which surface membrane glycoproteins are embedded. These proteins may not

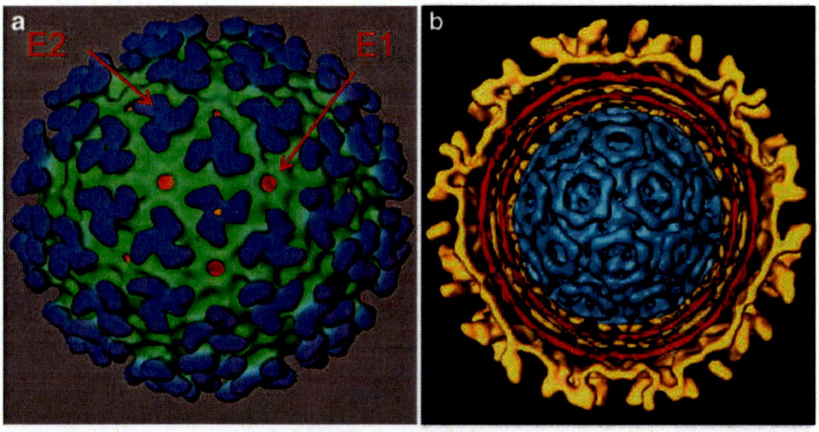

Fig. 8.1 Reconstruction of the surface (**a**) and interior (**b**) of Sindbis virus produced from electron cryomicrographs. The *arrows* in (**a**) indicate the proposed positions of proteins E1 and E2. (**b**) Shows the outer shell of the virus cut away to revile the membrane bilayer in *red* and the inner core structure in *blue* (provided by Dr. Angel Paredes, University of Texas Health sciences Center)

have significant lateral associations and may be free to move/float in the fluid lipid environment. The prototype of this type of virus is influenza virus. In this case, it is well established that the flu glycoproteins, HA1 and HA2, drive the fusion of the virus membrane with the host cell membrane in a low-pH environment (Skehel and Wiley 2000). In the case of viruses such as flu, the virus membrane is the form-determining structure, and the virus is structurally a membrane with associated glycoproteins. This configuration may prove optimal for a virus which gains entry by membrane fusion by allowing the exposure of significant virus membrane surface.

The alphaviruses are structurally distinct (Fig. 8.1). They have a precise size, shape, and composition. The alphaviruses are composed of 240 copies each of membrane glycoproteins E1 and E2 matched to 240 copies of the core "capsid" (Paredes et al. 1993). The proteins are organized into two nested protein shells between which is sandwiched a membrane bilayer. The glycoproteins E1 and E2 are embedded in the membrane bilayer by virtue of transmembrane domains and the E2 protein has a 33-amino-acid internal "endodomain" (Hernandez et al. 2000, 2003, 2005; Lee et al. 1996). E1 and E2 form the outer protein shell and are organized into a $T=4$ icosahedral lattice. In that protein shell, the E1 and E2 proteins are organized into 80 trimeric aggregates. The E2 proteins form the 80 prominent spike structures seen on the virus surface, while the E1 glycoproteins hold the surface lattice in its icosahedral geometry (Mukhopadhyay et al. 2006; Pletnev et al. 2001; Zhang et al. 2002). Beneath the internally situated membrane bilayer, 240 copies of the capsid protein form a $T=4$ icosahedral protein shell surrounding the virus plus polarity RNA (Paredes et al. 1993). The outer protein shell is linked to the inner protein shell by interaction of the endodomain of the E2 glycoprotein with a hydrophobic cleft in the capsid protein (Lee et al. 1996). This interaction occurs 240 times with these multiple interactions creating a rigid structure. The membrane of the alphaviruses is not exposed significantly on the surface of the virion, and it is clear that the geometry of the virus is determined by the component proteins. It has been demonstrated that treatment with a mild reducing agent can cause the outer protein shell to open releasing the internal nucleocapsid (Anthony et al. 1992). When this happens, the membrane does not reseal, enclosing space, but remains open indicating that the protein shell dictates the configuration of the membrane bilayer. Thus, in contrast to flu virus, the alphaviruses are better described as protein shells with an associated membrane bilayer. This type of structure presents significant consequences for the process of cell penetration.

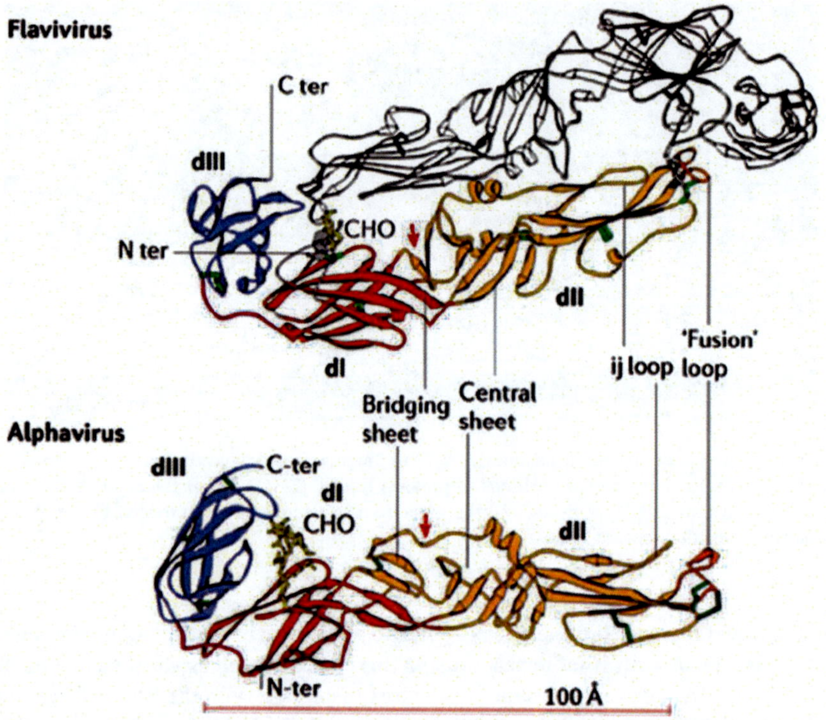

Fig. 8.2 Class II virus membrane fusion proteins. Structure and sequence diagram of class II virus membrane fusion proteins. The tick-borne encephalitis (TBE) flavivirus glycoprotein E and the Semliki Forest alphavirus (SFV) E1 protein are shown, with domains I, II, and III represented in *red*, *yellow*, and *blue*, respectively (dI, dII, and dIII). Common elements and important loops in domain II are indicated. *Red arrowheads* point to the hinge region between domains I and II. The fusion loop is shown in *orange*, disulfide bonds as *green tubes*, and carbohydrates (CHO) in *pale yellow*. The flavivirus E protein dimer is indicated with one subunit in *white*. Reprinted by permission from Macmillan Publishers Ltd. from Kielian and Rey (2006)

A crystal structure has been produced in detergent of a truncated form of the E1 glycoprotein (Gibbons et al. 2004) (Fig. 8.2). This structure is divided into three domain designated dI, dII, and dIII. dIII is the membrane proximal domain and dII is the most distal domain which contains the hydrophobic "fusion loop." The dII domain is situated between the two. This structure has been fitted into the electron density map produced by cryoelectron microscopy (Mukhopadhyay et al. 2006; Zhang et al. 2002). This is a very difficult exercise because the structure of the E2 glycoprotein is not known and the portion of the electron density map occupied by E2 cannot be determined. Thus, one cannot determine if the fit clashes with the presence of E2. It has also been shown that the E1 glycoprotein matures through a series of disulfide-bridged intermediates into a metastable structure which when released from the virus by detergent treatment reshuffles into several nonnative structures which can be separated by nondenaturing gel electrophoresis (Mulvey and Brown 1994). It is unclear which of these forms would form a crystal. It has also been shown by chemical protein modification of the intact virus that certain structures present in the crystal structure of E1 do not exist in E1 in the intact virion (Whitehurst et al. 2007). More recently, crystal structures of E1–E2 dimers have been produced by cloning the ectodomains of the two proteins into a chimeric construct linked together by either a Strep-tag or a glycine–serine linker (Li et al. 2010; Voss et al. 2010). These constructs eliminated the 6k spacer protein that exists in the wild-type virus sequence. Although crystal structures were obtained and were found to be similar to previous structures produced of E1, the assumption is that the proteins are native in the conformation that exists in the intact virion and that the membrane plays no defining role if the assembly of these proteins remains untested. The high-resolution structure of the virus surface remains unclear. Recent data using SANS on virus particles from BHK or C7-10 cells has revealed that Sindbis virus particles from mammalian or insect cells are not structurally equivalent (He et al. 2010). This finding was unexpected and suggests that mature Sindbis virus can exist in two separate metastable infectious intermediates. This finding imposes an additional level of structural complexity which will require further investigation.

8.5 Alphaviruses as Molecular Motors

Virus particles are molecular motors designed to encapsulate and protect their genomic material until a suitable host is encountered and a subsequent infection can ensue. During the process of infection, these particles undergo multiple rearrangements and conformational changes to breach the host cell membrane and inject their genetic material. Sindbis virus (SVHR strain), which demonstrates a particle to pfu of 1, is a symmetrical structure that is structurally stable and is uniquely poised to study the properties of complex molecular protein motors. For a protein complex to perform "work" it must use energy, which is hypothesized to be stored in the conformation of the individual proteins and within the protein–protein interactions that comprise the protomers that form the virus structure. This stored energy is accumulated during the process of particle assembly (Carleton and Brown 1996; Mulvey and Brown 1994). Much more is known about the assembly mechanisms used for Sindbis virus particle formation in mammalian cells. In this host system, it is known that E1 undergoes a complex set of disulfide-bonded folding intermediates prior to dimerization with PE2 and egress from the endoplasmic reticulum (Carleton and Brown 1996; Carleton et al. 1997; Mulvey and Brown 1995, 1996). Further conformational changes occur as the particles mature through the exocytic pathway and become glycosylated and proteolytically processed. During the late stages of maturation, the E2 tail is withdrawn from the host membrane, and association with the nucleocapsid is initiated (Liu and Brown 1993a, b). It is hypothesized that burial of the E2 tail into the hydrophobic pocket of the capsid protein is sufficient to drive the envelopment of 240 copies of each E2 1:1 with capsid (Ferreira et al. 2003). A final conformational change of the nucleocapsid occurs during budding and is one of the final adjustments of the structure (Coombs et al. 1984). Envelopment is completed with fission of the virus particle membrane which is mediated by interactions with the E2 tail (Brown et al. 1972; Hernandez et al. 2005). Once envelopment is accomplished, the mature particle has completed all the changes that are required to produce an energy-rich metastable structural intermediate. This "intermediate" is an infectious virus particle specifically programmed to probe its environment for a suitable host cell. Once a host cell is encountered, this complex structure will recognize a receptor(s) to initiate the process of cell penetration.

8.5.1 Receptor Recognition

The first critical event in the process of virus infection is the attachment of the virion to an appropriate cell receptor. This interaction must result in the penetration event that will deliver the virus genome to its site of replication within the host cell. The critical process of attachment may not happen as a one-step event. The process might be rendered very inefficient if interaction with the host cell could only occur by the specific interaction of the virus protein attachment domain with the specific site on the cell receptor which may not be present in high copy number. It is possible that attachment is preceded by a less specific "sticking" event that retains the virus in contact with the cell until the receptor is encountered. For some viruses the receptor has been identified, e.g., for flu the receptor has been shown to be sialic acid (Skehel and Wiley 2000). For the alphaviruses the situation remains unclear in spite of several published reports on the subject. Molecules implicated as participating in the receptor complex include the major histocompatibility receptor (Helenius et al. 1978; Maassen and Terhorst 1981; Oldstone et al. 1980), the major laminin receptor (Wang et al. 1992), DC sign L sign (Klimstra et al. 2003), heparan sulfate (Klimstra et al. 1998), the heat shock 70 protein (Ren et al. 2007), and an unidentified 110-kDa nerve cell–associated protein (Ubol and Griffin 1991). A cryo-EM image of the virus with attached heparin sulfate has been produced (Zhang et al. 2005). Whatever molecular structure constitutes the receptor, it is either ubiquitous in nature or more than one receptor is used by the alphaviruses. In the laboratory, alphaviruses can be shown to infect any tissue cultured cells (e.g., insect, mammals, bird) except those derived from plants. This enormous host range implies that a common set of cellular attributes which are employed in the penetration process exists in the biochemically and genetically distinct hosts.

8.5.2 Penetration of Cell Membranes

When an appropriate interaction between the virus and receptor is established, the process of transferring the virus genome to the cell interior can take place. To assay that infection has taken place has been determined by the detection of virus-specific RNA or protein synthesis. These are both relatively late events and occur in a time frame of 1–2 h postinfection. This time frame is well past RNA penetration since the earliest event that can be detected in alphavirus infection is the establishment of homologous interference (the inability to establish an infection with a superinfecting homologous virus) which can be detected in 15 min (Johnston et al. 1974). This event requires the initial translation of the incoming RNA and suggests that infection is rapidly established in this virus system. Many studies using inhibitors of endocytosis or acidification of endosomes have resulted in the conclusion that virus infection/penetration does not occur in the absence of endosome acidification (Glomb-Reinmund and Kielian 1998; Helenius et al. 1980, 1982; Helenius and Marsh 1982; Marsh et al. 1982, 1983; White and Helenius 1980; White et al. 1980). This approach is valid if the secondary effects of the chemical inhibitors on the cell and the infection are known. In the alphavirus system, it would be necessary to demonstrate that treatment of the infected cells with chemical inhibitors does not have any effect on RNA translation, RNA transcription, or protein processing.

8.5.2.1 Evidence for Membrane Fusion

It has long been a paradigm in animal virology that viruses that have membranes employ that membrane as a tool to penetrate host cells. Two mechanisms have been proposed for the role of the membrane in cell penetration both implicating fusion of the virus membrane with a cell membrane as the critical event. In one model, the virus engages the receptor complex and this contact initiates conformational changes in the virus membrane proteins. These conformational changes serve to draw the membranes into close proximity resulting in the disordering of the lipid bilayers and resulting in the fusion of the virus and cell membranes. Human immunodeficiency virus has been proposed to enter cells by this mechanism (Eckert and Kim 2001; Kielian and Jungerwirth 1990).

A second model implicates the endocytic pathway as a critical participant in the penetration process. In this model, the virus engages the receptor complex as described above. This attachment event is followed by the incorporation of the virus receptor aggregate into a developing endosome. As formation of the endosome vesicle is completed, the activity of vesicle-associated hydrogen ion pumps (vacuolar ATPase) lowers the pH of the vesicle contents. At some critical pH value the ionization state of the virus membrane proteins is changed leading to dramatic conformational rearrangements. The structural changes drive the process of moving and merging the virus and cell membranes producing the fusion event and releasing the nucleic acid. Influenza virus achieves cell penetration by this mechanism (Bullough et al. 1994; Skehel et al. 1995). Elegant models have been presented which illustrate mechanistically how these fusion events may occur although the mechanism by which the repulsive forces of the opposing phosphate head groups on the lipid bilayer are overcome remains unclear (Kielian and Rey 2006).

It has been proposed and widely accepted that the alphaviruses enter cells by the second of these mechanisms (Kielian 1995; Kielian and Rey 2006). Evidence supporting this model is provided by two basic experiments showing (1) physical demonstration of fusion. That alphavirus can cause membrane fusion upon exposure to low pH. This observation was made initially with living cells in experiments in which large numbers of virus were added to cells at neutral pH and then the monolayers were bathed in media at acid pH before subsequent return to neutral pH where fusion was evaluated (fusion from without) (White et al. 1980). Thus, for living cells, fusion does not occur at acid pH rather conditions are established at acid pH which favor the fusion event upon return to neutral pH. It has been subsequently shown that fusion with living cells requires return to neutral pH after exposure to acidic medium (Edwards and Brown 1986; Paredes et al. 2004). The experiments with living cells were replaced with experiments examining the interaction and fusion of virus with artificial membranes, liposomes, to facilitate a physical and chemical dissection of the fusion process (White and Helenius 1980). Another shows (2) chemical disruption of endosome acidification. Agents that prevent the acidification or formation of endosomes prevent the establishment of infection as determined by assay for the presence of newly synthesized virus RNA or protein (Helenius et al. 1982; Marsh et al. 1982).

Studies Using Artificial Membranes

Experiments involving the interaction of virus with model liposomes have provided a large body of information regarding the fusion of alphavirus with artificial membranes. In stating this, one must be mindful of the fact that such artificial membranes lack many of the attributes of a living cell including the virus receptor. As was the case with living cells, it was found that fusion of alphaviruses with liposomes required exposure to acid pH; however, unlike experiments with living cells, exposure to acid pH sufficed to drive the fusion event and return to neutral pH was not required. Fusion required significant concentrations of cholesterol (25–50 mol%) in the target membrane. The 3β-hydroxyl form of cholesterol was found to be particularly important for membrane fusion (Kielian and Helenius 1984; White and Helenius 1980). This also distinguished fusion of virus with liposomes from penetration of living cells such as insects (the alternate host), which are cholesterol auxotrophs and have very low levels of cholesterol (Clayton 1964; Cleverley et al. 1997; Hafer et al. 2009; Mitsuhashi et al. 1983). It has also been shown that infection of insect cells is not cholesterol dependent (Hafer et al. 2009). That some alphaviruses have been shown to be transmitted vertically from infected female mosquitoes to progeny males (transovarial transmission) argues against a major role of cholesterol in the virus life cycle because males do not take a blood meal (Dhileepan et al. 1996; Fulhorst et al. 1994). The arboviruses (alpha- and flaviviruses) are believed to be insect viruses which have extended their host range to include vertebrates (Schlesinger 1971). It is hard to understand how they would have developed a requirement for high levels of cholesterol in the membranes of cells they infect when they originated in a system that did not provide it.

The case for a critical role of cholesterol in the infection of living cells has been made in experiments in which insect cells were depleted of cholesterol by growing them in media containing serum which had been depleted of cholesterol by treatment with Cab-O-sil, a silicate (Lu et al. 1999).

While cells grown under these conditions were shown to be resistant to infection, it was subsequently shown that Cab-O-sil treatment does not specifically reduce cholesterol but depleted the serum of many lipids and other essential components (Hafer et al. 2009; Weinstein 1979). Other protocols for growing cells in cholesterol-free environments indicated that cholesterol depletion had little effect on the ability of the cells to become infected (Hafer et al. 2009).

The process of fusion of the virus membrane with the liposome membrane has been shown to be nonleaky, that is, the fusion event seals the mixed membrane product (Smit et al. 2002). By contrast the infection of living cells results in the production of pores in the membrane large enough to be penetrated by the drug alpha-sarcin (Madan et al. 2005) and which leaks ions and can be plugged with transitional metal ions (Koschinski et al. 2003; Wengler et al. 2003).

The experiments examining the fusion of alphavirus with liposomes are carried out by mixing virus with a membrane-associated dye which does not fluoresce at high concentrations but fluoresces when diluted as fusion and lipid mixing occur (fluorescence dequenching). The pH of the mixture is lowered and the increase in fluorescence is measured with time. The same experiment can be carried out with labeled virus and living cells using flu virus as a positive control (Keith Weninger, to be published elsewhere). These experiments are carried out by attaching virus to cells grown on coverslips in a chamber that allows the media to be exchanged. The attached virus fluoresces at a low level, and as the media is changed to acid pH the flu virions are seen to produce a distinct flash indicating that fusion has taken place. When this experiment is done with Sindbis, the virus is seen to be very mobile on the cell surface at neutral pH. When the media is changed for acid medium the viruses freeze in place but no flash of fluorescence can be seen indicating that lipid mixing has not taken place.

We have found that Sindbis virus membrane fusion has a critical pH of 5.8 which is the acid threshold required to establish conditions for fusion. In suspension the virus becomes very hydrophobic at pH lower than this and aggregates. This fact has made obtaining structural information on virus exposed to the critical fusion pH very difficult. We have hypothesized that exposure of the Sindbis virions attached to living cells results in the exposure of hydrophobic domains in the envelope protein which bind the virus to the lipid bilayer without the fusion event. In the liposome system, the exposure of the virus to acid pH allows the virus to bind to the receptor-free membrane initiating the fusion event.

Experiments have been reported that separate the ability of virus to fuse living cells in the two-step fusion process from their ability to penetrate cells. It has been shown that at low temperatures Sindbis virus can infect cells but cannot produce fusion. By contrast it has been shown that alkylation of two cysteine residues in the Sindbis virus E1 glycoprotein blocks the ability of the virus to infect cells but does not block the ability to produce fusion.

The data presented above raises questions regarding the structural changes that occur in the alphavirion as pH is lowered to the acid pH which establishes fusion in the liposome system and sets the conditions for fusion in the cell system upon return to neutral pH. Paredes et al. (2004) solved the problem of virus aggregation at the acid pH required for fusion and have produced three-dimensional reconstructions of the virus at pH 7.2, pH 5.3, and pH 5.3 returned to pH 7.2. The results are striking and have important implications for understanding the process of virus fusion with artificial membranes and living cells. As shown in Fig. 8.3, the transition from neutral to acid pH produces a dramatic reorganization of the virus surface. The virus swells from a diameter of 680 to 705 Å. The trimers located on the strict threefold axis extend their density of the tips of the trimers toward the twofold axis and make contact with the tips of the trimers on the neighboring threefold trimers. The trimers surrounding the threefold axis become distorted and form a ring of protein density around the fivefold axis (Fig. 8.3b, e). The most dramatic change at pH 5.3 is the appearance of a protruding structure at the center of the strict fivefold axis. The structure is 52 Å in width and protrudes 60 Å from the surface of the virus surface making it the most extended structure on the virus surface. Its position on the surface at the center of the fivefold axis implies that this structure has a

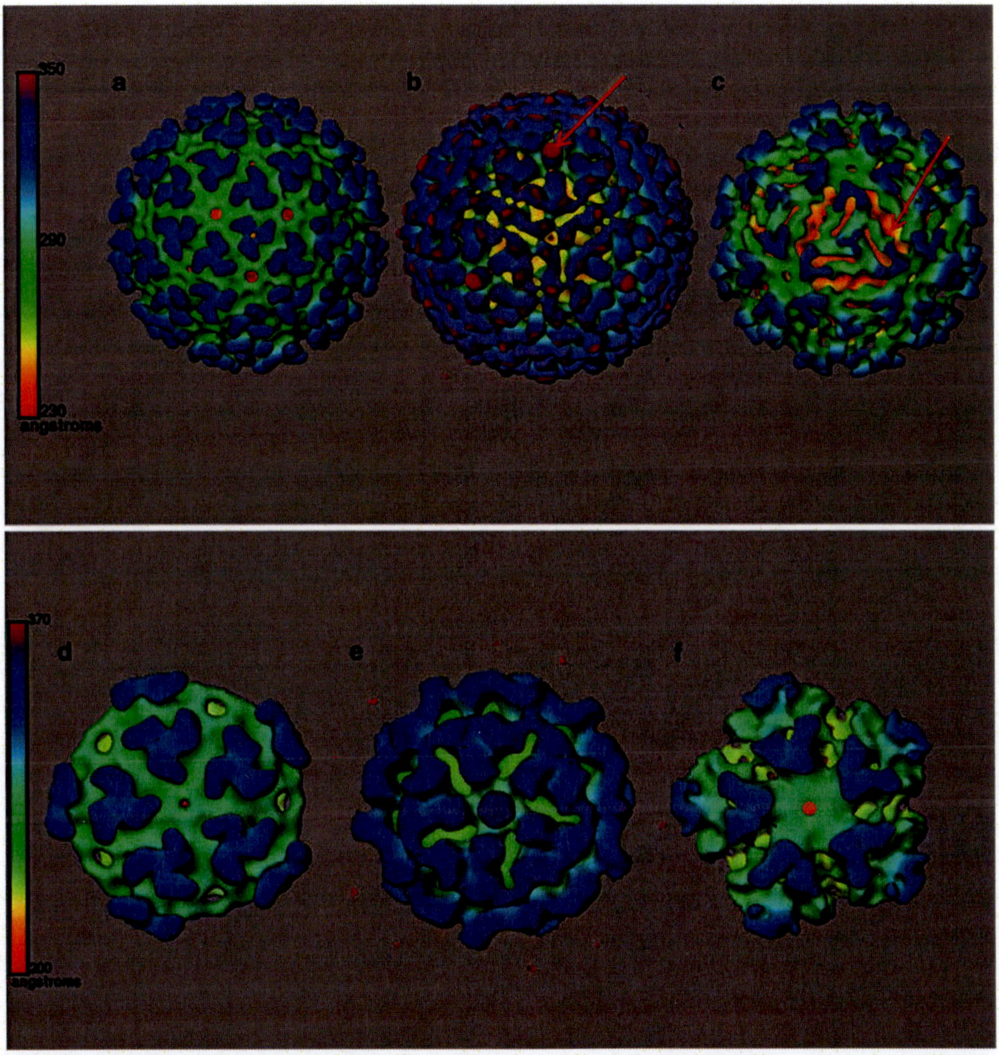

Fig. 8.3 Conformational changes in Sindbis virus after exposure to various pH conditions. The three-dimensional structures of Sindbis virus surface at 28 Å resolution viewed along icosahedral threefold (**a–c**). The reconstructions are colored according to a range of radii (key displayed) at different pH conditions. (**a**) Sindbis virus at pH 7.2, (**b**) Sindbis virus at pH 5.3, and (**c**) Sindbis virus exposed to pH 5.3 (5 min) and returned to pH 7.2. *Arrows*: (**b**) protrusion at the fivefold axis; (**c**) fissure at twofold axis; (**d–f**). The structure of Sindbis outer protein shell restricted to the fivefold axis in three different pH environments at 28 Å. (**d**) pH 7.2, (**e**) pH 5.3, and (**f**) exposed to pH 5.3 and returned to pH 7.2

fivefold rotational symmetry. It is very likely that this structure is a pentameric aggregate of five E1 glycoproteins. As the pH is returned from acid to neutral conditions, the structure of the virus surface reverts to a form that is almost, but not completely, like that of virus retained at neutral pH (Fig. 8.3c, f). The virion diameter reduces from 705 to 674 Å. The protruding density around the fivefold axis disappears and the skirt region is restored to an appearance similar to that seen in virus maintained at pH 7.2. The trimers appear deformed, although the strict threefold trimer appears as a tripartite ridge. The fenestration at the fivefold axis, present in virus maintained at neutral pH, reappears in this structure slightly larger than before. The quasi-threefold trimers have prominent clefts between two of the three appendages that make up the trimer (Fig. 8.3c, f). These clefts extend from the top of the timers toward the strict threefold axis into the skirt region. The appendages of the trimers are much shorter and are distorted compared to virus maintained at neutral pH with the fivefold portion of the spike not reaching into the fivefold axis to the same extent as control virus. The skirt region is also altered in that the twofold region has a large fissure caused by the redistribution of mass density that has not returned to its original position at pH 7.2 (Fig. 8.3c). These clefts would significantly weaken the virus protein shell. The loss of the protruding structure at the fivefold axis is particularly interesting as the return to neutral pH also eliminates the tendency for the virus to aggregate suggesting that hydrophobic domains exposed at pH 5.3 are buried upon return to pH 7.2. It is possible that the protruding structure contains the hydrophobic domains.

While the report by Paredes et al. contains the only image of an alphavirus at the pH that triggers fusion, the resolution of this structure is low (28 Å), the resolution limitation coming from the protocol required to prevent the virus from clumping. At the level of atomic resolution, the only data available come from studies on the E1 glycoprotein. When the E1 crystal structure is "fit" into the electron density map produced from the cryomicroscopy of the virus at neutral pH and 20 Å resolution, it is proposed that the E1 protein lies flat on the surface of the virus membrane as head to tail dimers forming the $T=4$ lattice which holds the icosahedral structure together (Zhang et al. 2002). In this process of fitting crystal structure to electron density, there is no information regarding the structure or position of glycoprotein E2. It is therefore assumed that any electron density not occupied by E1 constitutes E2. Using this assumption, it was concluded that E2 rested on top of E1 covering the hydrophobic loop required for initiating the fusion event. In vitro data have shown that when alphavirus is exposed to acid pH in the presence of cholesterol-containing liposomes, the E1 glycoprotein aggregates as trimers. In this model, it is postulated that these trimers function in the fusion event in a manner similar to that proposed for influenza virus (presented elsewhere) (see Fig. 8.4). The hypothesis continues stating that in the presence of acid pH and cholesterol-containing membranes, the E1–E2 heterodimers dissociate and reorganize into E1 trimers. It has only been possible to recover these trimers from membranes by detergent treatment which is also required to crystallize the E1 trimers themselves. The crystal structure of the trimer represents the structure of the E1 protein "postfusion," while the crystal of the neutral pH dimer is the structure of E1 "prefusion." These structures are dramatically different and imply a complex rearrangement of the virus surface during fusion in which E1 dimer interactions are broken and new trimer associations are formed. The interaction of the developing trimer with the target cell membrane is proposed to be mediated by the dII domain through an undefined interaction of the fusion loop with the target. E2 binding with the host receptor must precede the following reorganization of E1. Subsequently, the trimer is proposed to fold bringing domains dIII and dII together as the liposome and the virus membrane fuse (Kielian and Rey 2006). The energy for this event is provided by conformational changes in E1 in response to the change in the pH environment. The part of this model difficult to reconcile with all the available data is the part which relates to the anchoring of the E1 protein dII domain in the target membrane. The hydrophobic loop is rather short at 17 AA and cannot by itself anchor the E1 protein in the target membrane. As a protruding loop at the distal end of the E1 glycoprotein, it is difficult to understand how it could overcome the hydrophilic barrier created by the polar phosphate lipid head groups. It is proposed that the dII domain crosses the membrane bilayer completely or half way to tightly anchor it in the target (Kielian and Rey 2006). This would bring some very hydrophilic regions of the protein into the lipid interior of the membrane.

The development of the E1 trimer at low pH in the presence of cholesterol-rich membranes is different from the structural changes seen in the intact virion by electron cryomicroscopy. The most dramatic structural change seen in EM is the appearance of a prominent protruding structure at the strict fivefold axis indicating that this structure has a fivefold rotational symmetry. It is important to point out that this structure is produced by exposure of the virus to acid pH in the absence of membranes raising the possibility that there are multiple configurations that the alphavirus is capable of assuming in acid pH environments.

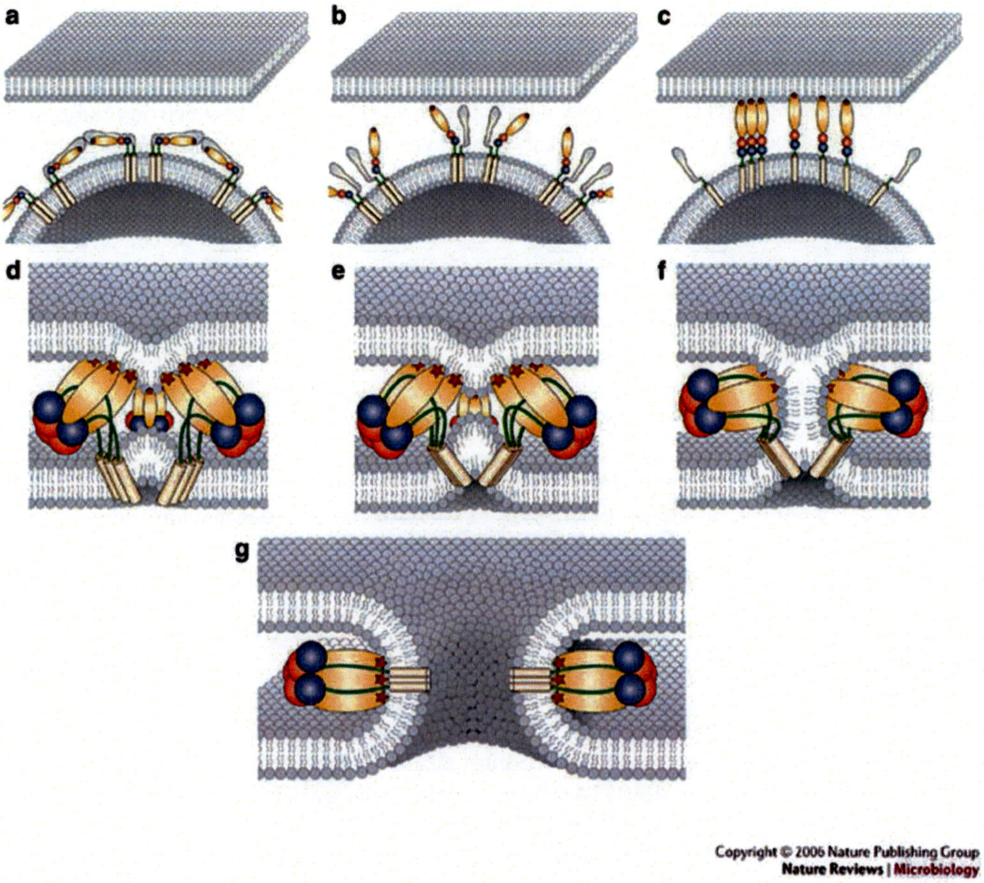

Fig. 8.4 (**a**) Native virions, showing E1 with domains colored as in Fig. 8.2, and E2 in *light gray* interacting with E1 and covering the fusion loop. Only the target-cell-proximal side of the virion represented in Fig. 8.2b, *right panel*, is shown. (**b**) Trigger 1: low pH triggers E1–E2 dissociation and exposure of the fusion loop. (**c**) Trigger 2: low-pH- and cholesterol-dependent insertion of the fusion loop, leading to the alignment of E1 subunits parallel to each other, favoring trimerization. (**d**) Fold-back process: domain III and the stem region move toward the fusion loop. Cooperative interactions between trimers through their fusion loops distort the target membrane. (**e**) Folding of the domain III and stem segments against the body of the trimer pulls transmembrane (TM) segments against each other, distorting the viral membrane. (**f**) Opposing dome-like deformations in the two membranes lead to mixing of the outer leaflets (hemifusion). (**g**) To reach the final stable conformation, the TM segments have to be closely juxtaposed to the fusion loop. This is only possible by opening an initial fusion pore

Studies Using Inhibitors of Vacuolar Acidification

The initial proposal that endocytosis and exposure to acid pH were required for infection by alphavirus came from studies employing lysosomotropic weak bases (Helenius et al. 1982). These basic compounds accumulate in intracellular compartments and neutralize the acid environment. Compounds such as ammonium chloride and chloroquine were applied to cells prior to infection and the event of infection was determined by assaying for newly synthesized virus RNA. Virus RNA synthesis, however, is not the next step in the infectious process. Prior to the synthesis of progeny RNA, the genomic RNA must be translated and transcribed. Both chloroquine and ammonium chloride were shown to block virus RNA synthesis when vertebrate cells were treated prior to the addition of virus. The weak bases are very nonspecific in activity and RNA synthesis is not a measure of penetration and is only a measure of infection if RNA is produced. It was also shown that while treatment of vertebrate cells with chloroquine blocked the production of virus RNA, it had no effect on the outcome of infection in insect cells (Coombs et al. 1981). It could be shown that chloroquine neutralized the endocytic compartments in the insect cell as it protected the cell from diphtheria toxin which is known to require an acid endosome for entry into cells (Edwards and Brown 1991). While ammonium chloride could block virus production in insect cells, it was shown that the block occurred at the level of proteolytic processing of the nonstructural polyprotein precursor, a process required for the production of the RNA polymerase (Hernandez et al. 2001). The nonspecific nature of the weak bases made them very poor tools for determining the role of acid pH in the infection process. More recently, the antibiotic Balfinomycin A has been shown to inhibit the V-ATPase (proton pump) responsible for maintaining cytoplasmic homeostasis and for the acidification of endosomes (Crider et al. 1994). Again, this drug was shown to prevent the synthesis of alphavirus RNA when cells were treated prior to or during infection (Glomb-Reinmund and Kielian 1998). We have recently reexamined the effect of this drug on alphavirus infection using Sindbis virus which has in its genome a fluorescent reporter protein under control of a separate subgenomic Sindbis virus promoter. The expression of this protein causes cells to fluoresce and indicates that infection and translation of the incoming RNA have taken place. In experiments examining the expression of the reporter, infection in the presence of drug only slightly reduced cell fluorescence while fluorescence was dramatically reduced if the drug was added after infection. Identical results were obtained when the virus RNA was transfected into the cells by electroporation (Hunt et al. 2011). Transfection bypasses the infection process, and this result suggests that drug treatment blocked translation of the Sindbis message. We also found that RNA entered the cell and was translated in the presence of Balfinomycin, but correct folding of the GFP protein required removal of the drug. These results with inhibitors underscore the caution that one must take in using them to study a particular event. The drugs may have secondary effects on cells which render the assay for direct effect on a particular event untrustworthy (Ferreira et al. 2000). The broad effects of these agents are underscored by the observation that weak bases can block alphavirus maturation when added after infection is established.

Evidence for Penetration in the Absence of Membrane Fusion

The problems with the indirect approach of investigating the penetration of cells by alphaviruses have produced a confusing and contradictory collection of results. Direct observation of alphavirus interaction with living cells by light and electron microscopy has been difficult because of the poor particle to plaque forming unit (infectivity ratio) of virus preparations. We have addressed this problem and produced pure preparations of Sindbis virus with infectivity ratios approaching unity. These virus preparations have been used in direct observation of virus–cell interaction (Paredes et al. 2004) (Fig. 8.5). As described above, Sindbis virus preparations determined to have a particle to pfu ratio of 1 were labeled with fluorescent dyes and used in experiments confirming the fusion of alphavirus with liposomes of the formulations described in the literature. However, it was not possible to demonstrate fusion of virus with living cells using fluorescence dequenching upon exposure to low pH using the same preparation of virus. Using electron microscopic methods, it was determined that when these viruses are added to living cells and examined by transmission electron microscopy of thin sectioned preparations, no evidence of membrane fusion or of involvement of endocytosis could be detected (Paredes et al. 2004). Virus appeared to attach to cell plasma membranes and lose their electron-dense RNA core. The attachment appeared to be mediated by the formation of a tubular, pore-like structure that engaged and penetrated the cell membrane. The loss of the core resulted in a loss of the tight spherical structure of the virus. The

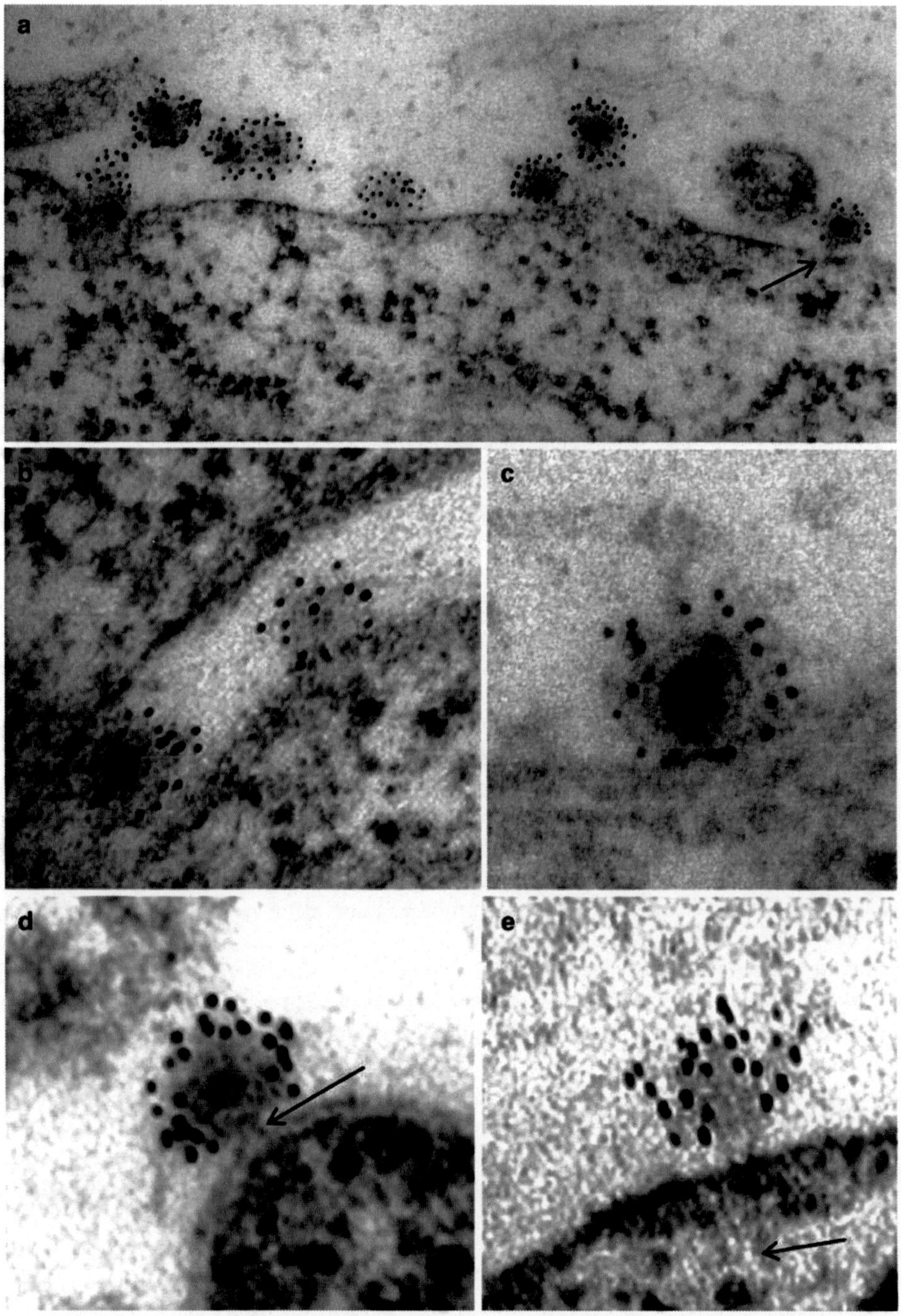

Fig. 8.5 Electron micrographs of thin sections of Sindbis virus–cell complexes at pH 7.2. (**a**) Low magnification showing "full" and "empty" particles and a particle attached by a pore to the cell surface (*arrow*). (**b**) An "empty" and "full" (RNA-containing) virions. The empty virion can only be identified as virus by antibody binding. (**c**) Reorganization of virus RNA into the developing pore. (**d**) The pore at the vertex of the icosahedron is seen to engage and penetrate the cell membrane (*arrow*). The virion has reduced electron density in the core region. (**e**) An empty particle with a possible RNA molecule entering the cell (*arrow*)

resulting empty particles could not have been identified as virus except that they were specifically labeled with antivirus antibody. That major structural changes in the structure of the virion take place during the penetration process is supported by the observation that attachment to the cell surface results in the exposure of otherwise buried virus antibody epitopes (Flynn et al. 1990). The virus appeared to break up completely after loss of the electron-dense core and was lost from the cell surface indicating that the integrity of the outer icosahedral protein shell depended on an intact core structure. The process by which the virus became empty was very fast and could only be effectively seen by attaching virus to cells at 4° and then washing the monolayer with warm (37°) media containing glutaraldehyde fixative. In these experiments, 26% of the attached viruses became empty when the virus/cell complex was warmed to 37°, and 4% became empty when the monolayer was warmed to 15°. This observation underscores the rapid nature of the penetration process and indicates again that penetration can take place at temperatures which block cell membrane fusion. These morphological studies are very similar to those published by Hase et al. (1989a, b) with the Flavivirus Dengue. These authors also found that the virus appeared to empty and disassemble when contacting the cell surface. They also found no involvement of endocytosis in the penetration process.

The nature of the pore-like structure that is formed as the virus contacts the cell surface is interesting. The pore seems to develop at the fivefold vertex of the icosahedron. The pore is 100 Å wide and about 90 Å long. The structure that develops at the fivefold axis of the free virus on exposure to acid pH is 52 Å wide and 62 Å long. The difference in size may be an artifact of the embedding process. The structure seen in the cryo-EM images may be a part of the pore structure, the rest being contributed by the cell receptor complex. Several pieces of evidence may support the formation of a pore in the cell plasma membrane as penetration takes place. It has been shown that when Sindbis virus infects cells they become permeable to the drug alpha-sarcin (Madan et al. 2005) and develop pores in the plasma membrane (Koschinski et al. 2003; Wengler et al. 2003, 2004). If virus penetrated by membrane fusion at low pH, as is the case in interaction with liposomes, this would not happen as the liposome fusion process has been shown to be nonleaky (Smit et al. 2002). The Wengler research group has shown that alphavirus infection creates pores of different sizes, one early and one later during infection (Koschinski et al. 2003). The appearance of the early pore causes the cell to become leaky, and the pore can be effectively plugged with transition metal salts. It may be that the attachment of the virus results in the development of a pore that is continuous from the interior of the virus core crossing the plasma membrane to the cell cytoplasm. The collapse of the virus structure after release of the RNA may leave the open pore structure embedded in the plasma membrane. It has been shown that exposure to acid pH results in the formation of a pore that crosses the virus membrane into the virus interior (Lanzrein et al. 1993). This latter observation suggests that in some ways the conformational changes induced by low pH may mimic those produced by receptor interaction by releasing the same switch required to engage the conformational changes required for RNA penetration into the cell.

8.6 Reconciling Contradictory Observations

The model favoring the entry of alphaviruses by low-pH-mediated membrane fusion is supported primarily by examining the interaction of virus with protein-free artificial membrane and examining the effect of inhibitors of endosome acidification on the expression of virus functions, post penetration. The model favoring entry by direct penetration of cell plasma membranes is supported by the observation of virus cell complexes by electron microscopy. This latter model is also consistent with reports of cell permeability changes occurring early in the infection process.

The alphaviruses are clearly capable of fusing with artificial membranes upon exposure to acid pH. They are also capable of fusing with membranes of living cells if exposed to low pH briefly and then returned to neutral pH. The ability of these viruses to execute these fusion events may be explained by the structural changes that are seen upon exposure to acidic conditions and upon return to neutral pH. As shown in Fig. 8.3 exposure to acid pH equivalent to that initiating the fusion event the surface of the virus undergoes dramatic reorganization. This rearrangement renders the virus sticky resulting in aggregation and falling out of solution. This implies that hydrophobic domains in the envelope proteins have been exposed by this treatment. The most prominent change in the surface is the appearance of a new protruding structure at the strict fivefold vertex of the icosahedrons. It is likely that this pentameric structure is presenting hydrophobic domains and that these domains insert into the membranes of the liposome or the living cell. The rearrangement of the virus surface may also weaken the protein–protein interactions responsible for maintaining the icosahedral structure. The highly curved and constrained radius of the liposome may provide the energy to dissociate the compromised virus protein shell permitting lipid fusion. The high concentration of cholesterol in the liposome may contribute to the expansive force required to absorb the virus membrane into the constrained liposome membrane. In the living cell, there is little curvature at the point of virus binding, so as the pH is lowered, the virus is able to extend its hydrophobic domains and freeze in position on the cell surface as these domains insert into the cell membrane without distortion of the virus particle. The lack of curvature of the plasma membrane may not provide the energy for the fusion event. As the pH is returned to neutrality in the virus structures prepared for cryo-EM, the hydrophobic pentameric protrusions are withdrawn back into the virus structure. In solution the virus loses its hydrophobic property and does not aggregate. Though the virion recovers much of its original structure upon return to neutral pH, the surface is further compromised by the appearance of a long fissure in the surface along the twofold axis. The withdrawal of the pentameric protrusion back into the virus structure may draw lipid of the plasma membrane toward the virus membrane. The combination of these events may allow lipid mixing to proceed. It is possible that treatment of the virus particles to low pH in solution and return to neutral pH causes conformational changes at each of the fivefold axes. This may not be the case when virus has associated with the viral receptor. In this situation, a more specific trigger such as binding with the cell receptor may specifically initiate the conformational rearrangement of the specific vertex from which RNA may be extruded.

Membrane fusion can be separated from infection as described above and was not seen in the electron microscopy studies of Paredes et al. Yet the changes in structure induced by acid pH are dramatic and significant. Does a role for acid pH exist in the infection of cells by alpha and by extension flaviviruses? The alphaviruses have a very broad ability to infect a variety of cell types. Fish, snake, mammalian, bird, and various insect cells, virtually any cell that can be cultured with the exception of plant protoplasts, can be infected by alphaviruses. This implies that the cell receptor complex must be available on these diverse cell types or that multiple receptors can be used. This fact and the effect of pH on virus structure led us to propose that a likely candidate for the virus receptor complex found on all cell types may be the vacuolar ATPase. This assembly of proteins is present in the membranes of every living cell from the *Archea* to mammals. It is possible that binding of the alphavirus by the pentameric ring of E2 proteins at the vertex of the icosahedron initiates the infection process. Action of the V-ATPase could produce a localized microenvironment at acid pH within the receptor virus protein complex. These conditions might expose the pentameric structure which together with the V-ATPase may produce the protein pore (Fig. 8.6) that is continuous from the interior RNA-containing virus core to the cell cytoplasm. Upon injection of the virus RNA, the breakdown of the virus membrane may leave the pore imbedded in the cell plasma membrane resulting in the observed increase in cell permeability seen to accompany alphavirus infection. This proposal would be strongly supported by the observation that Bafilomycin A1 (BAF), a

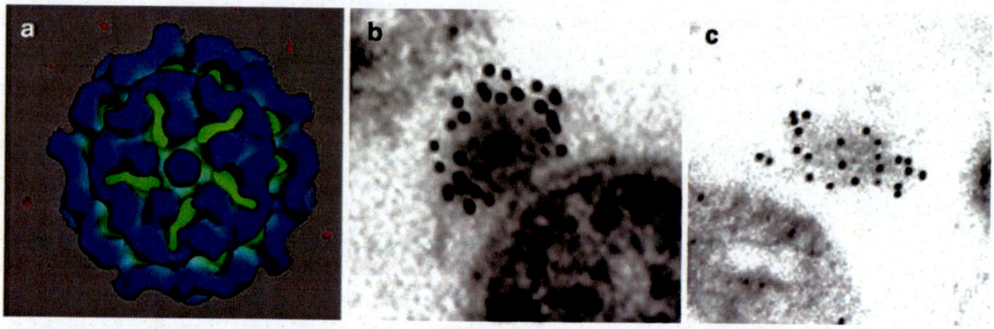

Fig. 8.6 Model for the direct penetration of cell membranes by alphaviruses. (**a**) The interaction of E2 proteins at the fivefold vertex with the cell receptor complex results in tight binding of E2 and the reorganization of E2 into a ring of protein subunits. The receptor-binding event results in the projection of E1 hydrophobic domains into the cell membrane/receptor complex. (**b**) These structural changes produce a pore which is likely composed of both host and virus proteins and forms a channel which is continuous from the interior of the virus to the interior of the cell. The virus RNA passes from the interior of the virion to the cytoplasm of the cell. (**c**) Transfer of the RNA to the host cell results in structural instability in the virus particle. The particles break up leaving the pore in the cell membrane. The presence of the pore makes the surface permeable to small molecules

presumed specific inhibitor of the V-ATPase, blocked virus infection as determined by failure to express a reporter gene (described above) in the presence of the drug. This possibility was diminished but not ruled out completely by the finding that BAF also blocked the translation of the incoming virus RNA.

8.7 Summary and Future Directions

The model for infection of cells by alphaviruses presented here is a significant departure from the existing paradigm describing the interaction of membrane-containing viruses with cells (Fig. 8.6). The process by which alphaviruses are proposed to infect cells here is very similar in many regards to the entry mechanism proposed by Hogle for poliovirus (Belnap et al. 2000). Hogle has hypothesized that polio interacts with cell membranes through the exposure of hydrophobic helices at the fivefold vertex of the virion upon attachment to the cell receptor. The exposed protein domains then form a pore in the plasma membrane through which the viral RNA passes. It is possible that in its function as a molecular machine the alphaviruses function as do nonmembrane-containing viruses to gain entry into a cell. This is supported by the contention presented above that alphaviruses are protein-enclosed viruses with an associated membrane. The function of the virus membrane may be to serve as a scaffold upon which the virus glycoproteins are folded into their high-energy metastable configuration. The membrane may play no functional role in the infection process. A single report exists in the literature suggesting that removal of the membrane of an alphavirus did not eliminate infectivity (Omar and Koblet 1988).

The important next step in the investigation of the interaction of alphaviruses with host cells is to unambiguously identify the host cell receptor complex and the domains of that complex that interact with the virus glycoproteins (E1 and E2) as penetration takes place. The identification of these structures may facilitate the activation of the RNA injection mechanism, which could be examined in solution allowing further elucidation of the penetration process.

Acknowledgments Dennis Brown and Raquel Hernandez are supported by "The Foundation for Research" Carson City, NV. The authors thank Sabrina Hunt for editing this manuscript.

References

Anthony RP, Paredes AM, Brown DT (1992) Disulfide bonds are essential for the stability of the Sindbis virus envelope. Virology 190(1):330–336

Belnap DM, Filman DJ, Trus BL, Cheng N, Booy FP, Conway JF, Curry S, Hiremath CN, Tsang SK, Steven AC, Hogle JM (2000) Molecular tectonic model of virus structural transitions: the putative cell entry states of poliovirus. J Virol 74(3):1342–1354

Brown DT, Condreay LD (1986) Replication of alphaviruses in mosquito cells. In: Schlesinger MJ (ed) The Togaviridae and Flaviviridae. New York, Plenum Publishing Corporation, pp 171–207

Brown DT, Waite MRF, Pfefferkorn ER (1972) Morphology and morphogenesis of Sindbis virus as seen with freeze-etching techniques. J Virol 10:534–536

Bullough PA, Hughson FM, Skehel JJ, Wiley DC (1994) Structure of influenza haemagglutinin at the pH of membrane fusion. Nature 371(6492):37–43

Carleton M, Brown DT (1996) Disulfide bridge-mediated folding of Sindbis virus glycoproteins. J Virol 70(8):5541–5547

Carleton M, Lee H, Mulvey M, Brown DT (1997) Role of glycoprotein PE2 in formation and maturation of the Sindbis virus spike. J Virol 71(2):1558–1566

Clayton RB (1964) The utilization of sterols by insects. J Lipid Res 5:3–19

Cleverley D, Geller H, Lenard J (1997) Characterization of cholesterol-free insect cells infectible by baculoviruses: effects of cholesterol on VSV fusion and infectivity and on cytotoxicity induced by influenza M2 protein. Exp Cell Res 233(2):288–296

Coombs K, Mann E, Edwards J, Brown DT (1981) Effects of chloroquine and cytochalasin B on the infection of cells by Sindbis virus and vesicular stomatitis virus. J Virol 37(3):1060–1065

Coombs K, Brown B, Brown DT (1984) Evidence for a change in capsid morphology during Sindbis virus envelopment. Virus Res 1(4):297–302

Crider BP, Xie XS, Stone DK (1994) Bafilomycin inhibits proton flow through the H+ channel of vacuolar proton pumps. J Biol Chem 269(26):17379–17381

Dhileepan K, Azuolas JK, Gibson CA (1996) Evidence of vertical transmission of Ross River and Sindbis viruses (Togaviridae: Alphavirus) by mosquitoes (Diptera: Culicidae) in southeastern Australia. J Med Entomol 33(1): 180–182

Eckert DM, Kim PS (2001) Mechanisms of viral membrane fusion and its inhibition. Annu Rev Biochem 70: 777–810

Edwards J, Brown DT (1986) Sindbis virus-mediated cell fusion from without is a two-step event. J Gen Virol 67 (Pt 2):377–380

Edwards J, Brown DT (1991) Sindbis virus infection of a Chinese hamster ovary cell mutant defective in the acidification of endosomes. Virology 182(1):28–33

Ferreira DF, Santo MP, Rebello MA, Rebello MC (2000) Weak bases affect late stages of Mayaro virus replication cycle in vertebrate cells. J Med Microbiol 49(4):313–318

Ferreira DF, Hernandez R, Horton M, Brown DT (2003) Morphological variants of Sindbis virus produced by a mutation in the capsid protein. Virology 307(1):54–66

Flynn DC, Meyer WJ, Mackenzie JM Jr, Johnston RE (1990) A conformational change in Sindbis virus glycoproteins E1 and E2 is detected at the plasma membrane as a consequence of early virus-cell interaction. J Virol 64(8):3643–3653

Fulhorst CF, Hardy JL, Eldridge BF, Presser SB, Reeves WC (1994) Natural vertical transmission of western equine encephalomyelitis virus in mosquitoes. Science 263(5147):676–678

Gibbons DL, Reilly B, Ahn A, Vaney MC, Vigouroux A, Rey FA, Kielian M (2004) Purification and crystallization reveal two types of interactions of the fusion protein homotrimer of Semliki Forest virus. J Virol 78(7):3514–3523

Glomb-Reinmund S, Kielian M (1998) The role of low pH and disulfide shuffling in the entry and fusion of Semliki Forest virus and Sindbis virus. Virology 248(2):372–381

Hafer A, Whittlesey R, Brown DT, Hernandez R (2009) Differential incorporation of cholesterol by Sindbis virus grown in mammalian or insect cells. J Virol 83:9113–9121

Hase T, Summers PL, Cohen WH (1989a) A comparative study of entry modes into C6/36 cells by Semliki Forest and Japanese encephalitis viruses. Arch Virol 108(1–2):101–114

Hase T, Summers PL, Eckels KH (1989b) Flavivirus entry into cultured mosquito cells and human peripheral blood monocytes. Arch Virol 104(1–2):129–143

He L, Piper A, Meilleur F, Myles DA, Hernandez R, Brown DT, Heller WT (2010) The structure of Sindbis virus produced from vertebrate and invertebrate hosts determined by small-angle neutron scattering. J Virol 84: 5270–5276

Helenius A (1984) Semliki Forest virus penetration from endosomes: a morphological study. Biol Cell 51(2):181–185

Helenius A, Marsh M (1982) Endocytosis of enveloped animal viruses. Ciba Found Symp 92:59–76
Helenius A, Morein B, Fries E, Simons K, Robinson P, Schirrmacher V, Terhorst C, Strominger JL (1978) Human (HLA-A and HLA-B) and murine (H-2K and H-2D) histocompatibility antigens are cell surface receptors for Semliki Forest virus. Proc Natl Acad Sci USA 75(8):3846–3850
Helenius A, Kartenbeck J, Simons K, Fries E (1980) On the entry of Semliki Forest virus into BHK-21 cells. J Cell Biol 84:404–420
Helenius A, Marsh M, White J (1982) Inhibition of Semliki forest virus penetration by lysosomotropic weak bases. J Gen Virol 58(Pt 1):47–61
Hernandez R, Lee H, Nelson C, Brown DT (2000) A single deletion in the membrane-proximal region of the Sindbis virus glycoprotein E2 endodomain blocks virus assembly. J Virol 74(9):4220–4228
Hernandez R, Luo T, Brown DT (2001) Exposure to low pH is not required for penetration of mosquito cells by Sindbis virus. J Virol 75(4):2010–2013
Hernandez R, Sinodis C, Horton M, Ferreira D, Yang C, Brown DT (2003) Deletions in the transmembrane domain of a Sindbis virus glycoprotein alter virus infectivity, stability, and host range. J Virol 77(23):12710–12719
Hernandez R, Ferreira D, Sinodis C, Litton K, Brown DT (2005) Single amino acid insertions at the junction of the Sindbis virus E2 transmembrane domain and endodomain disrupt virus envelopment and alter infectivity. J Virol 79(12):7682–7697
Hernandez R et al (2006) Sindbis virus: propagation, quantification and storage. In: Coico R et al (eds) Current protocols in microbiology, vol 1. Wiley, New York, p 15B.1
Hunt SR, Hernandez R et al. (2011) Role of the vacuolar-ATPase in Sindbis virus infection. Journal of virology 85(3):1257–1266
Johnston RE, Wan K, Bose HR (1974) Homologous interference induced by Sindbis virus. J Virol 14(5):1076–1082
Kielian M (1995) Membrane fusion and the alphavirus life cycle. Adv Virus Res 45:113–151
Kielian MC, Helenius A (1984) Role of cholesterol in fusion of Semliki Forest virus with membranes. J Virol 52(1):281–283
Kielian M, Jungerwirth S (1990) Mechanisms of enveloped virus entry into cells. Mol Biol Med 7(1):17–31
Kielian M, Rey FA (2006) Virus membrane-fusion proteins: more than one way to make a hairpin. Nat Rev Microbiol 4(1):67–76
Kielian MC, Keranen S, Kaariainen L, Helenius A (1984) Membrane fusion mutants of Semliki Forest virus. J Cell Biol 98(1):139–145
Klimstra WB, Ryman KD, Johnston RE (1998) Adaptation of Sindbis virus to BHK cells selects for use of heparan sulfate as an attachment receptor. J Virol 72(9):7357–7366
Klimstra WB, Nangle EM, Smith MS, Yurochko AD, Ryman KD (2003) DC-SIGN and L-SIGN can act as attachment receptors for alphaviruses and distinguish between mosquito cell- and mammalian cell-derived viruses. J Virol 77(22):12022–12032
Koschinski A, Wengler G, Repp H (2003) The membrane proteins of flaviviruses form ion-permeable pores in the target membrane after fusion: identification of the pores and analysis of their possible role in virus infection. J Gen Virol 84(Pt 7):1711–1721
Lanzrein M, Weingart R, Kempf C (1993) pH-dependent pore formation in Semliki forest virus-infected Aedes albopictus cells. Virology 193(1):296–302
Lee S, Owen KE, Choi HK, Lee H, Lu G, Wengler G, Brown DT, Rossmann MG, Kuhn RJ (1996) Identification of a protein binding site on the surface of the alphavirus nucleocapsid and its implication in virus assembly. Structure 4(5):531–541
Li L, Jose J et al. (2010) Structural changes of envelope proteins during alphavirus fusion. Nature 468(7324): 705–708
Liu N, Brown DT (1993a) Phosphorylation dephosphorylation events play critical roles in Sindbis virus maturation. Virology 196:703–711
Liu N, Brown DT (1993b) Transient translocation of the cytoplasmic (endo) domain of a type I membrane glycoprotein into cellular membranes. J Cell Biol 120(4):877–883
Lu YE, Cassese T, Kielian M (1999) The cholesterol requirement for Sindbis virus entry and exit and characterization of a spike protein region involved in cholesterol dependence. J Virol 73(5):4272–4278
Maassen JA, Terhorst C (1981) Identification of a cell-surface protein involved in the binding site of Sindbis virus on human lymphoblastic cell lines using a heterobifunctional cross-linker. Eur J Biochem 115(1):153–158
Madan V, Sanz MA, Carrasco L (2005) Requirement of the vesicular system for membrane permeabilization by Sindbis virus. Virology 332:307–315
Marsh M, Wellsteed J, Kern H, Harms E, Helenius A (1982) Monensin inhibits Semliki Forest virus penetration into culture cells. Proc Natl Acad Sci USA 79(17):5297–5301
Marsh M, Bolzau E, Helenius A (1983) Penetration of Semliki Forest virus from acidic prelysosomal vacuoles. Cell 32(3):931–940

Mitsuhashi J, Nakasone S, Horie Y (1983) Sterol-free eukaryotic cells from continuous cell lines of insects. Cell Biol Int Rep 7(12):1057–1062
Mukhopadhyay S, Zhang W, Gabler S, Chipman PR, Strauss EG, Strauss JH, Baker TS, Kuhn RJ, Rossmann MG (2006) Mapping the structure and function of the E1 and E2 glycoproteins in alphaviruses. Structure 14(1):63–73
Mulvey M, Brown DT (1994) Formation and rearrangement of disulfide bonds during maturation of the Sindbis virus E1 glycoprotein. J Virol 68(2):805–812
Mulvey M, Brown DT (1995) Involvement of the molecular chaperone BiP in maturation of Sindbis virus envelope glycoproteins. J Virol 69(3):1621–1627
Mulvey M, Brown DT (1996) Assembly of the Sindbis virus spike protein complex. Virology 219(1):125–132
Oldstone MB, Tishon A, Dutko FJ, Kennedy SI, Holland JJ, Lampert PW (1980) Does the major histocompatibility complex serve as a specific receptor for Semliki Forest virus? J Virol 34(1):256–265
Omar A, Koblet H (1988) Semliki Forest virus particles containing only E1 envelope glycoprotein are infectious and can induce cell-cell fusion. Virology 166:17–23
Paredes AM, Brown DT, Rothnagel R, Chiu W, Schoepp RJ, Johnston RE, Prasad BV (1993) Three-dimensional structure of a membrane-containing virus. Proc Natl Acad Sci USA 90(19):9095–9099
Paredes AM, Ferreira D, Horton M, Saad A, Tsuruta H, Johnston R, Klimstra W, Ryman K, Hernandez R, Chiu W, Brown DT (2004) Conformational changes in Sindbis virions resulting from exposure to low pH and interactions with cells suggest that cell penetration may occur at the cell surface in the absence of membrane fusion. Virology 324(2):373–386
Pletnev SV, Zhang W, Mukhopadhyay S, Fisher BR, Hernandez R, Brown DT, Baker TS, Rossmann MG, Kuhn RJ (2001) Locations of carbohydrate sites on alphavirus glycoproteins show that E1 forms an icosahedral scaffold. Cell 105(1):127–136
Ren J, Ding T, Zhang W, Song J, Ma W (2007) Does Japanese encephalitis virus share the same cellular receptor with other mosquito-borne flaviviruses on the C6/36 mosquito cells? Virol J 4:83
Schlesinger RW (1971) Some speculations on the possible role of arthropods in the evolution of arboviruses. Curr Top Microbiol Immunol 55:241–245
Skehel JJ, Wiley DC (2000) Receptor binding and membrane fusion in virus entry: the influenza hemagglutinin. Annu Rev Biochem 69:531–569
Skehel JJ, Bizebard T, Bullough PA, Hughson FM, Knossow M, Steinhauer DA, Wharton SA, Wiley DC (1995) Membrane fusion by influenza hemagglutinin. Cold Spring Harb Symp Quant Biol 60:573–580
Smit JM, Li G, Schoen P, Corver J, Bittman R, Lin KC, Wilschut J (2002) Fusion of alphaviruses with liposomes is a non-leaky process. FEBS Lett 521(1–3):62–66
Strauss JH, Strauss EG (1994) The alphaviruses: gene expression, replication, and evolution. Micro Rev 58:491–562
Ubol S, Griffin DE (1991) Identification of a putative alphavirus receptor on mouse neural cells. J Virol 65(12): 6913–6921
Voss JE, Vaney MC et al. (2010) Glycoprotein organization of Chikungunya virus particles revealed by X-ray crystallography. Nature 468(7324):709–712
Wang KS, Kuhn RJ, Strauss EG, Ou S, Strauss JH (1992) High-affinity laminin receptor is a receptor for Sindbis virus in mammalian cells. J Virol 66(8):4992–5001
Weinstein DB (1979) A single-step adsoption method for romoval of lipiproteins and preparation of cholesterol-free serum. Circulation 59–60(supplement II-54), abstract 204
Wengler G, Koschinski A, Dreyer F (2003) Entry of alphaviruses at the plasma membrane converts the viral surface proteins into an ion-permeable pore that can be detected by electrophysiological analyses of whole-cell membrane currents. J Gen Virol 84(Pt 1):173–181
Wengler G, Koschinski A, Repp H (2004) During entry of alphaviruses, the E1 glycoprotein molecules probably form two separate populations that generate either a fusion pore or ion-permeable pores. J Gen Virol 85(Pt 6): 1695–1701
White J, Helenius A (1980) pH-dependent fusion between the Semliki Forest virus membrane and lipsoomes. Proc Natl Acad Sci USA 77(6):3273–3277
White J, Kartenbeck J, Helenius A (1980) Fusion of Semliki Forest virus with the plasma membrane can be induced by low pH. J Cell Biol 87:264–272
Whitehurst CB, Soderblom EJ, West ML, Hernandez R, Goshe MB, Brown DT (2007) Location and role of free cysteinyl residues in the Sindbis virus E1 and E2 glycoproteins. J Virol 81(12):6231–6240
Zhang W, Mukhopadhyay S, Pletnev SV, Baker TS, Kuhn RJ, Rossmann MG (2002) Placement of the structural proteins in Sindbis virus. J Virol 76(22):11645–11658
Zhang W, Heil M, Kuhn RJ, Baker TS (2005) Heparin binding sites on Ross River virus revealed by electron cryo-microscopy. Virology 332(2):511–518

Chapter 9
Influenza Virus Entry

Ming Luo

Abstract As all the enveloped viruses, the entry of influenza viruses includes a number of steps in host cell infection. This chapter summarizes the current knowledge of the entry pathway and the role of the fusion protein of influenza virus, hemagglutinin, in this process. Hemagglutinin (HA) is a trimeric glycoprotein that is present in multiple copies in the membrane envelope of influenza virus. HA contains a fusion peptide, a receptor binding site, a metastable structural motif, and the transmembrane domain. The first step of influenza virus entry is the recognition of the host cell receptor molecule, terminal α-sialic acid, by HA. This multivalent attachment by multiple copies of trimetric HA triggers endocytosis of influenza virus that is contained in the endosome. The endosome-trapped virus traffics via a unidirectional pathway to near the nucleus. At this location, the interior pH of the endosome becomes acidic that induces a dramatic conformational change in HA to insert the fusion peptide into the host membrane, induce juxtaposition of the two membranes, and form a fusion pore that allows the release of the genome segments of influenza virus. HA plays a key role in the entire entry pathway. Inhibitors of virus entry are potentially effective antiviral drugs of influenza viruses.

9.1 Introduction

Influenza virus belongs to the family of *Orthomyxoviridae*. Infection of influenza virus causes a disease in humans with symptoms including high fever, cough, body ache, and runny nose. The word "influenza" was originated from the Italian language, meaning "influence." Since the disease caused by influenza virus infection occurs more often in winter seasons, it was thought in the ancient world to be caused by the influence of astrological movements. The true reason that influenza returns each year is because of the unique structure of the influenza virus. The influenza virion contains an envelope of lipid membrane that is derived from the plasma membrane of the infected host cell, the site of influenza virus assembly (Fig. 9.1). The membrane envelope forms a barrier inside of which are the viral components protected from the environment when the virus particle is in circulation. Immediately underneath the membrane envelope there is a layer of matrix formed by the matrix protein. The matrix protein layer encloses the viral genome that is composed of eight segments of

M. Luo (✉)
Department of Microbiology, University of Alabama at Birmingham,
Birmingham, AL 35294, USA
e-mail: mingluo@uab.edu

M.G. Rossmann and V.B. Rao (eds.), *Viral Molecular Machines*,
Advances in Experimental Medicine and Biology 726, DOI 10.1007/978-1-4614-0980-9_9,

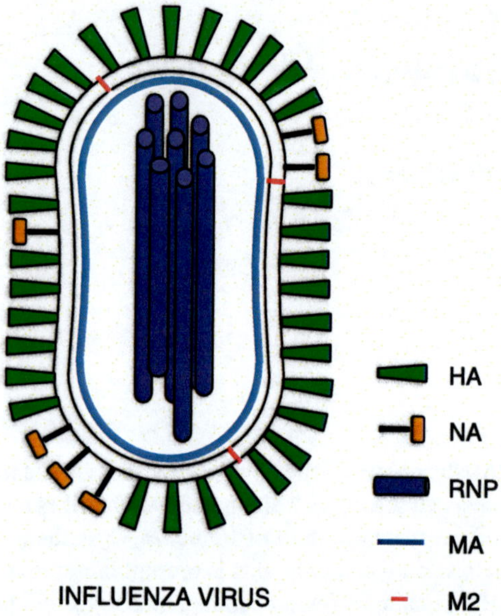

Fig. 9.1 Cartoon showing the architecture of influenza virus

nucleoprotein-enwrapped single-stranded RNA in the negative sense. These eight genome segments encodes ten viral proteins, including three subunits, PA, PB1, and PB2, of the viral-specific RNA polymerase, two surface glycoproteins, hemagglutinin (HA) and neuraminidase (NA), the nucleoprotein (NP), the matrix protein (M), the proton channel protein M2 that is translated from the spliced mRNA of M, and two nonstructural proteins NS1 and NS2 that are the products of the two alternatively spliced eighth viral mRNA. NS2 is also known as nuclear export protein (O'Neill et al. 1998). The outcome of the influenza virus entry into the host cell is to release the eight viral genome segments into the nucleus to initiate virus transcription and replication. The entire virus entry process may be divided into a number of steps. First, the influenza virus particle needs to recognize a specific receptor molecule on the surface of the target host cell to allow the virus particle to gain access to the specific host cell in which influenza virus may replicate. The receptor molecule for influenza virus is a terminal α-sialic acid that is linked to saccharides anchored on the host cell surface by various mechanisms. In the second step, the tight association of the influenza virus particle with the host cell induces endocytosis that generates an endosome to encapsulate the entering virus particle. The endosome is then translocated in the next step to the site near the nucleus. The fourth step is to fuse the viral membrane with the host membrane that forms the endosome, which is mediated by the HA glycoprotein embedded in the virus surface. This membrane fusion step releases all eight segments of the RNA genome into the nucleus where the initiation of transcription of viral genes and the later viral replication takes place.

The key player in virus entry is the surface glycoprotein HA that contains the host receptor binding site to allow the virus particle to attach to specific host cells, the fusion peptide that is inserted into the target cellular membrane during membrane fusion, and other structural elements that may refold during the membrane fusion process. Since HA is a surface glycoprotein on the virus particle, it is readily recognized by host antibodies when influenza virus infects the host. However, influenza virus is quite capable of escaping the host defense system by unique mechanisms. During the replication cycle, the viral-specific RNA polymerase makes errors with a high frequency in its synthesis of viral RNAs. The errors during replication will generate a large number of variants of the HA glycoprotein. Some of them will allow the influenza virus to become resistant to neutralization by the existing host antibodies. By this selection mechanism, a new influenza virus that can escape the host immune system is generated and can replicate readily in the infected host. The novel influenza virus variant can spread to other hosts if they have not been exposed to this variant previously either through vaccination or through natural infection. This is the main reason that influenza virus comes back year after year because in each season new influenza virus strains become the circulating variants to which the human population has not acquired sufficient immunity. The same variations also occur by the same mechanism in the NA glycoprotein, the other surface glycoprotein in influenza virus. Such changes in the HA and NA glycoproteins are termed "antigenic drift" (Hay et al. 2001). During antigenic drifts, amino acid sequences are replaced in areas that are exposed to antibody recognition, and at the same time, the amino acid substitutions are restricted from locations that are required for receptor binding, maintaining a function fusion peptide, and proper structural changes for mediating membrane fusion. When the HA amino acid substitutions are accumulated in the repertoire of influenza virus strains, they are mostly on the exposed surface of the HA glycoprotein except for the receptor binding site, the fusion peptide, and the amino acids that accommodate the fusion peptide before HA structural changes for membrane fusion (Wilson et al. 1981). Besides antigenic drifts, there is also occasionally a dramatic change in HA antigenicity of influenza A virus. The HA glycoprotein of influenza A virus has been classified in 20 subtypes (H1–H20) based on a hemagglutination-inhibition test (Yen and Webster 2009). Subtypes H3 and H1 have been circulating globally in recent years. However, there could be a sudden change of HA subtype in a given year, termed "antigenic shift." This type of change is the result of replacing one gene segment of a circulating influenza virus with another gene segment, a process called reassortment. Reassortment could occur in influenza virus because its genome prepresent in the virion is segmented. The HA glycoprotein must be compatible with other segments in the emerging influenza virus derived from the reassortment. If two influenza virus strains with totally different segmented genomes coinfect a common host, a novel influenza virus could emerge by regrouping a set of genome segments from the two sources. An antigenic shift usually leads to a global influenza pandemic, such as the pandemic of 1918 influenza that was the result of introducing a new H1N1 avian influenza virus in the human population (Compans et al. 1970). Throughout the antigenic changes over the many years, the HA-mediated virus entry remains robust in all influenza virus strains. The HA glycoprotein is fully functional despite many substitutions in its amino acid sequence.

9.2 Structure of Hemagglutinin

Influenza virus is divided into three types: influenza A virus, influenza B virus, and influenza C virus. For virus entry, hemagglutinin, the major glycoprotein on the surface of influenza virion, plays multiple roles in each step of the virus entry pathway. For influenza A virus, HA has been classified in 20 subtypes (H1–H20). The subtype of an influenza A virus is determined if the antisera corresponding to that subtype could inhibit the hemagglutination by the virus. Influenza B and C viruses have only one subtype and these two viruses have only been found circulating in humans.

The first glimpse of influenza virus HA was from electron micrographs. HA molecules appear as spikes covering the surface of the influenza virion (Compans et al. 1970). In the recent higher resolution three-dimensional structure reconstructed from tomographic images of influenza virion (Harris et al. 2006), the trimeric glycoprotein of HA is present as clearly distinct spikes scattered on the virion surface. The HA trimers are anchored through a transmembrane domain that penetrates the membrane envelope, which allows the C-terminal cytoplasmic tail of HA to interact directly with the matrix protein layer. The lollipop-shaped tetrameric NA is present among HA trimers in a small number (Fig. 9.1).

HA is inserted into the lumen of the ER during translation. The trimeric HA precursor travels through the secretory pathway from ER, Golgi apparatus, and secretory vesicles to the plasma membrane. Along the secretory pathway, HA becomes glycosylated and is processed into HA1 and HA2

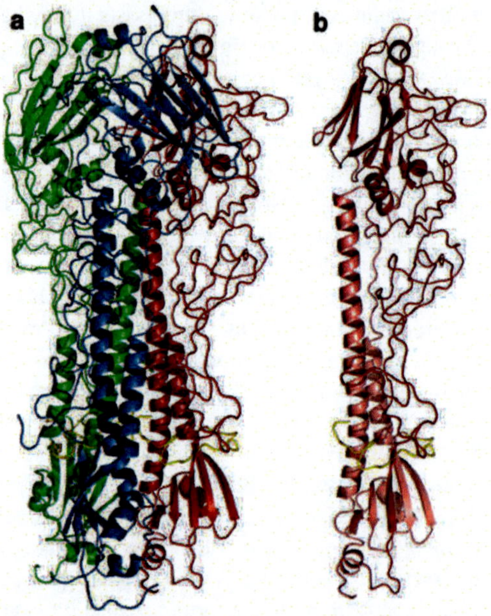

Fig. 9.2 Structure of hemagglutinin. (**a**) The crystal structure of the trimeric ectodomain of hemagglutinin (Wilson et al. 1981). Each subunit of HA is colored *red*, *green*, and *blue*, respectively, with all three fusion peptides colored *yellow*. The ribbon drawings in this and subsequent figures were prepared with the program PyMol (DeLano 2002). (**b**) A single subunit of the ectodomain is presented for clarity. The HA1 polypeptide is colored *red*, and the HA2, *lighter red*

polypeptides by the host proteinase. The cleavage of HA into HA1 and HA2 generates the fusion peptide at the N terminus of HA2. Uncleaved HA molecules are incompetent to mediate membrane fusion. The ectodomains of the mature HA trimers can be released from the surface of influenza virion by a proteinase, bromelain. The crystal structure of the HA ectodomain from an influenza A virus revealed the features that are important for the functions of HA (Fig. 9.2) (Wilson et al. 1981). The HA1 molecule is primarily on the outside of the HA trimer. The first 18 amino acids participate in a β-sheet formed by amino acids in the N-terminal and C-terminal regions of the HA2 molecule. This motif appears to be the structure that tethers the HA1 molecule to the membrane-anchored HA2 molecule. The next 18 amino acids form part of the pocket that accommodates the hydrophobic fusion peptide in the metastable HA structure on the influenza virion prior to virus entry. The rest of the HA1 molecule continues to extend to the top of the trimetric HA spike where a ligand-binding domain of a β-fold is located at the extreme end. The conformation of the ligand-binding domain at the top of the trimeric HA spike is partially stabilized by the contacts between themselves around the threefold symmetry axis. The ligand-binding domain is similar to the common protein fold that binds saccharides. It is composed of two β-sheets folded as a barrel. Since the host cell receptor for influenza virus HA is sialic acid, a saccharide moiety, it is not surprising that HA structure contains such a protein fold at the strategic location where it can recognize the host receptor molecule. Following the ligand-binding domain, the HA1 molecule extends downward along the HA2 molecule to end near the fusion peptide at the N terminus of the HA2 molecule. This conformation of the HA1 C terminus is the result of the cleavage between HA1 and HA2. In the uncleaved precursor, this region forms a surface loop accessible to protease (Chen et al. 1998). The size of the loop and the number of positively charged amino acids in the loop determine how easy the HA molecule is processed. When there are more than three positively charged amino acids in this loop, the cleavage at this site is more efficient and the influenza virus becomes more virulent, often highly pathogenic (Zambon 2001). The HA2 molecule forms the core of the HA trimer. The fusion peptide leads to the amino acids that participate in the anchoring β-sheet involving the N terminus of the HA1 molecule. After the β-sheet, the first α-helix of the HA2 molecule stands up roughly parallel to the threefold symmetry axis of the HA trimer. A large loop underneath the HA1 structural motif that contains the ligand-binding domain links the first α-helix to the second long α-helix that forms a coiled coil with the other two HA2 molecules to form a three-helix bundle toward the membrane envelope. The three-helix bundle is firmly anchored by the transmembrane domain in the membrane envelope. The three-helix bundle is a common structural motif in all the type I fusion proteins of enveloped viruses, including human immunodeficiency virus (HIV) and SARS virus (Colman and Lawrence 2003). The length of the helices may vary from virus to virus, but the three-helix configuration is the same in all viruses with type I fusion proteins. The three-helix bundle coiled coil is the key structural element that keeps the trimeric association of HA molecules throughout its structural changes during the virus entry process. The overall structure of the trimeric HA molecule could be viewed as a crown of the three HA1 molecules that caps the HA2 trimer tied together by the three-helix bundle. The HA1 crown is essential in keeping the HA structure in the metastable conformation prior to membrane fusion. The interactions between the top structural motifs and those between HA1 and HA2 molecules all contribute to the metastability, which is required for storage of energy that will be needed for the structural changes of HA during membrane fusion.

Both HA1 and HA2 proteins are glycosylated at N-linked glycosylation sites, with the HA1 molecule having more sites glycosylated near the top of the HA spike. There are a lot of variations between influenza virus strains in terms of the number of the N-linked sequences, ranging from three to ten sites (Bragstad et al. 2008). The presence of a sequence for N-linked glycosylation is not sufficient for this site being glycosylated. For a given influenza virus strain, the degree of glycosylation may depend on what types of cells the virus infects. For instance, there are five N-linked sequences in HA of subtype H1. When the virus was propagated in human cells, four of five sites were glycosylated (Gamblin et al. 2004). When the virus of an avian origin was propagated in chicken embryos, only three of the five sites were glycosylated (Lin et al. 2009). The degree and location of glycosylation has an effect on the entry of influenza virus. In the first place, proper glycosylation may be required for proper folding of the HA molecule (Segal et al. 1992). For instance, glycosylation at Asn8 and Asn22 after the cleavage of signal peptide helps recruitment of chaperone proteins that facilitate HA folding. In other cases, the glycosylation near the receptor binding site on the top of the HA spike has a direct influence on the affinity of the HA molecule to the virus receptor sialic acid. As shown in the case of influenza virus A/Vietnam/1203/04, a highly pathogenic strain of avian influenza A virus, the HA of this virus recognizes an α2,3-linked sialyl receptor and can spread systematically in mice. Removal of the glycosylation sequence, Asn158, near the receptor binding site, contributed to the increase of HA affinity to an α2,6-linked sialyl receptor, which resulted in reduction in systematic virus spread (Wang et al. 2007). Similar increases of receptor affinities after removal of N-linked glycosylation sites, and subsequent reduction in virus growth and spread, were also observed in other studies involving a H7N1 influenza virus (Wagner et al. 2000).

In addition to glycosylation, formation of disulfide bonds in HA is also important for the structure of HA. There are six strictly conserved disulfide bonds in HA in all subtypes of influenza viruses, one of which is between HA1 and HA2 molecules (Segal et al. 1992). This high conservation indicates that the pattern of disulfide bonds plays a critical role in the structure and function of HA. Studies showed that without proper formation of the disulfide bonds, the folding of HA was greatly impaired. Similarly as glycosylation, disulfide bonds are also additional factors that facilitate proper folding of HA during translation.

Among different HA subtypes, the structural differences are mainly located in three regions. First, the loops on the surface of the spike crown formed by the HA1 molecule are among the most variable structural elements. These variations are probably mostly responsible for the antigenic differences among the HA subtypes, which might not have significant effects on virus entry. The other two regions, however, are probably directly related to the process of HA-mediated virus entry. One of the regions is the receptor binding site. The structure of each HA subtype in this region appears to favor one type or the other α-linked sialic acid, which may be a determinant in host specificity of influenza A viruses. Another region of structural differences is near the fusion peptide. As discussed above, the N-terminal and the C-terminal sequences of the HA1 molecule are part of the hydrophobic pocket that accommodates the fusion peptide at the N terminus of the HA2 molecule. The fusion peptide adopts the same structure among different HA subtypes, but the sequence and structure of the HA molecule surrounding the fusion peptide changes between the HA subtypes. More details about the two regions will be discussed in later sections with regard to receptor binding and membrane fusion.

For influenza B virus, there is no subtype of HA. The single serotype of influenza B virus HA has a similar overall structure as that of influenza A virus HA. The HA1 molecule in influenza B virus also forms a crown structure on top of the trimeric HA2 molecules. However, the exact three-dimensional structure of influenza B virus HA is quite different. For the HA1 molecule, the ligand-binding domain that contains the receptor binding site has a conformation that requires a significant rotation up to 60° in order to be superimposed with that in the HA1 molecule of influenza A virus HA (Wang et al. 2007). Such a large degree of rotation may reflect two aspects of the HA molecule. First, the crown-like structure formed by HA1 molecules must disassemble during HA-mediated fusion. One of the requirements is that the structure of the HA1 molecule must have a hinge region that allows the rotation of the structural motif at the top of the spike in order to disassemble from the metastable trimeric conformation. It is therefore an intrinsic property in HA that the structural motif at the top of the spike is allowed to rotate without any blockage during HA-mediated membrane fusion. It is not surprising that the structural motif at the top of the spike of the influenza B virus HA adopts a conformation that is, although different from that in influenza A virus HA, within the spatial range of the HA structure.

Influenza C virus is also a virus that is only found in humans, and it has a surface glycoprotein that is quite different from HA in other influenza viruses. This surface glycoprotein in influenza C virus is called the hemagglutinin-esterase-fusion (HEF) protein. In addition to the ability to bind the host receptor molecule and to mediate membrane fusion as influenza virus HA, HEF has an additional enzymatic activity that destroys the HEF receptor molecule, 9-*O*-acetylated sialic acid, by enzymatic removal of the acetyl group. Since the receptor destroying enzyme (RDE) and the receptor-binding activities are both in HEF, influenza C virus does not therefore have any NA activity or the gene. The overall three-dimensional structure of HEF is very similar to that of HA when the two structures are superimposed, except that there are additional structural motifs in HEF corresponding to the esterase (Rosenthal et al. 1998). The detailed structures are, however, very different between HEF and HA. HEF is also cleaved into HEF1 and HEF2 molecules in the mature influenza C virion. The N terminus of HEF1 participates in the similar β-sheet at the anchoring region, but the sequence following that does not have a structural element that forms part of the hydrophobic pocket accommodating the fusion peptide. Instead, the polypeptide directly extends to the structural motif at the top of the spike. Comparing to HA, there is one structural insertion prior to the ligand-binding domain, and one after, which are two of the three segments in the esterase. It appears that HA deleted two of the three segments in the esterase during evolution. The remaining structural element in HA helps to keep the ligand-binding domain at the same position as in HEF. The esterase in HEF is located below the receptor binding site. The C terminus of HEF1 is closely located near the N terminus of HEF1 that contains six amino acids in front of the fusion peptide. The fusion peptide in HEF is therefore not accommodated in a hydrophobic pocket as in HA. The location of the fusion peptide in HEF cannot be superimposed with that in HA. Another significant difference is in the loop that links the two long α-helices in the HEF2 molecule. This loop is much shorter in HEF, as well as the second α-helix that forms the coiled coil in the trimer. This difference makes HEF look hollow underneath the crown formed by the HEF1 trimer. The significance of this structural difference is not yet clear.

9.3 Receptor Binding Site

As many animal viruses, influenza virus attaches to the host cell surface by recognizing a specific receptor. The commonly known receptor molecule for influenza virus is a sialic acid (also known as *N*-acetylneuraminic acid) that is covalently linked to the end of oligosaccharides on glycoproteins or glycolipids (Schauer 2009). Sialic acid is present on the surface of many cell types and also on cell-secreted products. Sialylation is an important mechanism of masking cells or microorganisms, as well

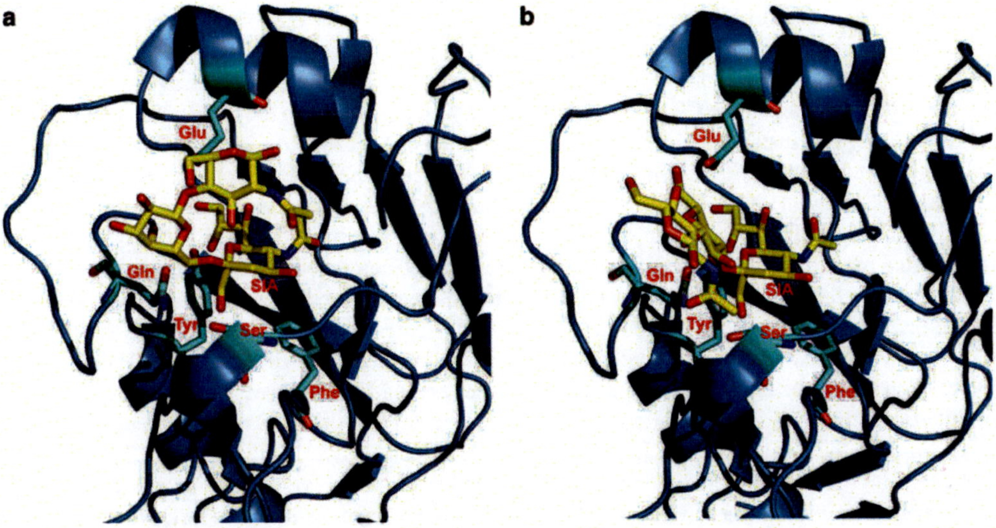

Fig. 9.3 The receptor binding site of an avian influenza virus hemagglutinin (Liu et al. 2009). (**a**) LSTc (α2,6-linked galactose) mimicking the human receptor bound in the binding site. (**b**) LSTa (α2,3-linked galactose) mimicking the avian receptor bound in the binding site. Critical residues for binding the receptor are presented as stick models with labels

as products they produce, to avoid being recognized by the host immune system. At the same time, the sialyl moiety is often used as ligands for attachment proteins that have important cellular functions, or as receptors (co-receptors) for microorganisms including viruses, bacteria, fungi, and parasites. Influenza virus is one of the best known example that uses sialic acid as the receptor for cell entry.

The sialic acid binding site of influenza virus is located in each subunit of the trimeric HA (Fig. 9.3). Since there are many copies of HA on each influenza virus particle, the attachment of influenza virus to the host cell surface may be considered as a multivalent binding event. As a result, the viral attachment is fairly tight even though the affinity of each individual sialyl moiety to its binding site on HA may be weak (Takemoto et al. 1996). The estimated dissociation constant (K_d) for sialic acid binding by influenza virus HA is about 3 mM (Hanson et al. 1992; Sauter et al. 1989). The precise interactions between sialic acid and HA have been described by cocrystal structures (Lin et al. 2009; Takemoto et al. 1996; Ibricevic et al. 2006; Tumpey et al. 2007). The binding site is a depression on the surface of a structural domain at the top of the HA trimer. The structural domain is composed of a β-barrel motif and α-helices. This structural motif is similar to other lectin proteins in shape, but topologically different. At the bottom of the receptor binding site there are two aromatic residues, a tyrosine (Tyr98) and a tryptophan (Trp153), in influenza A virus (A/Aichi/2/1968 H3N2). The pyranose ring of sialic acid sits on top of the aromatic residues. The four different functional groups linked to the pyranose ring form specific interactions with residues surrounding the depression. The carboxyl group in its α-configuration at the C2 position of the pyranose forms three hydrogen bonds with the side chain of Gln226 and main-chain atoms. The hydroxyl group at the C4 position does not seem to be involved in binding to HA. The acetylamido group at the C5 position, on the other hand, forms both hydrophobic and hydrophilic interactions. The amide group in the acetylamido group forms a hydrogen bond with a main-chain carbonyl oxygen atom, whereas the methyl group in the acetylamido group is within van der Waals contact with the side chain of Trp153. Finally, the last two hydroxyl groups in the glycerol moiety at the C6 position form hydrogen bonds with the side chains of Gln226 and Glu190, and a main-chain carbonyl oxygen atom. The network of these interactions renders the specificity for sialic acid recognition by HA. However, there are variations in the sialic acid binding site among different subtypes of influenza virus. For instance (Ha et al. 2001), residue Leu226 is replaced by Gln226 in an avian influenza A virus H5 HA. The side chain of Gln226 in H5 appears to form a hydrogen bond with one of the carboxylate oxygen atoms and more importantly, a hydrogen bond with the oxygen atom at the C2 position that is in the glycosidic bond linking the sialic acid moiety to the oligosaccharides. The hydrogen bond formed by the hydroxyl group at the C9 position is through a water molecule to the main-chain amide group of residue Gly228, instead of directly with the side chain of Ser228 in the H3 HA. In an H9 HA, the hydrogen bond formed with the water molecule as observed in the H5 HA is conserved, but that with the side chain of Glu190 as observed in both H3 and H5 HA is lost because this residue is a valine (Val190) in the H9 HA. In addition, the left-side loop in which residues Leu/Gln226 and Ser/Gly228 are located has a more opened conformation in H3 and H9, compared to that in H5.

Changes in the sialic acid binding site are related to the receptor preference of different subtypes of influenza virus HA. Currently, human influenza A viruses contain H1, H2, and H3 subtypes of influenza virus HA, whereas the avian and other animal influenza A viruses contain H1–H16 subtypes (Yen and Webster 2009). One of the differences among HA subtypes is that the HA of human influenza A viruses prefers a sialic acid that has an α2,6 glycosidic linkage to glycoproteins and glycolipids as the receptor molecule, but HA of avian influenza A viruses prefers an α2,3 glycosidic linkage. The receptor preference may be one of the barriers to prevent the wide spread of highly pathogenic avian influenza A viruses in humans to cause a severe pandemic (Ibricevic et al. 2006; Stevens et al. 2006; Yamada et al. 2006). Sialic acids with an α2,6 glycosidic linkage are abundant on the human airway epithelial cells (Stray et al. 2000; Thompson et al. 2006). It has been postulated that an avian influenza A virus must adapt to the human sialic acid receptor in order to transmit widely from humans to humans. The binding of HA to different receptor molecules may explain the difference in receptor preference. The avian H5 HA has a glutamine residue at position 226. In the crystal structure of H5 in complex with an α2,3 glycosidic linkage to a galactose (Ha et al. 2001), the glycosidic bond has a *trans* conformation that allows the side chain of Gln226 to make hydrogen bonds with both the glycosidic oxygen and the hydroxyl group at the C4 position of the galactose. The receptor binding site in H5 HA is also narrower so that it can only accommodate a sialic acid with the α2,3 glycosidic linkage in *trans* conformation. In the case of the avian H9 HA, a sialic acid with the α2,3 glycosidic linkage or the α2,6 glycosidic linkage may fit the receptor binding site, and the interactions with sialic acid by H9 HA are similar in both cases. However, the H9 HA can form hydrogen bonds with sugar moieties as the fourth or fifth molecule in oligosaccharides. The oligosaccharide chain adopts different conformation when bound to each individual HA even though the terminal sialic acid is bound in the same position.

In influenza B virus, the sialic acid binding site in its HA appears to be very different. First of all, the left-side loop homologous to that in influenza A virus HA is substantially more closed so the sialic acid moiety is no longer above a depressed site. The carboxylate group forms hydrogen bonds similar to the main-chain amide nitrogen and the side chain of a serine as in influenza A virus HA. However, the acetylamido group now forms hydrogen bonds with the side chain of Thr139, and the hydroxyl group at the C9 position forms hydrogen bonds with the side chain of Asp193 and Ser240. The narrower sialic acid binding site seems to be responsible for discriminating against the avian α2,3 glycosidic linkage by reducing two hydrogen bonds. Influenza B virus is found only circulating in humans.

In influenza C virus, the receptor binding site is very different. In fact, the receptor molecule is no longer the sialic acid, but 9-*O*-acetylsialic acid. The surface glycoprotein in influenza C virus is actually a HEF protein (Rosenthal et al. 1998). By comparison with the structure of HA, there is an inserted esterase with structural motifs flanking the receptor binding domain. The esterase may perform the receptor destroying function of NA, which is missing in influenza C virus. The main difference in the receptor binding site of the HEF protein is that the left-side loop is further opened, and the 9-*O*-acetyl group as well as the hydroxyl groups at the C8 and C7 positions all participate in forming hydrogen bonds with side chains of three different residues. The acetylamido group at the C5 position also forms new interactions with side chains of two different residues.

The receptor preference of a sialic acid with the α2,6 glycosidic linkage may be one of the requirements for influenza virus to infect the human airway efficiently, but may not be the only requirement. The 1918 pandemic influenza A virus has been shown to be originated from an avian influenza A virus (Compans et al. 1970; Lin et al. 2009; Nicholls et al. 2008; Shen et al. 2009; Hayden 2009). Its receptor preference is the α 2,6-linked sialic acid. As shown in (Liu et al. 1995), the change of one amino acid from aspartic acid to glutamate at position 190 would switch the receptor preference from the α2,6 glycosidic linkage to α2,3 glycosidic linkage. An additional change from aspartic acid to glycine at position 225 would diminish the transmission of the pandemic virus between model animals such as ferrets (Tumpey et al. 2007). However, the influenza virus carrying HA that has the α2,3 glycosidic linkage preference could cause the same lethal infection in mice as the 1918 pandemic influenza virus (Qi et al. 2009), suggesting that viral factors in addition to HA also contribute to the pandemic virulence of the 1918 influenza virus. For instance, the polymerase subunit gene, PB2, from the 1918 influenza virus was also required for transmission between ferrets (Lakadamyali et al. 2003). The NS1 gene was also shown to be an important virulence factor (Konig et al. 2010). The adaptation of the avian influenza viruses may begin with the avian influenza virus strain that could bind both types of receptors (Matlin et al. 1981). Additional adaptation may be required in other influenza virus genes in order for an influenza virus strain to be transmitted efficiently among humans or to increase virulence. This adaptation may take place in animal hosts, such as pigs or birds, or humans. The most threatening pandemic influenza virus strain is likely to have the receptor preference for the α2,6 glycosidic linkage and contain highly virulent genes in addition to HA.

The exact role of sialic acid binding by HA, however, may not be as clear cut as most literature described (Nicholls et al. 2008). First of all, avian H5N1 influenza viruses that prefer the α2,3 glycosidic linkage infect humans frequently since 1997, even though the transmission from humans to humans has not been confirmed for this type of influenza virus. Furthermore, it has been shown by a number of laboratories that cells could still be infected by influenza virus after the sialic acid was removed by sialidase treatment (Kogure et al. 2006; Stray et al. 2000; Thompson et al. 2006). One of the possible explanations is that the entry of influenza virus into the host cell employs a sialic acid independent mechanism, such as an alternative primary receptor molecule. The binding to sialic acid by HA may be a pre-entry step to attach the virus particle to the host cell surface. The properties of binding to sialic acid with a specific glycosidic linkage appear to be associated more closely with production of infectious progeny viruses and the ability of transmission from one host to another by influenza virus (Qi et al. 2009; Giannecchini et al. 2007; Nunes-Correia et al. 2004). This suggests the potential that a novel influenza virus variant may not need to bind sialic acid at all to become transmissible among humans.

Amino acid sequence changes in HA happen frequently because the changes allow influenza virus to escape host immune responses. However, these changes must not compromise the entry function carried by HA. It has been observed that when influenza virus was passaged in mice that had been immunized with inactivated wild-type virus, the selected escape mutants also exhibited increased cellular receptor binding avidity by mutated HA (Rust et al. 2004). Mutation sites could be near the sialic acid binding site or not close at all. Some escape mutations did not diminish polyclonal antibody binding, but simply increased virus binding avidity for cellular glycan receptors. These mutant viruses could still effectively agglutinate erythrocytes treated with *Vibrio cholerae* NA RDE that removes terminal sialic acids. The increase in receptor avidity by HA was toward sialyl glycan molecules with both the α2,3 and α2,6 glycosidic linkages. When the escape mutants were passaged in naïve mice, the receptor avidity was restored to the normal level, but the resistance to polyclonal antibody neutralization was retained. However, additional mutations were also identified in some cases that render the new mutants further resistance to polyclonal antibody neutralization. These studies suggest that under neutralizing immune pressure, influenza virus increases its receptor avidity by changing amino acid sequences throughout the HA globular domain. Additional amino acid changes may accumulate in naïve hosts in which high receptor avidity is not required for influenza virus infection. These further mutations compound on antibody resistance. Indeed, the sequence data of isolates from human and animal hosts since 1918 revealed a close relationship between receptor binding and escaping host immune response (Shen et al. 2009).

Currently, the popular antiviral drugs for therapeutic treatment of influenza virus infection are NA inhibitors (NAIs) (Hayden 2009). It is inevitable that drug resistance variants would arise after a period of time, some of which may have mutations in HA. The key function of NA is to remove sialic acid moieties from oligosaccharides attached to HA on newly assembled influenza virus particles when they bud out of the infected cells. If the NA activity is not present, the progeny virus particles form aggregates on the cell surface because the sialylated HA binds to each other between two influenza virus particles, and to cell surface (Liu et al. 1995). The aggregation prevents virus release, which is the mechanism used by NAIs to stop influenza virus infection. When influenza virus was passaged in cell culture in the presence of NAIs mutations in HA were identified in some drug-resistant mutants, in addition to drug-resistant mutants that had mutations in NA (Giannecchini et al. 2007; Hurt et al. 2009). It is obvious that influenza virus can replicate in the presence of NAIs if the sialic acid binding properties of HA have been changed. NAI-resistant mutants with mutations in HA have not been reported from clinical isolates, but the potential still exists.

9.4 Entry Pathway

After the virus is attached to the host cell surface via binding to sialic acid by HA, the process of importing the incoming virus particle to a location inside the cell begins. There is a large number of HA molecules on influenza virus particles, which may be required for adequate binding to the cell surface to trigger the internalization process. The entry process follows a specific pathway with a number of cellular factors involved (Fig. 9.4). At the cell plasma membrane, the influenza virus is taken up by endocytosis (Bottcher et al. 1999; Daniels et al. 1985). The virus particle quickly enters coated vesicles (endosomes) inside the cell. The internalized virus particles are transported inside the endosomes to locations near the nucleus (Lakadamyali et al. 2003). The fusion of viral mem-

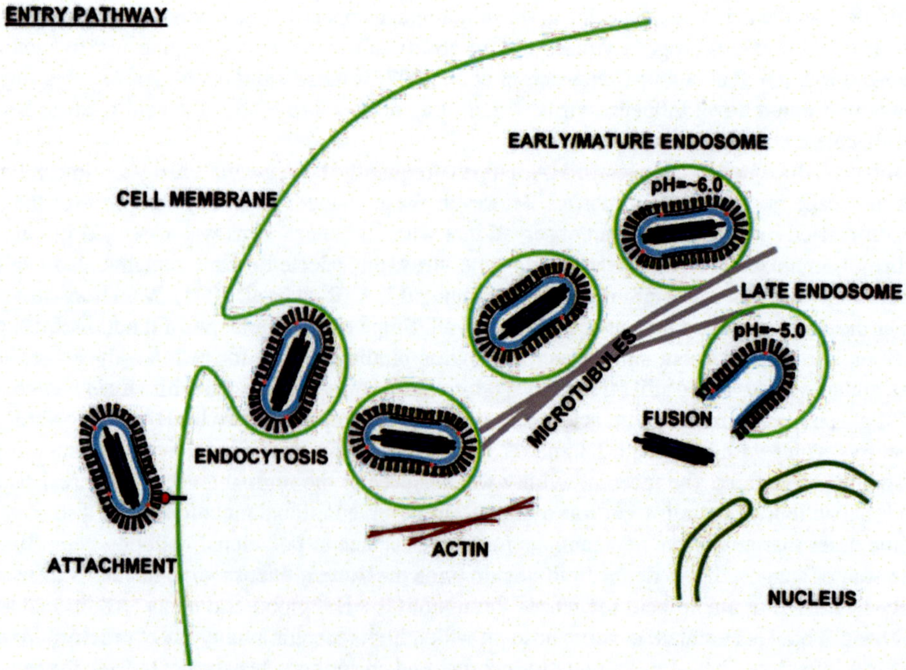

Fig. 9.4 A model for the entry pathway of influenza virus

branes with endosomes is triggered by acidification of late endosomes. The result of fusion is the release of viral RNA-dependent RNA polymerase-associated nucleocapsid (RNP) into the nucleus where viral transcription is initiated.

Studies by Lakadamyali et al. show that the trafficking pathway of influenza virus particles to the nucleus has three distinct stages (Lakadamyali et al. 2003). In Stage I, the endocytic vesicles are transported in an actin-dependent manner within the cytoplasm leading to the early endosomes. In Stage II, the early endosomes containing virus particles are rapidly transported by dynein-directed movement on microtubules toward the perinuclear region. At this location, the first acidification (pH = ≈6.0) takes place and the endosomes mature. The maturing endosomes continues to move on microtubules in Stage III. Finally, the second acidification (pH ≈ 5.0) of matured endosomes occurs and HA-mediated membrane fusion results in release of RNP into the nucleus.

The description of the entry pathway clearly indicates that the entry of influenza virus is a process directed by cellular endocytosis and transportation mechanisms. These functions are essential for the cell life cycle. Influenza virus utilizes some of the endocytosis and transportation mechanisms to support its own entry to infect the host cell. By systematic RNAi knockdown of host factors, up to 23 proteins have been shown to be related to the entry of influenza virus (Konig et al. 2010). One of the most obvious protein groups is the host proteins that form the endosomal coat protein complex (Chen and Zhuang 2008; Doxsey et al. 1987; Matlin et al. 1981; Roy et al. 2000). The clathrin-coated pits are formed around the attached virus particle once it is attached and the particle is then internalized. A number of other host factors are also involved in the initial internalization. Dynamin, a GTPase that play an important role in clathrin-mediated endocytosis, was shown to facilitate formation of virus-loaded coated vesicles (Roy et al. 2000). In the absence of dynamin's functions, the pinch off of the coated pits was incomplete. Epsin 1 is also a host factor that has been identified to be involved in influenza virus endocytosis via coated pits (Chen and Zhuang 2008). Epsin 1 is an adaptor protein that interacts with proteins such as clathrin, AP-2, and Eps15, all of which play important roles in assembly of the coated pits. Association of epsin 1 induces membrane curvature. On the other hand, influenza virus could also be endocytosed by other clathrin-independent pathways, even though clathrin-dependent endocytosis appeared to be the main route used by influenza virus entry (Nunes-Correia et al. 2004; Rust et al. 2004; Sieczkarski and Whittaker 2002). Analogous to using multiple receptor molecules, endocytosis of influenza virus may also utilize multiple routes.

Trafficking inside the host cell begins after the internalization of the influenza virus particle. Actin was shown to be required for the initial movement of the virus particle in polarized epithelial cells (Daniels et al. 1985; Bullough et al. 1994; Floyd et al. 2008; Kemble et al. 1994). When cytochalasin D, an inhibitor that disrupts actin polymerization, was added, influenza virus entry into the apical phase of polarized epithelial cells was inhibited. This inhibitor, however, has no effects on influenza virus entry into non-polarized cells or through the basolateral phase of polarized epithelial cells. Inhibition of myosin VI function, an actin motor protein, also decreased influenza virus infection. The coated vesicles containing influenza virus were arrested at the plasma membrane. These observations suggest that an active actin skeleton is required for directing influenza virus entry, which transports the coated vesicles to microtubules where the early endosomes are found (Lakadamyali et al. 2003). The movement of the virus-containing early endosome is rapid in Stage II before the first acidification (pH ≈ 6.0) occurs. The host cell protein sorting system appears to be selectively required for directional movement of early endosomes to late endosomes along microtubules. The virus-containing endosomes were shown to be colocalized with Rab9 and VPS4, suggesting that the ubiquitin/vacuolar protein sorting pathway was utilized by influenza virus (Khor et al. 2003). The lysosomal pathway for recycling could not be used for influenza virus infection. This specificity may be related to the functional requirement that virus-containing endosomes need to traffic through the early endosomes to the late endosomes along microtubules before the final membrane fusion takes place near the nucleus (Lakadamyali et al. 2003; Sieczkarski and Whittaker 2003). Before the final membrane fusion, the integrity of HA is not grossly affected during trafficking from endocytosis to the late endosomes through a directed, specific pathway to near the nucleus (Sieczkarski and Whittaker 2003). Since HA is the major surface glycoprotein that makes contact with the endosome membrane, it may very well play a role in selecting the specific trafficking pathway. This specificity may be related to recruitment of specific host proteins by HA or the unique envelope structure organized by HA.

9.5 Low pH-Induced Conformational Change in HA

In the late endosome, the HA mediates the fusion of the viral membrane with the endosome when the pH inside the endosome drops before 5.0. HA undergoes an extended conformational change at the acidic pH to insert the fusion peptide into the target membrane (Fig. 9.5). This is a common process used by many viral fusion proteins to mediate membrane fusion for viral entry. What has been learnt for influenza virus HA may have a broad application.

As discussed above, the native structure of HA is stabilized by trimeric association of HA1 and its interactions with HA2. Conformational changes of HA as well as membrane fusion could be prevented by introducing additional disulfide bonds at the top to stabilize the trimeric HA1 (Wharton et al. 1986), suggesting that the conformational change of the trimeric HA1 induced by low pH is required for HA-mediated membrane fusion. Upon binding to sialic acid, a step before the virus is exposed to low pH in the entry pathway, the HA1 trimer is stabilized and the conformational change of HA at low pH is actually restricted by sialic acid binding (Leikina et al. 2000). Since the virus particle is inside the endosome after endocytosis, there are multivalent interactions between trimeric

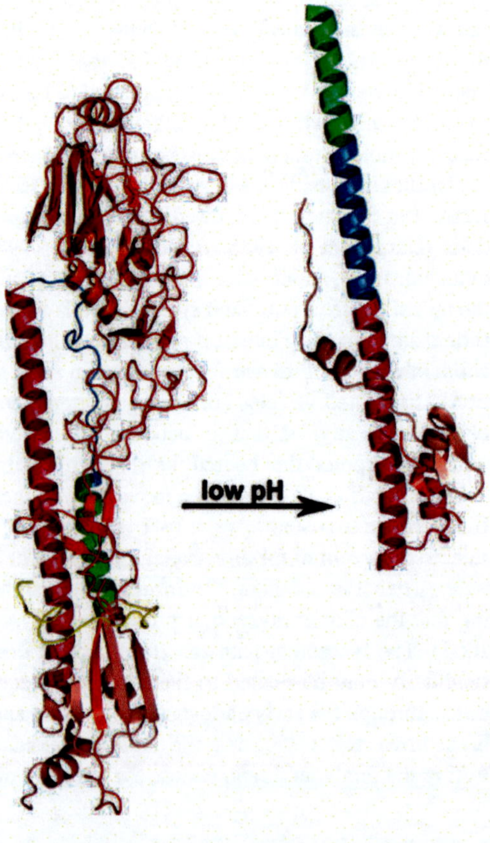

Fig. 9.5 A ribbon drawing to compare the metastable conformation of a HA subunit (*left*) with the refolded HA subunit at low pH. HA1 is colored *red*, and HA2, *lighter red*. The fusion peptide is colored *yellow*. The helix-loop-helix region in HA2 is colored *green*, *blue*, and *magenta*, for the conformation at neural and low pH. The *magenta* helix region at low pH is leveled with that in the metastable HA

HA and the receptor molecules on the endosome membranes. In such a structure, the fusion peptide at the N terminus of HA2 may be released when pH is lowered prior to HA refolding. In order to allow full refolding of HA2, it may be necessary for a step to dissociate HA from binding to sialic acid on the endosome membrane. Lowering pH may actually be a mechanism for such dissociation. Further downstream of the refolding pathway, there may be stepwise conformational changes that proceed sequentially at low pH. Before protonation of HA surface residues when the pH is lowered, the crown formed by the HA1 trimer has a positively charged surface both inside and outside (Huang et al. 2002). On the contrary, the HA2 trimer in its metastable conformation has a negatively charged surface. This negatively charged surface of the HA2 trimer electrostatically matches the positively charged interior surface of the trimeric HA1 crown. These attractive interactions allow HA to maintain its metastable conformation. When the pH is lowered, some surface residues in HA1 become more positively charged, such as residues His75, Lys259, and Lys299 (Huang et al. 2003). These extra charges increase the force of repulsion among HA1 subunits so the association within the HA1 trimer is weakened. Indeed, an obvious expansion of the HA1 trimer was observed at pH 4.9 by cryo-EM imaging (Bottcher et al. 1999). Furthermore, the structure corresponding to HA2 also appeared to move away from the native position, leaving an appearance of a channel through the center of the HA trimer. This may suggest that there is also significant destabilization of the HA2 trimer at low pH before the extended refolding. Mutagenesis studies support this notion (Daniels et al. 1985). Mutations that rearrange the interface of the HA2 molecule resulted in a higher pH needed for triggering HA2 refolding. These mutations are located in the helix that forms a coiled coil to hold the tree HA2 subunits together. This is consistent with the suggestion that there is a threshold of structural destabilization of the HA trimer before refolding. Protonation of surface residues, especially those on HA1, induces structural changes that begin to destabilize the HA trimer. The loosen-up of the HA trimer may allow more residues to be protonated until the complete HA2 refolding.

The refolding of HA2 under low pH is a remarkable structural phenomenon observed for any protein structure (Bullough et al. 1994). In the metastable native structure of HA, the central region of HA2 forms a helix-loop-helix structure, with residues 40–55 corresponding to the first helix; residues 56–74, the loop; and residues 75–125, the second helix. The second helix forms a coiled coil in HA2 trimer. The fusion peptide, the first 20 residues of HA2, is in front of the first helix, but oriented near the viral membrane envelope. At acidic pH, residues 40–105, which cover the first helix, the loop, and the first half of second helix in the native helix-loop-helix motif, become a continuous helix of about 100 Å in length. This extended refolding will bring the fusion peptide from its location near the viral envelope to the top of the refolded HA2 trimer, allowing it to reach the target endosome membrane. This may be the most important event during HA-mediated membrane fusion. Many viral fusion proteins have similar structural refolding to connect the viral membrane with the target membrane in the host cell. The second half of the second helix in the native helix-loop-helix motif undergoes a different conformational change. Residues 106–112 switch from a helical structure to an extended loop, and residues 113–125, although still exiting as a helix, has a 180° reorientation from its original orientation. This new orientation was suggested to represent the final collapsed "hairpin" structure of HA2 to bring the two merging membranes close to each other.

In addition to triggering the extended refolding of HA2, low pH also induces other critical structural changes that affect the fusion process. The release of the fusion peptide from the native pocket near the viral envelope is pH related and could influence the HA2 refolding. For instance, the carboxylate side chains of Asp109 and Asp112 form hydrogen bonds with residues 2–6 of the fusion peptide. Mutating either Asp109 or Asp112 to Ala increased the pH required for HA-mediated membrane fusion, suggesting that deionization of Asp109 and Asp112 at low pH could contribute to destabilization of HA (Heider et al. 1985). Biochemical analyses conformed the extended refolding of mutant HA at the elevated pH. Other residues in the vicinity of the fusion peptide binding pocket have similar effects when mutated (Heider et al. 1985; Ghendon et al. 1986). The fusion peptide appears to play a major role in stabilizing the metastable conformation of HA trimer and participates in low-pH triggering of HA2 refolding.

In addition to structural changes of HA, low pH also induces structural changes in the matrix protein (M1) that may be required for completion of membrane fusion and RNP release. It was proposed that the matrix protein, which constitutes a protein layer beneath the viral membrane envelope to anchor HA and NA (Harris et al. 2006), is the last barrier that was broken at low pH in the fusion process (Lee 2010). Furthermore, low pH dissociates the matrix protein from RNP to unveil the nuclear import mechanism (Bui et al. 1996). The synchronized structural changes induced by low pH in the endosomes make up the process to complete the ultimate task to release RNP into the nucleus for initiation of virus replication.

9.6 HA-Mediated Membrane Fusion

The low pH-triggered conformational refolding of HA projects the fusion peptide to the target endosomal membrane. This is the first step in HA-mediated membrane fusion (Fig. 9.6). The fusion peptide is generated after the HA precursor, HA0, is cleaved into HA1 and HA2 during influenza virus assembly. Before the cleavage, an extended loop consisting of residues 323–329 in HA1 and 1–12 in HA2 is exposed on the outer surface of the HA trimer (Chen et al. 1998). Residue 319 is usually a positively charged residue that allows the site to be cleaved by a trypsin-like host protease. Residues preceding 329, usually including a number of positively charged residues, would enlarge the loop to facilitate protease cleavage as found in case of highly pathogenic avian influenza viruses (Nicholls et al. 2008). The polybasic residues that precede the cleavage site are postulated to allow a broad spectrum of host proteases to carry out the cleavage so that the virus gains increased virulence. However, the loop did not have an exposed conformation in HA of the 1918 H1N1 virus, a highly pathogenic human influenza virus (Luo et al. 1996). Other structural features in the pocket

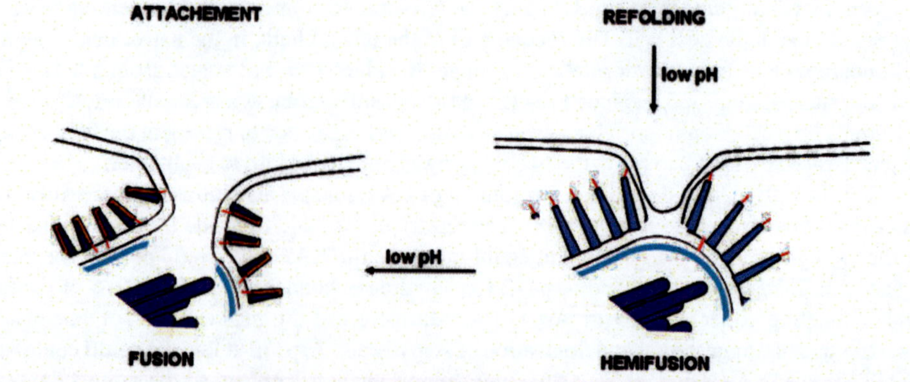

Fig. 9.6 A model for the membrane fusion process. Possible steps are attachment of the fusion peptide at initial refolding of HA2, hemifusion mediated by the fusion peptide and HA2, completion of membrane fusion with final refolding of HA2

where the fusion peptide is tucked in after cleavage also contribute to the pathogenicity of influenza virus. Normally, the side chains of ionizable residues in the pocket provide a number of hydrogen bond interactions with main chain amides of the fusion peptide to stabilize it in the pocket (Wilson et al. 1981). In highly pathogenic strains, the structural changes near the pocket may allow the fusion peptide to be released from the pocket more easily.

The released fusion peptide is capable of penetrating into the target membrane, especially when it is lifted by the refolded HA. To fuse the two membranes, the fusion peptide first catalyses the hemifusion, mixing the outer leaflets of the two lipid bilayers, followed by formation of a fusion pore where both leaflets are fused. The fusion peptide forms an α-helix and inserts into the lipid bilayer at an angle (Lear and DeGrado 1987; Luneberg et al. 1995). The fusion peptide helix in the lipid bilayer has a kink that divides the structure into two halves (Yoshimoto et al. 1999; Bodian et al. 1993). At the fusion pH of 5.0, residues 2–10 form an α-helix in the first half and residues 13–18 form a 3_{10}-helix in the second half. The two helices make almost a right angle. Such a tight structure allows the fusion peptide to insert deeper in the lipid bilayer with charged side chains in the fusion peptide left out of the membrane (Bodian et al. 1993). Insertion of the tightly folded fusion peptide likely induces membrane perturbation to facilitate lipid exchanges between juxtaposed membranes. Studies showed that the insertion of the fusion peptide into the lipid bilayer is reversible at neutral pH at which the fusion peptide could not form tightly folded two helices (Chang and Cheng 2006). Only at low pH, insertion of the fusion peptide would lead to the following steps toward irreversible membrane fusion to form a functional fusion pore. Prior to hemifusion of the lipid bilayer, the fusion peptide undergoes a process involving diffusion in the lipid bilayer in order to become self-associated, which may be a rate-limiting step and require the participation of multiple copies of HA (Chernomordik et al. 1998; Floyd et al. 2008).

In the in vitro experiments, the fusion peptide alone could facilitate the complete fusion of two membranes. However, the fusion peptide is presumably covalently linked to the refolded HA2 at the fusogenic pH during influenza virus infection. At this stage, it is conceivable that the viral and endosomal membranes are still apart, but hemifusion could already occur (Chernomordik et al. 1998; Kemble et al. 1994). Studies suggest that both the fusion peptide and the exposed part of HA2 after low pH refolding contribute to hemifusion under low pH (Leikina et al. 2001). First of all, the polypeptide following the fusion peptide is responsible for the formation of HA2 trimer. Fusion activities are dependent upon effective concentrations of HA. The contribution of HA2 trimer to fusion can at least in part attribute to concentrating the fusion peptide. Second, the low pH conformation of HA2 prior to the final refolded hairpin structure is required for catalysis of hemifusion. In fact, the whole HA1–HA2 ectodomain, released from the influenza virus particle by bromelain treatment, could not induce lipid mixing or membrane fusion (Leikina et al. 2001; Wharton et al. 1986; White et al. 1982). In addition, deletion of residues 91–127 or mutations in this region also diminished its activities to induce hemifusion (Leikina et al. 2001). These observations suggest that the entire fusogenic HA2 molecule plays a role to induce hemifusion in addition to the fusion peptide. Furthermore, it has been found that small pores are induced by insertion of the fusion peptide. They flicker at this stage, i.e., they are open or closed reversibly. The HA2 molecules could aggregate membrane-inserted fusion peptides to enlarge the pores, but the system did not proceed to complete fusion in the absence of the viral membrane. This leads to a hypothetical model that HA induces hemifusion and large pores in the target endosomal membrane before final refolding (Bonnafous and Stegmann 2000).

Expansion of the initial fusion pore leads to complete fusion of the two membranes and release of the contents in the viral particle. This last step requires the participation of the transmembrane domain in HA. The main function of the transmembrane domain is to package and anchor HA in the viral membrane envelope. In addition, studies showed that the transmembrane domain also participates in the final step of membrane fusion (Kemble et al. 1994). If HA is anchored to a cell membrane through GI-linked lipids, it may induce hemifusion with other cell membranes, but not complete fusion. This shows that only a full-length HA with the transmembrane domain could function to mediate membrane fusion. The transmembrane domain of HA could be replaced by a different transmembrane domain derived from another virus, and the complete fusion could still be induced by the chimeric HA. This suggests that the role of transmembrane domain is more biophysical in membrane fusion. The specific amino acid sequence of the transmembrane domain is not required for membrane fusion as long as its biophysical properties are suitable for this process. It was shown that the transmembrane domain forms helices that can aggregate in lipids and increases the order of acyl chains in lipid bilayers, which may be required for both targeting HA to the lipid "raft" during assembly and complete fusion (Tatulian and Tamm 2000). Mutations in the transmembrane domain not only reduced the copy number of HA in virus particles, but also the efficiency of HA-mediated membrane fusion (Takeda et al. 2003). Changes of the amino acid sequence altered the interactions of the transmembrane domain with lipid bilayer in terms of biophysical interactions, which is the reason that the functions of HA were affected in both the anchoring and membrane fusion. The transmembrane domain forms tight oligomers in the lipid bilayer, and also insert deeper into the lipid bilayer at low pH (Chang et al. 2008). However, the transmembrane domain itself could not disturb membrane structure or induce hemifusion. Its contribution to membrane fusion must be in association with the fusion peptide. As the result of refolding, the fusion peptide and the transmembrane domain are brought close to each other. The residues at the N terminus of HA2 in the refolded structure form a cap connecting the ends of the three helices to close off the helices in the refolded HA2 trimer (Chen et al. 1999). The residues preceding the transmembrane domain are also structured next to the cap. The region between the cap and the fusion peptide, as well as the region between the end of the structured C-terminal part of HA2 and the transmembrane domain, appear to be very flexible, which allows the fusion peptide and the transmembrane domain to assume their conformation independent of the rigid HA2 trimer. Evidence suggested that the fusion peptide could associate with the oligomerized transmembrane domain in lipid bilayer (Chang et al. 2008). The association of the fusion peptide with the transmembrane domain oligomer may be one mechanism that the transmembrane domain facilitates fusion pore expansion and complete fusion since the transmembrane domain can insert deeper in the lipid bilayer at low pH, which brings the fusion peptide further into the lipid bilayer because of the association between the two sequences.

The complete process of HA-mediated membrane fusion was captured as snapshots using EM cryo-tomography (Lee 2010). Liposomes were incubated with influenza virus particles at different pH for different time periods. Images showed that the fusion peptide was first released and inserted into the target liposomal membrane as the pH was lowered. The insertion of the fusion peptide from multiple copies of HA glycoproteins on the virus surface deformed the target membrane, but the membrane structure of the virus particle was still intact at this stage. Following the initial insertion of the fusion peptide, formation of pores was observed in the target membrane and the refolding of HA brought the target membrane in touch with the viral membrane. Opening in the membranes was observed while the fusion process was in progress. However, a protection layer was still present to protect the content of the virus particle. This layer was identified as the matrix protein layer, which gave way as the fusion process went further at low pH. The layer of the matrix protein provided not only the last protection of viral contents, but also a platform for HA to anchor on during the initial deformation of the target membrane and subsequent joining of the two membrane parties for fusion. This may be another reason that the transmembrane domain of HA is required for complete fusion because the anchoring of HA on the layer of the matrix protein underneath the viral membrane envelope is by interactions of the transmembrane domain with the matrix protein. The final stage would require expansion of pores at low pH and release of the viral contents. The EM cryo-tomography snapshots could only resolve certain steps in the membrane fusion process. The limitation of the image resolution does not allow the fine details of membrane structures to be fully revealed. The exact details of lipid exchange, hemifusion, and fusion pores may only be found out with images of higher resolution.

9.7 Entry Inhibitors as Antiviral Drugs

Since entry of influenza virus is an essential step in its replication cycle, inhibitors that block this step could be effective antiviral drugs. An entry inhibitor of HIV, Enfuvirtide, is a drug currently used for treatment of HIV infection (Kilby et al. 1998). There is no reason why entry inhibitors of influenza virus would not be effective antiviral drugs. In light of emergence of mutant viruses resistant to current drugs, such as NAIs, the need for novel antiviral drugs is obvious for treatment of influenza virus infection, especially as combination therapy.

The efforts to develop effective entry inhibitors of influenza virus have been attempted previously by a number of groups. One of the early reports is on triperiden (Fig. 9.7) that was shown to inhibit influenza virus replication at a concentration of 20 μg/ml (Heider et al. 1985). This compound was later shown to inhibit hemolysis of red blood cells and the sensitivity of HA1 to trypsin after low pH treatment of HA (Ghendon et al. 1986). Reassortment of the drug-sensitive strain with a strain that was not sensitive confirmed that the gene sensitive to triperiden was HA. Triperiden-resistant mutations were mapped to HA as well (Prosch et al. 1988). In a more recent report, it was shown that inhibition of influenza virus replication by triperiden may be due to its ability to lower the internal pH in the prelysosomal compartment (Ott and Wunderli-Allenspach 1994). In a later effort, a group in BMS Pharmaceutical Research Institute discovered a novel HA inhibitor (BMY-27709) that has an EC_{50} of 6–8 μM against influenza viruses that have HA subtypes H1 and H2, but not H3 (Luo et al. 1996). The compound was shown to inhibit virus replication at an early stage and the inhibition was reversible. These data suggest that the inhibitor target is HA. This was further confirmed by the inhibitor-resistant mutations found in HA and inhibition of hemolysis by the compound (Luo et al. 1997). Similarly, a group in Lilly Research Laboratories found a novel inhibitor, a podocarpic acid derivative (180299), by screening a chemical library (Staschke et al. 1998). The EC_{50} of 180299 is 0.01 μg/ml against A/Kawasaki/86 but ≥10 μg/ml against other trains. The target of action was also confirmed to be HA by inhibition of cell fusion and positions of inhibitor-resistant mutations. By screen, another inhibitor was also discovered, Stachyflin, that has an EC_{50} in the μM range against H1 and H2 viruses, but not H3 (Yoshimoto et al. 1999). HA as the target for Stachyflin was also confirmed by time of addition, inhibition of hemolysis, and reassortment between subtype H1 and H3. Another group initiated a structure-aided approach to identify inhibitors of HA-mediated membrane fusion

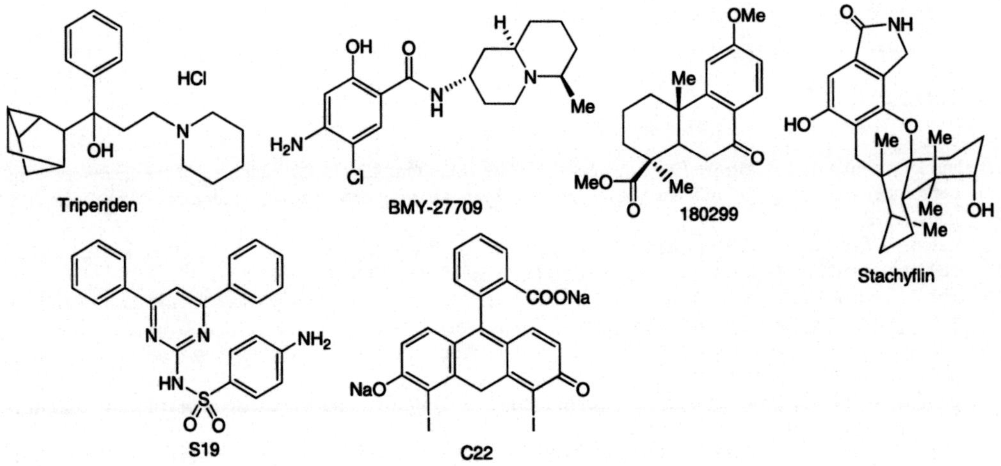

Fig. 9.7 The chemical structure of some published fusion inhibitors of influenza virus

(Bodian et al. 1993; Hoffman et al. 1997). The most effective compound identified (S19) has an EC_{50} of 0.8 μM against influenza virus X-31, and activities on other strains were not reported. What was remarkable about their work is that they found another moderate inhibitor (C22). Unlike other inhibitors that prevented the conformational change of HA, C22 facilitated the conformational change at fusion pH and its effect was irreversible. C22 destabilizes HA and also inhibits hemolysis, fusion, and viral infectivity. The authors concluded that since C22 does not induce the conformational change at neutral pH, it was conceivable that it might facilitate fusion by destabilizing HA as an effector.

The mutations identified in the process of verifying the mechanism of action by the fusion inhibitors of HA may provide insights on how HA fusion may be inhibited. The mutations may occur throughout HA1 and HA2 molecules, but most of them are in HA2. They can generally be classified in two groups. The first group includes the mutations that destabilize HA so the conformational change of HA2 can still occur even in the presence of the inhibitor. These mutations mostly were found in HA2 and occurred more often as found in resistant mutants to BMS-27709, 180299, and S19 (Hoffman et al. 1997; Luo et al. 1997; Staschke et al. 1998). Mutations reduce either the stabilizing interactions with the fusion peptide or the interfaces that stabilize the trimer association of HA. The mutant HAs are therefore more fusogenic at a pH higher than that required by the fusion of wild-type HA. Such mutations in HA were also identified in mutant viruses that are resistant to high concentrations of amantadine, an antiviral compound that elevates intracellular pH (Hoffman et al. 1997; Staschke et al. 1998). These data indicate that the fusion inhibitors discovered previously primarily stabilize HA when binding to HA. The stabilization by these inhibitors makes the conformational change of HA more energetically unfavorable so the chance of fusion is reduced, which leads to reduction of virus replication. The second group of mutations was identified by selection of resistant mutants to C22 that is an effector of the conformational change in HA, instead of an inhibitor (Hoffman et al. 1997). As expected, the mutant HAs have a downward shift in fusion pH, and the inhibitor resistant viruses in the first group are actually more sensitive to C22.

The results support the notion that entry inhibitors can inhibit influenza virus replication effectively. The challenge for developing an antiviral drug that can be used clinically is to discover entry inhibitors that have a high potency against all strains of human influenza virus. Since the conservation in amino acid sequence is low among influenza virus HA, a clear picture on how to discover a commonly efficacious entry inhibitor has not yet emerged. Since HA-mediated membrane fusion is the common mechanism for influenza virus entry, there is the potential to discover an inhibitor that blocks influenza virus entry through the common mechanism, if not a common inhibitor binding site in HA per se.

References

Bodian DL, Yamasaki RB, Buswell RL, Stearns JF, White JM, Kuntz ID (1993) Inhibition of the fusion-inducing conformational change of influenza hemagglutinin by benzoquinones and hydroquinones. Biochemistry 32: 2967–2978

Bonnafous P, Stegmann T (2000) Membrane perturbation and fusion pore formation in influenza hemagglutinin-mediated membrane fusion. A new model for fusion. J Biol Chem 275:6160–6166

Bottcher C, Ludwig K, Herrmann A, van Heel M, Stark H (1999) Structure of influenza haemagglutinin at neutral and at fusogenic pH by electron cryo-microscopy. FEBS Lett 463:255–259

Bragstad K, Nielsen LP, Fomsgaard A (2008) The evolution of human influenza A viruses from 1999 to 2006: a complete genome study. Virol J 5:40

Bui M, Whittaker G, Helenius A (1996) Effect of M1 protein and low pH on nuclear transport of influenza virus ribonucleoproteins. J Virol 70:8391–8401

Bullough PA, Hughson FM, Skehel JJ, Wiley DC (1994) Structure of influenza haemagglutinin at the pH of membrane fusion. Nature 371:37–43

Chang DK, Cheng SF (2006) pH-dependence of intermediate steps of membrane fusion induced by the influenza fusion peptide. Biochem J 396:557–563

Chang DK, Cheng SF, Kantchev EA, Lin CH, Liu YT (2008) Membrane interaction and structure of the transmembrane domain of influenza hemagglutinin and its fusion peptide complex. BMC Biol 6:2

Chen C, Zhuang X (2008) Epsin 1 is a cargo-specific adaptor for the clathrin-mediated endocytosis of the influenza virus. Proc Natl Acad Sci USA 105:11790–11795

Chen J, Lee KH, Steinhauer DA, Stevens DJ, Skehel JJ, Wiley DC (1998) Structure of the hemagglutinin precursor cleavage site, a determinant of influenza pathogenicity and the origin of the labile conformation. Cell 95: 409–417

Chen J, Skehel JJ, Wiley DC (1999) N- and C-terminal residues combine in the fusion-pH influenza hemagglutinin HA(2) subunit to form an N cap that terminates the triple-stranded coiled coil. Proc Natl Acad Sci USA 96: 8967–8972

Chernomordik LV, Frolov VA, Leikina E, Bronk P, Zimmerberg J (1998) The pathway of membrane fusion catalyzed by influenza hemagglutinin: restriction of lipids, hemifusion, and lipidic fusion pore formation. J Cell Biol 140:1369–1382

Colman PM, Lawrence MC (2003) The structural biology of type I viral membrane fusion. Nat Rev Mol Cell Biol 4:309–319

Compans RW, Klenk HD, Caliguiri LA, Choppin PW (1970) Influenza virus proteins. I. Analysis of polypeptides of the virion and identification of spike glycoproteins. Virology 42:880–889

Daniels RS, Downie JC, Hay AJ, Knossow M, Skehel JJ, Wang ML, Wiley DC (1985) Fusion mutants of the influenza virus hemagglutinin glycoprotein. Cell 40:431–439

DeLano WL (2002) The PyMOL Molecular Graphics System. DeLano Scientific, Palo Alto, CA

Doxsey SJ, Brodsky FM, Blank GS, Helenius A (1987) Inhibition of endocytosis by anti-clathrin antibodies. Cell 50:453–463

Floyd DL, Ragains JR, Skehel JJ, Harrison SC, van Oijen AM (2008) Single-particle kinetics of influenza virus membrane fusion. Proc Natl Acad Sci USA 105:15382–15387

Gamblin SJ, Haire LF, Russell RJ, Stevens DJ, Xiao B, Ha Y, Vasisht N, Steinhauer DA, Daniels RS, Elliot A, Wiley DC, Skehel JJ (2004) The structure and receptor binding properties of the 1918 influenza hemagglutinin. Science 303:1838–1842

Ghendon Y, Markushin S, Heider H, Melnikov S, Lotte V (1986) Haemagglutinin of influenza A virus is a target for the antiviral effect of Norakin. J Gen Virol 67(Pt 6):1115–1122

Giannecchini S, Campitelli L, Bandini G, Donatelli I, Azzi A (2007) Characterization of human H1N1 influenza virus variants selected in vitro with zanamivir in the presence of sialic acid-containing molecules. Virus Res 129:241–245

Ha Y, Stevens DJ, Skehel JJ, Wiley DC (2001) X-ray structures of H5 avian and H9 swine influenza virus hemagglutinins bound to avian and human receptor analogs. Proc Natl Acad Sci USA 98:11181–11186

Hanson JE, Sauter NK, Skehel JJ, Wiley DC (1992) Proton nuclear magnetic resonance studies of the binding of sialosides to intact influenza virus. Virology 189:525–533

Harris A, Cardone G, Winkler DC, Heymann JB, Brecher M, White JM, Steven AC (2006) Influenza virus pleiomorphy characterized by cryoelectron tomography. Proc Natl Acad Sci USA 103:19123–19127

Hay AJ, Gregory V, Douglas AR, Lin YP (2001) The evolution of human influenza viruses. Philos Trans R Soc Lond B Biol Sci 356:1861–1870

Hayden F (2009) Developing new antiviral agents for influenza treatment: what does the future hold? Clin Infect Dis 48(Suppl 1):S3–S13

Heider H, Markushin S, Schroeder C, Ghendon Y (1985) The influence of Norakin on the reproduction of influenza A and B viruses. Arch Virol 86:283–290

Hoffman LR, Kuntz ID, White JM (1997) Structure-based identification of an inducer of the low-pH conformational change in the influenza virus hemagglutinin: irreversible inhibition of infectivity. J Virol 71:8808–8820

Huang Q, Opitz R, Knapp EW, Herrmann A (2002) Protonation and stability of the globular domain of influenza virus hemagglutinin. Biophys J 82:1050–1058

Huang Q, Sivaramakrishna RP, Ludwig K, Korte T, Bottcher C, Herrmann A (2003) Early steps of the conformational change of influenza virus hemagglutinin to a fusion active state: stability and energetics of the hemagglutinin. Biochim Biophys Acta 1614:3–13

Hurt AC, Holien JK, Barr IG (2009) In vitro generation of neuraminidase inhibitor resistance in A(H5N1) influenza viruses. Antimicrob Agents Chemother 53:4433–4440

Ibricevic A, Pekosz A, Walter MJ, Newby C, Battaile JT, Brown EG, Holtzman MJ, Brody SL (2006) Influenza virus receptor specificity and cell tropism in mouse and human airway epithelial cells. J Virol 80:7469–7480

Kemble GW, Danieli T, White JM (1994) Lipid-anchored influenza hemagglutinin promotes hemifusion, not complete fusion. Cell 76:383–391

Khor R, McElroy LJ, Whittaker GR (2003) The ubiquitin-vacuolar protein sorting system is selectively required during entry of influenza virus into host cells. Traffic 4:857–868

Kilby JM, Hopkins S, Venetta TM, DiMassimo B, Cloud GA, Lee JY, Alldredge L, Hunter E, Lambert D, Bolognesi D, Matthews T, Johnson MR, Nowak MA, Shaw GM, Saag MS (1998) Potent suppression of HIV-1 replication in humans by T-20, a peptide inhibitor of gp41-mediated virus entry. Nat Med 4:1302–1307

Kogure T, Suzuki T, Takahashi T, Miyamoto D, Hidari KI, Guo CT, Ito T, Kawaoka Y, Suzuki Y (2006) Human trachea primary epithelial cells express both sialyl(alpha2-3)Gal receptor for human parainfluenza virus type 1 and avian influenza viruses, and sialyl(alpha2-6)Gal receptor for human influenza viruses. Glycoconj J 23:101–106

Konig R, Stertz S, Zhou Y, Inoue A, Hoffmann HH, Bhattacharyya S, Alamares JG, Tscherne DM, Ortigoza MB, Liang Y, Gao Q, Andrews SE, Bandyopadhyay S, De Jesus P, Tu BP, Pache L, Shih C, Orth A, Bonamy G, Miraglia L, Ideker T, Garcia-Sastre A, Young JA, Palese P, Shaw ML, Chanda SK (2010) Human host factors required for influenza virus replication. Nature 463:813–817

Lakadamyali M, Rust MJ, Babcock HP, Zhuang X (2003) Visualizing infection of individual influenza viruses. Proc Natl Acad Sci USA 100:9280–9285

Lear JD, DeGrado WF (1987) Membrane binding and conformational properties of peptides representing the NH2 terminus of influenza HA-2. J Biol Chem 262:6500–6505

Lee KK (2010) Architecture of a nascent viral fusion pore. EMBO J 29:1299–1311

Leikina E, LeDuc DL, Macosko JC, Epand R, Shin YK, Chernomordik LV (2001) The 1-127 HA2 construct of influenza virus hemagglutinin induces cell-cell hemifusion. Biochemistry 40:8378–8386

Leikina E, Markovic I, Chernomordik LV, Kozlov MM (2000) Delay of influenza hemagglutinin refolding into a fusion-competent conformation by receptor binding: a hypothesis. Biophys J 79:1415–1427

Lin T, Wang G, Li A, Zhang Q, Wu C, Zhang R, Cai Q, Song W, Yuen KY (2009) The hemagglutinin structure of an avian H1N1 influenza A virus. Virology 392:73–81

Liu C, Eichelberger MC, Compans RW, Air GM (1995) Influenza type A virus neuraminidase does not play a role in viral entry, replication, assembly, or budding. J Virol 69:1099–1106

Liu J, Stevens DJ, Haire LF, Walker PA, Coombs PJ, Russell RJ, Gamblin SJ, Skehel JJ (2009) Structures of receptor complexes formed by hemagglutinins from the Asian Influenza pandemic of 1957. Proc Natl Acad Sci USA 106:17175–17180

Luneberg J, Martin I, Nussler F, Ruysschaert JM, Herrmann A (1995) Structure and topology of the influenza virus fusion peptide in lipid bilayers. J Biol Chem 270:27606–27614

Luo G, Colonno R, Krystal M (1996) Characterization of a hemagglutinin-specific inhibitor of influenza A virus. Virology 226:66–76

Luo G, Torri A, Harte WE, Danetz S, Cianci C, Tiley L, Day S, Mullaney D, Yu KL, Ouellet C, Dextraze P, Meanwell N, Colonno R, Krystal M (1997) Molecular mechanism underlying the action of a novel fusion inhibitor of influenza A virus. J Virol 71:4062–4070

Matlin KS, Reggio H, Helenius A, Simons K (1981) Infectious entry pathway of influenza virus in a canine kidney cell line. J Cell Biol 91:601–613

Nicholls JM, Chan RW, Russell RJ, Air GM, Peiris JS (2008) Evolving complexities of influenza virus and its receptors. Trends Microbiol 16:149–157

Nunes-Correia I, Eulalio A, Nir S, Pedroso de Lima MC (2004) Caveolae as an additional route for influenza virus endocytosis in MDCK cells. Cell Mol Biol Lett 9:47–60

O'Neill RE, Talon J, Palese P (1998) The influenza virus NEP (NS2 protein) mediates the nuclear export of viral ribonucleoproteins. EMBO J 17:288–296

Ott S, Wunderli-Allenspach H (1994) Effect of the virostatic Norakin (triperiden) on influenza virus activities. Antiviral Res 24:37–42

Prosch S, Heider H, Schroeder C, Kruger DH (1988) Mutations in the hemagglutinin gene associated with influenza virus resistance to norakin. Arch Virol 102:125–129

Qi L, Kash JC, Dugan VG, Wang R, Jin G, Cunningham RE, Taubenberger JK (2009) Role of sialic acid binding specificity of the 1918 influenza virus hemagglutinin protein in virulence and pathogenesis for mice. J Virol 83:3754–3761

Rosenthal PB, Zhang X, Formanowski F, Fitz W, Wong CH, Meier-Ewert H, Skehel JJ, Wiley DC (1998) Structure of the haemagglutinin-esterase-fusion glycoprotein of influenza C virus. Nature 396:92–96

Roy AM, Parker JS, Parrish CR, Whittaker GR (2000) Early stages of influenza virus entry into Mv-1 lung cells: involvement of dynamin. Virology 267:17–28

Rust MJ, Lakadamyali M, Zhang F, Zhuang X (2004) Assembly of endocytic machinery around individual influenza viruses during viral entry. Nat Struct Mol Biol 11:567–573

Sauter NK, Bednarski MD, Wurzburg BA, Hanson JE, Whitesides GM, Skehel JJ, Wiley DC (1989) Hemagglutinins from two influenza virus variants bind to sialic acid derivatives with millimolar dissociation constants: a 500-MHz proton nuclear magnetic resonance study. Biochemistry 28:8388–8396

Schauer R (2009) Sialic acids as regulators of molecular and cellular interactions. Curr Opin Struct Biol 19:507–514

Segal MS, Bye JM, Sambrook JF, Gething MJ (1992) Disulfide bond formation during the folding of influenza virus hemagglutinin. J Cell Biol 118:227–244

Shen J, Ma J, Wang Q (2009) Evolutionary trends of A(H1N1) influenza virus hemagglutinin since 1918. PLoS One 4:e7789

Sieczkarski SB, Whittaker GR (2002) Influenza virus can enter and infect cells in the absence of clathrin-mediated endocytosis. J Virol 76:10455–10464

Sieczkarski SB, Whittaker GR (2003) Differential requirements of Rab5 and Rab7 for endocytosis of influenza and other enveloped viruses. Traffic 4:333–343

Staschke KA, Hatch SD, Tang JC, Hornback WJ, Munroe JE, Colacino JM, Muesing MA (1998) Inhibition of influenza virus hemagglutinin-mediated membrane fusion by a compound related to podocarpic acid. Virology 248:264–274

Stevens J, Blixt O, Tumpey TM, Taubenberger JK, Paulson JC, Wilson IA (2006) Structure and receptor specificity of the hemagglutinin from an H5N1 influenza virus. Science 312:404–410

Stray SJ, Cummings RD, Air GM (2000) Influenza virus infection of desialylated cells. Glycobiology 10:649–658

Takeda M, Leser GP, Russell CJ, Lamb RA (2003) Influenza virus hemagglutinin concentrates in lipid raft microdomains for efficient viral fusion. Proc Natl Acad Sci USA 100:14610–14617

Takemoto DK, Skehel JJ, Wiley DC (1996) A surface plasmon resonance assay for the binding of influenza virus hemagglutinin to its sialic acid receptor. Virology 217:452–458

Tatulian SA, Tamm LK (2000) Secondary structure, orientation, oligomerization, and lipid interactions of the transmembrane domain of influenza hemagglutinin. Biochemistry 39:496–507

Thompson CI, Barclay WS, Zambon MC, Pickles RJ (2006) Infection of human airway epithelium by human and avian strains of influenza a virus. J Virol 80:8060–8068

Tumpey TM, Maines TR, Van Hoeven N, Glaser L, Solorzano A, Pappas C, Cox NJ, Swayne DE, Palese P, Katz JM, Garcia-Sastre A (2007) A two-amino acid change in the hemagglutinin of the 1918 influenza virus abolishes transmission. Science 315:655–659

Wagner R, Wolff T, Herwig A, Pleschka S, Klenk HD (2000) Interdependence of hemagglutinin glycosylation and neuraminidase as regulators of influenza virus growth: a study by reverse genetics. J Virol 74:6316–6323

Wang Q, Tian X, Chen X, Ma J (2007) Structural basis for receptor specificity of influenza B virus hemagglutinin. Proc Natl Acad Sci USA 104:16874–16879

Wharton SA, Skehel JJ, Wiley DC (1986) Studies of influenza haemagglutinin-mediated membrane fusion. Virology 149:27–35

White J, Helenius A, Gething MJ (1982) Haemagglutinin of influenza virus expressed from a cloned gene promotes membrane fusion. Nature 300:658–659

Wilson IA, Skehel JJ, Wiley DC (1981) Structure of the haemagglutinin membrane glycoprotein of influenza virus at 3 A resolution. Nature 289:366–373

Yamada S, Suzuki Y, Suzuki T, Le MQ, Nidom CA, Sakai-Tagawa Y, Muramoto Y, Ito M, Kiso M, Horimoto T, Shinya K, Sawada T, Usui T, Murata T, Lin Y, Hay A, Haire LF, Stevens DJ, Russell RJ, Gamblin SJ, Skehel JJ, Kawaoka Y (2006) Haemagglutinin mutations responsible for the binding of H5N1 influenza A viruses to human-type receptors. Nature 444:378–382

Yen HL, Webster RG (2009) Pandemic influenza as a current threat. Curr Top Microbiol Immunol 333:3–24

Yoshimoto J, Kakui M, Iwasaki H, Fujiwara T, Sugimoto H, Hattori N (1999) Identification of a novel HA conformational change inhibitor of human influenza virus. Arch Virol 144:865–878

Zambon MC (2001) The pathogenesis of influenza in humans. Rev Med Virol 11:227–241

Chapter 10
Molecular Mechanisms of HIV Entry

Craig B. Wilen, John C. Tilton, and Robert W. Doms

Abstract Human immunodeficiency virus (HIV) entry is a complex and intricate process that facilitates delivery of the viral genome to the host cell. The only viral surface protein, Envelope (Env), is composed of a trimer of gp120 and gp41 heterodimers. It is essentially a fusion machine cloaked in a shroud of carbohydrate structures and variable loops of amino acids that enable it to evade the humoral immune response. For entry to occur gp120 sequentially engages the host protein CD4 and then one of two chemokine coreceptors, either CCR5 or CXCR4. CD4 binding facilitates exposure and formation of the coreceptor-binding site, and coreceptor binding then triggers the membrane fusion machinery in the gp41 subunit. Our understanding of HIV entry has led to the development of successful small molecule inhibitors for the clinical treatment of HIV infection as well as insights into viral tropism and pathogenesis.

Abbreviations

6HB	Six-helix bundle
AIDS	Acquired immunodeficiency syndrome
CD4bs	CD4-binding site
Cryo-EM	Cryo-electron microscopy
ECL	Extracellular loop
Env	Envelope
ER	Endoplasmic reticulum
HIV	Human immunodeficiency virus
HR-C	C-terminal heptad repeat
HR-N	N-terminal heptad repeat
R5 HIV	CCR5-tropic HIV
R5X4 HIV	Dual-tropic HIV
SIV	Simian immunodeficiency virus
V3	Variable loop 3
X4 HIV	CXCR4-tropic HIV

C.B. Wilen • J.C. Tilton • R.W. Doms (✉)
Department of Microbiology, University of Pennsylvania,
3610 Hamilton Walk, Johnson Pavilion, Philadelphia, PA 19104, USA
e-mail: doms@mail.med.upenn.edu

M.G. Rossmann and V.B. Rao (eds.), *Viral Molecular Machines*,
Advances in Experimental Medicine and Biology 726, DOI 10.1007/978-1-4614-0980-9_10,

10.1 Introduction

Acquired immunodeficiency syndrome (AIDS) is a devastating disease caused by human immunodeficiency virus (HIV)-mediated destruction of CD4+ T lymphocytes (Barre-Sinoussi et al. 1983; Gallo et al. 1983). Since its emergence over 25 years ago, HIV/AIDS has killed more than 25 million people, and another 33 million are currently infected (UNAIDS 2009). The profound effect of this pandemic has led to tremendous research efforts to elucidate the mechanisms of the HIV life cycle and identify susceptible targets for therapeutic intervention. Here we will focus on the molecular mechanisms of the first stage of the HIV life cycle: viral entry.

HIV is an enveloped, single-strand RNA virus that probably first infected humans via a cross-species transmission from SIV-infected (simian immunodeficiency virus) chimpanzees in the first half of the twentieth century (Gao et al. 1999). For HIV and other known non-human primate lentiviruses to infect and replicate inside their hosts, the viral surface protein Envelope (Env) must first bind the host receptor CD4 and subsequently a coreceptor, most commonly CCR5 or CXCR4 (Alkhatib et al. 1996; Choe et al. 1996; Dalgleish et al. 1984; Deng et al. 1996; Doranz et al. 1996; Dragic et al. 1996; Feng et al. 1996; Oberlin et al. 1996). Coreceptor binding then triggers a conformational change in Env that mediates pH-independent membrane fusion and delivery of the viral payload (Fig. 10.1). Once inside the cell, the viral capsid is uncoated, revealing the viral genome. The viral RNA is then reverse transcribed via error-prone reverse transcriptase into double-strand DNA that is transported into the nucleus of the host cell where it integrates into a host chromosome. The integrated provirus is then transcribed and translated by host machinery to generate a polyprotein that is autocatalytically cleaved and processed to form nascent virions that bud from the host cell, enabling additional rounds of replication (Freed 2001).

Advances in understanding the HIV entry pathway have directly led to the development of specific and effective antiviral agents that prevent virus entry, including the first antiviral agent that targets a host cell molecule needed for virus infection. This provides proof of principle that virus entry in general, including the viral membrane fusion machines, can be targeted. Currently, two small molecule HIV entry inhibitors, the fusion inhibitor enfuvirtide and the CCR5 antagonist maraviroc, are FDA approved for the treatment of HIV, while others are currently under various stages of development (Dwyer et al. 2007; He et al. 2008a, b; Schurmann et al. 2007; Strizki et al. 2005; Tilton and Doms 2010). The success of these compounds emphasizes the potential therapeutic benefit in

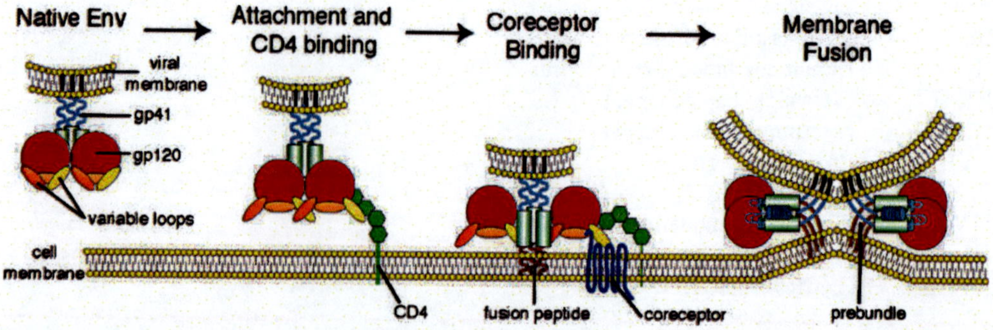

Fig. 10.1 HIV entry schematic. The HIV entry process can be divided into three key steps, all mediated by the viral surface protein Env, which is a trimer of gp120 and gp41 heterodimers. First, the virion binds to the host cell which can be facilitated by nonspecific cell attachment factors and allows Env to bind CD4. Second, CD4 binding induces conformational changes in Env that enable engagement of a coreceptor, most commonly CCR5 or CXCR4. This step triggers membrane fusion. Third, the gp41 subunits of Env enable fusion of the viral and host membranes allowing delivery of the viral payload

elucidating the mechanisms of viral entry. In addition, these drugs serve as critical tools in studying the mechanics of entry machines.

In addition to the identification of new drug targets, characterizing the HIV entry pathway has led to the identification of genetic polymorphisms in the human population that help account for the variable progression to AIDS seen after HIV infection (Gonzalez et al. 2005; McDermott et al. 1998). The most significant of these polymorphisms is a 32 base-pair deletion in *ccr5* called *ccr5Δ32*. Homozygosity for this mutation renders individuals highly resistant to HIV infection due to the functional loss of this viral coreceptor (Dean et al. 1996; Liu et al. 1996; Samson et al. 1996). People heterozygous for the *ccr5Δ32* mutation have somewhat reduced susceptibility to infection, and if they do get infected, they tend to progress to AIDS more slowly (Blanpain et al. 2002; Dean et al. 1996; Hoffman et al. 1997; Huang et al. 1996). In addition to the *ccr5Δ32* mutation, segmental duplications of a natural CCR5 ligand, CCL3L1 (MIP1α), may be associated with both decreased risk of HIV acquisition and disease progression. The gene duplications are thought to result in increased CCL3L1 expression and thus reduced CCR5 cell surface expression (Gonzalez et al. 2005). While genetic polymorphisms influencing CCR5 expression contribute significantly in explaining the variable disease progression observed in HIV-infected individuals, much of this variability cannot yet be explained and thus calls for additional investigation.

Despite the tremendous advances in understanding HIV entry, many questions remain unanswered. To penetrate cells, enveloped viruses must fuse with cells either directly at the plasma membrane or in intracellular compartments (Marsh and Helenius 2006). Historically, many have assumed that HIV fuses at the plasma membrane due to early studies demonstrating that Env can mediate cell–cell fusion both in vitro and in vivo, and because HIV entry can be augmented, but not inhibited, by lysomotropic agents (Maddon et al. 1988; McClure et al. 1988). However, the fact that HIV entry is pH independent provides no spatial information regarding the site of entry: pH-independent viruses can conceivably fuse at the cell surface or from within endocytic vesicles, albeit in a pH-independent fashion (Stein et al. 1987). Identifying the cellular site at which viral membrane fusion occurs is not a trivial process, but recent evidence using sophisticated live cell imaging to follow single virions suggests that productive membrane fusion may actually occur in endosomal compartments (Miyauchi et al. 2009; Uchil and Mothes 2009). Regardless of where fusion occurs, the mechanics of entry will likely be the same, and entry inhibitors clearly function both in vivo and in vitro.

In addition to questions regarding the site of membrane fusion during infection of primary cells and the implications this may hold for viral entry, several additional questions remain unanswered. Our knowledge of Env structure remains incomplete, and this complicates our understanding of how receptor binding induces such dramatic structural rearrangements in Env. It is also not clear how the fusion pore forms and expands, or how many trimers are needed for fusion. This chapter will review what is known about the mechanisms of the HIV fusion machine and how this informs our understanding of viral tropism, pathogenesis, and the development of antiviral therapies.

10.2 Key Players in HIV Entry

10.2.1 Env

The Env protein is a type I integral membrane protein first synthesized in the endoplasmic reticulum (ER) as a precursor termed gp160. While in the ER, gp160 folds, forms non-covalently associated trimers, is extensively glycosylated, and establishes critical disulfide bonds (Earl et al. 1991; Land and Braakman 2001; Land et al. 2003). The protein then traffics to the Golgi where the host protease furin cleaves it into gp120 and gp41 subunits. This cleavage event renders the protein fusion-competent by generating the N-terminus of gp41 that is composed of a stretch of hydrophobic amino acids termed the fusion peptide (Hallenberger et al. 1992). After cleavage, the soluble surface gp120 and the transmembrane gp41 subunits continue to interact noncovalently forming native Env, a trimer of heterodimers.

Each gp120 subunit contains a highly conserved inner domain and a more variable outer domain. Within these are five highly conserved regions (C1–C5) and five hypervariable loops (V1–V5) with all but V5 defined by a disulfide bridge at its base. The five conserved regions are predominantly located in the inner domain, or gp120 core, while the variable loops predominate at the protein surface. The conserved domains are critical to gp120 folding and function, while the surface-exposed variable loops play roles in immune evasion by presenting a constantly moving target for the host immune system. In addition, the V3 region plays a critical role in coreceptor binding, as will be discussed later (Hartley et al. 2005; Poignard et al. 2001).

The gp41 subunit contains an ectodomain followed by a transmembrane domain and C-terminal cytoplasmic tail. The ectodomain encodes an N-terminal, hydrophobic fusion peptide that inserts into the target cell membrane and two heptad repeat regions (HR-N and HR-C) separated by a hinge region. The transmembrane domain anchors Env in the viral membrane, and the very long cytoplasmic tail plays a complex role in packaging, infectivity, and cell pathogenicity (Yang et al. 2010).

10.2.2 CD4

CD4 is an immunoglobulin superfamily member that contains four extracellular immunoglobulin domains (D1–D4), a single transmembrane domain, and a short intracellular cytoplasmic tail. CD4 plays a key signaling role that assists the T-cell receptor in activating cells in response to immunologic stimuli. It is expressed on macrophages, dendritic cells, and the eponymous CD4+ T cells, which include naive, central, and effector memory subtypes. Depletion of CD4+ T cells, and in particular the memory subsets, is indicative of disease progression and a poor clinical outcome.

10.2.3 Coreceptor

Most Env strains utilize one of two chemokine receptors, either CCR5 or CXCR4, for viral entry. Both CCR5 and CXCR4 are seven-transmembrane G-protein coupled receptors. Each has an extracellular N-terminus, three extracellular loops (ECLs), three intracellular loops, and a cytoplasmic C-terminus.

10.2.3.1 CCR5

CCR5 has three known natural ligands: CCL3, CCL4, and CCL5 formerly known as MIP-1α, MIP-1β, and RANTES, respectively. CCR5 is expressed on macrophages, microglia, and central and effector memory CD4+ T cells (Combadiere et al. 1996; Raport et al. 1996; Zaitseva et al. 1997). Of note, CCR5 is broadly expressed in the gut lamina propria contributing to destruction of the gut integrity during infection (Brenchley and Douek 2008; Kotler et al. 1984). Before CCR5 was discovered as the predominant HIV coreceptor, it was demonstrated that increased levels of its three chemokine ligands reduced susceptibility to HIV infection (Cocchi et al. 1995). This is due to both direct competition with Env and ligand-induced receptor internalization (Alkhatib et al. 1997). As previously discussed, variation in ligand expression may significantly affect disease progression (Gonzalez et al. 2005).

10.2.3.2 CXCR4

Compared to CCR5, CXCR4 is broadly expressed on the vast majority of hematopoietic cells and on many parenchymal cells as well. Genetic disruption of either CXCR4 or its only known natural ligand, CXCL12 (SDF-1), is embryonic lethal in mice and leads to defects in B-cell, cardiac, and cerebellar development, leukocyte migration, and bone marrow colonization by hematopoietic stem cells (Bleul et al. 1996; Nagasawa et al. 1996; Oberlin et al. 1996; Zou et al. 1998). Small molecule inhibitors of CXCR4 advanced to clinical trials for the treatment of HIV when it was noticed that the compounds induced peripheral mobilization of hematopoietic stem cells (Hendrix et al. 2004). Recently, the CXCR4 antagonist plerixafor was FDA approved for this very purpose: to harvest hematopoietic stem cells prior to bone marrow transplantation. Unlike for CCR5, there are no known CXCR4 mutations that contribute to HIV resistance, presumably due to the embryonic lethality of such defects. However, a gain-of-function mutation in CXCR4 has been reported that causes excessive signaling by preventing ligand-induced receptor internalization. This genetic lesion leads to WHIM syndrome, characterized by warts, hypogammaglobulinemia, increased susceptibility to infections, and myelokathexis, or retention of neutrophils in the bone marrow (Gorlin et al. 2000; Hernandez et al. 2003).

While the vast majority of transmitted virions use CCR5 (R5 HIV), viruses capable of using CXCR4 either alone (X4 HIV) or in combination with CCR5 (R5X4 HIV) emerge in approximately 50% of infected individuals in the developed world, typically 5–7 years after infection (Connor et al. 1997). Such a coreceptor switch expands viral target cell tropism and heralds a poor prognosis; however, the reasons it occurs and why only in later stages of disease remain unclear. It is of note that some exceedingly uncommon HIV isolates have been reported to use alternative coreceptors; however, the clinical relevance of alternative coreceptor use remains to be elucidated (Choe et al. 1996; Deng et al. 1997; Doranz et al. 1996; Edinger et al. 1998; Farzan et al. 1997).

10.3 The Structure of Env

Since HIV establishes a chronic infection, the viral Env protein must employ mechanisms to evade the ever-evolving humoral immune response. First, the shroud of N-linked carbohydrate structures and flexible variable loops act as a malleable shield masking more conserved and potentially neutralizable epitopes. Second, as a result of the error-prone reverse transcriptase, the HIV genome is highly diverse (Saag et al. 1988). On average, between one and ten of the 10,000 nucleotides comprising the HIV genome are mutated during each round of replication, and with some individuals producing as many as ten billion new virions per day, each nucleotide site in the viral genome can be theoretically mutated to each of the other three nucleotides at every site in the genome each day (Drosopoulos et al. 1998; Keulen et al. 1997; Lal et al. 2005). This variability enables Env to escape potentially neutralizing antibodies, partially explaining the difficulty in developing an effective neutralizing antibody vaccine (Stamatatos et al. 2009).

10.3.1 Env Crystal Structures

Several crystal structures have revealed detailed pictures of the HIV-1 gp120 monomer in complex with soluble CD4 (sCD4) and various antibodies (Huang et al. 2005; Kwong et al. 1998), though in all cases portions of gp120 were removed and the resulting purified protein deglycosylated to assist in crystallization efforts. An unliganded, glycosylated form of SIV gp120 has also been solved

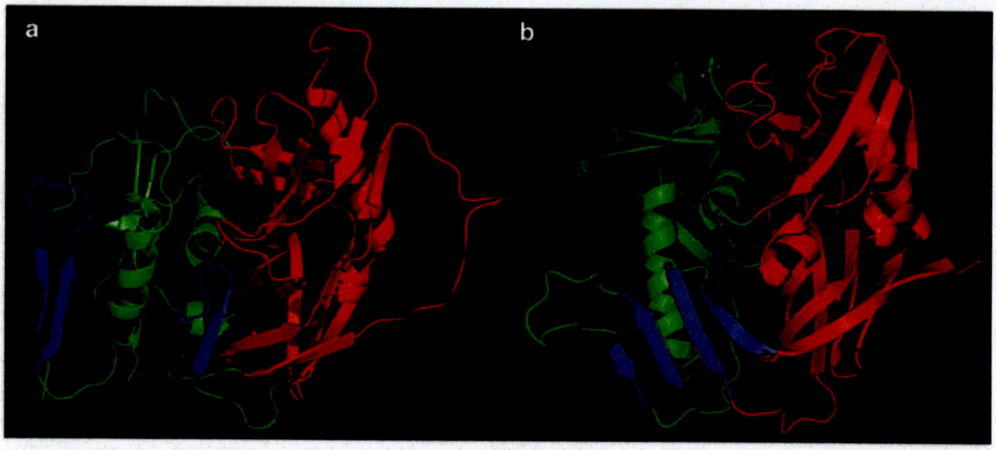

Fig. 10.2 Liganded and unliganded gp120 structures. CD4 binding induces massive structural rearrangements in gp120 that result in exposure and formation of the coreceptor-binding site. (**a**) Depicts the unliganded SIV gp120 monomer structure. The inner domain is in *green*, the outer domain is in *red*, and the bridging sheet is in *blue* (PDB ID: 2BF1) (Chen et al. 2005). (**b**) Depicts the CD4-bound HIV gp120 structure. The color scheme is the same as in (**a**). CD4 binding induces movement of the V3 loop that partially reveals the coreceptor-binding site. In addition, it causes the two pairs of β-strands to form the four-stranded β-sheet that enables coreceptor binding (PDB ID: 1GC1) (Kwong et al. 1998)

(Chen et al. 2005) (Fig. 10.2). In addition, portions of gp41 in its postfusion state have been solved (Chan et al. 1997; Weissenhorn et al. 1997). However, no crystal structures of trimeric Env have yet been reported.

Each gp120 monomer is comprised of a relatively conserved inner, or core, domain and a more variable outer domain. In addition, the outer domain contains the stems of the V3, V4, and V5 loops. While all the variable loops are thought to confer some degree of protection against neutralizing antibodies, the V3 loop is of particular interest due to its interactions with coreceptor. The V3 loop is almost always 35 residues in length and extends 30 Å from the gp120 surface (Huang et al. 2005, 2007). It is comprised of a relatively conserved disulfide-linked base, a long flexible stem, and a tip defined by a β-hairpin.

10.3.2 Cryo-Electron Microscopy

While crystallography has yielded atomic snapshots into Env subunit structure, cryo-electron tomography (see chapter 4) has revealed intriguing and though at times somewhat controversial structural properties of native Env trimers. By fitting known crystal structures into density maps of native Env obtained by cryo-electron tomography, detailed insights can be made into the native structure. The viral spike contains three propeller-like globular domains displaying threefold symmetry. The globular domains share a common density at their apex and are perched on a 50-Å stalk, most likely composed of gp41. The entire structure extends 120–140 Å from the viral membrane and is 110–120 Å in diameter (Sougrat et al. 2007; Zanetti et al. 2006; Zhu et al. 2006, 2008).

The abundance of structural studies has at times yielded conflicting results about several aspects of Env, with most of these revolving around how to place the known crystal structures within the tomographic densities. For instance, the position of the V1/V2 loop at the apex of Env proposed by Liu et al. (2008) is in contrast to that proposed using the unliganded SIV gp120 monomer, which

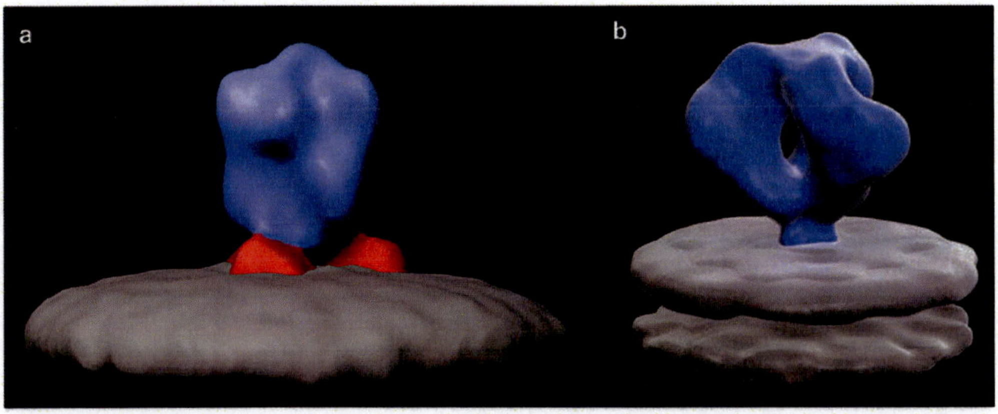

Fig. 10.3 Cryo-EM structures of the gp41 stalk. (**a**) Depicts the tripod-like stalk (highlighted in *red*) of native Env. Image courtesy of Kenneth Roux, Florida State University, 2010. In contrast, (**b**) Depicts native Env with a compact stalk. Reprinted by permission from Macmillan Publishers Ltd: Nature (Liu et al. 2008), copyright 2008

suggested that the V1/V2 loop is located nearer to the base of the trimer (Zanetti et al. 2006; Zhu et al. 2006). Two possible but not mutually exclusive explanations for this discrepancy are (1) that the V1/V2 loops of HIV and SIV are structurally different and (2) that the crystallization of the modified SIV gp120 resulted in a nonphysiologic location of V1/V2 (Liu et al. 2008). Further structural studies are needed to illuminate the detailed interactions that maintain and form stable trimers due to therapeutic and vaccine implications.

In addition, conflicting structural studies have described the native gp41 conformation as forming both a compact stalk and tripod-like structure (Liu et al. 2008; Zanetti et al. 2006; Zhu et al. 2008) (Fig. 10.3). Zhu et al. first reported a tripod-like stalk structure of a cytoplasmic tail-truncated SIV that had an increased number of Env spikes in the viral membrane (Zhu et al. 2006). They later reached a similar conclusion using native HIV trimers. However, Zanetti et al. and Liu et al. independently concluded that, like other class I viral fusion proteins studied to date, the viral stalk is compact and devoid of tripod-like legs (Liu et al. 2008; Zanetti et al. 2006). While it is possible that the discrepancy is due to inherent structural differences between the highly similar viral strains, it is more likely due to differential image collection and analysis. Specifically, electron tomography studies are limited by the "missing wedge" problem arising from the fact that the maximal angle between the microscope stage and image detector is typically 70°. This results in a "missing wedge," which the groups accounted for differently. In addition, the authors used different reference images to compensate for noise, which may result in downstream bias (Subramaniam 2006). Finally, all groups assumed threefold symmetry when solving the native Env structure, which may introduce a common bias.

10.4 The Entry Pathway

10.4.1 Attachment and CD4 Binding

HIV entry follows an intricate and sequential sequence of events whose primary mission is to deliver the viral payload. Our understanding of this complex pathway has been facilitated by the development of small molecule and protein-based inhibitors that target the key steps of the pathway. Entry begins with virion binding to the cell surface. This is often, but not always, facilitated by noncovalent interactions between Env and cell attachment factors. While not required for virus infection, the presence of such attachment factors can enhance infection of virus particles by accelerating virus–cell interactions, typically the rate-limiting step of virus infection in vitro. Examples of such molecules include heparan sulfate and mannose-binding lectins such as DC-SIGN (Baribaud et al. 2001; Geijtenbeek et al. 2000; Mondor et al. 1998; Pohlmann et al. 2001a, b). The role of attachment factors likely differs between cell types and viral strains, and its significance in vivo where direct cell-to-cell spread via virological synapses can occur remains unclear.

While a single receptor is sufficient for most enveloped viruses, the sequential receptor-binding strategy utilized by HIV has several benefits. Initial binding of CD4 induces exposure and creation of the coreceptor-binding site and thus may serve to protect this conserved region from host immune recognition (Chen et al. 2005; Kwong et al. 1998; Lal et al. 2005). In addition, the requirement for CD4 binding may reduce inappropriate and irreversible triggering of the fusion machinery, thus maximizing the infectivity of each virion. Finally, the CD4 molecule protrudes substantially further from the cell surface than the chemokine receptors and thus may also serve to facilitate virus binding.

The CD4-binding site (CD4bs), located at the interface of the outer and inner domains of gp120, is depressed 20 Å from the apex of the trimer, surrounded by carbohydrate moieties, and is likely partially occluded by the flexible V1/V2 loops (Kwong et al. 1998). CD4 binding to this highly conserved site initiates a sequence of structural rearrangements in Env defined both biochemically (Myszka et al. 2000) and structurally (Chen et al. 2005; Kwong et al. 1998) that reveal and create the coreceptor-binding site.

First, CD4 binding is thought to induce movement of the V1/V2 stem from the central axis of symmetry toward the lateral aspect of the trimer. This leads to V3 loop rearrangement and exposure of part of the coreceptor-binding site. Furthermore, the V3 stem shift toward the distal end of the trimer places it in direct apposition to the host target membrane, thus priming it for coreceptor engagement. Second, the bridging sheet, the second coreceptor-binding site, is formed by bringing together four antiparallel beta sheets, two each from the inner and outer domains of gp120. Third, the CD4-induced gp120 movement is thought to result in an outward rotation of each gp120 monomer that partially reveals the gp41 stalk. Finally, a recent model suggests that CD4–Env interactions cause CD4 to fold like a hinge between its second and third immunoglobulin-like domains. This folding brings the viral and host membranes into closer proximity and may serve to further limit antibody exposure of the conserved coreceptor-binding site (Liu et al. 2008).

In total, the aforementioned rearrangements culminate in coreceptor accessibility to its two binding sites, the V3 loop and the bridging sheet. The N-terminus of coreceptor engages the base of the V3 loop and the newly formed bridging sheet, while the ECLs, particularly ECL2, bind to the crown or tip of the V3 loop (Rizzuto and Sodroski 2000; Rizzuto et al. 1998). Most, but not all, Env strains depend upon both N-terminal and ECL2 interactions for entry.

10.4.1.1 Thermodynamics of CD4-Induced Changes

Thermodynamic studies complement the CD4-induced changes revealed by structural studies. Myszka et al. carried out a series of elegant experiments assessing the binding interactions between sCD4 and gp120 (Myszka et al. 2000). Both a full-length form of gp120 as well as a gp120 core devoid of N-linked glycans and containing truncations in the V1–3 loops, N-, and C-termini were compared. Both core and full-length gp120 had similar thermodynamic profiles suggesting that the CD4-induced changes between the two structures are conserved. Titration microcalorimetry revealed a surprisingly favorable CD4–gp120 binding enthalpy (−63 kcal/mol) suggesting that substantial, novel molecular interactions form, specifically van der Waals forces and hydrogen bonding. The magnitude of enthalpy is significantly greater than other common protein–protein interactions such as antibody–antigen and T cell receptor–major histocompatibility complex binding. In addition, there is a dramatic and unfavorable change in entropy upon CD4 binding. This demonstrates a reduction in randomness and is further evidence of significant structural rearrangements. The profound magnitudes of both entropy and enthalpy suggest that nearly 100 amino acids in the full length gp120 change conformation upon CD4 binding – greater than nearly all other reported binary protein–protein interactions (Myszka et al. 2000).

10.4.2 Coreceptor Binding

CD4 binding is thought to induce movement of the V1/V2 loop to expose the V3 loop. This allows the tip, or crown, of V3 to interact with the ECL2 of coreceptor. A second key gp120–coreceptor interaction involves sulfated tyrosines present in the CCR5 N-terminal domain with the base of the V3 loop and the four-stranded bridging sheet, which is formed by CD4-induced rearrangement of a pair of spatially separated two-stranded beta sheets. Extracellular loops 1 and 3 (ECL1 and ECL3) may also interact with Env, although the strength and significance of such interactions are uncertain.

Evidence for the interactions between the tip of V3 and ECL2 comes from a V3-containing gp120 crystal structure revealing that the V3 tip protrudes 30 Å from the gp120 core and is directed toward the target cell membrane allowing V3 to act as a "molecular hook" (Huang et al. 2005). Additional evidence for this interaction comes from the mechanism in which small molecule coreceptor antagonists inhibit HIV entry. A number of compounds have been designed that bind to a hydrophobic pocket within the transmembrane helices of both CCR5 and CXCR4. These molecules inhibit coreceptor engagement by causing a stable conformational shift of the ECLs. This suggests that the orientation of the ECLs is critical for coreceptor function. In addition, antibodies directed at the ECL2 are sufficient to prevent entry of most HIV isolates (Trkola et al. 2001). These antibodies likely preclude Env binding by both steric hindrance and alteration of ECL2 conformation. Further, site-directed mutations in ECL2 can abrogate infection (Doranz et al. 1997; Quinonez et al. 2003). Finally, a laboratory-engineered virus containing a V3 loop truncation that preserves the V3 base and eliminates the crown results in broad resistance to a number of CCR5 inhibitors. This virus relies solely on the CCR5 N-terminus as opposed to both the ECL2 and N-terminus for entry (Laakso et al. 2007).

The CCR5 N-terminus (residues 2–31) binds the base of V3 and the bridging sheet (Choe and Farzan 2009; Cormier et al. 2000; Huang et al. 2007). The CCR5 N-terminus contains four sulfotyrosine residues at positions 3, 10, 14, and 15; only that at position 15 is dispensible without reducing viral entry efficiency (Cormier et al. 2000; Farzan et al. 1998; Rabut et al. 1998). In addition, the N-terminus contains acidic residues which further facilitate binding to the negatively charged site in Env (Rizzuto and Sodroski 2000). Substantial evidence supports these interactions. First, sulfated N-terminal CCR5 peptides are sufficient to block entry, albeit at high concentrations (Cormier et al. 2000). Second, inhibition of tyrosine sulfation decreases the ability to engage CCR5 (Farzan et al. 1999). Third, several gp120 neutralizing antibodies have been reported that contain sulfated tyrosines at the site of antigen contact. These antibodies require sulfation for activity, are more effective in the presence of CD4, and compete with N-terminal-sulfated CCR5 peptides for gp120 binding (Choe et al. 2003). Recently, a crystal structure of one of these antibodies (412d) and CD4 bound to gp120 core containing V3 was solved (Huang et al. 2007). This structure demonstrated that sulfated tyrosine residues formed salt bridges and numerous hydrogen bonds with gp120. The molecular rearrangements in gp120 needed to form the tyrosine-binding domain for CCR5 Tyr14 and 412d Tyr100 result in the creation of a β-hairpin and stabilization of the previously flexible V3 loop.

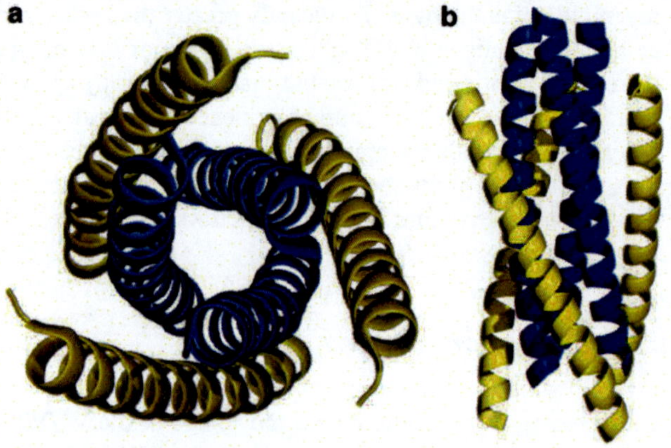

Fig. 10.4 Six-helix bundle structure (6HB). The three central HR-N domains are shown in *blue*, and the three HR-C domains are shown in *yellow*. The HR-C helices are antiparallel to the HR-N helices and pack into exterior grooves formed by the HR-N central coils. **(a)** Top view and **(b)** side view (PDB ID: 3F4Y) (Horne et al. 2009)

10.4.2.1 Env Determinants of CCR5 and CXCR4 Use

Elucidating the mechanistic differences governing CCR5 or CXCR4 use has significant clinical implications. The emergence of X4 viruses heralds progression to AIDS, and while causation has not been proven, it is known that X4 HIV is more pathogenic to CD4+ T cells in vitro. At present it is unclear whether the outgrowth of X4 HIV results from a depletion of CCR5-positive cells and thus exhaustion of the R5 HIV reservoir, or whether X4 viruses stochastically emerge and preferentially expand over R5 viruses. Being able to predict such a R5 to X4 coreceptor switch may enable clinicians to better predict a patient's disease progression and response to R5 antagonists.

In general, R5 viruses depend more upon the N-terminal interactions than X4 viruses which favor ECL2 interactions (Rizzuto and Sodroski 2000). The most significant Env contributor to coreceptor tropism is the V3 loop with the degree of charge playing a significant role in determining whether a virus uses CCR5 or CXCR4 (Hartley et al. 2005; Hwang et al. 1991). Most R5 viruses have a net V3 charge of +3 to +5, while most X4 viruses have a charge of +7 to +10, consistent with the increased negative charge on the CXCR4 N-terminus and ECLs compared to that of CCR5 (Jensen et al. 2003; Kwong et al. 2000). More specifically, positively charged residues at positions 11 and 24/25 in V3 are highly predictive of X4 tropism, while the absence of positive charges at these locations does not preclude CXCR4 use (Fouchier et al. 1992; Hartley et al. 2005).

10.4.3 Fusion: Formation and Opening of the Fusion Pore

Coreceptor binding triggers a "cast-and-fold" mechanism of membrane fusion (Melikyan 2008). The hydrophobic, N-terminal gp41 fusion peptide is exposed and likely projected toward the target cell membrane into which it inserts, tethering the viral and host membranes and destabilizing the host lipid bilayer (Tamm and Han 2000). The gp41 ectodomain contains two helical regions, both approximately 36 residues in length. The N-terminal helical regions (HR-N) from each gp41 subunit form a triple-stranded, coiled-coil structure. During the process of membrane fusion, the three C-terminal helical regions (HR-C) fold back and pack into exterior grooves formed at the external interface of the three HR-N domains in an antiparallel fashion. This results in a highly stable six-helix bundle (6HB) in which the fusion peptide and gp41 transmembrane domains, along with their respective membranes, are brought into close spatial proximity (Fig. 10.4) (Chan et al. 1997; Weissenhorn et al. 1997).

Several lines of evidence suggest that 6HB formation is critically important for membrane fusion. First, purified HR-N and HR-C peptides spontaneously fold into a 6HB in the absence of gp120.

Second, the 6HB is relatively thermostable compared to native gp41 suggesting it has adopted the lowest energy conformation. Third, both 6HB formation and fusion can be inhibited with peptides that mimic the 6HB N-terminal helices, even after CD4 binding. Fourth, the most robust amino acid conservation in gp41 between HIV and SIV is seen at the residues predicted to interact in the 6HB (Chan et al. 1997; Furuta et al. 1998; Weissenhorn et al. 1997). Finally, similar 6HB structures are seen in postfusion influenza hemagglutinin, Ebola glycoprotein, Moloney murine leukemia virus, and numerous other viral fusion machines suggesting this is a commonly utilized mechanism of membrane fusion (Bullough et al. 1994; Fass et al. 1996; Weissenhorn et al. 1997, 1998a, b).

Historically, it has been assumed that the highly conserved 6HB was responsible for the initiation of pore formation (Chan et al. 1997; Weissenhorn et al. 1997). However, recent evidence suggests that the prebundle complex actually initiates early pore formation and that the 6HB stabilizes and facilitates expansion of the nascent fusion pore (Markosyan et al. 2003). The prebundle state can be divided into an early prebundle, or temperature-arrested stage in which HR-N and HR-C are maximally separated, and a late prebundle that is present after hinge folding, but immediately before 6HB formation (Markosyan et al. 2003). The late prebundle can revert to the early prebundle in vitro by lowering the temperature to 4°C resulting in pore closure; however, the 6HB structure is the lowest energy state and cannot revert to the late prebundle, and thus it locks the pore open. Inhibitors of 6HB formation, such as the peptide enfuvirtide, allow pore formation, but prevent its expansion and stabilization.

10.5 Env Stoichiometry

While a tremendous amount has been learned about the molecular mechanisms of HIV entry and fusion, the overall stoichiometry of such entry events remains elusive. At present, it is still unclear how many Env trimers are required for entry and how many gp120 subunits per trimer are needed to productively engage CD4 and coreceptor. Resolving these questions may have implications for drug and vaccine design and may further reveal common principles of membrane fusion.

It has been reported that most wild-type HIV virions contain 10–15 randomly distributed Env spikes per virion (Liu et al. 2008; Yuste et al. 2004; Zhu et al. 2006), placing an upper limit on the number of spikes needed for membrane fusion. The trimeric Env spikes are likely mobile within the plane of the viral membrane, since an electron tomography study analyzing HIV and CD4+ T cells revealed between five and seven electron dense structures consistent in size and shape with HIV spikes at the point of virus–cell contact (Sougrat et al. 2007). Such structures were not detected in the presence of a small molecule CCR5 inhibitor or fusion inhibitor, consistent with the densities playing a role in entry. However, the presence of five to seven Env trimers at the site of virus–cell contact does not necessarily mean that all are involved in the membrane fusion process.

A variety of modeling approaches have been taken to estimate the number of Env trimers needed for membrane fusion, though these have yielded different answers, sometimes even from the same datasets. Yang et al. expressed in cells different ratios of wild-type Env proteins along with very similar Envs that, due to one or more amino acid changes, could not be neutralized by a specific monoclonal antibody (Yang et al. 2005). Since coexpression of similar Envs in the same cell results in the production of mixed trimers, varying the ratio of wild-type to mutant Env would in turn vary the proportion of wild-type to mixed heterotrimers. Using this approach, and assessing the efficiency with which virus particles could be neutralized, Yang et al. concluded that a single Env spike is sufficient for HIV entry. However, several other groups analyzed the data generated by Yang et al. and reached a different conclusion with $T > 1$. Klasse concluded that virions containing nine spikes require a T of five (Klasse 2007). Later, Magnus et al. generated several different mathematical models to determine a range for T between 2 and 19. The discrepancy between the analyses results from the different assumptions including how to account for imperfect transfections, nonrandom heterotrimer formation among the mutant and wild-type Env subunits, and the distribution of spike number among different virions (Magnus et al. 2009). Furthermore, Kuhmann et al. (2000) created stable cell lines expressing wild-type, N-terminal, and ECL2 mutants of CCR5 at various levels and then determined the relative infectivity of each cell population. Employing mathematical modeling, they determined that entry requires four to six coreceptor–gp120 interactions suggesting that two to six spikes are required per virion.

In addition to the uncertainty regarding the number of spikes needed for entry, it is at present uncertain how many CD4 and coreceptor molecules are needed to activate an Env trimer. In vivo, CCR5 expression is likely limiting on most target cells, compared to both CD4 and CXCR4. As a result, viruses bearing Env proteins that can be activated by one or two CCR5-binding events rather than three might be better able to infect cells expressing low levels of CCR5. In fact, it is commonly observed that some Envs are more "fusogenic" than others even when expressed at similar levels, though the mechanism(s) that accounts for these functional differences is not known. Conceivably more fusogenic Envs could be triggered by a lower number of receptor-binding events, though they could also undergo the conformational changes needed for membrane fusion with a higher degree of fidelity. A fuller understanding of how receptor and Env cooperativity govern membrane fusion will likely provide greater insight into viral tropism, and why some virus strains are not able to infect some primary cell types even though the relevant viral receptors are expressed (Salazar-Gonzalez et al. 2009).

10.6 Inhibitors of HIV Entry

While studying HIV entry has informed us about basic principles of virology and cell biology, the overarching goal should be to develop novel and effective therapeutics to limit the morbidity and mortality of the HIV/AIDS pandemic. Currently, two entry inhibitors, the CCR5 inhibitor maraviroc and the fusion inhibitor enfuvirtide, are FDA approved for the treatment of HIV infection. A number of other compounds that have targeted nearly every step of the entry pathway have also been tested in the clinic. For purposes of this discussion on therapeutics, HIV entry will be divided into three components: (1) attachment and CD4 binding, (2) coreceptor binding, and (3) membrane fusion (Table 10.1).

Table 10.1 HIV entry inhibitors

Class	Compound	Development status	Reference
Attachment inhibitor	PRO 2000	Phase III trial showed no efficacy	Rusconi et al. (1996)
Attachment inhibitor	Carraguard	Phase III trial showed no efficacy	Skoler-Karpoff et al. (2008)
Attachment inhibitor	Cellulose sulfate	Phase III trial showed no efficacy	Halpern et al. (2008) and Van Damme et al. (2008)
CD4-binding site inhibitor	BMS-378806	Preclinical development	Lin et al. (2003)
CD4 mimetic	Soluble CD4	Phase I study showed no efficacy	Daar et al. (1990) and Schacker et al. (1994)
CD4 downmodulators	CADA derivatives	Preclinical development	Vermeire et al. (2002, 2007, 2008)
CCR5 antagonist	Maraviroc	FDA approved	(2009)
CCR5 antibody	PRO 140	Phase II trials	Jacobson et al. (2008)
CXCR4 antagonist	Plerixafor	FDA approved for hematopoietic stem cell mobilization	Flomenberg et al. (2010)
Fusion inhibitor	Enfuvirtide	FDA approved	Wild et al. (1992, 1993)
Fusion inhibitor	D-peptides	Preclinical development	Welch et al. (2007)

10.6.1 Attachment and CD4-Binding Inhibitors

Attachment and CD4-binding inhibitors include relatively nonspecific anionic polymers, CD4bs inhibitors, sCD4 mimetics, and CD4 downmodulators. Anionic polymers, which act by preventing the favorable electrostatic interactions between the positively charged Env and negatively charged cell surface, have been predominantly studied for use in vaginal microbicides. PRO 2000, a naphthalene sulfonate polymer, inhibits soluble gp120 and CD4 binding in vitro (Rusconi et al. 1996). However, in a phase III clinical microbicide trial, PRO 2000 demonstrated no efficacy (CONRAD 2009, 14 December). Other anionic polymers including cellulose sulfate and Carraguard, derived from seaweed, demonstrated no efficacy, and cellulose sulfate may have actually increased the risk of HIV transmission (Halpern et al. 2008; Skoler-Karpoff et al. 2008; Van Damme et al. 2008).

Another approach involves targeting the CD4bs on gp120. Several small molecules that bind to gp120, such as BMS-378806 and BMS-488043, have antiviral activity in vitro and prevent CD4–gp120 binding (Ho et al. 2006; Lin et al. 2003). However, for at least BMS-378806, HIV quickly adapts resistance limiting its therapeutic potential (Lin et al. 2003). Furthermore, sCD4 demonstrated anti-HIV activity in vitro, but early-stage clinical trials were unable to demonstrate antiviral activity, most likely due to insufficient circulating concentrations of sCD4 (Daar et al. 1990). An additional class of compounds inhibits Env–CD4 interactions by downregulating CD4. These drugs, derivatives of cyclotriazadisulfonamide (CADA) (Vermeire et al. 2002), reduce CD4 expression by an unknown mechanism. However, they do not alter CD4 mRNA levels suggesting they exert their function in a posttranscriptional manner (Vermeire et al. 2007).

10.6.2 Coreceptor-Binding Inhibitors

Discovery of the *ccr5Δ32* mutation demonstrated that CCR5 is not essential for normal human growth and development, suggesting that it could be safely targeted by small molecule inhibitors (Dean et al. 1996; Liu et al. 1996; Samson et al. 1996). Several small molecule CCR5 antagonists have been developed and shown to have antiviral activity in vivo, including maraviroc, which received FDA approval in 2007 for use in treating HIV-infected individuals. Most small molecule CCR5 inhibitors, including maraviroc, function by binding to a hydrophobic pocket within the transmembrane domains of the protein – a region of the receptor not thought to directly interact with the viral Env protein (Baba et al. 1999; Tilton et al. 2010; Tilton and Doms 2010). As a result, CCR5 antagonists likely function by an allosteric mechanism, inducing conformational changes in the ECL domains of the receptor that subsequently prevent Env binding. Viral resistance to such compounds occurs by one of two pathways. In vivo, it appears that the most common resistance pathway is outgrowth of CXCR4-using viruses, even when present below the limit of detection in standard assays at the initiation of therapy. A second, less common pathway results from mutations in Env that enable it to utilize the drug-bound conformation of the receptor (Berro et al. 2009). In at least some cases, it appears that enhanced utilization of the CCR5 N-terminal domain is associated with this phenotype. In addition to traditional small molecule inhibitors, CCR5 blocking antibodies are being explored for therapeutic purposes. One such antibody, PRO 140, blocks HIV utilization of CCR5 while preserving CCR5 ligand function. PRO 140 is currently in phase II clinical trials (Jacobson et al. 2008; Tilton and Doms 2010).

Unlike for CCR5, inhibiting CXCR4 has been met with limited success primarily due to the concerns of systemic toxicity. Several CXCR4 inhibitors advanced to early-stage clinical trials, but none are currently ongoing for the treatment of HIV. One CXCR4 inhibitor, plerixafor, was recently FDA approved to mobilize hematopoietic stem cells to the peripheral blood for harvesting prior to bone marrow transplantation (Flomenberg et al. 2010).

10.6.3 *Fusion Inhibitors*

Membrane fusion is the net result of Env–receptor interactions and is the target of the first entry inhibitor ever approved, enfuvirtide. Enfuvirtide, previously known as T20, is a 36-amino-acid mimetic of the HR-C domain. The peptide binds the central coiled coils comprised of three HR-N molecules and inhibits 6HB formation, thus preventing fusion (Wild et al. 1992, 1993). Despite the in vivo efficacy of enfuvirtide, resistance mutations in a ten-amino-acid region of HR-N that prevent enfuvirtide binding have been well documented (Greenberg and Cammack 2004). Importantly though, resistance to enfuvirtide does not confer cross-resistance to other classes of entry inhibitors (Reeves et al. 2005). Novel fusion peptide-based inhibitors have been designed to combat enfuvirtide-resistant viruses, and some of these molecules have synergistic effects with enfuvirtide (Dwyer et al. 2007; Pan et al. 2009a, b).

While these peptide-based fusion inhibitors exhibit efficacy, they are limited by the fact that they are not orally bioavailable and therefore must be injected, a significant hindrance in maintaining patient adherence. One potential solution is the development of orally bioavailable small molecules that recapitulate enfuvirtide's mechanism of action by blocking the hydrophobic "knob-into-hole" interactions. The knobs are hydrophobic HR-C residues, specifically tryptophans and isoleucines, that pack into large hydrophobic holes present in the HR-N central coil (Chan et al. 1997). D-peptide inhibitors of gp120 represent one such exciting new class of compounds. They have potent in vitro activity and are not degraded by intestinal proteases and thus have the potential to be orally bioavailable (Welch et al. 2007).

10.7 Conclusion

Work on the mechanisms of HIV entry has led to the discovery of human mutations affecting HIV susceptibility and disease progression as well as the development of new antiviral agents, such as enfuvirtide and maraviroc. In addition to the continued development of entry inhibitors, a critical future challenge is translating our molecular understanding of HIV entry into therapeutically useful information. Will our knowledge of HIV entry make it possible to predict which patients stand most to benefit from the use of entry inhibitors? More generally, the discovery of the HIV coreceptors revealed a new mechanism for triggering conformational changes in viral membrane fusion proteins – that of sequential receptor engagement. In addition, the ability of attachment factors such as DC-SIGN to enhance HIV-1 infection has led to similar discoveries for other viruses. Finally, the ability of HIV-1 to enter particular T-cell subsets is still not fully understood. As our understanding of HIV entry becomes more complete, so too should our understanding of HIV tropism and pathogenesis.

References

(2009). FDA advisory committee approves Selzentry. AIDS Patient Care STDS 23, 987

Alkhatib G, Ahuja SS, Light D, Mummidi S, Berger EA, Ahuja SK (1997) CC chemokine receptor 5-mediated signaling and HIV-1 Co-receptor activity share common structural determinants. Critical residues in the third extracellular loop support HIV-1 fusion. J Biol Chem 272:19771–19776

Alkhatib G, Combadiere C, Broder CC, Feng Y, Kennedy PE, Murphy PM, Berger EA (1996) CC CKR5: a RANTES, MIP-1alpha, MIP-1beta receptor as a fusion cofactor for macrophage-tropic HIV-1. Science 272:1955–1958

Baba M, Nishimura O, Kanzaki N, Okamoto M, Sawada H, Iizawa Y, Shiraishi M, Aramaki Y, Okonogi K, Ogawa Y, Meguro K, Fujino M (1999) A small-molecule, nonpeptide CCR5 antagonist with highly potent and selective anti-HIV-1 activity. Proc Natl Acad Sci USA 96:5698–5703

Baribaud F, Pohlmann S, Doms RW (2001) The role of DC-SIGN and DC-SIGNR in HIV and SIV attachment, infection, and transmission. Virology 286:1–6

Barre-Sinoussi F, Chermann JC, Rey F, Nugeyre MT, Chamaret S, Gruest J, Dauguet C, Axler-Blin C, Vezinet-Brun F, Rouzioux C, Rozenbaum W, Montagnier L (1983) Isolation of a T-lymphotropic retrovirus from a patient at risk for acquired immune deficiency syndrome (AIDS). Science 220:868–871

Berro R, Sanders RW, Lu M, Klasse PJ, Moore JP (2009) Two HIV-1 variants resistant to small molecule CCR5 inhibitors differ in how they use CCR5 for entry. PLoS Pathog 5:e1000548

Blanpain C, Libert F, Vassart G, Parmentier M (2002) CCR5 and HIV infection. Receptors Channels 8:19–31

Bleul CC, Fuhlbrigge RC, Casasnovas JM, Aiuti A, Springer TA (1996) A highly efficacious lymphocyte chemoattractant, stromal cell-derived factor 1 (SDF-1). J Exp Med 184:1101–1109

Brenchley JM, Douek DC (2008) HIV infection and the gastrointestinal immune system. Mucosal Immunol 1:23–30

Bullough PA, Hughson FM, Skehel JJ, Wiley DC (1994) Structure of influenza haemagglutinin at the pH of membrane fusion. Nature 371:37–43

Chan DC, Fass D, Berger JM, Kim PS (1997) Core structure of gp41 from the HIV envelope glycoprotein. Cell 89:263–273

Chen B, Vogan EM, Gong H, Skehel JJ, Wiley DC, Harrison SC (2005) Structure of an unliganded simian immunodeficiency virus gp120 core. Nature 433:834–841

Choe H, Farzan M (2009) Chapter 7. Tyrosine sulfation of HIV-1 coreceptors and other chemokine receptors. Methods Enzymol 461:147–170

Choe H, Farzan M, Sun Y, Sullivan N, Rollins B, Ponath PD, Wu L, Mackay CR, LaRosa G, Newman W, Gerard N, Gerard C, Sodroski J (1996) The beta-chemokine receptors CCR3 and CCR5 facilitate infection by primary HIV-1 isolates. Cell 85:1135–1148

Choe H, Li W, Wright PL, Vasilieva N, Venturi M, Huang CC, Grundner C, Dorfman T, Zwick MB, Wang L, Rosenberg ES, Kwong PD, Burton DR, Robinson JE, Sodroski JG, Farzan M (2003) Tyrosine sulfation of human antibodies contributes to recognition of the CCR5 binding region of HIV-1 gp120. Cell 114:161–170

Cocchi F, DeVico AL, Garzino-Demo A, Arya SK, Gallo RC, Lusso P (1995) Identification of RANTES, MIP-1 alpha, and MIP-1 beta as the major HIV-suppressive factors produced by CD8+ T cells. Science 270: 1811–1815

Combadiere C, Ahuja SK, Tiffany HL, Murphy PM (1996) Cloning and functional expression of CC CKR5, a human monocyte CC chemokine receptor selective for MIP-1(alpha), MIP-1(beta), and RANTES. J Leukoc Biol 60:147–152

Connor RI, Sheridan KE, Ceradini D, Choe S, Landau NR (1997) Change in coreceptor use correlates with disease progression in HIV-1-infected individuals. J Exp Med 185:621–628

CONRAD (2009) CONRAD statement on MDP 301 results. December 14

Cormier EG, Persuh M, Thompson DA, Lin SW, Sakmar TP, Olson WC, Dragic T (2000) Specific interaction of CCR5 amino-terminal domain peptides containing sulfotyrosines with HIV-1 envelope glycoprotein gp120. Proc Natl Acad Sci USA 97:5762–5767

Daar ES, Li XL, Moudgil T, Ho DD (1990) High concentrations of recombinant soluble CD4 are required to neutralize primary human immunodeficiency virus type 1 isolates. Proc Natl Acad Sci USA 87:6574–6578

Dalgleish AG, Beverley PC, Clapham PR, Crawford DH, Greaves MF, Weiss RA (1984) The CD4 (T4) antigen is an essential component of the receptor for the AIDS retrovirus. Nature 312:763–767

Dean M, Carrington M, Winkler C, Huttley GA, Smith MW, Allikmets R, Goedert JJ, Buchbinder SP, Vittinghoff E, Gomperts E, Donfield S, Vlahov D, Kaslow R, Saah A, Rinaldo C, Detels R, O'Brien SJ (1996) Genetic restriction of HIV-1 infection and progression to AIDS by a deletion allele of the CKR5 structural gene. Hemophilia Growth and Development Study, Multicenter AIDS Cohort Study, Multicenter Hemophilia Cohort Study, San Francisco City Cohort, ALIVE Study. Science 273:1856–1862

Deng H, Liu R, Ellmeier W, Choe S, Unutmaz D, Burkhart M, Di Marzio P, Marmon S, Sutton RE, Hill CM, Davis CB, Peiper SC, Schall TJ, Littman DR, Landau NR (1996) Identification of a major co-receptor for primary isolates of HIV-1. Nature 381:661–666

Deng HK, Unutmaz D, KewalRamani VN, Littman DR (1997) Expression cloning of new receptors used by simian and human immunodeficiency viruses. Nature 388:296–300

Doranz BJ, Lu ZH, Rucker J, Zhang TY, Sharron M, Cen YH, Wang ZX, Guo HH, Du JG, Accavitti MA, Doms RW, Peiper SC (1997) Two distinct CCR5 domains can mediate coreceptor usage by human immunodeficiency virus type 1. J Virol 71:6305–6314

Doranz BJ, Rucker J, Yi Y, Smyth RJ, Samson M, Peiper SC, Parmentier M, Collman RG, Doms RW (1996) A dual-tropic primary HIV-1 isolate that uses fusin and the beta-chemokine receptors CKR-5, CKR-3, and CKR-2b as fusion cofactors. Cell 85:1149–1158

Dragic T, Litwin V, Allaway GP, Martin SR, Huang Y, Nagashima KA, Cayanan C, Maddon PJ, Koup RA, Moore JP, Paxton WA (1996) HIV-1 entry into CD4+ cells is mediated by the chemokine receptor CC-CKR-5. Nature 381:667–673

Drosopoulos WC, Rezende LF, Wainberg MA, Prasad VR (1998) Virtues of being faithful: can we limit the genetic variation in human immunodeficiency virus? J Mol Med 76:604–612
Dwyer JJ, Wilson KL, Davison DK, Freel SA, Seedorff JE, Wring SA, Tvermoes NA, Matthews TJ, Greenberg ML, Delmedico MK (2007) Design of helical, oligomeric HIV-1 fusion inhibitor peptides with potent activity against enfuvirtide-resistant virus. Proc Natl Acad Sci USA 104:12772–12777
Earl PL, Moss B, Doms RW (1991) Folding, interaction with GRP78-BiP, assembly, and transport of the human immunodeficiency virus type 1 envelope protein. J Virol 65:2047–2055
Edinger AL, Hoffman TL, Sharron M, Lee B, Yi Y, Choe W, Kolson DL, Mitrovic B, Zhou Y, Faulds D, Collman RG, Hesselgesser J, Horuk R, Doms RW (1998) An orphan seven-transmembrane domain receptor expressed widely in the brain functions as a coreceptor for human immunodeficiency virus type 1 and simian immunodeficiency virus. J Virol 72:7934–7940
Farzan M, Choe H, Martin K, Marcon L, Hofmann W, Karlsson G, Sun Y, Barrett P, Marchand N, Sullivan N, Gerard N, Gerard C, Sodroski J (1997) Two orphan seven-transmembrane segment receptors which are expressed in CD4-positive cells support simian immunodeficiency virus infection. J Exp Med 186:405–411
Farzan M, Choe H, Vaca L, Martin K, Sun Y, Desjardins E, Ruffing N, Wu L, Wyatt R, Gerard N, Gerard C, Sodroski J (1998) A tyrosine-rich region in the N terminus of CCR5 is important for human immunodeficiency virus type 1 entry and mediates an association between gp120 and CCR5. J Virol 72:1160–1164
Farzan M, Mirzabekov T, Kolchinsky P, Wyatt R, Cayabyab M, Gerard NP, Gerard C, Sodroski J, Choe H (1999) Tyrosine sulfation of the amino terminus of CCR5 facilitates HIV-1 entry. Cell 96:667–676
Fass D, Harrison SC, Kim PS (1996) Retrovirus envelope domain at 1.7 angstrom resolution. Nat Struct Biol 3:465–469
Feng Y, Broder CC, Kennedy PE, Berger EA (1996) HIV-1 entry cofactor: functional cDNA cloning of a seven-transmembrane, G protein-coupled receptor. Science 272:872–877
Flomenberg N, Comenzo RL, Badel K, Calandra G (2010) Plerixafor (Mozobil((R))) alone to mobilize hematopoietic stem cells from multiple myeloma patients for autologous transplantation. Biol Blood Marrow Transplant 16:695–700
Fouchier RA, Groenink M, Kootstra NA, Tersmette M, Huisman HG, Miedema F, Schuitemaker H (1992) Phenotype-associated sequence variation in the third variable domain of the human immunodeficiency virus type 1 gp120 molecule. J Virol 66:3183–3187
Freed EO (2001) HIV-1 replication. Somat Cell Mol Genet 26:13–33
Furuta RA, Wild CT, Weng Y, Weiss CD (1998) Capture of an early fusion-active conformation of HIV-1 gp41. Nat Struct Biol 5:276–279
Gallo RC, Sarin PS, Gelmann EP, Robert-Guroff M, Richardson E, Kalyanaraman VS, Mann D, Sidhu GD, Stahl RE, Zolla-Pazner S, Leibowitch J, Popovic M (1983) Isolation of human T-cell leukemia virus in acquired immune deficiency syndrome (AIDS). Science 220:865–867
Gao F, Bailes E, Robertson DL, Chen Y, Rodenburg CM, Michael SF, Cummins LB, Arthur LO, Peeters M, Shaw GM, Sharp PM, Hahn BH (1999) Origin of HIV-1 in the chimpanzee *Pan troglodytes troglodytes*. Nature 397:436–441
Geijtenbeek TB, Kwon DS, Torensma R, van Vliet SJ, van Duijnhoven GC, Middel J, Cornelissen IL, Nottet HS, KewalRamani VN, Littman DR, Figdor CG, van Kooyk Y (2000) DC-SIGN, a dendritic cell-specific HIV-1-binding protein that enhances trans-infection of T cells. Cell 100:587–597
Gonzalez E, Kulkarni H, Bolivar H, Mangano A, Sanchez R, Catano G, Nibbs RJ, Freedman BI, Quinones MP, Bamshad MJ, Murthy KK, Rovin BH, Bradley W, Clark RA, Anderson SA, O'Connell RJ, Agan BK, Ahuja SS, Bologna R, Sen L, Dolan MJ, Ahuja SK (2005) The influence of CCL3L1 gene-containing segmental duplications on HIV-1/AIDS susceptibility. Science 307:1434–1440
Gorlin RJ, Gelb B, Diaz GA, Lofsness KG, Pittelkow MR, Fenyk JR Jr (2000) WHIM syndrome, an autosomal dominant disorder: clinical, hematological, and molecular studies. Am J Med Genet 91:368–376
Greenberg ML, Cammack N (2004) Resistance to enfuvirtide, the first HIV fusion inhibitor. J Antimicrob Chemother 54:333–340
Hallenberger S, Bosch V, Angliker H, Shaw E, Klenk HD, Garten W (1992) Inhibition of furin-mediated cleavage activation of HIV-1 glycoprotein gp160. Nature 360:358–361
Halpern V, Ogunsola F, Obunge O, Wang CH, Onyejepu N, Oduyebo O, Taylor D, McNeil L, Mehta N, Umo-Otong J, Otusanya S, Crucitti T, Abdellati S (2008) Effectiveness of cellulose sulfate vaginal gel for the prevention of HIV infection: results of a Phase III trial in Nigeria. PLoS One 3:e3784
Hartley O, Klasse PJ, Sattentau QJ, Moore JP (2005) V3: HIV's switch-hitter. AIDS Res Hum Retroviruses 21:171–189
He Y, Cheng J, Lu H, Li J, Hu J, Qi Z, Liu Z, Jiang S, Dai Q (2008a) Potent HIV fusion inhibitors against enfuvirtide-resistant HIV-1 strains. Proc Natl Acad Sci USA 105:16332–16337
He Y, Xiao Y, Song H, Liang Q, Ju D, Chen X, Lu H, Jing W, Jiang S, Zhang L (2008b) Design and evaluation of sifuvirtide, a novel HIV-1 fusion inhibitor. J Biol Chem 283:11126–11134
Hendrix CW, Collier AC, Lederman MM, Schols D, Pollard RB, Brown S, Jackson JB, Coombs RW, Glesby MJ, Flexner CW, Bridger GJ, Badel K, MacFarland RT, Henson GW, Calandra G (2004) Safety, pharmacokinetics,

and antiviral activity of AMD3100, a selective CXCR4 receptor inhibitor, in HIV-1 infection. J Acquir Immune Defic Syndr 37:1253–1262

Hernandez PA, Gorlin RJ, Lukens JN, Taniuchi S, Bohinjec J, Francois F, Klotman ME, Diaz GA (2003) Mutations in the chemokine receptor gene CXCR4 are associated with WHIM syndrome, a combined immunodeficiency disease. Nat Genet 34:70–74

Ho HT, Fan L, Nowicka-Sans B, McAuliffe B, Li CB, Yamanaka G, Zhou N, Fang H, Dicker I, Dalterio R, Gong YF, Wang T, Yin Z, Ueda Y, Matiskella J, Kadow J, Clapham P, Robinson J, Colonno R, Lin PF (2006) Envelope conformational changes induced by human immunodeficiency virus type 1 attachment inhibitors prevent CD4 binding and downstream entry events. J Virol 80:4017–4025

Hoffman TL, MacGregor RR, Burger H, Mick R, Doms RW, Collman RG (1997) CCR5 genotypes in sexually active couples discordant for human immunodeficiency virus type 1 infection status. J Infect Dis 176:1093–1096

Horne WS, Johnson LM, Ketas TJ, Klasse PJ, Lu M, Moore JP, Gellman SH (2009) Structural and biological mimicry of protein surface recognition by alpha/beta-peptide foldamers. Proc Natl Acad Sci USA 106:14751–14756

Huang CC, Lam SN, Acharya P, Tang M, Xiang SH, Hussan SS, Stanfield RL, Robinson J, Sodroski J, Wilson IA, Wyatt R, Bewley CA, Kwong PD (2007) Structures of the CCR5 N terminus and of a tyrosine-sulfated antibody with HIV-1 gp120 and CD4. Science 317:1930–1934

Huang CC, Tang M, Zhang MY, Majeed S, Montabana E, Stanfield RL, Dimitrov DS, Korber B, Sodroski J, Wilson IA, Wyatt R, Kwong PD (2005) Structure of a V3-containing HIV-1 gp120 core. Science 310:1025–1028

Huang Y, Paxton WA, Wolinsky SM, Neumann AU, Zhang L, He T, Kang S, Ceradini D, Jin Z, Yazdanbakhsh K, Kunstman K, Erickson D, Dragon E, Landau NR, Phair J, Ho DD, Koup RA (1996) The role of a mutant CCR5 allele in HIV-1 transmission and disease progression. Nat Med 2:1240–1243

Hwang SS, Boyle TJ, Lyerly HK, Cullen BR (1991) Identification of the envelope V3 loop as the primary determinant of cell tropism in HIV-1. Science 253:71–74

Jacobson JM, Saag MS, Thompson MA, Fischl MA, Liporace R, Reichman RC, Redfield RR, Fichtenbaum CJ, Zingman BS, Patel MC, Murga JD, Pemrick SM, D'Ambrosio P, Michael M, Kroger H, Ly H, Rotshteyn Y, Buice R, Morris SA, Stavola JJ, Maddon PJ, Kremer AB, Olson WC (2008) Antiviral activity of single-dose PRO 140, a CCR5 monoclonal antibody, in HIV-infected adults. J Infect Dis 198:1345–1352

Jensen MA, Li FS, van't Wout AB, Nickle DC, Shriner D, He HX, McLaughlin S, Shankarappa R, Margolick JB, Mullins JI (2003) Improved coreceptor usage prediction and genotypic monitoring of R5-to-X4 transition by motif analysis of human immunodeficiency virus type 1 env V3 loop sequences. J Virol 77:13376–13388

Keulen W, Nijhuis M, Schuurman R, Berkhout B, Boucher C (1997) Reverse transcriptase fidelity and HIV-1 variation. Science 275:229, author reply 230–221

Klasse PJ (2007) Modeling how many envelope glycoprotein trimers per virion participate in human immunodeficiency virus infectivity and its neutralization by antibody. Virology 369:245–262

Kotler DP, Gaetz HP, Lange M, Klein EB, Holt PR (1984) Enteropathy associated with the acquired immunodeficiency syndrome. Ann Intern Med 101:421–428

Kuhmann SE, Platt EJ, Kozak SL, Kabat D (2000) Cooperation of multiple CCR5 coreceptors is required for infections by human immunodeficiency virus type 1. J Virol 74:7005–7015

Kwong PD, Wyatt R, Robinson J, Sweet RW, Sodroski J, Hendrickson WA (1998) Structure of an HIV gp120 envelope glycoprotein in complex with the CD4 receptor and a neutralizing human antibody. Nature 393:648–659

Kwong PD, Wyatt R, Sattentau QJ, Sodroski J, Hendrickson WA (2000) Oligomeric modeling and electrostatic analysis of the gp120 envelope glycoprotein of human immunodeficiency virus. J Virol 74:1961–1972

Laakso MM, Lee FH, Haggarty B, Agrawal C, Nolan KM, Biscone M, Romano J, Jordan AP, Leslie GJ, Meissner EG, Su L, Hoxie JA, Doms RW (2007) V3 loop truncations in HIV-1 envelope impart resistance to coreceptor inhibitors and enhanced sensitivity to neutralizing antibodies. PLoS Pathog 3:e117

Lal RB, Chakrabarti S, Yang C (2005) Impact of genetic diversity of HIV-1 on diagnosis, antiretroviral therapy & vaccine development. Indian J Med Res 121:287–314

Land A, Braakman I (2001) Folding of the human immunodeficiency virus type 1 envelope glycoprotein in the endoplasmic reticulum. Biochimie 83:783–790

Land A, Zonneveld D, Braakman I (2003) Folding of HIV-1 envelope glycoprotein involves extensive isomerization of disulfide bonds and conformation-dependent leader peptide cleavage. FASEB J 17:1058–1067

Lin PF, Blair W, Wang T, Spicer T, Guo Q, Zhou N, Gong YF, Wang HG, Rose R, Yamanaka G, Robinson B, Li CB, Fridell R, Deminie C, Demers G, Yang Z, Zadjura L, Meanwell N, Colonno R (2003) A small molecule HIV-1 inhibitor that targets the HIV-1 envelope and inhibits CD4 receptor binding. Proc Natl Acad Sci USA 100: 11013–11018

Liu J, Bartesaghi A, Borgnia MJ, Sapiro G, Subramaniam S (2008) Molecular architecture of native HIV-1 gp120 trimers. Nature 455:109–113

Liu R, Paxton WA, Choe S, Ceradini D, Martin SR, Horuk R, MacDonald ME, Stuhlmann H, Koup RA, Landau NR (1996) Homozygous defect in HIV-1 coreceptor accounts for resistance of some multiply-exposed individuals to HIV-1 infection. Cell 86:367–377

Maddon PJ, McDougal JS, Clapham PR, Dalgleish AG, Jamal S, Weiss RA, Axel R (1988) HIV infection does not require endocytosis of its receptor, CD4. Cell 54:865–874

Magnus C, Rusert P, Bonhoeffer S, Trkola A, Regoes RR (2009) Estimating the stoichiometry of human immunodeficiency virus entry. J Virol 83:1523–1531

Markosyan RM, Cohen FS, Melikyan GB (2003) HIV-1 envelope proteins complete their folding into six-helix bundles immediately after fusion pore formation. Mol Biol Cell 14:926–938

Marsh M, Helenius A (2006) Virus entry: open sesame. Cell 124:729–740

McClure MO, Marsh M, Weiss RA (1988) Human immunodeficiency virus infection of CD4-bearing cells occurs by a pH-independent mechanism. EMBO J 7:513–518

McDermott DH, Zimmerman PA, Guignard F, Kleeberger CA, Leitman SF, Murphy PM (1998) CCR5 promoter polymorphism and HIV-1 disease progression. Multicenter AIDS Cohort Study (MACS). Lancet 352: 866–870

Melikyan GB (2008) Common principles and intermediates of viral protein-mediated fusion: the HIV-1 paradigm. Retrovirology 5:111

Miyauchi K, Kim Y, Latinovic O, Morozov V, Melikyan GB (2009) HIV enters cells via endocytosis and dynamin-dependent fusion with endosomes. Cell 137:433–444

Mondor I, Ugolini S, Sattentau QJ (1998) Human immunodeficiency virus type 1 attachment to HeLa CD4 cells is CD4 independent and gp120 dependent and requires cell surface heparans. J Virol 72:3623–3634

Myszka DG, Sweet RW, Hensley P, Brigham-Burke M, Kwong PD, Hendrickson WA, Wyatt R, Sodroski J, Doyle ML (2000) Energetics of the HIV gp120-CD4 binding reaction. Proc Natl Acad Sci USA 97:9026–9031

Nagasawa T, Hirota S, Tachibana K, Takakura N, Nishikawa S, Kitamura Y, Yoshida N, Kikutani H, Kishimoto T (1996) Defects of B-cell lymphopoiesis and bone-marrow myelopoiesis in mice lacking the CXC chemokine PBSF/SDF-1. Nature 382:635–638

Oberlin E, Amara A, Bachelerie F, Bessia C, Virelizier JL, Arenzana-Seisdedos F, Schwartz O, Heard JM, Clark-Lewis I, Legler DF, Loetscher M, Baggiolini M, Moser B (1996) The CXC chemokine SDF-1 is the ligand for LESTR/fusin and prevents infection by T-cell-line-adapted HIV-1. Nature 382:833–835

Pan C, Cai L, Lu H, Qi Z, Jiang S (2009a) Combinations of the first and next generations of human immunodeficiency virus (HIV) fusion inhibitors exhibit a highly potent synergistic effect against enfuvirtide-sensitive and -resistant HIV type 1 strains. J Virol 83:7862–7872

Pan C, Lu H, Qi Z, Jiang S (2009b) Synergistic efficacy of combination of enfuvirtide and sifuvirtide, the first- and next-generation HIV-fusion inhibitors. AIDS 23:639–641

Pohlmann S, Baribaud F, Lee B, Leslie GJ, Sanchez MD, Hiebenthal-Millow K, Munch J, Kirchhoff F, Doms RW (2001a) DC-SIGN interactions with human immunodeficiency virus type 1 and 2 and simian immunodeficiency virus. J Virol 75:4664–4672

Pohlmann S, Soilleux EJ, Baribaud F, Leslie GJ, Morris LS, Trowsdale J, Lee B, Coleman N, Doms RW (2001b) DC-SIGNR, a DC-SIGN homologue expressed in endothelial cells, binds to human and simian immunodeficiency viruses and activates infection in trans. Proc Natl Acad Sci USA 98:2670–2675

Poignard P, Saphire EO, Parren PW, Burton DR (2001) gp120: biologic aspects of structural features. Annu Rev Immunol 19:253–274

Quinonez R, Sinha I, Singh IR, Sutton RE (2003) Genetic footprinting of the HIV co-receptor CCR5: delineation of surface expression and viral entry determinants. Virology 307:98–115

Rabut GE, Konner JA, Kajumo F, Moore JP, Dragic T (1998) Alanine substitutions of polar and nonpolar residues in the amino-terminal domain of CCR5 differently impair entry of macrophage- and dualtropic isolates of human immunodeficiency virus type 1. J Virol 72:3464–3468

Raport CJ, Gosling J, Schweickart VL, Gray PW, Charo IF (1996) Molecular cloning and functional characterization of a novel human CC chemokine receptor (CCR5) for RANTES, MIP-1beta, and MIP-1alpha. J Biol Chem 271:17161–17166

Reeves JD, Lee FH, Miamidian JL, Jabara CB, Juntilla MM, Doms RW (2005) Enfuvirtide resistance mutations: impact on human immunodeficiency virus envelope function, entry inhibitor sensitivity, and virus neutralization. J Virol 79:4991–4999

Rizzuto C, Sodroski J (2000) Fine definition of a conserved CCR5-binding region on the human immunodeficiency virus type 1 glycoprotein 120. AIDS Res Hum Retroviruses 16:741–749

Rizzuto CD, Wyatt R, Hernandez-Ramos N, Sun Y, Kwong PD, Hendrickson WA, Sodroski J (1998) A conserved HIV gp120 glycoprotein structure involved in chemokine receptor binding. Science 280:1949–1953

Rusconi S, Moonis M, Merrill DP, Pallai PV, Neidhardt EA, Singh SK, Willis KJ, Osburne MS, Profy AT, Jenson JC, Hirsch MS (1996) Naphthalene sulfonate polymers with CD4-blocking and anti-human immunodeficiency virus type 1 activities. Antimicrob Agents Chemother 40:234–236

Saag MS, Hahn BH, Gibbons J, Li Y, Parks ES, Parks WP, Shaw GM (1988) Extensive variation of human immunodeficiency virus type-1 in vivo. Nature 334:440–444

Salazar-Gonzalez JF, Salazar MG, Keele BF, Learn GH, Giorgi EE, Li H, Decker JM, Wang S, Baalwa J, Kraus MH, Parrish NF, Shaw KS, Guffey MB, Bar KJ, Davis KL, Ochsenbauer-Jambor C, Kappes JC, Saag MS, Cohen MS, Mulenga J, Derdeyn CA, Allen S, Hunter E, Markowitz M, Hraber P, Perelson AS, Bhattacharya T, Haynes BF, Korber BT, Hahn BH, Shaw GM (2009) Genetic identity, biological phenotype, and evolutionary pathways of transmitted/founder viruses in acute and early HIV-1 infection. J Exp Med 206:1273–1289

Samson M, Libert F, Doranz BJ, Rucker J, Liesnard C, Farber CM, Saragosti S, Lapoumeroulie C, Cognaux J, Forceille C, Muyldermans G, Verhofstede C, Burtonboy G, Georges M, Imai T, Rana S, Yi Y, Smyth RJ, Collman RG, Doms RW, Vassart G, Parmentier M (1996) Resistance to HIV-1 infection in Caucasian individuals bearing mutant alleles of the CCR-5 chemokine receptor gene. Nature 382:722–725

Schacker T, Coombs RW, Collier AC, Zeh JE, Fox I, Alam J, Nelson K, Eggert E, Corey L (1994) The effects of high-dose recombinant soluble CD4 on human immunodeficiency virus type 1 viremia. J Infect Dis 169:37–40

Schurmann D, Fatkenheuer G, Reynes J, Michelet C, Raffi F, van Lier J, Caceres M, Keung A, Sansone-Parsons A, Dunkle LM, Hoffmann C (2007) Antiviral activity, pharmacokinetics and safety of vicriviroc, an oral CCR5 antagonist, during 14-day monotherapy in HIV-infected adults. AIDS 21:1293–1299

Skoler-Karpoff S, Ramjee G, Ahmed K, Altini L, Plagianos MG, Friedland B, Govender S, De Kock A, Cassim N, Palanee T, Dozier G, Maguire R, Lahteenmaki P (2008) Efficacy of Carraguard for prevention of HIV infection in women in South Africa: a randomised, double-blind, placebo-controlled trial. Lancet 372:1977–1987

Sougrat R, Bartesaghi A, Lifson JD, Bennett AE, Bess JW, Zabransky DJ, Subramaniam S (2007) Electron tomography of the contact between T cells and SIV/HIV-1: implications for viral entry. PLoS Pathog 3:e63

Stamatatos L, Morris L, Burton DR, Mascola JR (2009) Neutralizing antibodies generated during natural HIV-1 infection: good news for an HIV-1 vaccine? Nat Med 15:866–870

Stein BS, Gowda SD, Lifson JD, Penhallow RC, Bensch KG, Engleman EG (1987) pH-independent HIV entry into CD4-positive T cells via virus envelope fusion to the plasma membrane. Cell 49:659–668

Strizki JM, Tremblay C, Xu S, Wojcik L, Wagner N, Gonsiorek W, Hipkin RW, Chou CC, Pugliese-Sivo C, Xiao Y, Tagat JR, Cox K, Priestley T, Sorota S, Huang W, Hirsch M, Reyes GR, Baroudy BM (2005) Discovery and characterization of vicriviroc (SCH 417690), a CCR5 antagonist with potent activity against human immunodeficiency virus type 1. Antimicrob Agents Chemother 49:4911–4919

Subramaniam S (2006) The SIV surface spike imaged by electron tomography: one leg or three? PLoS Pathog 2:e91

Tamm LK, Han X (2000) Viral fusion peptides: a tool set to disrupt and connect biological membranes. Biosci Rep 20:501–518

Tilton JC, Amrine-Madsen H, Miamidian JL, Kitrinos KM, Pfaff J, Demarest JF, Ray N, Jeffrey JL, Labranche CC, Doms RW (2010) HIV type 1 from a patient with baseline resistance to CCR5 antagonists uses drug-bound receptor for entry. AIDS Res Hum Retroviruses 26:13–24

Tilton JC, Doms RW (2010) Entry inhibitors in the treatment of HIV-1 infection. Antiviral Res 85:91–100

Trkola A, Ketas TJ, Nagashima KA, Zhao L, Cilliers T, Morris L, Moore JP, Maddon PJ, Olson WC (2001) Potent, broad-spectrum inhibition of human immunodeficiency virus type 1 by the CCR5 monoclonal antibody PRO 140. J Virol 75:579–588

Uchil PD, Mothes W (2009) HIV entry revisited. Cell 137:402–404

UNAIDS (2009) AIDS epidemic update: December 2009. World Health Organization, Geneva

Van Damme L, Govinden R, Mirembe FM, Guedou F, Solomon S, Becker ML, Pradeep BS, Krishnan AK, Alary M, Pande B, Ramjee G, Deese J, Crucitti T, Taylor D (2008) Lack of effectiveness of cellulose sulfate gel for the prevention of vaginal HIV transmission. N Engl J Med 359:463–472

Vermeire K, Brouwers J, Van Herrewege Y, Le Grand R, Vanham G, Augustijns P, Bell TW, Schols D (2008) CADA, a potential anti-HIV microbicide that specifically targets the cellular CD4 receptor. Curr HIV Res 6:246–256

Vermeire K, Lisco A, Grivel JC, Scarbrough E, Dey K, Duffy N, Margolis L, Bell TW, Schols D (2007) Design and cellular kinetics of dansyl-labeled CADA derivatives with anti-HIV and CD4 receptor down-modulating activity. Biochem Pharmacol 74:566–578

Vermeire K, Zhang Y, Princen K, Hatse S, Samala MF, Dey K, Choi HJ, Ahn Y, Sodoma A, Snoeck R, Andrei G, De Clercq E, Bell TW, Schols D (2002) CADA inhibits human immunodeficiency virus and human herpesvirus 7 replication by down-modulation of the cellular CD4 receptor. Virology 302:342–353

Weissenhorn W, Calder LJ, Wharton SA, Skehel JJ, Wiley DC (1998a) The central structural feature of the membrane fusion protein subunit from the Ebola virus glycoprotein is a long triple-stranded coiled coil. Proc Natl Acad Sci USA 95:6032–6036

Weissenhorn W, Carfi A, Lee KH, Skehel JJ, Wiley DC (1998b) Crystal structure of the Ebola virus membrane fusion subunit, GP2, from the envelope glycoprotein ectodomain. Mol Cell 2:605–616

Weissenhorn W, Dessen A, Harrison SC, Skehel JJ, Wiley DC (1997) Atomic structure of the ectodomain from HIV-1 gp41. Nature 387:426–430

Welch BD, VanDemark AP, Heroux A, Hill CP, Kay MS (2007) Potent D-peptide inhibitors of HIV-1 entry. Proc Natl Acad Sci USA 104:16828–16833

Wild C, Greenwell T, Matthews T (1993) A synthetic peptide from HIV-1 gp41 is a potent inhibitor of virus-mediated cell-cell fusion. AIDS Res Hum Retroviruses 9:1051–1053

Wild C, Oas T, McDanal C, Bolognesi D, Matthews T (1992) A synthetic peptide inhibitor of human immunodeficiency virus replication: correlation between solution structure and viral inhibition. Proc Natl Acad Sci USA 89:10537–10541

Yang P, Ai LS, Huang SC, Li HF, Chan WE, Chang CW, Ko CY, Chen SS (2010) The cytoplasmic domain of human immunodeficiency virus type 1 transmembrane protein gp41 harbors lipid raft association determinants. J Virol 84:59–75

Yang X, Kurteva S, Ren X, Lee S, Sodroski J (2005) Stoichiometry of envelope glycoprotein trimers in the entry of human immunodeficiency virus type 1. J Virol 79:12132–12147

Yuste E, Reeves JD, Doms RW, Desrosiers RC (2004) Modulation of Env content in virions of simian immunodeficiency virus: correlation with cell surface expression and virion infectivity. J Virol 78:6775–6785

Zaitseva M, Blauvelt A, Lee S, Lapham CK, Klaus-Kovtun V, Mostowski H, Manischewitz J, Golding H (1997) Expression and function of CCR5 and CXCR4 on human Langerhans cells and macrophages: implications for HIV primary infection. Nat Med 3:1369–1375

Zanetti G, Briggs JA, Grunewald K, Sattentau QJ, Fuller SD (2006) Cryo-electron tomographic structure of an immunodeficiency virus envelope complex in situ. PLoS Pathog 2:e83

Zhu P, Liu J, Bess J Jr, Chertova E, Lifson JD, Grise H, Ofek GA, Taylor KA, Roux KH (2006) Distribution and three-dimensional structure of AIDS virus envelope spikes. Nature 441:847–852

Zhu P, Winkler H, Chertova E, Taylor KA, Roux KH (2008) Cryoelectron tomography of HIV-1 envelope spikes: further evidence for tripod-like legs. PLoS Pathog 4:e1000203

Zou YR, Kottmann AH, Kuroda M, Taniuchi I, Littman DR (1998) Function of the chemokine receptor CXCR4 in haematopoiesis and in cerebellar development. Nature 393:595–599

Part II
Genome Replication Machines

Chapter 11
Bunyavirus: Structure and Replication

Tom S.Y. Guu, Wenjie Zheng, and Yizhi J. Tao

Abstract The *Bunyaviridae* family is comprised of a large number of negative-sense, single-stranded RNA viruses that infect animals, insects, and plants. The tripartite genome of bunyaviruses, encapsidated in the form of individual ribonucleoprotein complexes, encodes four structural proteins, the glycoproteins Gc and Gn, the nucleoprotein N, and the viral polymerase L. Some bunyaviruses also use an ambi-sense strategy to encode the nonstructural proteins NSs and NSm. While some bunyaviruses have a $T=12$ icosahedral symmetry, others only have locally ordered capsids, or capsids with no detectable symmetry. Bunyaviruses enter cells through clathrin-mediated endocytosis or phagocytosis. In endosome, viral glycoproteins facilitate membrane fusion at acidic pH, thus allowing bunyaviruses to uncoat and deliver their genomic RNA into host cytoplasm. Bunyaviruses replicate in cytoplasm where the viral polymerase L catalyzes both transcription and replication of the viral genome. While transcription requires a cap primer for initiation and ends at specific termination signals before the 3′ end of the template is reached, replication copies the entire template and does not depend on any primer for initiation. This review will discuss some of the most interesting aspects of bunyavirus replication, including L protein/N protein-mediated cap snatching, prime-and-realign for transcription and replication initiation, translation-coupled transcription, sequence/secondary structure-dependent transcription termination, ribonucleoprotein encapsidation, and N protein-mediated initiation of viral protein translation. Recent developments on the structure and functional characterization of the bunyavirus capsid and the RNA synthesis machineries (including both protein L and N) will also be discussed.

Abbreviations

BDV	Borna disease virus
BUNV	Bunyamwera virus
CCHFV	Crimean–Congo hemorrhagic fever virus
cRNA	Complementary RNA or antigenomic RNA
EM	Electron microscopy
FluA	Influenza A virus

T.S.Y. Guu • W. Zheng • Y.J. Tao (✉)
Department of Biochemistry and Cell Biology, Rice University,
6100 Main Street, MS140, Houston, TX 77005, USA
e-mail: ytao@rice.edu

M.G. Rossmann and V.B. Rao (eds.), *Viral Molecular Machines*,
Advances in Experimental Medicine and Biology 726, DOI 10.1007/978-1-4614-0980-9_11,

JCV	Jamestown Canyon virus
LACV	La Crosse virus
nt	Nucleotide
NTR	Nontranslated terminal region
OTU	Ovarian tumor-like protease motif
PTV	Punta Toro virus
RdRp	RNA-dependent RNA polymerase
RNP	Ribonucleoprotein complex
RSV	Respiratory syncytial virus
RV	Rabies virus
RVFV	Rift Valley fever virus
SNV	Sin Nombre virus
TSWV	Tomato spotted wilt virus
UUKV	Uukuniemi virus
vRNA	Viral RNA or genomic RNA
VSV	Vesicular stomatitis virus

11.1 Introduction

Bunyaviridae is a large family of enveloped, negative-stranded RNA viruses that encompasses five genera, *Orthobunyavirus*, *Hantavirus*, *Nairovirus*, *Phlebovirus*, and *Tospovirus*, totaling approximately 350 members (Schmaljohn and Nichol 2007). The first four genera are animal-infecting viruses that include many serious human pathogens, such as the Crimean–Congo hemorrhagic fever virus (CCHFV), La Crosse virus (LACV), Rift Valley fever virus (RVFV), and Sin Nombre virus. Members of the *Tospovirus* genus infect plants only. Most viruses in the *Bunyaviridae* family are carried and transmitted by arthropods (e.g., mosquitoes, ticks, and other insects). Hantaviruses, however, are rodent-borne and are transmitted by rodent bites or by direct contact with the excreta of infected rodents.

11.2 Genome Organization

Bunyavirus genome contains three negative-sense, single-stranded RNA segments, designated as large (L), medium (M), and small (S). Unlike typical negative-sense RNA viruses, some bunyaviruses use an ambi-sense strategy to encode their nonstructural proteins. The L segment of all bunyaviruses encodes the L protein, or the viral RNA-dependent RNA polymerase (RdRp), in negative-sense (i.e., the mRNA is copied from the viral RNA template). As a multifunctional protein, L varies from ~250 kDa in *Orthobunyavirus*, *Hantavirus*, and *Phlebovirus* to ~450 kDa in *Nairovirus*. The M segment also uses the negative-sense strategy to encode the glycoprotein precursor which is subsequently processed to yield two glycoproteins, Gn and Gc. The M segment of some bunyaviruses also encodes a nonstructural protein NSm. Interestingly, the protein NSm is encoded in ambi-sense in *Orthobunyavirus* and *Phlebovirus* but positive-sense in *Tospovirus* (i.e., the mRNA is copied from the antigenomic cRNA template). The segment S encodes the protein N, and in some bunyaviruses, also the nonstructural protein NSs. However, while the protein N is always encoded in negative-sense, NSs is encoded in negative-sense in *Orthobunyavirus* but in ambi-sense at the 3′ end of the segment S in *Tospovirus*. The size of the protein NSs varies profoundly from 10 kDa in *Orthobunyavirus*, ~30 kDa in *Phlebovirus*, to ~50 kDa in *Tospovirus*. *Hantavirus* and *Nairovirus* do not appear to encode NSm and NSs. In addition to the negative-sense vRNA, some bunyaviruses are known to package antigenomic cRNA in their virions so that messengers for nonstructural proteins can be made early during infection prior to protein synthesis and genome replication (Ikegami et al. 2005).

11.3 Virus Structure

Bunyaviruses are enveloped viruses that are generally spherical in shape, with a diameter of 80–120 nm. Due to the lack of a matrix protein layer, bunyaviruses are very flexible and can be easily distorted in negative-staining electron microscopy (EM) images. The surfaces of bunyaviruses are decorated by glycoprotein spikes assembled from the glycoproteins Gn and Gc. Phleboviruses have been shown to possess a $T=12$ icosahedral symmetry with pentameric and hexameric capsomeres. Hantaviruses, however, are pleomorphic with either spherical or rod-shaped particles. Their glycoprotein spikes form patches of square-shaped lattices which are locally ordered on the viral surface, and each spike adopts a fourfold symmetry. Orthobunyaviruses, nairoviruses, and tospoviruses do not appear to have any particular order in the arrangement of their glycoprotein spikes. Viral RNAs are packaged in the form of ribonucleoprotein complexes (RNPs), each of which contains one viral RNA molecule that is extensively coated with the viral protein N. The viral polymerase protein L (RdRp) is also associated with RNP. EM studies show that the RNPs of Bunyaviruses assume a closed, circular-shaped structure that is often loosely coiled.

The first three-dimensional structure of a bunyavirus was that of the Phlebovirus Uukuniemi virus (UUKV) reported by Overby et al. (2008) (Fig. 11.1a, b). The cryo-electron tomography reconstruction of UUKV showed that the lipid bilayer shell was almost fully occupied by spike-shaped structures, presumably glycoproteins, in a $T=12$ icosahedral symmetry. The UUKV surface glycoproteins

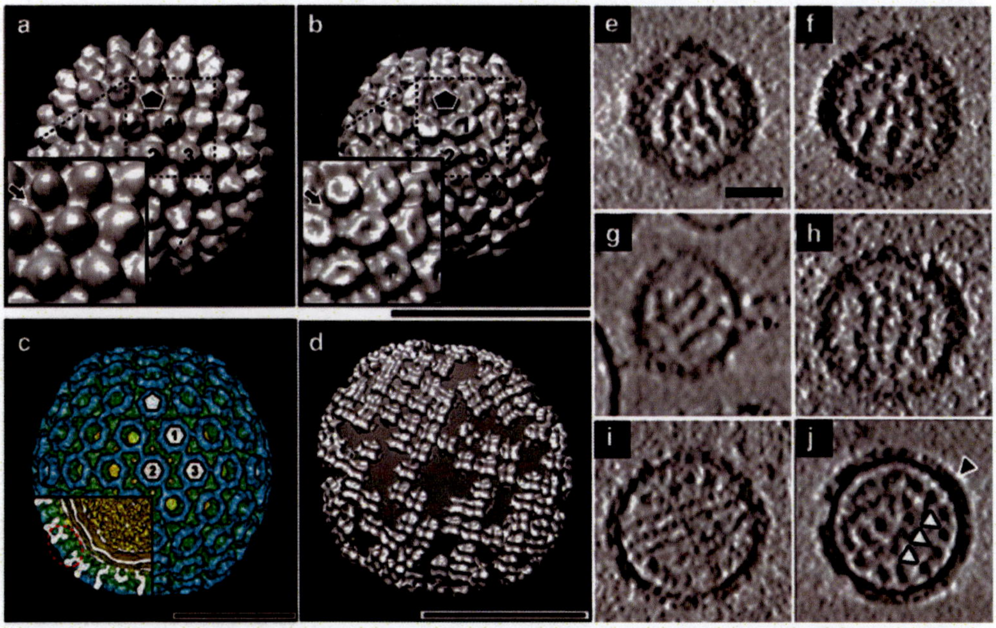

Fig. 11.1 Structure of bunyaviruses. **(a, b)** Icosahedrally symmetrized electron tomography reconstruction of UUKV at pH7 **(a)** and pH6 **(b)**. *Insets* show close-ups of the area indicated with *dashed lines*. Bridging densities (indicated by *arrows*) are present between capsomeres at every position and in both conformations. Scale bar, 100 nm. **(c)** Three-dimensional icosahedral reconstruction of RVFV in radial color representation. Glycoprotein cylinders are in *blue*, the glycoprotein base layer is in *green*, and the membrane and RNP density are in *brown*. In the *lower left* part of the panel, the front half of the reconstruction is removed to reveal the capsomers from the side (a single capsomer is indicated with a *red oval*). Scale bar, 50 nm. **(d)** Surface models of the Tula hantavirus spike lattices. Glycoprotein spikes form ordered patches on the viral membrane. Scale bar, 100 nm. **(e–j)** Tomographic sections of Hantaan virus. Rod-like densities are presumably RNP complexes. White triangles in **(f)** indicate cross-section views of the RNP complexes, and the *black triangle* indicates the Gn–Gc spikes. Bar, 50 nm. Figures are modified from Overby et al. (2008), Huiskonen et al. (2009, 2010), and Battisti et al. (2011)with permission

underwent conformational changes in a pH-dependent manner: most virions showed "flat" barrel-like spikes (~8 nm in height) at pH 6.0, while the "tall" spikes (~13 nm in height) were observed on most virions at pH 7.0. Cavities were found at the base of the spikes in both "tall" and "flat" conformations, but the spike was closed at the top only in the "tall" conformation. The different pH-dependent conformations of the glycoprotein spikes were possibly due to the Gc glycoprotein, a putative class II fusion protein (Garry and Garry 2004; Plassmeyer et al. 2005, 2007; Tischler et al. 2005), as the structural conformation of Gc was much more sensitive to pH compared to Gn (Ronka et al. 1995). This observation is consistent with an earlier study in which the Orthobunyavirus LACV was shown to aggregate at pH 6.2 and below (Wang et al. 1991). Therefore, it was proposed that low pH could induce conformational change to expose the hydrophobic parts of the glycoproteins, most likely in the Gc protein. In addition to the surface spikes, the UUKV lipid bilayer and inner RNP densities have also been revealed in the tomograms. The RNPs almost fully occupied the interior of the virion as "densely packed, threadlike" structures (Overby et al. 2008).

Subsequently, the three-dimensional structure of another Phlebovirus RVFV was obtained by cryo-electron tomography (Fig. 11.1c). Similar to UUKV, RVFV was found to have an icosahedral structure with a $T=12$ triangulation number, suggesting that other members of the *Phlebovirus* genus, and perhaps other members of the *Bunyaviridae* family, may also adopt the icosahedral symmetry (Freiberg et al. 2008). Using single-particle averaging, the three-dimensional structure of the RVFV virion was established at 2.2-nm resolution (Huiskonen et al. 2009; Sherman et al. 2009). The structure revealed a "multilayered structure," including the glycoprotein shell, the lipid bilayer, and the RNP layer. The transmembrane densities of the glycoproteins were observed to modulate the curvature of the lipid bilayer. Molecular-mass measurements led to the conclusion that Gn and Gc were organized in the virion as heterodimers. These heterodimers formed pentons and hexons on a $T=12$ icosahedral lattice. Without a structural matrix protein (Elliott 1990), the bridging densities observed among the capsomeres of both RVFV and the aforementioned UUKV were speculated to function in the virus assembly as well as the structure stabilization. Mutagenesis studies of tomato spotted wilt virus (TSWV), UUKV, and Bunyamwera virus (BUNV) have shown that the cytoplasmic tails of the glycoproteins directly interact with the RNP complexes (Overby et al. 2007; Shi et al. 2007; Snippe et al. 2007). Indeed, the RVFV RNPs colocalize with the transmembrane densities of the glycoproteins in the tomography reconstruction, suggesting specific interactions between the RNP complexes and the cytoplasmic tails of the glycoproteins (Huiskonen et al. 2009).

In spite of the overall similarity, some differences are observed in the two Phlebovirus structures, such as the morphologies of glycoprotein capsomers, pH sensitivities, and RNP–glycoprotein interactions (Fig. 11.1a–c). The shape of RVFV capsomers resembles hollow cylinders, whereas the shape of UUKV capsomers resembles spikes. The RVFV capsomers protrude 9 nm from the membrane, shorter than the 13 nm for UUKV capsomers. This difference could be explained by the larger molecular mass of the glycoproteins of UUKV (Gn = 75 kDa; Gc = 65 kDa) compared to those of RVFV (Gn = 54 kDa; Gc = 56 kDa). The RVFV glycoprotein showed little conformational change in acidic pH (Huiskonen et al. 2009), in contrast to the pH sensitivity exhibited by the glycoproteins of UUKV as discussed above. UUKV glycoproteins were shown to self-assemble to form empty virus-like particles without the N protein (Overby et al. 2006). For RVFV, however, the N protein and possibly also RNAs were required along with the glycoproteins to generate virus-like particles (Habjan et al. 2009). It was suggested that the RNPs of RVFV facilitate virion assembly and promote virion stability by interacting with the glycoproteins (Huiskonen et al. 2009). This hypothesis was supported by the observation that the positions of the membrane-proximal RNP densities colocalize with the transmembrane densities of the glycoproteins (Huiskonen et al. 2009).

Detailed EM analyses have also been reported for two *Hantaviruses*, the Tula virus and the Hantaan virus. The heterogeneous nature of Tula virus virions has impeded successful single-particle analysis of whole virions. Nevertheless, cryo-EM micrographs revealed regular arrays of the glycoprotein spikes, suggesting that the spikes on the virion surface were ordered in patches (Fig. 11.1d). The surfaces of the virus particles were covered by many such patches of protruding structures (the "spikes"), with only small areas of the membrane exposed. This observation is consistent with the grid-like patterns observed in negatively stained EM images (Martin et al. 1985). Huiskonen et al. solved the structure of the Gn–Gc spike complex to 3.6-nm resolution by averaging tomographic subvolumes. Each of the spikes has a four-lobed, square-shaped structure that can be conceptually divided into five different parts: the four globular densities, the central stalk, the four peripheral stalks, the membrane part, and the internal density under the membrane. The four globular "head" domains, shown to be a square-shaped and possess a fourfold rotationally symmetric assembly, were distal to the membrane. Within the ordered patches on the viral membrane, spike complexes interact with each other through specific lateral interactions. These interactions may be sufficient for creating membrane curvature during virus budding. The RNPs, identified as either thread-shaped or rod-shaped, can be seen almost evenly distributed in the virion and occasionally connected to the membrane (Huiskonen et al. 2010).

The structure of Hantaan virus reported by Battisti et al. also shows ordered patches of Gn–Gc spike complex on the virus surface (Battisti et al. 2011) (Fig. 11.1e–j). Like the Tula virus, the spikes on Hantaan virus have a fourfold symmetry. The fourfold averaged Gn–Gc spike complex of the Hantaan virus closely resembles that of Tula virus with a correlation of 0.92. The molecular weight of the complex was estimated to be 392 kDa, which corresponds to four copies each of Gn and Gc. Considering that the four globular densities on top are very similar in shape and size, these densities are more likely to represent four heterodimer of Gn–Gc rather than two different homodimers. Rod-shaped structures, presumably RNPs, are distributed in parallel within the virion, and sometimes in close proximity to the membrane, suggesting there might be interactions with the Gn and Gc cytoplasmic tails, similar to which was aforementioned in Phleboviruses. Hantaviruses enter cells via β3 integrins, presumably through interactions with the glycoproteins (Gavrilovskaya et al. 1999).

11.4 Viral RNA Transcription

Bunyaviruses replicate in the host cell cytoplasm. The RNPs released from the infecting virus support primary transcription, a reaction that is catalyzed by the viral polymerase L associated with the RNPs. Viral protein synthesis then allows viral RNA replication, thus generating cRNAs and vRNAs, both encapsidated in the form of RNP by interacting with the viral protein N as well as L. Such newly assembled RNPs are used for further rounds of mRNA and viral protein synthesis. We will discuss bunyavirus RNA replication and transcription in separate sections, as these two processes are characterized by different initiation and termination mechanisms.

The termini of the three viral RNA segments of bunyaviruses are surrounded by nontranslated regions (NTRs). The NTRs contain conserved sequences that exhibit nucleotide complementarity of various lengths for each respective S, M, and L segment (Barr and Wertz 2004; Kukkonen et al. 1998) (Fig. 11.2a). For the *Orthobunyavirus* BUNV, Barr et al. demonstrated that base-pairing between the 5′ and 3′ NTRs is required for RNA synthesis. Optimal replication could only be achieved by perfect complementarity between the 5′ and 3′ NTRs, except a U–G mismatched pair that is conserved among all three segments of viral RNA (Barr and Wertz 2004). While a U is strictly needed at the position 9 of the 3′ UTR, the corresponding nucleotide at the 5′ UTR may be varied without affecting transcription initiation from the vRNA template (Barr and Wertz 2005). Such terminal nucleotide complementarity allows the 5′ and 3′ NTRs of each viral genomic segment to fold into a characteristic "panhandle" hairpin structure (Kohl et al. 2004; Mir et al. 2006; Mir and Panganiban 2004). The panhandle structure distinguishes between viral and nonviral RNAs (Mir and Panganiban 2004), and thus may be important for not just transcription initiation but also viral RNA replication, viral RNA encapsidation, and viral genome packaging.

In addition to the secondary structures, several studies indicated that the primary sequence of NTRs is also crucial in determining the promoter strength (Flick et al. 2002; Kohl et al. 2004; Mir

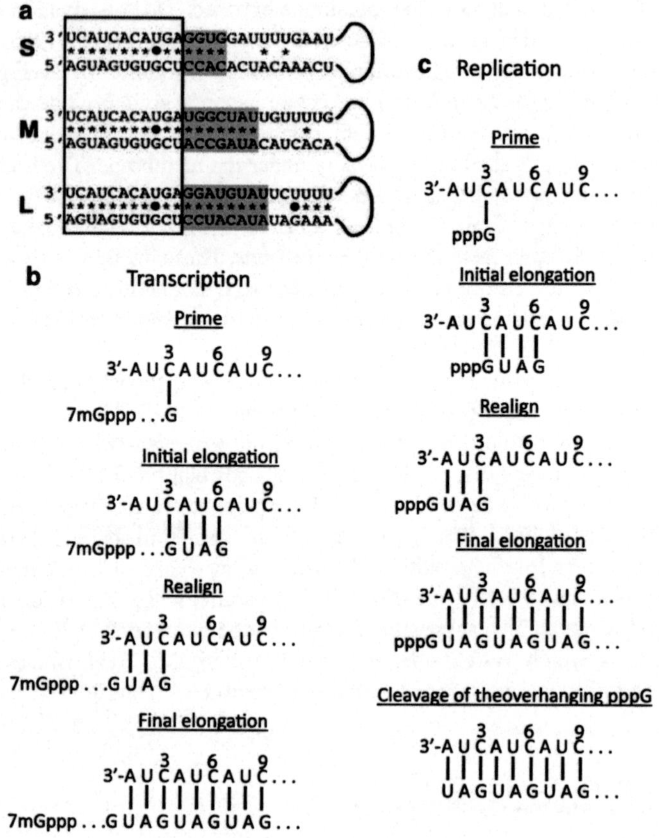

Fig. 11.2 The NTRs of bunyaviruses. (**a**) BUNV S, M, and L genomic RNAs. Only the first 25 nts of both 5′ and 3′ termini of each segment are shown. The first 11 nts of both 5′ and 3′ termini of all three segments are base-paired and conserved (*boxed* sequence), except for a conserved mismatch U–G. The variable complementary region of each segment is shown in *shaded* sequence. (**b, c**) The prime-and-realign mechanism for *Hantavirus* transcription (mRNA) and genome replication (vRNA and cRNA). The 3′ end is shown as the *top* strand. (**b**) The capped nonviral primer with a terminal G initiates transcription by first forming a base pair with the third C on the template. After synthesis of a few oligonucleotides, the nascent RNA slips three nucleotides backward. The original priming G now becomes an overhang, and the newly synthesized G forms a base pair with the third C of the template. This slipping process is dubbed as "realignment," which creates repetitive sequences at the 5′ end. (**c**) Similar prime-and-realign mechanism is also proposed for genomic and antigenomic syntheses, except that the genome chain is initiated with a GTP and a cleavage step occurs to remove the overhanging pppG, leaving a monophosphorylated U at the 5′ end. Figures are adapted and modified from Barr and Wertz (2004), Garcin et al. (1995), and Kukkonen et al. (2005) with permission

and Panganiban 2005). For instance, Flick et al. created complementary double nucleotide exchanges in the NTRs of the *Phlebovirus* UUKV that presumably preserved base-paring between the 5′ and 3′ ends and thus should have maintained the panhandle structure, yet the RNA synthesis of such mutant promoters was abolished. Lastly, the internal nonconserved regions of both UTRs also play a role in regulating viral RNA synthesis (Lowen et al. 2005).

Considering that the viral genomic RNAs (and also the antigenomic RNAs) are encapsidated in the form of RNPs and that the N protein has rather strong specific/nonspecific RNA-binding activities, the exact structure and property of the panhandle in the context of the RNP is not known and may deviate from those structures observed in vitro in the absence of N. It is also possible that the nonspecific RNA chaperone activity of N allows transient dissociation of misfolded RNA structures, thus facilitating the formation of correct higher-order RNA structures that are needed for viral RNA transcription/replication (Mir and Panganiban 2006a). It has been shown that the hantavirus N protein, in the form of a trimer, specifically recognizes and unwinds the terminal panhandle structure in the viral RNA (Mir and Panganiban 2004, 2006b). The stoichiometry calculation indicated that a single N trimer interacts with a single panhandle (Mir and Panganiban 2005). Such panhandle–N protein interactions are also genus-specific, that is, hantavirus N bound to hantavirus panhandle with high affinity but with much reduced affinity to panhandles from other genera (Mir et al. 2006). Following panhandle dissociation, the nucleocapsid protein may remain associated with the unwound 5′ terminus of the viral RNA, allowing transcription to start at the 3′ end (Mir and Panganiban 2006b).

Like influenza A virus (FluA) messengers, the mRNAs of all members of *Bunyaviridae* contain a nonviral heterogeneous sequence at the 5′ end that varies from 5 to 20 nucleotides (nts) in length (Bishop et al. 1983; Bouloy et al. 1990; Collett 1986; Eshita et al. 1985; Ihara et al. 1985; Jin and Elliott 1993a, b; Kormelink et al. 1992; Patterson and Kolakofsky 1984; Simons and Pettersson 1991; Vialat and Bouloy 1992). These heterogeneous oligonucleotide sequences, together with the 5′ cap, are derived from host mRNAs through a process called cap snatching, by the endonuclease activity associated with the RdRp/L protein (Patterson et al. 1984), to initiate viral mRNA synthesis during transcription initiation. Examination of various 5′-capped primers revealed that the polymerase preferably cleaves at certain nucleotides with specificity varying among the five virus genera (Jin and Elliott 1993a, b; Simons and Pettersson 1991). For instance, the capped primer of the *Tospovirus* TSWV is cleaved preferentially at an A residue, which is then base-paired to a U residue at the 3′ end of the viral template RNA (Duijsings et al. 2001). Unlike the FluA, which scavenges capped primers from cellular pre-mRNAs in the nucleus, bunyaviruses replicate in cytoplasm and thus cleave the 5′ cap from host cell mRNAs in the cytoplasm (Rossier et al. 1986). Since cellular mRNA degradation also occurs in the cytoplasm, specifically in the cytoplasmic processing bodies (P bodies), how bunyaviruses protect mRNAs from decay and preserve them for cap snatching is an important question. Mir et al. discovered that hantavirus N protein accumulates in the P bodies and that N protein binds to the 5′ cap with high affinity (Mir et al. 2008). The cap-binding site and the RNA-binding site of N are located in separate regions of the polypeptide, and N can simultaneously bind to both the mRNA cap and vRNA. N undergoes distinct conformational changes after binding to either the mRNA cap or vRNA or both mRNA cap and vRNA simultaneously (Mir et al. 2010). The conformationally altered N with a capped primer loaded at the cap-binding site specifically binds the nine conserved 3′ nucleotides of vRNA and assists the bound primer to anneal at the 3′ terminus, thus promoting transcription initiation by the hantavirus RdRp. Therefore, the bunyavirus N protein perhaps assumes a functional role similar to that of PB2 of the FluA (Panganiban and Mir 2009).

Another notable feature of bunyavirus viral mRNAs is the presence of repetitive sequences at its 5′ end (Garcin et al. 1995; Jin and Elliott 1993a, b). This partial reiteration of the 5′-terminal sequence could be explained by the so-called prime-and-realign mechanism. This model was first proposed by Garcin et al. from studying RNA synthesis by Hantaan virus (Fig. 11.2b) (Garcin et al. 1995). According to this model, the endonuclease of Hantaan virus cuts the host mRNA preferentially after the G nucleotide, which then base pairs with the third nucleotide, C^3, of the viral RNA template. The capped primer undergoes only a brief elongation, from C^3 to C^6. This nascent RNA slips three nucleotides backward and "realigns" with the original C^3 nucleotide before the completion of final elongation. A similar slippage mechanism is also proposed to describe the initiation of RNA synthesis for arenaviruses (Garcin and Kolakofsky 1990, 1992).

Unlike many other negative-stranded RNA viruses, the mRNAs of bunyaviruses do not contain a poly(A) tail that is generated by reiterative copying of a poly(U) tract near the 5′ end of the viral RNA template. Indeed, no such U-rich sequences have been universally identified in bunyaviruses. The transcription signals of bunyaviruses have been found to assume many different forms that are either sequence-dependent or structure-dependent. These termination signals are generally located some 100 nts upstream of the 3′ end of the genomic templates, depending on the RNA segments (Bouloy et al. 1990; Collett 1986; Eshita et al. 1985; Jin and Elliott 1993a). Some of the sequence-induced terminations include a U-rich motif for the M mRNA and a C-rich motif for the S mRNA of SNV (a hantavirus) (Hutchinson et al. 1996), and a purine-rich region for M mRNA of RVFV (a Phlebovirus) (Collett 1986). The termination signal for Bunyamwera virus S mRNA is a 33-nt sequence residing in the 5′ NTR of the S segment. Two nucleotide motifs within this region, 3′-GUCGAC-5′ and 3′-UGUCG-5′, assume critical roles in signaling termination (Barr et al. 2006). RNA secondary structures are also believed to be involved in the transcription termination of the ambi-sense gene segments of Phleboviruses and Tospoviruses. For example, in the intergenic region of Punta Toro virus (PTV), a Phlebovirus, S segment contains a long, inverted complementary sequence that has been identified to be an important region in terminating mRNA transcription (Emery and Bishop 1987). It is predicted that this inverted complementary sequence could potentially fold into a hairpin structure, causing the viral polymerase to stall and then dissociate from the template. Similar stable hairpin structures are also predicted to exist in the intergenic A/U-rich region of the S segment of TSWV, a Tospovirus (de Haan et al. 1990).

The mechanism for bunyavirus transcription initiation and termination is summarized below, using hantavirus as an example. At the onset of transcription, the trimeric N protein not only mediates the formation of but also directly binds to the panhandle structure through the base paring of the 5′ and 3′ NTRs of vRNA. Concurrently, other N protein also binds to the 5′ cap of the cellular mRNA, which is then cleaved by the polymerase through its endonuclease activity. Subsequently, the N protein unwinds and dissociates the panhandle, while remains attached to the 5′ end of the viral RNA. The cleaved nonviral capped primer is then base-paired to the viral RNA template, and the viral polymerase L proceeds with elongation until it reaches the transcription termination signal (Panganiban and Mir 2009).

11.5 Translation-Coupled Transcription

A unique feature of bunyavirus replication is the coupling of its transcription and translation processes. A number of studies have reported that the mRNA synthesis of several bunyaviruses bunyavirus was abolished in the presence of protein-synthesis inhibitors, such as puromycin and cycloheximide (Abraham and Pattnaik 1983; Patterson and Kolakofsky 1984; Vialat and Bouloy 1992). Bellocq et al. first proposed a possible mechanism for the two concurrent processes (Fig. 11.3) (Bellocq et al. 1987). Multiple transcriptional termination signals reside within the genomic coding sequence. The nascent mRNA can hybridize to the template at these sites, causing the polymerase to stall, and thus premature termination occurs. However, translocating ribosomes, which trail behind polymerase, prevent RNA interactions during transcription until they reach the authentic termination sequence located in the 5′ NTR (Fig. 11.3a). Therefore, translating ribosomes suppress these spurious termination signals. In the absence of translocating ribosomes, as induced by puromycin and cycloheximide, the premature termination signals become active, resulting in the synthesis of truncated transcripts and a fatal loss of gene expression (Fig. 11.3b). Interestingly, edeine, another protein-synthesis inhibitor, did not cause premature termination of mRNA synthesis (Vialat and Bouloy 1992). This was presumably because edeine was not able to block the migration of 40S ribosomal subunit along the mRNA chain (Kozak and Shatkin 1978). This translation-coupled transcription model was recently verified by Barr (Barr 2007). Using reverse genetics, translation stop codons were introduced into the gene segment to prevent ribosome translocation along nascent mRNAs. As expected, mRNA transcription was prohibited. However, when additional nucleotides were inserted or deleted prior to the translation stop codon, rendering the reading sequence "out-of-frame," the transcription activity was restored. This suggests that sequence modification allows the ribosomes to translocate across the translation stop codons (Barr 2007). In conclusion, the transcription of bunyavirus relies on ongoing protein synthesis due to the translocation of ribosomes. Although such coupling between transcription and translation is well documented in prokaryotic systems (Roland et al. 1988), this was the first report that this type of obligatory coupling also exists in eukaryotic cells.

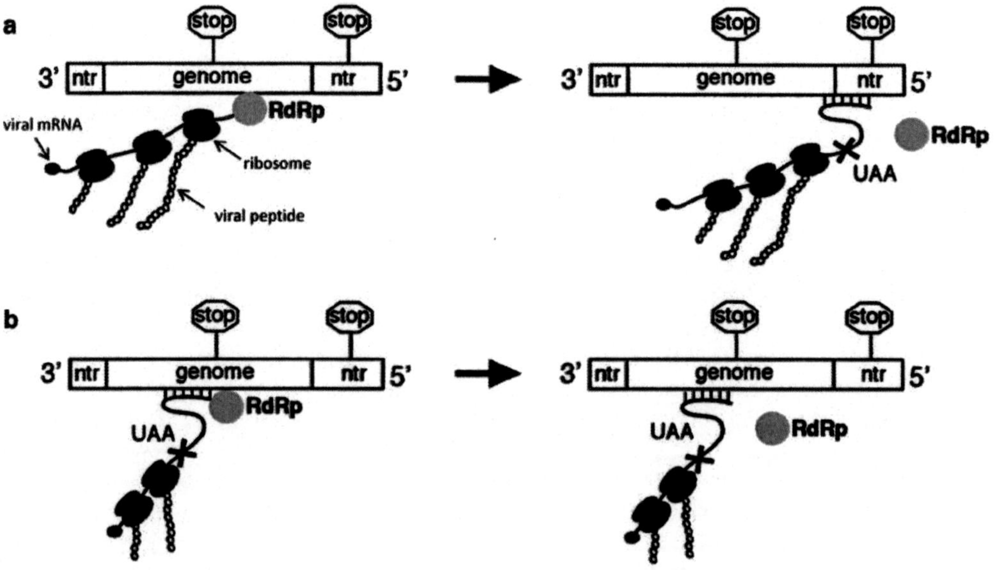

Fig. 11.3 Model for *Bunyaviridae* translation-coupled transcription. The transcriptional termination signals in the coding region and in the NTRs are shown as *stop signs*. (**a**) A typical translation-coupled transcription. While RdRp are synthesizing mRNA, ribosomes are trailing behind to synthesize viral proteins. The translocating ribosomes prevent nascent mRNA from interacting with template. Once reaching the termination signal in the NTR, translation stops (represented by a UAA stop codon), allowing RNA interactions to occurs (shown as a *loop*). Consequently, RdRp disengages. (**b**) When translocation of ribosomes are inhibited by chemicals such as puromycin and cycloheximide (represented by a UAA stop codon), the nascent mRNA hybridizes to its template at the internal transcriptional termination sites, causing the polymerase to stall and thus resulting in premature transcription termination. Figures are adapted and modified from Barr (2007) with permission

11.6 Viral RNA Replication

During replication, antigenomic RNAs (cRNAs) are first synthesized and then used as template to generate progeny genomic RNAs (vRNAs). Replication efficiency of the three RNA segments appears to vary. Studies on the *Orthobunyavirus* BUNV have indicated that the relative level of replication of the three segments is M>L>S. Through direct RNA analysis, Barr et al. was able to distinguish between template activities that affected either RNA replication or mRNA transcription (Barr et al. 2003). Exchange of segment-specific terminal nucleotides identified that a 12-nt region located within both the 3′ and 5′ termini of the M segment plays a major role in determining its high replication ability. A comparison of the nucleotide sequences of the first 23 nts of the M and S segments indicates that the M segment termini exhibit a greater extent of terminal complementarity than between those of the S segment (18 and 16 nt, respectively). Therefore, it is possible that this increased complementarity influences the replication ability of these segments. In addition, sequences outside the exchanged region likely contribute to the greater replication ability of a gene segment as well (Barr et al. 2003), a phenomenon that has also been observed for the transcription of the BUNV genes (Lowen et al. 2005).

Analyses of both cRNAs and vRNAs of bunyaviruses show the lack of cap structure and nonviral sequences at the 5′ terminus (Bishop et al. 1983; Eshita et al. 1985; Garcin et al. 1995; Obijeski et al. 1980). This indicates a switch from the primer-dependent mRNA transcription to the nonprimed full-length cRNA and vRNA replication, although the mechanism still remains undefined. This switch presumably involves some viral and host factors that are needed to suppress transcription initiation as well as premature termination. It has been observed that the nonstructural protein NSs of RVFV binds to the p44 subunit of TFIIH, a transcription factor, causing the attenuation of host mRNA transcription. The lack of cellular primers due to strong transcription inhibition, as speculated by Le May et al., would drive nonprimed RNA replication of RVFV. In other words, TFIIH concentration may modulate the balance between transcription and replication (Le May et al. 2004). Bunyavirus genome replication is sensitive to translational inhibitors such as cycloheximide (Eshita et al. 1985; Ihara et al. 1985), indicating that continuous protein synthesis is required. It is likely that the synthesis of the viral N protein is required for genome replication, as described for other negative-strand RNA viruses such as the rhabdovirus and FluA (Schmaljohn and Nichol 2007). Encapsidation by N may help to keep templates/products in a linear, nonstructured form, thus allowing full-length genome synthesis.

A "prime-and-realign" model is also proposed for the nonprimed replication of hantavirus vRNA and cRNA, similar to that for transcription. Sequence studies have shown that vRNAs contain uridine monophosphates and not triphosphates at the 5′ termini (Garcin et al. 1995). This is explained by the similar polymerase slippage mechanism (Fig. 11.2c). First, a guanosine triphosphate (pppG) base-pairs with the third nucleotide, C^3, of the template. After synthesizing for a few nucleotides, the polymerase slips backward and realigns the nascent RNA. As a result, the priming pppG overhangs the template. The endonuclease activity of polymerase then removes the overhanging pppG, leaving a monophosphorylated uridine at the 5′ end (Garcin et al. 1995). The elongation continues to generate full-length copies of genomic or antigenomic segments.

Preserving the sequence integrity of the terminal sequence during genome replication, especially the NTRs, is vital since NTRs contain important signals for RNA and protein syntheses. Walter and Barr recently discovered that the RdRp of BUNV is capable of correcting sequence alternations to the NTRs (Walter and Barr 2010). When an intact NTR was introduced into the antigenomic 3′ end, it was precisely removed, and the impaired RNA synthesis that resulted from the NTR addition was restored. Likewise, when nucleotides were deleted from the antigenomic 3′ end and replaced with a duplicate and intact NTR, the inserted NTR was removed and the missing nucleotides were reinstated. Furthermore, Walter and Barr demonstrated that the repair of NTRs is likely accomplished by direct entry of the RdRp into the internal NTRs during replication (Walter and Barr 2010). However, the precise repair mechanism employed by RdRp remains unclear at this point.

11.7 The RNA Synthesis Machinery

It is well established that both the L protein and the N protein of bunyavirus are essential components for viral RNA synthesis, that is, replication and transcription (Accardi et al. 2001; Blakqori et al. 2003; Dunn et al. 1995; Flick and Pettersson 2001; Jin and Elliott 1993a; Kariwa et al. 2003; Lopez et al. 1995). Using reporter gene expression, it was found that heterogeneous L and N proteins are capable of transcribing heterologous RNA templates from related viruses from the same genera, although the efficiency may be reduced (Accardi et al. 2001; Dunn et al. 1995). Blakqori et al. also found that the ratio between L and N proteins is critical to the replication activity, with too much of either protein resulting in the reduction of polymerase activity (Blakqori et al. 2003).

Conflicting results have been reported on the involvement of the nonstructural protein NSs in the RNA synthesis. While some showed that the co-expression of NSs protein had little effect on RNA replication or transcription when using the *Phlebovirus* RVFV and UUKV minigenome system (Flick and Pettersson 2001; Lopez et al. 1995), others noted a significant enhancement of minigenome RNA replication and transcription following NSs coexpression (Ikegami et al. 2005). On the other hand, inhibitory viral RNA synthesis resulting from NSs coexpression has been observed for BUNV and LACV from the *Orthobunyavirus* genus (Blakqori et al. 2003; Weber et al. 2001).

This may imply different biological roles for NSs proteins from different genera of *Bunyaviridae* (Ikegami et al. 2005). Indeed, the size of NSs ranges from 10 kDa for *Orthobunyavirus* to more than 50 kDa for *Tospovirus*; hence, the different activities reported for this protein are perhaps not surprising. Host cell factors may also contribute to bunyavirus RNA synthesis. For *Tospovirus* TSWV, a host protein called FoTF (*Frankliniella occidentalis* transcription factor) from its insect vector is able to bind to both the viral RdRp and to viral RNA. Using in vitro RNA synthesis assays, it was shown that addition of purified FoTF improves viral replication but not transcription (de Medeiros et al. 2005).

11.7.1 Protein L: The Viral RNA-Dependent RNA Polymerase

The protein L of bunyaviruses is a multifunctional protein consisting of a single polypeptide that varies from ~2,000 amino acid residues (in *Hantavirus* and *Phlebovirus*) to ~4,000 residues (in *Nairovirus*) in length. A core polymerase domain has been identified near the middle of the L polypeptide with all five polymerase motifs that are known to conserve among viral RdRps (Kukkonen et al. 2005). In addition to the core polymerase domain, an endonuclease domain has also been identified in all bunyavirus L proteins (Reguera et al. 2010). As in influenza viruses, the endonuclease activity is used for cap snatching during transcription initiation. The endonuclease domain has been shown to locate at the N-terminus of the L protein of all bunyaviruses except for nairoviruses. In nairoviruses, which have an L protein nearly twice as other known L proteins, an ovarian tumor (OTU)-like protease motif has been identified at the N-terminus of their L proteins (Honig et al. 2004). The OTU domain of the *Nairovirus* CCHFV has been shown to be a functional protease (Bergeron et al. 2010). However, no evidence of L protein autoproteolytic processing is found, and the OTU protease activity is dispensable for virus RNA replication (Bergeron et al. 2010). Bunyavirus L protein may also harbor a cap-binding activity like the influenza virus polymerase protein PB2. However, the cap-binding domain, which is needed for the recognition of mRNA cap for cap snatching, has not yet been mapped to the L polypeptide sequence. Alternatively, for hantaviruses, the cap-binding activity may be supplied by the N protein, which has been shown to bind the cap with high affinity (Mir et al. 2008; Mir and Panganiban 2008). It has been shown that the *Phlebovirus* RVFV L protein formed an oligomer, and the N- and C-terminal regions of L are involved in L oligomerization (Zamoto-Niikura et al. 2009). In general, biochemical characterization of the L protein activities is lacking presumably due to its large size and hence the associated challenges in expressing and isolating the protein in large quantities.

Recently, the crystal structure of an endonuclease domain from a bunyavirus L protein was determined (Reguera et al. 2010). The endonuclease activity site of the *Orthobunyavirus* LACV resides in the N-terminal region of its RdRP/L protein. The topology of the LACV endonuclease domain, encompassing residues 1–183, bears close resemblance to that of the FluA PA (Dias et al. 2009; Yuan et al. 2009), with a twisted plane of antiparallel β-strands surrounded by α-helices (Fig. 11.4). Two manganese ions are found in the 2,4-dioxo-4-phenylbutanoic acid (DPBA, a known endonuclease inhibitor)-bound protein (Fig. 11.4b). These two ions are coordinated by two sets of residues: His34, Asp79, and Asp92 for one, and Asp52 and Asp79 for the other (Fig. 11.4b). In the presence of divalent cations (e.g., Mn, Co, Zn, and Ni), LACV N-terminal domain is capable of digesting ssRNA, ssDNA, and even some highly structured RNA. Mutation of amino acid residues at the active site resulted in the loss of nuclease activity for the isolated LACV endonuclease domain. Similarly, a loss of transcriptional activity was observed in an in vivo RNP reconstitution system when the identical point mutations were made in the context of the full-length LACV L protein. Structure-based sequence alignment reveals that a similar endonuclease motif exists at the N-terminus of L or PA proteins of essentially all known negative-strand and cap-snatching segmented RNA

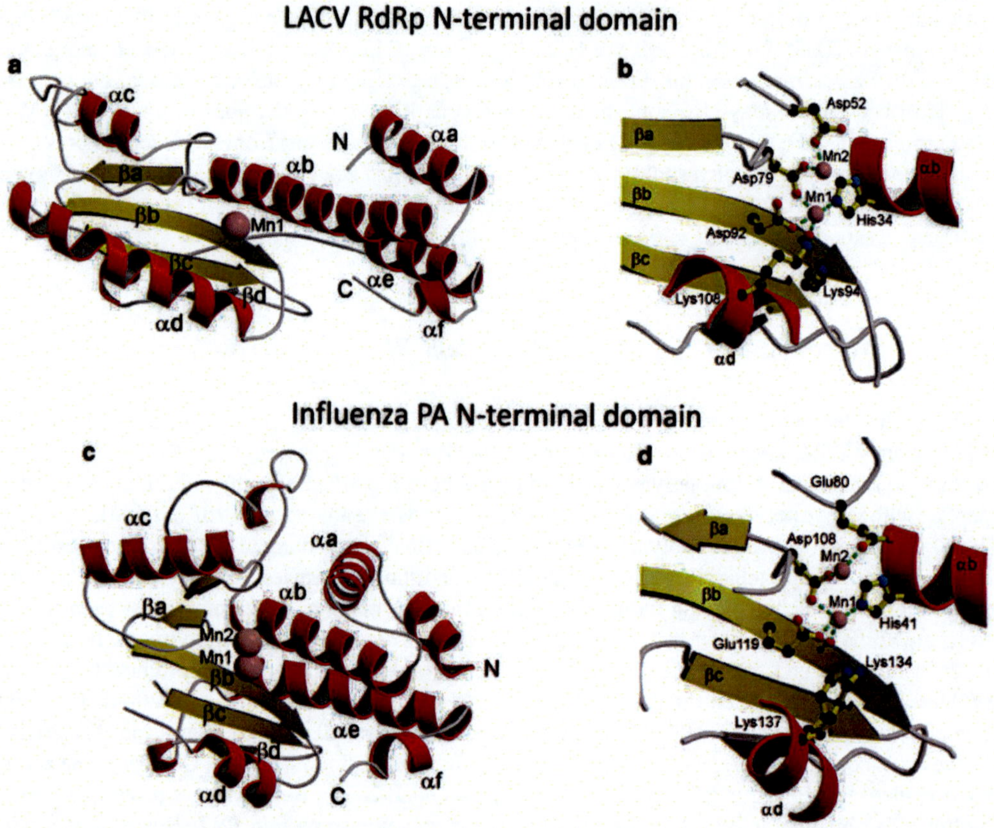

Fig. 11.4 Crystal structure of the N-terminal domain of the La Crosse virus (LACV) L protein and the influenza A virus PA. (**a**, **c**) *Ribbon diagram* showing the endonuclease domain structure of LACV (PDB ID 2XI5) and influenza A/H3N2 (PDB ID 2W69). Manganese ions are shown as *pink spheres*. Corresponding helices and strands are labeled. (**b**, **d**) Endonuclease active sites of both proteins. DPBA-bound LACV N-terminal L protein is depicted in (**b**). For LACV L, the first manganese ion is coordinated by His34, Asp79, and Asp92; the second ion is coordinated by Asp52 and Asp79. For Influenza A PA, the first manganese ion is coordinated by His 41, Asp 108, Glu119, and Ile120; the second ion is coordinated by Glu80 and Asp108. The catalytic lysine residue is Lys94 for LACV L and Lys134 for influenza A PA. Figures are adapted from Reguera et al. (2010) with permission

viruses, including arenaviruses (2 segments), bunyaviruses (3 segments), Tenuiviruses (4–6 segments), and orthomyxoviruses (6–8 segments) (Reguera et al. 2010).

11.7.2 *Protein N: The Viral Nucleoprotein*

The size of the N proteins from different genera of the *Bunyaviridae* family varies from 19–30 kDa (*Orthobunyavirus*, *Phlebovirus*, and *Tospovirus*) to 48–54 kDa (*Hantavirus* and *Nairovirus*). Although the sequences of N proteins from the same genera are clearly related, the sequence identity is generally poor among different genera. Despite their sequence variations, all bunyavirus N proteins possess two universal properties: homotypic oligomerization and nonspecific RNA binding, both required for their ability to encapsidate viral genomic and antigenomic RNAs.

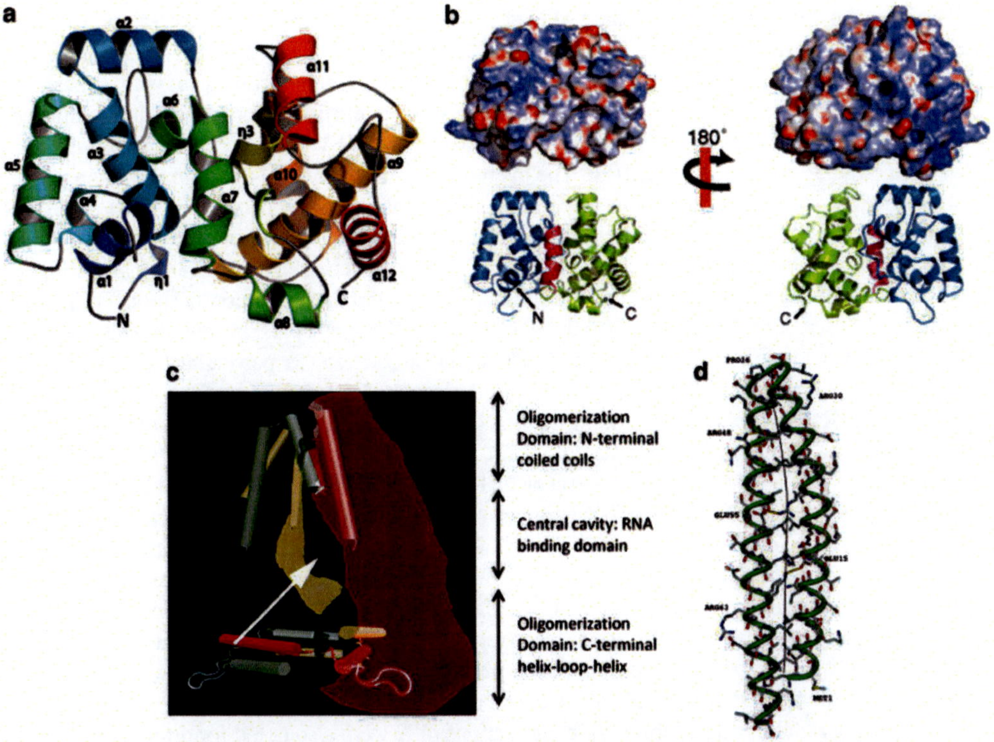

Fig. 11.5 (**a**) Crystal structure of the Rift Valley fever virus (RVFV) N. The *ribbon diagram* is rainbow-colored from *blue* at the N terminus to *red* at the C terminus. Helix α7 bridges the N-terminal and C-terminal domains. (**b**) Electrostatic surface potential of RVFV N. The negatively charged surface is colored in *red*, and positively charged surface is colored in *blue*. The orientation of the surface representations corresponds to that of the *ribbon diagrams*, with N-terminal domain colored in *blue*, C-terminal domain colored in *green*, and the linking helix in *pink*. (**c**) Three-dimensional model of a hantavirus N trimer EM reconstruction from a negatively stained sample. Three N monomers (in *red*, *blue*, and *yellow*) interact through oligomerization domains in the N-terminal region (coiled coil) and C-terminal region (helix–loop–helix). The central cavity (indicated by a *white arrow*) is speculated to be the RNA-binding domain. (**d**) Crystal structure of the N-terminal fragment (residues 1–75) of the Sin Nombre virus (SNV). The fragment adopts an α-helical coiled coil structure. Figures are adapted from Raymond et al. (2010), Kaukinen et al. (2004, 2005), and Boudko et al. (2007) with permission

The crystal structure of the protein N of the RVFV (genus *Phlebovirus*), the first atomic structure of any of the bunyavirus N proteins, was solved to 1.9 Å recently (Raymond et al. 2010). The full-length RVFV N is a 27-kDa protein of 245 residues. Overexpressed N was purified in complex with host nucleic acids and had to be denatured to obtain RNA-free N for crystallization. The refolded RVFV N protein was crystallized as pairs of symmetric dimers (i.e., tetramers). N protein tetramers have also been previously observed for the *Orthobunyavirus* BUNV (Mohl and Barr 2009; Rodgers et al. 2006). The RVFV N protein adopts a compact helical structure with 12 helices (Fig. 11.5a). Similar to other –ssRNA virus N protein, each RVFV N protein contains two lobes, the N-terminal and C-terminal domains, each containing one central helix surrounded by either four or five other helices. The two domains are connected by a linker helix, α7 (residues 112–121). Interestingly, the protein fold of these two domains is novel, different from the N protein folds of all other segmented or nonsegmented –ssRNA viruses [e.g., respiratory syncytial virus (RSV) (Tawar et al. 2009), FluA (Ye et al. 2006), rabies virus (RV) (Albertini et al. 2006), vesicular stomatitis virus (VSV) (Green et al. 2006), and Borna disease virus (BDV) (Rudolph et al. 2003)].

The oligomerization and RNA binding of N proteins of RVFV and members of *Orthomyxoviridae* (FluA) and *Mononegavirales* (BDV, RV, RSV, and VSV) are achieved via different mechanisms. The observation that RVFV N protein crystallized into symmetrical dimers is consistent with its behavior in solution, as the oligomeric species for *Phlebovirus* N protein is predominantly homodimer or multimer of two, for example, UUKV (Katz et al. 2010) and RVFV (Le May et al. 2005), although other high oligomers were also observed. The overall structure of RVFV N protein shows no prominent protrusions at both termini, and there is no structural exchange between the two RVFV N monomers, distinctly different from other –ssRNA viruses. The dimeric interface is mainly sustained by hydrophobic interactions between helices from the N-terminal domain and a salt bridge at the C terminus. Indeed, mutation of the hydrophobic Trp125 and the disruption of the salt bridge led to impaired transcriptional activities (Raymond et al. 2010). Le May et al. also demonstrated that the first 71 amino acids are involved in the RVFV N–N interaction, in particular Tyr4 and Phe11 (Le May et al. 2005).

In addition, Raymond et al. postulate that the N-terminal domain of RVFV may be involved in RNA binding, as it has a higher isoelectric point and a more positively charged surface than its C-terminal domain (Fig. 11.5b) (Raymond et al. 2010). No cleft is found between the N- and C-lobes of the RVFV N protein. How the RVFV N dimer relates to high-order oligomers, where the exact RNA-binding site is located, and whether RNA binding would induce conformational change of N protein remain to be determined. To date, the only three atomic structures of N in complex with RNA are from the rhabdoviruses RV and VSV and the paramyxovirus RSV. Each N protein of RV, RSV, and VSV associates with 9, 7, and 9 nts, respectively. These N–RNA complex structures reveal that the RNA chain wrapped tightly around the deep positively charged grooves, in between the N- and C-lobes of N proteins (Albertini et al. 2006; Green et al. 2006; Tawar et al. 2009).

In addition to Phleboviruses, homotypic interactions of N were also noted for orthobunyaviruses, for example, BUNV (Eifan and Elliott 2009; Leonard et al. 2005; Mohl and Barr 2009; Rodgers et al. 2006), and Tospoviruses, for example, TSWV (Kainz et al. 2004; Snippe et al. 2005; Uhrig et al. 1999). The N–N interaction interface of BUNV has been mapped to N-terminal region (residues 1–10), a small middle region (residues 94–158), and C-terminal region (residues 217–233) (Eifan and Elliott 2009; Leonard et al. 2005). Similar findings have been reported for TSWV: both N- and C-termini (residues 1–39 and residues 233–248) together with some discrete middle regions are involved in the homotypic interaction or multimerization of the N protein (Kainz et al. 2004; Uhrig et al. 1999). A "head-to-head-and-tail-to-tail" type of interaction and a "head-to-tail" model have been proposed for the BUNV and the TWSV N proteins, respectively.

Hantavirus N protein, which is perhaps the most extensively studied of this group, exhibits several unique features compared to those discussed above. First, the size of the Hantavirus N protein is approximately twice that of *Phlebovirus* and *Orthobunyavirus* with little sequence homology. Second, chemical cross-linking studies showed that the predominant oligomeric species of *Hantavirus* N protein is a trimer (Alfadhli et al. 2001; Kaukinen et al. 2001), which likely serves as the assembly intermediate of the RNP. In contrast, the main oligomeric states for *Phlebovirus* and *Orthobunyavirus* N are dimers and tetramers, respectively. Nevertheless, like N proteins from other bunyaviruses, the N–N interface of Hantavirus has been shown to reside on both terminal ends. The current understanding of hantavirus N is that trimerization is achieved by "head-to-head" and "tail-to-tail" interactions, as supported by a three-dimensional EM reconstruction of a trimer, in which the three subunits are attached in parallel to each other on both ends (Fig. 11.5c) (Kaukinen et al. 2004). Hantavirus N protein has a curved structure, and the central cavity formed between the N- and C-terminal domains of the N monomers was postulated to bind viral RNA (Xu et al. 2002). The crystal structure of a hantavirus N-terminal peptide (residues 1–75/93) shows that each polypeptide consists of two α-helices (residues 1–34 and 38–75) that form an antiparallel coiled coil, with two N-terminal polypeptides related by noncrystallographic symmetry forming a dimer (Fig. 11.5d)

(Boudko et al. 2007). While the crystal structure of this N-terminal peptide supports the "head-to-head" interaction model, it does not describe how the peptides interact with each other in the trimeric association.

11.8 Ribonucleoprotein Assembly

The genomic and antigenomic segments of bunyaviruses are encapsidated by the virus-encoded nucleocapsid (N) protein and associated with the RdRp or L protein to form RNPs, which are responsible for viral RNA synthesis and genome packaging. EM has been used to study the morphology of RNPs from various members of the *Bunyaviridae* family (Booth et al. 1991; Fontana et al. 2008; Huiskonen et al. 2010; Mohamed 1981; Obijeski et al. 1976; Pettersson and von Bonsdorff 1975; Raymond et al. 2010). Regardless of the virus origin, bunyavirus RNP seems to share very similar morphology: circular, convoluted, and supercoiled (Fig. 11.6). The irregular-shaped RNPs have string- or thread-like appearances and lack any well-defined symmetry. The length of RNPs also varies significantly, presumably depending on the length of associated RNA. At high magnification, the RNP filaments appear as "strings of beads" (Fig. 11.6c). Since no high resolution EM reconstruction is yet available, it is difficult to discern what oligomerization states the N protein adopts in the RNP structure.

Questions associated with RNP encapsidation include how N differentiates between viral versus nonviral RNAs, what regions of the viral RNAs are involved in encapsidation, what elements confer N–RNA binding specificity, and which part of N is essential in interacting with viral RNAs. As previously mentioned, the Hantavirus N trimer is able to discriminate between viral and nonviral RNAs (Mir and Panganiban 2004). Several studies have attributed the 5′ end of genomic and antigenomic

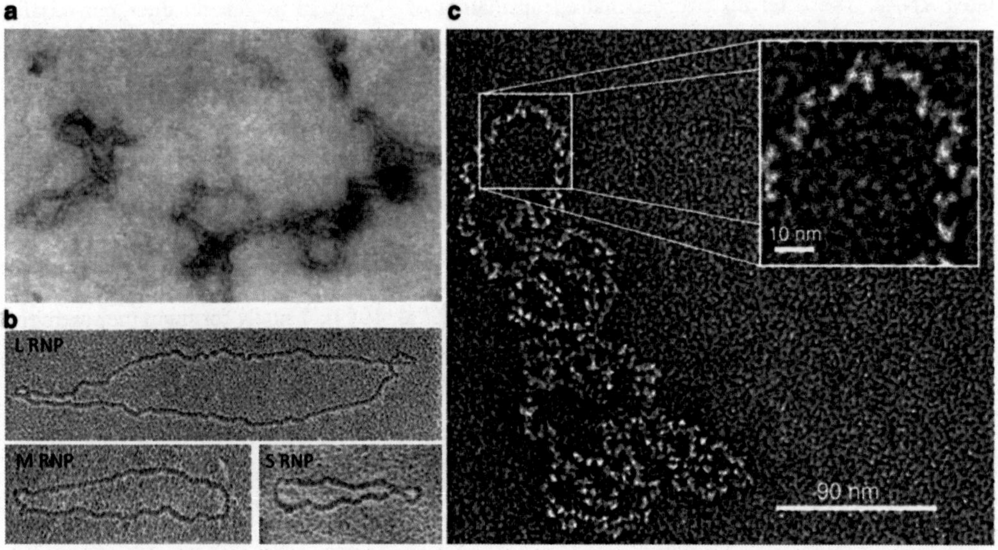

Fig. 11.6 Electron micrographs of RNP from (**a**) La Crosse virus (LACV), an Orthobunyavirus, (**b**) Uukuniemi virus (UUKV), a Phlebovirus, and (**c**) Rift Valley fever virus (RVFV), a Phlebovirus. All RNPs have irregular-shaped, circular, convoluted, and string/thread-like appearance. Figures are adapted from Obijeski et al. (1976), Pettersson and von Bonsdorff (1975), and Raymond et al. (2010) with permission

RNAs to be the region responsible for specific binding to N (Mohl and Barr 2009; Ogg and Patterson 2007; Osborne and Elliott 2000; Severson et al. 2001). Using competitive RNA-binding assays, Osborne et al. showed that orthobunyavirus BUNV N exhibits preferential binding toward the first 32 nts at the 5′ terminus of its S segment (Osborne and Elliott 2000). Likewise, Mohl et al. determined that the minimal sequence requirement for BUNV encapsidation is 17 nts within the genomic 5′ NTR and 18 nts within the antigenomic 3′ NTR (Mohl and Barr 2009). Consistent findings are also reported for another member of the *Orthobunyavirus* genus, Jamestown Canyon virus (JCV). Interestingly, when placing portions of 5′ vRNA or 3′ vRNA sequence in the nonviral RNAs, the modified nonviral RNAs would attain comparable binding potential as their wild-type counterpart (Ogg and Patterson 2007). In addition, similar encapsidation signals are found in Hantaan virus S-segment vRNA, whose 5′ end is essential for N binding, with the greatest affinity measured for the first 39 nts (Severson et al. 2001). Some of these 5′ end sequences were predicted to form secondary structure, for example, stem-loop, which may contribute to preferential binding to N (Mohl and Barr 2009; Osborne and Elliott 2000; Severson et al. 2001). In fact, a purified hantavirus N protein trimer was shown to bind tightly to the panhandle structure formed by the complementary 5′ and 3′ termini of vRNA (Mir and Panganiban 2004). Mir and Panganiban further demonstrated that the minimal size recognized by trimeric N is 23 nts long (Mir and Panganiban 2005). These high-order RNA structures are regarded as the encapsidation signals, as both the genomic and antigenomic segments are base-paired at the two termini (Raju and Kolakofsky 1989). The common consensus is that only vRNAs and cRNAs are encapsidated, but not mRNAs (Kukkonen et al. 2005; Schmaljohn and Nichol 2007). Since encapsidation depends on both the 5′ viral sequence and the formation of "panhandle" between the two complementary termini, the existence of 5′-capped nonviral primer and the truncated 3′ tail may prevent mRNAs from being encapsidated (Mir and Panganiban 2004; Osborne and Elliott 2000).

In order to probe the binding specificity between BUNV N and RNA, Mohl and Barr stripped bound RNAs from the N–RNA complex to analyze their primary sequences and secondary structures. However, no common sequence motifs and secondary structures were identified in the isolated RNAs. These results suggest that encapsidation of N protein in general does not require particular linear sequences or secondary structures (Mohl and Barr 2009). Therefore, a two-step model has been proposed for the encapsidation of genomic and antigenomic segments. In the first step, N makes specific interactions with the 5′ end of vRNA through the stem-loop structure or the panhandle structure. Following the initial binding, oligomerization of N takes place. In the second step, N oligomers bind nonspecifically to the remaining vRNA template (Osborne and Elliott 2000; Severson et al. 2001). Consequently, the encapsidated genomes and antigenomes are organized into RNPs with N proteins bound to RNA at a certain interval. For BUNV, the stoichiometry of the N–NRA complex is ~12 nts per N monomer (Mohl and Barr 2009). This N–RNA stoichiometry, nevertheless, is different from that of other negative-sense single-stranded RNA viruses, for example, 6 nts/N for the paramyxovirus Sendai virus (Egelman et al. 1989), 7 nts/N for the paramyxovirus RSV (Tawar et al. 2009), 9 nts/N for the rhabdovirus rabies virus (Albertini et al. 2006) and VSV (Green et al. 2006), and 24 nts/N for the orthomyxovirus influenza A virus 24 nts/N (Ortega et al. 2000).

The RNA-binding domain of Hantaan virus N protein was first mapped to a central and conserved region, amino acids 175–217 (Xu et al. 2002). This RNA-binding domain was later refined, and amino acids 194–204 were identified as the minimal region for binding (Severson et al. 2005). Site-directed mutagenesis showed that residues Glu192, Tyr206, and Ser217 are essential for RNA recognition. Since the crystal structures of the N–NRA complexes of rabies virus, RSV, and VSV showed that the positively charged side-chains interact with the phosphate backbone of RNA, the positively charged side-chains of Hantaan virus N protein were chemically modified, and it was found that positively charged residues, specifically lysine residues, distributed from amino acids 175–429 of the N protein, are required for the N protein–RNA interaction (Severson et al. 2005).

The involvement of positively charged residues of the N protein on RNA binding has also been noted in other genera of *Bunyaviridae* (Walter et al. 2011). In a recent study, a systematic single mutagenesis was conducted on the 15 positively charged and highly conserved residues of *Orthobunyavirus* BUNV N protein. As a result, Walter et al. identified that a single mutation, R94A, of N resulted in the complete loss of its RNA-binding ability (Walter et al. 2011). When tested in a minireplicon assay, the transcriptional activity of the R94A mutant was severely impaired while the replication remained unaffected. The recombinant virus bearing the R94A mutation was not viable (Eifan and Elliott 2009).

11.9 N Protein–Mediated Translation Initiation

An unusual translation initiation mechanism was recently elucidated for *Hantavirus* soon after the surprising discovery of a novel cap-binding activity for its N protein. This mechanism resembles the translation initiation found in eukaryotic cells, as illustrated in Fig. 11.7a (Panganiban and Mir 2009). In eukaryotic cells, at the onset of translation, the methionine-charged initiator tRNA is loaded onto the small ribosomal unit (40S) along with eIF2, GTP, and eIF3 cluster of peptides to form a 43S pre-initiation complex. The eukaryotic mRNA then binds to the small ribosomal subunit through the mediation of a cap-binding complex, namely eIF4F, in order to initiate translation. eIF4F consists of three proteins, eIF4E, eIF4G, and eIF4A. eIF4E binds to the 5′ cap of the mRNA. eIF4G links the 5′ cap to the 43S pre-initiation complex through interactions with eIF4E and eIF3, respectively. eIF4G also binds to poly(A)-binding protein (PABP), which attaches to the 3′ poly(A) tail of the mRNA. The 5′-to-3′ circularized arrangement of mRNA enhances translational efficiency as well as protects both termini from degradation. Lastly, eIF4A is a RNA helicase that presumably promotes scanning of the small ribosomal subunit through RNA secondary structure in order to find the first start (AUG) codon and allow elongation to commence.

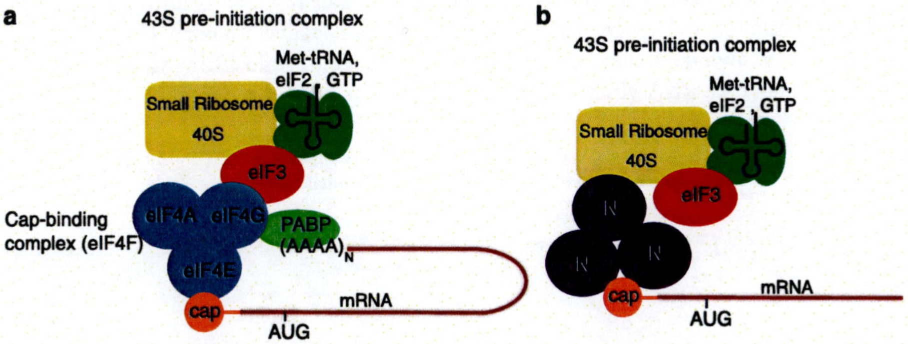

Fig. 11.7 Comparison of the translation initiation machinery between (**a**) a eukaryote and (**b**) a hantavirus. The 43S preinitiation complex consists of the small ribosomal subunit 40S, the eIF3 cluster of peptides, and met-tRNA/eIF2/GTP. This 43S pre-initiation complex is recruited to 5′ cap of mRNA through the mediation of a cap-binding complex (eIF4F). eIF4F has three components: eIF4A, eIF4G, and eIF4E. eIF4G also interacts with poly(A)-binding protein (PABP), allowing the circularization of mRNA. Once the 43S pre-initiation complex is docked onto the 5′ cap, the small ribosomal subunit 40S starts to search for the first start codon, AUG, to initiate protein translation. In Hantavirus, studies have shown that its N possesses functions analogous to eIF4A, eIF4G, and eIF4E. Therefore, hantavirus N can replace the entire cellular eIF4F complex to directly mediate translation initiation. Figures are adapted from Panganiban and Mir (2009) with permission

In a serendipitous discovery, Mir and Panganiban determined that the N protein of *Hantavirus* can replace the entire cap-binding eIF4F complex (Mir and Panganiban 2008) (Fig. 11.7b). They noted that the co-expression of N with various reporter mRNAs resulted in augmented protein expression. Subsequent experiments revealed that bunyavirus N possesses intrinsic properties of each of the three components of eIF4F. First, N binds to 5′ cap with high affinity, analogous to eIF4E. Although N is capable of facilitating nonviral mRNA translation, it exhibits preferential translation for viral mRNA. The cap-binding site of N is distinct from its RNA-binding site (Mir et al. 2010). Next, N is shown to directly interact with purified small ribosomal subunit and increases the rate of recruitment of the 43S pre-initiation complex onto mRNAs, thus replacing the role of eIF4G. Specifically, N interacts with the ribosomal protein S19 (RPS19), located at the head region of the 40S subunit (Haque and Mir 2010). Lastly, N can functionally substitute eIF4A by restoring the translational activity of a defective eIF4A mutant (Mir and Panganiban 2008). Taken together, hantavirus N is considered a surrogate of the entire eIF4F complex (Panganiban and Mir 2009). However, the N-mediated translation initiation does not seem to support the 5′-to-3′ closed-loop formation of the mRNA, unlike its eIF4F-mediated counterpart (Fig. 11.7). Indeed, hantavirus N protein does not appear to form a stable association with PABP (Mir and Panganiban), and the translation of BUNV-like mRNA is independent of PABP (Blakqori et al. 2009). Furthermore, as previously described, the transcription of bunyavirus is coupled with its translation. Hence, the circularization of viral mRNA is unlikely as these two processes cannot be accommodated simultaneously. Thus far, the cap-binding activity has only been demonstrated for *Hantavirus* N. It remains to be seen whether the N proteins of other bunyaviruses assume a similar role in translation initiation.

11.10 Summary

Recent advances in the bunyavirus field are starting to project a fascinating picture about their unique mechanisms for RNP encapsidation as well as viral RNA replication and transcription supported by the viral polymerase L and nucleoprotein N. Due to the substantial diversities in the capsid structure and genome coding strategies, it remains to be seen how viruses from the five genera of the *Bunyaviridae* family define their replication strategies. Further research is also likely to highlight similarities and differences between bunyaviruses and other negative-stranded RNA viruses of both nonsegmented (e.g., rhabdoviruses) and segmented genome (e.g., influenza viruses) in additional to those typical ambi-sense RNA viruses (e.g., arenaviruses).

Acknowledgments We thank Aaron Collier for critical reading of the manuscript. The authors are supported by grants C-1565 from the Welch Foundation and AI077785 from the National Institutes of Health.

References

Abraham G, Pattnaik AK (1983) Early RNA synthesis in Bunyamwera virus-infected cells. J Gen Virol 64(Pt 6):1277–1290

Accardi L, Prehaud C, Di Bonito P, Mochi S, Bouloy M, Giorgi C (2001) Activity of Toscana and Rift Valley fever virus transcription complexes on heterologous templates. J Gen Virol 82:781–785

Albertini AA, Wernimont AK, Muziol T, Ravelli RB, Clapier CR, Schoehn G, Weissenhorn W, Ruigrok RW (2006) Crystal structure of the rabies virus nucleoprotein-RNA complex. Science 313:360–363

Alfadhli A, Love Z, Arvidson B, Seeds J, Willey J, Barklis E (2001) Hantavirus nucleocapsid protein oligomerization. J Virol 75:2019–2023

Barr JN (2007) Bunyavirus mRNA synthesis is coupled to translation to prevent premature transcription termination. RNA 13:731–736

Barr JN, Wertz GW (2004) Bunyamwera bunyavirus RNA synthesis requires cooperation of 3′- and 5′-terminal sequences. J Virol 78:1129–1138
Barr JN, Wertz GW (2005) Role of the conserved nucleotide mismatch within 3′- and 5′-terminal regions of Bunyamwera virus in signaling transcription. J Virol 79:3586–3594
Barr JN, Elliott RM, Dunn EF, Wertz GW (2003) Segment-specific terminal sequences of Bunyamwera bunyavirus regulate genome replication. Virology 311:326–338
Barr JN, Rodgers JW, Wertz GW (2006) Identification of the Bunyamwera bunyavirus transcription termination signal. J Gen Virol 87:189–198
Battisti AJ, Chu YK, Chipman PR, Kaufmann B, Jonsson CB, Rossmann MG (2011) Structural studies of Hantaan virus. J Virol 85:835–841
Bellocq C, Raju R, Patterson J, Kolakofsky D (1987) Translational requirement of La Crosse virus S-mRNA synthesis: in vitro studies. J Virol 61:87–95
Bergeron E, Albarino CG, Khristova ML, Nichol ST (2010) Crimean-Congo hemorrhagic fever virus-encoded ovarian tumor protease activity is dispensable for virus RNA polymerase function. J Virol 84:216–226
Bishop DH, Gay ME, Matsuoko Y (1983) Nonviral heterogeneous sequences are present at the 5′ ends of one species of snowshoe hare bunyavirus S complementary RNA. Nucleic Acids Res 11:6409–6418
Blakqori G, Kochs G, Haller O, Weber F (2003) Functional L polymerase of La Crosse virus allows in vivo reconstitution of recombinant nucleocapsids. J Gen Virol 84:1207–1214
Blakqori G, van Knippenberg I, Elliott RM (2009) Bunyamwera orthobunyavirus S-segment untranslated regions mediate poly(A) tail-independent translation. J Virol 83:3637–3646
Booth TF, Gould EA, Nuttall PA (1991) Structure and morphogenesis of Dugbe virus (Bunyaviridae, Nairovirus) studied by immunogold electron microscopy of ultrathin cryosections. Virus Res 21:199–212
Boudko SP, Kuhn RJ, Rossmann MG (2007) The coiled-coil domain structure of the Sin Nombre virus nucleocapsid protein. J Mol Biol 366:1538–1544
Bouloy M, Pardigon N, Vialat P, Gerbaud S, Girard M (1990) Characterization of the 5′ and 3′ ends of viral messenger RNAs isolated from BHK21 cells infected with germiston virus (bunyavirus). Virology 175:50–58
Collett MS (1986) Messenger RNA of the M segment RNA of Rift Valley fever virus. Virology 151:151–156
de Haan P, Wagemakers L, Peters D, Goldbach R (1990) The S RNA segment of tomato spotted wilt virus has an ambisense character. J Gen Virol 71(Pt 5):1001–1007
de Medeiros RB, Figueiredo J, Resende Rde O, De Avila AC (2005) Expression of a viral polymerase-bound host factor turns human cell lines permissive to a plant- and insect-infecting virus. Proc Natl Acad Sci USA 102:1175–1180
Dias A, Bouvier D, Crepin T, McCarthy AA, Hart DJ, Baudin F, Cusack S, Ruigrok RW (2009) The cap-snatching endonuclease of influenza virus polymerase resides in the PA subunit. Nature 458:914–918
Duijsings D, Kormelink R, Goldbach R (2001) In vivo analysis of the TSWV cap-snatching mechanism: single base complementarity and primer length requirements. EMBO J 20:2545–2552
Dunn EF, Pritlove DC, Jin H, Elliott RM (1995) Transcription of a recombinant bunyavirus RNA template by transiently expressed bunyavirus proteins. Virology 211:133–143
Egelman EH, Wu SS, Amrein M, Portner A, Murti G (1989) The Sendai virus nucleocapsid exists in at least four different helical states. J Virol 63:2233–2243
Eifan SA, Elliott RM (2009) Mutational analysis of the Bunyamwera orthobunyavirus nucleocapsid protein gene. J Virol 83:11307–11317
Elliott RM (1990) Molecular biology of the Bunyaviridae. J Gen Virol 71(Pt 3):501–522
Emery VC, Bishop DH (1987) Characterization of Punta Toro S mRNA species and identification of an inverted complementary sequence in the intergenic region of Punta Toro Phlebovirus ambisense S RNA that is involved in mRNA transcription termination. Virology 156:1–11
Eshita Y, Ericson B, Romanowski V, Bishop DH (1985) Analyses of the mRNA transcription processes of snowshoe hare bunyavirus S and M RNA species. J Virol 55:681–689
Flick R, Pettersson RF (2001) Reverse genetics system for Uukuniemi virus (Bunyaviridae): RNA polymerase I-catalyzed expression of chimeric viral RNAs. J Virol 75:1643–1655
Flick R, Elgh F, Pettersson RF (2002) Mutational analysis of the Uukuniemi virus (Bunyaviridae family) promoter reveals two elements of functional importance. J Virol 76:10849–10860
Fontana J, Lopez-Montero N, Elliott RM, Fernandez JJ, Risco C (2008) The unique architecture of Bunyamwera virus factories around the golgi complex. Cell Microbiol 10:2012–2028
Freiberg AN, Sherman MB, Morais MC, Holbrook MR, Watowich SJ (2008) Three-dimensional organization of Rift Valley fever virus revealed by cryoelectron tomography. J Virol 82:10341–10348
Garcin D, Kolakofsky D (1990) A novel mechanism for the initiation of Tacaribe arenavirus genome replication. J Virol 64:6196–6203
Garcin D, Kolakofsky D (1992) Tacaribe arenavirus RNA synthesis in vitro is primer dependent and suggests an unusual model for the initiation of genome replication. J Virol 66:1370–1376

Garcin D, Lezzi M, Dobbs M, Elliott RM, Schmaljohn C, Kang CY, Kolakofsky D (1995) The 5′ ends of Hantaan virus (Bunyaviridae) RNAs suggest a prime-and-realign mechanism for the initiation of RNA synthesis. J Virol 69:5754–5762

Garry CE, Garry RF (2004) Proteomics computational analyses suggest that the carboxyl terminal glycoproteins of Bunyaviruses are class II viral fusion protein (beta-penetrenes). Theor Biol Med Model 1:10

Gavrilovskaya IN, Brown EJ, Ginsberg MH, Mackow ER (1999) Cellular entry of hantaviruses which cause hemorrhagic fever with renal syndrome is mediated by beta3 integrins. J Virol 73:3951–3959

Green TJ, Zhang X, Wertz GW, Luo M (2006) Structure of the vesicular stomatitis virus nucleoprotein-RNA complex. Science 313:357–360

Habjan M, Penski N, Wagner V, Spiegel M, Overby AK, Kochs G, Huiskonen JT, Weber F (2009) Efficient production of Rift Valley fever virus-like particles: the antiviral protein MxA can inhibit primary transcription of bunyaviruses. Virology 385:400–408

Haque A, Mir MA (2010) Interaction of hantavirus nucleocapsid protein with ribosomal protein S19. J Virol 84:12450–12453

Honig JE, Osborne JC, Nichol ST (2004) Crimean-Congo hemorrhagic fever virus genome L RNA segment and encoded protein. Virology 321:29–35

Huiskonen JT, Overby AK, Weber F, Grunewald K (2009) Electron cryo-microscopy and single-particle averaging of Rift Valley fever virus: evidence for GN-GC glycoprotein heterodimers. J Virol 83:3762–3769

Huiskonen JT, Hepojoki J, Laurinmaki P, Vaheri A, Lankinen H, Butcher SJ, Grunewald K (2010) Electron cryotomography of Tula Hantavirus suggests a unique assembly paradigm for enveloped viruses. J Virol 84:4889–4897

Hutchinson KL, Peters CJ, Nichol ST (1996) Sin Nombre virus mRNA synthesis. Virology 224:139–149

Ihara T, Matsuura Y, Bishop DH (1985) Analyses of the mRNA transcription processes of Punta Toro Phlebovirus (Bunyaviridae). Virology 147:317–325

Ikegami T, Peters CJ, Makino S (2005) Rift Valley fever virus nonstructural protein NSs promotes viral RNA replication and transcription in a minigenome system. J Virol 79:5606–5615

Jin H, Elliott RM (1993a) Characterization of Bunyamwera virus S RNA that is transcribed and replicated by the L protein expressed from recombinant vaccinia virus. J Virol 67:1396–1404

Jin H, Elliott RM (1993b) Non-viral sequences at the 5′ ends of Dugbe nairovirus S mRNAs. J Gen Virol 74(Pt 10):2293–2297

Kainz M, Hilson P, Sweeney L, Derose E, German TL (2004) Interaction between tomato spotted wilt virus N protein monomers involves nonelectrostatic forces governed by multiple distinct regions in the primary structure. Phytopathology 94:759–765

Kariwa H, Tanabe H, Mizutani T, Kon Y, Lokugamage K, Lokugamage N, Iwasa MA, Hagiya T, Araki K, Yoshimatsu K, Arikawa J, Takashima I (2003) Synthesis of Seoul virus RNA and structural proteins in cultured cells. Arch Virol 148:1671–1685

Katz A, Freiberg AN, Backstrom V, Schulz AR, Mateos A, Holm L, Pettersson RF, Vaheri A, Flick R, Plyusnin A (2010) Oligomerization of Uukuniemi virus nucleocapsid protein. Virol J 7:187

Kaukinen P, Koistinen V, Vapalahti O, Vaheri A, Plyusnin A (2001) Interaction between molecules of hantavirus nucleocapsid protein. J Gen Virol 82:1845–1853

Kaukinen P, Kumar V, Tulimaki K, Engelhardt P, Vaheri A, Plyusnin A (2004) Oligomerization of Hantavirus N protein: C-terminal alpha-helices interact to form a shared hydrophobic space. J Virol 78:13669–13677

Kaukinen P, Vaheri A, Plyusnin A (2005) Hantavirus nucleocapsid protein: a multifunctional molecule with both housekeeping and ambassadorial duties. Arch Virol 150:1693–1713

Kohl A, Dunn EF, Lowen AC, Elliott RM (2004) Complementarity, sequence and structural elements within the 3′ and 5′ non-coding regions of the Bunyamwera orthobunyavirus S segment determine promoter strength. J Gen Virol 85:3269–3278

Kormelink R, van Poelwijk F, Peters D, Goldbach R (1992) Non-viral heterogeneous sequences at the 5′ ends of tomato spotted wilt virus mRNAs. J Gen Virol 73(Pt 8):2125–2128

Kozak M, Shatkin AJ (1978) Migration of 40S ribosomal subunits on messenger RNA in the presence of edeine. J Biol Chem 253:6568–6577

Kukkonen SK, Vaheri A, Plyusnin A (1998) Completion of the Tula hantavirus genome sequence: properties of the L segment and heterogeneity found in the 3′ termini of S and L genome RNAs. J Gen Virol 79(Pt 11):2615–2622

Kukkonen SK, Vaheri A, Plyusnin A (2005) L protein, the RNA-dependent RNA polymerase of Hantaviruses. Arch Virol 150:533–556

Le May N, Dubaele S, Proietti De Santis L, Billecocq A, Bouloy M, Egly JM (2004) TFIIH transcription factor, a target for the Rift Valley hemorrhagic fever virus. Cell 116:541–550

Le May N, Gauliard N, Billecocq A, Bouloy M (2005) The N terminus of Rift Valley fever virus nucleoprotein is essential for dimerization. J Virol 79:11974–11980

Leonard VH, Kohl A, Osborne JC, McLees A, Elliott RM (2005) Homotypic interaction of Bunyamwera virus nucleocapsid protein. J Virol 79:13166–13172
Lopez N, Muller R, Prehaud C, Bouloy M (1995) The L protein of Rift Valley fever virus can rescue viral ribonucleoproteins and transcribe synthetic genome-like RNA molecules. J Virol 69:3972–3979
Lowen AC, Boyd A, Fazakerley JK, Elliott RM (2005) Attenuation of bunyavirus replication by rearrangement of viral coding and noncoding sequences. J Virol 79:6940–6946
Martin ML, Lindsey-Regnery H, Sasso DR, McCormick JB, Palmer E (1985) Distinction between Bunyaviridae genera by surface structure and comparison with Hantaan virus using negative stain electron microscopy. Arch Virol 86:17–28
Mir MA, Panganiban AT (2004) Trimeric Hantavirus nucleocapsid protein binds specifically to the viral RNA panhandle. J Virol 78:8281–8288
Mir MA, Panganiban AT (2005) The Hantavirus nucleocapsid protein recognizes specific features of the viral RNA panhandle and is altered in conformation upon RNA binding. J Virol 79:1824–1835
Mir MA, Panganiban AT (2006a) The bunyavirus nucleocapsid protein is an RNA chaperone: possible roles in viral RNA panhandle formation and genome replication. RNA 12:272–282
Mir MA, Panganiban AT (2006b) Characterization of the RNA chaperone activity of Hantavirus nucleocapsid protein. J Virol 80:6276–6285
Mir MA, Panganiban AT (2008) A protein that replaces the entire cellular eIF4F complex. EMBO J 27:3129–3139
Mir MA, Brown B, Hjelle B, Duran WA, Panganiban AT (2006) Hantavirus N protein exhibits genus-specific recognition of the viral RNA panhandle. J Virol 80:11283–11292
Mir MA, Duran WA, Hjelle BL, Ye C, Panganiban AT (2008) Storage of cellular 5′ mRNA caps in P bodies for viral cap-snatching. Proc Natl Acad Sci USA 105:19294–19299
Mir MA, Sheema S, Haseeb A, Haque A (2010) Hantavirus nucleocapsid protein has distinct m7G cap- and RNA-binding sites. J Biol Chem 285:11357–11368
Mohamed NA (1981) Isolation and characterization of subviral structures from tomato spotted wilt virus. J Gen Virol 53:197–206
Mohl BP, Barr JN (2009) Investigating the specificity and stoichiometry of RNA binding by the nucleocapsid protein of Bunyamwera virus. RNA 15:391–399
Obijeski JF, Bishop DH, Palmer EL, Murphy FA (1976) Segmented genome and nucleocapsid of La Crosse virus. J Virol 20:664–675
Obijeski JF, McCauley J, Skehel JJ (1980) Nucleotide sequences at the terminal of La Crosse virus RNAs. Nucleic Acids Res 8:2431–2438
Ogg MM, Patterson JL (2007) RNA binding domain of Jamestown Canyon virus S segment RNAs. J Virol 81:13754–13760
Ortega J, Martin-Benito J, Zurcher T, Valpuesta JM, Carrascosa JL, Ortin J (2000) Ultrastructural and functional analyses of recombinant influenza virus ribonucleoproteins suggest dimerization of nucleoprotein during virus amplification. J Virol 74:156–163
Osborne JC, Elliott RM (2000) RNA binding properties of Bunyamwera virus nucleocapsid protein and selective binding to an element in the 5′ terminus of the negative-sense S segment. J Virol 74:9946–9952
Overby AK, Popov V, Neve EP, Pettersson RF (2006) Generation and analysis of infectious virus-like particles of Uukuniemi virus (Bunyaviridae): a useful system for studying bunyaviral packaging and budding. J Virol 80:10428–10435
Overby AK, Pettersson RF, Neve EP (2007) The glycoprotein cytoplasmic tail of Uukuniemi virus (Bunyaviridae) interacts with ribonucleoproteins and is critical for genome packaging. J Virol 81:3198–3205
Overby AK, Pettersson RF, Grunewald K, Huiskonen JT (2008) Insights into bunyavirus architecture from electron cryotomography of Uukuniemi virus. Proc Natl Acad Sci USA 105:2375–2379
Panganiban AT, Mir MA (2009) Bunyavirus N: eIF4F surrogate and cap-guardian. Cell Cycle 8:1332–1337
Patterson JL, Kolakofsky D (1984) Characterization of La Crosse virus small-genome transcripts. J Virol 49:680–685
Patterson JL, Holloway B, Kolakofsky D (1984) La Crosse virions contain a primer-stimulated RNA polymerase and a methylated cap-dependent endonuclease. J Virol 52:215–222
Pettersson RF, von Bonsdorff CH (1975) Ribonucleoproteins of Uukuniemi virus are circular. J Virol 15:386–392
Plassmeyer ML, Soldan SS, Stachelek KM, Martin-Garcia J, Gonzalez-Scarano F (2005) California serogroup Gc (G1) glycoprotein is the principal determinant of pH-dependent cell fusion and entry. Virology 338:121–132
Plassmeyer ML, Soldan SS, Stachelek KM, Roth SM, Martin-Garcia J, Gonzalez-Scarano F (2007) Mutagenesis of the La Crosse Virus glycoprotein supports a role for Gc (1066–1087) as the fusion peptide. Virology 358:273–282
Raju R, Kolakofsky D (1989) The ends of La Crosse virus genome and antigenome RNAs within nucleocapsids are base paired. J Virol 63:122–128

Raymond DD, Piper ME, Gerrard SR, Smith JL (2010) Structure of the Rift Valley fever virus nucleocapsid protein reveals another architecture for RNA encapsidation. Proc Natl Acad Sci USA 107:11769–11774

Reguera J, Weber F, Cusack S (2010) Bunyaviridae RNA polymerases (L-protein) have an N-terminal, influenza-like endonuclease domain, essential for viral cap-dependent transcription. PLoS Pathog 6:e1001101

Rodgers JW, Zhou Q, Green TJ, Barr JN, Luo M (2006) Purification, crystallization and preliminary X-ray crystallographic analysis of the nucleocapsid protein of Bunyamwera virus. Acta Crystallogr Sect F Struct Biol Cryst Commun 62:361–364

Roland KL, Liu CG, Turnbough CL Jr (1988) Role of the ribosome in suppressing transcriptional termination at the pyrBI attenuator of *Escherichia coli* K-12. Proc Natl Acad Sci USA 85:7149–7153

Ronka H, Hilden P, Von Bonsdorff CH, Kuismanen E (1995) Homodimeric association of the spike glycoproteins G1 and G2 of Uukuniemi virus. Virology 211:241–250

Rossier C, Patterson J, Kolakofsky D (1986) La Crosse virus small genome mRNA is made in the cytoplasm. J Virol 58:647–650

Rudolph MG, Kraus I, Dickmanns A, Eickmann M, Garten W, Ficner R (2003) Crystal structure of the Borna disease virus nucleoprotein. Structure 11:1219–1226

Schmaljohn CS, Nichol ST (2007) Bunyaviridae. In: Knipe D, Howley P (eds) Fields virology, vol 2. Lippincott Williams & Wilkins, Philadelphia, PA, pp 1741–1790

Severson WE, Xu X, Jonsson CB (2001) cis-Acting signals in encapsidation of Hantaan virus S-segment viral genomic RNA by its N protein. J Virol 75:2646–2652

Severson W, Xu X, Kuhn M, Senutovitch N, Thokala M, Ferron F, Longhi S, Canard B, Jonsson CB (2005) Essential amino acids of the Hantaan virus N protein in its interaction with RNA. J Virol 79:10032–10039

Sherman MB, Freiberg AN, Holbrook MR, Watowich SJ (2009) Single-particle cryo-electron microscopy of Rift Valley fever virus. Virology 387:11–15

Shi X, Kohl A, Li P, Elliott RM (2007) Role of the cytoplasmic tail domains of Bunyamwera orthobunyavirus glycoproteins Gn and Gc in virus assembly and morphogenesis. J Virol 81:10151–10160

Simons JF, Pettersson RF (1991) Host-derived 5′ ends and overlapping complementary 3′ ends of the two mRNAs transcribed from the ambisense S segment of Uukuniemi virus. J Virol 65:4741–4748

Snippe M, Borst JW, Goldbach R, Kormelink R (2005) The use of fluorescence microscopy to visualise homotypic interactions of tomato spotted wilt virus nucleocapsid protein in living cells. J Virol Methods 125:15–22

Snippe M, Willem Borst J, Goldbach R, Kormelink R (2007) Tomato spotted wilt virus Gc and N proteins interact in vivo. Virology 357:115–123

Tawar RG, Duquerroy S, Vonrhein C, Varela PF, Damier-Piolle L, Castagne N, MacLellan K, Bedouelle H, Bricogne G, Bhella D, Eleouet JF, Rey FA (2009) Crystal structure of a nucleocapsid-like nucleoprotein-RNA complex of respiratory syncytial virus. Science 326:1279–1283

Tischler ND, Gonzalez A, Perez-Acle T, Rosemblatt M, Valenzuela PD (2005) Hantavirus Gc glycoprotein: evidence for a class II fusion protein. J Gen Virol 86:2937–2947

Uhrig JF, Soellick TR, Minke CJ, Philipp C, Kellmann JW, Schreier PH (1999) Homotypic interaction and multimerization of nucleocapsid protein of tomato spotted wilt Tospovirus: identification and characterization of two interacting domains. Proc Natl Acad Sci USA 96:55–60

Vialat P, Bouloy M (1992) Germiston virus transcriptase requires active 40 S ribosomal subunits and utilizes capped cellular RNAs. J Virol 66:685–693

Walter CT, Barr JN (2010) Bunyamwera virus can repair both insertions and deletions during RNA replication. RNA 16:1138–1145

Walter CT, Bento DF, Alonso AG, Barr JN (2011) Amino acid changes within the Bunyamwera virus nucleocapsid protein differentially affect the mRNA transcription and RNA replication activities of assembled ribonucleoprotein templates. J Gen Virol 92:80–84

Wang GJ, Hewlett M, Chiu W (1991) Structural variation of La Crosse virions under different chemical and physical conditions. Virology 184:455–459

Weber F, Dunn EF, Bridgen A, Elliott RM (2001) The Bunyamwera virus nonstructural protein NSs inhibits viral RNA synthesis in a minireplicon system. Virology 281:67–74

Xu X, Severson W, Villegas N, Schmaljohn CS, Jonsson CB (2002) The RNA binding domain of the Hantaan virus N protein maps to a central, conserved region. J Virol 76:3301–3308

Ye Q, Krug RM, Tao YJ (2006) The mechanism by which influenza A virus nucleoprotein forms oligomers and binds RNA. Nature 444:1078–1082

Yuan P, Bartlam M, Lou Z, Chen S, Zhou J, He X, Lv Z, Ge R, Li X, Deng T, Fodor E, Rao Z, Liu Y (2009) Crystal structure of an avian influenza polymerase PA(N) reveals an endonuclease active site. Nature 458:909–913

Zamoto-Niikura A, Terasaki K, Ikegami T, Peters CJ, Makino S (2009) Rift Valley fever virus L protein forms a biologically active oligomer. J Virol 83:12779–12789

Chapter 12
Viral Polymerases

Kyung H. Choi

Abstract Viral polymerases play a central role in viral genome replication and transcription. Based on the genome type and the specific needs of particular virus, RNA-dependent RNA polymerase, RNA-dependent DNA polymerase, DNA-dependent RNA polymerase, and DNA-dependent RNA polymerases are found in various viruses. Viral polymerases are generally active as a single protein capable of carrying out multiple functions related to viral genome synthesis. Specifically, viral polymerases use variety of mechanisms to recognize initial binding sites, ensure processive elongation, terminate replication at the end of the genome, and also coordinate the chemical steps of nucleic acid synthesis with other enzymatic activities. This review focuses on different viral genome replication and transcription strategies, and the polymerase interactions with various viral proteins that are necessary to complete genome synthesis.

12.1 Introduction

Viral polymerases play a central role in viral genome replication and transcription. Due to the limitations in the genome size that can be packaged in the virus shell, viral polymerases are generally active as a single protein capable of carrying out multiple functions related to viral genome synthesis. Based on genome type and the specific needs of particular virus families, viral polymerases utilize a variety of mechanisms to ensure complete copy of the genome. In particular, they have to recognize initial binding sites, ensure processive elongation, terminate replication at the end of the genome, and also coordinate the chemical steps of nucleic acid synthesis with other modifications of the genome. This chapter focuses on different viral genome replication and transcription strategies and the polymerase interactions with various viral proteins that are necessary to complete genome synthesis.

K.H. Choi (✉)
Department of Biochemistry and Molecular Biology,
University of Texas Medical Branch, Galveston, TX 77555, USA
e-mail: kychoi@utmb.edu

M.G. Rossmann and V.B. Rao (eds.), *Viral Molecular Machines*,
Advances in Experimental Medicine and Biology 726, DOI 10.1007/978-1-4614-0980-9_12,

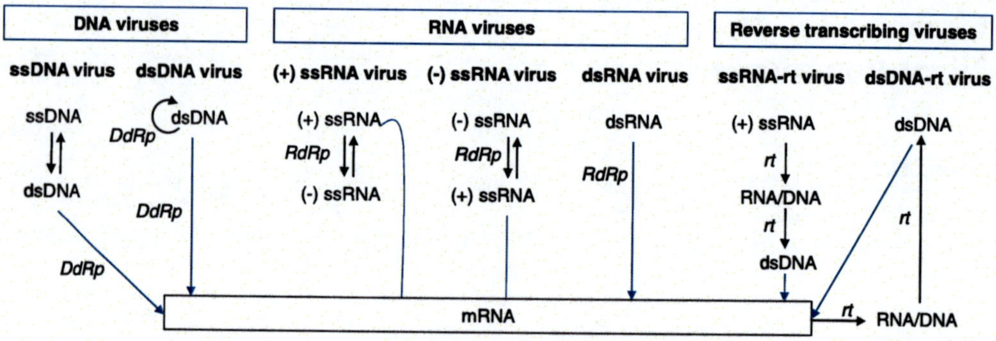

Fig. 12.1 Baltimore virus classification scheme based on genome type and method of mRNA synthesis (Baltimore 1971). The seven virus classes are DNA (ssDNA and dsDNA), RNA (plus-sense (+) ssRNA, minus-sense (–) ssRNA, and dsRNA), and reverse-transcribing (ssRNA-reverse transcribing (rt) and dsDNA-rt) viruses. Viral genome transcription and replication steps are shown as *blue* and *black arrows*, respectively. The steps that viral polymerases catalyze are indicated with RNA-dependent RNA polymerase (RdRp), reverse transcriptase (rt), DNA-dependent DNA polymerase (DdDp), and DNA-dependent RNA polymerase (DdRp). Viral plus-sense RNA can serve as an mRNA and does not require modification for transcription. Note that many eukaryotic dsDNA viruses do not encode a viral polymerase and instead use cellular enzymes in the nucleus for viral genome transcription and replication

12.1.1 Viral Polymerases

Several steps in the virus life cycle require the activity of a polymerase. Viruses first need to convert their genomes into mRNA that can be translated into viral proteins. Viruses must then synthesize identical copies of their genomes for encapsidation into newly assembled virus capsids. The requirement for a virus-encoded polymerase for either transcription and/or replication depends on the cellular location of genome replication and the nature of the viral genome, i.e., whether it is DNA or RNA, double-stranded or single-stranded, or whether it is plus- or minus-sense. Viruses that spend their entire life cycle in the cytoplasm do not have access to host polymerases and thus need to encode their own polymerases for transcription and replication. In contrast, some small DNA viruses that spend their entire life cycle in the nucleus do not need to encode their own polymerase, instead relying entirely on host cell machinery for genome replication and mRNA synthesis. Based on the chemical nature of the viral genome, the need for mRNA synthesis, and the particular genome replication strategy used, viruses are grouped into seven different classes (Baltimore 1971). Figure 12.1 summarizes the seven classes of viruses along with their mode of polymerase usage.

DNA viruses use DNA as their genetic material, which is generally double-stranded (dsDNA) but may also be single-stranded (ssDNA). DNA viruses replicate their genomes using DNA-dependent DNA polymerases (also called DNA polymerases) and transcribe mRNA using DNA-dependent RNA polymerases (also called RNA polymerases). RNA viruses have RNA genomes, which can also be either double-stranded (dsRNA) or single-stranded (ssRNA). Single-stranded RNA genomes can be either plus (+) sense or minus (–) sense. RNA viruses replicate and transcribe their genomes using RNA-dependent RNA polymerases. Reverse-transcribing viruses replicate using reverse transcription, a process for making DNA from RNA templates. Their genomes can be either RNA or DNA. Retroviruses are ssRNA viruses that replicate through a DNA intermediate, thus requiring a virus-encoded RNA-dependent DNA polymerase (reverse transcriptase). Hepatitis B virus is a dsDNA virus that replicates through an RNA intermediate using a virally encoded reverse transcriptase. Thus, all four possible types of polymerases, i.e., RNA-dependent

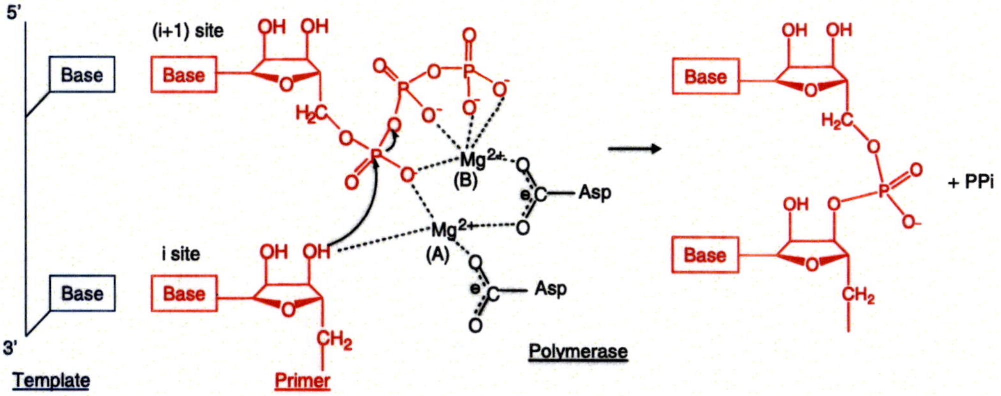

Fig. 12.2 Two-metal mechanism used by polymerases to catalyze the nucleotidyl transfer reaction. Template, primer, and polymerase residues are shown in *blue*, *red*, and *black*, respectively. The polymerase active site contains the binding sites for a template, the RNA or DNA terminus (the initiation, "i", or priming site), and the incoming NTP (the "i+1" site). Two metal ions, A and B, are coordinated by the conserved Asp residues in the polymerase active site. Metal ion A binds to the 3′-OH of the primer terminus and lowers the affinity of the 3′-OH for its hydrogen, facilitating a nucleophilic attack of 3′-O⁻ on the α-phosphate of the incoming NTP (Steitz 1998). Metal ion B stabilizes the incoming NTP and pyrophosphate (ppi) leaving group

RNA polymerases, RNA-dependent DNA polymerases, DNA-dependent RNA polymerases, and DNA-dependent DNA polymerases, are found in viruses. RNA-dependent RNA polymerases and reverse transcriptases are unique to viruses since the host cell does not require the RNA replication or reverse transcription.

12.1.2 Catalytic Mechanism

The main function of a polymerase is to copy a template nucleic acid strand to produce a daughter strand. Hence, the primary catalytic activity of a polymerase is to transfer a nucleotidyl moiety of an incoming nucleoside triphosphate (NTP) that is complementary to the template strand to the 3′-hydroxyl end of a growing daughter strand of RNA or DNA (Throughout the remainder of the text, NTP refers to either rNTP or dNTP unless specified). Two divalent metal ions, such as Mg^{2+} or Mn^{2+}, are required for polymerase activity. The polymerase active site thus must have binding sites for a template strand, the DNA or RNA primer terminus (the initiation "i" site), and the incoming NTP (the "i+1" site). The catalytic reaction steps consist of (1) binding of template–primer and NTP, (2) incorporation of nucleoside monophosphate into the growing daughter strand, (3) release of pyrophosphate, and (4) translocation along the template. Upon binding of the template and primer, the correct NTP is incorporated into the daughter strand via Watson–Crick base-pairing with the template base (Fig. 12.2). The polymerase active site provides conserved Asp residues that coordinate the two metal ions (either Mg^{2+} or Mn^{2+}). One metal ion, metal ion A in Fig. 12.2, binds the 3′-hydroxyl (3′-OH) of the terminal RNA or DNA primer at the "i" site and lowers the pK_a of the 3′-OH, thus facilitating nucleophilic attack of 3′-O⁻ on the α-phosphate of the incoming nucleotide (Steitz 1998). The other metal ion, metal B, binds the phosphates of the incoming NTP at the "i+1" site, positioning the incoming NTP and stabilizing the pyrophosphate leaving group. The 3′-OH of the RNA or DNA primer terminus then attacks the α-phosphate of the NTP, and a new phosphodiester bond is formed with a release of pyrophosphate. Both metal ions would stabilize the charge and

a

			A		B		C		D		E
RdRp	Polio	230	AFDYTG.Y	46	GMPSGCSGT.SI	25	YGDDV	20	TPADKSA		
	WN	534	ADDTAG.W	59	QRGSGQVVT.YA	54	SGDDC	16	NAMSKVR		
	Influ	304	TGDLKKYCL	91	SLSPGMMMG..N	31	SSDDF	27	.SKKKSY		
	LCMV	1169	SMDHSK.WG	93	DMGQGILHN..A	25	SSDDQ	44	.EAFKSR		
	BTV	635	IIDFGY.GE	74	THLSGENST.LI	33	VGDDT	31	.SPSKTM		
	IBDV	413	SIDLEK.GE	58	GQGSGNAAT.FI	42	SIDDI	37	.TYSKDL		
RdDp	HIV-1	108	VLDVGDAYF	31	VLPQGWKCSPAI	23	YMDDI	29	.PDKKHQ	4	FLWMGYELHPD
	HBV	249	SLDVSAAFY	78	KIPMGVGLSPFL	23	YMDDV	26	.NPNKTK	6	LNFMGYVIGCY

b

			A		B		C
DdRp	T7	536	AFDGSCSGI	103	KRSVMTLAYG	187	ALIHDS
	N4	1554	EADGVTNGP	104	KNPLTITIYG	267	LKIFDG
DdDp	ϕ29	248	FDVNSLYP	143	KLMLNSLYG	77	YCDTDS
	Adeno	542	YDICGMYA	145	KLLSNALYG	163	YGDTDS

Fig. 12.3 Polymerase sequence motifs in RNA-dependent and DNA-dependent polymerases. (**a**) Alignment of the RNA-dependent polymerase motifs A, B, C, D, and E. The alignment includes RdRps encoded by plus-sense ssRNA viruses (poliovirus and West Nile virus), minus-sense ssRNA viruses (influenza and Lassa viruses), dsRNA viruses (bluetongue and infectious bursal disease viruses), and DdRps (reverse transcriptases) encoded by a ssRNA virus (human immune deficiency virus type 1) and a dsDNA virus (hepatitis B virus). The numbers of residues separating the motifs are indicated. (**b**) Alignment of the DNA-dependent polymerase motifs A, B, and C. The alignment includes DdDps encoded by bacteriophages T7 and N4, and DdRps encoded by bacteriophages ϕ29 and adenovirus. DNA-dependent polymerase motifs A and C and RNA-dependent polymerase motifs A and C coordinate divalent metal ions for catalytic activity

geometry of the pentavalent transition state during the nucleotidyl transfer reaction. Upon release of a pyrophosphate product, the newly formed RNA or DNA primer terminus translocates by one base from the i+1 site to the i site, and the nucleotidyl transfer reaction is repeated until the template strand is completely copied or a termination signal is encountered.

12.1.3 Sequence Motifs of Viral Polymerases

Although overall sequence identity among viral polymerase is very low, even within the same family of viruses, different families of polymerases share a characteristic arrangement of consensus sequences that are involved in binding template, positioning of the priming nucleotide, the nucleotidyl transfer reaction, and maintaining structural integrity of the protein (Poch et al. 1989). These polymerase sequence motifs also provide the basis for identification of new polymerase genes in newly sequenced virus genomes. Conserved sequence motifs have been described for all four types of viral polymerases. RNA-dependent RNA polymerases share four major conserved sequence motifs, A, B, C, and D (Fig. 12.3a). Motif A and Motif C contain the catalytic Asp residues that bind two metal ions in the active site; in particular, motif C contains the characteristic Gly-Asp-Asp (GDD) sequence embedded in hydrophobic residues. Reverse transcriptases (RNA-dependent DNA polymerases) have five conserved sequence motifs, four of which can be aligned with those from RNA-dependent RNA polymerases. Similar to RNA-dependent RNA polymerases, motifs A and C are most strictly conserved.

Viral DNA-dependent RNA and DNA polymerases carry out similar functions to those of the multisubunit polymerases in prokaryotic and eukaryotic organisms and thus share similar sequences. The single subunit DNA-dependent RNA polymerases from bacteriophages T3 and T7 are more closely related to mitochondrial RNA polymerases than to any *Escherichia coli* (*E. coli*) polymerases. The multisubunit poxvirus RNA polymerase resembles polymerases from eukaryotic and prokaryotic organisms (Masters et al. 1987). Both DNA-dependent RNA polymerases and DNA-dependent DNA polymerases share the three conserved sequence motifs A, B, and C (Delarue et al. 1990). Motif A and C contain the essential Asp residues that bind divalent metal ions necessary for catalytic activity and can be aligned with RNA-dependent polymerase motifs (Fig. 12.3b).

12.1.4 General Features of Viral Polymerase Structures

All polymerases have roughly the shape of a "cupped" right hand, consisting of fingers, palm and thumb domains, based on an analogy used to describe the structure of the Klenow fragment of DNA polymerase (Ollis et al. 1985). The palm domain is the most highly conserved domain among different types of polymerases. Most polymerase sequence motifs, including the motifs A and C, are located in the palm domain. This domain consists of a four- to six-stranded β-sheet flanked on one side by two α-helices, with the essential Asp residues in motif C located in a hairpin between the two central β-strands. The fingers and thumb domains vary significantly in both size and secondary structure and reflect the specific requirements of replicating single- or double-stranded RNA/DNA genomes. The inner surfaces of the fingers and thumb domains form a template-binding channel that leads into the catalytic palm domain. Template and primer strands bind in the template-binding channel, positioning the primer terminus near the active site (see Figs. 12.6, 12.8, and 12.11). In the active site, the correct NTP to be added to the daughter strand is selected by Watson–Crick base-pairing with the template base. The selectivity of ribose versus deoxyribose NTPs is also provided by the interaction of the polymerase with the 2′-OH of the NTP. In general, DNA polymerases that incorporate dNTP in the growing daughter strand have a large side chain that prevents binding of an rNTP with a 2′-OH. In contrast, RNA polymerases directly interact with the 2′-OH of an rNTP via hydrogen bonding and utilize amino acids with a small side chain to accommodate the 2′-OH moiety. The polymerase active site binds the correct NTP with 10–1,000-fold higher affinity than incorrect NTPs (Gao et al. 1997).

In addition to fingers, palm, and thumb domains, viral polymerases often have other domains that carry out functions related to other respects of viral genome transcription and replication. For example, flavivirus RNA-dependent RNA polymerases have a methyltransferase domain that catalyzes methylations of a 5′-RNA cap. Retrovirus reverse transcriptases have an additional ribonuclease H domain that catalyzes degradation of the RNA strand in the RNA–DNA hybrid during genome replication. Some viral DNA-dependent polymerases have a nuclease domain with a proofreading activity to correct nucleotides incorrectly incorporated during genome synthesis. The arrangement of different domains and their functions in viral replication will be discussed in the following sections.

12.1.5 Polymerase Reaction Steps

Nucleic acid synthesis by polymerases can be divided into initiation, elongation, and termination phases. Unlike cellular DNA and RNA polymerases, which require oligonucleotides to initiate nucleic acid synthesis, viral polymerases can initiate genome replication using a variety of mechanisms reflecting their adaptation to the host cell.

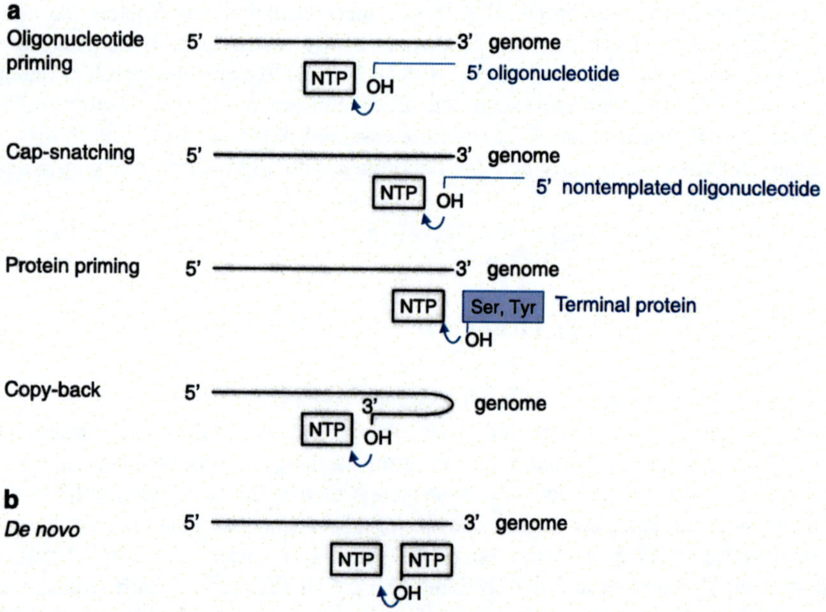

Fig. 12.4 Schematic diagram of initiation mechanisms of viral polymerases. (**a**) Primer-dependent initiation. The primers can be an oligonucleotide, a snatched cap from host mRNA, a terminal protein, or the 3′-terminus of the template strand that folds back onto itself. (**b**) De novo initiation. The 3′-hydroxyl group of the initiation NTP attacks the α-phosphate of the incoming NTP

12.1.5.1 Initiation

All known polymerases synthesize nucleic acid in the 5′→3′ direction and thus initiate the nucleotidyl transfer reaction at the 3′-end of the template strand. Polymerases use one of two initiation mechanisms, either a primer-dependent or a de novo (primer-independent) mechanism (Fig. 12.4). Primer-dependent initiation requires a primer, either in an oligonucleotide or a protein form, which provides the initial 3′-hydroxyl for addition of the first incoming nucleotide. Viral polymerases use one of the four following strategies to initiate nucleic acid synthesis via the primer-dependent mechanism (van Dijk et al. 2004).

1. *Oligonucleotide primer*. All DNA-dependent DNA polymerases initiate DNA synthesis using a short RNA primer. They can only add new NTPs to the preexisting strand. Reverse transcriptase also uses an RNA primer to initiate DNA synthesis. Some RNA-dependent RNA polymerases synthesize multiple copies of a two to five nucleotide-long RNA that is subsequently used as a primer in the elongation phase.
2. *Cap-snatching*. RNA-dependent RNA polymerases from minus-sense RNA viruses use a fragment of cellular-capped mRNA to make viral plus-sense mRNA. These viruses encode an endonuclease that cleaves capped oligonucleotides from the 5′-end of a cellular-capped mRNA and use it as a primer for the minus-sense RNA template. As a result, the viral mRNA contains a 5′-cap and short nontemplated sequences prior to the beginning of a viral transcript sequence (see Sect. 12.2.2).
3. *Protein-priming*. Similar to the 3′-hydroxyl group of an oligonucleotide primer, a terminal protein can provide a hydroxyl group to which the first nucleotide can be linked via a phosphodiester bond. Viral polymerases catalyze the addition of the first nucleotide to the terminal protein and then use the protein-linked primer for further elongation. As a result, a terminal protein is

covalently linked to the 5′-end of the genome. Picornavirus (ssRNA virus), adenoviruses (dsDNA virus), bacteriophage ϕ29 (dsDNA virus), and reverse transcribing hepatitis B virus (dsDNA virus) encode a terminal protein that provides the hydroxyl group from a Tyr or Ser residue for protein priming. The size of terminal proteins varies from small 20 amino-acid peptides in picornavirus to a 30-kDa protein in bacteriophage ϕ29 (see Sects. 2.1.1 and 4.1).

4. *Copy-back mechanism.* A special case of a primer-dependent RNA initiation is 3′-end elongation by a "copy-back" mechanism, where the 3′-end of the template folds back upon itself to serve as its own primer. The resulting product is a single chain nucleotide that has twice the length of the template. This type of chain elongation has been observed for recombinant RNA-dependent RNA polymerases from flavivirus (Ackermann and Padmanabhan 2001). However, this is unlikely to be a productive initiation mechanism in infected cells since the doubling of the genome would require a larger protein shell for encapsidation. De novo initiation is the likely mechanism for flavivirus RNA replication in infected cells.

Polymerases can also initiate nucleic acid synthesis without the need of a primer. De novo initiation requires two NTPs, i.e., the initiation NTP at the "i" site and the incoming NTP at the "i+1" site (Fig. 12.2). The initiation NTP provides a 3′-hydroxyl group for the addition of the next nucleotide and thus functions similarly to the 3′-hydroxyl of a primer terminus in a primer-dependent mechanism. Precise initiation at the 3′-terminus of the genome is necessary for preservation of the entire genome sequence. RNA-dependent RNA polymerases from flaviviruses and bacteriophage ϕ6 use a de novo mechanism to initiate RNA synthesis at the 3′-terminus of the template and continue to elongate the entire length of RNA. DNA-dependent RNA polymerases can also initiate RNA synthesis de novo. During the abortive initiation phase, the RNA polymerases initiate RNA synthesis at a promoter DNA sequence and produce short RNA oligonucleotides (3 to 8 nucleotide long). Once the nascent RNA chain becomes approximately nine-nucleotides-long, the polymerase forms a stable elongation complex and synthesizes the entire RNA transcript (see Sect. 12.5.1).

12.1.5.2 Elongation

Once a stable polymerase–nucleic acid complex is formed, the polymerase consecutively adds many nucleotides without dissociation from the template. Most polymerases can add thousands of nucleotides during processive elongation. For example, poliovirus RNA-dependent RNA polymerase can add approximately 5,000 nucleotides and 18,000 nucleotides in the absence and presence of its accessory protein 3AB, respectively (Rodriguez-Wells et al. 2001). This indicates that the poliovirus polymerase can synthesize the entire 7.5-kb genome in a single binding event in infected cells. The ϕ29 DNA-dependent DNA polymerase can also synthesize the entire 20-kb DNA genome upon a single binding event without dissociation (Meijer et al. 2001). High processivity requires a stable association of template and primer in the polymerase that prevents dissociation but which should also be nonspecific and weak enough to facilitate movement upon each nucleotidyl transfer reaction steps.

12.1.5.3 Termination

It is poorly understood how viral polymerases terminate nucleic acid synthesis at the 5′-end of the genome. Ending precisely at the 5′-terminus of the genome would be particularly challenging for viruses with linear genomes such as single-stranded RNA or DNA viruses. If the end of the genome is not copied during successive rounds of replication, the genome will become progressively shorter or longer. An exception is DNA-dependent RNA polymerases that terminate RNA transcription at a terminal signal. A newly synthesized RNA forms a GC-rich hairpin followed by a stretch of uridine nucleotides at the terminal signal, and the polymerase–RNA complex is released from the DNA template.

12.1.6 Polymerases as Molecular Motors

Polymerases are not only catalytic enzymes but are also mechanical machines that use the energy released from nucleotide transfer to move along the DNA or RNA strand. The movement of the *E. coli* RNA polymerase along its DNA template has been observed by single-molecule experiments, including microscopy and use of an optical trap to measure force generation during the polymerase reaction. When high mechanical force is applied in opposition to the movement of the polymerase, the rate of replication decreases and finally stops. This "stall" force that is necessary to block the polymerization reaction varies for different polymerases but is in the range of 15–40 pN (Abbondanzieri et al. 2005; Rothwell and Waksman 2005). For example, the force generated by the *E. coli* RNA polymerase is approximately 20 pN, exerting five to six times higher forces than those measured for cytoskeletal motors of myosin or kinesin (Wang et al. 1998; Yin and Steitz 2004). In addition, an optical trap experiment showed that the *E. coli* RNA polymerase translocates along its DNA template in discreet steps of 3.7 ± 0.6 Å. This 3.7 Å step corresponds to the spacing of dsDNA in a B-form helix, indicating that the polymerase moves along the DNA template one base at a time, which had been proposed for all polymerases but never measured (Abbondanzieri et al. 2005).

The mechanism that couples chemical reaction of polymerases to mechanical movement is not known. Two models have been proposed for nucleic acid translocation. In the active "power-stroke" model, the movement is powered by the free energy liberated from phosphodiester bond cleavage, i.e., release of pyrophosphate drives a conformational change that "pushes" the polymerase forward on the template (Yin and Steitz 2004). This model is suggested by structural studies of T7 RNA polymerases in successive steps of nucleotide addition. In brief, the polymerase has the same conformation in both the catalytic NTP substrate complex, i.e., prior to the nucleotidyl transfer reaction, and the pyrophosphate product complex, i.e., after the nucleotidyl transfer reaction and prior to pyrophosphate release. This suggests that the nucleotidyl transfer reaction does not require a conformational change in the polymerase and does not promote translocation, and thus, it is likely that the dissociation of pyrophosphate leads to a conformational change that is used to translocate the template–primer (Sect. 12.5.1.3).

Alternatively, in the "Brownian-ratchet" model, or "translocational equilibrium" model, the polymerase oscillates back and forth along the template among number of states that are in equilibrium, including the pre- and posttranslocated states (Guajardo and Sousa 1997). Binding of incoming NTPs stabilizes the posttranslocated state so that at high concentration of NTPs, the translocational equilibrium will favor the posttranslocated state. Binding of NTP acts as a ratchet type of device and prevents back-sliding of the polymerase (Brownian-ratchet model). Biochemical and single-molecule kinetic studies support the Brownian-ratchet model and suggest that translocation can be explained in terms of the affinity of NTP and the relative free energies of the polymerase–template–NTP complexes at each step of the polymerase reaction (Abbondanzieri et al. 2005). Whether pyrophosphate release is coupled to a conformational change is a key distinction between the power-stroke and Brownian-ratchet models, and is a question that remains unresolved. Although no conformational changes are involved in the Brownian-ratchet model, it does not exclude a conformational change. Understanding the mechanochemistry of polymerases will require detailed studies on the biochemical, kinetic, and mechanical pathways of the polymerase reaction.

12.2 RNA-Dependent RNA Polymerase

12.2.1 Plus-Sense ssRNA Viruses

Viruses that possess plus-sense ssRNA genomes include poliovirus and foot-and-mouth virus (*Picornaviridae*), Norwalk virus (*Caliciviridae*), Sindbis virus (*Togaviridae*), and flavivirus (*Flaviviridae*). RNA genome replication in ssRNA viruses takes place in a membrane-bound replication complex consisting of the viral polymerase and other viral proteins, cellular proteins, and viral RNA. The exact composition and nature of the replication complex is still unknown. Plus-sense genomic RNA serves as an mRNA in infected cells to produce viral proteins without any modification of the viral genome, and thus, viral RNA-dependent RNA polymerases are responsible for replication of the viral RNA genome. The RNA-dependent RNA polymerases first transcribe genomic plus-strand RNA into a complementary minus-strand RNA to form a dsRNA intermediate (Uchil and Satchidanandam 2003). The newly synthesized minus-strand of the dsRNA then serves as a template to generate multiple copies of plus-strand RNA. RNA synthesis occurs in an asymmetric replication cycle in which plus-strand RNA is synthesized in large excess over minus-strand, and all minus-strands exist as a part of the dsRNA intermediate. Viral polymerases copy only virus-specific RNA species in the infected cell, while isolated recombinant polymerases lack specificity and use both viral and nonviral templates for RNA synthesis. Thus, the replication complex may provide selectivity for the viral template by concentrating various substrates in the vicinity of the polymerase and thus creating an appropriate environment for selective and efficient genome replication.

Structures of RNA-dependent RNA polymerases have been determined from plus-sense ssRNA and dsRNA viruses. The first structure of an RNA-dependent RNA polymerase was from poliovirus, and several polymerase structures followed in either an unliganded state or complexed with various templates and NTPs (Hansen et al. 1997; Ng et al. 2002; Zamyatkin et al. 2008). A unique feature of the RNA-dependent RNA polymerases is an extensive interaction between the fingers and thumb domains; the fingers domain contains a long insert of several strands, called the "fingertips," that interacts with the thumb domain (Figs. 12.6 and 12.8). The fingertips contribute to formation of a well-defined template-binding channel in the "front" of the polymerase and an NTP channel in the "back" of the polymerase. It was proposed that NTPs would enter the active site through the NTP channel. Because of fingertip-mediated interactions between the fingers and thumb domains, large movements of the thumb domain relative to the fingers domain are restricted, and hence there are no significant conformational changes upon binding of template, primer, or NTP in the template-binding channel in RNA-dependent RNA polymerases. The size of thumb domains vary significantly among RNA-dependent RNA polymerases and seem to play a role in determining whether primer-dependent versus de novo initiation mechanisms are used. Polymerases from *Picornaviridae* and *Caliciviridae* that use a primer to initiate nucleotide synthesis have small thumb domains and thus have wider template-binding channels necessary to accommodate both a template and a primer (Fig. 12.6). In contrast, *Flaviviridae* polymerases that use de novo initiation have large thumb domains and thus have narrower template-binding channels that can accommodate only single-stranded RNA and NTPs (Fig. 12.8).

Viral RNA-dependent RNA polymerases do not have a proofreading nuclease activity, which leads to high rates of misincorporation of nucleotides during genome replication. The mutation rates are estimated to be on the order of one mutation per 10^3–10^7 nucleotides, resulting in approximately one error per replicated genome (Drake 1991). In comparison, the mutation rate in *E. coli* or *S. cerevisiae*, where cellular polymerases benefit from an error-correcting mechanism, is on the order of one mutation per 10^9–10^{10} nucleotides. Diverse viral genomes resulting from high mutation rates by viral RNA-dependent RNA polymerases may provide an evolutionary advantage for virus survival. For example, mutations in the viral surface protein could lead to new viruses that can evade the host antiviral responses or resist antiviral compounds.

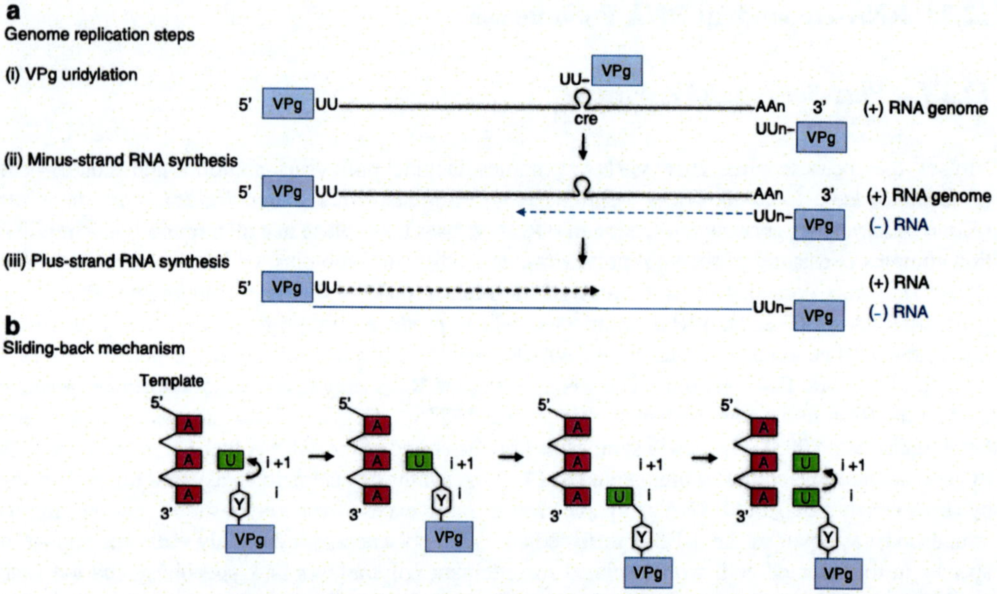

Fig. 12.5 Poliovirus genome replication. (**a**) Genome replication steps. (i) VPg uridylation. Terminal protein VPg is covalently attached to the 5′-end of plus-sense viral genome. The hydroxyl group of VPg Tyr3 acts analogously to the free 3′-OH of the primer terminus to form an uridylated form of VPg, VPg-pU, which is then further elongated to VPg-pUpU. The VPg-pUpU can be synthesized from either the poly(A) tail or conserved *cis*-acting replication element (CRE) as a template. (ii) Minus-strand RNA synthesis. VPg-pUpU is used as a primer to base-pair with a plus-strand genome and elongated to form a VPg-linked minus-strand RNA (*blue strand*). (iii) Plus-strand RNA synthesis. The VPg-pUpU made using the CRE as a template is used as a primer for plus-strand RNA synthesis in the subsequent step. Thus, both plus- and minus-strand RNAs are VPg-linked at their 5′-ends in vivo. (**b**) Schematic of the sliding-back mechanism. The polymerase initiates VPg uridylation at the second A on the 3′-end template. The VPg-pU then slides back one base to base-pair with the 3′-terminal A, and the polymerase begins second uridylation reaction. Thus, the sliding-back mechanism requires two sequential adenosines in the template

12.2.1.1 Picornavirus and Calicivirus Initiate RNA Synthesis via a Primer-Dependent Mechanism

Picornaviridae (poliovirus, foot-and-mouth disease virus (FMDV), and coxsackievirus) and *Caliciviridae* (rabbit hemorrhagic disease virus and Norwalk virus) initiate RNA synthesis via a primer-dependent mechanism. Poliovirus, for example, has a small 7.5-kb ssRNA genome which consists of 5′-nontranslated region (NTR), a single open-reading frame, and a 3′-NTR with a poly-adenylated [poly(A)] tail. The 5′-terminus of the RNA genome is covalently linked to a terminal protein VPg (virion protein genome-linked), a 22-amino acid peptide. The viral RNA-dependent RNA polymerase "3D" is synthesized as a 3CD precursor protein that is cleaved into a protease 3C and a polymerase 3D. The 3CD precursor protein does not have polymerase activity, and rather, it functions as a protease with cleavage specificity that differs from that of the cleaved 3C protease. The cleaved viral polymerase 3D catalyzes both the initial attachment VPg to the genome and subsequent elongation to produce a new genomic RNA.

3D polymerase initially catalyzes the uridylation of VPg using the 3′ poly(A) tail as a template (step (i) in Fig. 12.5a). RNA synthesis is initiated at the second nucleotide A rather than the 3′-terminal A of the template and covalently attaches UTP to VPg at a conserved Tyr residue (Tyr3). The hydroxyl group of Tyr3 acts analogously to the free 3′-hydroxyl of the primer terminus and forms a phosphodiester bond with the first UTP to form VPg-pU (Fig. 12.5b). This VPg-pU then "slides back" one base and base-pairs with the 3′-terminal A of the template, and 3D polymerase adds the second UTP to form VPg-pUpU (Paul et al. 2003b). The polymerase then uses the VPg-pUpU as a protein-linked primer and the plus-strand genomic RNA as a template to synthesize the entire length of minus-strand RNA (step (ii) in Fig. 12.5a). Next, the 3D polymerase synthesizes the plus-strand RNA using the VPg-pUpU as a primer and the newly synthesized minus strand as a template (step (iii) in Fig. 12.5a). The VPg-pUpU used for plus-strand RNA synthesis is suggested to be synthesized from a conserved *cis*-acting replication element (CRE) as a template; CRE is an RNA stem-loop containing a 5′-AAAC-3′ sequence in the coding region. It is not clear how the polymerase discriminates between the two types of VPg-pUpU, but mutations in the CRE region prevent the viral genome replication (Murray and Barton 2003; Paul et al. 2003a). The structures of 3D polymerase and its complexes that depict successive steps in viral genome replication are described below.

VPg Uridylation by 3D Polymerase (Initiation)

Crystal structures of 3D polymerases from *Picornaviridae* and *Caliciviridae* are all similar and have the characteristic features of RNA-dependent RNA polymerases such as fingertips that connect the fingers domain to the thumb domain and well-defined template-binding and NTP channels (Thompson and Peersen 2004). These polymerases have small thumb domains, consisting of only several helices, and thus a large template-binding channel that can accommodate an RNA template and a VPg-linked primer (Fig. 12.6a). Additionally, the 3D polymerases complexed with a template, NTP, and VPg, respectively, are very similar to the unbound polymerase, indicating that there are no large conformational changes or domain movements upon binding of VPg, template–primer or template–primer–NTP.

The polymerase initiation complex of VPg-uridylation could be formed by VPg, template $(rA)_{10}$, UTP, and metal ions (Mg^{2+} and Mn^{2+}). However, crystallization conditions containing this combination of chemicals produced a VPg-pU complex, indicating that the uridylation reaction occurred prior to crystallization (Ferrer-Orta et al. 2006). VPg-pU binds in the template-binding channel with the N terminus of VPg located in the NTP entry channel and the C terminus directed toward the template-binding channel (Fig. 12.6a, b). The hydroxyl group of VPg Tyr3 is covalently attached to the α-phosphate of UMP in VPg-pU and interacts with a divalent metal ion that binds to the catalytic Asp of the GDD motif (Asp338) in the active site. This arrangement of VPg-pU is similar to that of the primer terminus in the nucleotidyl transfer reaction, indicating that 3D polymerase catalyzes VPg uridylation using the same two-metal mechanism as does the nucleotidyl transfer reaction (Fig. 12.2).

The predominant forms of VPg-linked primers in infected cells are VPg-pU and VPg-pUpU, indicating that VPg-pUpU dissociates from the polymerase active site after VPg uridylation and then reassociates as a protein-linked primer for elongation. If VPg uridylation begins at the 3′ poly(A) tail of the template, then the polymerase may continue nucleotidyl transfer reactions to replicate the entire genome. However, if uridylation begins from the CRE, positioned in the middle of the genome, it needs to dissociate and be transferred to the 3′-end of the minus-strand RNA. It is not clear how 3D polymerase terminates RNA synthesis at VPg-pUpU, how the VPg-pUpU is transferred to the 3′-end of the genome, or whether the reassociated VPg-pUpU would bind the same way as in the VPg uridylation reaction.

Nucleotidyl Transfer Reaction

A general model of nucleotide incorporation is that polymerases catalyze the nucleotidyl transfer reaction using an ordered mechanism. Polymerase first binds a template–primer (step 1), followed by binding of an NTP complementary to the template base (step 2). This polymerase complex is then converted into an activated form, which includes a conformational change from an "open" to a

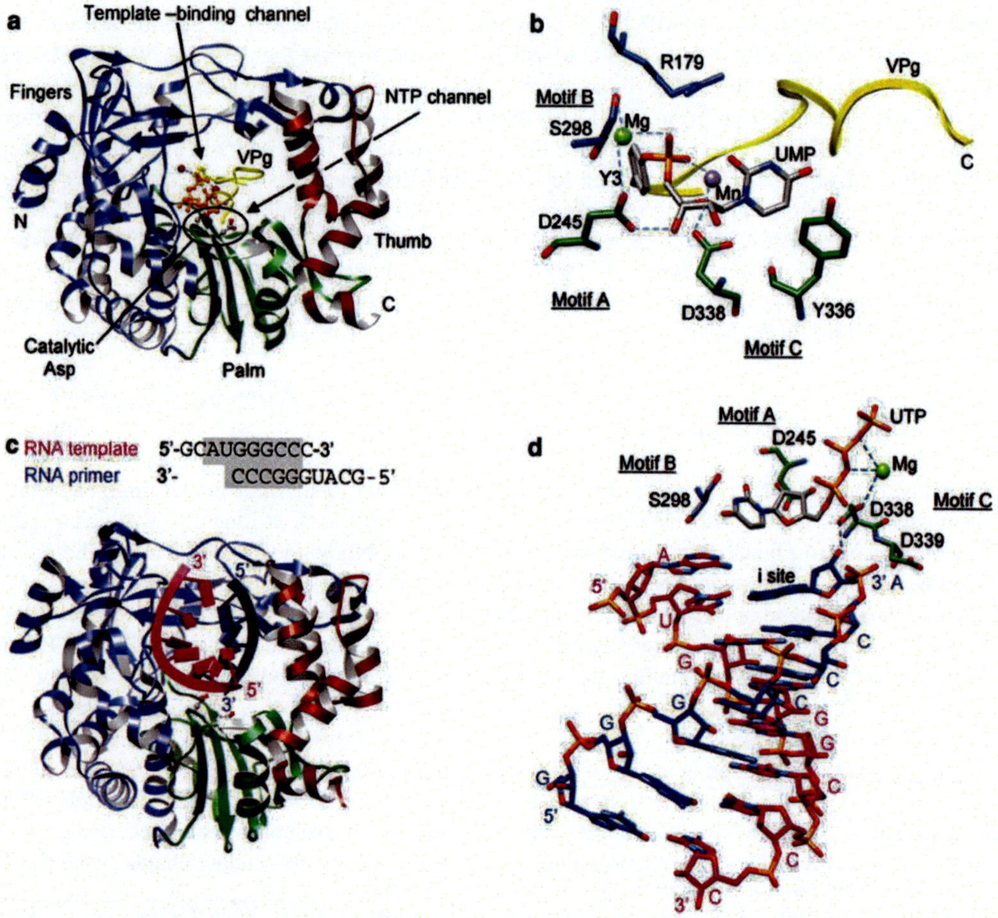

Fig. 12.6 Crystal structures of 3D polymerase that use VPg-linked primers. (a) FMDV 3D polymerase and VPg-pU complex. The fingers, palm, and thumb domains are shown in *blue*, *green*, and *red*, respectively. The active site residues in the GDD motif are depicted as ball-and-stick models. The UMP moiety in VPg-pU is shown as an orange ball-and-stick model. (b) Active site in a VPg-pU complex. Tyr3 and the UMP moiety of VPg-pU are shown as *gray sticks*. The active site residues are colored according to their domains as in (A), and polymerase sequence motifs are labeled. (c) 3D polymerase and template–primer complex. The template and primer strands are shown in *magenta* and *blue*, respectively. (d) 3D polymerase and template–primer–NTP complex (open form). Note that the incoming NTP (UTP) does not base-pair with the template base at the "i+1" site, although the triphosphate moiety interacts with a metal ion in the active site

"closed" form that is competent to undergo the chemistry step (step 3). Next, the nucleotidyl transfer reaction occurs (step 4), the pyrophosphate product is released, and template–primer translocates by one base (steps 5 and 6). The 3D polymerase structures have been captured in template–primer, open NTP, and closed NTP complexes (steps 1, 2, and 3) and are described below.

1. *Template–primer binding*. In a FMDV polymerase complex containing a self-complementary RNA, the template–primer RNA duplex binds in a right-hand A-form helix in the template-binding channel with the 3′-end of the template exiting the active site through the "front" of the polymerase (Ferrer-Orta et al. 2004) (Fig. 12.6c). The template strand in the RNA duplex interacts predominantly with the fingers domain, while the primer strand interacts at the thumb and palm domains. The template-binding channel accommodates seven nucleotides in the template strand. In the active site, the 3′-hydroxyl of the primer terminus forms a hydrogen bond with the catalytic

Asp residue in motif C (Asp338), which also binds to a metal ion. Only one metal ion, metal B in Fig. 12.2, is observed in the active site.

2. *NTP binding (open conformation)*. The template–primer–NTP complex in FMDV polymerase has been captured in an "open" conformation, in which the polymerase structure remains unchanged from the template–primer bound form (Ferrer-Orta et al. 2007). In this conformation, the incoming NTP does not base-pair with the template base at the "i+1" site although the phosphate group of the incoming NTP binds a metal ion near the essential Asp in motif C (Asp338) and the 3′-OH of the NTP forms a hydrogen bond with the essential Asp in motif A (Asp245) (Fig. 12.6d). This conformation is unable to carry out catalytic nucleotidyl transfer reactions and thus suggests that the elongation complex assembles in a sequential order, i.e., the second metal ion that binds to the 3′-hydroxyl of a primer terminus would bind in the polymerase active site only after the incoming NTP is correctly assembled in the active site.
3. *NTP binding (closed conformation)*. An elongation complex from Norwalk virus, containing self-complementary RNA, incoming NTP, Mg^{2+}, and Mn^{2+}, represents a perfectly assembled intermediate poised for the nucleotidyl transfer reaction (Zamyatkin et al. 2008). The binding of the template–primer–NTP complex is accompanied by rotations of the helices in the thumb domain, which provides a binding groove for a primer strand, and the loop containing the motif B in the fingers domain relative to the template–primer–NTP. In this "closed" form, the incoming NTP forms a Watson–Crick base pair with the template base, and the triphosphate group of the NTP binds two Mn^{2+} ions in the active site. The essential Asp residues in motif A (Asp242) and in the GDD motif (Asp343 and Asp344) also coordinate the divalent metal ions. The correct incoming NTP to be added to the primer strand also seems to be selected in this step by the interaction of polymerase with the 2′-hydroxyl of the rNTP. The 2′-hydroxyl of the rNTP forms hydrogen bonds with Ser and Asn residues in the motif B (Ser303 and Asn309) in the fingers domain. This interaction is missing in the "open" conformation where the incoming NTP is not correctly assembled for catalysis. Thus, this conformation represents a closed complex trapped immediately prior to chemical catalysis. These open and closed conformations described by polymerase complex structures are subtle changes in the active site and should be distinguished from the global domain rearrangements seen in other types of polymerases.

12.2.1.2 *Flaviviridae* RdRps Initiate RNA Synthesis via a De Novo Mechanism

Flaviviridae consists of hepacivirus (hepatitis C virus), pestivirus (bovine viral diarrhea virus, BVDV), and flavivirus (dengue and West Nile viruses). These viruses have approximately 10–12-kb plus-sense ssRNA genomes, which consists of a 5′-NTR, a single ORF, and a 3′-NTR lacking a poly(A) tail. Unique to flavivirus within *Flaviviridae*, the 5′-end of the viral genome has a cap structure and thus requires additional capping enzymes to carry out genome replication. The single ORF is translated into a polyprotein that is cleaved into individual proteins by cellular and viral proteases. Viral RNA-dependent RNA polymerases are synthesized at the C terminus of the polyprotein. *Flaviviridae* RNA-dependent RNA polymerases vary in size. The ~60-kDa hepatitis C virus "NS5B" polymerase is the smallest *Flaviviridae* polymerase and contains only a polymerase domain. Pestivirus "NS5B" polymerase has an additional small N-terminal domain whose function is unknown. Flavivirus "NS5" polymerase consists of an N-terminal methyltransferase domain (residues 1–270) that is involved in RNA cap methylation, and a C-terminal polymerase domain.

RNA-dependent RNA polymerases of *Flaviviridae* initiate RNA synthesis via a de novo mechanism, in which the second NTP is added directly to the 3′-hydroxyl of the first initiation NTP without the need of a primer. The viral polymerases bind at the 3′-end of the plus-strand genomic RNA and copy an entire genome without dissociation (Fig. 12.7). The newly synthesized, negative-sense RNA then serves as a template for synthesizing multiple copies of plus-sense genomic RNA. De novo

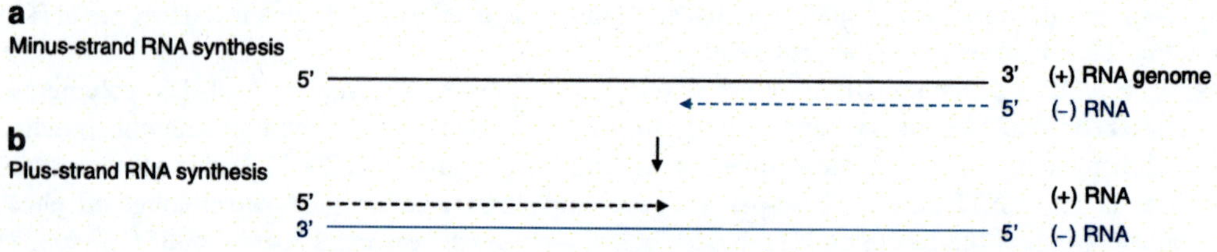

Fig. 12.7 Flavivirus genome replication. (**a**) Minus-strand RNA synthesis. The viral RdRp uses a plus-sense viral genome as a template to synthesize minus-strand RNA without a primer (de novo). Minus-strand RNA forms a dsRNA with plus-sense RNA. (**b**) Plus-strand RNA synthesis. Plus-sense RNA is synthesized using a minus-strand RNA in a dsRNA intermediate as a template via de novo initiation. In flavivirus, the newly synthesized plus-strand RNA is capped at its 5′-end by the viral replication complex

initiation has an advantage in genome replication in that no additional enzymes are required for synthesizing primers, and RNA synthesis can start at the very end of the genome, ideal for copying the entire length of the genome without loss of information (van Dijk et al. 2004). However, formation of a stable de novo initiation complex is challenging. Polymerases must selectively accommodate template, an initiation NTP, an incoming NTP, and two metal ions with precise positioning, which might require additional cofactors. Flavivirus polymerases require a high concentration of GTP as a cofactor for de novo initiation (Kao et al. 1999; Luo et al. 2000). Although de novo initiation is the likely mechanism of genome replication in infected cells, recombinant viral polymerases are capable of RNA synthesis via both de novo and primer-dependent initiation mechanisms.

De Novo Initiation by Flavivirus Polymerases

Structures of the polymerase domains from all three members of the *Flaviviridae* family have been determined (Ago et al. 1999; Bressanelli et al. 1999; Lesburg et al. 1999; Choi et al. 2004; Malet et al. 2007; Yap et al. 2007). Compared to the RNA-dependent RNA polymerases from *Picornaviridae* or *Caliciviridae*, all three polymerases possess a large thumb domain and a C-terminal motif within the same domain that encircles the active site (Fig. 12.8a). The larger thumb domain reduces the size of the template-binding channel. The C-terminal motif in the thumb domain (a β-hairpin in hepatitis C virus NS5B, a β-strand and connecting loop in pestivirus NS5B, and a loop in flavivirus NS5) occludes the active site and further reduces the size of the template-binding channel. This arrangement of the thumb domain restricts access to the template-binding channel, thus allowing only single-stranded RNA to access the active site during de novo initiation. When the C-terminal motif (β-hairpin) is replaced with a shorter turn in hepatitis C virus NS5B polymerase, the template-binding channel widens, and the mutant polymerase is able to use a dsRNA template–primer complex for a primer-dependent initiation, which wild-type polymerase is unable to use (Hong et al. 2001). Hence, the thumb domain and the C-terminal motif play a role in discriminating between de novo and primer-dependent initiation mechanisms.

A high concentration of GTP (but not any other nucleotide) is required for de novo initiation of RNA synthesis in all three members of the *Flaviviridae* family, regardless of the nucleotide at the 3′-end of the RNA template (Kao et al. 1999; Luo et al. 2000). The role of GTP has been proposed to stabilize the initiation NTP in the de novo initiation complex. The BVDV polymerase, complexed with GTP, shows that GTP binds 6 Å away from the catalytic GDD motif inside the template-binding channel and interacts with residues from all three polymerase domains (Choi et al. 2004) (Fig. 12.8a). When the GTP-binding site residues in BVDV polymerase are individually mutated to Ala, their ability to direct de novo RNA synthesis is almost completely abolished, whereas primer-dependent

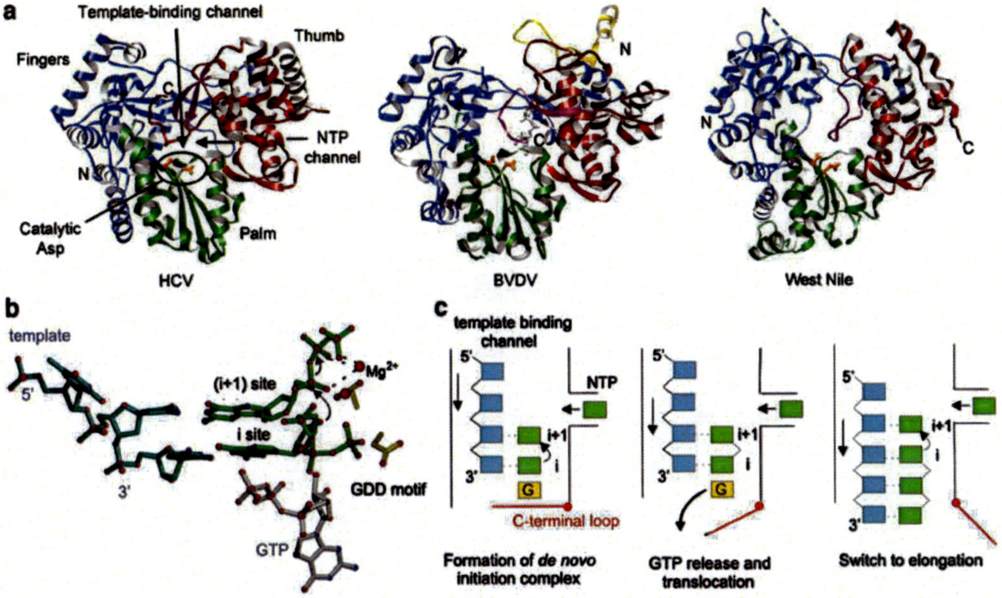

Fig. 12.8 The *Flaviviridae* RNA-dependent RNA polymerases that use a de novo initiation mechanism. (**a**) Hepatitis C virus (HCV), Bovine viral diarrhea virus (pestivirus, BVDV), and West Nile virus (flavivirus) are shown with their fingers, palm, and thumb domains colored in *blue*, *green*, and *red*, respectively. The *yellow* region of BVDV polymerase is a part of the N-terminal domain. The C-terminal motif in the thumb domain that occludes the active site is shown in purple, and active site residues are shown as *orange sticks*. (**b**) Model of initiation complex illustrating the proposed role of the de novo GTP. The role of GTP in de novo initiation may be to stabilize the initiating NTP at the "i" site. (**c**) Schematic of de novo initiation in *Flaviviridae* polymerases. The initiation complex, consisting of template (*blue*), two NTPs (*green*), and GTP (*yellow*), is assembled in the active site. The C-terminal loop is closed to further stabilize the initiation complex. After the nucleotidyl transfer reaction, the GTP molecule required for stabilizing the initiation NTP is released and the template–primer translocates by one base. The C-terminal loop in the thumb domain opens up to allow dsRNA to move out of the active site and the polymerase transitions from the initiation to the elongation phase (Choi and Rossmann 2009)

RNA elongation is reduced by only two- to tenfold (Ranjith-Kumar et al. 2002). This suggests that the GTP-binding site is essential for de novo initiation of RNA synthesis. Modeling the de novo initiation complex by superposition of a ϕ6 polymerase–template–NTP complex with the BVDV polymerase–GTP complex suggests that the ribose triphosphate moiety of the GTP would be positioned as might be expected for the "i – 1" site of already synthesized RNA strand (Fig. 12.8b). The 3′-hydroxyl of the GTP ribose is positioned near the α- and β-phosphates of the initiation NTP at the "i" site. Thus, the GTP molecule may stabilize the initiation NTP by helping to position the 3′-hydroxyl of the initiation NTP for nucleophilic attack on the α-phosphate of the incoming nucleotide (Choi et al. 2004). Once the nucleotidyl transfer reaction occurs, the GTP molecule required for stabilizing the initiation NTP would be released from the active site and the template RNA and product RNA translocated by one base (Fig. 12.8c).

Elongation

Currently, there is no structural information available for initiation or elongation complexes that contain template–NTPs or template–primer–NTP. Modeling of template–primer complexes into the active site of *Flaviviridae* polymerases suggests that the C-terminal motif in the thumb domain and the enclosed active site would block the path of elongating dsRNA. Thus, as the template-product RNA grows following initiation, a conformational change in the thumb domain to an "open" form is required to allow the dsRNA product to exit the active site (Fig. 12.8c). All *Flaviviridae* polymerases thus far have been crystallized in the closed form, where the template-binding channel is occluded by the C-terminal motif.

Other Functions of *Flavivirus* Polymerases

Pestivirus and flavivirus polymerases have an additional domain at their N-termini. The BVDV polymerase is approximately 130 residues longer at its N terminus than the hepatitis C virus polymerase. The function of this N-terminal domain is not known, and up to 90 residues can be truncated from the polymerase without loss of polymerase activity. However, deletion of first 106 residues causes a 90% loss of activity, indicating that residues 91–106 of the N-terminal domain may be involved in the polymerase reaction (Lai et al. 1999). The effects of the deletions are similar for both de novo and 3′-elongative RNA synthesis, and thus, the N-terminal domain may be involved in a step common to both modes of synthesis. The part of the N-terminal region that is ordered in the crystal structure (residues 92–138) folds into a separate domain (Fig. 12.8a). This domain is positioned over the thumb domain, interacting with the fingertips and the fingers domain either from the same polypeptide chain or from the neighboring protein (Choi et al. 2004, 2006). This domain-swapping of the N-terminal domain and its location near the entrance to the template-binding channel suggests an intrinsic movement of the domain that, along with the fingertips, might drive translocation of the template.

Flavivirus NS5 polymerase has an N-terminal 5′-RNA methyltransferase domain that catalyzes methylations of the N7 position of guanine in the cap and the ribose 2′-OH position of the first nucleotide of the RNA (Egloff et al. 2007). Presence of a 5′-RNA methyltransferase and polymerase in a single polypeptide chain suggests that RNA synthesis and 5′-capping might be coupled during the initial stages of RNA elongation. The coupled process differs from an uncoupled one in that the modified RNA is growing and progressively folding, rather than first synthesizing a full-length RNA before initiating the capping reactions (Gu and Lima 2005). The coupling of the two processes would be beneficial for genomic replication because capping enzymes would be concentrated near the nascent RNA, and interactions between RNA capping enzymes and the RNA polymerase elongation complex may allosterically influence the activity of other proteins in the replicase complex. Little is known as to how 5′-RNA methyltransferase and polymerase activities are coordinated in genome replication.

12.2.2 Minus-Sense ssRNA Virus

Viruses that possess minus-sense ssRNA genomes include influenza (*Orthomyxoviridae*), measles and mumps viruses (*Paramyxoviridae*), hantavirus (*Bunyaviridae*), rabies virus (*Rhabdoviridae*), Ebola and Marburg virus (*Filoviridae*), and borna disease virus (*Bornaviridae*). Their genomes can be either nonsegmented or segmented. Segmented genomes are viral genomes that are divided into two or more segments of nucleic acid, all of which are then packaged into a single virus particle. For example, the influenza virus genome is divided into eight viral RNA segments, and the arenavirus genome consists of two RNA segments. The viral minus-sense RNA genome cannot be used for protein synthesis and needs first be transcribed into a plus-sense RNA. Host polymerases cannot carry out RNA-dependent RNA polymerization, and thus, viral RNA-dependent RNA polymerases are packaged in the virion along with the genome and carry out both transcription and replication of the genome (Fig. 12.1). The plus-sense mRNA of these viruses contains a cap at 5′-end and an additional 5–15 nontemplated nucleotides, i.e., a heterogeneous sequence prior to the beginning of viral

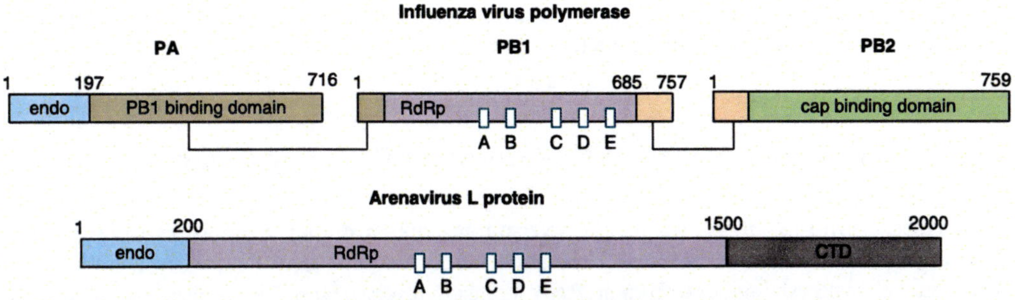

Fig. 12.9 Domain arrangement of RNA-dependent RNA polymerases from negative-sense ssRNA viruses. Influenza virus polymerase consists of three subunits, PA, PB1, and PB2 (*top*). All RNA polymerase motifs are found in PB1. The N-terminal domain of PA has an endonuclease (endo) activity that is proposed to be involved in the cap-snatching mechanism. Domains known to interact with each other are shown in the same color and are connected with lines (Boivin et al. 2010). Arenavirus L protein consists of a single polypeptide chain of 2,000 amino acids (*bottom*). Endonuclease (endo), RNA-dependent RNA polymerase (RdRp), and C-terminal domains (CTD) are labeled. Sequence motifs common to all RdRps (boxed and labeled A–E) are shown with the active site Ser-Asp-Asp (GDD) motif located in box C

transcript sequence (Plotch et al. 1981; Lelke et al. 2010). Since the virus does not encode capping enzymes, it is believed that the 5′-cap and the nontemplated nucleotides are derived from cellular mRNA via a "cap-snatching" mechanism. During transcription, the viral polymerase binds a cellular mRNA cap and cleaves it 10–15 nucleotides downstream from the 5′-end via an endonuclease activity. The cleaved, capped RNA is then used as a primer for viral mRNA synthesis. This cap-snatching mechanism was first suggested for influenza virus and later for bunyavirus and arenavirus as well (Bishop et al. 1983; Polyak et al. 1995; Morin et al. 2010).

Functional polymerases for minus-sense RNA viruses are either multiprotein or multidomain complexes. Influenza virus polymerase is a heterotrimer consisting of PA, PB1, and PB2 proteins, and bunyavirus and arenavirus polymerase is the large "L" protein composed of multiple domains (Fig. 12.9). The N-terminal domain of influenza virus PA protein and the N-terminal domain of bunyavirus and arenavirus L protein have endonuclease activities that are likely to be involved in the cap-snatching mechanism. The polymerase sequence motifs containing the essential GDD motif are located in PB1 of influenza virus and in the central region of the L protein. The enzymatic function of the C terminus of L protein is unknown. The crystal structures of endonuclease domains have been determined from influenza virus, La Crosse orthobunyavirus (*Bunyaviridae*), and lymphocytic choriomeningitis virus (*Arenaviridae*); all possess a common type II nuclease fold, consisting of a mixed β-sheet with four strands flanked on both sides by several α-helices (Orlowski and Bujnicki 2008; Morin et al. 2010; Reguera et al. 2010). Structures of neither the polymerase domain nor the full-length polymerase are available for polymerases from minus-sense ssRNA viruses. It is likely that the RNA polymerase domain is similar to other RNA-dependent RNA polymerases, but the details of domain arrangements and mechanism of coordinated cap-snatching and RNA transcription reaction await further structural and functional studies. For a detailed description of bunyavirus transcription and replication, please refer to chapter 11.

12.2.3 Double-Stranded RNA Virus

Viruses with segmented dsRNA genomes include reovirus and rotavirus (*Reoviridae*), birnavirus (*Birnaviridae*), and bacteriophage ϕ6 (*Cystoviridae*). The dsRNA virus genomes are always segmented. Because dsRNA cannot be used as mRNA for transcription and translation by host enzymes, these viruses also bring their own viral polymerases to synthesize mRNA and replicate their genomes. These viral polymerases are located inside the viral capsid and transcribe the packaged segmented dsRNA genomes within the virion.

12.3 Reverse Transcriptase (RNA-Dependent DNA Polymerase)

Two types of viruses are known to use reverse-transcription during their life cycle: retrovirus (*Retroviridae*) and hepatitis B virus (*Hepadnaviridae*). Although they both carry out the same reaction, i.e., synthesizing DNA from an RNA template, reverse transcriptases from retrovirus and hepatitis B virus are involved in different steps in the viral life cycle (Fig. 12.1). Retrovirus is a ssRNA virus that uses reverse transcriptase to convert its genome into a dsDNA form that can be integrated into the host genome. Once the retroviral genome is integrated into the host's DNA, mRNA synthesis and its translation are carried out by cellular enzymes. Because cellular polymerases are unable to carry out reverse transcription, the retrovirus encapsidates its own reverse transcriptase in the virion. In contrast, hepatitis B virus is a dsDNA virus that replicates through an RNA intermediate rather than a DNA intermediate. Transcription and protein translation are carried out by cellular polymerases. Hepatitis B virus uses reverse transcriptase to synthesize DNA from the progeny RNA that is synthesized by the cellular polymerase.

12.3.1 Retrovirus Reverse Transcriptase

The most studied retrovirus is human immunodeficiency virus (HIV). HIV packages two duplicate copies of a 9.2-kb ssRNA genome, reverse transcriptase, integrase, and protease within each viral particle. HIV uses reverse transcriptase to convert the ssRNA genome into a dsDNA that is subsequently incorporated into the host's DNA genome by the integrase. HIV reverse transcriptase consists of polymerase and ribonuclease (RNase) H domains. RNase H is an endonuclease that hydrolyzes the RNA strand in an RNA–DNA hybrid and generates 5′-phosphate and 3′-hydroxyl ends (Krug and Berger 1989). The RNase H activity is not coordinated with the polymerase activity, and the rate of cleavage is seven to ten times slower than the polymerase synthesis rate (DeStefano et al. 1994; Basu et al. 2008).

Reverse transcriptase first uses the viral genomic RNA as a template and a cellular tRNA as a primer and initiates minus-strand DNA synthesis at the *p*rimer *b*inding *s*ite (PBS) near the 5′-end of the genome (step A in Fig. 12.10) (Sarafianos et al. 2009; Wendeler et al. 2009). During the synthesis of minus-strand DNA, the RNase H cleaves the RNA strand of the resulting RNA–DNA duplex at multiples sites. When the reverse transcriptase reaches the 5′-end of the RNA genome, the minus-strand DNA is transferred and annealed to the *r*epeat "R" region at the 3′-end of the RNA genome; the RNA genome has identical R regions at each end (step B in Fig. 12.10). Reverse transcriptase continues to synthesize the minus-strand DNA resulting in a RNA–DNA hybrid, and RNase H continues to degrade the viral RNA in the RNA–DNA hybrid. RNase H is unable to cleave two short purine-rich RNA segments, called *p*oly*p*urine *t*racks (PPT), and the PPT stay bound to the newly synthesized minus-strand DNA. Next, reverse transcriptase synthesizes the plus-strand DNA using the PPT as primers and the minus-strand DNA as a template (step C in Fig. 12.10). The polymerase stops plus-strand DNA synthesis after copying the PBS region from tRNA, and RNase H cleaves the 3′-end of the tRNA to create a single-stranded PBS region in the plus-strand DNA. This plus-strand PBS is complementary to the minus strand PBS, and the newly synthesized plus-strand DNA is transferred and annealed to the minus strand PBS (step D in Fig. 12.10). Viral DNA synthesis continues via bidirectional DNA synthesis through the circular intermediate to generate a dsDNA form of the viral genome (step E in Fig. 12.10). The dsDNA is subsequently inserted into the host DNA

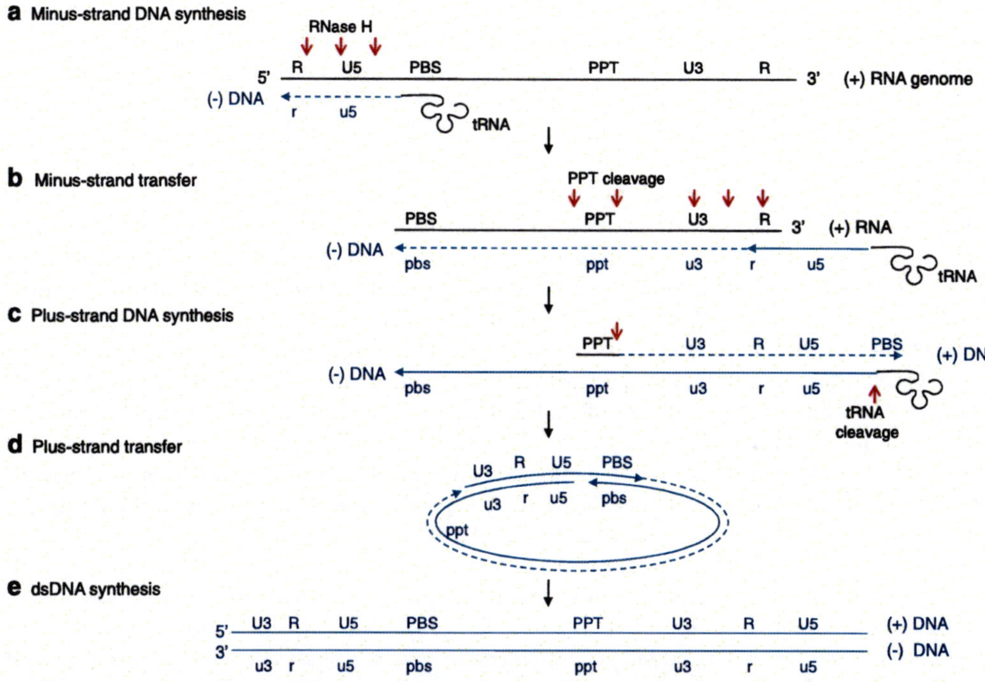

Fig. 12.10 Reverse transcription of the HIV-1 genome. **(a)** Minus-strand DNA synthesis. Minus-sense DNA synthesis is initiated near the 5′-end of the RNA genome using a tRNA as a primer that anneals at the primer-binding site (PBS). During the DNA synthesis, RNase H cleaves RNA segments of RNA–DNA hybrid (*red arrows*). **(b)** Minus-strand transfer. The single-stranded minus-strand DNA anneals with the repeat R region at the 3′-end of RNA genome and continues to elongate. **(c)** Plus-strand DNA synthesis. The polypurine track (PPT), which is resistant to RNase H, is used as a primer to initiate plus-sense DNA synthesis. The synthesis stops after copying the PBS region from tRNA, and RNase H cleaves tRNA. **(d)** Plus-strand transfer. The plus-strand DNA is transferred and annealed with the PBS region of the minus-strand DNA, and DNA synthesis continues via bidirectional DNA synthesis. **(e)** dsDNA synthesis. Ligation of a circular intermediate produces a linear dsDNA with terminal repeats containing R and promoter regions (U3 and U5) at both ends (Wendeler et al. 2009)

genome by an integrase in the host cell nucleus, and the viral genome (called provirus) is replicated and transcribed as part of the host's DNA.

The complexity of the retrovirus replication strategy is due to the requirement of synthesizing a DNA that can produce viral RNA genome during transcription in the host. Because the cellular RNA polymerases require a promoter DNA sequence for initiation that is not copied into the RNA transcript, some information in DNA will not be copied to mRNA during transcription. To overcome this problem, retrovirus copies their promoter sequences and transfers them to the opposite ends during reverse transcription. The resulting DNA is thus longer than the RNA template and has identical sequences in both ends. Thus, even if the promoter sequences are lost during transcription, the newly synthesized RNA would be identical to the original RNA genome.

12.3.1.1 DNA Synthesis by Reverse Transcriptase

Reverse transcriptase is synthesized from the "pol" gene product as a 66-kDa protein (p66) that forms a homodimer. P66 is organized as an N-terminal reverse transcriptase domain, a connection domain, and a C-terminal RNase H domain. The viral protease subsequently cleaves the 66-kDa homodimer and removes the RNase H domain from one monomer, making a heterodimer consisting of a 66-kDa protein (p66) and a 51-kDa protein (p51) (Kohlstaedt et al. 1992). Only p66 is active as a polymerase and an RNase H. Structures of the p66–p51 heterodimer explain why polymerase activity is observed only for p66 (Fig. 12.11a). Although the individual domains of p66 and p51 subunits are very similar to each other, the arrangements of domains are significantly different between the p66 and p51 subunits. P66 polymerase has the classical fingers, palm, and thumb domain configuration where the template-binding channel is formed by internal surfaces of all three domains. The connection and RNase H domains of p66 are positioned away from the template-binding channel. In contrast, the polymerase domain of p51 does not have a template-binding channel. The connection domain occupies the space between the palm and thumb domains and thus blocks the access to the active site residues, making p51 inactive as a polymerase.

Several steps in the nucleotidyl transfer reaction have been captured by structural studies. Binding of nucleic acid or an inhibitor in the polymerase induces a large movement of the thumb domain in respect to the fingers domain (Kohlstaedt and Steitz 1992; Rodgers et al. 1995). In unbound polymerase, the thumb domain contacts a portion of the fingers domain, which "opens" as much as 34° in relation to the fingers domain to accommodate the template–primer duplex in the polymerase complexes (Fig. 12.11a). This is in contrast to RNA-dependent RNA polymerases, where the fingertips in the fingers domain interact extensively with the thumb domain.

Reverse transcriptase, complexed with dsDNA or a PPT-containing RNA–DNA hybrid, shows that the enzyme interacts with nucleic acid through both p66 and p51 domains (Jacobo-Molina et al. 1993; Huang et al. 1998; Sarafianos et al. 2001). The polymerase and the RNase H domain of the p66 subunit is arranged such that the p66 polymerase active site binds to the primer terminus, and the RNase H active site contacts the template–primer duplex 17 nucleotides downstream (Fig. 12.11b). The connecting and thumb domains of p51 form a floor in the nucleic acid binding cleft between the polymerase and RNase H domains. The dsDNA and RNA–DNA hybrid have an A-form helix near the polymerase active site that gradually becomes B-form near the RNase H active site. The p66 polymerase active site binds to a 3′-end of the primer mainly through a β-hairpin in the palm domain, named a "primer grip." The primer grip orients the 3′-hydroxyl of a primer terminus for nucleophilic attack on the α-phosphate of the incoming NTP. Similarly, template strand binds in the polymerase active site through a "template grip" consisting of residues from the palm and fingers domains (Fig. 12.11b).

An elongation complex of reverse transcriptase containing template, primer, and incoming NTP was obtained by cross-linking the polymerase and template–primer dsDNA and by incorporating dideoxynucleotide at the 3′-end of the primer (Huang et al. 1998). Although the primer terminus lacks the 3′-hydroxyl group and was thus unable to attack the incoming NTP, the deoxyribose is ideally positioned to attack the incoming NTP via its missing 3′-OH. The NTP forms a Watson–Crick base pair with the template base, and the triphosphate group of NTP binds two metal ions that are coordinated by the essential Asp residues in motifs A and C. Compared to the template–primer complex, the binding of the incoming NTP induces a movement of the fingers domain closer to the thumb domain and thus closes down on the NTP binding site.

12.3.1.2 RNA Degradation by RNase H

While minus-strand DNA is synthesized from the RNA template, RNase H degrades most of the RNA in the RNA–DNA hybrid nonspecifically. In particular, RNase H cleaves 17–18 nucleotides downstream from the polymerase active site (Furfine and Reardon 1991; Basu et al. 2008). As described earlier, the distance between the polymerase and RNase active sites is ~60 Å, which corresponds to 17 base pairs for the duplex DNA and 18 base pairs for an RNA–DNA hybrid, consistent with the biochemical data. The RNase H is proposed to control the cleavage specificity by reducing

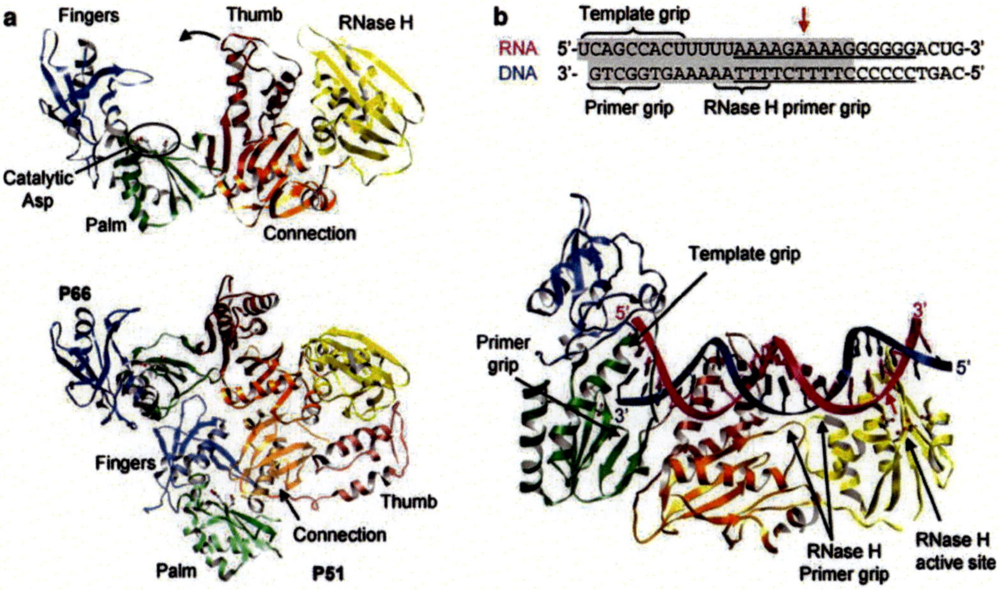

Fig. 12.11 Structures of HIV reverse transcriptase. (**a**) p66 subunit (*top*) and p66–p51 heterodimer (*bottom*). The fingers (*blue*), palm (*green*), thumb (*red*), connection (*orange*), and RNase H (*yellow*) domains of p66 are in solid colors and domains of p51 are in transparent colors. Two aspartic residues in the YMDD motif (motif C) are shown as ball-and-stick models. The polymerase is in an "open" conformation due to binding of substrate, which is omitted in the figure for clarity. Note the connecting domain in p51 is inserted between the palm and the thumb domains, making p51 inactive as a polymerase. (**b**) Structure of p66–p51 heterodimer complexed with a PPT-containing RNA–DNA hybrid. Only the p66 subunit is shown for clarity. The sequence of the RNA–DNA hybrid used in the crystallization is shown with the PPT *underlined* and the portion visible in the crystal structure shaded in *gray* (*top*). Template RNA and primer DNA are shown in *magenta* and *blue*, respectively, and the RNase H cleavage site is indicated with a *red arrow*. The primer grip, template grip, and RNase H primer grip are indicated

the width of minor groove and interacting with the DNA primer strand near the RNase H active site. The RNA–DNA hybrid typically adopts a conformation close to an A-form helix, in which the minor groove is 9–10 Å on average. RNase H cleaves substrates with different minor groove widths much less efficiently. In addition, residues near the RNase H active site, named the "RNase H primer grip," bind the DNA primer strand four to nine nucleotides downstream to the scissile phosphate base pair (Fig. 12.11b). Mutations of residues in the RNase H primer grip affect the specificity of RNase H cleavage in vitro and in vivo, suggesting that the RNase H primer grip might control the specificity of the RNase H (Julias et al. 2002; Rausch et al. 2002). This idea is further supported by a structure of reverse transcriptase bound with PPT-containing RNA–DNA, which is resistant to RNase H cleavage. In this complex, the RNA–DNA hybrid has unpaired and mispaired bases in the PPT region [$(rA)_4$:$(dT)_4$] and a significantly narrower minor groove of ~7 Å, less than the typical 9–10 Å groove. Although it is not clear whether this conformation of adenosine-rich PPT is an intrinsic property of the helix, this conformation places the scissile phosphate ~3 Å away from the RNase H active site, thus preventing cleavage.

12.3.1.3 Coordination of Multiple Activities in Reverse Transcriptase

During genome replication, reverse transcriptase catalyzes RNA-dependent DNA synthesis, RNA cleavage in an RNA–DNA hybrid, DNA strand transfer, and DNA-dependent DNA synthesis. Thus, reverse transcriptase has to recognize and orient different primer–template pairs in the active sites of the polymerase and RNase H domains for appropriate activities. How does reverse transcriptase determine the correct orientation of the polymerase domain or RNase H domain on the template–primer complex for its activity? Recent single-molecule fluorescence resonance energy transfer (FRET) measurements suggest that the binding orientation of reverse transcriptase on the template–primer complex determines the enzymatic activity (Abbondanzieri et al. 2008). Specifically, interaction between the primer backbone near the 5′-end of the primer, i.e., DNA versus RNA, and the "RNase H primer grip" determines the binding orientation of reverse transcriptase. When DNA primer binds to the DNA template (which would be the case for plus-sense DNA synthesis from minus-strand DNA), the polymerase domain binds to the 3′-end of the DNA primer, and the RNase H domain binds near the 5′-end of the primer. This is the correct orientation of the polymerase domain to initiate DNA synthesis. In contrast, when an RNA primer with a random sequence binds to the DNA template (which would be a case for nonspecific cleavage of the RNA template in a RNA–DNA hybrid), the polymerase domain binds to the 5′-end of the RNA primer, and RNase H domain is close to the 3′-end of the primer. Since the polymerase active site does not bind to the 3′-end of a primer, the polymerase would not elongate the primer. Instead, the RNase H domain would cleave RNA at 18 nt from the 5′-end of template. However, when the polypurine PPT-containing RNA is used as a primer (which would be the case for minus-strand DNA synthesis, step C in Fig. 12.10), the reverse transcriptase slides on the template–primer complex, rapidly shuttling between the 3′- and 5′-ends of the primer, and thus, the polymerase can use PPT as a primer for a DNA synthesis.

12.3.2 Hepatitis B Virus Reverse Transcriptase

Hepatitis B virus (hepadna virus) is a dsDNA virus, but its transcription and replication mechanisms are distinct from those of other dsDNA viruses. The 3.2-kb genome of hepatitis B is a mostly double-stranded circular DNA, but is not covalently closed, and is thus termed a *r*elaxed *c*ircular DNA (rcDNA). The minus-sense DNA strand is a genome-length, and the viral polymerase "P" protein is covalently linked at its 5′-end. The plus-sense DNA strand is variable in length but generally shorter than the full-length genome and has an RNA primer attached at the 5′-end (Nassal 2008; Sohn et al. 2009). Upon entry into cells, the genomic DNA is transported to the nucleus and converted into a plasmid-like *c*ovalently *c*losed *c*ircular DNA (cccDNA) by cellular enzymes. Subgenomic RNA is then synthesized from the cccDNA by cellular RNA polymerase II. One of the subgenomic RNAs, called the *pregenomic* RNA (pgRNA), has a full viral DNA sequence and is encapsidated into the viral capsid with the polymerase "P" protein. Only a single pgRNA and a single P protein are packaged per particle. Inside the capsid, plus-sense pgRNA is reverse-transcribed to minus-sense DNA by the P polymerase. During minus-sense DNA synthesis, most pgRNA of the RNA–DNA hybrid is degraded by the RNase H of the P protein. The last cleavage by RNase H occurs 11–16 nt from the 5′-end of pgRNA. The P polymerase subsequently uses the remaining ~15 RNA nucleotides as a primer and the minus-strand DNA as a template to synthesize plus-sense DNA and generate partially double-stranded rcDNA.

The 90-kDa polymerase P protein from hepatitis B virus consists of terminal protein, spacer, reverse transcriptase, and RNase H domains (Radziwill et al. 1990). The terminal protein domain (~180 amino acids) at the N terminus of P protein acts as a protein primer, similar to VPg in poliovirus and the terminal protein in bacteriophage ϕ29. Tyr63 provides a hydroxyl group to which the first nucleotide of the minus-sense DNA is added. Because the terminal protein is a part of reverse transcriptase P, the polymerase domain is also covalently linked to the 5′-end of minus-sense DNA. Hepatitis B virus is the only example of a viral polymerase that is covalently attached to the viral genome (Nassal 2008). The next ~150 amino acids of the spacer region are highly diverse among hepadna viruses, and the majority of the region can be deleted without loss of polymerase activity (Radziwill et al. 1990). Thus, the function of the spacer region is suggested to tether the terminal protein domain at the N terminus and the C-terminal polymerase domain. Recently, a zinc finger motif has been identified within the spacer region that is critical for pgRNA binding and encapsidation (Kim et al. 2009). The reverse transcriptase domain carries out minus-sense DNA synthesis using pgRNA as a template and the terminal protein as a protein primer. The C-terminal RNase H domain degrades pgRNA during minus-strand DNA synthesis, similar to RNase H of retrovirus reverse transcriptase. Currently, no structure is available for any of the hepadna virus reverse transcriptases. Future structural studies of P protein will help illuminate the strategy that hepadna viruses use for genome transcription and replication.

12.4 DNA-Dependent DNA Polymerases

DNA viruses replicate using either host or viral DNA-dependent DNA polymerases, also called DNA polymerases. Viral DNA-dependent DNA polymerases are similar to host DNA polymerases, and some of these virus polymerases even have a proofreading activity. Most DNA viruses have dsDNA genomes that can be in either a linear or circular form. Viruses with linear dsDNA genomes include adenovirus (*Adenoviridae*), herpesvirus (*Herpesviridae*), small pox or vaccinia virus (*Poxviridae*), and tailed bacteriophages such as ϕ29 and T7 (*Podoviridae*). Viruses with circular dsDNA genomes include papillomavirus (*Papillomaviridae*) and polyomavirus (*Polyomaviridae*). Eukaryotic DNA viruses with small genomes (4–9 kb), such as papillomavirus and polyomavirus, use cellular enzymes in the nucleus for both genome transcription and replication and thus do not encode a viral polymerase (Mesters et al. 2006). In contrast, cytoplasmic DNA viruses with large genomes, which spend most of their life cycles in the cytoplasm, are unable to access host DNA and RNA polymerases and thus encode their own DNA-dependent DNA and RNA polymerases. For example, vaccinia virus (*Poxviridae*), with an approximately 190-kb-long linear dsDNA genome, encodes its own DNA and RNA polymerase as well as other enzymes that are responsible for replication and the transcription. Since the replication and transcription machineries are needed immediately upon infection, they are packaged in viral capsids along with the DNA genome (Broyles 2003).

All DNA-dependent DNA polymerases require a primer for nucleic acid synthesis. Thus, viruses with linear genomes face the challenge of copying the 5′-end of the genome during replication. If the end of the genome is not copied during successive rounds of replication, the genome will become progressively shorter. To overcome this problem, some bacteriophages release a linear genome into the cell that is then converted into a circular form prior to replication or is integrated into the host chromosome (prophage) (Weigel and Seitz 2006). Cellular DNA polymerases can then elongate either the 3′-OH end of nicked dsDNA as a primer using the unnicked strand as a template in a rolling circle mechanism.

12.4.1 dsDNA Bacteriophage ϕ29 DNA-Dependent DNA Polymerase

Bacteriophages ϕ29, Cp-1, PRD-1, and adenovirus have a linear dsDNA genome and use a protein-primed initiation mechanism to copy the very end of the genome (Meijer et al. 2001). *Bacillus subtilis* phage ϕ29 has a 19.3-kb dsDNA genome with an origin of replication at each end. Both 5′-termini of the dsDNA genome are covalently attached to a terminal protein "gp3." The ϕ29 DNA-dependent

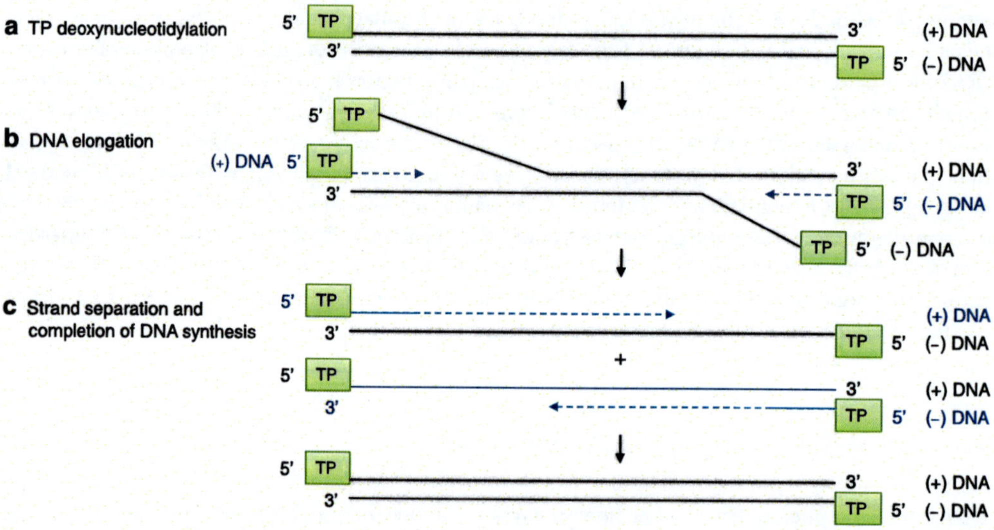

Fig. 12.12 ϕ29 protein-primed DNA replication. The linear dsDNA genome of ϕ29 has a terminal protein (TP) attached at both 5′-ends. (**a**) TP deoxynucleotidylation. The ϕ29 polymerase uses TP to initiate DNA synthesis from either end of the linear DNA genome. (**b**) DNA elongation. The intermediate contains a full-length dsDNA with one or more short ssDNA. (**c**) Strand separation and completion of DNA synthesis. When the two converging DNA polymerases merge, the templates become separated into two replication intermediates. Each of these consists of a full-length genome, one end of which is double-stranded and the other is single-stranded. Continuous DNA synthesis completes the replication of the ~20-kb DNA genome

DNA polymerase "gp2" initiates DNA synthesis using the terminal protein gp3 as a protein primer via the sliding-back mechanism (Fig. 12.5b). The polymerase begins DNA synthesis at the second T in the 3′-end of the ϕ29 genome and covalently attaches dATP to the hydroxyl group of a conserved Ser residue of the terminal protein to form a terminal protein (TP)-dAMP link (step A in Fig. 12.12). This TP-dAMP then slides back one base and forms a base pair with the 3′-terminal T of the template. The polymerase then starts DNA synthesis at the +2 position and elongates the rest of the template (Blanco and Salas 1996; Meijer et al. 2001). Because the gp2 polymerase can initiate DNA synthesis from each terminus of the dsDNA genome, there is no equivalent of lagging strand synthesis seen in cellular DNA polymerases (Blanco and Salas 1996). Instead, the DNA replication intermediate is a dsDNA containing single-stranded DNAs (step B in Fig. 12.12). When two DNA polymerases pass by each other, the templates separate into two duplex DNAs containing one full-length DNA and one shorter DNA (step C in Fig. 12.12). DNA synthesis continues until the polymerase reaches the 5′-end of the template, resulting in a DNA duplex with terminal proteins at each 5′-end of the genome. The gp2 polymerase also has a DNA helicase activity to unwind duplex DNA (strand separation) and an exonuclease activity to correct misincorporated nucleotides during the DNA replication. In addition to gp2 polymerase and gp3 terminal protein, the replication of ϕ29 in vitro requires ssDNA binding protein gp5 and dsDNA binding protein gp6. Gp5 binds and protects the single-stranded region of DNA from degradation during replication, and gp6 forms a nucleoprotein complex with the dsDNA genome and helps open the ends of the dsDNA (Blanco and Salas 1996).

12.4.1.1 Protein-Primed Initiation by ϕ29 Polymerase

The ϕ29 gp2 polymerase belongs to the B-type superfamily of DNA-dependent DNA polymerases that also includes DNA polymerases of eukaryotic polymerase α and *E. coli* polymerase II. The gp2 polymerase consists of an N-terminal exonuclease domain (residues 1–189) and a C-terminal DNA polymerase domain (190–575). The exonuclease domain catalyzes 3′→5′ cleavage of a single-stranded DNA and is responsible for the proofreading activity. In addition to the classical fingers, palm, and thumb domains, the ϕ29 polymerase domain has two insertions in the palm domain called *t*erminal *p*rotein *r*egions 1 and 2 (TPR1 and TPR2). The TPR1 and TPR2 are only present in the B-type family of DNA-polymerases that replicate via a protein-priming mechanism (Meijer et al. 2001). TPR1 forms a small domain next to the palm domain and TPR2, together with the thumb domain, forms an arch over the template-binding channel (Fig. 12.13a). The exonuclease domain is positioned between the fingers and thumb domains and encircles the template-binding channel together with the palm, thumb, TPR1, and TPR2 domains. Mutations of residues in TPR1 and TPR2 greatly reduce protein-primed DNA synthesis and exonuclease activities, consistent with the observed structure (Dufour et al. 2003).

The structure of the terminal protein-priming initiation complex consisting of ϕ29 DNA polymerase and the terminal protein suggests how the large terminal protein is used to initiate DNA synthesis (Kamtekar et al. 2006). The terminal protein consists of three domains, the N-terminal (residues 1–73), the intermediate (74–172), and the C-terminal priming domains (173–266) (Fig. 12.13b). The intermediate and priming domains interact with the TPR1 and TPR2 of the polymerase, respectively. In particular, the priming domain of the terminal protein binds in the template-binding channel, positioning the priming loop containing the conserved Ser required for initiation (Ser232) near the catalytic site of polymerase. Unfortunately, the density for Ser232 is not observed in the structure, suggesting that the priming loop might be flexible without a template strand. The interaction between the priming domain and polymerase is mostly electrostatic, similar to the interaction of the template–primer complex in the polymerase active site. The priming domain occupies the most of the template-binding channel, and thus, only a short template could bind in the channel during the initiation of DNA synthesis. The first addition of dATP to the terminal protein requires precise positioning of the template terminus, dATP, and terminal protein in the polymerase active site. By excluding most of the volume in the template-binding channel, the terminal protein could provide specific interaction with the template terminus to form a stable initiation complex. As the template and protein-linked primer extend, the priming domain of the terminal protein would move out of the template-binding channel to accommodate the longer template–primer duplex.

The terminal protein has high affinity (80 nM) for polymerase even in the absence of nucleic acid but dissociates from the polymerase after synthesis of six to ten nucleotides (Lazaro et al. 1995; Mendez et al. 1997). Modeling of template–primer in the polymerase active site suggests that the intermediate domain of terminal protein could maintain the interaction with TPR1 of the polymerase, while the priming domain is pushed out of the template-binding channel at each addition up to six or seven nucleotides (Kamtekar et al. 2006). The terminal protein would dissociate from the polymerase after seven nucleotide synthesis steps, consistent with the biochemical data. The release of the terminal protein from the polymerase would trigger a transition from the initiation to the elongation phase in which a stable template–primer DNA complex is formed, and DNA synthesis proceeds with high processivity.

12.4.1.2 Elongation

DNA-dependent DNA polymerases are highly processive, and the ϕ29 polymerase can synthesize the entire 20-kb DNA genome in a single binding event without dissociation (Meijer et al. 2001). High processivity requires a tight polymerase–DNA interaction to prevent dissociation but also should allow relative movements between the polymerase and template at each step of the polymerization. The ϕ29-type DNA polymerases accomplish this by treading the template DNA under a tunnel, which would prevent the dissociation of the template DNA from the polymerase during

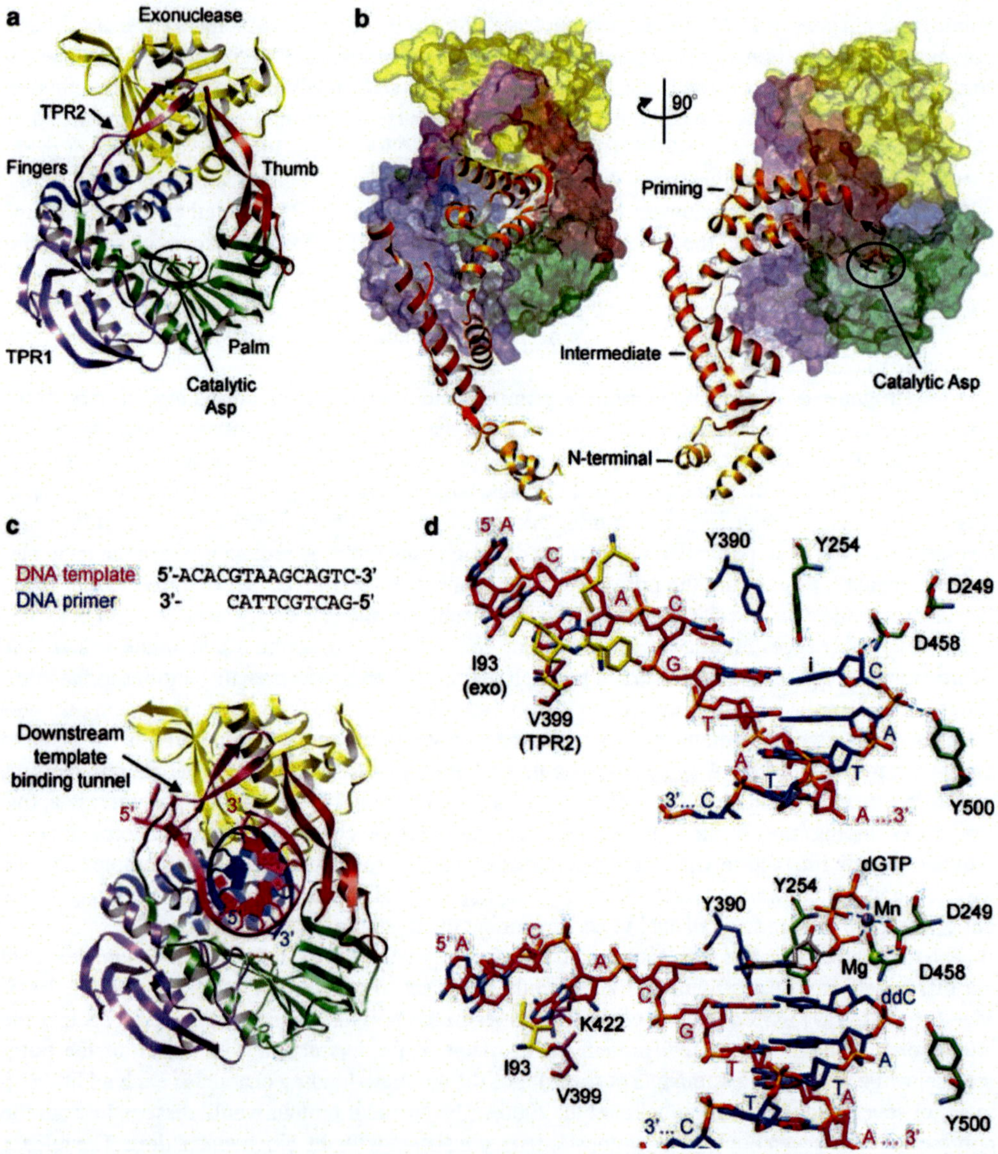

Fig. 12.13 Structures of ϕ29 DNA-dependent DNA polymerase. **(a)** ϕ29 polymerase. The N-terminal exonuclease, fingers, palm, and thumb domains are shown in *yellow*, *blue*, *green*, and *red*, respectively. The TPR1 and TPR2 regions are shown in *purple* and *magenta*. Two catalytic Asp residues (Asp249 and Asp458) in the motifs A and C are shown as ball-and-stick models. **(b)** ϕ29 polymerase-terminal protein (TP) heterodimer. The polymerase is shown as a transparent surface, and the TP is in ribbons. The N-terminal, intermediate, and priming domains of TP are labeled and shown in *gold*, *dark orange*, and *light orange*, respectively. The polymerase on the *left* is in a same orientation as in **(a)**. **(c)** ϕ29 polymerase complexed with template–primer. The template and primer strands are shown in *magenta* and *blue*. The sequences of template and primer strands are shown. Note that the 5′-end template is threaded under the downstream template-binding tunnel. **(d)** Active site in template–primer (*top*) and template–primer–NTP (*bottom*) complexes. The template–primer–NTP complex was trapped by incorporating a 2′,3′-dideoxynucleotide into the primer terminus and thus preventing a further nucleotidyl transfer reaction. The active site residues are colored according to their domains as in **(a)**

translocation and nucleotide addition steps. The structure of the ϕ29 polymerase complexed with a DNA duplex shows that the duplex region of the template–primer is encircled by the thumb domain and TPR2 (Fig. 12.13c) (Berman et al. 2007). The 5′-end single-stranded template is threaded under the "downstream template tunnel" formed from the exonuclease domain, TPR2, palm, and fingers domain. The downstream template tunnel is only large enough to accommodate single-stranded DNA and may play a role in strand separation. Deletion of TPR2, which forms an arch over the template-binding channel, reduces processivity and strand separation of the mutant enzyme, consistent with the model (Rodriguez et al. 2005; Kamtekar et al. 2006).

The overall polymerase structure is similar among unbound, terminal protein complex, template–primer complex, and template–primer–NTP complexes, and there is no large movement of domains. However, one distinctive difference between the template–primer complex and the template–primer–NTP complex is that the incoming NTP binding site is occupied by the side chain of a conserved Tyr residue (Tyr254) in the fingers domain in the absence of the incoming NTP (Fig. 12.13d top). Upon binding of incoming NTP to the template–primer, the fingers domain rotates 14° toward the polymerase active site and "closes" onto the incoming NTP. The Tyr254 which occupies the NTP binding site in the absence of incoming NTP now rotates to stack on the incoming dNTP (Fig. 12.13d bottom). The dNTP base-pairs with the template base, and the triphosphate moiety interacts with the conserved Lys in motif B (Lys383) and the two metal ions that are coordinated by the catalytic Asp residues in polymerase motif A (Asp249) and C (Asp458).

12.4.1.3 Proofreading by the ϕ29 Polymerase

The ϕ29 polymerase contains an exonuclease domain that provides a proofreading activity during DNA synthesis. The RB69 polymerase (a ϕ29-like polymerase) and DNA complex shows that the exonuclease domain contacts only three nucleotides at the 3′-end of the ssDNA, suggesting that primer transfer from the polymerase to the exonuclease active site would require an opening of at least three to four base pairs in template–primer duplex (Hogg et al. 2004). During replication, a primer terminus containing an error will be transferred from the polymerase to the exonuclease active site for cleavage and then be transferred back to the polymerase active site for elongation. The major determinant of this switch coordinating DNA synthesis and degradation has been proposed to be the strength with which the primer terminus binds to the respective active sites. Mutational studies show that a single mutation in a conserved motif (YxGG/A), located between the exonuclease and polymerase domains, results in polymerases that exhibit different ratios of the polymerase activity versus the exonuclease activity due to loss of DNA binding in either the polymerase or the exonuclease active site. For example, the loss of DNA binding in the polymerase active site results in the increased activity of the exonuclease (Truniger et al. 1996, 1999). Consistent with this idea, a single-molecule experiment showed that the mechanical tension between the polymerase and the mismatched DNA duplex promotes the transfer of the primer terminus to the exonuclease active site (Ibarra et al. 2009). Mismatched dsDNA induces a long-range tension on the dsDNA helix, and weakens the binding of the template–primer in the polymerase active site. The polymerase subsequently stalls, which facilitates the transfer of the primer terminus to the exonuclease active site.

12.4.2 *Eukaryotic ssDNA Viruses Do Not Encode a Polymerase Gene*

A small number of eukaryotic viruses are ssDNA viruses that have either linear or circular DNA genomes. Smaller-sized viral genomes indicate a higher dependence of the viral life cycle on the host cellular proteins for transcription and replication of the viral genome. Parvovirus B19

(*Parvoviridae*) has a small linear ssDNA genome of ~5 kb in size and does not encode a viral polymerase (Kasamatsu and Nakanishi 1998). In the nucleus, the single-stranded genome is converted into a dsDNA intermediate by host DNA polymerases. This dsDNA is used for both transcription and replication of the genome. Host DNA polymerases synthesize more plus- and minus-sense ssDNA by a rolling-circle mechanism. Either form can be encapsidated into the viral capsid.

12.5 DNA-Dependent RNA Polymerase

Viral DNA-dependent RNA polymerases share similar functions and specificities with cellular RNA polymerases. Many DNA viruses rely on cellular RNA polymerases for transcription and therefore carry out part of their infection cycle in the nucleus. An exception is cytoplasmic dsDNA viruses, which carry out viral transcription in the cytoplasm, and thus encode DNA-dependent RNA polymerases. DNA-dependent RNA polymerases are different from DNA-dependent DNA polymerases in that they bind at a specific promoter duplex DNA sequence to initiate RNA synthesis, and they are able to carry out de novo synthesis without a primer. The newly synthesized RNA is generally released from the DNA template, and thus, the double-stranded DNA helix is maintained except for a short region called a transcription bubble within which a transient DNA–RNA hybrid is formed. DNA-dependent RNA polymerases also exhibit a distinct transition from initiation to elongation. Following the initial RNA synthesis, RNA polymerases release many short (<10) nucleotides that failed to be converted into a full-length transcript (abortive initiation). Only after the polymerase synthesizes ~10 nucleotides does the polymerase begin processive elongation.

12.5.1 dsDNA Bacteriophage T7 DNA-Dependent RNA Polymerase

Bacteriophage T7 has a linear dsDNA genome of ~40 kb in size. The T7 DNA-dependent RNA polymerase is related to the mitochondrial RNA polymerases. Unlike multisubunit cellular RNA polymerases, the T7 RNA polymerase is a single-subunit protein and performs the complete transcription cycle without any additional proteins. Thus, it has been used as a model system to study multisubunit RNA polymerases. The T7 RNA polymerase is extremely promoter-specific and only transcribes the DNA downstream of a 17 base-pair T7 promoter. The polymerase first binds at a promoter site and separates the dsDNA into single strands (Fig. 12.14). The coding strand then serves as a template for the RNA synthesis. While bound at the promoter sequence, the T7 polymerase first undergoes abortive initiation, where the polymerase synthesizes and releases many short (three to eight nucleotide-long) RNA transcripts from the polymerase–promoter complex (step A in Fig. 12. 14). When the nascent RNA transcript reaches eight to nine nucleotides, the RNA stably associates with the transcription complex, and the polymerase proceeds to synthesize the entire RNA transcript until it reaches a termination signal (step B in Fig. 12. 14). When the polymerase reaches the termination signal, it stops, and the polymerase and the RNA transcript dissociate from the dsDNA template (step C in Fig. 12. 14).

12.5.1.1 Promoter Recognition and Abortive Initiation

T7 RNA polymerase consists of an N-terminal domain (35 kDa) unique to DNA-dependent RNA polymerases and a C-terminal polymerase domain (60 kDa) that is homologous to the *E. coli* DNA

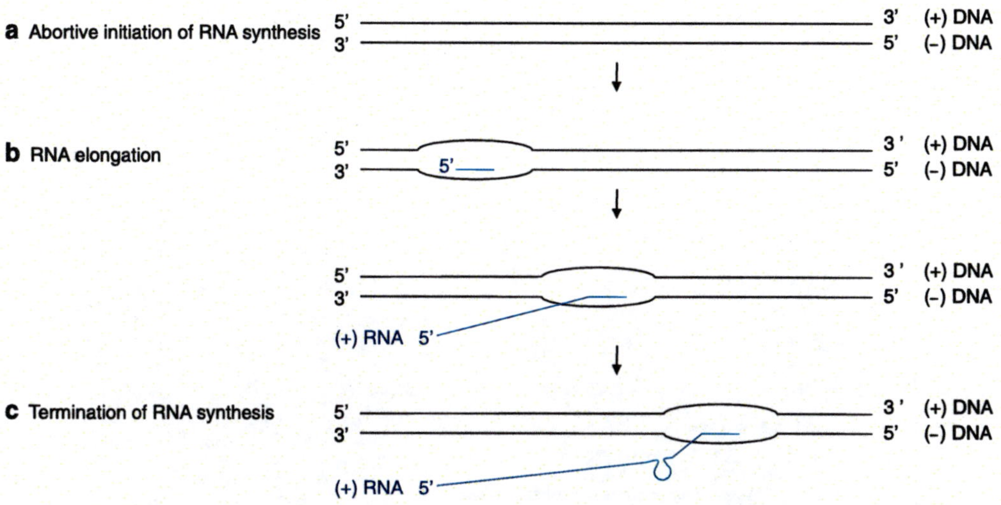

Fig. 12.14 RNA transcription by T7 DNA-dependent RNA polymerase. (**a**) Initiation of RNA synthesis. The T7 RNA polymerase binds at the promoter site, separates the dsDNA genome, and initiates RNA transcription. The polymerase synthesizes short three- to eight-nucleotide long RNA oligomers de novo that are released from the polymerase–promoter complex. Unlike the DNA replication seen in Fig. 12.11, only a small portion of dsDNA is separated to form a DNA–RNA hybrid (transcription bubble). Two DNA strands reanneal around the transcription bubble as the polymerase moves along the template. (**b**) RNA elongation. Once the RNA chain reaches about nine nucleotides, the polymerase elongates the RNA transcript with high processivity. (**c**) Termination. Once the polymerase encounters a termination signal, the polymerase and the RNA transcript are released from the nucleic acid

polymerase I (pol I) family of polymerases (Jeruzalmi and Steitz 1998). The N-terminal domain, positioned in front of the polymerase, interacts with the palm and thumb domains and partially occludes the polymerase active site (Fig. 12.15a) (Sousa et al. 1993; Cheetham and Steitz 1999; Tahirov et al. 2002). The T7 polymerase initiation complex, containing a three-nucleotide RNA transcript, shows the structural basis for specific recognition of the promoter DNA and abortive initiation. The T7 polymerase initiation complex was produced by addition of a 17 base-pair promoter DNA duplex, which contains a single-stranded region at the 5′-end, and a three-nucleotide RNA that can base-pair with the ssDNA region (Cheetham and Steitz 1999). In the crystal structure of the initiation complex, the polymerase separated the –4 to –1 region of the promoter DNA duplex near the RNA synthesis initiation site (+1 site) to form a transcription bubble. The +1 to +3 sites of the single-stranded DNA template base-pair with the RNA transcript in an A-form helix in the active site (Fig. 12.15a). The promoter DNA binds mainly to the N-terminal domain and a β-hairpin (residues 739–770) in the fingers domain, named the "specificity loop"; the N-terminal domain binds the duplex region (–17 to –5) of the promoter sequence, while the specificity loop interacts with the major groove of the promoter DNA in a sequence-specific manner.

DNA protection and cross-linking experiments suggest that the initiation complex of T7 RNA polymerase accommodates up to eight base pairs of the DNA–RNA hybrid in the polymerase active site while bound in the promoter site (Temiakov et al. 2000). However, the T7 initiation complex shows that the template-binding channel is just large enough to accommodate three to four base pairs, and a DNA–RNA hybrid longer than four base pairs would clash with the N-terminal domain. This apparent discrepancy suggests that the existence of additional conformations of the polymerase initiation complex that would accommodate longer RNA transcripts. Recent structural studies show one such intermediate, where the promoter binding site rotates along with the bound promoter DNA to incorporate a longer DNA–RNA hybrid in the active site (Durniak et al. 2008).

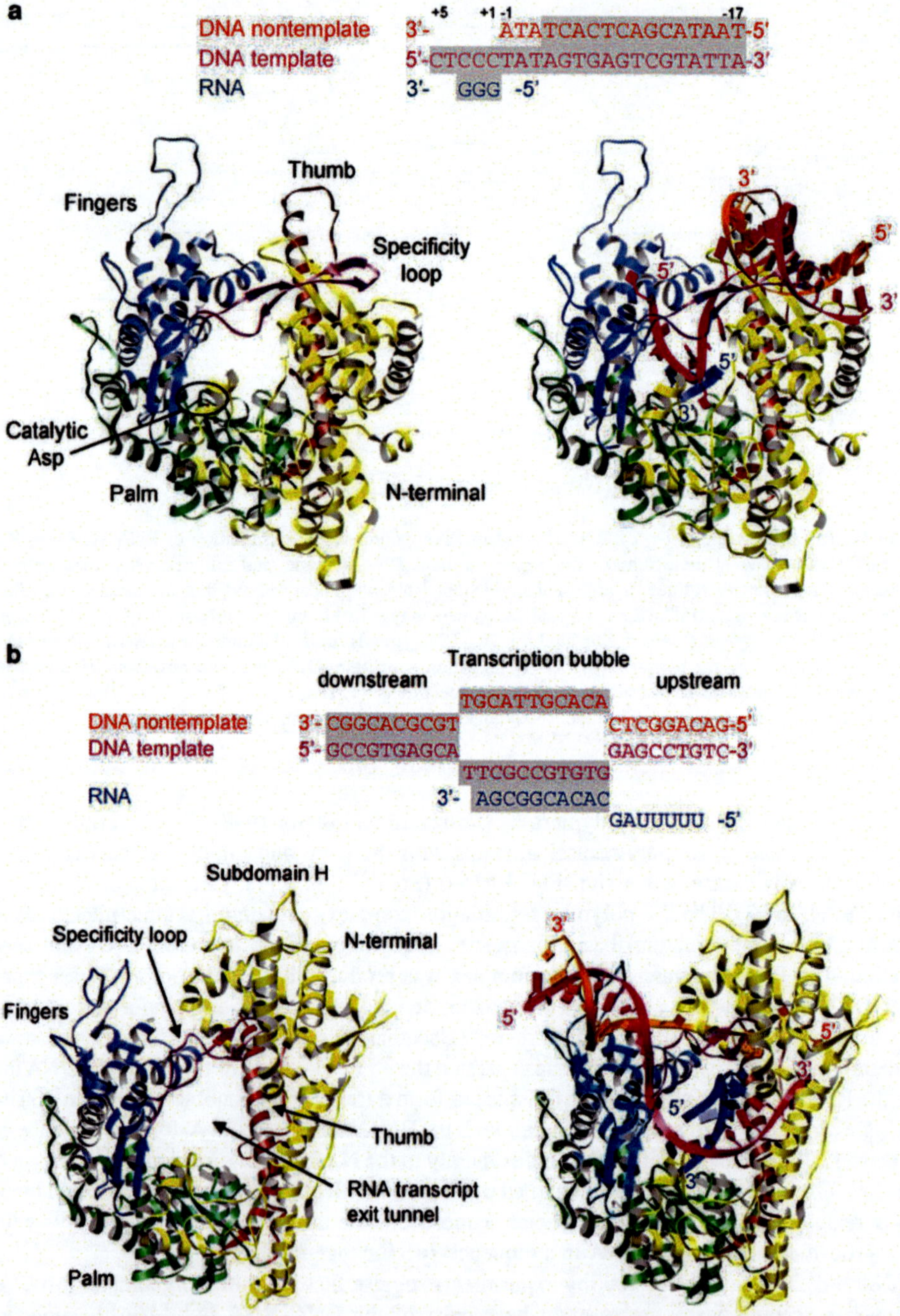

Fig. 12.15 T7 RNA polymerase initiation and elongation complexes. (**a**) Initiation complex containing the promoter dsDNA and a three-nucleotide ssRNA transcript. The sequences of the nucleic acids are shown with the portions visible in the crystal structure shaded in *gray* (*top*). RNA synthesis starts at the +1 position. Polymerase is shown in the absence and presence of the bound nucleotides for clarity (*bottom*). The N-terminal domain, fingers, palm, and thumb domains are shown in *yellow*, *blue*, *green*, and *red*, respectively. The specificity loop in the fingers domain is shown in *purple*. The template (coding) strand, nontemplate strand, and the RNA transcript are shown in *magenta*, *orange*, and *blue* ribbons, respectively. (**b**) Elongation complex containing dsDNA and a ten-nucleotide RNA transcript. Note three nucleotides at the 5′-end of the RNA transcript are separated from the template and are exiting the active site

Thus, transition from initiation to elongation requires successive conformational changes in the N-terminal domain. Many of these intermediate conformations are not faithful, leading to abortive initiations.

12.5.1.2 Processive Elongation

When the nascent RNA transcript reaches eight to nine nucleotides, the polymerase releases the promoter DNA and starts synthesizing downstream DNA sequences. T7 RNA polymerase is highly processive and can synthesize 200 nucleotides per second and the entire 40-kb genome within minutes (Golomb and Chamberline 1974). DNA-dependent RNA polymerases require specific interaction with the promoter in the initiation phase but need high processivity and nonspecific DNA interactions in the elongation phase. How does the T7 RNA polymerase accomplish these seemingly exclusive tasks? T7 RNA polymerase undergoes structural reorganization that removes the promoter binding site in the N-terminal domain and encircles the RNA transcript in a tunnel that prevents dissociation of DNA–RNA hybrid from the polymerase.

A T7 RNA polymerase elongation complex was obtained using a duplex DNA containing 11 noncomplementary bases in the middle to mimic a transcription bubble, and a ten-nucleotide RNA that would base-pair with the coding DNA strand in the transcription bubble (Yin and Steitz 2002). Among the ten-nucleotide RNA, seven nucleotides at the 3′-end RNA base-pair with the template DNA in the active site, and the other three nucleotides are separated from the template strand, exiting the polymerase active site (Fig. 12.15b). Compared to the initiation complex, the polymerase undergoes a major rearrangement of the N-terminal domain and the specificity loop in the elongation phase. The N-terminal domain, which forms a part of the promoter binding site in the initiation complex, now forms a part of the "RNA transcript exit tunnel" together with the thumb domain and the specificity loop (Fig. 12.15b). In particular, residues 160–190 in the N-terminal domain move approximately 70 Å from the bottom to the top of the thumb domain and form a helix-turn-helix. This refolded region of the N-terminal domain, named "subdomain H," or "flap," creates binding sites for the 5′-end of the RNA on one side and the single-stranded nontemplate DNA on the other. These structural changes result in a larger template-binding channel that can accommodate up to five to eight base pairs of DNA–RNA duplex. The specificity loop, which makes sequence-specific interactions with the promoter DNA in the initiation complex, also moves away from the template-binding channel to open space in the RNA exit tunnel. The RNA exit tunnel would thread an RNA transcript that is longer than eight nucleotides, consistent with the transition to elongation at ~8 nucleotides derived by biochemical studies. Similar to the TPR2 in ϕ29 DNA polymerases (Fig. 12.13c) and the sliding clamp of DNA replication, T7 RNA polymerase threads the RNA transcript through an RNA exit tunnel created in the elongation complex, making the DNA–RNA hybrid unable to dissociate from the polymerase.

12.5.1.3 Nucleotidyl Transfer Cycle in DNA-Dependent RNA Polymerases

The structures of the T7 polymerase have been determined in successive steps of nucleotidyl transfer cycle including (1) template–primer binding (posttranslocated state), (2) NTP binding in noncatalytic site ("open" template–primer–NTP complex), (3) NTP binding in a catalytic site ("closed" template–primer–NTP complex), and (4) pyrophosphate binding (pretranslocated state). After pyrophosphate is released and translocation occurs, the polymerases cycle back to the template–primer binding state (Yin and Steitz 2004).

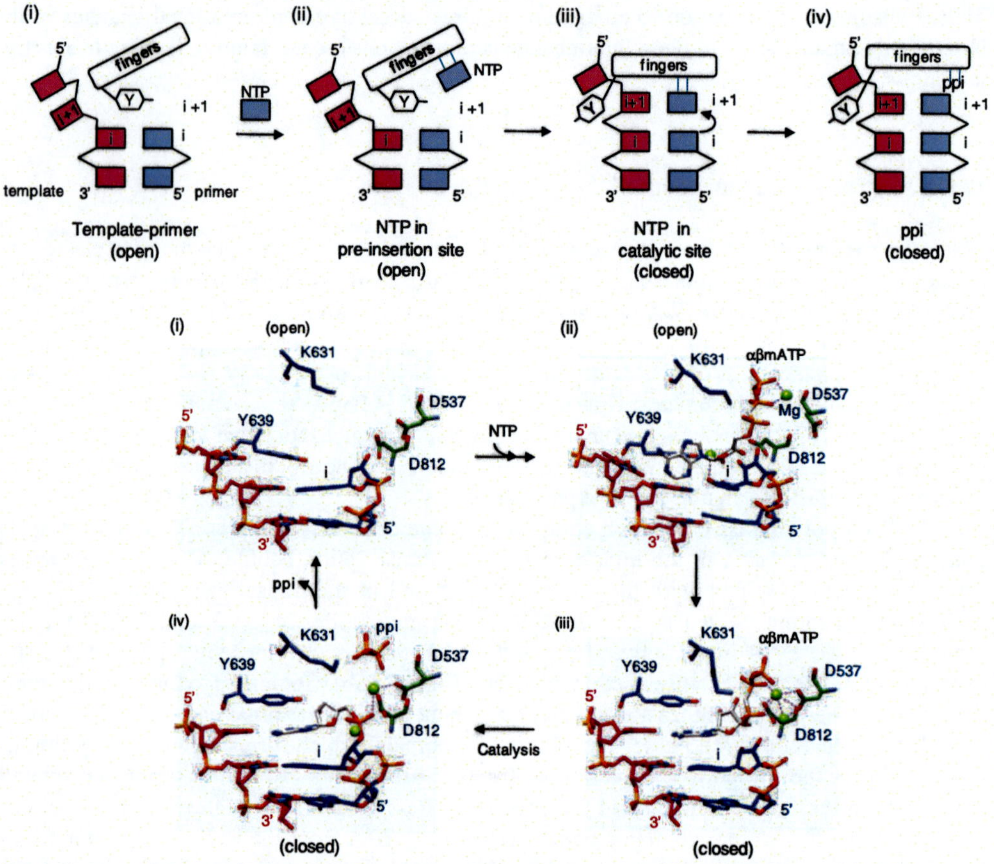

Fig. 12.16 Schematic of T7 polymerase reaction cycle (*top*) and structures of T7 RNA polymerase complexes in each step of the cycle (*bottom*). (i) Template–primer binding (post-translocated state). DNA template (*magenta*) and RNA primer bind in the active site. Tyr639, shown as a *hexagon*, occupies the NTP binding site in the absence of incoming NTP. (ii) NTP binding the "pre-insertion" site. The incoming NTP binds in the pre-insertion site because the binding (insertion) site for NTP is partially occupied by Tyr639. The triphosphate of the incoming NTP interacts with residues in the fingers domain. (iii) Binding of the NTP in the catalytic site. Closing of the fingers domain onto the incoming NTP positions the NTP ideally for the nucleotidyl transfer reaction. (iv) Catalysis (pre-translocated state). Nucleotidyl transfer reaction occurs. Pyrophosphate (ppi) leaving group binds in the active site in the closed conformation of the fingers domain

1. *Template–primer binding (posttranslocated state)*. Template and primer bind in the template-binding channel, and the primer terminus near the catalytic metal binding sites (Fig. 12.16; (i)). In the absence of the incoming NTP, the side chain of a conserved Tyr in the fingers domain (Tyr639) occupies the NTP binding site and stacks with the base at the primer terminus. This configuration would sterically block binding of an incoming NTP. In addition, the template nucleotide at the "i + 1" site (that would base-pair with incoming NTP) "flips out" to a position that is incapable of forming a base pair with the incoming NTP. It has been suggested that this template-binding site could function as a "ratchet" to prevent a template strand slippage, until the correct NTP is assembled in the active site. As mentioned earlier, ϕ29 DNA polymerase also has a similar Tyr residue occupying the incoming NTP binding site in the absence of NTP (Fig. 12.13d).
2. *NTP binding in the noncatalytic site ("open" form)*. The incoming NTP binds in a noncatalytic ("pre-insertion") site that is incapable of participating in a nucleotidyl transfer reaction (Fig. 12.16; (ii)). This T7 polymerase complex is formed by an addition of a noncleavable α,β-methylene

ATP (in which an oxygen that links α- and β-phosphate in ATP is replaced with a methyl group) to the template–primer. The template nucleotide at the "i + 1" site is still in a "flipped-out" position that is not competent to form a base pair with incoming NTP. The NTP binding site is partially occupied by Tyr639. Although the ATP analog does not base-pair with the template base at the "i + 1" site, the phosphate moiety of the ATP analog binds near the catalytic Asp residues in motif C and interacts with the basic residues in the α-helix in the fingers domain (Arg627 and Lys631) that also contains Tyr639 at the distal end. This mode of binding has not been described in ϕ29 DNA polymerase.

3. *NTP binding in the catalytic site* (*"closed" form*). The incoming NTP, the template at the "i + 1" site, and the primer terminus are ideally positioned for nucleotidyl transfer (Fig. 12.16; (iii)). This T7 polymerase complex, also formed by addition of an α,β-methylene ATP, shows that the template nucleotide at the i + 1 site now binds in a position that is expected for the nucleotidyl transfer reaction. The adenosine moiety of α,β-methylene ATP forms a base pair with the template nucleotide thymidine at the "i + 1" site, and the triphosphate moiety binds two catalytic metal ions that are coordinated by the catalytic Asp residues in polymerase motifs A and C (Asp537 and Asp811). Compared to the template–primer complex or the template–primer–ATP analog complex bound in the noncatalytic site, the α-helix in the fingers domain that contains Tyr639, Arg627, and Lys631 rotates approximately 20° to interact with correctly bound incoming NTP. In this "closed" form, the triphosphate of the ATP analog maintains interactions with basic residues Arg627 and Lys631. The correct incoming NTP to be added to the primer strand would be selected in this step by the interaction of polymerase with the 2′-OH of the rNTP.
4. *Catalysis and product release* (*pretranslocated state*). This is the conformation following chemical catalysis but prior to pyrophosphate product release (Fig. 12.16; (iv)). The T7 RNA polymerase–pyrophosphate complex is formed by addition of 3′-deoxyATP and pyrophosphate. The structure shows that the 3′-deoxyATP has been added to the growing RNA and that the pyrophosphate moiety binds where the β- and γ-phosphate of the catalytic NTP bind. The pyrophosphate maintains interactions with Arg627 and Lys631 in the α-helix of the fingers domain and with the metal ion and thus has the same conformation as in the catalytic NTP substrate complex. This suggests that the nucleotidyl transfer reaction does not require a conformational change in the polymerase and does not promote translocation prior to pyrophosphate release. Thus, it is likely that the dissociation of pyrophosphate leads to a conformational change in the fingers domain to "open" and allows translocation of the template–primer (Yin and Steitz 2004). This would support a power-stroke model of polymerase translocation over a Brownian-ratchet model because the energy from nucleotide incorporation, i.e., pyrophosphate release, drives a conformational change that "pushes" a polymerase forward on the template.

12.6 Summary and Prospectus

Viral polymerases have evolved to utilize replication strategies that are uniquely adapted to their hosts. They use a variety of mechanisms to initiate nucleic acid synthesis, prevent dissociation from nucleic acid during processive elongation, and coordinate multiple enzymatic activities. Compared to multisubunit cellular polymerases, many single-unit viral polymerases have all synthesizing activities and other functions necessary for genome replication. Unlike cellular host polymerases that require a strict regulation of transcription and replication, viral polymerases generally lack proofreading abilities and thus explore large areas of mutational space that could facilitate viral evolution. Although detailed knowledge of replication pathways, involvement of viral and cellular proteins in genome synthesis, and structural and functional characterization of viral polymerases is still limited to a few virus families, some common features of viral polymerases can be derived.

Viral RNA-dependent RNA polymerases and DNA-dependent DNA polymerases use several different mechanisms to ensure the initiation at the template terminus. The polymerase active site can generally accommodate five to seven nucleotides of template and primer strands; thus, it may not provide the interactions necessary to precisely position a template terminus and an NTP. Viruses overcome this problem by reducing the effective size of the template-binding channel using either a separate protein or a domain within a polymerase protein. The picornaviruses and the dsDNA phage ϕ29 utilize a terminal protein that occupies most of the template-binding channel and mimics the primer terminus in the active site. Alternatively, flavivirus polymerases use a motif within the polymerase to occlude the active site so that only the template terminus can bind in the active site for initiation. In both cases, a terminal protein or the polymerase motif would need to move out of the template-binding channel after several cycles of nucleotide additions to accommodate the template and the growing chain.

The hallmark of the elongation phase is its high processivity. Although the rate and processivity have not been measured for many vial polymerases, it is likely that they synthesize the entire genome in one or a few trials. High processivity not only requires a tight polymerase–nucleic acid interaction to prevent dissociation, but should also allow relative movements between the polymerase and template at each step of polymerization. DNA-dependent DNA and RNA polymerases, which exhibit high processivity without any accessory proteins, encircle a template and growing chain within the polymerase and thus prevent dissociation from the polymerase. RNA-dependent RNA or DNA polymerases with wide open template-binding channels do not seem to have such motifs. In picornaviruses, the processivity of the polymerase increases in the presence of an accessory protein 3AB. Thus, additional proteins in a replication complex may provide a similar function for RNA-dependent polymerases.

Since the first structure of the Klenow fragment of DNA polymerase was solved, we have learned a great detail about the universal aspects of nucleic acid polymerase reactions. In particular, the two-metal mechanism used to stabilize the transition state of the nucleotidyl transfer step is common to all nucleic acid–synthesizing polymerases. Furthermore, selection of the correct NTP and catalysis are linked to a conformational change in the fingers domain that orients the incoming NTP in the catalytic site. A remaining fundamental question regarding the catalytic mechanism is how the chemical steps are linked to actual translocation of the template and daughter strands. Another active area of research is to understand how viral polymerases coordinate their nucleotide synthesizing activity with other enzymatic activities in the overall scheme of genome synthesis. For example, proofreading activities in DNA polymerases, 5′-capping of mRNA in RNA-dependent RNA polymerases, RNase H cleavage in reverse transcriptase, and protein-priming in hepatitis B virus reverse transcriptase are carried out by separate domains within a single polymerase protein, and hence, the mechanism of coordination of these additional catalytic activities with the basic polymerase reaction provides another layer of complexity that will need to be addressed.

Acknowledgments I would like to thank Drs. Michael Rossmann, James Groarke, Marc Morais, Lucia Rothman-Denes, Peter Mason, and many colleagues who shared their passion for viruses with me throughout the years. The work is supported by NIH grants AI057156 and AI087856.

References

Abbondanzieri EA, Greenleaf WJ, Shaevitz JW, Landick R, Block SM (2005) Direct observation of base-pair stepping by RNA polymerase. Nature 438:460–465

Abbondanzieri EA, Bokinsky G, Rausch JW, Zhang JX, Le Grice SF, Zhuang X (2008) Dynamic binding orientations direct activity of HIV reverse transcriptase. Nature 453:184–189

Ackermann M, Padmanabhan R (2001) De novo synthesis of RNA by the dengue virus RNA-dependent RNA polymerase exhibits temperature dependence at the initiation but not elongation phase. J Biol Chem 276: 39926–39937

Ago H, Adachi T, Yoshida A, Yamamoto M, Habuka N, Yatsunami K, Miyano M (1999) Crystal structure of the RNA-dependent RNA polymerase of hepatitis C virus. Structure 7:1417–1426
Baltimore D (1971) Expression of animal virus genomes. Bacteriol Rev 35:235–241
Basu VP, Song M, Gao L, Rigby ST, Hanson MN, Bambara RA (2008) Strand transfer events during HIV-1 reverse transcription. Virus Res 134:19–38
Berman AJ, Kamtekar S, Goodman JL, Lazaro JM, de Vega M, Blanco L, Salas M, Steitz TA (2007) Structures of phi29 DNA polymerase complexed with substrate: the mechanism of translocation in B-family polymerases. EMBO J 26:3494–3505
Bishop DH, Gay ME, Matsuoko Y (1983) Nonviral heterogeneous sequences are present at the 5' ends of one species of snowshoe hare bunyavirus S complementary RNA. Nucleic Acids Res 11:6409–6418
Blanco L, Salas M (1996) Relating structure to function in phi29 DNA polymerase. J Biol Chem 271:8509–8512
Boivin S, Cusack S, Ruigrok RW, Hart DJ (2010) Influenza A virus polymerase: structural insights into replication and host adaptation mechanisms. J Biol Chem 285(37):28411–28417
Bressanelli S, Tomei L, Roussel A, Incitti I, Vitale RL, Mathieu M, De Francesco R, Rey FA (1999) Crystal structure of the RNA-dependent RNA polymerase of hepatitis C virus. Proc Natl Acad Sci USA 96:13034–13039
Broyles SS (2003) Vaccinia virus transcription. J Gen Virol 84:2293–2303
Cheetham GM, Steitz TA (1999) Structure of a transcribing T7 RNA polymerase initiation complex. Science 286: 2305–2309
Choi KH, Groarke JM, Young DC, Kuhn RJ, Smith JL, Pevear DC, Rossmann MG (2004) The structure of the RNA-dependent RNA polymerase from bovine viral diarrhea virus establishes the role of GTP in de novo initiation. Proc Natl Acad Sci USA 101:4425–4430
Choi KH, Gallei A, Becher P, Rossmann MG (2006) The structure of bovine viral diarrhea virus RNA-dependent RNA polymerase and its amino-terminal domain. Structure 14:1107–1113
Choi KH, Rossmann MG (2009) RNA-dependent RNA polymerases from Flaviviridae. Curr Opin Struct Biol 19(6):746–751
Delarue M, Poch O, Tordo N, Moras D, Argos P (1990) An attempt to unify the structure of polymerases. Protein Eng 3:461–467
DeStefano JJ, Mallaber LM, Fay PJ, Bambara RA (1994) Quantitative analysis of RNA cleavage during RNA-directed DNA synthesis by human immunodeficiency and avian myeloblastosis virus reverse transcriptases. Nucleic Acids Res 22:3793–3800
Drake JW (1991) A constant rate of spontaneous mutation in DNA-based microbes. Proc Natl Acad Sci USA 88:7160–7164
Dufour E, Rodriguez I, Lazaro JM, de Vega M, Salas M (2003) A conserved insertion in protein-primed DNA polymerases is involved in primer terminus stabilisation. J Mol Biol 331:781–794
Durniak KJ, Bailey S, Steitz TA (2008) The structure of a transcribing T7 RNA polymerase in transition from initiation to elongation. Science 322:553–557
Egloff MP, Decroly E, Malet H, Selisko B, Benarroch D, Ferron F, Canard B (2007) Structural and functional analysis of methylation and 5'-RNA sequence requirements of short capped RNAs by the methyltransferase domain of dengue virus NS5. J Mol Biol 372:723–736
Ferrer-Orta C, Arias A, Perez-Luque R, Escarmis C, Domingo E, Verdaguer N (2004) Structure of foot-and-mouth disease virus RNA-dependent RNA polymerase and its complex with a template-primer RNA. J Biol Chem 279:47212–47221
Ferrer-Orta C, Arias A, Agudo R, Perez-Luque R, Escarmis C, Domingo E, Verdaguer N (2006) The structure of a protein primer-polymerase complex in the initiation of genome replication. EMBO J 25:880–888
Ferrer-Orta C, Arias A, Perez-Luque R, Escarmis C, Domingo E, Verdaguer N (2007) Sequential structures provide insights into the fidelity of RNA replication. Proc Natl Acad Sci USA 104:9463–9468
Furfine ES, Reardon JE (1991) Reverse transcriptase.RNase H from the human immunodeficiency virus. Relationship of the DNA polymerase and RNA hydrolysis activities. J Biol Chem 266:406–412
Gao G, Orlova M, Georgiadis MM, Hendrickson WA, Goff SP (1997) Conferring RNA polymerase activity to a DNA polymerase: a single residue in reverse transcriptase controls substrate selection. Proc Natl Acad Sci USA 94:407–411
Golomb M, Chamberlin M (1974) Characterization of T7-specific ribonucleic acid polymerase. IV. Resolution of the major in vitro transcripts by gel electrophoresis. J Biol Chem 249(9):2858–2863
Gu M, Lima CD (2005) Processing the message: structural insights into capping and decapping mRNA. Curr Opin Struct Biol 15:99–106
Guajardo R, Sousa R (1997) A model for the mechanism of polymerase translocation. J Mol Biol 265:8–19
Hansen JL, Long AM, Schultz SC (1997) Structure of the RNA-dependent RNA polymerase of poliovirus. Structure 5:1109–1122
Hogg M, Wallace SS, Doublie S (2004) Crystallographic snapshots of a replicative DNA polymerase encountering an abasic site. EMBO J 23:1483–1493

Hong Z, Cameron CE, Walker MP, Castro C, Yao N, Lau JY, Zhong W (2001) A novel mechanism to ensure terminal initiation by hepatitis C virus NS5B polymerase. Virology 285:6–11
Huang H, Chopra R, Verdine GL, Harrison SC (1998) Structure of a covalently trapped catalytic complex of HIV-1 reverse transcriptase: implications for drug resistance. Science 282:1669–1675
Ibarra B, Chemla YR, Plyasunov S, Smith SB, Lazaro JM, Salas M, Bustamante C (2009) Proofreading dynamics of a processive DNA polymerase. EMBO J 28:2794–2802
Jacobo-Molina A, Ding J, Nanni RG, Clark AD Jr, Lu X, Tantillo C, Williams RL, Kamer G, Ferris AL, Clark P et al (1993) Crystal structure of human immunodeficiency virus type 1 reverse transcriptase complexed with double-stranded DNA at 3.0 A resolution shows bent DNA. Proc Natl Acad Sci USA 90:6320–6324
Jeruzalmi D, Steitz TA (1998) Structure of T7 RNA polymerase complexed to the transcriptional inhibitor T7 lysozyme. EMBO J 17:4101–4113
Julias JG, McWilliams MJ, Sarafianos SG, Arnold E, Hughes SH (2002) Mutations in the RNase H domain of HIV-1 reverse transcriptase affect the initiation of DNA synthesis and the specificity of RNase H cleavage in vivo. Proc Natl Acad Sci USA 99:9515–9520
Kamtekar S, Berman AJ, Wang J, Lazaro JM, de Vega M, Blanco L, Salas M, Steitz TA (2006) The phi29 DNA polymerase:protein-primer structure suggests a model for the initiation to elongation transition. EMBO J 25: 1335–1343
Kao CC, Del Vecchio AM, Zhong W (1999) De novo initiation of RNA synthesis by a recombinant flaviviridae RNA-dependent RNA polymerase. Virology 253:1–7
Kasamatsu H, Nakanishi A (1998) How do animal DNA viruses get to the nucleus? Annu Rev Microbiol 52:627–686
Kim S, Lee J, Ryu WS (2009) Four conserved cysteine residues of the hepatitis B virus polymerase are critical for RNA pregenome encapsidation. J Virol 83:8032–8040
Kohlstaedt LA, Steitz TA (1992) Reverse transcriptase of human immunodeficiency virus can use either human tRNA(3Lys) or Escherichia coli tRNA(2Gln) as a primer in an in vitro primer-utilization assay. Proc Natl Acad Sci USA 89:9652–9656
Kohlstaedt LA, Wang J, Friedman JM, Rice PA, Steitz TA (1992) Crystal structure at 3.5 A resolution of HIV-1 reverse transcriptase complexed with an inhibitor. Science 256:1783–1790
Krug MS, Berger SL (1989) Ribonuclease H activities associated with viral reverse transcriptases are endonucleases. Proc Natl Acad Sci USA 86:3539–3543
Lai VC, Kao CC, Ferrari E, Park J, Uss AS, Wright-Minogue J, Hong Z, Lau JY (1999) Mutational analysis of bovine viral diarrhea virus RNA-dependent RNA polymerase. J Virol 73:10129–10136
Lazaro JM, Blanco L, Salas M (1995) Purification of bacteriophage phi 29 DNA polymerase. Methods Enzymol 262:42–49
Lelke M, Brunotte L, Busch C, Gunther S (2010) An N-terminal region of Lassa virus L protein plays a critical role in transcription but not replication of the virus genome. J Virol 84:1934–1944
Lesburg CA, Cable MB, Ferrari E, Hong Z, Mannarino AF, Weber PC (1999) Crystal structure of the RNA-dependent RNA polymerase from hepatitis C virus reveals a fully encircled active site. Nat Struct Biol 6:937–943
Luo G, Hamatake RK, Mathis DM, Racela J, Rigat KL, Lemm J, Colonno RJ (2000) De novo initiation of RNA synthesis by the RNA-dependent RNA polymerase (NS5B) of hepatitis C virus. J Virol 74:851–863
Malet H, Egloff MP, Selisko B, Butcher RE, Wright PJ, Roberts M, Gruez A, Sulzenbacher G, Vonrhein C, Bricogne G, Mackenzie JM, Khromykh AA, Davidson AD, Canard B (2007) Crystal structure of the RNA polymerase domain of the West Nile virus non-structural protein 5. J Biol Chem 282:10678–10689
Masters BS, Stohl LL, Clayton DA (1987) Yeast mitochondrial RNA polymerase is homologous to those encoded by bacteriophages T3 and T7. Cell 51:89–99
Meijer WJ, Horcajadas JA, Salas M (2001) Phi29 family of phages. Microbiol Mol Biol Rev 65:261–287, second page, table of contents
Mendez J, Blanco L, Salas M (1997) Protein-primed DNA replication: a transition between two modes of priming by a unique DNA polymerase. EMBO J 16:2519–2527
Mesters JR, Tan J, Hilgenfeld R (2006) Viral enzymes. Curr Opin Struct Biol 16:776–786
Morin B, Coutard B, Lelke M, Ferron F, Kerber R, Jamal S, Frangeul A, Baronti C, Charrel R, de Lamballerie X, Vonrhein C, Lescar J, Bricogne G, Gunther S, Canard B (2010) The N-terminal domain of the arenavirus L protein is an RNA endonuclease essential in mRNA transcription. PLoS Pathog 6:e1001038
Murray KE, Barton DJ (2003) Poliovirus CRE-dependent VPg uridylylation is required for positive-strand RNA synthesis but not for negative-strand RNA synthesis. J Virol 77:4739–4750
Nassal M (2008) Hepatitis B viruses: reverse transcription a different way. Virus Res 134:235–249
Ng KK, Cherney MM, Vazquez AL, Machin A, Alonso JM, Parra F, James MN (2002) Crystal structures of active and inactive conformations of a caliciviral RNA-dependent RNA polymerase. J Biol Chem 277:1381–1387
Ollis DL, Brick P, Hamlin R, Xuong NG, Steitz TA (1985) Structure of large fragment of *Escherichia coli* DNA polymerase I complexed with dTMP. Nature 313:762–766

Orlowski J, Bujnicki JM (2008) Structural and evolutionary classification of Type II restriction enzymes based on theoretical and experimental analyses. Nucleic Acids Res 36:3552–3569

Paul AV, Peters J, Mugavero J, Yin J, van Boom JH, Wimmer E (2003a) Biochemical and genetic studies of the VPg uridylylation reaction catalyzed by the RNA polymerase of poliovirus. J Virol 77:891–904

Paul AV, Yin J, Mugavero J, Rieder E, Liu Y, Wimmer E (2003b) A "slide-back" mechanism for the initiation of protein-primed RNA synthesis by the RNA polymerase of poliovirus. J Biol Chem 278:43951–43960

Plotch SJ, Bouloy M, Ulmanen I, Krug RM (1981) A unique cap(m7G pppXm)-dependent influenza virion endonuclease cleaves capped RNAs to generate the primers that initiate viral RNA transcription. Cell 23:847–858

Poch O, Sauvaget I, Delarue M, Tordo N (1989) Identification of four conserved motifs among the RNA-dependent polymerase encoding elements. EMBO J 8:3867–3874

Polyak SJ, Zheng S, Harnish DG (1995) 5' termini of Pichinde arenavirus S RNAs and mRNAs contain nontemplated nucleotides. J Virol 69:3211–3215

Radziwill G, Tucker W, Schaller H (1990) Mutational analysis of the hepatitis B virus P gene product: domain structure and RNase H activity. J Virol 64:613–620

Ranjith-Kumar CT, Kim YC, Gutshall L, Silverman C, Khandekar S, Sarisky RT, Kao CC (2002) Mechanism of de novo initiation by the hepatitis C virus RNA-dependent RNA polymerase: role of divalent metals. J Virol 76:12513–12525

Rausch JW, Lener D, Miller JT, Julias JG, Hughes SH, Le Grice SF (2002) Altering the RNase H primer grip of human immunodeficiency virus reverse transcriptase modifies cleavage specificity. Biochemistry 41:4856–4865

Reguera J, Weber F, Cusack S (2010) Bunyaviridae RNA polymerases (L-protein) have an N-terminal, influenza-like endonuclease domain, essential for viral cap-dependent transcription. PLoS Pathog 6:e1001101

Rodgers DW, Gamblin SJ, Harris BA, Ray S, Culp JS, Hellmig B, Woolf DJ, Debouck C, Harrison SC (1995) The structure of unliganded reverse transcriptase from the human immunodeficiency virus type 1. Proc Natl Acad Sci USA 92:1222–1226

Rodriguez I, Lazaro JM, Blanco L, Kamtekar S, Berman AJ, Wang J, Steitz TA, Salas M, de Vega M (2005) A specific subdomain in phi29 DNA polymerase confers both processivity and strand-displacement capacity. Proc Natl Acad Sci USA 102:6407–6412

Rodriguez-Wells V, Plotch SJ, DeStefano JJ (2001) Primer-dependent synthesis by poliovirus RNA-dependent RNA polymerase (3D(pol)). Nucleic Acids Res 29:2715–2724

Rothwell PJ, Waksman G (2005) Structure and mechanism of DNA polymerases. Adv Protein Chem 71:401–440

Sarafianos SG, Das K, Tantillo C, Clark AD Jr, Ding J, Whitcomb JM, Boyer PL, Hughes SH, Arnold E (2001) Crystal structure of HIV-1 reverse transcriptase in complex with a polypurine tract RNA:DNA. EMBO J 20: 1449–1461

Sarafianos SG, Marchand B, Das K, Himmel DM, Parniak MA, Hughes SH, Arnold E (2009) Structure and function of HIV-1 reverse transcriptase: molecular mechanisms of polymerization and inhibition. J Mol Biol 385: 693–713

Sohn JA, Litwin S, Seeger C (2009) Mechanism for CCC DNA synthesis in hepadnaviruses. PLoS One 4:e8093

Sousa R, Chung YJ, Rose JP, Wang BC (1993) Crystal structure of bacteriophage T7 RNA polymerase at 3.3 A resolution. Nature 364:593–599

Steitz TA (1998) A mechanism for all polymerases. Nature 391:231–232

Tahirov TH, Temiakov D, Anikin M, Patlan V, McAllister WT, Vassylyev DG, Yokoyama S (2002) Structure of a T7 RNA polymerase elongation complex at 2.9 A resolution. Nature 420:43–50

Temiakov D, Mentesana PE, Ma K, Mustaev A, Borukhov S, McAllister WT (2000) The specificity loop of T7 RNA polymerase interacts first with the promoter and then with the elongating transcript, suggesting a mechanism for promoter clearance. Proc Natl Acad Sci USA 97:14109–14114

Thompson AA, Peersen OB (2004) Structural basis for proteolysis-dependent activation of the poliovirus RNA-dependent RNA polymerase. EMBO J 23:3462–3471

Truniger V, Lazaro JM, Salas M, Blanco L (1996) A DNA binding motif coordinating synthesis and degradation in proofreading DNA polymerases. EMBO J 15:3430–3441

Truniger V, Blanco L, Salas M (1999) Role of the "YxGG/A" motif of Phi29 DNA polymerase in protein-primed replication. J Mol Biol 286:57–69

Uchil PD, Satchidanandam V (2003) Architecture of the flaviviral replication complex. Protease, nuclease, and detergents reveal encasement within double-layered membrane compartments. J Biol Chem 278:24388–24398

van Dijk AA, Makeyev EV, Bamford DH (2004) Initiation of viral RNA-dependent RNA polymerization. J Gen Virol 85:1077–1093

Wang HY, Elston T, Mogilner A, Oster G (1998) Force generation in RNA polymerase. Biophys J 74:1186–1202

Weigel C, Seitz H (2006) Bacteriophage replication modules. FEMS Microbiol Rev 30:321–381

Wendeler M, Miller JT, Le Grice SFJ (2009) Human immunodeficiency virus reverse transcriptase. Cameron, Craig E.; Gotte, Matthias; Raney, Kevin D. (Eds.) In: Viral genome replication, Springer Science + Business Media, New York, pp 403–427

Yap TL, Xu T, Chen YL, Malet H, Egloff MP, Canard B, Vasudevan SG, Lescar J (2007) Crystal structure of the dengue virus RNA-dependent RNA polymerase catalytic domain at 1.85-angstrom resolution. J Virol 81: 4753–4765

Yin YW, Steitz TA (2002) Structural basis for the transition from initiation to elongation transcription in T7 RNA polymerase. Science 298:1387–1395

Yin YW, Steitz TA (2004) The structural mechanism of translocation and helicase activity in T7 RNA polymerase. Cell 116:393–404

Zamyatkin DF, Parra F, Alonso JM, Harki DA, Peterson BR, Grochulski P, Ng KK (2008) Structural insights into mechanisms of catalysis and inhibition in Norwalk virus polymerase. J Biol Chem 283:7705–7712

Part III
Capsid Assembly and Maturation

Chapter 13
Chaperonin-Mediated Folding of Viral Proteins

Zacariah L. Hildenbrand and Ricardo A. Bernal

Abstract Chaperonins are universally conserved molecular machines that facilitate the proper folding of nascent and partially folded polypeptides into their respective three-dimensional structures. These multimeric protein complexes utilize the energy derived from ATP hydrolysis to fuel a protein-folding mechanism that consists of multiple rounds of substrate binding, encapsulation, and eventual expulsion back into the cytosolic environment. In this portion of the chapter, the structure and function of group I and group II chaperonins are discussed. Furthermore, the general mechanism of chaperonin-mediated protein folding is addressed in addition to illustrating how viral phages such as Lambda, T4, and RB49 exploit the host machinery for the proper folding of viral gene products. Lastly, the phiEL chaperonin from phage EL is revealed to be the first virally encoded chaperonin and is proposed to function independently of the host chaperonin machinery. The molecular architecture of the phiEL chaperonin, coupled with its unique functional abilities, renders its characterization a challenge and further highlights its novelty as a potentially whole new class of chaperonins.

13.1 Introduction

The folding and assembly of newly synthesized polypeptides into their stable three-dimensional states is one of the least understood processes of protein synthesis (Booth et al. 2008; Carrascosa et al. 2001). The transition of a nascent polypeptide into its respective native conformation is a complex process of intermolecular interactions that is largely dictated by the primary amino acid sequence (Anfinsen 1973). The major driving force behind the protein-folding process involves an increased entropy at the solvent interface by way of sequestering hydrophobic residues within a hydrophobic core (Pace et al. 1996). Additionally, the formation of hydrogen bonds, charge–charge salt-bridge interactions, and disulfide bonds each contribute to the stability of folded protein as it enters the lowest free energy state. In some instances the complete folding of a given protein can be thermodynamically unfavorable in which the relative activation energies required to reach the fully folded state are insurmountable. Furthermore, the in vivo folding of nascent polypeptides is often faced with adverse intracellular conditions such as high overall protein concentration and the potential accumulation of hydrophobic interactions (Xu and Sigler 1998). In such cases, productive protein folding is facilitated by the distinct family of ubiquitous sequence-related proteins known as chaperonins (Nitsch et al. 1998).

Z.L. Hildenbrand • R.A. Bernal (✉)
Chemistry Department, University of Texas at El Paso, El Paso, TX 79968, USA
e-mail: rbernal@utep.edu

M.G. Rossmann and V.B. Rao (eds.), *Viral Molecular Machines*,
Advances in Experimental Medicine and Biology 726, DOI 10.1007/978-1-4614-0980-9_13,

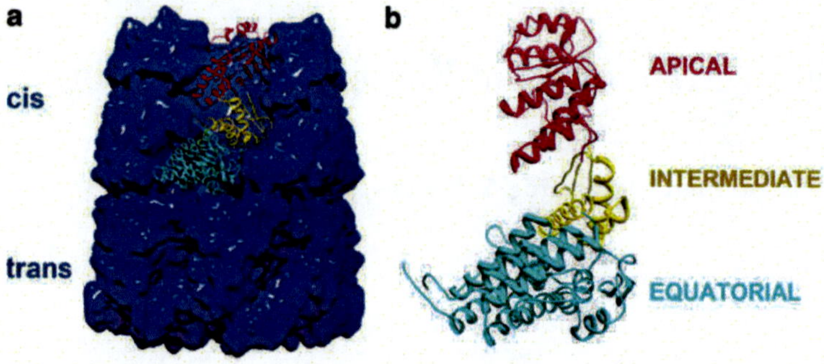

Fig. 13.1 General chaperonin architecture. (**a**) A side view of the extensively studied GroEL chaperonin highlighted by the nucleotide-bound and active folding *cis* ring (*top*) and the inactive *trans* ring (*bottom*) (PDB 1PF9). (**b**) The apical (*pink*), intermediate (*yellow*), and equatorial (*cyan*) domain boundaries within an individual GroEL subunit

Chaperonins are large, multi-subunit assemblies that function by mediating the ATP-dependent protein folding of nascent polypeptides into their native state. Chaperonins utilize the energy derived from ATP hydrolysis to undergo large conformational changes that guide substrate proteins along a refined folding pathway within a well-sequestered and encapsulated environment (Ellis 1994). By providing kinetic assistance, chaperonins prevent or reverse protein misfolding that can lead to irreversible protein aggregation (Horwich et al. 2007). The accumulation of protein aggregates is associated with the development of a number of neurodegenerative conditions including Alzheimer's and Parkinson's diseases (Gadad et al. 2011). The importance of chaperonins is best exemplified by knock-out experiments revealing the deletion of the host chaperonin gene is lethal in both bacteria and yeast (Fayet et al. 1989; Rospert et al. 1993; Reading et al. 1989). Furthermore, mutations in the human mitochondrial chaperonin Hsp60 are responsible for a number of hereditary diseases including spastic paraplegia (Hansen et al. 2002). As a result of the omniscient role in a variety of cellular compartments and the prevention of numerous pathologies, the host chaperonin operates beyond the machineries of transcription and translation to bridge the final gap between DNA and biologically active protein.

The general architecture of chaperonins is characterized by the back-to-back stacking of homo- or hetero-oligomers, each containing seven to nine individual subunits (Nitsch et al. 1998; Roseman et al. 1996; Waldmann et al. 1995) (Fig. 13.1a). Each subunit is organized into separate equatorial, intermediate, and apical domains that execute distinct functions respectively (Schoehn et al. 2000a, b; Spiess et al. 2004; Iizuka et al. 2004) (Fig. 13.1b). Of particular interest, the equatorial domains house the nucleotide binding pockets and are responsible for maintaining the interring contacts (Braig et al. 1994; Ditzel et al. 1998; Zhang et al. 2010). Additionally, the intermediate domains tether the equatorial and apical domains together as the apical domain is responsible for interacting with both the unfolded substrate protein and a co-chaperonin, if required (McCormack et al. 2001).

Chaperonins are characterized as molecular ATPases based on a universal requirement for ATP and the energy derived from its hydrolysis to drive the protein-folding mechanism (Sigler et al. 1998). Furthermore, chaperonins are specifically classified into two subgroups according to their origin and structural characteristics. Group I chaperonins are found in eubacteria, chloroplasts, and mitochondria, while group II chaperonins are found in archaea and eukaryotes (Frydman 2001). Group I chaperonins occupy a homo-oligomeric quaternary assembly and employ a detachable lid-like co-chaperone structure that binds in an ATP-dependent fashion (Roseman et al. 1996) (Fig. 13.2a). Group II chaperonins generally display a hetero-oligomeric subunit organization with exception to the homo-oligomeric archaeal mm-cpn chaperonin (Kusmierczyk and Martin 2003a). Furthermore, group II chaperonins make use of a built-in structure that protrudes upward during the substrate-accepting state and then retracts into a closed conformation upon ATP binding, producing

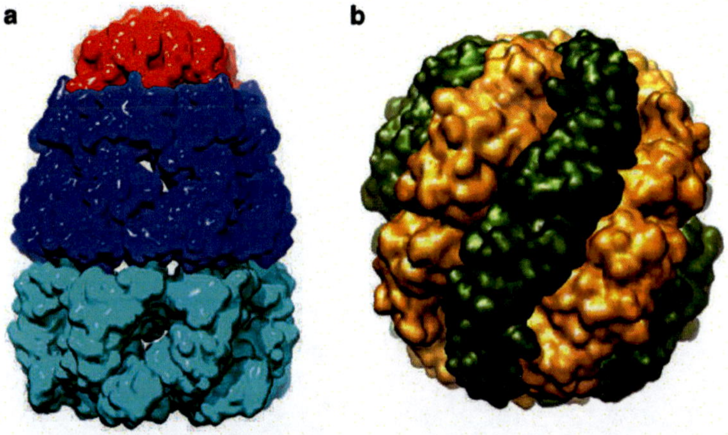

Fig. 13.2 Structural differences between type I and II chaperonins. (**a**) Type I chaperonins like the homo-oligomeric GroEL–GroES (*cis* and *trans* rings are represented in *blue* and *cyan*, respectively) require a co-chaperonin (*red*) to assist a two-stroke protein-folding mechanism (Kafri et al. 2001; Yifrach and Horovitz 1995) (PDB 1PF9). (**b**) On the contrary, type II chaperonins like the hetero-oligomeric thermosome (α and β subunits are outlined in *yellow* and *green*, respectively) utilize a built-in lid structure in their cooperative binding mechanism (Valpuesta et al. 2002; Bigotti and Clarke 2005) (PDB 1A6D)

an active-folding, encapsulated state (Klumpp et al. 1997; Ditzel et al. 1998; Pappenberger et al. 2002; Meyer et al. 2003) (Fig. 13.2b). Despite the different encapsulation mechanisms, the role that ATP plays in group I and II chaperonins is very similar in that ATP hydrolysis is directed by identical ATP-binding motifs within the equatorial domains (Horwich et al. 2007).

In this portion of the chapter, members of the group I and group II chaperonin families are introduced and characterized with respect to their molecular architecture and functional idiosyncrasies. Furthermore, the methods by which viral phages exploit infected host chaperonin machinery are addressed, in addition to revealing a distinct Myoviridae phage that encodes for its own unique chaperonin.

13.2 Group I Chaperonins

13.2.1 GroEL/ES

Of all chaperonins, the type I bacterial GroEL/ES has been the most extensively studied chaperonin system and serves as the basis from which the general mechanism of chaperonin-mediated protein folding was first elucidated (Booth et al. 2008; Saibil et al. 1993; Langer et al. 1992a). The three-dimensional structure of GroEL from *Escherichia coli* is characterized as a porous cylinder of 14 identical 57-kDa subunits that are organized into two nearly sevenfold rotationally symmetrical rings stacked back-to-back with dyad symmetry (Braig et al. 1994). Each ring is distinguished by either a *cis* or *trans* conformation in which ATP, substrate, and co-chaperonin binding occur in the *cis* ring (Fig. 13.1a). Each individual subunit within the rings consists of three distinct domains that each plays an important role in the GroEL-mediated protein-folding process (Fig. 13.1b). The mostly helical equatorial domain houses the ATP-binding pocket in addition to holding the apposing rings together through multiple of intra- and interring contacts (Xu and Sigler 1998). Together the seven equatorial domains provide a structural barricade obstructing interring substrate exchange across the

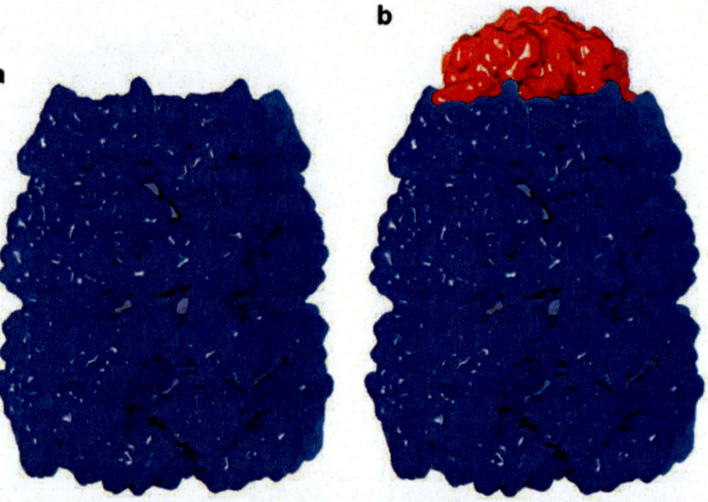

Fig. 13.3 The GroEL–GroES complex. (**a**) The double-ring GroEL chaperonin (*blue*). (**b**) The binding of the heptameric GroES lid (*red*) to the *cis* ring of GroEL

equatorial plane thus restricting substrate entry and escape to the apical domains (Rye et al. 1997). The loosely structured apical domain surrounds the opening to the central cavity and is responsible for the binding of substrate and the GroEL co-chaperonin (Braig et al. 1994; Fenton et al. 1994). Large conformation flexibility within the apical domains contributes to the opening and closing of the ring cavity allowing for substrate encapsulation (Zhang et al. 2010). Lastly, the hinge-like intermediate domain tethers the equatorial and apical domains together and is regarded as the central gear in the ATP-mediated allosteric mechanism of GroEL (Xu and Sigler 1998).

GroEL functions in close coordination with its GroES co-chaperonin counterpart in an ATP-dependent fashion (Tilly et al. 1981). GroES monomers assemble into a dome-shaped heptameric quaternary structure that contributes to a bullet-like structure when bound to GroEL (Ishii et al. 1992; Langer et al. 1992b; Saibil et al. 1991) (Fig. 13.3). GroES binds to GroEL through aliphatic residues within its mobile loop motif which interact with a hydrophobic patch atop the apical domain (Landry et al. 1993; Xu et al. 1997). Upon forming the GroEL–GroES complex, the internal volume of the GroES-bound *cis* ring is more than doubled from that of the unbound *trans* ring, potentially accommodating a molten globule intermediate with a molecular mass greater than 70 kDa (Xu and Sigler 1998). Once formed, the GroEL–GroES complex provides the respective substrate with an isolated environment in which unimpeded folding is segregated away from other folding intermediates, limiting the incidence of unwanted aggregation (Xu and Sigler 1998).

13.2.2 Hsp60

The group I chaperonin Hsp60 and its respective co-chaperonin Hsp10 are structurally and functionally homologous to the bacterial GroEL and GroES proteins, respectively (Nielsen and Cowan 1998). Hsp60 is abundantly scattered throughout the mitochondria where it catalyzes the folding of proteins destined for the mitochondrial matrix (Koll et al. 1992). Hsp60 subunits organize themselves into heptameric single-ring structures as indicated by size-exclusion chromatography experiments (Picketts et al. 1989). While the general ring architecture of Hsp60 is similar to that of GroEL, the two differ drastically in the schematics of their respective protein-folding mechanisms.

Hsp60–Hsp10 complexes facilitate protein folding as a single-ring species in an ATP-dependent fashion (Viitanen et al. 1992). This is in stark contrast to the GroEL–GroES complex that utilizes negatively cooperative allosteric binding between its two rings to facilitate protein folding (Rye et al. 1997). Transient double-ring intermediates are believed to be part of the Hsp60-mediated protein-folding pathway; however, this remains to be fully understood (Todd et al. 1995; Viitanen et al. 1998). Mutagenesis results have denounced the need for a double-ring Hsp90–Hsp10 structure, indicating that Hsp60 functions as an efficient "one-stroke" engine (Nielsen and Cowan 1998).

In GroEL–GroES, single-ring assemblies formed by the mutational disruption of the equatorial interface between the two rings rendered the bacterial chaperonin incapable of releasing its encapsulated substrate (Weissman et al. 1996). Interestingly, the single-ring Hsp60–Hsp10 has the ability to replace GroEL–GroES as indicated by a proliferation of a GroEL–GroES-deficient *E. coli* strain supplemented with the genes encoding for Hsp60 and Hsp10 (Nielsen et al. 1999). Chimeric Hsp-GroE strains were also found to be viable, albeit to a lesser extent. These results conclude that the single-ring method of action used by the mammalian Hsp60–Hsp10 can replace the double-ring "two-stroke" mechanism of GroEL–GroES in maintaining the viability of bacterial cells (Nielsen et al. 1999). However, Hsp60–Hsp10 cannot functionally substitute for the host chaperonin machinery in the proliferation of the bacteriophages lambda and T4, illustrating a level of specificity phages have developed for the host GroEL–GroES (Nielsen et al. 1999). In the case of the T4 phage, its inability to propagate in the absence of GroEL–GroES is likely due to the inability of Hsp60 to interact with the virally encoded gp31 co-chaperonin that is required for the successful folding of the respective capsid proteins (Hunt et al. 1997; Nielsen et al. 1999).

13.3 Group II Chaperonins

13.3.1 TRiC/CCT

The eukaryotic chaperonin TRiC (TCP1-Ring Complex or CCT) plays an essential role in the folding of numerous cytosolic proteins including cytoskeletal components, cell cycle regulators, and tumor suppressor proteins (Frydman 2001; Hartl and Hayer-Hartl 2002; Meyer et al. 2003; Yam et al. 2008; Thulasiraman et al. 1999). Substrates such as actin cannot be folded by other chaperonin systems, suggesting that TRiC possesses unique mechanistic features that are absent in other chaperonins (Spiess et al. 2004; Douglas et al. 2011). Furthermore, TRiC is believed to have a more substrate-specific role in protein folding on the basis of its unique subunit composition (Munoz et al. 2011). Specialized substrate binding is believed to occur through hydrophobic interactions; however, the extent of these interactions in a structural context remains unclear (Kubota et al. 2006; Yam et al. 2008).

Of the group II chaperonins, TRiC is the most structurally complex. Each of the two rings contains eight individual subunits and is related by pseudo-twofold symmetry (Munoz et al. 2011) (Fig. 13.4). ATP binding and hydrolysis drive TRiC through a conformational cycle that contains multiple transient states. In the absence of nucleotide, the lid-containing apical domains are extended in an open conformation awaiting substrate binding (Booth et al. 2008). Subsequent ATP binding induces a closed conformation where portions of the apical domains retract to form a beta-stranded iris that closes over the cavity of the complex (Douglas et al. 2011; Pereira et al. 2010). The formation of the lid structure confers allosteric subunit coupling within the complex and is essential for successful substrate folding to occur (Kanzaki et al. 2008; Meyer et al. 2003; Reissmann et al. 2007). Despite the recent unveilings of multiple conformational states within the TRiC-mediated protein-folding pathway, little is known about the trajectory through the conformational states or how substrate folding is achieved in group II chaperonins (Bigotti and Clarke 2008).

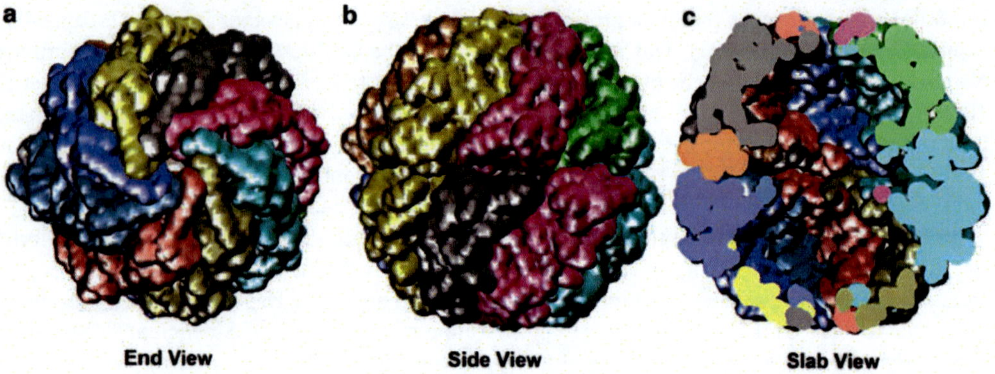

Fig. 13.4 The eukaryotic TRiC/CCT chaperonin. (**a–c**) End, side, and center-slabbed views of the hetero-oligomeric double-ringed TRiC/CCT complex in the closed conformation (PDB 3IYG). The eight different subunits within each ring are colored accordingly to illustrate the intra-ring subunit heterogeneity

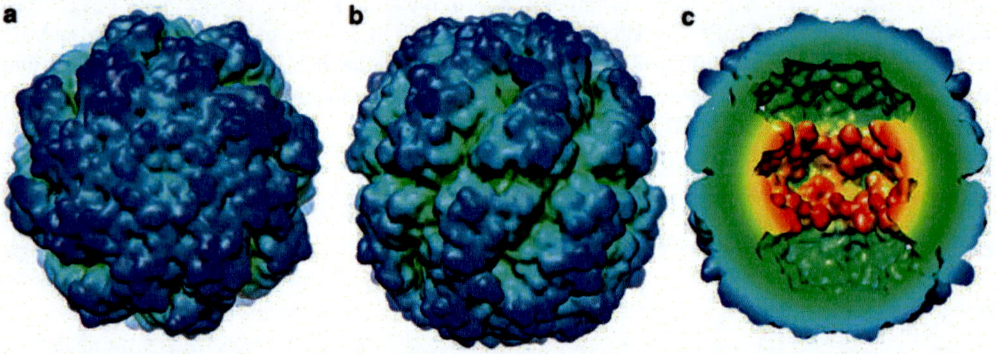

Fig. 13.5 Thermosome from archaebacteria. (**a–c**) End, side, and center-slabbed views of the thermosome chaperonin from *Thermoplasma acidophilum* (PDB 1A6D). Each structure is colored using a radial gradient from the center out to the distal regions of the external surface (*red* to *yellow* to *green* to *cyan* to *blue*)

13.3.2 Thermosome

The group II chaperonin, thermosome, found in thermophilic archaea, is structurally related to TF55 from *Sulfolobus solfataricus* and represents a functional homolog of the eukaryotic TRiC/CCT (Kim et al. 2002; Ditzel et al. 1998) (Fig. 13.5a, b). The structure of thermosome consists of two eight-membered rings stacked together with twofold symmetry, each formed by alternating α and β subunits (Ditzel et al. 1998). The α and β subunits exhibit approximately 60% primary sequence identity with one another and are virtually identical in regard to their fold and domain arrangements (Ditzel et al. 1998). Furthermore, the thermosome α subunit displays roughly 46% sequence identity with that of the bacterial GroEL monomer in which the similarity between the two is characterized by limited domain movements (Ditzel et al. 1998; Xu et al. 1997). Thermosome subunits each contain β-sheet protrusions at the distal ends of the apical domains that act as a built-in lid structure that can open and close in an ATP-dependent manner (Gutsche et al. 1999; Horwich and Saibil 1998). The presence of a built-in lid is consistent with all other group II chaperonins and alleviates the need for binding with an external co-chaperonin in order to successfully encapsulate the substrate.

Thermosome-mediated protein folding is thought to be regulated by an iterative two-step process driven by binding and hydrolysis of ATP. Initially, ATP binding would produce an asymmetric complex, where the upper ring adopts an open conformation facilitating substrate entry, while the built-in lid structure closes in the bottom ring. Upon ATP hydrolysis, the apical domains of the upper ring would extend forming the lid structure, sequestering the substrate in a thermodynamically favorable environment (Kim et al. 2002) (Fig. 13.5c). Despite previous work, the exact mechanism by which thermosome elicits its response to nonnative proteins remains unclear. However, the ability of the thermosome to hydrolyze ATP at temperatures beyond 80°C, provides an insight into the physical stability of the thermosome structure that is a likely representative of chaperonin structure in general (Kim et al. 2002).

13.3.3 mm-cpn

The mm-cpn chaperonin from the mesophilic archaea *Methanococcus maripaludis* is an eightfold double-ring structure that is unique to the group II subfamily of chaperonins (Kusmierczyk and Martin 2003a). The homomeric subunit organization of mm-cpn suggests a similarity to GroEL–GroES; however, structural investigations insist that mm-cpn is more closely related to TRiC/CCT (Zhang et al. 2010). Furthermore, mm-cpn employs a characteristic group II chaperonin built-in lid structure and exhibits an allosteric regulation that is similar to that of eukaryotic TRiC (Reissmann et al. 2007; Kusmierczyk and Martin 2003b). A lidless variant of mm-cpn maintains the ability to bind substrate and hydrolyze ATP, albeit in a less selective manner (Reissmann et al. 2007). Consistent with group II chaperonins, the mm-cpn has a natural affinity for substrates containing β-sheet motifs that are generally prone to thermodynamically unfavorable folding and aggregation (Knee et al. 2011; Yam et al. 2008). This affinity of group II chaperonins for β-sheet structure is thought to be the result of the subunit heterogeneity (Kubota et al. 2006). Such a conjecture is not applicable to the homo-oligomeric mm-cpn, and as a result further work is required to unveil the cause of this phenomenon.

During the ATPase cycle of mm-cpn, the barrel-shaped structure can adapt into different conformational states as illustrated by recent structural investigations (Reissmann et al. 2007; Zhang et al. 2010; Douglas et al. 2011) (Fig. 13.6). An open conformation is maintained in the absence of bound nucleotide whereas ATP binding triggers a closed conformational state (Zhang et al. 2010). The transition between the open and closed states is thought to involve a rocking motion that dramatically reconfigures the interring contacts as the result of tilting within the equatorial domains (Douglas et al. 2011). The conformational movements surrounding the equatorial plane likely contribute to the self-closure of the encapsulated chamber in the absence of an external co-chaperonin (Zhang et al. 2010).

13.4 GroEL-Mediated Protein-Folding Mechanism

13.4.1 ATP Binding and Hydrolysis

In the GroEL/ES system, there is positive cooperativity of intra-ring ATP binding within each ring by way of a concerted mechanism (Horwich et al. 2007). Structural studies suggest that movement within the intermediate domains of GroEL dictates communication between the ATP binding sites that is crucial for the propagation of positively cooperative ATP binding (Ranson et al. 2001). Rotation of the intermediate domain of the initial ATP-bound subunits leads to a new contact with the equatorial domain of a neighboring subunit near the ATP-binding site. Through this contact, information pertaining to the ATP-binding state of a single GroEL subunit is passed on to the ATP-binding site of the neighboring subunit and onward around the GroEL ring, thus increasing the ATP-binding affinity of successive sites (Ranson et al. 2001).

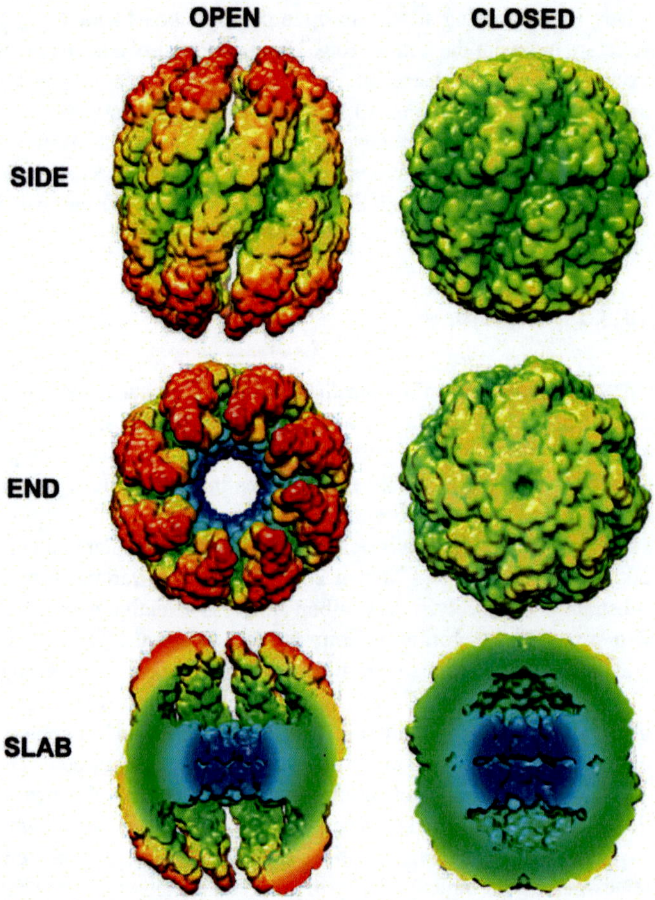

Fig. 13.6 Views of the mm-cpn chaperonin in both the open and closed conformations. Each structure is colored using a radial gradient away from the center (*blue* to *cyan* to *green* to *yellow* to *red* at the highest radius). In the open conformation, the apical domains are extended away from the center of the complex (*red*) (PDB 3IZI). Upon ATP binding and hydrolysis, mm-cpn adopts a closed conformation in which the previously extended apical domains retract toward the equatorial plan and contribute to ring closure (PDB 3IZH)

In conjunction with the positive cooperativity of intra-ring ATP binding, there is also negatively cooperative and sequential ATP binding between the two rings (Yifrach and Horovitz 1995; Rivenzon-Segal et al. 2005). ATP binding within one ring appears to alter the interring contacts such that the interface is distorted triggering the transmission of an allosteric signal between the two rings (Roseman et al. 2001; Ranson et al. 2001). Thus, the two rings function out of phase of one another, a phenomenon that is best exemplified by the binding of the GroES lid to only one of the two rings in an ATP-dependent manner (Horwich et al. 2007).

ATP binding within the equatorial domains of GroEL surrounds a well-conserved asparagine residue, Asp87, and is stabilized by the divalent magnesium cation (Fenton et al. 1994). Upon nucleotide loading, the ATPase cycle begins with a nascent or misfolded protein having a high affinity for the chaperonin in the ATP-bound conformation (Hayer-Hartl et al. 1996) (Fig. 13.7). In this state, the apical domains lining the distal end of the open conformation attract the substrate through hydrophobic interactions with exposed hydrophobic residues (Brinker et al. 2001; Saibil et al. 1993; Sigler et al. 1998). ATP is believed to bind to GroEL prior to substrate binding; however, this remains to be observed directly. ATP binding to each of the seven equatorial domains of the *cis* ring produces an allosterically mediated conformational change within the apical domains that primes the nucleotide-bound ring to accept the GroES co-chaperonin (Ranson et al. 2001). GroES binding induces further conformational changes altering the hydrophobic polypeptide-binding sites from their internal orientation to release the substrate into an encapsulated hydrophilic *cis* ring cavity where viable protein folding is isolated and thermodynamically favorable (Brinker et al. 2001; Saibil et al. 1993; Weissman et al. 1995). Folding continues in the *cis* cavity for the longest phase of the mechanistic

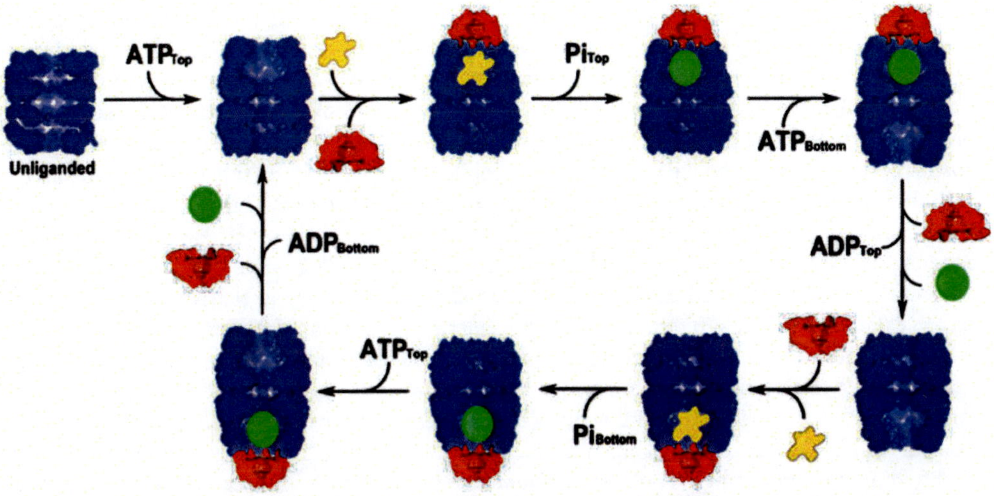

Fig. 13.7 The ATPase cycle of the GroEL–GroES complex. The hydrolytic reaction is divided into two phases. The first phase illustrates a pre-steady-state behavior of unbound GroEL prior to binding with ATP and the GroES co-chaperone (PDB 3C9V). Initially, the apo-GroEL *cis* ring (ATP_{TOP}) fills cooperatively with seven molecules of ATP, triggering the binding of substrate protein (*yellow star*) and the GroES co-chaperonin (*red*) (Lin and Rye 2006; Fenton and Horwich 2003). The capping of GroES onto the apical domain of the *cis* ring releases the bound substrate into the internal cavity where the folding process commences. The resulting GroEL–ATP–GroES complex hydrolyzes the ATP with an intrinsic turnover rate of 0.12 s^{-1} (Burston et al. 1995) (PDB 1PF9). The sequential binding of ATP to the open *trans* ring (ATP_{BOTTOM}) initiates the disassembly of the *cis* complex, leading to the release of GroES, ADP, and the folded protein product (*green*) (Rye et al. 1997, 1999; Todd et al. 1994). The steady-state cycling of the ATP- and ADP-bound conformations between the two negatively cooperative rings constitutes the second phase of the nucleotide-driven mechanism. In vivo the GroEL–GroES system persists almost exclusively within the latter phase (Madan et al. 2008; Lin et al. 2008)

cycle, approximately 10 s, upon which ATP hydrolysis within the equatorial domains of the *cis* ring weakens its affinity for GroES, concurrently allowing ATP to enter into the binding sites of the *trans* ring (Rye et al. 1997). Binding of ATP to the *trans* ring sends an allosteric signal that ejects the *cis* ring contents (co-chaperonin, partially or fully folded protein product, and ADP), while simultaneously converting the once *trans* ring into the *cis* ring conformation (Todd et al. 1994; Weissman et al. 1994; Rye et al. 1997). As a result, the rings of this "two-stroke engine" switch back and forth between the *cis* and *trans* conformations to employ a highly efficient protein-folding operation that is tightly regulated by ATP (Todd et al. 1994; Yifrach and Horovitz 1995).

13.4.2 Encapsulation and Release

Early gel filtration experiments have illustrated the ability GroEL has to bind promiscuously with a wide range of small and large nonnative proteins (Viitanen et al. 1992). Subsequently, it was determined that GroEL has little or no affinity for nondenatured proteins that have sequestered their hydrophobic residues, giving rise to the notion that GroEL–substrate interactions are likely initiated through hydrophobic interactions (Viitanen et al. 1992).

The transition between the substrate-bound open conformation of GroEL through to the active folding state where the substrate is sequestered within the *cis* cavity and protein folding occurs is difficult to understand and is yet to be fully elucidated. For instance, it remains unclear as to how the substrate is encapsulated upon GroES binding instead of escaping into the bulk solution (Horwich et al. 2007). Various biochemical analyses have shown that during a single round of ATP and GroES binding, all substrate molecules can reach their native state within the *cis* chamber implying that substrate escape of encapsulation is a rare occurrence (Weissman et al. 1996). Substrate loss is believed to be averted by the formation of a transient complex in which GroES and the substrate are simultaneously bound to the ATP-mobilized apical domains of GroEL (Miyazaki et al. 2002). In this complex, portions of the GroEL apical domain surface would be occupied by the polypeptide substrate, while others are bound to the β-hairpin mobile loops of the GroES subunits (Richardson et al. 2001). Furthermore, fluorescence-labeling studies measuring apical domain movement have revealed that substrate binding applies a load to the apical domains that resists their elevation. It was also found that the rate of apical movement is slower than that of GroES binding (Motojima et al. 2004). Collectively these results indicate that GroES binds to GroEL prior to the completion of apical domain extension thus guiding the substrate into entering the *cis* ring chamber.

In an Anfinsen cage model describing chaperonin-mediated protein folding, the substrate protein is encapsulated within the hydrophilic ring, separated from the outer bulk solution and protected from aggregation (Ellis 1994). Additionally, a confinement model exists describing how spatial confinement within the encapsulated chamber increases the entropic slope, which in turn drives the rapid folding process (Brinker et al. 2001; Baumketner et al. 2003). A recent cryo-EM reconstruction of a GroEL–substrate complex corroborates both of the aforementioned models in revealing an encapsulated substrate being held in the center of the *cis* cavity by repulsive charge interactions between the interior walls of GroEL and exposed residues of the substrate (Kanno et al. 2009). The importance of the electrostatic character within the *cis* ring cavity is further demonstrated by mutagenesis experiments in which single and multiple substitutions altering the overall charge at the *cis* ring cavity interface fail to rescue a GroEL-deficient *E. coli* strain (Farr et al. 2007). Additionally, the presence of hydrophobic residues at the *cis* cavity interface may also be important to the functioning of GroEL. Simulation experiments have suggested that residual hydrophobicity at the *cis* ring cavity interface may lower the transition states between folding intermediates, assisting in the folding process (Jewett et al. 2004). Alteration of hydrophobic residues facing the cavity has proven to be lethal, impairing GroES encapsulation, ATP turnover, and GroES release (Fenton et al. 1994; Koike-Takeshita et al. 2006; Farr et al. 2007). However, the promiscuous behavior of the GroEL chaperonin indicates that the internal *cis* cavity of the GroEL–GroES complex has likely been evolutionarily optimized to accommodate a broad range of nonnative substrate proteins irrespective of specific biochemical interactions (Horwich et al. 2007).

In regard to substrate release following encapsulation, the mechanism of GroEL-mediated protein folding is likely to be an iterative process. Thermodynamic analyses have indicated that time-limited attempts at folding within the encapsulated environment are followed by the release of the polypeptide into the bulk solution, regardless of whether or not the substrate has reached its native state (Craig et al. 1994). As a result, partially folded proteins likely encounter multiple rounds of chaperonin encapsulation, folding, and release. Of particular interest the iterative chaperonin-mediated protein folding is regarded as an energetically inexpensive event in that the complete folding of nascent polypeptides requires approximately 10% of the amount of ATP as is consumed in the translation of mRNA to polypeptide chains (Martin et al. 1991).

13.5 Viral Protein Folding

13.5.1 Lambda and T5

Chaperonin-mediated protein folding also plays an important role in the posttranslational processing of viral gene products. The genes encoding for the bacterial GroEL–GroES chaperonin machine were first established as host factors for bacteriophage morphogenesis, playing an essential role in

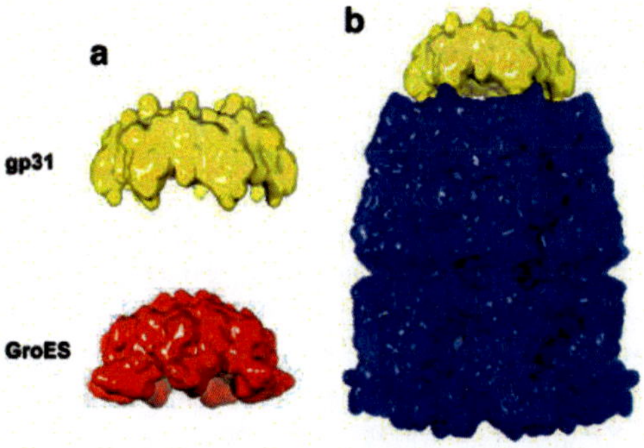

Fig. 13.8 Structural similarities between the virally encoded gp31 co-chaperonin and GroES. (**a**) The gp31 (*yellow*) and GroES (*red*) co-chaperonins share very little primary sequence identity with one another but are each assembled into structurally homologous heptameric lid formations. (**b**) gp31 can be incorporated into the GroEL system as a functional chimera for the folding of bacterial proteins; however, GroES is unable to substitute for gp31 during T4 capsid assembly (van der Vies et al. 1994) (PDB 2CGT)

the correct assembly of λ proheads and T5 tails (Georgopoulos et al. 1973; Zweig and Cummings 1973). Normal head morphogenesis of the viral phage λ is disrupted in selective *groE*-deficient bacteria as the result of the failed occurrence of required proteolytic cleavages (Georgopoulos et al. 1973; Sternberg 1973; Hendrix and Casjens 1975; Murialdo 1979). Additionally, point mutations altering the flexibility of the equatorial domains within GroEL block the formation of complete λ and T5 phage structural assemblies (van der Vies et al. 1994; Landry et al. 1993). Both phages require the GroES co-chaperonin to assist in the folding of their respective structural proteins; however, the T4-encoded gp31 co-chaperonin can substitute for the host co-chaperonin, as indicated by mutagenesis studies (van der Vies et al. 1994).

13.5.2 T4 and RB49

The bacteriophage phage T4 and the coliphage RB49 each encode for their own respective co-chaperonin that is functionally coupled with the host chaperonin to facilitate the proper folding of large viral proteins (Richardson and Georgopoulos 1999; Ang et al. 2001) (Fig. 13.8a). For example, the host GroEL, but not GroES, is required for T4 propagation as indicated by mutagenesis studies revealing *groEL* mutants that were unable to support T4 proliferation (Zeilstra-Ryalls et al. 1993). However, phage-encoded gene product 31 (gp31) acts during the early stages of the T4 life cycle and is required for the proper folding of the major capsid protein, gp23 (Keppel et al. 2002). Genetic and biochemical evidence indicate that the viral gp31 and the bacterial GroEL interact extensively where gp31 can compensate for the growth defect of *groES* temperature-sensitive mutations at nonpermissive temperatures (van der Vies et al. 1994). Additionally, gp31 can substitute for the host GroES in the refolding of prokaryotic rubisco and citrate synthase, irrespective of the two co-chaperonins possessing only 14% sequence identity with one another (van der Vies et al. 1994; Richardson and Georgopoulos 1999; Keppel et al. 1990; Koonin and van der Vies 1995) (Fig. 13.8b). Interestingly, the reverse is not true in regard to viral protein folding as GroES is unable to substitute for gp31 during T4 capsid assembly (van der Vies et al. 1994). Structural investigations of gp31 and GroES indicate that both co-chaperonins possess a mobile loop motif that interacts with the apical domains of GroEL, both playing an important regulatory role in the encapsulation step of GroEL–GroES-mediated protein folding (Landry et al. 1993; Richardson et al. 1999). However, gp31 is more specialized to facilitate the folding of large capsid proteins as is evident by a longer, more flexible mobile loop when compared to that of GroES (Hunt et al. 1997). As a result the gp31 heptameric lid structure protrudes further away from GroEL than GroES does, leading to the formation of a larger internal cavity.

The phage RB49 also encodes for a gp31-like homolog termed *co-c*haperonin *co*gnate (CocO) (Ang et al. 2001). CocO is distantly related to GroES and carries 35% primary sequence identity with gp31 yet can be incorporated into the folding of citrate synthase or the T4-encoded gp23 (Ang et al. 2001; Richardson et al. 1999). However, the reverse is not true as GroES cannot substitute for CocO in bacteriophage assembly and proliferation (Keppel et al. 2002). Furthermore, partial cross-complementation between gp31 and CocO was observed to be temperature sensitive, indicating that the two co-chaperonins exhibit a preference in function and prefer to fold their own respective capsid proteins (Keppel et al. 2002).

Collectively the dependence of the T4 and RB49 phages on their respective co-chaperonins is best epitomized by mutagenesis studies (Klein and Georgopoulos 2001). A mutation in the region responsible for modulating the en bloc movements of the apical domain rendered the GroEL mutant incompetent of co-chaperonin binding ultimately resulting in failed phage growth (Braig et al. 1994; Xu et al. 1997; Klein and Georgopoulos 2001).

13.5.3 Potato Leafroll Virus

The potato leafroll virus (PLRV) is a positive strand RNA virus that is transmitted to the phloem tissue of plants by way of the salivary glands of infected aphids, *Myzus persicae* (Harrison 1985; Gray and Gildow 2003). The aphid hemolymph acts as the reservoir in which PLRV is retained in the insect without replication (Eskandari et al. 1979). Of particular interest, the persistent nature of PLRV in the aphid hemolymph is dependent upon endosymbiotic *Buchnera* sp. bacteria that are harbored in specialized mycetocyte cells in the abdomen of the aphid (van den Heuvel et al. 1994). *Buchnera* spp. produce the MpB GroEL chaperonin which is highly homologous to the bacterial GroEL, sharing approximately 92% primary sequence similarity with the *E. coli* GroEL and other group I chaperonins (Clark et al. 1998; Hogenhout et al. 1998). MpB GroEL is required to prevent the proteolytic breakdown of PLRV viral proteins as indicated by antibiotic treatments of *M. persicae*, resulting in the absence of detectable MpB GroEL in the hemolymph in addition to a massive reduction in the transmission efficiency of PLRV (Eskandari et al. 1979).

Mutational analysis of the MpB GroEL gene revealed that the PLRV-binding sites are located in the equatorial domain as demonstrated by mutants lacking varying portions of the equatorial domain, losing the ability to bind PLRV proteins (Hogenhout et al. 1998). This particular substrate-binding modality exhibited by the MpB GroEL chaperonin is in stark contrast to its sequence-related homolog, GroEL, which targets substrate binding through its apical domain (Braig et al. 1994; Fenton et al. 1994). It is unclear as to why PLRV binding occurs at the equatorial domains of the *Buchnera* chaperonin; however, it is postulated that this form of substrate binding allows for large multimeric proteins, such as those of the PLRV capsid, to overcome the size limitations dictated by the spatial volume inside central cavity of the cylinder (Hogenhout et al. 1998).

While MpB GroEL is abundantly produced by *Buchnera* spp. in aphids, the respective GroES co-chaperonin is difficult to detect (Kakeda and Ishikawa 1991). The gene encoding for the MpB GroEL is located on the same operon that houses the *groES* gene; however, *Buchnera* GroEL is likely more stable and accumulates more readily, while the GroES co-chaperonin is rapidly degraded (Hogenhout et al. 1998). The *Buchnera* GroES is likely to be an important cofactor in the cellular processing of PLRV, yet the extent of its role in extracellular protein interactions within the aphid hemolymph remains to be determined (Ohtaka et al. 1992).

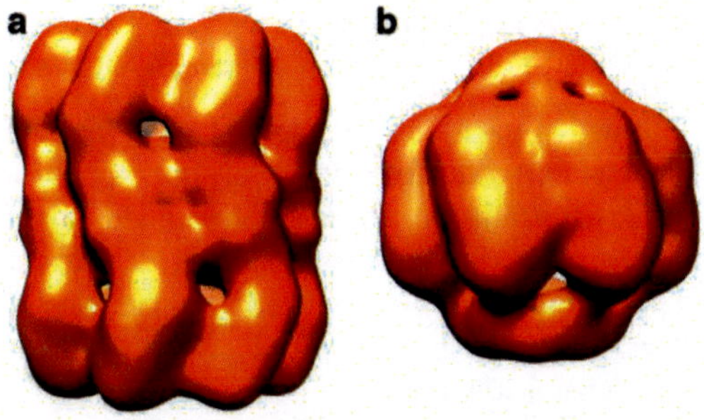

Fig. 13.9 The virally encoded φEL chaperonin. (**a**) In the ATP-bound conformation, the φEL chaperonin exhibits a double-ring structure with each ring occupying an open state. (**b**) ATP hydrolysis triggers the separation of the two rings, resulting in a single-ring structure with an enlarged internal cavity that is likely required to accommodate large viral proteins

13.5.4 φEL

The method of utilizing the host chaperonin to process nascent viral proteins is energetically favorable and alleviates the need for phages to carry their own protein-folding machinery. However, there is an exception to this phenomenon. Recently, an 860-kDa GroEL-like chaperonin complex has been identified and isolated from the bacteriophage φEL, a virus that infects the gram-negative bacterium *Pseudomonas aeruginosa* (Hertveldt et al. 2005). Of particular interest, the predicted gene products of the dsDNA φEL phage share moderate levels of primary sequence identity with that of the giant φKZ-like Myoviridae virus, yet the φEL is the only known phage that encodes for its own full chaperonin (Burkal'tseva et al. 2002). Much like the bacterial GroEL chaperonin described earlier, the φEL chaperonin is a protein assembly characterized by the back-to-back stacking of two heptameric rings comprised of identical subunits. In the ATP-bound open conformation, the φEL chaperonin occupies an overall cylindrical shape with two conspicuous openings at the distal portions of the apical domains that allow for unfolded protein substrate to enter either of the internal cavities (Fig. 13.9a). Substrate binding and ATP hydrolysis trigger encapsulation resulting in the partition of the chaperonin into two rings each occupying a closed conformation (Fig. 13.9b). In this state, the intermediate and equatorial domains of the single-ring structure extend away from the center, giving rise to a larger internal cavity that is likely designed to accommodate the spatial occupancy of particularly large φEL proteins. The mechanism through which the two rings reunite following the disposal of folded protein and bound nucleotide remains mysterious however, it likely involves the formation and a transient complex.

Of particular interest, the φEL chaperonin does not require a co-chaperonin to facilitate successful protein folding. Chaperonin function independent of a co-chaperonin is characteristic of a group II chaperonins; however, the φEL chaperonin is structurally related to the group I chaperonin GroEL. Furthermore, the functioning of the φEL chaperonin in the absence of a co-chaperonin starkly contrasts chaperonin-mediated protein folding observed in the T4 and RB49 phages, which utilize the host GroEL while encoding for their own co-chaperonin (van der Vies et al. 1994; Ang et al. 2001). The unique conformations of the φEL chaperonin in addition to its ability to function in the absence of an external co-chaperonin make its classification as a group I chaperonin difficult. Additionally, the homo-oligomeric subunit organization of the φEL chaperonin and its ability to partition itself into single-ring structures makes it equally difficult to distinguish as a group II chaperonin. Primary sequence-based phylogenetic analyses indicate that the φEL chaperonin is evolutionary independent of all other known chaperonins, further illustrating the potential emergence of a new distinctive group of protein-folding molecular machines.

13.6 Summary and Future Directions

It is understood that the information regarding the intermolecular interactions involved in a protein reaching its native state is maintained within the respectively primary amino acid sequence (Anfinsen 1973). However, the exact mechanism by which freshly translated polypeptides fold to reach their respective native conformations remains to be determined and requires further elucidation. The discovery of the chaperonin family of proteins has revealed a unique class of molecular machines that function to provide kinetic assistance for the proper folding of nascent or misfolded polypeptides within an isolated and thermodynamically favorable environment. Chaperonins have been identified in all three life kingdoms where they have been found to function as single- and double-ring structures, with and without the use of an external co-chaperonin (Ellis and Hartl 1996; Fenton and Horwich 1997; Hartl 1996). Additionally, a chaperonin has also been found to be encoded by the nonliving phage φEL, the structures of which provide insight into the folding mechanism of large viral gene products. Further work is required to greater understand the chaperonin-mediated protein-folding process as it pertains to the prevention of protein aggregation, the development of neurodegenerative diseases and the proliferation of viral phages. It is necessary that future efforts be spent unveiling the innate biophysical and biochemical interactions that take place within the encapsulated "chamber of secrets" (Spiess et al. 2004), to complete our understanding of the transition from DNA to functional protein.

Acknowledgments We would like to thank our collaborators and colleagues for their tremendous contribution to the chaperonin research field and for inspiring us to continue our pursuit of greater scientific understanding. We would also like to thank lab members for their hard work and insight toward our respective projects, which are supported by the Welch Foundation (AH-1649 to RAB) and the NSF (083295 to ZLH).

References

Anfinsen CB (1973) Principles that govern the folding of protein chains. Science 181(96):223–230

Ang D, Richardson A, Mayer MP, Keppel F, Krisch H, Georgopoulos C (2001) Pseudo-T-even bacteriophage RB49 encodes CocO, a cochaperonin for GroEL, which can substitute for *Escherichia coli's* GroES and bacteriophage T4's Gp31. J Biol Chem 276(12):8720–8726

Baumketner A, Jewett A, Shea JE (2003) Effects of confinement in chaperonin assisted protein folding: rate enhancement by decreasing the roughness of the folding energy landscape. J Mol Biol 332(3):701–713. doi:S002228360300929X [pii]

Bigotti MG, Clarke AR (2005) Cooperativity in the thermosome. J Mol Biol 348(1):13–26. doi:S0022-2836(05)00126-9 [pii] 10.1016/j.jmb.2005.01.066

Bigotti MG, Clarke AR (2008) Chaperonins: the hunt for the group II mechanism. Arch Biochem Biophys 474(2):331–339. doi:S0003-9861(08)00157-4 [pii] 10.1016/j.abb.2008.03.015

Booth CR, Meyer AS, Cong Y, Topf M, Sali A, Ludtke SJ, Chiu W, Frydman J (2008) Mechanism of lid closure in the eukaryotic chaperonin TRiC/CCT. Nat Struct Mol Biol 15(7):746–753

Braig K, Otwinowski Z, Hegde R, Boisvert DC, Joachimiak A, Horwich AL, Sigler PB (1994) The crystal structure of the bacterial chaperonin GroEL at 2.8 A. Nature 371(6498):578–586. doi:10.1038/371578a0

Brinker A, Pfeifer G, Kerner MJ, Naylor DJ, Hartl FU, Hayer-Hartl M (2001) Dual function of protein confinement in chaperonin-assisted protein folding. Cell 107(2):223–233. doi:S0092-8674(01)00517-7 [pii]

Burkal'tseva MV, Krylov VN, Pleteneva EA, Shaburova OV, Krylov SV, Volkart G, Sykilinda NN, Kurochkina LP, Mesianzhinov VV (2002) Phenogenetic characterization of a group of giant Phi KZ-like bacteriophages of *Pseudomonas aeruginosa*. Genetika 38(11):1470–1479

Burston SG, Ranson NA, Clarke AR (1995) The origins and consequences of asymmetry in the chaperonin reaction cycle. J Mol Biol 249(1):138–152. doi:S0022-2836(85)70285-9 [pii] 10.1006/jmbi.1995.0285

Carrascosa JL, Llorca O, Valpuesta JM (2001) Structural comparison of prokaryotic and eukaryotic chaperonins. Micron 32(1):43–50

Clark MA, Baumann L, Baumann P (1998) Sequence analysis of a 34.7-kb DNA segment from the genome of *Buchnera aphidicola* (endosymbiont of aphids) containing groEL, dnaA, the atp operon, gidA, and rho. Curr Microbiol 36(3):158–163

Craig EA, Weissman JS, Horwich AL (1994) Heat shock proteins and molecular chaperones: mediators of protein conformation and turnover in the cell. Cell 78(3):365–372. doi:0092-8674(94)90416-2 [pii]
Ditzel L, Lowe J, Stock D, Stetter KO, Huber H, Huber R, Steinbacher S (1998) Crystal structure of the thermosome, the archaeal chaperonin and homolog of CCT. Cell 93(1):125–138. doi:S0092-8674(00)81152-6 [pii]
Douglas NR, Reissmann S, Zhang J, Chen B, Jakana J, Kumar R, Chiu W, Frydman J (2011) Dual action of ATP hydrolysis couples lid closure to substrate release into the group II chaperonin chamber. Cell 144(2):240–252. doi:S0092-8674(10)01435-2 [pii] 10.1016/j.cell.2010.12.017
Ellis RJ (1994) Molecular chaperones. Opening and closing the Anfinsen cage. Curr Biol 4(7):633–635
Ellis RJ, Hartl FU (1996) Protein folding in the cell: competing models of chaperonin function. FASEB J 10(1):20–26
Eskandari F, Sylvester ES, Richardson J (1979) Evidence for lack of propagation of potato roll virus in its aphid vector, *Myzus persicae*. Phytopathology 69:45–47
Farr GW, Fenton WA, Horwich AL (2007) Perturbed ATPase activity and not "close confinement" of substrate in the cis cavity affects rates of folding by tail-multiplied GroEL. Proc Natl Acad Sci USA 104(13):5342–5347. doi:0700820104 [pii] 10.1073/pnas.0700820104
Fayet O, Ziegelhoffer T, Georgopoulos C (1989) The groES and groEL heat shock gene products of *Escherichia coli* are essential for bacterial growth at all temperatures. J Bacteriol 171(3):1379–1385
Fenton WA, Horwich AL (1997) GroEL-mediated protein folding. Protein Sci 6(4):743–760. doi:10.1002/pro.5560060401
Fenton WA, Horwich AL (2003) Chaperonin-mediated protein folding: fate of substrate polypeptide. Q Rev Biophys 36(2):229–256
Fenton WA, Kashi Y, Furtak K, Horwich AL (1994) Residues in chaperonin GroEL required for polypeptide binding and release. Nature 371(6498):614–619
Frydman J (2001) Folding of newly translated proteins in vivo: the role of molecular chaperones. Annu Rev Biochem 70:603–647
Gadad BS, Britton GB, Ks R (2011) Targeting oligomers in neurodegenerative disorders: lessons from alpha-synuclein, tau, and amyloid-beta peptide. J Alzheimers Dis. doi:RV33683375572T66 [pii] 10.3233/JAD-2011-110182
Georgopoulos CP, Hendrix RW, Casjens SR, Kaiser AD (1973) Host participation in bacteriophage lambda head assembly. J Mol Biol 76(1):45–60
Gray S, Gildow FE (2003) Luteovirus-aphid interactions. Annu Rev Phytopathol 41:539–566. doi:10.1146/annurev.phyto.41.012203.105815 012203.105815 [pii]
Gutsche I, Essen LO, Baumeister W (1999) Group II chaperonins: new TRiC(k)s and turns of a protein folding machine. J Mol Biol 293(2):295–312
Hansen JJ, Durr A, Cournu-Rebeix I, Georgopoulos C, Ang D, Nielsen MN, Davoine CS, Brice A, Fontaine B, Gregersen N, Bross P (2002) Hereditary spastic paraplegia SPG13 is associated with a mutation in the gene encoding the mitochondrial chaperonin Hsp60. Am J Hum Genet 70(5):1328–1332. doi:S0002-9297(07)62525-7 [pii] 10.1086/339935
Harrison BD (1985) Usefulness and limitations of the species concept for plant viruses. Intervirology 24(2):71–78
Hartl FU (1996) Molecular chaperones in cellular protein folding. Nature 381(6583):571–579
Hartl FU, Hayer-Hartl M (2002) Molecular chaperones in the cytosol: from nascent chain to folded protein. Science 295(5561):1852–1858
Hayer-Hartl MK, Weber F, Hartl FU (1996) Mechanism of chaperonin action: GroES binding and release can drive GroEL-mediated protein folding in the absence of ATP hydrolysis. EMBO J 15(22):6111–6121
Hendrix RW, Casjens SR (1975) Assembly of bacteriophage lambda heads: protein processing and its genetic control in petit lambda assembly. J Mol Biol 91(2):187–199
Hertveldt K, Lavigne R, Pleteneva E, Sernova N, Kurochkina L, Korchevskii R, Robben J, Mesyanzhinov V, Krylov VN, Volckaert G (2005) Genome comparison of *Pseudomonas aeruginosa* large phages. J Mol Biol 354(3): 536–545
Hogenhout SA, van der Wilk F, Verbeek M, Goldbach RW, van den Heuvel JF (1998) Potato leafroll virus binds to the equatorial domain of the aphid endosymbiotic GroEL homolog. J Virol 72(1):358–365
Horwich AL, Saibil HR (1998) The thermosome: chaperonin with a built-in lid. Nat Struct Biol 5(5):333–336
Horwich AL, Fenton WA, Chapman E, Farr GW (2007) Two families of chaperonin: physiology and mechanism. Annu Rev Cell Dev Biol 23:115–145. doi:10.1146/annurev.cellbio.23.090506.123555
Hunt JF, van der Vies SM, Henry L, Deisenhofer J (1997) Structural adaptations in the specialized bacteriophage T4 co-chaperonin Gp31 expand the size of the Anfinsen cage. Cell 90(2):361–371
Iizuka R, So S, Inobe T, Yoshida T, Zako T, Kuwajima K, Yohda M (2004) Role of the helical protrusion in the conformational change and molecular chaperone activity of the archaeal group II chaperonin. J Biol Chem 279(18):18834–18839. doi:10.1074/jbc.M400839200 M400839200 [pii]
Ishii N, Taguchi H, Sumi M, Yoshida M (1992) Structure of holo-chaperonin studied with electron microscopy. Oligomeric cpn10 on top of two layers of cpn60 rings with two stripes each. FEBS Lett 299(2):169–174. doi:0014-5793(92)80240-H [pii]

Jewett AI, Baumketner A, Shea JE (2004) Accelerated folding in the weak hydrophobic environment of a chaperonin cavity: creation of an alternate fast folding pathway. Proc Natl Acad Sci USA 101(36):13192–13197. doi:10.1073/pnas.0400720101 0400720101 [pii]

Kafri G, Willison KR, Horovitz A (2001) Nested allosteric interactions in the cytoplasmic chaperonin containing TCP-1. Protein Sci 10(2):445–449. doi:10.1110/ps.44401

Kakeda K, Ishikawa H (1991) Molecular chaperon produced by an intracellular symbiont. J Biochem 110(4):583–587

Kanno R, Koike-Takeshita A, Yokoyama K, Taguchi H, Mitsuoka K (2009) Cryo-EM structure of the native GroEL-GroES complex from *Thermus thermophilus* encapsulating substrate inside the cavity. Structure 17(2):287–293. doi:S0969-2126(09)00028-8 [pii] 10.1016/j.str.2008.12.012

Kanzaki T, Iizuka R, Takahashi K, Maki K, Masuda R, Sahlan M, Yebenes H, Valpuesta JM, Oka T, Furutani M, Ishii N, Kuwajima K, Yohda M (2008) Sequential action of ATP-dependent subunit conformational change and interaction between helical protrusions in the closure of the built-in lid of group II chaperonins. J Biol Chem 283(50):34773–34784. doi:M805303200 [pii] 10.1074/jbc.M805303200

Keppel F, Lipinska B, Ang D, Georgopoulos C (1990) Mutational analysis of the phage T4 morphogenetic 31 gene, whose product interacts with the *Escherichia coli* GroEL protein. Gene 86(1):19–25. doi:0378-1119(90)90109-5 [pii]

Keppel F, Rychner M, Georgopoulos C (2002) Bacteriophage-encoded cochaperonins can substitute for *Escherichia coli's* essential GroES protein. EMBO Rep 3(9):893–898. doi:10.1093/embo-reports/kvf176 kvf176 [pii]

Kim SK, Kim SR, Kim YH, Kwack P, Son D, Cheong GW (2002) Structural analysis of thermosome from hyperthermophilic archaeon Thermococcus. Mol Cells 14(1):85–92

Klein G, Georgopoulos C (2001) Identification of important amino acid residues that modulate binding of Escherichia coli GroEL to its various cochaperones. Genetics 158(2):507–517

Klumpp M, Baumeister W, Essen LO (1997) Structure of the substrate binding domain of the thermosome, an archaeal group II chaperonin. Cell 91(2):263–270. doi:S0092-8674(00)80408-0 [pii]

Knee KM, Goulet DR, Zhang J, Chen B, Chiu W, King JA (2011) The group II chaperonin Mm-Cpn binds and refolds human gammaD crystallin. Protein Sci 20(1):30–41. doi:10.1002/pro.531

Koike-Takeshita A, Shimamura T, Yokoyama K, Yoshida M, Taguchi H (2006) Leu309 plays a critical role in the encapsulation of substrate protein into the internal cavity of GroEL. J Biol Chem 281(2):962–967. doi:M506298200 [pii] 10.1074/jbc.M506298200

Koll H, Guiard B, Rassow J, Ostermann J, Horwich AL, Neupert W, Hartl FU (1992) Antifolding activity of hsp60 couples protein import into the mitochondrial matrix with export to the intermembrane space. Cell 68(6): 1163–1175. doi:0092-8674(92)90086-R [pii]

Koonin EV, van der Vies SM (1995) Conserved sequence motifs in bacterial and bacteriophage chaperonins. Trends Biochem Sci 20(1):14–15. doi:S0968000400889410 [pii]

Kubota S, Kubota H, Nagata K (2006) Cytosolic chaperonin protects folding intermediates of Gbeta from aggregation by recognizing hydrophobic beta-strands. Proc Natl Acad Sci USA 103(22):8360–8365. doi:0600195103 [pii] 10.1073/pnas.0600195103

Kusmierczyk AR, Martin J (2003a) Nested cooperativity and salt dependence of the ATPase activity of the archaeal chaperonin Mm-cpn. FEBS Lett 547(1–3):201–204. doi:S0014579303007221 [pii]

Kusmierczyk AR, Martin J (2003b) Nucleotide-dependent protein folding in the type II chaperonin from the mesophilic archaeon *Methanococcus maripaludis*. Biochem J 371(Pt 3):669–673

Landry SJ, Zeilstra-Ryalls J, Fayet O, Georgopoulos C, Gierasch LM (1993) Characterization of a functionally important mobile domain of GroES. Nature 364(6434):255–258

Langer T, Pfeifer G, Martin J, Baumeister W, Hartl FU (1992) Chaperonin-mediated protein folding – GroES binds to one end of the GroEL cylinder, which accommodates the protein substrate within its central cavity. EMBO J 11(13):4757–4765

Lin Z, Rye HS (2006) GroEL-mediated protein folding: making the impossible, possible. Crit Rev Biochem Mol Biol 41(4):211–239. doi:M315627H22375T33 [pii] 10.1080/10409230600760382

Lin Z, Madan D, Rye HS (2008) GroEL stimulates protein folding through forced unfolding. Nat Struct Mol Biol 15(3):303–311. doi:nsmb.1394 [pii] 10.1038/nsmb.1394

Madan D, Lin Z, Rye HS (2008) Triggering protein folding within the GroEL-GroES complex. J Biol Chem 283(46):32003–32013. doi:M802898200 [pii] 10.1074/jbc.M802898200

Martin J, Langer T, Boteva R, Schramel A, Horwich AL, Hartl FU (1991) Chaperonin-mediated protein folding at the surface of groEL through a 'molten globule'-like intermediate. Nature 352(6330):36–42. doi:10.1038/352036a0

McCormack EA, Llorca O, Carrascosa JL, Valpuesta JM, Willison KR (2001) Point mutations in a hinge linking the small and large domains of beta-actin result in trapped folding intermediates bound to cytosolic chaperonin CCT. J Struct Biol 135(2):198–204. doi:10.1006/jsbi.2001.4385 S1047-8477(01)94385-0 [pii]

Meyer AS, Gillespie JR, Walther D, Millet IS, Doniach S, Frydman J (2003) Closing the folding chamber of the eukaryotic chaperonin requires the transition state of ATP hydrolysis. Cell 113(3):369–381

Miyazaki T, Yoshimi T, Furutsu Y, Hongo K, Mizobata T, Kanemori M, Kawata Y (2002) GroEL-substrate-GroES ternary complexes are an important transient intermediate of the chaperonin cycle. J Biol Chem 277(52):50621–50628. doi:10.1074/jbc.M209183200 M209183200 [pii]

Motojima F, Chaudhry C, Fenton WA, Farr GW, Horwich AL (2004) Substrate polypeptide presents a load on the apical domains of the chaperonin GroEL. Proc Natl Acad Sci USA 101(42):15005–15012
Munoz IG, Yebenes H, Zhou M, Mesa P, Serna M, Park AY, Bragado-Nilsson E, Beloso A, de Carcer G, Malumbres M, Robinson CV, Valpuesta JM, Montoya G (2011) Crystal structure of the open conformation of the mammalian chaperonin CCT in complex with tubulin. Nat Struct Mol Biol 18(1):14–19. doi:nsmb.1971 [pii] 10.1038/nsmb.1971
Murialdo H (1979) Early intermediates in bacteriophage lambda prohead assembly. Virology 96(2):341–367
Nielsen KL, Cowan NJ (1998) A single ring is sufficient for productive chaperonin-mediated folding in vivo. Mol Cell 2(1):93–99
Nielsen KL, McLennan N, Masters M, Cowan NJ (1999) A single-ring mitochondrial chaperonin (Hsp60-Hsp10) can substitute for GroEL-GroES in vivo. J Bacteriol 181(18):5871–5875
Nitsch M, Walz J, Typke D, Klumpp M, Essen LO, Baumeister W (1998) Group II chaperonin in an open conformation examined by electron tomography. Nat Struct Biol 5(10):855–857
Ohtaka C, Nakamura H, Ishikawa H (1992) Structures of chaperonins from an intracellular symbiont and their functional expression in *Escherichia coli* groE mutants. J Bacteriol 174(6):1869–1874
Pace CN, Shirley BA, McNutt M, Gajiwala K (1996) Forces contributing to the conformational stability of proteins. FASEB J 10(1):75–83
Pappenberger G, Wilsher JA, Roe SM, Counsell DJ, Willison KR, Pearl LH (2002) Crystal structure of the CCTgamma apical domain: implications for substrate binding to the eukaryotic cytosolic chaperonin. J Mol Biol 318(5): 1367–1379. doi:S0022-2836(02)00190-0 [pii]
Pereira JH, Ralston CY, Douglas NR, Meyer D, Knee KM, Goulet DR, King JA, Frydman J, Adams PD (2010) Crystal structures of a group II chaperonin reveal the open and closed states associated with the protein folding cycle. J Biol Chem 285(36):27958–27966. doi:M110.125344 [pii] 10.1074/jbc.M110.125344
Picketts DJ, Mayanil CS, Gupta RS (1989) Molecular cloning of a Chinese hamster mitochondrial protein related to the "chaperonin" family of bacterial and plant proteins. J Biol Chem 264(20):12001–12008
Ranson NA, Farr GW, Roseman AM, Gowen B, Fenton WA, Horwich AL, Saibil HR (2001) ATP-bound states of GroEL captured by cryo-electron microscopy. Cell 107(7):869–879
Reading DS, Hallberg RL, Myers AM (1989) Characterization of the yeast HSP60 gene coding for a mitochondrial assembly factor. Nature 337(6208):655–659. doi:10.1038/337655a0
Reissmann S, Parnot C, Booth CR, Chiu W, Frydman J (2007) Essential function of the built-in lid in the allosteric regulation of eukaryotic and archaeal chaperonins. Nat Struct Mol Biol 14(5):432–440
Richardson A, Georgopoulos C (1999) Genetic analysis of the bacteriophage T4-encoded cochaperonin Gp31. Genetics 152(4):1449–1457
Richardson A, van der Vies SM, Keppel F, Taher A, Landry SJ, Georgopoulos C (1999) Compensatory changes in GroEL/Gp31 affinity as a mechanism for allele-specific genetic interaction. J Biol Chem 274(1):52–58
Richardson A, Schwager F, Landry SJ, Georgopoulos C (2001) The importance of a mobile loop in regulating chaperonin/co-chaperonin interaction: humans versus *Escherichia coli*. J Biol Chem 276(7):4981–4987. doi: 10.1074/jbc.M008628200 M008628200 [pii]
Rivenzon-Segal D, Wolf SG, Shimon L, Willison KR, Horovitz A (2005) Sequential ATP-induced allosteric transitions of the cytoplasmic chaperonin containing TCP-1 revealed by EM analysis. Nat Struct Mol Biol 12(3): 233–237. doi:nsmb901 [pii] 10.1038/nsmb901
Roseman AM, Chen S, White H, Braig K, Saibil HR (1996) The chaperonin ATPase cycle: mechanism of allosteric switching and movements of substrate-binding domains in GroEL. Cell 87(2):241–251
Roseman AM, Ranson NA, Gowen B, Fuller SD, Saibil HR (2001) Structures of unliganded and ATP-bound states of the *Escherichia coli* chaperonin GroEL by cryoelectron microscopy. J Struct Biol 135(2):115–125
Rospert S, Junne T, Glick BS, Schatz G (1993) Cloning and disruption of the gene encoding yeast mitochondrial chaperonin 10, the homolog of *E. coli* groES. FEBS Lett 335(3):358–360. doi:0014-5793(93)80419-U [pii]
Rye HS, Burston SG, Fenton WA, Beechem JM, Xu Z, Sigler PB, Horwich AL (1997) Distinct actions of cis and trans ATP within the double ring of the chaperonin GroEL. Nature 388(6644):792–798. doi:10.1038/42047
Rye HS, Roseman AM, Chen S, Furtak K, Fenton WA, Saibil HR, Horwich AL (1999) GroEL-GroES cycling: ATP and nonnative polypeptide direct alternation of folding-active rings. Cell 97(3):325–338. doi:S0092-8674(00)80742-4 [pii]
Saibil H, Dong Z, Wood S, auf der Mauer A (1991) Binding of chaperonins. Nature 353(6339):25–26. doi: 10.1038/353025b0
Saibil HR, Zheng D, Roseman AM, Hunter AS, Watson GM, Chen S, Auf Der Mauer A, O'Hara BP, Wood SP, Mann NH, Barnett LK, Ellis RJ (1993) ATP induces large quaternary rearrangements in a cage-like chaperonin structure. Curr Biol 3(5):265–273. doi:0960-9822(93)90176-O [pii]
Schoehn G, Hayes M, Cliff M, Clarke AR, Saibil HR (2000a) Domain rotations between open, closed and bullet-shaped forms of the thermosome, an archaeal chaperonin. J Mol Biol 301(2):323–332. doi:10.1006/jmbi.2000.3952 S0022-2836(00)93952-4 [pii]

Schoehn G, Quaite-Randall E, Jimenez JL, Joachimiak A, Saibil HR (2000b) Three conformations of an archaeal chaperonin, TF55 from *Sulfolobus shibatae*. J Mol Biol 296(3):813–819. doi:10.1006/jmbi.2000.3505 S0022-2836(00)93505-8 [pii]

Sigler PB, Xu Z, Rye HS, Burston SG, Fenton WA, Horwich AL (1998) Structure and function in GroEL-mediated protein folding. Annu Rev Biochem 67:581–608. doi:10.1146/annurev.biochem.67.1.581

Spiess C, Meyer AS, Reissmann S, Frydman J (2004) Mechanism of the eukaryotic chaperonin: protein folding in the chamber of secrets. Trends Cell Biol 14(11):598–604

Sternberg N (1973) Properties of a mutant of *Escherichia coli* defective in bacteriophage lambda head formation (groE). I. Initial characterization. J Mol Biol 76(1):1–23. doi:0022-2836(73)90078-8 [pii]

Thulasiraman V, Yang CF, Frydman J (1999) In vivo newly translated polypeptides are sequestered in a protected folding environment. EMBO J 18(1):85–95. doi:10.1093/emboj/18.1.85

Tilly K, Murialdo H, Georgopoulos C (1981) Identification of a second *Escherichia coli* groE gene whose product is necessary for bacteriophage morphogenesis. Proc Natl Acad Sci USA 78(3):1629–1633

Todd MJ, Viitanen PV, Lorimer GH (1994) Dynamics of the chaperonin ATPase cycle: implications for facilitated protein folding. Science 265(5172):659–666

Todd MJ, Walke S, Lorimer G, Truscott K, Scopes RK (1995) The single-ring *Thermoanaerobacter brockii* chaperonin 60 (Tbr-EL7) dimerizes to Tbr-EL14.Tbr-ES7 under protein folding conditions. Biochemistry 34(45):14932–14941

Valpuesta JM, Martin-Benito J, Gomez-Puertas P, Carrascosa JL, Willison KR (2002) Structure and function of a protein folding machine: the eukaryotic cytosolic chaperonin CCT. FEBS Lett 529(1):11–16. doi:S0014579302031800 [pii]

van den Heuvel JF, Verbeek M, van der Wilk F (1994) Endosymbiotic bacteria associated with circulative transmission of potato leafroll virus by *Myzus persicae*. J Gen Virol 75(Pt 10):2559–2565

van der Vies SM, Gatenby AA, Georgopoulos C (1994) Bacteriophage T4 encodes a co-chaperonin that can substitute for *Escherichia coli* GroES in protein folding. Nature 368(6472):654–656

Viitanen PV, Gatenby AA, Lorimer GH (1992) Purified chaperonin 60 (groEL) interacts with the nonnative states of a multitude of *Escherichia coli* proteins. Protein Sci 1(3):363–369. doi:10.1002/pro.5560010308

Viitanen PV, Lorimer G, Bergmeier W, Weiss C, Kessel M, Goloubinoff P (1998) Purification of mammalian mitochondrial chaperonin 60 through in vitro reconstitution of active oligomers. Methods Enzymol 290:203–217

Waldmann T, Nimmesgern E, Nitsch M, Peters J, Pfeifer G, Muller S, Kellermann J, Engel A, Hartl FU, Baumeister W (1995) The thermosome of *Thermoplasma acidophilum* and its relationship to the eukaryotic chaperonin TRiC. Eur J Biochem 227(3):848–856

Weissman JS, Kashi Y, Fenton WA, Horwich AL (1994) Groel-mediated protein-folding proceeds by multiple rounds of binding and release of nonnative forms. Cell 78(4):693–702

Weissman JS, Hohl CM, Kovalenko O, Kashi Y, Chen S, Braig K, Saibil HR, Fenton WA, Horwich AL (1995) Mechanism of GroEL action: productive release of polypeptide from a sequestered position under GroES. Cell 83(4):577–587. doi:0092-8674(95)90098-5 [pii]

Weissman JS, Rye HS, Fenton WA, Beechem JM, Horwich AL (1996) Characterization of the active intermediate of a GroEL-GroES-mediated protein folding reaction. Cell 84(3):481–490. doi:S0092-8674(00)81293-3 [pii]

Xu Z, Sigler PB (1998) GroEL/GroES: structure and function of a two-stroke folding machine. J Struct Biol 124 (2–3):129–141. doi:S1047-8477(98)94060-6 [pii] 10.1006/jsbi.1998.4060

Xu Z, Horwich AL, Sigler PB (1997) The crystal structure of the asymmetric GroEL-GroES-(ADP)7 chaperonin complex. Nature 388(6644):741–750

Yam AY, Xia Y, Lin HT, Burlingame A, Gerstein M, Frydman J (2008) Defining the TRiC/CCT interactome links chaperonin function to stabilization of newly made proteins with complex topologies. Nat Struct Mol Biol 15(12):1255–1262

Yifrach O, Horovitz A (1995) Nested cooperativity in the ATPase activity of the oligomeric chaperonin GroEL. Biochemistry 34(16):5303–5308

Zeilstra-Ryalls J, Fayet O, Baird L, Georgopoulos C (1993) Sequence analysis and phenotypic characterization of groEL mutations that block lambda and T4 bacteriophage growth. J Bacteriol 175(4):1134–1143

Zhang J, Baker ML, Schroder GF, Douglas NR, Reissmann S, Jakana J, Dougherty M, Fu CJ, Levitt M, Ludtke SJ, Frydman J, Chiu W (2010) Mechanism of folding chamber closure in a group II chaperonin. Nature 463(7279):379–383

Zweig M, Cummings DJ (1973) Cleavage of head and tail proteins during bacteriophage T5 assembly: selective host involvement in the cleavage of a tail protein. J Mol Biol 80(3):505–518

Chapter 14
Building the Machines: Scaffolding Protein Functions During Bacteriophage Morphogenesis

Peter E. Prevelige and Bentley A. Fane

Abstract For a machine to function, it must first be assembled. The morphogenesis of the simplest icosahedral virus would require only 60 copies of a single capsid protein to coalesce. If the capsid protein's structure could be entirely dedicated to this endeavor, the morphogenetic mechanism would be relatively uncomplicated. However, capsid proteins have had to evolve other functions, such as receptor recognition, immune system evasion, and the incorporation of other structure proteins, which can detract from efficient assembly. Moreover, evolution has mandated that viruses obtain additional proteins that allow them to adapt to their hosts or to more effectively compete in their respective niches. Consequently, genomes have increased in size, which has required capsids to do likewise. This, in turn, has lead to more complex icosahedral geometries. These challenges have driven the evolution of scaffolding proteins, which mediate, catalyze, and promote proper virus assembly. The mechanisms by which these proteins perform their functions are discussed in this review.

14.1 Introduction

The so-called spherical viral capsids have underlying icosahedral symmetry. While a true icosahedron is composed of 60 strictly equivalent subunits, viral capsids with true icosahedral or $T=1$ symmetry are relatively rare, presumably because the internal volume of $T=1$ particles is constrained. Instead, most "spherical" viruses are subtriangulated icosahedra with triangulation (T) numbers corresponding to a mathematical progression put forth by Caspar and Klug (Caspar and Klug 1962). For capsids with T numbers greater than 1, the subunits do not occupy identical positions within the lattice. This, in turn, leads to pathway-dependent self-assembly (Rossmann 1984). For many smaller viruses, all the information required for high fidelity assembly can be encoded, or self-contained, entirely in the coat protein subunits. However, larger viruses or small viruses upon which unique evolutionary constraints have been imposed frequently require additional proteins to insure robust assembly. Among those proteins are the "scaffolding" proteins, a class of auxiliary proteins that are present

P.E. Prevelige (✉)
Department of Microbiology, University of Alabama at Birmingham,
BBRB 416/6, 845 19th St. South, Birmingham, AL 35294-2170, USA
e-mail: prevelig@uab.edu

B.A. Fane
Department of Plant Sciences, The BIO5 Institute, University of Arizona,
Keating Building, 1657 E. Helen Street, Tucson, AZ 85721, USA

M.G. Rossmann and V.B. Rao (eds.), *Viral Molecular Machines*,
Advances in Experimental Medicine and Biology 726, DOI 10.1007/978-1-4614-0980-9_14,

transiently during assembly and are not part of the final structure. Although common, scaffolding proteins are not ubiquitous. Most plant viruses and small, enveloped animal viruses do not require them. In general, these coat proteins have not been constrained by the evolution of a receptor-binding domain or domains that recruit or interact with other structural proteins. As coat proteins evolved to perform additional functions, efficient assembly may have become compromised. Thus, additional adaptations may have become required. Scaffolding proteins constitute only one of these adaptations; coat proteins have evolved as well. In some systems, best represented by the parvoviruses, maximal infectivity and/or fitness most likely requires constructing capsids with two or three coat protein variants (Hernando et al. 2000). The $P=3$ picornavirus capsids represent a more complex example: capsids contain three unique coat proteins, which share similar structures and a common evolutionary origin (Hogle et al. 1985; Verlinden et al. 2000). And finally, the nodaviruses and caliciviruses coat proteins have evolved large exterior domains that may function in receptor binding and immune evasion (Chen et al. 2006; Tang et al. 2006).

Several factors may have driven the evolution of scaffolding proteins. These include increases in genome size and the need to interact with minor capsid components. As genomes enlarged, capsids enlarged as well, creating virions with larger T numbers and/or prolate morphologies. However, the coat proteins of large icosahedral viruses can still form capsids of smaller T numbers. Similarly, those found in prolate heads can form icosahedral structures. Moreover, most coat proteins will form aberrant capsid-like structures if left to their own devices. Therefore, a mechanism to ensure morphogenetic fidelity, vis-à-vis proper capsid size and shape formation, had to evolve, and this mechanism most often included scaffolding proteins. The need to rapidly assemble capsids before cell death and/or programmed cell lysis may also have contributed to their evolution. This contribution of this factor is most apparent with the microviruses, ostensibly simple viruses that accomplish an almost unimaginable fast replication cycle by employing two-scaffolding proteins.

14.2 Bacteriophage P22

14.2.1 The Assembly Pathway of Bacteriophage P22

The assembly pathway of the *Salmonella typhimurium* bacteriophage P22 is typical of the double-stranded (ds) DNA bacteriophages (Fig. 14.1). The first identifiable structural intermediate is a "procapsid" (King et al. 1973) composed of an outer shell of 415 molecules of the 47-kDa coat protein (the product of gene 5), arranged with $T=7$ symmetry. The procapsid does not contain nucleic acid. Instead, it contains a core composed of ~300 molecules of the 33-kDa scaffolding protein (encoded by gene 8). Biochemical and genetic studies demonstrated that in addition to scaffolding protein, the procapsid contains approximately 12 copies of the portal protein (the product of gene 1) and 12–20 copies of each of the pilot and ejection proteins (the products of genes 7, 16, and 20). All of these proteins are required for productive infection (Casjens and King 1974). In addition to promoting the fidelity of coat protein assembly, the results of genetic studies implicate the P22 scaffolding protein in the incorporation of these minor capsid proteins (Weigele et al. 2005). Scaffolding-dependent minor capsid protein incorporation is observed in many assembly systems, suggesting a common force driving their evolution.

One fivefold symmetrical vertex of the icosahedron is differentiated from the other 11 by the presence of a dodecameric portal protein complex. Structural studies indicate that the core of the portal protein is conserved among phages P22, Phi29, and SPP1 (Kang et al. 2008; Lander et al. 2009; Lebedev et al. 2007; Simpson et al. 2000). This conservation appears to extend even to the herpesviruses (Baker et al. 2005). DNA is packaged through this portal vertex. A terminase complex

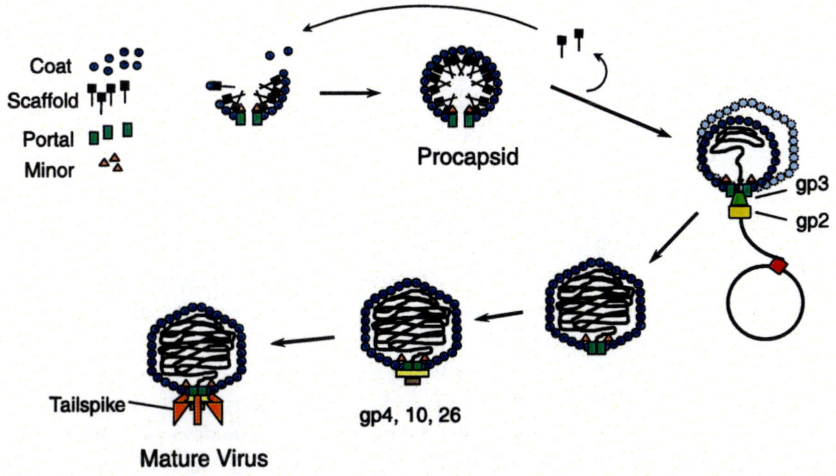

Fig. 14.1 The assembly pathway of bacteriophage P22. The initial structure formed, the procapsid, is assembled by the copolymerization of 415 molecules of the coat protein with approximately 300 molecules of the scaffolding protein. The portal and minor proteins are incorporated at this stage. The DNA is replicated as a concatemer. The terminase proteins (gp2 and gp3) bind to the DNA to deliver it to the portal vertex. The DNA is packaged by a headful mechanism in an ATP-dependent process. DNA packaging results in scaffolding protein exit. The released scaffolding protein can function in additional rounds of assembly. Packaging results in an expansion of the capsid and an increase in angularity. The portal vertex is closed by the binding of the products of gene 4, 10, and 26. Finally, attachment of the tailspike trimers renders the phage infectious

composed of multiple copies of two proteins is responsible for recognizing a "pac" sequence on the DNA, delivering the DNA to the portal vertex, and driving chemomechanical translocation through ATP hydrolysis (Rao and Feiss 2008). DNA packaging results in an approximately 10% expansion of the $T=7$ lattice, a pronounced increase in stability, and the egress of the scaffolding protein (Jiang et al. 2003; Kang and Prevelige 2005). In P22 and the *Bacillus subtilis* phage Phi29, the scaffolding protein exits intact and can be recycled in further rounds of assembly (Bjornsti et al. 1983; King and Casjens 1974). In most other dsDNA-containing bacteriophage and in herpesviruses, cleavage of the scaffolding protein by a virally encoded protease facilitates its removal.

14.2.2 The Role of the P22 Scaffolding Protein

In wild-type P22 infections, approximately 30 min postinfection, the cells burst and typically release approximately 500 newly formed virus particles. In cells infected with mutants that block DNA packaging, such as those with defective terminase proteins, the cells accumulate large numbers of procapsids within this time frame without evidence of partially formed shells (Lenk et al. 1975). In contrast, when cells are infected with mutants that prevent the synthesis of functional scaffolding protein, very few particles are produced. If infections are allowed to proceed long past the normal lysis time, aberrant particles, including small shells and spirals, accumulate (Earnshaw and King 1978). The morphology of the spirals, which constitute ~45% of the population by mass, suggests that they have an incorrect radius of curvature, which may be due to the improper positioning of the pentameric vertices. From these simple observations, two of the key functions of scaffolding protein can be discerned. Scaffolding protein promotes assembly, somehow lowering the concentration of coat protein required for polymerization, as evinced by the slower in vivo kinetics of aberrant particle formation, and ensures the fidelity of form determination.

14.2.3 Structural Studies of P22 Scaffolding Function

Biochemical and structural investigations have focused on two general phenomena: (1) the mechanism by which the P22 scaffolding protein imparts form determining information and (2) the mechanism by which it catalyzes assembly. The results of small angle X-ray diffraction (SAXS) experiments with procapsids indicated that the scaffolding protein density can be modeled as a thick shell or solid ball contained within the procapsid (Earnshaw et al. 1976). However, the results of subsequent biochemical experiments demonstrated that this core was formed during assembly and did not act as a template onto which the coat protein polymerized (Prevelige et al. 1988). Three-dimensional cryo-electron image reconstructions of P22 proheads elucidated the structure of these cores, revealing that the scaffolding protein was not packed with overall icosahedral symmetry (Thuman-Commike et al. 1996). Rather scaffolding protein interacts with the coat protein in a defined way only at the inside edge of the coat protein shell and the regions of ordered density falls off rapidly, presumably due to scaffolding flexibility. A similar situation is seen in herpesviruses where the scaffolding appears to be icosahedrally ordered only at the points of contact with the coat protein shell (Zhou et al. 1998). Thus, scaffolding proteins may impart form determining information through local interactions, for it is difficult to imagine how icosahedral information can pass through randomly oriented molecules, and indeed, a "local rules"-based model of how scaffolding protein can direct assembly has been proposed (Berger et al. 1994).

14.2.4 Kinetic Studies of P22 Scaffolding Function

Dissecting the assembly pathway demands a kinetic approach. Toward this end substantial progress has been made with a P22 in vitro assembly system in which purified coat and scaffolding protein subunits form procapsid-like particles with high yield and fidelity (Prevelige et al. 1988). The results of in vitro assembly experiments have demonstrated that (1) procapsid morphogenesis is a nucleation-limited process (Prevelige et al. 1993) and (2) that the pathway of assembly is well directed (Prevelige et al. 1993). Approximately 120 molecules of scaffolding protein are required for procapsid assembly (Prevelige et al. 1988). Scaffolding protein dimers are the dominant active form in assembly (Parker et al. 1997). However, monomers are required for completion of assembly (Tuma et al. 2008). Thus, scaffolding is required not just to nucleate assembly but throughout the assembly process. In the absence of monomeric scaffolding protein, assembly appears to become kinetically trapped leading to the production of partially formed shells. Full elongation can be achieved by the subsequent addition of monomeric scaffolding protein. Kinetic trapping can also be achieved by decreasing the ionic strength which favors the electrostatic coat/scaffolding interaction (Parent et al. 2005). In this case, completion can be achieved by increasing the salt concentration (Parent et al. 2005). Collectively, these experimental results fit nicely with the observation that the scaffolding protein is a weak monomer–dimer–tetramer association system (Parker et al. 1997) and suggest that the proper balance between nucleation and growth is maintained through the distribution of scaffolding oligomers.

14.2.5 The Structure of the P22 Scaffolding Protein

Although attempts to crystallize the P22 scaffolding protein have not been fruitful, other biophysical characterizations have provided some structural insights. The results of hydrodynamic studies indicate it is an extended, rod-like molecule (Fuller and King 1980), and the results of circular dichroism

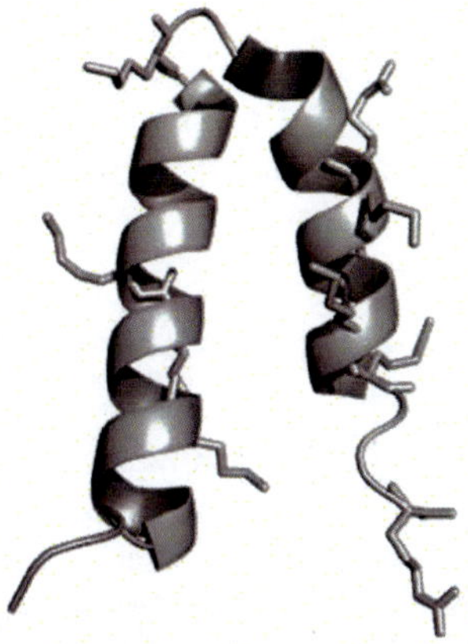

Fig. 14.2 The structure of the coat protein-binding domain of the P22 scaffolding protein. The NMR structure of the C-terminal coat protein-binding domain of P22 scaffolding protein (residues 264–303) is shown as a *ribbon diagram* (PDB 1gp8). The lysine and arginine side chains are shown in stick representation

studies indicate that it is highly α-helical (Tuma et al. 1998). Analytical ultracentrifuge analyses under assembly conditions have demonstrated that it is a monomer–dimer–tetramer self-associating system in solution (Parker et al. 1997). The Phi29 and SPP1 internal scaffolding proteins also dimerize (Morais et al. 2003; Poh et al. 2008) as do the øX174 and P4 external scaffolding proteins (Marvik et al. 1995; Morais et al. 2004). Evidence for a physiological role for oligomerization comes from in vitro assembly experiments. Kinetic analysis of the rate of assembly as a function of scaffolding protein concentration suggested that assembly displayed a kinetic order of between 1.7 and 3.5 with regard to scaffolding protein (Prevelige et al. 1993) and the isolation of a naturally occurring mutant (R74C/L177I) that spontaneously formed disulfide cross-linked dimers allowed for the direct testing of dimers in assembly reactions (Parker et al. 1997). The rate of assembly with covalently dimeric scaffolding protein was significantly faster than with an equivalent concentration of wild-type scaffolding protein suggesting an active role for dimers in assembly. While the regions of the scaffolding protein responsible for oligomerization remain undefined, there is some mutational data that bears on this question. The fact that the R74C mutation leads to the spontaneous formation of disulfide cross-linked dimers suggests that the dimers are likely symmetrical with amino acid 74 located near the dimer interface. A scaffolding protein mutant in which the N-terminal 140 amino acids were deleted was capable of promoting assembly. Analytical ultracentrifugation experiments indicated that this protein was capable of dimerization but not tetramerization, suggesting that tetramerization may not be critical during assembly (Tuma et al. 1998; Weigele et al. 2005).

Protein dissection experiments indicated that the C-terminus of P22 scaffolding protein was responsible for coat protein binding (Tuma et al. 1998). An NMR structure of the C-terminal peptide revealed that a relatively flexible region spanning 27 amino acids between residues 240–267 was followed by a well-defined helix-loop-helix motif spanning residues 268–303 (Fig. 14.2) (Sun et al. 2000). The two helices (helix I – residues 268–283; helix II – residues 289–303) were amphipathic in nature. Hydrophobic residues on each helix are packed together to form a hydrophobic core. The connecting loop consists of five residues (284–288) of which the first four form a type I β-turn. The fifth residue extends the loop allowing the necessary freedom for the helices to associate along their lengths. One striking feature of the structure is the high density of charged residues on the outside of the coat protein-binding domain. Five basic residues (R293, K294, K296, K298, and K300) are located on one side of the outer face of the 12-residue C-terminal helix. The highly charged nature of this surface may form the basis of the observed salt sensitivity of the scaffolding/coat protein interaction.

Directly measuring the strength of the interaction between coat and scaffolding is challenging because rapid assembly complicates the analysis. However, the scaffolding protein can be reversibly removed from and reenter the procapsid. Thus, the binding strength can be measured. Isothermal titration calorimetry provided direct evidence for two classes of scaffolding protein within the procapsid, a tightly bound class and a weakly bound class (Parker et al. 2001). These two classes roughly correlated with the two rates of scaffolding reentry seen in turbidity-based kinetic assays. The apparent Kd for binding to the high affinity sites was 100–300 nM and was almost entirely enthalpy driven between 10 and 37°C. A more negative than expected ΔCp suggests a conformational change upon binding or that bridging water molecules may be involved.

14.2.6 Auxiliary Functions of P22 Scaffolding Protein

Evidence that the P22 scaffolding protein may facilitate the incorporation of the pilot and portal proteins was first provided by the results of early biochemical characterizations of mutant infections. Procapsid-like particles assembled in the absence of scaffolding protein contained neither of these minor capsid proteins (Earnshaw and King 1978). The isolation of a temperature-sensitive scaffolding protein (S242F), which could drive the assembly of procapsids lacking both portal and pilot proteins, and the isolation of a Y214W mutant, which incorporated pilot proteins but not the portal protein, suggested the central region of the scaffolding protein was involved in these functions. Further support for the role of the central region in minor protein incorporation came from studies utilizing scaffolding protein fragments. Fragments missing the N-terminal 228 amino acids can incorporate portal protein, but the deletion of nine additional residues ablates incorporation (Weigele et al. 2005). Deletion analyses have also defined portal recruitment domains in the HSV-1 (Singer et al. 2005) and Phi29 (Fu et al. 2010) scaffolding proteins suggesting that this may be a general function of scaffolding proteins. Although these findings suggest that the P22 scaffolding protein directly interacts with the portal and minor proteins, no such interactions have been demonstrated biochemically.

14.2.7 The Mechanism of Scaffolding-Assisted Assembly

Any mechanistic model for scaffolding assisted assembly has to accommodate both its role in promoting assembly and ensuring proper form determination. In the assembly of the multishelled bluetongue virus, it has been suggested that the inner core forms a template upon which the outer shell polymerizes (Grimes et al. 1998). While such a mechanism is attractive, it most likely does not operate in the P22 system: no preformed cores of scaffolding protein could be identified under assembly conditions. Initiation of assembly is a critical control point, and it is possible that scaffolding protein might play a role in proper initiation, but it may not be required for further polymerization. However, in vitro experiments with limiting scaffolding protein demonstrated that a minimum of approximately 120 scaffolding proteins is required for assembly, suggesting that it is required throughout the assembly reaction.

Given that scaffolding protein is required for continued polymerization, there are two mechanisms by which it might promote assembly. Scaffolding binding might switch the coat protein subunit from an inactive, unassociable state to one competent for assembly. Alternatively, scaffolding protein might act as an entropy sink. In this model, each of the two coat-binding domains of a scaffolding protein dimer binds one coat protein molecule resulting in the formation of a heterotetramer (two each of scaffolding and coat subunits). Consequently, the effective concentration of coat protein is greatly increased. Thus, the binding energy for scaffolding dimerization is being used to increase the effective concentration of the coat protein. Of course, both mechanisms could be operative at the same time.

The fact that truncated scaffolding protein molecules such as the C-terminal 38 amino acid helix-loop-helix can activate the coat protein for assembly (Tuma et al. 2008) yet do not dimerize suggests that scaffolding binding can activate the coat protein for assembly. However, the breakdown of fidelity observed, with predominately aberrant particles being formed, suggests that the information required for form determination is not locally transmitted. Based upon a series of N-terminal deletion mutants (WT, Δ1-140, Δ1-237), which show a decreased ability to dimerize, fidelity seems to correlate with dimerization potential though the molecular details of this phenomenon remain obscure (Parker et al. 1998).

14.2.8 Scaffolding Exit from Procapsids

Although scaffolding-containing procapsids are stable structures in which the scaffolding protein can remain bound for prolonged periods of time, they are programmed to release the scaffolding protein intact during DNA packaging. Exit presumably occurs through ~25 Å holes present at the center of the hexameric capsomeres. Whether it is necessary to package some fraction of the DNA to drive egress – as is the case for bacteriophages T4 (Jardine et al. 1998), lambda (Hohn 1983), T3 (Shibata et al. 1987a, b), and T7 (Masker and Serwer 1982) – or whether only DNA/terminase complex docking at the portal vertex is required is unknown. However, circumstantial evidence suggests the former (Poteete et al. 1979). In the case of bacteriophage lambda, there appear to be specific DNA interaction sites within the capsid. Although similar sites have not been identified in P22, their existence seems likely. Parker and Prevelige proposed that scaffolding protein was bound through charge/charge interactions. In this model, the positive charges on the scaffolding proteins' basic C-terminal domain would interact with negatively charged patches on the coat protein subunits (Parker and Prevelige 1998). During DNA packaging, the negative charged DNA would drive the charged patches away from each other, thus promoting capsid expansion and destroying the scaffolding protein binding site. Based on high-resolution structural data, a similar model has recently been proposed for bacteriophage HK97 (Conway et al. 2001).

The NMR structure of the P22 scaffolding protein represented the unbound state, and it is also possible that a conformational change in the scaffolding protein itself contributes to its release. In accord with the suggestion of an active role for scaffolding protein in release, mutational studies have identified regions involved in this function. A deletion of the N-terminal 58 amino acids allows for the production of infectious virions; however, a deletion of an additional five residues halts assembly at the procapsid step. This suggests that residues 58–63 are involved in scaffolding exit. Characterization of the R74C mutant, which confers a temperature-sensitive phenotype, provides further support for an active role in scaffolding egress. At nonpermissive temperatures, the mutant protein halts assembly at the procapsid stage due to the inability of the scaffolding protein to exit. Mutations in herpesvirus scaffolding protein which block scaffolding exit also prevent DNA packaging and maturation (Newcomb et al. 2000).

14.3 The Bacteriophage Phi29 System

The Phi29 system is analogous to the P22 system: its overall life cycle involves prohead assembly, DNA packaging, and scaffolding exit without proteolysis. However, there are two significant differences: (1) five copies of a small RNA molecule (pRNA) must associate with the portal protein for DNA packaging (Guo et al. 1987) and (2) the capsid is elongated or prolate rather than icosahedral procapsid. In the case of Phi29, crystallography has been fruitful, and crystal structures of an assembly naive and released (postassembly) form of the scaffolding protein have been obtained (Morais et al. 2003). The 11-KDa predominately α-helical scaffolding protein forms parallel dimers. The N terminus is folded as a helix-loop-helix, strikingly similar to that seen at the C terminus

Fig. 14.3 The crystal structure of the Phi29 scaffolding protein dimer. The crystal structure of the preassembly form of the Phi29 scaffolding protein dimer (1NO4) is shown as a *ribbon diagram*. Note the structural similarity of the N terminus of the Phi29 scaffolding protein to the C terminus of the P22 scaffolding protein. The last 15 residues of the Phi29 protein are not seen in the crystal structure

of the P22 scaffolding protein (Fig. 14.3). The helix-loop-helix motifs interact to form an arrowhead-like structure. The shaft of the arrow is formed by a coiled-coil region, and the C-terminal 15 residues are disordered. Despite the structural similarity of the N-terminal region of Phi29 to the C-terminal coat protein binding region of P22, protein dissection experiments indicate that it is the unstructured C terminus of the scaffolding protein that interacts with the coat protein (Fu et al. 2007). Three-dimensional reconstructions of electron micrographs of the procapsid suggest that the scaffolding protein is arranged as a series of four concentric shells that have different symmetries when viewed down the long axis of the prolate icosahedron (Morais et al. 2003). Merging the EM density with the crystal structure suggests that scaffolding protein in the outermost shell selectively interacts with the equatorial coat protein hexamers, which are responsible for the prolate morphology of the procapsid. Evidence for multiple classes of scaffolding protein interactions within the procapsid comes from the observation that the hydrogen/deuterium exchange kinetics are perturbed for some but not all scaffolding protein molecules (Fu and Prevelige 2006). The crystal structure of the released form of the scaffolding protein is very similar to that of the bound form except for a change in the orientation of the last helix relative to the arrowhead (Morais et al. 2003). It is possible that this reorientation of the tail is coupled to scaffolding release.

An in vitro assembly system has also been developed for Phi29. Purified coat and scaffolding subunits can assemble into procapsids (Fu et al. 2007). Strikingly, when portal is included in the reactions, it is also incorporated (Fu and Prevelige 2009). Protein dissection, chemical cross-linking, and mutagenesis experiments have demonstrated that the region of scaffolding protein between amino acids 70–74, which resides adjacent to the unstructured coat-binding region (amino acids 75–98), is responsible for interacting with the portal protein (Fu et al. 2010). Noncovalent mass spectrometry experiments indicate that the scaffolding protein binds to the portal as a dimer and that a maximum of six dimers binds noncooperatively with a Kd of approximately 50 mM (Fu et al. 2010). This complex appears to be capable of nucleating assembly, suggesting a mechanism by which a single portal vertex can be defined during assembly (Fu and Prevelige 2009).

14.4 Scaffolding/Coat Protein Fusions

In contrast to the phages discussed above, some phages encode a scaffolding function located at the N terminus of the coat protein. This domain, known as a delta domain, is required for procapsid assembly but is cleaved by a virus-encoded protease during the transformation to the mature, DNA-containing phage.

14.4.1 HK79

HK97 is a double-stranded DNA-containing bacteriophage whose host is *Escherichia coli* and related bacteria. The structure of the phage capsid was first determined for a tailed bacteriophage (Wikoff et al. 2000), and the subunit fold, which has become known as the HK97 fold, has been found repeatedly in a wide variety of viruses suggesting an evolutionary relationship (Bamford et al. 2005). A remarkable feature of the HK97 capsid is the staged formation of stabilizing isopeptide bonds that covalently cross-link the coat protein subunits into protein chainmail (Wikoff et al. 2000). Similar to P22, the procapsid and capsid are $T=7$ levo structures composed of 415 coat protein subunits and a dodecamer of portal protein at one of the vertices. HK97 does not have a distinct scaffolding protein; instead a 103-amino-acid residue delta domain, located at the N-terminus of the coat protein, fulfills this function (Hendrix and Duda 1998). The delta domain is proteolytically removed following procapsid assembly but prior to DNA packaging. The delta domain itself is highly α-helical in solution. In contrast to many other scaffolding proteins, it does not have a tendency to oligomerize (Nemecek et al. 2009). In the procapsid, the delta domains form inward-projecting clusters, a single cluster can be visualized under the pentamers, while a doublet can be seen under the hexamers (Conway et al. 2007).

14.4.2 T5

The *Siphoviridae* coliphage T5 is distinguished by the large size of its genome and $T=13$ capsid. Like HK97 the genome does not encode a discrete scaffolding protein but rather a delta domain of 159 residues is fused to the N terminus of the coat protein. The delta domain is predicted to be highly α-helical and predominately coiled coil (Effantin et al. 2006). The intact fusion protein, known as pb8p, can assemble in vitro into $T=13$ particles (Huet et al. 2010). Similar to the case with HK97, these particles display inward projecting bundles whose length corresponds to the predicted length of the delta domain. The delta domain occupies most of the internal volume of the procapsid and may act in that manner to fix the size during assembly (Huet et al. 2010).

14.5 The *Microviridae*

14.5.1 The Evolutionary Context of a Dual Scaffolding Protein System

Dual scaffolding protein systems with internal and external species have evolved in both satellite and nonsatellite assembly systems. In satellite systems, typified by bacteriophages P2 and P4 (see below), the external scaffolding protein plays a parasitic role. Encoded by the satellite virus, it directs helper virus's coat and internal scaffolding proteins into smaller T number capsids, which can only accommodate the smaller satellite virus genome. However, the external scaffolding does not confer new properties to the helper virus coat protein. Most coat proteins that form capsids larger than $T=3$ have retained the ability to form capsids of smaller T numbers (Earnshaw and King 1978; Katsura and Kobayashi 1990; Kellenberger 1990; Newcomb et al. 2001; Stonehouse and Stockley 1993). Thus, the external scaffolding protein alters the thermodynamic barriers that govern the probability of two possible events.

The $T=1$ microviruses, which comprise a subfamily of the *Microviridae*, are nonsatellite viruses that evolved a dual scaffolding protein system. As discussed in the introduction, the necessity of recruiting minor structural proteins to growing capsids may have helped drive the evolution of scaffolding proteins. While this may explain the need for the internal scaffolding protein, which facilitates the incorporation of the minor vertex or DNA pilot protein H, an examination of the niche microviruses occupy is required to understand the evolution of the external scaffolding protein. Microviruses comprise only a small fraction, approximately 3%, of the phages infecting free-living bacteria. They must compete with a vast pool of large dsDNA viruses. The results of evolutionary analyses indicate that the external scaffolding gene, or its most current incarnation, is the most recently acquired microvirus gene (Rokyta et al. 2006). As discussed in greater detail below, the two-scaffolding proteins facilitate a very rapid life cycle (Chen et al. 2007). New progeny appears 5 min postinfection, a time when many dsDNA phages are still in middle gene expression. In contrast, members of the other *Microviridae* subfamily, the one-scaffolding protein gokushoviruses (Clarke et al. 2004), infect obligate intracellular parasitic bacteria. The only known viruses of Chlamydia, these coelacanths take 48 h to replicate (Salim et al. 2008), an affordable luxury when there is no competition.

14.5.2 The Early Stages of øX174 Morphogenesis: Internal Scaffolding Protein-Mediated Reactions

The internal scaffolding protein B mediates the early stages of øX174 morphogenesis. It performs functions reminiscent of other scaffolding proteins, such as ensuring fidelity and minor vertex protein recruitment. As observed in other systems (Desai et al. 1994; Matusick-Kumar et al. 1994, 1995), internal scaffolding–coat protein interactions prevent the aggregation of viral coat proteins into aberrant structures (Siden and Hayashi 1974; Tonegawa and Hayashi 1970). Thus, it confers fidelity to a thermodynamically favorable reaction: the inherent ability of coat proteins to self-associate into curve-surfaced structures.

After coat protein pentamer (9S) formation (Fig. 14.4), five copies of protein B bind to the pentamer's underside and facilitate the incorporation of one copy of the DNA pilot protein H, producing the 9S* particle (Cherwa et al. 2008). As evinced by cryo-EM and crystal models in three rather diverse procapsids, HSV, P22, and øX174, interactions with coat proteins are mostly mediated by the C-termini, which tend to be very ordered and/or exhibit local icosahedral symmetry (Dokland et al. 1997, 1999; Thuman-Commike et al. 1999; Trus et al. 1996; Zhou et al. 1998). Scaffolding protein densities become more diffuse toward the N-termini, suggesting that N-terminal-mediated interactions are much more flexible and variable. Genetic and biochemical data also underscores the critical nature of the C-termini interactions and the flexible nature of the N-termini. Cross-functional analyses have been conducted with scaffolding proteins in both the herpesviruses (BHV-1 and HSV-1) and microviruses (øX174, G4, and α3), and scaffolding proteins can cross-function despite sequence identities as low as 30% (Burch et al. 1999; Haanes et al. 1995). The activity of chimeric scaffolding proteins has also been characterized in both families. For the microviruses, optimal fitness is observed when the C-terminus of the scaffolding protein is of the same origin as the viral coat protein (Burch

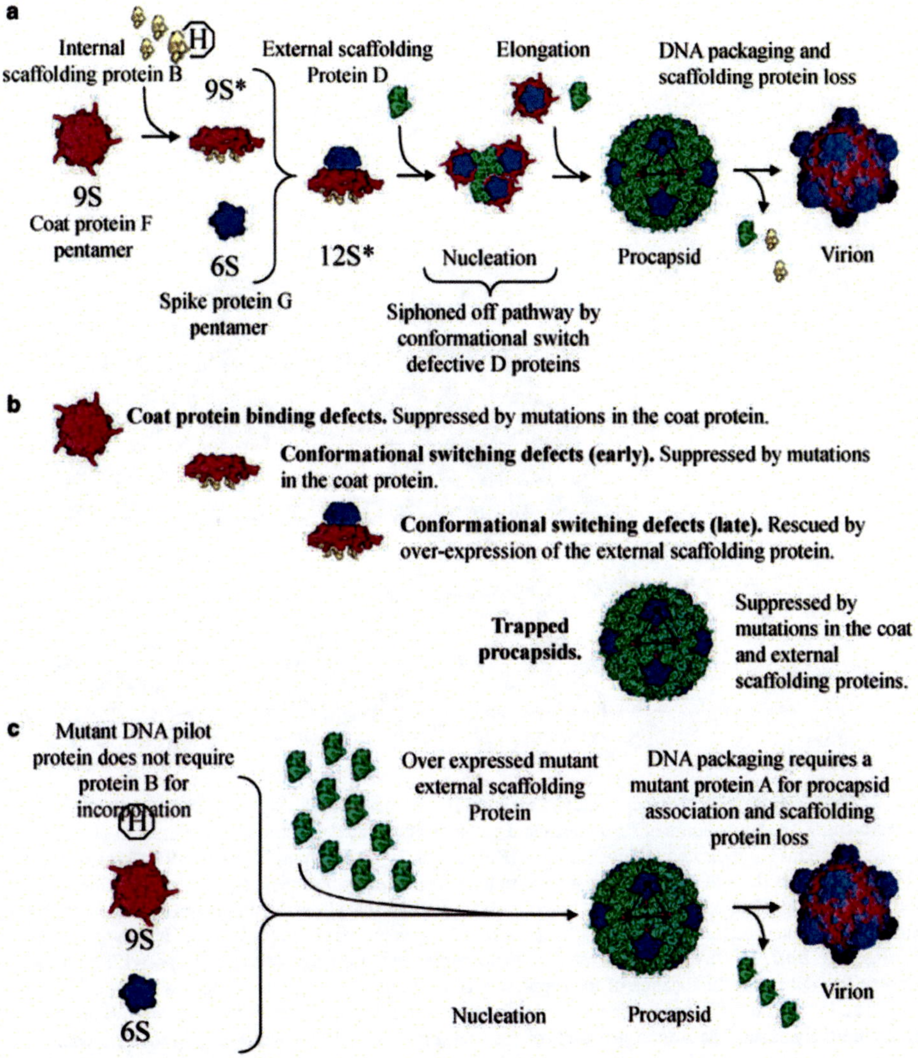

Fig. 14.4 Wild-type, mutant, and B-free øX174 assembly. (**a**) Wild-type assembly pathway. The first identifiable assembly intermediates are pentamers of the viral coat F and major spike G proteins: the respective 9S and 6S particles, which form independently of both scaffolding proteins (Tonegawa and Hayashi 1970). Five internal scaffolding B proteins bind to the underside of the 9S particle, yielding the 9S* intermediate (Cherwa et al. 2008). As *am*(*G*) infections yield particles containing one copy of the DNA pilot protein associated with five copies each of the major capsids F and internal scaffolding B proteins, the DNA pilot protein H may be incorporated at this stage. Protein F–B interactions induce conformational changes on the top of the particle that allows 9S*-6S particle associations, forming the 12S* intermediate (Siden and Hayashi 1974). Twenty external scaffolding D proteins, most likely in the form of five tetramers or ten asymmetric dimers (Morais et al. 2004), interact with the 12S* particle. D–D contacts mediate the construction of the procapsid (108S), an immature virus particle, presumably by allowing pentamers to interact across twofold axes of symmetry (Bernal et al. 2003; Dokland et al. 1997, 1999). (**b**) Particles isolated in cells infected with various B^- mutants. In cells infected with mutants that either do not produce B protein or produce a B protein that cannot bind to coat protein pentamers, 9S* intermediates accumulate. Two phenotypically different classes of B^- conformational switch mutants have been characterized: mutants of one class arrest assembly after the formation of the 9S* particle (early), mutants in the other class after the formation of the 12S* particle. Finally, one mutation has been isolated that produces defective procapsids that cannot be filled. (**c**) The B-free assembly pathway. A sextuple mutant strain of øX174 that can produce particles in the absence of protein B has been isolated (Chen et al. 2007). The 12S* intermediate is circumvented. Procapsid assembly requires the overexpression of a mutant external scaffolding protein. Mutations in protein A and H are also required to produce an infectious virion

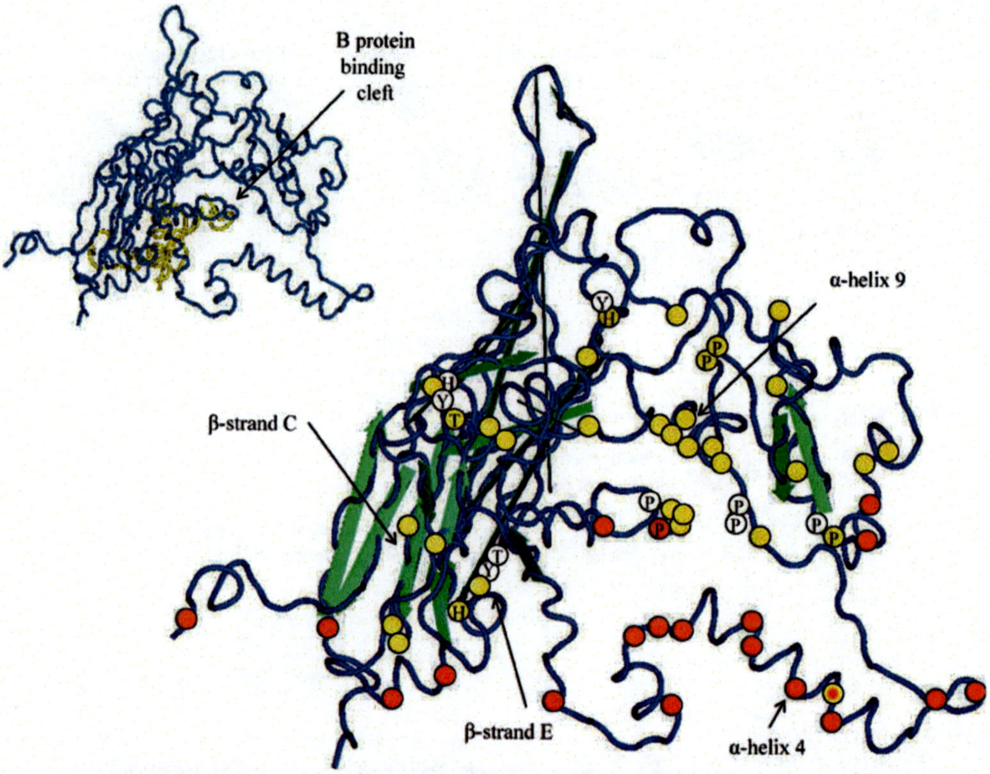

Fig. 14.5 The location of suppressors of mutant external D and internal scaffolding B proteins within the coat protein atomic structure. The suppressors of B^- mutants and mutations that allow the productive utilization of foreign internal scaffolding proteins are indicated with *yellow dots* (Burch et al. 1999; Fane and Hayashi 1991; Gordon, Knuff, Chen, and Fane, unpublished results). All of the parental mutants confer early morphogenetic defects, before the formation of the 12S* particle. The suppressors of D^- mutants and mutations that allow the productive utilization of N-terminal chimeric and deletion scaffolding proteins are indicated with *orange dots* (Fane et al. 1993; Uchiyama and Fane 2005; Uchiyama et al. 2007, 2009). All of the parental mutants confer late morphogenetic defects, post-12S* formation. The *orange–yellow dot* marks the location of the suppressor of a unique B^- protein that produces kinetically trapped procapsids. The location of adjacent proline residues and Thr-X-Tyr-X-X-His motifs is depicted with the letters P and TYH, respectively. Inset: The atomic structures of the coat (*blue*) and internal scaffolding (*yellow*) proteins

and Fane 2000b). Similar results were obtained with chimeric VZV–HSV-1 and CMV–HSV-1 proteins (Oien et al. 1997; Preston et al. 1997). And lastly, procapsid-like structures can be assembled with only C-terminal fragments of the øX174, P22, and P2 scaffolding proteins (Chang et al. 2009; Chen et al. 2007; Tuma et al. 1998; Weigele et al. 2005).

In the absence of the internal scaffolding B protein, coat protein pentamers do not interact with spike protein pentamers (6S particles). Thus, B protein binding to the underside of a 9S particle induces a conformational switch, which alters the pentamer's upper surface. The results of several genetic analyses have identified distinct regions of the coat protein that may participate in these switches. The first analyses, which were conducted before the atomic structure of the procapsid was solved, made use of extant N-terminal missense mutations. These substitutions conferred only weak *cs* (cold sensitive) phenotypes. Their second-site suppressors were located on the coat protein's outer surface and defined five distinct regions, which were characterized by two different primary structure motifs (Fane and Hayashi 1991). One of the motifs was characterized by a Thr-X-Tyr-X-X-His sequence, the other by adjacent proline residues (Fig. 14.5). These sequences may play a key role, perhaps as hinges, in mediating pentamer conformational switches. Mutations in the adjacent proline residues were also recovered as suppressors that allowed bacteriophage G4 to productively utilize the øX174 scaffolding (Burch et al. 1999) and the øX174 major spike protein (Li and Fane, unpublished results). This latter finding indicates that the ability of coat and spike protein pentamers to associate across species lines and considerable sequence variation during morphogenesis is a function of coat–scaffolding protein interactions.

The inability to isolate missense mutations with nonwild-type phenotypes has hindered investigations of early øX174 morphogenesis. The scarcity of nonwild-type phenotypes most likely reflects the proteins flexible nature, as evinced by the ability of related scaffolding protein to cross-function. "Foreign" scaffolding proteins can be regarded as "multiple mutants." Having diverged as much as 70% on the amino acid level, the dearth of single amino acid substitutions that obliterate function becomes apparent. However, the atomic structure revealed that the vast majority of coat–scaffolding interactions were mediated by just six aromatic amino acids, tyrosines and phenylalanines, in the C terminus of the protein (Dokland et al. 1997, 1999). Using the atomic structure as a guide, a more targeted genetic analysis has begun (Gordon, Knuff, and Fane, unpublished results). Amber mutations were placed at all these sites and four additional sites that do not encode aromatic amino acids. Mutants are then plated on *sup*$^+$ hosts that insert serine, glutamine, or tyrosine during protein synthesis. Thus, the ability to form plaque formation on the various hosts can serve as an indicator of the properties required at a specific site in the protein.

Substitutions of nonaromatic amino acids at the aromatic sites are not tolerated, conferring absolute lethal phenotypes. However, phenotypes differ on the molecular level and affect three distinct stages of early assembly, 9S→9S* transition, 9S*→12S* transition, and the 12S*→procapsid transition, suggesting defects in coat protein binding or complex stability. The latter two stages can be regarded as defects in conformational switching. The one mutant that is unable to support the 12S*→procapsid transition is rescued by the exogenous overexpression of the external scaffolding protein, indicating that a higher critical concentration of the external scaffolding protein is required to nucleate procapsids assembly. Second-site suppressors have been isolated for the remaining five of the "missense-suppressed" amber mutations. These suppressors define two hot spots in the coat protein atomic structure (Fig. 14.5), one residing in or adjacent to β-strands C and E and the second throughout α-helix #9, which is adjacent to the B protein binding cleft. In general, most substitutions are tolerated at nonaromatic amino acids. However, the introduction of a tyrosine residue at one of them leads the assembly of kinetically trapped procapsids, which cannot be filled. Unlike the suppressors of other B-mutants, which map to the upper half of the coat protein (Fig. 14.5), its suppressors map to a region of the coat protein rich in mutant external scaffolding protein suppressors (orange–yellow dot in Fig. 14.5).

14.5.3 Functional Redundancy in a Two-Scaffolding Protein System

A more evolutionary approach has been used to elucidate the functions of the internal scaffolding protein. Several lines of evidence suggest that the internal scaffolding protein may be less critical than the external scaffolding protein during morphogenesis. Unlike the internal scaffolding protein, the external scaffolding proteins are highly conserved between microvirus species, more highly ordered in atomic structures, and extremely sensitive to mutation (Dokland et al. 1997, 1999) (Burch and Fane 2000a, 2003; Cherwa et al. 2008). In the two-scaffolding protein P2/P4 satellite system, P4 procapsid-like structures can be generated in vitro from purified external scaffolding and coat proteins (Wang et al. 2000), circumventing the requirement for the internal scaffolding protein, which is required in vivo (Christie and Calendar 1990; Lengyel et al. 1973). However, morphogenesis is not efficient and the reaction is strongly dependent on PEG (Wang et al. 2000), which would effectively raise reactant concentrations via molecular crowding.

After the acquirement of the external scaffolding protein, which appears to be the newest of the microvirus genes (Rokyta et al. 2006), the internal scaffolding protein may have evolved into an "efficiency protein," aiding morphogenetic processes but not strictly required for any one reaction. While its gene was retained in microvirus genomes in an overlapping reading frame (Rokyta et al. 2006), it no longer had a genomic region solely dedicated to its coding as is seen in the gokushovirus genomes (Brentlinger et al. 2002; Liu et al. 2000). Thus, an inherent plasticity would allow other proteins to compensate for reduced or absent B protein function. To test this hypothesis, sequential-targeted selections were performed to reduce the requirement of B protein until it could be completely circumvented (Chen et al. 2007; Novak and Fane 2004). A series of N-terminal deletion B proteins was generated from cloned 5′ deletion genes and used to complement a *nullB* mutant (a nonsense mutation in codon 3). Since the C-terminus is known to mediate most of the known contacts with the viral coat protein (Dokland et al. 1997, 1999), the N-terminus was progressively removed. Proteins with progressively large N-terminal deletions were assayed for complementation on the level of plaque formation, which indicates that the strain could complete the viral life cyle. As the deletions increased in length, complementation became less efficient until it ceased. The deletion protein that failed to complement was then used to select a mutant strain capable of utilizing it. This utilizer mutant was then used for complementation assays with more truncated B proteins, and so forth. Thus, strains have acquired mutations in a sequential manner.

Internal scaffolding proteins lacking the first 53 (Δ53B) of its 120 amino acids can support morphogenesis. The ability to tolerate such a large deletion in a relatively small protein is further evidence that N-terminal functions are moderately expendable and mediate variable interactions with the viral coat protein. However, assembly is very cold sensitive and, most particles produced at permissive temperatures display a lower specific infectivity than wild-type particles due to the inefficient incorporation of the DNA pilot protein H. The recruitment of minor vertex proteins may be a general property of internal scaffolding proteins as similar observations have been made in the P22 system (Weigele et al. 2005). At low temperatures, no procapsids are produced. Substitutions in the external scaffolding protein D suppress the cold-sensitive assembly defect, but do not appear to increase the efficiency of DNA pilot protein H incorporation, which can be overcome by mutations in gene H. The procapsids produced by the Δ58B protein are altered and unable to be filled due to the inability to interact with the DNA-packaging complex. A substitution in the DNA-packaging protein A alleviates this defect. The overexpression of the mutant external scaffolding protein, via a promoter-up mutation, allows procapsids to be assembled with a Δ67B protein. The addition of a coat protein mutation that strengthens coat–external scaffolding proteins is both necessary and sufficient to allow assembly with a Δ97B protein. And finally, the creation of another promoter-up mutation allows plaque formation in the complete absence of protein B, the B-free phenotype. Due to the promoter mutations, viral protein expressions differ greatly in B-free infected cells from that typically observed in wild-type infections. The external scaffolding protein is vastly overproduced relative to the other four structural proteins, which appear to be downregulated. Thus, the ratio between interacting components is dramatically altered to drive procapsid assembly in the absence of the B protein. A summary of B-independent morphogenesis is given in Fig. 14.4.

The B-free phenotype was defined as the ability to form plaques, not the ability to produce progeny, which must occur before programmed cell lysis. It is possible to uncouple plaque formation from virion production by examining progeny production as a function of time in lysis-resistant cells. As can be seen in Fig. 14.6, only the first four of the six mutations found in the B-free mutant are required for progeny production. However, the lag phase before progeny production in vivo was 50 min, long after programmed cell lysis would occur in a lysis-sensitive cell. With the addition of the subsequent mutations, which included more promoter mutations, lag phases became progressively shorter. Thus, one of the primary functions of the internal scaffolding protein is to lower the critical concentration of the external scaffolding protein required to nucleate procapsid morphogenesis. A separate question involves the number of mutations that is required to form an assembled particle, regardless of

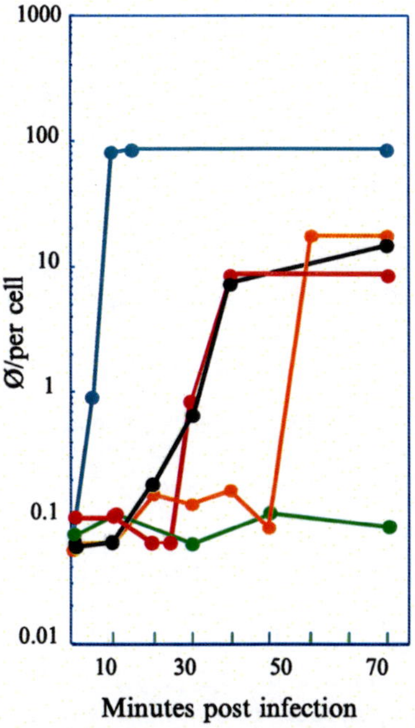

Genotypes and phenotypes of strains

Strain	Mutations1	Plaque Assay Viability	Time of Virion Production
Wild-type		YES	5 min p.i.
NullB Strains:			
Generation 2	A, H, D	$<10^{-6}$	None
Generation 3	A, H, D, P_{D1}	$<10^{-6}$	50 min p.i.
Generation 4	A, H, D, P_{D1}, F	$<10^{-6}$	25 min p.i.
Generation 5	A, H, D, P_{D1}, F, P_{D2}	YES	15 min p.i.

A: DNA packaging and replication protein
H: minor spike protein
D: external scaffolding protein
P_{D1} & P_{D2}: external scaffolding gene promoter
F: capsid protein

Fig. 14.6 Growth kinetics of B-free lineage strains in lysis-resistant cells. The strains used in these experiments were generated in targeted selections designed to produce a multiple mutant that does not require the internal scaffolding protein. Phenotypes and genotypes are given in the figure. Only the generation 5 mutant is viable in plaque assays with lysis-sensitive cells. However, the generation 3 and 4 mutants can produce viable progeny in lysis-resistant cells

infectivity. This may be as low as two. Two of the four mutations in the progeny producing the generation 3 mutant (Fig. 14.6) do not directly involve procapsid proteins. The mutation in protein A is not required to form procapsids, but to package the genome into an altered one. The other mutation is in the D gene promoter. Thus, it is likely that B-independent procapsid assembly only requires a mutation in the DNA pilot protein and the overexpression of a mutant external scaffolding protein.

14.5.4 The External Scaffolding Protein

14.5.4.1 Structure of the Assembled and Unassembled Protein

After 12S* particle formation, 240 copies of the external scaffolding protein D organize 12 12S* particles into a procapsid. The 240 D proteins form an exterior lattice, which bears no resemblance to a $T=4$ capsid. There are four structurally distinct external scaffolding proteins (Dokland et al. 1997, 1999) per asymmetric unit (Fig. 14.7). The four D proteins form dimers of asymmetric dimers (D_1D_2, D_3D_4). Although each subunit has a unique structure, which is partly determined by its unique interactions with the underlying coat protein and neighboring D proteins within and across asymmetric units, there are structurally similarities. The structures of D_1 and D_3 are more similar to each other than to either D_2 or D_4, and the converse is true with the D_2 and D_4 proteins. D_4 has the most unique structure, especially in the arrangement of the N- and C-termini, which appear to be critical for assembly and coat protein recognition (see below). The atomic structure of assembly naive

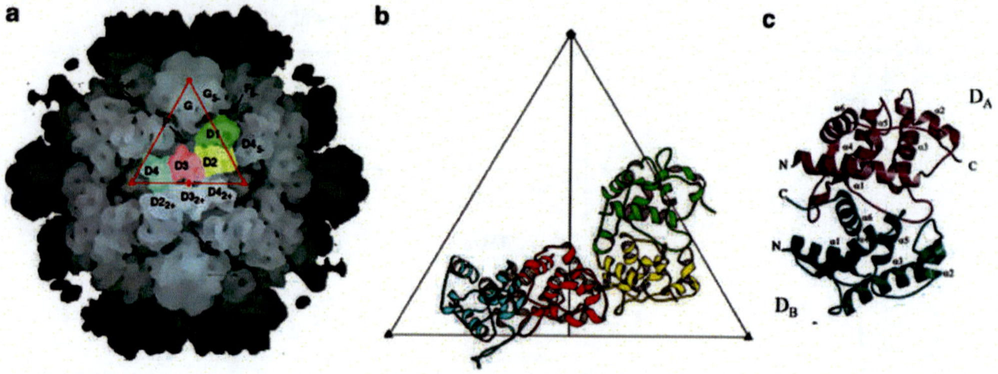

Fig. 14.7 The structure of the øX174 external scaffolding protein D. (**a**) Isosurface tracing of the procapsid illustrating the position of the D_1, D_2, D_3, and D_4 subunits. (**b**) The atomic structure of the D_1, D_2, D_3, and D_4 subunits. The color scheme is identical to that given in part (**a**). (**c**) The atomic structure of the assembly naive D protein dimer, D_AD_B

external scaffolding protein dimer (D_AD_B) has also been solved (Morais et al. 2004). While the D_AD_B dimer structure is unique, it bears a resemblance to the procapsid D_1D_3 and D_2D_4 dimers (Fig. 14.7). The D_A is structurally most similar to D_1 and D_3, while D_B shares a fold closely related to D_2 and D_4. Moreover, the relationship between neighboring asymmetric units in the D_AD_B crystal is similar to that found between the two asymmetric D_1D_2 and D_3D_4 dimers. Thus, the assembly naive subunits are poised to assume the $D_1D_2D_3D_4$ configuration found in the procapsid.

To achieve the unique dimer and tetramer arrangements found in the naive and assembled states, one monomer in each asymmetric dimer (D_1, D_3, or D_A) must be bent. If there were no conformational switch between the monomers, the intra- and interdimer interfaces would be identical. Consequently, the subunits would form a never-ending helix. The conformational switch occurs at glycine residue 61 in α-helix #3, which is bent by 30° in the D_1 and D_3 subunits (Fig. 14.8). Only a glycine residue can accommodate the torsional angle necessary for the D protein to properly adopt one of these two conformations, as the other amino acid side chains occupy forbidden regions of the Ramachandran plot. The organization of the genome further emphasizes the evolutionary importance of maintaining the G61 residue in the external scaffolding protein. Gene E, which encodes the lysis protein, resides entirely within the gene D sequence. Both proteins are translated from the same mRNA, and the G61 and gene E start codons overlap in all known microvirus sequences (Godson et al. 1978; Kodaira et al. 1992; Rokyta et al. 2006; Sanger et al. 1978).

14.5.4.2 Conformational Switching of the External Scaffolding Protein During Assembly

Extensive genetic analyses have been conducted with mutant proteins with substitutions of A, S, T, V, D, K, and P for glycine 61 (Cherwa et al. 2008, 2009; Cherwa and Fane 2009). Mutant genes were constructed on plasmids and assayed for the ability to complement a *nullD* mutant and to inhibit wild-type progeny production. As expected none of the mutant proteins could replace wild-type D protein function in *nullD* infections. In wild-type infections, the mutant proteins inhibit virion morphogenesis, with severity being a function of the substituted amino acid size.

The assembly intermediates made in wild-type and *nullD* infected cells expressing the strongly inhibitory G61D, G61V, and G61P external scaffolding proteins were characterized. In *nullD* infected cells, only the mutant D protein was present and 12S* particles readily accumulated. As the 12S* particle is the last intermediate before the first D protein mediated step in the assembly pathway

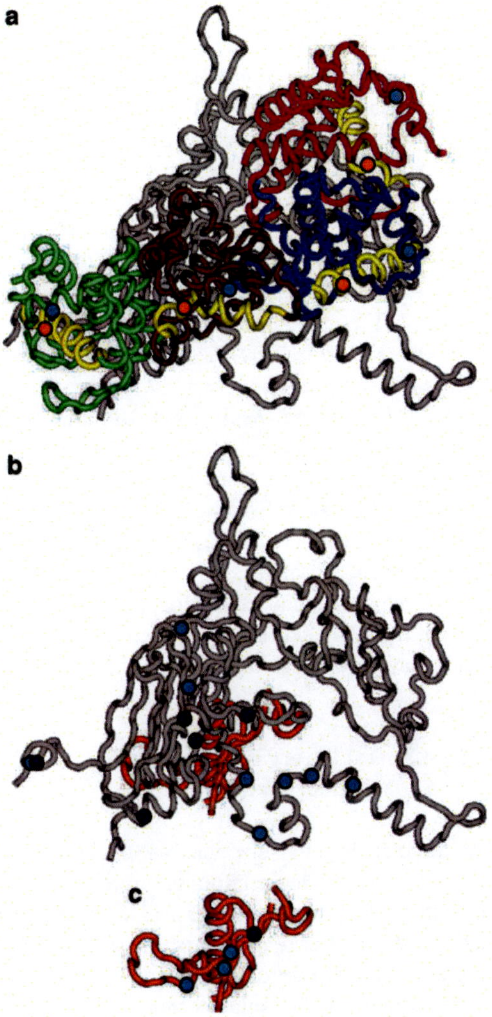

Fig. 14.8 Conformational switch mutants in the external scaffolding protein. (**a**) The location of G61 and D34 residues within the atomic structure of the procapsid. The coat protein is depicted in *gray*, the internal scaffolding protein in peach, and the D_1, D_2, D_3, and D_4 subunits in *magenta*, *blue*, *brown*, and *chartreuse*, respectively. α-helix #3, in which G61 resides (*orange dot*), is colored *yellow*. The helix assumes a straight (D_2 and D_4) or kinked conformation (D_1 and D_3). The location of residue D34 is indicated with a *blue dot*. (**b**) The location of resistance and utilizer mutations in the coat protein. Resistance and utilizer mutations are depicted with *light* and *dark blue dots*, respectively. (**c**) The location of resistance and utilizer mutations in the internal scaffolding protein. All structures are rendered to the same scale and orientation

(Fig. 14.4), the inhibitory proteins behave as if they have no activity vis-à-vis the ability to interact with other assembly intermediates. In wild-type infected cells, both the wild-type and inhibitory proteins would be present. In those infections, 12S* particles do not accumulate but appear to be channeled into off-pathway abominations (Fig. 14.4). As the inhibitory proteins require the wild-type protein to enter into the assembly pathway, the actual inhibitor is most likely a heterodimer.

Mutants resistant to the inhibitory external scaffolding proteins were isolated via direct genetic selections. The resistance mutations conferred amino acid substitution in the coat and internal scaffolding proteins. The affected amino acids cluster beneath or surround the D_3 subunit in the asymmetric unit (Fig. 14.8). The graphics in Fig. 14.8 depict atomic structure of the closed procapsid, which most likely represents an off-pathway product. During crystallization, the coat protein underwent a limited maturation event. In the cryo-electron microscopy reconstruction, which most likely represents the native species, large pores are present at the threefold axes of symmetry (Bernal et al. 2003; Ilag et al. 1995). Thus, the two α-helices that transverse the twofold axis of symmetry would be shifted upward, moving the resistance mutations at amino acids 212, 214, and 227 toward the center of the asymmetric unit. The clustering of the resistance mutations suggests models in which they operate to exclude or accommodate a heterodimer in the D_3D_4 position.

In a structure-based accommodation model, the resistance mutants would be expected to exhibit allele specificity, only accommodating the inhibitory protein for which they were selected. However, they do not exhibit strict allele specificity and even confer resistance against the G61P protein. The G→P substitution would affect the structure of α-helix #3 in a fundamentally different manner than the other substitutions. While the other substitutions would inhibit helix kinking, the proline substitution would lock the helix into a bent position.

While the expression of proteins with more conservative substitutions for G61 (serine and alanine) did not lower wild-type plating efficiencies, wild-type plaque morphologies and burst sizes were reduced. Morphogenesis in *nullD*-infected cells expressing these proteins was examined. Unlike assembly in the absence of D protein, which leads to a pronounced accumulation of 12S* assembly intermediates, particle recovery was skewed toward the recovery of the 9S* intermediate (Cherwa and Fane 2009). This observation may indicate that the G61A and S proteins do not need wild-type subunits to interact with other viral components, as was observed with the G61P, G61V, and G61D proteins. While these data are suggestive, a more compelling argument for G61A and G61S homodimer activity could be made if full complementation could be achieved. Thus, *nullD/utilizer* mutants were isolated. The seven *utilizer* mutations resided in the same regions of the coat and internal scaffolding proteins defined by the resistance mutations (Fig. 14.8). Some mutations were identical. *utilizer/nullD* virion production was investigated in cells expressing the wild-type and G61S proteins. G61S protein complementation resulted in a longer lag phase. As øX174 has no temporal gene expression, the longer lag phase may indicate that a higher critical concentration of the G61S protein is required to nucleate procapsid morphogenesis. Lower yields and slower rates may reflect secondary defects in the capsid elongation.

14.5.4.3 Inhibitory Proteins and the Evolution of a Three-Scaffolding Protein System

As targeted genetic selections with deletions of the internal scaffolding protein converted øX174 from a two- to a one-scaffolding protein system, the results of experimental evolution studies indicate that it may be possible to evolve a three-scaffolding protein system: one that requires the internal protein and two species of the external protein. The resistance phenotypes conferred by the mutations isolated in one-step genetic selections with the lethal dominant external scaffolding proteins were weak. To isolate a more robust phenotype, which may be predictive of resistant phenotypes that could be selected by prolonged exposure to antiviral agents that specifically target capsid assembly, wild-type øX174 was continually cultured through exponential phase cells expressing an inhibitory D protein (Cherwa et al. 2009). Initially the induction of the inhibitory gene was low, to allow low levels of virus replication. As the viral population began to exhibit increased fitness, indicative of the accumulation of mutant strains, induction levels were steadily increased. After approximately 180 life cycles, a quintuple-mutant resistant strain was isolated from the population. The recovered mutations were very similar to those isolated in one-step genetic selections with the exception of one mutation, which conferred a D→G substitution at amino 34 in α-helix #2 of the external scaffolding protein. As this particular helix mediates no contacts with other structural or scaffolding proteins, the substitution may provide an alternative locus for conformational flexibility, which compensates for the steric hindrance introduced into α-helix #3.

As can be seen in Table 14.1, acquiring resistance to an antiviral agent lowered strain fitness vis-à-vis the uninhibited wild-type levels, a phenomenon that has been observed in other studies (Collins et al. 2004; Paredes et al. 2009; Zhou et al. 2008). Moreover, the resistance strain's replication capacity is higher in the presence of inhibitor. Thus, as observed with other antiviral agents (Baldwin and Berkhout 2007; Salvati et al. 2004), the selection for resistance co-selected for a level of dependence. However, the virus appears to have evolved a mechanism to productively utilize the once

Table 14.1 Fitness (doublings/h) and virion yield (progeny/h/input phage) in continuous culture

	Inhibitory protein expression as percentage of total cellular D protein					
	0%		20%		50%	
Strain	Fitness	Yield	Fitness	Yield	Fitness	Yield
Wild type	9.4±0.3	675	7.4±0.2	170	2.0±0.1	4
Evolved	6.4±0.1	85	12.4±0.1	5,400	5.6±0.5	50

Experiments conducted in replicates of five

inhibitory protein to stimulate its fitness substantially higher than the uninhibited wild-type level, which is a novel phenomenon. Thus, optimal fitness now requires three scaffolding proteins.

The resistant strain was exhaustively characterized to determine the molecular basis of this effect. Although no significant differences in either attachment or eclipse kinetics were observed when compared to the ancestral wild-type strain, the inhibitory protein led to a dramatically shortened lag phase and increase in resistant strain progeny production. The most dramatic effect was observed when the inhibitory protein was present at a 1:3 ratio with the wild-type species. Under these conditions, progeny appears 8 min postinfection, as opposed to 14 min postinfection in the absence of the inhibitor, and progeny production is increased by approximately tenfold. These observations are best explained by the prudent particle – promiscuous protein hypothesis. The inhibitory proteins are known to promiscuously promote off-pathway reactions (Cherwa et al. 2008; Cherwa and Fane 2009); the resistance mutations may produce 12S* intermediates that are less prone to associate with both the inhibitory and wild-type external scaffolding protein. As the dependency phenomenon appears to operate on the level of capsid nucleation, efficient nucleation may now depend on a small concentration of the more promiscuous inhibitory protein.

14.5.4.4 Coat Protein Substrate Specificity Domains in the External Scaffolding Protein

The full extent of coat–external scaffolding protein interactions is not apparent in the atomic structure of the closed procapsid. During crystallization, the particles underwent a limited maturation event, which included a reorganization of the coat protein at the threefold axes of symmetry to generate a closed structure – the native particle contains 30 Å pores in these locations. Moreover, the coat protein pentamers moved inward in a radial manner, forming two- and threefold interactions similar to those observed in the virion, and contacts with the external scaffolding protein are minimal. In contrast, there are very few, if any, cross-pentamer contacts in the cryo-EM model of the native structure, and coat–external scaffolding protein interactions are much more extensive (Bernal et al. 2003; Dokland et al. 1997, 1999; Ilag et al. 1995). However, the results of genetic experiments conducted with chimeric scaffolding proteins have helped elucidate external scaffolding protein domains that confer coat protein specificity (Burch and Fane 2000a, 2003; Uchiyama and Fane 2005; Uchiyama et al. 2007, 2009).

Although most of the primary sequences of microvirus external scaffolding proteins are highly conserved (Godson et al. 1978; Kodaira et al. 1992; Rokyta et al. 2006; Sanger et al. 1978), a considerable degree of sequence divergence is localized to the N- and C-termini of the proteins, which constitute α-helices #1, #7, and loop #6 in the atomic structure. External scaffolding proteins do not cross-function between the related G4, α3, and øX174 species, nor does the expression of a foreign scaffolding protein dramatically interfere with cross-species assembly. However, chimeric øX174/α3 scaffolding proteins with interchanged first α-helices inhibit virion formation in a species-specific manner. Inhibition is observed when the first helix is of the same origin as the viral coat protein (Burch and Fane 2000a). Thus, the incorporation of the inhibitory chimera was a function of the first

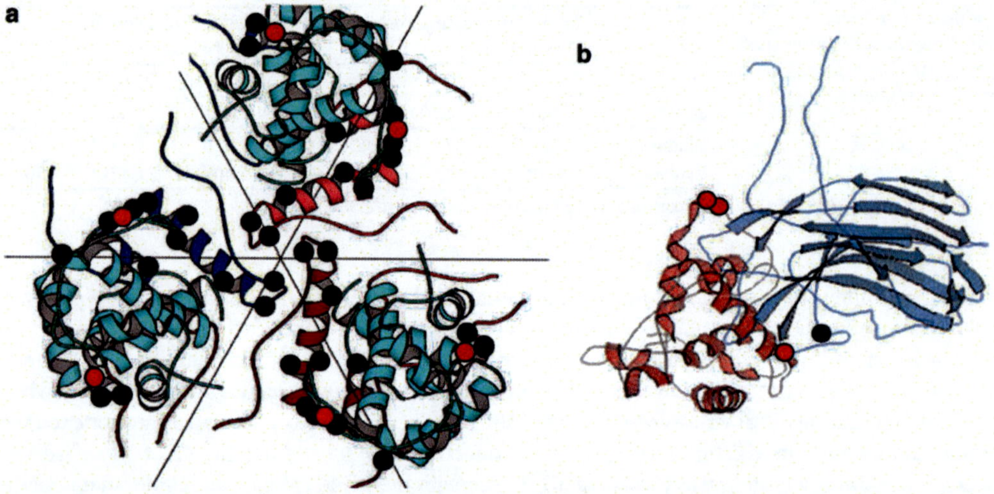

Fig. 14.9 The location of suppressors of chimeric and N-terminal deleted external scaffolding proteins. (**a**) The threefold axis of symmetry in the atomic structure of the closed procapsid. Coat protein sequences are depicted in *dark blue*, *pink*, and *purple*. The external scaffolding protein D_4 subunit is depicted in *light blue*. *Black dots* mark the location of suppressors in the coat protein. *Red dots* indicate the location of the suppressors within the external scaffolding protein. (**b**) Interactions between the external scaffolding protein D_1 subunit and the major spike protein G

α-helix. The results of further biochemical and genetic analyses indicate that the actual inhibitory effects were conferred by the foreign α-helix #7, which resulted in the generation of procapsids that could not be filled (Burch and Fane 2003).

Although the results of the above analysis suggested that α-helix #1 was required for the initial coat–external scaffolding protein recognition, sequence divergence between the øX174 and α3 proteins was too great to obtain mutants that could productively use the chimeric scaffolding proteins. Thus, chimeras between the more closely related G4 and øX174 phages were generated. While the G4/øX174 chimeric scaffolding proteins could not support øX174 *nullD* plaque formation, progeny was produced in lysis-resistant cell after an extended lag phase (Uchiyama and Fane 2005). However, once progeny production initiates, virions appear to be produced at a wild-type rate. Similar results were obtained when the chimeric gene was constructed directly into the viral genome (Uchiyama et al. 2007). These observations suggest that a higher critical concentration of the chimeric protein is required to nucleate virion morphogenesis. Plaque-forming mutants of the chimeric strains were isolated. The mutations shorten the lag phase, allowing progeny production before programmed cell lysis in lysis-sensitive cell. The mutations, along with previous suppressors of cold-sensitive α-helix-1 mutations (Fane et al. 1993), map to two coat protein α-helices that directly surround the threefold axes of symmetry (Figs. 14.5 and 14.9a). Strains with small deletions within the first helix also delay the timing of progeny production. These mutants were also suppressed by substitutions in the external scaffolding as well as by regulatory mutations that elevate the expression level of the N-terminal deleted D proteins (Uchiyama et al. 2009). Again, these results suggest a connection between the lag phase length and the critical concentration needed to nucleate assembly.

While the clustering of the suppressor mutations suggests that interactions between α-helix #1 of the external scaffolding protein and the coat protein helices near the threefold axes of symmetry (Fig. 14.9a) nucleate coat–external scaffolding protein interactions, it is not possible to ascertain which D subunit is involved in this interaction. However, two key structural features suggest the D_4 subunit. It is located at threefold axis of symmetry, and it is the only subunit in which the helix is pointed inward, toward the coat protein. These observations and the atomic structure of the assembly naive D protein dimer suggest a model in which a D_B subunit makes the initial contact with the 12S* particle, interacting with a coat protein in an adjacent fivefold-related asymmetric unit, and becomes the D_4 protein, as visualized in the procapsid structure. The associated D_A subunit would then sit in the D_3 position. The next D_AD_B dimer to attach would then occupy the D_1D_2 location. However, a secondary site for nucleation may exist. The second-site genetic analyses also identified mutations in the major spike and external scaffolding proteins. The suppressing D protein residues cluster in a loop in the D_1 subunit that extends toward the major spike protein (Fig. 14.9b). The suppressing residue in the major spike protein is in close proximity to α-helix 1 of the D_1 subunit. These suppressors may act by creating an alternate site for initial 12S*–D_AD_B dimer interactions.

14.6 Viruses with Satellites: Reprogramming Size Determination and Molecular Piracy

14.6.1 Bacteriophages P2 and P4

The P2/P4 bacteriophage system represents an example of size determination by the satellite virus P4 encoded external scaffolding protein gpSid (*size determination*). Like many dsDNA viruses, the assembly of the *T*=7 P2 coat protein (gpN) is mediated by its internal scaffolding proteins (gpO) (Christie and Calendar 1990). GpO shares many features with other internal scaffolding proteins. For example, C-terminal fragments are both necessary and sufficient to promote capsid assembly (Chang et al. 2009). However, a small proteolytic fragment of gpO, called O*, remains in the virion (Lengyel et al. 1973; Rishovd and Lindqvist 1992). Like the øX174 internal scaffolding protein (Bernal et al. 2003), gpO is an autoprotease, and the region of the protein responsible for enzymatic activity has been identified (Chang et al. 2009; Wang et al. 2006). It is also likely that gpO is responsible for proteolytic processing of the coat protein (Wang et al. 2006). The satellite bacteriophage encodes the external scaffolding protein sid (*size determination*). In P2/P4 co-infected cells, sid protein alters the curvature of the elongating capsids, which results in a *T*=4 particles. The smaller procapsids cannot accommodate the P2 genome. Thus, they are only filled with the P4 genome.

Structural and genetic studies have somewhat defined the molecular basis of the P4 piracy. The sid–coat protein interactions lead to the formation of a size-restricting cage by influencing the bend of the capsid protein at a flexible hinge region, which was genetically identified by *sir* mutations (*sid responsiveness*). Sir mutations render the capsid protein resistant to sid protein action (Six et al. 1991). In the cryo-EM model, the sid protein forms a lattice of 12 open pentagons, surrounding the vertices and bifurcating the surrounding hexameric capsomers (Dokland et al. 2002; Marvik et al. 1995) (Wang et al. 2000). The C-termini of sid proteins form trimeric structures at junctions containing three hexamers. Structurally distinct sid dimers interconnect the C-terminal trimer formations. Dimer–dimer and dimer–trimer interactions are most likely mediated via N-termini sid–sid interactions. Extragenic second-site suppressors of *sir* mutations (*nms*), which restore sid sensitivity to *sir* capsid proteins, are located in the C-terminus of the sid protein (Kim et al. 2001). These super-sid proteins most likely form stronger sid trimers, producing a more rigid external lattice that overrides the less flexible hinge regions created by the *sir* mutations. If P4 capsid assembly nucleates with a sid trimer at the junction of three hexamers, the concurrent elongation of the P2 capsid protein with the less flexible and size constraining sid lattice would direct assembly into the smaller *T*=4 capsid.

14.6.2 Phage 80a and Its Parasite SaPI1

In a manner analogous to the P2/P4 system, the *Staphylococcus aureus* pathogenicity island SaPI1, which carries the gene for the toxic shock syndrome toxin, is capable of redirecting the assembly of the phage 80a from a *T*=7 to a smaller *T*=4 procapsids (Spilman et al. 2011; T. Dokland, personal communication). Mass spectroscopic analysis of the protein composition of the SaPI1 procapsids detected the presence of two SaPI1 proteins, gp6 and gp7 (Poliakov et al. 2008), both of which had previously been implicated in form determination (Ubeda et al. 2008). Both gp6 and gp7 are absent from the mature phage. The NMR structure of purified gp6 indicates that it is structurally similar to the scaffolding protein of Phi29 in that it is highly α-helical, contains an N-terminal helix-loop-helix motif, and is dimeric in solution (T. Dokland, personal communication).

14.7 Concluding Remarks

All scaffolding proteins perform the same basic function: catalyzing and directing capsid assembly, a process that becomes more complicated as icosahedral geometries become more complex. However, as is clearly evident from the case of the $T=1$ microviruses, their performance is finely tuned and optimized by the competitive arena of natural selection. Moreover, many scaffolding proteins have extended their functional repertoire to include controlling the incorporation of other structural proteins, such as portal and pilot proteins, necessary for productive virus infection. Presumably, encoding additional functionality into an existing protein is a cost-effective way to expand function with a minimal increase in genetic load.

Although scaffolding proteins are found in both eukaryotic and prokaryotic viral assembly systems, most mechanistic insights have come from the study of bacteriophages, which are more readily adaptable to laboratory settings. The number of tailed bacteriophages on earth has been estimated to be 10^{31}, and there are an estimated 10^{23} bacterial infections per second (Hendrix 2003). To date we have sampled a small fraction of the continually evolving viruses in the biosphere. It is almost certain that as we sample more deeply, we will find additional and unexpected mechanisms and functions of scaffolding proteins.

References

Baker ML, Jiang W, Rixon FJ, Chiu W (2005) Common ancestry of herpesviruses and tailed DNA bacteriophages. J Virol 79:14967–14970

Baldwin C, Berkhout B (2007) HIV-1 drug-resistance and drug-dependence. Retrovirology 4:78

Bamford DH, Grimes JM, Stuart DI (2005) What does structure tell us about virus evolution? Curr Opin Struct Biol 15:655–663

Berger B, Shor PW, Tucker-Kellogg L, King J (1994) Local rule-based theory of virus shell assembly. Proc Natl Acad Sci USA 91:7732–7736

Bernal RA, Hafenstein S, Olson NH, Bowman VD, Chipman PR, Baker TS, Fane BA, Rossmann MG (2003) Structural studies of bacteriophage alpha3 assembly. J Mol Biol 325:11–24

Bjornsti MA, Reilly BE, Anderson DL (1983) Morphogenesis of bacteriophage phi 29 of *Bacillus subtilis*: oriented and quantized in vitro packaging of DNA protein gp3. J Virol 45:383–396

Brentlinger KL, Hafenstein S, Novak CR, Fane BA, Borgon R, McKenna R, Agbandje-McKenna M (2002) Microviridae, a family divided: isolation, characterization, and genome sequence of phiMH2K, a bacteriophage of the obligate intracellular parasitic bacterium *Bdellovibrio bacteriovorus*. J Bacteriol 184:1089–1094

Burch AD, Fane BA (2000a) Foreign and chimeric external scaffolding proteins as inhibitors of Microviridae morphogenesis. J Virol 74:9347–9352

Burch AD, Fane BA (2000b) Efficient complementation by chimeric Microviridae internal scaffolding proteins is a function of the COOH-terminus of the encoded protein. Virology 270:286–290

Burch AD, Fane BA (2003) Genetic analyses of putative conformation switching and cross-species inhibitory domains in Microviridae external scaffolding proteins. Virology 310:64–71

Burch AD, Ta J, Fane BA (1999) Cross-functional analysis of the Microviridae internal scaffolding protein. J Mol Biol 286:95–104

Casjens S, King J (1974) P22 morphogenesis. I: Catalytic scaffolding protein in capsid assembly. J Supramol Struct 2:202–224
Caspar DLD, Klug A (1962) Physical principles in the construction of regular viruses. Cold Spring Harb Symp Quant Biol 27:1–24
Chang JR, Spilman MS, Rodenburg CM, Dokland T (2009) Functional domains of the bacteriophage P2 scaffolding protein: identification of residues involved in assembly and protease activity. Virology 384:144–150
Chen R, Neill JD, Estes MK, Prasad BV (2006) X-ray structure of a native calicivirus: structural insights into antigenic diversity and host specificity. Proc Natl Acad Sci USA 103:8048–8053
Chen M, Uchiyama A, Fane BA (2007) Eliminating the requirement of an essential gene product in an already very small virus: scaffolding protein B-free oX174, B-free. J Mol Biol 373:308–314
Cherwa JE Jr, Fane BA (2009) Complete virion assembly with scaffolding proteins altered in the ability to perform a critical conformational switch. J Virol 83:7391–7396
Cherwa JE Jr, Uchiyama A, Fane BA (2008) Scaffolding proteins altered in the ability to perform a conformational switch confer dominant lethal assembly defects. J Virol 82:5774–5780
Cherwa JE Jr, Sanchez-Soria P, Wichman HA, Fane BA (2009) Viral adaptation to an antiviral protein enhances the fitness level to above that of the uninhibited wild type. J Virol 83:11746–11750
Christie GE, Calendar R (1990) Interactions between satellite bacteriophage P4 and its helpers. Annu Rev Genet 24:465–490
Clarke IN, Cutcliffe LT, Everson JS, Garner SA, Lambden PR, Pead PJ, Pickett MA, Brentlinger KL, Fane BA (2004) Chlamydiaphage Chp2, a skeleton in the phiX174 closet: scaffolding protein and procapsid identification. J Bacteriol 186:7571–7574
Collins JA, Thompson MG, Paintsil E, Ricketts M, Gedzior J, Alexander L (2004) Competitive fitness of nevirapine-resistant human immunodeficiency virus type 1 mutants. J Virol 78:603–611
Conway JF, Wikoff WR, Cheng N, Duda RL, Hendrix RW, Johnson JE, Steven AC (2001) Virus maturation involving large subunit rotations and local refolding. Science 292:744–748
Conway JF, Cheng N, Ross PD, Hendrix RW, Duda RL, Steven AC (2007) A thermally induced phase transition in a viral capsid transforms the hexamers, leaving the pentamers unchanged. J Struct Biol 158:224–232
Desai P, Watkins SC, Person S (1994) The size and symmetry of B capsids of herpes simplex virus type 1 are determined by the gene products of the UL26 open reading frame. J Virol 68:5365–5374
Dokland T, McKenna R, Ilag LL, Bowman BR, Incardona NL, Fane BA, Rossmann MG (1997) Structure of a viral procapsid with molecular scaffolding. Nature 389:308–313
Dokland T, Bernal RA, Burch A, Pletnev S, Fane BA, Rossmann MG (1999) The role of scaffolding proteins in the assembly of the small, single-stranded DNA virus phiX174. J Mol Biol 288:595–608
Dokland T, Wang S, Lindqvist BH (2002) The structure of P4 procapsids produced by coexpression of capsid and external scaffolding proteins. Virology 298:224–231
Earnshaw W, King J (1978) Structure of phage P22 coat protein aggregates formed in the absence of the scaffolding protein. J Mol Biol 126:721–747
Earnshaw W, Casjens S, Harrison SC (1976) Assembly of the head of bacteriophage P22: x-ray diffraction from heads, proheads and related structures. J Mol Biol 104:387–410
Effantin G, Boulanger P, Neumann E, Letellier L, Conway JF (2006) Bacteriophage T5 structure reveals similarities with HK97 and T4 suggesting evolutionary relationships. J Mol Biol 361:993–1002
Fane BA, Hayashi M (1991) Second-site suppressors of a cold-sensitive prohead accessory protein of bacteriophage phi X174. Genetics 128:663–671
Fane BA, Shien S, Hayashi M (1993) Second-site suppressors of a cold-sensitive external scaffolding protein of bacteriophage phi X174. Genetics 134:1003–1011
Fu CY, Prevelige PE Jr (2006) Dynamic motions of free and bound O29 scaffolding protein identified by hydrogen deuterium exchange mass spectrometry. Protein Sci 15:731–743
Fu CY, Prevelige PE Jr (2009) In vitro incorporation of the phage Phi29 connector complex. Virology 394:149–153
Fu CY, Morais MC, Battisti AJ, Rossmann MG, Prevelige PE Jr (2007) Molecular dissection of o29 scaffolding protein function in an in vitro assembly system. J Mol Biol 366:1161–1173
Fu CY, Uetrecht C, Kang S, Morais MC, Heck AJ, Walter MR, Prevelige PE Jr (2010) A docking model based on mass spectrometric and biochemical data describes phage packaging motor incorporation. Mol Cell Proteomics 9:1764–1773
Fuller MT, King J (1980) Regulation of coat protein polymerization by the scaffolding protein of bacteriophage P22. Biophys J 32:381–401
Godson GN, Barrell BG, Staden R, Fiddes JC (1978) Nucleotide sequence of bacteriophage G4 DNA. Nature 276:236–247
Grimes JM, Burroughs JN, Gouet P, Diprose JM, Malby R, Zientara S, Mertens PP, Stuart DI (1998) The atomic structure of the bluetongue virus core. Nature 395:470–478

Guo PX, Erickson S, Anderson D (1987) A small viral RNA is required for in vitro packaging of bacteriophage phi 29 DNA. Science 236:690–694

Haanes EJ, Thomsen DR, Martin S, Homa FL, Lowery DE (1995) The bovine herpesvirus 1 maturational proteinase and scaffold proteins can substitute for the homologous herpes simplex virus type 1 proteins in the formation of hybrid type B capsids. J Virol 69:7375–7379

Hendrix RW (2003) Bacteriophage genomics. Curr Opin Microbiol 6:506–511

Hendrix RW, Duda RL (1998) Bacteriophage HK97 head assembly: a protein ballet. Adv Virus Res 50:235–288

Hernando E, Llamas-Saiz AL, Foces-Foces C, McKenna R, Portman I, Agbandje-McKenna M, Almendral JM (2000) Biochemical and physical characterization of parvovirus minute virus of mice virus-like particles. Virology 267:299–309

Hogle JM, Chow M, Filman DJ (1985) Three-dimensional structure of poliovirus at 2.9 A resolution. Science 229:1358–1365

Hohn B (1983) DNA sequences necessary for packaging of bacteriophage lambda DNA. Proc Natl Acad Sci USA 80:7456–7460

Huet A, Conway JF, Letellier L, Boulanger P (2010) In vitro assembly of the T=13 procapsid of bacteriophage T5 with its scaffolding domain. J Virol 84:9350–9358

Ilag LL, Olson NH, Dokland T, Music CL, Cheng RH, Bowen Z, McKenna R, Rossmann MG, Baker TS, Incardona NL (1995) DNA packaging intermediates of bacteriophage phi X174. Structure 3:353–363

Jardine PJ, McCormick MC, Lutze-Wallace C, Coombs DH (1998) The bacteriophage T4 DNA packaging apparatus targets the unexpanded prohead. J Mol Biol 284:647–659

Jiang W, Li Z, Zhang Z, Baker ML, Prevelige PE Jr, Chiu W (2003) Coat protein fold and maturation transition of bacteriophage P22 seen at subnanometer resolutions. Nat Struct Biol 10:131–135

Kang S, Prevelige PE Jr (2005) Domain study of bacteriophage p22 coat protein and characterization of the capsid lattice transformation by hydrogen/deuterium exchange. J Mol Biol 347:935–948

Kang S, Poliakov A, Sexton J, Renfrow MB, Prevelige PE Jr (2008) Probing conserved helical modules of portal complexes by mass spectrometry-based hydrogen/deuterium exchange. J Mol Biol 381:772–784

Katsura I, Kobayashi H (1990) Structure and inherent properties of the bacteriophage lambda head shell. VII. Molecular design of the form-determining major capsid protein. J Mol Biol 213:503–511

Kellenberger E (1990) Form determination of the heads of bacteriophages. Eur J Biochem 190:233–248

Kim KJ, Sunshine MG, Lindqvist BH, Six EW (2001) Capsid size determination in the P2-P4 bacteriophage system: suppression of sir mutations in P2's capsid gene N by supersid mutations in P4's external scaffold gene sid. Virology 283:49–58

King J, Casjens S (1974) Catalytic head assembling protein in virus morphogenesis. Nature 251:112–119

King J, Lenk EV, Botstein D (1973) Mechanism of head assembly and DNA encapsulation in Salmonella phage P22. II. Morphogenetic pathway. J Mol Biol 80:697–731

Kodaira K, Nakano K, Okada S, Taketo A (1992) Nucleotide sequence of the genome of the bacteriophage alpha 3: interrelationship of the genome structure and the gene products with those of the phages, phi X174, G4 and phi K. Biochim Biophys Acta 1130:277–288

Lander GC, Khayat R, Li R, Prevelige PE, Potter CS, Carragher B, Johnson JE (2009) The p22 tail machine at subnanometer resolution reveals the architecture of an infection conduit. Structure 17:789–799

Lebedev AA, Krause MH, Isidro AL, Vagin AA, Orlova EV, Turner J, Dodson EJ, Tavares P, Antson AA (2007) Structural framework for DNA translocation via the viral portal protein. EMBO J 26:1984–1994

Lengyel JA, Goldstein RN, Marsh M, Sunshine MG, Calendar R (1973) Bacteriophage P2 head morphogenesis: cleavage of the major capsid protein. Virology 53:1–23

Lenk E, Casjens S, Weeks J, King J (1975) Intracellular visualization of precursor capsids in phage P22 mutant infected cells. Virology 68:182–199

Liu BL, Everson JS, Fane B, Giannikopoulou P, Vretou E, Lambden PR, Clarke IN (2000) Molecular characterization of a bacteriophage (Chp2) from *Chlamydia psittaci*. J Virol 74:3464–3469

Marvik OJ, Dokland T, Nokling RH, Jacobsen E, Larsen T, Lindqvist BH (1995) The capsid size-determining protein Sid forms an external scaffold on phage P4 procapsids. J Mol Biol 251:59–75

Masker WE, Serwer P (1982) DNA packaging in vitro by an isolated bacteriophage T7 procapsid. J Virol 43: 1138–1142

Matusick-Kumar L, Hurlburt W, Weinheimer SP, Newcomb WW, Brown JC, Gao M (1994) Phenotype of the herpes simplex virus type 1 protease substrate ICP35 mutant virus. J Virol 68:5384–5394

Matusick-Kumar L, Newcomb WW, Brown JC, McCann PJ 3rd, Hurlburt W, Weinheimer SP, Gao M (1995) The C-terminal 25 amino acids of the protease and its substrate ICP35 of herpes simplex virus type 1 are involved in the formation of sealed capsids. J Virol 69:4347–4356

Morais MC, Kanamaru S, Badasso MO, Koti JS, Owen BA, McMurray CT, Anderson DL, Rossmann MG (2003) Bacteriophage phi29 scaffolding protein gp7 before and after prohead assembly. Nat Struct Biol 10: 572–576

Morais MC, Fisher M, Kanamaru S, Przybyla L, Burgner J, Fane BA, Rossmann MG (2004) Conformational switching by the scaffolding protein D directs the assembly of bacteriophage phiX174. Mol Cell 15:991–997
Nemecek D, Overman SA, Hendrix RW, Thomas GJ Jr (2009) Unfolding thermodynamics of the Delta-domain in the prohead I subunit of phage HK97: determination by factor analysis of Raman spectra. J Mol Biol 385:628–641
Newcomb WW, Trus BL, Cheng N, Steven AC, Sheaffer AK, Tenney DJ, Weller SK, Brown JC (2000) Isolation of herpes simplex virus procapsids from cells infected with a protease-deficient mutant virus. J Virol 74: 1663–1673
Newcomb WW, Homa FL, Thomsen DR, Brown JC (2001) In vitro assembly of the herpes simplex virus procapsid: formation of small procapsids at reduced scaffolding protein concentration. J Struct Biol 133:23–31
Novak CR, Fane BA (2004) The functions of the N terminus of the phiX174 internal scaffolding protein, a protein encoded in an overlapping reading frame in a two scaffolding protein system. J Mol Biol 335:383–390
Oien NL, Thomsen DR, Wathen MW, Newcomb WW, Brown JC, Homa FL (1997) Assembly of herpes simplex virus capsids using the human cytomegalovirus scaffold protein: critical role of the C terminus. J Virol 71:1281–1291
Paredes R, Sagar M, Marconi VC, Hoh R, Martin JN, Parkin NT, Petropoulos CJ, Deeks SG, Kuritzkes DR (2009) In vivo fitness cost of the M184V mutation in multidrug-resistant human immunodeficiency virus type 1 in the absence of lamivudine. J Virol 83:2038–2043
Parent KN, Doyle SM, Anderson E, Teschke CM (2005) Electrostatic interactions govern both nucleation and elongation during phage P22 procapsid assembly. Virology 340:33–45
Parker MH, Prevelige PE Jr (1998) Electrostatic interactions drive scaffolding/coat protein binding and procapsid maturation in bacteriophage P22. Virology 250:337–349
Parker MH, Stafford WF 3rd, Prevelige PE Jr (1997) Bacteriophage P22 scaffolding protein forms oligomers in solution. J Mol Biol 268:655–665
Parker MH, Casjens S, Prevelige PE Jr (1998) Functional domains of bacteriophage P22 scaffolding protein. J Mol Biol 281:69–79
Parker MH, Brouillette CG, Prevelige PE Jr (2001) Kinetic and calorimetric evidence for two distinct scaffolding protein binding populations within the bacteriophage P22 procapsid. Biochemistry 40:8962–8970
Poh SL, el Khadali F, Berrier C, Lurz R, Melki R, Tavares P (2008) Oligomerization of the SPP1 scaffolding protein. J Mol Biol 378:551–564
Poliakov A, Chang JR, Spilman MS, Damle PK, Christie GE, Mobley JA, Dokland T (2008) Capsid size determination by *Staphylococcus aureus* pathogenicity island SaPI1 involves specific incorporation of SaPI1 proteins into procapsids. J Mol Biol 380:465–475
Poteete AR, Jarvik V, Botstein D (1979) Encapsulation of phage P22 DNA in vitro. Virology 95:550–564
Preston VG, Kennard J, Rixon FJ, Logan AJ, Mansfield RW, McDougall IM (1997) Efficient herpes simplex virus type 1 (HSV-1) capsid formation directed by the varicella-zoster virus scaffolding protein requires the carboxy-terminal sequences from the HSV-1 homologue. J Gen Virol 78(Pt 7):1633–1646
Prevelige PE Jr, Thomas D, King J (1988) Scaffolding protein regulates the polymerization of P22 coat subunits into icosahedral shells in vitro. J Mol Biol 202:743–757
Prevelige PE Jr, Thomas D, King J (1993) Nucleation and growth phases in the polymerization of coat and scaffolding subunits into icosahedral procapsid shells. Biophys J 64:824–835
Rao VB, Feiss M (2008) The bacteriophage DNA packaging motor. Annu Rev Genet 42:647–681
Rishovd S, Lindqvist B (1992) Bacteriophage P2 and P4 morphogenesis: protein processing and capsid size determination. Virology 187:548–554
Rokyta DR, Burch CL, Caudle SB, Wichman HA (2006) Horizontal gene transfer and the evolution of microvirid coliphage genomes. J Bacteriol 188:1134–1142
Rossmann MG (1984) Constraints on the assembly of spherical virus particles. Virology 134:1–11
Salim O, Skilton RJ, Lambden PR, Fane BA, Clarke IN (2008) Behind the chlamydial cloak: the replication cycle of chlamydiaphage Chp2, revealed. Virology 377:440–445
Salvati AL, De Dominicis A, Tait S, Canitano A, Lahm A, Fiore L (2004) Mechanism of action at the molecular level of the antiviral drug 3(2H)-isoflavene against type 2 poliovirus. Antimicrob Agents Chemother 48:2233–2243
Sanger F, Coulson AR, Friedmann T, Air GM, Barrell BG, Brown NL, Fiddes JC, Hutchison CA 3rd, Slocombe PM, Smith M (1978) The nucleotide sequence of bacteriophage phiX174. J Mol Biol 125:225–246
Shibata H, Fujisawa H, Minagawa T (1987a) Early events in DNA packaging in a defined in vitro system of bacteriophage T3. Virology 159:250–258
Shibata H, Fujisawa H, Minagawa T (1987b) Characterization of the bacteriophage T3 DNA packaging reaction in vitro in a defined system. J Mol Biol 196:845–851
Siden EJ, Hayashi M (1974) Role of the gene beta-product in bacteriophage phi-X174 development. J Mol Biol 89:1–16
Simpson AA, Tao Y, Leiman PG, Badasso MO, He Y, Jardine PJ, Olson NH, Morais MC, Grimes S, Anderson DL, Baker TS, Rossmann MG (2000) Structure of the bacteriophage phi29 DNA packaging motor. Nature 408: 745–750

Singer GP, Newcomb WW, Thomsen DR, Homa FL, Brown J (2005) Identification of a region in the herpes simplex virus scaffolding protein required for interaction with the portal. J Virol 79:132–139

Six EW, Sunshine MG, Williams J, Haggard-Ljungquist E, Lindqvist BH (1991) Morphopoietic switch mutations of bacteriophage P2. Virology 182:34–46

Spilman MS, Dearborn AD, Chang JR, Damle PK, Christie GE, Dokland TA (2011) Conformational switch involved in maturation of *Staphylococcus aureus* bacteriophage 80 alpha capsids. J Mol Biol 405:863–876

Stonehouse NJ, Stockley PG (1993) Effects of amino acid substitution on the thermal stability of MS2 capsids lacking genomic RNA. FEBS Lett 334:355–359

Sun Y, Parker MH, Weigele P, Casjens S, Prevelige PE Jr, Krishna NR (2000) Structure of the coat protein-binding domain of the scaffolding protein from a double-stranded DNA virus. J Mol Biol 297:1195–1202

Tang J, Johnson JM, Dryden KA, Young MJ, Zlotnick A, Johnson JE (2006) The role of subunit hinges and molecular "switches" in the control of viral capsid polymorphism. J Struct Biol 154:59–67

Thuman-Commike PA, Greene B, Jakana J, Prasad BV, King J, Prevelige PE Jr, Chiu W (1996) Three-dimensional structure of scaffolding-containing phage p22 procapsids by electron cryo-microscopy. J Mol Biol 260:85–98

Thuman-Commike PA, Greene B, Malinski JA, Burbea M, McGough A, Chiu W, Prevelige PE Jr (1999) Mechanism of scaffolding-directed virus assembly suggested by comparison of scaffolding-containing and scaffolding-lacking P22 procapsids. Biophys J 76:3267–3277

Tonegawa S, Hayashi M (1970) Intermediates in the assembly of phi X174. J Mol Biol 48:219–242

Trus BL, Booy FP, Newcomb WW, Brown JC, Homa FL, Thomsen DR, Steven AC (1996) The herpes simplex virus procapsid: structure, conformational changes upon maturation, and roles of the triplex proteins VP19c and VP23 in assembly. J Mol Biol 263:447–462

Tuma R, Parker MH, Weigele P, Sampson L, Sun Y, Krishna NR, Casjens S, Thomas GJ Jr, Prevelige PE Jr (1998) A helical coat protein recognition domain of the bacteriophage P22 scaffolding protein. J Mol Biol 281:81–94

Tuma R, Tsuruta H, French KH, Prevelige PE (2008) Detection of intermediates and kinetic control during assembly of bacteriophage P22 procapsid. J Mol Biol 381:1395–1406

Ubeda C, Maiques E, Barry P, Matthews A, Tormo MA, Lasa I, Novick RP, Penades JR (2008) SaPI mutations affecting replication and transfer and enabling autonomous replication in the absence of helper phage. Mol Microbiol 67:493–503

Uchiyama A, Fane BA (2005) Identification of an interacting coat-external scaffolding protein domain required for both the initiation of phiX174 procapsid morphogenesis and the completion of DNA packaging. J Virol 79:6751–6756

Uchiyama A, Chen M, Fane BA (2007) Characterization and function of putative substrate specificity domain in microvirus external scaffolding proteins. J Virol 81:8587–8592

Uchiyama A, Heiman P, Fane BA (2009) N-terminal deletions of the phiX174 external scaffolding protein affect the timing and fidelity of assembly. Virology 386:303–309

Verlinden Y, Cuconati A, Wimmer E, Rombaut B (2000) Cell-free synthesis of poliovirus: 14S subunits are the key intermediates in the encapsidation of poliovirus RNA. J Gen Virol 81:2751–2754

Wang S, Palasingam P, Nokling RH, Lindqvist BH, Dokland T (2000) In vitro assembly of bacteriophage P4 procapsids from purified capsid and scaffolding proteins. Virology 275:133–144

Wang S, Chang JR, Dokland T (2006) Assembly of bacteriophage P2 and P4 procapsids with internal scaffolding protein. Virology 348:133–140

Weigele PR, Sampson L, Winn-Stapley D, Casjens SR (2005) Molecular genetics of bacteriophage P22 scaffolding protein's functional domains. J Mol Biol 348:831–844

Wikoff WR, Liljas L, Duda RL, Tsuruta H, Hendrix RW, Johnson JE (2000) Topologically linked protein rings in the bacteriophage HK97 capsid. Science 289:2129–2133

Zhou ZH, Macnab SJ, Jakana J, Scott LR, Chiu W, Rixon FJ (1998) Identification of the sites of interaction between the scaffold and outer shell in herpes simplex virus-1 capsids by difference electron imaging. Proc Natl Acad Sci USA 95:2778–2783

Zhou Y, Bartels DJ, Hanzelka BL, Muh U, Wei Y, Chu HM, Tigges AM, Brennan DL, Rao BG, Swenson L, Kwong AD, Lin C (2008) Phenotypic characterization of resistant Val36 variants of hepatitis C virus NS3-4A serine protease. Antimicrob Agents Chemother 52:110–120

Chapter 15
Bacteriophage HK97 Capsid Assembly and Maturation

Roger W. Hendrix and John E. Johnson

Abstract The *Escherichia coli* phage HK97 has provided a productive experimental system for investigating how virus capsids are assembled from their protein components and how the assembled capsids mature to their final form. Aspects of the process for which the HK97 system has been particularly informative include assembly of the icosahedral capsid shell from the component proteins, structure of the capsid subunits and of the entire capsid as it progresses through its maturation, and the mechanism of the covalent cross-linking that links the subunits together into viral chain mail. The structural dynamics of the maturation as well as the energetics that drives maturation forward are beginning to be understood.

15.1 Introduction

There are at any one time an estimated 10^{31} tailed bacteriophages on Earth (Hendrix 2002; Suttle 2007), probably more than all other organisms combined. Recent evidence from both structural and genomic studies of capsid proteins argues that viruses resembling many of the contemporary viruses, including the tailed phages, were already present when the three domains of cellular life diverged approximately three billion years ago and must have been evolving for some time before that (Bamford et al. 2005; Hendrix 1999). The apparently ancient origin of tailed phage capsids implies that they have had as much opportunity as almost any other biological structure to evolve sophisticated and intricately coordinated mechanisms of assembly leading to mature structures finely adapted to their biological roles. In doing so, the virion proteins have diverged in sequence to an extreme degree, and we might wonder whether the mechanisms and pathways of capsid assembly have diverged to a similar extent, making the assembly process of each phage chosen for investigation a special case. Happily, studies of tailed phage capsid assembly extending over most of the past 50 years have amply demonstrated that all the tailed phages use fundamentally the same process to accomplish that assembly, even in the face of sequence divergence that erases any detectable trace

R.W. Hendrix (✉)
Department of Biological Sciences, Pittsburgh Bacteriophage Institute,
University of Pittsburgh, Pittsburgh, PA 15260, USA
e-mail: rhx@pitt.edu

J.E. Johnson
Department of Molecular Biology, MB31, The Scripps Research Institute,
10550 N. Torrey Pines Road, La Jolla, CA 92037, USA

M.G. Rossmann and V.B. Rao (eds.), *Viral Molecular Machines*,
Advances in Experimental Medicine and Biology 726, DOI 10.1007/978-1-4614-0980-9_15,

of similarity, and that that same common assembly process extends even to the tailed phages' distant cousins the herpesviruses (Baker et al. 2005; Duda et al. 2006; Johnson 2010). This means that new insights into capsid assembly that come from study of one phage typically illuminate the corresponding process in the other phages (and herpesviruses) under study. At the same time, each different phage that has been subjected to investigation shows its own variations, sometimes quite flamboyant, on the basic assembly process. As a result, work on understanding capsid assembly in several different tailed phage systems not only provides a multifaceted view of the basic assembly process they share but also gives a richly detailed picture of how a group of biological organisms can invent variations to ornament that basic process.

Bacteriophage HK97, the subject of this article, has provided significant contributions to understanding both aspects of capsid assembly. For example, the determination of the high-resolution structure of the capsid protein and the elucidation of the structural, dynamic, and energetic dimensions of the dramatic dance the protein goes through during capsid maturation are directly applicable to the corresponding processes in other tailed phages, while HK97's use of autocatalytic covalent bonds among the capsid subunits to stabilize the mature capsid is unique among the phages whose assembly mechanisms have been well studied. We describe these features and more below.

HK97 is an *Escherichia coli* phage that was isolated in about 1977, from a pigsty in Hong Kong (Dhillon et al. 1980). It originally came to the attention of one of us because of its long, elegant tail, which played a minor role in establishing that tail length in long-tailed phages is determined through the agency of what is now known as the tape measure protein (Katsura and Hendrix 1984). A surprising side observation from the tail length experiments was that there appeared to be no band on a sodium dodecyl sulfate (SDS) polyacrylamide gel corresponding to the major capsid protein. However, there was a large amount of protein that failed to enter even the stacking gel (Popa et al. 1991). Attempts to solve this conundrum led to the studies reported here.

15.2 The Assembly Pathway

The outlines of the HK97 head assembly pathway were established through a series of genetic and biochemical experiments (Duda et al. 1995a, b, c) together with some initial cryo-EM structures (Conway et al. 1995) (Fig. 15.1). Two of a collection of amber (conditional lethal) mutants of the phage failed to make heads but produced two different varieties of spherical structures that appeared similar to the proheads that had been identified in phages like λ and P22 (Duda et al. 1995c). Subsequent sequencing of the HK97 genome and biochemical and structural characterization of the heads and proheads led to the understanding that one of the proheads, named Prohead I (PH I), is made primarily of a full-length version of the major capsid protein (also known as gp5 = "*gene product* [of gene] *5*") (Duda et al. 1995b; Juhala et al. 2000). The other prohead, Prohead II (PH II), has a version of gp5 that is missing 102 aa from the N terminus of the 384 aa protein (Duda et al. 1995b), and this form of the protein is called gp5*. The mature head, called Head II (H II), has the same processed form of the protein as PH II, but it is in addition covalently cross-linked into the high-molecular-weight material that fails to enter the SDS gel. All three of these capsid-related forms made during phage infection also contain a small amount of the portal protein, gp3.

The cryo-EM structures show that the heads and proheads have icosahedral symmetry and a triangulation number (*T*) of 7 (Conway et al. 1995), which implies that there are 420 copies of the major capsid protein per shell, or 415 copies if one pentamer is replaced by the dodecameric portal ring. The H II shell has an angular, icosahedron-like shape, with a thin shell of uniform thickness. The hexamers appear to have nearly perfect sixfold symmetry, and the structure as a whole is a faithful example of quasi-equivalent packing of the subunits, as predicted in the 1962 Caspar and Klug paper (Caspar and Klug 1962). In contrast, the proheads have a more spherical shape, a smaller diameter, and a more corrugated surface with prominent protrusions on the outer surface, corre-

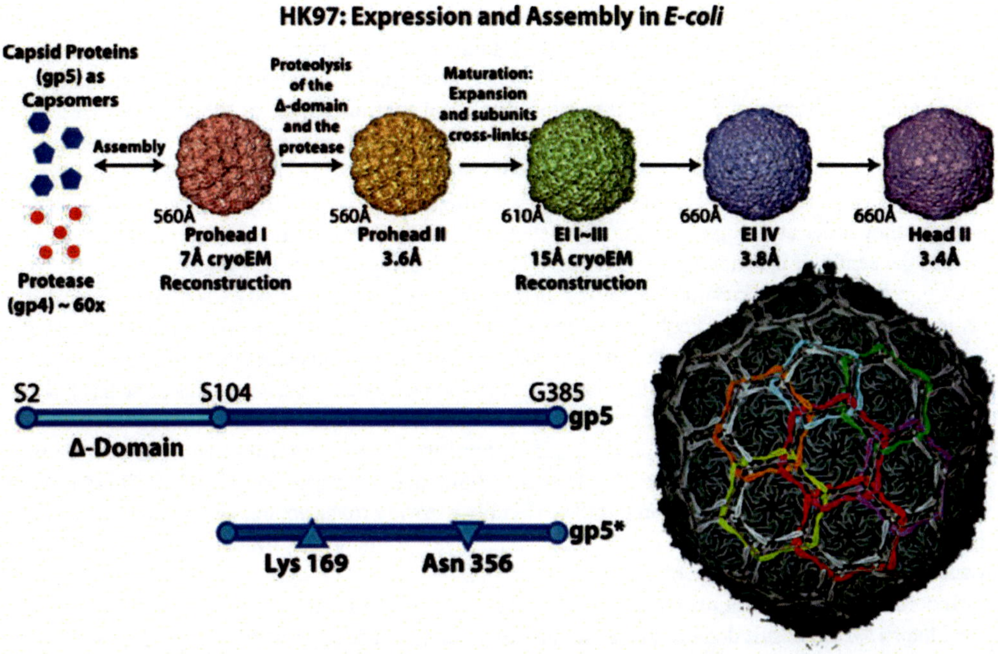

Fig. 15.1 The HK97 virus-like particle assembly and maturation pathway that is followed when only the coat protein and protease are co-expressed in *E. coli*. At the *top* are all the intermediates that have been characterized by crystallography and/or cryo-EM. Below is shown the processing that occurs to the capsid protein, gp5, and the residues that form the autocatalytic cross-link. At *right* is an enlargement of the final mature particle indicating that the cross-linked gp5 proteins mechanically chain-link the particle together. Each ring of the same color corresponds to five or six subunits chain-linked together by the isopeptide bond formed by the side chains of Asn356 and Lys169. Assembly and maturation: The capsid protein (gp5) immediately assembles into skewed hexamers and pentamers (generically called capsomers) shown in *blue*. The protease (gp4) is shown in *red*. The capsomers (60 hexamers and 12 pentamers) and approximately 60 copies of the protease co-assemble to form Prohead I, a roughly spherical particle, 51 nm in diameter. The hexamers in this particle are not symmetric, but "skewed," and correspond roughly to two shifted trapezoids (each composed of three subunits) related by twofold symmetry (see Fig. 15.3). Under normal circumstances, Prohead I is transient. The protease becomes active upon assembly and cleaves residues 2–103 (the delta domain) from all the gp5 subunits, creating gp5* as shown. The protease autodigests and all of the fragments diffuse from openings in the capsid. This creates Prohead II (51 nm) which is morphologically closely similar to Prohead I (including the skewed hexamers and virtually the same diameter); however, the mass of this particle is 13 MDa, compared to 17 MDa for Prohead I. Prohead I can be stabilized for study by not expressing the protease, or by co-expressing a mutant, inactive, protease. Either of these PH I particles can be disassembled and reassembled under mild conditions as indicated by the arrow from capsomers to Prohead I. Prohead II is a metastable particle that can only transition to EI-I. Conditions that disassemble Prohead I either cause nothing to happen to Prohead II or cause the transition to EI-I. There are many conditions that cause the transition, but dropping the pH from 7 to 4.0 was used for most of the in vitro studies of maturation. The transition to EI-I (56.0 nm), triggered by the pH drop, has a half-life of ~3 min and is stochastic, without populated intermediates. EI-I hexamers are sixfold symmetric, and the particle is cross-link competent, with cross-link initiation commencing virtually immediately after this particle is formed. Approximately 60% of the cross-links form before the morphology of EI-I changes (EI-II,III have essentially the same morphology as EI-I with increasing numbers of cross-links) to EI-IV. The process resembles a Brownian ratchet (a process by which thermal energy is captured by driving a process in only one direction) in which the loop containing Lys169 fluctuates until the covalent bond with Asn356 forms, locking it down and incrementally raising the energy of the particle. When a sufficient number form (~60%), the particle crosses the energy barrier and transitions to EI-IV, again without populated intermediates, a round (~62.5 nm), thin-shelled particle eventually forming all but 60 cross-links. This particle is the end point at pH 4.0 and was studied by crystallography and cryo-EM. These studies showed that subunits in the pentamers had not formed their final cross-link and that they were dynamic, fluctuating by 1.4 nm along the fivefold particle axes. Neutralizing EI-IV completes the cross-links with pentamer subunits, forming the fully mature, ~65 nm, faceted particle shown in the *lower right*

sponding to the hexameric and pentameric capsomers. Their most striking feature is a marked twofold skew to the hexamers, implying that the subunits of the hexamers are located in three different nonequivalent environments (Conway et al. 1995). The dramatic morphological difference between PH II and H II bespeaks a cataclysmic conformational rearrangement of the subunits within the structure; this morphological change is called "expansion" because it results in a ~20% increase in diameter and a roughly twofold increase in the interior volume of the shell (Conway et al. 1995).

The two prohead forms, PH I and PH II, are indistinguishable on their outer surfaces at low resolution, but they differ at the inner surface of the shell, where PH I has densities extending a short distance toward the center of the shell that are missing in PH II (Conway et al. 2007; Huang et al. 2011). These extra densities in PH I correspond to the 102 aa that have been removed from each subunit by proteolysis in PH II. This 102 aa proteolyzed segment of the protein is called the "delta domain," and it is becoming increasingly clear that the delta domain has crucial roles in assembly of the shell and, most likely, in driving the subsequent maturation of the shell. Exactly how it accomplishes these things is not yet entirely clear, and this is a topic of ongoing research (Gertsman et al. 2009; Johnson 2010).

The major capsid protein, gp5, can be produced in copious amounts from an overexpression plasmid, driven by a phage T7 promoter (Duda et al. 1995b), and this has greatly facilitated both production of structures from different parts of the assembly pathway for structural and biochemical studies as well as genetic analysis of the protein and the pathway. Remarkably, overexpression of gp5 alone leads to abundant production of correctly assembled T=7 PH I (Conway et al. 1995). Although correct folding of gp5 requires the host GroEL/S chaperonin (Ding et al. 1995; Popa et al. 1991; also see Chap. 13), this result demonstrates that no other phage-encoded proteins beyond gp5 are needed for efficient and accurate assembly of PH I. The next gene upstream from gene *5* encodes the maturation protease, gp4. If both genes *4* and *5* are included in the expression plasmid, PH II, with the delta domain removed from each subunit, is produced. In a HK97 infection, PH II is thought to be the capsid form that packages DNA, and by analogy with what is known for phage λ, we believe that it is the act of DNA packaging that triggers the shell expansion reaction in the infected cell. Purified PH II in Tris buffer can be triggered to go through the expansion reaction (in the absence of DNA) by a number of different manipulations of the solvent conditions, including addition of small amounts of organic solvents such as chloroform or butanol, or most conveniently, lowering the pH to about 4 (Duda et al. 1995a). Cross-linking occurs during the course of capsid expansion as catalytic sites form at the interfaces among subunits as a result of the conformational rearrangements of the subunits (Duda et al. 1995a).

15.3 Assembly of the Shell

PH I disassembles efficiently to capsomers (hexamers and pentamers) in high salt. The two capsomer forms are produced in the ratio expected for a T=7 structure (a 6:1 mass ratio, hexamers:pentamers). Remarkably, the two forms can be interconverted in vitro by adjusting solvent polarity in the presence of a small amount of denaturant (Xie and Hendrix 1995). The mechanism of this interconversion is unexplored, but it has allowed an experiment described below that asks what effect the hexamer:pentamer ratio has on shell assembly.

Capsomers dissociated in vitro from purified PH I can be reassembled in vitro by concentrating the protein by what is essentially a PEG precipitation (phase partition), with the assembly taking place in the "solid" phase (Xie and Hendrix 1995). This assembly is reasonably efficient (>50%), and strikingly accurate, with virtually all of the structures made being apparently complete, isometric, T=7 shells, indistinguishable from PH I made from plasmid expression of the same protein in the cell. Because these assembly reactions contain no protein other than gp5, we conclude that no other proteins, either phage- or host-encoded, are required for accurate assembly of T=7 PH I. However, this leaves open the question of whether the form (conformational or oligomeric) of gp5

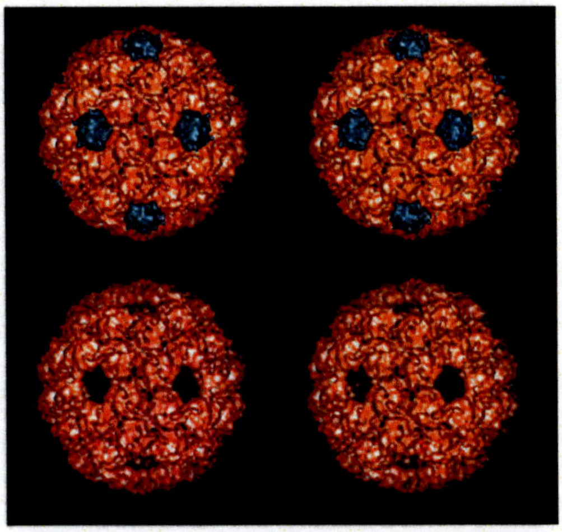

Fig. 15.2 Cryo-EM images of Whiffleballs. The *upper* structure is Prohead I produced in the cell by the Whiffleball mutant of gp5 (E219K). The *lower* structure is produced by disassembly and reassembly of the Prohead I in vitro. The pentamers are calculated as the difference between the two structures and are colored *blue*. The images are stereo pairs

is important for assembly. In particular, it could be imagined that the hexamer to pentamer ratio of the input protein could influence the T number of the assembled structure, inasmuch as icosahedral capsids of any T number have the same number of pentamers (one for each of the 12 vertices) but a variable number of hexamers depending on the T number. To test this question, we took advantage of the ability, mentioned above, to interconvert hexamers and pentamers and set up a series of in vitro capsid assembly reactions, all containing the same total mass of gp5 but differing in how that gp5 was partitioned between hexamers and pentamers. All of the structures produced were $T=7$ proheads, arguing that gp5 can accurately assemble into the correct size capsid regardless of the hexamer to pentamer ratio. On the other hand, there was a dramatic effect of this ratio on the efficiency with which the $T=7$ shells were made, with pure hexamers and pure pentamers both failing to make any detectable structures and the efficiency of assembly increasing from either extreme to a maximum at the ratio corresponding to a $T=7$ shell (Xie and Hendrix 1995). These results argue that hexamers and pentamers are the true precursors to shells in this assembly reaction, because if they were able, for example, to disassemble into monomers in the reaction conditions and then assemble directly into shells, all the different hexamer to pentamer ratios examined would have been expected to assemble with the same efficiency. The experiment also implies that neither pentamers nor hexamers can assemble into larger structures without the presence of the other flavor of capsomer.

The situation becomes both more complex and more interesting when we consider the properties of a point mutant of the major capsid protein, E219K (Li et al. 2005). This mutant protein is defective in a genetic complementation assay, but only slightly so. When the E219K protein is expressed from a plasmid, PH I is produced that is morphologically indistinguishable from the wild-type structure. However, the only unassembled capsomers detected in the lysate are hexamers, in contrast to the mixture of hexamers and pentamers found in a comparable wild-type lysate. Similarly, when the E219K PH I is dissociated in high salt, the only capsomers observed are hexamers, this despite the fact that the PH I had a normal complement of pentamers before dissociation. We speculate that for this mutant protein the equilibrium in the conversion between hexamers and pentamers is shifted in the direction to favor hexamers. Sufficient pentamers must be made in the host cytoplasm to support an abundant crop of PH I, but free E219K pentamers are evidently unstable in Tris buffer, possibly converting into hexamers. In any case, the real surprise from the E219K mutant comes when the hexamers produced by dissociation of mutant PH I are put under reassembly conditions, where they assemble into "Whiffleballs," structures that are indistinguishable from PH I except that they have holes where there should be pentamers (Fig. 15.2) (Li et al. 2005). On the face of it, this result appears to contradict the results from the earlier experiment described above, which indicated that hexamers could not assemble into shells in the absence of pentamers. We reconcile these results by postulating two conformational forms of hexamers, an "assembly competent" form and an "assembly naïve" form. We suggest that hexamers initially acquire their assembly competent form by an interaction with pentamers and that the assembly competent form is retained upon high-salt dissociation of PH I, as in the experiment with the E219K Whiffleball mutant. However, if the dissociated capsomers are subjected to interconversion between hexamers and pentamers prior to reassembly, as in the earlier experiment, we suggest the hexamers revert to the assembly naïve conformation and are unable to reassemble into shells until they have once again interacted with a pentamer. The most obvious candidate for the physical identity of the assembly competent form of the hexamer is the strongly twofold skewed hexamer that is seen in proheads but not in the mature capsid. It has not yet been possible to confirm this identification directly, but some of the results described below lend it support.

Genetic and biochemical experiments have identified a particular salt bridge between capsomers as having an essential role in assembly of capsomers into PH I (D. Tso, R. Duda & R. Hendrix, unpublished). The salt bridge forms between an aspartic acid (D) on one subunit and a lysine (K) on the subunit lying on the other side of the capsomer–capsomer interface. Each subunit contributes a D to one salt bridge and a K to a second salt bridge; this means there are 12 half salt bridges arrayed around the periphery of each hexamer and ten around each pentamer, each aligned precisely in the prohead structure with its partner on the adjacent capsomer. Although these KD salt bridges are unequivocal in both the PH I and PH II structures, the alignment of partners is completely lost in the mature H II structure, with former salt bridge partners separated by as much as 2.0 nm. In other words, if hexamers in the symmetrical (nonskewed) form that is found in H II had attempted to assemble into shells, they would have been unable to obtain a functional alignment of the crucial KD salt bridge partners. This argues that the skewed form of the hexamers is indeed the assembly competent form.

The role of the delta domain – the N-terminal 102 amino acids of each subunit that is removed by proteolysis in the transition from PH I to PH II – is still largely obscure but seems to be coming gradually into focus. The delta domain can be visualized in PH I structures as a poorly defined blob of density floating above the inner face of each capsomer and connected to it by thin threadlike densities. The delta domain is less well ordered than the main part of the subunit, and it has not been possible to trace the path of the polypeptide backbone of the delta domain. However, the amino acid sequence predicts that the delta domain should be almost entirely α-helix, with significant portions of coiled-coil structure (Conway et al. 1995), and the α-helical nature of the delta domain has been confirmed by Raman spectroscopy (Benevides et al. 2004; Nemecek et al. 2009). From the cryo-EM and X-ray structures of PH I, it appears that interactions between delta domains occur exclusively, or nearly so, between the delta domains of subunits of the same hexamer or pentamer and not across the boundary between capsomers (Conway et al. 2007; Huang et al. 2011).

We believe that the interactions among the delta domains of a hexamer have a role in allowing the hexamer to assume the skewed, assembly competent conformation. The most compelling evidence for this view comes from experiments combining differential scanning calorimetry with cryo-EM. Calorimetric experiments showed that PH I undergoes a reversible thermal transition at 53°C (Conway et al. 2007), and to determine the nature of the transition, grids with PH I being held at 60°C were directly plunge-frozen for examination in the microscope (Conway et al. 2007). A fraction of the proheads had reversed the thermal transition during freezing, and these were structurally indistinguishable from the PH I that had not been heated. On the other hand, reconstruction of the proheads that retained the high-temperature configuration differed from the low-temperature form in two regards: first, the delta domains were no longer visible, arguing that they were disordered and presumably not interacting with each other; second, the hexamers had lost their skew. Thus, in this experiment, there is a correlation between the interactions among the delta domains and the skewed conformation of the hexamers. The arrows of causation that relate these conformational states have not yet been determined experimentally, but a current working model (Gertsman et al. 2009; Johnson 2010) is as follows: (1) Energetically favorable interactions among the delta domains force energetically unfavorable changes in the main body of the subunit. These changes are manifested as the overall skewing of the hexamer and also as a twisting of each individual subunit (discussed in more detail below). (2) As a result of these changes, the hexamer is now in its assembly competent form, and capsid assembly can take place. (3) Once PH I is assembled, the delta domains are removed by proteolysis, but the skewed and twisted forms of the hexamers are maintained by quaternary interactions within the particle. (4) The conformational energy in the delta-less hexamer subunits is now available to drive the initial stages of capsid maturation in an exothermic process similar to that proposed for bacteriophage P22 (Galisteo and King 1993).

The different particle forms characterized both structurally and biochemically are described below. We start with the mature Head II because it was the first determined by crystallography and is the highest resolution structure. We then discuss each of the maturation intermediate structures and the details of the transitions.

15.4 Mature HK97 Virus-Like Particle Crystal Structure

The structure of the mature HK97 VLP capsid (H II) was determined at 3.5 Å resolution (Wikoff et al. 2000) (Fig. 15.1). This was the first dsDNA bacteriophage capsid structure determined, and it revealed the anticipated $T=7$ quasi-equivalent surface lattice composed of 12 pentamers and 60 hexamers of the single subunit type corresponding to gp5*, the proteolytically processed form of gp5 containing residues 104–385. The fold of gp5* was visualized in seven different subunits within the icosahedral asymmetric unit, assisting in the chain tracing and model building at the moderate resolution of the crystal structure. The independent subunits were closely similar when superimposed. The fold was unique when compared with other protein tertiary structures available at the time (Fig. 15.3). It was described in terms of an A (axial) domain, consisting of an α/β structure, that formed interactions near five and quasi-sixfold symmetry axes and a P (peripheral) domain, consisting of an extended two strand (E) loop at one end a tight (P) loop connecting two strands at the other. The two ends of the P domain participated in threefold and quasi-threefold interactions. A 4.3-nm "spine" helix supports the association of the A and P domains.

Since the HK97 structure was solved, subunit structures of six different bacteriophage capsids and the shell-forming domain of the herpes simplex virus subunit have been solved by either crystallography or cryo-EM at sufficient resolution to discern their tertiary structure organization. All of these tertiary structures appear to be related to the HK97 fold. This suggests that this tertiary structure is the canonical fold of dsDNA, tailed bacteriophages and, given their extraordinary abundance, may be the most common protein fold on the planet.

Previous studies showed that the subunits were covalently linked together. Mass spectrometry studies of protease digests revealed that the chemical cross-link occurred between Asn356 and Lys169 (Duda et al. 1995a). It was proposed that the side chains of these residues reacted to form an isopeptide bond with the loss of ammonia. This reaction has now been inferred in other cross-linked proteins (Kang et al. 2007), but at the time it was a novel observation. Further studies indicated that it was not a simple cross-link between subunits, but that the subunits were joined into catamers in which all the subunits associated in a fashion that could not be disrupted by treatment with strong denaturants such as SDS (Duda 1998). As described above, when run on an SDS gel, mature HK97 would not enter the gel, indicating an enormous molecular weight. Pretreatment with a protease, however, would release linear chains of hexamers and pentamers that did enter the gel. A proposal that explained this behavior invoked a chain-mail model similar to the armor worn by medieval knights (Duda 1998). The crystal structure proved this paradigm and showed that the chemical cross-link resulted in the formation of covalently joined pentameric and hexameric rings that were chain-linked together to form an integrated fabric that maintained continuity over the entire capsid shell (Fig. 15.1). A remarkable feature of the cross-linking reaction is that it is autocatalytic. The catalytic mechanism is discussed below.

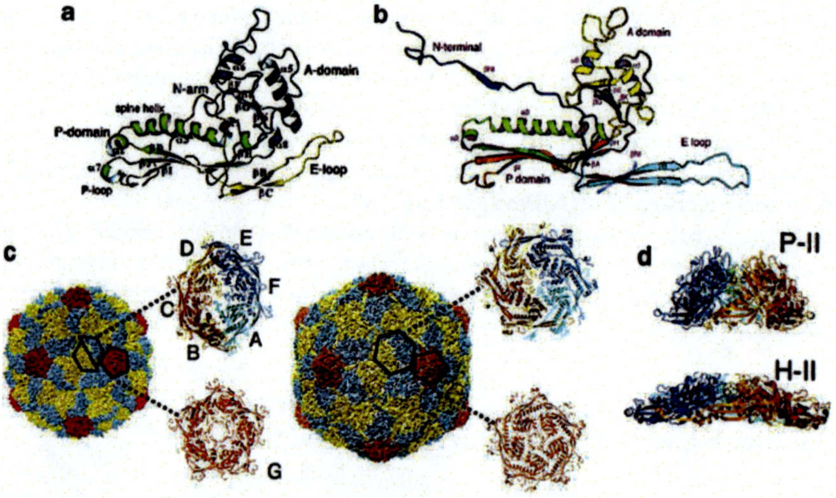

Fig. 15.3 Comparison of the Prohead II and Head II structures of HK97. (**a**) An isolated subunit from the 3.65 Å capsid structure of PH II showing the distortions of the tertiary structure proposed to generate the PH II metastability. The distortion is most apparent in the long "spine helix" (*green*) that is bent and partially unwound when compared to the same helix in panel (**b**). (**b**) An isolated subunit from the 3.4 Å structure of the fully mature H II capsid where the "spine helix" is straight and there are no unusually strained features of the tertiary structure. (**c**) Comparison of the PH II and H II capsids and hexamer and pentamer capsomers. The PH II particle is 51 nm in diameter, is composed of skewed hexamers, best described as twofold related trapezoids, has a shell thickness of ~3.5 nm, and is metastable. The mature H II particle is 65 nm in diameter with sixfold symmetric hexamer capsomers and a shell thickness of 1.8 nm. (**d**) A side view of hexamer capsomers in PH II (*top*) and H II (*bottom*). Subunits in PH II are in a more radial orientation relative to the particle center, increasing the thickness of the shell, while the subunits are more tangential in the H II particle, thinning the shell and increasing the diameter

15.5 Implication of the Mature Crystal Structure

The capsid structures of viruses determined prior to HK97 showed a packing of subunits that was rather typical of any protein oligomer. With the exception of short extensions at the N and C termini, the individual polypeptide chains remained within a discrete envelope that did not overlap significantly with neighboring subunits. For the most part subunits joined at continuous surfaces, defined by discrete protein envelopes that created the interactions that stabilized the particle. This is in fact a reasonable description of the HK97 *prohead*, as shown by the observation that if pentamers of wild-type protein are added to Whiffleballs (essentially PH I without the pentamers), the pentamers will join the Whiffleballs efficiently and close the holes (Li et al. 2005), into which the structure predicts they should fit exactly with no need for any conformational rearrangements (Huang et al. 2011). The quaternary association of *mature* HK97 is significantly different, as illustrated in Fig. 15.3. Each subunit contacts between five and nine neighbors, with the polypeptide chains extending great distances from the individual protein's center of gravity, overlapping in different layers with adjacent subunits.

The structure of the mature particle explains one aspect of why complex maturation pathways are required for dsDNA bacteriophages. The subunits in the final structure are so integrated that it is impossible to picture an assembly process that could reach that level of complexity in a single step. A logical sequence of assembly is envisioned in which an initial particle would have the protein composition and quasi-symmetry of the final product and the appearance of previously studied capsids with conventional protein–protein interactions that are relatively weak, bordering on reversible, in which subunits can anneal and self-correct their positions in the shell to form the "prohead." Encoded within the prohead is a maturation program that guides the protein quaternary structure through one or more intermediate structures that lead to the robust fabric observed in the mature particle. Deciphering the code of this program and its execution at the chemical level is a focus of effort that has produced some insights into the maturation process.

15.6 Prohead I

Normally PH I is transient and not detected. The PH I to PH II transition occurs with the proteolytic removal of the delta domain, accomplished by the phage-encoded protease, gp4. A homology model of gp4 has been proposed (Huang et al. 2011), based on the crystal structures of herpesvirus protease (reviewed in Tong 2002), which plays a similar role in herpesvirus maturation. Gp4 is a serine protease homolog as confirmed by mutating one of the predicted active site residues (His65) that resulted in loss of activity without affecting its folding and incorporation into the PH I particles. Prohead I particles can also be made without the protease if it is not co-expressed with the capsid protein. These particles have the characteristic skewed hexamers seen in PH II by cryo-EM. As described above, PH I can be disassembled into hexamers and pentamers and reassembled under mild conditions. Prohead I, containing an inactive mutant protease, can also be prepared by co-expressing mutant gp4 and gp5. These particles are designated Prohead I^{+P} and are more difficult to disassemble than PH I; however, they clearly distort under disassembly conditions, but the protease–protein complex largely maintains its integrity. A crystal structure at 5.2 Å resolution of Prohead I^{+P} was determined (Huang et al. 2011), and while the resolution was insufficient to observe side chains, it was possible to follow the main chain when guided by the subunit fold previously determined in the H II (Wikoff et al. 2003) and PH II (Gertsman et al. 2009) crystal structures. Figure 15.3 shows the distribution of subunits in the HK97 proheads. An interior view showed that the protease is associated with the delta domains of gp5 indicating molecular recognition between delta and gp4 (Huang et al. 2011). Wild-type gp4 efficiently proteolyses the delta domain and then autodigests into small peptides that diffuse from the particle (Conway et al. 1995). It is not known what the recognition elements are between the delta domain and protease, but it is likely that this localization is critical to the proper maturation of the particle.

15.7 Prohead II

Prohead II is the particle purified from the expression system when the wild-type gp4 is co-expressed with gp5. It has a mass of 13 MDa, a diameter of 56.0 nm, displays the skewed hexamers, and has no trace of the delta domain or protease when analyzed by SDS gel electrophoresis. It is a metastable particle poised for maturation (Li et al. 2005). The 3.65 Å resolution crystal structure of PH II was recently reported and revealed a distortion of the tertiary structures of the subunits when compared with the H II subunit structures (Gertsman et al. 2009, 2010). While the seven subunits in the icosahedral asymmetric unit of H II are virtually superimposable except for the N-terminal tails and the E-loops, each subunit of PH II has a different structure that depends on its location in the skewed hexamer. Striking differences in tertiary structure, when compared with the H II subunits, are observed in the long "spine" helix that is bent and unfolded to various degrees in different subunits. The "sixfold axis" in the skewed hexamers is closer to a twofold axis relating two trapezoids. This leads to differences in the distance of the subunits from the symmetry axis. As shown in Fig. 15.3, subunits B, C, and D are related to E, F, and A, respectively, with B and E furthest from the symmetry axis, D and A closest, and C and F intermediate. The further the subunit is from the symmetry axis, the more distorted the helix. The helix distortion is accompanied by a comparable degree of twist of the subunit tertiary structure about 3 β strands, as shown in Fig. 15.3. The subunit at the pentamer has a degree of distortion comparable to the D and A subunits at the hexamer. The entire particle appears to be poised for a transition that would place the subunit tertiary structures into their canonical folds observed in H II.

15.8 Expansion Intermediate I

When the transition is initiated by lowering the pH, expansion intermediate I (EI-I) is formed. The transition at pH 4.1 has a half-life of 1.6 min and is essentially a binary event, with no populated intermediates between PH II and EI-I, consistent with the strained appearance of PH II (Lee et al. 2005). This particle is 61.0 nm in diameter (~9% larger than PH II), has symmetric hexamers, and is cross-link competent (Lata et al. 2000; Lee et al. 2005, 2008). When EI-I forms, the end of the E-loop of the subunits moves approximately 4.0 nm, bringing Lys169 and Asn356 on neighboring subunits into proximity (Conway et al. 2001; Gertsman et al. 2009). As soon as EI-I forms, cross-links can be detected by SDS gel analysis. Cross-links form progressively into hexamers and pentamers. The quaternary structure of this particle remains the same until ~60% of the subunits are cross-linked. At that point, another binary transition occurs to the so-called balloon particle (Gan et al. 2004; Li et al. 2005). This suggested that the cross-links were driving the structural change from EI-I to the balloon, and this was further supported by the reduced rate of the transition when the cross-linking lysine residue was mutated to a tyrosine, inhibiting the covalent bond formation. A further mutation in which six residues were removed from the E-loop underwent no change beyond EI-I after lowering the pH, indicating that EI-I was at the energy minimum of the maturation landscape and that the balloon was generated through the action of a "Brownian ratchet" in which thermal energy was captured by the cross-links when the reactive residues were brought into proximity by Brownian fluctuations, a conclusion also inferred by a different route from calorimetric experiments (Ross et al. 2005). The fact that the Lys156→Tyr mutation could progress to the balloon, albeit at a much slower rate, without cross-link formation indicates that the noncovalent interactions of the loop were sufficient to drive further maturation.

15.9 Balloon

The penultimate intermediate in HK97 has an average diameter of 66 nm, the same as the mature H II. Its morphology, however, is spherical and not faceted like the mature H II. This intermediate is stable indefinitely at low pH and was purified and analyzed biochemically (Gan et al. 2004) and by crystallography and cryo-EM (Conway et al. 2007; Gan et al. 2006). The particle lacks a category of cross-links contributed by the pentamer subunits. The study demonstrated that balloon is a dynamic particle with the pentamers sampling a 1.4-nm trajectory in a pistonlike motion. When the pH is changed to 7 the final transition to H II occurs with the pentamers at their maximal radial position and fully cross-linked.

15.10 Cross-Linking Reaction

Just as there are seven different quasi-equivalent positions occupied by capsid subunits in a $T=7$ capsid, there are seven different quasi-equivalent sites of cross-linking in HK97's $T=7$ capsid. As stated above, cross-linking takes place throughout most of the course of expansion, starting when the

EI-I structure is reached. The cross-links can form only after the reacting side chains have moved into the correct positions adjacent to each other, and it is perhaps no surprise that this would happen with different kinetics at the different quasi-equivalent sites. Current data derived from biochemical and cryo-EM experiments (Lee et al. 2008) show that two of the cross-link classes (joining subunits A → A and E → D) form early and one (G → F) forms last. These results increase the precision with which we can describe the "dance" of the subunits during capsid maturation.

The cross-linking lysine (K169) is near the outer end of the E-loop, which in the prohead is oriented away from the surface of the capsid and disordered over the outer end. A prerequisite for the cross-linking reaction to occur is for the E-loop to descend and bring K169 next to the cross-linking asparagine (N356) on the adjacent subunit. It was apparent from the H II crystal structure that a glutamate (E363) from a third subunit is positioned ~ 0.2 nm from the cross-link bond, suggesting that it has a catalytic role, and this has been confirmed experimentally (Dierkes et al. 2009). Another important feature of the reaction site is that the three residues enumerated above are surrounded by three hydrophobic residues, a leucine, a methionine, and a valine (Helgstrand et al. 2003), the last of which is located in the E-loop near the cross-linking lysine, K169. When the E-loop descends and inserts K169 into the reaction center, it must simultaneously place the valine in position to complete the hydrophobic environment enclosing the reaction center. These considerations lead to a proposed reaction mechanism: (1) the positively charged cross-linking lysine (K169) and the negatively charged catalytic glutamate (E363) are enclosed together in a hydrophobic environment as a result of conformational changes in the capsid structure; (2) a proton moves from the lysine to the glutamate, neutralizing both buried charges (or in other words, the glutamate acts as a general base); and (3) the de-protonated amino group of the lysine, now a good nucleophile, attacks the amide carbon of the cross-linking asparagine (N356), forming the cross-link and releasing ammonia.

If an aspartate is substituted for the catalytic glutamate, the cross-linking reaction can still take place, and the cross-linking of a capsid can go to completion, but the rate of cross-linking is roughly 100-fold slower for the mutant protein than for wild type (Dierkes et al. 2009). This mutant protein does not support production of viable phages, arguing not only that cross-linking is essential for viability, but that it must happen rapidly to be effective. An important biological function of the cross-linking appears to be to stabilize the capsid, as implied by calorimetric experiments that show that almost all of the excess stability of the mature H II structure over PH II is due to the cross-links (Ross et al. 2005). By analogy with what is known for the role of phage lambda's capsid-stabilizing decoration protein, gpD (Sternberg and Weisberg 1977), we hypothesize that HK97 cross-linking is essential for allowing the capsid to resist the tightly packed genomic DNA and, therefore, that cross-linking must be completed before packaging is completed.

15.11 Conclusions and Prospects

We describe here the outlines of the developmental biology of phage HK97 heads – the assembly of several hundred protein subunits into a precisely specified icosahedrally symmetric protein shell and the maturation of that shell into the tough, highly cross-linked capsid that protects the phage's genomic DNA from environmental insult before delivering it to the next host. The HK97 system has proved to be particularly well suited to elucidating particular aspects of this developmental process. Among these is the initial assembly of the capsid shell from its subunits. Although many details remain to be worked out, we are able to present a scenario in which the proteins join together with sufficient precision to uniquely specify capsid size for this rather simple $T = 7$ structure, and they do so in a way that stores energy in the structure to drive the initial stages of capsid maturation. The maturation of the assembled shell has also yielded up a gratifying amount of useful information, largely due to the fact that it has been possible to produce several different structures from along the maturation pathway and to obtain structures of them at resolutions that have increased over the history of the project and continue to do so. We have made a movie of the maturation process, based on those structures, which is not only entertaining but is also a rich source of suggestions for experiments to test hypotheses about details of the maturation process. It has also begun to reveal in increasing detail dynamical aspects of the process, both at the level of individual protein subunits and of the whole capsid shell, and how the dynamics correlates with energetic changes in the proteins that drive maturation forward.

Progress on understanding the assembly and maturation of HK97 capsids has benefitted immeasurably from our being able to apply several complementary experimental approaches (genetics, biochemistry, structural biology, and other biophysical techniques) to the problem. For example, structural work suggested how the catalytic site for cross-linking forms and what the chemical mechanism might be, and we were able to test these with genetic and biochemical experiments. Our understanding of the detailed aspects of the protein dynamics and of the energetics is still at an early stage, and we anticipate that these will be areas of continuing progress for the HK97 system. Finally, we note that the delta domain, which is certainly the most enigmatic component of the whole pathway, appears to be central to almost every part of the process, particularly the dynamics and the energetics. Clarifying how this curious bit of protein carries out its roles will be a significant advance in understanding HK97 assembly and maturation, and may well be more generally informative about the different ways proteins can accomplish their jobs.

Acknowledgments The work described here is the product of a multilaboratory collaboration involving not only the two authors of this chapter but also Alasdair Steven, Bob Duda, and James Conway, as well as numerous graduate and undergraduate students, postdocs, and technical staff from all of our labs. There have been many important contributions as well from colleagues not as directly associated with the central "HK97 Cabal," including Lars Liljas, Philip Ross, George Thomas, Hiro Tsuruta, and their associates. Credit for our current understanding of HK97 capsid assembly and maturation is jointly owned. Work on this topic in the labs of the two authors of this chapter has been supported by NIH grants R01-GM47795 (RWH) and R01-AI40101 and R01-GM54076 (JEJ).

References

Baker ML, Jiang W, Rixon FJ, Chiu W (2005) Common ancestry of herpesviruses and tailed DNA bacteriophages. J Virol 79:14967–14970

Bamford DH, Grimes JM, Stuart DI (2005) What does structure tell us about virus evolution? Curr Opin Struct Biol 15:655–663

Benevides JM, Bondre P, Duda RL, Hendrix RW, Thomas GJ Jr (2004) Domain structures and roles in bacteriophage HK97 capsid assembly and maturation. Biochemistry 43:5428–5436

Caspar DL, Klug A (1962) Physical principles in the construction of regular viruses. Cold Spring Harb Symp Quant Biol 27:1–24

Conway JF, Duda RL, Cheng N, Hendrix RW, Steven AC (1995) Proteolytic and conformational control of virus capsid maturation: the bacteriophage HK97 system. J Mol Biol 253:86–99

Conway JF, Wikoff WR, Cheng N, Duda RL, Hendrix RW, Johnson JE, Steven AC (2001) Virus maturation involving large subunit rotations and local refolding. Science 292:744–748

Conway JF, Cheng N, Ross PD, Hendrix RW, Duda RL, Steven AC (2007) A thermally induced phase transition in a viral capsid transforms the hexamers, leaving the pentamers unchanged. J Struct Biol 158:224–232

Dhillon EK, Dhillon TS, Lai AN, Linn S (1980) Host range, immunity and antigenic properties of lambdoid coliphage HK97. J Gen Virol 50:217–220

Dierkes LE, Peebles CL, Firek BA, Hendrix RW, Duda RL (2009) Mutational analysis of a conserved glutamic acid required for self-catalyzed cross-linking of bacteriophage HK97 capsids. J Virol 83:2088–2098

Ding Y, Duda RL, Hendrix RW, Rosenberg JM (1995) Complexes between chaperonin GroEL and the capsid protein of bacteriophage HK97. Biochemistry 34:14918–14931

Duda R (1998) Protein chainmail: catenated protein in viral capsids. Cell 94:55–60

Duda RL, Hempel J, Michel H, Shabanowitz J, Hunt D, Hendrix RW (1995a) Structural transitions during bacteriophage HK97 head assembly. J Mol Biol 247:618–635

Duda RL, Martincic K, Hendrix RW (1995b) Genetic basis of bacteriophage HK97 prohead assembly. J Mol Biol 247:636–647
Duda RL, Martincic K, Xie Z, Hendrix RW (1995c) Bacteriophage HK97 head assembly. FEMS Microbiol Rev 17:41–46
Duda RL, Hendrix RW, Huang WM, Conway JF (2006) Shared architecture of bacteriophage SPO1 and herpesvirus capsids. Curr Biol 16:R11–R13
Galisteo ML, King J (1993) Conformational transformations in the protein lattice of phage P22 procapsids. Biophys J 65:227–235
Gan L, Conway JF, Firek BA, Cheng N, Hendrix RW, Steven AC, Johnson JE, Duda RL (2004) Control of crosslinking by quaternary structure changes during bacteriophage HK97 maturation. Mol Cell 14:559–569
Gan L, Speir JA, Conway JF, Lander G, Cheng N, Firek BA, Hendrix RW, Duda RL, Liljas L, Johnson JE (2006) Capsid conformational sampling in HK97 maturation visualized by X-ray crystallography and cryo-EM. Structure 14:1655–1665
Gertsman I, Gan L, Guttman M, Lee K, Speir JA, Duda RL, Hendrix RW, Komives EA, Johnson JE (2009) An unexpected twist in viral capsid maturation. Nature 458:646–650
Gertsman I, Fu CY, Huang R, Komives EA, Johnson JE (2010) Critical salt bridges guide capsid assembly, stability, and maturation behavior in bacteriophage HK97. Mol Cell Proteomics 9:1752–1763
Helgstrand C, Wikoff WR, Duda RL, Hendrix RW, Johnson JE, Liljas L (2003) The refined structure of a protein catenane: the HK97 bacteriophage capsid at 3.44 A resolution. J Mol Biol 334:885–899
Hendrix RW (1999) Evolution: the long evolutionary reach of viruses. Curr Biol 9:R914–R917
Hendrix RW (2002) Bacteriophages: evolution of the majority. Theor Popul Biol 61:471–480
Huang RK, Khayat R, Lee KK, Gertsman I, Duda RL, Hendrix RW, Johnson JE (2011) The prohead-I structure of bacteriophage HK97: implications for scaffold-mediated control of particle assembly and maturation. J Mol Biol 408:541–554
Johnson JE (2010) Virus particle maturation: insights into elegantly programmed nanomachines. Curr Opin Struct Biol 20:210–216
Juhala RJ, Ford ME, Duda RL, Youlton A, Hatfull GF, Hendrix RW (2000) Genomic sequences of bacteriophages HK97 and HK022: pervasive genetic mosaicism in the lambdoid bacteriophages. J Mol Biol 299:27–51
Kang HJ, Coulibaly F, Clow F, Proft T, Baker EN (2007) Stabilizing isopeptide bonds revealed in gram-positive bacterial pilus structure. Science 318:1625–1628
Katsura I, Hendrix RW (1984) Length determination in bacteriophage lambda tails. Cell 39:691–698
Lata R, Conway JF, Cheng N, Duda RL, Hendrix RW, Wikoff WR, Johnson JE, Tsuruta H, Steven AC (2000) Maturation dynamics of a viral capsid: visualization of transitional intermediate states. Cell 100:253–263
Lee KK, Tsuruta H, Hendrix RW, Duda RL, Johnson JE (2005) Cooperative reorganization of a 420 subunit virus capsid. J Mol Biol 352:723–735
Lee KK, Gan L, Tsuruta H, Moyer C, Conway JF, Duda RL, Hendrix RW, Steven AC, Johnson JE (2008) Virus capsid expansion driven by the capture of mobile surface loops. Structure 16:1491–1502
Li Y, Conway JF, Cheng N, Steven AC, Hendrix RW, Duda RL (2005) Control of virus assembly: HK97 "Whiffleball" mutant capsids without pentons. J Mol Biol 348:167–182
Nemecek D, Overman SA, Hendrix RW, Thomas GJ Jr (2009) Unfolding thermodynamics of the delta-domain in the prohead I subunit of phage HK97: determination by factor analysis of Raman spectra. J Mol Biol 385:628–641
Popa MP, McKelvey TA, Hempel J, Hendrix RW (1991) Bacteriophage HK97 structure: wholesale covalent crosslinking between the major head shell subunits. J Virol 65:3227–3237
Ross PD, Cheng N, Conway JF, Firek BA, Hendrix RW, Duda RL, Steven AC (2005) Crosslinking renders bacteriophage HK97 capsid maturation irreversible and effects an essential stabilization. EMBO J 24:1352–1363
Sternberg N, Weisberg R (1977) Packaging of coliphage lambda DNA II. The role of the D protein. J Mol Biol 117:733–759
Suttle CA (2007) Marine viruses – major players in the global ecosystem. Nat Rev Microbiol 5:801–812
Tong L (2002) Viral proteases. Chem Rev 102:4609–4626
Wikoff WR, Liljas L, Duda RL, Tsuruta H, Hendrix RW, Johnson JE (2000) Topologically linked protein rings in the bacteriophage HK97 capsid. Science 289:2129–2133
Wikoff WR, Che Z, Duda RL, Hendrix RW, Johnson JE (2003) Crystallization and preliminary analysis of a dsDNA bacteriophage capsid intermediate: prohead II of HK97. Acta Crystallogr D Biol Crystallogr 59:2060–2064
Xie Z, Hendrix RW (1995) Assembly in vitro of bacteriophage HK97 proheads. J Mol Biol 253:74–85

Chapter 16
Lipid-Containing Viruses: Bacteriophage PRD1 Assembly

Sarah J. Butcher, Violeta Manole, and Nelli J. Karhu

Abstract PRD1 is a tailless icosahedrally symmetric virus containing an internal lipid membrane beneath the protein capsid. Its linear dsDNA genome and covalently attached terminal proteins are delivered into the cell where replication occurs via a protein-primed mechanism. Extensive studies have been carried out to decipher the roles of the 37 viral proteins in PRD1 assembly, their association in virus particles and lately, especially the functioning of the unique packaging machinery that translocates the genome into the procapsid. These issues will be addressed in this chapter especially in the context of the structure of PRD1. We will also discuss the major challenges still to be addressed in PRD1 assembly.

16.1 Introduction

The archetypal member of the Tectiviridae family is PRD1, shown schematically in Fig. 16.1. It infects Gram-negative bacteria that contain a conjugative IncP, IncN or IncW plasmid which codes for the receptor. Viruses of this family typically have a spike-decorated, tailless, icosahedrally symmetric capsid approximately 65 nm in diameter, surrounding a lipid membrane which in turn encapsulates the linear, double-stranded DNA genome (Bamford and Ackermann 2000; Saren et al. 2005). There are four major reasons why PRD1 is interesting: First, the genome is replicated using a protein-primed, sliding-back mechanism utilising inverted terminal repeats on the genome and a covalently bound terminal protein at the 5′ ends (Bamford et al. 1983; Caldentey et al. 1992; Salas 1991; Savilahti and Bamford 1986; Savilahti et al. 1989, 1991). Second, it can be used to study viral membrane biogenesis in a well-characterised genetic background as it grows, for example, in *Escherichia coli* (Laurinavicius et al. 2004, 2007). Third, structure-based phylogeny of viruses has been shaken out of its slumber by the observation that the major capsid protein fold of PRD1 resembles that found in certain viruses infecting archaea and eukaryotes, possibly signifying descent from

S.J. Butcher (✉) • V. Manole
Structural Biology and Biophysics Programme, Institute of Biotechnology,
University of Helsinki, P.O. Box 65, 00014 Helsinki, Finland
e-mail: sarah.butcher@helsinki.fi

N.J. Karhu
Next Biomed Technologies NBT Oy, Biomedicum 2U, Tukholmankatu 8,
00290 Helsinki, Finland

M.G. Rossmann and V.B. Rao (eds.), *Viral Molecular Machines*,
Advances in Experimental Medicine and Biology 726, DOI 10.1007/978-1-4614-0980-9_16,

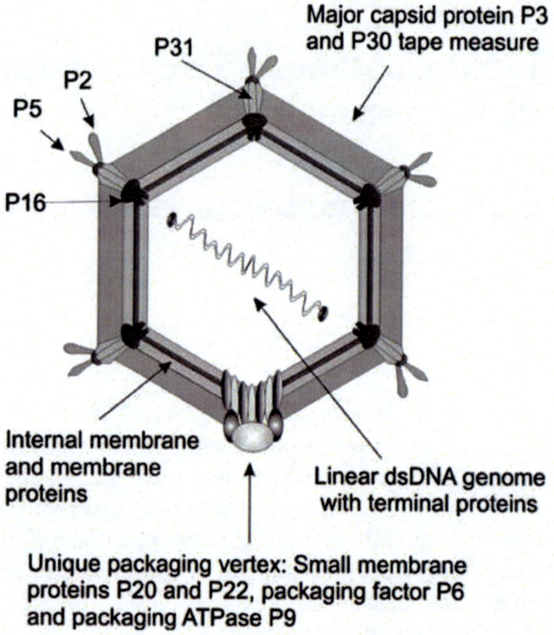

Fig. 16.1 Schematic representation of the PRD1 virion including a model for the unique vertex organisation

a common ancestor. This idea was originally formulated when the first picornavirus structures were determined and has recently been expanded to other capsid folds (Abad-Zapatero et al. 1980; Abrescia et al. 2010; Bamford 2000, 2003; Bamford et al. 2002a, 2005; Benson et al. 1999; Hogle et al. 1985; Nandhagopal et al. 2002; Rice et al. 2004; Rossmann et al. 1985; Rossmann and Johnson 1989; Simpson et al. 2003). Finally, and of most relevance to this work, the structure of the virion is known to atomic resolution; extensive genetic and biochemical analyses have been carried out on the system and an in vitro packaging system is available for PRD1, thus many of the steps in virus assembly have been characterised (Abrescia et al. 2004; Bamford et al. 1995; Caldentey et al. 1990; Cockburn et al. 2004; Huiskonen et al. 2007; Karhu et al. 2007; Merckel et al. 2005; Mindich et al. 1982b; Strömsten et al. 2005; Xu et al. 2003; Ziedaite et al. 2009).

16.2 Virion Structure and Properties

Extensive genetic and structural studies have been carried out on the virion and subviral particles of PRD1, using, for example, Raman spectroscopy, electron microscopy and X-ray crystallography. The constituents of the virion are approximately 70% protein, 15% lipid and 15% DNA, with a total mass of about 66 MDa. The virion contains about 18 protein species split roughly evenly between the capsid and the membrane (Fig. 16.1 and Table 16.1). Some of the structural proteins have been characterised individually, some in the context of the capsid or subviral complexes. X-ray crystallography of the major capsid protein, P3, showed that it consists of two eight-stranded β-barrels or jelly rolls (Fig. 16.2). The topology in both barrels is similar even though there is little sequence conservation. Each jelly roll contains a short α-helix inserted between the F and G strands. The first helix (FG1-α) locks the two jelly rolls together, the "trimerisation" loop (the FG1 loop) following FG1-α forms interactions with two other subunits to make a stable trimer. The barrels thus give the trimer pseudo-hexagonal symmetry that is well suited to packing on a spherical surface (Benson et al. 1999).

Table 16.1 PRD1 genes, open reading frames (ORF) and corresponding proteins

Gene[a]	ORF[a]	Coordinates in PRD1 genome[b]	Protein	Mass kDa[c]	Description
VIII		233–1012	P8	29.5	Genome terminal protein
I		1,016–2,677	P1	63.3	DNA polymerase
	(ORF a)	2,415–2,495		3.1	
XV		2,679–3,128	P15	17.3	Muramidase
II		3,128–4,903	P2	63.7	Receptor binding
	(ORF b)	3,453–3,587		5.1	
XXXI	(ORF c)	4,907–5,287	P31	13.7	Pentameric base of spike
	(ORF d)	5,103–5,294		7.2	
V		5,287–6,309	P5	34.2	Trimeric spike protein
XVII		6,328–6,588	P17	9.5	Assembly
XXXIII	(ORF f)	6,578–6,784	P33	7.5	Assembly
VI		6,784–7,284	P6	17.6	Minor capsid protein. DNA packaging
X		7,029–7,640	P10	20.6	Assembly
IX		7,637–8,320	P9	25.8	Minor capsid protein. DNA packaging ATPase
	(ORF i)	8,332–8,460		4.5	
XX	(ORF j)	8,460–8,588	P20	4.7	DNA packaging
III		8,595–9,782	P3	43.1	Major capsid protein
	(ORF h)	9,427–9,681		9.2	
XXII	(ORF k)	9,801–9,944	P22	5.5	DNA packaging
	(ORF l)	10,044–10,166		4.4	
XVIII	(ORF m)	10168–10440	P18	9.8	DNA delivery
XXXII	(ORF n)	10,440–10,604	P32	5.4	DNA delivery
XXXIV	(ORF o)	10,617–10,823	P34	6.7	
XXX	(ORF p)	10,833–11,087	P30	9.0	Minor capsid protein
	(ORF q)	11,090–11,200		4.2	
XI		11,202–11,825	P11	22.2	DNA delivery
XVI	(ORF s)	11,836–12,189	P16	12.6	Infectivity
VII		12,190–12,987	P7	27.1	DNA delivery, transglycosylase
XIV		12,535–12,987	P14	15.0	DNA delivery
XXXV	(ORF t)	12,984–13,337	P35	12.8	Holin
XXXVI	(ORF u)	13,330–13,692	P36	12.7	Lysis protein, Rz1-like
XXXVII	(ORF v)	13,616–13,888	P37	10.2	Lysis protein, Rz-like
XIX		14,132–13,848[d]	P19	10.5	ssDNA-binding protein
XII		14,687–14,205[d]	P12	16.6	ssDNA-binding protein

[a]ORF shown to code for functional proteins are classified as genes and are given Roman numerals
[b]GenBank accession AY848689
[c]The mass does only include the initial methionine if present in the mature protein
[d]Gene transcription in the opposite direction to that of the rest of the genes

Three-dimensional electron microscopy and X-ray crystallography of virions, and P2-deficient particles, respectively, have shown that the capsid is organised on a pseudo $T=25$ surface lattice. The major capsid protein is arranged in 240 trimers forming the facets, with pentamers of P31 at the vertices. There is an underlying framework of dimers of the minor coat protein P30 adjacent to the membrane, and the transmembrane protein P16 is found under the vertices as illustrated in Fig. 16.2 (Abrescia et al. 2004; Butcher et al. 1995; Cockburn et al. 2004). In the capsid, the termini of P3 face the membrane and have different conformations depending on the location in the surface lattice. In half of the P3 subunits, the first five amino-terminal residues are embedded in the outer leaflet of the

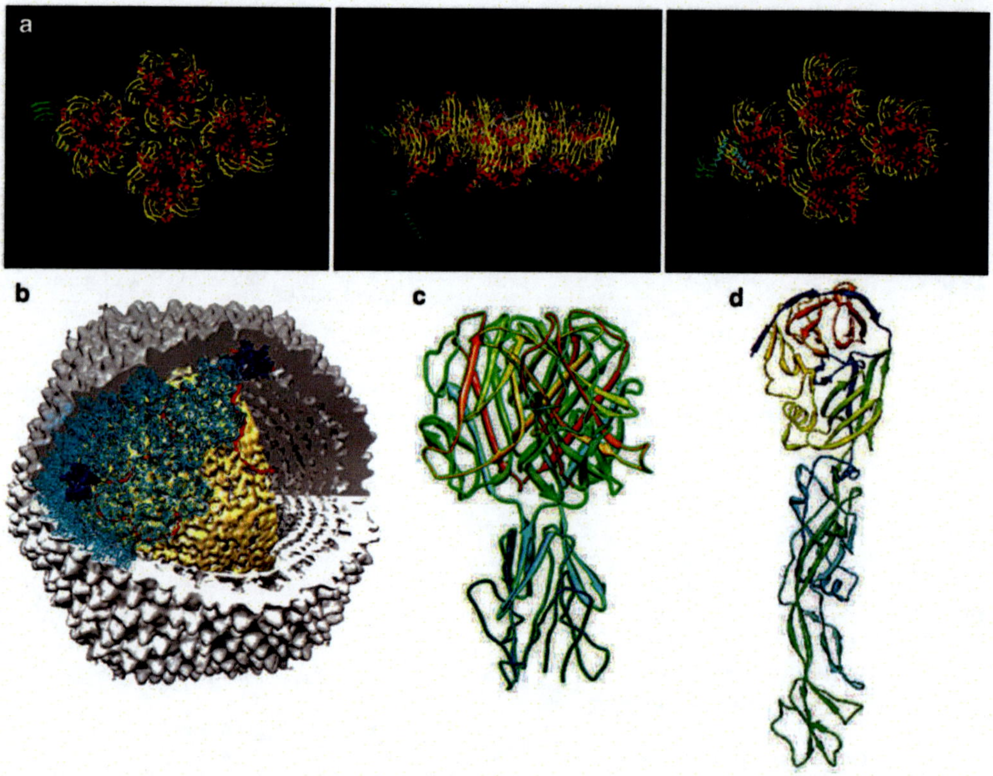

Fig. 16.2 Structure of PRD1. (**a**) Ribbon model of the asymmetric unit: P3 sheet in *yellow*, helices in *red*, loops in *grey*. P16 in *cyan*, P30 in *blue* and P31 in *green*, from the top, side and bottom. The polarity of the structure is very obvious, the helices control the interaction with the membrane, both in P3 and in P16, with the majority interacting with the phospholipid head groups, and only one transmembrane helix in P16 locking the membrane in place, at the position of greatest curvature of the membrane (Abrescia et al. 2004). (**b**) Isosurface representation of an icosahedral reconstruction of the particle (*grey*) with a segment cut away to reveal the underlying outer leaflet of the membrane (*yellow*), the inner leaflet and the organised DNA. Several copies of the atomic models of P3 (*blue ribbon*), P30 (*red ribbon*) and P31 (*navy blue ribbon*), are shown superposed on top of the outer leaflet showing how P30 stretches along the facet edges (San Martín et al. 2002; Abrescia et al. 2004). (**c**) Ribbon model of P5, rainbow coloured from the N terminus in *blue* to the C terminus in *red* (Merckel et al. 2005). (**d**) Ribbon model of P2, rainbow coloured from the N terminus in *blue* to the C terminus in *red* (Xu et al. 2003)

membrane, making P3 a peripheral membrane protein. These are the P3 subunits arranged along the edge of the facet, and may help to stabilise the membrane curvature at these points. The other termini switch the subunit conformation depending on the position in the lattice (Abrescia et al. 2004; Cockburn et al. 2004).

P31 is a single eight-stranded β-barrel which also shares the same topology as the P3 β-barrels (Abrescia et al. 2004). P31 pentamers are stabilised by β-strand A interacting with β-strand B of the adjacent P31 monomer and by side-to-side interactions between the BIDG sheet of one subunit and the neighbour's strands G and F. P31 pentamers are intricately linked to the peripentonal P3 trimers, to the C terminus of P30 and to the minor integral membrane protein, P16. P30 is an intrinsically disordered protein in solution, but in the capsid it plays an essential role in a very extended, yet distinct conformation, forming dimers along the facet edges, stabilising the junction of the facets and anchoring the vertices in place through interaction with the integral membrane protein P16 (Abrescia et al. 2004; Cockburn et al. 2004; Jaatinen et al. 2004; Rydman et al. 2001). The P30 dimers are formed by amino-terminal hooks. Together, 30 such dimers form a P30 network over the membrane (Fig. 16.2). The structures of the other membrane proteins are unresolved in the atomic model. Raman spectroscopy has shown that the lipids are in a liquid crystalline state, with the head groups tightly interacting with the DNA (Cockburn et al. 2004; Tuma et al. 1996b). The effective DNA concentration has been estimated to be about 515 mg/ml and the apparent pressure of the DNA within the capsid has been estimated to be around 45 atm. The spacing between the outermost DNA layer and the inner leaflet is about 2 nm, the spacing between the second, third and fourth layers is about 2.2 nm. However, the spacing between the first and second layers is about 2.6 nm (Cockburn et al. 2004).

Two types of spike proteins, the adsorption protein, P2, and a second trimeric spike protein, P5, are attached to P31 as shown at low resolution by averaging vertices from electron micrographs of virions, P2-deficient and P5-deficient particles (Grahn et al. 1999; Huiskonen et al. 2007; Rydman et al. 1999). The exact interaction is averaged out in the electron density of the P2-deficient virus particle and thus it is still not clear how the symmetry mismatches between P31, P5 and P2 are accommodated in the particle (Abrescia et al. 2004). P2 is an extended monomer with a pseudo β-propeller at the distal end to the capsid which most likely contains the receptor-binding site (Xu et al. 2003). P5 is a trimeric protein with the N terminus apparently inserted into the P31 pentamer (Caldentey et al. 2000; Huiskonen et al. 2007; Sokolova et al. 2001). The P5 stalk resembles the fold of the adenovirus fibre (Merckel et al. 2005; van Raaij et al. 1999) and the head is a ten-stranded β-barrel with a TNF-like fold (Huiskonen et al. 2007; Merckel et al. 2005) (Fig. 16.2).

16.3 Assembly

After the viral genome has entered the cell, transcription begins (Fig. 16.3). The genome of 14,927 bp with 37 identified genes (GenBank accession number AY848689) is organised into five operons as illustrated in Fig. 16.4. (Bamford et al. 1991, 2002b; Grahn et al. 1994; Saren et al. 2005). The properties of the identified ORFs in the genome are summarised in Table 16.1. The early operons, OE1 and OE2, express the terminal protein, the DNA polymerase and the two single-stranded DNA-binding proteins that take part in the replication reaction that is illustrated in the overall life cycle of PRD1 in Fig. 16.3. The three late operons mainly contain the genes for the structural components of the virion as shown in Fig. 16.4. PRD1 uses protein-primed replication to initiate DNA synthesis. Replication starts with the recognition of the terminal protein by the PRD1 DNA polymerase and free terminal protein. The covalently-linked terminal protein on each 5′ end of the linear dsDNA molecule serves as a primer for initiation instead of the more usual RNA priming mechanism. The polymerase then catalyses covalent phosphodiester bond formation between the 5′-terminal nucleotide (dGMP) and Tyr190 in the terminal protein (Savilahti et al. 1991). DNA polymerisation is then initiated in the inverted terminal repeats from the complementary base to the fourth nucleotide in the 3′-terminal end of the genome (Caldentey et al. 1993). The polymerase's "sliding-back" mechanism maintains the 5′ end of the sequence. Replication from both DNA ends is coupled to strand displacement, resulting in duplication of the parental strands (Salas 1999; Yoo and Ito 1989). Early on in this replication process, the terminal protein recruits the viral polymerase to the bacterial nucleoid, thus helping to organise the DNA replication machinery (Munoz-Espin et al. 2010). The single-stranded DNA-binding proteins P12 and P19 (Table 16.1) stimulate replication in a model system (Pakula et al. 1990, 1993).

The first DNA-containing viral particles are seen about 40 min in to the infection, and cell lysis occurs after 60 min. Late protein synthesis starts about 15 min post infection in EM, the first precursors produced in the cytoplasm are a number of soluble capsid proteins including P3 trimers, P5 trimers and P31 pentamers, reflecting their final state in the assembled capsid. Phage-encoded

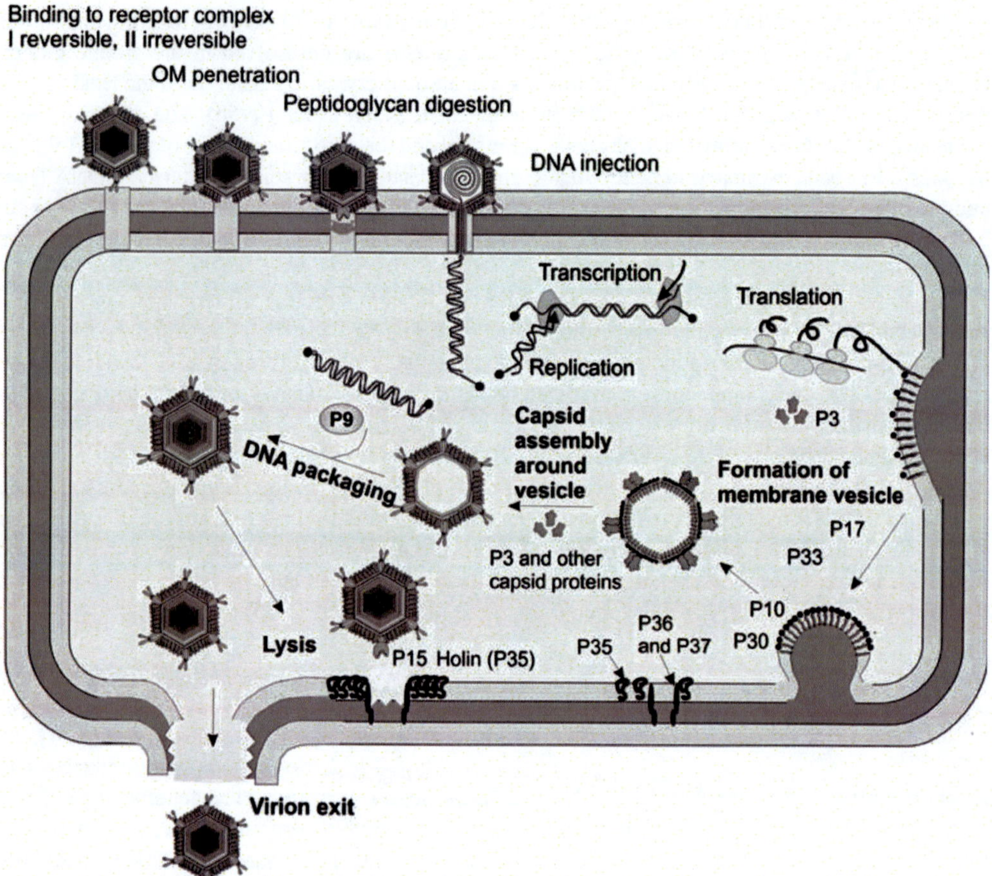

Fig. 16.3 Schematic representation of the life cycle of PRD1

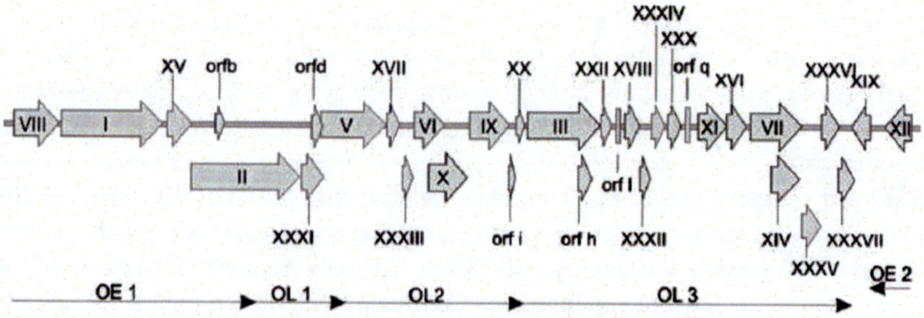

Fig. 16.4 Genome organisation of PRD1. ORFs have a roman numeral if they have been shown to encode a protein. The two early (OE) and the three late (OL) operons of PRD1 are shown at the *bottom*

membrane proteins such as P18, P7, P11 and P14 are found in the cytoplasmic membrane of the host (Mindich et al. 1982b; Rydman et al. 1999). The next observed intermediate is the procapsid which lacks the viral DNA. No other assembly intermediates have been directly observed. The procapsids are subsequently packaged with the linear dsDNA before the host cell is lysed. The cellular chaperones, GroEL and GroES, are required for the folding of the major capsid protein, although the viral chaperone P33 can substitute for GroES (Bamford et al. 2002b; Hänninen et al. 1997). The membrane-bound P10 is also a viral chaperone aiding the formation of the procapsid, potentially by budding off phage-specific membrane patches from the cytoplasmic membrane (Rydman et al. 2001). This process could involve the interaction with the essential major and minor capsid proteins P3, P31 and P30, and the integral membrane proteins of the virion, like P16. Budding is probably thus driven by multivalent interactions and by the extensive lateral interactions of P3 (Butcher et al. 1995; Mindich et al. 1982b; Rydman et al. 2001). The soluble chaperone P17 is also known to play an essential role in particle assembly and can interact with positively charged lipids (Holopainen et al. 2000; Mindich et al. 1982b). Particle assembly coincides with additional folding of the P3 N- and C-termini from random coil into short α-helices and β-sheets, detected by Raman spectroscopy (Tuma et al. 1996a). This agrees with the comparison of the atomic models of the major capsid protein in solution (disordered termini) and in the assembled capsid where folding of the termini into different conformational states dictates the quasi-equivalent interactions in the capsid and interaction with proteins P30, P31, P16 and the membrane (Abrescia et al. 2004; Benson et al. 1999; Cockburn et al. 2004; San Martín et al. 2002). Curvature is induced into what otherwise might be a planar P3 lattice by the introduction of the vertex protein P31 and the membrane curvature is maintained at this point by the integral membrane protein P16 (Abrescia et al. 2004; Cockburn et al. 2004; Jaatinen et al. 2004). It is thought that the length of each facet is dictated by the length of the essential P30 dimer that runs down the edges of the facets, organising the P3 trimers (Abrescia et al. 2004).

What role(s) does the membrane have in assembly? One could consider it initially as an assembly platform, where the viral membrane proteins can congregate prior to capsid formation. Once the vesicle has formed, it can affect both the DNA packaging and the assembly of additional capsid proteins. Careful characterisation of the lipid content of both the host and the phage has shown that the virus membranes contain significantly more phosphatidylglycerol and less phosphatidylethanolamine than the host membranes. The distribution within the membrane leaflets is skewed as phosphatidylglycerol and cardiolipin are enriched in the outer membrane leaflet and the zwitterionic phosphatidylethanolamine is enriched in the inner one (Cockburn et al. 2004; Laurinavicius et al. 2004, 2007). The membrane also undergoes an approximately 5% expansion in the virion compared to empty, packaging ATPase P9-deficient particles (Butcher et al. 1995). It is likely that the shape and charge of the phospholipid molecules and specific lipid–viral protein interactions are the driving forces for this asymmetric distribution of phospholipids in the membrane. As there is a very close interaction with the viral DNA and the inner leaflet, this distribution may be important in condensing the DNA during packaging (Tuma et al. 1996b). The negative charge of the outer leaflet could help the assembly of P3 which has a positively charged base (Cockburn et al. 2004; San Martín et al. 2002).

The assembly of the spike proteins P5 and P2 onto the virion is dependent on the incorporation of P31 (Rydman et al. 1999). Genetically, P2 assembly is dependent on P5, as mutants in P5 lack P2 (Bamford and Bamford 2000). However, P5 and P2 do not interact in solution and image reconstruction of the vertices has shown that P5 and P2 both interact with the vertex protein P31 (Caldentey et al. 2000; Huiskonen et al. 2007; Sokolova et al. 2001). This assembly occurs prior to packaging the DNA into the procapsid.

There is evidence for a unique packaging vertex in PRD1 from antibody labelling studies and mutational studies (Gowen et al. 2003; Strömsten et al. 2003), probably lacking proteins P2, P5 and P31 (Gowen et al. 2003; Huiskonen et al. 2007). How could this anomaly occur? The unique vertex might either be the initiation point for assembly as in tailed phages, or, due to the incorporation of the membrane, it could also be the last vertex to be formed, where pinching off from the host membrane has occurred. As the unique vertex is intact in P16-, P2-, P5- and P31-deficient particles, the 11 adsorption vertices and the 1 portal vertex are functionally and structurally distinct. Hence, their assembly is apparently independent (Jaatinen et al. 2004; Strömsten et al. 2003). It has been shown that the packaging vertex includes proteins P6, P9 (the packaging ATPase) and the membrane proteins P20 and P22 (Gowen et al. 2003; Strömsten et al. 2003). The tractability of the PRD1 system

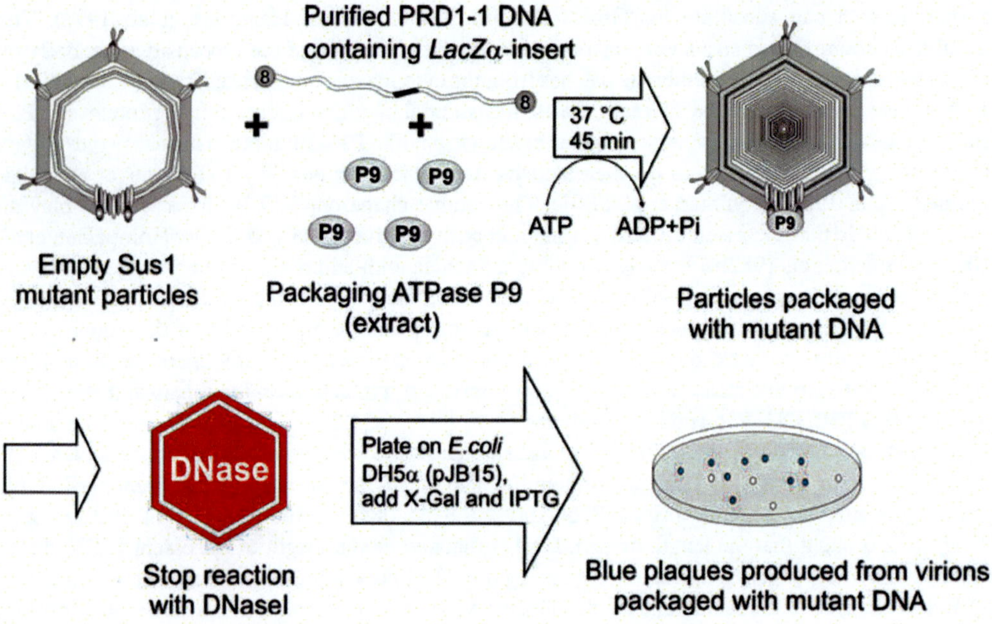

Fig. 16.5 PRD1 in vitro packaging assay

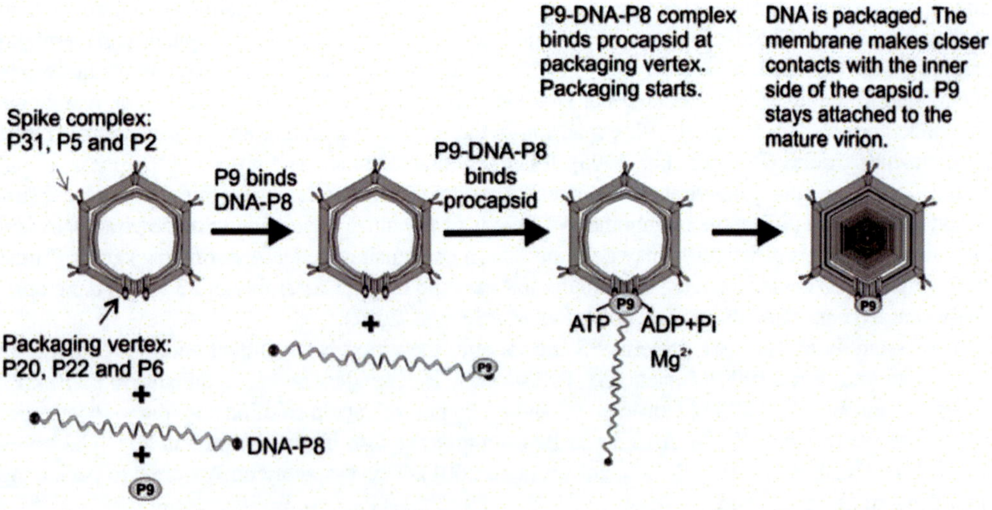

Fig. 16.6 A model for the events of PRD1 DNA packaging

as a molecular machine has been greatly aided by the setting up of an in vitro packaging system (Fig. 16.5) using purified mutant procapsids, a cell lysate containing recombinant P9 extract, and purified PRD1 DNA containing a lacZ marker and the covalently linked terminal proteins (Strömsten et al. 2005; Ziedaite et al. 2009). Using this system, it has been shown that the packaging ATPase can provide the energy required for DNA translocation from the breakdown of ATP, deoxy ATP and dideoxy ATP, and that it remains associated with the capsid after packaging has finished (Fig. 16.6) (Mindich et al. 1982b; Strömsten et al. 2005). In addition, sequence analysis of P9-type putative

ATPases in other viruses and mutation of conserved residues in P9 were used to identify a number of critical residues in three conserved motives that prevented packaging in vitro and in vivo complementation assays (Strömsten et al. 2003, 2005). Subsequently, it has been shown that P6 increases the packaging efficiency (Karhu et al. 2007). The genome is thought to pass through a pore in the membrane formed from the membrane proteins P20 and P22. The role of the terminal proteins in the packaging was elucidated using an optimised packaging reaction that allows the recovery of packaged particles (Ziedaite et al. 2009). PRD1 DNA with the covalently linked terminal proteins was shown to be essential for DNA packaging. Hence, it is likely that as in φ29, the packaging ATPase recognises the terminal protein linked to the genome (Guo et al. 1987). One of the consequences of this packaging pathway in assembly is that the terminal protein P8 and the packaging ATPase P9 are the last proteins to be added to the assembled particle.

The first infectious particles were detected after only 45 s in the optimised packaging reaction, the minimum time that could be measured experimentally. This gives a lower estimate of the packaging rate as 340 bp/s (Ziedaite et al. 2009) in comparison to a rate of up to 170 bp/s measured for φ29 using optical tweezers (Fuller et al. 2007). No large structural rearrangements are seen in the PRD1 procapsid after packaging, although there is an expansion and an increase in the order of the membrane (Butcher et al. 1995; Tuma et al. 1996b).

Like most dsDNA phages, PRD1 uses a two-component lysis (holin–endolysin) system (Mindich et al. 1982a) for host cell lysis (Fig. 16.3). The bacteriophage genome encodes a muraminidase, P15, which is virion-associated (Rydman and Bamford 2002), a holin P35 (Rydman and Bamford 2003) and two additional lysins, the proteins P36 and P37 (Krupovic et al. 2008). In concert, these proteins act on the host cytoplasmic membrane and peptidoglycans (Caldentey et al. 1994; Krupovic et al. 2008; Ziedaite et al. 2005). Monitoring of ion fluxes and the ATP content of the infected cells has revealed a sequence of lysis-related physiological changes. A decrease in the intracellular level of ATP is the earliest indicator of cell lysis, followed by the leakage of K+ from the cytosol approximately 20 min prior to the decrease in culture turbidity. However, the K+ efflux does not immediately lead to the depolarisation of the cytoplasmic membrane or leakage of intracellular ATP. These effects are only observed approximately 10 min prior to cell lysis (Ziedaite et al. 2005). The pair of accessory lysis proteins, P36 and P37, has been shown by complementation analysis to be functional analogues of the Rz and Rz1, respectively, of bacteriophage lambda. They ensure efficient disruption of the infected cell and consequent release of the phage progeny under less favourable growth conditions. A model has been proposed where complexes of P36 and P37 transform the mechanical stress caused by holin lesions at the cell membrane to the outer membrane leading to its disintegration (Krupovic et al. 2008).

16.4 Summary and Future Prospectives

There are still many open questions in the assembly of the PRD1 molecular machine. It is striking that so few intermediates in PRD1 assembly have been isolated. This really emphasises that in the future one should try to address how the capture of the viral membrane actually occurs. It is possible that there are host proteins involved in assembly, such as cell division proteins, and this area is as yet totally unexplored for PRD1. It is especially noteworthy that the particles seem to assemble in the centre of the cells, rather than at the periphery. It has become evident from extensive cryo-electron tomography work on many different bacterial cells (reviewed in Morris and Jensen 2008) that the organisation of the bacterial cytoplasm is much richer than previously thought, and that the cytoplasmic membrane can make extensive invaginations. Hence, in PRD1-infected cells perhaps the sites of DNA replication and the procapsid formation are actually juxtaposed. The role of some of the viral chaperones like P17 may actually be in remodelling the cytoplasmic membrane to allow this to occur. One possibility to explore this would be to couple fluorescence light microscopy of green-fluorescent-protein-labelled proteins with whole cell cryo-electron tomography of infected cells, using different host cell and virus phenotypes to explore the interplay of viral and host factors in more detail.

Assembly of the particle is tightly linked to the function of delivering the genomic cargo to the next host cell. One must also bear in mind that a viral particle is a dynamic assembly, whose structure does not reveal the whole truth. Instead, many stages present in the life cycle need to be considered when linking structure to function. In PRD1, P5 probably helps to roll the virus around the surface of the cell so that the primary adsorption protein, P2, can find its target on the cell, and this is still a reversible interaction. Under laboratory conditions using laboratory hosts, P5 is not terribly significant, but in the wild P5 may be much more important. There is still a further uncharacterised interaction that leads to irreversible attachment of the particle to the host cell. Delivery of the genome probably utilises the packaging vertex, with sequential release of proteins allowing penetration of the outer membrane and peptidoglycan possibly involving the transformation of the viral membrane in to a tube (Butcher et al. 1995; Grahn et al. 2002). It has been proposed that the genome is under pressure, leading to a passive release of the genome in the final stages of entry (Cockburn et al. 2004). To understand genome release, one should look carefully at the roles of the many membrane proteins in PRD1, how they are ordered, how the tube is organised and what components are in it. One should also investigate if the translocation of DNA in to the cell is really passive like in φ29 or is active like in T7 requiring transcription (Fuller et al. 2007; Garcia and Molineux 1995, 1996; Smith et al. 2001). In order to better understand both entry and packaging, the molecular organisation of the packaging vertex in capsids and the DNA should be explored, for instance, by orienting the capsids with respect to the receptor, or labelling the P9 in capsids to enable detailed structure determination without resorting to icosahedral symmetry, as has been done with, for example, MS2 and P-SSP7 (Liu et al. 2010; Toropova et al. 2011).

One can only really understand a car motor by looking under the hood. In order to really understand the detailed mechanism behind PRD1 packaging, one needs to move to single-particle experiments with laser tweezers as has been done so elegantly with φ29 (Smith et al. 2001). Currently, the frequency of packaging is still probably so low that this will need a lot of patience and optimisation.

One reason that entry should be from the same vertex where packaging occurs is that we expect the linear genome has one end at least close to the entry point, topologically, it would seem most likely that the DNA would start to exit from the end that was packaged last. Hence, it may be that even the terminal protein has a crucial role in DNA entry into the cell. We would like to leave you with one last thought, how does the terminal protein P8 get into the capsid through a channel designed for dsDNA?

Acknowledgements We would like to thank all our colleagues in the greasy phage field for their useful discussions over the years. This work was supported by the Academy of Finland Centre of Excellence Programme in Virus Research (2006–2011; 1129684 to SJB). V.M. is a fellow of the VGSB.

References

Abad-Zapatero C, Abdel-Meguid SS, Johnson JE, Leslie AG, Rayment I, Rossmann MG, Suck D, Tsukihara T (1980) Structure of southern bean mosaic virus at 2.8 A resolution. Nature 286:33–39

Abrescia NG, Cockburn JJ, Grimes JM, Sutton GC, Diprose JM, Butcher SJ, Fuller SD, San Martin C, Burnett RM, Stuart DI et al (2004) Insights into assembly from structural analysis of bacteriophage PRD1. Nature 432:68–74

Abrescia NG, Grimes JM, Fry EE, Ravantti JJ, Bamford DH, Stuart DI (2010) What does it take to make a virus: the concept of the viral 'self'. In: Stockley P, Twarock R (eds) Emerging topics in physical virology. Imperial College Press, London

Bamford DH (2000) Virus structures: those magnificent molecular machines. Curr Biol 10:R558–R561

Bamford DH (2003) Do viruses form lineages across different domains of life? Res Microbiol 154:231–236

Bamford DH, Ackermann H-W (2000) Family *Tectiviridae*. In: van Regenmortel MHV, Fauquet CM, Bishop DHL, Carstens EB, Estes MK, Lemon SM, Maniloff J, Mayo MA, McGeoch DJ, Pringle CR et al (eds) Virus taxonomy: classification and nomenclature of viruses. Academic, San Diego, pp 111–116

Bamford JKH, Bamford DH (2000) A new mutant class, made by targeted mutagenesis, of phage PRD1 reveals that protein P5 connects the receptor binding protein to vertex. J Virol 74:7781–7786

Bamford DH, McGraw T, Mackenzie G, Mindich L (1983) Identification of a protein bound to the termini of bacteriophage PRD1 DNA. J Virol 47:311–316

Bamford JKH, Hänninen A-L, Pakula TM, Ojala PM, Kalkkinen N, Frilander M, Bamford DH (1991) Genome organization of membrane-containing bacteriophage PRD1. Virology 183:658–676

Bamford DH, Caldentey J, Bamford JK (1995) Bacteriophage PRD1: a broad host range DSDNA tectivirus with an internal membrane. Adv Virus Res 45:281–319

Bamford DH, Burnett RM, Stuart DI (2002a) Evolution of viral structure. Theor Pop Biol 61:461–470

Bamford JK, Cockburn JJ, Diprose J, Grimes JM, Sutton G, Stuart DI, Bamford DH (2002b) Diffraction quality crystals of PRD1, a 66-MDa dsDNA virus with an internal membrane. J Struct Biol 139:103–112

Bamford DH, Grimes JM, Stuart DI (2005) What does structure tell us about virus evolution? Curr Opin Struct Biol 15:655–663

Benson SD, Bamford JKH, Bamford DH, Burnett RM (1999) Viral evolution revealed by bacteriophage PRD1 and human adenovirus coat protein structures. Cell 98:825–833

Butcher SJ, Bamford DH, Fuller SD (1995) DNA packaging orders the membrane of bacteriophage PRD1. EMBO J 14:6078–6086

Caldentey J, Bamford JK, Bamford DH (1990) Structure and assembly of bacteriophage PRD1, and *Escherichia coli* virus with a membrane. J Struct Biol 104:44–51

Caldentey J, Blanco L, Savilahti H, Bamford DH, Salas M (1992) *In vitro* replication of bacteriophage PRD1 DNA. Metal activation of protein-primed initiation and DNA elongation. Nucleic Acids Res 20:3971–3976

Caldentey J, Blanco L, Bamford DH, Salas M (1993) *In vitro* replication of bacteriophage PRD1 DNA. Characterization of the protein-primed initiation site. Nucleic Acids Res 21:3725–3730

Caldentey J, Hänninen A-L, Bamford DH (1994) Gene *XV* of bacteriophage PRD1 encodes a lytic enzyme with muramidase activity. Eur J Biochem 225:341–346

Caldentey J, Tuma R, Bamford DH (2000) Assembly of bacteriophage PRD1 spike complex: role of the multidomain protein P5. Biochemistry 39:10566–10573

Cockburn JJ, Abrescia NG, Grimes JM, Sutton GC, Diprose JM, Benevides JM, Thomas GJ Jr, Bamford JK, Bamford DH, Stuart DI (2004) Membrane structure and interactions with protein and DNA in bacteriophage PRD1. Nature 432:122–125

Fuller DN, Rickgauer JP, Jardine PJ, Grimes S, Anderson DL, Smith DE (2007) Ionic effects on viral DNA packaging and portal motor function in bacteriophage phi 29. Proc Natl Acad Sci USA 104:11245–11250

Garcia LR, Molineux IJ (1995) Rate of translocation of bacteriophage T7 DNA across the membranes of *Escherichia coli*. J Bacteriol 177:4066–4076

Garcia LR, Molineux IJ (1996) Transcription-independent DNA translocation of bacteriophage T7 DNA into *Escherichia coli*. J Bacteriol 178:6921–6929

Gowen B, Bamford JKH, Bamford DH, Fuller SD (2003) The tailless, icosahedral membrane virus PRD1 localizes the proteins involved in genome packaging and injection at a unique vertex. J Virol 77:7863–7871

Grahn AM, Bamford JKH, O'Neill MC, Bamford DH (1994) Functional organization of the bacteriophage PRD1 genome. J Bacteriol 176:3062–3068

Grahn AM, Caldentey J, Bamford JKH, Bamford DH (1999) Stable packaging of phage PRD1 DNA requires adsorption protein P2, which binds to the IncP plasmid-encoded conjugative transfer complex. J Bacteriol 181:6689–6696

Grahn AM, Daugelavicius R, Bamford DH (2002) Sequential model of phage PRD1 DNA delivery: active involvement of the viral membrane. Mol Microbiol 46:1199–1209

Guo P, Peterson C, Anderson D (1987) Prohead and DNA-gp3-dependent ATPase activity of the DNA packaging protein gp16 of bacteriophage phi 29. J Mol Biol 197:229–236

Hänninen A-L, Bamford DH, Bamford JKH (1997) Assembly of membrane-containing bacteriophage PRD1 is dependent on GroEL and GroES. Virology 227:207–210

Hogle JM, Chow M, Filman DJ (1985) Three-dimensional structure of poliovirus at 2.9 A resolution. Science 229:1358–1365

Holopainen JM, Säily M, Caldentey J, Kinnunen PK (2000) The assembly factor P17 from bacteriophage PRD1 interacts with positively charged lipid membranes. Eur J Biochem 267:6231–6238

Huiskonen JT, Manole V, Butcher SJ (2007) Tale of two spikes in bacteriophage PRD1. Proc Natl Acad Sci USA 104:6666–6671

Jaatinen ST, Viitanen SJ, Bamford DH, Bamford JK (2004) Integral membrane protein P16 of bacteriophage PRD1 stabilizes the adsorption vertex structure. J Virol 78:9790–9797

Karhu NJ, Ziedaite G, Bamford DH, Bamford JK (2007) Efficient DNA packaging of bacteriophage PRD1 requires the unique vertex protein P6. J Virol 81:2970–2979

Krupovic M, Cvirkaite-Krupovic V, Bamford DH (2008) Identification and functional analysis of the Rz/Rz1-like accessory lysis genes in the membrane-containing bacteriophage PRD1. Mol Microbiol 68:492–503

Laurinavicius S, Kakela R, Somerharju P, Bamford DH (2004) Phospholipid molecular species profiles of tectiviruses infecting Gram-negative and Gram-positive hosts. Virology 322:328–336

Laurinavicius S, Bamford DH, Somerharju P (2007) Transbilayer distribution of phospholipids in bacteriophage membranes. Biochim Biophys Acta 1768:2568–2577

Liu X, Zhang Q, Murata K, Baker ML, Sullivan MB, Fu C, Dougherty MT, Schmid MF, Osburne MS, Chisholm SW et al (2010) Structural changes in a marine podovirus associated with release of its genome into Prochlorococcus. Nat Struct Mol Biol 17:830–836

Merckel MC, Huiskonen JT, Bamford DH, Goldman A, Tuma R (2005) The structure of the bacteriophage PRD1 spike sheds light on the evolution of viral capsid architecture. Mol Cell 18:161–170

Mindich L, Bamford D, Goldthwaite C, Laverty M, Mackenzie G (1982a) Isolation of nonsense mutants of lipid-containing bacteriophage PRD1. J Virol 44:1013–1020

Mindich L, Bamford D, McGraw T, Mackenzie G (1982b) Assembly of bacteriophage PRD1: particle formation with wild-type and mutant viruses. J Virol 44:1021–1030

Morris DM, Jensen GJ (2008) Toward a biomechanical understanding of whole bacterial cells. Annu Rev Biochem 77:583–613

Munoz-Espin D, Holguera I, Ballesteros-Plaza D, Carballido-Lopez R, Salas M (2010) Viral terminal protein directs early organization of phage DNA replication at the bacterial nucleoid. Proc Natl Acad Sci USA 107: 16548–16553

Nandhagopal N, Simpson AA, Gurnon JR, Yan X, Baker TS, Graves MV, Van Etten JL, Rossmann MG (2002) The structure and evolution of the major capsid protein of a large, lipid-containing DNA virus. Proc Natl Acad Sci USA 99:14758–14763

Pakula TM, Caldentey J, Serrano M, Gutierrez C, Hermoso JM, Salas M, Bamford DH (1990) Characterization of a DNA binding protein of bacteriophage PRD1 involved in DNA replication. Nucleic Acids Res 18:6553–6557

Pakula TM, Caldentey J, Gutiérrez C, Olkkonen VM, Salas M, Bamford DH (1993) Overproduction, purification, and characterization of DNA-binding protein P19 of bacteriophage PRD1. Gene 126:99–104

Rice G, Tang L, Stedman K, Roberto F, Spuhler J, Gillitzer E, Johnson JE, Douglas T, Young M (2004) The structure of a thermophilic archaeal virus shows a double-stranded DNA viral capsid type that spans all domains of life. Proc Natl Acad Sci USA 101:7716–7720

Rossmann MG, Johnson JE (1989) Icosahedral RNA virus structure. Annu Rev Biochem 58:533–573

Rossmann MG, Arnold E, Erickson JW, Frankenberger EA, Griffith JP, Hecht HJ, Johnson JE, Kamer G, Luo M, Mosser AG et al (1985) Structure of a human common cold virus and functional relationship to other picornaviruses. Nature 317:145–153

Rydman PS, Bamford DH (2002) The lytic enzyme of bacteriophage PRD1 is associated with the viral membrane. J Bacteriol 184:104–110

Rydman PS, Bamford DH (2003) Identification and mutational analysis of bacteriophage PRD1 holin protein P35. J Bacteriol 185:3795–3803

Rydman PS, Caldentey J, Butcher SJ, Fuller SD, Rutten T, Bamford DH (1999) Bacteriophage PRD1 contains a labile receptor-binding structure at each vertex. J Mol Biol 291:575–587

Rydman PS, Bamford JK, Bamford DH (2001) A minor capsid protein P30 is essential for bacteriophage PRD1 capsid assembly. J Mol Biol 313:785–795

Salas M (1991) Protein-priming of DNA replication. Annu Rev Biochem 60:39–71

Salas M (1999) Mechanisms of initiation of linear DNA replication in prokaryotes. Genet Eng (N Y) 21:159–171

San Martín C, Huiskonen JT, Bamford JK, Butcher SJ, Fuller SD, Bamford DH, Burnett RM (2002) Minor proteins, mobile arms and membrane-capsid interactions in bacteriophage PRD1 capsid. Nat Struct Biol 9:756–763

Saren AM, Ravantti JJ, Benson SD, Burnett RM, Paulin L, Bamford DH, Bamford JK (2005) A snapshot of viral evolution from genome analysis of the tectiviridae family. J Mol Biol 350:427–440

Savilahti H, Bamford DH (1986) Linear DNA replication: inverted terminal repeats of the five closely related *Escherichia coli* bacteriophages. Gene 49:199–205

Savilahti H, Caldentey J, Bamford DH (1989) Bacteriophage PRD1 terminal protein: expression of gene *VIII* in *Escherichia coli* and purification of the functional P8 product. Gene 85:45–51

Savilahti H, Caldentey J, Lundström K, Syväoja JE, Bamford DH (1991) Overexpression, purification and characterization of *Escherichia coli* bacteriophage PRD1 DNA polymerase. *In vitro* synthesis of full-length PRD1 DNA with purified proteins. J Biol Chem 266:18737–18744

Simpson AA, Nandhagopal N, Van Etten JL, Rossmann MG (2003) Structural analyses of Phycodnaviridae and Iridoviridae. Acta Crystallogr D Biol Crystallogr 59:2053–2059
Smith DE, Tans SJ, Smith SB, Grimes S, Anderson DL, Bustamante C (2001) The bacteriophage phi29 portal motor can package DNA against a large internal force. Nature 413:748–752
Sokolova A, Malfois M, Caldentey J, Svergun DI, Koch MH, Bamford DH, Tuma R (2001) Solution structure of bacteriophage PRD1 vertex complex. J Biol Chem 276:46187–46195
Strömsten N, Bamford DH, Bamford JKH (2003) The unique vertex of bacterial virus PRD1 is connected to the viral internal membrane. J Virol 77:6314–6321
Strömsten NJ, Bamford DH, Bamford JK (2005) In vitro DNA packaging of PRD1: a common mechanism for internal-membrane viruses. J Mol Biol 348:617–629
Toropova K, Stockley PG, Ranson NA (2011) Visualising a viral RNA genome poised for release from its receptor complex. J Mol Biol 408:408–419
Tuma R, Bamford JH, Bamford DH, Russell MP, Thomas GJ Jr (1996a) Structure, interactions and dynamics of PRD1 virus I. Coupling of subunit folding and capsid assembly. J Mol Biol 257:87–101
Tuma R, Bamford JH, Bamford DH, Thomas GJ Jr (1996b) Structure, interactions and dynamics of PRD1 virus II. Organization of the viral membrane and DNA. J Mol Biol 257:102–115
van Raaij MJ, Mitraki A, Lavigne G, Cusack S (1999) A triple beta-spiral in the adenovirus fibre shaft reveals a new structural motif for a fibrous protein. Nature 401:935–938
Xu L, Benson SD, Butcher SJ, Bamford DH, Burnett RM (2003) The receptor binding protein P2 of PRD1, a virus targeting antibiotic-resistant bacteria, has a novel fold suggesting multiple functions. Structure 11:309–322
Yoo S-K, Ito J (1989) Protein-primed replication of bacteriophage PRD1 genome in vitro. Virology 170:442–449
Ziedaite G, Daugelavicius R, Bamford JK, Bamford DH (2005) The Holin protein of bacteriophage PRD1 forms a pore for small-molecule and endolysin translocation. J Bacteriol 187:5397–5405
Ziedaite G, Kivela HM, Bamford JK, Bamford DH (2009) Purified membrane-containing procapsids of bacteriophage PRD1 package the viral genome. J Mol Biol 386:637–647

Chapter 17
Assembly of Large Icosahedral Double-Stranded RNA Viruses

Minna M. Poranen and Dennis H. Bamford

Abstract Double-stranded RNA (dsRNA) viruses are a diverse group of viruses infecting hosts from bacteria to higher eukaryotes. Among the hosts are humans, domestic animals, and economically important plant species. Fine details of high-resolution virion structures have revealed common structural characteristics unique to these viruses including an internal icosahedral capsid built from 60 asymmetric dimers (120 monomers!) of the major coat protein. Here we focus mainly on the structures and assembly principles of large icosahedral dsRNA viruses belonging to the families of *Cystoviridae* and *Reoviridae*. It is obvious that there are a variety of assembly pathways utilized by different viruses starting from similar building blocks and reaching in all cases a similar capsid architecture. This is true even with closely related viruses indicating that the assembly pathway per se is not an indicator of relatedness and is achieved with minor changes in the interacting components.

Abbreviations

dsRNA	Double-stranded RNA
mRNA	Messenger RNA
ssRNA	Single-stranded RNA
Cryo-EM	Cryo-electron microscopy

17.1 Double-Stranded RNA Viruses

Double-stranded RNA (dsRNA) viruses represent a large and diverse group of viruses (Table 17.1). The hosts range from prokaryotes (e.g., *Pseudomonas* phage φ6, family *Cystoviridae*) to eukaryotes, including insect, plant, as well as animal hosts (e.g., rotavirus, family *Reoviridae*). In addition, cytoplasmic genetic elements, replicating dsRNA within virus-like particles which lack functions involved in host cell entry and exit, have been isolated from unicellular organisms such as yeast and

M.M. Poranen (✉) • D.H. Bamford
Department of Biosciences and Institute of Biotechnology, University of Helsinki,
Viikki Biocenter, 00014, Helsinki, Finland
e-mail: minna.poranen@helsinki.fi

M.G. Rossmann and V.B. Rao (eds.), *Viral Molecular Machines*,
Advances in Experimental Medicine and Biology 726, DOI 10.1007/978-1-4614-0980-9_17,

Table 17.1 The families of dsRNA viruses

Virus family	Size of the virion (nm)	Genome segments	Hosts
Birnaviridae	~65[a]	2 (co-encapsidated)	Fish, insects, birds, mollusks
Chrysoviridae	~40	4 (in separate capsids)	Fungi
Cystoviridae	~85	3 (co-encapsidated)	Bacteria
Hypoviridae	~50–80	1	Fungi
Partitiviridae	~30–40	2 (in separate capsids)	Plants, fungi
Picobirnaviridae[b]	~30–40	2 (co-encapsidated)	Mammals, birds, reptiles
Reoviridae	~70–90	10–12 (co-encapsidated)	Arachnids, arthropods, birds, crustaceans, fish, fungi, insects, mammals, plants, reptiles
Totiviridae	~30–40	1	Fungi

Updated from Mertens (2004)
[a]Bottcher et al. (1997)
[b]See Chandra (1997) and Fregolente et al. (2009)

protozoa (e.g., *Saccharomyces cerevisiae* virus L-A, family *Totiviridae*). So far no dsRNA archeal viruses have been described.

The vast majority of known dsRNA viruses have an icosahedral capsid structure, although pleomorphic dsRNA viruses have also been described (family *Hypoviridae*). Viruses with icosahedrally symmetric capsids can roughly be divided into two categories: those having 30–40 nm virion diameter (families *Chrysoviridae*, *Partitiviridae*, *Picobirnaviridae*, and *Totiviridae*) and those with 65–90 nm virion diameter (families *Cystoviridae*, *Birnaviridae, and Reoviridae*; Table 17.1). The large icosahedral virions typically comprise two to three distinct structural layers, except in birnaviruses (family *Birnaviridae*; Table 17.1; Bottcher et al. 1997) and cypoviruses (genera *Cypovirus*, family *Reoviridae*; Hill et al. 1999) which only have a single capsid layer. We will focus here on the virion assembly principles of selected dsRNA viruses with large multilayered virion structure of the families *Reoviridae* and *Cystoviridae* as most of the information has accumulated about these viruses.

17.1.1 Large Icosahedral dsRNA Viruses

The most prominent family among dsRNA viruses with 15 recognized genera is *Reoviridae* with several viruses of significant medical, veterinary, or agricultural importance. These include rotavirus (genus *Rotaviruses*) which is a major global cause of infant mortality, bluetongue virus (genus *Orbivirus*) causing acute disease of ruminants, and rice dwarf virus (genus *Phytoreovirus*) causing diseases in major crop species like rice and wheat. Because of their medical importance and economic significance these viruses have been subjected to extensive structural and biochemical studies. The type organism of *Reoviridae* is mammalian orthoreovirus of *Orthoreovirus* genus. Currently, the members of the *Reoviridae* family (reoviruses) are among the best structurally characterized viruses (see Sects. 17.2.1, 17.2.3, and 17.4).

All cystoviruses (family *Cystoviridae*) isolated so far belong to a single viral genus, *Cystovirus*. They all infect gram-negative bacteria, mainly plant-pathogenic *Pseudomonas* species. Among bacteriophages, cystoviruses represent a unique group having enveloped, nontailed virions and a segmented dsRNA genome. For a long time the type organism *Pseudomonas* phage φ6 was the only known dsRNA virus infecting prokaryotic hosts, but more recently additional cystoviruses have been isolated (Mindich et al. 1999; Qiao et al. 2010). Cystoviruses have been extensively studied as model systems for virus assembly, genome encapsidation, and replication. The structural and functional similarities to reoviruses and other dsRNA viruses suggest that cystoviruses may share a common evolutionary origin with eukaryotic dsRNA viruses (Bamford 2003; Mertens 2004).

17.1.1.1 Overall Virion Structure

The virions of large dsRNA viruses are typically multilayered consisting of two or three icosahedrally symmetrical concentric protein shells. The two innermost ones are structurally most conserved so that it is possible to identify cognate proteins with similar functions or structures even from distantly related dsRNA viruses as exemplified for simian rotavirus A/SA11, bluetongue virus, mammalian reovirus-3, rice dwarf virus, and *Pseudomonas* phage φ6 in Table 17.2 (see also Mertens 2004). The conserved viral core capsid, with the associated viral enzymes, mediates the replication and transcription of the viral dsRNA. This is a common feature for all icosahedral dsRNA viruses as well as the dsRNA virus-like genetic elements, regardless of the host. The outer virion layer (exposed during the extracellular phase of the viral life cycle) has adapted to functions involved in the initiation of infection in different host systems and shows much greater diversity in structural organization compared to the inner layers.

Depending on the history of the nomenclature for the particular genera of dsRNA viruses the different subviral particles and capsid shells have various designations. To avoid confusion we assign the innermost protein layer as T2 capsid and the protein layer surrounding the T2 capsid as T13 shell (Table 17.2). These acronyms arise from the overall organization of the corresponding capsid layers (see below Sects. 17.2.1 and 17.4). The corresponding subviral particles are assigned as a single-layered particle (SLP), that is T2 capsid enclosing the dsRNA genome, and a double-layered particle (DLP), that is SLP enclosed within the T13 shell. The possible additional outer capsid or envelope surrounding the DLP is designated as an outer virion layer (see Sect. 17.5).

17.1.1.2 Replication Cycle

The most distinctive feature in dsRNA virus life cycle is that the viral genome is always replicated and transcribed within an icosahedral protein shell. This icosahedral nucleoprotein structure also encompasses the core of the multilayered mature virions. Thus, the replication of the dsRNA genome is closely associated with the assembly of the virion.

The assembly and viral genome replication of reoviruses take place in discrete cellular structures termed viral inclusion bodies or viroplasm (see, e.g., Altenburg et al. 1980; Brookes et al. 1993; Miller et al. 2010; Shimizu et al. 2009). These perinuclear structures are rich in viral messenger RNA (mRNA), major capsid proteins, fully or partially assembled viral particles, and viral nonstructural proteins. The inclusion bodies serve as virus assembly and replication sites producing DLPs which subsequently mature to virions within the cell cytoplasm. Expression of nonstructural proteins in uninfected cells leads to the formation of viroplasm-like structures even in the absence of other viral components (Becker et al. 2003; Fabbretti et al. 1999; Thomas et al. 1990; Wei et al. 2006). Thus, viral capsid proteins or a full complement of viral RNA is not necessary for the formation of the inclusion bodies. On the other hand, the assembly of empty virus-like particles may take place in the absence of viral nonstructural proteins when viral structural proteins are co-expressed in insect, mouse, or plant cells implying that the viroplasm environment and the nonstructural proteins are not essential for the formation of viral shell structures (French and Roy 1990; Xu et al. 1993; Labbe et al. 1991; Zheng et al. 2000). This also implies that the viroplasm has mainly a role in the encapsidation and replication of viral RNA.

The replication and assembly of cystoviruses do not involve the formation of a viroplasm and therefore it has been possible to detect assembly intermediates for DLPs in infected cells. The first assembly intermediate detected is an empty T2 capsid which matures to dsRNA-filled SLP, to DLP, and subsequently to a complete enveloped triple-layered virion which is released by host cell lysis (Bamford and Mindich 1980). DLPs are also assembled in vitro in the absence of viral nonstructural proteins (Poranen et al. 2001).

Table 17.2 Comparable proteins of selected large dsRNA viruses

	T2 capsid					T13 capsid	Outer layer	
Virus family, subfamily Genus Species	T2 protein	Polymerase	Capping enzyme	Helicase	Other T2 capsid-associated proteins	T13 protein	Proteins	Lipids
Cystoviridae								
Cystovirus					P7, assembly cofactor		P6	Phosphatidylethanolamine
Pseudomonas phage $\phi 6$	P1	P2	–	P4		P8	P9	Phosphatidylglycerol
							P10	Cardiolipin[a]
							P13	
Reoviridae, Spinareovirinae								
Orthoreovirus					$\sigma 2$, core clamp		$\sigma 1$	–
Mammalian reovirus-3	$\lambda 1$	$\lambda 3$	$\lambda 2$	$\lambda 1$	$\mu 2$, putative polymerase cofactor	$\mu 1$[b]	$\sigma 3$	
Reoviridae, Sedoreovirinae								
Orbivirus							VP2	–
Bluetongue virus	VP3	VP1	VP4	VP6	–	VP7	VP5	
Phytoreovirus					P7, nucleic acid binding		P2 (P9)	–
Rice dwarf virus	P3	P1	P5	–		P8		
Rotavirus							VP4[d]	–
Simian rotavirus A/SA11	VP2	VP1	VP3	–[c]	–	VP6	VP7	

Adapted from Mertens (2004)
[a]Laurinavicius et al. (2004)
[b] Cleaved to $\mu 1C$ and $\mu 1N$
[c]Helicase activity assigned to rotavirus nonstructural protein NSP2
[d]Trypsin cleaved into VP5* and VP8*

17.2 T2 Capsid

Based on the overall morphology of the T2 capsid, reoviruses are divided into two subfamilies: *Spinareovirinae* (genuses *Aquareovirus*, *Coltivirus*, *Cypovirus*, *Dinovernavirus*, *Fijivirus*, *Idnoreovirus*, *Mycoreovirus*, *Orthoreovirus*, *Oryzavirus*) and *Sedoreovirinae* (genuses *Cardoreovirus*, *Mimoreovirus*, *Orbivirus*, *Phytoreovirus*, *Rotavirus*, *Seadornavirus*; Carstens 2010). The T2 capsids of the members of *Spinareovirinae* are decorated with 12 distinct turret-like structures extending from the fivefold vertices, while the members of *Sedoreovirinae* do not have such features. Turret-like structures are also evident in the T2 core capsid of cystoviruses (Butcher et al. 1997; Huiskonen et al. 2006; see Chap 27, Fig. 27.1). In this case, there is biochemical evidence that one of the vertices differs structurally and functionally from the other 11 vertices (Pirttimaa et al. 2002; see Sect. 17.3.1).

Channels close or at the icosahedral fivefold vertices are the exit points for viral RNA transcripts produced in the interior of the T2 capsid (Diprose et al. 2001; Lawton et al. 1997a; Mendez et al. 2008). Accordingly, the main enzymatic components involved in RNA metabolisms (polymerase, helicase, and mRNA-capping enzyme; see Sect. 17.2.4) are located in the proximity of the fivefold vertices inside the T2 capsid shell (Dryden et al. 1998; Miyazaki et al. 2010; Nason et al. 2004; Prasad et al. 1996; Sen et al. 2008) or in the turret structures exterior of the shell (de Haas et al. 1999; Luongo et al. 2002; Pirttimaa et al. 2002; see Chap 27, Fig. 27.1).

17.2.1 Architecture of the T2 Shell

The structural organization of the T2 capsid is the hallmark of icosahedral dsRNA viruses and dsRNA virus-like genetic elements. It is composed of 120 subunits of a large, relatively flat protein (designated here as T2 protein; see Table 17.2), with a dimer (subunits A and B) occupying the icosahedral asymmetric unit of a $T=1$ capsid (T for triangulation number). Such capsid organization, also referred as "$T=2$" capsid (Grimes et al. 1998), is unique for dsRNA viruses (see Fig. 17.1) and has not been described for any other virus group studied so far or in cellular environments.

This type of viral capsid organization was first discovered in cryo-electron microscopy (cryo-EM) reconstructions (Cheng et al. 1994; Huiskonen et al. 2006; Lawton et al. 1997b; Lu et al. 1998; Shaw et al. 1996; Yu et al. 2008; Zhang et al. 2010) and subsequently from X-ray structures of various large and small dsRNA virus virions or viral core particles (Duquerroy et al. 2009; Grimes et al. 1998; McClain et al. 2010; Naitow et al. 2002; Nakagawa et al. 2003; Pan et al. 2009; Reinisch et al. 2000; Fig. 17.1). Chrysoviruses (family *Chrysoviridae*; Table 17.1) with a small 60-subunit $T=1$ single-layered capsid are exceptions among dsRNA viruses. However, in this case, the capsid protein appears to have two genetically duplicated domains (Caston et al. 2003; Luque et al. 2010; Fig. 17.1).

In the high-resolution T2 capsid structures, first described for bluetongue virus (an orbivirus; Grimes et al. 1998), the approximately triangular subunits form asymmetric, relatively parallel dimers in which the two identical subunits or conformers, A and B, are located in distinctly nonequivalent structural environments, making nonequivalent contacts with each other. Five A subunits are arranged around an icosahedral fivefold axis of symmetry with five B subunits interdigitating between (see Fig. 17.1). The A subunits within neighboring fivefold axes of symmetry make pairs which span the icosahedral twofold axes so that a continuous scaffold of A subunits is formed encircling the triangular plugs formed by three B subunits at the icosahedral threefold axis. This type of overall organization is clearly recognizable in the high-resolution T2 capsid structures of all reoviruses (family *Reoviridae*; Grimes et al. 1998; McClain et al. 2010; Nakagawa et al. 2003; Reinisch et al. 2000; Yu et al. 2008; Zhang et al. 2010) as well as in the yeast L-A particle (family *Totiviridae*; Naitow et al. 2002) and *Penicillium chrysogenum* virus (*Chrysoviridae*; Luque et al. 2010; Fig. 17.1).

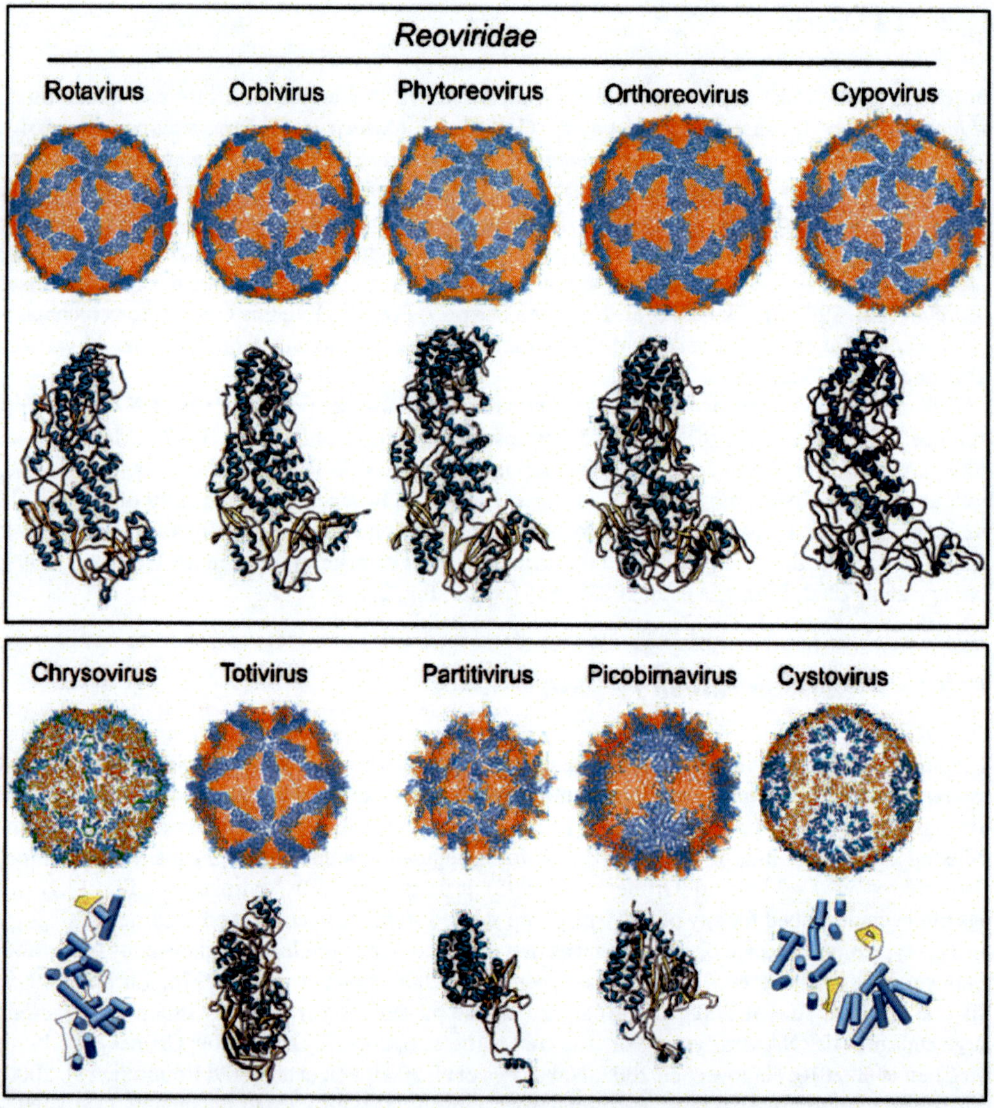

Fig. 17.1 Comparison of the T2 capsids of dsRNA viruses and virus-like particles based on X-ray or cryo-EM data. The arrangement of 120 copies of the T2 protein subunits into $T=1$ icosahedral lattice in selected dsRNA viruses (*upper rows*) and corresponding T2 protein structures shown from a top view (*lower rows*). The two conformers of the T2 protein, A and B (domains of a single protein in chrysoviruses), are shown in *blue* and *red*, respectively (*upper rows*). Reproduced/amended with permission from American Society for Microbiology from J Virol 2010 84, 7256–7266, doi: 10.1128/JVI.00432-10 (Luque et al. 2010)

A slightly different type of subunit organization is observed in the T2 capsid of cystoviruses (Huiskonen et al. 2006; Jaalinoja et al. 2007; Fig. 17.1). The quadrangular-shaped class A subunits encircle the fivefold axes, and the B subunits form a continuous net around the pentamers of A subunits by interacting both at the twofold and threefold axes of icosahedral symmetry (Fig. 17.1). Thus there are no direct interactions between A subunits in the neighboring fivefold axes as is seen in the T2 capsids of reoviruses, totiviruses, and chrysoviruses (Fig. 17.1). This cystovirus-type capsid organization resembles the capsid architecture recently described for rabbit picobirnavirus (family *Picobirnaviridae*; Duquerroy et al. 2009) and *Penicillium stoloniferum* virus F (family

Partitiviridae; Pan et al. 2009) although the contacts between the B subunits (at the twofold axes of icosahedral symmetry) are not as prominent in these cases (Fig. 17.1).

17.2.2 Structural Stability of the T2 Shell

Isolated T2 capsids of reoviruses and cystoviruses have compromised stability. Accordingly, the X-ray structures of the T2 capsids have revealed that the protein–protein interactions in the T2 shells of dsRNA viruses are relatively weak compared to other viral capsids (e.g., to the picornaviral capsids; see Reddy and Johnson 2005). Such weak interactions may be essential to allow the dynamic processes required for dsRNA replication and transcription. During the extracellular phase of the viral life cycle, the T2 capsid is typically protected by additional virion layer(s) stabilizing the structure, the only exception among reoviruses being the single-shelled cypovirus virion. In this case, the turret protein of the T2 capsid contains a polyhedron-binding domain inducing virion embedding in polyhedron matrix which forms micrometer-sized protein crystals called polyhedra (see Coulibaly et al. 2007; Yu et al. 2008). Such structures are remarkably stable, protecting the extracellular virions, but dissolve in the alkaline midgut of insect larvae (Rohrmann 1986).

17.2.3 T2 Protein

The T2 protein of reoviruses is thin, highly helical protein with crescent-like triangular shape (Grimes et al. 1998; McClain et al. 2010; Nakagawa et al. 2003; Reinisch et al. 2000). Despite the low sequence identity, the overall appearance and polypeptide chain fold are similar in all the reovirus T2 proteins for which high-resolution structural information is available (Cheng et al. 2010; Grimes et al. 1998; McClain et al. 2010; Nakagawa et al. 2003; Reinisch et al. 2000; Yu et al. 2008). Three distinct regions or domains have been recognized in the T2 proteins of bluetongue virus (an orbivirus), rotavirus (a rotavirus), and rice dwarf virus (a phytoreovirus): the apical region pointing toward the fivefold vertices, the central region (also known as carapace domain), and the dimer-forming region (Grimes et al. 1998; McClain et al. 2010; Nakagawa et al. 2003). In the case of mammalian reovirus (an orthoreovirus) and grass carp reovirus (an aquareovirus) only two subdomains can be identified (Cheng et al. 2010; Reinisch et al. 2000). The polypeptide chains of orbivirus, rotavirus, and orthoreovirus T2 proteins can be aligned for most parts of the proteins, the apical and the central regions being structurally most conserved (McClain et al. 2010). The interconnecting boundaries between the regions of the T2 proteins enable slight hinged motions needed to form the two conformers (A and B) observed in the T2 capsids. Thus the subdomains behave as rigid bodies when the T2 protein transforms between A and B conformers.

There is no high-resolution structural information on the T2 protein of cystoviruses. However, based on the cryo-EM reconstructions (Huiskonen et al. 2006; Jaalinoja et al. 2007), the overall shape of the protein is somewhat different from that observed for reoviruses (see Fig. 17.1). The protein is predicted to have two distinct clusters of α-helices, separated by two β-sheet regions (Huiskonen et al. 2006; Luque et al. 2010). The purified T2 protein of cystoviruses can be isolated as a monomer or a tetramer (Kainov et al. 2003a; Poranen et al. 2001). The monomer is elongated and exhibits limited solubility (Benevides et al. 2002).

The T2 proteins of large dsRNA viruses have RNA-binding activity (Labbe et al. 1991; Loudon and Roy 1992; Qiao et al. 2003). In the case of cystoviruses RNA-binding sites have been identified exterior of the T2 shell and are involved in the specific recognition of the viral mRNA for packaging (Qiao et al. 2003). The T2 protein of orthoreoviruses also contains RNA helicase activity (Bisaillon et al. 1997). The T2 capsids of cypoviruses and certain cystoviruses (*Pseudomonas* phage φ8) are apparently also active in the host plasma membrane penetration during entry into the host cell (Sun et al. 2003; Tan et al. 2003).

17.2.4 T2 Capsid-Associated Proteins

Several minor protein components are associated with the T2 capsid of dsRNA viruses (see Table 17.2). Most of these have enzymatic activities required for the transcription or replication of the dsRNA genome. Also particle-stabilizing minor capsid proteins, which have no detected catalytic activity, have been identified in the T2 capsids of cystoviruses and orthoreoviruses (see below Sect. 17.2.4.4).

17.2.4.1 Polymerase

The most conserved minor protein subunit found in the T2 capsids of all dsRNA viruses is the viral RNA-dependent RNA polymerase (Table 17.2) which is required both for the transcription (dsRNA to mRNA) of the dsRNA genome and its replication (mRNA to dsRNA). This globular protein has been localized at the fivefold symmetry position, inside the T2 capsid or close to that position (Sen et al. 2008; Zhang et al. 2003; see Chap. 26, Fig. 26.3). Direct protein contacts between the polymerase and the T2 shell are evident (Sen et al. 2008; Zhang et al. 2003). In the small icosahedral dsRNA viruses the polymerase may be the only minor protein component associated with the T2 capsid. In yeast L-A virus (*Totiviridae*) the polymerase (Pol) is covalently linked to the T2 capsid; such a polypeptide (known as Gag-Pol) is formed by a −1 ribosomal frameshift during translation (Dinman et al. 1991). In this case, and probably also in other dsRNA viruses, the polymerase is stably anchored to the T2 shell and consequently the RNA has to move relative to the viral polymerase during RNA replication and transcription.

The fold of the polymerase follows the canonical right-hand motif, common to many DNA and RNA polymerases, with finger, palm, and thumb subdomains (Butcher et al. 2001; Lu et al. 2008; Pan et al. 2007; Tao et al. 2002). Details of this type of polymerase structure and mechanism of catalysis are described elsewhere in this book (see Chap. 12 by Kyung H. Choi). The polymerases of rotavirus and reovirus, which apply conservative mechanism of transcription (Banerjee and Shatkin 1970), have four distinct tunnels leading from the surface of the molecule to the central catalytic site (Lu et al. 2008; Tao et al. 2002), while only three such tunnels exist in the polymerases of cystoviruses and birnaviruses (Butcher et al. 2001; Pan et al. 2007), which apply semiconservative transcription (Spies et al. 1987; Usala et al. 1980). Interestingly, birnaviruses initiate RNA polymerization using a protein primer (Calvert et al. 1991) while the other dsRNA viruses predominantly apply de novo initiation mechanism.

17.2.4.2 RNA-Capping Enzyme

The T2 capsid of reoviruses is capable of capping viral mRNA molecules. The capping enzyme (Table 17.2), guanylyl transferase, catalyzes the stepwise enzymatic processes required for the addition of the cap structure to the 5′ end of the viral mRNA. Such enzymes have been identified in a number of reoviruses (see, e.g., Chen et al. 1999; Cleveland et al. 1986; Hsiao et al. 2002; Le Blois et al. 1992; Liu et al. 1992; Ramadevi et al. 1998a). However, the structure and location of this enzyme are not conserved. In orthoreoviruses (and aquareoviruses) the capping enzyme forms distinct pentameric, turret-like structures at the fivefold vertices of the T2 capsid (Luongo et al. 2002; Nason et al. 2000), while in orbiviruses and rotaviruses the capping enzyme is located in the interior of the T2 capsid in close proximity of the polymerase (Nason et al. 2004; Prasad et al. 1996) and is a dimer in the case of orbiviruses (Ramadevi et al. 1998b). Cystoviruses do not have capping activity as they infect prokaryotic hosts where mRNA maturation does not involve addition of the 5′ cap.

Interestingly, there is no evidence of RNA capping among birnaviruses (although infecting eukaryotes). Instead the 5′ termini of the viral RNA are covalently bound to the viral polymerase (Calvert et al. 1991).

17.2.4.3 Helicase

Proteins with RNA helicase activity are common in large dsRNA viruses (Table 17.2). Such activity is required either during transcription to unwind dsRNA molecules or for translocation of single-stranded RNA (ssRNA) molecules across the T2 capsid shell during RNA packaging or mRNA extrusion (Kainov et al. 2004; Matsuo and Roy 2009; Pirttimaa et al. 2002). In bluetongue virus (an orbivirus), helicase activity has been assigned to a minor structural component (VP6) of the T2 capsid (Stauber et al. 1997). This 35.7-kDa protein readily assembles into hexamers (Kar and Roy 2003), which is a property shared with the approximately 35-kDa hexameric helicase (P4) of cystoviruses (Juuti et al. 1998; Kainov et al. 2003b). However, unlike the bluetongue virus helicase, the cystovirus helicase forms turret-like structures extending from the capsid vertices (de Haas et al. 1999; Huiskonen et al. 2007). The location of the hexameric structure at the fivefold vertex of the T2 capsid creates a symmetry mismatch thus being analogous to the DNA packaging devices of tailed dsDNA bacteriophages (reviewed by Rao and Feiss 2008; see also Chaps. 22 and 23).

RNA helicase activity has not been detected in any of the structural proteins of rotaviruses or phytoreoviruses. Instead, the nonstructural octameric protein NSP2 of rotavirus displays helix-destabilizing activity as well as NTPase activity suggesting that NSP2 may facilitate the encapsidation of mRNA into preformed T2 capsids or assembly intermediates (Jayaram et al. 2002; Schuck et al. 2001; Taraporewala et al. 1999, 2006; Taraporewala and Patton 2001). In orthoreoviruses, RNA unwinding activity has been assigned to the major capsid protein (T2 protein; Bisaillon et al. 1997). Details of the ATP hydrolysis-driven RNA translocation process by the cystoviral hexameric helicase are described elsewhere in this book by Erika Mancini and Roman Tuma (see Chap. 27).

17.2.4.4 Other T2 Capsid-Associated Proteins

The innermost layers of cystovirus, phytoreovirus, and orthoreovirus virions contain additional proteins with no apparent counterparts in orbivirus or rotavirus (see Table 17.2).

The orthoreovirus T2 capsid is decorated with a nodular clamp protein σ2. The compact globular σ2 monomer binds at three distinct positions within each icosahedral asymmetric unit (see Reinisch et al. 2000) reinforcing the T2-protein lattice. It apparently has an important role in the assembly and stability of the orthoreovirus T2 capsid as stable recombinant T2 shells are not formed in the absence of σ2 (Kim et al. 2002; Xu et al. 1993). Using electron microscopy similar nodular structures at positions corresponding to two of the three σ2 sites of orthoreoviruses have been described for the turreted T2 capsids of aquareoviruses and cypoviruses (Cheng et al. 2010; Hill et al. 1999).

Orthoreovirus protein μ2, cystovirus protein P7, and phytoreovirus protein P7 are T2 capsid components (Table 17.2) functionally associated with the RNA polymerization and assembly. Protein P7 of phytoreovirus is located close to the fivefold vertex, beneath the T2 capsid layer (Miyazaki et al. 2010) and a similar position has also been proposed for μ2 of orthoreovirus and P7 of cystovirus (Huiskonen et al. 2006; Jaalinoja et al. 2007; Zhang et al. 2003; Dryden et al. 1998), both of which appear to be dimers (Juuti and Bamford 1997; Kim et al. 2004). Orthoreovirus μ2 is a nonspecific NTPase which also possesses RNA triphosphatase activity. Its exact role in the viral RNA metabolism is not known but a temperature-sensitive mutation in the corresponding gene results defects in viral morphogenesis and genome replication (Coombs 1996). No enzymatic activity has been assigned for the cystoviral protein P7. P7-deficient T2 capsid of *Pseudomonas* phage φ6 (cystovirus) has compromised RNA encapsidation activity in an ssRNA packaging and protection assay (Juuti and Bamford 1995) and it functions as an assembly cofactor during T2 capsid assembly (Poranen et al. 2001, 2008; see Sect. 17.2.5). Such observations suggest that P7 could function as a particle reinforcing hinge or glue protein stabilizing the assembly intermediates as well as the end product, the T2 capsid. The phytoreovirus protein P7 is a nucleic acid-binding protein which also has affinity to the T2 protein as well as to the polymerase subunit suggesting a role in the T2 capsid assembly (Ueda et al. 1997).

17.2.5 Assembly of the T2 Capsid

When expressed within a heterologous protein expression cell line, the T2 capsid components of cystoviruses and several reoviruses assemble into T2 capsid-like structures (Hagiwara et al. 2003; Gottlieb et al. 1988b; Juuti and Bamford 1995; Kim et al. 2002; Labbe et al. 1991; Pirttimaa et al. 2002; Roy 2005; Xu et al. 1993; Zeng et al. 1996). Thus the assembly of such structures is not dependent on viral nonstructural proteins, other viral structural proteins (e.g., the T13 capsid protein), or factors specific for the host organism although such proteins may be used to boost the assembly. The formation of the T2 capsid-like particles is also independent of native viral RNA and the particles formed are empty (no ssRNA or dsRNA inside).

The T2 proteins of dsRNA viruses typically have a high intrinsic tendency to assemble into capsid-like structures, and in the case of rotavirus and orbivirus the T2 proteins can self-assemble into icosahedral particles also in the absence of any other viral proteins (Hagiwara et al. 2003; Labbe et al. 1991; Moss and Nuttall 1994). The orthoreovirus T2 shell assembly requires co-expression of the T2 protein and the nodular σ2 protein stabilizing the structure (Xu et al. 1993). The *Pseudomonas* phage φ6 T2-protein self-assembly products formed in *Escherichia coli* are aggregative and partially unfolded spherical particles (Benevides et al. 2002; Gottlieb et al. 1988b; Paatero et al. 1998) implying that the correct assembly of the precursor capsid of this cystovirus is dependent on the other protein components of the T2 capsid.

The different recombinant expression systems established for dsRNA viruses allow the production of a variety of incomplete T2 capsid structures missing different minor protein components (see Table 17.2). Such expression systems can reveal the order of dependency of the component proteins and have also been important in assigning functions and locations for individual proteins. However, it is evident that the structures formed do not necessarily present true assembly intermediates of the virion but rather demonstrate the tendency of the component proteins to make interactions with each other. For example, although particles are formed in the absence of the polymerase subunit (see, e.g., Juuti and Bamford 1995; Xu et al. 1993), it is obvious that the incorporation of the internal enzymes has to precede the enclosure of the T2 shell as the limited size of the pores in the assembled T2 capsid would prevent the entry of the subunits located in the interior of the capsid.

The in vitro assembly systems for cystoviruses using purified proteins and defined assembly conditions have revealed important aspects of the T2 capsid assembly process (Kainov et al. 2003a; Poranen et al. 2001, 2008). The incorporation of the enzymatic components seems to be ensured via kinetic control. The nucleation of the T2 shell assembly is dependent on one of the key enzymatic subunits, the hexameric helicase (P4, *Pseudomonas* phage φ6; Poranen et al. 2001) or the polymerase subunit (P2, *Pseudomonas* phage φ8; Kainov et al. 2003a). Kinetic analyses of phage φ6 assembly have also identified a role for the minor T2 capsid component protein P7 (Paatero et al. 1998; Poranen et al. 2001, 2008); The rate of T2 capsid assembly in vitro increases with rising P7 concentration, implying that P7 acts as an assembly cofactor. Consequently, the incorporation of

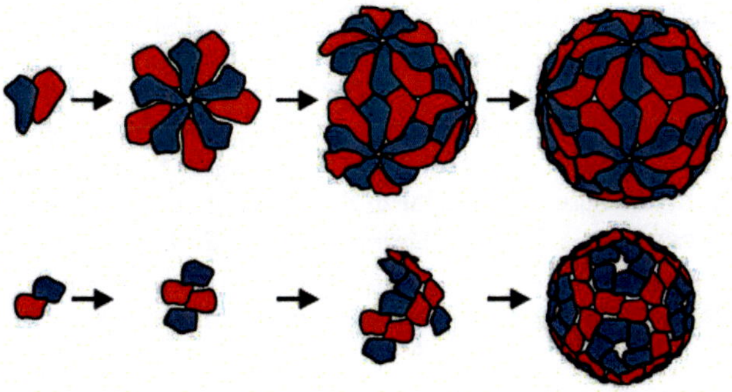

Fig. 17.2 Proposed assembly pathway for T2 capsid shells of reoviruses (*above*) and cystoviruses (*below*). The two different conformers of the T2 protein are shown in *blue* and *red*. Only the final expanded conformation for the cystoviral T2 capsid is depicted

P7 into the growing T2 shell structure is also kinetically favored. In line with these observations, the current models of rotavirus and orbivirus assembly pathways suggest that the viral polymerase-capping complex associated with several copies of the T2 protein is a likely assembly intermediate in the pathway of T2 capsid assembly (Guglielmi et al. 2010; Roy 2005).

17.2.5.1 T2 Shell Assembly Intermediates: Dimers of Dimers or Pentamers of Dimers?

Although the information on the assembly pathway for reovirus T2 capsid assembly is still scattered, it is likely that the assembly pathways for T2 capsids of cystoviruses and reoviruses are distinct. Thus, the slight differences detected in the organization of the T2 capsid architecture (Fig. 17.1) may also be reflected in the assembly process.

The *Pseudomonas* phage φ8 T2 protein is isolated as a stable tetramer (Kainov et al. 2003a) and kinetic analyses of *Pseudomonas* phage φ6 assembly have revealed a complex containing a tetramer of the T2 protein bridging two hexamers of the helicase protein (P4) between neighboring fivefold vertices as the rate-limiting, nucleation step (Poranen et al. 2001). This suggests that the assembly of cystoviral T2 capsid (Fig. 17.2) proceeds via a twofold centered assembly intermediate composed of dimers of A–B dimers (A and B being the different conformers of the T2 protein, see Sect. 17.2.1). Interestingly, recent structural analyses of *Penicillium stoloniferum* virus (a partitivirus) also suggested a tetrameric twofold centered capsid assembly intermediate (Pan et al. 2009). The strongest interactions between T2 protein A–B dimers in the *P. stoloniferum* virus T2 capsid occur at the twofold axes of icosahedral symmetry; the buried surface area between the dimer pair related by the twofold axis is about 3,500 Å^2 which is almost twice that observed between pairs of the dimers related by the icosahedral threefold or fivefold axes of symmetry (approximately 2,000 and 1,900 Å^2, respectively). In addition, the picobirnavirus crystal structure reveals T2 protein dimers with an extensive buried surface areas (6,200 Å^2 per subunit) that connect the two decamer structures at the icosahedral fivefold symmetry position (Duquerroy et al. 2009). Consequently, an assembly pathway proceeding via dimers of T2 protein A–B dimers (Fig. 17.2) seems to be common for the T2 capsids of several dsRNA viruses.

The strength of the interaction between protein subunits in the final assembly product has also been used to infer the assembly pathway for reovirus T2 capsids. The high-resolution structures of different reoviruses (Grimes et al. 1998; McClain et al. 2010; Reinisch et al. 2000) have revealed extensive interactions between the subunits within the T2 capsid protein decamers and a pathway from A–B dimers to pentamers of A–B dimers (i.e., decamers) and subsequently to the complete T2 shell has been proposed (Fig. 17.2). Further support for this hypothesis was obtained in a study in which the expression of a modified bluetongue virus (an orbivirus) T2 protein (with deleted dimerization domain responsible of the interdecamer contacts) resulted in the formation of dimers, pentamers, and decamers but prevented the T2 capsid assembly (Kar et al. 2004). This suggests that a decamer of T2 protein could be a true assembly intermediate of the bluetongue virus inner capsid.

Although assembly pathway models involving the formation of a T2 protein decamer are favored for different reoviruses, the relative strength of the T2 protein dimer interactions in rice dwarf virus (phytoreovirus) around the threefold icosahedral symmetry axis is stronger than around the fivefold icosahedral axis (Nakagawa et al. 2003; 24 hydrogen bonds and 10 salt bridges vs. 18 hydrogen bonds and 11 salt bridges for three- and fivefold, respectively). Thus, a trimer of dimers is another potential assembly intermediate during the assembly pathway of this reovirus.

17.3 Encapsidation of the Viral Genome

The assembly of the dsRNA virus SLP is tightly connected to the replication of the viral genome. The encapsidation of the single-stranded genome precursor molecules, which also serve as mRNAs, precedes the replication which always takes place within the T2 capsid. Consequently, there are specific packaging and replication signals in the mRNA molecules (Frilander et al. 1992; Gottlieb et al. 1994; Pirttimaa and Bamford 2000; Tortorici et al. 2003, 2006) which enable encapsidation of viral RNA from the pool of viral and cellular RNAs.

The assembly pathway of cystoviruses involves the formation of empty T2 capsids which have a distinct compact conformation compared to the dsRNA-filled mature T2 capsid (Butcher et al. 1997; Chap. 26, Fig. 26.2). The empty capsid, also known as a procapsid, has a dodecahedral appearance with fivefold vertices recessed to form inwardly oriented "cups," whereas the dsRNA-filled T2 capsid is roughly spherical and also slightly larger in diameter than the empty capsid (50 vs. 46 nm; Butcher et al. 1997). In addition, a third conformation, known as expanded procapsid, has been detected in preparations of empty T2 capsids as well as in virion-derived SLPs that have lost the genome. The expanded procapsid can be considered as an intermediate conformation between the empty and dsRNA-filled forms being a dodecahedron with shallow depressions at the fivefold vertices (Butcher et al. 1997). Based on these findings it is evident that the T2 capsid of cystoviruses undergoes substantial conformational changes during its maturation from empty to dsRNA-filled mature T2 capsid. The procapsid expansion leads to approximately 2.4-fold increase in the interior volume of the T2 capsid (Huiskonen et al. 2006).

Empty T2 capsids have not been detected in cells infected by reoviruses and it is unclear whether empty capsids are intermediates of virion assembly. Although the empty recombinant particles of reoviruses are indistinguishable from the dsRNA-filled SLPs (Kim et al. 2002; Zeng et al. 1996), it has been suggested that the radial plasticity between the A and B subunits within the capsid could allow collapse into a conformation similar to that described for empty T2 capsids of cystoviruses (Grimes et al. 1998). Accordingly, recombinant empty T2 capsid of rotavirus adopts a conformation resembling the intermediate particle of cystoviruses if incubated in the presence of viral nonstructural protein NSP5 (Berois et al. 2003).

In the large dsRNA viruses, the genome comprises multiple linear dsRNA molecules co-encapsidated in a single virus particle (Table 17.1), and apparently, only a single copy of each genome segment is present in each virion. The number of genome segments among reoviruses varies from 10 (orbiviruses, orthoreoviruses) to 12 (phytoreoviruses), and most of the segments encode a single gene and short noncoding regions (Mertens 2004). In cystoviruses the genome segments are polycistronic and each of the three genome segments encodes four to six distinct polypeptides (Hoogstraten et al. 2000; Poranen and Bamford 2011; Qiao et al. 2010).

The electron density maps of reoviruses have revealed that a significant proportion of RNA encapsidated within the T2 shell is ordered and appear as concentric layers of density in the icosahedrally averaged DLP structures (Gouet et al. 1999; Prasad et al. 1996; Reinisch et al. 2000). These observations support the proposal that the genomic dsRNA strands make a spiral around the enzymatic complexes at the icosahedral vertices (Gouet et al. 1999). It has been proposed that each genome segment is associated with a single replicase complex which repeatedly transcribes the segment (Mertens 2004). This would restrict the maximum number of genome segments to 12 (as there are 12 vertices in the icosahedrons). In the case of cystoviruses there is evidence that multiple polymerases can simultaneously transcribe a single genome segment (Usala et al. 1980).

The concentration of RNA in an orbivirus T2 capsid is approximately 400 g/l which corresponds to the concentration in which DNA forms liquid crystalline arrays (Gouet et al. 1999). Based on the estimated volume for cystovirus SLP (Huiskonen et al. 2006) it appears that the RNA concentration in cystoviruses is about 340 g/l. In spite of the high genome density the simultaneous transcription of multiple genomic segments can be efficiently accomplished.

The empty T2 capsid seems to be a true assembly intermediate for *Pseudomonas* phage φ6 and related cystoviruses (Bamford and Mindich 1980), but not necessarily for the different reoviruses (see Patton and Spencer 2000). The empty cystoviral capsids package the three viral mRNA molecules and replicate these to form dsRNA inside the capsid (Ewen and Revel 1988, 1990; Frilander and Bamford 1995; Gottlieb et al. 1992). Each genome segment has a unique packaging site at the 5′ end which is recognized by the T2 capsid (Gottlieb et al. 1994; Pirttimaa and Bamford 2000; Qiao et al. 2003). The translocation of the RNA into the capsid is dependent on NTP hydrolysis (Frilander and Bamford 1995; Gottlieb et al. 1992) and is driven by the hexameric helicase forming the turrets (Huiskonen et al. 2007; Pirttimaa et al. 2002; see Chap. 27 by Erica Mancini and Roman Tuma). To ensure the packaging of full complement of the three genome segments, the viral mRNA molecules are packaged in a specific order and the replication initiates only after the packaging site of the last RNA segment has been recognized within the particle (Frilander and Bamford 1995; Frilander et al. 1995; Poranen and Bamford 1999; Qiao et al. 1995). The initiation of replication is also dependent on a specific conserved 3′-terminal sequence in the viral RNAs (Frilander et al. 1992; Mindich et al. 1994; see also Chap. 26 by Leonard Mindich).

The mechanism by which the genome is encapsidated and how the different viral RNA segments are selected for encapsidation is not well understood for any member of the *Reoviridae* family. The current models of rotavirus assembly suggest that specific interactions between the polymerase and the conserved 3′ end of the genome segments would enable specific encapsidation of viral RNAs (Lu et al. 2008). The polymerase together with the bound RNA and the capping enzyme would recruit T2 proteins to form complete fivefold vertices (including a decamer of T2 protein; McClain et al. 2010). Twelve such structures would then come together to form the T2 capsid. The specific interactions between the RNA molecules would drive the formation of structures containing one copy of each of the genome segment thus ensuring encapsidation of a full complement of the segments (Patton and Spencer 2000). The properties of the replication intermediates suggest that the viral ssRNAs are largely exposed on the T2 capsid surface and are reeled into the capsid during replication (Patton and Gallegos 1990). The interaction between the T2 proteins with the polymerase is required to switch on the genome replication (Tortorici et al. 2003). The genome encapsidation and replication of reoviruses take place in the viroplasm and it is evident that the nonstructural proteins, many of which have RNA-binding activity, play a role in these processes (reviewed by Butan and Tucker 2010; Taraporewala and Patton 2004).

17.3.1 Single or Multiple Packaging Vertices

It is well established that the packaging of dsDNA bacteriophage capsid occurs via a portal vertex defining a unique capsid vertex (see Chaps. 22 and 23). Although the 12 icosahedral vertices in the dsRNA virus T2 capsid appear similar, there is biochemical evidence obtained for *Pseudomonas* phage φ6 (cystovirus) that one of the vertices could be structurally and functionally unique. A single amino acid substitution in the helicase protein of φ6 (the hexameric packaging NTPase P4 located in the capsid vertices; see Table 17.2 and Sect. 17.2.4.3) results in the formation of recombinant T2 capsids with a greatly reduced amount of P4 (Pirttimaa et al. 2002). Similar particles, with approximately one P4 hexamer per capsid, can also be obtained by gentle detergent treatment of complete T2 capsids (Pirttimaa et al. 2002). This suggests that one of the vertices in the T2 capsid is physically different from the other ones. Such P4-deficient capsids have the same RNA packaging and replication efficiencies and rates as the capsids with full complement of P4, whereas P4-null particles (recombinant particles with no P4) are completely inactive in RNA packaging (Pirttimaa et al. 2002). This indicates that cystovirus T2 capsids may have a special vertex that is necessary and sufficient for genome encapsidation. The P4 hexamers at the other vertices are required during transcription and act as passive conduits for mRNA produced in the capsid interior (Kainov et al. 2004; Pirttimaa et al. 2002). Thus the entry and exit of the genome appear to occur through different vertices in contrast to the dsDNA bacteriophages in which the genome packaging and ejection are accomplished via a single vertex.

17.4 T13 Shell

In the virions of large dsRNA viruses the T2 capsid is enclosed, in most cases, into a protein shell organized into $T=13$ *levo* icosahedral lattice (designated here as the T13 shell). In birnaviruses the T13 capsid comprises the only capsid layer (Bottcher et al. 1997; Coulibaly et al. 2005, 2010) making them unique among dsRNA viruses. The T13 shell of orbiviruses, rotaviruses, aquareovirus, and phytoreoviruses contains 780 copies of the T13 protein (Table 17.2), arranged as 260 trimers on the T2 shell scaffold (Grimes et al. 1998; Cheng et al. 2010; McClain et al. 2010; Nakagawa et al. 2003). In orthoreoviruses and cystoviruses (except in *Pseudomonas* phage φ8; Jaalinoja et al. 2007), there are only 600 copies (200 trimers) of the T13 protein as the $T=13$ icosahedral lattice is interrupted at the vertices by the turrets extending from the underlying T2 shell (Dryden et al. 1993; Huiskonen et al. 2006). Thus, ten capsid protein monomers contribute to one asymmetric unit in the T13 capsids of orthoreoviruses and cystoviruses, while 13 monomers are present in the asymmetric units of dsRNA viruses with complete $T=13$ shells. The inequality in the number of subunits in the T2 and T13 capsid layers (120 vs. 780 or 600) creates a symmetry mismatch.

The T13 protein trimers are located in four or five quasi-equivalent environments. The corresponding icosahedrally independent positions are designated as P, Q, R, S, and T (Grimes et al. 1998; Fig. 17.3). Position P, which is closest to the fivefold symmetry axis, is present only in the virions which have nonturreted T2 capsids (sedoreoviruses; Grimes et al. 1998; Cheng et al. 2010; McClain et al. 2010; Nakagawa et al. 2003). The T positions are centered on the threefold axes and this is the only position where the symmetries of the two protein layers coincide. The S subunits occupy positions adjacent to the twofold axis, and the Q and R subunits are located between the P and T sites (Fig. 17.3). In orthoreoviruses and cystoviruses, where P sites are not present, the Q sites are closest to the fivefold vertices, encircling the turrets (Dryden et al. 1993; Huiskonen et al. 2006).

17.4.1 T13 Protein

The T13 proteins of rotaviruses, orbiviruses, aquareovirus, orthoreoviruses, and phytoreoviruses have noticeable similar folds (Grimes et al. 1995; Liemann et al. 2002; Mathieu et al. 2001;

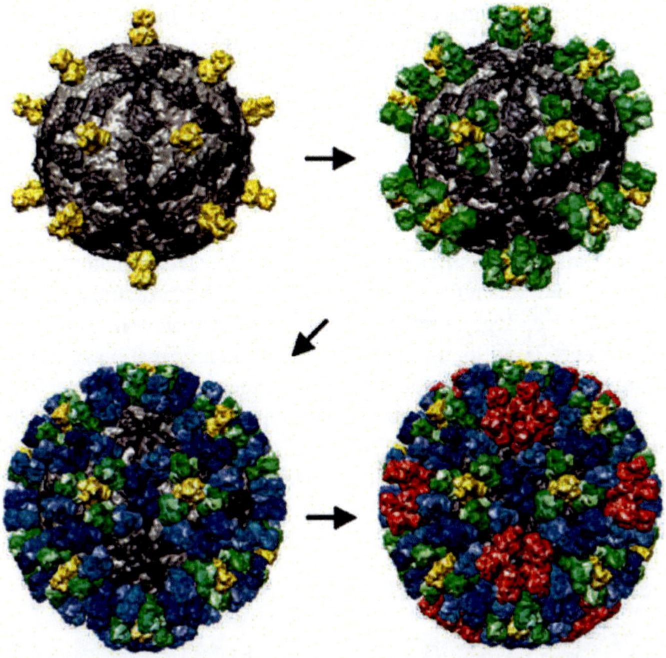

Fig. 17.3 Proposed assembly pathway for the T13 capsid and its structural organization. Five T13 protein trimers, P, Q, R, S, and T colored in *red*, *blue*, *green*, *dark blue*, and *yellow* can be identified in the T13 capsids of dsRNA viruses with nonturreted T2 capsids (*below*, *right*). The two T2 protein conformers A and B of the T2 shell are colored in *dark* and *light gray*, respectively (*up*, *left*). T13 protein T-trimer first associates with the T2 capsid shell at the icosahedral threefold symmetry positions. R-trimers are then attached, followed by Q- and S-trimers and finally by the P-trimers at the icosahedral fivefold positions. Molecular graphics images were produced using the UCSF Chimera package from the Resource for Biocomputing, Visualization, and Informatics at the University of California, San Francisco (supported by NIH P41 RR001081; Pettersen et al. 2004). The X-ray structure of bluetongue virus core (protein data bank accession number 2btv) was used as a template

Nakagawa et al. 2003; Zhang et al. 2010). The lower part, making contacts with the T2 shell, is α-helical and the upper part, exposed on the surface of the DLP, consists mainly of β-sheets with a jelly roll β-barrel topology frequently found in viral capsid proteins (Rossmann and Johnson 1989). The T13 proteins typically form stable trimers in solution (Gorziglia et al. 1985; Basak et al. 1992; Poranen et al. 2001). Within the trimer each monomer wraps around another in a right-handed conformation about the molecular threefold axis (Grimes et al. 1995; Liemann et al. 2002; Mathieu et al. 2001; Nakagawa et al. 2003; Zhang et al. 2010).

The T13 protein of cystoviruses is relatively small (approximately 16 kDa) compared to the corresponding proteins of reoviruses (38.5–76 kDa for bluetongue virus and mammalian reovirus, respectively; Mertens 2004). Consequently, to cover the underlying T2 capsid, the T13 protein trimer of cystoviruses has to be significantly flat (height about 2.5 nm, diameter about 7.5 nm; Huiskonen et al. 2006) compared to the corresponding proteins in reoviruses (e.g., orthoreovirus height about 10 nm, diameter about 7.5 nm; Liemann et al. 2002). Like the lower domain in the T13 protein of reoviruses the cystoviral T13 protein is mostly α-helical (Bamford et al. 1993; Huiskonen et al. 2006).

The T13 protein is important for the virion stability but in most cases it is also directly involved in the host plasma membrane penetration during virus entry (Chandran et al. 1999; Cvirkaite-Krupovic et al. 2010; Nibert and Fields 1992; Poranen et al. 1999; Tan et al. 2001; Xu et al. 1997). The entry-related functions in orthoreoviruses are triggered by specific protein cleavages which apparently take place during the virus entry (Nibert et al. 2005). In rotavirus the presence of the T13 layer is probably needed for the viral transcription (Bican et al. 1982; Charpilienne et al. 2002; Lawton et al. 1997a).

17.4.2 T13 Capsid-Associated Proteins

Cystoviruses have a small protein component P5 which is located beneath the envelope and loosely associated with the T13 shell (Hantula and Bamford 1988). This protein has murein peptidase activity and is essential for the peptidoglycan penetration during virus entry into its bacterial host (Mindich and Lehman 1979). There is no high-resolution structural information on this protein. However, image reconstructions of *Pseudomonas* phage φ12 have identified 60 symmetrically orientated densities on the T13 shell surface that could correspond to protein P5 (Wei et al. 2009).

To facilitate the comparison of dsRNA viruses, the different T13 shell-associated virion proteins that are exposed on the virion surface are discussed under Sect. 17.5.

17.4.3 Assembly of the T13 Capsid

In vitro reconstitution of DLP from SLPs and T13 protein has been demonstrated for cystoviruses (Olkkonen et al. 1990), rotaviruses (Charpilienne et al. 2002), and orthoreoviruses (T13 protein/σ3 heterohexamers; Chandran et al. 1999) either using purified constituents or cell extract. In the absence of the SLP, the T13 proteins of several dsRNA viruses self-assemble into different nonicosahedral structures such as two-dimensional hexagonal arrays, helical tubes, or polydisperse mixtures of aberrant spiral-like shells (Burroughs et al. 1994; Iwasaki et al. 2008; Lepault et al. 2001; Olkkonen et al. 1991; Zhu et al. 1997). Such observations suggest that the underlying T2 capsid template plays a critical role in organizing the surrounding $T=13$ icosahedral lattice.

The high-resolution structure of phytoreovirus, orbireovirus, and rotavirus DLPs has revealed major interaction sites between the T2 and T13 layers suggesting a pathway for T13 shell assembly. The first T13 protein trimer interacting with the underlying T2 shell is most likely the T-trimer located at the icosahedral threefold axis of the SLP (Fig. 17.3) as the interactions between T13 protein trimer and T2 capsid are strongest at these positions (Grimes et al. 1998; McClain et al. 2010; Nakagawa et al. 2003). Phytoreovirus SLPs decorated with the T13 protein trimers at the icosahedral threefold axis of symmetry have been isolated confirming that such structures are stable and thus potential assembly intermediates (Wu et al. 2000). The T13 protein T trimers likely nucleate the extension of the T13 lattice through strong side-to-side interactions with neighboring T13 protein trimers. The R-trimers of T13 protein first pack around the T-trimers followed by Q- and S-trimers. The assembly is finished by the incorporation of P-trimers which make the least favorable contacts with the T2 layer (Fig. 17.3). The interactions with the underlying T2 capsid are relatively weak at positions of symmetry mismatches between the $T=1$ and $T=13$ layers. Consistent with this model, particles with incomplete T13 shell, in which the five P-trimers around the fivefold vertices are absent, are produced in a recombinant expression system of orbivirus T2 and T13 proteins (Hewat et al. 1992; Nason et al. 2004).

The assembly pathway for the T13 shell presented above does not necessarily apply for dsRNA viruses with turreted T2 capsids. In orthoreoviruses, the capping enzyme forming the turrets is required for the assembly of the T13 shell (Kim et al. 2002) suggesting that interactions between the T13 protein and the capping enzyme at the icosahedral fivefold vertices are essential for the formation of the T13 shell, probably due to the need to nucleate the T13 shell assembly. Such hypothesis is supported by a recent cryo-EM-based model of aquareovirus virion, which shows clear areas of interactions between the turret and the Q-trimers (Cheng et al. 2010). Consequently, the incorporation of the T13 protein trimers on turreted and nonturreted T2 capsids may be accomplished in opposite orders.

The assembly of the orthoreovirus T13 shell is also dependent on the outer capsid proteins σ3 which forms heterohexameric complexes with the T13 protein (Liemann et al. 2002). This interaction has been shown to be essential for the assembly of the T13 capsid around the orthoreovirus SLPs (Chandran et al. 1999; Shing and Coombs 1996). Consequently, σ3 could be considered as to be an integral part of the T13 shell. Nevertheless, σ3 can be uncoated from the virion resulting in stable particles (Jane-Valbuena et al. 1999). Thus the integrity of the T13 layer is not dependent on σ3.

17.5 Outer Virion Layers

The outermost virion layer in large icosahedral dsRNA viruses is the most variable in structure, organization, and composition (Table 17.2) being a lipid–protein envelope in cystoviruses or composed of variety of unrelated proteins in different reoviruses (Table 17.2). The components external to the T13 capsid may either form a distinct structural shell (outer capsid in rotaviruses and envelope in cystoviruses) or a noncontiguous layer as observed in orthoreoviruses and orbiviruses.

The outer virion layers of rotaviruses and orthoreoviruses follow the $T=13$ icosahedral symmetry of the underlying T13 capsid. In orthoreoviruses, the T13 protein layer is decorated with σ3 protein (see Sect. 17.4.3) and the trimeric cell attachment protein σ1, with a distinct head and tail morphology (Chappell et al. 2002; Fraser et al. 1990) attached to the turrets of the underlying T2 capsid at the icosahedral vertices of the DLP. The rotavirus outer protein layer is composed of 260 trimers of the outer capsid protein VP7 located at the icosahedral and local threefold axes of the $T=13$ icosahedral lattice and 60 trimers of spike protein VP4 inserted into specific channels of the $T=13$ lattices, around the icosahedral fivefold axes (Li et al. 2009; Prasad et al. 1988, 1990; Settembre et al. 2011).

The outer capsid of orbiviruses is made up of two distinct structures, triskelions (trimers of VP2) located on the Q-type T13 protein trimers close to the icosahedral fivefold axes of the $T=13$ lattice and globular multimers (trimers of VP5) located at the quasi-sixfold axes of the underlying $T=13$ shell (Hassan et al. 2001; Hassan and Roy 1999; Zhang et al. 2011). The organization of proteins in the outer surface layer of phytoreovirus virions is not currently known but it contains at least protein P2, a minor virion component with membrane-fusogenic activity, which can be specifically removed from the surface of the virion without loss of other viral proteins (Yan et al. 1996; Zhou et al. 2007). Protein P9, which was initially considered to be a viral nonstructural protein, is likely also associated on the phytoreovirus virion surface (Zhong et al. 2003).

The envelope of cystoviruses is derived from the bacterial plasma membrane and it contains three to four integral membrane proteins (including protein P6 which mediates fusion between the bacterial outer membrane and the virion envelope; Bamford et al. 1987) and one to three peripheral membrane proteins forming the spikes (see, e.g., Gottlieb et al. 1988a, 2002; Hoogstraten et al. 2000). The membrane proteins in the cystoviral envelope are not arranged on an icosahedral lattice (Jaalinoja et al. 2007).

17.6 Conclusions

A wealth of exciting information on the structure and assembly of large icosahedral dsRNA viruses has been obtained during the past decades. The members of *Reoviridae* have been especially amiable for structural studies and the research has been pushing the structural biology forward in order to obtain high-resolution structural information on these complex multifunctional molecular machines (see, e.g., Grimes et al. 1998; Prasad et al. 1996; Settembre et al. 2011; Zhang et al. 2010). Such information has also led to the hypotheses on the assembly pathways for different representatives of this large viral family (see Sects. 17.2.5 and 17.4.3).

The unique in vitro assembly system for *Pseudomonas* phage φ6 has allowed analyses of a dsRNA virus assembly from purified protein and RNA components to infectious biologically functional DLPs (Poranen et al. 2001). All the information for the formation of the complex multifunctional φ6 DLP is carried within the structures of the components; no additional viral or cellular factors are required in the process. Consequently, it has been possible to reveal several fine details of the assembly and replication of cystoviruses (see Sects. 17.2.5 and 17.3; Poranen et al. 2005a, b; Chap 27, Fig. 27.3a).

Unlike in cystoviruses, the nonstructural proteins have a major role in the morphogenesis of reovirus DPLs. They induce the formation of viral inclusion bodies in infected cells and apparently are important in the encapsidation of viral RNAs into the T2 capsids (see Sects. 17.1.1.2 and 17.3). The structural proteins alone have also a strong inherent capacity to self-assemble into defined empty virus-like particles (Sect. 17.2.5). The details of the genome encapsidation processes are still obscure, and further research is required to get insights to these events.

The different approaches including analyses of mutant virus infections, recombinant expression systems for viral proteins, medium and high-resolution structural analyses, and in vitro assembly systems have been elementary in defining the assembly pathway for the large multilayered dsRNA viruses (see Sects. 17.2.5 and 17.4.3; Figs. 17.2 and 17.3). It appears that there is certain degree of flexibility in the way that principally similar virion structures (T2 and T13 capsids) are built up among the different dsRNA viruses.

Acknowledgments This work was supported by the Finnish Center of Excellence Program (2006–2011) of the Academy of Finland (grant 1129684) and Academy of Finland grants 250113 and 256069.

References

Altenburg BC, Graham DY, Estes MK (1980) Ultrastructural study of rotavirus replication in cultured cells. J Gen Virol 46:75–85

Bamford DH (2003) Do viruses form lineages across different domains of life? Res Microbiol 154:231–236

Bamford DH, Mindich L (1980) Electron microscopy of cells infected with nonsense mutants of bacteriophage φ6. Virology 107:222–228

Bamford DH, Romantschuk M, Somerharju PJ (1987) Membrane fusion in prokaryotes: bacteriophage φ6 membrane fuses with the *Pseudomonas syringae* outer membrane. EMBO J 6:1467–1473

Bamford JK, Bamford DH, Li T, Thomas GJ Jr (1993) Structural studies of the enveloped dsRNA bacteriophage φ6 of *Pseudomonas syringae* by Raman spectroscopy. II. Nucleocapsid structure and thermostability of the virion, nucleocapsid and polymerase complex. J Mol Biol 230:473–482

Banerjee AK, Shatkin AJ (1970) Transcription in vitro by reovirus-associated ribonucleic acid-dependent polymerase. J Virol 6:1–11

Basak AK, Stuart DI, Roy P (1992) Preliminary crystallographic study of bluetongue virus capsid protein, VP7. J Mol Biol 228:687–689

Becker MM, Peters TR, Dermody TS (2003) Reovirus σNS and μNS proteins form cytoplasmic inclusion structures in the absence of viral infection. J Virol 77:5948–5963

Benevides JM, Juuti JT, Tuma R, Bamford DH, Thomas GJ Jr (2002) Characterization of subunit-specific interactions in a double-stranded RNA virus: Raman difference spectroscopy of the φ6 procapsid. Biochemistry 41:11946–11953

Berois M, Sapin C, Erk I, Poncet D, Cohen J (2003) Rotavirus nonstructural protein NSP5 interacts with major core protein VP2. J Virol 77:1757–1763

Bican P, Cohen J, Charpilienne A, Scherrer R (1982) Purification and characterization of bovine rotavirus cores. J Virol 43:1113–1117

Bisaillon M, Bergeron J, Lemay G (1997) Characterization of the nucleoside triphosphate phosphohydrolase and helicase activities of the reovirus λ1 protein. J Biol Chem 272:18298–18303

Bottcher B, Kiselev NA, Stel'Mashchuk VY, Perevozchikova NA, Borisov AV, Crowther RA (1997) Three-dimensional structure of infectious bursal disease virus determined by electron cryomicroscopy. J Virol 71:325–330

Brookes SM, Hyatt AD, Eaton BT (1993) Characterization of virus inclusion bodies in bluetongue virus-infected cells. J Gen Virol 74:525–530

Burroughs JN, O'Hara RS, Smale CJ, Hamblin C, Walton A, Armstrong R, Mertens PP (1994) Purification and properties of virus particles, infectious subviral particles, cores and VP7 crystals of African horsesickness virus serotype 9. J Gen Virol 75:1849–1857

Butan C, Tucker P (2010) Insights into the role of the non-structural protein 2 (NS2) in Bluetongue virus morphogenesis. Virus Res 151:109–117

Butcher SJ, Dokland T, Ojala PM, Bamford DH, Fuller SD (1997) Intermediates in the assembly pathway of the double-stranded RNA virus φ6. EMBO J 16:4477–4487

Butcher SJ, Grimes JM, Makeyev EV, Bamford DH, Stuart DI (2001) A mechanism for initiating RNA-dependent RNA polymerization. Nature 410:235–240

Calvert JG, Nagy E, Soler M, Dobos P (1991) Characterization of the VPg-dsRNA linkage of infectious pancreatic necrosis virus. J Gen Virol 72:2563–2567

Carstens EB (2010) Ratification vote on taxonomic proposals to the International Committee on Taxonomy of Viruses (2009). Arch Virol 155:133–146

Caston JR, Ghabrial SA, Jiang D, Rivas G, Alfonso C, Roca R, Luque D, Carrascosa JL (2003) Three-dimensional structure of *Penicillium chrysogenum* virus: a double-stranded RNA virus with a genuine $T = 1$ capsid. J Mol Biol 331:417–431

Chandra R (1997) Picobirnavirus, a novel group of undescribed viruses of mammals and birds: a minireview. Acta Virol 41:59–62

Chandran K, Walker SB, Chen Y, Contreras CM, Schiff LA, Baker TS, Nibert ML (1999) In vitro recoating of reovirus cores with baculovirus-expressed outer-capsid proteins μ1 and σ3. J Virol 73:3941–3950

Chappell JD, Prota AE, Dermody TS, Stehle T (2002) Crystal structure of reovirus attachment protein σ1 reveals evolutionary relationship to adenovirus fiber. EMBO J 21:1–11

Charpilienne A, Lepault J, Rey F, Cohen J (2002) Identification of rotavirus VP6 residues located at the interface with VP2 that are essential for capsid assembly and transcriptase activity. J Virol 76:7822–7831

Chen D, Luongo CL, Nibert ML, Patton JT (1999) Rotavirus open cores catalyze 5′-capping and methylation of exogenous RNA: evidence that VP3 is a methyltransferase. Virology 265:120–130

Cheng RH, Caston JR, Wang GJ, Gu F, Smith TJ, Baker TS, Bozarth RF, Trus BL, Cheng N, Wickner RB et al (1994) Fungal virus capsids, cytoplasmic compartments for the replication of double-stranded RNA, formed as icosahedral shells of asymmetric Gag dimers. J Mol Biol 244:255–258

Cheng L, Zhu J, Hui WH, Zhang X, Honig B, Fang Q, Zhou ZH (2010) Backbone model of an aquareovirus virion by cryo-electron microscopy and bioinformatics. J Mol Biol 397:852–863

Cleveland DR, Zarbl H, Millward S (1986) Reovirus guanylyltransferase is L2 gene product lambda 2. J Virol 60:307–311

Coombs KM (1996) Identification and characterization of a double-stranded RNA$^-$ reovirus temperature-sensitive mutant defective in minor core protein μ2. J Virol 70:4237–4245

Coulibaly F, Chevalier C, Gutsche I, Pous J, Navaza J, Bressanelli S, Delmas B, Rey FA (2005) The birnavirus crystal structure reveals structural relationships among icosahedral viruses. Cell 120:761–772

Coulibaly F, Chiu E, Ikeda K, Gutmann S, Haebel PW, Schulze-Briese C, Mori H, Metcalf P (2007) The molecular organization of cypovirus polyhedra. Nature 446:97–101

Coulibaly F, Chevalier C, Delmas B, Rey FA (2010) Crystal structure of an Aquabirnavirus particle: insights into antigenic diversity and virulence determinism. J Virol 84:1792–1799

Cvirkaite-Krupovic V, Poranen MM, Bamford DH (2010) Phospholipids act as secondary receptor during the entry of the enveloped, double-stranded RNA bacteriophage phi6. J Gen Virol 91:2116–2120

de Haas F, Paatero AO, Mindich L, Bamford DH, Fuller SD (1999) A symmetry mismatch at the site of RNA packaging in the polymerase complex of dsRNA bacteriophage φ6. J Mol Biol 294:357–372

Dinman JD, Icho T, Wickner RB (1991) A 1 ribosomal frameshift in a double-stranded RNA virus of yeast forms a gag-pol fusion protein. Proc Natl Acad Sci USA 88:174–178

Diprose JM, Burroughs JN, Sutton GC, Goldsmith A, Gouet P, Malby R, Overton I, Zientara S, Mertens PP, Stuart DI, Grimes JM (2001) Translocation portals for the substrates and products of a viral transcription complex: the bluetongue virus core. EMBO J 20:7229–7239

Dryden KA, Wang G, Yeager M, Nibert ML, Coombs KM, Furlong DB, Fields BN, Baker TS (1993) Early steps in reovirus infection are associated with dramatic changes in supramolecular structure and protein conformation: analysis of virions and subviral particles by cryoelectron microscopy and image reconstruction. J Cell Biol 122:1023–1041

Dryden KA, Farsetta DL, Wang G, Keegan JM, Fields BN, Baker TS, Nibert ML (1998) Internal/structures containing transcriptase-related proteins in top component particles of mammalian orthoreovirus. Virology 245:33–46

Duquerroy S, Da Costa B, Henry C, Vigouroux A, Libersou S, Lepault J, Navaza J, Delmas B, Rey FA (2009) The picobirnavirus crystal structure provides functional insights into virion assembly and cell entry. EMBO J 28:1655–1665

Ewen ME, Revel HR (1988) In vitro replication and transcription of the segmented double-stranded RNA bacteriophage φ6. Virology 165:489–498

Ewen ME, Revel HR (1990) RNA-protein complexes responsible for replication and transcription of the double-stranded RNA bacteriophage φ6. Virology 178:509–519

Fabbretti E, Afrikanova I, Vascotto F, Burrone OR (1999) Two non-structural rotavirus proteins, NSP2 and NSP5, form viroplasm-like structures *in vivo*. J Gen Virol 80:333–339

Fraser RD, Furlong DB, Trus BL, Nibert ML, Fields BN, Steven AC (1990) Molecular structure of the cell-attachment protein of reovirus: correlation of computer-processed electron micrographs with sequence-based predictions. J Virol 64:2990–3000

Fregolente MC, de Castro-Dias E, Martins SS, Spilki FR, Allegretti SM, Gatti MS (2009) Molecular characterization of picobirnaviruses from new hosts. Virus Res 143:134–136

French TJ, Roy P (1990) Synthesis of bluetongue virus (BTV) corelike particles by a recombinant baculovirus expressing the two major structural core proteins of BTV. J Virol 64:1530–1536

Frilander M, Bamford DH (1995) *In vitro* packaging of the single-stranded RNA genomic precursors of the segmented double-stranded RNA bacteriophage φ6: the three segments modulate each other's packaging efficiency. J Mol Biol 246:418–428

Frilander M, Gottlieb P, Strassman J, Bamford DH, Mindich L (1992) Dependence of minus-strand synthesis on complete genomic packaging in the double-stranded RNA bacteriophage φ6. J Virol 66:5013–5017

Frilander M, Poranen M, Bamford DH (1995) The large genome segment of dsRNA bacteriophage φ6 is the key regulator in the *in vitro* minus and plus strand synthesis. RNA 1:510–518

Gorziglia M, Larrea C, Liprandi F, Esparza J (1985) Biochemical evidence for the oligomeric (possibly trimeric) structure of the major inner capsid polypeptide (45K) of rotaviruses. J Gen Virol 66:1889–1900

Gottlieb P, Metzger S, Romantschuk M, Carton J, Strassman J, Bamford DH, Kalkkinen N, Mindich L (1988a) Nucleotide sequence of the middle dsRNA segment of bacteriophage φ6: placement of the genes of membrane-associated proteins. Virology 163:183–190

Gottlieb P, Strassman J, Bamford DH, Mindich L (1988b) Production of a polyhedral particle in *Escherichia coli* from a cDNA copy of the large genomic segment of bacteriophage φ6. J Virol 62:181–187

Gottlieb P, Strassman J, Qiao X, Frilander M, Frucht A, Mindich L (1992) *In vitro* packaging and replication of individual genomic segments of bacteriophage φ6 RNA. J Virol 66:2611–2616

Gottlieb P, Qiao X, Strassman J, Frilander M, Mindich L (1994) Identification of the packaging regions within the genomic RNA segments of bacteriophage φ6. Virology 200:42–47

Gottlieb P, Wei H, Potgieter C, Toporovsky I (2002) Characterization of φ12, a bacteriophage related to φ6: nucleotide sequence of the small and middle double-stranded RNA. Virology 293:118–124

Gouet P, Diprose JM, Grimes JM, Malby R, Burroughs JN, Zientara S, Stuart DI, Mertens PP (1999) The highly ordered double-stranded RNA genome of bluetongue virus revealed by crystallography. Cell 97:481–490

Grimes J, Basak AK, Roy P, Stuart D (1995) The crystal structure of bluetongue virus VP7. Nature 373:167–170

Grimes JM, Burroughs JN, Gouet P, Diprose JM, Malby R, Zientara S, Mertens PP, Stuart DI (1998) The atomic structure of the bluetongue virus core. Nature 395:470–478

Guglielmi KM, McDonald SM, Patton JT (2010) Mechanism of intraparticle synthesis of the rotavirus double-stranded RNA genome. J Biol Chem 285:18123–18128

Hagiwara K, Higashi T, Namba K, Uehara-Ichiki T, Omura T (2003) Assembly of single-shelled cores and double-shelled virus-like particles after baculovirus expression of major structural proteins P3, P7 and P8 of *Rice dwarf virus*. J Gen Virol 84:981–984

Hantula J, Bamford DH (1988) Chemical crosslinking of bacteriophage φ6 nucleocapsid proteins. Virology 165:482–488

Hassan SS, Roy P (1999) Expression and functional characterization of bluetongue virus VP2 protein: role in cell entry. J Virol 73:9832–9842

Hassan SH, Wirblich C, Forzan M, Roy P (2001) Expression and functional characterization of bluetongue virus VP5 protein: role in cellular permeabilization. J Virol 75:8356–8367

Hewat EA, Booth TF, Loudon PT, Roy P (1992) Three-dimensional reconstruction of baculovirus expressed bluetongue virus core-like particles by cryo-electron microscopy. Virology 189:10–20

Hill CL, Booth TF, Prasad BV, Grimes JM, Mertens PP, Sutton GC, Stuart DI (1999) The structure of a cypovirus and the functional organization of dsRNA viruses. Nat Struct Biol 6:565–568

Hoogstraten D, Qiao X, Sun Y, Hu A, Onodera S, Mindich L (2000) Characterization of φ8, a bacteriophage containing three double-stranded RNA genomic segments and distantly related to φ6. Virology 272:218–224

Hsiao J, Martinez-Costas J, Benavente J, Vakharia VN (2002) Cloning, expression, and characterization of avian reovirus guanylyltransferase. Virology 296:288–299

Huiskonen JT, de Haas F, Bubeck D, Bamford DH, Fuller SD, Butcher SJ (2006) Structure of the bacteriophage φ6 nucleocapsid suggests a mechanism for sequential RNA packaging. Structure 14:1039–1048

Huiskonen JT, Jaalinoja HT, Briggs JA, Fuller SD, Butcher SJ (2007) Structure of a hexameric RNA packaging motor in a viral polymerase complex. J Struct Biol 158:156–164

Iwasaki K, Miyazaki N, Hammar L, Zhu Y, Omura T, Wu B, Sjoborg F, Yonekura K, Murata K, Namba K, Caspar DL, Fujiyoshi Y, Cheng RH (2008) Pleomorphic configuration of the trimeric capsid proteins of *Rice dwarf virus* that allows formation of both the outer capsid and tubular crystals. J Mol Biol 383:252–265

Jaalinoja HT, Huiskonen JT, Butcher SJ (2007) Electron cryomicroscopy comparison of the architectures of the enveloped bacteriophages φ6 and φ8. Structure 15:157–167

Jane-Valbuena J, Nibert ML, Spencer SM, Walker SB, Baker TS, Chen Y, Centonze VE, Schiff LA (1999) Reovirus virion-like particles obtained by recoating infectious subvirion particles with baculovirus-expressed σ3 protein: an approach for analyzing σ3 functions during virus entry. J Virol 73:2963–2973

Jayaram H, Taraporewala Z, Patton JT, Prasad BV (2002) Rotavirus protein involved in genome replication and packaging exhibits a HIT-like fold. Nature 417:311–315

Juuti JT, Bamford DH (1995) RNA binding, packaging and polymerase activities of the different incomplete polymerase complex particles of dsRNA bacteriophage φ6. J Mol Biol 249:545–554

Juuti JT, Bamford DH (1997) Protein P7 of phage φ6 RNA polymerase complex, acquiring of RNA packaging activity by in vitro assembly of the purified protein onto deficient particles. J Mol Biol 266:891–900

Juuti JT, Bamford DH, Tuma R, Thomas GJ Jr (1998) Structure and NTPase activity of the RNA-translocating protein (P4) of bacteriophage φ6. J Mol Biol 279:347–359

Kainov DE, Butcher SJ, Bamford DH, Tuma R (2003a) Conserved intermediates on the assembly pathway of double-stranded RNA bacteriophages. J Mol Biol 328:791–804

Kainov DE, Pirttimaa M, Tuma R, Butcher SJ, Thomas GJ Jr, Bamford DH, Makeyev EV (2003b) RNA packaging device of double-stranded RNA bacteriophages, possibly as simple as hexamer of P4 protein. J Biol Chem 278:48084–48091

Kainov DE, Lisal J, Bamford DH, Tuma R (2004) Packaging motor from double-stranded RNA bacteriophage φ12 acts as an obligatory passive conduit during transcription. Nucleic Acids Res 32:3515–3521

Kar AK, Roy P (2003) Defining the structure-function relationships of bluetongue virus helicase protein VP6. J Virol 77:11347–11356

Kar AK, Ghosh M, Roy P (2004) Mapping the assembly pathway of Bluetongue virus scaffolding protein VP3. Virology 324:387–399

Kim J, Zhang X, Centonze VE, Bowman VD, Noble S, Baker TS, Nibert ML (2002) The hydrophilic amino-terminal arm of reovirus core shell protein λ1 is dispensable for particle assembly. J Virol 76:12211–12222

Kim J, Parker JS, Murray KE, Nibert ML (2004) Nucleoside and RNA triphosphatase activities of orthoreovirus transcriptase cofactor μ2. J Biol Chem 279:4394–4403

Labbe M, Charpilienne A, Crawford SE, Estes MK, Cohen J (1991) Expression of rotavirus VP2 produces empty corelike particles. J Virol 65:2946–2952

Laurinavicius S, Kakela R, Bamford DH, Somerharju P (2004) The origin of phospholipids of the enveloped bacteriophage phi6. Virology 326:182–190

Lawton JA, Estes MK, Prasad BV (1997a) Three-dimensional visualization of mRNA release from actively transcribing rotavirus particles. Nat Struct Biol 4:118–121

Lawton JA, Zeng CQ, Mukherjee SK, Cohen J, Estes MK, Prasad BV (1997b) Three-dimensional structural analysis of recombinant rotavirus-like particles with intact and amino-terminal-deleted VP2: implications for the architecture of the VP2 capsid layer. J Virol 71:7353–7360

Le Blois H, French T, Mertens PP, Burroughs JN, Roy P (1992) The expressed VP4 protein of bluetongue virus binds GTP and is the candidate guanylyl transferase of the virus. Virology 189:757–761

Lepault J, Petitpas I, Erk I, Navaza J, Bigot D, Dona M, Vachette P, Cohen J, Rey FA (2001) Structural polymorphism of the major capsid protein of rotavirus. EMBO J 20:1498–1507

Li Z, Baker ML, Jiang W, Estes MK, Prasad BV (2009) Rotavirus architecture at subnanometer resolution. J Virol 83:1754–1766

Liemann S, Chandran K, Baker TS, Nibert ML, Harrison SC (2002) Structure of the reovirus membrane-penetration protein, μ1, in a complex with its protector protein, σ3. Cell 108:283–295

Liu M, Mattion NM, Estes MK (1992) Rotavirus VP3 expressed in insect cells possesses guanylyltransferase activity. Virology 188:77–84

Loudon PT, Roy P (1992) Interaction of nucleic acids with core-like and subcore-like particles of bluetongue virus. Virology 191:231–236

Lu G, Zhou ZH, Baker ML, Jakana J, Cai D, Wei X, Chen S, Gu X, Chiu W (1998) Structure of double-shelled rice dwarf virus. J Virol 72:8541–8549

Lu X, McDonald SM, Tortorici MA, Tao YJ, Vasquez-Del Carpio R, Nibert ML, Patton JT, Harrison SC (2008) Mechanism for coordinated RNA packaging and genome replication by rotavirus polymerase VP1. Structure 16:1678–1688

Luongo CL, Zhang X, Walker SB, Chen Y, Broering TJ, Farsetta DL, Bowman VD, Baker TS, Nibert ML (2002) Loss of activities for mRNA synthesis accompanies loss of λ2 spikes from reovirus cores: an effect of λ2 on λ1 shell structure. Virology 296:24–38

Luque D, Gonzalez JM, Garriga D, Ghabrial SA, Havens WM, Trus B, Verdaguer N, Carrascosa JL, Caston JR (2010) The T=1 capsid protein of *Penicillium chrysogenum* virus is formed by a repeated helix-rich core indicative of gene duplication. J Virol 84:7256–7266

Mathieu M, Petitpas I, Navaza J, Lepault J, Kohli E, Pothier P, Prasad BV, Cohen J, Rey FA (2001) Atomic structure of the major capsid protein of rotavirus: implications for the architecture of the virion. EMBO J 20:1485–1497

Matsuo E, Roy P (2009) Bluetongue virus VP6 acts early in the replication cycle and can form the basis of chimeric virus formation. J Virol 83:8842–8848

McClain B, Settembre E, Temple BR, Bellamy AR, Harrison SC (2010) X-ray crystal structure of the rotavirus inner capsid particle at 3.8 Å resolution. J Mol Biol 397:587–599

Mendez II, Weiner SG, She YM, Yeager M, Coombs KM (2008) Conformational changes accompany activation of reovirus RNA-dependent RNA transcription. J Struct Biol 162:277–289

Mertens P (2004) The dsRNA viruses. Virus Res 101:3–13

Miller CL, Arnold MM, Broering TJ, Hastings CE, Nibert ML (2010) Localization of mammalian orthoreovirus proteins to cytoplasmic factory-like structures via nonoverlapping regions of μ NS. J Virol 84:867–882

Mindich L, Lehman J (1979) Cell wall lysin as a component of the bacteriophage φ6 virion. J Virol 30:489–496

Mindich L, Qiao X, Onodera S, Gottlieb P, Frilander M (1994) RNA structural requirements for stability and minus-strand synthesis in the dsRNA bacteriophage φ6. Virology 202:258–263

Mindich L, Qiao X, Qiao J, Onodera S, Romantschuk M, Hoogstraten D (1999) Isolation of additional bacteriophages with genomes of segmented double-stranded RNA. J Bacteriol 181:4505–4508

Miyazaki N, Wu B, Hagiwara K, Wang CY, Xing L, Hammar L, Higashiura A, Tsukihara T, Nakagawa A, Omura T, Cheng RH (2010) The functional organization of the internal components of *Rice dwarf virus*. J Biochem 147:843–850

Moss SR, Nuttall PA (1994) Subcore- and core-like particles of Broadhaven virus (BRDV), a tick-borne orbivirus, synthesized from baculovirus expressed VP2 and VP7, the major core proteins of BRDV. Virus Res 32:401–407

Naitow H, Tang J, Canady M, Wickner RB, Johnson JE (2002) L-A virus at 3.4 Å resolution reveals particle architecture and mRNA decapping mechanism. Nat Struct Biol 9:725–728

Nakagawa A, Miyazaki N, Taka J, Naitow H, Ogawa A, Fujimoto Z, Mizuno H, Higashi T, Watanabe Y, Omura T, Cheng RH, Tsukihara T (2003) The atomic structure of rice dwarf virus reveals the self-assembly mechanism of component proteins. Structure 11:1227–1238

Nason EL, Samal SK, Venkataram Prasad BV (2000) Trypsin-induced structural transformation in aquareovirus. J Virol 74:6546–6555

Nason EL, Rothagel R, Mukherjee SK, Kar AK, Forzan M, Prasad BV, Roy P (2004) Interactions between the inner and outer capsids of bluetongue virus. J Virol 78:8059–8067

Nibert ML, Fields BN (1992) A carboxy-terminal fragment of protein μ1/μ1C is present in infectious subvirion particles of mammalian reoviruses and is proposed to have a role in penetration. J Virol 66:6408–6418

Nibert ML, Odegard AL, Agosto MA, Chandran K, Schiff LA (2005) Putative autocleavage of reovirus μ1 protein in concert with outer-capsid disassembly and activation for membrane permeabilization. J Mol Biol 345:461–474

Olkkonen VM, Gottlieb P, Strassman J, Qiao XY, Bamford DH, Mindich L (1990) *In vitro* assembly of infectious nucleocapsids of bacteriophage φ6: formation of a recombinant double-stranded RNA virus. Proc Natl Acad Sci USA 87:9173–9177

Olkkonen VM, Ojala PM, Bamford DH (1991) Generation of infectious nucleocapsids by *in vitro* assembly of the shell protein on to the polymerase complex of the dsRNA bacteriophage φ6. J Mol Biol 218:569–581

Paatero AO, Mindich L, Bamford DH (1998) Mutational analysis of the role of nucleoside triphosphatase P4 in the assembly of the RNA polymerase complex of bacteriophage φ6. J Virol 72:10058–10065

Pan J, Vakharia VN, Tao YJ (2007) The structure of a birnavirus polymerase reveals a distinct active site topology. Proc Natl Acad Sci USA 104:7385–7390

Pan J, Dong L, Lin L, Ochoa WF, Sinkovits RS, Havens WM, Nibert ML, Baker TS, Ghabrial SA, Tao YJ (2009) Atomic structure reveals the unique capsid organization of a dsRNA virus. Proc Natl Acad Sci USA 106:4225–4230

Patton JT, Gallegos CO (1990) Rotavirus RNA replication: single-stranded RNA extends from the replicase particle. J Gen Virol 71:1087–1094

Patton JT, Spencer E (2000) Genome replication and packaging of segmented double-stranded RNA viruses. Virology 277:217–225

Pettersen EF, Goddard TD, Huang CC, Couch GS, Greenblatt DM, Meng EC, Ferrin TE (2004) UCSF Chimera – a visualization system for exploratory research and analysis. J Comput Chem 25:1605–1612

Pirttimaa MJ, Bamford DH (2000) RNA secondary structures of the bacteriophage φ6 packaging regions. RNA 6:880–889

Pirttimaa MJ, Paatero AO, Frilander MJ, Bamford DH (2002) Nonspecific nucleoside triphosphatase P4 of double-stranded RNA bacteriophage φ6 is required for single-stranded RNA packaging and transcription. J Virol 76:10122–10127

Poranen MM, Bamford DH (1999) Packaging and replication regulation revealed by chimeric genome segments of double-stranded RNA bacteriophage φ6. RNA 5:446–454

Poranen MM, Bamford DH (2011) Cystovirus, cystoviridae. In: Tidona C, Darai G (eds) Springer index of viruses, 2nd edn. Springer, New York

Poranen MM, Daugelavicius R, Ojala PM, Hess MW, Bamford DH (1999) A novel virus-host cell membrane interaction. Membrane voltage-dependent endocytic-like entry of bacteriophage straight φ6 nucleocapsid. J Cell Biol 147:671–682

Poranen MM, Paatero AO, Tuma R, Bamford DH (2001) Self-assembly of a viral molecular machine from purified protein and RNA constituents. Mol Cell 7:845–854

Poranen MM, Pirttimaa MJ, Bamford DH (2005a) Encapsidation of the segmented double-stranded RNA genome of bacteriophage φ6. In: Catalano C (ed) Viral genome packaging machines: genetics, structure, and mechanism. Kluwer, New York, pp 117–134

Poranen MM, Tuma R, Bamford DH (2005b) Assembly of double-stranded RNA bacteriophages. Adv Virus Res 64:15–43

Poranen MM, Butcher SJ, Simonov VM, Laurinmaki P, Bamford DH (2008) Roles of the minor capsid protein P7 in the assembly and replication of double-stranded RNA bacteriophage φ6. J Mol Biol 383:529–538

Prasad BV, Wang GJ, Clerx JP, Chiu W (1988) Three-dimensional structure of rotavirus. J Mol Biol 199: 269–275

Prasad BV, Burns JW, Marietta E, Estes MK, Chiu W (1990) Localization of VP4 neutralization sites in rotavirus by three-dimensional cryo-electron microscopy. Nature 343:476–479

Prasad BV, Rothnagel R, Zeng CQ, Jakana J, Lawton JA, Chiu W, Estes MK (1996) Visualization of ordered genomic RNA and localization of transcriptional complexes in rotavirus. Nature 382:471–473

Qiao X, Casini G, Qiao J, Mindich L (1995) In vitro packaging of individual genomic segments of bacteriophage φ6 RNA: serial dependence relationships. J Virol 69:2926–2931

Qiao X, Qiao J, Mindich L (2003) Analysis of specific binding involved in genomic packaging of the double-stranded-RNA bacteriophage φ6. J Bacteriol 185:6409–6414

Qiao X, Sun Y, Qiao J, Di Sanzo F, Mindich L (2010) Characterization of φ2954, a newly isolated bacteriophage containing three dsRNA genomic segments. BMC Microbiol 10:55

Ramadevi N, Burroughs NJ, Mertens PP, Jones IM, Roy P (1998a) Capping and methylation of mRNA by purified recombinant VP4 protein of bluetongue virus. Proc Natl Acad Sci USA 95:13537–13542

Ramadevi N, Rodriguez J, Roy P (1998b) A leucine zipper-like domain is essential for dimerization and encapsidation of bluetongue virus nucleocapsid protein VP4. J Virol 72:2983–2990

Rao VB, Feiss M (2008) The bacteriophage DNA packaging motor. Annu Rev Genet 42:647–681

Reddy VS, Johnson JE (2005) Structure-derived insights into virus assembly. Adv Virus Res 64:45–68

Reinisch KM, Nibert ML, Harrison SC (2000) Structure of the reovirus core at 3.6 Å resolution. Nature 404:960–967

Rohrmann GF (1986) Polyhedrin structure. J Gen Virol 67:1499–1513

Rossmann MG, Johnson JE (1989) Icosahedral RNA virus structure. Annu Rev Biochem 58:533–573

Roy P (2005) Bluetongue virus proteins and particles and their role in virus entry, assembly, and release. Adv Virus Res 64:69–123

Schuck P, Taraporewala Z, McPhie P, Patton JT (2001) Rotavirus nonstructural protein NSP2 self-assembles into octamers that undergo ligand-induced conformational changes. J Biol Chem 276:9679–9687

Sen A, Heymann JB, Cheng N, Qiao J, Mindich L, Steven AC (2008) Initial location of the RNA-dependent RNA polymerase in the bacteriophage φ6 procapsid determined by cryo-electron microscopy. J Biol Chem 283:12227–12231

Settembre EC, Chen JZ, Dormitzer PR, Grigorieff N, Harrison SC (2011) Atomic model of an infectious rotavirus particle. EMBO J 30:408–416

Shaw AL, Samal SK, Subramanian K, Prasad BV (1996) The structure of aquareovirus shows how the different geometries of the two layers of the capsid are reconciled to provide symmetrical interactions and stabilization. Structure 4:957–967

Shimizu T, Yoshii M, Wei T, Hirochika H, Omura T (2009) Silencing by RNAi of the gene for Pns12, a viroplasm matrix protein of *Rice dwarf virus*, results in strong resistance of transgenic rice plants to the virus. Plant Biotechnol J 7:24–32

Shing M, Coombs KM (1996) Assembly of the reovirus outer capsid requires μ1/σ3 interactions which are prevented by misfolded σ3 protein in temperature-sensitive mutant tsG453. Virus Res 46:19–29

Spies U, Muller H, Becht H (1987) Properties of RNA polymerase activity associated with infectious bursal disease virus and characterization of its reaction products. Virus Res 8:127–140

Stauber N, Martinez-Costas J, Sutton G, Monastyrskaya K, Roy P (1997) Bluetongue virus VP6 protein binds ATP and exhibits an RNA-dependent ATPase function and a helicase activity that catalyze the unwinding of double-stranded RNA substrates. J Virol 71:7220–7226

Sun Y, Qiao X, Qiao J, Onodera S, Mindich L (2003) Unique properties of the inner core of bacteriophage φ8, a virus with a segmented dsRNA genome. Virology 308:354–361

Tan BH, Nason E, Staeuber N, Jiang W, Monastryrskaya K, Roy P (2001) RGD tripeptide of bluetongue virus VP7 protein is responsible for core attachment to *Culicoides* cells. J Virol 75:3937–3947

Tan YR, Sun JC, Lu XY, Su DM, Zhang JQ (2003) Entry of *Bombyx mori* cypovirus 1 into midgut cells *in vivo*. J Electron Microsc (Tokyo) 52:485–489

Tao Y, Farsetta DL, Nibert ML, Harrison SC (2002) RNA synthesis in a cage – structural studies of reovirus polymerase λ3. Cell 111:733–745

Taraporewala ZF, Patton JT (2001) Identification and characterization of the helix-destabilizing activity of rotavirus nonstructural protein NSP2. J Virol 75:4519–4527

Taraporewala ZF, Patton JT (2004) Nonstructural proteins involved in genome packaging and replication of rotaviruses and other members of the Reoviridae. Virus Res 101:57–66

Taraporewala Z, Chen D, Patton JT (1999) Multimers formed by the rotavirus nonstructural protein NSP2 bind to RNA and have nucleoside triphosphatase activity. J Virol 73:9934–9943

Taraporewala ZF, Jiang X, Vasquez-Del Carpio R, Jayaram H, Prasad BV, Patton JT (2006) Structure-function analysis of rotavirus NSP2 octamer by using a novel complementation system. J Virol 80:7984–7994

Thomas CP, Booth TF, Roy P (1990) Synthesis of bluetongue virus-encoded phosphoprotein and formation of inclusion bodies by recombinant baculovirus in insect cells: it binds the single-stranded RNA species. J Gen Virol 71:2073–2083

Tortorici MA, Broering TJ, Nibert ML, Patton JT (2003) Template recognition and formation of initiation complexes by the replicase of a segmented double-stranded RNA virus. J Biol Chem 278:32673–32682

Tortorici MA, Shapiro BA, Patton JT (2006) A base-specific recognition signal in the 5′ consensus sequence of rotavirus plus-strand RNAs promotes replication of the double-stranded RNA genome segments. RNA 12:133–146

Ueda S, Masuta C, Uyeda I (1997) Hypothesis on particle structure and assembly of rice dwarf phytoreovirus: interactions among multiple structural proteins. J Gen Virol 78:3135–3140

Usala SJ, Brownstein BH, Haselkorn R (1980) Displacement of parental RNA strands during in vitro transcription by bacteriophage φ6 nucleocapsids. Cell 19:855–862

Wei T, Shimizu T, Hagiwara K, Kikuchi A, Moriyasu Y, Suzuki N, Chen H, Omura T (2006) Pns12 protein of *Rice dwarf virus* is essential for formation of viroplasms and nucleation of viral-assembly complexes. J Gen Virol 87:429–438

Wei H, Cheng RH, Berriman J, Rice WJ, Stokes DL, Katz A, Morgan DG, Gottlieb P (2009) Three-dimensional structure of the enveloped bacteriophage φ12: an incomplete T=13 lattice is superposed on an enclosed T=1 shell. PLoS One 4:e6850

Wu B, Hammar L, Xing L, Markarian S, Yan J, Iwasaki K, Fujiyoshi Y, Omura T, Cheng RH (2000) Phytoreovirus T=1 core plays critical roles in organizing the outer capsid of T=13 quasi-equivalence. Virology 271:18–25

Xu P, Miller SE, Joklik WK (1993) Generation of reovirus core-like particles in cells infected with hybrid vaccinia viruses that express genome segments L1, L2, L3, and S2. Virology 197:726–731

Xu G, Wilson W, Mecham J, Murphy K, Zhou EM, Tabachnick W (1997) VP7: an attachment protein of bluetongue virus for cellular receptors in *Culicoides variipennis*. J Gen Virol 78:1617–1623

Yan J, Tomaru M, Takahashi A, Kimura I, Hibino H, Omura T (1996) P2 protein encoded by genome segment S2 of rice dwarf phytoreovirus is essential for virus infection. Virology 224:539–541

Yu X, Jin L, Zhou ZH (2008) 3.88 Å structure of cytoplasmic polyhedrosis virus by cryo-electron microscopy. Nature 453:415–419

Zeng CQ, Wentz MJ, Cohen J, Estes MK, Ramig RF (1996) Characterization and replicase activity of double-layered and single-layered rotavirus-like particles expressed from baculovirus recombinants. J Virol 70:2736–2742

Zhang X, Walker SB, Chipman PR, Nibert ML, Baker TS (2003) Reovirus polymerase λ3 localized by cryo-electron microscopy of virions at a resolution of 7.6 Å. Nat Struct Biol 10:1011–1018

Zhang X, Jin L, Fang Q, Hui WH, Zhou ZH (2010) 3.3 Å cryo-EM structure of a nonenveloped virus reveals a priming mechanism for cell entry. Cell 141:472–482

Zhang X, Boyce M, Bhattacharya B, Schein S, Roy P, Zhou ZH (2011) Bluetongue virus coat protein VP2 contains sialic acid-binding domains, and VP5 resembles enveloped virus fusion proteins. Proc Natl Acad Sci USA 107:6292–6297

Zheng H, Yu L, Wei C, Hu D, Shen Y, Chen Z, Li Y (2000) Assembly of double-shelled, virus-like particles in transgenic rice plants expressing two major structural proteins of rice dwarf virus. J Virol 74:9808–9810

Zhong B, Kikuchi A, Moriyasu Y, Higashi T, Hagiwara K, Omura T (2003) A minor outer capsid protein, P9, of *Rice dwarf virus*. Arch Virol 148:2275–2280

Zhou F, Pu Y, Wei T, Liu H, Deng W, Wei C, Ding B, Omura T, Li Y (2007) The P2 capsid protein of the nonenveloped rice dwarf phytoreovirus induces membrane fusion in insect host cells. Proc Natl Acad Sci USA 104:19547–19552

Zhu Y, Hemmings AM, Iwasaki K, Fujiyoshi Y, Zhong B, Yan J, Isogai M, Omura T (1997) Details of the arrangement of the outer capsid of rice dwarf phytoreovirus, as visualized by two-dimensional crystallography. J Virol 71:8899–8901

Chapter 18
The Papillomavirus Virion: A Machine Built to Hide Molecular Achilles' Heels

Christopher B. Buck and Benes L. Trus

Abstract Papillomaviruses are a diverse group of DNA viruses that infect the skin and mucosal tissues of vertebrates. More than 100 distinct human papillomavirus (HPV) genotypes have so far been identified. A subset of HPVs is known to cause human cancer. Although recently developed vaccines protect vaccinated individuals from the two most carcinogenic HPV types, there is a pressing need for next-generation vaccines that might offer broad-spectrum protection against the full range of cancer-causing HPVs. The ongoing development of such vaccines will be facilitated by a deeper understanding of the mechanics of the assembly of the nonenveloped papillomavirus virion, as well as the machinelike structural changes that occur in the virion during the process of infectious entry into host cells. This chapter reviews the field's current knowledge of these two aspects of papillomavirus biology and speculates about areas where further work is needed.

18.1 Introduction

Papillomaviruses are icosahedral nonenveloped circular dsDNA viruses that infect a wide range of vertebrate hosts. Papillomaviruses that infect humans [human papillomavirus (HPVs)] are currently divided into 120 serologically distinct genotypes, and the list of newly identified types continues to expand (Bernard et al. 2010). Although a subset of HPV types can cause skin warts (papillomas), for which the family is named, a great majority of HPV infections are clinically asymptomatic. Infection with a subset of about a dozen sexually transmitted HPV types is a causal factor underlying nearly all cases of cervical cancer, as well as a substantial fraction of head/neck cancers and anal/genital cancers. Although a recently licensed vaccine against the two most common cancer-causing HPVs is expected to dramatically reduce the incidence of cervical cancer in vaccinated women, the development of improved vaccines that can protect against a wider range of disease-causing HPVs

C.B. Buck (✉)
Laboratory of Cellular Oncology, National Cancer Institute, National Institutes of Health, Bethesda, MD 20892-4263, USA
e-mail: buckc@nih.gov

B.L. Trus
Laboratory of Structural Biology, National Institute for Arthritis, Musculoskeletal and Skin Diseases, National Institutes of Health, Bethesda, MD 20892, USA

Imaging Sciences Laboratory, Center for Information Technology, National Institutes of Health, Bethesda, MD 20892, USA

M.G. Rossmann and V.B. Rao (eds.), *Viral Molecular Machines*,
Advances in Experimental Medicine and Biology 726, DOI 10.1007/978-1-4614-0980-9_18,

is critically dependent on an improved understanding of the various machine-like features of the papillomavirus virion.

Most nonenveloped viruses must destroy the infected host cell to allow the escape of new progeny virions. Papillomaviruses, in contrast, achieve release into the environment via the constitutive shedding of cellular material that naturally occurs at the surface of healthy skin and mucosal tissues. To exploit this unique vulnerability in vertebrate biology, papillomaviruses have adopted an exquisitely tissue-specific life cycle, replicating exclusively in the keratinocytes that differentiate to produce the tough, cross-linked outer surface of the skin. The biology of this virus family can thus only be fully understood when viewed through the prism of vertebrate skin biology.

After reviewing the tissue biological underpinnings of the papillomavirus life cycle, this chapter will focus on the machine-like aspects of virion assembly. This aspect of the chapter will cover recent progress toward understanding how the virus exploits the slow process of keratinocyte differentiation that occurs near the skin surface. Then we will review recent progress toward a more detailed understanding of the multiple structural rearrangements that allow the virion to deliver the viral DNA to the nucleus of new host cells. Exciting progress has been made in recent years toward understanding how these steps occur in live animal models. An important concluding theme of the chapter will be the concept that improved understanding of how the papillomavirus mousetrap is sprung should facilitate the development of next-generation vaccines that could offer full protection against all cancer-causing HPVs, as well as the wide variety of HPV types that cause skin warts.

18.2 History and Medical Significance of Papillomaviruses

The concept that skin warts could be experimentally transmitted between humans was first put forward in the late nineteenth century (Variot 1894). The viral nature of the transmissible agent(s) capable of causing warts in humans was established by the early twentieth century (Ciuffo 1907; Serra 1924; Ullmann 1923). This early period in papillomavirus research also saw the development of animal model systems for the study of skin warts and other benign transmissible tumors, using viruses now known as canine oral papillomavirus (COPV), cottontail rabbit papillomavirus (CRPV), and bovine papillomavirus type 1 (BPV1) (reviewed in Lowy 2007 and Christensen 2005).

In a landmark series of studies beginning in the mid-1930s, Rous and others showed that the benign skin lesions induced by experimental challenge of rabbits with CRPV can progress to malignant skin cancer (Rous and Beard 1935; Syverton and Berry 1935). This was the first demonstration that a virus could cause cancer in mammals, building on Rous' previous demonstration of a cancer-causing virus of chickens (the retrovirus known as Rous sarcoma virus). Although a variety of follow-up studies using animal model systems shed further light on the basic biology of papillomaviruses, the field was presented with a major obstacle in the 1950s, when it became apparent that papillomaviruses could not be propagated in standard monolayer cell culture systems. Another major technical barrier has been the fact that no papillomavirus tropic for *Mus musculus* has yet been identified. Research interest therefore oriented more toward the polyomaviruses, a different group of nonenveloped DNA viruses that are readily propagated in vitro and readily cause cancer in small rodent systems. The study of papillomaviruses languished somewhat until the dawn of recombinant DNA technology in the 1970s. This technical revolution made it possible to study HPV biology, even in the face of the relative intractability of in vitro and in vivo culture systems.

Application of molecular approaches to the study of HPVs quickly led to a variety of important discoveries. First, the use of low-stringency hybridization techniques and Sanger sequencing of cloned HPV genomes revealed the existence of multiple HPV types, with each type exhibiting an apparent preference for specific skin areas (Danos et al. 1983; Gissmann et al. 1977; Orth et al. 1977; Schwarz et al. 1983). Molecular cloning also facilitated investigation of the function of papillomavirus oncogenes in vitro (Groff and Lancaster 1986; Schiller et al. 1984; Yang et al. 1985). Studies of an unusual condition known as epidermodysplasia verruciformis (EV), in which subjects are susceptible to the development of flat warts over wide areas of skin, uncovered the first evidence that HPVs, like CRPV, can cause cancer. Although EV lesions are generally benign, they can progress to squamous cell carcinoma, particularly in sun-exposed skin areas. Molecular analysis showed that, while benign EV lesions contain a variety of HPV types, a majority of malignant lesions contain two related HPV genotypes, HPV5 and HPV8 (Orth 1986; Orth et al. 1978, 1979).

The recombinant DNA revolution in the study of papillomaviruses set the stage for solving a long-standing medical mystery: namely, the concept that cancer of the uterine cervix might be caused by a sexually transmitted infectious agent. The roots of this hypothesis stem from early findings indicating that the incidence of cervical cancer among prostitutes is much greater than among nuns (Rigoni-Stern 1842). A breakthrough series of studies by zur Hausen and colleagues uncovered the existence of HPV types 16 and 18 in cervical tumors and cervical cancer cell lines (Boshart et al. 1984; Durst et al. 1983). Low-stringency hybridization approaches showed that a very high proportion of cervical cancers contain HPVs related to HPV types 16 or 18. The findings clearly implicated HPVs as causal agents that etiologically underlie the development of cervical cancer, triggering an avalanche of basic science and epidemiological research into the role of papillomaviruses in human disease. zur Hausen's pioneering work in establishing the causal role of a subset of HPV types in nearly all cases of cervical cancers was recognized with a Nobel Prize in Medicine in 2008 (reviewed in zur Hausen 2009). More recent studies have demonstrated that HPVs, particularly HPV16, also cause a substantial fraction of several additional types of cancer, including anal, penile, and vulvar cancer, as well as cancers of the tonsils and oropharynx (reviewed in Chung and Gillison 2009 and Giuliano et al. 2008).

To date, 120 distinct HPV types have been cloned and characterized (Bernard et al. 2010). About a dozen HPV types are associated with a high risk of cervical cancer, with HPV16 and HPV18 being the two most common cancer-causing types (reviewed in Schiffman and Castle 2003). Two HPV types, HPV6 and HPV11, cause most cases of genital warts (also known as condyloma accuminata). A different subset of HPV types can cause benign skin warts, which commonly occur among children. Although skin warts are not life threatening, they can be painful when located on the soles of the feet, and physician visits for treatment of skin warts present a nontrivial cost to health-care systems worldwide. In addition to the various HPV types that cause disease symptoms, it is also well established that essentially all humans harbor lifelong, clinically inapparent infections with various skin-dwelling HPV types. Since some of these clinically occult HPV types are found in EV lesions, they are often generically referred to as "EV types." The possibility that a minority of EV-type HPV infections can lead to various forms of skin cancer in non-EV individuals is the subject of ongoing investigation (reviewed in Feltkamp et al. 2008).

In 2006, a preventive vaccine against HPV types 6, 11, 16, and 18 was approved for use in girls and young women (reviewed in Lowy et al. 2008). The vaccine has recently achieved approval for use in boys and young men as well. An additional vaccine targeting only types 16 and 18 achieved licensure in various countries in 2007–2009. Both vaccines make use of the fact that the L1 major capsid protein of papillomaviruses can spontaneously assemble into virus-like particles that closely resemble authentic virions. The nature of recombinant papillomavirus VLPs will be discussed in greater detail below. Recombinant HPV VLPs are potently immunogenic and elicit high-titer L1-specific antibody responses capable of neutralizing cognate HPV types and preventing initial infection. Although the current HPV vaccines are expected to confer lifelong protection against initial infection with the specific HPV types in the vaccine, they are unlikely to protect against the full range of HPV types capable of causing cervical cancer (reviewed in Lowy et al. 2008). Thus, although HPV16 and HPV18 are thought to cause roughly 70% of all cervical cancers, vaccinated individuals will still be at some risk of being infected with cancer-causing HPV types not covered by the vaccine.

18.3 The Papillomavirus Life Cycle

Papillomaviruses have evolved to exploit the biology of keratinocytes, which form the surface of the skin, as well as the surface of the genital, oral, and much of the airway epithelium. The viral life cycle can be broadly divided into three stages: (1) production and release of progeny virions; (2) virion penetration of the epithelial surface, followed by a series of steps leading up to infectious delivery of the viral DNA to the nucleus of a host keratinocyte; and (3) establishment and maintenance of the viral episome. Although each step of the life cycle has a variety of machine-like features, this review will focus specifically on the molecular mechanics of papillomavirus virion structures during the final steps of virion production and the initial conformational changes that lead up to internalization of the virion into a new host cell. For additional information on the later steps in the infectious entry process, as well as the establishment and maintenance phase of the life cycle, the reader is referred to recent reviews on these topics (Burk et al. 2009; Day and Schiller 2009; Howley and Livingston 2009; McBride 2008; Sapp and Bienkowska-Haba 2009).

The histological architecture of the skin is shown, in cartoon form, in Fig. 18.1 (Doorbar 2005). The dermis, portrayed at the base of the cartoon, is composed primarily of fibroblasts and is overlaid by a relatively acellular basement membrane. The basement membrane delineates the dermis from the epidermis, the latter being composed primarily of keratinocytes. The epidermis is subdivided into distinct layers that are dictated by the differentiation state of the keratinocytes. Mitotically active basal keratinocytes that reside along the basement membrane give rise to transit-amplifying cells that are gradually pushed up off the basement membrane. The progeny keratinocytes then undergo a complex series of differentiation steps as they are gradually pushed toward the outer surface of the

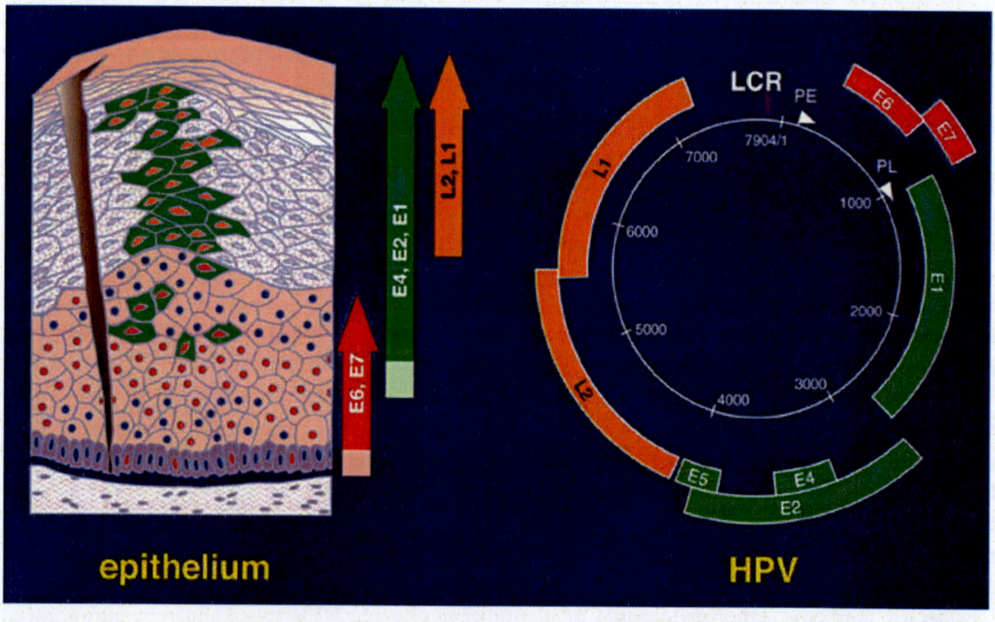

Fig. 18.1 The *left panel* shows a diagram of the skin. The fibroblast-rich dermis is portrayed at the bottom. This layer is overlaid with a basement membrane (*dark blue*) upon which mitotically active basal keratinocytes rest. As keratinocytes begin to differentiate to form the epidermis, they detach from the basement membrane and are pushed up toward the skin surface and are eventually shed through a process called desquamation. It is thought that papillomaviruses gain access to the basal keratinocytes that are the productive targets of infection via microtrauma (small wounds) in the epidermis. *Arrows* depict the expression patterns for papillomavirus genes. A diagram of the major genes common among papillomaviruses is shown on the *right* side of the figure

skin by the replicating cells underneath. The final stages in keratinocyte differentiation, which occur near the outer surface of the epidermis, include a series of biochemical reactions that form protein–protein cross-links and generate an unusual spectrum of lipids that form the tough exterior surface that prevents the loss of water from the underlying tissue and defends against the entry of microbes and chemical insults. Dead keratinocytes undergo a programmed form of cellular fragmentation and are shed into the environment through a process called desquamation. It is thought that the entire keratinocyte differentiation pathway takes roughly 2 weeks to complete in a typical skin surface.

While papillomaviruses can nonproductively infect a variety of cell types, including fibroblasts of the dermis, the full viral life cycle can only be completed in differentiating keratinocytes. The initial targets of productive infection are believed to be the relatively undifferentiated basal keratinocytes that reside along the basement membrane. Initially, the incoming viral genome is thought to express very low levels of the viral early genes, including the oncogenes E5, E6, and E7; the transcriptional regulator E2; and the DNA helicase E1. Collectively, these viral proteins drive the basal keratinocyte to divide and stimulate low levels of replication of the viral DNA. Several lines of indirect evidence support the concept that papillomaviruses can persist in a latent state in basal keratinocytes for the lifetime of the host. The fact that the latently infected cell expresses only trace amounts of the viral proteins is likely key to this remarkable ability to escape immune-mediated clearance. Thus, the virus is under a high degree of selective pressure to remain in a "quiet" state in cellular environments that resemble basal keratinocytes. This evolutionary pressure likely accounts for the failure of papillomaviruses to complete the late phase of the life cycle in conventional cultured cell lines.

Keratinocyte differentiation triggers a profound shift in the behavior of the virus. First, the viral DNA is amplified to much higher copy numbers (thousands of viral genomes per cell). This generally occurs in the middle "spinous" layer of the epidermis (reviewed in Doorbar 2005). The expression of the two viral capsid proteins, L1 and L2, occurs only in the outermost layers of the epidermis. It is thought that these tough, cross-linked layers of the epidermis are not generally subject to extensive immune surveillance, thus providing a cloak for the highly immunogenic capsid proteins. Assembled virions are ultimately shed into the environment through the normal process of desquamation. The fact that the papillomavirus does not need to risk triggering the proinflammatory signals associated with active virally induced cell lysis is believed to be a key aspect of the relatively quiet and languid lifestyle papillomaviruses have evolved to pursue.

In a technical breakthrough in the early 1990s, Meyers and colleagues showed that it is possible to induce HPV virion production by exposing infected keratinocytes to an air–liquid interface in organotypic "raft" culture systems (Meyers et al. 1992). The rafted keratinocytes undergo a differentiation process that closely resembles the normal process of differentiation in the epidermis. The raft culture system has been of great value for understanding the full papillomavirus life cycle (reviewed in Conway and Meyers 2009).

In recent years, systems have been developed for producing papillomavirus-based gene transfer vectors (commonly referred to as "pseudoviruses") using standard monolayer cell cultures (Buck et al. 2004, 2005b). In this system, plasmids encoding reporter genes are cotransfected into cells along with recombinant expression plasmids encoding papillomavirus L1 and L2 genes. A key element of the system is the use of L1 and L2 genes that carry an extensive series of silent mutations designed to eradicate the inhibitory RNA sequences that prevent the expression of L1 and L2 in basal keratinocytes and in cultured cell lines (Leder et al. 2001; Schwartz 2000; Zhou et al. 1999). It is possible to generate pseudovirion preparations with particle to infectivity ratios as low as 7:1 (Buck et al. 2004). This is a remarkably high degree of infectivity compared to most animal viruses. Thus far, no clear differences have been identified in comparisons between the infectious entry pathways used by authentic papillomavirus virions and the transduction pathways used by reporter pseudovirions in vitro. Although the pseudovirus systems are quite removed from many features of the natural papillomavirus life cycle, the ability to generate very high yields of reporter pseudovirions has helped shed light on virion structure, as well as the initial steps of papillomavirus infectious entry both in vitro and, more recently, in vivo (Johnson et al. 2009; Kines et al. 2009; Roberts et al. 2007).

18.4 Virion Maturation: Building a Brownian Mousetrap

Early negative stain electron microscopy (EM) analyses of virions extracted from skin warts revealed a generally icosahedral structure with regular knob-like protrusions (Finch and Klug 1965; Klug and Finch 1965, 1968). More detailed structures were eventually obtained by computerized image reconstruction based on electron micrographs of virions suspended in vitrified ice (Belnap et al. 1996; Hagensee et al. 1994; Trus et al. 1997). These cryo-electron microscopy (cryo-EM) studies and supporting biochemical analyses revealed that the knob-like capsomers were formed from pentamers of the L1 protein. The 72 pentameric L1 capsomers that form the virion are arranged on a $T=7$ icosahedral lattice. The atomic structure of the core of the capsomer knob was solved by X-ray crystallography of a 12 capsomer, $T=1$ subviral particle formed by a recombinant N- and C-truncated HPV16 L1 protein (Chen et al. 2000). The core fold of the capsomer knob is composed of an eight-stranded beta jellyroll. Modis and colleagues have fitted the atomic structure of HPV16 L1 into a high-resolution cryo-EM structure for BPV1 (Trus et al. 1997) to produce a quasi-atomic model of the structure of the full HPV16 virion (Modis et al. 2002). A very recent higher resolution cryo-EM image reconstruction study by Wolf and colleagues, using BPV1 virions, puts forth a revised quasi-atomic model (Wolf et al. 2010) for the full virion. Both the Modis and the revised Wolf model agree that a portion of L1 near the C terminus reaches out to invade neighboring pentameric capsomers (Fig. 18.2). In the model deduced by Wolf and colleagues, the L1 loops back to rejoin the original donor capsomer via a disconnected strut of density that travels through the canyon between the capsomer knobs.

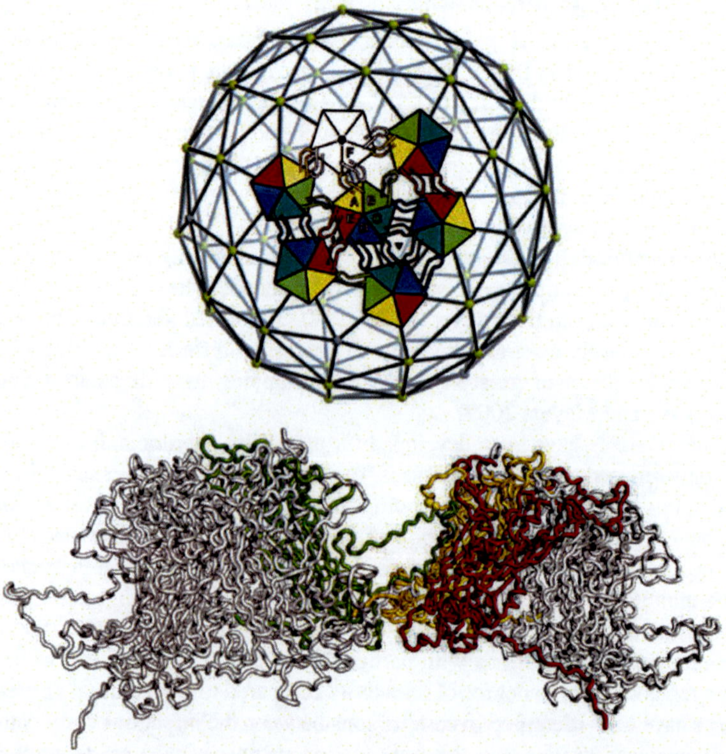

Fig. 18.2 The figure shows an atomic model of the arrangement of L1 based on high-resolution cryo-EM of BPV1 virions. The *top panel* shows the pattern in which the C-terminal arms of L1 invade neighboring capsomers. The *bottom panel* highlights the primary disulfide bond that stabilizes intercapsomeric contacts. In this construction, the invading arm of the *green chain* forms a disulfide bond near the upper cornice of a neighboring pentamer (invaded L1 is in *yellow*)

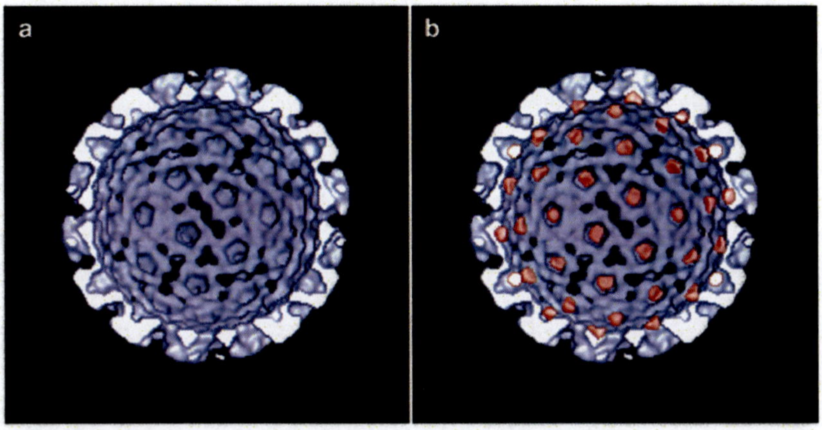

Fig. 18.3 (**a**) An inside view of an HPV16 capsid 3D reconstruction (3DR) that contains only the major capsid protein L1 is shown in *blue*. (**b**) After subtracting the L1-only 3DR from a 3D reconstruction containing both L1 and the minor capsid protein L2, the 3D difference map shows only the minor capsid protein L2 in *red*, which is superimposed on the L1-only 3D map in *blue*. The *red* density represents only about one-third of the protein from L2. The remainder is either disordered or not detected at this resolution. The capsid diameter is ~600 Å

Until recently, the abundance of a second capsid protein, L2, was not known, although older studies examining virions extracted from human skin warts suggested that the ratio of L1:L2 could be as low as 5:1 (i.e., 72 molecules of L2 per virion). We have recently confirmed that HPV1 and HPV2 virions extracted from skin warts display an L1:L2 ratio of about 5:1 (unpublished results). Cryo-EM comparison of L1+L2 pseudovirions to L1-only VLPs has revealed a button of L2-specific density at the base of the interior lumen of each capsomer knob (see Fig. 18.3). In order to visualize L2 within the capsid, two experiments were required (Buck et al. 2008). First, L1+L2 pseudovirions with an L1:L2 ratio approaching 5:1 were produced, imaged in a cryo-electron microscope, and analyzed with standard 3D reconstruction (3DR) techniques (Baker and Cheng 1996) to 22.7 Å resolution (Fourier Shell Correlation (FSC) = 0.3). Second, another pseudovirus system produced Ll-only capsids (Fig. 18.3a) which underwent the same analysis yielding 25.1 Å resolution (FSC = 0.3). Visually, the outside of the two 3DRs looked essentially the same. However, an additional density (red in Fig. 18.3b) present only in the L1+L2 reconstruction could be seen beneath the axial lumen of each L1 capsomer. This density (about 12 kDa) represents only a portion of L2, and the site through which L2 is thought to emerge from the virion (see below) has yet to be determined. The remaining L2 mass is either disordered or not sufficiently ordered to be visualized at the current resolution.

Although chaperones such as Hsc70 likely facilitate the virion assembly process in the setting of the cell nucleus, it is clear that purified L1 and L2 can spontaneously coassemble and even take up purified DNA without the assistance of any protein chaperones (Chromy et al. 2003; Florin et al. 2004) (Combita et al. 2001; Touze and Coursaget 1998). In the context of pseudovirion production in transfected cells, L1 and L2 appear to be highly promiscuous in their ability to take up various types of chromatinized DNA, including reporter plasmids (Buck et al. 2004), transfected recombinant HPV genomes (Pyeon et al. 2005), and even loops of cellular DNA (Buck et al. 2005a). The L1 and L2 of HPV16 appear to be particularly promiscuous, such that a majority of the DNA in pseudovirion preparations is composed of linear fragments of cellular DNA similar to the ~8 kb length of natural papillomavirus genomes. L2 appears to play a role in uptake of a complete "head full" of DNA, as L1 VLPs produced in the absence of L2 contain either no cellular DNA or much smaller fragments of cellular DNA, depending on the papillomavirus type. Although the pseudovirus production system can package essentially any plasmid less than 8 kb in size, some plasmids are taken up with very low efficiency (unpublished results). The pattern of packaging efficiency does not appear to have any clear correlates in terms of the primary DNA sequence, and fragments of cognate papillomavirus genomic DNA do not appear to improve the packaging efficiency for poorly packageable plasmids. A characteristic feature of papillomaviruses is that the viral genome is taken up fully decorated with cellular histones (Favre et al. 1975). Thus, one possibility is that L1 and L2 interact with a specific spectrum of posttranslationally modified histones that might be enriched on the viral minichromosome (or reporter plasmid) in vivo. Interestingly, the core fold of L1 closely resembles a family of pentameric histone-binding cellular proteins known as nucleoplasmins. Although papillomavirus L1 proteins share little primary amino acid homology with the nucleoplasmins, their common fold (and, conceivably, their ability to coordinate histones) might reflect common evolutionary origins.

Under the right conditions, purified L1 protein can assemble into pentameric capsomers that spontaneously coassociate to form virus-like particles with an exterior appearance that is essentially indistinguishable from natural virions. This remarkable feature of L1 made possible the current HPV VLP vaccines, which are generated by in vitro disassembly and reassembly of purified L1 protein (McCarthy et al. 1998). It is well established that papillomavirus virions and L1 VLPs are stabilized by disulfide bonds between cysteine side chains of neighboring L1 molecules (Li et al. 1998). The proper formation of disulfide bonds is likely important for the stability and immunogenicity of VLPs used in the current HPV vaccines (Schadlich et al. 2009; Zhao et al. 2002).

For HPV16, the stabilizing disulfide bonds involve cysteine residues 175 and 428 (Li et al. 1998; Modis et al. 2002; Sapp et al. 1998). L1 C175S mutants incapable of forming the 175–428 bond are able to form pseudovirions that protect encapsidated DNA from nucleases, but the mutant pseudovirions are fragile and easily disrupted (Buck et al. 2004). Interestingly, although C175S mutant virions are physically fragile, they remain capable of transducing reporter plasmids into cells at levels that approach wild-type mature pseudovirions. This suggests that virion maturation is not crucial for engagement of the various steps of infectious entry but primarily contributes to the environmental stability of the virus. Maturation may also serve to occlude conserved motifs on the virion surface from immunological recognition, although direct evidence for this hypothesis is currently lacking.

It is clear from biochemical analyses that disulfide cross-linking within the HPV virion results in dimeric and trimeric forms of L1. BPV1 and several other bovine papillomavirus types appear to be distinct from other papillomavirus types, in that the BPVs are stabilized by an additional cross-link between cysteines 22 and 473 in adjacent L1 chains (Wolf et al. 2010). It appears that these additional disulfide bonds in BPV1 covalently link the entire complement of L1 monomers in the virion into a single covalently closed molecule (Buck et al. 2004) (Wolf et al. 2010).

Pseudovirion preparations are generally subjected to an overnight incubation at 37°C to allow the formation of inter-L1 disulfide bonds (Buck et al. 2005b). In keratinocyte raft culture systems, it appears that establishment of an effective redox gradient across the thickness of the raft may take more than 2 weeks of culturing to achieve (Conway et al. 2009b). Interestingly, in the setting of normal skin, it appears that the redox gradient never reaches fully oxidized conditions, and the outermost surface of the undisturbed skin remains rich in free thiol groups, likely in the form of reduced glutathione (Ogawa et al. 1979). The thiol-rich layer may be important for the barrier function of the skin, serving as a sink for thiol-reactive chemicals (Pickard et al. 2009). Although the formation of inter-L1 disulfide bonds in cell lysates or in raft culture can be accelerated by the addition of oxidizing agents, such as oxidized glutathione, this forced acceleration results in the formation of virions with increased susceptibility to proteases (Buck et al. 2005b). The results suggest that virion maturation is generally a very slow process and indicate that a "musical chairs" effect may lock inappropriate L1 conformations into place if the maturation process is accelerated.

To examine the disulfide-driven maturation process in greater detail, we allowed pseudovirions to mature for different lengths of time (1, 4, or 18 h) and subjected them to cryo-EM. We also performed cryo-EM analyses on pseudovirions carrying the destabilizing C175S mutation in L1. A summary of these analyses is shown in Fig. 18.4. The C175S mutant (Fig. 18.4a), which is incapable of disulfide-driven maturation, displayed a larger diameter (640 Å) than the more mature forms

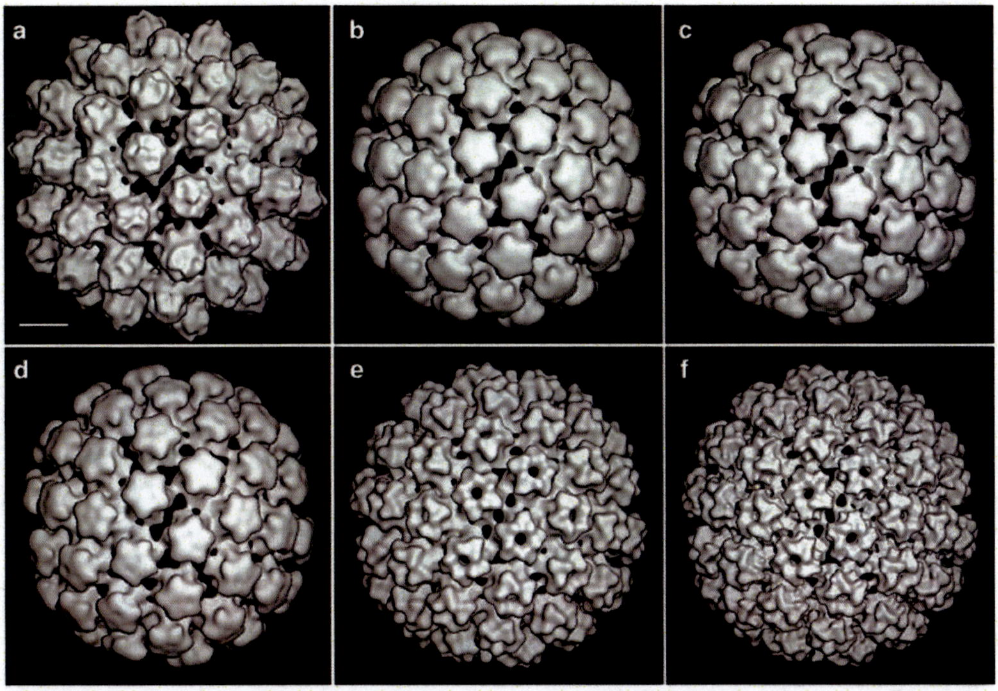

Fig. 18.4 (**a**) 3D Reconstruction (3DR) of a sample from lysates of mutant cells expressing L1 (Cys175Ser) and L2 at pH ~6. Because of the L1 mutation, the important Cys175 cross-link cannot form, and the capsids do not mature. The average diameter is about 640 Å. (**b–d**) 3D reconstructions of wild-type L1 + L2 allowed to mature for 18 h. Minor variation in diameter are observed as the HPV matures from 626 Å diameter (**b**) to 610 Å (**c**) and finally 600 Å (**d**) – the endpoint for maturation at slightly acidic pH. (**e, f**) When the experiment is repeated at pH ~7.5, the overwhelming majority of capsids are not only fully mature, but the 3DR can achieve significantly higher resolution (**e**) (~12 Å resolution). (**e**) Is a computationally band-limited version of (**e**) to bridge the gap between mature capsids at slightly acid pH (**d**) and ph ~7.5 (**f**). Bar = 100 Å

that predominated in the wild-type pseudovirion preparation allowed to mature for 18 h (Fig. 18.4b–d). Larger diameter pseudovirions were observed in the samples allowed to mature for only 1 or 4 h (data not shown). The less mature capsids showed a relative lack of density in the capsid floor between the capsomer knobs (e.g., at the twofold axis and around the threefold capsomers). A similar type of closing of the capsid floor upon maturation is also observed during the maturation of other viral capsids, including herpesvirus (HSV-1) (Heymann et al. 2003). The smallest diameter forms visualized in the pseudovirion preparation allowed to mature for 18 h could be reconstructed with only about 22 Å resolution (Fig. 18.4d), suggesting that even the most mature preparation may have contained a heterogeneous mixture of pseudovirion forms.

At the time of this analysis, we wondered why BPV1 virions could be reconstructed with such high resolution (Trus et al. 1997; Wolf et al. 2010) yet HPV16 pseudovirions allowed to reach an apparent endpoint disulfide cross-linking state yielded such mediocre resolution. The models put forward by Modis and colleagues and by Wolf and colleagues both postulate the existence of a topologically ring-shaped L1 trimer in the environments surrounding the vertex capsomer, as well as a dimer formed by reciprocal invasion of two L1 molecules (Fig. 18.2). Given these models, it was puzzling when nonreducing SDS-PAGE analysis of HPV16 pseudovirions allowed to mature overnight revealed an apparent endpoint cross-linking state with two distinct species of L1 trimer and two distinct species of L1 dimer in nonreducing SDS-PAGE analyses (Buck et al. 2005b). Although

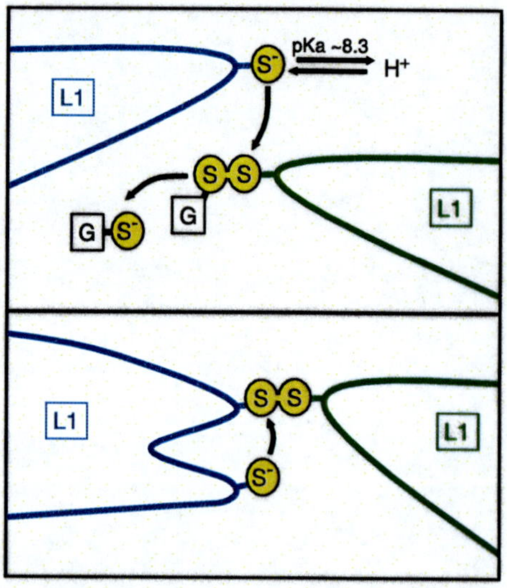

Fig. 18.5 The figure shows a hypothetical pathway for the formation of L1 disulfide bonds. In the *top panel*, a cysteine in the L1 chain in *green* is decorated with glutathione (GS). A free cysteine in a neighboring L1 chain ionizes to form a thiolate (S^-), which attacks the L1-glutathione disulfide bond. The reaction releases reduced glutathione. In the *bottom panel*, a thiolate within the *blue* L1 attacks the intermolecular disulfide bond. This results in an intramolecular disulfide bond within the *blue* L1

it has not been possible to isolate enough authentic HPV16 virions from natural infections or from organotypic raft culture systems to perform comparable biochemical analyses, it is possible to purify milligram amounts of HPV type 1 and type 2 virions from human skin wart specimens (Favre et al. 1975; Orth et al. 1977). Analysis of authentic HPV1 and HPV2 virions from skin warts revealed a single faster-migrating species of L1 dimer and a single slower-migrating species L1 trimer (unpublished results) consistent with the one-dimer one-trimer prediction of the Modis and Wolf models. Taken together, the results led us to hypothesize that HPV16 pseudovirions produced in vitro might not achieve the fully mature state observed in authentic virions. Specifically, we imagined that the second species of L1 trimer might reflect failure to close one leg of the predicted trimer ring. Likewise, the second species of L1 dimer might reflect a failure of reciprocal dimer formation, resulting in a topologically T-shaped dimer.

Although disulfide bonds can form through the direct attack of one free sulfhydryl group against another in the presence of molecular oxygen, the chemistry of this type of disulfide formation is very complex and is thought to require chemical catalysts, such as variable-valence metal ions (Bagiyan et al. 2003). A more common path to formation of disulfide bonds in proteins makes use of simple rearrangement of disulfide bonds, involving the attack of an existing disulfide by a free sulfhydryl group. Such disulfide shuffling reactions are often mediated by the cysteine-containing tripeptide glutathione, although enzymes such as protein disulfide isomerases (PDIs) and thioredoxin can also catalyze shuffling interchanges. The reaction is depicted in Fig. 18.5. An important first step in the disulfide shuffling reaction is the ionization of the free sulfhydryl group to thiolate (S^-). The pKa of this ionization is roughly 8.3. Thus, acidic conditions tend to inhibit the process of disulfide shuffling.

Analysis of the cell lysates in which the pseudovirion maturation process is performed showed an initial pH of around 6.0, with a gradual decline in pH during the maturation incubation. This led to the finding that by simply buffering the lysate to pH 7.5, a greater proportion of L1 molecules could achieve the slow trimer/fast dimer endpoint (data not shown). A complicating factor for analyses of L1 disulfide cross-linking is that dimers and trimers can undergo disulfide shuffling to yield L1 monomers in alkaline conditions in the presence of chaotropes (such as SDS or urea) or after heating to roughly 50°C (Fig. 18.6 and unpublished results). The collapse to monomers can occur in the absence of reducing agents. Thus, the most likely mechanism for breaking the intermolecular bond is via attack of the intermolecular disulfide bond by a free L1 sulfhydryl, resulting in a

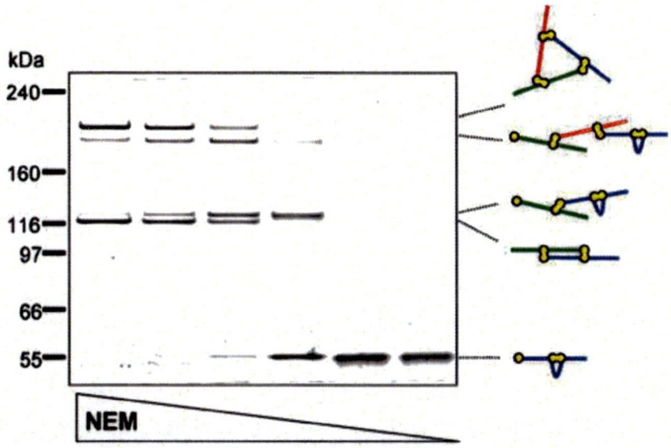

Fig. 18.6 HPV16 capsids were allowed to mature in crude cell lysate overnight then purified. The purified capsids were treated with alkaline SDS load dye (Nupage, Invitrogen) containing various concentrations of *N*-ethylmaleimide (NEM, a suflhydryl alkylating agent). The samples were then subjected to acrylamide gel electrophoresis. Neither the load dye nor the gel contained reducing agents. At low concentrations of NEM, disulfide shuffling reactions (see Fig. 18.5) resulted in collapse of L1 dimers and trimers to L1 monomers presumed to contain an intramolecular disulfide bond. *Cartoons* of the inferred topology of L1 dimers and trimers are shown to the *right* of the gel

rearrangement event that forms an L1 with an intramolecular disulfide (Fig. 18.5). The chaotrope-mediated collapse to monomers can be blocked by the addition of sulfhydryl-alkylating agents, such as *N*-ethyl maleimide (NEM). By varying the dose of NEM, thereby partially inhibiting the collapse of L1 to a monomeric form with an internal disulfide bond, we were able to deduce the topology of partially collapsed L1 dimers and trimers (Fig. 18.6).

Cryo-EM analysis of pseudovirions allowed to mature at neutral-buffered pH 7.5 produced only one 3DR (Fig. 18.4f). These more fully mature capsids were dramatically more regular, yielding a ~12 Å resolution map (FSC=0.3). The ability to generate more uniformly mature pseudovirion preparations should be of significant utility for achieving more detailed analyses of the structural dynamics of L1, and perhaps more detailed maps of the locations of L2 within the virion during various phases of virion assembly and infectious entry.

The redox environment of the maturing virion may also affect the biology of L2. Several recent reports have shown that formation of an intramolecular disulfide bond near the N terminus of L2 is critical for virion infectivity, at least for some papillomavirus species. It is not yet clear precisely how the intramolecular disulfide bond within L2 affects overall virion mechanics (Campos and Ozbun 2009; Conway et al. 2009a; Gambhira et al. 2009).

Currently available data suggest a model in which assembled virions undergo a slow process of disulfide bond formation through a mechanism that probably involves bond shuffling mediated by redox-active cellular components, such as glutathione. The fact that oxidative acceleration of the process appears to trap the virion in a kinetically unfavorable state suggests that the stabilizing disulfide bonds arise through a time-consuming process in which the L1 invading arms that connect the capsomers sample a variety of conformations prior to formation of inter-L1 disulfide bonds. This concept echoes a recently proposed "Brownian ratchet" model for virion maturation in the bacteriophage HK97. The model invokes the concept that random Brownian motion forces the HK97 gp5 capsid protein into various conformations. Once a particular conformation is achieved, the spontaneous formation of an intramolecular isopeptide cross-link within gp5 can fix the otherwise transient conformation in place. This provides a mechanism for a unidirectional evolution toward a stabilized chainmail topology for HK97 or the stabilized ring trimer/reciprocal dimer endpoint seen for mature papillomavirus virions. In the case of papillomaviruses, the languid pace of keratinocyte differentiation appears to allow for an exceptionally slow virion maturation process that is triggered by the gradual loss of cellular redox potential and driven by sequential chemical fixation (in the form of disulfide bonds) of otherwise transient L1 conformations.

18.5 Infectious Entry: Many Steps to Springing the Mousetrap

Papillomaviruses are generally thought to coevolve with their host species and tend to be capable of infecting only a narrow range of closely related animals. However, the primary barriers to cross-species infection appear to be at the level of regulation of the latent and/or productive phases of the viral life cycle, as opposed to defects in the initial process of infectious entry. For example, authentic BPV1 can induce immortalization of cultured mouse fibroblast lines in vitro (Dvoretzky et al. 1980). Likewise, reporter pseudovirions can transduce a wide variety of cell lines derived from distantly related animal species. The broad species promiscuity of the infectious entry process has recently made it possible to develop a pseudovirus-based mouse genital challenge model for infectious entry in vivo (Roberts et al. 2007). The mouse model and a variety of supporting in vitro lines of evidence have recently begun yielding new information about the machine-like behavior of the papillomavirus virion in the setting of the intact female genital tract of live mice.

The intact epidermis (or the closely related stratified epithelium of the mucosa) presents a substantial physical barrier for papillomaviruses. It is clear from studies with CRPV and COPV model systems that some form of active disruption of the natural structure of the epidermis is required for infection to occur (reviewed in Doorbar 2005 and Christensen 2005). The mouse genital challenge model likewise requires either mechanical or chemical disruption of the vaginal epithelium (Roberts et al. 2007). It has recently been shown that cultured cells must be mitotically active for HPV infection to occur (Pyeon et al. 2009). In the setting of the intact epidermis, suprabasal differentiating keratinocytes have typically exited the cell cycle and only basal keratinocytes and transit-amplifying cells resident along the basement membrane are mitotically active. Thus, the incoming papillomavirus virion must evidently bypass the full thickness of the epidermis to reach the target cells in which productive infection can occur.

It is well established that the initial interaction of papillomavirus virions with cultured cells is mediated primarily by a class of carbohydrates known as heparan sulfates (HS), a group of glycosaminoglycans that decorate the ectodomains of various cellular proteins. HS is composed of a long, unbranched chain of amine-decorated glucose units linked in an alternating series of α1-4/β1-4 saccharide bonds. Although the primary HS chain is chemically monotonous when first synthesized, enzyme complexes localized in the Golgi apparatus perform a coordinated series of chemical modifications, such as the addition of sulfate, acetyl, or carboxyl groups at various positions (reviewed in Esko et al. 2009; Esko and Selleck 2002). Papillomaviruses only effectively bind O-sulfated and, to a lesser extent, N-sulfated forms of HS (Johnson et al. 2009; Selinka et al. 2003). The less heavily modified forms of HS found on some cell types (particularly in vivo) do not support productive infection (Day et al. 2008b). Paradoxically, it appears that HS forms expressed at the plasma membrane of primary keratinocytes do not have high affinity for papillomavirus virions (Buck et al. 2006b; Johnson et al. 2009).

Although the interaction between the virion and heparan sulfate proteoglycans (HSPGs) has traditionally been postulated to occur at the cell surface (Joyce et al. 1999), recent evidence indicates that the virus can initially be recruited to HSPGs in extracellular matrix structures that are not closely associated with the cell surface (Culp et al. 2006). In the setting of the mouse genital challenge model, it appears that the initial HSPG interaction takes place not at the surface of the keratinocyte plasma membrane, but rather at the basement membrane upon which the less-differentiated basal keratinocytes rest (Fig. 18.7) (Johnson et al. 2009; Kines et al. 2009), reviewed in (Day and

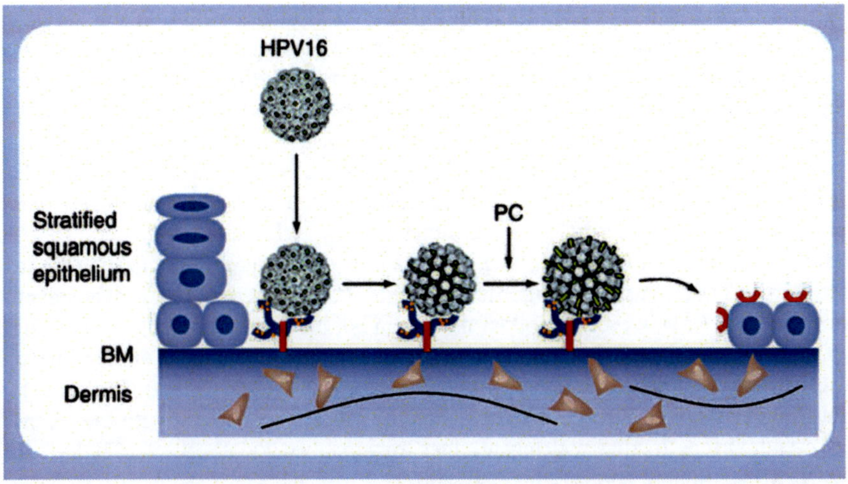

Fig. 18.7 The diagram depicts the initial binding of an HPV16 virion to a patch of denuded basement membrane in a disrupted mucosal surface. Initial interactions are mediated by heparan sulfate proteoglycans (HSPGs, branched *red/blue* structure). A conformational change then occurs that partially exposes the N terminus of L2. This exposure makes it possible for furin (or a related proprotein convertase, PC) to cleave an N-terminal segment of L2, resulting in a second conformational change. This second conformational change allows transfer of the virion from the HSPG to an unidentified entry receptor on basal keratinocytes (*red* crescent)

Schiller 2009). In the setting of a wound or microtrauma, the basement membrane may become denuded and exposed to the environment. Healing of the wound occurs when basal keratinocytes from adjacent areas migrate across the basement membrane and begin to proliferate. Thus, the specificity of the papillomavirus for basement membrane structures appears to be an elegant solution to the need for specific targeting of mitotically active basal keratinocytes.

Although the minor capsid protein L2 appears to be dispensable for formation of the basic structure of the virion, it plays a series of starring roles during the process of infectious delivery of the viral genome to new target cells. After the virion successfully traverses the epidermis and binds the underlying basement membrane, a conformational change must occur to allow transfer of the virion off the basement membrane and onto the keratinocyte cell surface (Kines et al. 2009). This conformational change appears to be triggered by proteolytic cleavage of L2. In cultured cell lines, the cleavage of L2 is mediated by furin, a serine protease that is a member of a family of proprotein convertases (Richards et al. 2006). A consensus furin cleavage motif (RXXR) near the N terminus of L2 is conserved among all known papillomaviruses. Furin is familiar to virologists as an essential factor for processing the surface proteins of a variety of enveloped virus types, for example, HIV-1 Env and dengue virus E proteins. For the enveloped viruses, furin processing generally occurs in the Golgi and is thought to prime the envelope protein to acquire a fusogenic conformation. However, furin is also present in limited amounts at the cell surface and in the extracellular environment, where it has been shown to activate the fusogenic form of a variety of bacterial toxin proteins, including *Pseudomonas* exotoxin and *Anthrax* toxin. In the case of papillomaviruses, furin cleavage of L2 appears to trigger a conformational change that allows the virion to dissociate from basement membrane HS and transfer onto an unidentified receptor on the surface of basal keratinocytes (Johnson et al. 2009; Richards et al. 2006), reviewed in (Day and Schiller 2009). It is not yet clear whether L2 is cleaved by furin or by a related proprotein convertase in the setting of the basement membrane in vivo. The process of transfer to the cell surface and internalization into cells is slow and asynchronous, with a half-time of roughly 14 h (reviewed in Sapp and Bienkowska-Haba 2009).

L2 is highly resistant to cleavage by furin in the context of the mature virion. After the virion binds the basement membrane, epitopes near the amino terminus of L2 become more exposed, allowing furin cleavage. It has recently been shown that cyclophilins, which serve as protein chaperones by mediating the isomerization of proline residues, facilitate the exposure of the N terminus of L2 on the virion surface (Bienkowska-Haba et al. 2009).

It is conceivable that the conformational changes associated with furin cleavage of L2 involve disulfide shuffling in a fashion reminiscent of the conversion of L1 dimers and trimers to monomers in the presence of chaotropes (Fig. 18.6). Cellular PDIs, which facilitate disulfide shuffling, have been shown to be important for the opening of the capsids of polyomaviruses (which are structurally similar to the papillomaviruses) during infectious entry (Gilbert et al. 2006; Magnuson et al. 2005; Schelhaas et al. 2007). However, inhibitors of PDI do not block the infectivity of papillomaviruses (Buck et al. 2006b; Ishii et al. 2007). Thus, disulfide shuffling would presumably have to occur without assistance from PDIs (but perhaps with assistance from cyclophilins). The concept that disulfide shuffling may be required during the infection process is consistent with the recent observation that pseudovirions treated with agents that irreversibly alkylate free sulfhydryl groups (for example, NEM) are noninfectious (Ishii et al. 2007).

After L2 becomes exposed on the surface of the virion, cleavage by furin leads to an additional conformational change associated with greater affinity for an unidentified cell-surface receptor for the furin-processed form of the virion. This change is also thought to reduce the affinity of the virion for HS. Furin cleavage and the opening events that precede it allow a greater degree of access for antibodies specific for L2 residues 17–36 (Day et al. 2008a). Some antibodies targeting this conserved region of L2 can neutralize the infectivity of a broad range of papillomavirus types (see below).

A series of recent findings by Day and colleagues show that, although L2 is not exposed to furin in the mature virion, a substantial fraction of L2 molecules can be cleaved by furin in the context of the immature virion. Thus, by simply adding recombinant furin to cell lysates early during the course of harvesting a pseudovirion preparation, it is possible to produce furin precleaved particles (Day et al. 2008b). Although mature L1/L2 pseudovirions do not readily bind HS-deficient cells (including primary keratinocytes), furin precleaved pseudovirions readily bind cells deficient in HS and can effectively transduce reporter genes in an HS-independent fashion. A surprising corollary finding was that L1-only VLPs also readily bind HS-deficient cells. The findings suggests that, prior to furin cleavage, L2 either covers up the binding site for the cell surface receptor or induces a conformational change in L1 that inhibits cell surface receptor binding. Thus, the function of furin may be merely to move L2 aside such that the cell surface receptor binding site formed primarily by L1 becomes dominant.

The various conformational changes that occur after the virion binds the cell surface are less well defined. An initial trafficking step may involve "surfing" of the virion along actin-rich plasma membrane extensions called filopodia (Schelhaas et al. 2008; Smith et al. 2008). The virion then traffics to endocytic compartments. A bewildering array of different internalization pathways have been proposed, and it may be that different papillomavirus types utilize different endocytic routes for productive infection of different cell types (reviewed in Sapp and Bienkowska-Haba 2009). Although papillomaviruses appear not to be capable of delivering their genomes to immune cell types, the Kast lab has shown that papillomavirus virions readily bind Langerhans cells of the epidermis and may mediate inactivation of their antigen-presenting function (reviewed in Horvath et al. 2010).

Although L1/L2 pseudovirions and L1-only VLPs are both internalized into endosomal compartments and both types of particles can eventually deform to expose the encapsidated DNA with the endosomal compartment, only L1/L2 particles appear to be capable of disrupting the endosomal membrane, allowing escape of the viral DNA into the cytoplasm. A hydrophobic motif near the C terminus of L2 is thought to play an essential role in endosome escape (Kämper et al. 2006). The escape process appears to depend on endosome acidification and may involve interactions between L2 and the transmembrane protein syntaxin-18 (Bossis et al. 2005). Endosome escape can be blocked by a range of different inhibitors, including a class of cationic cellular peptides known as defensins (Buck et al. 2006a). Although the mechanism by which this class of inhibitors act remains unclear, it has recently been shown that defensins block the infectivity of adenoviruses by directly binding the virion and interfering with conformational changes that expose membrane-disrupting motifs within the virion (Nguyen et al. 2010; Smith and Nemerow 2008), reviewed in (Buck 2008).

After escaping the endosome, L2 accompanies the viral genome to the nucleus, likely via interactions with dynein and with karyopherin nuclear import receptors (Bordeaux et al. 2006; Fay et al. 2004; Florin et al. 2006). Day and colleagues have shown that, after nuclear import, L2 is essential for directing the trafficking of the incoming viral genome to a subnuclear structure known as an ND10 domain that is rich in a variety of transcription factors (Day et al. 2004).

Although biochemical approaches clearly indicate that a mousetrap-like series of changes occur during the steps leading up to productive cell binding and infectious delivery of the viral DNA to the cell nucleus, high-resolution details about the nature of these conformational dynamics are currently lacking. Improved understanding of these functionally important conformational changes, as well as uncovering of the identity of the cell surface receptor, should be fertile areas of focus in coming years.

18.6 Practical Implications: Next-Generation HPV Vaccines

The current generation of HPV vaccines is based on VLPs assembled from purified L1. The vaccines elicit highly HPV type-specific responses and are therefore expected to protect vaccinated individuals against the cognate carcinogenic HPV types represented in the vaccine. However, vaccinated individuals are unlikely to be protected against an additional dozen or so sexually transmitted HPV types that collectively cause roughly 30% of all cervical cancer (Lowy et al. 2008). Epidemiological studies likewise suggest that natural infection generally elicits only HPV type-specific neutralizing antibody responses.

It is now widely believed that the virions of most or all viruses that infect vertebrates have "Achilles' heels" that must remain essentially constant to ensure appropriate interactions with cellular factors. Papillomaviruses appear to have at least two Achilles' heels. The first is the surface on L1 that mediates the crucial initial interaction between the virion and HS. The second Achilles' heel is composed, at least primarily, of a conserved motif formed by the N terminus of L2. The putative cell surface receptor binding site on L1 represents a possible third conserved patch of vulnerability.

Evidence that the HS binding site is highly conserved – and therefore presents a vulnerability to broad-spectrum neutralization – comes from studies showing that purified heparin, which shares a variety of chemical features with HS, can block the infectivity a wide variety of papillomavirus types (Giroglou et al. 2001; Joyce et al. 1999). Other types of sulfated polysaccharides can also block papillomavirus infectivity, in some instances with very high functional avidity (polysaccharide doses in the low ng/ml range) (Buck et al. 2006b; Johnson et al. 2009). Interestingly, some HS mimics are capable of neutralizing the infectivity of cell-bound virions, suggesting that they can also antagonize steps downstream of the initial virion–HS interaction (Buck et al. 2006b; Selinka et al. 2007). The fact that a wide variety of papillomavirus types are vulnerable to inhibition by sulfated carbohydrates suggests that an antibody with the ability to simply mimic HS could broadly cross-neutralize most or all papillomavirus types.

Although a heparin-binding motif has been shown to involve HPV16 L1 residues K278, K356, and K361 at the apex of the capsomer knob, the affinity of this motif for the less-sulfated HS on the basement membrane is not clear, and the motif is not well conserved among HPVs (Knappe et al. 2007). It has also been suggested that some papillomavirus types may display two distinct HS binding sites (Buck et al. 2006b; Johnson et al. 2009). Given these uncertainties, it is not clear whether the primary HS binding motif that might theoretically be the target of broad-spectrum neutralizing antibodies has been unambiguously identified.

Much clearer evidence supports the existence of the second highly conserved site of vulnerability that is influenced by motifs near the N terminus of L2. In contrast to L1-based vaccines, immunogens comprised of various fragments of the amino-terminal third of L2 have been shown to elicit antibody responses capable of neutralizing the infectivity of a broad range of different papillomavirus types (Roden et al. 2000). In particular, a recombinant immunogen composed of residues 1–88 of BPV1 L2 was shown to elicit antibody responses capable of neutralizing cancer-causing HPV types, such as HPV16 and HPV18, that are only very distantly related to BPV1 in phylogenetic terms (Pastrana et al. 2005). Although the overall neutralizing titers of animals given L2-based vaccine immunogens have typically been quite low compared to L1-based immunogens, the development of L2-based vaccines has seen rapid progress in recent years, and several initial candidate immunogens have recently been tested in human clinical trials (reviewed in Campo and Roden 2010 and Karanam et al. 2009). Recent studies using the neutralizing monoclonal antibody (mAb) RG-1 (Gambhira et al. 2007), which targets L2 residues 17–36, have shown that the cognate epitope is inaccessible in the context of mature pseudovirions but becomes exposed after the virion binds to the basement membrane (Day et al. 2008a). Exposure of the RG1 epitope is also substantially greater in pseudovirions precleaved with furin during the maturation process. The results indicate that the initial conformational shift that occurs on the basement membrane and the furin cleavage event that follows result in greater exposure of the L2 Achilles' heel. This concept is consistent with the hypothesis that highly conserved motifs on virions may be exposed only transiently during the infectious entry process. This may reflect a viral strategy for protecting conserved neutralization-vulnerable motifs from wholesale recognition by circulating B cells. Despite the relatively inaccessible nature of the neutralizable epitopes of L2 in the context of the mature virion, their transient exposure provides a window of opportunity for neutralizing antibodies to arrest the infection process in vivo (Alphs et al. 2008).

Papillomavirus vaccines capable of eliciting broadly cross-neutralizing antibody responses could theoretically protect vaccinated individuals from all cancer-causing HPVs, as well as the range of HPVs that cause skin warts. Although skin warts typically only present a nuisance, as opposed to a threat to life, they are very common among children and a source of substantial consternation among parents. Thus, a vaccine capable of preventing skin warts would likely be in great demand, facilitating vaccine uptake and thus resulting in a greater degree of herd immunity against HPVs. L2-based vaccines currently under study appear to be simpler to produce and would likely be cheaper to produce than the relatively complex VLP-based vaccines currently on the market. Thus, these next-generation vaccines might be more feasible for developing nations, where cervical cancer is the second most common cancer among women.

It is likely that improved understanding of details through which the papillomavirus molecular machine operates could facilitate the development of improved vaccine immunogens capable of eliciting more potent, broadly cross-neutralizing antibody responses. Given the tremendous potential benefits of broad-spectrum HPV vaccines, further study of the mechanical features of the papillomavirus virion should be an important area of focus in coming years.

Acknowledgments We would like to thank Drs Alasdair Steven and Naiqian Cheng for their ongoing participation in the electron microscopy of HPV and BPV. This research was supported by the Intramural Research Programs of CIT and NCI.

References

Alphs HH, Gambhira R, Karanam B, Roberts JN, Jagu S, Schiller JT et al (2008) Protection against heterologous human papillomavirus challenge by a synthetic lipopeptide vaccine containing a broadly cross-neutralizing epitope of L2. Proc Natl Acad Sci USA 105:5850–5855

Bagiyan GA, Koroleva IK, Soroka NV, Ufimstev AV (2003) Oxidation of thiol compounds by molecular oxygen in aqueous solutions. Russ Chem Bull 52:1135–1141

Baker TS, Cheng RH (1996) A model-based approach for determining orientations of biological macromolecules imaged by cryoelectron microscopy. J Struct Biol 116:120–130

Belnap DM, Olson NH, Cladel NM, Newcomb WW, Brown JC, Kreider JW et al (1996) Conserved features in papillomavirus and polyomavirus capsids. J Mol Biol 259:249–263

Bernard HU, Burk RD, Chen Z, van Doorslaer K, Hausen HZ, de Villiers EM (2010) Classification of papillomaviruses (PVs) based on 189 PV types and proposal of taxonomic amendments. Virology 401:70–79

Bienkowska-Haba M, Patel HD, Sapp M (2009) Target cell cyclophilins facilitate human papillomavirus type 16 infection. PLoS Pathog 5:e1000524

Bordeaux J, Forte S, Harding E, Darshan MS, Klucevsek K, Moroianu J (2006) The l2 minor capsid protein of low-risk human papillomavirus type 11 interacts with host nuclear import receptors and viral DNA. J Virol 80:8259–8262

Boshart M, Gissmann L, Ikenberg H, Kleinheinz A, Scheurlen W, zur Hausen H (1984) A new type of papillomavirus DNA, its presence in genital cancer biopsies and in cell lines derived from cervical cancer. EMBO J 3:1151–1157

Bossis I, Roden RB, Gambhira R, Yang R, Tagaya M, Howley PM et al (2005) Interaction of tSNARE syntaxin 18 with the papillomavirus minor capsid protein mediates infection. J Virol 79:6723–6731

Buck CB (2008) Defensins' offensive play: exploiting a viral Achilles' heel. Cell Host Microbe 3:3–4

Buck CB, Pastrana DV, Lowy DR, Schiller JT (2004) Efficient intracellular assembly of papillomaviral vectors. J Virol 78:751–757

Buck CB, Pastrana DV, Lowy DR, Schiller JT (2005a) Generation of HPV pseudovirions using transfection and their use in neutralization assays. Methods Mol Med 119:445–462

Buck CB, Thompson CD, Pang YY, Lowy DR, Schiller JT (2005b) Maturation of papillomavirus capsids. J Virol 79:2839–2846

Buck CB, Day PM, Thompson CD, Lubkowski J, Lu W, Lowy DR et al (2006a) Human alpha-defensins block papillomavirus infection. Proc Natl Acad Sci USA 103:1516–1521

Buck CB, Thompson CD, Roberts JN, Muller M, Lowy DR, Schiller JT (2006b) Carrageenan is a potent inhibitor of papillomavirus infection. PLoS Pathog 2:e69

Buck CB, Cheng N, Thompson CD, Lowy DR, Steven AC, Schiller JT et al (2008) Arrangement of L2 within the papillomavirus capsid. J Virol 82:5190–5197

Burk RD, Chen Z, Van Doorslaer K (2009) Human papillomaviruses: genetic basis of carcinogenicity. Public Health Genomics 12:281–290

Campo MS, Roden RB (2010) Papillomavirus prophylactic vaccines: established successes, new approaches. J Virol 84:1214–1220

Campos SK, Ozbun MA (2009) Two highly conserved cysteine residues in HPV16 L2 form an intramolecular disulfide bond and are critical for infectivity in human keratinocytes. PLoS One 4:e4463

Chen XS, Garcea RL, Goldberg I, Casini G, Harrison SC (2000) Structure of small virus-like particles assembled from the L1 protein of human papillomavirus 16. Mol Cell 5:557–567

Christensen ND (2005) Cottontail rabbit papillomavirus (CRPV) model system to test antiviral and immunotherapeutic strategies. Antivir Chem Chemother 16:355–362

Chromy LR, Pipas JM, Garcea RL (2003) Chaperone-mediated in vitro assembly of Polyomavirus capsids. Proc Natl Acad Sci USA 100:10477–10482

Chung CH, Gillison ML (2009) Human papillomavirus in head and neck cancer: its role in pathogenesis and clinical implications. Clin Cancer Res 15:6758–6762

Ciuffo G (1907) Innesto positivo con filtrate di verruca vulgare. Giornale Italiano delle Malattie Veneree 48:12–17

Combita AL, Touze A, Bousarghin L, Sizaret PY, Munoz N, Coursaget P (2001) Gene transfer using human papillomavirus pseudovirions varies according to virus genotype and requires cell surface heparan sulfate. FEMS Microbiol Lett 204:183–188

Conway MJ, Meyers C (2009) Replication and assembly of human papillomaviruses. J Dent Res 88:307–317

Conway MJ, Alam S, Christensen ND, Meyers C (2009a) Overlapping and independent structural roles for human papillomavirus type 16 L2 conserved cysteines. Virology 393:295–303

Conway MJ, Alam S, Ryndock EJ, Cruz L, Christensen ND, Roden RB et al (2009b) Tissue-spanning redox gradient-dependent assembly of native human papillomavirus type 16 virions. J Virol 83:10515–10526

Culp TD, Budgeon LR, Christensen ND (2006) Human papillomaviruses bind a basal extracellular matrix component secreted by keratinocytes which is distinct from a membrane-associated receptor. Virology 347:147–159

Danos O, Engel LW, Chen EY, Yaniv M, Howley PM (1983) Comparative analysis of the human type 1a and bovine type 1 papillomavirus genomes. J Virol 46:557–566

Day PM, Schiller JT (2009) The role of furin in papillomavirus infection. Future Microbiol 4:1255–1262

Day PM, Baker CC, Lowy DR, Schiller JT (2004) Establishment of papillomavirus infection is enhanced by promyelocytic leukemia protein (PML) expression. Proc Natl Acad Sci USA 101:14252–14257

Day PM, Gambhira R, Roden RB, Lowy DR, Schiller JT (2008a) Mechanisms of human papillomavirus type 16 neutralization by l2 cross-neutralizing and l1 type-specific antibodies. J Virol 82:4638–4646

Day PM, Lowy DR, Schiller JT (2008b) Heparan sulfate-independent cell binding and infection with furin-precleaved papillomavirus capsids. J Virol 82:12565–12568

Doorbar J (2005) The papillomavirus life cycle. J Clin Virol 32(Suppl 1):S7–S15

Durst M, Gissmann L, Ikenberg H, zur Hausen H (1983) A papillomavirus DNA from a cervical carcinoma and its prevalence in cancer biopsy samples from different geographic regions. Proc Natl Acad Sci USA 80:3812–3815

Dvoretzky I, Shober R, Chattopadhyay SK, Lowy DR (1980) A quantitative in vitro focus assay for bovine papilloma virus. Virology 103:369–375

Esko JD, Selleck SB (2002) Order out of chaos: assembly of ligand binding sites in heparan sulfate. Annu Rev Biochem 71:435–471

Esko JD, Kimata K, Lindahl U (2009) Proteoglycans and sulfated glycosaminoglycans. In: Varki A, Cummings RD, Esko JD, Freeze HH, Stanley P, Bertozzi CR et al (eds) Essentials of glycobiology, 2nd edn. The Consortium of Glycobiology Editors, La Jolla

Favre M, Breitburd F, Croissant O, Orth G (1975) Structural polypeptides of rabbit, bovine, and human papillomaviruses. J Virol 15:1239–1247

Fay A, Yutzy WH IV, Roden RB, Moroianu J (2004) The positively charged termini of L2 minor capsid protein required for bovine papillomavirus infection function separately in nuclear import and DNA binding. J Virol 78:13447–13454

Feltkamp MC, de Koning MN, Bavinck JN, Ter Schegget J (2008) Betapapillomaviruses: innocent bystanders or causes of skin cancer. J Clin Virol 43:353–360

Finch JT, Klug A (1965) The structure of viruses of the papilloma-polyoma type 3. Structure of rabbit papilloma virus, with an appendix on the topography of contrast in negative-staining for electron-microscopy. J Mol Biol 13:1–12

Florin L, Becker KA, Sapp C, Lambert C, Sirma H, Muller M et al (2004) Nuclear translocation of papillomavirus minor capsid protein L2 requires Hsc70. J Virol 78:5546–5553

Florin L, Becker KA, Lambert C, Nowak T, Sapp C, Strand D et al (2006) Identification of a dynein interacting domain in the papillomavirus minor capsid protein l2. J Virol 80:6691–6696

Gambhira R, Karanam B, Jagu S, Roberts JN, Buck CB, Bossis I et al (2007) A protective and broadly cross-neutralizing epitope of human papillomavirus L2. J Virol 81:13927–13931

Gambhira R, Jagu S, Karanam B, Day PM, Roden R (2009) Role of L2 cysteines in papillomavirus infection and neutralization. Virol J 6:176

Gilbert J, Ou W, Silver J, Benjamin T (2006) Downregulation of protein disulfide isomerase inhibits infection by the mouse polyomavirus. J Virol 80:10868–10870

Giroglou T, Florin L, Schafer F, Streeck RE, Sapp M (2001) Human papillomavirus infection requires cell surface heparan sulfate. J Virol 75:1565–1570

Gissmann L, Pfister H, Zur Hausen H (1977) Human papilloma viruses (HPV): characterization of four different isolates. Virology 76:569–580

Giuliano AR, Tortolero-Luna G, Ferrer E, Burchell AN, de Sanjose S, Kjaer SK et al (2008) Epidemiology of human papillomavirus infection in men, cancers other than cervical and benign conditions. Vaccine 26(Suppl 10): K17–K28

Groff DE, Lancaster WD (1986) Genetic analysis for the 3′ early region transformation and replication functions of bovine papillomavirus type 1. Virology 150:221–230

Hagensee ME, Olson NH, Baker TS, Galloway DA (1994) Three-dimensional structure of vaccinia virus-produced human papillomavirus type 1 capsids. J Virol 68:4503–4505

Heymann JB, Cheng N, Newcomb WW, Trus BL, Brown JC, Steven AC (2003) Dynamics of herpes simplex virus capsid maturation visualized by time-lapse cryo-electron microscopy. Nat Struct Biol 10:334–341

Horvath CA, Boulet GA, Renoux VM, Delvenne PO, Bogers JP (2010) Mechanisms of cell entry by human papillomaviruses: an overview. Virol J 7:11

Howley PM, Livingston DM (2009) Small DNA tumor viruses: large contributors to biomedical sciences. Virology 384:256–259

Ishii Y, Kondo K, Matsumoto T, Tanaka K, Shinkai-Ouchi F, Hagiwara K et al (2007) Thiol-reactive reagents inhibits intracellular trafficking of human papillomavirus type 16 pseudovirions by binding to cysteine residues of major capsid protein L1. Virol J 4:110

Johnson KM, Kines RC, Roberts JN, Lowy DR, Schiller JT, Day PM (2009) Role of heparan sulfate in attachment to and infection of the murine female genital tract by human papillomavirus. J Virol 83:2067–2074

Joyce JG, Tung JS, Przysiecki CT, Cook JC, Lehman ED, Sands JA et al (1999) The L1 major capsid protein of human papillomavirus type 11 recombinant virus-like particles interacts with heparin and cell-surface glycosaminoglycans on human keratinocytes. J Biol Chem 274:5810–5822

Kämper N, Day PM, Nowak T, Selinka HC, Florin L, Bolscher J et al (2006) A membrane-destabilizing peptide in capsid protein L2 is required for egress of papillomavirus genomes from endosomes. J Virol 80:759–768

Karanam B, Jagu S, Huh WK, Roden RB (2009) Developing vaccines against minor capsid antigen L2 to prevent papillomavirus infection. Immunol Cell Biol 87:287–299

Kines RC, Thompson CD, Lowy DR, Schiller JT, Day PM (2009) The initial steps leading to papillomavirus infection occur on the basement membrane prior to cell surface binding. Proc Natl Acad Sci USA 106:20458–20463

Klug A, Finch JT (1965) Structure of viruses of the papilloma-polyoma type. I. Human wart virus. J Mol Biol 11:403–423

Klug A, Finch JT (1968) Structure of viruses of the papilloma-polyoma type. IV. Analysis of tilting experiments in the electron microscope. J Mol Biol 31:1–12

Knappe M, Bodevin S, Selinka HC, Spillmann D, Streeck RE, Chen XS et al (2007) Surface-exposed amino acid residues of HPV16 L1 protein mediating interaction with cell surface heparan sulfate. J Biol Chem 282:27913–27922

Leder C, Kleinschmidt JA, Wiethe C, Muller M (2001) Enhancement of capsid gene expression: preparing the human papillomavirus type 16 major structural gene L1 for DNA vaccination purposes. J Virol 75:9201–9209

Li M, Beard P, Estes PA, Lyon MK, Garcea RL (1998) Intercapsomeric disulfide bonds in papillomavirus assembly and disassembly. J Virol 72:2160–2167

Lowy DR (2007) History of papillomavirus research. In: Garcea RL, DiMaio D (eds) The papillomaviruses. Springer, New York, NY, pp 13–28

Lowy DR, Solomon D, Hildesheim A, Schiller JT, Schiffman M (2008) Human papillomavirus infection and the primary and secondary prevention of cervical cancer. Cancer 113:1980–1993

Magnuson B, Rainey EK, Benjamin T, Baryshev M, Mkrtchian S, Tsai B (2005) ERp29 triggers a conformational change in polyomavirus to stimulate membrane binding. Mol Cell 20:289–300

McBride AA (2008) Replication and partitioning of papillomavirus genomes. Adv Virus Res 72:155–205

McCarthy MP, White WI, Palmer-Hill F, Koenig S, Suzich JA (1998) Quantitative disassembly and reassembly of human papillomavirus type 11 virus like particles in vitro. J Virol 72:32–41

Meyers C, Frattini MG, Hudson JB, Laimins LA (1992) Biosynthesis of human papillomavirus from a continuous cell line upon epithelial differentiation. Science 257:971–973

Modis Y, Trus BL, Harrison SC (2002) Atomic model of the papillomavirus capsid. EMBO J 21:4754–4762

Nguyen EK, Nemerow GR, Smith JG (2010) Direct evidence from single-cell analysis that human {alpha}-defensins block adenovirus uncoating to neutralize infection. J Virol 84:4041–4049

Ogawa H, Taneda A, Kanaoka Y, Sekine T (1979) The histochemical distribution of protein bound sulfhydryl groups in human epidermis by the new staining method. J Histochem Cytochem 27:942–946

Orth G (1986) Epidermodysplasia verruciformis: a model for understanding the oncogenicity of human papillomaviruses. Ciba Found Symp 120:157–174

Orth G, Favre M, Croissant O (1977) Characterization of a new type of human papillomavirus that causes skin warts. J Virol 24:108–120

Orth G, Jablonska S, Favre M, Croissant O, Jarzabek-Chorzelska M, Rzesa G (1978) Characterization of two types of human papillomaviruses in lesions of epidermodysplasia verruciformis. Proc Natl Acad Sci USA 75:1537–1541

Orth G, Jablonska S, Jarzabek-Chorzelska M, Obalek S, Rzesa G, Favre M et al (1979) Characteristics of the lesions and risk of malignant conversion associated with the type of human papillomavirus involved in epidermodysplasia verruciformis. Cancer Res 39:1074–1082

Pastrana DV, Gambhira R, Buck CB, Pang YY, Thompson CD, Culp TD et al (2005) Cross-neutralization of cutaneous and mucosal papillomavirus types with anti-sera to the amino terminus of L2. Virology 337:365–372

Pickard C, Louafi F, McGuire C, Lowings K, Kumar P, Cooper H et al (2009) The cutaneous biochemical redox barrier: a component of the innate immune defenses against sensitization by highly reactive environmental xenobiotics. J Immunol 183:7576–7584

Pyeon D, Lambert PF, Ahlquist P (2005) Production of infectious human papillomavirus independently of viral replication and epithelial cell differentiation. Proc Natl Acad Sci USA 102:9311–9316

Pyeon D, Pearce SM, Lank SM, Ahlquist P, Lambert PF (2009) Establishment of human papillomavirus infection requires cell cycle progression. PLoS Pathog 5:e1000318

Richards RM, Lowy DR, Schiller JT, Day PM (2006) Cleavage of the papillomavirus minor capsid protein, L2, at a furin consensus site is necessary for infection. Proc Natl Acad Sci USA 103:1522–1527

Rigoni-Stern D (1842) Fat ti statistici relativi alle malattie cancrose. Giornale Service Progr Pathol Terap Ser 2:507–517

Roberts JN, Buck CB, Thompson CD, Kines R, Bernardo M, Choyke PL et al (2007) Genital transmission of HPV in a mouse model is potentiated by nonoxynol-9 and inhibited by carrageenan. Nat Med 13:857–861

Roden RB, Yutzy WH IV, Fallon R, Inglis S, Lowy DR, Schiller JT (2000) Minor capsid protein of human genital papillomaviruses contains subdominant, cross-neutralizing epitopes. Virology 270:254–257

Rous P, Beard J (1935) The progression to carcinoma of virus induced rabbit papillomas (Shope). J Exp Med 62:523–545

Sapp M, Bienkowska-Haba M (2009) Viral entry mechanisms: human papillomavirus and a long journey from extracellular matrix to the nucleus. FEBS J 276:7206–7216

Sapp M, Fligge C, Petzak I, Harris JR, Streeck RE (1998) Papillomavirus assembly requires trimerization of the major capsid protein by disulfides between two highly conserved cysteines. J Virol 72:6186–6189

Schadlich L, Senger T, Gerlach B, Mucke N, Klein C, Bravo IG et al (2009) Analysis of modified human papillomavirus type 16 L1 capsomeres: the ability to assemble into larger particles correlates with higher immunogenicity. J Virol 83:7690–7705

Schelhaas M, Malmstrom J, Pelkmans L, Haugstetter J, Ellgaard L, Grunewald K et al (2007) Simian virus 40 depends on ER protein folding and quality control factors for entry into host cells. Cell 131:516–529

Schelhaas M, Ewers H, Rajamaki ML, Day PM, Schiller JT, Helenius A (2008) Human papillomavirus type 16 entry: retrograde cell surface transport along actin-rich protrusions. PLoS Pathog 4:e1000148

Schiffman MH, Castle P (2003) Epidemiologic studies of a necessary causal risk factor: human papillomavirus infection and cervical neoplasia. J Natl Cancer Inst 95:E2

Schiller JT, Vass WC, Lowy DR (1984) Identification of a second transforming region in bovine papillomavrirus DNA. Proc Natl Acad Sci USA 81:7880–7884

Schwartz S (2000) Regulation of human papillomavirus late gene expression. Ups J Med Sci 105:171–192

Schwarz E, Durst M, Demankowski C, Lattermann O, Zech R, Wolfsperger E et al (1983) DNA sequence and genome organization of genital human papillomavirus type 6b. EMBO J 2:2341–2348

Selinka HC, Giroglou T, Nowak T, Christensen ND, Sapp M (2003) Further evidence that papillomavirus capsids exist in two distinct conformations. J Virol 77:12961–12967

Selinka HC, Florin L, Patel HD, Freitag K, Schmidtke M, Makarov VA et al (2007) Inhibition of transfer to secondary receptors by heparan sulfate-binding drug or antibody induces noninfectious uptake of human papillomavirus. J Virol 81:10970–10980

Serra A (1924) Studi sul virus della verruca, del papilloma, del condiloma acuminato. Giornale Italiano delle Malattie Veneree e delle Pelle 65:1808–1814

Smith JG, Nemerow GR (2008) Mechanism of adenovirus neutralization by human alpha-defensins. Cell Host Microbe 3:11–19

Smith JL, Lidke DS, Ozbun MA (2008) Virus activated filopodia promote human papillomavirus type 31 uptake from the extracellular matrix. Virology 381:16–21

Syverton JT, Berry GP (1935) Carcinoma in the cottontail rabbit following spontaneous virus papilloma (Shope). Proc Soc Exp Biol Med 33:399–400

Touze A, Coursaget P (1998) In vitro gene transfer using human papillomavirus-like particles. Nucleic Acids Res 26:1317–1323

Trus BL, Roden RB, Greenstone HL, Vrhel M, Schiller JT, Booy FP (1997) Novel structural features of bovine papillomavirus capsid revealed by a three-dimensional reconstruction to 9 A resolution. Nat Struct Biol 4:413–420

Ullmann EV (1923) On the aetiology of the laryngeal papilloma. Acta Otolaryngol 5:317–334

Variot G (1894) Un cas d'incoluation experimentale des verrues de l'enfant a l'homme. J Clin Ther Infant 2:529–531

Wolf M, Garcea RL, Grigorieff N, Harrison SC (2010) Subunit interactions in bovine papillomavirus. Proc Natl Acad Sci USA 107:6298–6303

Yang YC, Okayama H, Howley PM (1985) Bovine papillomavirus contains multiple transforming genes. Proc Natl Acad Sci USA 82:1030–1034

Zhao Q, Wu S, Manger W, Gadam S (2002) Process for making human papillomavirus virus-like particles with improved properties. Merck & Co., Inc., Rahway, NJ

Zhou J, Liu WJ, Peng SW, Sun XY, Frazer I (1999) Papillomavirus capsid protein expression level depends on the match between codon usage and tRNA availability. J Virol 73:4972–4982

zur Hausen H (2009) Papillomaviruses in the causation of human cancers – a brief historical account. Virology 384:260–265

Chapter 19
Procapsid Assembly, Maturation, Nuclear Exit: Dynamic Steps in the Production of Infectious Herpesvirions

Giovanni Cardone, J. Bernard Heymann, Naiqian Cheng, Benes L. Trus, and Alasdair C. Steven

Abstract Herpesviruses, a family of animal viruses with large (125–250 kbp) linear DNA genomes, are highly diversified in terms of host range; nevertheless, their virions conform to a common architecture. The genome is confined at high density within a thick-walled icosahedral capsid with the uncommon (among viruses, generally) but unvarying triangulation number $T = 16$. The envelope is a membrane in which some 11 different viral glycoproteins are implanted. Between the capsid and the envelope is a capacious compartment called the tegument that accommodates ~20–40 different viral proteins (depending on which virus) destined for delivery into a host cell. A strong body of evidence supports the hypothesis that herpesvirus capsids and those of tailed bacteriophages stem from a distant common ancestor, whereas their radically different infection apparatuses – envelope on one hand and tail on the other – reflect subsequent coevolution with divergent hosts. Here we review the molecular components of herpesvirus capsids and the mechanisms that regulate their assembly, with particular reference to the archetypal alphaherpesvirus, herpes simplex virus type 1; assess their duality with the capsids of tailed bacteriophages; and discuss the mechanism whereby, once DNA packaging has been completed, herpesvirus nucleocapsids exit from the nucleus to embark on later stages of the replication cycle.

19.1 Introduction

Herpesviruses infect hosts throughout the animal kingdom, from molluscs to man. Eight family members cause diseases in humans, including two – Epstein–Barr virus and Kaposi's sarcoma-associated virus – that are tumorogenic. Here we review current understanding of herpesvirus capsid assembly, a largely conserved process. On the macromolecular scale, these are huge particles – 125 nm in diameter and ~200 MDa in mass – and composed, transiently or for the long term, of

G. Cardone • J.B. Heymann • N. Cheng • A.C. Steven (✉)
Laboratory of Structural Biology, National Institute for Arthritis,
Musculoskeletal and Skin Diseases, National Institutes of Health, Bethesda, MD 20892, USA
e-mail: stevena@mail.nih.gov

B.L. Trus
Laboratory of Structural Biology, National Institute for Arthritis,
Musculoskeletal and Skin Diseases, National Institutes of Health, Bethesda, MD 20892, USA

Imaging Sciences Laboratory, Center for Information Technology,
National Institutes of Health, Bethesda, MD 20892, USA

M.G. Rossmann and V.B. Rao (eds.), *Viral Molecular Machines*,
Advances in Experimental Medicine and Biology 726, DOI 10.1007/978-1-4614-0980-9_19,

~5,000 protein molecules of at least seven different kinds. We also discuss the hypothesis that the capsids of herpesviruses and tailed bacteriophages descend from a common ancestor. At first sight, this proposition may appear counterintuitive, in view of widely differing virion morphologies, disparate hosts, and the implication that elaborate assembly pathways evolved very early, prior to the point at which eukaryotic cells and bacteria diverged. In addition, we summarize some interactions between herpesvirus capsids and teguments, which are attracting increasing research interest. Compared with capsids, however, teguments are less conserved, less ordered, and less understood.

19.2 Herpesvirus Capsid Assembly

19.2.1 Assembly of the Procapsid

The sequence of events that take place as the herpesvirus nucleocapsid assembles is outlined in Fig. 19.1. First, a procapsid is produced which conforms to $T=16$ icosahedral symmetry, albeit in a particle that is almost perfectly spherical. One of its fivefold vertices is occupied by a dodecameric ring of portal protein subunits, giving a distinctive 5:12 symmetry mismatch. At the other 11 vertices are pentamers of UL19, the major capsid protein, with UL19 hexamers elsewhere in the surface lattice. It is plausible but not fully proven that the portal serves as the nucleation complex from which there is outgrowth of coassembling scaffolding protein and UL19 (Huffman et al. 2008; Yang and Baines 2008). In fact, HSV-1 employs two scaffolding systems – internal and external. The internal scaffold is a 25-nm thick shell of laterally packed dimers of an elongated protein, UL26.5. Approximately 10% of these subunits (UL26) have an additional protein, the maturational protease VP24, fused to the N terminus and positioned on the inside of the scaffold shell (Newcomb et al. 2000). The other end of each dimer, with the C-termini, contacts the inside of the surface shell. The external scaffold is contributed by "triplexes" which are $\alpha_2\beta$ heterotrimers of VP23 and VP19c. Remarkably, adjacent UL19 capsomers do not directly contact each other, but instead are coordinated around the sites of local threefold symmetry by their contacts with triplexes (Trus et al. 1995) (Fig. 19.2). Thus, the procapsid surface shell consists of one dodecamer of UL6, 150 hexamers and 11 pentamers of UL19, and 320 triplexes.

Biochemical quantitations estimate the scaffolding shell's complement of dimers of UL26.5 and/or UL26 as ~1,000 (Newcomb et al. 2000). It has yet to be established whether this shell conforms to a global symmetry – for example, icosahedral symmetry complementary to that of the outer shell – or whether it is only locally ordered and "micellar" in character. Of note, a similar dilemma persists concerning the packing of the retroviral polyprotein Gag in immature virions (Briggs et al. 2009; See Chapter 20). However, Gag shells, although also spherical, vary significantly in size whereas the diameter of the herpesvirus scaffold shell is fixed, as it tracks the inner surface of the precisely defined surface shell.

19.2.2 Procapsid Maturation

After the HSV-1 procapsid is complete, the protease is activated and performs two actions: detaching itself from the scaffolding protein and removing a 25-residue peptide from the C-termini of the scaffolding proteins (UL26.5 → VP22a; UL26 → VP24 + VP21) [The cytomegalovirus protease also cleaves at a third site (Welch et al. 1993)]. It is not clear how the protease is inhibited until the procapsid is complete. One possibility is kinetics, if assembly, once initiated, is fast and protease activation is relatively slow (herpesvirus proteases are known to be slow enzymes – Tong 2002).

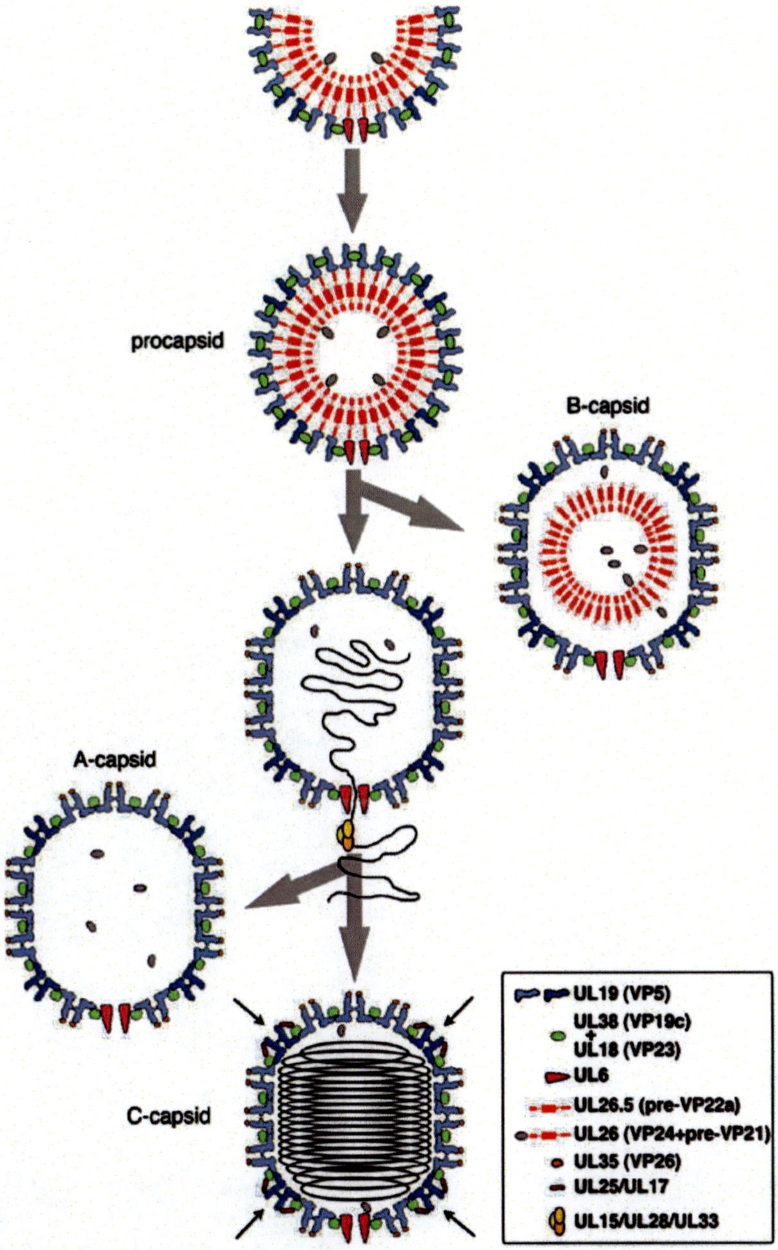

Fig. 19.1 Assembly and maturation of the HSV-1 nucleocapsid. Successive states of the capsid are shown schematically in cross-sections. This representation does not strictly comply with $T=16$ geometry (see Fig. 19.2) but is intended to convey key features, such as capsomer protrusions, the floor, triplexes, etc

Subsequently, the surface shell undergoes major alteration as it switches into its mature conformation which is more angular (polyhedral) and markedly more robust (Newcomb et al. 1996; Trus et al. 1996; Heymann et al. 2003). Contacts are established between neighboring capsomers via their innermost domains, forming the "floor" of the mature capsid (Fig. 19.2b, c). This transition is promoted by the packaging of DNA, which applies pressure on the inner surface of the capsid, and by

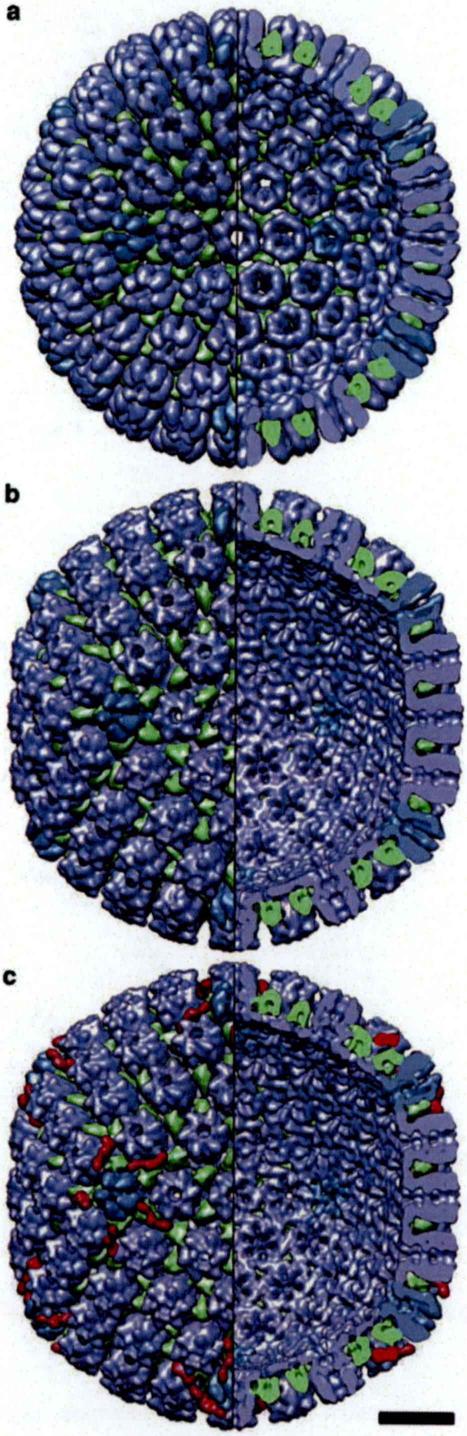

Fig. 19.2 Cryo-EM reconstructions of HSV-1 capsids. (**a**) The HSV-1 procapsid, (**b**) A-capsid, and (**c**) C-capsid, with DNA computationally removed. Left half of each reconstruction shows the outer surface, and right half, the inner surface. Hexameric capsomers are light blue; pentamers are mid-blue; triplexes are green; CCSCs are magenta. Bar, 20 nm

Table 19.1 Points of resemblance between phage capsids and herpesvirus capsids

Initial formation of a procapsid into which DNA is subsequently packaged
Packaging is accompanied by major structural changes in the procapsid (maturation)
Procapsid has an internal scaffolding protein that is expelled during maturation
Maturation is initiated by proteolytic processing of capsid protein and/or scaffolding protein[a]
During maturation, capsid protein subunits undergo large (~40°) rotations and other changes, re-assigning intersubunit and intercapsomer interactions
Procapsid is rounder; mature capsid is more polyhedral (flatter facets)
Procapsid has asymmetric hexamers that become symmetric on maturation
Capsid protein fold based on HK97 prototype, i.e., conserved core domain of ~300 aa plus additional domains on a case-by-case basis (Fig. 19.3)
Capsid geometry is based on hexamer/pentamer packing on icosahedral or quasi-icosahedral[b] foldings of a hexagonal surface lattice with 14 nm periodicity (mature capsid)
One capsid vertex is occupied by a 12-fold ring – the portal protein, conduit for DNA
Portal proteins may also conform to a conserved core structure plus additional domains on a case-by-case basis[c]
DNA is packaged into procapsid from a replicating concatemer
DNA is packaged by viral terminase, an ATPase/translocase/endonuclease
DNA is densely bundled with ~25 Å interduplex spacing, outer layers coaxially coiled[d], and no histone-like proteins

[a]Bioinformatic analyses suggest that herpesvirus proteases (four known structures) and some phage proteases have similar folds. Some phages (e.g., T7, P22) have no protease

[b]Term refers to nonspherical capsids, such as that of T4 which represents an icosahedron extended along a symmetry axis by insertion of additional rings of hexamers

[c]Simpson et al. (2001), Agirrezabala et al. (2005), Lebedev et al. (2007), and Olia et al. (2011)

[d]Booy et al. (1991) and Cerritelli et al. (1997)

detachment of the C-terminal peptides of the internal scaffolding proteins, which disrupts an interaction that restrains the surface shell from maturing. Concomitant with DNA packaging, the processed scaffolding proteins (VP22a, VP21) are expelled from the capsid but the protease (VP24) is retained. DNA packaging is effected by the viral terminase, a three-component ATPase complex that translocates DNA from the replicating concatemer into the maturing particle. DNA replication by phages and herpesviruses and, in particular, the roles played by helicases have been reviewed (Marintcheva and Weller 2001).

Two abortive capsids accumulate in the nuclei of infected cells as a result of miscarriage in DNA packaging. B-capsids have a mature surface shell, and their internal scaffolding protein is processed but not expelled, shrinking into a smaller and less regular structure. They contain no DNA and it appears that packaging has not been initiated. A-capsids, also with mature surface shells but retaining no scaffold, are viewed as particles on which DNA packaging was initiated but subsequently disrupted with loss of the packaged DNA. A third intranuclear capsid, the C-capsid, contains DNA and is an on-pathway intermediate. In a wild-type infection, the majority of intranuclear particles seen by thin section EM are B-capsids. A-capsids are less abundant, presumably because the event that produces them has lower incidence. Procapsids are few as, once formed, they rapidly convert to A-, B-, or C-capsids. C-capsids are also less abundant than B-capsids because they exit from the nucleus after packaging is complete. Below, in Sect. 19.3.1, we consider the events that regulate the exit of capsids from the nucleus.

19.2.3 Parallels in Bacteriophage Procapsid Assembly

Overall, the pathway in Fig. 19.1 is strikingly similar to those followed by tailed bacteriophages (Steven et al. 2005; See Chapter 15). This resemblance attracted comment as early as 1975 (Friedmann et al. 1975); since then, the number of properties recognized as shared has grown steadily (Table 19.1). A few aspects appear to be sticking points or at least bear further discussion.

1. All known herpesvirus capsids observe $T=16$ icosahedral symmetry, whereas tailed phage capsids exhibit a variety of T-numbers. This range does include $T=16$ (Parker et al. 1983; Duda et al. 2006; Weigele et al. 2007; Kuznetsov et al. 2010), and these are the only two families in which this capsid geometry has been documented. In view of the immense number of tailed phages in the biosphere (Hendrix 2003), it is hard to assess the extent to which the T-numbers reported to date afford a representative sampling. Given a common lattice constant of 14 nm arising from a shared capsid protein fold and packing arrangement (see below), it would be a straightforward undertaking to determine – or at least closely approximate – the T-numbers of many more tailed phage viruses by simply measuring capsid diameters by cryo-EM and hence to assess the overall incidence of $T=16$ geometry.
2. To date, no phages have been found to have a triplex-like external scaffold. Considering that scaffolding mechanisms are arguably the most diverse aspect of phage capsid assembly (Dokland 1999; See Chapter 14), it may be only a matter of time before one is discovered. However, we note that external scaffolding proteins have been observed in the phiX174 and P2/P4 phage systems - see Chapter 14. As the HSV procapsid matures, the role of its triplexes switches from morphogenesis to a clamp-like function at the trigonal lattice sites. In this respect, they resemble the trimers of clamping proteins centered on local threefold axes of phage capsids, such as those of T4 (Iwasaki et al. 2000; Fokine et al. 2004) and lambda (Yang et al. 2000). The latter proteins bind to their capsids only after maturation.
3. A strikingly distinctive behavior of phage procapsids is their maturation transformation, in which capsid protein subunits swivel through large angles and undergo some remodeling (Conway et al. 2001; Gertsman et al. 2009; Ionel et al. 2011; Chen et al. 2011). The main consequence of maturation is a stabilization that enables the capsid to resist the outward pressure of densely packed DNA; in some systems, further reinforcement follows from the binding of clamping proteins (see above) or the formation of covalent cross-links (Duda 1998). Maturation is generally accompanied by an increase in diameter by 15–20%, almost doubling the interior volume; hence, this transformation is often referred to as *expansion*. The HSV-1 procapsid also undergoes maturation (see above). Like phage capsids, it changes shape from round to polyhedral, but unlike them, it remains essentially the same size. However, this does not represent a fundamental discrepancy: the parts of UL19 subunits that form disconnected rings in the innermost part of the procapsid and which appear to be homologous to phage capsid proteins (Fig. 19.3a) reorganize so as to establish contacts between adjacent capsomers (Fig. 19.2). In this transition, the size of the per-subunit footprint in the floor increases by a similar margin as in maturing phage capsids. Thus, much the same allosteric mechanism is employed in both systems.

19.2.4 The Building Blocks and Their Roles in Assembly

19.2.4.1 Major Capsid Protein

UL19 (150 kDa). UL19 orthologs vary somewhat in size (~120–150 kDa) among different herpesviruses but in general are much larger than phage capsid proteins (~30–60 kDa). This discrepancy is attributable to their large C-terminal domains forming the external protrusions that account for the much greater thickness of herpesvirus capsids (~15 nm vs. ~4 nm). The HSV-1 protrusion consists of two domains, middle and upper (Fig. 19.3); a crystal structure has been determined for the upper domain (Bowman et al. 2003). The hexagonal lattice spacing in the floor (~14 nm) matches that of phage capsids, and cryo-EM data at relatively high resolution support the idea that the floor domain has a phage capsid protein-like fold (Zhou et al. 2000; Baker et al. 2005). Tailed phage MCPs consist of a basic core on to which additional domains may be grafted (Fig. 19.3a). It will be instructive to learn whether herpesvirus floor domains also have such domains. However, it is already clear that

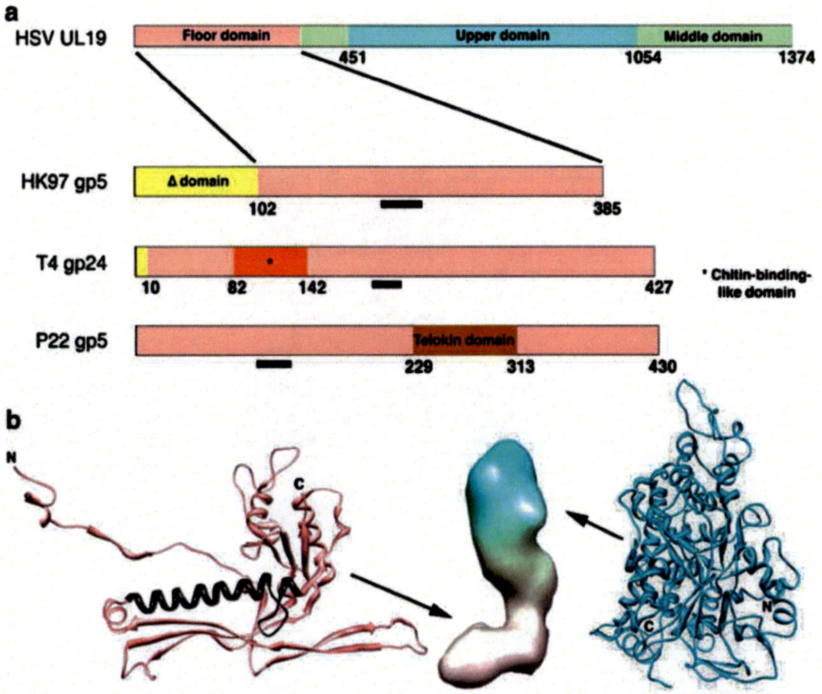

Fig. 19.3 Domain maps of capsid proteins of HSV-1 and three bacteriophages. (**a**) The phage proteins are ones for which high-resolution structures have been determined. The common core domain is colored *pink*. The Δ-domain of HK97 is equivalent to an internal scaffolding protein fused to the core domain and is proteolyzed during maturation (gp5→gp5*). The T4 capsid protein gp24 forms pentamers at all nonportal vertices of this elongated $T=13/Q=20$ capsid, whose other capsomers are hexamers of the homolog, gp23. Its short propeptide is removed during maturation (gp24→gp24*). P22 does not have a protease. Gp10A of T7 (not shown) appears to have a similar core domain plus an N-terminal domain of ~90 residues, which is not removed by proteolysis (Agirrezabala et al. 2007). The *black bars* mark the location of the long α-helix, a signature feature of this fold. (**b**) *Left*: fold of HK gp5* in its mature conformation (Wikoff et al. 2000); PDB: 1OHG. The long helix is colored in *black*. *Middle*: low-resolution surface rendering of a HSV-1 UL19 subunit in its B-capsid conformation. The floor domain (at *bottom*) is thought to have a HK97 gp5*-like fold. *Right*: upper domain of UL19 (Bowman et al. 2003); PDB: 1NO7

floor structures are markedly more conservative than protrusions (Booy et al. 1996; Trus et al. 2001; Davison et al. 2005).

19.2.4.2 Portal Protein

UL6 (74 kDa). When portal proteins of phages or herpesviruses are expressed at high levels, polymorphisms are commonly encountered in the number of subunits in the ring. However, it appears that, in both cases, only 12-mers are assembled into capsids (Lurz et al. 2001; Rochat et al. 2011). Crystal structures of several phage portals, of varying subunit size, have been determined (Simpson et al. 2001; Lebedev et al. 2007; Olia et al. 2011). As yet, the most detailed information on a herpesvirus portal is a cryo-EM reconstruction of the HSV-1 dodecamer at 16-Å resolution (Trus et al. 2004) (Fig. 19.4). The diameter of its wider end is compatible with fitting into the vertex space enclosed by five surrounding UL19 hexamers. There have been differing proposals, based on cryo-ET, as to whether the bulk of the portal complex faces outward (Cardone et al. 2007) or inward, as in phages (Deng et al. 2007; Chang et al. 2007; Rochat et al. 2011).

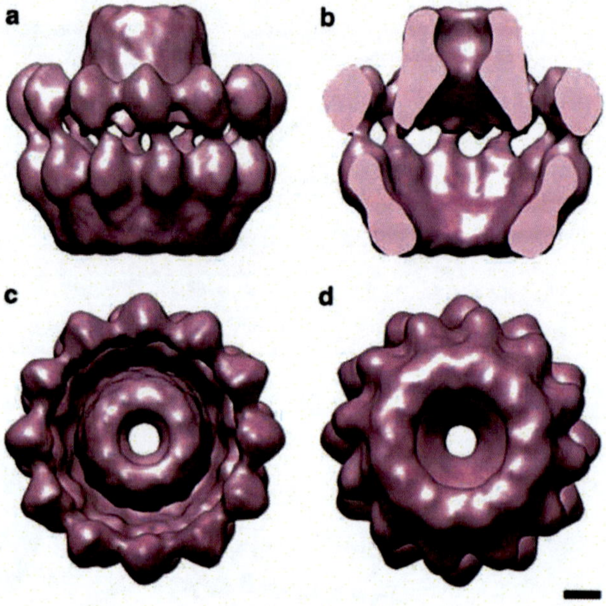

Fig. 19.4 The HSV1 portal protein, UL6. (**a**) Side view; (**b**) side view, cut; (**c**) top view; and (**d**) bottom view. Bar, 20 Å

19.2.4.3 Maturational Protease

VP24 (24 kDa). The structures of four herpesvirus proteases have been determined, exhibiting a common fold (Qiu et al. 1996, 1997; Newcomb et al 2000; Hoog et al. 1997). In the absence of crystal structures for phage proteases, bioinformatic analysis has identified a family that is predicted to share the canonical herpesvirus protease fold, with both sets of enzymes having conserved Ser and His residues in their active sites (Cheng et al. 2004). Of note, some phages (e.g., T7, P22) do not require a protease to mature. Thus, it would appear that the putatively shared structure represents the ancestral fold, with loss of the protease or its replacement representing more recent evolutionary developments.

19.2.4.4 Internal Scaffolding Proteins

UL26.5 (40 kDa) and UL26 (75 kDa). Two forms of this protein are expressed. The minor (less abundant) form, UL26, is initially extended at its N terminus by the protease and a linker, but its C-terminal part is identical to the major form, UL26.5. This fusion provides a convenient mechanism for incorporating the protease into the assembling procapsid (Sheaffer et al. 2000). Both forms are processed at their C-termini by the protease during maturation (see Sect. 19.2.2). The shared polypeptide appears to be a flexible α-helix-rich dimer that forms a segmented coiled coil (Pelletier et al. 1997). The crystal structure of the relatively short scaffolding protein of phage φ29 shows a dimeric coiled coil (Morais et al. 2003).

19.2.4.5 External Scaffolding Proteins: Triplex Proteins

UL18 (33 kDa) and UL38 (50 kDa). UL18 has been expressed in *Escherichia coli* (Kirkitadze et al. 1998). Assessment of its conformation by biophysical techniques led to the conclusion that the protein was not fully folded but in a "molten globule" state. It is plausible that UL38 is needed for UL18 to fold and enter functional trimers. UL38 has resisted expression until recently (Henson et al. 2011).

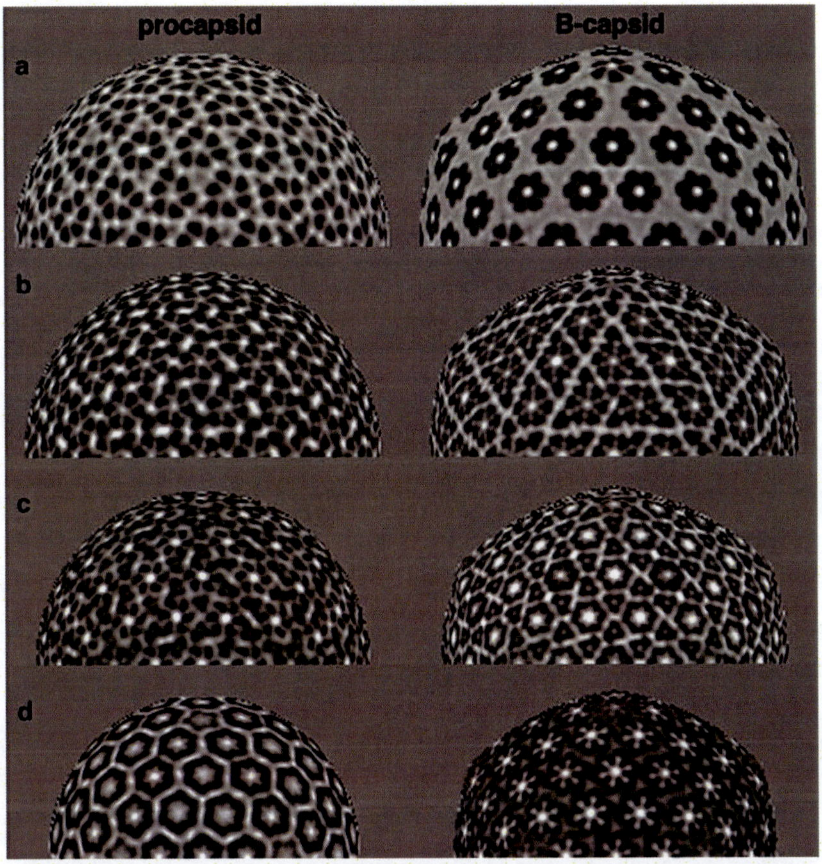

Fig. 19.5 Stratified representations of the HSV-1 procapsid and mature capsid. The procapsid is sampled on spherical sections one pixel thick, and the B-capsid on similarly sampled icosahedral sections. Protein density is dark. Half-sections are shown. (**a**) At a radius of 564 Å, the sections pass above the triplexes. The asymmetry of the procapsid hexons contrasts with the sixfold symmetry of the mature hexons. (**b**) At 536 Å radius, the triplexes present as triangular densities. (**c**) At 503 Å radius, the triplexes present as regular triplets of density. (**d**) At 470 Å radius, the floor domains form regular, noncontacting rings in the procapsid and the contiguous floor in the mature capsid

As yet, there is no high-resolution information on any triplex protein, but cryo-EM at moderate resolution shows well-defined densities, indicative of stably folded subunits. They show a triangular outer surface, underpinned by three equal and symmetrically distributed legs (Fig. 19.5). It may be, therefore, that the two proteins have similar folds, despite low sequence similarity.

19.2.4.6 Accessory Protein

UL35 (12 kDa). This small protein binds to the mature HSV-1 capsid at sites around the outer tips of hexons, but not to pentons (Trus et al. 1992; Zhou et al. 1995) nor to the procapsid (Newcomb et al. 2000). As UL35 is a monomer in solution (Wingfield et al. 1997), its specificity appears to reflect a bipartite binding site requiring appropriate juxtaposition of two adjacent UL19 protrusions. This is achieved in mature hexons but not in pentons nor in immature capsomers, where the protrusions are splayed apart (Figs. 19.2 and 19.5). A model has been proposed for part of UL35 based on *de novo* prediction and cryo-EM data (Baker et al. 2006). UL35 is dispensable for HSV-1, but proteins corresponding to it are needed in some other herpesvirus systems.

Survival of protein folds after sequences have diverged to unrecognizably low levels of similarity can provide telling clues as to shared antecedents. Evidence of this kind has provided a basis for the proposed common ancestry of other kinds of bacteriophages and animal viruses (Benson et al. 1999; Bamford et al. 2005; Krupovic and Bamford 2008). However, there is little scope for fold-based comparisons in the present context; of the seven herpesvirus capsid proteins, crystal structures have been determined only for the protease and major fragments of two other proteins, UL25 (Bowman et al. 2006) and the upper domain of UL19 (Bowman et al. 2003) (Fig. 19.3). As yet, there are no phage protease structures. Herpesvirus protrusions appear to be quite diverse (see above) and are not expected to have counterparts on phage capsid proteins. The same exclusion applies to UL25, which functions downstream from capsid assembly *per se* in phases of the replication cycle that are not shared by phages. Recognizing the substantial diversity that exists in the assembly strategies of phage capsids, we nevertheless anticipate structural similarity with herpesviruses in the following components: some proteases, portals, (possibly) internal scaffolding proteins, and terminases, for which a tract of significant sequence similarity has been detected between T4 and HSV-1. Further clarification should also come from more detailed information on floor domains.

Nevertheless, other less direct lines of evidence point to common evolutionary origins. There are so many highly distinctive molecular behaviors listed in Table 19.1 involving quaternary structure and other properties, which are common to phage capsids and herpesvirus capsids but in most cases shared by no other kind of virus, that the *a priori* probability of their having arisen by chance appears very small.

19.3 Later Capsid-Related Steps in the Replication Cycle

After completion of DNA packaging, the C-capsid has to exit the nucleus by primary envelopment at the inner nuclear membrane, followed by fusion of the membrane thus acquired with the outer nuclear membrane, delivering the nucleocapsid into the cytoplasm. It also has to acquire a tegument, traverse the cytoplasm, and bud into a *trans*-Golgi-derived compartment (secondary envelopment). Progeny virions are released by fusion of this compartment's membrane with the host cell plasma membrane. Virions may remain attached to the host cell for a while ("cell-associated virions") before being released as "extracellular virions." After entry into a host, the nucleocapsid is transported to the nucleus where it docks on to a nuclear pore complex to discharge its DNA, a mode of genome delivery reminiscent of tailed phages. These processes have been reviewed recently (Baines 2007; Mettenleiter et al. 2009; Guo et al. 2010), and the coverage given here focuses on some structural observations directly related to the nucleocapsid.

19.3.1 Nuclear Exit and the Roles of UL25 and UL17

As noted in Sect. 19.2.2, C-capsids migrate out of the nucleus more readily than B-capsids or A-capsids. This barrier is not absolute, as enveloped B-capsids have been observed, albeit rarely (unpublished results). There appears to be a pathway to tegumentation and envelopment on which C-capsids embark with high efficiency but which can also be entered by other particles, for example, small membrane vesicles nucleating the formation of L-particles, which resemble capsid-less virions (Szilagyi and Berriman 1994). The efficiency with which C-capsids exit the nucleus correlates with their contents of two viral proteins, UL25 and UL17, which form an elongated heterodimer called the CCSC (C-capsid-specific component) that binds to vertex-proximal sites on C-capsids (Figs. 19.2 and 19.6).

The penton-distal portion is contributed by UL25, as initially suggested by docking the crystal structure (Trus et al. 2007), leaving the penton-proximal portion to represent (part of) UL17. The UL25 assignment has since been proven by imaging C-capsids bearing UL25 with an N-terminal GFP fusion and Fab-labeling with an antibody whose epitope involves residues 99–111 (Conway et al. 2010). The

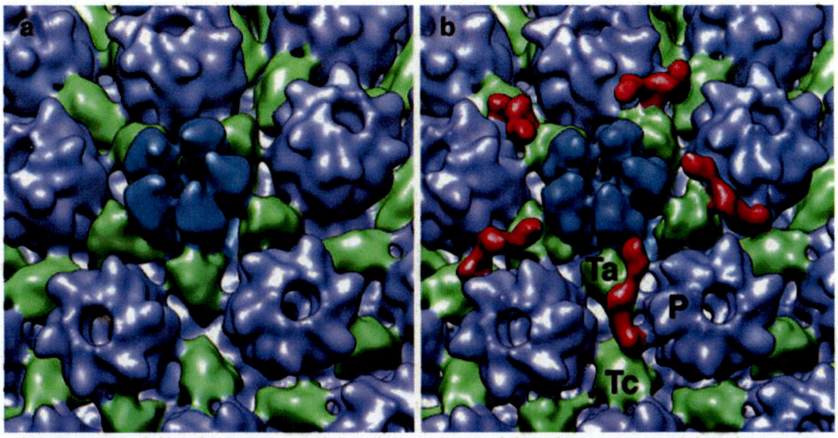

Fig. 19.6 Comparison of the penton-proximal regions of the A-capsid (**a**) and the C-capsid (**b**). The CCSC, a heterodimer of UL25 and UL17, is in *magenta*. It has multiple points of attachment to the capsid, on the a-triplex Ta, the c-triplex Tc, and the side of the P-hexon

CCSC density furthest from the penton was shown to be the N-terminal region of UL25 which has been shown to be important for capsid binding (Cockrell et al. 2009) but is not represented in the crystal structure, which starts at residue 134. More detailed mapping has assigned the P-hexon interaction to residues 1–26 and residues 28–39 as likely to make the Tc-triplex contact (Cockrell et al. 2011).

Cryo-EM of purified nuclear C-capsids found the CCSC sites to have an average occupancy of ~60% (Trus et al. 2007), whereas occupancy is undetectably low (i.e., <15% or so) on B-capsids and A-capsids. Biochemical studies have reported the presence of these proteins on all three capsids (Ogasawara et al. 2001; Thurlow et al. 2006; Newcomb et al. 2006; Cockrell et al. 2011), but their complements on A- and B-capsids are several-fold lower, and it is likely that these few molecules are scattered over other quasi-equivalent (not vertex-proximal) sites.

UL17 and UL25 are multifunctional proteins. The phenotype of a UL17-null mutant is the accumulation of B-capsids (Salmon and Baines 1998). In view of UL17's proximity to vertices, this phenotype may reflect the absence of its ability to engage terminase and hence DNA for packaging at the portal vertex. UL17 is also required for UL25 to associate with capsids (Thurlow et al. 2006). UL25 is needed for packaged genomes to remain stably encapsidated (McNab et al. 1998; Stow 2001; Cockrell et al. 2009) and is also involved in the interaction of incoming nucleocapsids with nuclear pore complexes (Preston et al. 2008). Here we focus on its role in nuclear exit (Klupp et al. 2006). Primary envelopment of the C-capsid is a complex process involving remodeling of the karyoskeleton and the inner nuclear membrane (Baines 2007). With that qualification, we envisage the following scenario.

CCSC binding is enhanced by a relatively subtle conformational change in the capsid, affecting its sites on the outer surface (Trus et al. 2007). Aspects of this change appear in a cryo-EM difference map between C-capsids and A-capsids (Fig. 19.7). Although similar, these structures are not identical as the difference map does not cleanly separate the CCSC; rather, small local changes affect various parts of the shell. Differences affecting the triplexes to which the CCSC binds are arrowed in Fig. 19.7b. Of note, small but significant structural differences have also been detected between DNA-filled capsids of phage HK97 and empty but nominally mature capsids (Duda et al. 2009).

This switch in the C-capsid is presumably elicited by mounting outward pressure as DNA packaging approaches completion. We posit that high occupancy of its CCSC sites enables the C-capsid to bind, via multiple low-affinity interactions, other protein(s) involved in primary envelopment. In this context, cryo-ET observations of vitrified sections of infected cells have detected an additional density layer between the INM and budding C-capsids arrow (Fig. 19.8a). If this layer represents

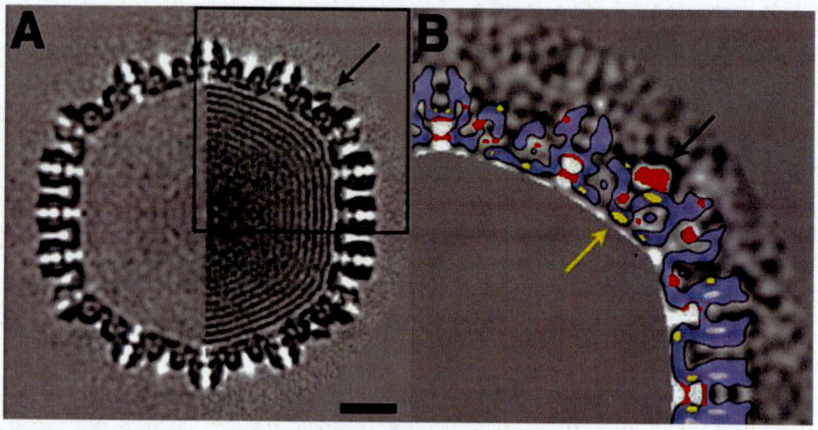

Fig. 19.7 Final switch in HSV-1 capsid maturation. This transition is shown in a comparison of A-capsid and C-capsid. (N.B. The A-capsid is not a direct precursor of the C-capsid; rather, it is an abortive capsid resulting from interrupted DNA packaging See Fig. 19.1. We take it to represent the mature capsid prior to the switch induced by late-stage DNA packaging that creates high-affinity sites for the CCSC). (**a**) Half-plane central sections of capsids viewed along a twofold axis: *left*: A-capsid and *right*: C-capsid. Scale bar, 20 nm. (**b**) Difference map (C-capsid minus A-capsid) of the boxed area in (**a**). Positive (negative) difference densities are light (dark). Positive densities above 1 standard deviation are colored *red*. Negative densities below 1 standard deviation are *yellow*. The strongest positive density are the CSSC [*arrows* in (**a**, **b**)] and density in the mouth of the axial channel through the floor. A single rigid-body shift of a capsid component would register as complementary *red* and *yellow* patches of the same size and shape, suitably offset; however, this difference map – and hence this transition – is more complex. The difference map is from density maps filtered at 30 Å. Outlines of capsomers and triplexes are contoured with *light blue* fill. DNA-related density inside the C-capsid is masked out in *gray*

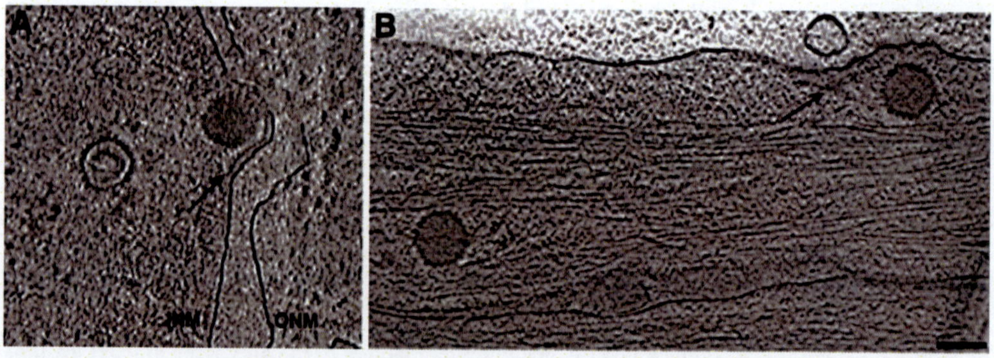

Fig. 19.8 Snapshots of the herpesvirus life cycle. (**a**) Nuclear egress. Tomographic slice from a vitrified section of a NIH 3T3 cell infected with murine cytomegalovirus. Inside the nucleus are a B-capsid and a C-capsid budding into the inner nuclear membrane. A layer of density between them (*arrow*) is likely to represent a layer of protein. Courtesy of Drs C. Hagen and K. Grünewald. For preparation details, see Scrivano et al. 2010. (**b**) Cell entry. Slice from a tomogram of a PtK2 cell, where two capsids of recently entered HSV-1 virions are visible. The strip of density along the cell membrane (*arrow*) appears to represent divested tegument. From Maurer et al. (2008), courtesy of Dr. K. Grünewald. Bar, 100 nm

viral proteins, UL34 and UL31 are plausible candidates (Baines 2007). If the envisaged interactions between the CCSC and its binding partner(s) are, as we suggest, of such low affinity as to need multivalency for their stable formation, the sparsity of CCSC complexes on the surfaces of A- and B-capsids would disallow embarkation on the nuclear exit pathway.

19.3.2 Tegumentation

In addition to delivering their genomes into host cells, tailed phages and herpesviruses both deliver proteins which perform functions needed early in the infection process. However, these proteins and their manner of delivery differ. In phage, they are incorporated into the procapsid and injected from the mature capsid into the host cell via the portal and tail-tube (Chaps. 7 and 21). Those of herpesviruses are incorporated into the tegument and released into the host cell cytoplasm when the viral envelope fuses with the host cell membrane. Whereas herpesvirus capsids are conservative in terms of their architecture and molecular components, teguments are more diverse (reviews: Britt 2007; Guo et al. 2010). Herpesviruses also vary in their teguments' point of attachment on the capsid; on simian cytomegalovirus, it is to triplexes as well as to capsomer protrusions (Trus et al. 1999) and on HSV-1, to pentons (Chen et al. 2001; Grünewald et al. 2003).

Moreover, teguments are markedly less regular in structure than capsids. Tegument proteins have been operationally distinguished into two classes in terms of their extractability from virions with the detergent Triton X-100 (Wolfstein et al. 2006). Those that are relatively easily extracted are called the outer tegument, and those that are more resistant, the inner tegument. The diversity of tegument components raises the challenging question of how they are selected for incorporation into nascent virions. A current working hypothesis is that a subset of inner tegment proteins, including the very large protein UL36 (~340 kDa), is anchored on the capsid and serves as a scaffold to which other tegument proteins attach (Coller et al. 2007; Newcomb and Brown 2010).

Cryo-ET of extracellular HSV-1 virions has demonstrated that their teguments are asymmetric, varying in thickness from ~5 nm, where the capsid is closely apposed to the envelope, to ~35 nm at the thickest part (Grünewald et al. 2003). Thus, the capsid is eccentrically positioned within an approximately spherical virion. Evidence from thin-sectioning EM is consistent with this observation while also showing that in cell-associated virions, the tegument is approximately uniform in thickness and the nucleocapsid is centered (Newcomb and Brown 2009). It follows that a substantial reorganization of the tegument takes place following exocytosis of progeny virions. It is not clear whether the inferred reorganization of tegument is accompanied by a redistribution of glycoproteins in the viral envelope or whether it may correlate with infectivity.

19.3.3 Capsid–Tegument Interactions in Host Cell Entry

Snapshots of HSV-1 nucleocapsids immediately after cell entry have been recorded by cryo-ET of PtK2 cells which are thin-enough specimens to allow this technique (Maurer et al. 2008). These data suggest that the nucleocapsid is released into the cytoplasm while the tegument remains largely connected and in contact with the viral envelope (now part of the host cell membrane) (Fig. 19.8b). This observation suggests that although tegument assembly is templated on the capsid, latterly it is more tightly bound to the envelope. The freed nucleocapsid, eventually retaining a few tegument proteins, is then transported along microtubules toward the nucleus by dynein/dynactin motors.

UL36 has been shown to play a role in the anterograde transport process but appears not to be the only player (Shanda and Wilson 2008); other inner tegument proteins are also involved (Radtke et al. 2010). Once the capsid reaches the nuclear envelope, interactions of UL25 (Pasdeloup et al. 2009) and UL36 (Radtke et al. 2010) with the nucleoporin Nup214 are important for its engagement with a nuclear pore complex.

Acknowledgment We thank Dr Robert Duda for helpful discussion and Drs Christoph Hagen and Kay Grünewald for the images in Fig. 19.8. This research was supported by the Intramural Research Programs of NIAMS and CIT.

References

Agirrezabala X, Martin-Benito J, Valle M, Gonzalez JM, Valencia A, Valpuesta JM, Carrascosa JL (2005) Structure of the connector of bacteriophage T7 at 8 Å resolution: structural homologies of a basic component of a DNA translocating machinery. J Mol Biol 347(5):895–902

Agirrezabala X, Velazquez-Muriel JA, Gomez-Puertas P, Scheres SH, Carazo JM, Carrascosa JL (2007) Quasi-atomic model of bacteriophage T7 procapsid shell: insights into the structure and evolution of a basic fold. Structure 15(4):461–472

Baines JD (2007) Chapter 11. Envelopment of herpes simplex virus nucleocapsids at the inner membrane. In: Arvin A, Campadelli-Fiume G, Mocarski E et al (eds) Human herpesviruses: biology, therapy, and immunoprophylaxis. Cambridge University Press, Cambridge

Baker ML, Jiang W, Rixon FJ, Chiu W (2005) Common ancestry of herpesviruses and tailed DNA bacteriophages. J Virol 79(23):14967–14970

Baker ML, Jiang W, Wedemeyer WJ, Rixon FJ, Baker D, Chiu W (2006) Ab initio modeling of the herpesvirus VP26 core domain assessed by CryoEM density. PLoS Comput Biol 2(10):e146

Bamford DH, Grimes JM, Stuart DI (2005) What does structure tell us about virus evolution? Curr Opin Struct Biol 15(6):655–663

Benson SD, Bamford JKH, Bamford DH, Burnett RM (1999) Viral evolution revealed by bacteriophage PRD1 and human adenovirus coat protein structures. Cell 98(6):825–833

Booy FP, Newcomb WW, Trus BL, Brown JC, Baker TS, Steven AC (1991) Liquid-crystalline, phage-like, packing of encapsidated DNA in herpes simplex virus. Cell 64:1007–1015

Booy FP, Trus BL, Davison AJ, Steven AC (1996) The capsid architecture of channel catfish virus, an evolutionarily distant herpesvirus, is largely conserved in the absence of discernible sequence homology with herpes simplex virus. Virology 215:134–141

Bowman BR, Baker ML, Rixon FJ, Chiu W, Quiocho FA (2003) Structure of the herpesvirus major capsid protein. EMBO J 22(4):757–765

Bowman BR, Welschhans RL, Jayaram H, Stow ND, Preston VG, Quiocho FA (2006) Structural characterization of the UL25 DNA-packaging protein from herpes simplex virus type 1. J Virol 80(5):2309–2317

Briggs JA, Riches JD, Glass B, Bartonova V, Zanetti G, Krausslich HG (2009) Structure and assembly of immature HIV. Proc Natl Acad Sci USA 106(27):11090–11095

Britt B (2007) Chapter 20. Maturation and egress. In: Arvin A, Campadelli-Fiume G, Mocarski E et al (eds) Human herpesviruses: biology, therapy and immunoprophylaxis. Cambridge University Press, Cambridge

Cardone G, Winkler DC, Trus BL, Cheng N, Heuser JE, Newcomb WW, Brown JC, Steven AC (2007) Visualization of the herpes simplex virus portal in situ by cryo-electron tomography. Virology 361(2):426–434

Cerritelli ME, Cheng N, Rosenberg AH, McPherson CE, Booy FP, Steven AC (1997) Encapsidated conformation of bacteriophage T7 DNA. Cell 91:271–280

Chang JT, Schmid MF, Rixon FJ, Chiu W (2007) Electron cryotomography reveals the portal in the herpesvirus capsid. J Virol 81(4):2065–2068

Chen DH, Jakana J, McNab D, Mitchell J, Zhou ZH, Dougherty M, Chiu W, Rixon FJ (2001) The pattern of tegument-capsid interaction in the herpes simplex virus type 1 virion is not influenced by the small hexon-associated protein VP26. J Virol 75(23):11863–11867

Chen DH, Baker ML, Hryc CF, DiMaio F, Jakana J, Wu W, Dougherty M, Haase-Pettingell C, Schmid MF, Jiang W, Baker D, King JA, Chiu W (2011) Structural basis for scaffolding-mediated assembly and maturation of a dsDNA virus. Proc Natl Acad Sci USA 108(4):1355–1360

Cheng H, Shen N, Pei J, Grishin NV (2004) Double-stranded DNA bacteriophage prohead protease is homologous to herpesvirus protease. Protein Sci 13(8):2260–2269

Cockrell SK, Sanchez ME, Erazo A, Homa FL (2009) Role of the UL25 protein in herpes simplex virus DNA encapsidation. J Virol 83(1):47–57

Cockrell SK, Huffman JB, Toropova K, Conway JF, Homa FL (2011) Residues of the UL25 protein of herpes simplex virus that are required for its stable interaction with capsids. J Virol 85(10):4875–4887

Coller KE, Lee JI, Ueda A, Smith GA (2007) The capsid and tegument of the alphaherpesviruses are linked by an interaction between the UL25 and VP1/2 proteins. J Virol 81(21):11790–11797

Conway JF, Wikoff WR, Cheng N, Duda RL, Hendrix RW, Johnson JE, Steven AC (2001) Virus maturation involving large subunit rotations and local refolding. Science 292(5517):744–748

Conway JF, Cockrell SK, Copeland AM, Newcomb WW, Brown JC, Homa FL (2010) Labeling and localization of the herpes simplex virus capsid protein UL25 and its interaction with the two triplexes closest to the penton. J Mol Biol 397(2):575–586

Davison AJ, Trus BL, Cheng N, Steven AC, Watson MS, Cunningham C, Le Deuff RM, Renault T (2005) A novel class of herpesvirus with bivalve hosts. J Gen Virol 86(Pt 1):41–53

Deng B, O'Connor CM, Kedes DH, Zhou ZH (2007) Direct visualization of the putative portal in the Kaposi's sarcoma-associated herpesvirus capsid by cryoelectron tomography. J Virol 81(7):3640–3644

Dokland T (1999) Scaffolding proteins and their role in viral assembly. Cell Mol Life Sci 56(7–8):580–603

Duda RL (1998) Protein chainmail: catenated protein in viral capsids. Cell 94(1):55–60

Duda RL, Hendrix RW, Huang WM, Conway JF (2006) Shared architecture of bacteriophage SPO1 and herpesvirus capsids. Curr Biol 16(1):R11–R13

Duda RL, Ross PD, Cheng N, Firek BA, Hendrix RW, Conway JF, Steven AC (2009) Structure and energetics of encapsidated DNA in bacteriophage HK97 studied by scanning calorimetry and cryo-electron microscopy. J Mol Biol 391(2):471–483

Fokine A, Chipman PR, Leiman PG, Mesyanzhinov VV, Rao VB, Rossmann MG (2004) Molecular architecture of the prolate head of bacteriophage T4. Proc Natl Acad Sci USA. 101(16):6003–6008

Friedmann A, Coward JE, Rosenkranz HS, Morgan C (1975) Electron microscopic studies on assembly of herpes simplex virus upon removal of hydroxyurea block. J Gen Virol 26:171–181

Gertsman I, Gan L, Guttman M, Lee K, Speir JA, Duda RL, Hendrix RW, Komives EA, Johnson JE (2009) An unexpected twist in viral capsid maturation. Nature 458(7238):646–650

Grünewald K, Desai P, Winkler DC, Heymann JB, Belnap DM, Baumeister W, Steven AC (2003) Three-dimensional structure of herpes simplex virus from cryo-electron tomography. Science 302(5649):1396–1398

Guo H, Shen S, Wang L, Deng H (2010) Role of tegument proteins in herpesvirus assembly and egress. Protein Cell 1(11):987–998

Hendrix RW (2003) Bacteriophage genomics. Curr Opin Microbiol 6(5):506–511

Henson BW, Johnson N, Bera A, Okoye ME, Desai KV, Desai PJ (2011) Expression of the HSV-1 capsid protein VP19C in *Escherichia coli*: a single amino acid change overcomes an expression block of the full-length polypeptide. Protein Expr Purif 77(1):80–85

Heymann JB, Cheng N, Newcomb WW, Trus BL, Brown JC, Steven AC (2003) Dynamics of herpes simplex virus capsid maturation visualized by time-lapse cryo-electron microscopy. Nat Struct Biol 10(5):334–341

Hoog SS, Smith WW, Qiu X, Janson CA, Hellmig B, McQueney MS, O'Donnell K, O'Shannessy D, DiLella AG, Debouck C, Abdel-Meguid SS (1997) Active site cavity of herpesvirus proteases revealed by the crystal structure of herpes simplex virus protease/inhibitor complex. Biochemistry 36(46):14023–14029

Huffman JB, Newcomb WW, Brown JC, Homa FL (2008) Amino acids 143 to 150 of the herpes simplex virus type 1 scaffold protein are required for the formation of portal-containing capsids. J Virol 82(13):6778–6781

Ionel A, Velazquez-Muriel JA, Luque D, Cuervo A, Caston JR, Valpuesta JM, Martin-Benito J, Carrascosa JL (2011) Molecular rearrangements involved in the capsid shell maturation of bacteriophage T7. J Biol Chem 286(1):234–242

Iwasaki K, Trus BL, Wingfield PT, Cheng N, Campusano G, Rao VB, Steven AC (2000) Molecular architecture of bacteriophage T4 capsid: vertex structure and bimodal binding of the stabilizing accessory protein, Soc. Virology. 271(2):321–333

Kirkitadze MD, Barlow PN, Price NC, Kelly SM, Boutell CJ, Rixon FJ, McClelland DA (1998) The herpes simplex virus triplex protein, VP23, exists as a molten globule. J Virol 72(12):10066–10072

Klupp BG, Granzow H, Keil GM, Mettenleiter TC (2006) The capsid-associated UL25 protein of the alphaherpesvirus pseudorabies virus is nonessential for cleavage and encapsidation of genomic DNA but is required for nuclear egress of capsids. J Virol 80(13):6235–6246

Krupovic M, Bamford DH (2008) Virus evolution: how far does the double beta-barrel viral lineage extend? Nat Rev Microbiol 6(12):941–948

Kuznetsov YG, Martiny JB, McPherson A (2010) Structural analysis of a Synechococcus myovirus S-CAM4 and infected cells by atomic force microscopy. J Gen Virol 91(Pt 12):3095–3104

Lebedev AA, Krause MH, Isidro AL, Vagin AA, Orlova EV, Turner J, Dodson EJ, Tavares P, Antson AA (2007) Structural framework for DNA translocation via the viral portal protein. EMBO J 26(7):1984–1994

Lurz R, Orlova EV, Gunther D, Dube P, Droge A, Weise F, van Heel M, Tavares P (2001) Structural organisation of the head-to-tail interface of a bacterial virus. J Mol Biol 310(5):1027–1037

Marintcheva B, Weller SK (2001) A tale of two HSV-1 helicases: roles of phage and animal virus helicases in DNA replication and recombination. Prog Nucleic Acid Res Mol Biol 70:77–118

Maurer UE, Sodeik B, Grünewald K (2008) Native 3D intermediates of membrane fusion in herpes simplex virus 1 entry. Proc Natl Acad Sci USA 105(30):10559–10564

McNab AR, Desai P, Person S, Roof LL, Thomsen DR, Newcomb WW, Brown JC, Homa FL (1998) The product of the herpes simplex virus type 1 UL25 gene is required for encapsidation but not for cleavage of replicated viral DNA. J Virol 72(2):1060–1070

Mettenleiter TC, Klupp BG, Granzow H (2009) Herpesvirus assembly: an update. Virus Res 143(2):222–234

Morais MC, Kanamaru S, Badasso MO, Koti JS, Owen BA, McMurray CT, Anderson DL, Rossmann MG (2003) Bacteriophage phi29 scaffolding protein gp7 before and after prohead assembly. Nat Struct Biol 10(7):572–576

Newcomb WW, Brown JC (2009) Time-dependent transformation of the herpesvirus tegument. J Virol 83(16):8082–8089

Newcomb WW, Brown JC (2010) Structure and capsid association of the herpesvirus large tegument protein UL36. J Virol 84(18):9408–9414

Newcomb WW, Homa FL, Thomsen DR, Booy FP, Trus BL, Steven AC, Spencer JV, Brown JC (1996) Assembly of the herpes simplex virus capsid: characterization of intermediates observed during cell-free capsid formation. J Mol Biol. 263(3):432–446

Newcomb WW, Trus BL, Cheng N, Steven AC, Sheaffer AK, Tenney DJ, Weller SK, Brown JC (2000) Isolation of herpes simplex virus procapsids from cells infected with a protease-deficient mutant virus. J Virol 74(4):1663–1673

Newcomb WW, Homa FL, Brown JC (2006) Herpes simplex virus capsid structure: DNA packaging protein UL25 is located on the external surface of the capsid near the vertices. J Virol 80(13):6286–6294

Ogasawara M, Suzutani T, Yoshida I, Azuma M (2001) Role of the UL25 gene product in packaging DNA into the herpes simplex virus capsid: location of UL25 product in the capsid and demonstration that it binds DNA. J Virol 75(3):1427–1436

Olia AS, Prevelige PE Jr, Johnson JE, Cingolani G (2011) Three-dimensional structure of a viral genome-delivery portal vertex. Nat Struct Mol Biol 18(5):597–603

Parker ML, Ralston EJ, Eiserling FA (1983) Bacteriophage SPO1 structure and morphogenesis. II. Head structure and DNA size. J Virol 46(1):250–259

Pasdeloup D, Blondel D, Isidro AL, Rixon FJ (2009) Herpesvirus capsid association with the nuclear pore complex and viral DNA release involve the nucleoporin CAN/Nup214 and the capsid protein pUL25. J Virol 83(13):6610–6623

Pelletier A, Do F, Brisebois JJ, Lagace L, Cordingley MG (1997) Self-association of herpes simplex virus type 1 ICP35 is via coiled-coil interactions and promotes stable interaction with the major capsid protein. J Virol 71:5197–5208

Preston VG, Murray J, Preston CM, McDougall IM, Stow ND (2008) The UL25 gene product of herpes simplex virus type 1 is involved in uncoating of the viral genome. J Virol 82(13):6654–6666

Qiu X, Culp JS, DiLella AG, Hellmig B, Hoog SS, Janson CA, Smith WW, Abdel-Meguid SS (1996) Unique fold and active site in cytomegalovirus protease. Nature 383:275–279

Qiu X, Janson CA, Culp JS, Richardson SB, Debouck C, Smith WW, Abdel-Meguid SS (1997) Crystal structure of varicella-zoster virus protease. Proc Natl Acad Sci USA 94(7):2874–2879

Radtke K, Kieneke D, Wolfstein A, Michael K, Steffen W, Scholz T, Karger A, Sodeik B (2010) Plus- and minus-end directed microtubule motors bind simultaneously to herpes simplex virus capsids using different inner tegument structures. PLoS Pathog 6(7):e1000991

Rochat RH, Liu X, Murata K, Nagayama K, Rixon FJ, Chiu W (2011) Seeing the portal in herpes simplex virus type 1 B capsids. J Virol 85(4):1871–1874

Scrivano L, Esterlechner J, Mühlbach H, Ettischer N, Hagen C, Grünewald K, Mohr CA, Ruzsics Z, Koszinowski U, Adler B (2010) The m74 gene product of murine cytomegalovirus (MCMV) is a functional homolog of human CMV gO and determines the entry pathway of MCMV. J Virol. 84(9):4469–4480

Salmon B, Baines JD (1998) Herpes simplex virus DNA cleavage and packaging: association of multiple forms of U(L)15-encoded proteins with B capsids requires at least the U(L)6, U(L)17, and U(L)28 genes. J Virol 72(4):3045–3050

Shanda SK, Wilson DW (2008) UL36p is required for efficient transport of membrane-associated herpes simplex virus type 1 along microtubules. J Virol 82(15):7388–7394

Sheaffer AK, Newcomb WW, Brown JC, Gao M, Weller SK, Tenney DJ (2000) Evidence for controlled incorporation of herpes simplex virus type 1 UL26 protease into capsids. J Virol 74(15):6838–6848

Simpson AA, Leiman PG, Tao Y, He Y, Badasso MO, Jardine PJ, Anderson DL, Rossmann MG (2001) Structure determination of the head-tail connector of bacteriophage phi29. Acta Crystallogr D Biol Crystallogr 57(Pt 9):1260–1269

Steven AC, Heymann JB, Cheng N, Trus BL, Conway JF (2005) Virus maturation: dynamics and mechanism of a stabilizing structural transition that leads to infectivity. Curr Opin Struct Biol 15(2):227–236

Stow ND (2001) Packaging of genomic and amplicon DNA by the herpes simplex virus type 1 UL25-null mutant KUL25NS. J Virol 75(22):10755–10765

Szilagyi JF, Berriman J (1994) Herpes simplex virus L particles contain spherical membrane-enclosed inclusion vesicles. J Gen Virol 75(Pt 7):1749–1753

Thurlow JK, Murphy M, Stow ND, Preston VG (2006) Herpes simplex virus type 1 DNA-packaging protein UL17 is required for efficient binding of UL25 to capsids. J Virol 80(5):2118–2126

Tong L (2002) Viral proteases. Chem Rev 102(12):4609–4626

Trus BL, Newcomb WW, Booy FP, Brown JC, Steven AC (1992) Distinct monoclonal antibodies separately label the hexons or the pentons of herpes simplex virus capsid. Proc Natl Acad Sci USA 89:11508–11512

Trus BL, Homa FL, Booy FP, Newcomb WW, Thomsen DR, Cheng N, Brown JC, Steven AC (1995) Herpes simplex virus capsids assembled in insect cells infected with recombinant baculoviruses: structural authenticity and localization of VP26. J Virol 69:7362–7366

Trus BL, Booy FP, Newcomb WW, Brown JC, Homa FL, Thomsen DR, Steven AC (1996) The herpes simplex virus procapsid: structure, conformational changes upon maturation, and roles of the triplex proteins VP19c and VP23 in assembly. J Mol Biol. 263(3):447–462

Trus BL, Gibson W, Cheng N, Steven AC (1999) Capsid structure of simian cytomegalovirus from cryoelectron microscopy: evidence for tegument attachment sites [published erratum appears in J Virol 1999 May;73(5):4530]. J Virol 73(3):2181–2192

Trus BL, Heymann JB, Nealon K, Cheng N, Newcomb WW, Brown JC, Kedes DH, Steven AC (2001) Capsid structure of Kaposi's sarcoma-associated herpesvirus, a gammaherpesvirus, compared to an alphaherpesvirus, HSV1, and a betaherpesvirus, CMV. J Virol 75(6):2879–2890

Trus BL, Cheng N, Newcomb WW, Homa FL, Brown JC, Steven AC (2004) Structure and polymorphism of the UL6 portal protein of herpes simplex virus type 1. J Virol 78(22):12668–12671

Trus BL, Newcomb WW, Cheng N, Cardone G, Marekov L, Homa FL, Brown JC, Steven AC (2007) Allosteric signaling and a nuclear exit strategy: binding of UL25/UL17 heterodimers to DNA-filled HSV-1 capsids. Mol Cell 26(4):479–489

Weigele PR, Pope WH, Pedulla ML, Houtz JM, Smith AL, Conway JF, King J, Hatfull GF, Lawrence JG, Hendrix RW (2007) Genomic and structural analysis of Syn9, a cyanophage infecting marine *Prochlorococcus* and *Synechococcus*. Environ Microbiol 9(7):1675–1695

Welch AR, McNally LM, Hall MRT, Gibson W (1993) Herpesvirus proteinase: site-directed mutagenesis used to study maturational, release, and inactivation cleavage sites of precursor and to identify a possible catalytic site serine and histidine. J Virol 67:7360–7372

Wikoff WR, Liljas L, Duda RL, Tsuruta H, Hendrix RW, Johnson JE (2000) Topologically linked protein rings in the bacteriophage HK97 capsid. Science 289(5487):2129–2133

Wingfield PT, Stahl SJ, Thomsen DR, Homa FL, Booy FP, Trus BL, Steven AC (1997) Hexon-only binding of VP26 reflects differences between the hexon and penton conformations of VP5, the major capsid protein of herpes simplex virus. J Virol 71:8955–8961

Wolfstein A, Nagel CH, Radtke K, Dohner K, Allan VJ, Sodeik B (2006) The inner tegument promotes herpes simplex virus capsid motility along microtubules in vitro. Traffic 7(2):227–237

Yang F, Forrer P, Dauter Z, Conway JF, Cheng N, Cerritelli ME, Steven AC, Plückthun A, Wlodawer A (2000) Novel Fold and Capsid-binding Properties of the λ Phage Display Platform Protein gpD. Nat. Struct. Biol. 7:230–237

Yang K, Baines JD (2008) Domain within herpes simplex virus 1 scaffold proteins required for interaction with portal protein in infected cells and incorporation of the portal vertex into capsids. J Virol 82(10):5021–5030

Zhou ZH, He J, Jakana J, Tatman JD, Rixon FJ, Chiu W (1995) Assembly of VP26 in herpes simplex virus-1 inferred from structures of wild-type and recombinant capsids. Nat Struct Biol 2:1026–1030

Zhou ZH, Dougherty M, Jakana J, He J, Rixon FJ, Chiu W (2000) Seeing the herpesvirus capsid at 8.5 Å. Science 288(5467):877–880

Chapter 20
Assembly and Architecture of HIV

Barbie K. Ganser-Pornillos, Mark Yeager, and Owen Pornillos

Abstract HIV forms spherical, membrane-enveloped, pleomorphic virions, 1,000–1,500 Å in diameter, which contain two copies of its single-stranded, positive-sense RNA genome. Virus particles initially bud from host cells in a noninfectious or immature form, in which the genome is further encapsulated inside a spherical protein shell composed of around 2,500 copies of the virally encoded Gag polyprotein. The Gag molecules are radially arranged, adherent to the inner leaflet of the viral membrane, and closely associated as a hexagonal, paracrystalline lattice. Gag comprises three major structural domains called MA, CA, and NC. For immature virions to become infectious, they must undergo a maturation process that is initiated by proteolytic processing of Gag by the viral protease. The new Gag-derived proteins undergo dramatic rearrangements to form the mature virus. The mature MA protein forms a "matrix" layer and remains attached to the viral envelope, NC condenses with the genome, and approximately 1,500 copies of CA assemble into a new cone-shaped protein shell, called the mature capsid, which surrounds the genomic ribonucleoprotein complex. The HIV capsid conforms to the mathematical principles of a fullerene shell, in which the CA subunits form about 250 CA hexamers arrayed on a variably curved hexagonal lattice, which is closed by incorporation of exactly 12 pentamers, seven pentamers at the wide end and five at the narrow end of the cone. This chapter describes our current understanding of HIV's virion architecture and its dynamic transformations: the process of virion assembly as orchestrated by Gag, the architecture of the immature virion, the virus maturation process, and the structure of the mature capsid.

Abbreviated Virus Names

HIV-1	Human immunodeficiency virus type 1
HTLV	Human T-cell leukemia virus
MLV	Murine leukemia virus
MPMV	Mason-Pfizer monkey virus
RSV	Rous sarcoma virus

B.K. Ganser-Pornillos (✉) • M. Yeager (✉) • O. Pornillos (✉)
Department of Molecular Physiology and Biological Physics, University of Virginia School of Medicine, P.O. Box 800736, Charlottesville, VA 22908, USA

Department of Cell Biology, The Scripps Research Institute, 10550 North Torrey Pines Road, La Jolla, CA 92037, USA
e-mail: bpornillos@virginia.edu; yeager@virginia.edu; opornillos@virginia.edu

M.G. Rossmann and V.B. Rao (eds.), *Viral Molecular Machines*,
Advances in Experimental Medicine and Biology 726, DOI 10.1007/978-1-4614-0980-9_20,

20.1 Introduction

The *Orthoretrovirinae* (orthoretroviruses) comprise the largest subfamily of *Retroviridae* and are subdivided into six genera (alpha, beta, gamma, delta, epsilon, and lentivirus). Orthoretroviruses are spherical membrane-enveloped viruses that bud from the plasma membrane of infected host cells (Fig. 20.1). The emergent virion is noninfectious, and is characterized by a protein shell located immediately beneath the envelope, which is composed of precursor Gag proteins bound to the inner bilayer leaflet of the viral membrane. During maturation, Gag is cleaved by the viral protease into three new proteins, called MA, CA, and NC. This disassembles the immature Gag shell and induces structural rearrangements that culminate in formation of the mature, infectious virion. The new MA proteins remain associated with the viral membrane, forming the matrix layer; a subset of the CA proteins assembles into a smaller, mature capsid; and NC condenses with the viral RNA and its associated enzymes into a compact ribonucleoprotein complex within the new capsid. The mature capsid and its contents are called the core.

To initiate replication, the core is introduced into the cytoplasm of a new host cell after receptor binding and fusion of the viral and cellular membranes. The capsid then undergoes a well-ordered but poorly understood uncoating process that culminates in a large enzymatic complex (the "preintegration complex"), which reverse transcribes the RNA genome into a double-stranded DNA copy called the provirus, and then integrates the provirus into the host's chromosomes. Once integrated, the provirus is transcribed by cellular machinery into various spliced and unspliced mRNA transcripts, which are then translated into the structural and enzymatic viral proteins required to maintain infection and propagate new virions.

Orthoretroviruses include important human pathogens (e.g., HTLV-1 and HTLV-2, which cause T-cell leukemias). The most significant of these has been HIV-1, which causes exceptionally high mortality rates through AIDS. The number of people living with HIV is currently estimated to be in excess of 33 million, with 2.7 million new infections and 2 million AIDS-related deaths in 2008

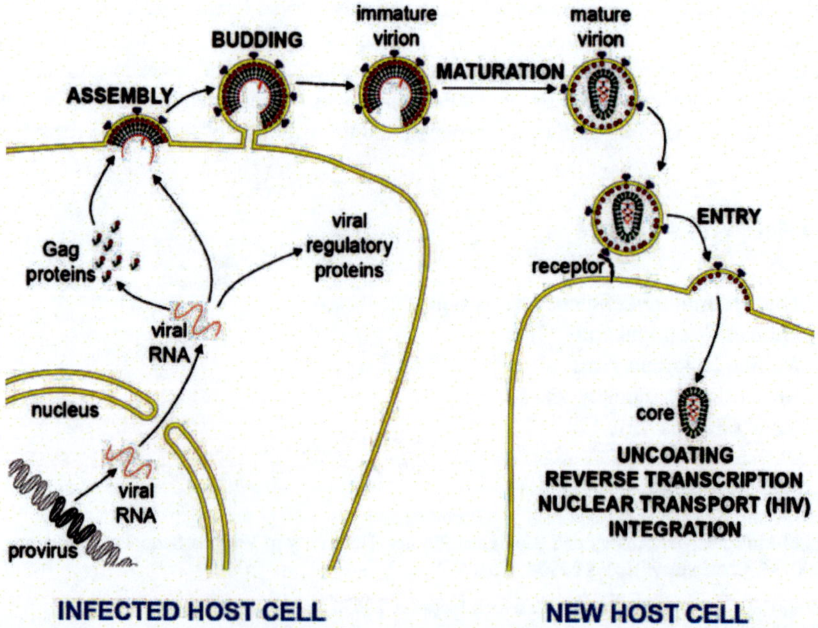

Fig. 20.1 Schematic of the HIV-1 replication cycle, emphasizing the stages discussed in this chapter. See text for details. Receptor-mediated virus entry is discussed in Chap. 10. (Adapted from Ganser-Pornillos et al. (2008), with permission from Elsevier)

(UNAIDS 2009). HIV-1 is, therefore, one of the most studied viruses and is discussed here as a model system for other orthoretroviruses. It is important to note, however, that valuable assembly principles have been discovered through studies of other orthoretroviruses (e.g., RSV, MLV). Furthermore, although the general principles of orthoretroviral structure and assembly are conserved, virus-specific variations also exist. The virus-specific features presumably reflect unique selective pressures or adaptations between a particular virus and its host.

20.2 Assembly and Architecture of the Immature Virion

Assembly of the immature virion is orchestrated by the precursor Gag protein (Fig. 20.2a). In fact, all the information necessary for assembly of a retrovirus is encoded in Gag. For instance, recombinant expression of Gag in the absence of all other viral proteins results in efficient release of immature particles (Gheysen et al. 1989). In addition, pure recombinant Gag proteins can assemble in vitro into spherical shells that resemble immature virions (Campbell and Vogt 1997; Campbell and Rein 1999; Gross et al. 2000; Campbell et al. 2001).

Gag is organized as a series of functional "modules" that work in concert to: (1) target Gag molecules to assembly sites at the plasma membrane, (2) create a membrane-bound protein shell composed of Gag hexamers, (3) specifically recognize and encapsulate the dimeric viral genome, (4) recruit and incorporate all other virus components into the assembling particle, and (5) conscript the

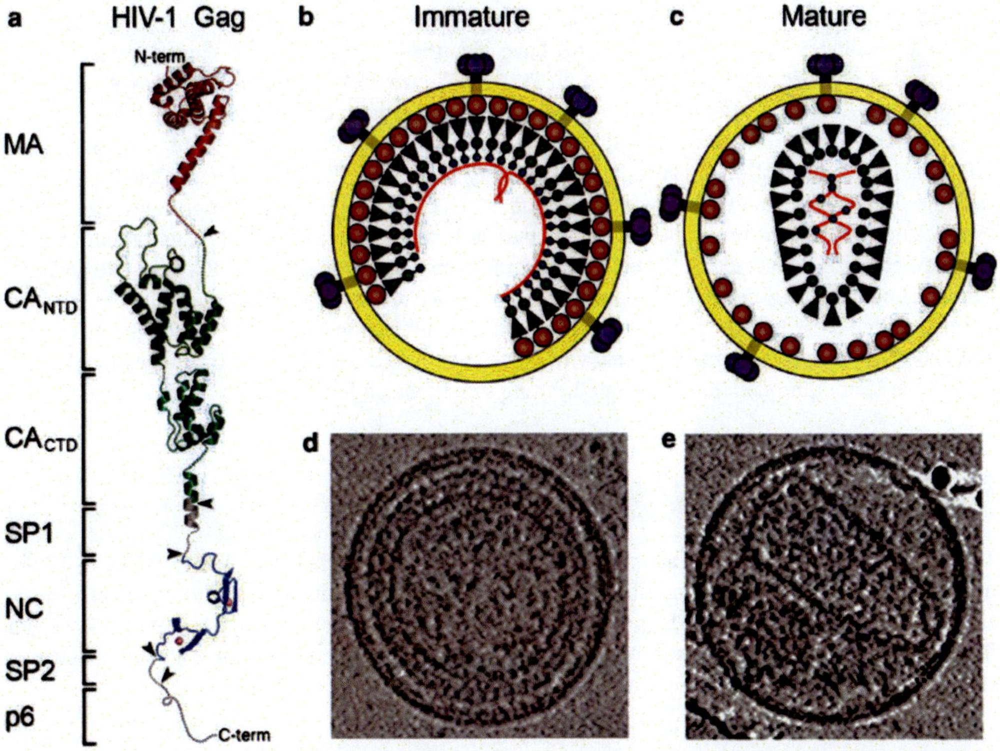

Fig. 20.2 Organization of the immature and mature HIV-1 virions. **(a)** Schematic tertiary structural model of full-length HIV-1 Gag. Individual domains are in different colors and are labeled on the *left*. This color scheme is maintained throughout the chapter. **(b)** Schematic model of the immature virion. **(c)** Schematic model of the mature virion. Images of **(d)** immature and **(e)** mature virions preserved in vitreous ice (Reprinted from Ganser-Pornillos et al. (2008), with permission from Elsevier)

required cellular factors and chaperones, such as the ESCRT budding machinery. The HIV-1 Gag protein is composed of four independently folded structural domains that are demarcated by flexible regions (Fig. 20.2a). The N-terminal domain corresponds to the mature MA protein, which is cotranslationally modified with a fatty acid chain (myristic acid) and facilitates Gag binding to the viral envelope. The MA domain (colored red in Fig. 20.2a) targets Gag to assembly sites at the plasma membrane and facilitates incorporation of the envelope glycoproteins into the assembling virion. The two central domains of Gag correspond to the mature CA protein, which is formed by independently folded N-terminal and C-terminal domains, which we designate as CA^{NTD} and CA^{CTD} (colored green and cyan in Fig. 20.2a). These domains mediate the principal Gag–Gag contacts in both the immature Gag shell and the mature capsid. The fourth domain corresponds to the mature NC protein and, in HIV-1, is composed of two zinc fingers (colored blue in Fig. 20.2a) that bind the RNA genome. High-affinity NC–RNA interactions enable specific recognition and packaging of two copies of the genome, whereas low-affinity interactions promote Gag assembly. HIV-1 Gag contains an additional region at its C terminus, called p6, which is specific to lentiviruses. The p6 sequence lacks a defined tertiary structure, but contains two peptide motifs (called "late domains") that function as docking sites for the cellular ESCRT machinery, which facilitates the release of newly assembled virions from the cell surface. (Other orthoretroviral Gag proteins harbor late domains with equivalent functions, but at different positions in the Gag primary sequence.) HIV-1 Gag also contains two spacer peptides, called SP1 and SP2, which respectively demarcate the CA/NC and NC/p6 domains. SP1 is a critical maturation switch, and equivalent regions are found in other orthoretroviruses.

Retrovirus assembly appears to be initiated by just a few Gag molecules that are specifically bound to two copies of the RNA genome (Jouvenet et al. 2009; Miyazaki et al. 2010). The molecular mechanism by which HIV-1 Gag specifically recognizes its genome is not yet fully elucidated, but studies of Moloney murine leukemia virus have identified an RNA switch that couples genomic dimerization and packaging (D'Souza et al. 2001; D'Souza and Summers 2004; Dey et al. 2005; Gherghe et al. 2010). Conserved sequence elements in the genomic packaging signal (called Ψ) are sequestered by intramolecular base pairing in the monomeric RNA and then become exposed upon dimerization, thereby allowing specific, high-affinity binding of the RNA dimer to the NC domain (D'Souza and Summers 2004). Upon nucleation, stochastic addition of Gag molecules extends the assembling lattice and curves the membrane outward, culminating in an immature particle whose envelope remains connected to the cellular plasma membrane by a narrow stalk. Live-cell imaging studies show that Gag proteins accumulate fairly rapidly at the membrane assembly site, and the immature shell is essentially complete about 10 min after nucleation (Jouvenet et al. 2008; Ivanchenko et al. 2009). However, the newly formed virion does not bud for another 15 min, suggesting that fission of the membrane stalk may be a rate-limiting step in virion formation (Ivanchenko et al. 2009).

20.2.1 Functions of the Gag MA Domain

The MA domain of HIV-1 Gag adopts a helical tertiary fold (Massiah et al. 1994; Hill et al. 1996), harbors the myristyl fatty acid modification at its N terminus (Göttlinger et al. 1989), directly mediates contacts between Gag and the plasma membrane (Spearman et al. 1994; Zhou et al. 1994), and facilitates incorporation of the envelope proteins into the assembling virion (Yu et al. 1992; Dorfman et al. 1994; Freed and Martin 1995, 1996) (Fig. 20.3).

20.2.1.1 Membrane Binding and Targeting

Low-angle X-ray solution scattering studies indicate that newly translated HIV-1 Gag proteins adopt a "folded-over" conformation, wherein the MA and NC domains are in proximity (Datta et al. 2007a) (Fig. 20.3a). In this soluble conformation, the N-terminal myristyl group (colored green in Fig. 20.3a)

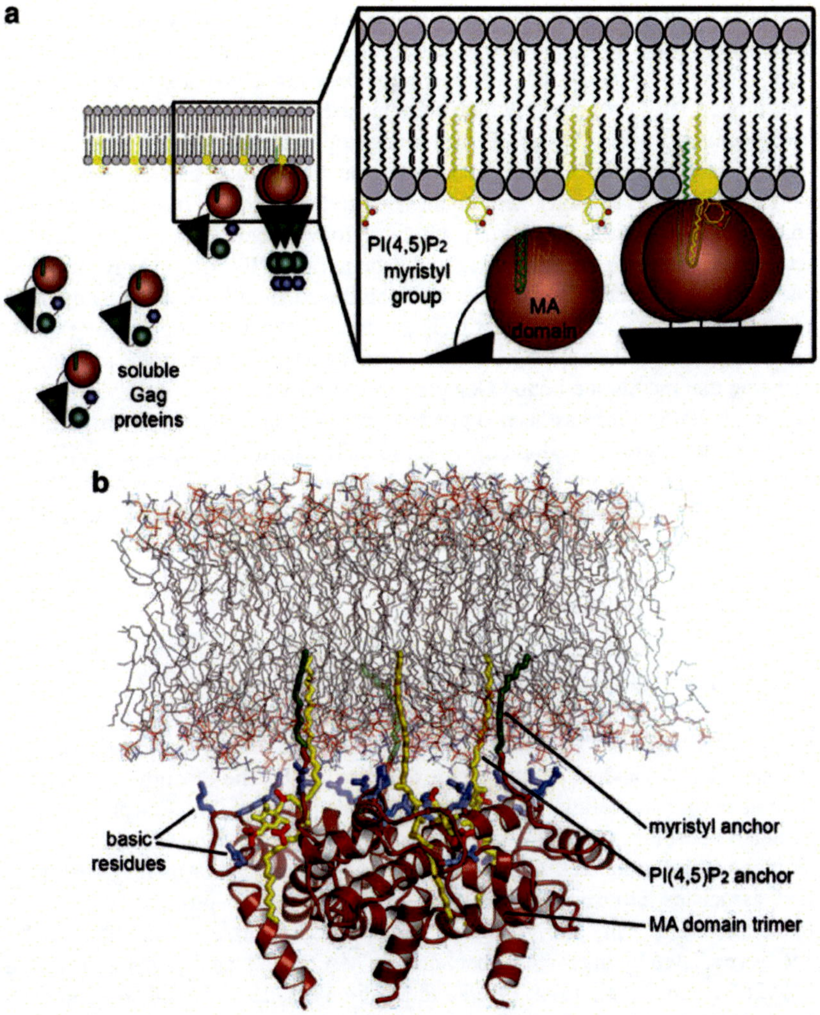

Fig. 20.3 The MA domain functions in binding and targeting of Gag to the plasma membrane. (**a**) Schematic showing soluble Gag proteins in a "folded-over" conformation, and membrane-bound assembling Gag molecules in a "beads-on-a-string" configuration. The boxed region illustrates the "myristyl switch" mechanism of membrane binding. (**b**) Structural model of the MA trimer bound to the lipid bilayer. [(**b**) Reprinted from Ganser-Pornillos et al. (2008), with permission from Elsevier]

is sequestered in a hydrophobic pocket in MA (Tang et al. 2004), which would presumably disfavor membrane binding. Stable attachment to the membrane requires three synergistic elements: (1) exposure of the myristyl group, (2) binding of the MA domain through nonspecific electrostatic interactions with negatively charged phospholipid headgroups and specific interactions with phosphatidylinositol-4,5-bisphosphate [PI(4,5)P_2], and (3) Gag oligomerization, wherein Gag adopts an extended conformation and begins to assemble into the immature shell (Zhou et al. 1994; Spearman et al. 1997; Paillart and Göttlinger 1999; Ono et al. 2000, 2004; Tang et al. 2004; Dalton et al. 2005, 2007; Saad et al. 2006; Datta et al. 2007b; Li et al. 2007). PI(4,5)P_2 (colored yellow in Fig. 20.3a) is enriched in the plasma membrane, and depletion of the cellular pools of this phosphoinositide incorrectly directs HIV-1 Gag to internal membranes (Ono et al. 2004). Thus, PI(4,5)P_2 binding simultaneously promotes stable membrane binding and proper targeting of Gag molecules to assembly sites at the plasma membrane.

The molecular mechanism by which PI(4,5)P_2 binding and myristyl exposure are coupled is now understood in detail (Saad et al. 2006) (Fig. 20.3). PI(4,5)P_2 belongs to a large class of membrane bilayer lipids (phosphatidylinositols), which are composed of a glycerol backbone, one saturated and one unsaturated fatty acid chain (which are attached to the 1′ and 2′ positions of the glycerol backbone, respectively), and a phosphoinositol headgroup (attached to the 3′ glycerol position). The phosphoinositol headgroup binds to a basic patch on the surface of MA and confers binding specificity. The 2′ unsaturated fatty acid chain is extruded from the lipid bilayer and binds in a hydrophobic groove adjacent to the myristyl binding groove. The 1′ saturated fatty acid chain does not interact with MA, but rather anchors Gag to the membrane. PI(4,5)P_2 and myristyl binding to the globular MA domain are mutually exclusive: upon binding, PI(4,5)P_2 repositions hydrophobic residues in the myristyl binding pocket, triggering exposure of the myristyl group. The exposed myristyl fatty acid chain then inserts into the lipid bilayer, forming a second membrane anchor.

It is intriguing that membrane-bound Gag proteins are anchored by two fully saturated fatty acid chains (Saad et al. 2006). This has been suggested as a potential mechanism for partitioning membrane-bound Gag into lipid microdomains (e.g., so-called lipid rafts) because rafts preferentially interact with saturated fatty acid chains (Brown and London 1997; Zacharias et al. 2002). Live-cell fluorescence imaging studies support this idea and show that Gag assembles at punctate sites at the plasma membrane (Hermida-Matsumoto and Resh 2000; Jouvenet et al. 2008; Ivanchenko et al. 2009), which appear to correspond to lipid microdomains (Aloia et al. 1993; Nguyen and Hildreth 2000; Ono and Freed 2001; Brügger et al. 2006).

20.2.1.2 Recruitment and Incorporation of Envelope Proteins

The exterior of HIV-1 virions is studded with trimeric envelope (Env) glycoprotein spikes composed of an ectodomain (gp120) and a transmembrane domain (gp41). The Env proteins mediate receptor-mediated fusion of viral and cellular membranes to initiate infection. The Env proteins are translated by ER-associated ribosomes as a single precursor polypeptide, called gp160, which is transported to the plasma membrane via the cellular secretory pathway. During transport, gp160 is extensively glycosylated, associates into trimers, and is cleaved by the cellular protease furin into the mature subunits (Swanstrom and Wills 1997). Incorporation of the Env proteins into the assembling virion is thought to be mediated by interactions between the MA domain and the cytoplasmic tail of gp41 (Yu et al. 1992; Dorfman et al. 1994; Freed and Martin 1995, 1996; Murakami and Freed 2000a), which may occur through a bridging cellular protein (Lopez-Vergès et al. 2006). Passive incorporation of Env may also occur through simple mass action (Murakami and Freed 2000b).

20.2.2 *Gag–Gag Lattice Interactions*

The immature virion contains around 2,500 copies of the Gag protein (Carlson et al. 2008), which assemble into a spherical shell immediately beneath the viral envelope. The assembled Gag molecules adopt an extended "beads on a string" configuration, wherein the sequential domains are arranged with the Gag N terminus on the outermost edge of the shell, and the C terminus points toward the center of the virion (Fuller et al. 1997; Yeager et al. 1998) (Fig. 20.2b). As discussed in the above section, the N terminus of Gag is tethered to the membrane by the MA domain. Although MA by itself can form trimers (Hill et al. 1996; Tang et al. 2004) (Fig. 20.3b), it does not form an extended lattice in the immature virion (Wright et al. 2007; Carlson et al. 2008; Briggs et al. 2009; de Marco et al. 2010; Keller et al. 2011) and is dispensable for particle assembly (Wang et al. 1993; Reil et al. 1998; Gross et al. 2000; Briggs et al. 2009). The C-terminal NC domain is also dispensable, provided it is replaced by a heterologous domain that is capable of self-association (e.g., leucine zipper motifs)

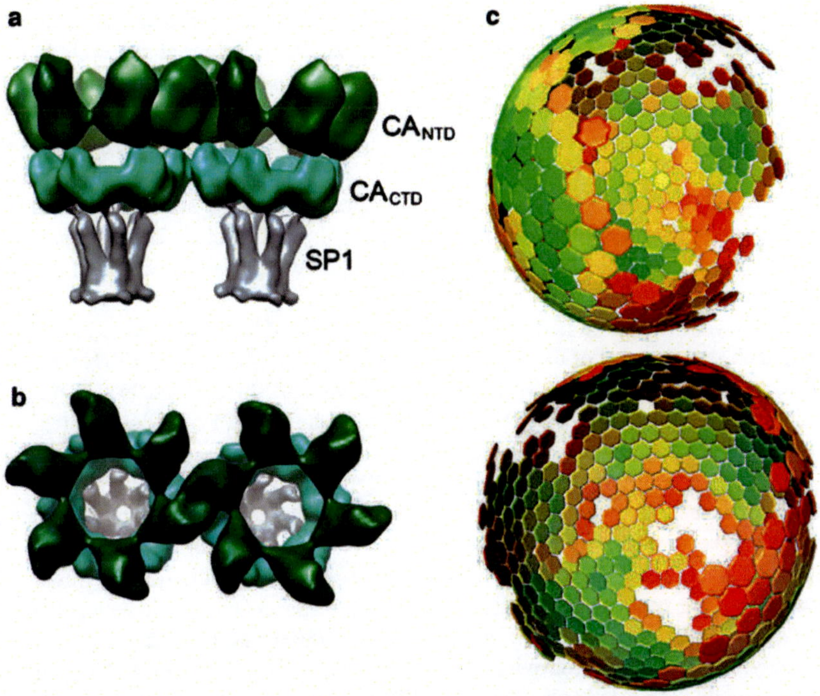

Fig. 20.4 The immature Gag lattice. (**a**) Side view and (**b**) top view of a low-resolution model of two interacting Gag hexamers (Wright et al. 2007). The CA^{NTD}, CA^{CTD}, and SP1 layers are colored *green*, *cyan*, and *gray*, respectively. (Model used to generate images courtesy of Elizabeth Wright.) (**c**) Global map of the immature HIV-1 lattice (Briggs et al. 2009). The Gag hexamers are represented by hexagons and colored according to symmetry cross-correlation on a scale from low (*red*) to high (*green*). (Lattice map images courtesy of John Briggs)

(Zhang and Barklis 1997). NC itself does not appear to form direct lattice-stabilizing contacts, but contributes to particle formation by tethering Gag to the RNA genome (via nonspecific contacts, as opposed to specific binding to the dimeric Ψ packaging element), thereby increasing the effective concentration of the assembling Gag molecules (Campbell and Vogt 1997; Gross et al. 2000).

The principal sites of lateral Gag–Gag interactions are located in the CA^{NTD} and CA^{CTD} domains, as well as the SP1 linker. These structural elements facilitate Gag assembly by forming a paracrystalline lattice of close-packed hexamers (Fig. 20.4).

The CA^{NTD} is an arrowhead-shaped domain composed of seven α-helices (numbered 1–7) (Gitti et al. 1996; Tang et al. 2002; Kelly et al. 2006) and directly contributes to Gag assembly by forming hexameric rings (Figs. 20.2a and 20.4). In reconstructed cryotomograms of immature particles, the CA^{NTD} hexamers are about 80 Å in diameter and interact across the threefold and/or twofold axes of the hexagonal lattice (Wright et al. 2007; Briggs et al. 2009; de Marco et al. 2010; Keller et al. 2011) (Fig. 20.4a, b, green). The current tomograms are not of sufficient resolution to unambiguously define domain orientations. However, biochemical analyses have identified two surfaces of CA^{NTD} that are important for Gag assembly: one encompassing helices 1 and 2, and a second near helices 4 and 7 (von Schwedler et al. 2003a; Lanman et al. 2004; Mortuza et al. 2008; Monroe et al. 2010). One suggested model is that the sixfold symmetric interactions are mediated by the first region, and interhexamer contacts are mediated by the second (Wright et al. 2007).

The CA^{CTD} is a globular domain composed of a short 3_{10} helix, a highly conserved sheet-turn-helix element termed the major homology region (MHR), and four α-helices (numbered 8–11) (Gamble et al. 1997; Worthylake et al. 1999) (Fig. 20.2a). Cryotomograms indicate that in the immature lattice, CA^{CTD} forms a second layer of hexamers directly below and with a similar (if slightly smaller) spacing as the CA^{NTD} layer (Wright et al. 2007; Briggs et al. 2009; de Marco et al. 2010) (Fig. 20.4a, b, cyan). The CA^{CTD} hexamers interact across the twofold symmetry axes of the hexagonal lattice, and this is consistent with biochemical data showing that the isolated CA^{CTD} is a dimer in solution (Gamble et al. 1997; Ivanov et al. 2007; Byeon et al. 2009). High-resolution X-ray and NMR structures of the isolated CA^{CTD} dimer have identified at least two dimerization modes (side-by-side and domain-swapped, with at least three possible side-by-side configurations) (Gamble et al. 1997; Worthylake et al. 1999; Ivanov et al. 2007; Bartonova et al. 2008; Byeon et al. 2009). However, the current cryotomograms cannot unambiguously discriminate which of these structures may exist in immature virions.

Cryotomograms of immature particles reveal a third layer of hexagonal order beneath the CA^{CTD} hexamers, which is composed of pillars of density that extend toward the NC/RNA region (Wright et al. 2007; Carlson et al. 2008; Briggs et al. 2009; de Marco et al. 2010; Keller et al. 2011) (Fig. 20.4a, b, gray). The pillars are interpreted as ordered residues at the CA/SP1 boundary, which is consistent with the finding that this region is essential for assembly of immature HIV-1 particles (Kräusslich et al. 1995; Accola et al. 1998). Residues at the CA/SP1 boundary have been previously predicted to adopt an α-helical secondary structure (Fig. 20.2a) to mediate quaternary Gag–Gag contacts in the immature shell (Accola et al. 1998; Worthylake et al. 1999; Liang et al. 2002; Newman et al. 2004; Morellet et al. 2005). Although the quaternary arrangement of the putative CA/SP1 helices is not yet firmly established, one proposal is that this region forms a sixfold symmetric helical bundle beneath the CA^{CTD} hexamer (Wright et al. 2007). Analogous regions have been identified in biochemical and structural analyses of other orthoretroviruses (Cheslock et al. 2003; Keller et al. 2008; de Marco et al. 2010), suggesting that structural roles of the CA/SP1 boundary residues are conserved.

Although our understanding of the architecture of the immature Gag shell is at an early stage, it is apparent that inherent curvature of the hexagonal Gag lattice likely arises from a systematic decrease in hexamer-to-hexamer packing distances across the sequential CA^{NTD}, CA^{CTD}, and SP1 (or equivalent) layers (Wright et al. 2007; Briggs et al. 2009) (Fig. 20.4a). Spherical hexagonal lattices cannot sustain long-range hexagonal order and must incorporate defects in order to maintain similar quaternary contacts across the different subunits. Quasi-equivalent icosahedral viruses and mature orthoretroviral capsids incorporate defects with local fivefold rotational symmetry (typically occupied by pentamers). In a hexagonal lattice, 12 such defects simultaneously accommodate lattice curvature and mediate shell closure. In contrast, the immature Gag lattice does not appear to contain pentamers and instead employs irregular defects of various geometries (Briggs et al. 2009) (Fig. 20.4c). Furthermore, the Gag shell is not completely closed, and surprisingly large regions (up to 50% of the virion) are devoid of ordered Gag molecules (Wright et al. 2007; Briggs et al. 2009; de Marco et al. 2010; Keller et al. 2011). It is not clear why orthoretroviruses do not form fully closed immature shells. One proposal is that Gag assembly is in kinetic competition with budding, and that membrane scission and virion release typically occur before complete Gag shells can be assembled (Carlson et al. 2008).

20.2.3 Budding

Virion release proceeds through the formation of a constriction in the connecting membrane, which produces both positive and negative curvatures in the lipid bilayer (Fig. 20.5). Membrane curvature during budding appears to not only be mediated by Gag but also by additional contributions from cellular proteins, and possibly, membrane lipids (Le Blanc et al. 2002; Wang et al. 2003; Carlson et al. 2008). Abscission of the connecting membrane and release of the nascent virion requires membrane fission, which is facilitated by cellular machinery that normally functions in the ESCRT budding pathway (Fig. 20.5).

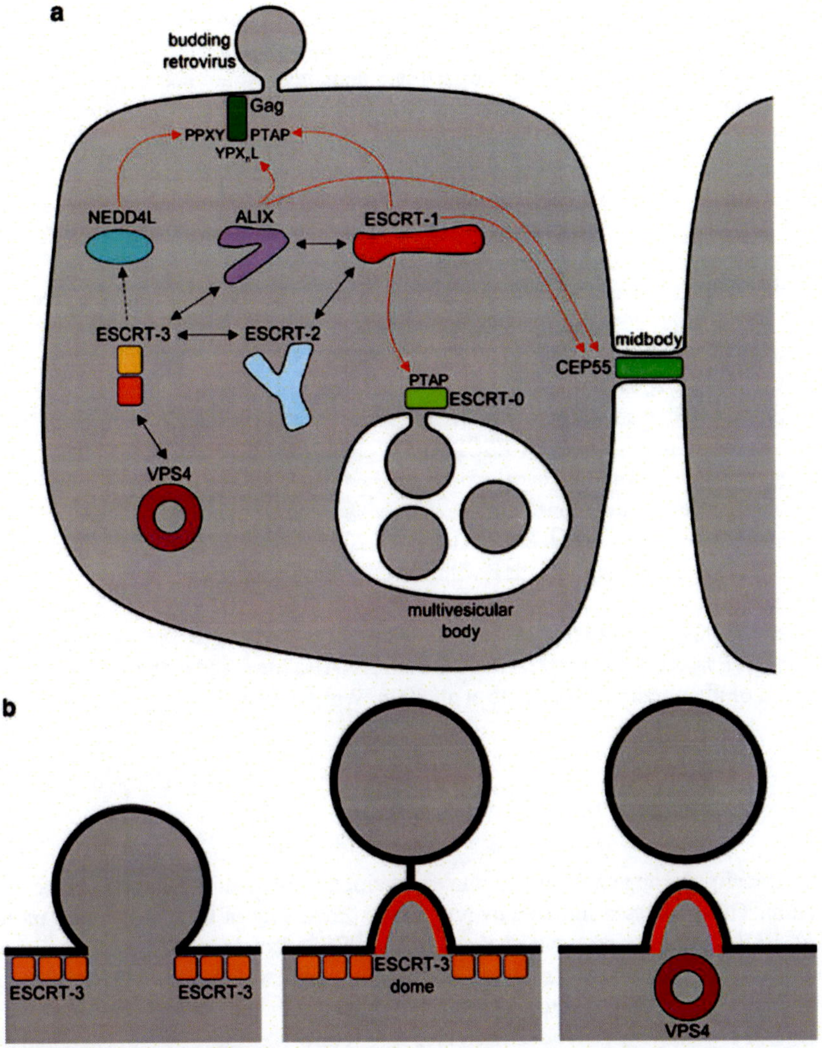

Fig. 20.5 Retrovirus budding. **(a)** Schematic representation of the cellular ESCRT pathway. Known interactions between ESCRT complexes and ESCRT-associated proteins are indicated by the *black double-headed arrows*. *Red arrows* connect viral late domains or native cellular recruitment factors with their corresponding complex. Recruitment factors (viral Gag protein, ESCRT-0, and CEP55) are colored in *green shades*. Although it is known that recruitment of NEDD4L to viral budding sites leads to the eventual recruitment of ESCRT-III, the precise mechanism by which this occurs is not known. **(b)** Schematic representation of the membrane fission mechanism catalyzed by ESCRT-III proteins and VPS4. (See text for details)

The ESCRT pathway comprises protein complexes (called ESCRT-0, ESCRT-I, ESCRT-II, ESCRT-III, and a dodecameric AAA ATPase called VPS4) that cycle between the cytoplasm and late endosomal membranes (Fig. 20.5a). On the surface of the endosome, ESCRT complexes have at least three functions: (1) sort protein cargos tagged with monoubiquitin into specific areas, (2) induce membrane invagination toward the lumen of the endosome and away from the cytoplasm (which results in formation of vesicles that are connected to the endosomal membrane, analogous to assembled immature virions), and (3) facilitate budding of the vesicles (Hurley and Emr 2006). During cell division, ESCRT complexes also facilitate abscission of the midbody, thereby separating mother and daughter cells (Carlton and Martin-Serrano 2007; Morita et al. 2007).

The mammalian ESCRT pathway is quite elaborate, and orthoretroviruses have evolved multiple short peptide motifs called "late domains" that mediate different entry points into the pathway (Fig. 20.5a). HIV-1 has two well-characterized late domains that are contained in the p6 region of Gag. These comprise the peptide sequences Pro-Thr/Ser-Ala-Pro (PTAP) and Tyr-Pro-X_n-Leu (YPXL), which respectively bind the proteins TSG101 (a component of the ESCRT-I complex) (Garrus et al. 2001; Martin-Serrano et al. 2001; VerPlank et al. 2001; Pornillos et al. 2002) or ALIX (which binds ESCRT-I and ESCRT-III proteins) (Strack et al. 2003; von Schwedler et al. 2003b; Zhai et al. 2008). HIV-1 and other orthoretroviruses also recruit members of the NEDD4 family of ubiquitin ligases, either via Pro-Pro-X-Tyr (PPXY) late domains (Kikonyogo et al. 2001; Yasuda et al. 2002; Bouamr et al. 2003) or via alternative interactions that have yet to be defined precisely (Chung et al. 2008; Usami et al. 2008). These late domain interactions lead to the eventual recruitment and conscription of the core fission machinery [composed of the ESCRT-III complex and the VPS4 enzyme (Wollert et al. 2009)], which then facilitates release of new virions from the cell surface.

One proposed mechanism of membrane fission at the neck of the budding virion has been inferred from the observation that a subset of ESCRT-III proteins (also called CHMPs) self-assemble into tubular structures with dome-like caps, whose external surfaces are lined with positively charged residues and, therefore, have considerable affinity for acidic phospholipid headgroups (Ghazi-Tabatabai et al. 2008; Hanson et al. 2008; Lata et al. 2008; Bajorek et al. 2009). The basic model is that assembly of the dome-like structure within the budding neck creates large elastic stress in the lipid bilayer, which drives membrane fission and separation of the viral and cellular membranes (Fabrikant et al. 2009) (Fig. 20.5b). Following release of the virion, the VPS4 enzyme uses energy derived from ATP hydrolysis to disassemble the ESCRT-III dome, and thereby resets the system for another round of membrane fission (Lata et al. 2008; Wollert et al. 2009).

20.3 Virion Maturation

Maturation transforms the noninfectious immature virion into a particle that is capable of transmitting infection. This process is initiated by proteolytic processing of Gag by the viral protease (PR) (Fig. 20.6). HIV-1 PR is encoded within the Gag–Pol fusion protein, which arises from a ribosomal frameshift during translation at the 3′ end of the *gag* gene (which occurs 5% of the time). The Gag–Pol proteins are packaged into virions through the immature lattice interactions mediated by the Gag region. The enzyme active site is created by PR dimerization, and the functional protease is liberated by autoproteolysis during or immediately after budding (Lapatto et al. 1989; Navia et al. 1989; Kräusslich 1991). Virions arrested at the late budding stage display immature morphology, implying that budding and maturation are intimately coupled events (Göttlinger et al. 1991).

The five proteolysis sites in HIV-1 Gag are cleaved at different rates, in the order: SP1/NC > MA/CA and SP2/p6 > NC/SP2 > CA/SP1 (Pettit et al. 1998) (Fig. 20.6a). This results in a defined sequential release of maturation intermediates. Mutations that disrupt cleavage of individual sites or alter the processing order result in particles that have severely reduced infectivity, indicating that both the accuracy and timing of Gag processing are essential for proper maturation (Tritch et al. 1991; Wiegers et al. 1998). Furthermore, doping the immature virion with Gag processing intermediates interferes with maturation, implying that individual Gag molecules are not cleaved at random (Müller et al. 2009). This is consistent with the idea that PR processing is initiated at the rim of the ordered Gag lattice in the incompletely closed immature shell and then proceeds as a "wave" into the interior of the lattice (Carlson et al. 2008; Müller et al. 2009).

Gag cleavage by PR disassembles the immature lattice and liberates the subunits of the mature capsid (the CA protein). CA is derived from the central region of the Gag precursor and is the last mature protein to be released as a result of sequential processing of Gag's proteolytic sites (Fig. 20.6a).

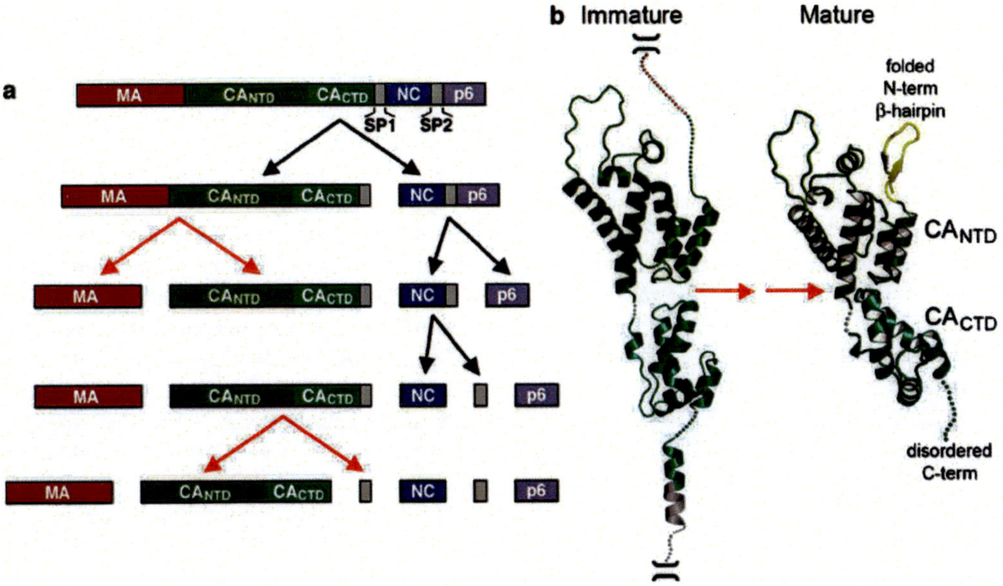

Fig. 20.6 Proteolytic processing of Gag during maturation. (**a**) Schematic showing the HIV-1 Gag proteolysis cascade. Cleavage events that generate the mature CA termini are indicated by the *red arrows*. (**b**) Secondary and tertiary structural changes at the N-terminal and C-terminal ends of the CA domain. The depicted conformations of the MA/CA and CA/SP1 junctions in Gag are based on a variety of structural, biochemical, and mutagenesis studies. In mature CA, the N-terminal 13 residues are folded into a β-hairpin (colored *yellow*), and the C-terminal 11 residues are disordered. (Structure images reprinted from Ganser-Pornillos et al. (2008), with permission from Elsevier)

Although a precise understanding of the molecular transformations that occur during maturation must await a high-resolution structure of the immature Gag lattice, deuterium exchange and mutagenesis studies indicate that the two CA domains (CA^{NTD} and CA^{CTD}) utilize highly overlapping interfaces to mediate interactions in both the immature and mature lattices (Lanman et al. 2003, 2004; von Schwedler et al. 2003a; Monroe et al. 2010). Indeed, the tertiary structures of these two domains are largely preserved in the Gag polyprotein and mature CA, except for residues in the immediate vicinity of the proteolytic sites (Worthylake et al. 1999; Tang et al. 2002; Newman et al. 2004; Kelly et al. 2006) (Fig. 20.6b).

Proteolytic cleavage at the N-terminal end of CA releases the protein from the upstream MA domain and results in refolding of the first 13 residues of CA from an extended conformation into a β-hairpin (colored yellow in Fig. 20.6b). The β-hairpin is stabilized by a salt bridge between the new Pro1 imino group and a conserved acidic residue (Asp51 in HIV-1) (Gitti et al. 1996; von Schwedler et al. 1998). Formation of the β-hairpin probably destabilizes the immature CA^{NTD} hexamer and facilitates proper juxtaposition of the adjacent domains in the mature hexamer conformation (Gross et al. 1998; von Schwedler et al. 1998; Kelly et al. 2006; Pornillos et al. 2010). Interestingly, it appears that folding of the β-hairpin does not proceed immediately after cleavage of the MA/CA junction, but after the final proteolytic processing step at the CA/SP1 junction (Monroe et al. 2010). Thus, cleavage at the CA C terminus appears to be the dominant maturation switch. Indeed, mutagenesis of this region is particularly detrimental to capsid assembly and virus infectivity (Kräusslich et al. 1995; Gross et al. 2000; Li et al. 2003; Müller et al. 2009). Proteolysis of the CA/SP1 junction is expected to unravel the putative helical bundle in this region that helps stabilize the immature lattice, and indeed, the last 11 residues of CA appear to be disordered in the mature capsid (Ganser-Pornillos et al. 2007; Pornillos et al. 2009).

Studies of other orthoretroviruses have revealed comparable structural switches at the CA termini. For example, folding of the β-hairpin during maturation appears to be a general switch because N-terminal extensions shift the in vitro assembly phenotypes of HIV-1, RSV, and MPMV CA proteins from mature-like particles to immature-like particles (Campbell and Vogt 1997; Gross et al. 1998; von Schwedler et al. 1998; Rumlova-Klikova et al. 2000). Interestingly, spacer peptides upstream of the CA domains of RSV and MPMV appear to make important protein–protein interactions in the immature lattice that are not present in HIV-1 (Joshi and Vogt 2000; Knejzlík et al. 2007; Phillips et al. 2008). The best-characterized example is the "p10" region immediately upstream of RSV CA, which folds into a helix and packs against the globular CA^{NTD} domain (Nandhagopal et al. 2004). This interaction is important for RSV Gag hexamerization and lattice assembly (Joshi and Vogt 2000; Phillips et al. 2008). Assembly regions downstream of CA, which are analogous to the SP1 spacer in HIV-1, have also been identified in MLV, RSV, and MPMV Gag, even though these proteins lack a bona fide spacer between their CA and NC domains (Cheslock et al. 2003; Keller et al. 2008; de Marco et al. 2010). Indeed, pillar-like densities below the CA^{CTD} hexamer layer are also visible in cryotomographic reconstructions of immature MPMV and RSV particles, although the MPMV pillar appears much smaller than in either HIV-1 or RSV (de Marco et al. 2010). These examples highlight both the degree of conservation in the assembly principles of orthoretroviruses and the structural/mechanistic variations that different viruses use to reach the same endpoint.

20.4 Assembly and Architecture of the Mature Capsid

In the infectious retrovirion, the ribonucleoprotein genomic complex is contained within the mature capsid, which is composed of around 1,500 copies of the mature CA protein (Fig. 20.2c). The architecture of the mature capsid of HIV-1 is described by the geometric principles of a fullerene cone (Ganser et al. 1999; Jin et al. 1999; Li et al. 2000; Heymann et al. 2008). The body of the capsid is a cone-shaped two-dimensional lattice of CA hexamers. By analogy to quasi-equivalent icosahedral viruses, the hexagonal capsid lattice incorporates 12 pentamers to form a closed shell. The pentamers are also composed of CA and are distributed asymmetrically across the capsid shell. Thus, the capsid itself is globally asymmetric even though it is composed of locally symmetric building blocks.

The mature capsids of orthoretroviruses adopt different preferred shapes and are primarily cones in HIV-1 and lentiviruses, cylinders in betaretroviruses (e.g., MPMV), and polyhedral or "spherical" in others (e.g., RSV, MLV) (Vogt 1997). The various shapes are a consequence of the distribution of the pentamers: "spherical" capsids have the 12 pentamers distributed randomly, cylinders have six pentamers at each end of a tube, and cones have five pentamers at the narrow end and seven at the wide end (Fig. 20.7) (Ganser-Pornillos et al. 2004; Benjamin et al. 2005; Briggs et al. 2006; Butan et al. 2008). This is in stark contrast to icosahedral capsids, wherein the 12 pentamers are arranged in a symmetric configuration.

20.4.1 CA–CA Lattice Interactions

Electron cryomicroscopy and cryotomography of mature virions and capsids of HIV-1 (Briggs et al. 2003, 2006; Benjamin et al. 2005) and RSV (Butan et al. 2008) confirm the essential tenets of the fullerene model and reveal that the hexamer-to-hexamer spacing in the mature lattice is 90–100 Å (Briggs et al. 2003). The mature CA hexamer is distinct from the immature Gag hexamer, and therefore the intersubunit interactions mediated by CA^{NTD} and CA^{CTD} must rearrange during maturation.

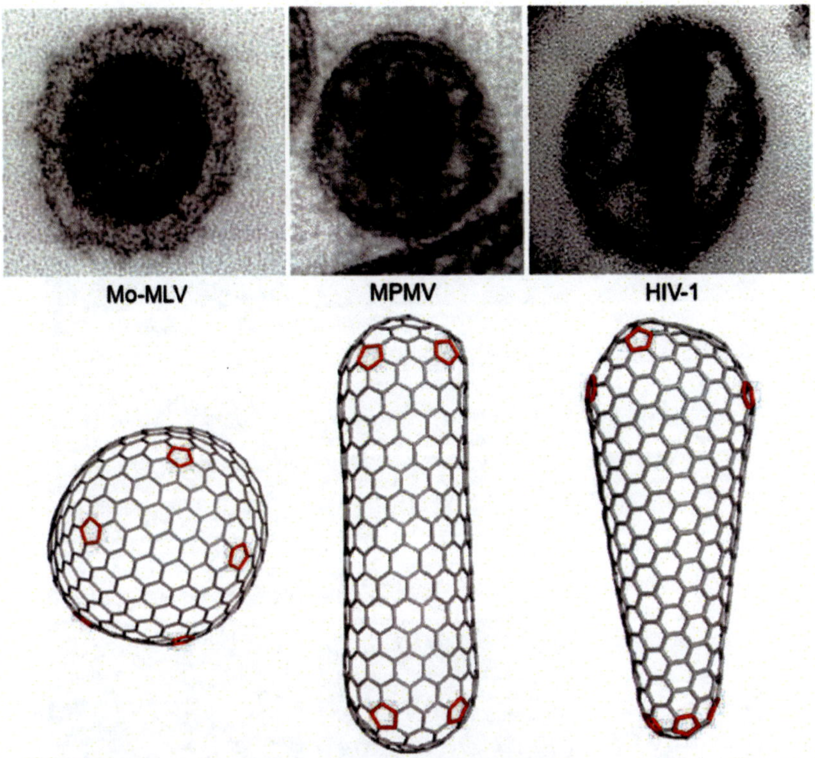

Fig. 20.7 Morphologies of representative mature capsids of different orthoretroviruses. Images and fullerene models of Moloney murine leukemia virus (Mo-MLV, a gammaretrovirus), Mason-Pfizer monkey virus (MPMV, a betaretrovirus), and HIV-1 (a lentivirus) are shown. In all cases, the capsids are organized as fullerene structures that incorporate 12 pentamers (*red*) to close the hexagonal lattice. (Virus images reproduced from Ganser-Pornillos et al. (2004), with permission from American Society for Microbiology)

To a large extent, our understanding of the quaternary interactions in the mature capsid is derived from analyses of in vitro model systems that are assembled from purified recombinant CA proteins (Ehrlich et al. 1992; Campbell and Vogt 1995; Gross et al. 1997; Kingston et al. 2000; Ganser-Pornillos et al. 2007; Purdy et al. 2008). These assemblies recapitulate the local symmetry of the building blocks but are also globally symmetric and are, thus, more amenable to biochemical and structural analyses (Fig. 20.8). For example, helical tubes and two-dimensional crystals composed of CA hexamers have provided low-resolution views of the hexagonal lattice (Li et al. 2000; Ganser-Pornillos et al. 2007; Byeon et al. 2009), whereas icosahedral assemblies of RSV CA allowed the first visualization of the CA pentamer (Cardone et al. 2009; Hyun et al. 2010). Biochemical and mutagenesis experiments have established that these in vitro assemblies faithfully recapitulate the lattice interactions of CA subunits in mature capsids (Gross et al. 1997; von Schwedler et al. 1998, 2003a; Lanman et al. 2003; Ganser-Pornillos et al. 2004; Purdy et al. 2008).

The assembly functions of the CA^{NTD} and CA^{CTD} domains in the mature capsid are clearly delineated: CA^{NTD} forms sixfold or fivefold symmetric rings, whereas side-by-side CA^{CTD} dimers connect each ring to its neighbors (Figs. 20.9, 20.10 and 20.11). Furthermore, the CA^{CTD} domains form a "belt" surrounding the central CA^{NTD} ring, with each CA^{CTD} packed against the CA^{NTD} of the adjacent subunit (Fig. 20.9). The quaternary intersubunit interactions in the CA hexamer, pentamer, and dimer have each now been visualized at atomic resolution. Thus, the architecture of the mature capsid is now understood in considerable detail (Fig. 20.9).

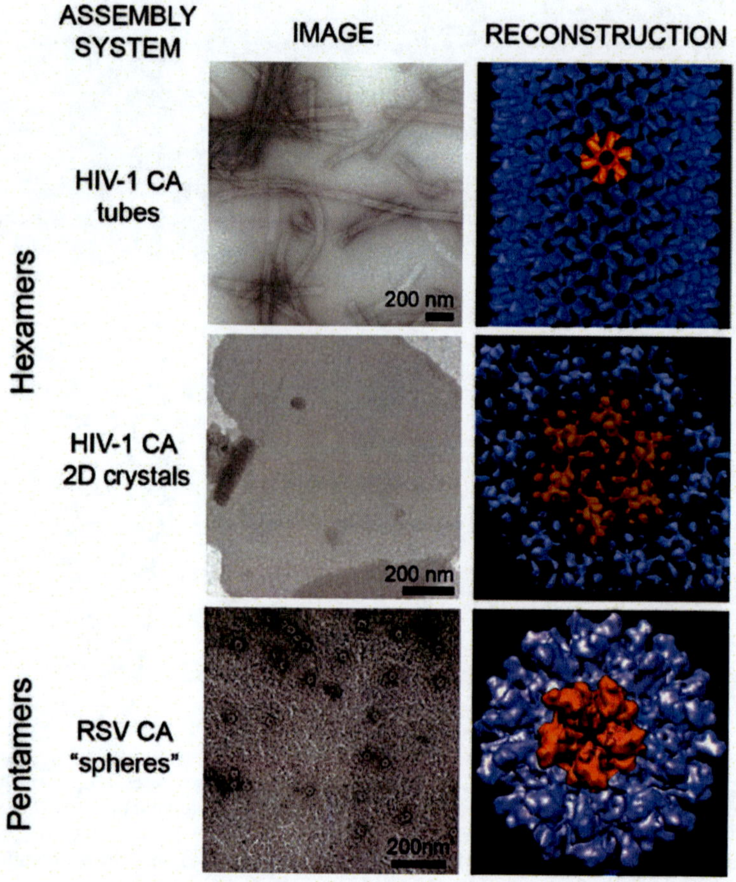

Fig. 20.8 Gallery of in vitro assembly systems for retroviral capsids: helical tubes, two-dimensional hexagonal crystals, and icosahedral particles. (Images of HIV-1 assemblies reprinted from (Ganser-Pornillos et al. 2008), with permission from Elsevier. Image of RSV particles courtesy of Rebecca Craven. Reconstructed images of HIV-1 tubes reprinted from (Li et al. 2000), with permission from Macmillan Publishers Ltd. Map used to generate T = 1 image courtesy of Alasdair Steven)

20.4.1.1 The CA^{NTD} Hexamer and Pentamer

The CA^{NTD} hexamer is a sixfold symmetric ring that is organized around an 18-helix barrel composed of the first three helices of each subunit (Mortuza et al. 2004; Ganser-Pornillos et al. 2007; Pornillos et al. 2009) (Figs. 20.9b and 20.10a). The repeating interaction is a three-helix bundle formed by lengthwise packing of helix 2 from one subunit against helices 1 and 3 of the adjacent subunit. In HIV-1, the center of the three-helix bundle contains a small hydrophobic core composed of aliphatic side chains (Fig. 20.10c), whereas interactions at the periphery are mediated by polar residues. Direct electrostatic protein–protein contacts are conspicuously absent, and essentially all the intersubunit hydrogen bonds are bridged by water molecules (Pornillos et al. 2009).

The quaternary organization of the CA^{NTD} pentamer is similar to the hexamer, except that the subunits form a fivefold symmetric ring organized around a 15-helix barrel (Cardone et al. 2009; Hyun et al. 2010; Pornillos et al. 2011) (Figs. 20.9c and 20.10b). Even though the packing angles

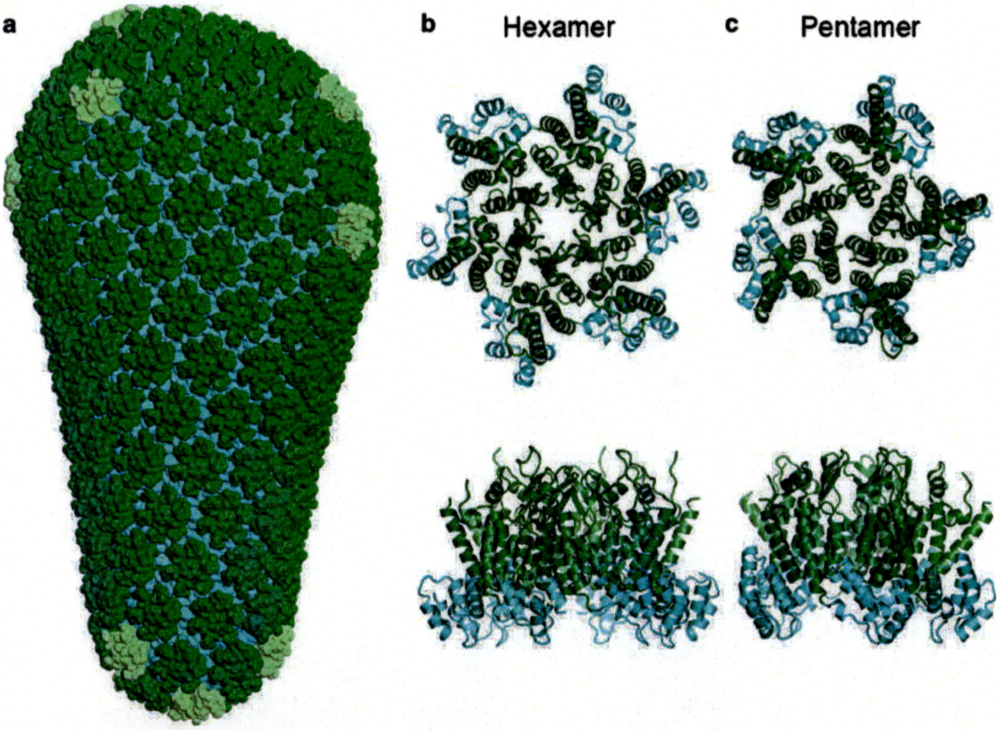

Fig. 20.9 Organization of the mature HIV-1 capsid. (**a**) Structural model of the complete capsid, with the CA^{NTD} hexamers in *green*, CA^{NTD} pentamers in *light green*, and CA^{CTD} dimers in *cyan*. (**b**) Two views of the X-ray structure of the HIV-1 CA hexamer (Pornillos et al. 2009). (**c**) Two views of the X-ray structure of the HIV-1 CA pentamer (Pornillos et al. 2011)

between adjacent subunits are different, the HIV-1 pentamer uses a similar three-helix repeat unit – stabilized by essentially identical hydrophobic interactions – as the hexamer (Pornillos et al. 2011) (Fig. 20.10c).

Switching between the CA^{NTD} hexamer and pentamer follows the general model for quasi-equivalence as originally proposed by Caspar and Klug (1962). Alternative packing of subunits in the two oligomers occurs by a simple rotation of adjacent subunits, about an axis that appears to coincide with the precise center of the three-helix repeat unit (Pornillos et al. 2011) (Fig. 20.10d). Residues at the center of the three-helix repeat unit can therefore maintain essentially the same hydrophobic packing interactions in both oligomers, and only subtle rearrangements in the hydrogen-bonding interactions at the inner and outer rims of the CA^{NTD} rings are required.

The energetic landscape of hexamer vs pentamer assembly appears to be controlled by an electrostatic switch (Cardone et al. 2009; Pornillos et al. 2011). In the case of HIV-1 CA, an almost universally conserved arginine residue (Arg18) occupies an annulus at the centers of both the hexamer and pentamer. Juxtaposition of like charges creates electrostatic repulsion, which is greater for the pentamer because the arginine residues are closer to each other in the fivefold ring (Pornillos et al. 2011). Indeed, substitution of Arg18 with uncharged residues promotes pentamer formation in vitro (Ganser-Pornillos et al. 2004, 2007). Electrostatic destabilization of the pentamer is most probably counterbalanced by local assembly rules and cooperativity, such that a pentamer is integrated into the assembling capsid lattice only at a position where incorporation of a hexamer is incompatible with the local lattice curvature.

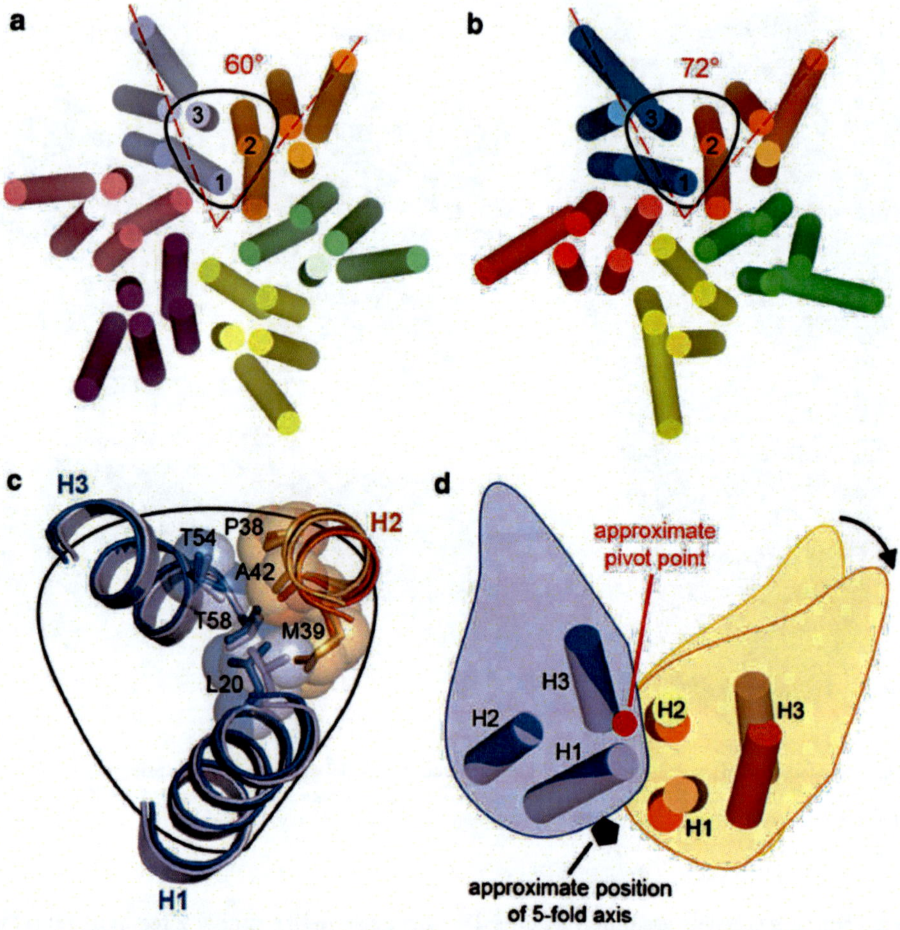

Fig. 20.10 Quasi-equivalence in the pentameric and hexameric CA^{NTD} rings. Top views of the (**a**) hexameric and (**b**) pentameric CA^{NTD} rings, with each subunit in a different color. Subunits in the pentamer and hexamer are shown in *darker* and *lighter* shades, respectively. The angles subtended by adjacent domains are shown explicitly for the *blue* and *orange* subunits. One of the repeating three-helix units is outlined in *black*. (**c**) Close-up view of the pentameric and hexameric repeat units, superimposed on helices 1 and 3 of the *blue* subunit. The aliphatic residues that form a small hydrophobic core are shown explicitly and labeled. (**d**) The "rotation" between adjacent subunits, in going from the hexamer to the pentamer. The approximate position of the rotation axis is indicated by the *red dot*. Note that this axis is parallel to neither the pentameric nor hexameric symmetry axes. (Reprinted from Pornillos et al. (2011), with permission from Macmillan Publishers Ltd)

20.4.1.2 The CA^{CTD} Dimer

The CA^{NTD} hexamers and pentamers are linked into a continuous lattice by symmetric CA^{CTD} dimers (Fig. 20.11a), with side-by-side interacting subunits (i.e., not domain-swapped). The affinity of the mature HIV-1 CA^{CTD} dimer is measurable in solution, with a dissociation constant (K_d) of ~10^{-5} M (Gamble et al. 1997; Byeon et al. 2009). Characterized CA proteins of other orthoretroviruses remain monomeric in solution (Khorasanizadeh et al. 1999; Campos-Olivas et al. 2000; Mortuza et al. 2009) but also form analogous dimers under conditions that promote capsid assembly (Bailey et al. 2009; Purdy et al. 2009). The CA^{CTD} dimer interface is mediated by symmetric packing of helix 9 across

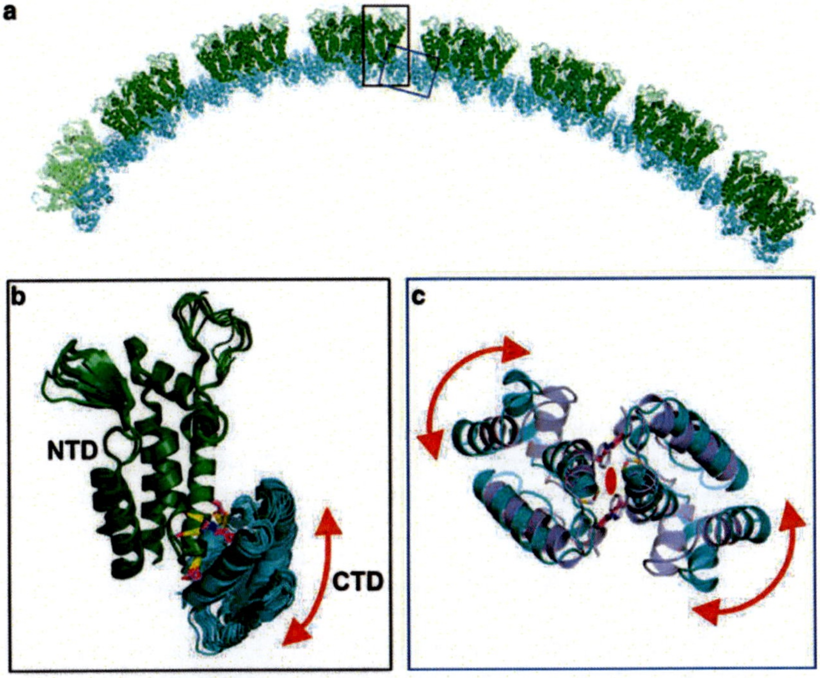

Fig. 20.11 Proposed mechanisms for generating curvature in the mature retroviral capsid. (**a**) Structural model of the capsid, showing a line of connected rings. The CA^{NTD} hexamers are colored in *green*, and CA^{CTD} dimers are in *cyan*. One pentamer is shown on the *left*, with the CA^{NTD} colored *light green*. The two regions that are proposed to contribute to curvature generation are indicated by the *black box* (NTD–CTD interface) and the *blue box* (CTD–CTD dimer interface). (**b**) Flexion at the intermolecular NTD–CTD interface occurs about molecular pivots composed of helix-capping hydrogen bonds. Shown are 17 independent X-ray structures of the NTD–CTD interface, superimposed on the CA^{NTD} domain. Rigid-body movement of CA^{CTD} relative to CA^{NTD} is indicated by the *red double-headed arrow*. Hydrogen-bonded side chains are shown explicitly. Hydrogen bonds are indicated as *yellow lines*. (**c**) Superposition of the X-ray structure (colored *light blue*) (Worthylake et al. 1999) and NMR structure (colored *cyan*) (Byeon et al. 2009) of the isolated full-affinity CA^{CTD} dimer. The structures show distinct configurations of the dimer. The *red double-headed arrows* indicate putative slippage or rotation about the dimer symmetry axis (*red oval*)

the dyad, and hydrophobic interactions between the 3_{10} helix of one subunit and helix 9 of the other (Gamble et al. 1997; Worthylake et al. 1999; Byeon et al. 2009). The lattice-stabilizing dimer interface is the same as the solution dimer interface (Ganser-Pornillos et al. 2007; Byeon et al. 2009), implying that the basic assembly unit of the mature capsid is composed of two CA subunits linked by CA^{CTD}.

20.4.2 Mechanisms of Capsid Lattice Curvature

A key feature of the HIV-1 fullerene cone capsid is that the curvature of the hexagonal CA lattice changes continuously (Ganser et al. 1999; Li et al. 2000) (Fig. 20.11). This is most easily appreciated by considering the bite angle between adjacent hexamers. The bite angles change continuously and range from around 135° between two hexamers connected to the same pentamer at the ends of the cone to around 180° in more flat regions at the body of the cone.

The CA protein has the requisite flexibility for creating such a variably curved lattice because the CA^{NTD} and CA^{CTD} domains are linked by a flexible stretch of four amino acid residues. Indeed, NMR analyses of full-length CA proteins in solution demonstrate that the two domains can rotate almost independently of each other (Khorasanizadeh et al. 1999; Campos-Olivas et al. 2000). In the assembled lattice, the relative orientations of the CA^{NTD} rings and CA^{CTD} dimers can also vary (Byeon et al. 2009; Cardone et al. 2009; Pornillos et al. 2009). Unlike the soluble CA protein, however, movement of the two domains in the assembled lattice is more restricted. One proposal is that capsid lattice curvature is mediated by defined modes of flexibility at two assembly interfaces: the NTD-CTD and CTD-CTD interfaces (Pornillos et al. 2011) (Fig. 20.11).

20.4.2.1 NTD–CTD Interactions

In the assembled CA lattice, each CA^{CTD} domain packs against the CA^{NTD} domain from the adjacent subunit, via an intermolecular set of interactions called the NTD–CTD interface (Bowzard et al. 2001; Lanman et al. 2003; Ganser-Pornillos et al. 2007) (Fig. 20.11b). A critical feature of the NTD–CTD interface is that it contains a set of direct protein–protein hydrogen bonds, each of which is a helix-capping interaction, and connects a flexible polar side chain in one domain with a helix terminus in the other domain (Pornillos et al. 2009, 2011). These helix-capping hydrogen bonds act as molecular pivots for limited rigid-body rotations of CA^{CTD} relative to CA^{NTD} (indicated by red double-headed arrow in Fig. 20.11b). In this manner, the NTD–CTD interface can restrict the movement of the CA^{CTD} dimers relative to the CA^{NTD} rings. Each dimer can only rotate about a fixed axis that is approximately parallel to the plane of the ring, and this flexion would allow each CA^{NTD} ring to adopt slightly different bite angles relative to its neighbors. The NTD–CTD interface, therefore, is likely to function by channeling the native flexibility of the CA protein into a mechanism for generating variable lattice curvature. This feature is common to the hexamer and pentamer since an identical set of interactions is found in both capsomers (Pornillos et al. 2009, 2011).

20.4.2.2 CTD–CTD Interactions

Biochemical and structural analyses also indicate that the CA^{CTD} domain is conformationally flexible (Ternois et al. 2005; Alcaraz et al. 2007; Ganser–Pornillos et al. 2007; Ivanov et al. 2007; Bartonova et al. 2008; Wong et al. 2008). For example, comparison of X-ray (Gamble et al. 1997; Worthylake et al. 1999) and NMR (Byeon et al. 2009) structures of full-affinity CA^{CTD} dimers reveals slight variations in the structure of the dimerization helix, which occur at both the tertiary level (in terms of helix 9 packing against other helices in the same domain) and quaternary level (in terms of the helix 9 crossing angle across the dimer dyad) (Fig. 20.11c). Flexibility of the CA^{CTD} dimer might have functional implications since twisting at the dimer interface may also dictate lattice curvature (indicated by red double-headed arrows in Fig. 20.11c) (Ternois et al. 2005; Ganser-Pornillos et al. 2007). Indeed, computational modeling of the HIV-1 capsid suggests that proper formation of the pentameric declinations requires a flexible dimer interface (Pornillos et al. 2011).

An electron cryomicroscopy reconstruction of helical tubes composed of HIV-1 CA hexamers indicates that adjacent CA^{CTD} domains surrounding the threefold axes of the hexagonal lattice may also interact via extended polar residues (Byeon et al. 2009). This interaction is not observed in structures of flattened lattices of CA (Ganser-Pornillos et al. 2007) and, therefore, appears to specifically stabilize subunit packing in a curved lattice. In the assembled capsid, it is likely that interactions across the threefold interface vary in response to different degrees of rotation at the NTD–CTD interface (Byeon et al. 2009; Pornillos et al. 2009).

20.5 Organization of the RNA Genome

The morphological and structural rearrangements that occur during retroviral maturation likely prepare the RNA genome for proper replication upon cell entry. Consistent with this idea, the genome itself undergoes structural rearrangements during maturation, in which the RNA condenses with the new NC proteins inside the capsid and adopts a more stable structure (Fu and Rein 1993). Furthermore, there appears to be a structural link between the genome and the capsid shell because alterations in capsid stability can significantly perturb reverse transcription (Forshey et al. 2002). Very little is currently known about the three-dimensional organization of the genome within the core, but significant progress has been made in delineating the secondary structural elements (Watts et al. 2009). This is an exciting unexplored area in retrovirus structure.

20.6 Conserved and Contrasting Features in the Design of Enveloped Spherical Viruses

As molecular machines, viruses have evolved efficient, elegant methods to perform their vital functions: (1) protect the genome upon exit from an infected cell, (2) recognize a susceptible host cell and trigger an entry mechanism, and (3) co-opt normal cellular machinery to replicate and assemble progeny. For retroviruses, and in fact most viruses, the first function is accomplished by the capsid, a protective protein shell with tightly packed subunits that exclude harmful proteins such as ribonucleases, and the second function is accomplished by trimeric Env glycoprotein spikes which recognize specific cellular receptors and mediate fusion of the viral envelope with the cell membrane.

The structures of the capsids and surface proteins from many enveloped viruses are now known and have begun to reveal both common and diverging features in virus architecture. As might be expected, many enveloped viruses have capsids and/or envelope proteins organized by the classical principles of icosahedral symmetry. The most striking example of this is the alphavirus family, where even the lipid bilayer of the virion is constrained to icosahedral symmetry (Zhang et al. 2002). On the other hand, hepatitis B virus does have an icosahedral capsid but also a pleomorphic envelope with a relatively random distribution of spikes (Dryden et al. 2006). Nevertheless, the envelope and capsid reside in close proximity, and the surface L protein mediates intimate interactions with the internal capsid. Like hepatitis B virus, herpesvirus has a pleomorphic envelope and an icosahedral inner capsid. However, a disordered tegument protein shell resides between the envelope and capsid (Grünewald et al. 2003). Finally, orthoretroviruses appear to be an extreme case of pleomorphism because both the capsids and surface glycoproteins are nonicosahedral.

This pleomorphism has made structural analyses of orthoretroviral capsids and envelope proteins very difficult. This chapter illustrates the "bottoms-up" approach that has been used to study these types of structures, wherein high-resolution information on the building blocks are coupled with lower-resolution electron microscopy structures and functional data to build models for the complete capsid. This approach requires close collaborations between structural biologists, biochemists, and virologists.

20.7 Future Directions

Future directions for studies of orthoretroviral capsid assembly include determination of higher-resolution structures of the immature Gag lattice, characterization of the molecular rearrangements (both within the RNA genome and the capsid shells) that occur during the regulated maturation of the virus, and dissection of the early events in the viral replication cycle. Of particular interest are the fate of the viral capsid once it enters the cell and how the capsid and its contents reorganize to form the enzymatic complex that reverse transcribes the genome and targets it to host chromosomes for integration.

Acknowledgments HIV research in the Yeager laboratory is supported by grants from the US National Institutes of Health (RO1-GM066087 and P50-GM082545). We thank John Briggs, Rebecca Craven, Alasdair Steven, and Elizabeth Wright for generously supplying materials for figures; Kelly Dryden and Jeong-Hyun Lee for assistance in preparing figures; and Wes Sundquist for helpful discussions. We apologize to colleagues whose works were not cited due to lack of space or inadvertent omission.

References

Accola MA, Höglund S, Göttlinger HG (1998) A putative α-helical structure which overlaps the capsid-p2 boundary in the human immunodeficiency virus type 1 Gag precursor is crucial for viral particle assembly. J Virol 72:2072–2078

Alcaraz LA, del Álamo M, Barrera FN et al (2007) Flexibility in HIV-1 assembly subunits: solution structure of the monomeric C-terminal domain of the capsid protein. Biophys J 93:1264–1276

Aloia RC, Tian H, Jensen FC (1993) Lipid composition and fluidity of the human immunodeficiency virus envelope and host cell plasma membranes. Proc Natl Acad Sci USA 90:5181–5185

Bailey GD, Hyun JK, Mitra AK et al (2009) Proton-linked dimerization of a retroviral capsid protein initiates capsid assembly. Structure 17:737–748

Bajorek M, Schubert HL, McCullough J et al (2009) Structural basis for ESCRT-III protein autoinhibition. Nat Struct Mol Biol 16:754–762

Bartonova V, Igonet S, Sticht J et al (2008) Residues in the HIV-1 capsid assembly inhibitor binding site are essential for maintaining the assembly-competent quaternary structure of the capsid protein. J Biol Chem 283:32024–32033

Benjamin J, Ganser-Pornillos BK, Tivol WF et al (2005) Three-dimensional structure of HIV-1 virus-like particles by electron cryotomography. J Mol Biol 346:577–588

Bouamr F, Melillo JA, Wang MQ et al (2003) PPPYEPTAP motif is the late domain of human T-cell leukemia virus type 1 Gag and mediates its functional interaction with cellular proteins Nedd4 and Tsg101. J Virol 77:11882–11895

Bowzard JB, Wills JW, Craven RC (2001) Second-site suppressors of Rous sarcoma virus CA mutations: evidence for interdomain interactions. J Virol 75:6850–6856

Briggs JAG, Wilk T, Welker R et al (2003) Structural organization of authentic, mature HIV-1 virions and cores. EMBO J 22:1707–1715

Briggs JA, Grünewald K, Glass B et al (2006) The mechanism of HIV-1 core assembly: insights from three-dimensional reconstructions of authentic virions. Structure 14:15–20

Briggs JAG, Riches JD, Glass B et al (2009) Structure and assembly of immature HIV. Proc Natl Acad Sci USA 106:11090–11095

Brown DA, London E (1997) Structure of detergent-resistant membrane domains: does phase separation occur in biological membranes? Biochem Biophys Res Commun 240:1–7

Brügger B, Glass B, Haberkant P et al (2006) The HIV lipidome: a raft with an unusual composition. Proc Natl Acad Sci USA 103:2641–2646

Butan C, Winkler DC, Heymann JB et al (2008) RSV capsid polymorphism correlates with polymerization efficiency and envelope glycoprotein content: implications that nucleation controls morphogenesis. J Mol Biol 376:1168–1181

Byeon I-JL, Meng X, Jung J et al (2009) Structural convergence between cryo-EM and NMR reveals intersubunit interactions critical for HIV-1 capsid function. Cell 139:780–790

Campbell S, Rein A (1999) *In vitro* assembly properties of human immunodeficiency virus type 1 Gag protein lacking the p6 domain. J Virol 73:2270–2279

Campbell S, Vogt VM (1995) Self-assembly *in vitro* of purified CA-NC proteins from Rous sarcoma virus and human immunodeficiency virus type 1. J Virol 69:6487–6497

Campbell S, Vogt VM (1997) *In vitro* assembly of virus-like particles with Rous sarcoma virus Gag deletion mutants: identification of the p10 domain as a morphological determinant in the formation of spherical particles. J Virol 71:4425–4435

Campbell S, Fisher RJ, Towler EM et al (2001) Modulation of HIV-like particle assembly *in vitro* by inositol phosphates. Proc Natl Acad Sci USA 98:10875–10879

Campos-Olivas R, Newman JL, Summers MF (2000) Solution structure and dynamics of the Rous sarcoma virus capsid protein and comparison with capsid proteins of other retroviruses. J Mol Biol 296:633–649

Cardone G, Purdy JG, Cheng N et al (2009) Visualization of a missing link in retrovirus capsid assembly. Nature 457:694–698

Carlson L-A, Briggs JAG, Glass B et al (2008) Three-dimensional analysis of budding sites and released virus suggests a revised model for HIV-1 morphogenesis. Cell Host Microbe 4:592–599

Carlton JG, Martin-Serrano J (2007) Parallels between cytokinesis and retroviral budding: a role for the ESCRT machinery. Science 316:1908–1912

Caspar DLD, Klug A (1962) Physical principles in the construction of regular viruses. Cold Spring Harb Symp Quant Biol 27:1–24

Cheslock SR, Poon DTK, Fu W et al (2003) Charged assembly helix motif in murine leukemia virus capsid: an important region for virus assembly and particle size determination. J Virol 77:7058–7066

Chung H-Y, Morita E, von Schwedler U et al (2008) NEDD4L overexpression rescues the release and infectivity of human immunodeficiency virus type 1 constructs lacking PTAP and YPXL late domains. J Virol 82: 4884–4897

Dalton AK, Murray PS, Murray D et al (2005) Biochemical characterization of rous sarcoma virus MA protein interaction with membranes. J Virol 79:6227–6238

Dalton AK, Ako-Adjei D, Murray PS et al (2007) Electrostatic interactions drive membrane association of the human immunodeficiency virus type 1 Gag MA domain. J Virol 81:6434–6445

Datta SAK, Curtis JE, Ratcliff W et al (2007a) Conformation of the HIV-1 Gag protein in solution. J Mol Biol 365:812–824

Datta SAK, Zhao Z, Clark PK et al (2007b) Interactions between HIV-1 Gag molecules in solution: an inositol phosphate-mediated switch. J Mol Biol 365:799–811

de Marco A, Davey NE, Ulbrich P et al (2010) Conserved and variable features of Gag structure and arrangement in immature retrovirus particles. J Virol 84:11729–11736

Dey A, York D, Smalls-Mantey A et al (2005) Composition and sequence-dependent binding of RNA to the nucleocapsid protein of Moloney murine leukemia virus. Biochemistry 44:3735–3744

Dorfman T, Mammano F, Haseltine WA et al (1994) Role of the matrix protein in the virion association of the human immunodeficiency virus type 1 envelope glycoprotein. J Virol 68:1689–1696

Dryden KA, Wieland SF, Whitten-Bauer C et al (2006) Native Hepatitis B virions and capsids visualized by electron cryomicroscopy. Mol Cell 22:843–850

D'Souza V, Summers MF (2004) Structural basis for packaging the dimeric genome of Moloney murine leukaemia virus. Nature 431:586–590

D'Souza V, Melamed J, Habib D et al (2001) Identification of a high affinity nucleocapsid protein binding element within the Moloney murine leukemia virus Ψ-RNA packaging signal: implications for genome recognition. J Mol Biol 314:217–232

Ehrlich LS, Agresta BE, Carter CA (1992) Assembly of recombinant human immunodeficiency virus type 1 capsid protein *in vitro*. J Virol 66:4874–4883

Fabrikant G, Lata S, Riches JD et al (2009) Computational model of membrane fission catalyzed by ESCRT-III. PLoS Comput Biol 5:e1000575

Forshey BM, von Schwedler U, Sundquist WI et al (2002) Formation of a human immunodeficiency virus type 1 core of optimal stability is crucial for viral replication. J Virol 76:5667–5677

Freed EO, Martin MA (1995) Virion incorporation of envelope glycoproteins with long but not short cytoplasmic tails is blocked by specific, single amino acid substitutions in the human immunodeficiency virus type 1 matrix. J Virol 69:1984–1989

Freed EO, Martin MA (1996) Domains of the human immunodeficiency virus type 1 matrix and gp41 cytoplasmic tail required for envelope incorporation into virions. J Virol 70:341–351

Fu W, Rein A (1993) Maturation of dimeric viral RNA of Moloney murine leukemia virus. J Virol 67:5443–5449

Fuller SD, Wilk T, Gowen BE et al (1997) Cryo-electron microscopy reveals ordered domains in the immature HIV-1 particle. Curr Biol 7:729–738

Gamble TR, Yoo S, Vajdos FF et al (1997) Structure of the carboxyl-terminal dimerization domain of the HIV-1 capsid protein. Science 278:849–853

Ganser BK, Li S, Klishko VY et al (1999) Assembly and analysis of conical models for the HIV-1 core. Science 283:80–83

Ganser-Pornillos BK, von Schwedler UK, Stray KM et al (2004) Assembly properties of the human immunodeficiency virus type 1 CA protein. J Virol 78:2545–2552

Ganser-Pornillos BK, Cheng A, Yeager M (2007) Structure of full-length HIV-1 CA: a model for the mature capsid lattice. Cell 131:70–79

Ganser-Pornillos BK, Yeager M, Sundquist WI (2008) The structural biology of HIV assembly. Curr Opin Struct Biol 18:203–217

Garrus JE, von Schwedler UK, Pornillos O et al (2001) Tsg101 and the vacuolar protein sorting pathway are essential for HIV-1 budding. Cell 107:55–65

Ghazi-Tabatabai S, Saksena S, Short JM et al (2008) Structure and disassembly of filaments formed by the ESCRT-III subunit Vps24. Structure 16:1345–1356

Gherghe C, Lombo T, Leonard CW et al (2010) Definition of a high-affinity Gag recognition structure mediating packaging of a retroviral RNA genome. Proc Natl Acad Sci USA 107:19248–19253

Gheysen D, Jacobs E, de Foresta F et al (1989) Assembly and release of HIV-1 precursor $Pr55^{gag}$ virus-like particles from recombinant baculovirus-infected insect cells. Cell 59:103–112

Gitti RK, Lee BM, Walker J et al (1996) Structure of the amino-terminal core domain of the HIV-1 capsid protein. Science 273:231–235

Göttlinger HG, Sodroski JG, Haseltine WA (1989) Role of capsid precursor processing and myristoylation in morphogenesis and infectivity of human immunodeficiency virus type 1. Proc Natl Acad Sci USA 86:5781–5785

Göttlinger HG, Dorfman T, Sodroski JG et al (1991) Effect of mutations affecting the p6 *gag* protein on human immunodeficiency virus particle release. Proc Natl Acad Sci USA 88:3195–3199

Gross I, Hohenberg H, Kräusslich H-G (1997) *In vitro* assembly properties of purified bacterially expressed capsid proteins of human immunodeficiency virus. Eur J Biochem 249:592–600

Gross I, Hohenberg H, Huckhagel C et al (1998) N-terminal extension of human immunodeficiency virus capsid protein converts the *in vitro* assembly phenotype from tubular to spherical particles. J Virol 72:4798–4810

Gross I, Hohenberg H, Wilk T et al (2000) A conformational switch controlling HIV-1 morphogenesis. EMBO J 19:103–113

Grünewald K, Desai P, Winkler DC et al (2003) Three-dimensional structure of herpes simplex virus from cryo-electron tomography. Science 302:1396–1398

Hanson PI, Roth R, Lin Y et al (2008) Plasma membrane deformation by circular arrays of ESCRT-III protein filaments. J Cell Biol 180:389–402

Hermida-Matsumoto L, Resh MD (2000) Localization of human immunodeficiency virus type 1 Gag and Env at the plasma membrane by confocal imaging. J Virol 74:8670–8679

Heymann JB, Butan C, Winkler DC et al (2008) Irregular and semi-regular polyhedral models for Rous sarcoma virus cores. Comput Math Methods Med 9:197–210

Hill CP, Worthylake D, Bancroft DP et al (1996) Crystal structures of the trimeric human immunodeficiency virus type 1 matrix protein: implications for membrane association and assembly. Proc Natl Acad Sci USA 93:3099–3104

Hurley JH, Emr SD (2006) The ESCRT complexes: structure and mechanism of a membrane-trafficking network. Annu Rev Biophys Biomol Struct 35:277–298

Hyun J-K, Radjainia M, Kingston RL et al (2010) Proton-driven assembly of the Rous Sarcoma virus capsid protein results in the formation of icosahedral particles. J Biol Chem 285:15056–15064

Ivanchenko S, Godinez WJ, Lampe M et al (2009) Dynamics of HIV-1 assembly and release. PLoS Pathog 5:e1000652

Ivanov D, Tsodikov OV, Kasanov J et al (2007) Domain-swapped dimerization of the HIV-1 capsid C-terminal domain. Proc Natl Acad Sci USA 104:4353–4358

Jin Z, Jin L, Peterson DL et al (1999) Model for lentivirus capsid core assembly based on crystal dimers of EIAV p26. J Mol Biol 286:83–93

Joint United Nations Programme on HIV/AIDS (2009) AIDS epidemic update. UNAIDS, Geneva

Joshi SM, Vogt VM (2000) Role of the Rous sarcoma virus p10 domain in shape determination of Gag virus-like particles assembled *in vitro* and within *Escherichia coli*. J Virol 74:10260–10268

Jouvenet N, Bieniasz PD, Simon SM (2008) Imaging the biogenesis of individual HIV-1 virions in live cells. Nature 454:236–240

Jouvenet N, Simon SM, Bieniasz PD (2009) Imaging the interaction of HIV-1 genomes and Gag during assembly of individual viral particles. Proc Natl Acad Sci USA 106:19114–19119

Keller PW, Johnson MC, Vogt VM (2008) Mutations in the spacer peptide and adjoining sequences in Rous sarcoma virus Gag lead to tubular budding. J Virol 82:6788–6797

Keller PW, Adamson CS, Heymann JB et al (2011) HIV-1 maturation inhibitor bevirimat stabilizes the immature Gag lattice. J Virol 85:1420–1428

Kelly BN, Howard BR, Wang H et al (2006) Implications for viral capsid assembly from crystal structures of HIV-1 Gag_{1-278} and $CA_{133-278}{}^{N}$. Biochemistry 45:11257–11266

Khorasanizadeh S, Campos-Olivas R, Summers MF (1999) Solution structure of the capsid protein from the human T-cell leukemia virus type-I. J Mol Biol 291:491–505

Kikonyogo A, Bouamr F, Vana ML et al (2001) Proteins related to the Nedd4 family of ubiquitin protein ligases interact with the L domain of Rous sarcoma virus and are required for gag budding from cells. Proc Natl Acad Sci USA 98:11199–11204

Kingston RL, Fitzon-Ostendorp T, Eisenmesser EZ et al (2000) Structure and self-association of the Rous sarcoma virus capsid protein. Structure 8:617–628

Knejzlík Z, Smékalová Z, Ruml T et al (2007) Multimerization of the p12 domain is necessary for Mason-Pfizer monkey virus Gag assembly *in vitro*. Virology 365:260–270

Kräusslich H-G (1991) Human immunodeficiency virus proteinase dimer as component of the viral polyprotein prevents particle assembly and viral infectivity. Proc Natl Acad Sci USA 88:3213–3217

Kräusslich H-G, Fäcke M, Heuser A-M et al (1995) The spacer peptide between human immunodeficiency virus capsid and nucleocapsid proteins is essential for ordered assembly and viral infectivity. J Virol 69:3407–3419

Lanman J, Lam TT, Barnes S et al (2003) Identification of novel interactions in HIV-1 capsid protein assembly by high-resolution mass spectrometry. J Mol Biol 325:759–772

Lanman J, Lam TT, Emmett MR et al (2004) Key interactions in HIV-1 maturation identified by hydrogen-deuterium exchange. Nat Struct Mol Biol 11:676–677

Lapatto R, Blundell T, Hemmings A et al (1989) X-ray analysis of HIV-1 proteinase at 2.7 Å resolution confirms structural homology among retroviral enzymes. Nature 342:299–302

Lata S, Schoehn G, Jain A et al (2008) Helical structures of ESCRT-III are disassembled by VPS4. Science 321:1354–1357

Le Blanc I, Prévost M-C, Dokhélar M-C et al (2002) The PPPY motif of human T-cell leukemia virus type 1 Gag protein is required early in the budding process. J Virol 76:10024–10029

Li S, Hill CP, Sundquist WI et al (2000) Image reconstructions of helical assemblies of the HIV-1 CA protein. Nature 407:409–413

Li F, Goila-Gaur R, Salzwedel K et al (2003) PA-457: a potent HIV inhibitor that disrupts core condensation by targeting a late step in Gag processing. Proc Natl Acad Sci USA 100:13555–13560

Li H, Dou J, Ding L et al (2007) Myristoylation is required for human immunodeficiency virus type 1 Gag-Gag multimerization in mammalian cells. J Virol 81:12899–12910

Liang C, Hu J, Russell RS et al (2002) Characterization of a putative α-helix across the capsid-SP1 boundary that is critical for the multimerization of human immunodeficiency virus type 1 Gag. J Virol 76:11729–11737

Lopez-Vergès S, Camus G, Blot G et al (2006) Tail-interacting protein TIP47 is a connector between Gag and Env and is required for Env incorporation into HIV-1 virions. Proc Natl Acad Sci USA 103:14947–14952

Martin-Serrano J, Zang T, Bieniasz PD (2001) HIV-1 and Ebola virus encode small peptide motifs that recruit Tsg101 to sites of particle assembly to facilitate egress. Nat Med 7:1313–1319

Massiah MA, Starich MR, Paschall C et al (1994) Three-dimensional structure of the human immunodeficiency virus type 1 matrix protein. J Mol Biol 244:198–223

Miyazaki Y, Garcia EL, King SR et al (2010) An RNA structural switch regulates diploid genome packaging by Moloney murine leukemia virus. J Mol Biol 396:141–152

Monroe EB, Kang S, Kyere SK et al (2010) Hydrogen/deuterium exchange analysis of HIV-1 capsid assembly and maturation. Structure 18:1483–1491

Morellet N, Druillennec S, Lenoir C et al (2005) Helical structure determined by NMR of the HIV-1 (345-392) Gag sequence, surrounding p2: implications for particle assembly and RNA packaging. Protein Sci 14:375–386

Morita E, Sandrin V, Chung H-Y et al (2007) Human ESCRT and ALIX proteins interact with proteins of the midbody and function in cytokinesis. EMBO J 26:4215–4227

Mortuza GB, Haire LF, Stevens A et al (2004) High-resolution structure of a retroviral capsid hexameric amino-terminal domain. Nature 431:481–485

Mortuza GB, Dodding MP, Goldstone DC et al (2008) Structure of B-MLV capsid amino-terminal domain reveals key features of viral tropism, Gag assembly and core formation. J Mol Biol 376:1493–1508

Mortuza GB, Goldstone DC, Pashley C et al (2009) Structure of the capsid amino-terminal domain from the betaretrovirus, Jaagsiekte sheep retrovirus. J Mol Biol 386:1179–1192

Müller B, Anders M, Akiyama H et al (2009) HIV-1 Gag processing intermediates trans-dominantly interfere with HIV-1 infectivity. J Biol Chem 284:29692–29703

Murakami T, Freed EO (2000a) Genetic evidence for an interaction between human immunodeficiency virus type 1 matrix and α-helix 2 of the gp41 cytoplasmic tail. J Virol 74:3548–3554

Murakami T, Freed EO (2000b) The long cytoplasmic tail of gp41 is required in a cell type-dependent manner for HIV-1 envelope glycoprotein incorporation into virions. Proc Natl Acad Sci USA 97:343–348

Nandhagopal N, Simpson AA, Johnson MC et al (2004) Dimeric Rous sarcoma virus capsid protein structure relevant to immature Gag assembly. J Mol Biol 335:275–282

Navia MA, Fitzgerald PMD, McKeever BM et al (1989) Three-dimensional structure of aspartyl protease from human immunodeficiency virus HIV-1. Nature 337:615–620

Newman JL, Butcher EW, Patel DT et al (2004) Flexibility in the P2 domain of the HIV-1 Gag polyprotein. Protein Sci 13:2101–2107

Nguyen DH, Hildreth JE (2000) Evidence for budding of human immunodeficiency virus type 1 selectively from glycolipid-enriched membrane lipid rafts. J Virol 74:3264–3272

Ono A, Freed EO (2001) Plasma membrane rafts play a critical role in HIV-1 assembly and release. Proc Natl Acad Sci USA 98:13925–13930

Ono A, Demirov D, Freed EO (2000) Relationship between human immunodeficiency virus type 1 Gag multimerization and membrane binding. J Virol 74:5142–5150

Ono A, Ablan SD, Lockett SJ et al (2004) Phosphatidylinositol (4,5) bisphosphate regulates HIV-1 Gag targeting to the plasma membrane. Proc Natl Acad Sci USA 101:14889–14894

Paillart J-C, Göttlinger HG (1999) Opposing effects of human immunodeficiency virus type 1 matrix mutations support a myristyl switch model of Gag membrane targeting. J Virol 73:2604–2612

Pettit SC, Sheng N, Tritch R et al (1998) The regulation of sequential processing of HIV-1 Gag by the viral protease. Adv Exp Med Biol 436:15–25

Phillips JM, Murray PS, Murray D et al (2008) A molecular switch required for retrovirus assembly participates in the hexagonal immature lattice. EMBO J 27:1411–1420

Pornillos O, Alam SL, Davis DR et al (2002) Structure of the Tsg101 UEV domain in complex with the PTAP motif of the HIV-1 p6 protein. Nat Struct Biol 9:812–817

Pornillos O, Ganser-Pornillos BK, Kelly BN et al (2009) X-ray structures of the hexameric building block of the HIV capsid. Cell 137:1282–1292

Pornillos O, Ganser-Pornillos BK, Banumathi S et al (2010) Disulfide bond stabilization of the hexameric capsomer of human immunodeficiency virus. J Mol Biol 401:985–995

Pornillos O, Ganser-Pornillos BK, Yeager M (2011) Atomic-level modelling of the HIV capsid. Nature 469:424–427

Purdy JG, Flanagan JM, Ropson IJ et al (2008) Critical role of conserved hydrophobic residues within the major homology region in mature retroviral capsid assembly. J Virol 82:5951–5961

Purdy JG, Flanagan JM, Ropson IJ et al (2009) Retroviral capsid assembly: a role for the CA dimer in initiation. J Mol Biol 389:438–451

Reil H, Bukovsky AA, Gelderblom HR et al (1998) Efficient HIV-1 replication can occur in the absence of the viral matrix protein. EMBO J 17:2699–2708

Rumlova-Klikova M, Hunter E, Nermut MV et al (2000) Analysis of Mason-Pfizer monkey virus Gag domains required for capsid assembly in bacteria: role of the N-terminal proline residue of CA in directing particle shape. J Virol 74:8452–8459

Saad JS, Miller J, Tai J et al (2006) Structural basis for targeting HIV-1 Gag proteins to the plasma membrane for virus assembly. Proc Natl Acad Sci USA 103:11364–11369

Spearman P, Wang J-J, Heyden NV et al (1994) Identification of human immunodeficiency virus type 1 Gag protein domains essential to membrane binding and particle assembly. J Virol 68:3232–3242

Spearman P, Horton R, Ratner L et al (1997) Membrane binding of human immunodeficiency virus type 1 matrix protein *in vivo* supports a conformational myristyl switch mechanism. J Virol 71:6582–6592

Strack B, Calistri A, Craig S et al (2003) AIP1/ALIX is a binding partner for HIV-1 p6 and EIAV p9 functioning in virus budding. Cell 114:689–699

Swanstrom R, Wills JW (1997) Synthesis, assembly and processing of viral proteins. In: Coffin JM, Hughes SH, Varmus HE (eds) Retroviruses. Cold Spring Harbor Laboratory Press, New York

Tang C, Ndassa Y, Summers MF (2002) Structure of the N-terminal 283-residue fragment of the immature HIV-1 Gag polyprotein. Nat Struct Biol 9:537–543

Tang C, Loeliger E, Luncsford P et al (2004) Entropic switch regulates myristate exposure in the HIV-1 matrix protein. Proc Natl Acad Sci USA 101:517–522

Ternois F, Sticht J, Duquerroy S et al (2005) The HIV-1 capsid protein C-terminal domain in complex with a virus assembly inhibitor. Nat Struct Mol Biol 12:678–682

Tritch RJ, Cheng Y-SE, Yin FH et al (1991) Mutagenesis of protease cleavage sites in the human immunodeficiency virus type 1 Gag polyprotein. J Virol 65:922–930

Usami Y, Popov S, Popova E et al (2008) Efficient and specific rescue of human immunodeficiency virus type 1 budding defects by a Nedd4-like ubiquitin ligase. J Virol 82:4898–4907

VerPlank L, Bouamr F, LaGrassa TJ et al (2001) Tsg101, a homologue of ubiquitin-conjugating (E2) enzymes, binds the L domain in HIV type 1 $Pr55^{Gag}$. Proc Natl Acad Sci USA 98:7724–7729

Vogt VM (1997) Retroviral virions and genomes. In: Coffin JM, Hughes SH, Varmus HE (eds) Retroviruses. Cold Spring Harbor Laboratory Press, New York

von Schwedler UK, Stemmler TL, Klishko VY et al (1998) Proteolytic refolding of the HIV-1 capsid protein amino-terminus facilitates viral core assembly. EMBO J 17:1555–1568

von Schwedler UK, Stray KM, Garrus JE et al (2003a) Functional surfaces of the human immunodeficiency virus type 1 capsid protein. J Virol 77:5439–5450

von Schwedler UK, Stuchell M, Müller B et al (2003b) The protein network of HIV budding. Cell 114:701–713

Wang C-T, Zhang Y, McDermott J et al (1993) Conditional infectivity of a human immunodeficiency virus matrix domain deletion mutant. J Virol 67:7067–7076

Wang MQ, Kim W, Gao G et al (2003) Endophilins interact with Moloney murine leukemia virus Gag and modulate virion production. J Biol 3:4

Watts JM, Dang KK, Gorelick RJ et al (2009) Architecture and secondary structure of an entire HIV-1 RNA genome. Nature 460:711–716

Wiegers K, Rutter G, Kottler H et al (1998) Sequential steps in human immunodeficiency virus particle maturation revealed by alterations of individual Gag polyprotein cleavage sites. J Virol 72:2846–2854

Wollert T, Wunder C, Lippincott-Schwartz J et al (2009) Membrane scission by the ESCRT-III complex. Nature 458:172–177

Wong HC, Shin R, Krishna NR (2008) Solution structure of a double mutant of the carboxy-terminal dimerization domain of the HIV-1 capsid protein. Biochemistry 47:2289–2297

Worthylake DK, Wang H, Yoo S et al (1999) Structures of the HIV-1 capsid protein dimerization domain at 2.6 Å resolution. Acta Crystallogr D Biol Crystallogr 55:85–92

Wright ER, Schooler JB, Ding HJ et al (2007) Electron cryotomography of immature HIV-1 virions reveals the structure of the CA and SP1 Gag shells. EMBO J 26:2218–2226

Yasuda J, Hunter E, Nakao M et al (2002) Functional involvement of a novel Nedd4-like ubiquitin ligase on retrovirus budding. EMBO Rep 3:636–640

Yeager M, Wilson-Kubalek EM, Weiner SG et al (1998) Supramolecular organization of immature and mature murine leukemia virus revealed by electron cryo-microscopy: implications for retroviral assembly mechanisms. Proc Natl Acad Sci USA 95:7299–7304

Yu X, Yuan X, Matsuda Z et al (1992) The matrix protein of human immunodeficiency virus type 1 is required for incorporation of viral envelope protein into mature virions. J Virol 66:4966–4971

Zacharias DA, Violin JD, Newton AC et al (2002) Partitioning of lipid-modified monomeric GFPs into membrane microdomains of live cells. Science 296:913–916

Zhai Q, Fisher RD, Chung H-Y et al (2008) Structural and functional studies of ALIX interactions with YPX$_n$L late domains of HIV-1 and EIAV. Nat Struct Mol Biol 15:43–49

Zhang Y, Barklis E (1997) Effects of nucleocapsid mutations on human immunodeficiency virus assembly and RNA encapsidation. J Virol 71:6765–6776

Zhang W, Mukhopadhyay S, Pletnev SV et al (2002) Placement of the structural proteins in Sindbis virus. J Virol 76:11645–11658

Zhou W, Parent LJ, Wills JW et al (1994) Identification of a membrane-binding domain within the amino-terminal region of human immunodeficiency virus type 1 Gag protein which interacts with acidic phospholipids. J Virol 68:2556–2569

Part IV
Genome Packaging Machines

Chapter 21
Condensed Genome Structure

Lindsay W. Black and Julie A. Thomas

Abstract Large, tailed dsDNA-containing bacteriophage genomes are packaged to a conserved and high density (~500 mg/ml), generally in ~2.5-nm, duplex-to-duplex, spaced, organized DNA shells within icosahedral capsids. Phages with these condensate properties, however, differ markedly in their inner capsid structures: (1) those with a naked condensed DNA, (2) those with many dispersed unstructured proteins embedded within the DNA, (3) those with a small number of localized proteins, and (4) those with a reduced or DNA-free internal protein structure of substantial volume. The DNA is translocated and condensed by a high-force ATPase motor into a procapsid already containing the proteins that are to be ejected together with the DNA into the infected host. The condensed genome structure of a single-phage type is unlikely to be precisely determined and can change without loss of function to fit an altered capsid size or internal structure. Although no such single-phage condensed genome structure is known exactly, it is known that a single general structure is unlikely to apply to all such phages.

21.1 Introduction

Viral genomes are enclosed within a capsid for protection, transfer, and infection of host cells. Most well-studied viruses with capsids of icosahedral, or similar, symmetry have genomes composed of either DNA or RNA [either single-stranded (ss) or double-stranded (ds) in both cases] that range in length from <4 to >1,200 kb. In some cases, the genome may consist of more than one nucleic acid segment. The dimensions of icosahedral capsids vary dramatically to accommodate such variations in nucleic acid content and may, in some viruses, be elongated, and thus depart from strict icosahedral symmetry. Despite wide variations in genome composition, there are numbers of virus components that are structurally conserved across the phylogenetic spectrum (Koonin et al. 2006). Notably, the translocation of DNA into an empty immature capsid (the procapsid) by a viral DNA-packaging machine is an ancient invention that is found in all kingdoms, and its mechanism appears to be highly conserved among many DNA viruses, including Herpes and related viruses (Salmon and Baines 1998; Sheaffer et al. 2001), as well as among most tailed dsDNA bacteriophages.

In most large (greater than ~10 kb) dsDNA viruses/phages, ATP-driven high force-generating molecular motors package the genome to high density within procapsids by a broadly conserved molecular

L.W. Black • J.A. Thomas (✉)
Department of Biochemistry and Molecular Biology,
University of Maryland School of Medicine, Baltimore, MD 21201-1503, USA
e-mail: lblack@umaryland.edu; JThomas@som.umaryland.edu

M.G. Rossmann and V.B. Rao (eds.), *Viral Molecular Machines*,
Advances in Experimental Medicine and Biology 726, DOI 10.1007/978-1-4614-0980-9_21,

mechanism (Smith et al. 2001). Generally, these molecular motors are composed of two main components, a portal-containing procapsid complexed with a multimeric ATPase enzyme, the terminase (Black 1989; Johnson and Chiu 2007; Rao and Feiss 2008; Rao and Black 2010). The portal is situated at a unique capsid vertex and contains a channel, through which DNA enters into the procapsid, and through which DNA and protein are ejected out of the capsid, via the tail (attached to the portal) into the host. DNA is often packaged in conjunction with formation of the mature genome ends by nucleolytic cleavage from a replicative concatemer (hence the name terminase) within bacteriophage or Herpes infected cells; in phages packaging is by a two component terminase, or as a protein terminated linear dsDNA by a single protein-component-packaging ATPase (e.g., Phi29). The terminase protein(s) in most phages/viruses dissociate from the virion after packaging and are not part of the mature virion (i.e., are not structural proteins). In contrast to this, and illustrating the amazing ability of viruses to create variations upon a theme, is the tailless phage PRD1 and its relatives, in which an icosahedral protein capsid encloses a lipid bilayer membrane packed with dsDNA. The packaging proteins of these dsDNA phages are permanently incorporated into a special vertex of the capsid and utilize this ATP driven motor to achieve a packaging rate and local packing density comparable to that of the tailed dsDNA phages (Gowen et al. 2003; Karhu et al. 2007). There are, of course, other icosahedral capsid phages with quite different packaging strategies that are often found to contain less densely packed genomes; e.g., circular supercoiled dsDNA containing PMS2; single-stranded DNA phage PhiX174; three-segment dsRNA containing outer lipid bilayer phage Phi6; not to mention filamentous phages and numerous others; however, phages of these families, which are very different from the tailed dsDNA phages utilizing molecular motors to densely fill procapsids with linear dsDNA, will not be reviewed here.

Most, if not all, tailed dsDNA containing bacteriophage genomes are highly condensed by the ATP-driven terminase complex to a remarkably characteristic and conserved concentration of ~500 mg/ml, near to a liquid crystalline DNA state. Phage DNA packaged within the icosahedral capsid is thus at a concentration at least fivefold higher than found in metaphase chromatin (Cerritelli et al. 1997). Capsid DNA condensates can be both packaged within the infected cell, and unpackaged upon delivery to infected bacteria, within several minutes or less. Nevertheless, despite such high mobility, the metabolically inactive condensed genomes are stable, infectious, and deliverable from the capsid after many years of storage (Ackermann et al. 2004).

The similar structural and functional properties of condensed genomes found among tailed dsDNA bacteriophages have been the subject of numerous reviews and structural models (e.g., Earnshaw and Harrison 1977; Black 1989). However, it has only recently been appreciated that the comparable condensed genomic DNA structures are actually found among tailed bacteriophages with fundamentally different inner capsid environments. Phages can be broadly categorized based on these inner capsid differences: (1) those containing only a naked condensed DNA within the capsid with few or no internal proteins (e.g., lambda, HK97); (2) those with more than a thousand dispersed unstructured proteins embedded within the DNA (e.g., T4 and other T-evens); (3) those with a small number of localized proteins (e.g., Phi29); and (4) those with a reduced, or DNA-free, internal structure composed of more than one type of protein that has a defined shape and volume and occupies an appreciable fraction of the total capsid volume (e.g., T7, PhiKZ). Examples of these capsid structure types can be seen in Fig. 21.1. As noted, despite such significant and fundamental internal structural differences, the degree of DNA compaction found among all four types of phages is comparable in the condensed DNA-containing portions in the capsid interior. Studies have shown that the capsid internal proteins, whether structured or unstructured, found together with the condensed DNA are an internal part of the DNA-free precursor procapsid and associate with the DNA only following packaging rather than being transferred into the procapsid together with the DNA. Indeed, proteins with high DNA-binding affinities found bound to the metabolically active cellular DNA are in fact actively excluded from being packaged together with the DNA. In all four types of bacteriophages, based on the T4 precedent, the capsid also most likely contains many small-molecule DNA counterions. These small molecules taken up into the procapsid can vary in composition due to mutational changes to the host which prevent normal polyamine synthesis but allow

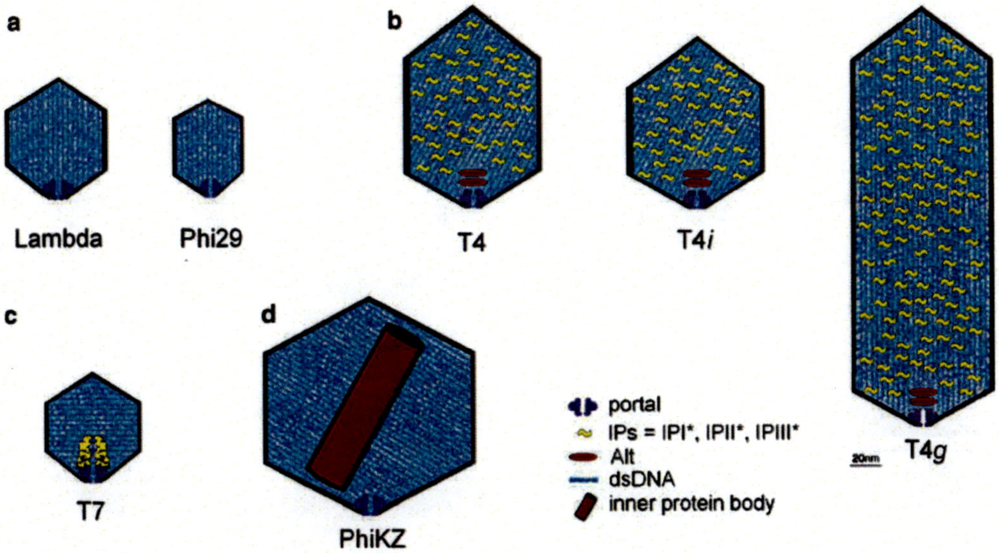

Fig. 21.1 Examples of different internal capsid arrangements found in tailed dsDNA bacteriophages. Tails on capsids are not shown. Portal complex depicted by *purple*; 2.5-nm spaced DNA depicted by *shades of blue*. (**a**) lambda, HK97, and Phi29 have only a "naked" condensed DNA within their capsids; (**b**) the T4 capsid contains more than a thousand dispersed internal proteins (*yellow*) embedded within its DNA, whose copy numbers vary relative to capsid dimensions, as well as several localized proteins whose copy numbers remain unchanged in different-sized capsids (*red*), such as icosahedral capsids (T4*i*) or giant capsids (T4*g*); (**c**) the T7 capsid has an inner core composed of three different proteins (*yellow*); and (**d**) the PhiKZ capsid contains an extremely large inner protein body (*purple*)

viral packaging with divalent cations, apparently without affecting the packaging or the overall packaged DNA structure (see Black 1989 for early references).

In this review, emphasis will be placed on recent advances and discoveries made among these four classes of highly condensed genomes found in tailed bacteriophages, especially among the more recently discovered "environmental" phages. Experimental work bearing on structure and function will be emphasized over theoretical structural and biophysical modeling of the condensed DNA. How modeling of condensed capsid DNA structure relates to DNA packaging force and velocity, and to DNA ejection from the capsid have been discussed recently (Fuller et al. 2007; Rickgauer et al. 2008).

21.2 Naked DNA Condensate Structure in Phage Capsids

The exact three-dimensional structural properties of the packaged double-stranded DNA in bacteriophage capsids remain controversial. The packaged condensate structure found among the four different types of phage capsids shown in Fig. 21.1 displays an ordered ~2.5-nm duplex-to-duplex DNA spacing that arises from the broadly conserved packaging motor mechanisms (Earnshaw and Casjens 1980). A number of models have been proposed to account for such broadly conserved structural features of these capsid condensates in which DNA is sometimes determined to be organized in concentric shells by X-ray scattering and cryo-EM (summarized in Mullaney and Black 1998; Comolli et al. 2008). However, definitive evidence to choose among these structural models (Fig. 21.2) appears to be lacking. In fact, despite some generally similar features such as the degree of packing density and the ordered 2.5-nm duplex spacings found among DNAs organized in shells, at least toward the periphery of the various phage capsids, other structural, chemical, and biological

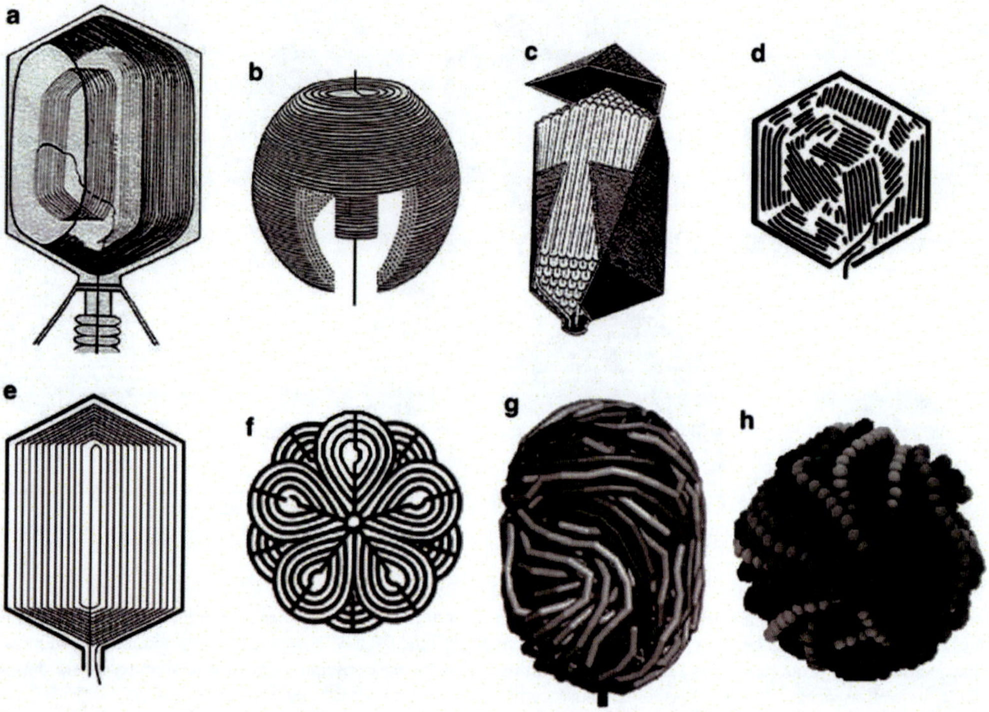

Fig. 21.2 Different models of packaged DNA structure. (**a**) The spool model containing longitudinally packaged DNA segments (Earnshaw et al. 1978); (**b**) spool or concentric shell models (Earnshaw and Harrison 1977; Cerritelli et al. 1997); (**c**) spiral-fold model (Black et al. 1985); (**d**) liquid crystal and hexagonal packing models (Lepault et al. 1987; Leforestier and Livolant 2010); (**e**) icosahedral-bend model which is the same as in (**a**) but sharp bends have been incorporated to follow the angular contours of the icosahedrons (Mullaney and Black 1998); (**f**) a model where the DNA forms 12 icosahedrally arranged *pear-shaped rings* was proposed for the large *Bacillus* phage G (Sun and Serwer 1997). This model is based on that proposed for lambda DNA (Witkiewicz and Schweiger 1985); (**g**) model for completion of genome packaging into shells in Phi29 based on a Monte Carlo simulation (Comolli et al. 2008); and (**h**) model for genome packaging with extensively knotted DNA that may arise following packaging (Marenduzzo et al. 2009). Nb. Capsid shell is not indicated in (**b, f, g, and h**)

features of packaged DNA condensates found among a number of phages argue against a single unitary and deterministic packaged capsid DNA structure. Significant differences found among a number of packaged phage capsids are the following: Depending upon the phage capsid, DNA is oriented transverse (T7) (Cerritelli et al. 1997) or more nearly parallel (T4) (Earnshaw et al. 1978; Lepault et al. 1987) to the head-tail axis of the phage particle; the first DNA end packaged is the last end out (lambda, T7) (Sternberg and Weisberg 1975; Hartman et al. 1979) or the first end packaged is the first end out (T4) (Black and Silverman 1978), and is packaged to a comparable 500 mg/ml density despite being found in the capsid as a protein-free condensate or together with unstructured or DNA-free protein structures (Cerritelli et al. 1997; Mullaney and Black 1998). Moreover, functional DNA-containing "giant" phage capsids (Earnshaw et al. 1978) are packed with DNA to the same 2.5-nm spacing and final DNA density as normal capsids. In these giant phages, the DNA is clearly oriented parallel to the extended long axis of the giant capsids that can be more than tenfold longer than wild-type T4, while maintaining the same width as normal capsids. However, a phage with the unusual highly positively charged base α-putrescinylthymine packages DNA to ~630 mg/ml, except when this modification is removed, showing that the DNA structure and charge critically determine packaging density (Scraba et al. 1983). Phages with unique sequence ends such as lambda significantly under-package deletion containing genomes of ~75% normal length to yield active phages with looser than 2.5-nm of duplex-to-duplex overall spacing. On the other hand, headful packaging phages such as T4, SPP1, or P22 that lack a sequence-specific terminase cutting site package DNA to a conserved ~500 mg/ml density irrespective of deletions because a headful cutting signal is communicated to the high-force packaging motor-terminase complex by the highly conserved portal DNA measuring device. Overall, functional fully packaged giant, more densely packaged positively charged base-containing DNA, and underpackaged short-genome phage capsids demonstrate how variable the capsid-condensate DNA structure can be among different phages.

DNA condensed within capsids of diverse size is often visualized by cryo-EM to be highly organized within shells with apparently gentle overall curvature in outer shells (Effantin et al. 2006; Xiang et al. 2006; Duda et al. 2009). An unresolved question is whether the DNA in such shells is uniformly bent in the wide outer shells even where contact is made with capsid angular icosahedral edges (e.g., Fig. 21.2e). Additional discontinuities in bending are expected in the inner shells near to the center of the icosahedron because packed DNA is observed to have essentially equal density in these portions of the capsid (Leforestier and Livolant 2010). This requires that DNA should be sharply bent below its persistence length in these regions. Overall, it appears uncertain whether DNA is arranged largely or exclusively with gentle bending (Fig. 21.2a, b), or with discontinuities – less gentle bends (Fig. 21.2e–g) or sharp kinks (Fig. 21.2c, d) – throughout the structure or only in those regions of the structure that apparently require this. Those DNA condensate models shown in Fig. 21.2 a–c, e, and g, whether or not they contain bends or kinks, would display DNA shells with generally similar overall structures in averaged cryo-EM reconstructions. It should be noted that DNA will readily undergo bending or kinking under certain constraints or interactions, even in the absence of high motor force, such as possibly with DNA-binding internal protein components of the capsid (see below) or domains of the main head protein itself which protrude into the capsid interior, as seen in the mature N4 capsid (Choi et al. 2008). Similarly, the T7 capsid apparently imparts angularity in portions of the outermost DNA shell (Agirrezabala et al. 2005).

Also, because of the highly energetic and rapid (up to 2,000 bp/s) packaging motor, it probably cannot be assumed a priori that minimum energy structures within the DNA must predominate in determining its structure. For example, it is frequently assumed in modeling (and other) studies that DNA is condensed within the capsid from outside to inside (Jiang et al. 2006; Petrov et al. 2007). However, in fact experimental evidence including ion etching of the intact fully packaged particle apparently establishes the opposite direction of packing condensation for phages T4 and lambda (Black and Silverman 1978; Black et al. 1985; Mendelson et al. 1992). EM observations of packaging intermediates in a number of phages also apparently show that the DNA is not initially preferentially packaged against the inner surface of the capsid, e.g., most recently in T3 (Fang et al. 2008).

It is unlikely that cryo-EM or other techniques presently have the resolution to resolve these questions of condensed DNA fine structure that requires that the entire length of an individual packaged duplex is resolved structurally at the nucleotide level. A number of techniques including Raman spectroscopy and solid-state NMR show that the condensed DNA is predominantly B form (Overman et al. 1998; Yu and Schaefer 2008). At the same time, evidence for some departure from uniform B-form structure comes from chemical evidence for kinks in the condensed DNA, in addition to known condensation below persistence length (see Black et al. 1985; Serwer 1986). Moreover, additional evidence against uniform B form comes from the observed hydrolysis of packaged T4 DNA in active phage particles by packaged staphylococcal nuclease whose enzymatic specificity is incompatible with B-form DNA cutting (see below). These experiments also show that many or a few molecules of the packaged nuclease can diffuse within the fully packaged DNA – condensed to ~500 mg/ml DNA contains 75–85% (vol/vol) water – to reduce it more or less rapidly, varying with the copy number, to ~160 bp limit digest lengths. Cryo-EM observations of such limit phage T4 head digests, however, show relatively small effects on DNA condensate structure. The observations also show the T4 DNA to be oriented differently from T7 in an elliptical coaxial spool whose long axes are oriented at approximately 30° tilt to the portal axis (N. Cheng, B. Mai, M.E. Cerritelli, L.W. Black, and A.C. Steven, personal communication).

A number of other fundamental properties of packaged DNA have recently been determined. It was known that one unique end of the packaged phage DNA descends through the portal to be injected into the infected host. Such a DNA duplex descending through the portal has recently been imaged by cryo-EM; the structure of the portal proximal DNA can also be visualized, and packaging is seen to change the portal structure (see below). Several studies had shown by classical electron microscopy that a single end is attached to the phage lambda (and other) tail(s) following disruption, thereby demonstrating that following packaging, but before ejection, DNA rearranges and descends through the portal and into the tail (Chattoraj and Inman 1974; Saigo and Uchida 1974; Thomas 1974; Saigo 1975). Recently, it has also been demonstrated by fluorescence correlation spectroscopy-Förster resonance energy transfer (FCS-FRET) and single-molecule FRET (smFRET) analysis of in vitro packaged Förster dye pair terminated 5- and 50-kb DNAs that the two DNA ends packaged into phage T4 capsids are held 80–90 Å apart; apparently both are within the portal channel of the T4 capsid (Ray et al. 2010). This structure suggests that a loop rather than an end of DNA is translocated by the T4 motor complex. Other less direct evidence (knotting and ion etching of DNA) is consistent with comparable colocalization of the two packaged DNA ends in other phages. However, colocalization of both packaged DNA ends such as in the phage T4 capsid also appears to place constraints on comparably packaged capsid DNA structures. How can this be reconciled with extensively knotted DNA derived from phage P4 capsids (Arsuaga et al. 2002)? DNA modeling suggests that the many knots found within DNA originating from the packaged phage P4 capsid would be delocalized and not interfere with DNA ejection from the capsid (see Fig. 21.2h) (Marenduzzo et al. 2009). However, if both P4 DNA ends are also anchored at the portal, it is not clear how extensive topological knotting could occur within the capsid; rather the knots may more likely arise following release and rearrangement of the condensate, or at least of its ends, from the native capsid structure.

21.3 Phage Capsids with DNA Condensed Together with Internal Proteins

In addition to the condensed genome, numbers of phage capsids contain a diversity of internal proteins, despite, as mentioned above, their DNA retaining conserved features, including a density of ~500 mg/ml and displaying the characteristic 2.5-nm duplex-to-duplex spacing. These proteins vary in length, copy number, function, and localization within the capsid, illustrating not only the remarkable diversity that occurs between different phages, but also highlighting the fact that genome packaging occurs successfully despite numerous internal capsid conditions/states.

21.3.1 Phage Capsids with DNA Condensed Together with Dispersed Unstructured Proteins

The most intensively studied phage capsid internal proteins (IPs) are those of phage T4 and relatives. In T4, more than 1,000 molecules of three different small (~8–20 kDa) proteins are packaged into DNA-free proheads by a capsid-targeting sequence (CTS) mechanism (Mullaney and Black 1996) (Fig. 21.3). IP incorporation into the procapsid is based on the affinity of a highly conserved consensus sequence found at the *N*-terminus of each IP (the CTS) that is responsible for incorporation of these proteins into the assembly and structure determining core of the procapsid. The affinity of the CTS for the core is such that it can be exploited to package diverse, nonphage proteins into the capsid (e.g., Fig. 21.3b, ii, iii, and iv). Processing of the assembly core by the morphogenetic protease (gp21) converts the essential shape-determining core protein (gp22) to small peptides, most of which exit the capsid, although a few of which are also retained within the mature capsid. Gp21 also modifies all of the IPs by removal of the 10–19-residue-long N-terminal CTS peptides. The internal proteins are unstructured and dispersed within the DNA condensate as shown by the ability of IP-nuclease

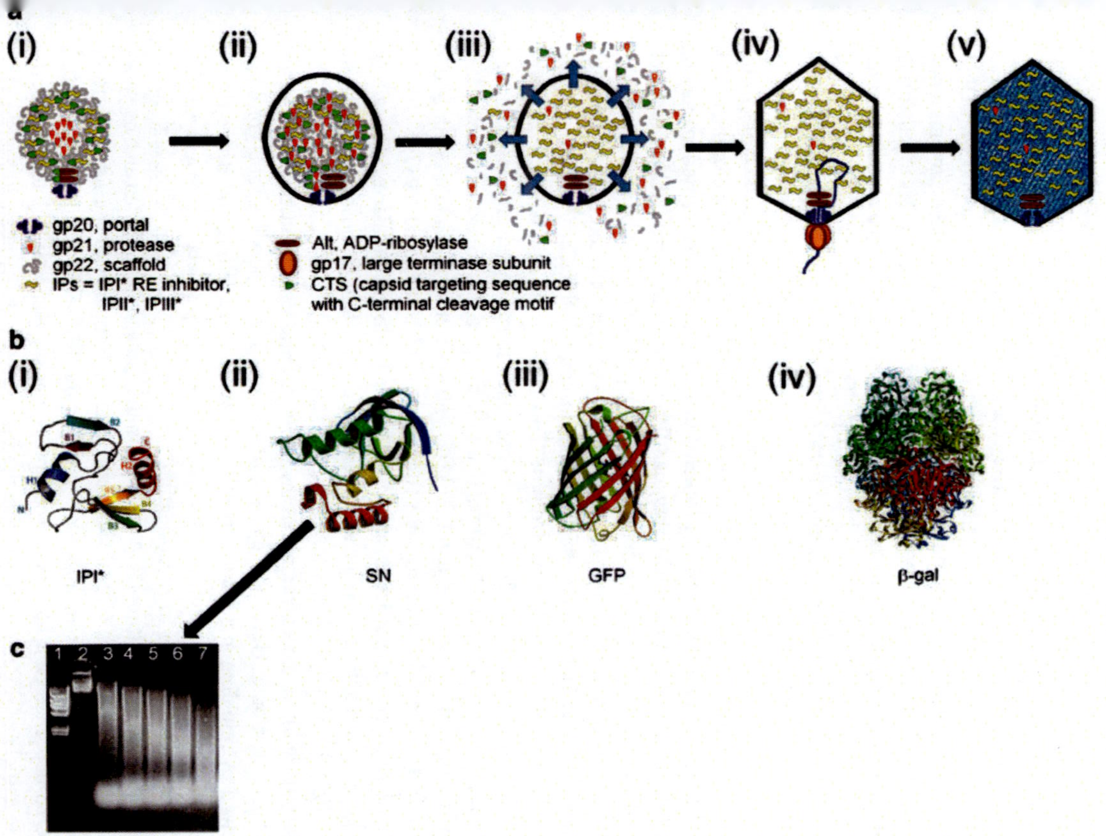

Fig. 21.3 **(a)** Overview of the packaging of proteins by the T4 capsid targeting sequence (CTS) system: (i) capsid-size prohead core, minus the outer shell and DNA, is seen to assemble in vivo; (ii) completion of the capsid shell triggers cleavage of prohead proteins; (iii) small fragments of processed prohead proteins (including scaffold protein, CTS sequences, and protease) exit the capsid presumably via small holes in the shell prior to maturation–expansion; (iv) phage DNA is packaged into the capsid by the terminase; (v) mature T4 head. **(b)** Examples of proteins packaged into the mature T4 phage capsid: (i) structure of restriction endonuclease inhibitor IPI* (9 kDa) (the processed form of IPI) found in wild-type T4; examples of proteins packaged into the phage capsid using the phage-derived expression, packaging, and processing (PEPP); (ii) SN nuclease (15 kDa) (Chen et al. 2000); (iii) GFP (27 kDa) (3ADF) (Ebisawa et al. 2010); (iv) β-galactosidase (540-kDa tetramer) (1DP0) (Juers et al. 2000) (Nb. Structures not to scale). **(c)** Genomic DNA of recombinant T4 phage digested by SN nuclease within the capsid. *Lane 1*, lambda *Hin*dIII standard; *lane 2*, control DNA from recombinant T4[CTS▼IPIII▼SN], not treated with Ca^{2+}; *lanes 3–7*, DNA from recombinant T4[CTS▼IPIII▼SN] incubated with Ca^{2+} for 1, 10, 30, 120 min, and 16 h, respectively (▼refers to a T4 protease processing site). Image reproduced from Mullaney and Black (1998)

fusion proteins to diffuse within and cut all the condensed DNA (Fig. 21.3c), IP protection from ion etching by the DNA (Black et al. 1985), and IP release and sedimentation together with packaged DNA released in condensed form by dissolution of the capsid (Zachary and Black 1991).

Thus, it appears that these proteins fit within the conserved DNA condensate structure without loss of function. In addition, all three IPs can be eliminated from the capsid by mutation with only minor effects on procapsid assembly and shape and without significantly affecting the amount of DNA packaged within the capsid. Packaging of each IP is independent of the presence of the two other IPs, and in giant particles, the IP copy numbers per capsid increase with the available volume, presumably because the interior of the giant procapsid is seen to be filled with a giant scaffolding-core (composed of gp22) that also contains increased numbers of IPs (see Fig. 3a in Black et al. 1993).

The three internal proteins of T4 all have a net positive charge (all have predicted pI>9) and are lysine rich relative to other T4 proteins, properties that might enhance their interaction with DNA. Solid-state NMR showed that there are specific IP-DNA contacts, and that these proteins could neutralize some (<5%) of the DNA charge balance (Yu and Schaefer 2008). Hence, this together with the fact that the small internal proteins are dispensable (Black et al. 1993) implies that their presence in the capsid is not essential for the stability of packaged DNA. This is supported by the fact that in other T-even phages the IPs are highly polymorphic and some of the internal proteins packaged by the CTS mechanism are small acidic or neutral proteins (Black et al. 1993; Repoila et al. 1994). Potentially, the main function of the IPs is to interact with the host in a manner that will enhance infection, as the T4 IPs are all injected into the host cell. Supporting this suggestion is the fact that one of them, IPI, protects the T4 DNA against a highly specific DNA modification dependent host restriction enzyme (Bair et al. 2007). Conditions or hosts in which the abundant IPII and IPIII are essential have not been found. The dimensions of IPI suggest it can be ejected through the portal and tail tube without unfolding thereby promoting immediate inhibition of its nuclease target in the infected host (Rifat et al. 2008). However, it appears probable that other such internal proteins (particularly larger ones) are ejected unfolded and are then refolded in the infected host. This is likely the case with *E. coli* phage P1 that may resemble the T-even phages with respect to packaged unstructured proteins, although little is known about their copy numbers and locations within the capsid. However, it is known that phage P1 injects two large and processed (60 and 200 kDa) anti-restriction endonuclease inhibitor proteins, dar A and dar B, to protect the injected DNA (Iida et al. 1987; Streiff et al. 1987).

That such permutations can occur was demonstrated by the packaging of β-galactosidase via an IPIII-β-galactosidase fusion and its subsequent ejection into, and activity within, the host cell (Fig. 21.3b, iv) (Hong and Black 1993). Physically, it is not feasible that packaged β-galactosidase could exit the capsid in its folded state. In addition, the appearance of the 540-kDa tetramer-dependent activity in the infected cell requires host chaperones. Hence, the ejection of the unfolded, or possibly semi-unfolded, protein must be followed by refolding and multimerization within the host (Hong and Black 1993). Possibly, unfolding of both unnatural and natural internal proteins, such as β-galactosidase and Alt, that are folded and active upon synthesis prior to encapsidation, is a secondary function of the high force-generating packaging motor and the resulting DNA pressure in the full capsid. GFP can also be packaged within viable T4 phage particles as CTS fusions and gradual formation of the fluorophore in mature phage shows the mobility of the encapsidated proteins; the IPIII-GFP fusions in DNA full heads, as opposed to CTS-deposited GFP, have fluorescent properties suggesting IPIII interacts specifically along the DNA to associate GFP monomers within the full head (Mullaney et al. 2000).

21.3.2 Phage Capsids with DNA Condensed with Localized Proteins

A number of unrelated phages contain proteins within their capsids that are injected into the host cell, ahead, or with, the phage DNA, but in contrast to the T4 IP proteins, these proteins have more specific locales, typically in close proximity to the portal complex. Examples of localized internal proteins are found in *E. coli* phages N4 and T4, and *Bacillus* phages Phi29 and SPP1. Despite having similar locales, these localized inner capsid proteins have a remarkably eclectic set of functions and several of them are apparently multifunctional. One localized protein that stands out for its unusualness is the N4 gp50, a virion RNA polymerase (vRNAP), the first RNAP to be identified as being packaged within the capsid of a tailed phage. Gp50 transcribes early phage promoters that are composed of four structural motifs at hairpin/stem structures (Gleghorn et al. 2008). The production of these structures and their binding by the vRNAP is assisted by host proteins (*E. coli* DNA gyrase and single-stranded DNA-binding protein) (Glucksmann et al. 1992; Gleghorn et al. 2008). The vRNAP is extremely large (~380 kDa) relative to its mitochondrial and T7-like RNAPs homologues (Cermakian et al. 1996) and comprises three main domains, a central domain responsible for RNAP activity, an *N*-terminal domain that is required for injection of the first 500 bp of the genome and a *C*-terminal domain that is required for its incorporation into the head (Kazmierczak et al. 2002). About four copies of the N4 vRNAP are present per virion, and cryo-EM comparison of wild-type versus gp50-minus mutants indicate that the vRNAP is located at the base of an inner core, above the portal in such a manner that at least part of the vRNAP is posited close to the internal entrance to the portal channel (Choi et al. 2008).

In addition to its dispersed high-copy IP proteins, T4 also has several proteins, Alt and gp2, whose positions within the capsid are localized. Alt is a large protein (~75 kDa) that is packaged into the capsids via its CTS. Alt is responsible for ribosylating host RNAP subunits (and other host proteins), resulting in preferential recognition of phage early promoters over host promoters (Depping et al. 2005). Alt is probably located above, possibly on, the portal complex as supported by the fact that in giant phages the copy number of Alt [about 40 copies (Black et al. 1993)] remains constant despite increases in the copy numbers of the three IP proteins proportional to increases in head size (Aebi et al. 1976; Bijlenga et al. 1976). Some finer details of this process remain to be clarified, such as why does Alt obtain such a very specific location within the head versus the dispersed locations of the IP proteins? Alt has the shortest CTS sequence, although comparable in sequence and also processing by the morphogenetic gp21 protease to the other IPs, leading to the question whether different portions of the Alt protein, other than the CTS sequence, may establish specific affinity for the portal protein.

N4 gp50 and Alt potentially have significant effect on the packaging of DNA and its final structure based on the volume they would occupy in the phage head, and the fact they are within the head prior to DNA packaging. Other proteins with specialized locations within the head potentially have less impact on the overall structure of packaged DNA as they are present in low copy number and/or have low molecular weights. However, their presence is likely to influence the structure of the packaged DNA in specific regions, particularly near or within the portal complex. Notable among such proteins are those that are noncovalently or covalently bound to the genome termini, such as T4 gp2 and Phi29 gp3, respectively. T4 gp2 prevents restriction of the phage genome in the host cell by exonuclease V (Lipinska et al. 1989) and likely also functions as a "genome plug" (see below) within the head as mutations in it produce unstable heads that leak DNA; however, the gp2 protein can be added to the already fully packaged head to complete active and stable phage particles (Wang et al. 2000). Since, as mentioned above, the ends of the T4 genome have been determined to be held 80–90 Å apart, it is likely that each copy of gp2 has a specialized location in close proximity to the portal. The main function of Phi29 gp3 is to act as a primer for DNA replication (Peñalva and Salas 1982; Watabe et al. 1983); however, it may also function as a genome plug (Xiang et al. 2006).

Genome plugs are proteins whose presence prevents the packaged genome from escaping through the portal vertex until the appropriate time for DNA ejection or descent into the tail tube. Genome plug proteins that are incorporated into the head are all likely to have some small effect on the structure of the packaged genome in the region close to where they reside. Due to the essential nature of such protein(s), all tailed phages are likely to have proteins with these functions, but as exemplified by T4 gp2 and Phi29 gp3, there is potentially a diversity of nonhomologous proteins that undertake this function. Conversely, it is also feasible that certain genome plug proteins may be conserved between related and even lesser-related phages. For instance, the minor capsid protein gp7 (NP_690662.1, also referred to as SPP1p011) of siphovirus SPP1 is a good candidate for a genome plug as it is present in one to two copies per virion, potentially injected into the host cell, and exhibits both DNA-binding capacity and affinity with the portal protein (Stiege et al. 2003). Homologues to gp7 exist in a wide diversity of phages (Stiege et al. 2003), as illustrated by its homology to the myovirus Mu gpF super family (cl10072) (e-30). However, for some potential genome plug proteins, such as T7 gp6.7 (9.2 kDa) (Kemp et al. 2005), determination of whether such a protein is widely conserved or not is complicated by their small size, which makes sequence comparisons difficult. It should be noted that the discussion above refers to genome plugs that are probably incorporated into the capsid, as this review focuses on DNA packaging and the factors that most immediately affect it. However, it should be noted that several phages, such as SPP1 and HK97, have neck or connector proteins that form rings under the portal structure which also function as genome plugs (and in these instances are homologous, belonging to large families) (Lhuillier et al. 2009; Cardarelli et al. 2010). Potentially, some phages, such as SPP1, have several proteins, in different locations, that participate in this function.

21.3.2.1 P22: A Localized Collection of Ejection Proteins?

The *Salmonella* podovirus P22 also potentially has localized proteins, one or more of which may act as a genome plug. These proteins, gps 7, 16, and 20 (whose genes are actually clustered), are known to be in the mature head and injected into the host cell, hence, they are referred to as pilot/injection or ejection proteins (Chang et al. 2006; Lander et al. 2006). Potentially, these ejection proteins are responsible for densities localized in the central channel of the portal, and/or in the channel of the tail hub (Chang et al. 2006); however, these densities observed by cryo-EM may, at least in part, be also due to the extended, large (12 × ~80 kDa) portal structure itself whose crown projects well into the interior of the capsid (Tang et al. 2011). The proteins within the P22 head are of particular interest as its genome is packaged as a coaxial spool (Zhang et al. 2000; Chang et al. 2006; Lander et al. 2006), apparently in a similar manner to that of the podoviruses below (e.g., T7) whose heads contain inner cores which are likely to have a major influence on the structure of packaged DNA. However, unlike those phages the P22 capsid apparently does not contain a large core, except for the large portal mentioned. Large portal proteins have been indicated as a cause of coaxial spooling in simulation studies (Petrov et al. 2007). Hence, it will be interesting to determine the impacts of the P22 portal and its ejection proteins on DNA packaging and structure.

21.3.3 Phage Capsids with Multiprotein Internal Structures

The capsids of some podoviruses and myoviruses have been determined to contain multiprotein core structures within the densely packaged DNA. These structures are roughly cylindrical in shape, positioned above the connector complex, and typically occupy a considerable volume in the central region of the capsid, normally occupied by dsDNA in other phages, such that they are large enough to be visualized by normal electron microscopy. Their presence raises numbers of questions as to their function(s), particularly their impact on packaged DNA structure, and the packaging and ejection processes. Of the known core structures, that of the *E. coli* podovirus, T7, is a paradigm, illustrating that these structures can be morphologically complex and have functions of major consequence to various stages of the phage lifecycle.

The 510-Å-diameter T7 capsid contains a 265-Å-high and 175-Å-wide inner core that is surrounded by about six coaxial/concentric rings of dsDNA (Cerritelli et al. 2003a; Agirrezabala et al. 2005). The T7 core encloses a channel/cavity (varying in diameter from 35 to 110 Å) that is continuous through to the outer side of the capsid as a consequence of its docking onto the portal complex. In mature capsids, this inner channel contains a density attributed to a terminal segment of DNA (Agirrezabala et al. 2005). The core structure is not a simple cylinder, but is composed of three domains, each with a different symmetry 12-, 8-, and 4-fold (Cerritelli et al. 2003b; Agirrezabala et al. 2005). The 12-fold symmetrical domain most probably represents the connector complex, whereas the other two domains are composed of three proteins gps 14 (21 kDa), 15 (84 kDa), and 16 (144 kDa), present in 10–12, 8, and 4 copies, respectively (Cerritelli et al. 2003a; Agirrezabala et al. 2005; Kemp et al. 2005). The genes encoding these three proteins are clustered, and they are transcribed together (McAllister and Wu 1978).

As the T7 core structure extends so far into the capsid, it is likely to have a significant influence on the packaging of DNA into coaxial spools (Cerritelli et al. 1997), possibly preventing the formation of concentric spools (Petrov et al. 2007). However, the T7 core is not only important for the structure of DNA in the mature head, it most likely acts as a conduit during DNA packaging and/or release (as implied by the presence of DNA within it) and may assist the assembly of proheads into correct dimensions (Agirrezabala et al. 2005). Indeed, evidence supports the core being multifunctional, playing different and complex roles in different parts of the phage life cycle. Its component proteins, known to be essential and injected into the host cell (Garcia and Molineux 1996), have been proposed to create a channel that spans the host cell envelope (i.e., become an extensible tail of sorts), thereby enabling the phage DNA to enter the host cell and overcome the obstacle posed by the short tail of the mature virion which prevents it from reaching the host cytoplasm (Molineux 2001). Support for this includes electron microscopy of partially emptied bacteriophage T7 capsids showing a needle-like extension, about twice the length of the normal T7 tail (Serwer et al. 2008), the *N*-terminal domain of gp16 has transglycosylase activity (Moak and Molineux 2000; Moak and Molineux 2004), and membrane localization experiments have shown that gp14 localizes to the outer membrane, whereas gp15 and gp16 are found predominantly in both soluble and cytoplasmic membrane-bound forms. Notably, recent studies have demonstrated that a DNA-gp15-gp16 motor is responsible for the translocation of the first 850 bp of the T7 genome into the host cell (Chang et al. 2010a). Another interesting feature of the three T7 core proteins is that their injection into the host implies that they would need to almost completely disassemble to be able to exit the head via the narrow connector channel (Kemp et al. 2005). Hence, the T7 core structure is likely to be intriguing, not only for its morphogenesis, but also for its disaggregation pathway.

Core structures have been identified in other podoviruses, such as *Prochlorococcus* phage P-SSP7, another member of the subfamily *Autographivirinae* to which T7 belongs, as well as in podoviruses that are more distantly related to T7 and belong to genera outside of its subfamily, such as *Salmonella* phage Epsilon15 and *E. coli* phage N4 (Lavigne et al. 2008). While the cores of these phages are not as well described as the T7 core, there is increasing evidence that they share a number of characteristics. Notably, the DNAs of all these phages, or at least the most outermost strands that have been resolved, have coaxial spools, or hexagonal packing (Leforestier and Livolant 2010) a formation that must most certainly be affected by the presence of the core in each phage. The core structures of these phages are all likely to contain an inner channel in which a terminal segment of DNA resides, either based on the observation of densities interpreted as DNA in Epsilon15 (Jiang et al. 2006) and P-SSP7 (Liu et al. 2010) or the presence of cylindrical structure as for N4. Interestingly, the extreme terminus of the DNA that descends into the portal in P-SSP7 is split into single-stranded DNA (Liu et al. 2010). The core proteins of Epsilon15 and P-SSP7 are also probably injected into the host cell, based on the absence of core density in empty particles (Chang et al. 2010b; Liu et al. 2010). In addition, Epsilon15 particles adsorbed to host cells gain a tubular density in the host periplasm (in conjunction with the disappearance of the core), suggesting that, as in T7, the Epsilon15 core protein(s) form a tunnel through which the phage DNA passes to the host cytoplasm (Chang et al. 2010b).

However, in addition to shared common features with the T7 core, the cores of Epsilon15, P-SSP7, and N4 also display dissimilarities to the T7 core. These differences include size differences, the Epsilon15 core is slightly smaller than that of T7 (200 Å high and about 180 Å) (Jiang et al. 2006), while the N4 core is much smaller (110 Å in height and ~80 Å in width) (Choi et al. 2008). The Epsilon15 core potentially lacks symmetry (Jiang et al. 2006), and no symmetry could be identified for the P-SSP7 core (Liu et al. 2010), although both these results could be a result of the difficulties of resolving the components of such small structures. The N4 core structure is apparently different from that of T7, and is composed of at least two components, an upper disk-like structure above a lower, longer cylindrical density that sits above a density attributed to the vRNAP (see above) (Choi et al. 2008).

Better comparisons of the cores of these phages with those of T7, and other phages, will be enhanced when the core proteins, and their copy numbers and functions are known. Currently, no protein in any of these three phages has been structurally allocated to the core, although candidates for core proteins exist in Epsilon15 and P-SSP7 based on other considerations. For instance, Epsilon15 gp17 has been named a core protein (Chang et al. 2010b), probably based on it being identified as a component of the phage particle by mass spectrometry (MS) (Jiang et al. 2006), and in an appropriate copy number (7.5 copies) (Kropinski et al. 2007). Epsilon15 gp15 is also a good candidate for a core protein, being identified by MS (Jiang et al. 2006) and present in the virion in about 20 copies (Kropinski et al. 2007). Epsilon gp14 is another potential core candidate having homology to the T7 core protein gp14 using Hidden Markov Modeling (2.5×10^{-12}) (our observation). Proteins identified by MS in P-SSP7 (gp15, gp16, and possibly gp14, equating to proteins 33, 34, and 35, respectively) are also good candidates for core components in this phage (Liu et al. 2010). It is also of note that the genes encoding the main candidates for inner core proteins in Epsilon15 and P-SSP7, like T7, are located within the same genome locale, a characteristic of phage virion genes with related functions (such as head genes). Such clustering of genes with related functions enables their transcription, and subsequent translation, to occur at a time appropriate for morphogenesis. Gene clustering may also protect essential structural protein interactions from being lost by recombinational exchanges with related coinfecting bacteriophages genomes. Currently, there is less indication of candidates for the components of the N4 core; however, it also has proteins identified as a component of the virion by MS whose functions are yet to be determined (e.g., gps 51, 52, 54, and 67) (Choi et al. 2008). Similarly, it will need to be determined whether the N4 vRNAP, which sits below the core, is technically a part of the core or not.

Of the proteins known to be components of a core structure in a mature capsid, there is no set of proteins, or even a single protein, that is strongly conserved at the sequence level (such as at the level of conservation seen between the major head proteins or terminase proteins of many tailed phages) between the different types of podoviruses with cores discussed above. This diversity of core proteins suggests that their current functions are also likely to be highly diversified. However, it is possible that some core proteins could share a distant ancestor, but have diverged greatly from that original state. This is reminiscent of scaffold proteins, found within the procapsid but not the mature capsid, between which there is typically little or no indication of a common ancestor at the sequence level in phages of different types. However, the fact that all share a conservation of structure (highly alpha helical) and function, argues otherwise (Steven et al. 2005). Considering the T7 precedent, where each of the core proteins has function(s) individualized to enable/enhance infection of its host, diversity in core proteins could potentially be akin to the diversity seen in phage tail fiber proteins, and in the T-even internal proteins. These proteins are under pressure constantly to adapt to changes in the host to preserve the ability of that phage to infect that host. Whether the core proteins in different phages have different impacts on the packaging, structure, and release of DNA, or whether they retain certain traits or structural elements to keep changes in DNA interactions to a minimum, remains to be determined.

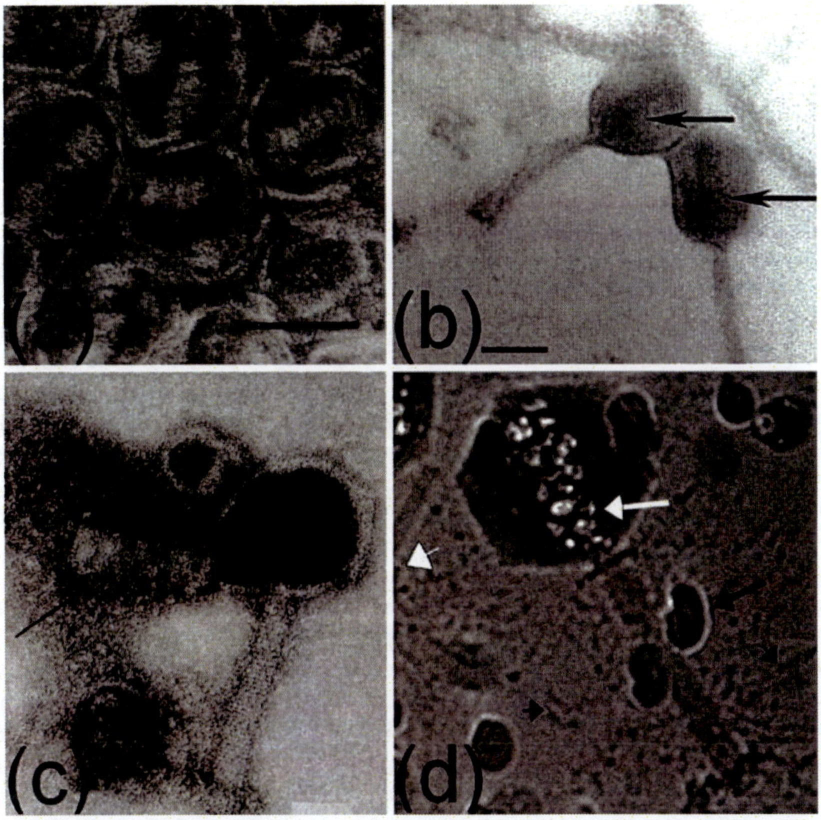

Fig. 21.4 The inner body found in the large myovirus, PhiKZ, and its relative 201phi2-1. (**a**) Electron micrograph of disrupted heads of the PhiKZ mutant ts759. Image reproduced from Krylov et al. (1984); (**b**) electron micrograph of partially disrupted PhiKZ particles. *Arrow* indicates the inner body. Image reproduced from Matsko et al. (2001); (**c**) electron micrograph of a PhiKZ particle disrupted by freeze–thawing. *Arrow* indicates inner body released from disrupted head. Image reproduced from Krylov et al. (1984); (**d**) cryo-electron micrograph of 201phi2-1. The *large white arrow* highlights bubbles which formed in the head apparently around the protein body. Image reproduced from Thomas et al. (2008)

21.3.3.1 Inner Bodies Found in Giant Myoviruses

A group of *Pseudomonas* phages, which includes PhiKZ, EL and 201phi2-1, also contains a multi-protein inner structure within their unusually large ($T=27$) capsids, based on electron microscopic observations (Krylov et al. 1984; Matsko et al. 2001) (Fig. 21.4). This structure, or inner body, is considerably larger than that of the T7 inner core, being roughly 90-nm in length and 35-nm in diameter in PhiKZ (Krylov et al. 1984). In fact, the volume occupied by this structure is potentially greater than the total capsid volume of some smaller phages. The PhiKZ inner body has a spring- or spool-like appearance and may disappear after phage particles adsorb to bacteria (Krylov et al. 1984) (Fig. 21.4c). However, whether this structure has an inner channel and/or is anchored to the portal complex, as in the T7 core, is yet to be determined. Recently, mass spectrometric studies of PhiKZ and EL (Lecoutere et al. 2009), and 201phi2-1 (Thomas et al. 2008, 2010), have supported earlier indications that these phages were structurally complex, based on the numbers of different proteins identified in their virions (>62). These phages also have an exceptionally large number of different proteins associated with their heads, even in comparison with T4 whose mature capsid is composed of 13 different proteins (Black et al. 1993). The latter observation is based on (1) the identification of 30 proteins in a tailless mutant of PhiKZ (Lecoutere et al. 2009) and (2) that nineteen 201phi2-1 (Thomas et al. 2010) and eighteen PhiKZ (Thomas and Black, unpublished) proteins are processed at a cleavage motif that ends in a glutamic acid residue comparable with the morphogenetic protease processing motif (LxE or IxE) in T4 head proteins. The latter suggests these proteins are processed by a T4-like prohead morphogenic protease, and are therefore head proteins. Overall, in these phages, there are enough different proteins to account for the capsid shell (including vertex and portal structures) as well as protein mass to account for the inner body structure.

The function(s) of the inner body of PhiKZ, and those of its relatives, is unknown, but potential functions include roles in DNA compaction, packaging and/or injection, or in strengthening of the large capsid (Krylov et al. 1984). Two other potential functions for the inner body are that it acts to regulate capsid size [i.e., analogous to the shape-determining T4 morphogenetic core (or to a tail tape measure protein)] and/or represents a reservoir of proteins that are injected into the host cell to aid the initial steps of infection and replication. A good candidate for a protein with the latter function is the multisubunit vRNAP conserved and packaged within each of these phages (Thomas et al. 2008). There are many unknowns regarding the inner bodies of PhiKZ and its relatives; however, based on the structural and functional complexity of the T7 core, the structure and function(s) of these large inner bodies represent fascinating subjects for further studies. One particularly intriguing aspect to resolve will be if the size of these inner bodies is constant in different members of this group of phages, or whether the size, composition, and even orientation within the capsid varies from phage to phage. Thus far, it appears there are variations, for instance between PhiKZ and EL, because not all structural proteins are identifiable as homologous between these two phages (Lecoutere et al. 2009). Also, it has been observed that the capsids of these two phages have apparently very similar dimensions (Hertveldt et al. 2005), whereas their genome lengths vary by nearly 70 kb. Since the density of packaged DNA appears to be similar in these phages, a larger inner body in EL, relative to that of PhiKZ, may play a role in determining packaged genome size.

21.4 Conclusions

Despite the generally common features of DNA packaging, such as the packaging motor itself and the resulting conserved density of DNA packaged within capsids, there are clearly different packaged DNA structures found in different large, tailed dsDNA phages. However, the reasons for these differences are in many cases still obscure. For instance, for certain phages, such as T7, the inner core is thought to have a major impact on coaxial spooling of the DNA within the capsid. While the large portal of P22 (gp3, 12 × 82 kDa) forms a complex that apparently extends well into the capsid and may similarly impart coaxial spooling (Tang et al. 2011), the core or portal is unlikely to be the only factor to result in this structure. Thus in the case of HK97, there does not appear to be an unusually large portal subunit (gp5, 12 × 47 kDa) for this to occur, and HK97 does not have a core, although its DNA is coaxially spooled (Duda et al. 2009). Overall, while the packaged DNA structure is necessarily expected to be significantly affected by the presence of inner head proteins, either structured or unstructured, often removal of these proteins (e.g., of over 1,000 T4 internal proteins) has no significant effect on either packaging or DNA ejection. Moreover, it is unknown whether observed differences in packaged DNA structure, such as the transverse packaging in T7 versus the more nearly parallel packaging found in T4 capsids, relate to the variable speeds and mechanisms (motor or nonmotor driven) DNA ejection into host bacteria that are observed; e.g., ≥10–15 min in T7 (Molineux 2001) to less than 2 min in T4 (Kalasauskaite et al. 1983) for complete ejection. For biophysical modeling studies of DNA condensation to prove more valuable for understanding both packaging and ejection, more experimental evidence from studies of different phages should be incorporated into the models or reasons should be given for excluding such evidence (e.g., see experimental determination of the direction of T4 and lambda DNA packaging condensation above).

In spite of conserved features of the packaged DNA structure, internal/injected capsid proteins have been observed in numbers of unrelated phage groups, suggesting that they may be even more common than currently appreciated. These capsid proteins are diverse, present in different copy numbers, found in discrete structures or dispersed within the DNA, and with different modes of entry into host cells and different functions, which are often only essential functions in a narrow set of specific hosts or conditions for a given phage [e.g., T4 IPI (Rifat et al. 2008)]. The diversity of inner capsid proteins in different phages with specialized locations shows some common themes, including that these proteins are apparently often packaged within close proximity to the portal channel for injection into the host cell. As already noted, these proteins enter the DNA-free precursor prohead before the DNA is packaged. The presence of portal-localized inner proteins raises a number of interesting questions, particularly as to how DNA packaging occurs around or through them. Do any of these proteins have any specific interactions with packaged DNA? Or conversely, do some of these proteins have limited DNA interactions so as to prevent any hindering of their injection into the host cell, possibly ahead of the DNA? Other questions regarding these proteins include the mechanisms by which they are targeted into their respective capsids. In the T4-like phages, it is clear that affinity of an N-terminal CTS peptide for the major shape-determining scaffolding core protein allows virtually any protein to which the CTS is joined to be incorporated into the prohead. But what of other phages, do they still utilize targeting affinity with the core or scaffold, or with other proteins? Supporting the use of other proteins was the finding by Stiege et al. (2003) that their candidate for the SPP1 genome plug (gp7) is apparently incorporated into the capsid via its affinity for the portal protein. A related intriguing feature of injected capsid proteins to be resolved is the mechanisms by which large proteins such as Alt, Dar, N4, and PhiKZ RNAPs, and even a nonphage protein such as β-galactosidase, are apparently unfolded within the capsid, probably while the DNA is being packaged, to allow their ejection through the narrow portal and tail tube channels. Could this unfolding be a secondary but essential function of the DNA packaging motor mechanism that, through high DNA pressure within the fully packaged capsid, promotes the ejection of unfolded or partially unfolded large proteins that are then refolded upon injection to display their sometimes essential enzymatic functions in the host cell?

Ejection of highly structured inner capsid core proteins, whose structure persists within the densely packaged DNA environment (e.g., T7 and possibly PhiKZ), raises the question of how such structures are disassembled to allow exit from the capsid. The role of the T4 core in determining head size suggests the possibility that a comparable sizing mechanism may operate in even larger capsids, e.g., may be one function of the capsid inner body in PhiKZ-related phages. However, it is likely that the proteins forming such bodies have multiple functions, including incorporation and ejection of proteins that enhance infection.

Acknowledgments The authors thank Julienne Mullaney, Ian Molineux, and Alasdair Steven for their reading of this manuscript and their helpful suggestions.

References

Ackermann H-W, Tremblay D, Moineau S (2004) Long-term bacteriophage preservation. WFCC Newslett 38:35–40

Aebi U, Bijlenga RKL, Bt H et al (1976) Comparison of the structural and chemical composition of giant T-even phage heads. J Supramol Struct 5:475–495

Agirrezabala X, Martin-Benito J, Caston JR et al (2005) Maturation of phage T7 involves structural modification of both shell and inner core components. EMBO J 24:3820–3829

Arsuaga J, Vázquez M, Trigueros S et al (2002) Knotting probability of DNA molecules confined in restricted volumes: DNA knotting in phage capsids. Proc Natl Acad Sci USA 99:5373–5377

Bair CL, Rifat D, Black LW (2007) Exclusion of glucosyl-hydroxymethylcytosine DNA containing bacteriophages is overcome by the injected protein inhibitor IPI*. J Mol Biol 366:779–789

Bijlenga RKL, Aebi U, Kellenberger E (1976) Properties and structure of a gene 24-controlled T4 giant phage. J Mol Biol 103:469–498

Black LW (1989) DNA packaging in dsDNA bacteriophages. Annu Rev Microbiol 43:267–292

Black LW, Silverman DJ (1978) Model for DNA packaging into bacteriophage T4 heads. J Virol 28:643–655

Black LW, Newcomb WW, Boring JW et al (1985) Ion etching bacteriophage T4: support for a spiral-fold model of packaged DNA. Proc Natl Acad Sci USA 82:7960–7964

Black LW, Showe MK, Steven AC (1993) Morphogenesis of the T4 head. In: Karam JD (ed) Molecular biology of bacteriophage T4. ASM, Washington, DC, pp 218–258

Cardarelli L, Lam R, Tuite A et al (2010) The crystal structure of bacteriophage HK97 gp6: defining a large family of head-tail connector proteins. J Mol Biol 395:754–768

Cermakian N, Ikeda TM, Cedergren R et al (1996) Sequences homologous to yeast mitochondrial and bacteriophage T3 and T7 RNA polymerases are widespread throughout the eukaryotic lineage. Nucleic Acids Res 24:648–654

Cerritelli ME, Cheng N, Rosenberg AH et al (1997) Encapsidated conformation of bacteriophage T7 DNA. Cell 91:271–280

Cerritelli ME, Conway JF, Cheng N et al (2003a) Molecular mechanisms in bacteriophage T7 procapsid assembly, maturation, and DNA containment. Adv Protein Chem 64:301–323

Cerritelli ME, Trus BL, Smith CS et al (2003b) A Second Symmetry Mismatch at the Portal Vertex of Bacteriophage T7: 8-fold Symmetry in the Procapsid Core. J Mol Biol 327:1–6

Chang J, Weigele P, King J et al (2006) Cryo-EM asymmetric reconstruction of bacteriophage P22 reveals organization of its DNA packaging and infecting machinery. Structure 14:1073–1082

Chang C-Y, Kemp P, Molineux IJ (2010a) Gp15 and gp16 cooperate in translocating bacteriophage T7 DNA into the infected cell. Virology 398:176–186

Chang JT, Schmid MF, Haase-Pettingell C et al (2010b) Visualizing the structural changes of bacteriophage Epsilon15 and its *Salmonella* host during infection. J Mol Biol 402:731–740

Chattoraj DK, Inman RB (1974) Location of DNA ends in P2, 186, P4 and lambda bacteriophage heads. J Mol Biol 87:11–22

Chen J, Lu Z, Sakon J et al (2000) Increasing the thermostability of staphylococcal nuclease: implications for the origin of protein thermostability. J Mol Biol 303:125–130

Choi KH, McPartland J, Kaganman I et al (2008) Insight into DNA and protein transport in double-stranded DNA viruses: the structure of bacteriophage N4. J Mol Biol 378:726–736

Comolli LR, Spakowitz AJ, Siegerist CE et al (2008) Three-dimensional architecture of the bacteriophage phi29 packaged genome and elucidation of its packaging process. Virology 371:267–277

Depping R, Lohaus C, Meyer HE et al (2005) The mono-ADP-ribosyltransferases Alt and ModB of bacteriophage T4: target proteins identified. Biochem Biophys Res Commun 335:1217–1223

Duda RL, Ross PD, Cheng N et al (2009) Structure and energetics of encapsidated DNA in bacteriophage HK97 studied by scanning calorimetry and cryo-electron microscopy. J Mol Biol 391:471–483

Earnshaw WC, Casjens SR (1980) DNA packaging by the double-stranded DNA bacteriophages. Cell 21:319–331

Earnshaw WC, Harrison SC (1977) DNA arrangement in isometric phage heads. Nature 268:598–602

Earnshaw WC, King J, Harrison SC et al (1978) The structural organization of DNA packaged within the heads of T4 wild-type, isometric and giant bacteriophages. Cell 14:559–568

Ebisawa T, Yamamura A, Kameda Y et al (2010) The structure of mAG, a monomeric mutant of the green fluorescent protein Azami-Green, reveals the structural basis of its stable green emission. Acta Crystallogr Sect F Struct Biol Cryst Commun 66:485–489

Effantin G, Boulanger P, Neumann E et al (2006) Bacteriophage T5 structure reveals similarities with HK97 and T4 suggesting evolutionary relationships. J Mol Biol 361:993–1002

Fang P-A, Wright ET, Weintraub ST et al (2008) Visualization of bacteriophage T3 capsids with DNA incompletely packaged in vivo. J Mol Biol 384:1384–1399

Fuller DN, Rickgauer JP, Jardine PJ et al (2007) Ionic effects on viral DNA packaging and portal motor function in bacteriophage phi 29. Proc Natl Acad Sci USA 104:11245–11250

Garcia L, Molineux I (1996) Transcription-independent DNA translocation of bacteriophage T7 DNA into *Escherichia coli*. J Bacteriol 178:6921–6929

Gleghorn ML, Davydova EK, Rothman-Denes LB et al (2008) Structural basis for DNA-hairpin promoter recognition by the bacteriophage N4 virion RNA polymerase. Mol Cell 32:707–717

Glucksmann MA, Markiewicz P, Malone C et al (1992) Specific sequences and a hairpin structure in the template strand are required for N4 virion RNA polymerase promoter recognition. Cell 70:491–500

Gowen B, Bamford JKH, Bamford DH et al (2003) The tailless icosahedral membrane virus PRD1 localizes the proteins involved in genome packaging and injection at a unique vertex. J Virol 77:7863–7871

Hartman PS, Eisenstark A, Pauw PG (1979) Inactivation of phage T7 by near-ultraviolet radiation plus hydrogen peroxide: DNA-protein crosslinks prevent DNA injection. Proc Natl Acad Sci USA 76:3228–3232

Hertveldt K, Lavigne R, Pleteneva E et al (2005) Genome comparison of *Pseudomonas aeruginosa* large phages. J Mol Biol 354:536–545

Hong Y-R, Black LW (1993) Protein folding studies in vivo with a bacteriophage T4 expression-packaging-processing vector that delivers encapsidated fusion proteins into bacteria. Virology 194:481–490

Iida S, Streiff MB, Bickle TA et al (1987) Two DNA antirestriction systems of bacteriophage P1, darA, and darB: characterization of darA- phages. Virology 157:156–166

Jiang W, Chang J, Jakana J et al (2006) Structure of epsilon15 bacteriophage reveals genome organization and DNA packaging/injection apparatus. Nature 439:612–616

Johnson JE, Chiu W (2007) DNA packaging and delivery machines in tailed bacteriophages. Curr Opin Struct Biol 17:237–243

Juers DH, Jacobson RH, Wigley D et al (2000) High resolution refinement of beta-galactosidase in a new crystal form reveals multiple metal-binding sites and provides a structural basis for alpha-complementation. Protein Sci 9:1685–1699

Kalasauskaite EV, Kadisaite DL, Daugelavicius RJ et al (1983) Studies on energy supply for genetic processes. Eur J Biochem 130:123–130

Karhu NJ, Ziedaite G, Bamford DH et al (2007) Efficient DNA packaging of bacteriophage PRD1 requires the unique vertex protein P6. J Virol 81:2970–2979

Kazmierczak KM, Davydova EK, Mustaev AA et al (2002) The phage N4 virion RNA polymerase catalytic domain is related to single-subunit RNA polymerases. EMBO J 21:5815–5823

Kemp P, Garcia LR, Molineux IJ (2005) Changes in bacteriophage T7 virion structure at the initiation of infection. Virology 340:307–317

Koonin EV, Senkevich TG, Dolja VV (2006) The ancient Virus World and evolution of cells. Biol Direct 1:29

Kropinski AM, Kovalyova IV, Billington SJ et al (2007) The genome of Epsilon15, a serotype-converting, Group E1 *Salmonella enterica*-specific bacteriophage. Virology 369:234–244

Krylov VN, Smirnova TA, Minenkova IB et al (1984) *Pseudomonas* bacteriophage phiKZ contains an inner body in its capsid. Can J Microbiol 30:758–762

Lander GC, Tang L, Casjens SR et al (2006) The structure of an infectious P22 virion shows the signal for headful DNA packaging. Science 312:1791–1795

Lavigne R, Seto D, Mahadevan P et al (2008) Unifying classical and molecular taxonomic classification: analysis of the *Podoviridae* using BLASTP-based tools. Res Microbiol 159:406–414

Lecoutere E, Ceyssens P-J, Miroshnikov K et al (2009) Identification and comparative analysis of the structural proteomes of phiKZ and EL, two giant *Pseudomonas aeruginosa* bacteriophages. Proteomics 9:3215–3219

Leforestier A, Livolant F (2010) The bacteriophage genome undergoes a succession of intracapsid phase transitions upon DNA ejection. J Mol Biol 396:384–395

Lepault J, Dubochet J, Baschong W et al (1987) Organization of double-stranded DNA in bacteriophages: a study by cryo-electron microscopy of vitrified samples. EMBO J 6:1507–1512

Lhuillier S, Gallopin M, Gilquin B et al (2009) Structure of bacteriophage SPP1 head-to-tail connection reveals mechanism for viral DNA gating. Proc Natl Acad Sci USA 106:8507–8512

Lipinska B, Rao AS, Bolten BM et al (1989) Cloning and identification of bacteriophage T4 gene 2 product gp2 and action of gp2 on infecting DNA *in vivo*. J Bacteriol 171:488–497

Liu X, Zhang Q, Murata K et al (2010) Structural changes in a marine podovirus associated with release of its genome into *Prochlorococcus*. Nat Struct Mol Biol 17:830–836

Marenduzzo D, Orlandini E, Stasiak A et al (2009) DNA-DNA interactions in bacteriophage capsids are responsible for the observed DNA knotting. Proc Natl Acad Sci USA 106:22269–22274

Matsko N, Klinov D, Manykin A et al (2001) Atomic force microscopy analysis of bacteriophages PhiKZ and T4. J Electron Microsc (Tokyo) 50:417–422

McAllister WT, Wu HL (1978) Regulation of transcription of the late genes of bacteriophage T7. Proc Natl Acad Sci USA 75:804–808

Mendelson EC, Newcomb WW, Brown JC (1992) Ar+ plasma-induced damage to DNA in bacteriophage lambda: implications for the arrangement of DNA in the phage head. J Virol 66:2226–2231

Moak M, Molineux IJ (2000) Role of the Gp16 lytic transglycosylase motif in bacteriophage T7 virions at the initiation of infection. Mol Microbiol 37:345–355

Moak M, Molineux I (2004) Peptidoglycan hydrolytic activities associated with bacteriophage virions. Mol Microbiol 51:1169–1183

Molineux IJ (2001) No syringes please, ejection of phage T7 DNA from the virion is enzyme driven. Mol Microbiol 40:1–8

Mullaney JM, Black LW (1996) Capsid targeting sequence targets foreign proteins into bacteriophage T4 and permits proteolytic processing. J Mol Biol 261:372–385

Mullaney JM, Black LW (1998) Activity of foreign proteins targeted within the bacteriophage T4 head and prohead: implications for packaged DNA structure. J Mol Biol 283:913–929

Mullaney JM, Thompson RB, Gryczynski Z et al (2000) Green fluorescent protein as a probe of rotational mobility within bacteriophage T4. J Virol Methods 88:35–40

Overman SA, Aubrey KL, Reilly KE et al (1998) Conformation and interactions of the packaged double-stranded DNA genome of bacteriophage T7. Biospectroscopy 4(5 Suppl):S47–S56

Peñalva MA, Salas M (1982) Initiation of phage phi 29 DNA replication in vitro: formation of a covalent complex between the terminal protein, p3, and 5′-dAMP. Proc Natl Acad Sci USA 79:5522–5526

Petrov AS, Boz MB, Harvey SC (2007) The conformation of double-stranded DNA inside bacteriophages depends on capsid size and shape. J Struct Biol 160:241–248

Rao V, Black L (2010) Structure and assembly of bacteriophage T4 head. Virol J 7:356

Rao VB, Feiss M (2008) The bacteriophage DNA packaging motor. Annu Rev Genet 42:647–681

Ray K, Ma JX, Oram M et al (2010) Single-molecule and FRET fluorescence correlation spectroscopy analyses of phage DNA packaging: colocalization of packaged phage T4 DNA ends within the capsid. J Mol Biol 395:1102–1113

Repoila F, Tetart F, Bouet J-Y et al (1994) Genomic polymorphism in the T-even phages. EMBO J 13:4181–4192

Rickgauer JP, Fuller DN, Grimes S et al (2008) Portal motor velocity and internal force resisting viral DNA packaging in bacteriophage phi29. Biophys J 94:159–167

Rifat D, Wright NT, Varney KM et al (2008) Restriction endonuclease inhibitor IPI* of bacteriophage T4: a novel structure for a dedicated target. J Mol Biol 375:720–734

Saigo K (1975) Tail-DNA connection and chromosome structure in bacteriophage T5. Virology 68:154–165

Saigo K, Uchida H (1974) Connection of the right-hand terminus of DNA to the proximal end of the tail in bacteriophage lambda. Virology 61:524–536

Salmon B, Baines JD (1998) Herpes simplex virus DNA cleavage and packaging: association of multiple forms of U (L) 15-encoded proteins with B capsids requires at least the U(L)6, U(L)17, and U(L)28 genes. J Virol 72:3045–3050

Scraba DG, Bradley RD, Leyritz-Wills M et al (1983) Bacteriophage phi W-14: the contribution of covalently bound putrescine to DNA packing in the phage head. Virology 124:152–160

Serwer P (1986) Arrangement of double-stranded DNA packaged in bacteriophage capsids: an alternative model. J Mol Biol 190:509–512

Serwer P, Wright ET, Hakala KW et al (2008) Evidence for bacteriophage T7 tail extension during DNA injection. BMC Res Notes 1:36

Sheaffer AK, Newcomb WW, Gao M et al (2001) Herpes simplex virus DNA cleavage and packaging proteins associate with the procapsid prior to its maturation. J Virol 75:687–698

Smith DE, Tans SJ, Smith SB et al (2001) The bacteriophage Phi29 portal motor can package DNA against a large internal force. Nature 413:748–752

Sternberg N, Weisberg R (1975) Packaging of prophage and host DNA by coliphage lambda. Nature 256:97–103

Steven AC, Heymann JB, Cheng N et al (2005) Virus maturation: dynamics and mechanism of a stabilizing structural transition that leads to infectivity. Curr Opin Struct Biol 15:227–236

Stiege AC, Isidro A, Dröge A et al (2003) Specific targeting of a DNA-binding protein to the SPP1 procapsid by interaction with the portal oligomer. Mol Microbiol 49:1201–1212

Streiff MB, Iida S, Bickle TA (1987) Expression and proteolytic processing of the darA antirestriction gene product of bacteriophage P1. Virology 157:167–171

Sun M, Serwer P (1997) The conformation of DNA packaged in bacteriophage G. Biophys J 72:958–963

Tang J, Lander Gabriel C, Olia A et al (2011) Peering down the barrel of a bacteriophage portal: the genome packaging and release valve in P22. Structure 19(4):496–502

Thomas JO (1974) Chemical linkage of the tail to the right-hand end of bacteriophage lambda DNA. J Mol Biol 87:1–9

Thomas JA, Rolando MR, Carroll CA et al (2008) Characterization of *Pseudomonas chlororaphis* myovirus 201phi2-1 via genomic sequencing, mass spectrometry, and electron microscopy. Virology 376:330–338

Thomas JA, Weintraub ST, Hakala K et al (2010) Proteome of the large *Pseudomonas* myovirus 201ϕ2-1: delineation of proteolytically processed virion proteins. Mol Cell Proteomics 9:940–951

Wang GR, Vianelli A, Goldberg EB (2000) Bacteriophage T4 self-assembly: in vitro reconstitution of recombinant gp2 into infectious phage. J Bacteriol 182:672–679

Watabe K, Shin M, Ito J (1983) Protein-primed initiation of phage phi 29 DNA replication. Proc Natl Acad Sci USA 80:4248–4252

Witkiewicz H, Schweiger M (1985) A model of lambda DNA arrangement in the viral particle. J Theor Biol 116:587–605

Xiang Y, Morais MC, Battisti AJ et al (2006) Structural changes of bacteriophage phi29 upon DNA packaging and release. EMBO J 25:5229–5239

Yu TY, Schaefer J (2008) REDOR NMR characterization of DNA packaging in bacteriophage T4. J Mol Biol 382:1031–1042

Zachary A, Black LW (1991) Isolation and characterization of a portal protein-DNA complex from dsDNA bacteriophage. Intervirology 33:6–16

Zhang Z, Greene B, Thuman-Commike PA et al (2000) Visualization of the maturation transition in bacteriophage P22 by electron cryomicroscopy. J Mol Biol 297:615–626

Chapter 22
The Bacteriophage DNA Packaging Machine

Michael Feiss and Venigalla B. Rao

Abstract Large dsDNA bacteriophages and herpesviruses encode a powerful ATP-driven DNA-translocating machine that encapsidates a viral genome into a preformed capsid shell or prohead. The key components of the packaging machine are the packaging enzyme (terminase, motor) and the portal protein that forms the unique DNA entrance vertex of prohead. The terminase complex, comprised of a recognition subunit (small terminase) and an endonuclease/translocase subunit (large terminase), cuts viral genome concatemers. The terminase–viral DNA complex docks on the portal vertex, assembling a motor complex containing five large terminase subunits. The pentameric motor processively translocates DNA until the head shell is full with one viral genome. The motor cuts the DNA again and dissociates from the full head, allowing head-finishing proteins to assemble on the portal, sealing the portal, and constructing a platform for tail attachment. A body of evidence from molecular genetics and biochemical, structural, and biophysical approaches suggests that ATP hydrolysis–driven conformational changes in the packaging motor (large terminase) power DNA motion. Various parts of the motor subunit, such as the ATPase, arginine finger, transmission domain, hinge, and DNA groove, work in concert to translocate about 2 bp of DNA per ATP hydrolyzed. Powerful single-molecule approaches are providing precise delineation of steps during each translocation event in a motor that has a speed as high as a millisecond/step. The phage packaging machine has emerged as an excellent model for understanding the molecular machines, given the mechanistic parallels between terminases, helicases, and numerous motor proteins.

22.1 Introduction

Large dsDNA viruses include the tailed bacteriophages and the eukaryotic herpes-, pox-, and adenoviruses. For the best understood of these viruses, the bacteriophages and the herpesviruses, viral DNA molecules are packaged into preformed empty shells, called procapsids or proheads. For many of these viruses, the viral DNA substrate for the packaging process is in the form of end-to-end

M. Feiss (✉)
Department of Microbiology, Roy J. and Lucille A. Carver College of Medicine,
University of Iowa, Iowa City, IA 52242, USA
e-mail: michael-feiss@uiowa.edu

V.B. Rao (✉)
Department of Biology, The Catholic University of America,
Washington, DC 20064, USA
e-mail: rao@cua.edu

M.G. Rossmann and V.B. Rao (eds.), *Viral Molecular Machines*,
Advances in Experimental Medicine and Biology 726, DOI 10.1007/978-1-4614-0980-9_22,

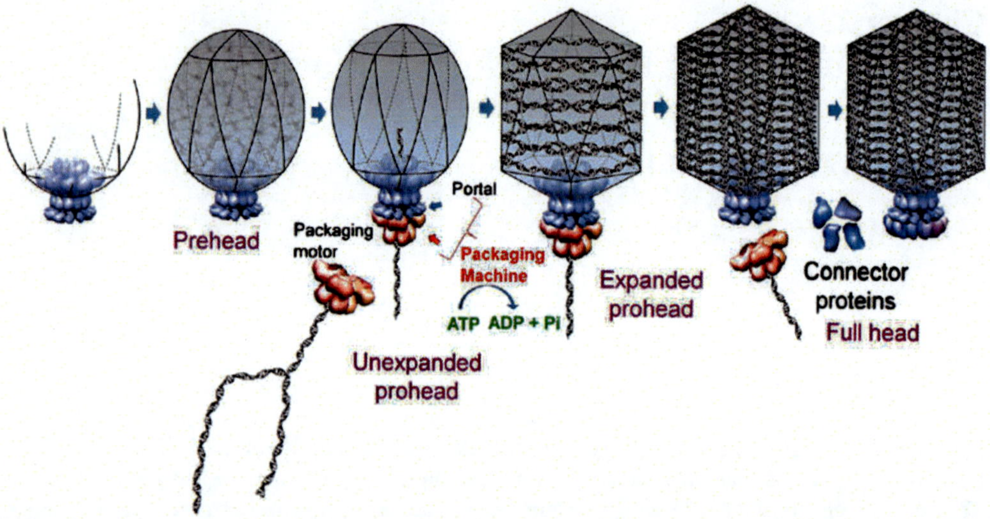

Fig. 22.1 Head assembly of large icosahedral bacteriophages. The dodecameric portal vertex acts as an initiator of head assembly. The major capsid protein polymerizes around a scaffolding core into a prehead. The core is removed to produce an empty unexpanded prohead. The packaging motor–DNA complex docks on the portal and initiates packaging. The prohead expands after about 10–25% of the DNA is packaged. After headful packaging that encapsidates one viral genome, the motor cuts the concatemeric DNA and dissociates from the DNA-full head. The head completion proteins assemble on the portal to seal off the DNA-full head and provide a platform for tail assembly

multimers (called concatemers) that must be cut to generate unit-length virus chromosomes. The cutting of concatemeric DNA occurs during DNA packaging. The tailed phages and herpesviruses produce a DNA packaging enzyme that contains the endonuclease activity that cuts concatemers into virion DNA molecules. These viral enzymes are called terminases because they produce virus chromosome ends. Terminases are complex, being composed of large and small subunits, which in turn form oligomers. The large subunit contains the endonuclease as well as an ATPase that powers DNA packaging. The small subunit contains a DNA-binding motif that targets the enzyme to a cleavage site on the concatemer. Although terminases have several other activities crucial to DNA packaging, the name terminase persists.

As late DNA replication produces concatemers, the procapsid components assemble an icosahedron that is generally formed of many copies of a major capsid protein. Of the 12 fivefold vertexes of the icosahedron, one is the unique portal vertex, which is occupied by a dodecameric portal protein. The portal protein assembles into a truncated cone-shaped structure, with 12 portal protein monomers arranged radially, creating a central channel through which DNA enters and exits the shell. The portal also serves as an initiator for assembly of the major capsid protein and the scaffolding protein(s) into an icosahedral lattice. In parallel, the terminase after cutting a concatemer remains bound to the newly created chromosome end and docks on the portal vertex of a procapsid. Formation of the ternary procapsid–terminase–DNA complex, or the packaging machine, activates terminase's powerful ATPase that fuels the machine to translocate the DNA into the procapsid (or prohead). After the interior of the shell is filled with DNA ("headful" packaging), the endonuclease cuts the DNA, completing DNA packaging. Terminase then undocks from the portal while remaining bound to the concatemer, poised to capture another prohead and package the next chromosome on the concatemer (Fig. 22.1). All these events occur precisely and efficiently in about 5 min during the late phase of phage life cycle.

Virus genome packaging thus is a complex and dynamic process, and the mechanisms are beginning to be understood. As genome packaging has been the subject of recent reviews (Rao and Feiss 2008; Sun et al. 2010), we focus here on the most recent progress while summarizing the previous work.

22.2 DNA Packaging by Large DNA Viruses: Variations on a Common Theme

22.2.1 Genome Recognition

Different phages use somewhat different strategies to process concatemers and assemble the translocation machine. Accordingly, some findings are phage specific, and others are general. For the related enteric phages P22 and Sf6, along with the *B. subtilis* phage SPP1, packaging is initiated when the small terminase recognition subunit binds to a specific site, called *pac*, and the endonuclease makes a cut nearby. Translocation ensues, and when the procapsid is "filled" with DNA, the concatemer is again cut to terminate the packaged chromosome. This downstream cut is made when ~105% of viral genome sequence has been packaged, which means that this sequence-nonspecific cut occurs at a sequence beyond the *pac* site. The resulting virion chromosomes have terminal redundancies, with only the first in the series along the concatemer having a cut near *pac*. Coliphage T4 also uses the headful packaging strategy, but the *pac* recognition sequence, which is not yet known, is even less stringent, and multiple chromosomal sites may be used (Lin and Black 1998). As a result, the ends of the terminally redundant genomes are nearly random (circularly permuted).

The terminase of λ, a temperate *E. coli* phage, introduces nicks, staggered by 12 bp at the *cosN* site, generating identical chromosomes with 5′-ended, 12-base-long cohesive ends that allow the molecule to cyclize upon infection. To ensure accurate nicking, *cosN* is flanked by *cosB*, a binding site for the recognition subunit. *cosB* thus is required for initiation. *cosQ* is a third subsite that is required for proper nicking at termination (Catalano et al. 1995). The chromosomes of related coliphages T3 and T7 have unique, terminally redundant DNA molecules and, like λ, require elaborate DNA signals for DNA processing and packaging (Fujisawa and Morita 1997). The direction of translocation is polarized, both for the phages and herpesviruses, reflecting the asymmetry of the DNA signals (Chattoraj and Inman 1974; Feiss and Bublitz 1975; Newcomb et al. 2009; Yang et al. 2008a).

The packaging strategies for *B. subtilis* phage phi29 and adenoviruses are similar to the extent that a terminal protein is covalently attached to the 5′ ends of the viral genome that primes DNA replication. Since phi29 does not produce concatemeric DNA, the gp16 "terminase" has the ATPase activity but lacks the cutting activity. Also, phi29's terminal protein, gp3, plays a role analogous to that of the recognition subunit, as it interacts with gp16 to escort the chromosome for packaging. An additional curiosity of phi29 is that a homo-oligomeric RNA ring docks on the portal protein of the procapsid and, in turn, provides a platform for assembly of the gp16 oligomer (Grimes et al. 2002).

22.2.2 Virions

Virion DNAs of tailed, dsDNA phages and herpesviruses are packed to the density of crystalline DNA, about 0.5 g/cc (Earnshaw and Casjens 1980). The internal shell pressure is calculated to be as high as 50 atmospheres due primarily to extensive DNA bending and electrostatic repulsion (Evilevitch et al. 2008; Qiu et al. 2011; Smith et al. 2001). The DNA inside the icosahedral shell is tightly packed in concentric layers, starting with the inner surface of the shell, with the layers toward the center being less ordered (Comolli et al. 2008). The last-packaged DNA segment is poised to be the first to exit through the portal; in many phages, it can be cross-linked to the tail. In cryo-EM reconstructions, this end of the DNA appears to be wrapped around the wide end of the portal P22 (Johnson and Chiu 2007; Lander et al. 2006). For many phages, following the undocking of terminase from the portal, additional proteins, the connector ("neck") proteins, assemble on the portal protein to assist in DNA retention and to create a platform for tail-binding (the term "connector" has often been used to indicate the portal, but here we adopt the recently proposed designation which refers to proteins that add to the portal and connect the filled head to the tail). These late steps in head assembly are discussed in detail by Tavares et al. in another chapter of this book.

22.2.3 *Energetics and Kinetics*

To provide the energy required to package DNA to such high density, phage packaging motors generate forces up to about 100 pN, as shown by optical tweezers experiments. This force ranks the DNA packaging motors as the most powerful biological motors yet described (Fuller et al. 2007a, b; Smith et al. 2001). The overall energy budget for packaging the phi29 and T3 genomes is about 2 bp packaged/ATP consumed (Guo et al. 1986; Morita et al. 1993). In addition to being powerful, the packaging motor translocates DNA at a high velocity: T4's 170-kb DNA is packaged at an average rate of about 900 bp/s; λ's 48.5-kb genome, at 600 bp/s; and phi29's 19-kb DNA, at about 125 bp/s. These observations lead to the suggestion that the packaging velocity is adjusted so that genome packaging can be completed in a few minutes. That the motor velocity is controlled is also supported by the isolation of mutants with altered translocation rates (Duffy and Feiss 2002; Tsay et al. 2009; Tsay et al. 2010). Fast packaging rates make sense for viruses with life cycles lasting about 30 min, which may also reduce host nuclease damage to concatemers in the process of being packaged. A comparison of the work done by the packaging motor (about 39 pN/nm) with the energy available from ATP hydrolysis (about 120 pN/nm) indicates that the translocation motor's efficiency is about 30% (Smith et al. 2001). The packaging motor encounters increasing resistance as the shell becomes increasingly packed with DNA. As a result, the velocity slows, and for phi29, the velocity approaches zero as the extent of filling approaches 100% (Smith et al. 2001).

22.2.4 *Promiscuity*

The phage packaging motor exhibits a certain degree of promiscuity with respect to the DNA substrate as well as the head intermediate on which it assembles. The motor is specific for double-stranded DNA; single-stranded DNA, DNA–RNA hybrids, and double-stranded RNA cannot be packaged. However, the motor tolerates significant abnormalities in the double helical structure (Moll and Guo 2005; Oram et al. 2008). Nicks, short single-strand gaps either at the ends (up to 12 nt) or in the middle (up to 10 nt), are tolerated. The motor seems to capture the 5′–3′ strand of DNA in the direction of translocation and track along its length. In the case of phi29 motor, the most important contacts are apparently made with two adjacent phosphates, as the motor loads ATP (Aathavan et al. 2009). Then the motor can traverse the next 10 bp, equivalent to one complete translocation cycle of the pentameric motor (see below), without necessarily making specific contacts with DNA. Surprisingly, in this 10-bp stretch, the motor can translocate even a single-stranded, abasic DNA, a three-stranded D-loop, and even a neutral non-DNA polymer. Whether this level of promiscuity is a common feature of all packaging motors remains to be shown.

An empty unexpanded prohead is considered to be the head intermediate in vivo on which the packaging motor–DNA complex normally assembles and initiates DNA translocation. In phage T4, this head intermediate would have undergone all the proteolytic head maturation cleavages that degrade the internal scaffold. However, in the T4 system, the expanded head and the partially full head or even the full head are as efficient as the unexpanded prohead in assembling the motor and driving DNA translocation (Rao and Black 1985; Zhang et al. 2011). In fact, the virion's packaged DNA can be emptied and refilled with DNA in vitro. Thus, the capsid shell might be a rather passive component, and the putative conformational changes in the portal are reversible, leading to promiscuous assembly of the packaging motor.

22.2.5 *Translocation Machine*

In a number of in vitro DNA packaging systems, terminase, proheads (shell and portal), DNA, and ATP are the minimal requirements for DNA translocation. For SPP1, both the terminase subunits are required, whereas for T4, the large terminase (gp17 ATPase/nuclease) alone is sufficient, although the small terminase (gp16) is essential in vivo.

22.2.5.1 The Shell

The icosahedral prohead shell, also known as the procapsid, uses the portal protein as the assembly initiator. Proper shell assembly by the major capsid protein generally requires the assistance of a scaffold protein. The scaffold assists in proper shell curvature and closure. Scaffold mutants show aberrant polyheads or spiral shells due to failure to close (Casjens and Hendrix 1988). Following completion of the shell, scaffold protein molecules exit the shell interior through openings in the portal shell and/or proteolysis by the virus-encoded protease. The prohead shell is an immature form of the virion head shell. Maturation of the shell accompanies DNA packaging and involves a local rearrangement of each major capsid protein in the lattice. The transitions accompanying prohead maturation are well understood for HK97 (Huang et al. 2011; Lee et al. 2008). Although the HK97 shell has the idiosyncrasy of forming covalent cross-links between major capsid protein monomers, the fold of the major capsid protein is evolutionarily conserved among many tailed dsDNA phages and the herpesviruses. After these local rearrangements, the shell is more angular, thinner, and expanded so that the interior volume is increased. The inter-subunit interactions are greatly strengthened, stabilizing the integrity of the capsid shell against the internal pressure of the tightly packed DNA. Shell expansion occurs early during DNA packaging. For example, in phage λ, a pausing of DNA translocation occurs when about 30% of the DNA has been packaged (Fuller et al. 2007b). The pause is argued to be due to increased resistance to translocation prior to shell expansion. Many phage head shells contain "decoration" proteins that fill lattice openings produced by shell expansion. The decoration proteins, like gpD of λ and gpSoc of T4, further strengthen the head shell. In the λ case, gpD is essential for viability unless the virus DNA is shorter than 78% the wild-type length. In the absence of gpD, packaging of full-length chromosomes leads to a shell failure presumably due to rupture of the lattice by the internal pressure of the packed DNA (Fuller et al. 2007b; Sternberg and Weisberg 1977; Wendt and Feiss 2004; Yang et al. 2008b).

Does the shell protein, particularly the sites lining the inner wall of capsid, participate in the organization and condensation of DNA? The cryo-EM structures of phage capsids show that the DNA strands that are in close proximity to the capsid wall are well ordered and the order progressively disappears toward the center of the shell. This suggests that the inner capsid surface might nucleate DNA condensation through interactions with the first translocated DNA. λ mutants that package but fail to retain full-length chromosomes are candidates for having defects in such inner wall–DNA contacts (Katsura 1989).

22.2.5.2 The Portal

In the prohead, the narrow end of the portal cone protrudes from the fivefold shell vertex, and the wide end is in the prohead's interior. Many viral portals have been studied by cryo-electron microscopy (see Rao and Feiss 2008 for references), and the atomic structures have been determined for the phi29, SPP1, and P22 portals (Lebedev et al. 2007; Olia et al. 2011; Simpson et al. 2000). Though there is little sequence identity among portal proteins, the overall shape and structure of the portals, including that of herpesviruses, are well conserved (Fig. 22.2).

The stem of portal contains two extended α-helices, called α3 and α5 in SPP1, angled about 40° from the channel axis. In the dodecamer, the α3- and α5-helices form a palisade lining the channel.

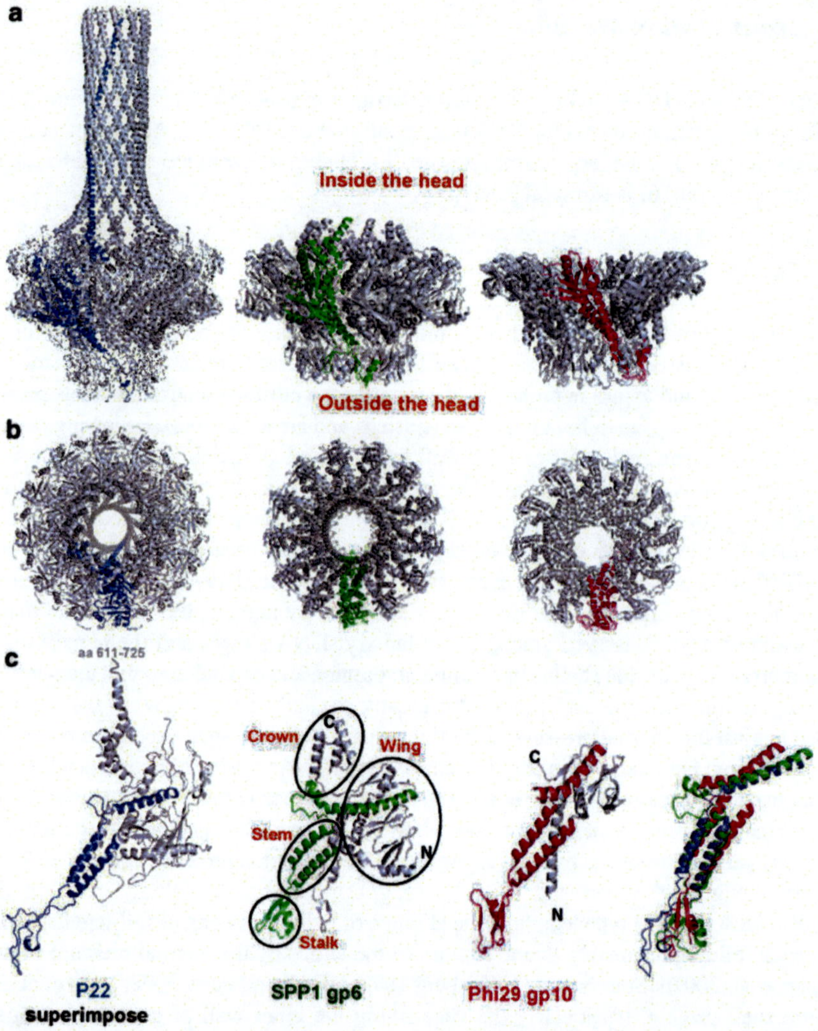

Fig. 22.2 Structure of phage portal. *Front* (**a**) and *top* (**b**) views of P22 (12-fold), SPP1 (13-fold), and phi29 (12-fold) portal structures. (**c**) *Left*: single portal subunits from each phage showing subdomains and the conserved core structure (subdomains are labeled using the SPP1 portal terminology; *blue*, P22 portal; *green*, SPP1 portal; *pink*, Phi29 portal). *Rightmost* picture shows superimposition of the core structures showing structural conservation of all the portals despite negligible sequence identity. The N- and C-termini of SPP1 and Phi29 portals are labeled

The wing, above the stem, contains a long curved α-helix, α6, one end of which extends into the channel. The wing domain contains the N-terminus, and the SPP1 portal's C-terminus forms a globular "crown" domain. The C- and N-termini of phi29's gp10 are similarly positioned, but these domains are much smaller. Below the stem is an αβ stalk structure, which protrudes from fivefold shell vertex. It is the portal's most solvent-exposed element. The terminase–DNA complex that is formed following concatemer cleavage docks on this exposed narrow end of the portal, though the molecular details of this interaction are not understood. In the phi29 motor, the pRNA ring is sandwiched between the portal protein and the gp16 ATPase (Ding et al. 2011; Guo et al. 1987; Zhao et al. 2008). The P22 portal has an additional α-helical domain at the C-terminus that is conserved among all the members of the P22 family phages (Olia et al. 2011). It consists of an ~120 amino-acid-long glutamine-rich α-helix that assembles into an α-helical barrel and extends into the interior of the capsid. Mutants that lack the barrel domain can package genome-length DNA, but the virus particles are noninfectious. Thus, the barrel domain is not essential for translocation, but it might help ordering the DNA inside the capsid and ejection of DNA into the host cell during infection.

The α6-helix connects to α5 via a 15-residue-long loop, the tunnel loop. The structure of the SPP1 gp6 channel loops is complicated by the fact that the gp6 atomic structure was of a 13-mer. In the 13-mer structure, the channel loops are ordered. To derive a structure for the dodecameric virion form of gp6, a pseudoatomic model was constructed by deleting one monomer. A consequence of the monomer deletion was that the channel at the level of the loops was constricted to ~18 Å, which means that different loops have to assume different conformations in order to accommodate dsDNA. In fact, the dodecameric pseudomolecular structure shows clashes at the channel loops, suggesting that the loops are not conformationally identical (Lebedev et al. 2007). Consistent with this hypothesis, the tunnel loops of the phi29 portal are not seen in the crystal structure, suggesting that the loops are flexible. Mutational changes of loop residues do not interfere with packaging but rather affect DNA retention in the head following completion of packaging (Grimes et al. 2011).

The phi29 portal can be inserted into planar lipid bilayers where it forms a pore through which DNA molecules can passively, but unidirectionally, translocate (Jing et al. 2010). Unidirectional DNA movement is consistent with the proposed role of the channel loops in DNA retention following completion of translocation. Saturation mutagenesis of the SPP1 portal identified mutants that were assembly competent but packaging defective (Isidro et al. 2004a, b). The mutations primarily affected residues lining the channel and included residues of the channel loops. Recent work shows that proheads with these mutant portals do not have a binding defect for the ATPase/nuclease (Oliveira et al. 2010). The SPP1's ATPase/nuclease, gp2, binds more strongly to gp6 in procapsids than to isolated gp6, suggesting that incorporation of gp6 into procapsid results in conformational changes.

Conformational changes in the portal, in large part, are thought to drive the sequential transitions that eventually produce a stable head that houses the viral genome as well as attaches to the tail injection machine (Fig. 22.1). In T4, heads in which about six of 12 portal subunits are replaced by gp20-GFP fusion protein are blocked in expansion (Baumann et al. 2006). In SPP1 and P22, portal conformational variants underpackage (~95% of genome per head) or overpackage (~105% of genome per head) DNA in the head (Casjens et al. 1992b; Orlova et al. 1999). In phage P22, a segment of packaged DNA spools around the portal, forcing a conformational change that apparently signals the motor to make the headful termination cut and disengage from the DNA-full head (Lander et al. 2006). Another portal conformational change primes DNA delivery following the binding of neck proteins (Zheng et al. 2008), although this has been refuted by recent structural comparisons made by Olia et al. (2011).

These portal conformational changes are not necessarily irreversible, as was previously thought. The promiscuous nature of packaging machine assembly indicates that the portal has unusual conformational plasticity. Other observations such as (1) the ectopically expressed and packaging-naïve P22 portal (gp10) interacting with the connector protein (gp4) which is supposed to be specific for the post-packaging state (Olia et al. 2006) and (2) phage lambda proheads reversibly interconverting between unexpanded and expanded conformations (Yang et al. 2008b; C. Catalano, personal communication) are consistent with the conformational flexibility of both the portal and the shell.

22.2.5.3 The Motor

The motor proteins were named as terminases because they were originally discovered as endonucleases that cut concatemeric DNA and generate termini of the packaged genome. The oligomeric holoenzyme is composed of a still unknown number of small terminase and large terminase subunits. We will also refer to these as the recognition and ATPase/nuclease subunits, respectively.

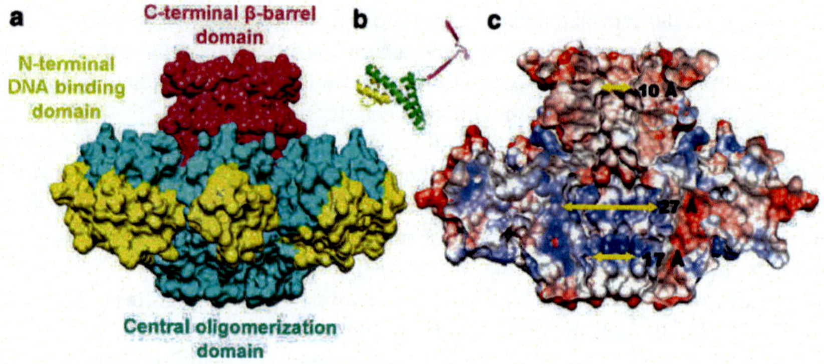

Fig. 22.3 Structure of phage SF6 small terminase. (**a**) Octamer ring structure of the small terminase showing three domains, N-terminal DNA-binding domain (*yellow*), central oligomerization domain (*cyan*), and C-terminal β-barrel domain (*magenta*). (**b**) Single subunit showing the three domains (the colors are matched with the oligomer). (**c**) Surface view cross section showing basic (*blue*) and acidic (*red*) residues and the central channel. The diameter at various positions of the central channel is shown

Recognition Subunit

The small terminase subunits selectively recognize viral DNA molecules in a pool that, in most cases, includes host nucleic acids. Small subunits position the large subunit endonuclease on the viral DNA for proper cutting.

Where information is available, the phage λ terminase being the best studied, the terminase–viral DNA complex is inherently asymmetric, making the direction of DNA polarized. Small terminase has an N-terminal domain (NTD) that is involved in DNA recognition (de Beer et al. 2000), a central domain for oligomerization (Kondabagil and Rao 2006), and a C-terminal domain that interacts with the large terminase subunit (Al-Zahrani et al. 2009; Frackman et al. 1985; Gao and Rao 2011). Oligomerization produces multimer rings that vary in size from octamers to decamers (or larger) (Gual and Alonso 1998; Lin et al. 1997; Nemecek et al. 2007; Zhao et al. 2010). The atomic structure of the Sf6 gp1 octamer is a vase-shaped molecule with a wide base and a central channel (Fig. 22.3; Zhao et al. 2010). Although the channel is proposed to be involved in DNA translocation, its 10-Å width at the narrowest point is too narrow to accommodate dsDNA, unless the native oligomeric state is different from that of the *E. coli*–expressed protein or the channel dynamically expands during packaging. For the known structures of the NTDs of the λ and Sf6 recognition subunits, the domains consist of three α-helices followed by two long α-helices that extend down and pack against each other to line the central channel. The C-terminal domain contains two β-strands that, with the strands of neighboring subunits, form a β-barrel or "neck," whereas the NTDs form appendages that hang down from the wider end of the vase.

For Sf6's gp1 and λ's gpNu1, DNA binding is specific for *pac* and *cosB* sequences, respectively. Specific binding in λ gpNu1 is due to interactions between gpNu1's winged helix-turn-helix (wHTH) motif and the target DNA sequences, called R sequences (de Beer et al. 2002; Feiss et al. 2010). gpNu1's NTD purifies as a dimer with C2 symmetry, allowing the helices of the HTH motifs to dock in the major grooves of the R3 and R2 binding sites of *cosB*, an interaction dependent on the presence of a sharp DNA bend at I1, the IHF binding site between R3 and R2 (de Beer et al. 2002). Whether the Sf6 gp1–*pac* interactions involve similar multiple binding sites remains to be determined.

Although the terminase recognition subunits of Sf6, T4, P22, SPP1, and λ apparently have a conserved quaternary structure, the details of concatemer cutting differ greatly, as exemplified by *pac* phages Sf6, P22, and SPP1. The *pac* site of phage P22 is perhaps the simplest yet documented – a 22-bp asymmetric sequence (Wu et al. 2002). The initiating *pac* cleavages occur at six different sites that are spaced at 20-bp intervals and extend in both directions (Casjens et al. 1987, 1992a). The initiating cleavages of Sf6, a relative of P22, are widely distributed over a 1,800-bp segment that likely contains *pac*. Such a broad distribution suggests that cleavage occurs after a short period of DNA translocation by the Sf6 terminase. SPP1's *pac*, like *cos*, is complex, consisting of two binding sites, *pacL* and *pacR*, for the gp1 recognition subunit flanking the *pacC* cleavage site (Chai et al. 1995). However, there is little to suggest that the *pac* and *cos* complexes are structurally related. For example, the gpNu1 binding sites of *cos* are all on one side of the *cosN* nicking site. Remarkably, *pacL* is large, approximately 100 bp, contains a sharp intrinsic bend, and does not show sequence identity with the smaller (30 bp) *pacR*. Chai et al. proposed that multiple gp1 oligomers assemble on *pacL* and *pacR* (Chai et al. 1995).

Do the oligomeric rings assemble to encircle the DNA prior to cleavage, or the DNA loops around the oligomer through interactions with NTDs? Is the small terminase central channel involved in active translocation? Does the small terminase assemble into rings in vivo, or does it assemble in association with the ATPase/nuclease subunit? On this latter point, the lambda terminase assembly studies are germane, as it appears that a $gpA_1/gpNu1_2$ heterotrimer is an active assembly species (Maluf et al. 2005). A lot need to be learned to answer these questions as well as define the nucleoprotein structures that assemble on *cos* and *pac* sites.

ATPase/Nuclease Subunit

The large terminases consist of two functional domains, an N-terminal ATPase domain and a C-terminal nuclease domain, which are connected by a flexible linker (Duffy and Feiss 2002; Kanamaru et al. 2004; Ortega et al. 2007). Early biochemical studies showed that the purified large terminase subunits from various phages possess an ATPase activity, and sequence gazing indicated the presence of one to three ATPase Walker A and B signature sequences, reviewed in Rao and Feiss (2008). Extensive sequence alignments of terminases and helicases later showed only one strictly conserved ATPase center near the amino-terminal end of essentially all large terminases (Fig. 22.4; Mitchell and Rao 2004; Mitchell et al. 2002). The alignments further showed that this ATPase center contained additional conserved signature motifs such as the adenine-binding motif, catalytic carboxylate, and coupling motifs (Goetzinger and Rao 2003; Mitchell and Rao 2006).

A search for the ATPase center in λ gpA, using the 8-azido-ATP cross-linking analogue, found that residues Y46 and K84 in the N-terminus are involved in ATP binding (Hang et al. 2000). Mutagenesis showed that codon changes at 46 and 84 produced defects in DNA packaging, consistent with the presence of an ATPase center involved in DNA packaging (Dhar and Feiss 2005; Hang et al. 2000; Tsay et al. 2009). A parallel genetic analysis of candidate Walker A and B sequences in T4 gp17 identified T4 gp17 ATPase center's catalytic residues (Mitchell and Rao 2006). Extensive combinatorial mutational analyses that followed led to the establishment of this ATPase center as the likely provider of energy for DNA translocation [Fig. 22.4; motifs (Goetzinger and Rao 2003; Mitchell and Rao 2006)].

Mutant terminases that are translocation competent but endonuclease defective have defects affecting the C-terminus of the large subunit (Alam et al. 2008; Arens et al. 1999; Davidson and Gold 1992; Hwang and Feiss 1996; Hwang and Feiss 2000; Kuebler and Rao 1998). Again, sequence alignments and extensive mutational analyses suggested that a Mg-coordinated triad of acidic residues in the C-domain is responsible for the DNA cleavage (Alam et al. 2008; Rentas and Rao 2003). Structural modeling suggested that the terminase nuclease center is related to that found in resolvase/RNAse H superfamily (Ponchon et al. 2006). The recently determined atomic structures of T4's gp17 and the nuclease domain structures of SPP1 and herpesvirus nuclease domains confirmed the above functional organization of the large terminase structure (Fig. 22.5; Nadal et al. 2010; Smits et al. 2009).

The ATPase Center

Useful crystals of the ATPase domain and intact gp17 were obtained using the ATP-binding but nonhydrolyzing mutant protein D255C/E256D (Fig. 22.4; Sun et al. 2007, 2008). The ATPase center

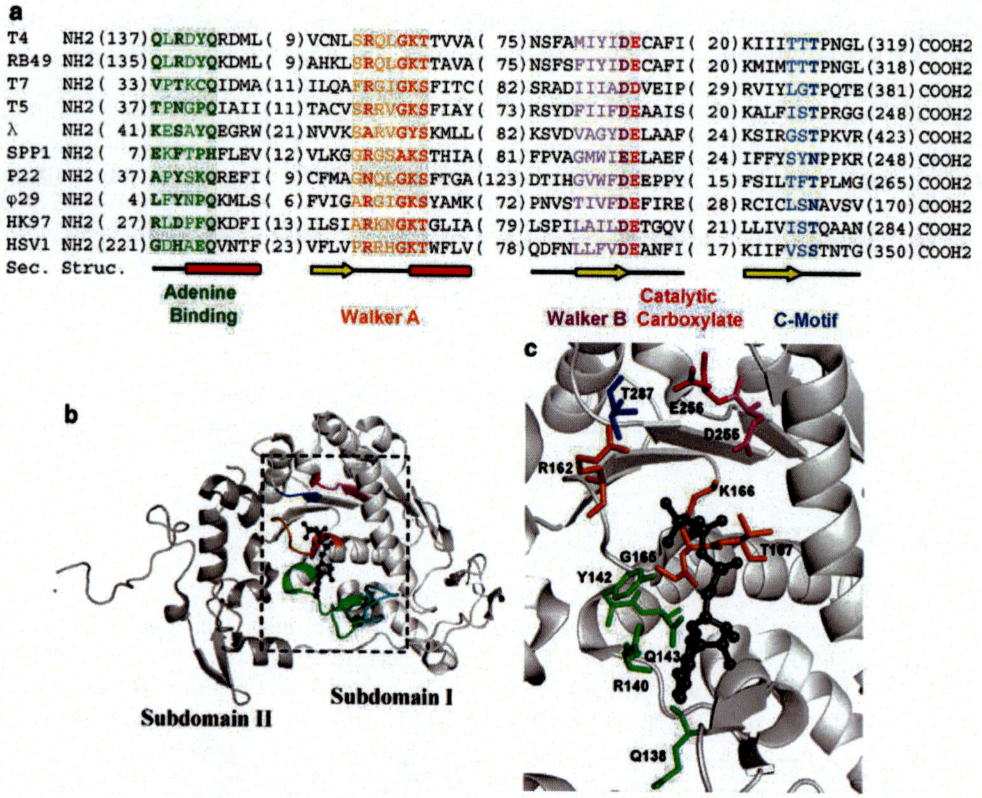

Fig. 22.4 The N-terminal ATPase center of the phage T4 DNA packaging motor. (**a**) The sequence signatures of the N-terminal ATPase center of phage T4 large terminase gp17 are conserved in other phage and viral large terminases. The secondary structure is shown at the *bottom* (*cylinder*, α-helix; *arrow*, β-strand; *line*, coil). (**b**) The structure of the T4 gp17-ATPase domain showing the classic NBD and RecA-type ATPase fold consisting of six parallel β-strands. The bound ATP molecule is shown in *black*. The adenine-binding loop (*green*) adopts a different conformation in the apo form (*cyan*). (**c**) The ATPase active center showing the catalytic residues mapped by mutagenesis studies (the colors of the residues match the colors shown in sequence alignments)

consists of all the basic features of a typical nucleotide-binding fold (Rossmann et al. 1974), a six-stranded β-sheet with interspersed α-helices that orient all the catalytic residues into an ATP-binding pocket (Iyer et al. 2004). The ATPase domain has two subdomains: subdomain I, consisting of the six β-strands arranged as β3-β2-β4-β5-β1-β6, with the Walker A motif at the start of the α1-helix connecting β1 and β2 and the Walker B motif at the C-terminus of β4 (Sun et al. 2007) and subdomain II, discussed below, comprising residues 1–58 and 314–360. The Walker A sequence [consensus = G/A-XXXXGK(T/S)] is highly conserved among the terminase ATPase centers. The Walker A segment forms a phosphate-binding loop in which the ∈-amino group of Lys coordinates the β and γ phosphates of ATP and Thr/Ser with the Mg–ATP complex. The Walker A motif of λ gpA is a variant Walker A sequence (76-KSARVGYS-83) in which Lys is replaced by a Tyr at position 82 and the first residue of the sequence (residue 76) is a Lys rather than a Gly or Ala. Modeling studies indicated that gpA Lys76 could coordinate the ATP phosphates (Draper and Rao 2007). Consistent with this proposal, a mutant enzyme with the highly conservative Lys to Arg change at 76 has a profoundly lethal defect in DNA translocation (Duffy and Feiss 2002; Tsay et al. 2009). The Walker B motif is a β-strand of four hydrophobic residues terminated by an Asp residue that coordinates the Mg^{2+} of the Mg–ATP complex. The Asp-Mg^{2+} positions the β and γ phosphates of ATP for

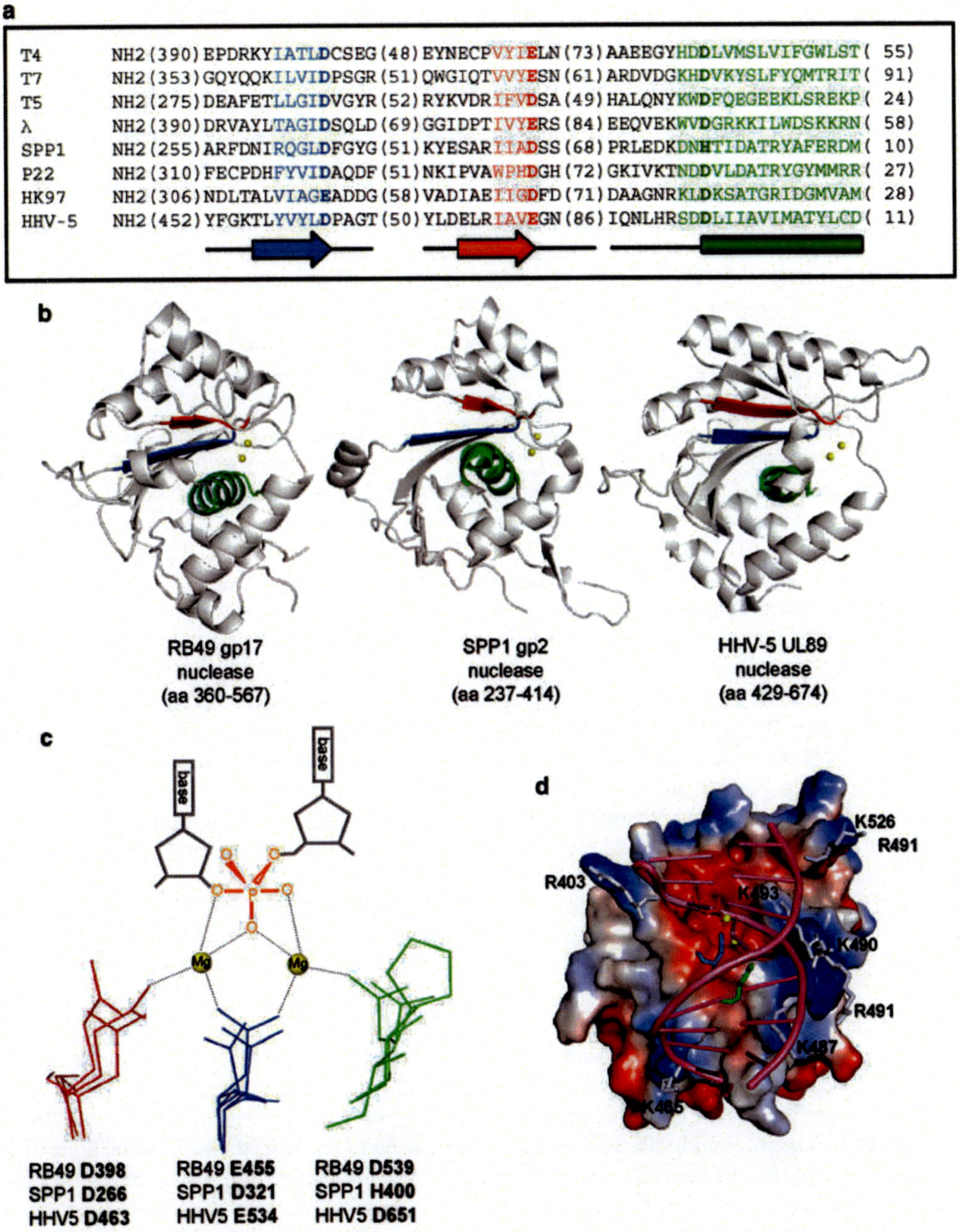

Fig. 22.5 The headful nuclease center in the C-domain of the DNA packaging motor. (**a**) The sequence signatures of the C-terminal nuclease center of large terminases. The secondary structure is shown at the *bottom* (*cylinder*, α-helix; *arrow*, β-strand; *line*, coil). (**b**) The X-ray structures of the C-terminal nuclease domains of three large terminases from RB49, SPP1, and human herpesvirus 5. All show a conserved structure with RNAse H-fold that is found in resolvases. The structural elements that form the nuclease catalytic center are shown in colors, which correspond to the colors of conserved residues shown in (**a**). Mg ions that coordinate with the catalytic residues are shown in *yellow*. (**c**) Schematic showing similar positioning of the catalytic triad residues (side chains shown in *red*, *blue*, and *green*) from all three structures. It shows the pentacovalent phosphorane intermediate formed during the cleavage reaction. (**d**) Surface view of the RB49 C-domain showing the DNA-binding groove. The acidic region (*dark red* at the *top*) is the catalytic center containing the acidic residues. A series of basic residues line the DNA groove with their side chains oriented in a position to interact with the backbone phosphates

nucleophilic attack by a water molecule that is coordinated in turn by the catalytic carboxylate of a Glu residue. In terminase ATPases, the catalytic Glu is adjacent to the Asp of the Walker B motif.

In SF2 helicases, there are additional motifs that are concerned with movement along the ssDNA or ssRNA substrate. One of these motifs, the C-motif (aka motif III), couples ATP hydrolysis to the nucleic acid binding and movement. The characteristic C-motif sequence contains a highly conserved amino acid triplet, with Thr or Ser most frequently found at the third position. SF2 helicases have two domains with a cleft between them. The N-terminal domain 1 contains the ATPase motifs including the C-motif, along with a nucleic acid–binding motif. Hydroxyls of motif C's Thr and Ser residues interact with the Walker B motif. The C-terminal domain 2 contains several helicase motifs including a nucleic acid–binding motif. Upon ATP binding, the cleft closes, forcing movement of the C-motif segment and, in turn, domain 2. ATP hydrolysis and dissociation of ADP and P_i returns the molecule to an open-cleft configuration. This conformational transition is accompanied by migration of the helicase along the nucleic acid. C-motif mutations generally uncouple ATP hydrolysis from translocation (Cordin et al. 2006).

The C-motif is found in terminase large subunits, 25–30 residues downstream of the Walker B (Fig. 22.4; Draper and Rao 2007). In T4's gp17, combinatorial mutagenesis studies of the 285-TTT-287 triplet show that a Thr or Ser in the third position is essential for enzyme function. Changing gp17's critical third Thr, Thr-287, to Ala or Asp abolishes the endonuclease and DNA translocation activities, the latter due to inability to turn over ATP (Draper and Rao 2007). The conserved triplet in λ gpA is 212-GST-214 and the gpA change G212S (in the first residue of the triplet) produces a lethal stalled translocation phenotype (Duffy and Feiss 2002).

The large terminase is also proposed to have an adenine-binding motif 10–25 residues upstream of Walker A (Mitchell et al. 2002), later named the Q-motif (Tanner et al. 2003) (Fig. 22.4). Mutant helicases in Q-motif residues lose affinity for ATP as well as the nucleic acid substrate (Cordin et al. 2004). GpA and gp17 have YQ residues appropriately positioned upstream of Walker A, and in gp17, these and additional residues form a "loop" that shows significant movement in the apo and ATP-bound states. Mutational studies in gpA suggest that this motif is important for DNA binding and translocation. The Y46F change translocates only part of the chromosome due to a greatly increased rate of DNA slipping and lowered ability to generate force to complete DNA filling (Dhar and Feiss 2005; Tsay et al. 2009).

The Endonuclease Center

The structural cores of various terminase nuclease domains, which can nearly be superimposed (although the amino acid sequences show no major conservation overall), is a five-stranded mostly anti-parallel β-sheet flanked by α-helices on both sides (Fig. 22.5). These elements form an RNAse H-fold that has been found in numerous RNAse Hs, resolvases, and integrases. There is a DNA-binding groove edged with basic residues that are proposed to have electrostatic interactions with backbone phosphates. At one end of this groove is a deep valley, the catalytic center in which a triad of acidic residues, plus a fourth acidic residue coordinate with two Mg^{2+} ions. This structure is similar to that found in RNase H nuclease active sites where four conserved acidic residues coordinate two Mg^{2+} ions that are positioned 4 Å apart (Fig. 22.5c). Mg^{2+} ion A is proposed to coordinate with the attacking water molecule and Mg^{2+} ion B to stabilize the transition state (Nowotny and Yang 2006).

It is crucial that the endonuclease of terminase be controlled such that it is inactive during translocation and active at the initiation and termination steps of DNA packaging. Unlike the RNase H enzymes, the gp17 and G2P nucleases contain a C-terminal β-hairpin structure. G2P's β-hairpin protrudes into the DNA groove, potentially clashing with the binding of dsDNA. This β-hairpin appears to be mobile, based on (1) comparison of the structures of Mn^{2+}-bound and metal-free G2P nucleases and (2) the different hairpin orientations in gp17 and G2P and the disordered nature of this region in RB49 and herpesvirus structures. Normal mode analysis shows several low-energy states with the β-hairpin showing significant movements (Smits et al. 2009). It was proposed that the β-hairpin movement controls endonuclease activity by governing the access to DNA groove (Smits et al. 2009). More recent analysis of the T4 structures, however, suggests that this model may be plausible only in the context of the isolated nuclease domain. In the full-length gp17, where the C and N domains are linked via a flexible hinge, normal mode analysis does not show such β-hairpin movement. Instead, a rather rigid body motion of the nuclease domain centered at the hinge is predicted.

Comparison of the X-ray structures of gp17 and cryo-EM reconstruction of prohead-docked gp17 suggested that gp17 exists in two conformational states, tensed and relaxed. As the full-length gp17 could be crystallized only in the tensed state, a relaxed state structure was modeled by molecular dynamics simulations (Ghosh-Kumar et al. 2011). Analysis of these two structures suggested that the β-hairpin movement does not clash with the bound DNA in either state. In fact, in the relaxed state, the β-hairpin moves away from the DNA groove and forms new contacts with the ATPase subdomains I and II. However, intriguingly, the topology of the DNA groove residues is different in the relaxed and tensed states. In the tensed state, K529 (β-hairpin residue) and the four positive residues, K490, K493, R494, and K496, that line one side of the DNA groove are ~21–23 Å from R406 or K468 that line the other side of the groove. In the relaxed state, however, K529 is predicted to move ~5 Å away, while the four positive residues move ~4 Å closer to R406, "compressing" the DNA groove. The orientation of the catalytic triad residues (D401, E458, and D542) do not, however, change. These analyses predict that the topology of DNA-binding residues is what is important for nuclease regulation, the tensed state groove allowing stable DNA binding whereas the compressed groove in the relaxed state being the least active. Biochemical results showing interconversion of tensed and relaxed states in the presence of different regulators, gp16, ATP, salt, and Mg, are consistent with this model.

The above analyses imply that one of the mechanisms by which the headful nuclease is regulated is by inter-domain communication (Ghosh-Kumar et al. 2011). A "communication track" consisting of residues from subdomain II, hinge, and β-hairpin is proposed to relay conformational signals from the ATPase center to the nuclease center. Different networks of interactions form when the protein interconverts between tensed and relaxed states, remodeling the nuclease groove into active or inactive states. This track also seems to be fundamentally important for driving DNA movement. The *cos* cleavage regulation in λ gpA might also be consistent with this model. Near the C-terminus of gpA is a basic leucine zipper (bZIP) DNA-binding motif, a dimerization interface with two α-helices forming a parallel coiled coil. The adjacent basic region, N-terminal to the coiled coil, is a 17-residue segment with two basic residue segments separated by a short sequence containing a highly conserved asparagine. This might make sequence specific contacts with DNA as mutant gpAs in this region are specifically nuclease-defective (Davidson and Gold 1992). One of the acidic residues, D579, which is predicted to coordinate Mg^{2+}-B (Alam et al. 2008), is located in the upstream basic segment of bZIP. It is conceivable that a similar regulatory mechanism via inter-domain communication alters the positioning of these nuclease center residues in coordination with DNA translocation.

The Translocation Center

The crystal structure of the full-length T4 gp17 shows additional features. First, as mentioned above, there is the predicted flexible "hinge" that connects the ATPase and nuclease domains. Previous biochemical studies showed that splitting gp17 into two domains at the linker retained the respective ATPase and nuclease functions, but DNA translocation activity was completely lost. Second, the

N- and C-domains have a >1,000-Å^2 complementary surface area consisting of an array of five charged pairs and hydrophobic patches. Third, the gp17 has a bound phosphate ion in the crystal structure. Docking of B-form DNA guided by shape and charge complementarity with one of the DNA phosphates superimposed on the bound phosphate aligns a number of basic residues, lining what appears to be a shallow translocation groove. Thus, the T4 gp17 C-domain appears to have two DNA grooves on different faces of the structure: one that aligns with the nuclease catalytic site and the second that aligns with the translocating DNA. Mutation of one of the groove residues (R406) showed loss of DNA translocation activity, but the ATPase and nuclease activities are retained. It is unclear if the SPP1 and herpesvirus C-domains contain this second translocation groove. While the deeper nuclease groove with the acidic catalytic triad patch is easily discernable from the surface views and DNA docking (Smits et al. 2009), the translocation groove is shallow and not so obvious. Therefore, further mutagenesis analyses need to be performed to confirm the dedicated nuclease and translocation DNA grooves in T4 gp17 and if a second translocation groove exists in other large terminases.

A functional T4 DNA packaging machine could be assembled by mixing proheads and purified gp17. The monomeric gp17 docks onto the portal and assembles into a pentameric translocation motor. The machine (motor plus portal) can now capture DNA and begin translocation and continue to do so until the head is full. Cryo-EM reconstruction of the packaging machine showed two extra rings of density at the portal vertex that are not seen with proheads alone (Fig. 22.6). The upper ring is flat, resembling the ATPase domain structure, and the lower ring is spherical, resembling the C-domain structure. This was confirmed by docking of the X-ray structures of the domains into the cryo-EM density (Fig. 22.6). The motor has pentamer stoichiometry, with the ATP-binding surface facing the portal and probably interacting with it. It has an open central channel that is in line with the portal channel and the translocation groove of the C-domain faces the channel. There are minimal contacts between the adjacent subunits suggesting that the ATPases may fire relatively independently during translocation. The same basic architecture is seen in the presence or absence of ATP or in the actively packaging machine when all the components including DNA are present (S. Sun and M. Rossmann, personal communication).

The phi29 packaging motor also shows pentameric stoichiometry (Fig. 22.6); however, it is more complex due to the presence of pRNA (Morais et al. 2008). The pRNA molecule first interacts with the portal causing a conformational change in the capsid protein that apparently exposes binding sites for pRNA. The pRNA molecules then interact with the capsid subunits, forming a pentameric ring, matching the fivefold symmetry of the icosahedral capsid vertex (Ding et al. 2011). In fact, extensive interaction between pRNA and capsid was evident in the 12.5-Å reconstruction. The appendages of the pRNA molecules that hang down serve as docking sites for the packaging ATPase molecules. As a result, the pentameric pRNA is sandwiched between the portal and ATPase subunits.

Hexameric and tetrameric packaging motor symmetries have been reported. In the case of phi29 motor, six prohead-bound fluorescent pRNA molecules were counted (Shu et al. 2007). A tetrameric holo-terminase complex was reported in the case of phage λ, which is active for *cos* cutting and DNA packaging (Maluf et al. 2006). However, it is not known if this complex assembles as a tetramer on the prohead. More work needs to be done to establish if the pentamer stoichiometry is universal and if it is absolutely essential for translocation mechanism.

22.2.6 Mechanism

Of several models proposed to explain the mechanism of viral DNA translocation (see review by Rao and Feiss 2008), the terminase-driven mechanism is the most supported by structural and biochemical evidence. In this model, the terminase not only provides energy but also actively translocates DNA. Conformational changes in the large terminase would cause changes in the

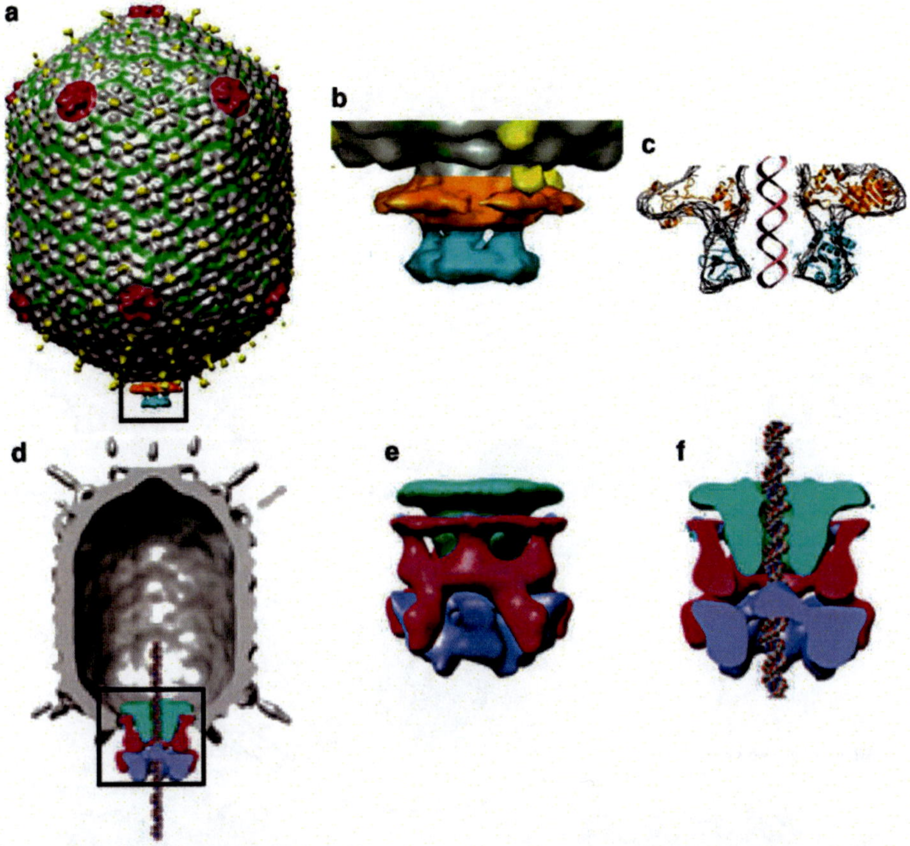

Fig. 22.6 Cryo-EM structure of the DNA packaging machine. (**a**) Cryo-EM reconstruction showing bound gp17 (*boxed*) at the portal vertex of phage T4. (**b**) A zoomed view of the cross section of the packaging motor. (**c**) The X-ray structures of the N-terminal ATPase domain (*yellow*) and the C-terminal nuclease domain (*cyan*) are fitted into the cryo-EM density and a piece of DNA is inserted into the center of the channel. (**d**), (**e**), and (**f**) are the cryo-EM reconstructions of the phi29 DNA packaging machine (portal, pRNA, and ATPase are shown in *cyan*, *pink*, and *purple*, respectively) (see Morais et al. 2008 for details)

DNA-binding affinity allowing binding and release of DNA coupled to movement, a mechanism reminiscent of the inchworm-type translocation by helicases. Most detailed mechanisms, though still speculative, have been postulated for the T4 and phi29 packaging machines.

The T4 gp17, as described above, consists of two domains: an N-terminal ATPase domain with two subdomains and a C-terminal nuclease/translocase domain. The ATPase center in subdomain I (NsubI) is the "engine" that provides energy by "burning" ATP fuel. The C-domain has a translocation groove, which is the "wheel" that moves DNA. The smaller ATPase subdomain II (NsubII) is the transmission domain that couples the engine to the wheel via a flexible hinge. The motor has an arginine finger, a "spark plug" that fires ATPase when the motor is locked in the firing mode, and a C-motif that is part of a communication track that couples ATP hydrolysis to DNA movement. The charged pairs at the domain interface act as an alternator, generating electrostatic force through conformational changes (Fig. 22.7).

The motor exists in two conformational states, relaxed and tensed. In the relaxed state, the hinge is extended (cryo-EM structure). Binding of ATP to NsubI increases the affinity of the translocation groove to bind DNA (a) and locks the motor in translocation mode and brings the arginine finger into

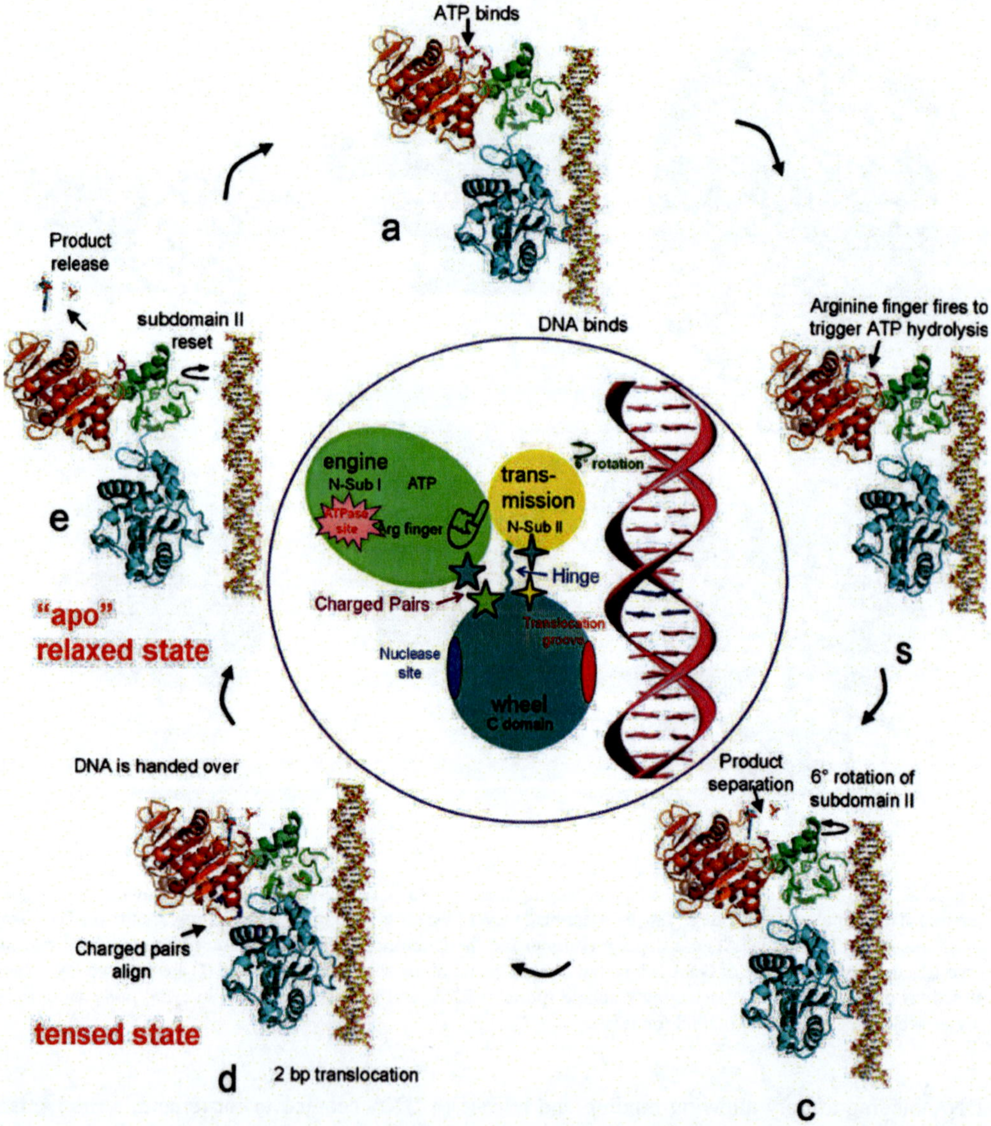

Fig. 22.7 A model for the electrostatic force–driven DNA packaging mechanism. A schematic representation showing the sequence of events that occur in a single gp17 molecule to translocate 2 bp of DNA [see text and Sun et al. (2008) for details]. Shown in the center is a schematic of gp17 molecule–DNA complex with various parts of the motor labeled

position, firing ATP hydrolysis (b). The repulsion between the negatively charged ADP(3–) and P_i(3–) drives them apart, causing NsubII to rotate by 6°, aligning the charge pairs between the N- and C-domains (c). This attracts the C-domain–DNA complex, causing 7 Å upward movement, attaining the tensed conformational state (X-ray structure) (d). As a result, the DNA is translocated into the capsid by 2 bp. Product release and loss of negative charges cause NsubII to rotate back to original position, misaligning the ion pairs and returning the C-domain to the relaxed state (e).

Translocation of 2 bp brings the translocation groove of an adjacent subunit into alignment with the backbone phosphates. DNA is then handed over to the next subunit by the matching motor and DNA symmetries. Thus, ATPase catalysis causes conformational changes which generate electrostatic force, which is then converted to mechanical force. The pentameric motor translocates 10 bp (one turn of the helix) when all five gp17 subunits fire in succession, bringing the first gp17 subunit once again in alignment with the DNA phosphates.

The basic features of the ATPase motor protein seems to be universally preserved in all phages and herpesviruses, even though there is no major overall sequence similarity. Alignments of signature motifs and structural modeling show that the domain architecture is very similar in all the large terminases, and most of the ATPase and nuclease functional motifs are present (Draper and Rao 2007; Mitchell et al. 2002). However, there is uncertainty about some of the motifs. For instance, although the crystal structures of the nuclease domains of phage SPP1 and CMV are very similar to that of the T4 and RB49 C-domains, the presence of charged pair residues and a separate DNA groove for translocation are not so obvious.

The single-molecule data from the phi29 packaging machine are also consistent with the above mechanism. However, high-resolution optical tweezers data suggest that only four of the five subunits fire in each cycle, translocating 2.5 bp per ATP hydrolyzed. The fifth subunit might be holding the DNA between translocation cycles when the subunits are being reloaded with ATP. In fact, when the ATP concentration is lowered to ~5 μM, the rate of reloading slows down considerably, resulting in a micropause, or "dwell," which is then followed by a "burst" of four rapid and successive ATPase firings. In this model, the motor is strictly coordinated with alternating dwell and burst phases; however, the mechanism is unknown. The hydrolytic phase is triggered only when the motor is fully loaded with ATP, resulting in 10 bp "steps" separated by pauses. Within each 10-bp step, four 2.5-bp substeps could be resolved, each presumably corresponding to a single ATP hydrolytic event.

Processive translocation requires that the DNA be held in place between successive translocation events. Otherwise, the translocated DNA could come out or, at late stages of packaging, even ejected out when the internal pressure mounts. There have been suggestions that portal might be involved in holding the DNA, perhaps by closure of the channel in coordination with the terminase pushing DNA into the capsid, one model suggesting compression of DNA (Oram et al. 2008). In the latter model, closure of the portal channel at the same time the terminase applies force "crunches" DNA in the motor channel. Opening of the channel relieves compression, causing movement of DNA into the capsid. There is some evidence suggesting that the DNA length might be decreased by ~20% when the motor tries to translocate a fluorescently labeled Y-DNA structure trapped in the motor channel (Ray et al. 2010). However, it is unclear if the apparent crunching is due to unsuccessful attempts by the packaging motor to force a branched DNA structure through a narrow translocation channel or if it is part of the normal mechanism that translocates a linear DNA substrate. Another aspect of the mechanism that has not yet been determined is the extent of DNA rotation during translocation. At minimum, rotation would be necessary to accommodate the winding of DNA inside the capsid to fit in the icosahedral capsid geometry.

22.3 Concluding Remarks

The virus packaging machine is one of the fastest and most powerful molecular machines known. More than 30 years of biochemical, molecular genetics, and structural analyses now lead researchers into an exciting phase to tease out the mechanistic details. The structures of all the components of the packaging machine, head, portal, large terminase, and small terminase, have now been determined. It is hoped that before long we will have all these structures from a single phage.

Central to the packaging mechanism are the conformational transitions in the large terminase that couple ATP hydrolysis by one of the domains to DNA movement by another domain. However simplified, this revelation allows us to focus on the right questions that might ultimately lead to reconstruction of the mechanism in near atomic detail. However, we have a long road to tread to realize this goal. Many basic questions still remain poorly understood. For instance, the terminase complexes at different stages of packaging must be different and dynamically change. What are the stoichiometries of these complexes and how are these modulated? What is the mechanism of packaging initiation which in large part governed by the small terminase? Does the small terminase have a mechanistic role in translocation? Where exactly is it located in the packaging machine? How are changes in ATPase site communicated to the translocation groove to modulate DNA-binding affinity? What are the precise distances the domains move during a single translocation event? How do the motor subunits communicate and coordinate ATPase firings? What is the structural basis for symmetry mismatch between portal and motor? Does the portal channel expand and contract? Is the DNA structure altered during translocation? Powerful biophysical strategies such as the optical tweezers and fluorescent techniques which can analyze single packaging motors, in combination with molecular genetics, biochemical, and structural analyses, are making it possible to address these fundamental questions. The results should unravel the mechanism in exquisite detail and keep the virus packaging field exciting for many years to come.

Acknowledgments The authors thank Bonnie Draper and Vishal Kottadiel for their assistance in preparing the figures. VBR thanks Michael Rossmann and Siyang Sun for many thoughtful discussions over the years. Special thanks for our present and former lab members for their contributions, and our apologies to colleagues whose work could not be cited due to space limitations. The research in the authors' laboratories has been funded by National Science Foundation (M.F. MCB-0717620 and VBR MCB-0110574, 423528, 0923873) and National Institutes of Health (MF GM-51611 and VBR NIAID-R56AI081726, NIBIB-EB009869).

References

Aathavan K, Politzer AT, Kaplan A, Moffitt JR, Chemla YR, Grimes S, Jardine PJ, Anderson DL, Bustamante C (2009) Substrate interactions and promiscuity in a viral DNA packaging motor. Nature 461:669–673

Alam TI, Draper B, Kondabagil K, Rentas FJ, Ghosh-Kumar M, Sun S, Rossmann MG, Rao VB (2008) The headful packaging nuclease of bacteriophage T4. Mol Microbiol 69:1180–1190

Al-Zahrani AS, Kondabagil K, Gao S, Kelly N, Ghosh-Kumar M, Rao VB (2009) The small terminase, gp16, of bacteriophage T4 is a regulator of the DNA packaging motor. J Biol Chem 284:24490–24500

Arens JS, Hang Q, Hwang Y, Tuma B, Max S (1999) Mutations that extend the specificity of the endonuclease activity of lambda terminase. J Bacteriol 181:218–224

Baumann RG, Mullaney J, Black LW (2006) Portal fusion protein constraints on function in DNA packaging of bacteriophage T4. Mol Microbiol 61:16–32

Casjens S, Hendrix R (1988) Control mechanisms in dsDNA bacteriophage assembly. In: Calendar R (ed) The bacteriophages, 1st edn. Plenum, New York, pp 15–91

Casjens S, Huang WM, Hayden M, Parr R (1987) Initiation of bacteriophage P22 DNA packaging series. Analysis of a mutant that alters the DNA target specificity of the packaging apparatus. J Mol Biol 194:411–422

Casjens S, Sampson L, Randall S, Eppler K, Wu H, Petri JB, Schmieger H (1992a) Molecular genetic analysis of bacteriophage P22 gene 3 product, a protein involved in the initiation of headful DNA packaging. J Mol Biol 227:1086–1099

Casjens S, Wyckoff E, Hayden M, Sampson L, Eppler K, Randall S, Moreno ET, Serwer P (1992b) Bacteriophage P22 portal protein is part of the gauge that regulates packing density of intravirion DNA. J Mol Biol 224:1055–1074

Catalano CE, Cue D, Feiss M (1995) Virus DNA packaging: the strategy used by phage lambda. Mol Microbiol 16:1075–1086

Chai S, Lurz R, Alonso JC (1995) The small subunit of the terminase enzyme of Bacillus subtilis bacteriophage SPP1 forms a specialized nucleoprotein complex with the packaging initiation region. J Mol Biol 252:386–398

Chattoraj DK, Inman RB (1974) Location of DNA ends in P2, 186, P4 and lambda bacteriophage heads. J Mol Biol 87:11–22

Comolli LR, Spakowitz AJ, Siegerist CE, Jardine PJ, Grimes S, Anderson DL, Bustamante C, Downing KH (2008) Three-dimensional architecture of the bacteriophage phi29 packaged genome and elucidation of its packaging process. Virology 371:267–277

Cordin O, Tanner NK, Doere M, Linder P, Banroques J (2004) The newly discovered Q motif of DEAD-box RNA helicases regulates RNA-binding and helicase activity. EMBO J 23:2478–2487

Cordin O, Banroques J, Tanner NK, Linder P (2006) The DEAD-box protein family of RNA helicases. Gene 367:17–37

Davidson A, Gold M (1992) Mutations abolishing the endonuclease activity of bacteriophage lambda terminase lie in two distinct regions of the *A* gene, one of which may encode a leucine zipper DNA binding domain. Virology 161:305–315

de Beer T, Ortega M, Berton N, Yang Q, Overduin M, Catalano CE (2000) Assignment of the 1H, 13C, and 15N resonances of the DNA binding domain of gpNu1, a genome packaging protein from bacteriophage lambda. J Biomol NMR 18:69–70

de Beer T, Fang J, Ortega M, Yang Q, Maes L, Duffy C, Berton N, Sippy J, Overduin M, Feiss M, Catalano CE (2002) Insights into specific DNA recognition during the assembly of a viral genome packaging machine. Mol Cell 9:981–991

Dhar A, Feiss M (2005) Bacteriophage lambda terminase: alterations of the high-affinity ATPase affect viral DNA packaging. J Mol Biol 347:71–80

Ding F, Lu C, Zhao W, Rajashankar KR, Anderson DL, Jardine PJ, Grimes S, Ke A (2011) Structure and assembly of the essential RNA ring component of a viral DNA packaging motor. Proc Natl Acad Sci USA 108(18):7357–7362

Draper B, Rao VB (2007) An ATP hydrolysis sensor in the DNA packaging motor from bacteriophage T4 suggests an inchworm-type translocation mechanism. J Mol Biol 369:79–94

Duffy C, Feiss M (2002) The large subunit of bacteriophage lambda's terminase plays a role in DNA translocation and packaging termination. J Mol Biol 316:547–561

Earnshaw WC, Casjens SR (1980) DNA packaging by the double-stranded DNA bacteriophages. Cell 21:319–331

Evilevitch A, Fang LT, Yoffe AM, Castelnovo M, Rau DC, Parsegian VA, Gelbart WM, Knobler CM (2008) Effects of salt concentrations and bending energy on the extent of ejection of phage genomes. Biophys J 94:1110–1120

Feiss M, Bublitz A (1975) Polarized packaging of bacteriophage lambda chromosomes. J Mol Biol 94:583–594

Feiss M, Reynolds E, Schrock M, Sippy J (2010) DNA packaging by lambda-like bacteriophages: mutations broadening the packaging specificity of terminase, the lambda-packaging enzyme. Genetics 184:43–52

Frackman S, Siegele DA, Feiss M (1985) The terminase of bacteriophage lambda. Functional domains for cosB binding and multimer assembly. J Mol Biol 183:225–238

Fujisawa H, Morita M (1997) Phage DNA packaging. Genes Cells 2:537–545

Fuller DN, Raymer DM, Kottadiel VI, Rao VB, Smith DE (2007a) Single phage T4 DNA packaging motors exhibit large force generation, high velocity, and dynamic variability. Proc Natl Acad Sci USA 104:16868–16873

Fuller DN, Raymer DM, Rickgauer JP, Robertson RM, Catalano CE, Anderson DL, Grimes S, Smith DE (2007b) Measurements of single DNA molecule packaging dynamics in bacteriophage lambda reveal high forces, high motor processivity, and capsid transformations. J Mol Biol 373:1113–1122

Gao S, Rao VB (2011) Specificity of interactions among the DNA-packaging machine components of T4-related bacteriophages. J Biol Chem 286:3944–3956

Ghosh-Kumar M, Alam TI, Draper B, Stack JD, Rao VB (2011) Regulation by interdomain communication of a headful packaging nuclease from bacteriophage T4. Nucleic Acids Res 39:2742–2755

Goetzinger K, Rao V (2003) Defining the ATPase center of bacteriophage T4 DNA packaging machine: requirement for a catalytic glutamate residue in the large terminase protein gp17. J Mol Biol 331:139–154

Grimes S, Jardine PJ, Anderson D (2002) Bacteriophage phi 29 DNA packaging. Adv Virus Res 58:255–294

Grimes S, Ma, S, Gao J, Atz R, Jardine P (2011) Role of the ø29 connector channel loops in late-stage DNA packaging. J Mol Biol 410:50–59

Gual A, Alonso JC (1998) Characterization of the small subunit of the terminase enzyme of the Bacillus subtilis bacteriophage SPP1. Virology 242:279–287

Guo P, Grimes S, Anderson D (1986) A defined system for in vitro packaging of DNA-gp3 of the Bacillus subtilis bacteriophage phi 29. Proc Natl Acad Sci USA 83:3505–3509

Guo PX, Erickson S, Anderson D (1987) A small viral RNA is required for in vitro packaging of bacteriophage phi 29 DNA. Science (New York, NY) 236:690–694

Hang Q, Tack B, Feiss M (2000) An ATPase center of bacteriophage lambda terminase involved in post-cleavage stages of DNA packaging: identification of ATP-interactive amino acids. J Mol Biol 302:777–795

Huang RK, Khayat R, Lee KK, Gertsman I, Duda RL, Hendrix RW, Johnson JE (2011) The Prohead-I structure of bacteriophage HK97: implications for scaffold-mediated control of particle assembly and maturation. J Mol Biol 408:541–554

Hwang Y, Feiss M (1996) Mutations affecting the high affinity ATPase center of gpA, the large subunit of bacteriophage lambda terminase, inactivate the endonuclease activity of terminase. J Mol Biol 261:524–535

Hwang Y, Feiss M (2000) The endonuclease and helicase activities of bacteriophage lambda terminase: changing nearby residue 515 restores activity to the gpA K497D mutant enzyme. Virology 277:204–214

Isidro A, Henriques AO, Tavares P (2004a) The portal protein plays essential roles at different steps of the SPP1 DNA packaging process. Virology 322:253–263

Isidro A, Santos MA, Henriques AO, Tavares P (2004b) The high-resolution functional map of bacteriophage SPP1 portal protein. Mol Microbiol 51:949–962

Iyer LM, Makarova KS, Koonin EV, Aravind L (2004) Comparative genomics of the FtsK-HerA superfamily of pumping ATPases: implications for the origins of chromosome segregation, cell division and viral capsid packaging. Nucleic Acids Res 32:5260–5279

Jing P, Haque F, Shu D, Guo P (2010) One-way traffic of a viral motor channel for double-stranded DNA translocation. Nano Lett 10(9):3620–3627

Johnson JE, Chiu W (2007) DNA packaging and delivery machines in tailed bacteriophages. Curr Opin Struct Biol 17:237–243

Kanamaru S, Kondabagil K, Rossmann MG, Rao VB (2004) The functional domains of bacteriophage T4 terminase. J Biol Chem 279:40795–40801

Katsura I (1989) Structure and inherent properties of the bacteriophage lambda head shell. VI. DNA-packaging-defective mutants in the major capsid protein. J Mol Biol 205:397–405

Kondabagil KR, Rao VB (2006) A critical coiled coil motif in the small terminase, gp16, from bacteriophage T4: insights into DNA packaging initiation and assembly of packaging motor. J Mol Biol 358:67–82

Kuebler D, Rao VB (1998) Functional analysis of the DNA-packaging/terminase protein gp17 from bacteriophage T4. J Mol Biol 281:803–814

Lander GC, Tang L, Casjens SR, Gilcrease EB, Prevelige P, Poliakov A, Potter CS, Carragher B, Johnson JE (2006) The structure of an infectious P22 virion shows the signal for headful DNA packaging. Science 312:1791–1795

Lebedev AA, Krause MH, Isidro AL, Vagin AA, Orlova EV, Turner J, Dodson EJ, Tavares P, Antson AA (2007) Structural framework for DNA translocation via the viral portal protein. EMBO J 26:1984–1994

Lee KK, Gan L, Tsuruta H, Moyer C, Conway JF, Duda RL, Hendrix RW, Steven AC, Johnson JE (2008) Virus capsid expansion driven by the capture of mobile surface loops. Structure 16:1491–1502

Lin H, Black LW (1998) DNA requirements in vivo for phage T4 packaging. Virology 242:118–127

Lin H, Simon MN, Black LW (1997) Purification and characterization of the small subunit of phage T4 terminase, gp16, required for DNA packaging. J Biol Chem 272:3495–3501

Maluf N, Yang Q, Catalano C (2005) Self-association properties of the bacteriophage lambda terminase holoenzyme: implications for the DNA packaging motor. J Mol Biol 347:523–542

Maluf N, Gaussier H, Bogner E, Feiss M, Catalano C (2006) Assembly of bacteriophage lambda terminase into a viral DNA maturation and packaging machine. Biochemistry 45:15259–15268

Mitchell MS, Rao VB (2004) Novel and deviant Walker A ATP-binding motifs in bacteriophage large terminase-DNA packaging proteins. Virology 321:217–221

Mitchell MS, Rao VB (2006) Functional analysis of the bacteriophage T4 DNA-packaging ATPase motor. J Biol Chem 281:518–527

Mitchell M, Matsuzaki S, Imai S, Rao V (2002) Sequence analysis of bacteriophage T4 DNA packaging/terminase genes 16 and 17 reveals a common ATPase center in the large subunit of viral terminases. Nucleic Acids Res 30:4009–4021

Moll WD, Guo P (2005) Translocation of nicked but not gapped DNA by the packaging motor of bacteriophage phi29. J Mol Biol 351:100–107

Morais MC, Koti JS, Bowman VD, Reyes-Aldrete E, Anderson DL, Rossmann MG (2008) Defining molecular and domain boundaries in the bacteriophage phi29 DNA packaging motor. Structure 16:1267–1274

Morita M, Tasaka M, Fujisawa H (1993) DNA packaging ATPase of bacteriophage T3. Virology 193:748–752

Nadal M, Mas PJ, Blanco AG, Arnan C, Sola M, Hart DJ, Coll M (2010) Structure and inhibition of herpesvirus DNA packaging terminase nuclease domain. Proc Natl Acad Sci USA 107:16078–16083

Nemecek D, Gilcrease EB, Kang S, Preveilige PE Jr, Casjens S, Thomas GJ Jr (2007) Subunit conformations and assembly states of a DNA-translocating motor: the terminase of bacteriophage P22. J Mol Biol 374:817–836

Newcomb WW, Cockrell SK, Homa FL, Brown JC (2009) Polarized DNA ejection from the herpesvirus capsid. J Mol Biol 392:885–894

Nowotny M, Yang W (2006) Stepwise analyses of metal ions in RNase H catalysis from substrate destabilization to product release. EMBO J 25:1924–1933

Olia AS, Al-Bassam J, Winn-Stapley DA, Joss L, Casjens SR, Cingolani G (2006) Binding-induced stabilization and assembly of the phage P22 tail accessory factor gp4. J Mol Biol 363:558–576

Olia AS, Prevelige PE Jr, Johnson JE, Cingolani G (2011) Three-dimensional structure of a viral genome-delivery portal vertex. Nat Struct Mol Biol 18:597–603

Oliveira L, Cuervo A, Tavares P (2010) Direct interaction of the bacteriophage SPP1 packaging ATPase with the portal protein. J Biol Chem 285:7366–7373

Oram M, Sabanayagam C, Black LW (2008) Modulation of the packaging reaction of bacteriophage T4 terminase by DNA structure. J Mol Biol 381:61–72

Orlova EV, Dube P, Beckmann E, Zemlin F, Lurz R, Trautner TA, Tavares P, van Heel M (1999) Structure of the 13-fold symmetric portal protein of bacteriophage SPP1. Nat Struct Biol 6:842–846

Ortega ME, Gaussier H, Catalano CE (2007) The DNA maturation domain of gpA, the DNA packaging motor protein of bacteriophage lambda, contains an ATPase site associated with endonuclease activity. J Mol Biol 373:851–865
Ponchon L, Boulanger P, Labesse G, Letellier L (2006) The endonuclease domain of bacteriophage terminases belongs to the resolvase/integrase/ribonuclease H superfamily: a bioinformatics analysis validated by a functional study on bacteriophage T5. J Biol Chem 281:5829–5836
Qiu X, Rau DC, Parsegian VA, Fang LT, Knobler CM, Gelbart WM (2011) Salt-Dependent DNA-DNA Spacings in Intact Bacteriophage lambda Reflect Relative Importance of DNA Self-Repulsion and Bending Energies. Phys Rev Lett 106:028102
Rao VB, Black LW (1985) DNA packaging of bacteriophage T4 proheads in vitro. Evidence that prohead expansion is not coupled to DNA packaging. J Mol Biol 185:565–578
Rao VB, Feiss M (2008) The bacteriophage DNA packaging motor. In: Annual review of genetics, Annual Reviews, Palo Alto, CA, pp 647–681
Ray K, Sabanayagam CR, Lakowicz JR, Black LW (2010) DNA crunching by a viral packaging motor: compression of a procapsid-portal stalled Y-DNA substrate. Virology 398:224–232
Rentas FJ, Rao VB (2003) Defining the bacteriophage T4 DNA packaging machine: evidence for a C-terminal DNA cleavage domain in the large terminase/packaging protein gp17. J Mol Biol 334:37–52
Rossmann MG, Moras D, Olsen KW (1974) Chemical and biological evolution of nucleotide-binding protein. Nature 250:194–199
Shu D, Zhang H, Jin J, Guo P (2007) Counting of six pRNAs of phi29 DNA-packaging motor with customized single-molecule dual-view system. EMBO J 26:527–537
Simpson A, Tao Y, Leiman P, Badasso M, He Y, Jardine P, Olson N, Morais M, Grimes S, Anderson D, Baker T, Rossmann M (2000) Structure of the bacteriophage phi29 DNA packaging motor. Nature 408:745–750
Smith D, Tans S, Smith S, Grimes S, Anderson D, Bustamante C (2001) The bacteriophage straight phi29 portal motor can package DNA against a large internal force. Nature 413:748–752
Smits C, Chechik M, Kovalevskiy OV, Shevtsov MB, Foster AW, Alonso JC, Antson AA (2009) Structural basis for the nuclease activity of a bacteriophage large terminase. EMBO Rep 10:592–598
Sternberg N, Weisberg R (1977) Packaging of coliphage lambda DNA. II. The role of the gene D protein. J Mol Biol 117:733–759
Sun S, Kondabagil K, Gentz PM, Rossmann MG, Rao VB (2007) The structure of the ATPase that powers DNA packaging into bacteriophage T4 procapsids. Mol Cell 25:943–949
Sun S, Kondabagil K, Draper B, Alam TI, Bowman VD, Zhang Z, Hegde S, Fokine A, Rossmann MG, Rao VB (2008) The structure of the phage T4 DNA packaging motor suggests a mechanism dependent on electrostatic forces. Cell 135:1251–1262
Sun S, Rao VB, Rossmann MG (2010) Genome packaging in viruses. Curr Opin Struct Biol 20:114–120
Tanner NK, Cordin O, Banroques J, Doere M, Linder P (2003) The Q motif: a newly identified motif in DEAD box helicases may regulate ATP binding and hydrolysis. Mol Cell 11:127–138
Tsay JM, Sippy J, Feiss M, Smith DE (2009) The Q motif of a viral packaging motor governs its force generation and communicates ATP recognition to DNA interaction. Proc Natl Acad Sci USA 106:14355–14360
Tsay JM, Sippy J, Deltoro D, Andrews BT, Draper B, Rao V, Catalano CE, Feiss M, Smith DE (2010) Mutations altering a structurally conserved loop-helix-loop region of a viral packaging motor change DNA translocation velocity and processivity. J Biol Chem 285(31):24282–24289
Wendt J, Feiss M (2004) A fragile lattice: replacing bacteriophage l's head stability gene *D* with the *shp* gene of phage 21 generates the Mg^{++}-dependent virus, lambda *shp*. Virology 326:41–46
Wu H, Sampson L, Parr R, Casjens S (2002) The DNA site utilized by bacteriophage P22 for initiation of DNA packaging. Mol Microbiol 45:1631–1646
Yang K, Poon AP, Roizman B, Baines JD (2008a) Temperature-sensitive mutations in the putative herpes simplex virus type 1 terminase subunits pUL15 and pUL33 preclude viral DNA cleavage/packaging and interaction with pUL28 at the nonpermissive temperature. J Virol 82:487–494
Yang Q, Maluf NK, Catalano CE (2008b) Packaging of a unit length viral genome: the role of nucleotides and the gpD decoration protein in stable nucleocapsid assembly in bacteriophage lambda. J Mol Biol 383:1037–1048
Zhang Z, Kottadiel VI, Vafabakhsh R, Dai L, Chemla YR, Ha T, Rao VB (2011) A promiscuous DNA packaging machine from bacteriophage T4. PLoS Biol 9:e1000592
Zhao W, Morais MC, Anderson DL, Jardine PJ, Grimes S (2008) Role of the CCA bulge of prohead RNA of bacteriophage φ29 in DNA packaging. J Mol Biol 383:520–528
Zhao H, Finch CJ, Sequeira RD, Johnson BA, Johnson JE, Casjens SR, Tang L (2010) Crystal structure of the DNA-recognition component of the bacterial virus Sf6 genome-packaging machine. Proc Natl Acad Sci USA 107:1971–1976
Zheng H, Olia AS, Gonen M, Andrews S, Cingolani G, Gonen T (2008) A conformational switch in bacteriophage P22 portal protein primes genome injection. Mol Cell 29:376–383

Chapter 23
The dsDNA Packaging Motor in Bacteriophage ø29

Marc C. Morais

Abstract The tailed dsDNA bacteriophage ø29 packages its 19.3-kb genome into a preassembled prolate icosahedral procapsid structure using a phage-encoded macromolecular motor. This process is remarkable considering that compaction of DNA to near crystalline densities within the confined space of the capsid requires that the motor work against considerable entropic, enthalpic, and DNA bending energies. The heart of the bacteriophage ø29 packaging motor consists of three macromolecular components: the connector protein, an RNA molecule known as the pRNA, and an ATPase. The pRNA is thus far unique to ø29, but the connector and ATPase are homologous to portal and terminase proteins, respectively, in other tailed dsDNA bacteriophages. Despite decades of effort and a wealth of genetic, biochemical, biophysical, structural, and single particle data, the mechanism of DNA packaging in bacteriophage ø29 remains elusive. In this chapter, we describe the development of a highly efficient in vitro DNA packaging system for ø29, review the data available for each individual macromolecular component in the packaging motor, and present and evaluate various packaging mechanisms that have been proposed to explain the available data.

23.1 Introduction

The ability to interconvert various forms of energy is an essential feature of living systems. Biological molecular motors accomplish this task by coupling the making and breaking of high-energy covalent bonds to conformational changes in large macromolecules. These conformational changes in turn drive a diverse array of biological phenomena, including muscle contraction, protein synthesis, nucleic acid manipulation, and the generation and maintenance of concentration gradients and electrostatic potentials. Understanding how molecular motors interconvert energy at the atomic scale provides insight into the fundamental question of how energy is harnessed in biological systems. Furthermore, characterizing the assembly pathways of molecular motors illuminates the mechanisms by which energy transduction systems are constructed and maintained in nature. There are only a few well-characterized biological molecular motors. These generally fall into two classes (1) rotary motors, such as the F_1 ATPase and bacterial flagellar motor, which turn a spindle by a rotary

M.C. Morais (✉)
Department of Biochemistry and Molecular Biology, Sealy Center for Structural and Computational Biology, University of Texas Medical Branch at Galveston, Galveston, TX 77555, USA
e-mail: mcmorais@utmb.edu

M.G. Rossmann and V.B. Rao (eds.), *Viral Molecular Machines*,
Advances in Experimental Medicine and Biology 726, DOI 10.1007/978-1-4614-0980-9_23,

conformational change mechanism and (2) linear motors, like those required for muscle contraction, where conformational fluctuations result in translation of the motor with each step. This chapter describes a third, hybrid class of molecular motors, viral DNA packaging motors, where rotary events are converted into translational movement. In this type of motor, several energy-generating ATPase subunits are arranged as a ring, and coordinated hydrolysis of ATP molecules around this ring induces a rotary pattern of conformational changes in the motor that are coupled to the translocation of double-stranded DNA (dsDNA) into preformed viral capsids.

The process of genome encapsidation is remarkable considering that the length of DNA to be packaged is approximately 150 times the width of the capsid itself (Earnshaw and Casjens 1980). Hence, considerable entropic, electrostatic, and bending energies must be overcome to package DNA to near-crystalline densities within the confines of the virus shell and against pressures estimated to exceed 6 Mpa (Smith et al. 2001). Indeed, viral DNA packaging motors are among the most powerful biological motors known, capable of generating forces up to nearly 100 pN (Smith et al. 2001; Fuller et al. 2007; Rickgauer et al. 2008). Furthermore, there is the nontrivial topological requirement that packaged DNA must also exit freely during genome ejection; knots and tangles cannot be tolerated since the portal for DNA entrance and exit is only large enough to accommodate passage of a single dsDNA helix. It is thus not surprising that early hypotheses regarding genome packaging proposed that the viral capsid must be assembled around condensed DNA (Kellenberger et al. 1959).

The virally encoded ATPases that power genome encapsidation are a subgroup of the large ASCE (*A*dditional *S*trand *C*atalytic *E* (glutamate)) ATPase family whose members are involved in various macromolecular remodeling tasks including: cell division; chromosome segregation; DNA recombination, strand separation, and conjugation; and the generation and maintenance of concentration gradients and electrostatic potentials (Mitchell et al. 2002; Burroughs et al. 2007; Singleton et al. 2007; Thomsen and Berger 2008). Thus, understanding the mechanisms by which these motors operate will elucidate the general principles of molecular partitioning and energy transformation in biological systems. Furthermore, although the oligomeric rings of virally encoded ATPases are similar to the ubiquitous ASCE macromolecular translocases, they are capable of generating considerably greater forces than other ASCE ATPase-based motors. Hence, insights gained from the study of viral packaging motors will not only shed light on the basic mechanistic principles of a broad class of macromolecular motors but can also illuminate how these principles have been adapted by viruses to generate and control the large molecular forces necessary for genome encapsidation.

Research on virus DNA packaging motors have converged on a few probable motor mechanisms that differ primarily in (1) the identity of the force-generating component, (2) the mechanics of the force-generating step, (3) the mode of coordination between motor components, and (4) the nature of substeps in one complete mechanical cycle of the motor. This chapter will examine genome packaging in dsDNA viruses based on available structural, biophysical, computational, genetic, and biochemical data that describe the self-assembly and mechanism of viral DNA packaging motors. These aspects of motor function will be examined in the bacteriophage ø29 DNA packaging motor, an outstanding model system for studying genome packaging due to its compositional simplicity, well-defined genetics, high in vitro packaging efficiency, and general experimental accessibility (Grimes et al. 2002).

A comprehensive understanding of the ø29 dsDNA packaging motor requires examining its mechanism at three levels (1) the chemical level, i.e., the mechanism of catalysis employed by the ATPase; (2) the mechanical level, i.e., the conformational changes of motor components that drive DNA translocation; and (3) the mechanochemical level, i.e., how the catalytic cycle of the ATPase is linked to coordinated mechanical movement of the various motor components. In addition to inspiring rationally designed molecular motors and illuminating basic biological questions regarding (1) molecular recognition and self-assembly; (2) coordinated protein–protein and protein–nucleic acid interactions; (3) nucleic acid translocation and compaction; and (4) conversion of chemical energy into mechanical motion, research on the ø29 packaging motor has the potential to inform rational design of therapeutics that inhibit genome packaging motors in medically important eukaryotic viruses such as herpesviruses, adenoviruses, and poxviruses.

23.2 Overview of Bacteriophage ø29

23.2.1 The ø29 Life Cycle and Assembly Pathways

Bacteriophage ø29 is a dsDNA bacteriophage belonging to the *Podoviridae* family. The mature phage consists of a prolate, $T=3$, $Q=5$ prolate icosahedral head elongated along a fivefold axis of symmetry and measuring 540 Å long by 450 Å wide with a shell thickness of approximately 16 Å. Attached to the head is a pseudo sixfold symmetric, 380-Å long tail (Tao et al. 1998; Morais et al. 2005; Xiang et al. 2006). The phage recognizes and binds to its *Bacillus subtilis* host by virtue of this short noncontractile tail. Hydrolytic enzymes incorporated into the tail structure help ø29 degrade and penetrate the approximately 250-Å thick external peptidoglycan cell wall and membrane of the Gram-positive *Bacillus* host (Xiang et al. 2008; Cohen et al. 2009). The phage then ejects its genome through the tail and into the bacterial cytoplasm where the genome is replicated, and the phage-encoded structural and nonstructural proteins are expressed, thus setting the stage for assembly of progeny ø29 virions (Fig. 23.1).

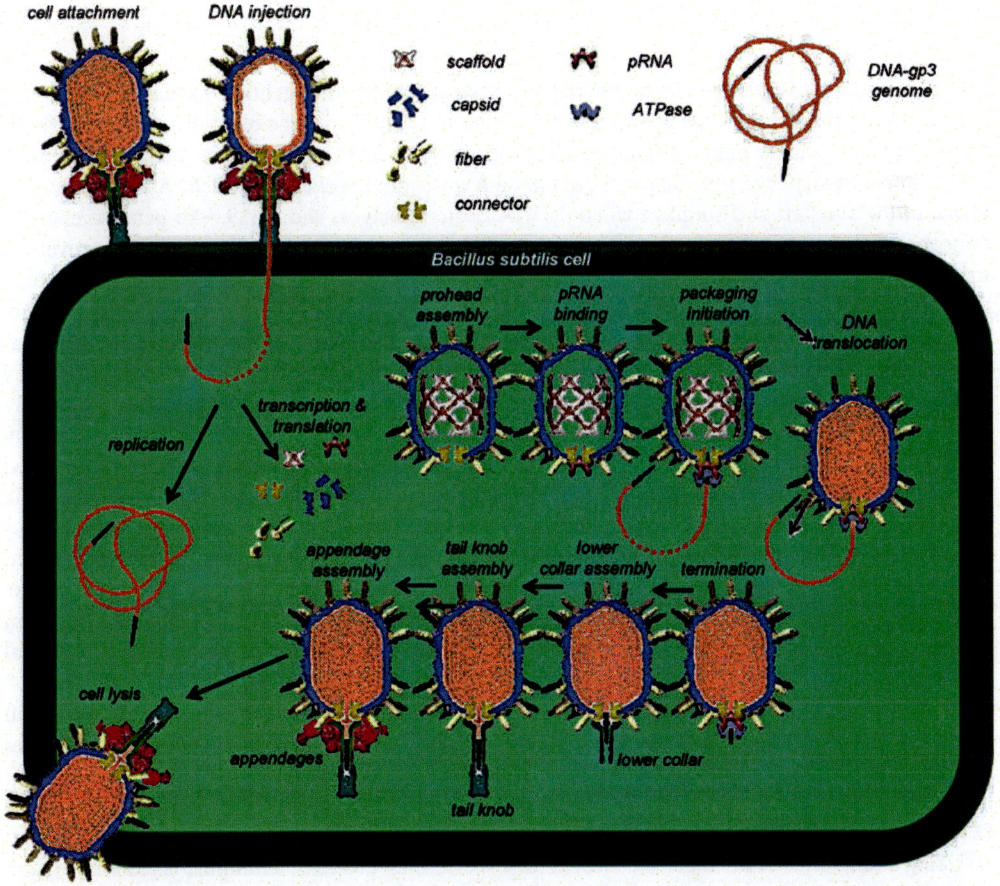

Fig. 23.1 Life cycle and assembly pathway of bacteriophage ø29. Sequential steps in the bacteriophage ø29 life cycle and assembly pathway are shown as a cartoon. A *Bacillus subtilis* host cell is represented as a *green rectangle*, various particles along the assembly pathway are shown as cross sections of cryoEM surface renderings, and individual phage components are either labeled directly or can be identified via the graphical key above the *B. subtilis* cell. Different steps in the assembly pathway are labeled accordingly, and the order of is events indicated by *black arrows*

The assembly of ø29 particles occurs by a well-ordered pathway (Anderson and Reilly 1993; Tao et al. 1998; Grimes et al. 2002). Proheads, the first particles assembled during morphogenesis, consist of a dodecameric head–tail connector protein, or gene product 10 (gp10), approximately 150 copies of the scaffolding protein (gp7), 235 copies of the major capsid protein (gp8), 55 trimeric head fiber proteins (gp8.5), and a pentameric phage-encoded 174 base RNA (pRNA). The 19.3-kb phage genome, covalently attached at either end to gene product 3 (DNA-gp3), is then threaded through a central channel of the connector and encapsidated within the prohead. Genome packaging is powered by a phage-encoded ATPase (gp16), analogous to the large terminase subunit in other phage systems, which transiently assembles at the connector/pRNA vertex. The scaffolding protein exits as DNA enters the shell, although the route and mechanism of exit remain unclear (Morais et al. 2003). Unlike many other tailed dsDNA bacteriophages, ø29 packages a unit-length genome and does not undergo significant structural changes during genome packaging (Xiang et al. 2006; Tang et al. 2008). Upon completion of packaging, the ATPase and pRNA detach from the head, and the lower collar (gp11), appendages (gp12), and tail knob (gp9) subsequently attachto form the phage tail and thus complete virus maturation.

23.2.2 *ø29 as a Model System for Genome Packaging*

There are several advantages inherent to the ø29 system for studying genome packaging. First, it is relatively easy to produce large quantities of ø29 particles. The burst size of wild-type ø29 is greater than 1,000, and mutant phage infections can produce up to 5,000 proheads per cell (Grimes et al. 2002). In contrast, phage lambda produces only a few hundred proheads per cell. Additionally, ø29 is one of the smallest and simplest tailed dsDNA bacteriophages, and its 19.3-kb genome encodes only about 20 genes. Hence, the small size of the ø29 genome affords the opportunity to uncover the structure and function of every gene. For comparison, consider that phage T4's 170,000-kb genome codes for approximately 200 genes. Similarly, the availability of conditional lethal mutants in every gene makes ø29 easy to manipulate and adapt to emerging technologies. Furthermore, the ø29 connector and ATPase components are small (~60% the size of other phages), suggesting that they represent the essential minimum for motor operation. Perhaps most importantly, an extraordinarily efficient in vitro packaging system has been developed for ø29 (Guo et al. 1986) in which nearly every prohead is active and every DNA can be packaged. Thus, compositional simplicity, combined with a large burst size and the high efficiency of in vitro packaging make the ø29 system amenable to complete dissection of the motor mechanism.

DNA packaging, defined as production of DNA filled heads, can be initiated by mixing purified proheads, DNA-gp3, recombinantly expressed ATPase, and ATP. The extent of packaging can be determined by assaying for head filling via DNase protection; addition of DNase I will result in digestion of unpackaged DNA (Guo et al. 1986), whereas packaged DNA is protected in the viral capsid. Hence, the efficiency of packaging can be determined by extracting the packaged DNA and quantifying it on an agarose gel. It was shown that packaging takes about 4 min to complete, and that the motor translocates approximately two base pairs of dsDNA per ATP hydrolyzed. By analyzing packaging of restriction digests, it was demonstrated that ø29 preferentially packages the left end of DNA-gp3, suggesting that the left end is the first end to enter the capsid (Bjornsti et al. 1983, 1984; Grimes and Anderson 1989).

Using fiberless proheads that have ejected their scaffolding proteins, a minimal packaging system can be constructed that consists of only five macromolecular components: (1) the capsid protein, (2) the connector, (3) the pRNA, (4) the ATPase, and (5) the DNA-gp3. The relative arrangement of these components in ø29 particles (Fig. 23.2) was initially proposed from cryoEM reconstructions of wild-type ø29 proheads (Tao et al. 1998; Simpson et al. 2000; Morais et al. 2001, 2005, 2008).

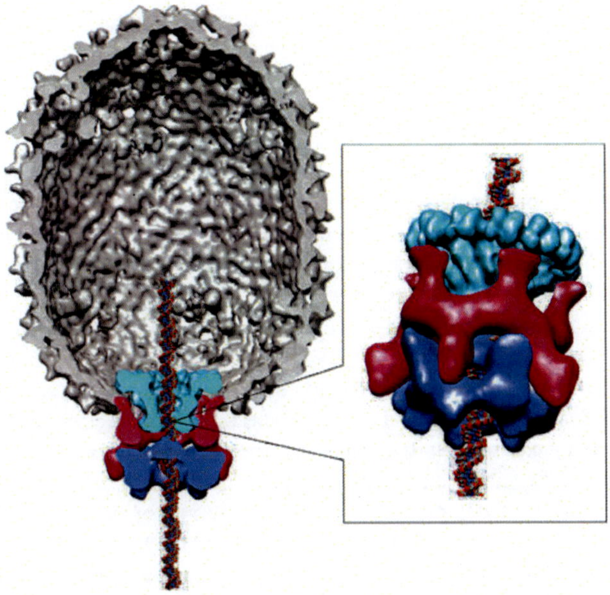

Fig. 23.2 Molecular envelopes of the bacteriophage ø29 motor components. Three-dimensional reconstruction of a ø29 prohead complexed with the packaging ATPase. The particle is shown in a cutaway view, where its front half has been removed to facilitate visualization of the dsDNA packaging motor. The capsid is shown in *gray*, the connector in *green*, the pRNA in *magenta*, and the ATPase in *blue*. The dsDNA genome, shown as *spheres* that have been colored by element, was modeled into the central channel of the connector. The inset shows a close-up of the packaging motor at a skewed angle. Molecular envelopes for various motor components were obtained via cryoEM difference maps (Morais et al. 2008)

The major capsid protein is arranged as a $T=3$, $Q=5$ prolate icosahedral lattice on the surface of the capsid. The dodecameric connector protein replaces a major capsid protein pentamer at a unique vertex situated in an icosahedral endcap along the long axis of the phage. The pRNA also binds this unique vertex, where it circumscribes the protruding narrow end of the connector as a five-membered ring. Five RNA spokes emanate from this central ring, extending away from the phage at a slight angle from the particle's central axis. The packaging ATPase binds the distal end of these pRNA spokes to form a presumably pentameric ring at the unique packaging vertex. These initial observations and predictions were later confirmed and refined in a series of cryoEM reconstructions that definitively located the different components of the packaging motor. Briefly, cryoEM reconstructions were determined for various ø29 particles that were engineered to lack different components of the motor such as the connector, the ATPase, full-length 174-base pRNA, or to lack different domains within the pRNA. The resulting reconstructions were then subtracted from reconstructions of particles containing an intact motor such that the residual density in the difference maps provided molecular envelopes for individual motor components and domains within components (Fig. 23.2) (Morais et al. 2008).

Hence, the connector, pRNA and ATPase form three concentric rings, and the DNA-gp3 is believed to be threaded through a continuous channel along their shared central axis and into the phage capsid. This overall picture of the motor not only provides a general context for understanding the structure and function of individual macromolecular motor components but also provides a framework for understanding how different motor components interact with each other to coordinate the mechanochemical cycle of the motor. Below, the structures and functions of individual motor components are examined in greater detail to illuminate their potential roles in both motor assembly and DNA translocation.

23.3 Components of the ø29 DNA Packaging Motor

23.3.1 The DNA-gp3

As mentioned above, the ø29 genome is one of the smallest and simplest dsDNA bacteriophage genomes, consisting of approximately 19.3 kb that code for about 20 genes. The early genes required for initial DNA replication are located at the ends of the genome, whereas the late genes coding for structural and morphogenetic proteins are expressed from the opposite strand in the middle of the genome (Mosharrafa et al. 1970; Schachtele et al. 1973). A distinguishing feature of ø29 is the requirement for a protein primer to initiate DNA replication. As a result, ø29 gene product 3, gp3 is covalently attached to the first adenine of each 5′ end of the ø29 genome via gp3 serine (Harding et al. 1978; Salas et al. 1978; Yehle 1978). Limited proteolysis of gp3 with trypsin or protease K reduces packaging two- and threefold, respectively, suggesting that gp3 is also necessary for efficient DNA packaging (Bjornsti et al. 1983; Grimes and Anderson 1989). Similarly, restriction fragments containing gp3 are packaged at least ten times more efficiently than fragments lacking gp3 (Grimes and Anderson 1989).

Available structural and biochemical data regarding genome organization within capsids (Xiang et al. 2006; Tang et al. 2008) indicate that the gp-3 attached to the left end of the genome is likely first to enter the capsid during packaging and last to leave upon ejection. Similarly, gp-3 bound to the right end should be last to enter the shell during encapsidation and first to leave upon genome ejection. The passage of gp3 through the central channel of the connector is remarkable in that the smallest dimension of gp3 is greater than the diameter of the connector channel (Kamtekar et al. 2006; Xiang et al. 2006). Presumably, gp3 must be at least partially unfolded to pass through the narrow channel of the connector. There is some controversy regarding the location of the right-end gp3 in mature phage; one group reports that gp3 is lodged in the central channel of the connector (Xiang et al. 2006), while another group reports that it is located in the central channel of the tail just below the connector (Tang et al. 2008). Regardless of the exact position of the right-end gp3 in mature phage, the ø29 packaging motor must at least recognize and package the left-end gp3 along with the phage DNA, and subsequently inject it into the host cell cytoplasm upon genome ejection to initiate a new replication cycle.

23.3.2 The Connector

The ø29 connector protein, analogous to the "portal" proteins in other phage systems, is remarkable in that it plays an essential role in nearly every step of the phage life cycle. In addition to providing a conduit for DNA entrance and egress, the connector nucleates capsid assembly, directs formation of an elongated prolate icosahedral shell (Hagen et al. 1976; Camacho et al. 1977; Guo et al. 1991a), and provides the site of attachment for the virus tail during phage maturation. As a consequence, the connector must selectively interact with the scaffolding protein, the capsid protein, the pRNA, the DNA, the lower collar, and the appendages at different stages of the assembly pathway (Guo et al. 1991a; Simpson et al. 2000; Morais et al. 2001, 2005, 2008 Xiao et al. 2005; Xiang et al. 2006; Atz et al. 2007; Tang et al. 2008; Fu and Prevelige 2009). Given the central importance of connector proteins in every aspect of the phage life cycle, it is not surprising that the structures of phage connector proteins are highly conserved in spite of often undetectable sequence homology (Valpuesta and Carrascosa 1994; Simpson et al. 2000; Lebedev et al. 2007; Zheng et al. 2008; Lander et al. 2009) (Fig. 23.3f). Indeed, cone-shaped connectors were some of the earliest defining morphological features observed in negative stained electron micrographs of tailed dsDNA bacteriophages.

The structure of isolated recombinantly expressed ø29 connectors has been studied by a variety of biophysical methods, including atomic force microscopy (Muller et al. 1997), cryoEM of two-dimensional arrays (Valpuesta et al. 1999), immune-electron microscopy (Valle et al. 1999), and

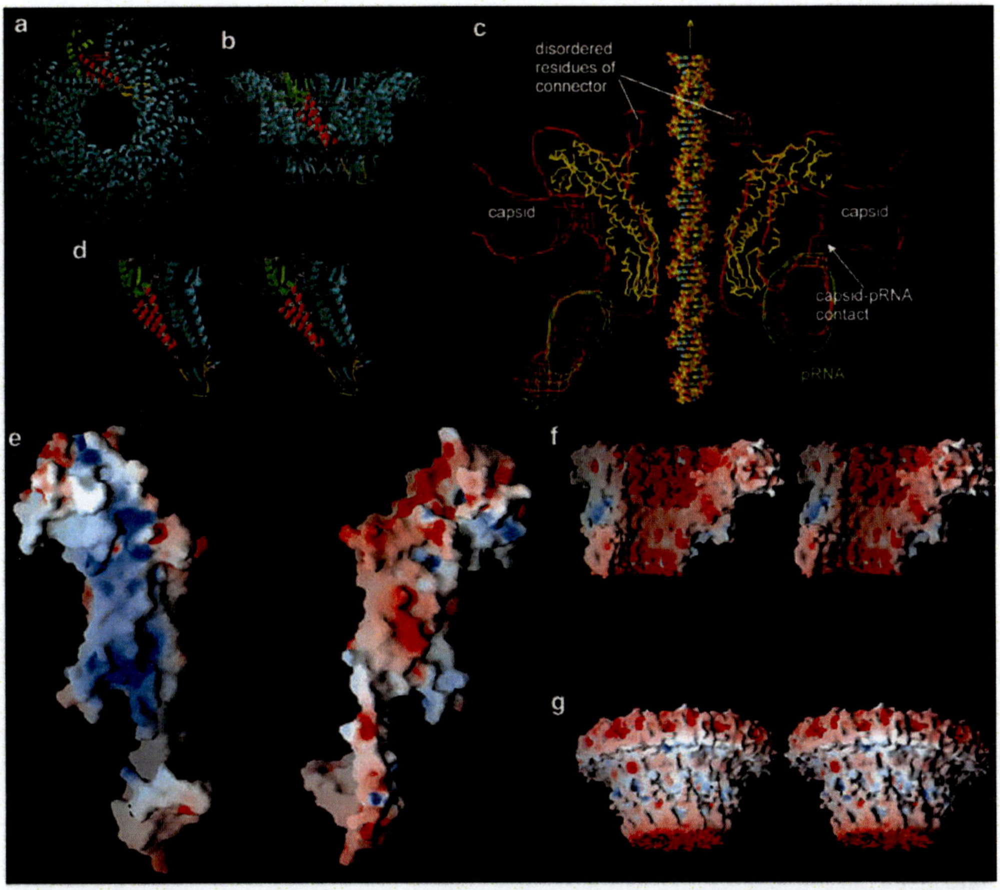

Fig. 23.3 Crystal structure of the bacteriophage ø29 connector protein. (**a**) Three-dimensional structure of the ø29 shown as a *ribbon diagram*. The direction of view is looking down the 12-fold symmetry axis from the narrow end of the connector. A single subunit is shown with its wide, central, and narrow domains in *green*, *red*, and *yellow*, respectively. (**b**) Side view of the connector with the central 12-fold symmetry axis shown as a white line. (**c**) Fit of the X-ray structure of the connector to cryoEM density of the prohead. The backbone of the connector is shown as *yellow wire*; a pRNA density is outlined with *green* mesh. A DNA molecule was modeled in the central channel of the connector. The capsid, one of the five contacts between the pRNA with the capsid, and the partially disordered residues 229–246 and 287–309 in the connector are indicated. (**d**) Stereodiagram of a pair of monomers. One monomer is in *blue* and the other as described in (**a**). Six α-helices are labeled, along with the N- and C-termini. (**e**) Electrostatic surface potential of opposite faces of a monomer showing the complementarily charged surfaces that interact between the subunits. (**f**) Internal electrostatic surface of the connector showing its primarily negative charge. (**g**) External electrostatic surface showing negatively charged regions at opposite ends of the connector and the relatively hydrophobic region in the center that interacts with the capsid. Parts (**a–c**) were modified from Simpson et al. (2000); parts (**d–f**) were modified from Simpson et al. (2001)

X-ray crystallography (Simpson et al. 2000, 2001; Guasch et al. 2002). Although 13-fold symmetric connectors have been reported (Tsuprun et al. 1994), both X-ray crystallography and cryoEM indicate that connectors form 12-fold symmetric dodecamers (Simpson et al. 2000; Morais et al. 2001; Xiang et al. 2006; Tang et al. 2008). The 12 monomers assemble a hollow cone-like structure where the 12-fold symmetry axis is coincident with the conical axis (Fig. 23.3a, b). The connector structure can be further divided into three approximately cylindrical regions that differ in height and radius: the narrow end, central part, and wide end have external radii of 33, 47, and 69 Å and heights of 25, 28, and 23 Å, respectively. Hence, the connector measures ~75 Å along its conical axis, and its width varies from ~140 Å at its wide end to ~65 Å at the narrow end. The diameter of the connector's central channel increases from approximately 36 Å at the narrow end to about 60 Å at the wide end, large enough to accommodate smooth passage of an extended dsDNA helix. Fitting the X-ray structure of the connector into cryoEM reconstructions of ø29 proheads shows that the wide end of the connector resides within the viral head, the central region is adjacent to the virus shell and the pRNA, and the narrow end protrudes from the head and interacts with the central ring of the pRNA (Simpson et al. 2000; Morais et al. 2005, 2008; Ding et al. 2011) (Fig. 23.3c).

The 309 residue connector monomer resembles a seahorse, where the wide domain corresponds to the head; the central domain, the body; and the narrow domain, the tail (Fig. 23.3d). Residues 158–202 comprise the narrow-end domain, forming mostly an antiparallel β-sheet that interacts with the central ring of the pRNA. The central domain (residues 11–14, 130–157, and 203–226) consists of an antiparallel three-helix bundle twisted in a left-handed superhelix around the central 12-fold axis at an angle of approximately 40. The wide-end domain (residues 1–10, 229–246, and 286–309) adopts a more globular mixed α/β-structure that is flexibly attached to the central domain, and contacts the inner surface of the head and the pRNA (Simpson et al. 2000; Ding et al. 2011) (Fig. 23.3c). Residues 1–10, 229–246, and 286–309 are not visible in the crystal structure, suggesting that these regions are disordered in recombinantly expressed connectors. However, the second and third disordered regions are visible in cryoEM maps where they are located on the inside of the channel at, and just above, the junction of the wide and narrow domains, respectively, consistent with the locations of analogous elements in the connector proteins from bacteriophages SPP1 and P22 (Orlova et al. 2003; Lebedev et al. 2007; Lander et al. 2009) (Fig. 23.3c).

Calculated electrostatic potentials show that the connector monomer has one predominantly negatively charged surface and a complementary net positively charged surface, promoting the circular interactions necessary for formation of a closed ring-like structure (Fig. 23.3e). The exterior surface of the central region is predominantly hydrophobic, whereas the exterior surfaces of the wide- and narrow ends have negatively charged patches (Fig. 23.3f, g). With the exception of two lysine rings, the surface corresponding to the central channel is negatively charged. It has been proposed that the net negative charge of this surface acts as a focus to center the opposing negatively charged DNA as it passes through the connector channel (Simpson et al. 2000), and that the positively charged lysine rings play a direct role in packaging by interacting with the negatively charged phosphate backbone of the translocating DNA (Guasch et al. 2002) (see also Sect. 23.5).

Comparison of connector structures fitted into cryoEM reconstructions of mature ø29, wild-type proheads, and particles lacking pRNA indicates that the conformation of the connector is strongly influenced by the attachment/detachment of the pRNA and the tail structure (Fig. 23.4a–c). For example, in the absence of pRNA, the outer and inner diameter of the central region and narrow end increases compared to connectors in particles bound to pRNA (Morais et al. 2008). The narrow domain of the connector also flares out in mature phage. It has been suggested that connector expansion occurs in response to high pressure in fully packaged proheads, and that this expansion forces the pRNA off the capsid and results in an optimal connector conformation for tail attachment (Xiang et al. 2006). However, it is not clear if this conformational change occurs prior to tail assembly, or if tail binding induces conformational change in the connector.

Additional information regarding conformational flexibility of the connector can be obtained by comparing connector structures from different crystal forms, and by analyzing structural differences among monomers within the asymmetric unit of a given crystal form (Simpson et al. 2001) (Fig. 23.4d, e). Differences between different crystal forms are localized in the wide-end domain. Similarly, superposition of different monomers within a crystallographic asymmetric unit shows that while the central and narrow domains strictly obey 12-fold symmetry, the wide end can undergo rigid body rotations of up to 5.4 relative to the central and narrow domains. These conformational fluctuations correspond to a "rolling" motion of the wide-end domain and were proposed as an essential component of a connector-based mechanism for DNA packaging (Simpson et al. 2000) (see also Sect. 23.3.2).

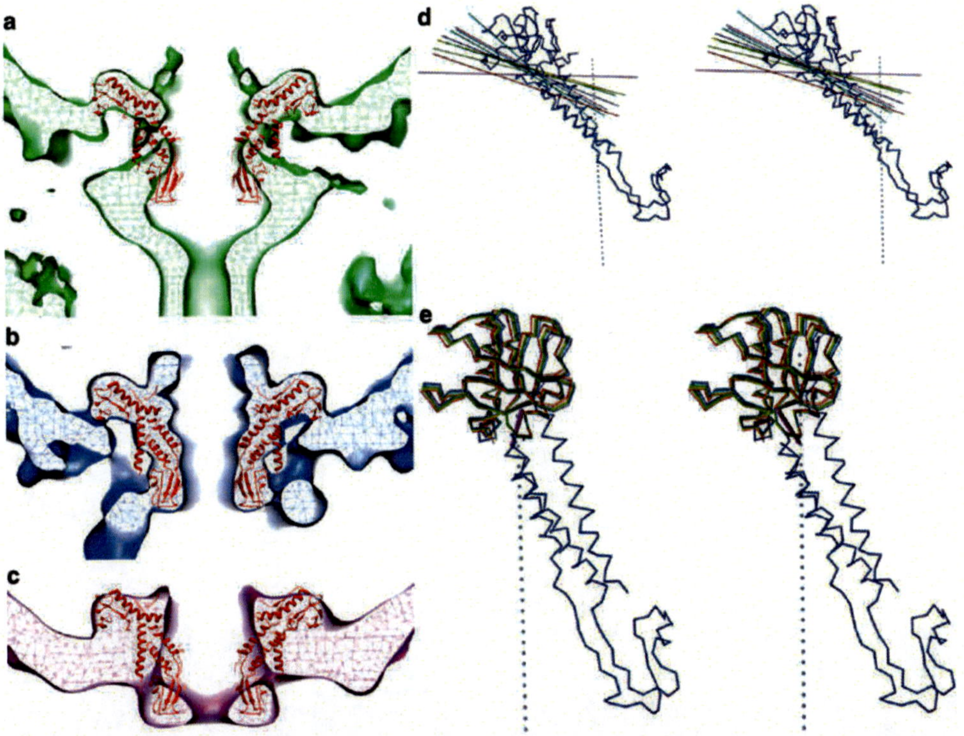

Fig. 23.4 Conformational flexibility of the connector. (**a**) Fit of the atomic coordinates of the connector (*red ribbon*, Protein Data Bank accession code: 1IJG) to cryoEM density (*green* density) corresponding to the connector in mature phage, (**b**) to prohead-bound phage (*blue* density), and (**c**) to pRNA-free proheads (*purple* density). Comparison of cryoEM densities indicates conformational changes involving the central and narrow domains during phage maturation. (**d**) Stereodiagram based on the superposition of the helical domains within the crystallographic asymmetric unit of the connector. Different colored lines represent the different axes of rotation relating any one wide-end domain to a reference monomer. (**e**) The variation in position of the wide-end domain in the crystallographic unit. Parts (**a–c**) were inspired by Xiang et al. (2006); parts (**d** and **e**) were adapted from Simpson et al. (2001)

23.3.3 The pRNA

One of the more surprising observations to emerge from early experiments on the ø29 packaging motor was the sensitivity of packaging activity to ribonuclease, as comparable ribonuclease sensitivity has not been observed in other phage systems. Perhaps equally surprising was the explanation for this sensitivity, i.e., the discovery of a 174-base phage-encoded RNA molecule that is absolutely required for packaging (Guo et al. 1987a). This RNA was shown to be a transcript from a promoter at the extreme left end of the genome (bases 320–147) (Guo et al. 1987a; Wichitwechkarn et al. 1989). Since the RNA was found to bind to proheads (Guo et al. 1987b; Garver and Guo 1997; Simpson et al. 2000), it was named "pRNA." The pRNA is not needed to assemble proheads (Guo et al. 1991b), suggesting that it attaches to proheads after capsid assembly. It was shown that pRNA binds to proheads in an Mg^{2+}-dependent reaction (Wichitwechkarn et al. 1989; Reid et al. 1994a; Trottier and Guo 1997). The role of pRNA is transitory and likely limited to DNA packaging, as pRNA is not present in mature ø29 (Guo et al. 1987a).

Bioinformatic analysis and secondary structure prediction indicated that the pRNA consists of two domains; domain I consists of the first 117 bases and is separated from domain II (bases 131–174) by

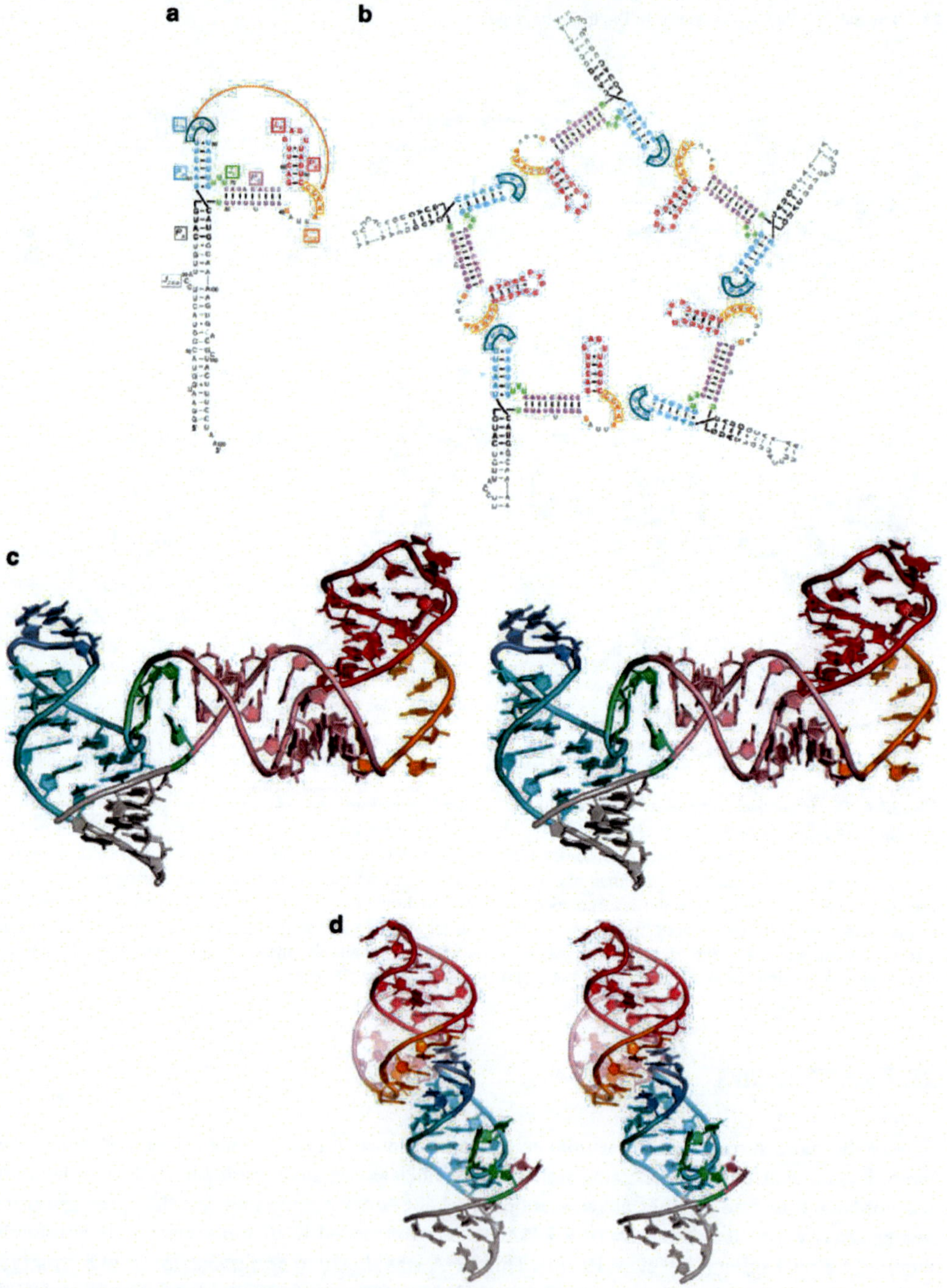

Fig. 23.5 Crystal structure of the pRNA prohead-binding domain. (**a**) Predicted secondary structure of pRNA domain I. Nucleotides are colored: *gray*, A-helix; *green*, three-way junction; *violet*, C-helix; *orange*, CE-loop; *magenta*, E-helix and E-loop; and *cyan*, D-helix and D-loop. In *light gray* are nucleotides deleted in the final crystallization construct. (**b**) Proposed cyclic oligomerization of the pRNA based on intermolecular base pairing at the pseudoknot. Although hexameric oligomerization was initially predicted, a pentameric structure is shown to reflect the likely oligomerization state on the prohead. (**c**) Stereodiagram of the crystal structure of the pRNA-binding domain. The phosphate backbone is shown as a coil, and the bases as filled slabs colored as above. (**d**) Stereoview of the intermolecular pseudoknot. The CE–D loop interaction mediates the continuous base stacking from the E-helix of one pRNA monomer to the A-helix of its neighbor, creating an RNA superhelix. This figure was adapted from Ding et al. (2011)

a 13 base single-stranded region (Bailey et al. 1990) (Fig. 23.5a). Although the full-length wild-type pRNA is a two-domain 174-base transcript, a 120-base construct encompassing domain I and lacking the 54 nucleotide 3′-domain II is fully competent for packaging and assembly in vitro. Although the function of domain II is not currently understood, its conservation in all ø29 relatives suggests that it has an essential function in vivo. Seven RNA helices, designated A–G, were also predicted, with helices B and F residing in domain II, and helices A, C, D, and E in domain I. Three conserved loops, CE, D, and E, named for their position relative to the similarly named helices, were also identified.

Packaging activity is eliminated upon removal of pRNA, but can be restored by reconstituting RNA-free particles with recombinantly expressed pRNA (Guo et al. 1987b). Reconstitution of RNA-free proheads with mutant pRNAs is a remarkably powerful technique, and in combination with bioinformatics and phylogenetic analysis, it has been used to exhaustively dissect structural and functional elements of the pRNA (for a comprehensive review, see Grimes et al. 2002). Early experiments distinguished two functional modules within pRNA domain I, one necessary for prohead binding (bases 22–84), the other required for DNA packaging activity (bases 1–20 and 92–117) (Reid et al. 1994a, b, c; Zhang et al. 1994, 1995). The prohead-binding domain contains the C, E, and D helices and the CE, E, and D loops; all of these structural elements were shown to be important for prohead binding, as were the base sequences within each element (Reid et al. 1994b). Phylogenetic analysis, combined with mutagenesis and biochemistry identified an essential U-rich three-way junction that was predicted to join helices A, D, and E in domain I (Bailey et al. 1990). The DNA packaging domain corresponds to the majority of the pRNA A-helix; mutants in this region bind proheads as tightly as wild type, but are defective in DNA packaging (Reid et al. 1994b, c; Zhang et al. 1994, 1995). Since many of these mutants were also defective in binding the ATPase, it was proposed that the pRNA A-helix provides the attachment site for the packaging ATPase.

Considering that quaternary structures larger than homodimers had never been observed for RNA (Ding et al. 2011), it was surprising that recombinantly expressed pRNA formed higher-order oligomers in solution (Guo et al. 1998; Zhang et al. 1998; Chen et al. 1999, 2000). This oligomerization was demonstrated to arise from a novel intermolecular RNA pseudoknot. From examination of the predicted secondary structure of the pRNA, it was clear that bases 45-AACC-48 from the C–E loop could form Watson–Crick base pairs with bases 85-UUGG-82 from the D loop (Reid et al. 1994c) (Fig. 23.5a, b). Furthermore, phylogenetic analysis indicated that comparable base-pairing potential is conserved in pRNAs from ø29 relatives. To test the biological significance of this potential pseudoknot, the CE-loop was mutated from AACC to GCGA in one mutant, and the D-loop mutated from UUGG to CGCU in another mutant. As expected, neither mutant alone could package DNA, yet a double mutant, which restores the base-pairing potential, regained full biological activity. Interestingly, a mixture of the two single mutants also restored packaging (Zhang et al. 1998). Since formation of intramolecular pseudoknots would not be possible in this experiment, it was concluded that the pseudoknot must be an intermolecular one between neighboring pRNAs. It was further proposed that formation of intermolecular pseudoknots between adjacent pRNAs would promote circular interactions resulting in an oligomeric ring-like structure (Fig. 23.5b).

Structural analysis has now confirmed many of the early predictions regarding pRNA structure and function. CryoEM analysis has definitively shown that (1) bases 22–84 form a pentameric ring-like structure that binds to the prohead; (2) five pRNA A-helices extend as spokes from this central ring; and (3) the packaging ATPase attaches to the distal end of the A-helix spokes (Morais et al. 2008) (Fig. 23.2). The recently determined X-ray structure of the prohead-binding domain of the pRNA confirms the presence of the U-rich junction and, via crystal contacts between pRNA monomers, the intermolecular pseudoknot (Ding et al. 2011) (Fig. 23.5c, d). That such sophisticated and prescient structure/function predictions could be made in the absence of any direct structural information is a testament to the power of carefully conceived genetic and biochemical experiments when applied to a well-characterized system.

Of course, structural data on the pRNA has also provided additional information that would not be available using other experimental approaches. The crystal structure of the prohead-binding domain shows that pRNA monomers adopt an extended conformation, and that the three major helical elements are organized around the U-rich three-way junction (J_u); helices A and D stack coaxially, whereas the C-helix extends outward at an approximately 70° angle (Ding et al. 2011) (Fig. 23.5c). At the distal end of the C-helix, the CE-loop causes the direction of the E-helix to change by about 110°. The conformation of the E-loop is similar to an NMR-derived structure of an E-loop hairpin (Harris and Schroeder 2010). Analysis of electron density and temperature factors in the crystal structure suggests the presence of two flexible joints at the J_u bulge and the CE-loop. In addition to confirming the presence of all four potential Watson–Crick base pairs in the intermolecular pseudoknot, the structure showed that there is an additional fifth interaction, a base triple between A_{44} in the CE-loop and G_{86} and U_{80} in the D-loop (Fig. 23.5d).

The crystal structure of the prohead-binding domain of the pRNA fits remarkably well into its corresponding density in a cryoEM reconstruction of the ø29 prohead (Fig. 23.6a). The excellent agreement between the two structures not only confirms that each is essentially correct but also provides insight as to how the pRNA interacts with the connector and the prohead. The structure fits into the density region corresponding to the central ring, where it makes contact with protruding narrow end of the connector (Fig. 23.6b). The pRNA bases adjacent to the connector are consistent with hydroxyl radical probing studies that showed increased cleavage sensitivity in the ring region of the pRNA when an Fe-BABE probe was tethered to the narrow end of the connector (Atz et al. 2007). Furthermore, the fitted pRNA is close to the expected position of the positively charged RKR residues at the N-terminus of the connector that were shown to be important for pRNA binding (Xiao et al. 2005; Atz et al. 2007). Additionally, this fitting placed the E-loop in density that connects the pRNA central ring to the prohead (Fig. 23.6a). This density had been identified as pRNA in previous cryoEM reconstructions (Morais et al. 2005, 2008), but existing pRNA models were unable to explain which bases of the pRNA accounted for this density. The current fitting clearly shows that E-loop bases G_{55} and A_{56} make contact with a unique-binding site on the capsid protein that is exposed as a result of rotation of the Ig-like C-terminal domains in capsid proteins circumscribing the unique motor vertex (Morais et al. 2005, 2008; Ding et al. 2011) (Fig. 23.6c). The importance of these bases in prohead binding was also confirmed via mutagenesis (Ding et al. 2011).

Although pRNA monomers were independently fitted into the cryoEM density without any prior assumptions, the intermolecular base-pairs were preserved in the resulting pRNA pentamer. This fitting also positions the small portion of the A-helix present in the truncated construct in cryoEM density identified as the A-helix by cryoEM (Morais et al. 2008; Ding et al. 2011), thus allowing modeling of the remainder of the A-helix and construction of pseudoatomic model of entire pRNA domain I. This model revealed a remarkable RNA superhelix that dominates the pRNA structure and arises from continuous stacking of multiple helical elements from adjacent pRNA monomers; an E-helix from one monomer shares a common superhelical axis with a D-helix (via the intermolecular pseudoknot) and an A-helix from a neighboring pRNA monomer (Fig. 23.6d). Hence, the basic structural unit of the pRNA spans two topological units, an arrangement that might mediate coordination between adjacent motor subunits.

The oligomerization state of the pRNA on the prohead has been a source of considerable controversy, and thus merits discussion here. Although the pRNA forms tetramers in the crystallographic unit cell (Ding et al. 2011), this arrangement is not believed to reflect the oligomerization state on the prohead. A truncated construct was crystallized, with additional mutations introduced to reduce molecular flexibility. Furthermore, the fourfold molecular symmetry axis is coincident with the crystallographic fourfold axis. Hence, since there is no evidence for wild-type pRNA tetramers, it seems likely that this packing arrangement might be an artifact of either/both crystallization and/or genetic manipulation of the construct.

Wild-type pRNA oligomerizes in solution, and both hexamers and dimers have been detected by analytical ultracentrifugation (Guo et al. 1998; Zhang et al. 1998; Chen et al. 1999, 2000). Initial genetic investigations into the nature of the intermolecular pseudoknots showed that two-, and three-way

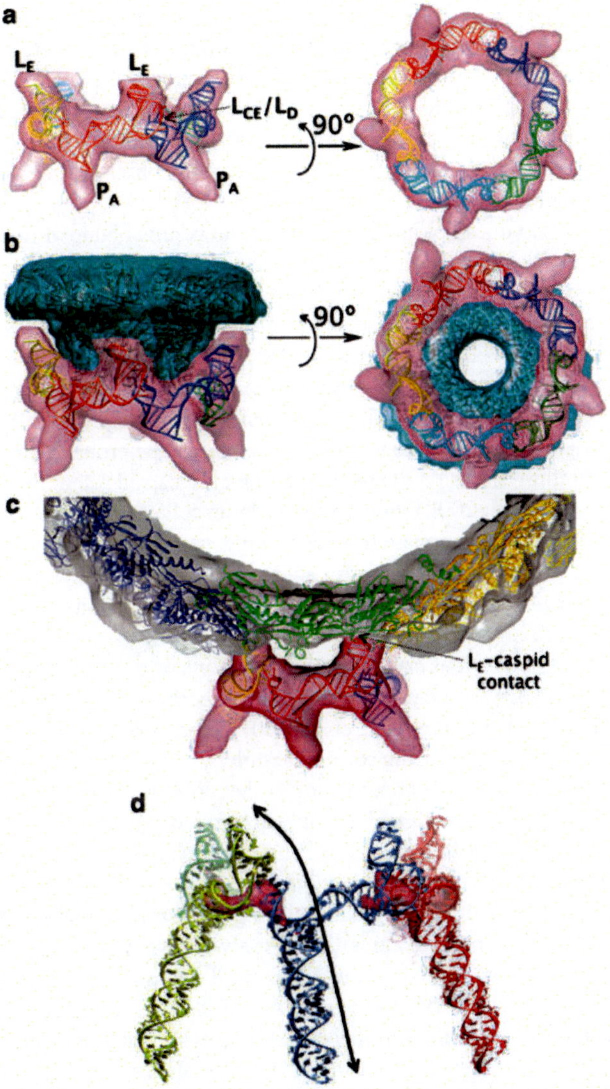

Fig. 23.6 Pseudoatomic structure of the pRNA on proheads. (**a**) Side and bottom views of the pentameric pRNA ring structure docked into its cryoEM envelope. (**b**) Side and bottom views of the pentameric pRNA ring and the dodecameric connector structures fit into their corresponding cryoEM envelopes. (**c**) Side view of the pRNA ring and a pseudoatomic model of the capsid protein fit into the cryoEM envelope showing that the pRNA contacts the capsid via the E-loop. The connector structure was omitted for clarity. (**d**) Pseudoatomic structure of full-length pRNA. Three of the five pRNA monomers are colored *blue*, *red*, and *yellow*, respectively, and part of a fourth monomer is shown in green. The A-helix was modeled as an ideal RNA helix and positioned via fitting to cryoEM density. The E-helix from one monomer shares a common super helical axis with a D-helix (via the intermolecular pseudoknot) and an A-helix from a neighboring pRNA monomer. The approximate direction of the superhelical axis is shown as a *double-headed arrow*. The strongest density in cryoEM maps is shown as a *magenta* isosurface; note that this density is near the intermolecular pseudoknot, demonstrating that this region is highly ordered in proheads. Parts (**a–c**) of this figure were modified from Ding et al. (2011)

complementation of CE/D-loop mutants could rescue packaging, suggesting that both multiples of two and three are compatible with an active pRNA oligomer. Hence, the pRNA was believed to be hexameric because (1) six is the smallest integer that is a multiple of both two and three, (2) hexamers were observed in analytical ultracentrifugation, and (3) the pRNA was initially believed to bind to the connector, an oligomer with 12-, and thus also sixfold symmetry. This view was recently reinforced by single-particle fluorescence quenching experiments where statistical analysis of photobleaching events was used to conclude that there is an average of six pRNAs per prohead (Shu et al. 2007). However, recent analytical ultracentrifugation experiments show that head-to-tail dimers dominate in solution, not hexamers (Chen et al. 2000), and it is likely these dimers are artifacts of recombinant overexpression since comparable dimers are not observed in X-ray or cryoEM structures of the pRNA (Ding et al. 2011).

A hexameric pRNA also does not agree with structural results obtained from either cryoEM or X-ray crystallography, which clearly show pentameric and tetrameric pRNAs, respectively. Fitting of the X-ray structure of a pRNA into cryoEM density results in a pentameric pRNA, even for carefully calculated asymmetric reconstructions where no symmetry is imposed (Simpson et al. 2000; Morais et al. 2001; Ding et al. 2011). This fitting also indicates that although the pRNA is close to the connector, it makes much more extensive contacts with the fivefold symmetric head than the 12-fold symmetric connector. Similarly, it is has been shown that pRNA binds to connectorless proheads with an occupancy comparable to that seen for wild-type prohead binding, indicating that the pRNA's primary-binding site is on the fivefold symmetric capsid rather than the 12/6-fold symmetric connector (Morais et al. 2008). Furthermore, the excellent fit of the X-ray structure of the pRNA to the cryoEM structure demonstrates that two independent methods give rise to essentially identical structures; it is difficult to account for such a result if the structures are not truly similar. Thus, it seems that the pentameric pRNA observed in cryoEM reconstructions reflects its true oligomeric state.

The argument has been made that cryoEM reconstructions of ø29 will always result in unintentional imposition of fivefold symmetry, and thus that the observed pentameric pRNA is an artifact of unintentional symmetry averaging (Shu et al. 2007). The underlying logic is that individual particles are aligned for averaging during reconstruction by virtue of their relatively massive fivefold symmetric capsids. Hence, small components of the motor that do not have the same symmetry as the prohead might not be aligned for meaningful averaging during the reconstruction procedure, see (Morais et al. 2001) for a more extensive discussion. However, it is important to point out that improper symmetry averaging, whether intentional or unintentional, results in smeared out uniform density without any distinguishing features. In contrast, the pRNA structure determined by cryoEM has distinct features that agree with the crystallographically determined structure of the pRNA (Ding et al. 2011). Furthermore, it was shown that subtle yet predictable structural changes can be introduced into the pRNA, and that these changes can then be readily observed in cryoEM reconstructions of proheads (Morais et al. 2008; Zhao et al. 2008), strongly indicating that the reconstruction procedure faithfully reproduces the molecular structure of the pRNA. Finally, the dodecameric connector provides an internal control to verify that symmetry mismatches have been resolved in cryoEM reconstructions of ø29 proheads. In particular, the ability to resolve 12 monomers, as well as secondary structural elements within monomers, in a recent ~8 Å asymmetric reconstruction of the packaging motor demonstrates that even relatively small nonpentameric components are properly aligned and averaged in asymmetric prohead reconstructions (Morais, unpublished).

The question of how pRNA assembles on the head also remains controversial. Based on the early analytical centrifugation results and pseudoknot complementarity experiments, it was assumed that the pRNA forms a hexameric ring which then docks onto the connector (Guo et al. 1998; Zhang et al. 1998). Later results suggested that pRNA forms head-to-tail dimers in solution, but that this dimer is dissembled as the pRNA binds to the connector to form a hand-to-hand circular ring. Additional evidence for connector-based docking came from experiments showing that the first 14 amino acids of the connector are important for pRNA binding, especially the short RKR sequence (Xiao et al. 2005; Atz et al. 2007). However, it was recently shown that pRNA can bind particles that lack connectors as well as bind wild-type particles (Morais et al. 2008).

An initial clue regarding pRNA assembly came from analysis of cryoEM reconstructions of the prohead, which show that the E-loop of the pRNA makes extensive contacts with the capsid that would be sterically impossible if the Ig-like domain of the capsid protein had not rotated ~180° from its expected position (Morais et al. 2005, 2008; Ding et al. 2011). Hence, an assembly scenario compatible with all the conflicting data is one in which the pRNA initially recognizes and binds to the RKR residues at the N-terminus of the connector, followed by a conformational change in the capsid proteins surrounding the motor vertex that creates a new, tight-binding site on the capsid for the pRNA. This proposed conformational change likely involves the previously described rotation of the Ig-like C-terminal domains in capsid proteins surrounding the pRNA (Morais et al. 2005), which would uncover the pRNA-binding site on the capsid. It was demonstrated that this conformational change is present in prohead particles whose pRNA has been removed by RNase digestion (Morais et al. 2008). Further support for this assembly scenario comes from cryoEM reconstructions of proheads that have never seen pRNA. It would be expected that the C-terminal Ig-like domain would not be rotated from its typical position since these particles are pRNA naive and thus would not have undergone the conformational change necessary for binding pRNA; this is indeed the case. Furthermore, it has been shown that when pRNA binds these particles, the Ig-like domain rotates from its expected position as in wild-type proheads (Morais, unpublished data). Thus, it may be that pRNA assembly proceeds via attachment to the procapsid first, followed by formation of intermolecular pseudoknots. However, since pseudoknots are absolutely required for packaging, they may play a role in the mechanism of translocation. Their location at the junction of helices contributed from neighboring monomers in the pRNA superhelix may provide a means of communication around the motor ring (Fig. 23.6d).

23.3.4 The Packaging ATPase

Packaging dsDNA into a preformed viral capsid is an inherently unfavorable energetic process. The dsDNA helix is a relatively rigid molecule, with a persistence length of approximately 50 nM. It is also highly negatively charged due to its phosphate backbone. Hence, there are considerable enthalpic and entropic costs associated with confining DNA within the limited space of the viral capsid. Coarse-grained simulations of DNA packaging and subsequent decomposition of packaging energy into its principle components suggests that electrostatic repulsions contribute most of the enthalpic penalty associated with packaging the ø29 DNA-gp3, whereas the elastic cost of bending the DNA helix is modest (Petrov and Harvey 2008). These analyses also indicate that the entropic cost of packaging accounts for approximately half of the total free energy cost of genome encapsidation. This is not surprising since ø29 DNA-gp3 has a radius of gyration of ~340 nm in solution compared to only ~15 nm inside proheads, corresponding to an ~10,000-fold volume compression; clearly, there must be a considerable entropic price associated with such a dramatic reduction of conformational space. Regardless of relative percentages, the enthalpic and entropic costs of genome encapsidation must be paid for via the input of energy.

ATP is the universal energy currency of the cell, and several protein families have evolved that couple ATP hydrolysis to otherwise thermodynamically unfavorable reactions. The net effect of this coupling is to drive the linked reactions in a direction necessary to do chemical or mechanical work. The dsDNA viruses, including tailed bacteriophages like ø29, have co-opted polymer-dependent ATPases from the ASCE P-loop NTPase superfamily to couple hydrolysis of host-supplied ATP to the translocation of the dsDNA viral genome into the preassembled procapsid (Mitchell et al. 2002; Burroughs et al. 2007; Singleton et al. 2007; Thomsen and Berger 2008) (Fig. 23.7). The energy released by hydrolysis of the high-energy phosphoanhydridic bond between the β- and γ-phosphates induces a conformational change in the ATPase that exerts a mechanical force that results in translocation of the DNA into the head.

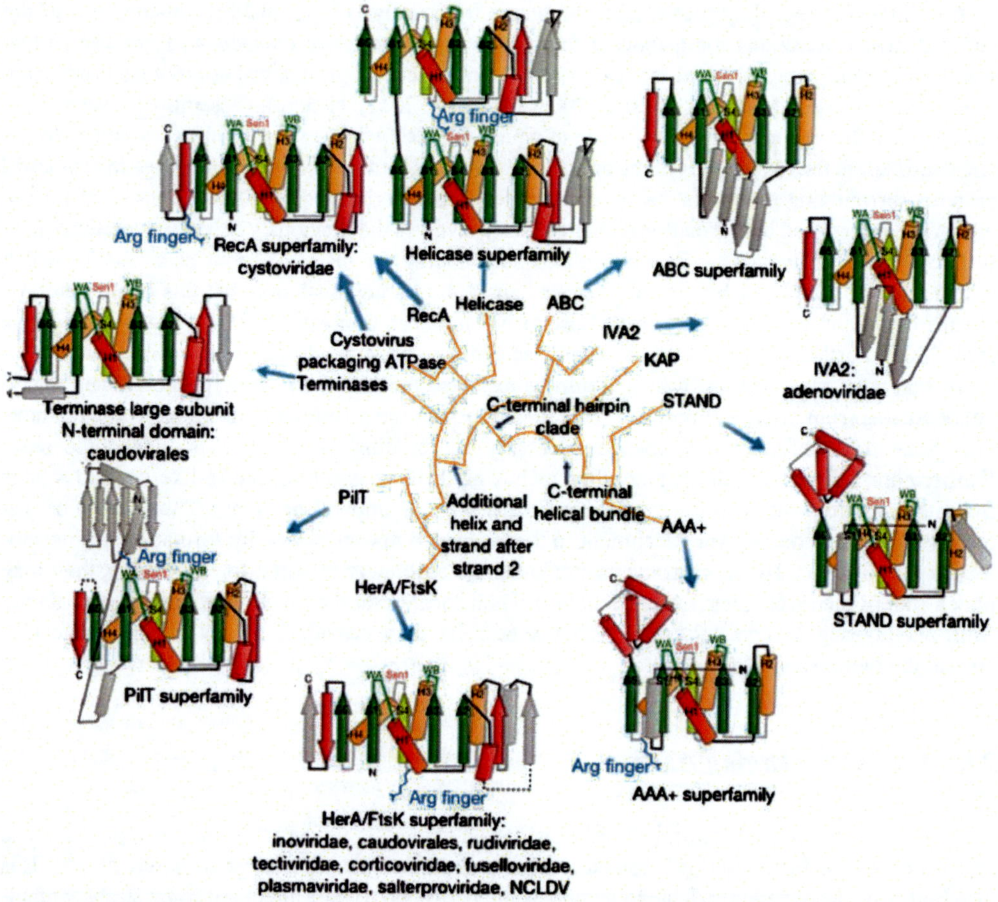

Fig. 23.7 ACSE superfamily. Proposed evolutionary relationships between ASCE P-loop NTPases are shown. Topology diagrams of different representative family members are superimposed on a cladogram depicting higher-order relationships. The highly conserved central strands and helices forming the core of the ASCE domain are numbered and colored. Strands are shown in *green* with the central strand S4 colored *yellow* and helices colored *orange*. Synapomorphies shared across different lineages are shown in *pink*, whereas nonconserved features are colored *gray* and outlined in *broken lines*. Lines connecting different lineages represent higher-order relationships constructed by comparison of shared structural and/or sequence similarities. *Broken lines* represent more tenuous relationships. Viral lineages with packaging ATPases from a specific superfamily are listed following a colon below the superfamily name. Abbreviations: *WA* Walker A, *WB Walker B*, and *Sen1* sensor-1. This figure is reproduced from Burroughs et al. (2007)

The 39-kDa 332-residue ø29 packaging ATPase consists of two domains (Koti et al. 2008). The first ~200 N-terminal amino acids correspond to the ASCE P-loop ATPase domain, and the last ~130 residues comprise the C-terminal domain. Sequence analysis suggests that the C-terminal domain adopts a five-stranded oligonucleotide/oligosaccharide-binding fold, possibly indicating that the C-terminal domain could either play a role in pRNA binding (Koti et al. 2008) or DNA translocation (Sun et al. 2007, 2008). Both dimers and trimers of recombinantly expressed ø29 ATPase have been reported (Lee and Guo 2006; Koti et al. 2008), although larger oligomeric rings are likely assembled on proheads (Simpson et al. 2000; Koti et al. 2008; Morais et al. 2008). Since the ATPase has been shown to bind to the A-helix of the pRNA, it seems likely that it would also form pentamers on the prohead. This prediction is supported by cryoEM reconstructions that show a pentameric ring of ATPases bound to distal ends of the pRNA A-helices (Koti et al. 2008;

Morais et al. 2008) (Fig. 23.2). It has been shown that the ATPase can either assemble on the pRNA and then bind the DNA-gp3 substrate, or first interact with DNA-gp3 and then attach to the pRNA (Koti et al. 2008). It is not clear which assembly pathway is utilized in infected *Bacillus* cells.

Similar to other phage packaging ATPases, the ø29 ATPase has only weak ATPase activity in the absence of other motor components (Grimes and Anderson 1990). This activity is modestly stimulated by the addition of pRNA and increases even further in the presence of pRNA–prohead complexes. There is no increase in activity when the ATPase is mixed with pRNA-free proheads, consistent with structural and biochemical results indicating that the ATPase binds to the A-helix of the pRNA (Reid et al. 1994b, 1994c; Zhang et al. 1994, 1995; Simpson et al. 2000; Morais et al. 2008). The increased activity observed for prohead–pRNA–ATPase mixtures compared to pRNA–ATPase mixtures indicates that the ring-like oligomerization of the ATPase upon prohead binding is important for catalysis. However, maximum activity is only observed in the presence of both pRNA–prohead complexes and the DNA-gp3 substrate. The mechanism of DNA-enhanced ATPase activity is not well understood, although it has been proposed that the translocation of the helically symmetric dsDNA substrate through the ring channel coordinates ATPase activity around the ring (Morais et al. 2008).

Although no high resolution structural information is available for the ø29 ATPase, considerable insight can be obtained from analysis of structurally homologous ATPases. The N-terminal domain of the ø29 ATPase contains several short sequence motifs characteristic of the ASCE P-loop NTPases, including an adenine-binding motif, Walker A and B motifs, a catalytic glutamate, a "phosphate sensor" motif, and an arginine finger (Fig. 23.7). The ASCE P-loop NTPases are built upon a Rossmann fold (Rossmann et al. 1974). First identified in dinucleotide-binding proteins, the Rossmann fold consists of three or more parallel β-strands interspersed by α-helices. This arrangement results in the C-terminal ends of each strand forming one edge of a β-sheet that is sandwiched between α-helices. The conserved nucleotide-binding site is located at this edge and consists of loops connecting β-strands to downstream α-helices. The ASCE P-loop NTPases adopt a modified version of this fold, with five conserved parallel β-strands forming a central β-sheet that contains all of the sequence signatures listed above (Story and Steitz 1992; Story et al. 1992). Consistent with the position of the nucleotide-binding site in the Rossmann fold, residues responsible for ATPase function are located almost exclusively on loops connecting the C-termini of the β-strands to α-helices. The position, sequence, and function of each of these motifs during catalysis are summarized below.

The Walker A and B motifs, named for their identification in the sequence of the bovine mitochondrial F_1-ATPase (Walker et al. 1982), are the most highly conserved sequences in the ASCE P-loop NTPase superfamily. Both are involved in nucleotide-binding and activation of the nucleophilic water molecule. The Walker A motif, also referred to as the P-loop, is located between strand β-1 and helix α-1 and is defined by the consensus sequence (G/A)XXXXGK(T/S), where X can be any amino acid. The backbone amides of the glycine residues coordinate the phosphate groups of the ATP, and the ε-amino group of the conserved lysine forms ion pairs with the β- and γ-phosphates. The hydroxyl group of the conserved threonine/serine coordinates a critical catalytic Mg^{2+} ion that coordinates the β- and γ-phosphates of ATP and activates the catalytic water nucleophile. The Walker B motif, located between strand β-3 and helix α-3, contains the consensus sequence ZZZZD/E, where Z is a hydrophobic residue. The conserved acidic group provides another coordinating ligand for the catalytic Mg^{2+}, thus also playing an essential role in positioning both the ATP substrate and the water nucleophile for in-line attack of the activated water molecule at the γ-phosphate. The adenine-binding motif, ZQ, is located ~15 amino acids upstream of the Walker lysine (Mitchell et al. 2002), where the conserved glutamine forms hydrogen bonds with the N6 and N7 nitrogens of the heterocyclic adenine ring (Sun et al. 2007).

Although the spatial position of the catalytic glutamate is highly conserved, its sequence position varies among different branches of the ASCE superfamily. For example, in the F_1/RecA, HerA/FtsK, and PilT ATPases, the catalytic glutamate is located on a loop between strand β-2 and helix α-2 (Story et al. 1992). In contrast, the catalytic glutamate in the helicase and ABC ATPases is positioned immediately after the aspartate of the Walker B motif (Subramanya et al. 1996; Hung et al. 1998; Geourjon et al. 2001). Sequence analysis of the ø29 ATPase and the bacteriophage terminase large subunits (TLS) ATPases indicates that their catalytic glutamates are also positioned immediately following the aspartate in the Walker B motif (Mitchell et al. 2002; Burroughs et al. 2007; Koti et al. 2008), a prediction confirmed by the X-ray crystal structure of the ATPase domain of the T4 large terminase domain (Sun et al. 2007). Regardless of the location of the catalytic glutamate in the primary amino acid sequence, its position in three-dimensional space is highly conserved among all known ASCE ATPase structures. In each case, the catalytic glutamate is positioned to act as a general base and activate a water molecule for nucleophilic attack on the γ-phosphate of ATP (Geourjon et al. 2001; Sun et al. 2007).

Another essential motif conserved in the active site of many ASCE ATPases is a polar residue at the C-terminal end of strand β-4. This residue, asparagine 159 in ø29, is proposed to function as a "γ-phosphate sensor" that plays a role in transmitting conformational changes to other parts of the protein upon ATP binding (Story and Steitz 1992; Yoshida and Amano 1995; Iyer et al. 2004a, b). A similar function was assigned to the switch II region of distantly related G proteins, members of the broader P-loop ATPase superfamily (Leipe et al. 2002). In ring-forming ATPases like the ø29 ATPase pentamer, this sensor might function to communicate the catalytic state of a particular monomer to its neighbors, and hence play an essential role in coordinating the activity of the ring.

Similarly, most ASCE P-loop ATPases also utilize an "arginine finger" to couple hydrolysis of ATP to large scale conformational changes between adjacent monomers and/or domains. The arginine finger from one monomer/domain is inserted into the active site of its neighbor, where the positively charged guanidinium group of the arginine stabilizes the pentavalent phosphoanhydride transition state that forms during ATP hydrolysis and promotes separation of the phosphate leaving group (Nadanaciva et al. 1999; Braig et al. 2000; Ogura et al. 2004). This in trans activation of substrate in a neighboring subunit/domain is reminiscent of the arginine (or lysine) fingers observed in GTPase-activating proteins (Wittinghofer et al. 1997).

As with the catalytic glutamate, the primary sequence position of the arginine finger varies in different branches of the ASCE superfamily, but the spatial position is highly conserved. In the HerA/FitsK superfamily, the arginine finger is at the N-terminal end of strand β-4 and is inserted into the active site of the adjacent monomer in an oligomeric ring. Similarly, the arginine finger in the Helicase superfamily is also located at the base of strand β-4 in one of the two nucleotide-binding domains that comprise the overall fold of this superfamily. In this case, the arginine finger participates in an interdomain rather than intersubunit engagement of a neighboring active site. In the TLS superfamily branch of the ASCE ATPases, the arginine finger is in the third position of the Walker B motif following strand β-3 (Burroughs et al. 2007; Sun et al. 2007). Interestingly, the crystal structure of the T4 large terminase domain indicates that this arginine finger acts *in cis*, via an intrasubunit interaction rather than in trans (Sun et al. 2007).

Bacteriophage ø29 and its relatives have conserved arginines at both positions, arginine 26 in the Walker A motif and arginine 146 at the base of strand β-4, an observation that has led to differing classifications of the ø29 ATPase within the ASCE superfamily (Mitchell et al. 2002; Burroughs et al. 2007). This distinction is significant, as it has important implications regarding the mechanism of the ø29 ATPase, i.e., whether it functions more like a monomeric helicase, or like an oligomeric ring helicase/DNA translocase. Based on secondary structure prediction and identification of the characteristic sequence motifs, it was initially assumed that the ø29 packaging ATPase belonged to the same superfamily as the large terminase proteins from other tailed dsDNA viruses (Mitchell et al. 2002). More recently, it was proposed that the ø29 ATPase is more closely related to the HerA/FtsK ATPases within the broader ASCE superfamily (Burroughs et al. 2007). This assignment is based on not only the position of the arginine finger but also the absence of a nuclease domain and the topology of the accessory domain inserted between strands β-2 and helix α-2. However, this classification is problematic in that all other viral ATPases in the HerA/FtsK superfamily come from viruses that have internal lipid membranes, lack portal/connector proteins, and are constructed from capsid proteins utilizing the jelly roll fold. Like all tailed dsDNA bacteriophages, ø29 lacks internal membranes, depends on portal/connector proteins for packaging, and utilizes the HK97-fold for their major capsid proteins (Wikoff et al. 2000; Fokine et al. 2005; Morais et al. 2005). Hence, the logical conclusion of grouping ø29 with the HerA/FitsK superfamily is that dsDNA packaging motors evolved from ancestral ASCE ATPases at two separate occasions during the evolution of tailed bacteriophages.

Regardless of the exact location of the arginine finger in the primary amino acid sequence, in three-dimensional space, it is typically positioned on the edge of the β-sheet opposite the edge where the ATP-binding loops are situated. This arrangement suggests an intriguing mechanism for coupling ATP hydrolysis to force generation in ring ATPases. In some members of the ASCE ATPase family, movement of the substrate-binding loops is coupled to twisting of the central β-sheet. This β-sheet deformation has been proposed to store elastic energy that can be used to power large motions in accessory domains. (Abrahams et al. 1994; Kabaleeswaran et al. 2006) Hence, controlling sheet deformation might provide a means for coordinating ATP hydrolysis and conformational change. Since the arginine finger and substrate-binding loops within a subunit are on opposite sides of the central β-sheet, the arginine finger might also help transmit information regarding the catalytic state of the neighboring active site across this sheet to the substrate-binding loops on the opposite edge, thus providing a route for communication between adjacent subunits.

Structural information regarding arginine finger-dependent coordination in ASCE ATPases comes from the ssRNA bacteriophage ø12 packaging motor P4 (related to SF4 helicases) (Lisal and Tuma 2005; Burroughs et al. 2007) and from PcrA and UvrD, members of the SF1 helicase branch of the ASCE superfamily (Velankar et al. 1999; Lee and Yang 2006). For these motors, structures of motor nucleic acid complexes in multiple ATP-bound states have been determined. These structures provide examples of how the relatively small intradomain conformational changes that occur upon ATP binding can be amplified in an oligomeric motor. In the empty state, the two ASCE nucleotide-binding domains are separated so that the arginine finger cannot reach its neighbor's active site. Upon binding ATP, one of these domains rotates toward the other, facilitating insertion of the arginine finger into the neighboring domain's active site and subsequent hydrolysis of the ATP. In P4, this movement is coupled to the movement of a luminal lever that interacts with the ssRNA to propel it into the capsid. In the SF1 helicases, this movement is coupled to unwinding the dsDNA duplex and translocation along the resulting ssDNA.

Several mechanistic generalizations can be drawn from the above analyses of the characteristic functional motifs in ASCE ATPases. When applied to ø29 packaging ATPase, a picture of its role in genome encapsidation begins to emerge. By analogy to other ASCE ring ATPases, ATP and Mg^{2+} are bound at an intersubunit interface such that adjacent monomers both contribute residues to the active site. Hence, ATP binding and hydrolysis directly affect adjacent subunits, and the motor can thus respond to allosteric interactions between multiple ATP and DNA-binding sites. In a ring-like structure, this arrangement can facilitate coordination of conformational changes required for DNA translocation. The glycine-rich Walker A P-loop and the PQ adenine-binding motif within the active site cleft recognize and bind the adenine base in ATP and thus help correctly orient the nucleotide substrate for hydrolysis. Walker A and Walker B loops both contribute to binding the catalytic Mg^{2+} ion. The catalytic glutamate immediately following the Walker B aspartate helps to activate the Mg^{2+}-bound water molecule for nucleophilic attack on the γ-phosphate of ATP. In order to couple coordinated conformational changes to particular nucleotide states, the ø29 ATPase likely uses the phosphate sensor at the tip of strand β-4 to distinguish ATP from its hydrolysis products. The arginine finger at the base of strand β-4 stabilizes the pentavalent transition state and phosphate leaving group during hydrolysis and likely plays a role in coupling hydrolysis of ATP to the conformational rearrangements that drive DNA translocation. Because the arginine finger belongs to a neighboring subunit, this in trans activation might also be important in initiating the next ATP hydrolysis event in a neighboring subunit in the oligomeric ring. Although clearly important for directional transport of DNA, the underlying mechanism for coordinating the sequential firing of ATP subunits around the ring is not well understood. It has been suggested that the inherent 10_1 helical symmetry present in B-form DNA might also play a role in determining the order of subunit firing, i.e., whichever subunit is aligned with a particular structural feature of DNA will be triggered to initiate ATP hydrolysis (Simpson et al. 2000; Morais et al. 2008; Sun et al. 2008).

23.4 Single Particle Analysis of the ø29 DNA Packaging Motor

A complete description of genome encapsidation by the ø29 dsDNA packaging motor remains elusive. In addition to capturing high-resolution structures of each intermediate step during DNA translocation, it is also necessary to correlate each catalytic step of ATP hydrolysis with the sequence of conformational changes that move the dsDNA substrate into the capsid. Although structural, genetic, and bulk biochemical assays have been instrumental in providing a framework for understanding the overall mechanochemistry of genome encapsidation, none of these techniques is particularly well suited for capturing the underlying dynamics of DNA packaging. Fortunately single-particle experiments have recently allowed researchers to directly examine the dynamic properties of individual ø29 motor complexes at extraordinarily high spatial and temporal resolutions (Smith et al. 2001; Chemla et al. 2005; Fuller et al. 2007; Hugel et al. 2007; Rickgauer et al. 2008; Aathavan et al. 2009; Moffitt et al. 2009).

In particular, sophisticated optical tweezer measurements have provided essential information regarding the mechanochemical cycle of the motor (Smith et al. 2001). In these experiments, force-measuring laser tweezers are used to follow packaging activity of a single complex in real time. Briefly, a stalled, partially packaged complex is attached to one polystyrene microsphere via the unpackaged end of the DNA, and this microsphere is captured in an optical trap. The phage capsid is then attached to a second bead, held by a pipette, which is coated with antibodies that selectively bind to the phage capsid. Hence, a stable tether is formed between the two beads; the virus head is bound to one bead, and the unpackaged end of the genome is bound to the other. Because the positions and forces applied to each bead can be precisely controlled, it is possible to monitor the changes in DNA length as a function of time, as well as quantitate the forces necessary to stall the motor. From this force–velocity data, it has been possible to extract detailed information regarding the mechanochemistry of the ø29 packaging motor.

Remarkably, it was found that the ø29 can generate approximately 100 pN of force, making it the most powerful biological motor reported to date (Rickgauer et al. 2008). Initial optical tweezer experiments showed that DNA shortened in bursts exceeding 5 μm, indicating that the motor is highly processive (Smith et al. 2001). Pauses and slips were also observed, occurring more frequently as the motor works against higher forces. From the force–velocity data, it was shown that the rate-limiting step of the motor's cycle is force dependent even at low loads; hence, the packaging rate decreases as the prohead is filled (Smith et al. 2001). Later experiments using higher resolution optical tweezers indicated that the motor moves in steps of ten base pairs with intervening dwells of varying intervals (Moffitt et al. 2009). Closer analysis indicates that each ten-base pair step consists of four 2.5 base-pair substeps. Single particle packaging experiments using chemically modified DNA substrates show that the motor is somewhat "promiscuous" and can tolerate short inserts of nonnucleic acid polymers in one of the two DNA strands in the double helix (Aathavan et al. 2009), suggesting that the motor tracks along the phosphate backbone of a single DNA strand. Optical tweezers experiments also suggest that the DNA is not simply translated into the capsid, but is also negatively rotated (in the underwinding direction) during packaging (Yu et al. 2010).

Force–velocity data was also obtained in the presence varying concentrations of ATP, ADP, inorganic phosphate, and various ATPase inhibitors, allowing for the determination of kinetic parameters for the motor as well as their dependence on external load (Chemla et al. 2005). The motor was shown to have a V_{max} of 103 bp/s, a K_m of ~30 μm, and a Hill coefficient of 1, showing that ATP binding is not cooperative. ADP was found to act as a competitive inhibitor for ATP, indicating that they share the same binding site. However, while affinity for DNA is high in the ATP-bound state, it is low in the ADP-bound state. In contrast to ADP, there is almost no rate dependence on the concentration of inorganic phosphate, reflecting the large equilibrium constant for product release and suggesting that this step is essentially irreversible. Based on these observations, the authors conclude that DNA translocation does not occur during ATP binding but rather is triggered by phosphate release (Chemla et al. 2005). Furthermore, although the motor has a Hill coefficient of 1, analysis of ATPase inhibitor data indicates that the motor subunits act in a coordinated, successive fashion with high processivity.

23.5 Proposed Packaging Mechanisms

Despite the wealth of genetic, biochemical, biophysical, structural, and single particle data accumulated regarding genome encapsidation by ø29, the fundamental mechanism of DNA translocation remains unknown. Proposed models typically fall into one of two basic categories (1) connector-driven models, where rotation of the connector is coupled to DNA translocation and (2) ATPase-driven models where different domains within the ATPase alternately grab and release DNA. It's important to note that in all proposed models, genome translocation depends on conformational changes in the ATPase induced by hydrolysis of ATP. In ATPase-driven models, these conformational changes directly translocate DNA, whereas in connector-driven models, conformational changes in the ATP modulate the connector structure to varying degrees depending on the proposed role of the connector in DNA translocation.

23.5.1 Connector Driven Packaging Models

Connector-driven models were initially based largely on the ideas of Hendrix (Hendrix 1978), who proposed that the symmetry mismatch between the fivefold symmetric prohead shell and 12-fold symmetric connector would facilitate rotation of the connector. Hendrix astutely realized that given a pair of concentric rings with differing symmetries, the energy minima for pairwise interactions between subunits in adjacent rings cannot add in phase, and hence only a small energy barrier prevents rotation of one ring with respect to the other. In contrast, if the number of subunits in the two rings is the same, then the energy minima for pairwise interactions between subunits in adjacent rings can add in phase, resulting in a deep energy minimum that essentially locks the two rings in place. Secondly, Hendrix recognized that as a helically symmetric molecule, dsDNA is a threaded object. Combining these observations led to the idea that the 12-fold symmetric connector would be relatively free to rotate in the fivefold symmetric prohead vertex where it sits, and that rotation of the connector around the helically threaded DNA would thus "screw" the DNA into the capsid much as a nut moves on a bolt. Of course, this type of model requires that the lumen of the connector is also threaded in a complementary way to the helical DNA. Since there is no apparent luminal threading observed in any connector structures determined so far, alternative connector-based packaging mechanisms have been developed.

One variation of the Hendrix model is based on observed flexibility of the connector protein and combined results from X-ray crystallography and cryoEM demonstrating that the DNA, connector,

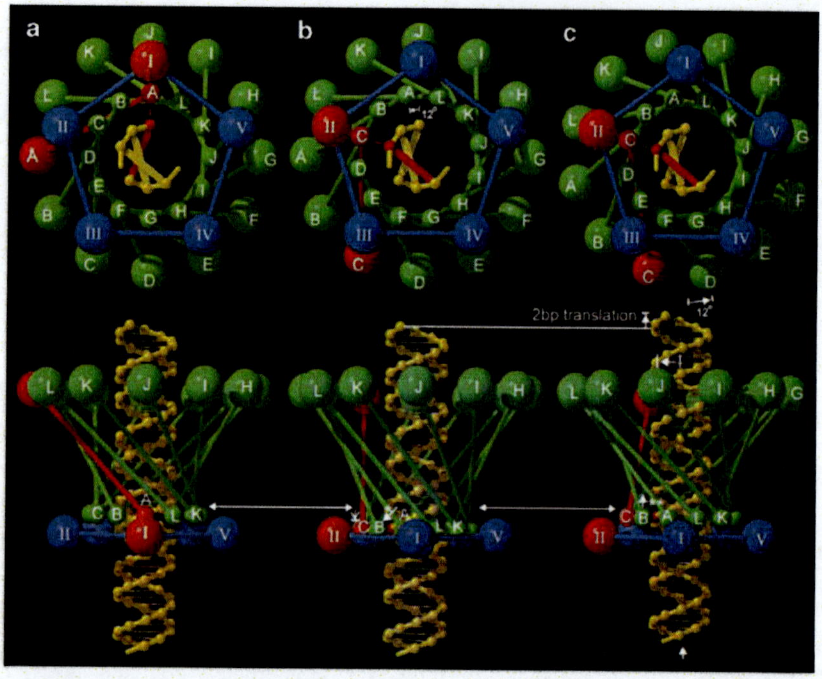

Fig. 23.8 Packaging mechanism based on connector expansion and contraction. A single cycle in the proposed packaging mechanism is depicted. The connector is shown as a set of *small spheres* at the narrow end and a set of *larger spheres* at the wide end joined by a thin tube corresponding to the central helical region. One "active" connector monomer is shown in *red*, the rest in *green*. The pRNA–ATPase complex is shown by a set of four *blue spheres* and one active *red sphere*. A DNA base that is aligned with the active connector monomer is also shown in *red*. The images in the *top* panel are end-on views looking down the wide end of the connector; the bottom images are side views corresponding to each top view. (**a**) The active pRNA–ATPase I interacts with the adjacent connector monomer that is properly aligned with the DNA. (**b**) The narrow end of the connector rotates anticlockwise by 12°, such that the narrow end of monomer C is aligned with ATPase II, the next ATPase to fire; this rotation causes the connector to expand lengthwise by slightly changing the angle of the helices in the central domain (*white arrow* with *asterisk*). (**c**) In the final step, to return to a low energy state, the wide end of the connector also rotates 12°, causing the connector to contract back to its original conformation, albeit rotated 12° (*white arrow* with two *asterisks*). Two base pairs of DNA are translated into the phage head during the contraction event. ATPase II and connector (C) are now position for the next cycle. This figure is reproduced from Simpson et al. (2000)

and prohead–pRNA–ATPase complex form a series of concentric structures with 10_1-, 12-, and 5-fold symmetries, respectively (Simpson et al. 2000, 2001; Morais et al. 2001). In this model, ATP hydrolysis promotes lengthwise expansion and contraction of the connector that is coupled to DNA translocation; the connector essentially acts as an oscillating helical spring that pumps DNA into the prohead (Fig. 23.8). To achieve this ATPase induced expansion, the antiparallel three-helix bundle that is twisted around the central 12-fold axis in the crystal structure is proposed to "straighten" out, thus extending its length and becoming more parallel with the 12-fold axis. The net result is that the central and narrow domains of the connector extend approximately two base pairs and rotate 12° relative to wide end. Upon extension and rotation, one connector monomer becomes aligned to bind DNA. During relaxation, the wide end of the connector follows the narrow end, rotating 12° and causing a contraction of the connector and subsequent translocation of two base pairs of DNA into the prohead. Because of the 10_1 helical symmetry of B-form DNA, translocation of two base pairs is equivalent to a 72° rotation; thus, the DNA would be aligned with the next ATPase in the pentameric ring. Furthermore, because the 12-fold symmetric connector has rotated 12°, it is also aligned with

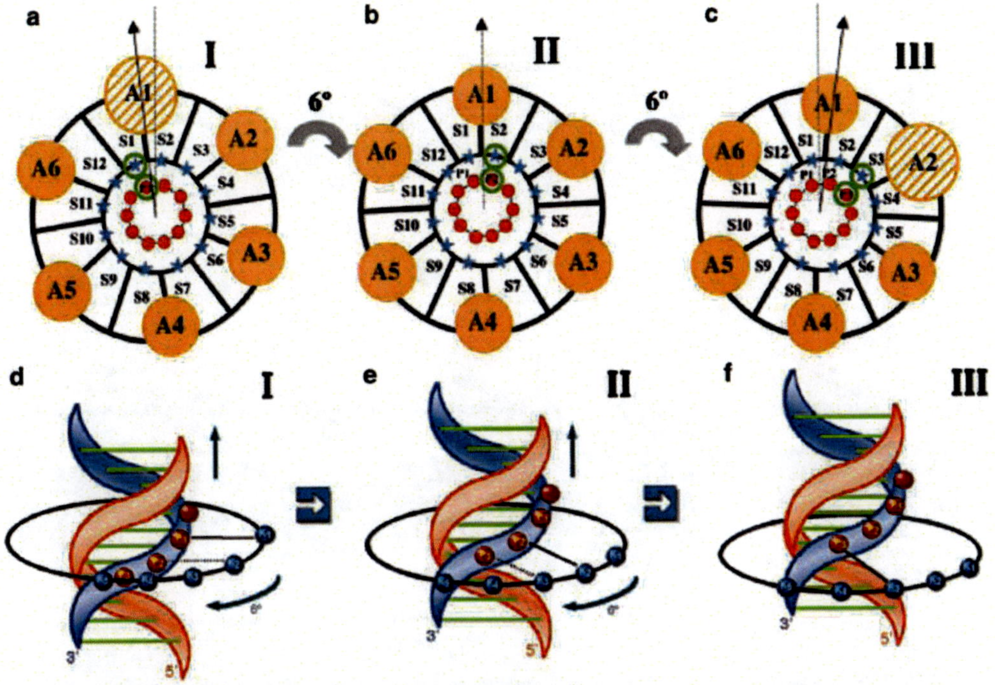

Fig. 23.9 Packaging mechanism based on lumenal lysine rings. A schematic of the packaging motor is shown looking down the packaging axis from inside the procapsid. A hexameric ATPase is depicted as six *yellow circles* numbered A1–A6; the connector is depicted as a disk with individual subunits numbered S1–S12 and corresponding lysines numbered K1–K12 (*blue star, top panel*; *blue ball, bottom panel*), and the DNA is represented as *red balls* (*top panel*) or a *cartoon ribbon* (*bottom panel*). (**a**) The cycle begins with a lysine aligned with connector subunit S1 and phosphate P1. Upon hydrolysis, ATPase A1 rotates the connector 12° clockwise while the DNA translocates along the connector axis by two base pairs. (**b**) At 6° rotation, there is a transient interaction between the lysine from S2 and the DNA P2. (**c**) An additional 6° rotation allows the lysine from S2 to interact with DNA P3. This movement aligns ATPase A2 for the next cycle. In (**a**) there is a strong interaction between lysine K1 and phosphate P1 (*continuous line*) and a weak interaction between K2 and P2 because P2 is still below the ring of lysine residues (*broken line*). In (**b**), the 6° rotation of the connector allows K2 to tightly bind P2 if DNA is translocated by one base pair. In (**c**), an additional 6° rotation of the connector and another one base pair movement of DNA aligns K3 with phosphate P3. Hence, the two-base pair step size is accompanied by 12° rotation of the connector. Panels (**d–f**) correspond to side views of (**a–c**). For simplicity, only a single DNA chain and one lysine ring are considered. This figure was modified from Guasch et al. (2002)

the DNA and the ATPase for the next mechanochemical cycle. Hence, the connector functions much like a ratchet in this two-state model, and overall rotation of the connector is not necessarily a driving force for translocation, but rather needs to occur to maintain alignment between the 12-fold symmetric connector, the fivefold symmetric ATPase, and the 10_1-helically symmetric B-form DNA. After all five ATPases have fired, the connector will have rotated 60°, and the DNA will have been translated the length of one pitch of its helix, thus packaging ten base pairs.

Another variation of connector-driven packaging mechanisms envisions that the two lysine rings in the connector lumen alternately grip and release DNA to ensure unidirectional transport into the phage capsid. These two positively charged lumenal lysine rings are formed by lysine 200 and lysine 209 (Fig. 23.9). The two rings are separated by approximately 20 Å, and neighboring lysines within the ring are about 10 Å apart (Guasch et al. 2002). Considering the symmetry mismatch between the connector and the DNA, it can be shown that the environment of any particular connector monomer is necessarily similar to that of another monomer five subunits around the ring. Hence, if lysine 200

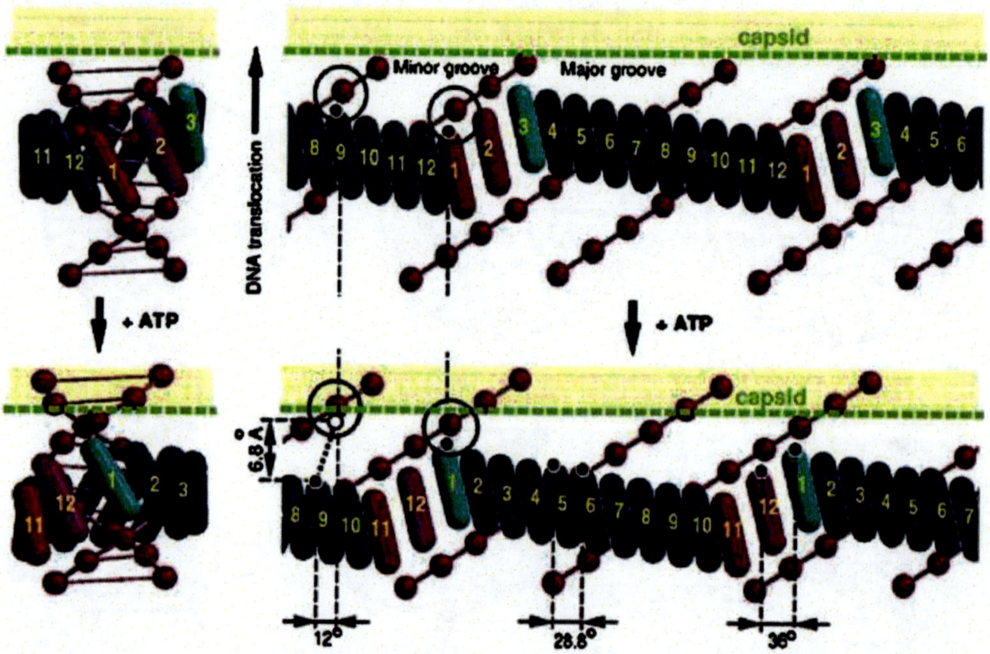

Fig. 23.10 SPP1 Packaging mechanism based on luminal loops. Pre- and post-ATP hydrolysis states (*top* and *bottom panels*, respectively) of the SPP1 portal are shown schematically. A three-dimensional conical rendering of the SPP1 connector (*left*) is opened (*right*) to provide a representation of packaging where the horizontal axis corresponds to the angular position inside the tunnel. Two helical repeats of B-form DNA are shown with two phosphates circled to provide reference points. Tilted tunnel loops are shown as *cylinders* and the DNA as *pink spheres* centered at phosphates. Specific loops are identified via numbers, and three specific conformational states are indicated via color. These three states, *red*, *magenta*, and *cyan*, propagate around the circle of loops. A larger angular separation for the three loops occupying these three states provides them with access to the major groove. Transition between the top and bottom states requires a 12° rotation of the SPP1 portal relative to the DNA. Loops 2–12 actively move with respect to the DNA during this transition, whereas loop 1 does not change its position with respect to the DNA, but instead, moves 6.8 Å along with the DNA. This figure was adapted from Lebedev et al. (2007)

in connector monomer 1 is aligned with the DNA phosphate backbone, then the lysine 200 in monomer 5 will be aligned with another phosphate on the other DNA strand. Similarly, 20 Å down the channel, two lysines in the lysine 209 ring can also interact with two phosphates. Hence, two portal subunits, four phosphates, and a pRNA/ATPase subunit can be brought into register, inducing ATP hydrolysis in the active ATPase subunit. Hydrolysis of ATP causes the connector to rotate 12° and translocate two base pairs of DNA. Presumably, rotation disrupts the lysine grip, allowing DNA to move and bringing another set of lysines into register with the DNA and the motor. In this model, the ATPase can either cause the portal to rotate, which pulls DNA into the head, or the ATPase can actively push DNA into the head, and the connector rotation is passive. This model requires that the ATPase is part of the rotor, and thus rotates with the connector to maintain motor alignment; hence, this model is more compatible with a hexameric ATPase rather than the pentameric ATPases that have been thus far observed in phage packaging systems (Morais et al. 2000; Simpson et al. 2000; Sun et al. 2008).

Yet, another variant of a connector-based packaging mechanism was proposed for bacteriophage SPP1, where the sequential movement of lumenal lever loops that project into the connector channel actively translocates dsDNA into the procapsid (Lebedev et al. 2007) (Fig. 23.10). The pseudo atomic structure of the dodecameric SPP1 portal indicates that 12 long flexible loops project into the central channel of the connector, and that these loops can assume at least two different conformations.

It was suggested that these lumenal loops interact directly with the DNA and undergo coordinated sequential conformational changes that are coupled to connector rotation and DNA translocation. Orchestration of these conformational changes gives rise to an "undulating belt" around the DNA with 12 connector–DNA contacts engaged in tight grip. Here again, ATP hydrolysis causes connector rotation, and sequential conformational rearrangements of the lumenal loops that translocate DNA and help align the motor for the next cycle. Symmetry mismatches between the connector, and the DNA are also accommodated by loop flexibility. Although there is considerable structural homology between phage connector/portal proteins, it is unlikely that this mechanism could apply to ø29 since the homologous loops in ø29 can be deleted with a minimal effect on packaging activity (S. Grimes, personal communication).

23.5.2 *ATPase-Driven Packaging Models*

In all the connector-based mechanisms described above, the inherent symmetry mismatch between the connector and the DNA requires that motor rotates in order maintain alignment of the motor and the DNA. However, recent experimental evidence from single particle and biochemical experiments indicates that the connector does not rotate during packaging (Baumann et al. 2006; Hugel et al. 2007). In light of this evidence, a connector/ATPase hybrid model was proposed that does not require connector rotation (Morais et al. 2008). In this model, conformational changes in the ATPase generate the power stroke that translocates DNA into the procapsid, and the connector provides a valve-like function holding the DNA in place while the ATPase reloads. The primary role of the connector in this third type of model is to prevent slippage of the DNA via nonspecific connector–DNA interactions and thus maintain unidirectional transport. It is not clear how this valve-like function of the connector would be regulated, but since the pRNA bridges the connector and ATPase, it also might also provide a means of communication between the two; one could imagine, for example, that the ATPase uses the pRNA superhelix as a lever to exert force on the connector.

In further contrast to rotary connector-based mechanisms, the ϕ29 motor was suggested to function more like a linear motor driven directly by conformational changes in the ATPase. Genetic analysis, sequence alignment, and structural homology between phage ATPases and monomeric helicases suggest that phage DNA packaging motors could be better explained by inchworm-type mechanisms, in which opening and closing of the ATP-binding cleft cause movement of a DNA-binding domain in the ATPase that is coupled to DNA translocation (Draper and Rao 2007; Sun et al. 2007, 2008). In these types of models, the connector functions as little more than a passive portal to provide a conduit for DNA to enter the phage shell.

A detailed mechanism for dsDNA packaging by bacteriophage T4 was recently proposed based on ATPase domain motions inferred from structural analysis of the T4 ATPase (Sun et al. 2008). The T4 ATPase consists of two functionally distinct domains. In addition to an N-terminal ASCE P-loop ATPase fold, the T4 ATPase also possesses a C-terminal nuclease domain necessary for cleavage of the concatenated T4 genome at the termination of packaging. The X-ray crystal structure of the N-terminal domain indicates that it is more closely related to the monomeric helicases than to hexameric ATPases, whereas the C-terminal nuclease domain is a member of the Rnase H/resolvase/integrase superfamily (Sun et al. 2007). The crystal structure of the full-length protein shows that the two domains pack tightly together, with extensive interactions between them including six charge pairs and a hydrophobic core. The structure of the ATPase bound to proheads was analyzed via cryoEM, which, like ø29, showed that five copies of the ATPase bind at the unique connector vertex of the phage. However, unlike ø29, there is very little contact between adjacent subunits, and thus the ATPase forms a pentameric ring only by virtue of binding to a pentameric vertex of the procapsid. In contrast to the crystal structure, where the two domains interact extensively to form a

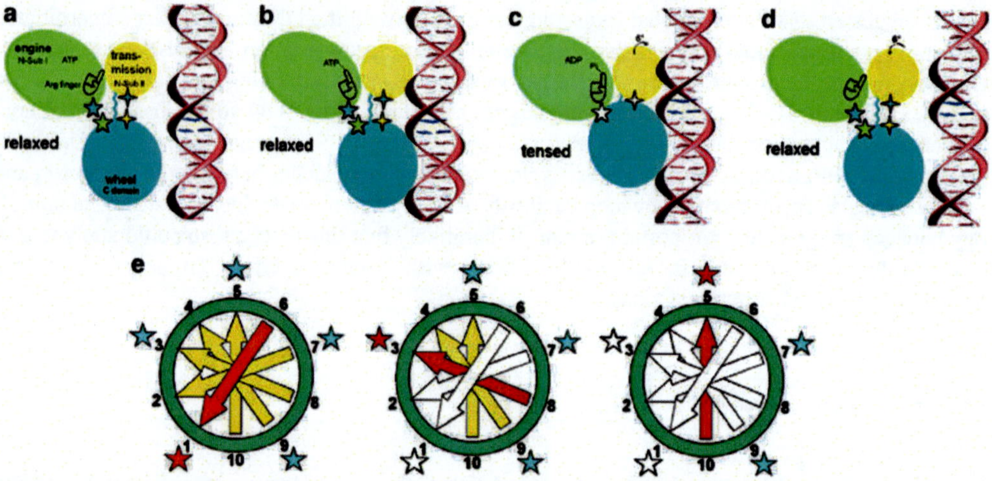

Fig. 23.11 T4 packaging mechanism based on electrostatic interactions. The T4 ATPase N-terminal subdomain I, subdomain II, and C-terminal domain are shown as *green*, *yellow*, and *cyan ovals*, respectively. Electrostatic interactions between the C-terminal domain and the N-terminal subdomains I and II are shown as five- and four-pointed stars, respectively. The linker between the N- and C-terminal domain is shown as a *cyan line*. (**a**) The C-terminal domain waits to bind DNA. (**b**) DNA binding causes the C-terminal domain to approach the N-terminal domain. A conformational change in the N-terminal domain inserts the arginine finger into the ATPase active site to trigger hydrolysis. (**c**) Hydrolysis of ATP causes the N-terminal subdomain II to rotate by about 6°, resulting in alignment of charge pairs and induction of an electrostatic attraction that moves the C-terminal domain and the bound DNA two base pairs. (**d**) Upon release of ADP and Pi, the C-terminal domain returns to its original position and DNA is released. The next subunit in the pentameric ring is now aligned for DNA translocation. (**e**) DNA base pairs outside, entering, and inside the procapsid are shown as *yellow*, *red*, and *white arrows*, respectively. Subunits in the pentameric ATPase are shown as *stars*, colored according to whether they are hydrolyzing ATP (*red*), ready to hydrolyze ATP (*blue*), or have already hydrolyzed ATP (*white*). Hydrolysis of ATP at position 1 translocates two base pairs of DNA (*left*), hydrolysis at position 3 translocates two more base pairs (*middle*), and the ATPase at position 5 is now ready to fire. This figure is reproduced from Sun et al. (2008)

relatively closed structure, the cryoEM structure suggested that the two domains are well separated, and thus each ATPase adopts a more open conformation.

Based on these observations, it was proposed that the two different conformations of the ATPase correspond to two distinct states in the DNA packaging pathway (Sun et al. 2008) (Fig. 23.11). The closed conformation present in the crystal structure is believed to correspond to the posttranslocation state, whereas the relaxed open state, obtained via independent fitting of the two ATPase domains into cryoEM density, represents the pretranslocation state. Comparison of the two structures shows that the C-terminal domain in the open state is translated approximately 7 Å along the packaging axis, similar to the two-base pair step size of the motor. Hence, if the C-terminal domain movements were coupled to binding and release of DNA, then this two state conformational switch might provide a means of translocating the dsDNA genome into the procapsid. It was proposed that the C-terminal domain has a high affinity for DNA in the open state. Upon binding DNA to the C-terminal nuclease domain, a cis-arginine finger located at the base of strand β-3 is inserted into the active site to trigger ATP hydrolysis. Hydrolysis is believed to be coupled to conformational change that aligns complementary charges in the N- and C-terminal domains; electrostatic forces then cause the subunit to close, and the DNA-bound C-terminal domain moves 7 Å along the packaging axis to package two base pairs of dsDNA. Since translocation of two base pairs is equivalent to a 72° rotation, the next subunit in the pentameric ATPase ring will be aligned with DNA for the next translocation cycle (Fig. 23.11b).

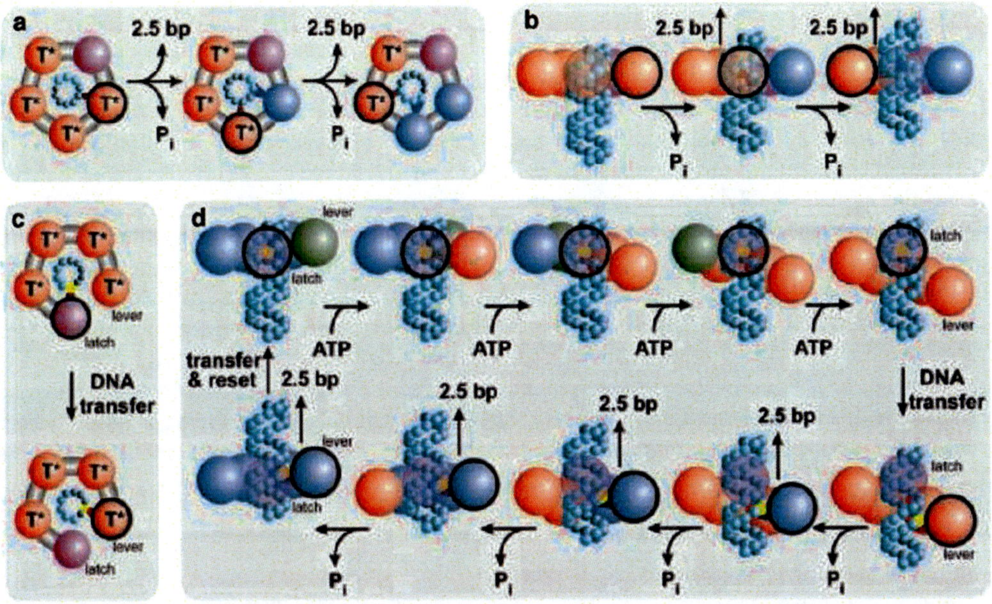

Fig. 23.12 Packaging mechanisms that accommodate a noninteger step size. (**a**) Schematic top view of a translocation model wherein every subunit interacts with the DNA (*cyan spheres*). The subunit presently interacting with DNA is outlined in *black* (*top view*). (**b**) Side view of (**a**) showing how the size of internal conformational changes determines the step size. (**c**) Top view of a schematic of a translocation model in which only two subunits contact the DNA (*black outline*). (**d**) Side view of the model shown in (**c**); one subunit maintains contact with the DNA (the latch) while binding of ATP induces relative subunit–subunit rotations in the other subunits that distort the ring. This distortion causes one subunit (the lever) to move along the DNA by ten base pairs. The DNA is then handed from the latch to the lever, and the release of hydrolysis products causes the ring to relax, retracting the lever and the DNA. The DNA contact is then handed back to the latch, the ring resets, and the cycle begins again. Because there are four pairs of subunits, the ring is retracted in four steps, dividing a ten-base pair step into four 2.5-base pair substeps. This figure is reproduced from Moffitt et al. (2009)

23.5.3 *Models with Noninteger Step Sizes*

Applying the T4 packaging model described above to ø29 is problematic for a number of reasons. Firstly, the ø29 ATPase does not have a C-terminal nuclease domain since it packages a unit-length genome. Given the central importance of this domain to the proposed packaging mechanism, it is hard to imagine how a homologous mechanism might work in ø29. It is possible that the small C-terminal oligonucleotide-binding domain of ø29 fulfills a function similar to that of the nuclease domain in the T4 ATPase, but additional research is required to refute or confirm this possibility. Secondly, single particle experiments have demonstrated that the ø29 packaging motor moves in ten-base pair bursts, where each burst consists of four noninteger 2.5-base pair steps (Moffitt et al. 2009). A step size that is a noninteger number of base pairs is incompatible with any mechanism wherein each subunit in the ATPase ring makes specific and interactions with the DNA. Thus, resolving the mechanistic details of ø29 DNA translocation may be more complex as it requires integration of a noninteger step size and the reconciliation of a four-step cycle within the structural framework of a pentameric motor.

Two alternative models incorporating four 2.5-base pair steps within a pentameric motor accompanied the initial report of a noninteger step size (Moffitt et al. 2009). The first model (Fig. 23.12a, b) proposes that the motor does not make any specific contacts with the DNA, but instead generates force via nonspecific steric interactions with the major groove of the DNA. Hence, the step size of the motor need not coincide with the helical symmetry of the DNA and is thus solely determined by the extent of the conformational changes that generate the power stroke. To accommodate four translocation steps, this model assumes asymmetry within the pentameric ring in that one of the five subunits is different from the other four. Because the nucleotide-free state has a low affinity for DNA (Chemla et al. 2005), the unique subunit may be required to hold the DNA at the end of each cycle, preventing slippage ATP during the subsequent dwell phase.

In the second model (Fig. 23.12c, d), specific interactions between the motor and the DNA occur but only for a select subset of the subunits. In the proposed model, two subunits make specific contact with the DNA, but a different number of DNA-binding subunits can also be accommodated in the same general framework. The DNA is pumped into the procapsid via an "inchworm-like" movement of these two subunits that is driven by sequential distortions around the ring. In short, the arrangement of the five ATPase subunits in the motor oscillates between a closed planar ring and open single turn helical structures. In this way, the ten-base pair burst can be divided into noninteger steps. This mechanism also suggests how a four-step cycle arises from a pentameric motor; one subunit interface sees a unique environment due to the accumulated distortion of the other four subunits, perhaps resulting in its inactivation. The proposed motor structures are similar to the quaternary structures of other ring ATPases, but were not implicated as part of any translocation mechanism (Singleton et al. 2000; Skordalakes and Berger 2003).

A comprehensive "push-and-roll" model was recently proposed that explicitly couples ATP hydrolysis with the coordinated conformational changes that drive packaging (Fig. 23.13). Although this model is primarily based on results from single particle experiments, it is also compatible with available structural, genetic, and biochemical data (Yu et al. 2010). In this model, genome encapsidation is not simply the result of translating DNA along the central axis of the phage, but rather requires that the DNA also roll around the inner lumen of the ATPase ring; hence, packaging combines a "push" and a "roll." Based on geometric arguments and packaging simulations, it was demonstrated that this push-and-roll mechanism can explain how a planar pentameric packaging motor translocates the DNA in bursts of four 2.5-base pair power strokes.

Lacking a high-resolution structure of the ATPase, the authors constructed a homology model based on the P4 ssRNA packaging motor in bacteriophage ø12 (Mancini et al. 2004; Lisal and Tuma 2005). Predicated on this model, it was suggested that during each power stroke, a positively charged lumenal lever loop from one subunit is electrostatically steered to the DNA backbone as in P4 (Fig. 23.13a). Upon engagement with the DNA, the lever then pushes sterically, orthogonal to the backbone axis; assuming B-form DNA geometry, this push will cause the right-handed DNA helix not only to be translated in the packaging direction but also to be rotated in a left-handed direction (Fig. 23.13b, c). Thus, the DNA movement consists of two steps, a vertical step along the packaging direction and a horizontal step in the tangential direction of the DNA cross section (Fig. 23.13b). It is not entirely clear how ATP hydrolysis is coupled to force generation by the lumenal levers, but it was proposed that, similar to the F_1-ATPase, movement of loops in the active site is coupled to twisting of the central β-sheet, and that this deformation is stored as elastic energy that can be used by the lumenal levers to apply a force on the DNA (Abrahams et al. 1994; Kabaleeswaran et al. 2006; Yu et al. 2010).

The proposed reaction scheme begins with one subunit bound to DNA (Fig. 23.14). Assuming that the subunit initially interacting with DNA is in an ATP-bound state, then the electrostatic coupling between the lever and the DNA will be tight (Chemla et al. 2005; Moffitt et al. 2009), thus holding the DNA in place and preventing slippage. It has been shown that affinity for DNA decreases in the ADP-bound state (Chemla et al. 2005). Hence the lever affinity for DNA would decrease upon ATP hydrolysis at end of the power stroke that accompanies Pi release (Chemla et al. 2005) causing the DNA to roll to the lever on the adjacent subunit. If the next subunit in the ring is in the ATP-bound

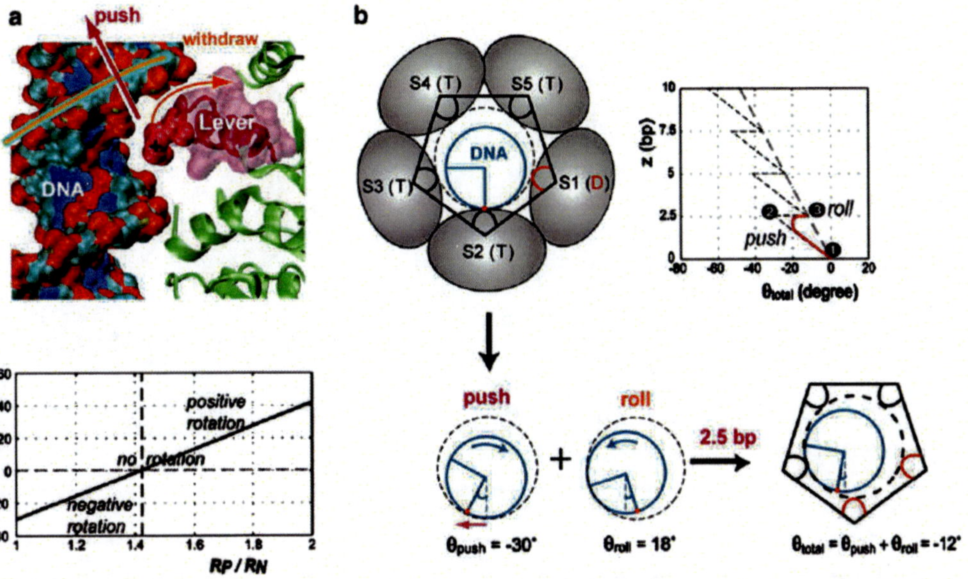

Fig. 23.13 Push-and-roll packaging mechanism. (**a**) Model of the proposed ATPase (*green ribbon*) lumenal lever (magenta isosurface) interacting with DNA; this model is based on the structure of the ssRNA packaging motor P4. (**b**) DNA movement generated by the power stroke. A schematic top view of the DNA and the motor ring is shown (*left*) with ATPase subunits labeled S1–5, and the hydrolysis state of each is in parentheses: T, ATP-bound; D, ADP-bound. During the power stroke (T→D), the motor lever pushes the DNA up by 2.5 base pairs (out of the paper) and rotates it by $\theta_{push} \approx -30°$. DNA rolls to the next subunit upon disengaging the lever resulting in an additional DNA rotation of $\theta_{roll} = +18°$ (when $R_p/R_n = 5/4$, see below); hence, each power stroke results in a total DNA rotation of −12°. A coordinate trace of a point on the DNA during packaging is shown on the right for the special case when the ratio of the radial cross section of the DNA (R_n) to the radius of the lumen of the ATPase ring (R_p) is 5/4: the DNA is rotated by $\theta_{push}(1 \rightarrow 2) + \theta_{roll}(2 \rightarrow 3) = \theta_{total} \approx -12°$ for every 2.5 base pairs translocated. The actual trajectory is likely to be more like the *red line* where the rolling commences before the push has completed. (**c**) The net DNA rotation predicted for each power stroke depends on the relative size of the motor ring cross section to that of the DNA (R_p/R_n): a negative (underwinding) rotation of the DNA occurs if $R_p/R_n < 1.4$, there is no rotation when $R_p/R_n \approx 1.4$, and a positive (over-winding) DNA rotation when $R_p/R_n > 1.4$. This figure was adapted from Yu et al. (2010)

state, it has a high affinity for DNA and can thus accept the incoming DNA. The proposed rolling motion thus carries the DNA from one subunit to the next. Similar to P4, each power stroke also causes the arginine finger in the active subunit to be inserted into the active site of its neighbor, thus promoting hydrolysis in the next subunit around the ring (Lisal and Tuma 2005; Kainov et al. 2008). However, because ADP release is slow, when it comes time for the fifth subunit to fire, subunit 1, the next subunit in a pentameric ring is still bound to ADP, thus having a low affinity for DNA and rendering it incapable of accepting the rolling DNA substrate. As a result, at the end of every four power strokes, the cycle pauses, constituting a dwell phase before the fifth subunit can fire, during which time four ATPs are loaded into the catalytic sites. The next burst phase of four power strokes starts once spontaneous ATP hydrolysis takes place in the fifth site without insertion of an arginine finger. A necessary consequence of this mechanism is that the order of subunit firing changes each cycle. For example, if the subunit firing order of a particular cycle is 1, 2, 3, 4 then the firing order in the subsequent cycle will be 5, 1, 2, 3, and so on. In summary, the push-and-roll model naturally accounts for the observed frequency and duration of the dwell phase, explains how a four 2.5-base pair step cycle arises from a planar pentameric motor, and provides a new perspective on how a multimeric ATPase might transport DNA.

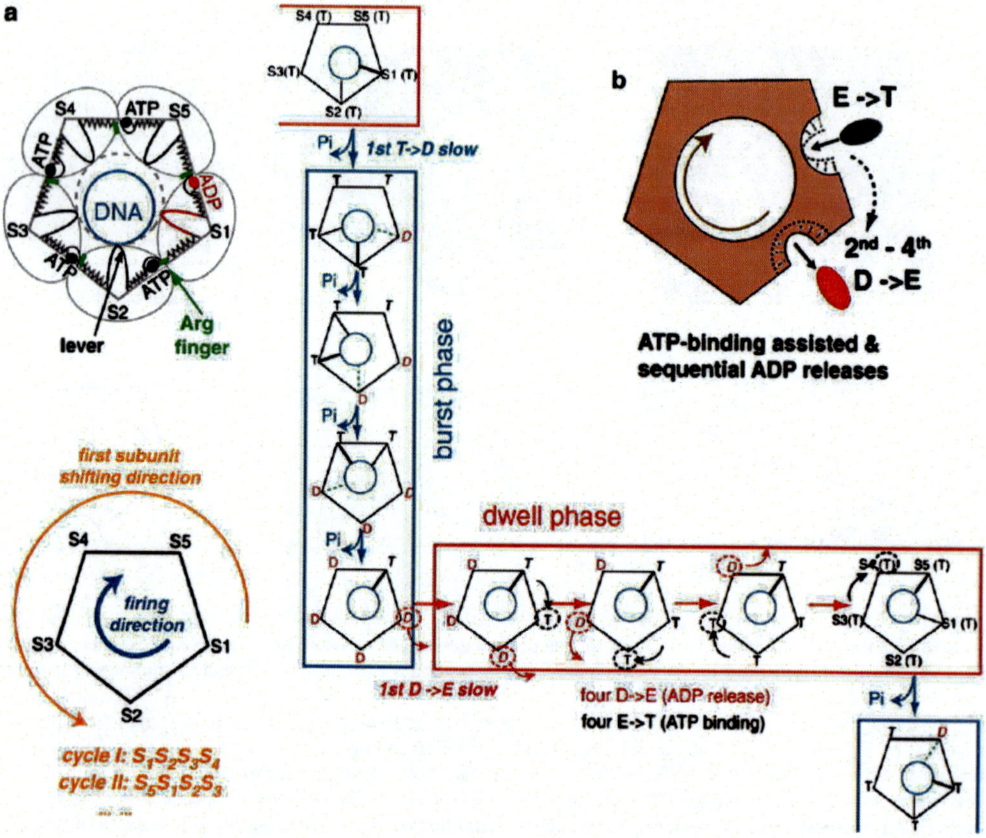

Fig. 23.14 Dominant reaction scheme of the push-and-roll model. (**a**) A schematic top view of the ATPase pentamer (*upper left*) and the direction of power stroke firing within a cycle versus the direction of the first subunit shifts in subsequent cycles (*lower left*) proposed in the model. (**b**) A cooperative ADP release mechanism would link four ADP release and ATP-loading events sequentially around the motor ring. The reaction cycle consists of a burst phase and a dwell phase. The burst phase includes four sequential 2.5-base pair power strokes at four consecutive catalytic sites. During the dwell phase, four consecutive ATP loadings and several non-ATP-binding events occur. A power stroke accompanies the T→D transition (Pi release) when the subunit is bound to DNA (*thick continuous line*). Each power stroke requires the next subunit (*thin continuous line*) is in the ATP-bound state (T) to receive the "rolling" DNA when it arrives at the end of the previous power stroke. After four rapid sequential power strokes in the burst phase, the motor must pause in the dwell phase because the next subunit is still in the low DNA affinity ADP-bound state (D). The first ADP release in the dwell phase is slow, but the following ADP releases are rapid if ATP binds quickly and accelerates ADP release at the next site (see **b**). Although the first power stroke after the four ATPs are loaded is also slow, the ensuing power strokes (second to fourth T→D) happen very quickly in the next burst phase since insertion of an arginine finger from the preceding subunit promotes catalysis. This figure was adapted from Yu et al. (2010)

23.6 Summary and Prospectus

The ø29 motor is arguably the best characterized DNA packaging motor, and new results regarding the mechanism of the motor can be interpreted in light of a rich background of sophisticated genetic, biochemical, structural, and single particle data. Research on the ø29 packaging motor is thus poised to move the DNA packaging field from speculation regarding the nature of the force-generating step to a detailed description of the coordinated conformational changes that drive DNA packaging. As our description of the mechanochemical cycle of the ø29 packaging motor becomes more and more complete, it is worth asking which aspects of the motor mechanism and assembly pathway are common to all dsDNA viruses and which are uniquely adapted to challenges faced by ø29. Ignoring close relatives, ø29 shares very little sequence similarity with other viruses or bacteriophages. Furthermore, interpreting bacteriophage relationships in the context of Darwinian evolution is complicated by the considerable influence of horizontal gene transfer on phage evolution (Hendrix 2002). Nonetheless, recent structural data shows that the tailed dsDNA bacteriophages and herpesviruses utilize the same basic protein folds for their capsid (Zhou et al. 2000; Jiang et al. 2003, 2008; Fokine et al. 2005; Morais et al. 2005; Agirrezabala et al. 2007; Dai et al. 2010) and portal proteins (Simpson et al. 2000; Orlova et al. 2003; Lebedev et al. 2007; Lander et al. 2009; Chen et al. 2011), whereas bioinformatics analysis indicates that the virus/phage-encoded ATPases that power dsDNA packaging all adopt the characteristic the ASCE superfamily fold (Mitchell et al. 2002; Burroughs et al. 2007). Hence, regardless of whether vertical or horizontal gene transfer has dominated phage evolution, it seems clear that all tailed phages utilize similar macromolecular components to package their DNA, pointing to a common mechanism of genome encapsidation.

There are, however, features of the ø29 packaging motor that are unique to ø29. These include the requirement for a phage-encoded pRNA, packaging of a unit-length genome, and the covalent attachment of gp3 to the ends of the ø29 genome. Clearly, other phages have evolved such that the role of the pRNA is carried out by proteins, or domains within proteins. Although it has been speculated that the pRNA is a vestige from an RNA-dominated world, it is not clear whether the pRNA evolved to replace a protein component, or whether the analogous protein components in other phages evolved to replace a pRNA. Similarly, regarding the requirement for gp3, it is not clear if the ø29 motor evolved to accommodate a polymerase that requires a protein primer (Blanco and Salas 1996), or if the genome replication machinery evolved to accommodate the specificity requirements of the packaging motor. Furthermore, because ø29 replicates nonconcatenated unit-length DNA, the ø29 ATPase does not require or possess any nuclease activity, and thus the ø29 motor must also utilize a different mechanism to terminate packaging than phages that encapsidate a concatenated multimeric genome. Indeed, this relative simplicity is a feature that makes ø29 an excellent model system for studying packaging; in addition to utilizing motor components that represent the bare minimum for a functional packaging motor, ø29 also carries out a simplified, and thus more easily dissected, packaging reaction.

As described above, genetic studies have identified all the components of the ø29 packaging motors (Grimes et al. 2002); structural, biochemical, and bioinformatics analyses have identified and characterized specific residues within these components that are critical for motor function, and biochemical, structural, and single particle analyses suggest ways that these components operate in a highly coordinated manner (Chemla et al. 2005; Aathavan et al. 2009; Moffitt et al. 2009; Yu et al. 2010). Yet, despite the wealth of accumulated data on genome packaging in ø29 and other dsDNA viruses, several aspects of DNA packaging remain obscure. Most importantly, it is still not known exactly how these molecular components interact with each other in a coordinated fashion to drive genome translocation.

As a first step toward understanding the conformational changes that drive packaging, it is necessary to determine high-resolution structures of the entire motor complex in different steps of the mechanochemical cycle. These structures would help identify the interactions between specific amino/nucleic acids within and between motor components. Since these contacts likely guide both motor assembly and coordination, this structure would provide valuable information regarding these two critical aspects of motor function. Due to the size and complexity of a multisubunit packaging motor, the most tractable approach to obtaining its atomic structure is to determine high-resolution structures of isolated motor components by X-ray crystallography and then fit these structures into subnanometer resolution cryoEM reconstructions of the intact motor complexes trapped in different stages of the packaging pathway.

As discussed above, atomic resolution structures of the ø29 connector and the prohead-binding domain of the ø29 pRNA have been determined (Simpson et al. 2000; Ding et al. 2011), and thus only the atomic structure of the ATPase remains unknown. Using cryoEM, the structure of a ø29 connector–pRNA complex has been determined to sub-nanometer resolution (Morais et al., in preparation), and thus it is likely that a similar complex that includes the ATPase can be reconstructed to a comparable resolution. At this resolution, secondary structural elements are easily recognized, and atomic resolution structures can be fit with great accuracy and reliability (Rossmann et al. 2005). Hence, should an atomic resolution structure of the ATPase become available, it will likely be possible to quantitatively fit the structures of the ATPase, connector, and pRNA into their corresponding densities in a cryoEM reconstruction of the entire motor complex for at least the ground state of the motor. The resulting pseudoatomic structure would map amino acid residues at component interfaces, identify interactions that guide motor assembly and coordinate its activity, suggest biochemical experiments to test specific packaging hypotheses, and serve as a reference for determining the conformational changes that drive packaging.

As with any mechanical device, the ø29 DNA packaging motor cannot be fully appreciated without some sense of how the different components and domains within components move relative to one another. The ability to visualize the packaging motor during DNA translocation would provide invaluable insight into the mechanism of packaging by showing how the different motor components move, interact with each other, and interact with the DNA. Direct visualization of conformational change has been extremely challenging. Although NMR provides information regarding molecular dynamics, it is limited to molecules with molecular weights less than ~30 kDa. Conventional X-ray crystallography and cryoEM are not restricted by size limitations, but produce only static images representing the average of a large ensemble of individual molecules. Thus, unless dynamic behavior can be synchronized in all molecules, it is not possible to simultaneously align all the moving parts in different individual molecules for averaging. Hence, macromolecules and their complexes are typically stalled and then imaged at different points in their enzymatic or mechanical cycle, and conformational changes are then inferred by interpolating between distinct static structures. For example, addition of γ-S ATP or AMP-PNP to ø29 particles would trap the motor in the ATP-bound, prehydrolysis state, whereas stalling with ADP-vanadate mimics the pentavalent transition state of ATP hydrolysis; comparing these structures would reveal any significant conformational differences between substrate and transition state-bound intermediates. An advantage of this approach is that by stopping packaging at defined points, it is possible to correlate specific structures with discrete steps in the mechanochemical pathway.

An alternative and particularly enticing approach for visualizing the conformational changes that drive packaging is to flash-freeze actively packaging particles, image them using cryoEM, and then computationally sort and reconstruct different motor conformations present in the sample. Upon freezing, motors in different particles would be trapped in different stages of the mechanochemical cycle, and thus the relative positions of the motor's moving parts will vary in different particles imaged. Until recently, this conformational heterogeneity would have been an obstacle in cryoEM analysis, i.e., unless dynamic behavior could be synchronized in all molecules (as described above), it would be impossible to simultaneously align all the moving parts in different individual molecules for image reconstruction. However, recent advances in image processing have demonstrated that it is possible to computationally sort different conformational species present in cryoEM micrographs and extract meaningful biological information from their differences (Leschziner and Nogales 2007). In these approaches, particles are initially classified according to direction of view by model-based projection matching and/or other well-established methods (Baker et al. 1999; Ludtke et al. 1999; van Heel et al. 2000; Frank et al. 2002). Each class of particles would then be analyzed for the presence of subclasses corresponding to different conformations of the various motor components.

Of course, structural data alone is not sufficient for a complete description of the motor, and a comprehensive understanding of the coordinated mechanochemical cycle of the bacteriophage ø29 DNA packaging motor will require a highly integrated genetic, biochemical, biophysical, computational, as well as structural approach. In addition to identifying mutations that completely abolish packaging, it would be informative to have a library of mutants that modulate, rather than eliminate, DNA packaging. Phage particles with altered activities could then be subjected to additional analysis via structural or single particle techniques. In this way, it may be possible to unravel more subtle aspects of the motor function and begin the task of determining the complete sequence of conformational changes that constitute the mechanochemical cycle of the packaging motor.

Insight provided by current and future results on ø29 and other model systems for genome encapsidation will inform us as to the fundamental mechanism of packaging in dsDNA viruses, as well as the operation of the broader class of ATP-driven, force-generating ring motors from which they evolved. Hence, future research on the ø29 DNA packaging motor promises to describe principles of the mechanochemistry of a broad range of molecular motors, provide insight into how these principles are optimized in viral packaging motors to generate large forces, inspire novel therapeutics that selectively target genome packaging to inhibit medically relevant dsDNA viruses, and stimulate work in the developing field of rational design of molecular machines.

Acknowledgments I am indebted to Dr. Michael Rossmann, who first introduced me to ø29, mentored much of my own research on the DNA packaging motor, and always graciously shared his considerable wisdom. I would also like to thank my collaborators: Drs. Dwight Anderson, Paul Jardine, Dr. Jaya Koti and Shelley Grimes for many thoughtful discussions regarding the mechanism of genome packaging in ø29, as well as for their invaluable contributions to the field of DNA packaging by dsDNA bacteriophages.

References

Aathavan K, Politzer AT, Kaplan A, Moffitt JR, Chemla YR, Grimes S, Jardine PJ, Anderson DL, Bustamante C (2009) Substrate interactions and promiscuity in a viral DNA packaging motor. Nature 461(7264):669–673

Abrahams JP, Leslie AG, Lutter R, Walker JE (1994) Structure at 2.8 A resolution of F1-ATPase from bovine heart mitochondria. Nature 370(6491):621–628

Agirrezabala X, Velazquez-Muriel JA, Gomez-Puertas P, Scheres SH, Carazo JM, Carrascosa JL (2007) Quasi-atomic model of bacteriophage t7 procapsid shell: insights into the structure and evolution of a basic fold. Structure 15(4): 461–472

Anderson DL, Reilly BE (1993) Morphogenesis of bacteriophage phi29. In: Hoch JA, Losick R, Sonenshein AL (eds) *Bacillus subtilis* and other gram positive bacteria: physiology, biochemistry and molecular genetics. ASM Press, Washington, DC, pp 859–867

Atz R, Ma S, Gao J, Anderson DL, Grimes S (2007) Alanine scanning and Fe-BABE probing of the bacteriophage ø29 prohead RNA-connector interaction. J Mol Biol 369(1):239–248

Bailey S, Wichitwechkarn J, Johnson D, Reilly BE, Anderson DL, Bodley JW (1990) Phylogenetic analysis and secondary structure of the *Bacillus subtilis* bacteriophage RNA required for DNA packaging. J Biol Chem 265(36): 22365–22370

Baker TS, Olson NH, Fuller SD (1999) Adding the third dimension to virus life cycles: three-dimensional reconstruction of icosahedral viruses from cryo-electron micrographs. Microbiol Mol Biol Rev 63(4):862–922, table of contents

Baumann RG, Mullaney J, Black LW (2006) Portal fusion protein constraints on function in DNA packaging of bacteriophage T4. Mol Microbiol 61(1):16–32

Bjornsti MA, Reilly BE, Anderson DL (1983) Morphogenesis of bacteriophage phi 29 of *Bacillus subtilis*: oriented and quantized in vitro packaging of DNA protein gp3. J Virol 45(1):383–396

Bjornsti MA, Reilly BE, Anderson DL (1984) Bacteriophage phi 29 proteins required for in vitro DNA-gp3 packaging. J Virol 50(3):766–772

Blanco L, Salas M (1996) Relating structure to function in phi29 DNA polymerase. J Biol Chem 271(15):8509–8512

Braig K, Menz RI, Montgomery MG, Leslie AG, Walker JE (2000) Structure of bovine mitochondrial F(1)-ATPase inhibited by Mg(2+) ADP and aluminium fluoride. Structure 8(6):567–573

Burroughs AM, Iyer LM, Aravind L (2007) Comparative genomics and evolutionary trajectories of viral ATP dependent DNA-packaging systems. Genome Dyn 3:48–65

Camacho A, Jimenez F, De La Torre J, Carrascosa JL, Mellado RP, Vasquez C, Vinuela E, Salas M (1977) Assembly of *Bacillus subtilis* phage phi29. 1. Mutants in the cistrons coding for the structural proteins. Eur J Biochem 73(1):39–55

Chemla YR, Aathavan K, Michaelis J, Grimes S, Jardine PJ, Anderson DL, Bustamante C (2005) Mechanism of force generation of a viral DNA packaging motor. Cell 122(5):683–692

Chen C, Zhang C, Guo P (1999) Sequence requirement for hand-in-hand interaction in formation of RNA dimers and hexamers to gear phi29 DNA translocation motor. RNA 5(6):805–818

Chen C, Sheng S, Shao Z, Guo P (2000) A dimer as a building block in assembling RNA. A hexamer that gears bacterial virus phi29 DNA-translocating machinery. J Biol Chem 275(23):17510–17516

Chen DH, Baker ML, Hryc CF, DiMaio F, Jakana J, Wu W, Dougherty M, Haase-Pettingell C, Schmid MF, Jiang W, Baker D, King JA, Chiu W (2011) Structural basis for scaffolding-mediated assembly and maturation of a dsDNA virus. Proc Natl Acad Sci USA 108(4):1355–1360

Cohen DN, Sham YY, Haugstad GD, Xiang Y, Rossmann MG, Anderson DL, Popham DL (2009) Shared catalysis in virus entry and bacterial cell wall depolymerization. J Mol Biol 387(3):607–618

Dai W, Hodes A, Hui WH, Gingery M, Miller JF, Zhou ZH (2010) Three-dimensional structure of tropism-switching Bordetella bacteriophage. Proc Natl Acad Sci USA 107(9):4347–4352

Ding F, Lu C, Zhao W, Rajashankar KR, Anderson DL, Jardine PJ, Grimes S, Ke A (2011) Structure and assembly of the essential RNA ring component of a viral DNA packaging motor. Proc Natl Acad Sci USA 108:7357–7362

Draper B, Rao VB (2007) An ATP hydrolysis sensor in the DNA packaging motor from bacteriophage T4 suggests an inchworm-type translocation mechanism. J Mol Biol 369(1):79–94

Earnshaw WC, Casjens SR (1980) DNA packaging by the double-stranded DNA bacteriophages. Cell 21(2):319–331

Fokine A, Leiman PG, Shneider MM, Ahvazi B, Boeshans KM, Steven AC, Black LW, Mesyanzhinov VV, Rossmann MG (2005) Structural and functional similarities between the capsid proteins of bacteriophages T4 and HK97 point to a common ancestry. Proc Natl Acad Sci USA 102(20):7163–7168

Frank J, Wagenknecht T, McEwen BF, Marko M, Hsieh CE, Mannella CA (2002) Three-dimensional imaging of biological complexity. J Struct Biol 138(1–2):85–91

Fu CY, Prevelige PE Jr (2009) In vitro incorporation of the phage Phi29 connector complex. Virology 394(1):149–153

Fuller DN, Rickgauer JP, Jardine PJ, Grimes S, Anderson DL, Smith DE (2007) Ionic effects on viral DNA packaging and portal motor function in bacteriophage phi 29. Proc Natl Acad Sci USA 104(27):11245–11250

Garver K, Guo P (1997) Boundary of pRNA functional domains and minimum pRNA sequence requirement for specific connector binding and DNA packaging of phage phi29. RNA 3(9):1068–1079

Geourjon C, Orelle C, Steinfels E, Blanchet C, Deleage G, Di Pietro A, Jault JM (2001) A common mechanism for ATP hydrolysis in ABC transporter and helicase superfamilies. Trends Biochem Sci 26(9):539–544

Grimes S, Anderson D (1989) In vitro packaging of bacteriophage phi 29 DNA restriction fragments and the role of the terminal protein gp3. J Mol Biol 209(1):91–100

Grimes S, Anderson D (1990) RNA dependence of the bacteriophage phi 29 DNA packaging ATPase. J Mol Biol 215(4):559–566

Grimes S, Jardine PJ, Anderson D (2002) Bacteriophage phi 29 DNA packaging. Adv Virus Res 58:255–294

Guasch A, Pous J, Ibarra B, Gomis-Ruth FX, Valpuesta JM, Sousa N, Carrascosa JL, Coll M (2002) Detailed architecture of a DNA translocating machine: the high-resolution structure of the bacteriophage phi29 connector particle. J Mol Biol 315(4):663–676

Guo P, Grimes S, Anderson D (1986) A defined system for in vitro packaging of DNA-gp3 of the *Bacillus subtilis* bacteriophage phi 29. Proc Natl Acad Sci USA 83(10):3505–3509

Guo P, Peterson C, Anderson D (1987a) Initiation events in in-vitro packaging of bacteriophage phi 29 DNA-gp3. J Mol Biol 197(2):219–228

Guo P, Peterson C, Anderson D (1987b) Prohead and DNA-gp3-dependent ATPase activity of the DNA packaging protein gp16 of bacteriophage phi 29. J Mol Biol 197(2):229–236

Guo PX, Erickson S, Xu W, Olson N, Baker TS, Anderson D (1991a) Regulation of the phage phi 29 prohead shape and size by the portal vertex. Virology 183(1):366–373

Guo PX, Rajagopal BS, Anderson D, Erickson S, Lee CS (1991b) sRNA of phage phi 29 of *Bacillus subtilis* mediates DNA packaging of phi 29 proheads assembled in Escherichia coli. Virology 185(1):395–400

Guo P, Zhang C, Chen C, Garver K, Trottier M (1998) Inter-RNA interaction of phage phi29 pRNA to form a hexameric complex for viral DNA transportation. Mol Cell 2(1):149–155

Hagen EW, Reilly BE, Tosi ME, Anderson DL (1976) Analysis of gene function of bacteriophage phi 29 of *Bacillus subtilis*: identification of cistrons essential for viral assembly. J Virol 19(2):501–517

Harding NE, Ito J, David GS (1978) Identification of the protein firmly bound to the ends of bacteriophage phi 29 DNA. Virology 84(2):279–292

Harris S, Schroeder SJ (2010) Nuclear magnetic resonance structure of the prohead RNA E-loop hairpin. Biochemistry 49(29):5989–5997

Hendrix RW (1978) Symmetry mismatch and DNA packaging in large bacteriophages. Proc Natl Acad Sci USA 75(10):4779–4783

Hendrix RW (2002) Bacteriophages: evolution of the majority. Theor Popul Biol 61(4):471–480

Hugel T, Michaelis J, Hetherington CL, Jardine PJ, Grimes S, Walter JM, Falk W, Anderson DL, Bustamante C (2007) Experimental test of connector rotation during DNA packaging into bacteriophage phi29 capsids. PLoS Biol 5(3):e59

Hung LW, Wang IX, Nikaido K, Liu PQ, Ames GF, Kim SH (1998) Crystal structure of the ATP-binding subunit of an ABC transporter. Nature 396(6712):703–707

Iyer LM, Leipe DD, Koonin EV, Aravind L (2004a) Evolutionary history and higher order classification of AAA + ATPases. J Struct Biol 146(1–2):11–31

Iyer LM, Makarova KS, Koonin EV, Aravind L (2004b) Comparative genomics of the FtsK-HerA superfamily of pumping ATPases: implications for the origins of chromosome segregation, cell division and viral capsid packaging. Nucleic Acids Res 32(17):5260–5279

Jiang W, Li Z, Zhang Z, Baker ML, Prevelige PE Jr, Chiu W (2003) Coat protein fold and maturation transition of bacteriophage P22 seen at subnanometer resolutions. Nat Struct Biol 10(2):131–135

Jiang W, Baker ML, Jakana J, Weigele PR, King J, Chiu W (2008) Backbone structure of the infectious epsilon 15 virus capsid revealed by electron cryomicroscopy. Nature 451(7182):1130–1134

Kabaleeswaran V, Puri N, Walker JE, Leslie AG, Mueller DM (2006) Novel features of the rotary catalytic mechanism revealed in the structure of yeast F1 ATPase. EMBO J 25(22):5433–5442

Kainov DE, Mancini EJ, Telenius J, Lisal J, Grimes JM, Bamford DH, Stuart DI, Tuma R (2008) Structural basis of mechanochemical coupling in a hexameric molecular motor. J Biol Chem 283(6):3607–3617

Kamtekar S, Berman AJ, Wang J, Lazaro JM, de Vega M, Blanco L, Salas M, Steitz TA (2006) The phi29 DNA polymerase: protein-primer structure suggests a model for the initiation to elongation transition. EMBO J 25(6): 1335–1343

Kellenberger E, Sechaud J, Ryter A (1959) Electron microscopical studies of phage multiplication. IV. The establishment of the DNA pool of vegetative phage and the maturation of phage particles. Virology 8:478–498

Koti JS, Morais MC, Rajagopal R, Owen BA, McMurray CT, Anderson DL (2008) DNA packaging motor assembly intermediate of bacteriophage phi29. J Mol Biol 381(5):1114–1132

Lander GC, Khayat R, Li R, Prevelige PE, Potter CS, Carragher B, Johnson JE (2009) The P22 tail machine at subnanometer resolution reveals the architecture of an infection conduit. Structure 17(6):789–799

Lebedev AA, Krause MH, Isidro AL, Vagin AA, Orlova EV, Turner J, Dodson EJ, Tavares P, Antson AA (2007) Structural framework for DNA translocation via the viral portal protein. EMBO J 26(7):1984–1994

Lee TJ, Guo P (2006) Interaction of gp16 with pRNA and DNA for genome packaging by the motor of bacterial virus phi29. J Mol Biol 356(3):589–599

Lee JY, Yang W (2006) UvrD helicase unwinds DNA one base pair at a time by a two-part power stroke. Cell 127(7): 1349–1360

Leipe DD, Wolf YI, Koonin EV, Aravind L (2002) Classification and evolution of P-loop GTPases and related ATPases. J Mol Biol 317(1):41–72

Leschziner AE, Nogales E (2007) Visualizing flexibility at molecular resolution: analysis of heterogeneity in single-particle electron microscopy reconstructions. Annu Rev Biophys Biomol Struct 36:43–62

Lisal J, Tuma R (2005) Cooperative mechanism of RNA packaging motor. J Biol Chem 280(24):23157–23164

Ludtke SJ, Baldwin PR, Chiu W (1999) EMAN: semiautomated software for high-resolution single-particle reconstructions. J Struct Biol 128(1):82–97

Mancini EJ, Kainov DE, Grimes JM, Tuma R, Bamford DH, Stuart DI (2004) Atomic snapshots of an RNA packaging motor reveal conformational changes linking ATP hydrolysis to RNA translocation. Cell 118(6):743–755

Mitchell MS, Matsuzaki S, Imai S, Rao VB (2002) Sequence analysis of bacteriophage T4 DNA packaging/terminase genes 16 and 17 reveals a common ATPase center in the large subunit of viral terminases. Nucleic Acids Res 30(18): 4009–4021

Moffitt JR, Chemla YR, Aathavan K, Grimes S, Jardine PJ, Anderson DL, Bustamante C (2009) Intersubunit coordination in a homomeric ring ATPase. Nature 457(7228):446–450

Morais MC, Baker AS, Dunaway-Mariano D, Allen KN (2000) Crystallization and preliminary crystallographic analysis of phosphonoacetaldehyde hydrolase. Acta Crystallogr D Biol Crystallogr 56(Pt 2):206–209

Morais MC, Tao Y, Olson NH, Grimes S, Jardine PJ, Anderson DL, Baker TS, Rossmann MG (2001) Cryoelectron-microscopy image reconstruction of symmetry mismatches in bacteriophage phi29. J Struct Biol 135(1):38–46

Morais MC, Kanamaru S, Badasso MO, Koti JS, Owen BA, McMurray CT, Anderson DL, Rossmann MG (2003) Bacteriophage phi29 scaffolding protein gp7 before and after prohead assembly. Nat Struct Biol 10(7):572–576

Morais MC, Choi KH, Koti JS, Chipman PR, Anderson DL, Rossmann MG (2005) Conservation of the capsid structure in tailed dsDNA bacteriophages: the pseudoatomic structure of phi29. Mol Cell 18(2):149–159

Morais MC, Koti JS, Bowman VD, Reyes-Aldrete E, Anderson DL, Rossmann MG (2008) Defining molecular and domain boundaries in the bacteriophage phi29 DNA packaging motor. Structure 16(8):1267–1274

Mosharrafa ET, Schachtele CF, Reilly BE, Anderson DL (1970) Complementary Strands of Bacteriophage phi29 Deoxyribonucleic Acid: Preparative Separation and Transcription Studies. J Virol 6(6):855–864
Muller DJ, Engel A, Carrascosa JL, Velez M (1997) The bacteriophage phi29 head-tail connector imaged at high resolution with the atomic force microscope in buffer solution. EMBO J 16(10):2547–2553
Nadanaciva S, Weber J, Wilke-Mounts S, Senior AE (1999) Importance of F1-ATPase residue alpha-Arg-376 for catalytic transition state stabilization. Biochemistry 38(47):15493–15499
Ogura T, Whiteheart SW, Wilkinson AJ (2004) Conserved arginine residues implicated in ATP hydrolysis, nucleotide-sensing, and inter-subunit interactions in AAA and AAA+ATPases. J Struct Biol 146(1–2):106–112
Orlova EV, Gowen B, Droge A, Stiege A, Weise F, Lurz R, van Heel M, Tavares P (2003) Structure of a viral DNA gatekeeper at 10 A resolution by cryo-electron microscopy. EMBO J 22(6):1255–1262
Petrov AS, Harvey SC (2008) Packaging double-helical DNA into viral capsids: structures, forces, and energetics. Biophys J 95(2):497–502
Reid RJ, Bodley JW, Anderson D (1994a) Characterization of the prohead-pRNA interaction of bacteriophage phi 29. J Biol Chem 269(7):5157–5162
Reid RJ, Bodley JW, Anderson D (1994b) Identification of bacteriophage phi 29 prohead RNA domains necessary for in vitro DNA-gp3 packaging. J Biol Chem 269(12):9084–9089
Reid RJ, Zhang F, Benson S, Anderson D (1994c) Probing the structure of bacteriophage phi 29 prohead RNA with specific mutations. J Biol Chem 269(28):18656–18661
Rickgauer JP, Fuller DN, Grimes S, Jardine PJ, Anderson DL, Smith DE (2008) Portal motor velocity and internal force resisting viral DNA packaging in bacteriophage phi29. Biophys J 94(1):159–167
Rossmann MG, Moras D, Olsen KW (1974) Chemical and biological evolution of nucleotide-binding protein. Nature 250(463):194–199
Rossmann MG, Morais MC, Leiman PG, Zhang W (2005) Combining X-ray crystallography and electron microscopy. Structure 13(3):355–362
Salas M, Mellado RP, Vinuela E (1978) Characterization of a protein covalently linked to the 5′ termini of the DNA of *Bacillus subtilis* phage phi29. J Mol Biol 119(2):269–291
Schachtele CF, De Sain CV, Anderson DL (1973) Transcription during the development of bacteriophage phi29: definition of "early" and "late" phi29 ribonucleic acid. J Virol 11(1):9–16
Shu D, Zhang H, Jin J, Guo P (2007) Counting of six pRNAs of phi29 DNA-packaging motor with customized single-molecule dual-view system. EMBO J 26(2):527–537
Simpson AA, Tao Y, Leiman PG, Badasso MO, He Y, Jardine PJ, Olson NH, Morais MC, Grimes S, Anderson DL, Baker TS, Rossmann MG (2000) Structure of the bacteriophage phi29 DNA packaging motor. Nature 408(6813): 745–750
Simpson AA, Leiman PG, Tao Y, He Y, Badasso MO, Jardine PJ, Anderson DL, Rossmann MG (2001) Structure determination of the head-tail connector of bacteriophage phi29. Acta Crystallogr D Biol Crystallogr 57(Pt 9): 1260–1269
Singleton MR, Sawaya MR, Ellenberger T, Wigley DB (2000) Crystal structure of T7 gene 4 ring helicase indicates a mechanism for sequential hydrolysis of nucleotides. Cell 101(6):589–600
Singleton MR, Dillingham MS, Wigley DB (2007) Structure and mechanism of helicases and nucleic acid translocases. Annu Rev Biochem 76:23–50
Skordalakes E, Berger JM (2003) Structure of the Rho transcription terminator: mechanism of mRNA recognition and helicase loading. Cell 114(1):135–146
Smith DE, Tans SJ, Smith SB, Grimes S, Anderson DL, Bustamante C (2001) The bacteriophage straight phi29 portal motor can package DNA against a large internal force. Nature 413(6857):748–752
Story RM, Steitz TA (1992) Structure of the recA protein-ADP complex. Nature 355(6358):374–376
Story RM, Weber IT, Steitz TA (1992) The structure of the E. coli recA protein monomer and polymer. Nature 355(6358):318–325
Subramanya HS, Bird LE, Brannigan JA, Wigley DB (1996) Crystal structure of a DExx box DNA helicase. Nature 384(6607):379–383
Sun S, Kondabagil K, Gentz PM, Rossmann MG, Rao VB (2007) The structure of the ATPase that powers DNA packaging into bacteriophage T4 procapsids. Mol Cell 25(6):943–949
Sun S, Kondabagil K, Draper B, Alam TI, Bowman VD, Zhang Z, Hegde S, Fokine A, Rossmann MG, Rao VB (2008) The structure of the phage T4 DNA packaging motor suggests a mechanism dependent on electrostatic forces. Cell 135(7):1251–1262
Tang J, Olson N, Jardine PJ, Grimes S, Anderson DL, Baker TS (2008) DNA poised for release in bacteriophage phi29. Structure 16(6):935–943
Tao Y, Olson NH, Xu W, Anderson DL, Rossmann MG, Baker TS (1998) Assembly of a tailed bacterial virus and its genome release studied in three dimensions. Cell 95(3):431–437
Thomsen ND, Berger JM (2008) Structural frameworks for considering microbial protein- and nucleic acid-dependent motor ATPases. Mol Microbiol 69(5):1071–1090

Trottier M, Guo P (1997) Approaches to determine stoichiometry of viral assembly components. J Virol 71(1): 487–494

Tsuprun V, Anderson D, Egelman EH (1994) The bacteriophage phi 29 head-tail connector shows 13-fold symmetry in both hexagonally packed arrays and as single particles. Biophys J 66(6):2139–2150

Valle M, Kremer L, Martinez AC, Roncal F, Valpuesta JM, Albar JP, Carrascosa JL (1999) Domain architecture of the bacteriophage phi29 connector protein. J Mol Biol 288(5):899–909

Valpuesta JM, Carrascosa JL (1994) Structure of viral connectors and their function in bacteriophage assembly and DNA packaging. Q Rev Biophys 27(2):107–155

Valpuesta JM, Fernandez JJ, Carazo JM, Carrascosa JL (1999) The three-dimensional structure of a DNA translocating machine at 10 A resolution. Structure 7(3):289–296

van Heel M, Gowen B, Matadeen R, Orlova EV, Finn R, Pape T, Cohen D, Stark H, Schmidt R, Schatz M, Patwardhan A (2000) Single-particle electron cryo-microscopy: towards atomic resolution. Q Rev Biophys 33(4):307–369

Velankar SS, Soultanas P, Dillingham MS, Subramanya HS, Wigley DB (1999) Crystal structures of complexes of PcrA DNA helicase with a DNA substrate indicate an inchworm mechanism. Cell 97(1):75–84

Walker JE, Saraste M, Runswick MJ, Gay NJ (1982) Distantly related sequences in the alpha- and beta-subunits of ATP synthase, myosin, kinases and other ATP-requiring enzymes and a common nucleotide binding fold. EMBO J 1(8):945–951

Wichitwechkarn J, Bailey S, Bodley JW, Anderson D (1989) Prohead RNA of bacteriophage phi 29: size, stoichiometry and biological activity. Nucleic Acids Res 17(9):3459–3468

Wikoff WR, Liljas L, Duda RL, Tsuruta H, Hendrix RW, Johnson JE (2000) Topologically linked protein rings in the bacteriophage HK97 capsid. Science 289(5487):2129–2133

Wittinghofer A, Scheffzek K, Ahmadian MR (1997) The interaction of Ras with GTPase-activating proteins. FEBS Lett 410(1):63–67

Xiang Y, Morais MC, Battisti AJ, Grimes S, Jardine PJ, Anderson DL, Rossmann MG (2006) Structural changes of bacteriophage phi29 upon DNA packaging and release. EMBO J 25(21):5229–5239

Xiang Y, Morais MC, Cohen DN, Bowman VD, Anderson DL, Rossmann MG (2008) Crystal and cryoEM structural studies of a cell wall degrading enzyme in the bacteriophage phi29 tail. Proc Natl Acad Sci USA 105(28): 9552–9557

Xiao F, Moll WD, Guo S, Guo P (2005) Binding of pRNA to the N-terminal 14 amino acids of connector protein of bacteriophage phi29. Nucleic Acids Res 33(8):2640–2649

Yehle CO (1978) Genome-linked protein associated with the 5′ termini of bacteriophage phi29 DNA. J Virol 27(3):776–783

Yoshida M, Amano T (1995) A common topology of proteins catalyzing ATP-triggered reactions. FEBS Lett 359(1):1–5

Yu J, Moffitt J, Hetherington CL, Bustamante C, Oster G (2010) Mechanochemistry of a viral DNA packaging motor. J Mol Biol 400(2):186–203

Zhang C, Lee CS, Guo P (1994) The proximate 5′ and 3′ ends of the 120-base viral RNA (pRNA) are crucial for the packaging of bacteriophage phi 29 DNA. Virology 201(1):77–85

Zhang C, Tellinghuisen T, Guo P (1995) Confirmation of the helical structure of the 5′/3′ termini of the essential DNA packaging pRNA of phage phi 29. RNA 1(10):1041–1050

Zhang F, Lemieux S, Wu X, St-Arnaud D, McMurray CT, Major F, Anderson D (1998) Function of hexameric RNA in packaging of bacteriophage phi 29 DNA in vitro. Mol Cell 2(1):141–147

Zhao W, Morais MC, Anderson DL, Jardine PJ, Grimes S (2008) Role of the CCA bulge of prohead RNA of bacteriophage ø29 in DNA packaging. J Mol Biol 383(3):520–528

Zheng H, Olia AS, Gonen M, Andrews S, Cingolani G, Gonen T (2008) A conformational switch in bacteriophage p22 portal protein primes genome injection. Mol Cell 29(3):376–383

Zhou ZH, Dougherty M, Jakana J, He J, Rixon FJ, Chiu W (2000) Seeing the herpesvirus capsid at 8.5 A. Science 288(5467):877–880

Chapter 24
Single-Molecule Studies of Viral DNA Packaging

Yann R. Chemla and Douglas E. Smith

Abstract Many double-stranded DNA bacteriophages and viruses use specialized ATP-driven molecular machines to package their genomes into tightly confined procapsid shells. Over the last decade, single-molecule approaches – and in particular, optical tweezers – have made key contributions to our understanding of this remarkable process. In this chapter, we review these advances and the insights they have provided on the packaging mechanisms of three bacteriophages: $\phi 29$, λ, and T4.

24.1 Introduction

A critical step in the assembly of many dsDNA viruses inside infected host cells is the packaging of newly synthesized viral genomes into procapsid shells. The procapsids have a small, ~3 nm diameter "portal" through which the ~2 nm diameter DNA must enter (Simpson et al. 2000). A molecular motor complex transiently assembles at the portal and converts chemical energy from ATP hydrolysis into mechanical work needed to translocate the DNA into the procapsid. This is a remarkable process from a biophysical point of view because the fully packed DNA attains very tight confinement, reaching crystalline density against huge resisting forces arising from electrostatic self-repulsion, bending rigidity, and entropic penalty.

This chapter reviews "single-molecule" studies of viral DNA packaging, which have been an exciting and powerful development during the last decade. In these approaches, the packaging of a single DNA molecule into a single viral procapsid can be measured in real time. Using "optical tweezers," in particular, one can directly measure binding of the procapsid–motor–DNA complex, initiation of packaging, and measure the DNA translocation dynamics with better than 1 nm displacement resolution and 0.1 s time resolution. One can also directly measure the forces exerted by the motor on the DNA with piconewton-level resolution. The optical tweezers method has now been applied to

Y.R. Chemla (✉)
Department of Physics, Center for the Physics of Living Cells, Center for Biophysics and Computational Biology, University of Illinois, Urbana-Champaign, 1110 W. Green St, Urbana, IL 61801, USA
e-mail: ychemla@illinois.edu

D.E. Smith
Department of Physics, University of California at San Diego, Mail Code 0379, 9500 Gilman Drive, La Jolla, San Diego, CA 92093, USA

M.G. Rossmann and V.B. Rao (eds.), *Viral Molecular Machines*, Advances in Experimental Medicine and Biology 726, DOI 10.1007/978-1-4614-0980-9_24,

the study of three systems – bacteriophages ϕ29, λ, and T4 – and this approach has provided a much more detailed picture of viral DNA packaging than traditional biochemical assays.

New findings have provided insight on motor force generation, velocity and processivity, internal forces resisting packaging, procapsid expansion and rupture, details of the mechanochemical kinetic cycle and motor stepping dynamics, functional roles of structural motifs, and conformational dynamics of motor units. These advances will be reviewed in detail below. Though other single-molecule techniques such as fluorescence imaging and spectroscopy have been applied to the study of viral DNA packaging, we limit the scope of this chapter predominantly to the advances made using optical tweezers due to space limitations. We also emphasize that single-molecule techniques will likely continue to have high applicability in the field of viral DNA packaging. Potential future directions, such as further extension of high-resolution and mutational studies in the ϕ29, λ, and T4 systems and studies of packaging initiation and termination mechanisms, will be discussed in the conclusion.

24.2 Single-Molecule Approaches

Viral DNA packaging has long been studied by traditional methods from biochemistry and structural biology. Structural methods such as cryo-electron microscopy and X-ray crystallography can provide detailed information on structures of viral proteins and assemblies with atomic or near-atomic resolution but do not provide information on the kinetics of packaging or conformational dynamics in protein subunits. Ensemble biochemical techniques, on the other hand, can provide information on molecular dynamics. However, to measure kinetics quantitatively and accurately requires synchronizing populations of complexes, which is often difficult or impossible due to heterogeneity in the ensemble. This heterogeneity stems both from the inherent stochasticity (i.e., randomness due to thermal fluctuations) in the molecular processes and from potential structural and conformational variability in individual complexes. Traditionally, viral DNA packaging has been assessed in bulk biochemical reactions by quantifying the amount of DNA protected by the capsid from degradation (Grimes and Anderson 1989). This is achieved by adding a nuclease (DNase or sometimes a restriction enzyme) to a packaging reaction to digest any unpackaged DNA, releasing any packed DNA by treatment with proteinase K, and quantifying it by gel electrophoresis. This assay can be used, for example, to determine the efficiency of packaging (the fraction of DNA successfully packaged) or estimate the total time to complete packaging. It is quite difficult, however, to go beyond these basic measurements and extract quantitative information on rates of packaging, how they vary in time, and how they vary among different complexes.

Single-molecule techniques provide a powerful complementary method for studying packaging kinetics in detail. Unlike traditional bulk methods, these techniques do not rely on temporal and population averaging, instead collecting statistics from individual packaging complexes. Moreover, conformational dynamics and force generation can be detected *directly* in real time with single-molecule techniques, providing a clear advantage over ensemble methods. Techniques such as single-molecule fluorescence, optical traps, and magnetic tweezers (reviewed in Neuman and Block 2004; Myong et al. 2006; Greenleaf et al. 2007; Moffitt et al. 2008; Neuman and Nagy 2008; Joo et al. 2008; Rickgauer and Smith 2008) have been instrumental in deciphering the mechanism of a wide range of biological phenomena and have been applied to the study of viral DNA packaging in recent years. While much of the work has been with optical tweezers, single-molecule fluorescence microscopy has also recently been employed. Single fluorophore imaging methods were used to search for conformational changes and subunit stoichiometries in the ϕ29 motor (Hugel et al. 2007; Shu et al. 2007), and fluorescence correlation spectroscopy and single-pair fluorescence resonance energy transfer were recently used to detect packaging and conformational changes in DNA substrates in T4 packaging (Sabanayagam et al. 2007; Oram et al. 2008; Ray et al. 2010a, b). This chapter, however, will focus primarily on reviewing findings in our main area of expertise: single DNA molecule packaging experiments using optical tweezers.

24.2.1 Optical Traps

Light carries momentum and thus can be used to exert forces on microscopic objects. Optical traps or "tweezers" use a tightly focused laser beam to trap a dielectric object such as a polystyrene microsphere in all three dimensions near the focal point (Ashkin 1986). The trap exerts a restoring force proportional to the displacement of the microsphere. Typically, the position of the microsphere in its trap can be detected to great precision by imaging the trap light scattered by the microspheres onto position-sensitive photodetectors (Neuman and Block 2004). This method can provide subnanometer (nm) resolution of displacements and subpiconewton resolution of forces (pN = 10^{-12} N).

Optical traps have been used to study the dynamics of a variety of different biomolecular processes, ranging from how individual macromolecules such as proteins, DNA, and RNA unfold under force (Bustamante et al. 2000; Cecconi et al. 2005; Tinoco et al. 2006) to how molecular motors translocate and exert forces (Ross et al. 2008; Michaelis et al. 2009). Of particular interest, single-molecule optical trap experiments have provided novel insights into the mechanism of nucleic acid translocases. In a typical measurement, an individual DNA molecule is tethered between a microsphere held in an optical trap and a second attachment point. This attachment point can be the surface of a sample chamber or a second microsphere suctioned onto the end of a micropipette or held in a second trap (reviewed in Moffitt et al. 2008; Chemla 2010; Fuller et al. 2006; Smith et al. 2003). The tethered DNA molecule may be stretched by an applied force, and displacements of the trapped microsphere report on the actions of the biological system under study. Alternately, the instrument can be controlled by a feedback loop in order to maintain a constant force on the trapped microsphere. In this *force feedback* mode, the position of the trap or attachment point is actively controlled to apply a steady tension to the tethering molecule (Neuman and Block 2004; Smith et al. 2001). Instead of detecting the displacement of the microsphere, which is constant, the readout is the change in separation between one trap and the second attachment point.

24.2.2 Single-Molecule Measurement of ϕ29 Packaging

Bacteriophage ϕ29 was the first viral DNA packaging system successfully studied using single-molecule optical tweezers measurements (Smith et al. 2001). The initial approach for carrying out this type of measurement was to create a "stalled complex" consisting of a prohead with partially packaged DNA hanging out. Prior to preparing the stalled complex, one end of the ϕ29 DNA was labeled with biotin by cutting the molecule with a restriction endonuclease and using DNA polymerase I to incorporate biotinylated nucleotides. Packaging was initiated in a bulk in vitro reaction and allowed to proceed for ~1 min until roughly 30–50% of the DNA was packaged. An excess of nonhydrolyzable ATP analog (γS-ATP) was then added, which caused the motors to stall with the unpackaged biotin-labeled end of the DNA dangling out of the prohead.

The single-molecule measurements were performed by carrying out the following steps. Stalled complexes were first attached to streptavidin-coated polystyrene microspheres via the biotin tag. Then, a second batch of microspheres coated with anti-ϕ29 antibodies was prepared. Both types of microspheres were injected into a microfluidic flow chamber in the optical tweezers instrument. In the initial work of Smith et al., the anti-ϕ29 microsphere was held by suction onto the end of a glass micropipette, and the microsphere carrying the ϕ29 packaging complex was trapped with an optical trap. The fluid chamber was filled with a solution containing ATP, such that the nonhydrolyzable ATP was eluted away as soon as the microspheres were injected, whereupon stalled complexes resumed packaging. When the two microspheres were brought into proximity by moving the pipette (attached to the chamber) with a piezo-actuated stage, the prohead bound the anti-ϕ29 microsphere, tethering the DNA between the two microspheres (Fig. 24.1a). Successful packaging events were observed from progressive shortening of the DNA tether as the motor reeled in its DNA against the force exerted by the trap,

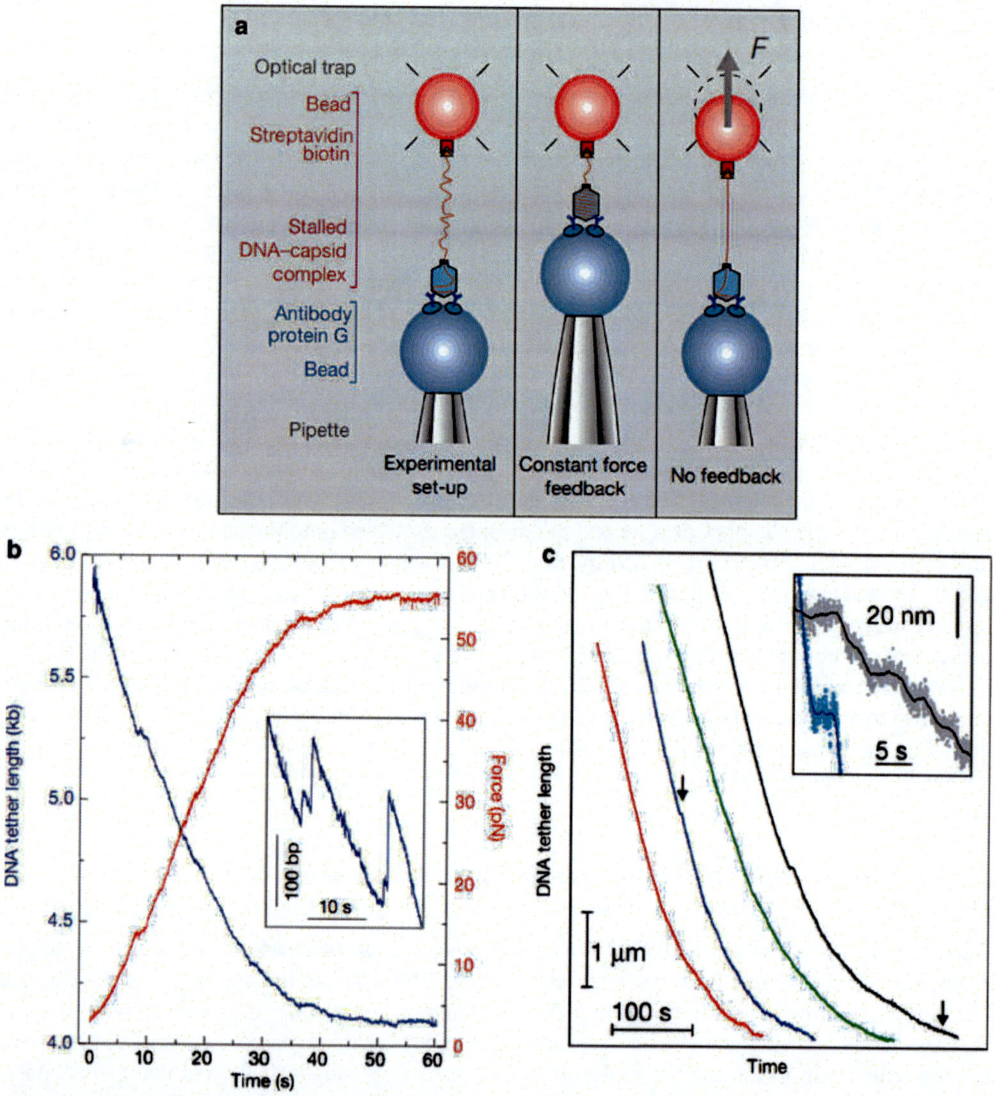

Fig. 24.1 Single-molecule viral DNA packaging assay. (**a**) Schematic of the experimental setup used in the earliest work. A single DNA molecule hanging out of a stalled ϕ29 packaging complex was tethered at one end to a microsphere held in an optical trap while the procapsid was bound to a second microsphere held by a micropipette. After initiating packaging with ATP, two measurement modes were used: "Constant force feedback," where the separation between the microspheres was adjusted to keep the DNA stretching force constant, or "No feedback" where the separation was fixed and the DNA stretching force was allowed to rise as packaging proceeded. (**b**) Force versus time (*red line*) for a packaging event measured without feedback, reaching ~55 pN before the motor paused or stalled, and corresponding tether length versus time (*blue line*). Inset is a zoomed view illustrating occasional slipping events where the DNA moved backward out of the capsid. (**c**) DNA tether length (i.e., unpackaged DNA length) versus time during packaging with 5 pN force feedback (the four different colored lines indicate four different single packaging events, shifted arbitrarily along the time axis for clarity). (**d**) Inset is a zoomed view of the regions marked with *arrows*, illustrating occasional pauses in translocation

pulling the two microspheres together. Packaging was monitored either by allowing the force to increase as the microspheres were pulled out of their traps by the phage (Fig. 24.1b) or by using force feedback and measuring the decreasing tether length as DNA was packaged (Fig. 24.1c).

This single-molecule approach has been further improved and also successfully extended to study the packaging motors of different viruses (bacteriophages λ and T4). In the following sections, we highlight the contributions made by this assay to our understanding of viral DNA packaging through its initiation (Sect. 24.3), its early stages (Sect. 24.4), and completion (Sect. 24.5).

24.3 Initiation of DNA Packaging

During their life cycles, viruses co-opt the host cell's own machinery to replicate their genomes and to express the proteins essential to their proliferation. In the case of many dsDNA viruses, including many tailed bacteriophages, expression of structural proteins coded by the viral genome leads to self-assembly of *empty* procapsid shells into which viral genomes must be packaged. Procapsids contain a portal ring through which DNA is imported. The first step in viral DNA packaging is the assembly of a molecular motor complex on the portal that can translocate the DNA into the procapsid (Catalano 2005). Packaging motors are multicomponent, multisubunit complexes. Unlike many cellular molecular motors such as myosins or kinesins, viral packaging motors assemble only transiently for the purpose of packaging the viral DNA and then disassemble prior to the formation of mature infectious viruses. The transient nature of these motor complexes is one feature that makes them challenging to study. Here, we review features of the initiation of packaging in the phage ϕ29, T4, and λ systems, focusing on recent findings from single-molecule studies.

24.3.1 Initiation in ϕ29

The *Bacillus subtilis* phage ϕ29 is one of the smallest tailed dsDNA phages, having a 19.3-kbp genome, and has been extensively characterized using genetic, biochemical, and structural methods (Grimes et al. 2002). The packaging motor is situated at a unique fivefold portal vertex in the 40-nm by 50-nm prolate (elongated) icosahedral prohead (Morais et al. 2005). The motor complex consists of a dodecameric head–tail connector ("portal ring") comprised of gene product 10 (gp10), a ring of RNA molecules ("prohead RNA" or "pRNA") attached to the narrow end of the connector, and five copies of the gp16 ATPase (Simpson et al. 2000; Morais et al. 2008). The ϕ29 genome, like that of human adenovirus, has a terminal protein covalently bound to each 5′ end (DNA–gp3) that primes DNA replication and enhances DNA packaging efficiency and selectivity (Grimes and Anderson 1989).

DNA can be efficiently packaged in vitro by mixing together ϕ29 DNA, gp16 motor protein, and empty ϕ29 proheads (procapsids) in a buffer containing ATP (Grimes and Anderson 1989), facilitating detailed biochemical and biophysical studies. Empty proheads can be produced with a mutant that lacks gp16 but contain the gp10 portal ring and pRNA. The original single-molecule experiments with stalled complexes were useful for initial studies (discussed in Sect. 24.4) but did not permit study of the initiation and early stages of packaging. Thus, an improved method facilitating measurement of packaging from initiation to completion was developed (Rickgauer et al. 2008).

24.3.1.1 Method for Tracking Single DNA Packaging from Initiation to Completion

In the improved method of Rickgauer et al., complexes consisting of proheads and the packaging ATPase gp16 were assembled with γS-ATP in the absence of DNA. An optical tweezers instrument consisting of two traps was also used, facilitating more accurate measurements of packaged DNA

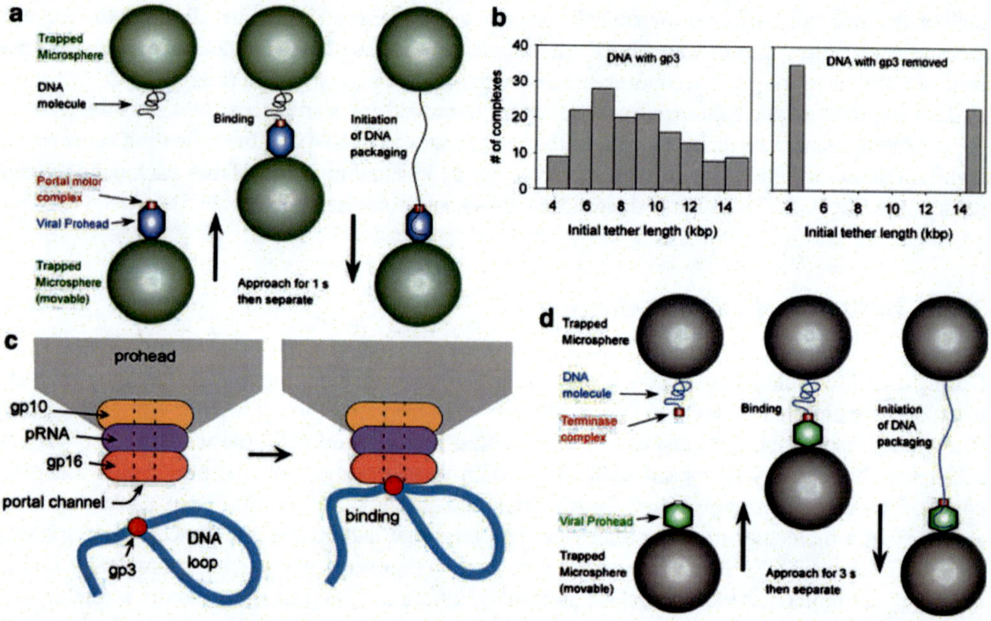

Fig. 24.2 Measurements of the initiation of viral DNA packaging. (**a**) Schematic of the experimental setup used to initiate ϕ29 packaging in real time. Instead of using a stalled, partially packaged complex as in Fig. 24.1, a preassembled prohead–motor complex is "fed" DNA. Each microsphere was held in a separate optical trap. The same approach was used for initiating phage T4 packaging. (**b**) Measured distribution of initial tether lengths for packaging native ϕ29 DNA with its gp3 terminal protein, or with gp3 removed by digestion with proteinase K. (**c**) Proposed model for gp3-mediated DNA looping at the initiation of ϕ29 packaging. (**d**) Modified approach used to initiate single-molecule phage λ packaging. A motor–DNA complex is preassembled and brought into proximity of a λ procapsid (instead of preassembling a motor–procapsid complex)

length (Fuller et al. 2006; Rickgauer et al. 2006). A microsphere carrying DNA was injected into the chamber and captured in one optical trap and a microsphere carrying prohead–gp16 complexes was injected and captured in a second trap. Packaging was initiated by bringing the two microspheres into near contact, allowing a DNA molecule to bind to a prohead–motor complex (Fig. 24.2a). Binding was observed to occur within seconds and DNA translocation within 1 s of binding.

While the order of assembly of the components was unclear in prior bulk packaging assays, these single-molecule studies demonstrate an assembly pathway wherein a motor–prohead complex forms *first* and then engages the DNA. In these studies, γS-ATP appeared to stabilize a packaging-competent conformation of the prohead–motor complex, and ATP appeared to destabilize it, but following initiation the translocating complex was highly stable in the presence of ATP. As with many DNA processing enzymes, Mg^{2+} was found to be a necessary cofactor; 1 mM Mg^{2+}, beyond that complexed with ATP, was necessary for initiation of packaging (Fuller et al. 2007b).

24.3.1.2 Effect on Initiation of Packaging of the ϕ29 gp3 Terminal Protein

Following initiation of packaging as described above, it was unexpectedly found that the initial end-to-end extension of stretched DNA tethers was highly variable, ranging from 30 to 100% of the full-length of the DNA substrate (Fig. 24.2b) (Rickgauer et al. 2008). This suggests that ϕ29 DNA has a complex structure that does not bind the motor by a free end. The variability was shown to be due to the gp3 terminal protein since it disappeared when the ϕ29 DNA was pretreated with proteinase K. Both left-end (15 kb) and right-end (4 kb) restriction fragments with proteolysed gp3 were readily packaged with full tether lengths at initiation (Fig. 24.2b). Arbitrary nonnative DNA molecules generated by PCR could also be packaged with no tether length variability. While non-gp3 DNA is not packaged in vivo and packaged inefficiently in bulk in vitro reactions compared with gp3-DNA, non-gp3 DNA was efficiently packaged in the optical tweezers assay (Rickgauer et al. 2008). The observed variability in DNA extension immediately following initiation of DNA packaging in the optical tweezers is most likely due to gp3-mediated DNA looping. Electron microscopy shows that ϕ29 DNA can form loops at its termini and that purified gp16 motor protein can bind at DNA–gp3 loop junctions (Grimes and Anderson 1997). The observed extension variability in the optical tweezers measurements further suggests that packaging initiates with binding of prohead–gp16 complexes to these loop junctions (Fig. 24.2c). A recent study of T4 packaging suggests, based on colocalization of the packaged DNA ends detected by single-pair FRET, that the T4 motor might also be capable of packaging looped DNA (Ray et al. 2010a). On the other hand, it is unclear how such looped DNA can fit through the portal channel since X-ray crystal structures suggest that it could not easily thread more than one segment of dsDNA at a time (Simpson et al. 2000).

After initiation, no abrupt increase in DNA extension was observed, indicating that the putative loop does not open prior to DNA translocation. One possibility is that the section of looped DNA is cleaved prior to DNA translocation. Although the ϕ29 genome is replicated as a monomer, related viral motors including λ and T4 have endonuclease activities that are needed to excise their genomes from concatemeric substrates (Rao and Feiss 2008). Alternatively, it has been suggested that gp16 may have gyrase activity, based on the observation that it appears to be capable of supercoiling DNA–gp3 loops (Grimes and Anderson 1997). It is therefore possible that cleavage of the DNA loop may result from stretching the DNA during gyrase action. A second possibility is that a DNA loop is present throughout packaging and the motor is capable of translocating DNA in one side of a loop while packaging the DNA from the other side of the loop.

24.3.2 Initiation in Phage T4 by Procapsid–Motor Complex Assembly

T4 is an important model for large tailed phages that exhibits distinct differences from ϕ29. Notably, T4 has a much larger capsid size (120×86 nm) and must package a 9× longer genome (Rao and Feiss 2008; Fokine et al. 2004). T4 is also a prototype for viruses that package DNA by a "headful" mechanism in which unit length DNA segments that fill the procapsid must be excised by the packaging motor from a concatenated string of multiple T4 genomes produced by rolling circle replication. The T4 packaging motor consists of a terminase complex comprised of a small subunit, gp16, and large subunit, gp17 (containing the packaging ATPase), which appears from cryo-EM studies to form a pentameric ring (Sun et al. 2008) connected to a dodecameric portal ring (gp20). Notably, T4 is the only virus for which an atomic structure of the ATPase subunit responsible for powering DNA packaging (gp17) has been determined by X-ray crystallography (Sun et al. 2008).

The single-molecule T4 packaging assay employed a similar strategy as used with ϕ29. An efficient, defined in vitro T4 packaging system consisting of only three components – empty proheads (containing a portal connector ring), the large terminase subunit (gp17, the packaging ATPase), and DNA – was developed by V. Rao and co-workers (Kondabagil et al. 2006), also building on work by L. Black and co-workers (Baumann and Black 2003). It was found that a packaging-competent prohead–motor complex could be prepared by incubating the T4 gp17 with empty procapsids in the presence of nonhydrolyzable γS-ATP (Fuller et al. 2007c). These complexes were then bound to microspheres, and DNA molecules were attached to separate microspheres. As in the method of initiation developed by Rickgauer et al. for ϕ29, DNA was "fed" to the T4 packaging motor by rapidly bringing both types of microspheres into close proximity inside a flow chamber containing ATP (Fig. 24.2a). Although a small terminase subunit (gp16) is needed for packaging in vivo, inclusion of only the large subunit (gp17) was found to be sufficient for efficient in vitro packaging.

24.3.3 *Initiation in Phage λ by Motor–DNA Complex Assembly*

Phage λ has been one of the most important model systems in molecular biology for over half a century (Gottesman and Weisberg 2004). This *Escherichia coli* virus has a 62-nm diameter icosahedral capsid containing a 48.5-kbp genome. Packaging is carried out by λ terminase, a hetero-oligomer composed of the viral gene products gpA (large terminase subunit) and gpNu1 (small subunit) (Catalano 2005). Like T4, the viral DNA is copied by rolling circle replication, yielding concatenated genomes that must be excised. However, unlike T4, λ terminase binds to and cleaves the DNA at a specific site (*cos* site) where packaging initiates. A stable terminase complex can be assembled onto DNA by adding recombinant gpA and gpNu1 proteins to form a stable intermediate referred to as Complex I. The gpNu1 subunit mediates the assembly of terminase at this site while gpA possesses endonuclease, strand separation, and ATPase/DNA translocation activities. Biochemical studies indicate that the proteins assemble into a stable $gpA_1/gpNu1_2$ heterotrimer, and these trimers can assemble into a homogeneous tetrameric ring of sufficient size to encircle dsDNA (Maluf et al. 2006). Presumably, the terminase ring assembles at a *cos* site, then binds to the portal of a procapsid (to form "Complex II"), and then initiates DNA translocation. At least two accessory proteins aid packaging: *E. coli* IHF appears to aid the formation of Complex I, and λ gpFI aids the formation of Complex II (Sippy and Feiss 2004; Gaussier et al. 2006).

The strategy used by Fuller et al. to develop a single-molecule packaging assay was opposite of that used with the ϕ29 and T4 systems. Instead of assembling a procapsid–motor complex and feeding it DNA, the terminase complex was assembled on a DNA substrate containing a *cos* site (required for λ packaging) near one terminus and biotin-labeled nucleotides at the opposite terminus (Fuller et al. 2007a). This DNA was tethered to a streptavidin-coated microsphere and attached λ procapsids to a separate batch of microspheres. Packaging was initiated by injecting the two microspheres into a microfluidic chamber containing ATP, trapping them in separate optical traps, and bringing the two microspheres into proximity such that the terminase–DNA complex could bind the procapsid (Fig. 24.2d). As in the ϕ29 and T4 systems, DNA binding and initiation of translocation occurred rapidly within seconds. Although different pathways for initiation, via either a motor–DNA complex (λ) or motor–prohead complex (ϕ29 and T4), have been observed in the single molecule studies, it is unclear whether multiple pathways may be followed in vivo.

24.4 Early Stages of DNA Packaging

Upon successful assembly of the packaging motor complex, the viral genome is internalized into the procapsid. This is accomplished in one continuous process in which the portal motor translocates the viral DNA using the energy of ATP hydrolysis. It is useful for our purposes to distinguish between "early" and "late" stages of packaging. During the initial stages, DNA encounters relatively little resistance to encapsidation. However, as the capsid is filled, internal forces resisting DNA confinement build, and the motor operates under a significant load. In this section, we will focus on motor function in the early stage of packaging, where there is little load on the motor, and how single-molecule studies have revealed key aspects of its mechanism. The late stages of packaging will be discussed in Sect. 24.5.

24.4.1 Insights on Packaging from Structural and Biochemical Studies

Prior to the single-molecule measurements highlighted in this chapter, years of extensive structural and biochemical studies provided important insights into the mechanism of viral DNA packaging. An early structural model for the mechanism of the packaging motor was provided by the observation of a mismatch between the sixfold symmetry of the portal connector and fivefold symmetry of the capsid vertex in which it resides. This observation led Hendrix (1978) to propose that packaging might be driven by a "nut-and-bolt" mechanism in which rotation of the connector (the nut) causes linear motion of the helical DNA (the bolt). More recently, the first X-ray structures of the ϕ29 connector by Simpson et al. (2000) verified this symmetry mismatch, though the connector was found to have 12-fold symmetry, rather than sixfold. The portal structure is remarkably conserved in other phages, hinting at a common mechanism. Based on this new structural information, several rotary motor models were proposed, involving such mechanisms as cyclic compression and relaxation of the connector (Simpson et al. 2000), electrostatic interaction with lysine rings inside the portal channel (Guasch et al. 2002), and sequential movement of loops (molecular levers) in the connector tunnel in kind of a "Mexican wave" (Lebedev et al. 2007) (reviewed in Rao and Feiss 2008). All of these proposed mechanisms necessitate connector rotation to align the proper structural motifs of the motor with the helical pitch of the DNA during translocation and also require coordinating this rotation with ATP hydrolysis in the ATPases.

Despite the attractiveness of these models, recent experiments have challenged these "connector-centric" mechanisms of packaging. Portal rotation was tested directly in T4 by cross-linking the connector to the capsid using the Hoc protein (Baumann et al. 2006). If rotation was essential for packaging, tethering of the connector would have abolished packaging. However, packaging efficiency was comparable to that in wild-type phages. Single-molecule fluorescence experiments in ϕ29 led to the same conclusion. Here, portals were labeled with a single fluorescent Cy3 dye whose polarized emission was used as a reporter for the orientation of the protein. In measurements of labeled complexes that were actively packaging, assessed by observing the translocation of tethered microspheres, no evidence was found for portal rotation (Hugel et al. 2007). Lastly, in ϕ29, mutations to the connector loops predicted in one model to interact with DNA did not lead to any observable decrease in packaging efficiency (R. Atz, S. Grimes, D.L. Anderson, personal communication). These studies together strongly suggest that the portal does not play an active role in DNA translocation. Interestingly, the ϕ29 loop mutation studies indicate that mutants are more prone to release packaged DNA than wild-type phages, implying that the connector may be important in *retention* of packaged DNA, rather than in its translocation.

The alternative to a portal-driven packaging mechanism is that the ATPase motor component directly translocates DNA (Black 1989; Fujisawa and Morita 1997). Compelling evidence supporting this mechanism has come from the recently solved structures of the T4 large terminase, gp17 (Sun et al. 2007). These studies, along with sequence homology analyses (Draper and Rao 2007), indicate a similar architecture in gp17 to monomeric SF1 and SF2 DNA (and RNA) helicases. Comparisons of X-ray crystal and cryo-electron microscopy structures of T4 gp17 monomers revealed two globular subdomains separated by a "hinge." These can adopt "relaxed" and "tensed" conformations separated by approximately 2 bp that correspond to the apo and nucleotide-bound states, respectively (Sun et al. 2008). These ATP-driven conformational rearrangements in the subdomains support an inchworm-type mechanism for DNA translocation in which a C-terminal DNA-binding domain ratchets the DNA into the prohead. Further supporting this "terminase-centric" model of packaging, the observed conformational switch in gp17 is consistent with estimations of the motor step size by biochemical assays. Measurements of ATP consumption in bulk, in vitro packaging with phage ϕ29 (Guo et al. 1987; Chemla et al. 2005) and T3 (Morita et al. 1993) have been used to determine the number of ATPs hydrolyzed per length of DNA packaged, a parameter called the "coupling ratio." These measurements revealed that on average, ~2 bp of DNA is packaged per 1 ATP hydrolyzed. These have been interpreted as indicating the motor steps in increments of 2 bp, consistent with the structural model of T4 gp17.

Despite the many insights provided by these structural and biochemical studies, many questions on viral DNA packaging remained open. Below, we discuss recent single-molecule optical trap experiments and how they have shaped our understanding of key aspects of this process: the kinetics of packaging, the mechanism of force generation, how ATP hydrolysis is coupled to translocation, how the ATPases are coordinated, the precise step size, and the interaction of the motor to DNA.

24.4.2 Single-Molecule Kinetics of Packaging

24.4.2.1 Motor Velocities with Saturating ATP at Low Force and Low Capsid Filling

The single-molecule optical trap assay allowed, for the first time, extensive and accurate characterizations of packaging kinetics in phages ϕ29, T4, and λ (Smith et al. 2001; Rickgauer et al. 2008; Fuller et al. 2007a, c; Chemla et al. 2005). In these three systems, the observed kinetics proved to be far more complex than anticipated. Packaging is interrupted by pauses and slips, where the motor transiently disengages and reengages its DNA substrate, occurs at a rate that depends on how much DNA is encapsidated and on the load force applied to the motor, and can vary for individual motor complexes.

Measurements of motor velocities in the early stages of packaging (near-zero capsid filling, where there is little resistance to packaging) and at a saturating ATP concentration were carried out in all three systems. Force feedback was used to apply a small constant force, typically ~5 pN, to keep the DNA stretched and allow continuous tracking of packaging in real time (Smith et al. 2001). Under these conditions, and in standard packaging buffer (25 mM Tris-HCl (pH 7.8), 50 mM NaCl, 5 mM $MgCl_2$), the ϕ29 motor translocates at an average rate of ~145 bp/s (Rickgauer et al. 2008). Solution conditions can affect motor velocity (Fuller et al. 2007b); average speeds were found to vary between 170 bp/s (at 5 pN load) with 100 mM NaCl and 114 bp/s with 0 NaCl. As discussed further below, velocities are also dependent on the load force applied by the optical trap. Using a ramped DNA stretching method, Rickgauer et al. were able to extract packaging rates at near-zero applied load, revealing a slightly higher average motor velocity of 165 bp/s (Rickgauer et al. 2008). Studies employing a temperature-controlled sample chamber also showed that increasing the temperature from 20 to 35°C (a more physiologically relevant temperature) increased the motor velocity two- to threefold (unpublished data, M. White and D. Smith). Phage λ has a 2.5× longer genome than ϕ29, and its average motor velocity was measured to be ~4.2× higher under similar conditions, 590 bp/s at a 5 pN load. T4, one of the largest phages, has an enormous 171 kbp genome, and its average motor velocity was measured to be ~5.4× higher than ϕ29, 770 bp/s at a 5 pN load (Fuller et al. 2007c). This observed trend of faster packaging rates for longer genome lengths is consistent with the need for each virus to package its complete genome in a limited time window of ~2–5 min to complete the infection cycle within 20–30 min. Kinetic parameters for ϕ29, T4, and λ from optical trap measurements are summarized in Table 24.1.

Another intriguing finding revealed by the single-molecule studies is that the velocity of individual motors can vary significantly in time, and the average velocity of different individual motors can also vary significantly. Such variability was most striking in T4 (Fig. 24.3a) (Fuller et al. 2007c). The average velocities of different individual motors ranged from as low as 70 bp/s to as high as 1,840 bp/s with a standard deviation of 320 bp/s, or 40% of the mean. By comparison, the average velocities of ϕ29 and λ motors were also variable, though by smaller amounts (~10 and 20% of the mean velocities, respectively) (Rickgauer et al. 2008; Fuller et al. 2007a) (Fig. 24.3b).

Table 24.1 Summary of kinetic parameters from phage ϕ29, λ, and T4

Phage	Genome length (kbp)	Maximal velocity[a] (bp/s)	Std/mean velocity (%)	Hydrolysis rate (ATP/s^{-1})	Processivity[b] (kbp)
ϕ29	19.3	145	10	>6	~20
λ	48.5	590	20	>25	>48.5
T4	171	770	40	>40	13

[a]Measured at room temperature, in standard packaging buffer containing saturating ATP, at low capsid filling, and under an externally applied load of 5 pN

[b]Defined as the distance between slipping events (scored as >50 bp backward movement of the DNA out of the capsid at a load of 5 pN)

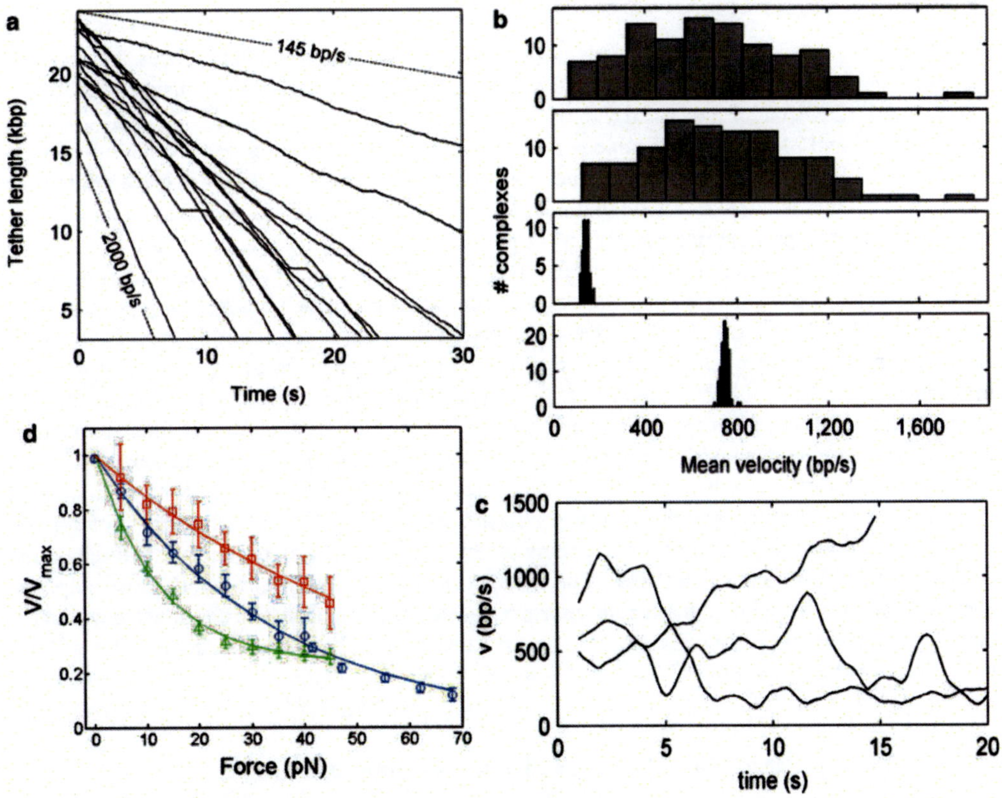

Fig. 24.3 Phage T4 DNA packaging dynamics. **(a)** Repeated measurements of DNA tether length versus time with 5 pN force feedback, showing variable packaging rates for individual complexes ranging from ~145 to 2,000 bp/s and showing occasional pauses in translocation (plateaus). **(b)** Histogram of average packaging rates measured for individual T4 motors (*top panel*); same histogram but with rates calculated not including pauses (*second panel*); histogram of average ϕ29 packaging rates for comparison (*third panel*); histogram showing stochastic variation in T4 packaging rates predicted by a simple Poisson stepper model if individual complexes are assumed to have uniform kinetics (*fourth panel*). **(c)** Examples of three packaging events where large variations in instantaneous motor velocity versus time were observed. **(d)** Dependence of motor velocity on applied load force for T4 (*red squares*), λ (*green triangles*), and ϕ29 (*blue circles*). Velocities are normalized to unity at zero load

Some individual T4 motors exhibited velocity changes in time ranging as wide as 500–1,500 bp/s (Fig. 24.3c). The variations in T4 motor velocities are too large to be reconciled in standard kinetic models as being due to inherent stochastic thermal fluctuations alone, suggesting that individual motors can adopt different active conformational states gearing different DNA translocation velocities and can switch between these states in time (Fuller et al. 2007c). Static and dynamic variability (also termed "disorder") in enzyme kinetics has been reported in single-molecule studies of several other simpler enzyme complexes including lactate dehydrogenase, cholesterol oxidase, λ exonuclease, and RecBCD helicase (Perkins et al. 2004; Xue and Yeung 1995; Lu et al. 1998; van Oijen et al. 2003). An understanding of the structural origin of these heterogeneities is currently lacking. They may be due, in part, to changes in conformation or chemical structure of individual motor subunits, variation in the number of active subunits in individual complexes, or conformational changes in the arrangement of subunits in the whole motor complex.

24.4.2.2 Pauses and Slips

Optical tweezers measurements revealed that packaging motors are remarkably processive, meaning that many base pairs of DNA are continuously packaged with only occasional pausing or backward slipping of the DNA out of the capsid (Smith et al. 2001; Rickgauer et al. 2008; Fuller et al. 2007a, c) (Figs. 24.1b, c and 24.3a). While the appearance of pauses and slips is intriguing and not fully understood, reversible pauses and slips only reduce the overall average motor velocity at low capsid filling by ~10% (Fuller et al. 2007c) (although the λ motors were occasionally observed to pause abruptly and never resumed packaging even after >1–2 min, after which data collection was stopped). The ϕ29 and λ motors exhibited less than one significant slip (>50 bp backward movement of the DNA out of the capsid) per genome length packaged. The T4 motor, which has the longest genome to package, exhibited the most slipping (one slip every 13 kbp) but is still a highly processive motor. The majority of slips were relatively small, less than a few hundred bp, meaning that once packaging initiates the probability that DNA slips out of the capsid completely appears to be quite small.

The mechanisms of pausing and slipping are not known, but the finding that their frequencies increase with applied load and that neither type of event correlates with particular positions along the DNA suggests that they may be off-pathway events that stochastically occur during normal DNA translocation. The dependence on load may indicate that force is able to disrupt the motor–DNA interactions made during translocation, leading to temporarily arrests in packaging (pauses) or disengagements with the molecule (slips). The capacity of these motors to reversibly pause, slip, and change velocity may be biologically relevant since packaging in vivo must be coordinated with other biochemical processes potentially ongoing on the same DNA substrate, including DNA transcription, recombination, and repair (Mosig and Eiserling 2006).

24.4.2.3 Dependence of Motor Velocity and Power on Applied Load

In all three systems studied, ϕ29, λ, and T4, the motor velocity with saturating ATP was found to decrease with applied force (Smith et al. 2001; Rickgauer et al. 2008; Fuller et al. 2007a, c). A general implication of this finding is that the rate-limiting step in packaging must involve DNA translocation (Wang et al. 1998). Displacement against an opposing force implies that mechanical work must be done by the motor, leading to an increased free-energy barrier for motor stepping and slowed reaction rate. In the simplest single-reaction energy barrier model, the extra work required to translocate DNA against the optical trap force is $F\Delta x_1$, where Δx_1 is the distance to the "transition state" ($\Delta x_1 \leq d$, the step size), defined along a reaction coordinate corresponding to the amount of DNA packaged. The translocation rate thus decreases as $V = V_{max} \exp(-F\Delta x_1 / k_B T)$, where k_B is the Boltzmann constant and T is the absolute temperature. As shown in Fig. 24.3d, the dependence of velocity on force for each system displayed quantitative differences. The T4 motor velocity dropped by ~40% as the force was increased from 5 to 40 pN, whereas the ϕ29 and λ motor velocities dropped ~60%. Neither the ϕ29 nor λ velocity versus force datasets could be well fit by a single force-dependent rate suggested by the simple single-reaction energy barrier model, indicating that different kinetic transitions become rate limiting at high force. A notable feature of the λ motor was that the motor velocity appeared to plateau at ~200 bp/s above ~50 pN, rather than approaching zero, suggesting that a purely chemical transition (i.e., one not involving DNA translocation) becomes rate-limiting at high force. In T4, the velocity decreased approximately linearly with force, over the range studied, possibly indicating a very small transition state distance for that system. Despite commonalities in behavior, these differences in force dependence may point to different mechanisms of translocation. Future work will be necessary to resolve these differences further.

By multiplying the average applied force by the corresponding average velocity, the mechanical power generated by the motor can be calculated. The maximum power observed to be generated by the λ motor occurred with a load of 45 pN, where the motor velocity was 208 bp/s, implying an average power of 9,400 pN·bp/s = 3,200 pN nm/s. This is ~4× higher than the maximum power detected for the ϕ29 motor. Assuming a free-energy release of 130 pN·nm per ATP (note that we express this free energy in units of pN·nm (force × distance) relevant in the single-molecule trap studies and ~130 pN nm = 73 kJ/mol), this implies an ATP hydrolysis rate of *at least* 3,200 pN nm/s ÷ 130 pN nm per ATP = 25 ATP/s for the λ motor. This figure is higher than the figure of ~10 ATP/s previously estimated in bulk biochemical assays (Yang and Catalano 2003), suggesting that those assays underreport the rate due to difficulties in accounting for "futile" ATP hydrolysis not associated with packaging. The maximum T4 mechanical power was observed with 40 pN load, where the velocity was 380 bp/s, yielding a power of 15,200 pN bp/s = 5,200 pN nm/s, about 7× higher than that detected for ϕ29 and implying an ATP hydrolysis rate of *at least* 5,200 pN nm/s ÷ 130 pN nm per ATP = 40 ATP/s. While these power figures may seem small, it must be kept in mind that the motor is a nanoscale device occupying a volume of only ~ (10 nm)^3, implying a *power density* on the order of 5,000 kW/m^3, which is roughly twice that generated by a high-performance automobile engine.

24.4.2.4 Motor Force Generation

One of the most striking features revealed by the single-molecule measurements was that viral DNA packaging motors generate very high forces, among the highest known for biomolecular motors (Ross et al. 2008; Michaelis et al. 2009; Oster and Wang 2003). Optical tweezers measurements allow one to measure directly the packaging force exerted by the motor on the DNA since that force is transmitted directly to the trapped microspheres. Measurements at low capsid filling, where internal forces resisting DNA confinement are small (discussed in Sect. 24.5), revealed that the ϕ29, λ, and T4 motors are all able to translocate DNA against externally applied load forces of >50–60 pN (Smith et al. 2001; Fuller et al. 2007a, c). These figures are lower bounds because most measurements ended with the DNA tether detaching from the microspheres, likely due to rupture of the prohead–antibody–microsphere linkage. While some motors did stall at forces <50 pN, others were still translocating when rupture occurred at >50 pN. Extrapolation of measurements made at high capsid filling, where forces resisting DNA confinement (discussed in Sect. 24.5) contribute a large additional load on the motor suggest that the ϕ29 motor can exert total forces as high as 110 pN (Rickgauer et al. 2008), strikingly large compared with many cellular molecular motors. For example, skeletal muscle myosin II, which powers skeletal muscle contraction, only generates 2–3 pN of force (Finer et al. 1994). It is likely that large forces are necessary to package DNA against the enormous internal forces generated from compacting the viral genome into the capsid. This issue will be discussed in depth in Sect. 24.5 on completion of packaging.

Current understanding dictates that molecular motors translocate in discrete-sized steps tightly coupled to the hydrolysis of ATP molecules. Energetic considerations then impose a tradeoff between a motor's mechanical step size and force generation. Each ATP hydrolysis releases a free energy on the order of 130 pN nm, depending on solution conditions (Lehninger et al. 1993). A motor, depending on its efficiency, can then convert up to ~130 pN nm of chemical energy into mechanical work to translocate one step. If the motor translocates a step size *d* against force *F*, the total mechanical work performed is *Fd*, which must be less than ~130 pN nm. Thus, a force of ~60 pN generated by the packaging motor places an upper bound of ~2 nm = 6 bp on its step size. For ϕ29, for which a maximum total force of 110 pN was reported, the step size must be smaller than ~1.2 nm, or 3.5 bp. (Precise direct measurements of motor step size are discussed below.) It is tempting to speculate that different types of motors may have evolved to generate more or less force depending on the physical bounds on their step size. Thus, cytoskeletal motors like myosin must take large steps, as dictated by the periodicity of the actin tracks on which they translocate, and thus generate small forces. On the other hand, viral DNA packaging motors, which are built to exert large forces to counteract the internal forces generated by compaction of the DNA in the capsid, must take relatively small steps.

24.4.3 Mechanochemistry of Packaging

The viral packaging machine requires ATP as a cofactor to drive processive encapsidation of DNA. A fundamental question is how the energy contained in a molecule of ATP is utilized by the motor to generate the large forces exerted during DNA translocation. The conversion of chemical energy into mechanical action – termed "mechanochemistry" – is a process carried out by all molecular motors and provides an important clue to their mechanism. In all nucleic acid translocases like DNA packaging motors, a mechanism couples ATP hydrolysis to DNA movement. ATP hydrolysis can be viewed as a multistep process involving docking of ATP into the catalytic cleft, accommodation of the nucleotide into the proper orientation, nucleophilic attack of the γ-phosphate, and release of the hydrolysis products phosphate and ADP, to name a few examples. These steps must somehow connect to the translocation cycle, during which the motor must engage with the DNA, translocate it by the motor step size, disengage the DNA, and reset the machinery for the subsequent cycle. Thus, one key goal in understanding this process is identifying which step or steps in the ATP hydrolysis cycle lead to DNA translocation. A second goal is to understand the structural specifics governing these steps.

Recent structural studies of T4 gp17 terminase have provided important clues on the mechanochemical coupling process in packaging motors. The observation of a "C motif," in which a network of hydrogen bonds connects the γ-phosphate of ATP to DNA (Draper and Rao 2007), provides the first structural basis for the communication between chemical and the mechanical activities of the protein. While instructive, molecular structures only provide static pictures of the mechanochemical conversion process. Single-molecule measurements provide complementary information in the form of real-time measurements of the DNA translocation dynamics. In particular, optical traps provide an ideal platform for investigating the mechanochemistry of molecular motors because force can be utilized as a probe for mechanical motion. In the single-molecule packaging assays described above, the phage translocates DNA against the force exerted by the optical trap; increasing this tension thus decreases the rate at which the translocation step occurs while leaving the other (force-independent) steps unperturbed. This feature of optical trap measurements is unique but loosely analogous to classical enzymology, where how an enzyme binds a substrate can be measured by varying the substrate concentration or by adding inhibitors that compete with the substrate. Here, rather than vary concentrations of a chemical species, force is used to modulate the translocation rate in the mechanochemical cycle of the protein.

24.4.3.1 Single-Molecule Measurements of Mechanochemistry in ϕ29

Chemla et al. (2005) investigated mechanochemical coupling in the DNA packaging motor of phage ϕ29 exploiting this unique ability of optical traps. Combined with methods of classical enzymology

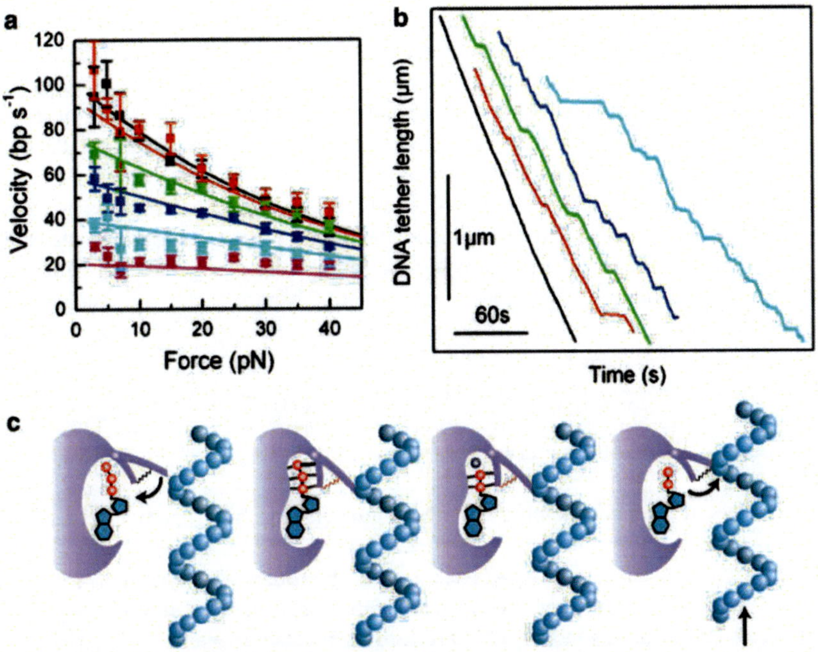

Fig. 24.4 Mechanochemistry of ϕ29. **(a)** Force–velocity behavior. As ATP was decreased from 500 µM (*black data*) to 10 µM (*magenta*), the packaging velocity displayed a decreasing dependence on force, indicating that ATP binding is not a force-generating kinetic event. **(b)** Analog-induced stalls. As the nonhydrolyzable ATP analog AMP-PNP was added to the packaging buffer (containing 100 µM ATP), packaging was interrupted by stalls. The frequency of the stalls increased as AMP-PNP was increased from 0 (*black data*) to 5 µM (*cyan*). **(c)** Model of mechanochemical coupling in ϕ29. ATP binding in the catalytic cleft of gp16 induces elastic strain that is stored as a compressed spring. This binding step occurs while the ATPase is disengaged from the DNA, such that no translocation occurs and binding is insensitive to tension. Upon hydrolysis and release of the cleaved phosphate, this elastic strain is relieved in a "recoil" step. Here, the gp16 is engaged to the DNA, and recoil drives the force-generating translocation

to understand the chemical aspects of packaging and force dependence to decipher the mechanical aspects, they were able to provide key insights on the coupling between the two. To achieve this understanding, Chemla et al. performed an extensive set of experiments measuring the generation of force by the ϕ29 motor as a function of ATP, non- (or slowly) hydrolyzable ATP analogs (AMP-PNP and γS-ATP), and product (ADP and P_i) concentrations. In the first set of measurements, ATP and force were varied, and their effect on the packaging rate was investigated. The packaging rate V was found to depend on ATP concentration according to the classical Michaelis–Menten equation: $V = V_{max}[\mathrm{ATP}]/([\mathrm{ATP}]+K_M)$. At low forces (~5 pN), a Michaelis–Menten constant K_M of ~30 µM was observed. More illuminating were measurements of the force dependence of the packaging rate at different ATP concentrations, as shown in Fig. 24.4a. At saturating ATP levels ($[\mathrm{ATP}] >> K_M$), the velocity depended strongly on force, indicating that the DNA translocation step was rate-limiting. In contrast, at low ATP levels ($[\mathrm{ATP}] < K_M$), where ATP binding was now rate-limiting, the packaging rate was largely independent of force. This observation revealed that the force-generating DNA translocation step must not occur during ATP binding.

Poisoning the ATP packaging buffer with nonhydrolyzable analogs was found to induce stalls in packaging lasting several seconds (Fig. 24.4b) and at a frequency (measured in number of induced stalls per length packaged) proportional to the analog concentration. Neither the frequency nor the duration of these stalls – kinetic parameters corresponding to the binding and unbinding of the analog to the motor, respectively – displayed any dependence on force. These results again indicated that the nucleotide-binding step in the mechanochemical cycle of the motor does not generate force. Measurements in the presence of hydrolysis products (ADP and P_i) illuminated the role of product release in the motor cycle. Though no stalls were detected, the packaging rate did decrease significantly with increasing ADP concentration, in a manner consistent with it acting as a competitive inhibitor to ATP (presumably, ADP binding and release were too rapid to observe stalls of significant duration). Moreover, similarly to the force dependence of the packaging velocity at varying ATP concentration described above, the pattern of ADP inhibition at various forces was inconsistent with ADP release being a force-generating step. Finally, increasing the inorganic phosphate level a 1,000-fold had no discernable effect on the packaging rate, indicating that the release of phosphate was highly irreversible. Taken together, these results led Chemla et al. to conclude that DNA translocation must occur at some step between hydrolysis (i.e., after ATP binding) and phosphate release (i.e., before ADP release and resetting of the cycle).

24.4.3.2 Model of ATP Coupling in ϕ29

Theoretical models of mechanochemistry in other molecular machines, particularly in other ringed ATPases such as the members of the AAA+ superfamily, provide a useful context for interpreting these optical trap measurements. In F_1-ATPase, part of the rotary ATP synthase motor, it has been proposed that the energy driving the mechanical rotation is ultimately derived from ATP binding (Oster and Wang 2000). In this model, ATP binds to each catalytic site in a zippering of hydrogen bonds that induces elastic strain and produces the first of two force-generating "power strokes" driving the motor. A second power stroke occurs when that elastic strain is relieved in a "recoil" step triggered by phosphate release and immediately preceding ADP release. The hydrolysis step itself is not believed to drive any mechanical motion because it involves only small rotation of the terminal phosphate, which is almost isoenergetic in this protein (Oster and Wang 2000).

The measurements of Chemla et al. on ϕ29 suggest a model for the packaging motor partly consistent with this mechanism. One prediction of the F_1-ATPase mechanochemical model is that both ATP binding and recoil steps should involve large changes in free energy in order to drive mechanical motion. This is consistent with two observations in ϕ29: (1) long stalls with ATP analogs, indicating that nucleotides are tightly bound to the motor and release at very slow rates and (2) the evidence for irreversible phosphate release. On the other hand, the F_1-ATPase model proposes that ATP binding should drive mechanical motion, yet this step was shown to be force-independent in the trap experiments. This suggests that in ϕ29, ATP binding occurs at a gp16 subunit that is disengaged with the DNA and thus insensitive to the forces applied by the optical trap. This is supported by the observation that the motor is more likely to slip, or disengage from DNA, at low ATP concentrations (Chemla et al. 2005) favoring the apo state. This suggests that elastic strain induced by ATP binding may cock the spring that drives translocation. After ATP binds, the gp16 engages the DNA and relieves this elastic strain, releasing the spring and translocating the DNA right before ADP is released. The data is thus consistent with a recoil step generating force, with phosphate release acting as an irreversible trigger committing the motor to the rest of the reaction cycle. This mechanism proposed by Chemla et al. is depicted schematically in Fig. 24.4c.

24.4.3.3 Coupling Mechanism and Motor Structure/Function Relationships in T4 and λ

Studies of λ and T4 phage are so far consistent with this mechanism of mechanochemical coupling. In T4, structures of gp17 in its apo and nucleotide-bound states by Sun et al. (2008) qualitatively fit with the model proposed by Chemla et al. In the apo state, gp17 is in a relaxed conformation and putatively disengaged from DNA, whereas the ATP-bound state is tensed and putatively engaged

Table 24.2 Summary of point mutations in phage λ gpA

Residue change	Putative location	Maximal velocity[a]	Force dependence[a]	Processivity[a]
Y46F	Q motif	Reduced 40%	Steep	Reduced >10×
K84A	Walker A	Reduced 40%	Wild type	Wild type
T194M	Loop–helix–loop[b]	Reduced 8×	Wild type	Wild type
T194V	Loop–helix–loop[b]	Wild type	Wild type	Wild type
G212S	C motif	Reduced 3×	Not yet measured	Reduced 6×

[a]Measured relative to wild-type gpA

[b]Modeling suggests this loop–helix–loop may position the Walker B and C motif in the ATPase center

with the DNA. This is consistent with the ATPase being "cocked" by nucleotide binding in the Chemla model. Upon hydrolysis, the C motif of gp17 is proposed to unlock the terminal phosphate of ATP and the catalytic cleft recoils to its open state, driving translocation of the DNA, consistent with the proposed ϕ29 mechanism.

In the phage λ system, optical tweezers studies of the effects of amino acid point mutations in the motor's large terminase subunit (gpA) on packaging dynamics recently shed light on structure–function relationships (summarized in Table 24.2) (Tsay et al. 2009, 2010). Photo-cross-linking with 8-azido-ATP showed that residues Y46 and K84 of gpA interact with ATP (Hang et al. 2000). On the basis of sequence homology with AAA+superfamily ATPases, gpA was predicted to have a "Walker A-like" phosphate-binding motif at 76-KSARVGYS-83 (Mitchell and Rao 2004). In support of this hypothesis, optical tweezers measurements indicated that change K76R abolished packaging and K84A, immediately adjacent to the motif, decreased motor velocity by ~40% but did not alter processivity or the steepness of the velocity–force dependence (Fig. 24.5a) (Tsay et al. 2009). This finding is consistent with the notion that the Walker A motif is involved in ATP binding but not coupling and supports a model in which ATP binding and hydrolysis are not the force-generating steps in λ as in the ϕ29 system discussed above.

Based on sequence comparisons with T4 gp17 and RNA helicases, gpA was predicted to have an adenine-binding "Q motif" at 46-YQ-47 involved in mechanochemical coupling (Draper and Rao 2007; Mitchell et al. 2002). This motif, located 17 residues upstream of the Walker A motif, was recently discovered in SF2 RNA helicases and observed to contain highly conserved aromatic residues proposed to aid in hydrophobic stacking interactions with adenine. Experiments with helicase mutants indicate that the Q motif plays a role in regulating nucleic acid affinity and conformational changes driven by nucleotide and ATP hydrolysis (Cordin et al. 2004). Consistent with those findings, optical tweezers studies of λ terminase showed that mutation Y46F in its putative Q motif decreased motor velocity 40% and increased the frequency of motor slipping during DNA translocation by >tenfold (Fig. 24.5a) (Tsay et al. 2009). This alteration in function is in sharp contrast to that observed with, which also exhibited a 40% reduction in velocity but no significant change in motor slipping. These findings support the hypothesis that viral DNA packaging motors contain an adenine-binding Q motif, that regulates substrate affinity, analogous to that found in RNA helicases. Such a feature qualitatively fits with the proposed translocation models for ϕ29 and T4. In addition, optical tweezers measurements revealed that change Y46F caused motor velocity to decrease more steeply with increasing load force relative to wild type, suggesting that the Q motif can also regulate motor power.

gpA is also hypothesized to have a "C motif" involved in coupling at 212-GST-214 (Draper and Rao 2007). In helicases, C motif residues hydrogen bond with the γ-PO_4 of ATP and the DNA. Mutants generally retain ATPase and DNA-binding activities but fail to translocate due to loss of coupling between the two activities (Banroques et al. 2010). In T4, certain mutants with changes in the putative C motif were shown to be able to hydrolyze one ATP but not turn over (Draper and Rao 2007). In λ, optical tweezers measurements found that change G212S in this motif caused a

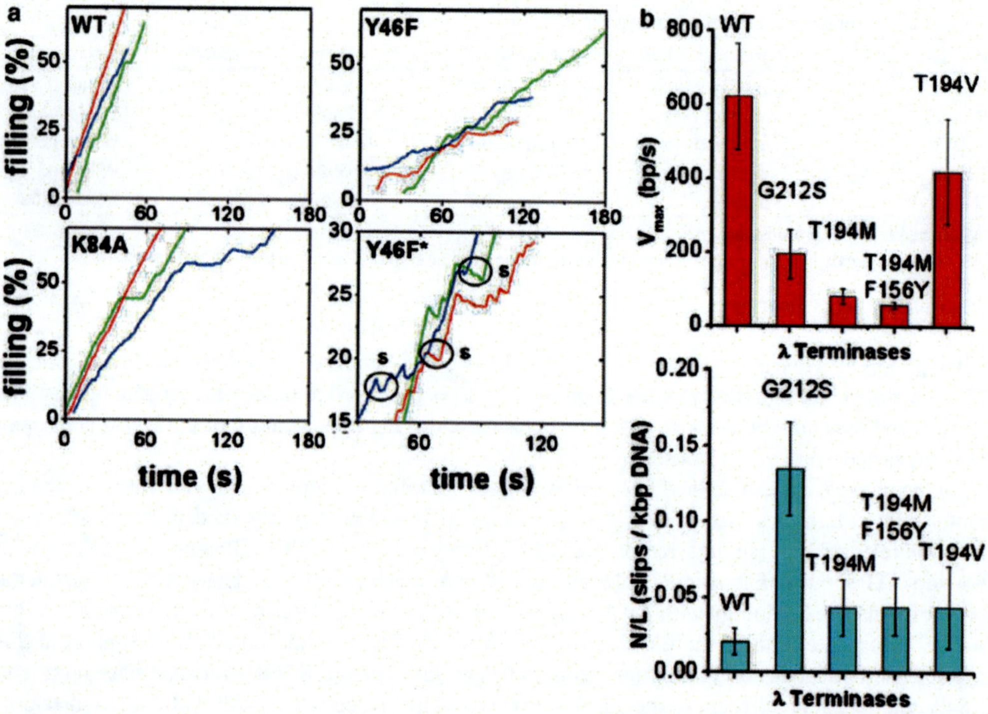

Fig. 24.5 Effects on phage λ packaging dynamics of mutations altering the gpA large terminase subunit of the λ packaging motor. (**a**) Representative recordings of length of DNA packaged (expressed as % of wild-type genome length) versus time for wild type (WT), Y46F mutants (residue change in the putative Q motif), and K84A mutants (residue change adjacent the putative Walker A motif). The panel labeled "Y46F*" is a zoomed view of the Y46F data illustrating frequent motor slipping events (labeled "s") observed with this mutant. (**b**) *Bar charts* comparing average motor velocity (*top chart*) and slipping frequency (*bottom chart*) for WT, G212S mutants (change in the putative C motif), T194M mutants (change in a structurally conserved loop–helix–loop), T194M/F156Y double mutants, and T194V pseudorevertants

threefold decrease in velocity and a sixfold reduction in processivity, again consistent with a coupling defect (Fig. 24.5b).

Genetic screening experiments also revealed several packaging defective mutants in a region of λ gpA outside any known functional motifs (Duffy and Feiss 2002). Change G191S resulted in no detectable packaging activity while change T194M caused an eightfold reduction in motor velocity without substantially changing motor processivity or force dependence (Fig. 24.5b) (Tsay et al. 2010). Structural modeling of gpA based on homology with T4 gp17 indicates that T194 is part of a loop–helix–loop region that connects the β4 (Walker B) and β5 (C motif) strands of the nucleotide-binding domain. This region has high structural similarity with analogous regions in T4 gp17, chromosome transport motor FtsK, and MjDEAD RNA helicase. Underscoring the importance of this region, change D584A in the proximal (predicted) loop segment of the prokaryotic SpoIIIE chromosome segregation motor (a close homolog of FtsK) was reported to reduce DNA translocation rate significantly (approximately threefold) (Burton et al. 2007). Together, these findings suggest the presence of a conserved structural region between the Walker B motif and C motif that may be part of a mechanism that governs motor velocity and processivity in several different types of nucleic acid translocases. Variations in this region may explain how the ϕ29, λ, and T4 motors evolved different packaging motor velocities that scale with viral genome size.

24.4.4 Subunit Coordination and Step Size

Though the optical trap measurements of Chemla et al. provided a framework for understanding the coupling of the chemical and mechanical reactions in each gp16 subunit of the ϕ29 motor, they did not fully address how the subunits coordinate their actions to package DNA. On one hand, the observation of Michaelis–Menten kinetics in this work suggests that there is no cooperativity in ATP binding in the pentameric ring. At the same time, the measurements of stalls with nonhydrolyzable ATP analogs indicate that the whole machinery is locked when one catalytic site is unable to complete its reaction and the subunits cannot act independently. According to the model proposed by Sun et al. based on structural data on the T4 large terminase, coordination between ATPases may be imposed by the DNA itself. Geometrical constraints based on the helical structure of B-DNA (10.5 bp per turn), the pentameric ring conformation of the packaging ATPases, and the putative 2-bp step size would dictate that each subunit must pass off DNA to its neighbor in ordinal sequence in order to remain in register with the molecule (Sun et al. 2008; Chemla et al. 2005). The alternative would be that coordination results from an underlying inherent communication between the subunits, independent of DNA geometry. Recent optical trap measurements have begun to address these issues in the ϕ29 system.

24.4.4.1 High-Resolution Optical Trap Measurements of ϕ29 Packaging

The development of *high-resolution* optical tweezers recently enabled the direct observation of viral DNA packaging step by step. Whereas the optical traps used in the packaging experiments described above were not sufficiently sensitive to monitor the stepwise encapsidation of DNA, these new instruments are capable of detecting motion at the scale of a single base pair (Abbondanzieri et al. 2005; Moffitt et al. 2006; Carter et al. 2009) (reviewed in Moffitt et al. 2008; Chemla 2010). With this technical breakthrough, Bustamante and co-workers (Moffitt et al. 2009) were able to detect directly, for the first time, the fundamental stepping motion of the ϕ29 packaging motor, revealing not only its step size but also its intersubunit coordination.

In a first set of experiments, packaging was monitored with the high-resolution optical traps at low force (~8 pN) and at varying ATP concentrations. Contrary to expectations, Moffitt et al. observed that packaging occurred in large 10-bp "bursts" in which the DNA was translocated rapidly, separated by "dwells," regions in which the motor remained at one position on the DNA (Fig. 24.6a). Further insight into this surprising result came from an analysis of the kinetics of the two classes of events. As ATP concentration was increased, the same 10-bp burst-dwell structure was observed. The duration of the dwells became shorter, while that of the bursts remained constant, indicating that ATP binding occurred solely during the dwells. Moreover, the dwell durations varied with ATP in a manner consistent with the Michaelis–Menten kinetics observed in the previous optical trap work. By measuring the durations of hundreds of dwells at each ATP concentration, accurate distributions of dwell times under each condition were compiled. If a single rate-limiting kinetic step governed the dwells, then an exponential distribution would have been expected. However, Moffitt et al. instead observed peaked distributions (Fig. 24.6b), better described by a *convolution* of multiple exponentials, consistent with multiple rate-limiting kinetic events during each dwell period. This key observation, in tandem with the ATP dependence of the dwell durations, led to the conclusion that several ATP molecules must bind to the motor during each dwell prior to translocation in 10-bp bursts.

These studies indicate a wholly unexpected mechanism of coordination in the pentameric gp16 ring. Rather than each subunit binding ATP, hydrolyzing it, and translocating DNA in succession, the high-resolution kinetics showed that the motor must wait to load multiple ATPs before translocation can occur. A question in this model is how the 10-bp bursts are generated by the motor. In the first

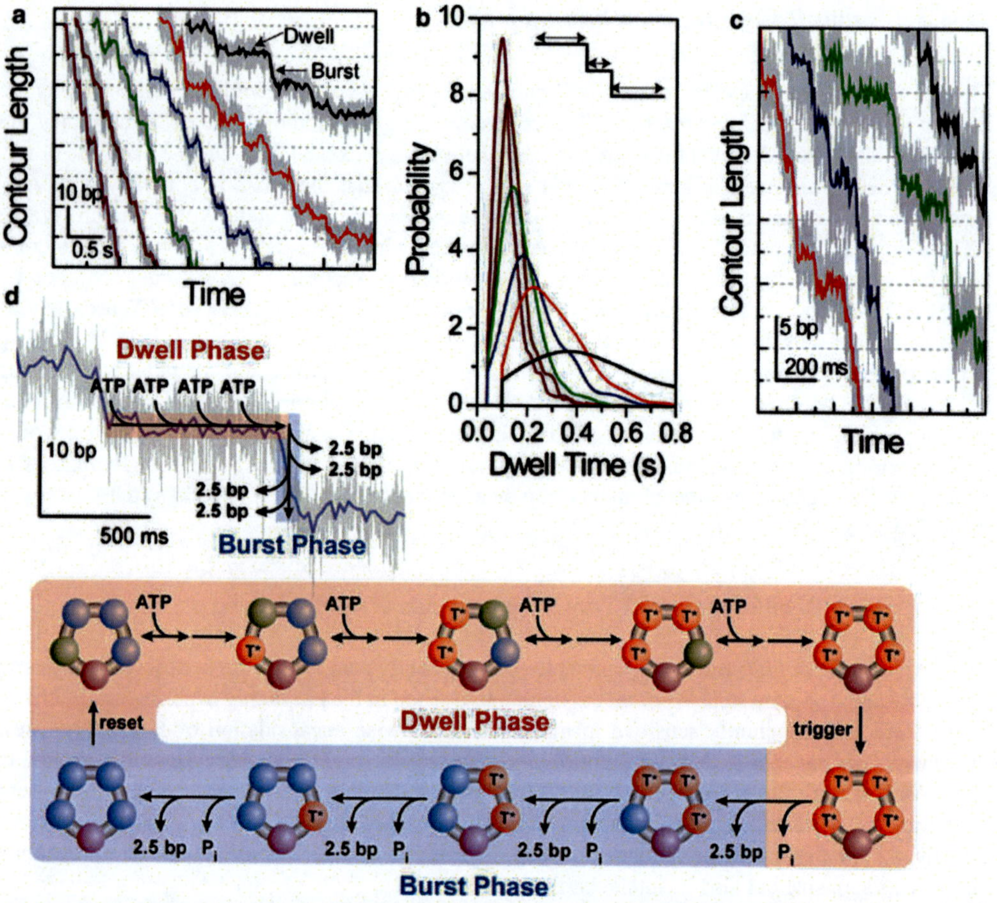

Fig. 24.6 High-resolution measurement of ϕ29 step size. (**a**) Burst-dwell behavior. At low tensions (~8 pN), packaging was observed to occur in large 10-bp "bursts" separated by flat "dwells." The behavior was seen across the full range of ATP, from 10 μM (*black data*) to 500 μM (*purple*). (**b**) Dwell time distributions. The duration of each dwell in the stepping traces was measured and used to compile a probability distribution at each ATP concentration (same color code as (**a**)). The distributions were peaked rather than exponential, indicating that multiple rate-limiting kinetic steps occurred during the dwells. (**c**). Substep size. At higher tensions (~40 pN) and at 250 μM ATP, conditions under which DNA translocation is rate-limiting, the 10-bp bursts were observed to consist of four 2.5-bp substeps. (**d**) Model of intersubunit coordination in ϕ29. Packaging occurs via a biphasic mechanism in which the gp16 ring loads multiple (most likely 4) ATPs during "dwells" and translocates the DNA in four rapid and successive 2.5-bp substeps during the 10-bp "bursts"

experiment of Moffitt et al., the finite duration of the bursts already provided a hint that these were themselves composite events made up of multiple substeps. At the low forces in these measurements, however, most bursts were too rapid to observe such substructure. Thus, the authors performed a second set of experiments exerting higher forces (~40 pN) with the optical traps to slow down each DNA translocation step. These measurements indeed revealed substeps within the 10-bp bursts. Unexpectedly, however, the substeps were a noninteger number of base pairs in size, 2.5 bp, as shown in Fig. 24.6c. An analysis of the durations of the "microdwells" preceding each 2.5-bp substep revealed that three in four were fast events and one in four was slow, corresponding to the long dwells taken every 10 bp as observed at low forces.

The surprising picture emerging from these experiments is that packaging in ϕ29 occurs via a biphasic mechanism in which ATP binding and DNA translocation are temporally segregated, as depicted in Fig. 24.6d. During a dwell, multiple ATPs load to the gp16 ring. Based on the observation of multiple rate-limiting kinetic events during the dwell at low forces and the four 2.5-bp substeps at high forces, the likeliest scenario is that four ATPs bind the motor during this phase. Then, following a dwell, the motor translocates DNA in a burst of four rapid and successive 2.5-bp substeps totaling 10 bp. At first glance, the requirement that multiple ATPs bind the motor prior to translocation may appear at odds with the observation of Michaelis–Menten kinetics, which would indicate a lack of cooperativity between subunits. However, as argued by Moffitt et al., the two results can be reconciled *if and only if* the binding of each ATP involves an irreversible step that commits it to the remainder of the cycle. This "commitment" step is consistent with the large free-energy change upon ATP binding in the mechanochemical model of Chemla et al., as discussed above. Thus, each gp16 subunit must bind ATP in succession. Since DNA is not translocated during the dwell, this time-ordered binding of ATP around the ring must indicate some level of communication between subunits, rather than a coordination imposed by the helical geometry of the DNA. The simplest model is that ATP binding at one subunit allosterically activates the catalytic site of its neighboring subunit, leading to successive binding around the ring. A more recent analysis of the data (Moffitt et al. 2010) further indicates that binding is progressively accelerated as each ATP loads to the motor.

24.4.4.2 Implications of High-Resolution Optical Trap Studies

Two key questions remain regarding this mechanism. The first is how to explain the measurement of a 2.5-bp step size. Assuming four ATPs are consumed during each burst-dwell cycle, a coupling ratio of 2.5 bp/ATP is obtained, which differs from the value of ~2 bp/ATP measured in bulk. However, the bulk measurements of ATP hydrolysis are often difficult to make precise because of basal ATPase activity of nonfunctional complexes in the ensemble or because of futile hydrolysis by active proteins. If hydrolysis of ATP does not necessarily lead to translocation, then the coupling ratio need not equal the step size of the motor. As discussed above, single-molecule experiments show that packaging is interrupted by slips in which DNA spills out of the capsid and is repackaged by the motor. These events lead to futile ATP hydrolysis and an underestimate of the coupling stoichiometry. Initiation of packaging may also require ATP hydrolysis. This overhead of ATP consumption not directly tied to translocation could also lead to an underestimate of the coupling ratio. Thus, the discrepancy between 2.5 and 2 bp estimated step sizes is likely consistent, given these systematic errors expected for the ensemble measurements. An interesting implication of a 2.5-bp step size is that it indicates that the motor is highly efficient. Assuming this step size persists at the highest force believed to be exerted by the motor, 110 pN (a caveat being that Moffitt et al. measured this step size out to only ~45 pN), then over 70% of the energy available from ATP hydrolysis can be converted into mechanical work driving DNA translocation. This would make ϕ29 one of the most efficient molecular machines known (Bustamante et al. 2004). Whether the remaining energy available from ATP hydrolysis is simply dissipated as heat or utilized to drive conformational changes unrelated to DNA translocation remains unclear.

A second question is why the motor must wait to bind the requisite number of ATPs (likely 4) before translocation can occur. A related issue is why only four of the five ATPases in the ring participate in each mechanochemical cycle. It is currently not understood why there is a "special" subunit in the pentameric ring that does not translocate DNA and whether it is the same subunit every 10-bp cycle or if its position rotates around the ring after each cycle. One attractive model is that after every cycle, the last subunit to step remains engaged with the DNA to keep it from slipping out of the capsid while the others begin loading with ATP. This subunit would most likely have to be bound to a nucleotide since the apo state is known to have a weak affinity for DNA, leaving only four available sites to bind ATP. A consequence of this model is that the identity of the special subunit would rotate around the ring after every 10-bp burst. What event triggers the switch from ATP-loading dwell phase to burst phase is not explained by this mechanism. One possibility is that once all available sites are occupied, ATP hydrolysis can occur spontaneously in one subunit, triggering a concerted wave of hydrolysis and translocation through the ring. Another is that the binding of each ATP induces strain in the ring and that a critical threshold is attained upon binding four molecules that triggers a conformational switch in the motor, leading to the translocation burst. An alternative mechanism proposed by Moffitt et al. to explain four translocation steps in a pentameric ring is that the ATPase ring could be broken at one subunit–subunit interface. Though "open" ATPase ring structures have been observed in ringed ATPase motors notably in ϕ8 bacteriophage (Lisal et al. 2005), there is no structural evidence to date for an *active* open conformation. Thus, we do not discuss this class of model presently.

The model described above implies a symmetry mismatch between the DNA helical pitch and the translocation by the ATPase ring. While B-form DNA makes one full turn per 10.5 bp, the motor translocates DNA by 10 bp using only four of five subunits, i.e., four-fifths of a complete turn around the ring. Thus, the motor subunits and DNA would have to rotate relative to each other to remain in register. Preliminary single-molecule measurements (C.L. Hetherington and C. Bustamante, personal communication) indicate that DNA indeed rotates during packaging in an underwinding direction. Encouragingly, this is so far consistent with the mechanism described above in which one special subunit remains engaged with the DNA after the translocation burst and its identity rotates in the ring after every mechanochemical cycle (in a counterclockwise direction around the ring, as viewed outward from inside the capsid). Further measurements will be needed to confirm this aspect of the model.

24.4.5 Motor–DNA Interaction

The apparent 2.5-bp step size of the ϕ29 motor has important implications regarding the interaction between the motor and DNA. Prior to the high-resolution studies, models of packaging explicitly or implicitly assumed that the motor translocates DNA through interactions with phosphates located at every base pair along the DNA backbone. The observed 2.5-bp step size, however, suggests that a different type of motor–DNA contact must be made. To better understand the nature of the motor–DNA interaction in ϕ29, Aathavan et al. (2009) performed extensive optical trap measurements challenging the ability of the motor to package modified substrates. To test the electrostatic nature of the motor–DNA contact, the authors incorporated methylphosphonate (i.e., neutral) DNA of varying lengths on each or both DNA strands; to reveal the structural requirements of this interaction, they utilized substrates such as abasic DNA, single-stranded DNA, DNA bulges, and even unstructured non-DNA linkers. These modified substrates were integrated within an 8-kb ordinary double-stranded DNA molecule to ensure normal initiation and start to packaging.

Faced with these modified inserts, the ϕ29 motors displayed a uniform pattern of pausing at the modification and either traversing it or dissociating completely (Fig. 24.7a). The probability of traversal depended not only on the type of modification but also on its length, which strand was modified, and the tension in the substrate. In general, inserts residing in the 5′–3′ strand (as measured along the direction of packaging) decreased traversal, whereas those on the 3′–5′ strand had little effect on packaging progress, indicating that the motor preferentially tracked the 5′–3′ strand. However, surprisingly, the motor displayed a remarkable ability to accommodate most modifications. Provided they were short (10 bp or less), neutral or abasic DNA, bulges, and even unstructured non-DNA linkers were traversed with over 80% probability. How the ϕ29 motor could package a substrate with no resemblance to canonical dsDNA is an intriguing question. One possibility is that the motor is able to "jump" over these barriers by diffusion, while another is that it can actively

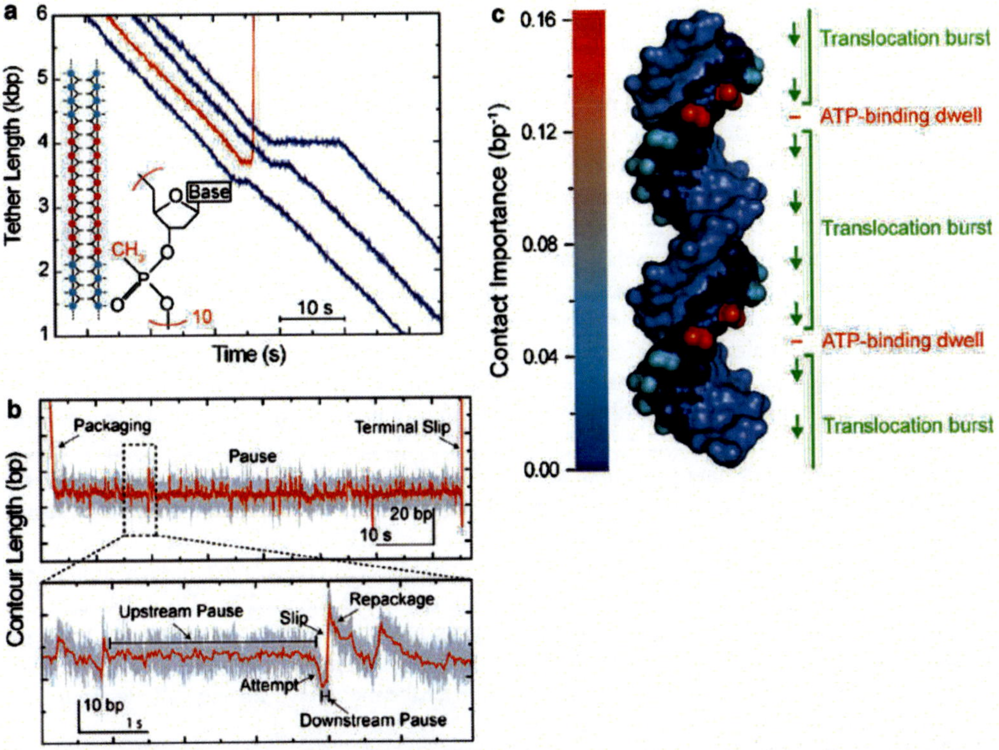

Fig. 24.7 Motor–DNA interactions in ϕ29. (**a**) Motor behavior at the site of modified DNA. The packaging complexes were challenged with modified DNA inserts of various lengths and compositions (inset; here, showing a 10-bp methylphosphonated dsDNA insert) at various tensions (here, ~5 pN) and ATP (1 mM). Upon reaching the modified inserts in the course of normal packaging, motors were observed to pause and either traverse the insert (*blue traces*) or dissociate completely from the DNA (*red*). Pause durations and the probabilities of traversal were dependent on the insert type and on tension. (**b**) High-resolution measurement of pause and dissociation. Pauses were composite events consisting of long "upstream" pauses, followed by either packaging attempts with short "downstream" pauses, small slips, or terminal dissociation. The two pause types are believed to occur during the dwell and burst phases of the motor, respectively. (**c**) "Heat map" of motor–DNA interactions in ϕ29. The biphasic coordination mechanism of ϕ29 dictates that two different types of contacts are made to DNA. During packaging, the motor tracks the 5′–3′strand, making strong ionic contacts to adjacent phosphates (*red*) during the dwell phase and making transient, promiscuous contacts along the backbone (*cyan*) during the translocation burst (The color map indicates the "contact importance" scale with *red* as the highest and *blue* as the lowest)

package through a variety of inserts. To test these possibilities, Aathavan et al. studied the effect of substrate tension on traversal and pause duration. For a diffusive model, they predicted that force would stretch the modified substrate, creating a larger barrier for the motor to jump over, leading to a lower traversal probability and longer pauses. While increased tension on the substrate did decrease the traversal probability and increase pause duration, the effect was much less dramatic than that predicted by diffusion alone. The authors also decreased the ATP concentration and observed a decrease in traversal probability and increase in pause duration, favoring an *active* – as opposed to purely diffusive – mechanism of traversal. Thus, the surprising conclusion is that over short length scales (≤10 bp), the motor is able to actively package through many types of substrates and neither phosphate charge nor DNA structure appears essential.

As the insert length is increased, however, this behavior changed remarkably. While double-stranded methylphosphonate DNA stretches 10 bp or shorter were packaged with 80% or better probability, increasing the length by a single base pair to 11 bp dramatically decreased the traversal probability to less than 50%. The shift in behavior between 10 and 11 bp is highly suggestive because this corresponds to the burst size of the motor, as determined by the high-resolution studies. Thus, Aathavan et al. argued that the shift may indicate that contacts made *within* a burst (i.e., on length scales < 10 bp) may be different than those made during dwells (which occur every 10 bp) in the mechanochemical phases of the motor. To probe this idea further, the authors carried out high-resolution optical trap experiments with a 10-bp methylphosphonate dsDNA insert. These measurements revealed that the pauses at the inserts were composite events with multiple packaging attempts, sub-pauses, and temporary disengagements from DNA or slips (Fig. 24.7b). They were able to identify two distinct types of subpauses termed "upstream" and "downstream" depending on whether they occurred at longer or shorter tether lengths, respectively. Upstream pauses were long lasting (1 s), occurred exclusively before slips or packaging attempts, and took place at the same position (within 1 bp) on the substrate, suggesting that they occurred at the boundary between normal and neutral DNA. Downstream pauses, on the other hand, were shorter in duration (80 ms) and occurred after packaging attempts and at a broader set of positions. Based on this observation and other evidence, Aathavan et al. proposed that upstream pauses may correspond to dwells and downstream pauses to the microdwells during bursts. According to this identification, dwells were more affected by neutral DNA, lasting much longer than their counterparts in normal DNA, compared to the bursts. The authors also predicted that if this model were correct, phages would be able to bypass neutral inserts shorter than one 10-bp burst as long as no dwells occurred within the modified DNA. Indeed, when phages were challenged with 5-bp methylphosphonate DNA insert, 50% of the phages traversed without exhibiting long pauses. This fraction is consistent with the phage arriving at the insert with an arbitrary "phase," such that half are able to burst over the modified 5 bp and half are required to dwell within this region, leading to a pause.

The data of Aathavan et al. paint a picture of motor–DNA interaction that is in concert with the biphasic mechanism proposed in Moffitt et al. Two types of contacts are made with the 5′–3′ DNA strand during packaging, corresponding to the two mechanochemical phases of the motor. During dwells, phosphate contacts are made every 10 bp. Since the traversal probability decreases at 11 bp, not 10 bp, the authors argued that motor may be interacting with adjacent phosphates on the DNA. These contacts are important; their absence stalls the motor, suggesting that they play a sensory role in the mechanochemical coordination of the motor. During the bursts, phosphate contacts are not essential, pointing to weaker "promiscuous" interactions with the substrate. These nonspecific contacts may explain why a noninteger base pair step size is observed on double-stranded DNA and how the motor can actively traverse short substrates structurally dissimilar to canonical B-form DNA. This model of motor–DNA interaction is summarized in Fig. 24.7c. Based on current understanding of the coordination of the motor subunits, strong ionic contacts might be required during dwells, when the motor must wait for the requisite number of ATP molecules to load. On the other hand, weak, transient contacts may be preferable during bursts to encapsidate DNA quickly. At present, there is no model for two different motor–DNA contacts in the available T4 gp17 structures, though it is possible that a single molecular lever could be responsible for both electrostatic and nonionic contacts. Crystal structures of ϕ29 gp16 and λ gpA would likely provide additional insight into this mechanism.

24.5 Completion of Packaging

The previous sections have reviewed studies probing the function of the packaging motor at low capsid filling where the forces resisting DNA confinement are negligible. To complete packaging, however, the whole viral genome must be translocated into the capsid, resulting in very high densities of DNA. Remarkably, the DNA is confined at ultrahigh densities of ~0.5 g/ml, meaning that roughly half the volume of the capsid is filled by DNA and half by closely associated water molecules and ions. This tight packing of DNA is a highly unfavorable conformation from the point of view of DNA bending energy, electrostatic self-repulsion, and conformational entropy loss. In addition to characterizing motor function, single-molecule studies have enabled the characterization of the forces resisting the tight confinement of the DNA in the capsid (Smith et al. 2001; Rickgauer et al. 2008; Fuller et al. 2007b). While many single-molecule studies involve isolating motor complexes away from their natural contexts (e.g., isolation of single myosin molecules from muscle fibers), the packaging motor dynamics can be measured during the natural biological task of translocating the viral DNA into the viral procapsid.

24.5.1 Internal Force Resisting DNA Confinement in Phage ϕ29

In solution, ϕ29 DNA forms a random coil of radius ~250 nm (Robertson et al. 2006). This DNA must be compressed by approximately 1,000-fold in volume to fit inside a ϕ29 capsid of average radius ~24 nm (Morais et al. 2005). The persistence length of DNA is 50 nm, meaning that significant energy is required to bend the DNA to fit in the capsid. DNA in solution is also highly charged, carrying one negative charge per phosphate group. While this charge is predicted to be 80–90% screened by mono- and divalent cations such as Na^+, Mg^{2+}, the residual charge would still create large electrostatic repulsion forces when the DNA is packaged (Riemer and Bloomfield 1978).

Thus, the motor must do mechanical work to translocate DNA against "internal forces" resisting DNA confinement. An approach for inferring these internal forces is based on the observation that the motor velocity steadily decreases with increasing amount of DNA packaged, slowing to nearly a stop when the full genome length is packaged (Fig. 24.8a–c) (Smith et al. 2001; Rickgauer et al. 2008). Since measurements at low capsid filling showed that the motor velocity decreases with an increasing applied load force, the progressive slowing with increasing capsid filling can be interpreted as arising from increasing load due to the internal force resisting DNA packaging. (Incidentally, accounting for this decrease in velocity with capsid filling means that the total time to package the full ϕ29 genome length is ~2–3 min on average). Careful measurements of the velocity versus force relationship at low capsid filling and the velocity versus filling relationship with low applied load allow one to deduce the internal force versus filling. Measurements with low (<10%) capsid filling were also made using the initiation method described above, and measurements with low force (<0.15 pN on average) were made using a ramped DNA stretching technique (Rickgauer et al. 2008) (not shown, see reference). Measurements made in this manner were more accurate than, while revealing similar trends as, earlier measurements (Smith et al. 2001). The earlier work used complexes with higher initial filling (>30%), a 5 pN force clamp, a less-accurate micropipette-trap system rather than a dual trap system, and DNA with the gp3 terminal protein (which appears to cause DNA looping, as described in the initiation section).

In the standard ϕ29 in vitro packaging buffer (25 mM Tris-HCl buffer (pH 7.8), 50 mM NaCl, 5 mM $MgCl_2$, 0.5 mM ATP), the internal force was found to rise to ~7 pN at one-third filling (~6% of the maximum), ~14 pN at half filling (~12% of the maximum), and increase sharply beyond 70 pN during the final stages of filling (Rickgauer et al. 2008). Extrapolation of the velocity versus load data to the low velocities observed in the final stages of packaging yielded an estimate of 110±9 pN for the maximum internal force. The presence of such high internal forces resisting DNA confinement rationalizes the observed capability of the viral packaging motor to exert very high forces, among the highest reported any molecular motors. The forces that build during packaging also likely play a critical biological role to drive DNA ejection when the virus infects a host cell (Smith et al. 2001; Kindt et al. 2001; Evilevitch et al. 2003).

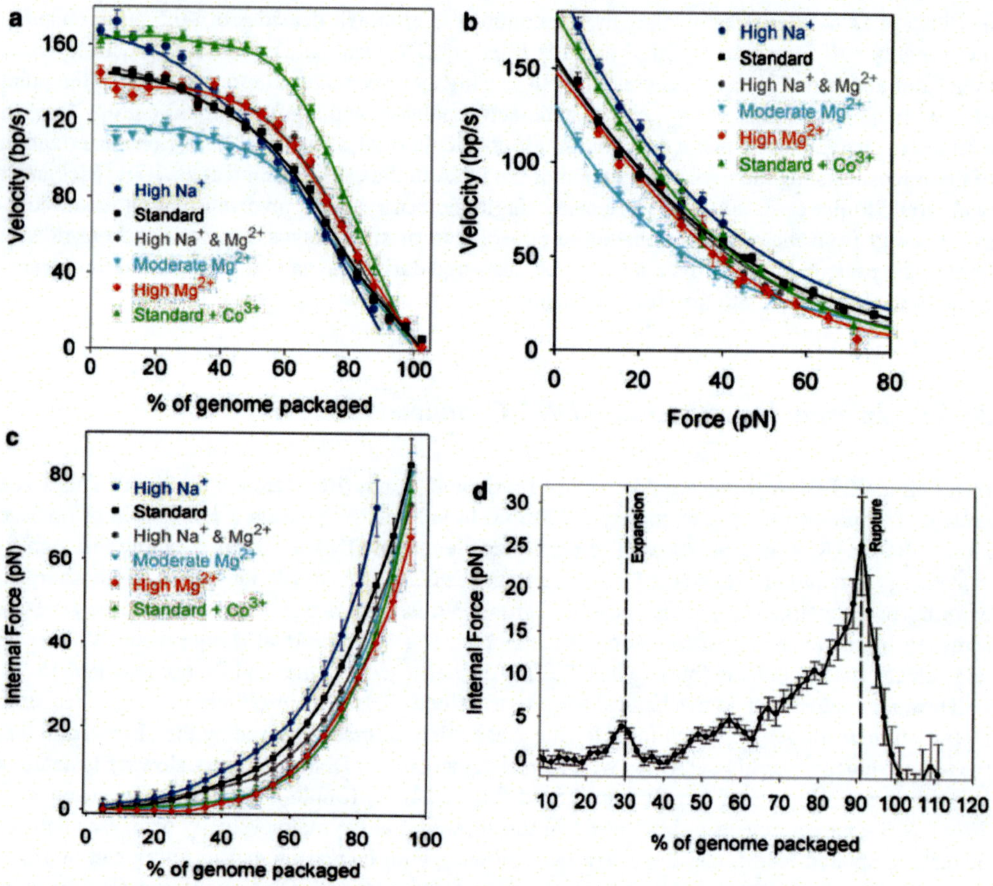

Fig. 24.8 Motor velocity and internal forces resisting DNA packaging. **(a)** ϕ29 motor velocity versus percentage of genome packaged (measured with 5 pN force feedback) in various ionic conditions (see legend). **(b)** ϕ29 motor velocity versus applied load force (measured at low capsid filling with no feedback). **(c)** ϕ29 internal force versus percentage of genome packaged deduced from the measurements in panels (**a** and **b**). (**d**) Internal force versus percentage of genome packaged measured for phage λ in the absence of gpD (a putative capsid-stabilizing protein). Proposed capsid expansion and capsid rupture events are marked by the *dashed vertical lines*

24.5.2 Effect of Ionic Screening on DNA Packaging Forces

The effect of varying ionic conditions on ϕ29 DNA packaging has also been studied by carrying out sets of packaging measurements with buffers containing different concentrations of Na^+, Mg^{2+}, and $CoHex^{3+}$ (Fuller et al. 2007b) (Fig. 24.8a–c). When Na^+ is the dominant ion screening the DNA, the effective DNA charge is predicted to be reduced by ~80%. Mg^{2+} is predicted to screen more strongly, reducing DNA charge by ~90% (Manning 1978). A smaller amount of a trivalent cation such as $CoHex^{3+}$ can screen more effectively than Mg^{2+}, and beyond a critical concentration trivalent cations or ions with higher valence can induce DNA condensation (Baumann et al. 2000). In the work of Fuller et al., measurements were carried out using a 5 pN force clamp to measure velocity versus filling, and a fixed trap position mode to measure velocity versus force (Fuller et al. 2007b). While this is a slightly less accurate method than the lower-force, lower-filling method described in the previous section, resulting in slightly lower estimates for internal force, it was adequate for discerning the trends of how packaging forces vary with ionic condition.

Under each ionic condition, the motor velocity decreased monotonically with capsid filling, and in all conditions, a sharp drop in velocity was observed above ~50% filling (Fig. 24.8a). The velocity converged toward zero at ~100% genome packaging in all cases except with a high Na^+ buffer, where the velocity approached zero at ~90% filling, indicating that packaging would not proceed to completion in this condition (of highest net DNA charge). The internal forces resisting DNA confinement, deduced as described in the previous section, followed the expected trend of decreasing with increasing ionic screening (Fig. 24.8c). Reduced forces similar to those with high Mg^{2+} were also observed with 1 mM $CoHex^{3+}$ added to the standard packaging buffer (less than that needed to induce DNA condensation), showing that trivalent cations can screen the DNA very effectively, even when in competition with higher concentrations of mono- and divalent cations. A plateau of near-zero internal force was observed during the first one-third of packaging with the high Mg^{2+} and $CoHex^{3+}$ buffers, but not with the lower-screening buffers. These measurements confirm the prediction that electrostatic self-repulsion is the dominant force resisting packaging (Riemer and Bloomfield 1978; Tzlil et al. 2003; Purohit et al. 2005), although agreement with theoretical predictions is not perfect, as discussed below.

24.5.3 Theoretical Models for DNA Packaging and Ejection Forces

Spurred by the experimental studies described above and complementary measurements of viral DNA ejection forces (Evilevitch et al. 2004), many investigators recently worked on theoretical models of the forces governing DNA confinement. Both analytical continuum-elastic models and dynamic simulation approaches were used. Analytical models predict specific packed DNA conformations that would minimize the free-energy change, while the simulations predict DNA conformations resulting from stochastic dynamics. Building on earlier work (Riemer and Bloomfield 1978; Odijk 1998), Gelbart and co-workers used Brownian dynamics simulations and analytical theory to predict the forces resisting DNA compaction (Kindt et al. 2001; Tzlil et al. 2003). Their calculations suggest that the DNA conformation would transition through toroidal and spool-like geometries with increasing length packaged, resulting in forces rising to tens of piconewtons, in the same range as inferred experimentally for ϕ29 with optical tweezers measurements (and for λ as well, as discussed further below). The Brownian dynamics simulations assumed a harmonic DNA bending potential and an empirical interaction potential deduced from X-ray diffraction data on the condensation of DNA by osmotic pressure, which reveal short-range interaxial DNA separations of 2–3 nm (Rau and Parsegian 1992). Assuming a purely repulsive potential (as with modest screening by Na^+ and Mg^{2+}), a disordered structure was observed in simulations, but it was speculated that this might equilibrate to a spool-like structure, as seen in simulations with an attractive–repulsive potential on longer time scales (Kindt et al. 2001). Phillips and co-workers further developed coaxial inverse spool models and made internal force predictions for various ionic conditions and various capsid shapes and sizes (Purohit et al. 2003, 2005). These models agreed well with measurements of phage λ DNA ejection forces (Evilevitch et al. 2003, 2005; Grayson et al. 2006), and the experimentally inferred packaging force in ϕ29 with Na^+ screening was within a factor of two of those predicted (2× higher than predicted). However, the experimentally inferred packaging forces with Mg^{2+} as the dominant screening were ~6× higher than predicted (Fuller et al. 2007b), a result that so far remains unexplained.

Equilibrium thermodynamic simulations also predicted a spool structure (Marenduzzo and Micheletti 2003), but stochastic rotation dynamics simulations predicted a more random conformation in the latter stages of packaging (Ali et al. 2006). Harvey and co-workers carried out an extensive series of molecular dynamics (MD) simulations (Harvey et al. 2009). They predicted concentric spooling with spherical capsids and folded toroidal conformations with prolate capsids, rather than coaxial spools. They also predicted, contrary to other investigators, that entropy plays a major role in the free energy of packaging. Other dynamic simulations also found evidence for folded toroids (Spakowitz and Wang 2005; Forrey and Muthukumar 2006) and evidence for nonequilibrium effects and molecular heterogeneity (i.e., different stochastic conformational dynamics of different individual packaging complexes, an effect observed in the optical tweezers experiments as well).

Several factors may potentially contribute to the discrepancy between the experimentally inferred ϕ29 packaging forces and theoretical predictions. First, ϕ29 DNA may not be uniformly packaged as an axially symmetric coaxial spool, as assumed in the analytical theoretical models. Notably, disordered structures are predicted to result in higher internal forces. Cryo-EM reconstructions of T7 phages suggest that the fully packed DNA is at least partially organized in a coaxial spool, although the degree of order toward the interior of the capsid remains unclear (Cerritelli et al. 1997). EM images of partially filled ϕ29 capsids, on the other hand, surprisingly revealed DNA mass density uniformly distributed throughout the capsid rather than progressive outside-in layering (Comolli et al. 2008). Second, the assumption that internal forces have the same effect on motor velocity as externally applied forces may not hold exactly, leading to inaccuracy in determining the former from measurements of motor velocity versus capsid filling. Third, energy dissipation due to friction may occur during packaging, such that the measured work done is higher than the theoretically predicted gain in potential energy of the packed DNA. However, observations of rapid slipping during ϕ29 and λ packaging and of rapid DNA ejection in the T5 and λ phages suggest that friction is negligible (Smith et al. 2001; Mangenot et al. 2005; Grayson et al. 2007). Fourth, as suggested by some simulations, the dynamic conformation adopted by the DNA during rapid packaging may not be exactly equal to the equilibrated free-energy minimum conformation. Fifth, uncertainty in the exact interior volumes of procapsids accessible to packed DNA assessed from cryo-EM data could lead to uncertainty in the theoretical predictions. Finally, DNA–DNA interaction potentials assumed in theoretical calculations may not be universally applicable in describing the DNA packaged in all types of phages in all ionic environments. The potentials used in many models were derived empirically from experiments in which straight DNA segments were condensed into hexagonally packed bundles in solution by applied osmotic pressure and probed by X-ray diffraction (Rau and Parsegian 1992).

24.5.4 Internal Force, Procapsid Expansion, and Procapsid Rupture in Phage λ

Using the approaches developed for ϕ29, Fuller et al. measured internal forces in phage λ. The average λ packaging rate decreased substantially with packaging, going from 580 to 240 bp/s as the procapsid filled from 20 to 90% of the genome length (Fuller et al. 2007a). Since the motor velocity decreases with increasing load, this decrease in velocity with capsid filling is indicative of a building internal force resisting DNA confinement in the procapsid, analogous to that observed with ϕ29. Extrapolation of the decreasing packaging rate trend versus length of DNA packaged to 100% filling suggests that the total time required to package the λ genome is 2–3 min, about the same as for ϕ29. In the λ system, however, the force reached 25 ± 6 pN with 90% of the genome length packaged (Fig. 24.8d), which is notably two- to threefold lower than that found for ϕ29 under a similar ionic condition. It is especially noteworthy, however, that this internal force figure is in excellent agreement with phage λ DNA ejection forces inferred by osmotic pressure experiments (Evilevitch et al. 2004) and DNA confinement forces predicted by theoretical calculations (Tzlil et al. 2003; Purohit et al. 2003). That multiple independent approaches to inferring internal force quantitatively agree strongly support the validity of these force determinations and the proposal that the internal force that builds during packaging is the primary force driving DNA ejection when the virus infects a host cell.

An intriguing behavior observed in the λ packaging dynamics, not seen in ϕ29, was a temporary dip in the packaging rate when ~30% of the λ genome length was packaged (Fuller et al. 2007a). Electron microscopy studies showed that λ procapsids undergo a dramatic conformational change that roughly doubles the internal volume of the capsid at some point during packaging (Dokland and Murialdo 1993). The dip observed in motor velocity corresponds to a 4 pN increase, then decrease in the internal force (Fig. 24.8d). The presence of this dip strongly supports the long-standing hypothesis that internal force builds in the unexpanded procapsid and triggers expansion, which temporarily reduces the internal force due to a reduction in DNA confinement. As packaging proceeds the motor velocity again slows significantly, consistent with subsequent re-buildup of internal force in the expanded procapsid. Notably, the location and magnitude of the dip was found to be variable in individual datasets, suggesting that stochastic dynamics are involved and individual procapsids expand at different internal force and filling levels.

In measurements with phage λ, a number of events were unexpectedly observed where greater than 100% of the genome length of DNA was translocated (Fuller et al. 2007a). In all of these events, after slowing dramatically, the motor abruptly accelerated to full speed, and this acceleration occurred at a distinct point between 90 and 100% packaging (Fig. 24.8d). Translocation then continued up to ~105–146% of the genome length, significantly more than expected for the assembly of a viable phage. These events strongly suggest that the building internal force at 90–100% of genome packaging causes rupture of the expanded procapsid, which releases the confined DNA, presumably through cracks in the capsid shell, and relieves the opposing load on the motor. Such rupture likely occurred because an accessory capsid protein, gpD, which binds the procapsid during packaging in vivo, was not included in the in vitro optical tweezers experiments. These findings provide strong support for the long-held notion (Sternberg and Weisberg 1977; Perucchetti et al. 1988; Yang et al. 2008) that gpD binding stabilizes the procapsid against a building internal force that accompanies DNA packaging. Recent higher resolution cryo-EM structures of λ capsids reveal the structural mechanism for this stabilization (Lander et al. 2008).

24.6 Conclusions and Future Prospects

As described above, single-molecule approaches developed over the last decade, particularly the optical tweezers method, have led to many advances in our understanding of the detailed biophysics of DNA packaging. These single-molecule studies strongly complement traditional biochemical studies, in vivo studies, and structural studies. Experiments have informed us about both the function and mechanisms of viral molecular motor complexes and the nature of the motor's task, which is to overcome the large forces resisting dense DNA confinement. Studies of three different virus systems, bacteriophage ϕ29, λ, and T4, have revealed some universal properties, such as very high force generation compared with other known motors, and some differences, such as motor velocities, capsid expansion effects, and internal forces. High-resolution optical tweezers measurements have revealed a surprising burst-dwell stepping behavior for the ϕ29 motor, a novel mode of operation that has never before been seen in any other molecular motor system. In the λ system, progress has been made in establishing motor structure–function relationships by identifying several mutants having specific alterations in packaging dynamics. Some progress has also been achieved in using single-molecule fluorescence measurements to characterize packaging and conformational dynamics of the motor and DNA substrate.

While much has been learned from single-molecule studies over the last several years, the findings also raise many new questions that will be the target of continuing future investigation.

There is much more work to be done on each of the three virus systems already studied (ϕ29, λ, and T4), and it is also hoped that eventually, these methods will be applicable to other virus systems that utilize analogous motor complexes for DNA packaging, such as the medically relevant herpes- and adenoviruses. Let us review some potential directions to be pursued:

Mechanisms of initiation of packaging: Optical tweezers experiments showed that packaging can initiate within seconds when DNA is brought into close proximity with procapsid–motor complexes (in the ϕ29 and T4 systems) or when DNA–motor complexes are brought into close proximity with procapsids (in the λ system) (Rickgauer et al. 2008; Fuller et al. 2007a, c). Studies suggest that this latter pathway of assembly is the prevalent one in λ in vivo, so it is of interest to investigate whether packaging in ϕ29 and T4 could also initiate via the formation of motor–DNA complexes. In ϕ29 initiation, optical tweezers measurements revealed an unusual DNA structure at the initiation of packaging, with a DNA loop apparently mediated by the gp3 terminal protein (Rickgauer et al. 2006). It is unclear how this looped structure is resolved during packaging, and combined optical tweezers and fluorescence measurements with labeled DNA could shed light on this. Optical tweezers can also be used to directly probe the formation of DNA loops (Gemmen et al. 2006). In the λ system, an accessory protein, gpFI, is essential for efficient viral assembly in vivo and has been implicated in influencing initiation of DNA packaging in vitro (Feiss and Catalano 2005). A possible role for gpFI in postcleavage steps is suggested by single-molecule packaging experiments, which thus far have lacked gpFI and demonstrate that many motor–procapsid binding events do not proceed to active translocation while others show a significant delay before translocation starts (Tsay et al. 2009, 2010). It has been proposed that gpFI may facilitate diffusion of the DNA motor complex over the procapsid surface to the portal, or may coat the procapsid surface to block nonproductive DNA–motor–procapsid binding, or may alter the interaction between the motor and DNA-binding site (*cos* site) to initiate translocation (Murialdo 1991). These models can potentially be discerned by comparing initiation dynamics and efficiency with and without gpFI.

High-resolution measurements of motor stepping in the λ and T4 systems: Having uncovered the unusual coordinated 2.5-bp burst-dwell stepping behavior in the ϕ29 system (Moffitt et al. 2009), an area of immediate interest for single-molecule experiments is to investigate whether such behavior is universal among other virus systems. It will hopefully be possible to extend high-resolution optical tweezers measurements to the λ and T4 systems, which do not have significant global sequence homology with the ϕ29 system (although they may have high structural similarity). Such measurements are challenging, however, due to the 4–10× higher packaging rate of these systems (Fuller et al. 2007a, c). Measurement resolution is ultimately limited by inherent Brownian motion of the trapped microspheres and can only be averaged down by reducing measurement bandwidth. However, the finding that the motors can be slowed by decreasing ATP, increasing load force, decreasing temperature, and/or by appropriate mutations (Smith et al. 2001; Chemla et al. 2005; Tsay et al. 2009, 2010) should make it possible to resolve the stepping dynamics of the λ and T4 motors.

Structure–function relationships: Studies of λ motor mutants exhibiting altered packaging dynamics have begun to inform us on the presence of various motifs and regions of the motor responsible for hydrolyzing ATP, converting the released energy into mechanical work, and/or gripping the DNA substrate (Tsay et al. 2009, 2010). Future studies using site-directed mutagenesis could more systematically map out the specific critical regions and residues. Of particular interest is to further explore the region between the Walker B motif and C motif, wherein change T194M was found to cause a sharp reduction in motor velocity without affecting processivity, pausing, or force generation (Tsay et al. 2010). Analogous mutational studies could also be extended to the ϕ29 and T4 systems, and high-resolution motor stepping measurements with mutants may be very informative. Mutant studies in ϕ29 revealed that deletion of residues of the portal ring, proposed in earlier models to drive DNA translocation, did not strongly affect packaging dynamics (R. Atz, S. Grimes, D.L. Anderson, personal communication). It is thus of interest to study the effects of changes in the ϕ29 gp16 ATPase, which is analogous to the λ gpA large terminase subunit that drives DNA translocation. The T4 system is also particularly attractive for studies of structure–function relationships because the X-ray crystal structure of the large terminase subunit has been determined and the first specific structural model for motor function has been proposed (Sun et al. 2008). Specifically, electrostatic interactions between ion pairs within two globular subdomains separated by a flexible hinge are proposed that produce force and DNA ratcheting motion. Studies with optical tweezers of the effect of site-directed mutations in these regions on packaging dynamics are underway. An ambitious future direction is to combine optical tweezers measurements of DNA translocation with single-molecule fluorescence measurements to probe conformational changes within the motor and the effects of site-directed mutations. One unsolved question is to understand why motor velocity can dramatically vary in time and for different individual complexes (Fuller et al. 2007c). One explanation could be that these motor complexes may have multiple different active conformational states gearing different translocation rates.

Further understanding of forces resisting DNA packaging: Studies of ϕ29 and λ revealed that large internal force resists packaging and depends on ionic screening of the DNA charge (Smith et al. 2001; Rickgauer et al. 2008; Fuller et al. 2007a, b). Such studies have also not yet been done with bacteriophage T4 because current optical tweezers instruments would have to be reconfigured to attain sufficient separation range to fully stretch the extremely long T4 genome (171 kbp). While studies in ϕ29 were able to follow packaging to completion, λ packaging could not be tracked beyond 90% of the genome length without capsid rupture, presumably due to the lack of the gpD protein putatively needed to stabilize the capsid (Fuller et al. 2007a). Optical tweezers measurements with added gpD would shed further light on the role of this protein. While inferred internal forces were in good agreement with theoretical predictions for λ packaging and measurements of ejection forces (Tzlil et al. 2003; Purohit et al. 2005; Evilevitch et al. 2004), they were not as in good agreement for ϕ29, particularly with Mg^{2+} as the dominant screening ion. Several issues that could explain these discrepancies remain to be investigated, such as the effect of prohead shape (ϕ29 and T4 have prolate capsids) and whether packaging involves nonequilibrium dynamics and/or dissipative effects that could make the transient force resisting packaging higher than the equilibrated force, later driving DNA ejection. Another unexplored issue is the effect on packaging of polyamine ions such as spermidine^{3+} and spermine^{4+}, which are present in host cells and have higher screening capacity than Mg^{2+}. At high enough concentrations, such polyamines are capable of completely neutralizing the DNA charge and inducing spontaneous DNA condensation (Baumann et al. 2000). While such an effect could reduce the internal packing forces, condensation of DNA outside the capsid could interfere with packaging and/or present an "external" load on the motor. In vivo, there are also DNA-binding proteins of interest to study, both viral and host gene products, as they would likely have to be stripped from the DNA during packaging and likely present another load on the motor that could affect packaging dynamics.

Mechanisms of termination of packaging: While ϕ29 packages a genome that is replicated as a monomer, λ and T4 and many other dsDNA virus motors utilize an endonuclease function to excise a unit length genome from a string of catenated genomes produced by rolling circle replication (Rao and Feiss 2008). The λ motor initiates packaging at a specific *cos* site in the concatemer, translocates DNA into the capsid, and finally recognizes the downstream *cos* site, which signals the end of the genome. Arrest of the translocating terminase complex at the terminal *cos* site requires not only interaction with the *cosQ* and *cosN* elements but also activation of a sensor that somehow detects the extent of DNA packaging in the capsid. The existence of such a "headful" sensor mechanism is indicated by experiments that show that λ terminase fails to arrest at an ectopic *cos* site placed earlier in the DNA substrate such that the site is reached with <80% of the wild-type genome length packaged (Cue and Feiss 1997). Bacteriophage T4 packaging also relies on a headful mechanism independent of a specific termination site.

Several models have been proposed for how an extent-of-packaging sensor may work (Casjens et al. 1992; Lander et al. 2006). One possibility is that the density of packaged DNA is monitored by the motor, perhaps via a domain of the portal that senses internal pressure of DNA confinement inside the capsid. When a critical point is reached, a conformational change of the portal protein activates terminase's endonuclease activity. A second possibility is that efficient binding of *cosQ* requires that motor velocity has slowed substantially, as occurs in the latter stages of packaging due to buildup of internal pressure resisting DNA confinement. A third possibility is that the packaging motor can directly measure the energy (or work = resisting force × distance translocated per motor step) required to package DNA and cleavage activity is enabled when a critical energy or resisting force has built up.

Single-molecule measurements should permit these models to be distinguished in λ packaging termination experiments. Specifically, Feiss, Catalano, and Smith have proposed experiments to probe for termination in a DNA substrate with an ectopic *cos* termination site placed at an early position that would be reached by the motor when only a small amount of DNA has been packaged. Measurements with reduced ATP, reduced temperature, and/or using the slow T194M mutant would examine whether reducing the motor velocity triggers early termination. Measurements with increasing load force applied with optical tweezers would test whether an "energy sensor" triggers early termination. If neither velocity reduction nor force increases triggers termination, this would suggest that the motor has a separate packaging sensor domain which functions independently of the DNA translocating motor domain.

References

Aathavan K, Politzer AT, Kaplan A, Moffitt JR, Chemla YR, Grimes S, Jardine PJ, Anderson DL, Bustamante C (2009) Substrate interactions and promiscuity in a viral DNA packaging motor. Nature 461:669–673

Abbondanzieri EA, Greenleaf WJ, Shaevitz JW, Landick R, Block SM (2005) Direct observation of base-pair stepping by RNA polymerase. Nature 438:460–465

Ali I, Marenduzzo D, Yeomans JM (2006) Polymer packaging and ejection in viral capsids: shape matters. Phys Rev Lett 96:208102

Ashkin A (1986) Observation of a single-beam gradient force optical trap for dielectric particles. Opt Lett 11:288–290

Banroques J, Doere M, Dreyfus M, Linder P, Tanner NK (2010) Motif III in superfamily 2 "helicase" helps convert the binding energy of ATP into a high-affinity RNA binding site in the yeast DEAD-box protein Ded1. J Mol Biol 396:949–966

Baumann RG, Black LW (2003) Isolation and characterization of T4 bacteriophage gp17 terminase, a large subunit multimer with enhanced ATPase activity. J Biol Chem 278:4618–4627

Baumann CG, Bloomfield VA, Smith SB, Bustamante C, Wang MD, Block SM (2000) Stretching of single collapsed DNA molecules. Biophys J 78:1965–1978

Baumann RG, Mullaney J, Black LW (2006) Portal fusion protein constraints on function in DNA packaging of bacteriophage T4. Mol Microbiol 61:16–32

Black LW (1989) DNA packaging in dsDNA bacteriophages. Annu Rev Microbiol 43:267–292

Burton BM, Marquis KA, Sullivan NL, Rapoport TA, Rudner DZ (2007) The ATPase SpoIIIE transports DNA across fused septal membranes during sporulation in *Bacillus subtilis*. Cell 131:1301–1312

Bustamante C, Smith SB, Liphardt J, Smith D (2000) Single-molecule studies of DNA mechanics. Curr Opin Struct Biol 10:279–285

Bustamante C, Chemla YR, Forde NR, Izhaky D (2004) Mechanical processes in biochemistry. Annu Rev Biochem 73:705–748

Carter AR, Seol Y, Perkins TT (2009) Precision surface-coupled optical-trapping assay with one-basepair resolution. Biophys J 96:2926–2934

Casjens S, Wyckoff E, Hayden M, Sampson L, Eppler K, Randall S, Moreno ET, Serwer P (1992) Bacteriophage P22 portal protein is part of the gauge that regulates packing density of intravirion DNA. J Mol Biol 224:1055–1074

Catalano CE (ed) (2005) Viral genome packaging machines: genetics, structure, and mechanism. Kluwer, New York

Cecconi C, Shank EA, Bustamante C, Marqusee S (2005) Direct observation of the three-state folding of a single protein molecule. Science 309:2057–2060
Cerritelli ME, Cheng NQ, Rosenberg AH, McPherson CE, Booy FP, Steven AC (1997) Encapsidated conformation of bacteriophage T7 DNA. Cell 91:271–280
Chemla YR (2010) Revealing the base pair stepping dynamics of nucleic acid motor proteins with optical traps. Phys Chem Chem Phys 12:3080–3095
Chemla YR, Aathavan K, Michaelis J, Grimes S, Jardine PJ, Anderson DL, Bustamante C (2005) Mechanism of force generation of a viral DNA packaging motor. Cell 122:683–692
Comolli LR, Spakowitz AJ, Siegerist CE, Jardine PJ, Grimes S, Anderson DL, Bustamante C, Downing KH (2008) Three-dimensional architecture of the bacteriophage phi29 packaged genome and elucidation of its packaging process. Virology 371:267–277
Cordin O, Tanner NK, Doere M, Linder P, Banroques J (2004) The newly discovered Q motif of DEAD-box RNA helicases regulates RNA-binding and helicase activity. EMBO J 23:2478–2487
Cue D, Feiss M (1997) Genetic evidence that recognition of cosQ the signal for termination of phage lambda DNA packaging, depends on the extent of head filling. Genetics 147:7–17
Dokland T, Murialdo H (1993) Structural transitions during maturation of bacteriophage lambda capsids. J Mol Biol 233:682–694
Draper B, Rao VB (2007) An ATP hydrolysis sensor in the DNA packaging motor from bacteriophage T4 suggests an inchworm-type translocation mechanism. J Mol Biol 369:79–94
Duffy C, Feiss M (2002) The large subunit of bacteriophage lambda's terminase plays a role in DNA translocation and packaging termination. J Mol Biol 316:547–561
Evilevitch A, Lavelle L, Knobler CM, Raspaud E, Gelbart WM (2003) Osmotic pressure inhibition of DNA ejection from phage. Proc Natl Acad Sci USA 100:9292–9295
Evilevitch A, Castelnovo M, Knobler CM, Gelbart WM (2004) Measuring the force ejecting DNA from phage. J Phys Chem B 108:6838–6843
Evilevitch A, Gober JW, Phillips M, Knobler CM, Gelbart WM (2005) Measurements of DNA lengths remaining in a viral capsid after osmotically suppressed partial ejection. Biophys J 88:751–756
Feiss M & Catalano C (2005) Bacteriophage lambda terminase and the mechanism of viral DNA packaging. in Viral Genome Packaging Machines: Genetics, Structure, and Mechanism", Springer US, pp. 5–39
Finer JT, Simmons RM, Spudich JA (1994) Single myosin molecule mechanics: piconewton forces and nanometre steps. Nature 368:113–119
Fokine A, Chipman PR, Leiman PG, Mesyanzhinov VV, Rao VB, Rossmann MG (2004) Molecular architecture of the prolate head of bacteriophage T4. Proc Natl Acad Sci USA 101:6003–6008
Forrey C, Muthukumar M (2006) Langevin dynamics simulations of genome packing in bacteriophage. Biophys J 91:25–41
Fujisawa H, Morita M (1997) Phage DNA packaging. Genes Cells 2:537–545
Fuller DN, Gemmen GJ, Rickgauer JP, Dupont A, Millin R, Recouvreux P, Smith DE (2006) A general method for manipulating DNA sequences from any organism with optical tweezers. Nucleic Acids Res 34:e15
Fuller DN, Raymer DM, Rickgauer JP, Robertson RM, Catalano CE, Anderson DL, Grimes S, Smith DE (2007a) Measurements of single DNA molecule packaging dynamics in bacteriophage lambda reveal high forces, high motor processivity, and capsid transformations. J Mol Biol 373:1113–1122
Fuller DN, Rickgauer JP, Jardine PJ, Grimes S, Anderson DL, Smith DE (2007b) Ionic effects on viral DNA packaging and portal motor function in bacteriophage phi 29. Proc Natl Acad Sci USA 104:11245–11250
Fuller DN, Raymer DM, Kottadiel VI, Rao VB, Smith DE (2007c) Single phage T4 DNA packaging motors exhibit barge force generation, high velocity, and dynamic variability. Proc Natl Acad Sci USA 104:16868–16873
Gaussier H, Yang O, Catalano CE (2006) Building a virus from scratch: assembly of an infectious virus using purified components in a rigorously defined biochemical assay system. J Mol Biol 357:1154–1166
Gemmen GJ, Millin R, Smith DE (2006) DNA looping by two-site restriction endonucleases: heterogeneous probability distributions for loop size and unbinding force. Nucleic Acids Res 34:2864–2877
Gottesman ME, Weisberg RA (2004) Little lambda, who made thee? Microbiol Mol Biol Rev 68:796–813
Grayson P, Evilevitch A, Inamdar MM, Purohit PK, Gelbart WM, Knobler CM, Phillips R (2006) The effect of genome length on ejection forces in bacteriophage lambda. Virology 348:430–436
Grayson P, Han L, Winther T, Phillips R (2007) Real-time observations of single bacteriophage lambda DNA ejections in vitro. Proc Natl Acad Sci USA 104:14652–14657
Greenleaf WJ, Woodside MT, Block SM (2007) High-resolution, single-molecule measurements of biomolecular motion. Annu Rev Biophys Biomol Struct 36:171–190
Grimes S, Anderson D (1989) In vitro packaging of bacteriophage phi 29 DNA restriction fragments and the role of the terminal protein gp3. J Mol Biol 209:91–100
Grimes S, Anderson D (1997) The bacteriophage phi29 packaging proteins supercoil the DNA ends. J Mol Biol 266:901–914

Grimes S, Jardine PJ, Anderson D (2002) Bacteriophage phi 29 DNA packaging. Adv Virus Res 58:255–294
Guasch A, Pous J, Ibarra B, Gomis-Ruth FX, Valpuesta JM, Sousa N, Carrascosa JL, Coll M (2002) Detailed architecture of a DNA translocating machine: the high-resolution structure of the bacteriophage phi29 connector particle. J Mol Biol 315:663–676
Guo P, Peterson C, Anderson D (1987) Prohead and DNA-gp3-dependent ATPase activity of the DNA packaging protein gp16 of bacteriophage phi 29. J Mol Biol 197:229–236
Hang JQ, Tack BF, Feiss M (2000) ATPase center of bacteriophage lambda terminase involved in post-cleavage stages of DNA packaging: identification of ATP-interactive amino acids. J Mol Biol 302:777–795
Harvey SC, Petrov AS, Devkota B, Boz MB (2009) Viral assembly: a molecular modeling perspective. Phys Chem Chem Phys 11:10553–10564
Hendrix RW (1978) Symmetry mismatch and DNA packaging in large bacteriophages. Proc Natl Acad Sci USA 75:4779–4783
Hugel T, Michaelis J, Hetherington CL, Jardine PJ, Grimes S, Walter JM, Falk W, Anderson DL, Bustamante C (2007) Experimental test of connector rotation during DNA packaging into bacteriophage phi29 capsids. PLoS Biol 5:e59
Joo C, Balci H, Ishitsuka Y, Buranachai C, Ha T (2008) Advances in single-molecule fluorescence methods for molecular biology. Annu Rev Biochem 77:51–76
Kindt J, Tzlil S, Ben-Shaul A, Gelbart WM (2001) DNA packaging and ejection forces in bacteriophage. Proc Natl Acad Sci USA 98:13671–13674
Kondabagil KR, Zhang Z, Rao VB (2006) The DNA translocating ATPase of bacteriophage T4 packaging motor. J Mol Biol 363:486–499
Lander GC, Tang L, Casjens SR, Gilcrease EB, Prevelige P, Poliakov A, Potter CS, Carragher B, Johnson JE (2006) The structure of an infectious P22 virion shows the signal for headful DNA packaging. Science 312: 1791–1795
Lander GC, Evilevitch A, Jeembaeva M, Potter CS, Carragher B, Johnson JE (2008) Bacteriophage lambda stabilization by auxiliary protein gpD: timing, location, and mechanism of attachment determined by cryo-EM. Structure 16:1399–1406
Lebedev AA, Krause MH, Isidro AL, Vagin AA, Orlova EV, Turner J, Dodson EJ, Tavares P, Antson AA (2007) Structural framework for DNA translocation via the viral portal protein. EMBO J 26:1984–1994
Lehninger AL, Nelson DL, Cox MM (1993) Principles of biochemistry. Worth Publishers, New York
Lisal J, Lam TT, Kainov DE, Emmett MR, Marshall AG, Tuma R (2005) Functional visualization of viral molecular motor by hydrogen-deuterium exchange reveals transient states. Nat Struct Mol Biol 12:460–466
Lu HP, Xun L, Xie XS (1998) Single-molecule enzymatic dynamics. Science 282:1877–1882
Maluf NK, Gaussier H, Bogner E, Feiss M, Catalano CE (2006) Assembly of bacteriophage lambda terminase into a viral DNA maturation and packaging machine. Biochemistry (N Y) 45:15259–15268
Mangenot S, Hochrein M, Radler J, Letellier L (2005) Real-time imaging of DNA ejection from single phage particles. Curr Biol 15:430–435
Manning GS (1978) The molecular theory of polyelectrolyte solutions with applications to the electrostatic properties of polynucleotides. Q Rev Biophys 11:179–246
Marenduzzo D, Micheletti C (2003) Thermodynamics of DNA packaging inside a viral capsid: the role of DNA intrinsic thickness. J Mol Biol 330:485–492
Michaelis J, Muschielok A, Andrecka J, Kugel W, Moffitt JR (2009) DNA based molecular motors. Phys Life Rev 6:250–255
Mitchell MS, Rao VB (2004) Novel and deviant Walker A ATP-binding motifs in bacteriophage large terminase-DNA packaging proteins. Virology 321:217–221
Mitchell MS, Matsuzaki S, Imai S, Rao VB (2002) Sequence analysis of bacteriophage T4 DNA packaging/terminase genes 16 and 17 reveals a common ATPase center in the large subunit of viral terminases. Nucleic Acids Res 30:4009–4021
Moffitt JR, Chemla YR, Izhaky D, Bustamante C (2006) Differential detection of dual traps improves the spatial resolution of optical tweezers. Proc Natl Acad Sci USA 103:9006–9011
Moffitt JR, Chemla YR, Smith SB, Bustamante C (2008) Recent advances in optical tweezers. Annu Rev Biochem 77:205–228
Moffitt JR, Chemla YR, Aathavan K, Grimes S, Jardine PJ, Anderson DL, Bustamante C (2009) Intersubunit coordination in a homomeric ring ATPase. Nature 457:446–450
Moffitt JR, Chemla YR, Bustamante C (2010) Mechanistic constraints from the substrate concentration dependence of enzymatic fluctuations. Proc Natl Acad Sci USA 107:15739–15744
Morais MC, Choi KH, Koti JS, Chipman PR, Anderson DL, Rossmann MG (2005) Conservation of the capsid structure in tailed dsDNA bacteriophages: the pseudoatomic structure of phi29. Mol Cell 18:149–159
Morais MC, Koti JS, Bowman VD, Reyes-Aldrete E, Anderson DL, Rossmann MG (2008) Defining molecular and domain boundaries in the bacteriophage phi29 DNA packaging motor. Structure 16:1267–1274

Morita M, Tasaka M, Fujisawa H (1993) DNA packaging ATPase of bacteriophage T3. Virology 193:748–752

Mosig G, Eiserling F (2006) T4 and related phages: structure and development. In: Calendar R, Abedon ST (eds) The bacteriophages. Oxford University Press, Oxford

Murialdo H (1991) Bacteriophage lambda DNA maturation and packaging. Annu Rev Biochem 60:125–153

Myong S, Stevens BC, Ha T (2006) Bridging conformational dynamics and function using single-molecule spectroscopy. Structure 14:633–643

Neuman KC, Block SM (2004) Optical trapping. Rev Sci Instrum 75:2787–2809

Neuman KC, Nagy A (2008) Single-molecule force spectroscopy: optical tweezers, magnetic tweezers and atomic force microscopy. Nat Methods 5:491–505

Odijk T (1998) Hexagonally packed DNA within bacteriophage T7 stabilized by curvature stress. Biophys J 75:1223–1227

Oram M, Sabanayagam C, Black LW (2008) Modulation of the packaging reaction of bacteriophage t4 terminase by DNA structure. J Mol Biol 381:61–72

Oster G, Wang H (2000) Reverse engineering a protein: the mechanochemistry of ATP synthase. Biochim Biophys Acta 1458:482–510

Oster G, Wang H (2003) Rotary protein motors. Trends Cell Biol 13:114–121

Perkins TT, Li HW, Dalal RV, Gelles J, Block SM (2004) Forward and reverse motion of single RecBCD molecules on DNA. Biophys J 86:1640–1648

Perucchetti R, Parris W, Becker A, Gold M (1988) Late stages in bacteriophage lambda head morphogenesis: in vitro studies on the action of the bacteriophage lambda D-gene and W-gene products. Virology 165:103–114

Purohit PK, Kondev J, Phillips R (2003) Mechanics of DNA packaging in viruses. Proc Natl Acad Sci USA 100:3173–3178

Purohit PK, Inamdar MM, Grayson PD, Squires TM, Kondev J, Phillips R (2005) Forces during bacteriophage DNA packaging and ejection. Biophys J 88:851–866

Rao VB, Feiss M (2008) The bacteriophage DNA packaging motor. Annu Rev Genet 42:647–681

Rau DC, Parsegian VA (1992) Direct measurement of the intermolecular forces between counterion-condensed DNA double helices – evidence for long-range attractive hydration forces. Biophys J 61:246–259

Ray K, Ma J, Oram M, Lakowicz JR, Black LW (2010a) Single-molecule and FRET fluorescence correlation spectroscopy analyses of phage DNA packaging: colocalization of packaged phage T4 DNA ends within the capsid. J Mol Biol 395:1102–1113

Ray K, Sabanayagam CR, Lakowicz JR, Black LW (2010b) DNA crunching by a viral packaging motor: compression of a procapsid-portal stalled Y-DNA substrate. Virology 398:224–232

Rickgauer, JP and Smith, DE (2008) Single-Molecule Studies of DNA: Visualization and Manipulation of Individual DNA Molecules with Fluorescence Microscopy and Optical Tweezers. In: Borsali R and Pecora R (ed) Soft Matter: Scattering, Imaging and Manipulation, Vol. 4, Springer

Rickgauer JP, Fuller DN, Smith DE (2006) DNA as a metrology standard for length and force measurements with optical tweezers. Biophys J 91:4253–4257

Rickgauer JP, Fuller DN, Grimes S, Jardine PJ, Anderson DL, Smith DE (2008) Portal motor velocity and internal force resisting viral DNA packaging in bacteriophage phi29. Biophys J 94:159–167

Riemer SC, Bloomfield VA (1978) Packaging of DNA in bacteriophage heads – some considerations on energetics. Biopolymers 17:785–794

Robertson RM, Laib S, Smith DE (2006) Diffusion of isolated DNA molecules: dependence on length and topology. Proc Natl Acad Sci USA 103:7310–7314

Ross JL, Ali MY, Warshaw DM (2008) Cargo transport: molecular motors navigate a complex cytoskeleton. Curr Opin Cell Biol 20:41–47

Sabanayagam CR, Oram M, Lakowicz JR, Black LW (2007) Viral DNA packaging studied by fluorescence correlation spectroscopy. Biophys J 93:L17–L19

Shu D, Zhang H, Jin J, Guo P (2007) Counting of six pRNAs of phi29 DNA-packaging motor with customized single-molecule dual-view system. EMBO J 26:527–537

Simpson AA, Tao Y, Leiman PG, Badasso MO, He Y, Jardine PJ, Olson NH, Morais MC, Grimes S, Anderson DL, Baker TS, Rossmann MG (2000) Structure of the bacteriophage phi29 DNA packaging motor. Nature 408:745–750

Sippy J, Feiss M (2004) Initial cos cleavage of bacteriophage lambda concatemers requires proheads and gpFI in vivo. Mol Microbiol 52:501–513

Smith DE, Tans SJ, Smith SB, Grimes S, Anderson DL, Bustamante C (2001) The bacteriophage phi29 portal motor can package DNA against a large internal force. Nature 413:748–752

Smith SB, Cui Y, Bustamante C (2003) Optical-trap force transducer that operates by direct measurement of light momentum. Methods Enzymol 361:134

Spakowitz AJ, Wang ZG (2005) DNA packaging in bacteriophage: is twist important? Biophys J 88:3912–3923

Sternberg N, Weisberg R (1977) Packaging of coliphage lambda DNA. II. The role of the gene D protein. J Mol Biol 117:733–759

Sun S, Kondabagil K, Gentz PM, Rossmann MG, Rao VB (2007) The structure of the ATPase that powers DNA packaging into bacteriophage t4 procapsids. Mol Cell 25:943–949

Sun S, Kondabagil K, Draper B, Alam TI, Bowman VD, Zhang Z, Hegde S, Fokine A, Rossmann MG, Rao VB (2008) The structure of the Phage T4 DNA packaging motor suggests a mechanism dependent on electrostatic forces. Cell 135:1251–1262

Tinoco I Jr, Li PT, Bustamante C (2006) Determination of thermodynamics and kinetics of RNA reactions by force. Q Rev Biophys 39:325–360

Tsay JM, Sippy J, Feiss M, Smith DE (2009) The Q motif of a viral packaging motor governs its force generation and communicates ATP recognition to DNA interaction. Proc Natl Acad Sci USA 106:14355–14360

Tsay JM, Sippy J, DelToro D, Andrews BT, Draper B, Rao V, Catalano CE, Feiss M, Smith DE (2010) Mutations altering a structurally conserved loop-helix-loop region of a viral packaging motor change DNA translocation velocity and processivity. J Biol Chem 285:24282–24289

Tzlil S, Kindt JT, Gelbart WM, Ben-Shaul A (2003) Forces and pressures in DNA packaging and release from viral capsids. Biophys J 84:1616–1627

van Oijen AM, Blainey PC, Crampton DJ, Richardson CC, Ellenberger T, Xie XS (2003) Single-molecule kinetics of lambda exonuclease reveal base dependence and dynamic disorder. Science 301:1235–1238

Wang MD, Schnitzer MJ, Yin H, Landick R, Gelles J, Block SM (1998) Force and velocity measured for single molecules of RNA polymerase. Science 282:902–907

Xue Q, Yeung ES (1995) Differences in the chemical reactivity of individual molecules of an enzyme. Nature 373:681–683

Yang Q, Catalano CE (2003) Biochemical characterization of bacteriophage lambda genome packaging in vitro. Virology 305:276–287

Yang Q, Maluf NK, Catalano CE (2008) Packaging of a unit-length viral genome: the role of nucleotides and the gpD decoration protein in stable nucleocapsid assembly in bacteriophage lambda. J Mol Biol 383:1037–1048

Chapter 25
Genome Gating in Tailed Bacteriophage Capsids

Paulo Tavares, Sophie Zinn-Justin, and Elena V. Orlova

Abstract Tailed bacteriophages use a portal system for genome entry and exit from viral capsids. Here, we review the mechanisms how these movements are controlled by the genome gatekeeper that assembles at the portal structure. Phage DNA is packaged at high pressure inside the viral capsid by a powerful motor. The viral genome is translocated through the central channel of the portal protein found at a single vertex of the capsid. Packaging is normally terminated by endonucleolytic cleavage of the substrate DNA followed by disassembly of the packaging motor and closure of the portal system, preventing leakage of the viral genome. This can be achieved either by conformational changes in the portal protein or by sequential addition of proteins that extend the portal channel (adaptors) and physically close it preventing DNA exit (stoppers). The resulting connector structure provides the interface for assembly of short tails (podoviruses) or for attachment of preformed long tails (siphoviruses and myoviruses). The connector maintains the viral DNA correctly positioned for ejection that is triggered by interaction of the phage particle with bacterial receptors. Recent exciting advances are providing new molecular insights on the mechanisms that ensure precise coordination of these critical steps required both for stable viral genome packaging and for its efficient release to initiate infection.

25.1 Introduction

Virus particles (virions) are designed to protect their viral genome from environmental insult and to ensure its efficient delivery to the host cell. Two major strategies were devised by viruses to achieve this goal. The first is the coassembly of nucleic acid and coat protein leading to genomes packed in

P. Tavares (✉)
Unité de Virologie Moléculaire et Structurale, UPR CNRS 3296 and IFR 115,
Bâtiment 14B, CNRS, 1 avenue de la Terrasse, 91198 Gif-sur-Yvette, France
e-mail: tavares@vms.cnrs-gif.fr

S. Zinn-Justin
Laboratoire de Biologie Structurale et Radiobiologie, iBiTec-S and URA CNRS 2096,
Bâtiment 144, CEA Saclay, 91190 Gif-sur-Yvette, France
e-mail: Sophie.ZINN@cea.fr

E.V. Orlova
Crystallography and Institute of Structural and Molecular Biology, Birkbeck College,
Malet Street, WC1E 7HX, London, UK
e-mail: e.orlova@mail.cryst.bbk.ac.uk

M.G. Rossmann and V.B. Rao (eds.), *Viral Molecular Machines*,
Advances in Experimental Medicine and Biology 726, DOI 10.1007/978-1-4614-0980-9_25,

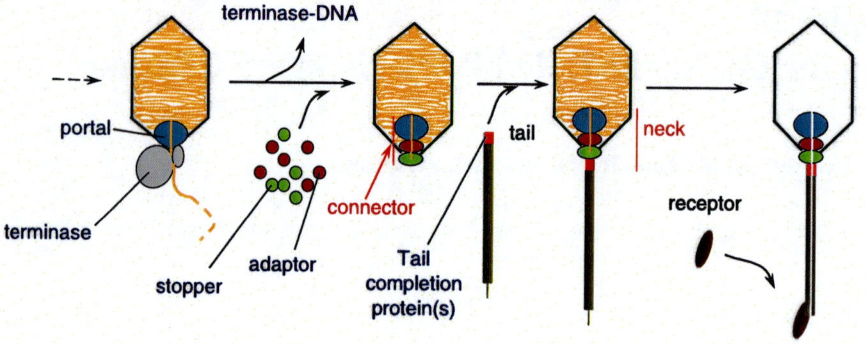

Fig. 25.1 Late steps of the siphovirus assembly pathway and DNA ejection. The effectors of DNA packaging (portal in *blue* and terminase in *gray*) and connector assembly are indicated. The adaptor (*magenta*) is a head completion protein that extends the portal channel and the stopper (*green*) closes this channel during connector assembly. The tail completion proteins are shown in *pink*. They form the neck structure together with the connector components upon head-to-tail joining. The bacterial receptor that triggers DNA release from the virion is depicted in *brown*

a protein-rich viral particle. Some families of these viruses are also characterized by presence of a lipid envelope. The resulting virus particle (partially) disassembles upon infection to expose the viral genome for its expression inside the host cell. A second strategy is the preassembly of a protein lattice called procapsid followed by packaging of the viral genome to the procapsid interior powered by a molecular motor. The energy consuming nucleic acid pumping leads to tight genome packing inside the viral particle, thus maximizing the amount of genetic information that is transported in the nucleic acid-rich virion.

Double-stranded DNA (dsDNA) viruses of the tailed bacteriophages–herpesviruses viral lineage (Bamford et al. 2005) provide a paradigm for viruses using a genome packaging motor. Their preassembled icosahedral procapsid is characterized by the presence of a specialized vertex formed by the dodecameric *portal protein*. The portal vertex provides the docking point for the viral ATPase complex (*terminase*). This complex contains normally multiple copies of two different subunits, a large two-domain subunit with ATPase and endonuclease activities (Sun et al. 2008), and a small subunit that recognizes viral DNA (Chap. 22). The terminase bound to the portal vertex forms the packaging motor that translocates the viral dsDNA to the procapsid interior through the central portal channel (Fig. 25.1). At late stages of encapsidation, the increasingly compact arrangement of the highly negatively charged DNA leads to a steep rise of pressure inside the capsid that can reach ~6 MPa (Smith et al. 2001; Chaps. 21 and 24).

Packaging termination and disassembly of the packaging motor are tightly coordinated with closure of the portal system to avoid leakage of the viral genome. In tailed bacteriophages, this is most frequently achieved through the binding of *head completion proteins*. The knob formed by the portal dodecamer and these proteins is named *connector*[1] (Lurz et al. 2001; Orlova et al. 2003). In functional terms, it is the structure present at the portal vertex of metastable DNA-filled capsids before attachment of the phage tail. This definition comprises most of the podoviruses small tails offering the advantage of a functional comparison between the structures found in the three morphotypes of tailed phages: podoviruses, siphoviruses, (phages with long noncontractile tails), and myoviruses (phages with long contractile tails). The connector provides a platform for completion of short-tail assembly in podoviruses (Johnson and Chiu 2007; Chap. 7) or for docking of preassembled long tails in myo- and siphoviruses (Chaps. 5 and 6). In phages with a long tail, the region between the portal protein and the helical tail tube is defined as the *head-to-tail interface* or *neck* (Fig. 25.1). The neck is composed of the connector and of the *tail completion proteins*. These proteins are added to the tail tube at a late step of assembly. They form the tail interface that will interact with the capsid connector to form mature virions (Fig. 25.1). The connector, possibly in coordination with other neck components, is the

[1] Please note that some authors use the term "connector" to designate exclusively the portal protein.

valve that opens in response to tail binding to the host cell for genome release from the capsid. DNA then exits through the tail tube to enter the bacterial cytoplasm, a process initially energized by the internal DNA pressure inside the phage capsid (Evilevitch et al. 2003; São-José et al. 2007). Here, we review current knowledge on the coordinated processes of genome packaging termination, connector assembly, and reversible closure of the capsid portal system in tailed bacteriophages.

25.2 DNA Packaging and Its Termination

The dsDNAs found in virions of tailed phages are linear molecules as found also in pox-, adeno-, and herpesviruses. However, the topology of the chromosome differs among viruses as a result of the substrate used for encapsidation or of different modes of packaging termination. In most phages, the dsDNA substrates for packaging are head-to-tail multimers of the viral genome (concatemers) generated by rolling circle replication or recombination. The chromosomes of many phages are terminally redundant [e.g., phages T4, P22, T1, P1, or SPP1 (reviewed in Black 1989; Chap. 22)]. Their encapsidation is normally initiated by an endonucleolytic cleavage at a defined sequence (*pac*) of the substrate DNA concatemer although some phages like T4 do not use a unique site for the initial cleavage. Packaging proceeds unidirectionally until a threshold amount of DNA is reached inside the viral capsid. Head filling triggers an apparently sequence-independent cleavage of the DNA concatemer defining the size of the packaged DNA. The headful cleavage is imprecise leading to variations in chromosome size of more than 1 kb (Casjens and Hayden 1988; Tavares et al. 1996). The mechanism requires a sensor that measures the level of DNA head filling and a headful nuclease. Single amino acid substitutions in the portal proteins of bacteriophages P22 (Casjens et al. 1992) and SPP1 (Tavares et al. 1992; Isidro et al. 2004a, b) were shown to affect the headful gauge. The SPP1 wild type and mutant proteins have structural differences in the region that they expose to the capsid interior (Orlova et al. 1999). This region in direct contact with the tightly packed DNA is ideally suited for a sensor function (Lander et al. 2006; Tang et al. 2011). The effector of the headful cut is most likely the terminase large subunit, the only component of the DNA packaging motor with known endonucleolytic activity (reviewed in Rao and Feiss 2008). Cleavage is probably achieved by a switch of its ATPase to endonuclease activity (Camacho et al. 2003). After termination of the first packaging cycle initiated at *pac* (initiation cycle), a second packaging event is initiated at the nonencapsidated DNA end created by the headful cleavage and additional cycles of encapsidation follow (Streisinger et al. 1967; Tye et al. 1974). Some packaging series can yield 12 encapsidation events or more revealing the high processivity of the packaging machinery (Adams et al. 1983; Tavares et al. 1996). Mechanistically, this implies that the viral ATPase/endonuclease remains associated to the substrate DNA concatemer end after headful cleavage and motor disassembly from the DNA-filled capsid. The nucleoprotein complex rebinds at a procapsid portal, and a new packaging cycle ensues.

Phages like lambda have their genome delimited by complementary cohesive ends (*cos*) in the concatemer substrate. The *cos* sequence is cleaved at the initiation step of the encapsidation cycle. Termination of packaging is also achieved by introducing staggered nicks at another *cos* generating a set of identical unit-genome length chromosomes found in mature virions (Feiss and Catalano 2005; Chap. 22). In spite of this sequence specificity, a requirement for head filling to achieve the terminal cos cleavage was reported for lambda (Feiss and Siegele 1979). Elegant genetic experiments also showed sequential processive packaging of lambda DNA (Feiss et al. 1985) highlighting mechanistic similarities with the headful packaging system described above.

Other phages, whose prototypes are T3 and T7, have unique ends with terminally redundant sequences. The packaged DNA molecules are thus identical. In this case, the concatemer DNA substrate being packaged is cleaved at a specific sequence, and local replication duplicates the terminal redundancy. One copy of the terminal redundancy is encapsidated to terminate the packaging cycle, while the other copy remains associated to the terminase. This nucleoprotein complex then binds a procapsid to initiate the following packaging cycle (Fujisawa and Morita 1997). The termination cleavage of the DNA substrate can also occur by a headful mechanism if defined packaging signals are absent in the substrate concatemer (Hashimoto and Fujisawa 1992).

Viruses with protein-primed DNA replication like phage ϕ29 or adenovirus package linear unit-genome length DNA molecules with a terminal protein (gp3 in case of ϕ29) covalently linked to each 5′ end (Salas 2006). Endonucleolytic cleavage is thus not required to terminate packaging. In this last case, no DNA headful requirement was reported for termination of DNA packaging and subsequent achievement of phage assembly.

25.3 Sealing of the Portal System: Components and Assembly Mechanism

Termination of phage genome packaging, frequently achieved by an endonucleolytic cleavage and marked by dissociation of the terminase motor from the portal protein, is followed by sequential binding of proteins to the portal. These assembly reactions gate the portal system and create the structural context for attachment of the machinery for delivery of the viral genome to the host bacterium. We describe below components of the genome gatekeeper system and current knowledge of its mechanisms of assembly.

25.3.1 The Portal Protein

The portal protein is the key component of capsids for DNA packaging and for initiation of connector formation at the end of the genome encapsidation process. The structural organization of portal proteins is highly conserved among tailed bacteriophages and herpesviruses (Bazinet and King 1985; Valpuesta and Carrascosa 1994; Orlova et al. 1999; Simpson et al. 2000; Newcomb et al. 2001; Guasch et al. 2002; Lebedev et al. 2007; Olia et al. 2011). They are cyclical oligomers organized around a central channel or tunnel through which dsDNA enters and exits from the viral capsid (Figs. 25.1 and 25.2). Isolated assembly naïve portals of different viruses represent homogeneous or mixed populations of oligomers comprising 11, 12, 13, or 14 subunits (Dube et al. 1993; van Heel et al. 1996; Orlova et al. 2003; Cingolani et al. 2002; Trus et al. 2004; our unpublished results). However, in viral particles, portals are invariably dodecamers, an association state that is strictly imposed during procapsid assembly.

The subunit of portal proteins can vary greatly in molecular mass, from the small gp10 of phage ϕ29 (~36 kDa) to the large gp1 of podovirus P22 (~83 kDa). In spite of these variations, all portals share a common fold revealed by the crystallographic structures of the ϕ29, SPP1, and P22 portals (Simpson et al. 2000; Lebedev et al. 2007; Olia et al. 2011; Fig. 25.3). A search using their subunit three-dimensional fold identified a fourth portal encoded by the *Corynebacterium diphtheriae* genome. Each portal subunit can be subdivided into the clip, stem, wing, and crown regions (Fig. 25.3; Lebedev et al. 2007). The clip represents the region of the portal exposed to the capsid exterior for sequential binding to the terminase for DNA packaging (Simpson et al. 2000; Sun et al. 2008) and later to a head completion protein for connector assembly. This part of the portal that provides extensive intersubunit contacts is linked to the wing and crown regions through the α-helical stem. The protein parts forming the clip and stem have highly conserved folds among portals, while the wing and crown vary significantly in size and complexity (Fig. 25.3). The wing irradiates outward from the stem to form a well-distinct radial domain with an α/β fold. It has a clear handedness conferring a turbine-like appearance to the portal. The α-helical crown that is very small in phage ϕ29 becomes more elaborated in SPP1 and features an impressive 213 amino acid long α-helical region in phage P22 (Fig. 25.3). This helix assembles into a coiled-coil barrel of 200 Å height that forms a

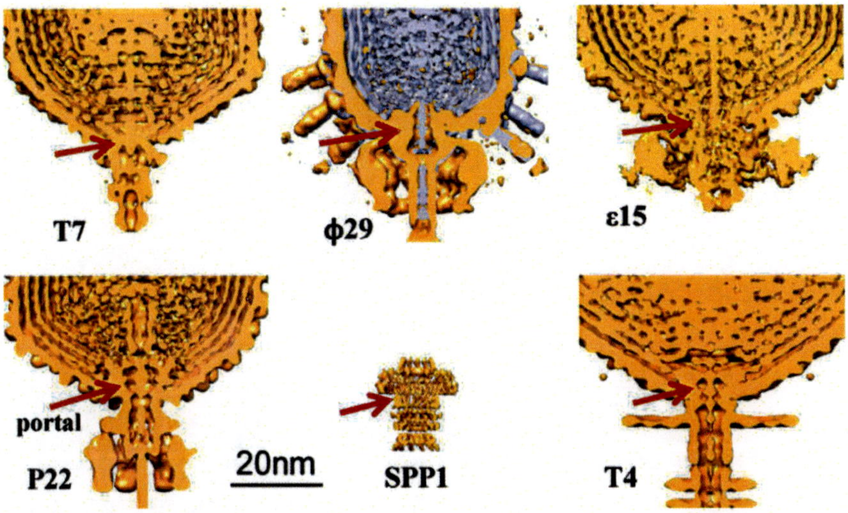

Fig. 25.2 Asymmetric reconstructions of the DNA-filled capsids of bacteriophages T7 (Agirrezabala et al. 2005; EMD-1164), ϕ29 (Xiang et al. 2006; EMD-1265 and EMD-1266), ε15 (Jiang et al. 2006; EMD-1175), P22 (Lander et al. 2006; EMD-1220), and T4 (Fokine et al. 2004; EMD-1075). In case of ϕ29 superimposition of the empty and DNA-filled capsid allowed to distinguish DNA (*violet*) from the capsid (*gold*). The reconstruction of the SPP1 connector purified from disrupted DNA-filled capsids is also shown (Orlova et al. 2003; EMD1021). The position of the portal protein, whose typical shape is easily identifiable at the capsid portal vertex, is shown by a *red arrow*

core protruding from the crown to near the capsid center (Olia et al. 2011; Tang et al. 2011). Sequence homology searches revealed the presence of putative C-terminal barrels in a large subgroup of podoviruses. P22 portals with long deletions of the barrel (>75 amino acids) were competent for assembly of mature virions. However, most of these particles were noninfectious suggesting that the major function of the barrel is exerted during initiation of host infection (Tang et al. 2011).

25.3.2 Head Completion Proteins: Adaptors and Stoppers

Head completion proteins bind sequentially to the portal vertex after DNA packaging termination. Their function is to seal the portal gate and to provide a hub for tail assembly in podoviruses or a platform for binding of a preformed long tail in siphoviruses and myoviruses. Mutants defective on one or both of these functions led to the identification of head completion proteins in a number of phages and provided information on their order of assembly at the portal vertex (Sect. 25.3.3). More recently, cryo-electron microscopy (cryo-EM) reconstructions of intact virions or of their subassemblies allowed correlating the mutant phenotypes with positioning of head completion proteins in the virion structure. These proteins can be distinguished according to their role as adaptors, when they extend the portal protein tunnel, or stoppers, if they plug reversibly the tunnel to prevent DNA exit.

Structural analyses of head completion proteins identified by now four families that have distinct folds. Bioinformatics searches showed that each family comprises a large number of proteins found in bacteriophages and prophages. The first family is presently represented by gp15 from bacteriophage SPP1, two proteins encoded by prophages of *Bacillus subtilis*, gp6 from phage HK97, and gp4 of phage P22. The first three structures were obtained by NMR of monomers (Lhuillier et al. 2009), while gp6 (HK97) and gp4 (P22) were determined by crystallographic analysis of a 13-mer ring (Cardarelli et al. 2010a) and of a cyclical dodecamer (Olia et al. 2011), respectively (Fig. 25.4).

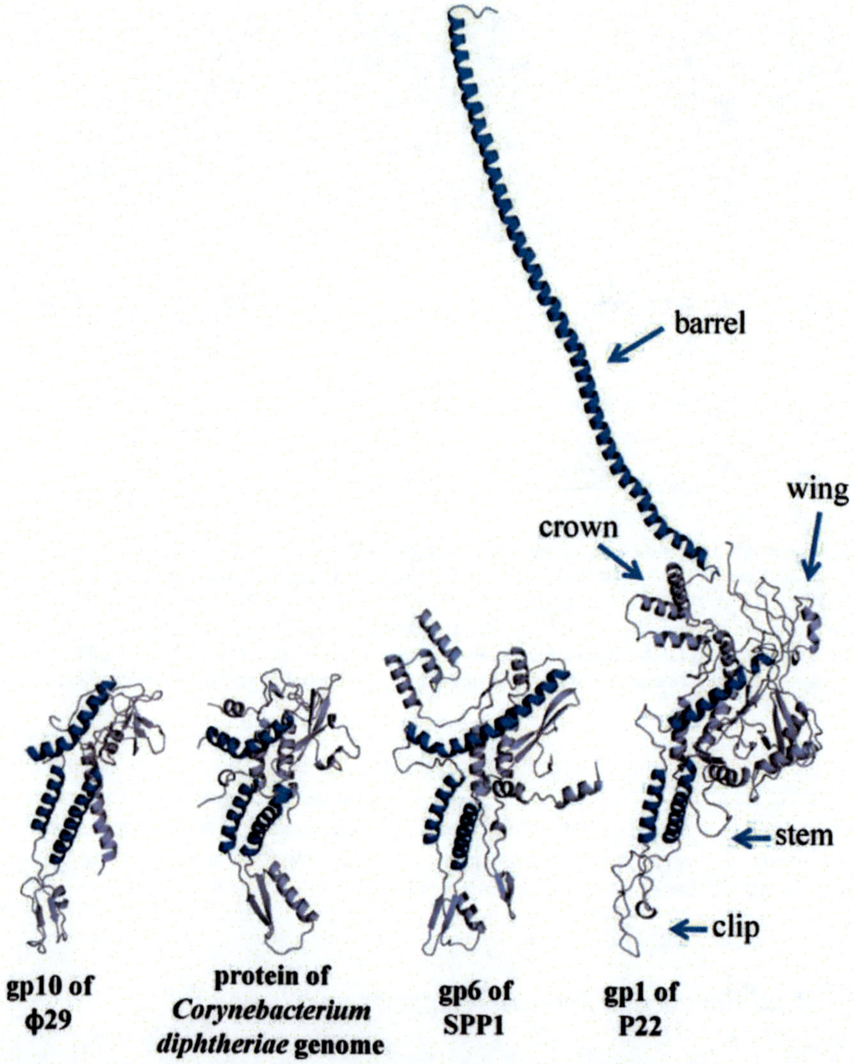

Fig. 25.3 Single subunits extracted from crystallographic structures of portal protein cyclical oligomers. The name of the protein and its origin are shown in the figure. The different regions of the portal structure are labeled in the subunit of gp1 from bacteriophage P22. The three α-helices that are conserved among portal structures and the helix forming the long barrel of the phage P22 portal are highlighted. The PDB codes are 1FOU/1HJW for gp10 of ϕ29, 3KDR for the protein of *Corynebacterium diphteriae* genome, 2JES for gp6 of SPP1, and 3LJ5 for gp1 of P22. Structures are ordered from left to right according to the increasing size of their amino acid sequence

The latter was obtained by assembly of gp4 monomers onto the portal dodecamer of phage P22, thus mimicking the first step of connector assembly (Olia et al. 2011). The three known phage proteins of this family operate as adaptors that bind directly to the portal protein during connector assembly suggesting that the prophage-encoded proteins have a similar function. All are formed by a 3- or 4-α-helix bundle stabilized by a small hydrophobic core. They are also characterized by the presence of highly mobile loops, as assessed by NMR for the α2α3 loop of SPP1 gp15 (Lhuillier et al. 2009), and/or flexible C-terminal regions, as suggested from the X-ray data on HK97 gp6 (Cardarelli et al. 2010a) and P22 gp4 (Olia et al. 2011).

The second family comprises small proteins whose prototype is gpW of bacteriophage λ. GpW is a component of the bacteriophage lambda connector (Gaussier et al. 2006; Cardarelli et al. 2010b).

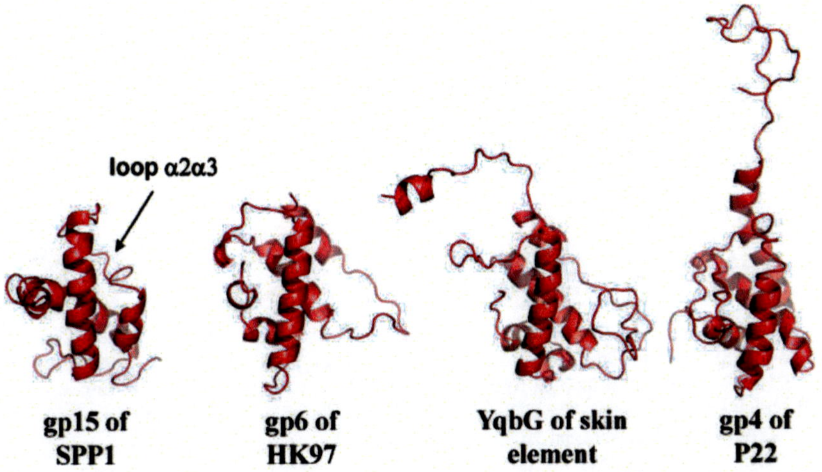

Fig. 25.4 Monomers (gp15 and YqbG) and single subunits (gp6 and gp4) of adaptors of the gp15 (SPP1) – gp6 (HK97) family. The structures were aligned relatively to gp4 of P22 which is the only protein whose orientation relative to the portal protein was established at atomic resolution (Fig. 25.7; Olia et al. 2011). The flexible loop α2α3 of gp15 (SPP1) is identified in the figure. The PDB codes are 2KBZ for gp15 of SPP1, 3JVO for gp6 of HK97, 1XN8 for YqbG of the skin element encoded by the *B. subtilis* genome, and 3LJ4 for gp4 of P22

Fig. 25.5 Monomer of the stopper gpW from phage λ (PDB code 1HYW)

It likely binds directly to the portal protein as found for the gp15 and gp4 adaptors of phages SPP1 and P22, respectively. Although these three proteins share an α-helical-rich structure, their fold is distinct (Lhuillier et al. 2009). The gpW basic protein is a monomer formed of two α-helices separated by a β-hairpin (Fig. 25.5; Maxwell et al. 2001). Its unstructured C terminus is essential for function (Maxwell et al. 2000, 2001). GpW acts as a stopper that prevents exit of the genome from λ DNA-filled capsids (Perucchetti et al. 1988).

The third family includes gpFII of phage λ and gp16 of phage SPP1 sharing a common β-barrel fold that is also found in proteins XkdH and SF1141 encoded by phage-related sequences from the *B. subtilis* and *Shigella flexneri* 2a genomes, respectively (Fig. 25.6; Maxwell et al. 2002; Lhuillier et al. 2009). The protein monomers are again characterized by large unstructured loops that are highly mobile. Gp16 was shown to be the stopper of the SPP1 connector (Lhuillier et al. 2009). In contrast, DNA is retained in λ capsids lacking gpFII. The possible function of this protein that binds after gpW to the portal vertex of the DNA-filled capsid could be providing an interface for tail attachment (Casjens et al. 1972; Cardarelli et al. 2010b).

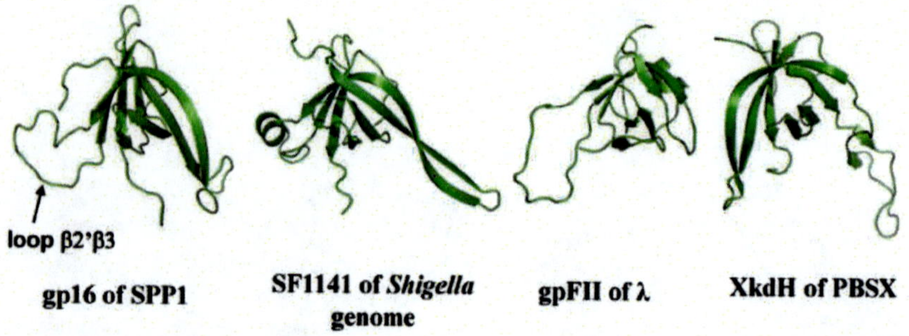

Fig. 25.6 Monomers from the gpFII (λ) – gp16 (SPP1) family. The structures were aligned relatively to gp16 whose orientation in the SPP1 viral particle was defined by docking its NMR structure into the connector EM map (Lhuillier et al. 2009). The flexible loop *β2′β3* of gp16 from SPP1 that folds to a *β*-strand during connector assembly (see text) is identified in the figure. The PDB codes are 2KCA for gp16 of SPP1, 2KZ4 for SF1141 of *Shigella* genome, 1K0H for gpFII of phage λ, and 3F3B for XkdH of PBSX

The fourth family is typified by gp26, the thin needle that closes the connector of bacteriophage P22 during assembly and penetrates the bacterial envelope at the beginning of infection (Fig. 25.2). These needle-like proteins are characteristic of a subgroup of podoviruses where the needle plays the double role of stopper and puncturing device (Bhardwaj et al. 2009; Chap. 7). Piercing of the cell envelope and opening of the conduit for DNA passage to the cytoplasm are thus intimately linked in these viruses. Gp26 is a homotrimeric 240-Å-long fiber formed by two coiled coils interrupted by a triple β-helix (Olia et al. 2007b). Consistent with its cell puncturing role, the gp26 needle has a slender rigid structure that can be unfolded only at 85°C or in the presence of ~6.4 M guanidine hydrochloride (Bhardwaj et al. 2007, 2009).

25.3.3 Mechanisms of Connector and Neck Assembly

Closure of the portal vertex is a late step in the obligate sequential program of interactions leading to assembly of infectious tailed phages. The interaction of head completion proteins with the portal vertex needs to be tightly regulated, both to avoid premature binding before DNA packaging completion and to ensure that departure of the terminase from the portal is coordinated with rapid connector assembly. The control of the correct order of assembly relies mostly on the conformational properties of viral proteins and their complexes because the phage building blocks are synthesized roughly at the same time in the infected cell (Casjens and Hendrix 1988). A detailed study of the sequence of assembly reactions leading to connector and phage assembly in an in vitro system with purified components was reported for phage λ (Gaussier et al. 2006).

There is little information on the mechanisms that ensure coordination of the endonucleolytic cleavage that terminates packaging with departure of the terminase–DNA complex from the portal to start the following packaging cycle (Sect. 25.2). No proteins other than those found in the packaging motor were reported to be required to achieve packaging termination. This step can be uncoupled from binding of head completion proteins that seal the portal channel (Orlova et al. 2003). However, during phage assembly, dissociation of the terminase–DNA complex from the portal shall be coordinated with the subsequent binding of head completion proteins, a requirement to produce metastable DNA-filled capsids in most phages. Conformational changes in the portal protein that precede or result from terminase departure were proposed to occur. Their role is to avoid packaged DNA exit from capsids as reported for phage T3 (Donate et al. 1988; Valpuesta et al. 1992) and/or to create the binding sites for the first head completion protein. These sites are expected to be exposed for binding only after DNA encapsidation to avoid that the head completion proteins compete with

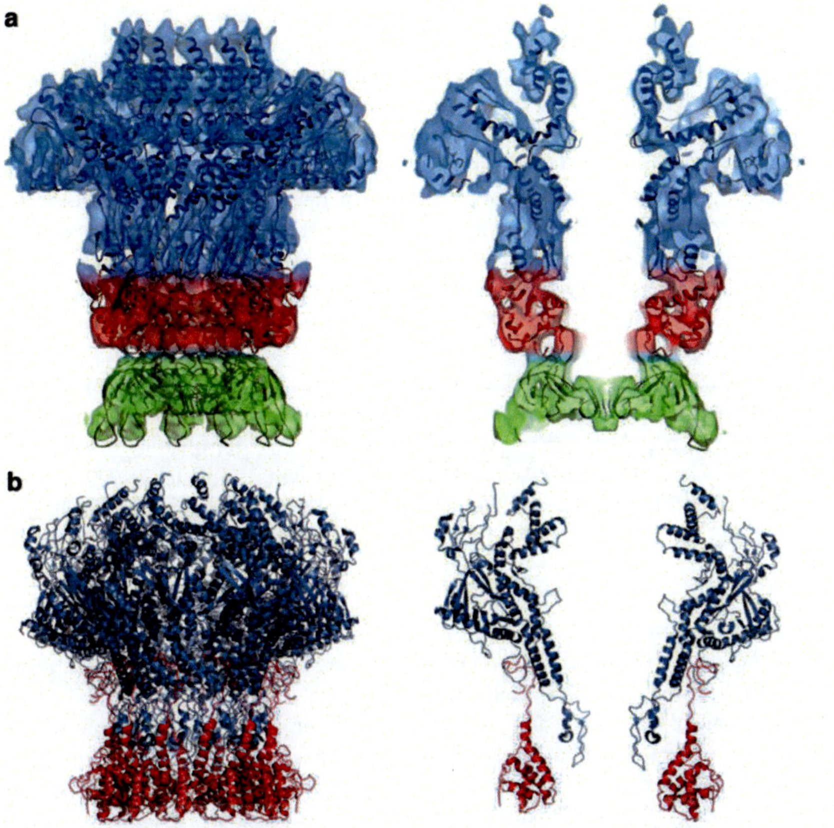

Fig. 25.7 SPP1 connector with the structures of its individual protein components (gp6, *blue* – portal; gp15, *magenta* – adaptor; gp16, *green* – stopper) fitted into the cryoEM map of the connector (Orlova et al. 2003; Lhuillier et al. 2009) (**a**) and structure of the P22 portal (gp1, *blue*) – adaptor (gp4, *magenta*) complex determined by X-ray crystallography (Olia et al. 2011) (**b**). Note that the large coiled-coil barrel of P22 gp1 (cf. Fig. 25.3) is not visible in this structure (Olia et al. 2011). Side views of the whole complex (*left*) and of a central slice of the EM map (SPP1) or of two opposite subunits (P22) (*right*) are shown

the terminase for assembly at the portal vertex, ensuring the correct order of assembly events for productive virus particles assembly. Interestingly, a single amino acid substitution in the SPP1 portal protein allows for headful cleavage but impairs retention of unit-length SPP1 chromosomes inside the capsid (Isidro et al. 2004a). The mutation is located in a portal region exposed to the capsid exterior where other mutations that block earlier DNA packaging steps were found (Isidro et al. 2004b; Lebedev et al. 2007; Oliveira et al. 2006, 2010) suggesting that the terminase and the adaptor gp15 bind sequentially to identical portal regions. Comparison between cryo-EM reconstructions of the isolated SPP1 portal and the portal embedded in the connector suggested structural changes in the portal clip that were interpreted to prime binding of gp15, but the boundaries between the two proteins could not be defined (Fig. 25.7a; Orlova et al. 2003; Lebedev et al. 2007). In contrast, the first atomic structure of a portal – head completion protein, the gp1–gp4 complex of phage P22 (Olia et al. 2011; Fig. 25.7b) – showed virtually no major differences between the portal in the complex and the structure of isolated portal protein gp1 at 7.5 Å resolution.

The P22 gp1–gp4 structure provides an instructive case for the mechanism of head completion proteins assembly at the portal protein (Olia et al. 2011). Gp4 (adaptor), gp10 (adaptor), and gp26 (stopper) bind sequentially to the portal vertex of P22 to seal the packaged DNA (Strauss and King

1984; Olia et al. 2007a). Gp4 is a monomer in solution at millimolar concentrations. It readily oligomerizes upon binding to the gp1 portal dodecamer (Olia et al. 2006; Zheng et al. 2008). In the crystallographic structure of the gp1–gp4 stacked dodecamers, each gp4 subunit binds between two gp1 subunits, and its C-terminal tail extends to sit at the interface between two other neighbor gp1 subunits. Each gp4 chain thus contacts a total of four different portal subunits explaining why gp4 does not bind to gp1 monomers. The binding interface between gp4 subunits in the complex is only ~900 Å^2 explaining also why gp4 does not oligomerize in the absence of portal oligomers. The structure reveals a mechanism for gp4 association driven by the portal but does not explain how the interaction is avoided earlier in assembly. This could be achieved if gp1 is a monomer when free in the infected cell (Moore and Prevelige 2001, 2002) and if the interfaces of gp1 where gp4 binds are shielded in the procapsid structure, becoming exposed upon departure of the terminase–DNA complex (Tang et al. 2011).

Gp4 of phage P22 is structurally similar to the adaptors gp15 of SPP1 and gp6 of HK97 (Fig. 25.4) suggesting that they share a common association mechanism to the portal. However, gp15 does not bind to isolated SPP1 portal protein oligomers, and it lacks the C-terminal region found in gp4 that is located in the groove of the interface between adjacent portal subunits in P22. It remains to be established if these differences account for the inability of gp15 to bind to the portal outside of the capsid context and if a similar portal–adaptor interaction interface is found in the P22 and SPP1 phage particle structures. Alternatively, a hinge-like structural rearrangement is proposed for gp15 stable assembly at the portal vertex on the basis of the docking of the gp15 structure into the cryo-EM map of the SPP1 connector (Lhuillier et al. 2009).

The gp6 adaptor of HK97 has the particularity to assemble cyclical oligomers at micromolar concentrations (Cardarelli et al. 2010a). It was observed that the gene coding for the adaptor protein has a weak ribosome binding site leading to production of very low amounts of gp6 which are probably monomers in infected bacteria. Overproduction of gp6 is highly detrimental for HK97 multiplication probably due to gp6 oligomerization, an effect that can be reversed by mutations that impair association of gp6 rings (Cardarelli et al. 2011). This study suggests that the requirement for adaptor monomers for connector assembly is fulfilled by a low-gene-dosage mechanism in case of phage HK97. It also shows that single amino acid substitutions can have a dramatic impact on the mechanisms of large macromolecular machines assembly.

Numerous nonassembled head completion proteins and components of long tails from many phages are characterized by the presence of large disordered regions that frequently account for more than 20% of the total protein amino acid sequence (Maxwell et al. 2001, 2002; Edmonds et al. 2007; Lhuillier et al. 2009; Pell et al. 2009). These regions are proposed to regulate assembly by folding upon interaction with their partners in a macromolecular complex and, in this way, creating the binding sites for the following component in the assembly pathway. A compelling case is illustrated by the gp16 stopper of SPP1. The gp16 monomer has two large unstructured loops (Fig. 25.6). When gp16 is in the SPP1 connector, both loops adopt well-defined positions that correspond to regions of electron density in the cryo-EM map of the connector suggesting that they acquired a defined fold (Lhuillier et al. 2009). One of these (β2′β3) fits into the stopper region that closes the connector (Figs. 25.6 and 25.7a). Modeling calculations and intersubunit cross-linking experiments with engineered disulfide bridges strongly suggest that the flexible loop of each subunit folds to form one of the 12 strands of the parallel β-sheet that closes the portal channel (Lhuillier et al. 2009). Studies with the structurally homologous protein gpFII indicated that this region corresponding to the gp16 β2′β3 loop is essential for tail attachment to DNA-filled capsids in phage λ (Cardarelli et al. 2010b).

In phages with long tails, the connector of the DNA-filled capsid binds to the tail end distal from the adsorption apparatus to form the neck. This end of the tail is formed by *tail completion proteins* which are required to stop assembly of the tail tube and to build a functional joint with their connector interaction partners (Chaps. 5 and 6).

25.3.4 DNA Priming for Ejection at the Head-to-Tail Interface

A critical requirement for phage viability is that the viral DNA can find its way out of the capsid at the beginning of infection. The observation that the last end of the dsDNA linear molecule that was packaged remains bound to the neck explains how polar exit of the viral chromosome from phage particles is ensured (Thomas 1974; Chattoraj and Inman 1974; Tavares et al. 1996). This DNA end was observed to exit first from virions when SPP1 DNA ejection is triggered in vitro (Tavares et al. 1996). Furthermore, λ phages defective for the neck–DNA-end interaction have very reduced infectivity in spite of the normal morphology of their viral particle (Thomas et al. 1978). The phenotype was attributed, at least partially, to inefficient delivery of viral DNA to the host bacterium. One of these mutants lacks the tail completion protein gpZ present in the neck structure of wild-type virions after the head-to-tail joining reaction. It was also proposed that λ DNA might penetrate the tail tube in mature virions (Thomas 1974). These observations suggest that λ tail binding to DNA-filled capsids leads to opening of the connector channel, previously closed by gpW (see Sect. 25.3.2; Perucchetti et al. 1988), allowing for DNA to enter the tail in the infectious phage. The situation is different in phage SPP1 where a 19–29 nm segment of the phage chromosome is firmly attached to the neck, a size that fits accurately to the height of the portal channel closed by the gp16 stopper (Fig. 25.7a; Tavares et al. 1996; Orlova et al. 2003).

The determination of asymmetric cryo-EM reconstructions from complete podovirus particles in recent years opened the way to direct visualization of DNA interactions in the connectors of phages T7 (Agirrezabala et al. 2005), ϕ29 (Xiang et al. 2006; Tang et al. 2008), ε15 (Jiang et al. 2006; Murata et al. 2010), P22 (Lander et al. 2006; Chang et al. 2006), and P-SSP7 (Liu et al. 2010) (Fig. 25.2; Chap. 7). DNA in the neck of myovirus T4 was also found (Leiman et al. 2004) (Fig. 25.2). Defining the boundaries between the averaged DNA molecule density and surrounding proteins is a challenging task. However, this problem is overcome by the improving resolution of the reconstructions and by the studies of structures with and without DNA that provide snapshots of the initial and final states of the viral genome release process (cf. structure of ϕ29 in Fig. 25.2).

25.4 Phage–Host Interaction: Trigger of the Portal System Opening and DNA Egress

Closure of the portal system preventing leakage of packaged DNA needs to be reversed for genome delivery to the host cell at the beginning of viral infection.

The phage tail adsorption apparatus of bacteriophages recognizes and binds specifically receptors at the bacterial surface (Vinga et al. 2006). These apparatus vary in their complexity from single tail fibers to intricate combinations of tail spikes and complex baseplates (Fig. 25.2; Rossmann et al. 2004; Agirrezabala et al. 2005; Tang et al. 2005; Xiang et al. 2006; Jiang et al. 2006; Lander et al. 2006; Plisson et al. 2007; Sciara et al. 2010; Veesler et al. 2010; Chaps. 5–7). Irreversible binding of the adsorption machinery to the bacterium commits the virus to infection. The tail structure then orchestrates the controlled transfer of the phage DNA from the viral capsid, across the bacterial envelope, to the bacterial cytoplasm (Vinga et al. 2006). The program of structural rearrangements that occur in the virus particle to fulfill these tasks is still poorly understood.

In phages with short tails, there are a limited number of tail effectors for infection, and the receptor-binding proteins (RBPs) are physically proximal or directly associated to the connector. This structural design requires a close coordination between RBPs binding to the receptors, attack of the bacterial membrane(s) and cell wall, and opening of the connector for DNA transfer to the cytoplasm. DNA ejection from phage P22 was triggered in vitro by aggregates of purified *Salmonella*

LPS (Andres et al. 2010). Multivalent binding of P22 tail spikes to LPS and O antigen hydrolysis was a prerequisite for DNA ejection suggesting a model in which the phage particle descends along the bacterial envelope through digestion of the envelope LPS chains. This movement facilitates possibly penetration of the gp26 tail needle through the cell envelope to reach the cytoplasmic membrane. Gp26, the phage P22 stopper (Sect. 25.3.2), would then open the connector for productive DNA release into the host cytoplasm (Andres et al. 2010) which requires also the pilot protein gp16 (Hoffman and Levine 1975). Cryo-electron tomography (cryo-ET) of podovirus P-SSP7 infecting its host cyanobacterium suggests also that a reorientation of the tail fibers precedes DNA release from viral particles (Liu et al. 2010). Cryo-ET observations of phage ε15 initiation of infection revealed that the phage short tail is prolonged by a tube that crosses the bacterial envelope establishing a direct connection between the phage capsid interior and the host cytoplasm (Chang et al. 2010). Future studies shall clarify the mechanism of ε15 connector opening and tube assembly.

In phages with long tails, the irreversible binding of the tail adsorption apparatus to bacterial receptors generates a signal to trigger DNA release from the viral capsid. In myoviruses, the interaction with the host surface causes a dramatic rearrangement of the tail shaft that surrounds the tail tube (Leiman et al. 2004; Rossmann et al. 2004). Contraction of the shaft leads to penetration of the tail tube across the cell envelope to access the bacterial membrane but can be uncoupled from DNA ejection at least in vitro (Leiman et al. 2004 and references therein). The signaling from the adsorption apparatus to trigger genome release is thus hypothesized to be communicated through the tail tube, possibly by a common mechanism in myo- and siphoviruses. The tube structure is built by a lattice of major tail protein(s) (MTPs) subunits organized in a sixfold symmetric helical arrangement, or threefold symmetric in a few phages (Effantin et al. 2006 and references therein), surrounding the internal tape measure protein (Kostyuchenko et al. 2005; Plisson et al. 2007 and references therein). The tape measure has an extended conformation that spans the length of the tube, from the adsorption apparatus to the neck, determining the length of the tail (Katsura 1987; Abuladze et al. 1994). The mechanism leading to DNA egress from the viral capsid can conceivably involve signaling through the helical tail structure and/or the tape measure protein. An informative approach to follow modifications in the virion upon interaction with the host cell is to reproduce the DNA ejection reaction in vitro as established for siphophages λ, T5, and SPP1 when incubated with LamB, FhuA, or YueB receptors, respectively (Roessner et al. 1983; Tosi et al. 1984; Boulanger et al. 1996; Mangenot et al. 2005; Böhm et al. 2001; São-José et al. 2006, 2007). Comparison of the phage SPP1 tail before and after DNA ejection showed that the tail tube internal wall undergoes extensive structural rearrangements (Plisson et al. 2007). These changes were proposed to be triggered by interaction of the tail adsorption with the YueB receptor. The resulting initial switch in the MTP subunits structure adjacent to the adsorption apparatus would be propagated as a domino-like cascade by a sequential rearrangement in the MTP hexameric rings along the 1,600-Å-long helical tail structure (Plisson et al. 2007). This signaling path could target directly the SPP1 neck leading to its opening or act indirectly, eliminating potential interactions of the MTP with the tape measure protein whose release from the tail tube lumen signals for aperture of the connector.

The mechanism of neck aperture for DNA release depends on the positioning of the DNA extremity bound at the head-to-tail interface at the end of virus particle assembly (Sect. 25.3.4). In case of phage λ, its DNA appears to penetrate partially the tail tube in mature virions (Thomas 1974) implying, in such case, that connector opening occurred already during viral assembly (Sect. 25.3.4). Signaling through the tail helical structure triggered by interaction of λ with the host would thus only play a role if it is necessary for tape measure release leaving the tube free for DNA exit. In case of phage SPP1 opening of the connector is required for DNA egress. Cysteine bridges engineered in the intersubunit β-sheet of the gp16 stopper that closes the connector central channel (Sect. 25.3.3) were shown to impair DNA ejection from SPP1 when exposed to its host receptor. The blocking effect was reversed when the disulfide bonds were removed by reduction (Lhuillier et al. 2009). These results indicate that viral DNA is retained at the level of the gp16 stopper in the mature virion. During infection, the specific interaction of the tail adsorption apparatus with the host triggers a signal transmitted to the connector that leads to unzipping of the gp16 intersubunit β-sheet, a key step of the mechanism of nucleocapsid gatekeeper opening for DNA release from the virus particle (Lhuillier et al. 2009).

25.5 Conclusions and Research Perspectives

The mechanisms of capsid assembly are highly conserved among viruses of the tailed bacteriophages–herpesviruses lineage. Functional and structural data reveal a number of strategies for closure of the portal DNA conduit during viral assembly and its aperture at the beginning of infection. They also suggest some variations on the mechanisms by which DNA is retained in the mature virion and thus primed for ejection. There is, at present, a wealth of knowledge on the effectors engaged in these processes that is complemented by information on their structure and organization in viral particles. In contrast, little is known about how DNA packaging termination is coordinated with the following steps of connector assembly, which precise molecular mechanisms ensure the correct sequence of protein–protein interactions, and what is(are) the trigger mechanism(s) for DNA release from virions. These questions are central to establish the unifying molecular strategies as well as the extent of diversity allowed by the tailed phages structure to retain their genome in the viral particle primed for infection. Their study shall also allow to pinpoint a component of the connector where the structural divergence between podoviruses and phages with long tails occurs. This might happen at the level of the β-strand-rich structures typified by the gpFII/gp16 family (Maxwell et al. 2002; Lhuillier et al. 2009), a fold also found among the different proteins building the tail tube (Cardarelli et al. 2010b).

Acknowledgments We are most thankful to Marie-Christine Vaney (Institut Pasteur, Paris) for the invaluable help to prepare Figs. 25.3–25.7. Work in our laboratories is supported by institutional funding from the CNRS (P.T and S.Z.-J.), the CEA (S.Z.-J.), the ANR grant "DNA Gating" (ANR-09-BLAN-0149-01), and the BBRSC (E.V.O.).

References

Abuladze NK, Gingery M, Tsai J, Eiserling FA (1994) Tail length determination in bacteriophage T4. Virology 199:301–310

Adams MB, Hayden M, Casjens S (1983) On the sequential packaging of bacteriophage P22 DNA. J Virol 46:673–677

Agirrezabala X, Martín-Benito J, Castón JR, Miranda R, Valpuesta JM, Carrascosa JL (2005) Maturation of phage T7 involves structural modification of both shell and inner core components. EMBO J 24:3820–3829

Andres D, Hanke C, Baxa U, Seul A, Barbirz S, Seckler R (2010) Tailspike interactions with lipopolysaccharide effect DNA ejection from phage P22 particles in vitro. J Biol Chem 285:36768–36775

Bamford DH, Grimes JM, Stuart DI (2005) What does structure tell us about virus evolution? Curr Opin Struct Biol 15:655–663

Bazinet C, King J (1985) The DNA translocating vertex of dsDNA bacteriophage. Annu Rev Microbiol 39:109–129

Bhardwaj A, Olia AS, Walker-Kopp N, Cingolani G (2007) Domain organization and polarity of tail needle GP26 in the portal vertex structure of bacteriophage P22. J Mol Biol 371:374–387

Bhardwaj A, Walker-Kopp N, Casjens SR, Cingolani G (2009) An evolutionarily conserved family of virion tail needles related to bacteriophage P22 gp26: correlation between structural stability and length of the alpha-helical trimeric coiled coil. J Mol Biol 391:227–245

Black LW (1989) DNA packaging in dsDNA bacteriophages. Annu Rev Microbiol 43:267–292

Böhm J, Lambert O, Frangakis AS, Letellier L, Baumeister W, Rigaud JL (2001) FhuA-mediated phage genome transfer into liposomes: a cryo-electron tomography study. Curr Biol 11:1168–1175

Boulanger P, le Maire M, Bonhivers M, Dubois S, Desmadril M, Letellier L (1996) Purification and structural and functional characterization of FhuA, a transporter of the *Escherichia coli* outer membrane. Biochemistry 35: 14216–14224

Camacho AC, Gual A, Lurz R, Tavares P, Alonso JC (2003) *Bacillus subtilis* bacteriophage SPP1 DNA packaging motor requires terminase and portal proteins. J Biol Chem 278:23251–23259

Cardarelli L, Lam R, Tuite A, Baker LA, Sadowski PD, Radford DR, Rubinstein JL, Battaile KP, Chirgadze N, Maxwell KL, Davidson AR (2010a) The crystal structure of bacteriophage HK97 gp6: defining a large family of head-tail connector proteins. J Mol Biol 395:754–768

Cardarelli L, Pell LG, Neudecker P, Pirani N, Liu A, Baker LA, Rubinstein JL, Maxwell KL, Davidson AR (2010b) Phages have adapted the same protein fold to fulfill multiple functions in virion assembly. Proc Natl Acad Sci USA 107:14384–14389

Cardarelli L, Maxwell KL, Davidson AR (2011) Assembly mechanism is the key determinant of the dosage sensitivity of a phage structural protein. Proc Natl Acad Sci USA 108(25):10168–10173

Casjens S, Hayden M (1988) Analysis *in vivo* of the bacteriophage P22 headful nuclease. J Mol Biol 199:467–474

Casjens S, Hendrix R (1988) Control mechanisms in dsDNA bacteriophage assembly. In: Calendar R (ed) The bacteriophages, vol 1. Plenum Press, New York

Casjens S, Horn T, Kaiser AD (1972) Head assembly steps controlled by genes F and W in bacteriophage lambda. J Mol Biol 64:551–563

Casjens S, Wyckoff E, Hayden M, Sampson L, Eppler K, Randall S, Moreno E, Serwer P (1992) Bacteriophage P22 portal protein is part of the gauge that regulates packing density of intravirion DNA. J Mol Biol 224:1055–1074

Chang J, Weigele P, King J, Chiu W, Jiang W (2006) Cryo-EM asymmetric reconstruction of bacteriophage P22 reveals organization of its DNA packaging and infecting machinery. Structure 14:1073–1082

Chang JT, Schmid MF, Haase-Pettingell C, Weigele PR, King JA, Chiu W (2010) Visualizing the structural changes of bacteriophage epsilon 15 and its Salmonella host during infection. J Mol Biol 402:731–740

Chattoraj DK, Inman RB (1974) Location of DNA ends in P2, 186, P4 and lambda bacteriophage heads. J Mol Biol 87:11–22

Cingolani G, Moore SD, Prevelige PE Jr, Johnson JE (2002) Preliminary crystallographic analysis of the bacteriophage P22 portal protein. J Struct Biol 139:46–54

Donate LE, Herranz L, Secilla JP, Carazo JM, Fujisawa H, Carrascosa JL (1988) Bacteriophage T3 connector: three-dimensional structure and comparison with other viral head-tail connecting regions. J Mol Biol 201:91–100

Dube P, Tavares P, Lurz R, van Heel M (1993) Bacteriophage SPP1 portal protein: a DNA pump with 13-fold symmetry. EMBO J 12:1303–1309

Edmonds L, Liu A, Kwan JJ, Avanessy A, Caracoglia M, Yang I, Maxwell KL, Rubenstein J, Davidson AR, Donaldson LW (2007) The NMR structure of the gpU tail-terminator protein from bacteriophage lambda: identification of sites contributing to Mg(II)-mediated oligomerization and biological function. J Mol Biol 365:175–186

Effantin G, Boulanger P, Neumann E, Letellier L, Conway JF (2006) Bacteriophage T5 structure reveals similarities with HK97 and T4 suggesting evolutionary relationships. J Mol Biol 361:993–1002

Evilevitch A, Lavelle L, Knobler CM, Raspaud E, Gelbart WM (2003) Osmotic pressure inhibition of DNA ejection from phage. Proc Natl Acad Sci USA 100:9292–9295

Feiss M, Catalano CE (2005) Bacteriophage lambda terminase and the mechanism of viral DNA packaging. In: Catalano CE (ed) Viral genome packaging machines: genetics, structure, and mechanism. Landes Bioscience, Georgetown, TX

Feiss M, Siegele DA (1979) Packaging of the bacteriophage lambda chromosome: dependence of cos cleavage on chromosome length. Virology 92:190–200

Feiss M, Sippy J, Miller G (1985) Processive action of terminase during sequential packaging of bacteriophage λ chromosomes. J Mol Biol 186:759–771

Fokine A, Chipman PR, Leiman PG, Mesyanzhinov VV, Rao VB, Rossmann MG (2004) Molecular architecture of the prolate head of bacteriophage T4. Proc Natl Acad Sci USA 101:6003–6008

Fujisawa H, Morita M (1997) Phage DNA packaging. Genes Cells 2:537–545

Gaussier H, Yang Q, Catalano CE (2006) Building a virus from scratch: assembly of an infectious virus using purified components in a rigorously defined biochemical assay system. J Mol Biol 357:1154–1166

Guasch A, Pous J, Ibarra B, Gomis-Rüth FX, Valpuesta JM, Sousa N, Carrascosa JL, Coll M (2002) Detailed architecture of a DNA translocating machine: the high-resolution structure of the bacteriophages phi29 connector particle. J Mol Biol 315:663–676

Hashimoto C, Fujisawa H (1992) DNA sequences necessary for packaging bacteriophage T3 DNA. Virology 187:788–795

Hoffman B, Levine M (1975) Bacteriophage P22 virion protein which performs an essential early function. II. Characterization of the gene *16* function. J Virol 16:1547–1559

Isidro A, Henriques AO, Tavares P (2004a) The portal protein plays essential roles at different steps of the SPP1 DNA packaging process. Virology 322:253–263

Isidro A, Santos MA, Henriques AO, Tavares P (2004b) The high-resolution functional map of bacteriophage SPP1 portal protein. Mol Microbiol 51:949–962

Jiang W, Chang J, Jakana J, Weigele P, King J, Chiu W (2006) Structure of epsilon 15 bacteriophage reveals genome organization and DNA packaging/injection apparatus. Nature 439:612–661

Johnson JE, Chiu W (2007) DNA packaging and delivery machines in tailed bacteriophages. Curr Opin Struct Biol 17:237–243

Katsura I (1987) Determination of bacteriophage lambda tail length by a protein ruler. Nature 327:73–75

Kostyuchenko VA, Chipman PR, Leiman PG, Arisaka F, Mesyanzhinov VV, Rossmann MG (2005) The tail structure of bacteriophage T4 and its mechanism of contraction. Nat Struct Mol Biol 12:810–813

Lander GC, Tang L, Casjens SR, Gilcrease EB, Prevelige P, Poliakov A, Potter CS, Carragher B, Johnson JE (2006) The structure of an infectious P22 virion shows the signal for headful DNA packaging. Science 312:1791–1795

Lebedev AA, Krause MH, Isidro AL, Vagin A, Orlova EV, Turner J, Dodson EJ, Tavares P, Antson AA (2007) Structural framework for DNA translocation via the viral portal protein. EMBO J 26:1984–1994

Leiman PG, Chipman PR, Kostyuchenko VA, Mesyanzhinov VV, Rossmann MG (2004) Three-dimensional rearrangement of proteins in the tail of bacteriophage T4 on infection of its host. Cell 118:419–429

Lhuillier S, Gallopin M, Gilquin B, Brasilès S, Lancelot N, Letellier G, Gilles M, Dethan G, Orlova EV, Couprie J, Tavares P, Zinn-Justin S (2009) Structure of bacteriophage SPP1 head-to-tail connection reveals mechanism for viral DNA gating. Proc Natl Acad Sci USA 106:8507–8512

Liu X, Zhang Q, Murata K, Baker ML, Sullivan MB, Fu C, Dougherty MT, Schmid MF, Osburne MS, Chisholm SW, Chiu W (2010) Structural changes in a marine podovirus associated with release of its genome into Prochlorococcus. Nat Struct Mol Biol 17:830–836

Lurz R, Orlova EV, Günther D, Dube P, Dröge A, Weise F, van Heel M, Tavares P (2001) Structural organisation of the head-to-tail interface of a bacterial virus. J Mol Biol 310:1027–1037

Mangenot S, Hochrein M, Radler J, Letellier L (2005) Real-time imaging of DNA ejection from single phage particles. Curr Biol 15:430–435

Maxwell KL, Davidson AR, Murialdo H, Gold M (2000) Thermodynamic and functional characterization of protein W from bacteriophage lambda. The three C-terminal residues are critical for activity. J Biol Chem 275:18879–18886

Maxwell KL, Yee AA, Booth V, Arrowsmith CH, Gold M, Davidson AR (2001) The solution structure of bacteriophage lambda protein W, a small morphogenetic protein possessing a novel fold. J Mol Biol 308:9–14

Maxwell KL, Yee AA, Arrowsmith CH, Gold M, Davidson AR (2002) The solution structure of the bacteriophage lambda head-tail joining protein, gpFII. J Mol Biol 318:1395–1404

Moore SD, Prevelige PE Jr (2001) Structural transformations accompanying the assembly of bacteriophage P22 portal protein rings in vitro. J Biol Chem 276:6779–6788

Moore SD, Prevelige PE Jr (2002) Bacteriophage P22 portal vertex formation in vivo. J Mol Biol 315:975–994

Murata K, Liu X, Danev R, Jakana J, Schmid MF, King J, Nagayama K, Chiu W (2010) Zernike phase contrast cryo-electron microscopy and tomography for structure determination at nanometer and subnanometer resolutions. Structure 18:903–912

Newcomb WW, Juhas RM, Thomsen DR, Homa FL, Burch AD, Weller SK, Brown JC (2001) The UL6 gene product forms the portal for entry of DNA into the herpes simplex virus capsid. J Virol 75:10923–10932

Olia AS, Al-Bassam J, Winn-Stapley DA, Joss L, Casjens SR, Cingolani G (2006) Binding-induced stabilization and assembly of the phage P22 tail accessory factor gp4. J Mol Biol 363:558–576

Olia AS, Bhardwaj A, Joss L, Casjens S, Cingolani G (2007a) Role of gene 10 protein in the hierarchical assembly of the bacteriophage P22 portal vertex structure. Biochemistry 46:8776–8784

Olia AS, Casjens S, Cingolani G (2007b) Structure of phage P22 cell envelope-penetrating needle. Nat Struct Mol Biol 14:1221–1226

Olia AS, Prevelige PE Jr, Johnson JE, Cingolani G (2011) Three-dimensional structure of a viral genome-delivery portal vertex. Nat Struct Mol Biol 18:597–603

Oliveira L, Henriques AO, Tavares P (2006) Modulation of the viral ATPase activity by the portal protein correlates with DNA packaging efficiency. J Biol Chem 281:21914–21923

Oliveira L, Cuervo A, Tavares P (2010) Direct interaction of the bacteriophage SPP1 packaging ATPase with the portal protein. J Biol Chem 285:7366–7373

Orlova EV, Dube P, Beckmann E, Zemlin F, Lurz R, Trautner TA, Tavares P, van Heel M (1999) Structure of the 13-fold symmetric portal protein of bacteriophage SPP1. Nat Struct Biol 6:842–846

Orlova EV, Gowen B, Dröge A, Stiege A, Weise F, Lurz R, van Heel M, Tavares P (2003) Structure of a viral DNA gatekeeper at 10 Å resolution by cryo-electron microscopy. EMBO J 22:1255–1262

Pell LG, Kanelis V, Donaldson LW, Howell PL, Davidson AR (2009) The phage lambda major tail protein structure reveals a common evolution for long-tailed phages and the type VI bacterial secretion system. Proc Natl Acad Sci USA 106:4160–4165

Perucchetti R, Parris W, Becker A, Gold M (1988) Late stages in bacteriophage lambda head morphogenesis: in vitro studies on the action of the bacteriophage lambda D-gene and W-gene products. Virology 165:103–114

Plisson C, White HE, Auzat I, Zafarani A, São-José C, Lhuillier S, Tavares P, Orlova EV (2007) Structure of bacteriophage SPP1 tail reveals trigger for DNA ejection. EMBO J 26:2728–3720

Rao VB, Feiss M (2008) The bacteriophage DNA packaging motor. Annu Rev Genet 42:647–681

Roessner CA, Struck DK, Ihler GM (1983) Injection of DNA into liposomes by bacteriophage lambda. J Biol Chem 258:643–648

Rossmann MG, Mesyanzhinov VV, Arisaka F, Leiman PG (2004) The bacteriophage T4 DNA injection machine. Curr Opin Struct Biol 14:171–180

Salas M (2006) Phage ϕ29 and its relatives. In: Calendar R (ed) The bacteriophages, 2nd edn. Oxford University Press, New York

São-José C, Lhuillier S, Lurz R, Melki R, Lepault J, Santos MA, Tavares P (2006) The ectodomain of the viral receptor YueB forms a fiber that triggers DNA ejection of bacteriophage SPP1 DNA. J Biol Chem 281:11464–11470

São-José C, de Frutos M, Raspaud E, Santos MA, Tavares P (2007) Pressure built by DNA packing inside virions: enough to drive DNA ejection in vitro, largely insufficient for delivery into the bacterial cytoplasm. J Mol Biol 374:346–355

Sciara G, Bebeacua C, Bron P, Tremblay D, Ortiz-Lombardia M, Lichière J, van Heel M, Campanacci V, Moineau S, Cambillau C (2010) Structure of lactococcal phage p2 baseplate and its mechanism of activation. Proc Natl Acad Sci USA 107:6852–6857

Simpson AA, Tao Y, Leiman PG, Badasso MO, He Y, Jardine PJ, Olson NH, Morais MC, Grimes S, Anderson DL, Baker TS, Rossmann MG (2000) Structure of the bacteriophage ϕ29 DNA packaging motor. Nature 408:745–750

Smith DE, Tans SJ, Smith SB, Grimes S, Anderson DL, Bustamante C (2001) The bacteriophage ϕ29 portal motor can package DNA against a large internal force. Nature 413:748–752

Strauss H, King J (1984) Steps in the stabilisation of newly packaged DNA during phage P22 morphogenesis. J Mol Biol 172:523–543

Streisinger G, Emrich J, Stahl MM (1967) Chromosome structure in phage T4, iii. Terminal redundancy and length determination. Proc Natl Acad Sci USA 57:292–295

Sun S, Kondabagil K, Draper B, Alam TI, Bowman VD, Zhang Z, Hegde S, Fokine A, Rossmann MG, Rao VB (2008) The structure of the phage T4 DNA packaging motor suggests a mechanism dependent on electrostatic forces. Cell 135:1251–1262

Tang L, Marion WR, Cingolani G, Prevelige PE, Johnson JE (2005) Three-dimensional structure of the bacteriophage P22 tail machine. EMBO J 24:2087–2095

Tang J, Olson N, Jardine PJ, Grimes S, Anderson DL, Baker TS (2008) DNA poised for release in bacteriophage phi29. Structure 16:935–943

Tang J, Lander GC, Olia A, Li R, Casjens S, Prevelige P Jr, Cingolani G, Baker TS, Johnson JE (2011) Peering down the barrel of a bacteriophage portal: the genome packaging and release valve in P22. Structure 19:496–502

Tavares P, Santos MA, Lurz R, Morelli G, de Lencastre H, Trautner TA (1992) Identification of a gene in *Bacillus subtilis* bacteriophage SPP1 determining the amount of packaged DNA. J Mol Biol 225:81–92

Tavares P, Lurz R, Stiege A, Rückert B, Trautner TA (1996) Sequential headful packaging and fate of the cleaved DNA ends in bacteriophage SPP1. J Mol Biol 264:954–967

Thomas JO (1974) Chemical linkage of the tail to the right-hand end of bacteriophage lambda DNA. J Mol Biol 87:1–9

Thomas JO, Sternberg N, Weisberg R (1978) Altered arrangement of the DNA in injection-defective lambda bacteriophage. J Mol Biol 123:149–161

Tosi F, Labedan B, Legault-Démare J (1984) Analysis of the coliphage T5 DNA ejection process with free and liposome-associated TonA protein. J Virol 50:213–219

Trus BL, Cheng N, Newcomb WW, Homa FL, Brown JC, Steven AC (2004) Structure and polymorphism of the UL6 portal protein of herpes simplex virus type 1. J Virol 78:12668–12671

Tye BK, Huberman JA, Botstein D (1974) Non-random circular permutation of phage P22 DNA. J Mol Biol 85:501–532

Valpuesta JM, Carrascosa JL (1994) Structure of viral connectors and their function in bacteriophage assembly and DNA packaging. Q Rev Biophys 27:107–155

Valpuesta JM, Fujisawa H, Marco S, Carazo JM, Carrascosa JL (1992) Three-dimensional structure of T3 connector purified from overexpressing bacteria. J Mol Biol 224:103–112

van Heel M, Orlova EV, Dube P, Tavares P (1996) Intrinsic versus imposed curvature in cyclical oligomers: the portal protein of bacteriophage SPP1. EMBO J 15:4785–4788

Veesler D, Robin G, Lichière J, Auzat I, Tavares P, Bron P, Campanacci V, Cambillau C (2010) Crystal structure of bacteriophage SPP1 distal tail protein (gp 19.1): a baseplate hub paradigm in gram-positive infecting phages. J Biol Chem 285:36666–36673

Vinga I, São-José C, Tavares P, Santos MA (2006) Bacteriophage entry in the host cell. In: Wegrzyn G (ed) Modern bacteriophage biology and biotechnology. Research Signpost, Kerala, India

Xiang Y, Morais MC, Battisti AJ, Grimes S, Jardine PJ, Anderson DL, Rossmann MG (2006) Structural changes of bacteriophage phi29 upon DNA packaging and release. EMBO J 25:5229–5239

Zheng H, Olia AS, Gonen M, Andrews S, Cingolani G, Gonen T (2008) A conformational switch in bacteriophage p22 portal protein primes genome injection. Mol Cell 29:376–383

Chapter 26
Packaging in dsRNA Viruses

Leonard Mindich

Abstract Several families of viruses have segmented genomes with 3–12 chromosomes. They are capable of packaging these segments in a precise manner so that each virus particle contains one each of the genomic segments. The Cystoviridae are a family of bacteriophages that contain three genomic segments of dsRNA. During infection, the virus produces empty dodecahedral core particles that have the ability to specifically package plus strand transcripts of the genomic segments. The program of packaging makes use of the conformational changes in the surface of the particle as each transcript is packaged. The particles have complexes of a hexameric NTPase that serve as motors to bring the transcripts into the particle, and they have polymerase molecules in the interior that synthesize minus and plus strand copies of the genomic segments.

26.1 Introduction

The cystoviruses are bacteriophages that contain a genome of three unique double-stranded molecules of RNA. The family of Reoviridae consists of viruses that have genomes of 10, 11, or 12 dsRNA segments. In all of these cases, the packaging of the RNA is precise, and each virus particle has one each of the genomic segments. The mechanism for accomplishing this is unknown for the Reoviridae but has been worked out for the Cystoviridae, primarily because of the fortunate availability of a functioning in vitro packaging system.

The core particle of the Cystoviridae is a complex structure that not only encloses the double-stranded RNA (dsRNA) genome of the virus but also plays a role in the packaging of the transcripts of genomic segments and the regulation of their replication. Empty cores are capable of packaging plus strand transcripts of the genomic segments in a serially dependent order (Qiao et al. 1995). Polymerase molecules, which are located inside the cores, are activated at a particular step in the packaging program to synthesize minus strand copies of the packaged RNA (Frilander et al. 1992). Polymerase is again activated to synthesize plus strand molecules when minus strand synthesis is completed. But this transcription is under yet another set of controls wherein specific host proteins bind to the exterior of the cores and activate the polymerases to transcribe one of the genomic segments.

L. Mindich (✉)
Public Health Research Institute Center, UMDNJ - New Jersey Medical School,
225 Warren Street, Newark, NJ 07103, USA
e-mail: mindicle@umdnj.edu

M.G. Rossmann and V.B. Rao (eds.), *Viral Molecular Machines*,
Advances in Experimental Medicine and Biology 726, DOI 10.1007/978-1-4614-0980-9_26,

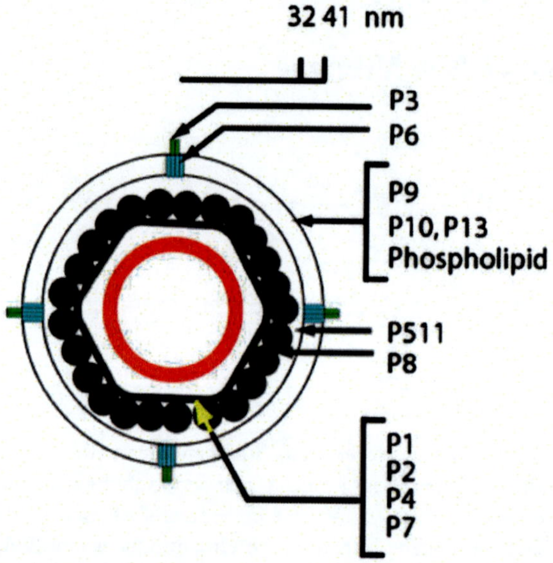

Fig. 26.1 Diagram of the structure of bacteriophage Φ6

In order to understand the mechanisms of these activities, we have to first describe the structure of the cores (Fig. 26.1). The cores are composed of 120 molecules of protein P1 that alone can form a dodecahedral structure. Dodecahedra are 12-sided solids with five-, three-, and twofold symmetry axes. The P1 molecules are arranged as pentamers at the fivefold vertices and trimers at the threefold vertices (Jaalinoja et al. 2007). Twelve hexameric structures of protein P4 are located at the exterior of the fivefold vertices. P4 is an NTPase and is the motor for the packaging of the plus strands into the core particles. Twelve molecules of the RNA-dependent RNA polymerase P2 are located at the interior of the fivefold vertices but are probably somewhat off center (Sen et al. 2008). There is a fourth protein, P7, that plays a role in genomic packaging and replication but whose exact location in the core is presently unknown. There are about 60 molecules of P7 per core. In filled core particles, there are three dsRNA genomic segments L, M, and S that code for about 14 genes depending upon the particular virus species. The filled particles are almost spherical in shape while empty cores have their fivefold vertices drawn into the interior of the particle with the result that the interior volume of the empty particle is much smaller than that of the filled one (Figs. 26.2 and 26.3). The structures of the empty and filled core particles have been investigated by cryoelectron microscopy, and high resolution images have been obtained (Butcher et al. 1997; de Haas et al. 1999; Sen et al. 2008). The polymerase of Φ6 has been crystallized, and X-ray diffraction studies have resulted in molecular models of the protein in several conformations (Butcher et al. 2001; Salgado et al. 2004). The NTPase P4 has been crystallized from bacteriophage Φ12, and its structure has been described in great detail (Mancini et al. 2004). Recently, the structure of P7 of Φ12 was also crystallized and described (Eryilmaz et al. 2008). Although the secondary structures of P1 have been determined by high resolution cryoelectron microscopy, it has not been possible to obtain structures at the level of atomic resolution (Huiskonen et al. 2006).

In order to place the operation of the cystovirus cores in context, we should describe the life cycle of these bacteriophages. The core particle is enclosed in a shell of protein P8 to form the nucleocapsid (Fig. 26.1). This structure is enclosed in a lipid-containing membrane that contains a major structural protein P9 and a host attachment protein P3 anchored by a membrane protein P6. In some cases, there are several attachment proteins 3a and 3b (Mindich et al. 1999). The phage attaches to a type IV pilus or to rough LPS (Mindich et al. 1999). The membrane fuses with the outer membrane of the host *Pseudomonas syringae* to place the nucleocapsid in the periplasmic space (Bamford et al. 1987). A muramidase P5 on the surface of the nucleocapsid digests a pathway through the cell wall

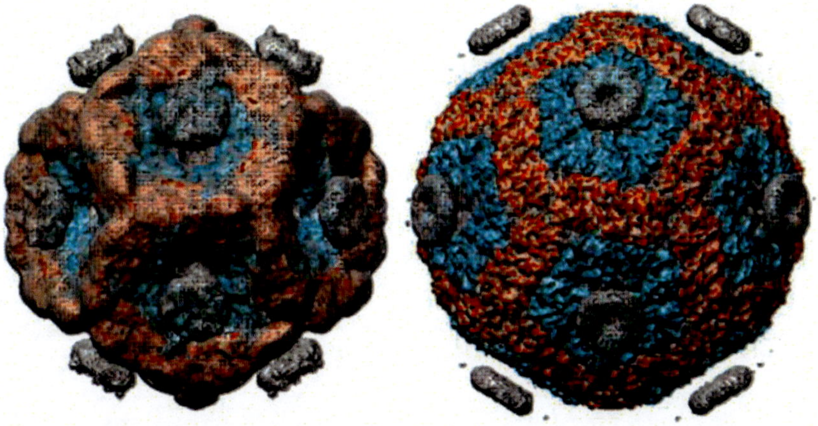

Fig. 26.2 Image reconstructions of an empty and a filled procapsid of bacteriophage Φ6. Discs at the fivefold vertices are hexamers of the packaging NTPase P4. Pentamers of P1 are shown in *blue* while trimers are shown in *brown* (Huiskonen et al. 2006)

Fig. 26.3 Section through a reconstruction of the Φ6 procapsid showing the position of the RNA-dependent polymerase molecules shown in *green* (Sen et al. 2008)

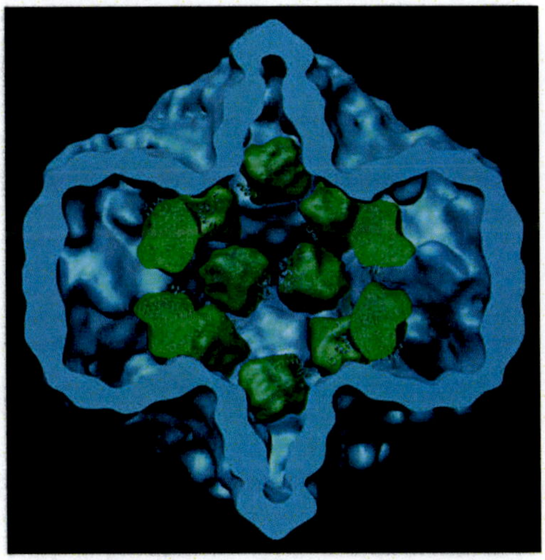

to allow the nucleocapsid to enter the cytoplasm (Mindich and Lehman 1979; Romantschuk et al. 1988). Inside the cell, the nucleocapsid loses the P8 shell, and the core particle is bound by a specific host protein that activates the transcription of genomic segment L. The transcription of segments S and M does not require the host protein activation. The L transcript codes for the four proteins of the core, and these are formed early in infection. The empty core particles are able to package the plus strand transcripts and to synthesize minus strand copies to form dsRNA. These particles are then able to transcribe plus strands which result in the expression of the late proteins of the virus and the formation of nucleocapsids, which are then enclosed in viral membranes, which bud off from the host cytoplasmic membrane. A combination of a holin P10 and the muramidase P5 leads to the lysis of the host cells, liberating several hundred mature virions (Mindich and Lehman 1979). Under some circumstances, the viral infection can lead to a carrier state wherein the virus replicates in concert with the host system and does not lyse the cell (Cuppels et al. 1979). In these cases, reporter genes in the virus genome can be expressed much as genes on plasmids (Onodera et al. 1992).

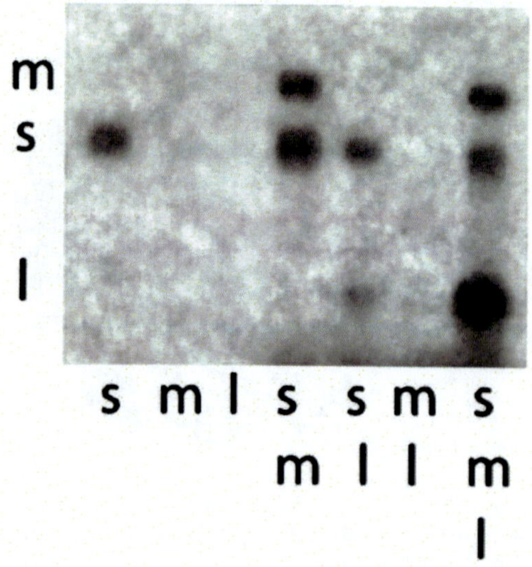

Fig. 26.4 Autoradiogram showing the results of in vitro packaging of labeled transcripts of S, M, and a truncated segment L (Qiao et al. 1995)

26.1.1 The Genomic Packaging Model

Empty particles can be formed by the expression of the genes in genomic segment L expressed in plasmids in *Escherichia coli*. These particles can package plus strand transcripts of the genomic segments in vitro in the presence of ATP or other nucleotide triphosphates (Qiao et al. 1995). In the case of Φ6, the most intensively studied member of the Cystoviridae, there are sequences of about 200 nucleotides about 50 bases from the 5′ end of the transcripts that are necessary for packaging (Fig. 26.4) (Gottlieb et al. 1994). In the case of Φ6, the distance from the *pac* sequences to the 5′ end is critical for packaging. A deletion of 11 bases allows binding but not packaging. A larger deletion prevents both binding and packaging. However, in some phages, the *pac* sequences are much closer to the 5′ end (Qiao et al. 2003b). The *pac* sequences contain a number of stem-loop structures, and these seem to be important. Closely related phages show similar stem-loop structures with base changes that conserve the structures (Mindich 1999). The *pac* sequences are different for each of the genomic segments. In Φ6, there are sequences of 18 bases at the 5′ ends that are identical for the three segments with the only difference being that segments S and M begin with GG while L begins with GU.

The transcript of segment S can be packaged alone; however, the transcript of M can only be packaged after the packaging of the S transcript. The transcript of L can only be packaged after the packaging of the M transcript. The phages have one segment of each of the three under normal conditions. However, if the transcript of S was made half its normal size, two transcripts would be packaged along with one of M and one of L. If the transcript was even smaller, multiple copies would be packaged so that the total amount of RNA for that segment would be equal to that of the normal transcript (Fig. 26.5). The empty particle has binding sites for the S transcript, but the binding sites for the transcripts of M and L are not exposed (Qiao et al. 2003b). The binding sites are on protein P1. Cross-linking of labeled RNA to particles shows binding to P1 (Qiao et al. 2003b). Mutations in the *pac* sequences have been made, and some of these result in the lack of packaging activity. However, suppressor mutations can be selected, and they turn out to be in the gene for protein P1 (Qiao et al. 2003a).

The Cystoviridae are able to establish carrier state infections in their host cells (Cuppels et al. 1979). These are fairly stable and can be maintained easily if drug resistance genes are incorporated into one of the genomic segments. The phage does not require all of its genes for maintenance of the carrier state since the phage does not have to reinfect cells. In some cases where a kanamycin resistance gene is located in segment M, the phages in this state lose genomic segment S. Since the packaging program normally is serially dependent with the packaging of the S transcript as the first

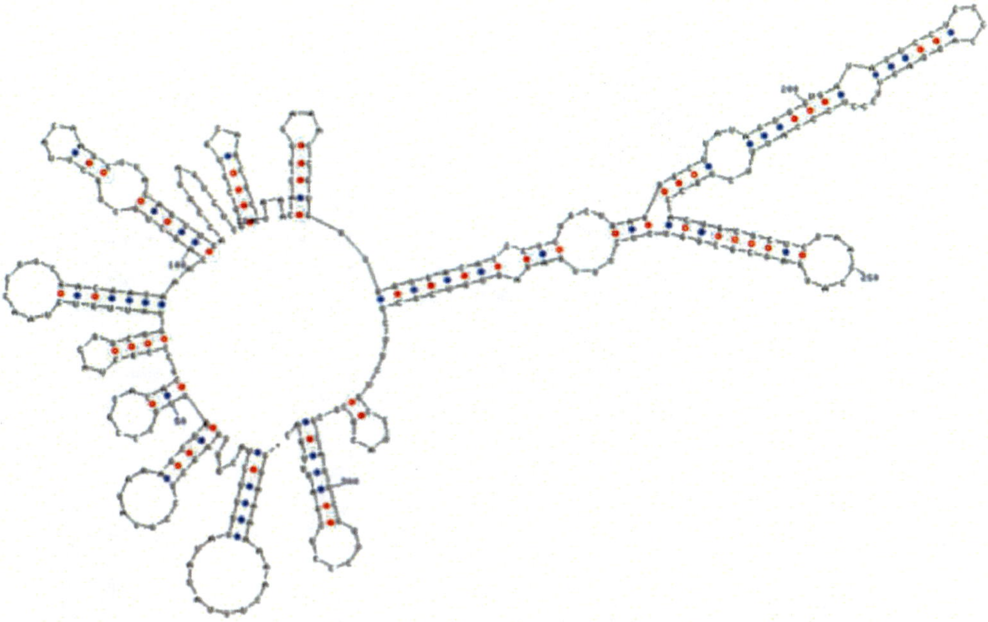

Fig. 26.5 Prediction of the folding pattern of the 5′ region of the Φ6 S transcript (Mindich 1999; Zuker 2003)

molecule to enter the particle, this finding seemed anomalous. However, it was found that these phages had produced mutations in the gene for protein P1. When empty core particles were produced from the cDNA copies of the mutant segment L, it was found that they would package transcripts of M and L without that of S. These results further supported the idea that the specificity of genomic packaging is manifested through interaction of the RNA with protein P1 (Qiao et al. 2003a).

A model was proposed to explain the results of the packaging experiments (Fig. 26.6) (Mindich 1999). It was stated that the empty core particles had binding sites for the transcript of S on its surface, and that these binding sites are in protein P1. RNA is bound to these sites such that the 5′ end is positioned on one of the hexameric NTPase complexes of protein P4. The details of the entry of RNA through the P4 hexamers are described fully by Mancini and Tuma in Chap. 27. The RNA is drawn into the core through a process involving the hydrolysis of ATP. The core particle expands as the S transcript is drawn in, and this process involves the rearranging of the surface of the particle so that the binding sites for S are lost and those for M appear. The process is then repeated for M. And finally, the process is repeated for the transcript of L. A model for the expansion of the core has been developed by the laboratory of Sarah Butcher (Huiskonen et al. 2006) where the molecules of P1 change their orientations but not their shapes. The nature of the binding sites is presently unknown. It is not known whether the conformation of P1 changes during expansion or whether the exposure of particular sites changes or that the RNA is binding to several molecules of P1 and that the distances between sites on different P1 molecules is the determinant of binding specificity. It is not known whether the binding sites for S involve the same amino acids as those for M or L or whether the different binding sites involve different domains of protein P1. However, there is a mutation in protein P1 that results in the packaging of a mutated S *pac* sequence but excludes S when M is present (Qiao et al. 2003a).

The model produces some predictions that have been tested. If the transcripts are smaller than normal, multiple molecules should be packaged to reach the content of a normal transcript. This has been shown to be true for transcripts of S, M, and L. If the transcript of S is made larger, so that it is equal to that of S and M combined, it should be possible to package L without M. This has been found to be true. In addition, the entire genome can be combined into a single transcript, and this can be packaged as long as it has the *pac* sequence of S at its 5′ end (Onodera et al. 1998).

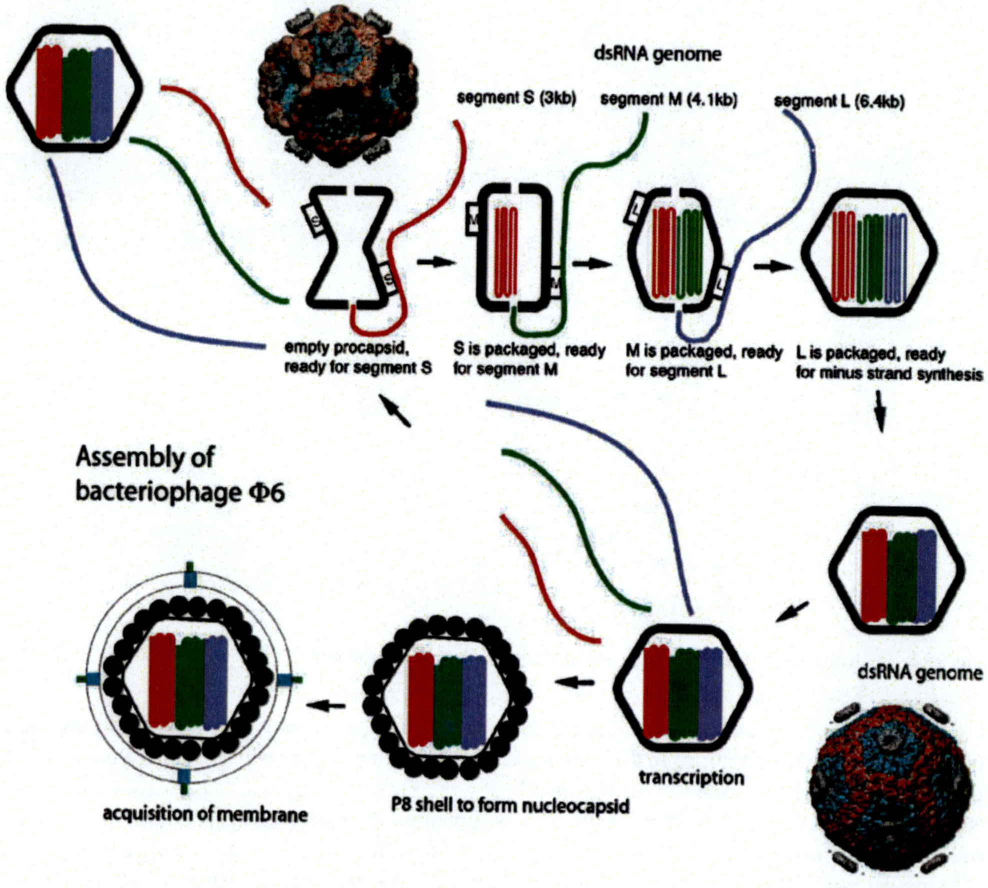

Fig. 26.6 The packaging model (Mindich 1999)

26.1.2 *Control of Polymerase Activity*

We see then, that the packaging process is very orderly and is under a stringent control. There is another level of control that is manifested in the packaging process. The polymerase molecules are capable of synthesizing the minus strand copies of the transcripts when incubated in vitro in the absence of the capsid structure (Makeyev and Bamford 2000). However, when inside the core particles, minus strand synthesis does not begin until the entire genome is packaged (Frilander et al. 1992). The explanation is that the polymerase molecules inside the core are in a conformation that prevents activity until the core particle attains its expanded state. It appears that the structure of the capsid shell determines the activity of the polymerase. It is also found that transcription of the genomic segments does not take place until minus strand synthesis is complete. This too appears to be a consequence of the further expansion of the capsid shell.

How the conformation of the capsid shell determines the activity of the polymerase is a mystery. It is of interest that the activity of the polymerase of the rotavirus is only activated when an additional shell is formed around the core capsid (Sandino et al. 1986). Rotavirus is rather similar to the cystoviruses in that it also has a capsid core with 120 molecules of the major structural protein and a genome of 11 dsRNA genomic segments packaged along with polymerase molecules inside the core particle.

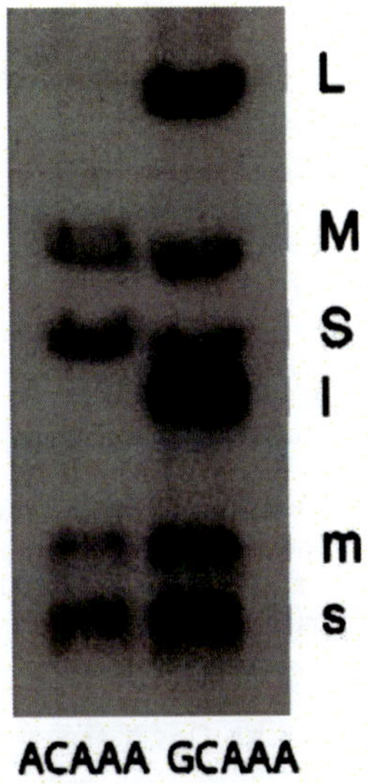

Fig. 26.7 In vitro transcription of Φ2954 nucleocapsids. ACAA is the sequence at the 5′ end of the wild-type L transcript. GCAA is the sequence at the 5′ end of a mutated L transcript. The transcripts of segments S and M begin with GCAA in both cases (Qiao et al. 2010b). L, M, and S designate dsRNA while l, m, and s designate ssRNA transcripts

Many members of the Cystoviridae exhibit another aspect of control of polymerase activity. Transcription activity is different for genomic segments S and M versus that of L. Core particles will transcribe S and M segments without intervention of additional factors; however, transcription of L requires the binding of specific host proteins to the outside of the core particle (Qiao et al. 2008). In the case of Φ6, it has been shown that the binding is to protein P1. Again, we see that the conformation of the capsid structure can influence the activity of the polymerase and, in this case, the template specificity of the polymerase.

The nature of the template differences between the L segment and those of S and M is known. In the case of Φ6, the L transcript starts with GU while those of S and M start with GG. It has been shown that the Φ6 polymerase prefers G as the second nucleotide, and that if the L segment sequence is changed to GG, the differences in transcription disappear (Butcher et al. 2001; Van Dijk et al. 1995). In the case of bacteriophage Φ2954, the L transcript starts with AC while those of S and M start with GC. Changing the sequence of L to GC results in the loss of the transcriptional differences (Fig. 26.7) (Qiao et al. 2010b). In the case of Φ6, the binding of host protein YajQ to the core results in transcription of the L segment. In the case of Φ2954, the binding of host protein glutaredoxin 3 (GrxC) to the core results in transcription of the L segment (Qiao et al. 2010a). In both the case of Φ6 and Φ2954, the virus can mutate to independence of the host protein. In each case, the mutations involve the gene for P1. The nature of the conformational change effected in the polymerase is not known.

References

Bamford DH, Romantschuk M, Somerharju PJ (1987) Membrane fusion in prokaryotes: bacteriophage Φ6 membrane fuses with the Pseudomonas syringae outer membrane. EMBO J 6:1467–1473

Butcher SJ, Dokland T, Ojala PM, Bamford DH, Fuller SD (1997) Intermediates in the assembly pathway of the double-stranded RNA virus φ6. EMBO J 16:4477–4487

Butcher SJ, Grimes JM, Makeyev EV, Bamford DH, Stuart DI (2001) A mechanism for initiating RNA-dependent RNA polymerization. Nature 410:235–240

Cuppels DA, Vidaver AK, Van Etten JL (1979) Resistance to bacteriophage Φ6 by Pseudomonas phaseolicola. J Gen Virol 44:493–504 de Haas F, Paatero AO, Mindich L, Bamford DH, Fuller SD (1999) A symmetry mismatch at the site of RNA packaging in the polymerase complex of dsRNA bacteriophage φ6. J Mol Biol 294:357–372

Eryilmaz E, Benach J, Su M, Seetharaman J, Dutta K, Wei H, Gottlieb P, Hunt JF, Ghose R (2008) Structure and dynamics of the P7 protein from the bacteriophage phi 12. J Mol Biol 382(2):402–422

Frilander M, Gottlieb P, Strassman J, Bamford DH, Mindich L (1992) Dependence of minus strand synthesis upon complete genomic packaging in the dsRNA bacteriophage Φ6. J Virol 66:5013–5017

Gottlieb P, Qiao X, Strassman J, Frilander M, Mindich L (1994) Identification of the packaging regions within the genomic RNA segments of bacteriophage Φ6. Virology 200:42–47

Huiskonen JT, de Haas F, Bubeck D, Bamford DH, Fuller SD, Butcher SJ (2006) Structure of the bacteriophage phi6 nucleocapsid suggests a mechanism for sequential RNA packaging. Structure 14(6):1039–1048

Jaalinoja HT, Huiskonen JT, Butcher SJ (2007) Electron cryomicroscopy comparison of the architectures of the enveloped bacteriophages phi6 and phi8. Structure 15(2):157–167

Makeyev EV, Bamford DH (2000) Replicase activity of purified recombinant protein P2 of double-stranded RNA bacteriophage phi6. EMBO J 19:124–131

Mancini EJ, Kainov DE, Grimes JM, Tuma R, Bamford DH, Stuart DI (2004) Atomic snapshots of an RNA packaging motor reveal conformational changes linking ATP hydrolysis to RNA translocation. Cell 118:743–755

Mindich L (1999) Precise packaging of the three genomic segments of the double-stranded-RNA bacteriophage Φ6. Microbiol Mol Biol Rev 63:149–160

Mindich L, Lehman J (1979) Cell wall lysin as a component of the bacteriophage Φ6 virion. J Virol 30:489–496

Mindich L, Qiao X, Qiao J, Onodera S, Romantschuk M, Hoogstraten D (1999) Isolation of additional bacteriophages with genomes of segmented double-stranded RNA. J Bacteriol 181:4505–4508

Onodera S, Olkkonen VM, Gottlieb P, Strassman J, Qiao X, Bamford DH, Mindich L (1992) Construction of a transducing virus from dsRNA bacteriophage Φ6: establishment of carrier states in host cells. J Virol 66:190–196

Onodera S, Qiao X, Qiao J, Mindich L (1998) Directed changes in the number of dsRNA genomic segments in bacteriophage Φ6. Proc Natl Acad Sci USA 95:3920–3924

Qiao X, Casini G, Qiao J, Mindich L (1995) In vitro packaging of individual genomic segments of bacteriophage Φ6 RNA: serial dependence relationships. J Virol 69:2926–2931

Qiao J, Qiao X, Sun Y, Mindich L (2003a) Isolation and analysis of mutants with altered packaging specificity in the dsRNA bacteriophage Φ6. J Bacteriol 185:4572–4577

Qiao X, Qiao J, Mindich L (2003b) Analysis of the specific binding involved in genomic packaging of the dsRNA bacteriophage Φ6. J Bacteriol 185(21):6409–6414

Qiao X, Sun Y, Qiao J, Mindich L (2008) The role of host protein YajQ in the temporal control of transcription in bacteriophage Phi6. Proc Natl Acad Sci USA 105:15956–15960

Qiao J, Qiao X, Sun Y, Mindich L (2010a) Role of host protein glutaredoxin 3 in the control of transcription during bacteriophage Φ2954 infection. Proc Natl Acad Sci USA 107(13):6000–6004

Qiao X, Sun Y, Qiao J, Di Sanzo F, Mindich L (2010b) Characterization of phi2954, a newly isolated bacteriophage containing three dsRNA genomic segments. BMC Microbiol 10:55

Romantschuk M, Olkkonen VM, Bamford DH (1988) The nucleocapsid of bacteriophage Φ6 penetrates the host cytoplasmic membrane. EMBO J 7:1821–1829

Salgado PS, Makeyev EV, Butcher SJ, Bamford DH, Stuart DI, Grimes JM (2004) The structural basis for RNA specificity and Ca2+ inhibition of an RNA-dependent RNA polymerase. Structure 12(2):307–316

Sandino AM, Jashes M, Faundez G, Spencer E (1986) Role of the inner protein capsid on in vitro human rotavirus transcription. J Virol 60(2):797–802

Sen A, Heymann JB, Cheng N, Qiao J, Mindich L, Steven AC (2008) Initial location of the RNA-dependent RNA polymerase in the bacteriophage phi 6 procapsid determined by cryo-electron microscopy. J Biol Chem 283:12227–12231

Van Dijk AA, Frilander M, Bamford DH (1995) Differentiation between minus- and plus-strand synthesis: polymerase activity of dsRNA bacteriophage Φ6 in an in vitro packaging and replication system. Virology 211:320–323

Zuker M (2003) Mfold web server for nucleic acid folding and hybridization prediction. Nucleic Acids Res 31(13): 3406–3415

Chapter 27
Mechanism of RNA Packaging Motor

Erika J. Mancini and Roman Tuma

Abstract P4 proteins are hexameric RNA packaging ATPases of dsRNA bacteriophages of the *Cystoviridae* family. P4 hexamers are integral part of the inner polymerase core and play several essential roles in the virus replication cycle. P4 proteins are structurally related to the hexameric helicases and translocases of superfamily 4 (SF4) and other RecA-like ATPases. Recombinant P4 proteins retain their 5' to 3' helicase and translocase activity in vitro and thus serve as a model system for studying the mechanism of action of hexameric ring helicases and RNA translocation. This review summarizes the different roles that P4 proteins play during virus assembly, genome packaging, and transcription. Structural and mechanistic details of P4 action are laid out to and subsequently compared with those of the related hexameric helicases and other packaging motors.

Abbreviations

AMPcPP	[[(2R,3S,4R,5R)-5-(6-aminopurin-9-yl)-3,4-dihydroxyoxolan-2-yl]methoxy-hydroxyphosphoryl]methylphosphonic acid
AMPPNP	Adenylyl-imidodiphosphate
ASCE	Additional strand conserved E
LTAg	Simian virus 40 large T antigen
PC	Procapsid
P_i	Inorganic phosphate
γP	Gamma phosphate in ATP
SF	Superfamily

E.J. Mancini (✉)
Division of Structural Biology, Wellcome Trust Centre for Human Genetics,
Oxford University, Oxford OX3 7BN, UK
e-mail: erika@strubi.ox.ac.uk

R. Tuma (✉)
The Astbury Centre for Structural Molecular Biology, Institute of Cellular
and Molecular Biology, University of Leeds, Leeds LS2 9JT, UK
e-mail: r.tuma@leeds.ac.uk

M.G. Rossmann and V.B. Rao (eds.), *Viral Molecular Machines*,
Advances in Experimental Medicine and Biology 726, DOI 10.1007/978-1-4614-0980-9_27,

27.1 Introduction

P4 is a 35-kDa ssRNA packaging ATPase from dsRNA bacteriophages belonging to the *Cystoviridae* family (ϕ6–ϕ14) (Mindich 2004; Poranen and Tuma 2004; see Chapter 26 of this volume). P4 proteins form hexameric rings which are integral part of the polymerase core and play an essential role in several key stages during virus assembly and replication (Poranen et al. 2008; see Chapter 26 of this volume). Belonging to the RecA-like family of ATPases, P4 proteins are structurally most similar to the hexameric ring helicases and translocases of superfamily 4 (SF4) (Kainov et al. 2006; Mancini et al. 2004a). Isolated recombinant P4 proteins retain their ability to form hexamers and exhibit RNA-stimulated ATPase activity in the absence of other virion proteins (Juuti et al. 1998; Kainov et al. 2003, 2004). Furthermore, recombinant P4 proteins from several cystoviruses (e.g., ϕ8, ϕ13) exhibit 5′ to 3′ helicase activity in vitro (Kainov et al. 2003). P4 proteins can be readily expressed and purified in quantities large enough to allow investigation by a wide variety of structural and biophysical methods. Hence, the P4 proteins constitute an important model system for studying the mechanism of action of hexameric helicases and other translocases.

This chapter summarizes the roles that P4 protein plays in the viral life cycle, specifically during RNA packaging into empty procapsids. Structural and functional relationships to hexameric helicases and dsDNA packaging motors are discussed next. Survey of high resolution structures, site-directed mutagenesis, and biochemical properties provides basis for a mechanistic model of RNA packaging. Other aspects of P4 proteins have been reviewed elsewhere (Kainov et al. 2006; Mancini and Tuma 2008; Rabhi et al. 2010; Tuma 2010).

27.2 Structure and Sequence Similarity to Other Molecular Motors

In their native state, P4 hexamers are integral part of the procapsid (PC) (Fig. 27.1, see also Chapter 17 of this volume). However, all P4 proteins characterized to date (phages ϕ6, ϕ8, ϕ12, ϕ13) remain hexameric in absence of the viral capsid and were successfully expressed in *E. coli*, purified and crystallized (Juuti et al. 1998; Kainov et al. 2003; Mancini et al. 2004b). The first high-resolution structure of ϕ12 P4 hexamer revealed six identical subunits arranged in a ring, resembling a domed toroid around a central channel (Fig. 27.1) (Mancini et al. 2004a) Similar arrangement has been

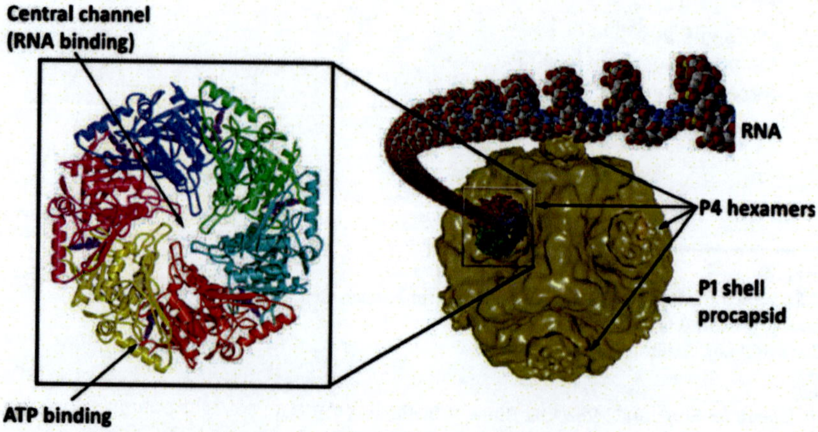

Fig. 27.1 Artistic impression of P4 pumping ssRNA into the PC (*right*). A ribbon diagram of ϕ12 P4 hexamer with nucleotide triphosphate binding sites located at subunit interfaces and RNA binding sites within the central channel (*left*)

subsequently found for P4 from a related bacteriophage ϕ13 (Meier et al. 2005). Six identical ATP binding sites are located toward the external perimeter of the ring at the interfaces between adjacent subunits and are formed by residues contributed by both subunits. The central channel hosts nucleic acid binding sites which are essential for packaging and helicase activity (Kainov et al. 2008; Lisal et al. 2005) (Fig. 27.1b). As originally proposed on the basis of low-resolution techniques (Juuti et al. 1998), the quaternary structure of P4 rings resembles that of other hexameric helicases. Indeed, recombinant P4 hexamers from bacteriophages ϕ8 and ϕ13 exhibit 5′ to 3′ RNA-specific helicase activity (Kainov et al. 2004). This relationship has been subsequently investigated and detailed by a combination of sequence alignments and high-resolution structural information.

27.2.1 Conserved Helicase Fold and Sequence Motifs

The atomic structure of P4 solved by X-ray crystallography (Mancini et al. 2004a) revealed three domains (Fig. 27.2a, b): an N-terminal domain which is important for hexamer assembly and stability, a central RecA-like ATPase domain, and an α/β C-terminal extension which serves as a platform for attachment to the virion (Lisal et al. 2006; Paatero et al. 1998). While the overall sequence identity between P4 proteins from different phages is low (<30%), five conserved sequence motifs (H1, H1a, H2, H3, H4) are clearly discernible within the RecA-like ATPase module of these molecular motors. These motifs are conserved among ATPases that belong to helicase superfamily 4 (SF4) (Fig. 27.2b). Proteins within is and related ATPase superfamilies (e.g., SF3 viral hexameric helicases, SF5 Rho transcription terminator) contain the conserved ASCE core topology (additional strand conserved E (Iyer et al. 2004)) (Fig 27.2c). SF4 encompasses hexameric bacterial and bacteriophage helicases (e.g., *E. coli* replicative helicase DnaB, bacteriophage T7 gp4 helicase–primase), most of which are involved in DNA replication and translocate in the 5′ to 3′ direction (Singleton et al. 2007) along one strand while splitting DNA duplex. P4 proteins are the only RNA-specific helicases within this superfamily.

Structurally, the closest relatives of the P4 subunit fold are the helicase domain of T7 gp4, RecA, and the F1-ATPase subunits (Kainov et al. 2008) (see also chapter 2 in this volume). However, the *E. coli* transcription terminator Rho (Skordalakes and Berger 2003), which is a hexameric RNA-specific helicase with 5′ to 3′ directionality, resembles P4 structurally (Fig. 27.2d) as well as functionally (Rabhi et al. 2010). Several structures of Rho in complex with RNA have been recently obtained and provided additional insights into the mechanism of RNA-specific translocation. For this reason, a detailed comparison between Rho and P4 is employed in this review.

Hexameric viral helicases belonging to SF3 (e.g., E1 from papilloma viruses or large T antigen from SV40) also form toroidal rings and thus are morphologically similar to P4. However, the fold of their ATPase domains is of AAA+ (*A*TPases associated with diverse cellular activities) type and exhibits only three conserved motifs: A, B, and C. It is interesting to note that while the P4 fold clearly belongs to the RecA type, the ATPase domain lacks the typical RecA insertion between the Walker A (H1 motif) and Walker B (H1a and H2) motifs within the ASCE core. Instead, two additional β-strands and a helix are appended at the C-terminus of the ATPase core (Fig. 27.2b, c). As a result, the P4 fold resembles that of the ASCE motif within AAA+ proteins while still superposing well onto the typical eight-stranded β-sheet RecA-like topology. Due to this structural and topological resemblance between the ASCE core in P4 and SF3 helicases, the three conserved motifs (A, B, C) of the latter superpose onto the corresponding residues in P4. While superposition of A (Walker A, H1 in SF4) and B (Walker B, H2 in SF4) is a rather trivial consequence of both proteins belonging to the P-loop ATPases, the structural conservation of the polar residue within motif C (H3 in P4) highlights its universal importance as a γ-phosphate (γP) sensor in diverse molecular motors (Enemark and Joshua-Tor 2006; Kainov et al. 2008).

P4 also exhibits H1a and H4 motifs that are typical of SF4 helicases. The former hosts a catalytic acidic residue and together with H2 residues acts as a split Walker B motif. H4 motif within the ASCE

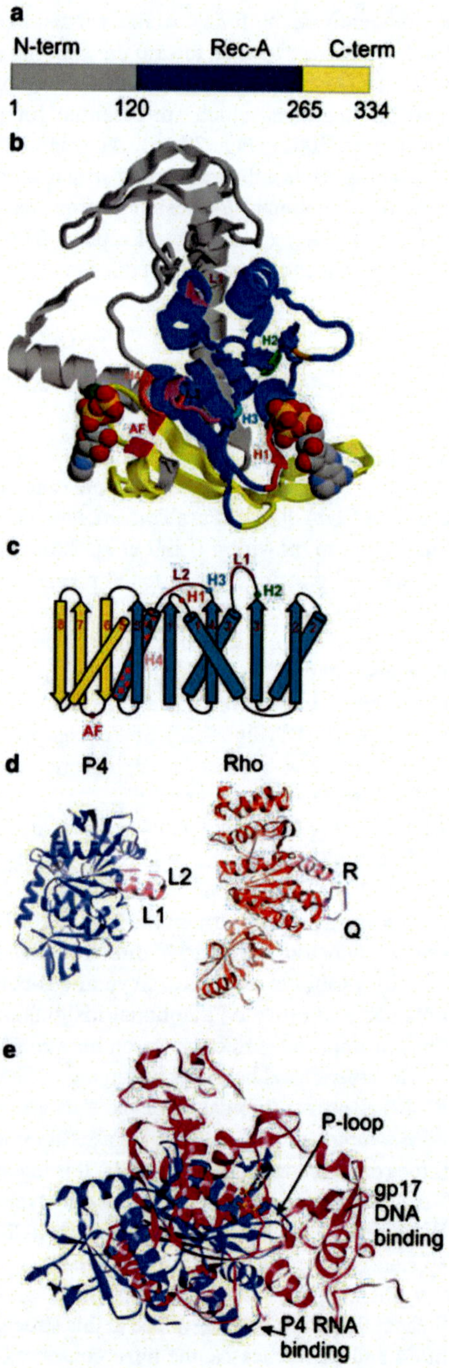

Fig. 27.2 (**a**) Domain structure of P4. (**b**) P4 subunit fold – domains are colored according to panel A, SF4 helicase motifs highlighted in different colors: H1 – *red*, H2 – *green*, H3 – *cyan*, H4 – *pink*. Additional structurally conserved regions are shown: arginine finger (AF, *magenta*) and L1 and L2 loops (*dark magenta*). (**c**) Topology of the ASCE core (*blue*) and C-terminal RecA-like extension (*yellow*). (**d**) RecA-like fold of P4 compared to Rho (PDB, 3ICE, Thomsen and Berger 2009). P4 loops L1 and L2 and the lever helix α6 are highlighted together with the corresponding loops Q and R and the equivalent "lever" helix α12 in Rho. Note that α12 does not act as the lever in Rho (see the text for more detail). (**e**) Comparison between P4 (*blue*) and the ATPase domain of phage T4 gp17 (*magenta*, PDB 2O0H, Sun et al. 2007) with the ATPase active site serving as the alignment reference

core usually contains a conserved polar residue (S252 in ϕ12 P4 (Mancini et al. 2004a), N494 in T7 phage helicase–primase gp4 (Singleton et al. 2000)) which contributes to the ATP binding site *in trans* (i.e., from a neighboring subunit within the hexamer) and is essential for activity (Kainov et al. 2008).

In addition to the conserved helicase motifs, there are other structurally conserved and catalytically important residues. These are difficult to identify solely on the basis of sequence. "Arginine finger" motifs are arginine residues inserted into the active site *in trans*. An arginine finger is thought to facilitate formation of the transition state by neutralizing the buildup of negative change on the γP (Scheffzek et al. 1997). In P4, there are two essential arginine residues, one of which (R272 in ϕ12 P4) remains stationary throughout the catalytic cycle (Kainov et al. 2008; Mancini et al. 2004a). The other (R279 in ϕ12 P4) is inserted into the catalytic pocket during hydrolysis (see below). In SF4 helicases, these residues are located downstream of H4 and outside of the conserved ASCE core. In contrast, arginine (and lysine) fingers in SF3 helicases are located within the ASCE core.

The RNA-interacting loops L1 and L2 (Fig. 27.2c, d) are also structurally conserved among P4 proteins from different cystoviruses (Kainov et al. 2006; Mancini et al. 2004a) and other SF4 and SF5 helicases (Sawaya et al. 1999). For example, Rho possesses two RNA binding loops, Q and R (structurally equivalent to L1 and L2 in P4, respectively) (Fig. 27.2c, d).

In summary, Rho is the closest functional relative of the P4 that also exhibits similar structure. Analysis of bacterial sequences suggests that Rho-like homologues are ubiquitous and highly conserved in the bacteria; therefore, it is plausible to speculate that dsRNA bacteriophages hijacked this essential host cell helicase to become an RNA-specific packaging motor.

27.2.2 P4 Is Only Remotely Related to Packaging ATPases of dsDNA Viruses

Recently, the first high-resolution structure of a dsDNA packaging ATPase, the gp17 terminase from bacteriophage T4, has been determined (Sun et al. 2007). This enzyme provides the energy for dsDNA translocation but, unlike P4, it only transiently associates as a pentamer (Sun et al. 2008) with the portal vertex of the phage procapsid (empty virion precursor, PC). In addition, its C-terminal domain exhibits endonuclease activity that is used to cut the concatenated genomic dsDNA precursor after a headful packaging (Rao and Feiss 2008).

A structural alignment of the RecA cores of P4 and gp17 reveals a very different disposition of nucleic acid binding sites with respect to the ATP binding site (Fig. 27.2e). While in gp17 one of the proposed DNA binding sites is located within an intervening domain adjacent to the ATP binding site (the second site being located within the C-terminal nuclease domain), the RNA binding sites in P4 are located much closer to the ATPase active site. The intervening domain in gp17 provides the arginine finger, which inserts *in cis* into the ATPase active site during catalysis. In P4 (and in other hexameric ring helicases), this residue is provided *in trans* by the neighboring subunit. The bacteriophage T4 ATPase has in fact the greater structural similarity to the monomeric helicases of the superfamilies 1 and 2. This extends to the presence of a catalytically important glutamate immediately following the Walker B motif (whereas this is upstream of the Walker B motif in P4) and the TTT sequence "motif III," which is important in monomeric helicases for sensing the state of ATP in the catalytic pocket (Mitchell et al. 2002). Taken together, these results suggest that the two classes of packaging motors are quite distinct and, as discussed later, their mechanisms are likely to be different.

27.3 P4 Is Essential for Several Steps During the Viral Life Cycle

27.3.1 Procapsid Assembly

The dsRNA bacteriophages have a 14-kbp genome composed of three segments, small (S), medium (M), and large (L) which are enclosed within an icosahedral protein core (Huiskonen et al. 2006; Jaalinoja et al. 2007, see also Chapter 17 of this volume). The core is further enclosed in a nucleocapsid shell and enveloped by a lipid bilayer to yield an infectious virus (for details and overview

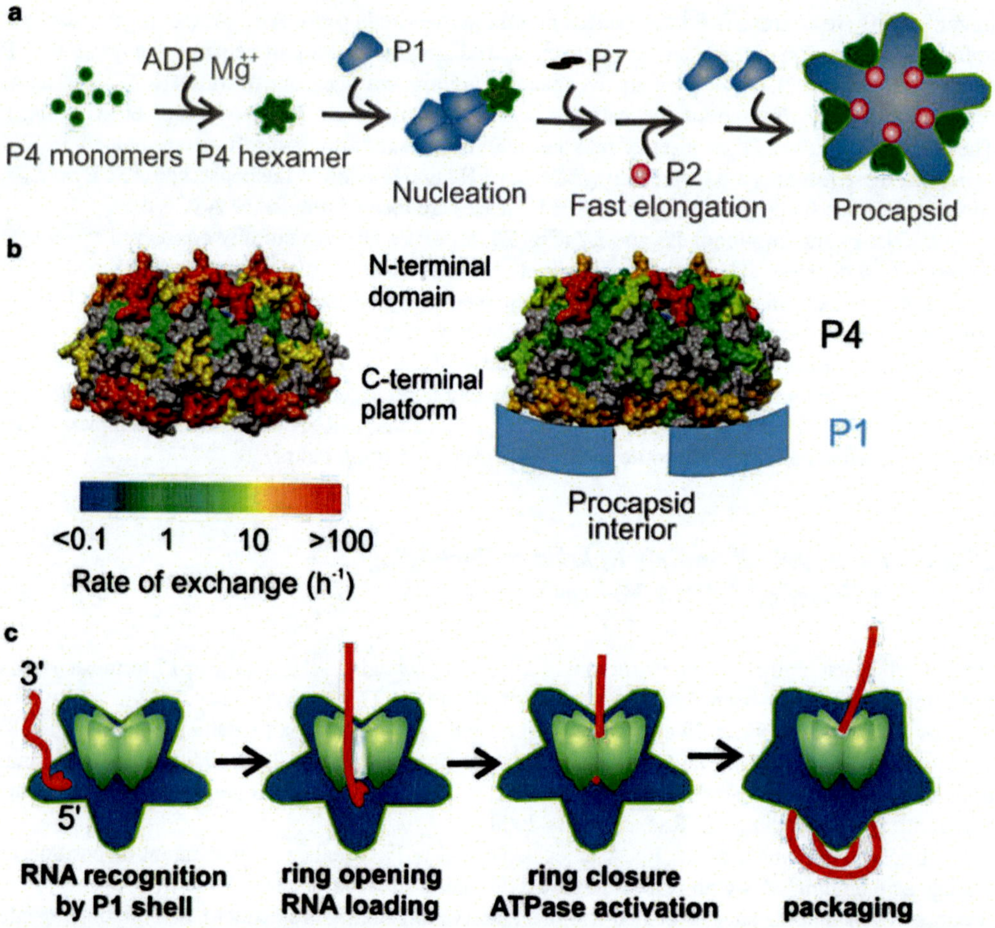

Fig. 27.3 (**a**) Role of P4 in assembly of dsRNA phages. P4 forms hexamers in the presence of divalent cations and/or nucleotide di/triphosphates and then associates with the major PC protein P1 and nucleates PC assembly in vitro. (**b**) Interaction of P4 hexamer with the viral PC as visualized by hydrogen–deuterium exchange. P4 is represented as a surface plot colored according to the backbone amide (NH) exchange rates. Isolated, PC-free hexamer is shown in the *left panel* together with the exchange *scale bar*. The *right panel* shows the exchange representation for the PC-bound P4. P1 shell (*blue*) is shown schematically to indicate the deduced relative orientation. (**c**) Proposed mechanism of RNA selection, loading, and packaging. RNA pac sites are specifically recognized by P1 shell (*blue*) and brought to the vicinity of the P4 ring (*green*). Upon contact with RNA, the ring opens at one interface to allow the RNA to slip into the central channel and activate the ATPase. Translocation ensues in 5′ to 3′ direction

see (Huiskonen and Butcher 2007; Tuma 2008). Capsid assembly proceeds in multiple steps with an intermediate dodecahedral PC (Poranen et al. 2008). Initially, PC is void of RNA (Butcher et al. 1997; Mindich et al. 1976) and is built by co-assembly of 120 copies of the major structural protein P1 with ~12 copies of the viral RNA-dependent RNA polymerase P2, 12 hexamers of P4, and 30 dimers of the assembly cofactor P7 (Poranen and Tuma 2004). P4 hexamers nucleate PC assembly in vitro and are essential for virion formation in cells (Juuti et al. 1998; Poranen et al. 2001) (Fig. 27.3a). Up to 12 hexamers occupy the fivefold facets of the dodecahedron (Huiskonen et al. 2007) (Fig. 27.1a), creating a symmetry mismatch. However, recent electron tomography revealed variable occupancy of ϕ6 P4, with only about five hexamers per procapsid (Nemecek et al. 2010). P4 hexamers also play a structural role in the PC, contributing up to 30% of the total protein mass. P4 interacts with the PC through its C-terminal domain (Lisal et al. 2006; Paatero et al. 1998), and the interaction further stabilizes the hexamer, preventing spontaneous opening of the ring (Fig. 27.3b).

27.3.2 RNA Packaging

RNA in dsRNA bacteriophages is packaged into preformed empty PCs in a form of single-stranded (plus sense) genomic precursors (Mindich 2004, see also Chapter 26 of this volume). The three ssRNA precursors (s, m, and l) are packaged in a sequential manner (Mindich 1999). The process encompasses three essential steps: recognition of the specific RNA, RNA loading into the P4 hexamer, and ATP-dependent RNA translocation. Genomic precursors contain unique recognition sequences (a *pac* sites) at 5′ end (Gottlieb et al. 1994; Qiao et al. 1997, see also Chapter 26 of this volume). The pac sequences fold into well-defined structures (Pirttimaa and Bamford 2000) that are recognized by the protein P1 within the PC (Qiao et al. 1997, 2003) (Fig. 27.3c). This recognition brings the correct RNA into the close proximity of P4 which alone is a nonspecific translocase that can turnover heterologous RNAs (Kainov et al. 2003).

In a second step, the bound RNA precursor is loaded into a P4 hexamer, most likely via a ring-opening mechanism. Evidence for ring opening was obtained for recombinant isolated ϕ8 P4, where the transient opening of one subunit interface during RNA loading was detected by hydrogen–deuterium exchange (Lisal et al. 2005). A snapshot of similar ring-opening process was captured in the crystal structure of *E. coli* Rho in complex with RNA fragments bound to the primary binding sites at the periphery of the hexamer (Skordalakes and Berger 2003). The ring-opening mechanism was also proposed for other hexameric helicases based on kinetic considerations (Levin and Patel 2003). However, the ring-opening mechanism of P4 in the context of the whole PC remains to be characterized.

The captured RNA stimulates the ATPase activity of P4 which translocates it inside the PC in the 5′ to 3′ direction (Gottlieb et al. 1991, 1992a). The packaged RNA is subsequently replicated by the viral polymerase P2 within the PC; at this point, the PC becomes the polymerase core, which is able to synthesize new plus-sense RNA transcripts (Bamford et al. 1995; Frilander et al. 1992; Gottlieb et al. 1992b).

27.3.3 Transcription

During infection, the polymerase core is released from the nucleocapsid shell into the cytoplasm. This activates the polymerase which starts to produce ssRNA transcripts (Mindich 2004). Also, the newly assembled polymerase cores transcribe the packaged genome in order to produce genomic precursors for assembly (see Chapter 17 of this volume for further details).

PC-associated P4 hexamers are indispensable for transcription and act as passive ports for ssRNA export, that is, translocation occurs in absence of catalytic activity (Kainov et al. 2004). During extrusion, RNA moves through the central channel in the direction opposite to that of packaging. This is consistent with the diminished ATPase activity of P4 within the core of the ϕ12 virus (Kainov et al. 2004) and with the structural changes observed for the P4 ring within the ϕ6 virus core upon maturation (Huiskonen et al. 2007).

27.4 Biochemical Properties and Enzyme Kinetics

27.4.1 Hexamer Assembly

Cystoviral packaging motors remain hexameric at low (submicromolar) protein concentrations in the presence of millimolar Mg^{2+} and ADP (Kainov et al. 2003; Juuti et al. 1998). Thus, in infected cells, P4 exists almost exclusively in the hexameric form, readily available for assembly. However, in vitro, (and in absence of nucleotides and divalent cations) P4 hexamers from phage ϕ12 can be slowly and reversibly dissociated, allowing assortment of subunits, for example, wild-type and mutant subunits (Telenius et al. 2008).

27.4.2 ATP Binding

The ATPase cycle can be divided into four essential steps: (1) ATP binding, (2) phosphodiester hydrolysis, (3) inorganic phosphate (P_i) release, and (4) ADP dissociation. All P4 motors studied thus far (ϕ6 P4 (Juuti et al. 1998), ϕ8 P4 (Lisal et al. 2004) and ϕ12 P4 (Lisal and Tuma 2005)) bind ATP and ADP noncooperatively with submillimolar affinities (K_d ~ 100–200 μM). All six ATP binding sites are equivalent, and concurrent RNA binding does not have significant effect on the nucleotide affinities. This is consistent with the comparison of ϕ12 P4 structures in the apo and ATP-/ADP-bound states. Apart from a slight closure of the ATP binding cleft at the subunit interfaces, there is little change that would indicate intersubunit communication and relative rotation of subunits (Mancini et al. 2004a). However, pre-steady-state kinetics revealed a rate-limiting conformational change associated with ATP binding to ϕ8 P4 hexamer. In the case of ϕ8 P4, the presence of RNA affected the ATP binding kinetics (Lisal et al. 2004), suggesting that a conformational change concomitant with ATP binding may be involved in modulating RNA binding and release. These conformational changes were subsequently mapped by hydrogen–deuterium exchange, suggesting that the same motifs that are involved in RNA translocation in ϕ12 P4 (i.e., H3 and H4) also modulate RNA affinity in ϕ8 P4 (Lisal et al. 2005). This alludes to a common translocation mechanism for P4 proteins from different phages.

27.4.3 Nucleic Acid Binding

P4 proteins from different phages bind both RNA and DNA in a sequence-independent fashion with affinities ranging from submicromolar (ϕ8, ϕ13) to millimolar (ϕ6, ϕ12) (Kainov et al. 2003; Lisal et al. 2004; Lisal and Tuma 2005). ϕ8 P4 binds both single-stranded as well as double-stranded nucleic acids (Kainov et al. 2003), while the central channel within the ϕ12 P4 hexamer is not large enough to accommodate a duplex nucleic acid (Mancini et al. 2004a). ATPase activity of ϕ12 P4 is stimulated by ssRNA. This suggests that single- and double-stranded nucleic acids bind in distinct ways, perhaps employing different binding sites. ϕ8 P4 has been shown to associate with ssDNA as well as dsRNA and dsDNA, but only in the presence of ssRNA will it attain ATPase activity, suggesting that the protein discriminates between RNA and DNA at the level of coupling between ATPase activity and translocation. Here, we focus on the biologically relevant binding to ssRNA.

Two loops, L1 and L2, which protrude into the central channel (Fig. 27.1b), constitute the RNA binding sites in P4 (Kainov et al. 2008; Lisal et al. 2005). These loops are structurally equivalent to

Fig. 27.4 Plausible model for RNA binding within P4 hexamer based on fitting of the "staircase" RNA structure from *E. coli* Rho (PDB, 3ICE; Thomsen and Berger 2009) into the central channel of P4 hexamer (PDB 1W49; Mancini et al. 2004a). *Side view* with only two subunits shown for clarity and loops L1, L2 highlighted (*left*). *Top view* shown in the *right panel*

RNA and DNA binding sites in the related hexameric helicases (e.g., R- and Q-loops in Rho, Fig. 27.2d). In ϕ8 P4, two positively charged residues within the L1 loop (motif LKK) were shown to be essential for RNA binding. In ϕ12 P4, which binds only weakly to RNA, a single residue within L2 (K241) seems essential. However, additional mutational analysis demonstrated that polar residues within the L1 loop of ϕ12 P4 (e.g., S203) are also essential and thus may interact with RNA (Mancini, Kainov, Purdy, Rong and Tuma, unpublished results). As seen in Rho, the two loops most likely form transient salt bridges and hydrogen bonds with the RNA backbone (Thomsen and Berger 2009). L1 loop in P4 may also contribute to the RNA specificity as proposed for the Q-loop in Rho (Thomsen and Berger 2009). A plausible configuration of RNA inside the central channel is illustrated in Fig. 27.4 in which the bound RNA adopts a "staircase" structure seen in a recent Rho crystal structure (Thomsen and Berger 2009). In this configuration, RNA snuggly fits within the channel, contacts the L1 loops, and certainly would come into contact with the L2 loops on its way down the channel during translocation. Note that the RNA orientation in Fig. 27.4 is derived from the Rho structure and has its 3′ pointing toward the C-terminal platform, that is, opposite to that expected for P4. This is because Rho appears to run "in reverse" with respect to the SF4 helicases (Thomsen and Berger 2009).

One important mechanistic issue is how many subunits are engaged with the RNA strand at any given time and whether this number depends on the nucleotide state. An estimate of the number of subunits that engage RNA during translocation was obtained using the dependence of ATPase activity on the oligoribonucleotide length (Lisal and Tuma 2005). This estimate was further confirmed from the minimum number of active subunits that cooperate during hydrolysis (Telenius et al. 2008). In ϕ12 P4, 3–4 subunits were estimated to interact with RNA. More subunits within the ϕ8 P4, which exhibits higher affinity for RNA, may interact with RNA at once. For comparison, the related DNA helicase gp4 from bacteriophage T7 has been shown to require binding of all six subunits to DNA for helicase activity (Crampton et al. 2006). Five subunits were seen to engage with RNA in the crystal structure of Rho, forming the staircase binding site configuration within the hexamer (Thomsen and Berger 2009) similar to that found in the structure of bovine papillomavirus helicase E1 with ssDNA (Enemark and Joshua-Tor 2006).

However, a previous Rho structure exhibited a very different binding pattern in which each pair of subunits was bound to a single strand (perhaps dictated by the overall threefold symmetry of the crystal). Recent chemical interference probing of Rho–RNA interactions in solution indicated that the protein makes a critical contact with the RNA strand on average at every seventh nucleotide. Hence, RNA binding may be more dynamic than suggested by the few crystallized complexes. Given the close structural and functional relationship between Rho and P4, it is likely that P4 binds RNA in a similar fashion, that is, utilizing the staircase configuration of binding sites, but may engage fewer subunits at a time. It should be noted that the topological enclosure of nucleic acid also contributes to the overall RNA binding process and increases the apparent affinity for long substrates by slowing down the dissociation rate which is governed by ring opening or, alternatively, by sliding off the strand end. Spontaneous ring opening is rare for ϕ8 P4 ((Lisal et al. 2005), see also translocation processivity below), while ϕ12 P4 ring is more dynamic unless it is bound to the procapsid (Lisal et al. 2006).

27.4.4 ATP Hydrolysis

In many cases, the rate of hydrolysis is measured indirectly by monitoring the rate of P_i released from the enzyme. In addition to the bond breaking, the resulting rate constant reflects any conformational changes that configure the active site for hydrolysis and also those leading to P_i release. Based on detailed kinetic characterization of many ATPases, the intrinsic bond-breaking step is expected to be much faster than the associated conformational changes or the subsequent P_i release, and thus, the latter two are considered the rate-limiting steps (Liao et al. 2005).

Under cellular conditions, that is, high [ATP]/[ADP] ratio, nucleotide exchange does not generally constitute the rate-limiting step, and the enzyme works close to the maximum turnover (k_{cat}=turnover number, i.e., the average number of ATP molecules hydrolyzed by one subunit per unit time). However, in vitro reaction conditions (e.g., ATP concentrations, presence of inhibitors) can be judiciously chosen to emphasize the contributions of the various steps, including ATP binding and ADP release, to the overall cycle, as this is reflected in the apparent Michaelis constant K_M (i.e., substrate concentration necessary to achieve 50% of k_{cat} activity; for monomeric enzymes, K_M is usually equal to the substrate binding dissociation constant K_d). Note that in the case of oligomeric proteins, such as P4, the apparent K_M does not necessarily correspond to K_d and instead reflects cooperativity (binding or kinetic) between multiple catalytic sites.

As stated above, ATPase activity of P4 is stimulated by ssRNA (Kainov et al. 2003) (Fig. 27.5a). ϕ8 P4 exhibits a tight coupling between ATP hydrolysis and translocation, that is, virtually no ATP is hydrolyzed in the absence of RNA. However, other P4 motors retain basal ATPase activity in the absence of RNA (Kainov et al. 2003; Lisal and Tuma 2005). Such imperfect coupling can be exploited to detail the mechanism, for example, by comparing effects of inhibitors or mutations in the presence and absence of RNA. This strategy has been employed in the study of ϕ12 P4 mechanism, and hence, we focus primarily on this model system.

Without affecting ATP/ADP association, RNA has profound effect on the steady-state kinetic parameters (K_M and k_{cat}, n=Hill coefficient which is related to number of cooperating subunits). In the absence of RNA, K_M (1.5 mM) is almost an order of magnitude higher than the corresponding K_d (0.2 mM) for ATP binding. Little cooperativity is observed. The high K_M value (higher than the corresponding K_d) reflects cooperation between several subunits within the hexamer that have to bind ATP simultaneously in order to trigger hydrolysis (Lisal and Tuma 2005). RNA binding causes a threefold decrease of the apparent K_M and a concomitant increase of k_{cat} and cooperativity (n=2.7) (Fig. 27.5a) (Lisal and Tuma 2005). These observations, together with inhibition studies, led to a simple sequential model of catalytic cooperativity (Lisal and Tuma 2005) (Fig. 27.5b) in which all subunits within hexamer are active, but only three neighboring subunits are cooperating in hydrolysis at a time. The mechanism is intrinsically the same in the absence or presence of RNA.

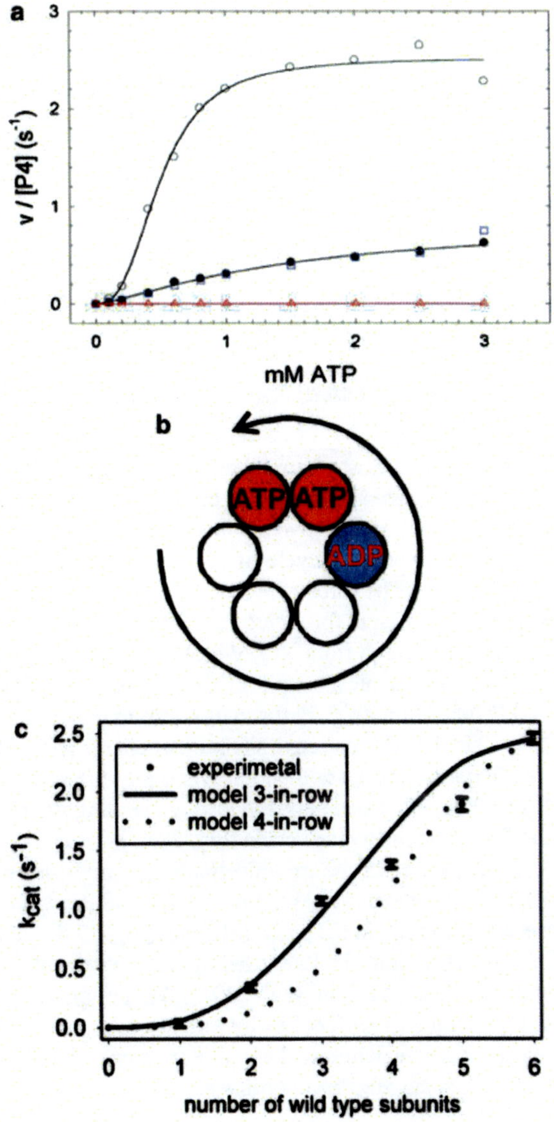

Fig. 27.5 (**a**) RNA-induced catalytic cooperativity as demonstrated by steady-state kinetics of ATP hydrolysis. Rate of P_i release was measured as a function of substrate concentration (*solid circles*, without nucleic acid; *empty circles*, in the presence of 1 mM poly(C); *squares*, in the presence of ssDNA). *Triangles* represent AMPPNP hydrolysis as a negative control. *Lines* represent fits to the Hill equation as described in Lisal and Tuma (2005). (**b**) A sequential model for ATP hydrolysis encompassing three subunits in a row, two of which are occupied by ATP while the remaining is in the ADP state. The *arrow* indicates the direction in which sequential ATP hydrolysis proceeds in an orientation with the hexameric ring viewed from the apical N-terminal facet along the central channel. (**c**) Incorporation of increasing number of arginine finger mutant (R279A) subunits into the hexamer. Dependence of the turnover number k_{cat} on the number of wild-type subunits. Computed data for the three and four cooperating subunit model are shown as *solid* and *dotted curves*, respectively. Adapted from Telenius et al. (2008)

In the absence of RNA, the hexamer stochastically attains the right configuration of the three subunits and becomes catalytically competent. In the presence of RNA, that is, during translocation, RNA binding coordinates configurations of the neighboring subunits and thus promotes sequential hydrolysis at ATP concentrations higher than the K_d, and the apparent K_M becomes closer to the K_d. This scheme is compatible with the structural features of the P4 hexamer (Mancini et al. 2004a) and was further supported by the enzymatic properties of hexamers with increasing number of mutant subunits (Telenius et al. 2008). In the latter case, two models, with three or four neighboring subunits cooperating, respectively, reproduced the experimentally determined dependence of ATPase activity on the number of wt subunits within the hexamer (Fig. 27.5c). Results of site-directed mutagenesis and structural analysis (Kainov et al. 2006, 2008; Mancini et al. 2004a) are discussed in the following sections.

In contrast to P4, cooperativity in other hexameric helicases, such as Rho and T7 gp4, is achieved by cooperative ATP binding in which the ATP affinity (K_d in the lower μM range) is modulated by RNA binding. Despite this, a sequential mechanism of hydrolysis similar to the one proposed for P4 has been put forward for these two helicases (Liao et al. 2005; Adelman et al. 2006). The proposed mechanism is also in agreement with rapid kinetics studies which showed that multiple ATPs need to be hydrolyzed before the first phosphate can be released from the Rho hexamer in a single turnover experiment (Adelman et al. 2006).

27.5 Structure-Based Mechanism

27.5.1 Mechanical Reaction Coordinate: Molecular Lever

Molecular motors are thermodynamic machines that convert chemical energy into mechanical work in a cyclical fashion. An analogy can be drawn between a standard combustion engine and a molecular motor but the respective thermodynamic operating principles differ considerably (Fig. 27.6). The former is driven by a cycle of gas expansion and cooling, utilizing reservoirs with two different temperatures. The force generation step is associated with gas expansion during combustion. The situation is different for molecular motors which operate at one temperature and harness the chemical energy released from ATP hydrolysis. Such energy conversion is often called mechanochemistry or mechano–chemical coupling. This process exploits concentration differences between ATP, ADP, and P_i in solution (e.g., in the cytoplasm) to create subtle structural changes (on the order of 0.1 nm or less) in the ATPase active site such as charge redistribution accompanying P_i release or ATP binding. These changes are then amplified into the nanometer-scale motion of the bound nucleic acid (Fig. 27.6b). Consequently, the force generating steps may be associated with injection or exhaust in the "combustion engine" cycle (Fig. 27.6a).

In many motors, the amplification is accomplished by a "molecular lever" (Fig. 27.6b). In the ϕ12 P4, the lever encompasses a rigid helix α6 which connects the ATP binding site with the RNA binding site (loop L2) in the central channel (Fig. 27.6c). Two positions of this lever have been identified in high-resolution structures, and motion between these two positions is controlled by ATP binding and hydrolysis (Mancini et al. 2004a). We designate this motion, which spans about 6–7 Å (the positively charged tip of K241 swings up to 11 Å), as the mechanical reaction coordinate. In the following sections, we delineate the key interactions within the ATP active site which control motion along the mechanical reaction coordinate.

27.5.2 ATP Binding and Hydrolysis Control Lever Motion

In absence of bound nucleotides (apo form), two lever positions, one designated as "up" and the other as "down," were observed in the crystal structure within the same hexamer (Fig. 27.6c) (Mancini et al. 2004a). This suggests that in the absence of ATP or ADP, the two positions are energetically equivalent, while intermediate positions are perhaps less favored. The two principal lever positions were also observed in structures with bound nucleotides (Fig. 27.6d). In presence of ATP (or an ATP analog), the loops and α6-helices adopt the "up" position, while in the ADP:Mg^{2+} state, they assume the "down" position. This is consistent with a model in which the lever moves down the central channel in response to ATP hydrolysis and/or P_i release (Mancini et al. 2004a).

The upstream end of the L2 loop reaches back into the active site of the same subunit where an essential asparagine (N234) within the H3 motif interacts with γP in the ATP-bound state (Fig. 27.7a).

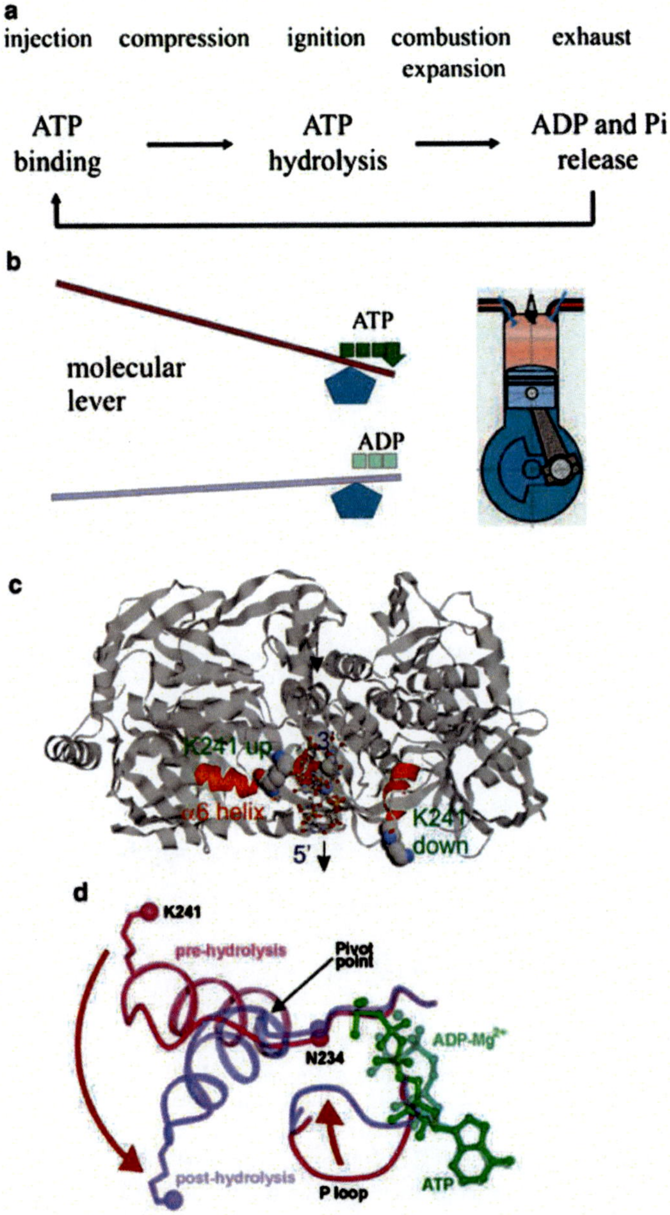

Fig. 27.6 (**a**) Schematic comparison between the cycle of a molecular motor and a typical combustion engine. (**b**) Molecular lever mechanism (*left*) compared to a combustion-driven piston engine (*right*). (**c**) Cross-section through the apo hexamer (*ribbon*) with a short RNA (*ball* and *stick*, taken from Rho structure PDB 3IEC) modeled inside the central channel. The two positions of α6 helix (*red*) and K241 (*spacefill*) are shown looking at three neighboring subunits from the center of the hexamer. The direction of RNA translocation is designated by *arrows* (**d**); α6 helix and L2 loop act as a molecular lever which connects the ATP binding site to the central channel and amplifies small changes into nanometer-scale swiveling motion

Structurally equivalent polar residues are found within the H3 motif (C motif in SF3) of other hexameric helicases. In these helicases, residues within the H3 motif usually serve as γP sensors. This is also the case for ϕ12 P4 residue N234, where the role was confirmed by structural analysis of the N234G mutant (Kainov et al. 2008). In this mutant, the lever adopts the posthydrolysis down position even when an ATP analog is bound in the active site (Fig. 27.7b). Thus, the H3 sensor motif conveys information from the active site to the lever within the same subunit.

The lever helix (α6) is positioned downstream of the RNA binding loop L2 and its C-terminus inserts into the ATP binding site of the next subunit. The loop at the C-terminal tip displays a conserved serine (S252 in ϕ12 sequence) which is part of H4 motif and is essential for catalysis (see below

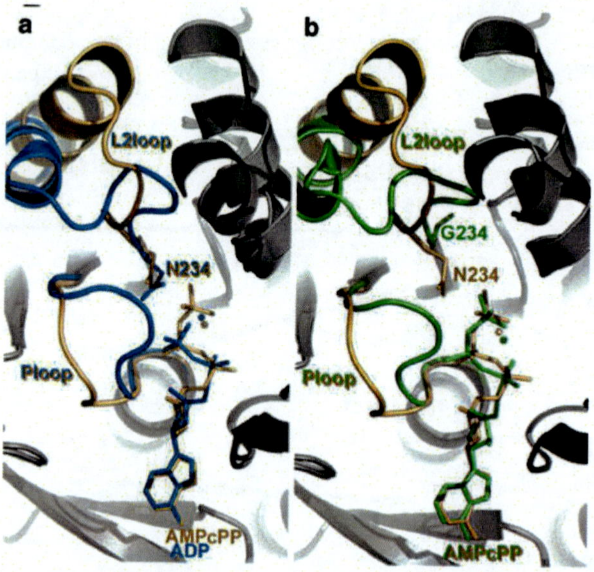

Fig. 27.7 Comparison of lever and ATP binding site conformations viewed from the C-terminal facet along the central channel. (**a**) Structures with ATP analog AMPcPP (*yellow*) and with ADP (*blue*) bound together with Mg^{2+} (*sphere*) in the active site. Note the different position of the P-loop and the L2 loop/α6 helix. (**b**) Comparison of wild-type (*yellow*) and N234G mutant (*green*) co-crystallized with AMPcPP and Mg^{2+} in the active site. Note that the position of the P-loop and the L2 loop/α6 helix in the mutant is almost identical to that in the ADP structure in (**a**)

and Kainov et al. 2008). This residue is part of the communication network between subunits and conveys information about the lever state to the neighbor.

27.5.3 *Lever Motion Is Essential for Hydrolysis*

As discussed above, RNA increases the catalytic rate, and thus, the presence of RNA in the central channel (L2 binding site) needs to be communicated to the ATPase active site. The ATP and RNA binding sites are more than 15 Å apart in the structure; hence, the mechanism is most likely indirect and perhaps involves the lever motion itself. One possibility is that the lever motion in one subunit triggers hydrolysis in the neighboring ATP site. Such "mechanical trigger" would be consistent with the sequential mechanism of hydrolysis. To test this hypothesis, a cysteine mutant of the RNA binding site was generated (K241C), and the levers from neighboring subunits were effectively immobilized by reversible cross-linking of their L2 loops via a disulfide bond. Immobilization of the L2 loops abolished the basal ATPase activity (Kainov et al. 2008) which could be restored upon reduction of the disulfide bonds. This suggests that the lever position is coupled to hydrolysis, that is, the chemical reaction coordinate. But what constitutes the chemical reaction coordinate?

In strictly chemical terms, the reaction coordinate usually represents progress along the reaction path and usually is characterized by a single parameter, for example, scissile bond length. However, in enzymes, the reaction path is seldom described by a single parameter since the contribution of the complex protein motions needs to be taken into account. This broader view is adopted below, with the aim of delineating the structural prerequisites for hydrolysis.

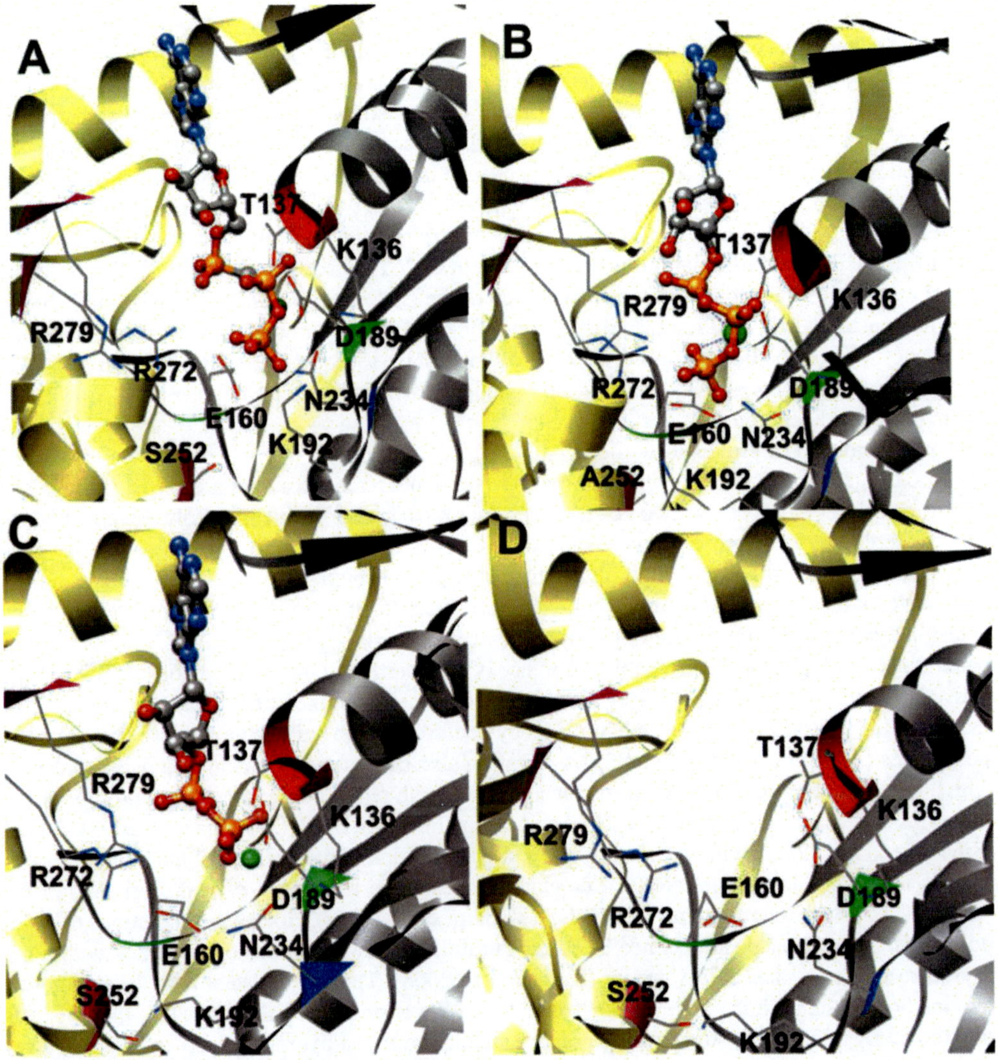

Fig. 27.8 Details of the ATP binding site along the chemical reaction coordinate. (**a**) Wild-type protein in the presence of AMPcPP (*ball* and *stick*) and Mg (*green sphere*). (**b**) S252A mutant in the presence of ATP and Mg. (**c**) Wild-type protein in the presence of ADP and Mg. (**d**) Apoprotein

27.5.4 Coupling Between Mechanical and Chemical Reaction Coordinate

The mechanical trigger may be related to the motion of the lever on the catalytic subunit which would be mediated by *cis*-acting residues. On the other hand, sequential hydrolysis would require residues which are contributed *in trans* by the neighboring subunit. Both mechanisms are plausible and may apply simultaneously. In order to identify the key residues which define the chemical reaction coordinate, it is instrumental to compare constellations of conserved side chains within the active site in different nucleotides states (Fig. 27.8).

The *cis*-acting residues are considered first. The active site coordinates the bound nucleotide di- or triphosphate together with a Mg^{2+} ion which is essential for hydrolysis. In the ATP-bound pre-hydrolysis state, the Mg^{2+} ion is bound to oxygens on β- and γP and coordinated by T137 in the P-loop (Walker A, H1 motif) and D189 (H2 motif). Extrapolating from other helicases, the conserved E160 (motif H1a) most likely acts as the catalytic base during hydrolysis.

The P-loop (Walker A, SF3 motif A, SF4 motif H1) constitutes a flexible link between strand 1 and helix 1 in the ASCE core (Fig. 27.2). In P4, the P-loop adopts different conformations in the ATP- and ADP-bound states (Fig. 27.7a). This change is characterized by a swinging motion of the flexible upstream portion (residues 130–135) which correlates with the position of the moving lever (Fig. 27.7a). The P-loop position does not necessarily reflect the nucleotide occupancy since it can adopt the ADP-like conformation in the presence of an ATP analog within the active site of the N234G mutant which adopts the "down" position of the L2 lever (Fig. 27.7b) (Kainov et al. 2008). This demonstrates that the P-loop conformation is linked to the lever position rather than nucleotide occupancy. Hence, the P-loop motion may act to synchronize lever motion with ATP hydrolysis, that is, formation of the transition state may only happen when both the levers push the P-loop into a certain threshold position.

Trans residues are provided by the neighboring subunit. Given that the mechanical motion is likely to trigger hydrolysis in a sequential manner (see above), these residues are expected to play an important role in this process. Two key trans residues have been identified, namely S252 and R279. The conserved S252 (motif H4) is located at the C-terminus of helix α6 (the lever) and has been shown to be essential for ATP hydrolysis, to the point that mutant protein S252A could be co-crystallized with ATP and Mg^{2+} in a "pre-hydrolysis" state (Kainov et al. 2008). However, this state is distinctly different from the "pre-hydrolysis" state in the wild-type protein (P4:AMPcPP:Mg^{2+}, cf panel a and b in Fig. 27.8). One notable difference is the insertion of the arginine finger R279 into the active site where it interacts with γP (Kainov et al. 2008). An arginine finger is thought to assist formation of the transition state during hydrolysis by neutralizing the buildup of negative charge on the phosphates (Allin et al. 2001). The S252A mutation also causes displacement of a water molecule which is held by Ser252 hydroxyl in an apical position with respect to the γP in the wild-type structure. This water is most likely involved in the nucleophilic attack on γP (Telenius et al. 2008). A final difference in the S252A mutant structure was a slight variation in the position of the L2 loop.

It is interesting to note that R279 side chain moves again as a result of hydrolysis and in the "product" structure interacts with βP of ADP (Mancini et al. 2004a). However, interactions of R279 do not affect the lever position since the latter faithfully follows nucleotide occupancy within the crystal structures of the R279A mutant (Kainov et al. 2008). This is consistent with the notion that the position of the arginine finger is controlled by the lever position. Taken together, this suggests that the S252A:ATP:Mg^{2+} structure represents a state close to the hydrolysis step and that the lever motion is necessary for arginine finger insertion which in turn mediates sequential hydrolysis.

27.5.5 Comparison Between P4 and Related Hexameric Molecular Motors

Comparison between molecular motors may reveal possible conserved structural and mechanistic features. Such comparisons have been done for AAA+ proteins (*A*TPases associated with diverse cellular activities) (Erzberger and Berger 2006) and hexameric helicases (Donmez and Patel 2006; Enemark and Joshua-Tor 2008). More recently, P4 has been compared with RNA-specific hexameric helicases, *E. coli* Rho (Rabhi et al. 2010) and viral helicases E1 and Simian virus 40 large T antigen (LTAg) (Tuma 2010). Here, we summarize the latter comparisons.

Domain and loop motions have been proposed to be involved in hexameric helicase action and nucleic acid translocation (Lisal et al. 2004; Singleton et al. 2000). However, the first structural correlations between conformational changes and ATP binding and hydrolysis were provided for ϕ12 P4 and SV40 LTAg hexamers (Gai et al. 2004; Mancini et al. 2004a). These structures revealed

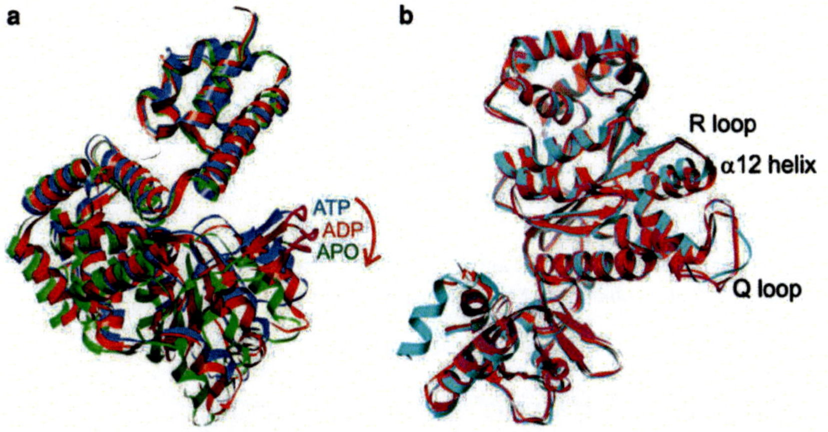

Fig. 27.9 Comparison of subunit conformations associated with different positions of the DNA/RNA binding sites along the central channel. (**a**) LTAg structures in the presence of ATP (*blue*), ADP (*red*), and in apo form (*green*) were aligned using the N-terminal domain (Gai et al. 2004). Motion of the DNA binding loops along the central channel is shown by the *arrow*. (**b**) Comparison of two different conformations of Rho (subunits F and A) within the same hexamer crystallized in the presence of RNA (Thomsen and Berger 2009)

motions of the nucleic acid binding sites and correlated them with the nucleotide occupancy of the ATPase catalytic pocket. Subsequently, similar motions were found in the E1 helicase (AAA+ motor closely related to LTAg) and Rho crystal structures in which the hexamers are bound to their respective DNA/RNA substrates (Enemark and Joshua-Tor 2006; Thomsen and Berger 2009). These later structures also revealed the staircase arrangement of the nucleic acid binding sites, that is, gradual movement of the lever along the mechanical reaction coordinate and subtle differences in ATP binding pockets associated with the movement.

The mechanical reaction coordinates of these molecular motors, encompassing simple loop motion along the central channel, are similar. RNA binding sites are structurally conserved between the RecA-like helicases P4, Rho, and T7 gp4, while those within the AAA+ group (LTAg and E1) differ from the former. The mechanism by which motion of these loops is accomplished differs. In the case of P4, motion is accomplished by the swiveling helix α6 while subunits remain virtually motionless. In contrast, in Rho, whole subunits undergo rigid body rotations within the hexamer, while there is little internal motion within subunits themselves as manifested by a negligible shift in the position of loops Q and R and helix α12 (equivalent of the P4 lever helix α6) with respect to the RecA core (Fig. 27.9b). Similarly to Rho, motion within the LTAg and E1 hexamers is accomplished by domain rotations with respect to the neighboring subunits, leaving the N-terminal domain stationary (Fig. 27.9a). Thus, in these motors, the whole domain or subunit acts as a lever, amplifying small changes in the active site at subunit interfaces into intersubunit motions.

Perhaps it is not surprising that the remotely related RecA-like P4 and the AAA+ viral helicases (LTAg, E1) employ different molecular lever mechanisms. However, in the light of the close structural and functional relationship between P4 and Rho and their possible evolutionary relationship, this difference is puzzling. One possibility is that P4 evolved to limit large domain motions within the hexamer in favor of motions within individual subunits. Unlike Rho and LTAg, which act alone during their motor action, P4 is an integral part of the viral capsid during packaging. P4 associates with the capsid via its C-terminal facet and these steric constraints may limit subunit rotations. Indeed, these interactions stabilize subunit interfaces within the hexamer, further limiting domain and subunit motions (Lisal et al. 2006) and limit spontaneous ring opening and consequently increase processivity. Indeed, P4 is a more processive translocase than Rho (Rabhi et al. 2010).

In summary, there are two principal ways to amplify small changes in the ATP binding site into nanometer-scale motions. One involves rigid body rotation of subunits or domains, while the other relies on internal motions of rigid elements (e.g., helices) within subunits.

27.6 Mechanism of Translocation

This section builds upon the structural and biochemical observations presented above and attempts to integrate them into a mechanistic model (schematically represented in Fig. 27.10).

27.6.1 Sequential ATP Hydrolysis Drives Translocation

In P4 and related hexameric motors, the movement along the mechanical reaction coordinate is controlled by ATP hydrolysis (Kainov et al. 2008; Thomsen and Berger 2009). While P_i release seems to drive the lever with the bound RNA down the central channel (i.e., power stroke), ATP binding most likely drives the reverse stroke and the detachment from the oligonucleotide. The biochemical properties and structural characteristics are compatible with a sequential hydrolysis scheme in which three consecutive subunits cooperate (either through binding cooperativity as in the case of Rho or via catalytic cooperativity in P4). Under this scheme, a perpetual wave of conformational changes propagates around the ring in one direction (a molecular version of Mexican wave during a football match). The direction is structurally dictated by the insertion of the arginine finger. When the P4 ring is viewed down from the N-terminal facet, then the direction of propagation of these changes is counterclockwise, as shown in Fig. 27.5b.

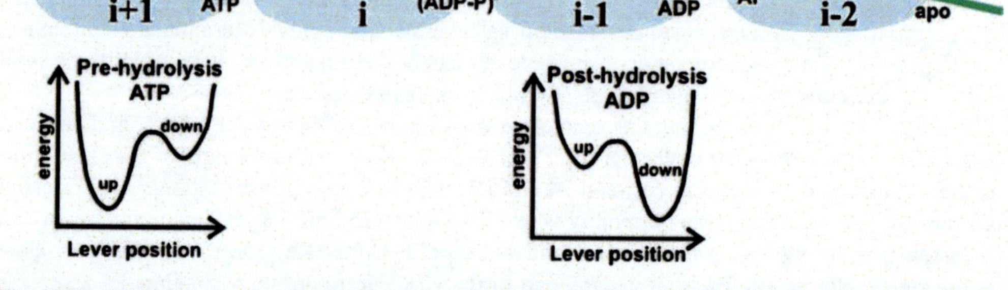

Fig. 27.10 Schematic representation of the P4 cooperative mechanism. The *top panel* depicts the outline of four neighboring subunits (*light gray*), with the lever helix depicted in *magenta* and L2 loop in *cyan* with the RNA binding site, K241, in *blue*. P-loop is *red*, while the bound nucleotide di/triphosphates are *yellow*. The conserved H3 motif (γP sensor) N234 is in *pink* and arginine finger (AF) is *green* and the H4 motif S252 is *orange*. *Lower panel*: simplified hypothetical energy landscapes for the lever motion along the mechanical reaction coordinate for the corresponding nucleotide states depicted in the *top panel*

27.6.2 RNA Stimulates Hydrolysis by Concerted Lever Movement

ATPase activity of hexameric helicases is stimulated by ssDNA/RNA which binds to several subunits within the central channel. The number of subunits engaged with the nucleic acid at a time may vary depending on the helicase and the method used in estimating it. For example, enzymatic and mutational analyses of P4 indicated that three to four subunits are engaged at once with RNA (Kainov et al. 2008; Telenius et al. 2008). In contrast, five loops were shown to interact with RNA in the most recent Rho crystal structure (Thomsen and Berger 2009). However, enzymatic studies and modeling suggest that approximately three subunits cooperate in Rho during translocation (Adelman et al. 2006).

In the simplest case, ATPase stimulation can involve just two neighboring subunits, both bound to RNA and with an ATP molecule bound at their interface. Since lever motion is essential in P4 for hydrolysis (e.g., arginine finger insertion), one plausible mechanism of stimulation is coordinated motion of two consecutive levers. In this model, the bound RNA provides a conformational link which facilitates concerted motion of the consecutive L2 loops (Fig. 27.10, subunits $i-1$, i). ATP hydrolysis at the preceding site (subunit $i-1$) forces the lever (L2 loop and α6) down, inserting at the same time Arg279 into the next active site for transition state formation (subunit i). The mechanical drag of the attached RNA on the L2 loop of the next subunit helps the lever to move down while configuring the P-loop into a catalytically competent conformation. If a third subunit is considered, then the attached RNA mechanically drags the L2 loop of the next subunit down during the next cycle, promoting sequential hydrolysis (Fig. 27.10, subunit $i+1$). After hydrolysis and P_i release on subunit i, the cycle repeats itself starting on position $i+1$. There is some evidence from Raman spectroscopy that the bound RNA becomes stretched during ATP binding and hydrolysis (Lisal et al. 2004), effectively becoming a "molecular bungee cord" linking the neighboring levers.

After reaching the lowest position in the apo state (e.g., $i-2$ in Fig. 27.10), energy associated with ATP binding is utilized to detach RNA from this subunit simply by lifting the lever up. Reverse motion is prevented by simultaneous interaction of RNA with several consecutive subunits (i and $i-1$) that are engaged in sequential hydrolysis.

27.7 Conclusions

Despite belonging to different helicase superfamilies, the bacteriophage packaging enzyme P4 (SF4), the viral helicases LTAg and E1 (SF3), and Rho transcription terminator (SF5) share many mechanistic features. These hexameric enzymes seem to employ a sequential scheme of hydrolysis that drives conformational changes around the hexameric ring. The ATP binding sites of these molecular motors contain structurally similar residues which control the motion of the RNA binding sites and/or trigger hydrolysis in response to translocation of RNA. The clear relationship between these mechanisms in such diverse systems suggests that their mechanistic features may reflect intrinsic chemical and physical constraints. However, other packaging motors that adopt different stoichiometry clearly differ from these hexamers and most likely adopt alternative scheme of translocation.

References

Adelman JL, Jeong YJ, Liao JC, Patel G, Kim DE, Oster G, Patel SS (2006) Mechanochemistry of transcription termination factor Rho. Mol Cell 22:611–621

Allin C, Ahmadian MR, Wittinghofer A, Gerwert K (2001) Monitoring the GAP catalyzed H-Ras GTPase reaction at atomic resolution in real time. Proc Natl Acad Sci USA 98:7754–7759

Bamford DH, Ojala PM, Frilander M, Walin L, Bamford JKH (1995) Isolation, purification, and function of assembly intermediates and subviral particles of bacteriophages PRD1 and phi6. In: Adolph KW (ed) Methods in molecular genetics. Academic, San Diego, pp 455–474

Butcher SJ, Dokland T, Ojala PM, Bamford DH, Fuller SD (1997) Intermediates in the assembly pathway of the double-stranded RNA virus phi6. EMBO J 16:4477–4487

Crampton DJ, Mukherjee S, Richardson CC (2006) DNA-induced switch from independent to sequential dTTP hydrolysis in the bacteriophage T7 DNA helicase. Mol Cell 21:165–174

Donmez I, Patel SS (2006) Mechanisms of a ring shaped helicase. Nucleic Acids Res 34:4216–4224

Enemark EJ, Joshua-Tor L (2006) Mechanism of DNA translocation in a replicative hexameric helicase. Nature 442:270–275

Enemark EJ, Joshua-Tor L (2008) On helicases and other motor proteins. Curr Opin Struct Biol 18:243–257

Erzberger JP, Berger JM (2006) Evolutionary relationship and structural mechanisms of AAA+ proteins. In: Rees DC (ed) Annual reviews of biophysics and biomolecular structure. Annual Reviews, Palo Alto, pp 93–114

Frilander M, Gottlieb P, Strassman J, Bamford DH, Mindich L (1992) Dependence of minus-strand synthesis on complete genomic packaging in the double-stranded RNA bacteriophage phi 6. J Virol 66:5013–5017

Gai D, Zhao R, Li D, Finkielstein CV, Chen XS (2004) Mechanisms of conformational change for a replicative hexameric helicase of SV40 large tumor antigen. Cell 119:47–60

Gottlieb P, Strassman J, Frucht A, Qiao XY, Mindich L (1991) In vitro packaging of the bacteriophage phi 6 ssRNA genomic precursors. Virology 181:589–594

Gottlieb P, Strassman J, Mindich L (1992a) Protein P4 of the bacteriophage phi 6 procapsid has a nucleoside triphosphate-binding site with associated nucleoside triphosphate phosphohydrolase activity. J Virol 66:6220–6222

Gottlieb P, Strassman J, Qiao X, Frilander M, Frucht A, Mindich L (1992b) In vitro packaging and replication of individual genomic segments of bacteriophage phi 6 RNA. J Virol 66:2611–2616

Gottlieb P, Qiao X, Strassman J, Frilander M, Mindich L (1994) Identification of the packaging regions within the genomic RNA segments of bacteriophage phi 6. Virology 200:42–47

Huiskonen JT, Butcher SJ (2007) Membrane-containing viruses with icosahedrally symmetric capsids. Curr Opin Struct Biol 17:229–236

Huiskonen JT, de Haas F, Bubeck D, Bamford DH, Fuller SD, Butcher SJ (2006) Structure of the bacteriophage phi6 nucleocapsid suggests a mechanism for sequential RNA packaging. Structure 14:1039–1048

Huiskonen JT, Jaalinoja HT, Briggs JA, Fuller SD, Butcher SJ (2007) Structure of a hexameric RNA packaging motor in a viral polymerase complex. J Struct Biol 158:156–164

Iyer LM, Leipe DD, Koonin EV, Aravind L (2004) Evolutionary history and higher order classification of AAA+ ATPases. J Struct Biol 146:11–31

Jaalinoja HT, Huiskonen JT, Butcher SJ (2007) Electron cryomicroscopy comparison of the architectures of the enveloped bacteriophages phi6 and phi8. Structure 15:157–167

Juuti JT, Bamford DH, Tuma R, Thomas GJ Jr (1998) Structure and NTPase activity of the RNA-translocating protein (P4) of bacteriophage phi 6. J Mol Biol 279:347–359

Kainov DE, Pirttimaa M, Tuma R, Butcher SJ, Thomas GJ Jr, Bamford DH, Makeyev EV (2003) RNA packaging device of double-stranded RNA bacteriophages, possibly as simple as hexamer of P4 protein. J Biol Chem 278:48084–48091

Kainov DE, Lisal J, Bamford DH, Tuma R (2004) Packaging motor from double-stranded RNA bacteriophage phi12 acts as an obligatory passive conduit during transcription. Nucleic Acids Res 32:3515–3521

Kainov DE, Tuma R, Mancini EJ (2006) Hexameric molecular motors: P4 packaging ATPase unravels the mechanism. Cell Mol Life Sci 63:1095–1105

Kainov DE, Mancini EJ, Telenius J, Lisal J, Grimes JM, Bamford DH, Stuart DI, Tuma R (2008) Structural basis of mechanochemical coupling in a hexameric molecular motor. J Biol Chem 283:3607–3617

Levin MK, Patel SS (2003) Helicases as molecular motors. In: Schliwa M (ed) Molecular motors. Wiley-VCH, Weinheim, pp 179–203

Liao JC, Jeong YJ, Kim DE, Patel SS, Oster G (2005) Mechanochemistry of T7 DNA helicase. J Mol Biol 350:452–475

Lisal J, Tuma R (2005) Cooperative mechanism of RNA packaging motor. J Biol Chem 280:23157–23164

Lisal J, Kainov DE, Bamford DH, Thomas GJ Jr, Tuma R (2004) Enzymatic mechanism of RNA translocation in double-stranded RNA bacteriophages. J Biol Chem 279:1343–1350

Lisal J, Lam T, Kainov DE, Emmett MR, Marshall AG, Tuma R (2005) Functional visualization of viral molecular motor by hydrogen-deuterium exchange reveals transient states. Nat Struct Mol Biol 12:460–466

Lisal J, Kainov DE, Lam TT, Emmett MR, Wei H, Gottlieb P, Marshall AG, Tuma R (2006) Interaction of packaging motor with the polymerase complex of dsRNA bacteriophage. Virology 351:73–79

Mancini EJ, Tuma R (2008) Structure and function of P4, a dsRNA virus packaging motor. In: Patton JT (ed) Segmented double-stranded RNA viruses: structure and molecular biology. Caister Academic, Norfolk, UK, pp 259–272

Mancini EJ, Kainov DE, Grimes JM, Tuma R, Bamford DH, Stuart DI (2004a) Atomic snapshots of an RNA packaging motor reveal conformational changes linking ATP hydrolysis to RNA translocation. Cell 118:743–755

Mancini EJ, Kainov DE, Wei H, Gottlieb P, Tuma R, Bamford DH, Stuart DI, Grimes JM (2004b) Production, crystallization and preliminary X-ray crystallographic studies of the bacteriophage phi 12 packaging motor. Acta Crystallogr D 60:588–590

Meier C, Mancini EJ, Bamford DH, Walsh MA, Stuart DI, Grimes JM (2005) Overcoming the false-minima problem in direct methods: structure determination of the packaging enzyme P4 from bacteriophage phi13. Acta Crystallogr D 61:1238–1244

Mindich L (1999) Precise packaging of the three genomic segments of the double-stranded RNA bacteriophage phi6. Microbiol Mol Biol Rev 63:149–160

Mindich L (2004) Packaging, replication and recombination of the segmented genome of bacteriophage Phi6 and its relatives. Virus Res 101:83–92

Mindich L, Sinclair JF, Cohen J (1976) The morphogenesis of bacteriophage phi6: particles formed by nonsense mutants. Virology 75:224–231

Mitchell MS, Matsuzaki S, Imai S, Rao VB (2002) Sequence analysis of bacteriophage T4 DNA packaging/terminase genes 16 and 17 reveals a common ATPase center in the large subunit of viral terminases. Nucleic Acids Res 30:4009–4021

Nemecek D, Heymann JB, Qiao J, Mindich L, Steven AC (2010) Cryo-electron tomography of bacteriophage phi6 procapsids shows random occupancy of the binding sites for RNA polymerase and packaging NTPase. J Struct Biol 171:389–396

Paatero AO, Mindich L, Bamford DH (1998) Mutational analysis of the role of nucleoside triphosphatase P4 in the assembly of the RNA polymerase complex of bacteriophage phi6. J Virol 72:10058–10065

Pirttimaa MJ, Bamford DH (2000) RNA secondary structures of the bacteriophage phi6 packaging regions. RNA 6:880–889

Poranen MM, Tuma R (2004) Self-assembly of double-stranded RNA bacteriophages. Virus Res 101:93–100

Poranen MM, Paatero AO, Tuma R, Bamford DH (2001) Self-assembly of a viral molecular machine from purified protein and RNA constituents. Mol Cell 7:845–854

Poranen MM, Tuma R, Bamford DH (2008) Dissecting the assembly pathway of bacterial dsRNA viruses: infectious nucleocapsids produced by self-assembly. In: Patton JT (ed) Segmented double-stranded RNA viruses: structure and molecular biology. Caister Academic, Norfolk, UK, pp 115–132

Qiao X, Qiao J, Mindich L (1997) Stoichiometric packaging of the three genomic segments of double-stranded RNA bacteriophage phi6. Proc Natl Acad Sci USA 94:4074–4079

Qiao X, Qiao J, Mindich L (2003) Analysis of specific binding involved in genomic packaging of the double-stranded-RNA bacteriophage phi6. J Bacteriol 185:6409–6414

Rabhi M, Tuma R, Boudvillain M (2010) RNA remodeling by hexameric RNA helicases. RNA Biol 7:16–27

Rao VB, Feiss M (2008) The bacteriophage DNA packaging motor. Annu Rev Genet 42:647–681

Sawaya MR, Guo S, Tabor S, Richardson CC, Ellenberger T (1999) Crystal structure of the helicase domain from the replicative helicase-primase of bacteriophage T7. Cell 99:167–177

Scheffzek K, Ahmadian MR, Kabsch W, Wiesmuller L, Lautwein A, Schmitz F, Wittinghofer A (1997) The Ras-RasGAP complex: structural basis for GTPase activation and its loss in oncogenic Ras mutants. Science 277:333–338

Singleton MR, Sawaya MR, Ellenberger T, Wigley DB (2000) Crystal structure of T7 gene 4 ring helicase indicates a mechanism for sequential hydrolysis of nucleotides. Cell 101:589–600

Singleton MR, Dillingham MS, Wigley DB (2007) Structure and mechanism of helicases and nucleic acid translocases. Annu Rev Biochem 76:23–50

Skordalakes E, Berger JM (2003) Structure of the Rho transcription terminator: mechanism of mRNA recognition and helicase loading. Cell 114:135–146

Sun S, Kondabagil K, Gentz PM, Rossmann MG, Rao VB (2007) The structure of the ATPase that powers DNA packaging into bacteriophage T4 procapsids. Mol Cell 25:943–949

Sun S, Kondabagil K, Draper B, Alam TI, Bowman VD, Zhang Z, Hegde S, Fokine A, Rossmann MG, Rao VB (2008) The structure of the phage T4 DNA packaging motor suggests a mechanism dependent on electrostatic forces. Cell 135:1251–1262

Telenius J, Wallin AE, Straka M, Zhang H, Mancini EJ, Tuma R (2008) RNA packaging motor: from structure to quantum mechanical modeling and sequential-stochastic mechanism. Comput Math Methods Med 9:351–369

Thomsen ND, Berger JM (2009) Running in reverse: the structural basis for translocation polarity in hexameric helicases. Cell 139:523–534

Tuma R (2008) Icosahedral enveloped dsRNA bacterial viruses. In: Mahy BWJ, Regenmortel MHVV (eds) Encyclopedia of virology. Elsevier, Oxford, pp 6–13

Tuma R (2010) Hexameric viral RNA helicases. In: Jankowsky E (ed) RNA helicases. RSC, Cambridge, UK, pp 213–242

Chapter 28
Helical Viruses

Gerald Stubbs and Amy Kendall

Abstract Virtually all studies of structure and assembly of viral filaments have been made on plant and bacterial viruses. Structures have been determined using fiber diffraction methods at high enough resolution to construct reliable molecular models or several of the rigid plant tobamoviruses (related to tobacco mosaic virus, TMV) and the filamentous bacteriophages including Pf1 and fd. Lower-resolution structures have been determined for a number of flexible filamentous plant viruses using fiber diffraction and cryo-electron microscopy. Virions of filamentous viruses have numerous mechanical functions, including cell entry, viral disassembly, viral assembly, and cell exit. The plant viruses, which infect multicellular organisms, also use virions or virion-like assemblies for transport within the host. Plant viruses are generally self-assembling; filamentous bacteriophage assembly is combined with secretion from the host cell, using a complex molecular machine. Tobamoviruses and other plant viruses disassemble concomitantly with translation, by various mechanisms and involving various viral and host assemblies. Plant virus movement within the host also makes use of a variety of viral proteins and modified host assemblies.

28.1 Introduction

28.1.1 Helical Viruses

The vast majority of helical viruses are either filamentous plant viruses or filamentous bacteriophages. The filoviruses (Ebola and Marburg viruses), which are among the agents that cause life-threatening viral hemorrhagic fevers in humans, are a rare exception; they are, however, pleomorphic membrane-enveloped assemblies, and neither their structures nor the molecular mechanisms that underlie their life cycles are well understood. In many other animal viruses, such as the retroviruses (including human immunodeficiency virus, HIV), the paramyxoviruses (measles, mumps), orthomyxoviruses (influenza A, B, and C), and rhabdoviruses (vesicular stomatitis virus and rabies, as well as many plant viruses), the genome, RNA or DNA, is packaged in a helical filamentous nucleocapsid assembly, but the nucleocapsid is highly condensed into a compact, coiled structure, which is in turn encapsidated by

---

G. Stubbs (✉) • A. Kendall
Department of Biological Sciences, Vanderbilt University,
Nashville, TN, USA
e-mail: gerald.stubbs@vanderbilt.edu

M.G. Rossmann and V.B. Rao (eds.), *Viral Molecular Machines*,
Advances in Experimental Medicine and Biology 726, DOI 10.1007/978-1-4614-0980-9_28,

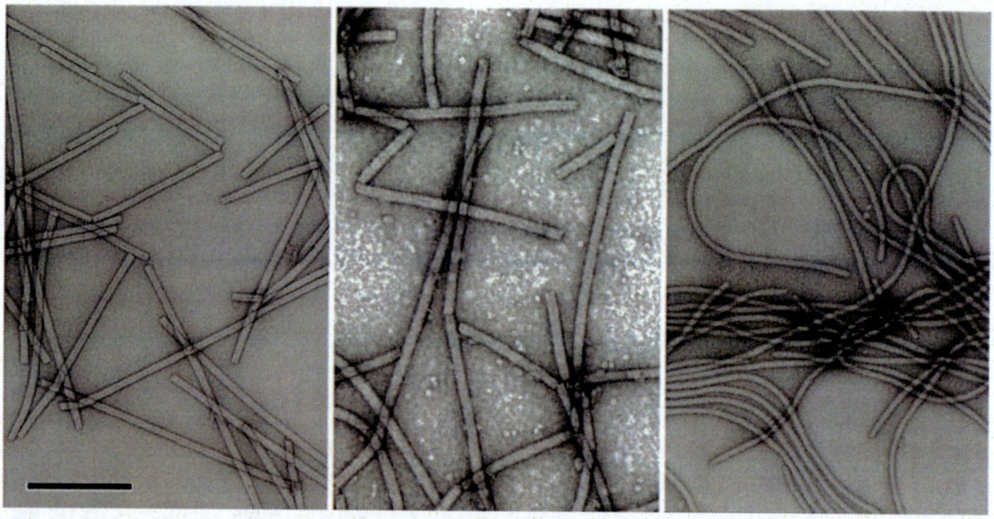

Fig. 28.1 Negative-stain electron micrographs of filamentous plant viruses. From *left*: tobacco mosaic virus, barley stripe mosaic virus, potato virus Y. TMV and BSMV are rigid *rod*-shaped viruses; PVY is flexible. TMV and PVY particles are all the same length; BSMV has three particle lengths. *Bar*=250 nM

other proteins, sometimes enveloped in a membrane, in various degrees of complexity. In this chapter, we will address only the simple helical filamentous plant viruses and bacteriophages. Only in these two groups does the filamentous nucleocapsid comprise essentially the entire virion. This simplicity of structure, together with the relative ease of obtaining and purifying them, has meant that almost all studies of structure and assembly of viral filaments have been made on plant and bacterial viruses.

Almost 500 individual species of filamentous plant viruses have been identified (Fauquet et al. 2005; Carstens 2010), making up about one-third of the plant virus genera and almost half of the species. The family *Potyviridae* alone, including about 20% of the known plant viruses, is responsible for more than half the viral crop damage in the world (López-Moya and García 2008) and constitutes a significant threat to agriculture and food security. The potyvirus cassava brown streak virus has been identified as one of Africa's biggest threats to food security (Pennisi 2010). A number of filamentous plant viruses appear to have significant potential in biotechnology (Cañizares et al. 2005; Denis et al. 2008; Lico et al. 2008), a result of the combination of ease of large-scale plant production and the ready accommodation of additional RNA in the filamentous viral genome.

Because of their unique combination of stability, physical manipulability, inherent mechanical function, and the wealth of biophysical information available, filamentous viruses, particularly the filamentous plant virus tobacco mosaic virus (TMV) (Dujardin et al. 2003) and the filamentous bacteriophage M13 (Hemminga et al. 2010) have attracted particular attention in the field of bionanotechnology.

Filamentous plant viruses are either rigid (rod-shaped) or flexible (Fig. 28.1). The International Committee on Taxonomy of Viruses currently recognizes eight genera and one family (the *Virgaviridae*, Adams et al. 2009) of rigid filamentous plant viruses, and 20 genera and four families of flexible filamentous plant viruses (Fauquet et al. 2005; Carstens 2010). Not all viruses have yet been assigned to genera and families. All of the filamentous plant viruses are RNA viruses, and most consist of a single type of coat protein encapsidating a single-stranded RNA molecule in a simple helical array. In some genera, for example the tobraviruses and hordeiviruses, the genome is divided among two or more RNA molecules; in these cases, the virus consists of multiple particles, typically of different lengths (Fig. 28.1).

Many filamentous plant viruses, even from different families, appear to be similar to each other at the electron microscopic level, but others exhibit large differences in morphology, and most families have little or no similarity in their coat protein amino acid sequences with other families. Some authors have suggested that the coat protein structures of all filamentous plant viruses are nevertheless similar; more convincingly, it has been argued that most fall into one of two groups, corresponding to the rigid rods and the flexible filaments (Dolja et al. 1991). Coat protein sequences suggest that that the tobamoviruses, tobraviruses, and furoviruses share a common protein fold, the four α-helix bundle (Goulden et al. 1992; Chen et al. 1997). The amino acid sequences and protein chain lengths of viruses in different groups are, however, so different that even in these cases there must be major structural differences, too great to predict without structure determinations using objective methods.

The filamentous bacteriophages (Marvin 1998) are members of the *Inovirus* and *Plectrovirus* genera (family *Inoviridae*) (Fauquet et al. 2005) and infect mostly Gram-negative bacteria. Virtually all structural studies have been carried out on members of the genus *Inovirus*. These viruses have been grouped into two classes on the basis of helical symmetry, class I and class II (Marvin et al. 1974a), but all appear to be very similar at the level of electron microscopy. The Inoviruses have been extremely useful in biotechnology and in the laboratory, because they are easily genetically modified, and accommodate very large inserts of foreign DNA without difficulty. The Ff viruses (the closely related bacteriophages fd, f1, and M13, which have almost identical coat protein amino acid sequences) are widely used as vectors for gene transfer, and have for many years occupied a particularly useful niche as vehicles for peptide display, presenting peptides on the virion surface. In this technique, the gene for a foreign peptide is fused with the gene for one of the viral coat proteins, providing a very useful way to investigate the interactions of the foreign peptide with other molecules (Russel and Model 2006).

Filamentous bacteriophage particles are long, flexible DNA viruses. Although the coat protein structures of the various bacteriophages (at least among the well-studied members of the genus *Inovirus*) are all similar (Sect. 28.3), the DNA content, both in total number of nucleotides and nucleotide to protein ratio, is very variable.

28.1.2 Structural Analysis of Filamentous Assemblies

Like most helical assemblies, filamentous viruses do not crystallize, and although there have been some solid-state nuclear magnetic resonance (ssNMR) studies, primarily of the filamentous bacteriophages (Cross et al. 1983; Marvin et al. 2006; Opella et al. 2008), they are too large for conventional NMR structure determination. The most informative structural studies have therefore come from X-ray fiber diffraction and electron microscopy.

Fiber diffraction (Stubbs 1999a; Stubbs 2001; Chandrasekaran and Stubbs 2006) is the scattering of radiation from specimens in which filamentous structural aggregates are aligned approximately parallel to each other but are randomly oriented about a common axis, the fiber axis. Because of the random orientation, fiber diffraction patterns represent the cylindrical average of the filament diffraction pattern. Cylindrical averaging reduces the information content of the patterns, but for filaments of high symmetry such as viruses, the reduction is much less than one might expect. For the plant virus TMV (the type member of the genus *Tobamovirus*), cylindrical averaging reduces the number of observable diffraction data at 3 Å resolution by a factor of about 2.5, while for the filamentous bacteriophage Pf1, the corresponding factor is only 1.7 (Makowski 1982). The reduction is negligible for TMV at 15 Å resolution, and for Pf1 at 7 Å. Fiber diffraction patterns (Fig. 28.2) are characterized by layer lines at spacings inversely proportional to the helical filament repeat distance.

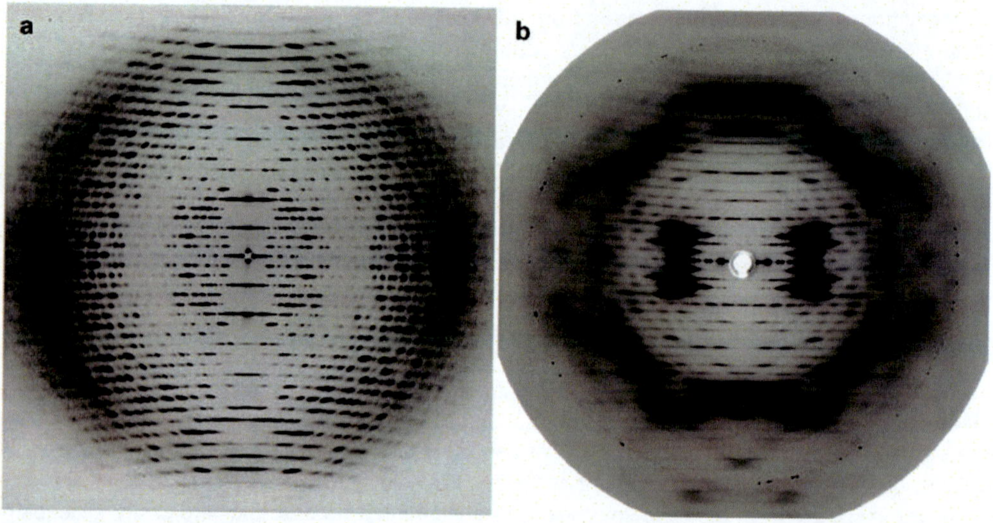

Fig. 28.2 X-ray fiber diffraction patterns from (**a**) an oriented sol of TMV (Namba et al. 1989) (**b**) a dried fiber of filamentous bacteriophage Pf1 (Welsh et al. 2000). The data lie on layer lines, characteristic of diffraction from helically repeating structures. Part (**b**) courtesy of Dr. Don Marvin

Fiber diffraction specimens may be oriented sols (Gregory and Holmes 1965; Kendall and Stubbs 2006) or dried fibers (McDonald et al. 2008). Magnetic fields (Torbet and Maret 1979; Welsh et al. 1998; Yamashita et al. 1998) have been used to improve alignment, particularly for filamentous bacteriophages. In order to resolve diffraction from large repeat distances and to reach reasonably high resolution, data collection methods for filamentous virus analysis have more in common with protein crystallography than with other fiber diffraction methods, but because of the difficulties associated with background subtraction and other data processing steps from layer line patterns, synchrotron beam lines specializing in fiber diffraction (Fischetti et al. 2004; Smolsky et al. 2007) have been useful.

The resolution of fiber diffraction analyses may be limited by imperfect alignment, but in favorable cases, patterns can contain sufficient data to determine structures, particularly when stereochemical restraints can be imposed. Initial structures may be determined by a multidimensional analog of isomorphous replacement (Stubbs and Diamond 1975) or by molecular replacement (Wang et al. 1997; Welsh et al. 1998). The filamentous bacteriophage structures are relatively simple, and models may be constructed directly from consideration of diffraction and spectroscopic data (Makowski et al. 1980). Cryo-electron microscopy (cryo-EM) is now providing models at moderate resolution (Egelman 2007; Kendall et al. 2008) that also appear to have potential as starting models for fiber diffraction analyses. Models have been refined by various methods; in recent years, molecular dynamics refinement has been effective (Wang and Stubbs 1993; Wang et al. 1997; Welsh et al. 1998).

Cryo-EM has achieved increasingly high resolutions in recent years (Zhang et al. 2008; Zhou 2008). In favorable cases, resolution may be better than 10 Å, sufficient to allow the visualization of secondary structures, particularly α-helices. The helical symmetry of filamentous viruses is a useful constraint in reconstructions and has allowed methods such as iterative helical real-space reconstruction (IHRSR) (Egelman 2007) to be used to advantage (Kendall et al. 2008). Although a relative newcomer to filamentous virus structure determination, this approach promises to be of increasing importance, both as a method of structure determination in itself and as a source of models with which to begin refinement against fiber diffraction data.

28.1.3 Structural Studies of Helical Viruses

Structural studies have been made of a number of filamentous plant viruses and bacteriophages. In rare cases, isolated coat proteins of viruses have been crystallized; the structure of the TMV coat protein was determined crystallographically (Bloomer et al. 1978; Bhyravbhatla et al. 1998). However, the natural tendency of filamentous virus coat protein subunits to form helical aggregates has prevented crystallization of the coat proteins of other filamentous viruses, including other tobamoviruses. Even when crystals can be grown, the protein–protein interactions do not correspond to the biologically significant interactions in the filaments, and there are of course no protein–nucleic acid interactions. Structural studies of filamentous viruses have therefore primarily used the methods of fiber diffraction and electron microscopy.

TMV was the first virus to be discovered (Beijerinck 1898), and the first to be studied by X-ray diffraction. Solution scattering was described by Wyckoff and Corey (1936), and Bawden et al. (1936) and Bernal and Fankuchen (1941) showed that TMV can form highly oriented sols, in which the rod-shaped particles are aligned to within about 1° of each other. The sols yield high-quality fiber diffraction patterns (Fig. 28.2a); the virions are randomly oriented about their long axes, but the patterns are nevertheless rich in information. Bernal and Fankuchen (1941) also obtained fiber diffraction patterns from the flexible filamentous potato virus X (PVX), although those patterns were not as well oriented as those of TMV.

The arrangement of layer lines in the TMV diffraction pattern showed that the virion structure was periodic, repeating every 69 Å. The intensity of every third layer line close to the meridian (the vertical axis of the pattern) further showed that there was an approximate repeat every 23 Å. After Cochran et al. (1952) had developed the theory of diffraction from a helix, Watson (1954) was able to show that the 69 Å repeating unit of TMV must contain $3n+1$ subunits in three turns of a helical structure, with n being an integer. Franklin and Holmes (1958) showed that $n=16$, that is, there are 49 subunits in three turns of the TMV helix.

The protein crystallographic method of isomorphous replacement, binding heavy atoms such as mercury and lead to the protein subunits, can be adapted for fiber diffraction. Caspar (1956) and Franklin (1956a) independently used this method to determine the radial density distribution of TMV. They showed that the virion was about 90 Å in radius, with a central hole about 20 Å in radius and the RNA located about 40 Å from the viral axis. The method was later adapted further (Stubbs and Diamond 1975) to determine the structure of TMV at 6.7 Å resolution (Holmes et al. 1975), and later, in conjunction with other developments in fiber diffraction structure determination, at 2.9 Å resolution (Namba et al. 1989). Structures of other tobamoviruses, discussed below, followed.

Fiber diffraction studies of other filamentous plant virus genera have not yet provided the detailed structural models that are available for the tobamoviruses, but there have been studies of other rigid rod-shaped viruses. More recently, there has been considerable progress in obtaining diffraction data from several families of flexible filamentous viruses (Sect. 28.2.2). Fiber diffraction data from these viruses have not been sufficient by themselves for structure determination, but the combination of fiber diffraction and cryo-EM is beginning to yield structural information about a number of filamentous plant virus families (Kendall et al. 2008).

Diffraction patterns from dried fibers of filamentous bacteriophage fd were described by Marvin (1966). Patterns from other class I viruses were very similar (Marvin et al. 1974b), but diffraction patterns from class II virions, while generally similar to class I, differed significantly in details (Marvin et al. 1974a). Strong magnetic fields were used to induce the bacteriophage particles to align (Torbet and Maret 1979), yielding diffraction patterns of exceptionally high quality (Fig. 28.2b). The simple α-helical structure of the coat protein (Marvin 1966) facilitated the construction of models that could be refined against the diffraction data. The structure of Pf1 was determined at 7 Å resolution (Makowski et al. 1980), and structures of a number of filamentous bacteriophages from both structural classes (Sect. 28.3) were later determined at resolutions as high as 3.0 Å (Welsh et al. 2000).

28.2 Filamentous Plant Virus Structure

28.2.1 Rigid Rod-Shaped Plant Viruses

28.2.1.1 Tobamoviruses

Perhaps more because of their ready availability than their agricultural importance, tobamoviruses have served as models in structural biology, biochemistry, virology, molecular biology, and cell biology (Harrison and Wilson 1999). They are easily purified in large quantities, physically and chemically stable, and particularly suited to fiber diffraction studies because they can be aligned with little difficulty. They have been a model fiber diffraction system for over 70 years and are the largest molecular assemblies for which complete three-dimensional structural models in atomic detail have been obtained by fiber diffraction alone. TMV was the first virus for which protein–nucleic acid interactions were directly visualized, and because of the simplicity of the nucleic acid binding, the tobamoviruses remain the only viruses for which the three-dimensional structure of essentially the entire nucleic acid component is known.

Early studies using fiber diffraction, electron microscopy, and other methods (reviewed by Caspar 1963) showed that TMV (as well as several other tobamoviruses) was about 3,000 Å long, 180 Å in diameter, and made up of about 2,000 identical coat protein subunits of molecular weight about 17,500, following a right-handed helix of pitch 23 Å with very close to 49 subunits in three turns of the helix. A central hole of about 40 Å diameter runs down the viral axis. A single RNA strand follows the viral helix at a radius of about 40 Å, with three nucleotides binding to each coat protein subunit (Fig. 28.3). Later, more detailed fiber diffraction analyses showed that the symmetry is close to 49.02 subunits in three turns (Stubbs and Makowski 1982), and that the helical pitch is close to 22.9 Å (Kendall et al. 2007b). There is very little variation in helical symmetry among the tobamoviruses; all measured tobamoviruses have 16.34 ± 0.01 protein subunits per turn of the viral helix (Franklin 1956b; Holmes and Franklin 1958; Stubbs and Makowski 1982).

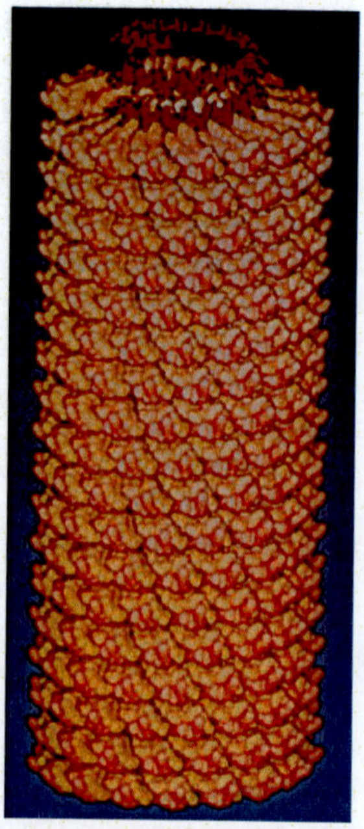

Fig. 28.3 Representation of about one-fifth of the TMV virion (Namba et al. 1988). Protein subunits are *yellow*; RNA nucleotides are *red*. The RNA is shown extending beyond the end of the protein helix for clarity

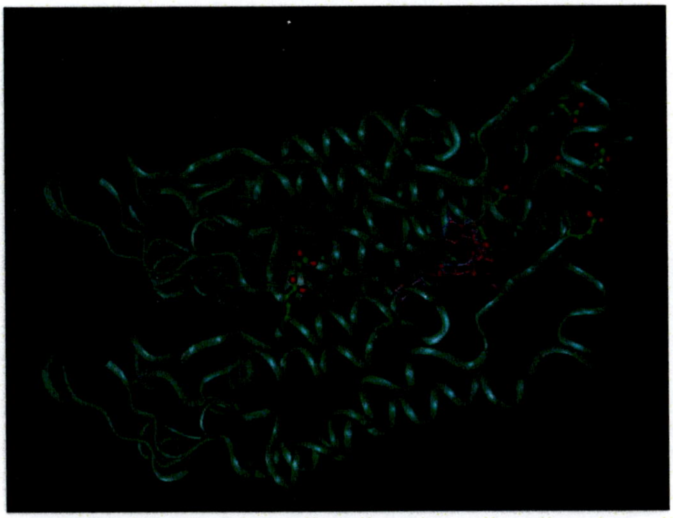

Fig. 28.4 Three coat protein subunits of TMV with associated RNA. Coat protein ribbon representation is *blue* with the carboxylate side chains involved in viral disassembly in *green* and *red*. RNA is *purple*. The viral axis is vertical, to the *right* of the figure. Figure by Dr. Hong Wang

TMV Structure

The complete structure of TMV was determined by fiber diffraction at 2.9 Å resolution (Namba et al. 1989). The structure was refined to an *R*-factor of 0.096; it should be noted that *R*-factors of structures determined by fiber diffraction are inherently lower than those determined by crystallography (Millane 1989; Stubbs 1989), but this is nevertheless an *R*-factor low enough to suggest that the model is comparable in quality to most icosahedral virus structures refined at similar resolution.

The structure of three adjacent TMV coat protein subunits and part of the bound RNA is shown in Fig. 28.4. The coat protein fold (Stubbs et al. 1977; Bloomer et al. 1978) does not resemble that of any icosahedral virus coat protein. As in many viruses, the N- and C-termini are on the surface of the virion. The protein is about 50% α-helical, and the subunit is built around a core of four α-helices, stabilized by closely packed hydrophobic interactions. The interactions between neighboring subunits make up a mosaic of hydrophobic and hydrophilic patches (Namba et al. 1989), stabilizing the overall virus structure. The intersubunit interactions are extensive, so much so that there are approximately the same number of intermolecular van der Waals interactions as intramolecular. A small β-sheet caps the four α-helices, and outside the sheet, numerous aromatic and other hydrophobic side chains interact within the subunit and between adjacent subunits, forming a continuous hydrophobic ribbon that follows the viral helix at high radius.

The RNA electron density in the TMV map represents an average of all of the bound trinucleotides in the viral genome, but it was known that GAA binds particularly strongly to the coat protein (Steckert and Schuster 1982), and that this binding has a function in viral assembly (Zimmern 1977). The RNA is therefore represented by a repeating GAA trinucleotide in Figs. 28.4 and 28.5, and this trinucleotide did in fact fit the electron density very well (Namba et al. 1989). The three bases lie flat against hydrophobic amino acid side chains on one of the α-helices, providing an interaction that does not depend on the nucleotide sequence. Two of the bases are in the *anti* conformation, and one is in the unusual *syn* conformation (Stubbs and Stauffacher 1981). The *syn* base and one of the *anti* bases stack together in a cavity between two intersubunit salt bridges. Although most of the protein–nucleic acid interactions are not base specific, there does appear to be a slight preference for hydrogen bonding to guanine in one of the three base-binding sites. The importance of this specificity is discussed in Sects. 28.4.1.1 and 28.4.2.1. The charges on the three nucleotide phosphate groups are offset by the charges on three arginine side chains, although not in the form of simple ion pairs.

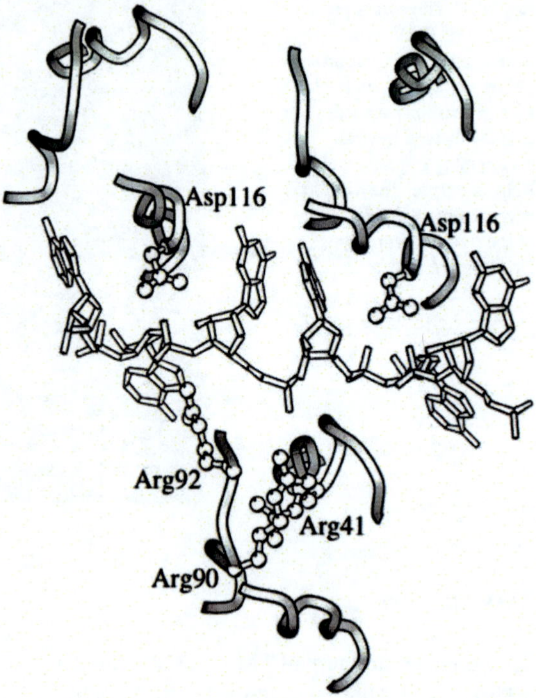

Fig. 28.5 The RNA-binding site of TMV (Stubbs 1999b). Parts of two coat protein subunits are shown above the RNA and one below. The view is from the interior of the virion, with the viral axis vertical

Electrostatic interactions between negatively charged groups are important in the coat protein subunit interfaces, as well as the protein–nucleic acid interaction. Carboxylic acid side chains from Glu50 and Asp77, from adjacent subunits separated in the axial direction, are in close proximity to each other (Namba and Stubbs 1986; Culver et al. 1995), as are the side chains in a complex interaction at low radius between Glu106 from one subunit and Glu95, Glu97, and Asp109 on the adjacent subunit in the lateral direction (Fig. 28.4) (Namba et al. 1989; Lu et al. 1996, 1998). A third close interaction of this type is between one of the RNA phosphates and Asp116 (Fig. 28.5). The Glu106/Glu95/Glu97/Asp109 and phosphate/Asp116 interactions appear to be calcium-binding sites (Namba et al. 1989; Pattanayek et al. 1992), and all three interactions are expected to bind protons with anomalous pK values. These electrostatic interactions and binding sites are important in the assembly and disassembly of the virion (Sects. 28.4.1.1 and 28.4.2.1).

Other Tobamoviruses

Structures have been determined for several other tobamoviruses: tobacco mild green mosaic virus (TMGMV, also called TMV-U2) (Pattanayek and Stubbs 1992), cucumber green mottle mosaic virus (CGMMV) (Lobert et al. 1987; Wang and Stubbs 1994), ribgrass mosaic virus (RMV) (Wang et al. 1997), and odontoglossum ringspot virus (ORSV) (Planchart 1995; Wang et al. 1998). Structures were determined by molecular replacement, refining models based on the TMV structure against fiber diffraction data. In the case of CGMMV, some information from isomorphous replacement was also incorporated (Lobert and Stubbs 1990). With the exception of RMV, the structures were determined at resolutions of about 3.5 Å, with *R*-factors in all cases a little less than 0.10. The RMV structure was determined at a resolution of 2.9 Å, and its refinement parameters suggest that it may be the most accurately determined of the tobamovirus structures. In any case, the relatively low resolution of most of these structure determinations is not a concern, because many of the most significant differences arise from the amino acid sequences rather than the details of chain folding, and the reliability of the

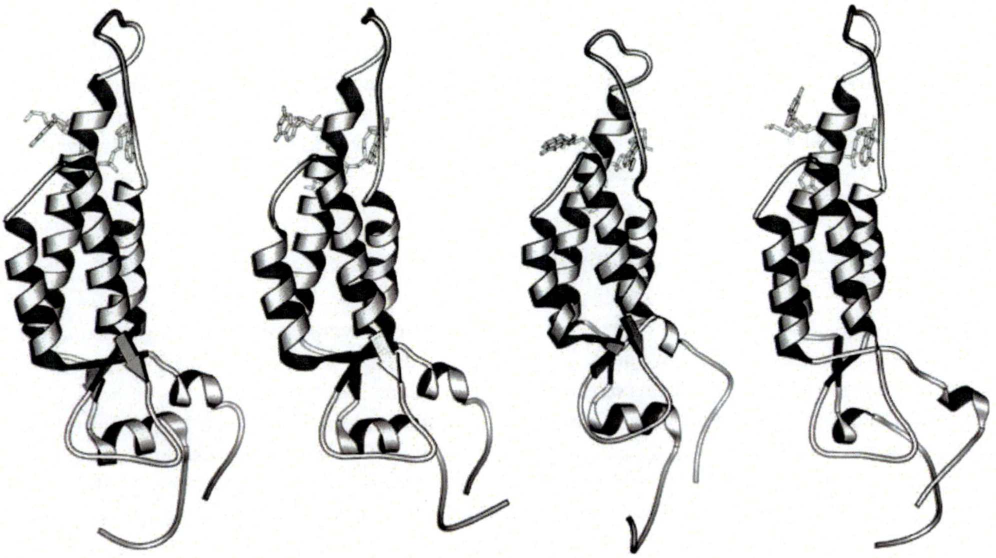

Fig. 28.6 Ribbon drawings of four tobamovirus coat proteins, with the three bound nucleotides of RNA in skeletal form. The viral axis is perpendicular to the page, at the top of each figure component. From *left*: TMV, TMGMV, CGMMV, RMV

TMV and RMV structures is sufficient to base conclusions on for the whole group. The degree of similarity among these viruses varies considerably; the coat protein amino acid sequences of TMVMG and ORSV, respectively, are 72 and 70% identical to TMV, but those of CGMMV and RMV are only 36 and 46% identical to TMV, and CGMMV and RMV are themselves only 37% identical (Altschuh et al. 1987; Dubs and van Regenmortel 1990; Wang et al. 1997).

All of the tobamovirus structures share the four-α-helix coat protein core, and the coat protein structures are generally similar (Fig. 28.6). As might be expected, the degree of difference reflects the degree of coat protein sequence identity to a considerable extent. Somewhat unexpectedly, the intersubunit electrostatic interactions are not conserved, although some corresponding interactions are seen in all of the tobamoviruses studied. The importance of carboxylate groups in viral disassembly was recognized very early (Caspar 1963), and the interacting groups are often called Caspar carboxylates, but attempts to identify the groups on the basis of sequence conservation (Butler and Durham 1972) were unsuccessful; we can now see why this would have been the case. The Asp116-phosphate interaction is conserved in all the viruses. In TMGMV, the Glu50–Asp77 interaction is retained (as Asp–Asp), but the low-radius Glu106/Glu95/Glu97/Asp109 interaction is absent (Pattanayek and Stubbs 1992); this absence may account for the exceptional stability of TMGMV at high pH. In CGMMV, the low-radius interaction is modified to become Glu106/Glu95/Asp98 and is across the axial rather than the lateral subunit interface (Wang and Stubbs 1994). The high-radius interaction is replaced by Glu46/Asp126, displaced one α-helical turn to lower radius from the Glu50/Asp77 interaction in TMV. RMV does not appear to have this high-radius interaction but has an extensive and strong interaction between Glu97/Glu98/Glu99 and the corresponding side-chains from the laterally adjacent subunit (Wang et al. 1997). These observations, combined with mutagenesis studies (Lu et al. 1996), prompted us to suggest that the simple electrostatic interactions that drive tobamovirus disassembly (Sect. 28.4.2.1) can "walk" (Wang and Stubbs 1994) or "slide" (Wang et al. 1997) across the subunit interface during viral evolution, maintaining functionality but altering the viral structure in order to evade host defensive responses.

There have been limited structural studies of other tobamoviruses. The Dahlemense strain of TMV exhibits an apparent doubled repeat distance of 138 Å, attributed to a deformation of the coat protein, suggesting that the high-radius part of the coat protein may be unusually flexible (Caspar and Holmes 1969). While in itself little more than a curiosity, this observation points to the significance of double-layered aggregates of coat protein, important in the assembly of tobamoviruses (Sect. 28.4.1). Sunn-hemp mosaic virus (SHMV) is a type II tobamovirus (discussed in the context of viral assembly) and appears from fiber diffraction data to have a structure similar to that of CGMMV (H. Wang and G. Stubbs, unpublished).

28.2.1.2 Other Rigid Rod-Shaped Plant Viruses

Structural studies of rigid filamentous plant viruses have been very limited and largely confined to the tobraviruses and the hordeiviruses. These studies, and our own unpublished studies by fiber diffraction and cryo-EM, have been somewhat ambivalent about the relationships among these viruses. On the one hand, as discussed in this section, there are many commonalities, for example, the diameters of the virions and their central holes, and modeling studies have supported a common coat protein fold, at least for the tobamoviruses and furoviruses and probably also for the tobraviruses. On the other hand, the helical symmetries of the different genera appear to differ; proposed values for the helical parameters have varied, and our own recent studies of the hordeiviruses certainly suggest that they do not share the tobamovirus helical parameters.

The tobraviruses (MacFarlane 1999, 2010) are bipartite viruses, with two types of particle, each with its characteristic length, encapsidating the two RNA molecules that make up the viral genome. Both particles use the same coat protein and have diameters of about 230 Å. One particle is consistently about 2,000 Å long, but the length of the other particle varies considerably among species, from 500 to 1,100 Å. The C terminus of the coat protein is exposed on the viral surface and is mobile (Brierley et al. 1993), as in the tobamoviruses. A model for the coat protein structure has been proposed from sequence comparisons with TMV (Goulden et al. 1992), but the tobravirus coat protein is about 40% longer than the TMV coat protein, the sequence identity is low, and this model has yet to be substantiated experimentally.

Fiber diffraction patterns from oriented sols have been published for tobacco rattle virus (Finch 1965) and pepper ringspot virus (known at that time as the Campinas strain of tobacco rattle virus) (Tollin and Wilson 1971). As in the tobamoviruses, the patterns show that there are close to $3n+1$ subunits in three turns of the viral helix; the helical pitch is about 25 Å. The value of the integer n is not clear; early electron microscopic observations suggested a value of 25 (Finch 1965; Tollin and Wilson 1971), but stoichiometric calculations (Kendall et al. 2008) favor a value closer to the 16 found for tobamoviruses. The central hole in the virion appears to be about 40 Å in diameter, again like the tobamoviruses.

Hordeiviruses (Jackson et al. 2009) are generally tripartite, with particle lengths varying from 1,000 to almost 2,000 Å depending on the particle and the species (Fauquet et al. 2005). They are believed to bind three nucleotides to each coat protein subunit, like the tobamoviruses. The only published diffraction pattern of a hordeivirus is from barley stripe mosaic virus (BSMV) (Finch 1965). The virus appears to have a maximum diameter of about 220 Å and a central hole about 40 Å in diameter. The pattern shows that there are close to $5n+1$ subunits in five turns of the viral helix ($5n-1$ is possible, but unlikely). We have confirmed these results and shown by cryo-EM and scanning transmission electron microscopy (STEM) that there are probably more than 20 subunits per turn of the helix.

There has been one modeling study of a furovirus (Chen et al. 1997); on the basis of sequence comparison with TMV, this study suggested that the structure of soilborne wheat mosaic virus (SBWMV) is extremely similar to that of TMV, with the addition of several loops projecting from the SBWMV surface.

28.2.2 *Flexible Filamentous Plant Viruses*

While there are not yet any detailed experimentally determined structural models for the flexible filamentous plant viruses, the considerable importance of these viruses in agriculture and biotechnology has led to increased interest in them in recent years. The potexviruses have been studied for many years. More recent results have shown that the potyviruses have structures similar to those of the potexviruses. There have also been a few limited structural studies of the closteroviruses.

28.2.2.1 Potexviruses

Members of the genus *Potexvirus* (Verchot-Lubicz et al. 2007) in the family *Alphaflexiviridae* are about 5,000 Å long and 130 Å in diameter. Following the early work of Bernal and Fankuchen (1941), Tollin et al. (1967, 1980) examined a number of potexviruses by fiber diffraction and a larger number by EM (Richardson et al. 1981). These studies were largely confined to the determination of helical parameters. The helical pitch varied between 32 and 36 Å, while the helical symmetry was between 8.7 and 8.9 subunits per turn, a substantially greater variation than had been seen in the tobamoviruses. There are probably five nucleotides bound to each protein subunit (Tollin et al. 1980).

Circular dichroism measurements (Homer and Goodman 1975; Wilson et al. 1991) showed that the potexviruses have an α-helical content of around 45%. Other spectroscopic studies showed that the virion surface (Baratova et al. 2004), including parts of the α-helices (Blanch et al. 2002), is highly hydrated, more so than in the tobamoviruses. Tritium bombardment of PVX (Baratova et al. 1992) suggested that the N-terminus of the coat protein is exposed on the viral surface, and that the C terminus is close to the surface.

More recent fiber diffraction experiments have shown that PVX has a helical symmetry of 8.9 subunits per turn and a pitch of 34.5 Å (Parker et al. 2002). The surface contains deep grooves, consistent with both flexibility and high surface hydration. Fiber diffraction patterns have also been obtained from narcissus mosaic virus (NMV). NMV has 8.8 subunits per turn, and the small difference between 8.8 and 8.9 is enough to separate the layer lines in the NMV patterns, allowing the determination of a low-resolution radial density distribution (Kendall et al. 2007a). However, more detailed analysis required more information than could be obtained from fiber diffraction alone. By combining fiber diffraction, cryo-EM, and STEM, we have constructed a model of PVX (Fig. 28.7) with a resolution of about 14 Å (Kendall et al. 2008). The protein subunits are compact and well defined, with a major domain at high radius and smaller regions of density reaching into the center of the virion. The subunits have their longest axes running approximately radially, as in TMV, but they are much shorter and wider; the PVX subunits are only about 45 Å long. Intersubunit contacts are mostly confined to the high-radius region; the viral structure is very open at low radius. The viral RNA would not be distinguishable from protein at this resolution, but it is likely to be at a radius of 30–35 Å.

In the absence of an experimentally determined atomic model, there have been several attempts to predict the three-dimensional structure of PVX. The predicted structures are variations and extensions of the four-helix bundle structure of TMV (Namba et al. 1989), consistent with the high α-helical content of the protein and the known location of the N- and C-termini. The most detailed models (Baratova et al. 2001; Nemykh et al. 2008) consist of two domains, an inner four-helix bundle and an outer α/β domain. These models are similar in overall shape to the subunits that we have observed, more so in a recent revision (Lukashina et al. 2009) to fit the subunit length in our reconstruction.

One unusual virus related to the potexviruses, Lolium latent virus (family *Alphaflexiviridae*, genus *Lolavirus*) appears to have two coat proteins, possibly arising from the use of an alternative in-frame start codon (Vaira et al. 2008). Our unpublished fiber diffraction patterns do not indicate

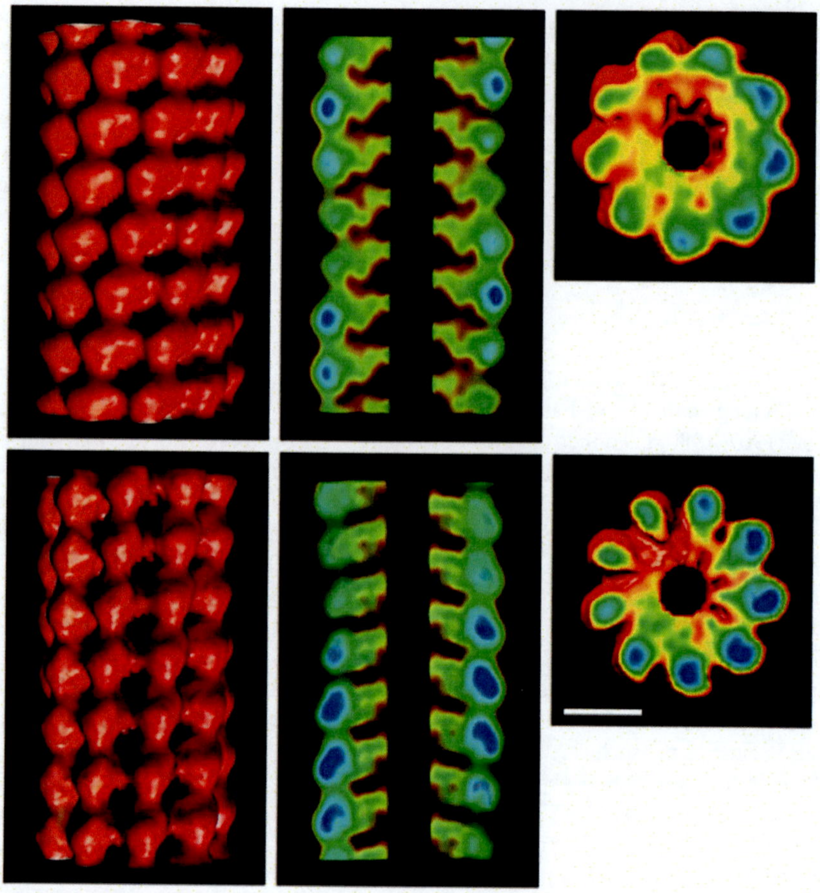

Fig. 28.7 Reconstructions of SMV (*top row*) and PVX (*bottom row*). From *left*: outside surface view, section through the viral axis, section normal to the viral axis. Color coding is from *red-orange* (low density) to *green-blue* (high density). *Bar*: 50 Å

any major structural difference (such as a heterodimer repeat) from the potexviruses, but these data are only preliminary. Fiber diffraction data from a still more distantly related virus, Heracleum latent virus (HLV) (family *Betaflexiviridae*, genus *Vitivirus*), including a helical pitch of 32.6 Å, were described by Tollin et al. (1992), although at that time HLV was believed to be a closterovirus.

28.2.2.2 Potyviruses

The family *Potyviridae* (López-Moya et al. 2009) includes the genera *Potyvirus*, *Rymovirus*, *Tritimovirus*, *Bymovirus*, *Macluravirus*, *Ipomovirus*, and *Brambyvirus* (Fauquet et al. 2005; Carstens 2010), making up almost a quarter of the known plant viruses.

The virions are approximately 7,500 Å long and 120 Å in diameter and have helical pitches of about 33 Å (Varma et al. 1968; McDonald and Bancroft 1977). Circular dichroism measurements and secondary structure predictions suggest that the coat proteins are about 50% α-helical, similar to the potexviruses (Shukla et al. 1988; Baratova et al. 2001). The N- and C-termini of the coat proteins are believed to be located near the surface of the virions (Shukla et al. 1988; Baratova et al.

2001). Models have been proposed for the protein structure (Dobrov et al. 2004), derived along similar lines to those of the potexviruses.

Despite their importance, very little was known until recently about the structure of the potyviruses beyond the level of gross morphology. There was one very limited fiber diffraction study of wheat streak mosaic virus (WSMV) (Parker et al. 2005). However, the combination of fiber diffraction, cryo-EM, and STEM that was effective for PVX has also been applied to the potyvirus soybean mosaic virus (SMV) (Kendall et al. 2008). At a resolution of about 14 Å, the structure is very similar to that of PVX at the same resolution (Fig. 28.7), supporting the suggestion that all flexible filamentous plant viruses are structurally related (Dolja et al. 1991; Kendall et al. 2008). Fiber diffraction data from a number of members of the family, including representatives of the genera *Rymovirus* (Agropyron mosaic virus) and *Tritimovirus* (WSMV), show that all of these related structures are similar, although there do appear to be some differences in the WSMV structure. Helical parameters vary among species, like the potexviruses but unlike the tobamoviruses (McDonald et al. 2010).

The potyvirus protein VPg is covalently linked to the 5′ end of the viral RNA, and therefore forms part of the intact virion. Atomic force microscopy (AFM) together with immunogold labeling EM shows that in PVY and PVA, some virions have an unusual structure at one end, which appears to contain the protein HC-Pro as well as VPg (Torrance et al. 2006).

28.2.2.3 Closteroviruses

The family *Closteroviridae* (Dolja et al. 2006; Moreno et al. 2008) includes three genera of flexible filamentous plant viruses of highly variable length (12,000–22,000 Å) and about 120 Å diameter. Studies of structure and assembly have been reported for beet yellows virus (BYV), citrus tristeza virus (CTV), grapevine leafroll-associated virus 2 (GLRaV-2), and Heracleum virus 6 (HV-6; tentative genus assignment) from the genus *Closterovirus*. Virions have been examined by EM and AFM (Tollin and Wilson 1988; Peremyslov et al. 2004). Unusually, closteroviruses have a major and a minor coat protein; the minor protein coats a short segment (less than 5% of the total virion length) at the 5′ end of the RNA, giving a "rattlesnake" appearance (Agranovsky et al. 1995; Alzhanova et al. 2007). The short 5′-end tail end also contains other protein molecules: the Hsp homolog Hsp70h and the 64-kDa protein p64 in CTV and BSV, and the 20-kDa protein p20, at least in BYV (Satyanarayana et al. 2004; Peremyslov et al. 2004). There have been no diffraction studies of closteroviruses reported, but unpublished data (M. McDonald, A. Prokhnevsky, V. Dolja, AK, GS) indicate a helical pitch of about 37 Å for BYV. A helical pitch of about 36 Å and an estimated symmetry of about 9 subunits per turn have been reported for HV-6 from EM studies (Tollin et al. 1992); other EM studies report pitches between 30 and 40 Å.

28.3 Filamentous Bacteriophage Structure

The filamentous bacteriophages are flexible, with diameters close to 60 Å (Fig. 28.8). Lengths are extremely variable, ranging from 8,000 (Ff) to 20,000 Å (Pf1). Three to five copies of two minor proteins are found at each end of the virion, p3 and p6 at one end and p7 and p9 at the other, but most of the virion capsid is made up of a single coat protein (p8) of about 50 amino acid residues (molecular weight about 6,000). Amino acid sequences vary greatly among different species, but the principal coat protein in all of the filamentous bacteriophages is believed to have a common structure, a simple extended α-helix (Fig. 28.8).

Filamentous bacteriophage particles contain one molecule of single-stranded circular DNA. Protein–nucleic acid interactions do not appear to be either well defined or conserved in detail

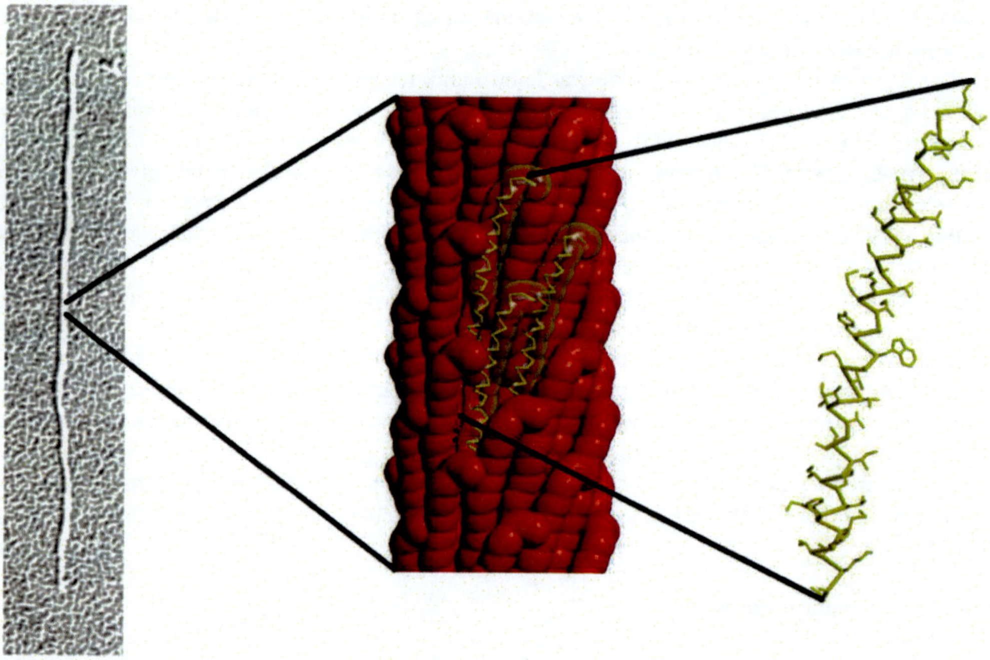

Fig. 28.8 Structure of filamentous bacteriophage fd (Marvin et al. 2006). From *left*: electron micrograph of a single virion, subunit structure of the virion, α-helical structure of an individual coat protein subunit

among members of the family; viral species differ considerably in length, and the nucleotide to protein ratio varies considerably. There are significant differences, even within the same class, for example, in the class II virus Pf1, the nucleotide to protein ratio is 1.0, whereas for both Pf3 (class II) and Ff (class I), it is 2.4 (Welsh et al. 1998). As a result of this ratio, Pf1 is more than twice as long as Ff or Pf3. As in most filamentous viruses, when additional genetic material (DNA in this case) is inserted into the genome, the length of the virion increases, being proportional to the total DNA length.

The helical symmetries of the two classes are actually small perturbations of each other (Fig. 28.9). The repeating unit of fd, the type member of class I, consists of two rings of five subunits, although the five subunits in a ring do not contact each other. The second ring is related to the first by a translation of about 16 Å along the viral axis and a rotation of 36° about that axis. The repeating unit of Pf1, the type member of class II, consists of 27 subunits in five turns of a helix, that is, there are 5.40 subunits in one helical turn. Within a symmetry class, there are small variations in these parameters, for example, in the class I bacteriophages IKe and If1, the rotation is 38° rather than 36°. Symmetries may depend on solution conditions; below 6°C, Pf1 changes slightly to a form with 5.46 rather than 5.40 subunits per turn (Wachtel et al. 1976; Nave et al. 1979). The Pf1 symmetry may be constructed by moving the fd rings about 2 Å closer together and translating each subunit about 2.8 Å relative to its neighbor (illustrated in Makowski 1984 and Stubbs 2001).

The α-helical conformation of the coat protein has been confirmed by solid-state NMR (Cross et al. 1983; Opella et al. 2008). The ssNMR results suggested that the coat protein of fd was a single continuous α-helix, while the coat protein of Pf1 consisted of two helices, the smaller one tilted at an angle of about 20° relative to the larger. In both the X-ray and the ssNMR models, the α-helices are interleaved to form two layers (Fig. 28.8), with the DNA in the center of the filament. The C terminus of the protein is near the center of the virion, and the C-terminal ends of the protein

Fig. 28.9 Subunit arrangement in the two classes of filamentous bacteriophage (after Marvin (1990); see also Welsh et al. 2000; Marvin et al. 2006). *Left*: fd (class I). *Right*: Pf1 (class II). Courtesy of Dr. Don Marvin

molecules form the helices of the inner layer. Protein–protein interactions in this layer tend to be hydrophobic. The N-terminal part of the protein is more hydrophilic, and the N-terminus is exposed at the virion surface. The α-helices in the two layers cross each other at a negative angle (Bryan et al. 1983), rather than at a positive angle as in most α-helical proteins.

Although the coat protein model for Pf1 shown in Fig. 28.9 is one continuous α-helix, an alternative model partly derived from low-resolution neutron diffraction data with selectively deuterated amino acids has two helices, abutted end-to-end to form a continuous helix, but with a seven-residue loop folded out of the helical structure (Stark et al. 1988; Nambudripad et al. 1991). This model is consistent with the ssNMR results, but the single-helix model has been refined against higher-resolution X-ray diffraction data (3 Å) using molecular dynamics methods (Welsh et al. 2000). Although the X-ray and ssNMR models of fd (Marvin et al. 1994; Zeri et al. 2003) were broadly similar, there were significant differences. These differences were to some extent resolved in a refinement against both X-ray and ssNMR data (Marvin et al. 2006) (Fig. 28.8). A structure has also been determined by cryo-EM (Wang et al. 2006); differences between this structure and the X-ray and ssNMR structures are not completely understood (Straus et al. 2008), but the cryo-EM structure is at significantly lower resolution than the X-ray structures.

The structure of Pf3 (class I) has been refined at 3.1 Å resolution, starting from the Pf1 model (Welsh et al. 1998). The Pf1 and Pf3 diffraction patterns and refined structures are very similar, despite the large differences in nucleotide/protein subunit ratio.

In all filamentous bacteriophage coat proteins, basic amino acid side chains near the C terminus are believed to neutralize the phosphate groups of the DNA. The DNA does not appear to conform to the helical symmetry (Welsh et al. 1998), and the ratio of nucleotides to coat protein subunits is often not integral. In fd, removal of one of the lysine residues near the C terminus results in a viable bacteriophage about 35% longer than the wild type, that is, the length is inversely proportional to the positive charge at the DNA-binding site. This observation suggests that simple neutralization of charge with no specific molecular interaction is all that is required to bind the DNA (Hunter et al. 1987).

28.4 Filamentous Viruses as Molecular Machines

In addition to its passive function in genome protection, the virion of a filamentous virus has numerous mechanical functions: cell entry, viral disassembly, viral assembly, and cell exit. In the case of those viruses (such as the plant viruses) that infect multicellular organisms, virions or virion-like assemblies are also essential for transport within the host. Early studies focused on assembly and disassembly, and these processes are now well understood for some viruses. We will therefore treat them first, and then consider the various membrane-crossing functions together.

28.4.1 Assembly

28.4.1.1 Tobamoviruses

Fraenkel-Conrat and Williams (1955) showed in reconstitution experiments that TMV is a self-contained assembly system: no external machinery such as a scaffolding protein is required.

TMV coat protein forms a number of aggregates, depending on the solution conditions (Butler and Durham 1977); the most important for viral assembly is a 20S aggregate, which forms near neutral pH at low ionic strength. Although there has been some disagreement about the nature of the 20S aggregate involved in assembly, specifically whether it is the 34-subunit disk seen in the coat protein crystal structure (Bloomer et al. 1978) or a short helix of a little over two turns (Correia et al. 1985), its precise nature is not important to the nature of the assembly process (Klug 1999). The aggregate is required for initiation of viral assembly (Butler and Klug 1971).

The 20S aggregate binds to a specific RNA sequence, the origin-of-assembly (OAS) sequence, located about one-sixth of the way along the RNA molecule from the 3′ end (Zimmern and Wilson 1976; Zimmern 1977). The sequence fits a proposed secondary structure in which the sequence GAAGAAGUU is in a loop at the end of a hairpin (Zimmern 1977) (Fig. 28.10). If the sequence is considered as a succession of trinucleotides, consistent with the binding of three nucleotides to each protein subunit, G occupies six consecutive equivalent positions. The base sequence and composition of the OAS are consistent with the trinucleotide binding studies of Steckert and Schuster (1982) and with the slight preference for guanine observed in one of the three base-binding sites in the virus structure (Namba et al. 1989). No other repeats in the complete nucleotide sequence of TMV RNA reflect the protein–RNA binding (Goelet et al. 1982).

The stable assembly nucleus formed by the 20S aggregate and the OAS rapidly binds approximately two more 20S aggregates, so that the RNA is bound to a short helix of coat protein, with the

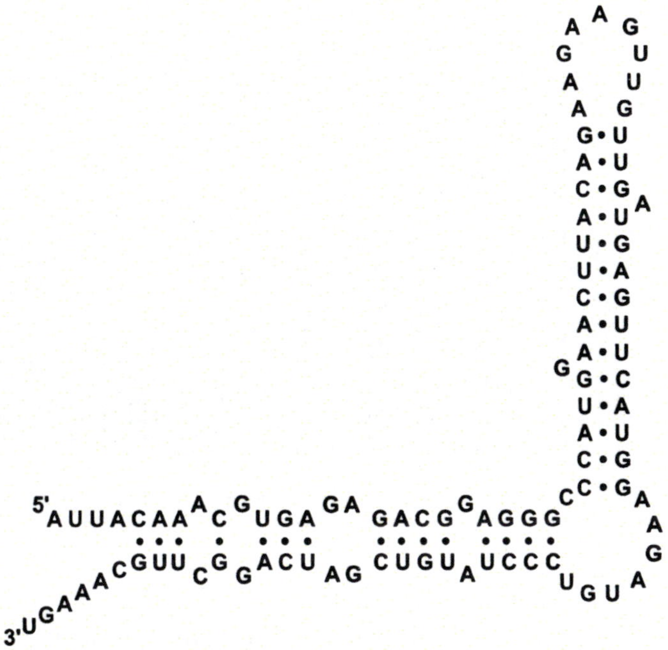

Fig. 28.10 Nucleotide sequence of the TMV origin of assembly, folded as proposed by Zimmern (1977)

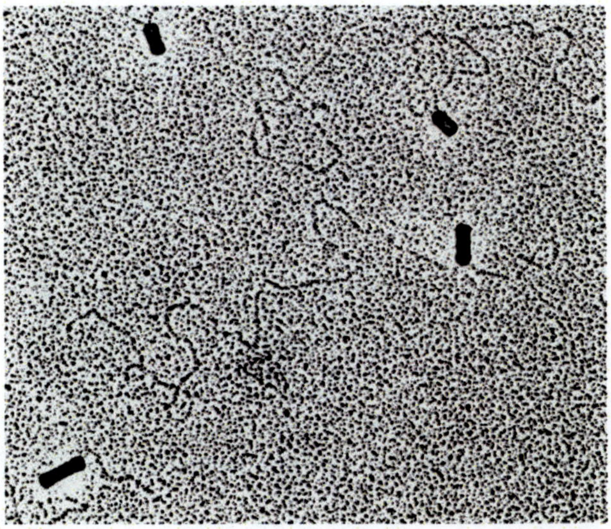

Fig. 28.11 Electron micrograph of partially assembled TMV with dimethyl sulfoxide (DMSO) present. The DMSO separates the two RNA tails, which are clearly coming from the same end of the growing virion. Courtesy of Dr. Genevieve Lebeurier

relatively short 3′ tail extending from one end of the helix, and the long 5′ tail threaded through the central hole of the growing virion so that both tails extend from the same end of the protein helix (Figs. 28.11 and 28.12). The existence of this topologically remarkable process was shown by Butler et al. (1977) and Lebeurier et al. (1977). The assembly process continues by pulling the 5′ end of the RNA through the hole, as more 20S aggregates (and possible smaller aggregates as well) bind to the growing end of the assembly. Incorporation of the 3′ end of the RNA requires a different mechanism, perhaps involving the smaller protein aggregates.

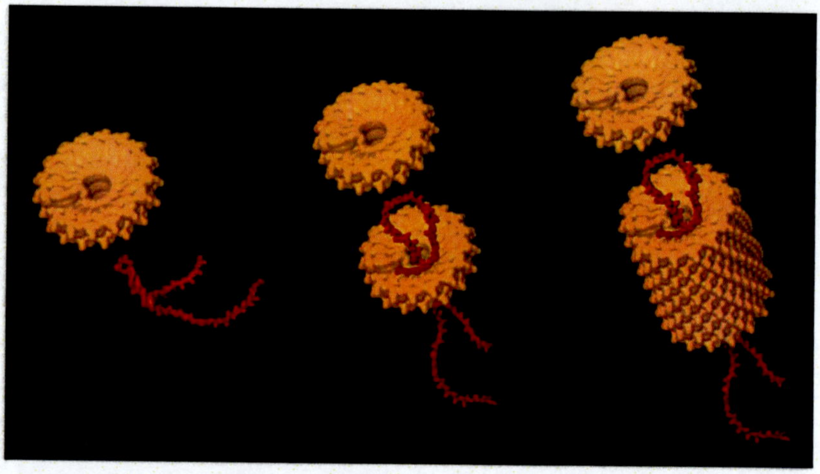

Fig. 28.12 Assembly of TMV. The origin of assembly (Fig. 28.10) of the RNA binds to a 20S aggregate, followed by the addition of more 20S aggregates. Because the OAS is at an internal site on the RNA, the 5′ end of the RNA passes through the axial hole in the center of the growing virion. Figure by Dr. Hong Wang

Tobamoviruses have been classified into two subgroups on the basis of the OAS (Okada 1986). Subgroup I viruses, including TMV and most other studied tobamoviruses, begin assembling at the TMV OAS, about 900 nucleotides from the 3′ end of the RNA in the cell-to-cell movement protein cistron. Subgroup II viruses, including CGMMV and SHMV, start assembling at a point only about 300 nucleotides from the 3′ end, in the coat protein cistron (Fukuda et al. 1980). The two subgroups are clearly separated in comparisons based on coat protein amino acid sequences. Subgroup II virus preparations characteristically include large numbers of short particles, formed when the coat protein mRNA is encapsidated by coat protein as a result of its internal OAS.

28.4.1.2 Other Filamentous Plant Viruses

The coat proteins of other filamentous viruses also form a variety of aggregates depending on solution conditions. Virus-like particles without RNA have been reconstituted from coat protein isolated from virions or expressed in prokaryotic or eukaryotic cells; these experiments have included members of the family *Potyviridae* (McDonald et al. 1976; Jagadish et al. 1991; Hammond et al. 1998; Joseph and Savithri 1999; Jacob and Usha 2002; Cerovská et al. 2002; Hema et al. 2008), potexviruses (Tremblay et al. 2006), and hordeiviruses (Atabekov et al. 1968). Although TMV forms virus-like particles at low pH (Schramm 1947) that are structurally extremely like the virions (Franklin 1955; Mandelkow et al. 1981), this is not necessarily the case for other viruses, and in many cases the particles are stacked disks rather than helical empty capsids. These aggregates are not, however, believed to be part of the assembly pathways.

Smaller aggregates, many of which appear to be double-layered and comparable to the TMV 20S aggregate (the disk or short helix), have been seen in the coat proteins of potexviruses (Erickson et al. 1983; Tremblay et al. 2006; and our own unpublished observations on PVX), hordeiviruses (Atabekov et al. 1968), and tobraviruses (Mayo et al. 1993). In at least some cases, these aggregates appear to be involved in the formation of TMV-like initiation complexes (Tremblay et al. 2006).

In contrast to TMV, the OAS is located at the 5′ end of the RNA in a number of filamentous plant viruses, including the tobravirus tobacco rattle virus (Hamilton et al. 1987), the potexvirus papaya mosaic virus (Sit et al. 1994), the potyvirus tobacco vein mottling virus (Wu and Shaw 1998), the pecluvirus Indian peanut clump virus (Bragard et al. 2000), and two closteroviruses, BYV (Peremyslov et al. 2004) and CTV (Satyanarayana et al. 2004). A stem-loop region at the 5′ end of the RNA of PVX potexvirus serves as a recognition site for the binding of coat protein subunits (Kwon et al. 2005). Most potexviruses contain an AC-rich region at the 5′ end of the viral RNA, and the first of several ACCA motifs is critical for PVX replication (Park et al. 2008). A sequence at the 5′ end of the RNA of citrus tristeza closterovirus that contains two stem-loops separated by a spacer region has also been shown to be important for virus replication (Gowda et al. 2003).

Closterovirus assembly requires several different structural proteins. The majority of the body of the closterovirus virion is encapsidated by a single coat protein, but the 5′-end tail (encapsidating about 650 bases) includes the minor coat protein, Hsp70h, p64, and p20 (Peremyslov et al. 2004; Satyanarayana et al. 2004). The minor coat protein, Hsp70h, and p64 coordinate to encapsidate the 5′ end of the RNA; the major coat protein and p20 are not required to form the tails (Alzhanova et al. 2007). The 5′ end of the closterovirus RNA contains the packaging signal that is recognized by the minor coat protein (Peremyslov et al. 2004; Satyanarayana et al. 2004).

Assembly, disassembly, and other molecular processes in the potexvirus life cycle have been reviewed by Verchot-Lubicz et al. (2007).

28.4.1.3 Filamentous Bacteriophages

Filamentous bacteriophages are not self-assembling in the sense applied to the plant viruses. Assembly is concomitant with secretion from the host cell through the plasma and outer membranes, using a mechanism very different from that of the filamentous plant viruses, carried out by a complex molecular machine (Marvin 1998; Russel and Model 2006; Opella et al. 2008). It will therefore be discussed in Sect. 28.4.3.2.

28.4.2 Disassembly

Little is known about the actual disassembly of filamentous bacteriophages, although a great deal is known about the membrane transport steps that immediately precede it (Sect. 28.4.3.2). This section will therefore consider only the filamentous plant viruses.

28.4.2.1 Tobamoviruses

By combining structural information with numerous other observations, it has been possible to propose a detailed mechanism for the disassembly of TMV (Namba et al. 1989; Stubbs 1999b), which can be applied to all tobamoviruses. A virion in a plant cell is subjected to much lower concentrations of both calcium and protons than it is in the external environment. Protons and calcium ions are removed from the intersubunit carboxylate and carboxylate–phosphate clusters (Sect. 28.2.1.1), and repulsive interactions between the resulting negatively charged groups destabilize the virion. This destabilization is not by itself sufficient to initiate disassembly, but the protein–nucleic acid interactions involving the 69 nucleotides at the 5′ end of the RNA are slightly weaker than in the rest of the genome because there are no guanine bases (Sect. 28.2.1.1) in this part of the RNA (Goelet et al. 1982). The protein subunits forming about 1½ turns of the viral helix at the end of the virion are therefore more easily lost. The first start codon is consequently exposed so that a cellular ribosome can bind. The ribosome moves toward the 3′ end of the RNA in the process of translation, competing with the coat protein to continue disassembly and begin the cycle of viral replication. This phenomenon, called co-translational disassembly, has been observed experimentally in vitro (Wilson 1984) and in vivo (Shaw et al. 1986).

28.4.2.2 Other Filamentous Plant Viruses

While much less is known about the details of disassembly for other viruses, there have been a number of observations that point to various steps in the process, particularly for the potexviruses. Phosphorylation of PVX coat protein by host kinases appears to be important, enhancing RNA translation, presumably by destabilizing the virions and promoting disassembly (Atabekov et al. 2001). Potexviruses and many other filamentous viruses encode three proteins involved in cell-to-cell movement of the virus; these are known as the triple gene block proteins TGBp1, TGBp2, and TGBp3. Binding of TGBp1 to the 5′ end of the PVX particle converts nontranslatable virions to translatable virions (Atabekov et al. 2000; Kiselyova et al. 2003; Zayakina et al. 2008), apparently also by destabilizing the particle. Disassembly after binding TGBp1 requires translation; to this extent, there are some parallels with the co-translational disassembly of TMV (Wilson 1984, Sect. 28.4.2.1), but whereas co-translational disassembly of TMV is gradual and proceeds along the virion with the ribosome, the disassembly of PVX associated with TGBp1 requires only the first few steps of translation to disassemble the entire virion (Rodionova et al. 2003). It has been suggested that when a plant is first infected, phosphorylation promotes disassembly, but as the infection spreads to other plant cells, TGBp1-induced disassembly may serve the same function, eliminating dependence on host kinases (Verchot-Lubicz et al. 2007).

In potyviruses, one molecule of the viral protein VPg is covalently attached to the 5′ end of the viral RNA (Hari 1981). It has been shown that the bound VPg can be phosphorylated by host kinases, and it is possible that this phosphorylation may trigger virus disassembly and RNA translation (Puustinen et al. 2002).

28.4.3 Host Cell Entry, Exit, and Membrane Transport

28.4.3.1 Filamentous Plant Viruses

Plant viruses enter and leave the host by poorly defined mechanisms, generally involving mechanical damage to the plant by an insect or other vector. However, they have sophisticated mechanisms for transport within the host, usually considered separately as cell-to-cell (meaning adjacent cells within a leaf or half-leaf) and long-distance (involving transport through the phloem) transport (Samuel 1934; Carrington et al. 1996; Scholthof 2005).

A number of recent reviews discuss movement of the filamentous plant viruses in plants (for example Morozov and Solovyev 2003; Verchot-Lubicz et al. 2007; Taliansky et al. 2008; Jackson et al. 2009; Epel 2009; Verchot-Lubicz et al. 2010). Taliansky et al. (2008) divide the filamentous plant viruses into three groups, depending on which proteins they require to move from cell to cell. Some viruses, such as TMV, require a single type of movement protein. A second group, which includes the potexviruses and the hordeiviruses, requires a triple gene block (Sect. 28.4.2.2) for movement, and the third group, including the potyviruses and the closteroviruses, requires multiple proteins for movement. Verchot-Lubicz et al. (2010) further divide the triple gene block-encoding viruses into three groups based on protein interactions occurring during infection and the types of proteins required for movement, and propose a model for virus movement in each group. Many plant viruses require the coat protein for systemic movement within plants, although the tobraviruses (Swanson et al. 2002), hordeiviruses (Petty et al. 1990), and several others do not appear to.

Very little is known about the detailed molecular mechanisms by which plant viruses move within plants. While it is clear that these viruses are able to override the normal intercellular trafficking mechanisms within plant cells (reviewed by Heinlein and Epel 2004; Epel 2009), the form that the virus takes while moving from cell to cell is not. In the potexviruses, it is possible that the particles that travel from cell to cell are the single-tailed particles described by Karpova et al. (2006) or another altered virion form (Verchot-Lubicz et al. 2007; Verchot-Lubicz et al. 2010). In the hordeiviruses, it appears that the movement complex is made up of TGBp1 and the viral genomic and subgenomic RNAs (Verchot-Lubicz et al. 2010); TGBp1 must interact with a TGBp2–TGBp3 complex in order to be targeted to the cell wall (Lim et al. 2008).

As early as 2001, Dolja and his colleagues had begun to speculate that the unusual tail structure in closteroviruses might serve as a specialized transport device (Alzhanova et al. 2001). They suggested that the tails help guide the virus to the plasmodesmata, and that the smaller diameter of the virion tails may even allow for insertion of the virus into the plasmodesmata (Alzhanova et al. 2007). This type of directional transport would allow the 5′ end of the RNA to enter the adjacent cell first to be recruited by ribosomes (Dolja et al. 2006). Tail structures at the end of some potyvirus particles (Torrance et al. 2006) and the TGBp1 complex at the end of PVX particles (Atabekov et al. 2000; Kiselyova et al. 2003) may serve a similar purpose (Taliansky et al. 2008).

28.4.3.2 Filamentous Bacteriophages

Disassembly is the final step of the process of cell entry for filamentous bacteriophages, and assembly is part of the process of secretion. Most work has been on the Ff group of viruses, so the description that follows is primarily taken from those studies. More details and accounts of other viruses are found in reviews by Marvin (1998) and Russel and Model (2006).

Cell Entry and Disassembly

Bacteriophage particles initially bind to the tip of a pilus (F pili in the case of *E. coli* phages) through the p3 protein at the end of the virion. The pilus disassembles into the plasma membrane, drawing the virion into contact with the membrane through interaction between p3 and the host membrane proteins TolA, TolQ, and TolR. The bacteriophage disassembles and the DNA is released into the host cytoplasm.

Assembly and Secretion

The phage DNA is prepackaged in the host cell by protein p5, forming a long flexible filamentous complex about 80 Å in diameter. Although p5 proteins vary greatly among the different bacteriophage species in both size and amino acid sequence, they all appear to be structurally similar.

The five virion coat proteins – the major coat protein and all four minor proteins – and two other bacteriophage-encoded proteins, p1 and p11, are inserted into the host plasma membrane as integral membrane proteins. In addition, phage protein p4 is an integral protein of the outer membrane. The major coat protein p8 has been particularly well studied; it takes up a tilted orientation in the membrane with the N-terminus in the periplasm and the C terminus in the cytoplasm. Varying views have been expressed concerning the structure of p8 in this environment; it appears to be either a single α-helix similar to the structure in the assembled virion (Hemminga et al. 2010) or two α-helices, one in the membrane hydrophobic core and a smaller N-terminal helix in the surface of the membrane (Opella et al. 2008). Some phages use a leader sequence to insert p8 into the membrane; some do not. The leader sequence is subsequently cleaved by a bacterial peptidase. Phage assembly and secretion takes place at sites where the plasma membrane is very close to the outer membrane. A complex of p1 and p11 in the plasma membrane interacts with p4, 12–14 subunits of which form a pore in the outer membrane (Opalka et al. 2003).

Assembly of the virion begins when the packaging signal, an imperfect but stable DNA hairpin having some analogies to the OAS of filamentous plant viruses, interacts with p7 and p9 in the membrane. The packaging signal protrudes from the end of the p5–DNA prepackaging complex. This interaction involves p1 and host-encoded thioredoxin, although the thioredoxin appears to be acting to manipulate the DNA, rather than utilizing its normal function as a redox enzyme. ATP hydrolysis, possibly by p1, is part of the packaging process. During the next step, phage elongation, p8 replaces p5 and the growing virion moves into the periplasm. Finally, the minor coat proteins p3 and p6 are added to the end of the virion. Addition of p3 and p6 is essential to the termination of assembly and detachment from the membrane; without either of these proteins, filaments continue to elongate rather than detach. Finally, the p4 pore complex translocates the virion across the periplasm and extrudes it through the outer membrane.

Acknowledgments We thank Dr. Don Marvin for much helpful advice and assistance with filamentous bacteriophage figures. The plant virus work from this laboratory was supported by NSF grant MCB-0743931.

References

Adams MJ, Antoniw JF, Kreuze J (2009) *Virgaviridae*: a new family of rod-shaped plant viruses. Arch Virol 154:1967–1972

Agranovsky AA, Lesemann DE, Maiss E, Hull R (1995) "Rattlesnake" structure of a filamentous plant RNA virus built of two capsid proteins. Proc Natl Acad Sci USA 92:2470–2473

Altschuh D, Lesk AM, Bloomer AC, Klug A (1987) Correlation of coordinated amino acid substitutions with function in viruses related to tobacco mosaic virus. J Mol Biol 193:693–707

Alzhanova DV, Napuli AJ, Creamer R, Dolja VV (2001) Cell-to-cell movement and assembly of a plant closterovirus: roles for the capsid proteins and Hsp70 homolog. EMBO J 20:6997–7007

Alzhanova DV, Prokhnevsky AI, Peremyslov VV, Dolja VV (2007) Virion tails of *Beet yellows virus*: coordinated assembly by three structural proteins. Virology 359:220–226

Atabekov JG, Novikov VK, Kiselev NA, Kaftanova AS, Egorov AM (1968) Stable intermediate aggregates formed by the polymerization of barley stripe mosaic virus protein. Virology 36:620–638

Atabekov JG, Rodionova NP, Karpova OV, Kozlovsky SV, Poljakov VY (2000) The movement protein-triggered *in situ* conversion of Potato virus X virion RNA from a nontranslatable into a translatable form. Virology 71:259–263

Atabekov JG, Rodionova NP, Karpova OV, Kozlovsky SV, Novikov VK, Arkhipenko MV (2001) Translational activation of encapsidated potato virus X RNA by coat protein phosphorylation. Virology 286:466–474

Baratova LA, Grebenshchikov NI, Dobrov EN, Gedrovich AV, Kashirin IA, Shishkov AV, Efimov AV, Järvekülg L, Radavsky YL, Saarma M (1992) The organization of potato virus X coat proteins in virus particles studied by tritium planigraphy and model building. Virology 188:175–180

Baratova L, Efimov AV, Dobrov EN, Fedorova NV, Hunt R, Badun GA, Ksenofontov AL, Torrance L, Järvekülg L (2001) In situ spatial organization of Potato virus A coat protein subunits as assessed by tritium bombardment. J Virol 75:9696–9702

Baratova LA, Fedorova NV, Dobrov EN, Lukashina EV, Kharlanov AN, Nasonov VV, Serebryakova MV, Kozlovsky SV, Zayakina OV, Rodionova NP (2004) N-terminal segment of potato virus X coat protein subunits is glycosylated and mediates formation of a bound water shell on the virion surface. Eur J Biochem 271:3136–3145

Bawden FC, Pirie NW, Bernal JD, Fankuchen I (1936) Liquid crystalline substances from virus-infected plants. Nature 138:1051–1052

Beijerinck MW (1898) Ueber ein *contagium vivum fluidum* als Ursache der Fleckenkrankheit der Tabaksblätter. Verh Kon Akad Wetensch 5:3–21

Bernal JD, Fankuchen I (1941) X-ray and crystallographic studies of plant virus preparations. J Gen Physiol 28:111–165

Bhyravbhatla B, Watowich S, Caspar DLD (1998) Refined atomic model of the four-layer aggregate of the tobacco mosaic virus coat protein at 2.4-Å resolution. Biophys J 74:604–615

Blanch EW, Robinson DJ, Hecht L, Syme CD, Nielsen K, Barron LD (2002) Solution structures of potato virus X and narcissus mosaic virus from Raman optical activity. J Gen Virol 83:241–246

Bloomer AC, Champness JN, Bricogne G, Staden R, Klug A (1978) Protein disk of tobacco mosaic virus at 2.8 Å resolution showing the interactions within and between subunits. Nature 276:362–368

Bragard C, Duncan GH, Wesley SV, Naidu RA, Mayo MA (2000) Virus-like particles assemble in plants and bacteria expressing the coat protein gene of *Indian peanut clump virus*. J Gen Virol 81:267–272

Brierley KM, Goodman BA, Mayo MA (1993) A mobile element on a virus particle surface identified by nuclear magnetic resonance spectroscopy. Biochem J 293:657–659

Bryan RK, Bansai M, Folkhard W, Nave C, Marvin DA (1983) Maximum-entropy calculation of the electron density at 4 Å resolution of Pf1 filamentous bacteriophage. Proc Natl Acad Sci USA 80:4728–4731

Butler PJG, Durham ACH (1972) Structures and roles of the polymorphic forms of tobacco mosaic virus. V. Conservation of the abnormally titrating groups in tobacco mosaic virus. J Mol Biol 72:19–24

Butler PJG, Durham ACH (1977) Tobacco mosaic virus protein aggregation and the virus assembly. Adv Protein Chem 31:187–251

Butler PJG, Klug A (1971) Assembly of the particle of tobacco mosaic virus from RNA and disks of protein. Nat New Biol 229:47–50

Butler PJG, Finch JT, Zimmern D (1977) Configuration of tobacco mosaic virus RNA during virus disassembly. Nature 265:217–219

Cañizares MC, Nicholson L, Lomonossoff GP (2005) Use of viral vectors for vaccine production in plants. Immunol Cell Biol 83:263–270

Carrington JC, Kasschau KD, Mahajan SK, Schaad MC (1996) Cell-to-cell and long-distance transport of viruses in plants. Plant Cell 8:1669–1681

Carstens EB (2010) Ratification vote on taxonomic proposals to the International Committee on Taxonomy of Viruses. Arch Virol 155:133–146

Caspar DLD (1956) The radial density distribution in the tobacco mosaic virus particle. Nature 177:928

Caspar DLD (1963) Assembly and stability of the tobacco mosaic virus particle. Adv Protein Chem 18:37–121

Caspar DLD, Holmes KC (1969) Structure of Dahlemense strain of tobacco mosaic virus: a periodically deformed helix. J Mol Biol 46:99–133

Cerovská N, Moravec T, Velemínsky J (2002) Polyclonal antibodies to a recombinant coat protein of Potato virus A. Acta Virol 46:147–151

Chandrasekaran R, Stubbs G (2006) Fibre diffraction. In: Rossmann MG, Arnold E (eds) International tables for crystallography, vol. F: Crystallography of biological macromolecules. Kluwer, The Netherlands. doi:10.1107/97809553602060000702

Chen J, Torrance L, Cowan GH, MacFarlane SA, Stubbs G, Wilson TMA (1997) Detection of a single amino acid difference between the coat proteins of soil-borne wheat mosaic virus isolates by monoclonal antibodies: implications for virus structure. Phytopathology 87:295–301

Cochran W, Crick FHC, Vand V (1952) The structure of synthetic polypeptides. I. The transform of atoms on a helix. Acta Crystallogr 5:581–586

Correia JJ, Shire S, Yphantis DA, Schuster TM (1985) Sedimentation equilibrium measurements of the intermediate-size tobacco mosaic virus protein polymers. Biochemistry 24:3292–3329

Cross TA, Opella SJ, Stubbs G, Caspar DLD (1983) ^{31}P NMR of the RNA in TMV. J Mol Biol 170:1037–1043

Culver JN, Dawson WO, Plonk K, Stubbs G (1995) Site-specific mutagenesis confirms the involvement of carboxylate groups in the disassembly of tobacco mosaic virus. Virology 206:724–730

Denis J, Acosta-Ramirez E, Zhao Y, Hameliní M-E, Koukavica I, Baz M, Abed Y, Savard C, Pare C, Macias CL, Boivin G, Leclerc D (2008) Development of a universal influenza A vaccine based on the M2e peptide fused to the papaya mosaic virus (PapMV) vaccine platform. Vaccine 26:3395–3403

Dobrov EN, Efimov AV, Baratova LA (2004) Investigation of helical plant virus ribonucleoprotein structures with the help of tritium planigraphy and theoretical modeling. Mol Biol (Mosk) 41:706–710

Dolja VV, Boyko VP, Agranovsky AA, Koonin EV (1991) Phylogeny of capsid proteins of rod-shaped and filamentous RNA plant viruses: two families with distinct patterns of sequence and probably structure conservation. Virology 184:79–86

Dolja VV, Kreuze JF, Valkonen JPT (2006) Comparative and functional genomics of closteroviruses. Virus Res 117:38–51

Dubs MC, van Regenmortel MH (1990) Odontoglossum ringspot virus coat protein: sequence and antigenic comparisons with other tobamoviruses. Arch Virol 115:239–249

Dujardin E, Peet C, Stubbs G, Culver JN, Mann S (2003) Organization of metallic nanoparticles using tobacco mosaic virus templates. Nano Lett 3:413–417

Egelman EH (2007) The iterative helical real space reconstruction method: surmounting the problems posed by real polymers. J Struct Biol 157:83–94

Epel BL (2009) Plant viruses spread by diffusion on ER-associated movement-protein-rafts through plasmodesmata gated by viral induced host β-1,3-glucanases. Semin Cell Dev Biol 20:1074–1081

Erickson JW, Hallett FR, Bancroft JB (1983) Subassembly aggregates of papaya mosaic virus protein. Virology 129:207–211

Fauquet CM, Mayo MA, Maniloff J, Desselberger U, Ball LA (eds) (2005) Virus taxonomy: VIIIth report of the International Committee on Taxonomy of Viruses. Elsevier/Academic, London

Finch JT (1965) Preliminary X-ray diffraction studies on tobacco rattle and barley stripe mosaic viruses. J Mol Biol 12:612–619

Fischetti R, Stepanov S, Rosenbaum G, Barrea R, Black E, Gore D, Heurich R, Kondrashkina E, Kropf AJ, Wang S, Zhang K, Irving TC, Bunker GB (2004) The BioCAT undulator beamline 18ID: a facility for biological non-crystalline diffraction and X-ray absorption spectroscopy at the Advanced Photon Source. J Synchrotron Rad 11:399–405

Fraenkel-Conrat H, Williams R (1955) Reconstitution of active tobacco mosaic virus from its inactive protein and nucleic acid components. Proc Natl Acad Sci USA 41:690–698

Franklin RE (1955) Structure of tobacco mosaic virus. Nature 175:379–381

Franklin RE (1956a) Location of the ribonucleic acid in the tobacco mosaic virus particle. Nature 177:928–930

Franklin RE (1956b) X-ray diffraction studies of cucumber virus 4 and three strains of tobacco mosaic virus. Biochim Biophys Acta 19:203–211

Franklin RE, Holmes KC (1958) Tobacco mosaic virus: application of the method of isomorphous replacement to the determination of the helical parameters and radial density distribution. Acta Crystallogr 11:213–220

Fukuda M, Okada Y, Otsuki Y, Takebe I (1980) The site of initiation of rod assembly on the RNA of a tomato and a cowpea strain of tobacco mosaic virus. Virology 101:493–502

Goelet P, Lomonossoff GP, Butler PJG, Akam ME, Gait MJ, Karn J (1982) Nucleotide sequence of tobacco mosaic virus RNA. Proc Natl Acad Sci USA 79:5818–5822

Goulden MG, Davies JW, Wood KR, Lomonossoff GP (1992) Structure of tobraviral particles: a model suggested from sequence conservation in tobraviral and tobamoviral coat proteins. J Mol Biol 227:1–8

Gowda S, Satyanarayana T, Ayllón MA, Moreno P, Flores R, Dawson WO (2003) The conserved structures of the 5′ nontranslated region of Citrus tristeza virus are involved in replication and virion assembly. Virology 317:50–64

Gregory J, Holmes KC (1965) Methods of preparing orientated tobacco mosaic virus sols for x-ray diffraction. J Mol Biol 13:796–801

Hamilton WDO, Boccara M, Robinson DJ, Baulcombe DC (1987) The complete nucleotide sequence of tobacco rattle virus RNA-1. J Gen Virol 68:2563–2575

Hammond JM, Sproat KW, Wise TG, Hyatt AD, Jagadish MN, Coupar BE (1998) Expression of the potyvirus coat protein mediated by recombinant vaccinia virus and assembly of potyvirus-like particles in mammalian cells. Arch Virol 143:1433–1439

Hari V (1981) The RNA of tobacco etch virus: further characterization and detection of protein linked to RNA. Virology 112:391–399

Harrison BD, Wilson TMA (1999) Milestones in research on tobacco mosaic virus. Phil Trans R Soc Lond B 354:521–529

Heinlein M, Epel BL (2004) Macromolecular transport and signaling through plasmodesmata. Int Rev Cytol 235:93–164

Hema M, Subba Reddy ChV, Savithri HS, Sreenivasulu P (2008) Assembly of recombinant coat protein of sugarcane streak mosaic virus into potyvirus-like particles. Indian J Exp Biol 46:793–796

Hemminga MA, Vos WL, Nazarov PV, Koehorst RBM, Wolfs CJAM, Spruijt RB, Stopar D (2010) Viruses: incredible nanomachines. New advances with filamentous phages. Eur Biophys J 39:541–550

Holmes KC, Franklin RE (1958) The radial density distribution in some strains of tobacco mosaic virus. Virology 6:328–336

Holmes KC, Stubbs GJ, Mandelkow E, Gallwitz U (1975) Structure of tobacco mosaic virus at 6.7 Å resolution. Nature 254:192–196

Homer RB, Goodman RM (1975) Circular dichroism and fluorescence studies on potato virus X and its structural components. Biochim Biophys Acta 378:296–304

Hunter GJ, Rowitch DH, Perham RN (1987) Interactions between DNA and coat protein in the structure and assembly of filamentous bacteriophage fd. Nature 327:252–254

Jackson A, Lim H-L, Bragg J, Ganesan U, Lee MY (2009) Hordeivirus replication, movement, and pathogenesis. Annu Rev Phytopathol 47:385–422

Jacob T, Usha R (2002) Expression of *Cardamom mosaic virus* coat protein in *Escherichia coli* and its assembly into filamentous aggregates. Virus Res 86:133–141

Jagadish MN, Ward CW, Gough KH, Tulloch PA, Whittaker LA, Shukla DD (1991) Expression of potyvirus coat protein in *Escherichia coli* and yeast and its assembly into virus-like particles. J Gen Virol 72:1543–1550

Joseph J, Savithri HS (1999) Determination of 3′-terminal nucleotide sequence of pepper vein banding virus RNA and expression of its coat protein in *Escherichia coli*. Arch Virol 144:1679–1687

Karpova OV, Zayakina OV, Arkhipenko MV, Sheval EV, Kiselyova OI, Poljakov VY, Yaminsky IV, Rodionova NP, Atabekov JG (2006) Potato virus X RNA-mediated assembly of single-tailed ternary 'coat protein-RNA-movement protein' complexes. J Gen Virol 87:2731–2740

Kendall A, Stubbs G (2006) Oriented sols for fiber diffraction from limited quantities or hazardous materials. J Appl Crystallogr 39:39–41

Kendall A, Bian W, Junn J, McCullough I, Gore D, Stubbs G (2007a) Radial density distribution and symmetry of a *Potexvirus*, narcissus mosaic virus. Virology 357:158–164

Kendall A, McDonald M, Stubbs G (2007b) Precise determination of the helical repeat of tobacco mosaic virus. Virology 369:226–227

Kendall A, McDonald M, Bian W, Bowles T, Baumgarten SC, Shi J, Stewart PL, Bullitt E, Gore D, Irving TC, Havens WM, Ghabrial SA, Wall JS, Stubbs G (2008) Structure of flexible filamentous plant viruses. J Virol 82:9546–9554

Kiselyova OI, Yaminsky IV, Karpova OV, Rodionova NP, Kozlovsky SV, Arkhipenko MV, Atabekov JG (2003) AFM study of potato virus X disassembly induced by movement protein. J Mol Biol 332:321–325

Klug A (1999) The tobacco mosaic virus particle: structure and assembly. Phil Trans R Soc Lond B 354:531–535

Kwon S-J, Park M-R, Kim K-W, Plante CA, Hemenway CL, Kim K-H (2005) *cis*-Acting sequences required for coat protein binding and in vitro assembly of *Potato virus X*. Virology 334:83–97

Lebeurier G, Nicolaieff A, Richards KE (1977) Inside-out model for self-assembly of tobacco mosaic virus. Proc Natl Acad Sci USA 74:149–153

Lico C, Chen Q, Santi L (2008) Viral vectors for production of recombinant proteins in plants. J Cell Physiol 216:366–377

Lim H-S, Bragg JN, Ganesan U, Lawrence DM, Yu J, Isogai M, Hammond J, Jackson AO (2008) Triple gene block protein interactions involved in movement of *Barley stripe mosaic virus*. J Virol 82:4991–5006

Lobert S, Stubbs G (1990) Fiber diffraction analysis of cucumber green mottle mosaic virus using limited numbers of heavy-atom derivatives. Acta Crystallogr A46:993–997

Lobert S, Heil P, Namba K, Stubbs G (1987) Preliminary x-ray fiber diffraction studies of cucumber green mottle mosaic virus, watermelon strain. J Mol Biol 196:935–938

López-Moya JJ, García JA (2008) Potyviruses. In: Mahy BWJ, van Regenmortel MHV (eds) Encyclopedia of virology, 3rd edn. Elsevier/Academic, London

López-Moya JJ, Valli A, García JA (2009) Potyviridae. In: Encyclopedia of life sciences (ELS). Wiley, Chichester. doi:10.1002/9780470015902.a0000755.pub2

Lu B, Stubbs G, Culver JN (1996) Carboxylate interactions involved in the disassembly of tobacco mosaic tobamovirus. Virology 225:11–20

Lu B, Stubbs G, Culver JN (1998) Coat protein interactions involved in tobacco mosaic tobamovirus cross-protection. Virology 248:188–198

Lukashina E, Badun G, Fedorova N, Ksenofontov A, Nemykh M, Serebyakova M, Mukhamedzhanova A, Karpova O, Rodionova N, Baratova L, Dobrov E (2009) Tritium planigraphy study of structural alterations in the coat protein of *Potato virus X* induced by binding of its triple gene block 1 protein to virions. FEBS J 276:7006–7015

Macfarlane S (1999) Molecular biology of the tobraviruses. J Gen Virol 80:2799–2807

Macfarlane S (2010) Tobraviruses - plant pathogens and tools for biotechnology. Mol Plant Pathol. doi:10.1111/j.1364-3703.2010.00617.x

Makowski L (1982) The use of continuous diffraction data as a phase constraint II. Application to fiber diffraction data. J Appl Crystallogr 15:546–557

Makowski L (1984) Structural diversity in filamentous bacteriophages. In: Jurnak FA, McPherson A (eds) Biological macromolecules and assemblies, vol 1. Wiley, New York

Makowski L, Caspar DLD, Marvin DA (1980) Filamentous bacteriophage Pf1 structure determined at 7 Å resolution by refinement of models for the α-helical subunit. J Mol Biol 140:149–181

Mandelkow E, Stubbs G, Warren S (1981) Structures of the helical aggregates of tobacco mosaic virus protein. J Mol Biol 152:375–386

Marvin DA (1966) X-ray diffraction and electron microscope studies on the structure of the small filamentous bacteriophage fd. J Mol Biol 15:8–17

Marvin DA (1990) Model-building studies of *Inovirus:* genetic variations on a geometric theme. Int J Biol Macromol 12:125–138

Marvin DA (1998) Filamentous phage structure, infection and assembly. Curr Opin Struct Biol 8:150–158

Marvin DA, Wiseman RL, Wachtel EJ (1974a) Filamentous bacterial viruses XI. Molecular architecture of the Class II (Pf1, Xf) virion. J Mol Biol 82:121–138

Marvin DA, Pigram WJ, Wiseman RL, Wachtel EJ, Marvin FJ (1974b) Filamentous bacterial viruses XII. Molecular architecture of the Class I (fd, If1, IKe) virion. J Mol Biol 88:581–600

Marvin DA, Hale RD, Nave C, Helmer-Citterich M (1994) Molecular models and structural comparisons of native and mutant class I filamentous bacteriophages Ff (fd, fl, M13), If1 and IKe. J Mol Biol 235:260–286
Marvin DA, Welsh LC, Symmons MF, Scott WRP, Straus SK (2006) Molecular structure of fd (f1, M13) filamentous bacteriophage refined with respect to X-ray fibre diffraction and solid-state NMR data supports specific models of phage assembly at the bacterial membrane. J Mol Biol 355:294–309
Mayo MA, Brierley KM, Goodman BA (1993) Developments in the understanding of the particle structure of tobraviruses. Biochimie 75:639–644
McDonald JG, Bancroft JB (1977) Assembly studies on Potato virus Y and its coat protein. J Gen Virol 35:251–263
McDonald JG, Beveridge TJ, Bancroft JB (1976) Self-assembly of a protein from a flexuous virus. Virology 69:327–331
McDonald M, Kendall A, Tanaka M, Weissman JS, Stubbs G (2008) Enclosed chambers for humidity control and sample containment in fiber diffraction. J Appl Crystallogr 41:206–209
McDonald M, Kendall A, Bian W, McCullough I, Lio E, Havens WW, Ghabrial SA, Stubbs G (2010) Architecture of the potyviruses. Virology 30:309–313
Millane RP (1989) R factors in x-ray fiber diffraction. II. Largest likely R factors for N overlapping terms. Acta Crystallogr A45:573–576
Moreno P, Ambrós S, Albiach-Martí MR, Guerri J, Peña L (2008) *Citrus tristeza virus*: a pathogen that changed the course of the citrus industry. Mol Plant Pathol 9:251–268
Morozov SY, Solovyev AG (2003) Triple gene block: modular design of a multifunctional machine for plant virus movement. J Gen Virol 84:1351–1366
Namba K, Stubbs G (1986) Structure of tobacco mosaic virus at 3.6 Å resolution: implications for assembly. Science 231:1401–1406
Namba K, Caspar DLD, Stubbs G (1988) Enhancement and simplification of macromolecular images. Biophys J 53:469–475
Namba K, Pattanayek R, Stubbs G (1989) Visualization of protein-nucleic acid interactions in a virus. Refined structure of intact tobacco mosaic virus at 2.9 Å resolution by x-ray fiber diffraction. J Mol Biol 208:307–325
Nambudripad R, Stark W, Opella SJ, Makowski L (1991) Membrane-mediated assembly of filamentous bacteriophage Pf1 coat protein. Science 252:1305–1308
Nave C, Fowler AG, Malsey S, Marvin DA, Siegrist H, Wachtel EJ (1979) Macromolecular structural transitions in Pf1 filamentous bacterial virus. Nature 281:232–234
Nemykh MA, Efimov AV, Novikov VK, Orlov VN, Arutyunyan AM, Drachev VA, Lukashina EV, Baratova LA, Dobrov EN (2008) One more probable structural transition in potato virus X virions and a revised model of the virus coat protein structure. Virology 373:61–71
Okada Y (1986) Molecular assembly of tobacco mosaic virus *in vitro*. Adv Biophys 22:95–149
Opalka N, Beckmann R, Boisset N, Simon MN, Russel M, Darst SA (2003) Structure of the filamentous phage pIV multimer by cryo-electron microscopy. J Mol Biol 325:461–470
Opella SJ, Zeri AC, Park SH (2008) Structure, dynamics, and assembly of filamentous bacteriophages by nuclear magnetic resonance spectroscopy. Annu Rev Phys Chem 59:635–657
Park M-R, Kwon S-J, Choi H-S, Hemenway CL, Kim K-H (2008) Mutations that alter a repeated ACCA element located at the 5′ end of the *Potato virus X* genome affect RNA accumulation. Virology 378:133–141
Parker L, Kendall A, Stubbs G (2002) Surface features of potato virus X from fiber diffraction. Virology 300:291–295
Parker L, Kendall A, Berger PH, Shiel PJ, Stubbs G (2005) Wheat streak mosaic virus – structural parameters for a *Potyvirus*. Virology 340:64–69
Pattanayek R, Stubbs G (1992) Structure of the U2 strain of tobacco mosaic virus refined at 3.5 Å resolution using x-ray fiber diffraction. J Mol Biol 228:516–528
Pattanayek R, Elrod M, Stubbs G (1992) Characterization of a putative calcium-binding site in tobacco mosaic virus. Proteins 12:128–132
Pennisi E (2010) Armed and dangerous. Science 327:802–803
Peremyslov VV, Andreev IA, Prokhnevsky AI, Duncan GH, Taliansky ME, Dolja VV (2004) Complex molecular architecture of beet yellows virus particles. Proc Natl Acad Sci USA 101:5030–5035
Petty IT, French R, Jones RW, Jackson AO (1990) Identification of barley stripe mosaic virus genes involved in RNA replication and systemic movement. EMBO J 9:3453–3457
Planchart A (1995) X-ray fiber diffraction studies of odontoglossum ringspot virus: lessons on how nature produces a virus with a different host specificity. Thesis, Vanderbilt University
Puustinen P, Rajamäki M-L, Ivanov KI, Valkonen JPT, Mäkinen K (2002) Detection of the potyviral genome-linked protein VPg in virions and its phosphorylation by host kinases. J Virol 76:12703–12711
Richardson JF, Tollin P, Bancroft JB (1981) The architecture of the potexviruses. Virology 112:34–39
Rodionova NP, Karpova OV, Kozlovsky SV, Zayakina OV, Arkhipenko MV, Atabekov JG (2003) Linear remodeling of a helical virus by movement protein binding. J Mol Biol 333:565–572

Russel M, Model P (2006) Filamentous phage. In: Calendar RL (ed) The bacteriophages, 2nd edn. Oxford University Press, New York

Samuel G (1934) The movement of tobacco mosaic virus within the plant. Ann Appl Biol 21:90–111

Satyanarayana T, Gowda S, Ayllón MA, Dawson WO (2004) Closterovirus bipolar virion: evidence for initiation of assembly by minor coat protein and its restriction to the genomic RNA 5′ region. Proc Natl Acad Sci USA 101:799–804

Scholthof HB (2005) Plant virus transport: motions of functional equivalence. Trends Plant Sci 10:376–382

Schramm G (1947) Uber die Spaltung des Tabakmosaikvirus und die Wiedervereinigung der Spaltstücke zu höhermolekularen Proteinen. Z Naturforsch 2b:249–257

Shaw JG, Plaskitt KA, Wilson TMA (1986) Evidence that tobacco mosaic virus particles disassemble cotranslationally *in vivo*. Virology 148:326–336

Shukla DD, Strike PM, Tracy SL, Gough KH, Ward CW (1988) The N and C termini of the coat proteins of potyviruses are surface-located and the N terminus contains the major virus specific epitopes. J Gen Virol 69: 1497–1508

Sit TL, Leclerc D, AbouHaidar MG (1994) The minimal 5′ sequence for *in vitro* initiation of papaya mosaic assembly. Virology 199:239–242

Smolsky IL, Liu P, Niebuhr M, Ito K, Weiss TM, Tsuruta H (2007) Biological small-angle X-ray scattering facility at the Stanford Synchrotron Radiation Laboratory. J Appl Cryst 40:s453–s458

Stark W, Glucksman MJ, Makowski L (1988) Conformation of the coat protein of the filamentous bacteriophage Pf1 determined by neutron diffraction from magnetically oriented gels of specifically deuterated virions. J Mol Biol 199:171–182

Steckert JJ, Schuster TM (1982) Sequence specificity of trinucleoside diphosphate binding to polymerized tobacco mosaic virus protein. Nature 299:32–36

Straus SK, Scott WRP, Symmons MF, Marvin DA (2008) On the structures of filamentous bacteriophage Ff (fd, f1, M13). Eur Biophys J 37:521–527

Stubbs G (1989) The probability distributions of x-ray intensities in fiber diffraction: largest likely values for fiber diffraction R factors. Acta Crystallogr A45:254–258

Stubbs G (1999a) Developments in fiber diffraction. Curr Opin Struct Biol 9:615–619

Stubbs G (1999b) Tobacco mosaic virus particle structure and the initiation of disassembly. Philos Trans R Soc B 354:551–557

Stubbs G (2001) Fibre diffraction studies of filamentous viruses. Rep Prog Phys 64:1389–1425

Stubbs G, Diamond R (1975) The phase problem for cylindrically averaged diffraction patterns. Solution by isomorphous replacement and application to tobacco mosaic virus. Acta Crystallogr A31:709–718

Stubbs G, Makowski L (1982) Coordinated use of isomorphous replacement and layer-line splitting in the phasing of fiber diffraction data. Acta Crystallogr A38:417–425

Stubbs G, Stauffacher C (1981) Structure of the RNA in tobacco mosaic virus. J Mol Biol 152:387–396

Stubbs G, Warren S, Holmes K (1977) Structure of RNA and RNA binding site in tobacco mosaic virus from a 4 Å map calculated from x-ray fibre diagrams. Nature 267:216–221

Swanson M, Barker H, Macfarlane S (2002) Rapid vascular movement of tobraviruses does not require coat protein: evidence from mutated and wild-type viruses. Ann Appl Biol 141:259–266

Taliansky M, Torrance L, Kalinina NO (2008) Role of plant virus movement proteins. Methods Mol Biol 451:33–54

Tollin P, Wilson HR (1971) Some observations on the structure of the Campinas strain of tobacco rattle virus. J Gen Virol 13:433–440

Tollin P, Wilson HR (1988) Particle structure. In: Milne RG (ed) The plant viruses, vol. 4: The filamentous plant viruses. Plenum, New York

Tollin P, Wilson HR, Young DW, Cathro J, Mowat WP (1967) X-ray diffraction and electron microscope studies of narcissus mosaic virus, and comparison with potato virus X. J Mol Biol 26:353–355

Tollin P, Wilson HR, Bancroft JB (1980) Further observations on the structure of particles of potato virus X. J Gen Virol 49:407–410

Tollin P, Wilson HR, Roberts IM, Murant AF (1992) Diffraction studies of the particles of two closteroviruses: heracleum latent virus and heracleum virus 6. J Gen Virol 73:3045–3048

Torbet J, Maret G (1979) Fibres of highly oriented Pf1 bacteriophage produced in a strong magnetic field. J Mol Biol 134:843–845

Torrance L, Andreev IA, Gabrenaite-Verhovskaya R, Cowan G, Mäkinen K, Taliansky M (2006) An unusual structure at one end of potato potyvirus particles. J Mol Biol 357:1–8

Tremblay M-H, Majeau N, Laliberté Gagné M-E, Lecours K, Morin H, Duvignaud J-B, Bolduc M, Chouinard N, Paré C, Gagné S, Leclerc D (2006) Effect of mutations K97A and E128A on RNA binding and self assembly of papaya mosaic potexvirus coat protein. FEBS J 273:14–25

Vaira AM, Maroon-Lango CJ, Hammond J (2008) Molecular characterization of Lolium latent virus, proposed type member of a new genus in the family *Flexiviridae*. Arch Virol 153:1263–1270

Varma A, Gibbs AJ, Woods RD, Finch JT (1968) Some observations on the structure of the filamentous particles of several plant viruses. J Gen Virol 2:107–114

Verchot-Lubicz J, Ye C-M, Bamunusinghe D (2007) Molecular biology of potexviruses: recent advances. J Gen Virol 88:1643–1655

Verchot-Lubicz J, Torrance L, Solovyev AG, Morozov SY, Jackson AO, Gilmer D (2010) Varied movement strategies employed by triple gene block-encoding viruses. Mol Plant Microbe Interact 23:1231–1247

Wachtel EJ, Marvin FJ, Marvin DA (1976) Structural transition in a filamentous protein. J Mol Biol 107:379–383

Wang H, Stubbs G (1993) Molecular dynamics in refinement against fiber diffraction data. Acta Cryst A49:504–513

Wang H, Stubbs G (1994) The structure of cucumber green mottle mosaic virus at 3.4 Å resolution by x-ray fiber diffraction: significance for the evolution of tobamoviruses. J Mol Biol 239:371–384

Wang H, Culver JN, Stubbs G (1997) Structure of ribgrass mosaic virus at 2.9 Å resolution: evolution and taxonomy of tobamoviruses. J Mol Biol 269:769–779

Wang H, Planchart A, Stubbs G (1998) Caspar carboxylates: the structural basis of tobamovirus disassembly. Biophys J 74:633–638

Wang YA, Yu X, Overman S, Tsuboi M, Thomas GJ Jr, Egelman EH (2006) The structure of a filamentous bacteriophage. J Mol Biol 361:209–215

Watson JD (1954) The structure of tobacco mosaic virus: I. X-ray evidence of a helical arrangement of subunits around the longitudinal axis. Biochim Biophys Acta 13:10–19

Welsh LC, Symmons MF, Sturtevant JM, Marvin DA, Perham RN (1998) Structure of the capsid of Pf3 filamentous phage determined from X-ray fibre diffraction data at 3.1 Å resolution. J Mol Biol 283:155–177

Welsh LC, Symmons MF, Marvin DA (2000) The molecular structure and structural transition of the α-helical capsid in filamentous bacteriophage Pf1. Acta Crystallogr D56:137–150

Wilson TM (1984) Cotranslational disassembly of tobacco mosaic virus *in vitro*. Virology 137:255–265

Wilson HR, Tollin P, Sawyer L, Robinsin DJ, Price NC, Kelly SM (1991) Secondary structures of narcissus mosaic virus protein. J Gen Virol 72:1479–1480

Wu X, Shaw JG (1998) Evidence that assembly of a potyvirus begins near the 5′ terminus of the viral RNA. J Gen Virol 79:1525–1529

Wyckoff RWG, Corey RB (1936) X-ray diffraction patterns of crystalline tobacco mosaic proteins. J Biol Chem 116:51–55

Yamashita I, Suzuki H, Namba K (1998) Multiple-step method for making exceptionally well-oriented liquid-crystalline sols of macromolecular assemblies. J Mol Biol 278:609–615

Zayakina O, Arkhipenko M, Kozlovsky S, Nikitin N, Smirnov A, Susi P, Rodionova N, Karpova O, Atabekov J (2008) (2008) Mutagenic analysis of *Potato Virus X* movement protein (TGBp1) and the coat protein (CP): *in vitro* TGBp1-CP binding and viral RNA translation activation. Mol Plant Pathol 9:37–44

Zeri AC, Mesleh MF, Nevzorov AA, Opella SJ (2003) Structure of the coat protein in fd filamentous bacteriophage particles determined by solid-state NMR spectroscopy. Proc Natl Acad Sci USA 100:6458–6463

Zhang X, Settembre E, Xu C, Dormitzer PR, Bellamy R, Harrison SC, Grigorieff N (2008) Near-atomic resolution using electron cryomicroscopy and single-particle reconstruction. Proc Natl Acad Sci USA 105:1867–1872

Zhou ZH (2008) Towards atomic resolution structural determination by single-particle cryoelectron microscopy. Curr Opin Struct Biol 18:218–228

Zimmern D (1977) The nucleotide sequence at the origin for assembly on tobacco mosaic virus RNA. Cell 11: 463–482

Zimmern D, Wilson TM (1976) Location of the origin for viral reassembly on tobacco mosaic virus RNA and its relation to stable fragment. FEBS Lett 71:294–298

Index

M.G. Rossmann and V.B. Rao (eds.), *Viral Molecular Machines*,
Advances in Experimental Medicine and Biology 726, DOI 10.1007/978-1-4614-0980-9,

G

H